C0-DXA-265

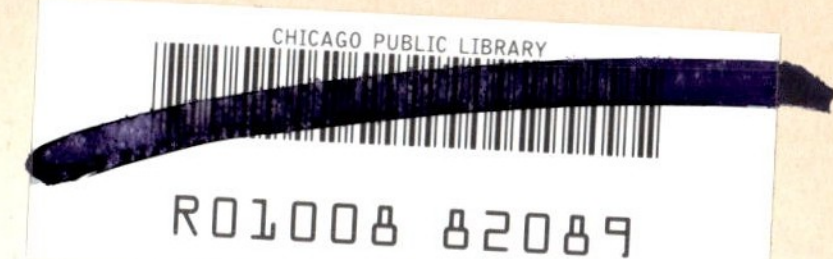


**CHILTON'S** GUIDE TO

# FUEL INJECTION & ELECTRONIC ENGINE CONTROLS–1988-90

## ACURA • DAIHATSU • HONDA • HYUNDAI ISUZU • MAZDA • MITSUBISHI



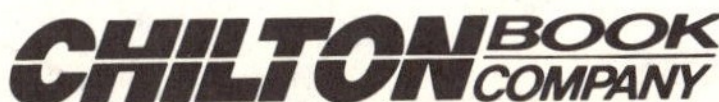


*ONE OF THE **ABC PUBLISHING COMPANIES,***
*A PART OF **CAPITAL CITIES/ABC, INC.***

Manufactured in USA

Chilton Way, Radnor, PA 19089
ISBN 0–8019–8043–7
1234567890 9876543210

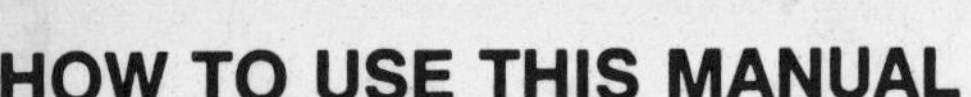


# HOW TO USE THIS MANUAL

For ease of use, this manual is divided into sections as follows:

**SECTION 1** Basic Electricity
**SECTION 2** Troubleshooting and Diagnosis
**SECTION 3** Self-Diagnostic Systems
**SECTION 4** Fuel Injection Systems

The **CONTENTS** summarize the subjects covered in each section.

To quickly locate the proper service section, use the application chart on the following pages. It references applicable **CAR AND TRUCK MODELS** and **SERVICE SECTIONS** for major electronic engine control systems.

It is recommended that the user be familiar with the applicable **GENERAL INFORMATION, SERVICE PRECAUTIONS** and **TROUBLESHOOTING AND DIAGNOSIS TECHNIQUES** before testing or servicing any engine control system.

Major service sections are grouped by vehicle manufacturer, with each engine control system subsection containing:

- **GENERAL INFORMATION** pertaining to the operation of the system, individual components and the overall logic by which components work together.

- **SERVICE PRECAUTIONS** (if any) of which the user should be aware to prevent injury or damage to the vehicle or components.

- **FAULT DIAGNOSIS** in the form of diagnostic charts or test procedures which lead the user through the various system circuit tests and explain the trouble codes stored in the computer memory.

## SAFETY NOTICE

Proper service and repair procedures are vital to the safe, reliable operation of all motor vehicles, as well as the personal safety of those performing service or repairs. This manual outlines procedures for servicing and repairing vehicles using safe, effective methods. The procedures contain many NOTES and CAUTIONS which should be followed along with standard safety procedures to eliminate the possibility of personal injury or improper service which could damage the vehicle or compromise its safety.

It is important to note that repair procedures and techniques, tools and parts for servicing motor vehicles, as well as the skill and experience of the individual performing the work vary widely. It is not possible to anticipate all of the hazards that may result. Standard and accepted safety precautions and equipment should be used when handling toxic or flammable fluids and safety goggles or other protection should be used during cutting, grinding, chiseling, prying or any other process that that can cause material removal or projectiles. Similar protection against the high voltages generated in all electronic ignition systems should be employed during service procedures.

Some procedures require the use of tools or test equipment specially designed for a specific purpose. Before substituting another tool or procedure, you must be completely satisfied that neither your personal safety, nor the performance of the vehicle will be endangered.

## PART NUMBERS

Part numbers listed in this reference are not recommendations by Chilton for any product by brand name. They are references that can be used with interchange manuals and aftermarket supplier catalogs to locate each brand supplier's discrete part number.

Although information in this manual is based on industry sources and is complete as possible at the time of publication, the possibilty exists that some car manufacturers made later changes which could not be included here. While striving for total accuracy, Chilton Book Company cannot assume responsibity for any errors, changes or omissions that may occur in the compilation of this data.

# Contents

# ENGINE CONTROL SYSTEM APPLICATION CHART
## ACURA, DAIHATSU AND HONDA

### ACURA

| Year | Model | Engine cc (liter) | Engine Serial Number | Fuel System | Ignition System |
|---|---|---|---|---|---|
| 1988 | Integra | 1590 (1.6) | D16A1 | PGM-FI | PGM-IG |
| | Legend | 2675 (2.7) | C27A1 | PGM-FI | PGM-IG |
| 1989 | Integra | 1590 (1.6) | D16A1 | PGM-FI | PGM-IG |
| | Legend | 2675 (2.7) | C27A1 | PGM-FI | PGM-IG |
| 1990 | Integra | 1834 (1.8) | BA18A1 | PGM-FI | PGM-IG |
| | Legend | 2675 (2.7) | C27A1 | PGM-FI | PGM-IG |

PGM-FI—Programmed fuel injection
PGM-IG—Programmed ignition

### DAIHATSU

| Year | Model | Engine cc (liter) | Engine Code | Fuel System | Ignition System |
|---|---|---|---|---|---|
| 1988-90 | Charade | 994 (1.0) | — | MPI | ETA |

MPI—Multiport injection
ETA—Electronic timing advance

### HONDA

| Year | Model | Engine cc (liter) | Engine Serial Number | Fuel System | Ignition System |
|---|---|---|---|---|---|
| 1988 | Accord DX, LX | 1955 (2.0) | A20A1 | 2bbl | Electronic |
| | Accord LXI | 1955 (2.0) | A20A3 | PGM-FI | Electronic |
| | Civic | 1493 (1.5) | D15B1 | PGM-FI | PGM-IG |
| | Civic | 1493 (1.5) | D15B2 | PGM-FI | PGM-IG |
| | Civic 4WD | 1590 (1.6) | D16A6 | PGM-FI | PGM-IG |
| | Civic CRX STD | 1488 (1.5) | D15B6 | PGM-FI | PGM-IG |
| | Civic CRX HF | 1488 (1.5) | D15B2 | PGM-FI | PGM-IG |
| | Civic CRX SI | 1590 (1.6) | D16A6 | PGM-FI | PGM-IG |
| | Prelude | 1958 (2.0) | B20A3 | Dual 1bbl ① | Electronic |
| | Prelude | 1958 (2.0) | B20A5 | PGM-FI | PGM-IG |
| 1989 | Accord DX, LX | 1955 (2.0) | A20A1 | 2bbl | Electronic |
| | Accord LXI | 1955 (2.0) | A20A3 | PGM-FI | Electronic |
| | Civic | 1493 (1.5) | D15B1 | PGM-FI | PGM-IG |
| | Civic | 1493 (1.5) | D15B2 | PGM-FI | PGM-IG |
| | Civic | 1590 (1.6) | D16A6 | PGM-FI | PGM-IG |
| | Civic CRX STD | 1493 (1.5) | D15B6 | PGM-FI | PGM-IG |
| | Civic CRX HF | 1493 (1.5) | D15B2 | PGM-FI | PGM-IG |
| | Civic CRX SI | 1590 (1.6) | D16A6 | PGM-FI | PGM-IG |
| | Prelude | 1958 (2.0) | B20A3 | Dual 1bbl ① | Electronic |
| | Prelude | 1958 (2.0) | B20A5 | PGM-FI | PGM-IG |

## HONDA

| Year | Model | Engine cc (liter) | Engine Serial Number | Fuel System | Ignition System |
|---|---|---|---|---|---|
| 1990 | Accord | 2156 (2.2) | F22A | PGM-FI | PGM-IG |
| | Civic | 1493 (1.5) | P15B1 | PGM-FI | PGM-IG |
| | Civic | 1493 (1.5) | D15B2 | PGM-FI | PGM-IG |
| | Civic | 1590 (1.6) | D16A6 | PGM-FI | PGM-IG |
| | Civic CRX STD | 1493 (1.5) | D15B6 | PGM-FI | PGM-IG |
| | Civic CRX HF | 1493 (1.5) | D15B2 | PGM-FI | PGM-IG |
| | Civic CRX SI | 1590 (1.6) | D16A6 | PGM-FI | PGM-IG |
| | Prelude | 1958 (2.0) | B20A5 | PGM-FI | PGM-IG |

BBL Barrel carburetor
PGM-FI Programmed fuel injection
① Side draft carburetor
PGM-IG Programmed ignition
PGM-IG

## HYUNDAI

| Year | Model | Engine cc (liter) | Engine VIN | Fuel System | Ignition System |
|---|---|---|---|---|---|
| 1988 | Excel | 1468 (1.5) | J | FBC | Electronic |
| 1989 | Excel | 1468 (1.5) | J | FBC | Electronic |
| | Sonata | 2351 (2.4) | S | MPI | Electronic |
| | | 2972 (3.0) | T | MPI | Electronic |
| 1990 | Excel | 1468 (1.5) | J | FBC | Electronic |
| | | 1468 (1.5) | J | MPI | Electronic |
| | Sonata | 2351 (2.4) | S | MPI | Electronic |
| | | 2972 (3.0) | T | MPI | Electronic |

## ISUZU

| Year | Model | Engine cc (liter) | Engine VIN | Fuel System | Ignition System |
|---|---|---|---|---|---|
| 1988 | I-Mark | 1471 (1.5) | 4XC1 | 2bbl | Electronic |
| | I-Mark (Turbo) | 1471 (1.5) | 4XC1-T | EFI | Electronic |
| | Impulse | 2254 (2.3) | 4Z01 | EFI | Electronic |
| | Impulse (Turbo) | 1994 (2.0) | 4ZC1-T | EFI | Electronic |
| | Pick-up | 2254 (2.3) | 4ZD1 | 2bbl | Electronic |
| | Pick-up | 2559 (2.6) | 4ZE1 | EFI | Electronic |
| | Trooper II | 2559 (2.6) | 4ZE1 | EFI | Electronic |
| 1989-90 | Amigo | 2254 (2.3) | 4ZD1 | 2bbl | Electronic |
| | Amigo | 2559 (2.6) | 4ZE1 | EFI | Electronic |
| | I-Mark | 1471 (1.5) | 4XC1 | 2bbl | Electronic |
| | I-Mark (Turbo) | 1471 (1.5) | 4XC1-T | EFI | Electronic |
| | I-Mark | 1588 (1.6) | 4XE1 | EFI | Electronic |
| | Impulse | 2254 (2.3) | 4ZD1 | EFI | Electronic |
| | Impulse (Turbo) | 1994 (2.0) | 4ZC1-T | EFI | Electronic |
| | Pick-up | 2254 (2.3) | 4ZD1 | 2bbl | Electronic |
| | Pick-up | 2559 (2.6) | 4ZE1 | EFI | Electronic |
| | Trooper, Trooper II | 2559 (2.6) | 4ZE1 | EFI | Electronic |
| | Trooper, Trooper II | 2837 (2.8) | LL2 | EFI | Electronic |

# ENGINE CONTROL SYSTEM APPLICATION CHART
## MAZDA AND MITSUBISHI

### MAZDA

| Year | Model | Engine cc (liter) | Engine Code | Fuel System | Ignition System |
|---|---|---|---|---|---|
| 1988 | Mazda RX-7 | 1308 (1.3) | Rotary | EGI | Electronic |
| | RX-7 (Turbo) | 1308 (1.3) | Rotary | EGI | Electronic |
| | 323 | 1597 (1.6) | B6 SOHC | EGI | Electronic |
| | 323 (Turbo) | 1597 (1.6) | B6 DOHC | EGI | Electronic |
| | 626, MX-6 | 2184 (2.2) | F2 | EFI | Electronic |
| | 626, MX-6 (Turbo) | 2184 (2.2) | F2 | EFI | Electronic |
| | B2200 | 2184 (2.2) | F2 | 2bbl | Electronic |
| | B2600 | 2555 (2.5) | G54B | 2bbl | Electronic |
| | 929 | 2954 (3.0) | JE | EFI | Electronic |
| 1989 | Mazda RX-7 | 1308 (1.3) | Rotary | EGI | Electronic |
| | Mazda RX-7 (Turbo) | 1308 (1.3) | Rotary | EGI | Electronic |
| | 323 | 1597 (1.6) | B6 SOHC | EGI | Electronic |
| | 323 (Turbo) | 1597 (1.6) | B6 DOHC | EGI | Electronic |
| | 626, MX-6 | 2184 (8.2) | F2 | EFI | Electronic |
| | 626, MX-6 (Turbo) | 2184 (2.2) | F2 | EFI | Electronic |
| | B2200 | 2184 (2.2) | F2 | 2bbl | Electronic |
| | B2600I | 2606 (2.5) | G6 | EFI | Electronic |
| | 929 | 2954 (3.0) | JE | EFI | Electronic |
| | MPV | 2954 (3.0) | JE | EFI | Electronic |
| | MPV | 2606 (2.5) | G6 | EFI | Electronic |
| 1990 | RX-7 | 1308 (1.3) | Rotary | EGI | Electronic |
| | RX-7 (Turbo) | 1308 (1.3) | Rotary | EGI | Electronic |
| | 323 | 1597 (1.6) | B6 SOHC | EFI | Electronic |
| | 323 | 1839 (1.9) | B6 DOHC | EFI | Electronic |
| | MX-5 | 1597 (1.6) | B6 DOHC | EFI | Electronic |
| | 626, MX-6 | 2184 (2.2) | F2 | EFI | Electronic |
| | 626, MX-6 (Turbo) | 2184 (2.2) | F2 | EFI | Electronic |
| 1990 | B2200 | 2184 (2.2) | F2 | 2bbl | Electronic |
| | B2600I | 2606 (2.5) | G6 | EFI | Electronic |
| | 929 | 2954 (3.0) | JE (SOHC) | EFI | Electronic |
| | 929 | 2954 (3.0) | JE (DOHC) | EFI | Electronic |
| | MPV | 2954 (3.0) | JE | EFI | Electronic |
| | MPV | 2606 (2.5) | G6 | EFI | Electronic |

EGI—Electronic gasoline injection

### MITSUBISHI

| Year | Model | Engine cc (liter) | Family | Fuel System | Ignition System |
|---|---|---|---|---|---|
| 1988 | Tredia | 1997 (2.0) | G63B | FBC | Electronic |
| | | 1795 (1.8) | G62B | ECI | ESC (HEI) |
| | Cordia | 1997 (2.0) | G63B | FBC | Electronic |
| | | 1795 (1.8) | G62B | ECI | ESC (HEI) |
| | Starion | 2555 (2.6) | G54B | ECI | ESC (HEI) |
| | Galant | 2972 (3.0) | 6G72 | MPI | ECIT |
| | Mirage | 1468 (1.5) | G15B | FBC | Electronic |
| | | 1597 (1.6) | G32B | ECI | ECIT |
| | Precis | 1468 (1.5) | G15B | FBC | Electronic |
| | Montero | 2555 (2.6) | G54B | FBC | Electronic |

## MITSUBISHI

| Year | Model | Engine cc (liter) | Family | Fuel System | Ignition System |
|---|---|---|---|---|---|
| 1988 | Truck | 1997 (2.0) | G63B | FBC | Electronic |
| | | 2555 (2.6) | G54B | FBC | Electronic |
| | Van/Wagon | 2350 (2.4) | G64B | MPI | ECIT |
| 1989 | Starion | 2555 (2.6) | G541B | ECI | ESC (HEI) |
| | Galant | 1997 (2.0) SOHC | 4G63 | MPI | ECIT |
| | | 1997 (2.0) DOHC | 4G63 | MPI | ECIT |
| | Mirage | 1468 (1.5) | 4G15 | MPI | ECIT |
| | | 1597 (1.6) | 4G61 | MPI | ECIT |
| | Precis | 1468 (1.5) | G15B | FBC | Electronic |
| | Montero | 2555 (2.6) | G54B | FBC | Electronic |
| | | 2972 (3.0) | 6G72 | MPI | ECIT |
| | Truck | 1997 (2.0) | G63B | FBC | Electronic |
| | | 2555 (2.6) | G54B | FBC | Electronic |
| | Van/Wagon | 2350 (2.4) | 4G64 | MPI | ECIT |
| 1990 | Galant | 1997 (2.0) SOHC | 4G63 | MPI | ECIT |
| | | 1997 (2.0) DOHC | 4G63 | MPI | ECIT |
| | Mirage | 1468 (1.5) | 4G15 | MPI | ECIT |
| | | 1597 (1.6) | 4G61 | MPI | ECIT |
| | Precis | 1468 (1.5) | G4DJ | MPI | ECIT |
| | Montero | 2972 (3.0) | 6G72 | MPI | ECIT |
| | Truck | 2350 (2.4) | 4G64 | MPI | ECIT |
| | | 2972 (3.0) | 6G72 | MPI | ECIT |
| | Van/Wagon | 2350 (2.4) | 4G64 | MPI | ECIT |
| | Eclipse | 1795 (1.8) | 4G37 | MPI | ECIT |
| | | 1997 (2.0) | 4G63 | MPI | ECIT |

ESC—Electronic spark control
ECIT—Electronic control ignition timing
FBC—Feedback carburetor
ECI—Electronic controlled injection (throttle body inj.)
MPI—Multiport injection
SOHC/DOHC—Single/double overhead cam

# Basic Electricity

SECTION 1

## INDEX

# FUNDAMENTALS OF ELECTRICITY

A good understanding of basic electrical theory and how circuits work is necessary to successfully perform the service and testing outlined in this manual. Therefore, this section should be read before attempting any diagnosis and repair.

All matter is made up of tiny particles called molecules. Each molecule is made up of two or more atoms. Atoms may be divided into even smaller particles called protons, neutrons and electrons. These particles are the same in all matter and differences in materials (hard or soft, conductive or non-conductive) occur only because of the number and arrangement of these particles. In other words, the protons, neutrons and electrons in a drop of water are the same as those in an ounce of lead, there are just more of them (arranged differently) in a lead molecule than in a water molecule. Protons and neutrons packed together form the nucleus of the atom, while electrons orbit around the nucleus much the same way as the planets of the solar system orbit around the sun.

The proton is a small positive natural charge of electricity, while the neutron has no electrical charge. The electron carries a negative charge equal to the positive charge of the proton. Every electrically neutral atom contains the same number of protons and electrons, the exact number of which determines the element. The only difference between a conductor and an insulator is that a conductor possesses free electrons in large quantities, while an insulator has only a few. An element must have very few free electrons to be a good insulator and vice-versa. When we speak of electricity, we're talking about these free electrons.

In a conductor, the movement of the free electrons is hindered by collisions with the adjoining atoms of the element (matter). This hindrance to movement is called **RESISTANCE** and it varies with different materials and temperatures. As temperature increases, the movement of the free electrons increases, causing more frequent collisions and therefore increasing resistance to the movement of the electrons. The number of collisions (resistance) also increases with the number of electrons flowing (current). Current is defined as the movement of electrons through a conductor such as a wire. In a conductor (such as copper) electrons can be caused to leave their atoms and move to other atoms. This flow is continuous in that every time an atom gives up an electron, it collects another one to take its place. This movement of electrons is called electric current and is measured in amperes. When 6.28 billion, billion electrons pass a certain point in the circuit in one second, the amount of current flow is called 1 ampere.

The force or pressure which causes electrons to flow in any conductor (such as a wire) is called **VOLTAGE.** It is measured in volts and is similar to the pressure that causes water to flow in a pipe. Voltage is the difference in electrical pressure measured between 2 different points in a circuit. In a 12 volt system, for example, the force measured between the two battery posts is 12 volts. Two important concepts are voltage potential and polarity. Voltage potential is the amount of voltage or electrical pressure at a certain point in the circuit with respect to another point. For example, if the voltage potential at one post of the 12 volt battery is 0, the voltage potential at the other post is 12 volts with respect to the first post. One post of the battery is said to be positive (+); the other post is negative (−) and the conventional direction of current flow is from positive to negative in an electrical circuit. It should be noted that the electron flow in the wire is opposite the current flow. In other words, when the circuit is energized, the current flows from positive to negative, but the electrons actually move from negative to positive. The voltage or pressure needed to produce a current flow in a circuit must be greater than the resistance present in the circuit. In other words, if the voltage drop across the resistance is greater than or equal to the voltage input, the voltage potential will be zero—no voltage will flow through the circuit. Resistance to the flow of electrons is measured in ohms. One volt will cause 1 ampere to flow through a resistance of 1 ohm.

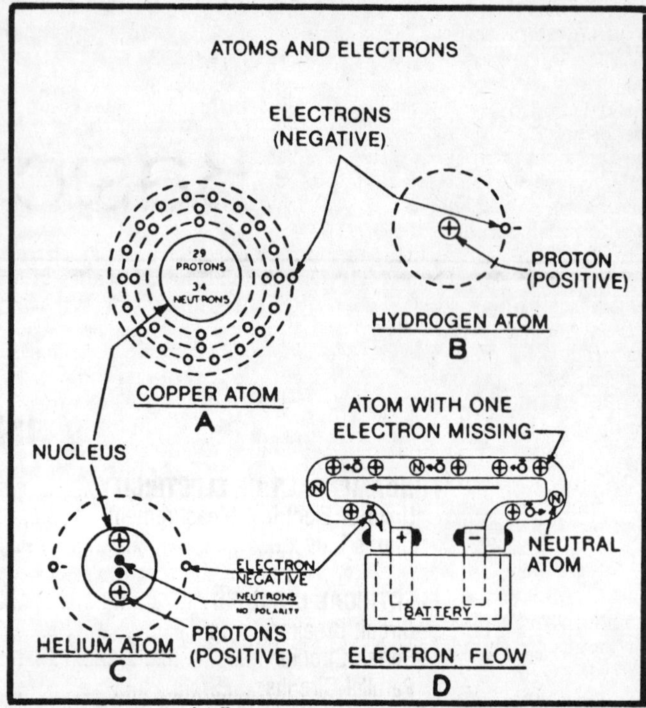


**Typical atoms of copper (A), hydrogen (B) and helium (C). Electron flow in battery circuit (D)**

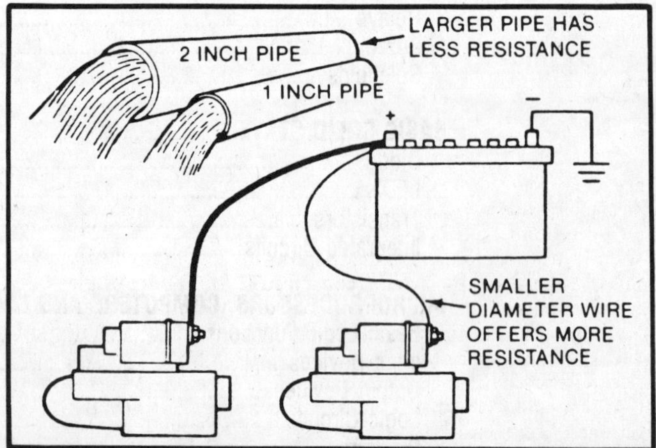


**Electrical resistance can be compared to water flow through a pipe. The smaller the wire (pipe), the more resistance to the flow of electrons (water)**

## Units Of Electrical Measurement

There are 3 fundamental characteristics of a direct-current electrical circuit: volts, amperes and ohms.

**VOLTAGE** in a circuit controls the intensity with which the loads in the circuit operate. The brightness of a lamp, the heat of an electrical defroster, the speed of a motor are all directly proportional to the voltage, if the resistance in the circuit and/or mechanical load on electric motors remains constant. Voltage available from the battery is constant (normally 12 volts), but as it operates the various loads in the circuit, voltage decreases (drops).

**AMPERE** is the unit of measurement of current in an electrical circuit. One ampere is the quantity of current that will flow through a resistance of 1 ohm at a pressure of 1 volt. The amount of current that flows in a circuit is controlled by the voltage and the resistance in the circuit. Current flow is directly proportional to resistance. Thus, as voltage is increased or decreased, current is increased or decreased accordingly. Current is decreased as resistance is increased. However, current is also increased as resistance is decreased. With little or no resistance in a circuit, current is high.

**OHM** is the unit of measurement of resistance, represented by the Greek letter Omega (Ω). One ohm is the resistance of a conductor through which a current of one ampere will flow at a pressure of one volt. Electrical resistance can be measured on an instrument called an ohmmeter. The loads (electrical devices) are the primary resistances in a circuit. Loads such as lamps, solenoids and electric heaters have a resistance that is essentially fixed; at a normal fixed voltage, they will draw a fixed current. Motors, on the other hand, do not have a fixed resistance. Increasing the mechanical load on a motor (such as might be caused by a misadjusted track in a power window system) will decrease the motor speed. The drop in motor rpm has the effect of reducing the internal resistance of the motor because the current draw of the motor varies directly with the mechanical load on the motor, although its actual resistance is unchanged. Thus, as the motor load increases, the current draw of the motor increases, and may increase up to the point where the motor stalls (cannot move the mechanical load).

Circuits are designed with the total resistance of the circuit taken into account. Troubles can arise when unwanted resistances enter into a circuit. If corrosion, dirt, grease, or any other contaminant occurs in places like switches, connectors and grounds, or if loose connections occur, resistances will develop in these areas. These resistances act like additional loads in the circuit and cause problems.

E
I | R

$$I = \frac{E}{R} \quad \text{or} \quad \text{AMPERES} = \frac{\text{VOLTS}}{\text{OHMS}}$$

$$R = \frac{E}{I} \quad \text{or} \quad \text{OHMS} = \frac{\text{VOLTS}}{\text{AMPERES}}$$

$$E = I \times R \quad \text{or} \quad \text{VOLTS} = \text{AMPERES} \times \text{OHMS}$$

**Ohms Law is the basis for all electrical measurements. By simply plugging in two values, the third can be calculated using the illustrated formula.**

$$R = \frac{E}{I} \qquad \text{Where: } E = 12 \text{ volts}, \quad I = 2.5 \text{ amperes}$$

$$R = \frac{12 \text{ volts}}{2.5 \text{ amps}} = 4.8 \text{ ohms}$$

**An example of calculating resistance (R) when the voltage (E) and amperage (I) is known.**

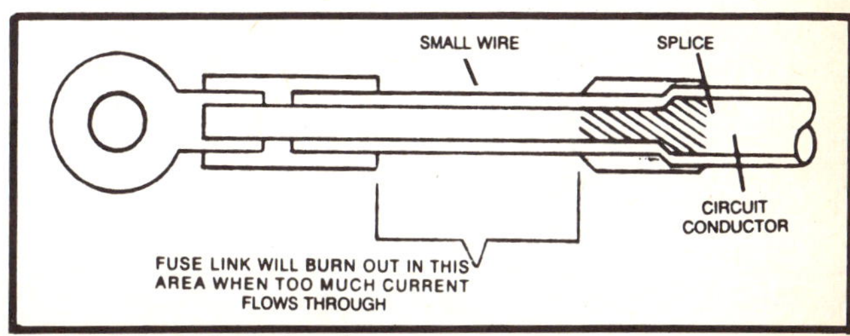


**Typical fusible link wire**

## OHM'S LAW

Ohm's law is a statement of the relationship between the 3 fundamental characteristics of an electrical circuit. These rules apply to direct current (DC) only.

Ohm's law provides a means to make an accurate circuit analysis without actually seeing the circuit. If, for example, one wanted to check the condition of the rotor winding in a alternator whose specifications indicate that the field (rotor) current draw is normally 2.5 amperes at 12 volts, simply connect the rotor to a 12 volt battery and measure the current with an ammeter. If it measures about 2.5 amperes, the rotor winding can be assumed good.

An ohmmeter can be used to test components that have been removed from the vehicle in much the same manner as an ammeter. Since the voltage and the current of the rotor windings used as an earlier example are known, the resistance can be calculated using Ohms law. The formula would be ohms equals volts divided by amperes.

If the rotor resistance measures about 4.8 ohms when checked with an ohmmeter, the winding can be assumed good. By plugging in different specifications, additional circuit information can be determined such as current draw, etc.

## Electrical Circuits

An electrical circuit must start from a source of electrical supply and return to that source through a continuous path. Circuits are designed to handle a certain maximum current flow. The maximum allowable current flow is designed higher than the normal current requirements of all the loads in the circuit. Wire size, connections, insulation, etc., are designed to prevent undesirable voltage drop, overheating of conductors, arcing of contacts and other adverse effects. If the safe maximum current flow level is exceeded, damage to the circuit components will result; it is this condition that circuit protection devices are designed to prevent.

Protection devices are fuses, fusible links or circuit breakers designed to open or break the circuit quickly whenever an overload, such as a short circuit, occurs. By opening the circuit quickly, the circuit protection device prevents damage to the wiring, battery and other circuit components. Fuses and fusible links are designed to carry a preset maximum amount of current and to melt when that maximum is exceeded, while circuit breakers merely break the connection and may be manually reset. The maximum amperage rating of each fuse is marked on the fuse body and all contain a see-through portion that shows the break in the fuse element when blown. Fusible link maximum amperage rating is indicated by gauge or thickness of the wire. Never replace a blown fuse or fusible link with one of a higher amperage rating.

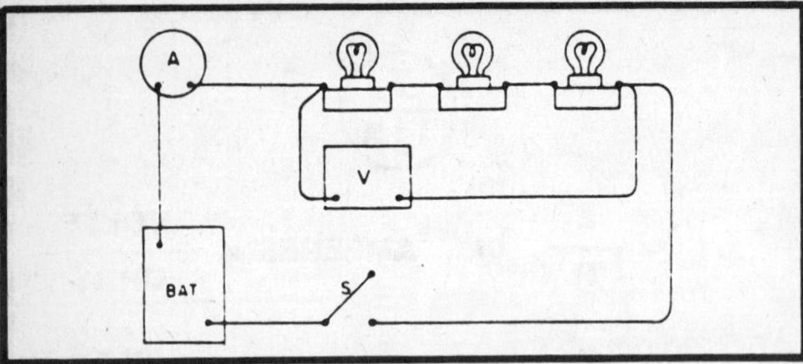

Example of a series circuit

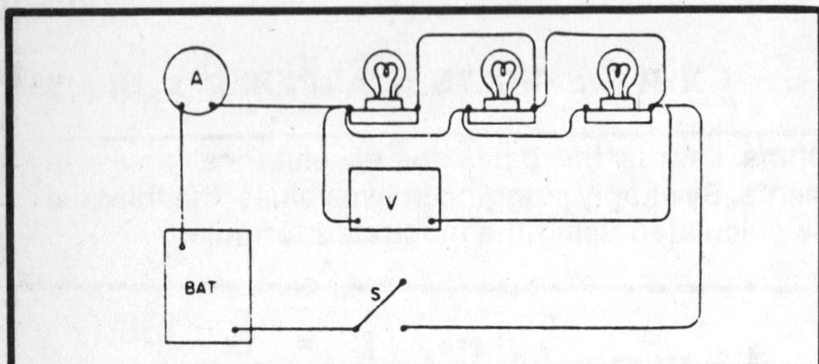

Example of a parallel circuit

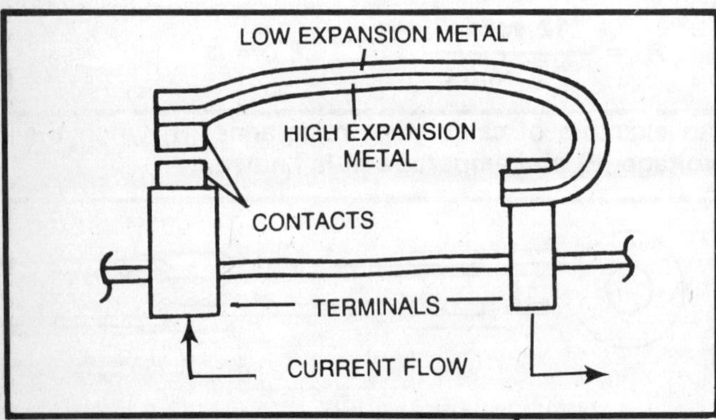

Typical circuit breaker construction

**CAUTION**

***Resistance wires, like fusible links, are also spliced into conductors in some areas. Do not make the mistake of replacing a fusible link with a resistance wire. Resistance wires are longer than fusible links and are stamped "RESISTOR-DO NOT CUT OR SPLICE."***

Circuit breakers consist of 2 strips of metal which have different coefficients of expansion. As an overload or current flows through the bimetallic strip, the high-expansion metal will elongate due to heat and break the contact. With the circuit open, the bimetal strip cools and shrinks, drawing the strip down until contact is re-established and current flows once again. In actual operation, the contact is broken very quickly if the overload is continuous and the circuit will be repeatedly broken and remade until the source of the overload is corrected.

The self-resetting type of circuit breaker is the one most generally used in automotive electrical systems. On manually reset circuit breakers, a button will pop up on the circuit breaker case. This button must be pushed in to reset the circuit breaker and restore power to the circuit. Always repair the source of the overload before resetting a circuit breaker or replacing a fuse or fusible link. When searching for overloads, keep in mind that the circuit protection devices protect only against overloads between the protection device and ground.

There are 2 basic types of circuit; Series and Parallel. In a series circuit, all of the elements are connected in chain fashion with the same amount of current passing through each element or load. No matter where an ammeter is connected in a series circuit, it will always read the same. The most important fact to remember about a series circuit is that the sum of the voltages across each element equals the source voltage. The total resistance of a series circuit is equal to the sum of the individual resistances within each element of the circuit. Using ohms law, one can determine the voltage drop across each element in the circuit. If the total resistance and source voltage is known, the amount of current can be calculated. Once the amount of current (amperes) is known, values can be substituted in the Ohms law formula to calculate the voltage drop across each individual element in the series circuit. The individual voltage drops must add up to the same value as the source voltage.

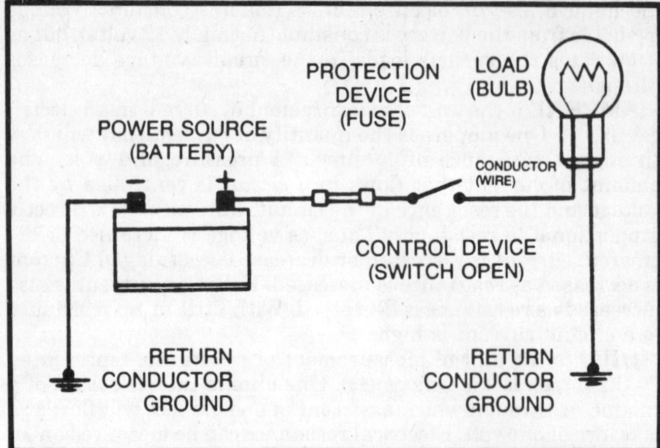

Typical circuit with all essential components

A parallel circuit, unlike a series circuit, contains 2 or more branches, each branch a separate path independent of the others. The total current draw from the voltage source is the sum of all the currents drawn by each branch. Each branch of a parallel circuit can be analyzed separately. The individual branches can be either simple circuits, series circuits or combinations of series-parallel circuits. Ohms law applies to parallel circuits just as it applies to series circuits, by considering each branch independently of the others. The most important thing to remember is that the voltage across each branch is the same as the source voltage. The current in any branch is that voltage divided by the resistance of the branch. A practical method of determining the resistance of a parallel circuit is to divide the product of the 2 resistances by the sum of 2 resistances at a time. Amperes through a parallel circuit is the sum of the amperes through the separate branches. Voltage across a parallel circuit is the same as the voltage across each branch.

By measuring the voltage drops the resistance of each element within the circuit is being measured. The greater the voltage drop, the greater the resistance. Voltage drop measurements are a common way of checking circuit resistances in automotive electrical systems. When part of a circuit develops excessive resistance (due to a bad connection) the element will show a higher than normal voltage drop. Normally, automotive wiring is selected to limit voltage drops to a few tenths of a volt. In parallel circuits, the total resistance is less than the sum of the individual resistances; because the current has 2 paths to take, the total resistance is lower.

## Magnetism and Electromagnets

Electricity and magnetism are very closely associated because when electric current passes through a wire, a magnetic field is created around the wire. When a wire carrying electric current

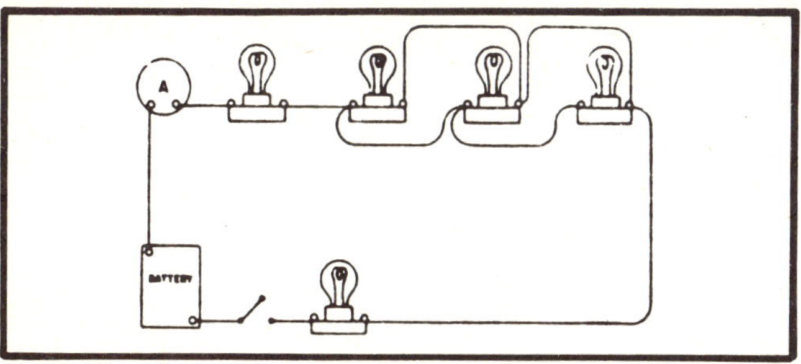
Example of a series-parallel circuit

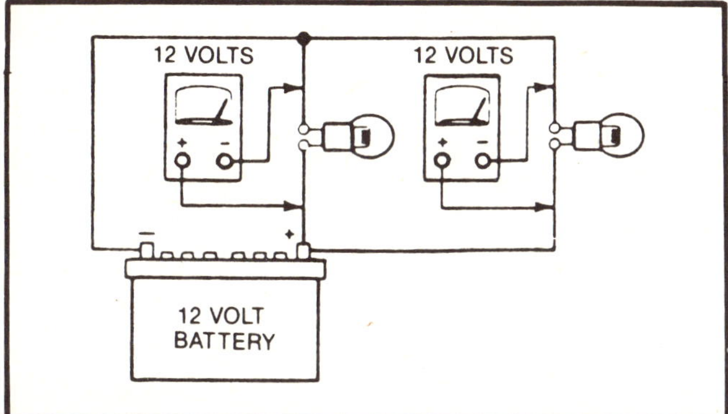

Voltage drop in a parallel circuit. Voltage drop across each lamp is 12 volts

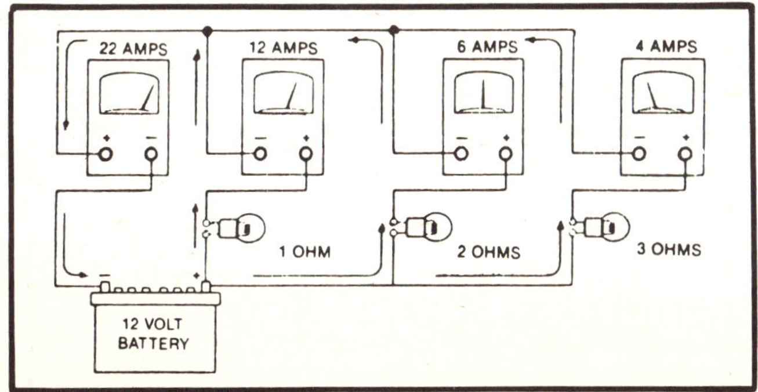

Total current in parallel circuit: 4 + 6 + 12 = 22 amps

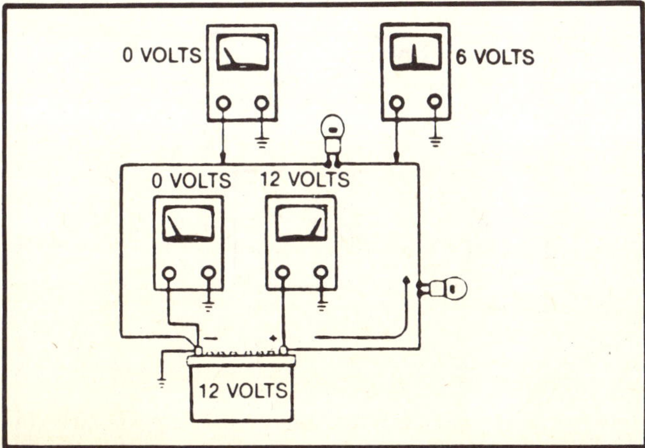

Voltage drop in a series circuit

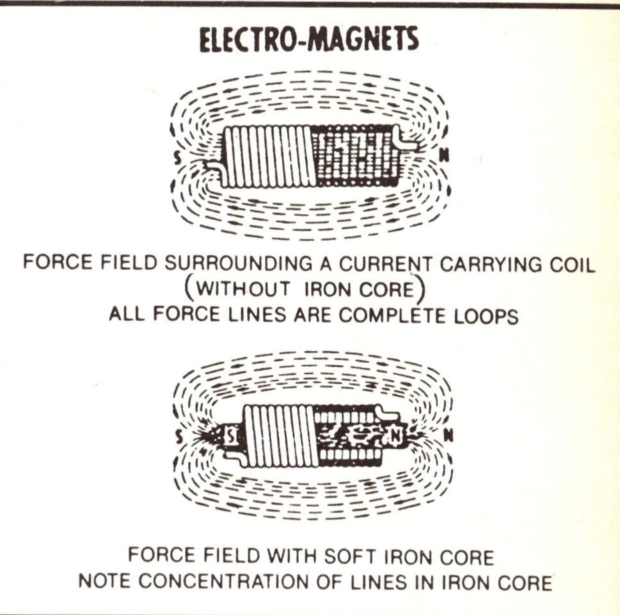

Magnetic field surrounding an electromagnet

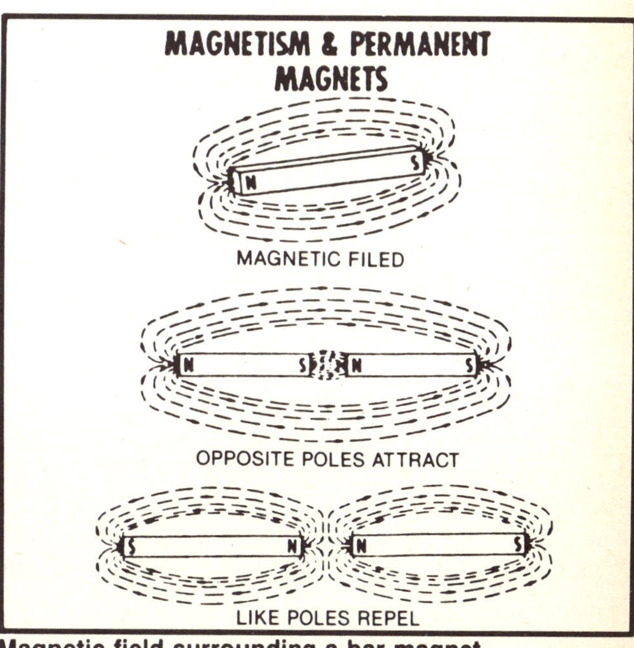

Magnetic field surrounding a bar magnet

is wound into a coil, a magnetic field with North and South poles is created just like in a bar magnet. If an iron core is placed within the coil, the magnetic field becomes stronger because iron conducts magnetic lines much easier than air. This arrangement is called an electromagnet and is the basic principle behind the operation of such components as relays, buzzers and solenoids.

A relay is basically just a remote-controlled switch that uses a small amount of current to control the flow of a large amount of current. The simplest relay contains an electromagnetic coil in series with a voltage source (battery) and a switch. A movable armature made of some magnetic material pivots at one end and is held a small distance away from the electromagnet by a spring or the spring steel of the armature itself. A contact point, made of a good conductor, is attached to the free end of the armature with another contact point a small distance away. When the relay is switched on (energized), the magnetic field created by the current flow attracts the armature, bending it until the contact points meet, closing a circuit and allowing current to flow in the second circuit through the relay to the load the circuit operates. When the relay is switched off (de-energized), the armature springs back and opens the contact points, cutting off the current flow in the secondary, or controlled, circuit. Relays can be designed to be either open or closed when energized, depending on the type of circuit control a manufacturer requires.

A buzzer is similar to a relay, but its internal connections are different. When the switch is closed, the current flows through the normally closed contacts and energizes the coil. When the coil core becomes magnetized, it bends the armature down and breaks the circuit. As soon as the circuit is broken, the spring-loaded armature remakes the circuit and again energizes the coil. This cycle repeats rapidly to cause the buzzing sound.

A solenoid is constructed like a relay, except that its core is allowed to move, providing mechanical motion that can be used to actuate mechanical linkage to operate a door or trunk lock or control any other mechanical function. When the switch is closed, the coil is energized and the movable core is drawn into the coil. When the switch is opened, the coil is de-energized and spring pressure returns the core to its original position.

## Basic Solid State

The term "solid state" refers to devices utilizing transistors, diodes and other components which are made from materials known as semiconductors. A semiconductor is a material that is neither a good insulator nor a good conductor; principally silicon and germanium. The semiconductor material is specially treated to give it certain qualities that enhance its function, therefore becoming either P-type (positive) or N-type (negative) material. Most semiconductors are constructed of silicon and can be designed to function either as an insulator or conductor.

### DIODES

The simplest semiconductor function is that of the diode or rectifier (the 2 terms mean the same thing). A diode will pass current in one direction only, like a one-way valve, because it has low resistance in one direction and high resistance on the other. Whether the diode conducts or not depends on the polarity of the voltage applied to it. A diode has 2 electrodes, an anode and a cathode. When the anode receives positive (+) voltage and the cathode receives negative (−) voltage, current can flow easily through the diode. When the voltage is reversed, the diode becomes non-conducting and only allows a very slight amount of current to flow in the circuit. Because the semiconductor is not a perfect insulator, a small amount of reverse current leakage will occur, but the amount is usually too small to consider. The application of voltage to maintain the current flow described is called "forward bias."

A light-emitting diode (LED) is made of a particular type of crystal that glows when current is passed through it. LED's are used in display faces of many digital or electronic instrument clusters. LED's are usually arranged to display numbers (digital readout), but can be used to illuminate a variety of electronic graphic displays.

Like any other electrical device, diodes have certain ratings that must be observed and should not be exceeded. The forward current rating (or bias) indicates how much current can safely pass through the diode without causing damage or destroying it. Forward current rating is usually given in either amperes or milliamperes. The voltage drop across a diode remains constant regardless of the current flowing through it. Small diodes designed to carry low amounts of current need no special provision for dissipating the heat generated in any electrical device, but large current carrying diodes are usually mounted on heat sinks to keep the internal temperature from rising to the point where the silicon will melt and destroy the diode. When diodes are operated in a high ambient temperature environment, they must be de-rated to prevent failure.

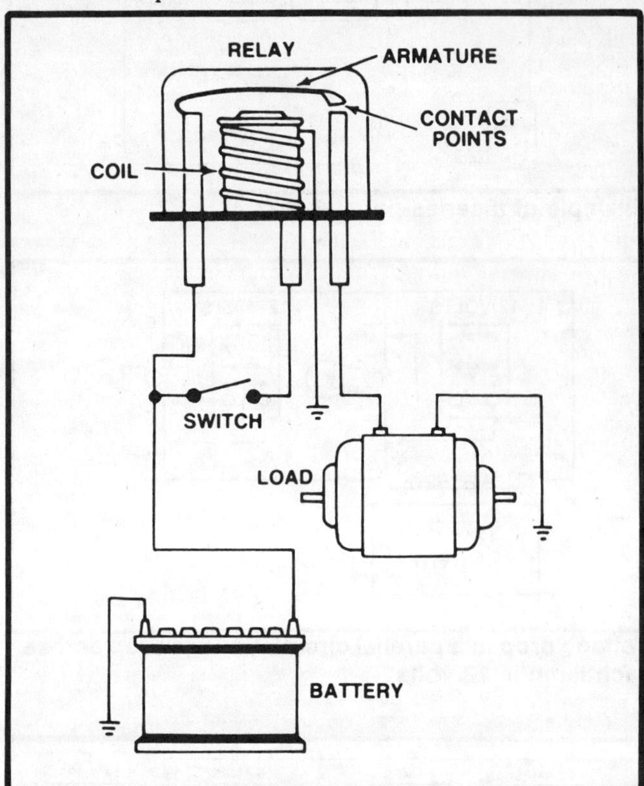

**Typical relay circuit with basic components**

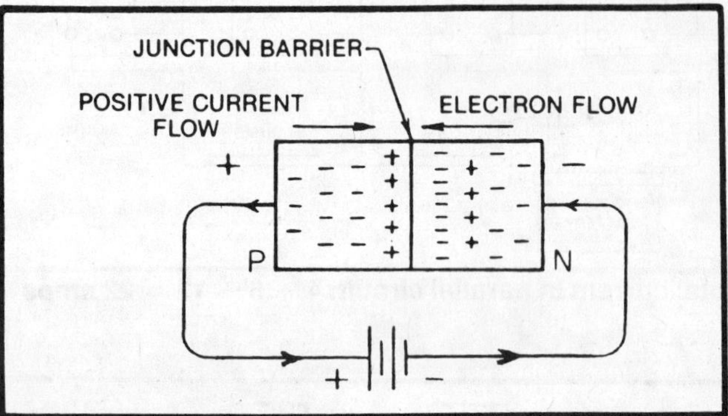

**Diode with forward bias**

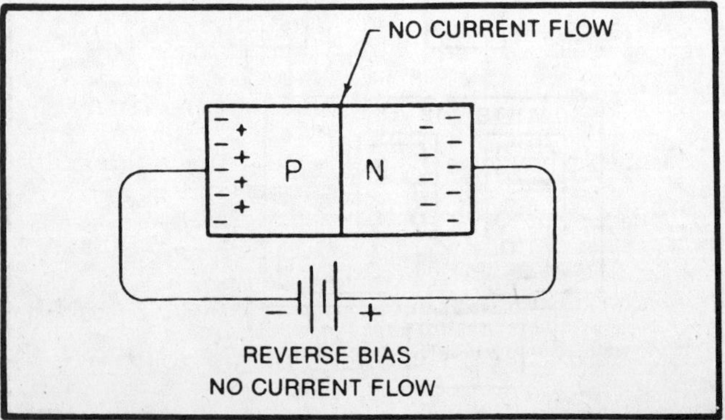

**Diode with reverse bias**

Another diode specification is its peak inverse voltage rating. This value is the maximum amount of voltage the diode can safely handle when operating in the blocking mode. This value can be anywhere from 50–1000 volts, depending on the diode. If voltage amount is exceeded, it will damage the diode just as too much forward current will. Most semiconductor failures are caused by excessive voltage or internal heat.

One can test a diode with a small battery and a lamp with the same voltage rating. With this arrangement one can find a bad diode and determine the polarity of a good one. A diode can fail and cause either a short or open circuit, but in either case it fails to function as a diode. Testing is simply a matter of connecting the test bulb first in one direction and then the other and making sure that current flows in one direction only. If the diode is shorted, the test bulb will remain on no matter how the light is connected.

## TRANSISTORS

The transistor is an electrical device used to control voltage within a circuit. A transistor can be considered a "controllable diode" in that, in addition to passing or blocking current, the transistor can control the amount of current passing through it. Simple transistors are composed of 3 pieces of semiconductor material, P and N type, joined together and enclosed in a container. If 2 sections of P material and 1 section of N material are used, it is known as a PNP transistor; if the reverse is true, then it is known as an NPN transistor. The 2 types cannot be interchanged.

Most modern transistors are made from silicon (earlier transistors were made from germanium) and contain 3 elements; the emitter, the collector and the base. In addition to passing or blocking current, the transistor can control the amount of current passing through it and because of this can function as an amplifier or a switch. The collector and emitter form the main current-carrying circuit of the transistor. The amount of current that flows through the collector-emitter junction is controlled by the amount of current in the base circuit. Only a small amount of base-emitter current is necessary to control a large amount of collector-emitter current (the amplifier effect). In automotive applications, however, the transistor is used primarily as a switch.

When no current flows in the base-emitter junction, the collector-emitter circuit has a high resistance, like to open contacts of a relay. Almost no current flows through the circuit and transistor is considered OFF. By bypassing a small amount of current into the base circuit, the resistance is low, allowing current to flow through the circuit and turning the transistor ON. This condition is known as "saturation" and is reached when the base current reaches the maximum value designed into the transistor that allows current to flow. Depending on various factors, the transistor can turn on and off (go from cutoff to saturation) in less than one millionth of a second.

Much of what was said about ratings for diodes applies to transistors, since they are constructed of the same materials. When transistors are required to handle relatively high currents, such as in voltage regulators or ignition systems, they are generally mounted on heat sinks in the same manner as diodes. They can be damaged or destroyed in the same manner if their voltage ratings are exceeded. A transistor can be checked for proper operation by measuring the resistance with an ohmmeter between the base-emitter terminals and then between the base-collector terminals. The forward resistance should be small, while the reverse resistance should be large. Compare the readings with those from a known good transistor. As a final check, measure the forward and reverse resistance between the collector and emitter terminals.

### INTEGRATED CIRCUITS

The integrated circuit (IC) is an extremely sophisticated solid state device that consists of a silicone wafer (or chip) which has been doped, insulated and etched many times so that it contains an entire electrical circuit with transistors, diodes, conductors and capacitors miniaturized within each tiny chip. Integrated circuits are often referred to as "computers on a chip" and are largely responsible for the current boom in electronic control technology.

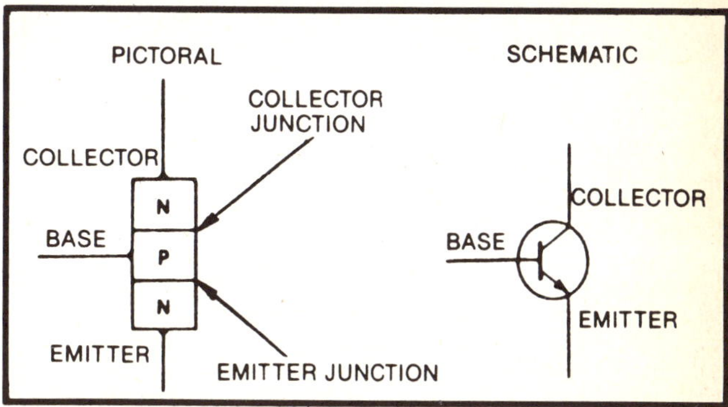

NPN transistor illustrations (pictorial and schematic)

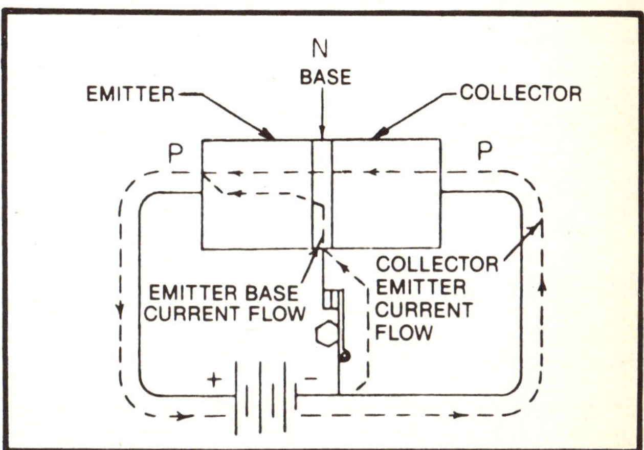

PNP transistor with base switch closed (base emitter and collector emitter current flow)

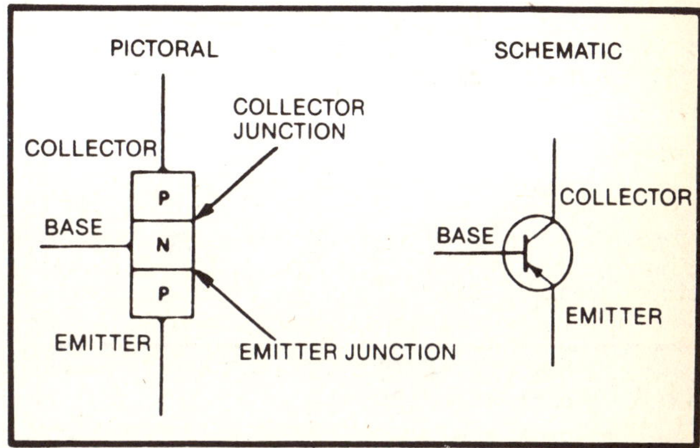

PNP transistor illustrations (pictorial and schematic)

## Microprocessors, Computers and Logic Systems

Mechanical or electromechanical control devices lack the precision necessary to meet the requirements of modern control standards. They do not have the ability to respond to a variety of

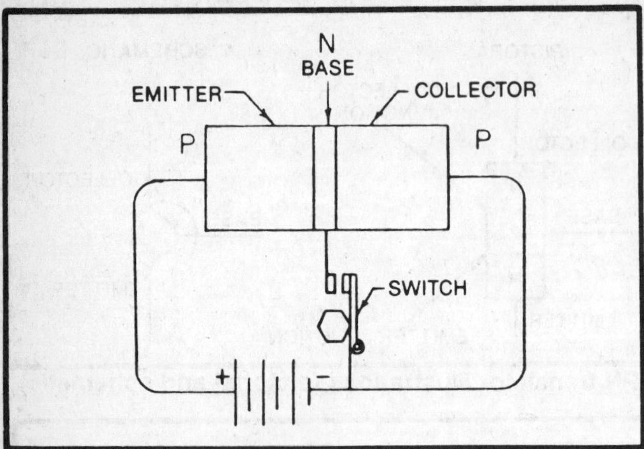


PNP transistor with base switch open (no current flow)

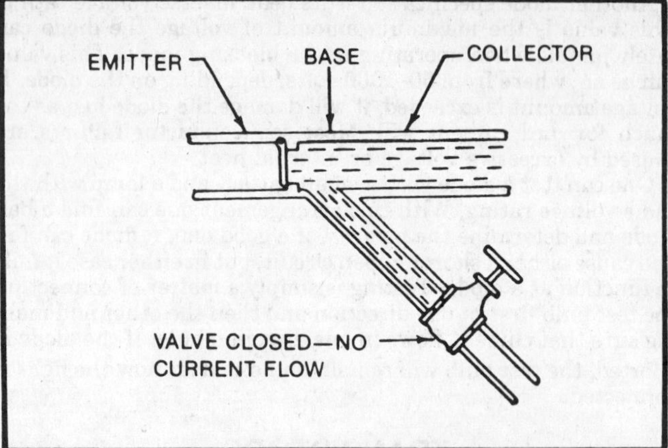


Hydraulic analogy to transistor function is shown with the base circuit shut off

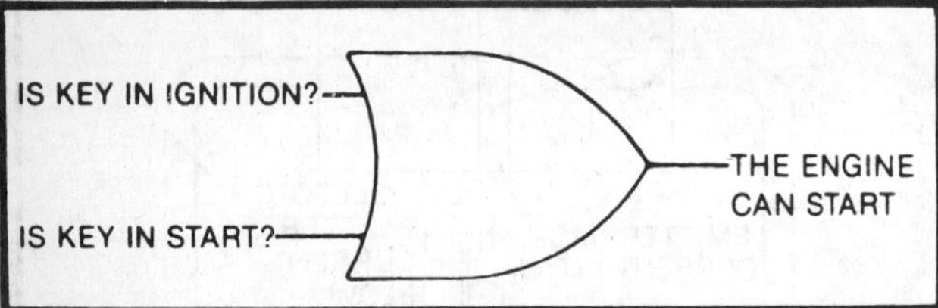


Typical two-input OR circuit operation

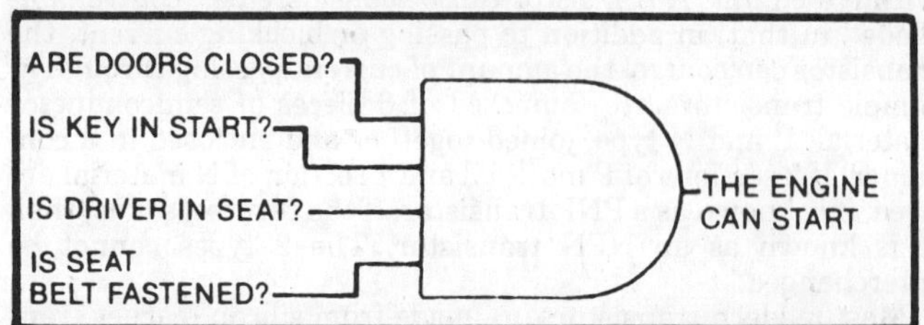


Multiple input AND operation in a typical automotive starting circuit

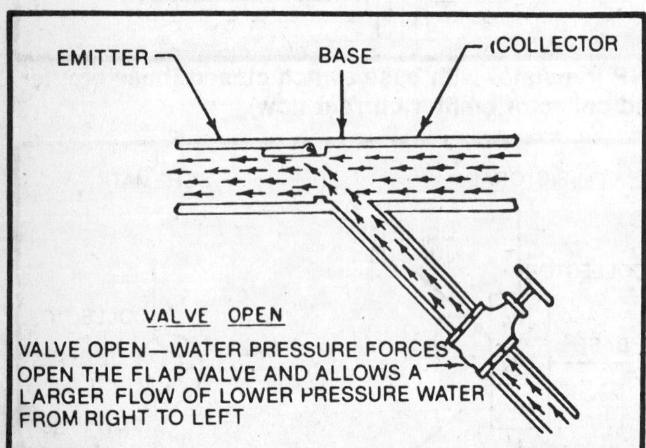


Hydraulic analogy to transistor function is shown with the base circuit energized

input conditions common to antilock brakes, climate control and electronic suspension operation. To meet these requirements, manufacturers have gone to solid state logic systems and microprocessors to control the basic functions of suspension, brake and temperature control, as well as other systems and accessories.

One of the more vital roles of microprocessor-based systems is their ability to perform logic functions and make decisions. Logic designers use a shorthand notation to indicate whether a voltage is present in a circuit (the number 1) or not present (the number 0). Their systems are designed to respond in different ways depending on the output signal (or the lack of it) from various control devices.

There are 3 basic logic functions or "gates" used to construct a microprocessor control system: the AND gate, the OR gate or the NOT gate. Stated simply, the AND gate works when voltage is present in 2 or more circuits which then energize a third (A and B energize C). The OR gate works when voltage is present at either circuit A or circuit B which then energizes circuit C. The NOT function is performed by a solid state device called an "inverter" which reverses the input from a circuit so that, if voltage is going in, no voltage comes out and vice versa. With these three basic building blocks, a logic designer can create complex systems easily. In actual use, a logic or decision making system may employ many logic gates and receive inputs from a number of sources (sensors), but for the most part, all utilize the basic logic gates discussed above.

Stripped to its bare essentials, a computerized decision-making system is made up of three subsystems:

a. Input devices (sensors or switches)
b. Logic circuits (computer control unit)
c. Output devices (actuators or controls)

The input devices are usually nothing more than switches or sensors that provide a voltage signal to the control unit logic circuits that is read as a 1 or 0 (on or off) by the logic circuits. The output devices are anything from a warning light to solenoid-operated valves, motors, linkage, etc. In most cases, the logic circuits themselves lack sufficient output power to operate these devices directly. Instead, they operate some intermediate device such as a relay or power transistor which in turn operates the appropriate device or control. Many problems diagnosed as computer failures are really the result of a malfunctioning intermediate device like a relay. This must be kept in mind whenever troubleshooting any microprocessor-based control system.

The logic systems discussed above are called "hardware" systems, because they consist only of the physical electronic components (gates, resistors, transistors, etc.). Hardware systems do not contain a program and are designed to perform specific or "dedicated" functions which cannot readily be changed. For many simple automotive control requirements, such dedicated logic systems are perfectly adequate. When more complex logic functions are required, or where it may be desirable to alter these functions (e.g. from one model vehicle to another) a true computer system is used. A computer can be programmed through its software to perform many different functions and, if that program is stored on a separate integrated circuit chip called a ROM (Read Only Memory), it can be easily changed simply by plugging in a different ROM with the desired program. Most on-board automotive computers are designed with this capability. The on-board computer method of engine control offers the manufacturer a flexible method of responding to data from a variety of input devices and of controlling an equally large variety of output controls. The computer response can be changed quickly and easily by simply modifying its software program.

The microprocessor is the heart of the microcomputer. It is the thinking part of the computer system through which all the data from the various sensors passes. Within the microprocessor, data is acted upon, compared, manipulated or stored for future use. A microprocessor is not necessarily a microcomputer, but the differences between the 2 are becoming very minor. Originally, a microprocessor was a major part of a microcomputer, but nowadays microprocessors are being called "single-chip microcomputers". They contain all the essential elements to make them behave as a computer, including the most important ingredient–the program.

All computers require a program. In a general purpose computer, the program can be easily changed to allow different tasks to be performed. In a "dedicated" computer, such as most on-board automotive computers, the program isn't quite so easily altered. These automotive computers are designed to perform one or several specific tasks, such as maintaining the passenger compartment temperature at a specific, predetermined level. A program is what makes a computer smart; without a program a computer can do absolutely nothing. The term "software" refers to the computer's program that makes the hardware preform the function needed.

The software program is simply a listing in sequential order of the steps or commands necessary to make a computer perform the desired task. Before the computer can do anything at all, the program must be fed into it by one of several possible methods. A computer can never be "smarter" than the person programming it, but it is a lot faster. Although it cannot perform any calculation or operation that the programmer himself cannot perform, its processing time is measured in millionths of a second.

Because a computer is limited to performing only those operations (instructions) programmed into its memory, the program must be broken down into a large number of very simple steps. Two different programmers can come up with 2 different programs, since there is usually more than one way to perform any task or solve a problem. In any computer, however, there is only so much memory space available, so an overly long or inefficient program may not fit into the memory. In addition to performing arithmetic functions (such as with a trip computer), a computer can also store data, look up data in a table and perform the logic functions previously discussed. A Random Access Memory (RAM) allows the computer to store bits of data temporarily while waiting to be acted upon by the program. It may also be used to store output data that is to be sent to an output device. Whatever data is stored in a RAM is lost when power is removed from the system by turning **OFF** the ignition key, for example.

Computers have another type of memory called a Read Only Memory (ROM) which is permanent. This memory is not lost when the power is removed from the system. Most programs for automotive computers are stored on a ROM memory chip. Data is usually in the form of a look-up table that saves computing time and program steps. For example, a computer designed to control the amount of distributor advance can have this information stored in a table. The information that determines distributor advance (engine rpm, manifold vacuum and temperature) is coded to produce the correct amount of distributor advance over a wide range of engine operating conditions. Instead of the computer computing the required advance, it simply looks it up in a pre-programmed table. However, not all electronic control functions can be handled in this manner; some must be

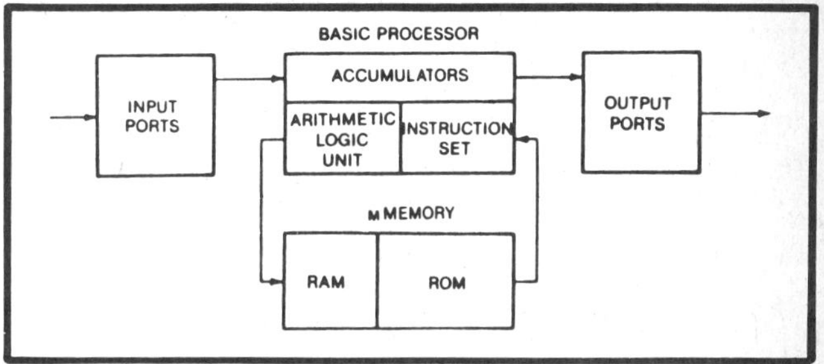


**Schematic of typical microprocessor based on-board computer showing essential components**

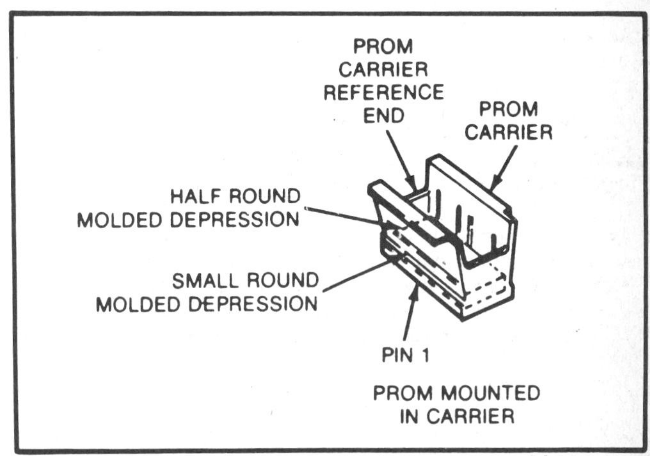


**Typical PROM showing carrier refernce markings**

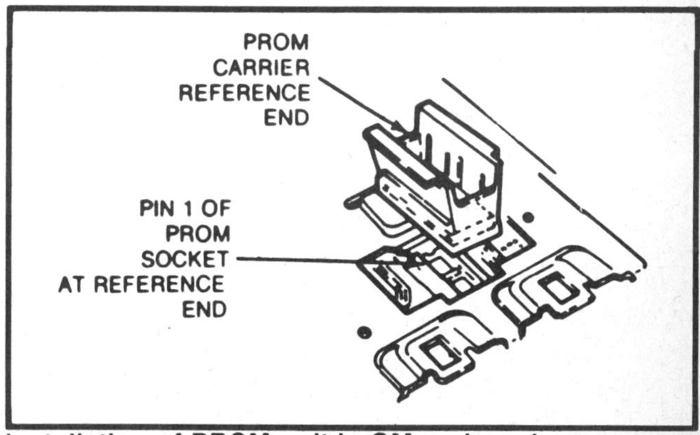


**Installation of PROM unit in GM on-board computer**

computed. On an antilock brake system, for example, the computer must measure the rotation of each separate wheel and then calculate how much brake pressure to apply in order to prevent one wheel from locking up and causing a loss of control.

There are several ways of programming a ROM, but once programmed the ROM cannot be changed. If the ROM is made on the same chip that contains the microprocessor, the whole computer must be altered if a program change is needed. For this reason, a ROM is usually placed on a separate chip. Another type of memory is the Programmable Read Only Memory (PROM) that has the program "burned in" with the appropriate programming machine. Like the ROM, once a PROM has been programmed, it cannot be changed. The advantage of the PROM is that it can be produced in small quantities economically, since it is manufactured with a blank memory. Program changes for various vehicles can be made readily. There is still another type of memory called an EPROM (Erasable PROM) which can be

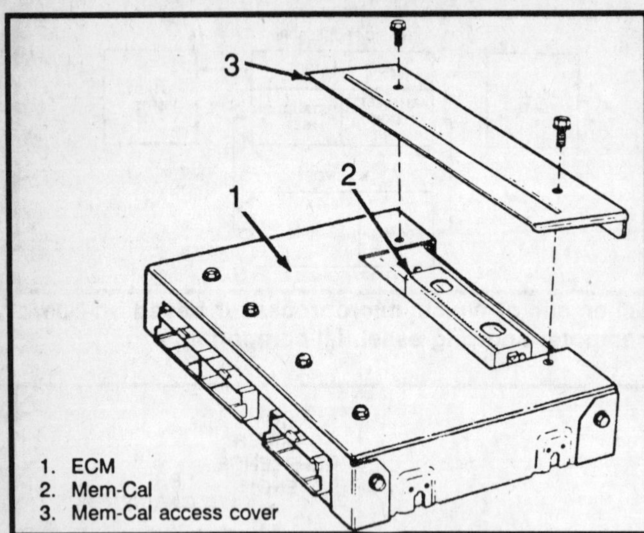


Electronic control module – with Mem-Cal

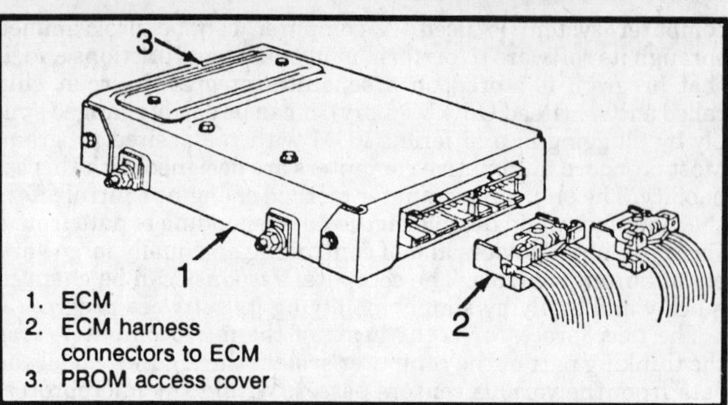


Electronic control module – with PROM and CalPak

erased and programmed many times. EPROM's are used only in research and development work, not on production vehicles.

General Motors refers to the engine controlling computer as an Electronic Control Module (ECM). The ECM contains the PROM necessary for all engine functions, it also contains a device called a CalPak. This allows the fuel delivery function should other parts of the ECM become damaged. It has an access door in the ECM, like the PROM has. There is a third type control module used in some ECMs called a Mem-Cal. The Mem-Cal contains the function of PROM, CalPak and Electronic Spark Control (EST) module. Like the PROM, it contains the calibrations needed for a specific vehicle, as well as the back-up fuel control circuitry required if the rest of the ECM should become damaged and the spark control. An ECM containing a PROM and CalPak can be identified by the 2 connector harnesses, while the ECM containing the Mem-Cal has 3 connector harnesses attached to it.

# Troubleshooting and Diagnosis

**2** SECTION

# TROUBLESHOOTING AND DIAGNOSIS

## Diagnostic Equipment and Special Tools

While we may think that with no moving parts, electronic components should never wear out, in the real world malfunctions do occur. The problem is that any computer-based system is extremely sensitive to electrical voltages and cannot tolerate careless or haphazard testing or service procedures. An inexperienced individual can literally do major damage looking for a minor problem by using the wrong kind of test equipment or connecting test leads or connectors with the ignition switch ON. Therefore, when selecting test equipment, make sure the manufacturers instructions state that the tester is compatible with whatever type of electronic control system is being serviced. Read all instructions carefully and double check all test points before installing probes or making any connections.

The following section outlines basic diagnosis techniques for dealing with computerized engine control systems. Along with a general explanation of the various types of test equipment available to aid in servicing modern electronic automotive systems, basic repair techniques for wiring harnesses and connectors is given. Read the basic information before attempting any repairs or testing on any computerized system, to provide the background of information necessary to avoid the most common and obvious mistakes that can cost both time and money. Likewise, the individual system sections for engine controls, fuel injection and feedback carburetors should be read from the beginning to the end before any repairs or diagnosis is attempted. Although the replacement and testing procedures are simple in themselves, the systems are not, and unless one has a thorough understanding of all components and their function within a particular fuel injection system (for example), the logical test sequence these systems demand cannot be followed. Minor malfunctions can make a big difference, so it is important to know how each component affects the operation of the overall electronic system to find the ultimate cause of a problem without replacing good components unnecessarily. It is not enough to use the correct test equipment; the test equipment must be used correctly.

## Safety Precautions

**CAUTION**

***Whenever working on or around any computer-based microprocessor control system, always observe these general precautions to prevent the possibility of personal injury or damage to electronic components:***

- Never install or remove battery cables with the key ON or the engine running. Jumper cables should be connected with the key OFF to avoid power surges that can damage electronic control units. Engines equipped with computer controlled systems should avoid both giving and getting jump starts due to the possibility of serious damage to components from arcing in the engine compartment when connections are made with the ignition ON.
- Always remove the battery cables before charging the battery. Never use a high-output charger on an installed battery or attempt to use any type of "hot shot" (24 volt) starting aid.
- Exercise care when inserting test probes into connectors to insure good connections without damaging the connector or spreading the pins. Always probe connectors from the rear (wire) side, NOT the pin side, to avoid accidental shorting of terminals during test procedures.
- Never remove or attach wiring harness connectors with the ignition switch ON, especially to an electronic control unit.
- Do not drop any components during service procedures and never apply 12 volts directly to any component (like a solenoid or relay) unless instructed specifically to do so. Some component electrical windings are designed to safely handle only 4 or 5 volts and can be destroyed in seconds if 12 volts are applied directly to the connector.
- Remove the electronic control unit if the vehicle is to be placed in an environment where temperatures exceed approximately 176°F (80°C), such as a paint spray booth or when arc- or gas-welding near the control unit location in the car.

## Organized Troubleshooting

When diagnosing a specific problem, organized troubleshooting is a must. The complexity of a modern automobile demands that you approach any problem in a logical, organized manner. There are certain troubleshooting techniques that are standard:

1. Establish when the problem occurs. Does the problem appear only under certain conditions? Were there any noises, odors, or other unusual symptoms? Make notes on any symptoms found, including warning lights and trouble codes, if applicable.
2. Isolate the problem area. To do this, make some simple tests and observations; then eliminate the systems that are working properly. Check for obvious problems such as broken wires or split or disconnected vacuum hoses. Always check the obvious before assuming something complicated is the cause.
3. Test for problems systematically to determine the cause once the problem area is isolated. Are all the components functioning properly? Is there power going to electrical switches and motors? Is there vacuum at vacuum switches and/or actuators? Is there a mechanical problem such as bent linkage or loose mounting screws? Doing careful, systematic checks will often turn up most causes on the first inspection without wasting time checking components that have little or no relationship to the problem.
4. Test all repairs after the work is done to make sure that the problem is fixed. Some causes can be traced to more than one component, so a careful verification of repair work is important to pick up additional malfunctions that may cause a problem to reappear or a different problem to arise. A blown fuse, for example, is a simple problem that may require more than just replacing a fuse. If you don't look for a problem that caused a fuse to blow, a shorted wire may go undetected.

The diagnostic tree charts are designed to help solve problems by leading the user through closely defined conditions and tests so that only the most likely components, vacuum and electrical circuits are checked for proper operation when troubleshooting a particular malfunction. By using the trouble trees to eliminate those systems and components which normally will not cause the condition described, a problem can be isolated within one or more systems or circuits without wasting time on unnecessary testing. Experience has shown that most problems tend to be the result of a fairly simple and obvious cause, such as loose or corroded connectors or air leaks in the intake system. A careful inspection of components during testing is essential to quick and accurate troubleshooting. Frequent references to special test equipment will be found in the text and in the diagnosis charts. These devices or compatible equivalents are necessary to perform some of the more complicated test procedures listed, but many components can be functionally tested with the quick checks outlined in the "On-Car Service" procedures. Aftermarket testers are available from a variety of sources, as well as from the vehicle manufacturer, but care should be taken that any test equipment being used is designed to diagnose that particular system accurately without damaging the control unit (ECU) or components being tested.

**NOTE: Pinpointing the exact cause of trouble in an electrical system can sometimes only be done using special test equipment. The following describes commonly used test equipment and explains how to put it to best use in diagnosis. In addition to the information covered below, the manufacturer's instructions booklet provided with the tester should be read and clearly understood before attempting any test procedures.**

## Jumper Wires

Jumper wires are simple, yet extremely valuable pieces of test equipment. Jumper wires are merely wires that are used to bypass sections of a circuit. The simplest type of jumper wire is merely a length of multistrand wire with an alligator clip at each end. Jumper wires are usually fabricated from lengths of standard automotive wire and whatever type of connector (alligator clip, spade connector or pin connector) that is required for the particular vehicle being tested. The well-equipped tool box will have several different styles of jumper wires in several different lengths. Some jumper wires are made with three or more terminals coming from a common splice for special-purpose testing. In cramped, hard-to-reach areas it is advisable to have insulated boots over the jumper wire terminals in order to prevent accidental grounding, sparks, and possible fire, especially when testing fuel system components.

Jumper wires are used primarily to locate open electrical circuits, on either the ground (–) side of the circuit or on the hot (+) side. If an electrical component fails to operate, connect the jumper wire between the component and a good ground. If the component operates only with the jumper installed, the ground circuit is open. If the ground circuit is good, but the component does not operate, the circuit between the power feed and component is open. You can sometimes connect the jumper wire directly from the battery to the hot terminal of the component, but first make sure the component uses 12 volts in operation. Some electrical components, such as fuel injectors, are designed to operate on about 4 volts and running 12 volts directly to the injector terminals can burn out the wiring. By inserting an in-line fuseholder between a set of test leads, a fused jumper wire can be used for bypassing open circuits. Use a 5 amp fuse to provide protection against voltage spikes. When in doubt, use a voltmeter to check the voltage input to the component and measure how much voltage is being applied normally. By moving the jumper wire successively back from the lamp toward the power source, you can isolate the area of the circuit where the open is located. When the component stops functioning, or the power is cut off, the open is in the segment of wire between the jumper and the point previously tested.

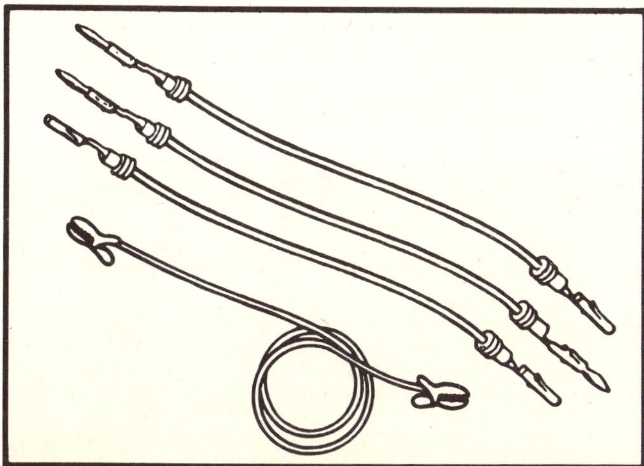

**Typical jumper wires with various terminal ends**

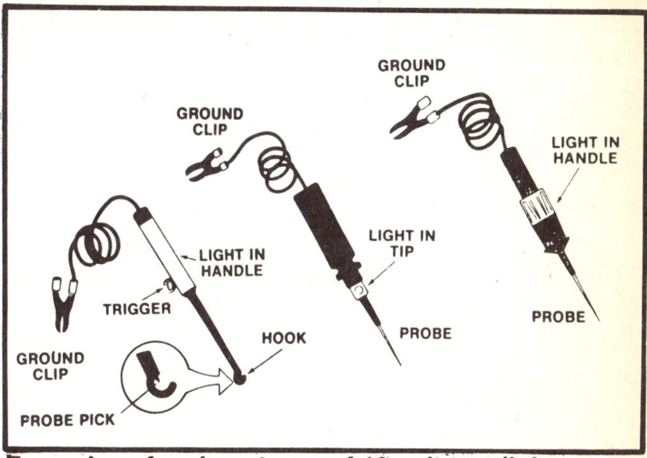


**Examples of various types of 12 volt test lights**

**CAUTION**

*Never use jumpers made from wire that is of lighter gauge than used in the circuit under test. If the jumper wire is of too small gauge, it may overheat and possibly melt. Never use jumpers to bypass high-resistance loads (such as motors) in a circuit. Bypassing resistances, in effect, creates a short circuit which may, in turn, cause damage and fire. Never use a jumper for anything other than temporary bypassing of components in a circuit.*

## 12 Volt Test Light

The 12 volt test light is used to check circuits and components while electrical current is flowing through them. It is used for voltage and ground tests. Twelve volt test lights come in different styles but all have three main parts; a ground clip, a probe, and a light. The most commonly used 12 volt test lights have pick-type probes. To use a 12 volt test light, connect the ground clip to a good ground and probe wherever necessary with the pick. The pick should be sharp so that it can penetrate wire insulation to make contact with the wire, without making a large hole in the insulation. The wrap-around light is handy in hard to reach areas or where it is difficult to support a wire to push a probe pick into it. To use the wrap around light, hook the wire to be probed with the hook and pull the trigger. A small pick will be forced through the wire insulation into the wire core.

**CAUTION**

*Do not use a test light to probe electronic ignition spark plug or coil wires. Never use a pick-type test light to probe wiring on computer controlled systems unless specifically instructed to do so.*

Like the jumper wire, the 12 volt test light is used to isolate opens in circuits. But, whereas the jumper wire is used to bypass the open to operate the load, the 12 volt test light is used to locate the presence of voltage in a circuit. If the test light glows, you know that there is power up to that point; if the 12 volt test light does not glow when its probe is inserted into the wire or connector, you know that there is an open circuit (no power). Move the test light in successive steps back toward the power source until the light in the handle does glow. When it does glow, the open is between the probe and point previously probed.

**NOTE: The test light does not detect that 12 volts (or any particular amount of voltage) is present; it only detects that some voltage is present. It is advisable before using the test light to touch its terminals across the battery posts to make sure the light is operating properly.**

### Self-Powered Test Light

The self-powered test light usually contains a 1.5 volt penlight battery. One type of self-powered test light is similar in design to the 12 volt test light. This type has both the battery and the light in the handle and pick-type probe tip. The second type has the light toward the open tip, so that the light illuminates the contact point. The self-powered test light is dual-purpose piece of test equipment. It can be used to test for either open or short circuits when power is isolated from the circuit (continuity test). A powered test light should not be used on any computer controlled system or component unless specifically instructed to do so. Many engine sensors can be destroyed by even this small amount of voltage applied directly to the terminals.

### Open Circuit Testing

To use the self-powered test light to check for open circuits, first isolate the circuit from the vehicle's 12 volt power source by disconnecting the battery or wiring harness connector. Connect the test light ground clip to a good ground and probe sections of the circuit sequentially with the test light. (start from either end of the circuit). If the light is out, the open is between the probe and the circuit ground. If the light is on, the open is between the probe and end of the circuit toward the power source.

### Short Circuit Testing

By isolating the circuit both from power and from ground, and using a self-powered test light, you can check for shorts to ground in the circuit. Isolate the circuit from power and ground. Connect the test light ground clip to a good ground and probe any easy-to-reach test point in the circuit. If the light comes on, there is a short somewhere in the circuit. To isolate the short, probe a test point at either end of the isolated circuit (the light should be on). Leave the test light probe connected and open connectors, switches, remove parts, etc., sequentially, until the light goes out. When the light goes out, the short is between the last circuit component opened and the previous circuit opened.

**NOTE: The 1.5 volt battery in the test light does not provide much current. A weak battery may not provide enough power to illuminate the test light even when a complete circuit is made (especially if there are high resistances in the circuit). Always make sure that the test battery is strong. To check the battery, briefly touch the ground clip to the probe; if the light glows brightly the battery is strong enough for testing. Never use a self-powered test light to perform checks for opens or shorts when power is applied to the electrical system under test. The 12-volt vehicle power will quickly burn out the 1.5 volt light bulb in the test light.**

## Voltmeter

A voltmeter is used to measure voltage at any point in a circuit, or to measure the voltage drop across any part of a circuit. It can also be used to check continuity in a wire or circuit by indicating current flow from one end to the other. Voltmeters usually have various scales on the meter dial and a selector switch to allow the selection of different voltages. The voltmeter has a positive and a negative lead. To avoid damage to the meter, always connect the negative lead to the negative (–) side of circuit (to ground or nearest the ground side of the circuit) and connect the positive lead to the positive (+) side of the circuit (to the power source or the nearest power source). Note that the negative voltmeter lead will always be black and that the positive voltmeter will always be some color other than black (usually red). Depending on how the voltmeter is connected into the circuit, it has several uses.

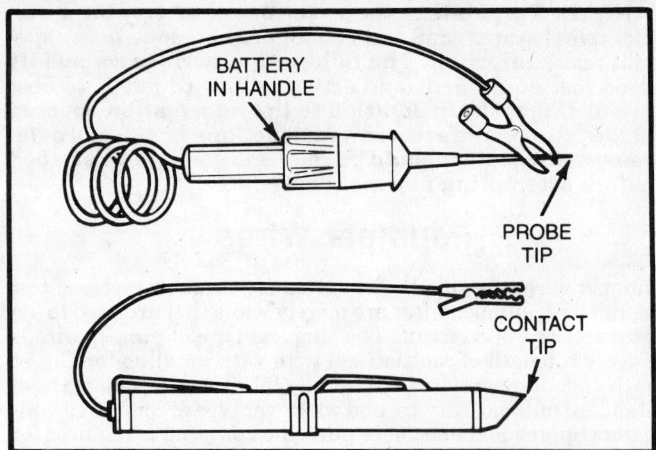

**Two types of self-powered test lights**

A voltmeter can be connected either in parallel or in series with a circuit and it has a very high resistance to current flow. When connected in parallel, only a small amount of current will flow through the voltmeter current path; the rest will flow through the normal circuit current path and the circuit will work normally. When the voltmeter is connected in series with a circuit, only a small amount of current can flow through the circuit. The circuit will not work properly, but the voltmeter reading will show if the circuit is complete or not.

### Available Voltage Measurement

Set the voltmeter selector switch to the 20V position and connect the meter negative lead to the negative post of the battery. Connect the positive meter lead to the positive post of the battery and turn the ignition switch ON to provide a load. Read the voltage on the meter or digital display. A well-charged battery should register over 12 volts. If the meter reads below 11.5 volts, the battery power may be insufficient to operate the electrical system properly. This test determines voltage available from the battery and should be the first step in any electrical trouble diagnosis procedure. Many electrical problems, especially on computer controlled systems, can be caused by a low state of charge in the battery. Excessive corrosion at the battery cable terminals can cause a poor contact that will prevent proper charging and full battery current flow.

Normal battery voltage is 12 volts when fully charged. When the battery is supplying current to one or more circuits it is said to be "under load". When everything is off the electrical system is under a "no-load" condition. A fully charged battery

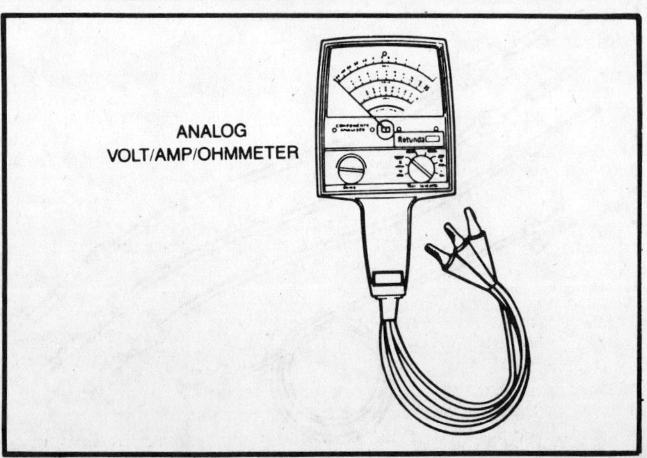

**Typical analog-type voltmeter**

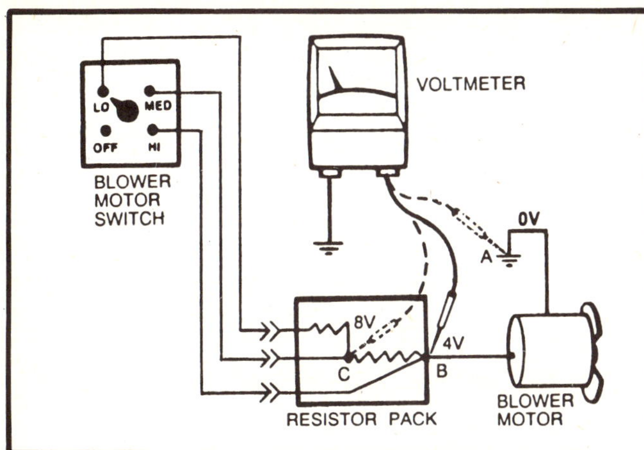

**Measuring available voltage in a blower circuit**

may show about 12.5 volts at no load; will drop to 12 volts under medium load; and will drop even lower under heavy load. If the battery is partially discharged the voltage decrease under heavy load may be excessive, even though the battery shows 12 volts or more at no load. When allowed to discharge further, the battery's available voltage under load will decrease more severely. For this reason, it is important that the battery be fully charged during all testing procedures to avoid errors in diagnosis and incorrect test results.

## VOLTAGE DROP

When current flows through a resistance, the voltage beyond the resistance is reduced (the larger the current, the greater the reduction in voltage). When no current is flowing, there is no voltage drop because there is no current flow. All points in the circuit which are connected to the power source are at the same voltage as the power source. The total voltage drop always equals the total source voltage. In a long circuit with many connectors, a series of small, unwanted voltage drops due to corrosion at the connectors can add up to a total loss of voltage which impairs the operation of the normal loads in the circuit.

### Indirect Computation of Voltage Drops

1. Set the voltmeter selector switch to the 20 volt position.
2. Connect the meter negative lead to a good ground.
3. Probe all resistances in the circuit with the positive meter lead.
4. Operate the circuit in all modes and observe the voltage readings.

### Direct Measurement of Voltage Drops

1. Set the voltmeter switch to the 20 volt position.
2. Connect the voltmeter negative lead to the ground side of the resistance load to be measured.
3. Connect the positive lead to the positive side of the resistance or load to be measured.
4. Read the voltage drop directly on the 20 volt scale.

Too high a voltage indicates too high a resistance. If, for example, a blower motor runs too slowly, you can determine if there is too high a resistance in the resistor pack. By taking voltage drop readings in all parts of the circuit, you can isolate the problem. Too low a voltage drop indicates too low a resistance. If, for example, a blower motor runs too fast in the MED and/or LOW position, the problem can be isolated in the resistor pack by taking voltage drop readings in all parts of the circuit to locate a possibly shorted resistor. The maximum allowable voltage drop under load is critical, especially if there is more than one high resistance problem in a circuit because all voltage drops are cumulative. A small drop is normal due to the resistance of the conductors.

### High Resistance Testing

1. Set the voltmeter selector switch to the 4 volt position.
2. Connect the voltmeter positive lead to the positive post of the battery.
3. Turn on the headlights and heater blower to provide a load.
4. Probe various points in the circuit with the negative voltmeter lead.
5. Read the voltage drop on the 4 volt scale. Some average maximum allowable voltage drops are:

FUSE PANEL – 7 volts
IGNITION SWITCH – 5volts
HEADLIGHT SWITCH – 7 volts
IGNITION COIL (+) – 5 volts
ANY OTHER LOAD – 1.3 volts

**NOTE: Voltage drops are all measured while a load is operating; without current flow, there will be no voltage drop.**

## Ohmmeter

The ohmmeter is designed to read resistance (ohms) in a circuit or component. Although there are several different styles of ohmmeters, all will usually have a selector switch which permits the measurement of different ranges of resistance (usually the selector switch allows the multiplication of the meter reading by 10, 100, 1000, and 10,000). A calibration knob al-

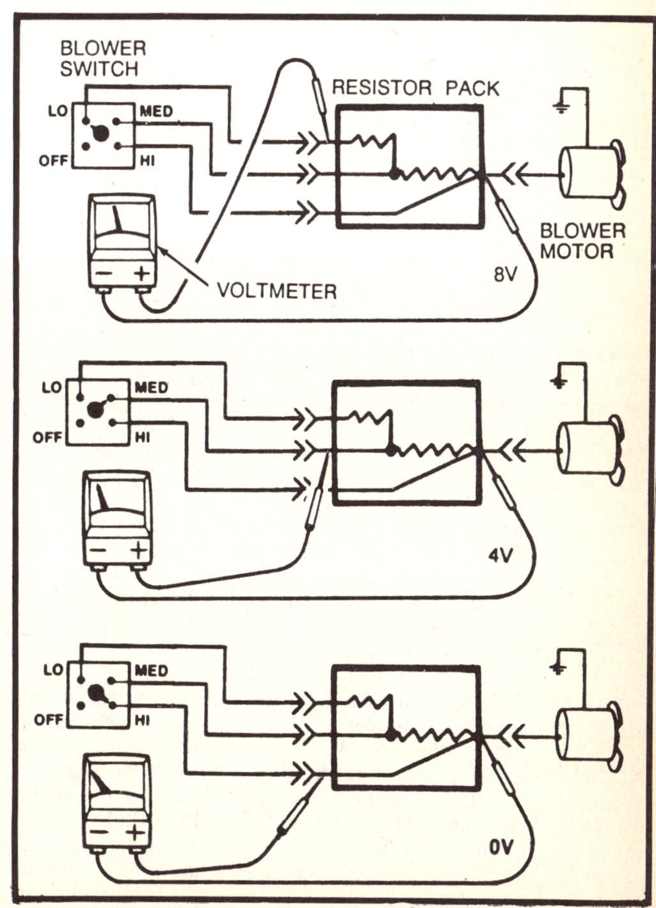

**Direct measurement of voltage drops in a circuit**

lows the meter to be set at zero for accurate measurement. Since all ohmmeters are powered by an internal battery (usually 9 volts), the ohmmeter can be used as a self-powered test light. When the ohmmeter is connected, current from the ohmmeter flows through the circuit or component being tested. Since the ohmmeter's internal resistance and voltage are known values, the amount of current flow through the meter depends on the resistance of the circuit or component being tested.

The ohmmeter can be used to perform continuity test for opens or shorts (either by observation of the meter needle or as a self-powered test light), and to read actual resistance in a circuit. It should be noted that the ohmmeter is used to check the resistance of a component or wire while there is no voltage applied to the circuit. Current flow from an outside voltage source (such as the vehicle battery) can damage the ohmmeter, so the circuit or component should be isolated from the vehicle electrical system before any testing is done. Since the ohmmeter uses its own voltage source, either lead can be connected to any test point.

**NOTE: When checking diodes or other solid state components, the ohmmeter leads can only be connected one way in order to measure current flow in a single direction. Make sure the positive (+) and negative (-) terminal connections are as described in the test procedures to verify the one-way diode operation.**

In using the meter for making continuity checks, do not be concerned with the actual resistance readings. Zero resistance, or any resistance readings, indicate continuity in the circuit. Infinite resistance indicates an open in the circuit. A high resistance reading where there should be none indicates a problem in the circuit. Checks for short circuits are made in the same manner as checks for open circuits except that the circuit must be isolated from both power and normal ground. Infinite resistance indicates no continuity to ground, while zero resistance indicates a dead short to ground.

### Resistance Measurement

The batteries in an ohmmeter will weaken with age and temperature, so the ohmmeter must be calibrated or "zeroed" before taking measurements. To zero the meter, place the selector switch in its lowest range and touch the two ohmmeter leads together. Turn the calibration knob until the meter needle is exactly on zero.

**NOTE: All analog (needle) type ohmmeters must be zeroed before use, but some digital ohmmeter models are automatically calibrated when the switch is turned on. Self-calibrating digital ohmmeters do not have an adjusting knob, but it's a good idea to check for a zero readout before use by touching the leads together. All computer controlled systems require the use of a digital ohmmeter with at least 10 megohms impedance for testing. Before any test procedures are attempted, make sure the ohmmeter used is compatible with the electrical system, or damage to the on-board computer could result.**

To measure resistance, first isolate the circuit from the vehicle power source by disconnecting the battery cables or the harness connector. Make sure the key is OFF when disconnecting any components or the battery. Where necessary, also isolate at least one side of the circuit to be checked to avoid reading parallel resistances. Parallel circuit resistances will always give a lower reading than the actual resistance of either of the branches. When measuring the resistance of parallel circuits, the total resistance will always be lower than the smallest resistance in the circuit. Connect the meter leads to both sides of the circuit (wire or component) and read the actual measured ohms on the meter scale. Make sure the selector switch is set to

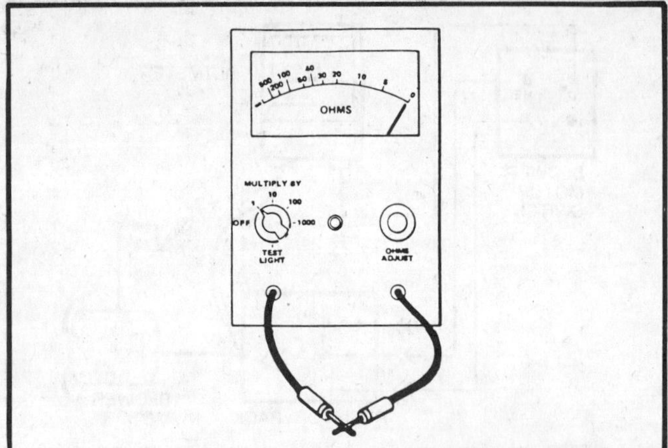


**Analog ohmmeters must be calibrated before use by touching the probes together and adjusting the knob**

the proper ohm scale for the circuit being tested to avoid misreading the ohmmeter test value.

**CAUTION**

*Never use an ohmmeter with power applied to the circuit. Like the self-powered test light, the ohmmeter is designed to operate on its own power supply. The normal 12 volt automotive electrical system current could damage the meter.*

## Ammeters

An ammeter measures the amount of current flowing through a circuit in units called amperes or amps. Amperes are units of electron flow which indicate how fast the electrons are flowing through the circuit. Since Ohm's Law dictates that current flow in a circuit is equal to the circuit voltage divided by the total circuit resistance, increasing voltage also increases the current level (amps). Likewise, any decrease in resistance will increase the amount of amps in a circuit. At normal operating voltage, most circuits have a characteristic amount of amperes, called "current draw" which can be measured using an ammeter. By referring to a specified current draw rating, measuring the amperes, and comparing the two values, one can determine what is happening within the circuit to aid in diagnosis. An open circuit, for example, will not allow any current to flow so the ammeter reading will be zero. More current flows through a heavily loaded circuit or when the charging system is operating.

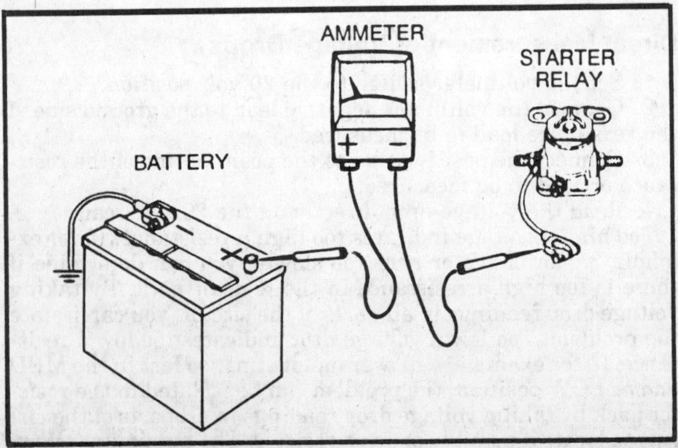


**Battery current drain test**

An ammeter is always connected in series with the circuit being tested. All of the current that normally flows through the circuit must also flow through the ammeter; if there is any other path for the current to follow, the ammeter reading will not be accurate. The ammeter itself has very little resistance to current flow and therefore will not affect the circuit, but it will measure current draw only when the circuit is closed and electricity is flowing. Excessive current draw can blow fuses and drain the battery, while a reduced current draw can cause motors to run slowly, lights to dim and other components not to operate properly. The ammeter can help diagnose these conditions by locating the cause of the high or low reading.

## Multimeters

Different combinations of test meters can be built into a single unit designed for specific tests. Some of the more common combination test devices are known as Volt-Amp testers, Tach-Dwell meters, or Digital Multimeters. The Volt-Amp tester is used for charging system, starting system or battery tests and consists of a voltmeter, an ammeter and a variable resistance carbon pile. The voltmeter will usually have at least two ranges for use with 6, 12 and 24 volt systems. The ammeter also has more than one range for testing various levels of battery loads and starter current draw and the carbon pile can be adjusted to offer different amounts of resistance. The Volt-Amp tester has heavy leads to carry large amounts of current and many later models have an inductive ammeter pickup that clamps around the wire to simplify test connections. On some models, the ammeter also has a zero-center scale to allow testing of charging and starting systems without switching leads or polarity. A digital multimeter is a voltmeter, ammeter and ohmmeter combined in an instrument which gives a digital readout. These are often used when testing solid state circuits because of their high input impedence (usually 10 megohms or more).

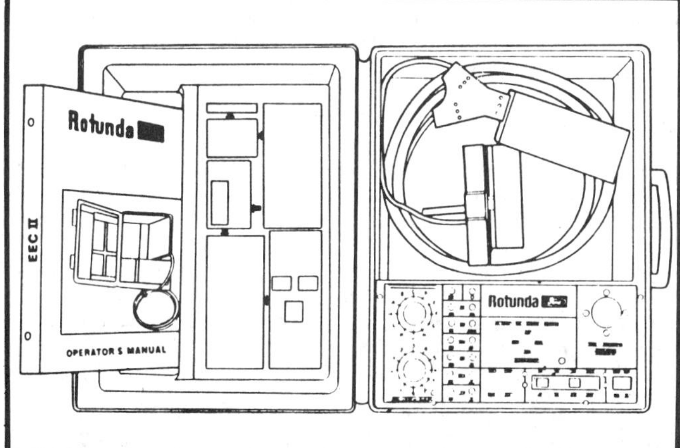

Typical electronic engine control tester

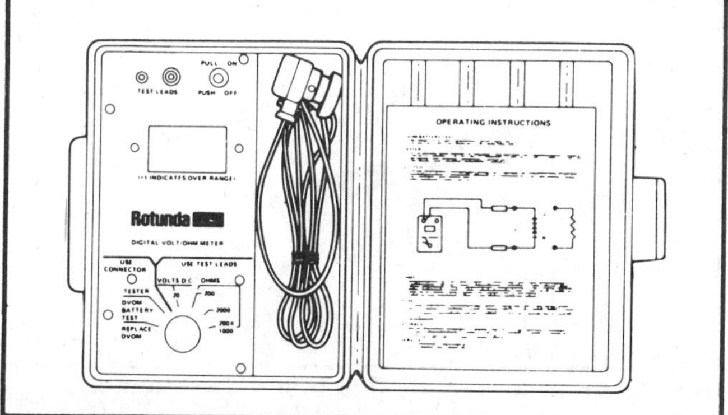

Digital volt-ohmmeter

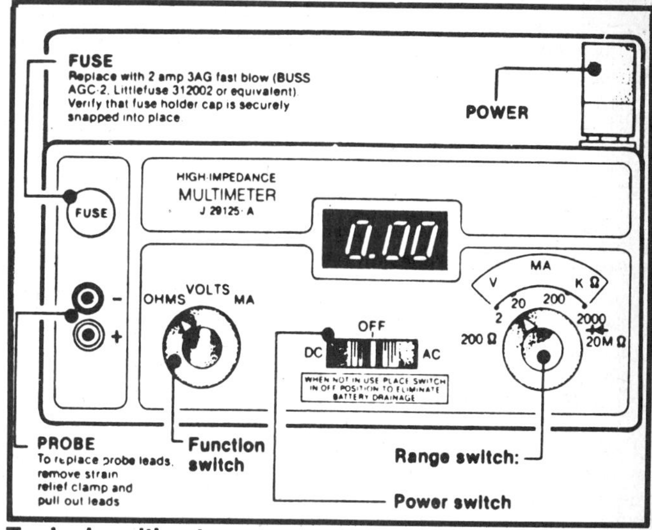

Typical multimeter

UF9

The tach-dwell meter combines a tachometer and a dwell (cam angle) meter and is a specialized kind of voltmeter. The tachometer scale is marked to show engine speed in rpm and the dwell scale is marked to show degrees of distributor shaft rotation. In most electronic ignition systems, dwell is determined by the control unit, but the dwell meter can also be used to check the duty cycle (operation) of some electronic engine control systems. Some tach-dwell meters are powered by an internal battery, while others take their power from the car battery in use. The battery powered testers usually require calibration much like an ohmmeter before testing.

## Special Test Equipment

A variety of diagnostic tools are available to help troubleshoot and repair computerized engine control systems. The most sophisticated of these devices are the console-type engine analyzers that usually occupy a garage service bay, but there are several types of aftermarket electronic testers available that will allow quick circuit tests of the engine control system by plugging directly into a special connector located in the engine compartment or under the dashboard. Several tool and equipment manufacturers offer simple, hand-held testers that measure various circuit voltage levels on command to check all system components for proper operation. Although these testers usually cost about $300–500, consider that the average computer control unit (or ECM) can cost just as much and the money saved by not replacing perfectly good sensors or components in an attempt to correct a problem could justify the purchase price of a special diagnostic tester the first time it's used.

These computerized testers can allow quick and easy test measurements while the engine is operating or while the car is being driven. In addition, the on-board computer memory can be read to access any stored trouble codes; in effect allowing the computer to tell you where it hurts and aid trouble diagnosis by pinpointing exactly which circuit or component is malfunctioning. In the same manner, repairs can be tested to make sure the problem has been corrected. The biggest advantage these special testers have is their relatively easy hookups that minimize or eliminate the chances of making the wrong connections and getting false voltage readings or damaging the computer accidentally.

**NOTE: It should be remembered that these testers check voltage levels in circuits; they don't detect mechanical problems or failed components if the circuit voltage falls within the preprogrammed limits stored in the tester PROM unit. Also, most of the hand-held testers are designed to work only on one or two systems made by a specific manufacturer.**

A variety of aftermarket testers are available to help diagnose different computerized control systems. Owatonna Tool Company (OTC), for example, markets a device called the OTC Monitor which plugs directly into the assembly line diagnostic link (ALDL). The OTC tester makes diagnosis a simple matter of pressing the correct buttons and, by changing the internal PROM or inserting a different diagnosis cartridge, it will work on any model from full size to subcompact, over a wide range of years. An adapter is supplied with the tester to allow connection to all types of ALDL links, regardless of the number of pin terminals used. By inserting an updated PROM into the OTC tester, it can be easily updated to diagnose any new modifications of computerized control systems.

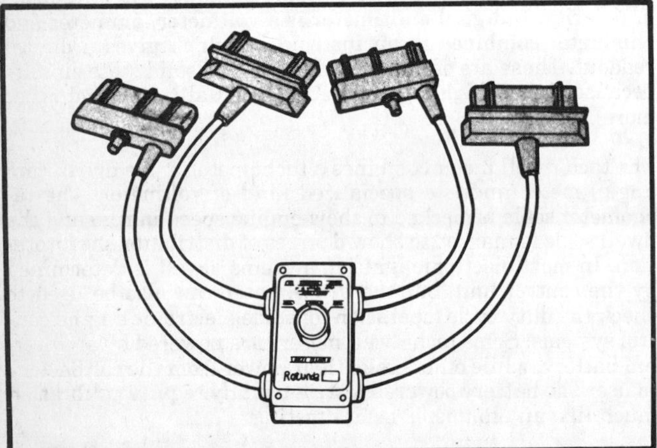


**Typical adapter wiring harness for connecting tester to diagnostic terminal**

## Wiring Diagrams

The average automobile contains about ½ mile of wiring, with hundreds of individual connections. To protect the many wires from damage and to keep them from becoming a confusing tangle, they are organized into bundles, enclosed in plastic or taped together and called wire harnesses. Different wiring harnesses serve different parts of the vehicle. Individual wires are color-coded to help trace them through a harness where sections are hidden from view.

A loose or corroded connection or a replacement wire that is too small for the circuit will add extra resistance and an additional voltage drop to the circuit. A ten percent voltage drop can result in slow or erratic motor operation, for example, even though the circuit is complete.

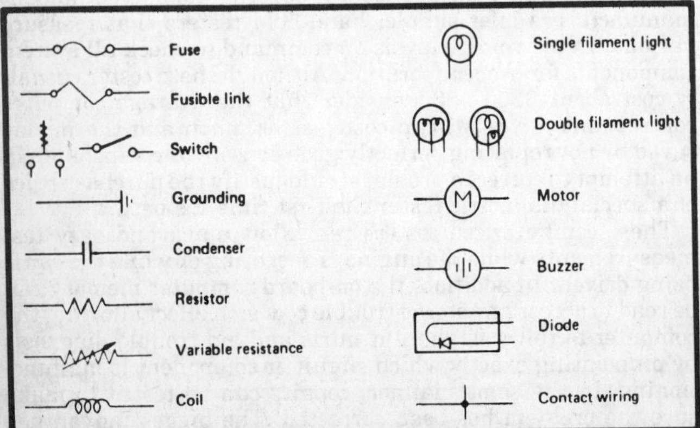


**Typical electrical symbols found on wiring diagrams**

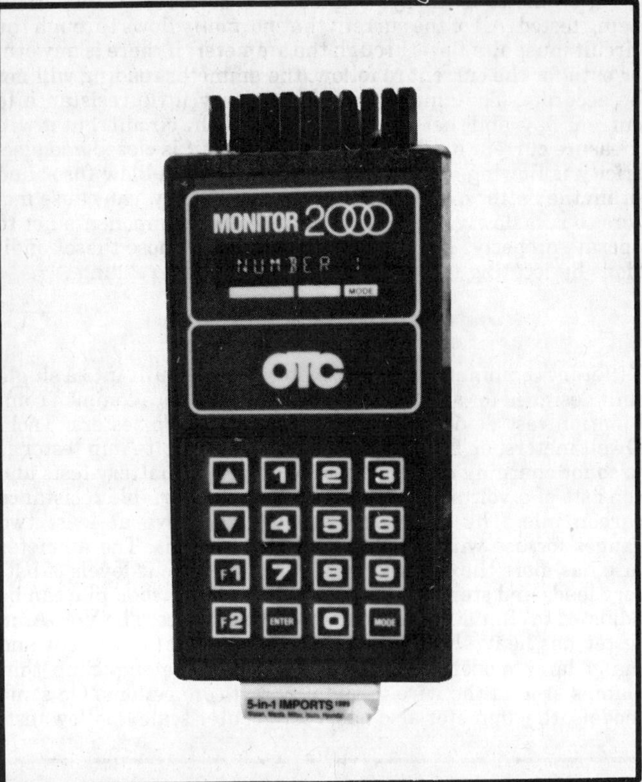


**Hand-held aftermarket tester used to diagnosis electronic engine control systems**

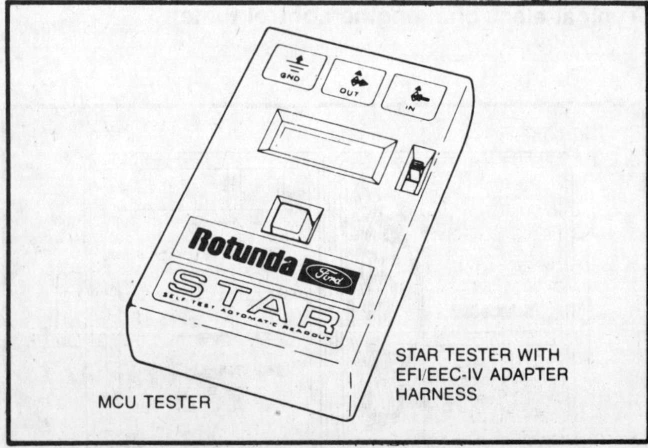


**Self-Test and Automatic Readout (STAR) tester**

# Self-Diagnostic Systems

# 3 SECTION

## INDEX

# ACURA PROGRAMMED FUEL INJECTION (PGM-FI)

## General Information

Programmed Fuel Injection (PGM-FI) System consists of 3 subsystems: air intake, electronic control and fuel supply. In order to get fuel into the cylinders at the correct instant and in correct amount, the control system must perform various separate functions. The Electronic Control Unit (ECU), the heart of the PGM-FI, uses an 8-bit microcomputer and consists of a CPU (Central Processing Unit), memories, and I/O (Input/Output) ports. Basic data stored in the memories are compensated by the signals sent from the various sensors to provide the correct air/fuel mixture for all engine needs.

### ELECTRONIC CONTROL UNIT (ECU)

The unit contains memories for the basic discharge duration at various engine speeds and manifold pressures. The basic discharge duration, after being read out from the memory, is further modified by signals sent from various sensors to obtain the final discharge duration. Other functions also include:

- Starting Control – The fuel system must vary the air/fuel ratio to suit different operating requirements. For example, the mixture must be rich for starting. The memories also contain the basic discharge durations to be read out by signals from the starter switch, and engine speed and coolant temperature sensors, thereby providing extra fuel needed for starting.
- Fuel Pump Control – When the speed of the engine falls below the prescribed limit, electric current to the fuel pump is cut off, preventing the injectors from discharging fuel.
- Safety – A fail-safe system monitors the sensors and detects any abnormality in the ECU, ensuring safe driving if a single or many sensors are faulty, or if the ECU malfunctions.

## Self-Diagnostic System

### ENTERING SELF-DIAGNOSTIC AND DIAGNOSTIC CODE DISPLAY

#### Self-Diagnosis Indicators

The quick reference chart covers the most common failure modes for the PGM-FI. The probable causes are listed in order of most-easily-checked first, then progressing to more difficult fixes. Run through all the causes listed. If problem is still unsolved, go on to the more detailed troubleshooting. Troubleshooting is divided into different LED displays. Find the correct light display and begin again.

For all the conditions listed, the PGM-FI warning light on the dashboard must be on (comes on and stays on). This indicates a problem in the electrical portion of the fuel injection system. At that time, check the LED display (self-diagnosis system) in the ECU.

On 1988–89 Acura Integra, there is a single LED display. The LED will blink consecutively to indicate the trouble code. The ECU is located under the passenger's seat.

On 1990 Acura Integra, there is a single LED display. The LED will blink consecutively to indicate the trouble code. The ECU is located behind the carpet, on the passenger's side under the dash.

On 1988–90 Acura Legend models, there are 2 LED displays. The yellow LED display is for idle speed adjustment check. The red LED display will blink consecutively to indicate the trouble code. The ECU is located under the passenger's seat on the Legend Sedan. On the Legend Coupe, the ECU is located under the floor kick panel in front of the passenger's seat.

Sometimes the dash warning light and/or ECU LED will come ON, indicating a system problem, when, in fact, there is only a bad or intermittent electrical connection. To troubleshoot a bad connection, note the ECU LED pattern that is lit, refer to the diagnosis chart and check the connectors associated with the items for that LED pattern (disconnect, clean or repair if necessary and reconnect those connections). Then, reset the ECU memory as described, restart the car and drive it for a few minutes and then recheck the car and drive it for a few minutes and then recheck the LED(s). If the same pattern lights up, begin system troubleshooting; if it does not light up, the problem was only a bad connection.

On the 1990 Legend, if the CHECK ENGINE and the S warning light simultaneously when the LED blinks Code 6, 7 or 17. Check the PGM-FI system according to troubleshooting charts and check the S warning light and circuit.

**NOTE: The CHECK ENGINE light does not come on when there is a malfunction in the A.T. FI signal. However, the ECU LED will indicate codes.**

### CLEARING TROUBLE CODES

The memory for the PGM-FI warning light on the dashboard will be erased when the ignition switch is turned **OFF**; however, the memory for the LED display will not be canceled. Thus, the warning light will not come ON when the ignition switch is again turned **ON** unless the trouble is once more detected. Troubleshooting should be done according to the LED display even if the warning light is OFF.

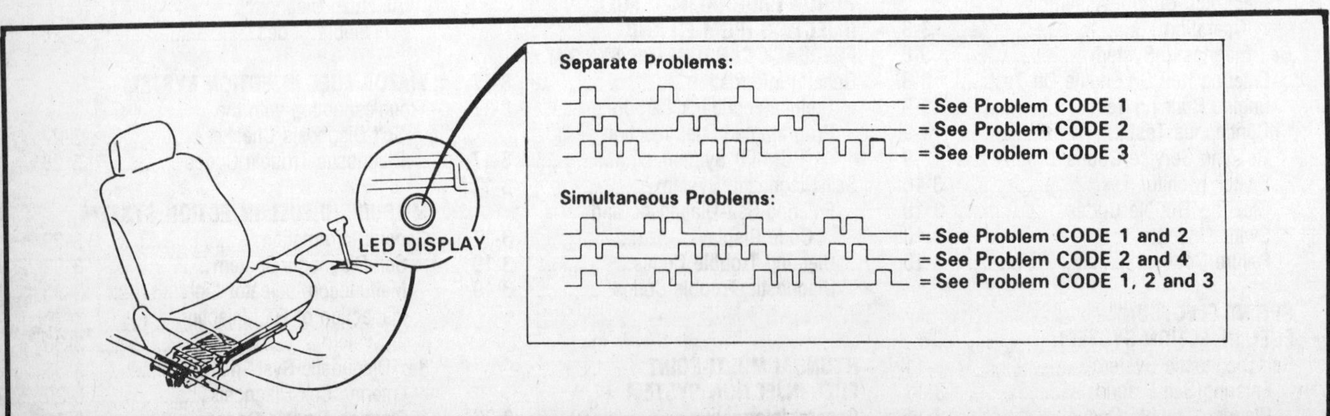


ECU location and diagnostic LED display – 1988–89 Integra

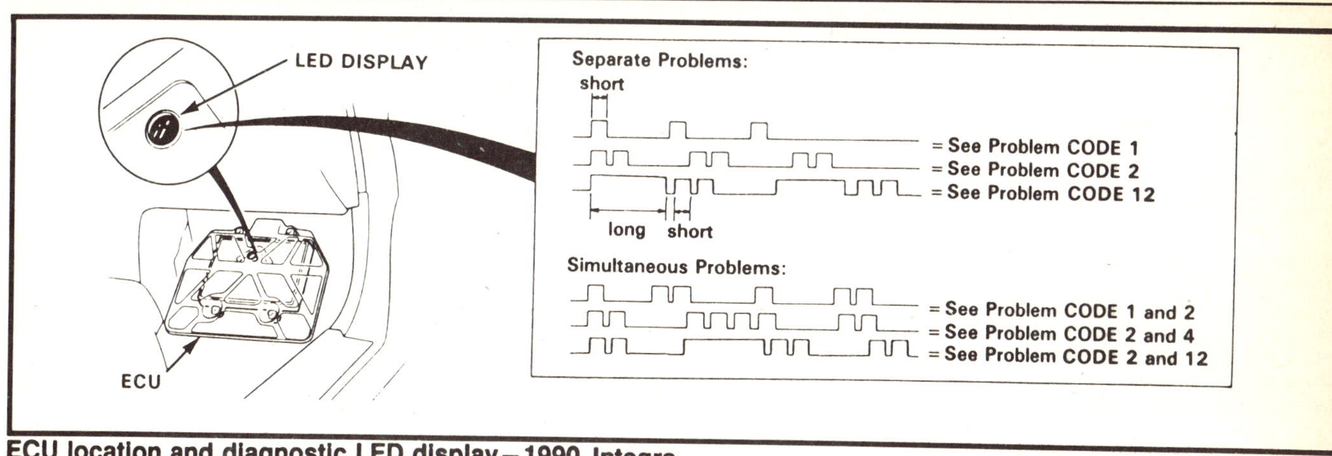


**ECU location and diagnostic LED display—1990 Integra**

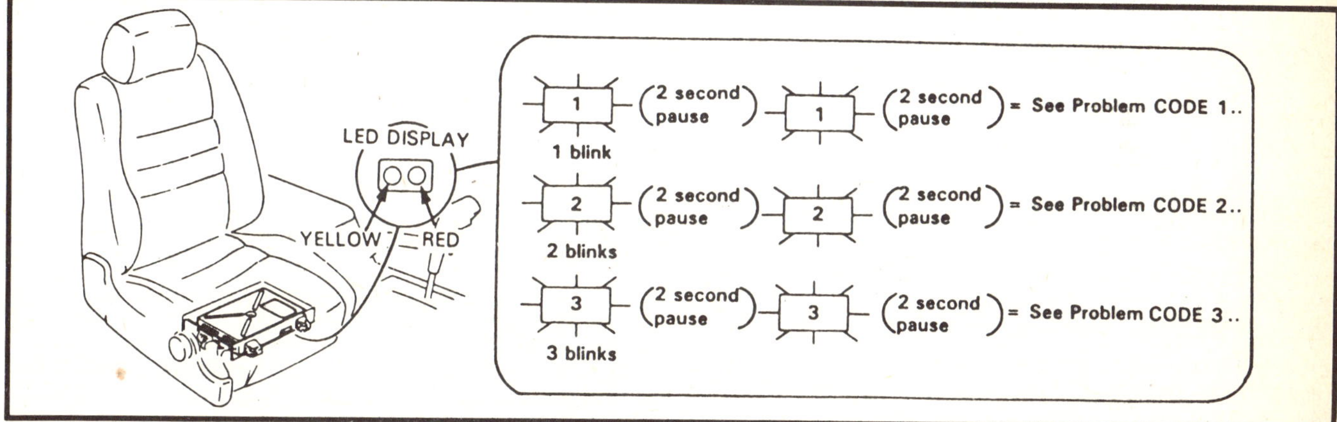


**ECU location and diagnostic LED display—1989–90 Legend**

To clear the ECU trouble code memory on the Intergra, remove the HAZARD fuse at the battery positive terminal for at least 10 seconds. To clear the ECU trouble code memory on the Legend, remove the ALTERNATOR SENSE fuse from the underhood relay box for at least 10 seconds.

**NOTE: Some vehicle computers have a learning ability. If a change is noted in vehicle performance after clearing codes, it may be due to the computers learning ability. To restore performance, warm vehicle to normal operating temperature and drive at part throttle with moderate acceleration, before performing any additional diagnosis.**

## 1988–90 ACURA DIAGNOSTIC TROUBLE CODES

| Trouble Codes | Integra | Legend |
|---|---|---|
| 0—Electronic control unit | X | X |
| 1—Oxygen content | X | ④ |
| 2—Rear oxygen content | | X |
| 3—Manifold absolute pressure | X | X |
| 4—Crank angle | X | X |
| 5—Manifold absolute pressure | X | X |
| 6—Coolant temperature | X | X |
| 7—Throttle angle | X | X |

## 1988–90 ACURA DIAGNOSTIC TROUBLE CODES

| Trouble Codes | Integra | Legend |
|---|---|---|
| 8—TDC position | X | X |
| 9—No. 1 cylinder position | X | X |
| 10—Intake air temperature | X | X |
| 12—Exhaust gas recirculation | ① ② | X |
| 13—Atmospheric pressure | X | X |
| 14—Electronic air control | X | X |
| 15—Ignition output signal | X | X |
| 16—Fuel injector | X | |
| 17—Vehicle speed sensor | X | X |
| 18—Ignition timing adjustment | | X |
| 19—Lock-up control solenoid valve | ③ | |
| 20—Electric load | ③ | |
| 30—A/T F1 signal A | | X |
| 31—A/T F1 signal B | | X |
| 43—Fuel supply system | ① | |

① 1990 only
② Auto Trans only
③ 1989–90 only
④ Front

# CHRYSLER FUEL INJECTION

## General Information

All Chrysler fuel injected engines systems include Electronic Controlled Injection (ECI) or Multi-Point Injection (MPI). Both systems contain a self-diagnosis system which monitors various input signals from the engine sensors and enters a trouble code in the on-board computer memory if a problem is detected. This constant checking by the ECU looks for both the presence of voltage and for the voltage to be within certain pre-determined limits; if either condition is not met, a fault code will be set in memory. There are several monitored items, including the "Normal Operation" code, all of which can be read either by using an analog (dial-type) voltmeter or the ECI/MPI tester or DRB–II tester and the appropriate adapters. The adapters connect the tester to the on-board diagnosis connector.

Depending on the year and model, the diagnostic connector may be found either behind the glove box, behind the right front kick panel or just above the fusebox on the left side of the dashboard. Certain ECI systems have a separate connector under the hood.

Because the computer memory draws its power directly from the battery, the trouble codes are not erased when the ignition is switched **OFF**. The memory can only be cleared (trouble codes erased) if a battery cable is disconnected or the ECU wiring harness connector is disconnected from the computer module.

**NOTE: ECI systems will not retain memory of oxygen sensor function after ignition is turned OFF. Check for this code with engine fully warmed and running at idle.**

If more than a single trouble code is stored in the memory, the computer will read out the codes in numerical order beginning with the lowest numbered code.

**NOTE: The order in which codes are displayed does NOT indicate the order of occurrence.**

### OPERATION OF ECI/MPI SYSTEM

The Electronic Control Unit (ECU), based on the information from various sensors, determines (computes) an optimum control for varying operating conditions and accordingly drives the output actuators. The ECU consists of an microprocessor, random access memory (RAM), read only memory (ROM) and input/output (I/O) interface.

### MALFUNCTION INDICATOR LIGHT

On some models, a dashboard light will illuminate when a code is stored during engine operation. If the irregular condition returns to normal during operation, the indicator light will go OFF. Once the ignition switch is turned **OFF**, the dash light will not come back ON until the malfunction is detected again. Every time the ignition is turned **ON**, the dash light will illuminate for 2.5 seconds as a test. If a fault code is detected, the light will stay ON beyond the test period.

## Self-Diagnostic System

### ENTERING SELF-DIAGNOSIS

Chrysler recommends the use of Chrysler DRB–II diagnostic tester with the MMC Mitsubishi adapter for accessing codes and testing the ECU components and circuits. In the event that the DRB–II tester is not available codes can still be obtained through the use of a voltmeter connected to the diagnostic terminal.

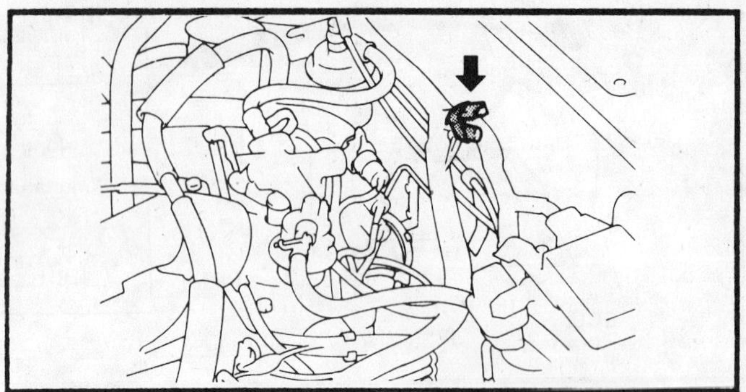

**Diagnostic terminal location – 1988 Colt 1.6L**

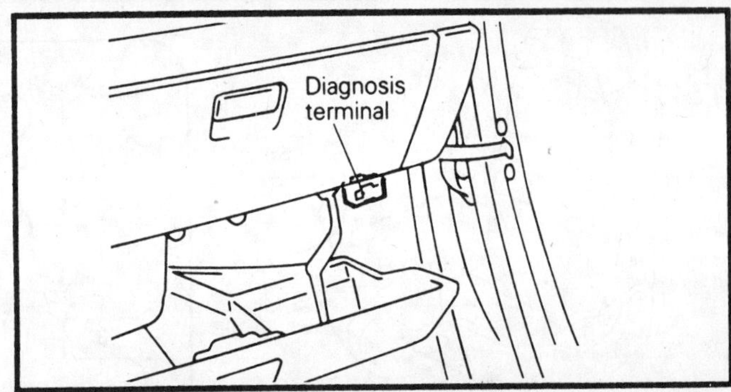


**Diagnostic terminal location – 1988–89 Colt Vista 2.0L**

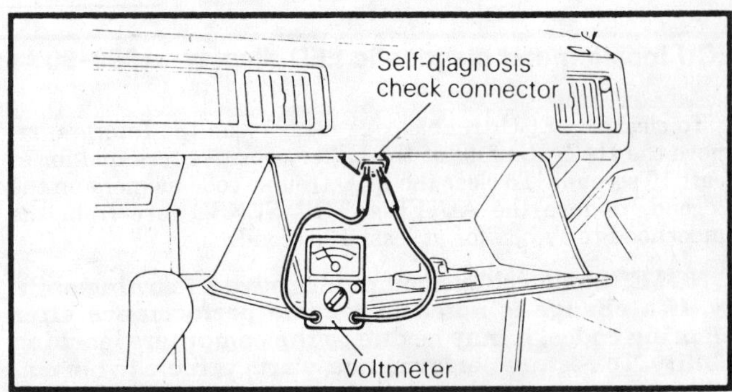


**Diagnostic terminal location – Conquest**

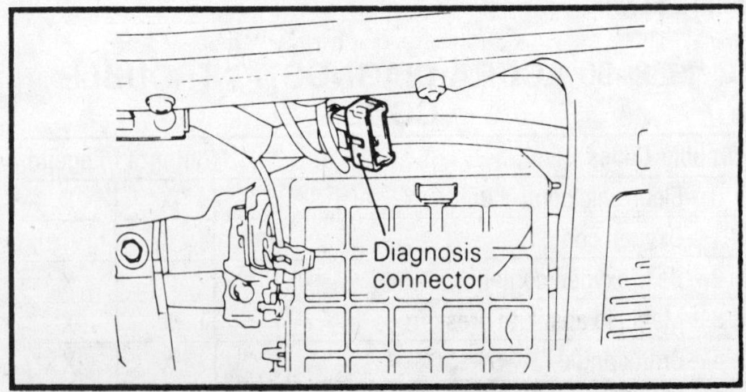


**Diagnostic terminal location – Raider**

**NOTE: If battery voltage is low, the self-diagnosis system will not operate properly. If the battery voltage is below 11 volts the ECU may not store codes.**

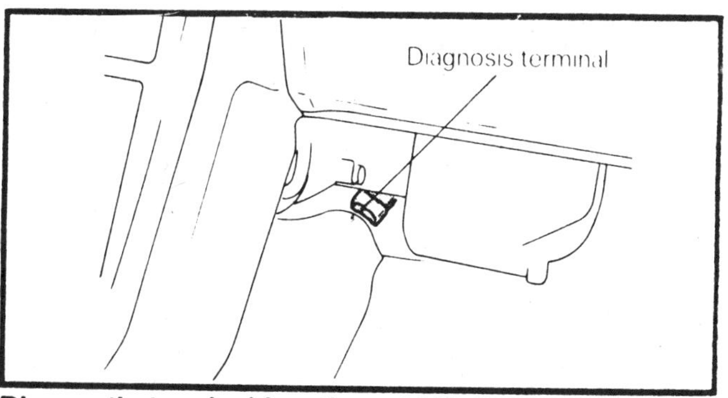

Diagnostic terminal location – 1988 Colt 1.5L, 1989–90 Colt 1.5L and 1.8L wagon and 1989–90 Vista 2.0L wagon

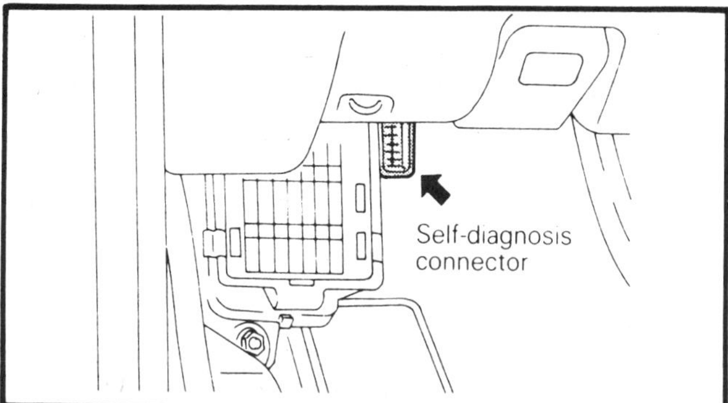

Diagnostic terminal location – 1989–90 Colt and Summit Canada

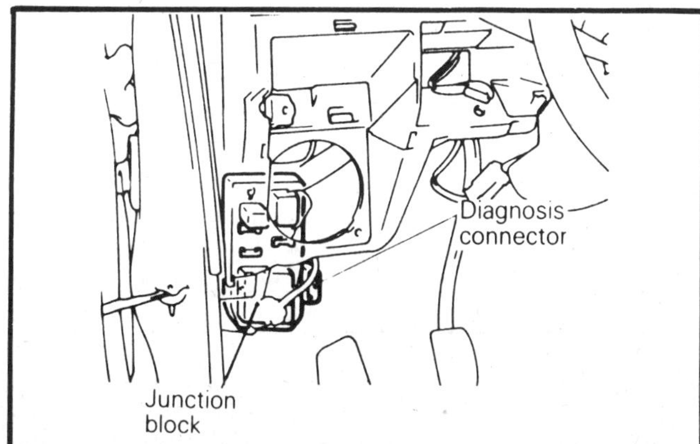

Diagnostic terminal location – 1989–90 Colt and Summit except Canada

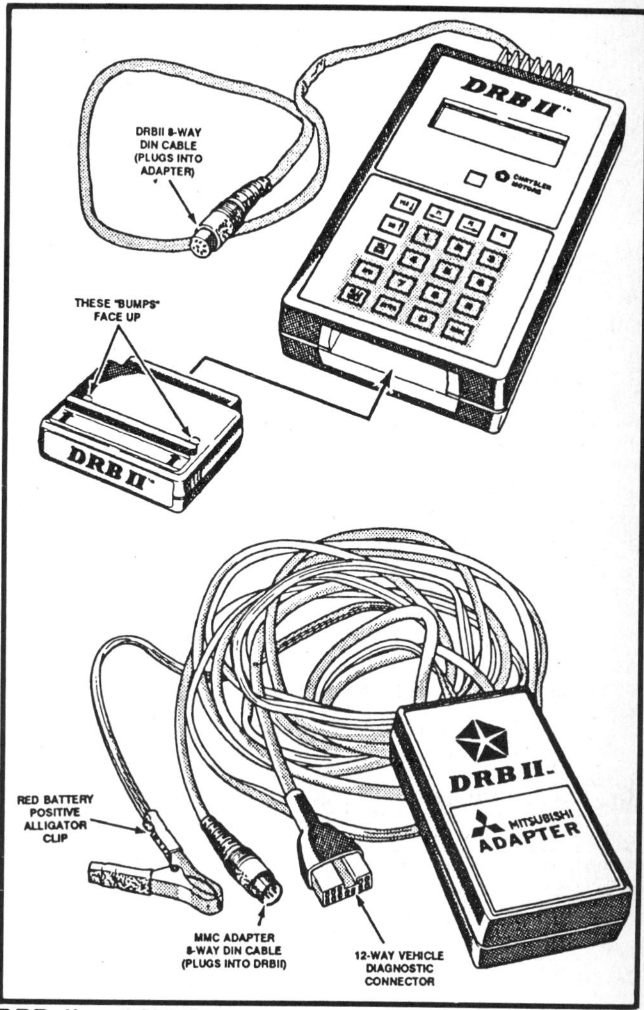

DRB–II and MMC test adapter

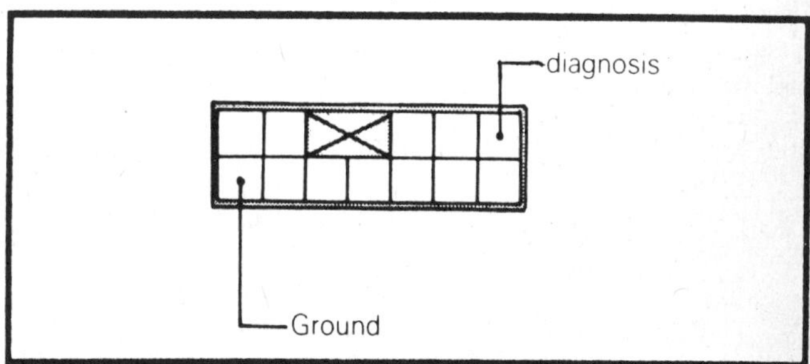

Diagnostic terminal identification

1. Turn ignition switch to **OFF**.
2. Either connect ECI or DRB–II engine tester, or using an analogue voltmeter. The following procedures are for use of an analogue voltmeter.
3. Connect a voltmeter between test terminal and terminal for ground.
4. Turn ignition switch to **ON**, and indication of ECU memory contents will start. If system is not storing any fault codes, the "normal operation" code is transmitted. If any codes are stored in memory, pointer of voltmeter or tester will deflect. (For example: Two ½–second sweeps followed by a 2 second pause followed by three ½–second sweeps might indicate Code 23 or Code 2 and Code 3, depending on the system.)
5. Record each code as it is transmitted. Once the stored code(s) have been transmitted, the system will pause and retransmit the code(s).
6. Turn ignition switch **OFF**.
7. After completing service procedures, the trouble code should be erased from the ECU.
8. After the engine has been operated, check again for presence of codes; if repair was successful, no code will be transmitted.

## CLEARING TROUBLE CODES

In all cases, the memory of the ECU is erased by disconnecting the negative battery cable for at least 30 seconds. Make certain the ignition switch is turned **OFF** before disconnecting the cable.

**NOTE: Some vehicle computers have a learning ability. If a change is noted in vehicle performance after clearing codes, it may be due to the computers learning ability. To restore performance, warm vehicle to normal operating temperature and drive at part throttle with moderate acceleration, before performing any additional diagnosis.**

### 1988-90 CHRYSLER IMPORT DIAGNOSTIC TROUBLE CODES

| Trouble Codes | Colt | 1988 Vista | 1989-90 Summit 2.0L | Raider | Conquest |
|---|---|---|---|---|---|
| 1—Oxygen sensor | ① ⑤ | X | | | X |
| 2—Ignition signal | ① ⑤ | X | | | X |
| 3—Airflow sensor | ① ⑤ | X | | | X |
| 4—Barometric pressure sensor | | X | | | |
| 5—Throttle position sensor | ① ⑤ | X | | | X |
| 6—Idle speed control Motor position sensor | ① ⑤ | X | | | X |
| 7—Coolant temperature sensor | ① ⑤ | X | | | X |
| 8—No. 1 cylinder TDC sensor | | X | | | |
| 11—Oxygen sensor | X | | X | X | |
| 12—Air flow sensor | X | | X | X | |
| 13—Intake air temperature sensor | X | | X | X | |
| 14—Throttle position sensor | X | | X | X | |
| 15—Motor position sensor | ④ | | ④ | | |
| 21—Coolant temperature sensor | X | | X | X | |
| 22—Crank angle sensor | X | | X | X | |
| 23—No. 1 cylinder TDC sensor | X | | X | X | |
| 24—Vehicle speed sensor | X | | X | X | |
| 25—Barometric pressure sensor | X | | X | X | |
| 31—Detonation sensor | ② ⑤ | | ⑤ | X | |
| 41—Injector | X | | X | X | |
| 42—Fuel Pump | X | | X | X | |
| 43—Exhaust gas recirculation (California only) | X | | X | X | |
| 44—Ignition coil | ② ⑤ | | X | X | |

① 1988 only
② 1989-90 only
④ 1.5 L engine only
④ 1.6 L engine only

# DAIHATSU ELECTRONIC FUEL INJECTION (EFI)

## General Information

Daihatsu electronic fuel injection system consists of 3 sub-systems: air intake, electronic control and fuel supply.

### ENGINE CONTROL SYSTEM

The Electronic Control Unit (ECU) controls the amount of fuel and mixture based on engine conditions and running conditions, as determined by signals input by various sensors. The ECU also controls electronic ignition timing, fuel pump operation, idle-up control, a self-diagnosis function and the fail safe and back-up functions.

### SELF-DIAGNOSIS

If any malfunction should occur in an input signal circuit, the ECU memorizes the fault. This malfunction is indicated during the trouble diagnosis, by codes through the checking lamps. If a major malfunction is encountered, the ECU turns ON the CHECK ENGINE lamp to warn the driver.

### Fail-Safe Function

When a malfunction is detected, the ECU determines, based on programmed standard values, whether the engine may be allowed to continue operation or should be stopped immediately.

### Back-Up Function

In the event the ECU detects a malfunction, the back-up function makes it possible to provide fuel injection amount and ignition timing that have been predetermined by the back-up data. This is to ensure safe driving if a single or many sensors are faulty, or if the ECU malfunctions.

# Self-Diagnostic System

## ENTERING SELF-DIAGNOSTICS

1. Make certain battery voltage is above 11 volts. The ECU may not store codes if the battery is low.
2. Make certain the throttle is closed.
3. Turn all accessories **OFF**.
4. Short the Brown T terminal and the Black ground terminal of the test connector.

**NOTE: The Check Connector is located at the upper section of the transmission. The use of a special test harness SST No. 09991–87702–000 or equivalent, connected to the Check Connector may be necessary.**

5. Turn the ignition switch to **ON**. Do not start the engine.

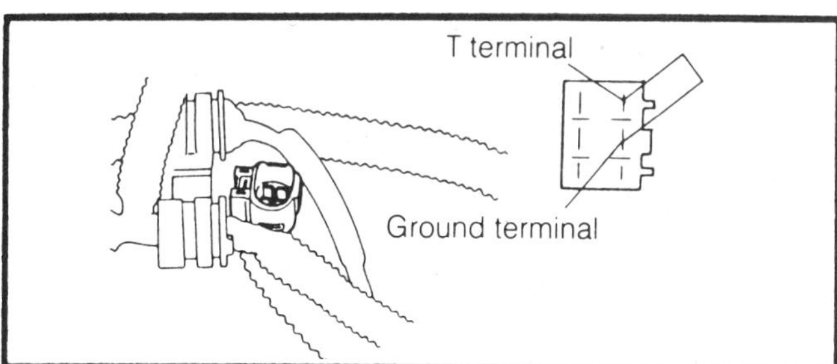


**Check Connector**

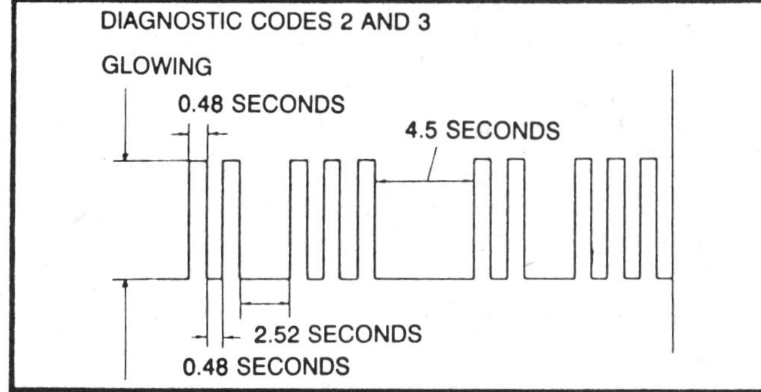


**Diagnostic code pattern—typical**

6. Read the diagnosis codes by counting the CHECK ENGINE lamp flashes.

## CLEARING TROUBLE CODES

To clear the diagnosis code from the ECU, disconnect the negative terminal from the battery or disconnect the BACK-UP fuse of the relay block assembly in the engine compartment with the ignition switch **OFF** for at least 30 seconds.

**NOTE: Some vehicle computers have a learning ability. If a change is noted in vehicle performance after clearing codes, it may be due to the computers learning ability. To restore performance, warm vehicle to normal operating temperature and drive at part throttle with moderate acceleration, before performing any additional diagnosis.**

| Code No | Number of glowing of check engine lamp | Diagnosis item | Diagrosis contents | Trouble area |
|---|---|---|---|---|
| 1 | | Normal | — | — |
| 2 | | Pressure sensor | When the signal from pressure sensor becomes open or shorted | 1. Pressure sensor circuit<br>2. Pressure sensor<br>3. ECU |
| 3 | | Ignition signal | • No ignition confirmation signal (IGf) is inputted. | 1. Ignition circuit (+B, IGF)<br>2. Ignitor<br>3. ECU |
| 4 | | Water temperature sensor | • When the signal from the water temperature sensor circuit becomes open or shorted: | 1. Water temperature sensor circuit<br>2. Water temperature sensor<br>3. ECU |
| 5 | | Oxygen sensor signal | • When the oxygen sensor signal circuit becomes open or shorted: | 1. Oxygen sensor circuit<br>2. Oxygen sensor<br>3. ECU |
| 6 | | Revolution signal | • When Ne and/or G signal is not inputted within a few seconds after starting of engine cranking:<br>• When the Ne signal of a few decade milliseconds is not inputted when the engine speed is 1000 rpm or more: | 1. Distributor circuit<br>2. Distributor<br>3. ECU |
| 7 | | Throttle position sensor signal | • When the throttle position sensor signal circuit becomes open or shorted: | 1. Throttle position sensor circuit<br>2. Throttle position sensor<br>3. ECU |
| 8 | | Intake air temperature sensor signal | • When the intake air temperature signal circuit becomes open or shorted: | 1. Air temperature sensor circuit<br>2. Air temperature sensor<br>3. ECU |
| 9 | | Vehicle speed sensor signal | • When the vehicle speed sensor signal circuit becomes open or shorted: | 1. Vehicle speed sensor circuit<br>2. Vehicle speed sensor<br>3. ECU |
| 10 | | Starter signal | When the starter signal becomes open or shorted:However, it should be noted that this code may be memorized when the vehicle is started by being pushed. | 1. Starter signal circuit<br>2. ECU |
| 11 | | Switch signal | When the air conditioner is turned ON or the idle switch is turned OFF with the terminal T shorted: However, no memorizing will take place. | 1. Air conditioner switch circuit<br>2. Idle switch circuit<br>3. Air conditioner switch<br>4. Throttle position sensor<br>5. ECU |
| 12 | | EGR control system | When the EGR system is not operating normally: | 1. EGR valve<br>2. Modulator<br>3. EVSV<br>4. Water temperature sensor<br>5. ECU |

**Diagnostic engine codes—Charade**

# FORD FESTIVA AND TRACER ELECTRONIC FUEL INJECTION (EFI)

## General Information

The Ford Tracer fuel injection system uses an electronic engine control system similar to previous Ford EEC-IV engine control systems. The heart of the system is a microcomputer called an Electronic Control Assembly (ECA). The ECA receives data (system inputs) from sensors, switches, relays and other electronic components and issues command signals (system outputs) to various devices in order to control engine operation under a variety of loads and ambient conditions. The ECA is calibrated according to the powertrain, axle ratio and vehicle weight and options to optimize fuel economy and driveability while minimizing harmful emissions.

The Ford Festiva fuel injection system uses an electronic engine control system identical to Ford Tracer engine control system. Both electronic fuel injection systems are classified as a multi-port, pulse time, mass airflow, fuel injection systems. The basic fuel requirement of the engine is determined from data supplied to the ECA by the vane airflow, air temperature, atmospheric pressure, coolant temperature, engine speed, exhaust oxygen, throttle position sensors, in addition to the common sensors used on closed loop feedback engines.

### ELECTRONIC CONTROL ASSEMBLY OPERATION

#### Cranking Mode

While the engine is being cranked, the ECA receives a signal from the ignition switch, causing the amount of fuel by the injectors to increase. This provides reliable engine starting. The fuel supply is also increase after starting to provide idle stability.

#### Warm-up Mode

The primary concern during this mode is for rapid and smooth engine warm-up. The ECA monitors the input signals from the engine coolant temperature sensor and the air charge temperature sensor and increases the amount of fuel to help warm up.

#### Open Loop

Before engine is warm, the ECA receives signals from the vane airflow sensor (intake air amount) and the ignition module (engine rpm). It uses this information to provide best fuel economy, performance and emissions.

#### Closed Loop

In closed-loop, the system uses the exhaust gas oxygen sensor to monitor the exhaust gas to determine the actual air/fuel ratio being delivered by the fuel injectors. This is sometimes referred to as the "feedback system".

#### High Altitude

When the vehicle is operated at high altitudes, the ECA will vary the air/fuel ratio to compensate for reduced air density. The barometric pressure sensor signals the ECA, which reduces the amount of fuel supplied by the injectors. In addition, the ECA also uses the signal to adjust ignition timing. At altitudes above 3280 ft., the ECA signals the ignition module to advance timing approximately 6 degrees. At altitudes above 6560 ft., the ECA signals the ISC solenoid to increase air flow and stabilize the idle speed.

#### Acceleration

While driving under heavy load or acceleration, more fuel is supplied. This mode is activated by the throttle position sensor and/or when intake manifold pressure drops below a specific point. Because a manifold vacuum sensor is not used, the ECA calculates low intake manifold vacuum using the input signals from the vane airflow sensor and the ignition module (rpm). Additional fuel is also added when the engine is above 4700 rpm.

#### Deceleration

On closed throttle deceleration, the engine requires a very lean mixture to reduce emissions and improve fuel economy. Deceleration is signaled by the closed throttle idle switch and throttle position sensor switch position. The ECA gradually reduces the amount of fuel supplied to the injectors until the engine speed falls below 2200 rpm. When the engine rpm is below 2200 rpm, the ECA gradually increase the fuel supply to normal.

#### Failure Mode

The ECA will initiate Failure Mode Effects Management (FMEM) if a failure in the vane airflow, air charge temperature, coolant temperature, barometric pressure or exhaust gas oxygen sensor is detected. The vehicle will still run, but poor performance or fuel economy may be experienced.

#### Malfunction Indicator Light (MIL)

The MIL (CHECK ENGINE light) will illuminate to warn the driver of an emission related problem, when a malfunction occurs in any input sensor. The MIL can also be used to read extracted fault codes from the ECA.

## Self-Diagnostic System

### ENTERING KEY ON ENGINE OFF TEST

#### MIL Equipped

On vehicle equipped with MIL on the dash, when in the Self-Test the CHECK ENGINE lamp will also flash the service codes.

1. Turn the ignition switch **OFF**.
2. Connect a jumper wire from the Self-Test Input (STI) connector to ground.
3. Turn the ignition switch **ON**.
4. To activate codes, disconnect and reconnect the STI ground jumper wire.
5. Record codes.
6. Clear codes and retest.

#### Using a Voltmeter

On vehicles not equipped with MIL, the codes can be obtained by the use of an analog voltmeter, by the same method used on other Fords with EEC-IV.

1. Set the analog voltmeter on a DC voltage range to read from 0 to 20 volts, except on non-MIL equipped Festiva, set on the 0 to 54 volt range.
2. On the Festiva, insert a voltmeter probe wire into the BL/W – EFI or GN/R – FBC terminal on the 6-pin Self-Test Output (STO) connector and the other probe to negative battery terminal. On the Tracer connect the voltmeter between the green and Black terminals of the under-dash 6–pin STO connector.
3. Jumper the Self-Test Input (STI) terminal to ground.
4. Turn the ignition switch **ON**.

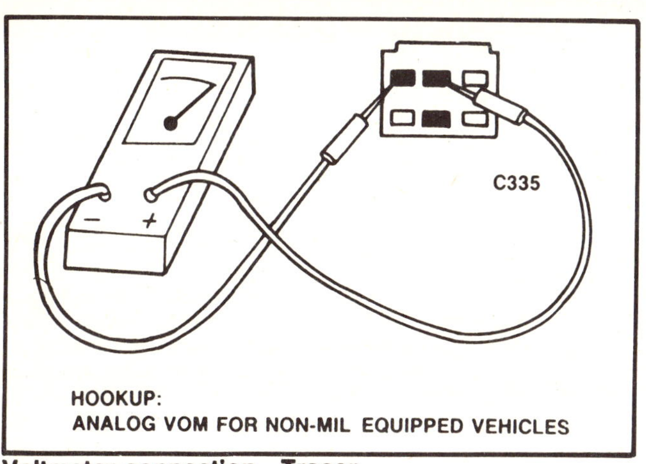


**Voltmeter connection – Tracer**

5. To activate codes, disconnect and reconnect the STI ground jumper wire.
6. Record codes.
7. Clear codes and retest.

## ENGINE RUNNING TEST

### MIL Equipped

1. Deactivate Self-Test and clear memory of codes.
2. Run engine at 2000 rpm for 15 minutes, to warm up the oxygen sensor.
3. Ground the STI connector.
4. Turn engine **OFF**, turn ignition switch **ON** and wait 10 seconds.
5. Run engine at idle.
6. To activate codes, disconnect and reconnect the STI ground jumper wire.
7. Record codes.

### Non-MIL Equipped

1. Deactivate Self-Test and clear memory of codes.
2. Run engine at 2000 rpm for 15 minutes, to warm up the oxygen sensor.
3. Ground the STI connector.
4. Turn engine **OFF**, turn ignition switch **ON** and wait 10 seconds.
5. Run engine at idle.
6. To activate codes, disconnect and reconnect the STI ground jumper wire.
7. Disconnect the STI ground wire.
8. Turn engine **OFF**, turn ignition **ON** and wait 10 seconds.
9. Run engine at idle.
10. Connect the STI ground wire.
11. Record codes.

## CONTINUOUS TEST

### Intermittent Fault Confirmation Check (Wiggle Test)

The ECA is continually looking for shorts, open circuits and other problems within the EEC–IV system and when noted, stores them in the memory, when they occur. Memory codes obtained during the Key On Engine Off segment of the self test, recalls intermittent faults from the ECA memory.

1. Repeat the self-test segment when any code other than "System Pass", is generated by the ECA.
2. Attempt to re-create intermittent faults while the test equipment is still connected to the system. This is called an intermittent fault confirmation check.
3. During the Key On Engine Off segment, perform this check with self-test sequence de-activated (ECA not in the self-test mode). A fault is indicated when the VOM deflection is 10.5 volts or greater, or the MIL flashes a code.
4. During the Key On Engine Running mode, perform this check with self-test sequence activated (ECA in self-test mode). A fault will be indicated in the same way as it is with the Key On Engine Off mode.

**RECREATING INTERMITTENT FAULTS**

Intermittent faults can generally be recreated by the following methods:

1. Wiggling connectors and harnesses
2. Manipulating moveable sensors and actuators
3. Heating thermistor type sensors with a heat gun

Suspected components, such as the sensors, actuators and harnesses, are identified by matching service codes obtained during the Key On Engine Off segment of the self-test.

When an intermittent fault is re-created, the voltmeter needle will sweep back and forth across the scale or sweep to the right and stay there.

Malfunctioning components identified with this procedure can be repaired or replaced without further diagnostic testing. Further testing must be done for hard faults and for intermittent faults that can not be re-created by the above method.

## READING SERVICE CODES

After the ECA system completes its self-test, it communicates with the service technician by way of the analog voltmeter needle and scale, or MIL on the dash.

When a service code is obtained by the use of the MIL on the dash the codes are read as a pattern of light flashes. The MIL displays the same pattern as the voltmeter display on Ford EEC–IV systems.

When a service code is reported on the analog meter, it will represent itself as a pulsing or sweeping movement of the voltmeter scale's needle across the dial face of the meter. Therefore, a single digit number of 3 will be indicated by 3 pulses (sweeps) of the needle across the dial. Since the service codes are indicated by a 2 digit number, such as 32, the self-test's service Code of 32 will be displayed as 3 pulses (sweeps) of the meter's needle with a 1½ second pause between pulses (sweeps) and a 2 second pause between digits. When more than a single service code is indicated, a 4 second pause between service codes will occur.

Separator and dynamic response codes (numeral 10 in both cases), are represented by a single pulse or sweep of the needle. There are no pulses or sweeps generated for the digit "0". There will be a 6 second pause before each of these.

The key to sorting out the service codes on the voltmeter is to keep the pulses and the pauses straight.

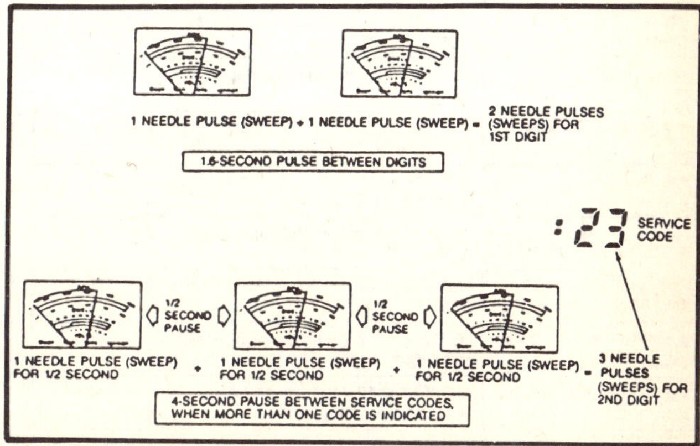


**Reading codes from a voltmeter**

1. Each digit is separated by a 2 second pause.
2. Each code is separated by a 4 second pause.
3. Separator and dynamic response codes are separated from previous and subsequent codes by 6 second (or longer) pauses.

## SWITCH MONITOR TEST

The switch monitor test checks the input signals from individual switches to the ECA. All switches must be tested individually, leaving a switch on while another is testing will lead to false results.

1. Deactivate the Self-Test.
2. Turn all accessories **OFF**.
3. Allow engine to completely cool.
4. Place the transaxle in **N**.
5. Ground the STI connector.
6. Turn the ignition switch **ON**.
7. Connect a voltmeter across the proper STO terminals:
   Festiva with FBC – BK/W and Y/BK
   Festiva with EFI – Y/BK and BL/GN
   Tracer – R/BU and Y/GR
8. Watch the output of the voltmeter as each switch is activated.

| Switch | LED or Analog VOM Condition | Pinpoint Test |
|---|---|---|
| Neutral Safety Switch (ATX only) | LED "ON" or 12v in D, 1, 2, R Range | FB |
| Clutch Engage switch and Neutral Gear Switch (MTX) | LED "ON" or 12v in gear and clutch pedal released | FA |
| Brake On/Off Switch (1.3L FBC) | LED "ON" or 12v with brake pedal depressed | FD |
| Idle Switch | LED "ON" or 12v with accelerator depressed | FE |
| WOT Switch (1.3L EFI) | LED "OFF" or 0v with accelerator fully depressed | FH |
| A/C Switch | LED "ON" or 12v with A/C on and blower in position 1 | FF |
| Headlamp Switch | LED "ON" or 12v with headlamps on | FJ |
| Rear Defroster | LED "ON" or 12v with rear defroster on | FK |
| A/C-Heater Blower Switch | LED "ON" or 12v with blower in position 2 or 3 | FL |
| Cooling Fan Temperature Switch | LED "ON" or 12v coolant temperature above 195°F. | FM |

Switch monitor test – Festiva

| Switch | Led or Analog VOM Condition | Pinpoint 1.6L EFI |
|---|---|---|
| Neutral Gear Switch and clutch engage switch (MTX) | LED "ON" or 12 V in gear and clutch pedal released (VOM) | FA1 |
| Inhibitor Switch (ATX) | LED "ON" or 12 V in D, 1, 2, R Range (VOM) | FB1 |
| Brake On-Off | LED "ON" or 12 V with Pedal Depressed (VOM) | FD1 |
| Throttle Position Sensor | LED "ON" or 12 V with Accelerator Depressed (VOM) | DH1 |
| A/C Switch | LED "ON" or 12 V with A/C on and blower in position 1 (VOM) | FF1 |
| Headlamp Switch | LED "ON" or 12 V with headlamps on (VOM) | FG8 |
| Rear Defroster | LED "ON" or 12 V with rear defroster on (VOM) | FG6 |
| A/C-Heater Blower Control Switch | LED "ON" or 12 V with blower in position 3 or 4 (VOM) | FG4 |
| Cooling Fan Temperature Switch | LED "ON" or 12 V with connector disconnected (VOM) | FG2 |

Switch monitor test – Tracer

## CLEARING TROUBLE CODES

1. Perform the Key On Engine Off test.
2. As soon as the first codes are received, disconnect the STI ground.
3. Disconnect the negative battery cable.
4. Depress the brake pedal from 10 seconds.
5. Reconnect the negative battery cable.
6. Perform the Key On Engine OFF test to confirm the codes have cleared.

**NOTE: Some vehicle computers have a learning ability. If a change is noted in vehicle performance after clearing codes, it may be due to the computers learning ability. To restore performance, warm vehicle to normal operating temperature and drive at part throttle with moderate acceleration, before performing any additional diagnosis.**

| Sensor or Sub-system | Engine | | FMEM | Malfunction Display | |
|---|---|---|---|---|---|
| | 1.3L FBC | 1.3L EFI | | Malfunction Code No. | MIL Output Signal Pattern |
| Ignition Diagnostic Monitor | X | X | — | 01 |  |
| Vane Airflow Meter | | X | Maintains basic signal at preset value of midrange vane position | 08 |  |
| Engine Coolant Temperature Sensor | X | X | EFI: Maintains constant command — 35°C (95°F) fuel. 50°C (122°F) for ISC control use FBC: Maintains constant 80°C (176°F) | 09 |  |

Failure Codes – Festiva

| Sensor or Sub-system | Engine | | FMEM | Malfunction Display | |
|---|---|---|---|---|---|
| | 1.3L FBC | 1.3L EFI | | Malfunction Code No. | MIL Output Signal Pattern |
| Vane Air Temperature Sensor | | X | Maintains constant 20°C (68°F) command | 10 | 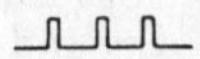 |
| Manifold Absolute Pressure Sensor | X | | Maintains constant command of 27.6-29.9 in. Hg manifold vacuum | 13 |  |
| Barometric Pressure Sensor | | X | Maintains constant command of sea level pressure | 14 |  |
| EGO Sensor | X | X | Cancels feedback operation | 15 |  |
| EGR Valve Position Sensor | California Only | | Cuts-off EGR | 16 |  |
| EGO Feedback | X | X | Cancels Feedback operation | 17 |  |
| Wide Open Throttle Vacuum Switch | X | | Maintains constant command of 27.6-29.9 in. Hg manifold vacuum | 70 | |

Failure Codes – Festiva

| Code No. | Input devices | Malfunction | Code Pattern | Fail-safe function | Code in constant memory |
|---|---|---|---|---|---|
| 01 | Ignition pulse | Broken wire, short circuit | ON OFF | — | Yes |
| 08 | Vane air flow meter | Broken wire, short circuit | ON OFF | Basic fuel injection amount fixed as for 2 driving modes 1) Idle switch: ON 2) Idle switch: OFF | Yes |
| 09 | Engine coolant temp. | Broken wire, short circuit | ON OFF | Coolant temp. input fixed at 80°C (176°F) | Yes |
| 10 | Vane air temp. sensor | Broken wire, short circuit | ON OFF | Intake air temp. input fixed at 20°C (68°F) | Yes |
| 14 | Barometric pressure sensor | Broken wire, short circuit | ON OFF | Atmospheric pressure input fixed at 760 mmHg (29.9 in Hg) | Yes |
| 15 | EGO sensor | Sensor output continues less than 0.55V 120 sec. after engine starts (1500 rpm) | ON OFF | Feedback system operation cancelled | Yes |
| 17 | Feedback system | Oxygen sensor output not changed 20 sec. after engine exceeds 1500 rpm | ON OFF | Feedback system operation cancelled | Yes |

**Failure Codes—Tracer**

| Code No. | Output devices | Pattern of output signals (Self-Diagnosis Checker or MIL) | Code in constant memory |
|---|---|---|---|
| 25 | Pressure regulator control solenoid valve | ON OFF | No |
| 26 | Canister purge regulator solenoid valve | ON OFF | No |
| 27 | Canister purge solenoid valve | ON OFF | No |
| 34 | Idle speed control solenoid valve | ON OFF | No |

**Failure Codes—Tracer**

# SPRINT ELECTRONIC FUEL INJECTION

## Self-Diagnostic System

If a malfunction exists in the input signal to the ECM as a result of some trouble in the EFI components, the CHECK ENGINE light on the instrument panel will turn ON or flash to inform a presence of trouble.

### ENTERING SELF-DIAGNOSTIC SYSTEM

1. With the ignition switch **ON** with the engine not running, the light turns ON to indicate the circuit is in good condition. The light will turn OFF when the engine is started.
2. If the light stays ON after the engine is started, a malfunction exists in the signal to the ECM. The ECM will store a malfunction even if it an intermittent one. A poor contact may cause the light to come ON and then turn OFF when the circuit is completed, but the code will be stored even with the light OFF.
3. To diagnose a problem, turn ON the diagnostic switch located under the instrument panel. With the engine not running and the ignition switch **ON**, the CHECK ENGINE light will flash to indicate the diagnostic code in the ECM memory.
4. The codes consist of flashes and pauses to indicate trouble codes. Each component has a code designation. The code represents a 2 digit number as shown in the diagnostic code chart.
5. If 2 or more areas are involved and that many codes are to be indicated, the CHECK ENGINE light indicates each code corresponding to the area of trouble 3 times in the increasing order of the code numbers and then repeats them.

**NOTE: Always write down the codes as they are flashing. This memory will be lost when the power to the ECM is disconnected.**

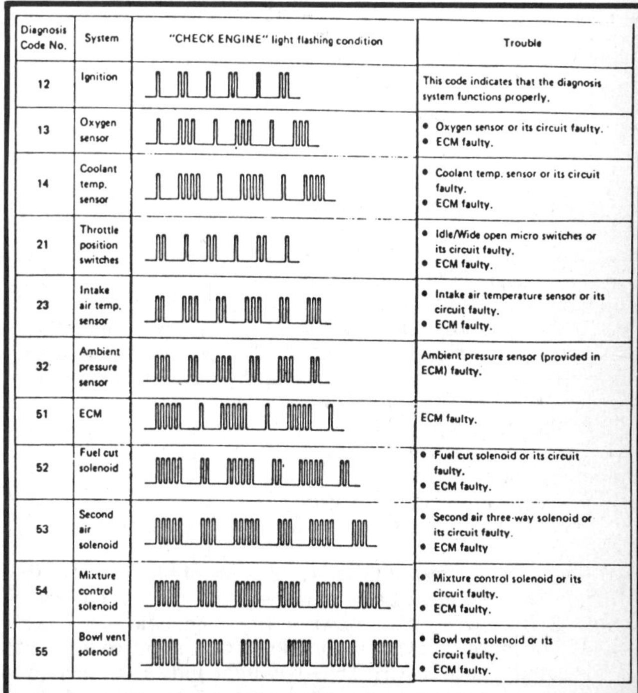

| Diagnosis Code No. | System | "CHECK ENGINE" light flashing condition | Trouble |
|---|---|---|---|
| 12 | Ignition | | This code indicates that the diagnosis system functions properly. |
| 13 | Oxygen sensor | | • Oxygen sensor or its circuit faulty. • ECM faulty. |
| 14 | Coolant temp. sensor | | • Coolant temp. sensor or its circuit faulty. • ECM faulty. |
| 21 | Throttle position switches | | • Idle/Wide open micro switches or its circuit faulty. • ECM faulty. |
| 23 | Intake air temp. sensor | | • Intake air temperature sensor or its circuit faulty. • ECM faulty. |
| 32 | Ambient pressure sensor | | Ambient pressure sensor (provided in ECM) faulty. |
| 51 | ECM | | ECM faulty. |
| 52 | Fuel cut solenoid | | • Fuel cut solenoid or its circuit faulty. • ECM faulty. |
| 53 | Second air solenoid | | • Second air three-way solenoid or its circuit faulty. • ECM faulty |
| 54 | Mixture control solenoid | | • Mixture control solenoid or its circuit faulty. • ECM faulty. |
| 55 | Bowl vent solenoid | | • Bowl vent solenoid or its circuit faulty. • ECM faulty. |

**Diagnostic trouble code chart—Sprint**

### CLEARING TROUBLE CODES

The code will be stored until the power from the battery to the ECM is cut off. Clear trouble codes by disconnecting the negative battery terminal for at least 30 seconds.

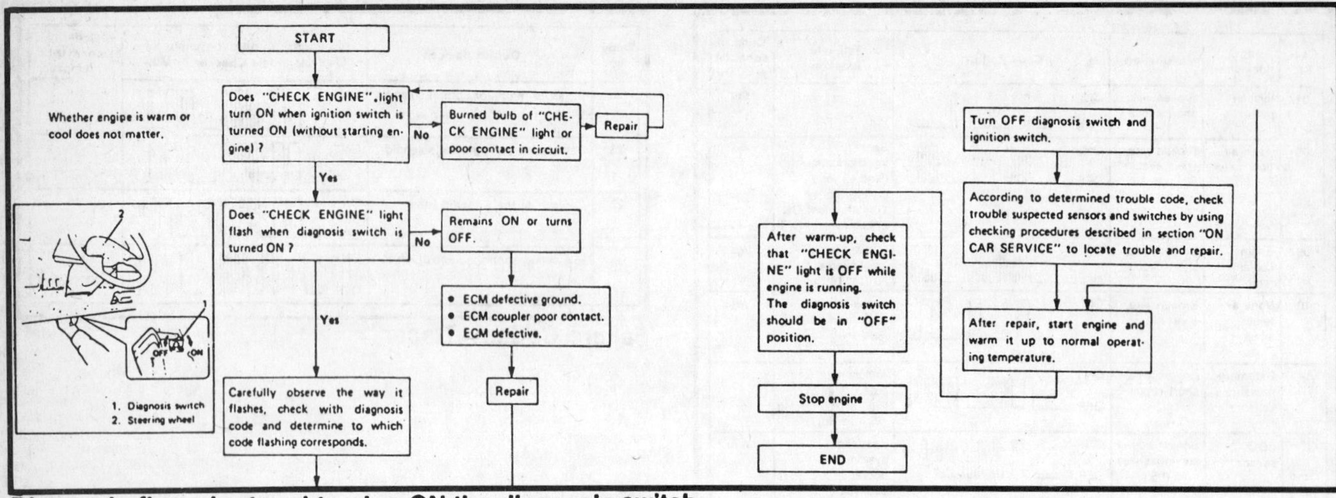

**Diagnosis flow chart and turning ON the diagnosis switch**

**NOTE: Some vehicle computers have a learning ability. If a change is noted in vehicle performance after clearing codes, it may be due to the computers learning ability. To restore performance, warm vehicle to normal operating temperature and drive at part throttle with moderate acceleration, before performing any additional diagnosis.**

# GEO PRIZM AND NOVA EFI AND CARBURETED

## Self-Diagnostic System

If a malfunction exists in the input signal to the ECM as a result of some trouble in the EFI components, the CHECK ENGINE light on the instrument panel will turn ON or flash to inform a presence of trouble.

### ENTERING SELF-DIAGNOSTIC SYSTEM

1. With the ignition switch **ON** and the engine not running, the light turns ON to indicate the circuit is in good condition. The light will turn OFF when the engine is started.
2. If the light stays ON after the engine is started, a malfunction exists in the signal to the ECM. The ECM will store a malfunction even if it an intermittent one. A poor contact may cause the light to come ON and then turn OFF when the circuit is completed, but the code will be stored even with the light OFF.
3. To diagnose a problem, ground the diagnostic terminal in the engine compartment with a jumper wire. With the engine not running and the ignition switch **ON**, the CHECK ENGINE light will flash to indicate the diagnostic code in the ECM memory.
4. The codes consist of flashes and pauses to indicate trouble codes. Each component has a code designation. The code represents a 2 digit number as shown in the diagnostic code chart.
5. If 2 or more areas are involved and that many codes are to be indicated, the CHECK ENGINE light indicates each code corresponding to the area of trouble 3 times in the increasing order of the code numbers and then repeats them.

**NOTE: Always write down the codes as they are flashing. This memory will be lost when the power to the ECM is disconnected.**

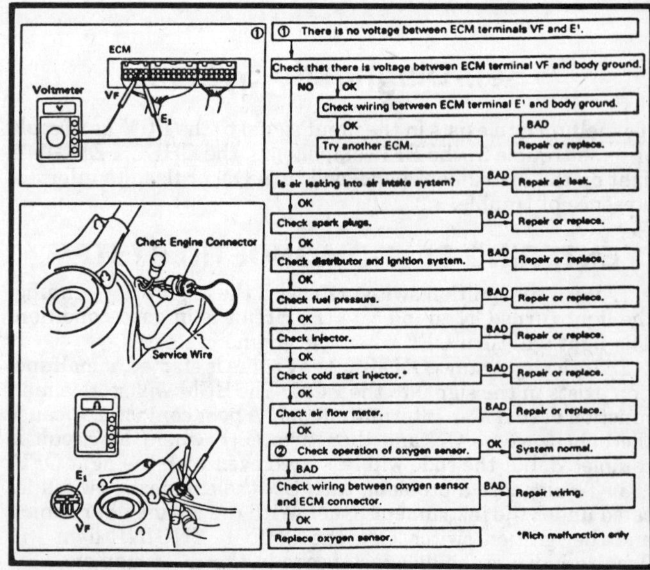

**Check engine connector – Nova**

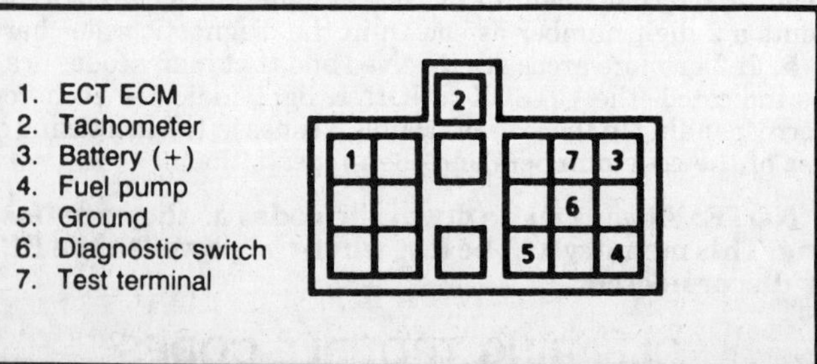

**Diagnostic link – Prizm**

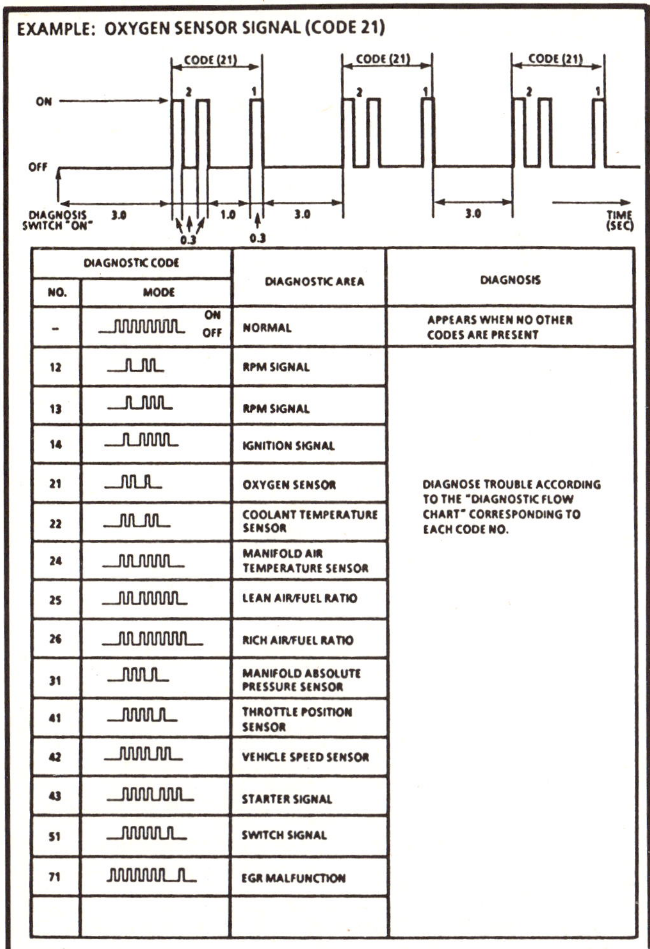


| DIAGNOSTIC CODE NO. | DIAGNOSTIC CODE MODE | DIAGNOSTIC AREA | DIAGNOSIS |
|---|---|---|---|
| – | ON OFF | NORMAL | APPEARS WHEN NO OTHER CODES ARE PRESENT |
| 12 | | RPM SIGNAL | DIAGNOSE TROUBLE ACCORDING TO THE "DIAGNOSTIC FLOW CHART" CORRESPONDING TO EACH CODE NO. |
| 13 | | RPM SIGNAL | |
| 14 | | IGNITION SIGNAL | |
| 21 | | OXYGEN SENSOR | |
| 22 | | COOLANT TEMPERATURE SENSOR | |
| 24 | | MANIFOLD AIR TEMPERATURE SENSOR | |
| 25 | | LEAN AIR/FUEL RATIO | |
| 26 | | RICH AIR/FUEL RATIO | |
| 31 | | MANIFOLD ABSOLUTE PRESSURE SENSOR | |
| 41 | | THROTTLE POSITION SENSOR | |
| 42 | | VEHICLE SPEED SENSOR | |
| 43 | | STARTER SIGNAL | |
| 51 | | SWITCH SIGNAL | |
| 71 | | EGR MALFUNCTION | |
| | | | |

Diagnostic trouble code chart—Geo Prizm and Nova

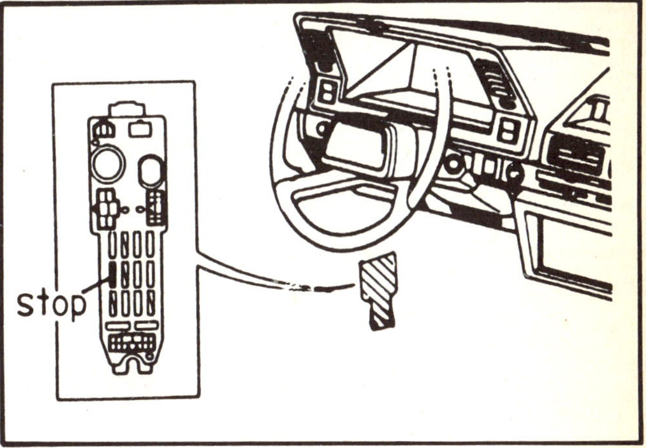


ECM power disconnect—Nova

## CLEARING TROUBLE CODES

The code will be stored until the power from the battery to the ECM is cut off. After correcting the problem, erase the memory by pulling the 15A STOP fuse located behind the drivers kick panel for 10 seconds or disconnect the battery negative cable for more than 30 seconds. Disconnecting the battery cable will cancel other memory functions such as the clock and radio.

**NOTE: Some vehicle computers have a learning ability. If a change is noted in vehicle performance after clearing codes, it may be due to the computers learning ability. To restore performance, warm vehicle to normal operating temperature and drive at part throttle with moderate acceleration, before performing any additional diagnosis.**

# PONTIAC LEMANS EFI SYSTEM

## Self-Diagnostic System

If a malfunction exists in the input signal to the ECM as a result of some trouble in the EFI components, the SERVICE ENGINE SOON light on the instrument panel will turn ON or flash to inform a presence of trouble.

### ENTERING SELF-DIAGNOSTIC SYSTEM

1. With the ignition switch **ON** and the engine not running, the light turns ON to indicate the circuit is in good condition. The light will turn OFF when the engine is started.
2. If the light stays ON after the engine is started, a malfunction exists in the signal to the ECM. The ECM will store a malfunction even if it an intermittent one. A poor contact may cause the light to come ON and then turn OFF when the circuit is completed, but the code will be stored even with the light OFF.
3. To diagnose a problem, ground the Assembly Line Diagnostic Link (ALDL) terminal A and B together with a jumper wire or connect a diagnostic tool to the ALDL. The ALDL is located on the passenger side near the kick panel. With the engine not running and the ignition switch **ON**, the SERVICE ENGINE SOON light will flash to indicate the diagnostic code in the ECM memory.

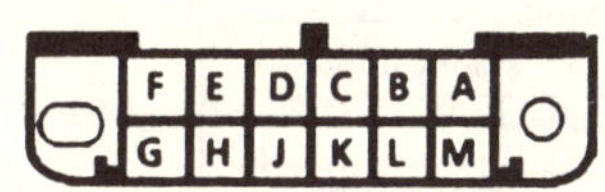


TERMINAL IDENTIFICATION

A. Ground
B. Diagnostic terminal
F. Torque converter clutch (if used)
G. Fuel pump
M. Serial data for special diagnostic tool

Diagnostic terminal—Pontiac Lemans

4. The codes consist of flashes and pauses to indicate trouble codes. Each component has a code designation. The code represents a 2 digit number code.
5. If 2 or more areas are involved and that many codes are to be indicated, the SERVICE ENGINE SOON light indicates each code corresponding to the area of trouble 3 times in the increasing order of the code numbers and then repeats them.

**NOTE: Always write down the codes as they are flashing. This memory will be lost when the power to the ECM is disconnected.**

## CLEARING TROUBLE CODES

The code will be stored until the power from the battery to the ECM is cut off. After correcting the problem, erase the memory by pulling the 20A ECM fuse (F-10) located in the fuse panel for 30 seconds or disconnect the negative battery cable for more than 30 seconds. Disconnecting the battery cable will cancel other memory functions such as the clock and radio.

**NOTE: Some vehicle computers have a learning ability. If a change is noted in vehicle performance after clearing codes, it may be due to the computers learning ability. To restore performance, warm vehicle to normal operating temperature and drive at part throttle with moderate acceleration, before performing any additional diagnosis.**

# GEO METRO EFI

## Self-Diagnostic System

If a malfunction exists in the input signal to the ECM as a result of some trouble in the EFI components, the CHECK ENGINE light on the instrument panel will turn ON or flash to inform a presence of trouble.

### ENTERING SELF-DIAGNOSTIC SYSTEM

1. With the ignition switch **ON** and engine not running, the light turns ON to indicate the circuit is in good condition. The light will turn OFF when the engine is started.

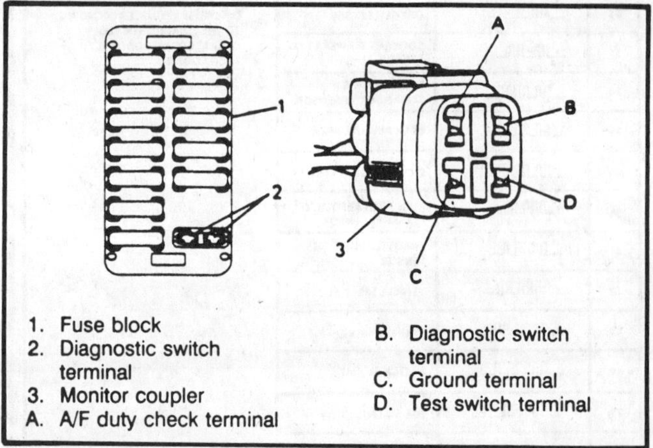


**Diagnostic monitor coupler and switch – Geo Metro**

2. If the light stays ON after the engine is started, a malfunction exists in the signal to the ECM. The ECM will store a malfunction even if it an intermittent one. A poor contact may cause the light to come ON and then turn OFF when the circuit is completed, but the code will be stored even with the light OFF.
3. To diagnose a problem, ground the monitor connector located in the engine compartment, driver's side, near the ECM. The trouble codes can be displayed by grounding the diagnostic terminal in the fuse block or grounding the B and C terminal in the monitor connector with a jumper wire. Use a spare fuse to ground the terminal in the fuse block. Connect a Scan tool to the monitor connector for diagnosis purposes. With the engine not running and the ignition switch **ON**, the CHECK ENGINE light will flash to indicate the diagnostic code in the ECM memory.
4. The codes consist of flashes and pauses to indicate trouble codes. Each component has a code designation. The code represents a 2 digit number as shown in the diagnostic code charts.
5. If 2 or more areas are involved and that many codes are to be indicated, the CHECK ENGINE light indicates each code corresponding to the area of trouble 3 times in the increasing order of the code numbers and then repeats them.

**NOTE: Always write down the codes as they are flashing. This memory will be lost when the power to the ECM is disconnected.**

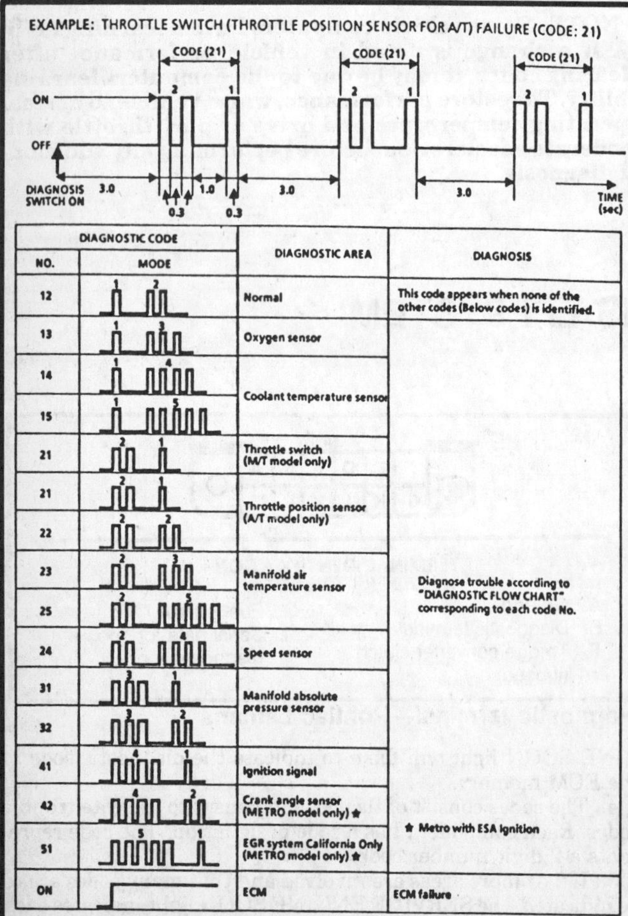


| DIAGNOSTIC CODE NO. | MODE | DIAGNOSTIC AREA | DIAGNOSIS |
|---|---|---|---|
| 12 | 1 2 | Normal | This code appears when none of the other codes (Below codes) is identified. |
| 13 | 1 3 | Oxygen sensor | Diagnose trouble according to "DIAGNOSTIC FLOW CHART" corresponding to each code No. |
| 14 | 1 4 | Coolant temperature sensor | |
| 15 | 1 5 | | |
| 21 | 2 1 | Throttle switch (M/T model only) | |
| 21 | 2 1 | Throttle position sensor (A/T model only) | |
| 22 | 2 2 | | |
| 23 | 2 3 | Manifold air temperature sensor | |
| 25 | 2 5 | | |
| 24 | 2 4 | Speed sensor | |
| 31 | 3 1 | Manifold absolute pressure sensor | |
| 32 | 3 2 | | |
| 41 | 4 1 | Ignition signal | |
| 42 | 4 2 | Crank angle sensor (METRO model only) ★ | ★ Metro with ESA Ignition |
| 51 | 5 1 | EGR system California Only (METRO model only) ★ | |
| ON | | ECM | ECM failure |

**Diagnostic trouble code chart – Geo Metro**

## CLEARING TROUBLE CODES

The code will be stored until the power from the battery to the

ECM is cut off. After correcting the problem, erase the memory by removing the 15A TAIL LAMP fuse located in the fuse panel for 30 seconds or disconnect the negative battery cable for more than 30 seconds. Disconnecting the battery cable will cancel other memory functions such as the clock and radio.

**NOTE: Some vehicle computers have a learning ability. If a change is noted in vehicle performance after clearing codes, it may be due to the computers learning ability. To restore performance, warm vehicle to normal operating temperature and drive at part throttle with moderate acceleration, before performing any additional diagnosis.**

# GEO STORM EFI SYSTEM

## Self-Diagnostic System

If a malfunction exists in the input signal to the ECM as a result of some trouble in the EFI components, the CHECK ENGINE light on the instrument panel will turn ON or flash to inform a presence of trouble.

### ENTERING SELF-DIAGNOSTIC SYSTEM

1. With the ignition switch **ON** and engine not running, the light turns ON to indicate the circuit is in good condition. The light will turn OFF when the engine is started.
2. If the light stays ON after the engine is started, a malfunction exists in the signal to the ECM. The ECM will store a malfunction even if it an intermittent one. A poor contact may cause the light to come ON and then turn OFF when the circuit is completed, but the code will be stored even with the light OFF.
3. To diagnose a problem, ground the diagnostic terminal near the kick panel on the passenger side of the vehicle with a jumper wire from terminal 1 to 3. With the engine not running and the ignition switch **ON**, the CHECK ENGINE light will flash to indicate the diagnostic code in the ECM memory.
4. The codes consist of flashes and pauses to indicate trouble codes. Each component has a code designation. The code represents a 2 digit number as shown in the diagnostic code chart.
5. If 2 or more areas are involved and that many codes are to be indicated, the CHECK ENGINE light indicates each code corresponding to the area of trouble 3 times in the increasing order of the code numbers and then repeats them.

**NOTE: Always write down the codes as they are flashing. This memory will be lost when the power to the ECM is disconnected.**

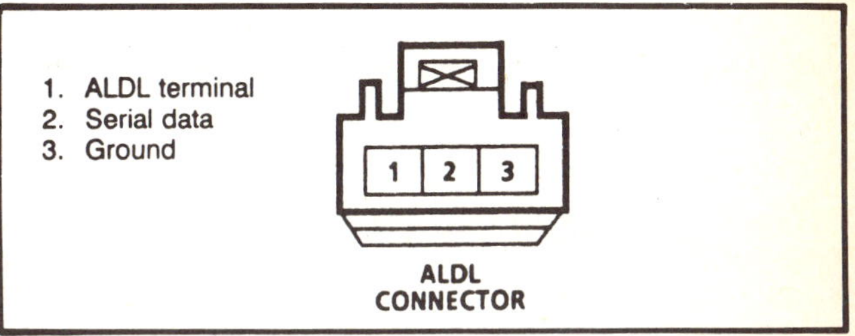


**ALDL (assembly line diagnostic link) – Geo Storm**

### CLEARING TROUBLE CODES

The code will be stored until the power from the battery to the ECM is cut off. After correcting the problem, erase the memory by removing the 30A MAIN fuse located behind the drivers instrument panel for 30 seconds or disconnect the negative battery cable for more than 30 seconds. Disconnecting the battery cable will cancel other memory functions such as the clock and radio.

**NOTE: Some vehicle computers have a learning ability. If a change is noted in vehicle performance after clearing codes, it may be due to the computers learning ability. To restore performance, warm vehicle to normal operating temperature and drive at part throttle with moderate acceleration, before performing any additional diagnosis.**

# GEO TRACKER EFI

## Self-Diagnostic System

If a malfunction exists in the input signal to the ECM as a result of some trouble in the EFI components, the CHECK ENGINE light on the instrument panel will turn ON or flash to inform a presence of trouble.

When the mileage reaches 50000, 80000 and 100000 miles respectively the mileage sensor turns ON the CHECK ENGINE light even while the engine is running with no problems. This is to warn the driver that it is time for periodic inspection. Turn the light OFF with its cancel switch located behind the access panel below the steering column. To reset, slide the switch right or left until the light goes OFF.

### ENTERING SELF-DIAGNOSTIC SYSTEM

1. With the ignition switch **ON** and the engine not running, the light turns ON to indicate the circuit is in good condition. The light will turn OFF when the engine is started.
2. If the light stays ON after the engine is started, a malfunction exists in the signal to the ECM or the mileage interval has been reached. The ECM will store a malfunction even if it an intermittent code. A poor contact may cause the light to come ON and then turn OFF when the circuit is completed, but the code will be stored even with the light OFF.
3. To diagnose a problem, ground the diagnostic terminal in the fuse block. Use a spare fuse to ground the terminal in the fuse block. Connect a Scan tool to the monitor connector for diagnosis purposes. With the engine not running and the ignition switch **ON**, the CHECK ENGINE light will flash to indicate the diagnostic code in the ECM memory.
4. The codes consist of flashes and pauses to indicate trouble codes. Each component has a code designation. The code represents a 2 digit number as shown in the diagnostic code charts.

5. If 2 or more areas are involved and that many codes are to be indicated, the CHECK ENGINE light indicates each code corresponding to the area of trouble 3 times in the increasing order of the code numbers and then repeats them.

**NOTE: Always write down the codes as they are flashing. This memory will be lost when the power to the ECM is disconnected.**

### CLEARING TROUBLE CODES

The code will be stored until the power from the battery to the ECM is cut off. After correcting the problem, erase the memory by removing the 15A TAIL LAMP fuse located in the fuse panel for 30 seconds or disconnect the negative battery cable for more than 30 seconds. Disconnecting the battery cable will cancel other memory functions such as the clock and radio.

**NOTE: Some vehicle computers have a learning ability. If a change is noted in vehicle performance after clearing codes, it may be due to the computers learning ability. To restore performance, warm vehicle to normal operating temperature and drive at part throttle with moderate acceleration, before performing any additional diagnosis.**

# SPECTRUM FEEDBACK CARBURETED

## Self-Diagnostic System

The ECM in the Closed Loop Emission Control system features a self-diagnosis function. The self-diagnosis function identifies troubles in the area related to the sensors, including wiring harnesses and ECM.

If a malfunction exists in the input signal to the ECM as a result of some trouble in the emission related components, the CHECK ENGINE light on the instrument panel will turn ON or flash to inform a presence of trouble. The system diagnosis should be performed in the following sequence:

1. Diagnostic circuit check
2. Driver Complaint
3. System Performance Check

### ENTERING SELF-DIAGNOSTIC SYSTEM

1. With the ignition switch **ON** and engine not running, the light turns ON to indicate the circuit is in good condition. The light will turn OFF when the engine is started.

| | |
|---|---|
| TROUBLE CODE 12 | No distributor reference pulses to the ECM. This code is not stored and will only flash while the fault is present. |
| TROUBLE CODE 13 | Oxygen Sensor circuit - The engine must run up to two minutes at part throttle. under road load, before this code will set. |
| TROUBLE CODE 14 | Shorted Coolant Sensor circuit - The engine must run up to two minutes before this code will set. |
| TROUBLE CODE 15 | Open Coolant Sensor circuit - The engine must run up to five minutes before this code will set. |
| TROUBLE CODE 21 | Idle Switch misadjusted and/or circuit open. This code will set if engine speed falls below 600 rpm for longer than 32 seconds, or if TPS and Idle Switch are faulty or misadjuted. This code will set if the ECM detects both idle and WOT condition at the same time. |
| TROUBLE CODE 22 | Fuel Cut-Off Relay and/or circuit open. |
| TROUBLE CODE 23 | Open or grounded M/C solenoid circuit. |
| TROUBLE CODE 42 | Fuel Cut-Off Relay and/or circuit shorted |
| TROUBLE CODE 44 | Lean Oxygen Sensor condition - The engine must run up to 2 minutes, in closed loop, at part throttle, under road load, before this code will set. This code will not set when the coolant temperature is below 70°C and/or the air temperature in air cleaner is below 0°C, in a "low altitude" condition. This code will not set in a "high altitude" condition. |
| TROUBLE CODE 45 | Rich system indication - The engine must run up to 2 minutes, in closed loop, at part throttle, under road load, before this code will set. This code will not set when the engine is not between 1500 to 2500 rpm, and/or the coolant temperature is below 70°C and/or at "high altitude condition". |
| TROUBLE CODE 51 | Faulty calibration unit (PROM) or installation. It takes up to 30 seconds before this code will set. |
| TROUBLE CODE 54 | Shorted M/C solenoid circuit and/or faulty ECM. |
| TROUBLE CODE 55 | Faulty ECM - problem in A/D converter in ECM. |

**Diagnostic trouble code chart—Spectrum carbureted**

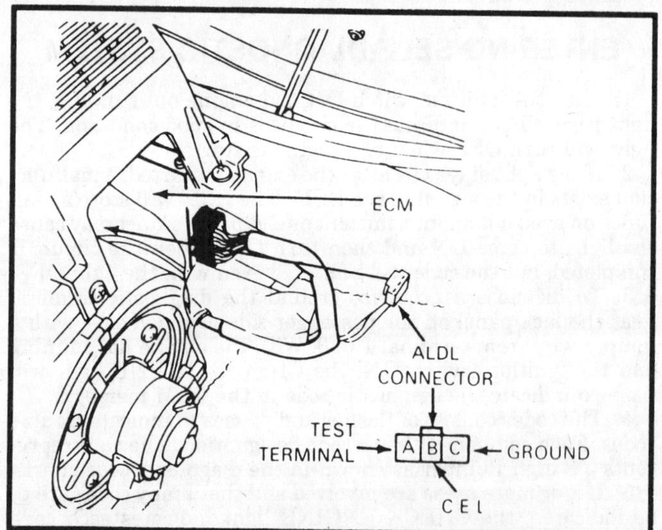


**ALDL (assembly line diagnostic link)—Spectrum carbureted**

2. If the light stays ON after the engine is started, a malfunction exists in the signal to the ECM. The ECM will store a malfunction even if it an intermittent one. A poor contact may cause the light to come ON and then turn OFF when the circuit is completed, but the code will be stored even with the light OFF.

3. To diagnose a problem, turn ON the diagnostic switch located under the instrument panel. With the engine not running and the ignition switch **ON**, the CHECK ENGINE light will flash to indicate the diagnostic code in the ECM memory.

4. The codes consist of flashes and pauses to indicate trouble codes. Each component has a code designation. The code represents a 2 digit number as shown in the diagnostic code chart. Code 12 is 1 flash, a short pause and 2 flashes. This code means the ECM is in the self-diagnosis mode and it is working. This code is not a problem code.

5. If 2 or more areas are involved and that many codes are to be indicated, the CHECK ENGINE light indicates each code corresponding to the area of trouble 3 times in the increasing order of the code numbers and then repeats them.

**NOTE: Always write down the codes as they are flashing. This memory will be lost when the power to the ECM is disconnected.**

## CLEARING TROUBLE CODES

The code will be stored until the power from the battery to the ECM is cut off. After correcting the problem, erase the memory by removing the ECM fuse in the fuse block or disconnect the negative battery cable for more than 20 seconds. Disconnecting the battery cable will cancel clock, radio and any other memory functions.

**NOTE: Some vehicle computers have a learning ability. If a change is noted in vehicle performance after clearing codes, it may be due to the computers learning ability. To restore performance, warm vehicle to normal operating temperature and drive at part throttle with moderate acceleration, before performing any additional diagnosis.**

# HONDA PROGRAMMED FUEL INJECTION (PGM-FI) AND FEEDBACK CARBURETOR

## General Information

### PRELUDE FEEDBACK CARBURETOR

The carburetor consists of 2 side draft carburetors, each having a variable venturi. The variable venturi carburetors allow a smooth increase of engine speed and output, due to the change in venturi area in proportion to the carburetor intake airflow rate.

#### Feedback control system

The feedback control system maintains the proper air/fuel mixture ratio by allowing air into the intake manifold, as is necessary, to adjust a temporarily fuel-rich condition. This system in made up of 4 subsystems: the air/fuel ratio control, shot of air control, deceleration air supply and hot engine start control.

**AIR/FUEL RATIO**

This system is designed to achieve a stoichimetric air/fuel ratio, needed for proper operation of the 3-way catalyst. The carburetor mixture is calibrated on the richer side. The air supply through the EACV dilutes the mixture for controlling the mixture as needed.

**SHOT OF AIR CONTROL**

This system provides air into the intake manifold to improve emissions performance and prevent afterburning due to over rich mixtures during deceleration.

The control unit receives signals from speed, coolant temperature, MAP and rpm sensors. The air shot is induced from the EACV when the manifold vacuum increases suddenly except when the vehicle is moving at very low speed with the coolant temperature below normal. The amount of air provides is based on the amount of manifold vacuum.

**DECELERATION AIR SUPPLY**

This system is designed to improve emission performance by supplying air into the intake manifold during deceleration in relatively high engine rpm.

The control unit receives signals from speed, coolant temperature, MAP, gear position and rpm sensors. The air is induced from the EACV, based on signals from the ECU.

**HOT ENGINE START CONTROL**

This system is designed to provide air into the intake manifold for engine starting when the coolant temperature is very high. The EACV provides the additional air flow based on signals from the ECU.

### PROGRAMMED FUEL INJECTION (PGM-FI) SYSTEM

Programmed Fuel Injection (PGM-FI) System consists of 3 subsystems: air intake, electronic control, and fuel. The Electronic Control System (ECU) in order to get fuel into the cylinders at the correct instant and in correct amount must perform various separate functions. The ECU (Electronic Control Unit), the heart of the PGM-FI, uses an 8-bit microcomputer and consists of a CPU (Central Processing Unit), memories, and I/O (Input/Output) ports. Basic data stored in the memories are compensated by the signals sent from the various sensors to provide the correct air/fuel mixture for all engine needs.

The unit contains memories for the basic discharge duration at various engine speeds and manifold pressures. The basic discharge duration, after being read out from the memory, is further modified by signals sent from various sensors to obtain the final discharge duration. Other functions also include:

- Starting Control – The fuel system must vary the air/fuel ratio to suit different operating requirements. For example, the mixture must be rich for starting. The memories also contain the basic discharge durations to be read out by signals from the starter switch, and engine speed and coolant temperature sensors, thereby providing extra fuel needed for starting.
- Injector Control – The ECU controls the discharge durations at various engine speeds and loads.
- Electronic Air Control – The ECU controls the EACV to maintain correct idle speed based on engine and accessories demand.
- Ignition Timing Control – The ECU controls the basic ignition timing based on engine load, engine rpm, vehicle speed and coolant temperature.
- Fuel Pump Control – When the speed of the engine falls below the prescribed limit, electric current to the fuel pump is cut off, preventing the injectors from discharging fuel.
- Fuel Cut-Off Control – During deceleration with the throttle valve nearly closed, electric current to the injectors is cut off at speeds over 900 rpm, contributing to improved fuel economy. Fuel cut-off action also takes place when engine speed exceeds approximately 7500 rpm regardless of the position of the throttle valve.
- Safety – A fail-safe system monitors the sensors and detects any abnormality in the ECU, ensuring safe driving even if one or more sensors are faulty, or if the ECU malfunctions.
- Self-Diagnosis – When a abnormality occurs, the ECU lights the engine warning light and stores the failure code in erasable memory. The ECU LED will display the code any time the ignition is turned **ON**.

## Self-Diagnostic System

### ENTERING SELF-DIAGNOSTIC AND DIAGNOSTIC CODE DISPLAY

#### Self-Diagnosis Indicators

The quick reference chart covers the most common failure

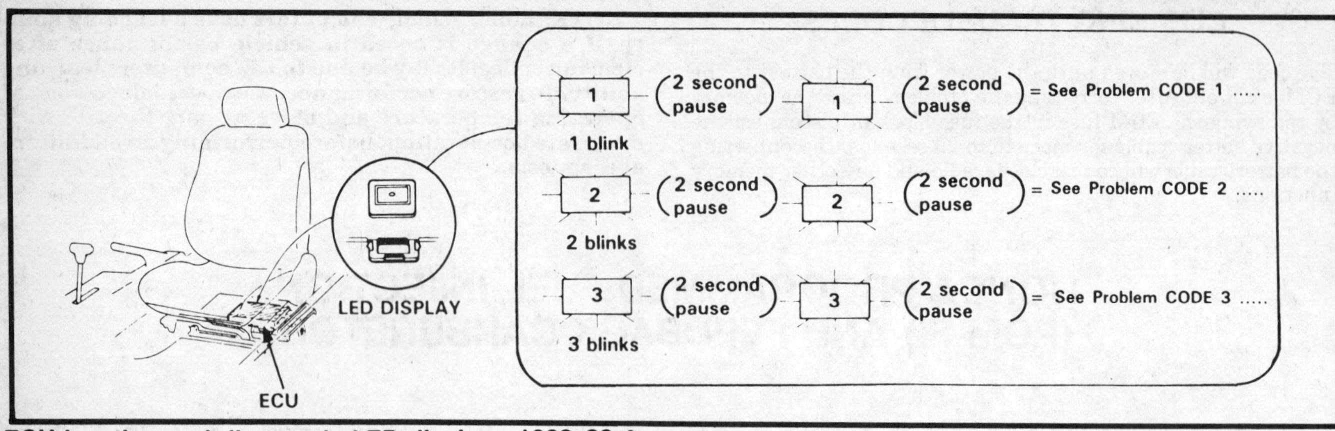


ECU location and diagnostic LED display—1988–89 Accord

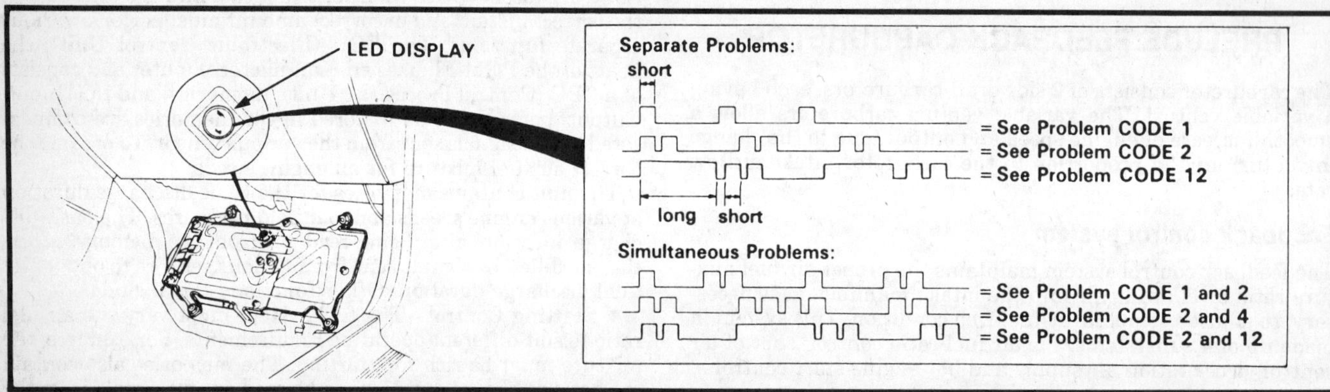


ECU location and diagnostic LED display—1990 Accord

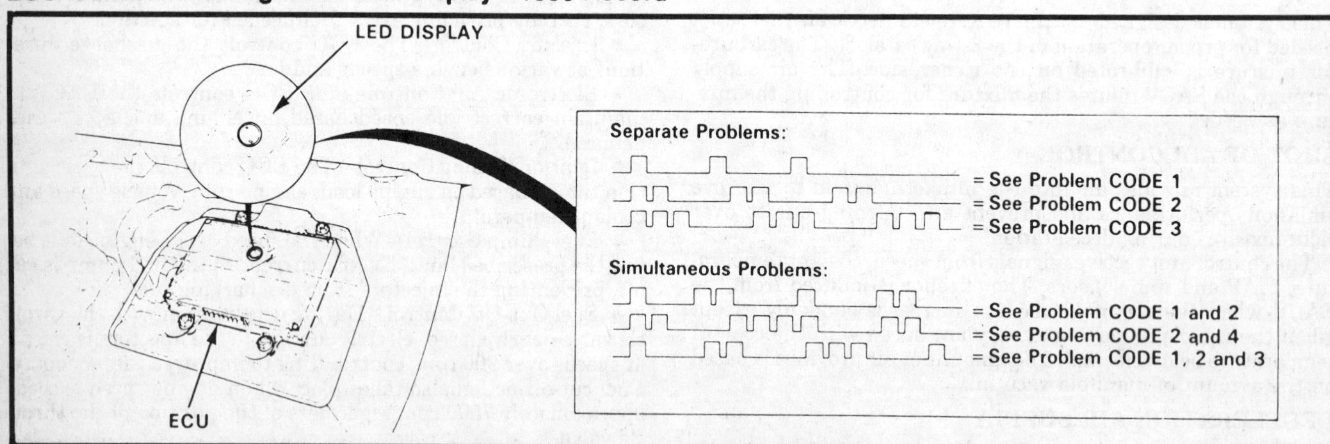


ECU location and diagnostic LED display—Civic

modes for the PRG-CARB and PGM-FI. The probable causes are listed in order of most-easily-checked first, then progressing to more difficult fixes. Run through all the causes listed. If problem is still unsolved, go on to the more detailed troubleshooting. Troubleshooting is divided into different LED displays. Find the correct light display and begin again.

For all the conditions listed, the PRG-CARB and PGM-FI warning light on the dashboard must be on (comes on and stays on). This indicates a problem in the electrical portion of the fuel injection system. At that time, check the LED display (self-diagnosis system) in the ECU.

There is only one LED display. The LED will blink consecutively to indicate the trouble code. The ECU is located under the driver's seat on the 1988–89 Accord or on other models beneath the carpet on the passenger's side under the dash.

Sometimes the dash warning light and/or ECU LED will come on, indicating a system problem, when, in fact, there is only a bad or intermittent electrical connection. To troubleshoot a bad connection, note the ECU LED pattern that is lit, refer to the diagnosis chart and check the connectors associated with the items mentioned in the "Possible Cause" column for that LED pattern (disconnect, clean or repair if necessary and reconnect those connections). Then, reset the ECU memory as described, restart the car and drive it for a few minutes and then recheck the car and drive it for a few minutes and then recheck the LED. If the same pattern lights up, begin system troubleshooting; if it does not light up, the problem was only a bad connection.

The memory for the PRG-CARB and PGM-FI warning light on the dashboard will be erased when the ignition switch is turned **OFF**; however, the memory for the LED display will not

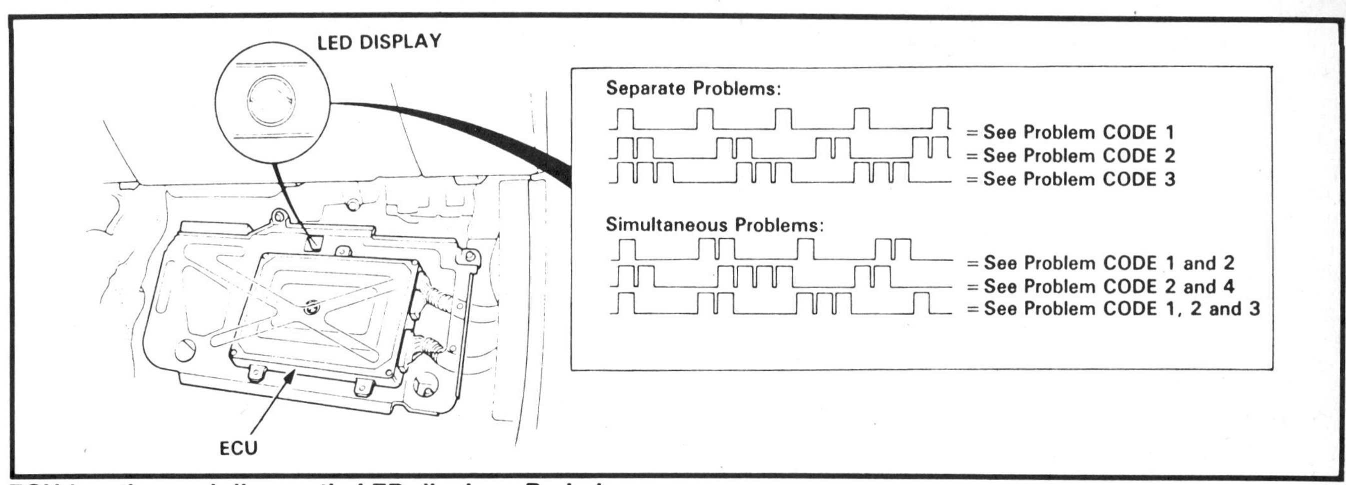

**ECU location and diagnostic LED display—Prelude**

be canceled. Thus, the warning light will not come on when the ignition switch is again turned **ON** unless the trouble is once more detected. Troubleshooting should be done according to the LED display even if the warning light is OFF.

**NOTE: On the 1990 Accord, if the terminals of the service connector are connected with a jumper wire the LED on the ECU and the CHECK ENGINE light will indicate the same codes.**

If the LED blinks codes that don't exist, clear the ECU, road test the vehicle and check them again. If the same codes repeat, replace the ECU.

## CLEARING TROUBLE CODES

The memory for the PRG-CARB and PGM-FI warning light on the dashboard may be erased when the ignition switch is turned **OFF**; however, the memory for the LED display will not be canceled. Thus, the warning light may not come ON when the ignition switch is again turned **ON** unless the trouble is once more detected. Troubleshooting should be done according to the LED display even if the warning light is OFF.

To clear the ECU trouble code memory remove the ECU memory power fuse for at least 10 seconds. The correct fuse to remove is:

Civic—HAZARD fuse at the main fuse box
Prelude with PRG-CARB—EFI-ECU fuse from the underhood relay box
Prelude with PRG-FI—CLOCK fuse from the underhood relay box
1988–89 Accord—No. 11 CLOCK fuse from the underhood relay box
1990 Accord—BACK UP fuse from the underhood relay box

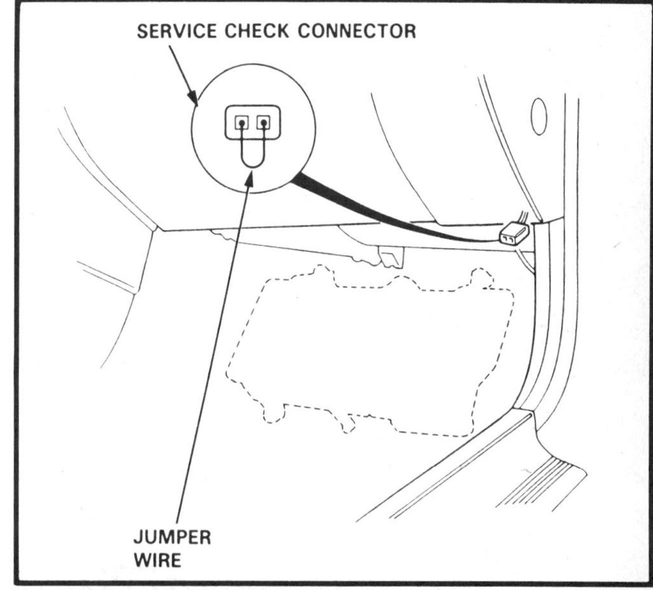

**Service check connector—1990 Accord**

**NOTE: Some vehicle computers have a learning ability. If a change is noted in vehicle performance after clearing codes, it may be due to the computers learning ability. To restore performance, warm vehicle to normal operating temperature and drive at part throttle with moderate acceleration, before performing any additional diagnosis.**

## 1988-90 HONDA DIAGNOSTIC TROUBLE CODES

| Trouble Codes | Accord | Civic | Civic CRX | Prelude PRG-CARB | Prelude PRG-FI |
|---|---|---|---|---|---|
| 0—Electronic control unit | X | X | X | | X |
| 1—Oxygen content | ② | X | X | X | |
| 1—Oxygen content A | ① | | | | X |
| 2—Oxygen content B | ① | | | | X |
| 2—Vehicle speed pulser | | | | X | |
| 3—Manifold absolute pressure | X | X | X | X | X |

## 1988-90 HONDA DIAGNOSTIC TROUBLE CODES

| Trouble Codes | Accord | Civic | Civic CRX | Prelude PRG-CARB | Prelude PRG-FI |
|---|---|---|---|---|---|
| 4—Crank angle | ② | X | X | | X |
| 4—Vacuum switch signal | | | | X | |
| 5—Manifold absolute pressure | X | X | X | X | X |
| 6—Coolant temperature | X | X | X | X | X |
| 7—Throttle angle | X | X | X | | X |
| 8—Crank angle TDC position | X | X | X | | X |
| 8—Ignition coil signal | | | | X | |
| 9—No. 1 cylinder position | X | ③ | ⑤ | | X |
| 10—Intake air temperature | X | X | X | X | X |
| 12—Exhaust gas recirculation system | X | ④ | ⑥ | | X |
| 13—Atmospheric pressure | X | X | X | | X |
| 14—Electronic air control | X | X | X | X | X |
| 15—Ignition output signal | ② | X | X | | X |
| 16—Fuel injector | ② | X | X | | X |
| 17—Vehicle speed sensor | X | X | X | | X |
| 19—Lock-up control solenoid valve (A/T) | | X | X | | |
| 20—Electric load detector | ② | X | X | | |
| 30—A/T F1 signal A | ② | | | | |
| 31—A/T F1 signal B | ② | | | | |
| 41—Oxygen sensor heater | ② | | | | |
| 43—Fuel supply system | ② | | | | |

① 1989–90 only
② 1990 only
③ 1.6 L engine only
④ 1.5 L engine with Auto Trans (Calif.) only
⑤ HF and Si only
⑥ HF and Standard California with Auto Trans only

# HYUNDAI MULTI-POINT FUEL INJECTION (MPI)

## General Information

The air/fuel ratio is controlled by varying the injector driving time. An electric fuel pump supplies sufficient fuel to the injection system and the pressure regulator maintains a constant pressure to the injectors. These injectors inject a metered quantity of fuel into the intake manifold in accordance with signals from the Electronic Control Unit (ECU) or engine computer. After pressure regulation excess fuel is returned to the fuel tank.

When the injector is energized, the valve inside the injector opens fully to inject the fuel. The injectors inject fuel to each manifold port, in the sequential firing order of each cylinder, as controlled by the ECU. The injectors fire simultaneous or non-synchronous injection, when during engine starting. During acceleration, additional fuel in a proportionate amount is injected for a fixed time of 10 milliseconds to 2 cylinders during intake and exhaust strokes. After starting and during normal engine operation, the ECU activates the injectors sequentially or as synchronous injection. Meaning that after each stroke the injectors are activated at the exhaust stroke of each cylinder.

Based on information from various sensors the ECU computes an optimum control for fuel mixture and ignition timing. The ECU consists of an 8-bit microprocessor, Random Access Memory (RAM), Read Only Memory (ROM) and input and output signal interface system.

### ELECTRONIC CONTROL SYSTEM

The ECU monitors the input and output signals of several sensors. Based on the sensor output the ECU controls fuel and ignition timing functions through the use of actuators.

When the ECU detects a problems for a specified time, the ECU memorizes the trouble code and output the signal to the self-diagnostic connector under the left side dash or kick panel.

## Self-Diagnostic System

### ENTERING SELF-DIAGNOSTIC AND DIAGNOSTIC CODE DISPLAY

**NOTE: If the battery voltage is low, no trouble codes will be stored by the ECU.**

| Trouble code | Diagnosis item | Trouble code | Diagnosis item |
|---|---|---|---|
| 11 | Oxygen sensor | 23 | No.1 cylinder top dead center sensor |
| 12 | Air-flow sensor | 24 | Vehicle-speed reed switch |
| 13 | Intake air temperature sensor | 25 | Barometric pressure sensor |
| 14 | Throttle position sensor | 41 | Injector |
| 15 | Motor position sensor | 42 | Fuel pump |
| 21 | Engine coolant temperature sensor | 43 | EGR temperature sensor (California Vehicles Only) |
| 22 | Crank angle sensor | | |

**Engine diagnostic codes**

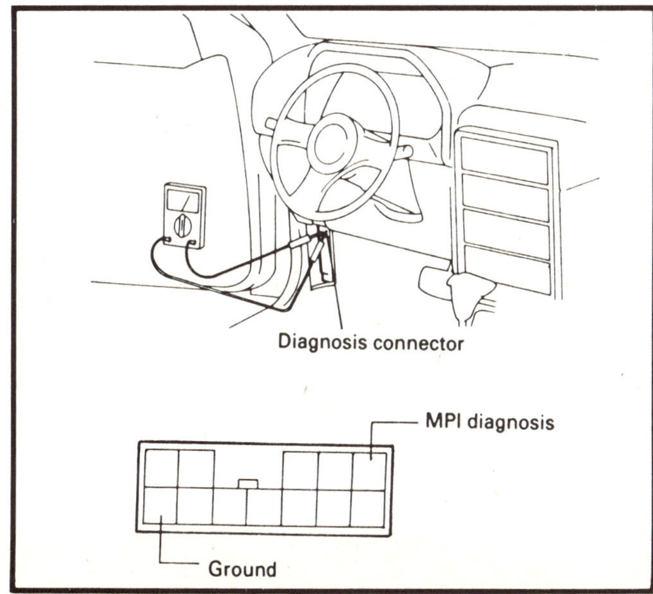


**Diagnosis connector and terminal location**

1. Connect the voltmeter to the self-diagnostic connector, across the MPI diagnosis and ground terminals.
2. Turn the ignition switch **ON**. The ECU diagnostics memory will immediately start.
3. If the voltmeter displays a steady needle sweep, the system is normal and no codes are in the memory.
4. If the voltmeter displays a steady **HIGH**, signal the ECU is damaged.
5. If the ECU has detected a malfunction, the voltmeter will deflect the indicating the diagnostic code.
6. Record the codes displayed by the voltmeter needle deflections.
7. The ECU will continue to send any memorized trouble codes to the self-diagnostic for as long as the ignition is **ON**, or until the codes have been cleared from memory.

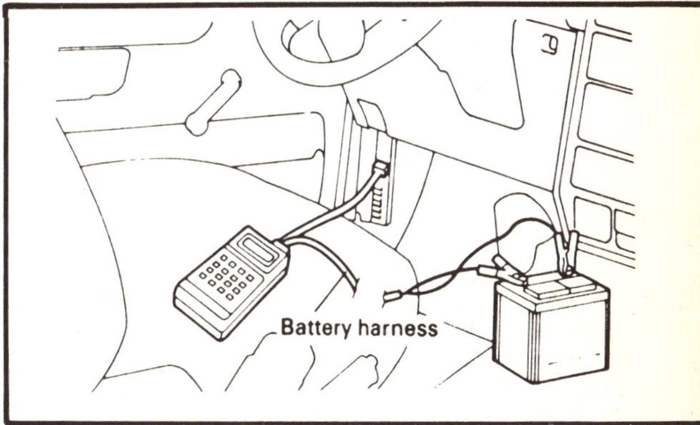


**Use separate battery during cranking test**

**NOTE: If using a multi-use tester to pull engine codes, it is necessary to power the tester from a separate battery if test is made during cranking. Power to the cigarette lighter is interrupted during cranking.**

## CLEARING TROUBLE CODES

1. Disconnect the negative battery cable for at least 15 seconds.
2. Reconnect the negative battery cable.
3. Start engine and allow to reach normal operating temperature.
4. Road test vehicle at speeds above 10 mph.
5. Check that no trouble codes have reset.

**NOTE: Some vehicle computers have a learning ability. If a change is noted in vehicle performance after clearing codes, it may be due to the computers learning ability. To restore performance, warm vehicle to normal operating temperature and drive at part throttle with moderate acceleration, before performing any additional diagnosis.**

# ISUZU WITH FEEDBACK CARBURETOR

## General Information

The FBC closed loop emission control system precisely controls the air/fuel ratio near the optimum mixture and allows the use of the 3-way catalyst to reduce the oxides of nitrogen and oxidize hydrocarbons and carbon monoxide. The essential components of the closed loop system are the coolant temperature sensor, oxygen sensor, Electronic Control Module (ECM), Feedback Carburetor (FBC), 3-way catalytic converter, idle and WOT switches (1988 Pick-Up), idle switch and MAP sensor (1989-90 Amigo and Pick-Up), duty solenoid, fuel cut solenoid (Amigo and Pick-up).

### ELECTRONIC CONTROL MODULE

The ECM generates a control signal to the vacuum controller solenoid which controls carburetor air/fuel ratio through vacuum signals. This control signal is continuously cycling the solenoid between on and off time (duty cycle) as a function of the input voltages from the sensors. The control signal generated by the ECM is selected from 4 operational modes as follows:

1. Inhibit Mode – no signal to the vacuum controller solenoid.
2. Enrichment Mode – a fixed pre-programmed duty cycle to the vacuum controller solenoid.
3. Open Loop Mode – a fixed pre-programmed duty cycle to the vacuum controller.
4. Close Loop Mode – a calculated duty cycle is generated based on oxygen sensor and other sensor outputs.

During closed loop operation, the ECM monitors the voltage output of the oxygen sensor. As the sensor voltage increases and passes through the ECM threshold set point, the proportional gain immediately changes the duty cycle of the output signal. The duty cycle is further changed at a constant rate (integral gain) until the sensor input voltage decreases and passes through the ECM threshold point.

The selection integral and proportional gain rates by the ECM is based on the engine operating conditions (idle or off-idle condition). At idle condition different gain rates are required for optimum air/fuel ratio control than those at partial load condition. The ECM also stores in an adaptive memory the current duty cycle being used for either idle or off-idle condition (below and above the adaptive switch point). When the ECM sees a transition from idle condition (as signaled by the vacuum switch) to off-idle condition, it immediately steps to the duty cycle last recorded for stoichiometric operation. From then on while at that engine operating condition the system uses the basic proportional and integral gain controls as previously described.

The ECM outputs the signal to control the slow cut solenoid valve incorporated in the carburetor. The ECM senses the coasting condition by means of the signals from the transmission gear position switch, clutch pedal position switch and idle position switch, cutting off the flow of electric current to slow the cut solenoid valve in such a instance when the engine running speed is beyond the limit of the speed specified, thus stopping the fuel flow to the carburetor. The flow of the electric current to slow the cut solenoid valve is cut off only at a time when the following 2 conditions exist simultaneously. They are:

1. The vacuum signal of the vacuum switch is below the specified vacuum.
2. The engine speed exceeds the limit of the specified speed.

### SYSTEM MALFUNCTION LAMP

This system utilizes a dashboard mounted malfunction indicator lamp which, for some failure modes will inform the driver of the need for unscheduled maintenance. In the event of a system malfunction the Legend CHECK ENGINE will light and remain ON as long as the fault function occurs and the engine is running. The electronic control unit incorporates a diagnostic program which will assist in diagnosing the closed loop control system faults. When activated the diagnostic program will flash a code through the malfunction lamp which isolates the source of the system fault.

### SELF-DIAGNOSTIC SYSTEM

The diagnostic circuit check makes sure that the self-diagnostic system works, determines that the trouble codes will display and guides diagnosis to other problem areas. When the engine is running and a problem develops in the system which the self diagnosis can evaluate, the CHECK ENGINE light will come on and a trouble code will be stored in the ECM trouble code memory. The light will remain ON with the engine running as long as there is a problem. If the problem is intermittent the CHECK ENGINE light will go OFF, but the trouble codes will be stored in the ECM trouble code memory.

With the ignition turned **ON** and the engine stopped, the CHECK ENGINE lamp should be ON. This is a bulb check to indicate the light is working properly. The trouble code test leads are located as follows:

On the I-Mark, a 3 terminal connector is located near the ECM connector, this connector is used to actuate the trouble code system in the ECM. This connector is also known as the Assembly Line Diagnostic Link (ALDL) or the Assembly Line Communications Link (ALCL). Terminals A and C of this connector are used to activate the trouble code system in the ECM.

On the Impulse, Amigo and Pick-Up a trouble code "TEST" lead (white cable) and a ground lead (black cable) are branched from a harness at a distance of 8 in. from the ECM connector (next to the clutch pedal).

## Self-Diagnostic System

### ENTERING SELF-DIAGNOSTIC AND DIAGNOSTIC CODE DISPLAY

With the ignition turned **ON** and the engine stopped, the CHECK ENGINE lamp should be ON. This is a bulb check to indicate the light is working properly.

Connect the ALDL leads, on the 3 terminal connector, jumper the A and C terminals (outer terminals) or on the Amigo and Pick-Up connect the trouble code "TEST" lead (white cable) and a ground lead (black cable), located 8 in. from the ECM connector (next to the clutch pedal).

On Amigo and Pick-Up, the trouble code memory is activated by placing the ignition switch in the **ON** position and connecting the trouble code "TEST" lead to the ground lead. On the I-Mark vehicles, the trouble code memory is activated by placing the ignition switch in the **ON** position and running a jumper wire between terminals **A** and C of the ALDL connector.

The CHECK ENGINE light will begin to flash a trouble Code 12. Code 12 consists of 1 flash, a short pause and then 2 more flashes. There will be a longer pause and a Code 12 will repeat 2 more times. The check indicates that the self-diagnostic system is working. This cycle will repeat itself until the engine is started or the ignition switch is turned **OFF**. If more than a single fault code is stored in the memory, the lowest number code will flash 3 times followed by the next highest code number until all the

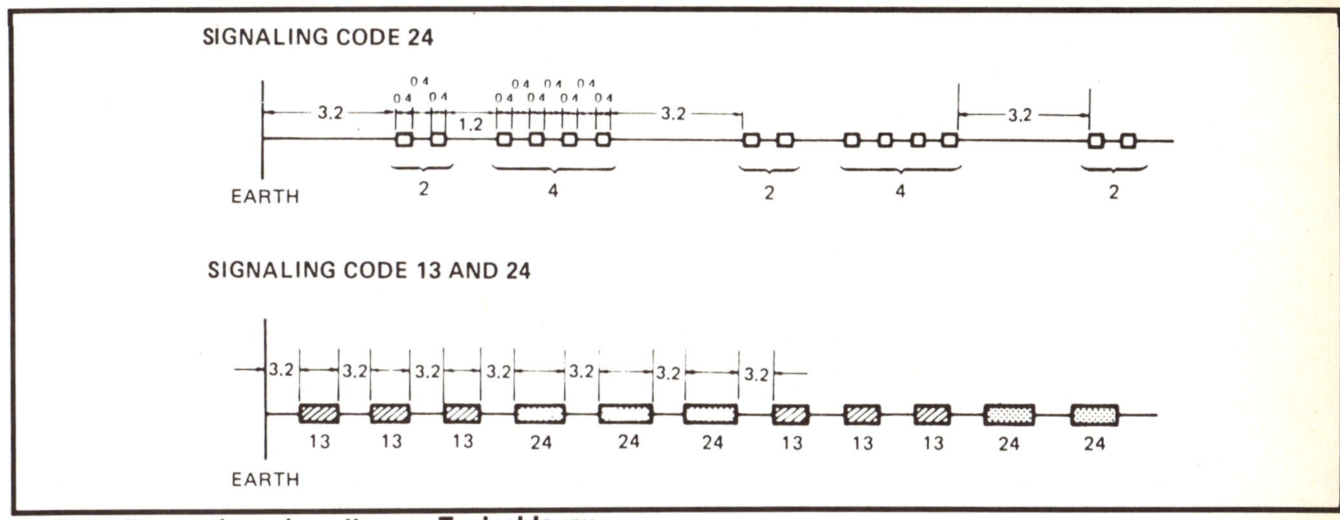

Reading diagnostic code patterns—Typical Isuzu

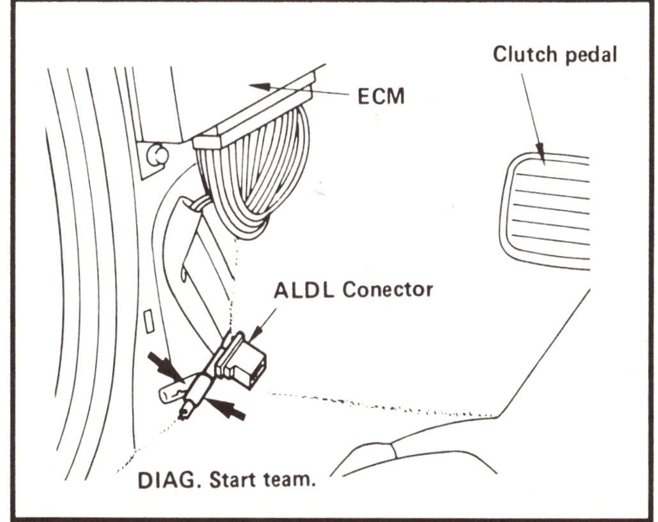

ALDL location—Amigo and Pick-up

codes have been flashed. The faults will then repeat in the same order. In most cases, codes will be checked with the engine running since no codes other than Codes 12 or 51 will be present on the initial "KEY ON". Remove the ground from the test terminal before starting the engine.

**NOTE: The fault indicated by trouble Code 15 takes 5 minutes of engine operation before it will display. The diagnostic charts for trouble Codes 13, 31, 44 and 45 should be used if any of these trouble codes can be obtained.**

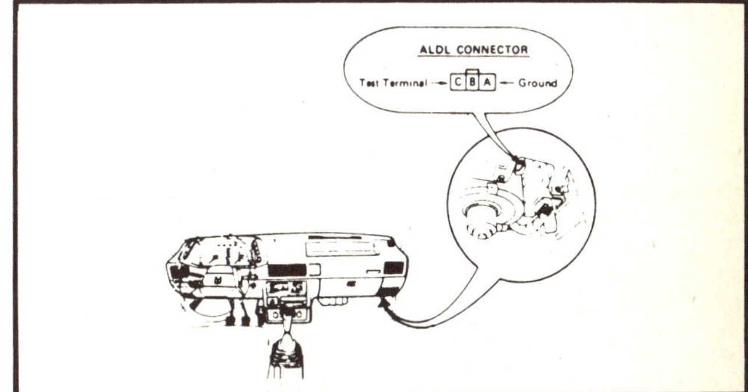

ALDL location—I-Mark

## CLEARING THE TROUBLE CODE MEMORY

The trouble code memory is fed a continuous 12 volts even with the ignition switch in the **OFF** position. After a fault has been corrected, it will be necessary to remove the voltage for 10 seconds to clear any stored codes. The quickest way to remove the voltage is to remove the "ECM" fuse from the fuse block or the MAIN 60A fuse for 10 seconds. The voltage can also be removed by disconnecting the negative battery cable, but this will mean if the vehicle is equipped with electronic instrumentation, such as a clock and radio, they would have to be reset.

**NOTE: Some vehicle computers have a learning ability. If a change is noted in vehicle performance after clearing codes, it may be due to the computers learning ability. To restore performance, warm vehicle to normal operating temperature and drive at part throttle with moderate acceleration, before performing any additional diagnosis.**

# ISUZU FUEL INJECTION SYSTEM EXCEPT 2.8L V6

## General Information

The fuel injection control system constantly monitors and controls engine operation, which in turn helps lower emissions while maintaining the fuel economy and driveability . The Electronic Control Unit (ECU) controls the fuel injection system, the ignition system and the turbocharger control system (if so equipped).

In general, the fuel injection system consists of the following components; the crank angle sensor, the throttle valve switch, the vehicle speed sensor, the coolant temperature sensor, air flow sensor, oxygen sensor, fuel injectors and the ECU.

The turbocharger system, if equipped, consists of the following components; the throttle valve switch, the throttle position sensor, the knock sensor, the coolant temperature sensor, air flow sensor, oxygen sensor, the stepping motor, the turbocharger controller and the ECU.

## ELECTRONIC CONTROL UNIT

The ECU is usually located under the instrument panel. The ECU analyzes all electrical data signals from the sensor. It controls the fuel injection system, the ignition system and the turbocharger control system (is so equipped). The ECU has a built in back-up, diagnostic and fail safe control systems.

The ECU contains a Back-Up System, which is used in case there is a malfunction with the microcomputer within the ECU, the back-up control system works to maintain the necessary functions of the control unit to permit continuous operation of the vehicle.

The sensors used in this system, provide electrical impulses to the ECM by monitoring the pressure, temperature, vacuum and other engine operating conditions.

## SELF-DIAGNOSTIC SYSTEM

In the event of a failure, the ECU stores the trouble code in memory and operates the CHECK ENGINE light on the instrument panel when the nature of trouble is important, to warn the operator of failure.

**NOTE: The self-diagnosis system is capable of troubleshooting the electrical circuits in the Closed Loop Emission Control system only and does not cover the trouble in the sensors, actuators or the engine itself.**

With the engine not running, first, a Code 12 is flashed 3 times, which indicates that the ECM is functioning normally. If there are other trouble codes in the ECU, it displays each code 3 times at 3.2 second intervals each, in ascending order of code numbers. When all the codes are displayed, the computer will go through the entire code display again.

The malfunction codes are also displayed when the engine is running, but a Code 12 will not be indicated.

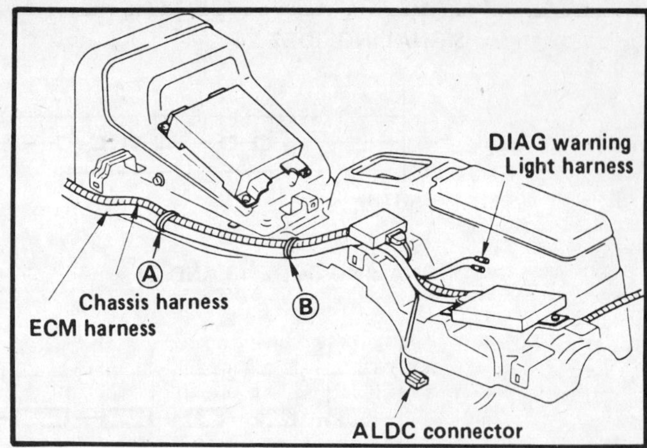

ALDL location—Trooper and Trooper II

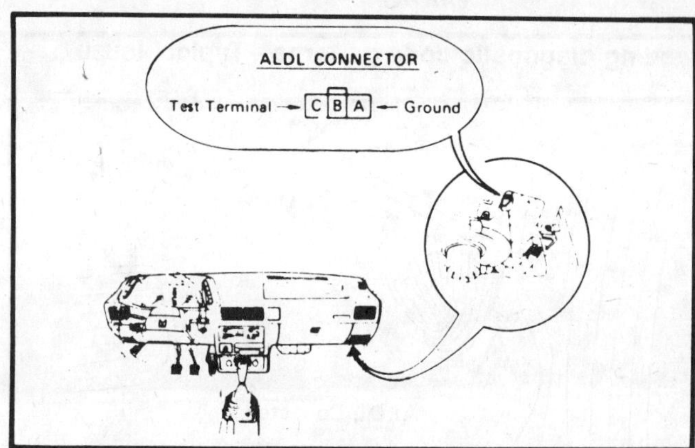

ALDL location—I-Mark

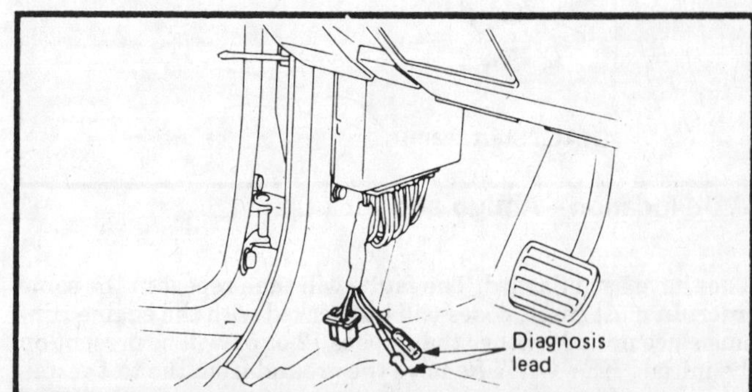

ALDL location—Amigo, Pick-up and Impulse

# Self-Diagnostic System

## ENTERING SELF-DIAGNOSTIC AND DIAGNOSTIC CODE DISPLAY

When the diagnosis lead in the vicinity of the control unit is connected with the ignition switch in the **ON** position (engine not running), the trouble codes stored in the memory will be displayed by the CHECK ENGINE light.

To display codes on vehicles with 2 separate terminals, connect the terminals together. On vehicles with a 3 terminal connector, jumper the A and C terminals (outer terminals) together. Codes are read by the CHECK ENGINE light flashes.

## CLEARING TROUBLE CODES

Clear the trouble codes stored in memory by disconnecting the memory fuse in the fuse junction block or by disconnecting the negative battery cable for at least 20 seconds, then check that only Code 12 is displayed.

Since all memory is cleared when disconnecting the battery cable or removing the fuse, the clock, radio and other electrical equipment will need to be reset.

**NOTE: Some vehicle computers have a learning ability. If a change is noted in vehicle performance after clearing codes, it may be due to the computers learning ability. To restore performance, warm vehicle to normal operating temperature and drive at part throttle with moderate acceleration, before performing any additional diagnosis.**

# TROOPER AND TROOPER II WITH 2.8L ENGINE

## General Information

### CHECK ENGINE LIGHT

The CHECK ENGINE light is located on the instrument panel and has 2 functions:

1. It alerts the driver that a problem has occurred in the system and that the car should be taken to the service station for evaluation as soon as possible.
2. Once in the shop, the light allows the technician to read the trouble codes to help diagnose system problems. Normally, the light will go ON with the ignition switch **ON** and the engine not running. If the ECM has detected a problem within the system, the light will remain ON. If the problem goes away, the light will go OFF in most cases but a trouble code will remain stored in the ECM.

#### Intermittent CHECK ENGINE light

Fault codes are generally divided into 2 categories: intermittent and hard.

An intermittent code is defined as a code that does not re-set itself, but the fault condition is not active when the technician is working on the vehicle. The most probable cause of intermittent faults are loose wiring connections.

A hard code is defined as present when the vehicle is brought in for service and that the ECM recognizes and displays to the technician.

### TROUBLE CODES

The codes stored in the ECM's memory can be read either through a hand held diagnostic scanner plugged into the ALDL connector or by counting the number of flashes on the "CHECK ENGINE" light when the diagnostic terminal of the ALCL connector is grounded. The ALDL connector terminal **B** (diagnostic terminal) is the second terminal from the right of the top row in the ALDL connector. The terminal is most easily grounded by connecting to terminal **A** (internal ECM ground), the terminal to the right of the terminal **B** on the top row in the ALDL connector.

Once terminals **A** and **B** are grounded, the ignition switch must be turned **ON** with the engine not running. At this point the CHECK ENGINE light should flash a Code 12, 3 times consecutively. This would be the following flash sequence; flash, pause, flash-flash, long pause, flash, pause, flash-flash, long pause, flash, pause, flash-flash.

Code 12 indicates that the ECM's diagnostic system is operating. If Code 12 is not indicated, a problem is present within the diagnostic system itself and should by addresses by using the appropriate diagnostic chart.

Following the output of Code 12, the CHECK ENGINE light will indicate a diagnostic trouble code 3 times, if a code is present or it will simply continue to output Code 12. If more than a single code has been stored in the ECM's memory, the codes will be output from the lowest to the highest, with each code being displayed 3 times.

#### Diagnostic Mode

**NOTE: In the Diagnostic mode, codes can be read only with the engine stopped. Grounding the Diagnostic terminal with the engine running sends the ECM into the "FIELD SERVICE MODE" described below.**

If the diagnostic terminal of the ALDL connector is grounded with the ignition switch in the **ON** position and the engine not running, the ECM will enter the Diagnostic mode. In this mode the ECM will perform the following functions:

1. Display a Code 12 by flashing the CHECK ENGINE light which indicates that the system is functioning properly. A Code 12 consists of 1 flash, followed by a short pause, then 2 quick flashes. The code will be repeated 3 times. If not other codes are stored in the system, code will continue to flash until the diagnostic terminal is ungrounded.
2. Display any stored trouble codes by flashing the CHECK ENGINE light. Each code will be code will be flashed 3 times, then Code 12 will be flashed again. If a trouble code is displayed, use the diagnostic code trouble code charts to troubleshoot the system.
3. Energize the all ECM relays and solenoids except the fuel pump relay.
4. Move the AIC valve to the fully extended position.

#### Field Service Mode

The Field Service Mode is entered by grounding the diagnostic terminal with the engine running and will indicate to the technician whether the ECM is in "Open" or "Closed" loop.

In "Open Loop" the CHECK ENGINE flashes 2½ time a second, in "Closed Loop" the CHECK ENGINE light once per second or will stay OFF most of the time if the system is too lean or ON most of the time if the system is too rich.

When the system is in the Field Service Mode, new trouble codes cannot be stored in the ECM and the closed loop timer is by-passed.

## Self-Diagnostic System

### ENTERING SELF-DIAGNOSTIC AND DIAGNOSTIC CODE DISPLAY

The ALDL diagnostic terminal **B** must be jumper to the ground terminal **A**, the ignition switch must be turned **ON** with the engine not running. At this point the CHECK ENGINE light should flash each code 3 times consecutively.

F E D C B A
G M

TERMINAL IDENTIFICATION

| | | | |
|---|---|---|---|
| A | GROUND | E | SERIAL DATA |
| B | DIAGNOSTIC TERMINAL | F | TCC (IF USED) |
| C | A.I.R. (IF USED) | G | FUEL PUMP (IF USED) |
| D | CHECK ENGINE LIGHT (IF USED) | M | SERIAL DATA (IF USED) |

**ALDL terminal identification—2.8 V6**

### CLEARING TROUBLE CODES

When the ECM sets a trouble code, the CHECK ENGINE light will illuminate and the code will be stored in the computer's memory. If the problem is intermittent, the light will go OFF in 10 seconds when the fault goes away. However, the code will remain in the ECU's memory until battery voltage to the ECU is removed. Removing battery voltage for 30 seconds will clear all stored trouble codes. Trouble codes will be cleared from the ECU after the fault is repaired.

**NOTE: To prevent damage to the ECM, the ignition switch must be turned to the OFF position when disconnecting voltage to the ECM (battery cable, ECM pigtail, ECM fuse, jumper cables). Some vehicle computers have a learning ability. If a change is noted in vehicle performance after clearing codes, it may be due to the computers learning ability. To restore performance, warm vehicle to normal operating temperature and drive at part throttle with moderate acceleration, before performing any additional diagnosis.**

## ENGINE DIAGNOSTIC CODES

**CODE 12**–ECU functioning
**CODE 13**–Oxygen sensor
**CODE 14**–Coolant temperature HIGH
**CODE 15**–Coolant temperature LOW
**CODE 21**–Throttle position sensor HIGH
**CODE 22**–Throttle position sensor LOW
**CODE 24**–Vehicle speed sensor
**CODE 32**–EGR system
**CODE 33**–Manifold absolute pressure voltage HIGH
**CODE 34**–Manifold absolute pressure voltage LOW
**CODE 42**–Electronic spark timing circuit
**CODE 43**–Electronic spark control circuit
**CODE 44**–Oxygen sensor, exhaust Lean
**CODE 45**–Oxygen sensor, exhaust Rich
**CODE 51**–Prom error
**CODE 52**–Calpak error
**CODE 54**–Fuel pump circuit
**CODE 55**–ECM error

## 1988–90 ISUZU DIAGNOSTIC TROUBLE CODES

| Trouble Codes | I-Mark (non-turbo) | I-Mark (Turbo) | Impulse | Trooper and Trooper II | Amigo and Pick-Up |
|---|---|---|---|---|---|
| 12—No distributor reference pulse | X | X | X | X | X |
| 13—Oxygen sensor circuit | X | X | X | X | X |
| 14—Shorted coolant sensor circuit | X | X | X | X | X |
| 15—Open coolant sensor circuit | X | X | X | X | X |
| 21—Idle switch circuit | X | | X | ② | X |
| 21—Manifold air pressure sensor circuit | | | | | ① ④ |
| 21—Throttle position sensor circuit (high) | | X | | ③ | |
| 22—Fuel cut off circuit | X | | | | ① |
| 22—Throttle position sensor circuit (low) | | X | | ③ | |
| 22—Starter signal circuit | | | X | ② | ② |
| 23—Mixture control solenoid circuit | X | | | | ① |
| 23—Manifold air temperature sensor circuit (low) | | X | | ③ | |
| 23—Power transistor ignition shorted | | | X | ② | ② |
| 24—Vehicle speed sensor | | X | | ③ | |
| 25—Vacuum switch valve | X | | X | ② | X |
| 25—Manifold air temperature sensor circuit (high) | | X | | | ① |
| 26—Canister vacuum switch valve circuit | | | X | ② | X |
| 27—Vacuum switch valve (high) | | | X | ② | X |
| 31—No reference pulse | | | | | ① |
| 31—Turbo wastegate control | | X | | | |
| 32—EGR system failure | | X | X | X | X |
| 33—Manifold air pressure sensor (high) | | X | | ③ | |
| 33—Fuel injector system circuit | | | X | ② | ② |
| 34—EGR gas sensor circuit | | | X | ② | X |
| 34—Manifold air pressure sensor (low) | | X | | ③ | |
| 35—Power transistor ignition open | | | X | ② | ② |
| 41—Crank angle sensor circuit | | | X | ② | ② |
| 42—Fuel cut off relay circuit | X | | | | |
| 42—Electronic spark timing circuit | | X | | ③ | |
| 43—Electronic spark control circuit | | X | | ③ | |
| 43—Throttle valve switch circuit | | | X | ② | ② |

## 1988–90 ISUZU DIAGNOSTIC TROUBLE CODES

| Trouble Codes | I-Mark (non-turbo) | I-Mark (Turbo) | Impulse | Trooper and Trooper II | Amigo and Pick-Up |
|---|---|---|---|---|---|
| 44—Oxygen sensor circuit (lean) | X | X | X | X | X |
| 45—Oxygen sensor circuit (rich) | X | X | X | X | X |
| 51—Faulty prom | X | | | ③ | |
| 51—Fuel cut solenoid circuit shorted | | | | | ① |
| 51—MEM-CAL error | | X | | | |
| 51—Micro-computer unit | | | X | ② | ② |
| 52—Calpak error | | | | ③ | |
| 52—Faulty ECM | | | | | ① |
| 52—Micro-computer unit | | | X | ② | ② |
| 53—Vacuum switching valve shorted or ECM | X | | X | ② | ① |
| 54—Mixture control solenoid shorted or ECM | X | | | | |
| 54—Fuel pump circuit | | | | ③ | |
| 54—Vacuum control solenoid shorted or ECM | | | | | ① |
| 54—Power transistor ignition circuit | | | X | ② | ② |
| 55—Faulty ECM | X | | | ③ | ① |
| 61—Air flow sensor circuit | | | X | ② | ② |
| 62—Air flow sensor circuit | | | X | ② | ② |
| 63—Vehicle speed sensor circuit | | | X | ② | ② |
| 64—Fuel injector system circuit | | | X | ② | ② |
| 65—Throttle valve switch full contact | | | X | ② | ② |
| 66—Knocking sensor system circuit | | | X | | |
| 71—Throttle position sensor circuit | | | X | | |
| 72—EGR Vacuum switching valve circuit | | | X | | |
| 73—EGR Vacuum switching valve circuit | | | X | | |

① 2.3L engine only
② 2.6 L engine only
③ 2.8 L engine only
④ 1989–90 only

# MAZDA FUEL INJECTION

**NOTE: For further diagnosis and testing, refer to the fuel injection section and/or the carburetor feedback section.**

## TROUBLESHOOTING WITH THE SELF-DIAGNOSIS CHECKER

The self diagnosis checker (49–H018–9A1) and System Selector (49–B019–9A0, used on 1990 Miata) are used to retrieve code numbers of malfunctions which have happened and were memorized or are continuing. The malfunction is indicated by the code number and a buzzer.

If there is more than 1 malfunction, the code numbers will display on the self diagnosis checker 1 by 1 in numerical order. In the case of malfunctions, 09, 13, and 01, the code numbers are displayed in a order of 01, 09 and then 13.

The memory of malfunctions is canceled by disconnecting the negative battery cable for at least 5 seconds.

The ECU has a built in fail-safe mechanism for the main input sensors. If a malfunction occurs, the emission control unit will substitute values. This will slightly effect the driving performance, but the vehicle may still be driven.

The ECU continuously checks for malfunctions of the input devices. But, the ECU checks checks for malfunctions of the output devices within 3 seconds after turning the ignition switch to the **ON** position and the test connector is grounded.

The malfunction indicator light indicates a pattern the same as the buzzer of the self-diagnosis checker when the self-diagnosis check connector is grounded. When the self-diagnosis check connector is not grounded, the lamp illuminates steady while malfunction of the main input sensor occurs and goes OFF if the malfunction recovers. However, the malfunction code is memorized in the emission control unit.

## Inspection Procedure

### EXCEPT RX-7

1. On all models except 1990 Miata, Connect the Self Diagnosis Checker (49–H018–9A1) or equivalent to the check connector. The check connector is usually located above the right side wheel housing.
2. On 1990 Miata, connect Self Diagnosis Checker (49–H018–9A1) and System Selector (49–B019–9A0) to the diagnosis connector and ground.
3. On all models, set the select switch on the Self Diagnosis Checker to the "A" position.
4. On 1990 Miata, set the System Selector (49–B019–9A0) to position 1 and the SELF TEST switch to the up position.
5. On all models except 1990 Miata, ground the test connector using a suitable jumper wire.
6. On all models, turn the ignition switch to the **ON** position. Check that the number "88" flashes on the digital display and the buzzer sounds for 3 seconds after turning the ignition switch **ON**.
7. If the number "88" does not flash, check the main relay, power supply circuit and the check the check connector wiring.
8. On all models except 1990 Miata, if the number "88" flashes and the buzzer sounds continuously for more than 20 seconds, replace the ECU and perform Steps 3 and 4 again.
9. On 1990 Miata, if the number "88" flashes and the buzzer sounds continuously for more than 20 seconds, check for a short circuit between the ECU and terminal 1F on the diagnosis connector. Replace the ECU if necessary.
10. On all models, note the code numbers and check the causes, repair as necessary. Be sure to recheck the code numbers by performing the "After Repair Procedure," after repairing.

### RX-7

1. Start and allow the engine to reach normal operating temperature. Stop the engine.
2. Connect the Self Diagnosis Checker (49–H018–9A1) or equivalent to the check connector and the battery ground cable.
3. Set the select switch on the Self Diagnosis Checker to the "B" position on 1988 models, and the "A" position on 1989–90 models.
4. On 1988 models, check the Self-Diagnosis checker for trouble codes, and proceed to appropriate trouble code diagnostic chart.
5. On 1989–90 models, connect a jumper wire between the test connector and ground.
6. Turn the ignition switch to the **ON** position. Check that the number "88" flashes on the digital display and the buzzer sounds for 3 seconds after turning the ignition switch **ON**.
7. If the number "88" does not flash, check the check connector wiring.
9. If the number "88" flashes and the buzzer sounds continuously for more than 20 seconds, check for a short circuit between terminal 1F on the ECU and the check connector. Check ECU terminals 3X and 3Z voltages. Replace the ECU if necessary.
10. Note the code numbers and check the causes, repair as necessary. Be sure to recheck the code numbers by performing the "After Repair Procedure," after repairing.

## After Repair Procedure

1. Cancel the memory of malfunctions by disconnecting the battery negative cable and depressing the brake pedal for at least 5 seconds, then reconnect the battery ground cable.
2. On all models except 1990 Miata, connect the Self-Diagnosis Checker 49–H018–9A1 to the check connector. Ground the test connector (green: 1 pin) using a suitable jumper wire.
3. On 1990 Miata, connect Self-Diagnosis Checker (49–H018–9A1) and System Selector (49–B019–9A0) to the diagnosis connector.
4. On all models, turn the ignition switch to the **ON** position, but do not start the engine for approximately 6 seconds.
5. Start the engine and allow it to reach normal operating temperature, the run the engine at 2000 rpm for 2 minutes. Check that no code numbers are displayed.

## 1988–90 MAZDA DIAGNOSTIC TROUBLE CODES

| Code | | 1988 328 | 1989–90 323 | 1988–90 626/MX-6 Non-Turbo | 1988–90 626/MX-6 Turbo | 1988–90 929 | 1990 Miata | 1988 RX-7 | 1989–90 RX-7 | 1989–90 mpr 4-cyl. | 1989–90 mrp 6-cyl. | 1989–90 B2200 Carb | 1989–90 B2200/B2600 EFI |
|---|---|---|---|---|---|---|---|---|---|---|---|---|---|
| 1 | Ignition Pulse | X | X | X | X | X | X | ⑦ | X | X | X | X | X |
| 2 | Crank angle sensor (NE) signal | — | ⑥ | — | X | X | X | — | X | — | — | — | — |
| 2 | Air flow meter (1988 RX-7) | — | — | — | — | — | — | X | — | — | — | — | — |
| 3 | $G^1$ signal | ④ | ④ | — | X | X | X | — | X | X | X | — | X |
| 3 | Water temp. sensor (1988 RX-7) | — | — | — | — | — | — | X | — | — | — | — | — |
| 4 | $G^2$ signal | — | — | — | X | X | — | — | — | — | — | — | — |
| 4 | Intake air temp. sensor (1988 RX-7) | — | — | — | — | — | — | X | — | — | — | — | — |
| 5 | Knock sensor | — | — | — | X | ⑩ | — | — | ④ | — | — | — | — |
| 5 | Oxygen sensor (1988 RX-7) | — | — | — | — | — | — | X | — | — | — | — | — |
| 6 | Throttle sensor (1988 RX-7) | — | — | — | — | — | — | X | — | — | — | — | — |
| 7 | Boost sensor (1988 RX-7) | — | — | — | — | — | — | ④ | — | — | — | — | — |
| 8 | Air flow meter | X | X | X | X | X | X | — | X | X | X | — | X |
| 9 | Water temp. sensor | X | X | X | X | X | X | — | X | X | X | X | X |
| 9 | Atmosphere pressure sensor (1988 RX-7) | — | — | — | — | — | — | X | — | — | — | — | — |
| 10 | Intake air temp. sensor | X | X | X | X | X | X | — | X | — | — | — | — |
| 11 | Intake air temp. sensor | — | — | — | — | X | — | — | X | X | X | — | X |
| 12 | Throttle sensor | ④ | ④ | X | X | X | — | — | X | X | X | — | X |
| 12 | Coil igniter (trailing side) (1988 RX-7) | — | — | — | — | — | — | X | — | — | — | — | — |
| 13 | Intake manifold pressure sensor | — | — | — | — | — | — | — | X | — | — | X | — |
| 14 | Atmosphere pressure sensor | X | X | X | X | X | X | — | X | X | X | X | X |
| 15 | Oxygen sensor | X | X | X | X | ③ | X | — | X | X | X | X | X |
| 15 | Intake air temp. sensor (1988 RX-7) | — | — | — | — | — | — | X | — | — | — | — | — |

## 1988–90 MAZDA DIAGNOSTIC TROUBLE CODES

| Code | 1988 328 | 1989–90 323 | 1988–90 626/MX-6 Non-Turbo | 1988–90 626/MX-6 Turbo | 1988–90 929 | 1990 Miata | 1988 RX-7 | 1989–90 RX-7 | 1989–90 mpr 4-cyl. | 1989–90 mrp 6-cyl. | 1989–90 B2200 Carb | 1989–90 B2200/B2600 EFI |
|---|---|---|---|---|---|---|---|---|---|---|---|---|
| 16 EGR position sensor | — | — | ⑥ | X | X | — | — | — | — | — | X | — |
| 17 Feedback system | X | X | X | X | X | X | — | X | X | X | X | X |
| 18 Air/fuel solenoid valve | — | — | — | — | — | — | — | — | — | — | X | — |
| 18 Throttle sensor | — | — | — | — | — | — | — | X | — | — | — | — |
| 20 Metering oil pump sensor (MOP) | — | — | — | — | — | — | — | X | — | — | — | — |
| 22 Slow fuel cut solenoid valve | — | — | — | — | — | — | — | — | — | — | X | — |
| 23 Coasting richer solenoid valve | — | — | — | — | — | — | — | — | — | — | X | — |
| 23 Oxygen senser—right side | — | — | — | — | ⑩ | — | — | — | — | — | — | — |
| 24 Feedback system—right side | — | — | — | — | ⑩ | — | — | — | — | — | — | — |
| 25 Pressure regulator solenoid valve | X | ④ | X | X | X | — | — | X | X | X | — | X |
| 26 Vacuum switch solenoid valve | X | — | ⑥ | X | ⑪ | — | — | — | — | — | — | — |
| 26 Purge solenoid valve | — | ① | X | — | ② | X | — | ⑫ | X | X | X | X |
| 27 Purge solenoid valve | ① | ② | — | — | ⑪ ① | — | — | ⑫ | — | — | — | — |
| 28 EGR solenoid | — | — | X | X | X | — | — | — | — | — | — | — |
| 28 Duty solenoid vacuum valve | — | — | — | — | — | — | — | — | — | — | X | — |
| 29 Duty solenoid vent valve | — | — | — | — | — | — | — | — | — | — | X | — |
| 29 EGR solenoid valve | — | — | — | X | X | — | — | — | — | — | — | — |
| 30 ACV solenoid valve (split air) | — | — | — | — | — | — | — | X | — | — | X | — |
| 31 Solenoid relief valve | — | — | — | — | — | — | — | X | — | — | — | — |
| 32 Solenoid valve (switch) | — | — | — | — | — | — | — | X | — | — | — | — |
| 33 Port air solenoid valve | — | — | — | — | — | — | — | X | — | — | — | — |
| 34 Idle speed control solenoid | X | X | X | X | X | X | — | ⑧ | X | ⑧ | ⑨ | X |
| 35 A/T idle-up solenoid | — | — | — | — | — | — | — | — | — | — | X | |
| 36 Oxygen sensor heater relay | — | — | — | — | ⑩ | — | — | — | — | — | — | |
| 37 Metering oil pump (MOP) | — | — | — | — | — | — | — | X | — | — | — | |
| 38 Accelerated warm-up solenoid valve | — | — | — | — | — | — | — | X | — | — | — | |
| 40 Triple induction control system solenoid | — | — | — | — | X | — | — | ⑤ | — | — | — | |
| 41 Variable induction control system | — | ⑥ | — | — | X | — | — | X | — | X | — | |
| 42 Turbo boost duty solenoid | — | — | — | X | — | — | — | ④ | — | — | — | |
| 45 Vacuum solenoid valve or waste gate | — | — | — | — | — | — | — | — | — | — | X | |
| 51 Fuel pump resistor relay | — | — | — | — | — | — | — | X | — | — | — | |
| 71 Injector—front | — | — | — | — | — | — | — | X | — | — | — | |
| 73 Injector—rear | — | — | — | — | — | — | — | X | — | — | — | |

① No. 1 purge solenoid
② No. 2 purge solenoid
③ Left side on V6
④ Turbo
⑤ Non-Turbo auxiliary port valve
⑥ 1990
⑦ Crank angle signal
⑧ BAC valve
⑨ A/C idle up solenoid
⑩ 1990 V6
⑪ 1988–89
⑫ Step motor (MOP)

# MITSUBISHI FUEL INJECTION

## General Information

### SELF-DIAGNOSIS SYSTEM

Each fuel injected engine (ECI or MPI) contains a self-diagnosis system which monitors various input signals from the engine sensors and enters a trouble code in the on-board computer memory if a problem is detected. This constant checking by the ECU looks for both the presence of voltage and for the voltage to be within certain pre-determined limits; if either condition is not met, a fault code will be set in memory. There are several monitored items, including the "normal operation" code, all of which can be read either by using an analog (dial – type) voltmeter or the ECI/MPI tester and adapters. The adapters connect the tester to the on-board diagnosis connector.

Depending on the year and model, the diagnostic connector may be found either behind the glove box, behind the right front kick panel or just above the fusebox on the left side of the dashboard. Certain ECI systems have a separate connector under the hood.

Because the computer memory draws its power directly from the battery, the trouble codes are not erased when the ignition is switched OFF. The memory can only be cleared (trouble codes erased) if a battery cable is disconnected or the main ECU wiring harness connector is disconnected from the computer module.

**NOTE: ECI systems will not retain memory of oxygen sensor function after ignition is turned OFF. Check for this code with engine fully warmed and running at idle.**

If 2 or more trouble codes are stored in the memory, the computer will read out the codes in numerical order beginning with the lowest numbered code.

**NOTE: The order in which codes are displayed does not indicate the order of occurrence.**

## MALFUNCTION INDICATOR LIGHT

On some models, a dashboard light will illuminate when a code is stored during engine operation. If the irregular condition returns to normal during operation, the indicator light will go OFF. Once the ignition switch is turned **OFF**, the dash light will not come back ON until the malfunction is detected again. Every time the ignition is turned **ON**, the dash light will illuminate for 2.5 seconds as a test. If a fault code is detected, the light will stay ON beyond the test period.

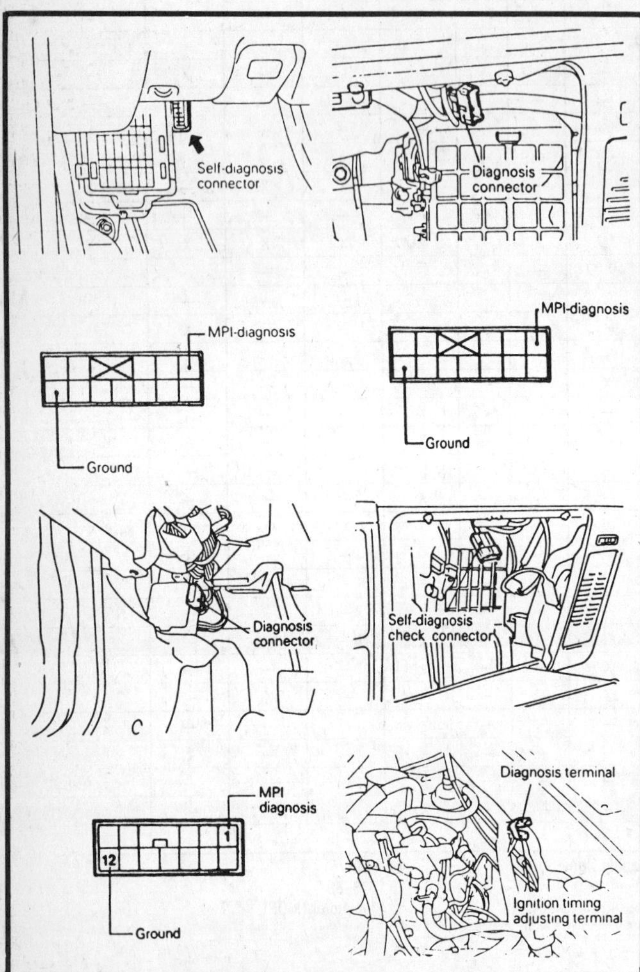


**Locations of diagnostic connector**

## INSPECTION OF FUEL INJECTION SYSTEM

If any system components (sensors, ECU, injector, etc.) fail, interruption of the correct fuel supply will result. When checking and correcting engine problems, it is important to start with inspection of the basic systems. Such conditions as failure to start, unstable idling or poor acceleration are often caused by items other than the fuel injection or engine management systems.

## Self-Diagnostic System

### ENTERING SELF-DIAGNOSIS

**NOTE: If battery voltage is low, the self-diagnosis system will not operate properly; charge should be checked before attempting any self-diagnosis procedures.**

1. Turn ignition switch to **OFF**.
2. Either connect ECI or MPI checker according to instructions given with the unit using a voltmeter. The following procedure is for testing using an analog voltmeter.
3. Connect a voltmeter between test terminal and terminal for ground.
4. Turn ignition switch to **ON**, and indication of ECU memory contents will start. If system is not storing any fault codes, the "normal operation" code is transmitted. If any codes are stored in memory, pointer of voltmeter or tester will deflect. (For example: 2 ½-second sweeps followed by a 2 second pause followed by 3 ½-second sweeps might indicate Code 23 or Code 2 and Code 3, depending on the system.) Record each code as it is transmitted. Once the stored code(s) have been transmitted, the system will pause and retransmit the code(s).
5. Turn ignition switch to **OFF**.
6. Each code, when interpreted, points to the unit or system which may be the problem. It must still be checked along with the attendant wiring, connectors and controls. A great number of fault codes are set because of loose or dirty connections in the wiring which fool the ECU into thinking the unit has failed.

Not every component can be tested with a voltmeter or ohmmeter; some require the use of the ECI checker. In each case remember that you are only reading voltage used to transmit a code; the actual voltage running within the system can only be checked with the factory diagnostic unit.

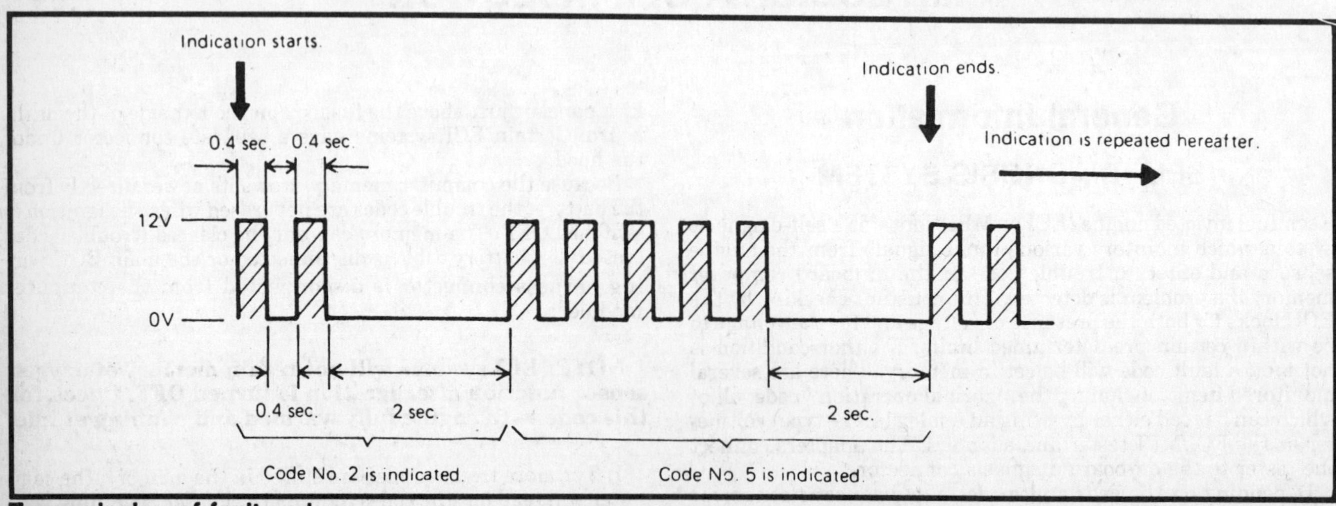


**Transmission of fault codes**

**NOTE: Any resistances given are for 68°F (20°C) unless otherwise stated. Remember that resistance will increase or decrease respectively as the temperature rises or falls. Use common sense in interpreting the readings.**

7. Before disconnecting any sensor or component within the system, make certain the ignition is **OFF** and disconnect the negative battery cable. Removal or connection of battery cable during engine operation or while the ignition switch is **ON** could cause faulty operation of the ECU or damage to semiconductors.
8. After completing service procedures, the trouble code should be erased from the ECU by disconnecting the negative battery cable for at least 15 seconds.
9. After the engine has been operated, check again for presence of codes; if repair was successful, no code will be transmitted.

## CLEARING TROUBLE CODES

In all cases, the memory of the ECU is erased by disconnecting the negative battery cable for at least 15 seconds. Make certain the ignition in in the **OFF** position before disconnecting the cable. Memory should be cleared only after system repairs have been performed.

**NOTE: Some vehicle computers have a learning ability. If a change is noted in vehicle performance after clearing codes, it may be due to the computers learning ability. To restore performance, warm vehicle to normal operating temperature and drive at part throttle with moderate acceleration, before performing any additional diagnosis.**

## 1988-90 MITSUBISHI DIAGNOSTIC TROUBLE CODES

| Trouble Codes | Cordia/ Tredia with ECI | Starion | Galant | Sigma | Mirage | 1990 Precis | Eclipse | V6 | Van/ Wagon/ | 1990 Truck |
|---|---|---|---|---|---|---|---|---|---|---|
| 1—Oxygen sensor | X | X | | | | | | | ① | |
| 2—Ignition signal | X | X | | | | | | | ① | |
| 3—Air flow sensor | X | X | | | | | | | ① | |
| 4—Barometric pressure sensor | X | | | | | | | | ① | |
| 5—Throttle position sensor | X | X | | | | | | | ① | |
| 6—Idle speed control Motor position sensor | X | X | | | | | | | | |
| 7—Coolent temperature sensor | X | X | | | | | | | ① | |
| 8—No. 1 cylinder TDC sensor | | | | | | | | | ① | |
| 11—Oxygen sensor | | | X | X | X | X | X | X | ② | X |
| 12—Air flow sensor | | | X | X | X | X | X | X | ② | X |
| 13—Intake air temperature sensor | | | X | X | X | X | X | X | ② | X |
| 14—Throttle position sensor | | | X | X | X | X | X | X | ② | X |
| 15—Motor position sensor | | | ② ⑩ | | ② ⑤ | X | ⑦ | | ② | ⑨ |
| 21—Coolant temperature sensor | | | X | X | X | X | X | X | ② | X |
| 22—Crank angle sensor | | | X | X | X | X | X | X | ② | X |
| 23—No. 1 cylinder TDC sensor | | | X | X | X | X | X | X | ② | X |
| 24—Vehicle speed sensor | | | X | X | X | X | X | X | ② | X |
| 25—Barometric pressure sensor | | | X | X | X | X | X | X | ② | X |
| 31—Detonation Sensor | | | | | ③ ⑥ | | ⑧ | | | |
| 41—Injector | | | X | X | X | X | X | X | ② | X |
| 42—Fuel pump | | | X | X | X | X | X | X | ② | X |
| 43—Exhaust gas recirculation | | | X | X | X | X | X | X | ② | X |
| 44—Ignition coil | | | ③ ⑪ | | ② ⑥ | | ⑧ | | | |

① 1988 only
② 1989–90 only
③ 1989 only
④ 1990 only
⑤ 1.5L engine only
⑥ 1.6L engine only
⑦ 1.8L engine only
⑧ 2.0L engine only
⑨ 2.4L engine only
⑩ SOHC engine only
⑪ DOHC engine only

## CONTROL FUNCTIONS

| | Related components \ Function | Air/fuel mixture control (ECI) | Ignition timing control | Idle speed control (ISC) | Air conditioner power relay control | Fuel pump drive control | Purge control | EGR control |
|---|---|---|---|---|---|---|---|---|
| Input | Power supply (ignition switch coupled) | × | × | × | × | × | × | × |
| | Power supply (battery backup) | × | × | × | × | × | × | × |
| | Air-flow sensor | × | × | | | | × | × |
| | Barometric pressure sensor | × | × | × | | | × | |
| | Intake air temperature sensor | × | × | × | | | × | |
| | Engine coolant temperature sensor | × | × | × | | | × | × |
| | Throttle position sensor | × | × | × | ×* | | | |
| | Idle switch | × | × | × | | | | |
| | TDC sensor | × | × | × | × | × | | × |
| | Crank angle sensor | × | × | | × | × | | |
| | Oxygen sensor | × | | | | | | |
| | Vehicle-speed sensor | | × | × | | | | |
| | Air conditioner switch | | | × | ×* | | × | |
| | Inhibitor switch <A/T> | | × | × | × | | | |
| | Power steering oil pressure switch | | | × | | | | |
| | Ignition switch | × | | × | | × | | |
| | Ignition switch-ST terminal (start signal) | × | × | × | | × | | |
| Output | Injector | × | | | | | | |
| | Idle speed control servo (Stepper motor) | | | × | | | | |
| | Power transistor | | × | | | | | |
| | Air conditioner power relay | | | | × | | | |
| | Control relay | | | | | × | | |
| | Purge control solenoid valve | | | | | | × | |
| | EGR control solenoid valve <California> | | | | | | | × |

NOTE
* : Vehicles with an automatic transaxle

# Fuel Injection Systems

# SECTION 4

## INDEX

### ACURA PROGRAMMED FUEL INJECTION (PGM-FI) SYSTEM



### DAIHATSU ELECTRONIC FUEL INJECTION SYSTEM



### HONDA PROGRAMMED FUEL INJECTION (PGM-FI) SYSTEM



### HYUNDAI MULTI-POINT FUEL INJECTION (MPI) SYSTEM

## ISUZU FUEL INJECTION SYSTEM



## MAZDA FUEL INJECTION SYSTEMS



## MAZDA RX- ELECTRONIC GASOLINE INJECTION (EGI) SYSTEMS



## MITSUBISHI ELECTRONICALLY CONTROLLED INJECTION (ECI) SYSTEM



## MITSUBISHI MULTI-POINT INJECTION (MPI) SYSTEM

# PROGRAMMED FUEL INJECTION (PGM-FI) SYSTEM

## ENGINE CONTROL SYSTEM APPLICATION CHART

| Year | Model | Engine cc (liter) | Engine Serial Number | Fuel System | Ignition System |
|---|---|---|---|---|---|
| 1988 | Integra | 1590 (1.6) | D16A1 | PGM-FI | PGM-IG |
| | Legend | 2675 (2.7) | C27A1 | PGM-FI | PGM-IG |
| 1989 | Integra | 1590 (1.6) | D16A1 | PGM-FI | PGM-IG |
| | Legend | 2675 (2.7) | C27A1 | PGM-FI | PGM-IG |
| 1990 | Integra | 1834 (1.8) | BA18A1 | PGM-FI | PGM-IG |
| | Legend | 2675 (2.7) | C27A1 | PGM-FI | PGM-IG |

## General Description

Programmed Fuel Injection (PGM-FI) System consists of 3 subsystems: Air intake, electronic control and fuel supply.

### SYSTEM OPERATION

#### Air Intake System

The system supplies air for all engine needs. It consists of the air cleaner, air intake pipe, throttle body, Electronic Idle Control Valve (EICV) system, fast idle mechanism, and intake manifold. A resonator in the air intake pipe provides additional silencing as air is drawn into the system.

#### THROTTLE BODY

The throttle body is a single-barrel side-draft type. The lower portion of the throttle valve is heated by engine coolant which is led from the cylinder head. The idle adjusting screw, which increases/decreases bypass air, and the canister/purge port are located on the top of the throttle body. On cars equipped with a manual transmission, a dashpot is used to slow the throttle as it approaches the closed position.

#### IDLE CONTROL SYSTEM

The idle speed of engine is controlled by the Electronic Idle Control Valve (EICV) and the fast idle valve. The valve changes the amount of air bypassing into the intake manifold in response to electric current sent from the ECU. When the EICV is activated, the valve opens to maintain the proper idle speed.

After the engine starts, the EICV opens for a certain time. The amount of air is increased to raise the idle speed about 0–150 rpm. When the coolant temperature is low, EICV is opened to obtain the proper fast idle speed. The amount of bypassed air is, thus controlled in relation to the coolant temperature. When the coolant temperature reaches 86°F (30°C), it also activates the fast idle valve to prevent the idle speed from dropping.

#### FAST IDLE CONTROL SYSTEM

To prevent erratic running when the engine is warming up, it is necessary to raise the idle speed. The air bypass valve is controlled by a thermowax plunger. When the thermowax is cold, the valve is open. When the thermowax is heated, the valve is closed. With the engine cold and the thermowax consequently cold, additional air is bypassed into the intake manifold so that the engine idles faster than normal. When the engine reaches

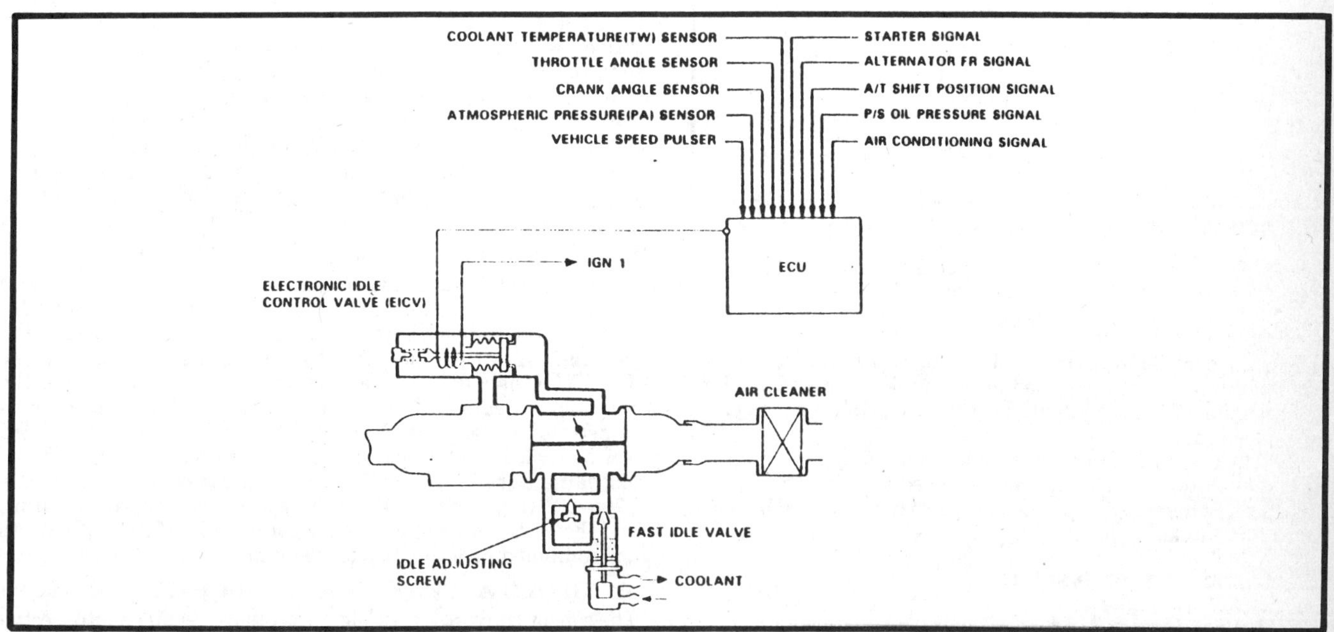


**Idle control system—Acura Legend Sedan and Sterling shown, others similar**

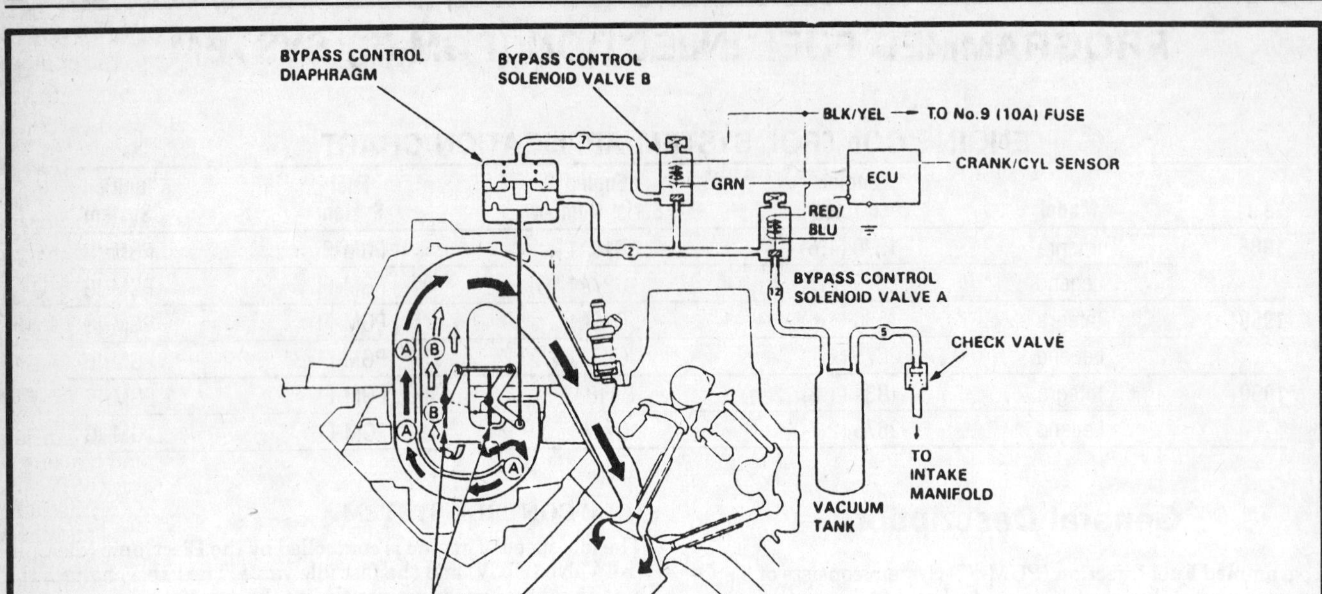


**Bypass control system – Legend**

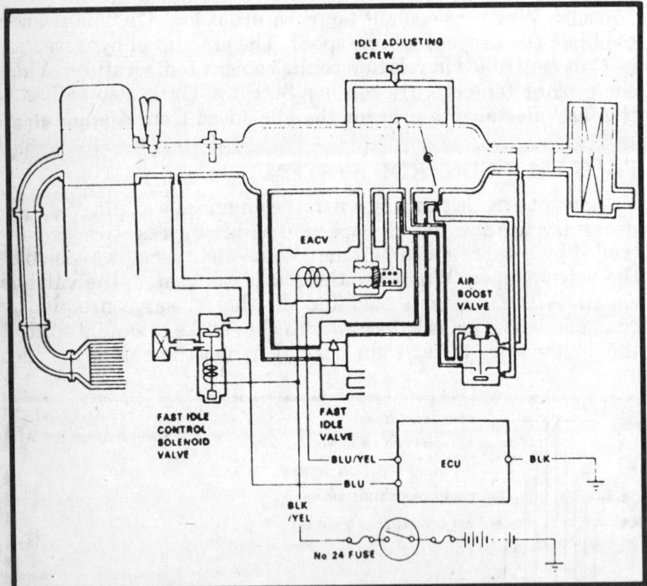


**Idle control system – 1990 Integra shown**

operating temperature, the valve begins to close, reducing the amount of air bypassing into the manifold.

### BYPASS CONTROL SYSTEM

The Legend uses a bypass control system, which utilizes 2 air intake paths which are provided in the intake manifold to allow the selection of the intake path length most favorable for a given engine speed.

Satisfactory power performance is achieved by switching the paths. High torque at low rpm is achieved by using the long intake path, whereas high power at high rpm is achieved by using the short intake path.

## Electronic Control System

### CONTROL SYSTEM

In order to get fuel into the cylinders at the correct instant and in correct amount, the control system must perform various separate functions. The ECU (Electronic Control Unit), the heart of the PGM-FI, uses an 8-bit microcomputer and consists of a CPU (Central Processing Unit), memories, and I/O (Input/Output) ports. Basic data stored in the memories are compensated by the signals sent from the various sensors to provide the correct air/fuel mixture for all engine needs.

### ELECTRONIC CONTROL UNIT (ECU)

The unit contains memories for the basic discharge duration at various engine speeds and manifold pressures. The basic discharge duration, after being read out from the memory, is further modified by signals sent from various sensors to obtain the final discharge duration. Other functions also include:

- Starting Control – The fuel system must vary the air/fuel ratio to suit different operating requirements. For example, the mixture must be rich for starting. The memories also contain the basic discharge durations to be read out by signals from the starter switch, and engine speed and coolant temperature sensors, thereby providing extra fuel needed for starting.
- Fuel Pump Control – When the speed of the engine falls below the prescribed limit, electric current to the fuel pump is cut off, preventing the injectors from discharging fuel.
- Safety – A fail-safe system monitors the sensors and detects any abnormality in the ECU, ensuring safe driving if a single or many sensors are faulty, or if the ECU malfunctions.

### CRANK ANGLE SENSOR (TDC/CYL SENSORS)

The sensors are designed as an assembly to save space and weight. The entire unit consist of a pair of rotors, TDC and CYL, and a pickup for each rotor. Since the rotors are coupled to the camshaft, they turn together as a unit as the camshaft rotates. The CYL sensor detects the position of the No. 1 cylinder as the base for the Sequential Injection whereas the TDC sensor serves to determine the injection timing for each cylinder. The TDC sensor is also used to detect engine speed to read out the basic discharge duration for different operating conditions. On the 1988–89 Integra the TDC/Crank sensor is serviced as a unit, and the CYL sensor is serviced separately. In 1990 all 3 sensors are combined into the distributor unit.

### MANIFOLD AIR PRESSURE SENSOR (MAP SENSOR)

The sensor converts manifold air pressure readings into electrical voltage signals and sends them to the ECU. This information with signals from the crank angle sensor is then used to read out the basic discharge duration from the memory.

**ATMOSPHERIC PRESSURE SENSOR (PA SENSOR)**

Like the MAP sensor, the unit converts atmospheric pressures into voltage signals and sends them to the ECU. The signals then modify the basic discharge duration to compensate for changes in the atmospheric pressure.

**COOLANT TEMPERATURE SENSOR (TW SENSOR)**

The sensor uses a temperature-dependent diode (thermistor) to measure differences in the coolant temperature. The basic discharge duration is read out by the signals sent from this sensor through the ECU. The resistance of the thermister decreases with a rise in coolant temperature.

**INTAKE AIR TEMPERATURE SENSOR (TA SENSOR)**

This device is also a thermistor and is placed in the intake manifold. It acts much like the water temperature sensor but with a reduced thermal capacity for for quicker response. The basic discharge duration read out from the memory is again compensated for different operating conditions by the signals sent from this sensor through the ECU.

**THROTTLE ANGLE SENSOR**

This sensor is essentially a variable resistor. In construction, the rotor shaft is connected to the throttle valve shaft such that, as the throttle valve is moved, the resistance varies, altering the output voltage to the control unit.

**OXYGEN SENSOR**

The oxygen sensor, by detecting the oxygen content in the exhaust gas, maintains the stoichiometric air/fuel ratio. In operation, the ECU receives the signals from the sensor and changes the duration during which fuel is injected. The oxygen sensor is located in the exhaust manifold.

The sensor is a hollow shaft of zirconia with a closed end. The inner and outer surfaces are plated with platinum, thus forming a platinum electrode. The inner surface or chamber is open to the atmosphere whereas the outer surface is exposed to the exhaust gas flow through the manifold.

Voltage is induced at the platinum electrode when there is any difference in oxygen concentration between the 2 layers of air over the surfaces. Operation of the device is dependent upon the fact that voltage induced changes sharply as the stoichiometric air/fuel ratio is exceeded when the electrode is heated above a certain temperature.

### Fuel System

**FUEL PUMP**

On 1988–89 Integra, the fuel pump is an in-line, direct drive type. Fuel is drawn into the pump through a filter, flows around the armature through the one-way valve and is delivered to the engine compartment. A baffle is provided to prevent fuel pulsation. The fuel pump has a relief valve to prevent excessive pressure. It opens if there is a blockage in the discharge side. When the relief valve opens, fuel flows from the high pressure to the low pressure side. A check valve is provided to maintain fuel pressure in the line after the pump is stopped. This is to ease restarting.

The pump section is composed of a rotor, rollers and pump spacer. When the rotor turns, the rollers turn and travel along the inner surface of the pump spacer by centrifugal force. The volume of the cavity enclosed by these 3 parts changes, drawing and pressurizing the fuel.

On the Legend and 1990 Integra, the fuel pump is a compact impeller design and is installed inside the fuel tank, thereby saving space and simplifying the fuel line system.

The fuel pump is comprised of a DC motor, a circumference flow pump, a relief valve for protecting the fuel line systems, a check valve for retaining residual pressure, an inlet port, and a discharge port. The pump assembly consists of the impeller (driven by the motor), the pump casing (which forms the pumping chamber), and cover of the pump.

**PRESSURE REGULATOR**

The fuel pressure regulator maintains a constant fuel pressure to the injectors. The spring chamber of the pressure regulator is connected to the intake manifold to constantly maintain the fuel pressure at 36–41 psi (250–279 kPa) higher than the pressure in the manifold. When the difference between the fuel pressure and manifold pressure exceeds 36–41 psi (250–279 kPa), the diaphragm is pushed upward, and the excess fuel is fed back into the fuel tank through the return line.

**INJECTOR**

The injector is of the solenoid-actuated constant-stroke pintle type consisting of a solenoid, plunger, needle valve and housing. When current is applied to the solenoid coil, the valve lifts up and pressurized fuel fills the inside of the injector and is injected close to the intake valve. Because the needle valve lifts and the fuel pressure is constant, the injection quantity is determined by the length of time that the valve is open, i.e., the duration the current is supplied to the solenoid coil. The injector is sealed by an O-ring and seal ring at the top and bottom. These seals also reduce operating noise.

**RESISTOR**

The injector timing, which controls the opening and closing intervals, must be very accurate since it dictates the air/fuel mixture ratio. The injector must also be durable. For the best possible injector response, it is necessary to shorten the current rise time when voltage is applied to the injector coil. Therefore, the number of windings of the injector coil is reduced to reduce the inductance in the coil. This, however, makes low resistance in the coil, allowing a large amount of current to flow through the coil. As a result, the amount of heat generated is high, which compromises the durability of the coil. Flow of current in the coil is therefore restricted by a resistor installed in series between the electric power source and the injector coil.

**MAIN RELAY**

The main relay is a direct coupler type which contains the relays for the electronic control unit power supply and the fuel pump power supply. This relay is installed behind the fuse box or at the left side cowl.

## SERVICE PRECAUTIONS

- Do not operate the fuel pump when the fuel lines are empty.
- Do not operate the fuel pump when removed from the fuel tank.
- Do not reuse fuel hose clamps.
- Make sure all ECU harness connectors are fastened securely. A poor connection can cause an extremely high surge voltage in the coil and condenser and result in damage to integrated circuits.
- Keep ECU all parts and harnesses dry during service.
- Before attempting to remove any parts, turn **OFF** the ignition switch and disconnect the battery ground cable.
- Always use a 12 volt battery as a power source.
- Do not attempt to disconnect the battery cables with the engine running.
- Do not depress the accelerator pedal when starting.
- Do not rev up the engine immediately after starting or just prior to shutdown.
- Do not apply battery power directly to injectors.

# Diagnosis and Testing

## SELF-DIAGNOSTIC SYSTEM

### Self-Diagnosis Indicators

The quick reference chart covers the most common failure modes for the PGM-FI. The probable causes are listed in order of most-easily-checked first, then progressing to more difficult fixes. Run through all the causes listed. If problem is still unsolved, go on to the more detailed troubleshooting. Troubleshooting is divided into different LED displays. Find the correct light display and begin again.

For all the conditions listed, the PGM-FI warning light on the dashboard must be on (comes on and stays on). This indicates a problem in the electrical portion of the fuel injection system. At that time, check the LED display (self-diagnosis system) in the ECU.

On 1988–89 Acura Integra, there is a single LED display. The LED will blink consecutively to indicate the trouble code. The ECU is located under the passenger's seat.

On 1990 Acura Integra, there is a single LED display. The LED will blink consecutively to indicate the trouble code. The ECU is located behind the carpet, on the passenger's side under the dash.

On 1988–90 Acura Legend models, there are 2 LED displays. The yellow LED display is for idle speed adjustment check. The red LED display will blink consecutively to indicate the trouble code. The ECU is located under the passenger's seat on the Legend Sedan. On the Legend Coupe, the ECU is located under the floor kick panel in front of the passenger's seat.

Sometimes the dash warning light and/or ECU LED will come ON, indicating a system problem, when, in fact, there is only a bad or intermittent electrical connection. To troubleshoot a bad connection, note the ECU LED pattern that is lit, refer to the diagnosis chart and check the connectors associated with the

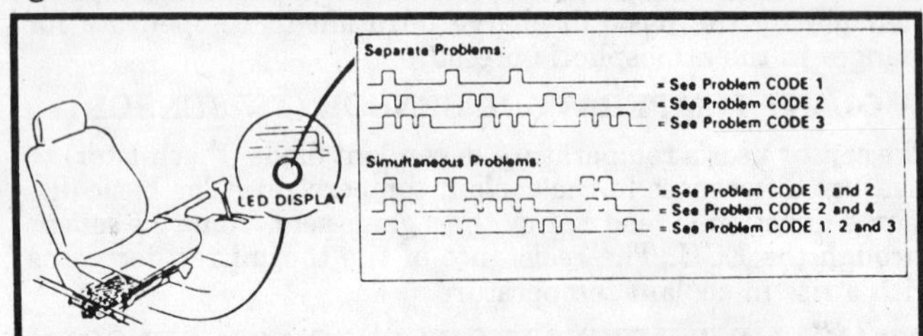


ECU location and LED display – 1988–89 Integra

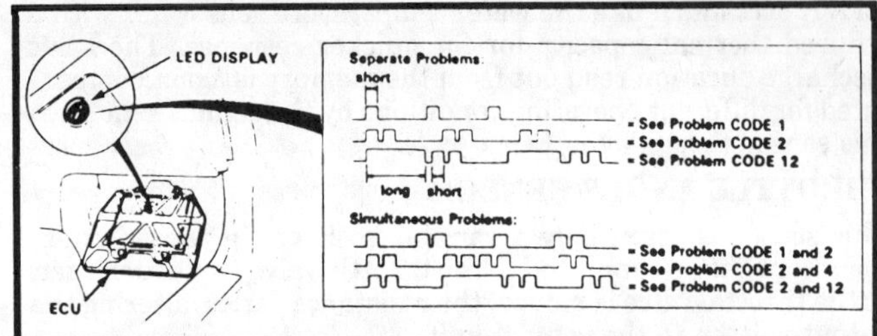


ECU location and LED display – 1990 Integra

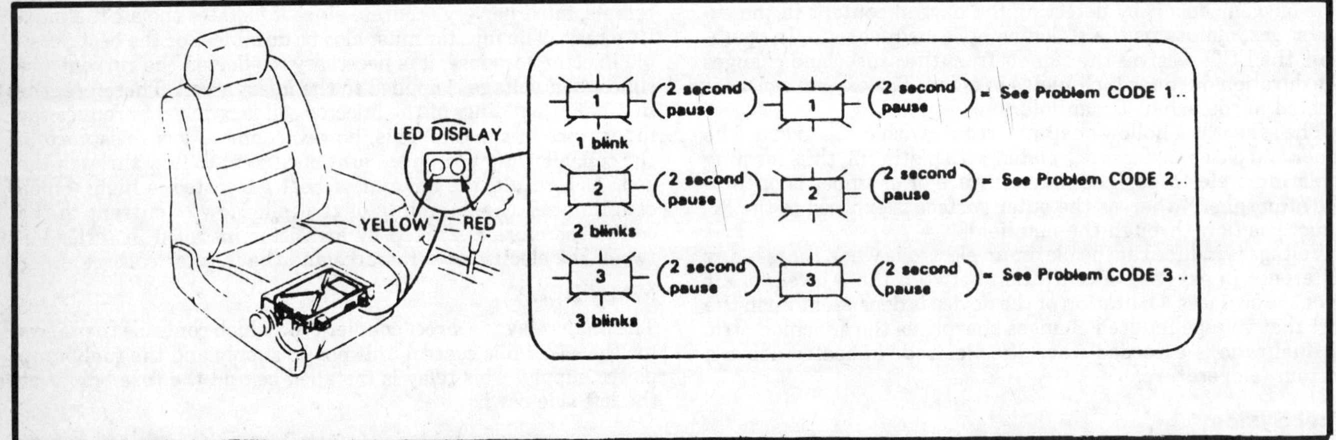


ECU location and self-diagnostic LED display – 1986–88 Acura Legend Sedan and Sterling

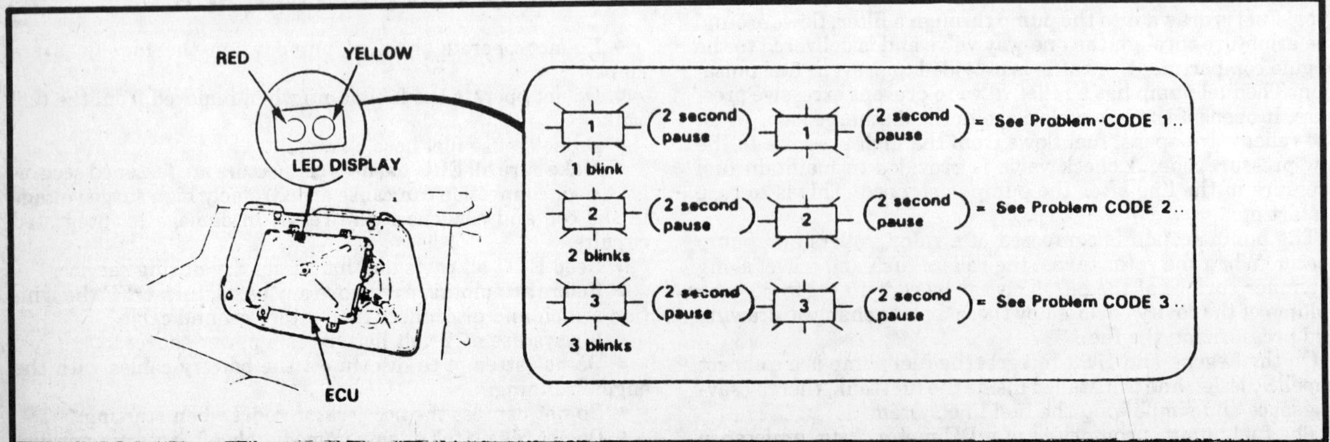


ECU location and self-diagnostic LED display – 1987–88 Acura Legend Coupe

| SELF-DIAGNOSIS INDICATOR BLINKS | SYSTEM INDICATED |
|---|---|
| 0 | ECU |
| 1 | FRONT OXYGEN CONTENT |
| 2 | REAR OXYGEN CONTENT |
| 3 | MANIFOLD ABSOLUTE PRESSURE |
| 5 | |
| 4 | CRANK ANGLE |
| 6 | COOLANT TEMPERATURE |
| 7 | THROTTLE ANGLE |
| 8 | TDC POSITION |
| 9 | No.1 CYLINDER POSITION |
| 10 | INTAKE AIR TEMPERATURE |
| 12 | EXHAUST GAS RECIRCULATION SYSTEM |
| 13 | ATMOSPHERIC PRESSURE |
| 14 | ELECTRONIC IDLE CONTROL |
| 15 | IGNITION OUTPUT SIGNAL |
| 17 | VEHICLE SPEED PULSER |
| 18 | IGNITION TIMING ADJUSTMENT |

**ECU self-diagnostic code identification – 1987–88 Acura Legend Coupe and 1988 Legend Sedan**

| SELF DIAGNOSIS INDICATOR BLINKS | SISTEM INDICATED |
|---|---|
| 0 | ECU |
| 1 | FRONT OXYGEN CONTENT |
| 2 | REAR OXYGEN CONTENT |
| 3 | MANIFOLD ABSOLUTE PRESSURE |
| 5 | |
| 4 | CRANK ANGLE |
| 6 | COOLANT TEMPERATURE |
| 7 | THROTTLE ANGLE |
| 8 | TDC POSITION |
| 9 | No. 1 CYLINDER POSITION |
| 10 | INTAKE AIR TEMPERATURE |
| 12 | EXHAUST GAS RECIRCULATION SYSTEM |
| 13 | ATMOSPHERIC PRESSURE |
| 14 | ELECTRONIC IDLE CONTROL |
| 15 | IGNITION OUTPUT SIGNAL |
| 17 | VEHICLE SPEED PULSER |
| 18 | IGNITION TIMING ADJUSTMENT |
| 30 | A/T FI SIGNAL A |
| 31 | A/T FI SIGNAL B |

**ECU Self-diagnosis code identification – Legend**

| SELF-DIAGNOSIS INDICATOR BLINKS | SYSTEM INDICATED |
|---|---|
| 0 | ECU |
| 1 | OXYGEN CONTENT |
| 3 | MANIFOLD ABSOLUTE PRESSURE |
| 5 | |
| 4 | CRANK ANGLE |
| 6 | COOLANT TEMPERATURE |
| 7 | THROTTLE ANGLE |
| 8 | TDC POSITION |
| 9 | No.1 CYLINDER POSITION |
| 10 | INTAKE AIR TEMPERATURE |
| 12 | EXHAUST GAS RECIRCULATION SYSTEM (A/T) |
| 13 | ATMOSPHERIC PRESSURE |
| 14 | ELECTRONIC AIR CONTROL |
| 15 | IGNITION OUTPUT SIGNAL |
| 16 | FUEL INJECTOR |
| 17 | VEHICLE SPEED SENSOR |
| 43 | FUEL SUPPLY SYSTEM |

**ECU Self-diagnosis code identification – 1990 Integra**

| SELF-DIAGNOSIS INDICATOR BLINKS | SYSTEM INDICATED |
|---|---|
| 0 | ECU |
| 1 | OXYGEN CONTENT |
| 3 | MANIFOLD ABSOLUTE PRESSURE |
| 5 | |
| 4 | CRANK ANGLE |
| 6 | COOLANT TEMPERATURE |
| 7 | THROTTLE ANGLE |
| 8 | TDC POSITION |
| 9 | No.1 CYLINDER POSITION |
| 10 | INTAKE AIR TEMPERATURE |
| 13 | ATMOSPHERIC PRESSURE |
| 14 | ELECTRONIC AIR CONTROL |
| 15 | IGNITION OUTPUT SIGNAL |
| 16 | FUEL INJECTOR |
| 17 | VEHICLE SPEED SENSOR |
| 19 | LOCK-UP CONTROL SOLENOID VALVE |
| 20 | ELECTRIC LOAD |

**ECU Self-diagnosis code identification – 1988–89 Integra**

items mentioned in the "Possible Cause" column for that LED pattern (disconnect, clean or repair if necessary and reconnect those connections). Then, reset the ECU memory as described, restart the car and drive it for a few minutes and then recheck the car and drive it for a few minutes and then recheck the LED(s). If the same pattern lights up, begin system troubleshooting; if it does not light up, the problem was only a bad connection.

The memory for the PGM-FI warning light on the dashboard will be erased when the ignition switch is turned **OFF**; however, the memory for the LED display will not be canceled. Thus, the warning light will not come ON when the ignition switch is again turned **ON** unless the trouble is once more detected. Troubleshooting should be done according to the LED display even if the warning light is OFF.

Other ECU information:

- After making repairs, disconnect the battery negative cable from the battery negative terminal for at least 10 seconds and reset the ECU memory. After reconnecting the cable, check that the LED display is turned OFF.
- Turn the ignition switch **ON**. The PGM-FI warning light should come on for about 2 seconds. If the warning light won't come ON, check for:
  – Blown warning light bulb
  – Blown fuse (causing faulty back up light, seat belt alarm, clock, memory function of the car radio)
  – Open circuit in Yellow wire between fuse and combination meter
- After the PGM-FI warning light and self-diagnosis indicators have been turned on, turn the ignition switch **OFF**. If the LED display fails to come ON when the ignition switch is turned **ON** again, check for:
  – Blown fuse
  – Open circuit in wire between ECU A17 terminal and fuse
- Replace the ECU only after making sure that all couplers and connectors are connected securely.

## TROUBLESHOOTING GUIDE – 1988–89 LEGEND

NOTE: Across each row in the chart, the systems that could be sources of a symptom are ranked in the order they should be inspected starting with ①. Find the symptom in the left column, read across to the most likely source, then refer to the page listed at the top of that column. If inspection shows the system is OK, try the next most likely system ②, etc.

| SYMPTOM | | PGM-FI: ELECTRONIC CONTROL UNIT | OXYGEN SENSOR | MANIFOLD ABSOLUTE PRESSURE SENSOR | CRANK/CYL SENSOR | COOLANT TEMPERATURE SENSOR | THROTTLE ANGLE SENSOR | TDC SENSOR | INTAKE AIR TEMPERATURE SENSOR |
|---|---|---|---|---|---|---|---|---|---|
| CHECK ENGINE WARNING LIGHT TURNS ON | | □ or ☼ | ☼ | ☼ | ☼ | ☼ | ☼ | ☼ | ☼ |
| SELF-DIAGNOSIS INDICATOR (RED LED) BLINKS | | 0 or ✳ | 1 or 2 | 3 or 5 | 4 or 9 | 6 | 7 | 8 | 10 |
| ENGINE WON'T START | | 2 | | 3 | | 3 | | | |
| DIFFICULT TO START ENGINE WHEN COLD | | BU | | 3 | | 1 | | | |
| IRREGULAR IDLING | WHEN COLD FAST IDLE OUT OF SPECIFIC | BU | | | | 3 | | | |
| | ROUGH IDLE | BU | | 2 | | | | | |
| | WHEN WARM RPM TOO HIGH | BU | | | | 3 | | | |
| | WHEN WARM RPM TOO LOW | BU | | 3 | | | | | |
| FREQUENT STALLING | WHILE WARMING UP | BU | | 3 | | | | | |
| | AFTER WARMING UP | BU | | | | | | | |
| POOR PERFORMANCE | MISFIRE OR ROUGH RUNNING | BU | | 3 | | | | | |
| | FAILS EMISSION TEST | BU | 2 | 1 | | 3 | | | |
| | LOSS OF POWER | BU | | 2 | | | | | |

- CODE 11, 16 or exceeds 18: count the number of blinks again. If the indicator is in fact blinking these codes, substitute a known–good ECU and recheck. If the indication goes away, replace the original ECU.

(BU): When the Check Engine warning light and the self-diagnosis indicator are on, the back-up system is in operation. Substitute a known-good ECU and recheck. If the indication goes away, replace the original ECU.

NOTE: If there is abnormal air intake noise, check the resonator control system

## TROUBLESHOOTING GUIDE – 1990 LEGEND

NOTE: Across each row in the chart, the systems that could be sources of a symptom are ranked in the order they should be inspected starting with ①. Find the symptom in the left column, read across to the most likely source, then refer to the page listed at the top of that column. If inspection shows the system is OK, try the next most likely system ②, etc.

| SYMPTOM | | PGM-FI: ELECTRONIC CONTROL UNIT | OXYGEN SENSOR | MANIFOLD ABSOLUTE PRESSURE SENSOR | CRANK/CYL SENSOR | COOLANT TEMPERATURE SENSOR | THROTTLE ANGLE SENSOR | TDC SENSOR | INTAKE AIR TEMPERATURE SENSOR |
|---|---|---|---|---|---|---|---|---|---|
| CHECK ENGINE WARNING LIGHT TURNS ON | | □ or ☼ | ☼ | ☼ | ☼ | ☼ | ☼ | ☼ | ☼ |
| SELF-DIAGNOSIS INDICATOR (RED LED) BLINKS | | 0 or ✳ | 1 or 2 | 3 or 5 | 4 or 9 | 6 | 7 | 8 | 10 |
| ENGINE WON'T START | | 2 | | 3 | | 3 | | | |
| DIFFICULT TO START ENGINE WHEN COLD | | BU | | 3 | | 1 | | | |
| IRREGULAR IDLING | WHEN COLD FAST IDLE OUT OF SPECIFIC | BU | | | | 3 | | | |
| | ROUGH IDLE | BU | | 2 | | | | | |
| | WHEN WARM RPM TOO HIGH | BU | | | | 3 | | | |
| | WHEN WARM RPM TOO LOW | BU | | 3 | | | | | |
| FREQUENT STALLING | WHILE WARMING UP | BU | | 3 | | | | | |
| | AFTER WARMING UP | BU | | | | | | | |
| POOR PERFORMANCE | MISFIRE OR ROUGH RUNNING | BU | | 3 | | | | | |
| | FAILS EMISSION TEST | BU | 2 | 1 | | 3 | | | |
| | LOSS OF POWER | BU | | 2 | | | | | |

- If codes other than those listed above are indicated, count the number of blinks again. If the indicator is in fact blinking these codes, substitute a known-good ECU and recheck. If the indication goes away, replace the original ECU.

(BU): When the Check Engine warning light and the self-diagnosis indicator are on, the back-up system is in operation. Substitute a known-good ECU and recheck. If the indication goes away, replace the original ECU.

NOTE: If there is abnormal air intake noise, check the resonator control system

## TROUBLESHOOTING GUIDE – 1988–90 LEGEND

| PGM-FI | | | | IDLE CONTROL | | | | |
|---|---|---|---|---|---|---|---|---|
| ATMOSPHERIC PRESSURE SENSOR | IGNITION OUTPUT SIGNAL | VEHICLE SPEED PULSER | IGNITION TIMING ADJUSTER | ELECTRONIC AIR CONTROL VALVE | OTHER IDLE CONTROLS | FUEL SUPPLY | AIR INTAKE | EGR CONTROL SYSTEM |
| 13 | 15 | 17 | 18 | 14 | | | | 12 |
| ③ | ③ | | | | | ① | | |
| | | | | | | ② | | |
| | | | | ② | ① | | | |
| | | | | ① | | | | ③ |
| | | | | ① | ② | | | |
| | | | | ① | ③ | | | ② |
| | | | | ① | | ② | | ③ |
| | | | | ② | | ③ | | ① |
| | | | | ② | | ① | | ③ |
| | | | | | | | | |
| | | | | | | ① | ③ | |

## TROUBLESHOOTING GUIDE – 1990 LEGEND

| PGM-R | | | | | | IDLE CONTROL | | | | |
|---|---|---|---|---|---|---|---|---|---|---|
| ATMOSPHERIC PRESSURE SENSOR | IGNITION OUTPUT SIGNAL | VEHICLE SPEED PULSER | IGNITION TIMING ADJUSTER | A/T FI SIGNAL A | A/T FI SIGNAL B | ELECTRONIC AIR CONTROL VALVE | OTHER IDLE CONTROLS | FUEL SUPPLY | AIR INTAKE | EGR CONTROL SYSTEM |
| 13 | 15 | 17 | 18 | 30 | 31 | 14 | | | | 12 |
| ③ | ③ | | | | | | | ① | | |
| | | | | | | | | ② | | |
| | | | | | | ② | ① | | | |
| | | | | | | ① | | | | ③ |
| | | | | | | ① | ② | | | |
| | | | | | | ① | ③ | | | ② |
| | | | | | | ① | | ② | | ③ |
| | | | | | | ② | | ③ | | ① |
| | | | | | | ② | | ① | | ③ |
| | | | | | | | | | | |
| | | | | | | | | ① | ③ | |

## ECU AND CHECK LIGHT – 1988–90 LEGEND

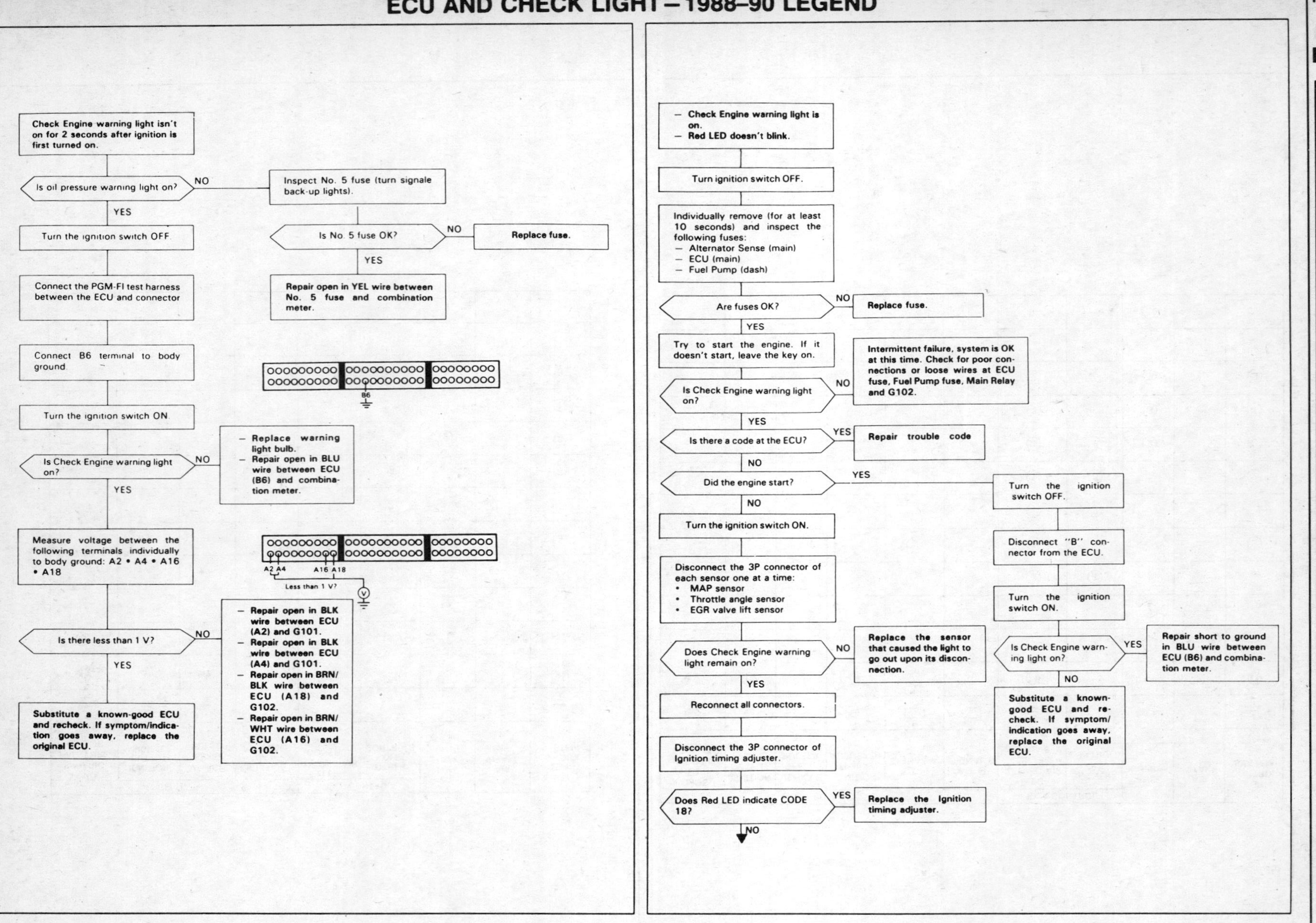

Check Engine warning light isn't on for 2 seconds after ignition is first turned on.
Is oil pressure warning light on?
NO
Inspect No. 5 fuse (turn signale back-up lights).
Is No. 5 fuse OK?
NO
Replace fuse.
YES
Repair open in YEL wire between No. 5 fuse and combination meter.
YES
Turn the ignition switch OFF.
Connect the PGM-FI test harness between the ECU and connector
Connect B6 terminal to body ground.
B6
Turn the ignition switch ON.
Is Check Engine warning light on?
NO
– Replace warning light bulb.
– Repair open in BLU wire between ECU (B6) and combination meter.
YES
Measure voltage between the following terminals individually to body ground: A2 • A4 • A16 • A18
A2 A4 A16 A18
Less than 1 V?
Is there less than 1 V?
NO
– Repair open in BLK wire between ECU (A2) and G101.
– Repair open in BLK wire between ECU (A4) and G101.
– Repair open in BRN/BLK wire between ECU (A18) and G102.
– Repair open in BRN/WHT wire between ECU (A16) and G102.
YES
Substitute a known-good ECU and recheck. If symptom/indication goes away, replace the original ECU.
– Check Engine warning light is on.
– Red LED doesn't blink.
Turn ignition switch OFF.
Individually remove (for at least 10 seconds) and inspect the following fuses:
– Alternator Sense (main)
– ECU (main)
– Fuel Pump (dash)
Are fuses OK?
NO
Replace fuse.
YES
Try to start the engine. If it doesn't start, leave the key on.
Is Check Engine warning light on?
NO
Intermittent failure, system is OK at this time. Check for poor connections or loose wires at ECU fuse, Fuel Pump fuse, Main Relay and G102.
YES
Is there a code at the ECU?
YES
Repair trouble code
NO
Did the engine start?
YES
Turn the ignition switch OFF.
Disconnect "B" connector from the ECU.
Turn the ignition switch ON.
Is Check Engine warning light on?
YES
Repair short to ground in BLU wire between ECU (B6) and combination meter.
NO
Substitute a known-good ECU and recheck. If symptom/indication goes away, replace the original ECU.
NO
Turn the ignition switch ON.
Disconnect the 3P connector of each sensor one at a time:
• MAP sensor
• Throttle angle sensor
• EGR valve lift sensor
Does Check Engine warning light remain on?
NO
Replace the sensor that caused the light to go out upon its disconnection.
YES
Reconnect all connectors.
Disconnect the 3P connector of Ignition timing adjuster.
Does Red LED indicate CODE 18?
YES
Replace the Ignition timing adjuster.
NO

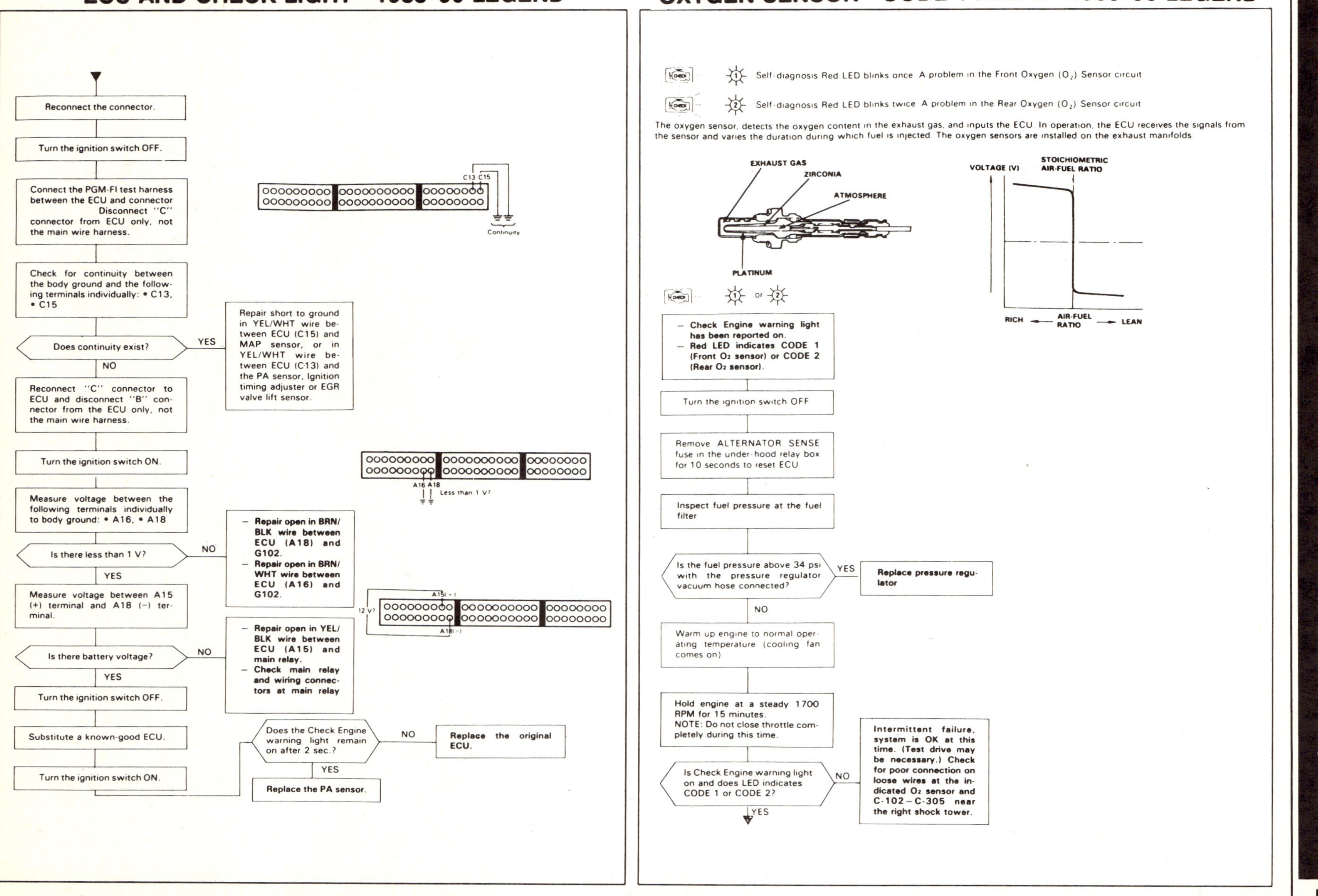
ECU AND CHECK LIGHT—1988–90 LEGEND
Reconnect the connector.
Turn the ignition switch OFF.
Connect the PGM-FI test harness between the ECU and connector Disconnect "C" connector from ECU only, not the main wire harness.
C13 C15
Continuity
Check for continuity between the body ground and the following terminals individually: • C13, • C15
Does continuity exist?
YES
Repair short to ground in YEL/WHT wire between ECU (C15) and MAP sensor, or in YEL/WHT wire between ECU (C13) and the PA sensor, Ignition timing adjuster or EGR valve lift sensor.
NO
Reconnect "C" connector to ECU and disconnect "B" connector from the ECU only, not the main wire harness.
Turn the ignition switch ON.
A16 A18
Less than 1 V?
Measure voltage between the following terminals individually to body ground: • A16, • A18
Is there less than 1 V?
NO
– Repair open in BRN/BLK wire between ECU (A18) and G102.
– Repair open in BRN/WHT wire between ECU (A16) and G102.
YES
Measure voltage between A15 (+) terminal and A18 (−) terminal.
12 V?
A15(+)
A18(−)
Is there battery voltage?
NO
– Repair open in YEL/BLK wire between ECU (A15) and main relay.
– Check main relay and wiring connectors at main relay
YES
Turn the ignition switch OFF.
Substitute a known-good ECU.
Turn the ignition switch ON.
Does the Check Engine warning light remain on after 2 sec.?
NO
Replace the original ECU.
YES
Replace the PA sensor.
OXYGEN SENSOR—CODE 1 AND 2—1988–90 LEGEND
Self-diagnosis Red LED blinks once A problem in the Front Oxygen ($O_2$) Sensor circuit
Self-diagnosis Red LED blinks twice A problem in the Rear Oxygen ($O_2$) Sensor circuit
The oxygen sensor, detects the oxygen content in the exhaust gas, and inputs the ECU. In operation, the ECU receives the signals from the sensor and varies the duration during which fuel is injected. The oxygen sensors are installed on the exhaust manifolds.
EXHAUST GAS
ZIRCONIA
ATMOSPHERE
PLATINUM
VOLTAGE (V)
STOICHIOMETRIC AIR-FUEL RATIO
RICH
AIR-FUEL RATIO
LEAN
or
– Check Engine warning light has been reported on.
– Red LED indicates CODE 1 (Front $O_2$ sensor) or CODE 2 (Rear $O_2$ sensor).
Turn the ignition switch OFF
Remove ALTERNATOR SENSE fuse in the under-hood relay box for 10 seconds to reset ECU
Inspect fuel pressure at the fuel filter
Is the fuel pressure above 34 psi with the pressure regulator vacuum hose connected?
YES
Replace pressure regulator
NO
Warm up engine to normal operating temperature (cooling fan comes on)
Hold engine at a steady 1700 RPM for 15 minutes.
NOTE: Do not close throttle completely during this time.
Is Check Engine warning light on and does LED indicates CODE 1 or CODE 2?
NO
Intermittent failure, system is OK at this time. (Test drive may be necessary.) Check for poor connection on loose wires at the indicated $O_2$ sensor and C-102 – C-305 near the right shock tower.
YES

## OXYGEN SENSOR—CODE 1 AND 2—1988–90 LEGEND

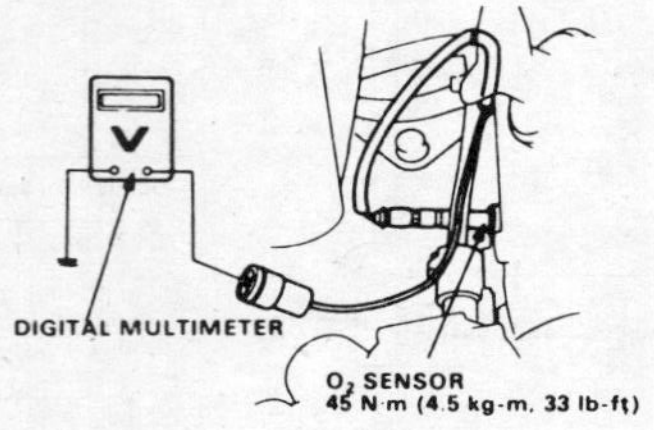

Disconnect engine wire harness from indicated $O_2$ sensor.

Warm up engine to normal operating temperature again.

Measure voltage between the connector terminal and body ground.

DIGITAL MULTIMETER

$O_2$ SENSOR
45 N·m (4.5 kg-m, 33 lb-ft)

Is voltage above 0.6 V at 5,000 rpm and below 0.4 V when the throttle is quickly released from 5,000 rpm?
- NO: Replace $O_2$ sensor.
- YES: Stop engine.

Reconnect the $O_2$ sensor.

Connect the PGM-FI test harness between the ECU and connector

Restart and warm up engine to normal operating temperature.

Measure voltage between
- Front $O_2$ sensor C12 (+) and A18 (-) terminals
- Rear $O_2$ sensor C10 (+) and A18 (-) terminals

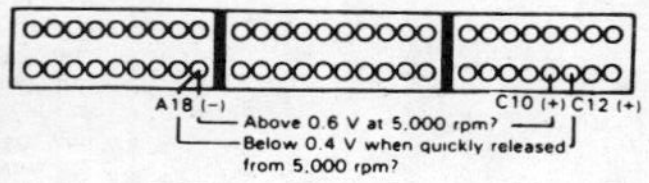

A18 (-)
C10 (+) C12 (+)
Above 0.6 V at 5,000 rpm?
Below 0.4 V when quickly released from 5,000 rpm?

Is voltage above 0.6 V at 5,000 rpm and below 0.4 V when the throttle is quickly released from 5,000 rpm?
- NO: **Repair wire between ECU and $O_2$ sensor for short or open:**
  - **Front $O_2$ sensor WHT wire to C12.**
  - **Rear $O_2$ sensor RED/BLU wire to C10.**
- YES: **Substitute a known-good ECU and recheck. If symptom/indication goes away, replace the original ECU.**

## MAP SENSOR—CODE 3—1988–90 LEGEND

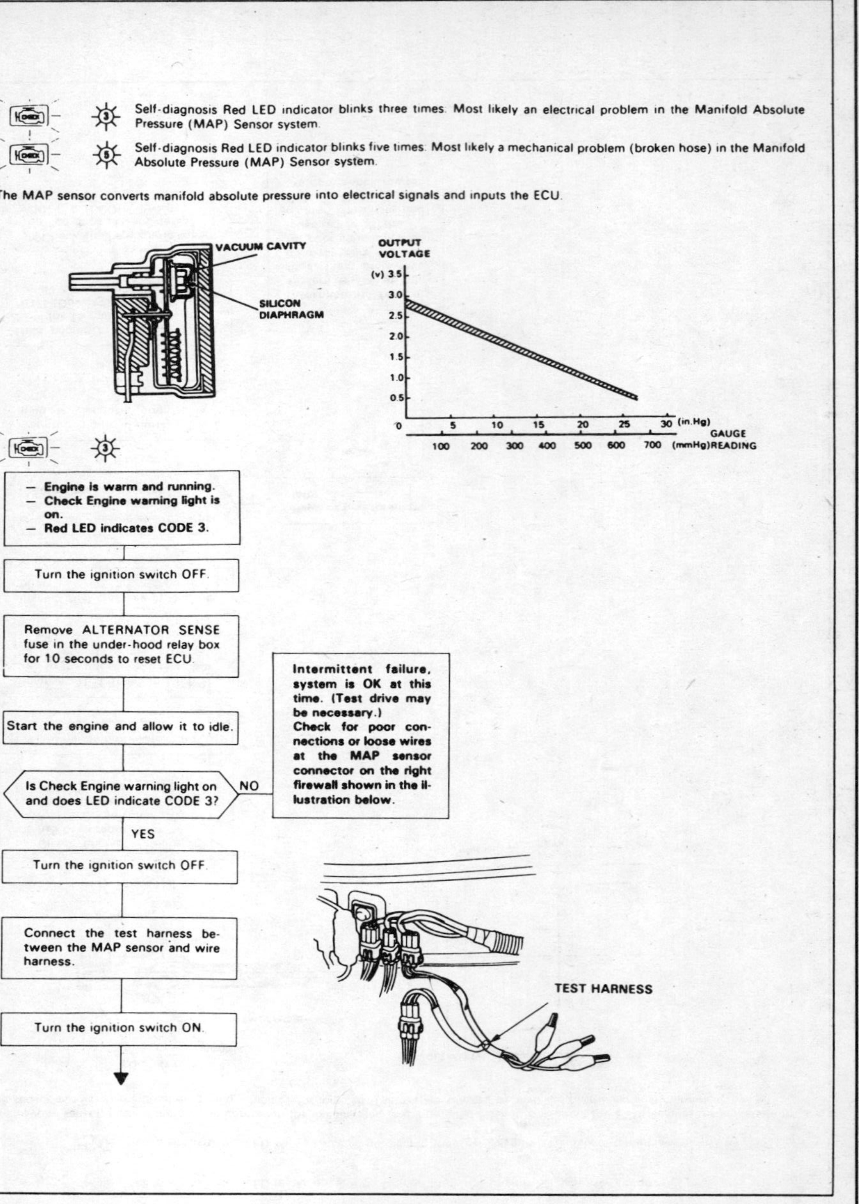


Self-diagnosis Red LED indicator blinks three times. Most likely an electrical problem in the Manifold Absolute Pressure (MAP) Sensor system.

Self-diagnosis Red LED indicator blinks five times. Most likely a mechanical problem (broken hose) in the Manifold Absolute Pressure (MAP) Sensor system.

The MAP sensor converts manifold absolute pressure into electrical signals and inputs the ECU.

- **Engine is warm and running.**
- **Check Engine warning light is on.**
- **Red LED indicates CODE 3.**

Turn the ignition switch OFF.

Remove ALTERNATOR SENSE fuse in the under-hood relay box for 10 seconds to reset ECU.

Start the engine and allow it to idle.

Is Check Engine warning light on and does LED indicate CODE 3?
- NO: **Intermittent failure, system is OK at this time. (Test drive may be necessary.) Check for poor connections or loose wires at the MAP sensor connector on the right firewall shown in the illustration below.**
- YES: Turn the ignition switch OFF.

Connect the test harness between the MAP sensor and wire harness.

Turn the ignition switch ON.

TEST HARNESS

## MAP SENSOR—CODE 3—1988–90 LEGEND

## MAP SENSOR—CODE 5—1988–90 LEGEND

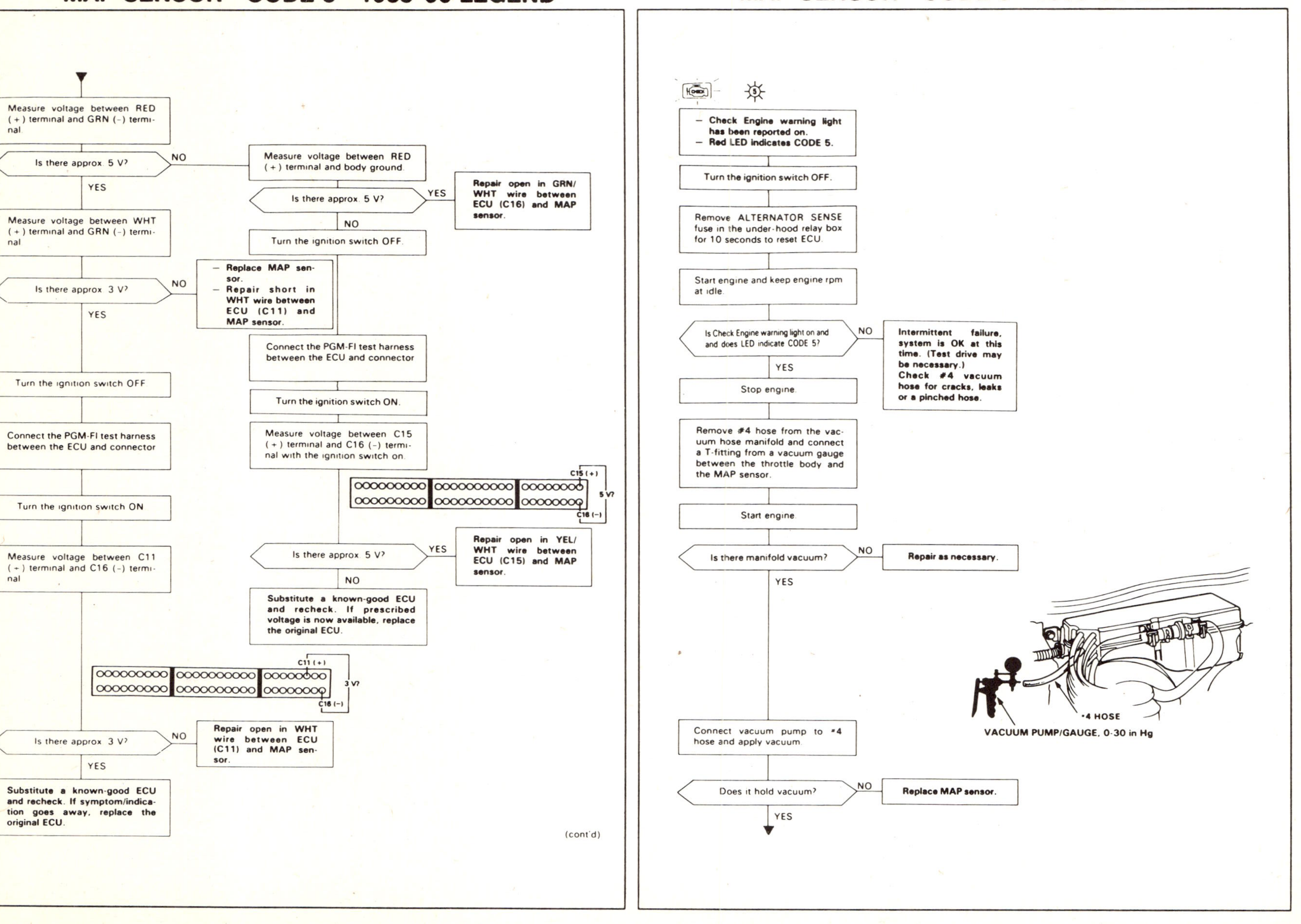

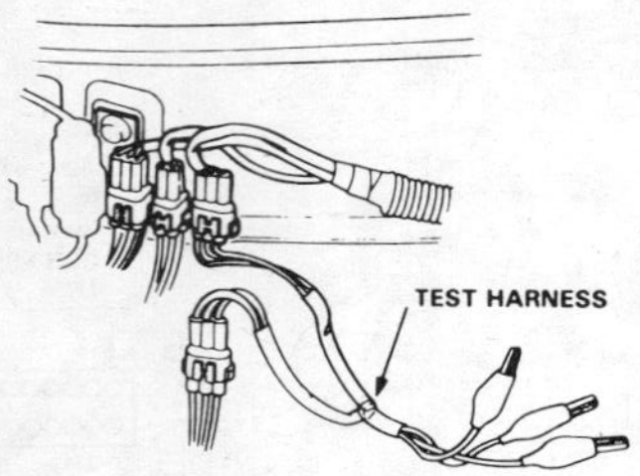

## CRANK/CYL SENSOR – CODE 4 – 1988–90 LEGEND

Self-diagnosis Red LED indicator blinks four times. A problem in the CRANK circuit of the CRANK/CYL Sensor.

Self-diagnosis Red LED indicator blinks nine times. A problem in the CYL circuit of the CRANK/CYL Sensor.

The CRANK sensor determines timing for fuel injection and ignition of each cylinder and also detects engine RPM. The CYL sensor detects the position of No.1 cylinder for sequential fuel injection to each cylinder.

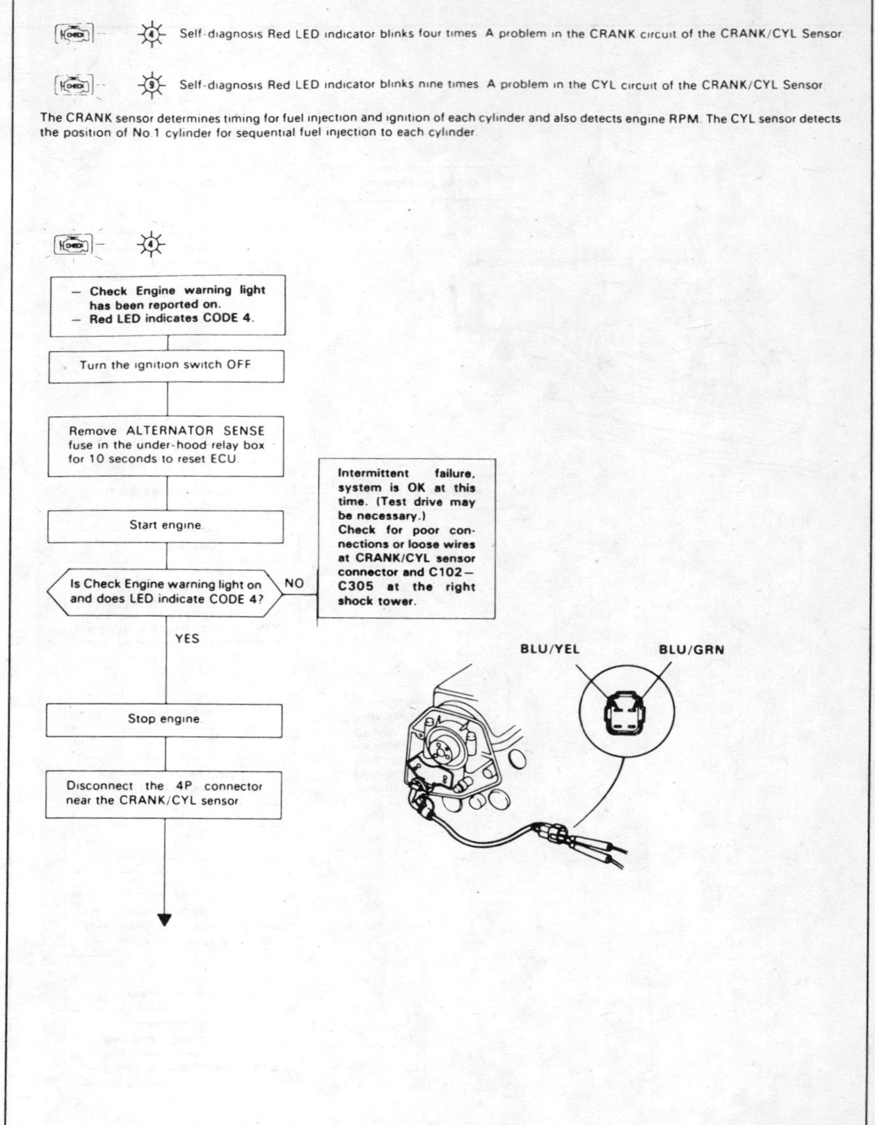


## MAP SENSOR – CODE 5 – 1988–90 LEGEND

Stop engine.

Connect the test harness between the MAP sensor and wire harness.

Turn the ignition switch ON.

Measure voltage between WHT (+) terminal and GRN (–) terminal.

Is there approx. 3V?

NO → **Replace MAP sensor.**

YES → **Substitute a known-good ECU and recheck. If symptom/indication goes away, replace the original ECU.**

TEST HARNESS

## CRANK/CYL SENSOR – CODE 9 – 1988–90 LEGEND

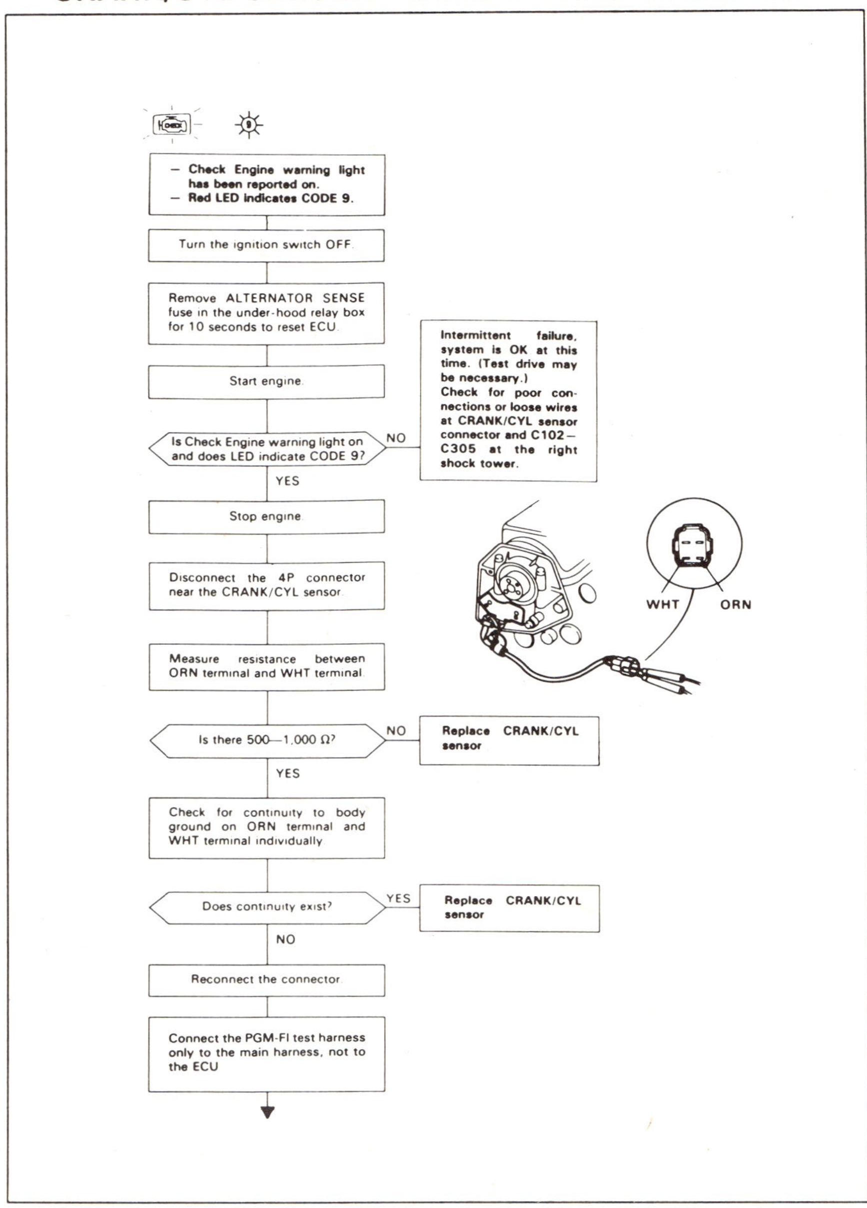


## CRANK/CYL SENSOR – CODE 4 – 1988–90 LEGEND

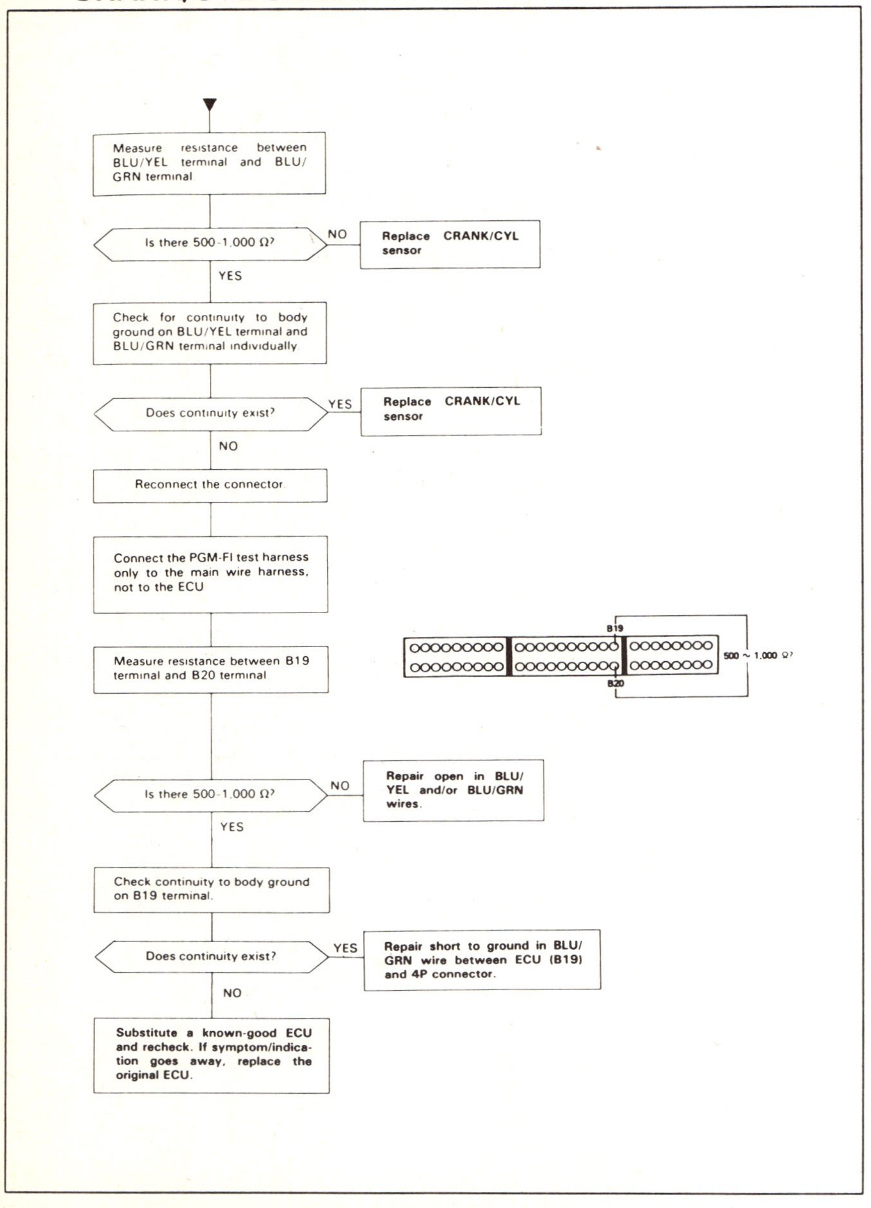

Measure resistance between BLU/YEL terminal and BLU/GRN terminal
Is there 500-1,000 Ω?
NO
Replace CRANK/CYL sensor
YES
Check for continuity to body ground on BLU/YEL terminal and BLU/GRN terminal individually.
Does continuity exist?
YES
Replace CRANK/CYL sensor
NO
Reconnect the connector.
Connect the PGM-FI test harness only to the main wire harness, not to the ECU
Measure resistance between B19 terminal and B20 terminal
B19
B20
500 ~ 1,000 Ω?
Is there 500-1,000 Ω?
NO
Repair open in BLU/YEL and/or BLU/GRN wires.
YES
Check continuity to body ground on B19 terminal.
Does continuity exist?
YES
Repair short to ground in BLU/GRN wire between ECU (B19) and 4P connector.
NO
Substitute a known-good ECU and recheck. If symptom/indication goes away, replace the original ECU.

## CRANK/CYL SENSOR – CODE 9 – 1988–90 LEGEND

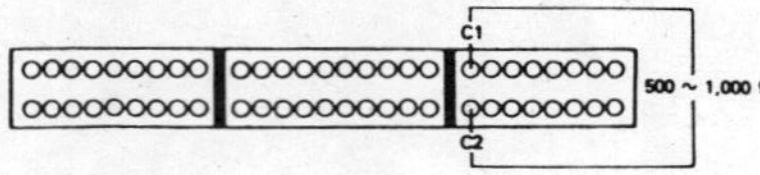

C1, C2: 500 ~ 1,000 Ω?

Measure resistance between C1 terminal and C2 terminal.

Is there 500-1,000Ω?
- NO → Repair open in ORN and/or WHT wires.
- YES → Check for continuity to body, ground on C1 terminal.

Does continuity exist?
- YES → Repair short to ground in ORN wire between ECU (C1) and 4P connector.
- NO → Substitute a known-good ECU and recheck. If symptom/indication goes away, replace the original ECU.

## COOLANT TEMPERATURE (TW) SENSOR – CODE 6 1988–90 LEGEND

Self-diagnosis Red LED indicator blinks six times. Most likely a problem in the Coolant Temperature (TW) Sensor circuit.

The TW sensor is a temperature dependant resistor (thermistor). The resistance of the thermistor decreases as the coolant temperature increases as shown below.

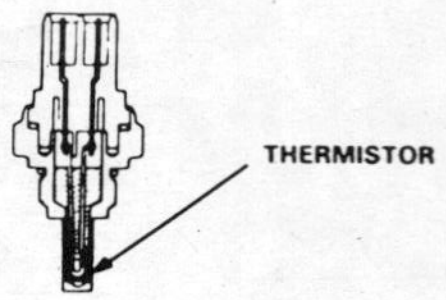

THERMISTOR

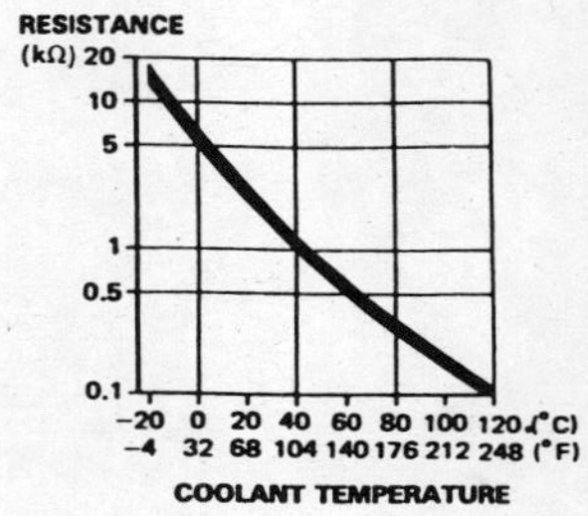

RESISTANCE (kΩ): 20, 10, 5, 1, 0.5, 0.1

COOLANT TEMPERATURE: −20, 0, 20, 40, 60, 80, 100, 120 (°C); −4, 32, 68, 104, 140, 176, 212, 248 (°F)

- Check Engine warning light is on.
- Red LED indicates CODE 6.

Turn the ignition switch OFF.

Remove ALTERNATOR SENSE fuse in the under-hood relay box for 10 seconds to reset ECU.

Turn the ignition switch ON.

Is Check Engine warning light on and does LED indicate CODE 6?
- NO → Intermittent failure, system is OK at this time. (Test drive may be necessary.) Check for poor connections or loose wires at TW sensor and RED/WHT wire at C303–C103 and GRN/WHT at C305–C102 at the right shock tower.
- YES → Turn the ignition switch OFF.

Connect the PGM-FI test harness between the ECU and connector.

Warm up engine to normal operating temperature (cooling fan comes on).

Turn the ignition switch OFF.

Disconnect "C" connector from ECU only, not the main wire harness.

Measure resistasnce between C6 terminal and C14 terminal.

Is there 200–400 Ω?
- NO → Inspect for an open circuit in GRN/WHT wire between ECU (C14) and TW sensor or open circuit in RED/WHT wire between ECU (C6) and TW sensor. If wires are OK, replace TW sensor.
- YES → Reconnect "C" connector to ECU and disconnect "C" connector from main wire harness.

## COOLANT TEMPERATURE (TW) SENSOR – CODE 6 1988–90 LEGEND

Turn the ignition switch ON.

Measure voltage between C6 (+) terminal and C14 (–) terminal.

Is there approx. 5 V?

- NO → Substitute a known-good ECU and recheck. If prescribed voltage is now available, replace the original ECU.
- YES ↓

Turn the ignition switch OFF.

Reconnect "C" connector to main wire harness.

Turn the ignition switch ON.

Measure voltage between C6 (+) terminal and C14 (–) terminal.

Is there less than 0.50 V?

- NO → Substitute a known-good ECU and recheck. If symptom/indication goes away, replace the original ECU.
- YES → AT ONLY: Disconnect "A" connector from Automatic Transmission control unit.
  - Is there less than 0.50 V?
    - NO → Replace Automatic Transmission control unit.
    - YES → Repair short in RED/WHT wire between ECU (C6) and TW sensor.

## THROTTLE ANGLE SENSOR – CODE 7 1988–90 LEGEND

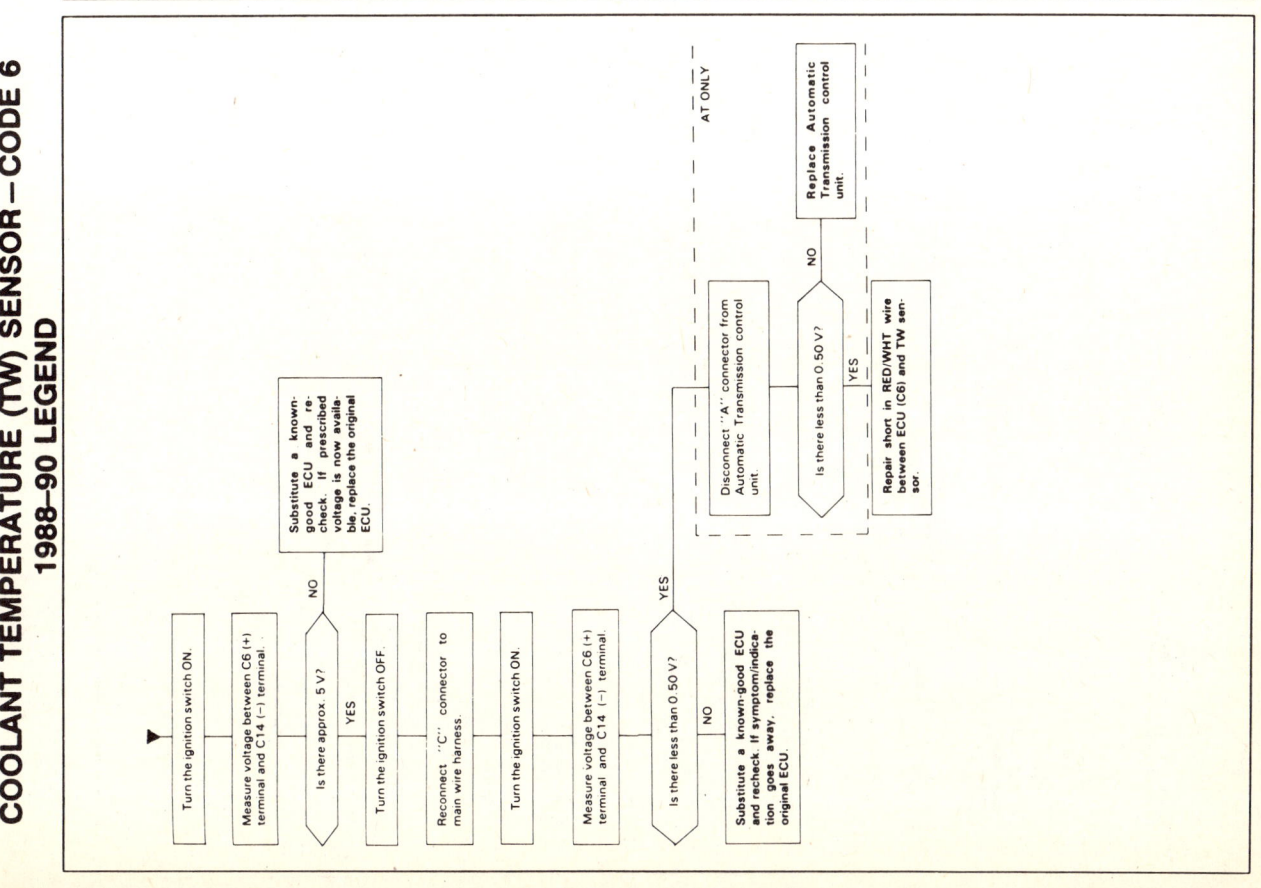

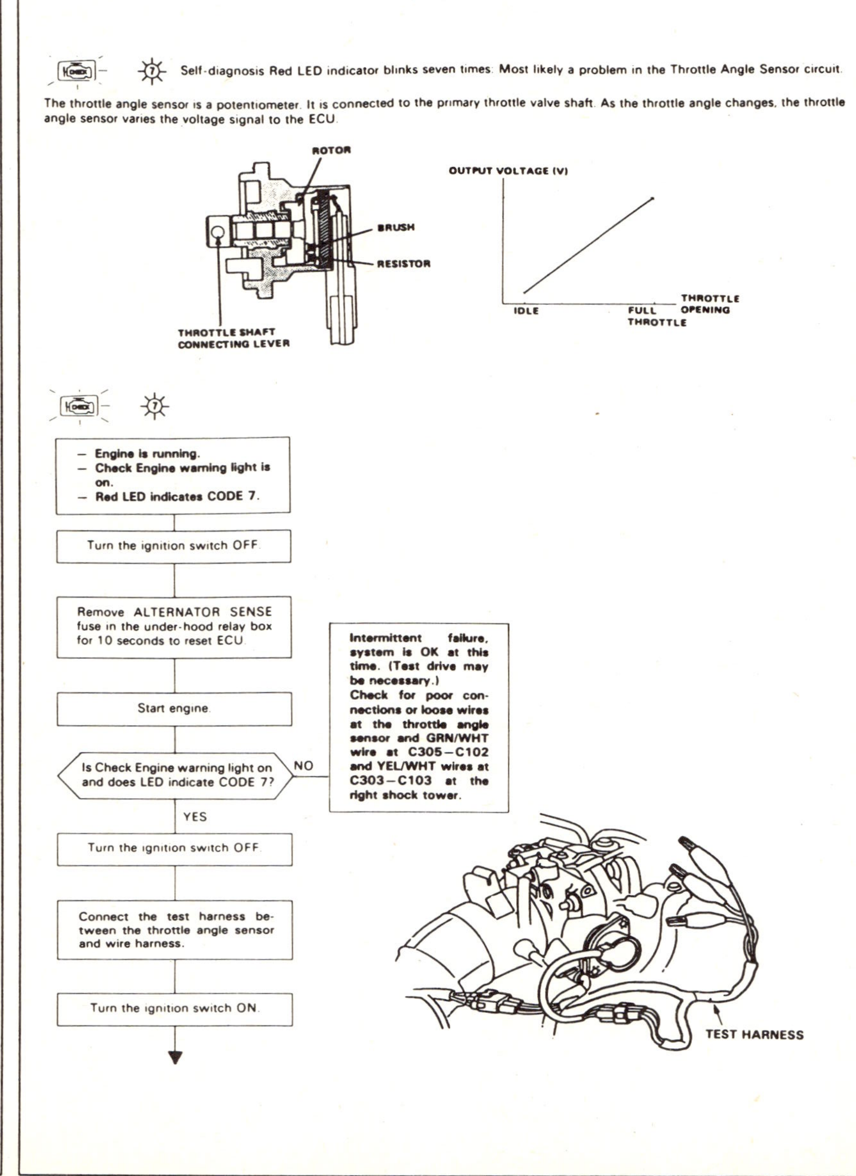

Self-diagnosis Red LED indicator blinks seven times. Most likely a problem in the Throttle Angle Sensor circuit.
The throttle angle sensor is a potentiometer. It is connected to the primary throttle valve shaft. As the throttle angle changes, the throttle angle sensor varies the voltage signal to the ECU.
ROTOR
BRUSH
RESISTOR
THROTTLE SHAFT CONNECTING LEVER
OUTPUT VOLTAGE (V)
IDLE
FULL THROTTLE
THROTTLE OPENING
– Engine is running.
– Check Engine warning light is on.
– Red LED indicates CODE 7.
Turn the ignition switch OFF.
Remove ALTERNATOR SENSE fuse in the under-hood relay box for 10 seconds to reset ECU.
Start engine.
Is Check Engine warning light on and does LED indicate CODE 7?
NO
Intermittent failure, system is OK at this time. (Test drive may be necessary.) Check for poor connections or loose wires at the throttle angle sensor and GRN/WHT wire at C305–C102 and YEL/WHT wires at C303–C103 at the right shock tower.
YES
Turn the ignition switch OFF.
Connect the test harness between the throttle angle sensor and wire harness.
Turn the ignition switch ON.
TEST HARNESS

## THROTTLE ANGLE SENSOR – CODE 7 1988–90 LEGEND

Measure voltage between RED (+) terminal and GRN (−) terminal

Is there approx 5V?

- NO → Measure voltage between RED (+) terminal and body ground.
  - Is there approx. 5V?
    - YES → Repair open in GRN/WHT wire between ECU (C14) and throttle angle sensor.
    - NO → Turn the ignition switch OFF.
      - Connect the PGM-FI test harness between the ECU and connector
      - Turn the ignition switch ON.
      - Measure voltage between C13 (+) terminal and C14 (−) terminal (C13 (+), C14 (−), 5 V?)
      - Is there approx 5 V?
        - YES → Repair open in YEL/WHT wire between ECU (C13) and throttle angle sensor.
        - NO → Substitute a known-good ECU and recheck. If prescribed voltage is now available, replace the original ECU.
- YES → Measure voltage between WHT (+) terminal and GRN (−) terminal at wide open throttle.

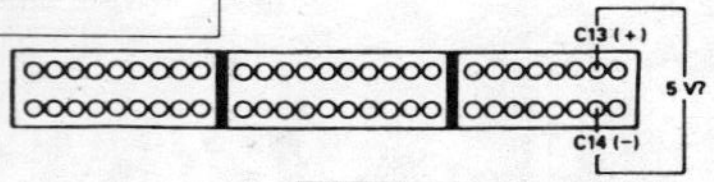

Is there approx 4 V?

- NO →
  - Replace throttle angle sensor.
  - Repair short in RED/YEL wire between ECU (C7) and throttle angle sensor.
  - Replace Automatic Transmission control unit.
- YES → Turn the ignition switch OFF

Connect the PGM-FI test harness between the ECU and connector

Turn the ignition switch ON

Measure voltage between C7 (+) terminal and C14 (−) terminal at wide open throttle. (C7 (+), C14 (−), 4 V?)

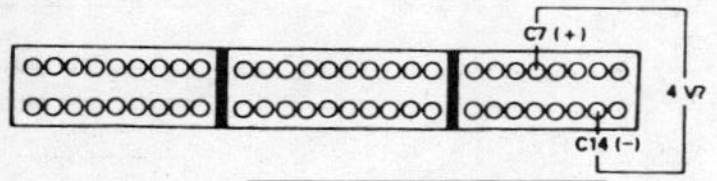

Is there approx 4 V?

- NO → Repair open in RED/YEL wire between ECU (C7) and throttle angle sensor.
- YES → Substitute a known-good ECU and recheck. If symptom/indication goes away, replace the original ECU.

## TDC SENSOR – CODE 8 – 1988–90 LEGEND

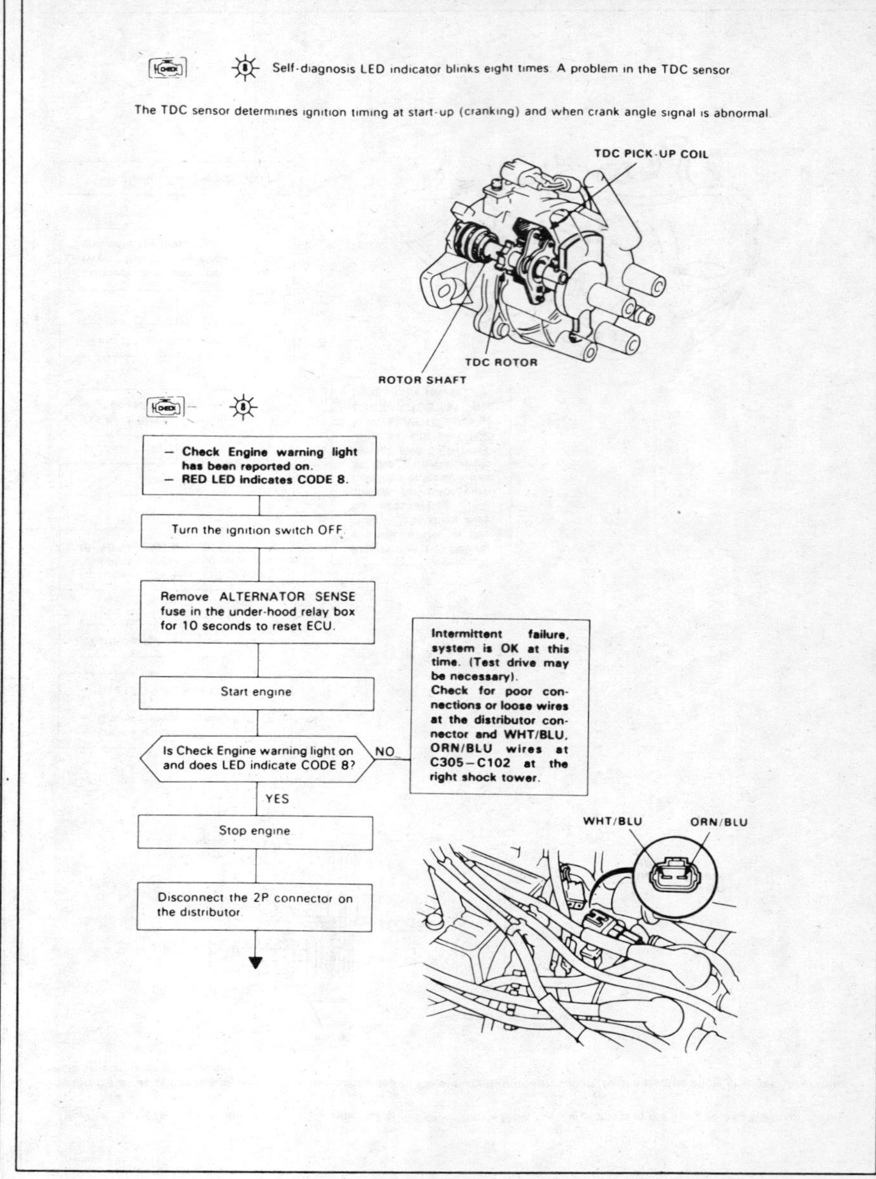

Self-diagnosis LED indicator blinks eight times. A problem in the TDC sensor
The TDC sensor determines ignition timing at start-up (cranking) and when crank angle signal is abnormal.
TDC PICK-UP COIL
TDC ROTOR
ROTOR SHAFT
– Check Engine warning light has been reported on.
– RED LED indicates CODE 8.
Turn the ignition switch OFF.
Remove ALTERNATOR SENSE fuse in the under-hood relay box for 10 seconds to reset ECU.
Start engine
Is Check Engine warning light on and does LED indicate CODE 8?
NO
Intermittent failure, system is OK at this time. (Test drive may be necessary). Check for poor connections or loose wires at the distributor connector and WHT/BLU, ORN/BLU wires at C305–C102 at the right shock tower.
YES
Stop engine
Disconnect the 2P connector on the distributor
WHT/BLU
ORN/BLU

## TDC SENSOR — CODE 8 — 1988–90 LEGEND

- Measure resistance between WHT/BLU terminal and ORN/BLU terminal.
- Is there 500 — 1,000 Ω?
  - NO → Replace TDC sensor
  - YES → Check for continuity to body ground on WHT/BLU terminal and ORN/BLU terminal individually
- Does continuity exist?
  - YES → Replace TDC sensor
  - NO → Reconnect the connector
- Connect the PGM-FI test harness only to the main wire harness, not to the ECU
- Measure resistance between C3 terminal and C4 terminal

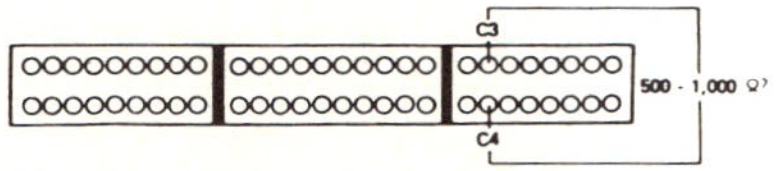

C3
C4
500 - 1,000 Ω?

- Is there 500 — 1,000 Ω?
  - NO → Repair open is ORN/BLU and/or WHT/BLU wires.
  - YES → Check for continuity to body ground on C3 terminal.
- Does continuity exist?
  - YES → Repair short in ORN/BLU wire between ECU (C3) and 29 connector.
  - NO → Substitute a known-good ECU and recheck. If symptom/indication goes away, replace the original ECU.

## AIR TEMPERATURE (TA) SENSOR — CODE 10 1988–90 LEGEND

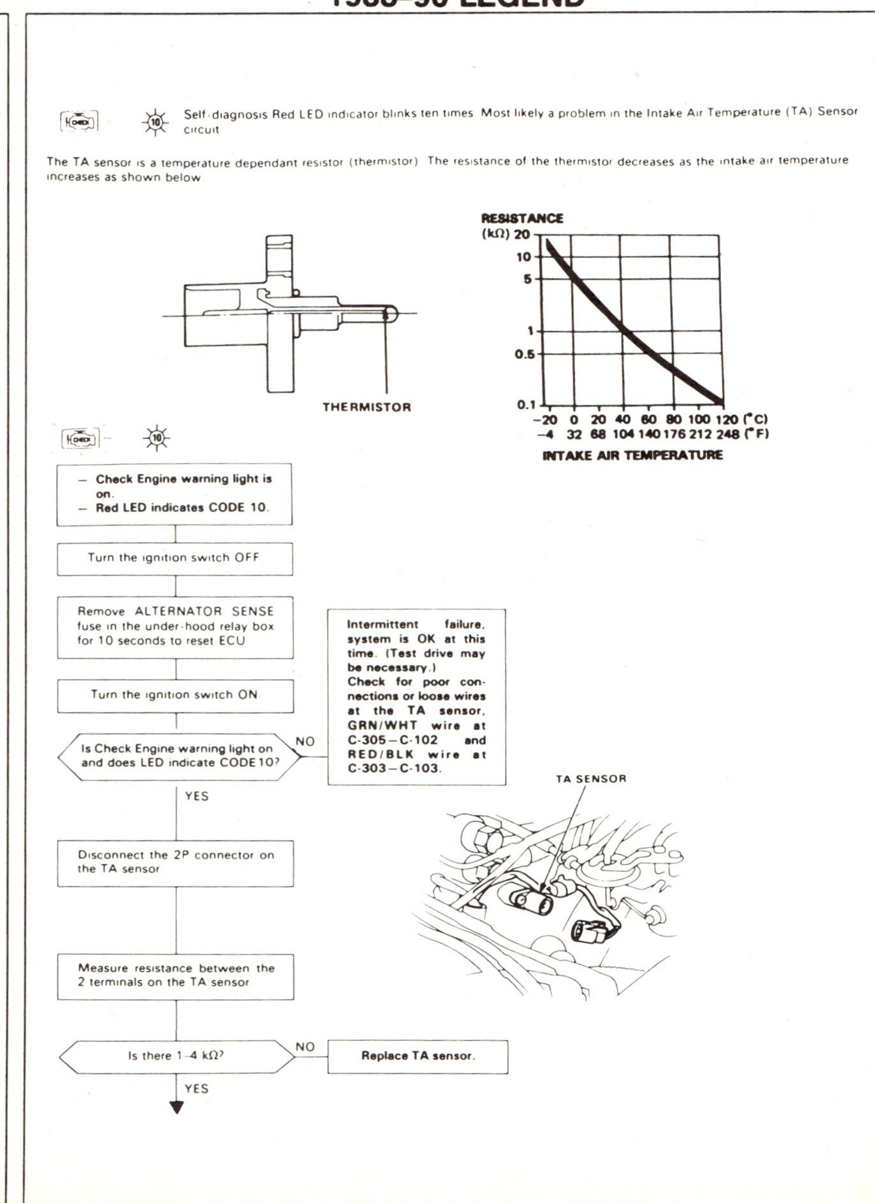

Self-diagnosis Red LED indicator blinks ten times. Most likely a problem in the Intake Air Temperature (TA) Sensor circuit
The TA sensor is a temperature dependant resistor (thermistor). The resistance of the thermistor decreases as the intake air temperature increases as shown below
THERMISTOR
RESISTANCE (kΩ)
20
10
5
1
0.5
0.1
-20 0 20 40 60 80 100 120 (°C)
-4 32 68 104 140 176 212 248 (°F)
INTAKE AIR TEMPERATURE
– Check Engine warning light is on.
– Red LED indicates CODE 10.
Turn the ignition switch OFF
Remove ALTERNATOR SENSE fuse in the under-hood relay box for 10 seconds to reset ECU
Turn the ignition switch ON
Is Check Engine warning light on and does LED indicate CODE 10?
NO
Intermittent failure, system is OK at this time. (Test drive may be necessary.) Check for poor connections or loose wires at the TA sensor, GRN/WHT wire at C-305—C-102 and RED/BLK wire at C-303—C-103.
YES
Disconnect the 2P connector on the TA sensor
Measure resistance between the 2 terminals on the TA sensor
Is there 1–4 kΩ?
NO
Replace TA sensor.
YES
TA SENSOR

## AIR TEMPERATURE (TA) SENSOR—CODE 10
## 1988–90 LEGEND

- Measure voltage between RED/BLK (+) terminal and body ground.
- Is there approx. 5 V?
  - NO → Turn the ignition switch OFF. → Connect the PGM-FI test harness between the ECU and connector → Turn the ignition switch ON. → Measure voltage between C5 (+) terminal and C14 (−) terminal. 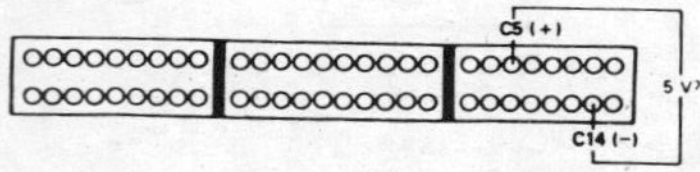
    - Is there approx. 5 V?
      - YES → Repair open in RED/BLK wire between ECU (C5) and TA sensor.
      - NO → Disconnect "C" connector from the main harness only, not the ECU. → Measure voltage between C5 (+) terminal and C14 (−) terminal.
        - Is there approx. 5 V?
          - YES → Repair short in RED/BLK wire between ECU (C5) and TA sensor.
          - NO → Substitute a known-good ECU and recheck. If prescribed voltage is now available, replace the original ECU.
  - YES → Measure voltage between RED/BLK (+) terminal and GRN/WHT (−) terminal.
    - Is there approx. 5 V?
      - NO → Repair open in GRN/WHT wire between ECU (C14) and TA sensor.
      - YES → Substitute a known-good ECU and recheck. If symptom/indication goes away, replace the original ECU.

C5 (+) C14 (−) 5 V?

## ATMOSPHERIC PRESSURE (PA) SENSOR—CODE 13
## 1988–90 LEGEND

Self-diagnosis Red LED indicator blinks thirteen times: A problem in the Atmospheric Pressure (PA) Sensor circuit.

The PA sensor converts atmospheric pressure into electrical signals and inputs the ECU.

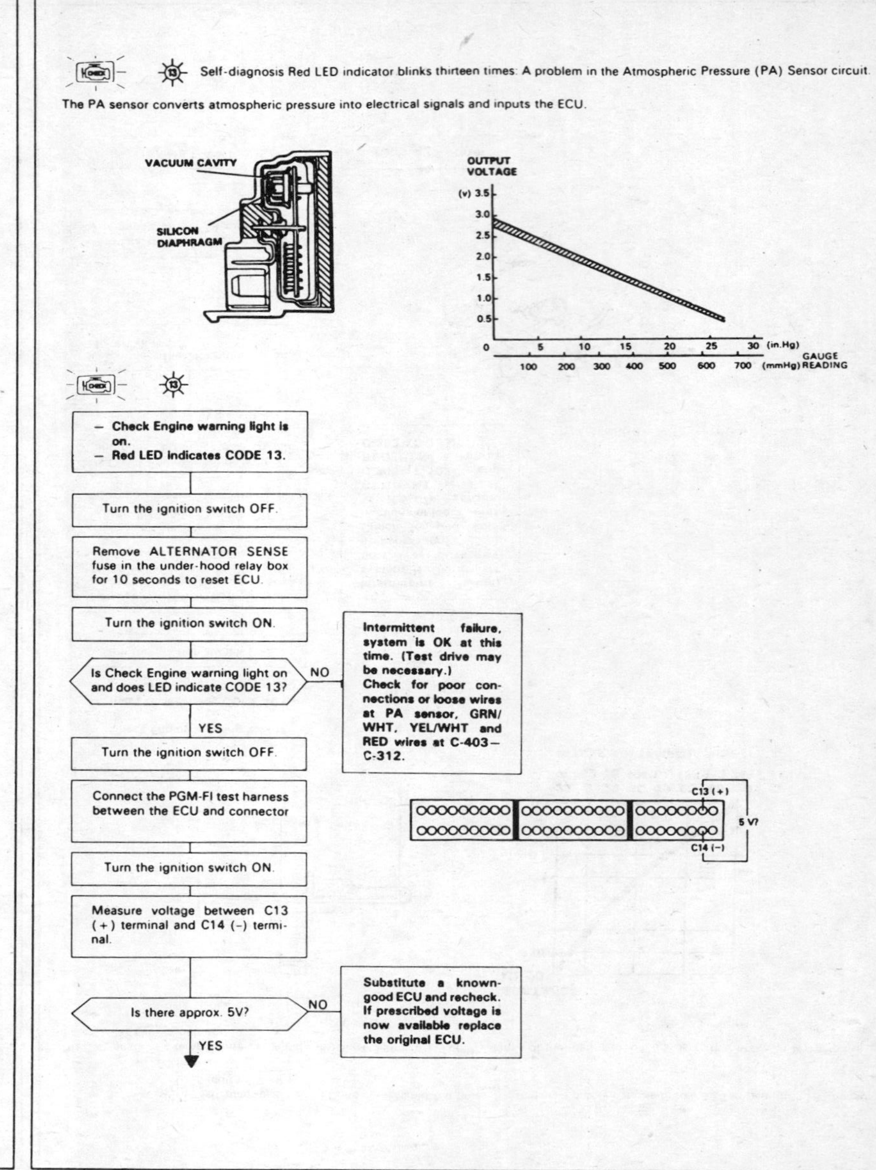


- Check Engine warning light is on.
- Red LED indicates CODE 13.

1. Turn the ignition switch OFF.
2. Remove ALTERNATOR SENSE fuse in the under-hood relay box for 10 seconds to reset ECU.
3. Turn the ignition switch ON.
4. Is Check Engine warning light on and does LED indicate CODE 13?
   - NO → Intermittent failure, system is OK at this time. (Test drive may be necessary.) Check for poor connections or loose wires at PA sensor, GRN/WHT, YEL/WHT and RED wires at C-403–C-312.
   - YES → Turn the ignition switch OFF. → Connect the PGM-FI test harness between the ECU and connector → Turn the ignition switch ON. → Measure voltage between C13 (+) terminal and C14 (−) terminal.
5. Is there approx. 5V?
   - NO → Substitute a known-good ECU and recheck. If prescribed voltage is now available replace the original ECU.
   - YES ▼

C13 (+) C14 (−) 5 V?

## ATMOSPHERIC PRESSURE (PA) SENSOR—CODE 13 1988–90 LEGEND

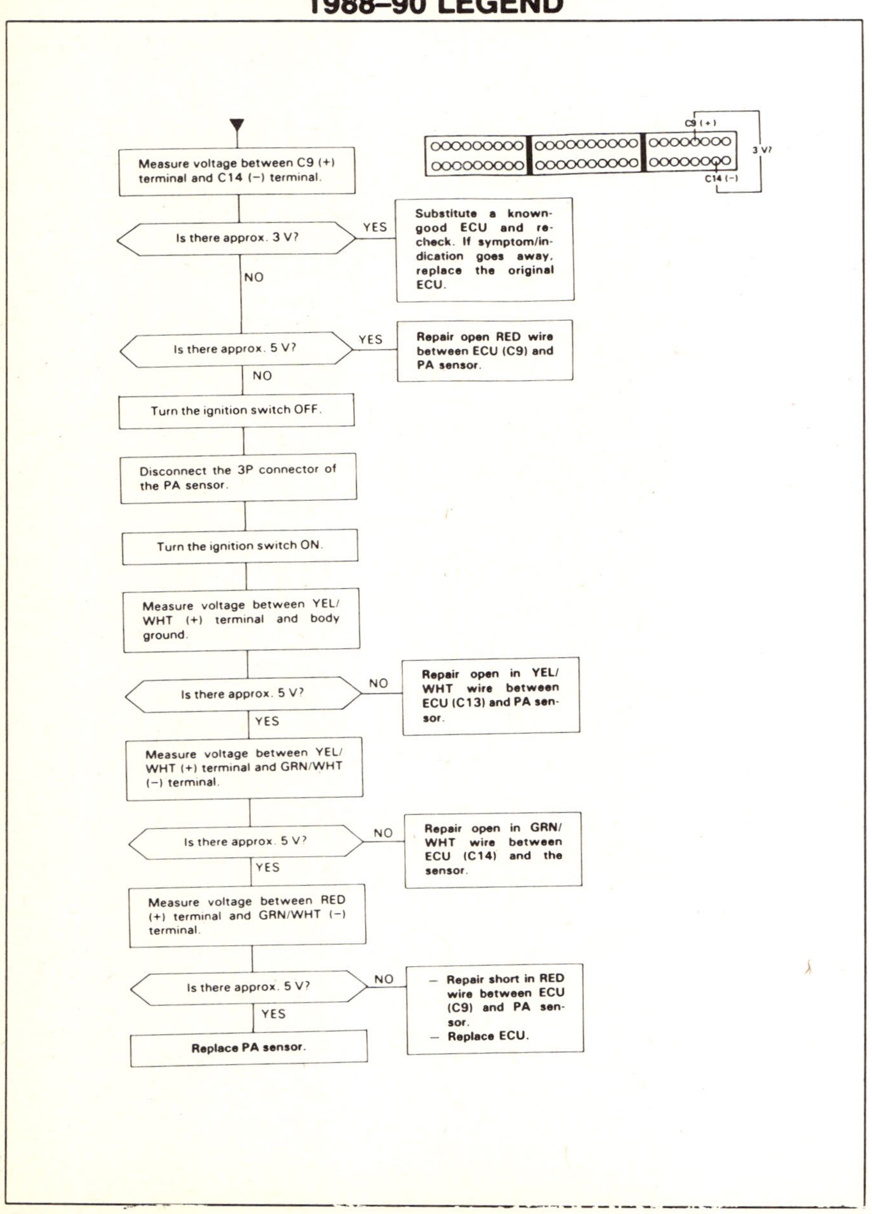


## IGNITION OUTPUT SIGNAL—CODE 15 1988–90 LEGEND

CHECK – 15 – Self-diagnosis Red LED indicator blinks fifteen times: A problem in the Ignition Output Signal circuit.

- Check Engine warning light is on.
- Red LED indicates CODE 15.

Turn the ignition switch OFF.

Remove ALTERNATOR SENSE fuse in the under-hood relay box for 10 seconds to reset ECU.

Turn the ignition switch ON.

Is Check Engine warning light on and does LED indicate CODE 15?

NO → **Intermittent failure, system is OK at this time. (Test drive may be necessary.) Check for poor connections or loose wires at Igniter and RED/BLU wire at C-439—C-257 and C-403—C-312.**

YES → Turn the ignition switch OFF.

Disconnect the 4P connector on the igniter unit.

Turn the ignition switch ON.

Measure voltage between the BLK/YEL (+) terminal and body ground.

Is there battery voltage?

NO → **Repair open in BLK/YEL wire between the igniter unit and ignition switch.**

YES → Turn the ignition switch OFF.

BLU (+)
RED/BLU
BLK (−)
BLK/YEL (+)
IGNITER UNIT

## IGNITION OUTPUT SIGNAL—CODE 15 1988–90 LEGEND

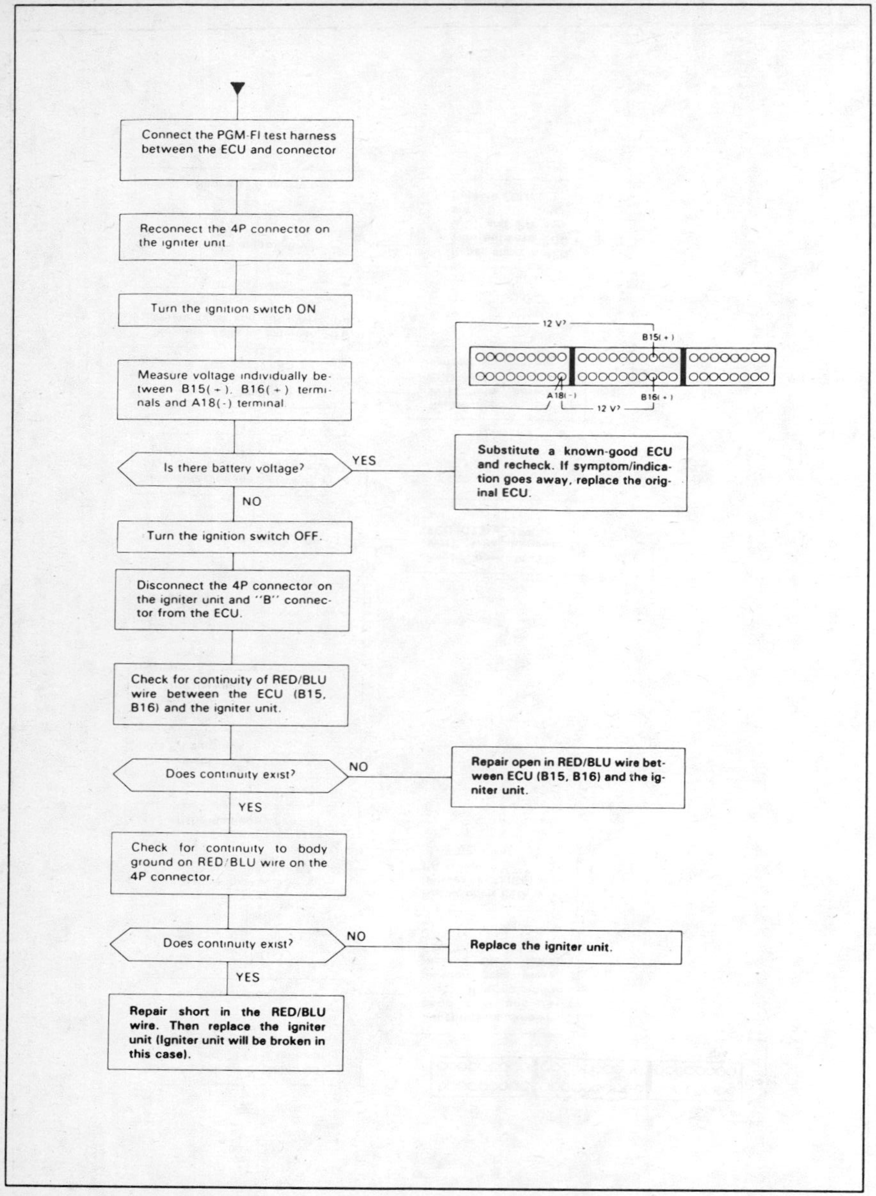

Connect the PGM-FI test harness between the ECU and connector

Reconnect the 4P connector on the igniter unit

Turn the ignition switch ON

Measure voltage individually between B15(+), B16(+) terminals and A18(-) terminal

12 V?
B15(+)
A18(-)
B16(+)
12 V?

Is there battery voltage?

YES — **Substitute a known-good ECU and recheck. If symptom/indication goes away, replace the original ECU.**

NO

Turn the ignition switch OFF.

Disconnect the 4P connector on the igniter unit and "B" connector from the ECU.

Check for continuity of RED/BLU wire between the ECU (B15, B16) and the igniter unit.

Does continuity exist?

NO — **Repair open in RED/BLU wire between ECU (B15, B16) and the igniter unit.**

YES

Check for continuity to body ground on RED/BLU wire on the 4P connector.

Does continuity exist?

NO — **Replace the igniter unit.**

YES

**Repair short in the RED/BLU wire. Then replace the igniter unit (Igniter unit will be broken in this case).**

## VEHICLE SPEED SENSOR—CODE 17—1988–90 LEGEND

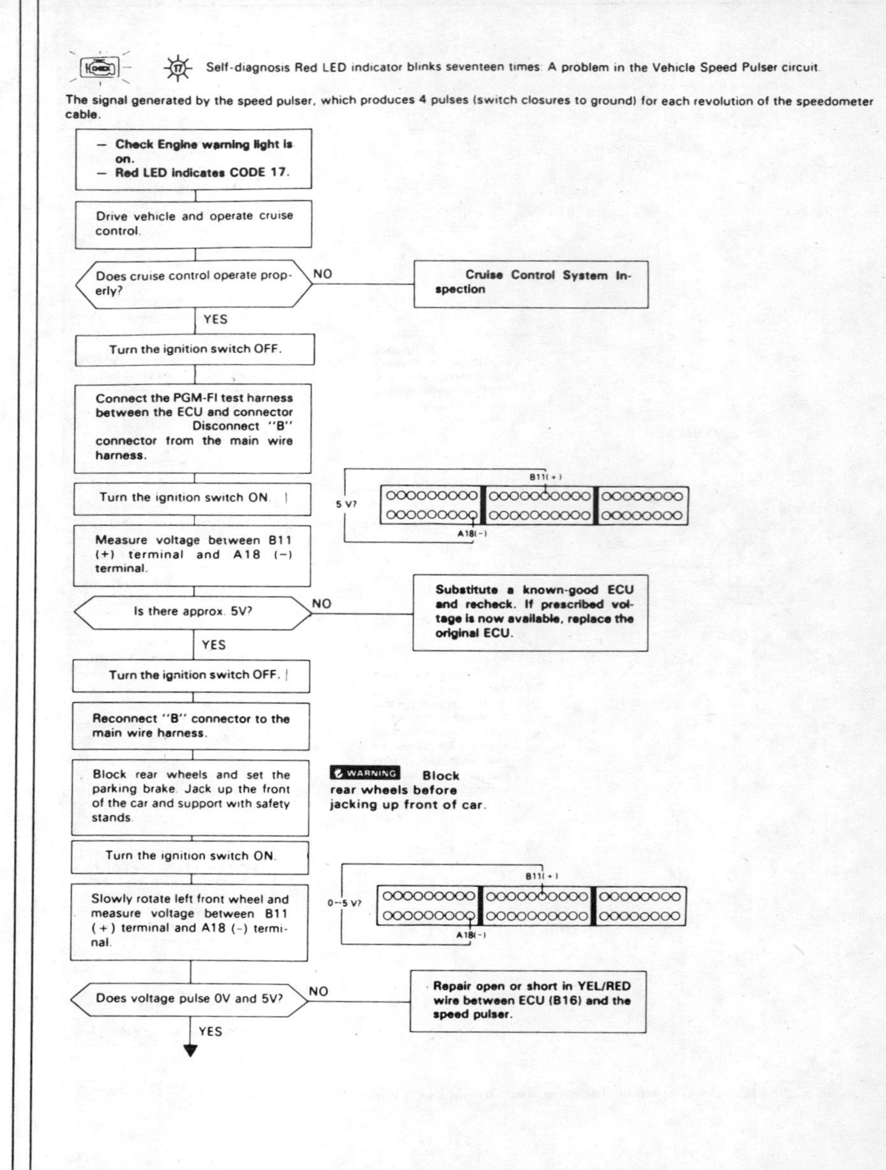

Self-diagnosis Red LED indicator blinks seventeen times. A problem in the Vehicle Speed Pulser circuit.
The signal generated by the speed pulser, which produces 4 pulses (switch closures to ground) for each revolution of the speedometer cable.
— Check Engine warning light is on.
— Red LED indicates CODE 17.
Drive vehicle and operate cruise control.
Does cruise control operate properly?
NO
Cruise Control System Inspection
YES
Turn the ignition switch OFF.
Connect the PGM-FI test harness between the ECU and connector Disconnect "B" connector from the main wire harness.
Turn the ignition switch ON.
5 V?
B11(+)
A18(-)
Measure voltage between B11 (+) terminal and A18 (−) terminal.
Is there approx. 5V?
NO
Substitute a known-good ECU and recheck. If prescribed voltage is now available, replace the original ECU.
YES
Turn the ignition switch OFF.
Reconnect "B" connector to the main wire harness.
Block rear wheels and set the parking brake. Jack up the front of the car and support with safety stands.
WARNING Block rear wheels before jacking up front of car.
Turn the ignition switch ON.
0–5 V?
B11(+)
A18(-)
Slowly rotate left front wheel and measure voltage between B11 (+) terminal and A18 (-) terminal.
Does voltage pulse 0V and 5V?
NO
Repair open or short in YEL/RED wire between ECU (B16) and the speed pulser.
YES

## VEHICLE SPEED SENSOR—CODE 17 1988–90 LEGEND

Does voltage hold 0 V or 5 V?

- 5 V: 
  - Repair open in YEL/RED wire between ECU (B11) and the speed pulser.
  - Faulty speed pulser.
- 0 V: Disconnect the connectors from the main wire harness only, not the ECU.

Measure voltage between B11 (+) terminal and A18 (-) terminal.

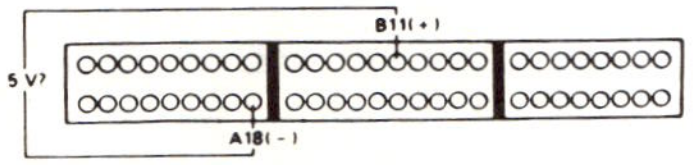

5 V? B11 (+) A18 (-)

Is there approx. 5V?

- YES:
  - Repair short in YEL/RED wire between ECU (B11) and the speed pulser.
  - Faulty speed pulser.
- NO: **Substitute a known-good ECU and recheck. If prescribed voltage is now available, replace the original ECU.**

## IGNITION TIMING ADJUSTER—CODE 18 1988–90 LEGEND

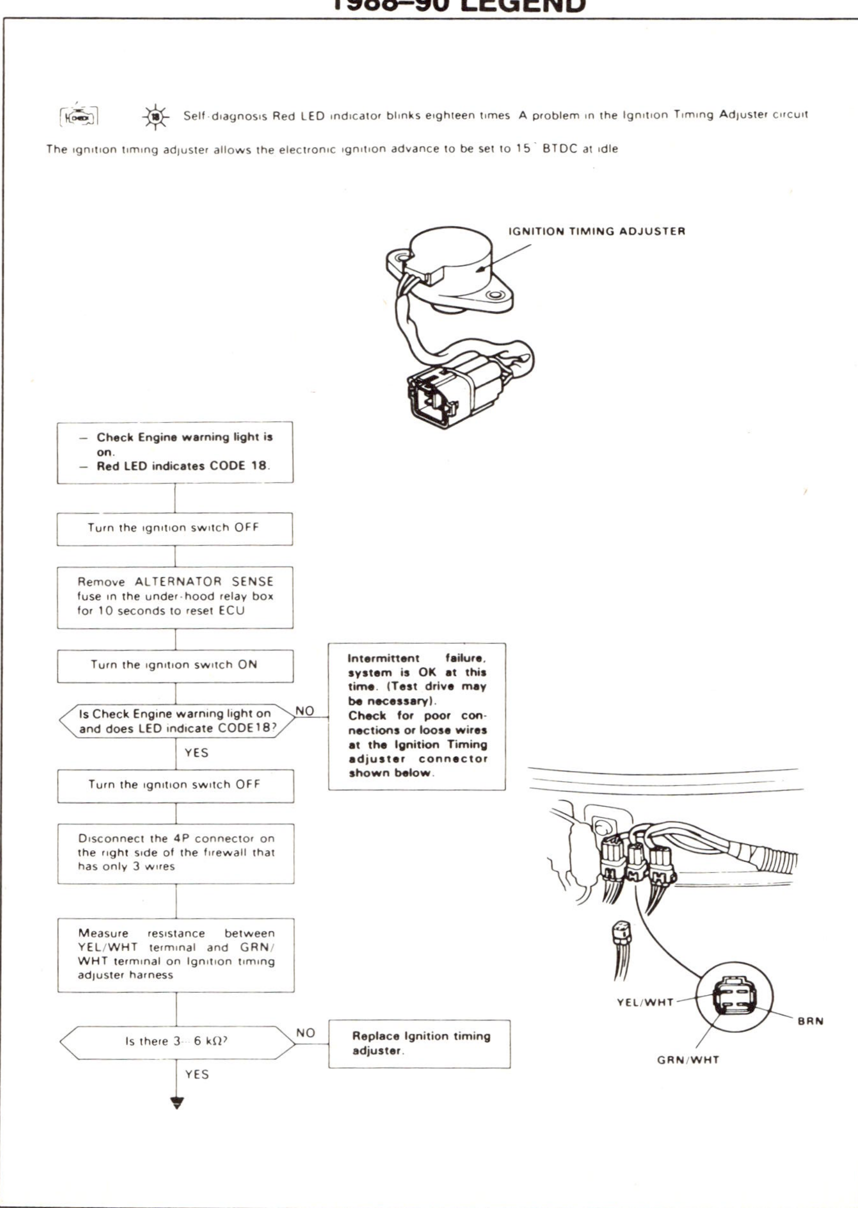


Self-diagnosis Red LED indicator blinks eighteen times. A problem in the Ignition Timing Adjuster circuit.

The ignition timing adjuster allows the electronic ignition advance to be set to 15° BTDC at idle.

- Check Engine warning light is on.
- Red LED indicates CODE 18.

Turn the ignition switch OFF.

Remove ALTERNATOR SENSE fuse in the under-hood relay box for 10 seconds to reset ECU.

Turn the ignition switch ON.

Is Check Engine warning light on and does LED indicate CODE18?

- NO: **Intermittent failure, system is OK at this time. (Test drive may be necessary). Check for poor connections or loose wires at the Ignition Timing adjuster connector shown below.**
- YES: Turn the ignition switch OFF.

Disconnect the 4P connector on the right side of the firewall that has only 3 wires.

Measure resistance between YEL/WHT terminal and GRN/WHT terminal on Ignition timing adjuster harness.

Is there 3–6 kΩ?

- NO: **Replace Ignition timing adjuster.**
- YES

## EACV – CODE 14 – 1988–90 LEGEND

Self diagnosis Red LED indicator blinks fourteen times: A problem in the Electronic Air Control Valve (EACV) circuit

The EACV changes the amount of air bypassing the throttle body in response to a current signal from the ECU in order to maintain the proper idle speed.

VALVE
SPRING
COIL
TO INTAKE MANIFOLD
FROM AIR CLEANER
VALVE SHAFT

– Engine is running.
– Check Engine warning light is on.
– Red LED indicates CODE 14.

Turn the ignition switch OFF

Remove ALTERNATOR SENSE fuse in the under-hood relay box for 10 seconds to reset ECU

Start engine

Is Check Engine warning light on and does LED indicate CODE 14?

NO → Intermittent failure system is OK. (test drive may be necessary). Check for poor connections or loose wires at the EACV, C228–C127, C418–C256, C402–C313 and C403–C312.

YES

Stop engine

Disconnect the 2P connector on the EACV.

Measure resistance between the 2 terminals on the EACV.

EACV

## IGNITION TIMING ADJUSTER – CODE 18 1988–90 LEGEND

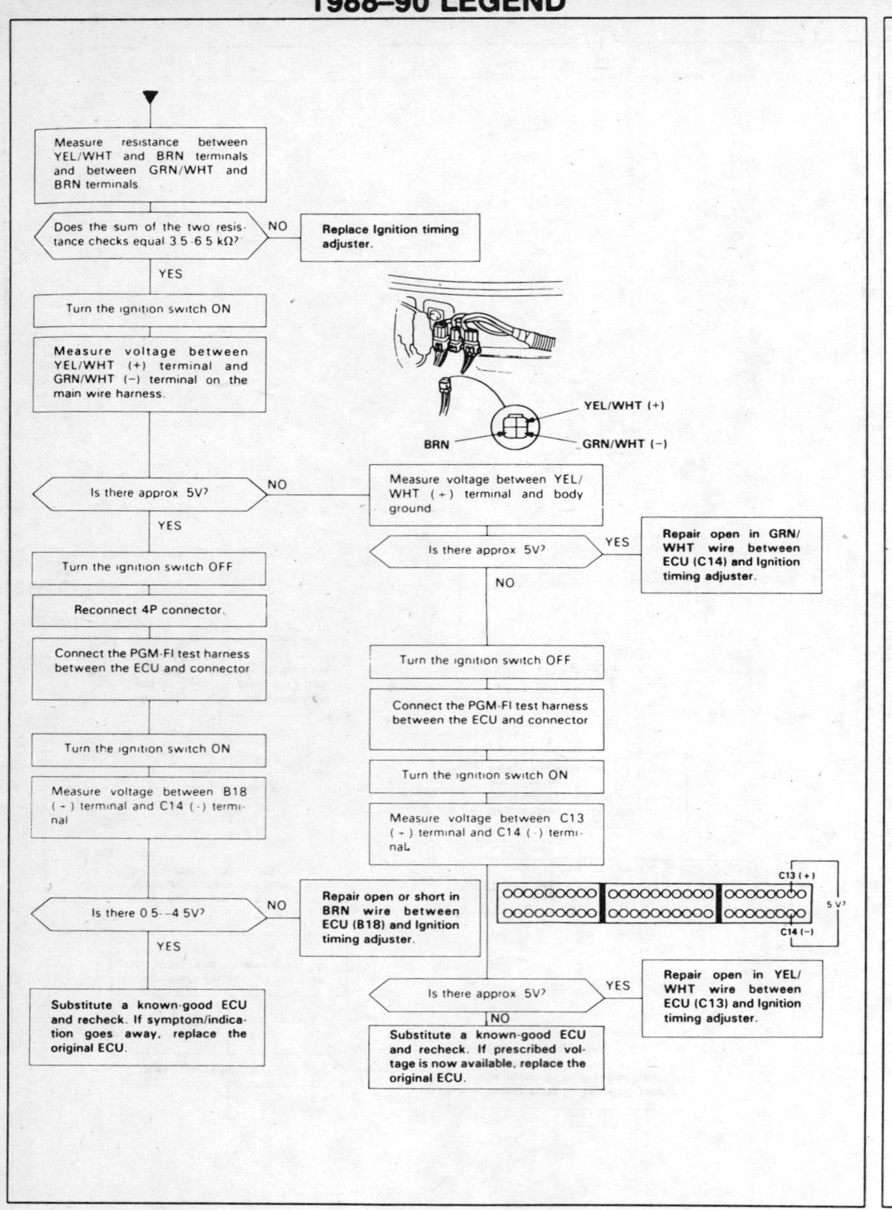

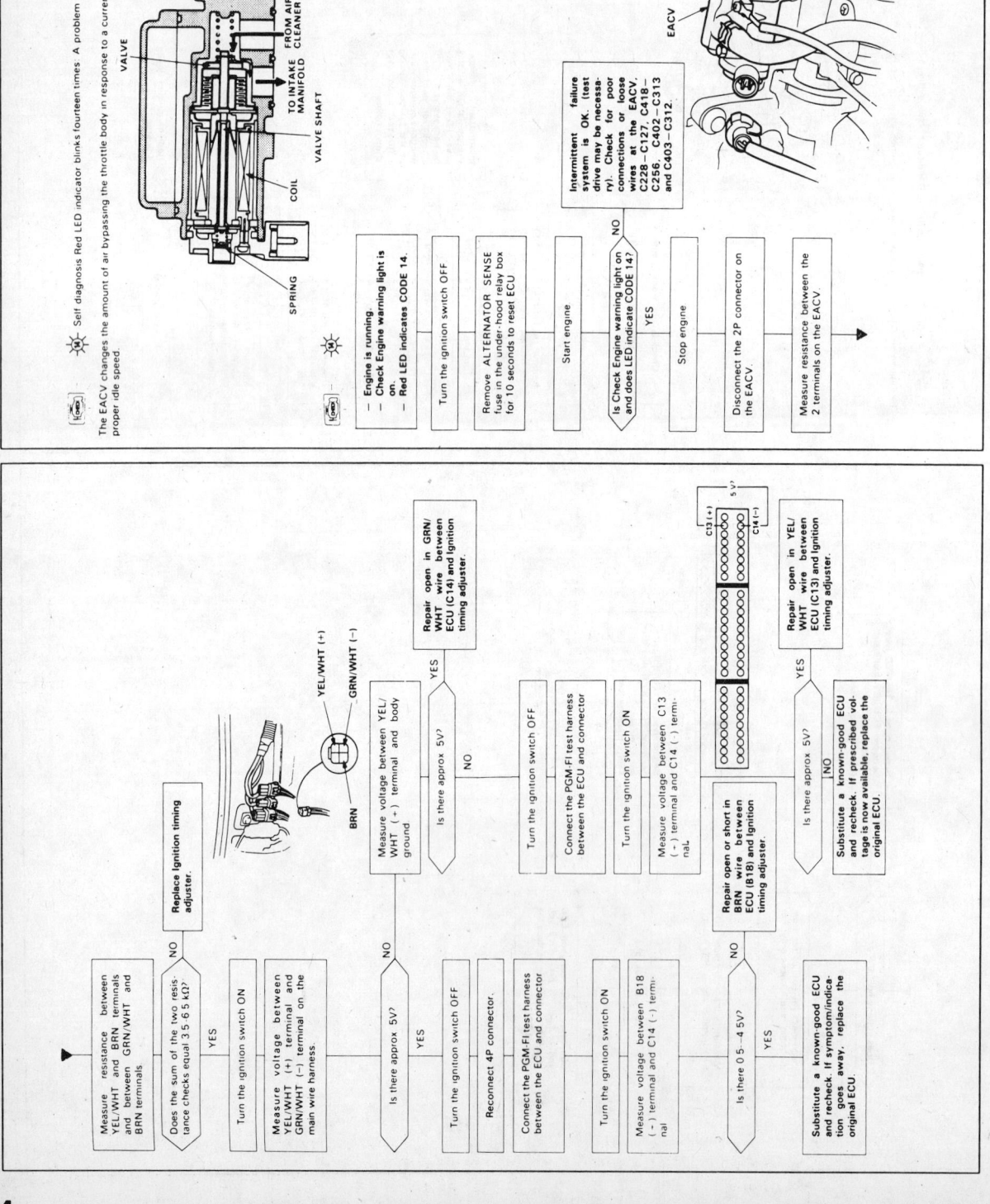

## EACV – CODE 14 – 1988–90 LEGEND

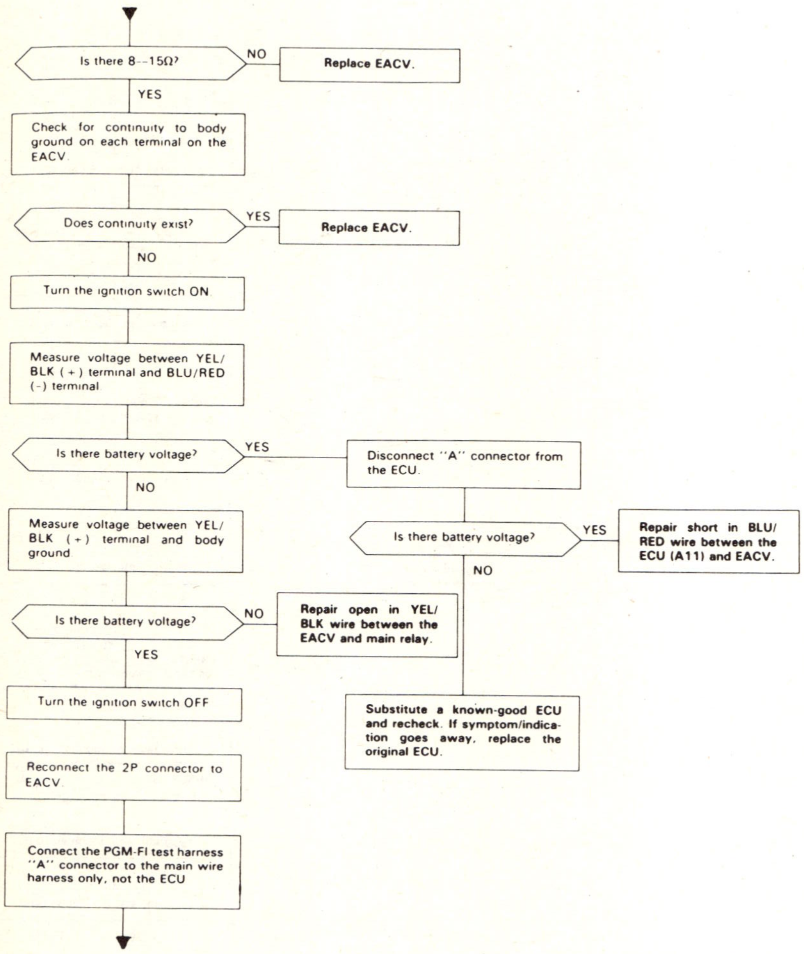
Is there 8–15Ω?
NO
Replace EACV.
YES
Check for continuity to body ground on each terminal on the EACV.
Does continuity exist?
YES
Replace EACV.
NO
Turn the ignition switch ON.
Measure voltage between YEL/BLK (+) terminal and BLU/RED (-) terminal.
Is there battery voltage?
YES
Disconnect "A" connector from the ECU.
Is there battery voltage?
YES
Repair short in BLU/RED wire between the ECU (A11) and EACV.
NO
Substitute a known-good ECU and recheck. If symptom/indication goes away, replace the original ECU.
NO
Measure voltage between YEL/BLK (+) terminal and body ground.
Is there battery voltage?
NO
Repair open in YEL/BLK wire between the EACV and main relay.
YES
Turn the ignition switch OFF.
Reconnect the 2P connector to EACV.
Connect the PGM-FI test harness "A" connector to the main wire harness only, not the ECU.

## EACV – CODE 14 – 1988–90 LEGEND

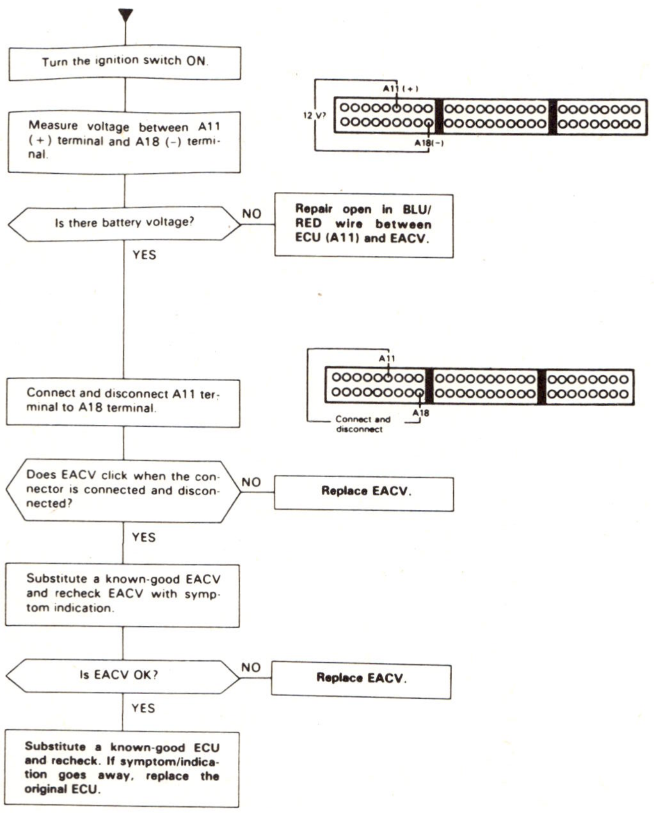
Turn the ignition switch ON.
Measure voltage between A11 (+) terminal and A18 (-) terminal.
A11 (+)
12 V?
A18 (-)
Is there battery voltage?
NO
Repair open in BLU/RED wire between ECU (A11) and EACV.
YES
Connect and disconnect A11 terminal to A18 terminal.
A11
A18
Connect and disconnect
Does EACV click when the connector is connected and disconnected?
NO
Replace EACV.
YES
Substitute a known-good EACV and recheck EACV with symptom indication.
Is EACV OK?
NO
Replace EACV.
YES
Substitute a known-good ECU and recheck. If symptom/indication goes away, replace the original ECU.

## A/T FI SIGNAL A–CODE 30–1990 LEGEND

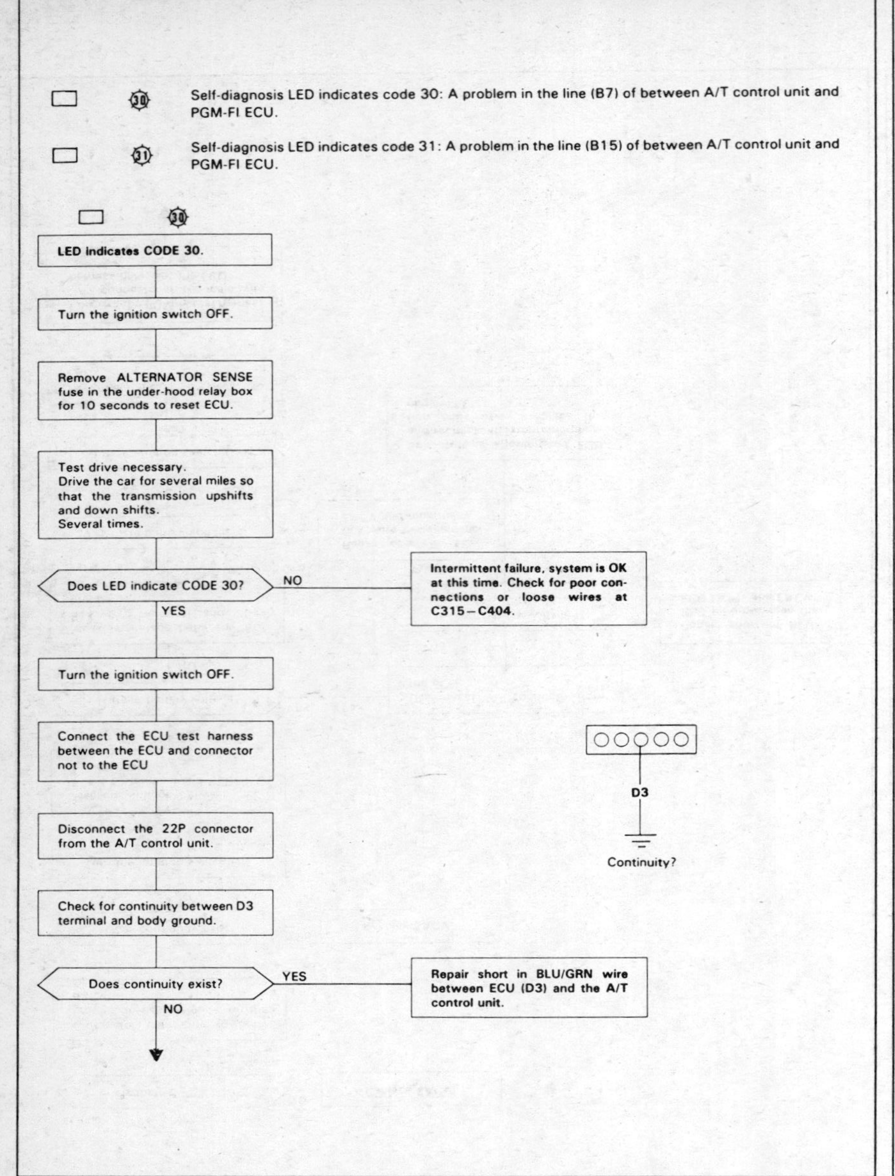


## A/T FI SIGNAL A–CODE 30–1990 LEGEND

Check for Continuity BLU/GRN wire between D3 and 22P connector of the A/T control unit.

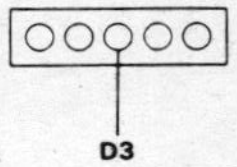

Does continuity exist?

NO — Repair open in BLU/GRN wire between ECU (D3) and the A/T control unit.

D3

YES — Substitute a known-good ECU and recheck. If symptom/indication goes away, replace the original ECU.

## A/T FI SIGNAL B—CODE 31—1990 LEGEND

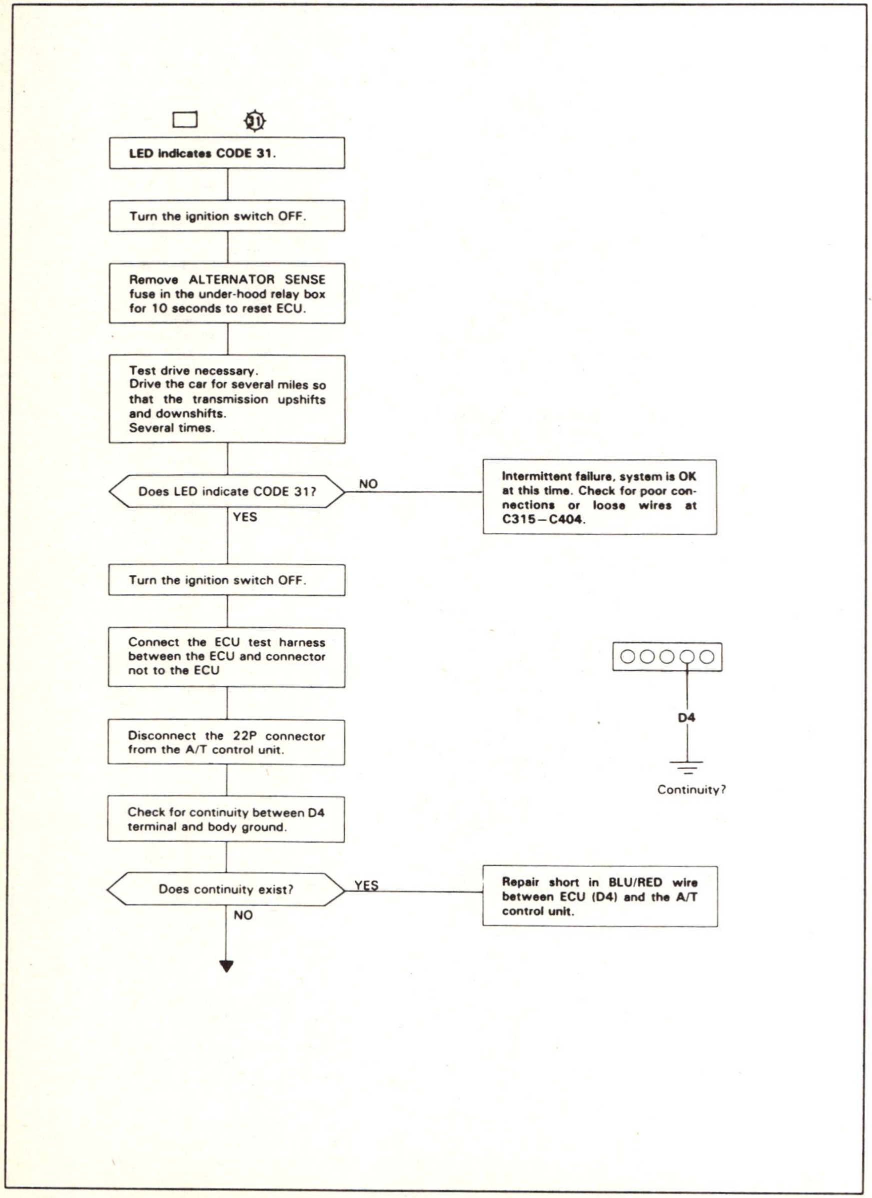

LED indicates CODE 31.

Turn the ignition switch OFF.

Remove ALTERNATOR SENSE fuse in the under-hood relay box for 10 seconds to reset ECU.

Test drive necessary.
Drive the car for several miles so that the transmission upshifts and downshifts.
Several times.

Does LED indicate CODE 31?

- NO: Intermittent failure, system is OK at this time. Check for poor connections or loose wires at C315–C404.
- YES: Turn the ignition switch OFF.

Connect the ECU test harness between the ECU and connector not to the ECU

Disconnect the 22P connector from the A/T control unit.

Check for continuity between D4 terminal and body ground.

D4

Continuity?

Does continuity exist?

- YES: Repair short in BLU/RED wire between ECU (D4) and the A/T control unit.
- NO: (continued)

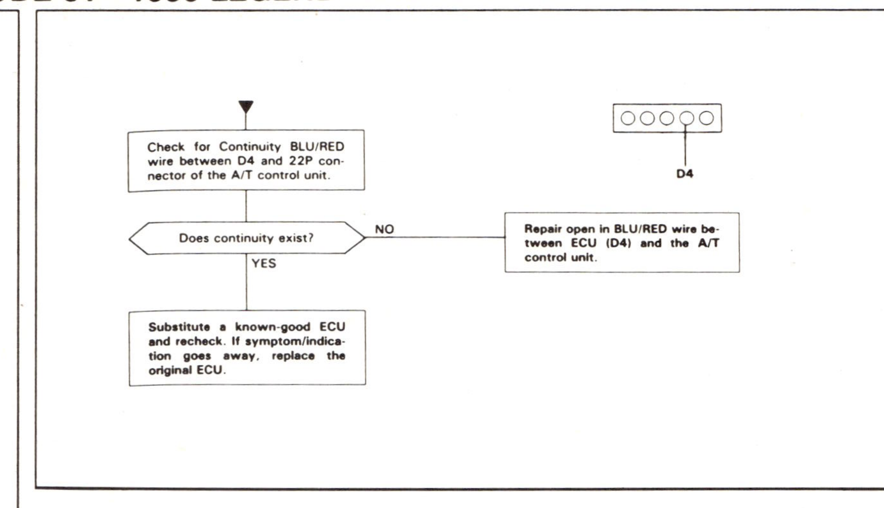


## STARTER SWITCH SIGNAL—1988–90 LEGEND

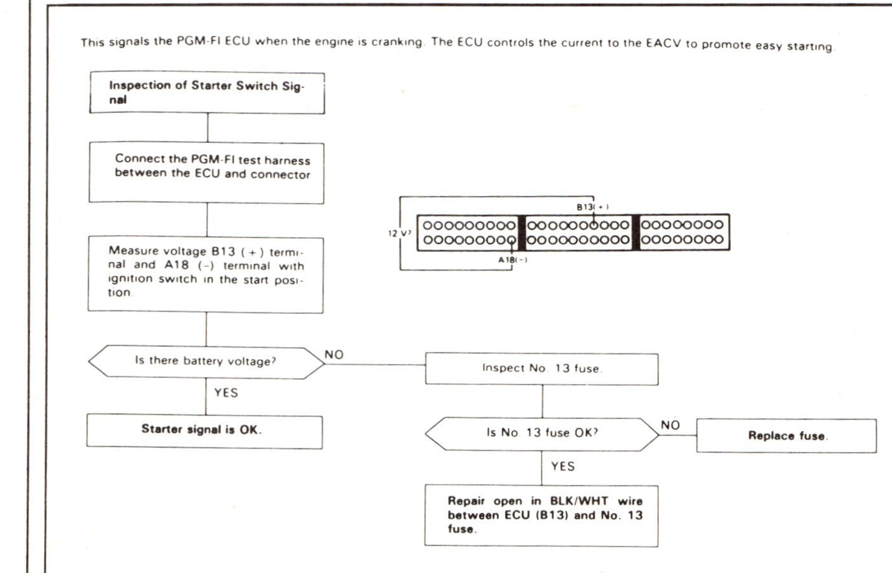

## ALTERNATOR FR SIGNAL – 1988–90 LEGEND

This signals the PGM-FI ECU when the alternator is charging

Note: Make sure the battery is fully charged before performing inspection of the FR terminal.

**Inspection of Alternator FR signal**

Connect the PGM-FI test harness between the ECU and connector. Disconnect "B" connector from the main wire harness only, not the ECU.

Turn the ignition switch ON

Measure voltage between B14 (+) terminal and A18 (-) terminal

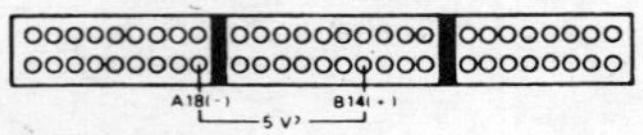

Is there approx 5V?

- NO: **Substitute a known-good ECU and recheck. If prescribed voltage is now available, replace the original ECU.**
- YES: Turn the ignition switch OFF

A18(-) — 5 V? — B14(+)

Reconnect "B" connector to the main wire harness.

Warm up engine to normal operating temperature (cooling fan comes on).

Measure voltage between B14 (-) terminal and A18 ( ) terminal

A18(-) — Voltage? — B14(+)

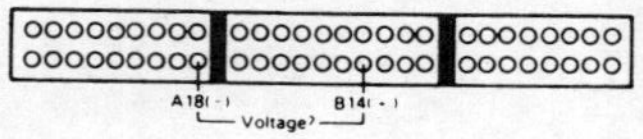

Does the voltage decrease when headlight and rear defogger are turned on?

- NO: Stop engine
- YES: **Alternator FR signal is OK. Turn the ignition switch off and remove ALTERNATOR SENSE fuse for at least 10 seconds to reset ECU.**

## ALTERNATOR FR SIGNAL – 1988–90 LEGEND

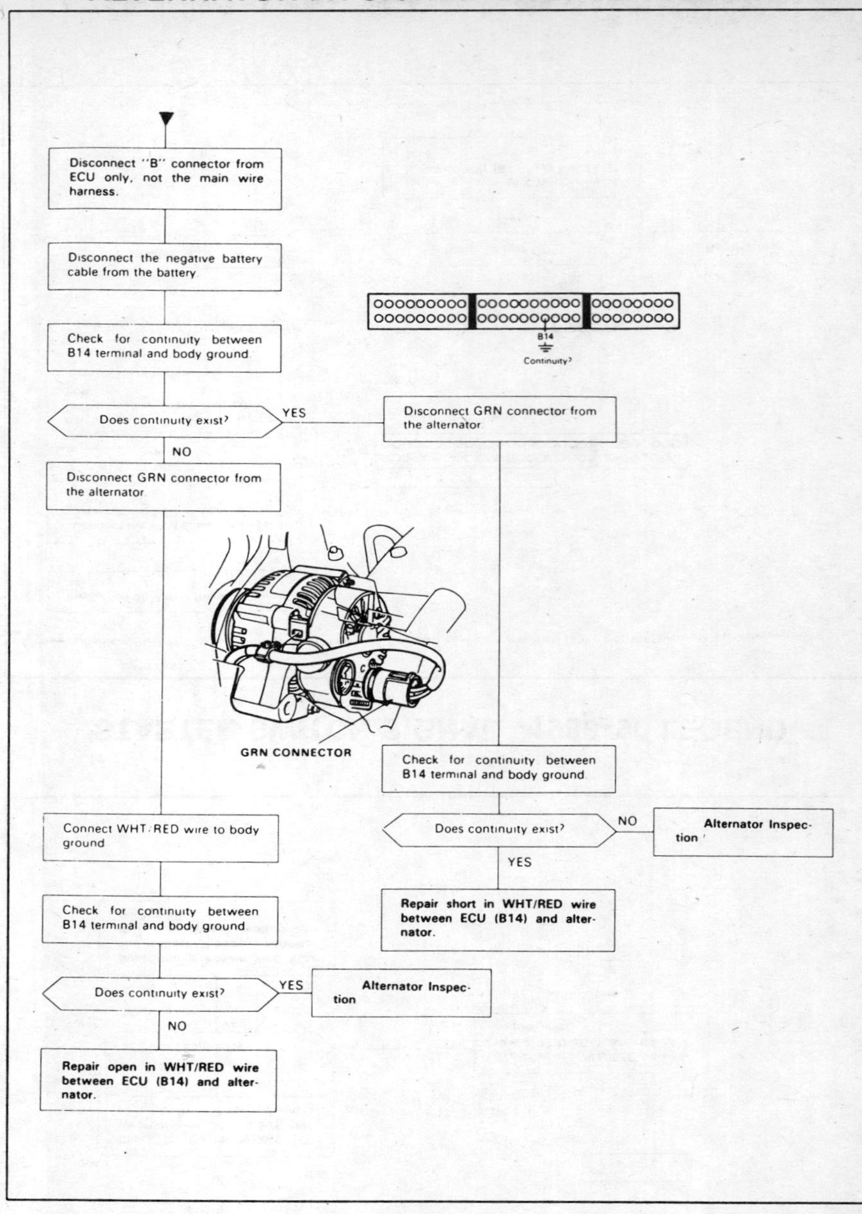

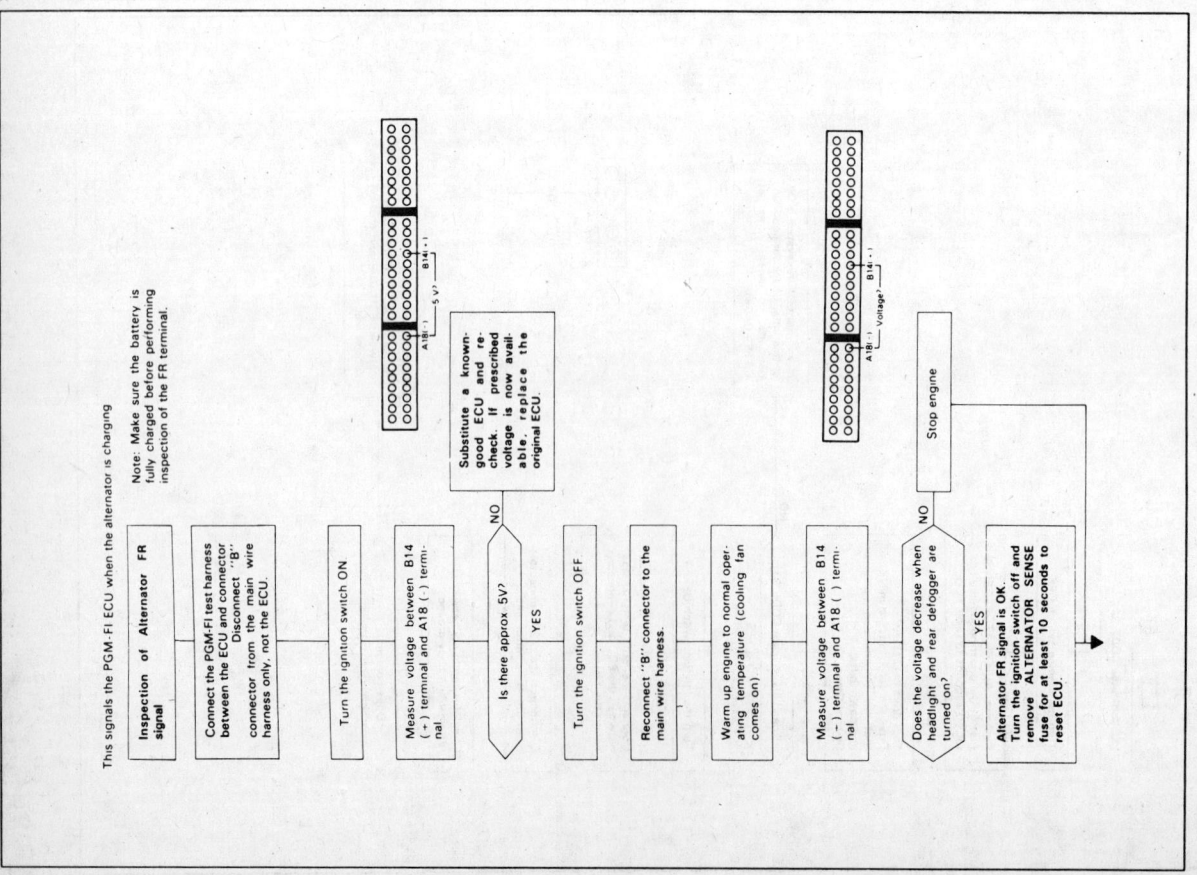

# A/T SHIFT POSITION SIGNAL – 1988–89 LEGEND

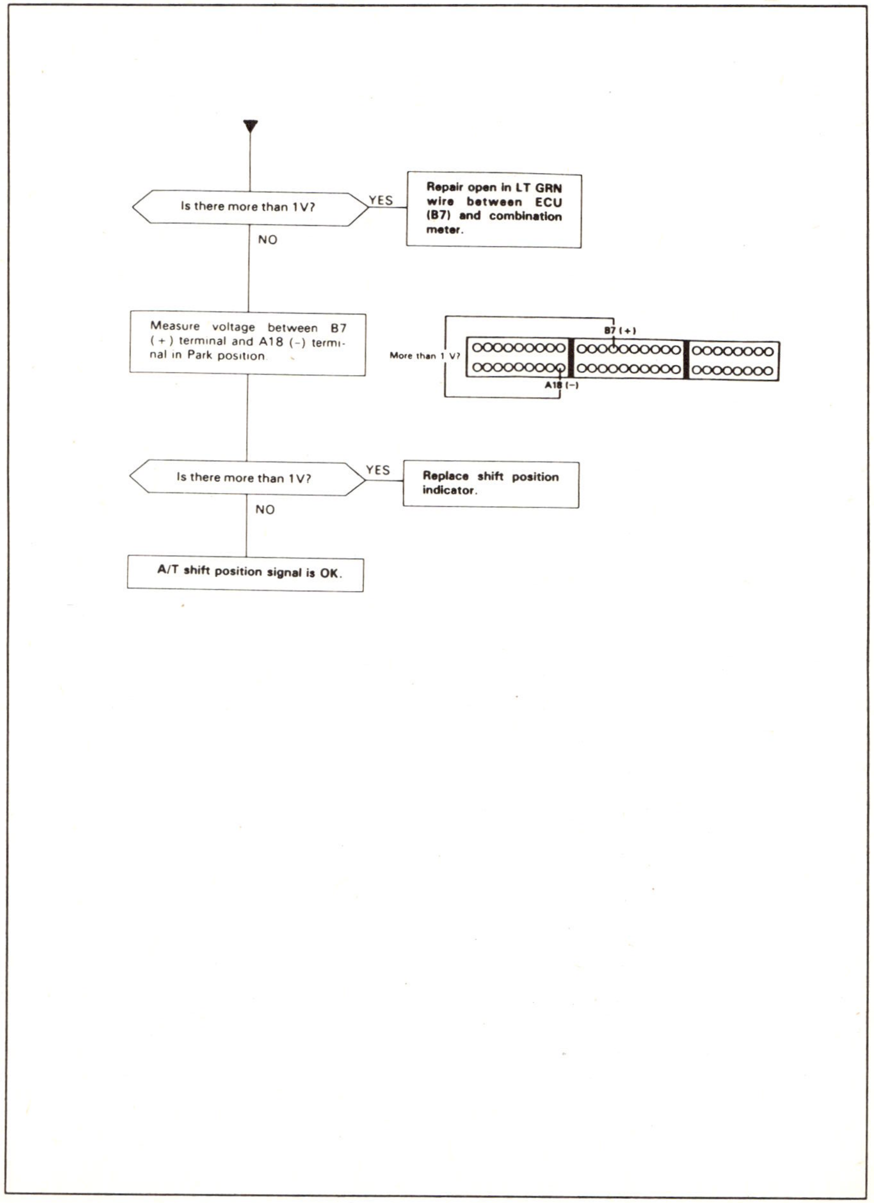

This signals the PGM-FI ECU when the transmission is in Neutral or Park

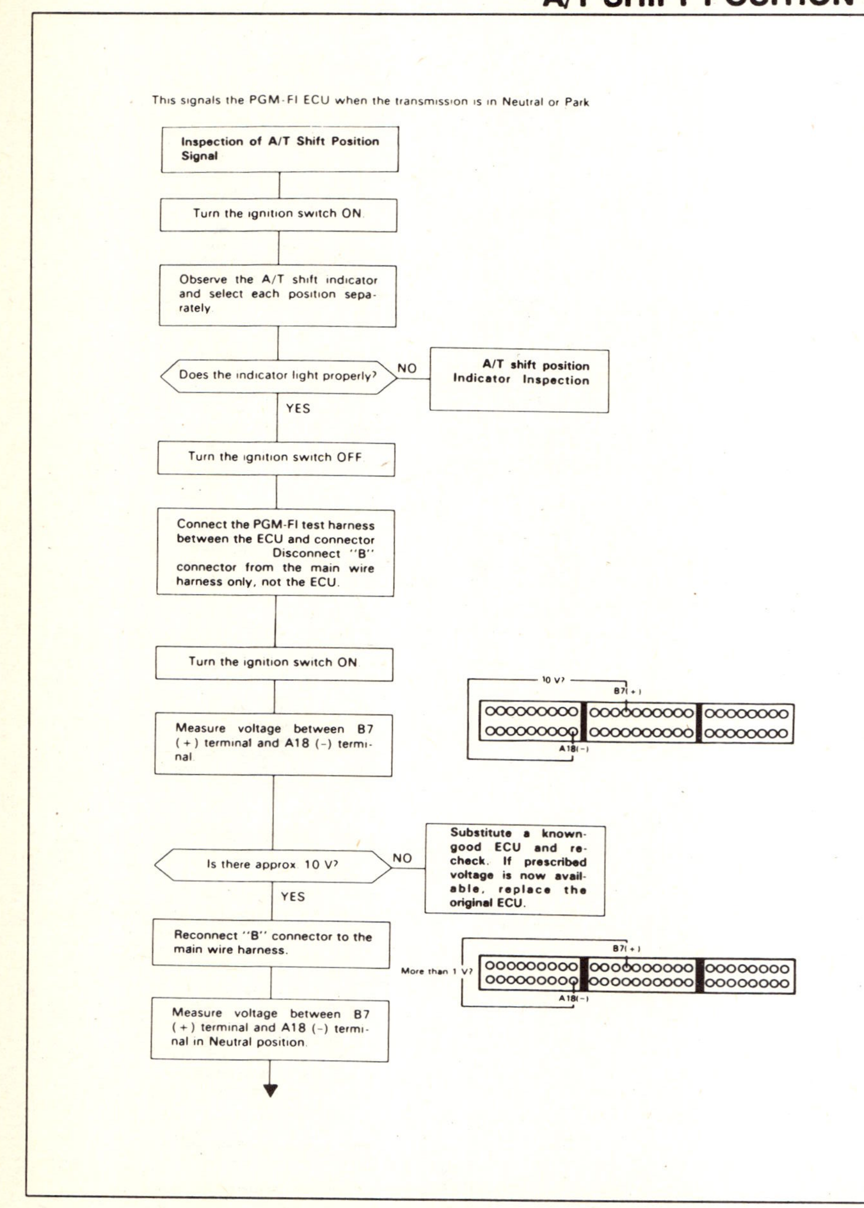


Is there more than 1V?

YES → Repair open in LT GRN wire between ECU (B7) and combination meter.

NO ↓

Measure voltage between B7 (+) terminal and A18 (-) terminal in Park position.

More than 1 V?
B7 (+)
A18 (-)

Is there more than 1V?

YES → Replace shift position indicator.

NO ↓

A/T shift position signal is OK.

# A/T SHIFT POSITION SIGNAL – 1990 LEGEND

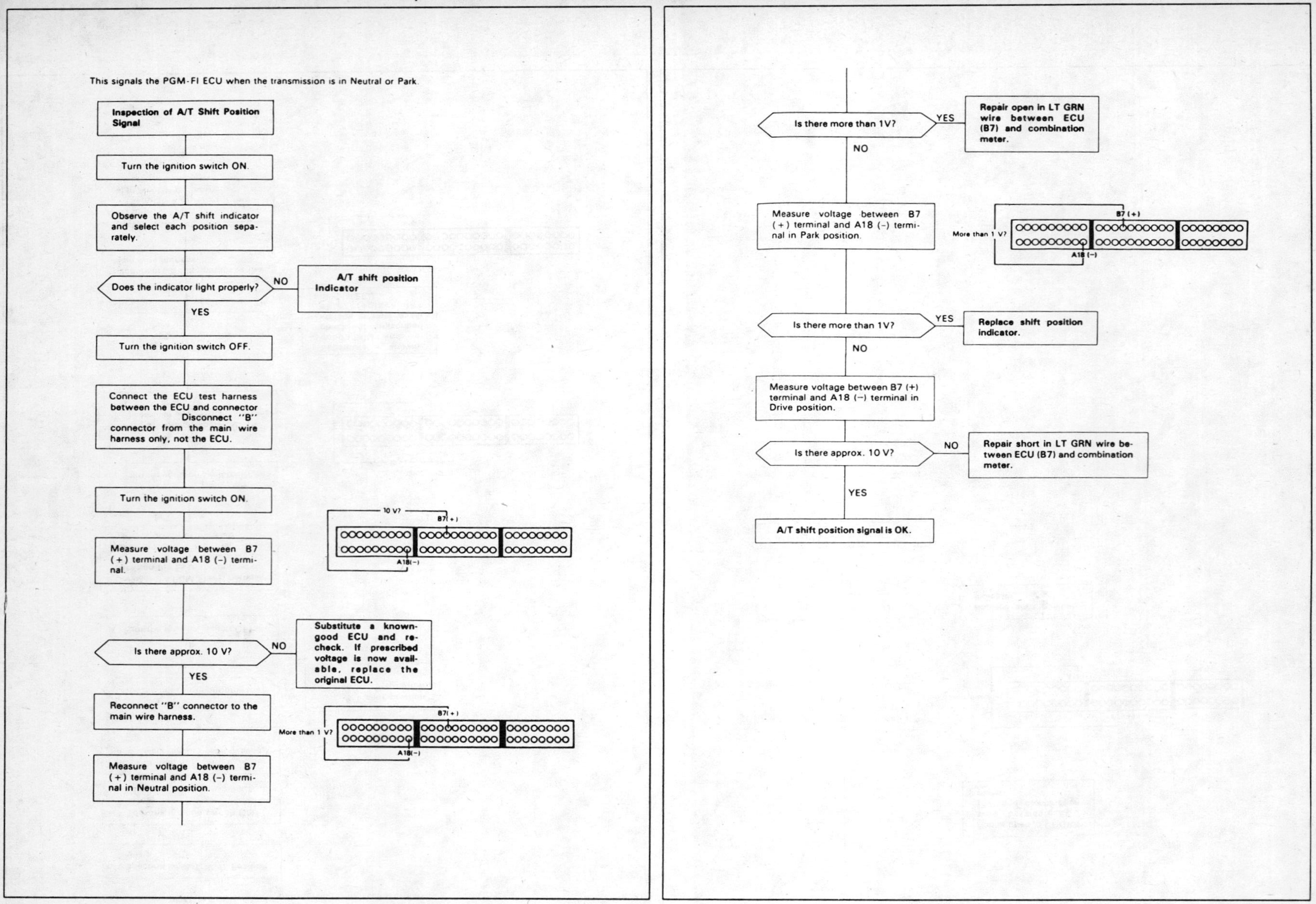

This signals the PGM-FI ECU when the transmission is in Neutral or Park.

**Inspection of A/T Shift Position Signal**

Turn the ignition switch ON.

Observe the A/T shift indicator and select each position separately.

Does the indicator light properly?
- NO: **A/T shift position Indicator**
- YES:

Turn the ignition switch OFF.

Connect the ECU test harness between the ECU and connector Disconnect "B" connector from the main wire harness only, not the ECU.

Turn the ignition switch ON.

Measure voltage between B7 (+) terminal and A18 (–) terminal.

10 V?
B7(+)
A18(–)

Is there approx. 10 V?
- NO: **Substitute a known-good ECU and re-check. If prescribed voltage is now available, replace the original ECU.**
- YES:

Reconnect "B" connector to the main wire harness.

Measure voltage between B7 (+) terminal and A18 (–) terminal in Neutral position.

More than 1 V?
B7(+)
A18(–)

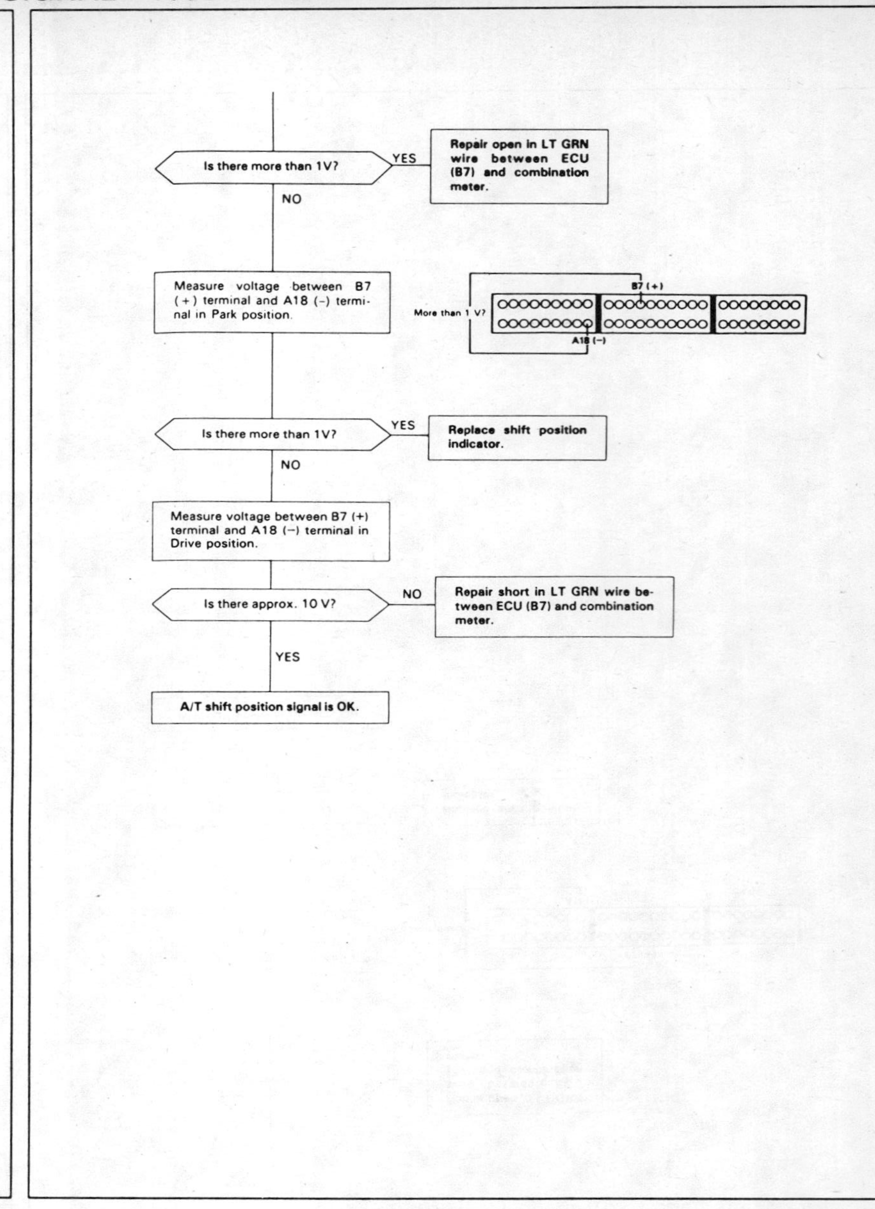

## POWER STEERING PRESSURE SIGNAL 1988–90 LEGEND

This signals the PGM-FI ECU when the power steering load is high

- **Inspection of P/S Oil Pressure Signal**
- Connect the PGM-FI test harness between the ECU and connector
- Turn the ignition switch ON
- Measure voltage between B12 (+) terminal and A18 (-) terminal (A18 (-), B12 (+): 12 V? 10 V?)
- Is there battery voltage?
  - YES: Disconnect the 2P connector on the P/S oil pressure switch → Connect RED terminal to BLK terminal → Is there approx 10 V?
    - NO: **Replace P/S oil pressure switch.**
    - YES: **Repair open in RED wire between ECU (B12) and P/S oil pressure switch or BLK wire between P/S oil pressure switch and G3.**
  - NO: Start engine → Turn steering wheel slowly → Measure voltage between B12 (+) terminal and A18 (-) terminal while steering wheel is turning → Is there approx 10 V?
    - YES: **P/S oil pressure signal is OK.**
    - NO: Disconnect "B" connector from main wire harness only, not the ECU → Is there approx 10 V?
      - NO: **Substitute a known-good ECU and recheck. If prescribed voltage is now available, replace the original ECU.**
      - YES: Reconnect "B" connector to main wire harness and disconnect 2P connector on the P/S oil pressure switch → Is there approx 10 V?
        - YES: **Replace P/S oil pressure switch.**
        - NO: **Repair short in RED wire between ECU (B12) and the P/S oil pressure switch.**

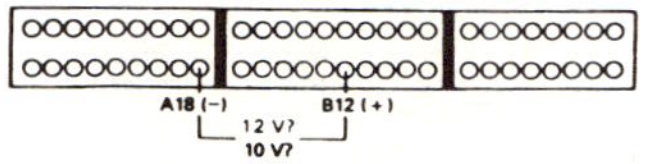

## M/T NEUTRAL SWITCH SIGNAL 1988–90 LEGEND

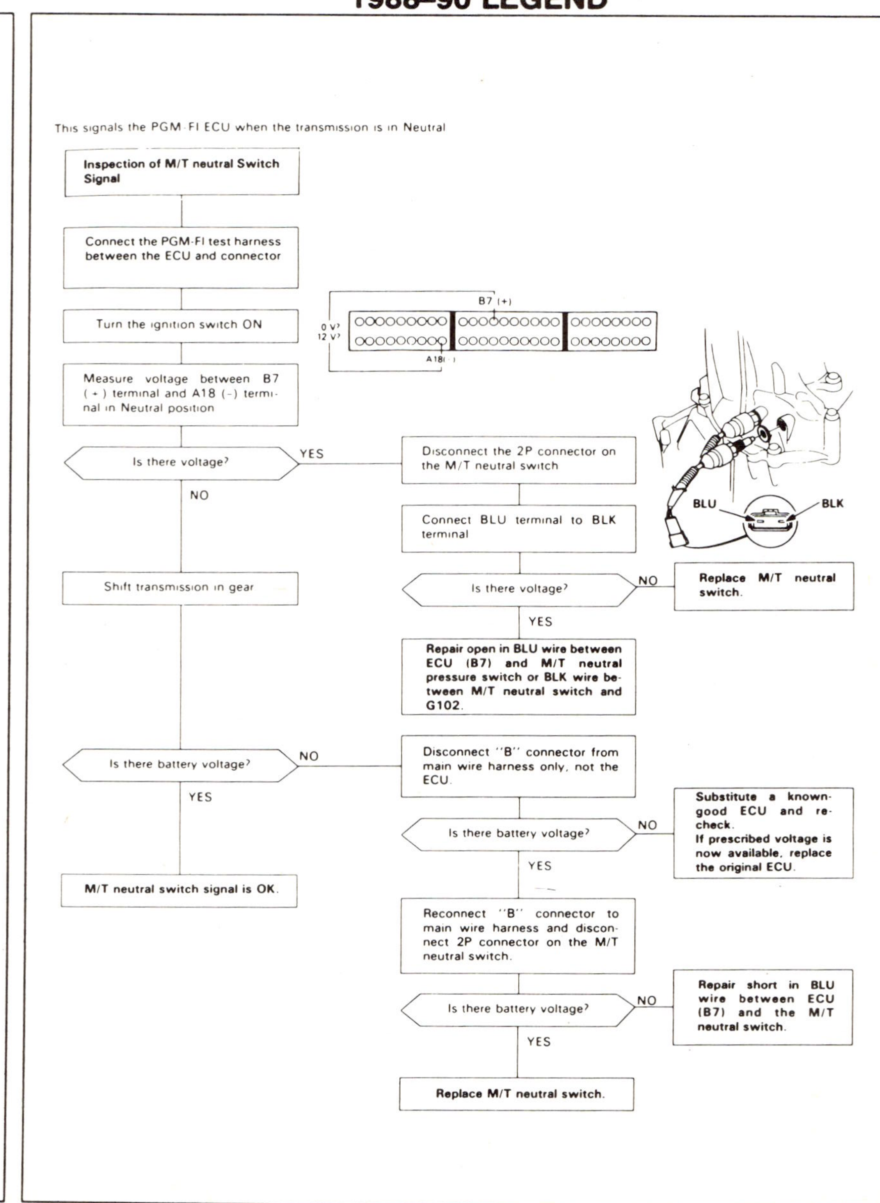

# CLUTCH SWITCH SIGNAL—1988–90 LEGEND

This signals the PGM-FI ECU when the clutch is engaged.

**Inspection of clutch switch signal**

Connect the PGM-FI test harness between the ECU and connector

Turn the ignition switch ON.

Measure voltage between B9 (+) terminal and A18 (-) terminal.

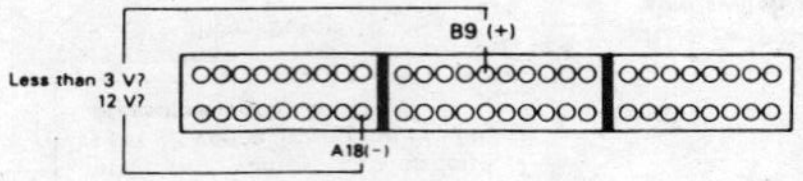

Less than 3 V?
12 V?
B9 (+)
A18(-)

Is voltage less than 3 V?

NO → Turn the ignition switch OFF.

Disconnect the 2P connector on the clutch switch.

Check for continuity between the 2 terminals on the clutch switch.

BLK (−)
PNK (+)

Does continuity exist? — NO → **Replace the clutch switch.**

YES

Turn the ignition switch ON.

Measure voltage between PNK (+) terminal and body ground.

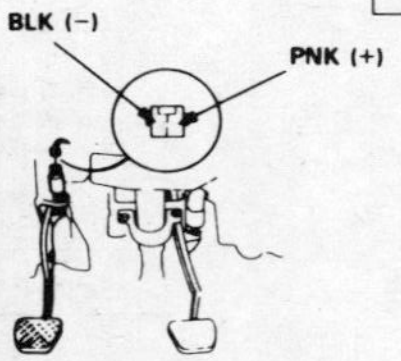

Is there battery voltage? — NO → **Repair open in PNK wire between ECU (B9) and the clutch switch.**

YES

**Repair open in BLK wire between the clutch switch and G401.**

YES (Is voltage less than 3 V?)

Depress the clutch pedal.

Measure voltage between B9 (+) terminal and A18 (-) termianl.

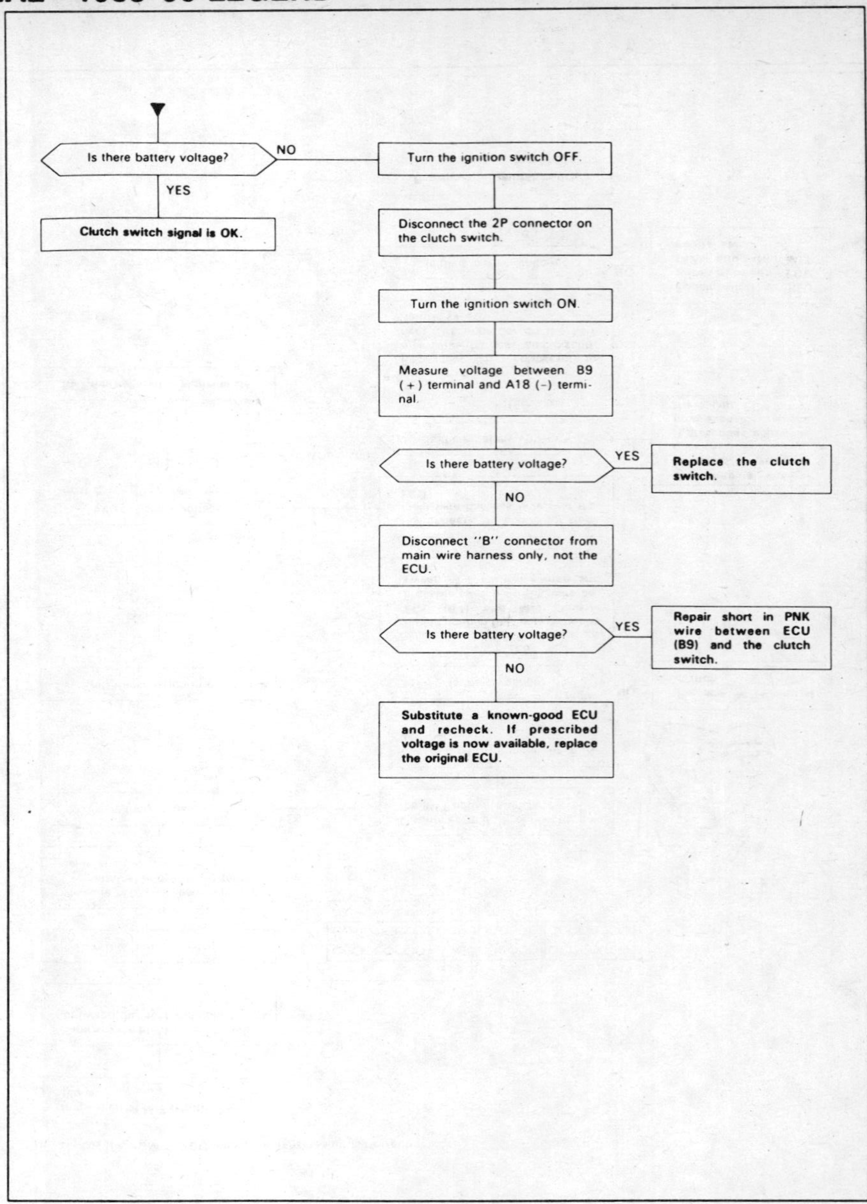

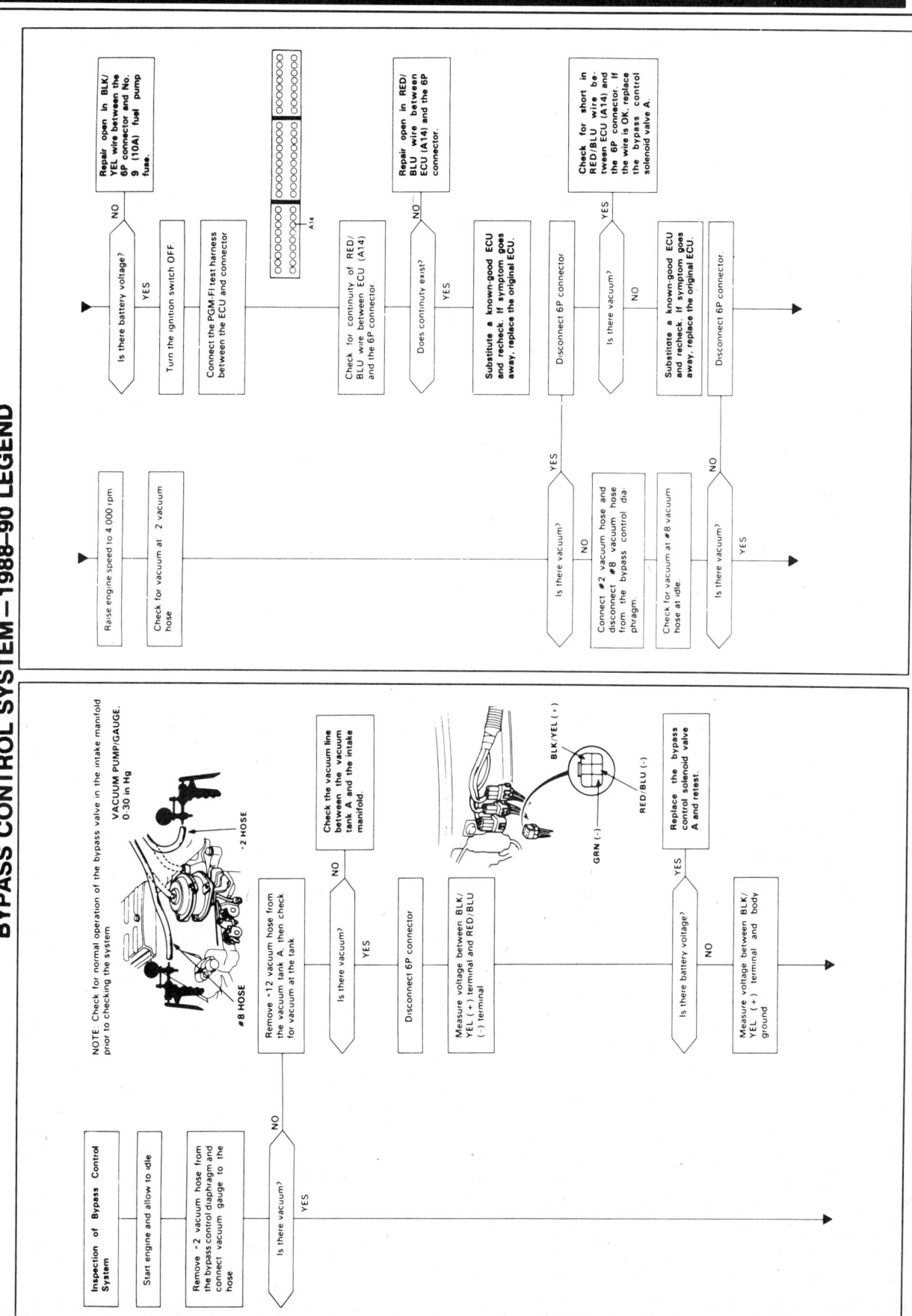

# BYPASS CONTROL SYSTEM—1988–90 LEGEND

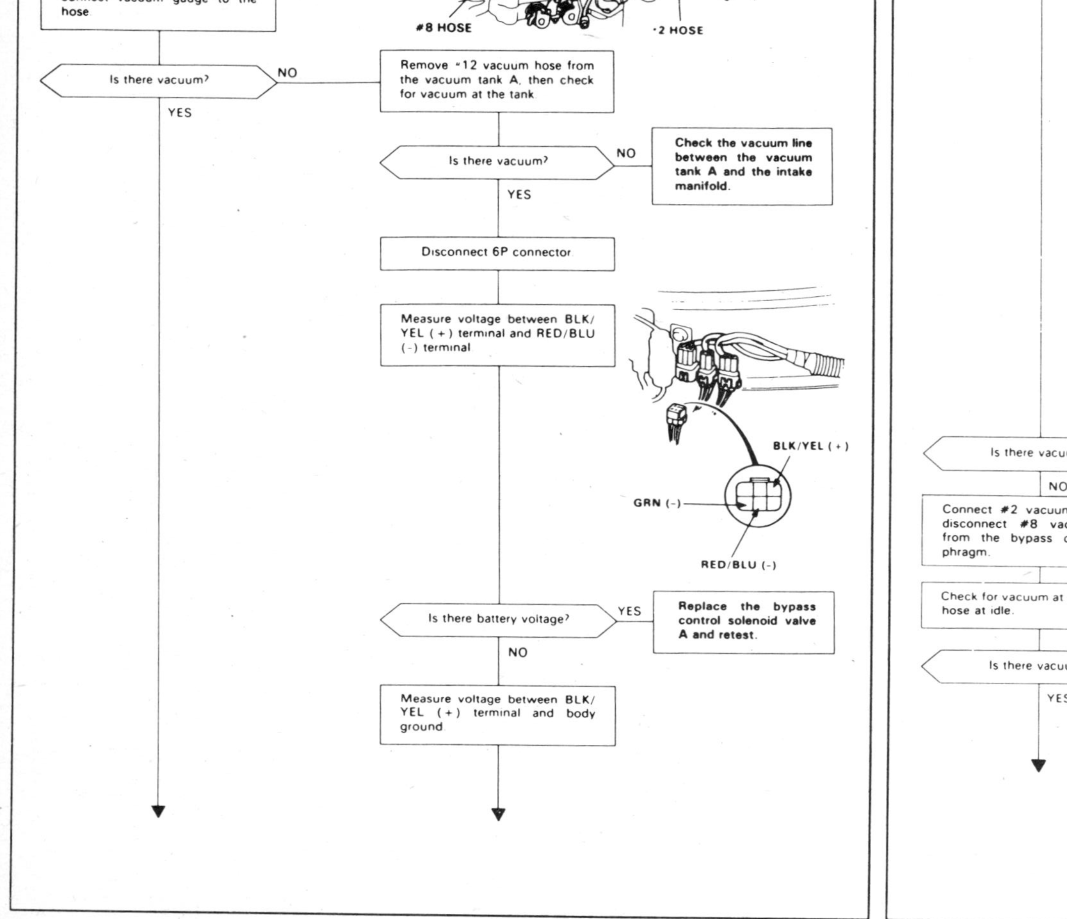


**Inspection of Bypass Control System**

Start engine and allow to idle.

Remove #2 vacuum hose from the bypass control diaphragm and connect vacuum gauge to the hose.

Is there vacuum?
- YES → (continued)
- NO → Remove #12 vacuum hose from the vacuum tank A, then check for vacuum at the tank.

NOTE: Check for normal operation of the bypass valve in the intake manifold prior to checking the system.

Is there vacuum?
- NO → **Check the vacuum line between the vacuum tank A and the intake manifold.**
- YES → Disconnect 6P connector.

Measure voltage between BLK/YEL (+) terminal and RED/BLU (–) terminal.

Is there battery voltage?
- YES → **Replace the bypass control solenoid valve A and retest.**
- NO → Measure voltage between BLK/YEL (+) terminal and body ground.

Is there battery voltage?
- NO → **Repair open in BLK/YEL wire between the 6P connector and No. 9 (10A) fuel pump fuse.**
- YES → Turn the ignition switch OFF.

Connect the PGM-FI test harness between the ECU and connector.

A14

Check for continuity of RED/BLU wire between ECU (A14) and the 6P connector.

Does continuity exist?
- NO → **Repair open in RED/BLU wire between ECU (A14) and the 6P connector.**
- YES → **Substitute a known-good ECU and recheck. If symptom goes away, replace the original ECU.**

Raise engine speed to 4,000 rpm.

Check for vacuum at #2 vacuum hose.

Is there vacuum?
- YES → Disconnect 6P connector.
- NO → Connect #2 vacuum hose and disconnect #8 vacuum hose from the bypass control diaphragm.

Disconnect 6P connector.

Is there vacuum?
- YES → **Check for short in RED/BLU wire between ECU (A14) and the 6P connector. If the wire is OK, replace the bypass control solenoid valve A.**
- NO → **Substitute a known-good ECU and recheck. If symptom goes away, replace the original ECU.**

Check for vacuum at #8 vacuum hose at idle.

Is there vacuum?
- NO → Disconnect 6P connector.
- YES → (continued)

## BYPASS CONTROL SYSTEM – 1988–90 LEGEND

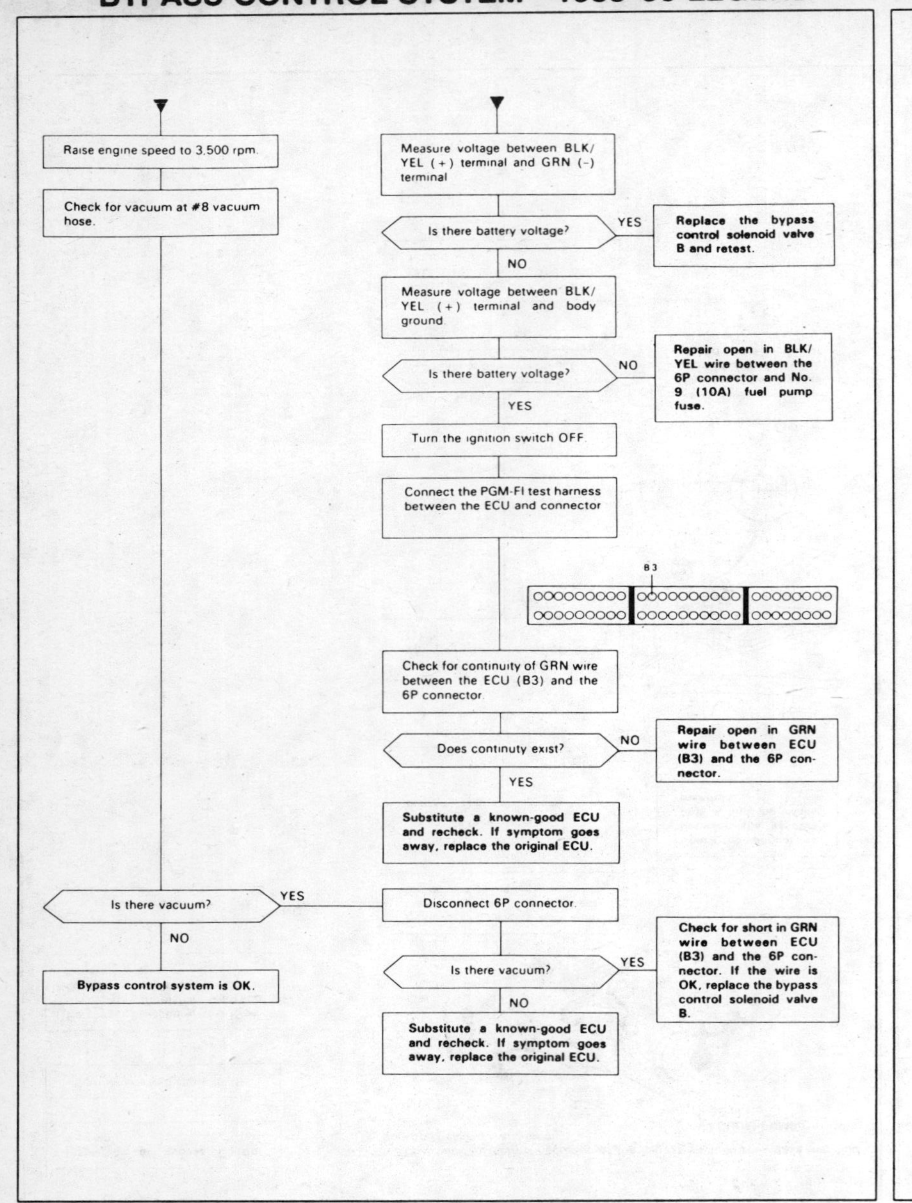


## EXHAUST GAS RECIRCULATION – CODE 12 1988–90 LEGEND

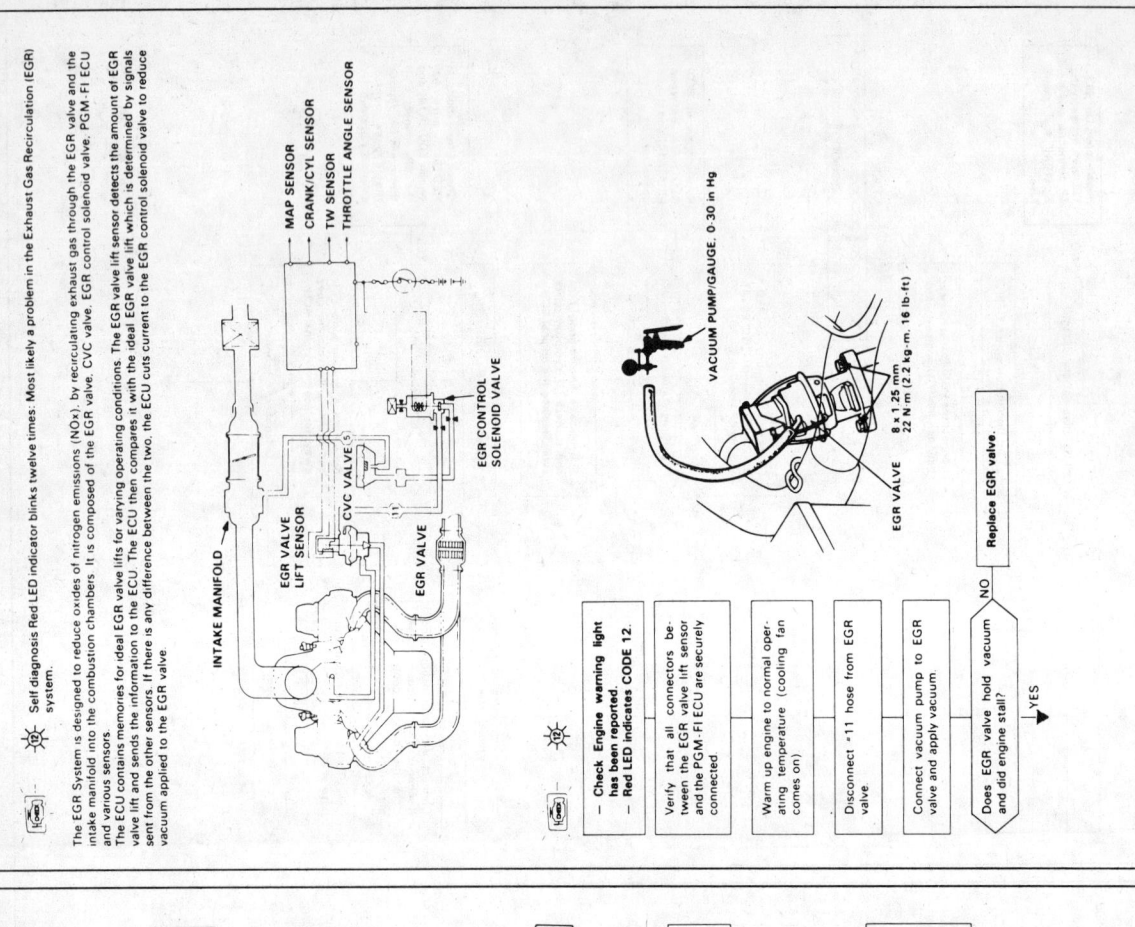

Self diagnosis Red LED indicator blinks twelve times: Most likely a problem in the Exhaust Gas Recirculation (EGR) system.

The EGR System is designed to reduce oxides of nitrogen emissions (NOx), by recirculating exhaust gas through the EGR valve and the intake manifold into the combustion chambers. It is composed of the EGR valve, CVC valve, EGR control solenoid valve, PGM-FI ECU and various sensors.

The ECU contains memories for ideal EGR valve lifts for varying operating conditions. The EGR valve lift sensor detects the amount of EGR valve lift and sends the information to the ECU. The ECU then compares it with the ideal EGR valve lift which is determined by signals sent from the other sensors. If there is any difference between the two, the ECU cuts current to the EGR control solenoid valve to reduce vacuum applied to the EGR valve.

INTAKE MANIFOLD
EGR VALVE LIFT SENSOR
CVC VALVE
EGR VALVE
EGR CONTROL SOLENOID VALVE
MAP SENSOR
CRANK/CYL SENSOR
TW SENSOR
THROTTLE ANGLE SENSOR

- Check Engine warning light has been reported.
- Red LED indicates CODE 12.

Verify that all connectors between the EGR valve lift sensor and the PGM-FI ECU are securely connected.

Warm up engine to normal operating temperature (cooling fan comes on).

Disconnect #11 hose from EGR valve.

Connect vacuum pump to EGR valve and apply vacuum.

Does EGR valve hold vacuum and did engine stall?

NO — Replace EGR valve.

YES

VACUUM PUMP/GAUGE, 0-30 in Hg
EGR VALVE
8 x 1.25 mm
22 N·m (2.2 kg-m, 16 lb-ft)

## EXHAUST GAS RECIRCULATION—CODE 12—1988–90 LEGEND

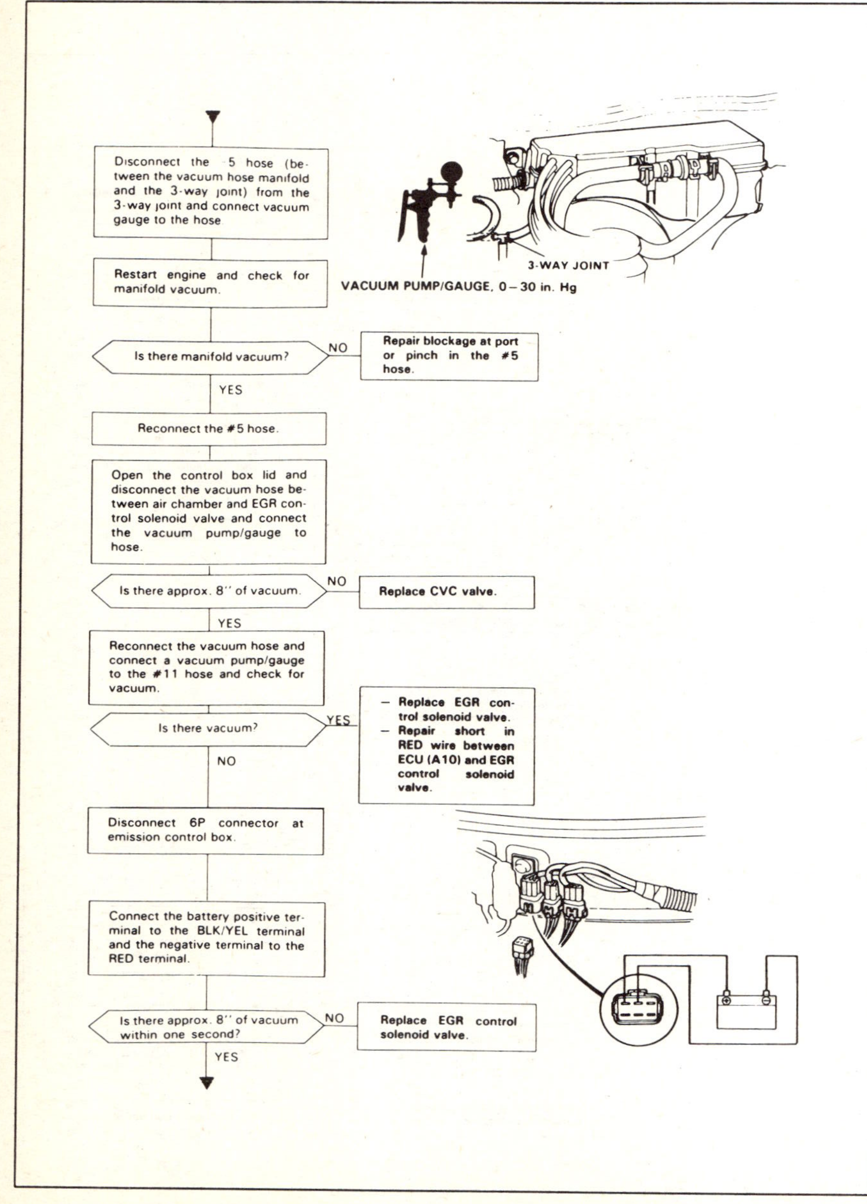

Disconnect the #5 hose (between the vacuum hose manifold and the 3-way joint) from the 3-way joint and connect vacuum gauge to the hose
3-WAY JOINT
VACUUM PUMP/GAUGE, 0–30 in. Hg
Restart engine and check for manifold vacuum.
Is there manifold vacuum?
NO
Repair blockage at port or pinch in the #5 hose.
YES
Reconnect the #5 hose.
Open the control box lid and disconnect the vacuum hose between air chamber and EGR control solenoid valve and connect the vacuum pump/gauge to hose.
Is there approx. 8'' of vacuum.
NO
Replace CVC valve.
YES
Reconnect the vacuum hose and connect a vacuum pump/gauge to the #11 hose and check for vacuum.
Is there vacuum?
YES
– Replace EGR control solenoid valve.
– Repair short in RED wire between ECU (A10) and EGR control solenoid valve.
NO
Disconnect 6P connector at emission control box.
Connect the battery positive terminal to the BLK/YEL terminal and the negative terminal to the RED terminal.
Is there approx. 8'' of vacuum within one second?
NO
Replace EGR control solenoid valve.
YES


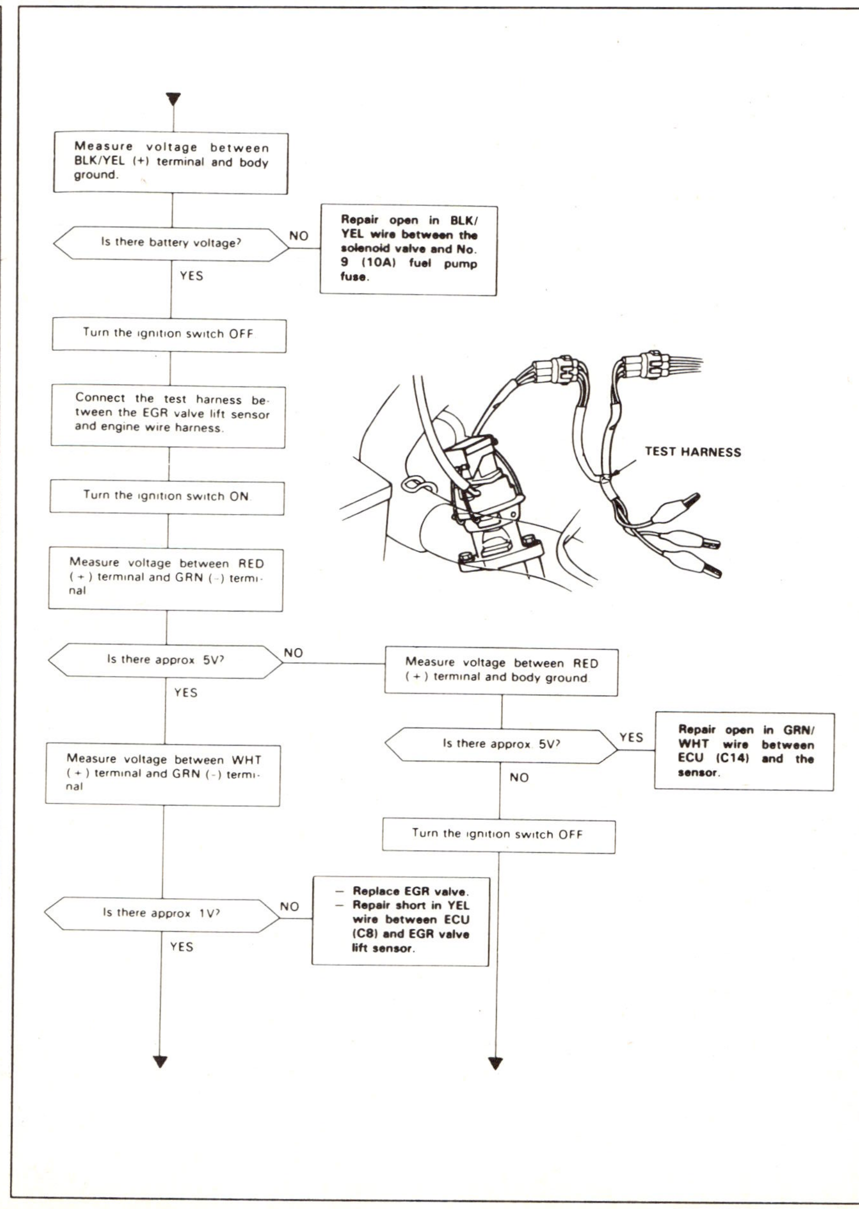

## EXHAUST GAS RECIRCULATION—CODE 12 1988–90 LEGEND

Turn the ignition switch OFF.

Connect the PGM-FI test harness between the ECU and connector

Turn the ignition switch ON.

Measure voltage between C8 (+) terminal and C14 (-) terminal.

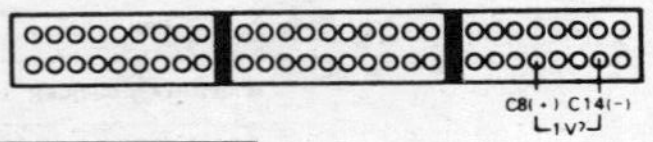


Is there approx. 1 V?
- NO: Repair open in WHT/GRN wire between ECU (C8) and the sensor.
- YES: Disconnect "A" connector from the ECU only, not the main wire harness.

Measure voltage between A10 (+) terminal and A18 (-) terminal.

Is there battery voltage?
- NO: Repair open in WHT wire between ECU (A10) and EGR control solenoid valve.
- YES: Substitute a known-good ECU and recheck. If symptom/indication goes away, replace the original ECU.

Connect the PGM-FI test harness between the ECU and connector

Turn the ignition switch ON.

Measure voltage between C13 (+) terminal and C14 (-) terminal.

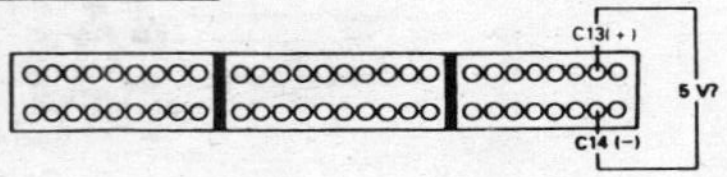


Is there approx. 5 V?
- YES: Repair open in YEL/WHT wire between ECU (C13) and the sensor.
- NO: Substitute a known-good ECU and recheck. If prescribed voltage is now available replace the original ECU.

## RESONATOR CONTROL SYSTEM—1988–90 LEGEND

Inspection of Resonator Control System

Start engine and allow it to idle.

Remove the vacuum hose from the resonator diaphragm and connect a vacuum gauge to the hose.

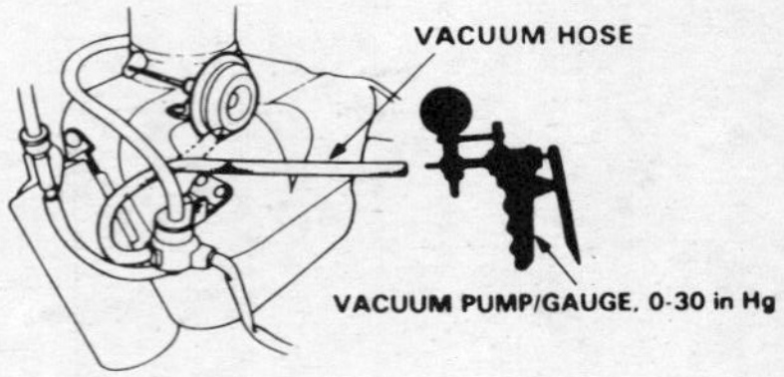


Is there vacuum?
- YES: (continue)
- NO: Remove resonator control vacuum hose from surge tank, and check for vacuum at the tank.

Is there vacuum?
- NO: Repair the blockage or vacuum leak between surge tank and intake manifold.
- YES: Disconnect the 2P connector from the resonator control solenoid valve.

Measure voltage between BLK/YEL (+) terminal and RED/BLU (-) terminal.

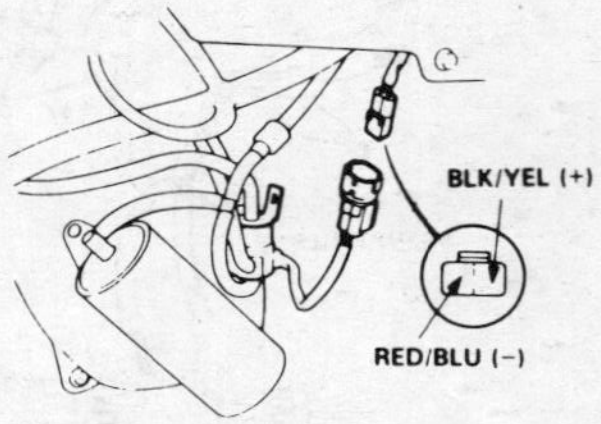


Is there battery voltage?
- YES: Replace the resonator control solenoid valve.
- NO: Measure voltage between BLK/YEL (+) terminal and body ground

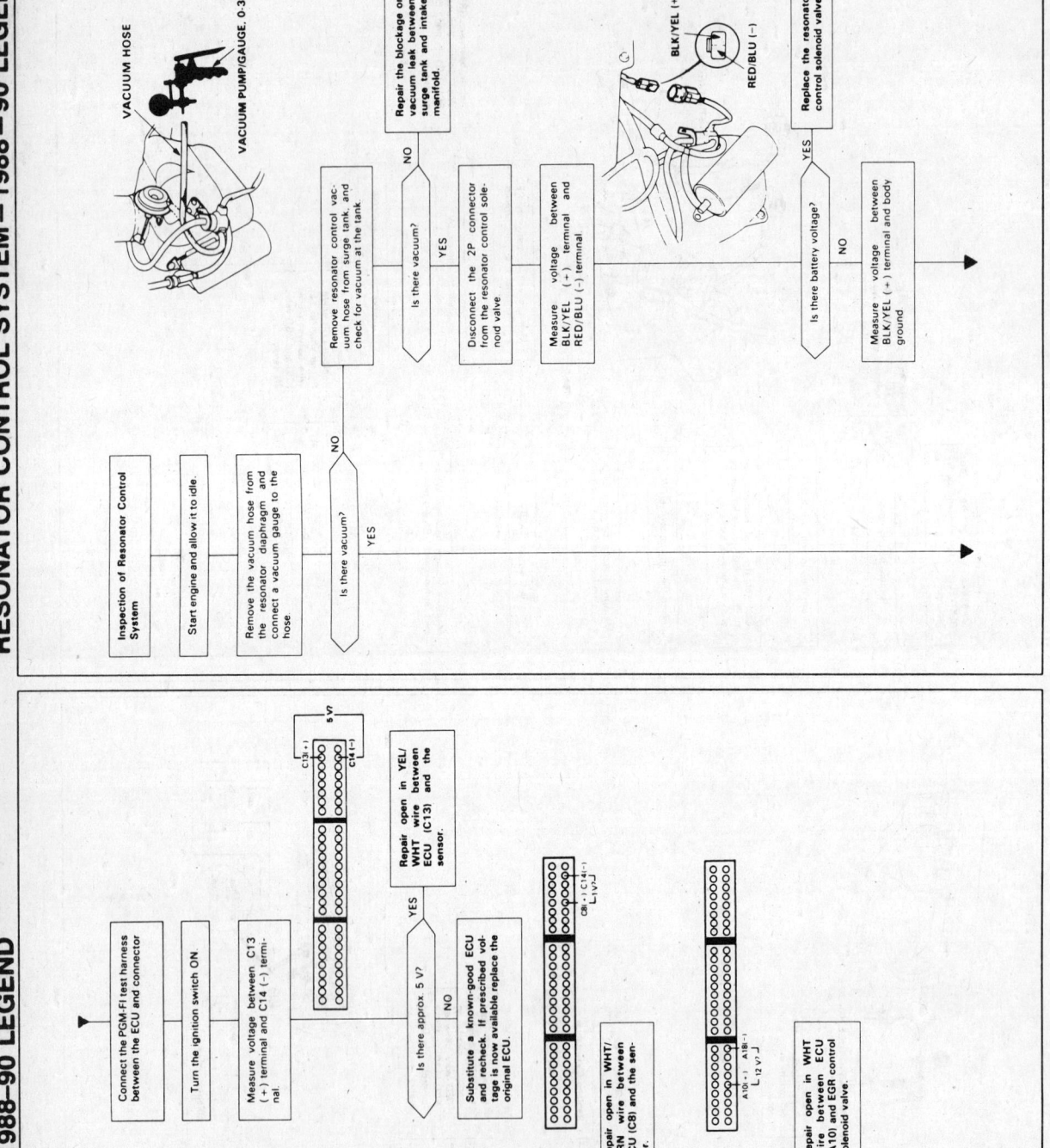

## RESONATOR CONTROL SYSTEM 1988–90 LEGEND

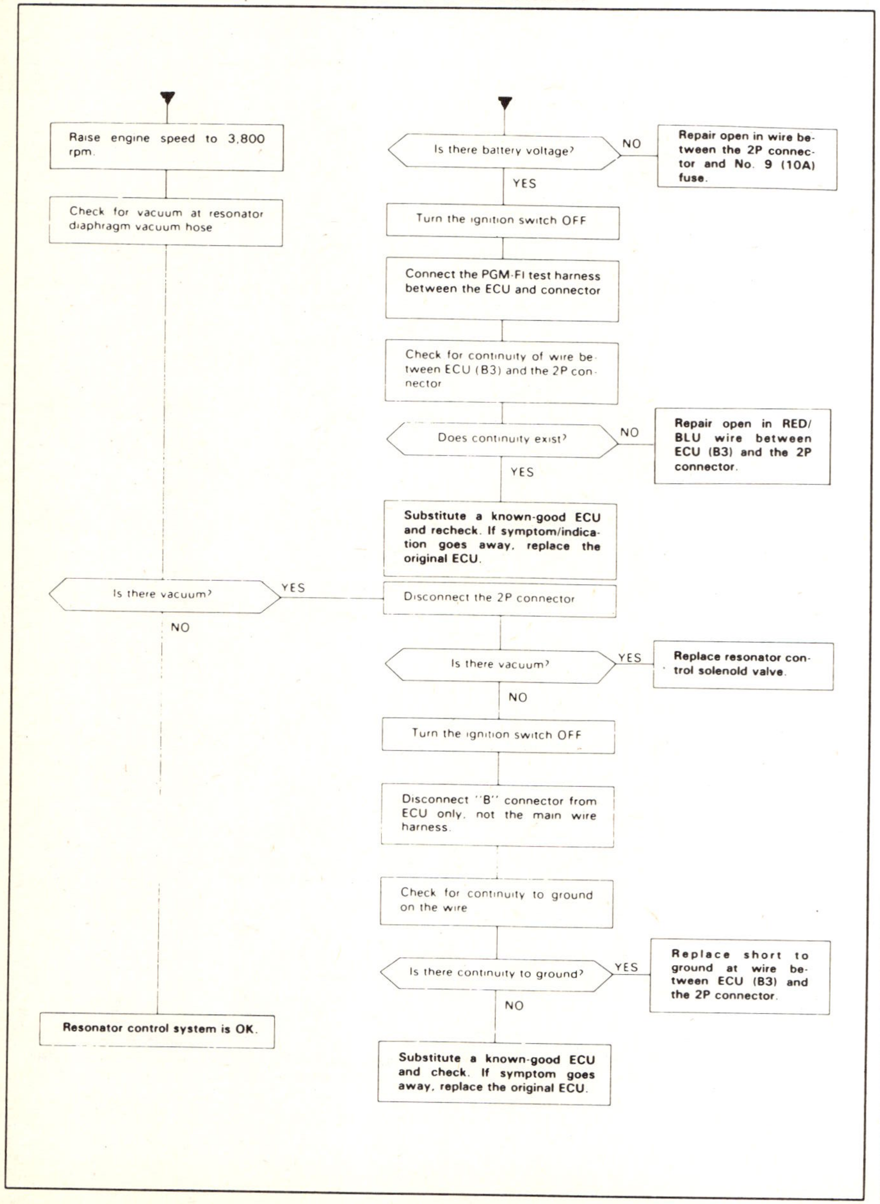
Raise engine speed to 3,800 rpm.
Check for vacuum at resonator diaphragm vacuum hose
Is there vacuum?
YES
NO
Resonator control system is OK.
Is there battery voltage?
NO
Repair open in wire between the 2P connector and No. 9 (10A) fuse.
YES
Turn the ignition switch OFF
Connect the PGM-FI test harness between the ECU and connector
Check for continuity of wire between ECU (B3) and the 2P connector
Does continuity exist?
NO
Repair open in RED/BLU wire between ECU (B3) and the 2P connector.
YES
Substitute a known-good ECU and recheck. If symptom/indication goes away, replace the original ECU.
Disconnect the 2P connector
Is there vacuum?
YES
Replace resonator control solenoid valve.
NO
Turn the ignition switch OFF
Disconnect "B" connector from ECU only, not the main wire harness.
Check for continuity to ground on the wire
Is there continuity to ground?
YES
Replace short to ground at wire between ECU (B3) and the 2P connector.
NO
Substitute a known-good ECU and check. If symptom goes away, replace the original ECU.

## Electrical circuits – 1990 Legend

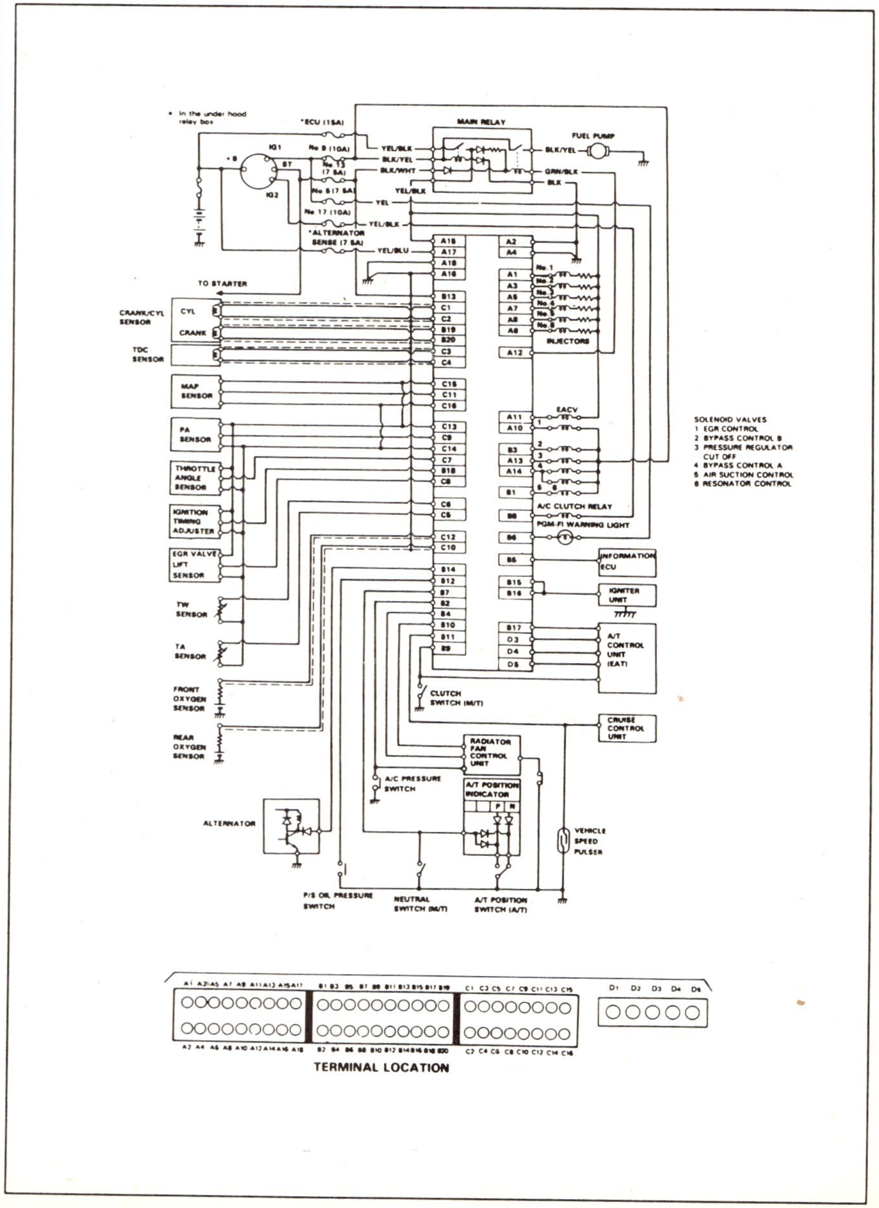
• In the under hood relay box
*ECU (15A)
MAIN RELAY
FUEL PUMP
No 9 (10A)
No 15 (7.5A)
No 6 (7.5A)
No 17 (10A)
*ALTERNATOR SENSE (7.5A)
YEL/BLK
BLK/YEL
BLK/WHT
GRN/BLK
BLK
YEL
YEL/BLU
TO STARTER
CRANK/CYL SENSOR
CYL
CRANK
TDC SENSOR
MAP SENSOR
PA SENSOR
THROTTLE ANGLE SENSOR
IGNITION TIMING ADJUSTER
EGR VALVE LIFT SENSOR
TW SENSOR
TA SENSOR
FRONT OXYGEN SENSOR
REAR OXYGEN SENSOR
ALTERNATOR
INJECTORS
EACV
A/C CLUTCH RELAY
PGM-FI WARNING LIGHT
INFORMATION ECU
IGNITER UNIT
A/T CONTROL UNIT (EAT)
CRUISE CONTROL UNIT
CLUTCH SWITCH (M/T)
RADIATOR FAN CONTROL UNIT
A/T POSITION INDICATOR
A/C PRESSURE SWITCH
VEHICLE SPEED PULSER
P/S OIL PRESSURE SWITCH
NEUTRAL SWITCH (M/T)
A/T POSITION SWITCH (A/T)
SOLENOID VALVES
1 EGR CONTROL
2 BYPASS CONTROL B
3 PRESSURE REGULATOR CUT OFF
4 BYPASS CONTROL A
5 AIR SUCTION CONTROL
6 RESONATOR CONTROL
TERMINAL LOCATION

# TROUBLESHOOTING GUIDE – 1988–89 INTEGRA

NOTE: Across each row in the chart, the systems that could be sources of a symptom are ranked in the order they should be inspected starting with ①. Find the symptom in the left column, read across to the most likely source, then refer to the page listed at the top of that column. If inspection shows the system is OK, try the next most likely system ②, etc.

| SYMPTOM \ SYSTEM | | PGM-FI: ECU | PGM-FI: OXYGEN SENSOR | PGM-FI: MANIFOLD ABSOLUTE PRESSURE SENSOR | PGM-FI: TDC/CRANK/CYL SENSOR | PGM-FI: COOLANT TEMPERATURE SENSOR | PGM-FI: THROTTLE ANGLE SENSOR | PGM-FI: INTAKE AIR TEMPERATURE SENSOR | PGM-FI: ATMOSPHERIC PRESSURE SENSOR |
|---|---|---|---|---|---|---|---|---|---|
| CHECK ENGINE WARNING LIGHT TURNS ON | | ☐ or ☼ | [CHECK] | [CHECK] | [CHECK] | [CHECK] | [CHECK] | [CHECK] | [CHECK] |
| SELF-DIAGNOSIS INDICATOR (LED) BLINKS | | 0 or ✻ | 1 | 3 or 5 | 4 or 8 or 9 | 6 | 7 | 10 | 13 |
| ENGINE WON'T START | | 2 | | | | | | | |
| DIFFICULT TO START ENGINE WHEN COLD | | BU | | 3 | | 1 | | | |
| IRREGULAR IDLING | WHEN COLD FAST IDLE OUT OF SPEC | BU | | | | 3 | | | |
| | ROUGH IDLE | BU | | 2 | | | | | |
| | WHEN WARM RPM TOO HIGH | BU | | | | | | | |
| | WHEN WARM RPM TOO LOW | BU | | | | | | | |
| FREQUENT STALLING | WHILE WARMING UP | BU | | 3 | | | | | |
| | AFTER WARMING UP | BU | | 3 | | | | | |
| POOR PERFORMANCE | MISFIRE OR ROUGH RUNNING | BU | | 3 | | | | | |
| | FAILS EMISSION TEST | BU | 1 | 2 | | 3 | | | |
| | LOSS OF POWER | BU | | 3 | | | 2 | | |

| SYMPTOM \ SYSTEM | | PGM-FI: IGNITION OUTPUT SIGNAL | PGM-FI: VEHICLE SPEED SENSOR | PGM-FI: LOCK-UP CONTROL SOLENOID VALVE | PGM-FI: ELECTRIC LOAD DETECTOR | IDLE CONTROL: ELECTRONIC AIR CONTROL VALVE | IDLE CONTROL: OTHER IDLE CONTROLS | FUEL SUPPLY: FUEL INJECTOR | FUEL SUPPLY: OTHER FUEL SUPPLY | AIR INTAKE | EMISSION CONTROL |
|---|---|---|---|---|---|---|---|---|---|---|---|
| CHECK ENGINE WARNING LIGHT TURNS ON | | [CHECK] | [CHECK] | [CHECK] | ☐ | [CHECK] | | [CHECK] | | | |
| SELF-DIAGNOSIS INDICATOR (LED) BLINKS | | 15 | 17 | 19 | 20 | 14 | | 16 | | | |
| ENGINE WON'T START | | 3 | | | | | | | 1 | | |
| DIFFICULT TO START ENGINE WHEN COLD | | | | | | 2 | | | | | |
| IRREGULAR IDLING | WHEN COLD FAST IDLE OUT OF SPEC | | | | | 1 | 2 | | | | |
| | ROUGH IDLE | | | | | 3 | | 1 | | | |
| | WHEN WARM RPM TOO HIGH | | | | 3 | 2 | 1 | | | | |
| | WHEN WARM RPM TOO LOW | | | | 3 | 2 | 1 | | | | |
| FREQUENT STALLING | WHILE WARMING UP | | | | | 1 | | 2 | 3 | 3 | |
| | AFTER WARMING UP | | 3 | | | 1 | | | 2 | | |
| POOR PERFORMANCE | MISFIRE OR ROUGH RUNNING | | | | | 2 | | 1 | | | |
| | FAILS EMISSION TEST | | | | | | | | | | |
| | LOSS OF POWER | | | | | | | | 1 | | |

If codes other than those listed above are indicated, count the number of blinks again. If the indicator is in fact blinking these codes, substitute a known-good ECU and recheck. If the indication goes away, replace the original ECU.

BU: When the Check Engine warning light and the self-diagnosis indicator are on, the back-up system is in operation. Substitute a known-good ECU and recheck. If the indication goes away, replace the original ECU.

## ECU AND CHECK LIGHT—1988–89 INTEGRA

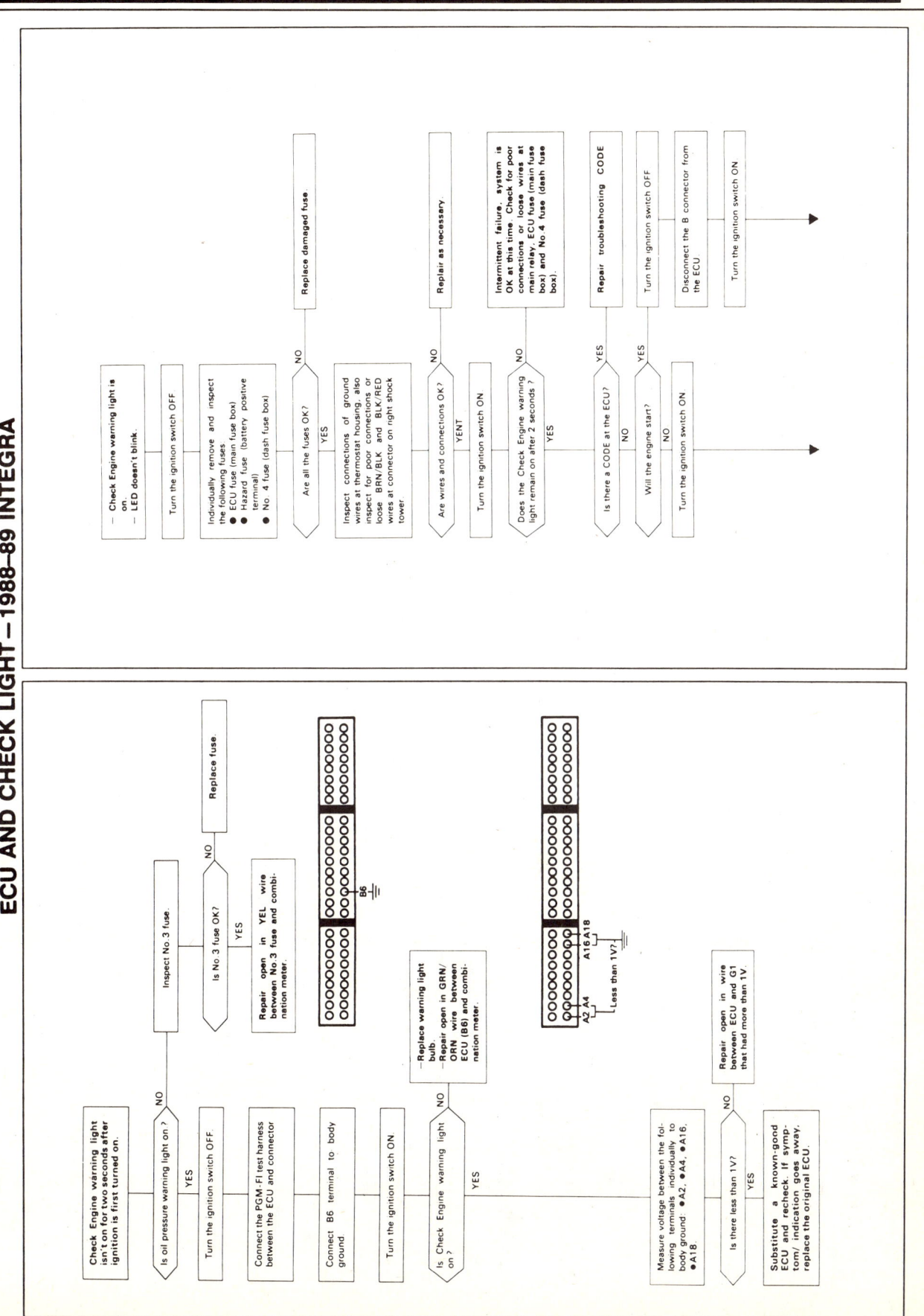

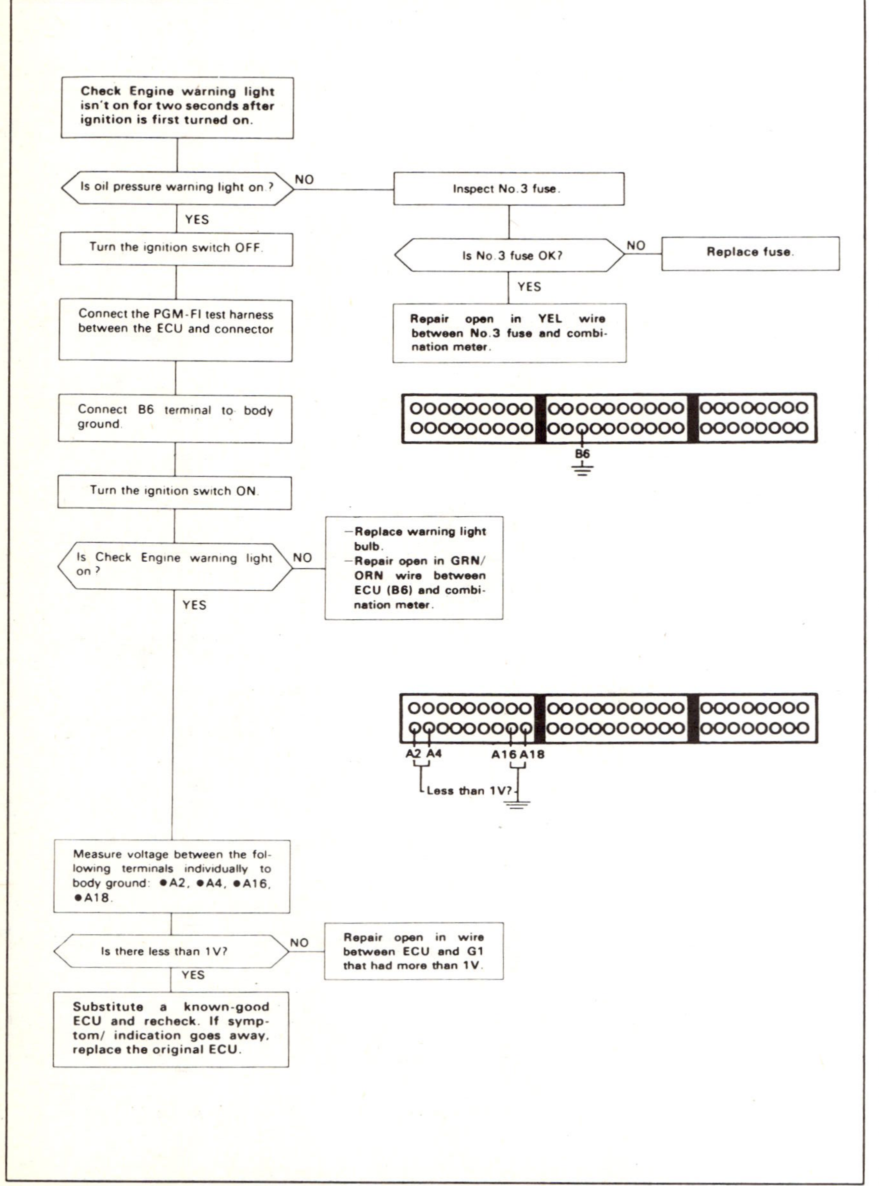


– Check Engine warning light is on.
– LED doesn't blink.

Turn the ignition switch OFF

Individually remove and inspect the following fuses:
- ECU fuse (main fuse box)
- Hazard fuse (battery positive terminal)
- No. 4 fuse (dash fuse box)

Are all the fuses OK? — NO → Replace damaged fuse.

YES

Inspect connections of ground wires at thermostat housing, also inspect for poor connections or loose BRN/BLK and BLK/RED wires at connector on right shock tower.

Are wires and connections OK? — NO → Replair as necessary.

YENT

Turn the ignition switch ON.

Does the Check Engine warning light remain on after 2 seconds ? — NO → Intermittent failure, system is OK at this time. Check for poor connections or loose wires at main relay, ECU fuse (main fuse box) and No.4 fuse (dash fuse box).

YES

Is there a CODE at the ECU? — YES → Repair troubleshooting CODE

NO

Will the engine start? — YES → Turn the ignition switch OFF → Disconnect the B connector from the ECU. → Turn the ignition switch ON.

NO

Turn the ignition switch ON.

## ECU AND CHECK LIGHT – CODE 13 1988–89 INTEGRA

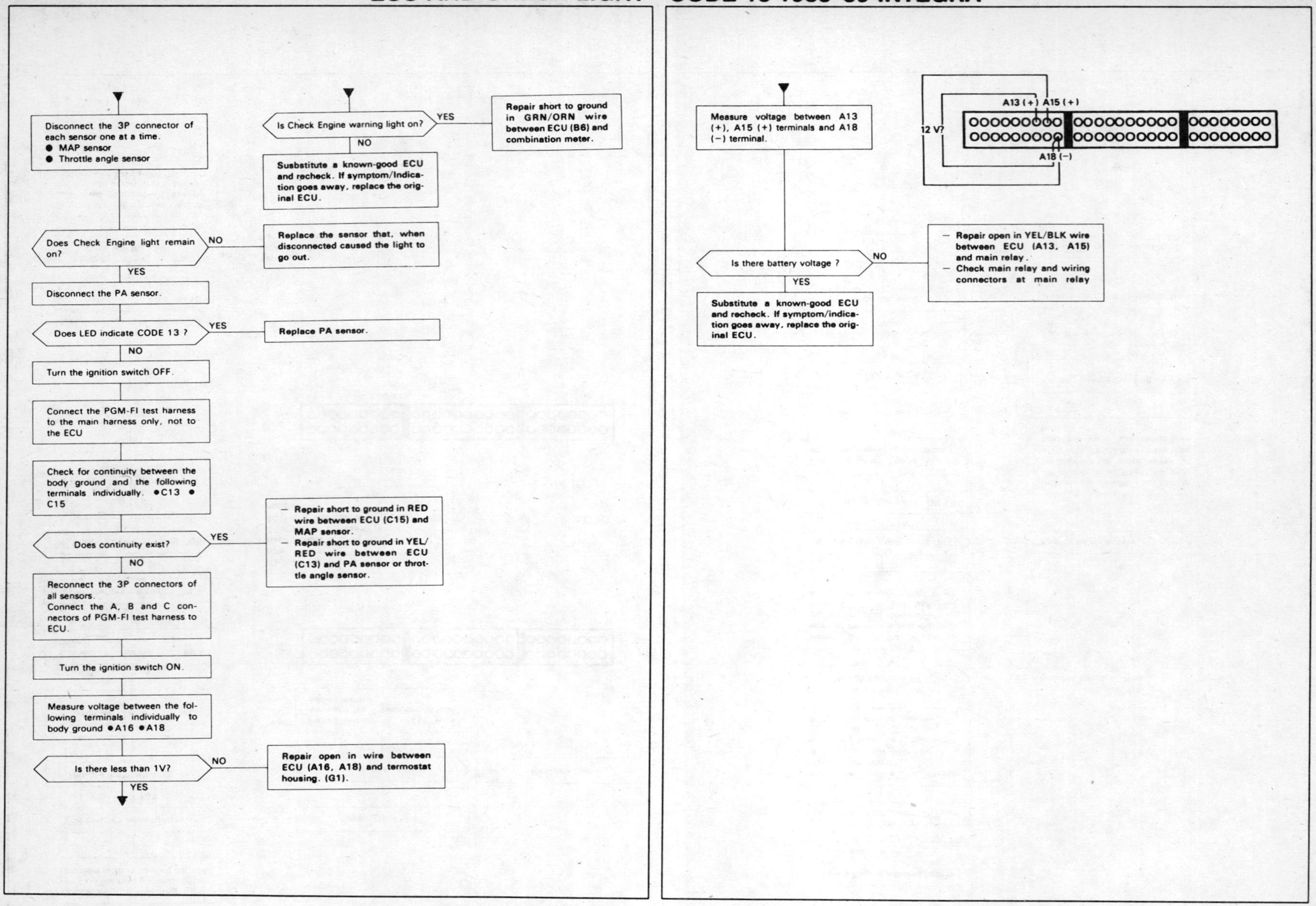

Disconnect the 3P connector of each sensor one at a time.
- MAP sensor
- Throttle angle sensor

Does Check Engine light remain on?
- NO → Replace the sensor that, when disconnected caused the light to go out.
- YES → Disconnect the PA sensor.

Does LED indicate CODE 13 ?
- YES → Replace PA sensor.
- NO → Turn the ignition switch OFF.

Connect the PGM-FI test harness to the main harness only, not to the ECU

Check for continuity between the body ground and the following terminals individually. ●C13 ● C15

Does continuity exist?
- YES →
  - Repair short to ground in RED wire between ECU (C15) and MAP sensor.
  - Repair short to ground in YEL/RED wire between ECU (C13) and PA sensor or throttle angle sensor.
- NO → Reconnect the 3P connectors of all sensors. Connect the A, B and C connectors of PGM-FI test harness to ECU.

Turn the ignition switch ON.

Measure voltage between the following terminals individually to body ground ●A16 ●A18

Is there less than 1V?
- NO → Repair open in wire between ECU (A16, A18) and termostat housing. (G1).
- YES ↓

Is Check Engine warning light on?
- YES → Repair short to ground in GRN/ORN wire between ECU (B6) and combination meter.
- NO → Susbstitute a known-good ECU and recheck. If symptom/Indication goes away, replace the original ECU.

Measure voltage between A13 (+), A15 (+) terminals and A18 (−) terminal.



Is there battery voltage ?
- NO →
  - Repair open in YEL/BLK wire between ECU (A13, A15) and main relay.
  - Check main relay and wiring connectors at main relay
- YES → Substitute a known-good ECU and recheck. If symptom/indication goes away, replace the original ECU.

# OXYGEN SENSOR—CODE 1—1988–89 INTEGRA

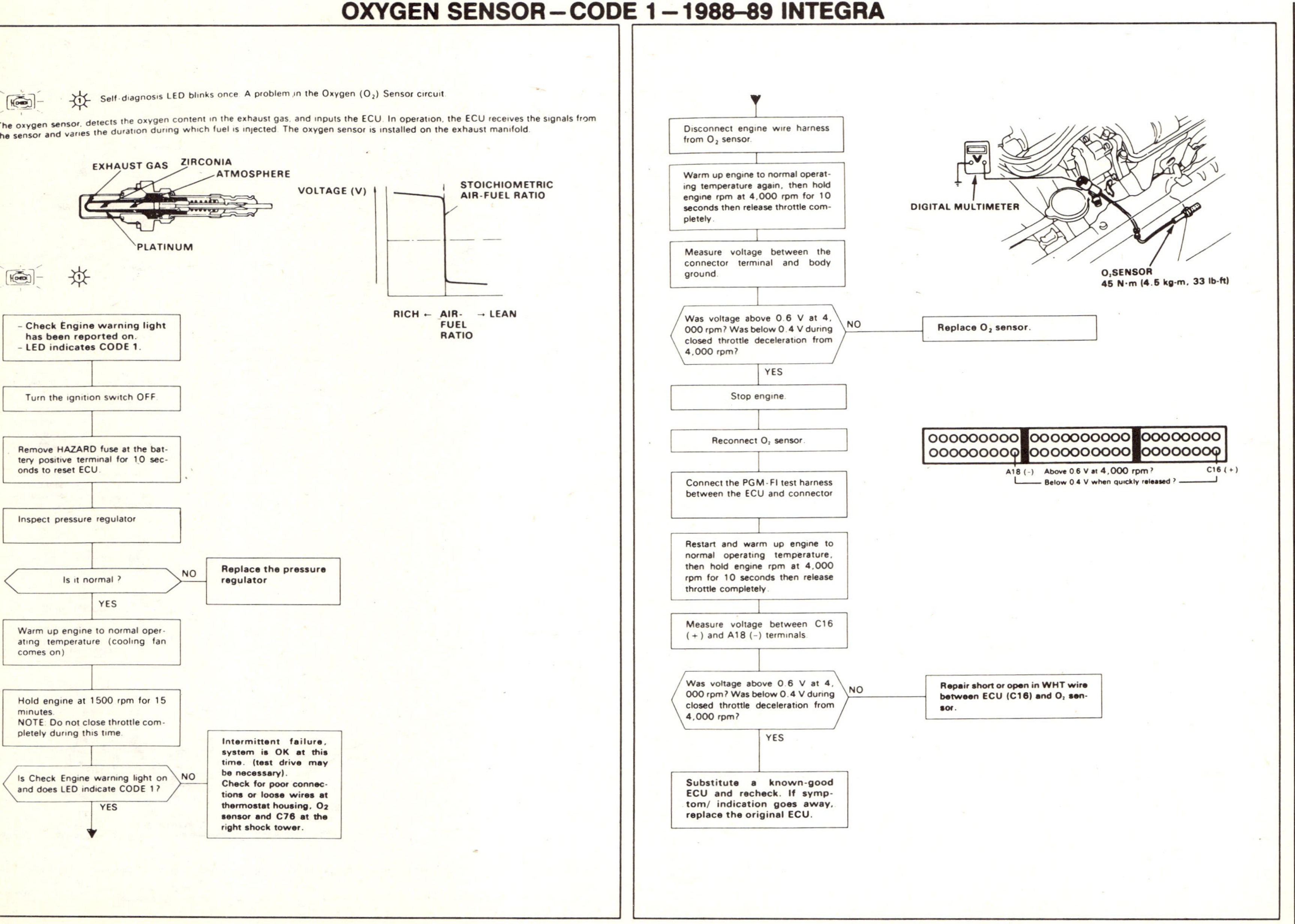

## MAP SENSOR—CODE 3—1988–89 INTEGRA

(3) Self-diagnosis LED indicator blinks three times: Most likely an electrical problem in the Manifold Absolute Pressure (MAP) Sensor system.

(5) Self-diagnosis LED indicator blinks five times: Most likely a mechanical problem (broken hose) in the Manifold Absolute Pressure (MAP) Sensor system.

The MAP sensor converts manifold absolute pressure into electrical signals and inputs the ECU.

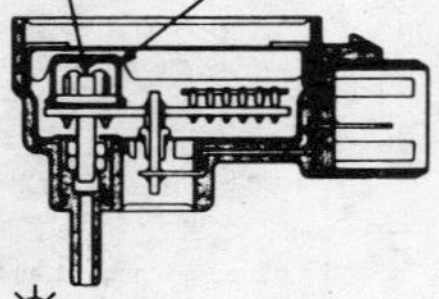

SILICON DIAPHARAGM
VACUUM CAVITY

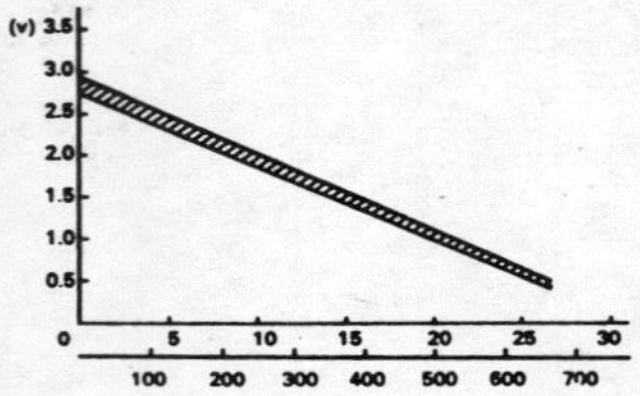

OUTPUT VOLTAGE
(v) 3.5, 3.0, 2.5, 2.0, 1.5, 1.0, 0.5, 0
0, 5, 10, 15, 20, 25, 30 (in. Hg) GAUGE
100, 200, 300, 400, 500, 600, 700 (mm Hg) READING

- Engine is warm and running.
- Check Engine warning light is on.
- LED indicates CODE 3.

Turn the ignition switch OFF.

Remove HAZARD fuse at the battery positive terminal for 10 seconds to reset ECU.

Warm up engine to normal operating temperature (cooling fan comes on).

Is Check Engine warning light on and does LED indicate CODE 3?

NO → Intermittent failure, system is OK at this time (test drive may be necessary). Check for poor connections or loose wires at MAP sensor connector and ECU.

YES ↓

Turn the ignition switch OFF.

Disconnect the 3P connector from the MAP sensor.

Turn the ignition switch ON.

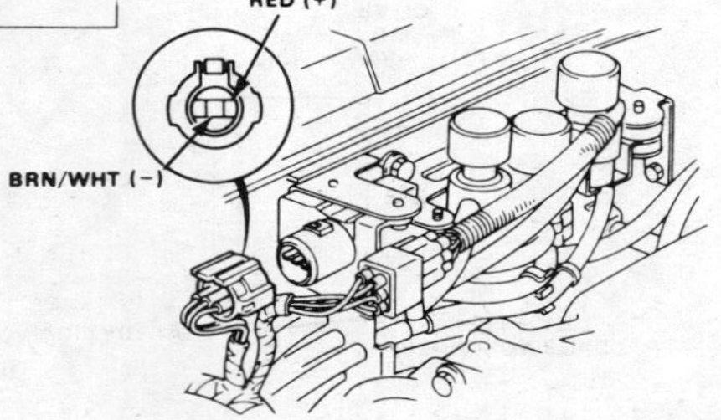

RED (+)
BRN/WHT (−)

Measure voltage between RED (+) terminal and body ground.

Is there approx. 5V ?

NO → Repair open in RED wire between ECU (C15) and MAP sensor. If wire is OK, substitute a known-good ECU and recheck. If prescribed voltage is new available, replace the original ECU.

YES → Measure voltage between RED (+) terminal and GRN/WHT (−) terminal.

Is there approx. 5V?

NO → Repair open in BRN/WHT wire between ECU (C14) and MAP sensor. If wire is OK, substitute a known-good ECU and recheck. If precribed voltage is new available, replace the original ECU.

YES ↓

Measure voltage between WHT (+) terminal and BRN/WHT (−) terminal.

Is there approx. 5V?

NO → Repair open in WHT wire between ECU (C11) and MAP sensor. If wire is OK, substitute a known-good ECU and recheck. If precribed voltage is new available, replace the original ECU.

YES ↓

Turn the ignition switch OFF.

Reconnect the 3P connector to the MAP sensor.

Connect the PGM-FI test harness between the ECU and connector

Turn the ignition switch ON.

## MAP SENSOR—CODE 3—1988–89 INTEGRA

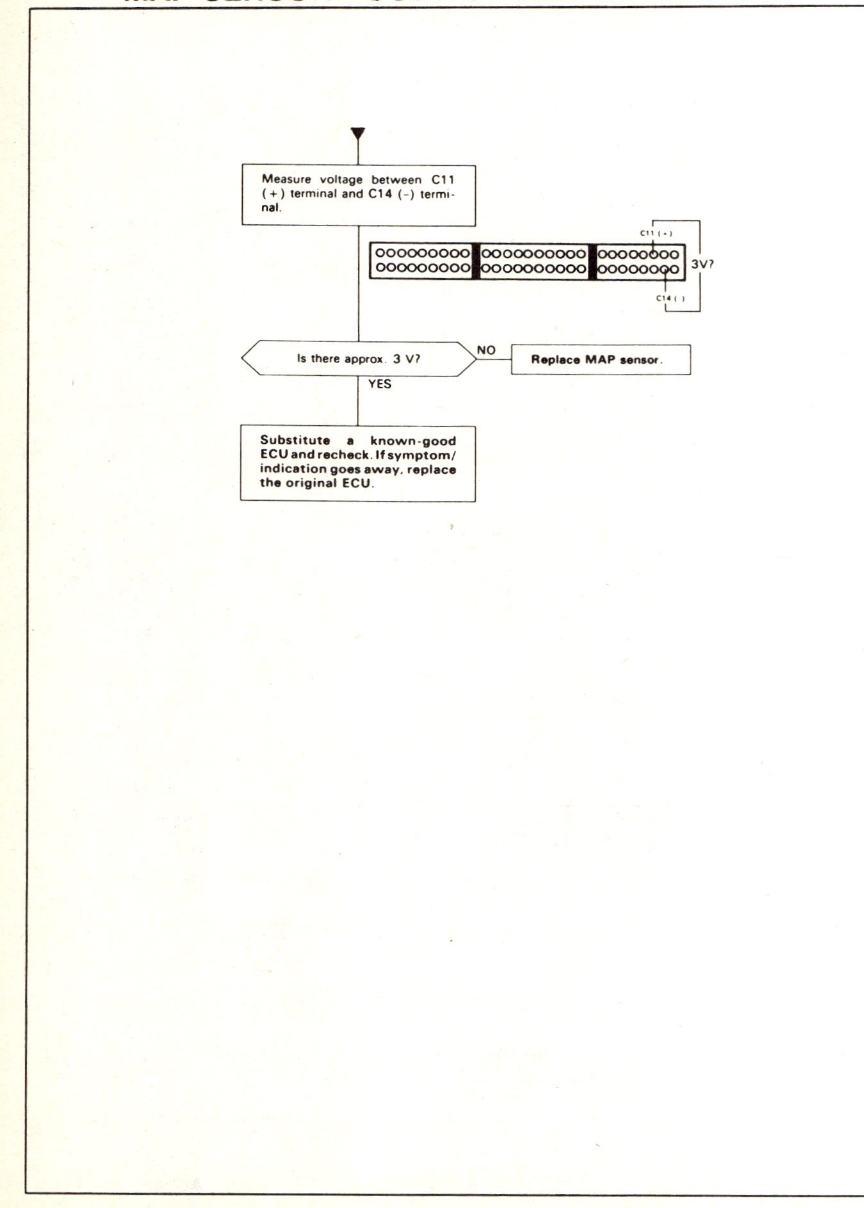


## MAP SENSOR—CODE 5—1988–89 INTEGRA

- Check Engine warning light has been reported on.
- LED indicates CODE 5.

Turn the ignition switch OFF.

Remove HAZARD fuse at the battery positive terminal for 10 seconds to reset ECU.

Start the engine.

Is Check Engine warning light on and does LED indicate CODE 5?

NO → —Intermittent failure, system is OK at this time (test drive may be necessary). —Check vacuum hoses, pipes, and connections. —Make sure all connectors are secure.

YES → Stop engine.

Connect vacuum pump to #21 hose and apply vacuum.

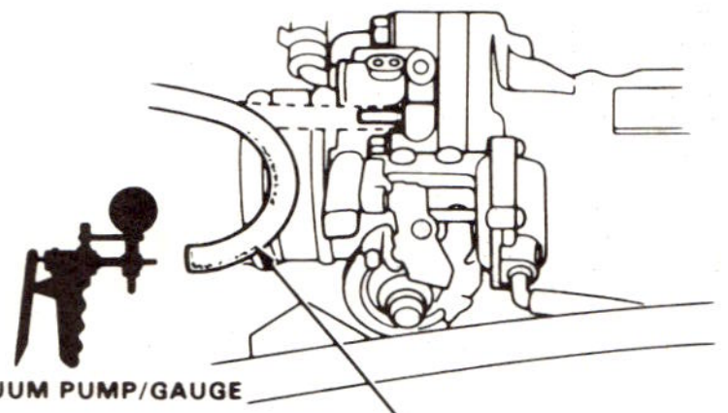

VACUUM PUMP/GAUGE

#21 HOSE

Does it hold vacuum?

NO → Connect a vacuum pump to the MAP sensor and apply vacuum.

Does it hold vacuum?

NO → Replace MAP sensor.

YES → Replace #21 hose.

YES → Disconnect #21 hose from the throttle body and connect a T-fitting from a vacuum gauge between the throttle body and MAP sensor.

## MAP SENSOR – CODE 5 – 1988–89 INTEGRA

- Start engine.
- Is there manifold vacuum?
  - NO → –Remove restriction from throttle body. –Replace throttle body.
  - YES ↓
- Stop engine.
- Connect the PGM-FI test harness between the ECU and connector
- Turn the ignition switch ON.
- Measure voltage between C11 (+) terminal and C14 (–) terminal.

C11 (+)
C14 (–)
3 V?

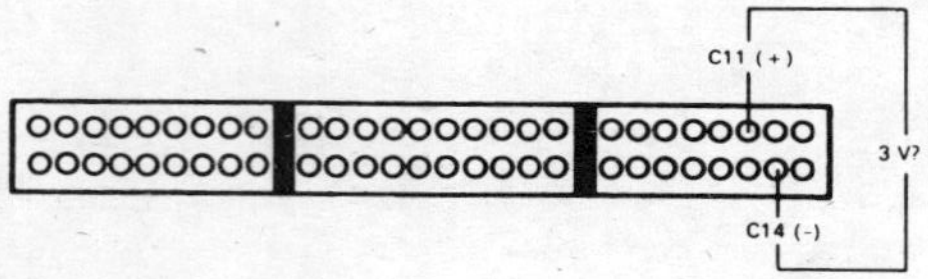

- Is there approx. 3 V ?
  - NO → Inspect open in WHT wire between the MAP sensor and ECU. If wire is OK, replace the MAP sensor.
  - YES ↓
- Start the engine and allow it to idle.
- Measure voltage between C11 (+) terminal and C14 (–) terminal.
- Is there approx. 1V?
  - NO → Replace MAP sensor.
  - YES ↓
- Substitute a known-good ECU and recheck. If symptom/indication goes away, replace the original ECU.

## TDC/CRANK SENSOR – CODE 4 – 1988–89 INTEGRA

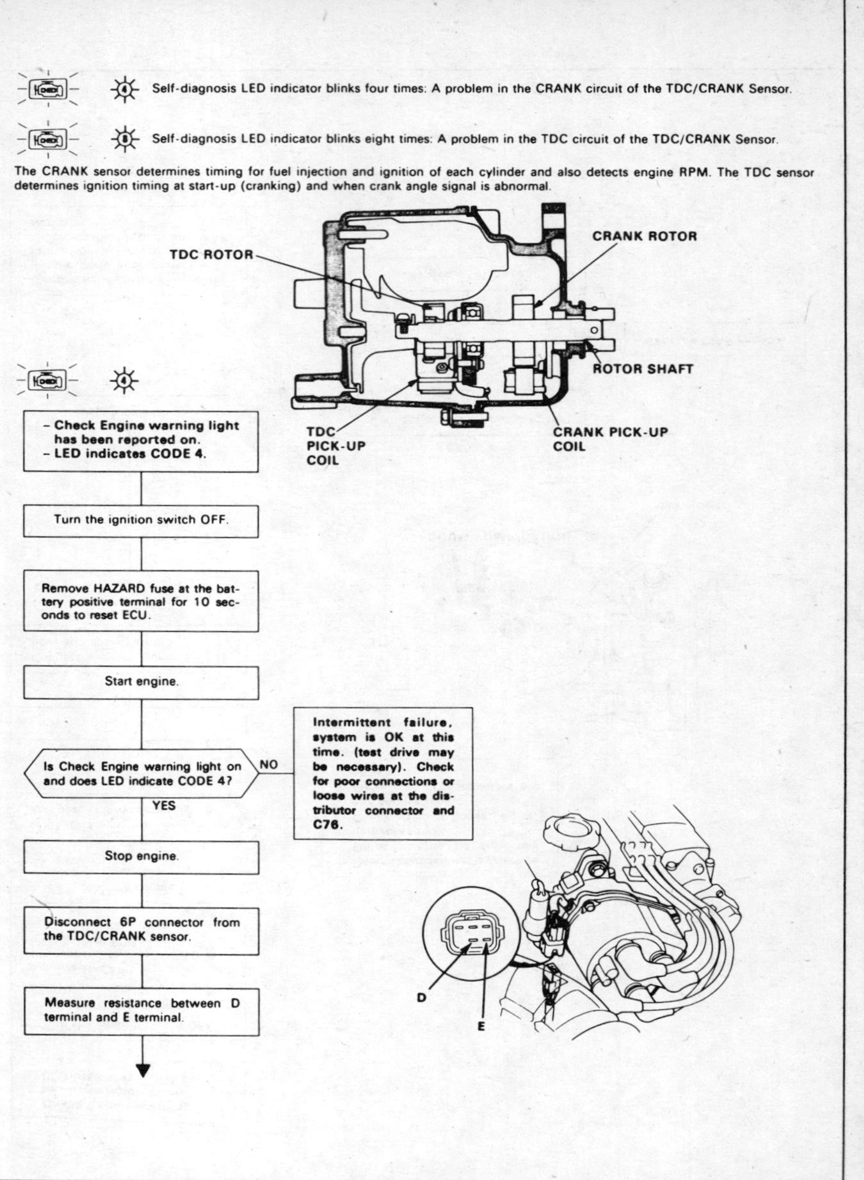

Self-diagnosis LED indicator blinks four times: A problem in the CRANK circuit of the TDC/CRANK Sensor.
Self-diagnosis LED indicator blinks eight times: A problem in the TDC circuit of the TDC/CRANK Sensor.
The CRANK sensor determines timing for fuel injection and ignition of each cylinder and also detects engine RPM. The TDC sensor determines ignition timing at start-up (cranking) and when crank angle signal is abnormal.
TDC ROTOR
CRANK ROTOR
ROTOR SHAFT
TDC PICK-UP COIL
CRANK PICK-UP COIL
– Check Engine warning light has been reported on.
– LED indicates CODE 4.
Turn the ignition switch OFF.
Remove HAZARD fuse at the battery positive terminal for 10 seconds to reset ECU.
Start engine.
Is Check Engine warning light on and does LED indicate CODE 4?
NO
Intermittent failure, system is OK at this time. (test drive may be necessary). Check for poor connections or loose wires at the distributor connector and C76.
YES
Stop engine.
Disconnect 6P connector from the TDC/CRANK sensor.
Measure resistance between D terminal and E terminal.
D
E

## TDC/CRANK SENSOR – CODE 4 – 1988–89 INTEGRA

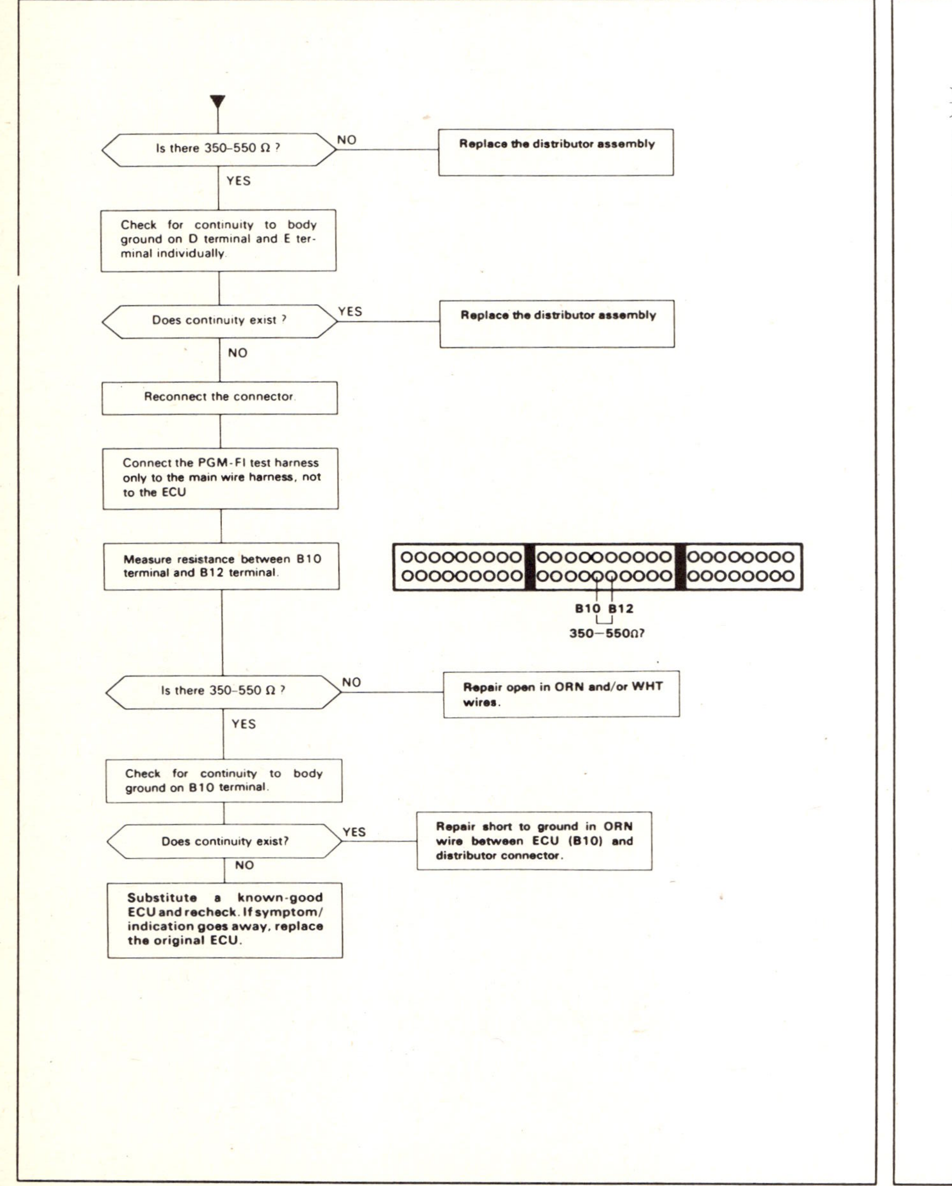


## TDC/CRANK SENSOR – CODE 8 – 1988–89 INTEGRA

- Check Engine warning light has been reported on.
- LED indicates CODE 8.

Turn the ignition switch OFF.

Remove HAZARD fuse at the battery positive terminal for 10 seconds to reset ECU.

Start engine.

Is Check Engine warning light on and does LED indicate CODE 8?

- NO → Intermittent failure, system is OK at this time (test drive may be necessary). Check for poor connections or loose wires at the distributor and C76 connector.
- YES ↓

Stop engine.

Disconnect the 6P connector from the TDC/CRANK sensor.

Measure resistance between B terminal and C terminal.

B C

Is there 350—550 Ω ?

- NO → Replace the distributor assembly
- YES ↓

Check for continuity to body ground on B terminal and C terminal individually.

Does continuity exist ?

- YES → Replace the distributor assembly
- NO ↓

Reconnect the connector.

## TDC/CRANK SENSOR – CODE 8 – 1988–89 INTEGRA

Connect the PGM-FI test harness only to the main wire harness, not to the ECU

Measure resistance between C3 terminal and C4 terminal.

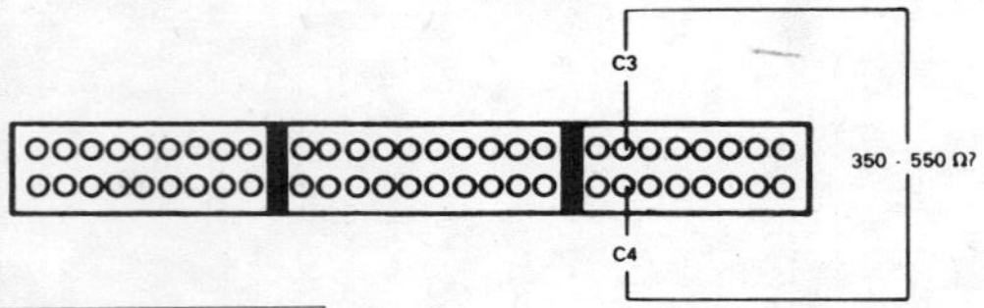

350 - 550 Ω?

Is there 350 — 550 Ω ?

- NO: Repair open in ORN/BLU and/or WHT/BLU wires.
- YES: Check for continuity to body ground on C3 terminal.

Does continuity exist?

- YES: Repair short in ORN/BLU wire between ECU (C3) and distributor connector.
- NO: Substitute a known-good ECU and recheck. If symptom/indication goes away, replace the original ECU.

## COOLANT TEMPERATURE (TW) SENSOR – CODE 6 1988–89 INTEGRA

Self-diagnosis LED indicator blinks six times: Most likely a problem in the Coolant Temperature (TW) Sensor circuit.

The TW sensor is a temperature dependant resistor (thermistor). The resistance of the thermistor decreases as the coolant temperature increases as shown below.

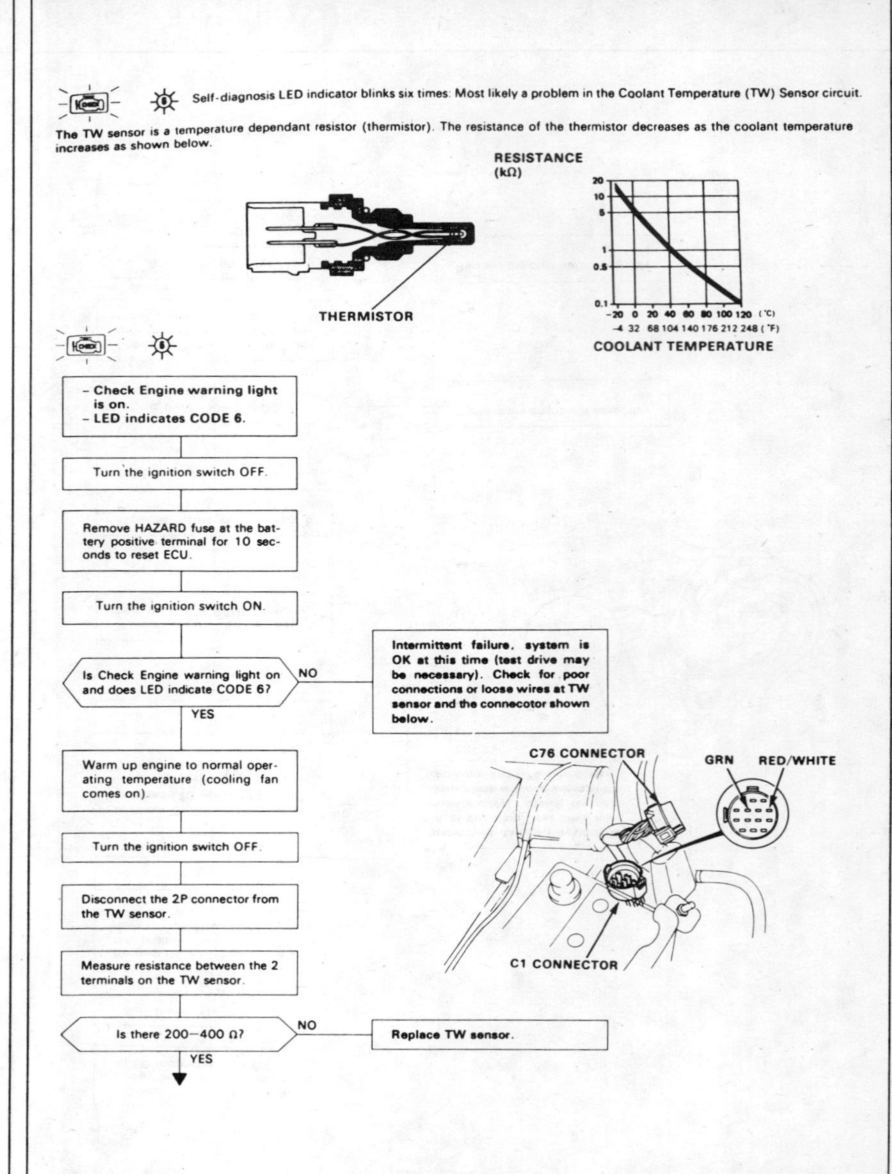


- Check Engine warning light is on.
- LED indicates CODE 6.

Turn the ignition switch OFF.

Remove HAZARD fuse at the battery positive terminal for 10 seconds to reset ECU.

Turn the ignition switch ON.

Is Check Engine warning light on and does LED indicate CODE 6?

- NO: Intermittent failure, system is OK at this time (test drive may be necessary). Check for poor connections or loose wires at TW sensor and the connecotor shown below.
- YES: Warm up engine to normal operating temperature (cooling fan comes on).

Turn the ignition switch OFF.

Disconnect the 2P connector from the TW sensor.

Measure resistance between the 2 terminals on the TW sensor.

Is there 200—400 Ω?

- NO: Replace TW sensor.
- YES

## COOLANT TEMPERATURE (TW) SENSOR – CODE 6 1988–89 INTEGRA



- Turn the ignition switch ON.
- Measure voltage between RED/WHT (+) terminal and body ground.
- Is there approx. 5V ?
  - NO → **Repair open in RED/WHT wire between ECU (C6) and TW sensor. If wire is OK, substitute a known-good ECU and recheck. If prescribed voltage is now available, replace the original ECU.**
  - YES ↓
- Measure voltage between RED/WHT (+) terminal and GRN/(−) terminal.
- Is there approx. 5V ?
  - NO → **Repair open in GRN and BRN/BLK wire between ECU (C12) and TW sensor.**
  - YES ↓
- **Substitute a known-good ECU and recheck. If symptom/indication goes away, replace the original ECU.**

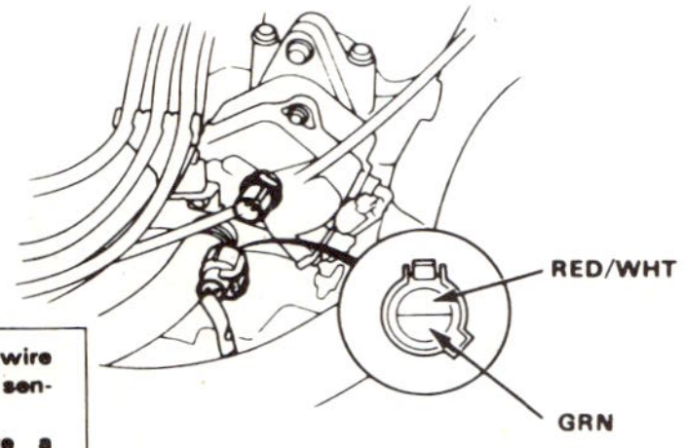

## THROTTLE ANGLE SENSOR – CODE 7 1988–89 INTEGRA

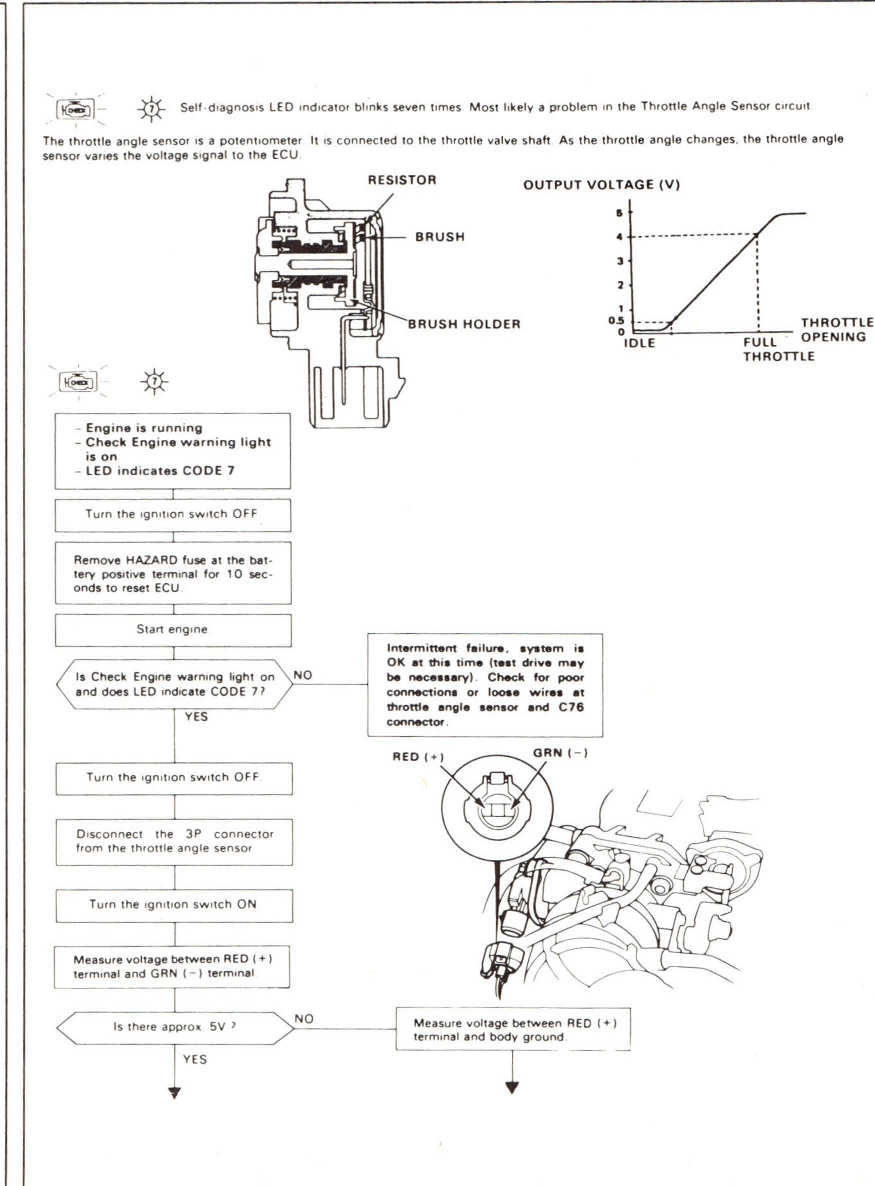

## THROTTLE ANGLE SENSOR—CODE 7 1988–89 INTEGRA

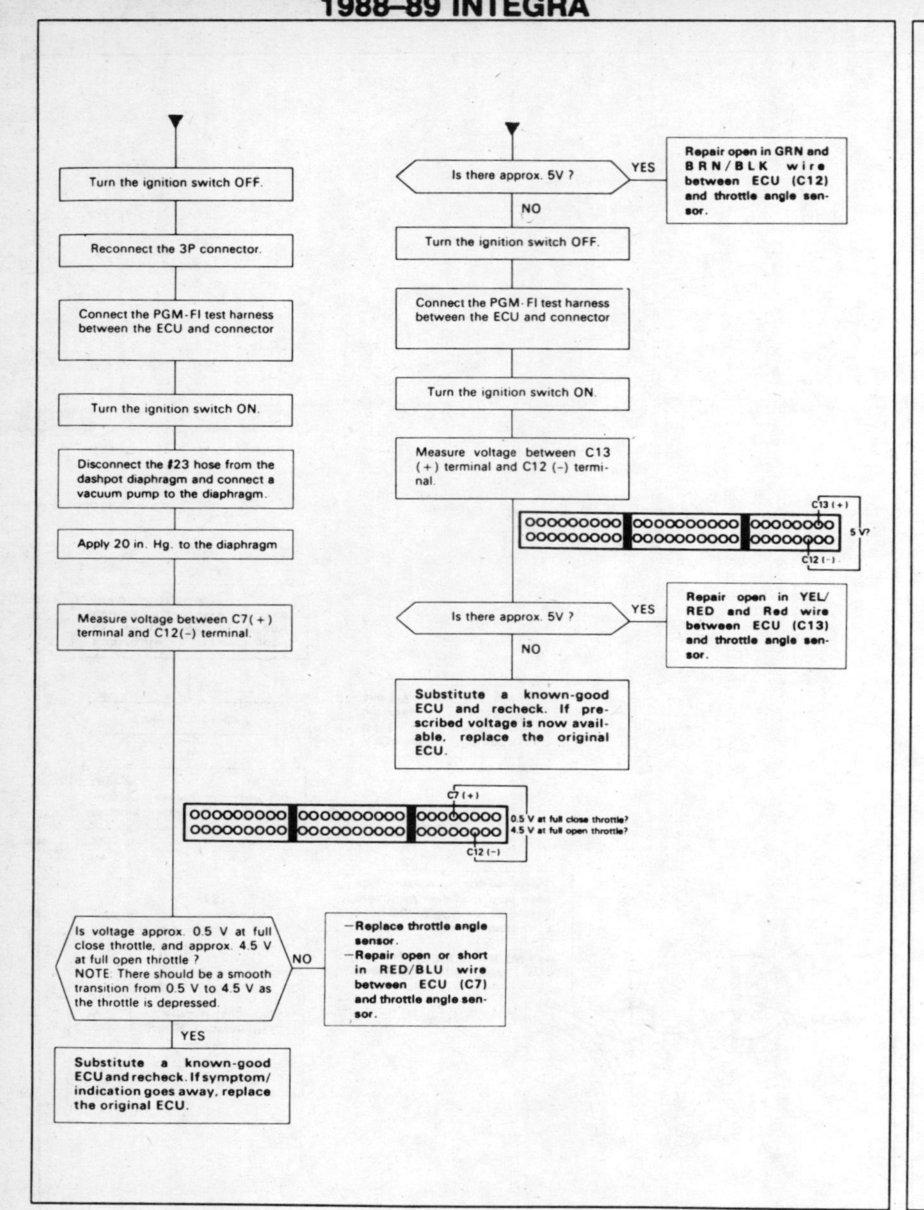


## CYL SENSOR—CODE 9—1988–89 INTEGRA

Self-diagnosis LED indicator blinks nine times: A problem in the CYL sensor.

The CYL sensor detects the position of NO.1 cylinder for sequential fuel injection to each cylinder.

CYL PICK-UP COIL

CYL ROTOR

ROTOR SHAFT

—Check Engine warning light has been reported on.
—LED indicates CODE 9.

Turn the ignition switch OFF.

Remove HAZARD fuse at the battery positive terminal for 10 seconds to reset ECU.

Start engine.

Is Check Engine warning light on and does LED indicate CODE 9?

NO: Intermittent failure, system is OK at this time (test drive may be necessary). Check for poor connections or loose wires at TA sensor and C76 connector.

YES: Stop engine.

Disconnect the 2P connector from the CYL sensor.

Measure resistance between 2 terminals on the CYL sensor.

CYL SENSOR

## CYL SENSOR—CODE 9—1988–89 INTEGRA

- Is there 700–1,000 Ω? — NO: Replace CYL sensor
- YES: Check for continuity to body ground on the 2 terminals individually.
- Does continuity exist? — YES: Replace CYL sensor
- NO: Reconnect the connector.
- Connect the PGM-FI test harness only to the main wire harness, not to the ECU
- Measure resistance between C1 terminal and C2 terminal.
- Is there 700–1,000 Ω? — NO: Repair open in ORN and/or WHT wires.
- YES: Check for continuity to body ground on C1 terminal.
- Does continuity exist? — YES: Repair short in ORN wire between ECU (C1) and cylinder connector.
- NO: Substitute a known-good ECU and recheck. If symptom/indication goes away, replace the original ECU.

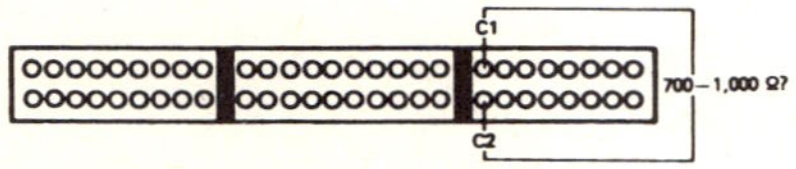

C1
C2
700–1,000 Ω?

## AIR TEMPERATURE (TA) SENSOR—CODE 10 1988–89 INTEGRA

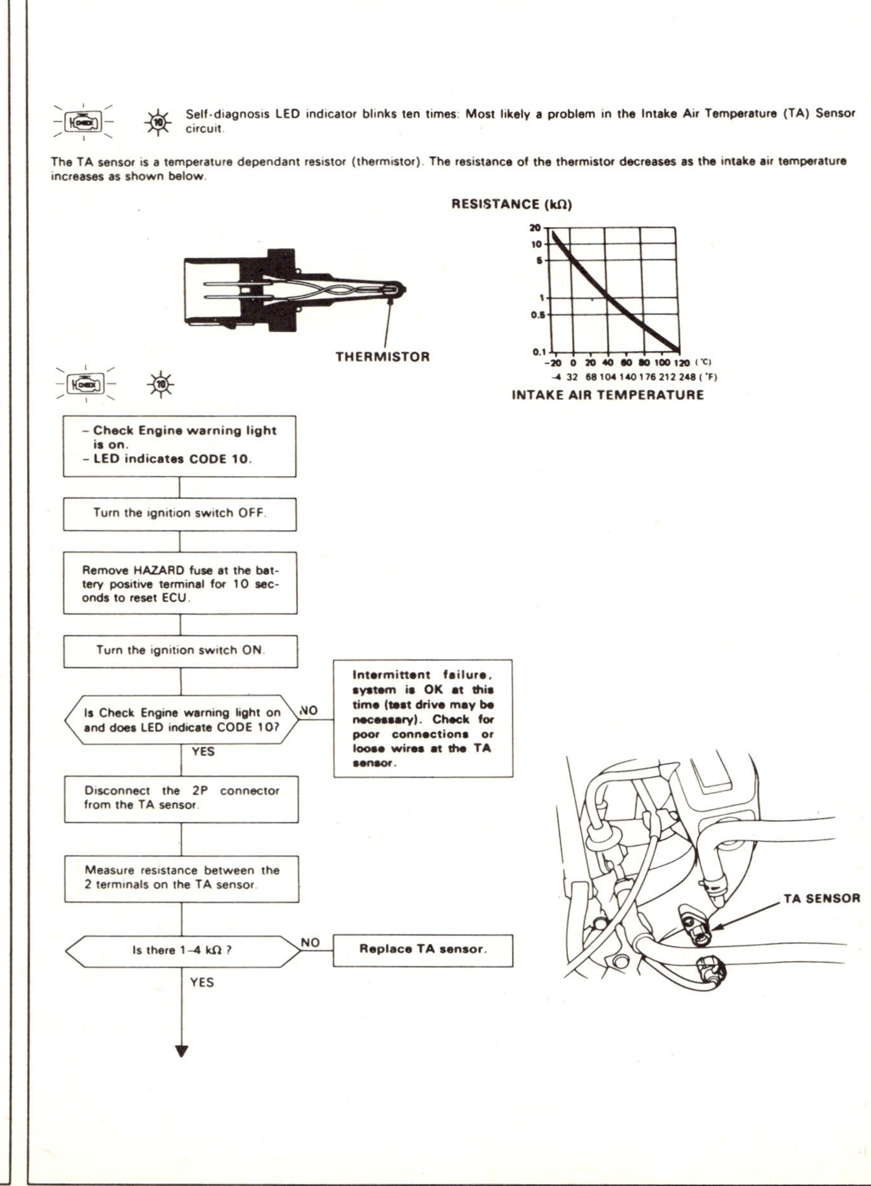

## AIR TEMPERATURE (TA) SENSOR—CODE 10 1988–89 INTEGRA

Measure voltage between WHT/BLK (+) terminal and body ground.

Is there approx. 5 V ?
- NO → Repair open or short in WHT/BLK and RED/YEL wire between TA sensor. If wire is OK, substitute a known-good ECU and recheck. If prescribed voltage is new available, replace the original ECU.
- YES ↓

Measure voltage between WHT/BLK (+) terminal and GRN (−) terminal.

Is there approx. 5 V ?
- NO → Repair open in GRN and BRN/BLK wire between ECU (C12) and TA sensor.
- YES ↓

Substitute a known-good ECU and recheck. If symptom/indication goes away, replace the original ECU.

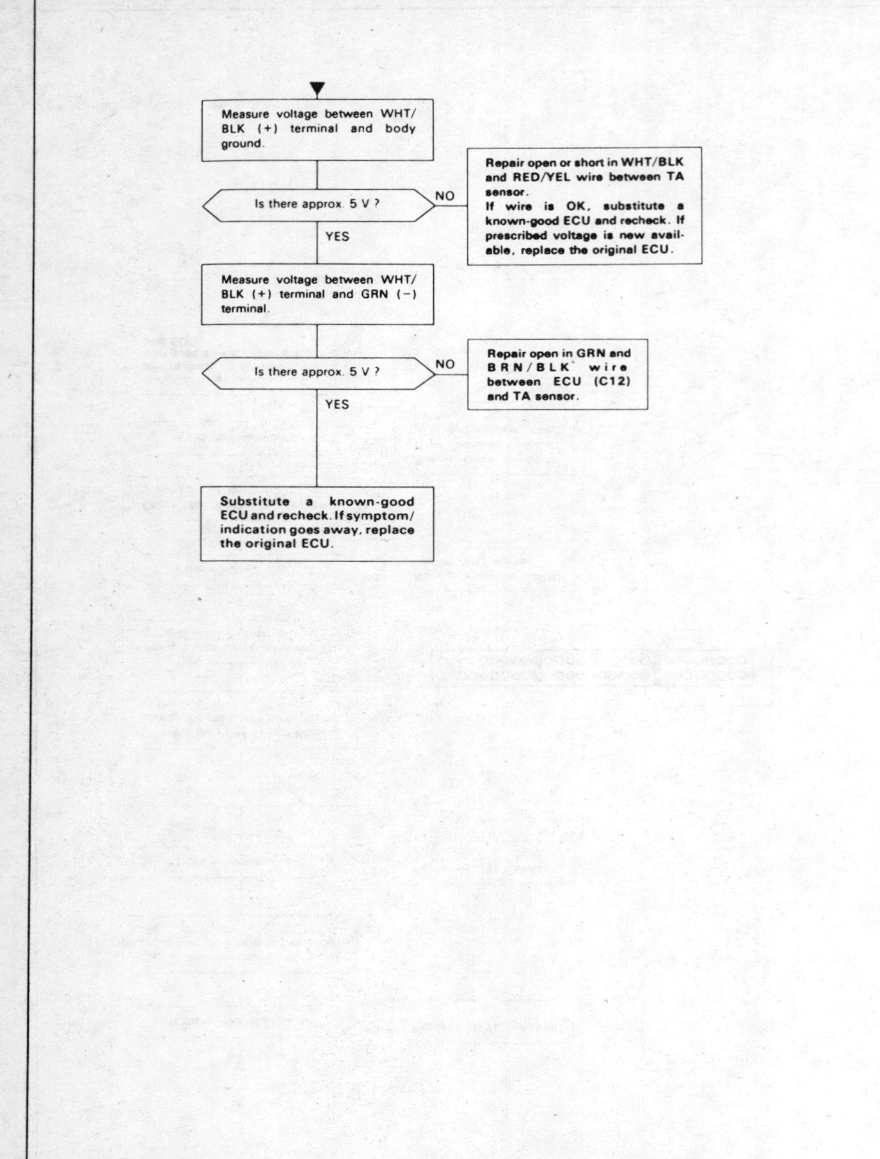

## ATMOSPHERIC PRESSURE (PA) SENSOR—CODE 13 1988–89 INTEGRA

Self-diagnosis LED indicator blinks thirteen times: A problem in the Atmospheric Pressure (PA) Sensor circuit.

The PA sensor converts atmospheric pressure into electrical signals and inputs the ECU.

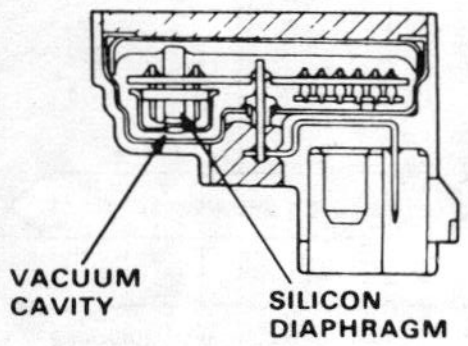

VACUUM CAVITY
SILICON DIAPHRAGM

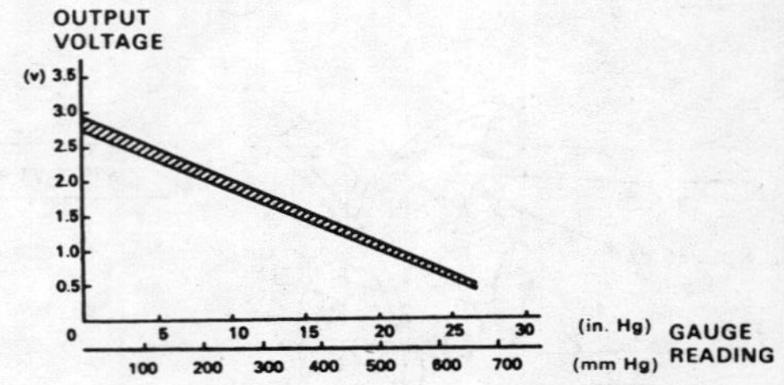

OUTPUT VOLTAGE
(v) 3.5
3.0
2.5
2.0
1.5
1.0
0.5
0
5 10 15 20 25 30 (in. Hg)
100 200 300 400 500 600 700 (mm Hg)
GAUGE READING

- Check Engine warning light is on.
- LED indicates CODE 13.

Turn the ignition switch OFF.

Remove HAZARD fuse at the battery positive terminal for 10 seconds to reset ECU.

Turn the ignition switch ON.

Is Check Engine warning light on and does LED indicate CODE 13?
- NO → Intermittent failure, system is OK at this time (test drive may be necessary). Check for poor connections or loose wires at the PA sensor.
- YES ↓

Turn the ignition switch OFF.

Disconnect the main wire harness from PA sensor.

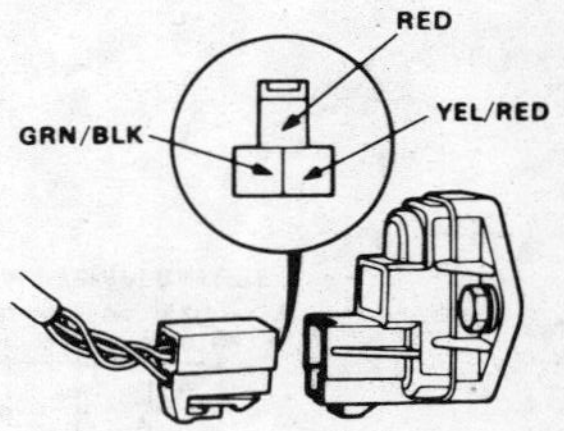

RED
GRN/BLK
YEL/RED

Measure voltage between YEL/RED (+) terminal and body ground.

Is there approx. 5V?
- NO → Repair open in YEL/RED wire between ECU (C13) and the sensor. If wire is OK, substitute a known-good ECU and recheck. If prescribed voltage is now available replace the original ECU.
- YES → Measure voltage between YEL/RED (+) terminal and GRN/BLK (−) terminal.

Is there approx. 5V?
- NO → Repair open or short in GRN/BLK wire between ECU (C12) and the sensor. If wire is OK, substitute a known-good ECU and recheck. If precribed voltage is now available replace the original ECU.
- YES ↓

## ATMOSPHERIC PRESSURE (PA) SENSOR—CODE 13 1988–89 INTEGRA

Measure voltage between RED (+) terminal and GRN/BLK (+) terminal.

Is there approx. 5V?

- NO → **Repair open or short in RED wire between ECU (C9) and PA sensor. If wire is OK, substitute a known-good ECU and recheck. If prescribed voltage is now available replace the original ECU.**
- YES ↓

Turn the ignition switch OFF.

Remove HAZARD fuse at the battery positive terminal for 10 seconds to reset ECU.

Substitute a known-good PA sensor.

Turn the ignition switch ON.

Is Check Engine warning light on and does LED indicate CODE 13?

- NO → **Replace original PA sensor.**
- YES ↓

**Substitute a known-good ECU and recheck. If prescribed voltage is now available replace the original ECU.**

## IGNITION OUTPUT SIGNAL—CODE 15 1988–89 INTEGRA

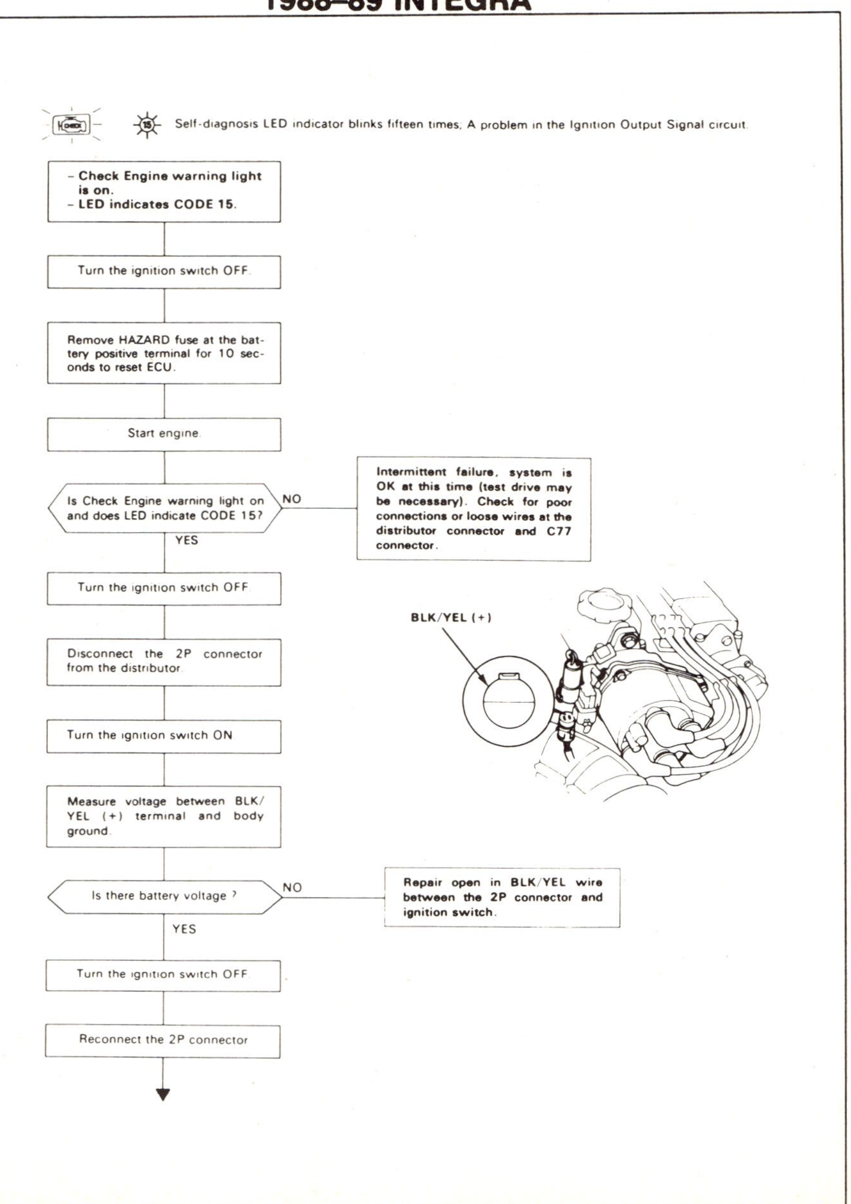

## IGNITION OUTPUT SIGNAL—CODE 15 1988–89 INTEGRA

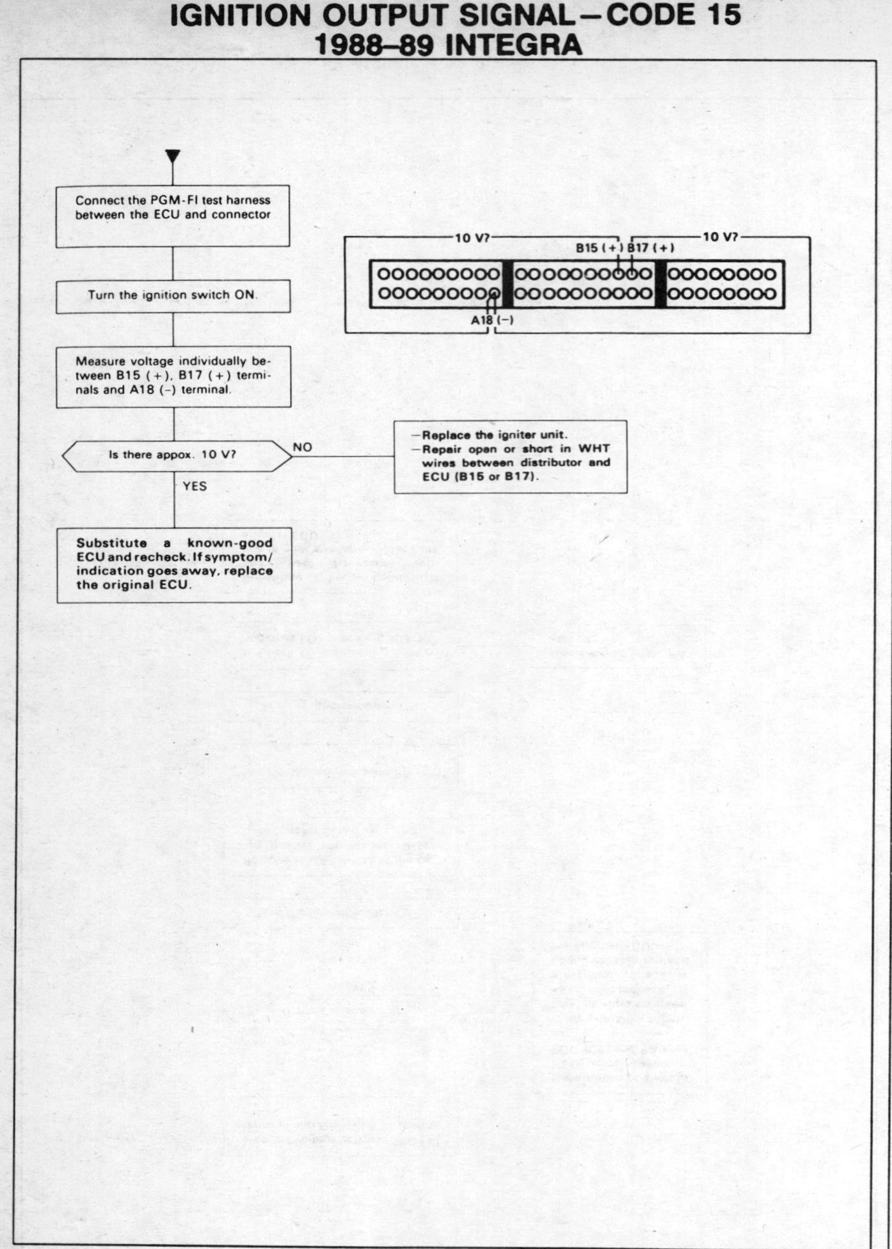


## VEHICLE SPEED SENSOR—CODE 17 1988–89 INTEGRA

Self-diagnosis LED indicator blinks seventeen times. A problem in the Vehicle Speed Sensor circuit.

The signal generated by the speed sensor, which produces 4 pulses (switch closures to ground) for each revolution of the speedometer cable.

- Check Engine warning light is on.
- LED indicates CODE 17.

Turn the ignition switch OFF.

Remove HAZARD fuse at the battery positive terminal for 10 seconds to reset ECU.

Road test necessary:
In 2nd gear accelerate to 3,500 rpm and decelerate to 1,500 rpm with throttle fully closed.

Is Check Engine warning light on and does LED indicate CODE 17?

NO → **Intermittent failure, system is OK at this time. Check for poor connections or loose wires at C202 and C208.**

YES ↓

Block rear wheels and set the parking brake. Jack up the front of the car and support with safety stands.

**WARNING Block rear wheels before jacking up front of car.**

Connect the PGM-FI test harness between the ECU and connector

Turn the ignition switch ON

Slowly rotate left front wheel and measure voltage between B16 (+) terminal and A18 (−) terminal.

A18 (−) B16 (+) 0 V↔5 V?

Does voltage pulse 0V and 5V?

NO →
- **Repair open or short in YEL/RED wire between ECU (B16) and the speed sensor.**
- **Faulty speed sensor.**
- **Faulty ECU.**

YES ↓

**Substitute a known-good ECU and recheck. If symptom/indication goes away, replace the original ECU.**

# LOCK-UP CONTROL SOLENOID VALVE (A/T)—CODE 19—1988—89 INTEGRA

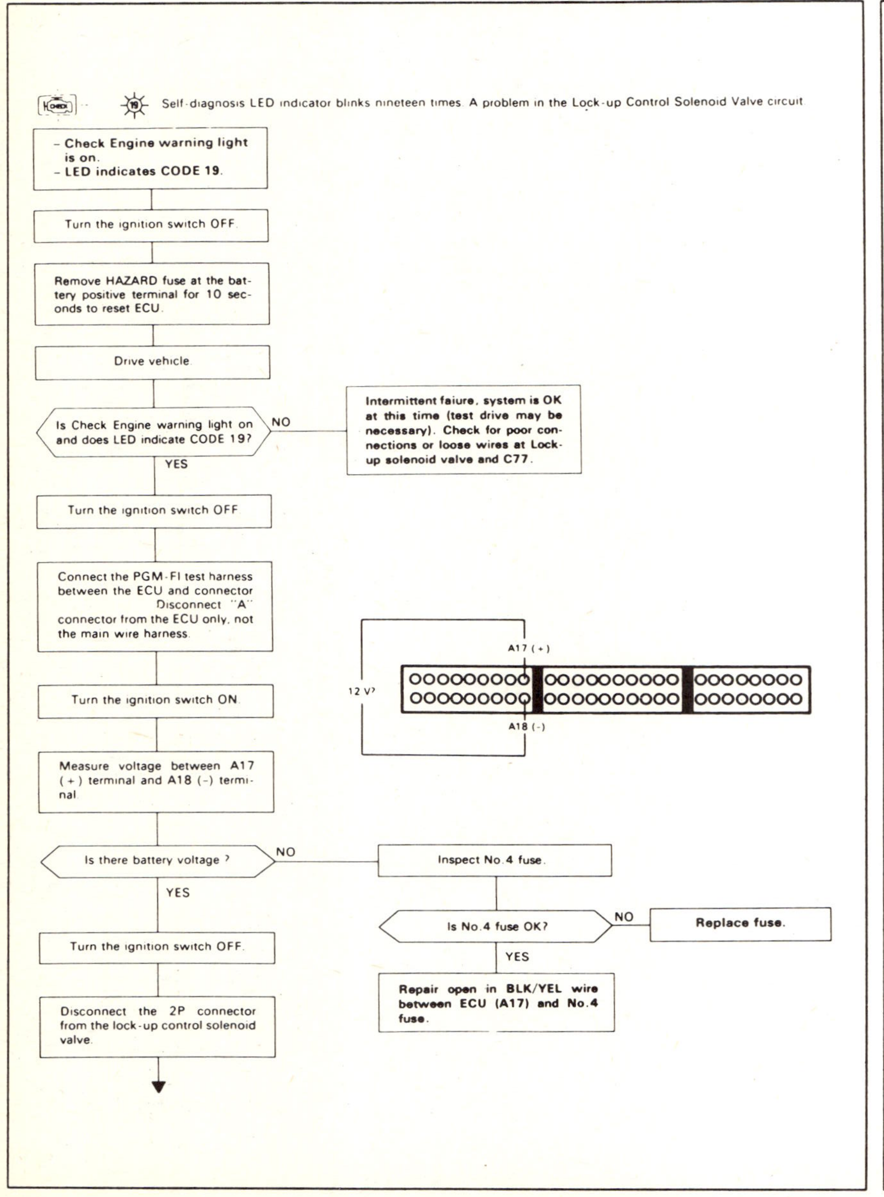

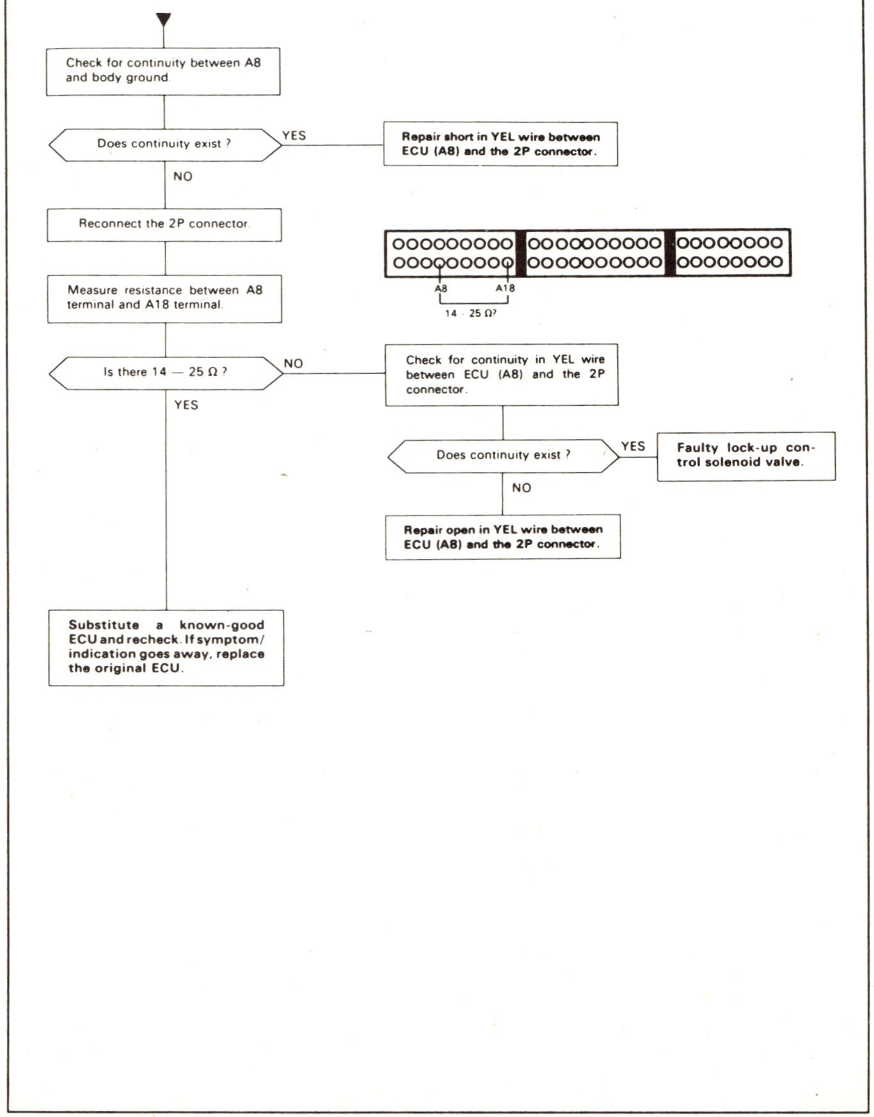

Check for continuity between A8 and body ground.

Does continuity exist?

YES → Repair short in YEL wire between ECU (A8) and the 2P connector.

NO

Reconnect the 2P connector.

Measure resistance between A8 terminal and A18 terminal.

A8 A18 14 - 25 Ω?

Is there 14 — 25 Ω?

NO → Check for continuity in YEL wire between ECU (A8) and the 2P connector.

Does continuity exist?

YES → Faulty lock-up control solenoid valve.

NO → Repair open in YEL wire between ECU (A8) and the 2P connector.

YES

Substitute a known-good ECU and recheck. If symptom/indication goes away, replace the original ECU.

# ELECTRIC LOAD DETECTOR—CODE 20—1988—89 INTEGRA

Self-diagnosis LED indicator blinks twenty times: A problem in the Electric Load Detector circuit.

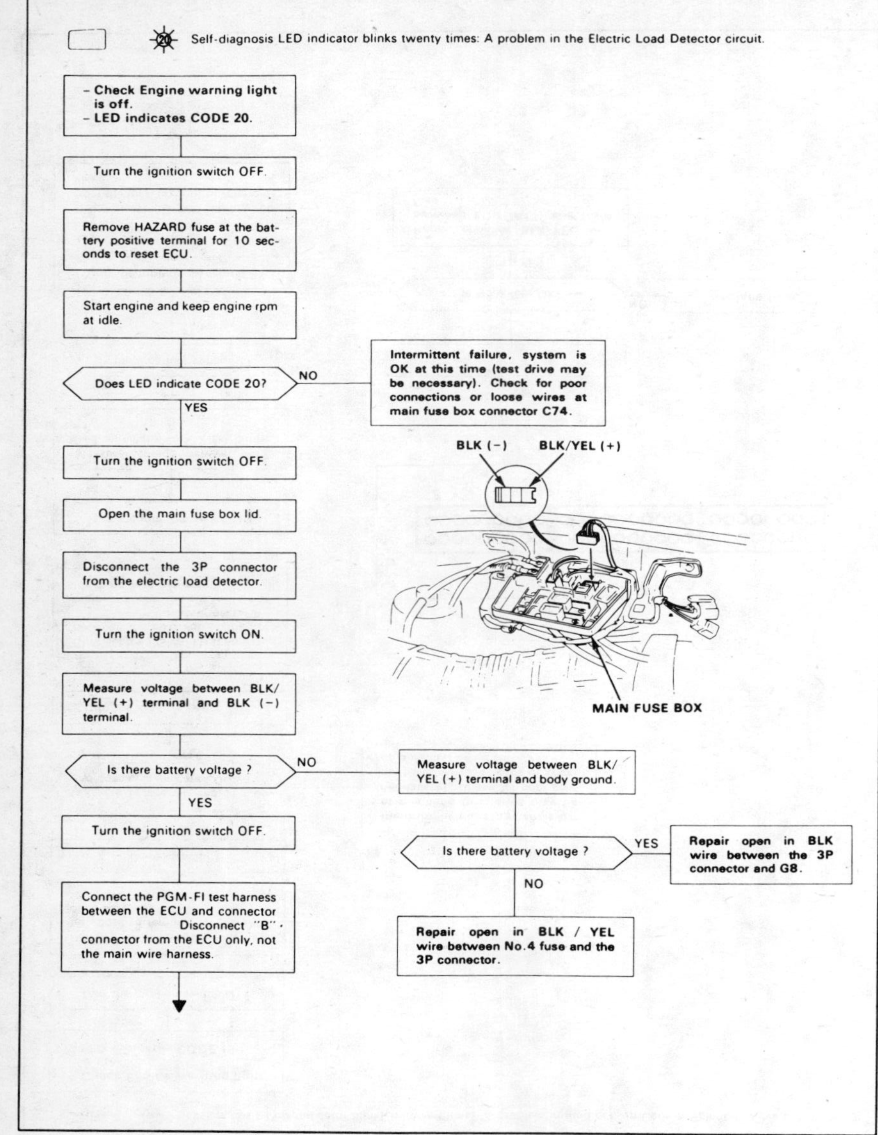


Connect GRN/RED terminal of the 3P connector to body ground.

Check for continuity in GRN/RED wire between ECU (B19) and body ground.

Does continuity exist ?

NO — Repair open in GRN/RED wire between ECU (B19) and the 3P connector.

BLK (−)
GRN/RED (+)

YES

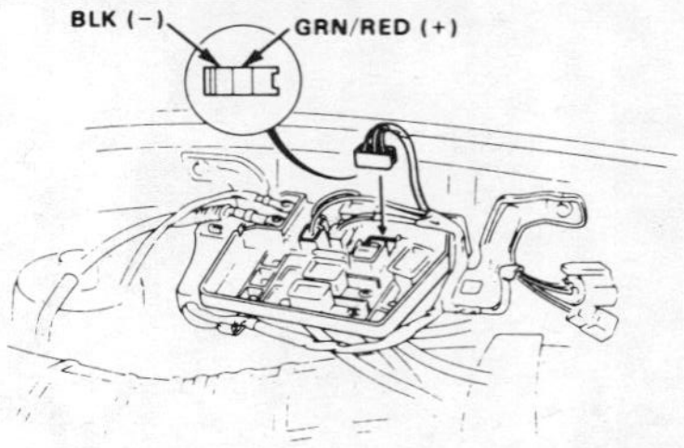

Connect "B" connector to the ECU.

Disconnect GRN/RED terminal of the 3P connector from body ground.

Turn the ignition switch ON.

Measure voltage between GRN/RED (+) terminal and BLK (−) terminal of the 3P connector.

Is there 4.5 — 5 V ?

NO — Disconnect "B" connector from the main wire harness only, not the ECU.

Measure voltage between B19 (+) terminal and A18 (−) terminal.

Is there 4.5 — 5 V ?

YES — Repair short in GRN/RED wire between ECU (B19) and the 3P connector.

NO — Substitute a known-good ECU and recheck. If prescribed voltage is now available, replace the original ECU.

YES

Connect the 3P connector to the electric load detector.

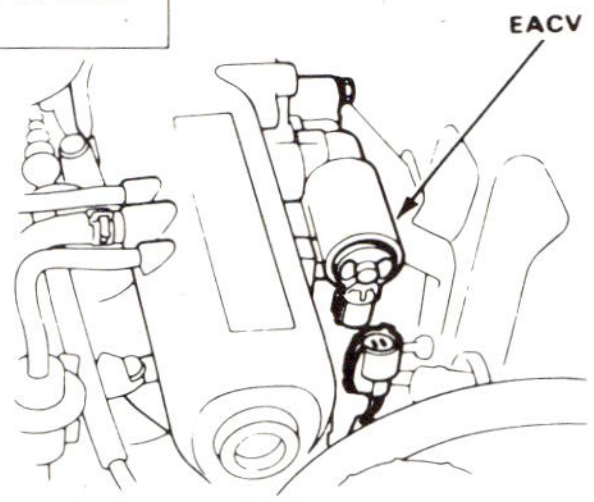

## ELECTRIC LOAD DETECTOR—CODE 20 1988—89 INTEGRA

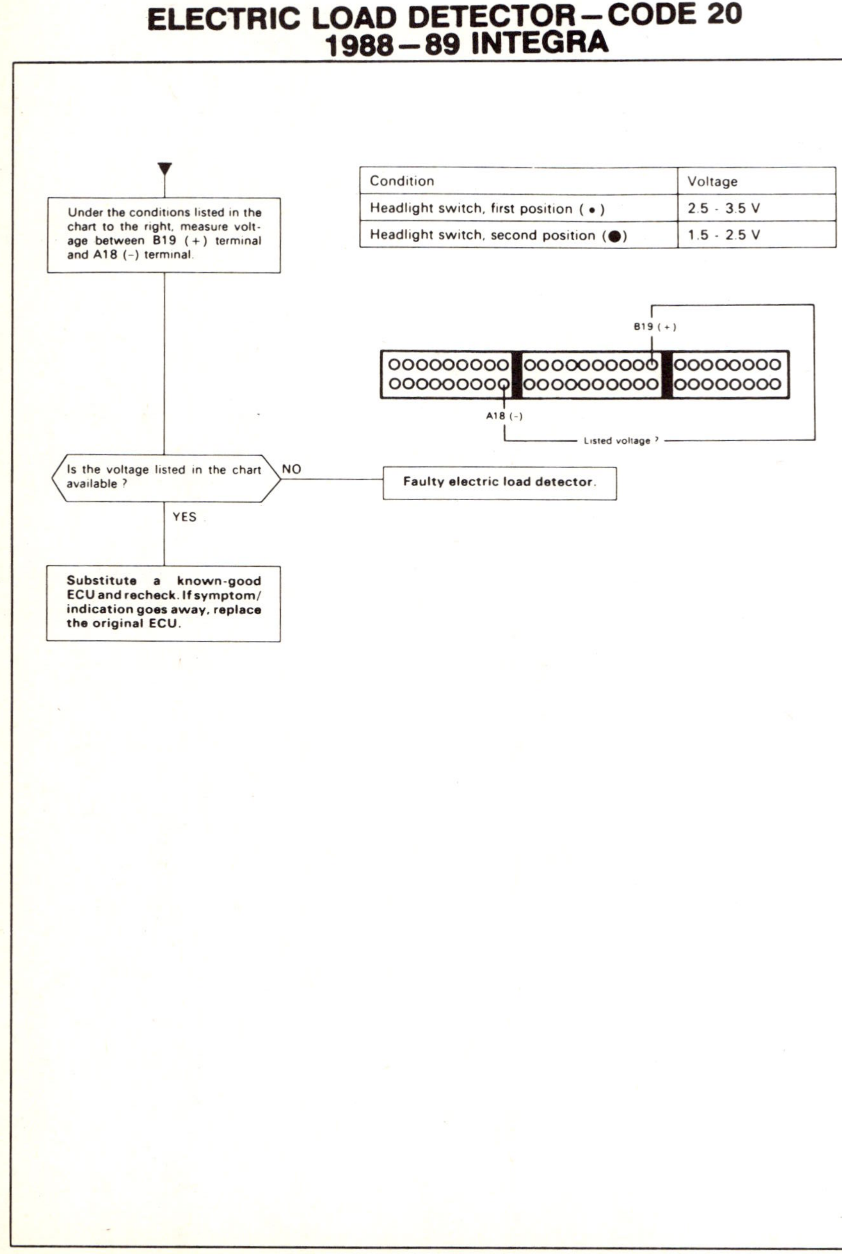


| Condition | Voltage |
|---|---|
| Headlight switch, first position ( • ) | 2.5 - 3.5 V |
| Headlight switch, second position (●) | 1.5 - 2.5 V |

## EACV—CODE 14—1988–89 INTEGRA

Self-diagnosis LED indicator blinks fourteen times. A problem in the Electronic Air Control Valve (EACV) circuit.

The EACV changes the amount of air bypassing the throttle body in response to a current signal from the ECU in order to maintain the proper idle speed.

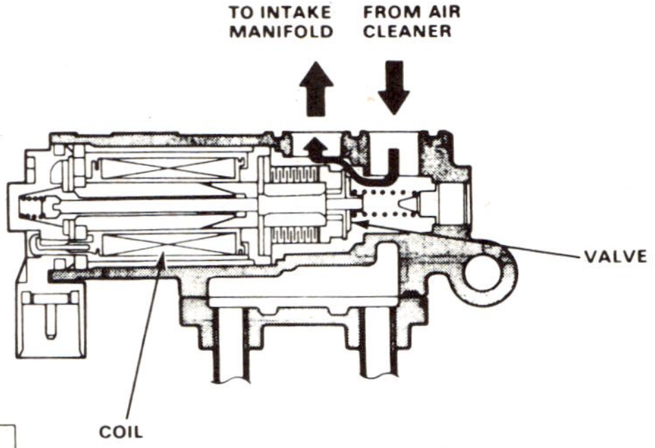

TO INTAKE MANIFOLD
FROM AIR CLEANER
VALVE
COIL

- Engine is running.
- Check Engine warning light is on.
- LED indicates CODE 14.

Turn the ignition switch OFF.

Remove HAZARD fuse at the battery positive terminal for 10 seconds to reset ECU.

Start engine

Is Check Engine warning light on and does LED indicate CODE 14?

NO → Intermittent failure, system is OK at this time (test driving may be necessary). Check for poor connections or loose wires at EACV and C78.

YES

Stop engine

Disconnect the 2P connector from the EACV.

Measure resistance between the 2 terminals on the EACV.

EACV

## EACV—CODE 14—1988–89 INTEGRA

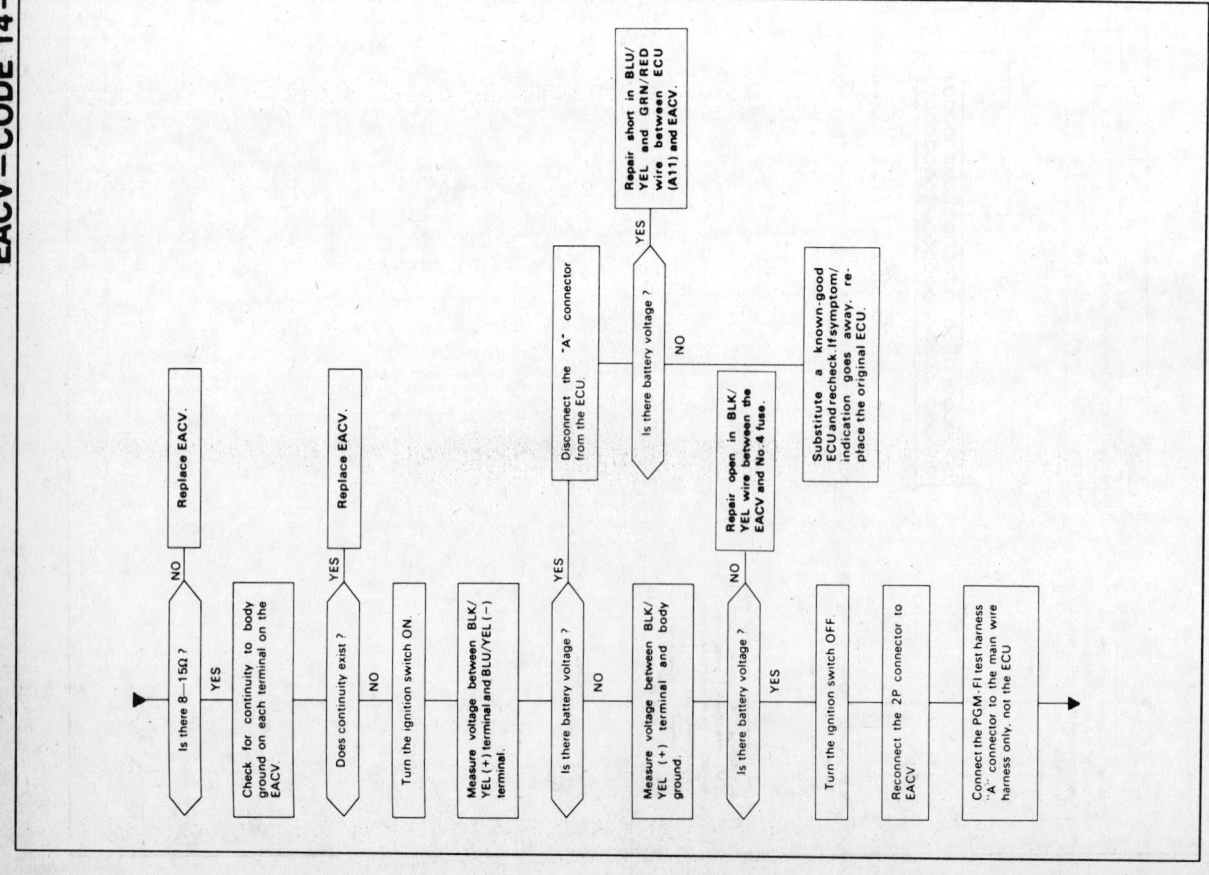

Is there 8—15Ω ?
- NO → Replace EACV.
- YES → Check for continuity to body ground on each terminal on the EACV.

Does continuity exist ?
- YES → Replace EACV.
- NO → Turn the ignition switch ON.

Measure voltage between BLK/YEL (+) terminal and BLU/YEL (−) terminal.

Is there battery voltage ?
- YES → Disconnect the "A" connector from the ECU.
  - Is there battery voltage ?
    - YES → Repair short in BLU/YEL and GRN/RED wire between ECU (A11) and EACV.
    - NO → Substitute a known-good ECU and recheck. If symptom/indication goes away, replace the original ECU.
- NO → Measure voltage between BLK/YEL (+) terminal and body ground.

Is there battery voltage ?
- NO → Repair open in BLK/YEL wire between the EACV and No.4 fuse.
- YES → Turn the ignition switch OFF.

Reconnect the 2P connector to EACV.

Connect the PGM-FI test harness "A" connector to the main wire harness only, not the ECU.

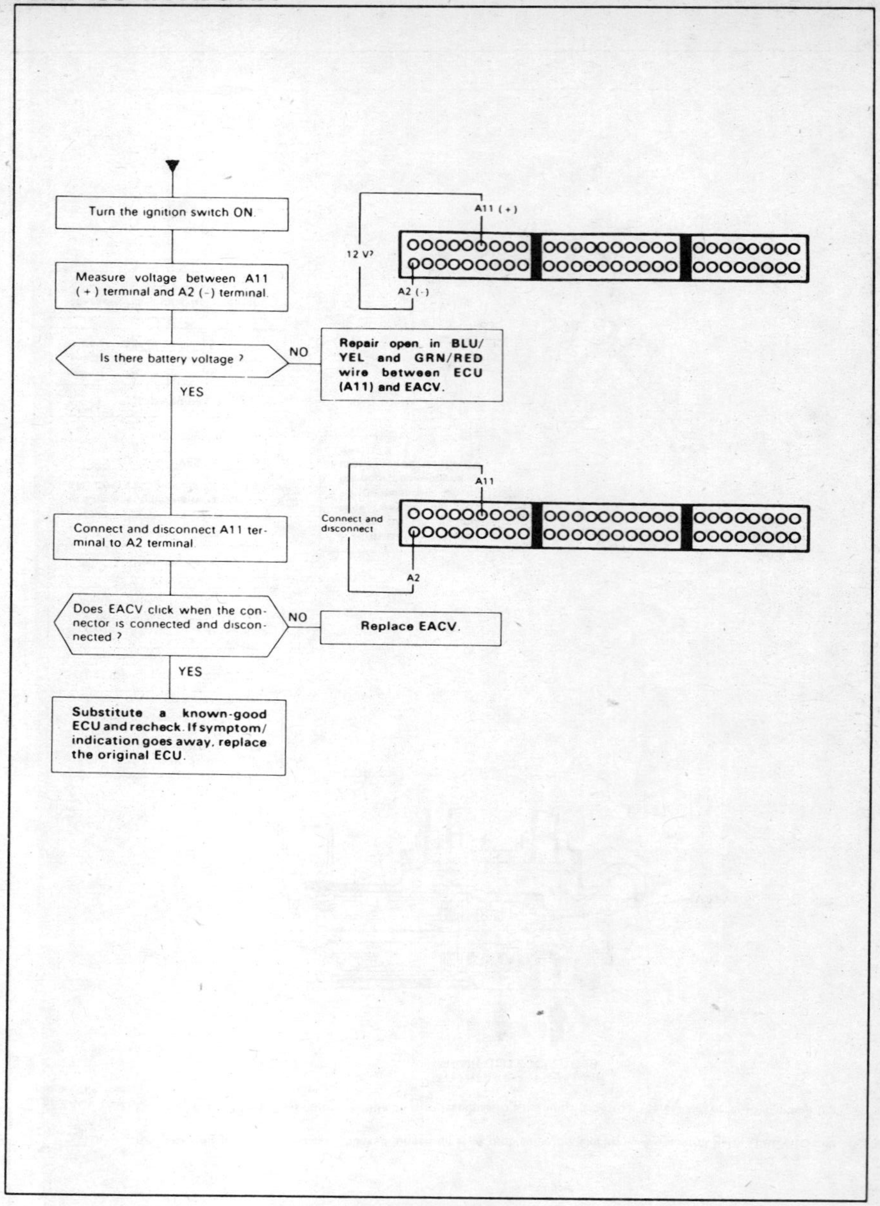

---

Turn the ignition switch ON.

Measure voltage between A11 (+) terminal and A2 (−) terminal.

12 V?
A11 (+)
A2 (−)

Is there battery voltage ?
- NO → Repair open in BLU/YEL and GRN/RED wire between ECU (A11) and EACV.
- YES → Connect and disconnect A11 terminal to A2 terminal.

Connect and disconnect
A11
A2

Does EACV click when the connector is connected and disconnected ?
- NO → Replace EACV.
- YES → Substitute a known-good ECU and recheck. If symptom/indication goes away, replace the original ECU.

# AIR CONDITIONING SIGNAL—1988–89 INTEGRA

This signals the PGM-FI ECU when there is a demand for cooling from the air conditioning circuits

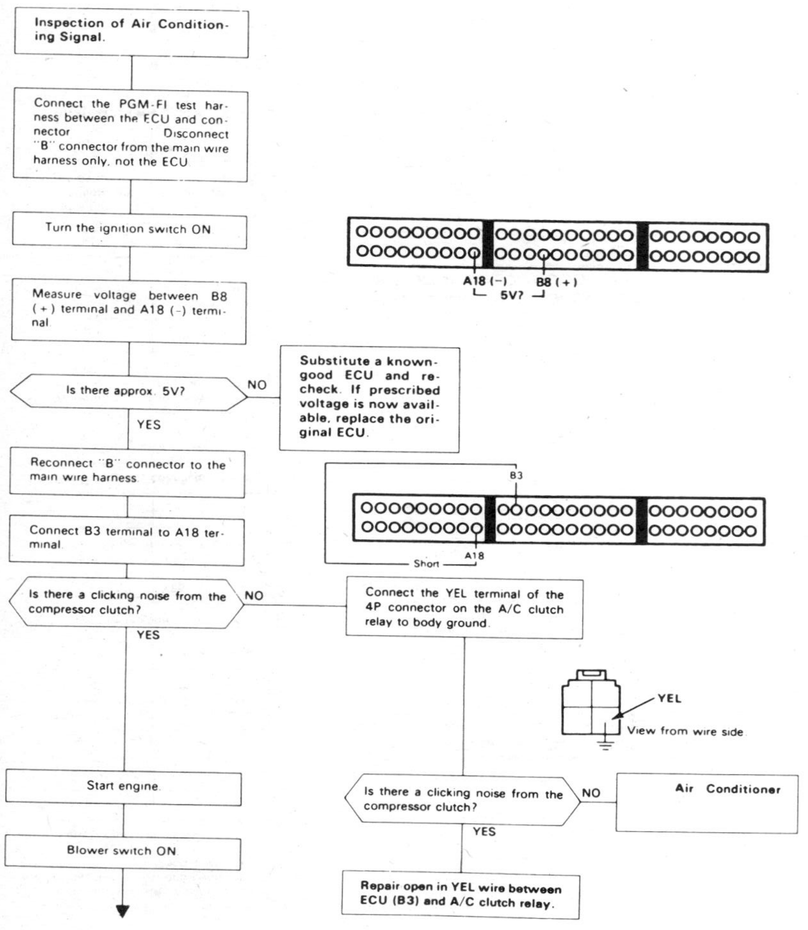

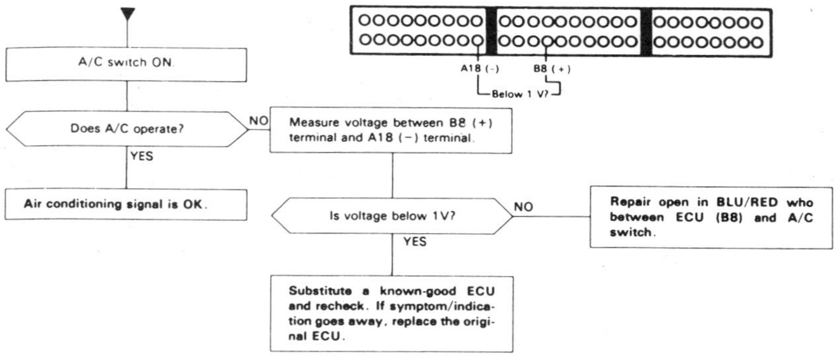

**Inspection of Air Conditioning Signal.**

Connect the PGM-FI test harness between the ECU and connector. Disconnect "B" connector from the main wire harness only, not the ECU.

Turn the ignition switch ON.

Measure voltage between B8 (+) terminal and A18 (-) terminal.

A18 (-) B8 (+) 5V?

Is there approx. 5V?

- NO: **Substitute a known-good ECU and recheck. If prescribed voltage is now available, replace the original ECU.**
- YES: Reconnect "B" connector to the main wire harness.

Connect B3 terminal to A18 terminal.

B3 — Short — A18

Is there a clicking noise from the compressor clutch?

- NO: Connect the YEL terminal of the 4P connector on the A/C clutch relay to body ground.

  YEL — View from wire side

  Is there a clicking noise from the compressor clutch?
  - NO: **Air Conditioner**
  - YES: **Repair open in YEL wire between ECU (B3) and A/C clutch relay.**
- YES: Start engine.

Blower switch ON.

A/C switch ON.

Does A/C operate?

- YES: **Air conditioning signal is OK.**
- NO: Measure voltage between B8 (+) terminal and A18 (−) terminal.

  A18 (-) B8 (+) Below 1 V?

  Is voltage below 1V?
  - NO: **Repair open in BLU/RED who between ECU (B8) and A/C switch.**
  - YES: **Substitute a known-good ECU and recheck. If symptom/indication goes away, replace the original ECU.**

## ALTERNATOR FR SIGNAL—1988–89 INTEGRA

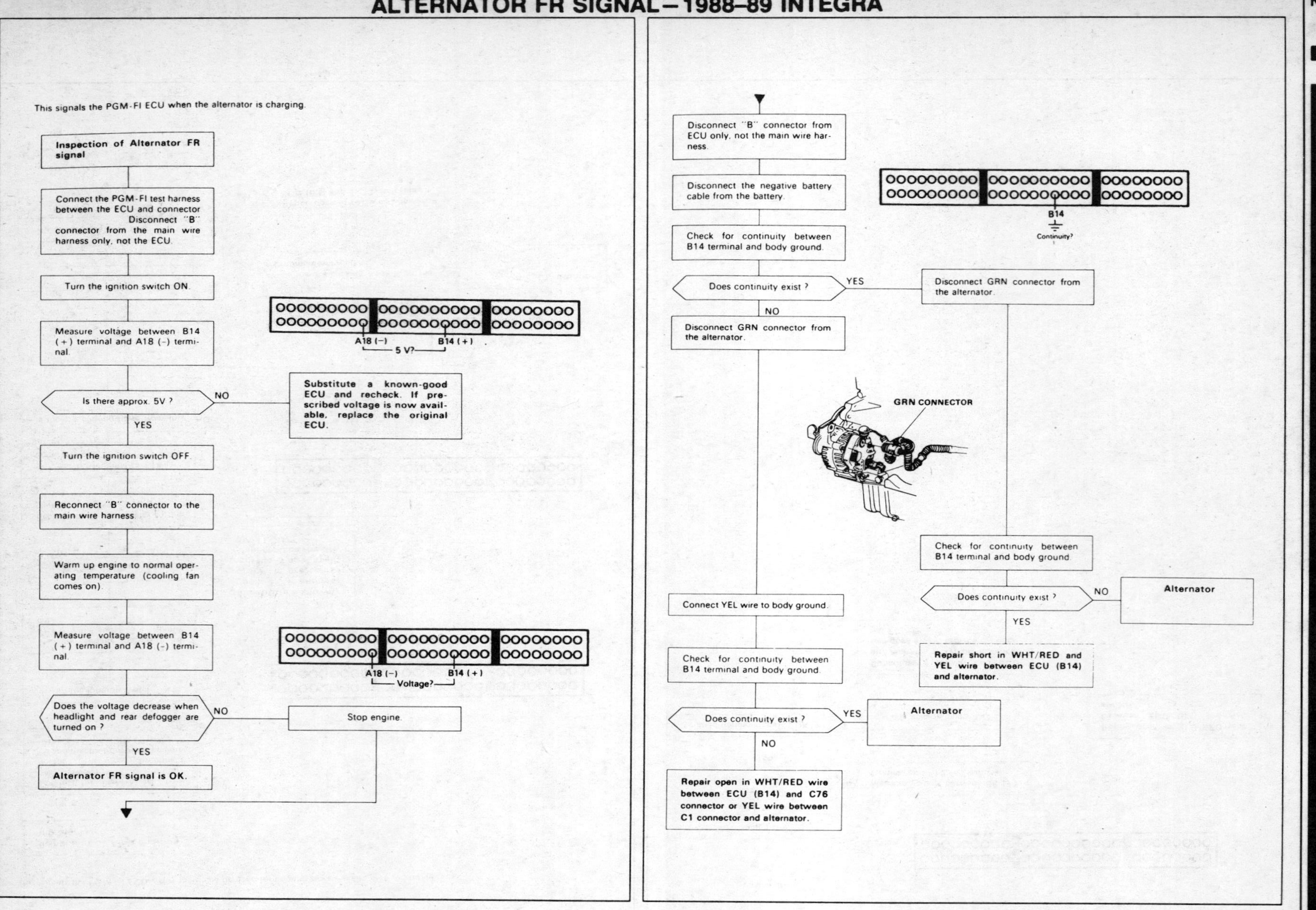

This signals the PGM-FI ECU when the alternator is charging.
Inspection of Alternator FR signal
Connect the PGM-FI test harness between the ECU and connector Disconnect "B" connector from the main wire harness only, not the ECU.
Turn the ignition switch ON.
Measure voltage between B14 (+) terminal and A18 (-) terminal.
A18 (–)
5 V?
B14 (+)
Is there approx. 5V ?
NO
Substitute a known-good ECU and recheck. If prescribed voltage is now available, replace the original ECU.
YES
Turn the ignition switch OFF.
Reconnect "B" connector to the main wire harness.
Warm up engine to normal operating temperature (cooling fan comes on).
Measure voltage between B14 (+) terminal and A18 (-) terminal.
A18 (–)
Voltage?
B14 (+)
Does the voltage decrease when headlight and rear defogger are turned on ?
NO
Stop engine.
YES
Alternator FR signal is OK.
Disconnect "B" connector from ECU only, not the main wire harness.
Disconnect the negative battery cable from the battery.
B14
Continuity?
Check for continuity between B14 terminal and body ground.
Does continuity exist ?
YES
Disconnect GRN connector from the alternator.
NO
Disconnect GRN connector from the alternator.
GRN CONNECTOR
Check for continuity between B14 terminal and body ground.
Does continuity exist ?
NO
Alternator
YES
Repair short in WHT/RED and YEL wire between ECU (B14) and alternator.
Connect YEL wire to body ground.
Check for continuity between B14 terminal and body ground.
Does continuity exist ?
YES
Alternator
NO
Repair open in WHT/RED wire between ECU (B14) and C76 connector or YEL wire between C1 connector and alternator.

## A/T SHIFT POSITION SIGNAL—1988–89 INTEGRA

This signals the PGM-FI ECU when the transmission is in Neutral, Park, D3 or D4

**Inspection of A/T Shift Position Signal**

Connect the PGM-FI test harness between the ECU and connector. Disconnect "B" connector from the main wire harness only, not the ECU.

Turn the ignition switch ON.

Measure voltage individually between B7 (+), B9 (+) B11 (+) terminals and A18 (−) terminal.

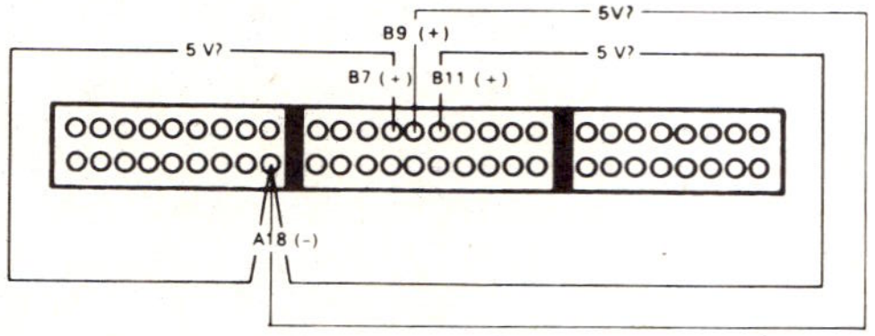

Is there approx. 5 V?

- NO → **Substitute a known-good ECU and recheck. If prescribed voltage is now available, replace the original ECU.**
- YES ↓



Reconnect "B" connector to the main wire harness.

Measure voltage between B7 (+) terminal and A18 (−) terminal in Neutral position.

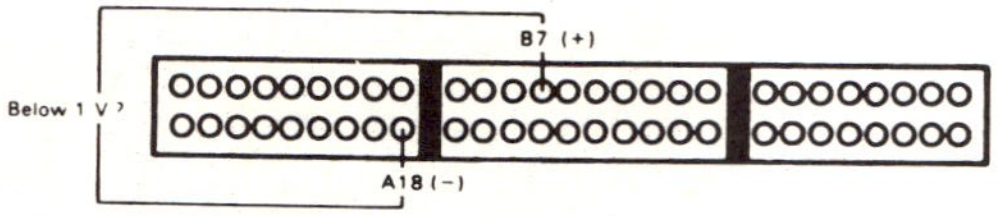

Is voltage below 1V?

- NO → **—Repair open in LT GRN wire between ECU (B7) and shift position console switch.**
  **—See shift position console switch inspection**
- YES ↓

Measure voltage between B7 (+) terminal and A18 (−) terminal in Park position.

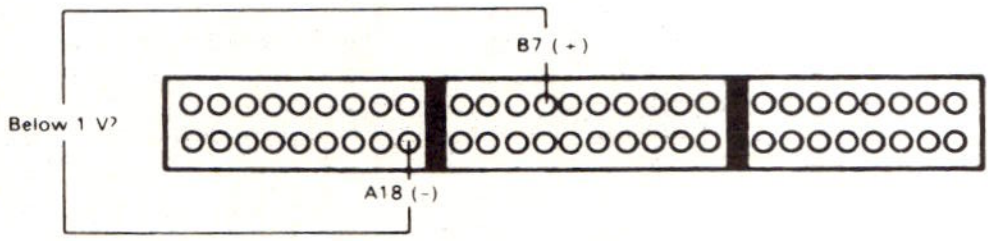

## A/T SHIFT POSITION SIGNAL—1988–89 INTEGRA

Is voltage below 1V?

- NO → **See shift position console switch inspection**
- YES ↓

Measure voltage between B9 (+) terminal and A18 (−) terminal in D3 position.

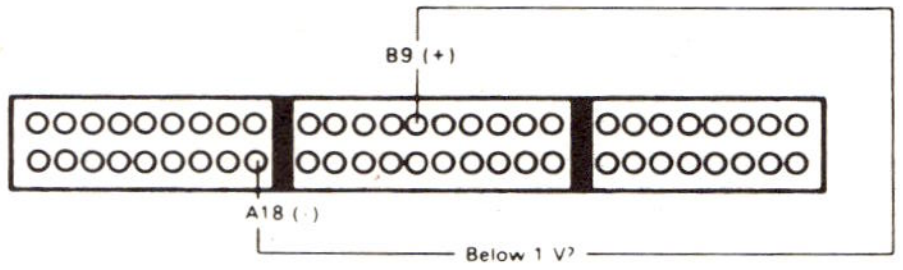

Is voltage below 1V?

- NO → **—Repair open in GRN/BLU wire between ECU (B9) and shift position console switch.**
  **— shift position console switch inspection**
- YES ↓

Measure voltage between B11 (+) terminal and A18 (−) terminal in D4 position.

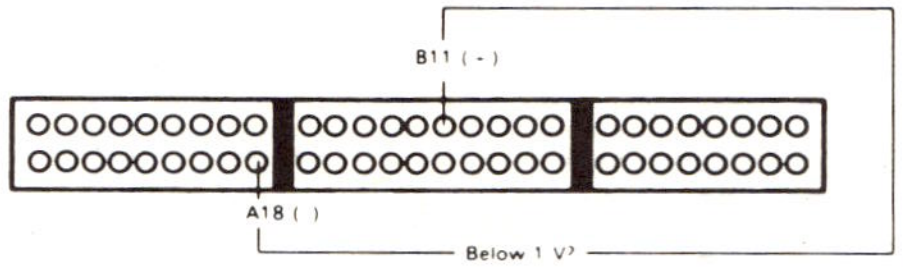

Is voltage below 1V?

- NO → **—Repair open in GRN/BLK wire between ECU (B11) and shift position console switch.**
  **— shift position console switch inspection**
- YES ↓

A/T shift position signal is OK.

## BRAKE SWITCH SIGNAL – 1988–89 INTEGRA

This signal the PGM-FI ECU when the brake pedal is depressed.

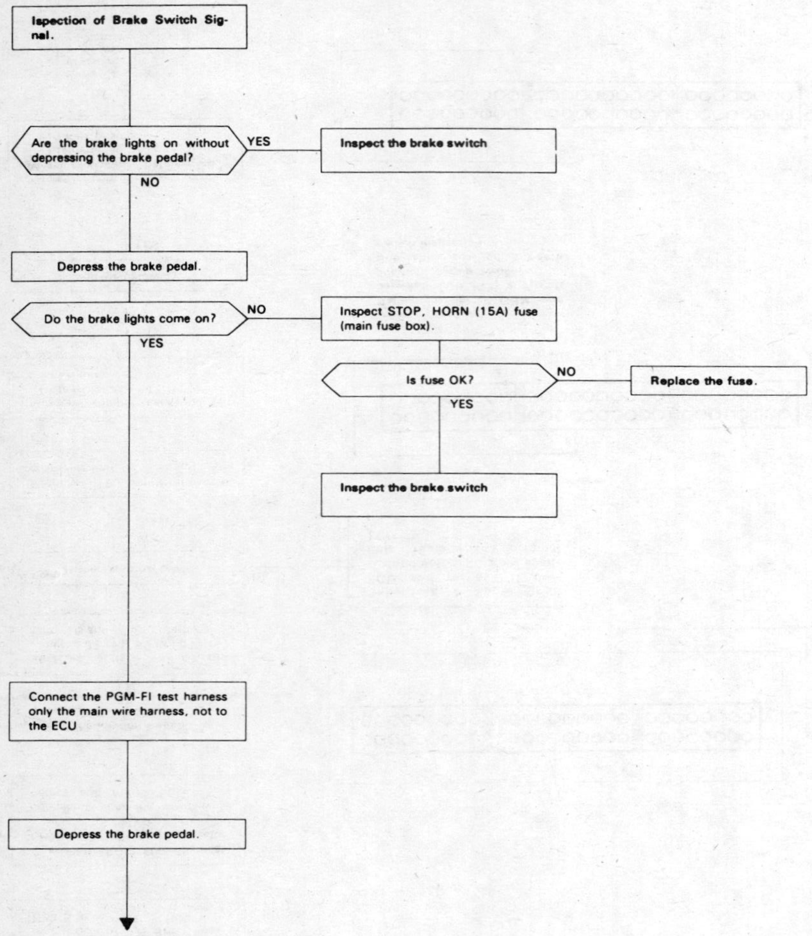

Ispection of Brake Switch Signal.

Are the brake lights on without depressing the brake pedal? — YES: Inspect the brake switch

NO:

Depress the brake pedal.

Do the brake lights come on? — NO: Inspect STOP, HORN (15A) fuse (main fuse box).

Is fuse OK? — NO: Replace the fuse. — YES: Inspect the brake switch

YES:

Connect the PGM-FI test harness only the main wire harness, not to the ECU

Depress the brake pedal.

## BRAKE SWITCH SIGNAL – 1988–89 INTEGRA

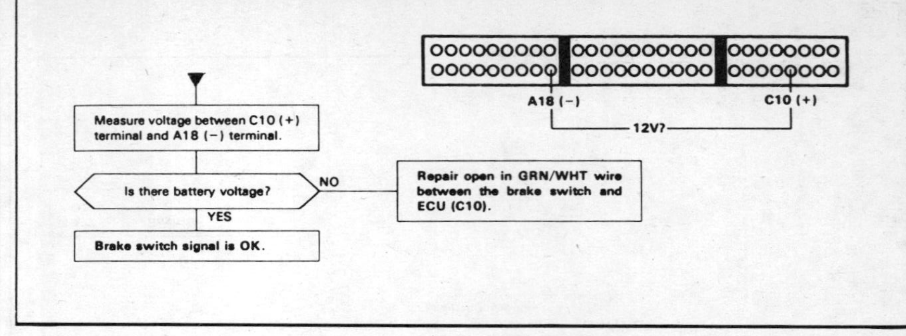


## STARTER SWITCH SIGNAL – 1988–89 INTEGRA

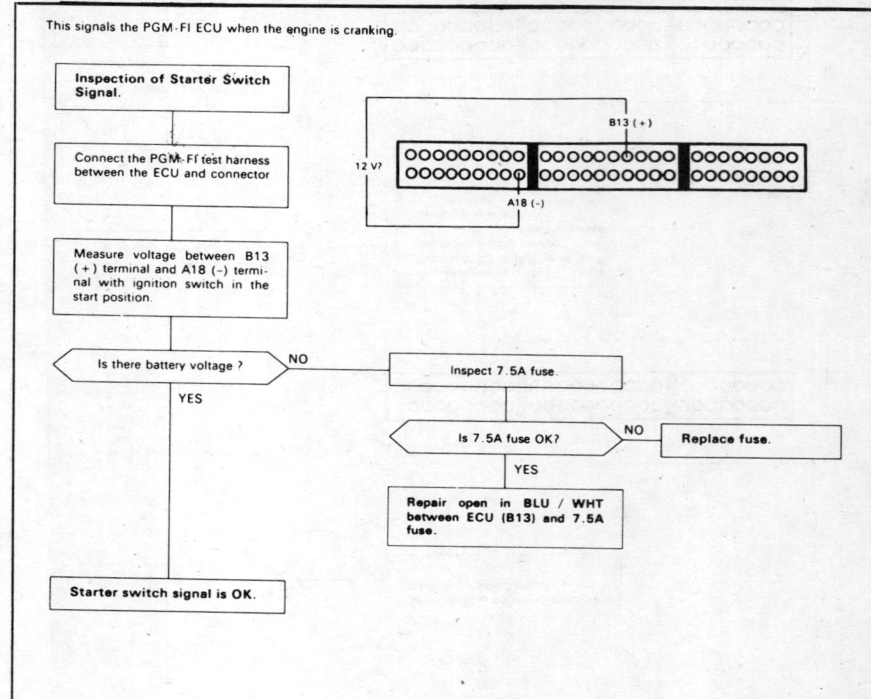

## FAST IDLE CONTROL—1988–89 INTEGRA

## FUEL INJECTORS—CODE 16—1988–89 INTEGRA

Self-diagnosis LED indicator blinks sixteen times: A problem in the fuel injector circuit.

The injectors are the solenoid-actuated constant-stroke pintle type consisting of a solenoid, plunger needle valve and housing. When current is applied to the solenoid coil, the valve lifts up and pressurized fuel is injected close to the intake valve. Because the needle valve lift and the fuel pressure are constant, the injection quantity is determined by the length of time that the valve is open (i.e., the duration the current is supplied to the solenoid coil). The injector is sealed by an O-ring and seal ring at the top and bottom. These seals also reduce operating noise.

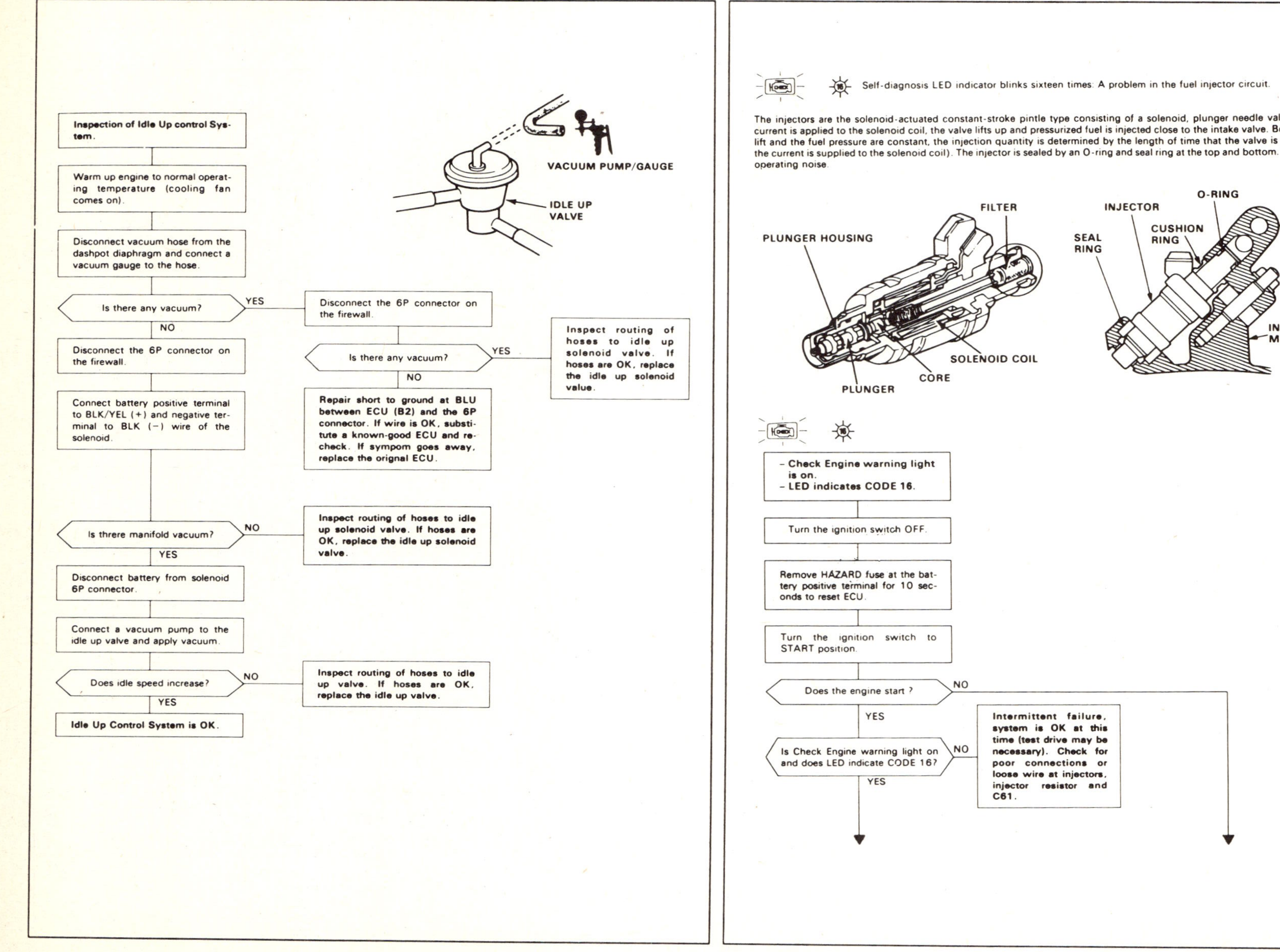

Inspection of Idle Up control System.
Warm up engine to normal operating temperature (cooling fan comes on).
Disconnect vacuum hose from the dashpot diaphragm and connect a vacuum gauge to the hose.
Is there any vacuum?
YES
Disconnect the 6P connector on the firewall.
Is there any vacuum?
YES
Inspect routing of hoses to idle up solenoid valve. If hoses are OK, replace the idle up solenoid value.
NO
Repair short to ground at BLU between ECU (B2) and the 6P connector. If wire is OK, substitute a known-good ECU and recheck. If sympom goes away, replace the orignal ECU.
NO
Disconnect the 6P connector on the firewall.
Connect battery positive terminal to BLK/YEL (+) and negative terminal to BLK (−) wire of the solenoid.
Is threre manifold vacuum?
NO
Inspect routing of hoses to idle up solenoid valve. If hoses are OK, replace the idle up solenoid valve.
YES
Disconnect battery from solenoid 6P connector.
Connect a vacuum pump to the idle up valve and apply vacuum.
Does idle speed increase?
NO
Inspect routing of hoses to idle up valve. If hoses are OK, replace the idle up valve.
YES
Idle Up Control System is OK.
VACUUM PUMP/GAUGE
IDLE UP VALVE
FILTER
PLUNGER HOUSING
SOLENOID COIL
CORE
PLUNGER
INJECTOR
O-RING
FUEL PIPE
CUSHION RING
SEAL RING
INTAKE MANIFOLD
– Check Engine warning light is on.
– LED indicates CODE 16.
Turn the ignition switch OFF.
Remove HAZARD fuse at the battery positive terminal for 10 seconds to reset ECU.
Turn the ignition switch to START position.
Does the engine start ?
NO
YES
Is Check Engine warning light on and does LED indicate CODE 16?
NO
Intermittent failure, system is OK at this time (test drive may be necessary). Check for poor connections or loose wire at injectors, injector resistor and C61.
YES

## FUEL INJECTORS—CODE 16—1988–89 INTEGRA

Check the clicking sound of each injector by means of a stethoscope when the engine is idling.

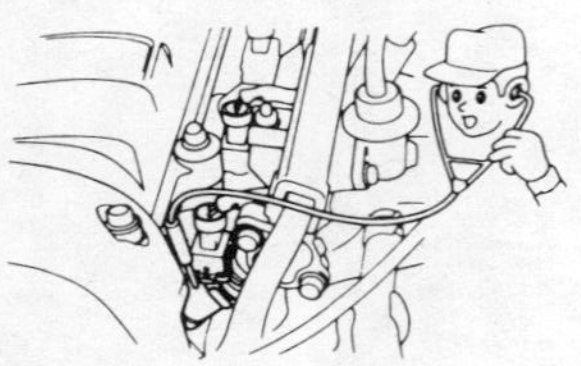

Do the injectors click ?

- YES → Substitute a known-good ECU and recheck. If symptom/indication goes away, replace the original ECU.
- NO ↓

Turn the ignition switch OFF.

Disconnect the 2P connector from the injector that does not click.

Measure resistance between the 2 terminals of the injector.

Is there 1.5—2.5Ω ?

- NO → Replace the injector.
- YES ↓

Turn the ignition switch ON.

(Right branch:)

Turn the ignition switch OFF.

Disconnect the 2P connector from each injector.

Measure resistance between the 2 terminals of the injector.

INJECTOR

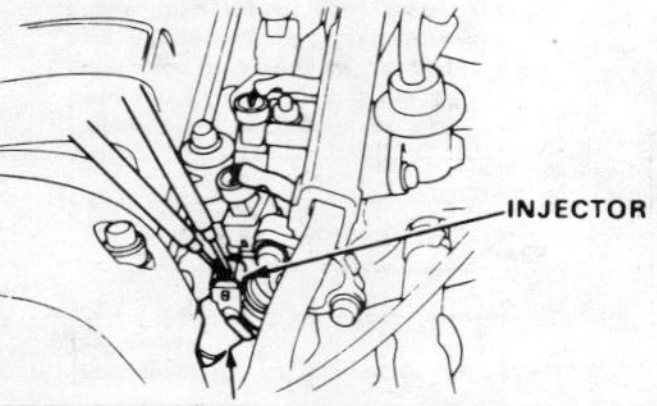

(cont'd)

Measure voltage between RED/BLK (+) terminal on the 2P connector and body ground.

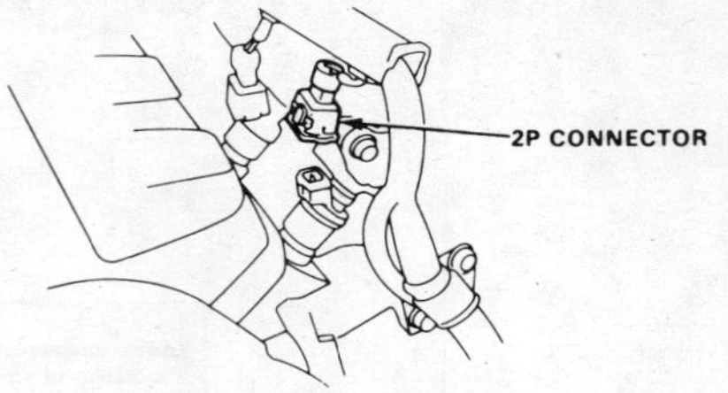

2P CONNECTOR

Is there battery voltage ?

- NO → Turn the ignition switch OFF. → Disconnect 6P connector on the injector resistor. → Turn the ignition switch ON. → Measure voltage between YEL/BLK (+) terminal and body ground. → Is there battery voltage ?
  - NO → Repair open in the YEL/BLK wire between the injector resistor and the main relay.
  - YES → —Replace the injector resistor. —Repair open in RED/BLK wire between 2P connector and resistor.
- YES ↓

Measure voltage between the following terminals,

- No.1 injector: RED/BLK (+) terminal and BRN (−) terminal.
- No.2 injector: RED/BLK (+) terminal and RED (−) terminal.
- No.3 injector: RED/BLK (+) terminal and LT BLU (−) terminal.
- No.4 injector: RED/BLK (+) terminal and YEL (−) terminal.

Is there battery voltage ?

- YES → Disconnect "A" connector from the ECU. → Is there battery voltage ?
  - YES → Repair short in the wire between the ECU (A1, A3, A5 or A7) and the injector.
  - NO → Substitute a known-good ECU and recheck. If prescribed voltage is now available, replace the original ECU.
- NO ↓

Reconnect the 2P connector to the injector.

Turn the ignition switch OFF.

Connect the PGM-FI test harness between the ECU and connector

Turn the ignition switch ON.

Measure voltage between A2 (−) terminal and following terminals,

- No.1 injector: A1 (+) terminal.
- No.2 injector: A3 (+) terminal.
- No.3 injector: A5 (+) terminal.
- No.4 injector: A7 (+) terminal.

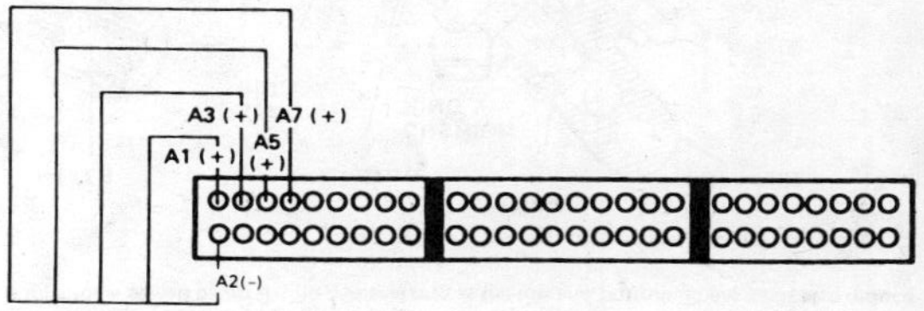

A3 (+) A7 (+) A5 (+) A1 (+) A2(−)

## FUEL INJECTORS—CODE 16—1988–89 INTEGRA

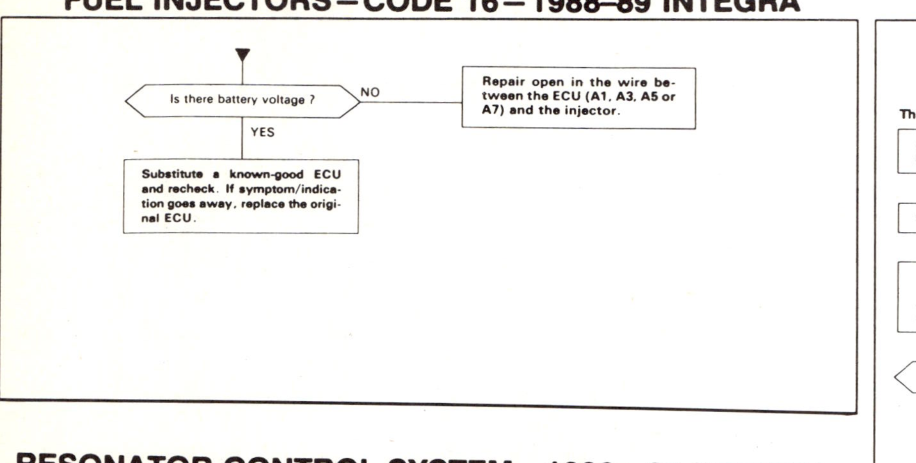


## RESONATOR CONTROL SYSTEM—1988–89 INTEGRA

**Description**
The resonator control system decreases air intake noise.

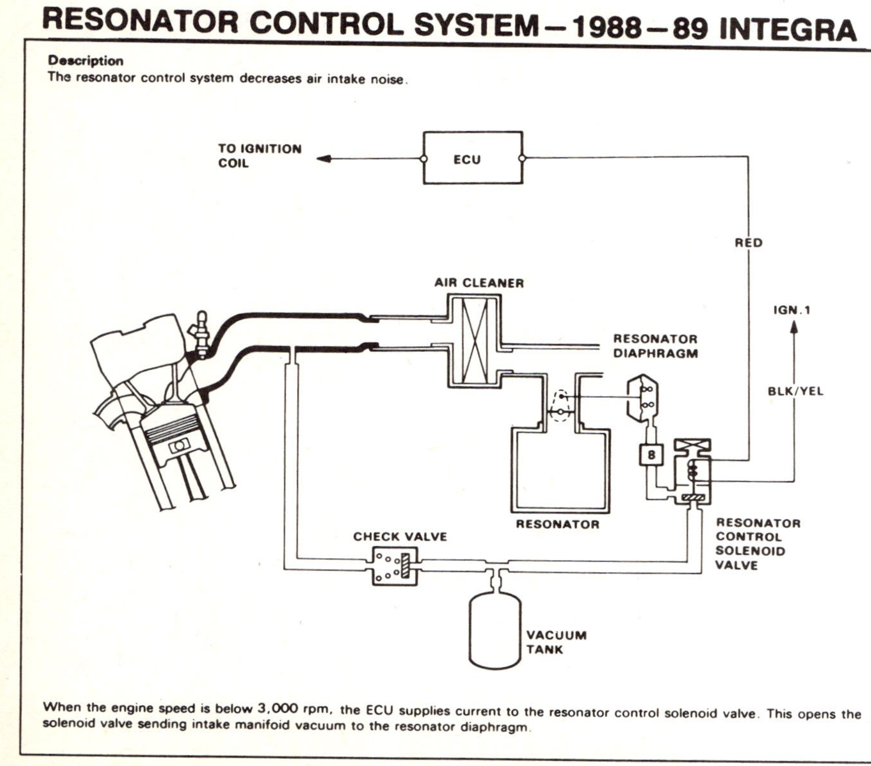


When the engine speed is below 3,000 rpm, the ECU supplies current to the resonator control solenoid valve. This opens the solenoid valve sending intake manifold vacuum to the resonator diaphragm.

## RESONATOR CONTROL SYSTEM 1988–89 INTEGRA

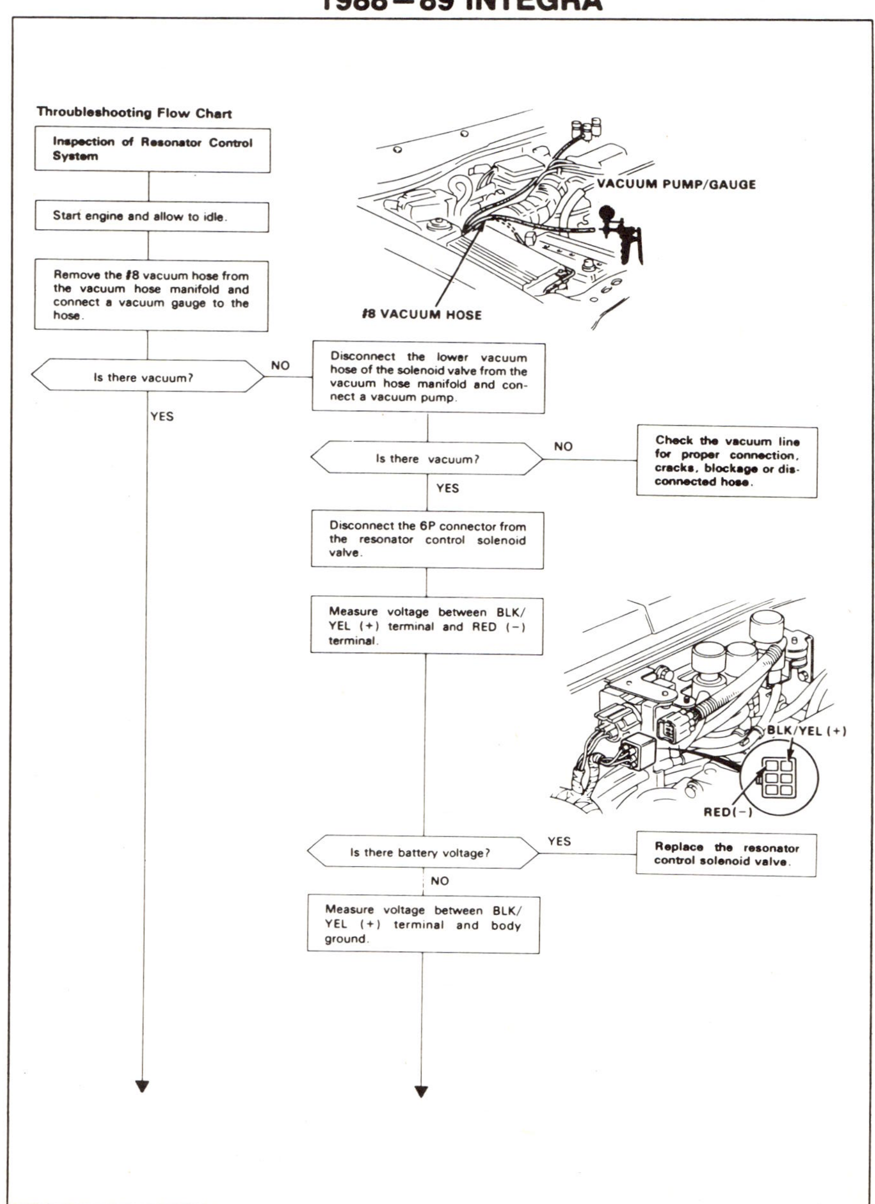

Troubleshooting Flow Chart
Inspection of Resonator Control System
Start engine and allow to idle.
Remove the #8 vacuum hose from the vacuum hose manifold and connect a vacuum gauge to the hose.
VACUUM PUMP/GAUGE
#8 VACUUM HOSE
Is there vacuum?
NO
Disconnect the lower vacuum hose of the solenoid valve from the vacuum hose manifold and connect a vacuum pump.
YES
Is there vacuum?
NO
Check the vacuum line for proper connection, cracks, blockage or disconnected hose.
YES
Disconnect the 6P connector from the resonator control solenoid valve.
Measure voltage between BLK/YEL (+) terminal and RED (−) terminal.
BLK/YEL (+)
RED(−)
Is there battery voltage?
YES
Replace the resonator control solenoid valve.
NO
Measure voltage between BLK/YEL (+) terminal and body ground.

# RESONATOR CONTROL SYSTEM – 1988 – 89 INTEGRA

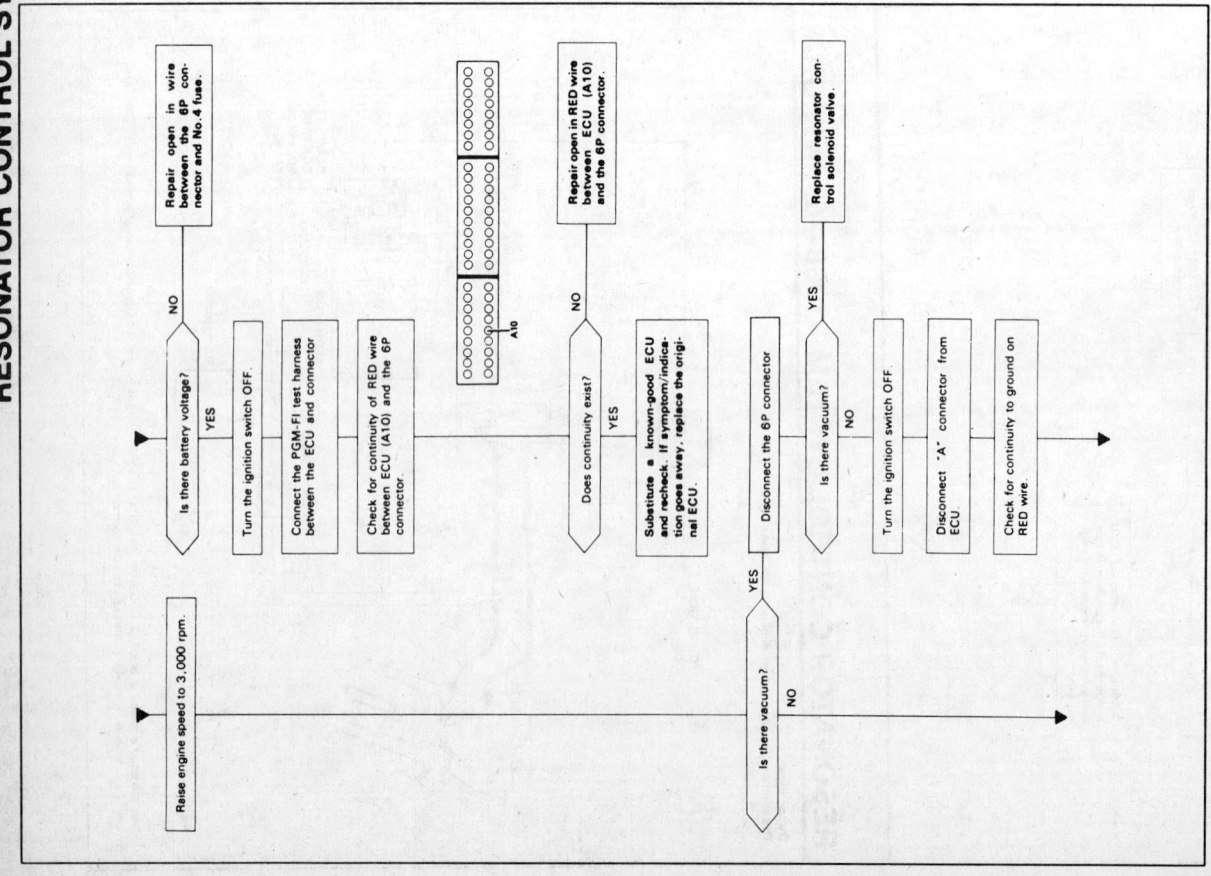

Raise engine speed to 3,000 rpm.

Is there vacuum?
- NO → (continue)
- YES → Disconnect the 6P connector

Is there vacuum?
- YES → Replace resonator control solenoid valve.
- NO → Turn the ignition switch OFF.

Disconnect "A" connector from ECU.

Check for continuity to ground on RED wire.

---

Is there battery voltage?
- NO → Repair open in wire between the 6P connector and No.4 fuse.
- YES → Turn the ignition switch OFF.

Connect the PGM-FI test harness between the ECU and connector

Check for continuity of RED wire between ECU (A10) and the 6P connector.



Does continuity exist?
- NO → Repair open in RED wire between ECU (A10) and the 6P connector.
- YES → Substitute a known-good ECU and recheck. If symptom/indication goes away, replace the original ECU.

---

Connect the vacuum hose to the vacuum hose manifold and connect a vacuum pump to the hose.

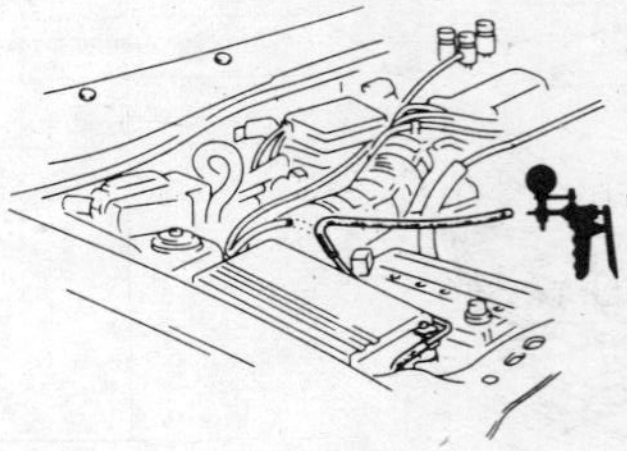

Apply vacuum.

Does it hold vacuum?
- NO → Check the vacuum line for proper connection or disconnected hose. If OK, replace the resonator diaphragm.
- YES → Resonator control system is OK.

---

Is there continuity to ground?
- YES → Repair short to ground in Red wire between ECU (A10) and the 6P connector.
- NO → Substitute a known good ECU and recheck. If symptom goes away, replace the original ECU.

# EVAPORATIVE EMISSIONS—1988–89 INTEGRA

### Description

The evaporative controls are designed to minimize the amount of fuel vapor escaping to the atmosphere. The system consists of the following components:

A. **Charcoal Canister**
A canister for the temporary storage of fuel vapor until the fuel vapor can be purged from the canister into the engine and burned.

B. **Vapor Purge Control System**
Canister purging is accomplished by drawing fresh air through the canister and into a port on the throttle body. The ported vacuum is controlled by the purge control diaphragm valve and the purge cut-off solenoid valve.

PURGE CUT-OFF SOLENOID VALVE OFF
5 seconds AFTER STARTING ENGINE

COOLANT TEMPERATURE ABOVE 60°C (140°F) AND THROTTLE VALVE NOT FULL CLOSE

C. **Fuel Tank Vapor Control System**
When fuel vapor pressure in the fuel tank is higher than the set value of the two-way valve, the valve opens and regulates the flow of fuel vapor to the caniser.

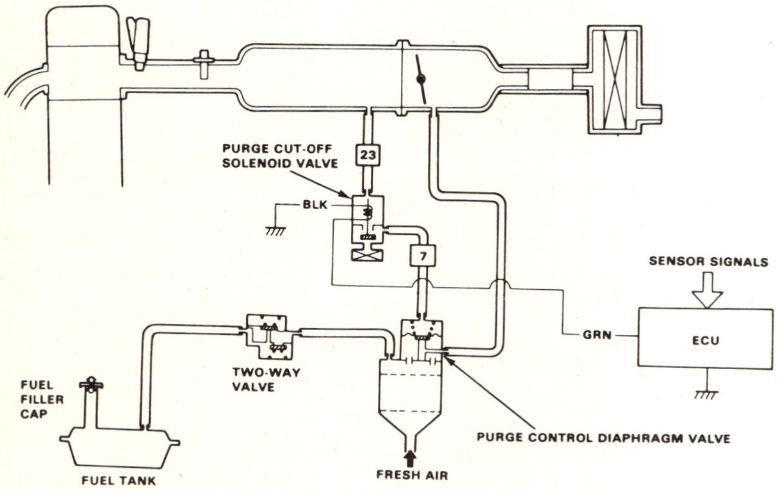

### Troubleshooting Flowchart

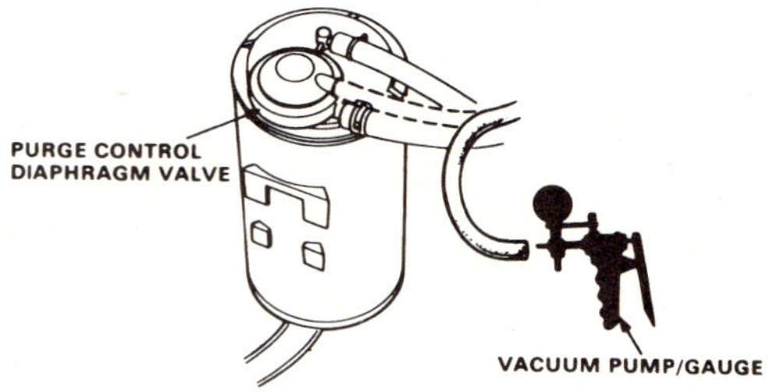

**Inspection of Evaporative Emission Controls**

Disconnect #7 hose from the purge control diaphragm valve (on the charcoal canister) and connect a vacuum gauge to the hose.

Start the engine and allow to idle. NOTE: Engine coolant temperature must be below 60°C (140°F).

Is there vacuum ?
- NO → (continued)
- YES → Disconnect the 6P connector.

Measure voltage between GRN (+) terminal and BLK (−) terminal.

Is there battery voltage ?
- YES → **Replace purge cut-off solenoid valve.**
- NO → Measure voltage between GRN (+) terminal and body ground.

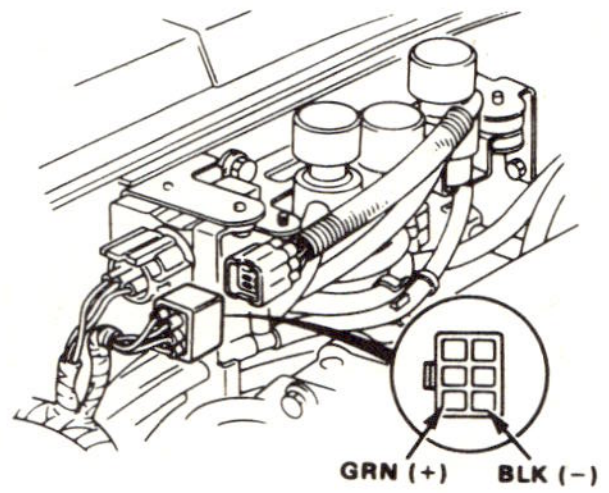

## EVAPORATIVE EMISSIONS—1988–89 INTEGRA

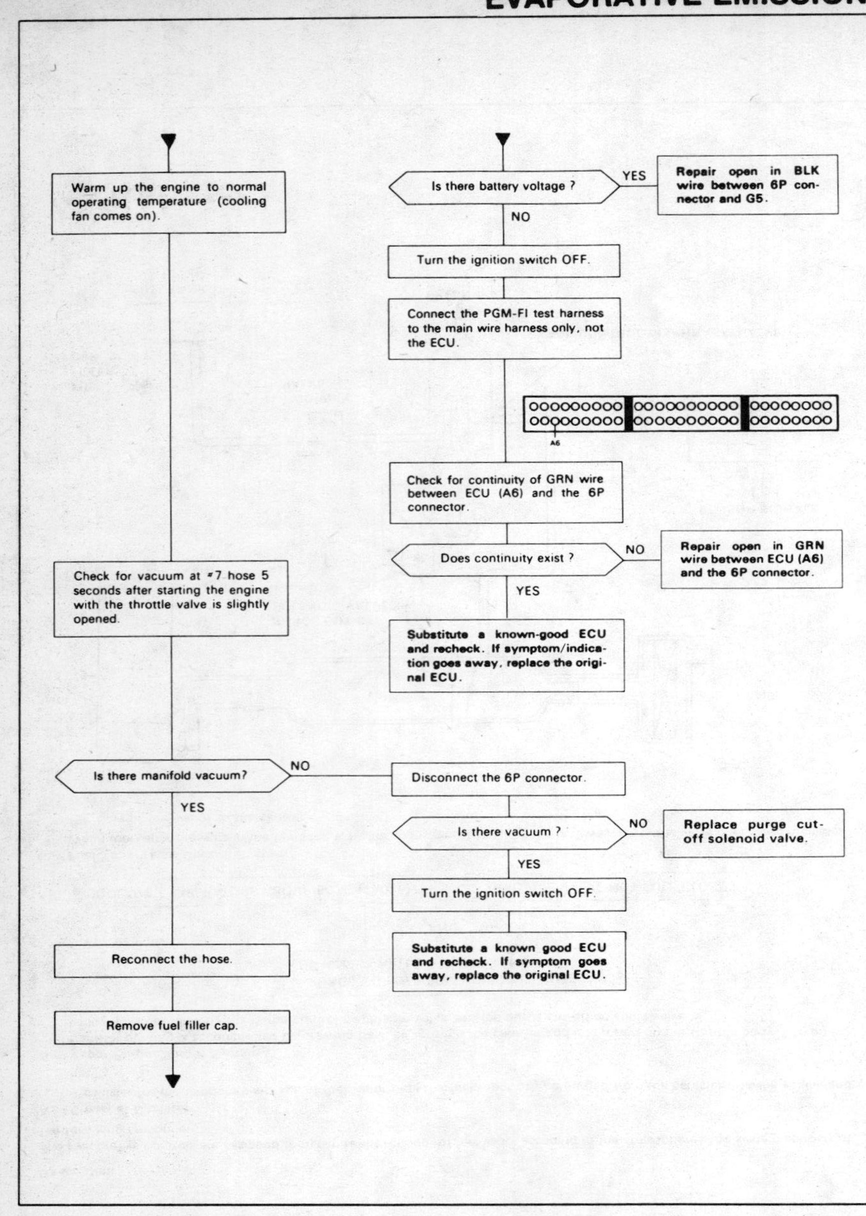


Connect a vacuum gauge to canister purge air hose.

Start the engine and rise speed to 3,500 rpm.

Does vacuum appear on gauge within 1 minute ?

NO — Replace the canister.

YES — See two-way valve to complete Evaporative.

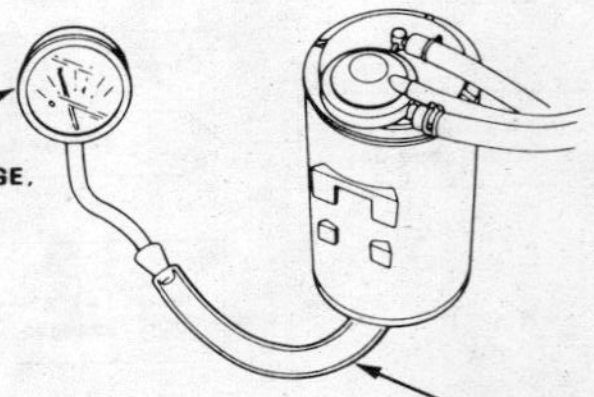

VACUUM/PRESSURE GAUGE, 0–4 in. Hg

PURGE AIR HOSE

# TROUBLESHOOTING GUIDE—1990 INTEGRA

NOTE: Across each row in the chart, the systems that could be sources of a symptom are ranked in the order they should be inspected starting with ①. Find the symptom in the left column, read across to the most likely source, then refer to the page listed at the top of that column. If inspection shows the system is OK, try the next most likely system ②, etc.

| SYMPTOM | | SYSTEM: ECU | PGM-FI: OXYGEN SENSOR | PGM-FI: MANIFOLD ABSOLUTE PRESSURE SENSOR | PGM-FI: TDC/CRANK/CYL SENSOR | PGM-FI: COOLANT TEMPERATURE SENSOR | PGM-FI: THROTTLE ANGLE SENSOR | PGM-FI: INTAKE AIR TEMPERATURE SENSOR | PGM-FI: ATMOSPHERIC PRESSURE SENSOR | PGM-FI: IGNITION OUTPUT SIGNAL | PGM-FI: VEHICLE SPEED SENSOR | IDLE CONTROL: ELECTRONIC AIR CONTROL VALVE | IDLE CONTROL: OTHER IDLE CONTROLS | FUEL SUPPLY: FUEL INJECTOR | FUEL SUPPLY: OTHER FUEL SUPPLY | AIR INTAKE | EMISSION CONTROL: EGR CONTROL SYSTEM (A/T only) | EMISSION CONTROL: OTHER EMISSION CONTROLS |
|---|---|---|---|---|---|---|---|---|---|---|---|---|---|---|---|---|---|---|
| CHECK ENGINE WARNING LIGHT TURNS ON | | ☐ or ✳ | ▣ | ▣ | ▣ | ▣ | ▣ | ▣ | ▣ | ▣ | ▣ | ▣ | | ▣ | | | ▣ | |
| SELF-DIAGNOSIS INDICATOR (LED) BLINKS | | ⓪ or ✱ | ① or 43 | ③ or ⑤ | ④ or ⑧ or ⑨ | ⑥ | ⑦ | ⑩ | ⑬ | ⑮ | ⑰ | ⑭ | | ⑯ | | | ⑫ | |
| ENGINE WON'T START | | ③ | | | | | | | | | | | | ② | ① | | | |
| DIFFICULT TO START ENGINE WHEN COLD | | BU | | ③ | | ① | | | | | | | ② | | | | | |
| IRREGULAR IDLING | WHEN COLD FAST IDLE OUT OF SPEC | BU | | | | ③ | | | | | | ① | ② | | | | | |
| | ROUGH IDLE | BU | | ③ | | | | | | | | ① | | ② | | | ③ | |
| | WHEN WARM RPM TOO HIGH | BU | | | | | | | | | | ② | ① | | | | | |
| | WHEN WARM RPM TOO LOW | BU | | | | | | | | | | ① | | ② | | | | |
| FREQUENT STALLING | WHILE WARMING UP | BU | | | | | | | | | | ① | ② | | ③ | | | |
| | AFTER WARMING UP | BU | | | | | | | | | | ① | ② | | ③ | | ③ | |
| POOR PERFORMANCE | MISFIRE OR ROUGH RUNNING | BU | | | | | | | | | | ② | | ① | | | ③ | |
| | FAILS EMISSION TEST | BU | ③ | ② | | | | | | | | | | ① | | | | |
| | LOSS OF POWER | BU | | ③ | | | ② | | | | | | | | ① | | | |

* If codes other than those listed above are indicated, count the number of blinks again. If the indicator is in fact blinking these codes, substitute a known-good ECU and recheck. If the indication goes away, replace the original ECU.

BU: When the Check Engine warning light and the self-diagnosis indicator are on, the back-up system is in operation. Substitute a known-good ECU and recheck. If the indication goes away, replace the original ECU.

## ECU AND CHECK LIGHT – 1990 INTEGRA

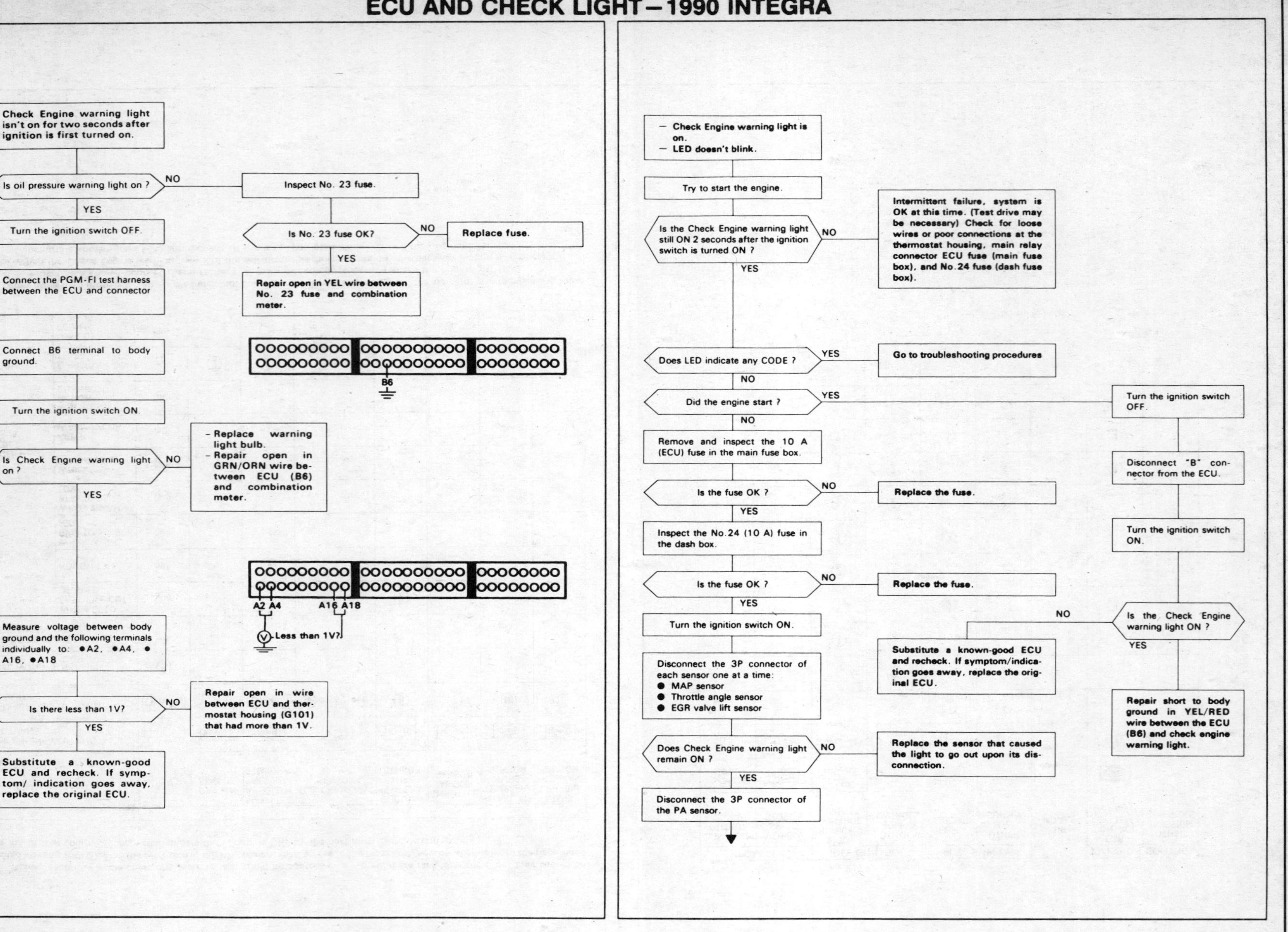

## ECU AND CHECK LIGHT—CODE 13—1990 INTEGRA

## OXYGEN SENSOR—CODE 1—1990 INTEGRA

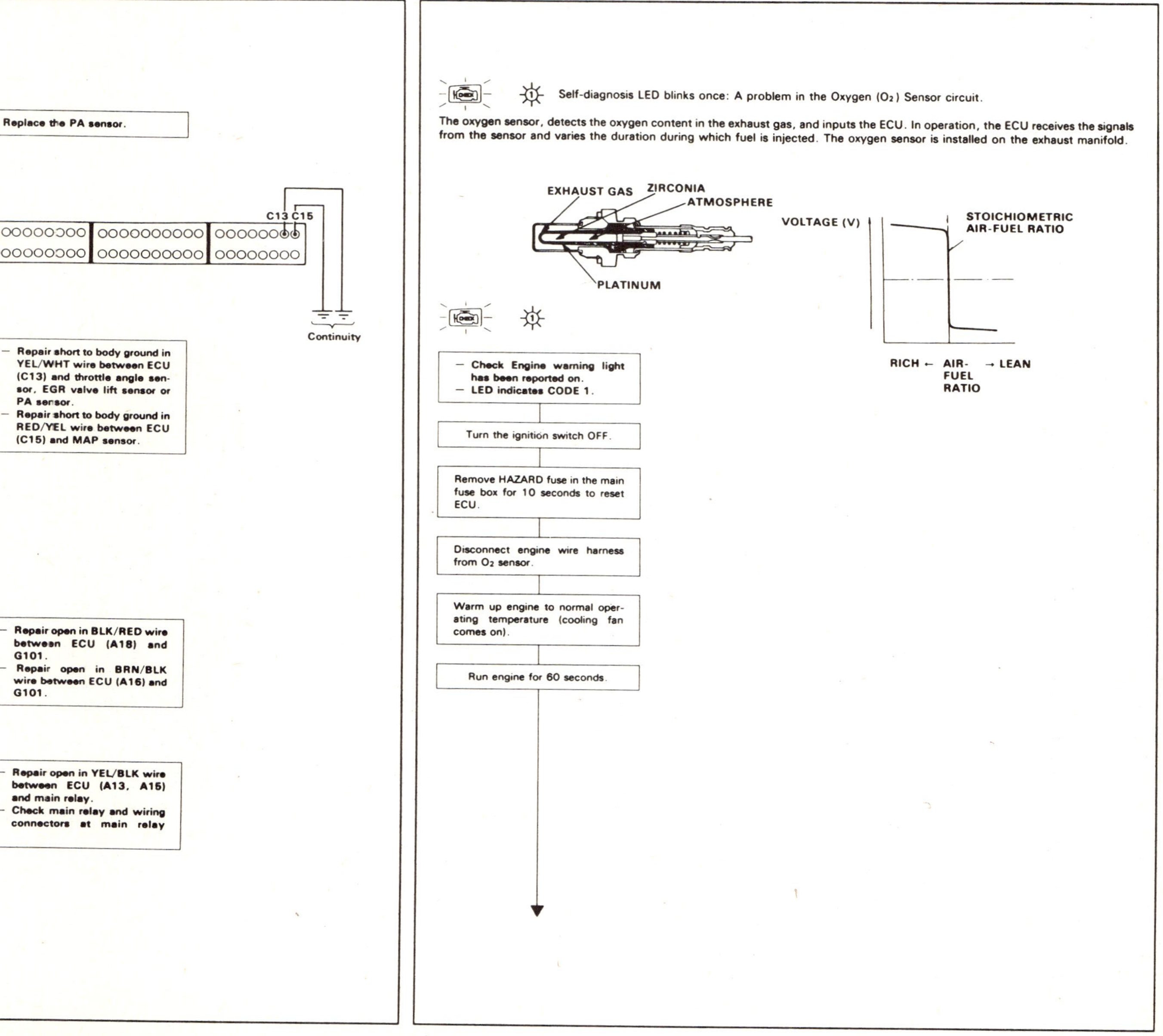

Does LED indicate CODE 13 ?
YES
Replace the PA sensor.
NO
Turn the ignition switch OFF.
Connect the PGM-FI test harness But disconnect the "C" connector from the ECU only, not the main wire harness.
C13 C15
Continuity
Check for continuity between body ground and the following terminals: C13, C15.
Does continuity exist ?
YES
– Repair short to body ground in YEL/WHT wire between ECU (C13) and throttle angle sensor, EGR valve lift sensor or PA sensor.
– Repair short to body ground in RED/YEL wire between ECU (C15) and MAP sensor.
NO
Reconnect all the connectors. Reconnect the "C" connector to the ECU.
Turn the ignition switch ON.
Individually connect the following terminal to Body Ground. A16 · A18
Is the Check Engine warning light still ON after 2 seconds ?
NO
– Repair open in BLK/RED wire between ECU (A18) and G101.
– Repair open in BRN/BLK wire between ECU (A16) and G101.
YES
Measure voltage between A18 (−) and the following: A13 (+) and A15 (+).
Is there battery voltage ?
NO
– Repair open in YEL/BLK wire between ECU (A13, A15) and main relay.
– Check main relay and wiring connectors at main relay
YES
Substitute a known-good ECU and recheck. If symptom/indication goes away, replace the original ECU.
Self-diagnosis LED blinks once: A problem in the Oxygen ($O_2$) Sensor circuit.
The oxygen sensor, detects the oxygen content in the exhaust gas, and inputs the ECU. In operation, the ECU receives the signals from the sensor and varies the duration during which fuel is injected. The oxygen sensor is installed on the exhaust manifold.
EXHAUST GAS
ZIRCONIA
ATMOSPHERE
PLATINUM
VOLTAGE (V)
STOICHIOMETRIC AIR-FUEL RATIO
RICH ← AIR-FUEL RATIO → LEAN
– Check Engine warning light has been reported on.
– LED indicates CODE 1.
Turn the ignition switch OFF.
Remove HAZARD fuse in the main fuse box for 10 seconds to reset ECU.
Disconnect engine wire harness from $O_2$ sensor.
Warm up engine to normal operating temperature (cooling fan comes on).
Run engine for 60 seconds.

## OXYGEN SENSOR – CODE 1 – 1990 INTEGRA

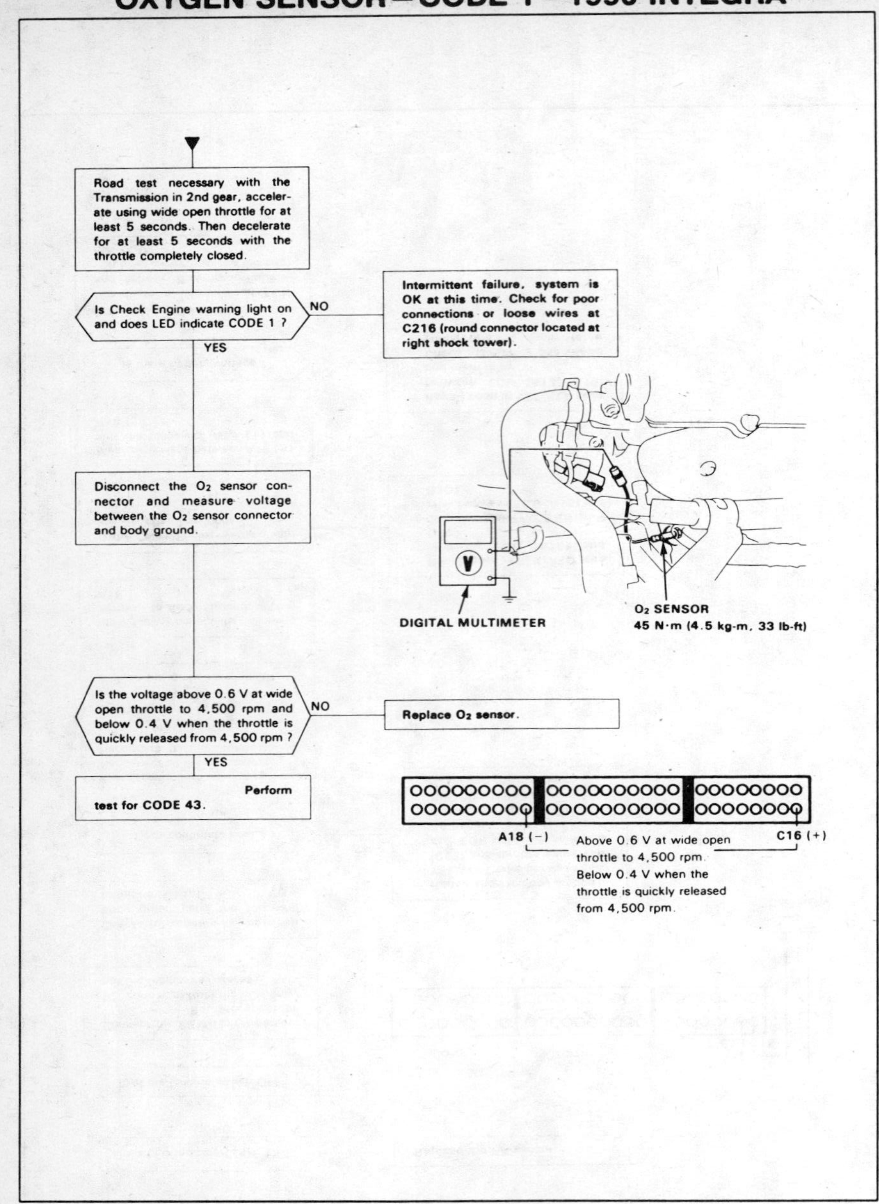

Road test necessary with the Transmission in 2nd gear, accelerate using wide open throttle for at least 5 seconds. Then decelerate for at least 5 seconds with the throttle completely closed.
Is Check Engine warning light on and does LED indicate CODE 1 ?
NO
Intermittent failure, system is OK at this time. Check for poor connections or loose wires at C216 (round connector located at right shock tower).
YES
Disconnect the O2 sensor connector and measure voltage between the O2 sensor connector and body ground.
DIGITAL MULTIMETER
O2 SENSOR
45 N·m (4.5 kg-m, 33 lb-ft)
Is the voltage above 0.6 V at wide open throttle to 4,500 rpm and below 0.4 V when the throttle is quickly released from 4,500 rpm ?
NO
Replace O2 sensor.
YES
Perform test for CODE 43.
A18 (–)
C16 (+)
Above 0.6 V at wide open throttle to 4,500 rpm.
Below 0.4 V when the throttle is quickly released from 4,500 rpm.


## FUEL SUPPLY SYSTEM – CODE 43 – 1990 INTEGRA

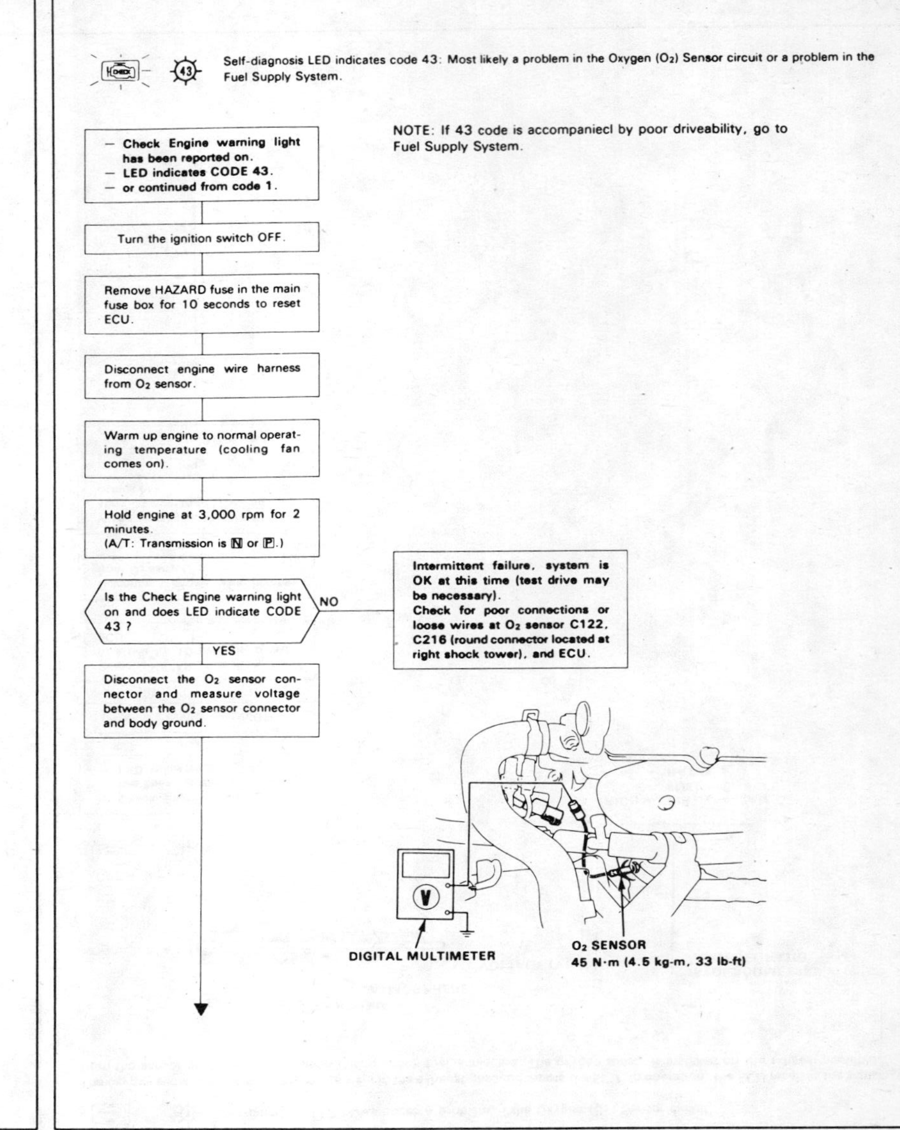

Self-diagnosis LED indicates code 43: Most likely a problem in the Oxygen (O2) Sensor circuit or a problem in the Fuel Supply System.
NOTE: If 43 code is accompaniecl by poor driveability, go to Fuel Supply System.
– Check Engine warning light has been reported on.
– LED indicates CODE 43.
– or continued from code 1.
Turn the ignition switch OFF.
Remove HAZARD fuse in the main fuse box for 10 seconds to reset ECU.
Disconnect engine wire harness from O2 sensor.
Warm up engine to normal operating temperature (cooling fan comes on).
Hold engine at 3,000 rpm for 2 minutes.
(A/T: Transmission is N or P.)
Is the Check Engine warning light on and does LED indicate CODE 43 ?
NO
Intermittent failure, system is OK at this time (test drive may be necessary).
Check for poor connections or loose wires at O2 sensor C122, C216 (round connector located at right shock tower), and ECU.
YES
Disconnect the O2 sensor connector and measure voltage between the O2 sensor connector and body ground.
DIGITAL MULTIMETER
O2 SENSOR
45 N·m (4.5 kg-m, 33 lb-ft)

## FUEL SUPPLY SYSTEM – CODE 43 – 1990 INTEGRA

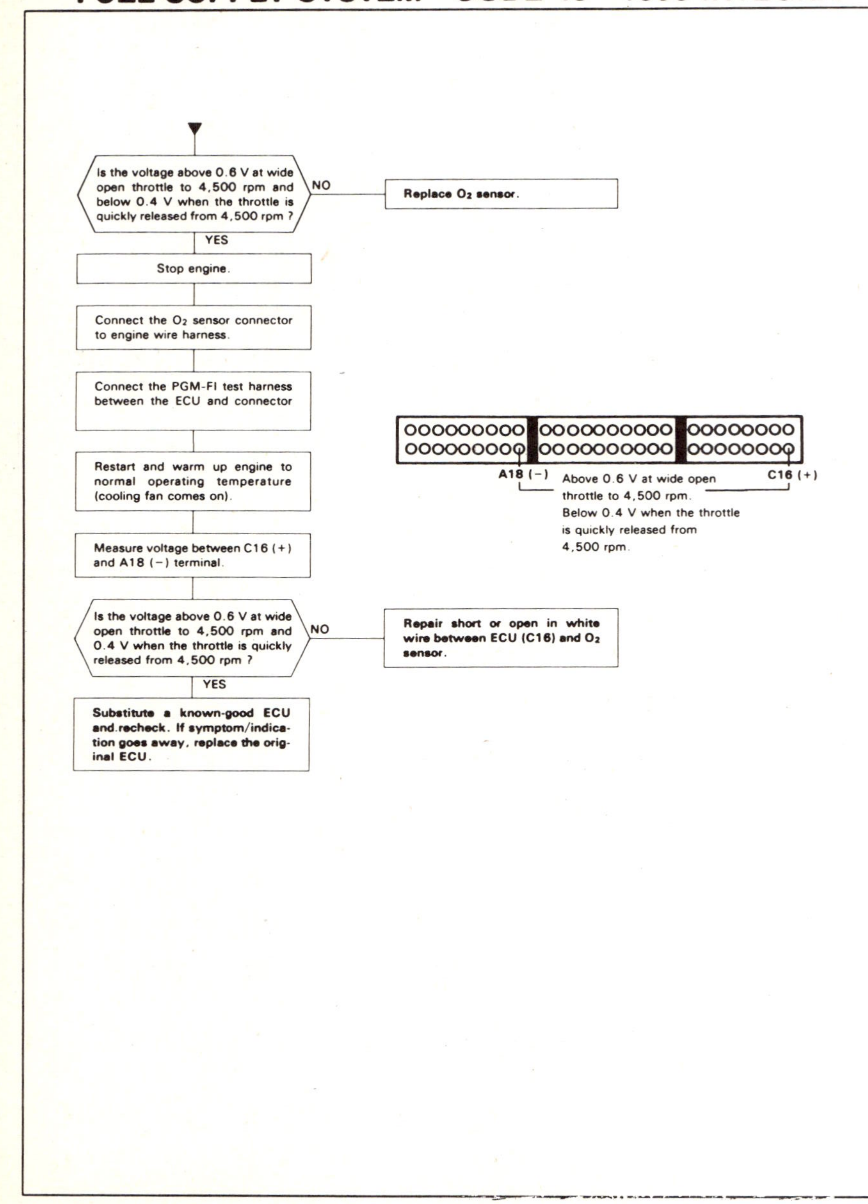


## MAP SENSOR – CODE 3 – 1990 INTEGRA

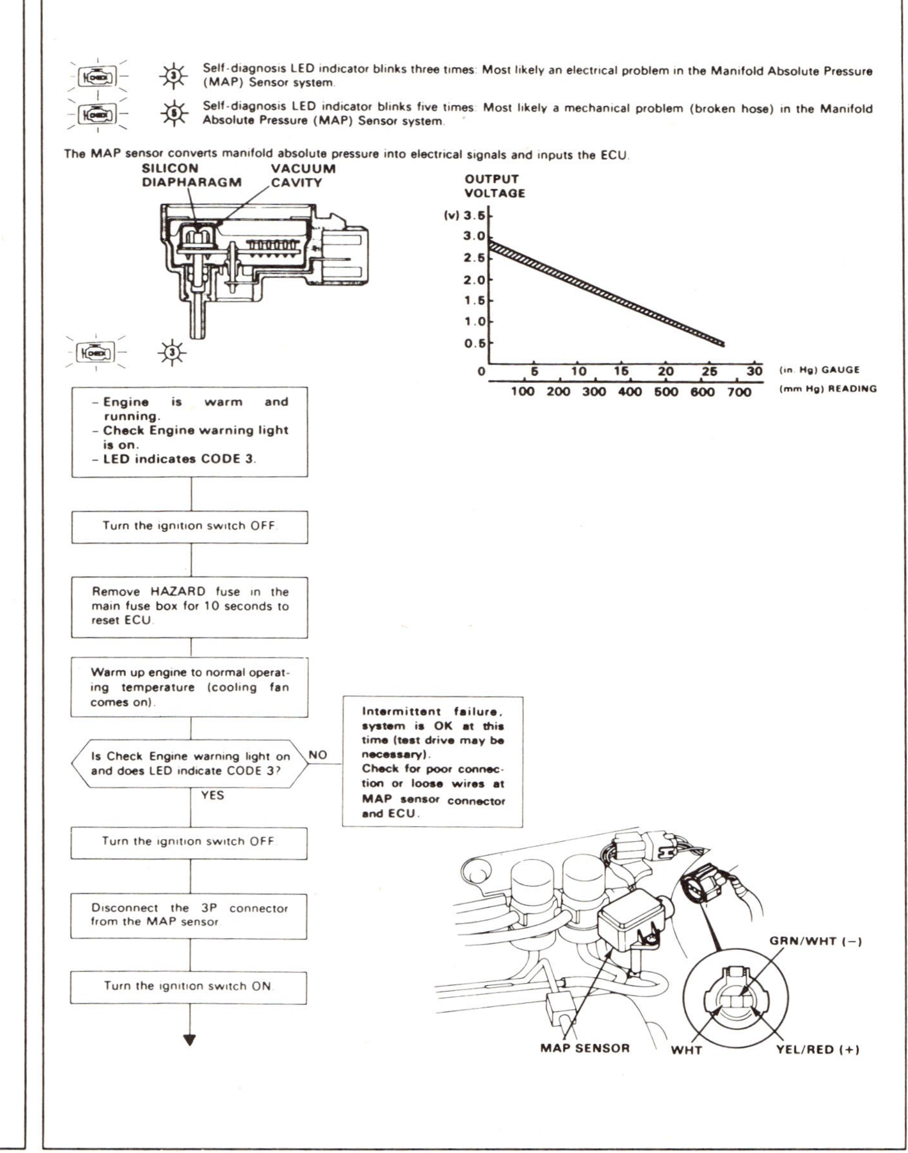

## MAP SENSOR – CODE 3 – 1990 INTEGRA

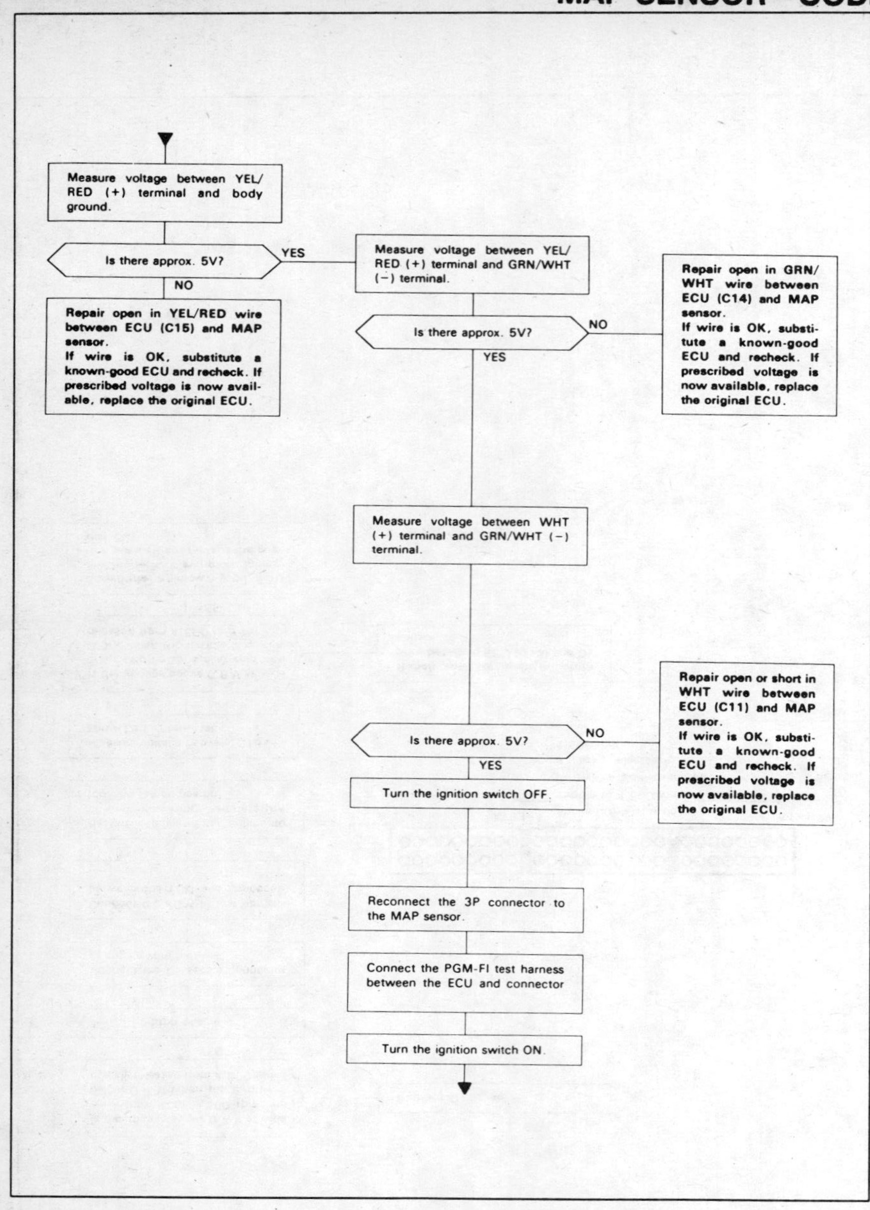


Measure voltage between C11 (+) terminal and C14 (–) terminal.

C11 (+)

3 V?

C14 (–)

Is there approx. 3 V ?

NO — Replace MAP sensor.

YES — Substitute a known-good ECU and recheck. If symptom/indication goes away, replace the original ECU.

# MAP SENSOR—CODE 5—1990 INTEGRA

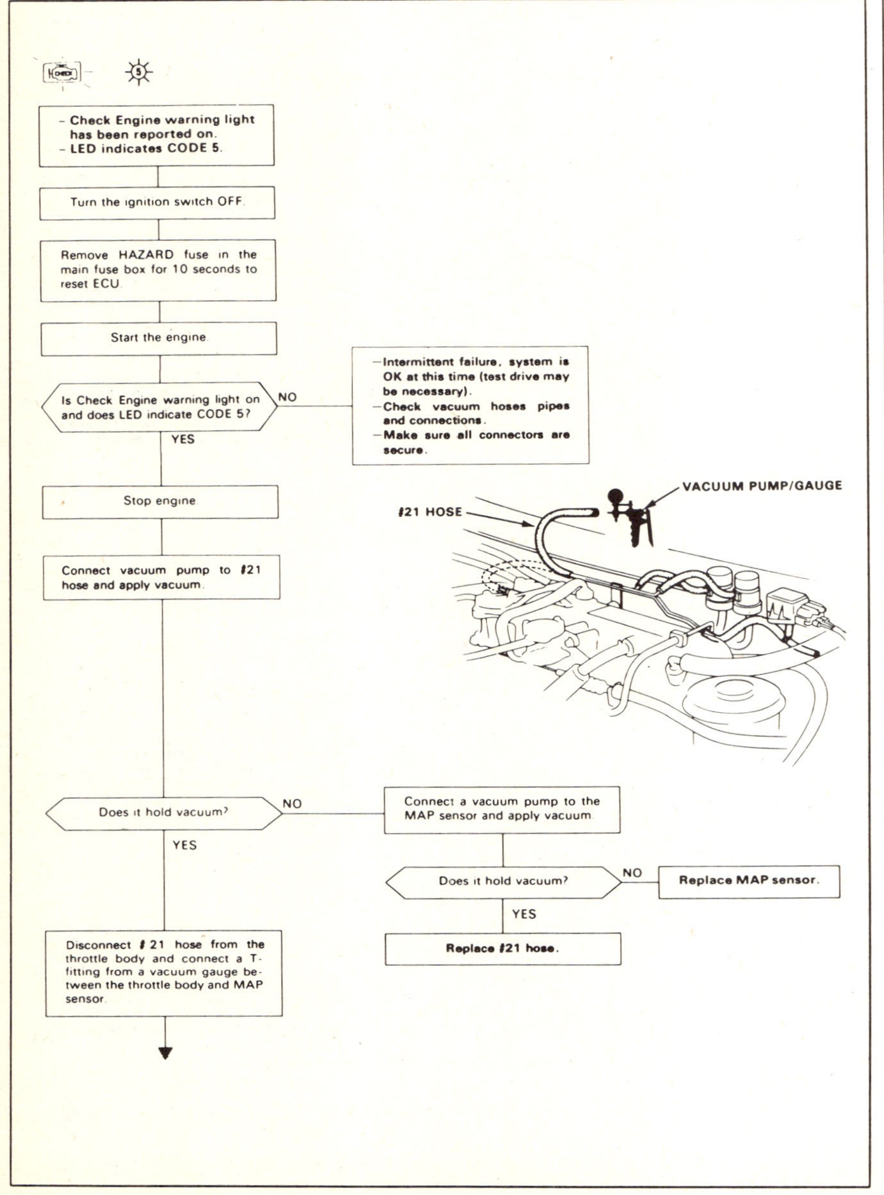


- Start engine.
- Is there manifold vacuum?
  - NO: –Remove restriction from throttle body. –Replace throttle body.
  - YES
- Stop engine.
- Connect the PGM-FI test harness between the ECU and connector.
- Turn the ignition switch ON.
- Measure voltage between C11 (+) terminal and C14 (-) terminal.

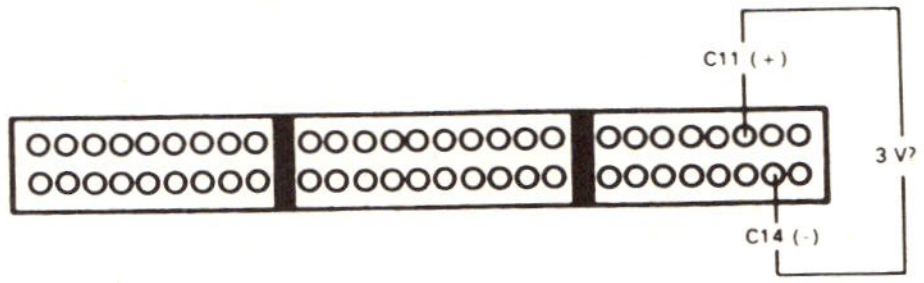

C11 (+)
C14 (-)
3 V?

- Is there approx. 3 V?
  - NO: **Replace the MAP sensor.**
  - YES
- Start the engine and allow it to idle.
- Is there approx. 1 V?
  - NO: **Replace MAP sensor.**
  - YES
- **Substitute a known-good ECU and recheck. If symptom/indication goes away, replace the original ECU.**

# TDC/CRANK/CYL SENSOR—CODE 4—1990 INTEGRA

Self-diagnosis LED indicator blinks four times: A problem in the circuit of the CRANK Sensor.

Self-diagnosis LED indicator blinks eight times: A problem in the circuit of the TDC Sensor.

Self-diagnosis LED indicator blinks nine times: A problem in the circuit of the CYL Sensor.

The CRANK sensor determines timing for fuel injection and ignition of each cylinder and also detects engine RPM. The TDC sensor determines ignition timing at start-up (cranking) and when crank angle is abnormal. The CYL sensor detects the position of No. 1 cylinder for sequential fuel injection to each cylinder.

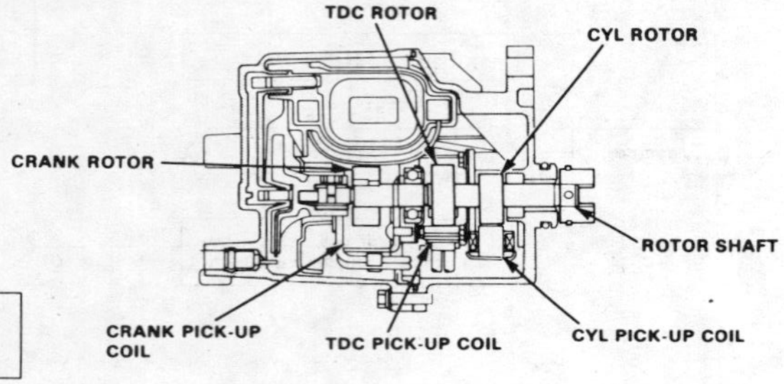

TDC ROTOR
CYL ROTOR
CRANK ROTOR
ROTOR SHAFT
CRANK PICK-UP COIL
TDC PICK-UP COIL
CYL PICK-UP COIL

- Check Engine warning light has been reported on.
- LED indicates CODE 4.

Turn the ignition switch OFF.

Remove HAZARD fuse in the main fuse box for 10 seconds to reset ECU.

Start engine.

Is Check Engine warning light on and does LED indicate CODE 4?

- NO: Intermittent failure, system is OK at this time (test drive may be necessary). Check for poor connections or loose wires at the distributor connector and C216 (round connector located at the right shock tower).
- YES: Stop engine.

Disconnect the 8P connector from the TDC/CRANK/CYL sensor.

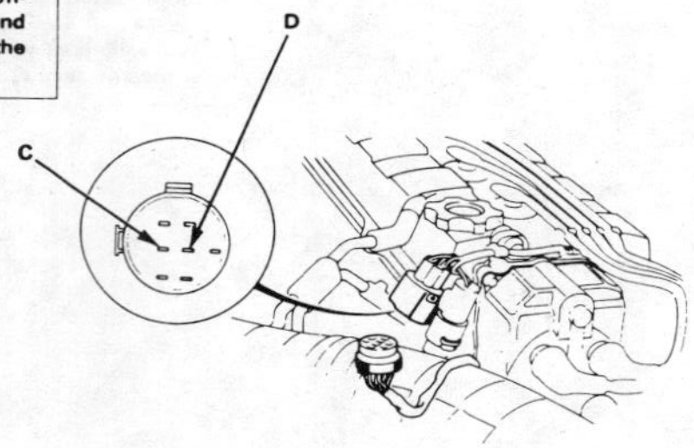

C
D

Measure resistance between C terminal and D terminal.

Is there 350–700 Ω ?

- NO: Replace the distributor assembly
- YES: Check for continuity to body ground on C terminal and D terminal individually.

Does continuity exist ?

- YES: Replace the distributor assembly
- NO: Reconnect the connector.

Connect the PGM-FI test harness only to the main wire harness, not to the ECU

Measure resistance between B10 terminal and B12 terminal.

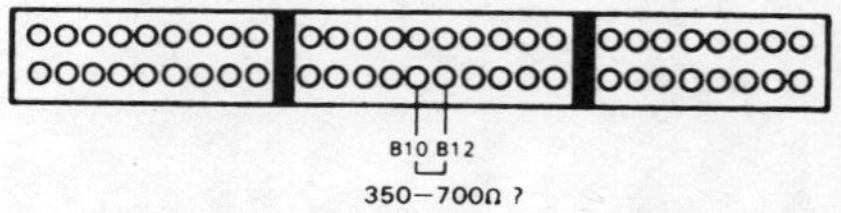

B10 B12
350–700Ω ?

Is there 350–700 Ω ?

- NO: Repair open in ORN and/or WHT wires.
- YES: Check for continuity to body ground on B10 terminal.

Does continuity exist?

- YES: Repair short in ORN wire between ECU (B10) and distributor connector.
- NO: Substitute a known-good ECU and recheck. If symptom/indication goes away, replace the original ECU.

# TDC/CRANK/CYL SENSOR—CODE 8—1990 INTEGRA

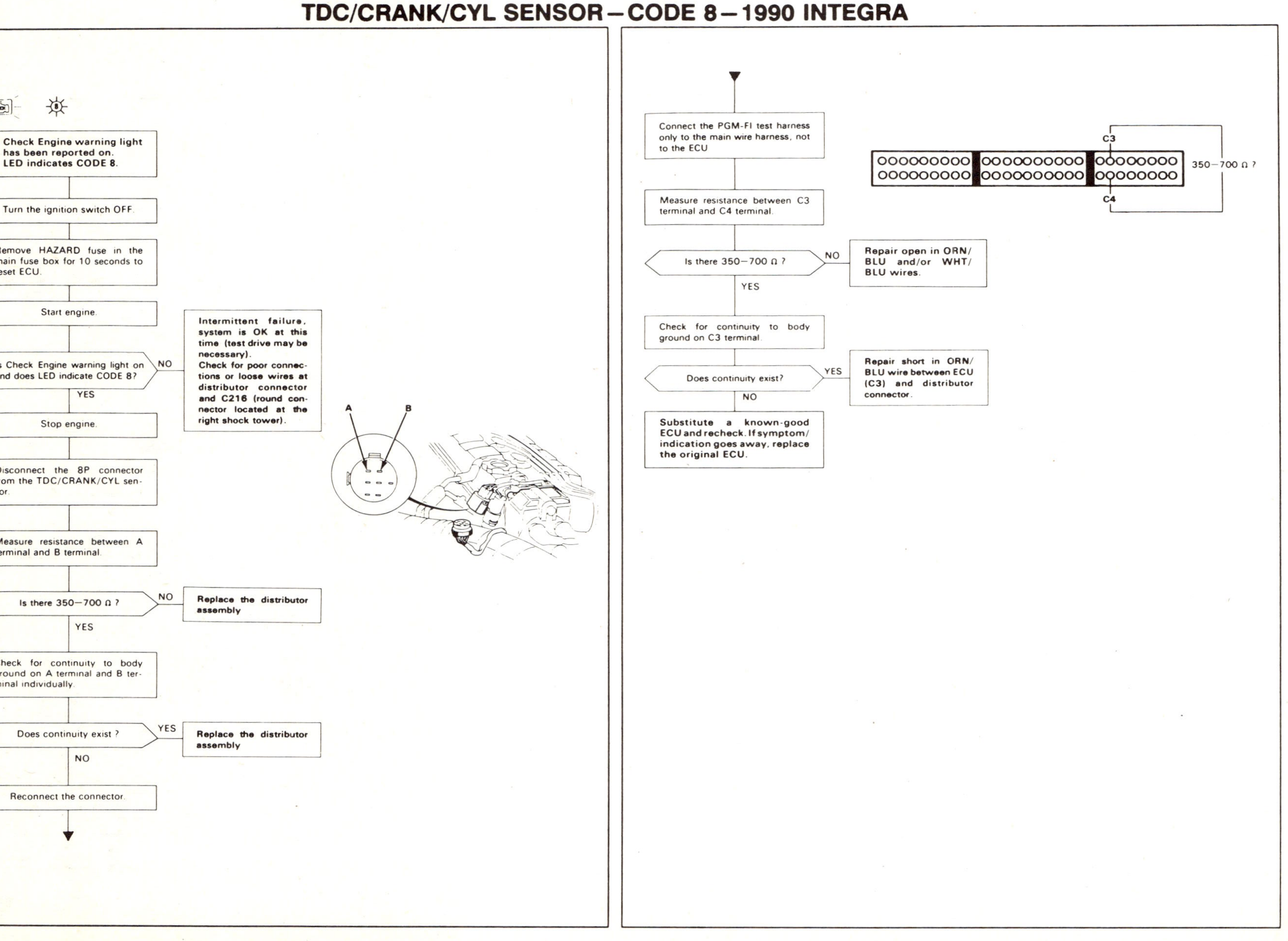

# TDC/CRANK/CYL SENSOR—CODE 9—1990 INTEGRA

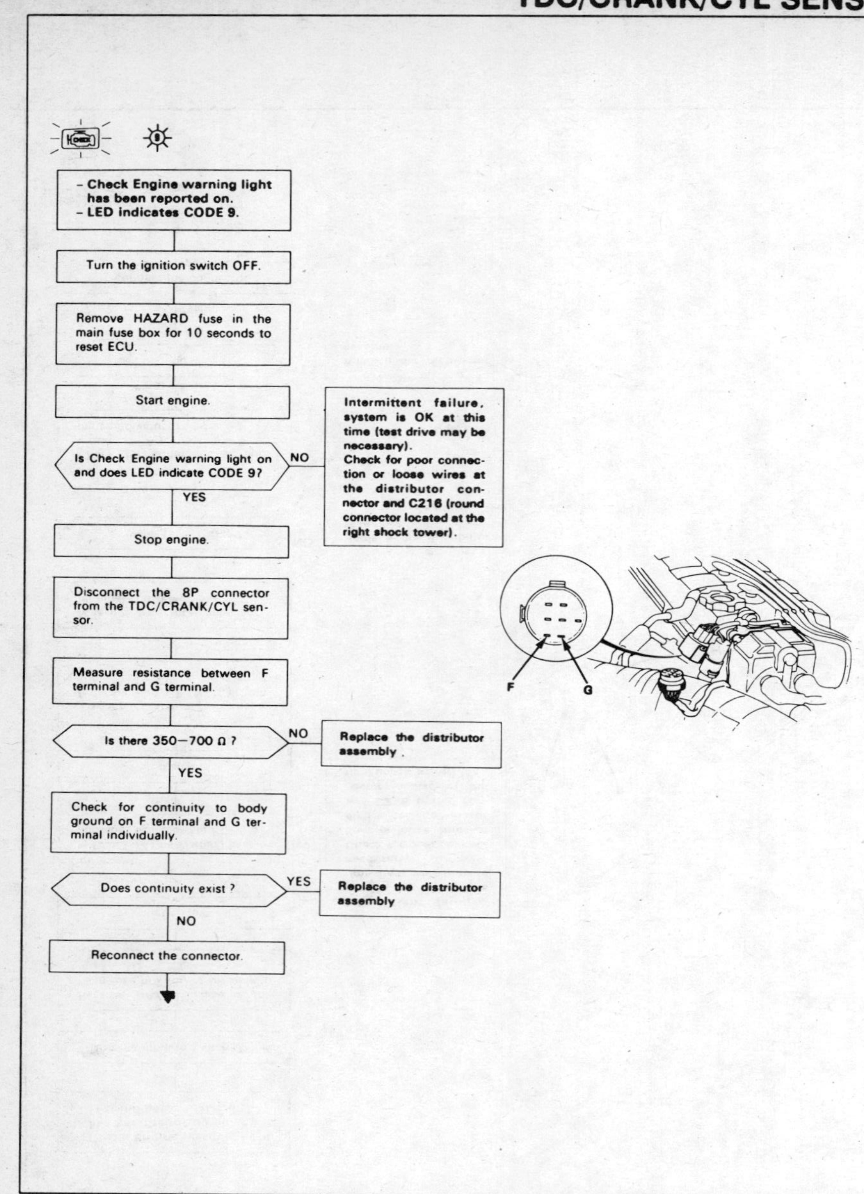


Connect the PGM-FI test harness only to the main wire harness, not to the ECU

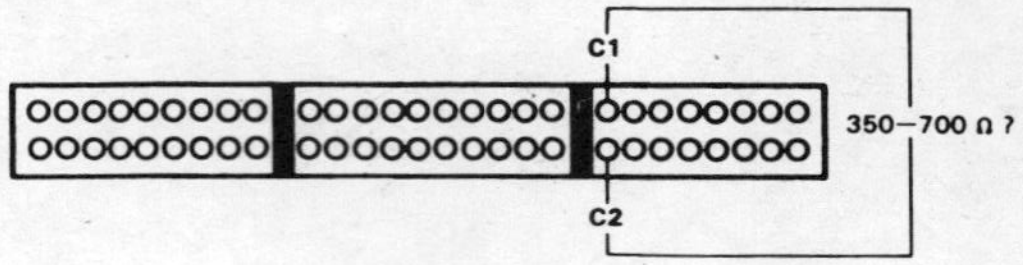

Measure resistance between C1 terminal and C2 terminal.

Is there 350—700 Ω ?

NO → Repair open in BLU/GRN and/or BLU/YEL wires.

YES

Check for continuity to body ground on C1 terminal.

Does continuity exist?

YES → Repair short in BLU/GRN wire between ECU (C1) and distributor connector.

NO

Substitute a known-good ECU and recheck. If symptom/indication goes away, replace the original ECU.

C1

C2

350—700 Ω ?

# COOLANT TEMPERATURE (TW) SENSOR – CODE 6 – 1990 INTEGRA

Self-diagnosis LED indicator blinks six times. Most likely a problem in the Coolant Temperature (TW) Sensor circuit.

The TW sensor is a temperature dependant resistor (thermistor). The resistance of the thermistor decreases as the coolant temperature increases as shown below.

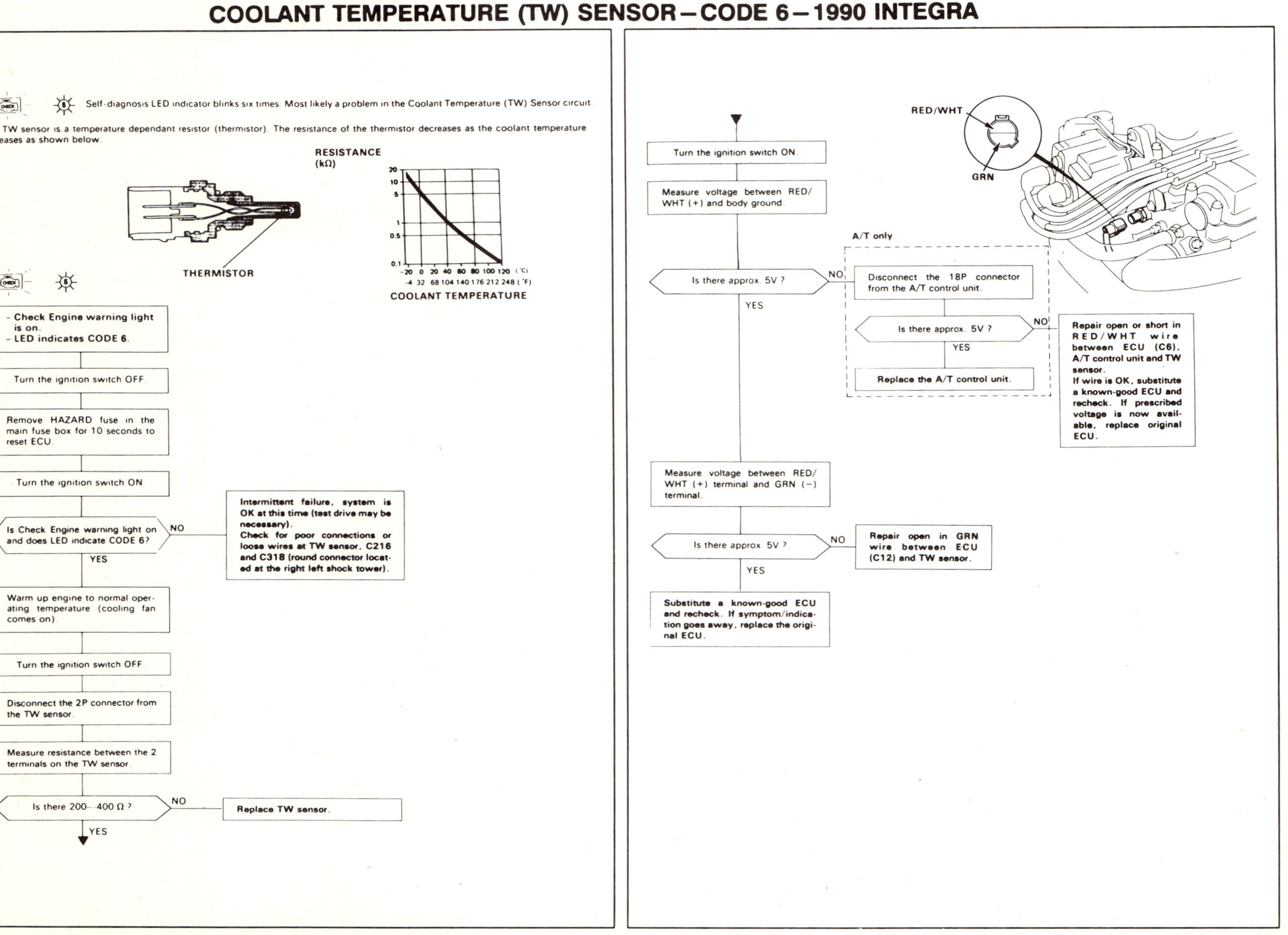

# THROTTLE ANGLE SENSOR—CODE 7—1990 INTEGRA

Self-diagnosis LED indicator blinks seven times. Most likely a problem in the Throttle Angle Sensor circuit.

The throttle angle sensor is a potentiometer. It is connected to the throttle valve shaft. As the throttle angle changes, the throttle angle sensor varies the voltage signal to the ECU.

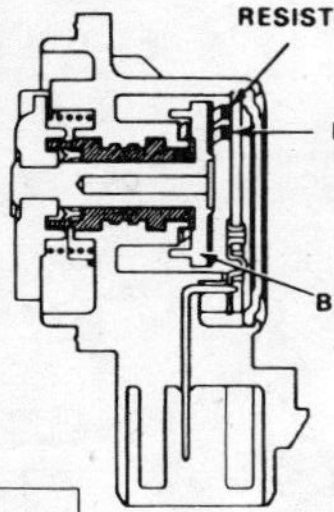

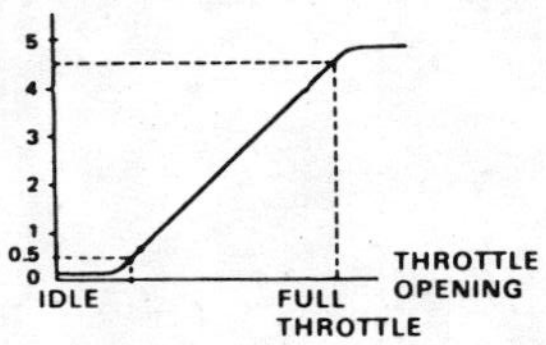

- Engine is running.
- Check Engine warning light is on.
- LED indicates CODE 7.

Turn the ignition switch OFF.

Remove HAZARD fuse in the main fuse box for 10 seconds to reset ECU.

Start engine.

Is Check Engine warning light on and does LED indicate CODE 7?

- NO: **Intermittent failure, system is OK at this time (test drive may be necessary). Check for poor connections or loose wires at throttle angle sensor, C216 (round connector located at the right shock tower) and C318 (round connector located at the left shock tower).**
- YES:

Turn the ignition switch OFF.

Disconnect the 3P connector from the throttle angle sensor.

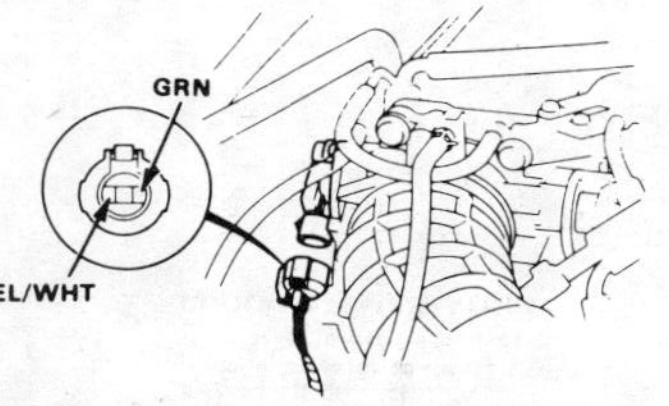

Turn the ignition switch ON.

Measure voltage between YEL/WHT (+) terminal and GRN (−) terminal.

Is there approx. 5V ?

- NO: Measure voltage between YEL/WHT (+) terminal and body ground. (continue to right column)
- YES: (continue)

**(YES branch)**

Turn the ignition switch OFF.

Reconnect the 3P connector.

Connect the PGM-FI test harness between the ECU and connector.

Turn the ignition switch ON.

Measure voltage between C7(+) terminal and C12(−) terminal.

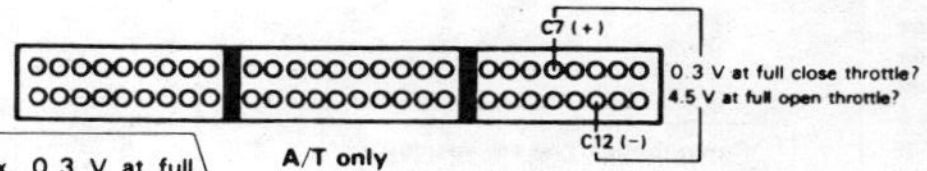

Is voltage approx. 0.3 V at full close throttle, and approx. 4.5 V at full open throttle ?
NOTE: There should be a smooth transmission from 0.3 V to 4.5 V as the throttle is depressed.

- YES: **Substitute a known-good ECU and recheck. If symptom/indication goes away, replace the original ECU.**
- NO: A/T only

Disconnect the 18 P connector from the A/T control unit.

Is voltage approx. 0.3 V at full close throttle, and approx. 4.5 V at full open throttle ?
NOTE: There should be a smooth transmission from 0.3 V to 4.5 V as the throttle is depressed.

- YES: **Replace the A/T control unit.**
- NO:
  - **Replace throttle angle sensor.**
  - **Repair open or short in RED/BLU wire between ECU (C7), A/T control unit and throttle angle sensor.**

**(NO branch)**

Is there approx. 5V ?

- YES: **Repair open in GRN wire between ECU (C12) and throttle angle sensor.**
- NO:

Turn the ignition switch OFF.

Connect the PGM-FI test harness between the ECU and connector.

Turn the ignition switch ON.

Measure voltage between C13 (+) terminal and C12 (−) terminal.

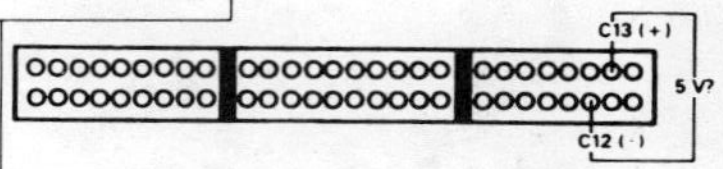

Is there approx. 5V ?

- YES: **Repair open in YEL/WHT wire between ECU(C13) and throttle angle sensor.**
- NO: **Substitute a known-good ECU and recheck. If prescribed voltage is now available, replace the original ECU.**

## AIR TEMPERATURE (TA) SENSOR—CODE 10—1990 INTEGRA

Self-diagnosis LED indicates code 10: Most likely a problem in the Intake Air Temperature (TA) Sensor circuit.

The TA sensor is a temperature dependant resistor (thermistor). The resistance of the thermistor decreases as the intake air temperature increases as shown below.

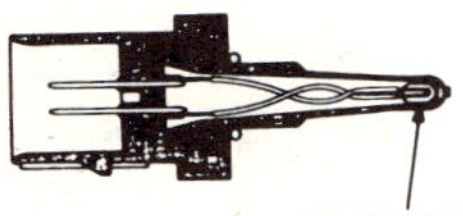

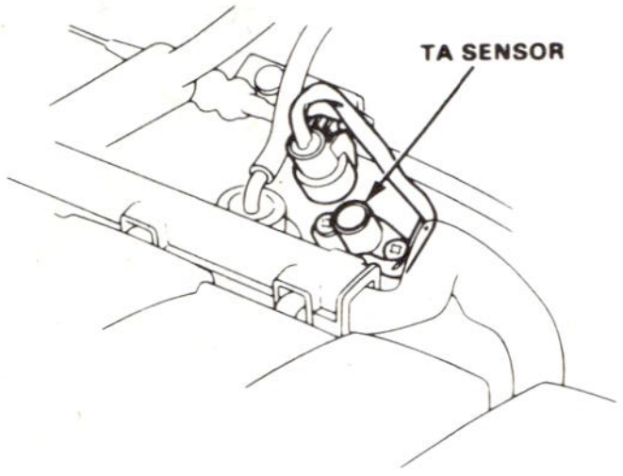

THERMISTOR

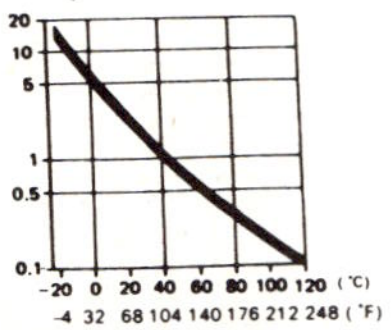

RESISTANCE (kΩ)

20, 10, 5, 1, 0.5, 0.1

-20 0 20 40 60 80 100 120 (°C)

-4 32 68 104 140 176 212 248 (°F)

INTAKE AIR TEMPERATURE

TA SENSOR

- Check Engine warning light is on.
- LED indicates CODE 10.

Turn the ignition switch OFF.

Remove HAZARD fuse in the main fuse box for 10 seconds to reset ECU.

Turn the ignition switch ON.

Is Check Engine warning light on and does LED indicate CODE 10?

NO → Intermittent failure, system is OK at this time (test drive may be necessary). Check for poor connections or loose wires at TA sensor and C318 (round connector located at left shock tower).

YES ↓

Turn the ignition switch OFF.

Disconnect the 2P connector from the TA sensor.

Measure resistance between the 2 terminals on the TA sensor.

Is there 1–4 kΩ ?

NO → Replace TA sensor.

YES ↓

Turn the ignition switch ON.

Measure voltage between RED/YEL (+) terminal and body ground.

Is there approx. 5 V ?

NO → Repair open or short in RED/YEL wire between ECU (C5) and TA sensor. If wire is OK, substitute a known-good ECU and recheck. If prescribed voltage is now available, replace original ECU.

YES ↓

Measure voltage between RED/YEL (+) terminal and GRN (−) terminal.

Is there approx. 5 V ?

NO → Repair open in GRN wire between ECU (C12) and TA sensor.

YES ↓

Substitute a known-good ECU and recheck. If symptom/indication goes away, replace the original ECU.

# ATMOSPHERIC PRESSURE (PA) SENSOR—CODE 13—1990 INTEGRA

Self-diagnosis LED indicates code 13: A problem in the Atmospheric Pressure (PA) Sensor circuit.

The PA sensor converts atmospheric pressure into electrical signals and inputs the ECU.

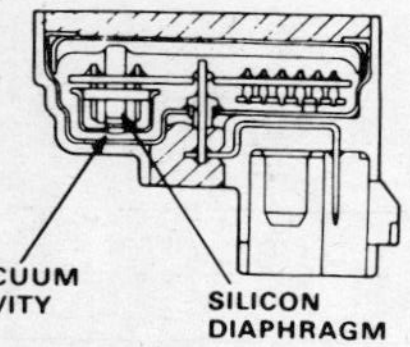

VACUUM CAVITY
SILICON DIAPHRAGM

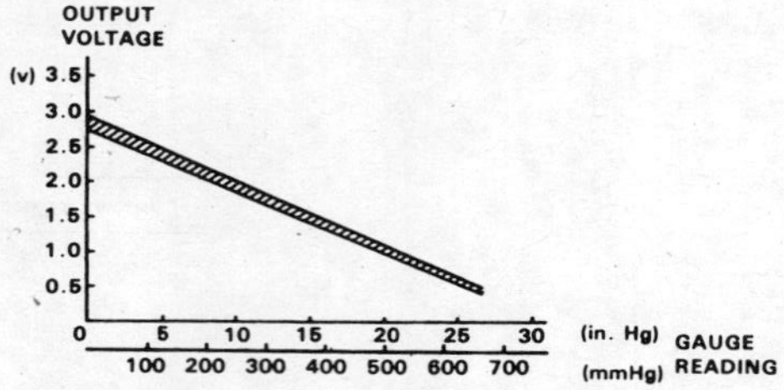

OUTPUT VOLTAGE (v): 3.5, 3.0, 2.5, 2.0, 1.5, 1.0, 0.5, 0
GAUGE READING (in. Hg): 5, 10, 15, 20, 25, 30
(mmHg): 100, 200, 300, 400, 500, 600, 700

- Check Engine warning light is on.
- LED indicates CODE 13.

Turn the ignition switch OFF.

Remove HAZARD fuse in the main fuse box for 10 seconds to reset ECU.

Turn the ignition switch ON.

Is Check Engine warning light on and does LED indicate CODE 13?

NO → Intermittent failure, system is OK at this time (test drive may be necessary). Check for poor connections or loose wires at the PA sensor.

YES → Turn the ignition switch OFF.

Disconnect the main wire harness from PA sensor.

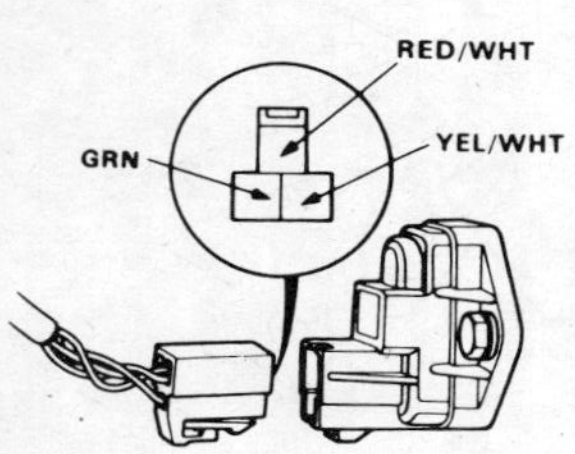

RED/WHT
YEL/WHT
GRN

Measure voltage between YEL/WHT (+) terminal and body ground.

Is there approx. 5V?

NO → Repair open in YEL/WHT wire between ECU (C13) and the sensor. If wire is OK, substitute a known-good ECU and recheck. If prescribed voltage is now available replace the original ECU.

YES → Measure voltage between YEL/WHT (+) terminal and GRN (−) terminal.

Is there approx. 5V?

NO → Repair open or short in GRN wire between ECU (C12) and the sensor. If wire is OK, substitute a known-good ECU and recheck. If prescribed voltage is now available, replace the original ECU.

YES ↓

Measure voltage between RED/WHT (+) terminal and GRN (−) terminal.

Is there approx. 5V?

NO → A/T only: Disconnect the 18 P connector from the A/T control unit.

Is there approx. 5 V ?

YES → Replace the A/T control unit.

NO → Repair open or short in RED/WHT wire between ECU (C9), A/T control unit and PA sensor. If wire is OK, substitute a known-good ECU and recheck. If prescribed voltage is now available, replace the original ECU.

YES → Turn the ignition switch OFF.

Remove HAZARD fuse in the main fuse box for 10 seconds to reset ECU.

Substitute a known-good PA sensor.

Turn the ignition switch ON.

Is Check Engine warning light on and does LED indicate CODE 13?

NO → Replace a original PA sensor.

YES → Substitute a known-good ECU and recheck. If prescribed voltage is now available replace the original ECU.

## IGNITION OUTPUT SIGNAL—CODE 15—1990 INTEGRA

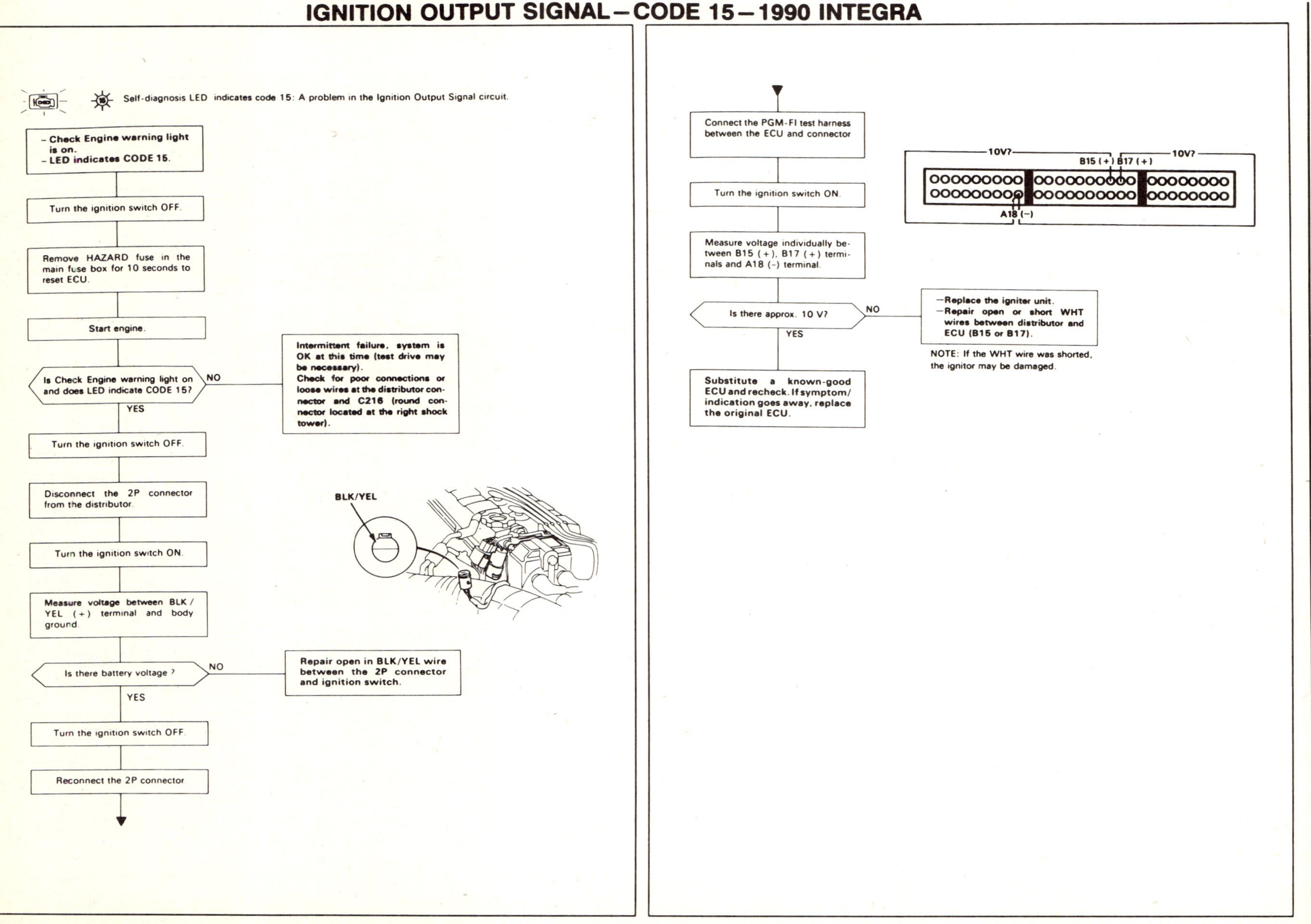

## VEHICLE SPEED SENSOR—CODE 17—1990 INTEGRA

[CHECK] [17] Self-diagnosis LED indicates code 17: A problem in the Vehicle Speed Sensor circuit.

The signal generated by the speed sensor, which produces 4 pulses (switch closures to ground) for each revolution of the speedometer cable.

- Check Engine warning light is on.
- LED indicates CODE 17.

↓

Turn the ignition switch OFF.

↓

Remove HAZARD fuse in the main fuse box for 10 seconds to reset ECU.

↓

Road test necessary.
In 2nd gear accelerate to 3,500 rpm and decelerate to 1,500 rpm with throttle fully closed.

↓

Is Check Engine warning light on and does LED indicate CODE 17?

- NO → **Intermittent failure, system is OK at this time. Check for poor connections or loose wires at C401 and C714.**
- YES ↓

Block rear wheels and set the parking brake. Jack up the front of the car and support with safety stands.

**WARNING Block rear wheels before jacking up front of car.**

↓

Connect the PGM-FI test harness between the ECU and connector

↓

Turn the ignition switch ON.

↓

Slowly rotate left front wheel and measure voltage between B16 (+) terminal and A18 (−) terminal.

A18 (−) B16 (+)
0 V↔5 V?

↓

Does voltage pulse 0V and 5V ?

- NO →
  - **Repair open or short in YEL/RED wire between ECU (B16) and the speed sensor.**
  - **Faulty speed sensor.**
  - **Substitute a known-good ECU and recheck. If symptom/indication goes away, replace the original ECU.**
- YES ↓

**Substitute a known-good ECU and recheck. If symptom/indication goes away, replace the original ECU.**

## IDLE CONTROL TROUBLESHOOTING GUIDE 1990 INTEGRA

NOTE:

- Across each row in the chart, the sub systems that could be sources of a symptom are ranked in the order they should be inspected, starting with ①. Find the symptom in the left column, read across to the most likely source, then refer to the page listed at the top of that column. If inspection shows the system is OK, try the next system ②, etc.
- If the idle speed still cannot be adjusted to specification (and LED does not blink CODE 14) after EACV replacement, substitute a known-good ECU and recheck. If symptom goes away, replace the original ECU.

| SYMPTOM \ SUB SYSTEM | | IDLE ADJUSTING SCREW | EACV | AIR CONDITIONING SIGNAL | ALTERNATOR FR SIGNAL | A/T SHIFT POSITION SIGNAL (A/T ONLY) | BRAKE SWITCH SIGNAL | STARTER SWITCH SIGNAL | P/S OIL PRESSURE SWITCH SIGNAL | FAST IDLE VALVE | AIR BOOST VALVE | FAST IDLE CONTROL SOLENOID VALVE (A/T ONLY) | HOSES AND CONNECTIONS |
|---|---|---|---|---|---|---|---|---|---|---|---|---|---|
| DIFFICULT TO START ENGINE WHEN COLD | | ③ | ② | | | | | | | ① | ③ | | |
| WHEN COLD FAST IDLE OUT OF SPEC (1,000—2,000 rpm) | | ③ | ② | | | | | | | ① | | ③ | |
| ROUGH IDLE | | | ② | | | | | | | | | | ① |
| WHEN WARM RPM TOO HIGH | | ③ | ① | | | | | | ③ | ② | | ③ | |
| WHEN WARM RPM TOO LOW | Idle speed is below specified rpm (no load) | ② | ① | | ③ | | | | | | | | |
| | Idle speed does not increase after initial start up. | | ① | | | | | ② | | | | | |
| | On models with automatic transmission, the idle speed drops in gear | | ② | | | ① | | | | | | | |
| | Idle speeds drops when air conditioner in ON | | ② | ① | | | | | | | | | |
| | Idle speed drops when steering wheel is turning | | ② | | | | | | ① | | | | |
| | Idle speed fluctuates with electrical local | ③ | ② | | ① | | | | | | | | |
| FREQUENT STALLING | WHILE WARMING UP | | ① | | | | | | | | | | |
| | AFTER WARMING UP | ② | ① | | | | | | | | | | |
| FAILS EMISSION TEST | | | | | | | | | | | | | ① |

# EACV—CODE 14—1990 INTEGRA

Self-diagnosis LED indicates code 14: A problem in the Electronic Air Control Valve (EACV) circuit.

The EACV changes the amount of air bypassing the throttle body in response to a current signal from the ECU in order to maintain the proper idle speed.

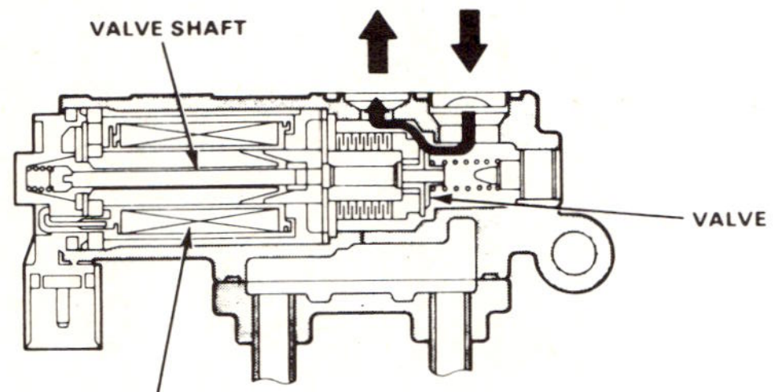

- Engine is running.
- Check Engine warning light is on.
- LED indicates CODE 14.

Turn the ignition switch OFF.

Remove HAZARD fuse in the main fuse box for 10 seconds to reset ECU.

Start engine.

Is Check Engine warning light on and does LED indicate CODE 14?
- NO → Intermittent failure, system is OK at this time (test driving may be necessary). Check for poor connections or loose wires at EACV and at C318.
- YES ↓

Stop engine.

Disconnect the 2P connector from the EACV.

Measure resistance between the 2 terminals on the EACV.

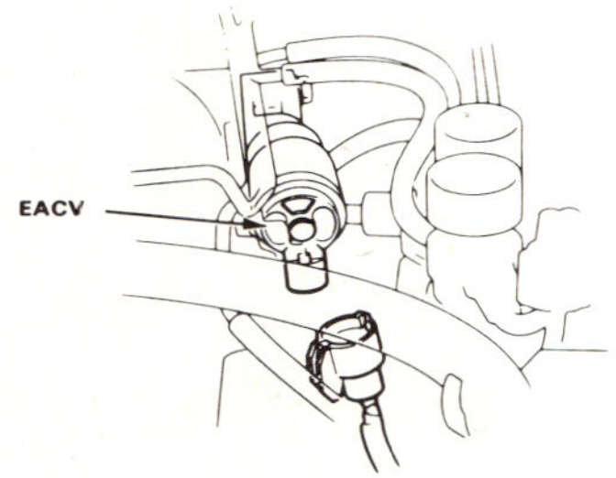

Is there 8—15Ω ?
- NO → Replace EACV.
- YES ↓

Check for continuity to body ground on each terminal on the EACV.

Does continuity exist ?
- YES → Replace EACV.
- NO ↓

Turn the ignition switch ON.

Measure voltage between BLK/YEL (+) terminal and BLU/YEL (-) terminal.

Is there battery voltage ?
- YES → Disconnect the "A" connector from the ECU. → Measure voltage between BLK/YEL (+) terminal and BLU/YEL (−) terminal. → Is there battery voltage ?
  - YES → Repair short in BLU/YEL wire between ECU (A11) and EACV.
  - NO → Substitute a known-good ECU and recheck. If symptom/indication goes away, replace the original ECU.
- NO ↓

Measure voltage between BLK/YEL (+) terminal and body ground.

Is there battery voltage ?
- NO → Repair open in BLK/YEL wire between the EACV and No. 24 fuse.
- YES ↓

Turn the ignition switch OFF.

Reconnect the 2P connector to EACV.

Connect the PGM-FI test harness "A" connector to the main wire harness only, not the ECU.

## EACV—CODE 14—1990 INTEGRA

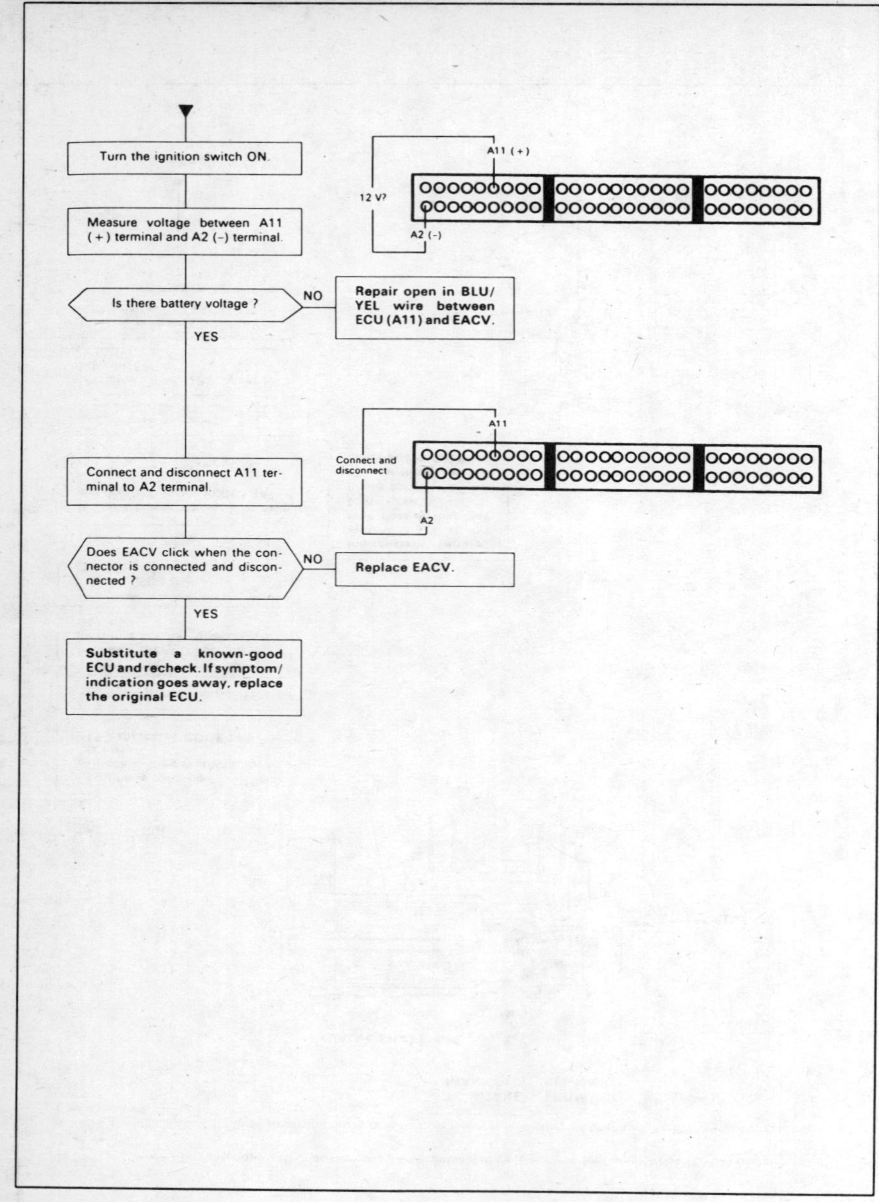


## AIR CONDITIONING SIGNAL—1990 INTEGRA

This signals the PGM-FI ECU when there is a demand for cooling from the air conditioning circuits.

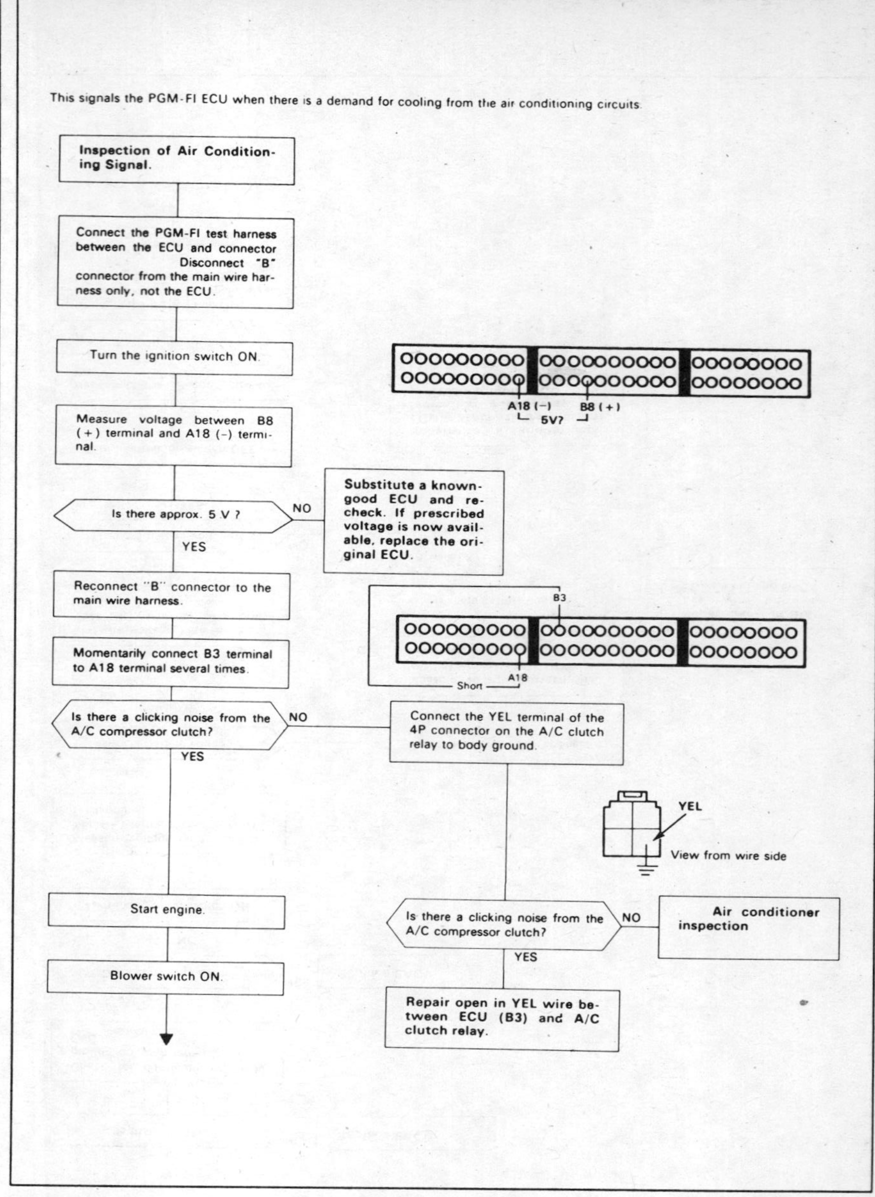

## AIR CONDITIONING SIGNAL—1990 INTEGRA

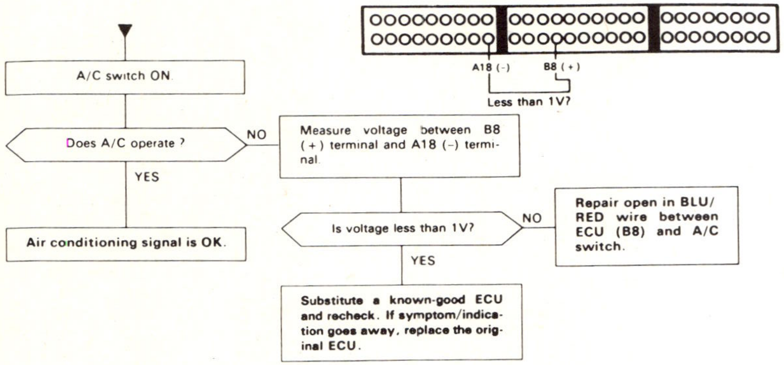

A/C switch ON

Does A/C operate ?

- YES: Air conditioning signal is OK.
- NO: Measure voltage between B8 (+) terminal and A18 (-) terminal.

A18 (-) B8 (+) Less than 1V?

Is voltage less than 1V?

- NO: Repair open in BLU/RED wire between ECU (B8) and A/C switch.
- YES: Substitute a known-good ECU and recheck. If symptom/indication goes away, replace the original ECU.

## ALTERNATOR FR SIGNAL—1990 INTEGRA

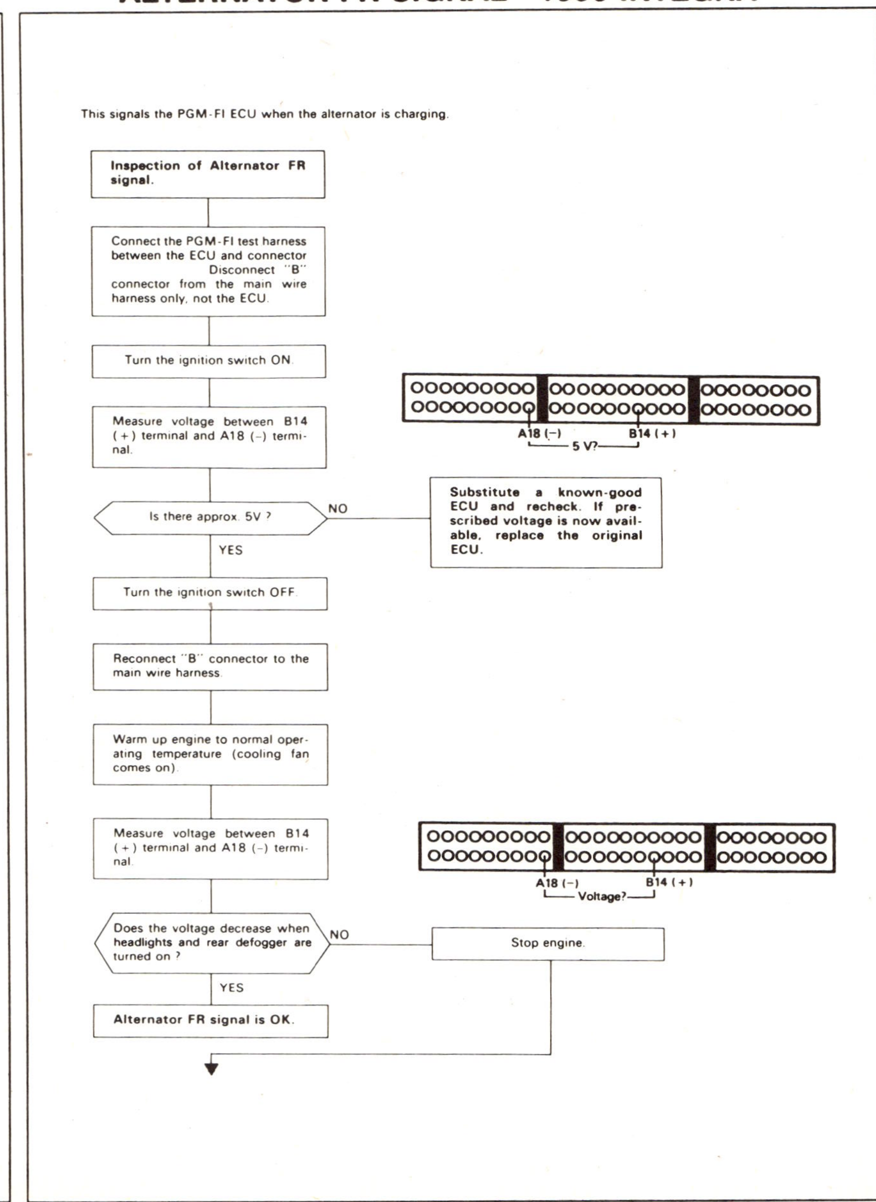

## ALTERNATOR FR SIGNAL—1990 INTEGRA

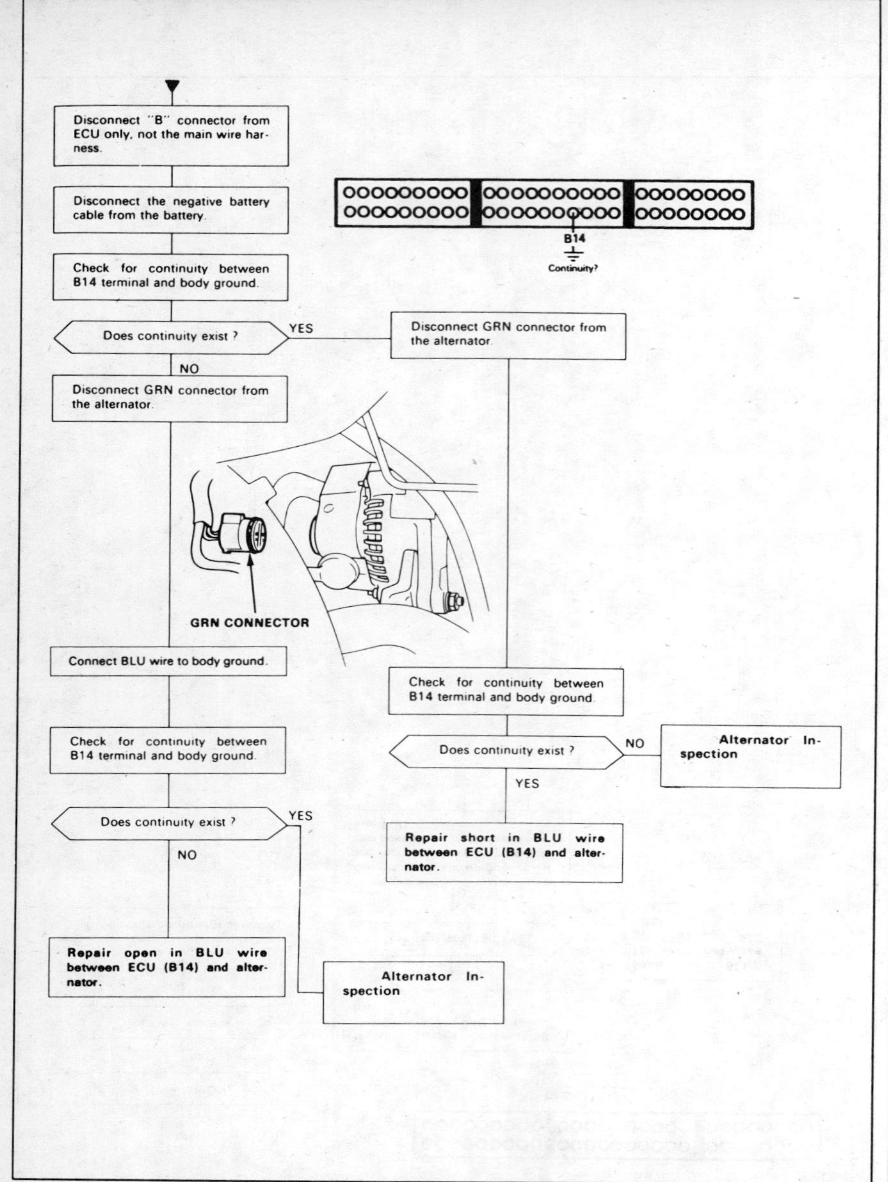


## A/T SHIFT POSITION SIGNAL—1990 INTEGRA

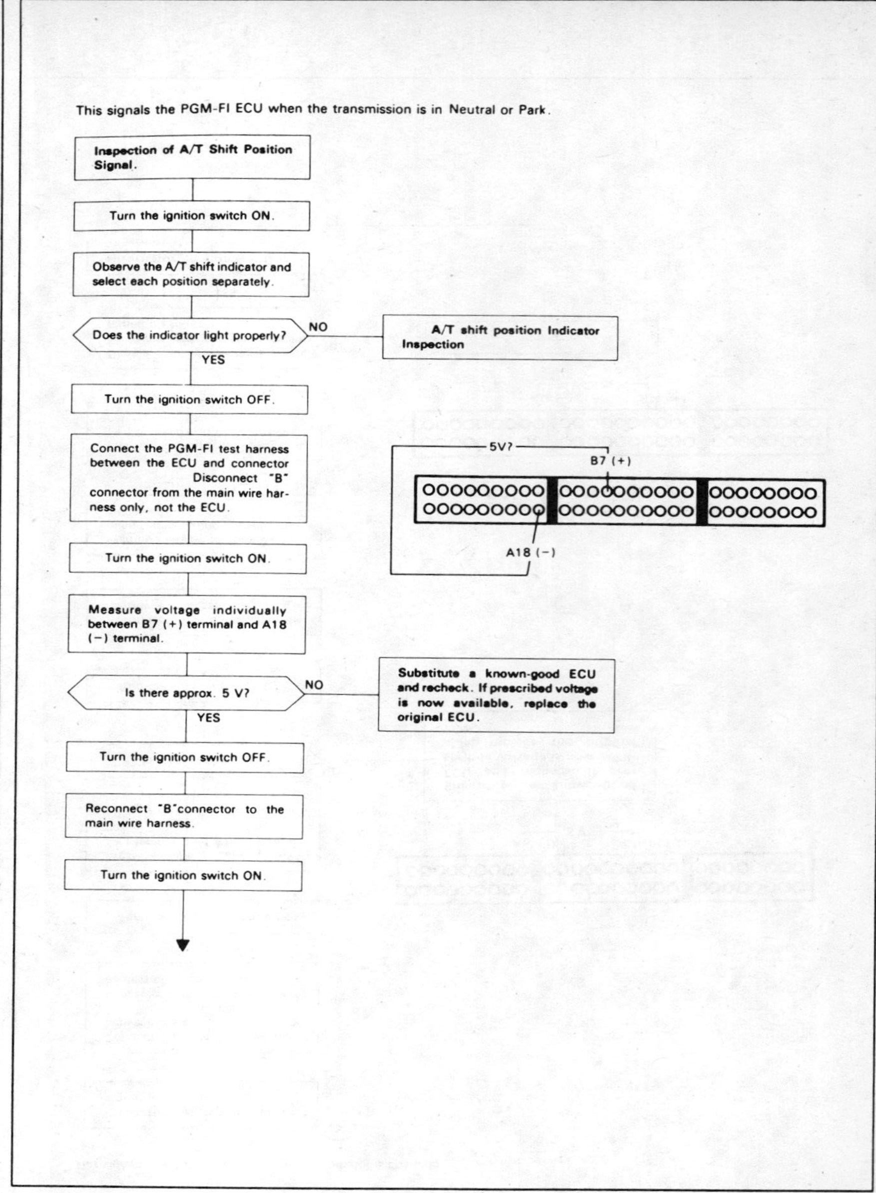

## A/T SHIFT POSITION SIGNAL—1990 INTEGRA

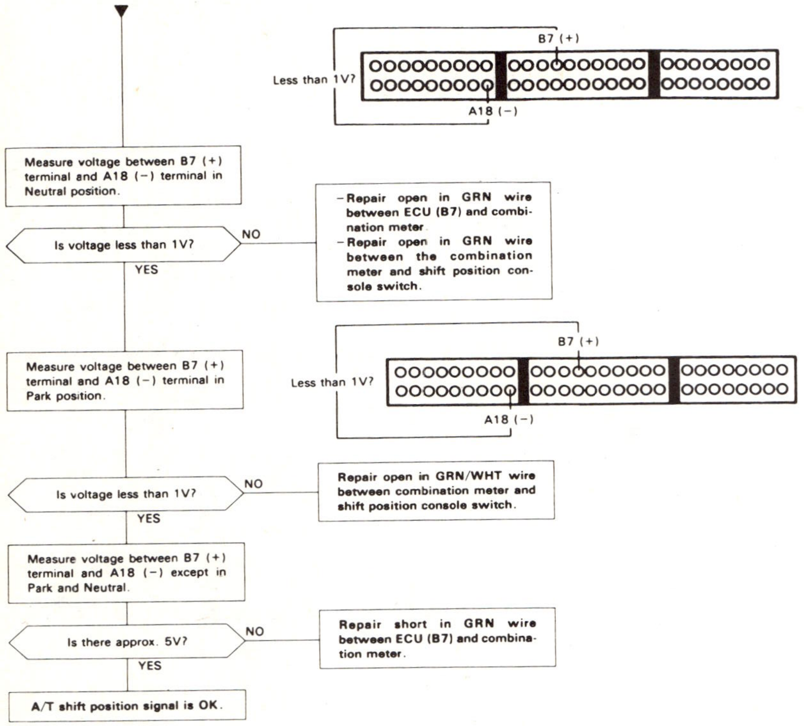

Less than 1V?
B7 (+)
A18 (−)

Measure voltage between B7 (+) terminal and A18 (−) terminal in Neutral position.

Is voltage less than 1V?

NO → –Repair open in GRN wire between ECU (B7) and combination meter.
–Repair open in GRN wire between the combination meter and shift position console switch.

YES ↓

Less than 1V?
B7 (+)
A18 (−)

Measure voltage between B7 (+) terminal and A18 (−) terminal in Park position.

Is voltage less than 1V?

NO → Repair open in GRN/WHT wire between combination meter and shift position console switch.

YES ↓

Measure voltage between B7 (+) terminal and A18 (−) except in Park and Neutral.

Is there approx. 5V?

NO → Repair short in GRN wire between ECU (B7) and combination meter.

YES ↓

A/T shift position signal is OK.

## BRAKE SWITCH SIGNAL—1990 INTEGRA

This signals the PGM-FI ECU when the brake pedal is depressed.

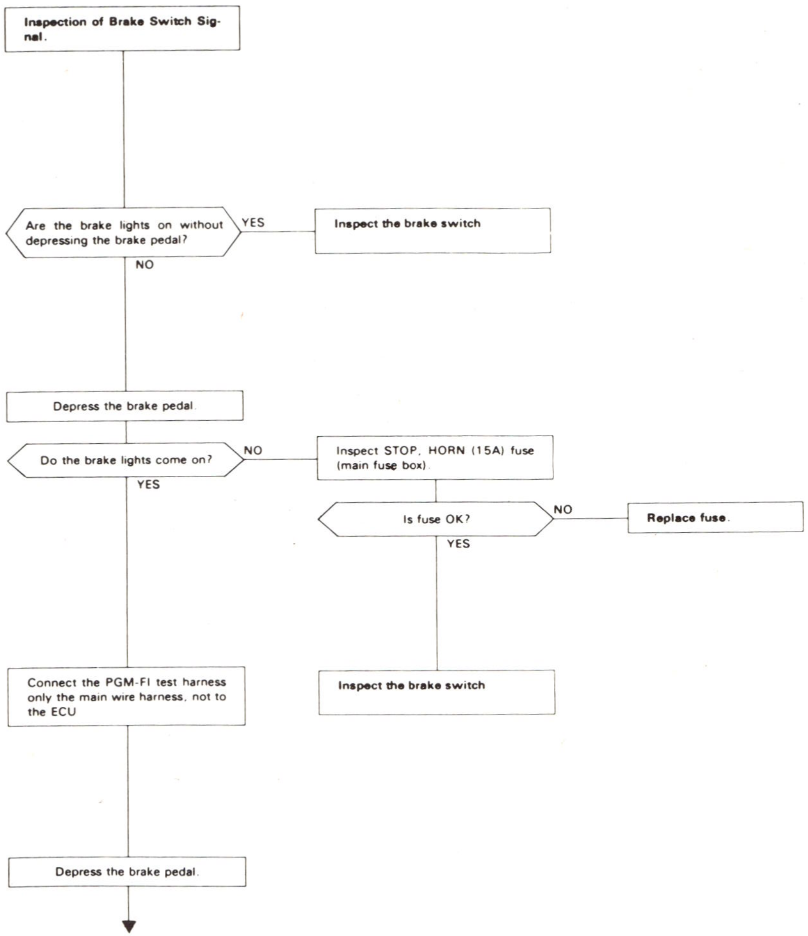

Inspection of Brake Switch Signal.

Are the brake lights on without depressing the brake pedal?

YES → Inspect the brake switch

NO ↓

Depress the brake pedal.

Do the brake lights come on?

NO → Inspect STOP, HORN (15A) fuse (main fuse box)

Is fuse OK?

NO → Replace fuse.

YES → Inspect the brake switch

YES ↓

Connect the PGM-FI test harness only the main wire harness, not to the ECU

Depress the brake pedal.

## BRAKE SWITCH SIGNAL—1990 INTEGRA

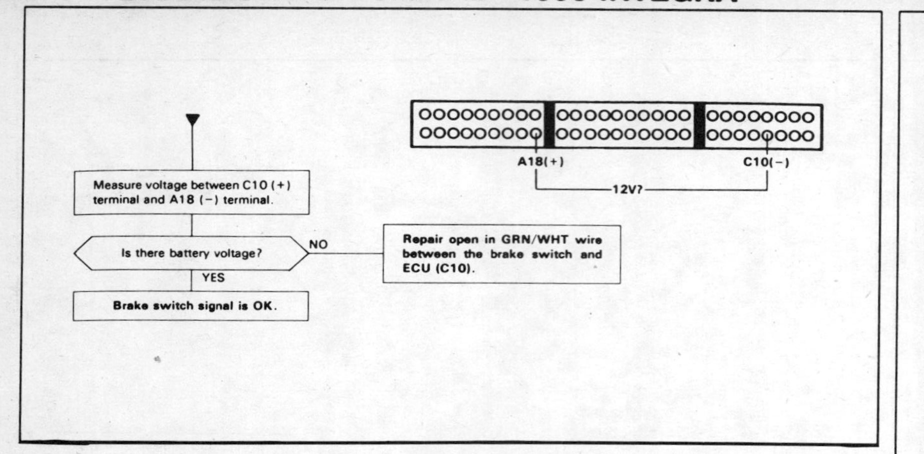


## STARTER SWITCH SIGNAL—1990 INTEGRA

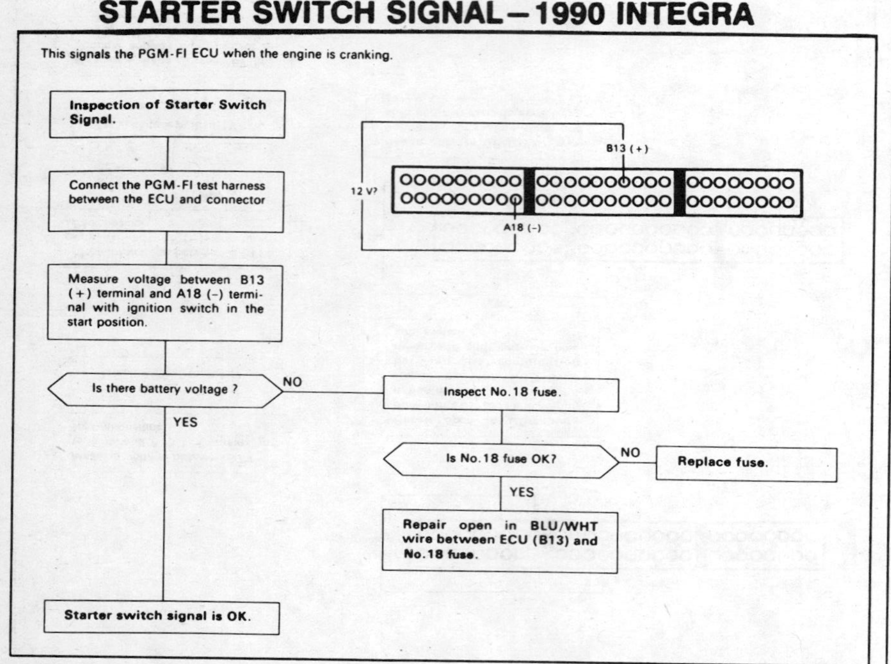


## POWER STEERING PRESSURE SIGNAL 1990 INTEGRA

This signals the PGM-FI ECU when the power steering load is high.

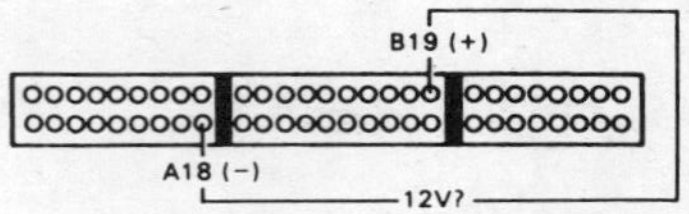

B19 (+)

A18 (−)

12V?

**Inspection of P/S Oil Pressure Signal**

Connect the PGM-FI test harness between the ECU and connector

Turn the ignition switch ON.

Measure voltage between B19 (+) terminal and A18 (−) terminal.

Is there battery voltage ?

YES → Disconnect the 2P connector on the P/S oil pressure switch.

Connect RED terminal to BLK terminal.

Is there battery voltage ?

NO → **Replace P/S oil pressure switch.**

YES → **Repair open in RED wire between ECU (B19) and P/S oil pressure switch or BLK wire between P/S oil pressure switch and G301.**

NO → Start engine.

Turn steering wheel slowly.

Measure voltage between B19 (+) terminal and A18 (−) terminal while steering wheel is turning.

Is there battery voltage ?

NO → Turn the ignition switch OFF.

Disconnect "B" connector from main wire harness only, not the ECU.

Turn the ignition switch ON.

Is there battery voltage ?

NO → **Substitute a known-good ECU and recheck. If prescribed voltage is now available, replace the original ECU.**

YES → Reconnect "B" connector to main wire harness and disconnect 2P connector on the P/S oil pressure switch.

Is there battery voltage ?

YES → **Replace P/S oil pressure switch.**

NO → **Repair short in RED wire between ECU (B19) and the P/S oil pressure switch.**

YES → **P/S oil pressure signal is OK.**

# FAST IDLE CONTROL SOLENOID VALVE (A/T) – 1990 INTEGRA

**Description**

The fast idle control solenoid valve is employed to increase the air flow rate for fast idling at extremely low ambient temperature

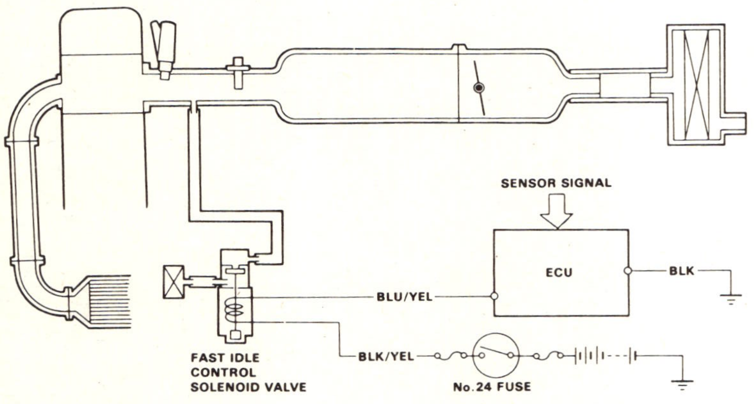

SENSOR SIGNAL
ECU
BLK
BLU/YEL
BLK/YEL
No.24 FUSE
FAST IDLE CONTROL SOLENOID VALVE

FAST IDLE CONTROL SOLENOID VALVE IS ON FOR 20 SECONDS

AND

COOLANT TEMPERATURE BELOW −10°C (14°F)

ENGINE SPEED BELOW 1,750 rpm

**Troubleshooting Flowchart**

**Inspection of Fast Idle Control Solenoid Valve.**

Warm up engine to normal operating temperature (cooling fan comes on).

Remove the control box cover, disconnect the vacuum hose from the air filter and connect a vacuum gauge to the hose.

Is there any vacuum?

- YES → Disconnect the 4P connector from the control box.
  - Is there any vacuum?
    - YES → **Replace the fast idle control solenoid valve.**
    - NO → **Repair short to ground at BLU wire between ECU (B2) and the 4P connector. If wire is OK, substitute a known-good ECU and recheck. If symptom goes away, replace the original ECU.**
- NO → Turn the ignition switch OFF.

Disconnect the 4P connector from the control box.

Connect battery positive to terminal A and battery negative to terminal C of the control box.

Start the engine.

Is there manifold vacuum?

- NO → **Replace the fast idle control solenoid valve.**
- YES → **Fast Idle Control Solenoid Valve is OK.**

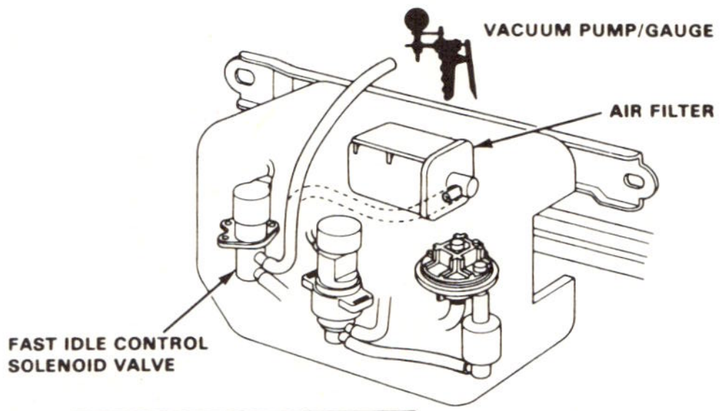

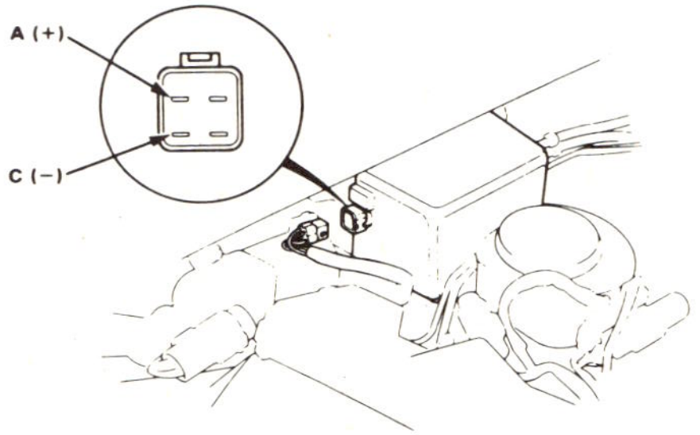

VACUUM PUMP/GAUGE
AIR FILTER
FAST IDLE CONTROL SOLENOID VALVE

A (+)
C (−)

# FUEL INJECTORS – CODE 16 – 1990 INTEGRA

**Troubleshooting Flowchart**

Self-diagnosis LED indicates code 16: A problem in the fuel injector circuit.

The injectors are the solenoid-actuated constant-stroke pintle type consisting of a solenoid, plunger needle valve and housing. When current is applied to the solenoid coil, the valve lifts up and pressurized fuel is injected close to the intake valve. Because the needle valve lift and the fuel pressure are constant, the injection quantity is determined by the length of time that the valve is open (i.e., the duration the current is supplied to the solenoid coil). The injector is sealed by an O-ring and seal ring at the top and bottom. These seals also reduce operating noise.

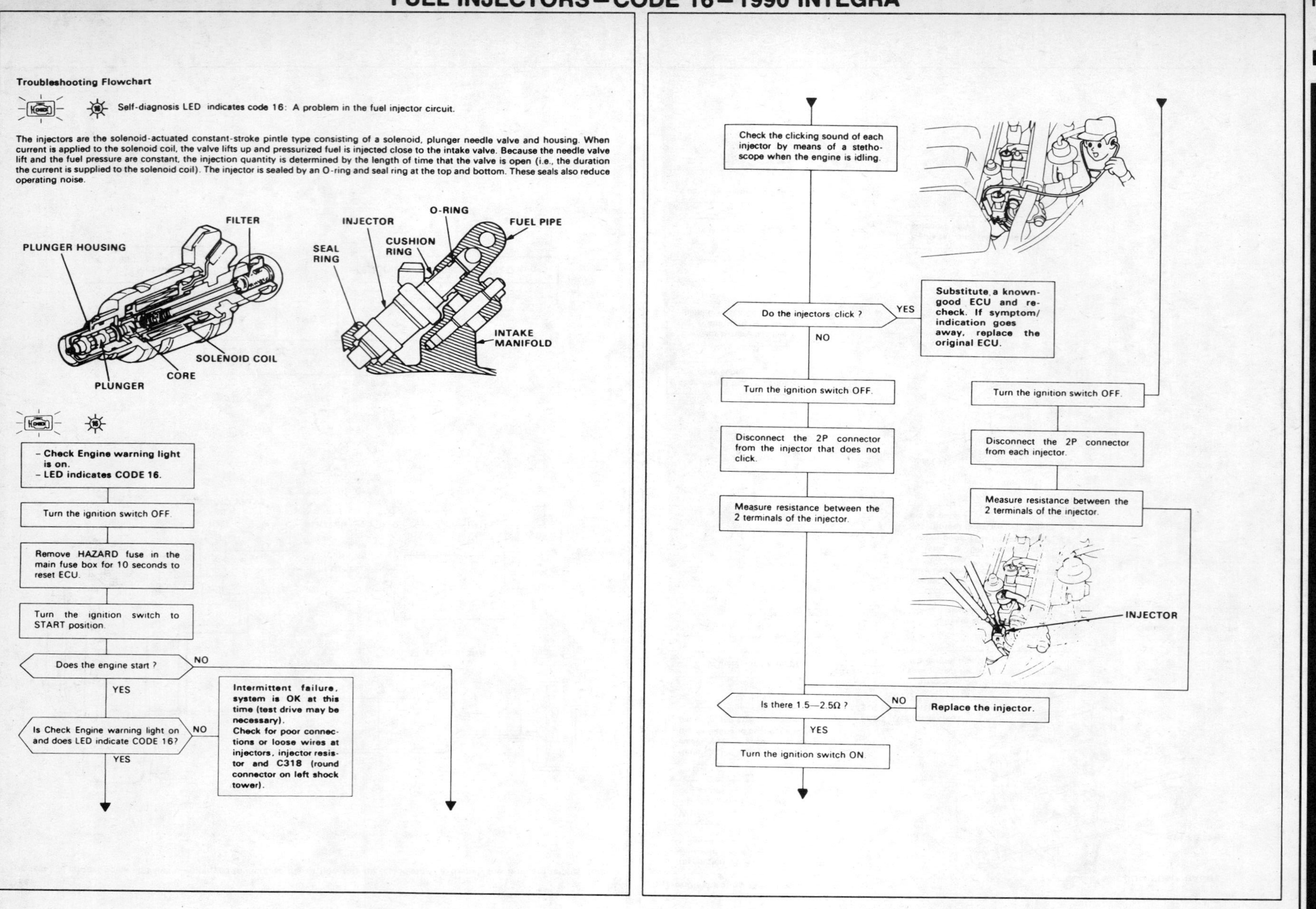

## FUEL INJECTORS—CODE 16—1990 INTEGRA

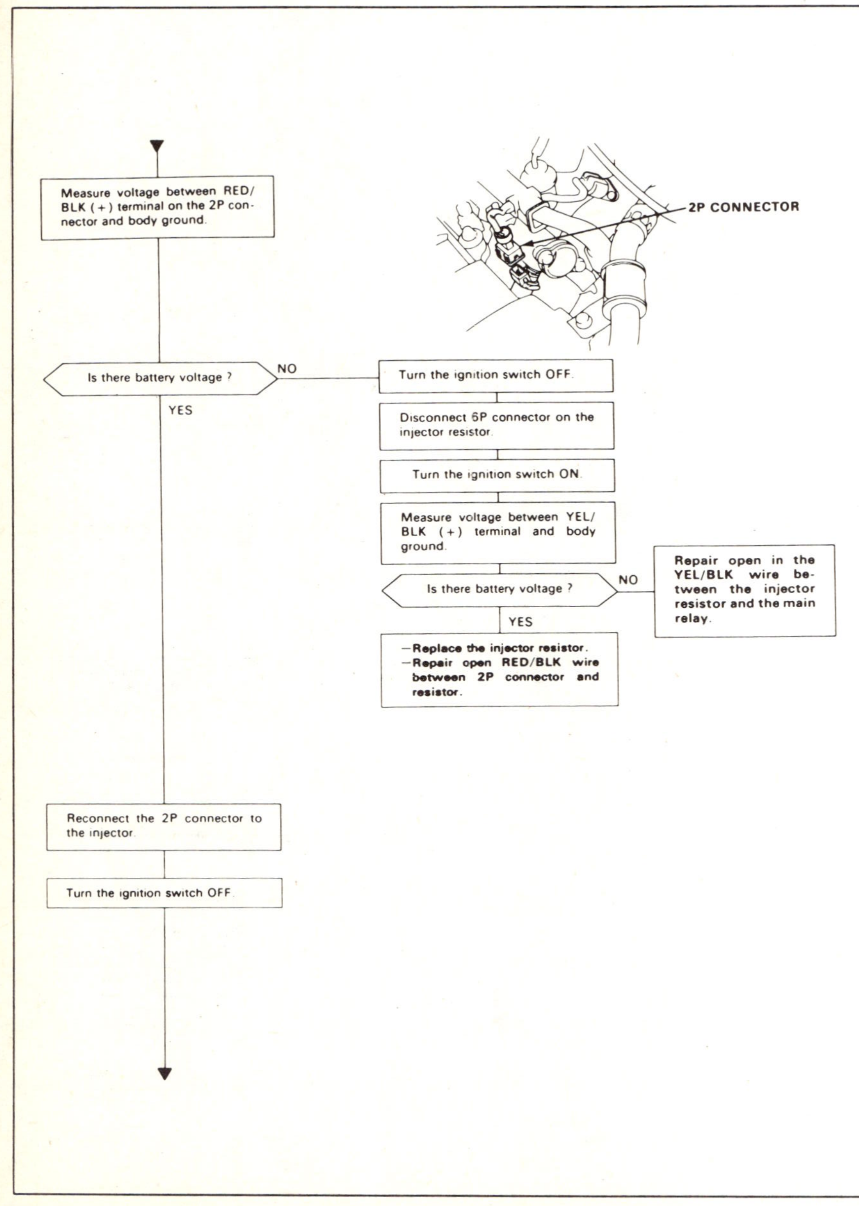


Connect the PGM-FI test harness between the ECU and connector

Turn the ignition switch ON.

Measure voltage between A2 (−) terminal and following terminals.
- No.1 injector: A1 (+) terminal.
- No.2 injector: A3 (+) terminal.
- No.3 injector: A5 (+) terminal.
- No.4 injector: A7 (+) terminal.

A3 (+) A7 (+) A5 (+) A1 (+) A2 (−)

Is there battery voltage ?

NO: **Repair open in the wire between the ECU (A1, A3, A5 or A7) and the injector.**

YES: **Substitute a known-good ECU and recheck. If symptom/indication goes away, replace the original ECU.**

# FUEL PRESSURE REGULATOR – 1990 INTEGRA

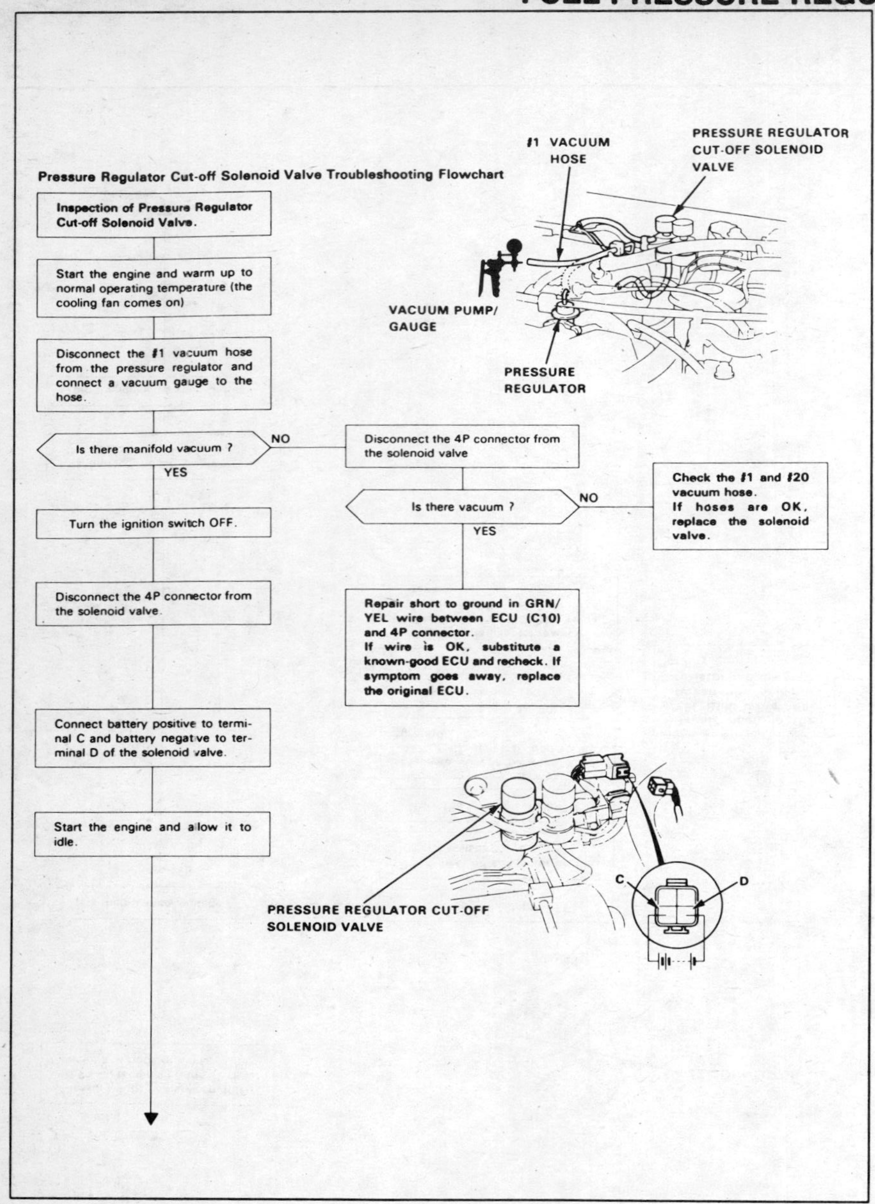


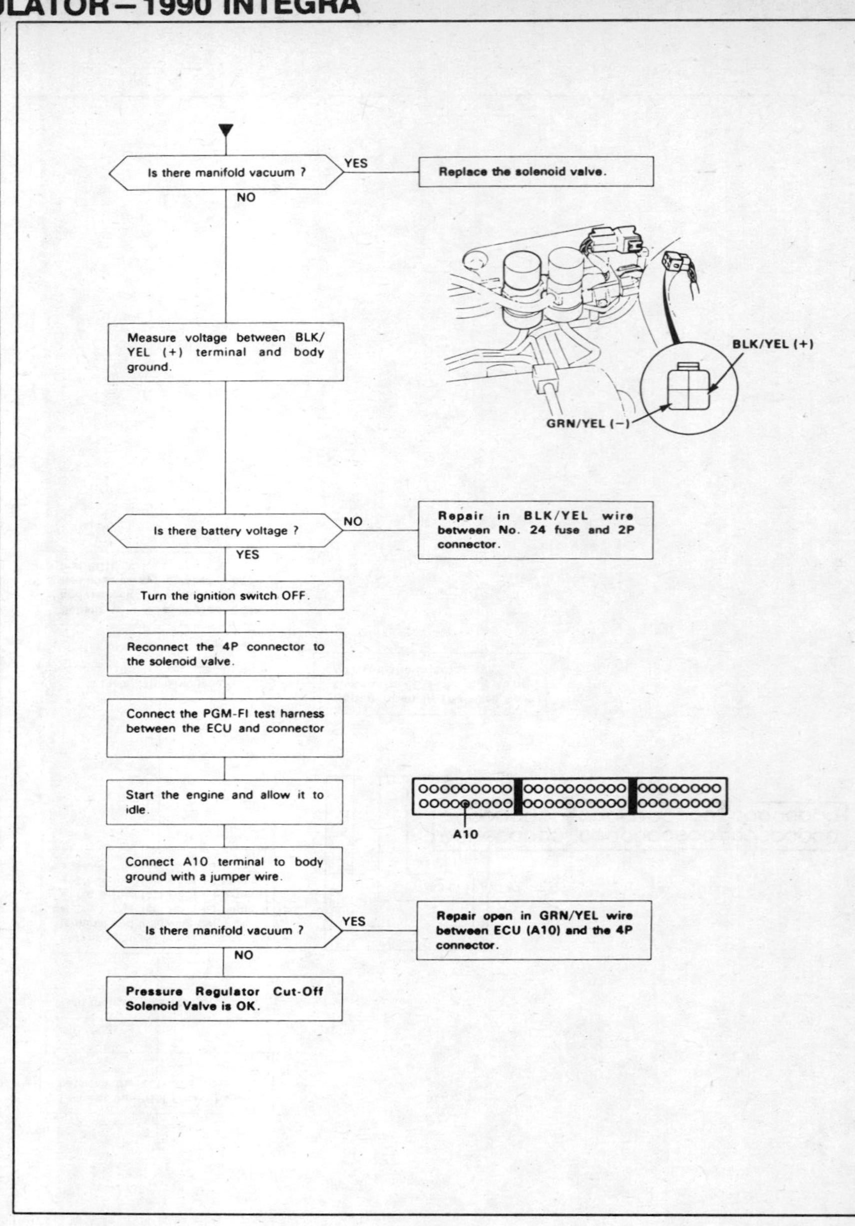

## EXHAUST GAS RECIRCULATION—CODE 12—1990 INTEGRA

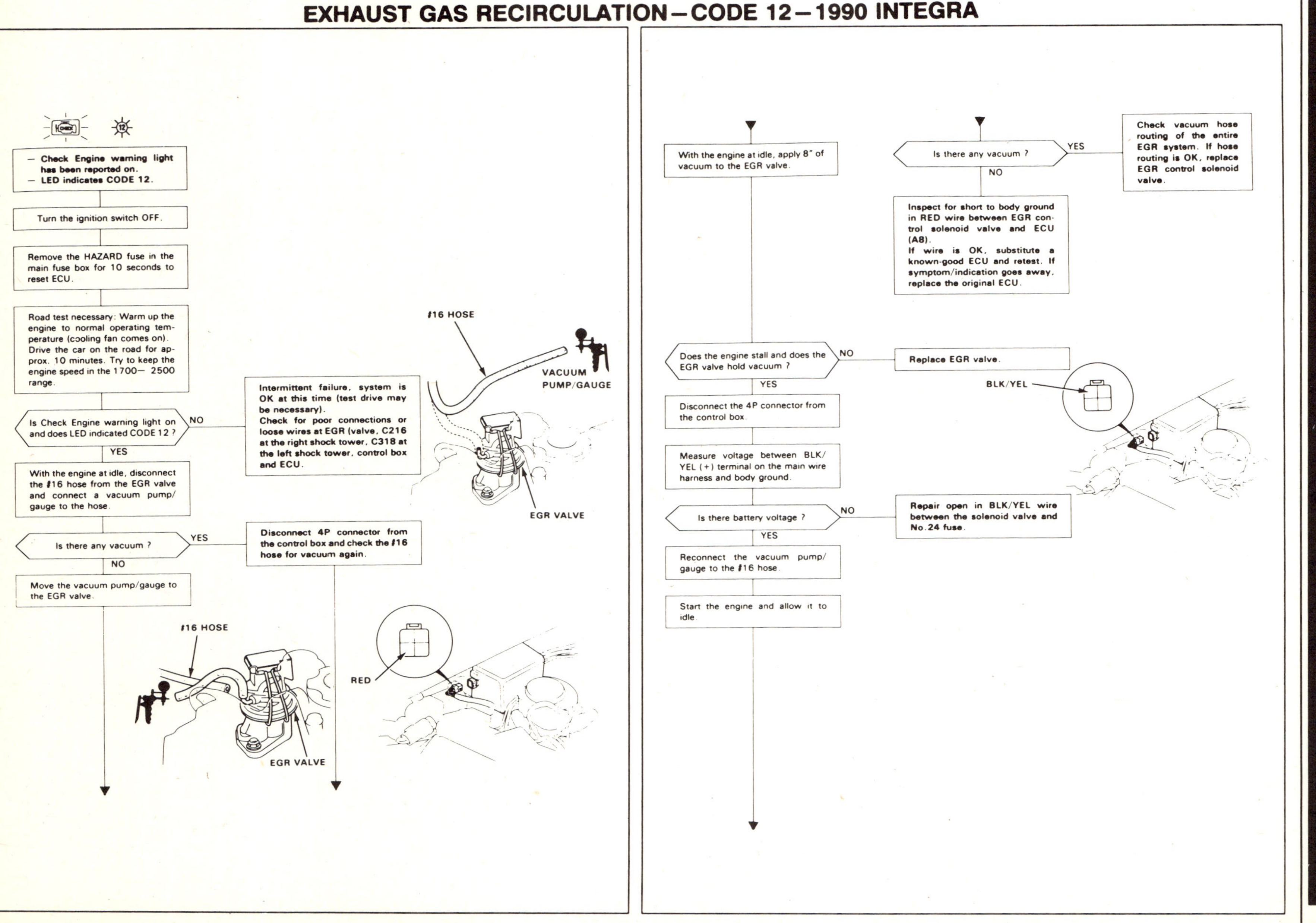

# EXHAUST GAS RECIRCULATION—CODE 12—1990 INTEGRA

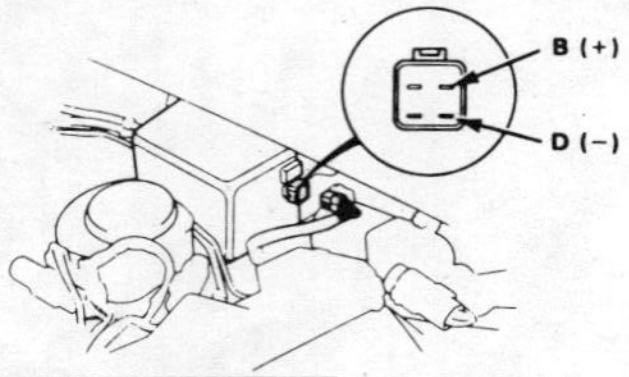

Connect the battery positive terminal to the B terminal of the 4P connector. While watching the vacuum gauge, connect the battery negative terminal to the D terminal.

Is there approx. 8" within 1 minute ?

- NO → Turn the ignition switch OFF and inspect the #16 and #20 hoses for leaks, restrictions, or mis-routing.
  - Are the hose OK ?
    - NO → Connect as necessary.
    - YES → Disconnect the lower lose on EGR control solenoid valve and connect a vacuum gauge to the hose.
      - Start the engine and allow it to idle.
      - Is there 6"— 10" of vacuum ?
        - NO → Replace CVC valve.
        - YES → Replace the EGR control solenoid valve.
- YES → Turn the ignition switch OFF and reconnect the 4P connector.

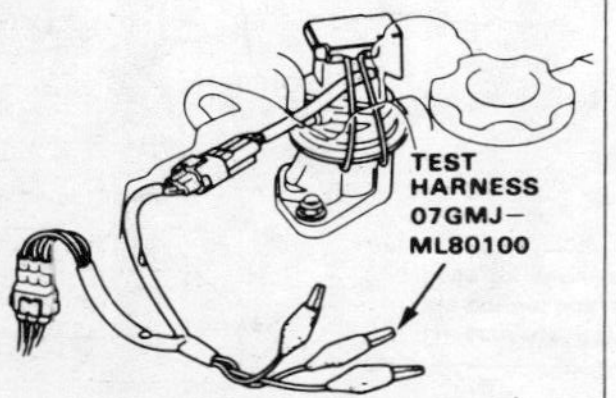

B (+)
D (−)

TEST HARNESS 07GMJ—ML80100

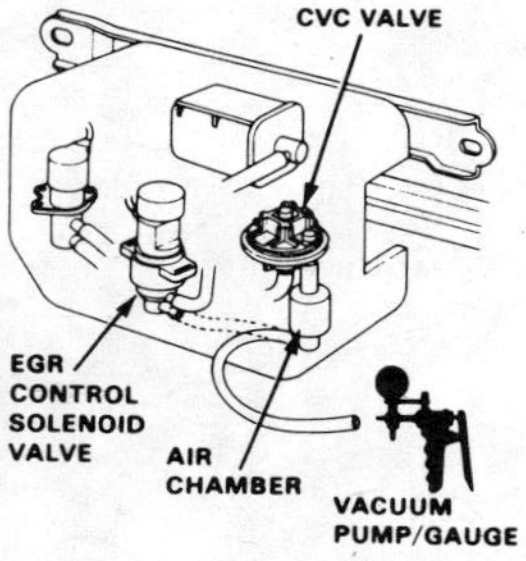

CVC VALVE
EGR CONTROL SOLENOID VALVE
AIR CHAMBER
VACUUM PUMP/GAUGE

Connect the test harness between the EGR valve lift sensor and engine wire harness.

Turn the ignition switch ON.

Measure voltage between RED (+) terminal and GRN (−) terminal.

Is there approx. 5 V ?

- NO → Repair open in YEL/WHT wire (C216) or GRN wire (C318) between EGR valve and ECU (C13).
- YES ↓

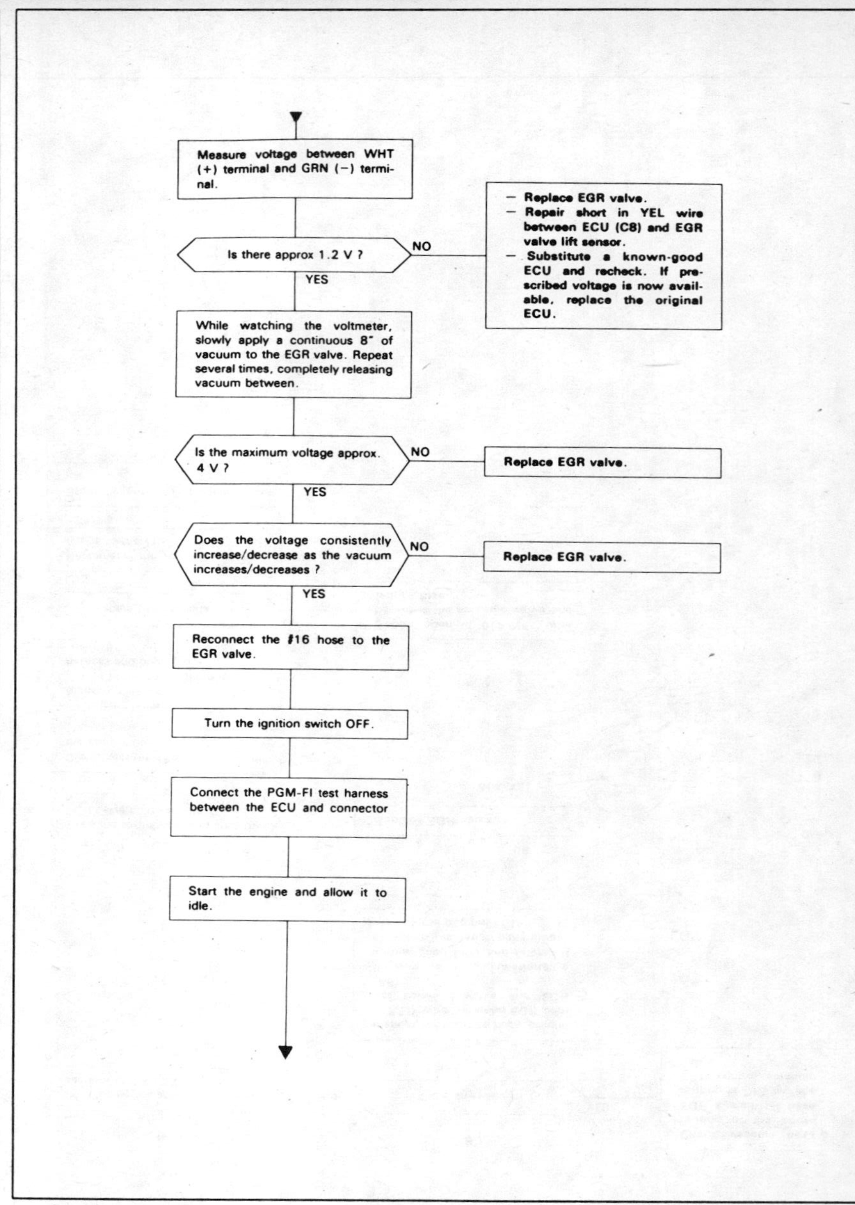

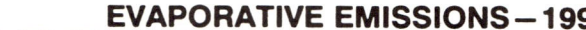

## EXHAUST GAS RECIRCULATION—CODE 12 1990 INTEGRA

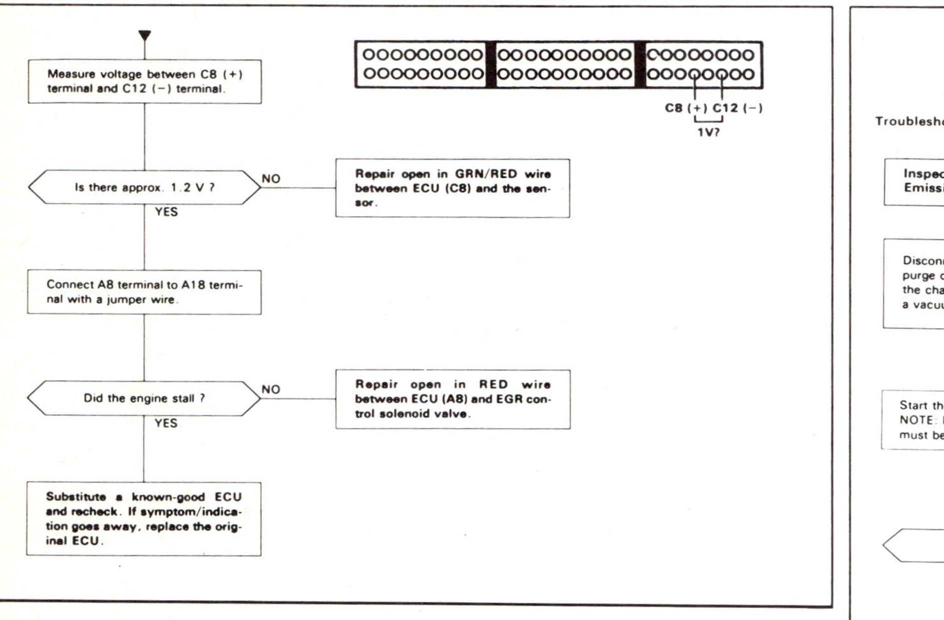

Measure voltage between C8 (+) terminal and C12 (−) terminal.
Is there approx. 1.2 V ?
NO
Repair open in GRN/RED wire between ECU (C8) and the sensor.
YES
Connect A8 terminal to A18 terminal with a jumper wire.
Did the engine stall ?
NO
Repair open in RED wire between ECU (A8) and EGR control solenoid valve.
YES
Substitute a known-good ECU and recheck. If symptom/indication goes away, replace the original ECU.
C8 (+) C12 (−)
1V?


## EVAPORATIVE EMISSIONS—1990 INTEGRA

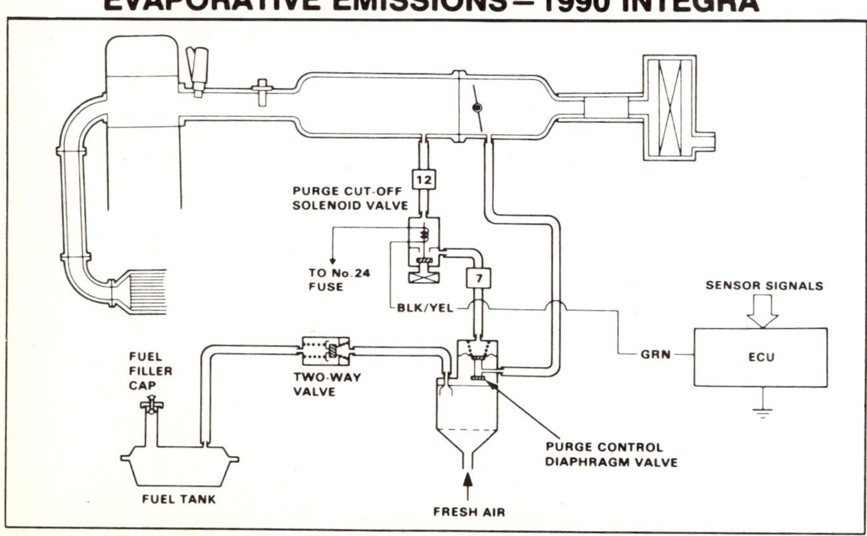


## EVAPORATIVE EMISSIONS—1990 INTEGRA

Troubleshooting Flowchart

Inspection of Evaporative Emission Controls

Disconnect vacuum hose from the purge control diaphragm valve (on the charcoal canister) and connect a vacuum gauge to the hose.

PURGE CONTROL DIAPHRAGM VALVE

VACUUM PUMP/GAUGE

Start the engine and allow to idle. NOTE: Engine coolant temperature must be below 80 °C (176°F).

Is there vacuum ?

NO →

YES → Disconnect the 4P connector.

GRN (−)

BLK/YEL (+)

Measure voltage between BLK/YEL (+) terminal and GRN (−) terminal.

Is there battery voltage ?

YES → Inspect vacuum hose routing. If OK, replace purge cut-off solenoid valve.

NO → Measure voltage between BLK/YEL (+) terminal and body ground. →

# EVAPORATIVE EMISSIONS—1990 INTEGRA

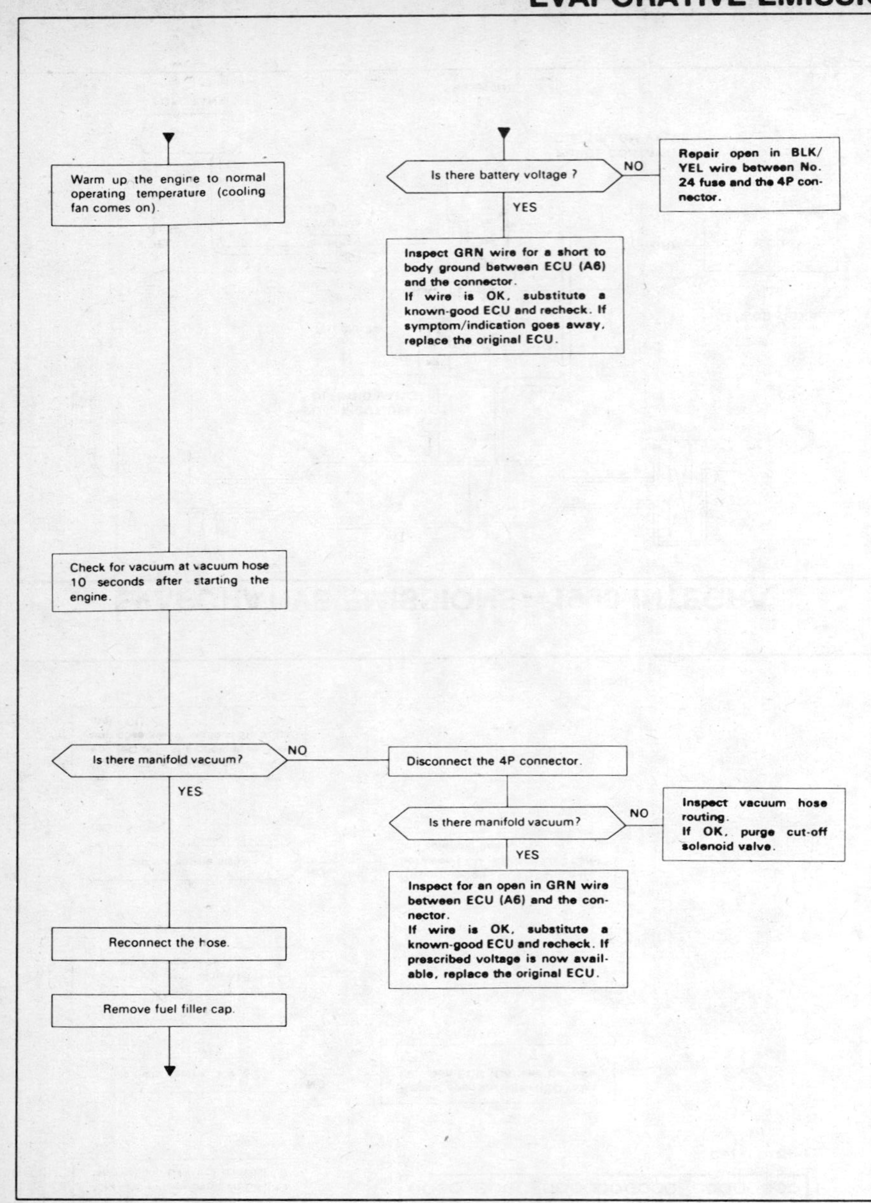

Warm up the engine to normal operating temperature (cooling fan comes on).
Check for vacuum at vacuum hose 10 seconds after starting the engine.
Is there manifold vacuum?
NO
YES
Reconnect the hose.
Remove fuel filler cap.
Is there battery voltage ?
NO
YES
Repair open in BLK/YEL wire between No. 24 fuse and the 4P connector.
Inspect GRN wire for a short to body ground between ECU (A6) and the connector.
If wire is OK, substitute a known-good ECU and recheck. If symptom/indication goes away, replace the original ECU.
Disconnect the 4P connector.
Is there manifold vacuum?
NO
YES
Inspect vacuum hose routing.
If OK, purge cut-off solenoid valve.
Inspect for an open in GRN wire between ECU (A6) and the connector.
If wire is OK, substitute a known-good ECU and recheck. If prescribed voltage is now available, replace the original ECU.


Connect a vacuum gauge to canister purge air hose.

VACUUM/PRESSURE GAUGE, 0—4 in. Hg

PURGE HOSE

PURGE AIR HOSE

Start the engine and raise speed to 3,500 rpm.

Does vacuum appear on gauge within 1 minute ?

YES: **See two-way valve test to complete. Evaporative emission control are OK.**

NO: Connect a vacuum gauge to the canister purge hose and raise the engine speed to 3,500 rpm.

Does vacuum appear on the gauge within 1 minute ?

NO: **Inspect the purge hose. If hose is OK, replace the throttle body.**

YES: **Replace the canister.**

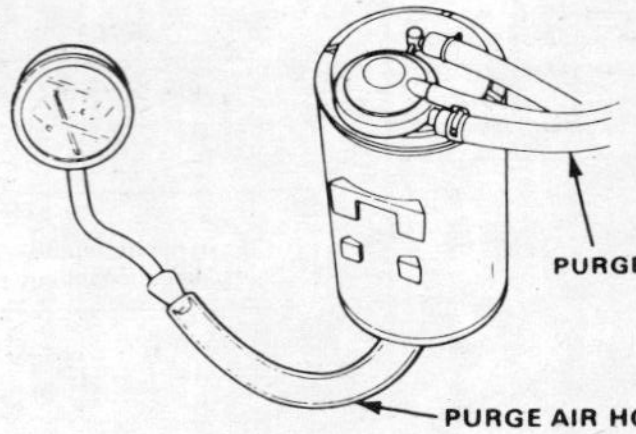

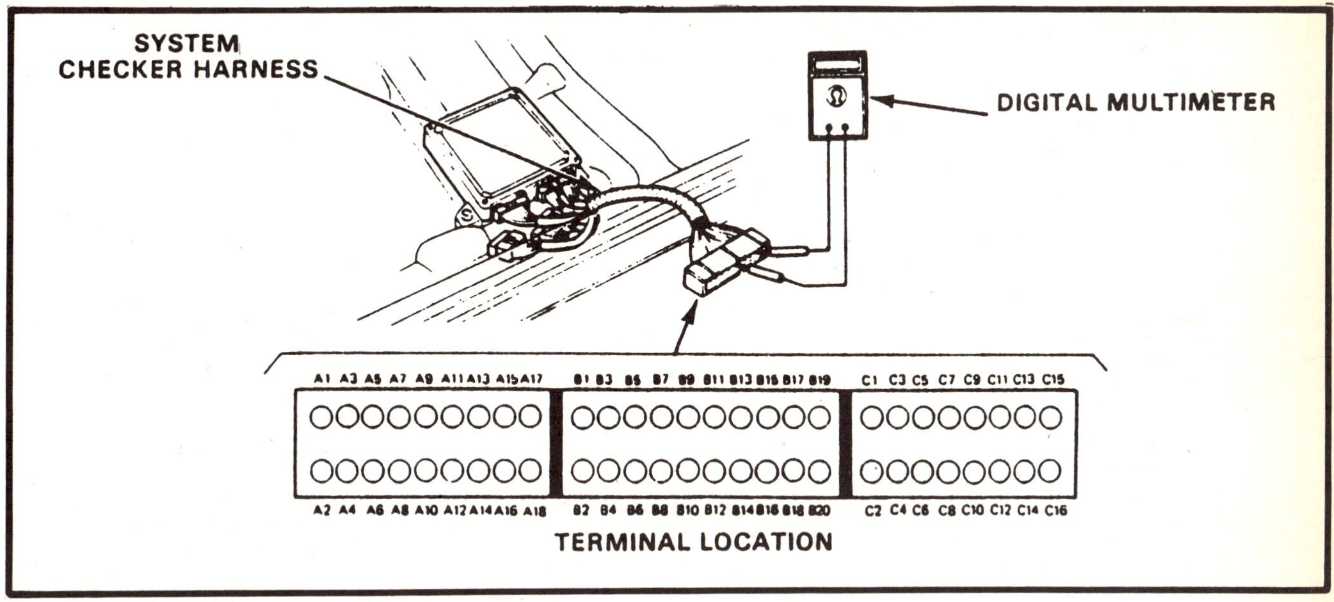


**ECU terminal identification and diagnostic tools**

## FAST IDLE VALVE

### Testing

#### ALL MODELS

**NOTE: Fast idle valve is factory adjusted, it should not be disassembled.**

1. Start engine.
2. Remove cover of fast idle valve.
3. Make sure that there is air flow with engine cold (coolant temperature below 86°F [30°C]). It can be detected by putting your finger on valve seat area.
4. If no air flow, replace fast idle valve.
5. Warm up to normal operating temperature (cooling fan comes on).
6. Check that valve is completely closed. If not, air is being sucked from valve seat area. It can be detected by putting your finger on valve seat area.
7. If any suction sound is heard, valve is leaking. Replace fast idle valve.

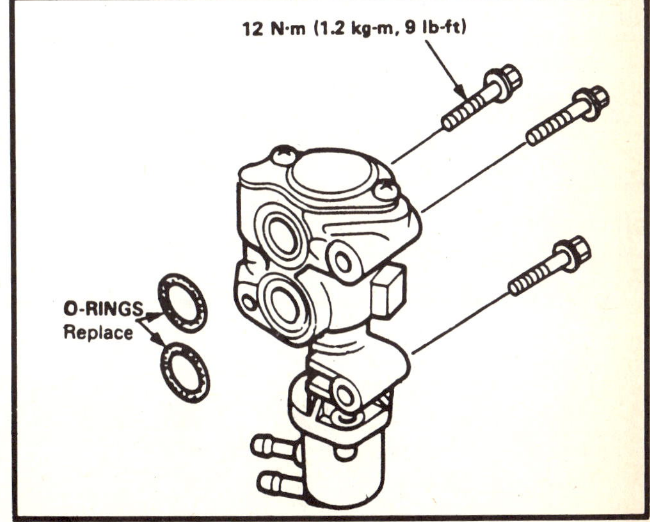


**Fast idle valve**

## DASHPOT

### Testing

#### INTEGRA

1. Check vacuum line for leaks, blockage or disconnected hose.
2. Disconnect vacuum hose from dashpot diaphragm, and connect a vacuum pump/gauge to hose.
3. Apply vacuum.
4. Rod should pull in and vacuum should hold.
5. If vacuum does not hold or rod does not move, replace dashpot diaphragm.

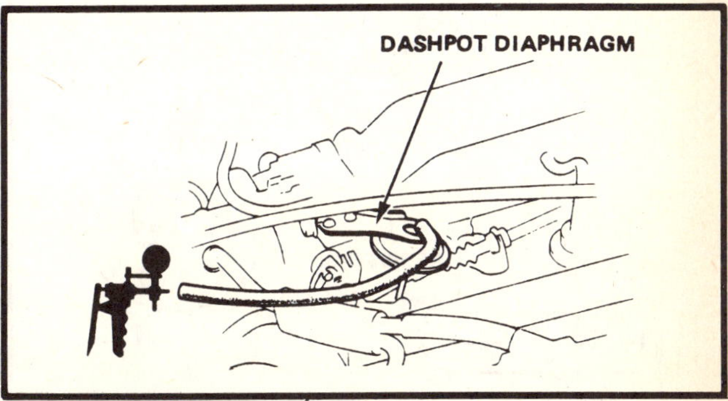


**Dashpot diaphragm**

## INJECTOR

### Testing

#### LEGEND

**NOTE: Check following items before testing idle speed, ignition timing, valve clearance and idle CO%.**

1. If engine will run, disconnect injector couplers with engine idling, and inspect change in idling speed. Idle should drop the same for each cylinder.

2. Check clicking sound of each injector by means of a stethoscope when engine is idling.

3. If any injector fails to make typical clicking sound, check wiring between ECU and injector. Voltage at injector coupler should fluctuate between 0–2 volts. If voltage is OK, replace injector.

4. If engine can not be started, remove coupler of injector, and measure resistance between terminals of injector. Resistance should be 1.5–2.5 ohms.

5. If resistance is not as specified, replace injector. If resistance is normal, check wiring between resistor and injector, wiring between resistors and control unit, and resistors.

**INTEGRA**

**Perform tests as listed in diagnosis chart under Code 16, for the proper year Integra.**

## FUEL SYSTEM RESISTOR

### Testing

**ALL MODELS**

1. Disconnect resistor coupler.

2. On Integra models, check for resistance between each of resistor terminals (E, D, C and B) and power terminal (A). Resistance should be 5–7 ohms.

3. On Legend and Sterling models, check for resistance between each of resistor terminals (g, f, e, d, c and b) and power terminal (a). Resistance should be 5–7 ohms.

4. Replace resistor if any of resistances are outside of specification.

## FUEL PRESSURE

### Testing

**ALL MODELS**

1. Relieve fuel pressure.

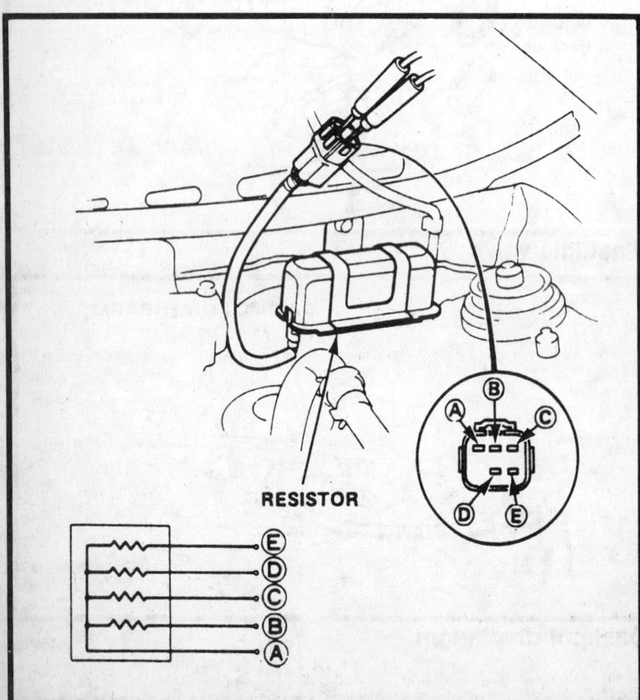


**Fuel injector resistor pin identification and location—1988–89 Integra shown**

2. Remove service bolt at the top of the fuel filter and attach fuel pressure gauge.

3. Start engine. Measure fuel pressure with engine idling and vacuum hose of pressure regulator disconnected. Pressure should be 36–41 psi (250–279 kPa).

4. If fuel pressure is not as specified, first check fuel pump. If pump is OK, check following:

a. If pressure is higher than specified, inspect for pinched or clogged fuel return hose or piping, and faulty pressure regulator.

b. If pressure is lower than specified, inspect for clogged fuel filter, pinched or clogged fuel hose from fuel tank to fuel pump, pressure regulator failure, leakage in fuel line, or pinched, broken or disconnected regulator vacuum hose.

## MAIN RELAY

### Testing

**ALL MODELS**

1. Remove main relay, near under-dash fuse box.

2. Connect battery positive terminal to No. 4 terminal and battery negative terminal to No. 8 terminal of main relay.

3. Check for continuity between No. 5 terminal and No. 7 terminal of main relay. If no continuity, replace main relay.

4. Connect battery positive terminal to No. 5 terminal and battery negative terminal to No. 2 terminal of main relay.

5. Check that there is continuity between No. 1 terminal and

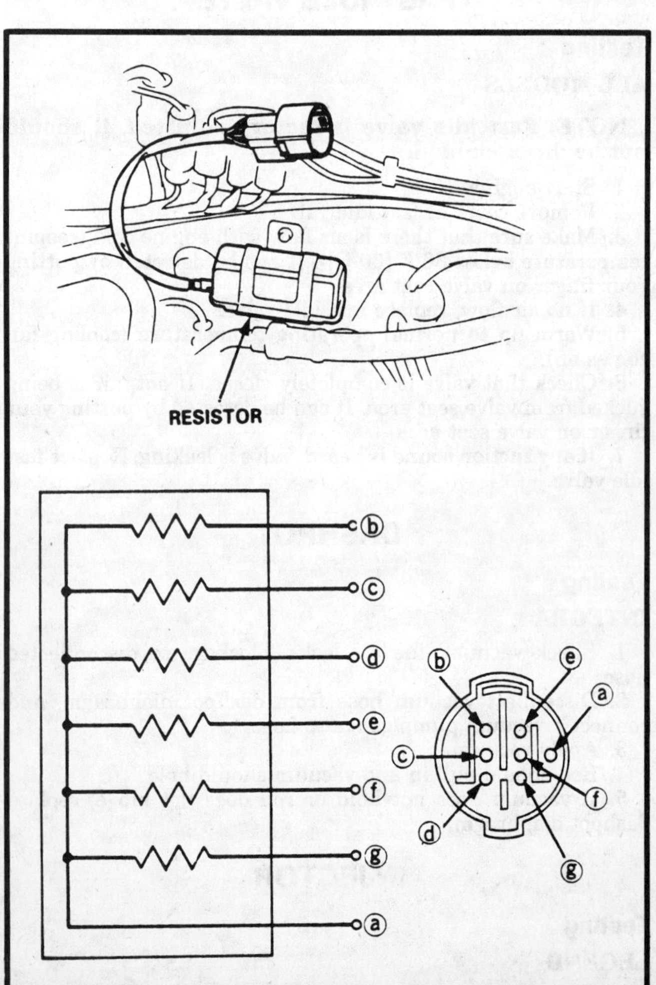


**Fuel injector resistor pin identification and location—Legend**

No. 3 terminal of main relay. If there is no continuity, replace main relay.

6. Connect battery positive terminal to No. 3 terminal and battery negative terminal to No. 8 terminal of main relay.

7. Check that there is continuity between No. 5 terminal and No. 7 terminal of main relay. If there is no continuity, replace main relay.

## HARNESS

### Testing

**LEGEND**

1. Keep ignition switch in **OFF** position.
2. Disconnect main relay connector.
3. Check continuity between Black wire in connector and body ground.
4. Connect positive probe of circuit tester to Yellow/Black wire and negative probe of tester to Black wire. Tester should read battery voltage.
5. If there is no voltage, check wiring between battery and main relay as well as ECU (15A) fuse.
6. Connect positive probe of tester to Black/Yellow wire of connector and ground negative probe of tester to Black wire.
7. Turn ignition switch **ON**. Tester should read battery voltage.
8. If no voltage, check wiring from ignition switch and main relay as well as fuse No. 9.
9. Connect positive probe of tester to Black/White wire and negative probe to Black wire.
10. Turn ignition switch to **START** position. Tester should read battery voltage.
11. If no voltage, check wiring between ignition switch and main relay as well as No. 13 (7.5A).
12. Connect a jumper wire between 2 Black/Yellow wires in connector. Turn ignition switch **ON**. Fuel pump should work.
13. If pump does not work, check wiring between battery and fuel pump and wiring from fuel pump to ground (Black wire).

**INTEGRA**

1. Keep ignition switch in **OFF** position.
2. Disconnect main relay connector.
3. Check continuity between Black wire in connector and body ground.
4. Connect positive probe of circuit tester to Yellow/White wire and negative probe of tester to Black wire. Tester should read battery voltage.
5. If there is no voltage, check wiring between battery and main relay as well as 15A fuse.
6. Connect positive probe of tester to Black/Yellow wire of connector and ground negative probe of tester to Black wire.
7. Turn ignition switch **ON**. Tester should read battery voltage.
8. If no voltage, check wiring from ignition switch and main relay as well as fuse No. 4 on 1988–89 models or fuse No. 24 on 1990.
9. Connect positive probe of tester to Blue/White wire and negative probe to Black wire.
10. Turn ignition switch to **START** position. Tester should read battery voltage.
11. If no voltage, check wiring between ignition switch and main relay as well as fuse (7.5A).
12. Connect a jumper wire between Black/Yellow and Yellow wire in connector. Turn ignition switch **ON**. Fuel pump should work.
13. If pump does not work, check wiring between battery and fuel pump and wiring from fuel pump to ground (Black wire).

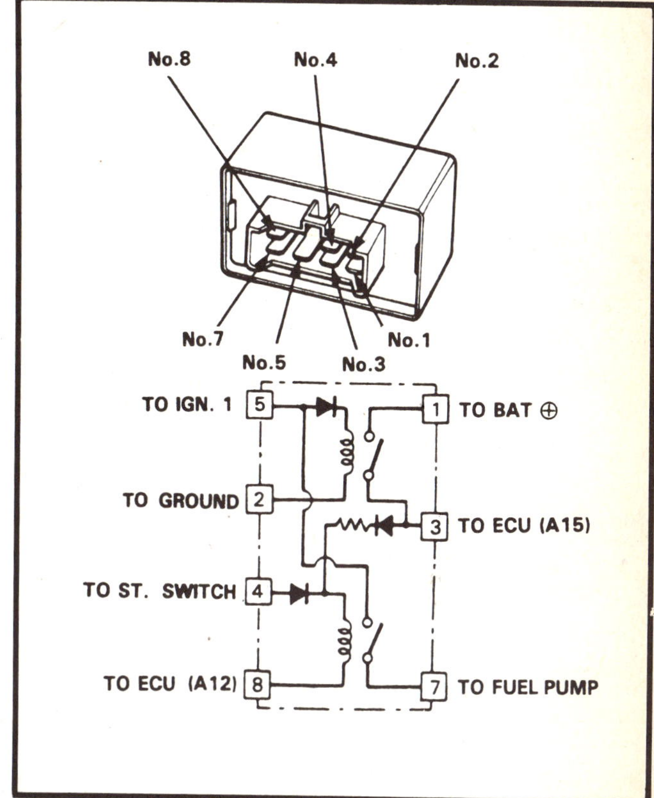


**Main relay pin identification**

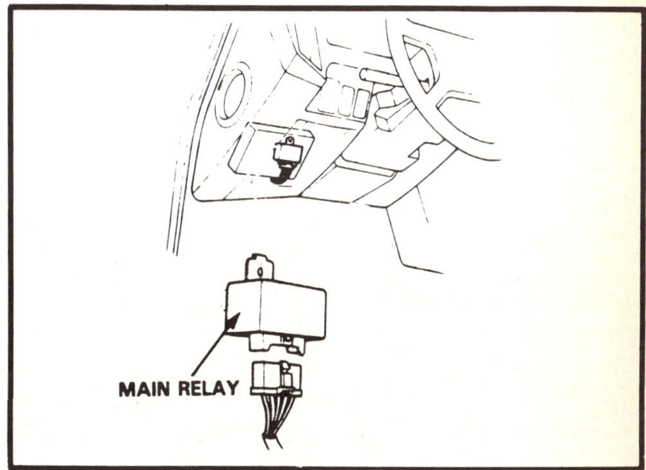


**Main relay location—Legend**

## BYPASS VALVE

### Testing

**LEGEND**

**NOTE: Do not adjust the bypass valve full-close screw. It was preset at the factory.**

1. Check bypass valve shaft for binding or sticking.
2. Check bypass valve for smooth movement.
3. Check that lever A of bypass valve is in close contact with stopper when bypass valve is fully open.
4. Check that lever B of bypass valve is in close contact with full-close screw when valve is fully closed.

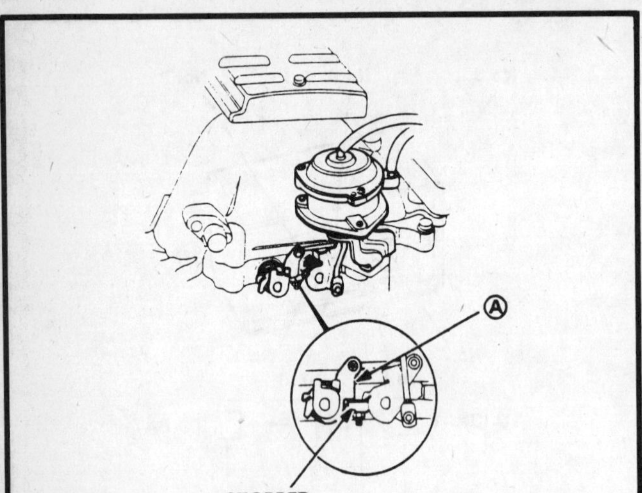

**Bypass valve lever A and lever stopper – Legend**

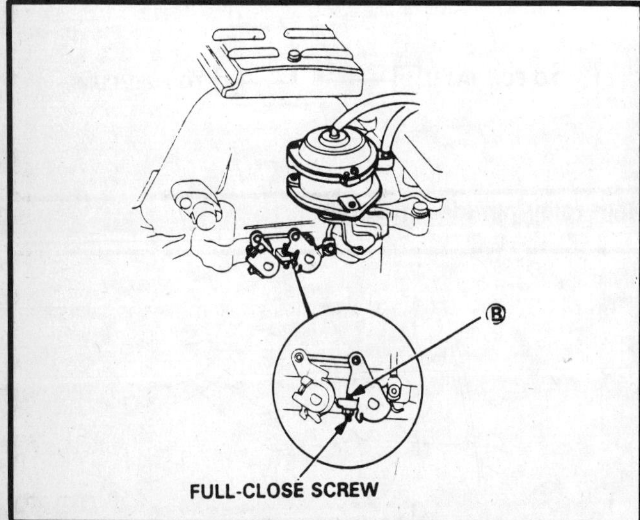

**Bypass valve lever B and full-close screw – Legend**

5. If any fault is found, clean linkage and shafts with carburetor cleaner. If problem still exists after cleaning, disassemble intake manifold and check bypass valve.

## CANISTER TWO-WAY VALVE

### Testing

**INTEGRA**

1. Remove the fuel tank cap.
2. Remove the vapor line from the fuel tank and connect to T-fitting from a vacuum test gauge and vacuum pump.
3. Apply vacuum slowly and continuously. Vacuum should stabilize momentarily at 0.2–0.6 in. Hg. If not install a new valve and retest.
4. Move vacuum pump hose from vacuum to pressure fitting, and move the gauge hose from the vacuum to pressure side.
5. Slowly pressurize the vapor line while watching the gauge. Pressure should stablize at 0.4–1.4 in. Hg. If not replace the valve and retest.

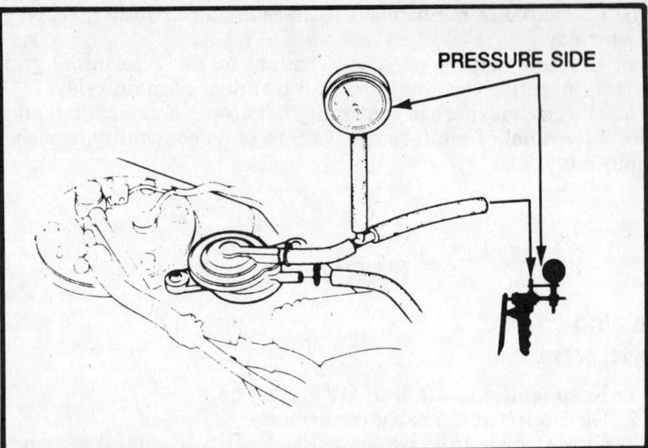

**Canister 2-way valve pressure test**

**LEGEND**

1. Remove the fuel tank cap.
2. Remove the vapor line from the fuel tank and connect to T-fitting from a vacuum test gauge and vacuum pump.
3. Apply vacuum slowly and continuously. Vacuum should stabilize momentarily at 0.2–0.6 in. Hg. If not install a new valve and retest.
4. Move vacuum pump hose from vacuum to pressure fitting, and move the gauge hose from the vacuum to pressure side.
5. Slowly pressurize the vapor line while watching the gauge. Pressure should stablize at 1.0–2.2 in. Hg. If not replace the valve and retest.

# Component Replacement

## FUEL SYSTEM

### Pressure Relieving

**ALL MODELS**

**CAUTION**

***Keep open flames or sparks from work area. Do not smoke while working on fuel system. Be sure to relieve fuel pressure while engine is off.***

**NOTE: Before disconnecting fuel pipes or hoses, release pressure from system by loosen 6mm service bolt at top of fuel filter.**

1. Disconnect battery negative cable from battery negative terminal.
2. Use a box end wrench on 6mm service bolt at top of fuel filter, while holding special Banjo bolt with another wrench.
3. Place a rag or a shop towel over 6mm service bolt.
4. Slowly loosen 6mm service bolt 1 complete turn.

**NOTE: A fuel pressure gauge can be attached at 6mm service bolt hole. Always replace washer between service bolt and special Banjo bolt, whenever service bolt is loosened to relieve fuel pressure. Replace all washers whenever bolts are removed to disassemble parts.**

## IDLE SPEED

### Adjustment

**LEGEND**

1. Start engine and warm up to normal operating temperature; cooling fan will come on.
2. Connect a tachometer.

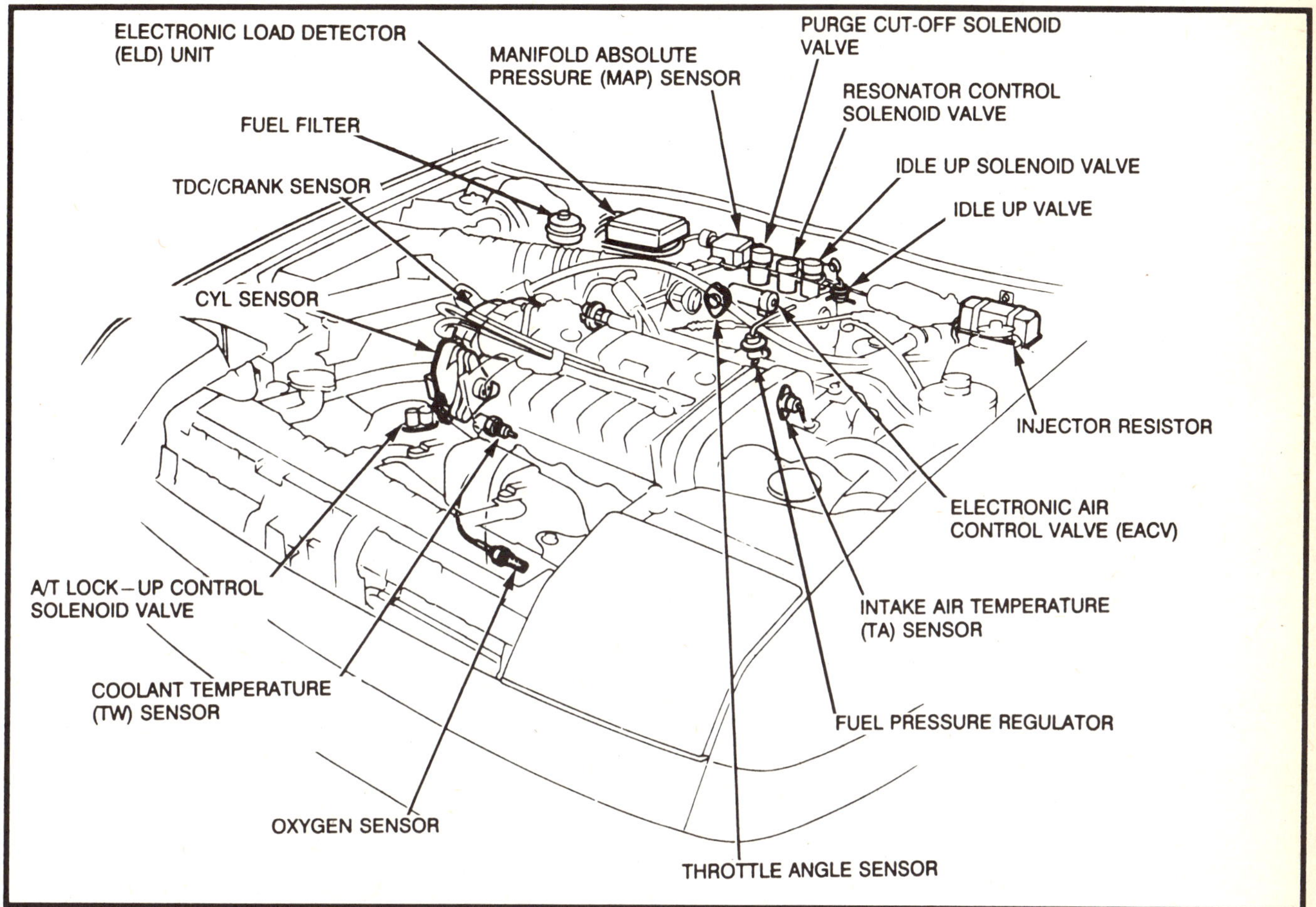

ELECTRONIC LOAD DETECTOR (ELD) UNIT
MANIFOLD ABSOLUTE PRESSURE (MAP) SENSOR
PURGE CUT-OFF SOLENOID VALVE
RESONATOR CONTROL SOLENOID VALVE
FUEL FILTER
IDLE UP SOLENOID VALVE
TDC/CRANK SENSOR
IDLE UP VALVE
CYL SENSOR
INJECTOR RESISTOR
ELECTRONIC AIR CONTROL VALVE (EACV)
A/T LOCK-UP CONTROL SOLENOID VALVE
INTAKE AIR TEMPERATURE (TA) SENSOR
COOLANT TEMPERATURE (TW) SENSOR
FUEL PRESSURE REGULATOR
OXYGEN SENSOR
THROTTLE ANGLE SENSOR

**Component locations – 1988–89 Integra shown**

SERVICE BOLT
12 N·m (1.2 kg-m, 9 lb-ft)
SHOP TOWEL
FUEL FILTER

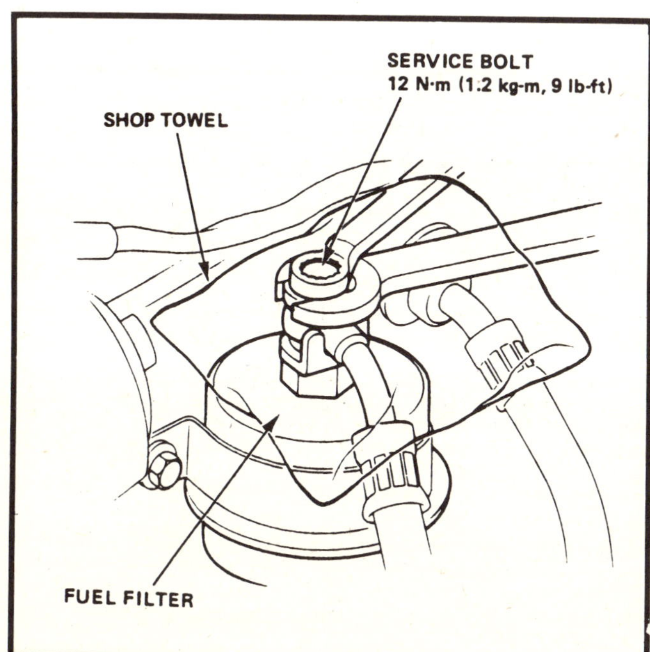

**Fuel pressure relief – Acura Integra shown, others similar**

3. Set steering in straight forward condition, and check idling in no-load conditions in which all accessories are turned **OFF** and/or not operating.

4. Check idle speed. Idle speed should be 680±50 rpm in **NEUTRAL**.

5. Check yellow LED display at ECU under passenger's seat.

a. If yellow LED is OFF, do not adjust idle adjusting screw.

b. If yellow LED is BLINKING, adjust idle screw ¼ turn clockwise.

c. If yellow LED is ON, adjust idle screw ¼ turn counterclockwise.

**NOTE: The yellow LED display may be lit at earlier stages, for example, when the distance covered is within 310 miles (500 km). However, no adjustment should be made. Check that the yellow LED goes OFF after approximately 30 seconds. If it does not go OFF, rotate the idle adjusting screw by ¼ turn in the same direction, and repeat the same operation until the yellow LED goes OFF.**

6. Check idle speed with highbeam headlights and rear window defogger **ON**. Idle should remain stable at specified idle speed.

7. Check idle speed with air conditioner compressor **ON**. Idle should remain stable at specified idle speed.

8. If equipped with automatic transmission, check idle speed with transmission in gear. Idle should remain stable at specified idle speed.

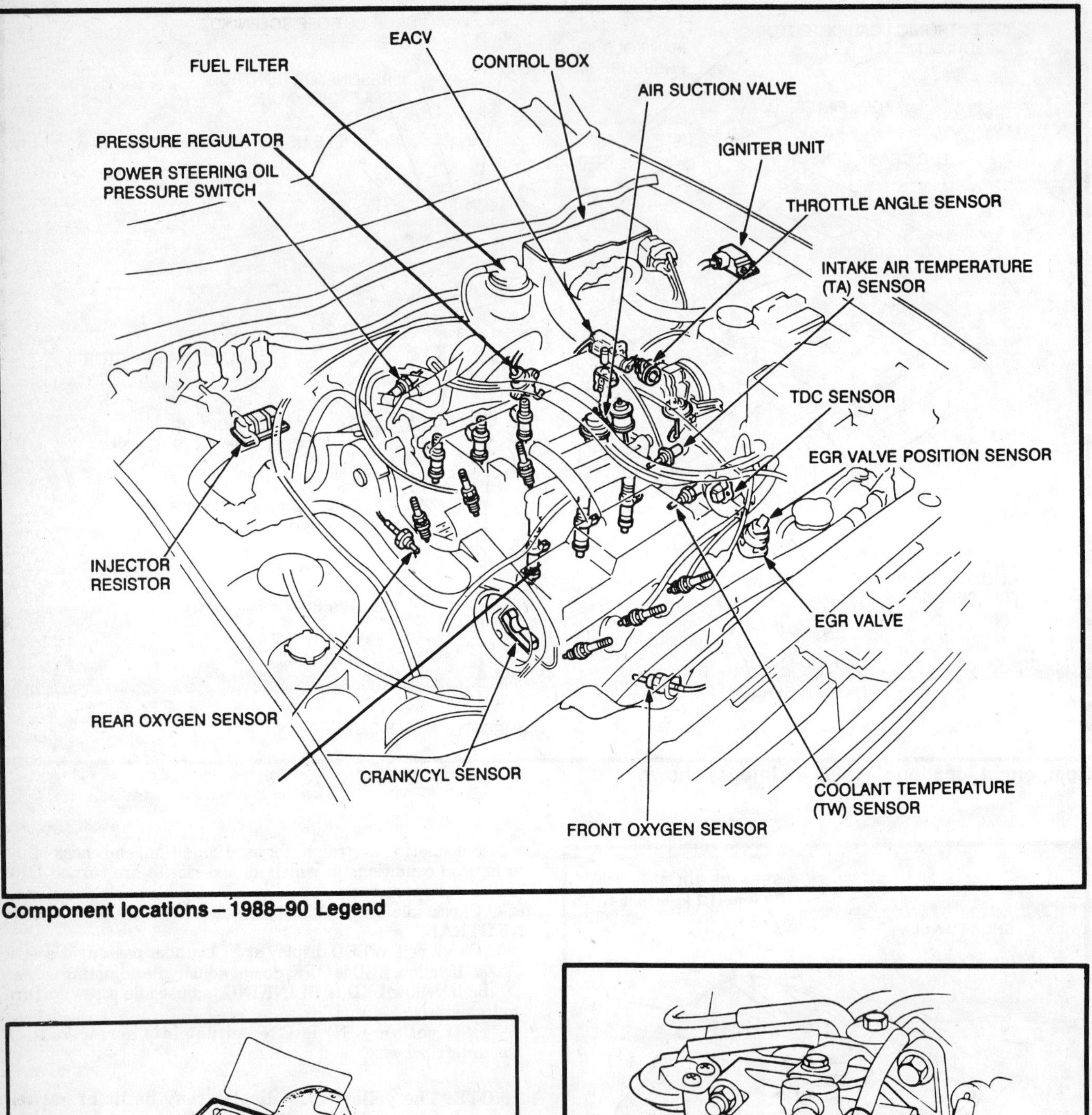

**Component locations – 1988–90 Legend**

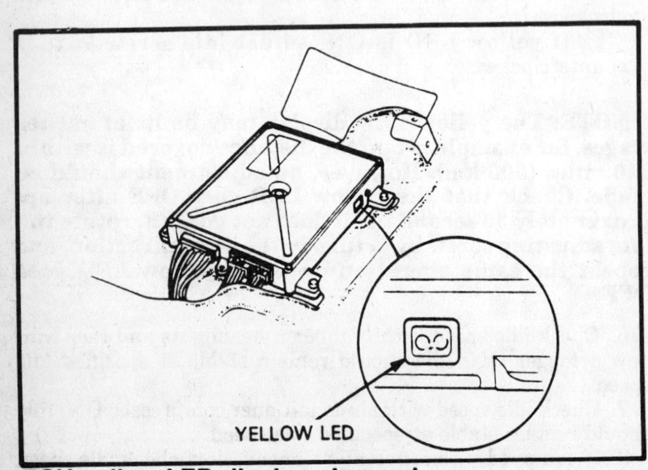


**ECU yellow LED display – Legend**

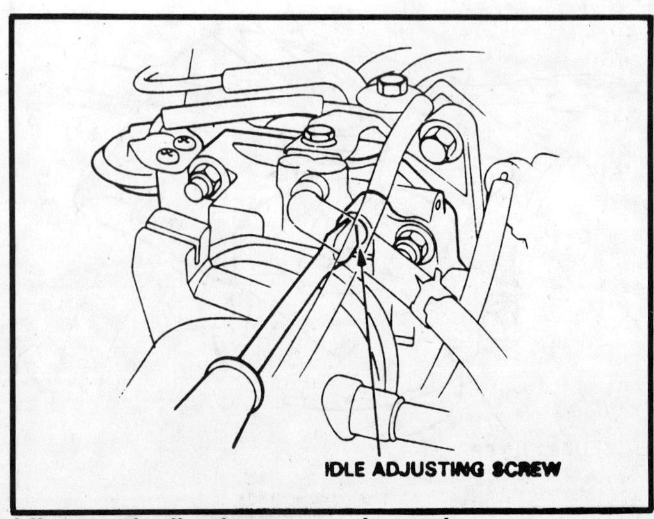


**Idle speed adjusting screw – Legend**

**1988–89 INTEGRA**

1. Start engine and warm up to normal operating temperature; cooling fan will come on.
2. Connect a tachometer.
3. With engine idling, disconnect connector at EICV.
4. With all accessories **OFF**, check idle speed. Idle speed should be 550±50 rpm (in **NEUTRAL**). Adjust idle speed, if necessary, by turning idle adjusting screw.

**NOTE: If the idle is excessively high, check the throttle control system.**

5. After adjustment, reconnect connector at EICV.
6. Disconnect hazard fuse at battery positive terminal for at least 10 seconds to reset ECU memory.
7. Start engine and warm it up to normal operating temperature (cooling fan comes on).
8. With all accessories **OFF**, check idle speed. Idle speed should be 700±50 rpm in **NEUTRAL.**
9. Check idle speed with highbeam headlights and rear window defogger **ON**. Idle should remain stabilize at 750±50 rpm with manual transmission or 700±50 rpm with automatic transmission after 1 minute.
10. Check idle speed with air conditioner compressor ON and blower on **HI**. Idle should remain stabilize 750±50 rpm with manual transmission or 730±50 rpm with automatic transmission after 1 minute.
11. If idle speed is out of specification and LED does not blinking a Code 14, check air conditioning, alternator, transmission shift, brake, starter and power steering signals.
12. Also check fast idle and idle boost valves, EACV O-rings and all hose and wire connections.
13. If idle speed is still out of specification and LED does not blink a Code 14, replace the EACV and readjust the idle.
14. If idle speed is still out of specification and LED still does not blink a Code 14, replace the ECU and readjust the idle.

**1990 INTEGRA**

1. Start engine and warm up to normal operating temperature; cooling fan will come on.
2. Connect a tachometer.
3. With engine idling, disconnect connector at EICV.
4. With all accessories **OFF**, check idle speed. Idle speed should be 650±50 rpm (in **NEUTRAL**). Adjust idle speed, if necessary, by turning idle adjusting screw.

**NOTE: If the idle is excessively high, check the throttle control system.**

5. Turn ignition switch **OFF**.
6. After adjustment, reconnect connector at EICV.

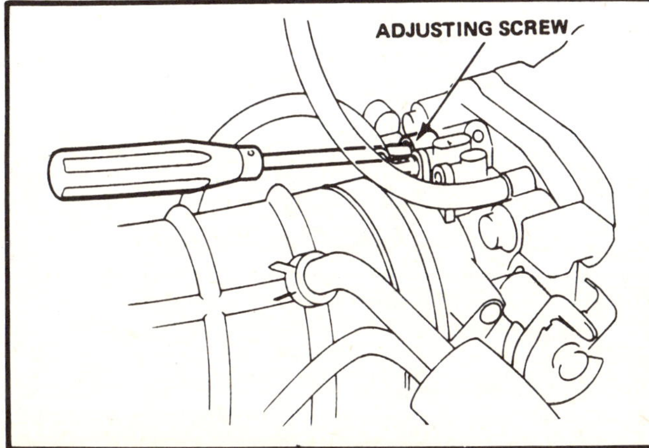

**Idle speed adjusting screw—Integra**

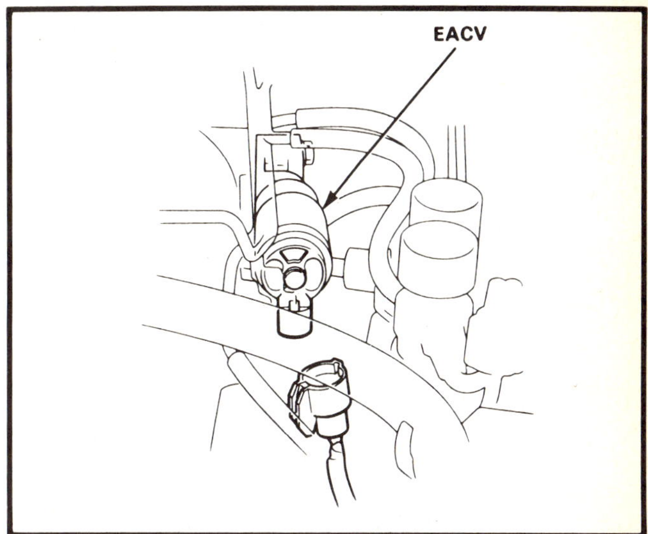

**EACV connector—1990 Integra shown**

7. Disconnect hazard fuse at battery positive terminal for at least 10 seconds to reset ECU memory.
8. Start engine and warm it up to normal operating temperature (cooling fan comes on).
9. With all accessories **OFF**, check idle speed. Idle speed should be 750±50 rpm in **NEUTRAL.**
10. Check idle speed with highbeam headlights and rear window defogger **ON**. Idle should remain stabilize at 750±50 rpm after 1 minute.
11. Check idle speed with air conditioner compressor ON and blower on **HI**. Idle should stabilize at 750±50 rpm after 1 minute.
12. If idle speed is out of specification and LED does not blinking Code 14, check air conditioning, alternator, transmission shift, brake, starter and power steering signals. Also check fast idle and idle boost valves, EACV O-rings and all hose and wire connections.
13. If idle speed is still out of specification and LED does not blink a Code 14, replace the EACV and readjust the idle.
14. If idle speed is still out of specification and LED still does not blink a Code 14, replace the ECU and readjust the idle.

## CRANK ANGLE SENSOR (CRANK/CYL SENSOR)

### Removal and Installation

**1988–89 ACURA INTEGRA**

1. Disconnect connector from crank/cyl sensor.
2. Remove 3 bolts and remove sensor.
3. To install, reverse removal procedure. Make sure to use new O-ring with installing.

**LEGEND**

1. Remove cruise control actuator.
2. Remove upper cover at front side of timing belt.
3. Remove timing belt.
4. Remove 3 bolts and detach pulley.
5. Remove 4 bolts and detach front side cover plate of timing belt.

**To install:**

6. Place sensor on cylinder head.
7. Align cam pulley pin with cam shaft hole to install cam pulley.
8. Complete installation by reversing removal procedure. Make sure cam pulley is aligned properly before install timing belt.

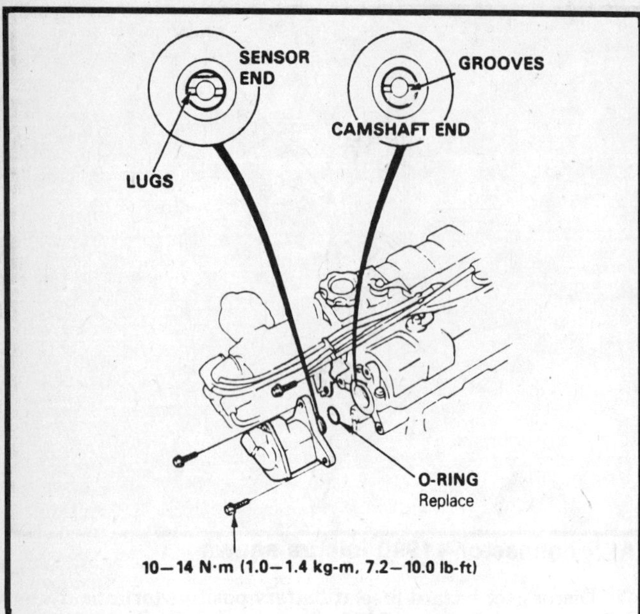


Crank angle sensor servicing – 1988–89 Integra

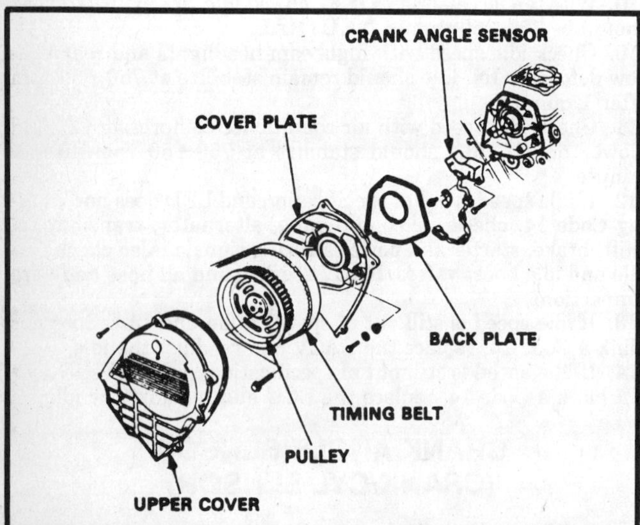


Crank angle sensor (Crank/CYL sensor) servicing – Legend

## Overhaul

### 1988–89 INTEGRA

1. Carefully pry up the CYL rotor by using equal pressure on both sides of rotor.
2. Pull the CYL coil assembly and mount rubber out from the sensor housing by removing the screws.
3. Remove the C-clip.
4. Slide off the pin retainer being careful not to stretch it.
5. Separate the coupling from the shaft by removing the roll pin.
6. Remove the rotor shaft.

**To assemble:**

7. Apply lubricate, install washers on shaft, install shaft into the housing and install a new C-clip.
8. Install the coupling with its index mark aligned with index on housing, install pin and retainer.

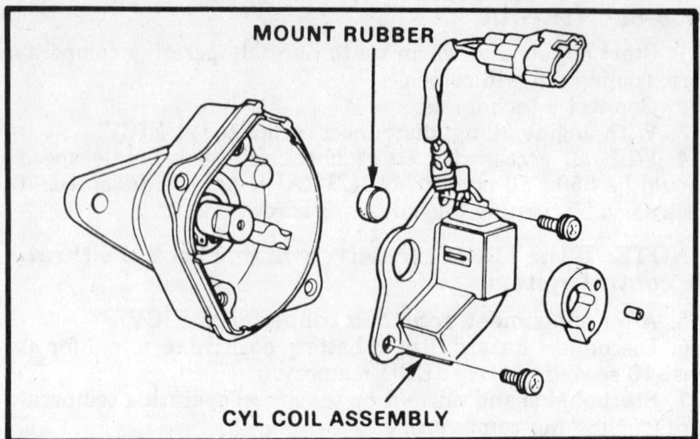


CYL assembly coil mounting

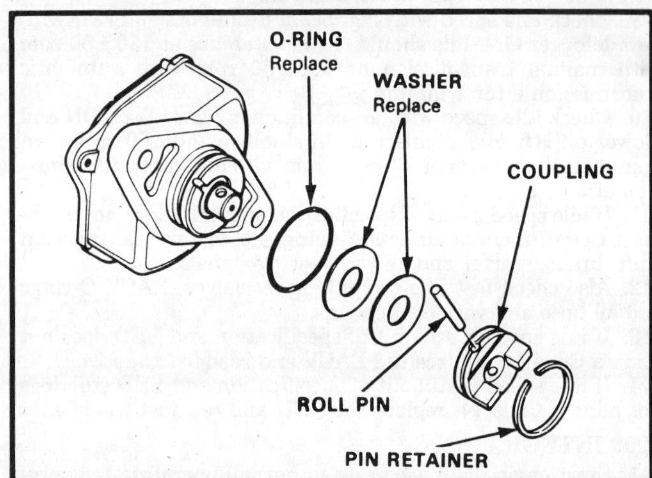


CYL assembly lower shaft

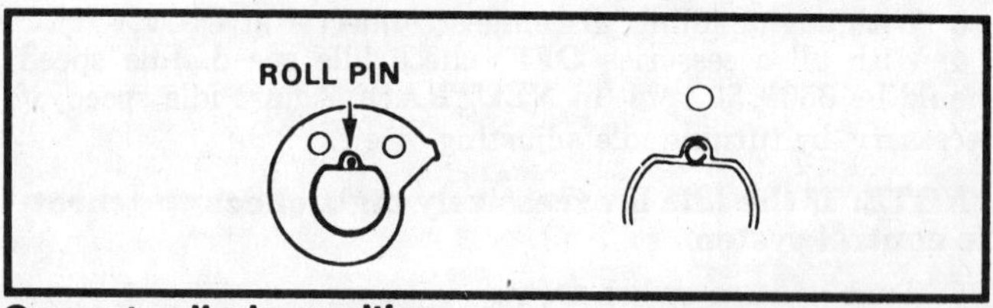


Correct roll pin position

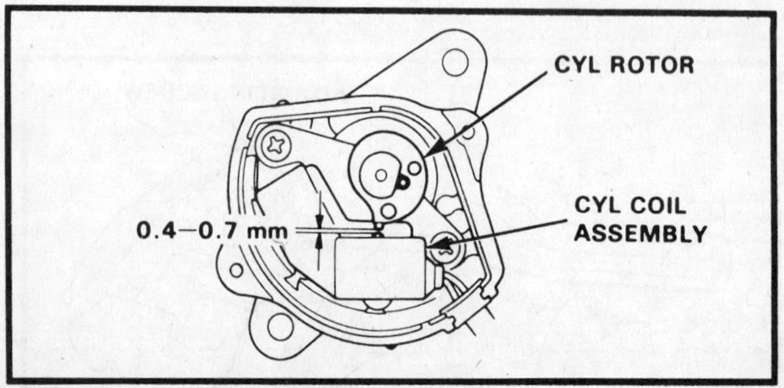


CYL sensor rotor air gap

9. Install the rubber mount, the install the CYL coil assembly and the CYL rotor. Adjust air gap to 0.01–0.02 in. (0.4–0.7mm).
10. Install the rotor with part number facing up.
11. Install O-ring onto housing and install sensor.

## TOP DEAD CENTER (TDC) SENSOR

### Removal and Installation

**ALL MODELS**

1. Remove the distributor.
2. Remove distributor cap from distributor.
3. Replace the distributor and TDC sensor as an assembly.
4. To install, reverse removal procedure. Be sure to use new O-ring.

## TDC/CRANK/CYL SENSOR

### Removal and Installation

**1990 INTEGRA**

1. Disconnect the 2 connectors at the distributor.
2. Remove the distributor assembly.
3. Remove distributor cap from distributor.
4. The TDC/CRANK/CYL sensor and the distributor are serviced as an assembly.
5. To install, reverse removal procedure. Be sure to use new O-ring.
6. Adjust timing.

## THROTTLE BODY

### Removal and Installation

1. Disconnect the negative battery cable.
2. Release clamp and remove air inlet from throttle body.
3. Tag and remove all electrical connectors at throttle body.
4. Tag and remove all hoses attached to throttle body.
5. Disconnect throttle cable and if equipped the automatic transmission cable from throttle body.
6. Remove throttle body from intake on Integra or insulator plate on Legend.
7. Service insulator plate, if equipped, as needed.

**To install:**

8. Replace gaskets and insulator plate if equipped and torque to 16 ft. lbs. (22 Nm).
9. Install throttle body and torque nuts or bolts to 16 ft. lbs. (22 Nm).
10. Connect all hoses, electrical connector and cables.
11. Adjust cables and make any necessary adjustments.
12. Install air inlet to throttle body and connect the negative battery cable.

## FUEL INJECTOR

### Removal and Installation

**INTEGRA**

1. Disconnect battery negative cable from battery negative.
2. Relieve fuel pressure.
3. Disconnect connector of injectors.
4. Disconnect vacuum hose and fuel return hose from pressure regulator.

**CAUTION**

***Place a rag or shop towel over hose and tube before disconnecting, to contain fuel spray.***

5. Remove pulsation damper.
6. Remove connector holder.
7. Loosen retainer nuts on fuel pipe.
8. Disconnect fuel pipe.
9. Remove injector from intake manifold.
10. Slide new cushion onto injector.
11. Coat new O-rings with clean engine oil and put O-rings on injectors.
12. Insert injectors into fuel pipe first.

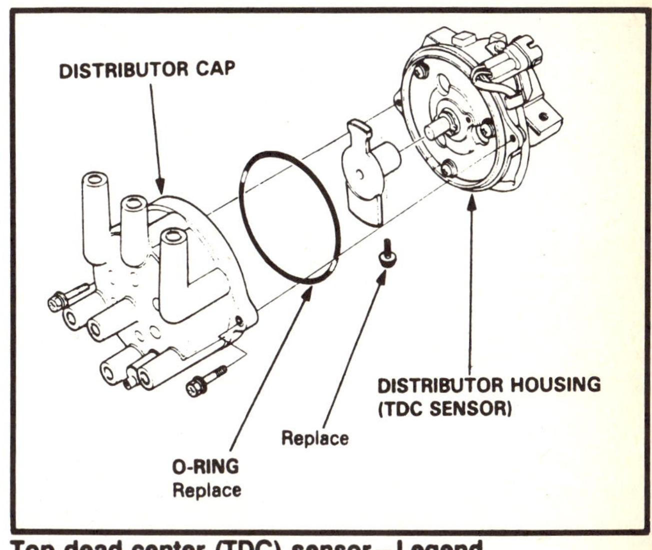


Top dead center (TDC) sensor—Legend

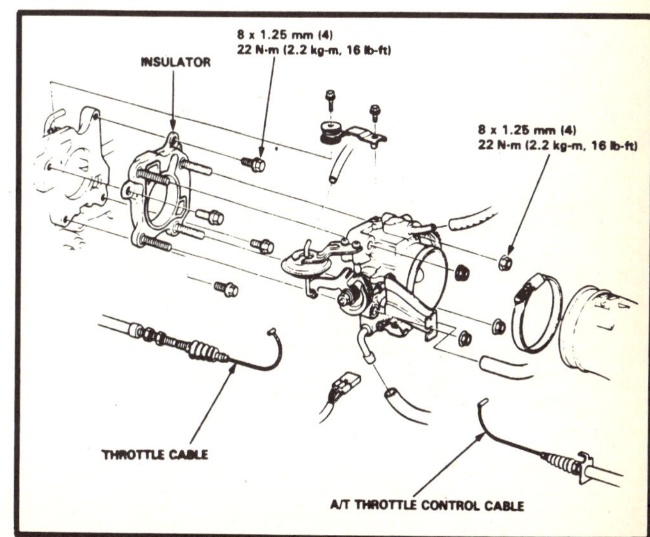


Throttle body servicing—Legend

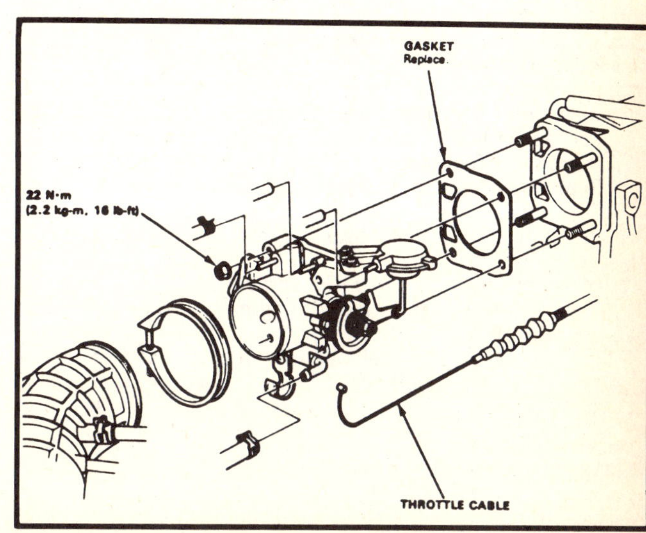


Throttle body—1990 Integra shown

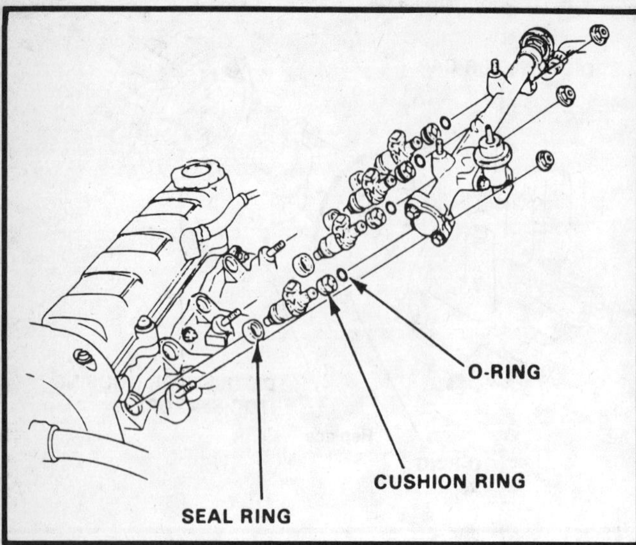

**Fuel injector service – Integra**

13. Coat new seal rings with clean engine oil and press into intake manifold.
14. Install injector and fuel pipe assembly in manifold.

**NOTE: To prevent damage to O-ring, install injectors in fuel pipe first, then install in intake manifold.**

15. Align center line on connector with mark on fuel pipe.
16. Install and tighten retainer nuts.
17. Install pulsation damper.
18. Connect vacuum hose ad fuel return hose to pressure regulator.
19. Install connectors on injectors.
20. Turn ignition switch **ON** but do not operate starter. After fuel pump runs for approximately 2 seconds, fuel pressure in fuel line rises. Repeat this 3 times, then check whether there is any fuel leakage.

**LEGEND**

1. Disconnect battery negative cable from battery negative.
2. Relieve fuel pressure.
3. Disconnect connector of injectors.
4. Disconnect vacuum hose and fuel return hose from pressure regulator.

**CAUTION**

*Place a rag or shop towel over hose and tube before disconnecting, to contain fuel spray.*

5. Disconnect fuel hose from fuel pipe.
6. Loosen retainer nuts on fuel pipe and harness holder.
7. Disconnect fuel pipe.
8. Remove injector from intake manifold.
9. Slide new cushion onto injector.
10. Coat new O-rings with clean engine oil and put O-rings on injectors.
11. Insert injectors into fuel pipe first.
12. Coat new seal rings with clean engine oil and press into intake manifold.
13. Install injector and fuel pipe assembly in manifold.

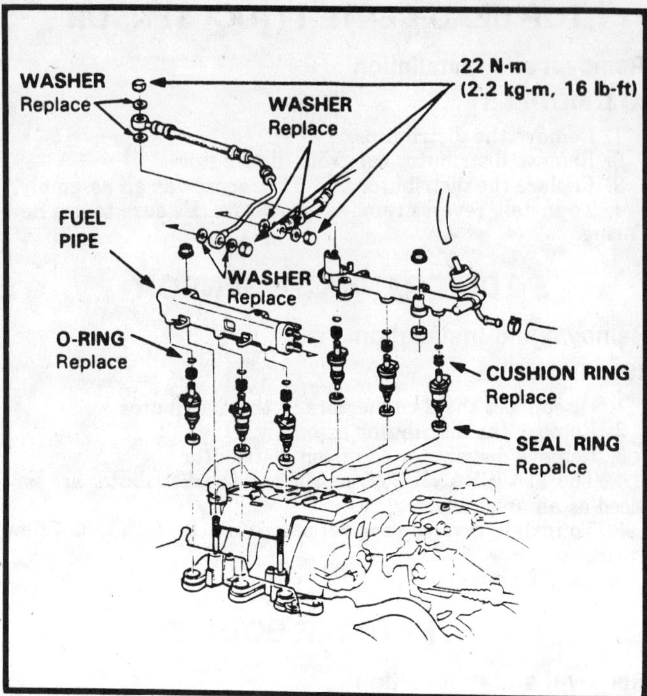

**Fuel injector servicing – Legend**

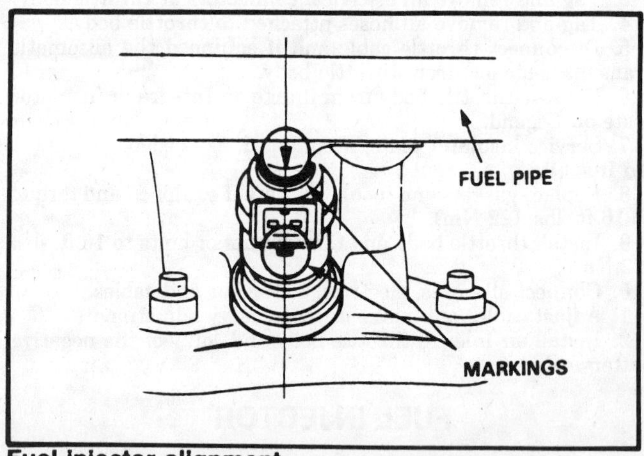

**Fuel injector alignment**

**NOTE: To prevent damage to O-ring, install injectors in fuel pipe first, then install in intake manifold.**

14. Align center line on connector with mark on fuel pipe.
15. Install and tighten retainer nuts.
16. Connect vacuum hose ad fuel return hose to pressure regulator.
17. Install connectors on injectors.
18. Turn ignition switch **ON** but do not operate starter. After fuel pump runs for approximately 2 seconds, fuel pressure in fuel line rises. Repeat this 3 times, then check whether there is any fuel leakage.

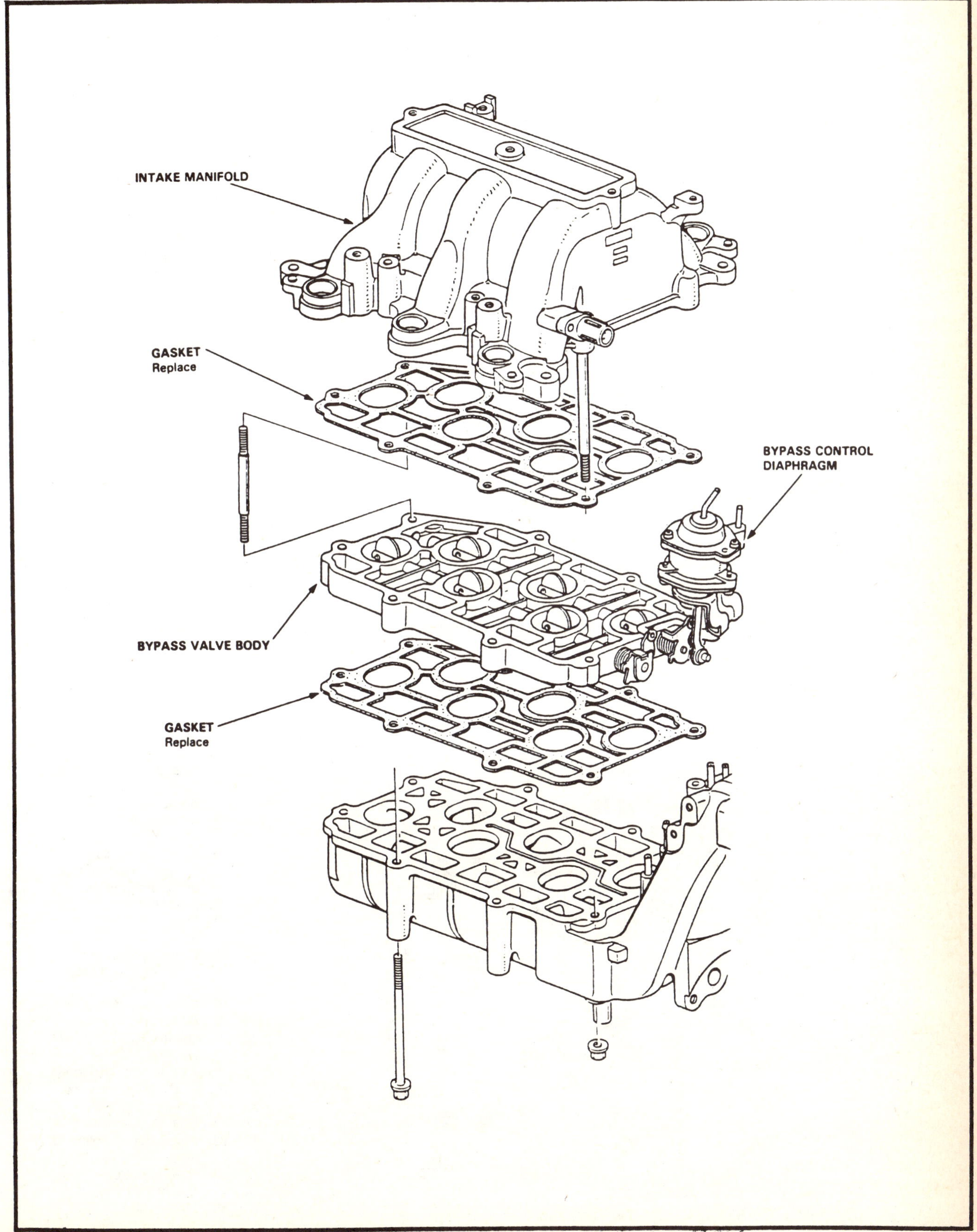


**Bypass valve body, exploded view—1987-88 Legend Coupe and 1988 Legend Sedan**

1. Manifold absolute pressure sensor
2. Pressure regulator
3. Fuel tank
4. Fuel pump
5. Fuel filter
6. Fuel injector
7. 2-way valve
8. Charcoal canister
9. Purge control valve
10. Dashpot diaphragm
11. Idle up solenoid valve
12. Resonator
13. Resonator diaphragm
14. Resonator control solenoid valve
15. Purge cut-off solenoid valve
16. Idle up valve
17. Electronic air control valve (EACV)
18. Vacuum tank

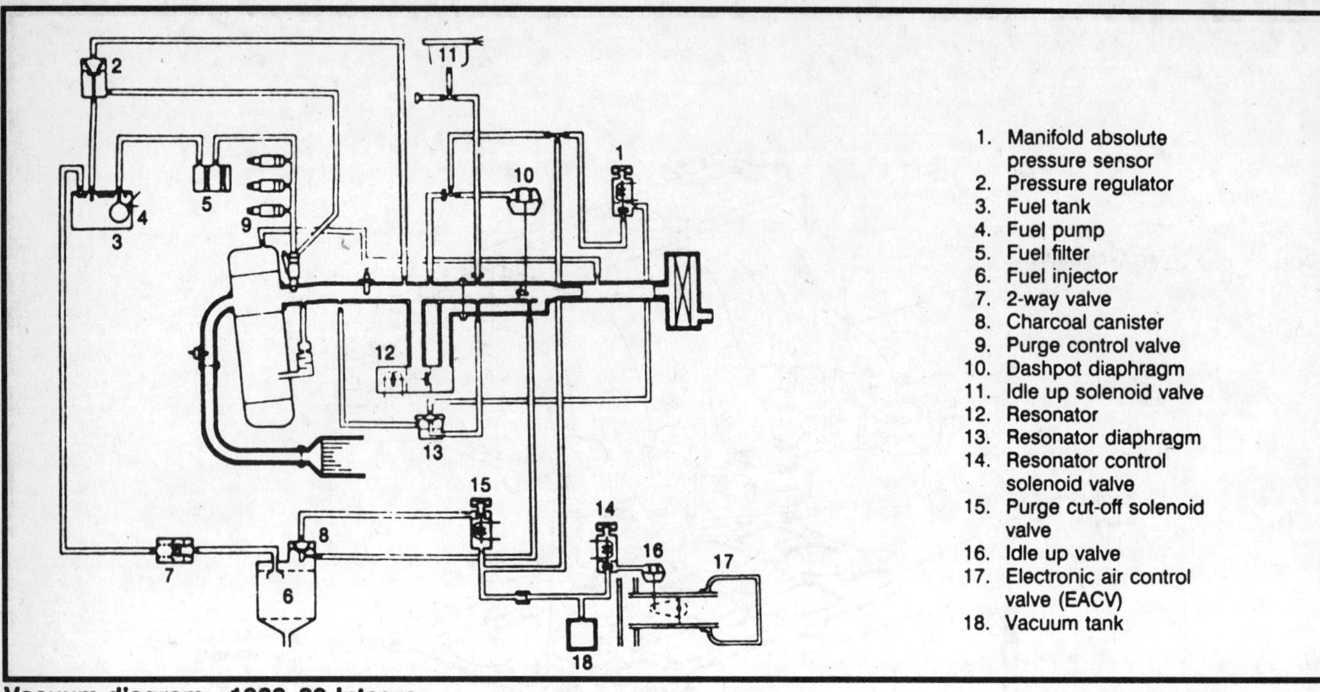

**Vacuum diagram – 1988–89 Integra**

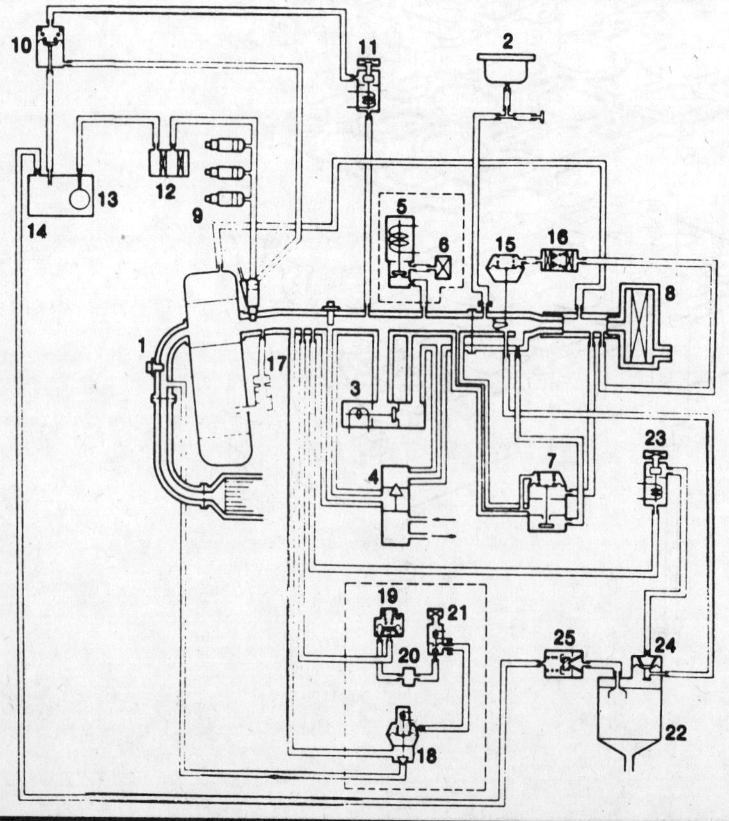

1. Oxygen sensor
2. Manifold Absolute Pressure (MAP) sensor
3. Electronic Air Control Valve (EACV)
4. Fast idle valve
5. Fast idle control solenoid valve (A/T only)
6. Air filter
7. Air boost valve
8. Air cleaner
9. Fuel injector
10. Pressure regulator
11. Pressure regulator cut-off solenoid valve
12. Fuel filter
13. Fuel pump
14. Fuel tank
15. Dashpot diaphragm
16. Dashpot check valve
17. PCV valve
18. EGR valve (A/T only)
19. Constant Vacuum Control (CVC) valve (A/T only)
20. Air chamber (A/T only)
21. EGR control solenoid valve (A/T only)
22. Charcoal canister
23. Purge cut-off solenoid valve
24. Purge control diaphragm valve
25. 2-way valve

**Vacuum diagram – 1990 Integra**

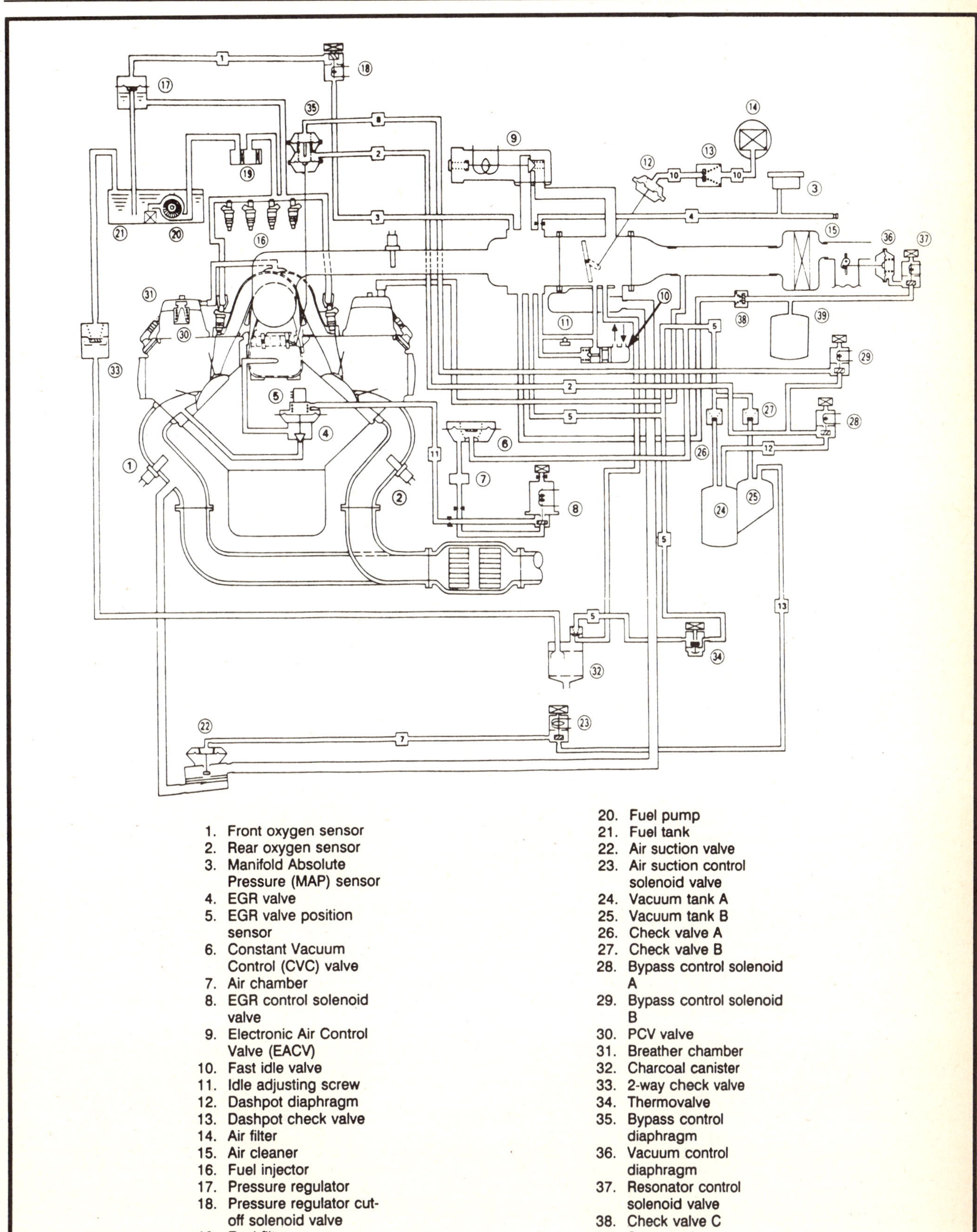

1. Front oxygen sensor
2. Rear oxygen sensor
3. Manifold Absolute Pressure (MAP) sensor
4. EGR valve
5. EGR valve position sensor
6. Constant Vacuum Control (CVC) valve
7. Air chamber
8. EGR control solenoid valve
9. Electronic Air Control Valve (EACV)
10. Fast idle valve
11. Idle adjusting screw
12. Dashpot diaphragm
13. Dashpot check valve
14. Air filter
15. Air cleaner
16. Fuel injector
17. Pressure regulator
18. Pressure regulator cut-off solenoid valve
19. Fuel filter
20. Fuel pump
21. Fuel tank
22. Air suction valve
23. Air suction control solenoid valve
24. Vacuum tank A
25. Vacuum tank B
26. Check valve A
27. Check valve B
28. Bypass control solenoid A
29. Bypass control solenoid B
30. PCV valve
31. Breather chamber
32. Charcoal canister
33. 2-way check valve
34. Thermovalve
35. Bypass control diaphragm
36. Vacuum control diaphragm
37. Resonator control solenoid valve
38. Check valve C
39. Surge tank

**Vacuum diagram – 1988–90 Legend**

ELECTRICAL CIRCUITS – 1988–89 INTEGRA

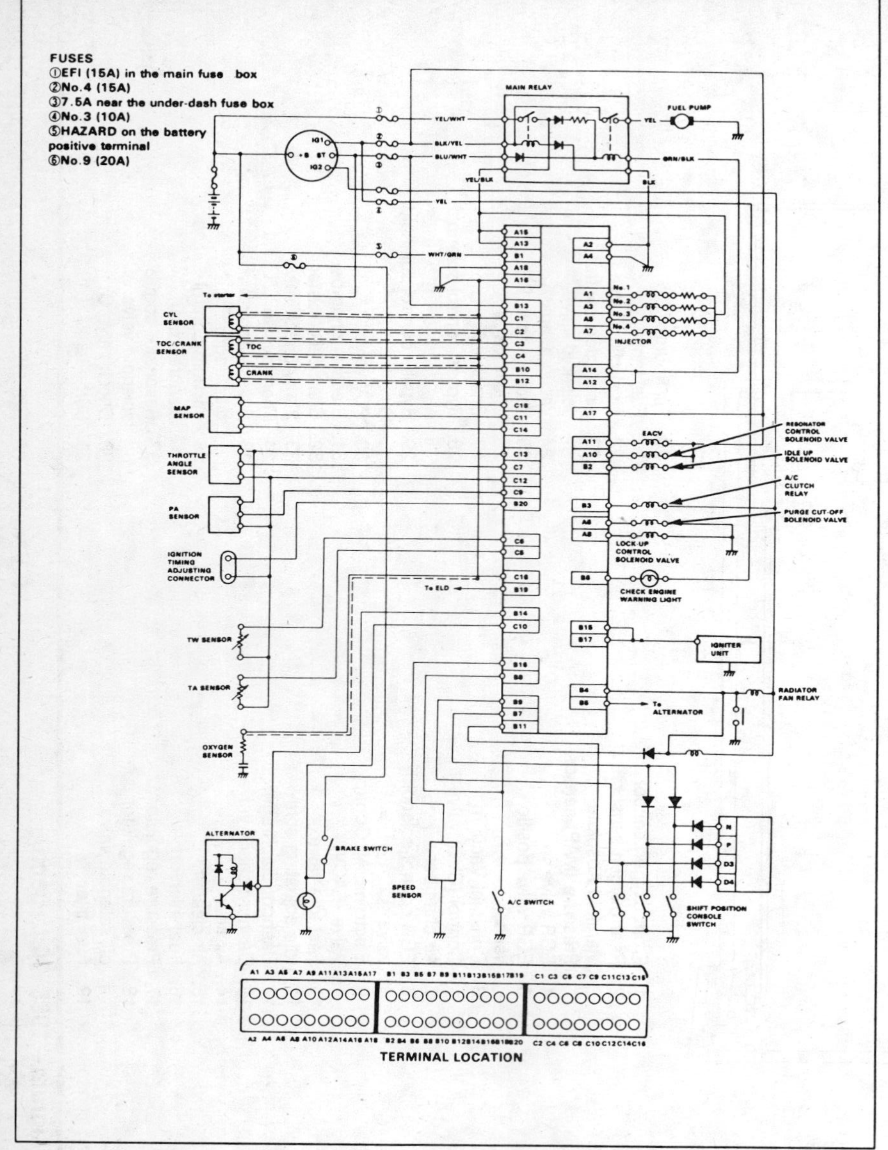


ELECTRICAL CIRCUITS – 1988–89 LEGEND

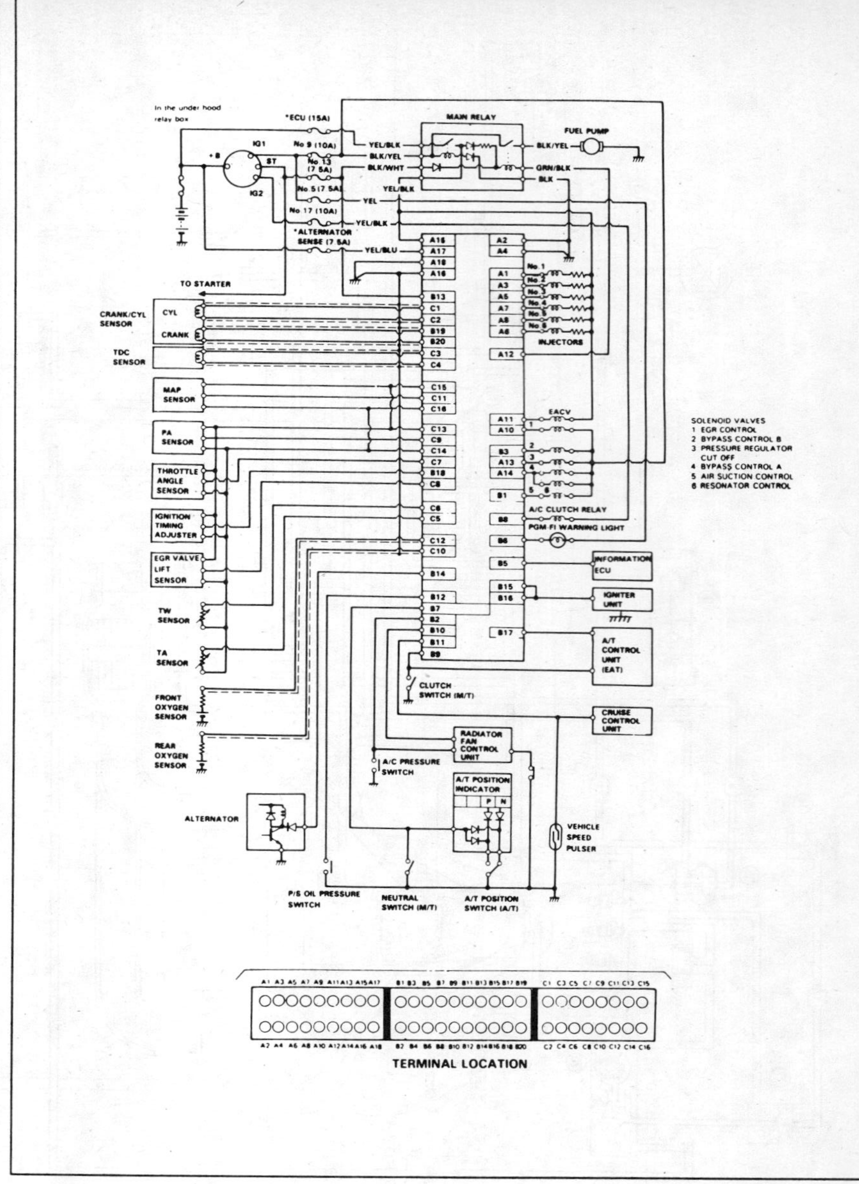

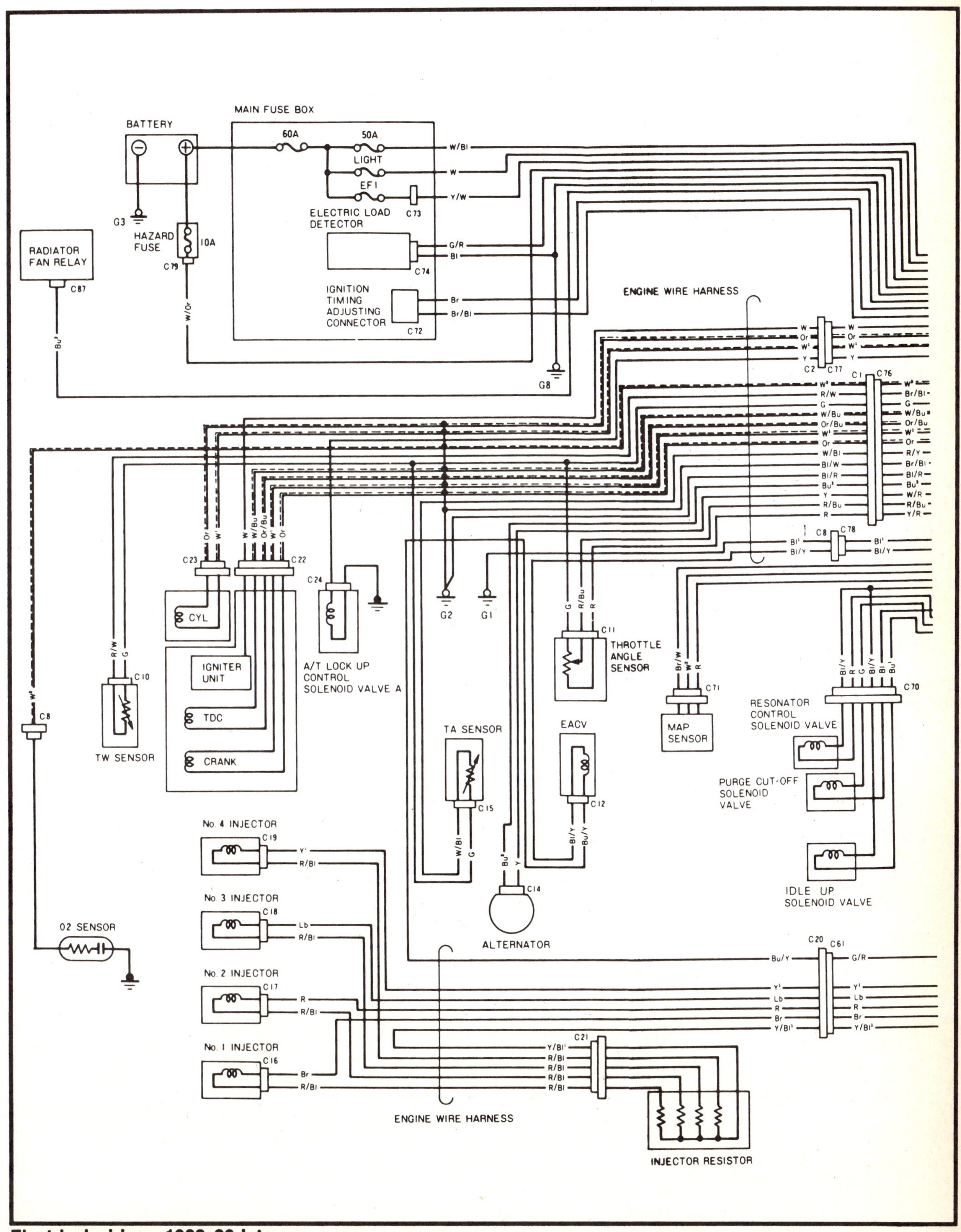


Electrical wiring—1988–89 Integra

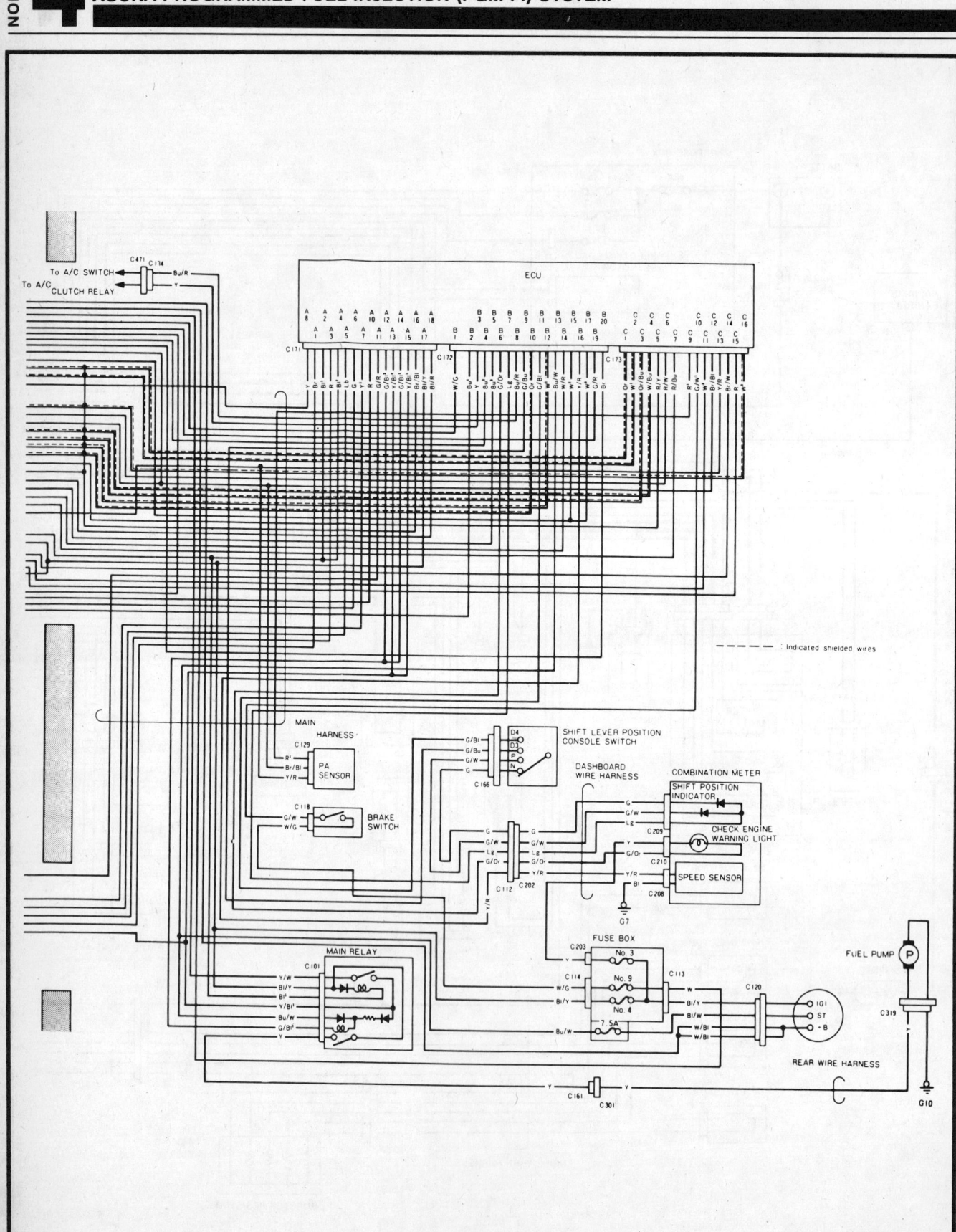

**Electrical wiring – 1988–89 Integra (cont.)**

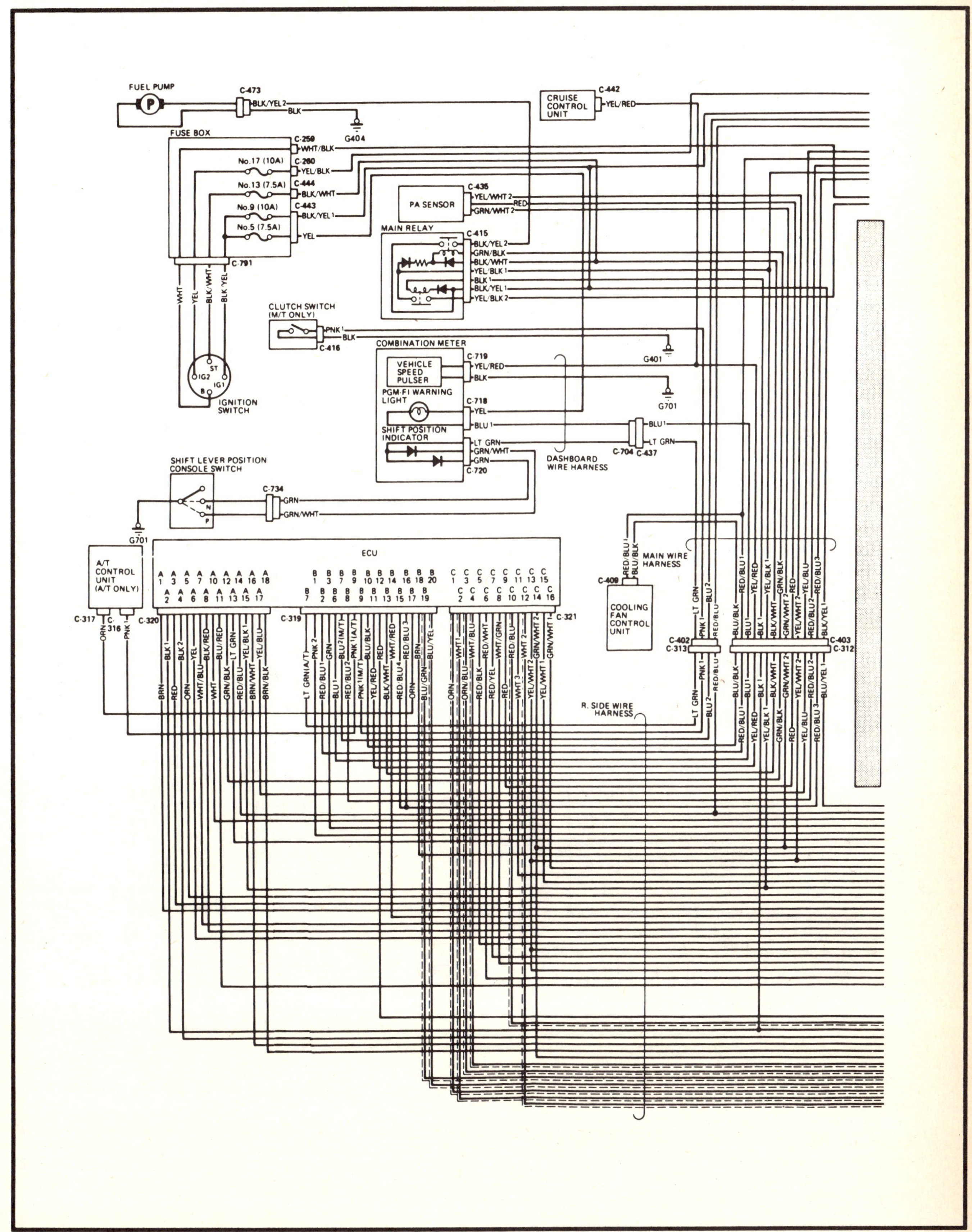


Electrical wiring—1988–89 Legend

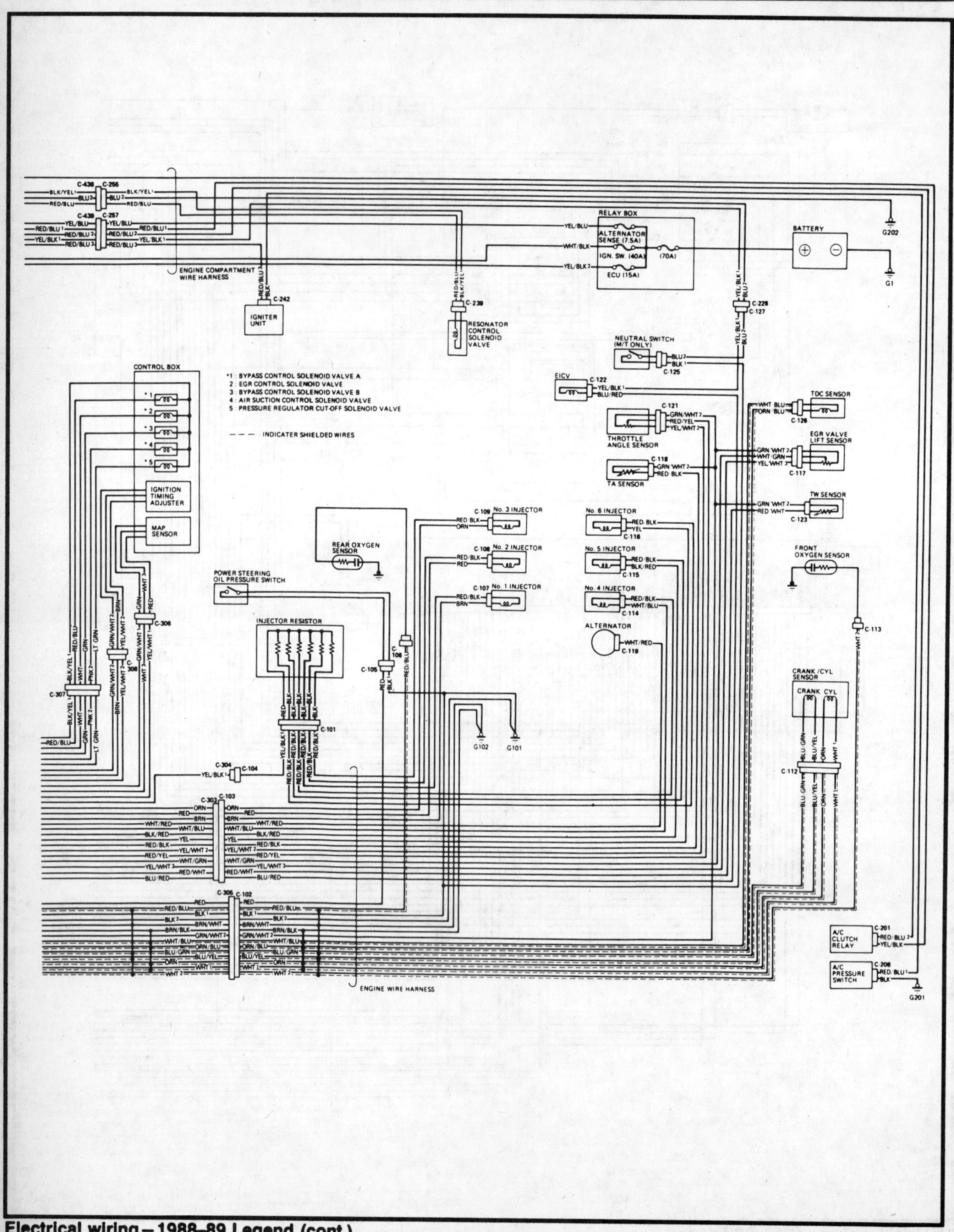


Electrical wiring – 1988–89 Legend (cont.)

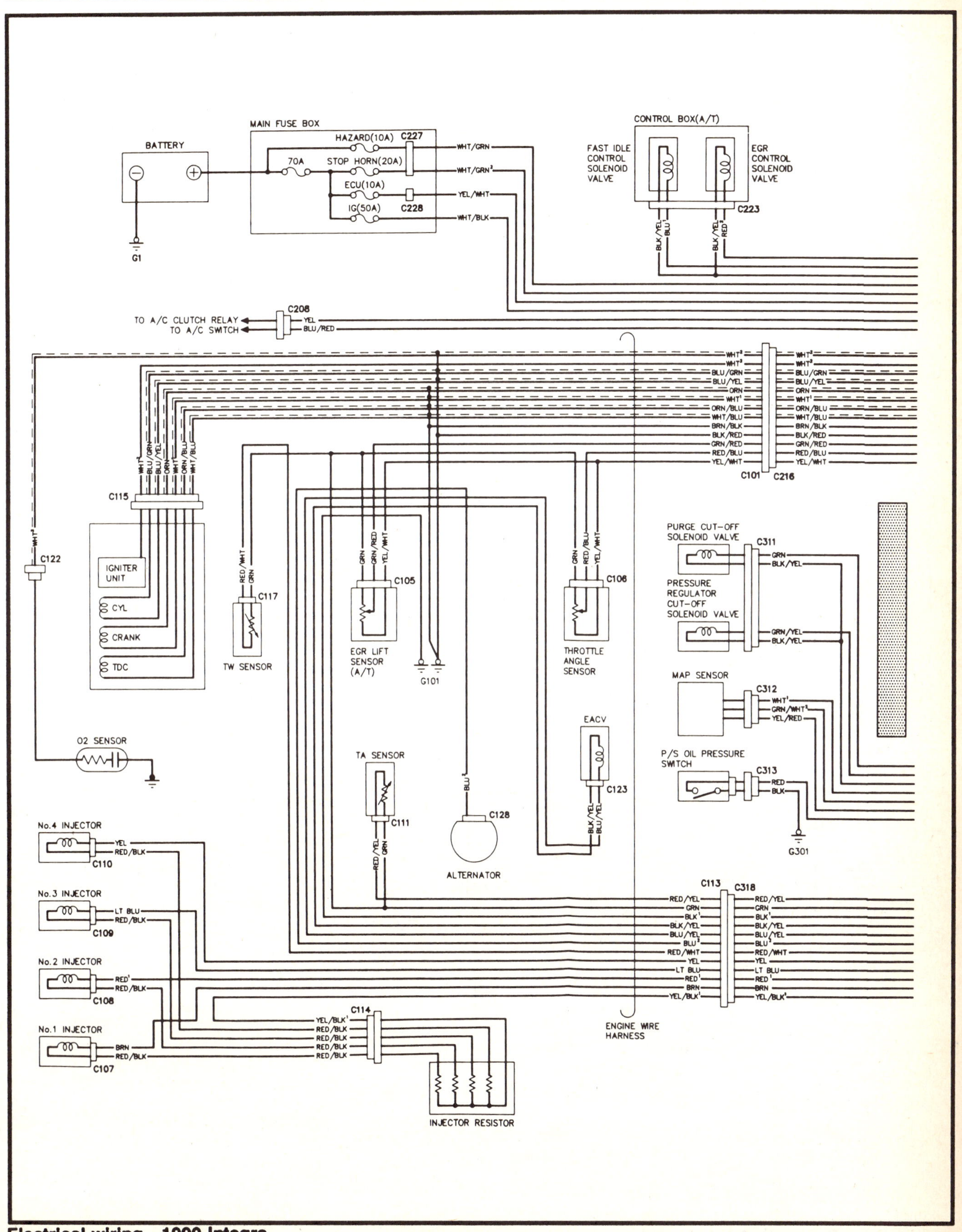


Electrical wiring – 1990 Integra

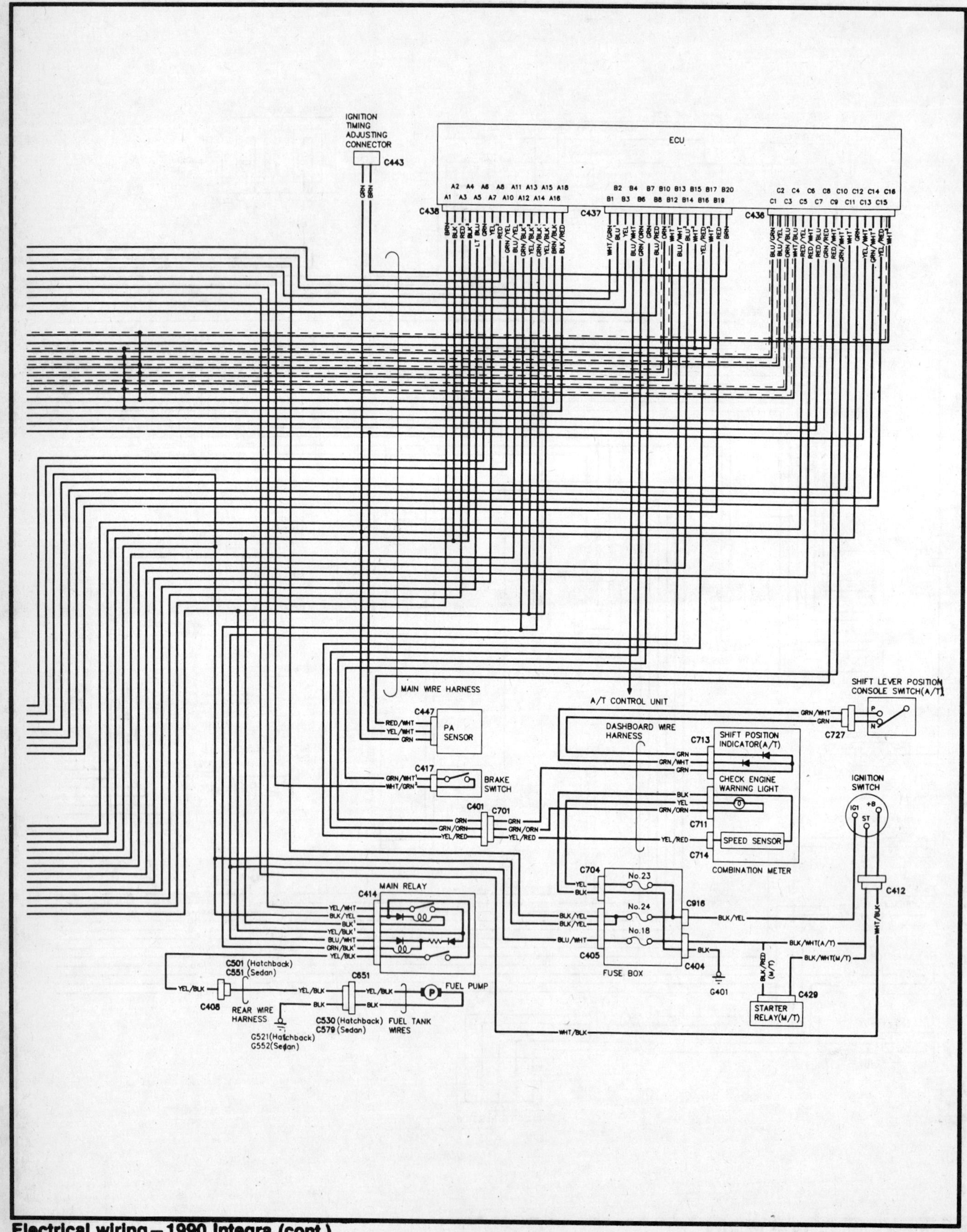


Electrical wiring – 1990 Integra (cont.)

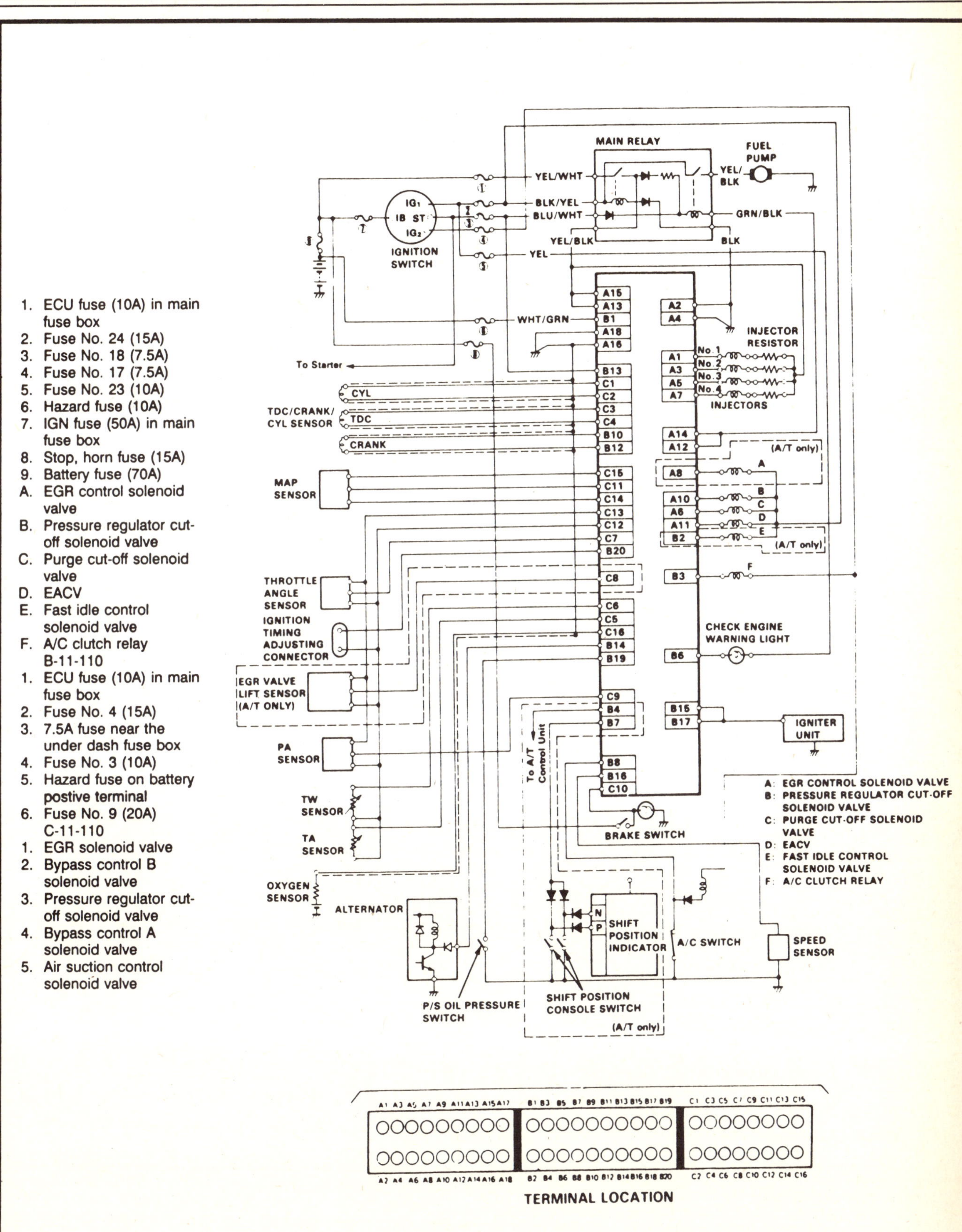


Electrical circuits—1990 Integra

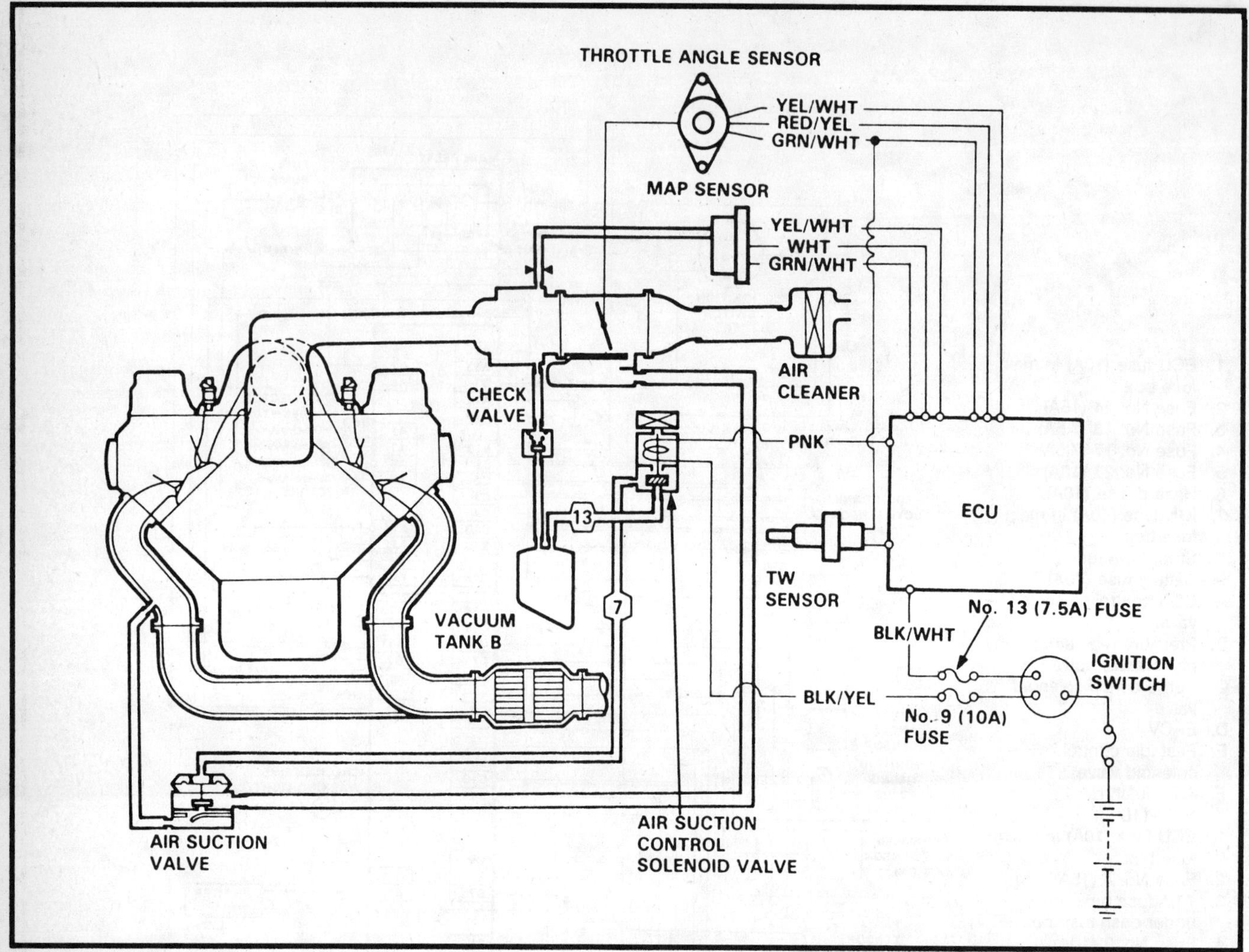


Air Injection System

# DAIHATSU ELECTRONIC FUEL INJECTION SYSTEM

## General Description

Daihatsu electronic fuel injection system consists of 3 sub-systems: air intake, electronic control and fuel supply.

### SYSTEM OPERATION

#### Fuel System

Fuel is supplied to the engine by an in-tank fuel pump. The fuel pump delivers fuel under a pressurized condition. The pressure regulator mounted at the delivery pipe keeps the fuel pressure at 36.7 psi higher that the surge tank inner pressure. The fuel injection amount for each injector is kept at a constant level by this method. The operation of the electric fuel pump is controlled by the Electronic Control Unit (ECU).

#### Air Intake System

The system supplies air for all engine needs. It consists of the air cleaner, air intake pipe, surge tank, throttle body, auxiliary air valve, fast idle mechanism, and intake manifold.

#### Engine Control System

The Electronic Control Unit (ECU) controls the amount of fuel and mixture based on engine conditions and running conditions, as determined by signals input by various sensors. The ECU also controls electronic ignition timing, fuel pump operation, idle-up control, a self-diagnosis function and the fail safe and back-up functions.

#### IGNITION TIMING CONTROL

Based on sensor input and engine load, the ECU controls the ignition timing. This maintains optimum engine operations.

#### SELF-DIAGNOSIS

If any malfunction should occur in an input signal circuit, the ECU memorizes the fault. This malfunction is indicated during the trouble diagnosis, by codes through the checking lamps. If a major malfunction is encountered, the ECU turns on the check engine lamp to warn the driver.

#### Fail-Safe Function

When a malfunction is detected, the ECU determines, based on programmed standard values, whether the engine may be allowed to continue operation or should be stopped immediately.

#### Back-Up Function

In the event the ECU detects a malfunction, the back-up function makes it possible to provide fuel injection amount and ignition timing that have been predetermined by the back-up data. This is to ensure safe driving if a single or many sensors are faulty, or if the ECU malfunctions.

#### IDLE-UP CONTROL

This system controls the idle-up Vacuum Switching Valve (VSV), which is necessary to increase engine idle to compensate for headlight or rear defogger operation.

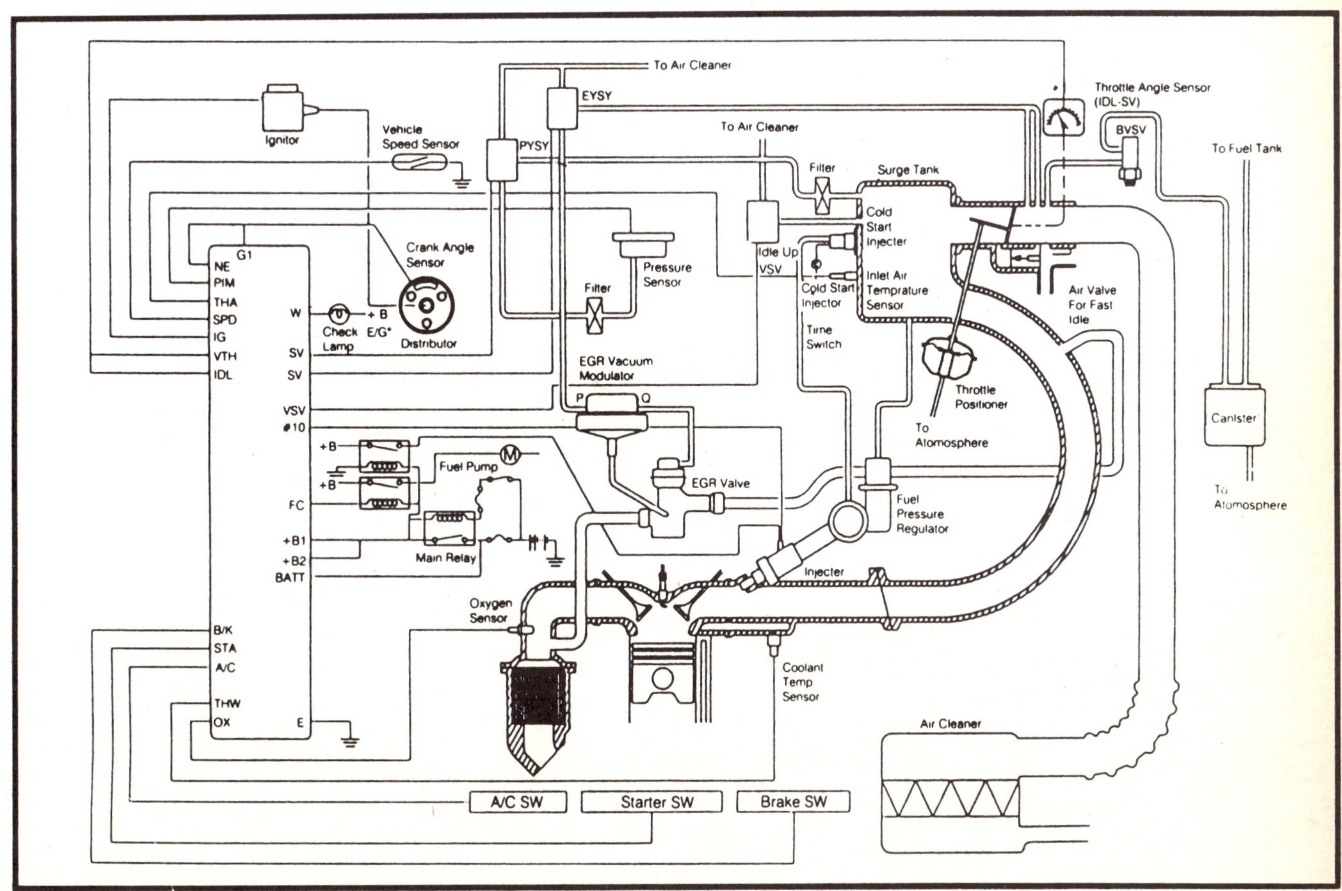

**FUEL PUMP CONTROL**

This system actuates the fuel pump by turning on the fuel pump relay when the ignition switch is turned to **START** or when the ECU detects the engine rotation.

## SERVICE PRECAUTIONS

- Disconnect the negative battery cable before disconnecting any fuel line.
- Do not operate the fuel pump when the fuel lines are empty.
- Do not operate the fuel pump when removed from the fuel tank.
- Make sure all ECU harness connectors are fastened securely. A poor connection can cause an extremely high surge voltage and could and result in damage to integrated circuits.
- Keep ECU all parts and harnesses dry during service.
- Before attempting to remove any parts, turn **OFF** the ignition switch and disconnect the battery ground cable.
- Do not attempt to disconnect the battery cables with the engine running.
- Do not depress the accelerator pedal when starting.
- Never open the cover to the ECU.

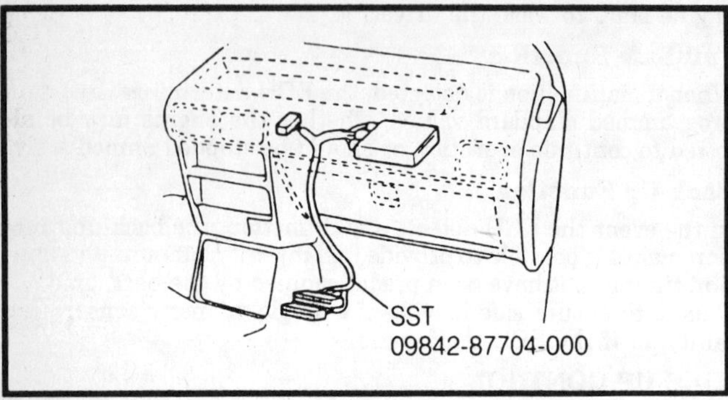


**Special Service Tool (SST) testing harness**

# Diagnosis and Testing

**NOTE: For Self-Diagnostic System and Accessing Trouble Code Memory, see Section 3 "SELF-DIAGNOSTIC SYSTEM".**

## SPECIAL SERVICE TOOL (SST)

Many ECU and sensor test require the installation of a special test wiring harness SST 09842–87704–000, or equivalent.

1. Disconnect the negative battery cable.
2. Remove the glove compartment box.
3. Disconnect the wiring harness from the ECU.
4. Connect the SST wiring harness between the ECU and the engine wiring harness. If performing resistance tests at the ECU connector terminals, never connect the SST to the ECU.
5. Reconnect the negative battery cable.
6. After testing, disconnect the negative battery cable and replace the engine harness to the ECU.

## TROUBLESHOOTING PROCEDURE—CHARADE

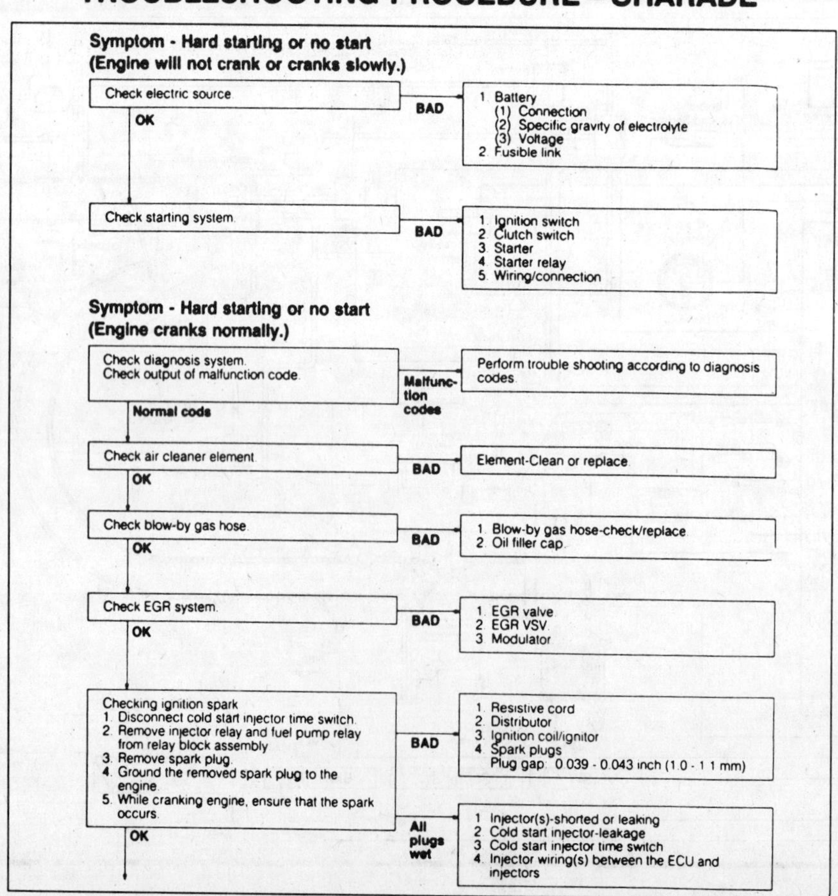

# TROUBLESHOOTING PROCEDURE—CHARADE

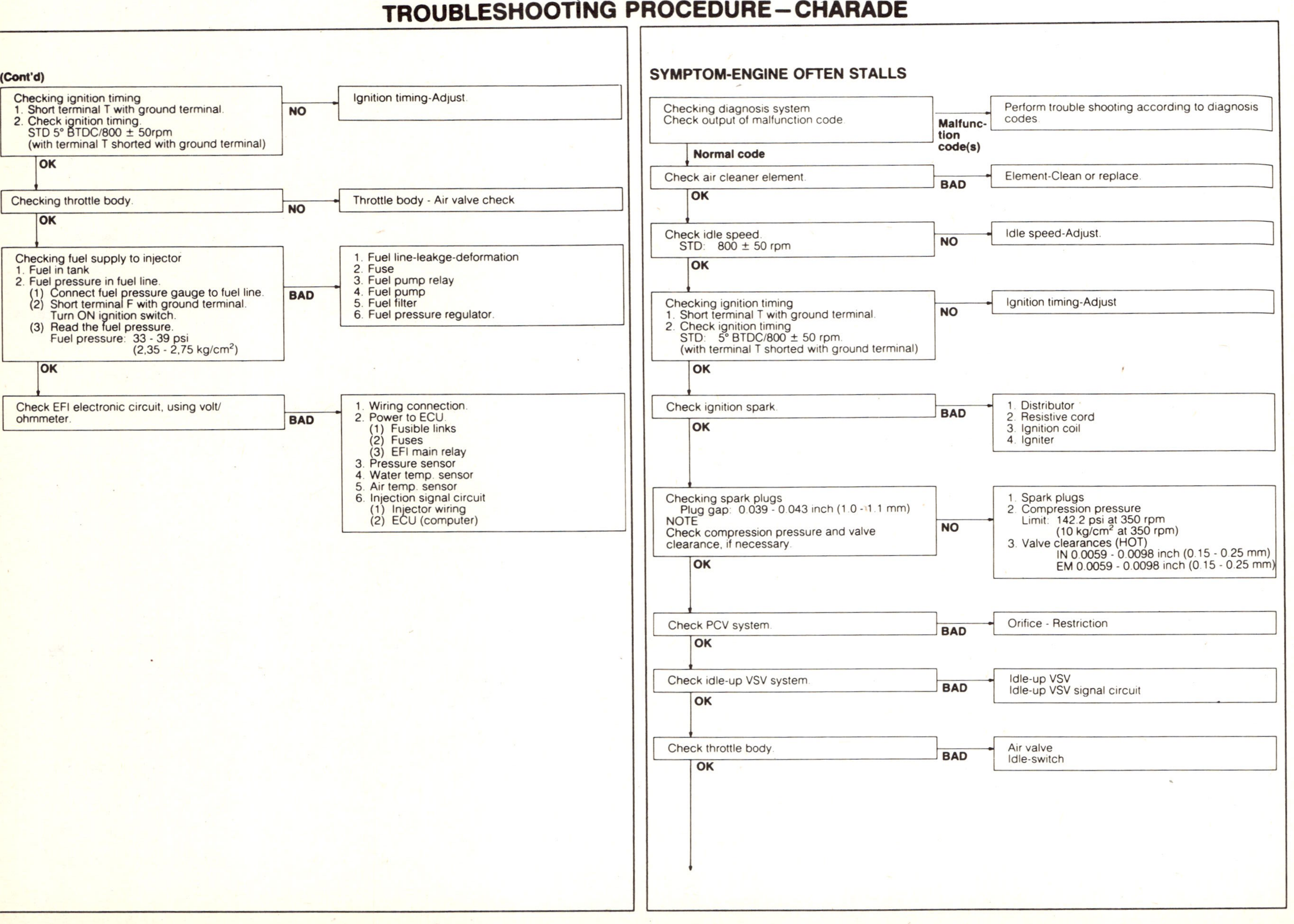

## TROUBLESHOOTING PROCEDURE—CHARADE

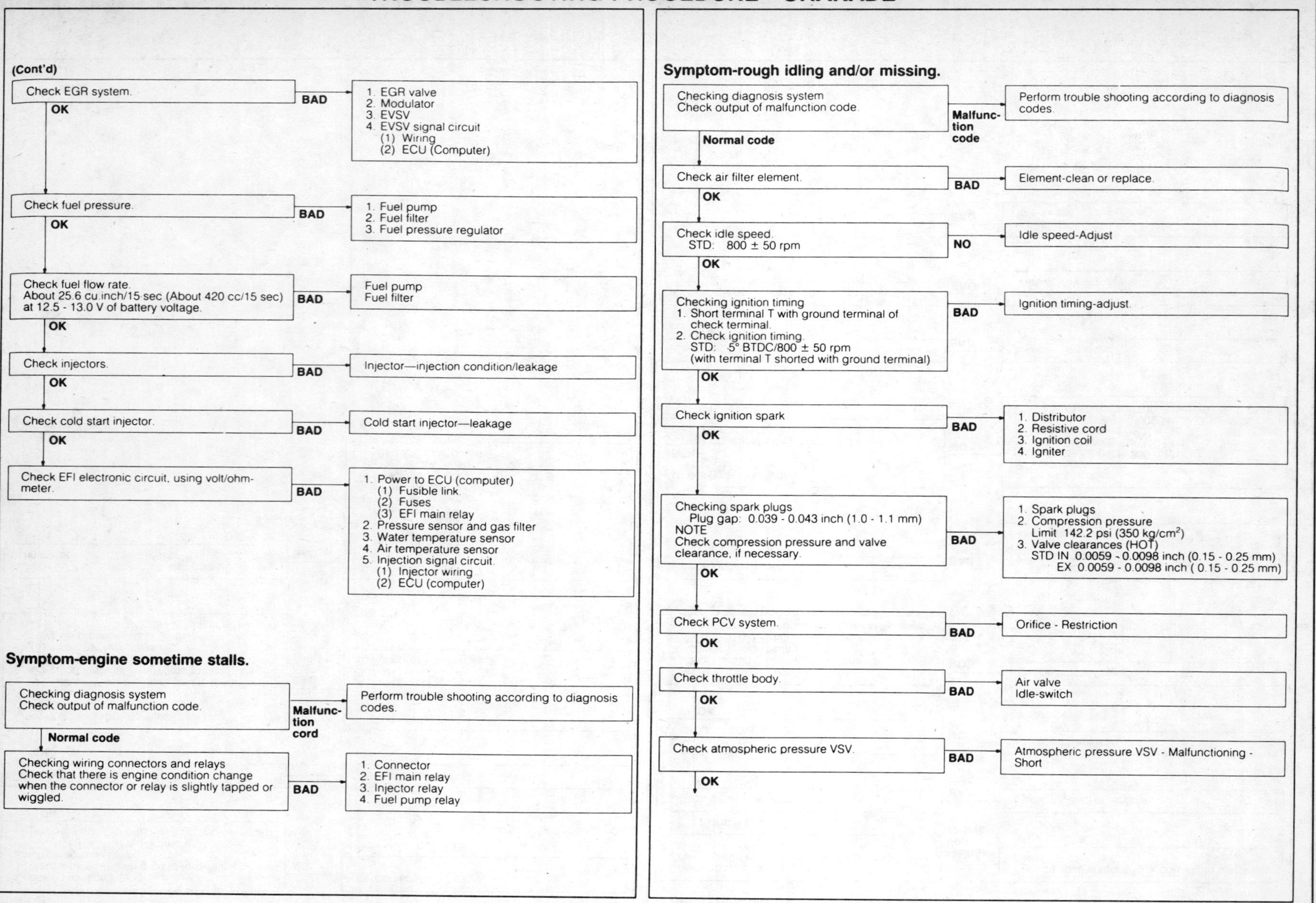

# TROUBLESHOOTING PROCEDURE—CHARADE

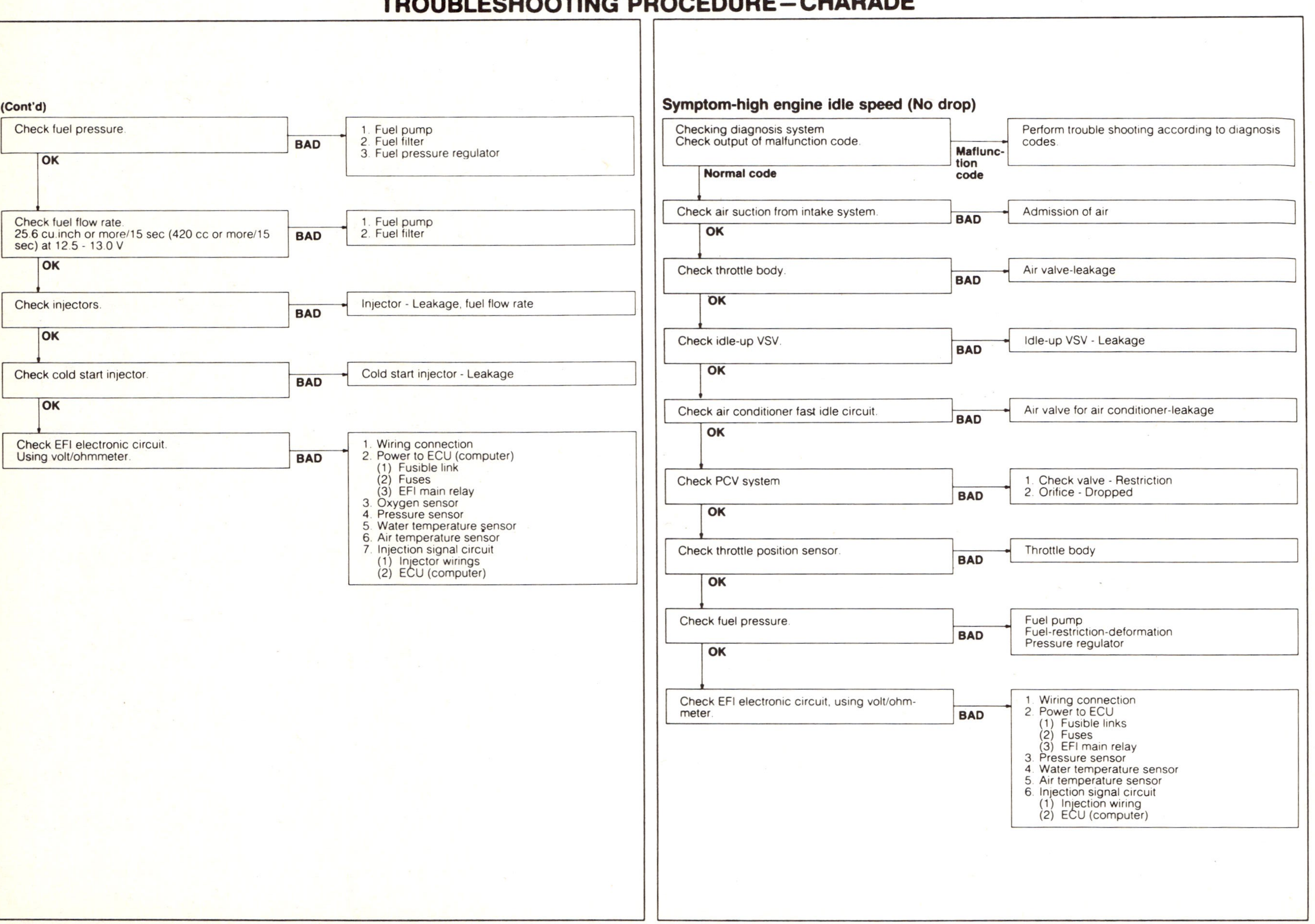

# TROUBLESHOOTING PROCEDURE—CHARADE

## Symptom-engine backfires

| Check | Result | Action |
|---|---|---|
| Checking diagnosis system<br>Check output of malfunction code. | Malfunction code | Perform trouble shooting according to diagnosis codes. |
| ↓ Normal code | | |
| Air cleaner element | BAD | Element—clean or replace |
| ↓ OK | | |
| Checking ignition timing<br>1. Short terminal T with ground terminal.<br>2. Check ignitiontiming<br>STD 5° BTDC/800 ± 50 rpm.<br>(with terminal T shorted with ground terminal) | NO | Ignition timing-adjust. |
| ↓ OK | | |
| Check idle speed.<br>STD 800 ± 50 rpm | NO | Idle speed-adjust |
| ↓ OK | | |
| Check valve clearances. (HOT)<br>STD IN 0.0059 - 0.0098 inch (0.15 - 0.25 mm)<br>EX 0.0059 - 0.0098 inch (0.15 - 0.25 mm) | NO | Valve clearance-adjust |
| ↓ OK | | |
| Check fuel pressure | BAD | 1. Fuel pump<br>2. Fuel filter<br>3. Fuel pressure regulator |
| ↓ OK | | |
| Check injectors. | BAD | Injector-clogged |
| ↓ OK | | |
| Check EFI electronic circuit, using volt/ohmmeter. | BAD | 1. Wiring connection<br>2. Power to ECU (computer)<br>(1) Fusible links<br>(2) Fuses<br>(3) EFI main relay<br>3. Pressure sensor<br>4. Water temperature sensor<br>5. Air temperature sensor<br>6. Throttle position sensor<br>7. Injection signal circuit<br>(1) Injection signal circuit<br>(2) ECU (computer)<br>8. Oxygen sensor |

## Symptom-engine afterfires

| Check | Result | Action |
|---|---|---|
| Checking diagnosis system<br>Check output of malfunction code. | Malfunction code | Perform trouble shooting according to diagnosis codes. |
| ↓ Normal code | | |
| Checking ignition timing<br>1. Short terminal T with ground terminal.<br>2. Check ignition timing<br>STD: 5° BTDC/800 ± 50 rpm<br>(with terminal T shorted with ground terminal) | NO | Ignition timing-adjust |
| ↓ OK | | |
| Check idle speed<br>STD: 800 ± 50 rpm | NO | Idle speed-adjust |
| ↓ OK | | |
| Check valve clearances<br>STD IN 0.0059 - 0.0098 inch (0.15 - 0.25 mm)<br>EX 0.0059 - 0.0098 inch (0.15 - 0.25 mm) | NO | Valve clearance-adjust |
| ↓ OK | | |
| Check spark plugs<br>Plug gap 0.039 - 0.043 inch (1.0 - 1.1 mm) | NO | Spark plug |
| ↓ OK | | |
| Check compression pressure. | BAD | Engine overhaul |
| ↓ OK | | |
| Check atmospheric pressure VSV. | BAD | Atmospheric pressure VSV |
| ↓ OK | | |
| Check fuel pressure. | BAD | 1. Fuel pump<br>2. Fuel filter<br>3. Fuel pressure regulator |
| ↓ OK | | |
| Check injector. | BAD | Injector-leakage |
| ↓ OK | | |
| Check cold start injector. | BAD | Cold start injector-leakage. |
| ↓ OK | | |

# TROUBLESHOOTING PROCEDURE—CHARADE

**(Cont'd)**

| | BAD |
|---|---|
| Check EFI electronic circuit, using volt/ohmmeter. | 1. Wiring connection<br>2. Power to ECU (computer)<br>(1) Fusible links<br>(2) Fuses<br>(3) EFI main relay<br>3. Oxygen sensor<br>4. Pressure sensor<br>5. Water temperature sensor<br>6. Air temperature sensor<br>7. Throttle position sensor<br>8. Injection signal circuit<br>(1) Injector wirings<br>(2) ECU (computer) |

## Symptom-Engine hesitation and/or poor acceleration

| | BAD |
|---|---|
| Check clutch or brake. | 1. Clutch-slips<br>2. Brakes-drag |
| OK ↓ | |
| Check air filter element. | Element-clean or replace. |
| OK ↓ | |
| Checking diagnosis system<br>Check output of malfunction code. | Perform trouble shooting according to diagnosis codes. |
| OK ↓ | |
| Checking ignition timing<br>1. Short terminal T with ground terminal.<br>2. Check ignition timing.<br>STD: 5° BTDC/800 ± 50 rpm<br>(with terminal T shorted with ground terminal) | Ignition timing-adjust |
| OK ↓ | |
| Check spark plugs.<br>Plug gap 0.039 - 0.043 inch (1.0 - 1.1 mm) | Spark plugs |
| OK ↓ | |
| Check compression poressure.<br>Limit: 142.2 psi at 350 rpm<br>(10 kg/cm² at 350 rpm) | Engine overhaul. |
| OK ↓ | |

**(Cont'd)**

| | BAD |
|---|---|
| Checking ignition spark<br>1. Disconnect cold start injector time switch.<br>2. Remove injector relay and fuel pump relay from relay block assembly.<br>3. Remove spark plug<br>4. Ground removed spark plug to engine by installing spark plug to resistive cord.<br>5. While cranking engine, ensure that spark occurs. | 1. Resistive cord<br>2. Distributor<br>3. Ignition coil/ignitor |
| OK ↓ | |
| Check idle-up VSV system. | Idle-up VSV<br>Idle-up VSV circuit |
| OK ↓ | |
| Check atmospheric pressure VSV system. | Atmospheric pressure VSV<br>Atmospheric pressure VSV signal circuit |
| OK ↓ | |
| Checking fuel pressure.<br>1. Fuel in tank<br>2. Fuel pressure in fuel line<br>(1) Connect fuel pressure gauge<br>(2) Short terminal T with ground terminal. Turn ON ignition switch.<br>(3) Read the fuel pressure<br>Fuel pressure 33 - 39 psi<br>(2.35 - 2.75 kg/cm²) | 1. Fuel line-leakage-deformation<br>2. Fuse<br>3. Fuel pump relay<br>4. Fuel pressure regulator |
| OK ↓ | |
| Check cold start injector. | Cold start injector-leakage |
| OK ↓ | |
| Checking fuel flow rate<br>25.6 ca. inch or more/15 sec<br>(420 cc or more/15 sec)<br>at 12.5 - 13.0 V | 1. Fuel pump<br>2. Fuel filter |
| OK ↓ | |
| Check injectors. | Injector-fuel flow amoant leakage |
| OK ↓ | |
| Check EFI electronic circuit, using volt/ohmmeter | 1. Wiring connection<br>2. Power to ECU (computer)<br>(1) Fusible links<br>(2) Fuses<br>(3) EFI main relay<br>3. Pressure sensor<br>4. Throttle position sensor<br>5. Water temperature sensor<br>6. Air temprerature sensor<br>7. Injection signal circuit<br>(1) Injector<br>(2) ECU (computer) |

## DIAGNOSTIC CODES—CHARADE

### DIAGNOSITIC CODE

| Code No. | Number of glowing of check engine lamp | Diagnosis item | Diagnosis contents | Trouble area |
|---|---|---|---|---|
| 1 | | Normal | — | — |
| 2 | | Pressure sensor | When the signal from pressure sensor becomes open or shorted | 1. Pressure sensor circuit<br>2. Pressure sensor<br>3. ECU |
| 3 | | Ignition signal | • No ignition confirmation signal (IGf) is inputted. | 1. Ignition circuit (+B, IGF)<br>2. Ignitor<br>3. ECU |
| 4 | | Water temperature sensor | • When the signal from the water temperature sensor circuit becomes open or shorted: | 1. Water temperature sensor circuit<br>2. Water temperature sensor<br>3. ECU |
| 5 | | Oxygen sensor signal | • When the oxygen sensor signal circuit becomes open or shorted: | 1. Oxygen sensor circuit<br>2. Oxygen sensor<br>3. ECU |
| 6 | | Revolution signal | • When Ne and/or G signal is not inputted within a few seconds after starting of engine cranking:<br>• When the Ne signal of a few decade milliseconds is not inputted when the engine speed is 1000 rpm or more: | 1. Distributor circuit<br>2. Distributor<br>3. ECU |
| 7 | | Throttle position sensor signal | • When the throttle position sensor signal circuit becomes open or shorted: | 1. Throttle position sensor circuit<br>2. Throttle position sensor<br>3. ECU |
| 8 | | Intake air temperature sensor signal | • When the intake air temperature signal circuit becomes open or shorted: | 1. Air temperature sensor circuit<br>2. Air temperature sensor<br>3. ECU |
| 9 | | Vehicle speed sensor signal | • When the vehicle speed sensor signal circuit becomes open or shorted: | 1. Vehicle speed sensor circuit<br>2. Vehicle speed sensor<br>3. ECU |
| 10 | | Starter signal | When the starter signal becomes open or shorted:However, it should be noted that this code may be memorized when the vehicle is started by being pushed. | 1. Starter signal circuit<br>2. ECU |
| 11 | | Switch signal | When the air conditioner is turned ON or the idle switch is turned OFF with the terminal T shorted: However, no memorizing will take place. | 1. Air conditioner switch circuit<br>2. Idle switch circuit<br>3. Air conditioner switch<br>4. Throttle position sensor<br>5. ECU |
| 12 | | EGR control system | When the EGR system is not operating normally: | 1. EGR valve<br>2. Modulator<br>3. EVSV<br>4. Water temperature sensor<br>5. ECU |

## DIAGNOSTIC CIRCUIT INSPECTION—CHARADE

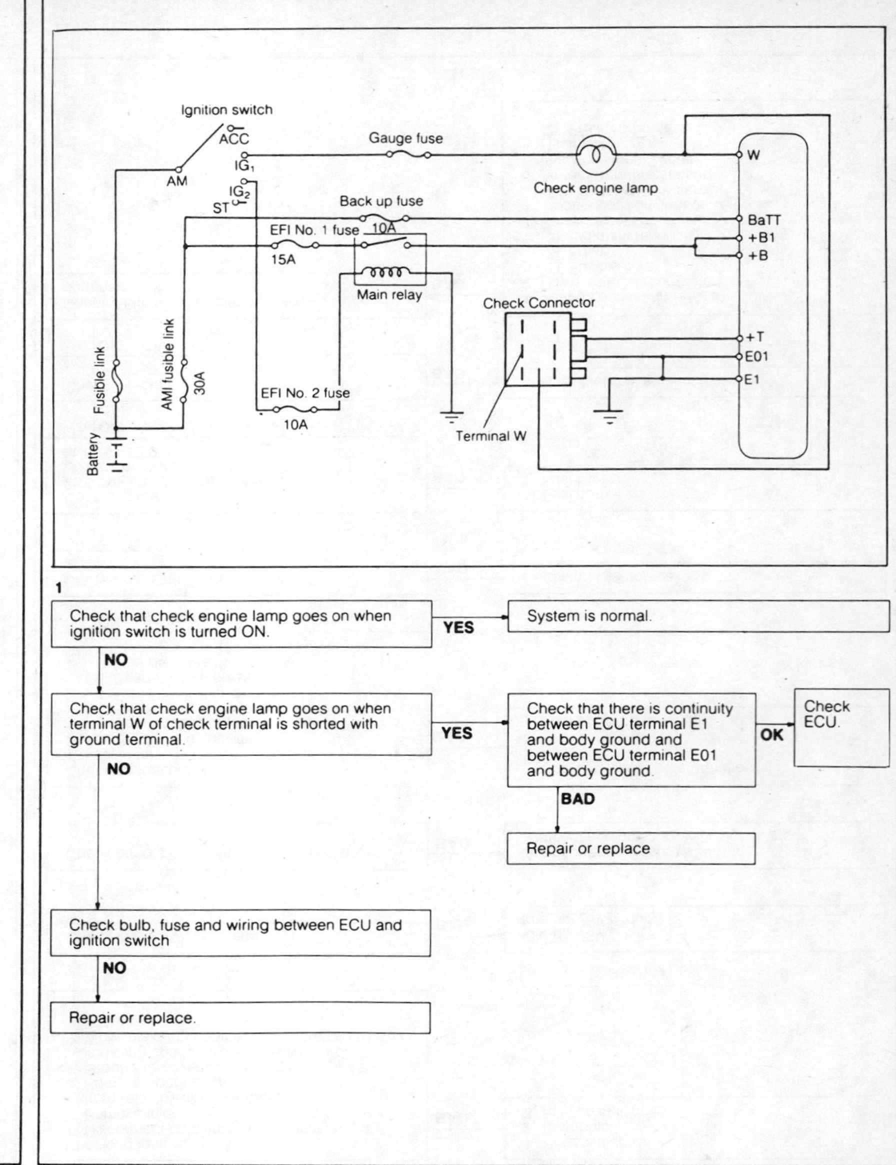

## DIAGNOSTIC CIRCUIT INSPECTION—CHARADE

2

- Check that check engine lamp goes off when engine starts.
  - YES → System is normal.
  - NO ↓
- Check wiring between ECU and check engine lamp.
  - BAD → Repair or replace
  - OK ↓
- Check that diagnosis code number is indicated when terminal T of check terminal is shorted with ground terminal.
  - NO → Check ECU.
  - YES ↓
- Perform trouble shooting according to diagnosis codes.
  - YES ↓
- After trouble shooting according to diagnosis codes has been performed, start engine. Check that check engine lamp goes off.
  - NO → Check diagnosis code again.
  - YES ↓
- System is normal. → Erase diagnosis code.

## TROUBLESHOOTING CHART 1—CHARADE

| NO | Terminals | Trouble | Conditions | STD voltage |
|---|---|---|---|---|
| 1 | BATT–E1 | NO voltage | At all time | 10 - 15.5 |
| | +B1 –E1<br>+B | NO voltage | Ignition switch ON | 10 - 15.5 |

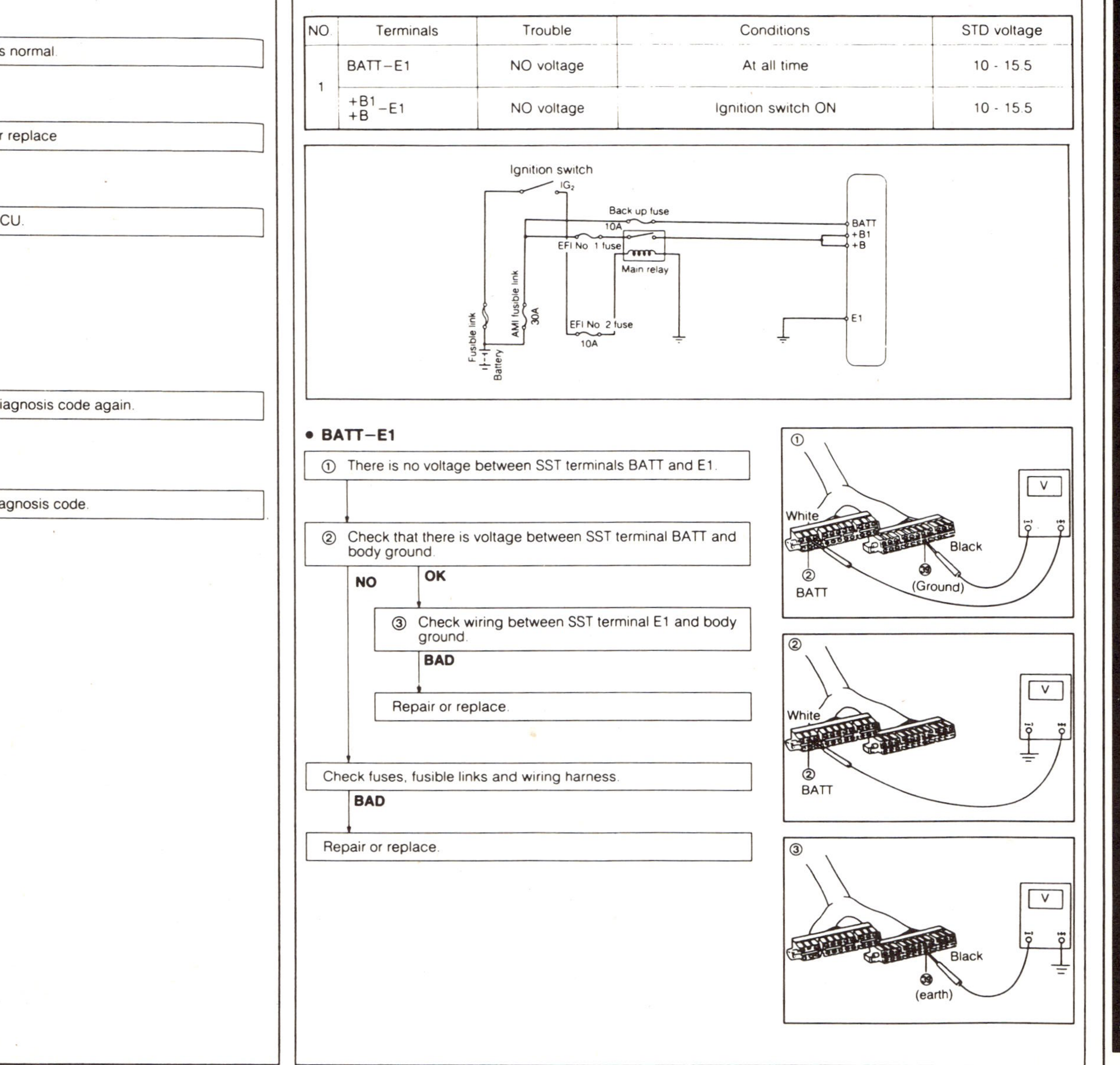


• **BATT–E1**

- ① There is no voltage between SST terminals BATT and E1.
- ② Check that there is voltage between SST terminal BATT and body ground.
  - OK → ③ Check wiring between SST terminal E1 and body ground.
    - BAD → Repair or replace.
  - NO → Check fuses, fusible links and wiring harness.
    - BAD → Repair or replace.

## TROUBLESHOOTING CHART 1—CHARADE

## TROUBLESHOOTING CHART 2—CHARADE

| NO. | Terminals | Trouble | Conditions | STD voltage |
|---|---|---|---|---|
| 2 | VCC–E21 | No voltage | Ignition switch ON | 4.5 - 5.5 |
| | Pim–E21 | | Ignition switch ON<br>At time of atmospheric pressure of 760 mmHg | 2.1 - 2.7 |

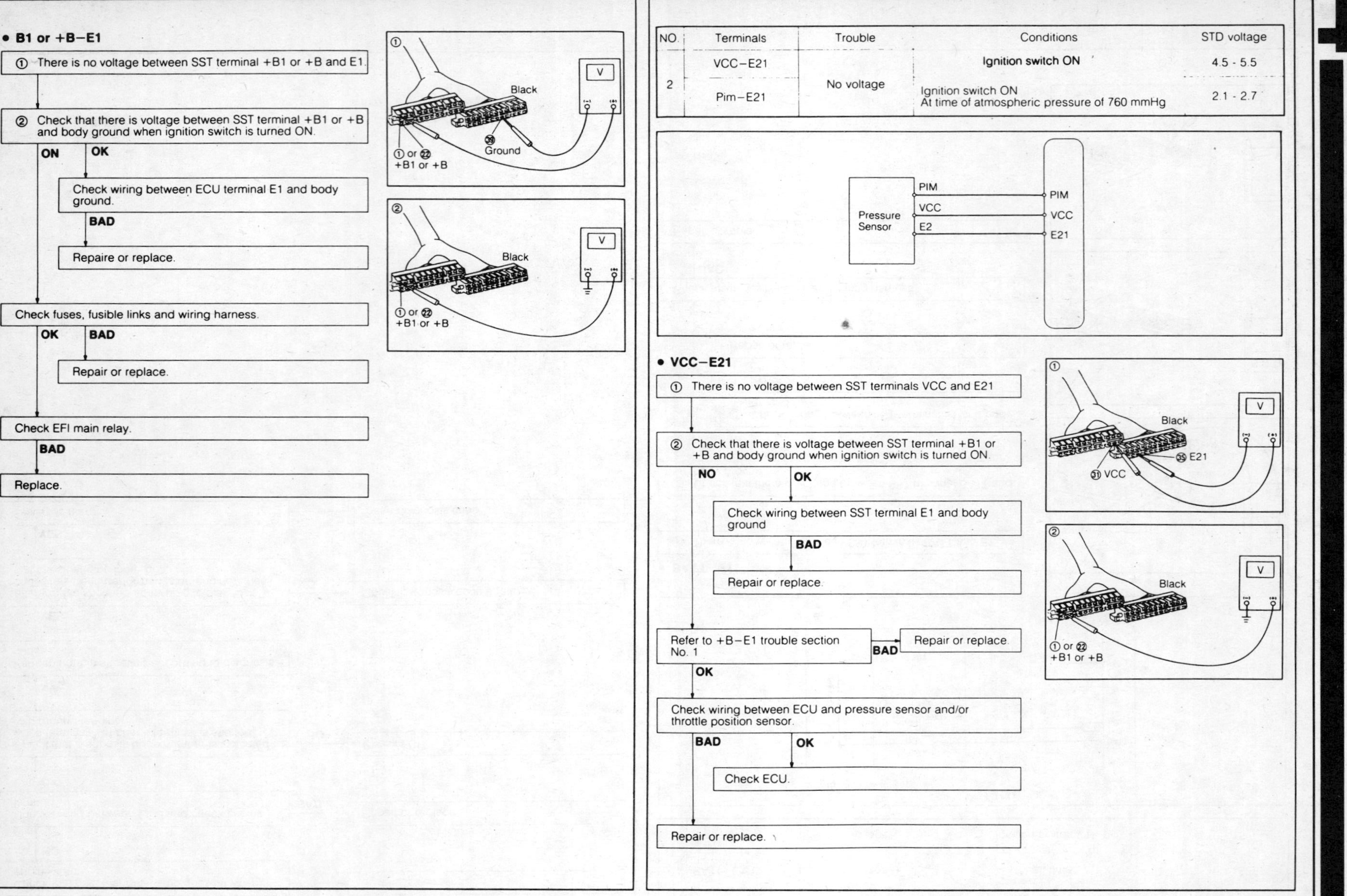

• B1 or +B–E1
① There is no voltage between SST terminal +B1 or +B and E1.
② Check that there is voltage between SST terminal +B1 or +B and body ground when ignition switch is turned ON.
ON
OK
Check wiring between ECU terminal E1 and body ground.
BAD
Repaire or replace.
Check fuses, fusible links and wiring harness.
OK
BAD
Repair or replace.
Check EFI main relay.
BAD
Replace.
①
Black
① or ㉒
+B1 or +B
㉟
Ground
V
②
Black
① or ㉒
+B1 or +B
V
Pressure Sensor
PIM
VCC
E2
PIM
VCC
E21
• VCC–E21
① There is no voltage between SST terminals VCC and E21
② Check that there is voltage between SST terminal +B1 or +B and body ground when ignition switch is turned ON.
NO
OK
Check wiring between SST terminal E1 and body ground
BAD
Repair or replace.
Refer to +B–E1 trouble section No. 1
BAD
Repair or replace.
OK
Check wiring between ECU and pressure sensor and/or throttle position sensor.
BAD
OK
Check ECU.
Repair or replace.
①
Black
㉟ E21
㉛ VCC
V
②
Black
① or ㉒
+B1 or +B
V

## TROUBLESHOOTING CHART 2—CHARADE

• **PIM—E21**

① There is no voltage between SST terminals Pim-E21

② Check that there is voltage between SST terminals VCC—E21

**OK** / **BAD**

BAD → Refer to VCC—E21 trouble section (No. 2)

**BAD** → Repair.

OK → Check wiring between ECU and pressure sensor.

**OK** / **BAD**

BAD → Repair or replace.

OK → Check pressur sensor.

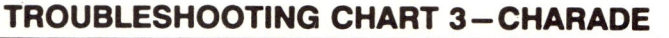

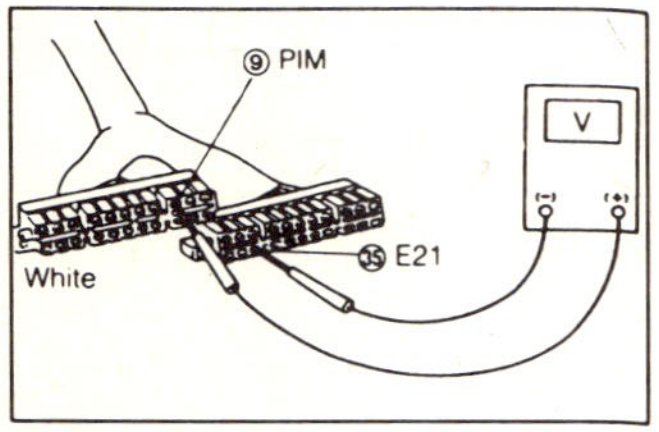

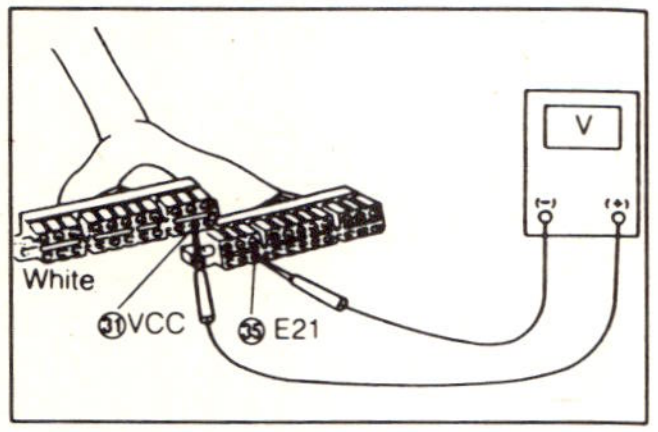

## TROUBLESHOOTING CHART 3—CHARADE

| No. | Terminals | Trouble | Conditions | STD voltage |
|---|---|---|---|---|
| 3 | IGF-E1 | 4.5 - 5.5 or 0 | Ignition switch ON | 0.5 - 1.5 (While engine is stopped) |

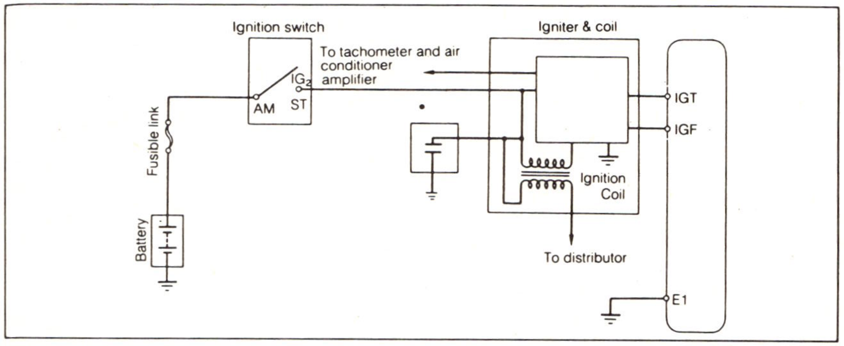

• **IGF—E1 4.5 - 5.5 V**

① Check that there is a voltage of 4.5 to 5.5 volts between SST terminals IGF and E1 when ignition switch is turned ON.

Check wiring between SST terminal IGT and igniter.

**OK** / **BAD**

BAD → Repair or replace.

OK → ② Check that voltage is applied to ignition coil.

**OK** / **BAD**

BAD → Repair or replace.

OK → Check that there is continuity between igniter body and body ground.

**OK** / **BAD**

BAD → Repair.

OK → Replace igniter.

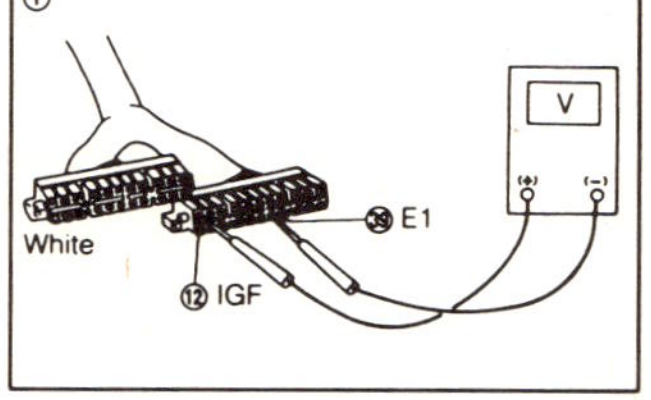

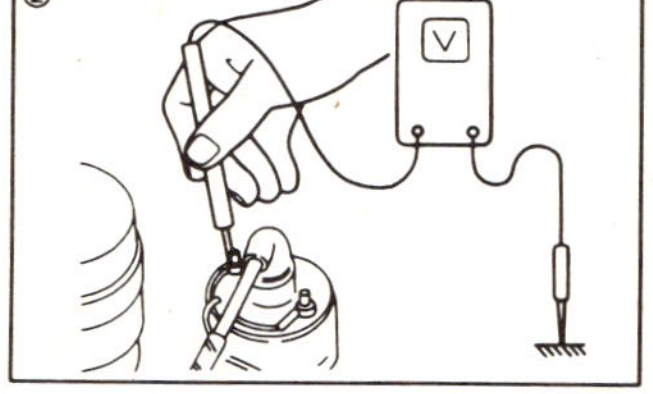

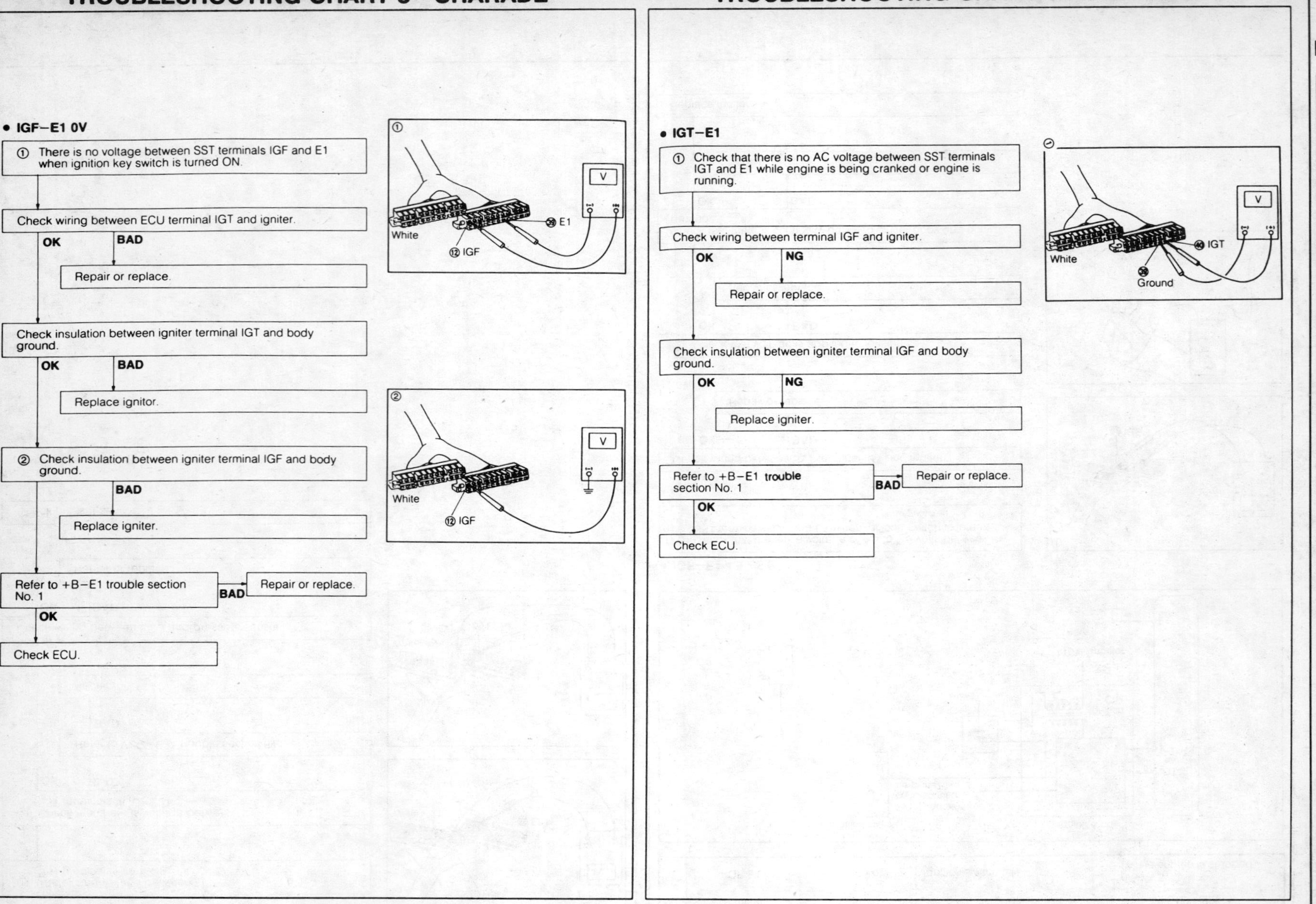
TROUBLESHOOTING CHART 3—CHARADE
• IGF—E1 0V
① There is no voltage between SST terminals IGF and E1 when ignition key switch is turned ON.
Check wiring between ECU terminal IGT and igniter.
OK
BAD
Repair or replace.
Check insulation between igniter terminal IGT and body ground.
OK
BAD
Replace ignitor.
② Check insulation between igniter terminal IGF and body ground.
BAD
Replace igniter.
Refer to +B—E1 trouble section No. 1
BAD
Repair or replace.
OK
Check ECU.
①
White
E1
⑫ IGF
V
②
White
⑫ IGF
V
TROUBLESHOOTING CHART 3—CHARADE
• IGT—E1
① Check that there is no AC voltage between SST terminals IGT and E1 while engine is being cranked or engine is running.
Check wiring between terminal IGF and igniter.
OK
NG
Repair or replace.
Check insulation between igniter terminal IGF and body ground.
OK
NG
Replace igniter.
Refer to +B—E1 trouble section No. 1
BAD
Repair or replace.
OK
Check ECU.
①
White
IGT
Ground
V

## TROUBLESHOOTING CHART 4—CHARADE

| No. | Terminals | Trouble | Conditions | | STD voltage |
|---|---|---|---|---|---|
| 4 | THW-E2 | No voltage | Ignition switch ON | Coolant temperature 176 °F (80 °C) | 0.1 - 0.7 |

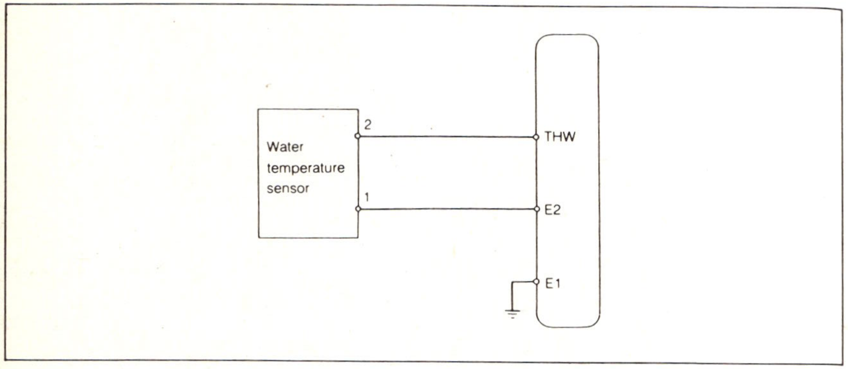

Water temperature sensor
2 — THW
1 — E2
E1

① There is no specified voltage between SST terminals THW and E2 when ignition switch is turned ON.

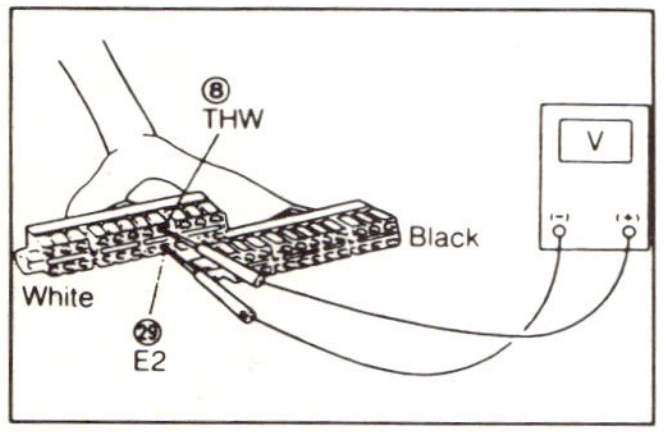

② Check that there is voltage between SST terminals +B1 or +B and body ground when ignition switch is turned ON.

- **OK** → Check water temperature sensor.
- **BAD** → Refer to +B–E1 trouble section No. 1.

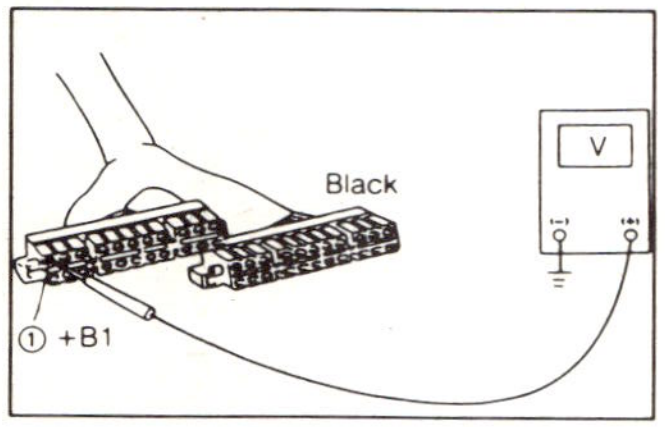

Check water temperature sensor.

- **OK** → Check wiring between ECU and water temperature sensor.
- **BAD** → Replace water temperature sensor.

Check wiring between ECU and water temperature sensor.

- **OK** → Check ECU.
- **BAD** → Repair or replace.

Check ECU.

⑧ THW
Black
White
㉙ E2
V

Black
① +B1
V

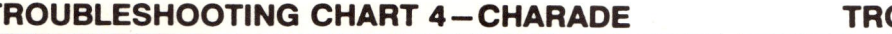

## TROUBLESHOOTING CHART 5—CHARADE

| No. | Terminals | Trouble | Conditions | | STD voltage |
|---|---|---|---|---|---|
| 5 | OX-E1 | No voltage changes | Ignition switch ON | When engine speed is held at 3000 rpm for two minutes after engine has been fully warmed up. | Voltage changes more than 8 times with in 10 seconds |

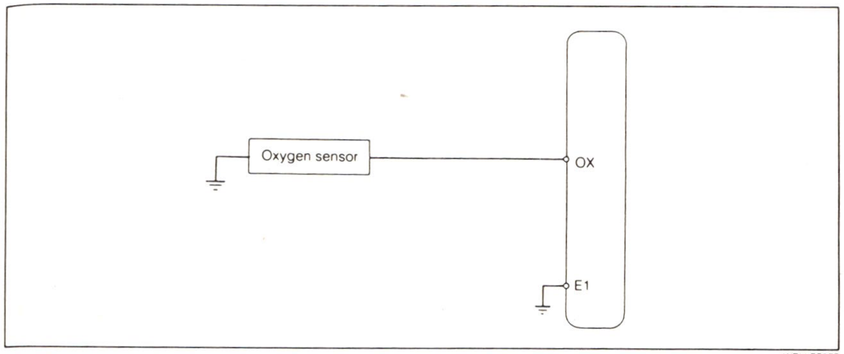

Oxygen sensor — OX
E1

WRU-EF077

① There is no change in voltage between SST terminals OX and E1 when engine speed is held at 3000 rpm for two minutes after engine has been fully warmed up.

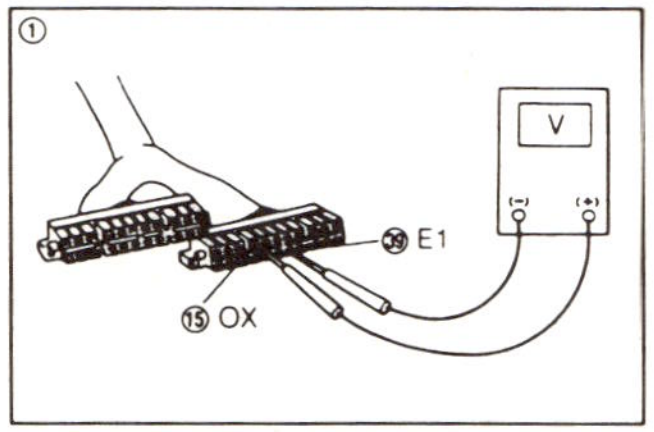

Check wiring between ECU and oxygen sensor.

- **BAD** → Repair or replace.

Check oxygen sensor.

- **BAD** → Check ECU.

①
㉚ E1
⑮ OX
V

## TROUBLESHOOTING CHART 6—CHARADE

| No. | Terminal | Trouble | Conditions | STD voltage |
|---|---|---|---|---|
| 6 | Ne−E1 | No voltage | Ignition switch ON | 0.5 - 1.5 |
| | GI−E1 | No voltage | Ignition switch ON | 0.5 - 1.5 |

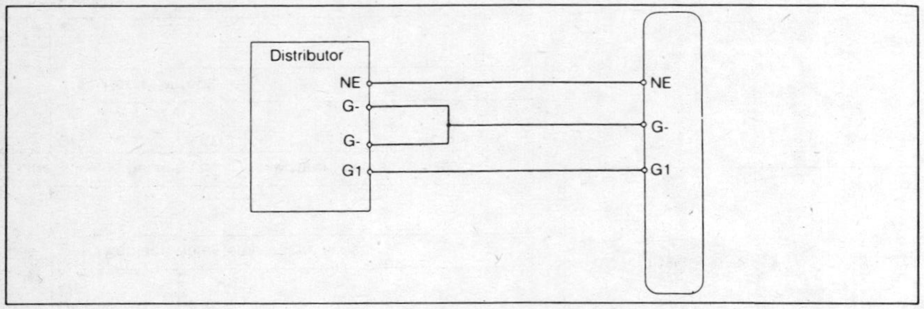

• **Ne−G-**

① There is no AC voltage between SST terminals Ne and G- while engine is being cranked or engine is running.

② Check that there is a voltage of about 0.3 - 0.9 Volt between the following terminals of SST terminal when ignition switch is turned ON.: Namely, between Ne and E1, between G1 and E1, and between G- and E1, respectively.

- **OK** → Check distributor.
  - **BAD** → Replace distributor.
- **BAD** → Check wiring between ECU and distributor.
  - **OK** → Repair or replace.
  - **OK** → Check distributor.
    - **OK** → Refer to +B−E1 trouble section No. 1
      - **BAD** → Repair or replace.
      - **OK** → Check ECU

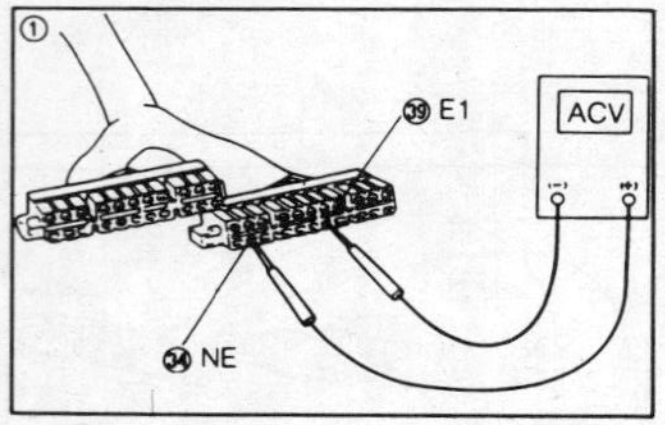

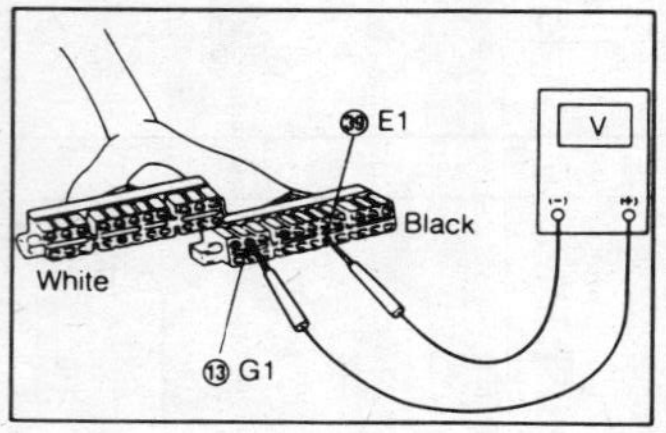

## TROUBLESHOOTING CHART 7—CHARADE

| No. | Terminal | Trouble | Conditions | | STD voltage |
|---|---|---|---|---|---|
| 7 | VCC−E2 | No voltage | Ignition switch ON | | 4.5 - 5.5 |
| | VTH−E2 | | Ignition switch ON | Throttle valve closed | About 0.55 |
| | | | | Throttle valve fully opened | About 3.91 |

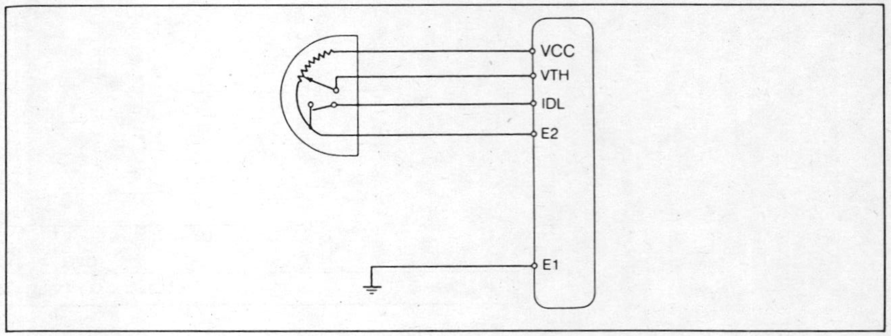

• **VCC−E2**

① There is no voltage between SST terminals VCC and E2 when ignition switch is turned ON.

② Check that there is voltage between SST terminal +B1 or +B and body ground when ignition switch is turned ON.

- **ON** → Refer to +B−E1 trouble section No. 1
- **OK** → Check throttle position sensor.
  - **BAD** → Repair or replace.
  - **OK** → Check wiring between ECU and throttle position sensor.
    - **BAD** → Repair or replace wiring.
    - **OK** → Check ECU.

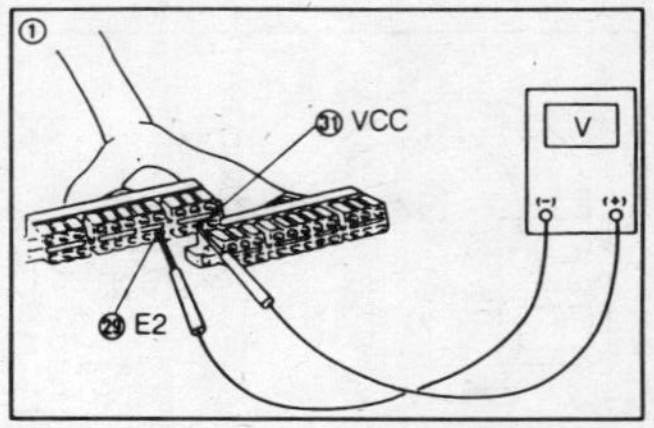

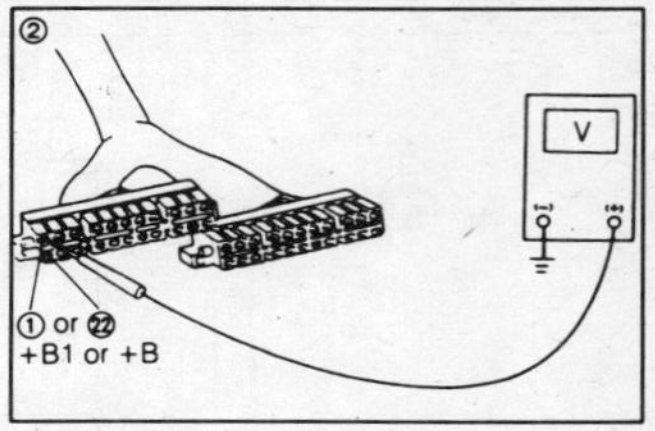

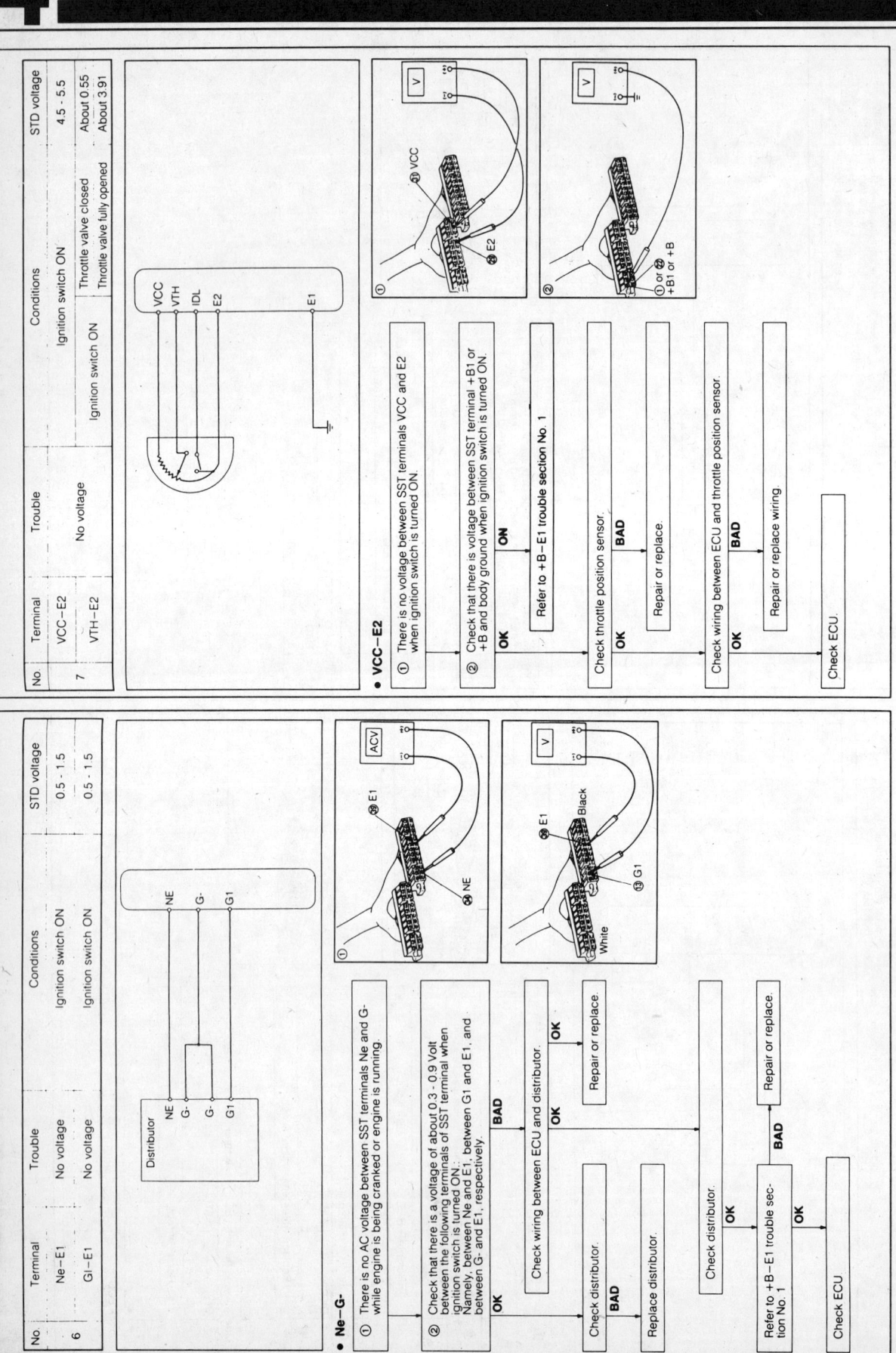

## TROUBLESHOOTING CHART 7—CHARADE

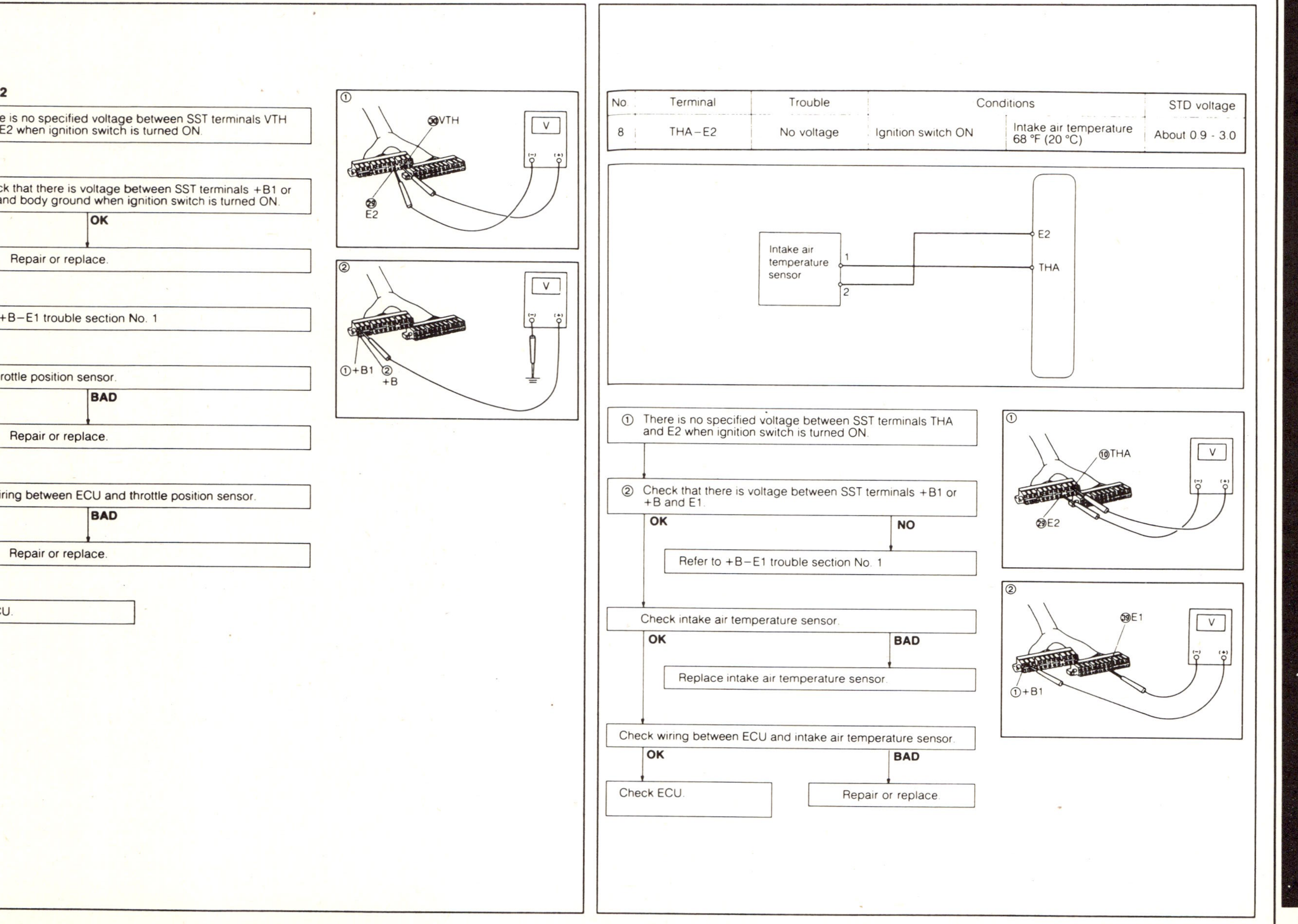


## TROUBLESHOOTING CHART 8—CHARADE

| No | Terminal | Trouble | Conditions | | STD voltage |
|---|---|---|---|---|---|
| 8 | THA–E2 | No voltage | Ignition switch ON | Intake air temperature 68 °F (20 °C) | About 0.9 - 3.0 |

Intake air temperature sensor
1
2
E2
THA
① There is no specified voltage between SST terminals THA and E2 when ignition switch is turned ON.
② Check that there is voltage between SST terminals +B1 or +B and E1.
OK
NO
Refer to +B–E1 trouble section No. 1
Check intake air temperature sensor.
OK
BAD
Replace intake air temperature sensor.
Check wiring between ECU and intake air temperature sensor.
OK
BAD
Check ECU.
Repair or replace.
①
⑩THA
㉙E2
V
②
㉙E1
①+B1
V

## TROUBLESHOOTING CHART 9—CHARADE

| No. | Terminal | Trouble | Conditions | | STD voltage |
|---|---|---|---|---|---|
| 9 | SPD–E1 | No voltage changes | Ignition switch ON | When vehicle is moved slowly. | 0 to 4.4 - 5.5 |

Speed sensor — SPD — Magnet — Speedometer Cable — E1

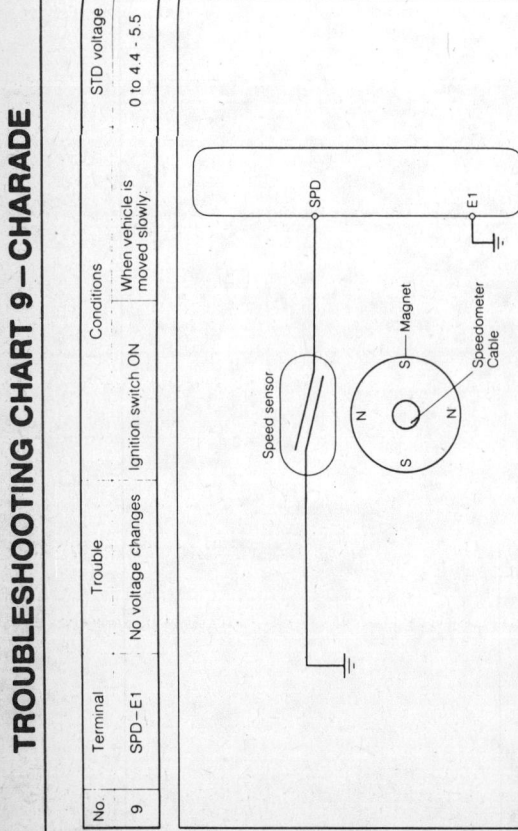

① No change in voltage occurs between SST terminals SPD and E1 when vehicle is moved slowly with ignition switch turned ON.

Check that there is voltage between SST terminal +B1 or +B and E1
- NO → Refer to +B–E1 trouble section No. 1
- OK ↓

Check speed sensor.
- → Replace speed sensor.
- OK ↓

Check wiring between ECU and speed sensor.
- BAD → Repair or replace.
- OK ↓

Check wiring between speed sensor and body ground.
- BAD → Repair or replace.
- OK → Check ECU.

① 26 SPD, 39 E1, V

## TROUBLESHOOTING CHART 10—CHARADE

| No. | Terminal | Trouble | Conditions | STD voltage |
|---|---|---|---|---|
| 10 | STA–E1 | No voltage | Ignition switch ST position | 6 - 15.5 |

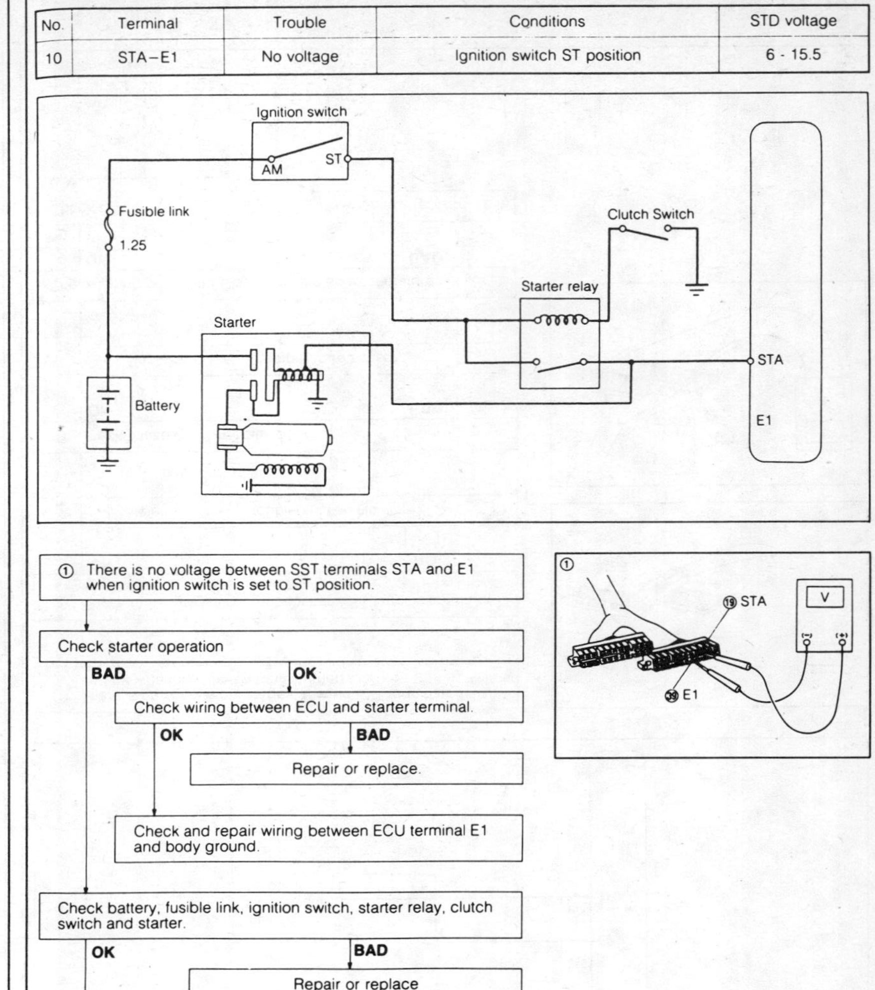

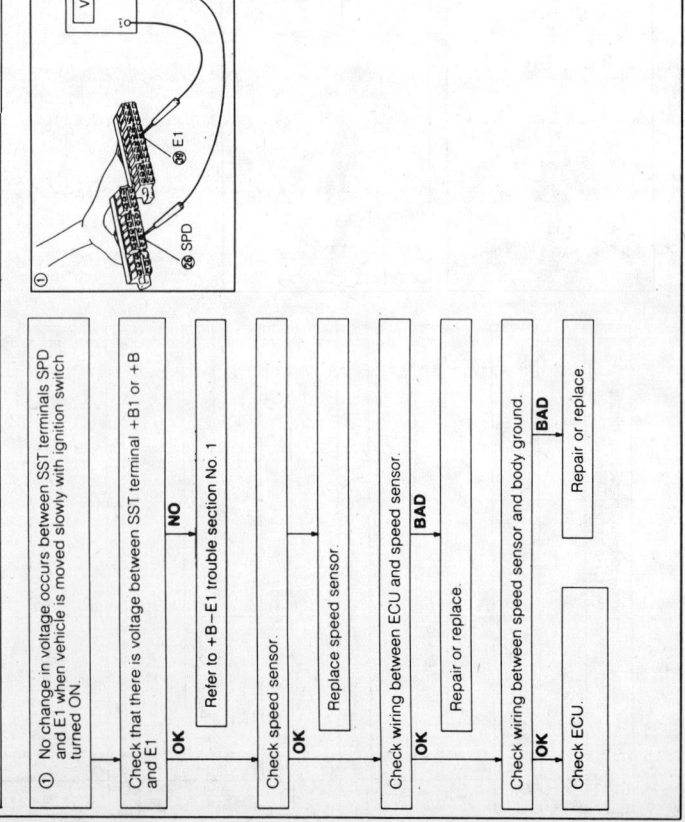


① There is no voltage between SST terminals STA and E1 when ignition switch is set to ST position.

Check starter operation
- OK → Check wiring between ECU and starter terminal.
  - BAD → Repair or replace.
  - OK → Check and repair wiring between ECU terminal E1 and body ground.
- BAD ↓

Check battery, fusible link, ignition switch, starter relay, clutch switch and starter.
- BAD → Repair or replace
- OK → Check and repair wiring between battery and ECU terminal STA.

① 19 STA, 39 E1, V

# TROUBLESHOOTING CHART 11—CHARADE

| No. | Terminals | Trouble | Conditions | | STD voltage |
|---|---|---|---|---|---|
| 11 | A/C–E1 | No voltage | When engine is idling | Air conditioner switch ON | 10 - 15.5 |
| | IDL–E2 | No specified voltage | Ignition switch ON | Throttle valve fully closed | 0 - 0.6 |
| | | | | Throttle valve opened | 4.5 - 5.5 |

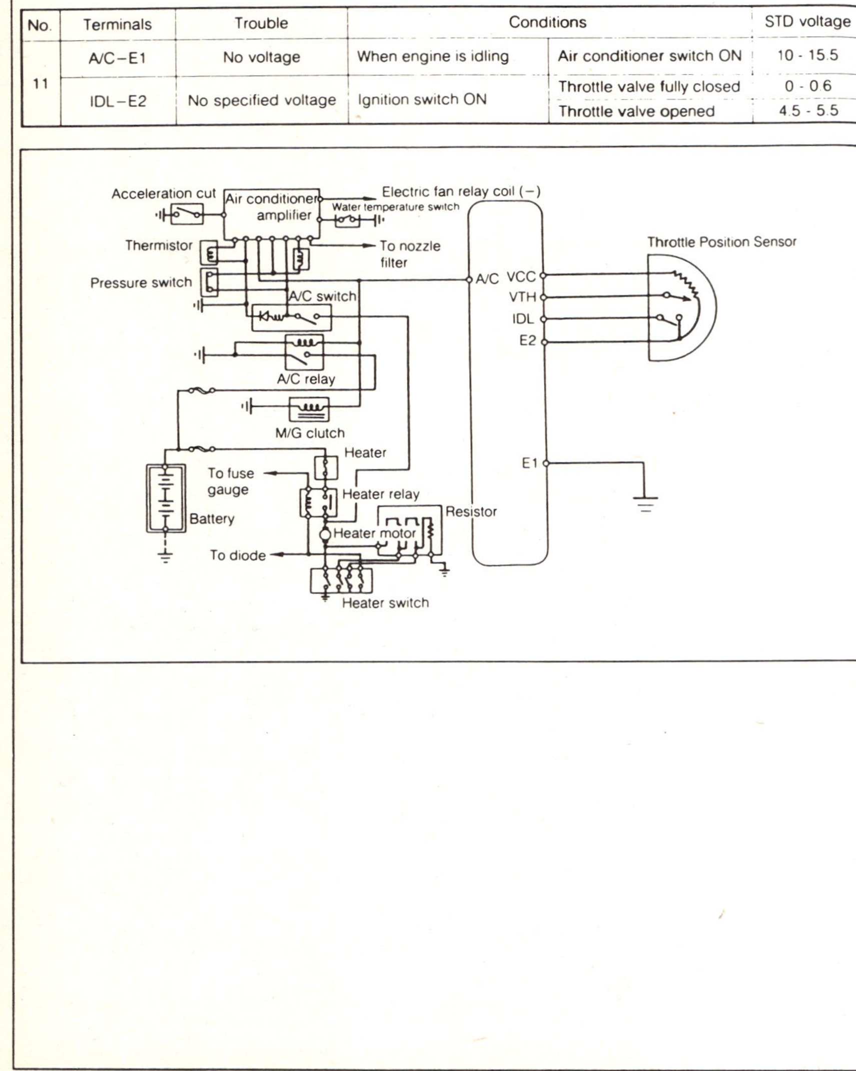


• **A/C–E1**

① There is no voltage at SST terminals A/C and E1 when air conditioner is operated.

② Check that there is voltage between SST terminal A/C and body ground.

- **OK** → Check wiring between ECU terminal E1 and body ground.
- **NO** → Check compressor cluch operation.
  - **OK** → Check and repair wiring between ECU terminal A/C and amplifier.
  - **BAD** → Check that there is voltage between amplifier terminal and body ground.
    - **OK** → Check fusible link, fuse, air conditioner switch, pressure switch, thermistor, acceleration cut switch and water temperature switch.
      - **OK** → Check wiring between battery and air conditioner amplifier.
        - **BAD** → Repair or replace.
        - **OK** → Replace air conditioner amplifier.
    - **OK** → Check and repair wiring between amplifier and ECU terminal A/C.

①

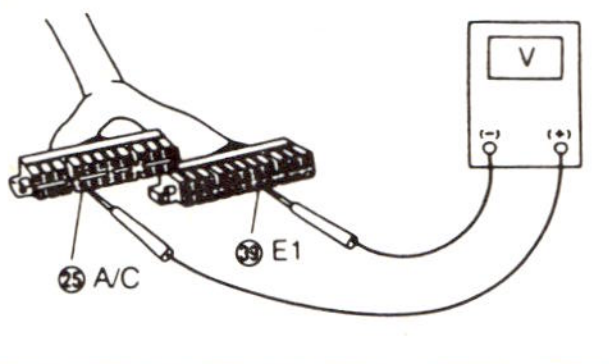

㉕ A/C
㉟ E1
V

②

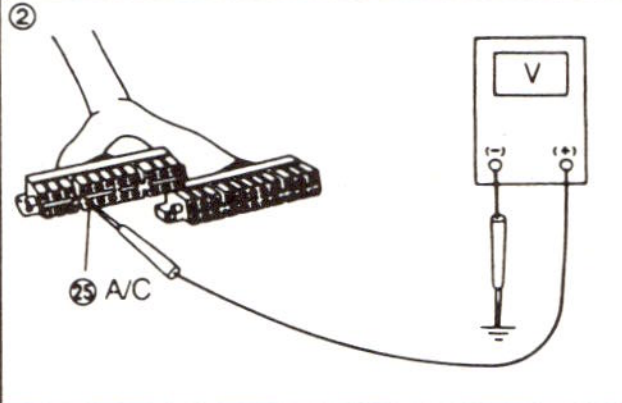

㉕ A/C
V

## TROUBLESHOOTING CHART 11 – CHARADE

• **IDL–E2**

① There is no voltage between SST terminals IDL and E2 when ignition switch is turned ON. (Throttle valve opened)

② Check that there is voltage between SST terminal +B1 or +B and E1.
- **OK** → Check throttle position sensor.
- **BAD** → Refer to +B–E1 trouble section No. 1

Check throttle position sensor.
- **OK** → Check wiring between ECU and throttle position sensor.
- **BAD** → Replace throttle position sensor.

Check wiring between ECU and throttle position sensor.
- **OK** → Check ECU.

Check ECU.

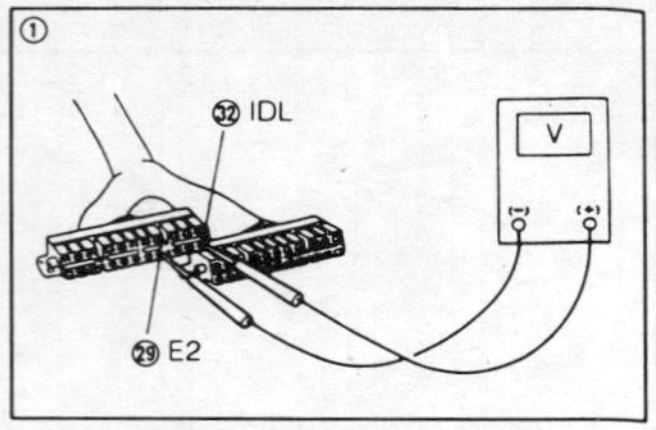

①: ㉜ IDL, ㉙ E2, V

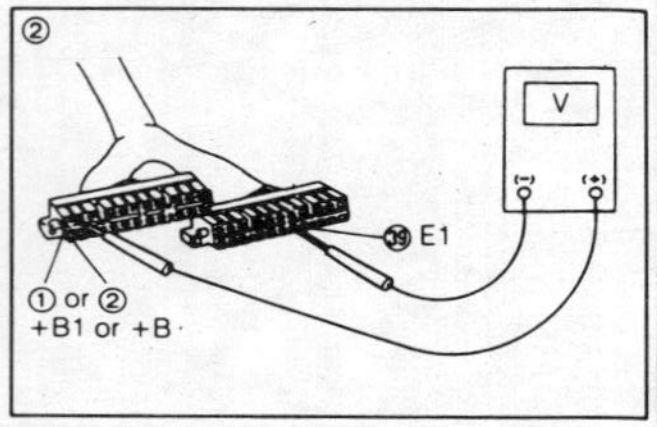

②: ㉙ E1, ① or ② +B1 or +B, V

## TROUBLESHOOTING CHART 12 – CHARADE

| No. | Terminal | Trouble | Conditions | | STD voltage |
|---|---|---|---|---|---|
| 12 | EVSV–E1 | No specified voltage | When engine is idling | Coolant temperature above 104 °F (40 °C) | 1.0 - 3.5 |

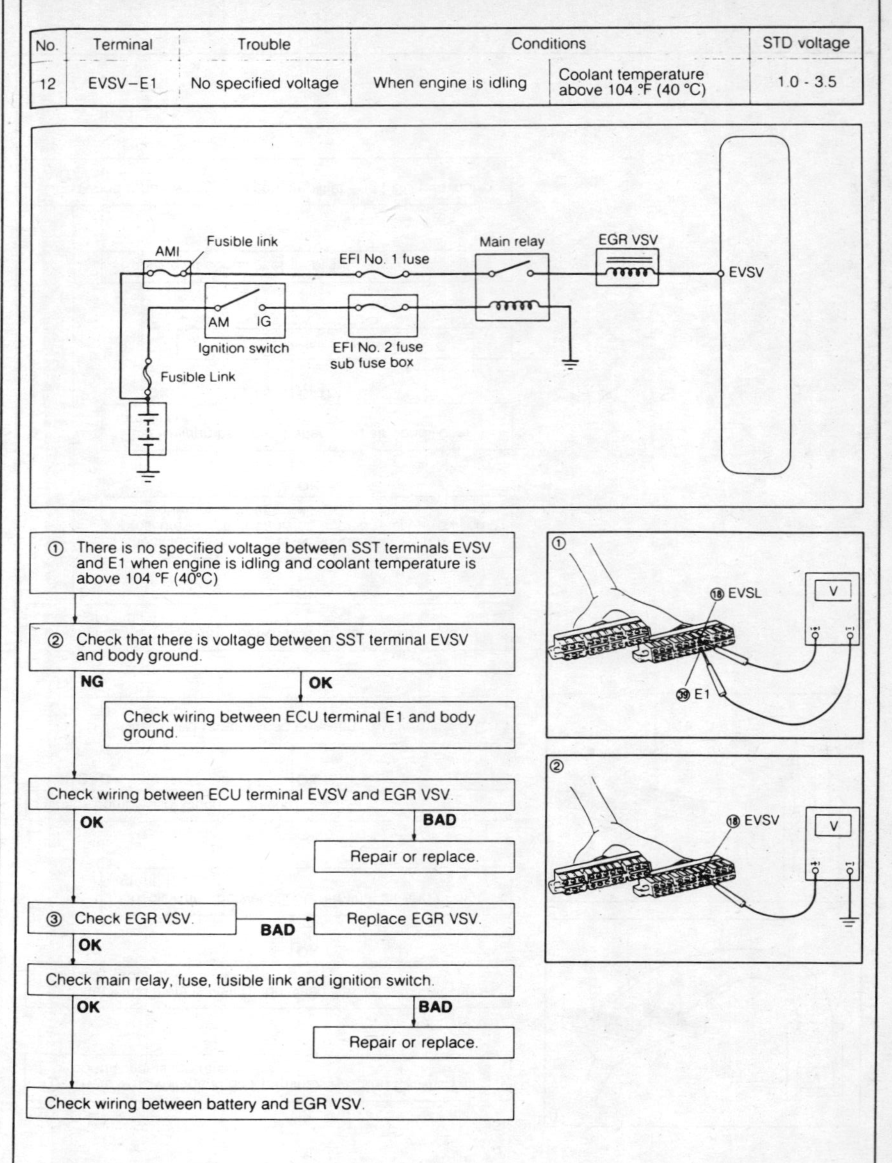


① There is no specified voltage between SST terminals EVSV and E1 when engine is idling and coolant temperature is above 104 °F (40°C)

② Check that there is voltage between SST terminal EVSV and body ground.
- **NG** → Check wiring between ECU terminal EVSV and EGR VSV.
- **OK** → Check wiring between ECU terminal E1 and body ground.

Check wiring between ECU terminal EVSV and EGR VSV.
- **OK** → ③ Check EGR VSV.
- **BAD** → Repair or replace.

③ Check EGR VSV.
- **OK** → Check main relay, fuse, fusible link and ignition switch.
- **BAD** → Replace EGR VSV.

Check main relay, fuse, fusible link and ignition switch.
- **OK** → Check wiring between battery and EGR VSV.
- **BAD** → Repair or replace.

Check wiring between battery and EGR VSV.

①: ⑱ EVSL, ㊴ E1, V

②: ⑱ EVSV, V

## FUEL PRESSURE

### Testing

1. Disconnect the negative battery cable.
2. Slowly remove the union bolt from the fuel filter using a cloth to catch excess fuel.
3. Connect pressure gauge to the fuel pipe.
4. Install the SST No. 09991–87702–000 test harness to the check terminal.
5. Ground the F terminal.
6. Reconnect the negative battery cable.
7. Fuel pressure should read 33–40 psi.
8. The fuel pressure should not drop below 25 psi after 3 or more minutes.
9. Remove the SST test harness, start the engine and check fuel pressure with engine running at normal operating temperature. The pressure should be 21–28 psi.
10. If fuel pressure is above specifications, check regulator vacuum line. If vacuum line and system check good, replace the fuel regulator.

## MAIN RELAY OR INJECTOR RELAY

### Testing

**CAUTION**

***Do not touch the relay if in operation. The relay becomes very hot during operation.***

1. Check that the relay clicks when the ignition switch is turned **ON**.
2. Check that there is only continuity between terminals 1 and 2.
3. The resistance between terminals 1 and 2 should be 60–80 ohms.
4. Apply battery voltage across terminals 1 and 2.
5. Check that there is continuity between terminals 3 and 4.
6. If any test fails, replace the relay.

## FUEL PUMP RELAY

### Testing

**CAUTION**

***Do not touch the relay if in operation. The relay becomes very hot during operation.***

1. Disconnect the ST terminal of the starter.
2. Check that the relay clicks when the ignition switch is turned to the **START** position.
3. Check that there is only continuity between terminals 1 and 2 of the relay.
4. The resistance between terminals 1 and 2 should be 70–90 ohms.
5. Apply battery voltage across terminals 1 and 2.
6. Check that there is continuity between terminals 3 and 4 of the relay.
7. If any test fails, replace the relay.

## WATER TEMPERATURE SENSOR

### Testing

1. Disconnect sensor.
2. Measure temperature of sensor from 0–180°F (–20–80°C).
3. Measure resistance between sensor terminals through temperature range and compare to graph.
4. If not within specification, replace sensor.

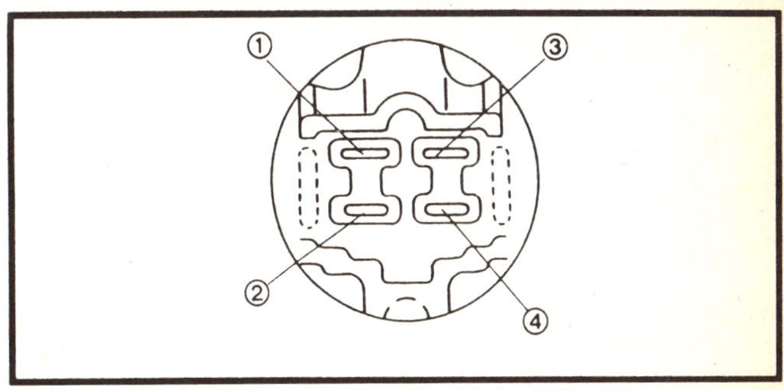


Main and injector relay terminal identification

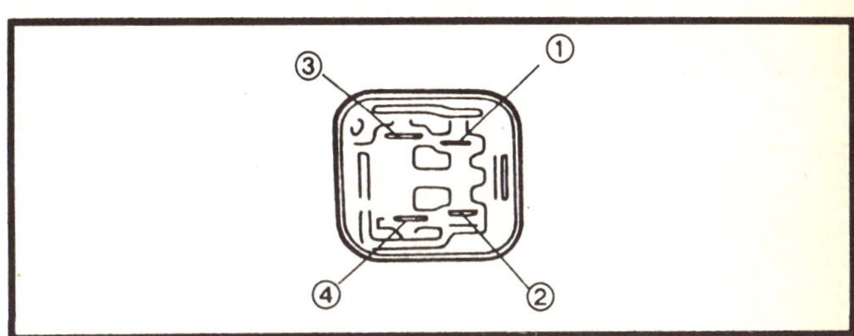


Fuel pump relay terminal identification

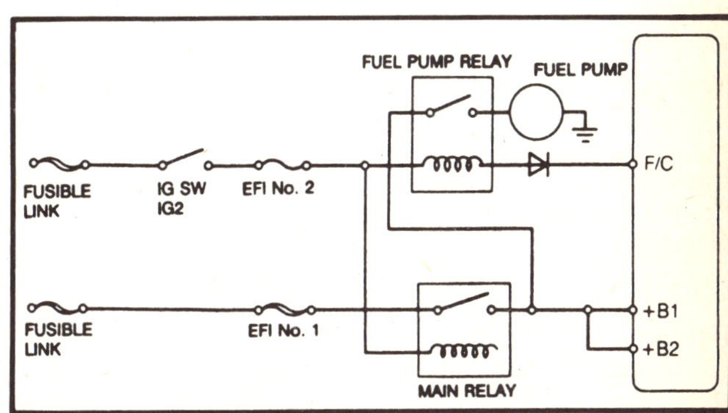


Fuel pump relay circuit

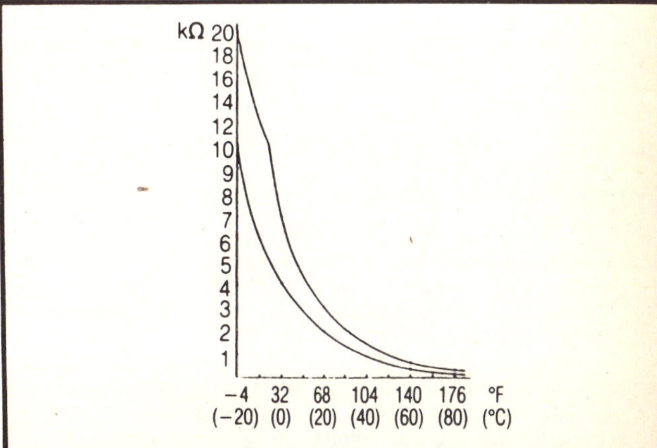


Water temperature and air temperature sensor graph

## AIR TEMPERATURE SENSOR

### Testing

1. Disconnect sensor and check that there is no continuity between the terminals and the sensor body.
2. Measure temperature inside the surge tank from 0–180°F (–20–80°C).

3. Measure resistance between sensor terminals through temperature range and compare to graph.
4. If not within specification, replace sensor.

## THROTTLE POSITION SENSOR (TPS)

### Testing

1. Disconnect the TPS.
2. Measure the resistance between the VCC and E2 terminals.
3. The resistance should be 4.25–8.25 ohms.
4. Measure the resistance between the VCC and VTH terminals.
5. The resistance should be 3.5–10.3 ohms at closed throttle and 0.3–6.3 ohms at full throttle. The resistance value should change smoothly through the complete range of the throttle valve.
6. Measure the resistance between the IDL and E2 terminals.
7. The resistance should be 15–35 ohms at closed throttle and over 1000 kilo ohms if the throttle is opened more than 1.5 degrees.
8. If any test fails, replace the throttle body.

## THROTTLE POSITIONER (TP)

### Testing

1. Allow engine to reach normal operating temperature.
2. Check that the idle speed is correct.
3. Raise engine to 3000 rpm and slowly close the throttle.
4. Observe the engine rpm at the time the dashpot lever comes in contact with the throttle lever.
5. If the engine speed is not 1500–1700 rpm, adjust by turning the TP adjusting screw.

## AUXILIARY AIR VALVE

### Testing

1. Disconnect the air cleaner hose from the throttle body.
2. Start the engine and check that when the cooling temperature is below 137°F (75°C) and the air valve port is blocked that the engine rpm drops.
3. Check that when the engine is above 146°F (80°C) and the air valve port is blocked that the engine rpm does not change.
4. If the auxiliary air valve malfunctions, replace the throttle body.

## PRESSURE SENSOR

### Testing

1. Disconnect the negative battery cable.
2. Connect the SST test harness to the ECU.
3. Reconnect the negative battery cable.
4. Measure the voltage between the PiM and E2 terminals of test harness with the ignition switch **ON**.
5. If not within specification and if there is 4.5–5.5 volts between the VCC and E2 terminals, replace the sensor.
6. Apply vacuum to the sensor and check that the voltage drops approximately 0.5–0.7 volts.
7. If the voltage does not drop, replace sensor. It is recommended that when the sensor is replaced that the air filter and fuel filter also be replaced.

## PRESSURE VACUUM SWITCHING VALVE (VSV)

### Testing

1. Disconnect the negative battery cable.

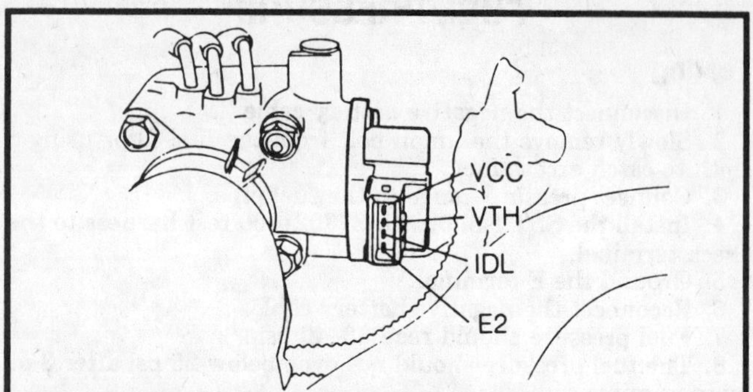

**Throttle position sensor terminals**

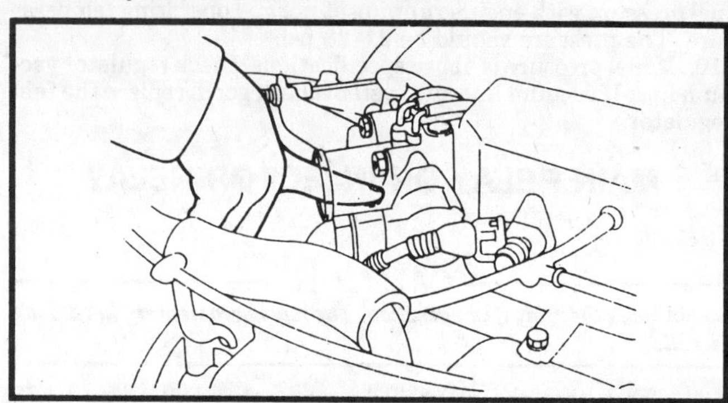
**Blocking auxiliary air valve port**

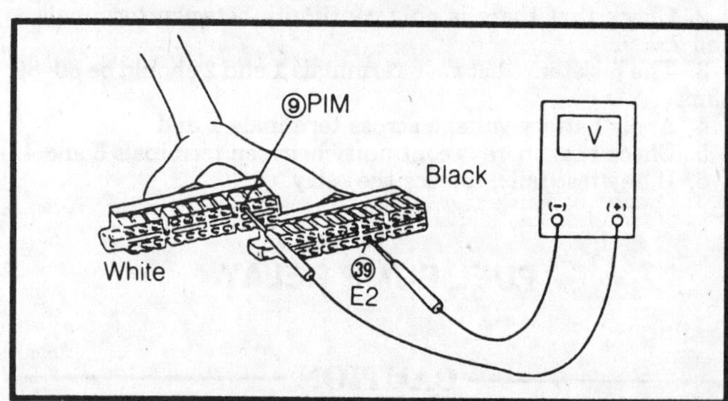

**Testing the pressure sensor**

| Measuring point Altitude (height above sea level) ft (m) | Atmospheric pressure inch Hg (mmHg) | Voltage V |
|---|---|---|
| 0 (0) | 29.92 (760) | 2.1 - 2.7 |
| 1640 (500) | 28.19 (716) | 1.9 - 2.5 |
| 3280 (1,000) | 26.54 (674) | 1.8 - 2.4 |

**Pressure sensor chart**

2. Connect the SST test harness to the ECU.
3. Reconnect the negative battery cable.
4. Start the engine and allow it to reach normal operating temperature.
5. Measure the voltage at terminal 38–PVSV and 39–E1.
6. Raise engine to 4000 rpm and release accelerator pedal; if the voltage drops momentarily, VSV is good; if not, proceed.
7. Disconnect the vacuum hose from the VSV at the air cleaner side and apply 4 in. Hg. If VSV leaks, replace it.

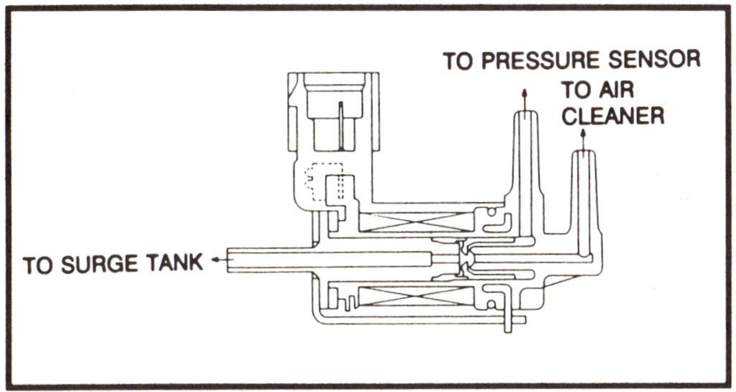


**Pressure VSV**

8. Turn the ignition switch **ON**.
9. **Short the PVSV and E1 terminals. The vacuum should drop to 0; if not, replace the VSV.**
10. Turn the ignition switch **OFF**.
11. Apply vacuum to the sensor side port of the VSV. It should not hold vacuum.

## EGR VSV

### Testing

1. Disconnect the negative battery cable.
2. Disconnect the EGR VSV connector and measure the resistance across the terminals. If the resistance is not approximately 44 ohms at 68°F (20°C), replace the EGR VSV.
3. Connect the SST test harness to the ECU.
4. Reconnect the negative battery cable.
5. Disconnect the hose connected to the EGR VSV.
6. Apply vacuum to the modulator side port of the VSV. It should not hold vacuum.
7. Apply vacuum to the surge tank side of the VSV and apply 4 in. Hg. It should hold vacuum.
8. EGR VSV should release this vacuum when the ignition switch is turned **ON**. If the vacuum is not released:
   a. Check for for voltage between the EVSV and E1 terminals of the SST harness.
   b. If there is voltage, check the ECU.
   c. If no voltage is applied, check the EGR VSV for power supply.
   d. Measure the resistance of the EGR VSV, the resistance should be 30–50 ohms.
   e. Replace defective parts, as necessary.
9. Start the engine, with the coolant temperature below 104°F (40°C). Apply vacuum to the surge tank port. If it does not hold vacuum, check the water temperature sensor.
10. With coolant temperature above 104°F (40°C), check that the engine stalls or runs poorly when vacuum is applied. If engine performance is not effected and vacuum is maintained, check the coolant temperature sensor. If engine performance is not affected without vacuum being held, check the EGR system.
11. Reconnect the vacuum hoses.
12. Start engine. While engine is warming up, measure the voltage between the EVSV and E1 terminals of the SST harness. Check that the voltage drops before the radiator cooling fan starts to rotate. If no voltage drop occurs, check the water temperature sensor. If the sensor is good, check the ECU.
13. Turn ignition **OFF** and disconnect the negative battery cable before removing the SST test harness.

## IDLE-UP VACUUM SWITCHING VALVE (VSV)

### Testing

1. Disconnect the negative battery cable.
2. Connect the SST test harness to the ECU.

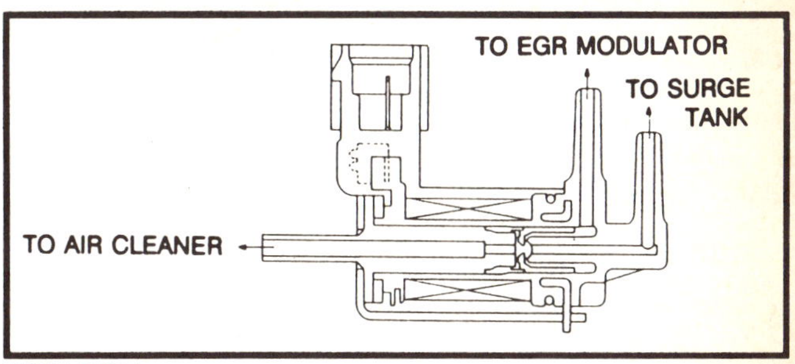


**EGR VSV**

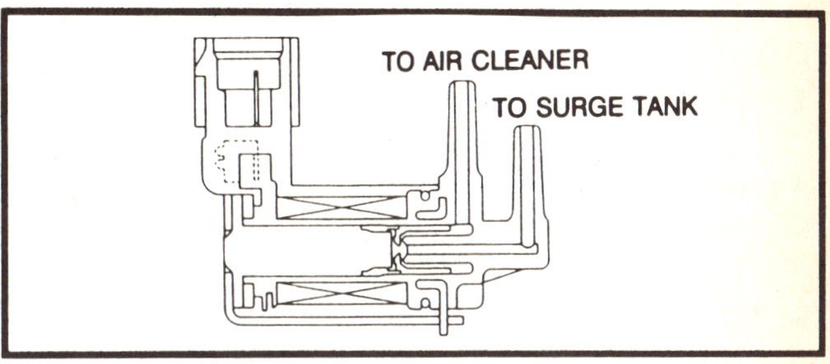


**Idle-up VSV**

3. Reconnect the negative battery cable.
4. Disconnect the rubber hose connected to the idle-up VSV.
5. Apply 4 in. Hg vacuum to the idle-up VSV; it should hold vacuum.
6. Turn **OFF** all accessory switches and turn **ON** the ignition switch.
7. The vacuum that was applied should drop to 0 when the ignition switch is turned **ON**. If the vacuum is not released:
   a. Check for for voltage between the VSV and E1 terminals of the SST harness.
   b. If there is voltage, check the the resistance between the VSV terminals. The resistance should be 30–50 ohms.
   c. If the resistance is correct, check the wiring between the ECU and the main relay.
   d. Replace defective parts, as necessary.
8. Start the engine and allow to idle for 1 minute.
9. Apply 4 in. Hg vacuum to the VSV.
10. If the VSV does not hold vacuum, check the wiring between the VSV and ECU. If wiring is good, check that there is voltage between the DSW and E1 terminals of the SST harness. If there is voltage, check the headlamp and defogger switches and circuits. If they check good, check the ECU and ECU circuit.
11. Turn **ON** the headlamp switch or defogger switch. Check that the vacuum applied in Step 9 drops to 0. If the vacuum does not drop to 0, check that there is voltage between the DSW and E1 terminals of the SST harness. If there is voltage, check the ECU; if there is no voltage, check the headlamp and defogger switches and circuits.
12. Turn **OFF** headlamp and defogger switches.
13. Apply 4 in. Hg vacuum to the VSV.
14. Turn **ON** the blower fan switch. The vacuum should drop to 0. If not check the blower switch and circuit.
15. Connect any vacuum lines that are disconnected.
16. Turn ignition **OFF** and disconnect the negative battery cable before removing the SST test harness.

## OXYGEN SENSOR

### Testing

1. Install test harness SST No. 09991-87702-000 or equivalent, to the check terminal.

2. Warm engine to normal operating temperature.
3. After warm, raise engine to 3000 rpm for 2 minutes.
4. Short the Brown T terminal and the Black ground terminal of the test connector.
5. Connect a voltmeter between the Green VF terminal and engine ground.
6. Hold the engine speed constant at 1500 rpm.
7. The voltmeter change 8 or more times a minute.
8. If the voltmeter changes less than 8 times, repeat Steps 3 to 7; if still less than 8 changes, replace sensor.
9. If no change at all, perform diagnosis check. Check and repair and trouble codes. When trouble Code 11 is obtained continue with Step 10.
10. Repeat Steps 2 to 7 if the voltmeter reads 5 volts; the mixture is too rich. Repair condition causing rich mixture.
11. If the voltmeter reads 0 volts and engine is idling normally, replace the oxygen sensor. If the engine is not idling normally, the mixture is too lean. Check for vacuum leak or repair other condition causing a lean mixture.

## FUEL INJECTOR

### Testing

1. Check injector for operating sound when engine is being cranked.
2. Remove surge tank.
3. Check resistance of injector.
4. Resistance should be 1.5–3.0 ohms.
5. If no operating sound or resistance not in range, replace injector.

## COLD START INJECTOR

### Testing

1. Disconnect the injector.
2. Measure the resistance between the terminals under the following conditions:
   - Terminals 1-2 – 67–79 ohms at 75°F (24°C)
   - Terminals 1-3 – 67–79 ohms at 75°F (24°C)
   - Terminals 2-3 – 140–155 ohms at 75°F (24°C)
   - Terminals 1-2 – 33–41 ohms at 68°F (20°C)
   - Terminals 1-3 – 33–41 ohms at 68°F (20°C)
   - Terminals 2-3 – 0 ohms at 68°F (20°C)
3. If any test fails, replace the cold start injector.

## FUEL CUT RPM

### Testing

1. Disconnect the negative battery cable.
2. Connect the SST test harness to the ECU.
3. Reconnect the negative battery cable.
4. Start engine and allow to reach normal operating temperature.
5. Short the IDL and E2 terminals of the SST harness.
6. Gradually raise the engine speed. Determine the rpm at which the engine speed ceases its increase because of fuel be cut off.
7. The fuel cut off rpm should be 1450 rpm with A/C **OFF** or 1750 rpm with the A/C **ON**.
8. The fuel return rpm should be 1200 rpm with A/C **OFF** or 1500 rpm with the A/C **ON**.

## ELECTRONIC CONTROL UNIT (ECU)

### Testing

1. Disconnect the negative battery cable.
2. Connect the SST test harness to the ECU.
3. Reconnect the negative battery cable.

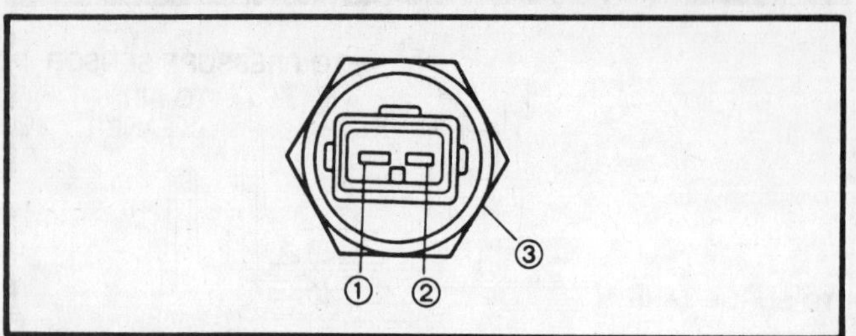

**Cold start injector terminals**

4. Measure the voltages at the test harness.

**NOTE: If measuring sensor resistances at the ECU connector, never connect the SST test harness to the ECU.**

5. If voltage inputs and outputs are within specification and if the sensor resistance values are correct and the engine still malfunctions, replace the ECU.

# Component Replacement

## COLD START INJECTOR

### Removal and Installation

1. Disconnect the negative battery cable.
2. Disconnect the connector of the cold start injector.
3. Place a cloth under the cold start injector and loosen the union bolt.
4. Remove the fuel hose and the cold start injector.

**To Install:**

5. Install the cold start injector to the surge tank using a new gasket and torque to 3–5 ft. lbs. (4–7 Nm).
6. Install the fuel hose with new gaskets to the cold start injector and torque the delivery pipe to in 3 stages to 7–11 ft. lbs. (10–15 Nm).
7. Connect the cold start injector connector.
8. Connect the negative battery cable and check for fuel leakage.

## SURGE TANK

### Removal and Installation

**NOTE: Tag all hoses and electrical connections to use as reference during reassembly.**

1. Disconnect the negative battery cable.
2. Drain the cooling system.
3. Remove the accelerator cable from the throttle body.
4. Remove the air cleaner assembly.
5. Remove oil cooler hose and water pump hose from the throttle body.
6. Remove the brake booster hose from the intake manifold.
7. Disconnect the vacuum hose connected to the air cleaner.
8. Disconnect the vacuum hose from the atmosphere pressure VSV.
9. Remove the vacuum hose from the pressure regulator.
10. Disconnect the vacuum hose connected to the intake manifold BVSV.
11. Disconnect the vacuum hose of the idle-up VSV, if equipped with air conditioning.
12. Disconnect the idle-up VSV, EGR VSV, intake air temperature sensor, throttle position sensor and cold start injector.
13. Disconnect the blow-by gas hose from the surge tank.
14. Remove the fuel hose from the cold start injector.
15. Remove the surge tank support.
16. Remove the wire retaining clamp from the surge tank.
17. Remove the EGR pipe.

## ECU and test harness connector identification

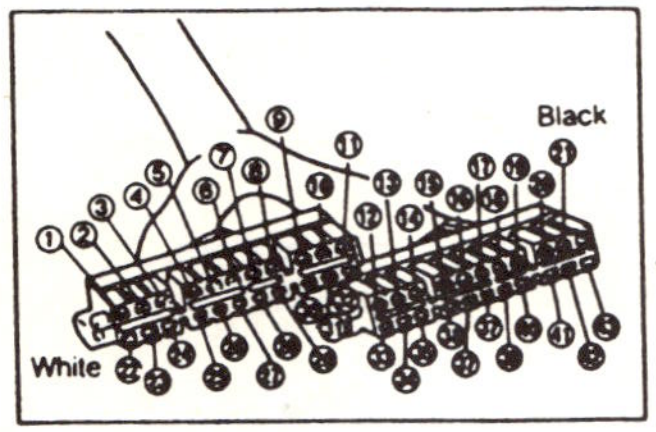

### ECU Connectors

| Code | | Contents of connection |
|---|---|---|
| 1 | IBI | Main relay (power supply) |
| 2 | BATT | Battery (back-up power supply) |
| 3 | | |
| 4 | FC | Fuel pump relay |
| 5 | | |
| 6 | | |
| 7 | | |
| 8 | THW | Water temperature sensor |
| 9 | PIM | Intake manifold pressure sensor |
| 10 | THA | Intake air temperature sensor |
| 11 | T | Check terminal |
| 12 | IGF | Ignition monitor |
| 13 | GI | Crank angle sensor |
| 14 | G– | Crank angle sensor ground |
| 15 | OX | O2 sensor |
| 16 | | |
| 17 | DSW | Electrical load signal |
| 18 | EVSV | VSV for EGR |
| 19 | STA | Starter switch |
| 20 | #10 | Injector |
| 21 | E01 | Power system ground (engine block) |

| Code | | Contents of connection |
|---|---|---|
| 22 | +B | Main relay (power supply) |
| 23 | W | Diagnosis lamp |
| 24 | | |
| 25 | A/C | Air conditioner amplifier |
| 26 | SPD | Vehicle speed sensor |
| 27 | | |
| 28 | | |
| 29 | E2 | Sensor system ground |
| 30 | VTH | Linear throttle sensor |
| 31 | VCC | Sensor power supply output (5 V) |
| 32 | IDL | Throttle sensor full closing signal |
| 33 | B/K | Brake signal |
| 34 | NE | Crank angle sensor revolution speed |
| 35 | E21 | Pressure sensor ground |
| 36 | VF | O2 sensor checker |
| 37 | VSV | Idle-up VSV |
| 38 | PVSV | VSV for introducing atmospheric pressure |
| 39 | EI | Operation system ground (engine block) |
| 40 | IGT | Ignitor control output |
| 41 | | |
| 42 | | |

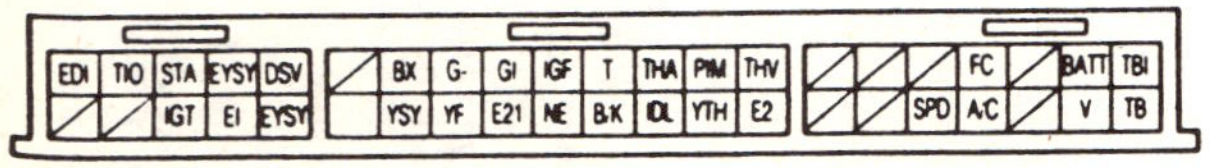

## Voltages at the ECU connector

| Terminals | STD voltage | Condition | |
|---|---|---|---|
| +B1–E1 | 10 - 15.5 | Ignition switch ON | |
| BATT–E1 | 10 - 15.5 | At all time | |
| FC–E1 | 10 - 15.5 | Ignition switch ON | When engine is stopped: |
| | 0.1 - 1.5 | Ignition switch ON | When engine is running: |
| THW–E2 | 0.1 - 0.7 | Ignition switch ON | Coolant temperature 176 °F (80 °C) |
| PIM–E2 | 2.1 - 2.7 | Ignition switch ON | When atmospheric pressure is 760 mmHg: |
| THA–E2 | About 0.9 - 3.0 | Ignition switch ON | Temperature of air in surge tank 68 °F (20 °C) |
| T–E1 | 4.5 - 5.5 | Ignition switch ON | T terminal OFF |
| IGF–E1 | 0.5 - 1.5 | Ignition switch ON | |
| GI–E1 | About 0.6 | Ignition switch ON | |
| G⊖–E1 | About 0.6 | | |
| OX–E1 | Change in output voltage | When engine speed is held at 3000 rpm for two minutes after engine has been fully warmed up: | |
| DSW–E1 | 10 - 15.5 | Ignition switch ON | When defogger and/or headlamp switch is turned ON: |
| EVSV–E1 | 10 - 15.5 | Ignition switch ON | Coolant temperature is below 104 °F (40 °C) |
| | 1.0 - 3.5 | | Coolant temperature is above 104 °F (40 °C) |
| STA–E1 | 0 | At all time | |
| | 6 - 15.5 | When ignition switch is set to ST position: | |
| #10–E1 | 10 - 15.5 | Ignition switch ON | |
| +B–E1 | 10 - 15.5 | Ignition switch ON | |
| W–E1 | 1.0 - 3.5 | Ignition switch ON | |
| | 10 - 15.5 | When engine is running with ignition switch turned ON and diagnosis code is normal during diagnosis code check: | |
| AC–E1 | 10 - 15.5 | When air conditioner switch is turned ON while engine is running: | |
| SPD–E1 | Change in voltage between 0 to 4.5–5.5 | Ignition switch ON | When vehicle is moved: |
| VTH–E2 | About 0.55 | Ignition switch ON | Throttle valve fully closed: |
| | About 3.91 | | Throttle valve fully opened: |
| VCC–E2 or E21 | 4.5 - 5.5 | Ignition switch ON | |
| IDL–E2 | 0 | Ignition switch ON | Throttle valve fully closed: |
| | 4.5 - 5.5 | | Throttle valve fully opened: |

### Voltages at the ECU connector

| Terminals | STD voltage | Condition | |
|---|---|---|---|
| B/K−E1 | 10 - 15.5 | Brake pedal is depressed: | |
| NE−E1 | About 0.6 | Ignition switch ON | |
| VSV−E1 | 1.0 - 3.5 | Ignition switch ON | |
| | 10 - 15.5 | After a lapse of at least 60 seconds after engine starting. | |
| PVSV−E1 | 10 - 15.5 | Ignition switch ON | |
| IGT−E1 | 0 | Ignition switch ON | |
| | AC 0.3 - 0.9 | When engine is running: | |
| VF−E1 | 1.5 - 3.5 | Ignition switch ON | |
| | 0 - 0.5 | Ignition switch ON | When diagnosis code is memorized while throttle valve is closed fully, SST (09991-87702-000) is installed on check terminal, and terminal T is shorted with ground terminal: |
| | 4.5 - 5.5 | | When no diagnosis code is memorized while throttle valve is closed fully, SST (09991-87702-000) is installed on check terminal, and terminal T is shorted with ground terminal: |
| | Change in voltage occurs eight times or more for 10 seconds between 0 - 5. | | When engine speed is held at 3000 rpm for two minutes after engine has been fully warmed up, SST (09991-87702-000) is installed on check terminal, and terminal T is shorted with ground terminal: |

### Resistance values at the ECU connector

| Terminals | Conditions | Resistance |
|---|---|---|
| THW−E2 | Coolant temperature 176 °F (80 °C) | About 0.32 kΩ |
| THA−E2 | Temperature of air in surge tank 68 °F (20 °C) | About 2.45 kΩ |
| VCC−VTH | Throttle valve fully closed: | 3.5 - 10.3 kΩ |
| | Throttle valve fully opened: | 0.3 - 6.3 kΩ |
| IDL−E2 | Throttle valve fully closed: | About 17Ω |
| | Throttle valve fully opened: | 1000 kΩ or more |
| NE−G- | — | 140 - 180Ω |
| G1−G- | — | 140 - 180Ω |

18. Remove the surge tank.

**To Install:**

19. Install the surge tank to the intake manifold with a new gasket.
20. Torque the bolts evenly in 3 stages to 11–16 ft. lbs. (15–21 Nm).
21. Reconnect the blow-by gas, air conditioner fast idle, BVSV vacuum, pressure regulator vacuum, atmosphere VSV, air cleaner and brake booster hoses.
22. Reconnect the cold start injector, throttle position sensor, intake air temperature sensor, EGR VSV and idle-up VSV electrical connectors.
23. Install the surge tank support.
24. Install the clamp for the wiring at the back of the surge tank.
25. Install the oil cooler hose and water pump hose to the throttle body.
26. Install the air hose to the throttle body.
27. Install the accelerator cable to the throttle body and torque the locknut to 7–12 ft. lbs. (10–15 Nm).
28. Check that the accelerator cable free-play is 0.12–0.31 in. (3–8mm).
29. Install the EGR pipe.
30. Fill the cooling system.
31. Connect the negative battery cable.
32. Start engine and check for fuel or coolant leaks.

## FUEL PRESSURE REGULATOR

### Removal and Installation

1. Disconnect the negative battery cable.
2. Drain the cooling system.
3. Remove the surge tank.
4. Place a cloth under the fuel hose and slowly disconnect the fuel hose from the delivery pipe.
5. Remove the fuel return hose from the pressure regulator.
6. Remove the delivery pipe from the intake manifold.
7. Clamp the hexagonal section of the delivery pipe in a vise and remove the fuel pressure regulator.

**To Install:**

8. Install a new O-ring on the fuel pressure regulator.
9. Clamp the hexagonal section of the delivery pipe in a vise and install the fuel pressure regulator. Back off the regulator so

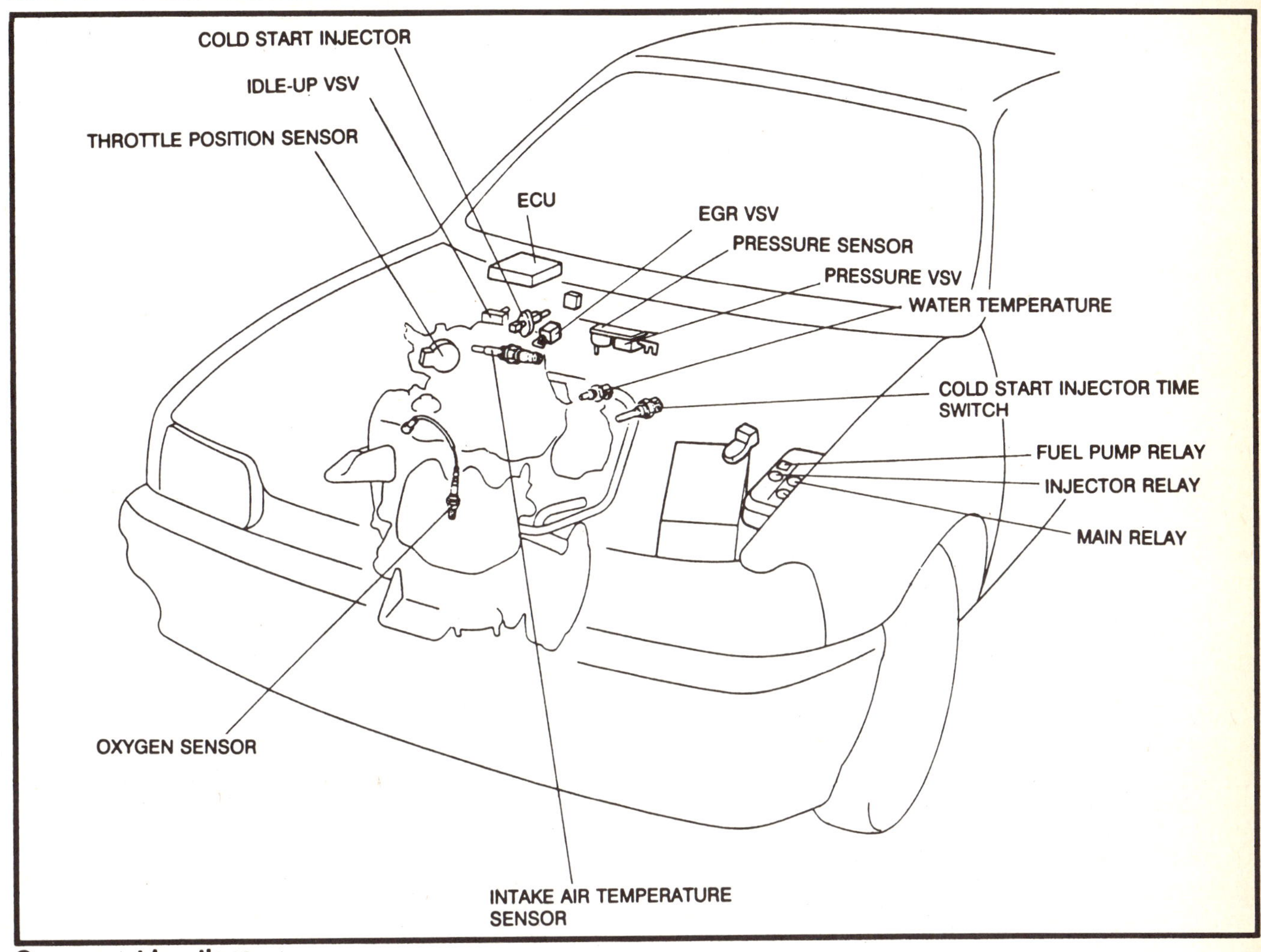


**Component locations**

the vacuum sensing pipe is 45 degrees from the horizontal plane, facing toward the rear of the engine. Torque the locknut to 25–32 ft. lbs. (34–44 Nm).

10. Install a new O-ring on the injector, apply gasoline to O-ring and install to the delivery pipe.
11. Torque the delivery pipe nuts in 3 stages to 7–11 ft. lbs. (10–15 Nm).
12. Check that the injector turns freely by hand; if not, replace the O-ring.
13. Install the fuel hose to the delivery pipe and torque to 25–33 ft. lbs. (34–44 Nm).
14. Install the fuel hose to the fuel filter with a new gasket and connect the fuel return pipe to the pressure regulator.
15. Install the surge tank.
16. Fill the cooling system.
17. Connect the negative battery cable.
18. Start engine and check for fuel or coolant leaks.

## FUEL INJECTOR

### Removal and Installation

1. Disconnect the negative battery cable.
2. Drain the cooling system.
3. Remove the surge tank.
4. Place a cloth under the fuel hose and slowly disconnect the fuel hose from the delivery pipe.
5. Disconnect the fuel return hose from the pressure regulator.
6. Remove 3 nuts and the fuel delivery pipe.
7. Remove the injectors from the intake manifold. Do not remove the injector cover.

**To Install:**

8. Install a new O-ring on the fuel injector.
9. Clamp the hexagonal section of the delivery pipe in a vise and install the fuel pressure regulator. Back off the regulator so the vacuum sensing pipe is 45 degrees from the horizontal plane, facing toward the rear of the engine. Torque the locknut to 25–32 ft. lbs. (34–44 Nm).
10. Check the injector vibration insulator for damage. Install the injector insulator to the intake manifold.
11. Remove the O-ring from the injector, check the grommet for damaged or flatten condition. Replace if necessary.
12. Install a new O-ring onto the injector.
13. Insert the injector into the intake manifold.
14. Check the delivery pipe heat insulator for damage. Install insulator, replace with new, if necessary.
15. Apply gasoline to the O-ring of the injector and install the delivery pipe. Make certain the delivery pipe is connected to the proper injector.
16. Torque the delivery pipe nuts in 3 stages to 7–11 ft. lbs. (10–15 Nm).
17. Check that the injector turns freely by hand; if not, replace the O-ring.

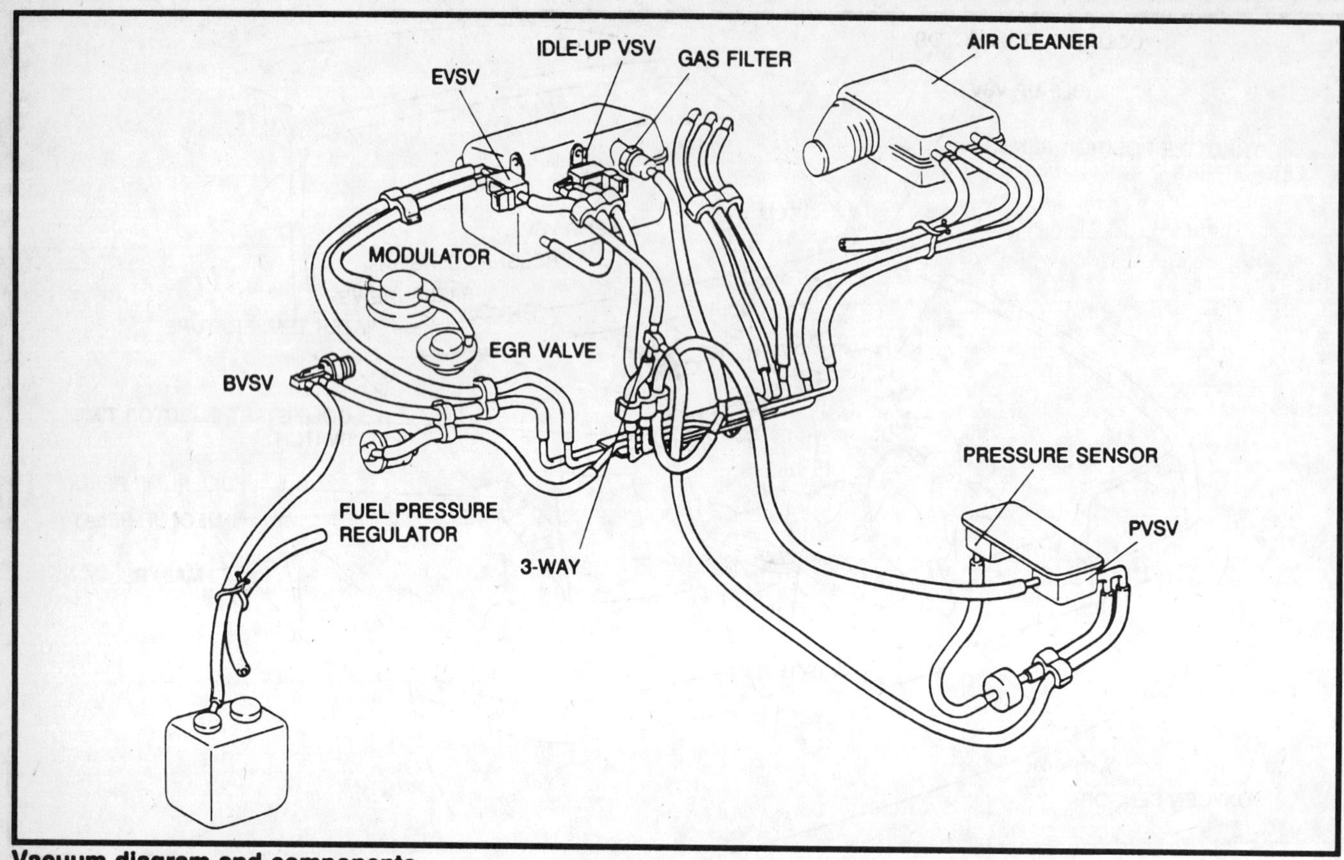


**Vacuum diagram and components**

18. Connect the injector connector.
19. Connect the return hose to the pressure regulator and return pipe. Attach the clip.
20. Install the fuel hose to the fuel delivery pipe with new gaskets and torque to 25–33 ft. lbs. 34–44 Nm).
21. Install the fuel hose to the fuel filter with new gaskets and torque to 25–32 ft. lbs. (34–44 Nm).
22. Install the surge tank.
23. Fill the cooling system.
24. Connect the negative battery cable.
25. Start engine and check for fuel or coolant leaks.

## THROTTLE BODY

### Removal and Installation

**NOTE: Tag all hoses and electrical connections to use as reference during reassembly.**

1. Disconnect the negative battery cable.
2. Drain the cooling system.
3. Remove the accelerator cable from the throttle body.
4. Disconnect the throttle position sensor.
5. Disconnect the air cleaner hose from the throttle body.
6. Disconnect the water hoses from the throttle body.
7. Remove 2 throttle body nuts and 2 throttle body bolts.
8. Remove the throttle body and gasket.

**To Install:**

9. Install the throttle body to the surge tank with a new gasket and torque the nuts and bolts to 11–16 ft. lbs. (15–21 Nm).
10. Connect the water hose to the throttle body and attach the clip.
11. Connect the air cleaner hose to the throttle body and attach the clip.
12. Connect the connector of the throttle position sensor.
13. Connect the vacuum hoses to the throttle body.
14. Install the accelerator cable to the throttle body and torque the locknut to 7–12 ft. lbs. (10–15 Nm).
15. Check that the accelerator cable free-play is 0.12–0.31 in. (3–8mm).
16. Install the EGR pipe.
17. Fill the cooling system.
18. Connect the negative battery cable.
19. Start engine and check for fuel or coolant leaks.

# PROGRAMMED FUEL INJECTION (PGM-FI) SYSTEM

## ELECTRONIC IGNITION SYSTEM ENGINE CONTROL SYSTEM APPLICATION CHART

| Year | Model | Engine cc (liter) | Engine Serial Number | Fuel System | Ignition System |
|---|---|---|---|---|---|
| 1988 | Accord DX,LX | 1955 (2.0) | A20A1 | 2bbl | Electronic |
| | Accord LXi | 1955 (2.0) | A20A3 | PGM-FI | Electronic |
| | Civic | 1493 (1.5) | D15B1 | PGM-FI | PGM-IG |
| | Civic | 1493 (1.5) | D15B2 | PGM-FI | PGM-IG |
| | Civic 4WD | 1590 (1.6) | D16A6 | PGM-FI | PGM-IG |
| | Civic CRX STD | 1488 (1.5) | D15B6 | PGM-FI | PGM-IG |
| | Civic CRX HF | 1488 (1.5) | D15B2 | PGM-FI | PGM-IG |
| | Civic CRX Si | 1590 (1.6) | D16A6 | PGM-FI | PGM-IG |
| | Prelude | 1958 (2.0) | B20A3 | Dual 1bbl ① | Electronic |
| | Prelude | 1958 (2.0) | B20A5 | PGM-FI | PGM-IG |
| 1989 | Accord DX, LX | 1955 (2.0) | A20A1 | 2bbl | Electronic |
| | Accord LXi | 1955 (2.0) | A20A3 | PGM-FI | Electronic |
| | Civic | 1493 (1.5) | D15B1 | PGM-FI | PGM-IG |
| | Civic | 1493 (1.5) | D15B2 | PGM-FI | PGM-IG |
| | Civic | 1590 (1.6) | D16A6 | PGM-FI | PGM-IG |
| | Civic CRX STD | 1493 (1.5) | D15B6 | PGM-FI | PGM-IG |
| | Civic CRX HF | 1493 (1.5) | D15B2 | PGM-FI | PGM-IG |
| | Civic CRX Si | 1590 (1.6) | D16A6 | PGM-FI | PGM-IG |
| | Prelude | 1958 (2.0) | B20A3 | Dual 1bbl ① | Electronic |
| | Prelude | 1958 (2.0) | B20A5 | PGM-FI | PGM-IG |
| 1990 | Accord | 2156 (2.2) | F22A | PGM-FI | PGM-IG |
| | Civic | 1493 (1.5) | P15B1 | PGM-FI | PGM-IG |
| | Civic | 1493 (1.5) | D15B2 | PGM-FI | PGM-IG |
| | Civic | 1590 (1.6) | D16A6 | PGM-FI | PGM-IG |
| | Civic CRX STD | 1493 (1.5) | D15B6 | PGM-FI | PGM-IG |
| | Civic CRX HF | 1493 (1.5) | D15B2 | PGM-FI | PGM-IG |
| | Civic CRX Si | 1590 (1.6) | D16A6 | PGM-FI | PGM-IG |
| | Prelude | 1958 (2.0) | B20A5 | PGM-FI | PGM-IG |

BBL Barrel carburetor
PGM-FI Programmed fuel injection
①Side draft carburetor
PGM-IG Programmed ignition
PGM-IG

## General Description

Programmed Fuel Injection (PGM-FI) System consists of 3 subsystems: air intake, electronic control, and fuel.

### AIR INTAKE SYSTEM

The system supplies air for all engine needs. It consists of the air cleaner, air intake pipe, throttle body, idle control system, fast idle mechanism, and intake manifold. A resonator in the air intake pipe provides additional silencing as air is drawn into the system.

#### Throttle Body

The throttle body, either a side-draft type or down-draft type, depending on the engine, with the primary air horn at the top. To prevent icing of the throttle valves and air horn walls, under certain atmospheric conditions of the throttle valves air horn walls, the lower portion of the throttle body is heated by engine coolant. A throttle sensor is attached to the primary throttle valve to sense changes in throttle opening. A dashpot is used to slow the throttle as it approaches the closed position.

### Idle Control System

The air/fuel ratio during idling is controlled by the electronic control unit and various solenoid valves such as Electronic Air Control Valve (EACV), fast idle valve and air boost valves. With the exception of the A/C idle control solenoid valve, these change the amounts of air bypassing into the air intake manifold. The A/C control solenoid valve opens the throttle when the air conditioner is turned on by signals sent from the ECU.

### Bypass Control Solenoid Valve (BPCSV)

When engine rpm is below 5000 rpm, the BPCSV directs air flow through the long intake path for higher torque. At higher speeds, intake air flow is through the shorter path to reduce resistance to air flow.

### Idle Adjuster (Bypass Circuit)

Fuel cut-off takes place at a set position or angle of the throttle valve. If the throttle valve is moved to adjust idle speed, this position or angle will be changed and the system may not cut off fuel supply. To solve this problem, the throttle body contains an adjustable bypass circuit. This circuit is designed to control the amount of air bypassing into the intake manifold without changing the position of throttle valve. The idle speed usually does not require adjustment. When the idle control system is in operation, the idle adjustment screw has no effect on the idle speed.

### Fast Idle Mechanism

To prevent erratic running when the engine is warming up, it is necessary to raise the idle speed. The air bypass valve is controlled by a thermowax plunger. When the thermowax is cold, the valve is open. When the thermowax is heated, the valve is closed. With the engine cold and the thermowax consequently cold, additional air is bypassed into the intake manifold so that the engine idles faster than normal. When the engine reaches operating temperature, the valve begins to close, reducing the amount of air bypassing into the manifold.

## ELECTRONIC CONTROL SYSTEM

### Control System

In order to get fuel into the cylinders at the correct instant and in correct amount, the control system must perform various separate functions. The ECU (Electronic Control Unit), the heart of the PGM-FI, uses an 8-bit microcomputer and consists of a CPU (Central Processing Unit), memories, and I/O (Input/Output) ports. Basic data stored in the memories are compensated by the signals sent from the various sensors to provide the correct air/fuel mixture for all engine needs.

### Electronic Control Unit (ECU)

The unit contains memories for the basic discharge duration at various engine speeds and manifold pressures. The basic discharge duration, after being read out from the memory, is further modified by signals sent from various sensors to obtain the final discharge duration. Other functions also include:

- Starting Control – The fuel system must vary the air/fuel ratio to suit different operating requirements. For example, the mixture must be rich for starting. The memories also contain the basic discharge durations to be read out by signals from the starter switch, and engine speed and coolant temperature sensors, thereby providing extra fuel needed for starting.
- Injector Control – The ECU controls the discharge durations at various engine speeds and loads.
- Electronic Air Control – The ECU controls the EACV to maintain correct idle speed based on engine and accessories demand.
- Ignition Timing Control – The ECU controls the basic ignition timing based on engine load, engine rpm, vehicle speed and coolant temperature.
- Fuel Pump Control – When the speed of the engine falls below the prescribed limit, electric current to the fuel pump is cut off, preventing the injectors from discharging fuel.
- Fuel Cut-Off Control – During deceleration with the throttle valve nearly closed, electric current to the injectors is cut off at speeds over 900 rpm, contributing to improved fuel economy. Fuel cut-off action also takes place when engine speed exceeds approximately 7500 rpm regardless of the position of the throttle valve.
- Safety – A fail-safe system monitors the sensors and detects any abnormality in the ECU, ensuring safe driving even if one or more sensors are faulty, or if the ECU malfunctions.
- Self-Diagnosis – When a abnormality occurs, the ECU lights the engine warning light and stores the failure code in erasable memory. The ECU LED will display the code any time the ignition is turned **ON**.

### Crank Angle Sensor (TDC/CYL Sensors)

The sensors and distributor are designed as an assembly to save space and weight. The entire unit consist of a pair of rotors, TDC and CYL, and a pickup for each rotor. Since the rotors are coupled to the camshaft, they turn together as a unit as the camshaft rotates. The CYL sensor detects the position of the No. 1 cylinder as the base for the Sequential Injection whereas the TDC sensor serves to determine the injection timing for each cylinder. The TDC sensor is also used to detect engine speed to read out the basic discharge duration for different operating conditions.

Most 1990 vehicles, incorporate the TDC sensor, CYL sensor and Crank sensor in a single assembly. The function of these components does not change, but if any single sensor is defective, the entire distributor must be replaced as an assembly.

### Manifold Air Pressure Sensor (MAP Sensor)

The sensor converts manifold air pressure readings into electrical voltage signals and sends them to the ECU. This information with signals from the crank angle sensor is then used to read out the basic discharge duration from the memory.

### Atmospheric Pressure Sensor (PA Sensor)

Like the MAP sensor, the unit converts atmospheric pressures into voltage signals and sends them to the ECU. The signals then modify the basic discharge duration to compensate for changes in the atmospheric pressure.

### Coolant Temperature Sensor (TW Sensor)

The sensor uses a temperature-dependent diode (thermistor) to measure differences in the coolant temperature. The basic discharge duration is read out by the signals sent from this sensor through the ECU. The resistance of the thermister decreases with a rise in coolant temperature.

### Intake Air Temperature Sensor (TA Sensor)

This device is also a thermistor and is placed in the intake manifold. It acts much like the water temperature sensor but with a reduced thermal capacity for for quicker response. The basic discharge duration read out from the memory is again compensated for different operating conditions by the signals sent from this sensor through the ECU.

### Throttle Angle Sensor

This sensor is essentially a variable resistor. In construction, the rotor shaft is connected to the throttle valve shaft such that, as the throttle valve is moved, the resistance varies, altering the output voltage to the control unit.

### Oxygen Sensor

The oxygen sensor, by detecting the oxygen content in the exhaust gas, maintains the stoichiometric air/fuel ratio. In operation, the ECU receives the signals from the sensor and changes the duration during which fuel is injected. The oxygen sensor is located in the exhaust manifold.

The sensor is a hollow shaft of zirconia with a closed end. The inner and outer surfaces are plated with platinum, thus forming a platinum electrode. The inner surface or chamber is open to the atmosphere whereas the outer surface is exposed to the exhaust gas flow through the manifold.

Voltage is induced at the platinum electrode when there is any difference in oxygen concentration between the 2 layers of air over the surfaces. Operation of the device is dependent upon the fact that voltage induced changes sharply as the stoichiometric air/fuel ratio is exceeded when the electrode is heated above a certain temperature.

Some 1990 models, use a heated oxygen sensor. The heater stabilizes the sensors outputs and allow the sensor to heat quicker after the engine as been started.

### Starter Switch

The air/fuel mixture must be rich for starting. During cranking, the ECU detects signal from the starter switch and increases the amount of fuel injected into the manifold according to the engine temperature. The amount of fuel injected is gradually reduced when the starter switch is turned **OFF**.

## FUEL SYSTEM

### Fuel Pump

The fuel pump is a compact impeller design and is installed inside the fuel tank, thereby saving space and simplifying the fuel line system.

The fuel pump is comprised of a DC motor, a circumference flow pump, a relief valve for protecting the fuel line systems, a check valve for retaining residual pressure, an inlet port, and a discharge port. The pump assembly consists of the impeller (driven by the motor), the pump casing (which forms the pumping chamber), and cover of the pump.

### Pressure Regulator

The fuel pressure regulator maintains a constant fuel pressure to the injectors. The spring chamber of the pressure regulator is connected to the intake manifold to constantly maintain the fuel pressure at 36 psi (2.55 kg/cm$^2$) higher than the pressure in the manifold. When the difference between the fuel pressure and manifold pressure exceeds 36 psi (2.55 kg/cm$^2$), the diaphragm is pushed upward, and the excess fuel is fed back into the fuel tank through the return line.

### Injector

The injector is of the solenoid-actuated constant-stroke pintle type consisting of a solenoid, plunger, needle valve and housing. When current is applied to the solenoid coil, the valve lifts up and pressurized fuel fills the inside of the injector and is injected close to the intake valve. Because the needle valve lifts and the fuel pressure are constant, the injection quantity is determined by the length of time that the valve is open, i.e., the duration the current is supplied to the solenoid coil. The injector is sealed by an O-ring and seal ring at the top and bottom. These seals also reduce operating noise.

### Resistor

The injector timing, which controls the opening and closing intervals, must be very accurate since it dictates the air/fuel mixture ratio. The injector must also be durable. For the best possible injector response, it is necessary to shorten the current rise time when voltage is applied to the injector coil. Therefore, the number of windings of the injector coil is reduced to reduce the inductance in the coil. This, however, makes low resistance in the coil, allowing a large amount of current to flow through the coil. As a result, the amount of heat generated is high, which compromises the durability of the coil. Flow of current in the coil is therefore restricted by a resistor installed in series between the electric power source and the injector coil.

### Main Relay

The main relay is a direct coupler type which contains the relays for the electronic control unit power supply and the fuel pump power supply. This relay is installed at the back of the fuse box.

## SERVICE PRECAUTIONS

- Do not operate the fuel pump when the fuel lines are empty.
- Do not operate the fuel pump when removed from the fuel tank.
- Do not reuse fuel hose clamps.
- Make sure all ECU harness connectors are fastened securely. A poor connection can cause an extremely high surge voltage in the coil and condenser and result in damage to integrated circuits.
- Keep ECU all parts and harnesses dry during service.
- Before attempting to remove any parts, turn **OFF** the ignition switch and disconnect the battery ground cable.
- Always use a 12 volt battery as a power source.
- Do not attempt to disconnect the battery cables with the engine running.
- Do not disconnect and wiring connector with the engine running, unless instructed to do so.
- Do not depress the accelerator pedal when starting.
- Do not rev up the engine immediately after starting or just prior to shutdown.
- Do not apply battery power directly to injectors.

# Diagnosis and Testing

## SELF-DIAGNOSTIC SYSTEM

### Self-Diagnosis Indicators

The quick reference chart covers the most common failure modes for the PGM-FI. The probable causes are listed in order of most-easily-checked first, then progressing to more difficult fixes. Run through all the causes listed. If problem is still unsolved, go on to the more detailed troubleshooting. Troubleshooting is divided into different LED displays. Find the correct light display and begin again.

For all the conditions listed, the PGM-FI warning light on the dashboard must be on (comes on and stays on). This indicates a problem in the electrical portion of the fuel injection system. At that time, check the LED display (self-diagnosis system) in the ECU.

There is only one LED display. The LED will blink consecutively to indicate the trouble code. The ECU is located under the driver's seat on the 1988–89 Accord or on other models beneath the carpet on the passenger's side under the dash.

Sometimes the dash warning light and/or ECU LED will come on, indicating a system problem, when, in fact, there is only a bad or intermittent electrical connection. To troubleshoot a bad connection, note the ECU LED pattern that is lit, refer to the diagnosis chart and check the connectors associated with the items mentioned in the "Possible Cause" column for that LED

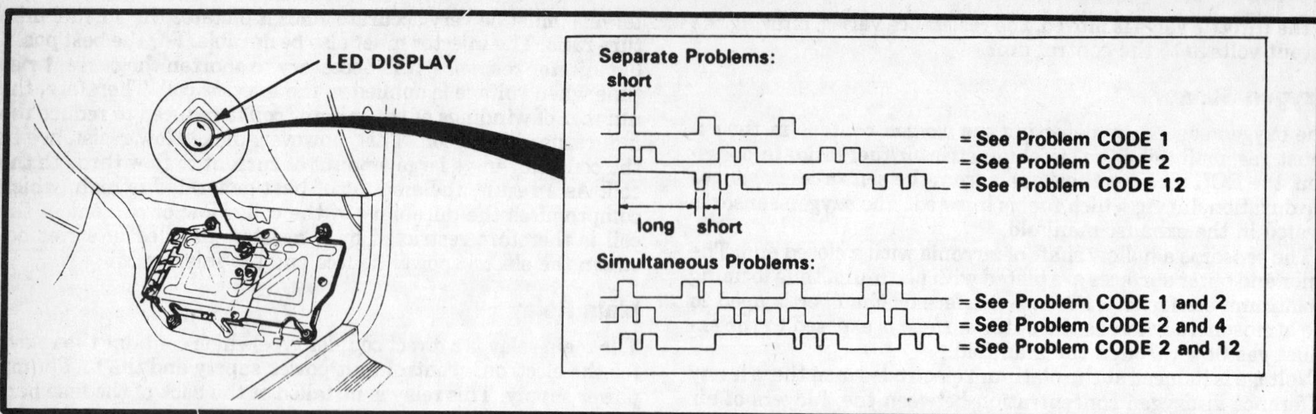


**PGM-FI LED display**

pattern (disconnect, clean or repair if necessary and reconnect those connections). Then, reset the ECU memory as described, restart the car and drive it for a few minutes and then recheck the car and drive it for a few minutes and then recheck the LED. If the same pattern lights up, begin system troubleshooting; if it does not light up, the problem was only a bad connection.

The memory for the PGM-FI warning light on the dashboard will be erased when the ignition switch is turned **OFF**; however, the memory for the LED display will not be canceled. Thus, the warning light will not come on when the ignition switch is again turned **ON** unless the trouble is once more detected. Troubleshooting should be done according to the LED display even if the warning light is off.

Other ECU information:

- After making repairs, disconnect the battery negative cable from the battery negative terminal for at least 10 seconds and reset the ECU memory. After reconnecting the cable, check that the LED display is turned off.
- Turn the ignition switch **ON**. The PGM-FI warning light should come on for about 2 seconds. If the warning light won't come on, check for:
  - –Blown warning light bulb
  - –Blown fuse (causing faulty back up light, seat belt alarm, clock, memory function of the car radio)
  - –Open circuit in Yellow wire
  - –Open circuit in wiring and control unit
- After the PGM-FI warning light and self-diagnosis indicators have been turned on, turn the ignition switch **OFF**. If the LED display fails to come on when the ignition switch is turned **ON** again, check for:
  - –Blown fuses, especially No. 10 fuse
  - –Open circuit in wire between ECU fuse
- Replace the ECU only after making sure that all couplers and connectors are connected securely.

| SELF-DIAGNOSIS INDICATOR BLINKS | SYSTEM INDICATED |
|---|---|
| 0 | ECU |
| 1 | OXYGEN CONTENT A |
| 2 | OXYGEN CONTENT B |
| 3<br>5 | MANIFOLD ABSOLUTE PRESSURE |
| 6 | COOLANT TEMPERATURE |
| 7 | THROTTLE ANGLE |
| 8 | CRANK ANGLE (TDC) |
| 9 | CRANK ANGLE (CYL) |
| 10 | INTAKE AIR TEMPERATURE |
| 12 | EXHAUST GAS RECIRCULATION SYSTEM |
| 13 | ATMOSPHERIC PRESSURE |
| 14 | ELECTRONIC IDLE CONTROL |
| 17 | VEHICLE SPEED SENSOR |

CODE 4, 11, 15, 16 or exceeds 17, count the number of blinks again. If the indicator is in fact, blinking these codes, substitute a known-good ECU and recheck. If the indication goes away, replace the original ECU.
The Check Engine warning light and ECU LED may come on, indicating a system problem, when, in fact, there is a poor or intermittent electrical connection. First, check the electrical connections, clean or repair connections if necessary.

**ECU Self-diagnosis code identification–1988–89 Accord**

| SELF-DIAGNOSIS INDICATOR BLINKS | SYSTEM INDICATED |
|---|---|
| 0 | ECU |
| 1 | OXYGEN CONTENT |
| 3 | MANIFOLD ABSOLUTE PRESSURE |
| 5 | |
| 4 | CRANK ANGLE |
| 6 | COOLANT TEMPERATURE |
| 7 | THROTTLE ANGLE |
| 8 | TDC POSITION |
| 9 | No.1 CYLINDER POSITION |
| 10 | INTAKE AIR TEMPERATURE |
| 12 | EXHAUST GAS RECIRCULATION SYSTEM |
| 13 | ATMOSPHERIC PRESSURE |
| 14 | ELECTRONIC AIR CONTROL |
| 15 | IGNITION OUTPUT SIGNAL |
| 16 | FUEL INJECTOR |
| 17 | VEHICLE SPEED SENSOR |
| 20 | ELECTRIC LOAD DETECTOR |
| 30 | A/T FI SIGNAL A |
| 31 | A/T FI SIGNAL B |
| 41 | OXYGEN SENSOR HEATER |
| 43 | FUEL SUPPLY SYSTEM |

- If codes other than those listed above are indicated, verify the code. If the code indicated is not listed above, replace the ECU.
- The Check Engine warning light may come on, indicating a system problem, when, in fact, there is a poor or intermittent electrical connection. First, check the electrical connections, clean or repair connections if necessary.
- The Check Engine warning light and S4 warning light may light simultaneously when the self-diagnosis indicator blinks 6, 7 and 17. Check the PGM-FI system according to the PGM-FI control system troubleshooting, then recheck the S4 warning light. If it lights, see page 14-36.
- The Check Engine warning light does not come on when there is a malfunction in the A/T FI signal or Electric Load Detector circuits. However, the ECU LED will indicate the codes.

**ECU Self-diagnosis code identification—1990 Accord**

| SELF-DIAGNOSIS INDICATOR BLINKS | SYSTEM INDICATED |
|---|---|
| 0 | ECU |
| 1 | OXYGEN CONTENT |
| 3 | MANIFOLD ABSOLUTE PRESSURE |
| 5 | |
| 4 | CRANK ANGLE |
| 6 | COOLANT TEMPERATURE |
| 7 | THROTTLE ANGLE |
| 8 | TDC POSITION |
| 9 | No.1 CYLINDER POSITION (1.6 ℓ) |
| 10 | INTAKE AIR TEMPERATURE |
| 12 | EXHAUST GAS RECIRCULATION SYSTEM (1.5 ℓ CAL: A T) |
| 13 | ATMOSPHERIC PRESSURE |
| 14 | ELECTRONIC AIR CONTROL |
| 15 | IGNITION OUTPUT SIGNAL |
| 16 | FUEL INJECTOR |
| 17 | VEHICLE SPEED SENSOR |
| 19 | LOCK-UP CONTROL SOLENOID VALVE (A/T) |
| 20 | ELECTRIC LOAD |

If codes other than those listed above are indicated, count the number of blinks again. If the indicator is in fact blinking these codes, substitute a known-good ECU and recheck. If the indication goes away, replace the original ECU.
The Check Engine warning light and ECU LED may come on, indicating a system problem, when, in fact, there is a poor or intermittent electrical connection. First, check the electrical connections, clean or repair connections if necessary.
If the Check Engine warning light is on and LED stays on, replace the ECU.

**ECU Self-diagnosis code identification—1988–90 Civic**

| SELF-DIAGNOSIS INDICATOR BLINKS | SYSTEM INDICATED |
|---|---|
| 0 | ECU |
| 1 | OXYGEN CONTENT A |
| 2 | OXYGEN CONTENT B |
| 3<br>5 | MANIFOLD ABSOLUTE PRESSURE |
| 4 | CRANK ANGLE |
| 6 | COOLANT TEMPERATURE |
| 7 | THROTTLE ANGLE |
| 8 | TDC POSITION |
| 9 | No.1 CYLINDER POSITION |
| 10 | INTAKE AIR TEMPERATURE |
| 12 | EXHAUST GAS RECIRCULATION SYSTEM |
| 13 | ATMOSPHERIC PRESSURE |
| 14 | ELECTRONIC IDLE CONTROL |
| 15 | IGNITION OUTPUT SIGNAL |
| 16 | FUEL INJECTOR |
| 17 | VEHICLE SPEED SENSOR |

If CODE 11, or more than 17, count the number of blinks again. If the indicator is in fact blinking these codes, substitute a known-good ECU and recheck. If the indication goes away, replace the original ECU.
The Check Engine dash warning light and ECU LED may come on, indicating a system problem, when, in fact, there is a poor or intermittent electrical conneciton. First, check the electrical connecitons, clean or repair connections if necessary.
NOTE: If the Check Engine dash warning light is on and the ECU LED is on but not blinking, substitute a known-good ECU and recheck.

**ECU Self-diagnosis code identification—1988–90 Prelude**

## DIAGNOSTIC TREE CHARTS

**NOTE: The 1.5L engine used in the CRX HF uses the same PGM-FI system as the 1.6L engine used in all Civics. For the CRX HF, use charts listed for the 1.6L Civic unless specifically instructed differently.**

# TROUBLESHOOTING GUIDE—1988–89 ACCORD

NOTE: Across each row in the chart, the systems that could be sources of a symptom are ranked in the order they should be inspected starting with ①. Find the symptom in the left column, read across to the most likely source, then refer to the page listed at the top of that column. If inspection shows the system is OK, try the next most likely system ②, etc.

| SYMPTOM \ SYSTEM | | PGM-FI: ELECTRONIC CONTROL UNIT | PGM-FI: OXYGEN SENSOR | PGM-FI: MANIFOLD ABSOLUTE PRESSURE SENSOR | PGM-FI: COOLANT TEMPERATURE SENSOR | PGM-FI: THROTTLE ANGLE SENSOR | PGM-FI: CRANK ANGLE SENSOR |
|---|---|---|---|---|---|---|---|
| CHECK ENGINE WARNING LIGHT TURNS ON | | □ or (light on) | (CHECK light) | (CHECK light) | (CHECK light) | (CHECK light) | (CHECK light) |
| SELF-DIAGNOSIS INDICATOR (LED) BLINKS | | 0 or (blinking) | 1 or 2 | 3 or 5 | 6 | 7 | 8 or 9 |
| ENGINE WON'T START | | ② | | ③ | | | ③ |
| DIFFICULT TO START ENGINE WHEN COLD | | (BU) | | | ① | | |
| IRREGULAR IDLING | WHEN COLD FAST IDLE OUT OF SPECIFICATION | (BU) | | ③ | ② | | |
| | ROUGH IDLE | (BU) | ② | ① | | | |
| | WHEN WARM RPM TOO HIGH | (BU) | (FS) | (FS) | (FS) | (FS) | (FS) |
| | WHEN WARM RPM TOO LOW | (BU) | | ② | | | |
| FREQUENT STALLING | WHILE WARMING UP | (BU) | | | ③ | | |
| | AFTER WARMING UP | (BU) | | | | | |
| POOR PERFORMANCE | MISFIRE OR ROUGH RUNNING | (BU) | ② | ① | | | |
| | FAILS TEST EMISSION | (BU) | ① | ② | | | |
| | LOSS OF POWER | (BU) | | ② | | | |

| SYMPTOM \ SYSTEM | | PGM-FI: INTAKE AIR TEMPERATURE SENSOR | PGM-FI: ATMOSPHERIC PRESSURE SENSOR | PGM-FI: VEHICLE SPEED SENSOR | IDLE CONTROL: ELECTRONIC AIR CONTROL VALVE | IDLE CONTROL: OTHER IDLE CONTROLS | FUEL SUPPLY | AIR INTAKE | EMISSION CONTROL: EGR CONTROL SYSTEM | EMISSION CONTROL: OTHER EMISSION CONTROLS |
|---|---|---|---|---|---|---|---|---|---|---|
| CHECK ENGINE WARNING LIGHT TURNS ON | | (CHECK light) | (CHECK light) | (CHECK light) | (CHECK light) | | | | (CHECK light) | |
| SELF-DIAGNOSIS INDICATOR (LED) BLINKS | | 10 | 13 | 17 | 14 | | | | 12 | |
| ENGINE WON'T START | | | ③ | | | | ① | | | |
| DIFFICULT TO START ENGINE WHEN COLD | | | | | | | | | | |
| IRREGULAR IDLING | WHEN COLD FAST IDLE OUT OF SPECIFICATION | | | | ① | | | | | |
| | ROUGH IDLE | | | | | | | | ③ | |
| | WHEN WARM RPM TOO HIGH | (FS) | (FS) | (FS) | (FS) | ① | | ② | | ③ |
| | WHEN WARM RPM TOO LOW | | | | ① | ③ | | | | |
| FREQUENT STALLING | WHILE WARMING UP | | | | ① | | | | ② | |
| | AFTER WARMING UP | | | | ② | | | | ① | |
| POOR PERFORMANCE | MISFIRE OR ROUGH RUNNING | | ③ | ③ | | | | | | |
| | FAILS TEST EMISSION | | ③ | | | | | | | |
| | LOSS OF POWER | | ③ | | | | ① | | | |

- CODE 4, 11, 15, 16 or exceeds 17: count the number of blinks again. If the indicator is in fact, blinking these codes, substitute a known-good ECU and recheck. If the indication goes away, replace the original ECU.
  (FS): When the Check Engine warning light is on the idle speed will increase due to fail-safe operation.
  (BU): When the Check Engine warning light is on with no blinks on the self-diagnosis indicator, the back-up system is in operation.

## ECU AND CHECK LIGHT—1988 ACCORD

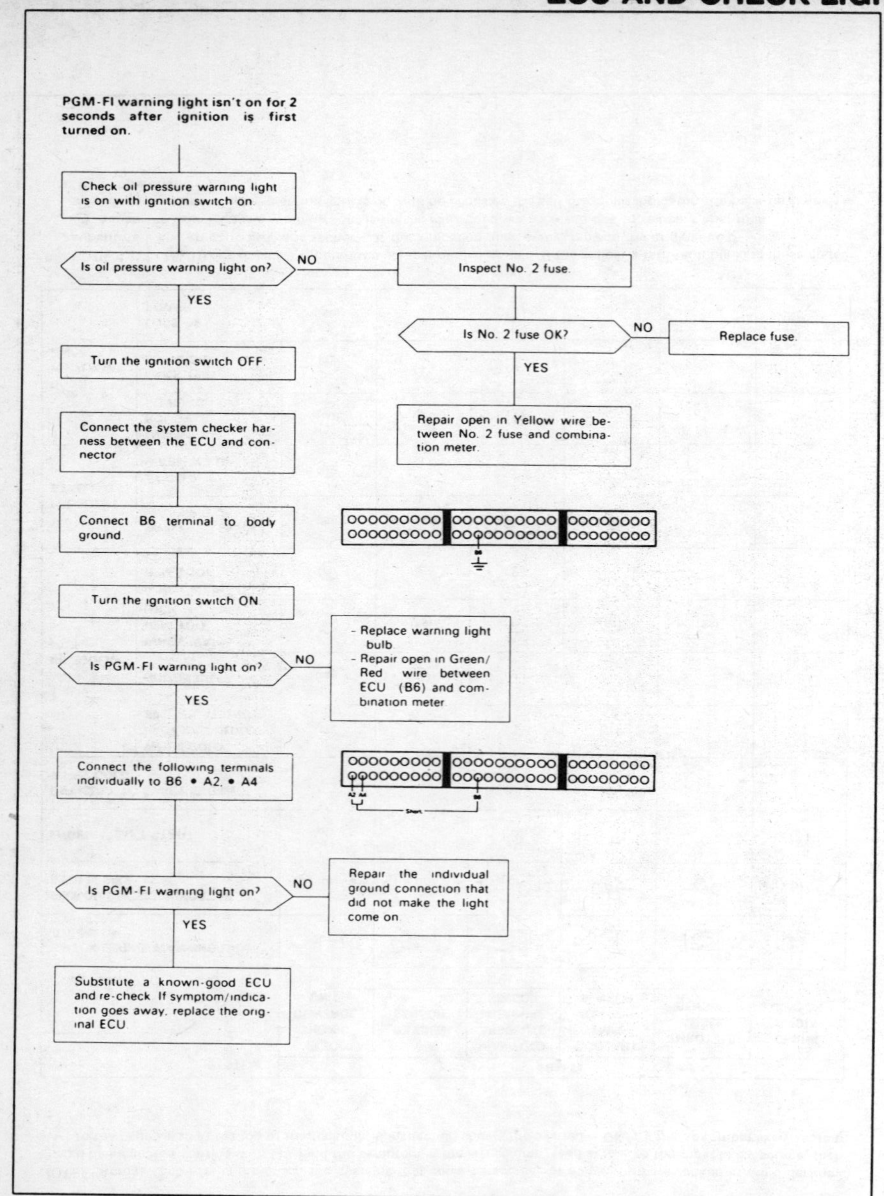

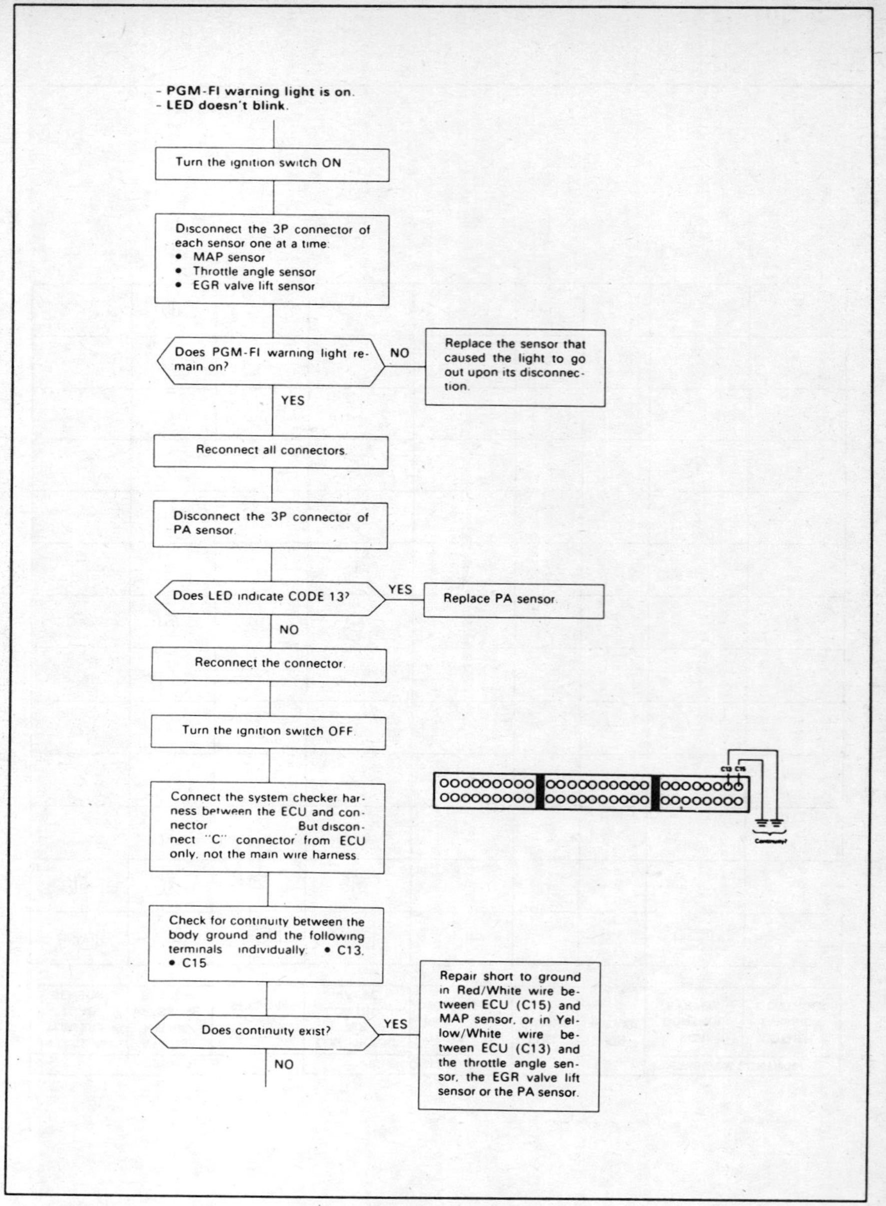

- PGM-FI warning light is on.
- LED doesn't blink.

Turn the ignition switch ON.

Disconnect the 3P connector of each sensor one at a time:
- MAP sensor
- Throttle angle sensor
- EGR valve lift sensor

Does PGM-FI warning light remain on?
- NO → Replace the sensor that caused the light to go out upon its disconnection.
- YES ↓

Reconnect all connectors.

Disconnect the 3P connector of PA sensor.

Does LED indicate CODE 13?
- YES → Replace PA sensor.
- NO ↓

Reconnect the connector.

Turn the ignition switch OFF.

Connect the system checker harness between the ECU and connector. But disconnect "C" connector from ECU only, not the main wire harness.

Check for continuity between the body ground and the following terminals individually. • C13, • C15

Does continuity exist?
- YES → Repair short to ground in Red/White wire between ECU (C15) and MAP sensor, or in Yellow/White wire between ECU (C13) and the throttle angle sensor, the EGR valve lift sensor or the PA sensor.
- NO

## ECU AND CHECK LIGHT – 1988 ACCORD

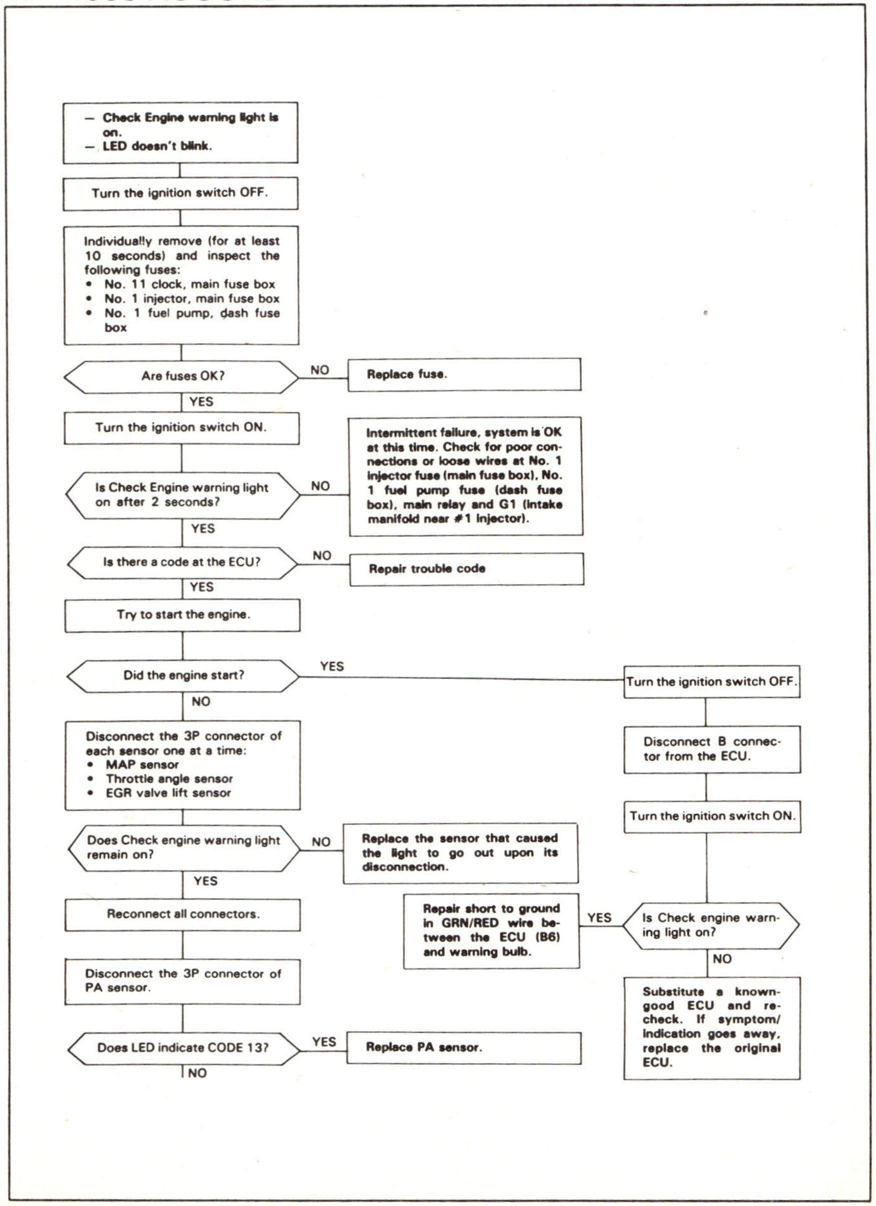

– Check Engine warning light is on.
– LED doesn't blink.

Turn the ignition switch OFF.

Individually remove (for at least 10 seconds) and inspect the following fuses:
- No. 11 clock, main fuse box
- No. 1 injector, main fuse box
- No. 1 fuel pump, dash fuse box

Are fuses OK? — NO → Replace fuse.

YES

Turn the ignition switch ON.

Is Check Engine warning light on after 2 seconds? — NO → Intermittent failure, system is OK at this time. Check for poor connections or loose wires at No. 1 injector fuse (main fuse box), No. 1 fuel pump fuse (dash fuse box), main relay and G1 (intake manifold near #1 injector).

YES

Is there a code at the ECU? — NO → Repair trouble code

YES

Try to start the engine.

Did the engine start? — YES → Turn the ignition switch OFF.

Disconnect B connector from the ECU.

Turn the ignition switch ON.

Is Check engine warning light on? — YES → Repair short to ground in GRN/RED wire between the ECU (B6) and warning bulb.

NO → Substitute a known-good ECU and re-check. If symptom/indication goes away, replace the original ECU.

Did the engine start? — NO

Disconnect the 3P connector of each sensor one at a time:
- MAP sensor
- Throttle angle sensor
- EGR valve lift sensor

Does Check engine warning light remain on? — NO → Replace the sensor that caused the light to go out upon its disconnection.

YES

Reconnect all connectors.

Disconnect the 3P connector of PA sensor.

Does LED indicate CODE 13? — YES → Replace PA sensor.

NO

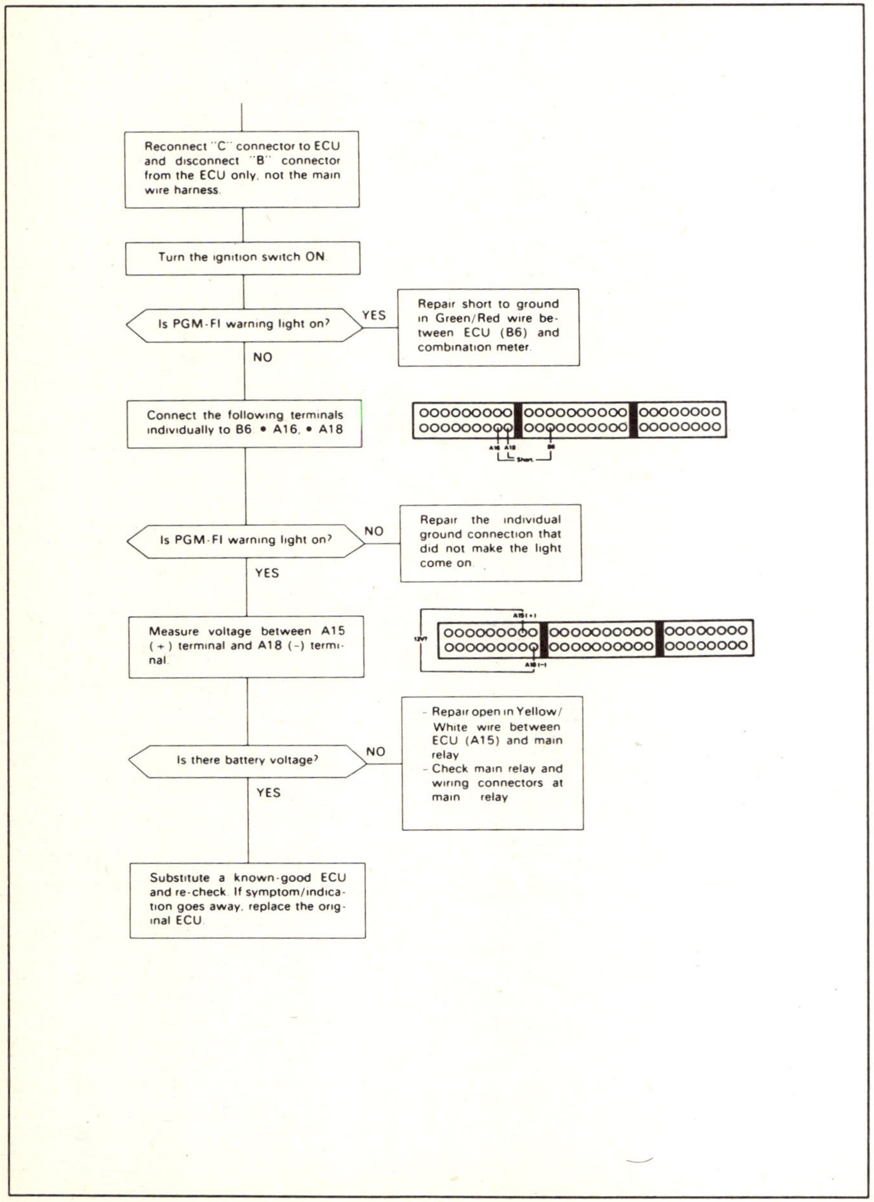

Reconnect "C" connector to ECU and disconnect "B" connector from the ECU only, not the main wire harness.

Turn the ignition switch ON

Is PGM-FI warning light on? — YES → Repair short to ground in Green/Red wire between ECU (B6) and combination meter.

NO

Connect the following terminals individually to B6 • A16, • A18

Is PGM-FI warning light on? — NO → Repair the individual ground connection that did not make the light come on.

YES

Measure voltage between A15 (+) terminal and A18 (–) terminal.

Is there battery voltage? — NO →
- Repair open in Yellow/White wire between ECU (A15) and main relay
- Check main relay and wiring connectors at main relay

YES

Substitute a known-good ECU and re-check. If symptom/indication goes away, replace the original ECU.

## ECU AND CHECK LIGHT – 1989 ACCORD

- Turn the ignition switch OFF.
- Connect the PGM-FI test harness between the ECU and connector But disconnect "C" connector from ECU only, not the main wire harness.
- Check for continuity between the body ground and the following terminals individually: • C13, • C15
- Does continuity exist?
  - YES → Disconnect C17–C93 connector located at the left shock tower.
    - Does continuity exist?
      - YES → Repair short to ground in RED/WHT wire between ECU (C15) and MAP sensor or short in YEL/WHT wire between ECU (C13) and C17 at the left shock tower.
      - NO → Repair short to body ground in YEL/WHT wire between C93 and throttle angle sensor or EGR valve lift sensor.
  - NO → Reconnect "C" connector to ECU
- Turn the ignition switch ON.
- Measure voltage between the following terminals individually to body ground • A16, • A18
- Is voltage less than 1 V?
  - NO → Repair the individual ground connection between ECU and G1 located on intake manifold near #1 injector.
  - YES → Measure voltage between A15 (+) terminal and A18 (–) terminal.
- Is there battery voltage?
  - NO → – Repair open in YEL/BLK wire between ECU (A15) and main relay. – Check main relay and wiring connectors at main relay
  - YES → Substitute a known-good ECU and recheck. If symptom/indication goes away, replace the original ECU.

## OXYGEN SENSOR – CODE 1 OR 2 – 1988–89 ACCORD

Self-diagnosis LED blinks once: A problem in the Oxygen ($O_2$) Sensor A circuit.

Self-diagnosis LED blinks twice: A problem in the Oxygen ($O_2$) Sensor B circuit.

The oxygen sensor, detects the oxygen content in the exhaust gas, and inputs the ECU. In operation, the ECU receives the signals from the sensor and varies the duration during which fuel is injected. The oxygen sensors are installed on the exhaust manifold.

1 or 2

- – Check Engine warning light has been reported on. – LED indicates CODE 1 ($O_2$ sensor A) or CODE 2 ($O_2$ sensor B).
- Turn the ignition switch OFF.
- Remove No. 11 CLOCK fuse in the under-hood relay box for 10 seconds to reset ECU.
- Inspect pressure regulator
- Is it normal ?
  - NO → Replace the pressure regulator
  - YES → Warm up engine to normal operating temperature (cooling fan comes on).
- Hold the engine at 1500 rpm for 15 minutes. NOTE: Do not close throttle completely during this time.
- Is Check Engine warning light on and does LED indicate CODE 1 or 2 ?
  - NO → Intermittent failure, system is OK at this time. Check for poor connections or loose wires at $O_2$ sensors, C93–C17, G1 (on intake manifold near #1 injector) and ECU.
  - YES

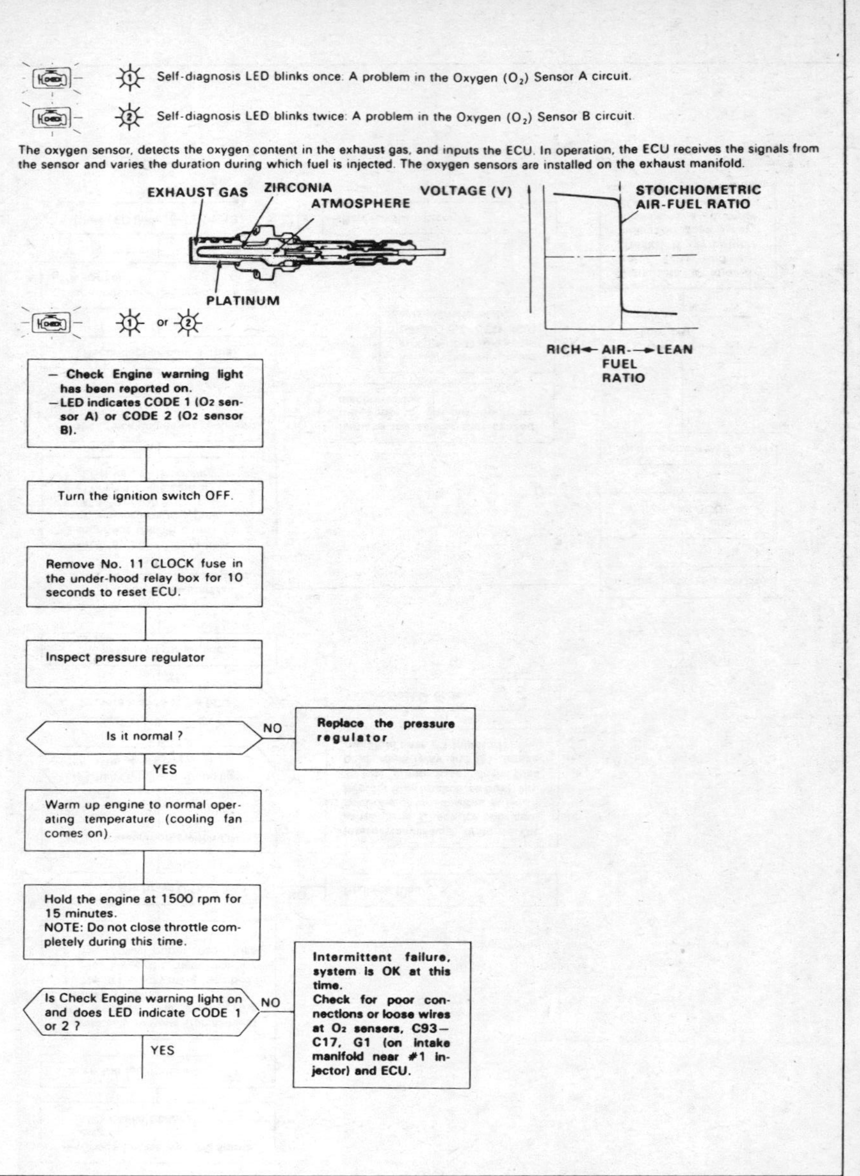

## OXYGEN SENSOR—CODE 1 OR 2—1988–89 ACCORD

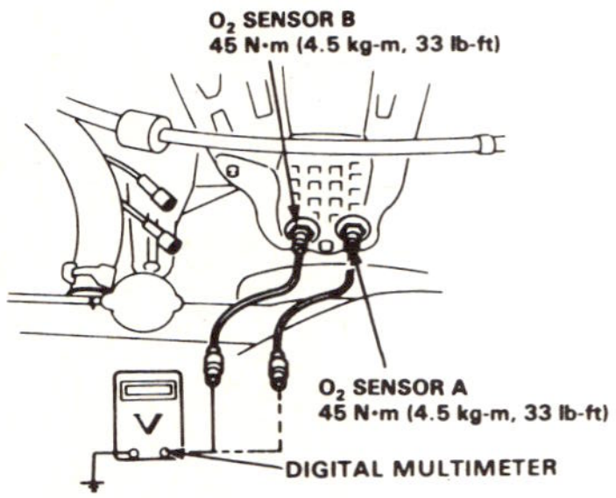

Disconnect engine wire harness from indicated $O_2$ sensor.

Warm up engine to normal operating temperature again.

Measure voltage between the connector terminal and body ground.

Is voltage above 0.6 V at 5,000 rpm and below 0.4 V when the throttle is quickly released from 5,000 rpm?

- NO → Replace $O_2$ sensor.
- YES ↓

$O_2$ SENSOR B
45 N·m (4.5 kg-m, 33 lb-ft)

$O_2$ SENSOR A
45 N·m (4.5 kg-m, 33 lb-ft)

DIGITAL MULTIMETER

Stop engine.

Reconnect $O_2$ sensor.

Connect the PGM-FI test harness between the ECU and connector

Restart and warm up engine to normal operating temperature.

Measure voltage between:

- $O_2$ sensor A
  C16 (+) and A18(−) terminals
- $O_2$ sensor B
  B11 (+) and A18 (−) terminals

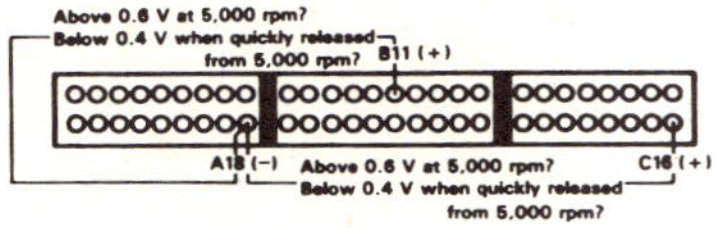

Above 0.6 V at 5,000 rpm?
Below 0.4 V when quickly released from 5,000 rpm? B11 (+)
A18 (−) Above 0.6 V at 5,000 rpm?
Below 0.4 V when quickly released from 5,000 rpm? C16 (+)

Is voltage above 0.6 V at 5,000 rpm and below 0.4 V when the throttle is quickly released from 5,000 rpm?

- NO → Repair open or short in wire between ECU and $O_2$ sensor:
  - $O_2$ sensor A
    WHT wire to C16
  - $O_2$ sensor B
    RED/BLU wire to C93 coupler and WHT wire to B11
- YES → Substitute a known-good ECU and recheck. If symptom/ indication goes away, replace the original ECU.

## MAP SENSOR—CODE 3—1988 ACCORD

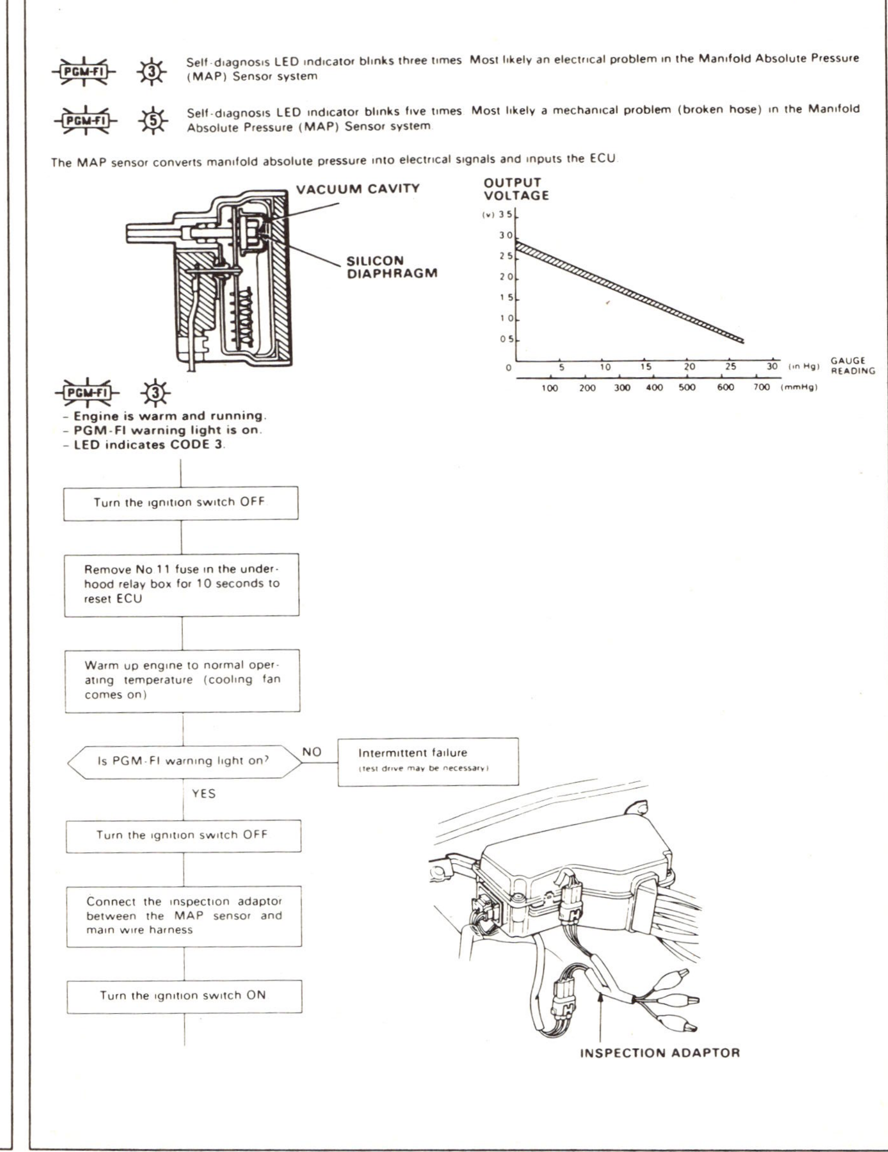

## MAP SENSOR—CODE 3—1988 ACCORD

- Measure voltage between Red (+) terminal and Green (-) terminal.
- Is there approx. 5 V?
  - NO → Measure voltage between Red (+) terminal and body ground.
    - Is there approx. 5 V?
      - YES → Repair open in Green/White wire between ECU (C14) and MAP sensor.
      - NO → Turn the ignition switch OFF.
        - Connect the system checker harness between the ECU and connector.
        - Measure voltage between C15 (+) terminal and C14 (−) terminal. 
        - Is there approx. 5 V?
          - YES → Repair open in Red/White wire between ECU (C15) and MAP sensor.
          - NO → Substitute a known-good ECU and re-check. If prescribed voltage is now available, replace the original ECU.
  - YES → Measure voltage between White (+) terminal and Green (-) terminal.
- Is there approx. 3 V?
  - NO → - Replace MAP sensor. - Repair short in White/Blue wire between ECU (C11) and MAP sensor.
  - YES → Turn the ignition switch OFF.
- Connect the system checker harness between the ECU and connector.
- Turn the ignition switch ON.
- Measure voltage between C11 (−) terminal and C14 (−) terminal. 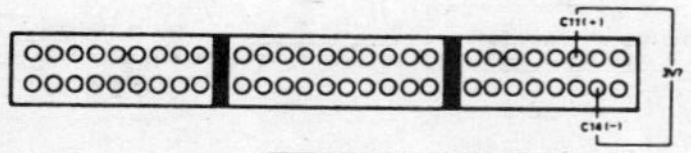
- Is there approx. 3 V?
  - NO → Repair open in White/Blue wire between ECU (C11) and MAP sensor.
  - YES → Substitute a known-good ECU and re-check. If symptom/indication goes away, replace the original ECU.

## MAP SENSOR—CODE 3—1989 ACCORD

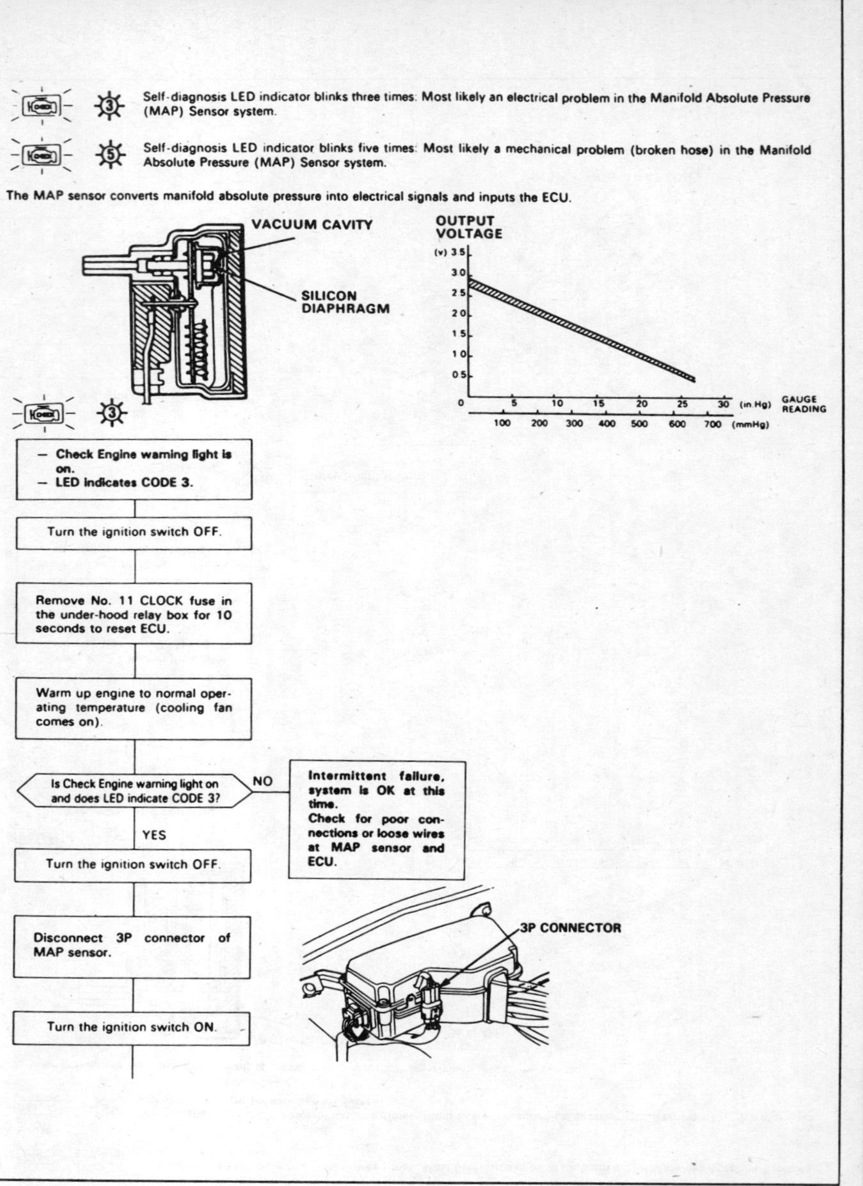

## MAP SENSOR – CODE 3 – 1989 ACCORD

Measure voltage between RED/WHT terminal and body ground.

Is there approx. 5 V?

- NO: Repair open in RED/WHT wire between ECU (C15) and MAP sensor. If wire is OK, substitute a known-good ECU and recheck. If prescribed voltage is now available, replace the original ECU.
- YES: Measure voltage between RED/WHT (+) terminal and BLU/WHT (−) terminal.

Is there approx. 5 V?

- NO: Repair open in BLU/WHT wire between ECU (C14) and 3P connector. If wire is OK, replace ECU and retest.
- YES: Measure voltage between WHT/BLU (+) terminal and BLU/WHT (−) terminal.

Is there approx. 5 V?

- NO: Repair open or short in WHT/BLU wire between ECU (C11) and 3P connector. If wire is OK, replace ECU and retest.
- YES: Install test harness between MAP sensor and main harness.

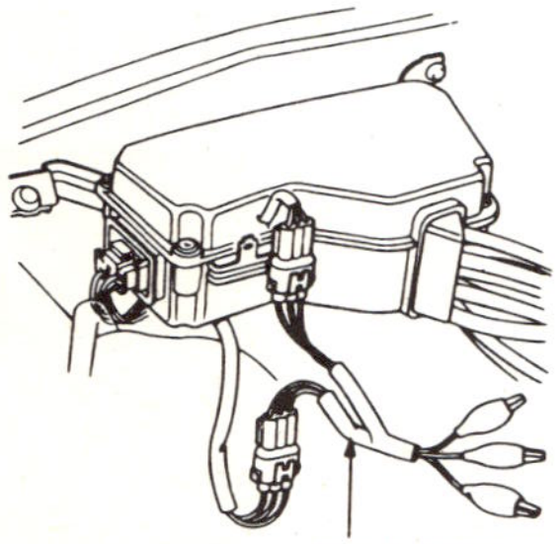

TEST HARNESS

Measure voltage between WHT (+) terminal and GRN (−) terminal.

Is there approx. 3 V?

- NO: Replace MAP sensor.
- YES: Replace ECU and retest.

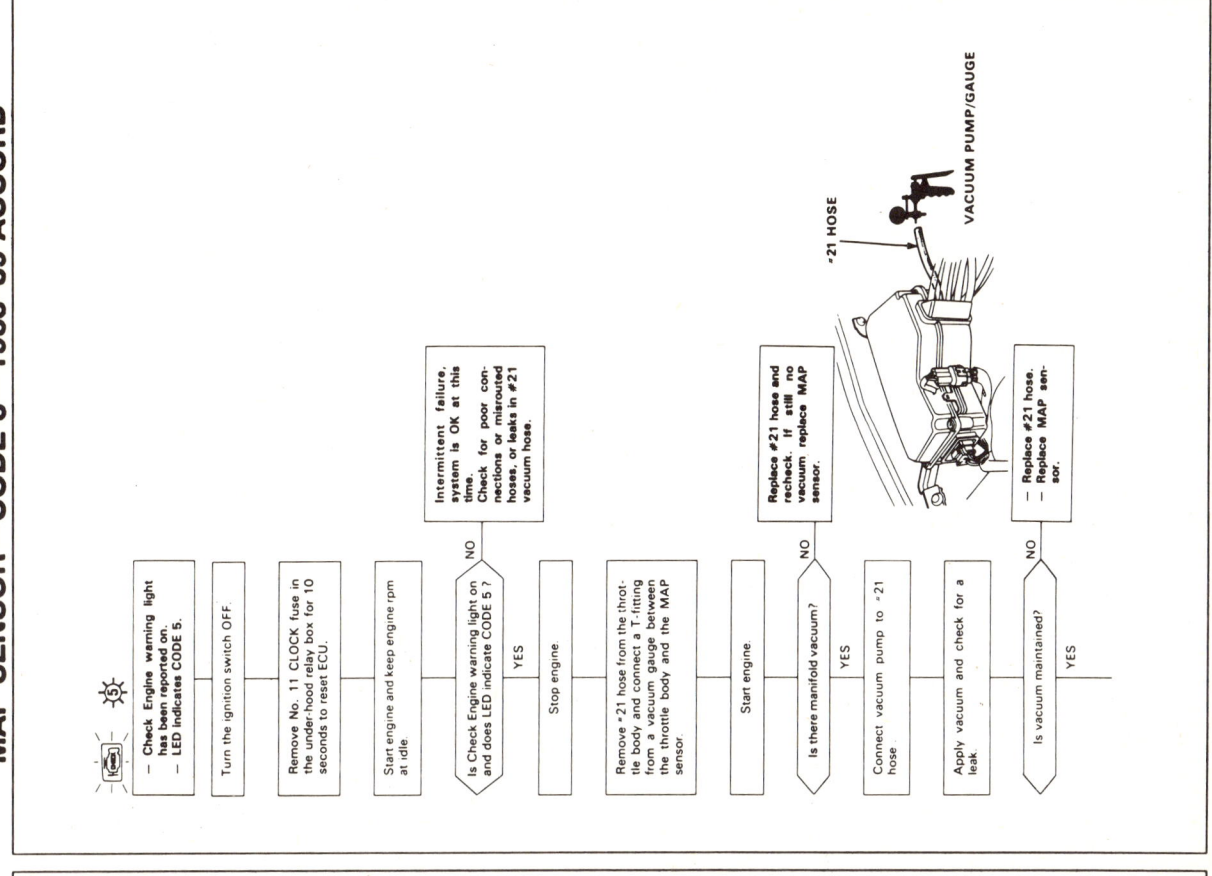

## MAP SENSOR – CODE 5 – 1988–89 ACCORD

- Check Engine warning light has been reported on.
- LED indicates CODE 5.

Turn the ignition switch OFF.

Remove No. 11 CLOCK fuse in the under-hood relay box for 10 seconds to reset ECU.

Start engine and keep engine rpm at idle.

Is Check Engine warning light on and does LED indicate CODE 5 ?

- NO: Intermittent failure, system is OK at this time. Check for poor connections or misrouted hoses, or leaks in #21 vacuum hose.
- YES: Stop engine.

Remove #21 hose from the throttle body and connect a T-fitting from a vacuum gauge between the throttle body and the MAP sensor.

Start engine.

Is there manifold vacuum?

- NO: Replace #21 hose and recheck. If still no vacuum replace MAP sensor.
- YES: Connect vacuum pump to #21 hose.

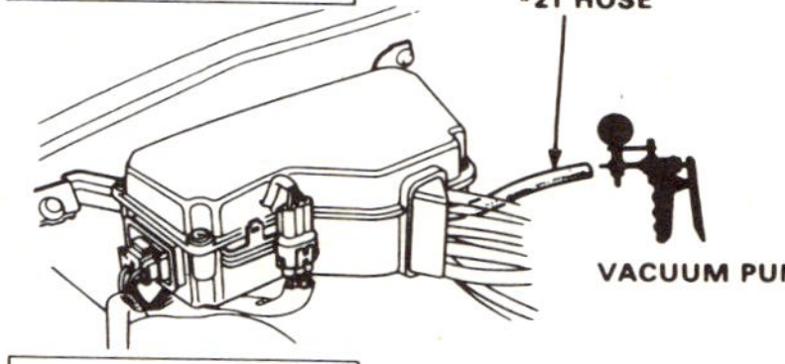

Apply vacuum and check for a leak.

#21 HOSE

VACUUM PUMP/GAUGE

Is vacuum maintained?

- NO:
  - Replace #21 hose.
  - Replace MAP sensor.
- YES

## MAP SENSOR—CODE 5—1988–89 ACCORD

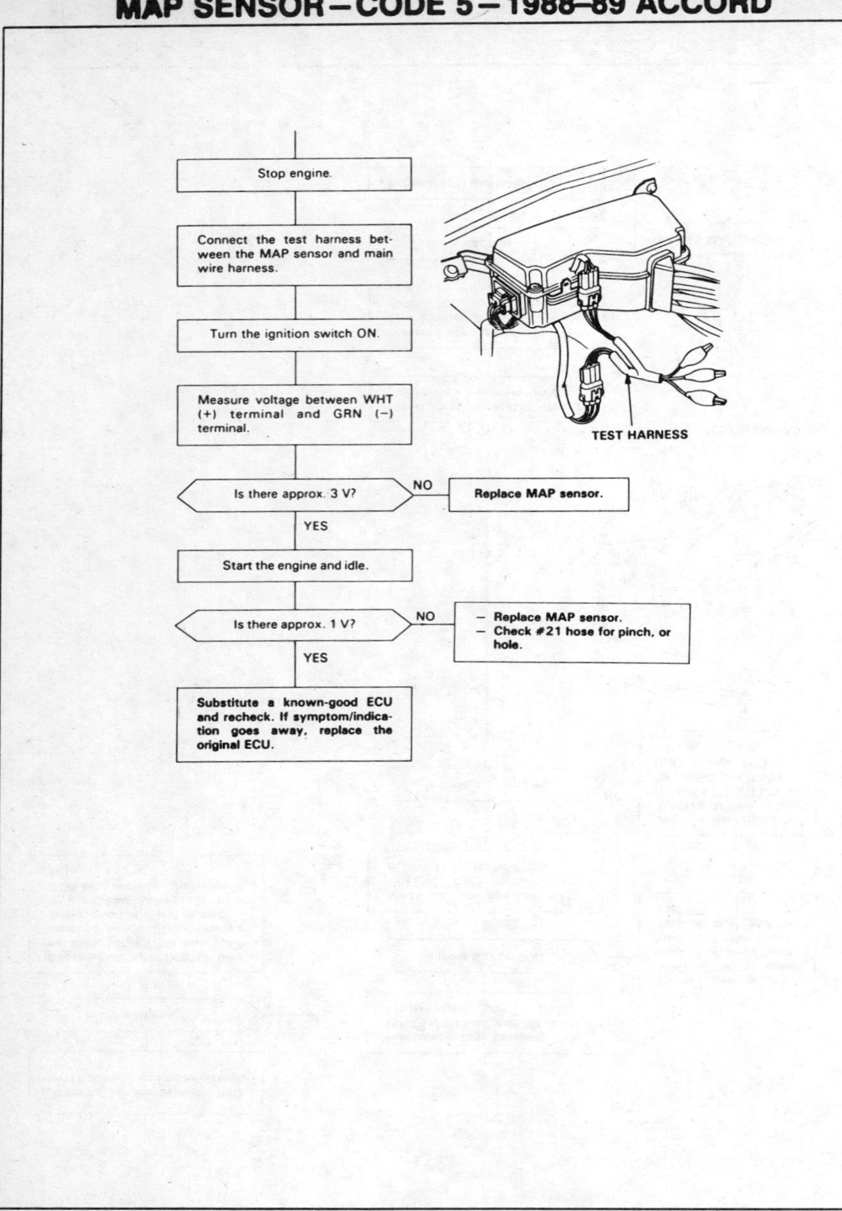


## COOLANT TEMPERATURE (TW) SENSOR—CODE 6—1988–89 ACCORD

PGM-FI (6) Self-diagnosis LED indicator blinks six times. Most likely a problem in the Coolant Temperature (TW) Sensor circuit.

The TW sensor is a temperature dependant resistor (thermistor). The resistance of the thermistor decreases as the coolant temperature increases as shown below.

THERMISTOR

RESISTANCE (kΩ): 20, 10, 5, 1, 0.5, 0.1

COOLANT TEMPERATURE: −20, 0, 20, 40, 60, 80, 100, 120 (°C) / −4, 32, 68, 104, 140, 176, 212, 248 (°F)

PGM-FI (6)
- PGM-FI warning light is on.
- LED indicates CODE 6.

Turn the ignition switch OFF.

Remove No.11 fuse in the under-hood relay box for 10 seconds to reset ECU.

Turn the ignition switch ON.

Is PGM-FI warning light on? — NO → Intermittent failure. (test drive may be necessary)

YES

Warm up engine to normal operating temperature (cooling fan comes on).

Disconnect the 2P connector on the TW sensor.

Measure resistance between the 2 terminals on the TW sensor.

Is there 200—400 Ω — NO → Replace TW sensor.

YES

TW SENSOR
28 N·m (2.8 kg-m, 20 lb-ft)

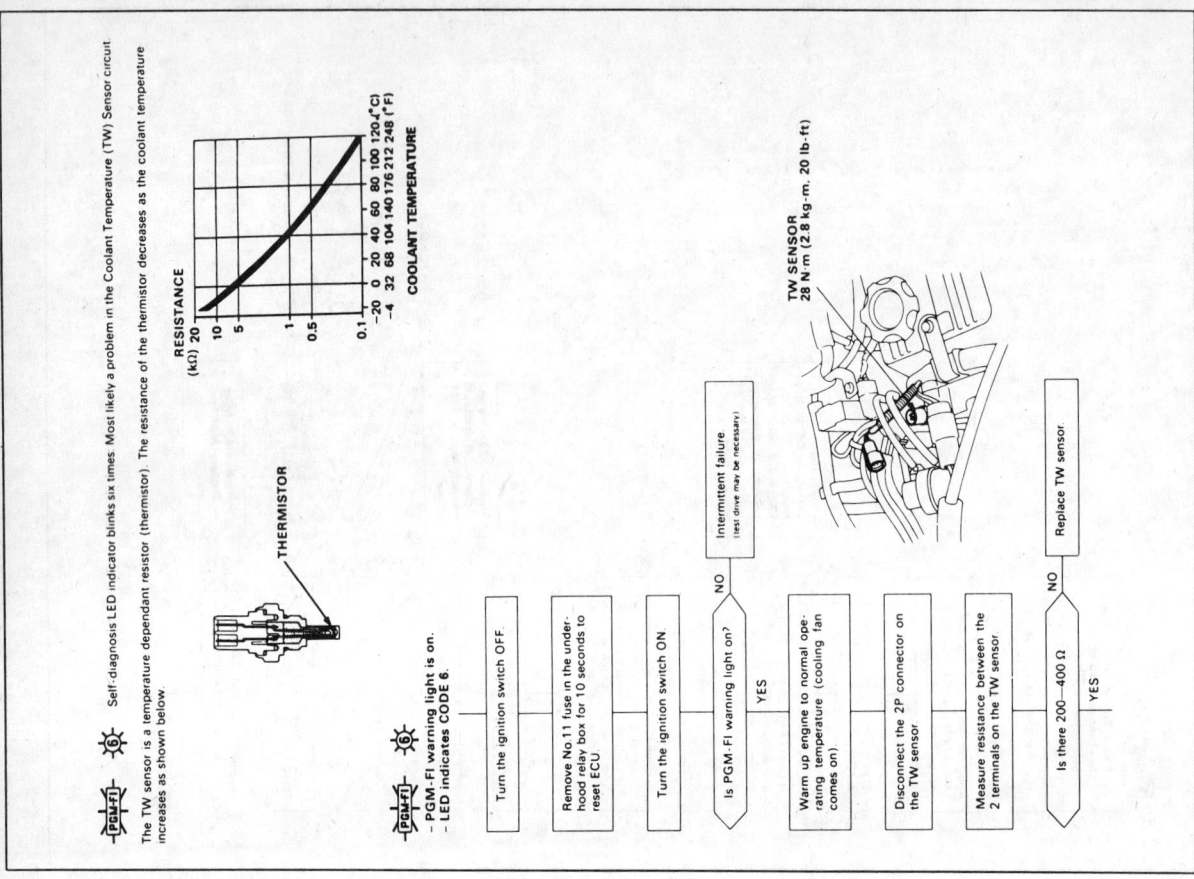

## COOLANT TEMPERATURE (TW) SENSOR – CODE 6 – 1988–89 ACCORD

Measure voltage between Yellow/Green (+) terminal and body ground.

Is there approx. 5 V?

- NO → Turn the ignition switch OFF. → Connect the system checker harness between the ECU and connector → Turn the ignition switch ON. → Measure voltage between C6 (+) terminal and C12 (–) terminal. 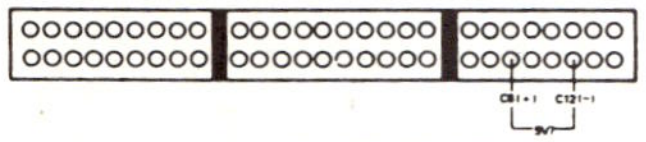
  - Is there approx. 5 V?
    - YES → Repair open in Yellow/Green wire between ECU (C6) and TW sensor.
    - NO → Disconnect "C" connector from the main wire harness only, not the ECU. → Measure voltage between C6 (+) terminal and C12 (–) terminal.
      - Is there approx. 5 V?
        - YES → Repair short in Yellow/Green wire between ECU (C6) and TW sensor.
        - NO → Substitute a known-good ECU and re-check. If prescribed voltage is now available, replace the original ECU.
- YES → Measure voltage between Yellow/Green (+) terminal and Green/White (–) terminal.
  - Is there approx. 5 V?
    - NO → Repair open in Green/White wire between ECU (C12) and TW sensor.
    - YES → Substitute a known-good ECU and re-check. If symptom/indication goes away, replace the original ECU.

## THROTTLE ANGLE SENSOR – CODE 7 – 1988–89 ACCORD

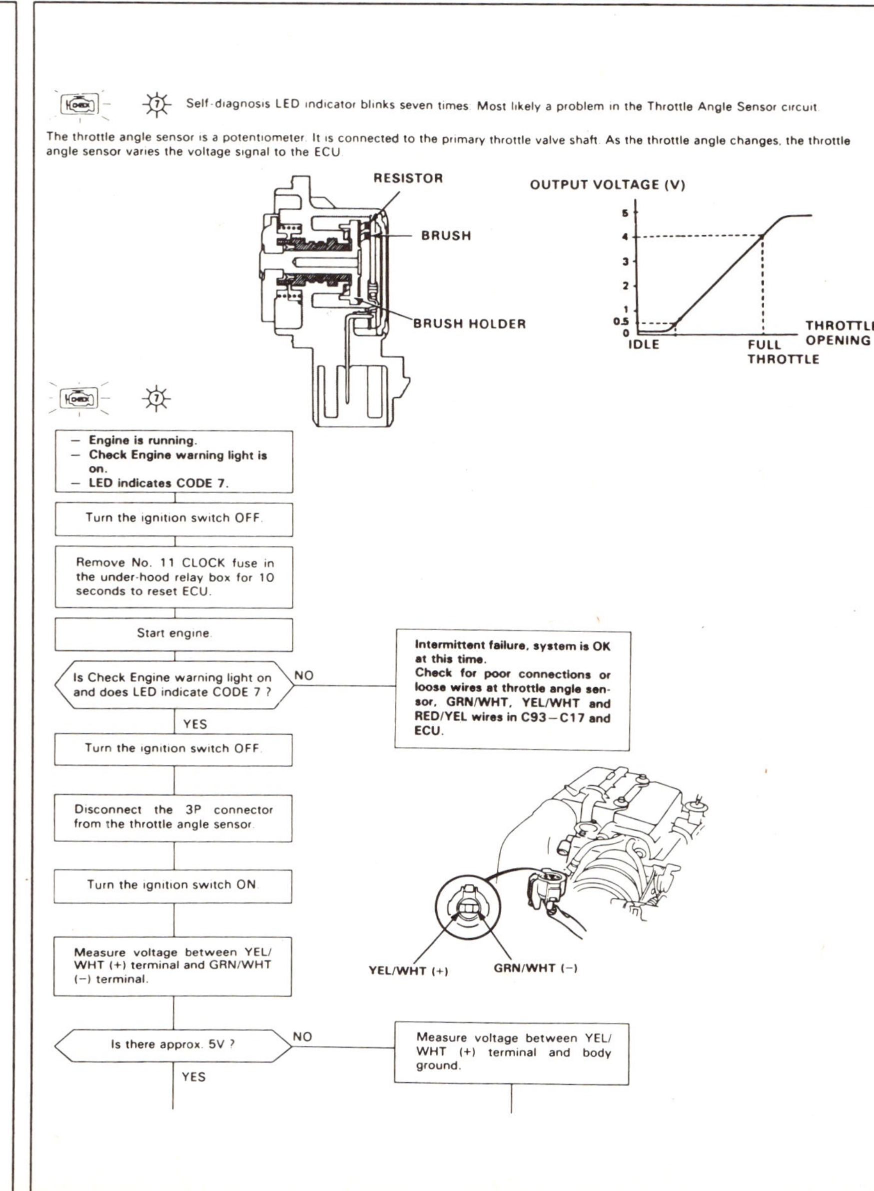

# THROTTLE ANGLE SENSOR CODE 7—1988–89 ACCORD

Turn the ignition switch OFF.

Reconnect the 3P connector.

Connect the PGM-FI test harness between the ECU and connector

Turn the ignition switch ON.

Measure voltage between C7(+) terminal and C12(−) terminal.

C7 (+)
C12 (−)
Approx. 0.5 V when throttle is closed?
Approx. 4.5 V at wide open throttle?

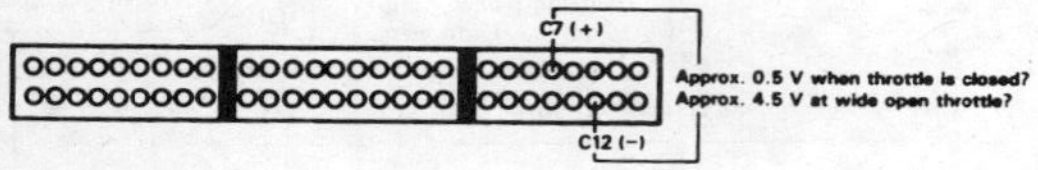

Is there approx. 0.5 V when throttle is closed and 4.5 V at wide open throttle?

- NO: 
  - Replace throttle angle sensor.
  - Repair open or short in RED/YEL wire between ECU (C7) and throttle angle sensor.
- YES: Substitute a known-good ECU and recheck. If symptom/indication goes away, replace the original ECU.

Is there approx. 5V ?

- YES: Repair open in GRN/WHT wire between ECU (C12) and throttle angle sensor.
- NO: Turn the ignition switch OFF.

Connect the PGM-FI test harness between the ECU and connector

Turn the ignition switch ON.

Measure voltage between C13(+) terminal and C12(−) terminal.

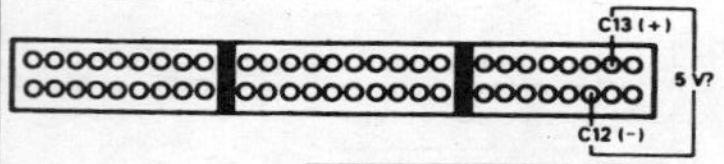

C13 (+)
5 V?
C12 (−)

Is there approx. 5V ?

- YES: Repair open in YEL/WHT wire between ECU (C13) and throttle angle sensor.
- NO: Substitute a known-good ECU and recheck. If prescribed voltage is now available, replace the original ECU.

# CRANK ANGLE SENSOR—CODE 8—1988–89 ACCORD

Self-diagnosis LED indicator blinks eight times: A problem in the TDC circuit of the Crank Angle Sensor.

Self-diagnosis LED indicator blinks nine times: A problem in the CYL circuit of the Crank Angle Sensor.

The sensor consists of a pair of rotors, TDC and CYL, and a pickup for each rotor. Since the rotors are coupled to the camshaft, they turn together as a unit as the camshaft rotates. The CYL sensor detects the position of the No. 1 cylinder as the base for the Sequential Injection whereas the TDC sensor serves to determine the injection timing for each cylinder. The TDC sensor also supplies the RPM signal to the ECU.

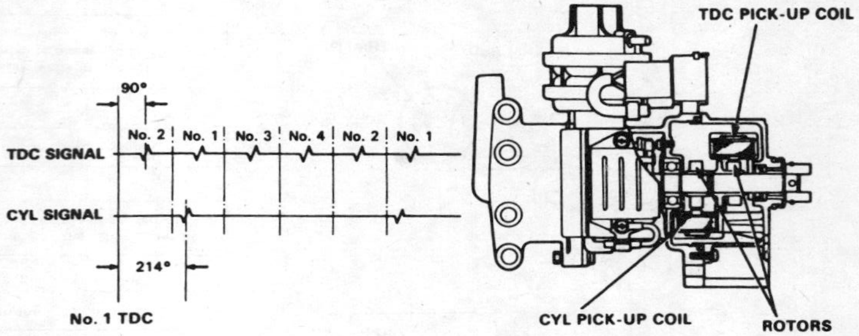

90°
No. 2 | No. 1 | No. 3 | No. 4 | No. 2 | No. 1
TDC SIGNAL
CYL SIGNAL
214°
No. 1 TDC
TDC PICK-UP COIL
CYL PICK-UP COIL
ROTORS

- Check Engine warning light has been reported on.
- LED indicates CODE 8.

Turn the ignition switch OFF.

Remove No. 11 CLOCK fuse in the under-hood relay box for 10 seconds to reset ECU.

Start engine.

Is Check Engine warning light on and does LED indicate CODE 8?

- NO: Intermittent failure, system is OK at this time. Check for poor connections or loose wires at distributor 4P connector, WHT/BLU and ORN/BLU wires in C93—C17 and ECU.
- YES: Stop engine.

Disconnect the 4P connector on the crank angle sensor.

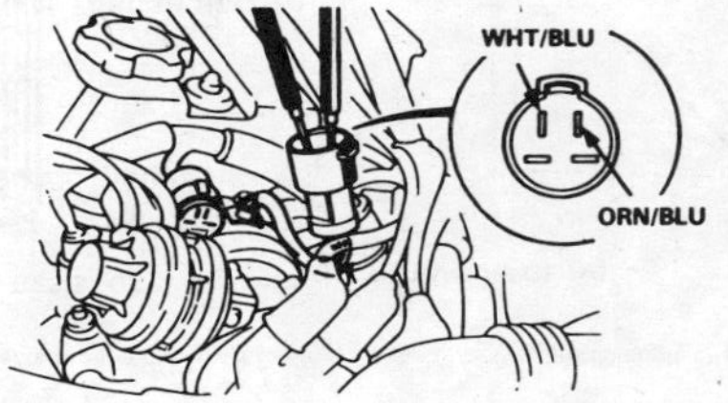

WHT/BLU
ORN/BLU

## CRANK ANGLE SENSOR—CODE 8—1988–89 ACCORD

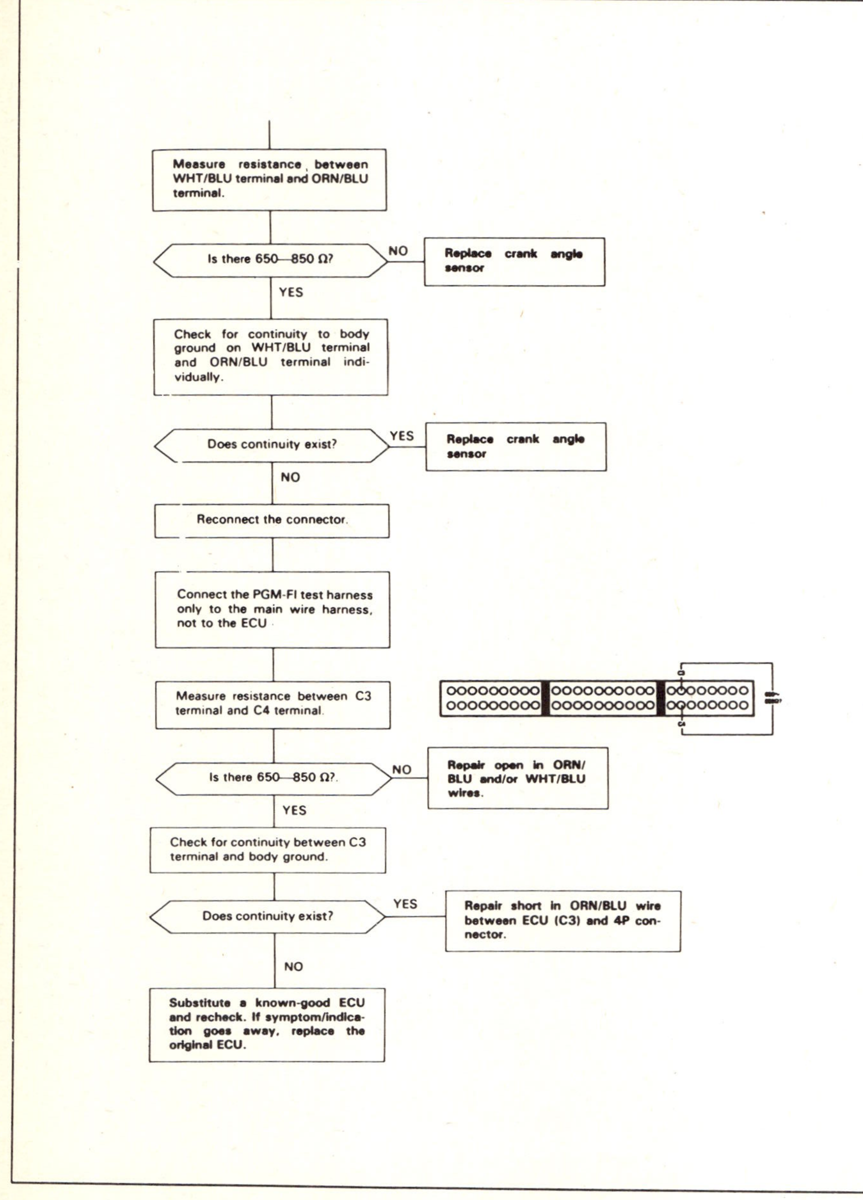


## CRANK ANGLE SENSOR—CODE 9—1988–89 ACCORD

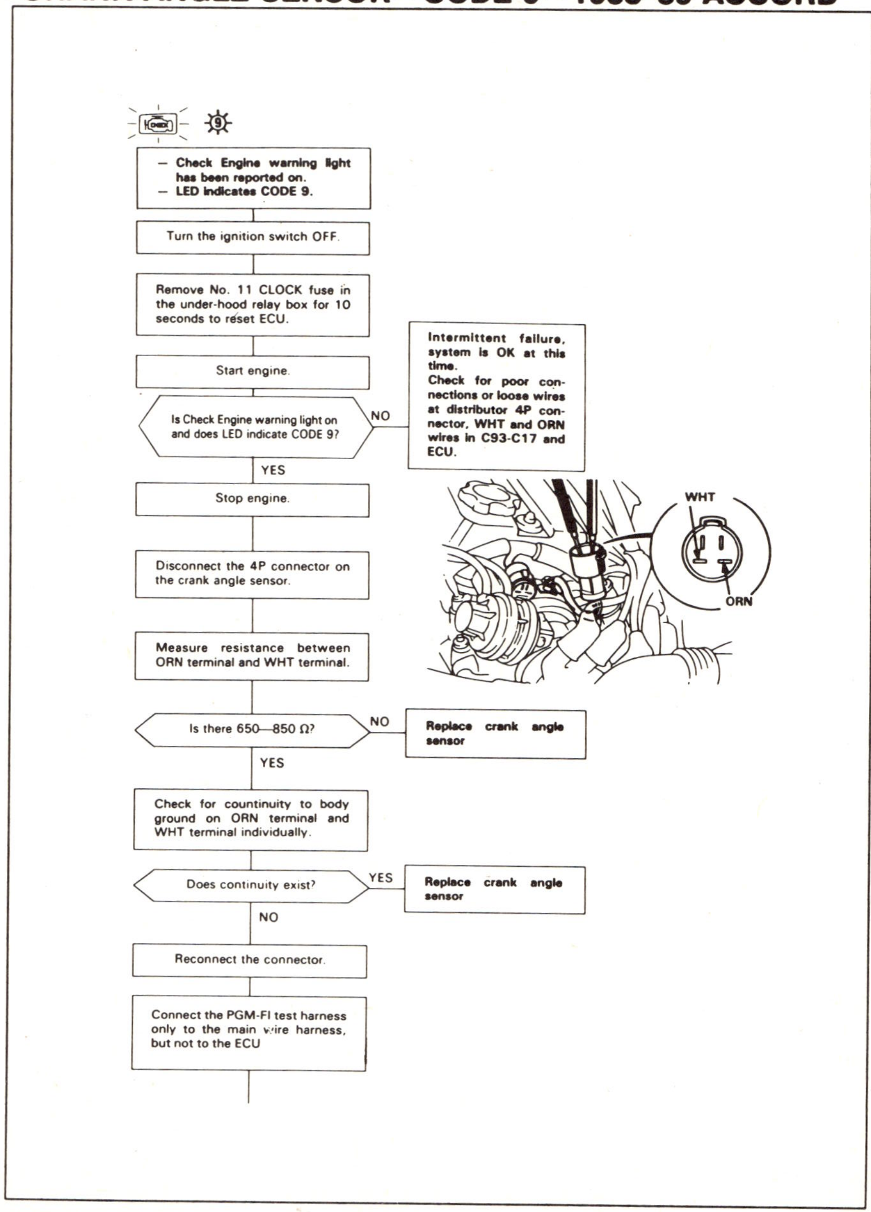

## CRANK ANGLE SENSOR—CODE 9—1988–89 ACCORD

Measure resistance between C1 terminal and C2 terminal.

Is there 650—850 Ω?
- NO: Repair open in ORN and/or WHT wires.
- YES: Check for continuity C1 to body ground.

Does continuity exist?
- YES: Reair short in ORN wire between ECU (C1) and 4P connector.
- NO: Substitute a known-good ECU and recheck. If symptom/indication goes away, replace the original ECU.

C1
C2
650~850 Ω?

Removal:
1. Remove the distributor
2. Remove the crank angle sensor assembly from the distributor.

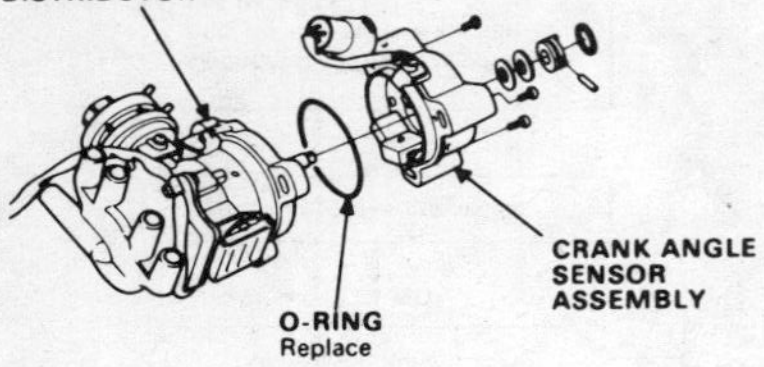

DISTRIBUTOR
CRANK ANGLE SENSOR ASSEMBLY
O-RING
Replace

Installation:
1. Install a new O-ring.
2. Install the sensor to the distributor.

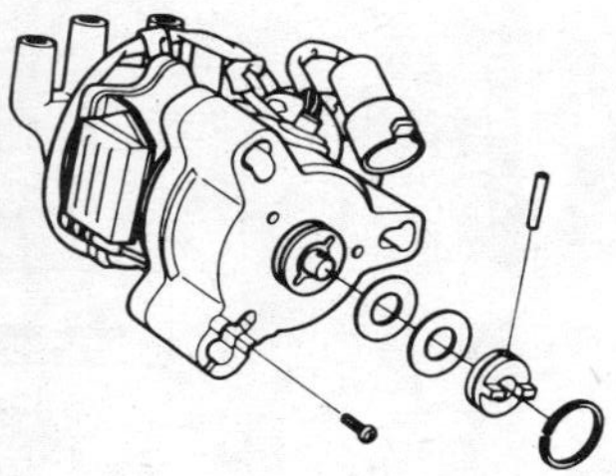

3. Install the distributor to the cylinder head

## AIR TEMPERATURE SENSOR CODE 10—1988–89 ACCORD

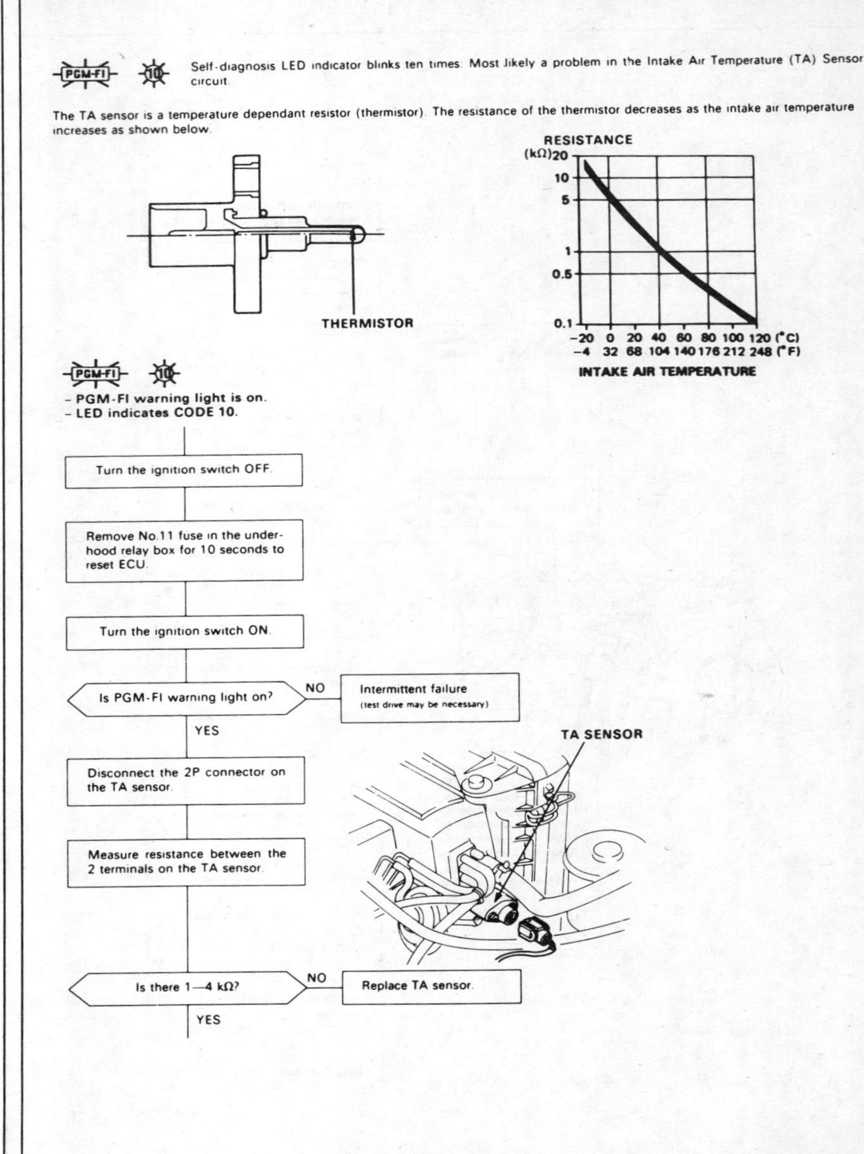

PGM-FI 10 Self-diagnosis LED indicator blinks ten times. Most likely a problem in the Intake Air Temperature (TA) Sensor circuit.

The TA sensor is a temperature dependant resistor (thermistor). The resistance of the thermistor decreases as the intake air temperature increases as shown below.



- PGM-FI warning light is on.
- LED indicates CODE 10.

Turn the ignition switch OFF.

Remove No.11 fuse in the under-hood relay box for 10 seconds to reset ECU.

Turn the ignition switch ON.

Is PGM-FI warning light on?
- NO: Intermittent failure (test drive may be necessary)
- YES: Disconnect the 2P connector on the TA sensor.

Measure resistance between the 2 terminals on the TA sensor.

Is there 1—4 kΩ?
- NO: Replace TA sensor.
- YES

TA SENSOR

## AIR TEMPERATURE SENSOR CODE 10—1988–89 ACCORD

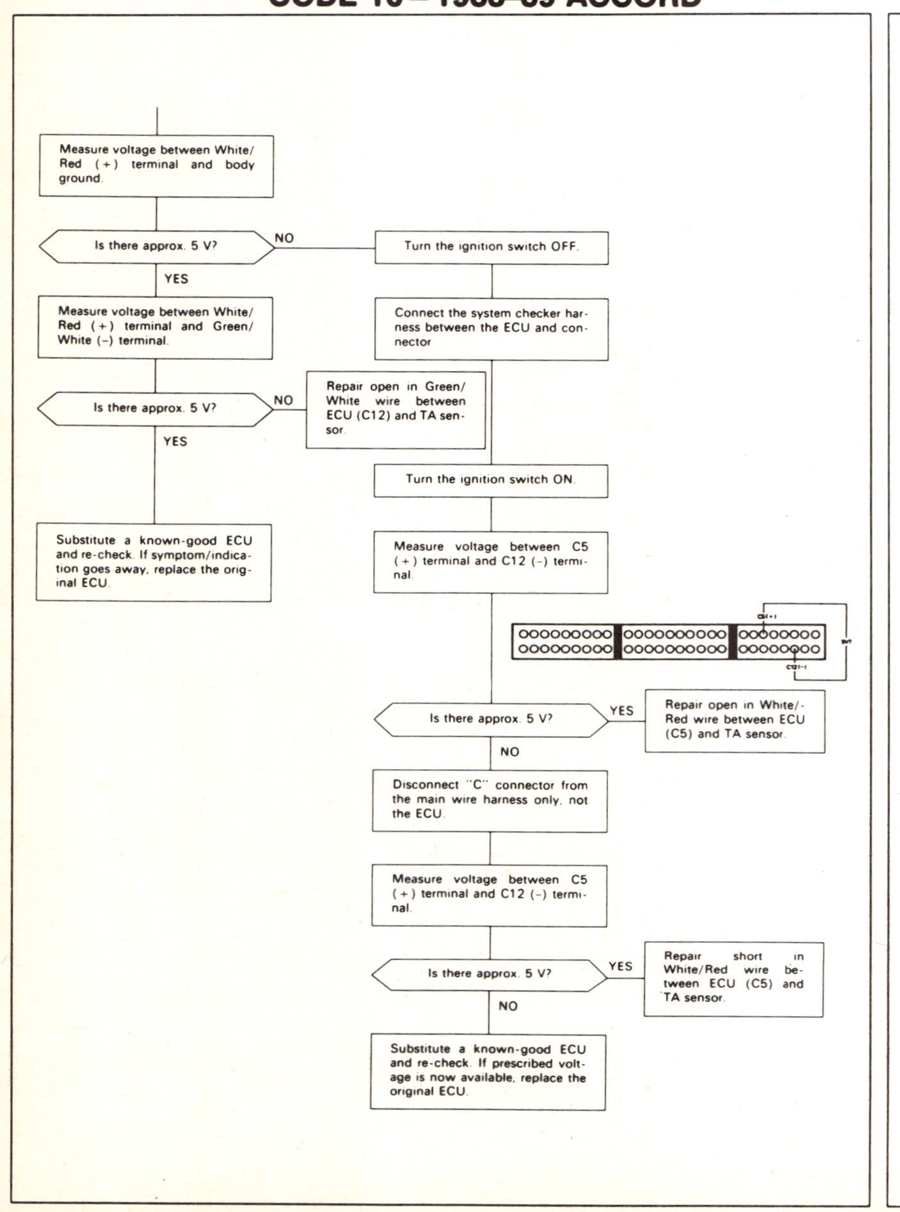

Measure voltage between White/Red (+) terminal and body ground.
Is there approx. 5 V?
NO
Turn the ignition switch OFF.
YES
Measure voltage between White/Red (+) terminal and Green/White (−) terminal.
Connect the system checker harness between the ECU and connector
Is there approx. 5 V?
NO
Repair open in Green/White wire between ECU (C12) and TA sensor.
YES
Turn the ignition switch ON.
Substitute a known-good ECU and re-check. If symptom/indication goes away, replace the original ECU.
Measure voltage between C5 (+) terminal and C12 (−) terminal.
Is there approx. 5 V?
YES
Repair open in White/Red wire between ECU (C5) and TA sensor.
NO
Disconnect "C" connector from the main wire harness only, not the ECU.
Measure voltage between C5 (+) terminal and C12 (−) terminal.
Is there approx. 5 V?
YES
Repair short in White/Red wire between ECU (C5) and TA sensor.
NO
Substitute a known-good ECU and re-check. If prescribed voltage is now available, replace the original ECU.


## ATMOSPHERIC PRESSURE (PA) SENSOR CODE 13—1988 ACCORD

PGM-FI 13 Self-diagnosis LED indicator blinks thirteen times. A problem in the Atmospheric Pressure (PA) Sensor circuit.

The PA sensor converts atmospheric pressure into electrical signals and inputs the ECU.

VACUUM CAVITY

SILICON DIAPHRAGM

OUTPUT VOLTAGE (v) 3.5, 3.0, 2.5, 2.0, 1.5, 1.0, 0.5

GAUGE READING 0, 5, 10, 15, 20, 25, 30 (in Hg); 100, 200, 300, 400, 500, 600, 700 (mmHg)

PGM-FI 13

- PGM-FI warning light is on.
- LED indicates CODE 13.

Turn the ignition switch OFF.

Remove No.11 fuse in the underhood relay box for 10 seconds to reset ECU.

Turn the ignition switch ON.

Is PGM-FI warning light on? — NO → Intermittent failure (test drive may be necessary)

YES

Turn the ignition switch OFF.

Connect the system checker harness between the ECU and connector

Turn the ignition switch ON.

Measure voltage between C13 (+) terminal and C12 (−) terminal.

Is there approx. 5 V? — NO → Substitute a known-good ECU and recheck. If prescribed voltage is now available replace the original ECU.

YES

## ATMOSPHERIC PRESSURE (PA) SENSOR—CODE 13—1988 ACCORD

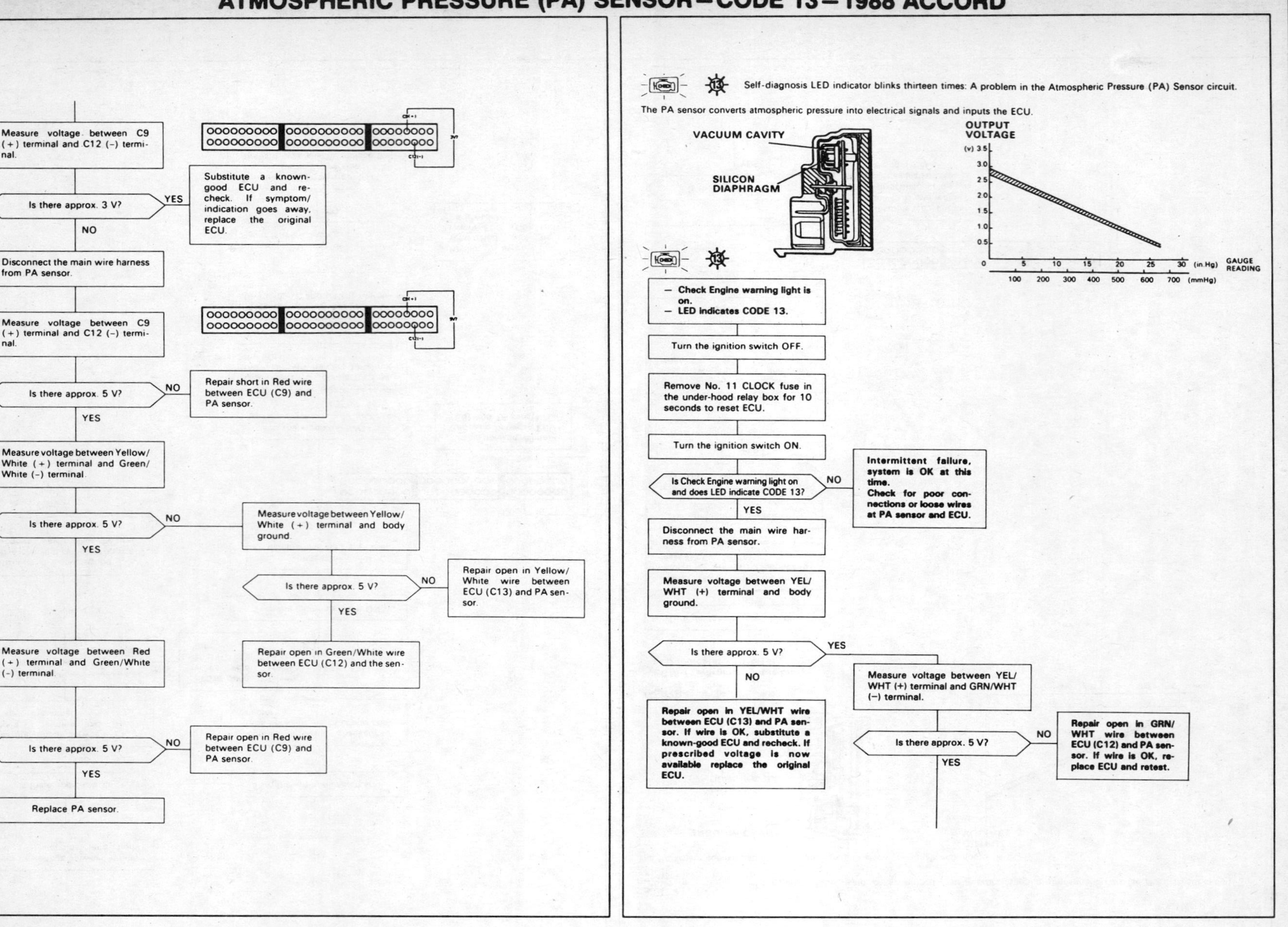

## ATMOSPHERIC PRESSURE (PA) SENSOR CODE 13—1989 ACCORD

- Turn the ignition switch OFF.
- Remove No. 11 CLOCK fuse in the underhood relay box for 10 seconds to reset ECU.
- Substitute a known-good PA sensor.
- Turn the ignition switch ON.
- Is Check Engine warning light on and does LED indicate CODE 13?
  - NO → **Replace original PA sensor.**
  - YES → Replace ECU and retest.

## ELECTRONIC AIR CONTROL VALVE (EACV) CODE 14—1988–89 ACCORD

Self-diagnosis LED indicator blinks fourteen times: A problem in the Electronic Air Control Valve (EACV) circuit.

The EACV changes the amount of air bypassing the throttle body in response to a current signal from the ECU in order to maintain the proper idle speed.

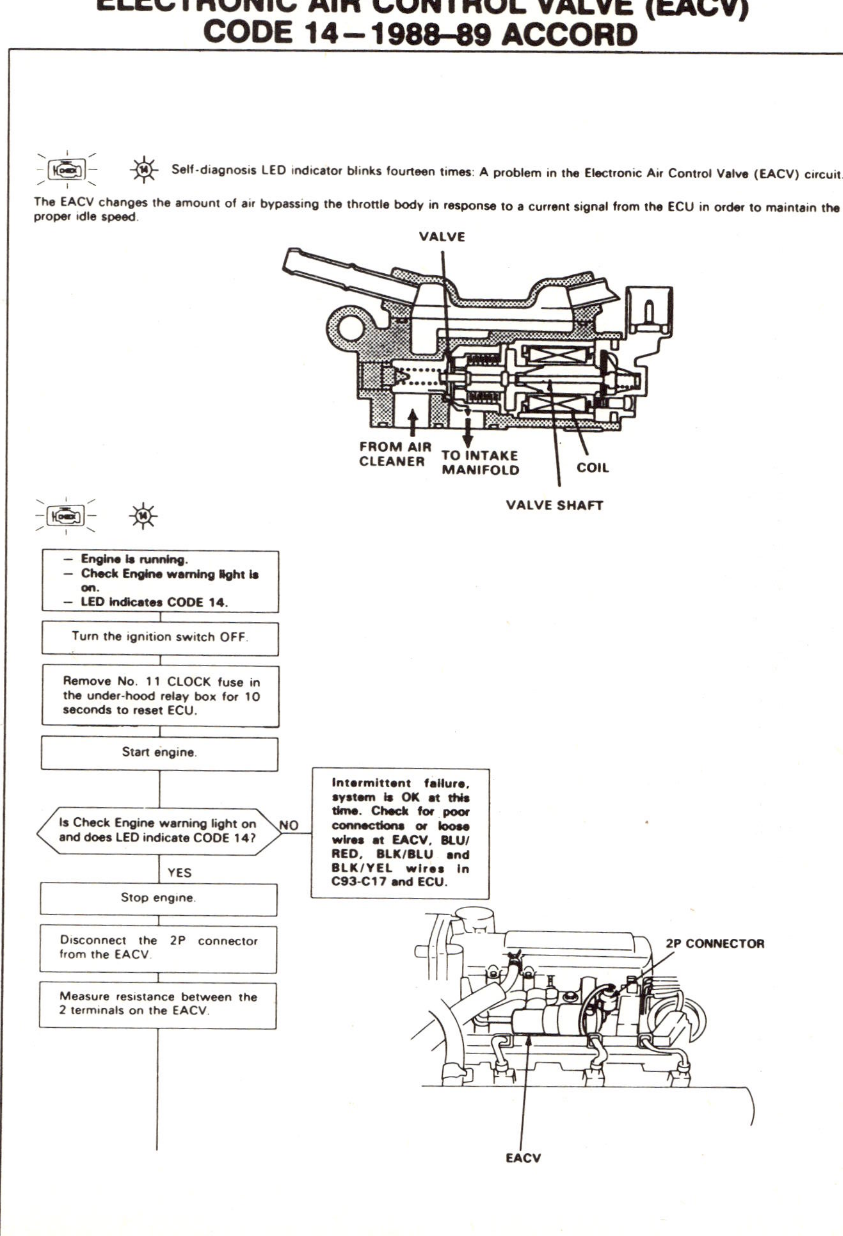


- **Engine is running.**
- **Check Engine warning light is on.**
- **LED indicates CODE 14.**

1. Turn the ignition switch OFF.
2. Remove No. 11 CLOCK fuse in the under-hood relay box for 10 seconds to reset ECU.
3. Start engine.
4. Is Check Engine warning light on and does LED indicate CODE 14?
   - NO → **Intermittent failure, system is OK at this time. Check for poor connections or loose wires at EACV, BLU/RED, BLK/BLU and BLK/YEL wires in C93-C17 and ECU.**
   - YES → Stop engine.
5. Disconnect the 2P connector from the EACV.
6. Measure resistance between the 2 terminals on the EACV.

## ELECTRONIC AIR CONTROL VALVE (EACV)—CODE 14—1988–89 ACCORD

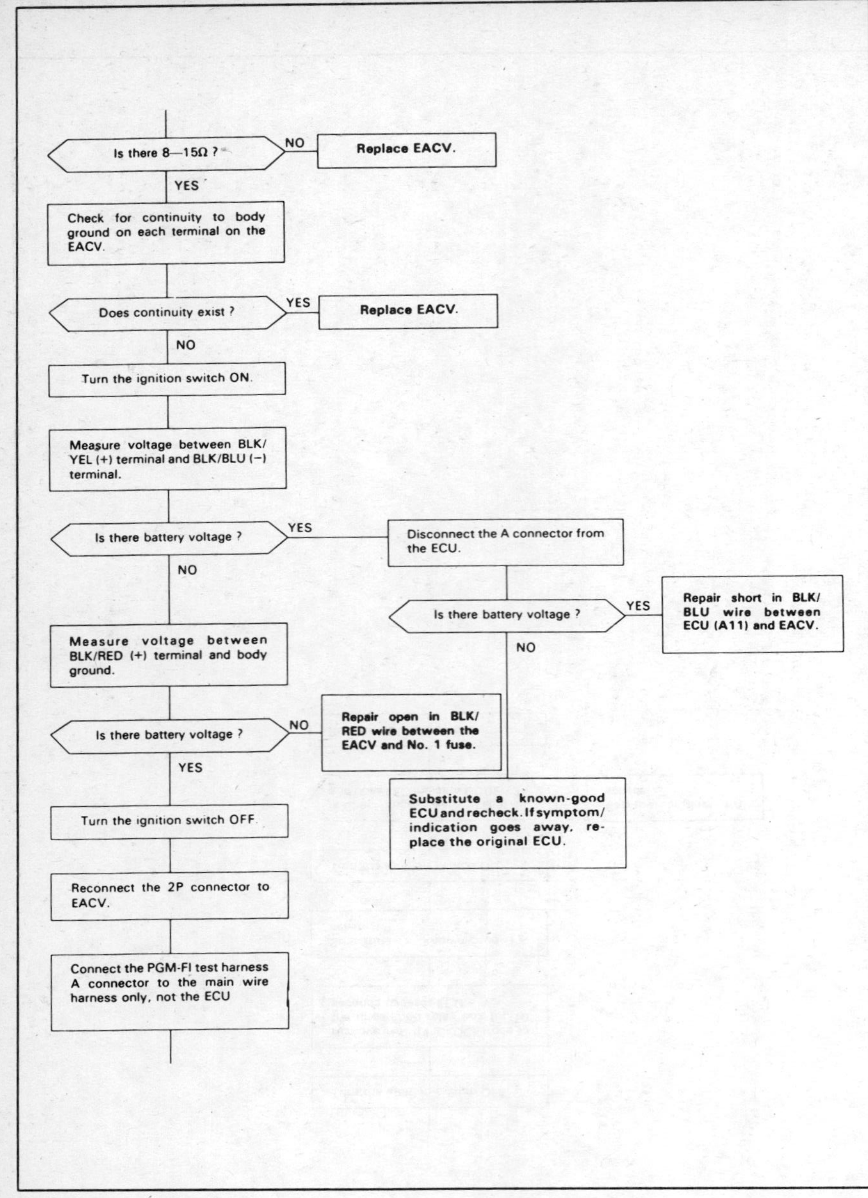


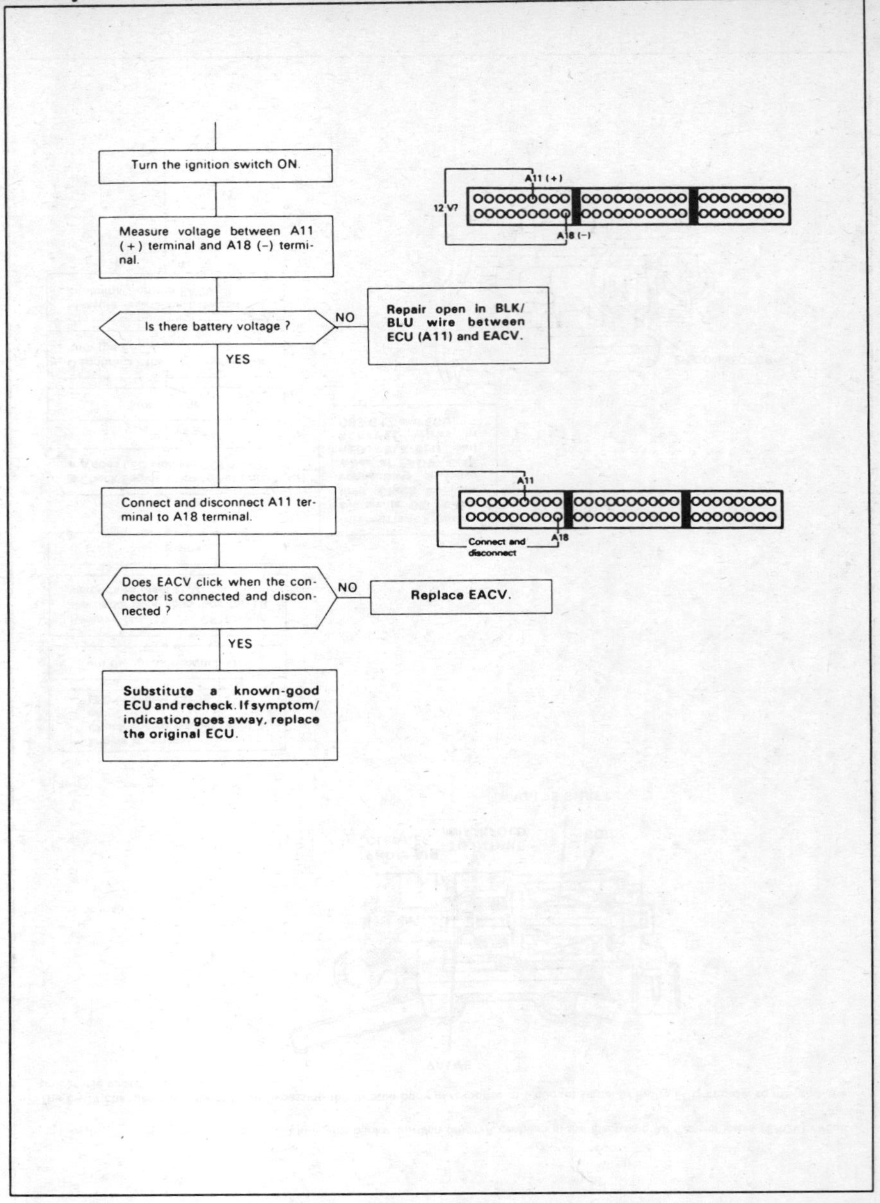

# VEHICLE SPEED SENSOR—CODE 17—1989 ACCORD

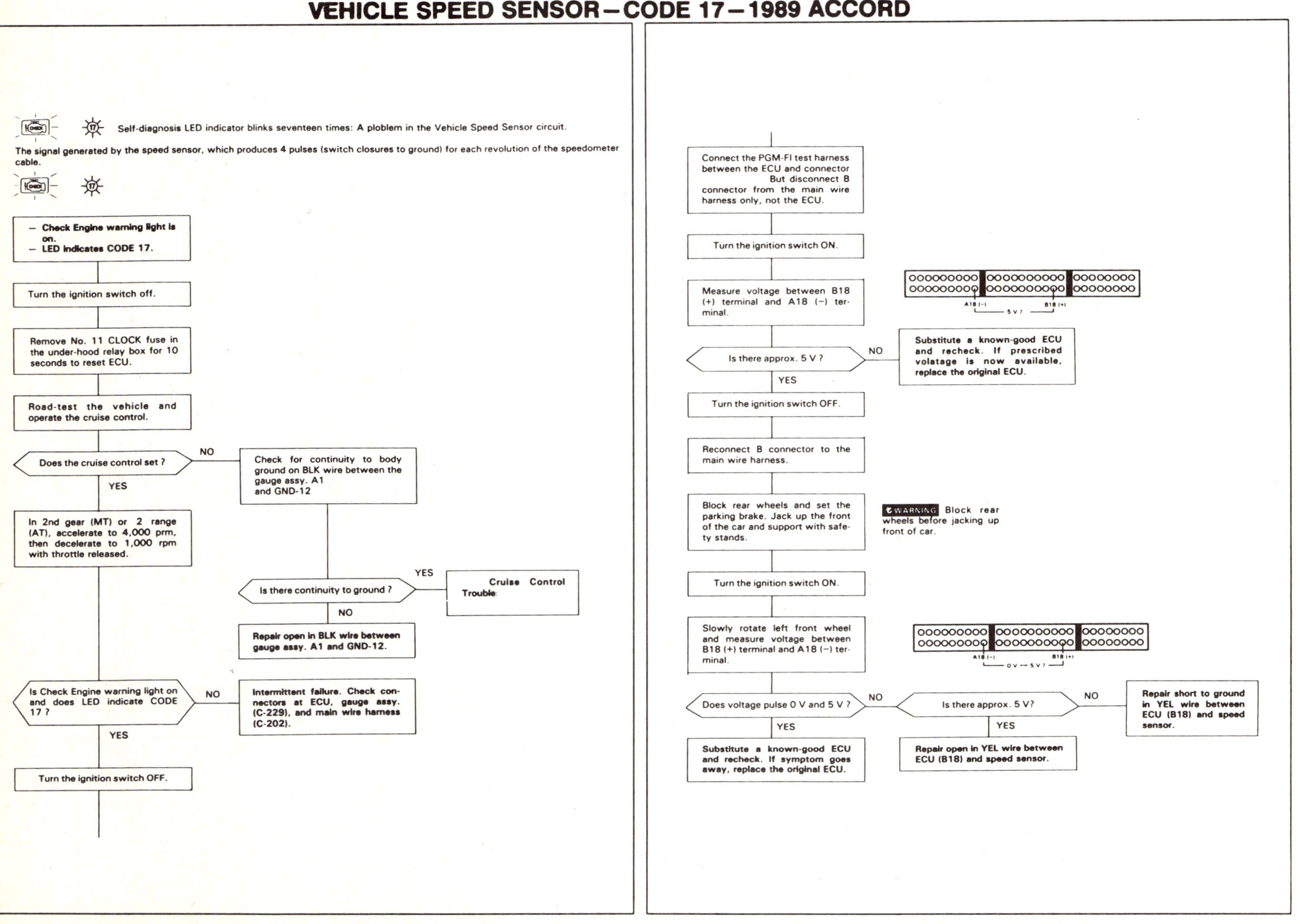

## VEHICLE SPEED SENSOR – 1988 ACCORD

This signals the PGM-FI ECU when the vehicle speed is above 10 mph.

**Inspection of Vehicle Speed Signal.**

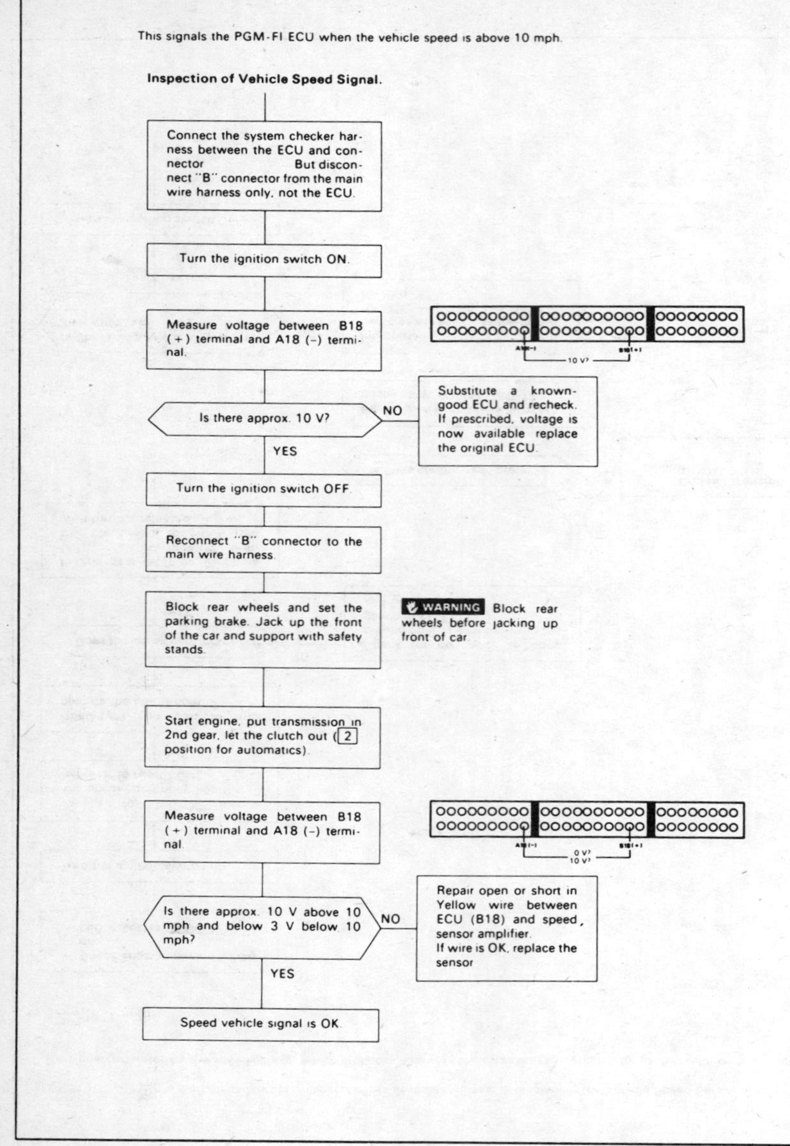


## IDLE CONTROL TROUBLESHOOTING – 1988–89 ACCORD

NOTE: Across each row in the chart, the sub systems that could be sources of a symptom are ranked in the order they should be inspected, starting with ①. Find the symptom in the left column, read across to the most likely source, then refer to the page listed at the top of that column. If inspection shows the system is OK, try the next system ②, etc.

| SYMPTOM | SUB SYSTEM | IDLE ADJUSTING SCREW | ELECTRONIC AIR CONTROL VALVE | STARTER SIGNAL | ALTERNATOR FR SIGNAL | A/T SHIFT POSITION SIGNAL |
|---|---|---|---|---|---|---|
| WHEN COLD | Fast idle speed is not as specified (1,200–2,000 rpm) | ③ | ② | | | |
| WHEN WARM RPM TOO HIGH | Idle speed is above specified rpm | | ③ | | | |
| WHEN WARM RPM TOO LOW | Idle speed is below specified rpm | | ① | | | |
| | Idle speed does not increase after initial start up. | | ① | ② | | |
| | Idle speed drops with electrical load | | | | ① | |
| | On models with automatic transmission, the idle speed drops in gear | | ① | | | ② |
| | Idle speed drops when air conditioner is ON | | ① | | | |
| | Idle speed drops when steering wheel is turning | | ② | | | |
| FREQUENT STALLING WHILE WARMING UP | | | | | | |

- – If by-pass passages are blocked, a low idle speed will result.
  – If hoses or by-pass passages are leaking, a high idle speed will result.

## IDLE CONTROL TROUBLESHOOTING—1988–89 ACCORD

| AIR CONDITIONING SIGNAL | P/S OIL PRESSURE SIGNAL | FAST IDLE VALVE | HOSES AND CONNECTIONS |
|---|---|---|---|
| | | | • |
| | | ① | |
| | | ② | ① |
| ② | | | |
| | | | |
| | | | |
| | | | |
| ② | | | |
| | ① | | |
| | | ① | |

## STARTER SIGNAL—1988–89 ACCORD

This signals the PGM-FI ECU when the engine is cranking. The ECU controls the idle control solenoid valve to promote easy starting.

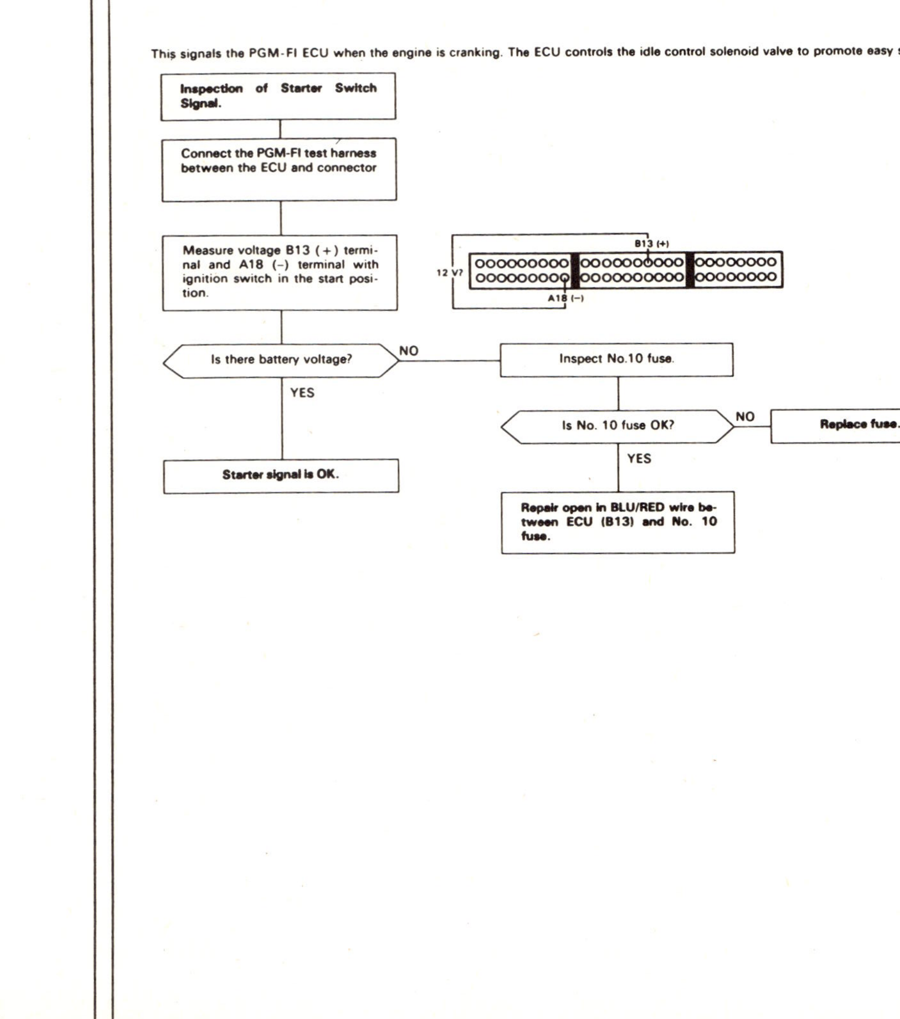

## ALTERNATOR FR SIGNAL—1988–89 ACCORD

This signals the PGM-FI ECU when the alternator is charging.

**Inspection of Alternator FR Signal.**

Connect the PGM-FI test harness between the ECU and connector But disconnect "B" connector from the main wire harness only, not the ECU.

Turn the ignition switch ON.

Measure voltage between B14 (+) terminal and A18 (–) terminal.

(A18 (–) — 5 V? — B14 (+))

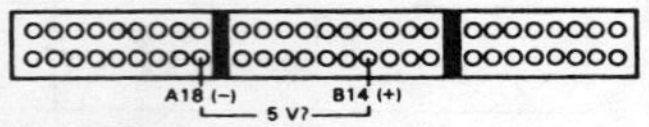

Is there approx. 5 V?

- NO: **Substitute a known-good ECU and recheck. If prescribed voltage is now available, replace the original ECU.**
- YES: Turn the ignition switch OFF.

Reconnect B connector to the main wire harness.

Warm up engine to normal operating temperature (cooling fan comes on).

Measure voltage between B14 (+) terminal and A18 (–) terminal.

(A18 (–) — Voltage? — B14 (+))

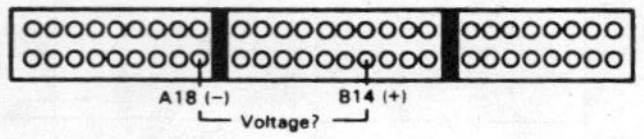

Does the voltage decrease when headlight and rear defogger are turned on?

- YES: **Alternator FR signal is OK.**
- NO: Stop the engine.

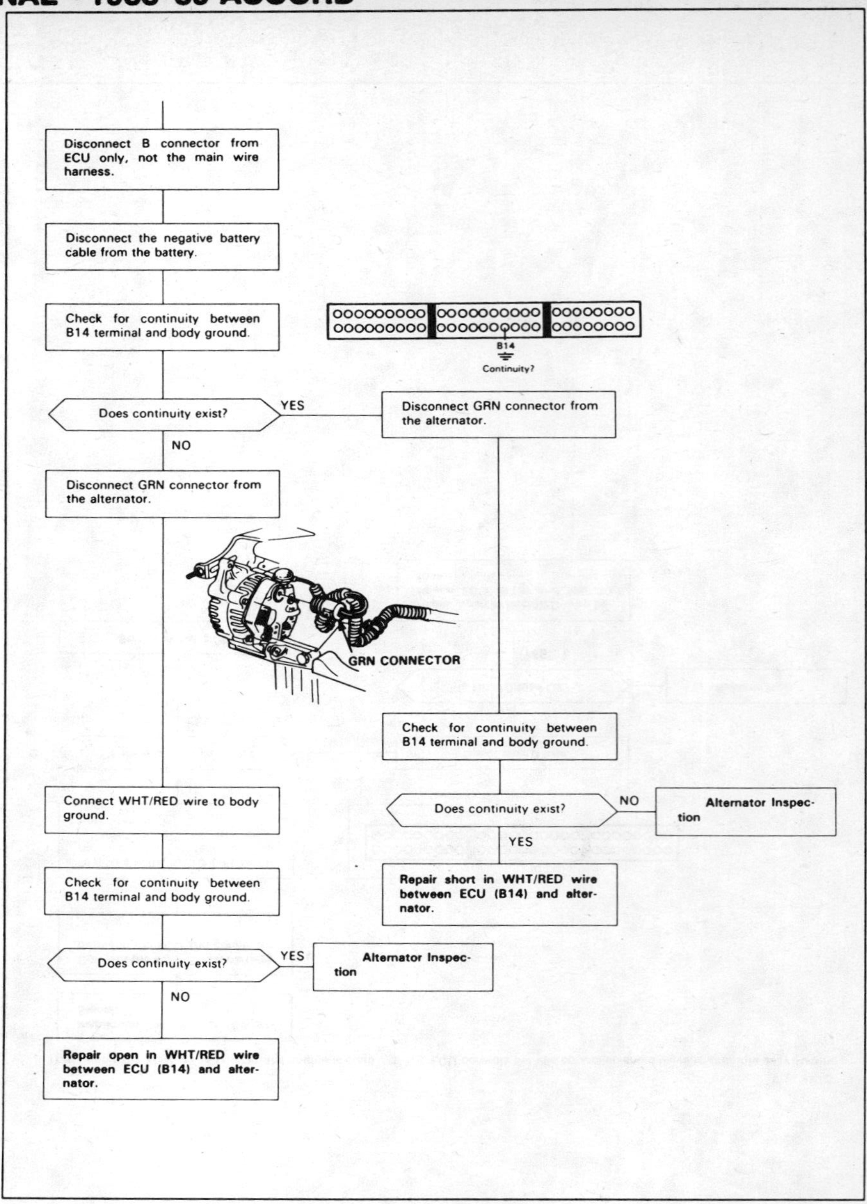

# A/T SHIFT POSITION SIGNAL—1988 ACCORD

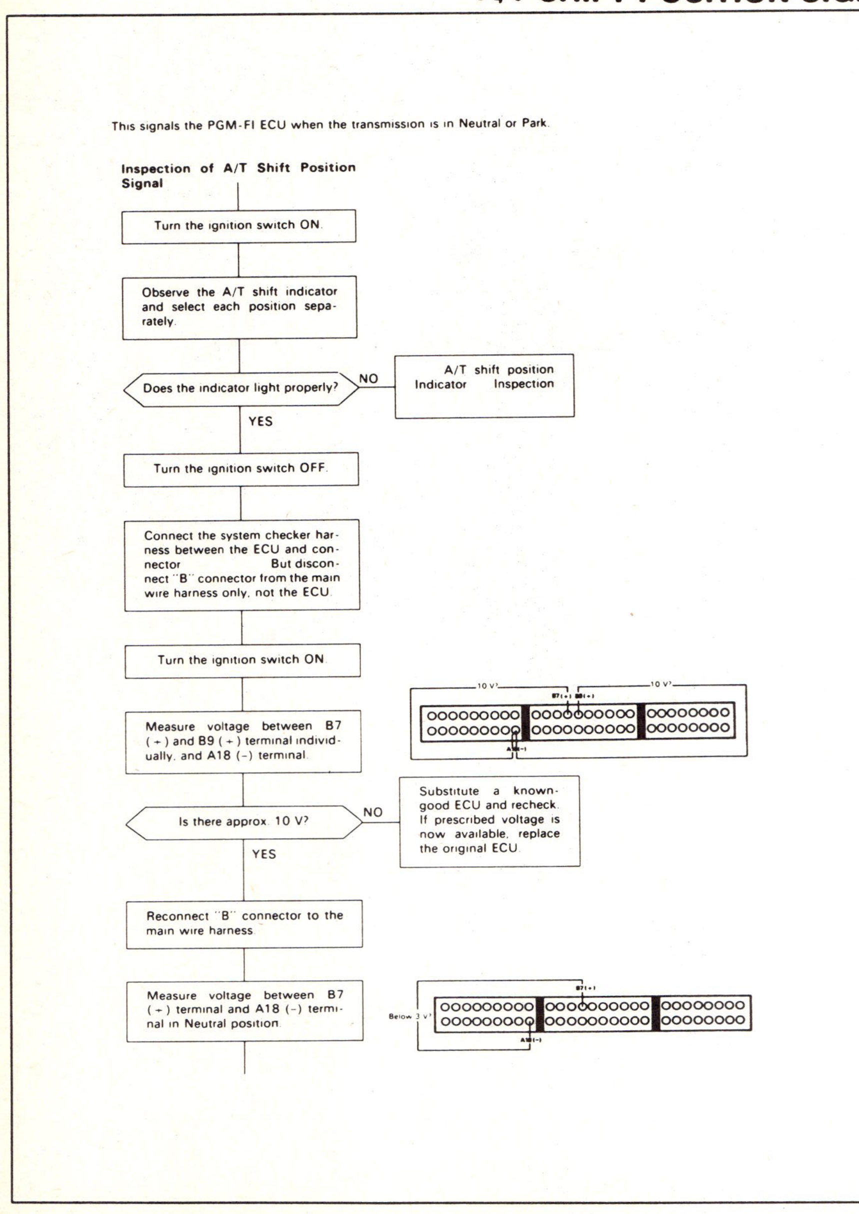


Is there below 3 V?

NO — Repair open Green wire between ECU (B7) and A/T shift position indicator switch

YES

Measure voltage between B9 (+) terminal and A18 (-) terminal in Park position.

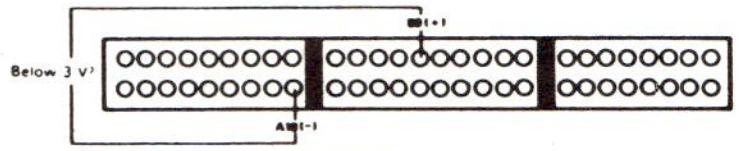

Below 3 V?

Is there below 3 V?

NO — Repair open Green/White wire between ECU (B9) and A/T shift position indicator switch

YES

A/T shift position signal is OK.

## A/T SHIFT POSITION SIGNAL—1989 ACCORD

This signals the PGM-FI ECU when the transmission is in Neutral or Park.

**Inspection of A/T Shift Position Signal**

1. Turn the ignition switch ON.
2. Observe the A/T shift indicator and select each position separately.
3. Does the indicator light properly?
   - NO: A/T shift position Indicator Inspection
   - YES: continue.
4. Turn the ignition switch OFF.
5. Connect the PGM-FI test harness between the ECU and connector . But disconnect B connector from the main wire harness only, not the ECU.
6. Turn the ignition switch ON.
7. Measure voltage between B7 (+) and A18 (−) terminal. (5 V? — B7 (+), A18 (−))
8. Is there approx. 5 V?
   - NO: **Substitute a known-good ECU and recheck. If prescribed voltage is now available, replace the original ECU.**
   - YES: continue.
9. Turn the ignition switch OFF.
10. Reconnect "B" connector to the main wire harness.
11. Turn the ignition switch ON.
12. Measure voltage between B7 (+) terminal and A18 (−) terminal in Neutral position. (Below 3 V? — B7 (+), A18 (−))
13. Is there less than 3 V?
    - NO: **Repair open in LT GRN wire between ECU (B7) and connector C226 at gauge assy.**
    - YES: **A/T shift position signal is OK.** (Less than 3 V? — B7 (+), A18 (−))

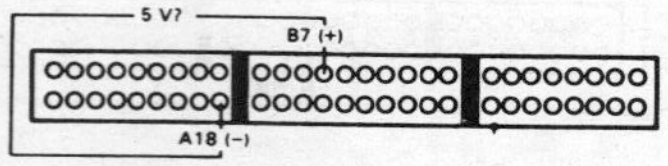

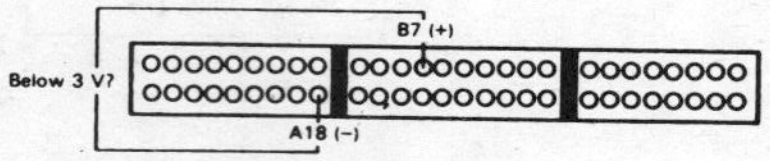

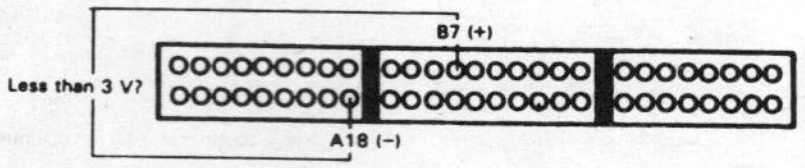

## AIR CONDITIONING SIGNAL—1988 ACCORD

This signals the PGM-FI ECU when there is a demand for cooling from the air conditioning circuits (A/C switch).

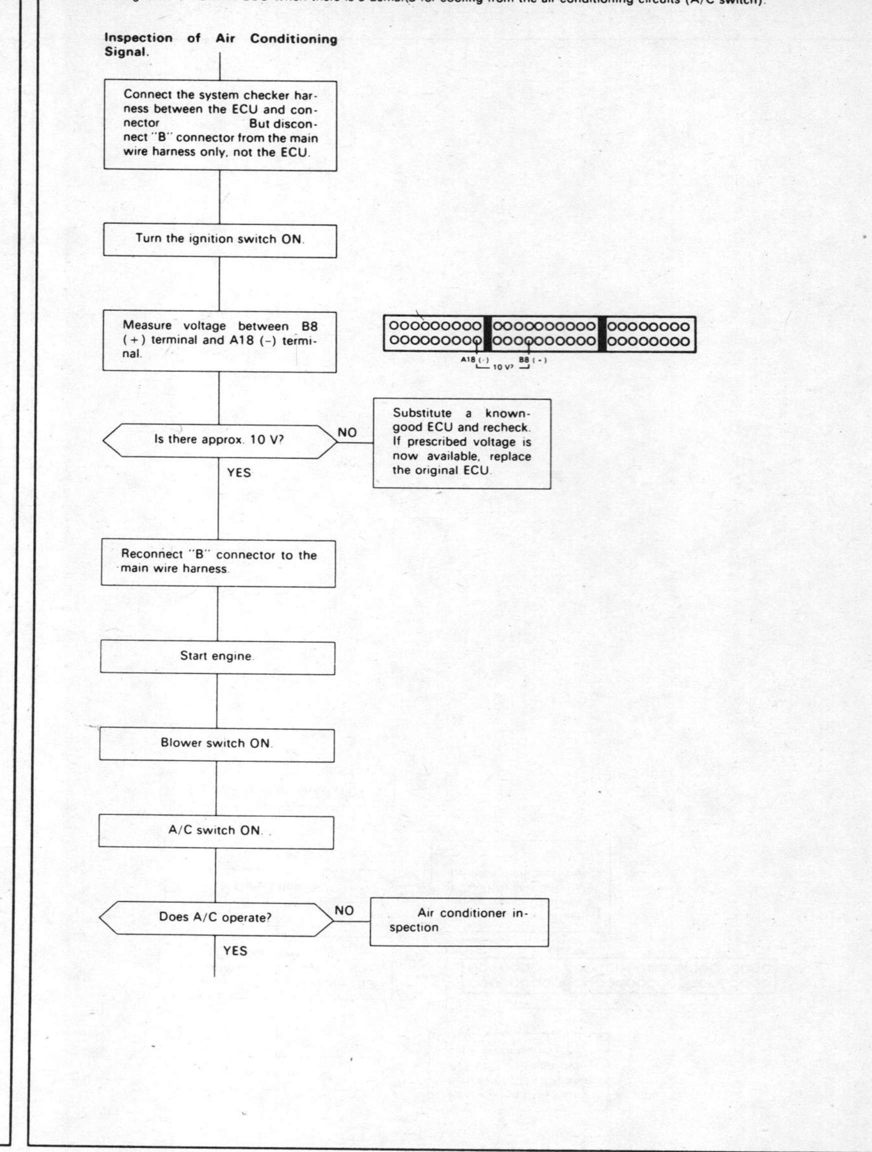

## AIR CONDITIONING SIGNAL—1988 ACCORD

This signals the PGM-FI ECU when there is a demand for cooling from the air conditioning circuits (A/C switch).

**Inspection of Air Conditioning Signal.**

Connect the PGM-FI test harness between the ECU and connector But disconnect B connector from the main wire harness only, not the ECU.

Turn the ignition switch ON.

Measure voltage between B8 (+) terminal and A18 (–) terminal.

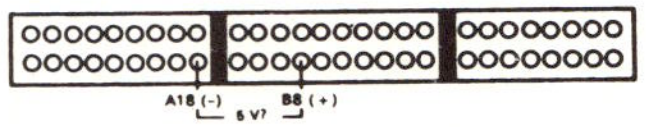

A18 (–) 5 V? B8 (+)

Is there approx. 5 V?

NO → **Substitute a known-good ECU and recheck. If prescribed voltage is now available, replace the original ECU.**

YES ↓

Reconnect B connector to the main wire harness.

Start engine.

Blower switch ON.

A/C switch ON.

Does A/C operate?

NO → **Air conditioner inspection**

YES ↓

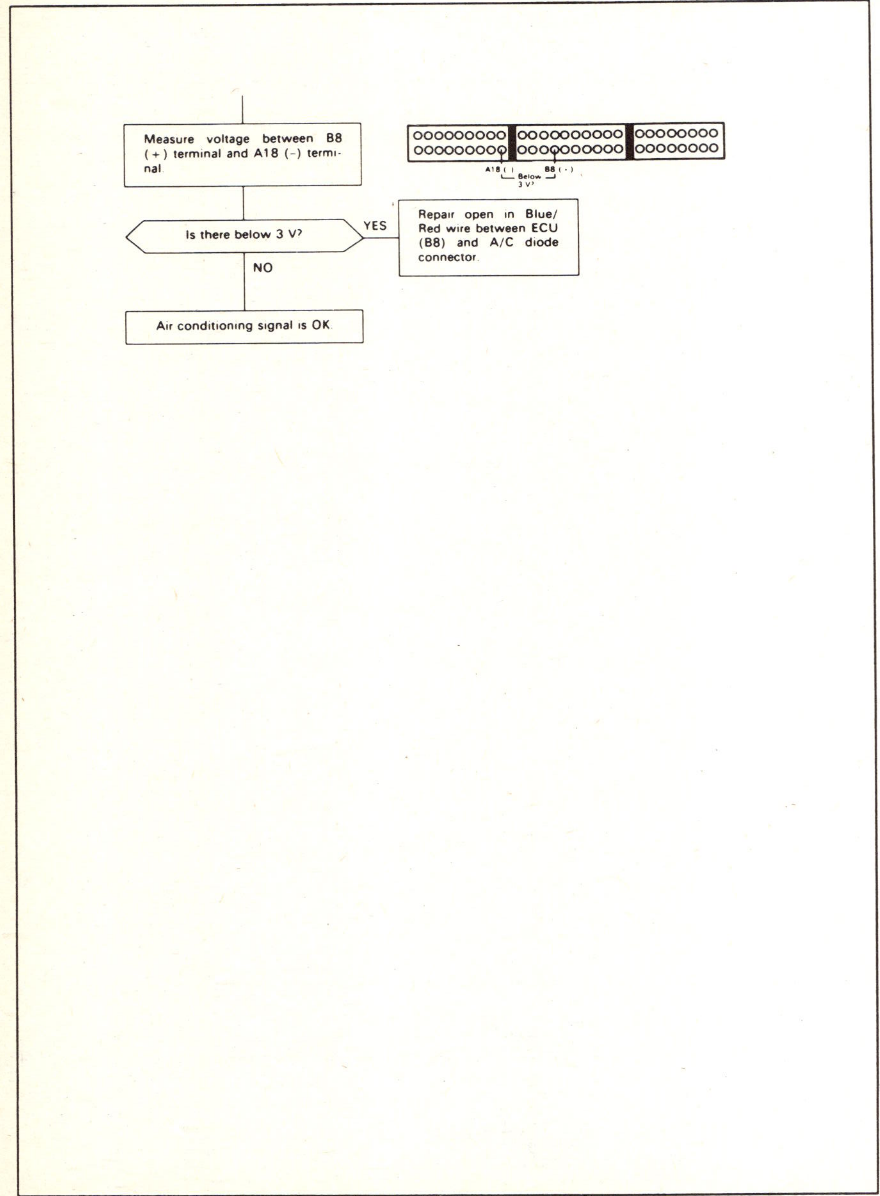

## POWER STEERING PRESSURE SIGNAL—1989 ACCORD

This signals the PGM-FI ECU when the power steering load is high.

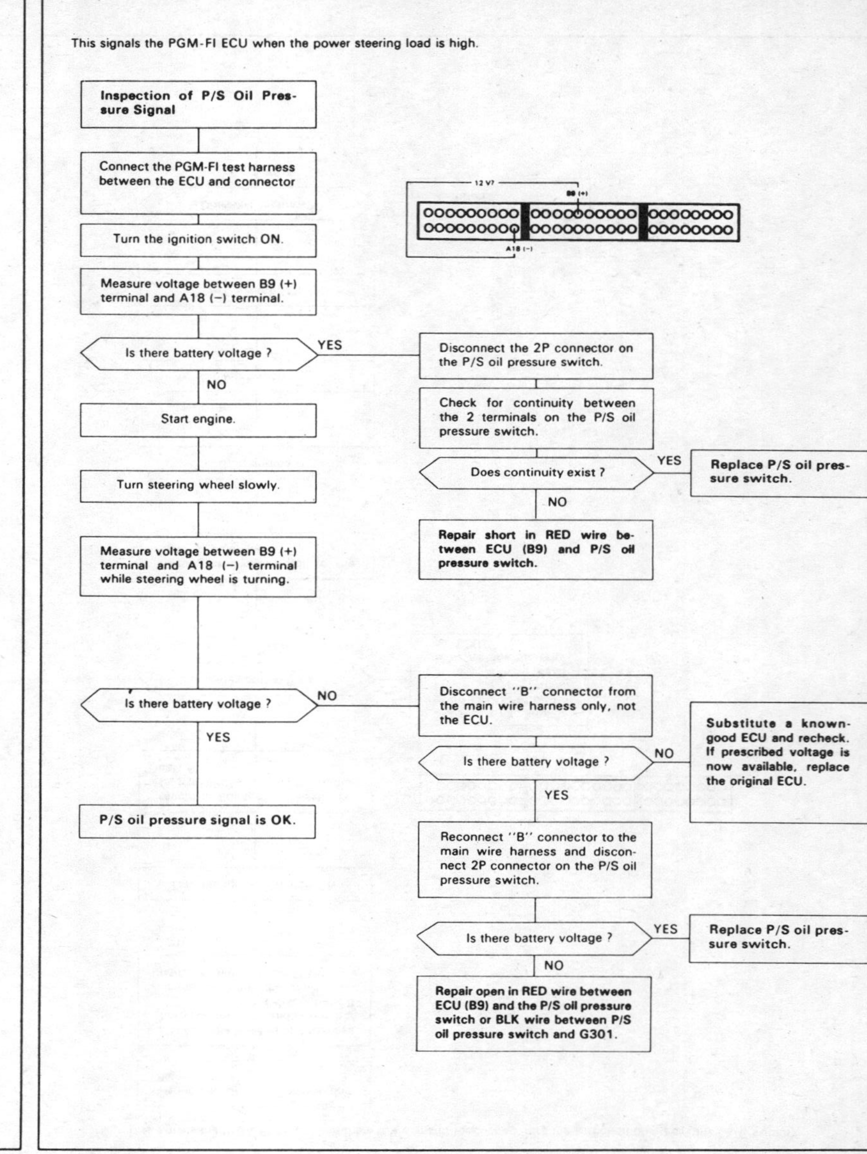


## AIR CONDITIONING SIGNAL—1989 ACCORD

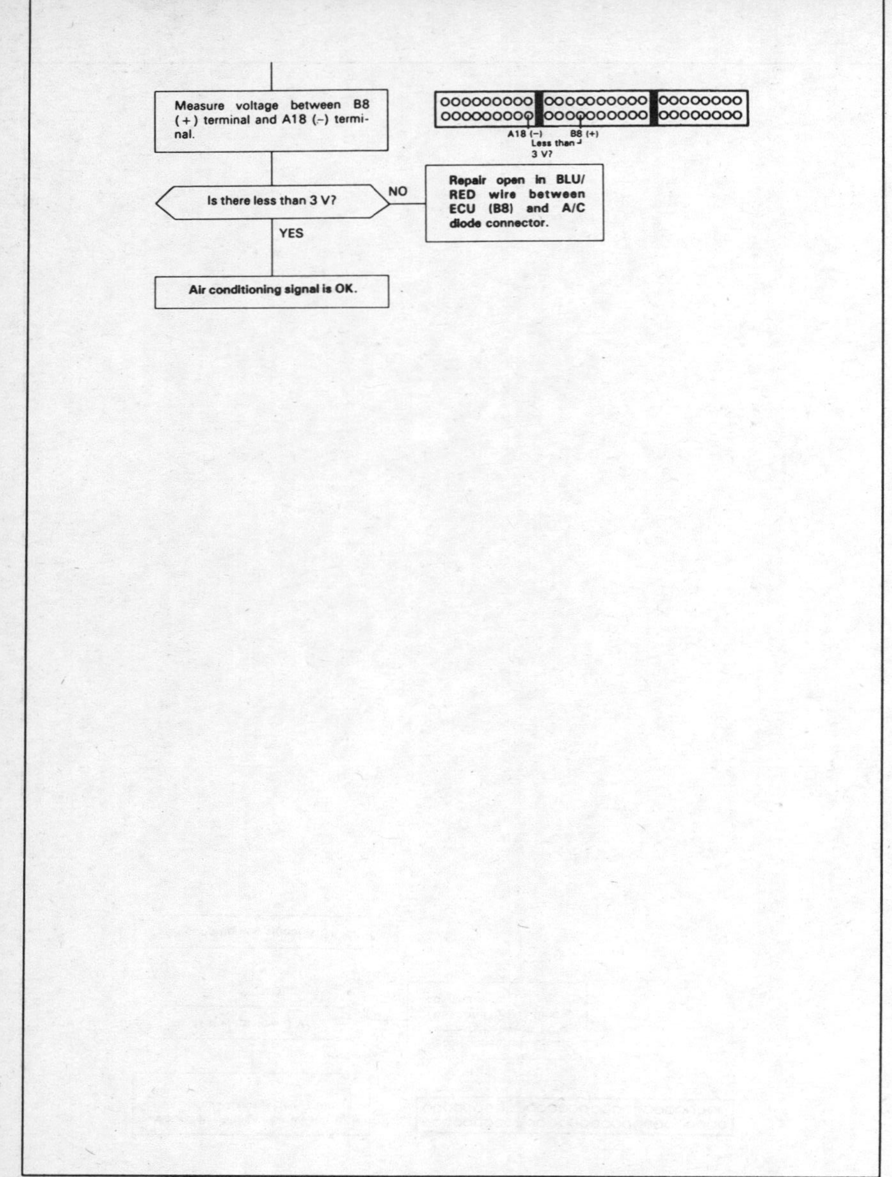

## BYPASS CONTROL SYSTEM—1988 ACCORD

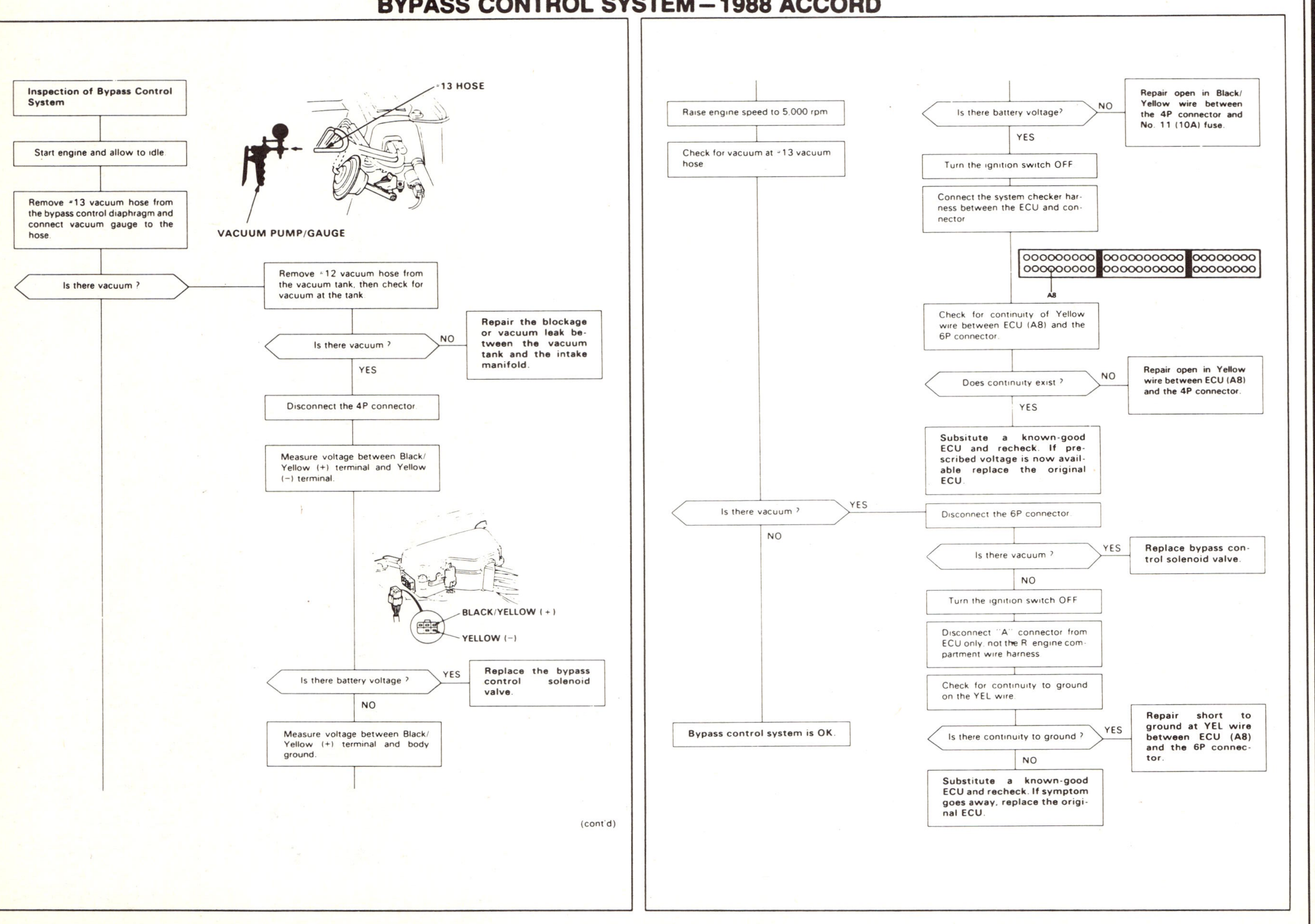

## BYPASS CONTROL SYSTEM – 1989 ACCORD

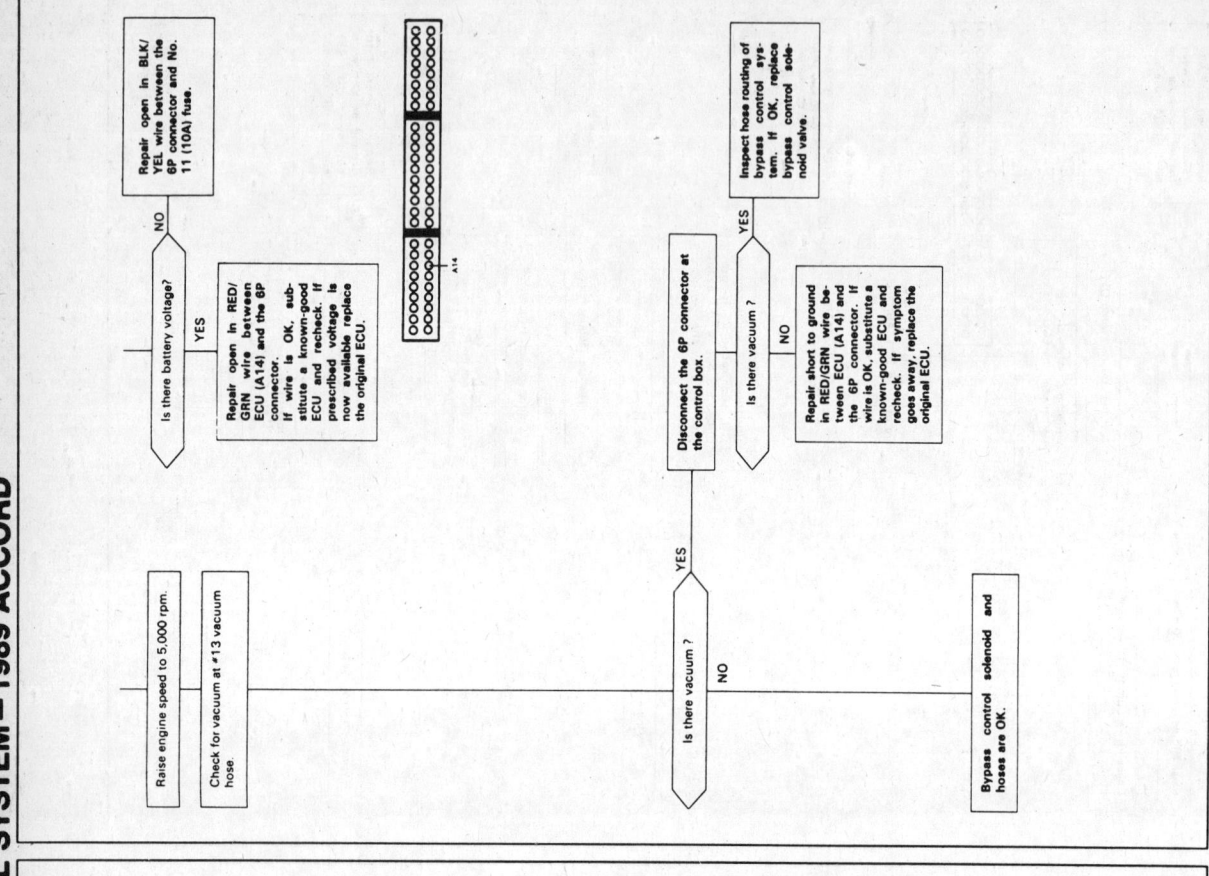

- Inspection of Bypass Control Solenoid
- Start engine and allow to idle.
- Remove #13 vacuum hose from the bypass control diaphragm and connect vacuum gauge to the hose.

*13 HOSE
VACUUM PUMP/GAUGE

**Is there manifold vacuum?**

- NO → Disconnect the 6P connector at the control box.
  - Measure voltage between BLK/YEL (+) terminal and RED/GRN (–) terminal.

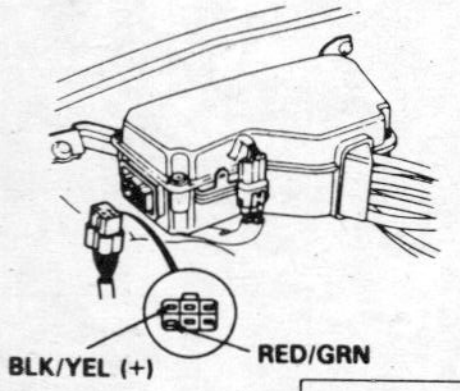

  BLK/YEL (+)
  RED/GRN

  **Is there battery voltage?**
  - YES → Inspect hose routing of bypass control system. If OK, replace the bypass control solenoid valve.
  - NO → Measure voltage between BLK/YEL (+) terminal and body ground.

    **Is there battery voltage?**
    - NO → Repair open in BLK/YEL wire between the 6P connector and No. 11 (10A) fuse.
    - YES → Repair open in RED/GRN wire between ECU (A14) and the 6P connector. If wire is OK, substitute a known-good ECU and recheck. If prescribed voltage is now available replace the original ECU.

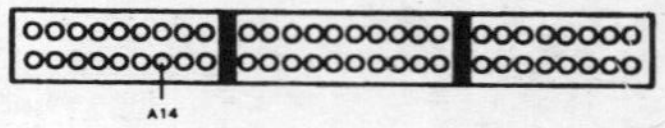

    A14

- YES → Raise engine speed to 5,000 rpm.
  - Check for vacuum at #13 vacuum hose.

  **Is there vacuum?**
  - NO → Bypass control solenoid and hoses are OK.
  - YES → Disconnect the 6P connector at the control box.

    **Is there vacuum?**
    - YES → Inspect hose routing of bypass control system. If OK, replace bypass control solenoid valve.
    - NO → Repair short to ground in RED/GRN wire between ECU (A14) and the 6P connector. If wire is OK, substitute a known-good ECU and recheck. If symptom goes away, replace the original ECU.

## TROUBLESHOOTING GUIDE – 1988–89 ACCORD

NOTE: Across each row in the chart, the sub systems that could be sources of a symptom are ranked in the order they should be inspected, starting with ①. Find the system in the left column, read across to the most likely source, then refer to the page listed at the top of that column. If inspection shows the system is OK, try the next system ②, etc.

| SUB SYSTEM / SYMPTOM | CATALYTIC CONVERTER | EGR SYSTEM | POSITIVE CRANKCASE VENTILATION SYSTEM | EVAPORATIVE EMISSION CONTROLS |
|---|---|---|---|---|
| | | | | |
| IRREGULAR IDLING | | ① | ② | |
| FREQUENT STALLING | | ① | | |
| FAILS EMISSION TEST | ① | ① | | ② |
| IDLE SPEED ABOVE SPECIFIED rpm | | | ① | |
| LOSS OF POWER | ① | ② | | |

## EGR SYSTEM – 1988–89 ACCORD

Self diagnosis indicator blinks twelve times: Most likely a problem in the Exhaust Gas Recirculation (EGR) system

The EGR System is designed to reduce oxides of nitrogen emissions (NOx), by recirculating exhaust gas through the EGR valve and the intake manifold into the combustion chambers. It is composed of the EGR valve, CVC valve, EGR control solenoid valve, PGM-FI ECU and various sensors.
The ECU contains memories for ideal EGR valve lifts for varying operating conditions. The EGR valve lift sensor detects the amount of EGR valve lift and sends the information to the ECU. The ECU then compares it with the ideal EGR valve lift which is determined by signals sent from the other sensors. If there is any difference between the two, the ECU cuts current to the EGR control solenoid valve to reduce vacuum applied to the EGR valve.

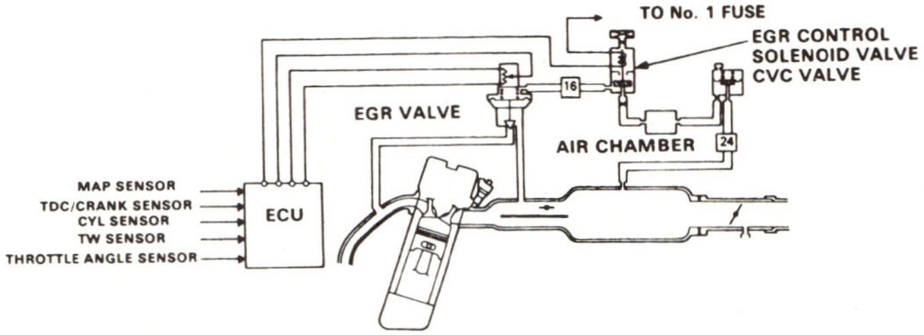

– Check Engine warning light has been reported on.
– LED indicates CODE 12.

Verify that all connectors between the EGR valve lift sensor and the PGM-FI ECU are securely connected.

Warm up engine to normal operating temperature (cooling fan comes on).

Disconnect #16 hose from EGR valve.

Connect a vacuum pump/gauge to EGR valve and apply vacuum.

Does EGR valve hold vaccum and did engine stall?

NO → Replace EGR valve.

YES

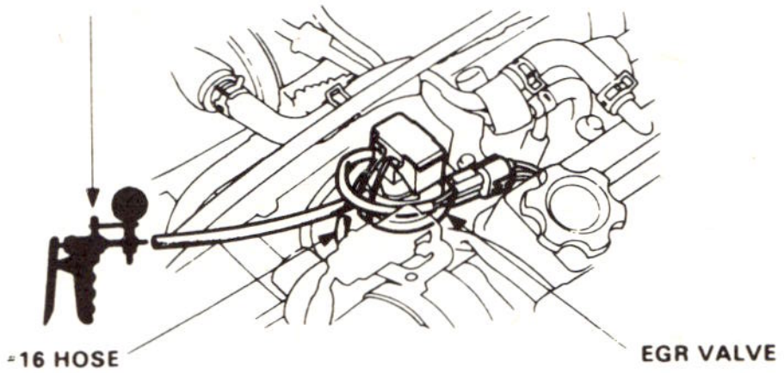

## EGR SYSTEM – 1988–89 ACCORD

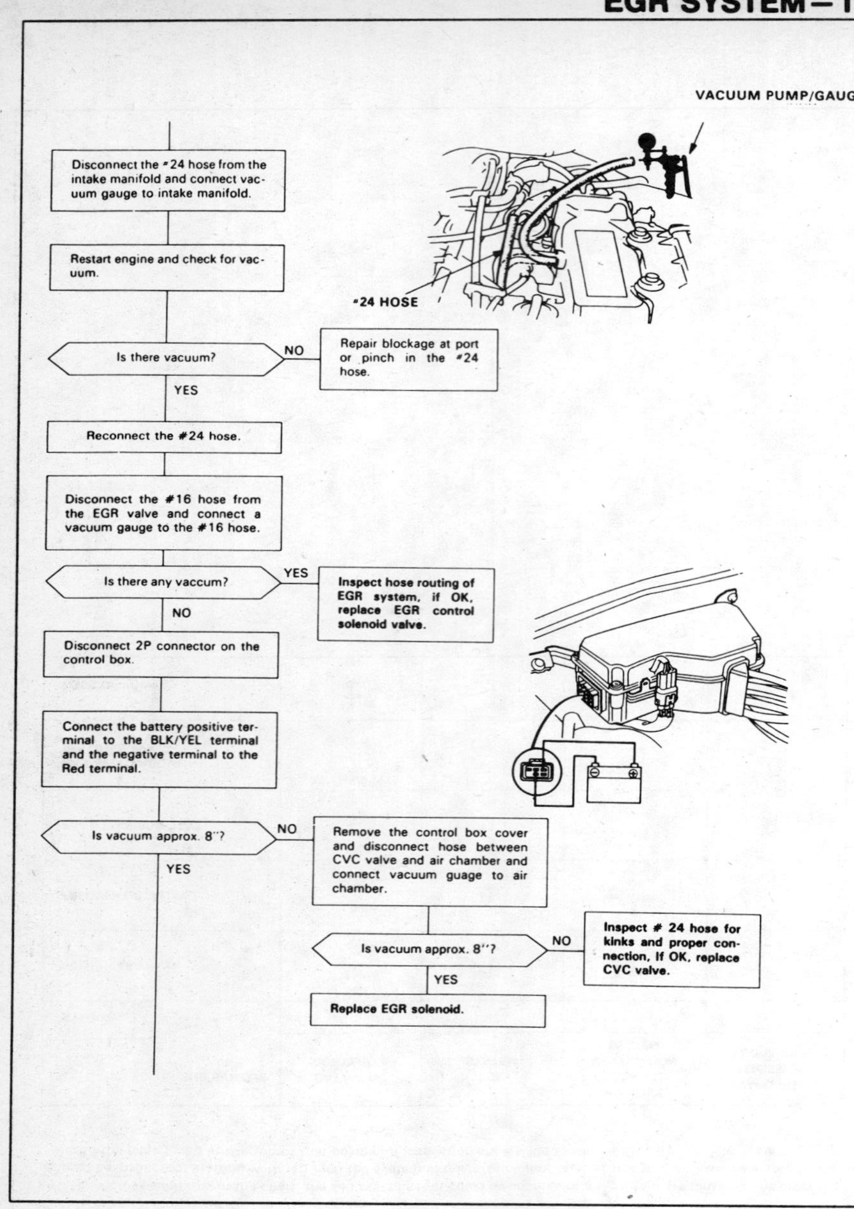

VACUUM PUMP/GAUGE
Disconnect the #24 hose from the intake manifold and connect vacuum gauge to intake manifold.
Restart engine and check for vacuum.
#24 HOSE
Is there vacuum?
NO
Repair blockage at port or pinch in the #24 hose.
YES
Reconnect the #24 hose.
Disconnect the #16 hose from the EGR valve and connect a vacuum gauge to the #16 hose.
Is there any vaccum?
YES
Inspect hose routing of EGR system, if OK, replace EGR control solenoid valve.
NO
Disconnect 2P connector on the control box.
Connect the battery positive terminal to the BLK/YEL terminal and the negative terminal to the Red terminal.
Is vacuum approx. 8''?
NO
Remove the control box cover and disconnect hose between CVC valve and air chamber and connect vacuum guage to air chamber.
Is vacuum approx. 8''?
NO
Inspect # 24 hose for kinks and proper connection, If OK, replace CVC valve.
YES
Replace EGR solenoid.
YES


- Measure voltage between BLK/YEL (+) terminal on the main wire harness and body ground.
- Is there battery voltage?
  - NO: Repair open in BLK/YEL wire between the solenoid valve and No. 1 fuse.
  - YES: Turn the ignition switch OFF.
- Connect the test harness between the EGR valve lift sensor and engine wire harness.
- Turn the ignition switch ON.
- Measure voltage between RED (+) terminal and GRN (–) terminal.
- Is there approx. 5 V?
  - NO: Measure voltage between RED (+) terminal and body ground.
    - Is there approx. 5 V?
      - YES: Repair open in GRN/WHT wire between ECU (C12) and the sensor.
      - NO: Turn the ignition switch OFF.
  - YES: Measure voltage between WHT (+) terminal and GRN (–) terminal.
- Is there approx. 1 V?
  - NO: –Replace EGR valve. –Repair short in YEL wire between ECU (C8) and EGR lift sensor
  - YES

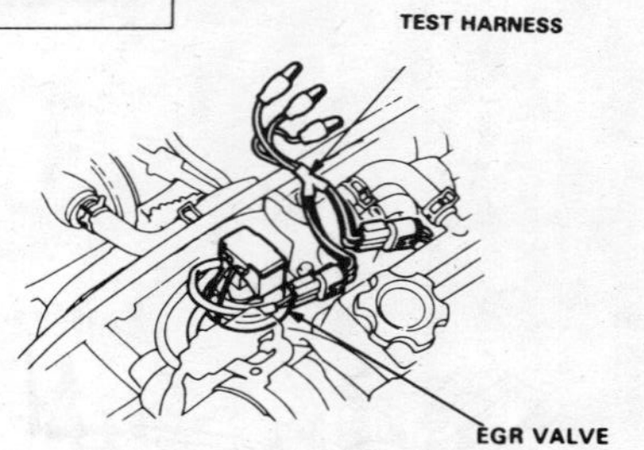

TEST HARNESS

EGR VALVE

## EGR SYSTEM—1988–89 ACCORD

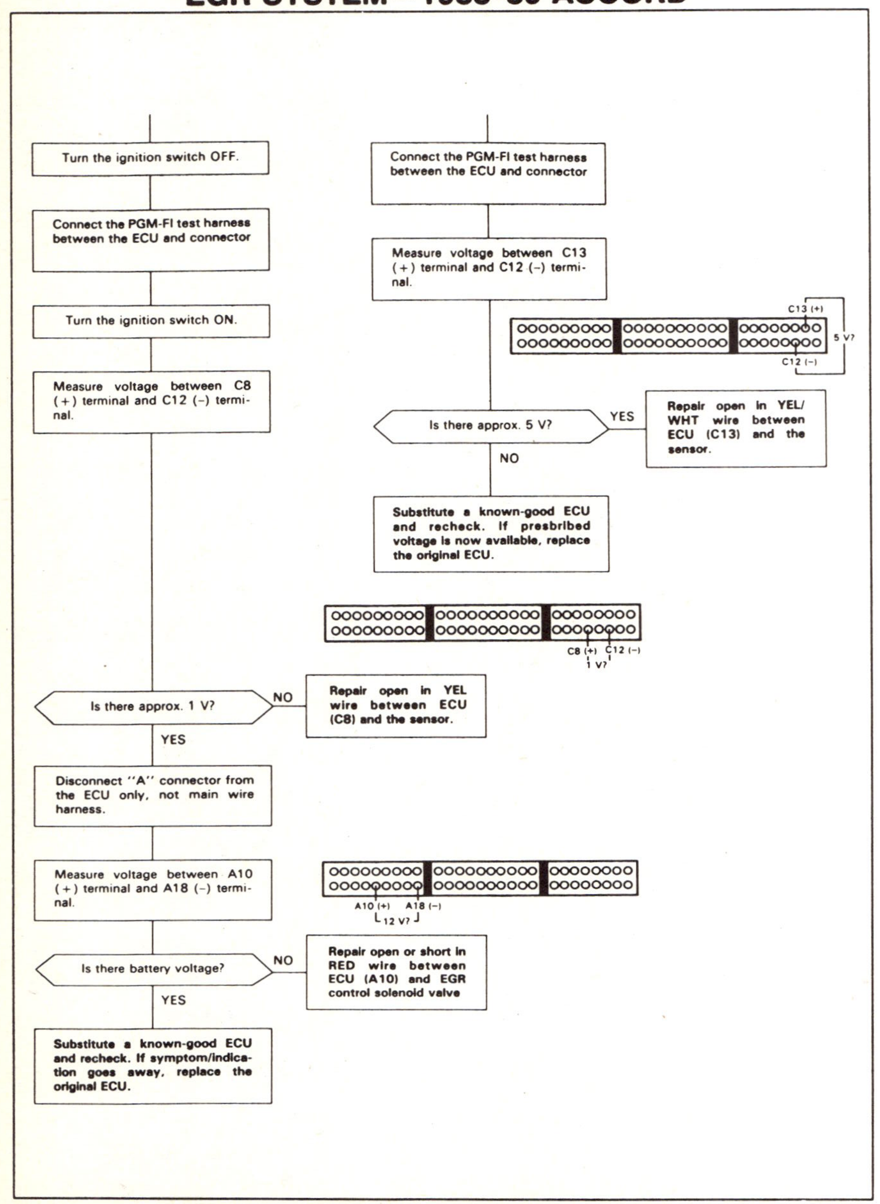


## PURGE CUT-OFF SOLENOID VALVE—1988 ACCORD

When the engine coolant temperature is above the set temperature of the TW sensor, the purge cut-off solenoid valve is activated by the control unit receiving signals from each sensor

**Inspection of Purge Cut-off Solenoid Valve**

Remove the control box and open the control box lid.

Disconnect the upper vacuum hose of the solenoid valve from the three-way joint.

VACUUM PUMP/GAUGE

Start the engine

NOTE Engine coolant temperature must be below 55°C (131°F)

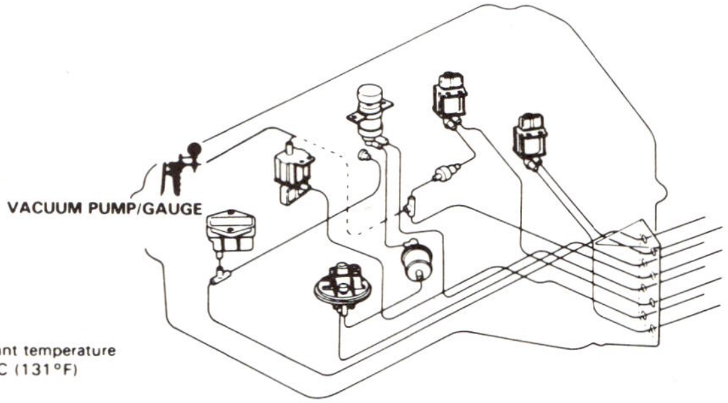

Apply vacuum

Does solenoid valve hold vacuum?

YES — Turn the ignition switch OFF

Disconnect the 6P connector on the control box

Start the engine.

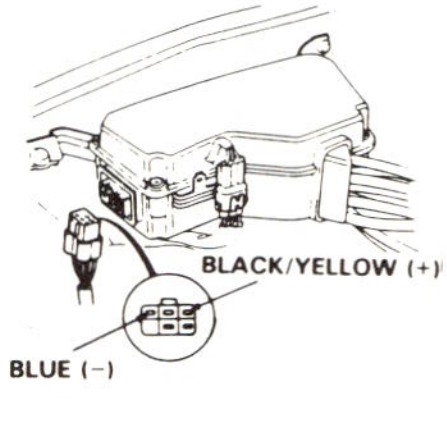

BLACK/YELLOW (+)

BLUE (−)

NO

## PURGE CUT-OFF SOLENOID VALVE—1988 ACCORD

When the engine coolant temperature is below the set temperature of the TW sensor, the purge cut-off solenoid valve is energized by the control unit to cut-off vacuum to the purge control diaphragm on the charcoal canister.

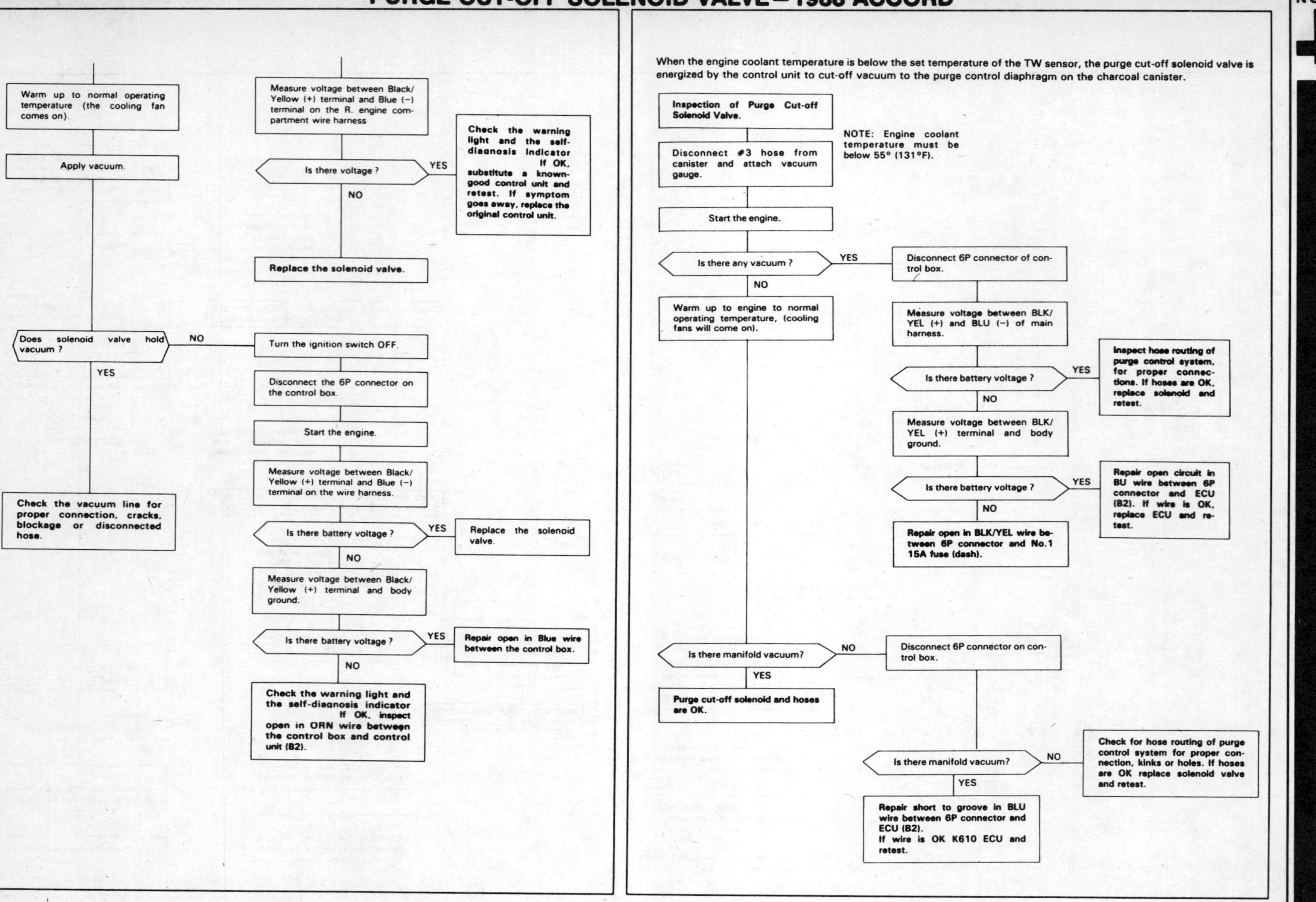

Inspection of Purge Cut-off Solenoid Valve.
Disconnect #3 hose from canister and attach vacuum gauge.
NOTE: Engine coolant temperature must be below 55° (131°F).
Start the engine.
Is there any vacuum ?
YES
Disconnect 6P connector of control box.
Measure voltage between BLK/YEL (+) and BLU (−) of main harness.
Is there battery voltage ?
YES
Inspect hose routing of purge control system, for proper connections. If hoses are OK, replace solenoid and retest.
NO
Measure voltage between BLK/YEL (+) terminal and body ground.
Is there battery voltage ?
YES
Repair open circuit in BU wire between 6P connector and ECU (B2). If wire is OK, replace ECU and retest.
NO
Repair open in BLK/YEL wire between 6P connector and No.1 15A fuse (dash).
NO
Warm up to engine to normal operating temperature, (cooling fans will come on).
Is there manifold vacuum?
YES
Purge cut-off solenoid and hoses are OK.
NO
Disconnect 6P connector on control box.
Is there manifold vacuum?
YES
Repair short to groove in BLU wire between 6P connector and ECU (B2). If wire is OK K610 ECU and retest.
NO
Check for hose routing of purge control system for proper connection, kinks or holes. If hoses are OK replace solenoid valve and retest.
Warm up to normal operating temperature (the cooling fan comes on).
Apply vacuum.
Does solenoid valve hold vacuum ?
YES
Check the vacuum line for proper connection, cracks, blockage or disconnected hose.
NO
Turn the ignition switch OFF.
Disconnect the 6P connector on the control box.
Start the engine.
Measure voltage between Black/Yellow (+) terminal and Blue (−) terminal on the wire harness.
Is there battery voltage ?
YES
Replace the solenoid valve.
NO
Measure voltage between Black/Yellow (+) terminal and body ground.
Is there battery voltage ?
YES
Repair open in Blue wire between the control box.
NO
Check the warning light and the self-diagnosis indicator If OK, inspect open in ORN wire between the control box and control unit (B2).
Measure voltage between Black/Yellow (+) terminal and Blue (−) terminal on the R. engine compartment wire harness
Is there voltage ?
YES
Check the warning light and the self-diagnosis indicator If OK, substitute a known-good control unit and retest. If symptom goes away, replace the original control unit.
NO
Replace the solenoid valve.

# TROUBLESHOOTING GUIDE—1990 ACCORD

NOTE: Across each row in the chart, the systems that could be sources of a symptom are ranked in the order they should be inspected starting with ①. Find the symptom in the left column, read across to the most likely source, then refer to the page listed at the top of that column. If inspection shows the system is OK, try the next most likely system ②, etc.

| SYMPTOM \ SYSTEM | | PGM-FI: ECU | PGM-FI: OXYGEN SENSOR | PGM-FI: MANIFOLD ABSOLUTE PRESSURE SENSOR | PGM-FI: TDC/CRANK/CYL SENSOR | PGM-FI: COOLANT TEMPERATURE SENSOR | PGM-FI: THROTTLE ANGLE SENSOR | PGM-FI: INTAKE AIR TEMPERATURE SENSOR | PGM-FI: ATMOSPHERIC PRESSURE SENSOR |
|---|---|---|---|---|---|---|---|---|---|
| CHECK ENGINE WARNING LIGHT TURNS ON | | □ or ☼ | CHECK (blinking) | CHECK (blinking) | CHECK (blinking) | CHECK (blinking) | CHECK (blinking) | CHECK (blinking) | CHECK (blinking) |
| SELF-DIAGNOSIS INDICATOR (LED) BLINKS | | ⓪ or ✱ | ① or ㊶ or ㊸ | ③ or ⑤ | ④ or ⑧ or ⑨ | ⑥ | ⑦ | ⑩ | ⑬ |
| ENGINE WON'T START | | ③ | | | ③ | | | | |
| DIFFICULT TO START ENGINE WHEN COLD | | BU | | ③ | ③ | ① | | | ③ |
| IRREGULAR IDLING | WHEN COLD FAST IDLE OUT OF SPEC | BU | | | | ③ | | | |
| | ROUGH IDLE | BU | | ③ | | | | | |
| | WHEN WARM RPM TOO HIGH | BU | | | | | | | |
| | WHEN WARM RPM TOO LOW | BU | | | | | | | |
| FREQUENT STALLING | WHILE WARMING UP | BU | | | | ③ | | | |
| | AFTER WARMING UP | BU | | | | | | | ③ |
| POOR PERFORMANCE | MISFIRE OR ROUGH RUNNING | BU | | | ③ | | | | |
| | FAILS EMISSION TEST | BU | ③ | ② | | | | | |
| | LOSS OF POWER | BU | | ③ | | | ② | | |

| SYMPTOM | | PGM-FI: IGNITION OUTPUT SIGNAL | PGM-FI: VEHICLE SPEED SENSOR | PGM-FI: ELECTRIC LOAD DETECTOR | PGM-FI: A/T FI SIGANL A | PGM-FI: A/T FI SIGNAL B | IDLE CONTROL: ELECTRONIC AIR CONTROL VALVE | IDLE CONTROL: OTHER IDLE CONTROLS | FUEL SUPPLY: FUEL INJECTOR | FUEL SUPPLY: OTHER FUEL SUPPLY | AIR INTAKE | EMISSION CONTROL: EGR CONTROL SYSTEM | EMISSION CONTROL: OTHER EMISSION CONTROLS |
|---|---|---|---|---|---|---|---|---|---|---|---|---|---|
| CHECK ENGINE WARNING LIGHT TURNS ON | | CHECK (blinking) | CHECK (blinking) | □ | □ | □ | CHECK (blinking) | | CHECK (blinking) | | | CHECK (blinking) | |
| SELF-DIAGNOSIS INDICATOR (LED) BLINKS | | ⑮ | ⑰ | ⑳ | ㉚ | ㉛ | ⑭ | | ⑯ | | | ⑫ | |
| ENGINE WON'T START | | ① | | | | | | | ② | ③ | | | |
| DIFFICULT TO START ENGINE WHEN COLD | | | | | | | | ② | | | | | |
| IRREGULAR IDLING | WHEN COLD FAST IDLE OUT OF SPEC | | | | | | ① | ② | | | | | |
| | ROUGH IDLE | | | | | | ① | | ② | | | ③ | |
| | WHEN WARM RPM TOO HIGH | | | | | | ① | ② | | | | | |
| | WHEN WARM RPM TOO LOW | | | ③ | | | ① | | ② | | | | |
| FREQUENT STALLING | WHILE WARMING UP | | | | | | ① | ② | | ③ | | | |
| | AFTER WARMING UP | | | | | | ③ | ① | | ② | | ③ | |
| POOR PERFORMANCE | MISFIRE OR ROUGH RUNNING | | | | | | ③ | | ① | | | ③ | |
| | FAILS EMISSION TEST | | | | | | | | ② | ③ | | | ① |
| | LOSS OF POWER | | | | | | | | ③ | ① | ③ | | ③ |

- If codes other than those listed above are indicated, count the number of blinks again. If the indicator is in fact blinking these codes, substitute a known-good ECU and recheck. If the indication goes away, replace the original ECU.

BU: When the Check Engine warning light and the self-diagnosis indicator are on, the back-up system is in operation. Substitute a known-good ECU and recheck. If the indication goes away, replace the original ECU.

# ECU AND CHECK LIGHT—1990 ACCORD

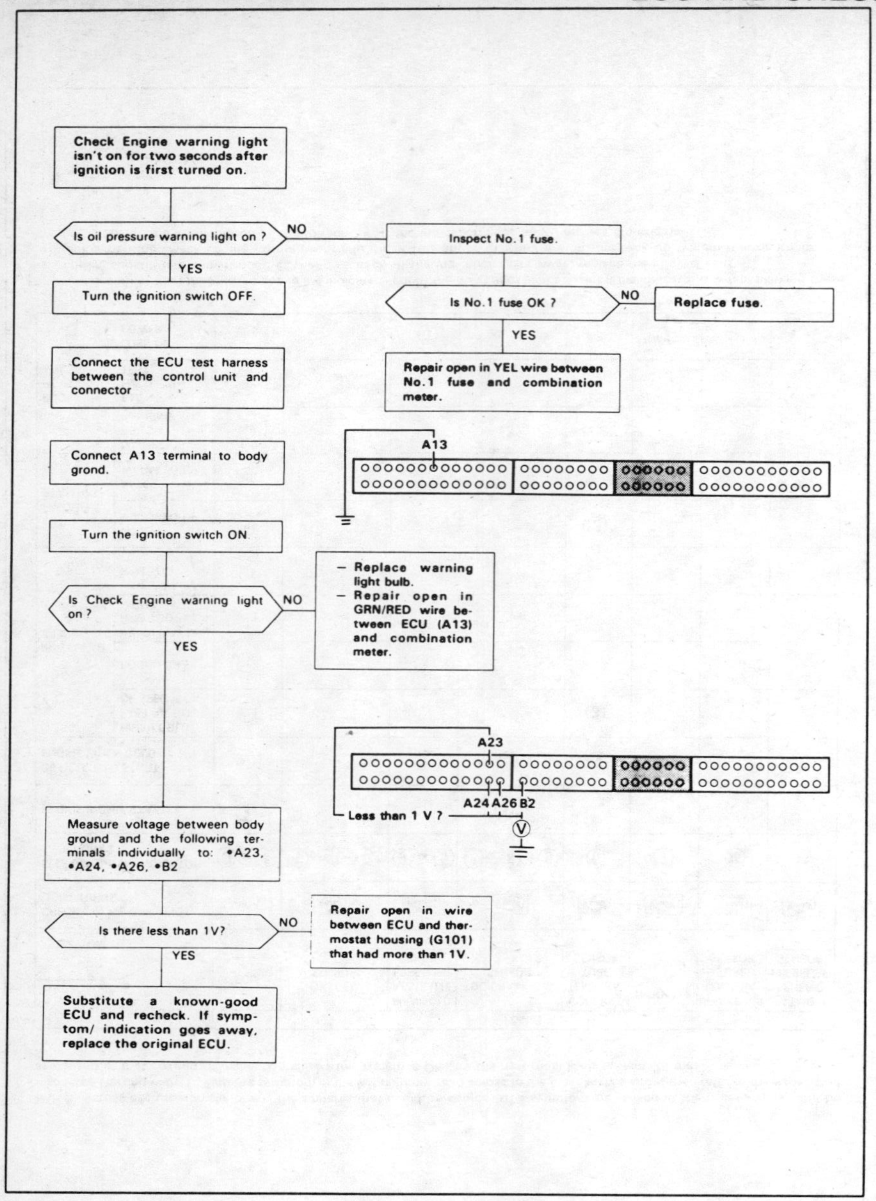

Check Engine warning light isn't on for two seconds after ignition is first turned on.

Is oil pressure warning light on ? NO → Inspect No.1 fuse.

Is No.1 fuse OK ? NO → Replace fuse.

YES → Repair open in YEL wire between No.1 fuse and combination meter.

YES → Turn the ignition switch OFF.

Connect the ECU test harness between the control unit and connector

Connect A13 terminal to body grond.

A13

Turn the ignition switch ON.

Is Check Engine warning light on ? NO → – Replace warning light bulb. – Repair open in GRN/RED wire between ECU (A13) and combination meter.

YES → Measure voltage between body ground and the following terminals individually to: •A23, •A24, •A26, •B2

A23 A24 A26 B2 Less than 1 V ?

Is there less than 1V? NO → Repair open in wire between ECU and thermostat housing (G101) that had more than 1V.

YES → Substitute a known-good ECU and recheck. If symptom/ indication goes away, replace the original ECU.

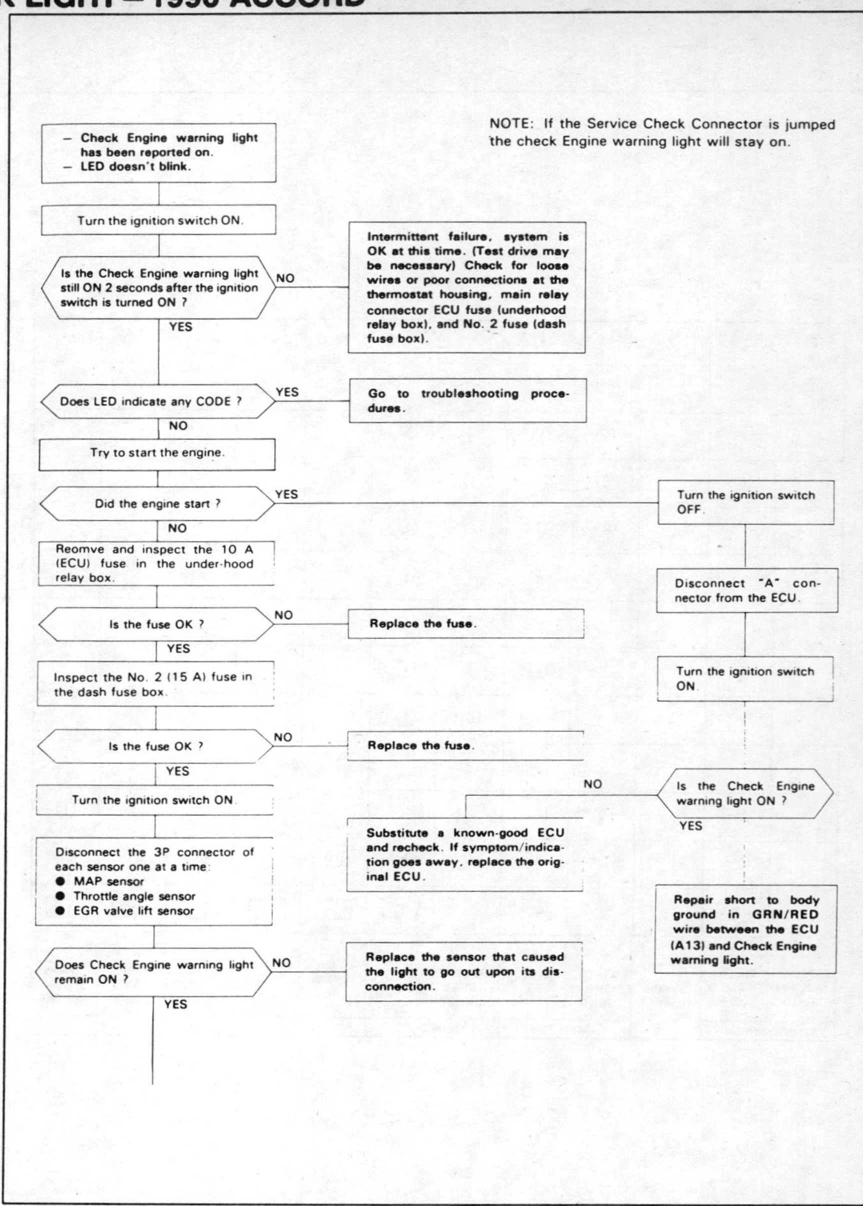

NOTE: If the Service Check Connector is jumped the check Engine warning light will stay on.
– Check Engine warning light has been reported on.
– LED doesn't blink.
Turn the ignition switch ON.
Is the Check Engine warning light still ON 2 seconds after the ignition switch is turned ON ?
NO
Intermittent failure, system is OK at this time. (Test drive may be necessary) Check for loose wires or poor connections at the thermostat housing, main relay connector ECU fuse (underhood relay box), and No. 2 fuse (dash fuse box).
YES
Does LED indicate any CODE ?
YES
Go to troubleshooting procedures.
NO
Try to start the engine.
Did the engine start ?
YES
Turn the ignition switch OFF.
Disconnect "A" connector from the ECU.
Turn the ignition switch ON.
Is the Check Engine warning light ON ?
NO
YES
Repair short to body ground in GRN/RED wire between the ECU (A13) and Check Engine warning light.
NO
Reomve and inspect the 10 A (ECU) fuse in the under-hood relay box.
Is the fuse OK ?
NO
Replace the fuse.
YES
Inspect the No. 2 (15 A) fuse in the dash fuse box.
Is the fuse OK ?
NO
Replace the fuse.
YES
Turn the ignition switch ON.
Disconnect the 3P connector of each sensor one at a time:
● MAP sensor
● Throttle angle sensor
● EGR valve lift sensor
Does Check Engine warning light remain ON ?
NO
Replace the sensor that caused the light to go out upon its disconnection.
YES
Substitute a known-good ECU and recheck. If symptom/indication goes away, replace the original ECU.

## ECU AND CHECK LIGHT—1990 ACCORD

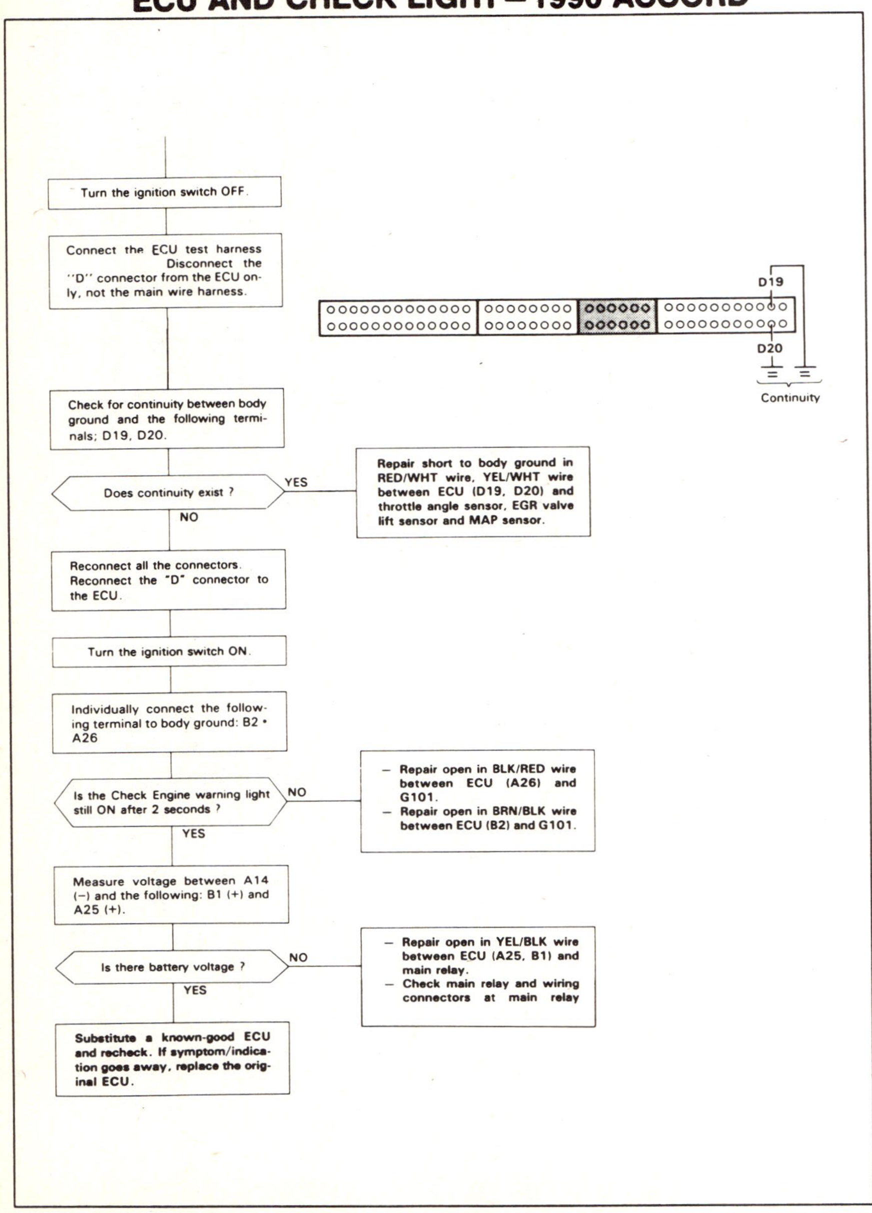

Turn the ignition switch OFF.
Connect the ECU test harness. Disconnect the "D" connector from the ECU only, not the main wire harness.
D19
D20
Continuity
Check for continuity between body ground and the following terminals; D19, D20.
Does continuity exist ?
YES
Repair short to body ground in RED/WHT wire, YEL/WHT wire between ECU (D19, D20) and throttle angle sensor, EGR valve lift sensor and MAP sensor.
NO
Reconnect all the connectors. Reconnect the "D" connector to the ECU.
Turn the ignition switch ON.
Individually connect the following terminal to body ground: B2 • A26
Is the Check Engine warning light still ON after 2 seconds ?
NO
– Repair open in BLK/RED wire between ECU (A26) and G101.
– Repair open in BRN/BLK wire between ECU (B2) and G101.
YES
Measure voltage between A14 (−) and the following: B1 (+) and A25 (+).
Is there battery voltage ?
NO
– Repair open in YEL/BLK wire between ECU (A25, B1) and main relay.
– Check main relay and wiring connectors at main relay
YES
Substitute a known-good ECU and recheck. If symptom/indication goes away, replace the original ECU.


## OXYGEN SENSOR—CODE 1—1990 ACCORD

Self-diagnosis LED indicates code 1: A problem in the Heated Oxygen ($O_2$) Sensor circuit.

The oxygen sensor, detects the oxygen content in the exhaust gas, and inputs the ECU. In operation, the ECU receives the signals from the sensor and varies the duration during which fuel is injected. The oxygen sensor is combined with a heater. The heater stabilizes the sensor output. The oxygen sensor is installed on the exhaust manifold (DX, LX) or in the exhaust pipe (EX).

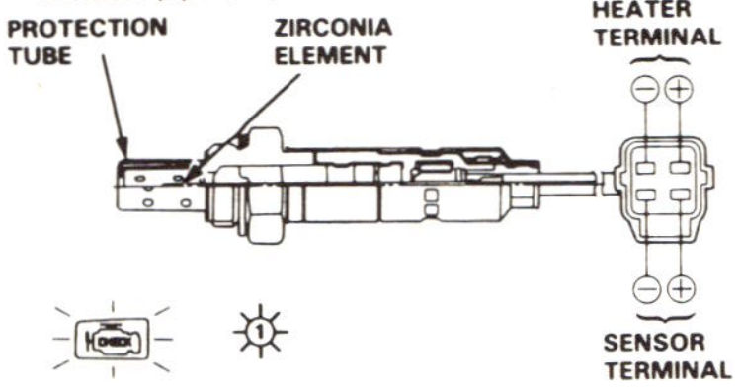

PROTECTION TUBE

ZIRCONIA ELEMENT

HEATER TERMINAL

SENSOR TERMINAL

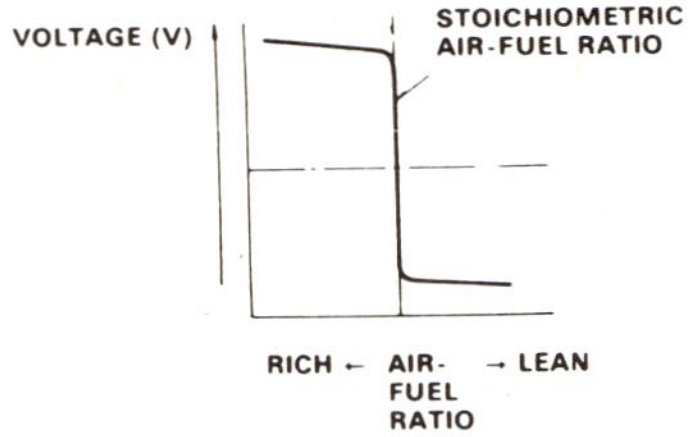

VOLTAGE (V)

STOICHIOMETRIC AIR-FUEL RATIO

RICH ← AIR-FUEL RATIO → LEAN

– Check Engine warning light has been reported on.
– LED indicates CODE 1.

Turn the ignition switch OFF.

Remove BACK UP fuse in the under-hood relay box for 10 seconds to reset ECU.

Warm up engine to normal operating temperature (cooling fan comes on).

Run engine for 60 seconds.

Road test with the transmission in 2nd gear, accelerate using wide open throttle for at least 5 seconds. Then decelerate for at least 5 seconds with the throttle completely closed.

Is Check Engine warning light on and does LED indicate CODE 1?

NO: Intermittent failure, system is OK at this time. Check for poor connections or loose wires at C210 (located at right shock tower) and at the $O_2$ sensor

YES: Go to CODE 43.

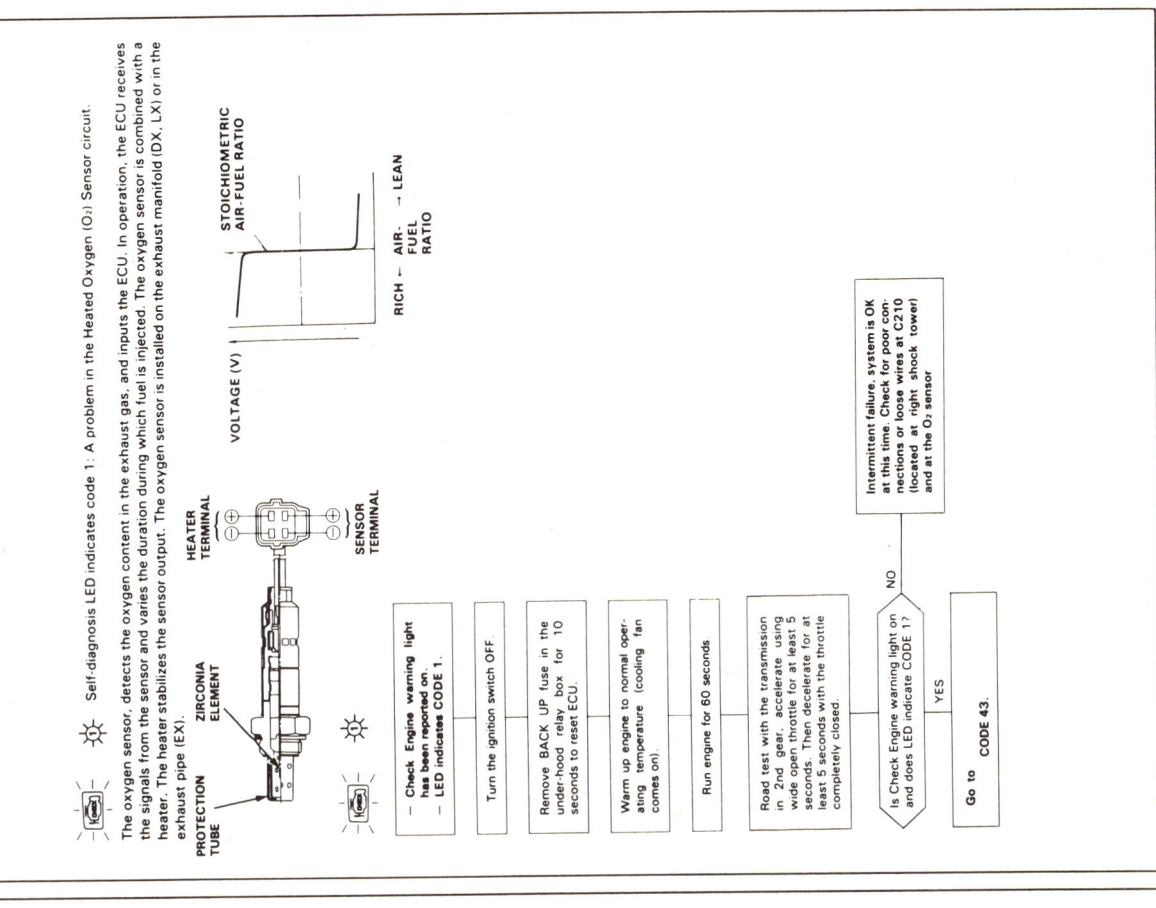

# OXYGEN SENSOR HEATER—CODE 41—1990 ACCORD

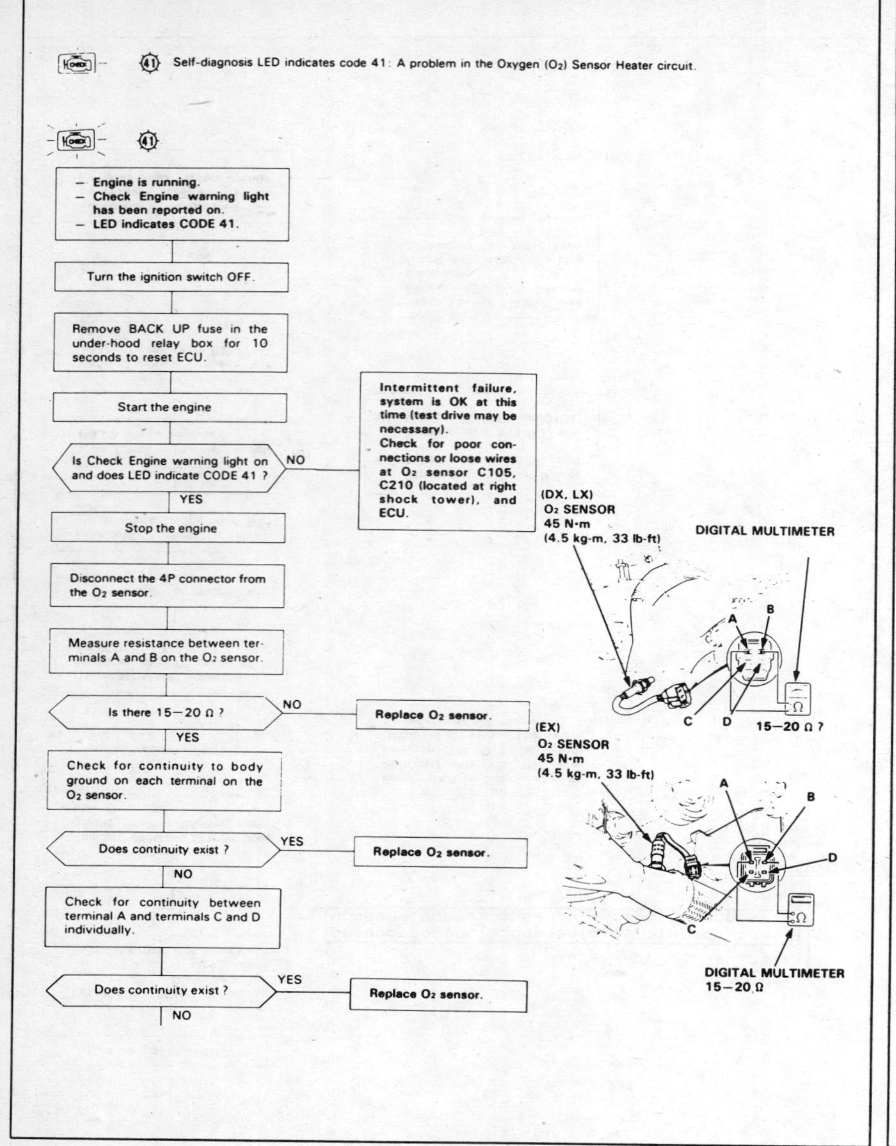


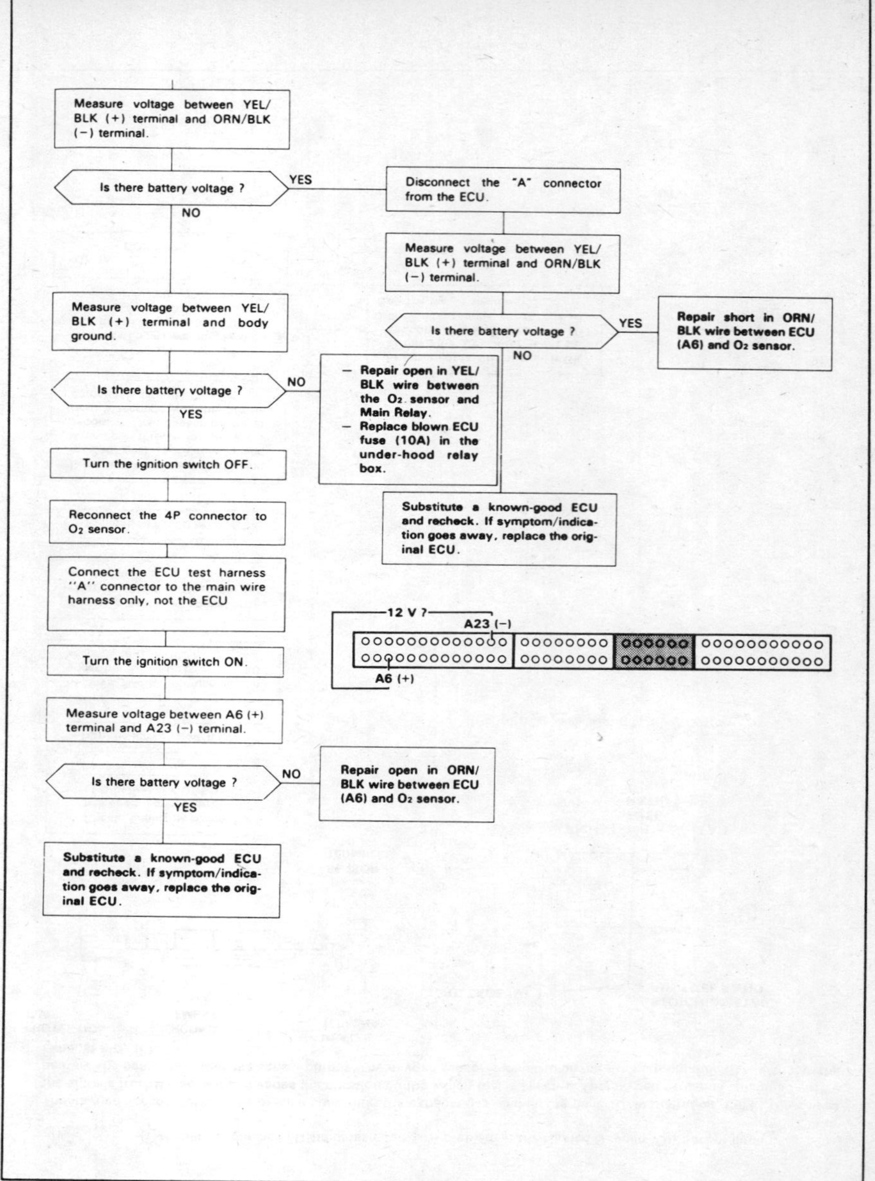

# OXYGEN SENSOR OR FUEL SUPPLY—CODE 43—1990 ACCORD

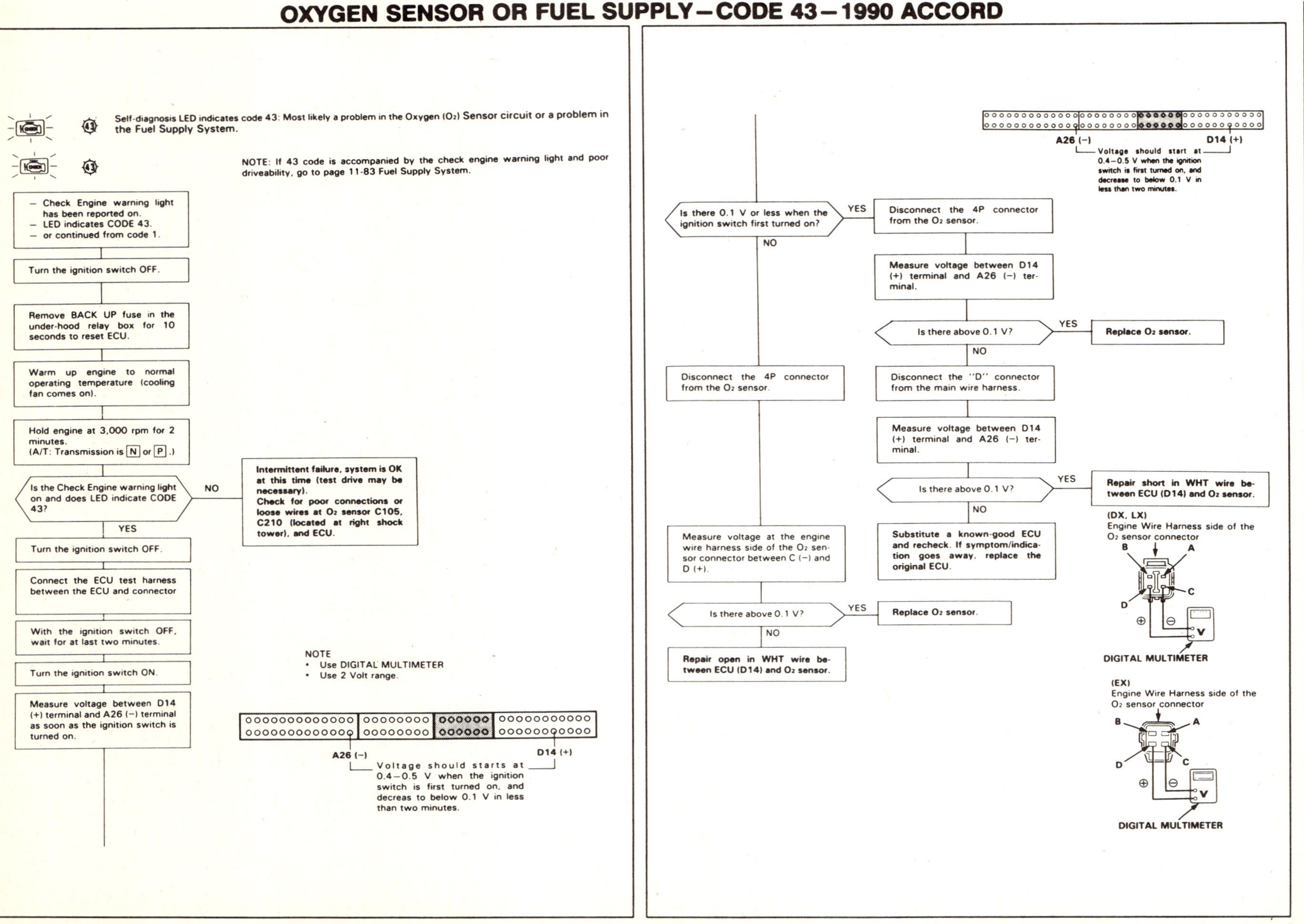

# MAP SENSOR—CODE 3—1990 ACCORD

Self-diagnosis LED indicates code 3: Most likely an electrical problem in the Manifold Absolute Pressure (MAP) Sensor system.

Self-diagnosis LED indicates code 5: Most likely a mechanical problem (broken hose) in the Manifold Absolute Pressure (MAP) Sensor System.

The MAP sensor converts manifold absolute pressure into electrical signals and inputs the ECU.

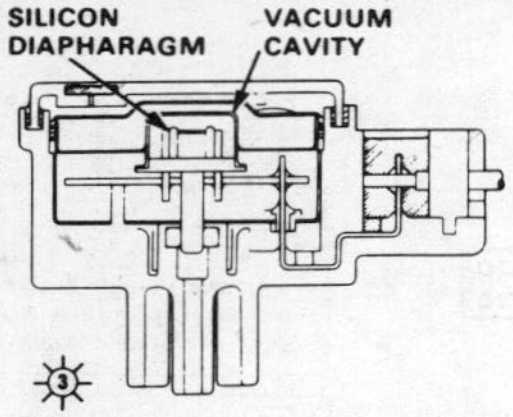

SILICON DIAPHARAGM
VACUUM CAVITY

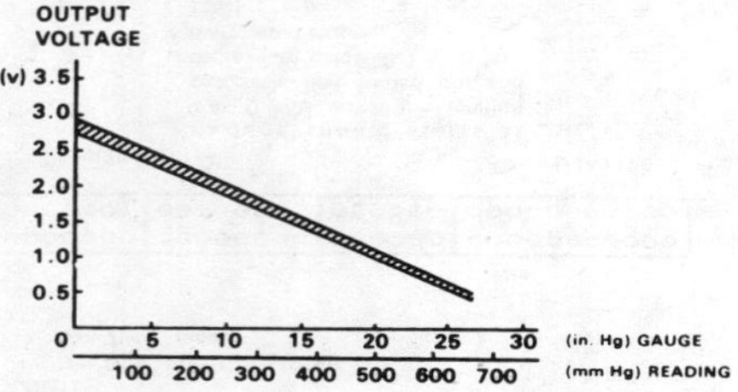

OUTPUT VOLTAGE
(v) 3.5, 3.0, 2.5, 2.0, 1.5, 1.0, 0.5, 0
(in. Hg) GAUGE: 5, 10, 15, 20, 25, 30
(mm Hg) READING: 100, 200, 300, 400, 500, 600, 700

- Engine is warm and running.
- Check Engine warning light has been reported on.
- LED indicates CODE 3.

Turn the ignition switch OFF.

Remove BACK UP fuse in the under-hood relay box for 10 seconds to reset ECU.

Warm up engine to normal operating temperature (cooling fan comes on).

Is Check Engine warning light on and does LED indicate CODE 3?

NO: **Intermittent failure, system is OK at this time (test drive may be necessary). Check for poor connection or loose wires at MAP sensor connector and ECU.**

YES:

Turn the ignition switch OFF.

Disconnect the 3P connector from the MAP sensor.

Turn the ignition switch ON.

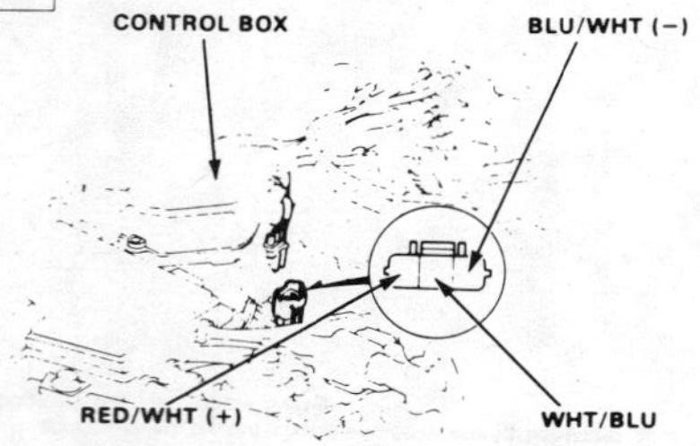

CONTROL BOX
BLU/WHT (−)
RED/WHT (+)
WHT/BLU

Measure voltage between RED/WHT (+) terminal and body ground.

Is there approx. 5V?

NO: **Repair open in RED/WHT wire between ECU (D19) and MAP sensor. If wire is OK, substitute a known-good ECU and recheck. If prescribed voltage is now available, replace the original ECU.**

YES: Measure voltage between RED/WHT (+) terminal and BLU/WHT (−) terminal.

Is there approx. 5V?

NO: **Repair open in BLU/WHT wire between ECU (D21) and MAP sensor. If wire is OK, substitute a known-good ECU and recheck. If prescribed voltage is now available, replace the original ECU.**

YES:

Measure voltage between WHT/BLU (+) terminal and BLU/WHT (−) terminal.

Is there approx. 5V?

NO: **Repair open or short in WHT/BLU wire between EUC (D17) and MAP sensor. If wire is OK, substitute a known-good ECU and recheck. If prescribed voltage is now available, replace the original ECU.**

YES:

Turn the ignition switch OFF.

Reconnect the 3P connector to the MAP sensor.

Connect the ECU test harness between the ECU and connector.

Turn the ignition switch ON.

## MAP SENSOR—CODE 3—1990 ACCORD

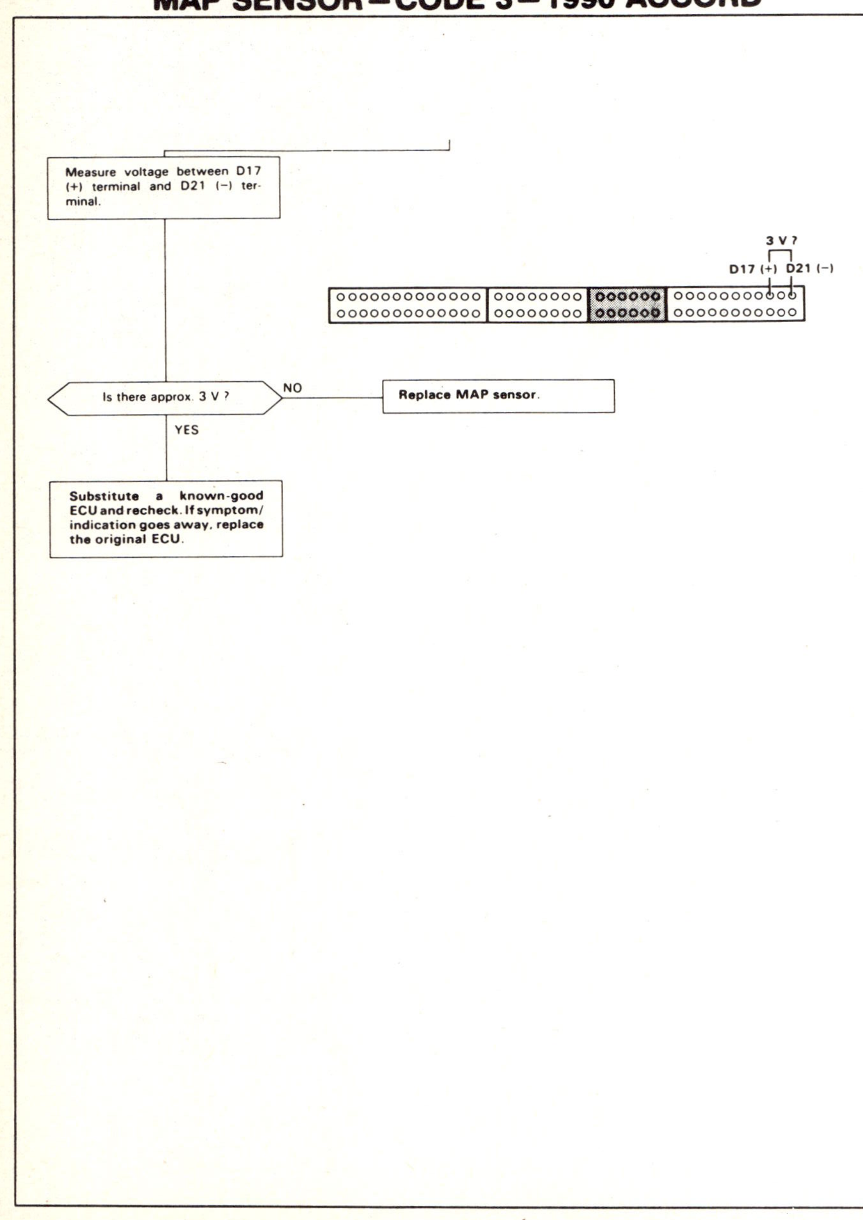


## MAP SENSOR—CODE 5—1990 ACCORD

- Check Engine warning light has been reported on.
- LED indicates CODE 5.

Turn the ignition switch OFF.

Remove BACK UP fuse in the under-hood relay box for 10 seconds to reset ECU.

Start the engine.

Is Check Engine warning light on and does LED indicate CODE 5?

NO → —Intermittent failure, system is OK at this time (test drive may be necessary).
—Check vacuum hoses, pipes and connections.
—Make sure all connectors are secure.

YES → Stop the engine.

Disconnect #21 hose from the throttle body, connect vacuum pump to the hose and apply vacuum.

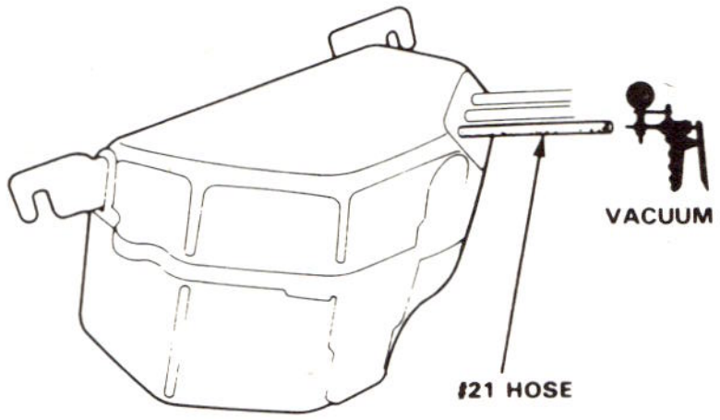

VACUUM PUMP/GAUGE

#21 HOSE

Does it hold vacuum?

NO → Connect a vacuum pump to the MAP sensor and apply vacuum.

Does it hold vacuum?

NO → Replace MAP sensor.

YES → Replace #21 hose.

YES → Connect a T-fitting from a vacuum gauge between the throttle body and MAP sensor.

## MAP SENSOR—CODE 5—1990 ACCORD

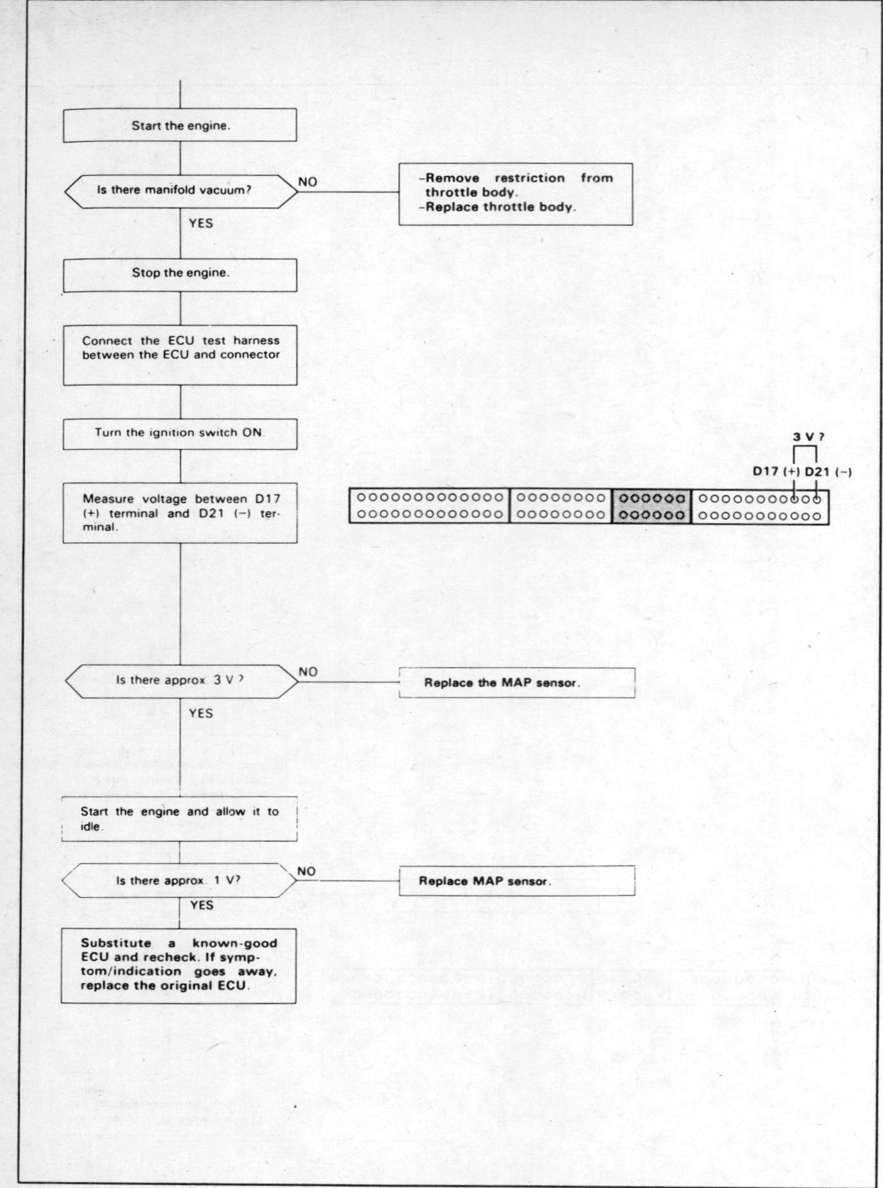


## TDC/CRANK/CYL SENSOR—CODE 4—1990 ACCORD

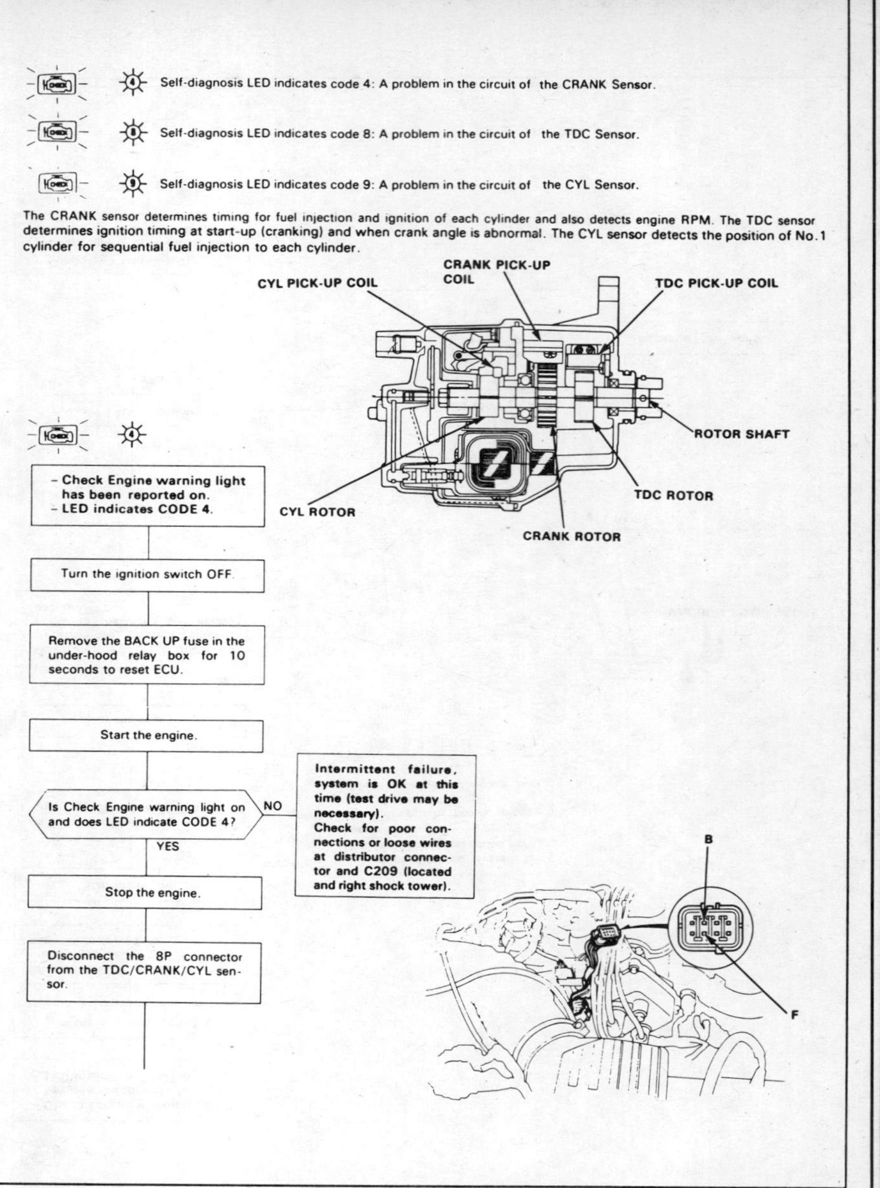

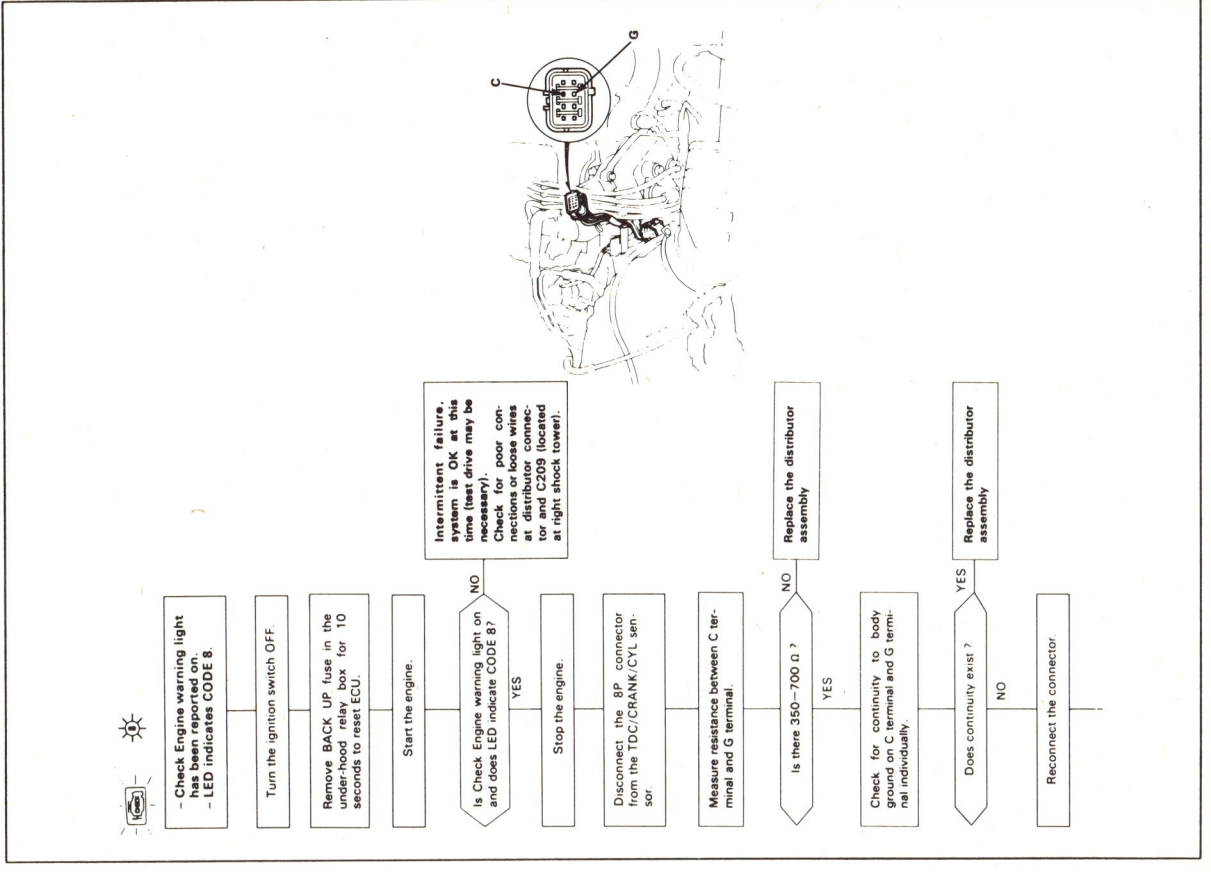

## TDC/CRANK/CYL SENSOR—CODE 4—1990 ACCORD

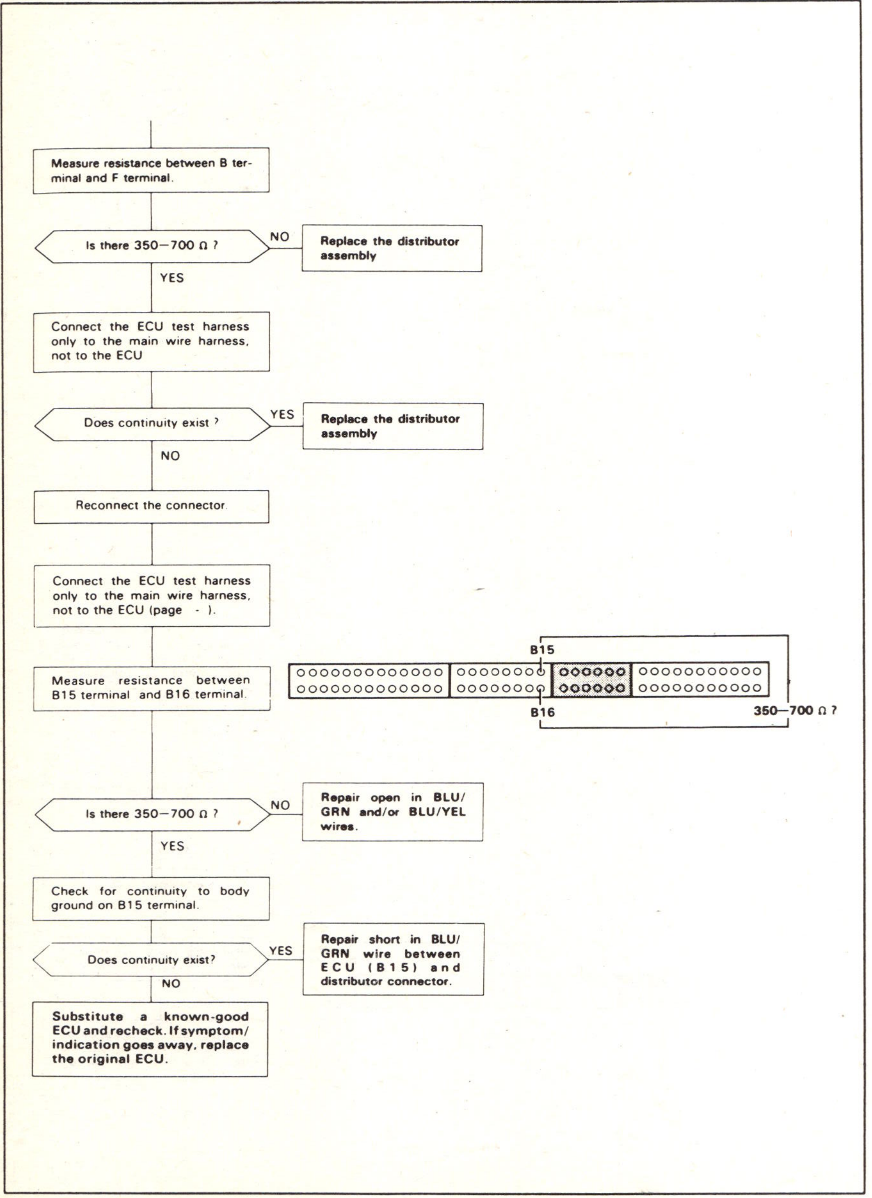


## TDC/CRANK/CYL SENSOR—CODE 8—1990 ACCORD

- Check Engine warning light has been reported on.
- LED indicates CODE 8.

Turn the ignition switch OFF.

Remove BACK UP fuse in the under-hood relay box for 10 seconds to reset ECU.

Start the engine.

Is Check Engine warning light on and does LED indicate CODE 8?

NO → **Intermittent failure, system is OK at this time (test drive may be necessary). Check for poor connections or loose wires at distributor connector and C209 (located at right shock tower).**

YES

Stop the engine.

Disconnect the 8P connector from the TDC/CRANK/CYL sensor.

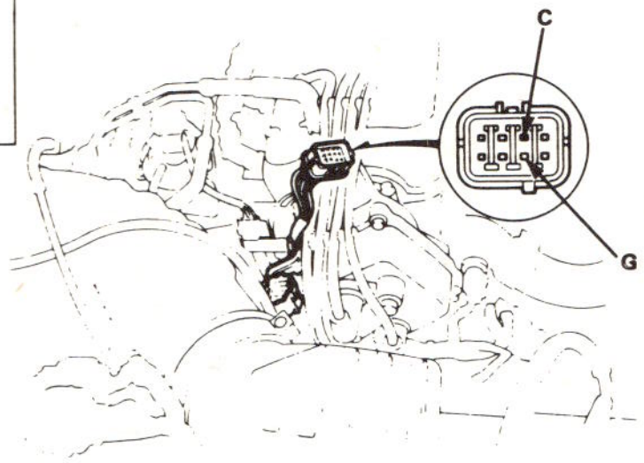

Measure resistance between C terminal and G terminal.

C

G

Is there 350—700 Ω ?

NO → **Replace the distributor assembly**

YES

Check for continuity to body ground on C terminal and G terminal individually.

Does continuity exist ?

YES → **Replace the distributor assembly**

NO

Reconnect the connector.

## TDC/CRANK/CYL SENSOR – CODE 8 – 1990 ACCORD

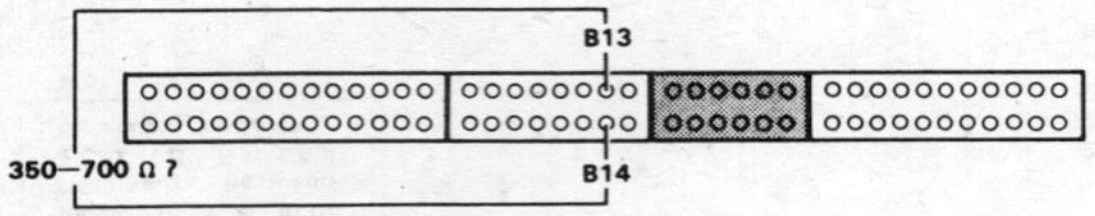

- Connect the ECU test harness only to the main wire harness, not to the ECU (page 11-18).
- Measure resistance between B13 terminal and B14 terminal.
- Is there 350–700 Ω ?
  - NO: Repair open in ORN/BLU and/or WHT/BLU wires.
  - YES: Check for continuity to body ground on B13 terminal.
- Does continuity exist?
  - YES: Repair short in ORN/BLU wire between ECU (B13) and distributor connector.
  - NO: Substitute a known-good ECU and recheck. If symptom/indication goes away, replace the original ECU.

B13
350–700 Ω ?
B14

## TDC/CRANK/CYL SENSOR – CODE 9 – 1990 ACCORD

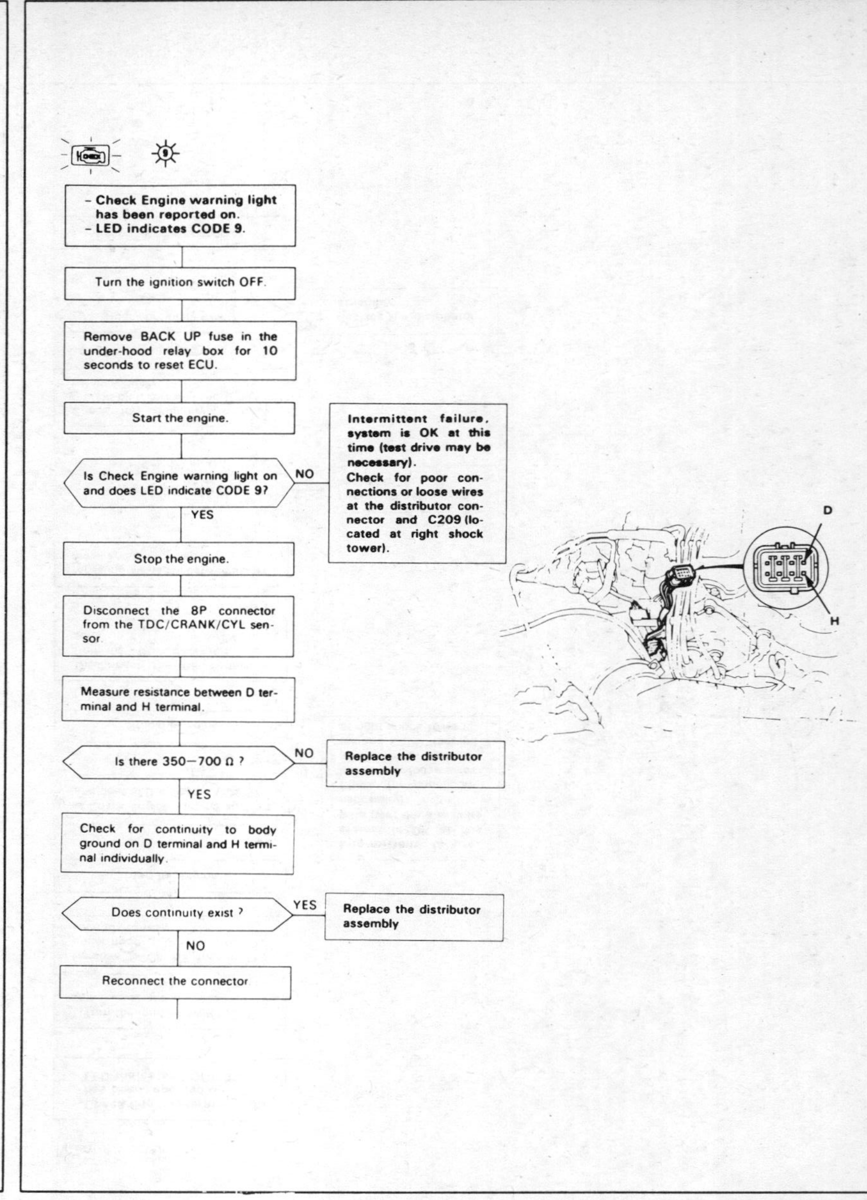

# TDC/CRANK/CYL SENSOR—CODE 9—1990 ACCORD

Connect the ECU test harness only to the main wire harness, not to the ECU

Measure resistance between B11 terminal and B12 terminal.

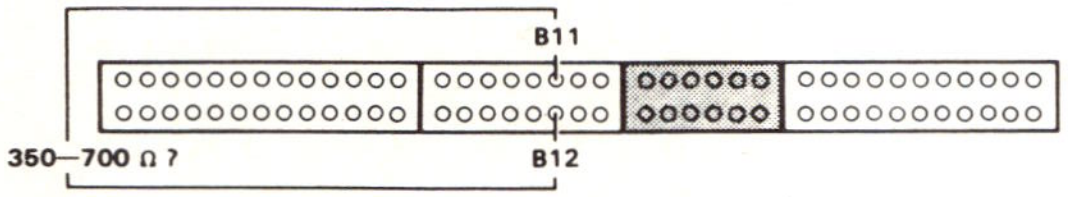

350—700 Ω ?

B11

B12

Is there 350—700 Ω ?

NO: **Repair open in ORN and/or WHT wires.**

YES:

Check for continuity to body ground on B11 terminal.

Does continuity exist?

YES: **Repair short in ORN wire between ECU (B11) and distributor connector.**

NO:

**Substitute a known-good ECU and recheck. If symptom/indication goes away, replace the original ECU.**

# COOLANT TEMPERATURE (TW) SENSOR—CODE 6—1990 ACCORD

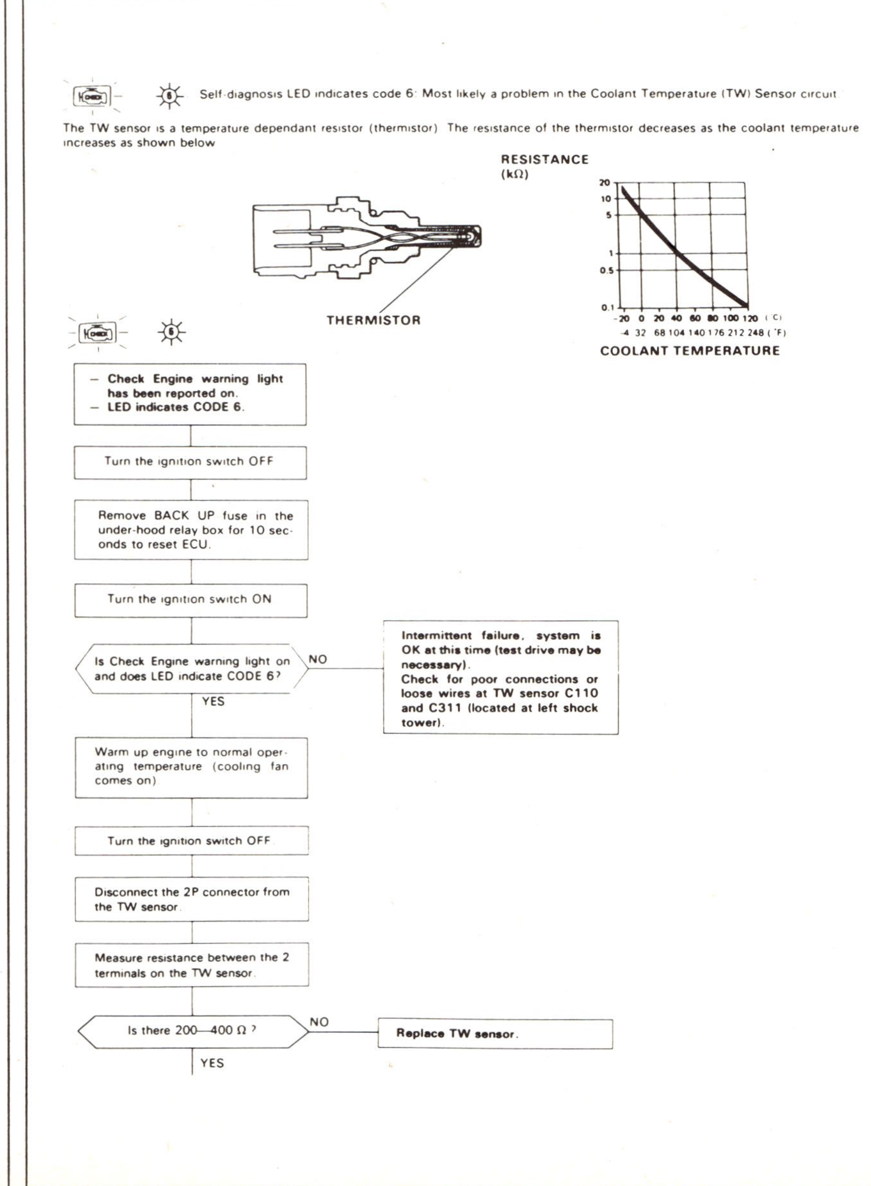

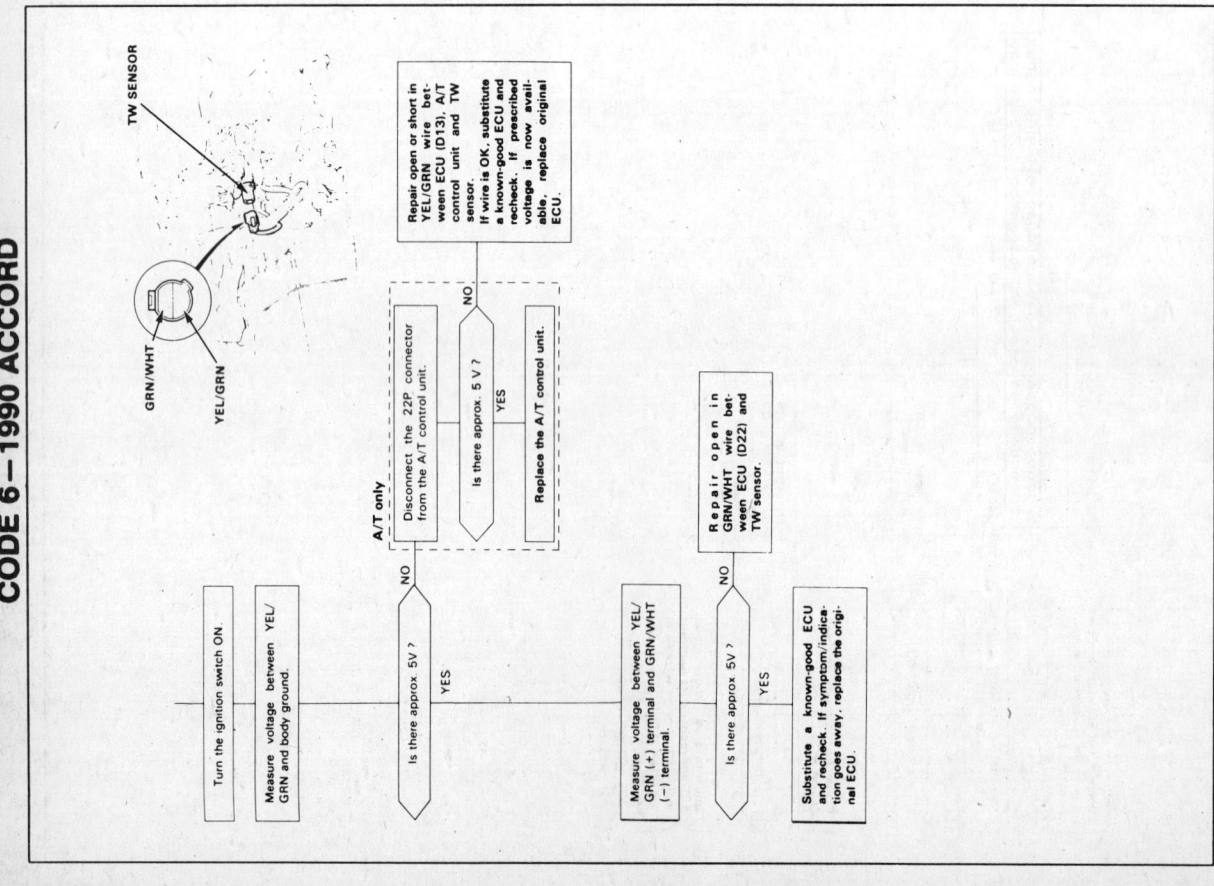

## COOLANT TEMPERATURE (TW) SENSOR CODE 6—1990 ACCORD

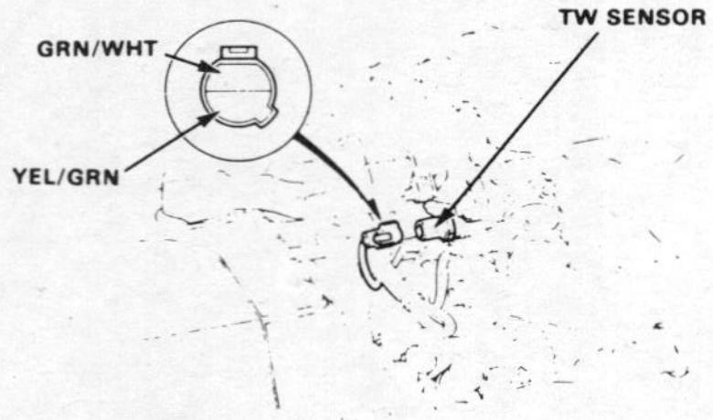

Turn the ignition switch ON.

Measure voltage between YEL/GRN and body ground.

Is there approx. 5V ?

- NO → **A/T only**
  - Disconnect the 22P connector from the A/T control unit.
  - Is there approx. 5 V ?
    - NO → **Repair open or short in YEL/GRN wire between ECU (D13), A/T control unit and TW sensor. If wire is OK, substitute a known-good ECU and recheck. If prescribed voltage is now available, replace original ECU.**
    - YES → **Replace the A/T control unit.**
- YES ↓

Measure voltage between YEL/GRN (+) terminal and GRN/WHT (−) terminal.

Is there approx. 5V ?

- NO → **Repair open in GRN/WHT wire between ECU (D22) and TW sensor.**
- YES ↓

**Substitute a known-good ECU and recheck. If symptom/indication goes away, replace the original ECU.**

## THROTTLE ANGLE SENSOR—CODE 7—1990 ACCORD

Self-diagnosis LED indicator blinks seven times: Most likely a problem in the Throttle Angle Sensor circuit.

The throttle angle sensor is a potentiometer. It is connected to the throttle valve shaft. As the throttle angle changes, the throttle angle sensor varies the voltage signal to the ECU.

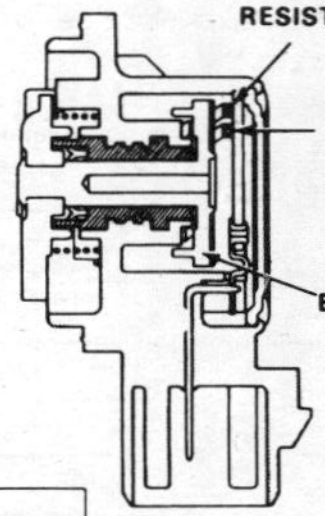

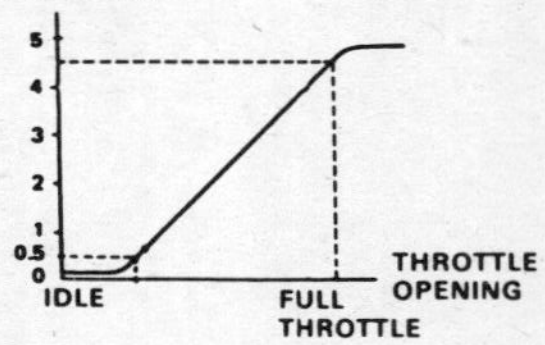

**- Engine is running.**
**- Check Engine warning light has been reported on.**
**- LED indicates CODE 7.**

Turn the ignition switch OFF.

Remove BACK UP fuse in the under-hood relay box for 10 seconds to reset ECU.

Start the engine.

Is Check Engine warning light on and does LED indicate CODE 7?

- NO → **Intermittent failure, system is OK at this time (test drive may be necessary). Check for poor connections or loose wires at throttle angle sensor C123 and C311 (located at left shock tower).**
- YES ↓

Turn the ignition switch OFF.

Disconnect the 3P connector from the throttle angle sensor.

Turn the ignition switch ON.

**Measure voltage between YEL/WHT (+) terminal and GRN/WHT (−) terminal.**

Is there approx. 5V ?

- NO → Measure voltage between YEL/WHT (+) terminal and body ground.
- YES ↓

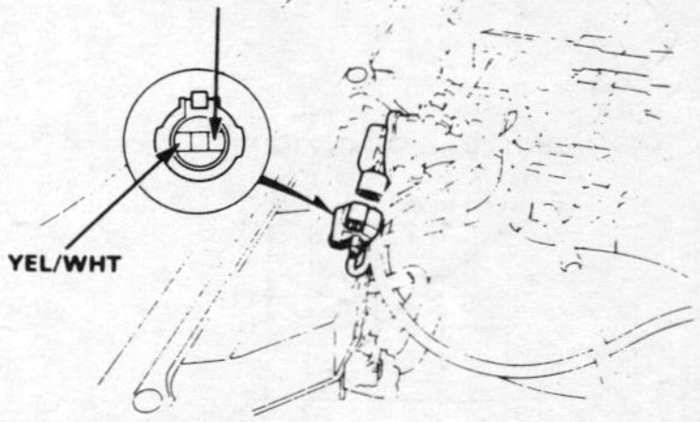

## THROTTLE ANGLE SENSOR—CODE 7—1990 ACCORD

Turn the ignition switch OFF.

Reconnect the 3P connector.

Connect the ECU test harness between the ECU and connector.

Turn the ignition switch ON.

Measure voltage between D11 (+) terminal and D22 (−) terminal.

Is voltage approx. 0.5 V at full close throttle, and approx. 4.5 V at full open throttle? NOTE: There should be a smooth transition from 0.5 V to 4.5 V as the throttle is depressed.

- YES: **Substitute a known-good ECU and recheck. If symptom/indication goes away, replace the original ECU.**
- NO: 

A/T only

Disconnect the 22P connector from the A/T control unit.

Is voltage approx. 0.5 V at full close throttle, and approx. 4.5 V at full open throttle? NOTE: There should be a smooth transition from 0.5 V to 4.5 V as the throttle is depressed.

- YES: **Replace the A/T control unit.**
- NO: **- Replace throttle angle sensor. - Repair open or short in RED/BLK wire between ECU (D11), A/T control unit and throttle angle sensor.**

Is there approx. 5V ?

- YES: **Repair open in GRN/WHT wire between ECU (D22) and throttle angle sensor.**
- NO:

Turn the ignition switch OFF.

Connect the ECU test harness between the ECU and connector.

Turn the ignition switch ON.

Measure voltage between D20 (+) terinal and D22 (−) terminal.

Is there approx. 5V ?

- YES: **Repair open in YEL/WHT wire between ECU (D20) and throttle angle sensor.**
- NO: **Substitute a known-good ECU and recheck. If prescribed voltage is now available, replace the original ECU.**

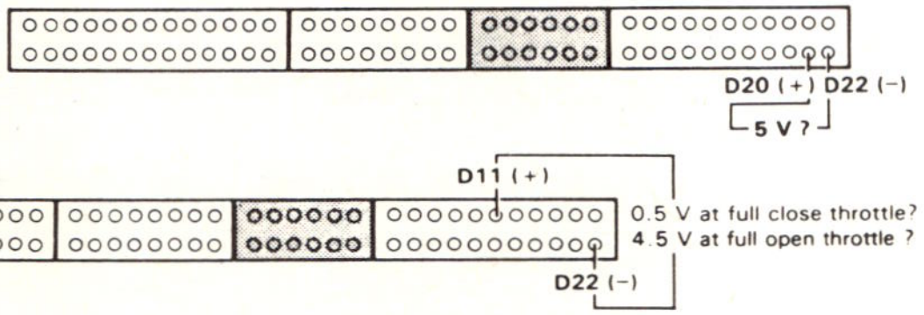

## AIR TEMPERATURE SENSOR CODE 10—1990 ACCORD

Self-diagnosis LED indicates code 10: Most likely a problem in the Intake Air Temperature (TA) Sensor circuit.

The TA sensor is a temperature dependant resistor (thermistor). The resistance of the thermistor decreases as the intake air temperature increases as shown below.

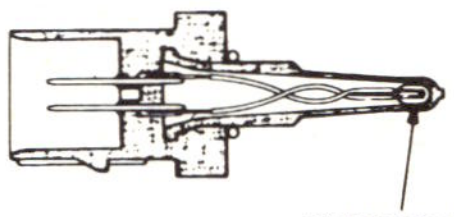

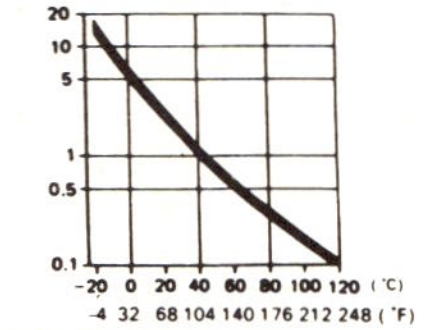

- **Check Engine warning light has beer reported on.**
- **LED indicates CODE 10.**

Turn the ignition switch OFF.

Remove BACK UP fuse in the under-hood relay box for 10 seconds to reset ECU.

Turn the ignition switch ON.

Is Check Engine warning light on and does LED indicate CODE 10?

- NO: **Intermittent failure, system is OK at this time (test drive may be necessary). Check for poor connections or loose wires at TA sensor C124 and C311 (located at left shock tower).**
- YES:

Turn the ignition switch OFF.

Disconnect the 2P connector from the TA sensor.

Measure resistance between the 2 terminals on the TA sensor.

Is there 1–4 kΩ ?

- NO: **Replace TA sensor.**
- YES

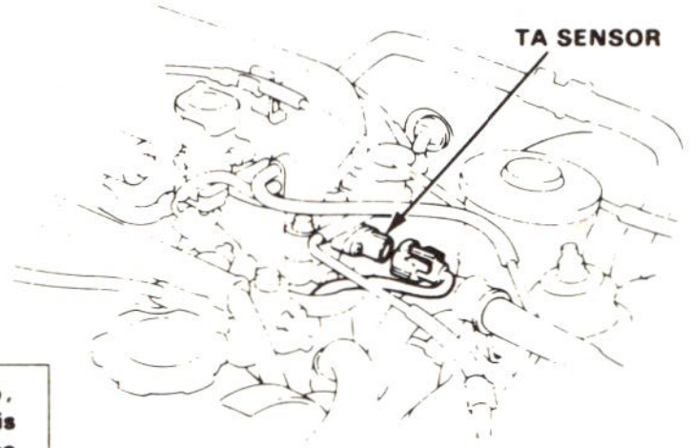

## AIR TEMPERATURE SENSOR 10—1990 ACCORD CODE

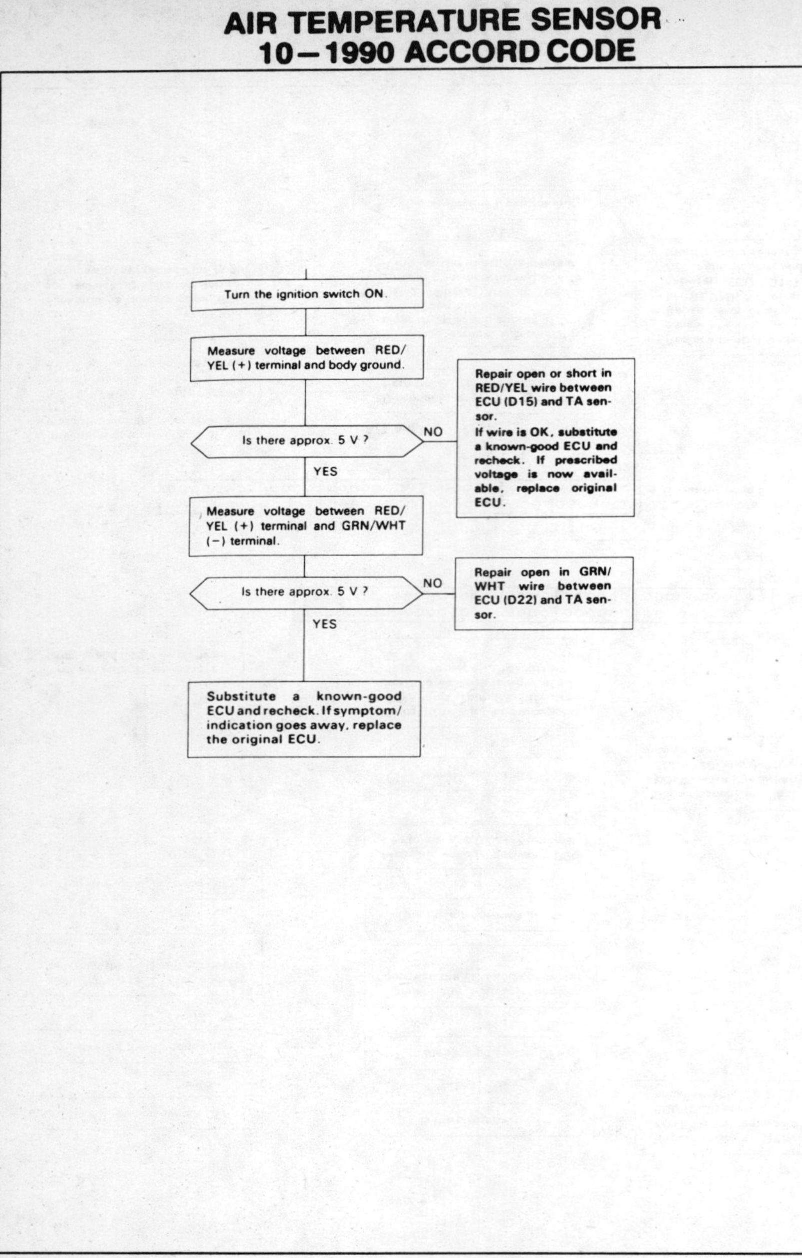


## ATMOSPHERIC PRESSURE (PA) SENSOR CODE 13—1990 ACCORD

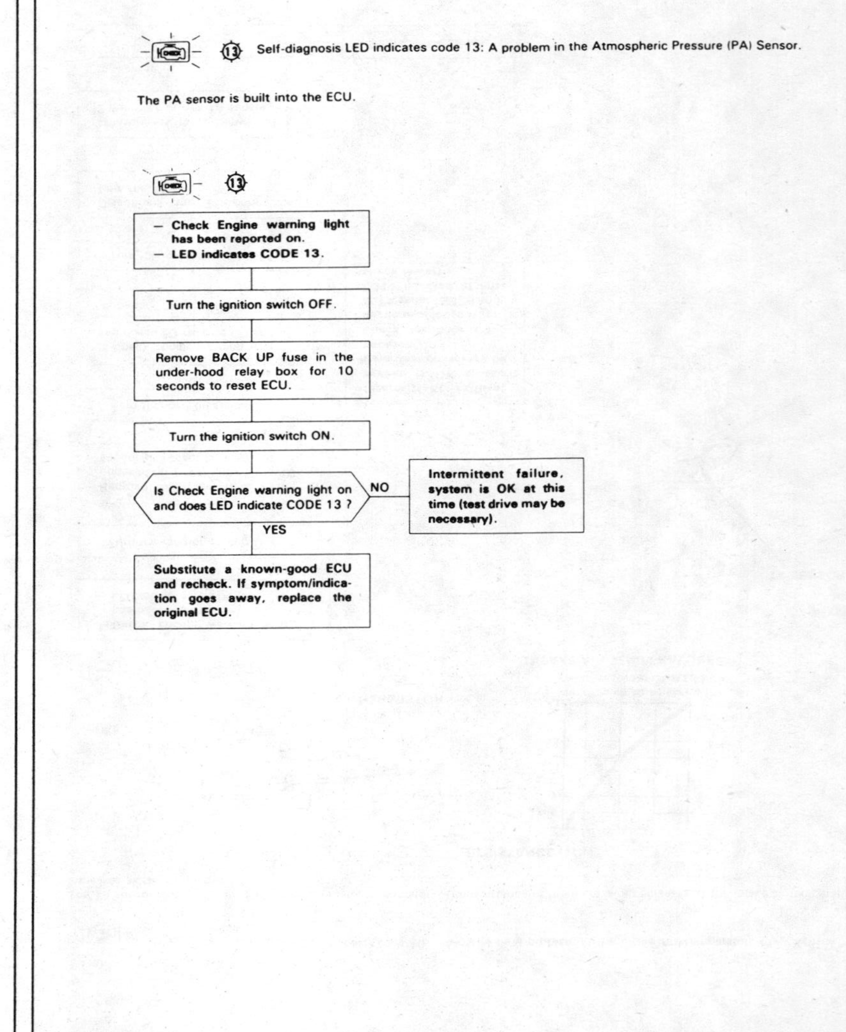

# IGNITION OUTPUT SIGNAL—CODE 15—1990 ACCORD

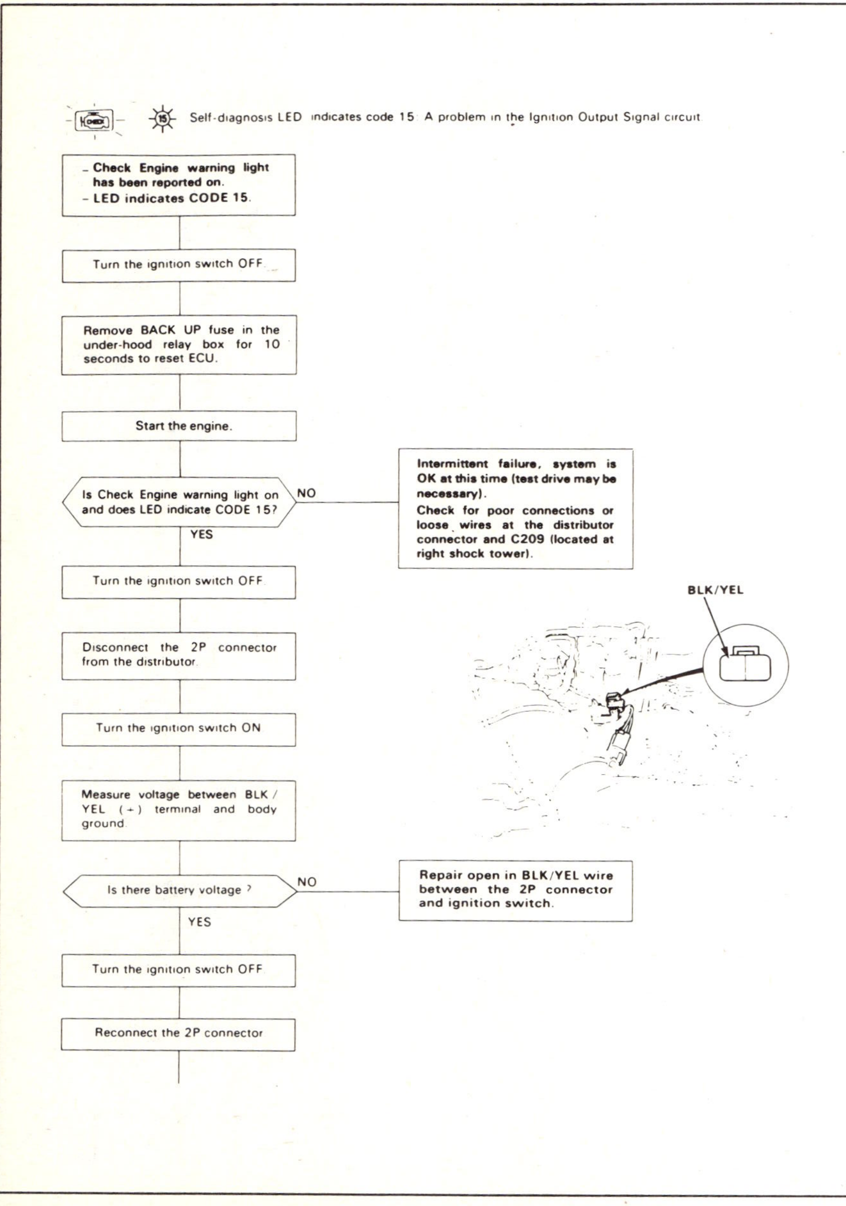


Self-diagnosis LED indicates code 15. A problem in the Ignition Output Signal circuit.

- **Check Engine warning light has been reported on.**
- **LED indicates CODE 15.**

Turn the ignition switch OFF.

Remove BACK UP fuse in the under-hood relay box for 10 seconds to reset ECU.

Start the engine.

Is Check Engine warning light on and does LED indicate CODE 15?

NO → **Intermittent failure, system is OK at this time (test drive may be necessary). Check for poor connections or loose wires at the distributor connector and C209 (located at right shock tower).**

YES ↓

Turn the ignition switch OFF.

Disconnect the 2P connector from the distributor.

Turn the ignition switch ON.

Measure voltage between BLK/YEL (+) terminal and body ground.

Is there battery voltage?

NO → Repair open in BLK/YEL wire between the 2P connector and ignition switch.

YES ↓

Turn the ignition switch OFF.

Reconnect the 2P connector.

Connect the ECU test harness between the ECU and connector.

Turn the ignition switch ON.

Measure voltage individually between A21 (+), A22 (+) terminals and A26 (−) terminal.

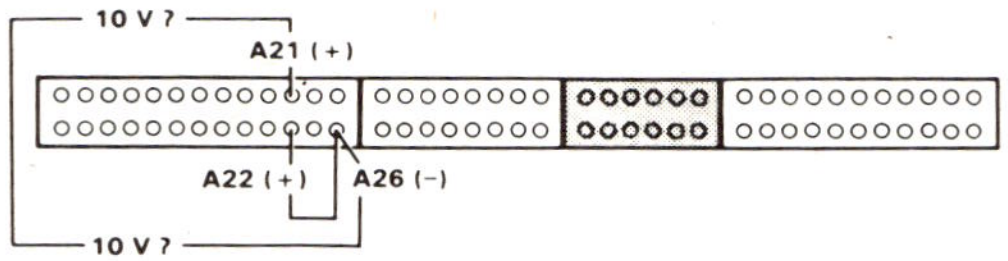

10 V ? A21 (+)
A22 (+) A26 (−) 10 V ?

Is there approx. 10 V?

NO →
- **Replace the igniter unit.**
- **Repair open or short YEL/GRN wires between distributor and ECU (A21 or A22).**

NOTE: If the YEL/GRN wire was shorted, the ignitor may be damaged.

YES ↓

Substitute a known-good ECU and recheck. If symptom/indication goes away, replace the original ECU.

## VEHICLE SPEED SENSOR—CODE 17—1990 ACCORD

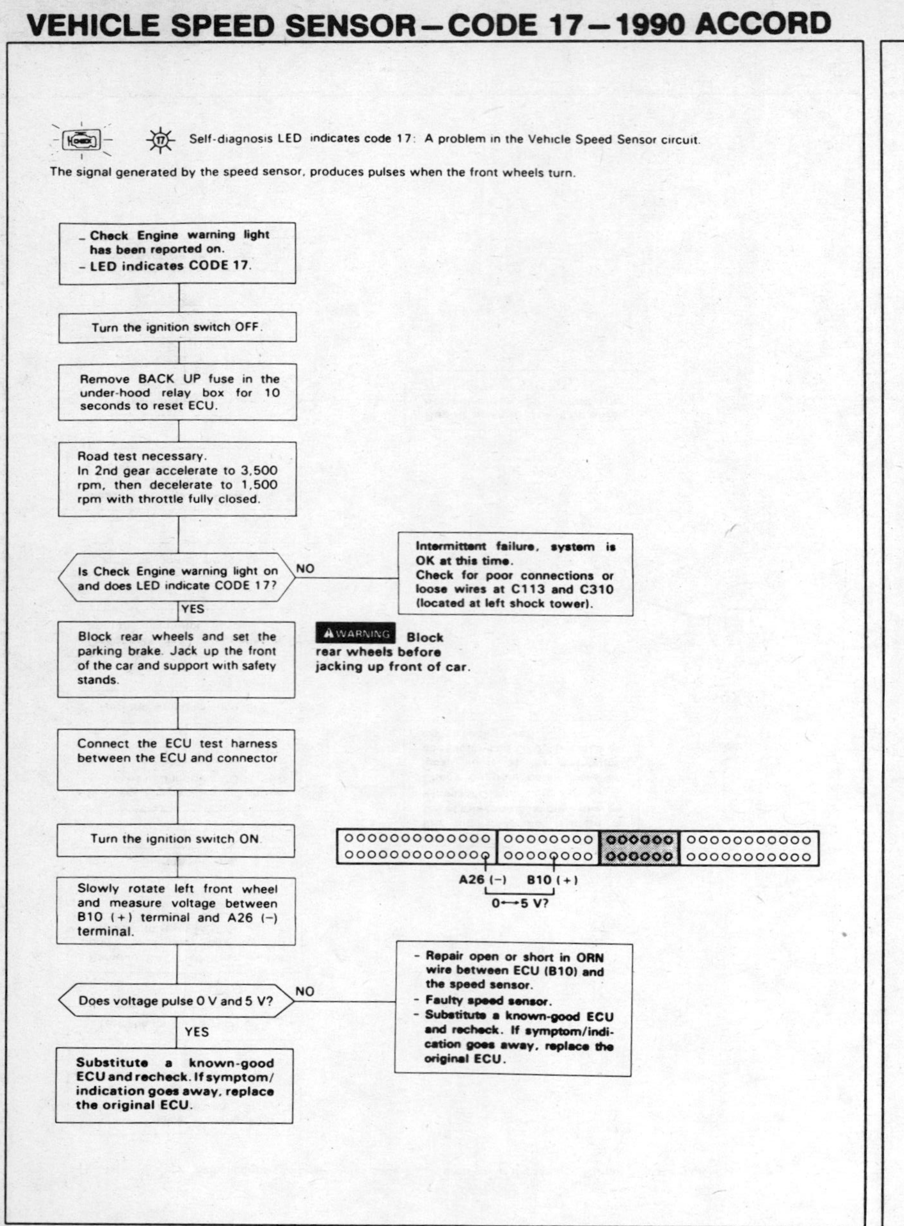

Self-diagnosis LED indicates code 17: A problem in the Vehicle Speed Sensor circuit.
The signal generated by the speed sensor, produces pulses when the front wheels turn.
- Check Engine warning light has been reported on.
- LED indicates CODE 17.
Turn the ignition switch OFF.
Remove BACK UP fuse in the under-hood relay box for 10 seconds to reset ECU.
Road test necessary. In 2nd gear accelerate to 3,500 rpm, then decelerate to 1,500 rpm with throttle fully closed.
Is Check Engine warning light on and does LED indicate CODE 17?
NO
Intermittent failure, system is OK at this time. Check for poor connections or loose wires at C113 and C310 (located at left shock tower).
YES
Block rear wheels and set the parking brake. Jack up the front of the car and support with safety stands.
WARNING Block rear wheels before jacking up front of car.
Connect the ECU test harness between the ECU and connector
Turn the ignition switch ON.
A26 (−)
B10 (+)
0↔5 V?
Slowly rotate left front wheel and measure voltage between B10 (+) terminal and A26 (−) terminal.
Does voltage pulse 0 V and 5 V?
NO
- Repair open or short in ORN wire between ECU (B10) and the speed sensor.
- Faulty speed sensor.
- Substitute a known-good ECU and recheck. If symptom/indication goes away, replace the original ECU.
YES
Substitute a known-good ECU and recheck. If symptom/indication goes away, replace the original ECU.


## ELECTRIC LOAD DETECTOR CODE 20—1990 ACCORD

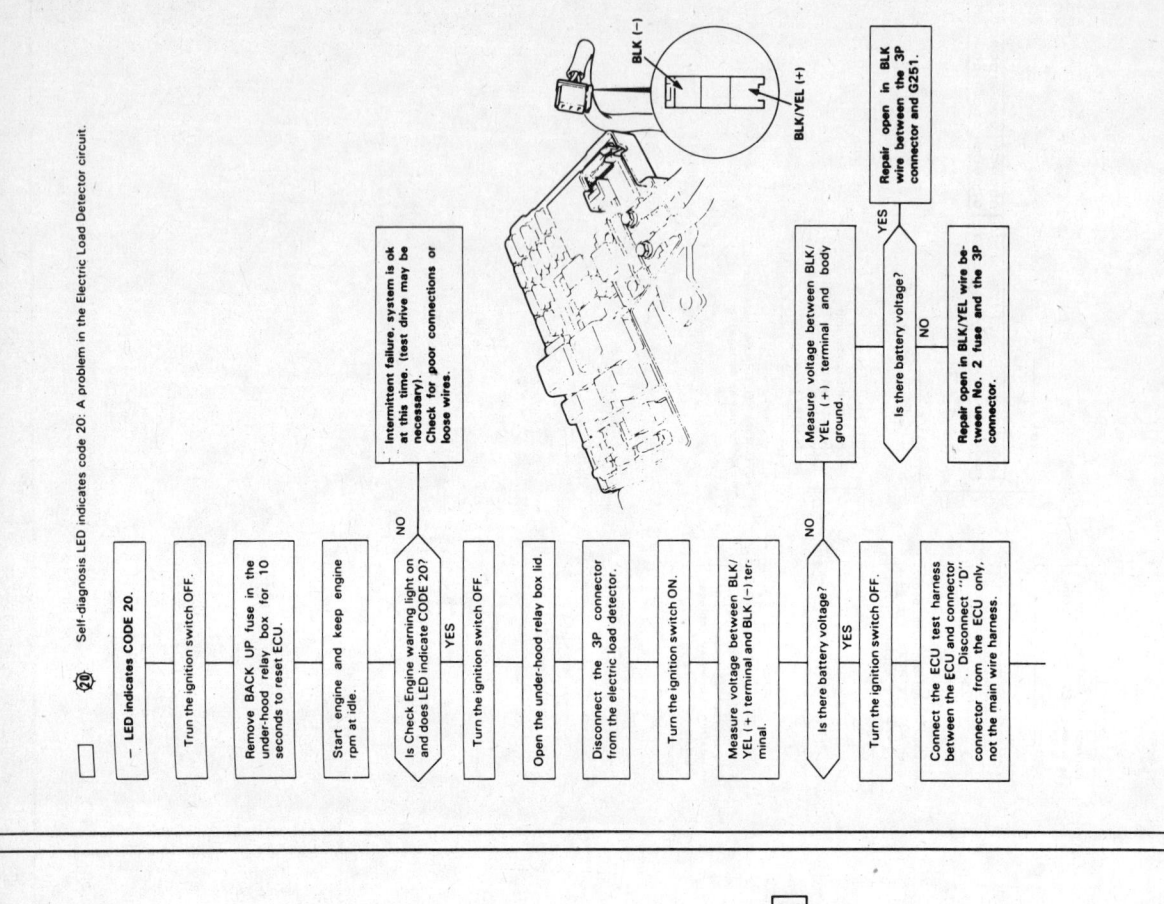

Self-diagnosis LED indicates code 20: A problem in the Electric Load Detector circuit.

- LED indicates CODE 20.

Trun the ignition switch OFF.

Remove BACK UP fuse in the under-hood relay box for 10 seconds to reset ECU.

Start engine and keep engine rpm at idle.

Is Check Engine warning light on and does LED indicate CODE 20?

NO → Intermittent failure, system is ok at this time. (test drive may be necessary). Check for poor connections or loose wires.

YES

Turn the ignition switch OFF.

Open the under-hood relay box lid.

Disconnect the 3P connector from the electric load detector.

Turn the ignition switch ON.

Measure voltage between BLK/YEL (+) terminal and BLK (−) terminal.

BLK (−)

BLK/YEL (+)

Is there battery voltage?

NO → Measure voltage between BLK/YEL (+) terminal and body ground.

Is there battery voltage?

YES → Repair open in BLK wire between the 3P connector and G251.

NO → Repair open in BLK/YEL wire between No. 2 fuse and the 3P connector.

YES

Turn the ignition switch OFF.

Connect the ECU test harness between the ECU and connector . Disconnect "D" connector from the ECU only, not the main wire harness.

# ELECTRIC LOAD DETECTOR—CODE 20—1990 ACCORD

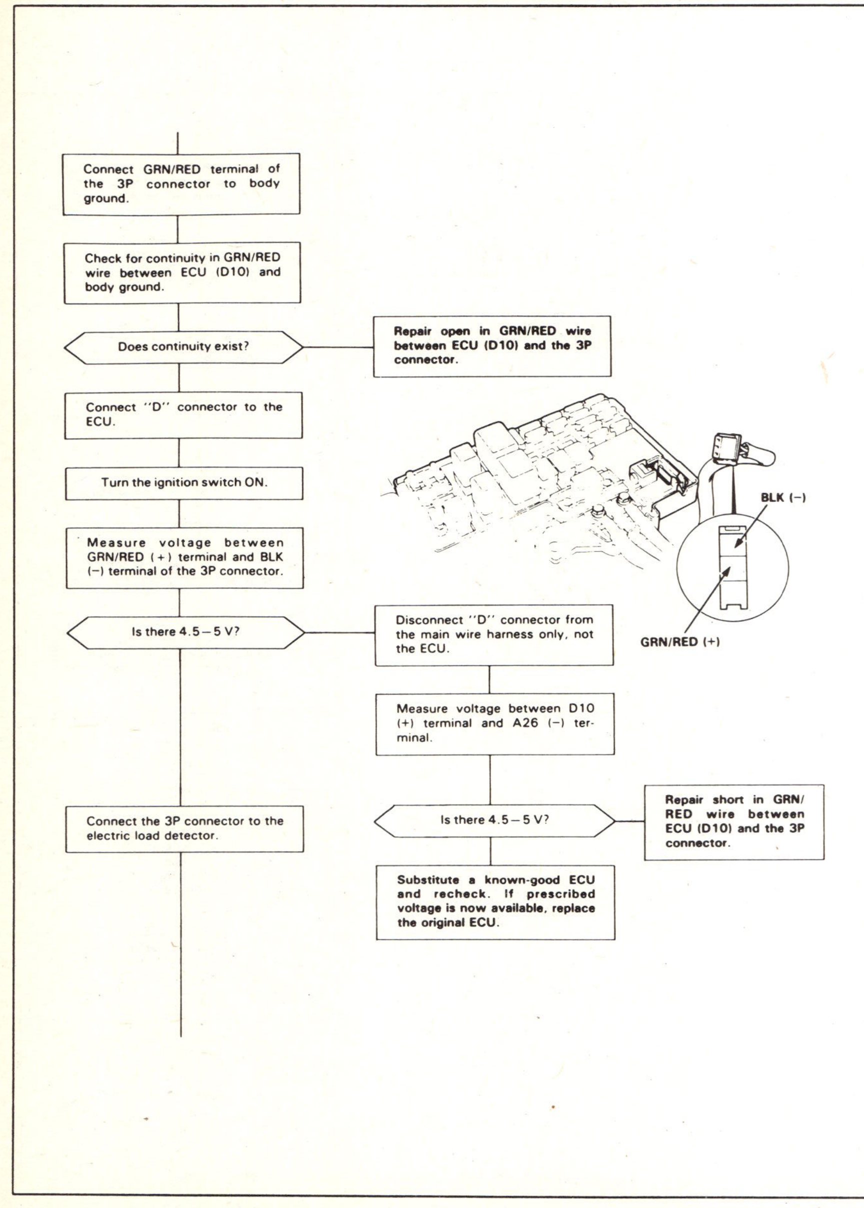


Under the conditions listed in the chart to the right, measure voltage between D10 (+) terminal and A26 (−) terminal.

| Condition | Voltage |
|---|---|
| Headlight switch, first position (•) | 2.5–3.5 V |
| Headlight switch, second position (●) | 1.5–2.5 V |

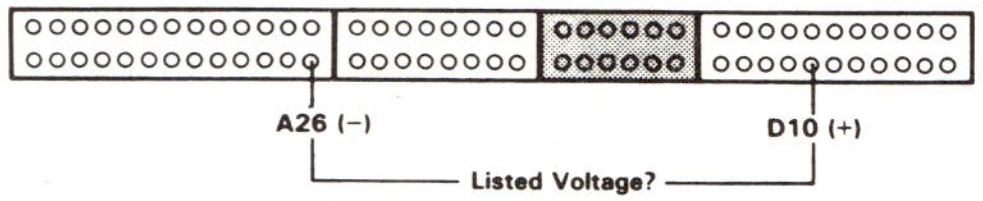

A26 (−)

D10 (+)

Listed Voltage?

Is the voltage listed in the chart available?

Faulty electric load detector.

Substitute a known-good ECU and recheck. If symptom/indication goes away, replace the original ECU.

## A/T FI SIGNAL A—CODE 30—1990 ACCORD

Self-diagnosis LED indicates code 30: A problem in the line (B3) between A/T control unit and PGM-FI ECU.

Self-diagnosis LED indicates code 31: A problem in the line (B4) between A/T control unit and PGM-FI ECU.

LED indicates CODE 30.

Turn the ignition switch OFF.

Remove BACK UP fuse in the under-hood relay box for 10 seconds to reset ECU.

Test drive necessary. Drive the car for several miles so that the transmission upshifts and downshifts several times.

Does LED indicate CODE 30?

NO — Intermittent failure, system is OK at this time.

YES — Turn the ignition switch OFF.

Connect the ECU test harness between the ECU and connector not to the ECU

Disconnect the 22P connector from the A/T control unit.

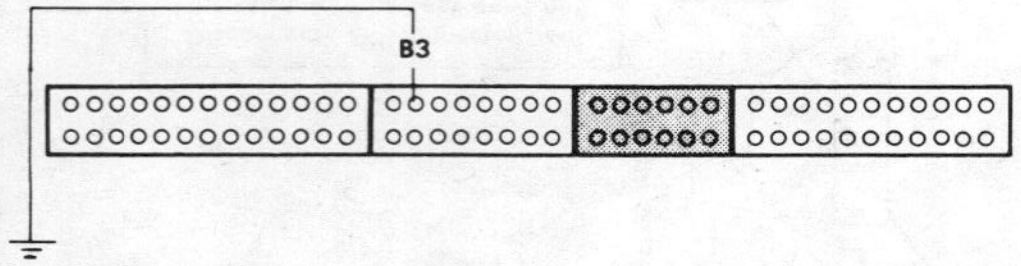

Check for continuity between B3, terminal and body ground.

B3

Continuity?

Does continuity exist?

YES — Repair short in WHT/RED wire between ECU (B3) and the A/T control unit.

NO — Check for continuity on WHT/RED wire between B3 and 22P connector of the A/T control unit.

B3

Does continuity exist?

NO — Repair open in WHT/RED wire between ECU (B3) and the A/T control unit.

YES — Substitute a known-good ECU and recheck. If symptom/indication goes away, replace the original ECU.

## A/T FI SIGNAL B—CODE 31—1990 ACCORD

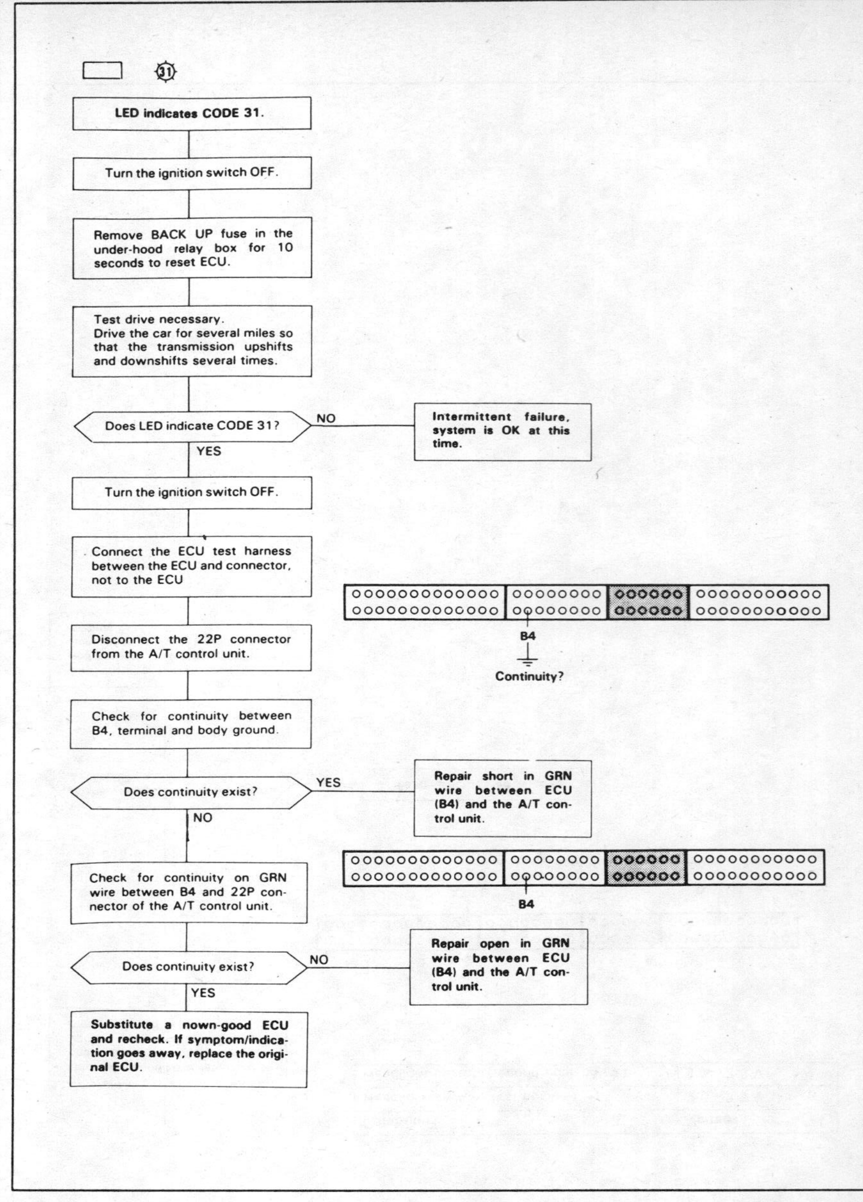

# IDLE CONTROL TROUBLESHOOTING—1990 ACCORD

NOTE:
- Across each row in the chart, the sub systems that could be sources of a symptom are ranked in the order they should be inspected, starting with ①. Find the symptom in the left column, read across to the most likely source, then refer to the page listed at the top of that column. If inspection shows the system is OK, try the next system ②, etc.

| SYMPTOM \ SUB SYSTEM | | IDLE ADJUSTING SCREW | EACV | AIR CONDITIONING SIGNAL | ALTERNATOR FR SIGNAL | A/T SHIFT POSITION SIGNAL (A/T ONLY) | BRAKE SWITCH SIGNAL | STARTER SWITCH SIGNAL | P/S OIL PRESSURE SWITCH SIGNAL | FAST IDLE VALVE | AIR BOOST VALVE | HOSES AND CONNECTIONS |
|---|---|---|---|---|---|---|---|---|---|---|---|---|
| DIFFICULT TO START ENGINE WHEN COLD | | | | | | | | | | ① | | |
| WHEN COLD FAST IDLE OUT OF SPEC (1,000–2,000 rpm) | | ③ | ② | | | | | | | ① | | |
| ROUGH IDLE | | | ② | | | | | | | | | ① |
| WHEN WARM RPM TOO HIGH | | ③ | ① | | | | | | ③ | ② | | ③ |
| WHEN WARM RPM TOO LOW | Idle speed is below specified rpm (no load) | ② | ① | | | | | | | | | |
| | Idle speed does not increase after initial start up | | ① | | | | | | | | | |
| | On models with automatic transmission, the idle speed drops in gear | | ② | | | ① | | | | | | |
| | Idle speeds drops when air conditioner in ON | | ② | ① | | | | | | | | |
| | Idle speed drops when steering wheel is turning | | ② | | | | | | ① | | | |
| | Idle speed fluctuates with electrical local | | ② | | | | | | | | | ① |
| FREQUENT STALLING | WHILE WARMING UP | | ① | | | | | | | | | |
| | AFTER WARMING UP | ① | | | | | | | | | | |
| FAILS EMISSION TEST | | | | | | | | | | | | ① |

- If the idle speed still cannot be adjusted to specification (and LED does not blink CODE 14) after EACV replacement, substitute a known-good ECU and recheck. If symptom goes away, replace the original ECU.

# ELECTRONIC AIR CONTROL VALVE (EACV) CODE 14—1990 ACCORD

Self-diagnosis LED indicates code 14. A problem in the Electronic Air Control Valve (EACV) circuit.

The EACV changes the amount of air bypassing the throttle body in response to a current signal from the ECU in order to maintain the proper idle speed.

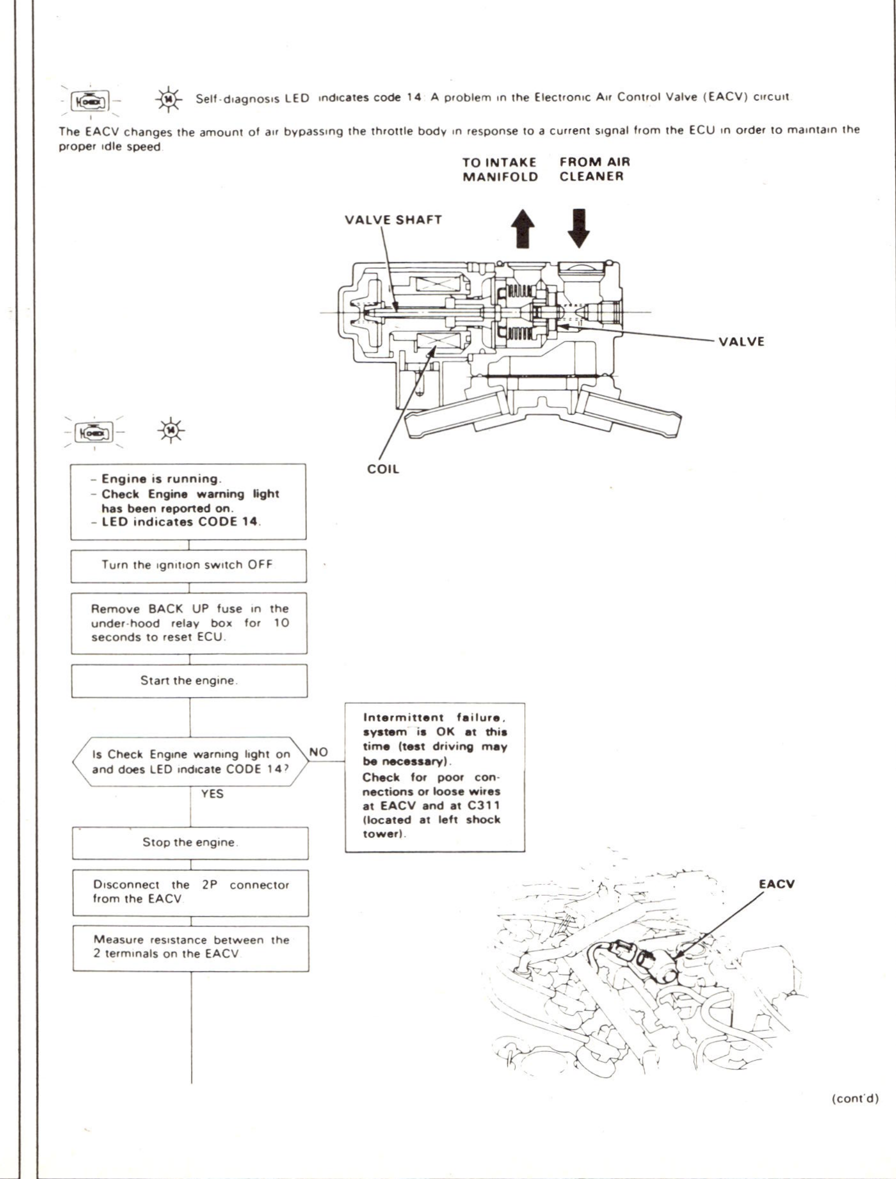


(cont'd)

# ELECTRONIC AIR CONTROL VALVE (EACV)—CODE 14—1990 ACCORD

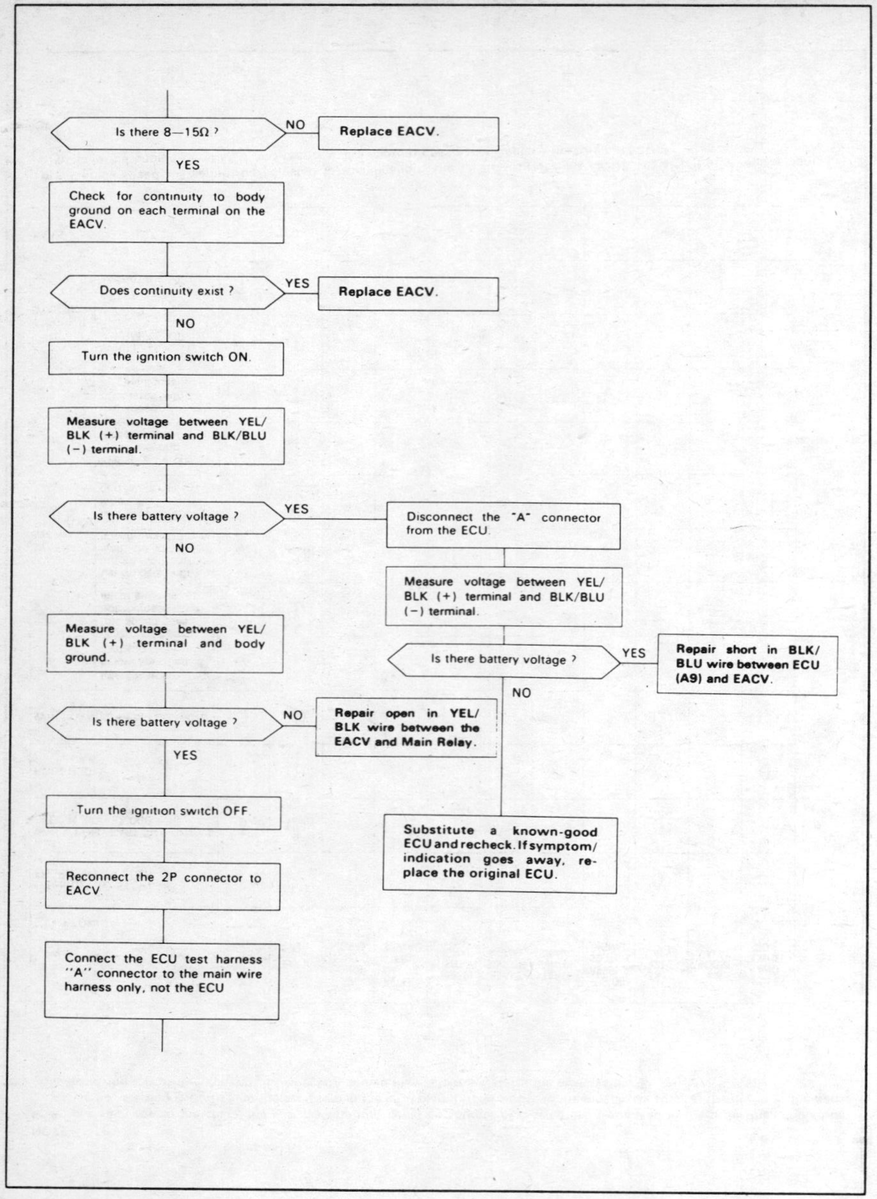

Is there 8—15Ω ? — NO → Replace EACV.

YES

Check for continuity to body ground on each terminal on the EACV.

Does continuity exist ? — YES → Replace EACV.

NO

Turn the ignition switch ON.

Measure voltage between YEL/BLK (+) terminal and BLK/BLU (−) terminal.

Is there battery voltage ? — YES → Disconnect the "A" connector from the ECU. → Measure voltage between YEL/BLK (+) terminal and BLK/BLU (−) terminal. → Is there battery voltage ? — YES → Repair short in BLK/BLU wire between ECU (A9) and EACV. — NO → Substitute a known-good ECU and recheck. If symptom/indication goes away, replace the original ECU.

NO

Measure voltage between YEL/BLK (+) terminal and body ground.

Is there battery voltage ? — NO → Repair open in YEL/BLK wire between the EACV and Main Relay.

YES

Turn the ignition switch OFF.

Reconnect the 2P connector to EACV.

Connect the ECU test harness "A" connector to the main wire harness only, not the ECU

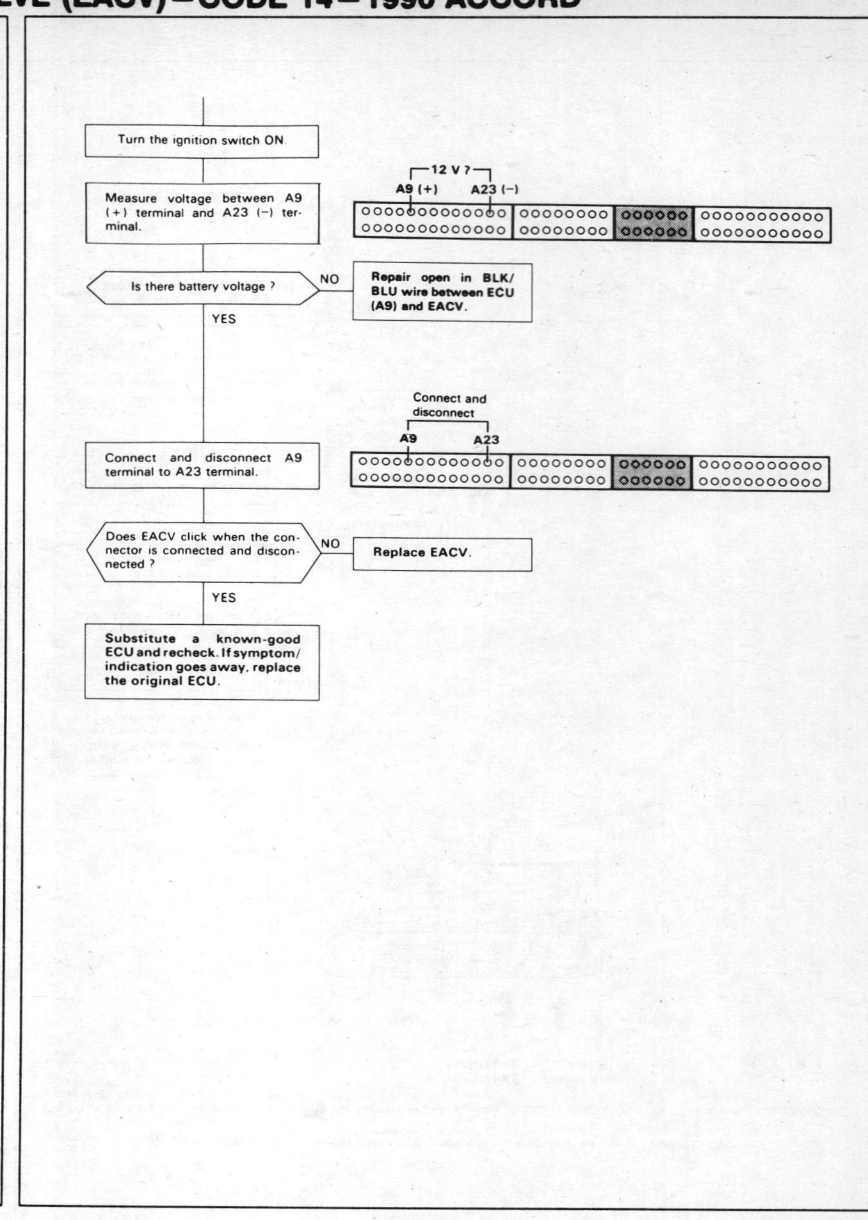

## AIR CONDITIONING SIGNAL—1990 ACCORD

This signals the PGM-FI ECU when there is a demand for cooling from the air conditioning circuits

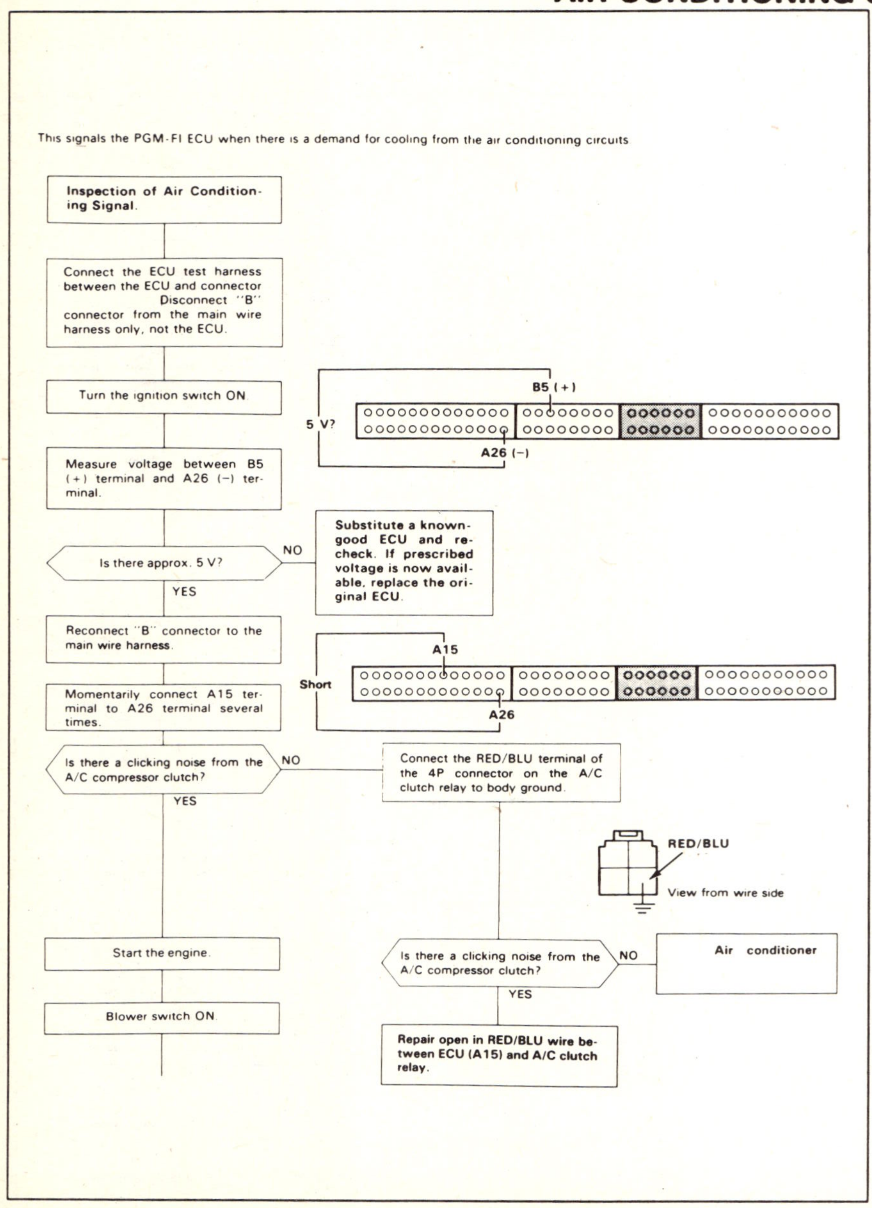

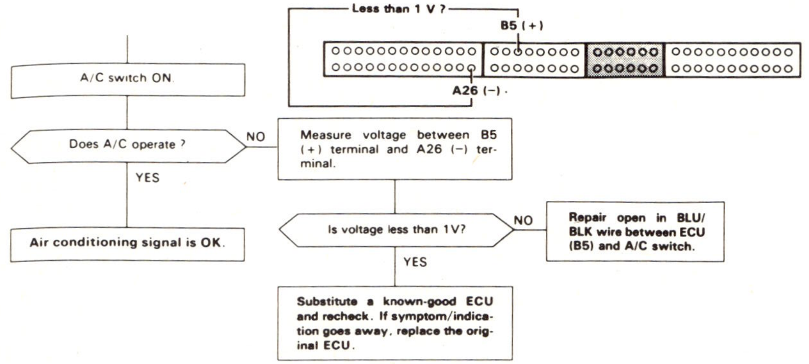

A/C switch ON.

Does A/C operate?

YES: Air conditioning signal is OK.

NO: Measure voltage between B5 (+) terminal and A26 (−) terminal.

Less than 1 V?

B5 (+)

A26 (−)

Is voltage less than 1V?

NO: Repair open in BLU/BLK wire between ECU (B5) and A/C switch.

YES: Substitute a known-good ECU and recheck. If symptom/indication goes away, replace the original ECU.

# ALTERNATOR FR SIGNAL—1990 ACCORD

This signals the PGM-FI ECU when the alternator is charging.

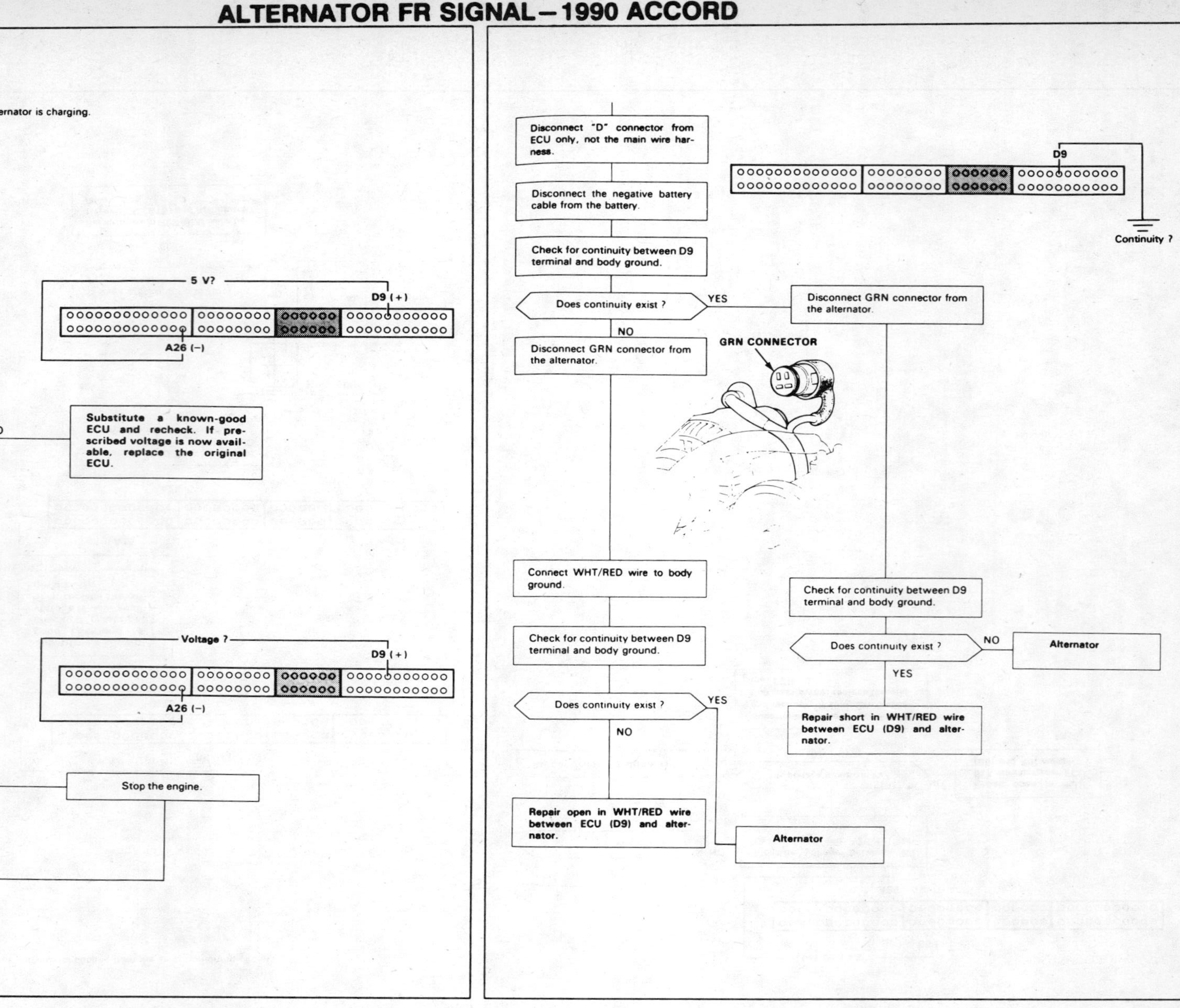

## A/T SHIFT POSITION SIGNAL—1990 ACCORD

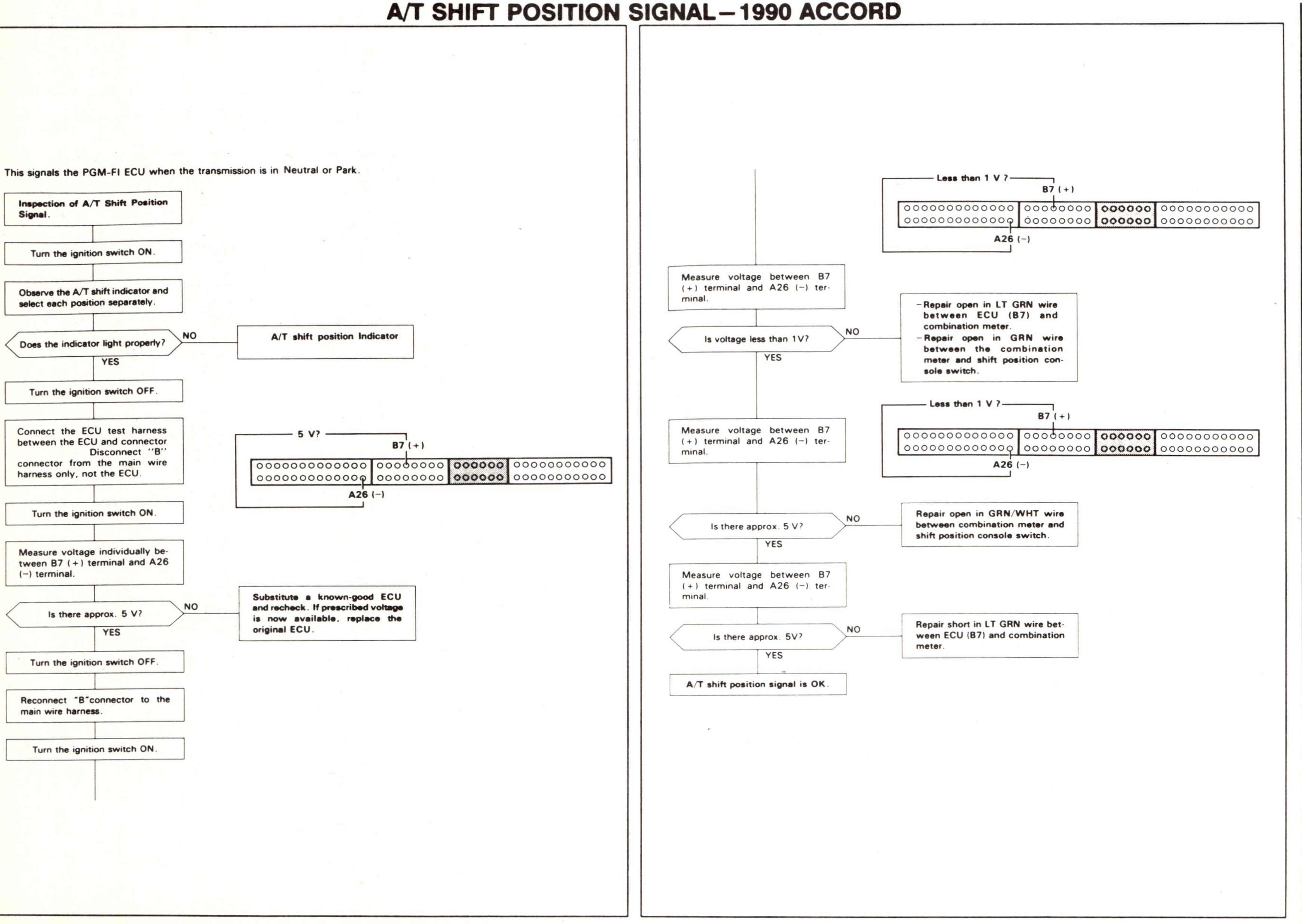

This signals the PGM-FI ECU when the transmission is in Neutral or Park.
Inspection of A/T Shift Position Signal.
Turn the ignition switch ON.
Observe the A/T shift indicator and select each position separately.
Does the indicator light properly?
NO
A/T shift position Indicator
YES
Turn the ignition switch OFF.
Connect the ECU test harness between the ECU and connector. Disconnect "B" connector from the main wire harness only, not the ECU.
5 V?
B7 (+)
A26 (−)
Turn the ignition switch ON.
Measure voltage individually between B7 (+) terminal and A26 (−) terminal.
Is there approx. 5 V?
NO
Substitute a known-good ECU and recheck. If prescribed voltage is now available, replace the original ECU.
YES
Turn the ignition switch OFF.
Reconnect "B" connector to the main wire harness.
Turn the ignition switch ON.
Less than 1 V ?
B7 (+)
A26 (−)
Measure voltage between B7 (+) terminal and A26 (−) terminal.
Is voltage less than 1V?
NO
–Repair open in LT GRN wire between ECU (B7) and combination meter.
–Repair open in GRN wire between the combination meter and shift position console switch.
YES
Less than 1 V ?
B7 (+)
A26 (−)
Measure voltage between B7 (+) terminal and A26 (−) terminal.
Is there approx. 5 V?
NO
Repair open in GRN/WHT wire between combination meter and shift position console switch.
YES
Measure voltage between B7 (+) terminal and A26 (−) terminal.
Is there approx. 5V?
NO
Repair short in LT GRN wire between ECU (B7) and combination meter.
YES
A/T shift position signal is OK.

## BRAKE SWITCH SIGNAL–1990 ACCORD

This signals the PGM-FI ECU when the brake pedal is depressed.

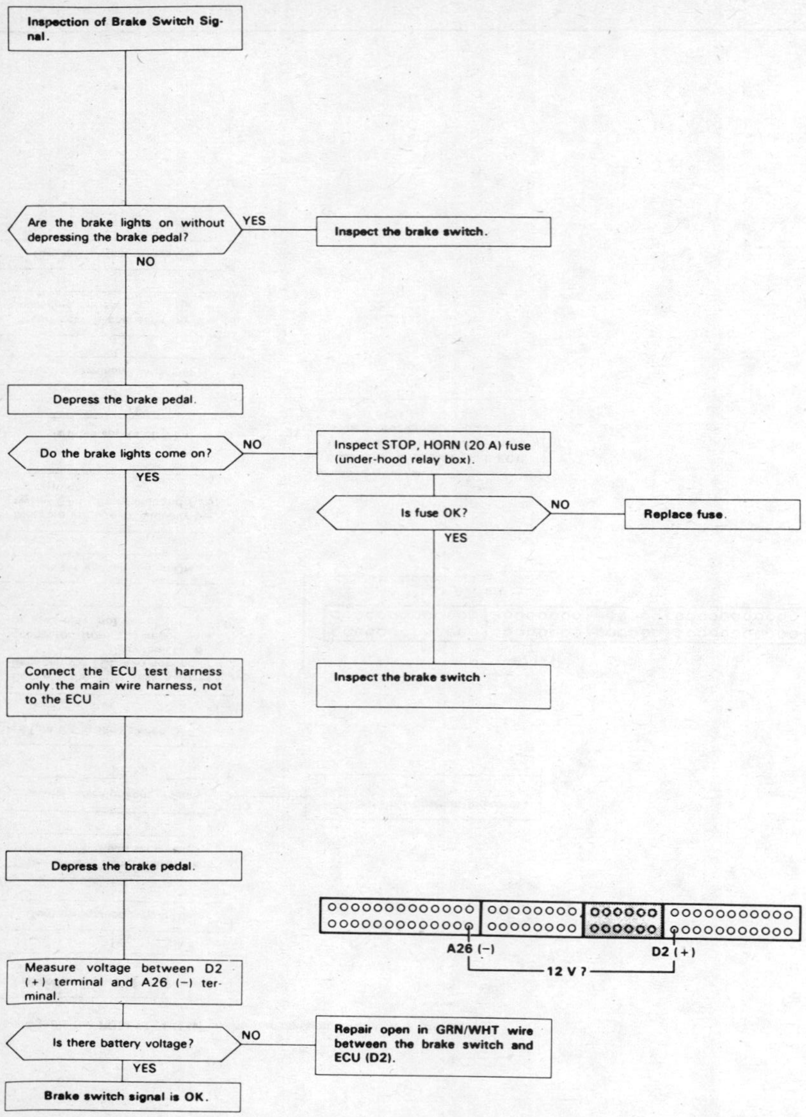

Inspection of Brake Switch Signal.
Are the brake lights on without depressing the brake pedal?
YES
Inspect the brake switch.
NO
Depress the brake pedal.
Do the brake lights come on?
NO
Inspect STOP, HORN (20 A) fuse (under-hood relay box).
Is fuse OK?
NO
Replace fuse.
YES
Inspect the brake switch
YES
Connect the ECU test harness only the main wire harness, not to the ECU
Depress the brake pedal.
A26 (−)
D2 (+)
12 V ?
Measure voltage between D2 (+) terminal and A26 (−) terminal.
Is there battery voltage?
NO
Repair open in GRN/WHT wire between the brake switch and ECU (D2).
YES
Brake switch signal is OK.

## STARTER SIGNAL–1990 ACCORD

This signals the PGM-FI ECU when the engine is cranking.

Inspection of Starter Switch Signal.
12 V ?
B9 (+)
A26 (−)
NOTE: Clutch pedal must be depressed on M/T models.
Connect the ECU test harness between the ECU and connector
Measure voltage between B9 (+) terminal and A26 (−) terminal.
Is there battery voltage ?
NO
Inspect No.9 fuse.
Is No.9 fuse OK ?
NO
Replace fuse.
YES
Repair open in BLU/RED wire between ECU (B9) and No. 9 fuse.
YES
Starter switch signal is OK.

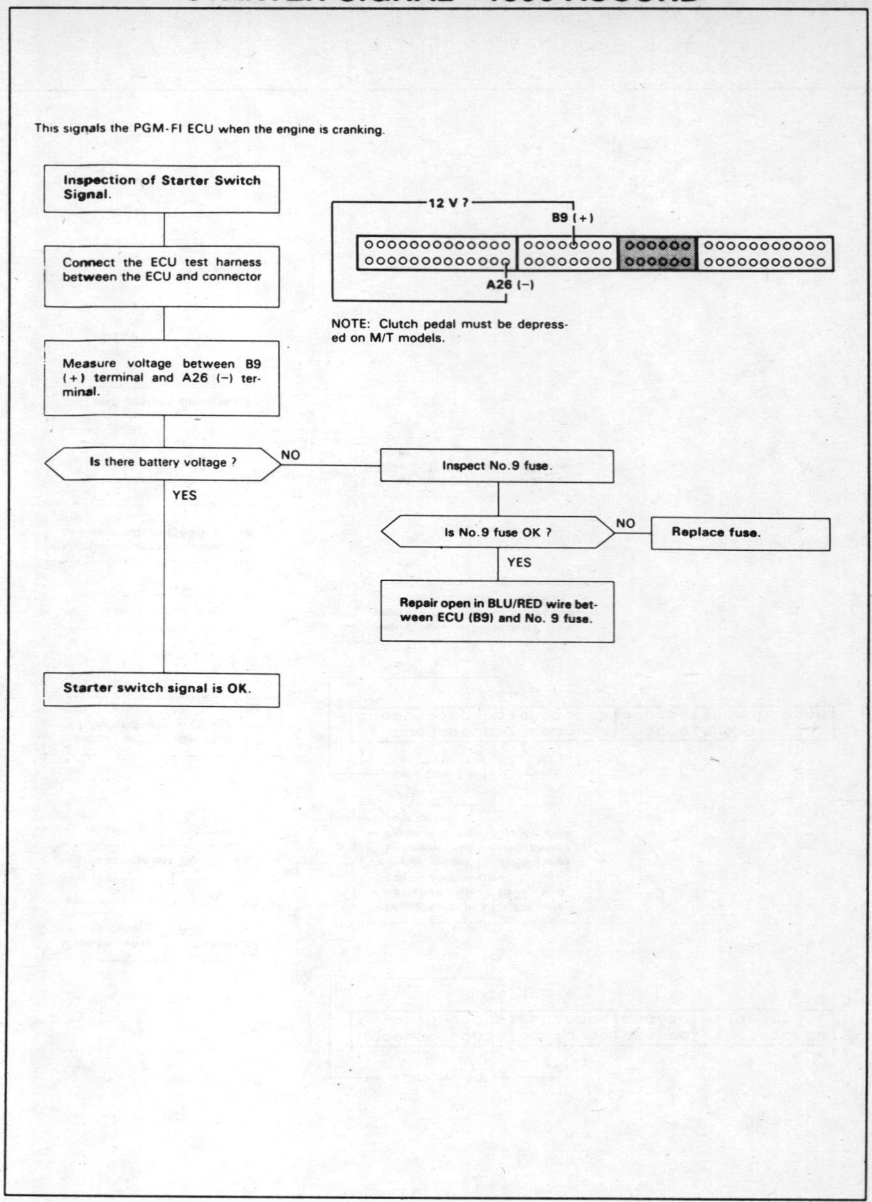

## POWER STEERING PRESSURE SIGNAL—1990 ACCORD

This signals the PGM-FI ECU when the power steering load is high.

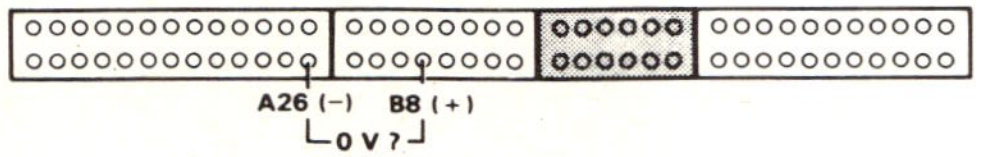

A26 (−)  B8 (+)
0 V ?

- **Inspection of P/S Oil Pressure Signal**
- Connect the ECU test harness between the ECU and connector
- Turn the ignition switch ON.
- Measure voltage between B8 (+) terminal and A26 (−) terminal.
- Is there voltage ?
  - YES → Disconnect the 2P connector on the P/S oil pressure switch.
    - Connect RED terminal to BLK terminal.
    - Is there voltage ?
      - NO → **Replace P/S oil pressure switch.**
      - YES → **Repair open in RED wire between ECU (B8) and P/S oil pressure switch or BLK wire between P/S oil pressure switch and G301.**
  - NO → Start the engine.
    - Turn steering wheel slowly.
    - Measure voltage between B8 (+) terminal and A26 (−) terminal while steering wheel is turning.
    - Is there battery voltage ?
      - YES → **P/S oil pressure signal is OK.**
      - NO → Turn the ignition switch OFF.
        - Disconnect "B" connector from main wire harness only, not the ECU.
        - Turn the ignition switch ON.
        - Is there battery voltage ?
          - NO → **Substitute a known-good ECU and recheck. If prescribed voltage is now available, replace the original ECU.**
          - YES → Reconnect "B" connector to main wire harness and disconnect 2P connector on the P/S oil pressure switch.
            - Is there battery voltage ?
              - YES → **Replace P/S oil pressure switch.**
              - NO → **Repair short in RED wire between ECU (B8) and the P/S oil pressure switch.**

## FUEL INJECTORS—CODE 16—1990 ACCORD

Troubleshooting Flowchart

Self-diagnosis LED indicates code 16: A problem in the fuel injector circuit.

The injectors are the solenoid-actuated constant-stroke pintle type consisting of a solenoid, plunger needle valve and housing. When current is applied to the solenoid coil, the valve lifts up and pressurized fuel is injected close to the intake valve. Because the needle valve lift and the fuel pressure are constant, the injection quantity is determined by the length of time that the valve is open (i.e., the duration the current is supplied to the solenoid coil). The injector is sealed by an O-ring and seal ring at the top and bottom. These seals also reduce operating noise.

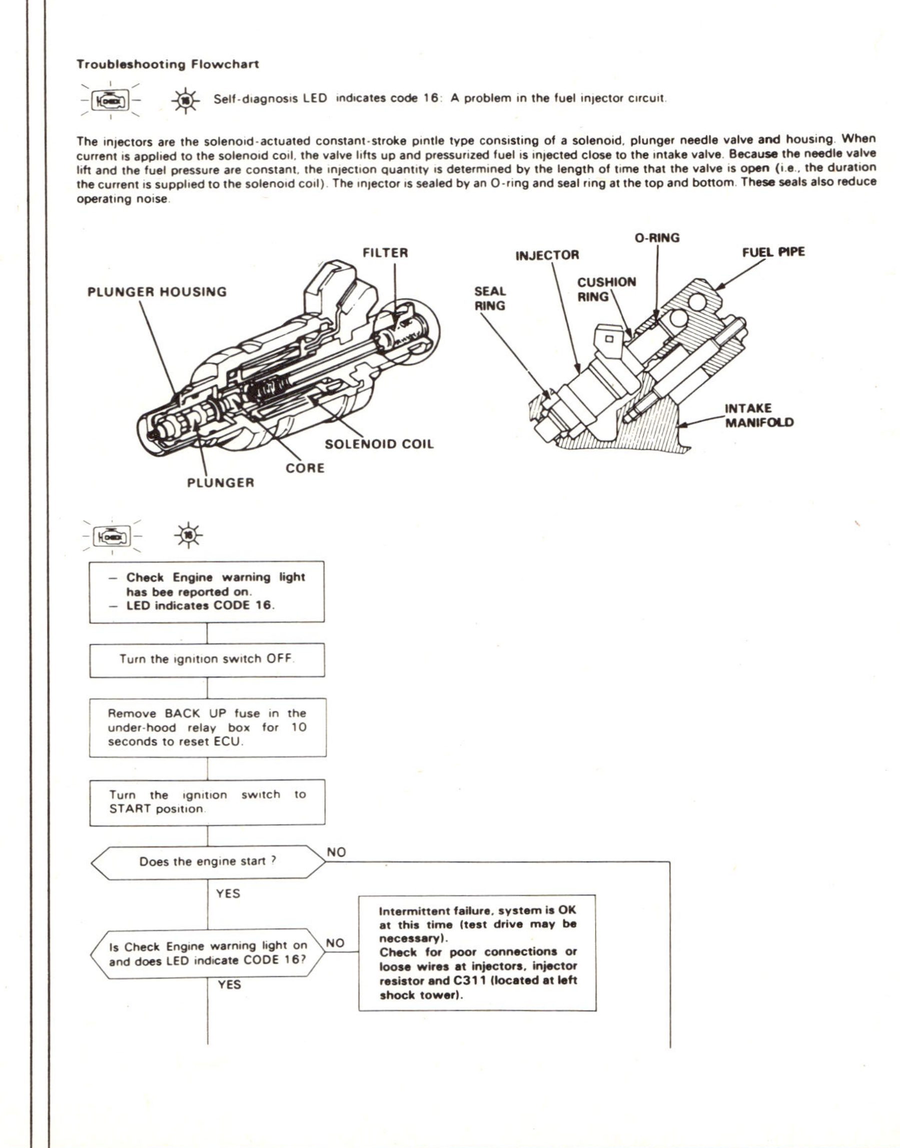


- **– Check Engine warning light has bee reported on.**
  **– LED indicates CODE 16.**
- Turn the ignition switch OFF.
- Remove BACK UP fuse in the under-hood relay box for 10 seconds to reset ECU.
- Turn the ignition switch to START position.
- Does the engine start ?
  - NO → (continued)
  - YES → Is Check Engine warning light on and does LED indicate CODE 16?
    - NO → **Intermittent failure, system is OK at this time (test drive may be necessary). Check for poor connections or loose wires at injectors, injector resistor and C311 (located at left shock tower).**
    - YES → (continued)

## FUEL INJECTORS—CODE 16—1990 ACCORD

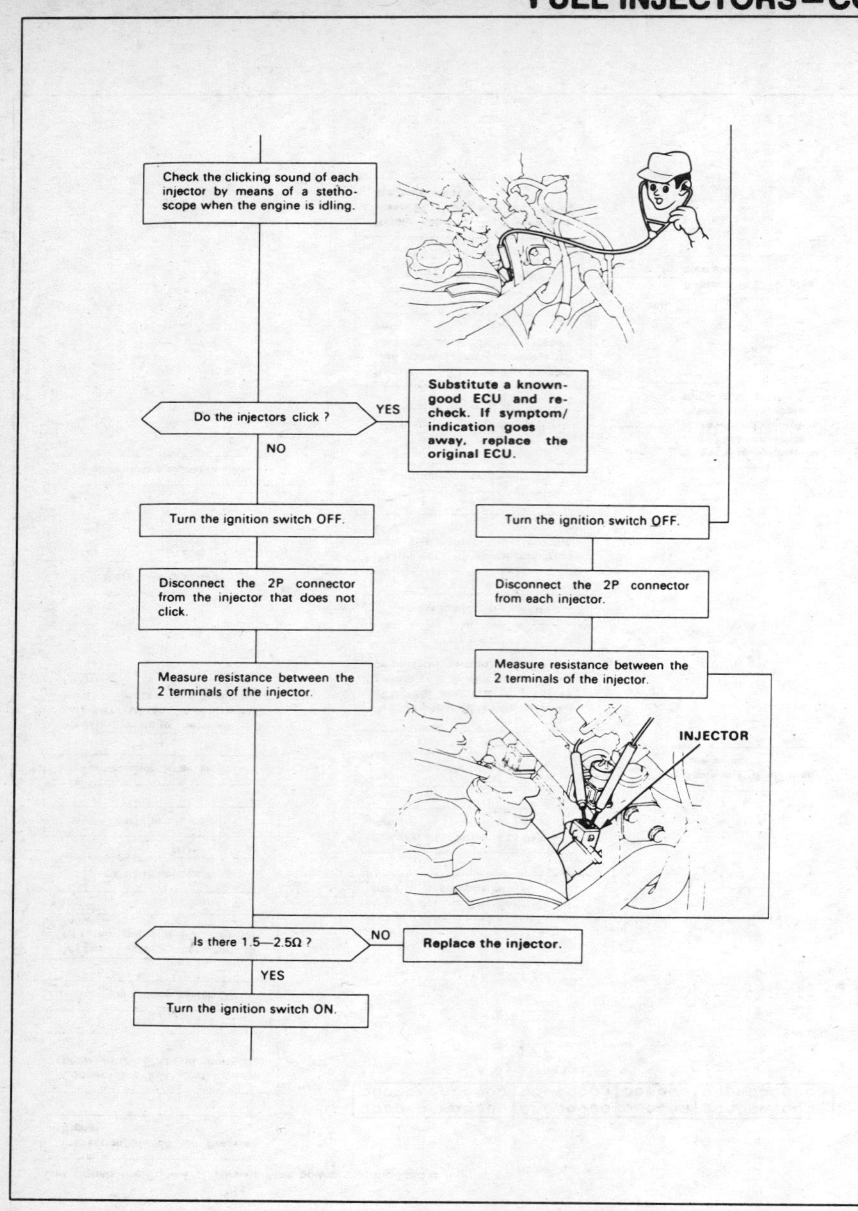


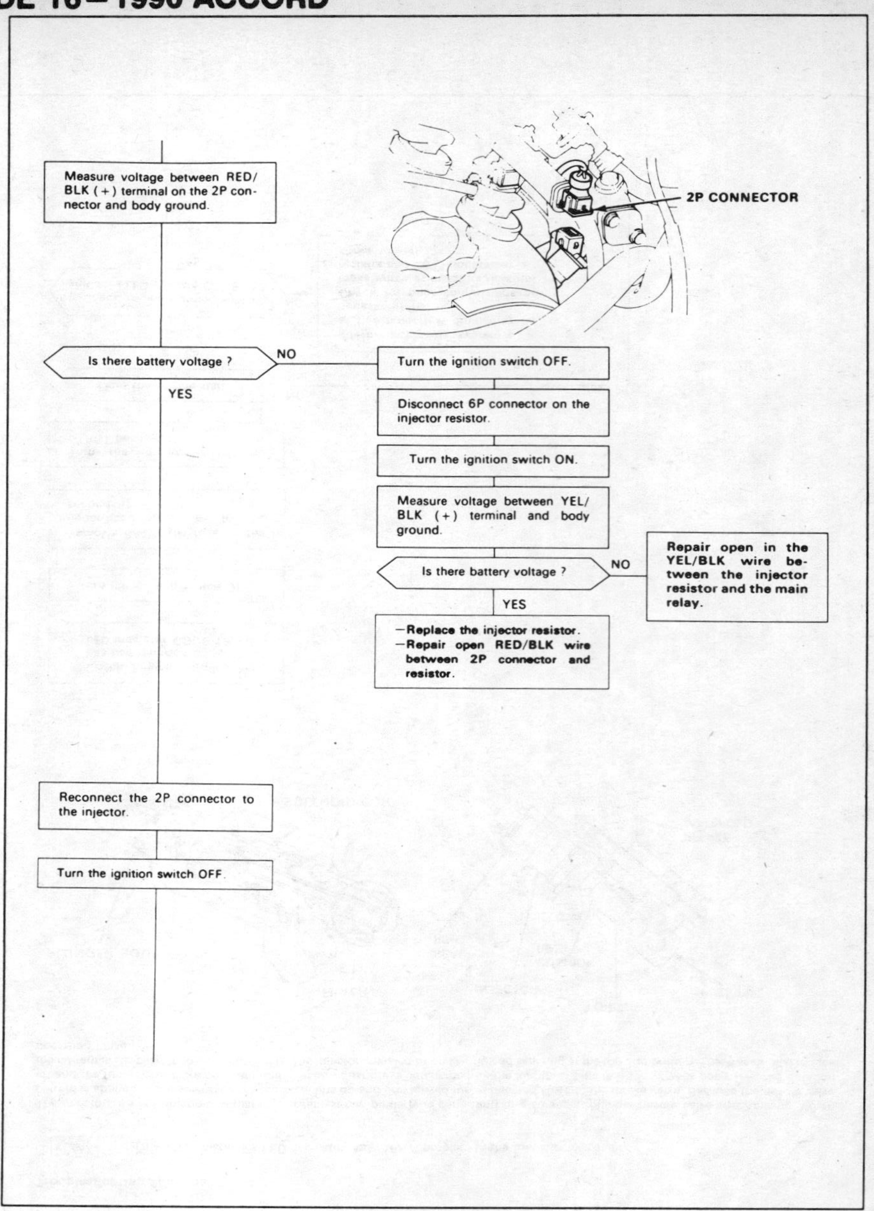

## FUEL INJECTORS—CODE 16—1990 ACCORD

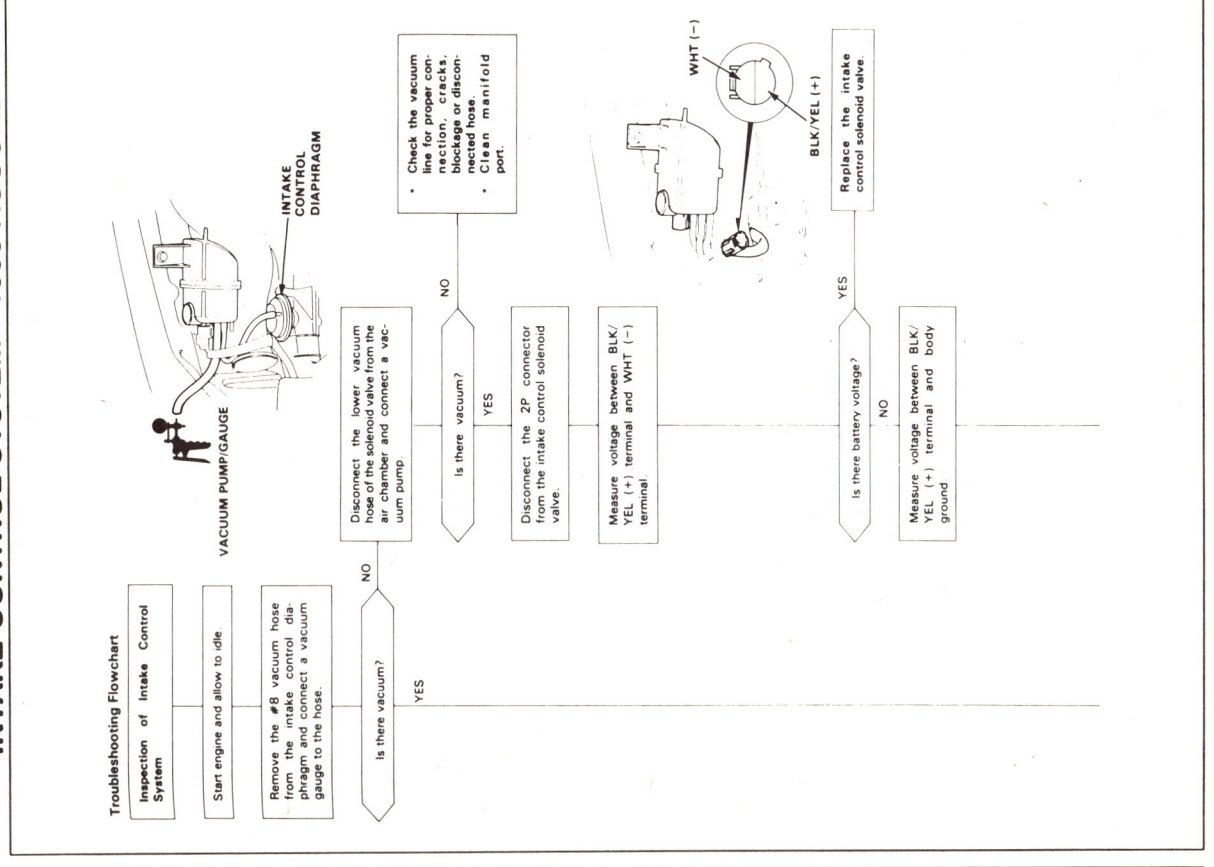

Connect the ECU test harness between the ECU and connector

Turn the ignition switch ON.

Measure voltage between A23(−) terminal and following terminals.
- No.1 injector: A1 (+) terminal.
- No.2 injector: A3 (+) terminal.
- No.3 injector: A5 (+) terminal.
- No.4 injector: A2 (+) terminal.

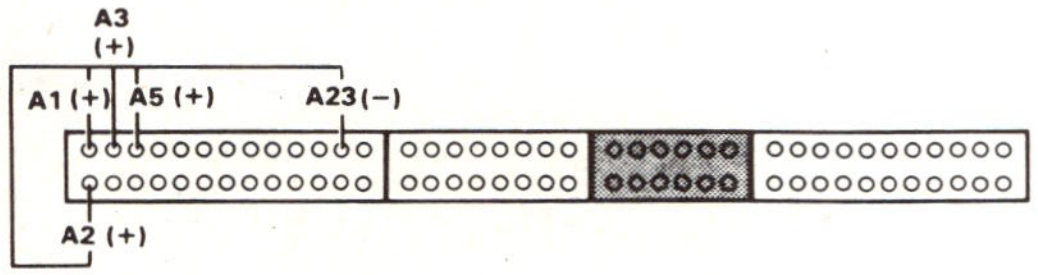

A3 (+)
A1 (+) A5 (+) A23(−)
A2 (+)

Is there battery voltage ?

NO — Repair open in the wire between the ECU (A1, A3, A5 or A2) and the injector.

YES — Substitute a known-good ECU and recheck. If symptom/indication goes away, replace the original ECU.

## INTAKE CONTROL SYSTEM—1990 ACCORD

Troubleshooting Flowchart

Inspection of Intake Control System

Start engine and allow to idle.

Remove the #8 vacuum hose from the intake control diaphragm and connect a vacuum gauge to the hose.

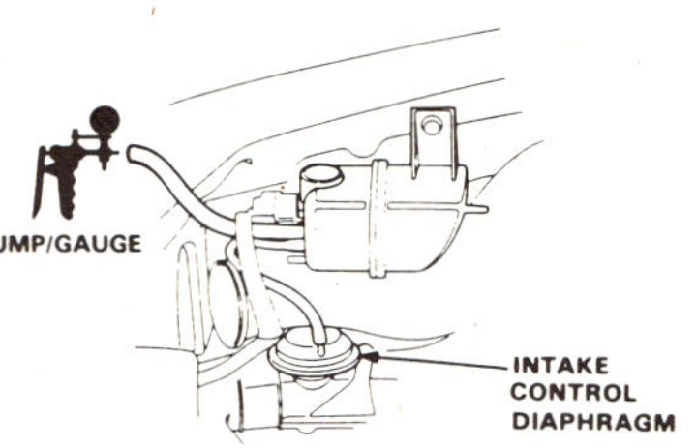

VACUUM PUMP/GAUGE

INTAKE CONTROL DIAPHRAGM

Is there vacuum?

YES

NO — Disconnect the lower vacuum hose of the solenoid valve from the air chamber and connect a vacuum pump.

Is there vacuum?

NO —
- Check the vacuum line for proper connection, cracks, blockage or disconnected hose.
- Clean manifold port.

YES — Disconnect the 2P connector from the intake control solenoid valve.

Measure voltage between BLK/YEL (+) terminal and WHT (−) terminal.

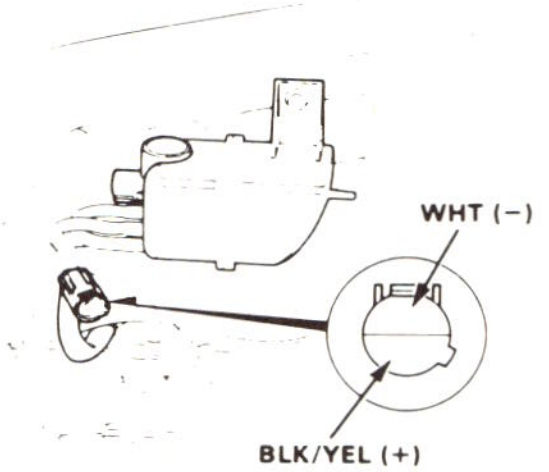

WHT (−)

BLK/YEL (+)

Is there battery voltage?

YES — Replace the intake control solenoid valve.

NO — Measure voltage between BLK/YEL (+) terminal and body ground.

# INTAKE CONTROL SYSTEM—1990 ACCORD

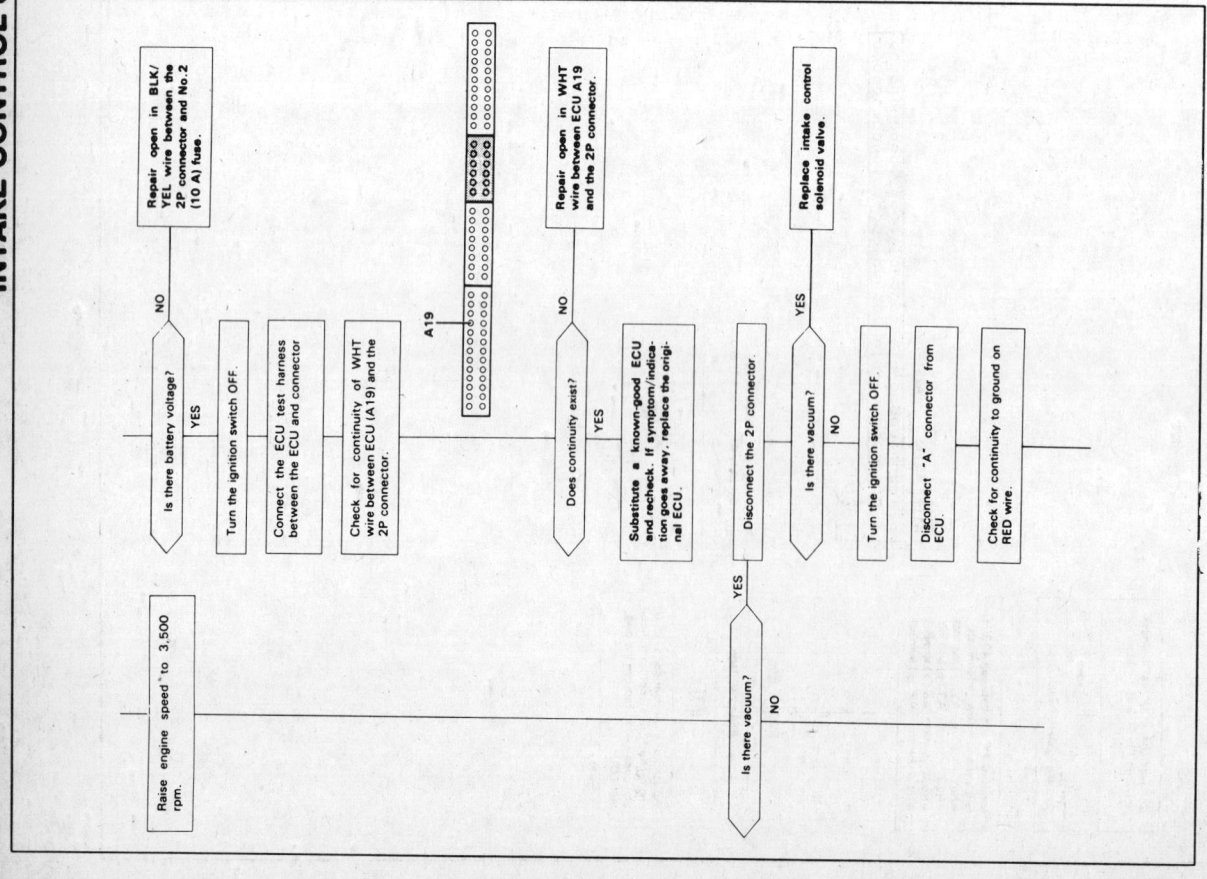

Raise engine speed to 3,500 rpm.

Is there battery voltage?
- NO: Repair open in BLK/YEL wire between the 2P connector and No.2 (10 A) fuse.
- YES:

Turn the ignition switch OFF.

Connect the ECU test harness between the ECU and connector

Check for continuity of WHT wire between ECU (A19) and the 2P connector.

A19

Does continuity exist?
- NO: Repair open in WHT wire between ECU A19 and the 2P connector.
- YES: Substitute a known-good ECU and recheck. If symptom/indication goes away, replace the original ECU.

Is there vacuum?
- YES: Disconnect the 2P connector.
- NO:

Is there vacuum?
- YES: Replace intake control solenoid valve.
- NO:

Turn the ignition switch OFF.

Disconnect "A" connector from ECU.

Check for continuity to ground on RED wire.

Is there continuity to ground?
- YES: Repair short to ground in WHT wire between ECU (A19) and the 2P connector.
- NO: Substitute a known good ECU and recheck. If symptom goes away, replace the original ECU.

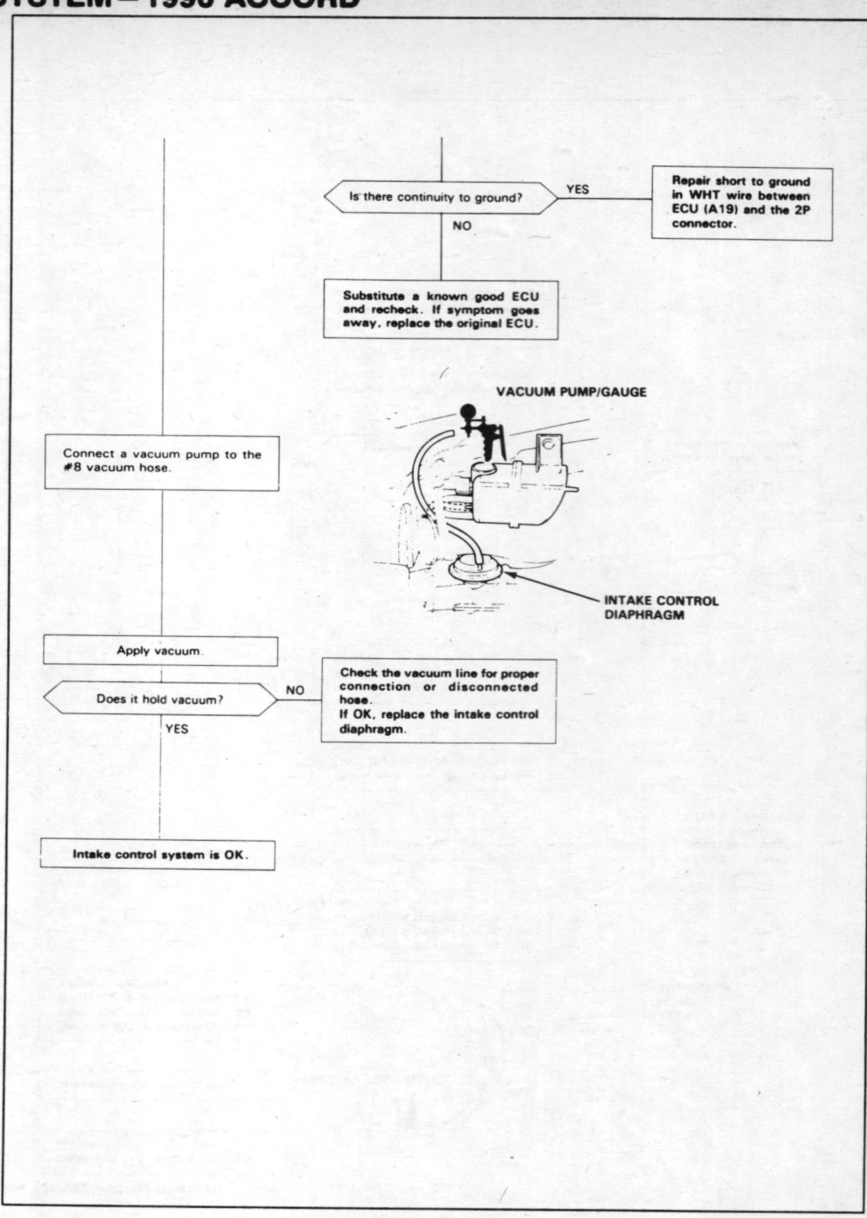


Connect a vacuum pump to the #8 vacuum hose.

Apply vacuum.

Does it hold vacuum?
- NO: Check the vacuum line for proper connection or disconnected hose. If OK, replace the intake control diaphragm.
- YES: Intake control system is OK.

# EGR SYSTEM—CODE 12—1990 ACCORD

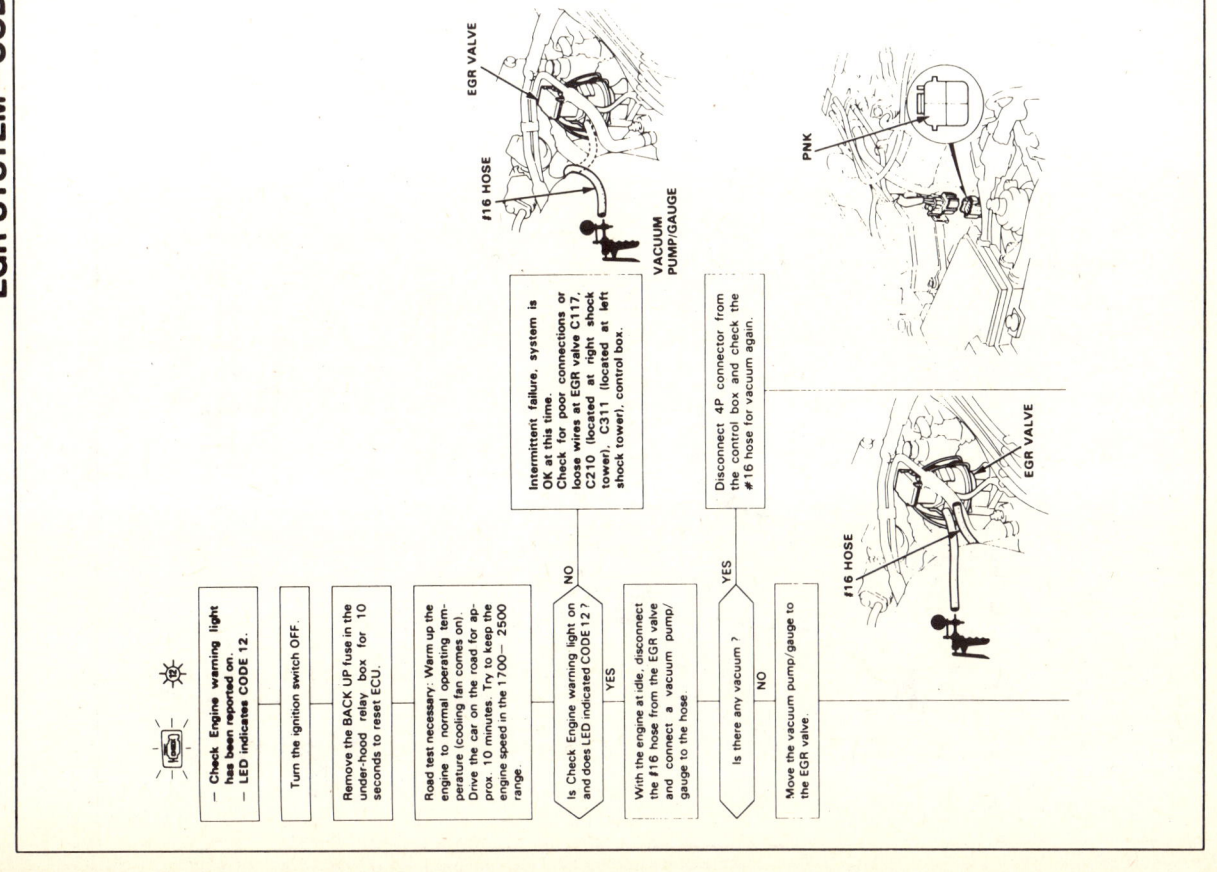

- Check Engine warning light has been reported on.
- LED indicates CODE 12.

Turn the ignition switch OFF.

Remove the BACK UP fuse in the under-hood relay box for 10 seconds to reset ECU.

Road test necessary: Warm up the engine to normal operating temperature (cooling fan comes on). Drive the car on the road for approx. 10 minutes. Try to keep the engine speed in the 1700– 2500 range.

Is Check Engine warning light on and does LED indicated CODE 12?

- NO: Intermittent failure, system is OK at this time. Check for poor connections or loose wires at EGR valve C117, C210 (located at right shock tower), C311 (located at left shock tower), control box.
- YES: With the engine at idle, disconnect the #16 hose from the EGR valve and connect a vacuum pump/gauge to the hose.

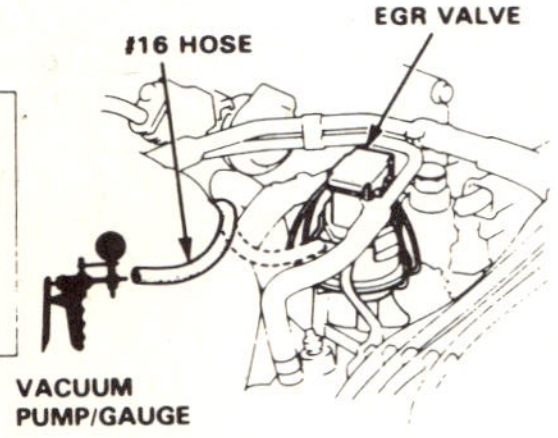

Is there any vacuum?

- YES: Disconnect 4P connector from the control box and check the #16 hose for vacuum again.

  Is there any vacuum?

  - YES: Check vacuum hose routing of the entire EGR system. If hose routing is OK, replace EGR control solenoid valve.
  - NO: Inspect for short to body ground in PNK wire between EGR control solenoid valve and ECU (A11). If wire is OK, substitute a known-good ECU and retest. If symptom/indication goes away, replace the original ECU.



- NO: Move the vacuum pump/gauge to the EGR valve.

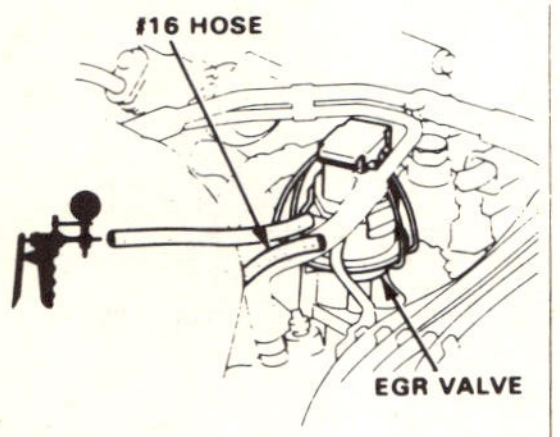

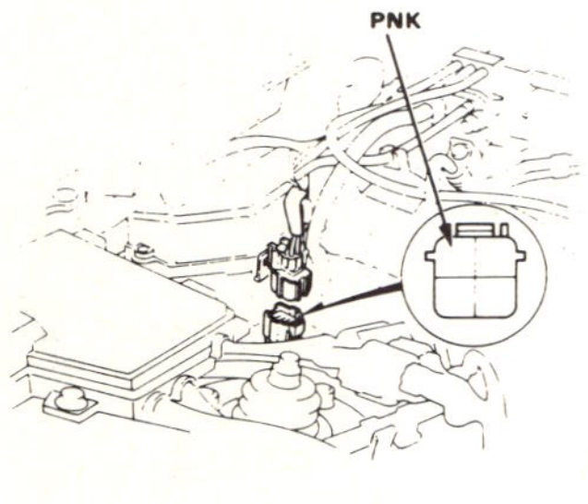

With the engine at idle, apply 8" of vacuum to the EGR valve.

Does the engine stall and does the EGR valve hold vacuum?

- NO: Replace EGR valve.
- YES: Disconnect the 4P connector from the control box.

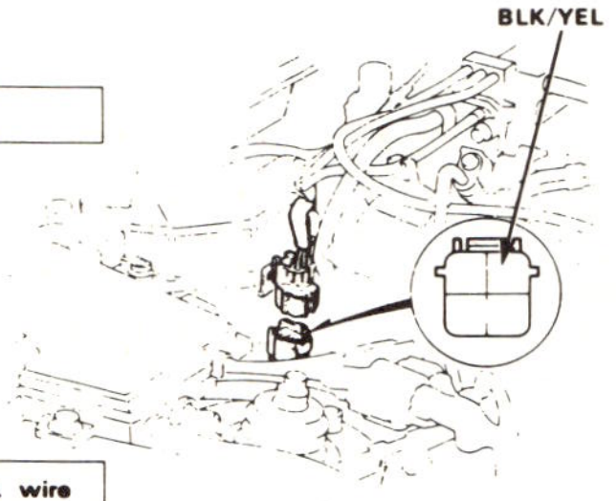

Measure voltage between BLK/YEL (+) terminal on the main wire harness and body ground.



Is there battery voltage?

- NO: Repair open in BLK/YEL wire between the solenoid valve and No. 2 fuse.
- YES: Reconnect the vacuum pump/gauge to the #16 hose.

Start the engine and allow it to idle.

## EGR SYSTEM – CODE 12 – 1990 ACCORD

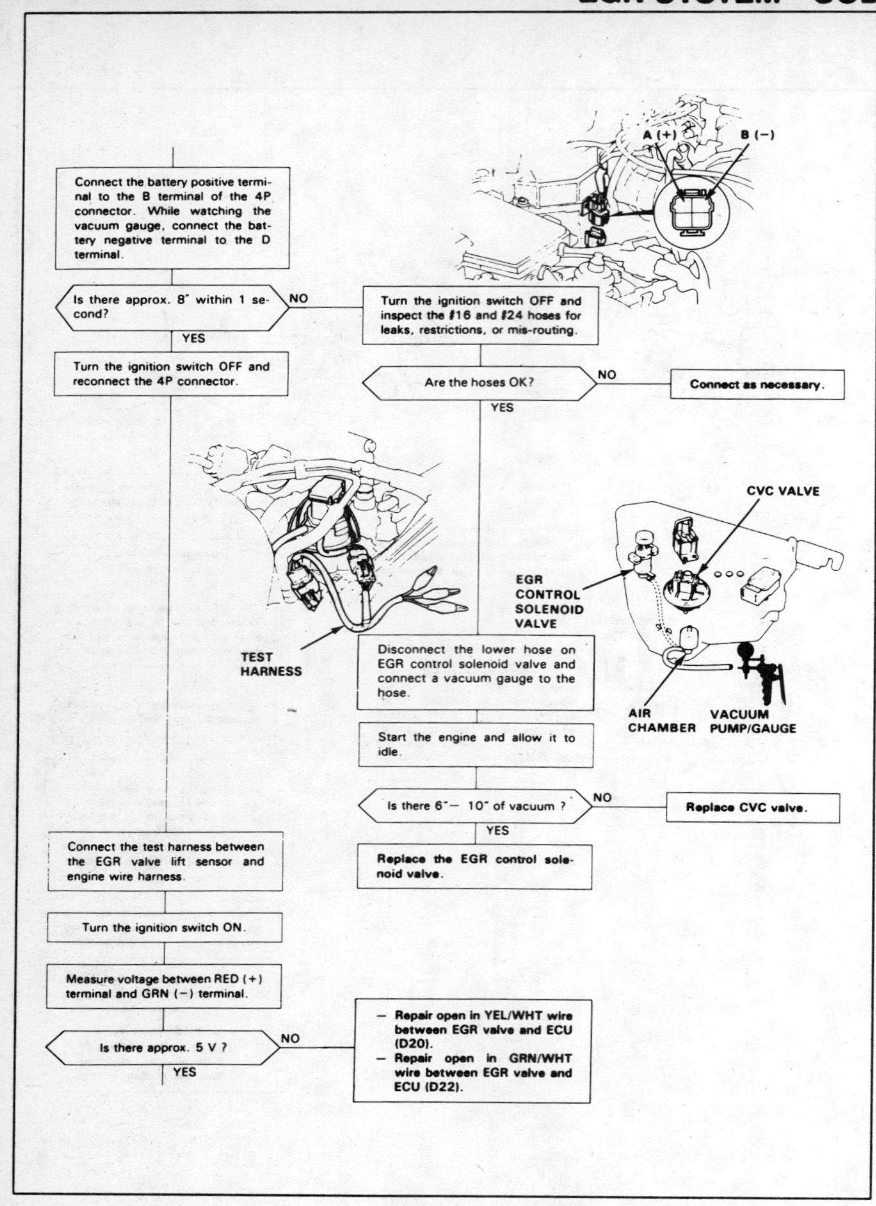


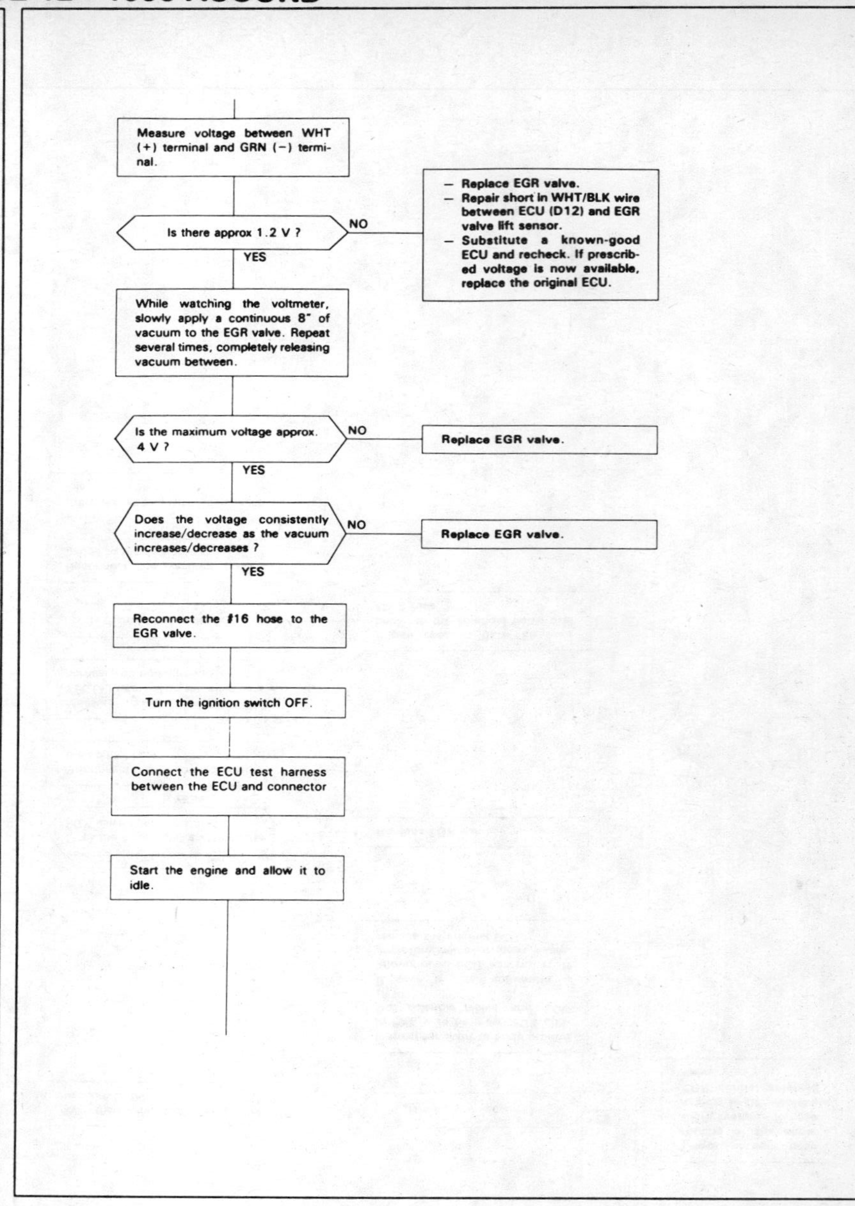

Measure voltage between WHT (+) terminal and GRN (–) terminal.
Is there approx 1.2 V ?
NO
– Replace EGR valve.
– Repair short in WHT/BLK wire between ECU (D12) and EGR valve lift sensor.
– Substitute a known-good ECU and recheck. If prescribed voltage is now available, replace the original ECU.
YES
While watching the voltmeter, slowly apply a continuous 8" of vacuum to the EGR valve. Repeat several times, completely releasing vacuum between.
Is the maximum voltage approx. 4 V ?
NO
Replace EGR valve.
YES
Does the voltage consistently increase/decrease as the vacuum increases/decreases ?
NO
Replace EGR valve.
YES
Reconnect the #16 hose to the EGR valve.
Turn the ignition switch OFF.
Connect the ECU test harness between the ECU and connector
Start the engine and allow it to idle.

## EGR SYSTEM—CODE 12—1990 ACCORD

Measure voltage between D12 (+) terminal and D22 (−) terminal.

D12 (+) D22 (−)
1 V ?

Is there approx. 1.2 V ?

NO → Repair open in WHT/BLK wire between ECU (D12) and the sensor.

YES ↓

Connect A11 terminal to A26 terminal with a jumper wire.

Did the engine stall ?

NO → Repair open in PNK wire between ECU (A11) and EGR control solenoid valve.

YES ↓

Substitute a known-good ECU and recheck. If symptom/indication goes away, replace the original ECU.

## EVAPORATIVE CONTROLS—1990 ACCORD

Troubleshooting Flowchart

Inspection of Evaporative Emission Controls

Disconnect vacuum hose from the purge control diaphragm valve (on the charcoal canister) and connect a vacuum gauge to the hose.

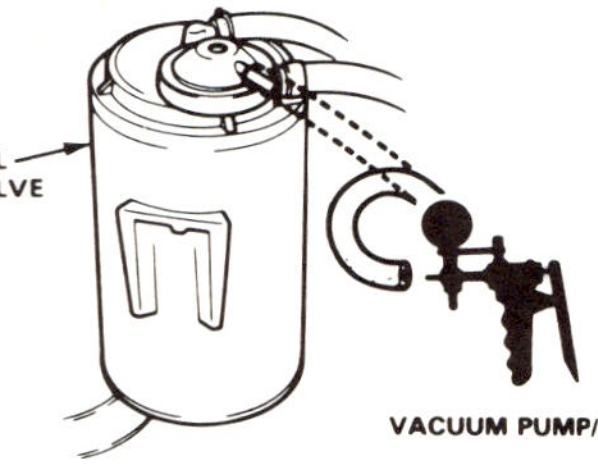

PURGE CONTROL DIAPHRAGM VALVE

VACUUM PUMP/GAUGE

Start the engine and allow to idle.
NOTE: Engine coolant temperature must be below 75°C (167°F).

Is there vacuum ?

NO ↓

YES → Disconnect the 4P connector.

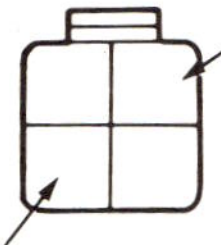

BLK/YEL (+)

RED/GRN (−)

Measure voltage between BLK/YEL (+) terminal and RED/GRN (−) terminal.

Is there battery voltage ?

YES → Inspect vacuum hose routing. If OK, replace purge cut-off solenoid valve.

NO ↓

Measure voltage between BLK/YEL (+) terminal and body ground.

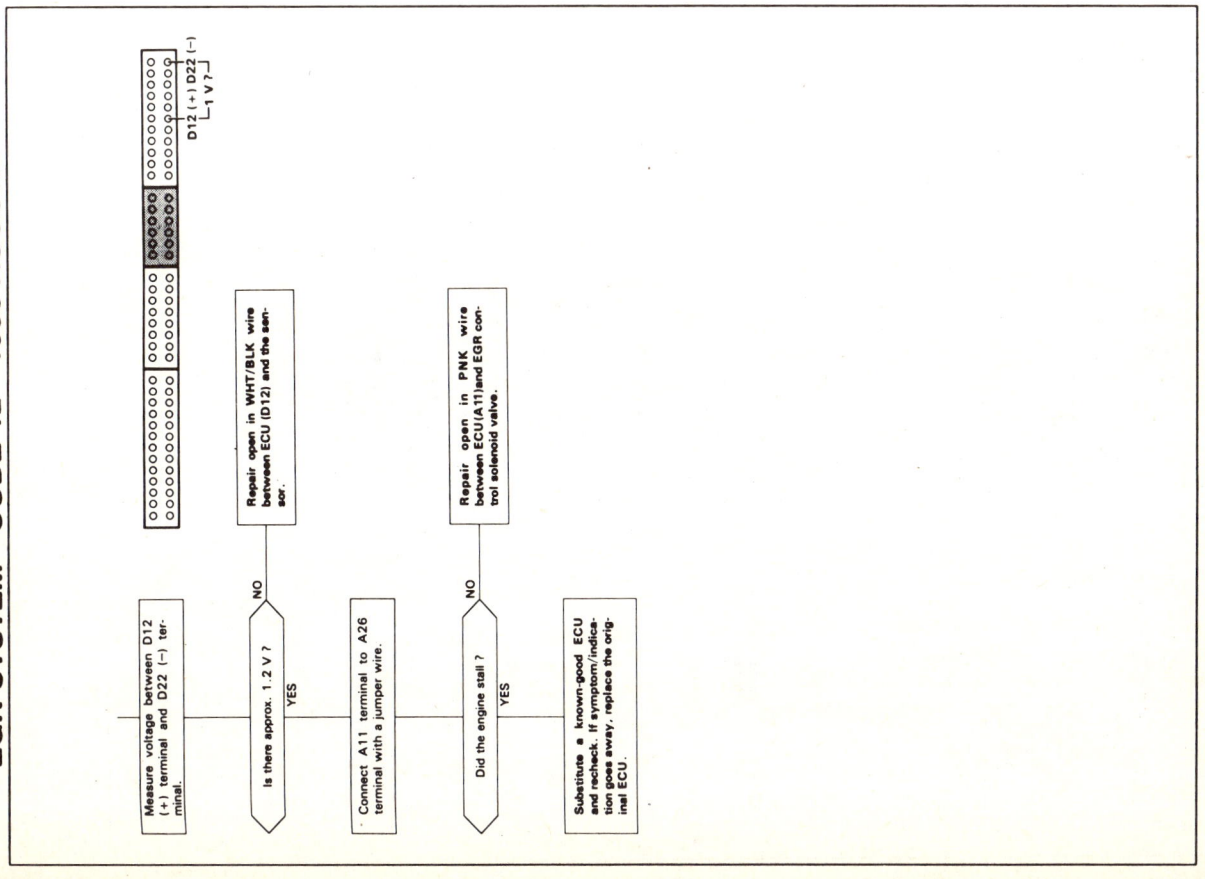

## EVAPORATIVE CONTROLS—1990 ACCORD

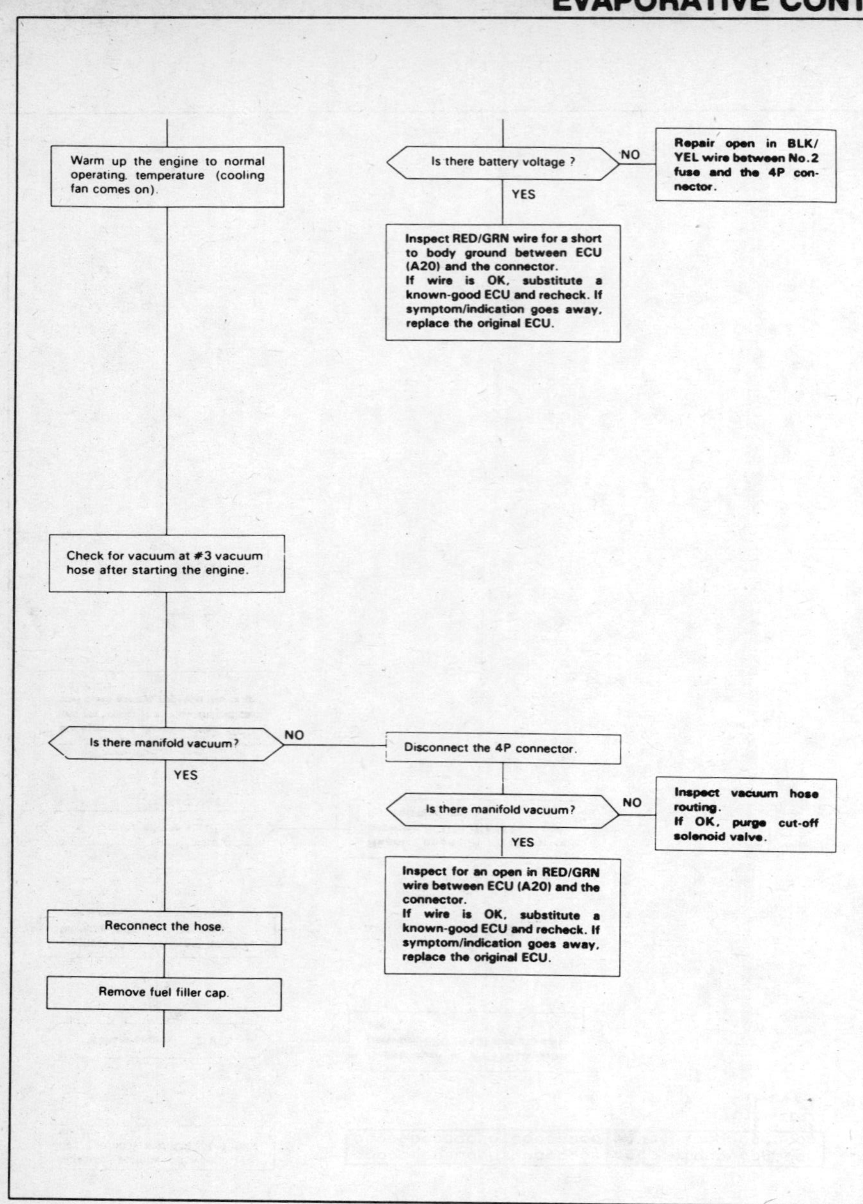


Connect a vacuum gauge to canister purge air hose.

VACUUM/PRESSURE GAUGE, 0–4 in. Hg 07JAZ–001000A

PURGE HOSE

PURGE AIR HOSE

Start the engine and raise speed to 3,500 rpm.

Does vacuum appear on gauge within 1 minute ?

YES — See two-way valve test to complete. Evaporative emission control are OK.

NO — Connect a vacuum gauge to the canister purge hose and raise the engine speed to 3,500 rpm.

Does vacuum appear on the gauge within 1 minute ?

NO — Inspect the purge hose. If hose is OK, replace the throttle body.

YES — Replace the canister.

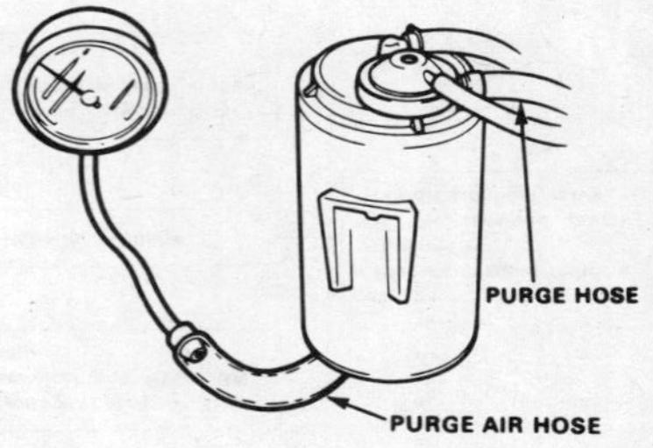

# TROUBLESHOOTING GUIDE – 1988–90 CIVIC 1.5L EXCEPT CRX HF

NOTE: Across each row in the chart, the systems that could be sources of a symptom are ranked in the order they should be inspected starting with (1). Find the symptom in the left column, read across to the most likely source, then refer to the page listed at the top of that column. If inspection shows the system is OK, try the next most likely system 2, etc.

| SYMPTOM \ SYSTEM | | PGM-FI: ECU | PGM-FI: OXYGEN SENSOR | PGM-FI: MANIFOLD ABSOLUTE PRESSURE SENSOR | PGM-FI: TDC/CRANK SENSOR | PGM-FI: COOLANT TEMPERATURE SENSOR | PGM-FI: THROTTLE ANGLE SENSOR | PGM-FI: INTAKE AIR TEMPERATURE SENSOR | PGM-FI: ATMOSPHERIC PRESSURE SENSOR |
|---|---|---|---|---|---|---|---|---|---|
| CHECK ENGINE WARNING LIGHT TURNS ON | | □ or ☼ | [CHECK] | [CHECK] | [CHECK] | [CHECK] | [CHECK] | [CHECK] | [CHECK] |
| SELF-DIAGNOSIS INDICATOR (LED) BLINKS | | (0) or (*) | *1* | *3* or *5* | *4* or *8* | *6* | *7* | *10* | *13* |
| ENGINE WON'T START | | (3) | | | | | | | |
| DIFFICULT TO START ENGINE WHEN COLD | | (BU) | | (3) | | (1) | | | |
| IRREGULAR IDLING | WHEN COLD FAST IDLE OUT OF SPEC | (BU) | | | | (3) | | | |
| | ROUGH IDLE | (BU) | | (3) | | | | | |
| | WHEN WARM RPM TOO HIGH | (BU) | | | | | | | |
| | WHEN WARM RPM TOO LOW | (BU) | | | | | | | |
| FREQUENT STALLING | WHILE WARMING UP | (BU) | | (3) | | | | | |
| | AFTER WARMING UP | (BU) | | | | | | | |
| POOR PERFORMANCE | MISFIRE OR ROUGH RUNNING | (BU) | | | | | | | |
| | FAILS EMISSION TEST | (BU) | (3) | (2) | | | | | |
| | LOSS OF POWER | (BU) | | | | | (2) | | |

| SYMPTOM | | PGM-FI: IGNITION OUTPUT SIGNAL | PGM-FI: VEHICLE SPEED SENSOR | PGM-FI: LOCK-UP CONTROL SOLENOID VALVE | PGM-FI: ELECTRIC LOAD DETECTOR | IDLE CONTROL: ELECTRONIC AIR CONTROL VALVE | IDLE CONTROL: OTHER IDLE CONTROLS | FUEL SUPPLY: FUEL INJECTOR | FUEL SUPPLY: OTHER FUEL SUPPLY | AIR INTAKE | EMISSION CONTROL: EGR CONTROL SYSTEM (CAL: A/T only) | EMISSION CONTROL: OTHER EMISSION CONTROLS |
|---|---|---|---|---|---|---|---|---|---|---|---|---|
| CHECK ENGINE WARNING LIGHT TURNS ON | | [CHECK] | [CHECK] | [CHECK] | □ | [CHECK] | | [CHECK] | | | [CHECK] | |
| SELF-DIAGNOSIS INDICATOR (LED) BLINKS | | *15* | *17* | *19* | *20* | *14* | | *16* | | | *12* | |
| ENGINE WON'T START | | | | | | | | (2) | (1) | | | |
| DIFFICULT TO START ENGINE WHEN COLD | | | | | | (2) | | | | | | |
| IRREGULAR IDLING | WHEN COLD FAST IDLE OUT OF SPEC | | | | | (1) | (2) | | | | | |
| | ROUGH IDLE | | | | | (1) | | (3) | | | (3) | |
| | WHEN WARM RPM TOO HIGH | | | | (3) | (2) | (1) | | | | | |
| | WHEN WARM RPM TOO LOW | | | | (2) | (1) | | (3) | | | | |
| FREQUENT STALLING | WHILE WARMING UP | (2) | | | | (1) | | | | | | |
| | AFTER WARMING UP | | | | | (1) | | (2) | (3) | | (3) | |
| POOR PERFORMANCE | MISFIRE OR ROUGH RUNNING | | | | | (3) | | (1) | (2) | | (3) | |
| | FAILS EMISSION TEST | | | | | | | | (1) | | | |
| | LOSS OF POWER | | | | | | | (3) | (1) | | | |

* If codes other than those listed above are indicated, count the number of blinks again. If the indicator is in fact blinking these codes, substitute a known-good ECU and recheck. If the indication goes away, replace the original ECU.

(BU): When the Check Engine warning light and the self-diagnosis indicator are on, the back-up system is in operation. Substitute a known-good ECU and recheck. If the indication goes away, replace the original ECU.

# TROUBLESHOOTING GUIDE – 1988–90 CIVIC 1.6L AND CRX HF

NOTE: Across each row in the chart, the systems that could be sources of a symptom are ranked in the order they should be inspected starting with 1. Find the symptom in the left column, read across to the most likely source, then refer to the page listed at the top of that column. If inspection shows the system is OK, try the next most likely system 2, etc.

| PAGE / SYSTEM | | | PGM-FI | | | | | | | |
|---|---|---|---|---|---|---|---|---|---|---|
| SYMPTOM | | ECU | OXYGEN SENSOR | MANIFOLD ABSOLUTE PRESSURE SENSOR | TDC/CRANK/CYL SENSOR | COOLANT TEMPERATURE SENSOR | THROTTLE ANGLE SENSOR | INTAKE AIR TEMPERATURE SENSOR | ATMOSPHERIC PRESSURE SENSOR |
| CHECK ENGINE WARNING LIGHT TURNS ON | | □ or ☼ | (CHECK) | (CHECK) | (CHECK) | (CHECK) | (CHECK) | (CHECK) | (CHECK) |
| SELF-DIAGNOSIS INDICATOR (LED) BLINKS | | ⓪ or ✱ | ① | ③ or ⑤ | ④ or ⑧ or ⑨ | ⑥ | ⑦ | ⑩ | ⑬ |
| ENGINE WON'T START | | ③ | | | | | | | |
| DIFFICULT TO START ENGINE WHEN COLD | | (BU) | | ③ | | ① | | | |
| IRREGULAR IDLING | WHEN COLD FAST IDLE OUT OF SPEC | (BU) | | | | ② | | | |
| | ROUGH IDLE | (BU) | | ③ | | | | | |
| | WHEN WARM RPM TOO HIGH | (BU) | | | | | | | |
| | WHEN WARM RPM TOO LOW | (BU) | | | | | | | |
| FREQUENT STALLING | WHILE WARMING UP | (BU) | | | | | | | |
| | AFTER WARMING UP | (BU) | | | | | | | |
| POOR PERFORMANCE | MISFIRE OR ROUGH RUNNING | (BU) | | | | | | | |
| | FAILS EMISSION TEST | (BU) | ③ | ② | | | | | |
| | LOSS OF POWER | (BU) | | ③ | | | ② | | |

| PAGE / SYSTEM | | PGM-FI | | | IDLE CONTROL | | FUEL SUPPLY | | | |
|---|---|---|---|---|---|---|---|---|---|---|
| SYMPTOM | | IGNITION OUTPUT SIGNAL | VEHICLE SPEED SENSOR | ELECTRIC LOAD DETECTOR | ELECTRONIC AIR CONTROL VALVE | OTHER IDLE CONTROLS | FUEL INJECTOR | OTHER FUEL SUPPLY | AIR INTAKE | EMISSION CONTROL |
| CHECK ENGINE WARNING LIGHT TURNS ON | | (CHECK) | (CHECK) | □ | (CHECK) | | (CHECK) | | | |
| SELF-DIAGNOSIS INDICATOR (LED) BLINKS | | ⑮ | ⑰ | ⑳ | ⑭ | | ⑯ | | | |
| ENGINE WON'T START | | | | | | | ② | ① | | |
| DIFFICULT TO START ENGINE WHEN COLD | | | | | | ② | | | | |
| IRREGULAR IDLING | WHEN COLD FAST IDLE OUT OF SPEC | | | | ① | ② | | | | |
| | ROUGH IDLE | | | | ① | | ② | | | |
| | WHEN WARM RPM TOO HIGH | | | ③ | ② | ① | | | | |
| | WHEN WARM RPM TOO LOW | | | ③ | ① | | ② | | | |
| FREQUENT STALLING | WHILE WARMING UP | | | | ① | ② | | ③ | | |
| | AFTER WARMING UP | | | | ① | ② | | ③ | | |
| POOR PERFORMANCE | MISFIRE OR ROUGH RUNNING | | | | ② | | ① | | | |
| | FAILS EMISSION TEST | | | | | | ① | | | |
| | LOSS OF POWER | | | | | | | ① | | |

* If codes other than those listed above are indicated, count the number of blinks again. If the indicator is in fact blinking these codes, substitute a known-good ECU and recheck. If the indication goes away, replace the original ECU.

(BU): When the Check Engine warning light and the self-diagnosis indicator are on, the back-up system is in operation. Substitute a known-good ECU and recheck. If the indication goes away, replace the original ECU.

## ECU AND CHECK LIGHT—1988 CRX

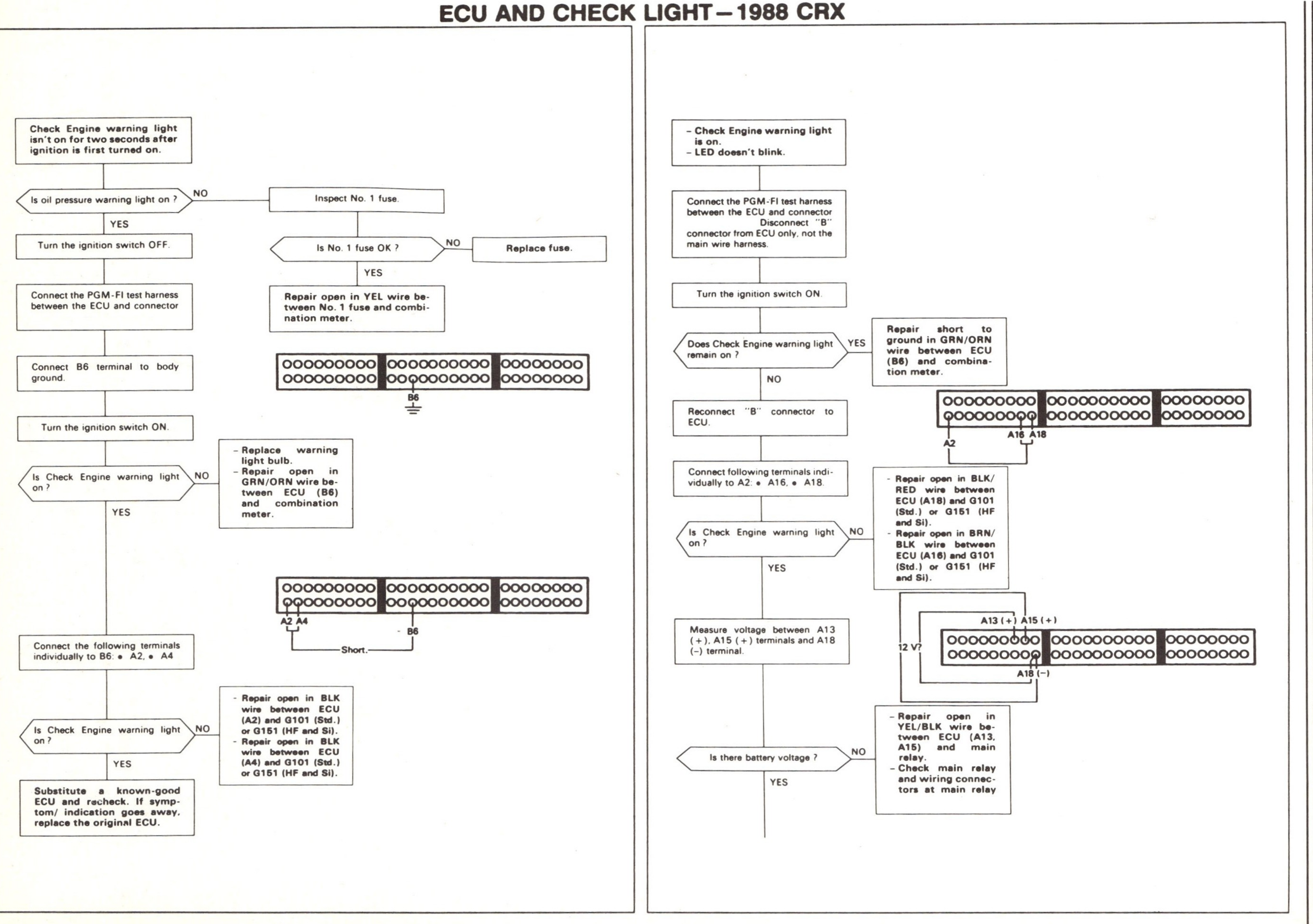

Check Engine warning light isn't on for two seconds after ignition is first turned on.
Is oil pressure warning light on ?
NO
Inspect No. 1 fuse.
Is No. 1 fuse OK ?
NO
Replace fuse.
YES
Repair open in YEL wire between No. 1 fuse and combination meter.
YES
Turn the ignition switch OFF.
Connect the PGM-FI test harness between the ECU and connector
Connect B6 terminal to body ground.
B6
Turn the ignition switch ON.
Is Check Engine warning light on ?
NO
- Replace warning light bulb.
- Repair open in GRN/ORN wire between ECU (B6) and combination meter.
YES
A2 A4
B6
Short.
Connect the following terminals individually to B6: • A2, • A4
Is Check Engine warning light on ?
NO
- Repair open in BLK wire between ECU (A2) and G101 (Std.) or G151 (HF and Si).
- Repair open in BLK wire between ECU (A4) and G101 (Std.) or G151 (HF and Si).
YES
Substitute a known-good ECU and recheck. If symptom/ indication goes away, replace the original ECU.
- Check Engine warning light is on.
- LED doesn't blink.
Connect the PGM-FI test harness between the ECU and connector Disconnect "B" connector from ECU only, not the main wire harness.
Turn the ignition switch ON.
Does Check Engine warning light remain on ?
YES
Repair short to ground in GRN/ORN wire between ECU (B6) and combination meter.
NO
A2
A16 A18
Reconnect "B" connector to ECU.
Connect following terminals individually to A2: • A16, • A18.
Is Check Engine warning light on ?
NO
- Repair open in BLK/RED wire between ECU (A18) and G101 (Std.) or G151 (HF and Si).
- Repair open in BRN/BLK wire between ECU (A16) and G101 (Std.) or G151 (HF and Si).
YES
A13 (+) A15 (+)
12 V?
A18 (−)
Measure voltage between A13 (+), A15 (+) terminals and A18 (−) terminal.
Is there battery voltage ?
NO
- Repair open in YEL/BLK wire between ECU (A13, A15) and main relay.
- Check main relay and wiring connectors at main relay
YES

## ECU AND CHECK LIGHT 1988–90 CIVIC EXCEPT 1988 CRX

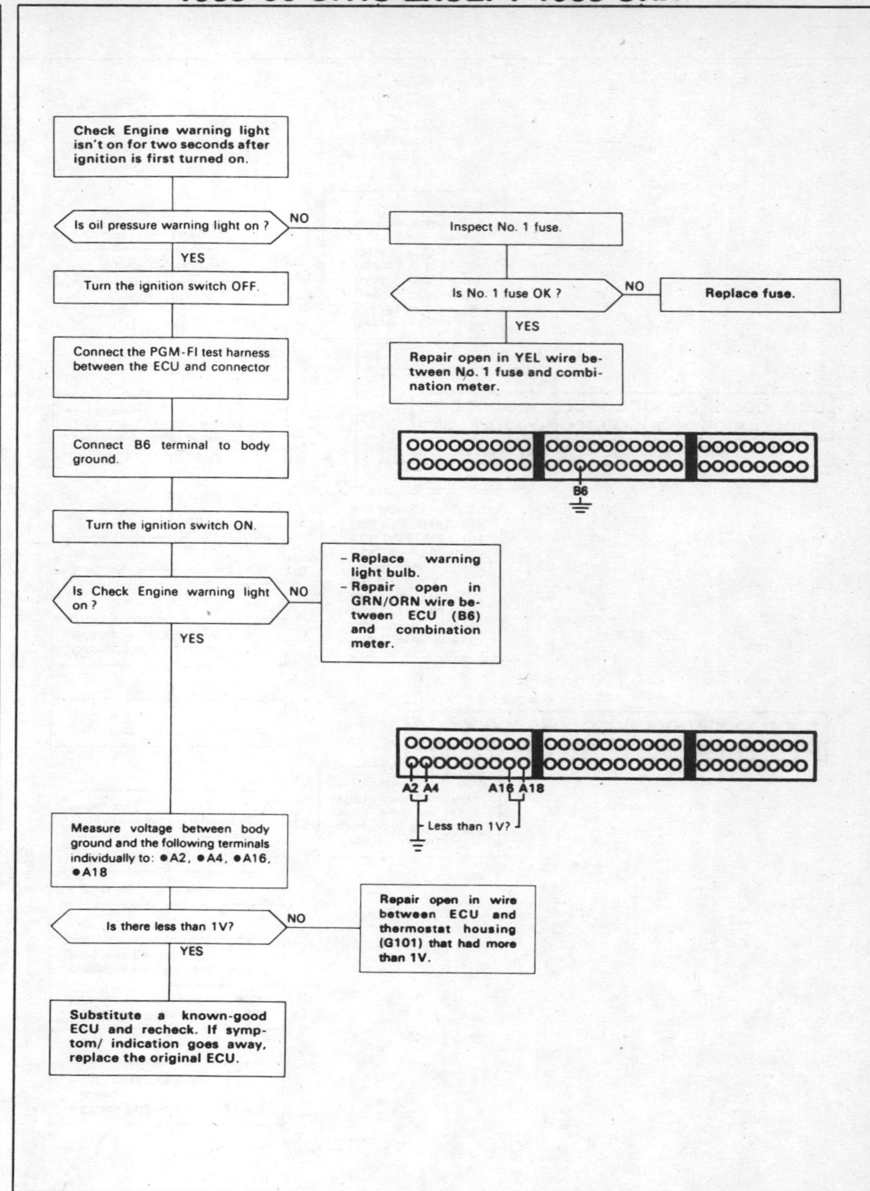


## ECU AND CHECK LIGHT – 1988 CRX

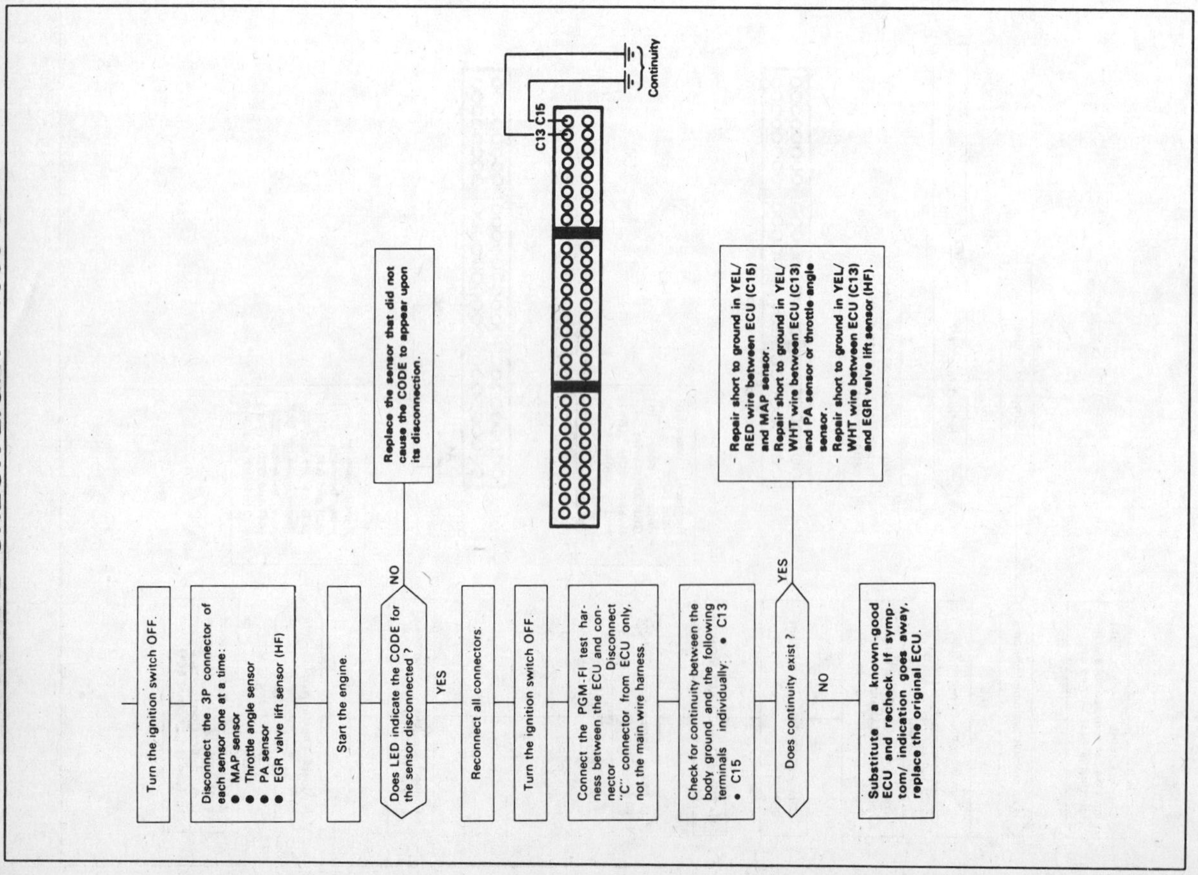

Turn the ignition switch OFF.

Disconnect the 3P connector of each sensor one at a time:
- MAP sensor
- Throttle angle sensor
- PA sensor
- EGR valve lift sensor (HF)

Start the engine.

Does LED indicate the CODE for the sensor disconnected ?

NO → **Replace the sensor that did not cause the CODE to appear upon its disconnection.**

YES

Reconnect all connectors.

Turn the ignition switch OFF.

Connect the PGM-FI test harness between the ECU and connector Disconnect "C" connector from ECU only, not the main wire harness.

C13 C15

Continuity

Check for continuity between the body ground and the following terminals individually: • C13 • C15

Does continuity exist ?

YES →
- **Repair short to ground in YEL/RED wire between ECU (C15) and MAP sensor.**
- **Repair short to ground in YEL/WHT wire between ECU (C13) and PA sensor or throttle angle sensor.**
- **Repair short to ground in YEL/WHT wire between ECU (C13) and EGR valve lift sensor (HF).**

NO

**Substitute a known-good ECU and recheck. If symptom/ indication goes away, replace the original ECU.**

## ECU AND CHECK LIGHT – 1988–90 CIVIC EXCEPT 1988 CRX

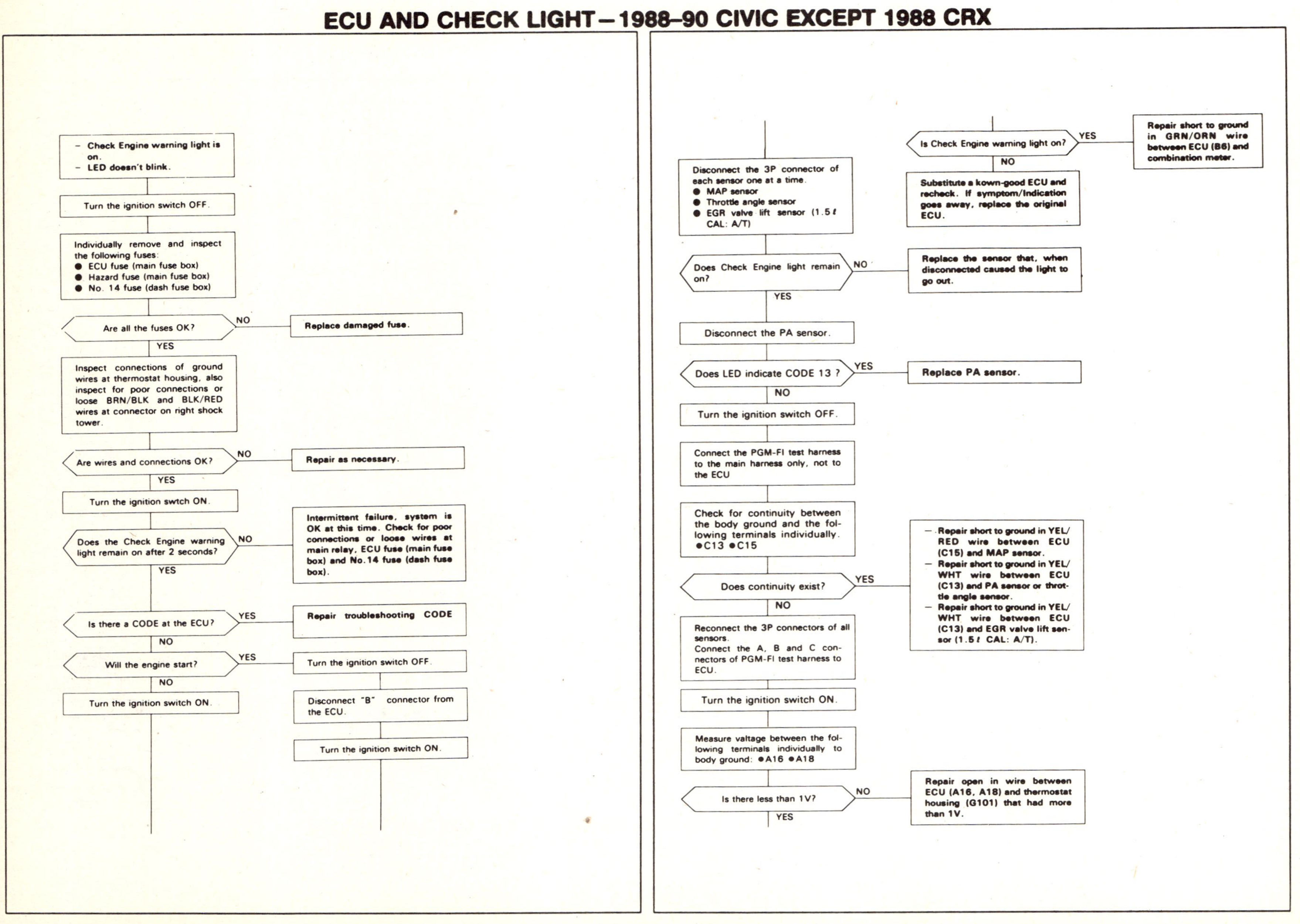

– Check Engine warning light is on.
– LED doesn't blink.
Turn the ignition switch OFF.
Individually remove and inspect the following fuses:
● ECU fuse (main fuse box)
● Hazard fuse (main fuse box)
● No. 14 fuse (dash fuse box)
Are all the fuses OK?
NO
Replace damaged fuse.
YES
Inspect connections of ground wires at thermostat housing, also inspect for poor connections or loose BRN/BLK and BLK/RED wires at connector on right shock tower.
Are wires and connections OK?
NO
Repair as necessary.
YES
Turn the ignition swtch ON.
Does the Check Engine warning light remain on after 2 seconds?
NO
Intermittent failure, system is OK at this time. Check for poor connections or loose wires at main relay, ECU fuse (main fuse box) and No. 14 fuse (dash fuse box).
YES
Is there a CODE at the ECU?
YES
Repair troubleshooting CODE
NO
Will the engine start?
YES
Turn the ignition switch OFF.
Disconnect "B" connector from the ECU.
Turn the ignition switch ON.
NO
Turn the ignition switch ON.
Is Check Engine warning light on?
YES
Repair short to ground in GRN/ORN wire between ECU (B6) and combination meter.
NO
Substitute a kown-good ECU and recheck. If symptom/indication goes away, replace the original ECU.
Disconnect the 3P connector of each sensor one at a time.
● MAP sensor
● Throttle angle sensor
● EGR valve lift sensor (1.5 ℓ CAL: A/T)
Does Check Engine light remain on?
NO
Replace the sensor that, when disconnected caused the light to go out.
YES
Disconnect the PA sensor.
Does LED indicate CODE 13 ?
YES
Replace PA sensor.
NO
Turn the ignition switch OFF.
Connect the PGM-FI test harness to the main harness only, not to the ECU
Check for continuity between the body ground and the following terminals individually.
●C13 ●C15
Does continuity exist?
YES
– Repair short to ground in YEL/RED wire between ECU (C15) and MAP sensor.
– Repair short to ground in YEL/WHT wire between ECU (C13) and PA sensor or throttle angle sensor.
– Repair short to ground in YEL/WHT wire between ECU (C13) and EGR valve lift sensor (1.5 ℓ CAL: A/T).
NO
Reconnect the 3P connectors of all sensors.
Connect the A, B and C connectors of PGM-FI test harness to ECU.
Turn the ignition switch ON.
Measure valtage between the following terminals individually to body ground: ●A16 ●A18
Is there less than 1V?
NO
Repair open in wire between ECU (A16, A18) and thermostat housing (G101) that had more than 1V.
YES

## ECU AND CHECK LIGHT 1988–90 CIVIC EXCEPT 1988 CRX

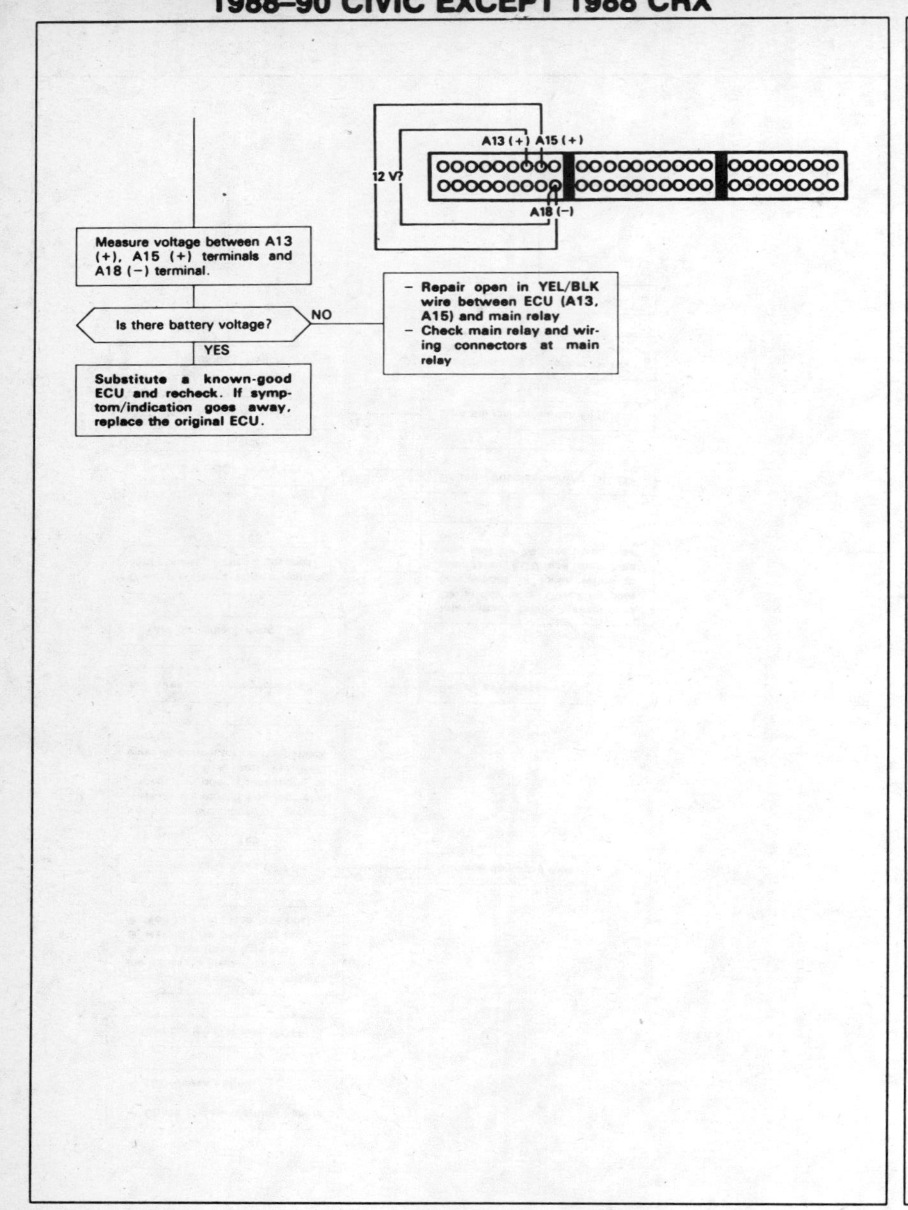


## OXYGEN SENSOR – CODE 1 – 1988 CRX

Self-diagnosis LED blinks once: A problem in the Oxygen ($O_2$) Sensor circuit.

The oxygen sensor, detects the oxygen content in the exhaust gas, and inputs the ECU. In operation, the ECU receives the signals from the sensor and varies the duration during which fuel is injected. The oxygen sensor is installed on the exhaust manifold.

EXHAUST GAS
ZIRCONIA
ATMOSPHERE
PLATINUM
VOLTAGE (V)
STOICHIOMETRIC AIR-FUEL RATIO
RICH ← AIR-FUEL RATIO → LEAN

- Check Engine warning light has been reported on.
- LED indicates CODE 1.

Turn the ignition switch OFF.

Remove HAZARD fuse in the main fuse box for 10 seconds to reset ECU.

Inspect pressure regulator (page

Is it normal?
- NO → Replace the pressure regulator
- YES ↓

Warm up engine to normal operating temperature (cooling fan comes on).

Hold engine at 1500 rpm for 15 minutes.
NOTE: Do not close throttle completely during this time.

Is Check Engine warning light on?
Does LED indicate CODE 1?
- NO → Intermittent failure (test drive may be necessary).
- YES

## OXYGEN SENSOR—CODE 1 1988–90 CIVIC EXCEPT 1988 CRX

Self-diagnosis LED blinks once: A problem in the Oxygen ($O_2$) Sensor circuit.

The oxygen sensor, detects the oxygen content in the exhaust gas, and inputs the ECU. In operation, the ECU receives the signals from the sensor and varies the duration during which fuel is injected. The oxygen sensor is installed on the exhaust manifold.

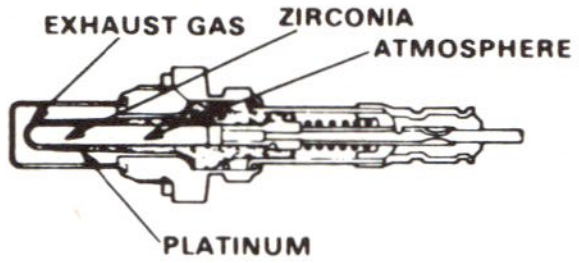

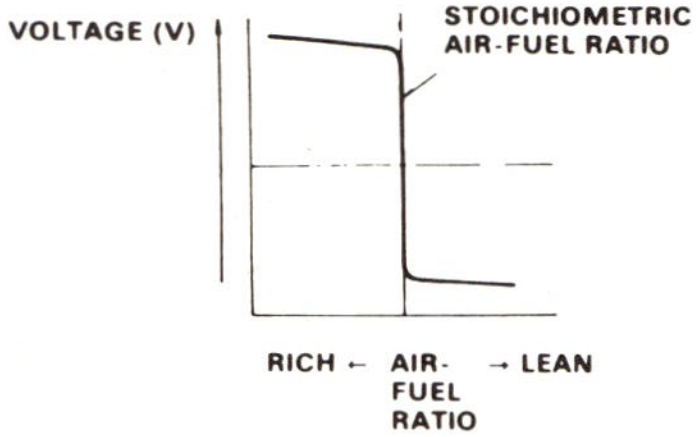

EXHAUST GAS
ZIRCONIA
ATMOSPHERE
PLATINUM
VOLTAGE (V)
STOICHIOMETRIC AIR-FUEL RATIO
RICH ← AIR-FUEL RATIO → LEAN

- Check Engine warning light has been reported on.
- LED indicates CODE 1.

Turn the ignition switch OFF.

Remove HAZARD fuse in the main fuse box for 10 seconds to reset ECU.

Inspect pressure regulator.

Is it normal?
- NO → **Replace the pressure regulator**
- YES ↓

Warm up engine to normal operating temperature (cooling fan comes on).

**1.5 ℓ**

Block rear wheels and set the parking brake. Jack up the front of car and support with safety stands.

**WARNING Block rear wheels before jacking up front of car.**

Warm up engine to normal operating temperature again, then put the transmission into second gear and run the engine at 2,000 rpm for 15 minutes.
NOTE: Do not close throttle completely during this time.

**HF and 1.6 ℓ**

Hold engine at 1500 rpm for 15 minutes.
NOTE: Do not close throttle completely during this time.

Is Check Engine warning light on and does LED indicate CODE 1?
- NO → **Intermittent failure, system is OK at this time (test drive may be necessary). Check for poor connections or loose wires at the thermostat housing, $O_2$ sensor and C210 (round connector located at the right shock tower).**
- YES ↓

## OXYGEN SENSOR—CODE 1—1988 CRX

Disconnect engine wire harness from $O_2$ sensor.

Warm up engine to normal operating temperature again, then open the throttle wide open then close it.

Measure voltage between the connector terminal and body ground.

**Is voltage above 0.6V during wide open acceleration? Is voltage below 0.4V during closed throttle deceleration from 5,000 rpm?**
- NO → **Replace $O_2$ sensor.**
- YES ↓

$O_2$ SENSOR
DIGITAL MULTIMETER

Stop engine.

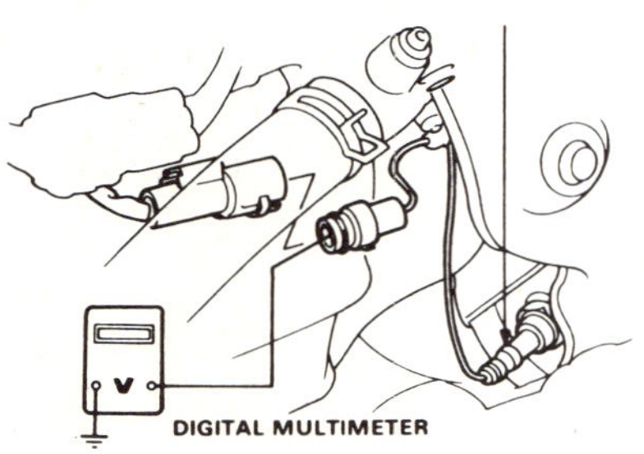

Connect the PGM-FI test harness between the ECU and connector (page 11-21).

Restart and warm up engine to normal operating temperature, then open the throttle wide open then close it.

Measure voltage between C16 (+) and A18 (−) terminals.

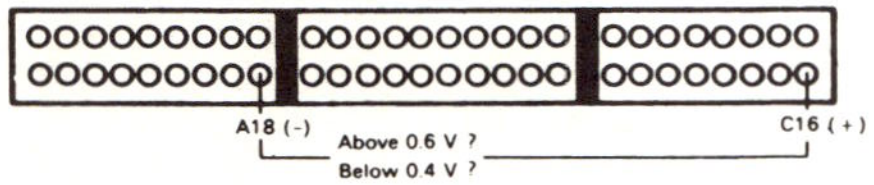

A18 (−)
C16 (+)
Above 0.6 V ?
Below 0.4 V ?

**Is voltage above 0.6V during wide open acceleration? Is voltage below 0.4V during closed throttle deceleration from 5,000 rpm?**
- NO → **Repair short or open in WHT wire between ECU (C16) and $O_2$ sensor.**
- YES ↓

**Substitute a known-good ECU and recheck. If symptom/ indication goes away, replace the original ECU.**

## OXYGEN SENSOR—CODE 1
## 1988–90 CIVIC EXCEPT 1988 CRX

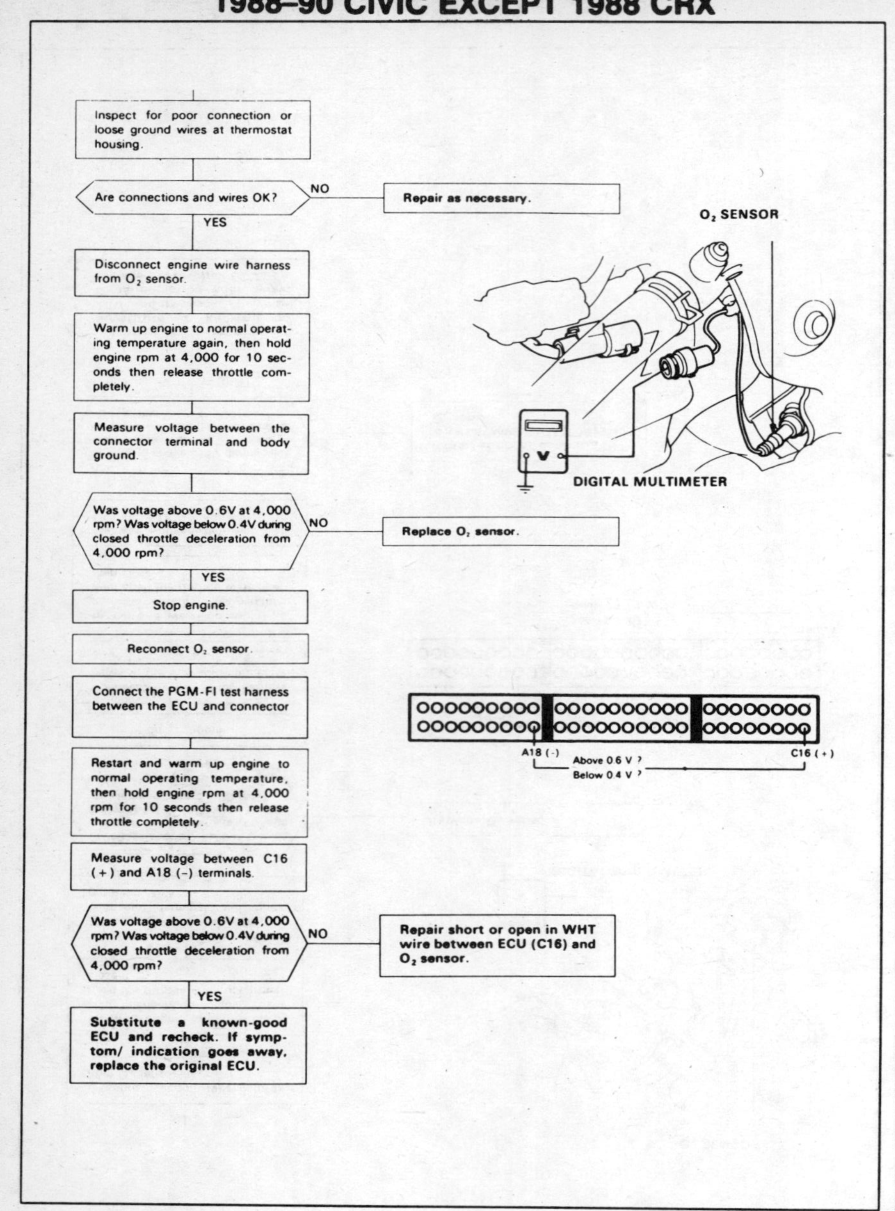


## MAP SENSOR—CODE 3—1988–90 CIVIC

Self-diagnosis LED indicator blinks three times: Most likely an electrical problem in the Manifold Absolute Pressure (MAP) Sensor system.

Self-diagnosis LED indicator blinks five times: Most likely a mechanical problem (broken hose) in the Manifold Absolute Pressure (MAP) Sensor system.

The MAP sensor converts manifold absolute pressure into electrical signals and inputs the ECU.

SILICON DIAPHARAGM

VACUUM CAVITY

OUTPUT VOLTAGE

(v) 3.5, 3.0, 2.5, 2.0, 1.5, 1.0, 0.5, 0

5, 10, 15, 20, 25, 30 (in. Hg) GAUGE

100, 200, 300, 400, 500, 600, 700 (mm Hg) READING

- Engine is warm and running.
- Check Engine warning light is on.
- LED indicates CODE 3.

Turn the ignition switch OFF.

Remove HAZARD fuse in the main fuse box for 10 seconds to reset ECU.

Warm up engine to normal operating temperature (cooling fan comes on).

Is Check Engine warning light on and does LED indicate CODE 3?

NO

**Intermittent failure, system is OK at this time. (test drive may be necessary). Check for poor connection or loose wires at MAP sensor connector and ECU.**

YES

Turn the ignition switch OFF.

Disconnect the 3P connector from the MAP sensor.

Turn the ignition switch ON.

(1.5 ℓ CAL: A/T)

YEL/RED (+)

WHT

GRN/WHT (−)

(Ex. 1.5 ℓ CAL: A/T)

YEL/RED (+)

WHT

GRN/WHT (−)

## MAP SENSOR—CODE 3—1988–90 CIVIC

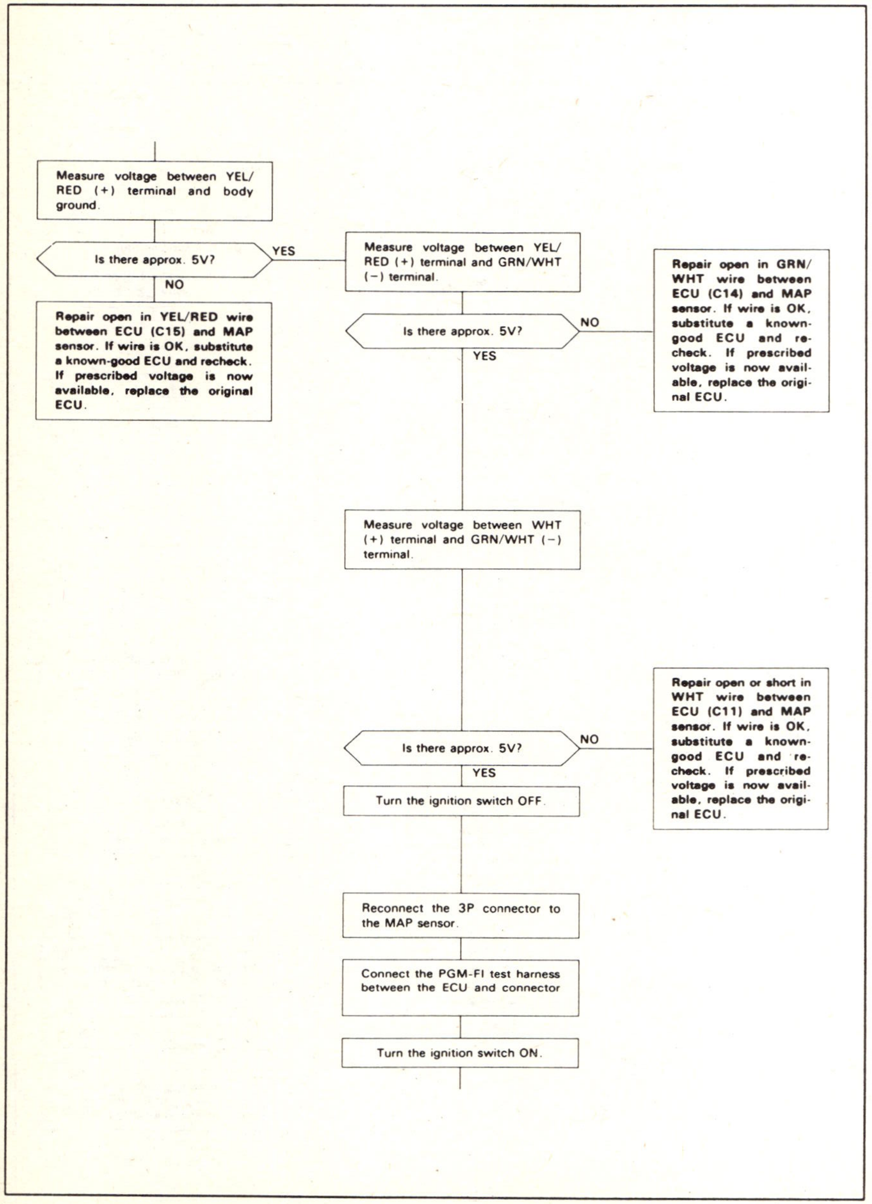

Measure voltage between YEL/RED (+) terminal and body ground.

Is there approx. 5V?

- NO: Repair open in YEL/RED wire between ECU (C15) and MAP sensor. If wire is OK, substitute a known-good ECU and recheck. If prescribed voltage is now available, replace the original ECU.
- YES: Measure voltage between YEL/RED (+) terminal and GRN/WHT (−) terminal.

Is there approx. 5V?

- NO: Repair open in GRN/WHT wire between ECU (C14) and MAP sensor. If wire is OK, substitute a known-good ECU and recheck. If prescribed voltage is now available, replace the original ECU.
- YES: Measure voltage between WHT (+) terminal and GRN/WHT (−) terminal.

Is there approx. 5V?

- NO: Repair open or short in WHT wire between ECU (C11) and MAP sensor. If wire is OK, substitute a known-good ECU and recheck. If prescribed voltage is now available, replace the original ECU.
- YES: Turn the ignition switch OFF.

Reconnect the 3P connector to the MAP sensor.

Connect the PGM-FI test harness between the ECU and connector

Turn the ignition switch ON.

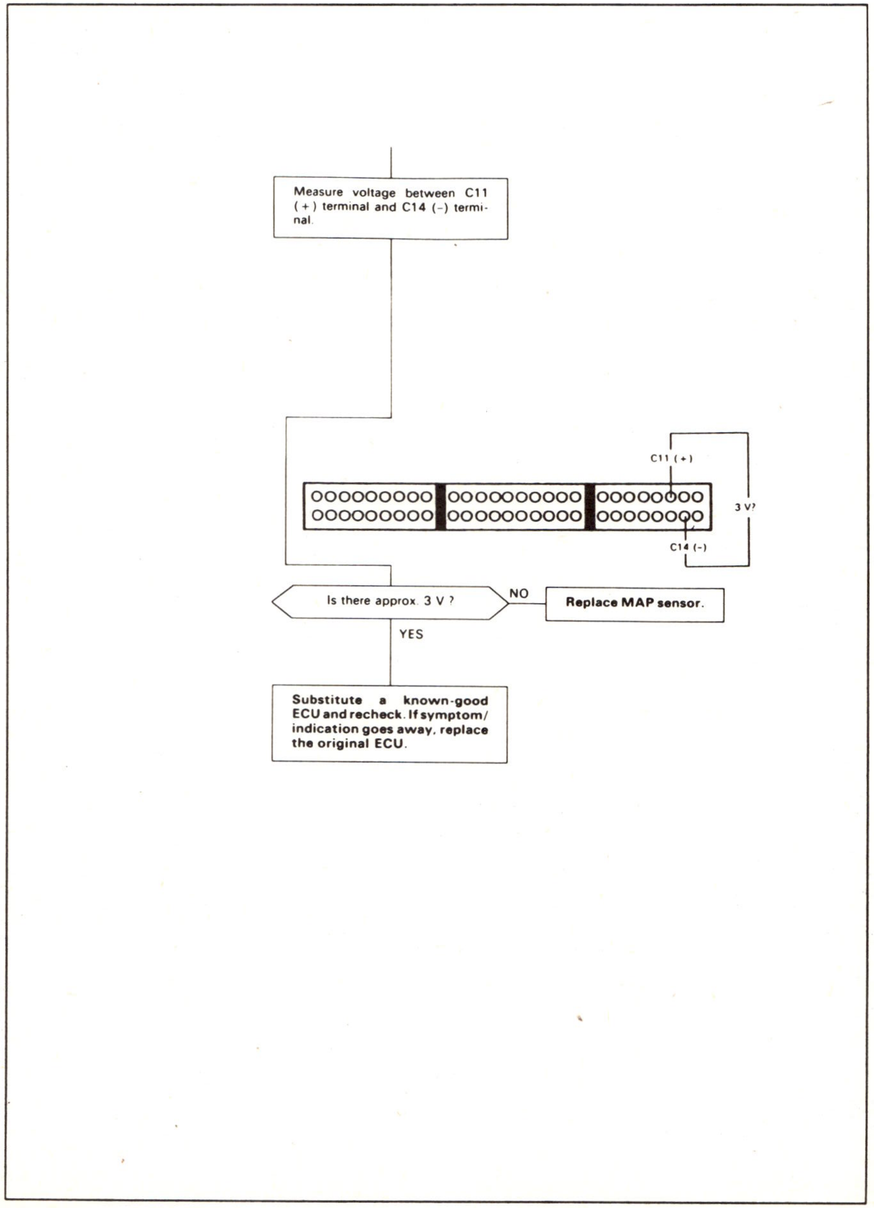

Measure voltage between C11 (+) terminal and C14 (−) terminal.

C11 (+)
3 V?
C14 (−)

Is there approx. 3 V?

- NO: Replace MAP sensor.
- YES: Substitute a known-good ECU and recheck. If symptom/indication goes away, replace the original ECU.

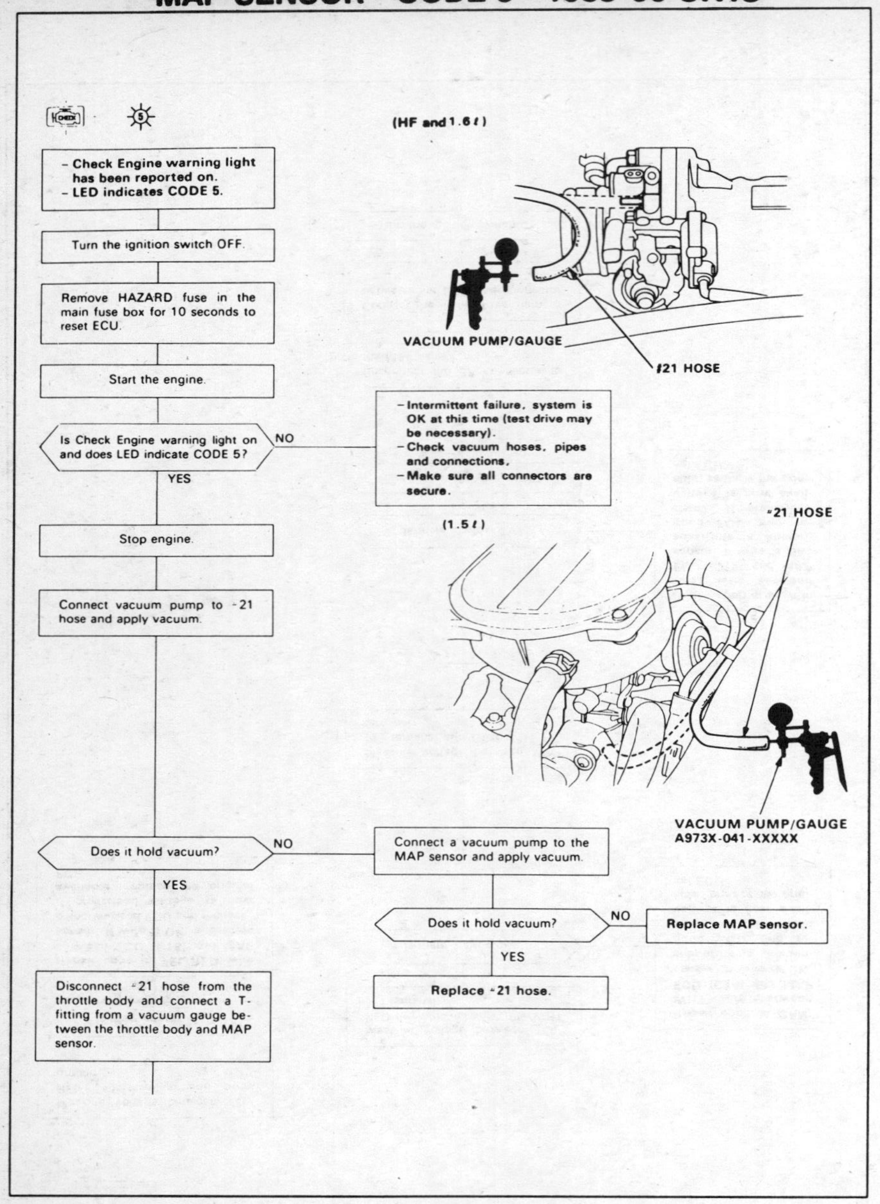
MAP SENSOR—CODE 5—1988–90 CIVIC
- Check Engine warning light has been reported on.
- LED indicates CODE 5.
Turn the ignition switch OFF.
Remove HAZARD fuse in the main fuse box for 10 seconds to reset ECU.
Start the engine.
Is Check Engine warning light on and does LED indicate CODE 5?
NO
YES
- Intermittent failure, system is OK at this time (test drive may be necessary).
- Check vacuum hoses, pipes and connections.
- Make sure all connectors are secure.
Stop engine.
Connect vacuum pump to #21 hose and apply vacuum.
Does it hold vacuum?
NO
YES
Disconnect #21 hose from the throttle body and connect a T-fitting from a vacuum gauge between the throttle body and MAP sensor.
Connect a vacuum pump to the MAP sensor and apply vacuum.
Does it hold vacuum?
NO
YES
Replace MAP sensor.
Replace #21 hose.
(HF and 1.6 l)
VACUUM PUMP/GAUGE
#21 HOSE
(1.5 l)
#21 HOSE
VACUUM PUMP/GAUGE
A973X-041-XXXXX

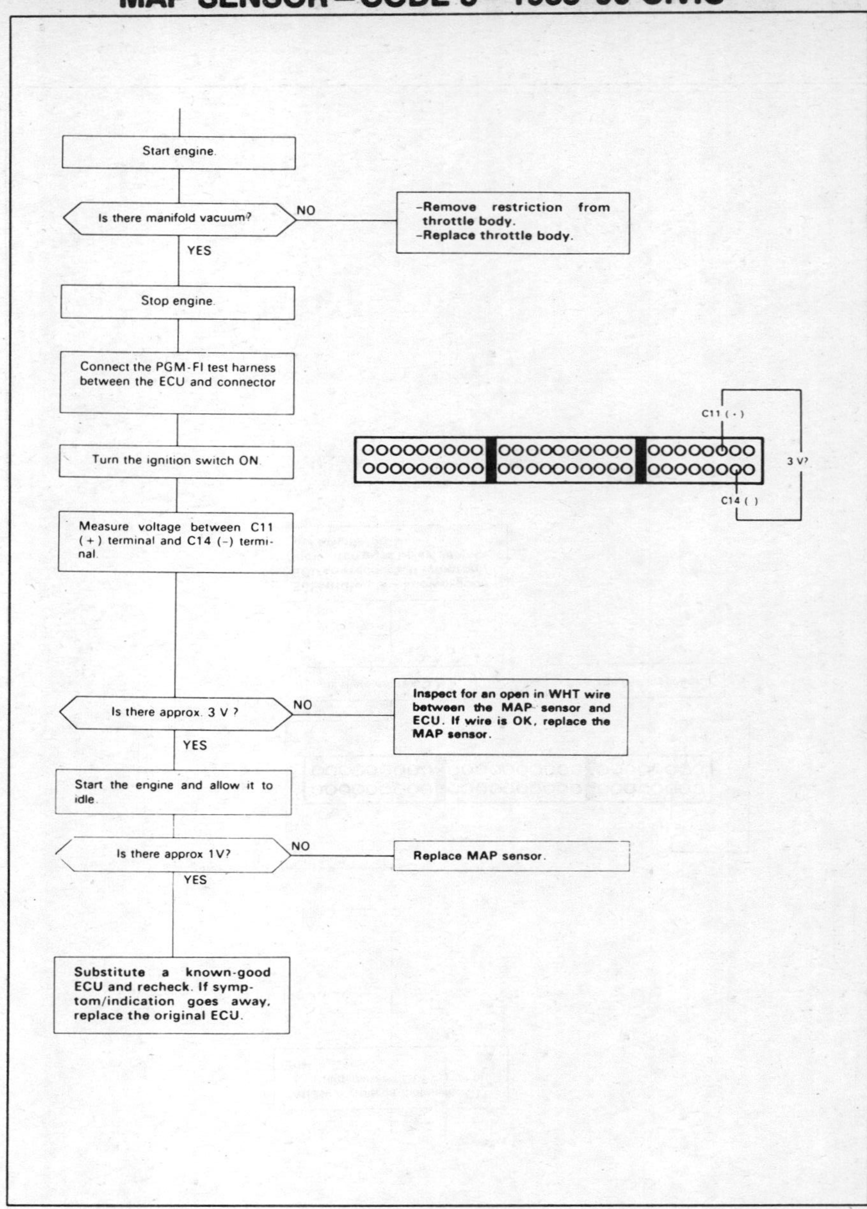
MAP SENSOR—CODE 5—1988–90 CIVIC
Start engine.
Is there manifold vacuum?
NO
YES
-Remove restriction from throttle body.
-Replace throttle body.
Stop engine.
Connect the PGM-FI test harness between the ECU and connector
Turn the ignition switch ON.
Measure voltage between C11 (+) terminal and C14 (-) terminal.
C11
3 V?
C14
Is there approx. 3 V ?
NO
YES
Inspect for an open in WHT wire between the MAP sensor and ECU. If wire is OK, replace the MAP sensor.
Start the engine and allow it to idle.
Is there approx 1V?
NO
YES
Replace MAP sensor.
Substitute a known-good ECU and recheck. If symptom/indication goes away, replace the original ECU.

# TDC/CRANK SENSOR—CODE 4—1988–90 CIVIC EXCEPT CRX HF

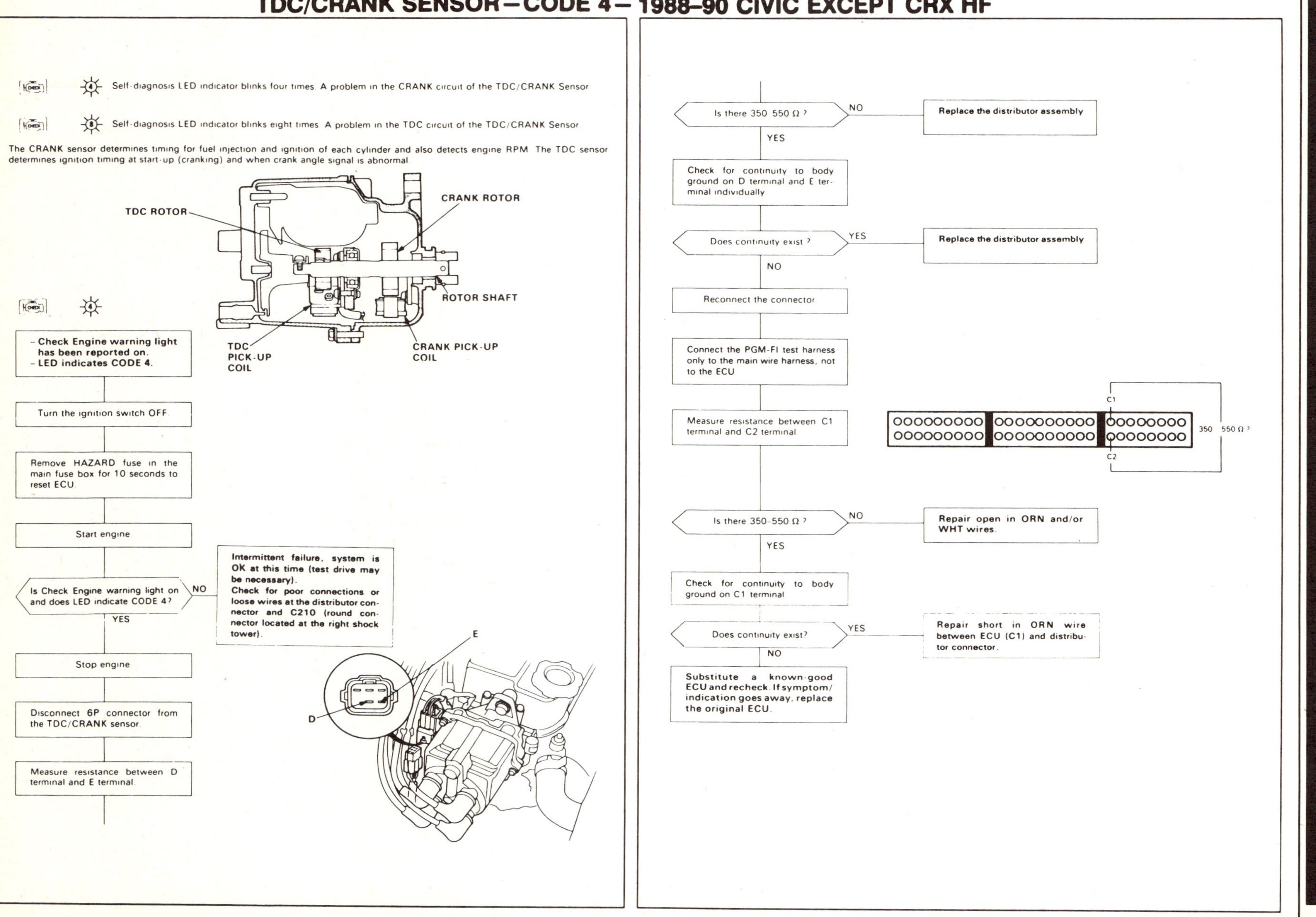

## TDC/CRANK SENSOR—CODE 8—1988–90 CIVIC EXCEPT CRX HF

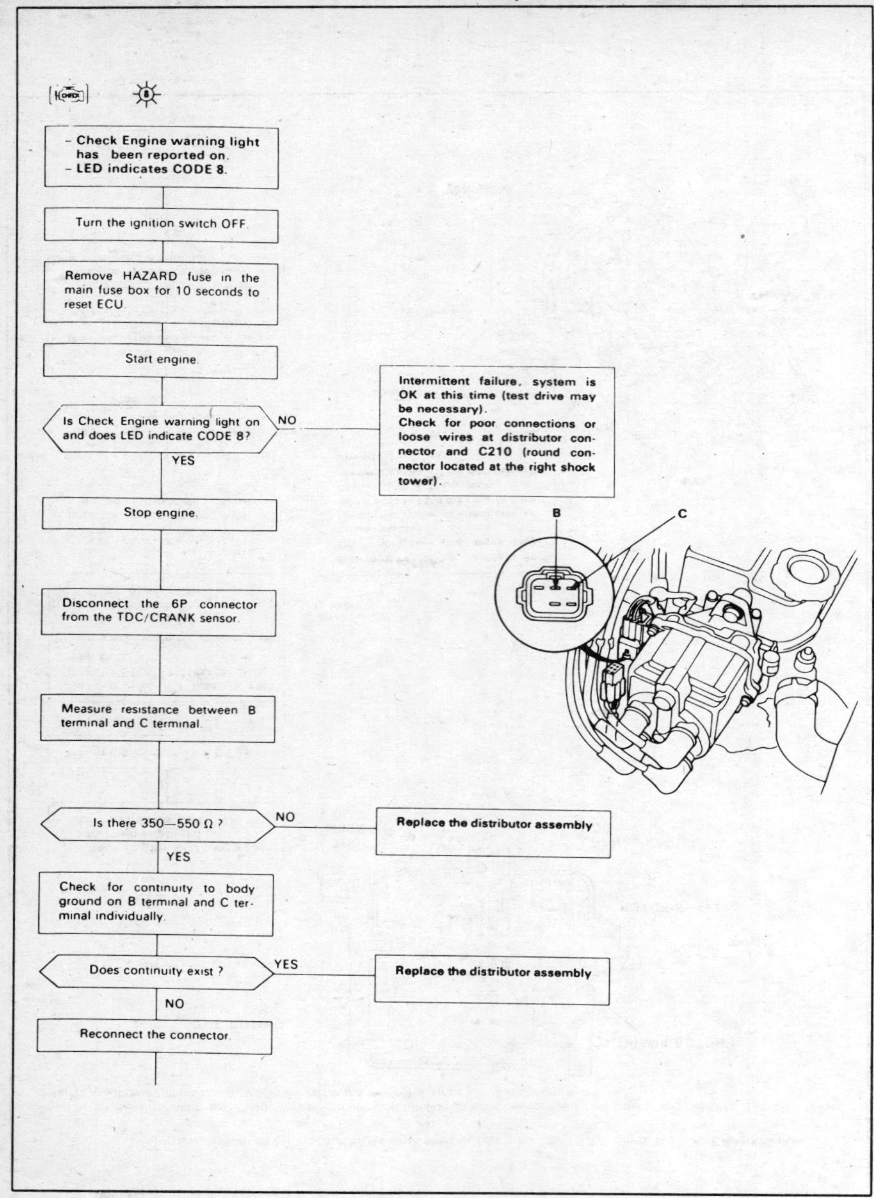


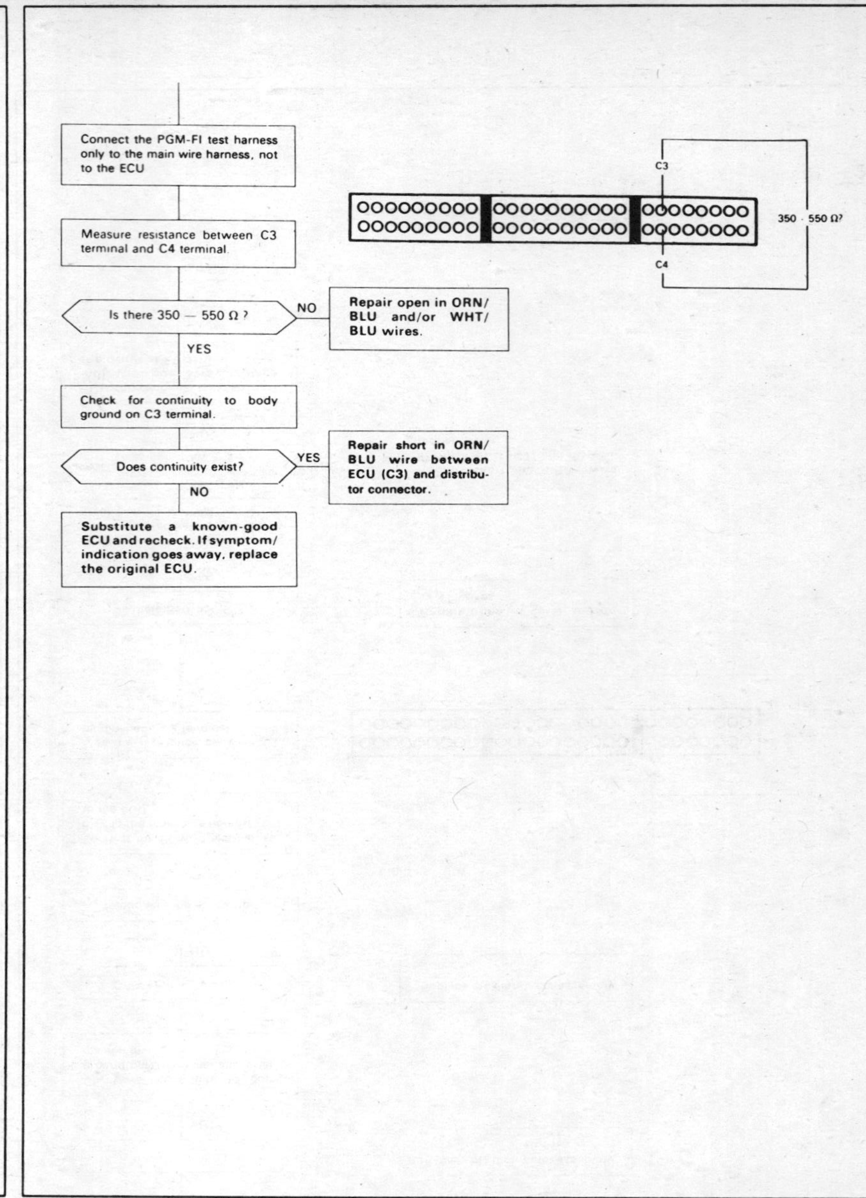

# TDC/CRANK/CYL SENSOR—CODE 4—1988–90 CIVIC 1.6L AND CRX HF

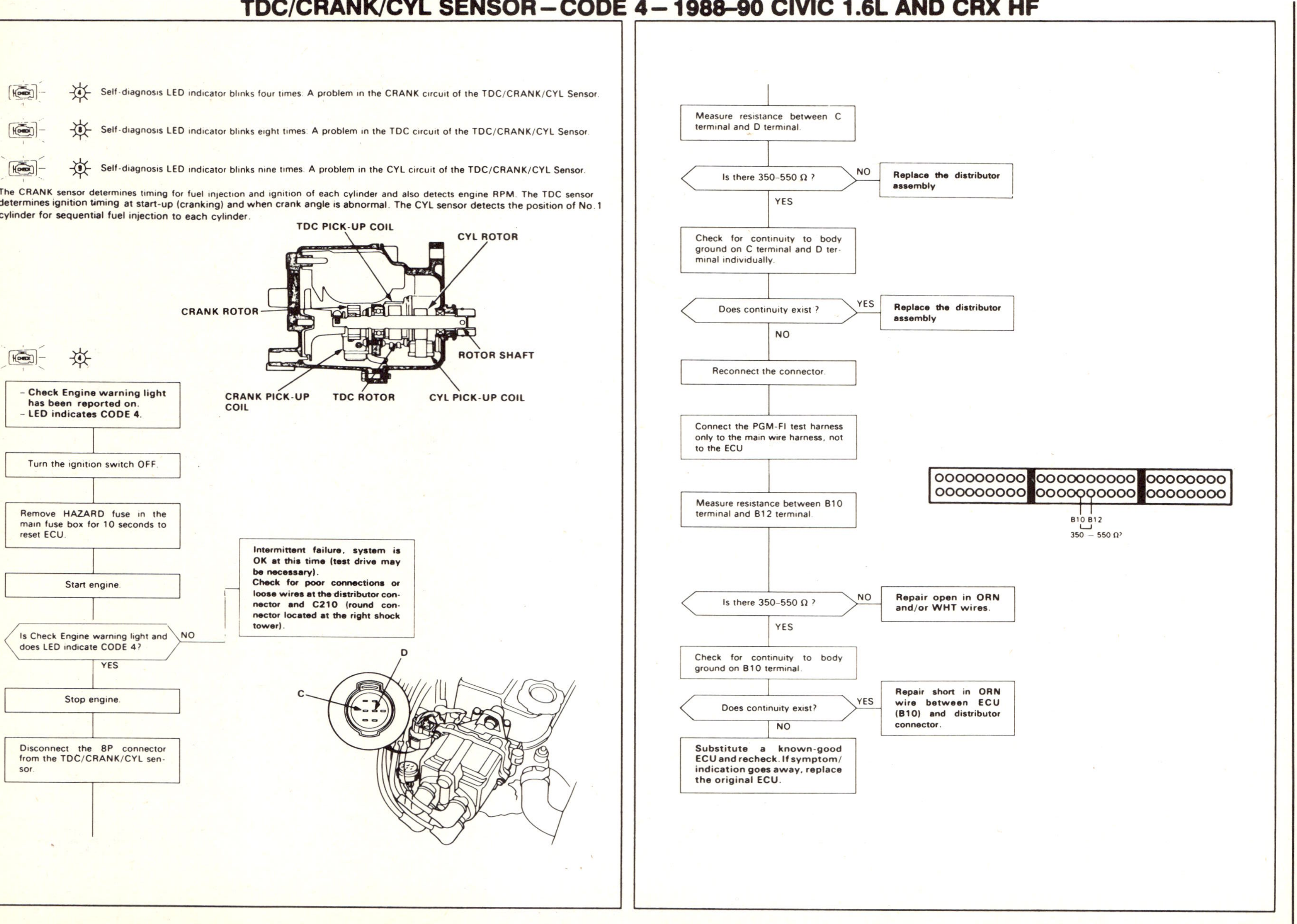

## TDC/CRANK/CYL SENSOR—CODE 8—1988–90 CIVIC 1.6L AND CRX HF

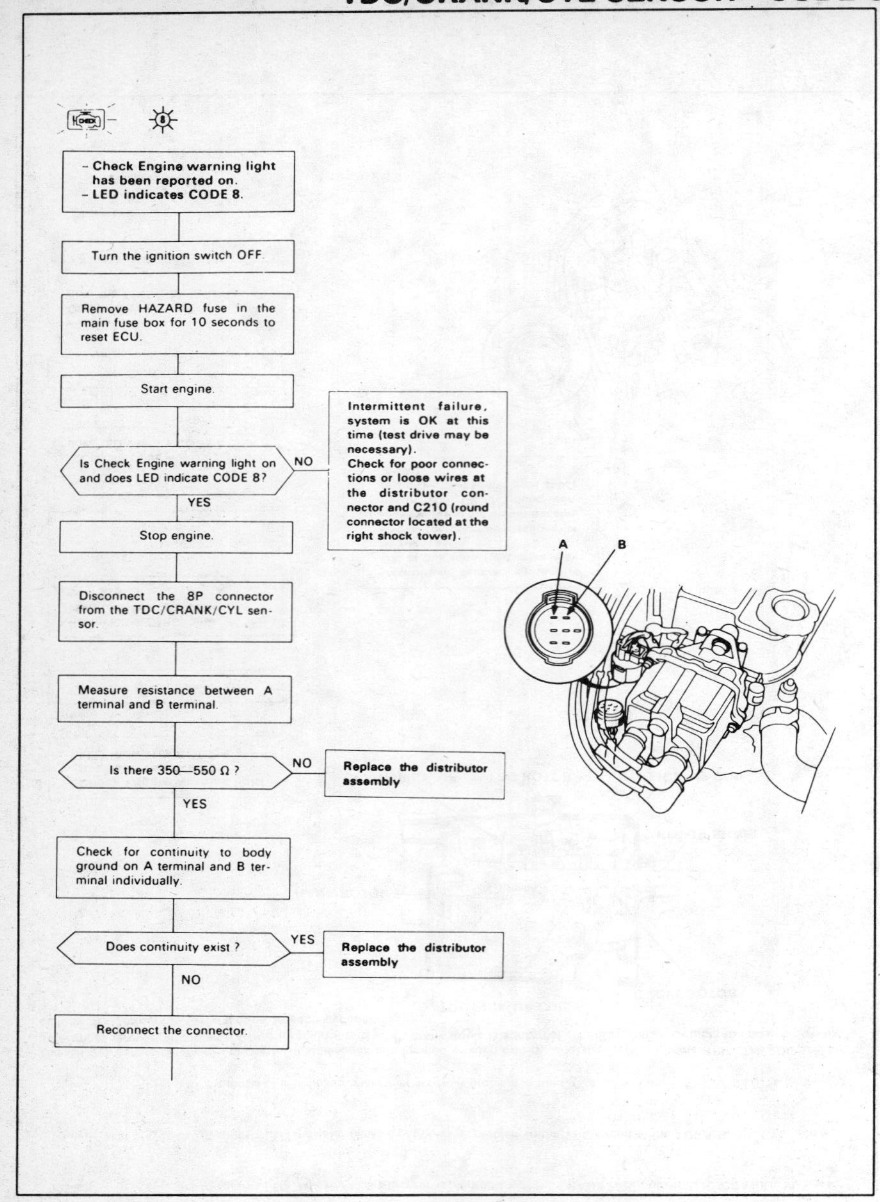


Connect the PGM-FI test harness only to the main wire harness, not to the ECU

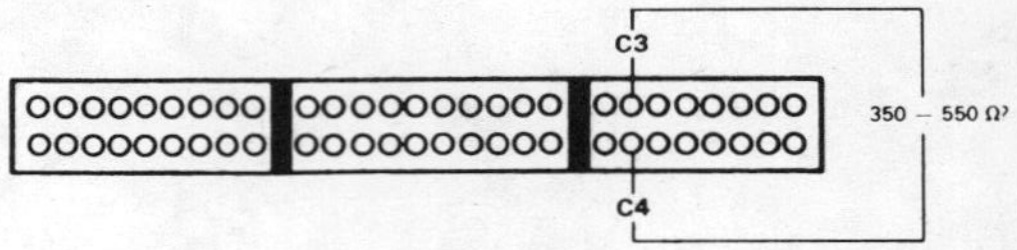

Measure resistance between C3 terminal and C4 terminal.

Is there 350 — 550 Ω ?

- NO → Repair open in ORN/BLU and/or WHT/BLU wires.
- YES ↓

Check for continuity to body ground on C3 terminal.

Does continuity exist?

- YES → Repair short in ORN/BLU wire between ECU (C3) and distributor connector.
- NO ↓

Substitute a known-good ECU and recheck. If symptom/indication goes away, replace the original ECU.

C3
C4
350 — 550 Ω?

# TDC/CRANK/CYL SENSOR—CODE 9—1988–90 CIVIC 1.6L AND CRX HF

- Check Engine warning light has been reported on.
- LED indicates CODE 9.

Turn the ignition switch OFF.

Remove HAZARD fuse in the main fuse box for 10 seconds to reset ECU.

Start engine.

Is Check Engine warning light on and does LED indicate CODE 9?

- NO → **Intermittent failure, system is OK at this time (test drive may be necessary). Check for poor connections or loose wires at the distributor connector and C210 (round connector located at the right shock tower).**
- YES ↓

Stop engine.

Disconnect the 8P connector from the TDC/CRANK/CYL sensor.

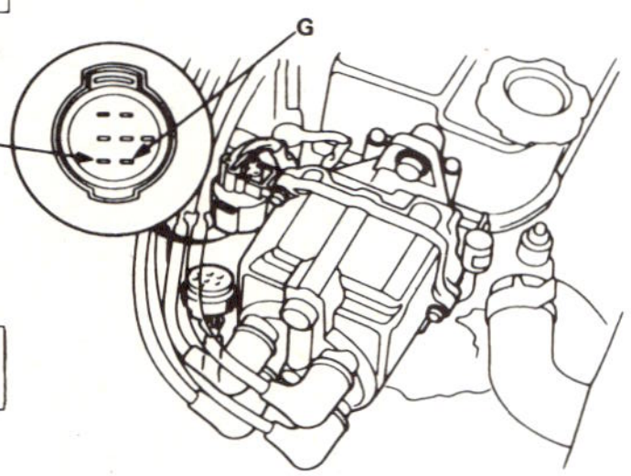

Measure resistance between F terminal and G terminal.

Is there 350—550 Ω ?

- NO → **Replace the distributor assembly**
- YES ↓

Check for continuity to body ground on F terminal and G terminal individually.

Does continuity exist ?

- YES → **Replace the distributor assembly**
- NO ↓

Reconnect the connector.



Connect the PGM-FI test harness only to the main wire harness, not to the ECU

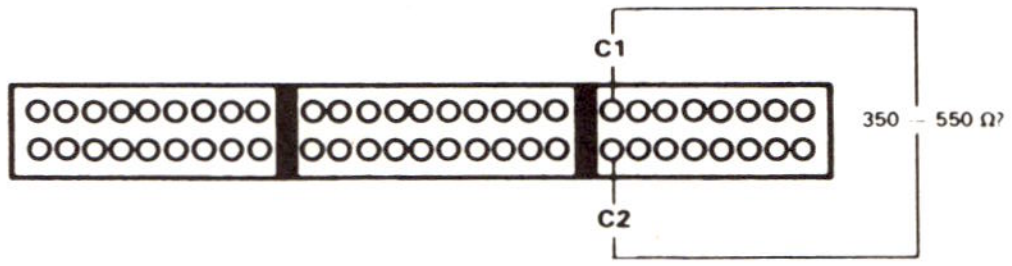

Measure resistance between C1 terminal and C2 terminal.

Is there 350 — 550 Ω ?

- NO → **Repair open in BLU/GRN and/or BLU/YEL wires.**
- YES ↓

Check for continuity to body ground on C1 terminal.

Does continuity exist?

- YES → **Repair short in BLU/GRN wire between ECU (C1) and distributor connector.**
- NO ↓

**Substitute a known-good ECU and recheck. If symptom/indication goes away, replace the original ECU.**

# COOLANT TEMPERATURE (TW) SENSOR—CODE 6—1988–90 CIVIC EXCEPT 1988 CRX

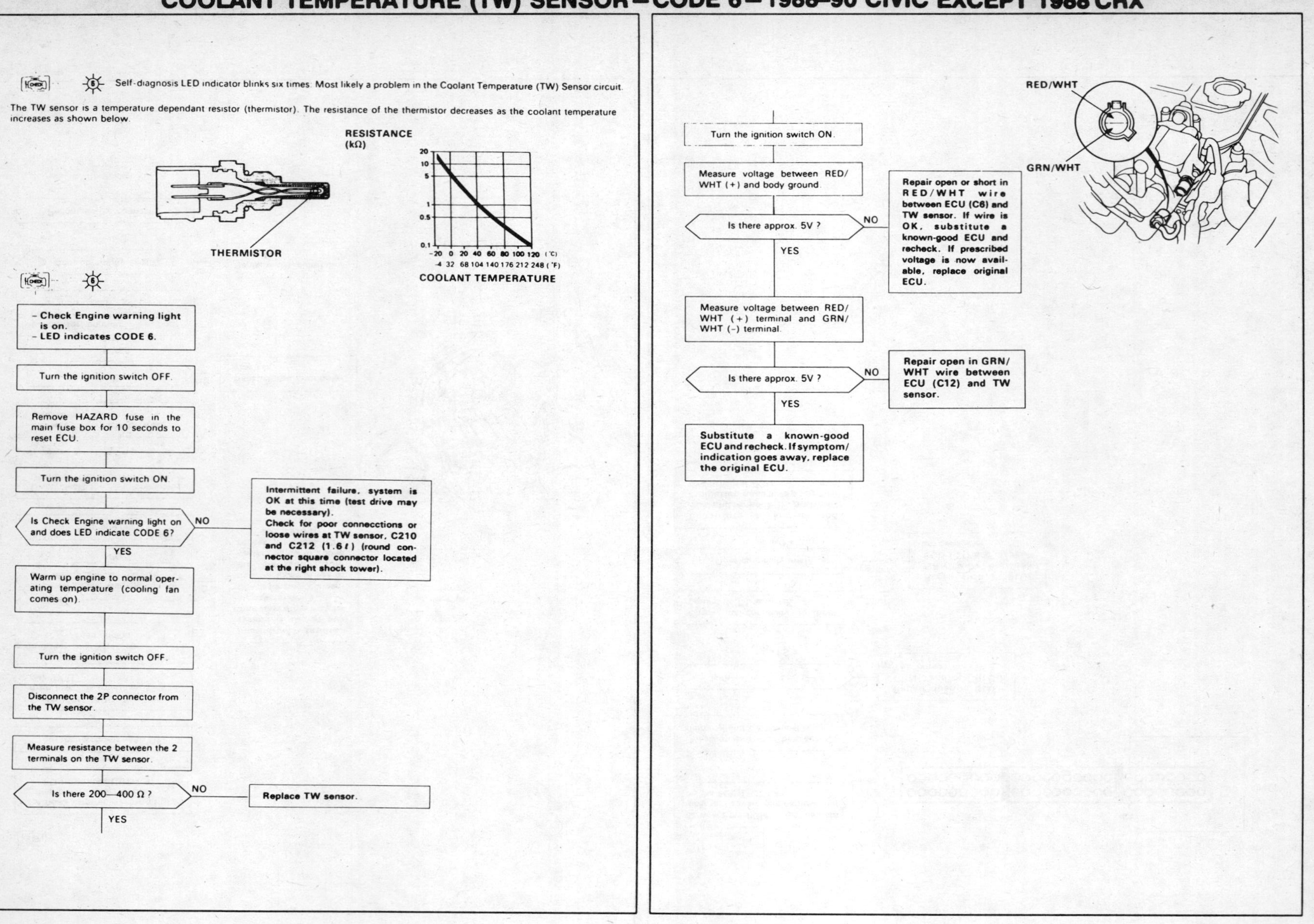

# COOLANT TEMPERATURE (TW) SENSOR – CODE 6 – 1988 CRX

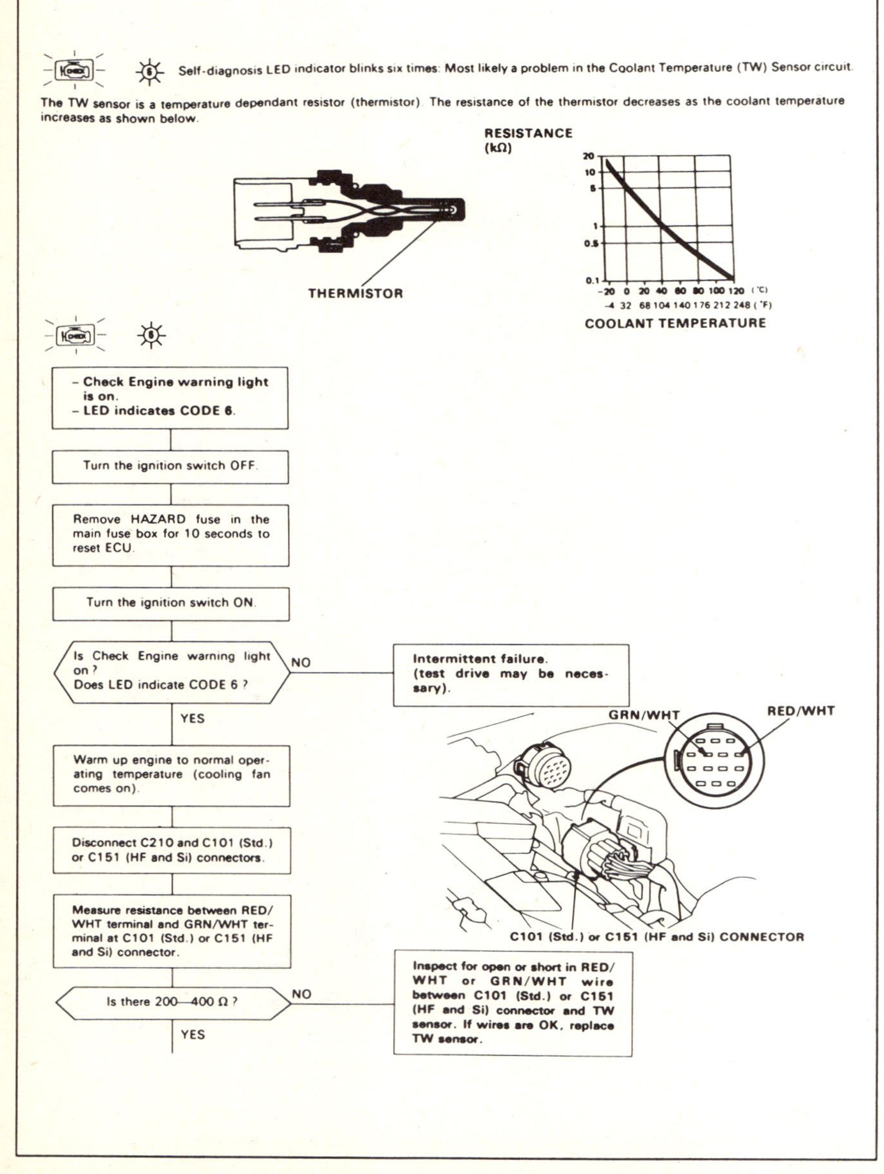


RED/WHT

GRN/WHT

- Reconnect C210 and C101 (Std.) or C151 (HF and Si) connectors, then disconnect the 2P connector from TW sensor.
- Is there approx. 5V?
  - NO: Turn the ignition switch OFF. → Connect the PGM-FI test harness between the ECU and connector → Turn the ignition switch ON. → Measure voltage between C6 (+) terminal and C12 (–) terminal.
    - C6 (+) C12 (–) 5 V
    - Is there approx. 5V?
      - YES: **Repair open in RED/WHT wire between ECU (C6) and TW sensor.**
      - NO: Disconnect "C" connector from the main wire harness only, not the ECU. → Measure voltage between C6 (+) terminal and C12 (–) terminal. → Is there approx. 5V?
        - YES: **Repair short in RED/WHT wire between ECU (C6) and TW sensor.**
        - NO: **Substitute a known-good ECU and recheck. If prescribed voltage is now available, replace the original ECU.**
  - YES: Measure voltage between RED/WHT (+) terminal and GRN/WHT (–) terminal.
    - Is there approx. 5V?
      - NO: **Repair open in GRN/WHT wire between ECU (C12) and TW sensor.**
      - YES: **Substitute a known-good ECU and recheck. If symptom/indication goes away, replace the original ECU.**

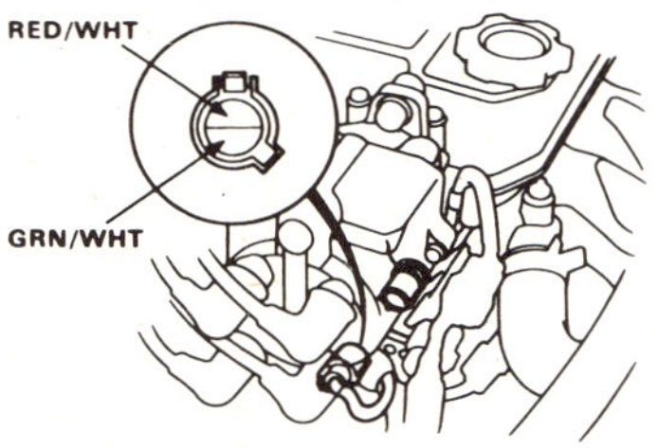

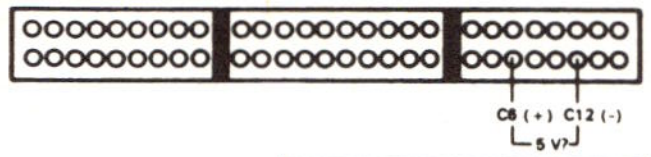

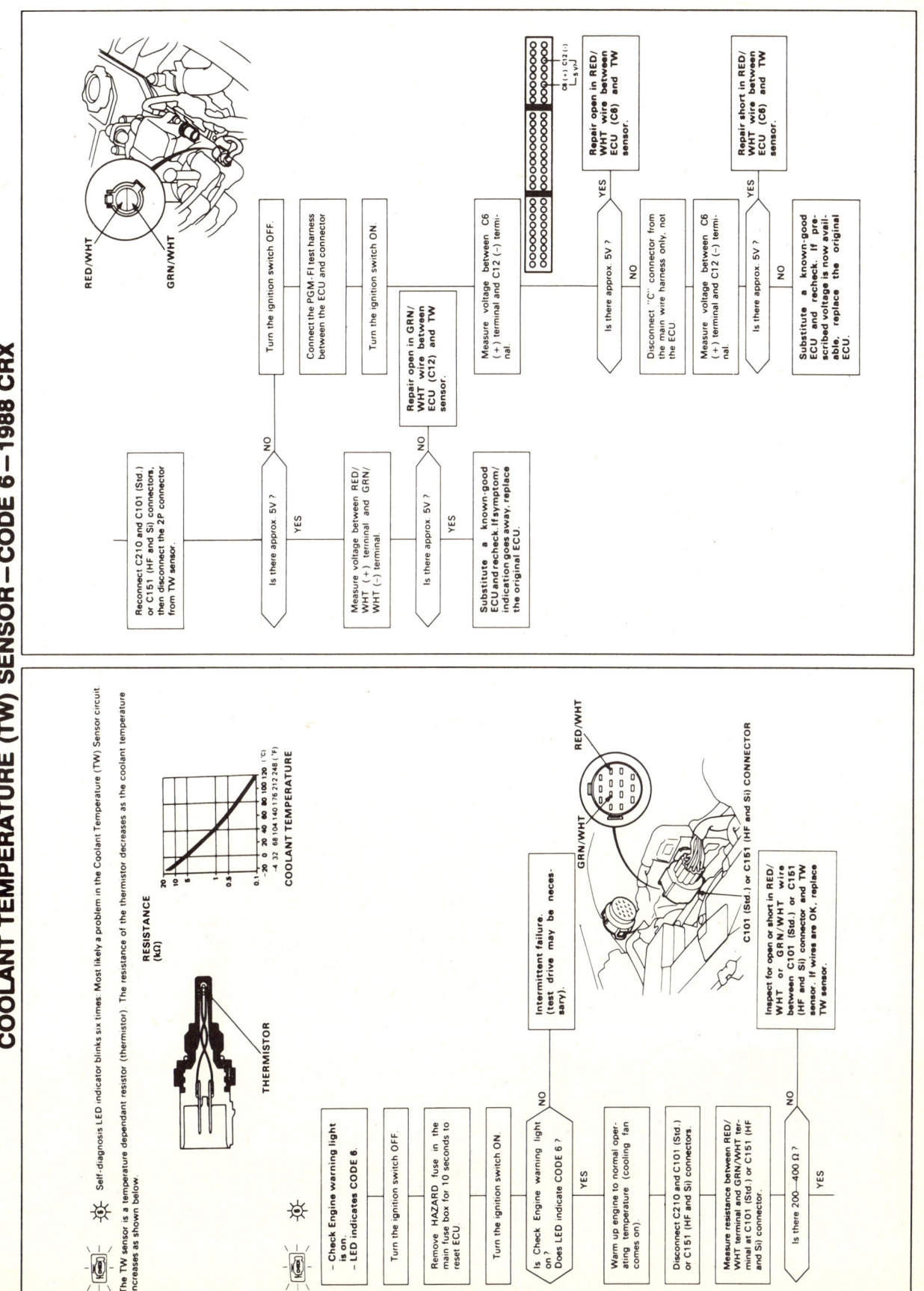

# THROTTLE ANGLE SENSOR—CODE 7—1988–90 CIVIC

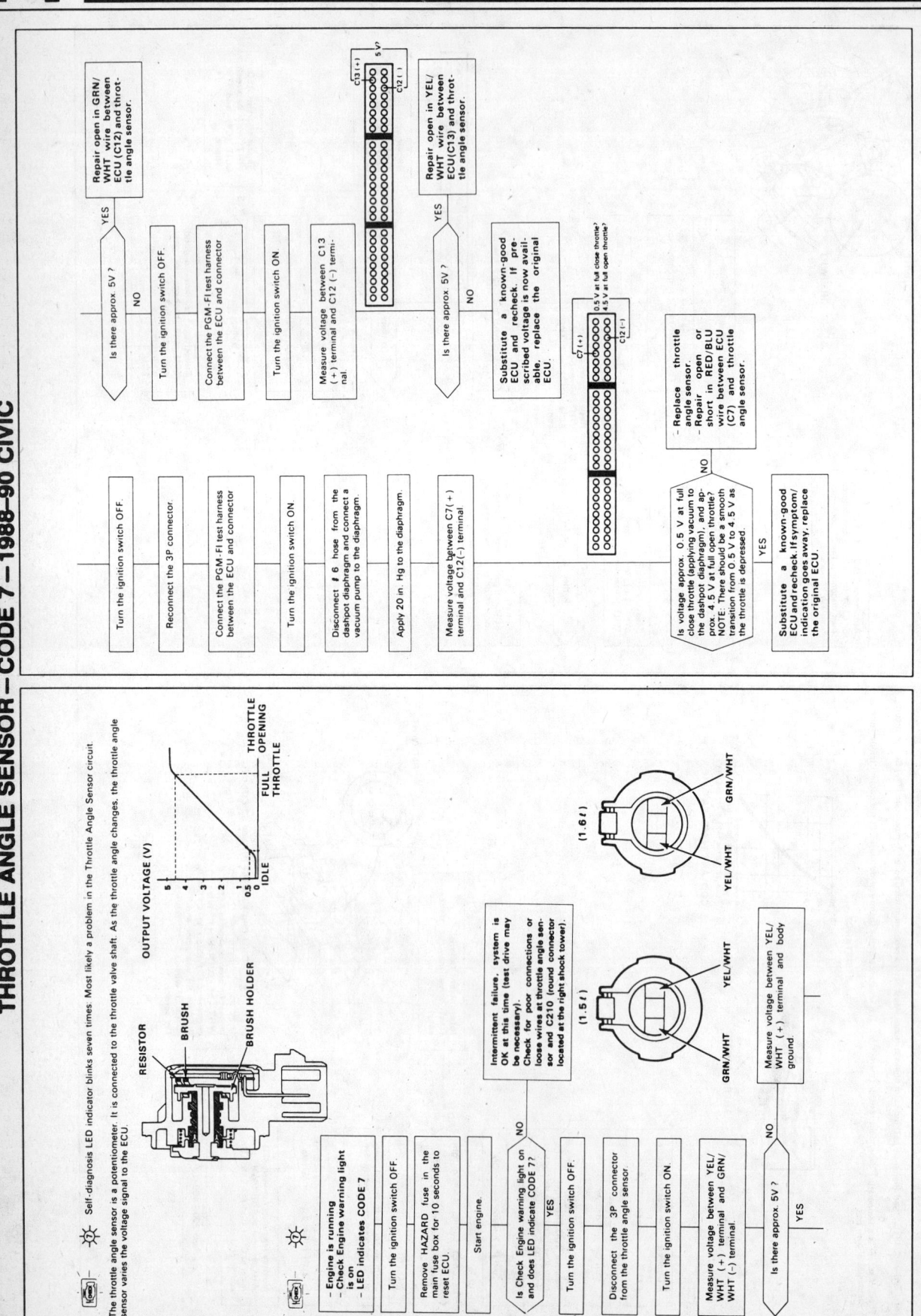

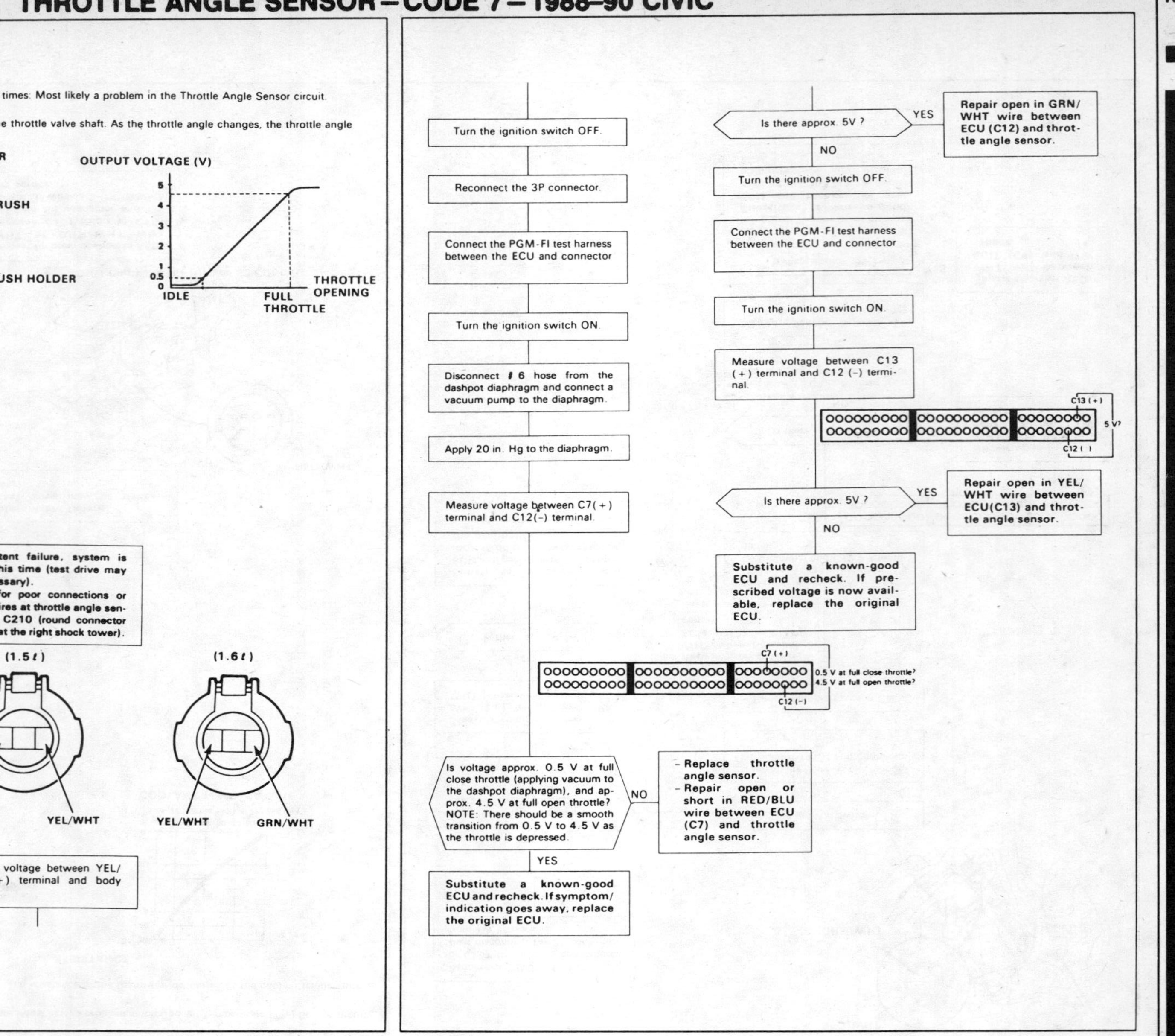

# AIR TEMPERATURE SENSOR—CODE 10—1988–90 CIVIC

Self-diagnosis LED indicator blinks ten times. Most likely a problem in the Intake Air Temperature (TA) Sensor circuit.

The TA sensor is a temperature dependant resistor (thermistor). The resistance of the thermistor decreases as the intake air temperature increases as shown below.

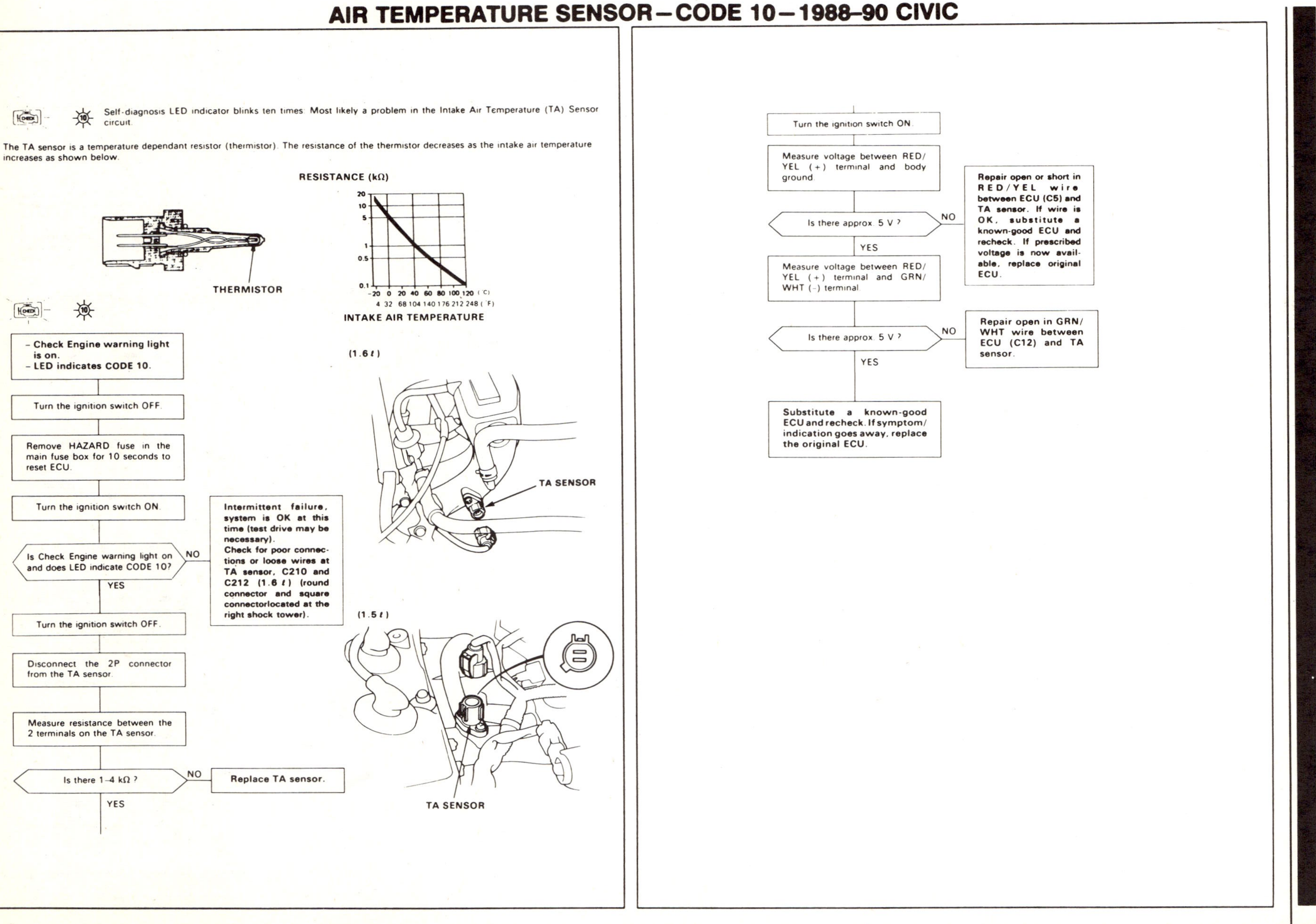

# ATMOSPHERIC PRESSURE (PA) SENSOR – CODE 13 – 1988–90 CIVIC

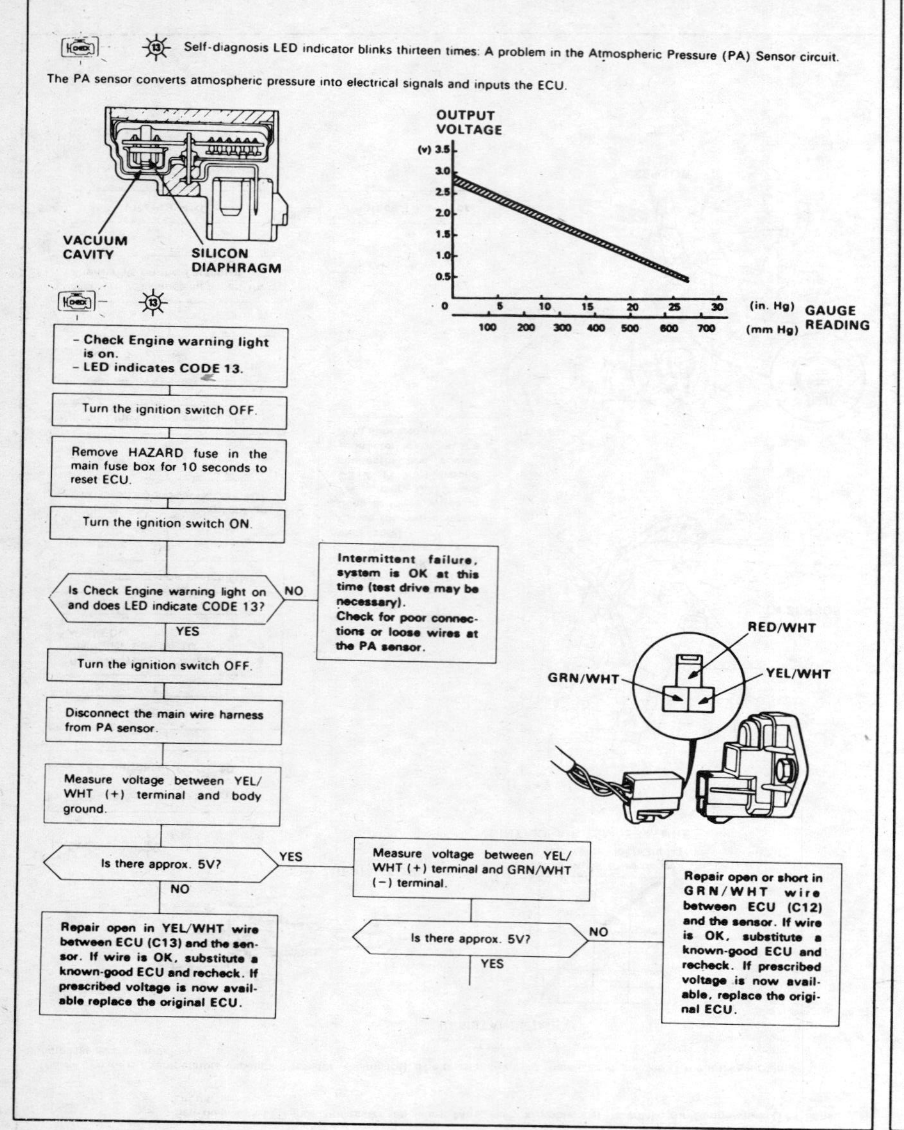


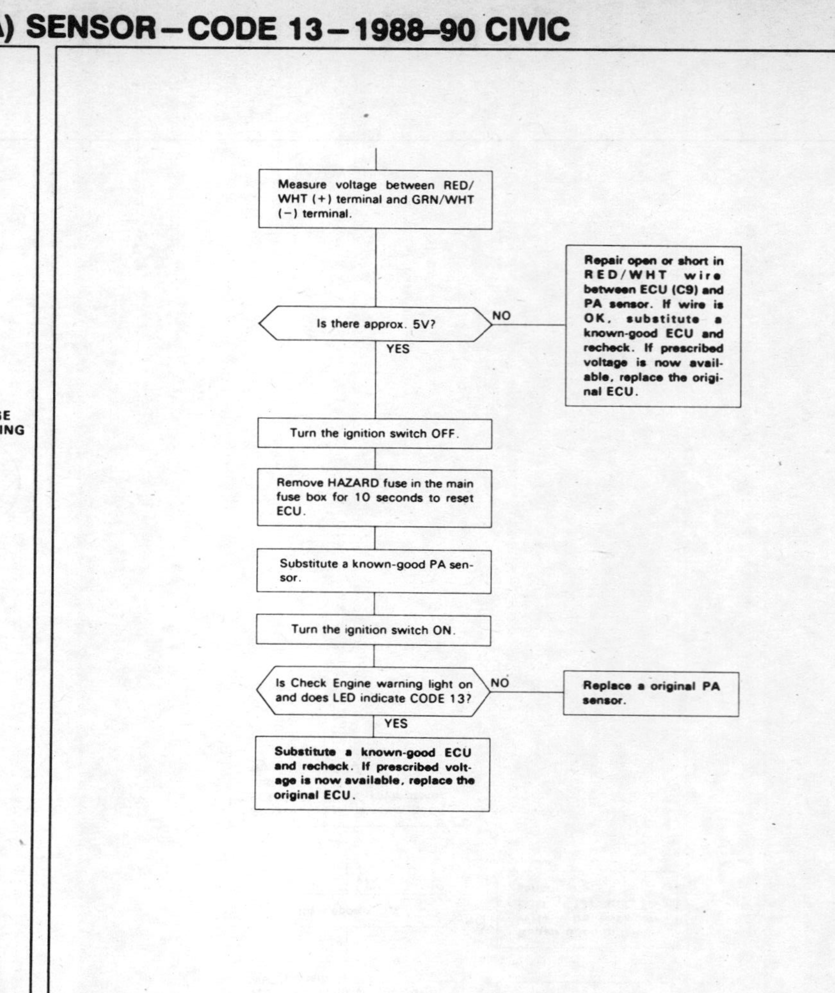

# IGNITION OUTPUT SIGNAL—CODE 15—1988–90 CIVIC

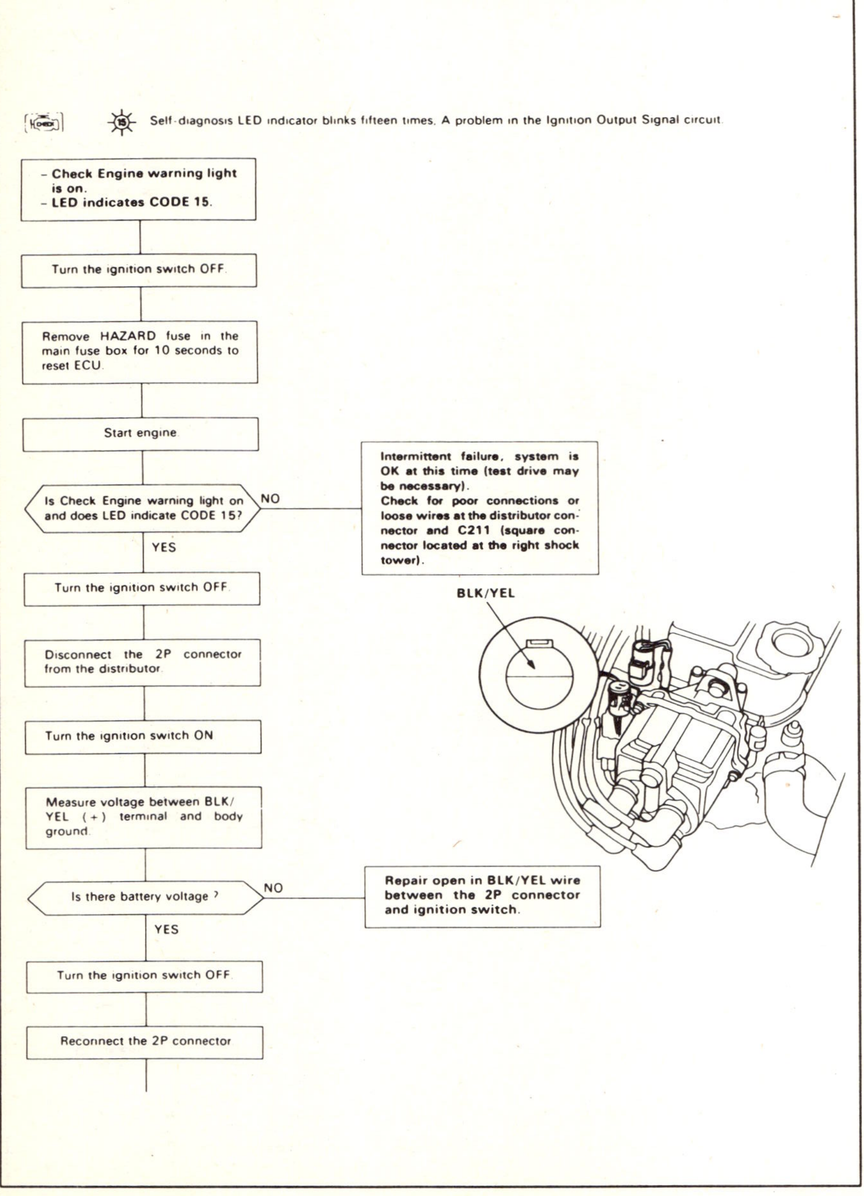


Connect the PGM-FI test harness between the ECU and connector.

Turn the ignition switch ON.

Measure voltage individually between B15 (+), B17 (+) terminals and A18 (-) terminal.

Is there approx. 10V?

NO → – Replace the igniter unit.
– Repair open or short in WHT wires between distributor and ECU (B15 or B17).

YES → Substitute a known-good ECU and recheck. If symptom/indication goes away, replace the original ECU.

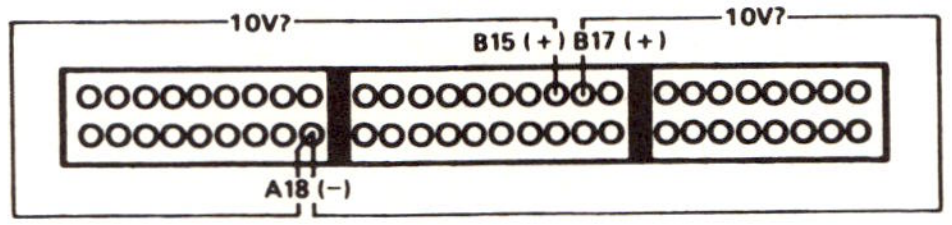

10V? B15 (+) B17 (+) 10V? A18 (–)

## VEHICLE SPEED SENSOR—CODE 17—1988–90 CIVIC

Self-diagnosis LED indicator blinks seventeen times. A problem in the Vehicle Speed Sensor circuit.

The signal generated by the speed sensor, which produces 4 pulses (switch closures to ground) for each revolution of the speedometer cable.

- Check Engine warning light is on.
- LED indicates CODE 17.

Turn the ignition switch OFF.

Remove HAZARD fuse in the main fuse box for 10 seconds to reset ECU.

Road test necessary: In 2nd gear accelerate to 3,500 rpm and decelerate to 1,500 rpm with throttle fully closed.

Is Check Engine warning light and does LED indicate CODE 17?

- NO → Intermittent failure, system is OK at this time. Check for poor connections or loose wires at C405 and C710.
- YES ↓

Block rear wheels and set the parking brake. Jack up the front of the car and support with safety stands.

**WARNING** Block rear wheels before jacking up front of car.

Connect the PGM-FI test harness between the ECU and connector

Turn the ignition switch ON

Slowly rotate left front wheel and measure voltage between B16 (+) terminal and A18 (-) terminal.

A18 (−) B16 (+) 0 V↔5 V?

Does voltage pulse 0V and 5V ?

- NO →
  - Repair open or short in YEL/RED wire between ECU (B16) and tthe speed sensor.
  - Faulty speed sensor.
  - Substitute a known-good ECU and recheck. If symptom/indication goes away, replace the original ECU.
- YES ↓

Substitute a known-good ECU and recheck. If symptom/indication goes away, replace the original ECU.

## A/T LOCK UP CONTROL SOLENOID VALVE CODE 19—1988–90 CIVIC

Self-diagnosis LED indicator blinks nineteen times: A problem in the Lock-up Control Solenoid Valve circuit.

- Check Engine warning light is on.
- LED indicates CODE 19.

Turn the ignition switch OFF.

Remove HAZARD fuse in the main fuse box for 10 seconds to reset ECU.

Drive vehicle.

Is Check Engine warning light on and does LED indicate CODE 19?

- NO → Intermittent failure, system is OK at this time (test drive may be necessary). Check for poor connections or loose wires at lock-up solenoid valve and C313 (round connector located at the left shock tower).
- YES ↓

Turn the ignition switch OFF.

Connect the PGM-FI test harness between the ECU and connector. Disconnect "A" connector from the ECU only, not the main wire harness.

Disconnect the 2P connector from the lock-up control solenoid valve.

Check for continuity between A8 and body ground.

Does continuity exist ?

- YES → Repair short in YEL wire between ECU (A8) and the 2P connector.
- NO ↓

Reconnect the 2P connector.

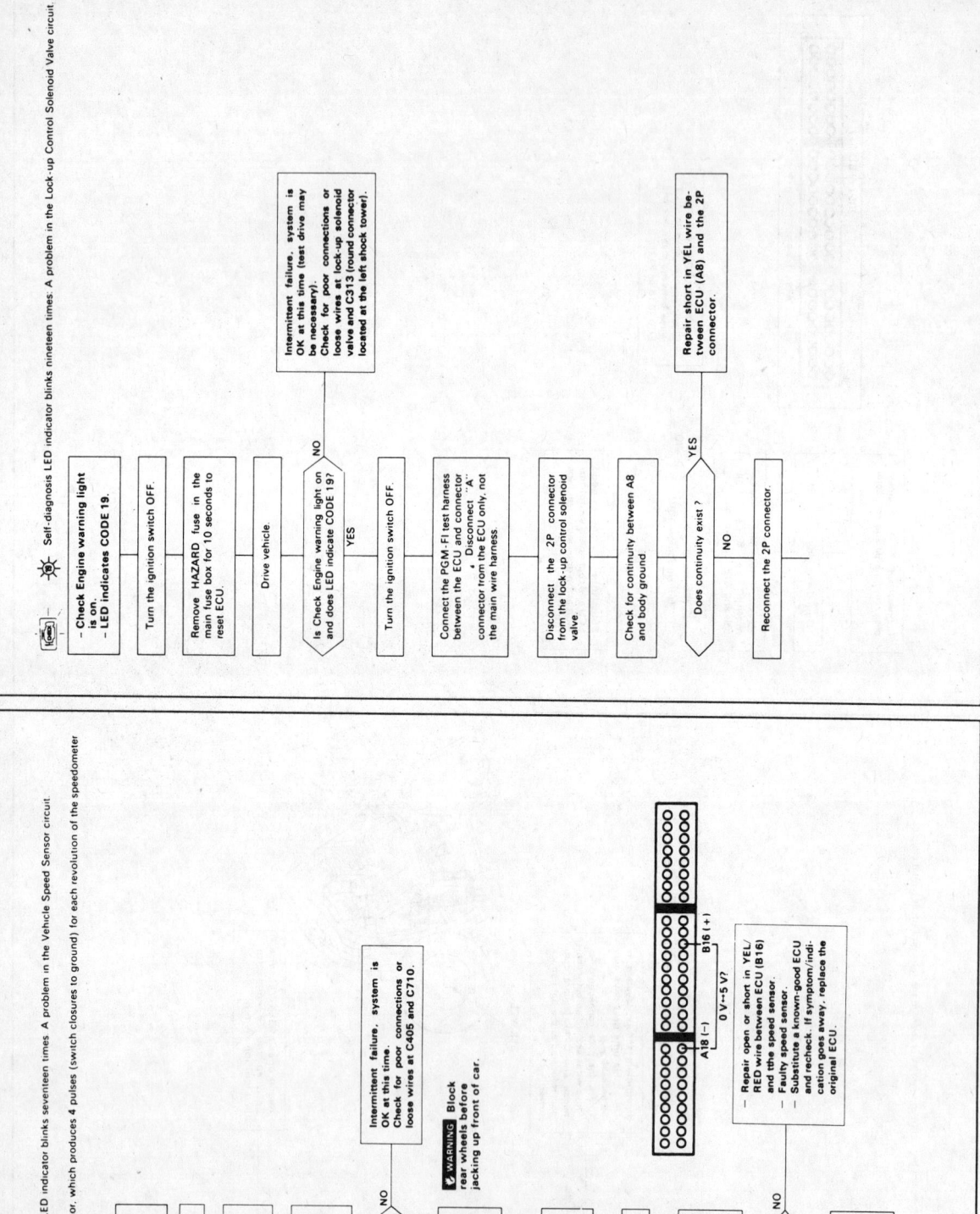

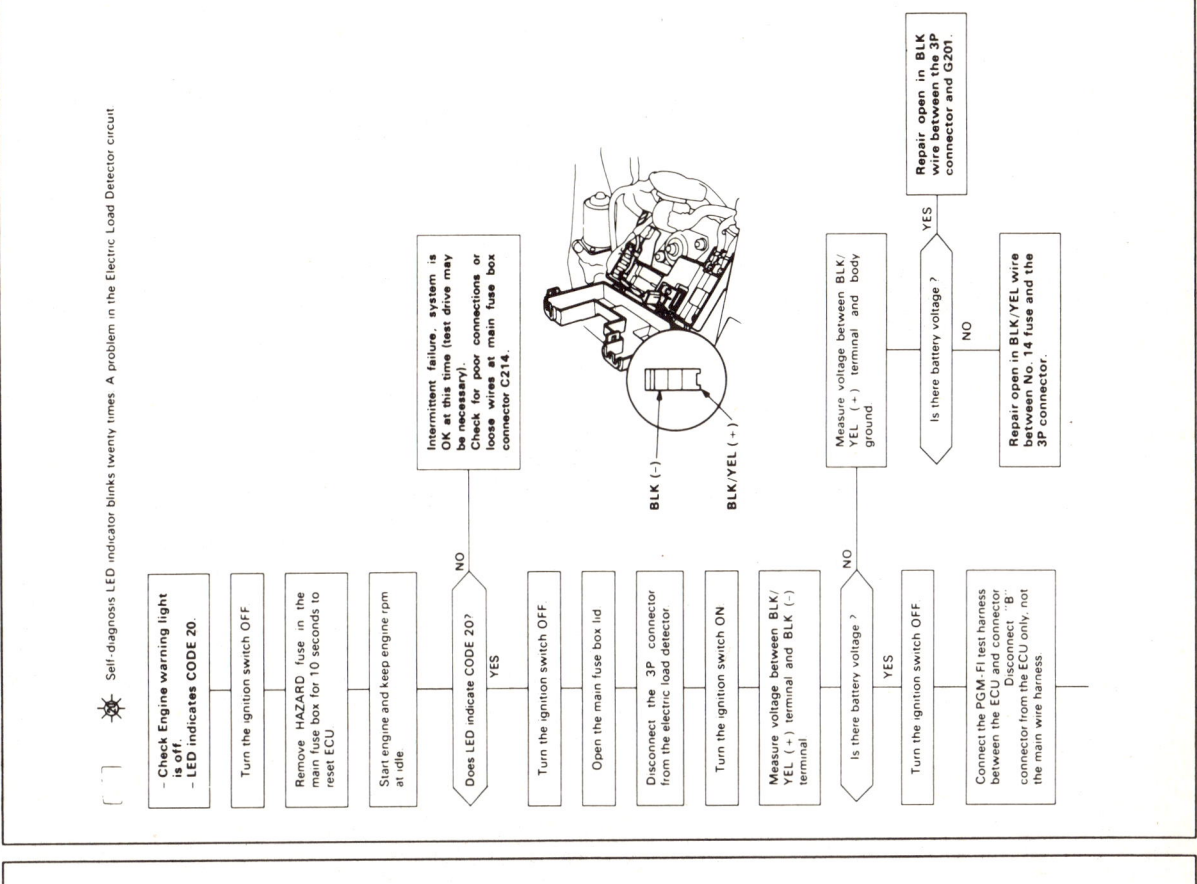

## A/T LOCK UP CONTROL SOLENOID VALVE CODE 19—1988–90 CIVIC

A8 · A18 · 14 · 25 Ω?

- Measure resistance between A8 terminal and A18 terminal.
- Is there 14 — 25 Ω ?
  - NO → Check for continuity in YEL wire between ECU (A8) and the 2P connector.
    - Does continuity exist ?
      - YES → **Faulty lock-up control solenoid valve.**
      - NO → **Repair open in YEL wire between ECU (A8) and the 2P connector.**
  - YES → **Substitute a known-good ECU and recheck. If symptom/indication goes away, replace the original ECU.**

## ELECTRIC LOAD DETECTOR—CODE 20—1988–90 CIVIC

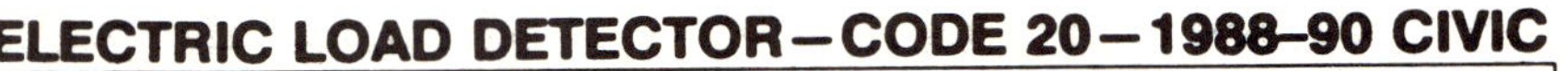

Self-diagnosis LED indicator blinks twenty times. A problem in the Electric Load Detector circuit.

- **– Check Engine warning light is off.**
  **– LED indicates CODE 20.**
- Turn the ignition switch OFF.
- Remove HAZARD fuse in the main fuse box for 10 seconds to reset ECU.
- Start engine and keep engine rpm at idle.
- Does LED indicate CODE 20?
  - NO → **Intermittent failure, system is OK at this time (test drive may be necessary). Check for poor connections or loose wires at main fuse box connector C214.**
  - YES → Turn the ignition switch OFF.
- Open the main fuse box lid.
- Disconnect the 3P connector from the electric load detector.
- Turn the ignition switch ON.
- Measure voltage between BLK/YEL (+) terminal and BLK (–) terminal.
- Is there battery voltage ?
  - NO → Measure voltage between BLK/YEL (+) terminal and body ground.
    - Is there battery voltage ?
      - YES → **Repair open in BLK wire between the 3P connector and G201.**
      - NO → **Repair open in BLK/YEL wire between No. 14 fuse and the 3P connector.**
  - YES → Turn the ignition switch OFF.
- Connect the PGM-FI test harness between the ECU and connector. Disconnect "B" connector from the ECU only, not the main wire harness.

# ELECTRIC LOAD DETECTOR—CODE 20—1988–90 CIVIC

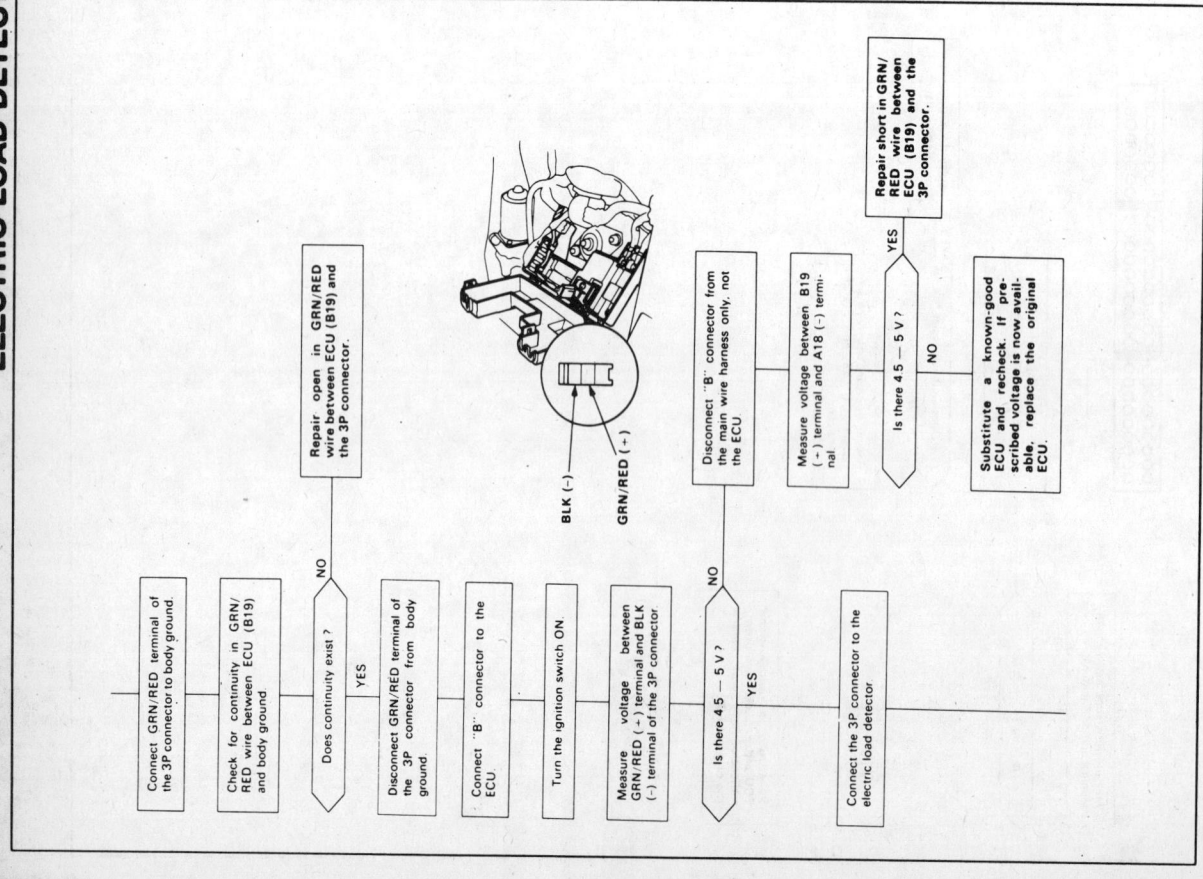

Connect GRN/RED terminal of the 3P connector to body ground.

Check for continuity in GRN/RED wire between ECU (B19) and body ground.

Does continuity exist ?

- NO → Repair open in GRN/RED wire between ECU (B19) and the 3P connector.
- YES → Disconnect GRN/RED terminal of the 3P connector from body ground.

Connect "B" connector to the ECU.

Turn the ignition switch ON.



Measure voltage between GRN/RED (+) terminal and BLK (-) terminal of the 3P connector.

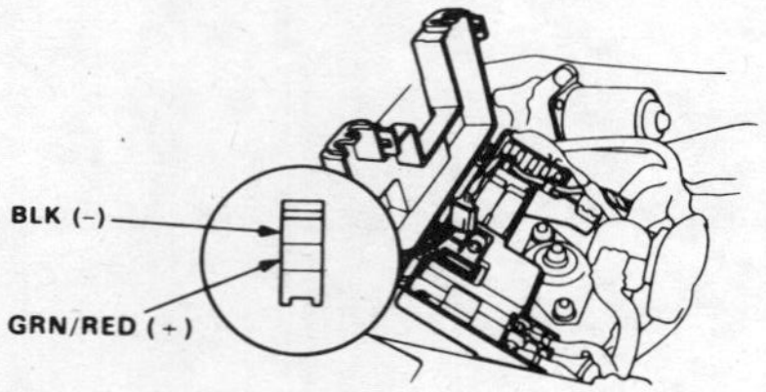

Is there 4.5 — 5 V ?

- YES → Connect the 3P connector to the electric load detector.
- NO → Disconnect "B" connector from the main wire harness only, not the ECU.

Measure voltage between B19 (+) terminal and A18 (-) terminal.

Is there 4.5 — 5 V ?

- YES → Repair short in GRN/RED wire between ECU (B19) and the 3P connector.
- NO → Substitute a known-good ECU and recheck. If prescribed voltage is now available, replace the original ECU.

Under the conditions listed in the chart to the right, measure voltage between B19 (+) terminal and A18 (-) terminal.

| Condition | Voltage |
|---|---|
| Headlight switch, first position ( • ) | 2.5 - 3.5 V |
| Headlight switch, second position (●) | 1.5 - 2.5 V |

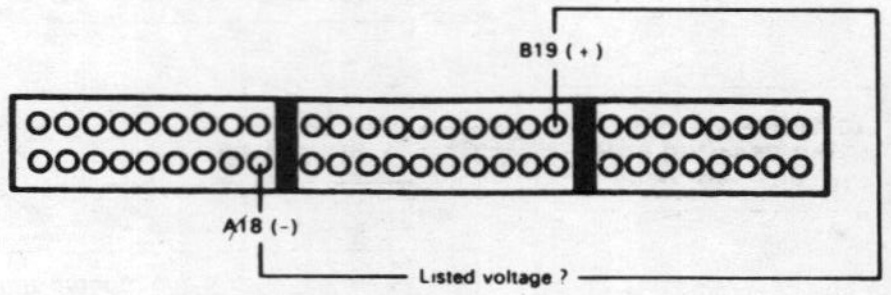

Is the voltage listed in the chart available ?

- NO → Faulty electric load detector.
- YES → Substitute a known-good ECU and recheck. If symptom/indication goes away, replace the original ECU.

# IDLE CONTROL TROUBLESHOOTING – 1988–90 CIVIC

NOTE:
- Across each row in the chart, the sub systems that could be sources of a symptom are ranked in the order they should be inspected, starting with ①. Find the symptom in the left column, read across to the most likely source, then refer to the page listed at the top of that column. If inspection shows the system is OK, try the next system ②, etc.

| SYMPTOM \ SUB SYSTEM | | IDLE ADJUSTING SCREW | EACV | AIR CONDITIONING SIGNAL | ALTERNATOR FR SIGNAL | A/T SHIFT POSITION SIGNAL (Automatic) | BRAKE SWITCH SIGNAL | STARTER SWITCH SIGNAL | FAST IDLE CONTROL (1.6 ℓ) | HOSES AND CONNECTIONS |
|---|---|---|---|---|---|---|---|---|---|---|
| DIFFICULT TO START ENGINE WHEN COLD | | ② | ① | | | | | | | |
| WHEN COLD FAST IDLE OUT OF SPEC (1,000–2,000 rpm) | | ② | ① | | | | | | | ③ |
| ROUGH IDLE | | ③ | ② | | | | | | | ① |
| WHEN WARM RPM TOO HIGH | | ③ | ② | | | | | | | ① |
| WHEN WARM RPM TOO LOW | Idle speed is below specified rpm (no load) | ② | ① | | ③ | | | | | |
| | Idle speed does not increase after initial start up. | | ① | | | | | ② | | |
| | On models with automatic transmission, the idle speed drops in gear | ③ | ② | | | ① | | | | |
| | Idle speeds drops when air conditioner in ON | ③ | ② | ① | | | | | | |
| FREQUENT STALLING | WHILE WARMING UP | ② | ① | | | | | | | |
| | AFTER WARMING UP | ② | ① | | | | | | | |
| FAILS EMISSION TEST | | | | | | | | | | ① |

When the idle speed is out of specification and LED does not blink CODE 14, check the following items:
- Adjust the idle speed
- Air conditioning signal
- Alternator FR signal
- A/T shift position signal
- Brake switch signal
- Starter switch signal
- Fast idle control
- Hoses and connections
- EACV and its mounting O-rings.

If the above items are normal, substitute a known-good EACV and readjust the idle speed

- If the idle speed still cannot be adjusted to specification (and LED does not blink CODE 14) after EACV replacement, substitute a known-good ECU and recheck. If symptom goes away, replace the original ECU.

# ELECTRONIC AIR CONTROL VALVE (EACV) CODE 14 – 1988–90 CIVIC

Self-diagnosis LED indicator blinks fourteen times: A problem in the Electronic Air Control Valve (EACV) circuit.

The EACV changes the amount of air bypassing the throttle body in response to a current signal from the ECU in order to maintain the proper idle speed.

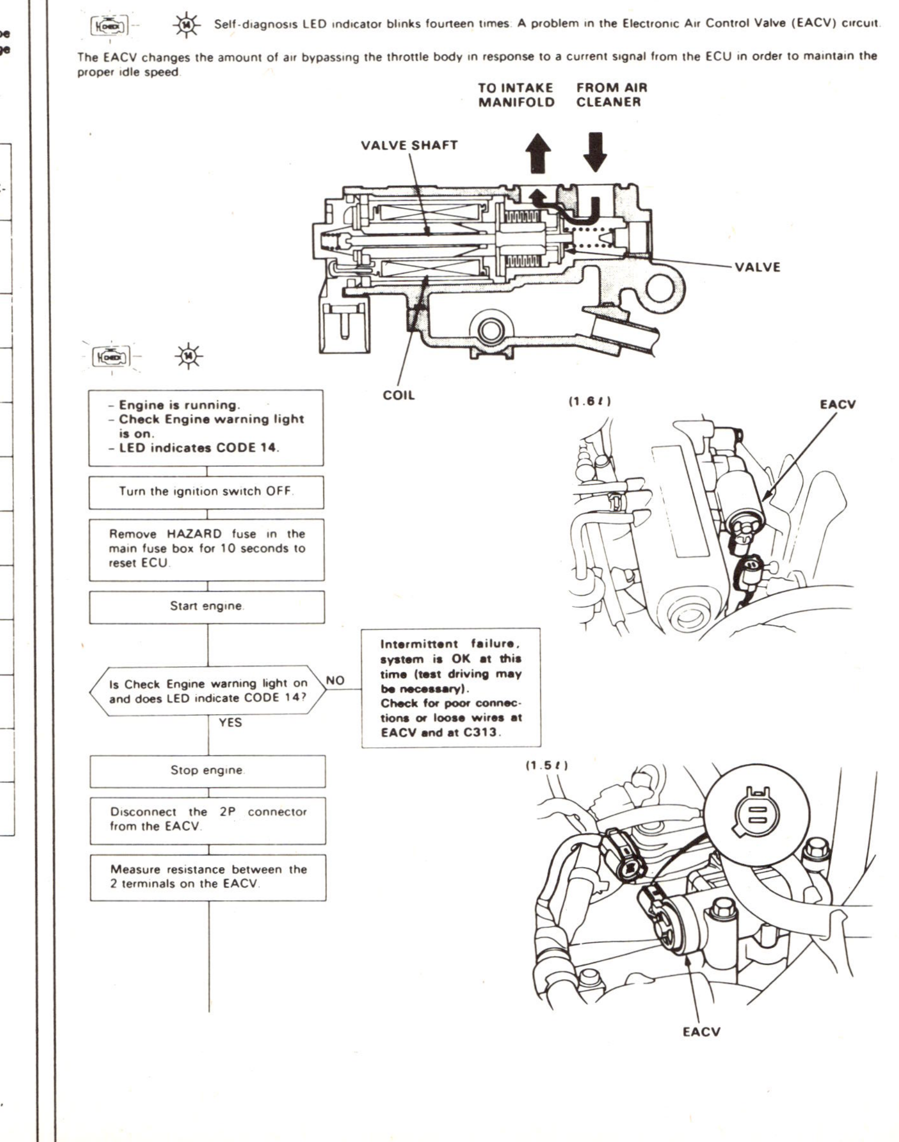

# ELECTRONIC AIR CONTROL VALVE (EACV)—CODE 14—1988–90 CIVIC

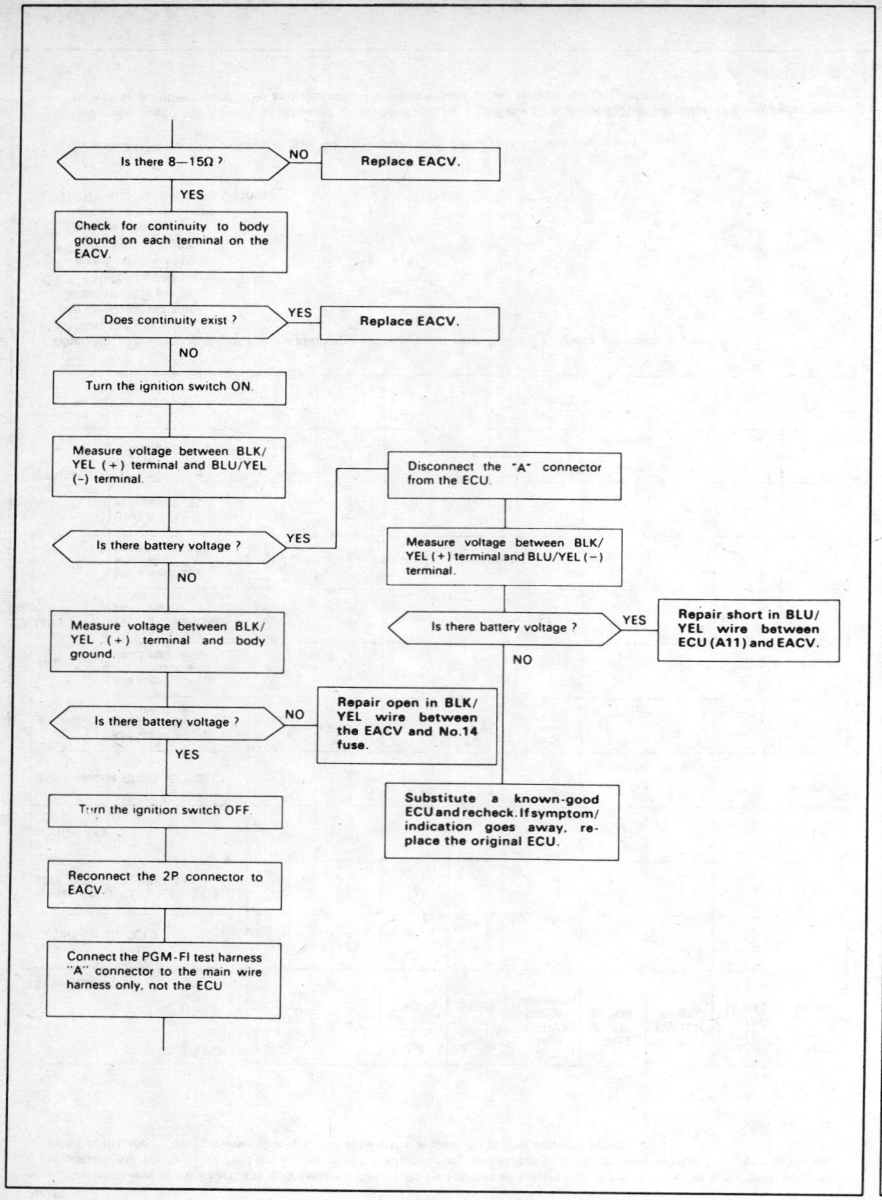

Is there 8—15Ω ?
NO → Replace EACV.
YES ↓

Check for continuity to body ground on each terminal on the EACV.

Does continuity exist ?
YES → Replace EACV.
NO ↓

Turn the ignition switch ON.

Measure voltage between BLK/YEL (+) terminal and BLU/YEL (-) terminal.

Is there battery voltage ?
YES → Disconnect the "A" connector from the ECU. → Measure voltage between BLK/YEL (+) terminal and BLU/YEL (−) terminal. → Is there battery voltage ?
- YES → Repair short in BLU/YEL wire between ECU (A11) and EACV.
- NO → Substitute a known-good ECU and recheck. If symptom/indication goes away, replace the original ECU.

NO ↓

Measure voltage between BLK/YEL (+) terminal and body ground.

Is there battery voltage ?
NO → Repair open in BLK/YEL wire between the EACV and No.14 fuse.
YES ↓

Turn the ignition switch OFF.

Reconnect the 2P connector to EACV.

Connect the PGM-FI test harness "A" connector to the main wire harness only, not the ECU

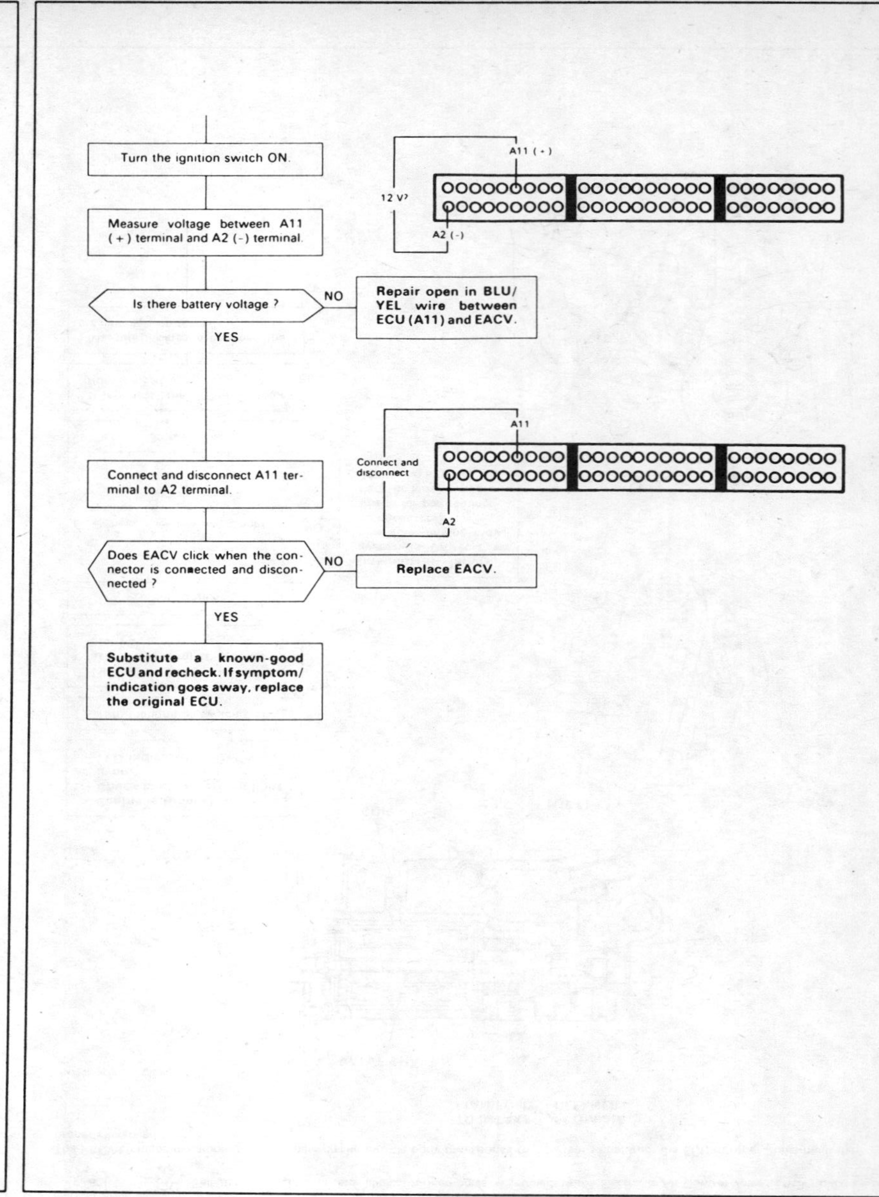

# AIR CONDITIONING SIGNAL—1988–90 CIVIC

This signals the PGM-FI ECU when there is a demand for cooling from the air conditioning circuits.

Inspection of Air Conditioning Signal.
Connect the PGM-FI test harness between the ECU and connector . Disconnect "B" connector from the main wire harness only, not the ECU.
Turn the ignition switch ON.
Measure voltage between B8 (+) terminal and A18 (–) terminal.
Is there approx. 4.5V?
NO
Substitute a known-good ECU and recheck. If prescribed voltage is now available, replace the original ECU.
YES
A18 (–)
B8 (+)
4.5V?
Reconnect "B" connector to the main wire harness.
Momentarily connect B3 terminal to A18 terminal several times.
B3
A18
Short
Is there a clicking noise from the A/C compressor clutch?
NO
Connect the YEL terminal of the 4P connector on the A/C clutch relay to body ground.
YEL
View from wire side.
Is there a clicking noise from the A/C compressor clutch?
NO
Air conditioner
YES
Repair open in YEL wire between ECU (B3) and A/C clutch relay.
YES
Start engine.
Blower switch ON.

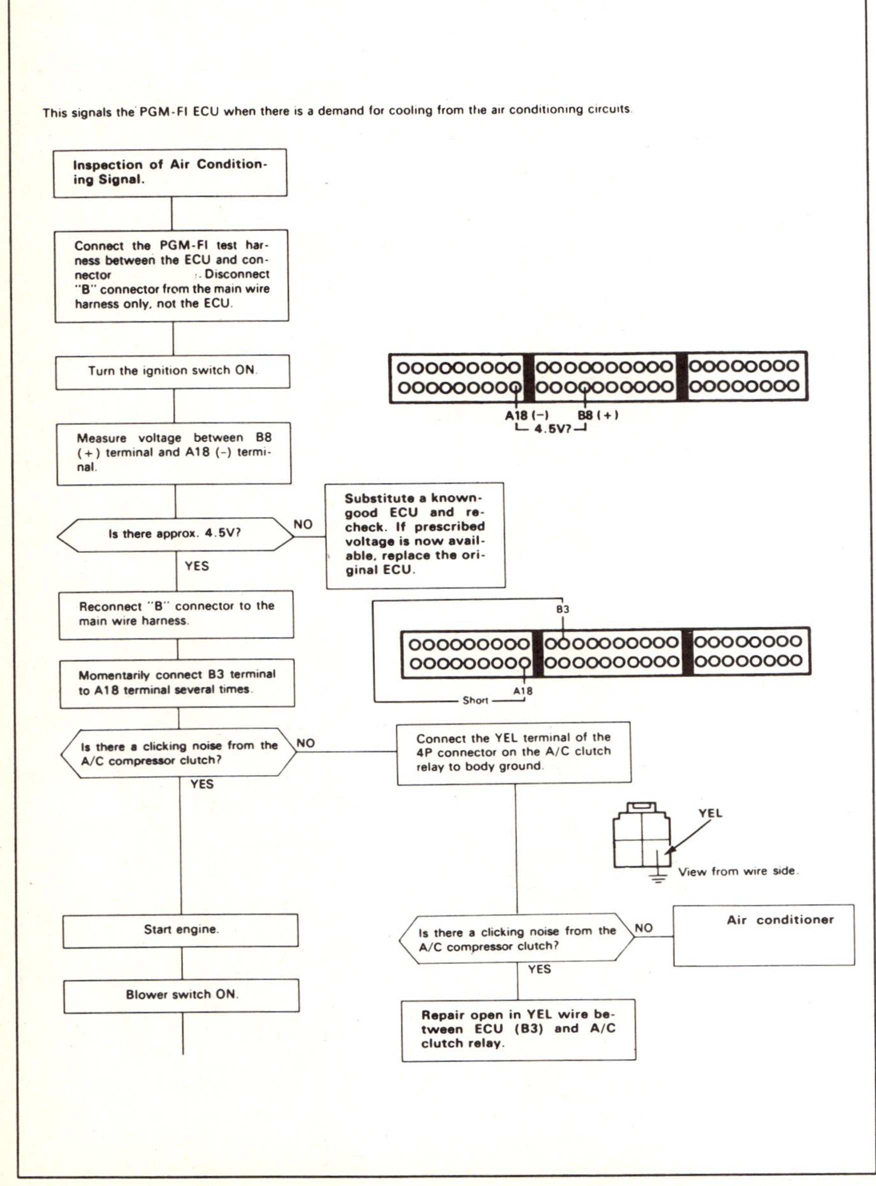

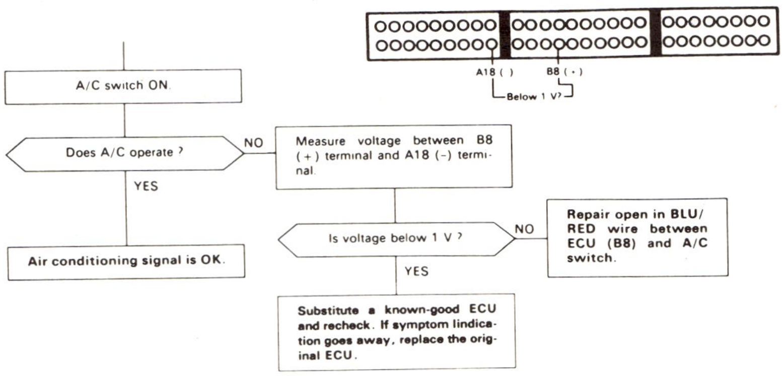

A/C switch ON.
Does A/C operate?
YES
Air conditioning signal is OK.
NO
Measure voltage between B8 (+) terminal and A18 (–) terminal.
A18 (–)
B8 (+)
Below 1 V?
Is voltage below 1 V?
NO
Repair open in BLU/RED wire between ECU (B8) and A/C switch.
YES
Substitute a known-good ECU and recheck. If symptom indication goes away, replace the original ECU.

## ALTERNATOR FR SIGNAL—1988–90 CIVIC

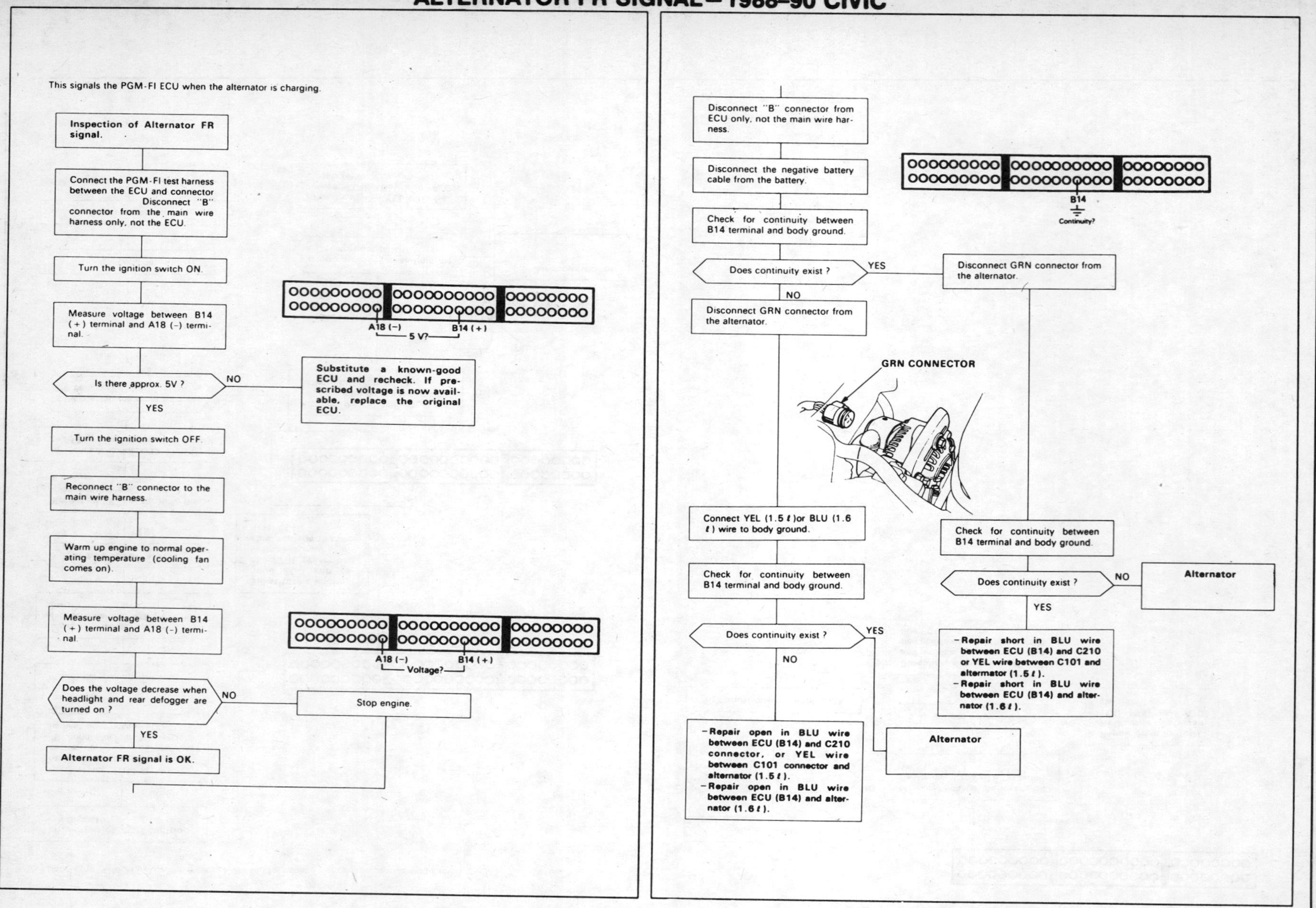

## A/T SHIFT POSITION SIGNAL – 1988–90 CIVIC

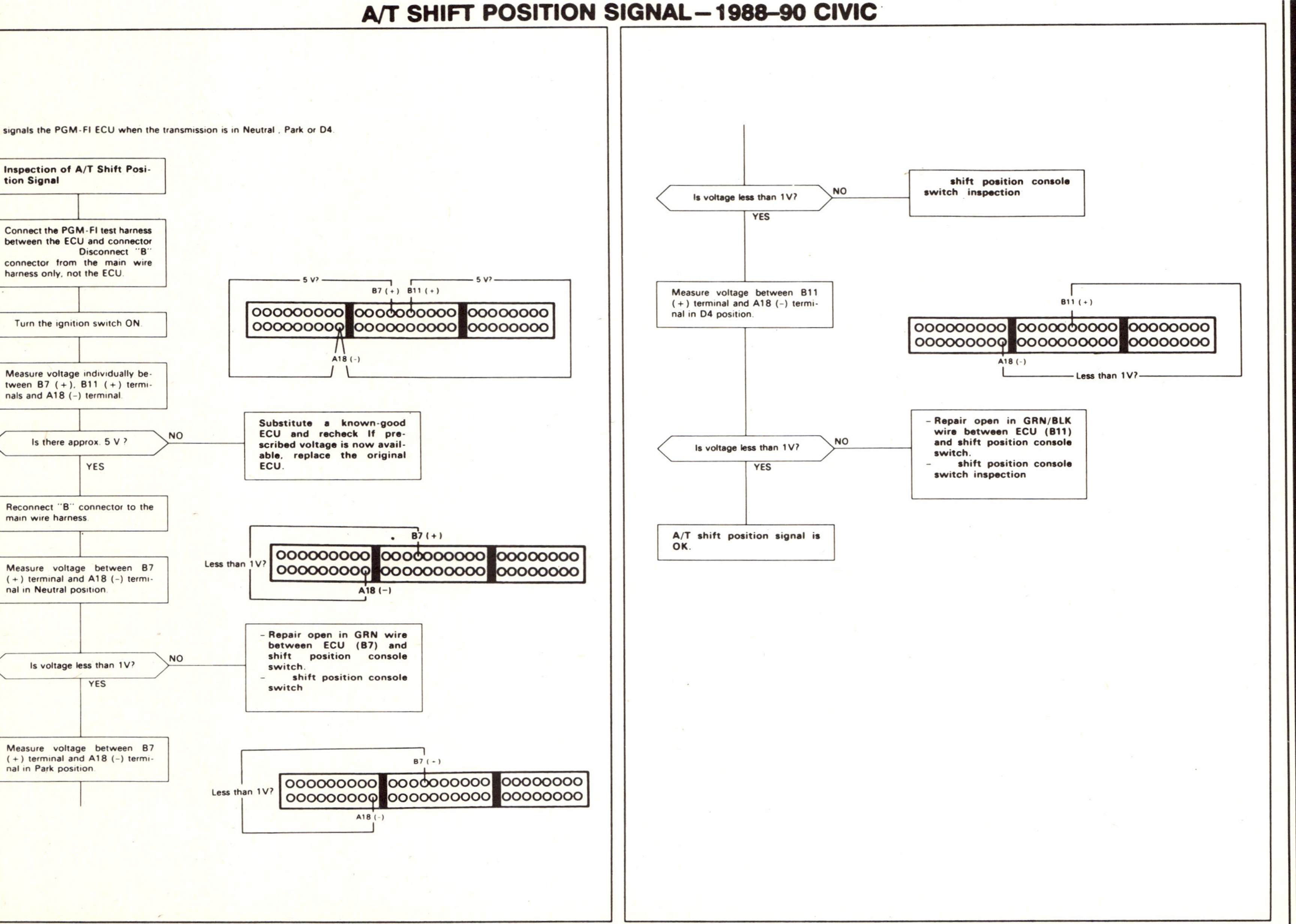

## BRAKE SWITCH SIGNAL—1988–90 CIVIC

This signals the PGM-FI ECU when the brake pedal is depressed.

Inspection of Brake Switch Signal
Are the brake lights on without depressing the brake pedal?
YES
Inspect the brake switch
NO
Depress the brake pedal.
Do the brake lights come on?
NO
Inspect STOP, HORN (15A) fuse (main fuse box).
Is fuse OK?
NO
Replace fuse.
YES
YES
Connect the PGM-FI test harness only the main wire harness, not to the ECU
Inspect the brake switch
Depress the brake pedal.
A18(−)
C10(+)
12V?
Measure voltage between C10 (+) terminal and A18 (−) terminal.
Is there battery voltage?
NO
Repair open in GRN/WHT wire between the brake switch and ECU (C10).
YES
Brake switch signal is OK.

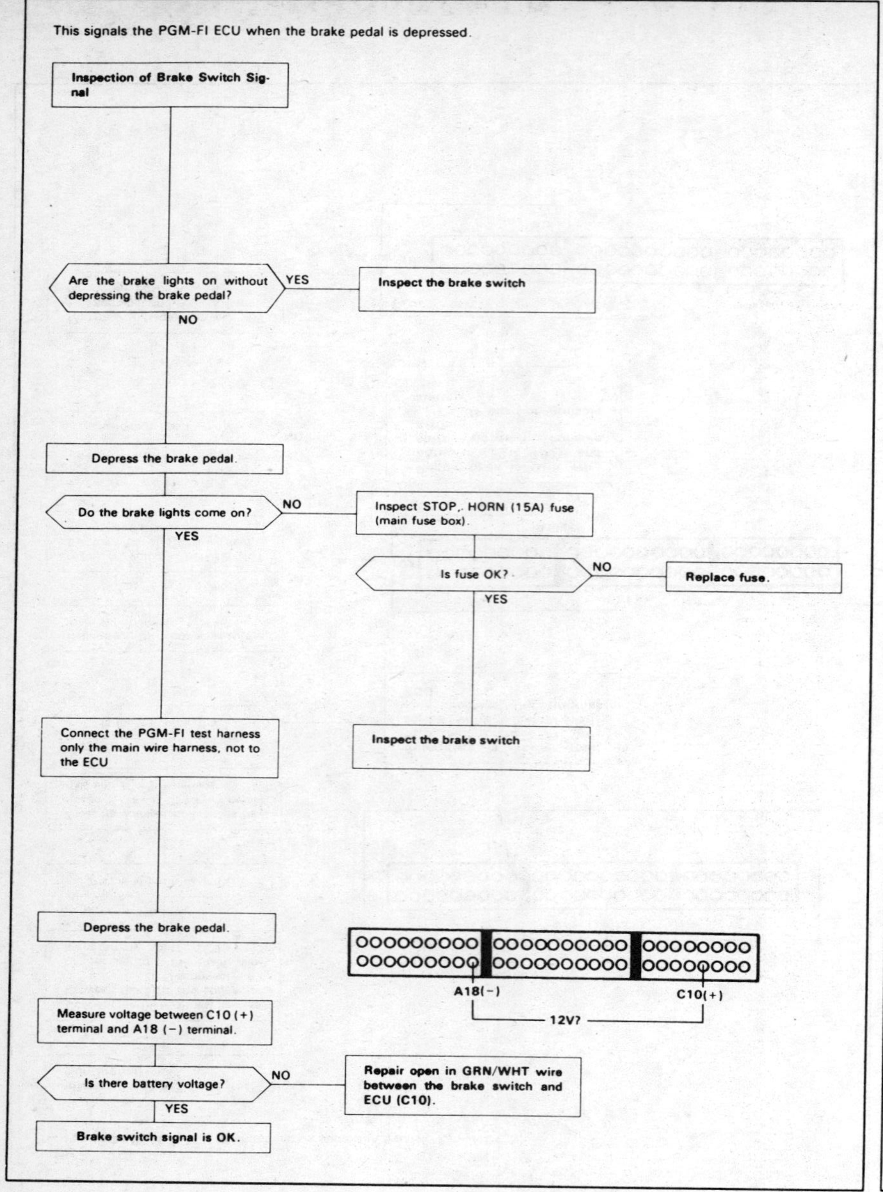

## STARTER SIGNAL—1988–90 CIVIC

This signals the PGM-FI ECU when the engine is cranking.

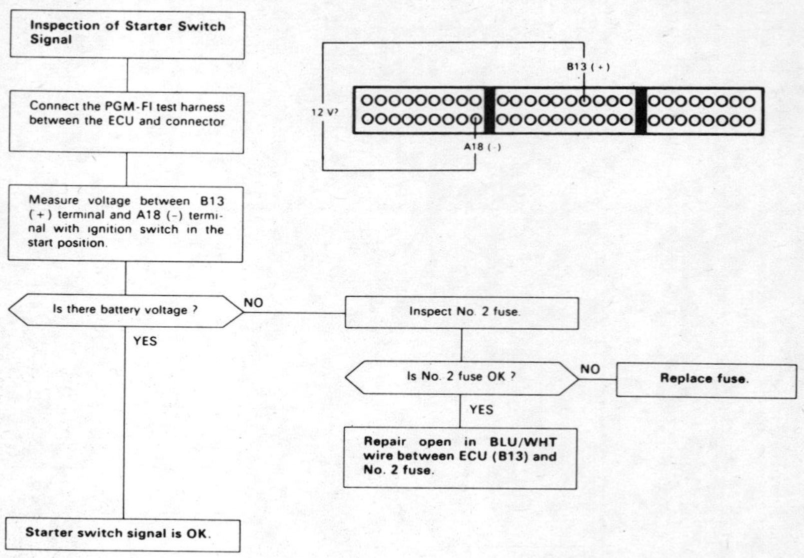

Inspection of Starter Switch Signal
B13 (+)
12 V?
A18 (−)
Connect the PGM-FI test harness between the ECU and connector
Measure voltage between B13 (+) terminal and A18 (−) terminal with ignition switch in the start position.
Is there battery voltage ?
NO
Inspect No. 2 fuse.
Is No. 2 fuse OK ?
NO
Replace fuse.
YES
Repair open in BLU/WHT wire between ECU (B13) and No. 2 fuse.
YES
Starter switch signal is OK.

## HEATER FAN SWITCH SIGNAL—1988–90 CRX HF

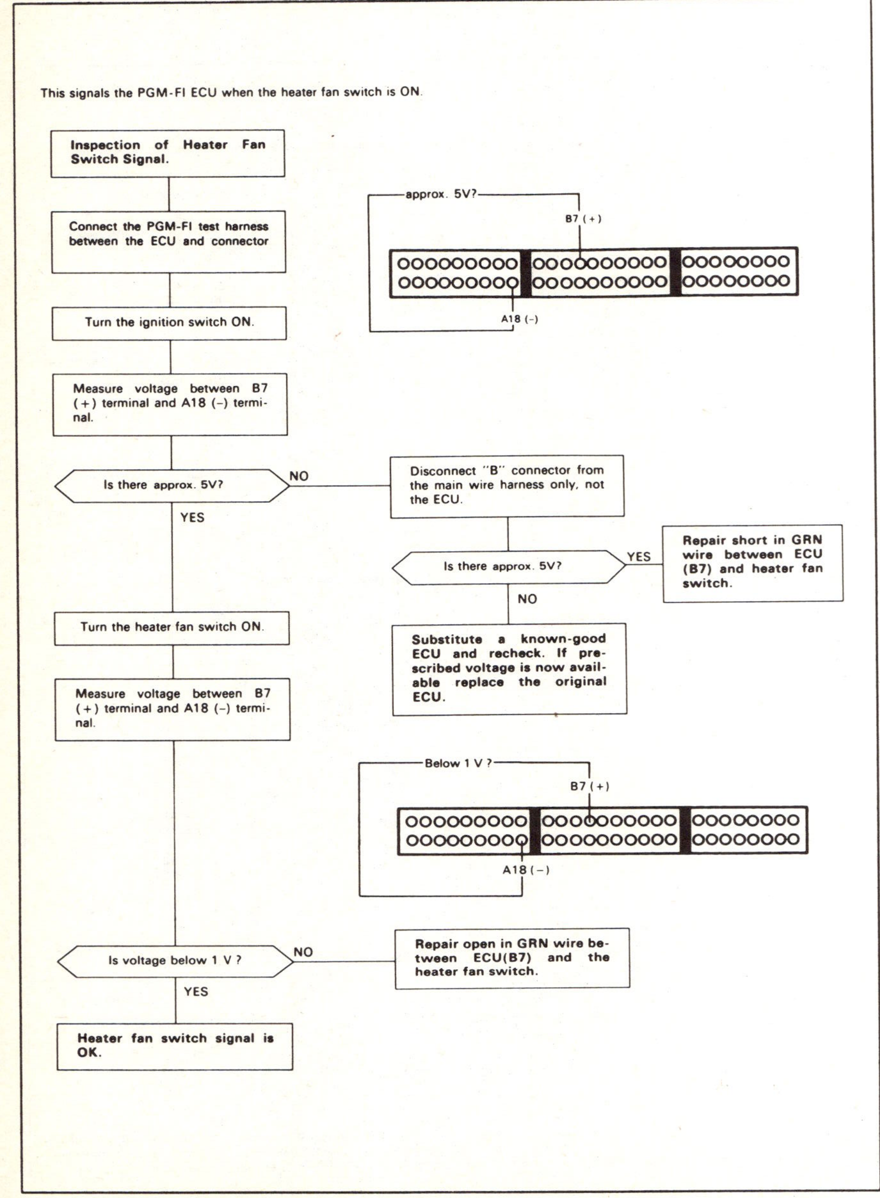

This signals the PGM-FI ECU when the heater fan switch is ON.
Inspection of Heater Fan Switch Signal.
Connect the PGM-FI test harness between the ECU and connector
Turn the ignition switch ON.
Measure voltage between B7 (+) terminal and A18 (–) terminal.
Is there approx. 5V?
NO
YES
approx. 5V?
B7 (+)
A18 (–)
Disconnect "B" connector from the main wire harness only, not the ECU.
Is there approx. 5V?
YES
NO
Repair short in GRN wire between ECU (B7) and heater fan switch.
Substitute a known-good ECU and recheck. If prescribed voltage is now available replace the original ECU.
Turn the heater fan switch ON.
Measure voltage between B7 (+) terminal and A18 (–) terminal.
Below 1 V?
B7 (+)
A18 (–)
Is voltage below 1 V?
NO
YES
Repair open in GRN wire between ECU(B7) and the heater fan switch.
Heater fan switch signal is OK.


## FAST IDLE CONTROL—1988–90 CIVIC 1.6L

Troubleshooting Flow Chart

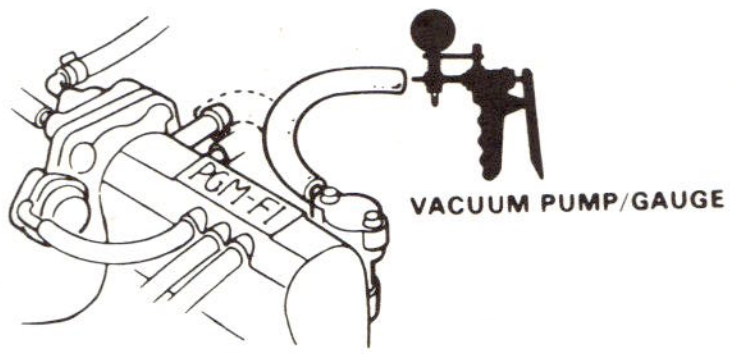

VACUUM PUMP/GAUGE

**Inspection of Fast Idle Control Solenoid Valve**

Warm up engine to normal operating temperature (cooling fan comes on).

Disconnect the upper vacuum hose from the fast idle control solenoid valve and connect a vacuum gauge to the valve.

Is there any vacuum?

- YES → Disconnect the 2P connector from the solenoid valve.
  - Is there any vacuum?
    - YES → **Replace the fast idle control solenoid valve.**
    - NO → **Repair short to ground at BLU wire between ECU (B2) and the 2P connector. If wire is OK, substitute a known-good ECU and recheck. If symptom goes away, replace the original ECU.**
- NO → Turn the ignition switch OFF.

Disconnect the 2P connector from the solenoid valve.

Connect battery positive to terminal A and battery negative to terminal B of the solenoid valve.

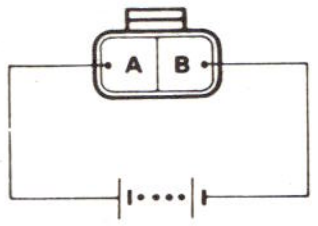

A B

Is there manifold vacuum?

- NO → **Replace the fast idle control solenoid valve.**
- YES → **Fast Idle control Solenoid Valve is OK.**

## FUEL SYSTEM TROUBLESHOOTING—1988–90 CIVIC

NOTE: Across each row in the chart, the systems that could be sources of a symptom are ranked in the order they should be inspected starting with 1. Find the symptom in the left column, read across to the most likely source, then refer to the page listed at the top of that column. If inspection shows the system is OK, try the next most likely system 2, etc.

**1.5ℓ**

| SYMPTOM | | FUEL INJECTOR | PRESSURE REGULATOR | FUEL FILTER | FUEL PUMP | MAIN RELAY | CONTAMINATED FUEL |
|---|---|---|---|---|---|---|---|
| ENGINE WON'T START | | ③ | | ④ | ① | ② | |
| DIFFICULT TO START ENGINE WHEN COLD | | ② | ③ | ① | | | |
| ROUGH IDLE | | ① | ② | | | | |
| FREQUENT STALLING | WHILE WARMING UP | ① | | ② | ③ | | |
| | AFTER WARMING UP | ① | ⑤ | ③ | ② | | ④ |
| POOR PERFORMANCE | MISFIRE OR ROUGH RUNNING | ① | ② | | | | ③ |
| | FAILS EMISSION TEST | ① | ② | | | | |
| | LOSS OF POWER | ④ | | ① | ③ | | ② |

**HF and 1.6ℓ**

| SYMPTOM | | FUEL INJECTOR | INJECTOR RESISTOR | PRESSURE REGULATOR | FUEL FILTER | FUEL PUMP | MAIN RELAY | CONTAMINATED FUEL |
|---|---|---|---|---|---|---|---|---|
| ENGINE WON'T START | | | ③ | | ④ | ① | ② | |
| DIFFICULT TO START ENGINE WHEN COLD | | ③ | | | ② | ① | | |
| ROUGH IDLE | | ① | | ② | | | | |
| FREQUENT STALLING | WHILE WARMING UP | ① | | | ② | ③ | | |
| | AFTER WARMING UP | ① | | ④ | ③ | ② | | |
| POOR PERFORMANCE | MISFIRE OR ROUGH RUNNING | ① | | ② | | | | |
| | FAILS EMISSION TEST | ① | | ② | | | | |
| | LOSS OF POWER | | | | ① | ③ | | |

• Fuel with dirt, water or a high percentage of alcohol is considered contaminated.

## FUEL INJECTORS—CODE 16
## 1988–90 CIVIC 1.5L EXCEPT CRX HF

**Troubleshooting Flowchart**

(CHECK) – (16) Self-diagnosis LED indicator blinks sixteen times: A problem in the fuel injector circuit.

The injectors are the solenoid-actuated constant-stroke pintle type consisting of a solenoid, plunger needle valve and housing. When current is applied to the solenoid coil, the valve lifts up and pressurized fuel is injected. Because the needle valve lift and the fuel pressure are constant, the injection quantity is determined by the length of time that the valve is open (i.e., the duration the current is supplied to the solenoid coil). The injector is sealed by two O-rings and a mount rubber. This also reduces operating noise.

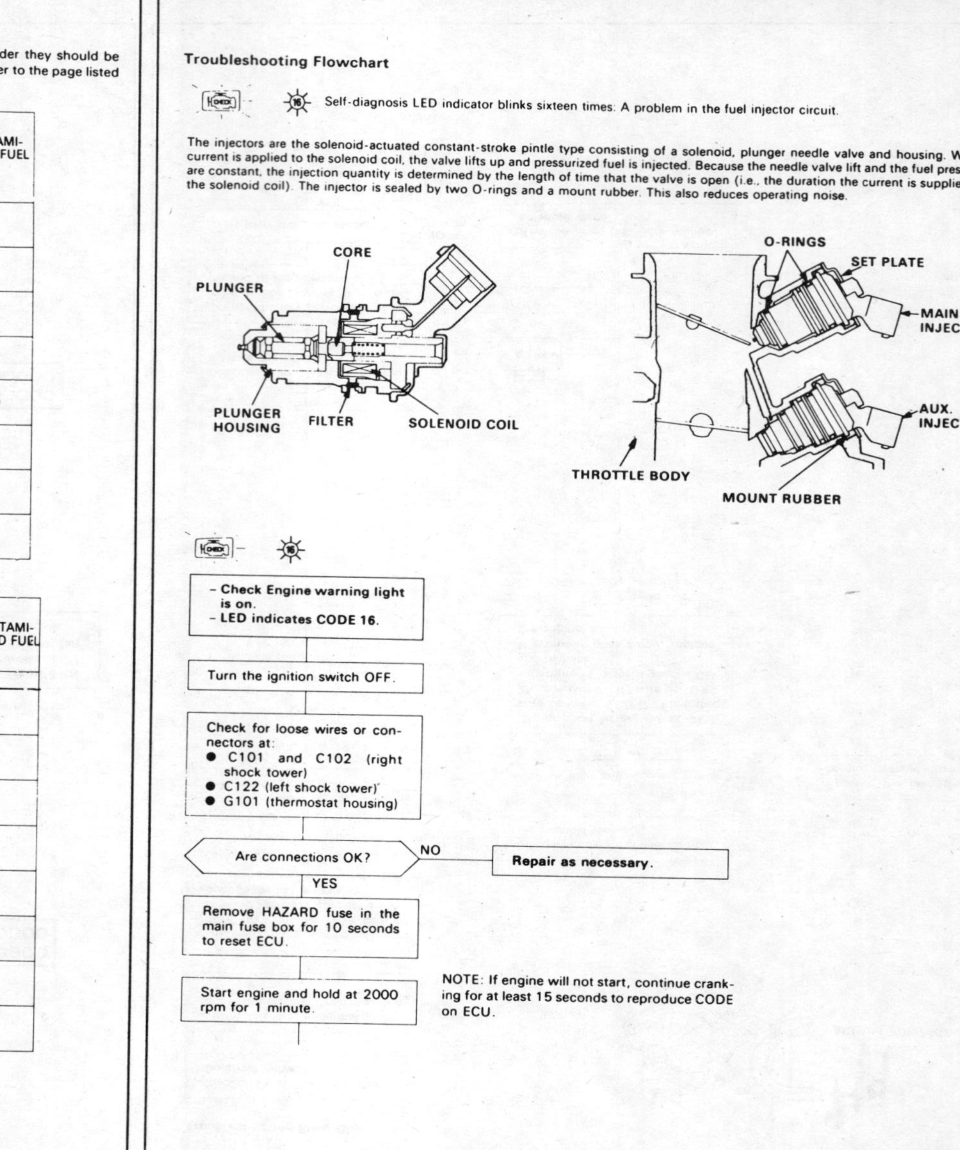


(CHECK) – (16)

- Check Engine warning light is on.
- LED indicates CODE 16.

↓

Turn the ignition switch OFF.

↓

Check for loose wires or connectors at:
- C101 and C102 (right shock tower)
- C122 (left shock tower)
- G101 (thermostat housing)

↓

Are connections OK? — NO → **Repair as necessary.**

YES ↓

Remove HAZARD fuse in the main fuse box for 10 seconds to reset ECU.

↓

Start engine and hold at 2000 rpm for 1 minute.

NOTE: If engine will not start, continue cranking for at least 15 seconds to reproduce CODE on ECU.

# FUEL INJECTORS—CODE 16—1988–90 CIVIC 1.5L EXCEPT CRX HF

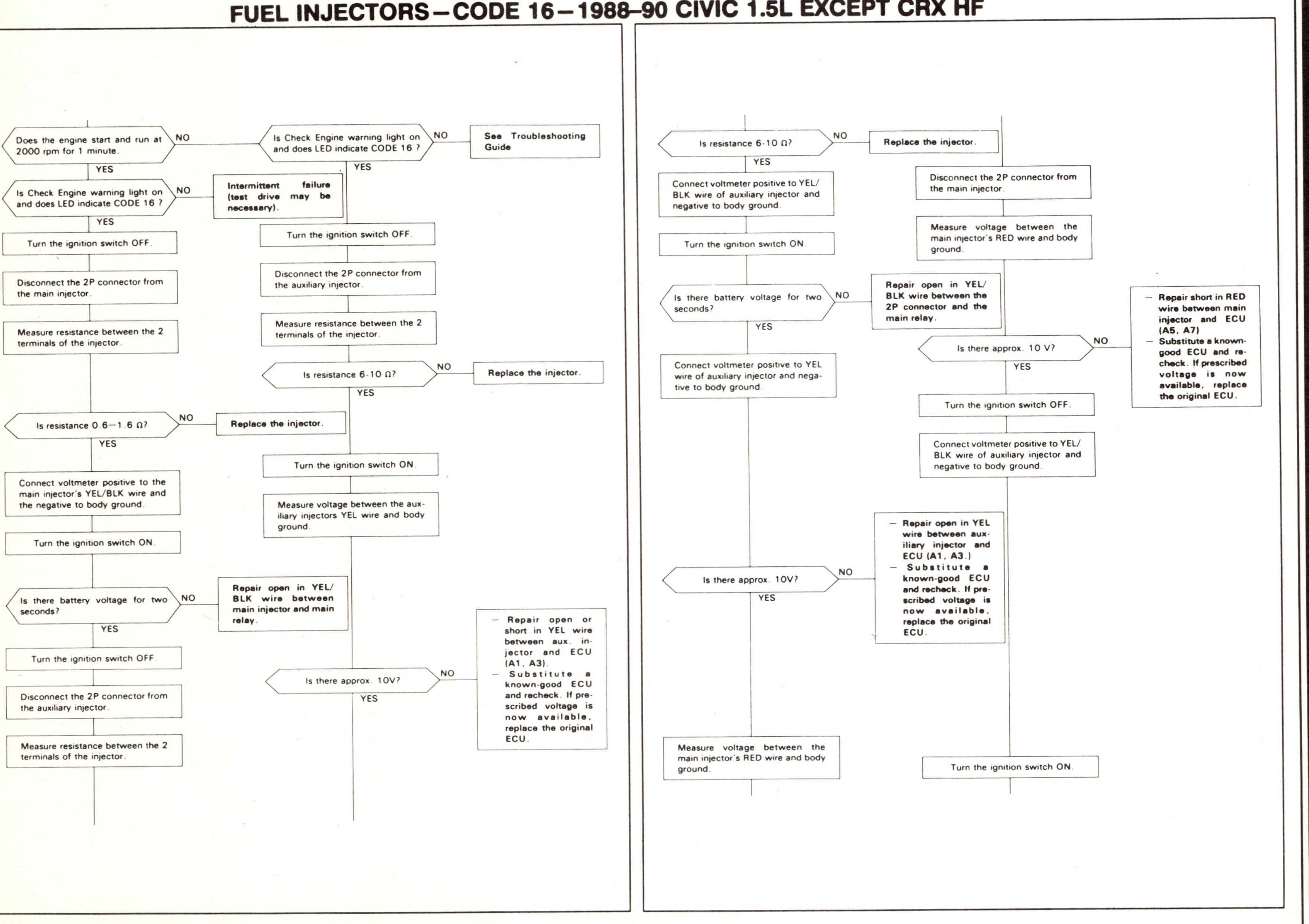

## FUEL INJECTORS—CODE 16
## 1988–90 CIVIC 1.5L EXCEPT CRX HF

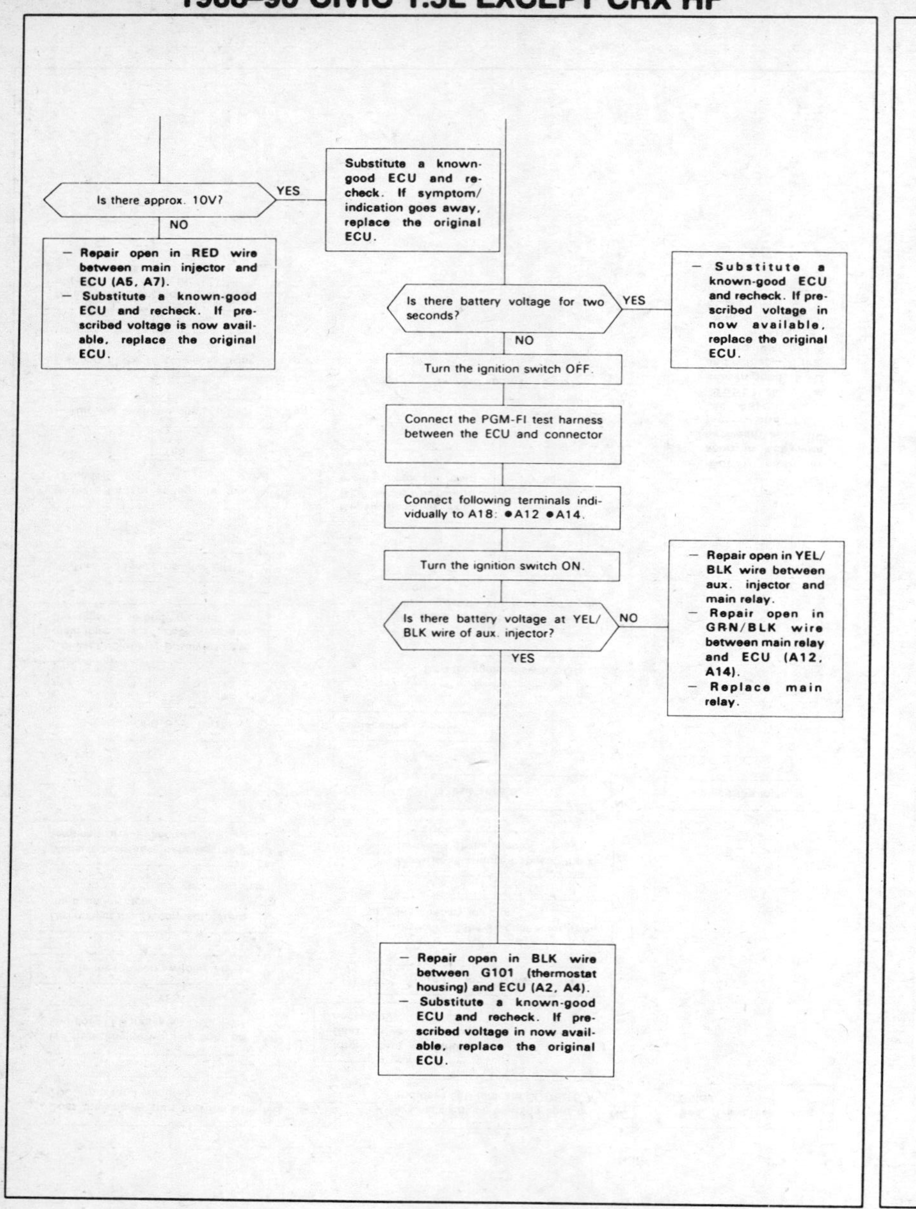


## FUEL INJECTORS—CODE 16
## 1988–90 CIVIC 1.6L AND CRX HF

**Troubleshooting Flowchart**

Self-diagnosis LED indicator blinks sixteen times: A problem in the fuel injector circuit.

The injectors are the solenoid-actuated constant-stroke pintle type consisting of a solenoid, plunger needle valve and housing. When current is applied to the solenoid coil, the valve lifts up and pressurized fuel is injected close to the intake valve. Because the needle valve lift and the fuel pressure are constant, the injection quantity is determined by the length of time that the valve is open (i.e., the duration the current is supplied to the solenoid coil). The injector is sealed by an O-ring and seal ring at the top and bottom. These seals also reduce operating noise.

PLUNGER HOUSING
FILTER
SOLENOID COIL
CORE
PLUNGER
SEAL RING
INJECTOR
CUSHION RING
O-RING
FUEL PIPE
INTAKE MANIFOLD

- Check Engine warning light is on.
- LED indicates CODE 16.

Turn the ignition switch OFF.

Remove HAZARD fuse in the main fuse box for 10 seconds to reset ECU.

Turn the ignition switch to START position.

Does the engine start?

NO — Intermittent failure, system is OK at this time (test drive may be necessary). Check for poor connections or loose wires at injectors, injector resistor and C313 (round connector on left shock tower).

YES — Is Check Engine warning light on and does LED indicate CODE 16?

NO — Intermittent failure, system is OK at this time (test drive may be necessary). Check for poor connections or loose wires at injectors, injector resistor and C313 (round connector on left shock tower).

YES

# FUEL INJECTORS—CODE 16—1988–90 CIVIC 1.6L AND CRX HF

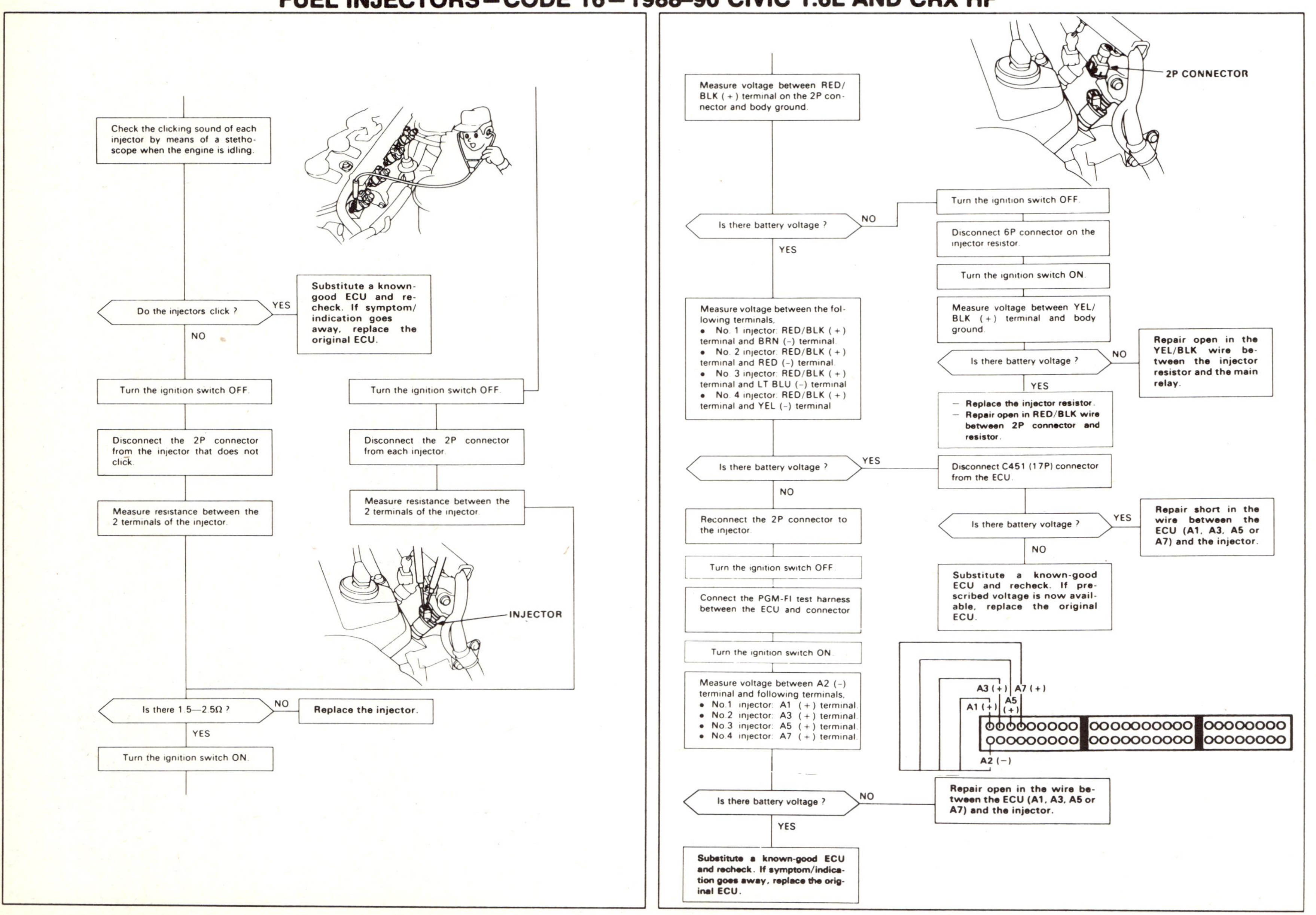

## TANDEM CONTROL SYSTEM – 1988–90 CIVIC 1.5L EXCEPT CRX HF

### Troubleshooting Flowchart

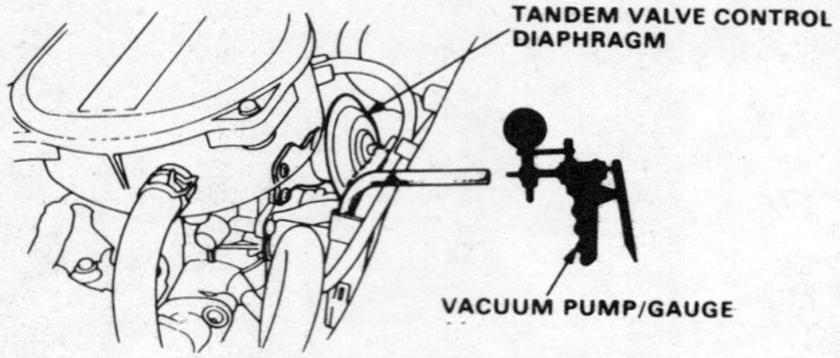

Inspection of Tandem Control System

Disconnect the vacuum hose from the tandem valve control diaphragm and connect a vacuum gauge to the hose.

TANDEM VALVE CONTROL DIAPHRAGM

VACUUM PUMP/GAUGE

Start engine and allow to idle. NOTE: Coolant temperature must be below 70°C (160°F).

Is there vacuum ?

- YES → Disconnect the 2P connector from the tandem valve control solenoid valve.
  - Is there vacuum ?
    - YES → **Replace the solenoid valve.**
    - NO → Turn the ignition switch OFF.
      - Disconnect "B" connector from ECU.
      - Check for continuity to ground on the ORN wire.
      - Is there continuity to ground ?
        - YES → **Repair short to ground at ORN wire between ECU (B2) and the 2P connector.**
        - NO → **Substitute a known-good ECU and recheck. If symptom goes away, replace the original ECU.**
- NO → Raise engine speed to.
  - Manual: 3,000 rpm
  - Automatic: 2,000 rpm

  Is there vacuum ?
  - YES
  - NO → Disconnect the 2P connector from the tandem valve control solenoid valve.

Measure voltage between BLK/YEL (+) terminal and ORN (–) terminal at 3,000 rpm.

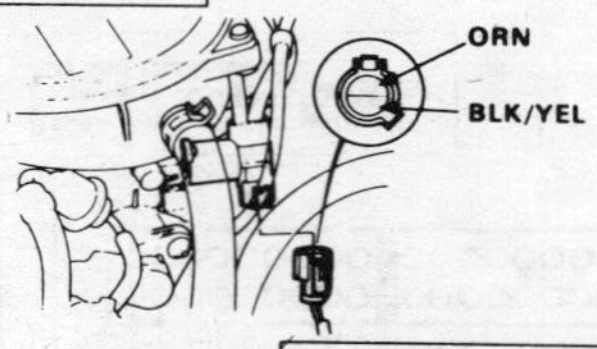

ORN

BLK/YEL

Is there battery voltage ?

- YES → **Remove the solenoid valve from the throttle body and check the port for blockage. If the port is OK, replace the solenoid valve.**
- NO → Measure voltage between BLK/YEL (+) terminal and body ground.
  - Is there battery voltage ?
    - NO → **Repair open in BLK/YEL wire between the 2P connector and No. 14 fuse.**
    - YES → Turn the ignition switch OFF.
      - Connect the PGM-FI test harness between the ECU and connector

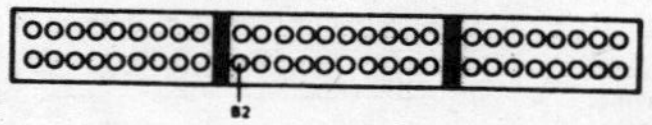

        B2
      - Check for continuity of ORN wire between ECU (B2) and the 2P connector.
      - Does continuity exist ?
        - NO → **Repair open in ORN wire between ECU (B2) and the 2P connector.**
        - YES → **Substitute a known-good ECU and recheck. If symptom goes away, replace the original ECU.**

## TANDEM CONTROL SYSTEM 1988–90 CIVIC 1.5L EXCEPT CRX HF

Warm up engine to normal operating temperature (the cooling fan comes on).

Slowly open the throttle.

Is there vacuum ?

- NO → **Substitute a known-good ECU and recheck. If symptom goes away, replace the original ECU.**
- YES → **Tandem control system is OK.**

## THROTTLE CONTROL SYSTEM—1988 CRX SI

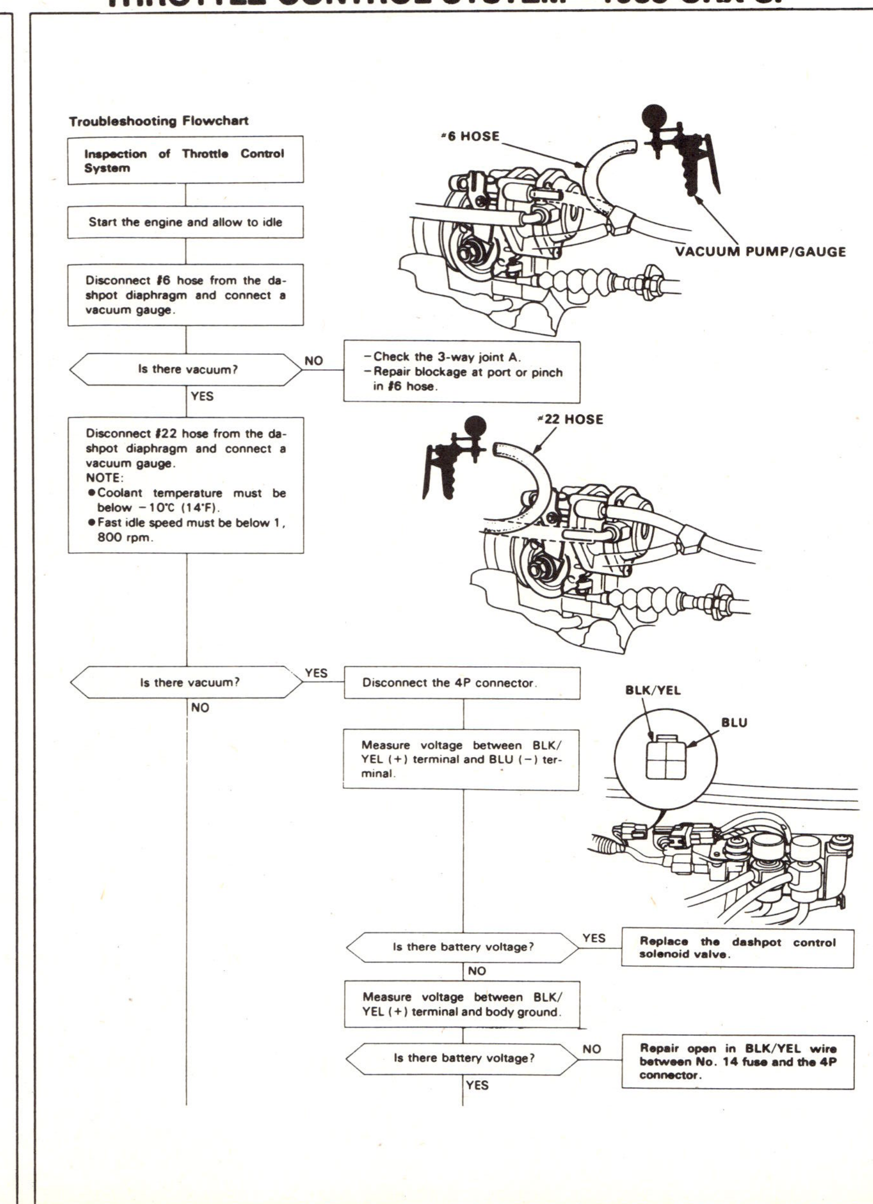

## THROTTLE CONTROL SYSTEM – 1988 CRX SI

**Troubleshooting Flowchart** (cont'd)

Warm up engine to normal operating temperature (cooling fan comes on).

Check for vacuum at #22 vacuum hose.

Is there vacuum?
- YES → Throttle control system is OK.
- NO → Disconnect the 4P connector.

Is there vacuum?
- NO → Replace the dashpot control solenoid valve.
- YES → Turn the ignition switch OFF.

Disconnect "B" connector from ECU.

Check for continuity to ground in the BLU (to ECU, B2) wire.

Is there continuity to ground?
- YES → **Repair short to ground at BLU wire between ECU (B2) and the 4P connector.**
- NO → **Substitute a known-good ECU and recheck. If symptom goes away, replace the original ECU.**

Turn the ignition switch OFF.

Connect the PGM-FI test harness between the ECU and connector

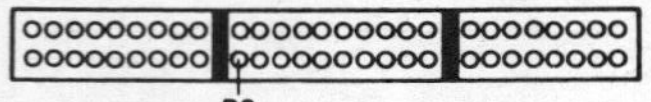

B2

Check for continuity of BLU wire between ECU (B2) and the 4P connector.

Does continuity exist?
- NO → **Repair open in BLU wire between ECU (B2) and the 4P connector.**
- YES → **Subsitute a known-good ECU and recheck. If symptom goes away replace the original ECU.**

## EGR SYSTEM – CODE 12
## 1988–90 CIVIC CALIF. A/T EXCEPT CRX

**Troubleshooting Flowchart**

Self diagnosis indicator blinks twelve times. Most likely a problem in the Exhaust Gas Recirculation (EGR) system.

The EGR System is designed to reduce oxides of nitrogen emissions (NOx) by recirculating exhaust gas through the EGR valve and the intake manifold into the combustion chambers. It is composed of the EGR valve, CVC valve, EGR control Solenoid valve, PGM-FI ECU and various sensors.
The ECU contains memories for ideal EGR valve lifts for varying operating conditions. The EGR valve lift sensor detects the amount of EGR valve lift and sends the information to the ECU. The ECU then compares it with the ideal EGR valve lift which is determined by signals sent from the other sensors. If there is any difference between the two, the ECU cuts current to the EGR control solenoid valve to reduce vacuum applied to the EGR valve.

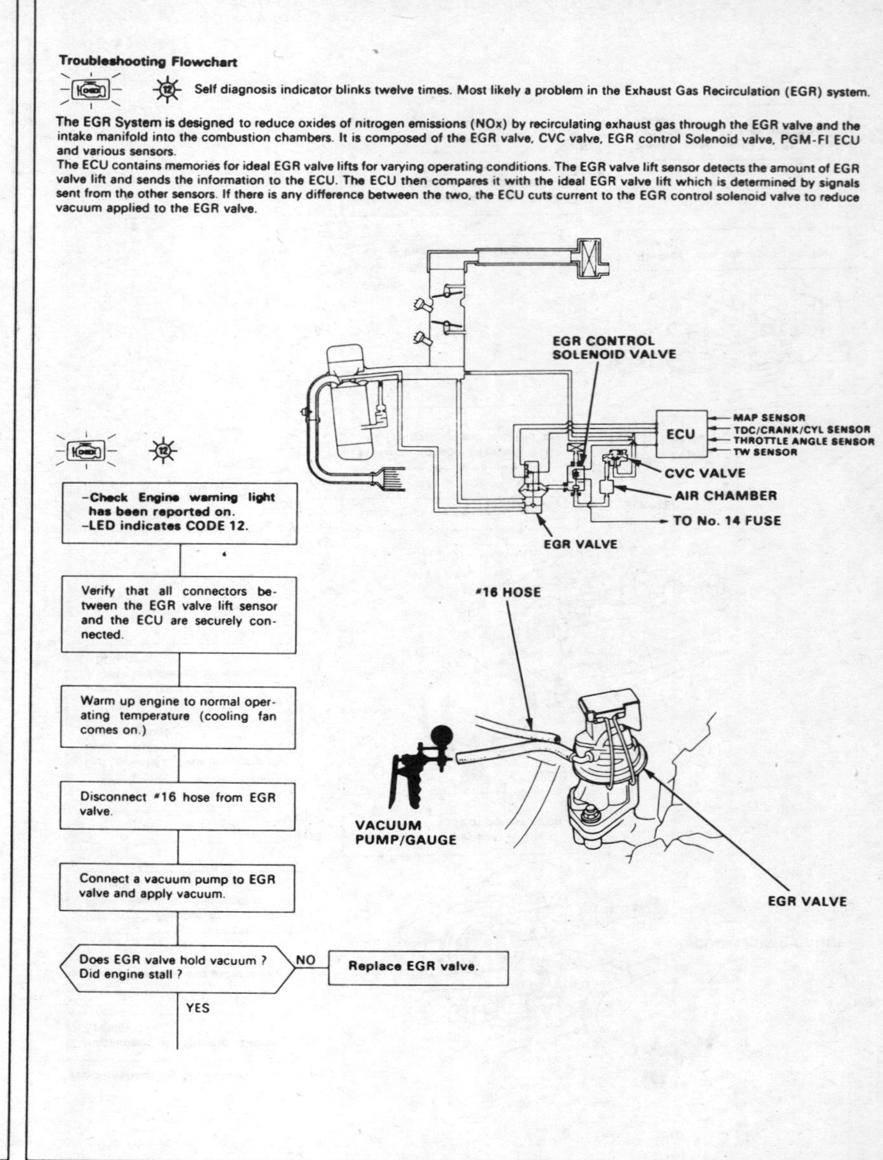


**–Check Engine warning light has been reported on.**
**–LED indicates CODE 12.**

Verify that all connectors between the EGR valve lift sensor and the ECU are securely connected.

Warm up engine to normal operating temperature (cooling fan comes on.)

Disconnect #16 hose from EGR valve.

Connect a vacuum pump to EGR valve and apply vacuum.

Does EGR valve hold vacuum ? Did engine stall ?
- NO → **Replace EGR valve.**
- YES

# EGR SYSTEM—CODE 12—1988–90 CIVIC CALIF. A/T EXCEPT CRX

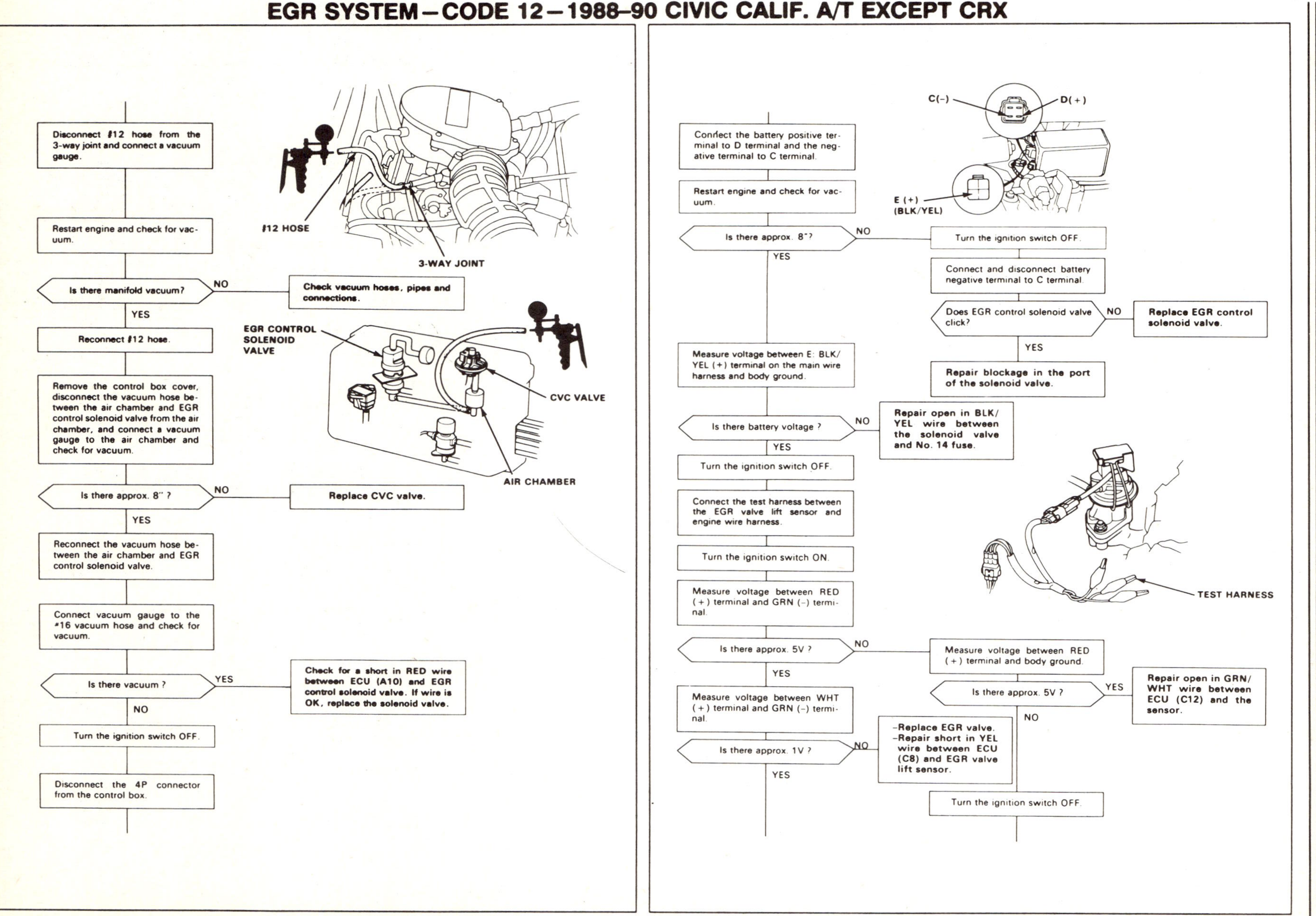

Disconnect #12 hose from the 3-way joint and connect a vacuum gauge.
#12 HOSE
3-WAY JOINT
Restart engine and check for vacuum.
Is there manifold vacuum?
NO
Check vacuum hoses, pipes and connections.
YES
Reconnect #12 hose.
EGR CONTROL SOLENOID VALVE
CVC VALVE
AIR CHAMBER
Remove the control box cover, disconnect the vacuum hose between the air chamber and EGR control solenoid valve from the air chamber, and connect a vacuum gauge to the air chamber and check for vacuum.
Is there approx. 8" ?
NO
Replace CVC valve.
YES
Reconnect the vacuum hose between the air chamber and EGR control solenoid valve.
Connect vacuum gauge to the #16 vacuum hose and check for vacuum.
Is there vacuum ?
YES
Check for a short in RED wire between ECU (A10) and EGR control solenoid valve. If wire is OK, replace the solenoid valve.
NO
Turn the ignition switch OFF.
Disconnect the 4P connector from the control box.
C(–)
D(+)
E(+) (BLK/YEL)
Connect the battery positive terminal to D terminal and the negative terminal to C terminal.
Restart engine and check for vacuum.
Is there approx. 8"?
NO
Turn the ignition switch OFF.
Connect and disconnect battery negative terminal to C terminal.
Does EGR control solenoid valve click?
NO
Replace EGR control solenoid valve.
YES
Repair blockage in the port of the solenoid valve.
YES
Measure voltage between E: BLK/YEL (+) terminal on the main wire harness and body ground.
Is there battery voltage ?
NO
Repair open in BLK/YEL wire between the solenoid valve and No. 14 fuse.
YES
Turn the ignition switch OFF.
Connect the test harness between the EGR valve lift sensor and engine wire harness.
TEST HARNESS
Turn the ignition switch ON.
Measure voltage between RED (+) terminal and GRN (–) terminal.
Is there approx. 5V ?
NO
Measure voltage between RED (+) terminal and body ground.
Is there approx. 5V ?
YES
Repair open in GRN/WHT wire between ECU (C12) and the sensor.
NO
Turn the ignition switch OFF.
YES
Measure voltage between WHT (+) terminal and GRN (–) terminal.
Is there approx. 1V ?
NO
–Replace EGR valve.
–Repair short in YEL wire between ECU (C8) and EGR valve lift sensor.
YES

## EGR SYSTEM—CODE 12
## 1988–90 CIVIC CALIF. A/T EXCEPT CRX

Turn the ignition switch OFF.

Connect the PGM-FI test harness between the ECU and connector

Turn the ignition switch ON.

Measure voltage between C8 (+) terminal and C12 (–) terminal.

C8 (+) C12 (–) 1 V?

Is there approx. 1V ?

NO — Repair open in YEL wire between ECU (C8) and the sensor.

YES — Disconnect "A" connector from the ECU only, not the main wire harness.

Connect the PGM-FI test harness between the ECU and connector

Turn the ignition switch ON.

Measure voltage between C13 (+) terminal and C12 (–) terminal.

C13 (+) C12 (–) 5 V?

Is there approx. 5 V ?

YES — Repair open in YEL/WHT wire between ECU (C13) and the sensor.

NO — Substitute a known-good ECU and recheck. If prescribed voltage is now available, replace the original ECU.

Measure voltage between A10 (+) terminal and A2 (–) terminal.

A2 (–) A10 (+) 12 V?

Is there battery voltage ?

NO — Repair open in RED wire between ECU (A10) and EGR control solenoid valve.

YES — Substitute a known-good ECU and recheck. If symptom/indication goes away, replace the original ECU.

## EGR SYSTEM—CODE 12
## 1988–90 CRX HF AND Std. CALIF. A/T

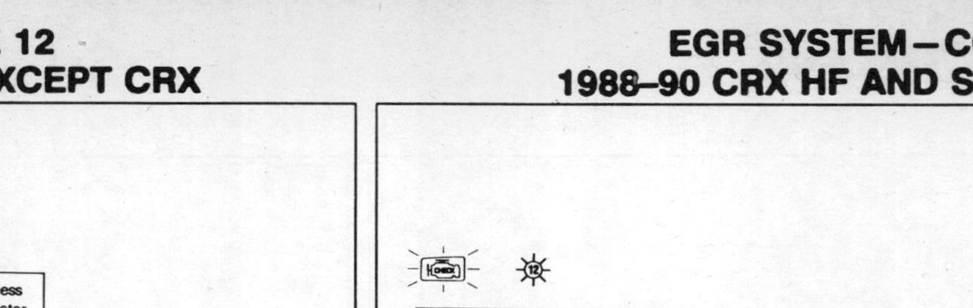

—Check Engine warning light has been reported on.
–LED indicates CODE 12.

Verify that all connectors between the EGR valve lift sensor and the ECU are securely connected.

Warm up engine to normal operating temperature (cooling fan comes on.)

Disconnect #16 hose from EGR valve.

Connect a vacuum pump to EGR valve and apply vacuum.



Does EGR valve hold vacuum ? Did engine stall ?

NO — Replace EGR valve.

YES — Std. CAL (A/T): Disconnect #7 hose from the 3-way joint and connect a vacuum gauge.
HF: Disconnect #10 hose from the intake manifold and connect a vacuum gauge to the port.



Restart engine and check for vacuum.

Is there manifold vacuum?

NO — Check vacuum hoses, pipes and connecting.

YES — Reconnect hose.

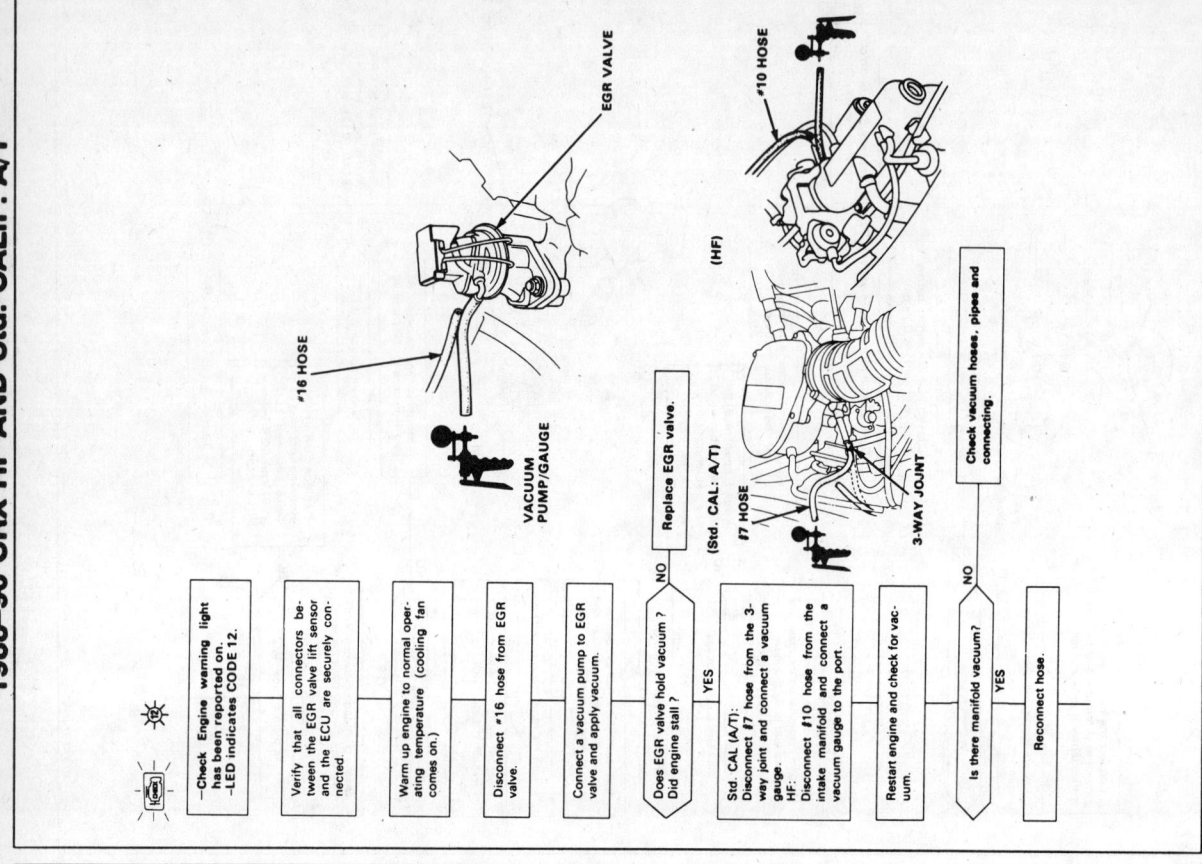

# EGR SYSTEM—CODE 12—1988–90 CRX HF AND Std. CALIF. A/T

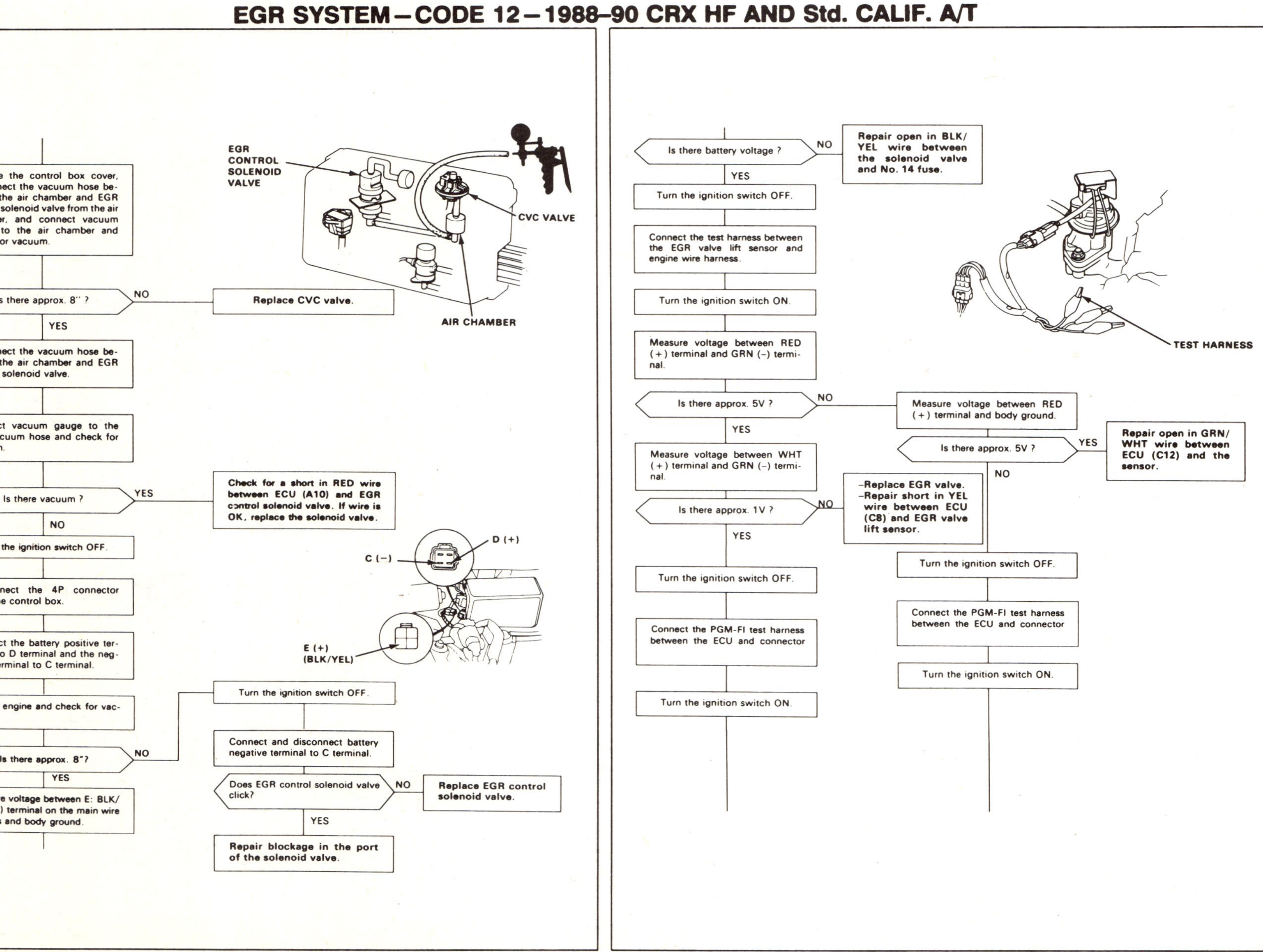

## EGR SYSTEM—CODE 12
## 1988–90 CRX HF AND Std. CALIF. A/T

Measure voltage between C13 (+) terminal and C12 (–) terminal.

C13 (+) / C12 (–) / 5 V?

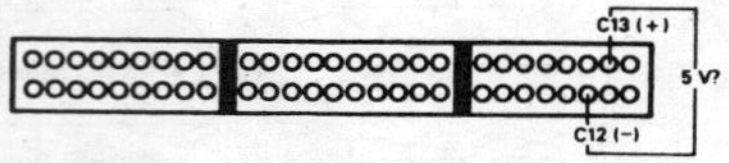

Is there approx. 5 V ?

- YES → Repair open in YEL/WHT wire between ECU (C13) and the sensor.
- NO → Substitute a known-good ECU and recheck. If prescribed voltage is now available, replace the original ECU.

Measure voltage between C8 (+) terminal and C12 (–) terminal.

C8 (+) / C12 (–) / 1 V?

Is there approx. 1V ?

- NO → Repair open in YEL wire between ECU (C8) and the sensor.
- YES → Disconnect "A" connector from the ECU only, not the main wire harness.

Measure voltage between A10 (+) terminal and A2 (–) terminal.

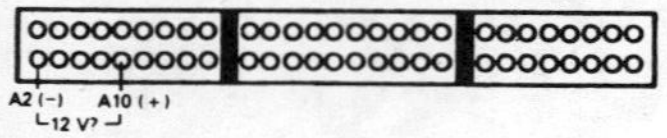

A2 (–) / A10 (+) / 12 V?

Is there battery voltage ?

- NO → Repair open in RED wire between ECU (A10) and EGR control solenoid valve.
- YES → Substitute a known-good ECU and recheck. If symptom/indication goes away, replace the original ECU.

## EVAPORATIVE CONTROLS
## 1988–90 CIVIC 1.5L EXCEPT CRX HF

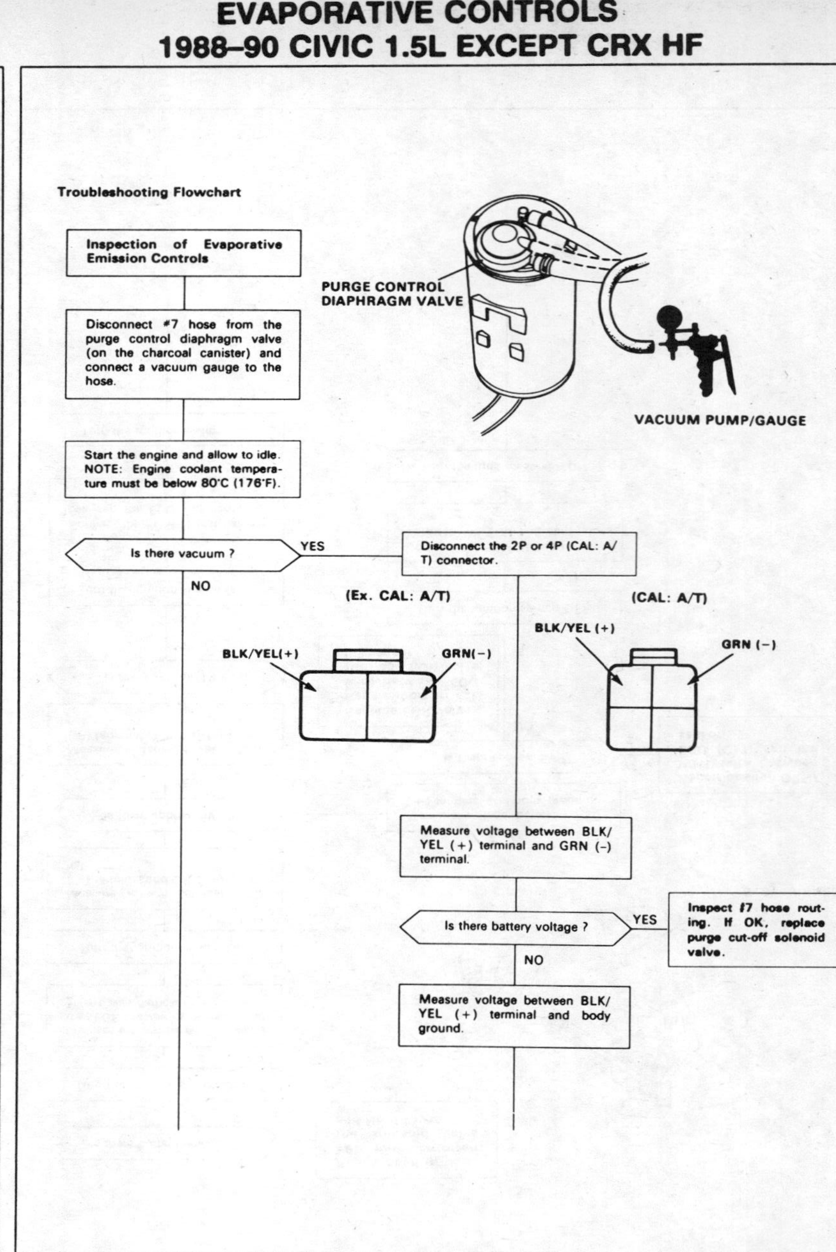

# EVAPORATIVE CONTROLS—1988–90 CIVIC 1.5L EXCEPT CRX HF

Warm up the engine to normal operating temperature (cooling fan comes on).

Check for vacuum at #7 hose 10 seconds after starting the engine NOTE: On the 1.5ℓ, check with the throttle valve sligtly opend.

Is there manifold vacuum?
- NO → Disconnect the 2P or 4P (CAL: A/T) connector.
  - Is there manifold vacuum?
    - NO → Inspect #7 hose routing. If OK, purge cut-off solenoid valve.
    - YES → Turn the ignition switch OFF.
  - Disconnect "A" connector from ECU.
  - Check for continuity to ground on GRN wire.
    - Is there continuity to ground ?
      - YES → Repair short to ground in GRN wire between ECU (A6) and the connector.
      - NO → Substitute a known-good ECU and recheck. If symptom goes away, replace the original ECU.
- YES → Reconnect the hose.
  - Remove fuel filler cap.
  - Is there battery voltage ?
    - NO → Repair in BLK/YEL wire between No.14 fuse and the 2P or 4P (CAL: A/T) connector
    - YES → Turn the ignition switch OFF.
  - Connect the PGM-FI test harness to the main wire harness only, not the ECU.

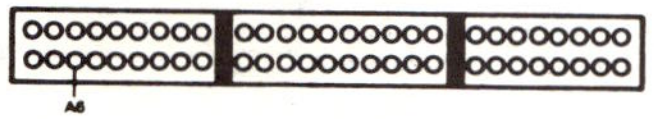

A6

- Check for continuity of GRN wire between ECU (A6) and the 2P or 4P connector.
- Does continuity exist ?
  - NO → Repair open in GRN wire between ECU (A6) and the connector.
  - YES → Substitute a known-good ECU and recheck. If symptom goes away, replace the original ECU.

Connect a vacuum gauge to canister purge air hose.

Start the engine and raise speed to 3,500 rpm.

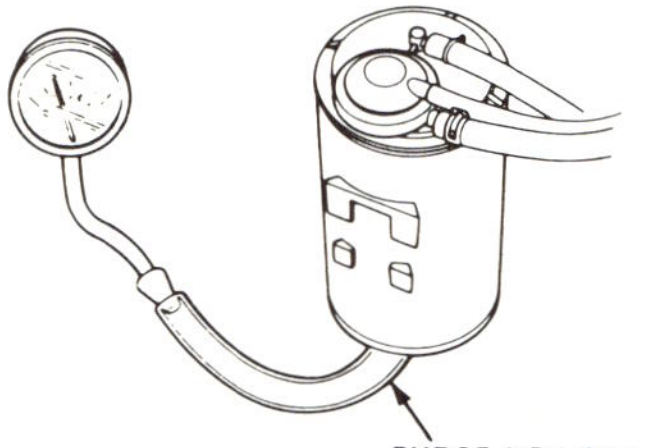

Does vacuum appear on gauge within 1 minute ?
- NO → Replace the canister.
- YES → See two-way valve test to complete. Evaporative emission conrtols are OK.

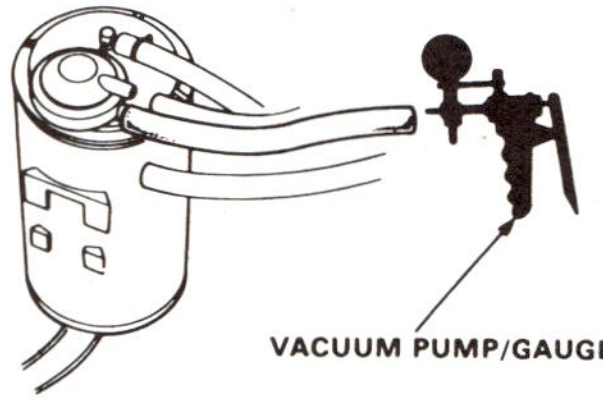

VACUUM/PRESSURE GAUGE, 0— 4 in. Hg

PURGE AIR HOSE

VACUUM PUMP/GAUGE

# EVAPORATIVE CONTROLS—1988–90 CIVIC 1.6L AND CRX HF

Troubleshooting Flowchart

Inspection of Evaporative Emission Controls

Disconnect #7 hose from the purge control diaphragm valve (on the charcoal canister) and connect a vacuum gauge to the hose.

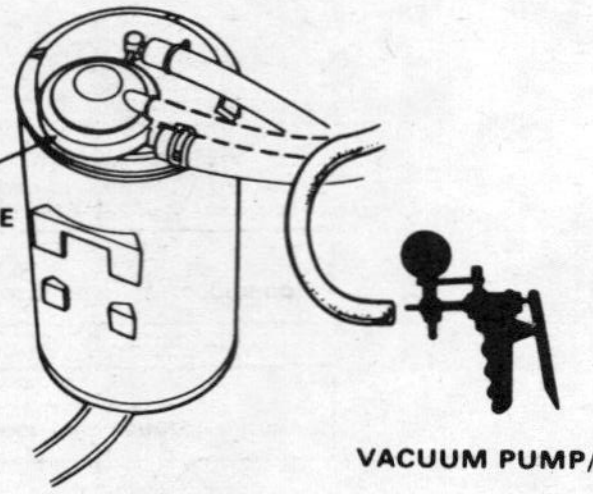

PURGE CONTROL DIAPHRAGM VALVE

VACUUM PUMP/GAUGE

Start the engine and allow to idle. NOTE: Engine coolant temperature must be below 57 °C (135 °F).

Is there vacuum ?

- YES → Disconnect the 2P connector.

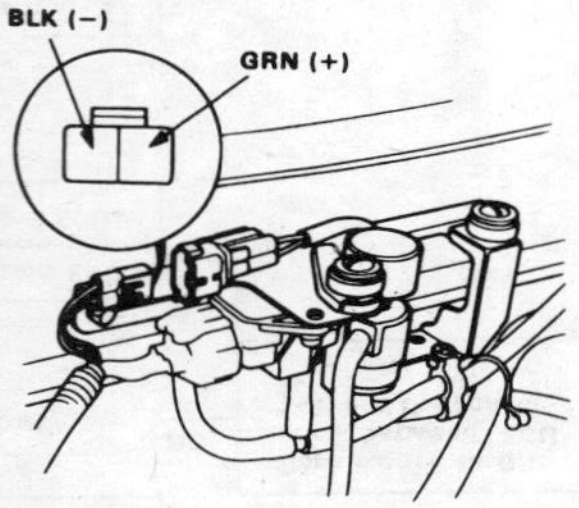

  BLK (–)

  GRN (+)

  Measure voltage between GRN (+) terminal and BLK (–) terminal.

  Is there battery voltage ?
  - YES → Replace purge cut-off solenoid valve.
  - NO → Measure voltage between GRN (+) terminal and body ground.

    Is there battery voltage ?
    - YES → Repair open in BLK wire between the 2P connector and G201.
    - NO → Turn the ignition switch OFF.

      Connect the PGM-FI test harness between the ECU and connector

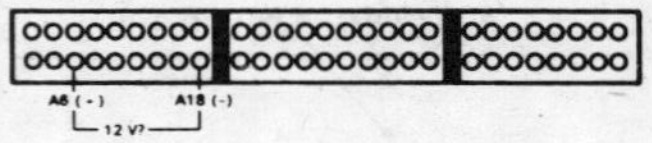

      A6 (+) A18 (–) 12 V?

      Turn the ignition switch ON.

      Measure voltage between A6 (+) terminal and A18 (–) terminal.

      Is there battery voltage?
      - YES → Repair open in GRN wire between ECU (A6) and the 2P connector.
      - NO → Inspect for a short on GRN wire between ECU (A6) and 2P connector. If wire is OK, substitute a known-good ECU and recheck. If prescribed voltage is now available, replace the original ECU.

- NO → Warm up the engine to normal operating temperature (cooling fan comes on).

  Check for vacuum at #7 hose 5 seconds after starting the engine.

## EVAPORATIVE CONTROLS—1988–90 CIVIC 1.6L AND CRX HF

Is there manifold vacuum?

- NO → Disconnect the 2P connector.
  - Is there manifold vacuum?
    - NO → **Replace purge cut-off solenoid valve.**
    - YES → Turn the ignition switch OFF.
      - Disconnect "A" connector from ECU.
      - Check for continuity to ground on GRN wire.
      - Is there continuity to ground ?
        - YES → **Repair short to ground in GRN wire between ECU (A6) and the 2P connector.**
        - NO → **Substitute a known good ECU and recheck. If symptom goes away, replace the original ECU.**
- YES → Remove fuel filter cap.
  - Connect a vacuum gauge to canister purge air hose.
  - Start the engine and raise speed to 3,500 rpm

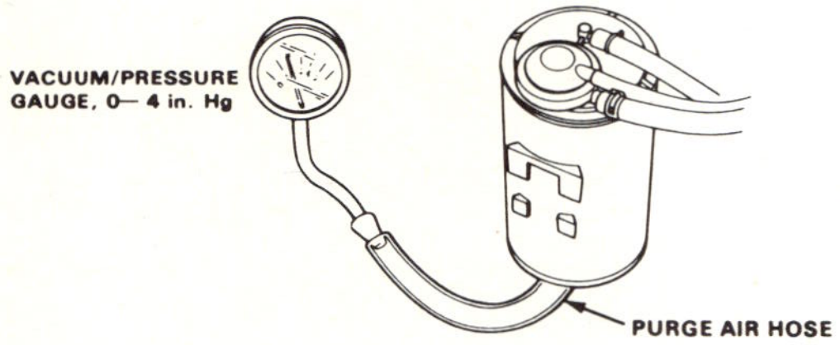

VACUUM/PRESSURE GAUGE, 0— 4 in. Hg

PURGE AIR HOSE

Does vacuum appear on gauge within 1 minute ?

- NO → **Replace the canister.**
- YES → **Evaporative emission controls are OK.**

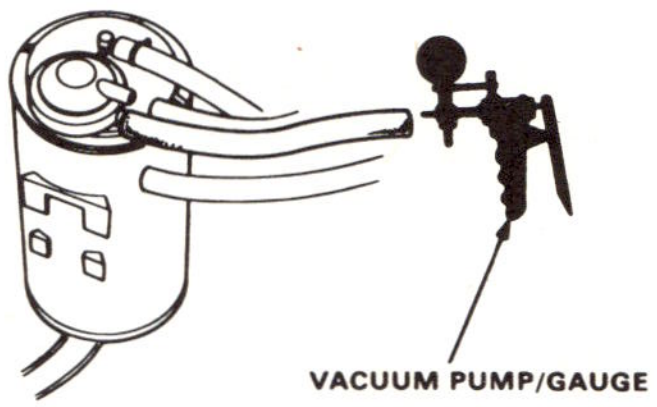

VACUUM PUMP/GAUGE

# TROUBLESHOOTING GUIDE—PRELUDE

NOTE: Across each row in the chart, the systems that could be sources of a symptom are ranked in the order they should be inspected starting with ①. Find the symptom in the left column, read across to the most likely source, then refer to the page listed at the top of that column. If inspection shows the system is OK, try the next most likely system ②, etc.

| SYMPTOM \ SYSTEM | | PGM-FI: ECU | PGM-FI: OXYGEN SENSOR | PGM-FI: MANIFOLD ABSOLUTE PRESSURE SENSOR | PGM-FI: TDC/CRANK SENSOR | PGM-FI: COOLANT TEMPERATURE SENSOR | PGM-FI: THROTTLE ANGLE SENSOR | PGM-FI: CYL SENSOR | PGM-FI: INTAKE AIR TEMPERATURE SENSOR |
|---|---|---|---|---|---|---|---|---|---|
| CHECK ENGINE WARNING LIGHT TURNS ON | | (lamp off) or (lamp lit) | (lamp lit) | (lamp lit) | (lamp lit) | (lamp lit) | (lamp lit) | (lamp lit) | (lamp lit) |
| SELF-DIAGNOSIS INDICATOR (LED) BLINKS | | 0 or * | 1 or 2 | 3 or 5 | 4 or 8 | 6 | 7 | 9 | 10 |
| ENGINE WON'T START | | ③ | | | | | | | |
| DIFFICULT TO START ENGINE WHEN COLD | | BU | | | | ① | | | |
| IRREGULAR IDLING | WHEN COLD FAST IDLE OUT OF SPECIFIC | BU | | | | ② | ③ | ③ | ③ |
| | ROUGH IDLE | BU | ③ | ② | | | ③ | | ③ |
| | WHEN WARM RPM TOO HIGH | BU | | | | | | | |
| | WHEN WARM RPM TOO LOW | BU | | | | | | | |
| FREQUENT STALLING | WHILE WARMING UP | BU | | ② | | ③ | ③ | | |
| | AFTER WARMING UP | BU | | ② | | | ② | | |
| POOR PERFORMANCE | MISFIRE OR ROUGH RUNNING | BU | ③ | ② | | | ③ | ③ | |
| | FAILS EMISSION TEST | BU | ② | ① | | | | | |
| | LOSS OF POWER | BU | | ③ | | | | | |

| SYMPTOM \ SYSTEM | | PGM-FI: ATMOSPHERIC PRESSURE SENSOR | PGM-FI: IGNITION OUTPUT SIGNAL | PGM-FI: VEHICLE SPEED SENSOR | IDLE CONTROL: ELECTRONIC AIR CONTROL VALVE | IDLE CONTROL: OTHER IDLE CONTROLS | FUEL SUPPLY: FUEL INJECTOR | FUEL SUPPLY: OTHER FUEL SUPPLY | AIR INTAKE | EMISSION CONTROL: EGR CONTROL SYSTEM | EMISSION CONTROL: OTHER EMISSION CONTROLS |
|---|---|---|---|---|---|---|---|---|---|---|---|
| CHECK ENGINE WARNING LIGHT TURNS ON | | (lamp lit) | (lamp lit) | (lamp lit) | (lamp lit) | | (lamp lit) | | | (lamp lit) | |
| SELF-DIAGNOSIS INDICATOR (LED) BLINKS | | 13 | 15 | 17 | 14 | | 16 | | | 12 | |
| ENGINE WON'T START | | | ② | | | | ③ | ① | | | |
| DIFFICULT TO START ENGINE WHEN COLD | | | | | ② | ③ | ③ | | | | |
| IRREGULAR IDLING | WHEN COLD FAST IDLE OUT OF SPECIFIC | ③ | | | | ① | | | | | |
| | ROUGH IDLE | ③ | | | ① | | ③ | | | ③ | ③ |
| | WHEN WARM RPM TOO HIGH | | | | | ① | | | ② | | ③ |
| | WHEN WARM RPM TOO LOW | | | | ① | ③ | | | ② | | |
| FREQUENT STALLING | WHILE WARMING UP | | | | ① | ③ | | ③ | ② | ③ | |
| | AFTER WARMING UP | | | | ② | | | ③ | | ① | |
| POOR PERFORMANCE | MISFIRE OR ROUGH RUNNING | | | ③ | | | ③ | ① | | | |
| | FAILS EMISSION TEST | | | | | | | | | ② | ② |
| | LOSS OF POWER | | | | | | ③ | ① | ③ | ③ | ② |

- CODE 11, or exceeds 17: count the number of blinks again. If the indicator is in fact blinking these codes, substitute a known—good ECU and recheck. If the indication goes away, replace the original ECU.

BU: When the Check Engine warning light and the self-diagnosis indicator are on, the back-up system is in operation. Substitute a known-good ECU and recheck. If the indication goes away, replace the original ECU.

# ECU AND CHECK LIGHT—PRELUDE

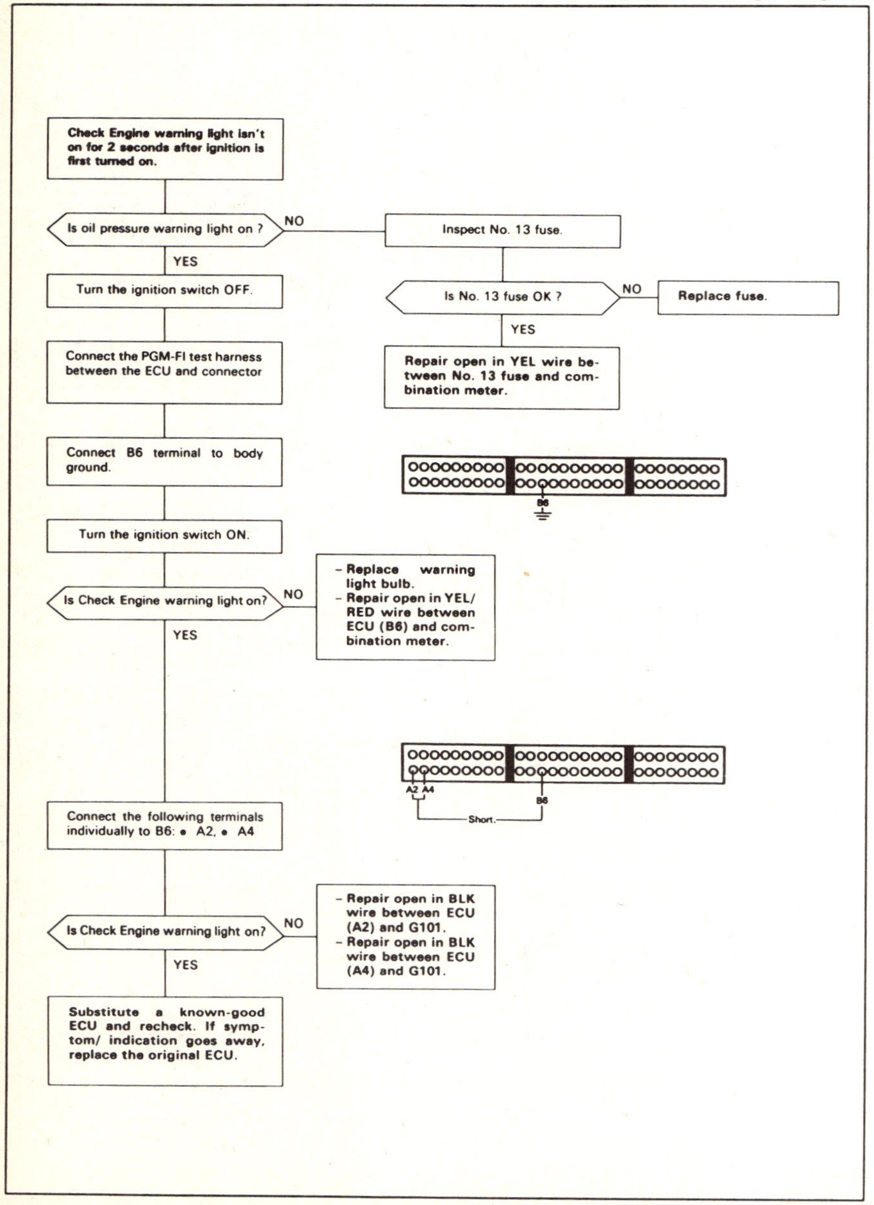

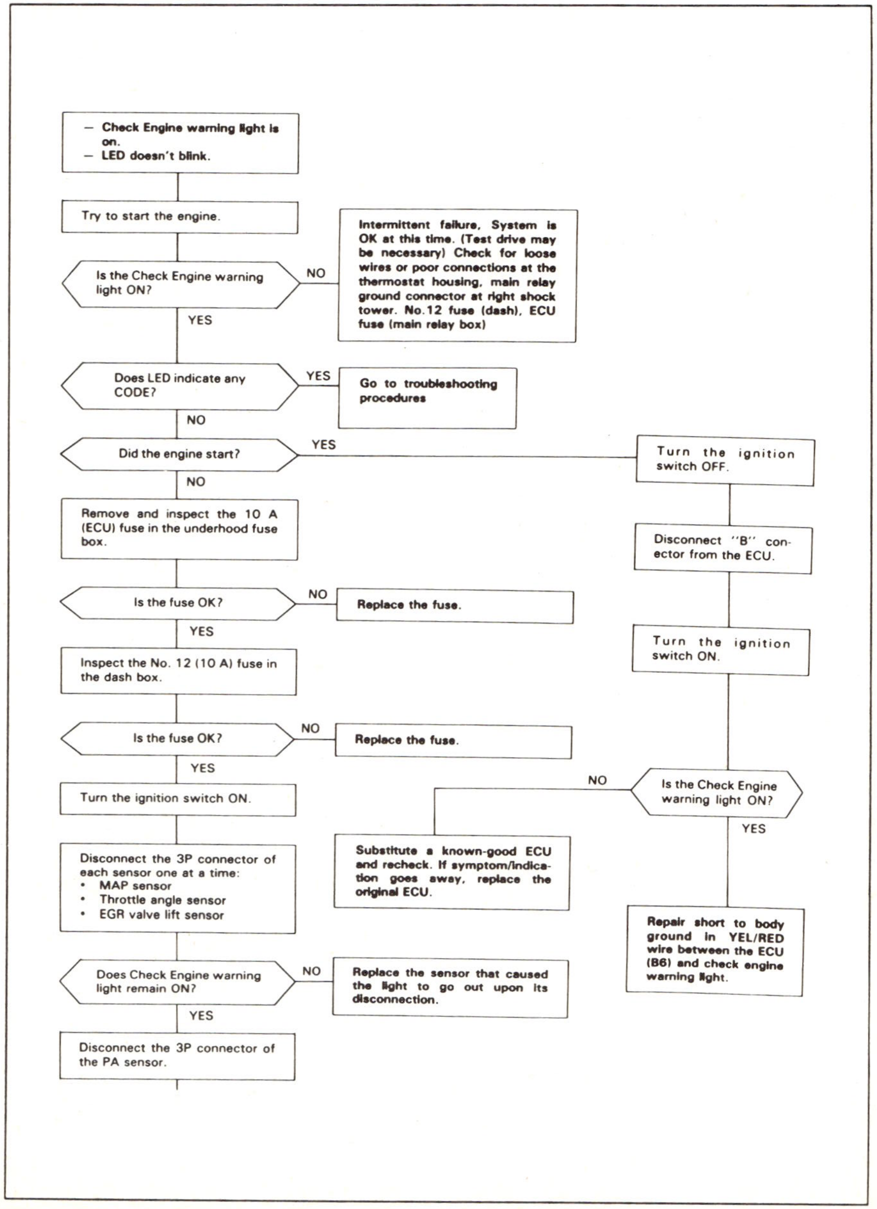

**Check Engine warning light isn't on for 2 seconds after ignition is first turned on.**

- Is oil pressure warning light on?
  - NO → Inspect No. 13 fuse. → Is No. 13 fuse OK?
    - NO → **Replace fuse.**
    - YES → **Repair open in YEL wire between No. 13 fuse and combination meter.**
  - YES → Turn the ignition switch OFF. → Connect the PGM-FI test harness between the ECU and connector → Connect B6 terminal to body ground. → Turn the ignition switch ON. → Is Check Engine warning light on?
    - NO → **– Replace warning light bulb. – Repair open in YEL/RED wire between ECU (B6) and combination meter.**
    - YES → Connect the following terminals individually to B6: • A2, • A4 → Is Check Engine warning light on?
      - NO → **– Repair open in BLK wire between ECU (A2) and G101. – Repair open in BLK wire between ECU (A4) and G101.**
      - YES → **Substitute a known-good ECU and recheck. If symptom/ indication goes away, replace the original ECU.**



**– Check Engine warning light is on.**
**– LED doesn't blink.**

- Try to start the engine. → Is the Check Engine warning light ON?
  - NO → **Intermittent failure. System is OK at this time. (Test drive may be necessary) Check for loose wires or poor connections at the thermostat housing, main relay ground connector at right shock tower. No.12 fuse (dash), ECU fuse (main relay box)**
  - YES → Does LED indicate any CODE?
    - YES → **Go to troubleshooting procedures**
    - NO → Did the engine start?
      - YES → Turn the ignition switch OFF. → Disconnect "B" connector from the ECU. → Turn the ignition switch ON. → Is the Check Engine warning light ON?
        - NO → **Substitute a known-good ECU and recheck. If symptom/indication goes away, replace the original ECU.**
        - YES → **Repair short to body ground in YEL/RED wire between the ECU (B6) and check engine warning light.**
      - NO → Remove and inspect the 10 A (ECU) fuse in the underhood fuse box. → Is the fuse OK?
        - NO → **Replace the fuse.**
        - YES → Inspect the No. 12 (10 A) fuse in the dash box. → Is the fuse OK?
          - NO → **Replace the fuse.**
          - YES → Turn the ignition switch ON. → Disconnect the 3P connector of each sensor one at a time:
            - MAP sensor
            - Throttle angle sensor
            - EGR valve lift sensor

            → Does Check Engine warning light remain ON?
            - NO → **Replace the sensor that caused the light to go out upon its disconnection.**
            - YES → Disconnect the 3P connector of the PA sensor.

## ECU AND CHECK LIGHT – PRELUDE

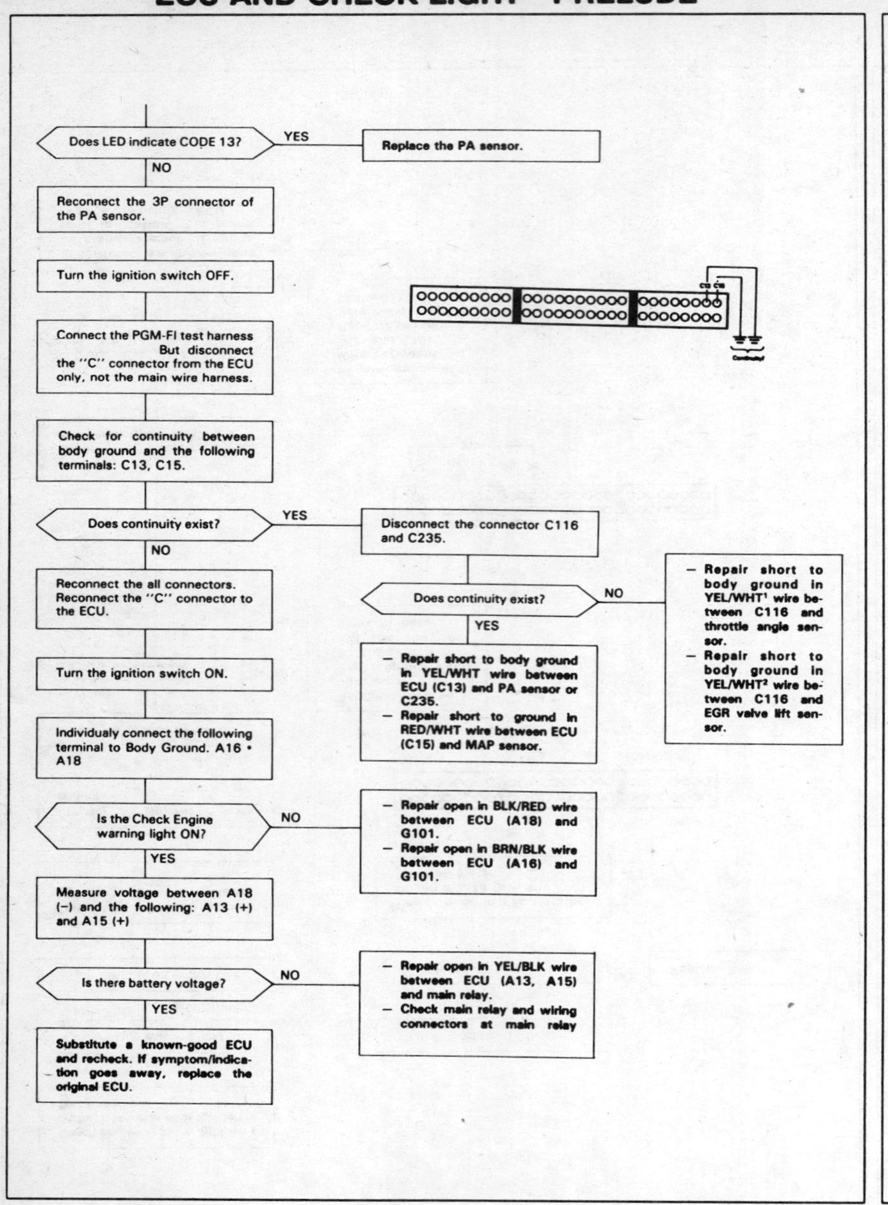

Does LED indicate CODE 13?
YES
Replace the PA sensor.
NO
Reconnect the 3P connector of the PA sensor.
Turn the ignition switch OFF.
Connect the PGM-FI test harness But disconnect the "C" connector from the ECU only, not the main wire harness.
Check for continuity between body ground and the following terminals: C13, C15.
Does continuity exist?
YES
Disconnect the connector C116 and C235.
Does continuity exist?
NO
– Repair short to body ground in YEL/WHT¹ wire between C116 and throttle angle sensor.
– Repair short to body ground in YEL/WHT² wire between C116 and EGR valve lift sensor.
YES
– Repair short to body ground in YEL/WHT wire between ECU (C13) and PA sensor or C235.
– Repair short to ground in RED/WHT wire between ECU (C15) and MAP sensor.
NO
Reconnect the all connectors. Reconnect the "C" connector to the ECU.
Turn the ignition switch ON.
Individualy connect the following terminal to Body Ground. A16 • A18
Is the Check Engine warning light ON?
NO
– Repair open in BLK/RED wire between ECU (A18) and G101.
– Repair open in BRN/BLK wire between ECU (A16) and G101.
YES
Measure voltage between A18 (–) and the following: A13 (+) and A15 (+)
Is there battery voltage?
NO
– Repair open in YEL/BLK wire between ECU (A13, A15) and main relay.
– Check main relay and wiring connectors at main relay
YES
Substitute a known-good ECU and recheck. If symptom/indication goes away, replace the original ECU.


## OXYGEN SENSOR (LATE MODEL) – CODE 1 – 1988 PRELUDE

Self-diagnosis LED blinks once: A problem in the Oxygen ($O_2$) Sensor circuit.

The oxygen sensor, detects the oxygen content in the exhaust gas, and inputs the control unit. In operation, the control unit receives the signals from the sensor and varies the duration during which fuel is injected. The oxygen sensor is installed on the exhaust manifold.

EXHAUST GAS — ZIRCONIA — ATMOSPHERE — PLATINUM

VOLTAGE (V) — STOICHIOMETRIC AIR-FUEL RATIO — RICH — AIR-FUEL RATIO — LEAN

– Check Engine warning light has been reported on.
– LED indicates CODE 1.

Warm up engine to normal operating temperature (the cooling fan comes on).

Turn the ignition switch OFF.

Remove EFI,ECU fuse in the under-hood relay box for 10 seconds to reset control unit.

Remove #2 hose from the vacuum hose manifold and connect a vacuum pump. Plug the vacuum hose manifold.

Start the engine and slowly apply 13" Hg. of vacuum.

Hold engine at 2500 rpm for 3 minutes.

Is Check Engine warning light on and does LED indicate CODE 1?

NO — Intermittent failure check connectors at the $O_2$ sensor, the battery, and the control unit.

YES

VACUUM PUMP/ GAUGE — #2 HOSE — PLUG

# OXYGEN SENSOR (EARLY MODEL)—CODE 1—1988 PRELUDE

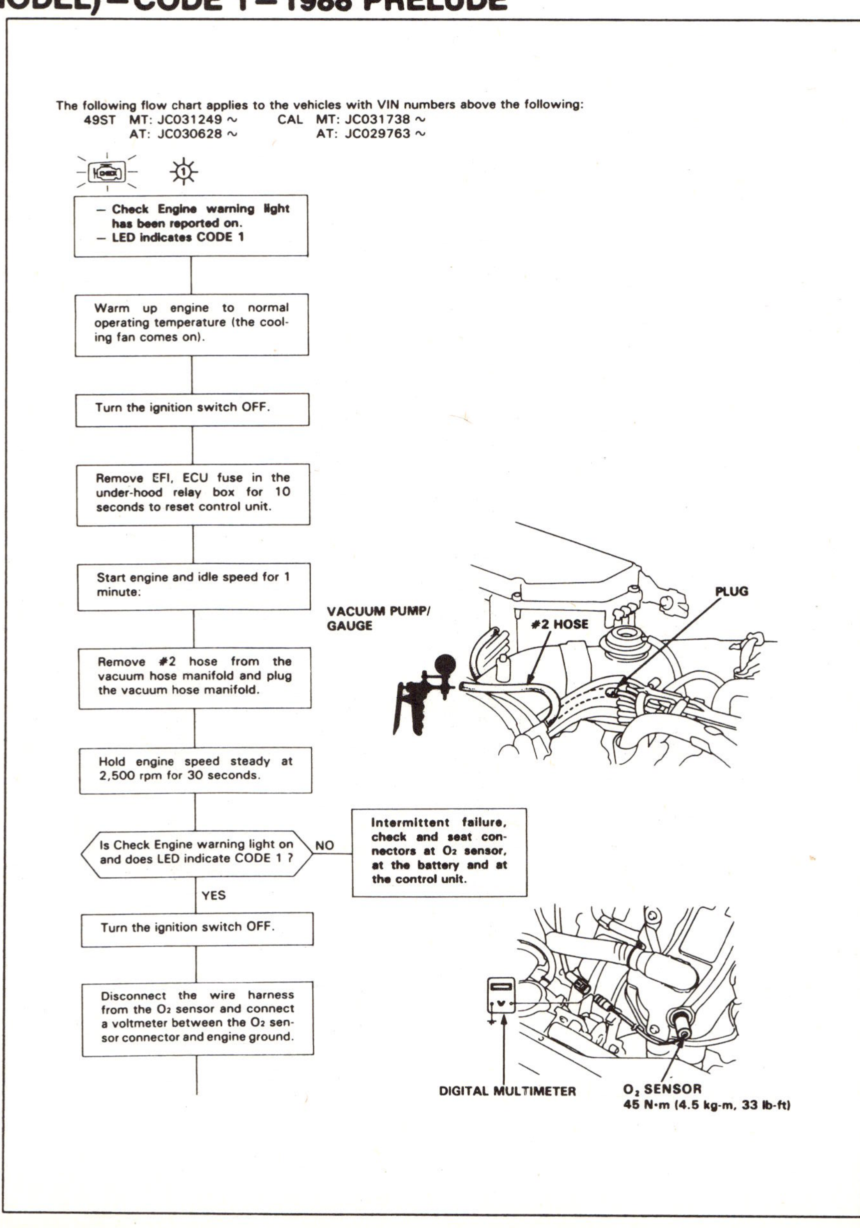

The following flow chart applies to the vehicles with VIN numbers above the following:
49ST MT: JC031249 ~ CAL MT: JC031738 ~
AT: JC030628 ~ AT: JC029763 ~
— Check Engine warning light has been reported on.
— LED indicates CODE 1
Warm up engine to normal operating temperature (the cooling fan comes on).
Turn the ignition switch OFF.
Remove EFI, ECU fuse in the under-hood relay box for 10 seconds to reset control unit.
Start engine and idle speed for 1 minute.
Remove #2 hose from the vacuum hose manifold and plug the vacuum hose manifold.
Hold engine speed steady at 2,500 rpm for 30 seconds.
Is Check Engine warning light on and does LED indicate CODE 1 ?
NO
Intermittent failure, check and seat connectors at O2 sensor, at the battery and at the control unit.
YES
Turn the ignition switch OFF.
Disconnect the wire harness from the O2 sensor and connect a voltmeter between the O2 sensor connector and engine ground.
VACUUM PUMP/ GAUGE
#2 HOSE
PLUG
DIGITAL MULTIMETER
O2 SENSOR
45 N·m (4.5 kg-m, 33 lb-ft)


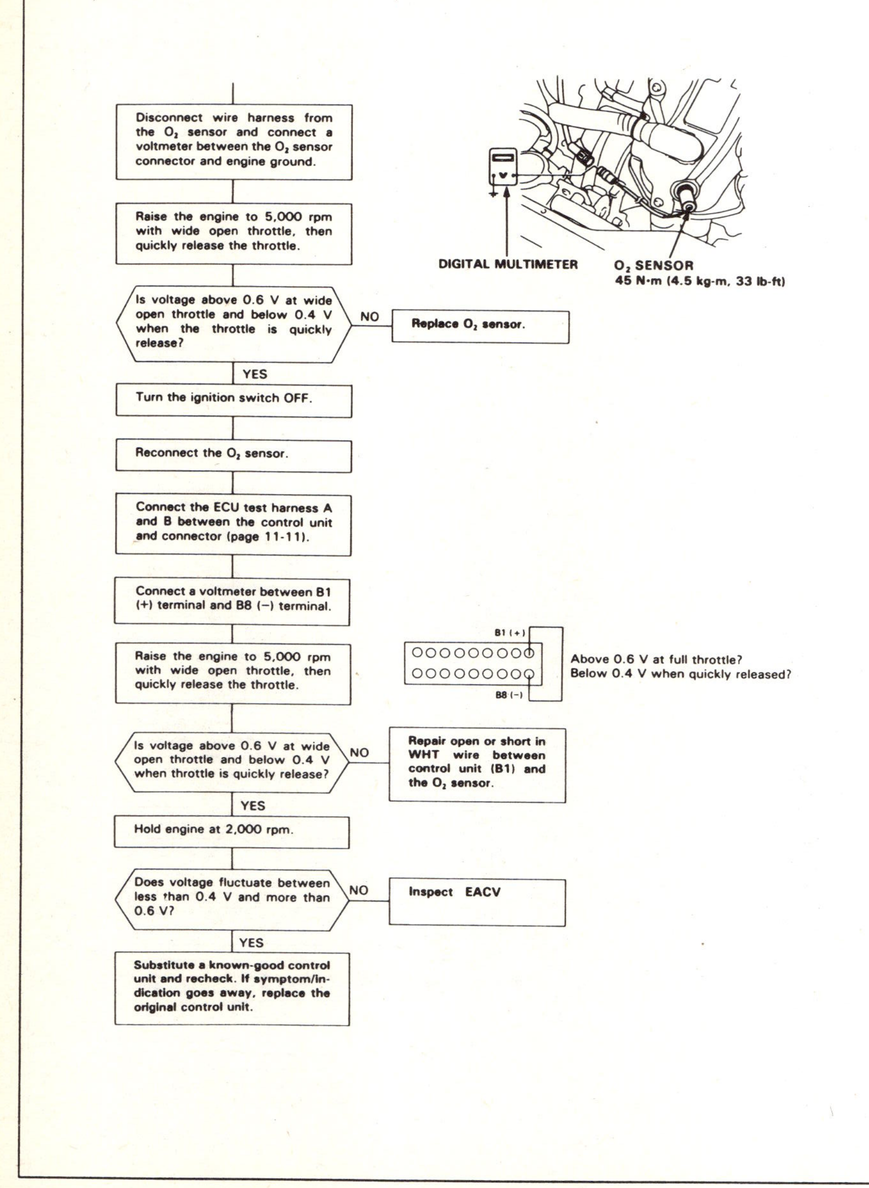

Disconnect wire harness from the O2 sensor and connect a voltmeter between the O2 sensor connector and engine ground.
Raise the engine to 5,000 rpm with wide open throttle, then quickly release the throttle.
Is voltage above 0.6 V at wide open throttle and below 0.4 V when the throttle is quickly release?
NO
Replace O2 sensor.
YES
Turn the ignition switch OFF.
Reconnect the O2 sensor.
Connect the ECU test harness A and B between the control unit and connector (page 11-11).
Connect a voltmeter between B1 (+) terminal and B8 (−) terminal.
Raise the engine to 5,000 rpm with wide open throttle, then quickly release the throttle.
B1 (+)
B8 (−)
Above 0.6 V at full throttle?
Below 0.4 V when quickly released?
Is voltage above 0.6 V at wide open throttle and below 0.4 V when throttle is quickly release?
NO
Repair open or short in WHT wire between control unit (B1) and the O2 sensor.
YES
Hold engine at 2,000 rpm.
Does voltage fluctuate between less than 0.4 V and more than 0.6 V?
NO
Inspect EACV
YES
Substitute a known-good control unit and recheck. If symptom/indication goes away, replace the original control unit.
DIGITAL MULTIMETER
O2 SENSOR
45 N·m (4.5 kg-m, 33 lb-ft)

## OXYGEN SENSOR (LATE MODEL) CODE 1—1988 PRELUDE

Start the engine and measure the voltage between O2 sensor connector and engine ground.

Is voltage above 0.6 at wide open throttle and below 0.4 on when the throttle is quickly released ?
- NO → Replace O2 sensor.
- YES ↓

Turn the ignition switch OFF.

Reconnect O2 sensor.

Connect the ECU test harness B between the control unit and connector

Start the engine.

Measure voltage between B1 (+) terminal and B8 (−) teminal.

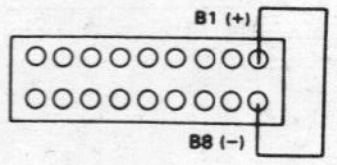
B1 (+)
B8 (−)

Above 0.6 V at full throttle ?
Below 0.4 V when quickly released ?

Is voltage above 0.6 V at wide open throttle and below 0.4 V when the throttle is quickly released ?
- NO → Repair open or short in WHT wire between the control unit (B1) and O2 sensor.
- YES ↓

**Substitute a known-good control unit and recheck. If symptom/indication goes away, replace the original control unit.**

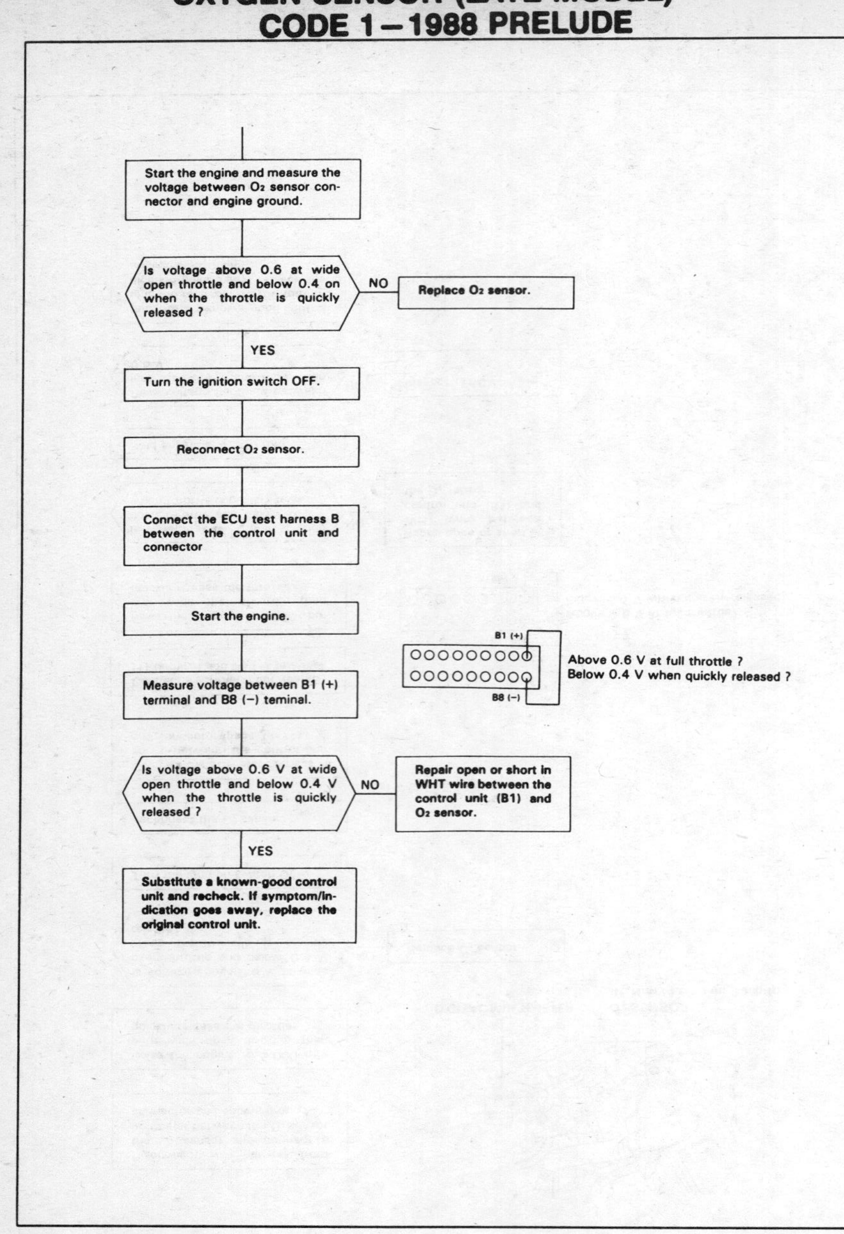

## OXYGEN SENSOR—CODE 1—1989–90 PRELUDE

Self-diagnosis LED blinks once: A problem in the Oxygen ($O_2$) Sensor A circuit.

Self-diagnosis LED blinks twice: A problem in the Oxygen ($O_2$) Sensor B circuit.

The oxygen sensor, detects the oxygen content in the exhaust gas, and inputs the ECU. In operation, the ECU receives the signals from the sensor and varies the duration during which fuel is injected. The oxygen sensors are installed on the exhaust manifold.

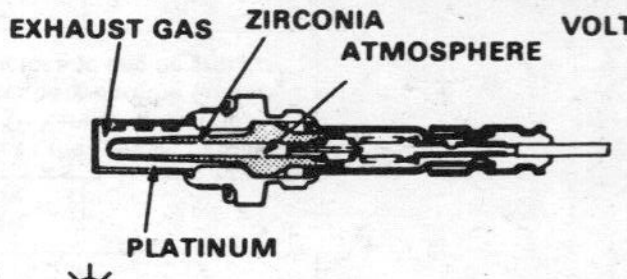
EXHAUST GAS
ZIRCONIA
ATMOSPHERE
PLATINUM

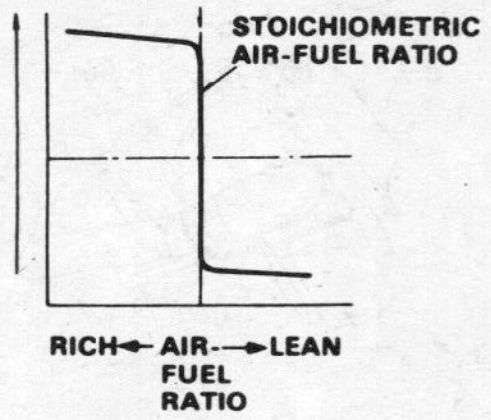
VOLTAGE (V)
STOICHIOMETRIC AIR-FUEL RATIO
RICH← AIR- →LEAN
FUEL
RATIO

1 or 2

- **Check Engine warning light has been reported on.**
- **LED indicates CODE 1 (O2 sensor A) or CODE 2 (O2 sensor B).**

Turn the ignition switch OFF.

Remove CLOCK fuse in the under-hood relay box for 10 seconds to reset ECU.

Inspect pressure regulator

Is it normal ?
- NO → **Replace the pressure regulator**
- YES ↓

Warm up engine to normal operating temperature (cooling fan comes on).

Hold engine at 1500 rpm for 15 minutes.
NOTE: Do not close throttle completely during this time.

Is Check Engine warning light on and does LED indicate CODE 1 or 2 ?
- NO → **Intermittent failure (test drive may be necessary).**
- YES ↓

## OXYGEN SENSOR—CODE 1—1989–90 PRELUDE

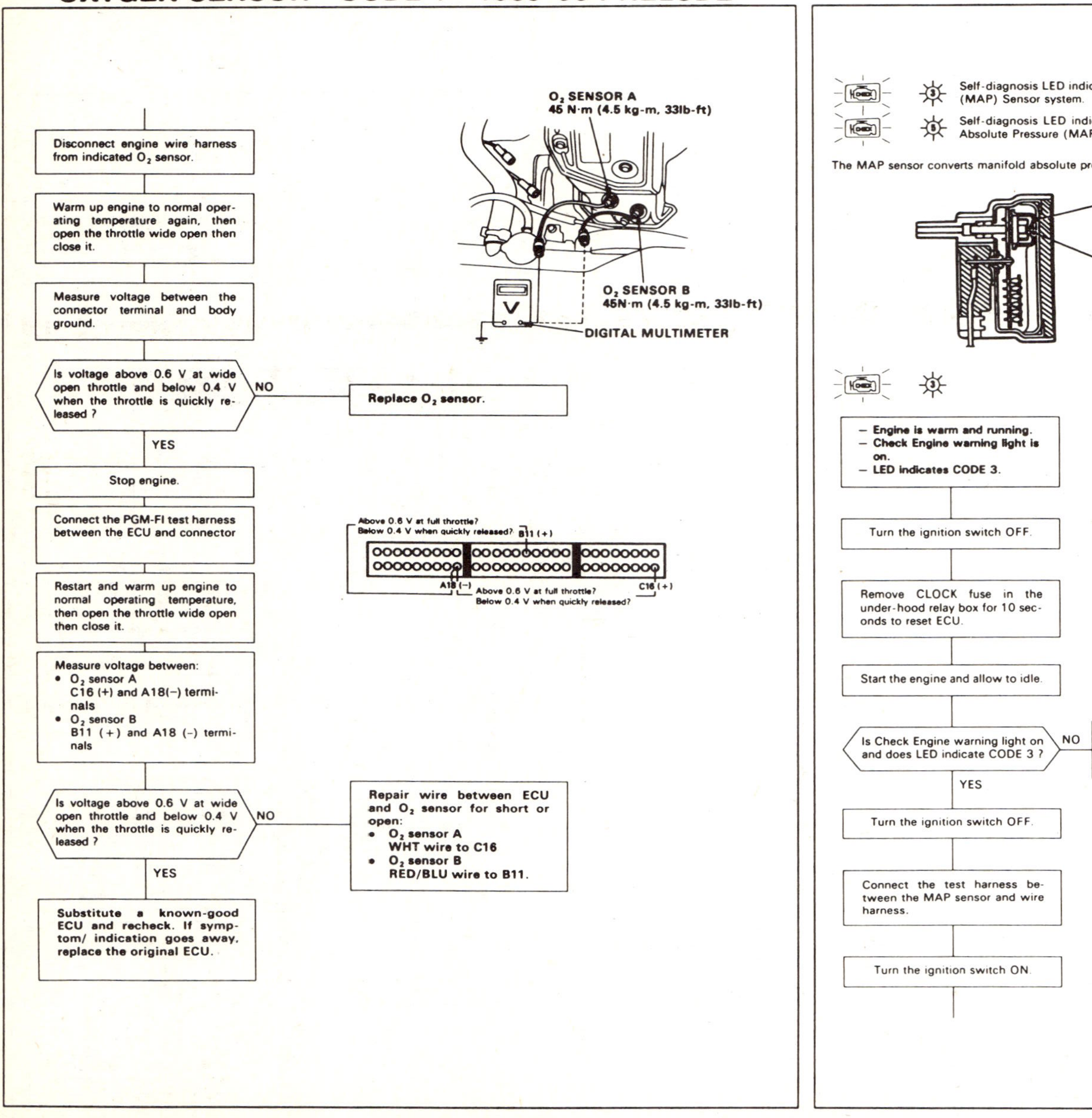


- Disconnect engine wire harness from indicated $O_2$ sensor.
- Warm up engine to normal operating temperature again, then open the throttle wide open then close it.
- Measure voltage between the connector terminal and body ground.
- Is voltage above 0.6 V at wide open throttle and below 0.4 V when the throttle is quickly released ?
  - NO → **Replace $O_2$ sensor.**
  - YES ↓
- Stop engine.
- Connect the PGM-FI test harness between the ECU and connector
- Restart and warm up engine to normal operating temperature, then open the throttle wide open then close it.
- Measure voltage between:
  - $O_2$ sensor A C16 (+) and A18(−) terminals
  - $O_2$ sensor B B11 (+) and A18 (−) terminals
- Is voltage above 0.6 V at wide open throttle and below 0.4 V when the throttle is quickly released ?
  - NO → **Repair wire between ECU and $O_2$ sensor for short or open:**
    - **$O_2$ sensor A WHT wire to C16**
    - **$O_2$ sensor B RED/BLU wire to B11.**
  - YES ↓
- **Substitute a known-good ECU and recheck. If symptom/ indication goes away, replace the original ECU.**

## MAP SENSOR—CODE 3—PRELUDE

(3) Self-diagnosis LED indicator blinks three times: Most likely an electrical problem in the Manifold Absolute Pressure (MAP) Sensor system.

(5) Self-diagnosis LED indicator blinks five times: Most likely a mechanical problem (broken hose) in the Manifold Absolute Pressure (MAP) Sensor system.

The MAP sensor converts manifold absolute pressure into electrical signals and inputs the ECU.

VACUUM CAVITY
SILICON DIAPHARAGM

OUTPUT VOLTAGE
(v) 3.5, 3.0, 2.5, 2.0, 1.5, 1.0, 0.5, 0
5, 10, 15, 20, 25, 30 (in. Hg) GAUGE
100, 200, 300, 400, 500, 600, 700 (mm Hg) READING

(3)

- **— Engine is warm and running.**
  **— Check Engine warning light is on.**
  **— LED indicates CODE 3.**
- Turn the ignition switch OFF.
- Remove CLOCK fuse in the under-hood relay box for 10 seconds to reset ECU.
- Start the engine and allow to idle.
- Is Check Engine warning light on and does LED indicate CODE 3 ?
  - NO → Intermittent failure (test drive may be necessary).
  - YES ↓
- Turn the ignition switch OFF.
- Connect the test harness between the MAP sensor and wire harness.
- Turn the ignition switch ON.

TEST HARNESS

## MAP SENSOR—CODE 3—PRELUDE

- Measure voltage between RED (+) terminal and GRN (–) terminal.
- Is there approx. 5 V?
  - NO → Measure voltage between RED (+) terminal and body ground.
    - Is there approx. 5 V?
      - YES → Repair open in BLU/WHT, wire between ECU (C14) and MAP sensor.
      - NO → Turn the ignition switch OFF.
    - Connect the PGM-FI test harness between the ECU and connector
    - Measure voltage between C15 (+) terminal and C14 (–) terminal with the ignition switch on. 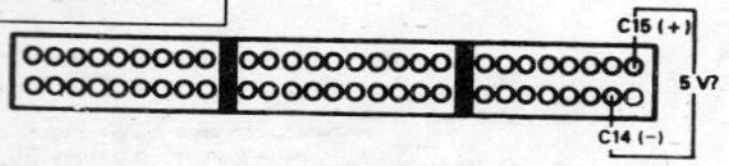
    - C15 (+) 5 V? C14 (–)
    - Is there approx. 5 V?
      - YES → Repair open in RED/WHT wire between ECU (C15) and MAP sensor.
      - NO → Substitute a known-good ECU and recheck. If prescribed voltage is now available, replace the original ECU.
  - YES
- Measure voltage between WHT (+) terminal and GRN (–) terminal.
- Is there approx. 3 V?
  - NO → - Replace MAP sensor.
    - Repair short in WHT/BLU wire between ECU (C11) and MAP sensor.
  - YES
- Turn the ignition switch OFF.
- Connect the PGM-FI test harness between the ECU and connector
- Turn the ignition switch ON.
- Measure voltage between C11 (+) terminal and C14 (–) terminal. 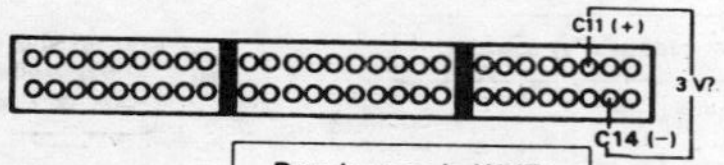
- C11 (+) 3 V? C14 (–)
- Is there approx. 3 V?
  - NO → Repair open in WHT/BLU wire between ECU (C11) and MAP sensor.
  - YES → Substitute a known-good ECU and recheck. If symptom/indication goes away, replace the original ECU.

## MAP SENSOR—CODE 5—PRELUDE

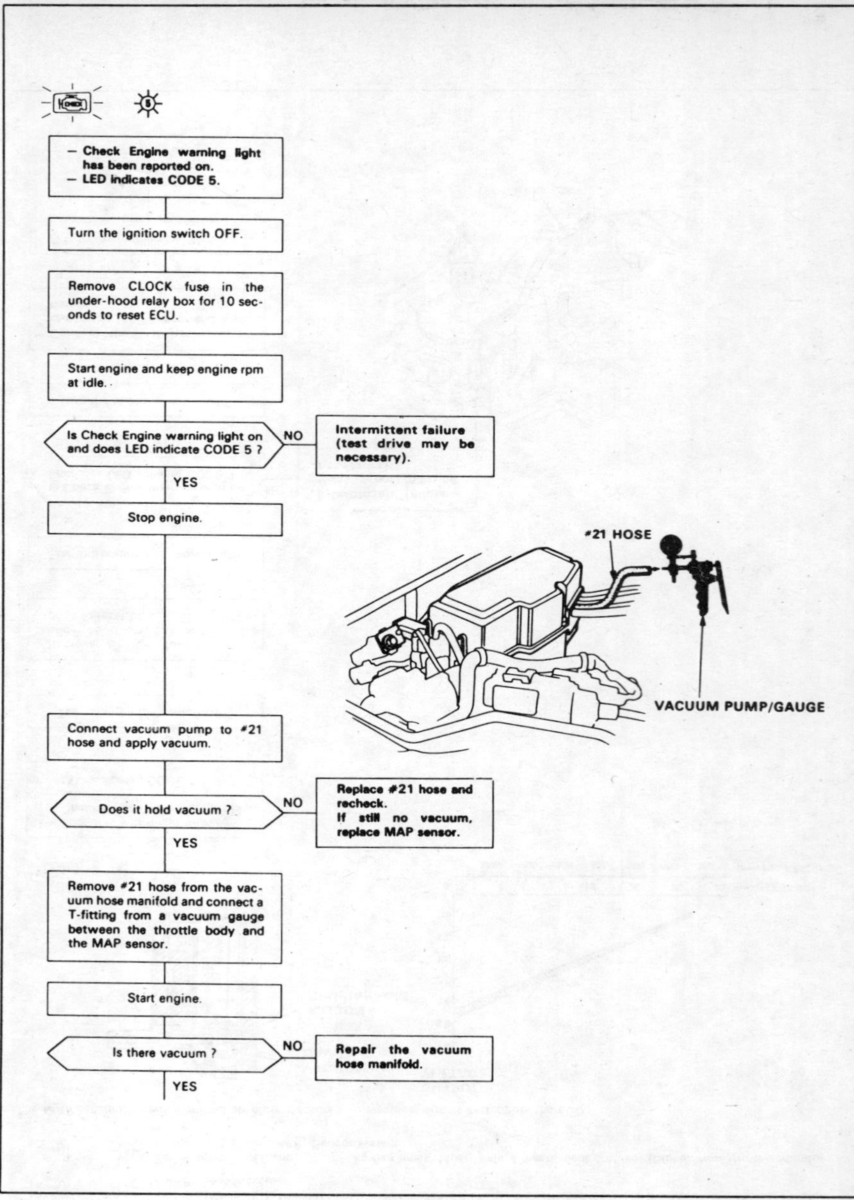


- – Check Engine warning light has been reported on.
  – LED indicates CODE 5.
- Turn the ignition switch OFF.
- Remove CLOCK fuse in the under-hood relay box for 10 seconds to reset ECU.
- Start engine and keep engine rpm at idle.
- Is Check Engine warning light on and does LED indicate CODE 5?
  - NO → Intermittent failure (test drive may be necessary).
  - YES
- Stop engine.
- Connect vacuum pump to #21 hose and apply vacuum.
- Does it hold vacuum?
  - NO → Replace #21 hose and recheck. If still no vacuum, replace MAP sensor.
  - YES
- Remove #21 hose from the vacuum hose manifold and connect a T-fitting from a vacuum gauge between the throttle body and the MAP sensor.
- Start engine.
- Is there vacuum?
  - NO → Repair the vacuum hose manifold.
  - YES

## MAP SENSOR—CODE 5—PRELUDE

Stop engine.

Connect the test harness between the MAP sensor and wire harness.

Turn the ignition switch ON.

Measure voltage between WHT (+) terminal and GRN (−) terminal.

Is there approx. 3V ?

- NO → **Replace MAP sensor.**
- YES → **Substitute a known-good ECU and recheck. If symptom/indication goes away, replace the original ECU.**

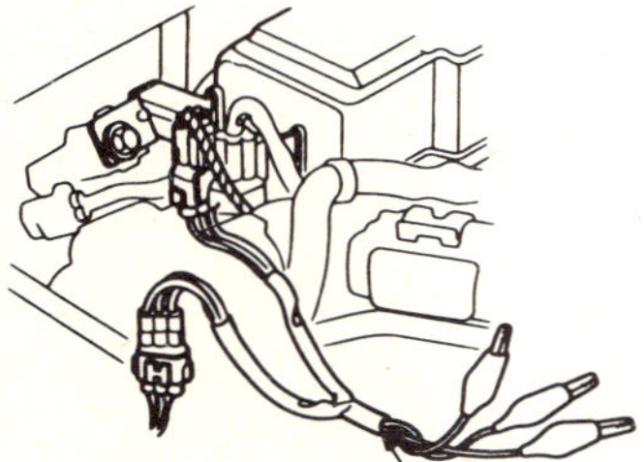

TEST HARNESS

## TDC/CRANK SENSOR—CODE 4—PRELUDE

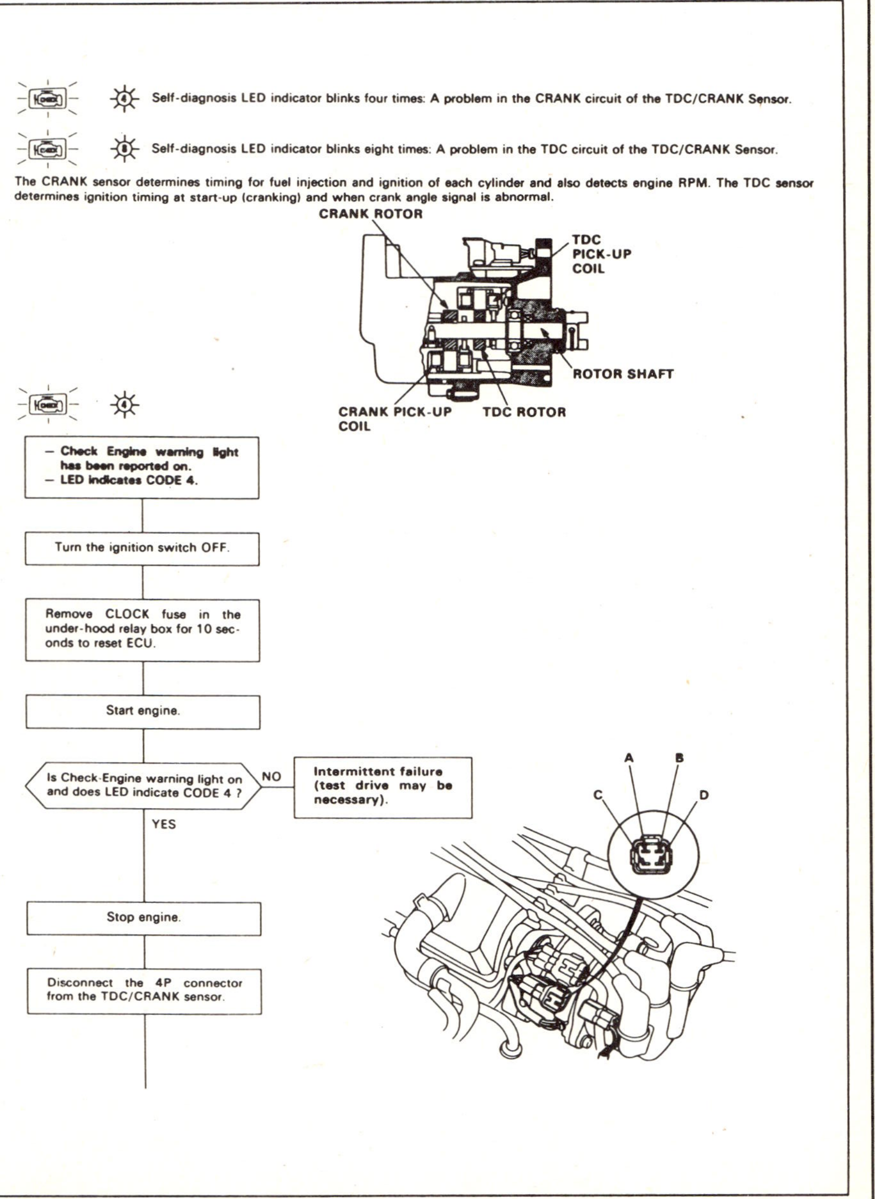


Self-diagnosis LED indicator blinks four times: A problem in the CRANK circuit of the TDC/CRANK Sensor.

Self-diagnosis LED indicator blinks eight times: A problem in the TDC circuit of the TDC/CRANK Sensor.

The CRANK sensor determines timing for fuel injection and ignition of each cylinder and also detects engine RPM. The TDC sensor determines ignition timing at start-up (cranking) and when crank angle signal is abnormal.

**– Check Engine warning light has been reported on.**
**– LED indicates CODE 4.**

Turn the ignition switch OFF.

Remove CLOCK fuse in the under-hood relay box for 10 seconds to reset ECU.

Start engine.

Is Check-Engine warning light on and does LED indicate CODE 4 ?

- NO → **Intermittent failure (test drive may be necessary).**
- YES ↓

Stop engine.

Disconnect the 4P connector from the TDC/CRANK sensor.

## TDC/CRANK SENSOR—CODE 8—PRELUDE

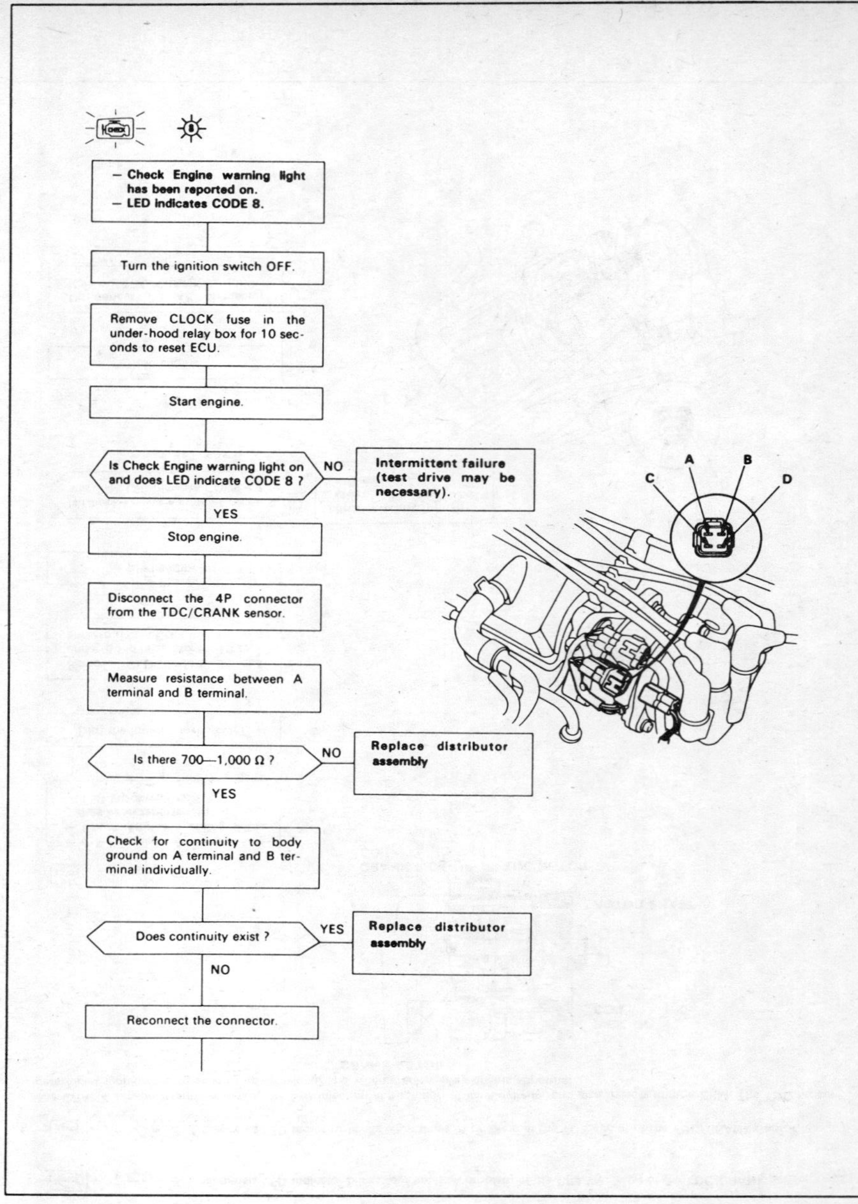


## TDC/CRANK SENSOR—CODE 4—PRELUDE

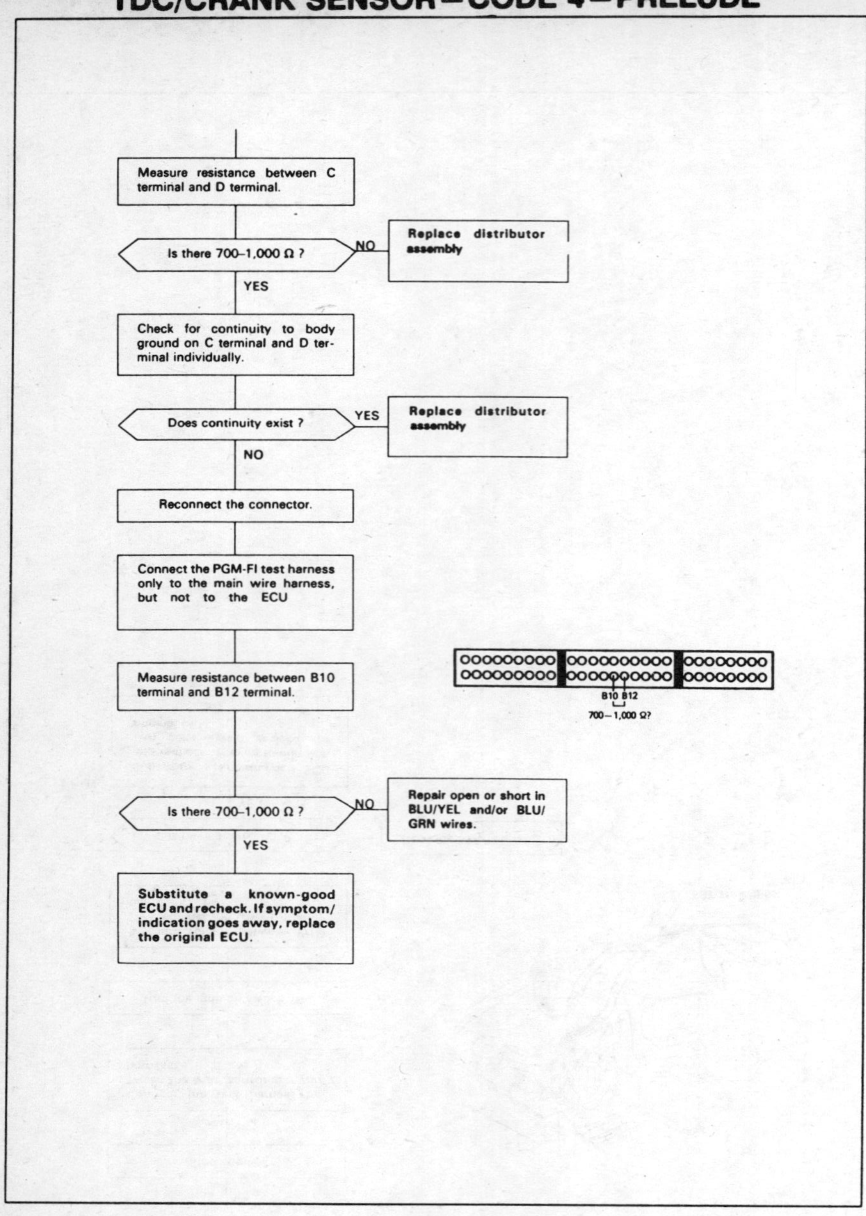

## TDC/CRANK SENSOR—CODE 8—PRELUDE

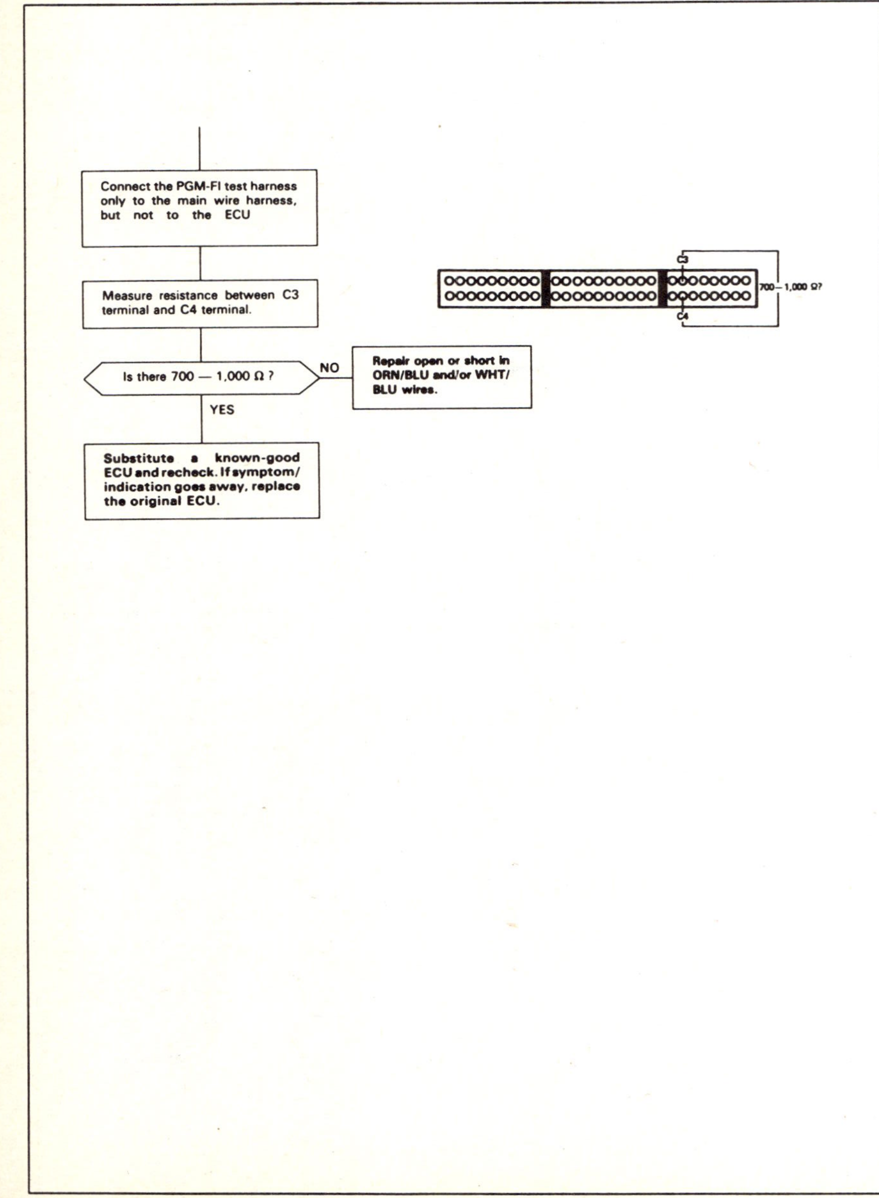


## COOLANT TEMPERATURE (TW) SENSOR CODE 6—PRELUDE

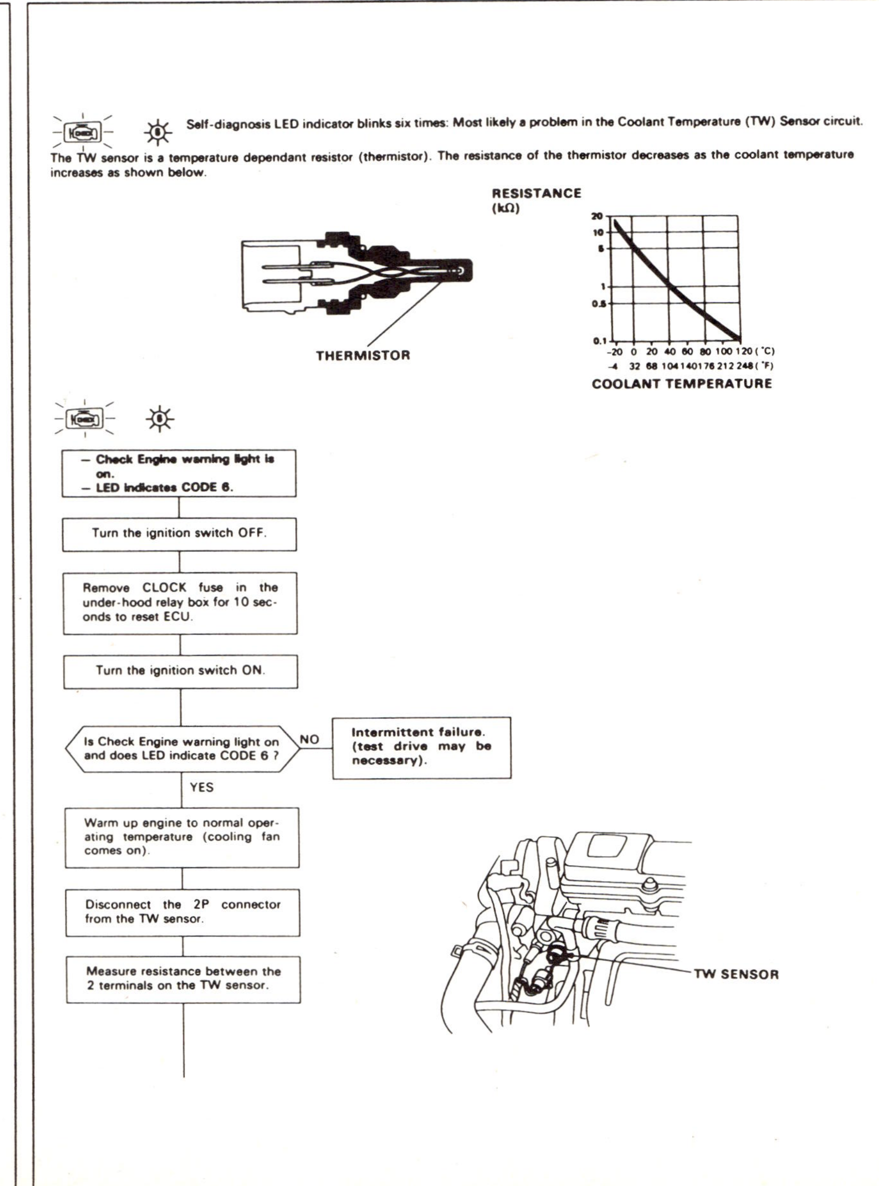

## COOLANT TEMPERATURE (TW) SENSOR CODE 6—PRELUDE

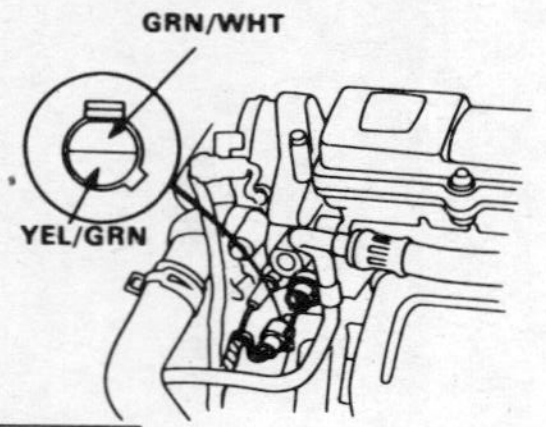

GRN/WHT

YEL/GRN

Is there 200—400 Ω ?

NO → Replace TW sensor.

YES

Measure voltage between YEL/GRN(+) terminal and body ground.

Is there approx. 5V ?

NO → Turn the ignition switch OFF.

Connect the PGM-FI test harness between the ECU and connector

Turn the ignition switch ON.

Measure voltage between C6 (+) terminal and C12 (–) terminal.

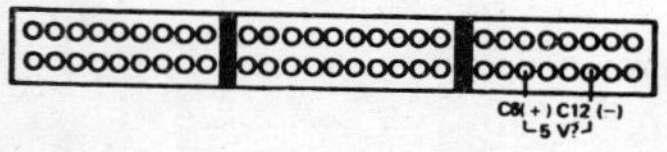

C6(+) C12 (–) 5 V?

Is there approx. 5V ?

YES → Repair open in YEL/GRN wire between ECU (C6) and TW sensor.

NO

Disconnect "C" connector from the main wire harness only, not the ECU.

Measure voltage between C6 (+) terminal and C12 (–) terminal.

Is there approx. 5V ?

YES → – Repair short in YEL/GRN wire between ECU (C6) and TW sensor.
– Repair short in YEL/GRN wire between TW sensor and A/T control unit (A11).

NO

Substitute a known-good ECU and recheck. If prescribed voltage is now available, replace the original ECU.

YES (from "Is there approx. 5V ?" after body ground measurement)

Measure voltage between YEL/GRN (+) terminal and GRN/WHT (–) terminal.

Is there approx. 5V ?

NO → Repair open in GRN/WHT wire between ECU (C12) and TW sensor.

YES

Substitute a known-good ECU and recheck. If symptom/indication goes away, replace the original ECU.

## THROTTLE ANGLE SENSOR—CODE 7—PRELUDE

Self-diagnosis LED indicator blinks seven times: Most likely a problem in the Throttle Angle Sensor circuit.

The throttle angle sensor is a potentiometer. It is connected to the primary throttle valve shaft. As the throttle angle changes, the throttle angle sensor varies the voltage signal to the ECU.

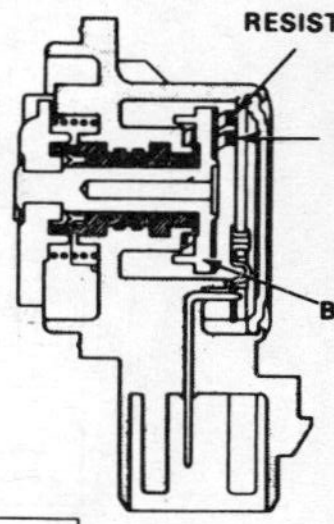

RESISTOR

BRUSH

BRUSH HOLDER

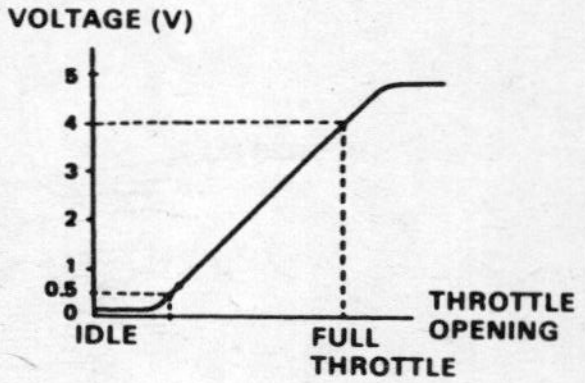

OUTPUT VOLTAGE (V)

5
4
3
2
1
0.5
0

IDLE

FULL THROTTLE

THROTTLE OPENING

– Engine is running.
– Check Engine warning light is on.
– LED indicates CODE 7.

Turn the ignition switch OFF.

Remove CLOCK fuse in the under-hood relay box for 10 seconds to reset ECU.

Start engine.

Is Check Engine warning light on and does LED indicate CODE 7 ?

NO → Intermittent failure. (test drive may be necessary).

YES

Turn the ignition switch OFF.

Disconnect the 3P connector from the throttle angle sensor.

Turn the ignition switch ON.

Measure voltage between YEL/WHT (+) terminal and GRN/WHT (–) terminal.

GRN/WHT (–)

YEL/WHT (+)

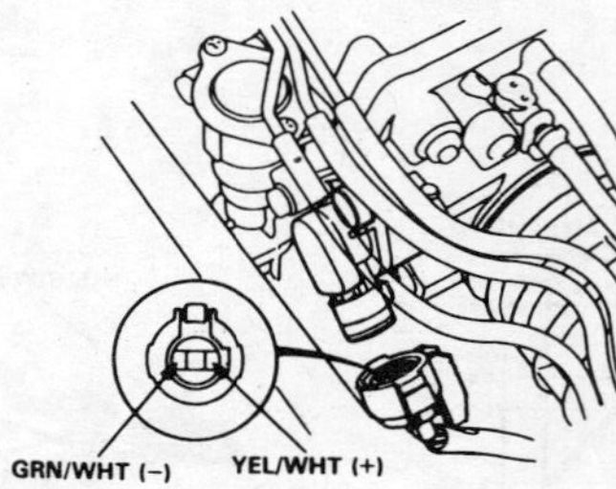

Is there approx. 5V ?

NO → Measure voltage between YEL/WHT (+) terminal and body ground.

YES

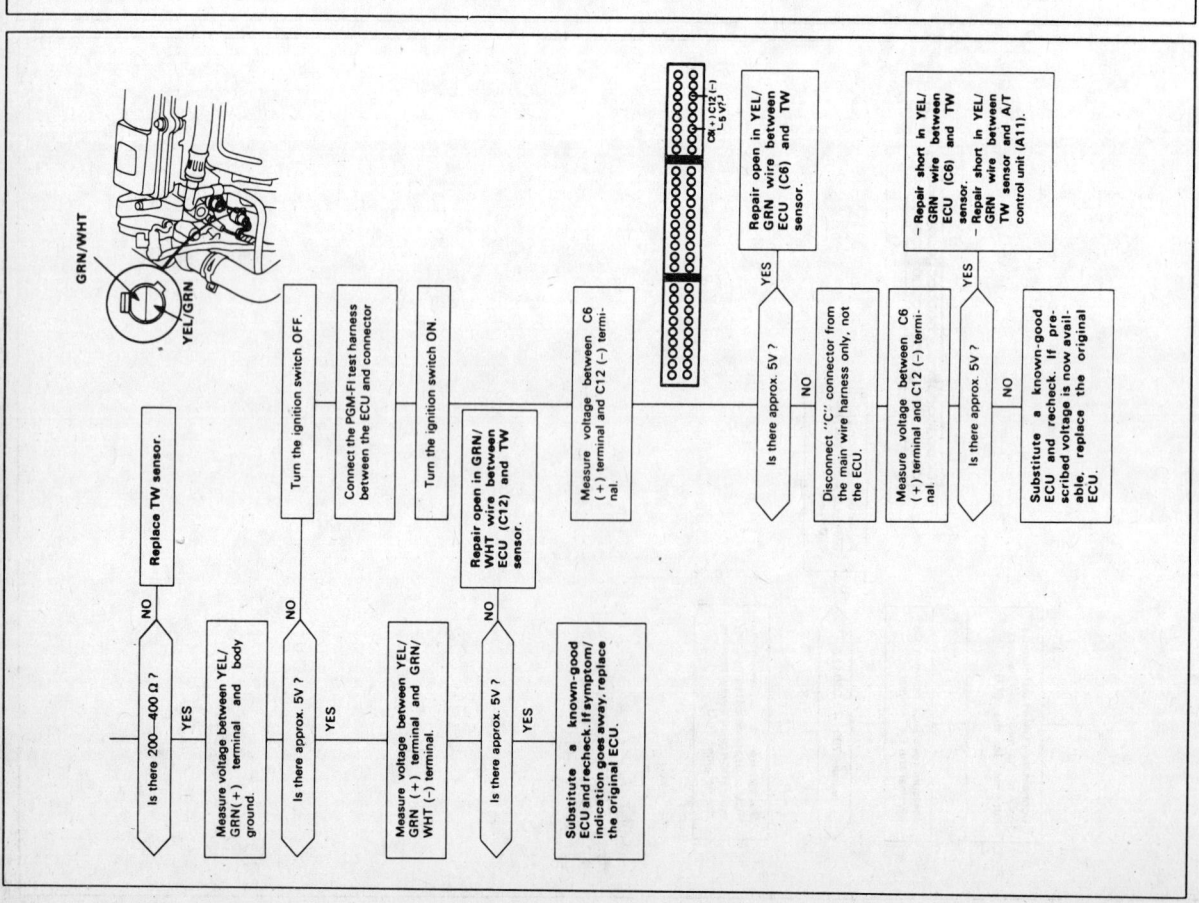

## THROTTLE ANGLE SENSOR—CODE 7—PRELUDE

Turn the ignition switch OFF.

Reconnect the 3P connector.

Connect the PGM-FI test harness between the ECU and connector

Turn the ignition switch ON.

Measure voltage between C7(+) terminal and C12(–) terminal.

C7 (+) / C12 (–): 0.5 V at full close throttle? 4.5 V at full open throttle?

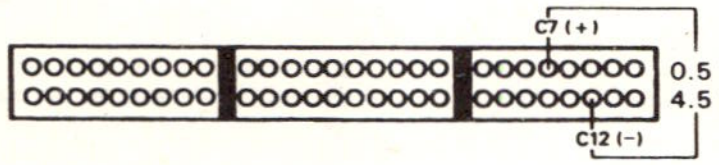

Is voltage approx. 0.5 V at full close throttle, and approx. 4.5 V at full open throttle?

- YES → Substitute a known-good ECU and recheck. If symptom/indication goes away, replace the original ECU.
- NO →
  - Replace throttle angle sensor.
  - Repair open or short in RED/YEL wire between ECU (C7) and throttle angle sensor.
  - Repair short in RED/YEL wire between throttle angle sensor and A/T control unit (A16).
  - NO → Is there approx. 5V ?
    - YES → Repair open in GRN/WHT wire between ECU (C12) and throttle angle sensor.
    - NO → Turn the ignition switch OFF.

Connect the PGM-FI test harness between the ECU and connector

Turn the ignition switch ON.

Measure voltage between C13(+) terminal and C12(–) terminal.

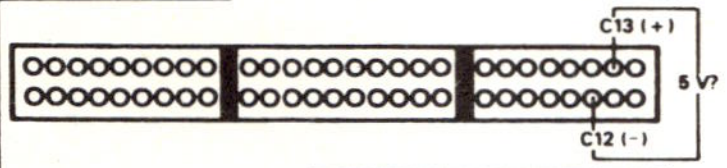

C13 (+) / C12 (–): 5 V?

Is there approx. 5V ?

- YES → Repair open in YEL/WHT wire between ECU(C13) and throttle angle sensor.
- NO → Substiture a known-good ECU and recheck. If prescribed voltage is now available, replace the original ECU.

## CYL SENSOR—CODE 9—PRELUDE

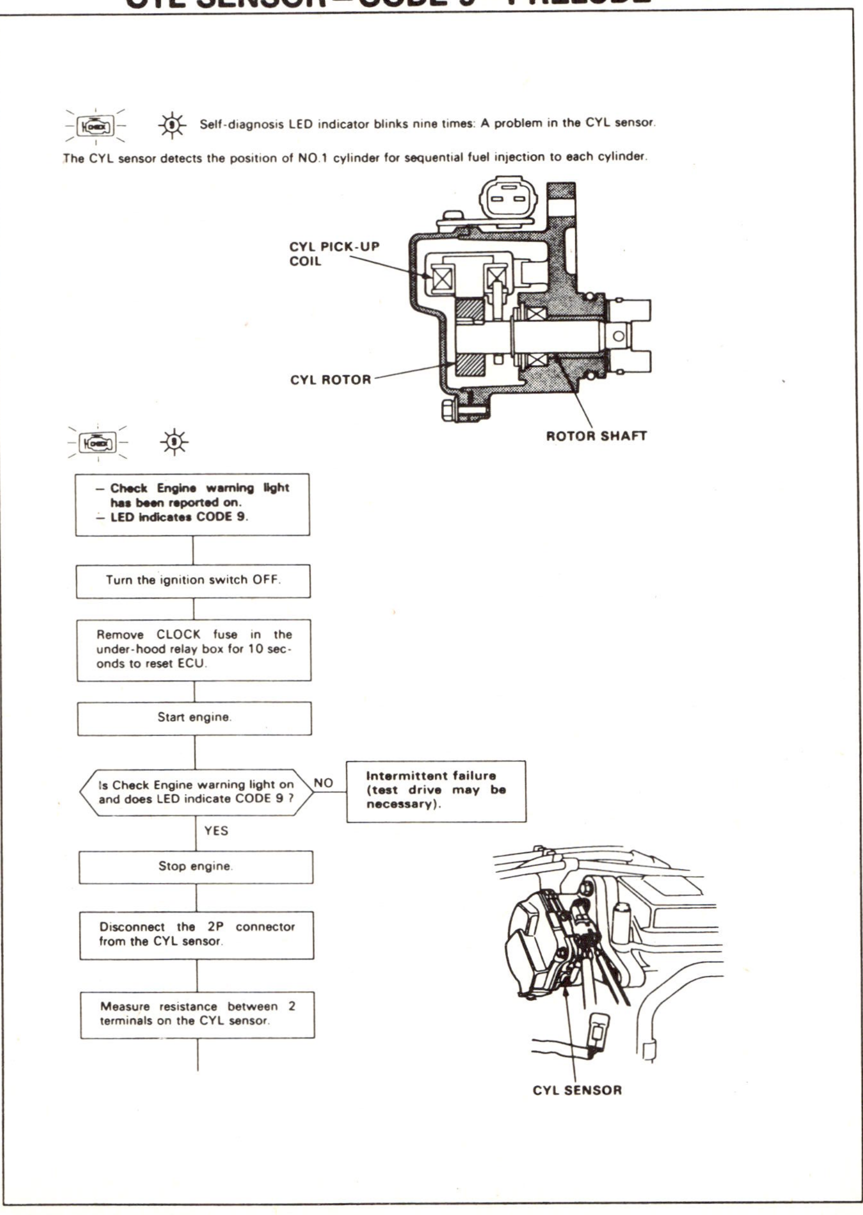


Self-diagnosis LED indicator blinks nine times: A problem in the CYL sensor.

The CYL sensor detects the position of NO.1 cylinder for sequential fuel injection to each cylinder.

- Check Engine warning light has been reported on.
- LED indicates CODE 9.

Turn the ignition switch OFF.

Remove CLOCK fuse in the under-hood relay box for 10 seconds to reset ECU.

Start engine.

Is Check Engine warning light on and does LED indicate CODE 9 ?

- NO → Intermittent failure (test drive may be necessary).
- YES → Stop engine.

Disconnect the 2P connector from the CYL sensor.

Measure resistance between 2 terminals on the CYL sensor.

## CYL SENSOR—CODE 9—PRELUDE

- Is there 700—1,000 Ω ?
  - NO: Replace CYL sensor assembly
  - YES: Check for continuity to body ground on the 2 terminals individually.
- Does continuity exist ?
  - YES: Replace CYL sensor assembly
  - NO: Reconnect the connector.
- Connect the PGM-FI test harness only to the main wire harness, but not to the ECU
- Measure resistance between C1 terminal and C2 terminal.
- Is there 700 — 1,000 Ω ?
  - NO: Repair open or short in ORN and/or WHT wires.
  - YES: Substitute a known-good ECU and recheck. If symptom/indication goes away, replace the original ECU.

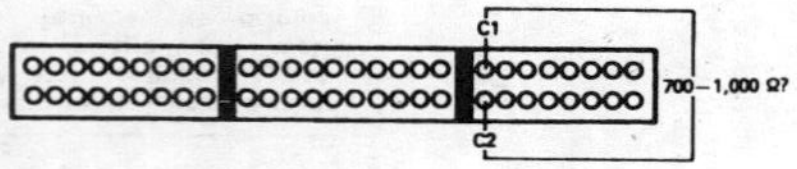

C1
C2
700—1,000 Ω?

## AIR TEMPERATURE SENSOR CODE 10—PRELUDE

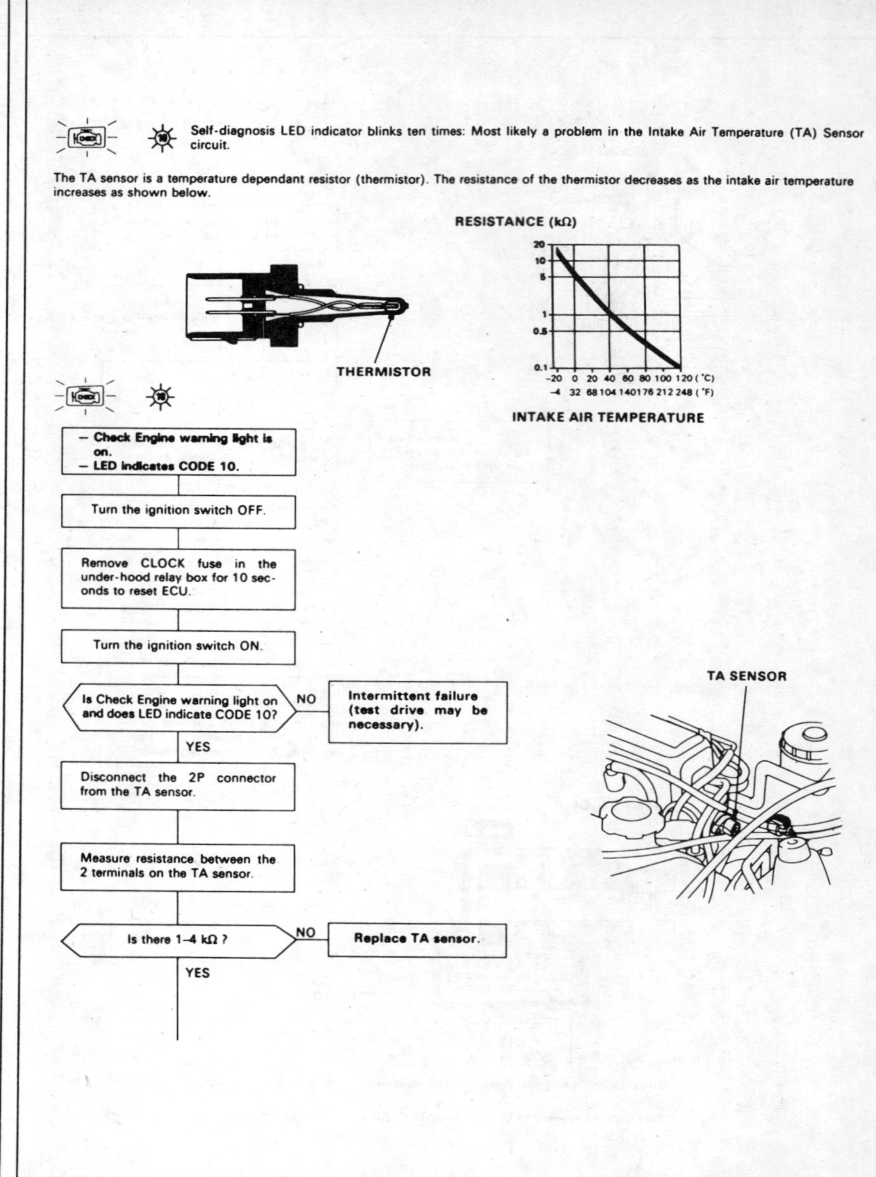


Self-diagnosis LED indicator blinks ten times: Most likely a problem in the Intake Air Temperature (TA) Sensor circuit.

The TA sensor is a temperature dependant resistor (thermistor). The resistance of the thermistor decreases as the intake air temperature increases as shown below.

- **– Check Engine warning light is on.**
- **– LED indicates CODE 10.**
- Turn the ignition switch OFF.
- Remove CLOCK fuse in the under-hood relay box for 10 seconds to reset ECU.
- Turn the ignition switch ON.
- Is Check Engine warning light on and does LED indicate CODE 10?
  - NO: **Intermittent failure (test drive may be necessary).**
  - YES: Disconnect the 2P connector from the TA sensor.
- Measure resistance between the 2 terminals on the TA sensor.
- Is there 1–4 kΩ ?
  - NO: **Replace TA sensor.**
  - YES

## ATMOSPHERIC PRESSURE (PA) SENSOR – CODE 13 – PRELUDE

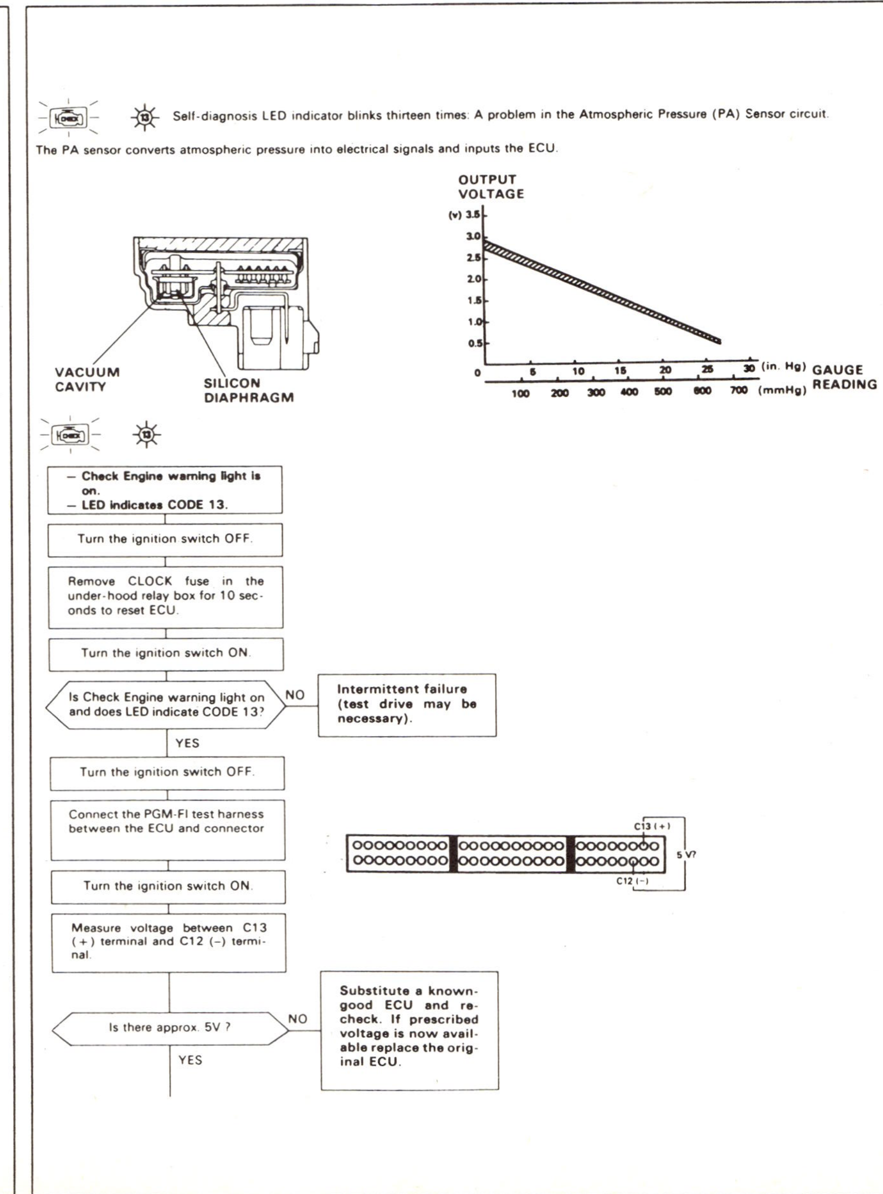


## AIR TEMPERATURE SENSOR CODE 10 – PRELUDE

Measure voltage between WHT/RED (+) terminal and body ground.

Is there approx. 5 V ?

- NO → Turn the ignition switch OFF.
  - Connect the PGM-FI test harness between the ECU and connector
  - Turn the ignition switch ON.
  - Measure voltage between C5 (+) terminal and C12 (–) terminal. 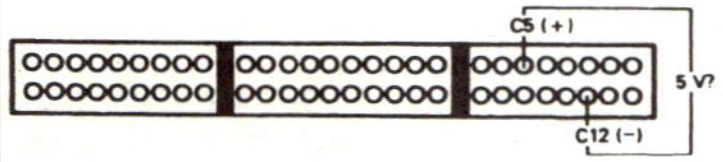
  - C5 (+) / C12 (–) / 5 V?
  - Is there approx. 5 V ?
    - YES → Repair open in WHT/RED wire between ECU (C5) and TA sensor.
    - NO → Disconnect "C" connector from the main wire harness only, not the ECU.
      - Measure voltage between C5 (+) terminal and C12 (–) terminal.
      - Is there approx. 5 V ?
        - YES → Repair short in WHT/RED wire between ECU (C5) and TA sensor.
        - NO → Substitute a known-good ECU and recheck. If prescribed voltage is now available, replace the original ECU.
- YES → Measure voltage between WHT/RED (+) terminal and GRN/WHT (–) terminal.
  - Is there approx. 5 V ?
    - NO → Repair open in GRN/WHT wire between ECU (C12) and TA sensor.
    - YES → Substitute a known-good ECU and recheck. If symptom/indication goes away, replace the original ECU.

## ATMOSPHERIC PRESSURE (PA) SENSOR CODE 13—PRELUDE

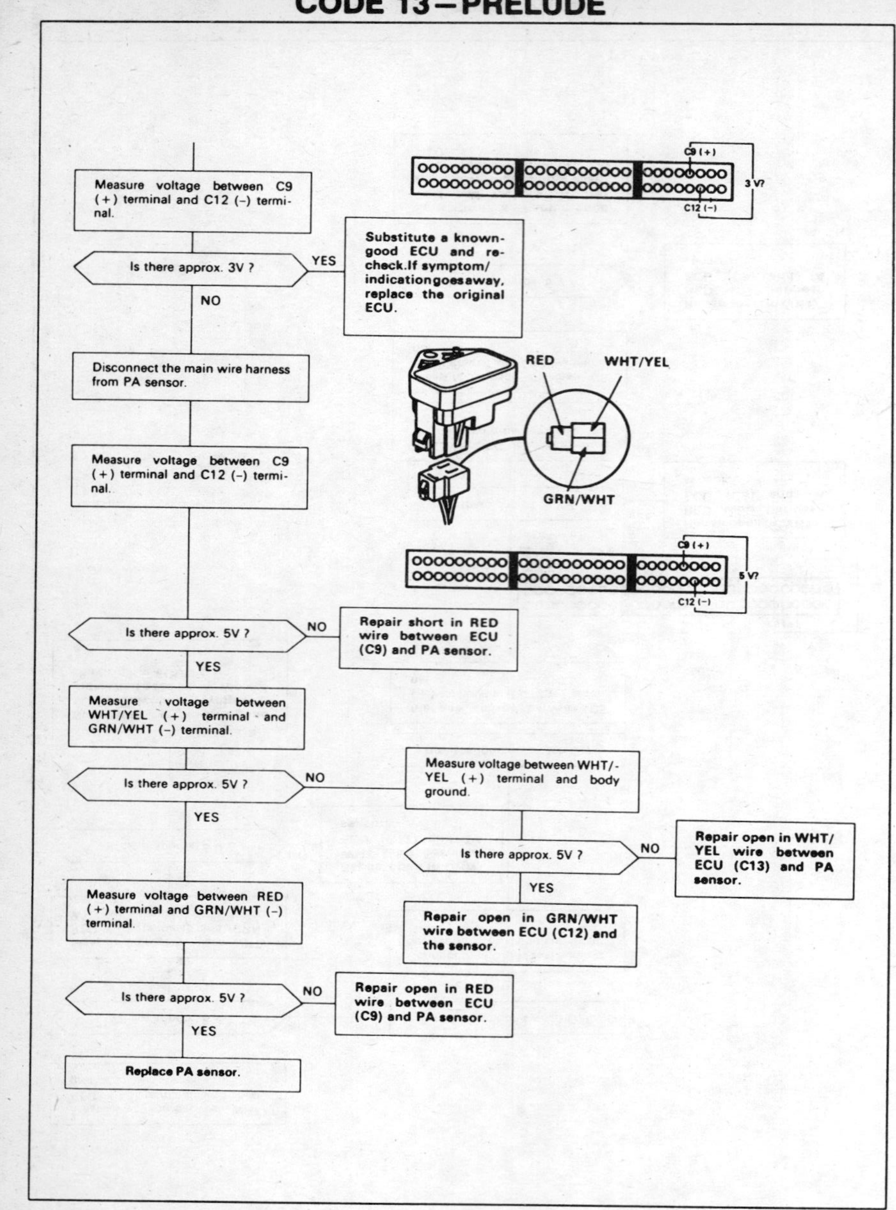


## IGNITION OUTPUT SIGNAL—CODE 15—PRELUDE

Self-diagnosis LED indicator blinks fifteen times: A problem in the Ignition Output Signal circuit.

- Check Engine warning light is on.
- LED indicates CODE 15.

Turn the ignition switch OFF.

Remove CLOCK fuse in the under-hood relay box for 10 seconds to reset ECU.

Turn the ignition switch ON.

Is Check Engine warning light on and does LED indicate CODE 15?

NO → Intermittent failure (test drive may be necessary).

YES

Turn the ignition switch OFF.

Disconnect the 6P connector on the igniter unit.

Turn the ignition switch ON.

Measure voltage between the BLK/YEL (+) terminal and body ground.

Is there battery voltage ?

NO → Repair open in the BLK/YEL wire between the igniter unit and ignition switch.

YES

Turn the ignition switch OFF.

Connect the PGM-FI test harness between the ECU and connector

BLK/YEL
BLK
WHT

## IGNITION OUTPUT SIGNAL—CODE 15—PRELUDE

Reconnect the 6P connector on the igniter unit.

Turn the ignition switch ON.

Measure voltage individually between B15 (+), B17 (+) terminals and A18 (−) terminal.

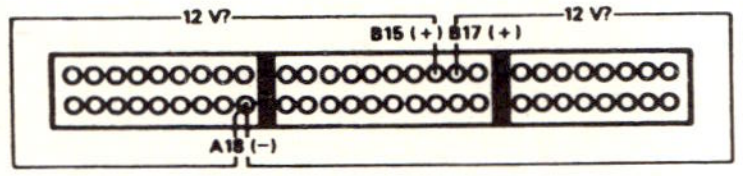

12 V? — B15 (+) B17 (+) — 12 V? — A18 (−)

Is there battery voltage ?

- YES: Substitute a known-good ECU and recheck. If symptom/indication goes away, replace the original ECU.
- NO: Turn the ignition switch OFF.

Disconnect the 6P connector on the igniter unit and the system checker harness from the ECU.

Check for continuity of WHT wires between the ECU (B15, B17) and the igniter unit.

Does continuity exist ?

- NO: Repair open in WHT wires between the ECU (B15, B17) and the igniter unit.
- YES: Check for continuity between white terminal of 6P connector and body ground.

Does continuity exist ?

- YES: Repair short in WHT wire.
- NO: Replace the igniter unit.

## VEHICLE SPEED SENSOR—CODE 17—PRELUDE

Self-diagnosis LED indicator blinks seventeen times: A problem in the Vehicle Speed Sensor circuit.

The signal generated by the speed sensor, which produces 4 pulses (switch closures to ground) for each revolution of the speedometer cable.

- Check Engine warning light is on.
- LED indicates CODE 17.

Turn the ignition switch OFF.

Remove CLOCK fuse in the under-hood relay box seconds to reset ECU.

Connect the PGM-FI test harness between the ECU and connector

Block rear wheels and set the parking brake. Jack up the front of the car and support with safety stands.

**WARNING** **Block rear wheels before jacking up front of car.**

Turn the ignition switch ON.

Slowly rotate left front wheel and measure voltage between B16 (+) terminal and A18 (−) terminal.

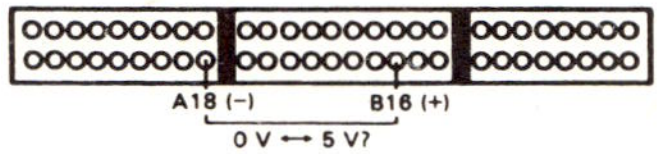

A18 (−) B16 (+) 0 V ↔ 5 V?

Does voltage pulse 0V and 5V?

- YES: Road test necessary. Place the shift or selector lever in 2nd gear (MT) or [2] position (AT), accelerate to 4,000 rpm and decelerate to 1,200 rpm with foot off accelerator.

  NOTE: The decel from 4,000 rpm to 1,200 rpm must be at least 6 seconds with your foot off the accelerator pedal.

  Is Check Engine warning light is on and does LED indicate CODE 17?

  - NO: Intermitten failure seat connectors C252, C263, C417 and C710.
  - YES: Substitute a known-good ECU and recheck. If prescribed voltage is now available replace the original ECU.
- NO: Is there approx. 5V?
  - YES:
    - Repair open in WHT/BLU wire between ECU (B16) and the speed sensor.
    - Replace the speed sensor.
  - NO

## VEHICLE SPEED SENSOR—CODE 17—PRELUDE

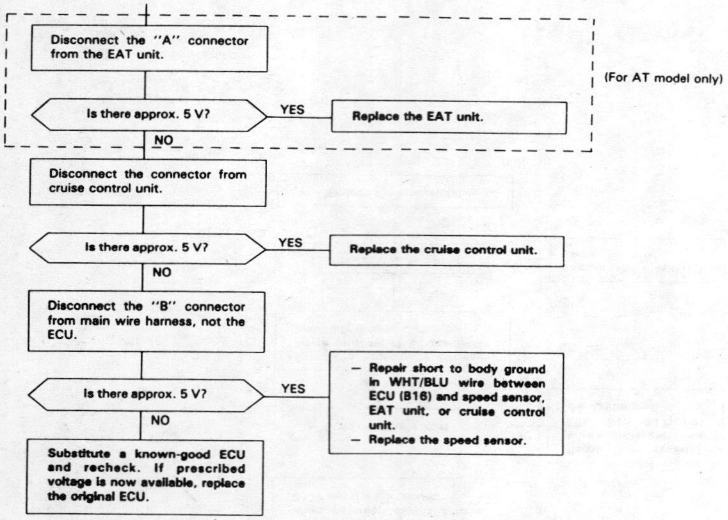

(For AT model only)

1. Disconnect the "A" connector from the EAT unit.
2. Is there approx. 5 V?
   - YES: Replace the EAT unit.
   - NO: go to step 3.

3. Disconnect the connector from cruise control unit.
4. Is there approx. 5 V?
   - YES: Replace the cruise control unit.
   - NO: go to step 5.
5. Disconnect the "B" connector from main wire harness, not the ECU.
6. Is there approx. 5 V?
   - YES:
     - Repair short to body ground in WHT/BLU wire between ECU (B16) and speed sensor, EAT unit, or cruise control unit.
     - Replace the speed sensor.
   - NO: Substitute a known-good ECU and recheck. If prescribed voltage is now available, replace the original ECU.

## IDLE CONTROL TROUBLESHOOTING—PRELUDE

NOTE: Across each row in the chart, the sub systems that could be sources of a symptom are ranked in the order they should be inspected, starting with ①. Find the symptom in the left column, read across to the most likely source, then refer to the page listed at the top of that column. If inspection shows the system is OK, try the next system ②, etc.

| SYMPTOM \ SUB SYSTEM | | IDLE ADJUSTING SCREW | ELECTRONIC AIR CONTROL VALVE | AIR CONDITIONING SIGNAL | ALTERNATOR FR SIGNAL | A/T SHIFT POSITION SIGNAL | P/S OIL PRESSURE SIGNAL | FAST IDLE VALVE | HOSES AND CONNECTIONS |
|---|---|---|---|---|---|---|---|---|---|
| DIFFICULT TO START ENGINE WHEN COLD | | | | | | | | ① | |
| WHEN COLD | Fast idle speed is not as specified (1,100–1,900 rpm) | | ② | | | | | ① | |
| WHEN WARM RPM TOO HIGH | Idle speed is above specified rpm | ② | ② | | | | | ② | ① |
| WHEN WARM RPM TOO LOW | Idle speed is below specified rpm | ② | ① | | | | | | ② |
| | Idle speed does not increase after initial start up. | | ① | | | | | | |
| | Idle speed drops when blipping throttle with electrical load | | ② | | ① | | | | |
| | On models with automatic transmission, the idle speed drops in gear | | ② | | | ① | | | |
| | Idle speed drops when steering wheel is turning | | ② | | | | ① | | |
| | Idle speed drops when air conditioner is ON. | | ② | ① | | | | | |
| FREQUENT STALLING WHILE WARMING UP | | | | | | | | ① | |

- – If bypass passages are blocked, a low idle speed will result.
  – If hoses or bypass passages are leaking, a high idle speed will result.

# ELECTRONIC AIR CONTROL VALVE (EACV)—CODE 14—PRELUDE

Self-diagnosis LED indicator blinks fourteen times: A problem in the Electronic Air Control Valve (EACV) circuit.

The EACV changes the amount of air bypassing the throttle body in response to a current signal from the ECU in order to maintain the proper idle speed.

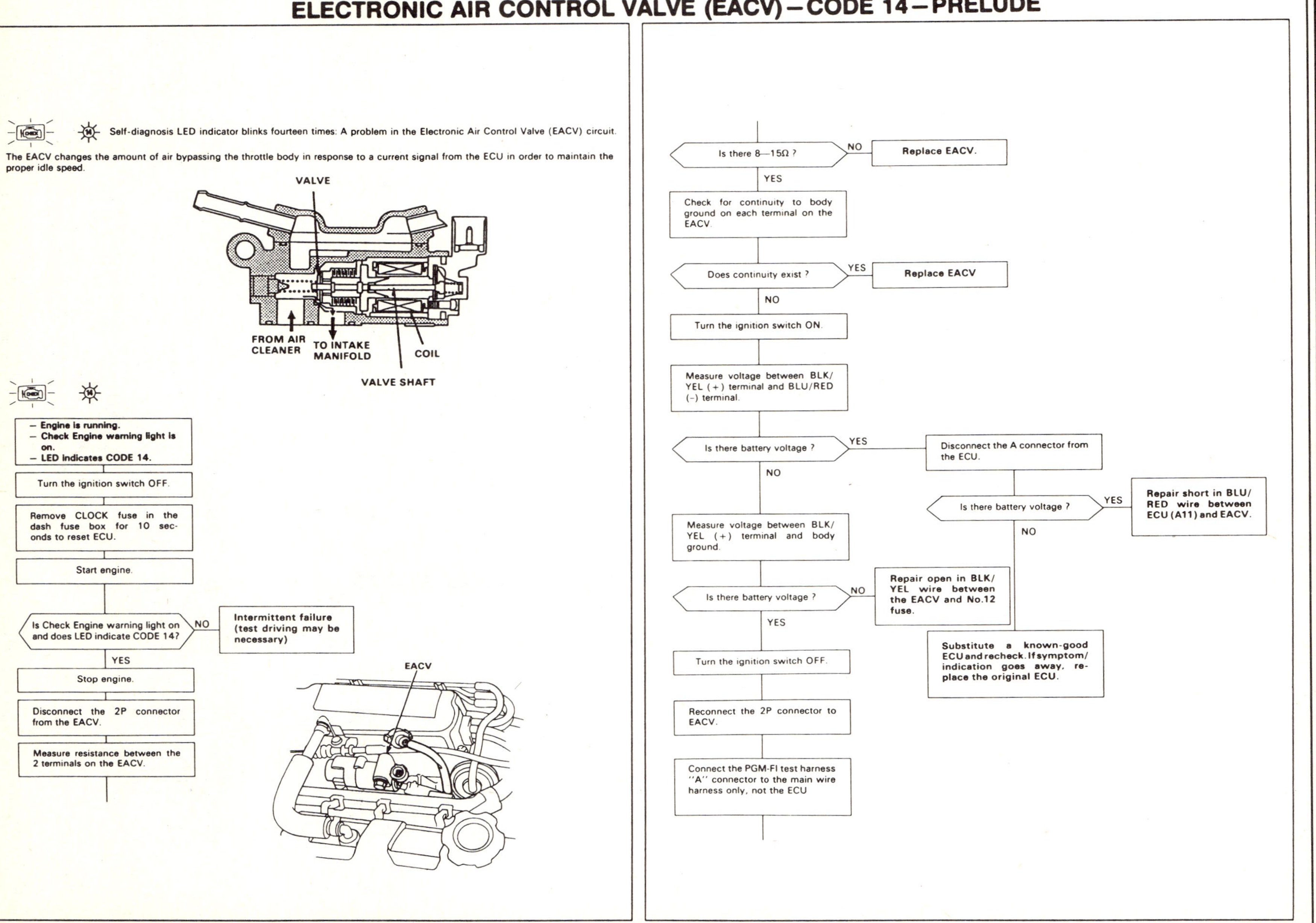

## ELECTRONIC AIR CONTROL VALVE (EACV) CODE 14—PRELUDE

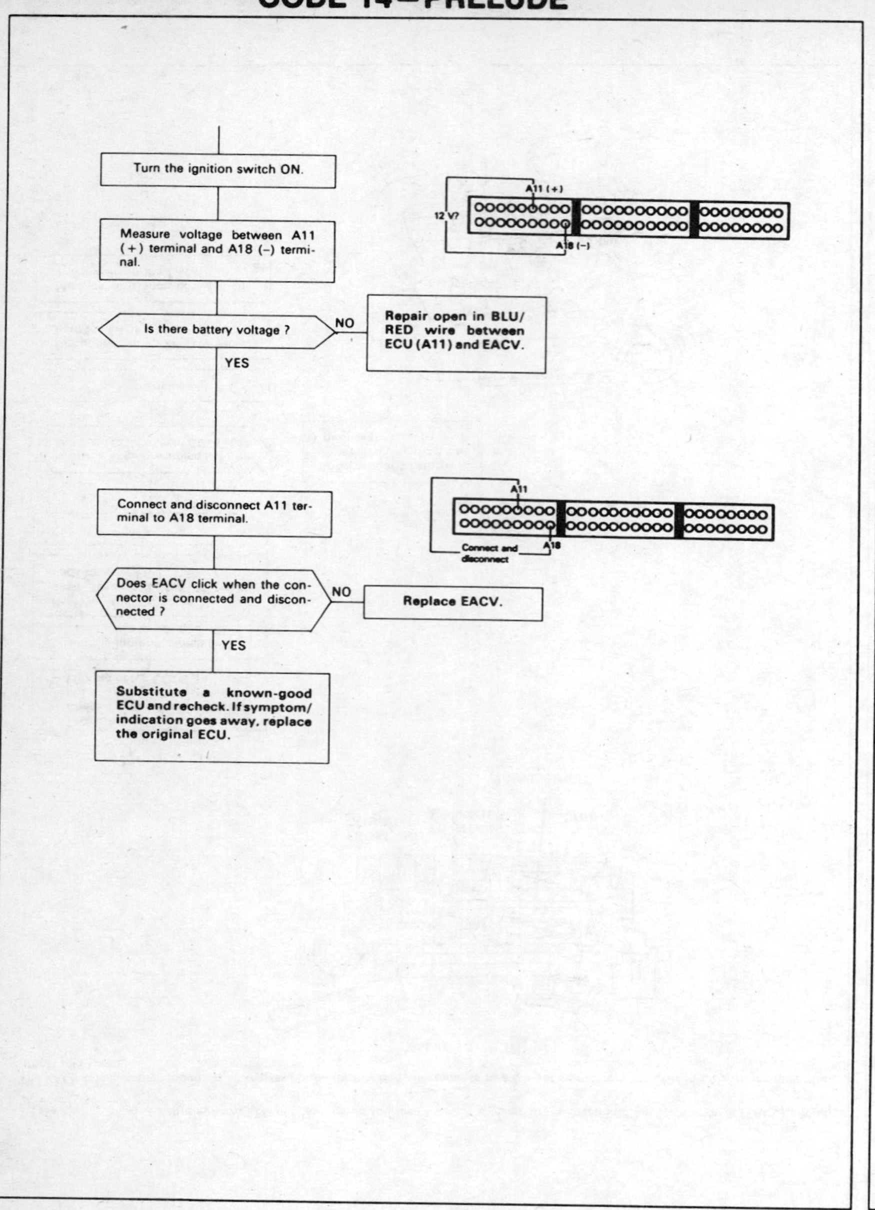


## AIR CONDITIONING SIGNAL—PRELUDE

This signals the PGM-FI ECU when there is a demand for cooling from the air conditioning circuits.

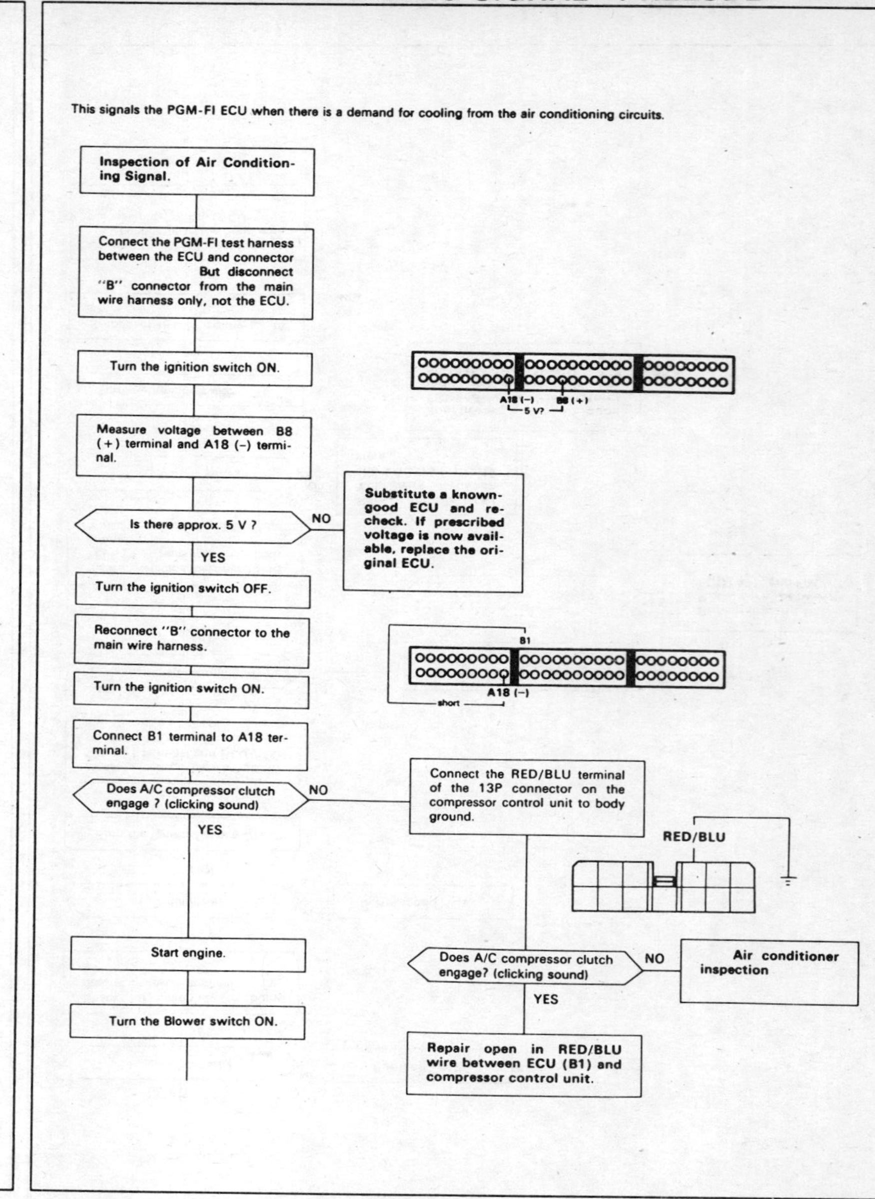

Inspection of Air Conditioning Signal.
Connect the PGM-FI test harness between the ECU and connector But disconnect "B" connector from the main wire harness only, not the ECU.
Turn the ignition switch ON.
A18 (−)
B8 (+)
5 V?
Measure voltage between B8 (+) terminal and A18 (−) terminal.
Is there approx. 5 V ?
NO
Substitute a known-good ECU and recheck. If prescribed voltage is now available, replace the original ECU.
YES
Turn the ignition switch OFF.
Reconnect "B" connector to the main wire harness.
Turn the ignition switch ON.
B1
A18 (−)
short
Connect B1 terminal to A18 terminal.
Does A/C compressor clutch engage ? (clicking sound)
NO
Connect the RED/BLU terminal of the 13P connector on the compressor control unit to body ground.
RED/BLU
YES
Start engine.
Turn the Blower switch ON.
Does A/C compressor clutch engage? (clicking sound)
NO
Air conditioner inspection
YES
Repair open in RED/BLU wire between ECU (B1) and compressor control unit.

## AIR CONDITIONING SIGNAL—PRELUDE

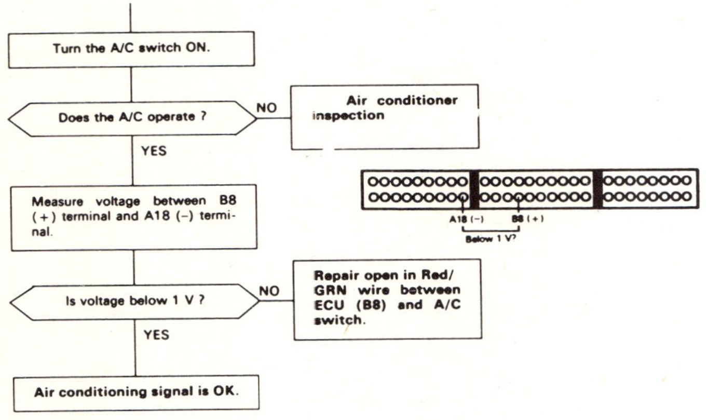

Turn the A/C switch ON.

Does the A/C operate ? — NO: Air conditioner inspection

YES

Measure voltage between B8 (+) terminal and A18 (−) terminal.

A18 (−) B8 (+) Below 1 V?

Is voltage below 1 V ? — NO: Repair open in Red/GRN wire between ECU (B8) and A/C switch.

YES

Air conditioning signal is OK.

## ALTERNATOR FR SIGNAL—PRELUDE

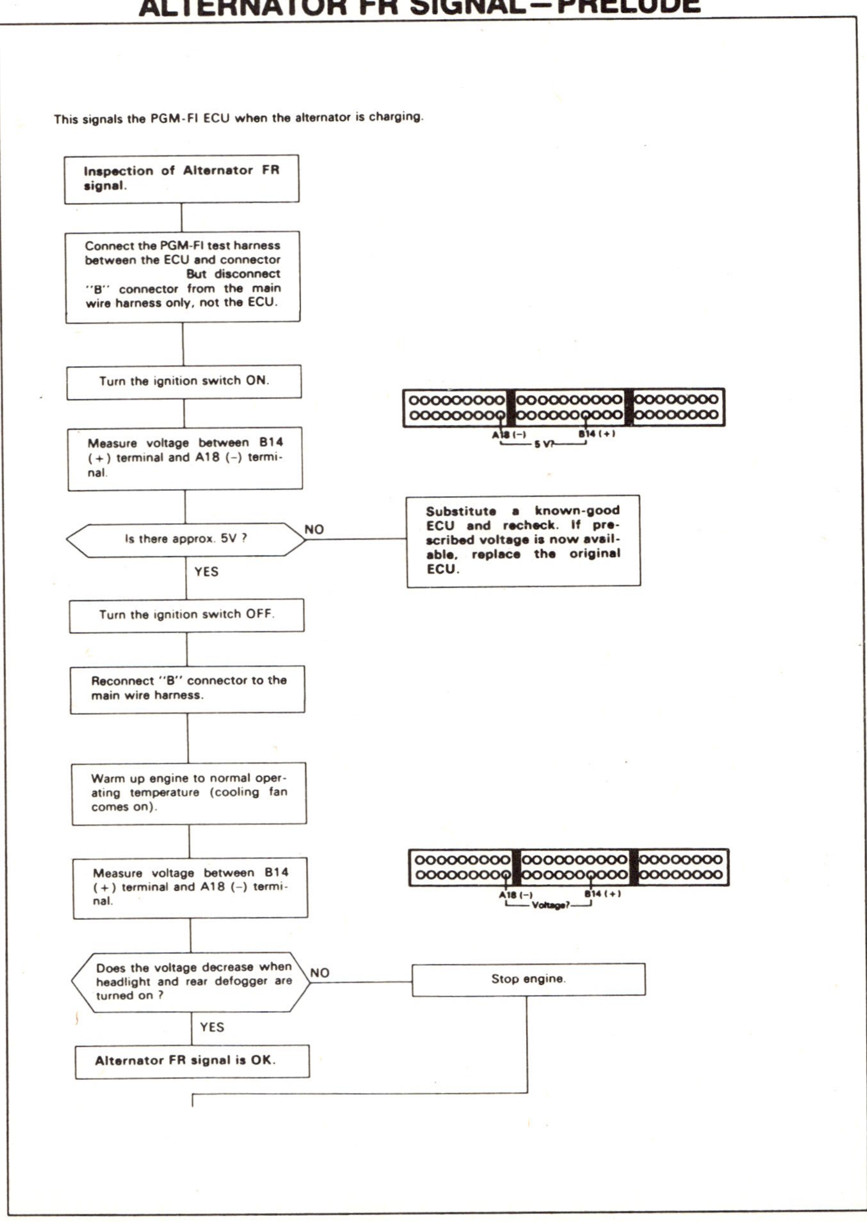

## ALTERNATOR FR SIGNAL—PRELUDE

Disconnect "B" connector from ECU only, not the main wire harness.

Disconnect the negative battery cable from the battery.

Check for continuity between B14 terminal and body ground.

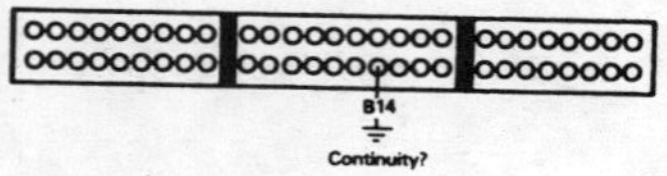

B14
Continuity?

Does continuity exist ?

YES → Disconnect GRN connector from the alternator.

- Check for continuity between B14 terminal and body ground.
- Does continuity exist ?
  - NO → Alternator Inspection
  - YES → Repair short in WHT/RED wire between ECU (B14) and alternator.

NO → Disconnect GRN connector from the alternator.

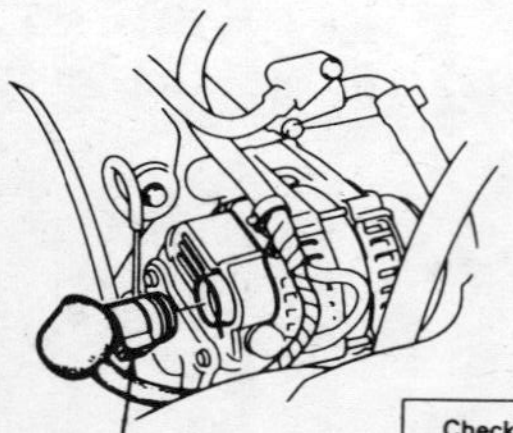

GRN CONNECTOR

Connect WHT/RED wire to body ground.

Check for continuity between B14 terminal and body ground.

Does continuity exist ?

- YES → Alternator Inspection
- NO → Repair open in WHT/RED wire between ECU (B14) and alternator.

## A/T SHIFT POSITION SIGNAL—PRELUDE

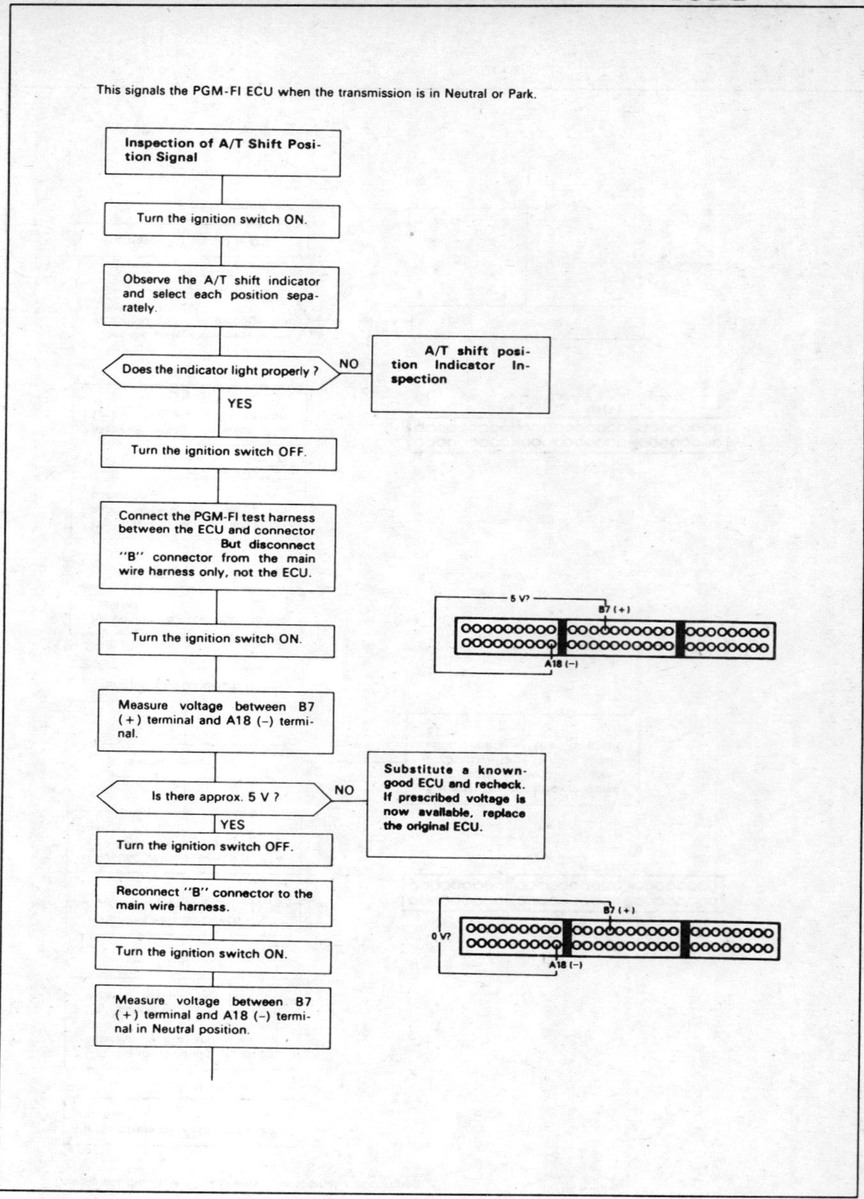

## A/T SHIFT POSITION SIGNAL—PRELUDE

Is voltage below 1V ?

NO → – Repair open in LT GRN wire between ECU (B7) and combination meter.
– Repair open in GRN wire between the combination meter and shift position console switch.

YES ↓

Measure voltage between B7 (+) terminal and A18 (–) terminal in Park position.

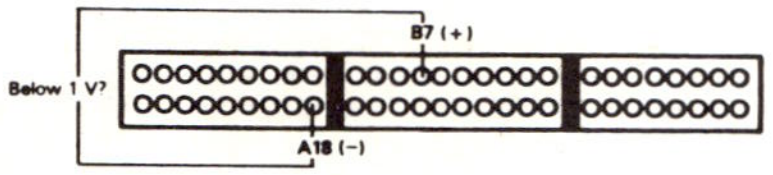

Below 1 V?  B7 (+)  A18 (–)

Is voltage below 1V ?

NO → Repair open in GRN/WHT wire between combination meter and shift position console switch.

YES ↓

Measure voltage between B7 (+) terminal and A18 (–) except in Park and neutral.

Is there approx. 5 V ?

NO → Repair short in LT GRN wire between ECU (B7) and combination meter.

YES ↓

A/T shift position signal is OK.

## POWER STEERING PRESSURE SIGNAL—PRELUDE

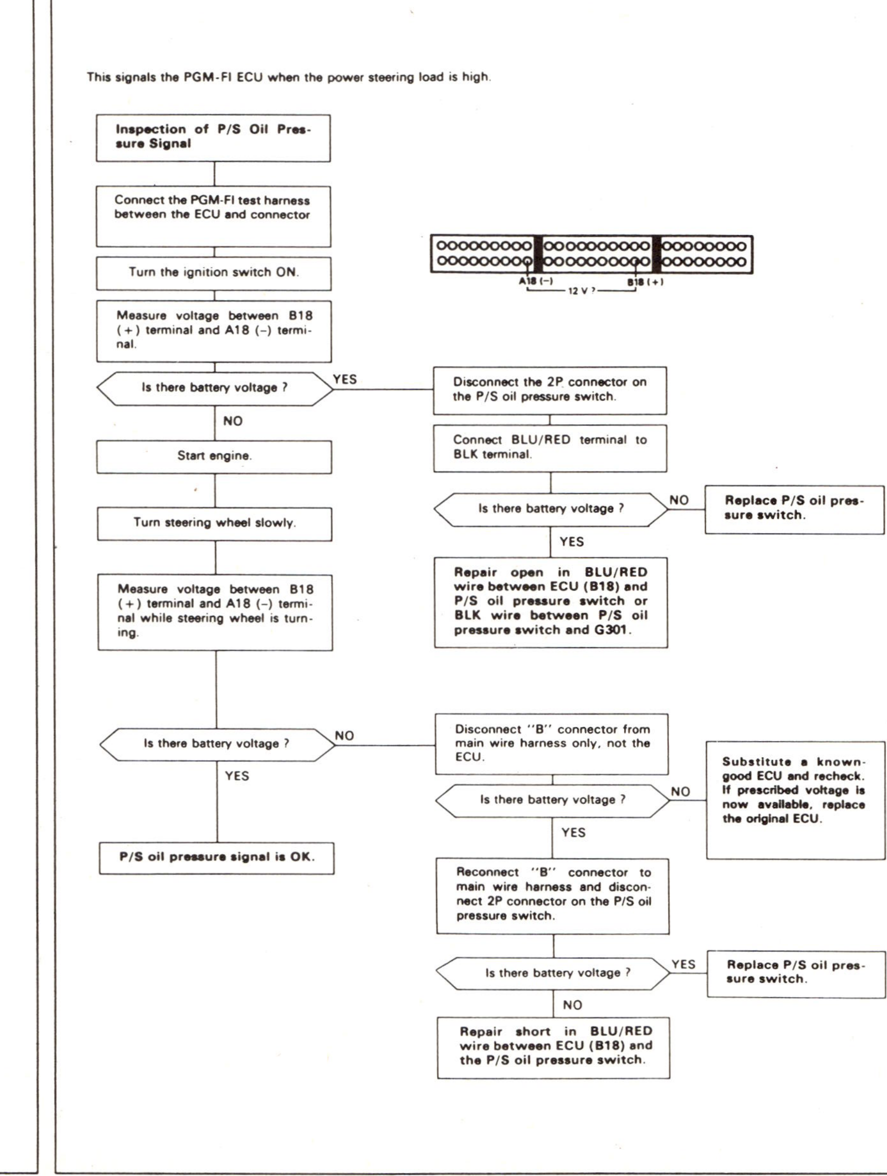

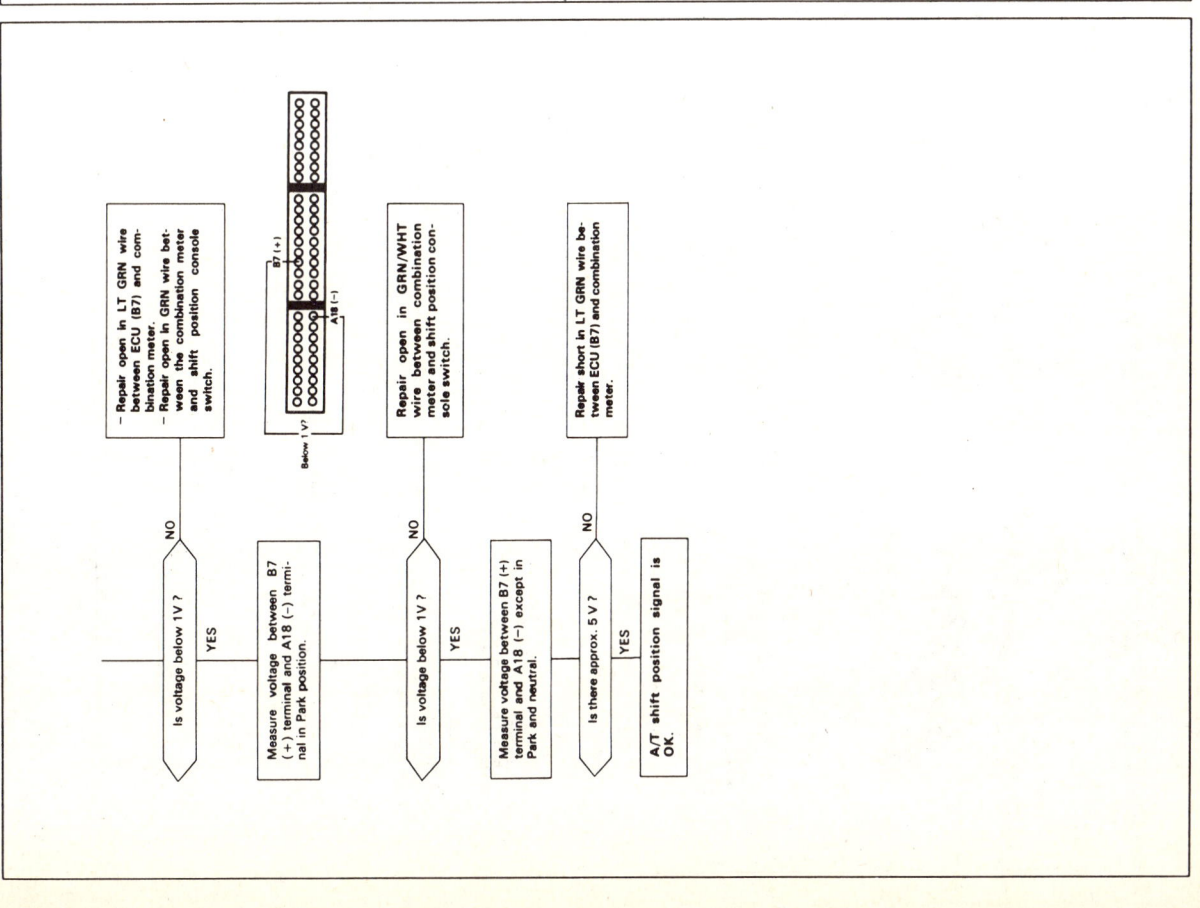

# FUEL INJECTORS—CODE 16—PRELUDE

Self-diagnosis LED indicator blinks sixteen times: A problem in the fuel injector circuit.

The injectors are the solenoid-actuated constant-stroke pintle type consisting of a solenoid, plunger needle valve and housing. When current is applied to the solenoid coil, the valve lifts up and pressurized fuel is injected close to the intake valve. Because the needle valve lift and the fuel pressure are constant, the injection quantity is determined by the length of time that the valve is open (i.e., the duration the current is supplied to the solenoid coil). The injector is sealed by an O-ring and seal ring at the top and bottom. These seals also reduce operating noise.

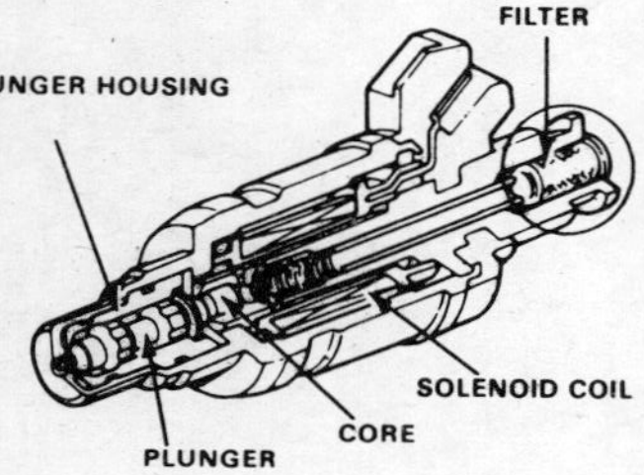

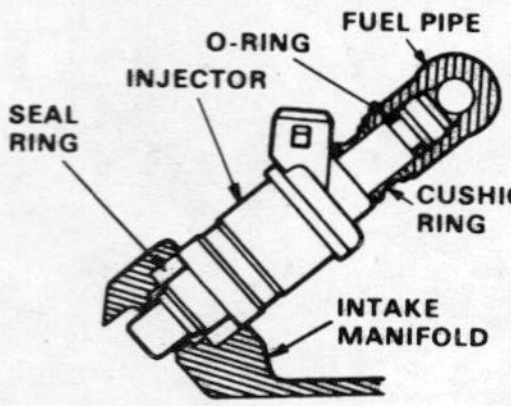

- Check Engine warning light is on.
- LED indicates CODE 16.

Turn the ignition switch OFF.

Remove CLOCK fuse in the underhood relay box for 10 seconds to reset ECU.

Turn the ignition switch to START position.
NOTE: It may be necessary to crank for a minimum of 15 seconds to reset failure CODES.

Does the engine start ? — NO: (continue to next chart) — YES:

Is Check Engine warning light on and does LED indicate CODE 16? — NO: Intermittent failure (Test drive may be necessary.) — YES: (continue)

Check the clicking sound of each injector by means of a stethoscope when the engine is idling.

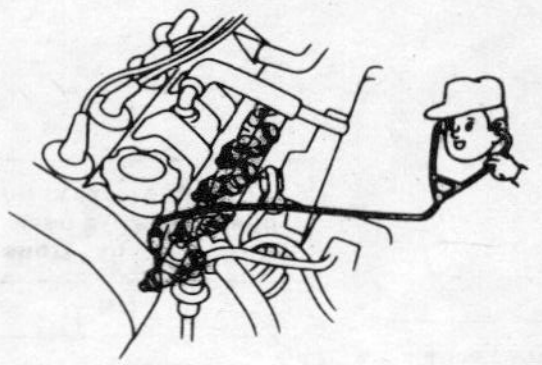

Does the injector click ? — YES: **Substitute a known-good ECU and recheck. If symptom/indication goes away, replace the original ECU.** — NO:

Turn the ignition switch OFF.

Disconnect the 2P connector from the injector that does not click.

Measure resistance between the 2 terminals of the injector.

(From engine does not start:) Turn the ignition switch OFF.

Disconnect the 2P connector from each injector.

Measure resistance between the 2 terminals of the injector.

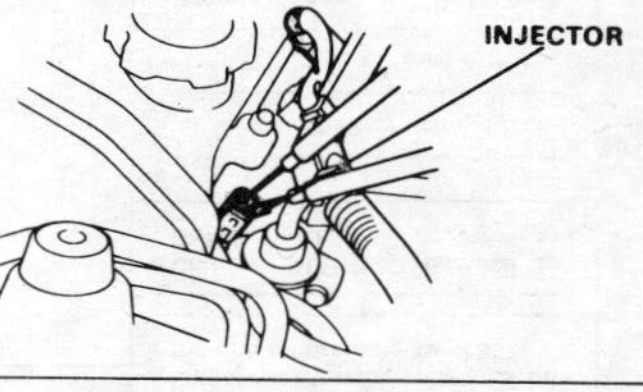

Is there 1.5—2.5Ω ? — NO: **Replace the injector.** — YES:

Turn the ignition switch ON.

## FUEL INJECTORS—CODE 16—PRELUDE

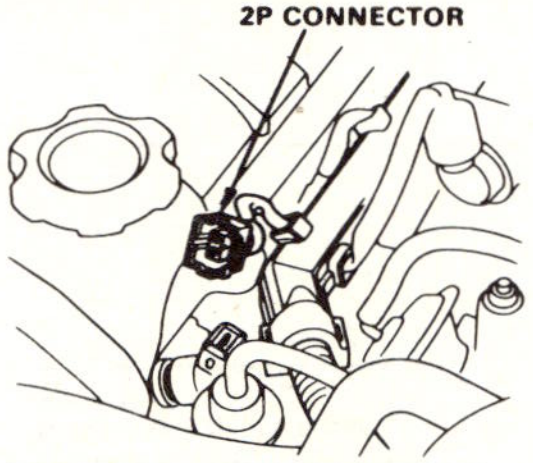

2P CONNECTOR

Measure voltage between RED/BLK (+) terminal on the 2P connector and body ground.

Is there battery voltage ?

NO → Turn the ignition switch OFF. → Disconnect 6P connector on the injector resistor. → Turn the ignition switch ON. → Measure voltage between YEL/BLK (+) terminal and body ground. → Is there battery voltage ?

- NO → **Repair open in YEL/BLK wire between the injector resistor and the main relay.**
- YES → **Replace the injector resistor.**

YES → Measure voltage between the following terminals.
- No. 1 injector: RED/BLK (+) terminal and BRN (-) terminal.
- No. 2 injector: RED/BLK (+) terminal and RED (-) terminal.
- No. 3 injector: RED/BLK (+) terminal and BLU (-) terminal
- No. 4 injector: RED/BLK (+) terminal and YEL (-) terminal

Is there battery voltage ?

YES → Disconnect "A" connector from the ECU only, not the main wire harness. → Is there battery voltage ?

- YES → **Repair short in the wire between the ECU (A1, A3, A5 or A7) and the injector.**
- NO → **Substitute a known-good ECU and recheck. If prescribed voltage is now available, replace the original ECU.**

NO → Reconnect the 2P connector to the injector. → Connect the PGM-FI test harness between the ECU and connector → Measure voltage between A18 (-) terminal and following terminals.
- No.1 injector: A1 (+) terminal.
- No.2 injector: A3 (+) terminal.
- No.3 injector: A5 (+) terminal.
- No.4 injector: A7 (+) terminal.

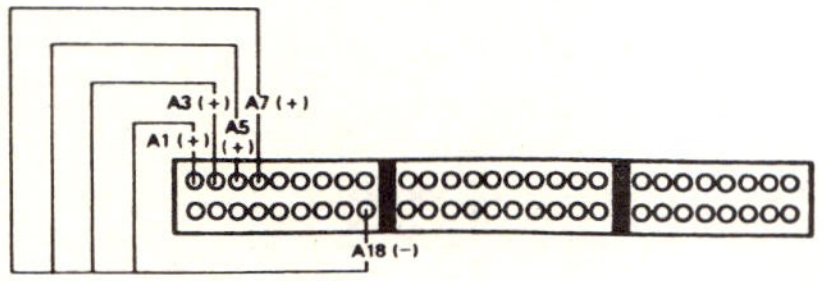

A3 (+) A7 (+) A5 (+) A1 (+) A18 (-)

Is there battery voltage ?

- NO → **Repair open in the wire between the ECU (A1, A3, A5 or A7) and the injector.**
- YES → **Substitute a known-good ECU and recheck. If prescribed voltage is now available, replace the original ECU.**

## BYPASS CONTROL SYSTEM—PRELUDE

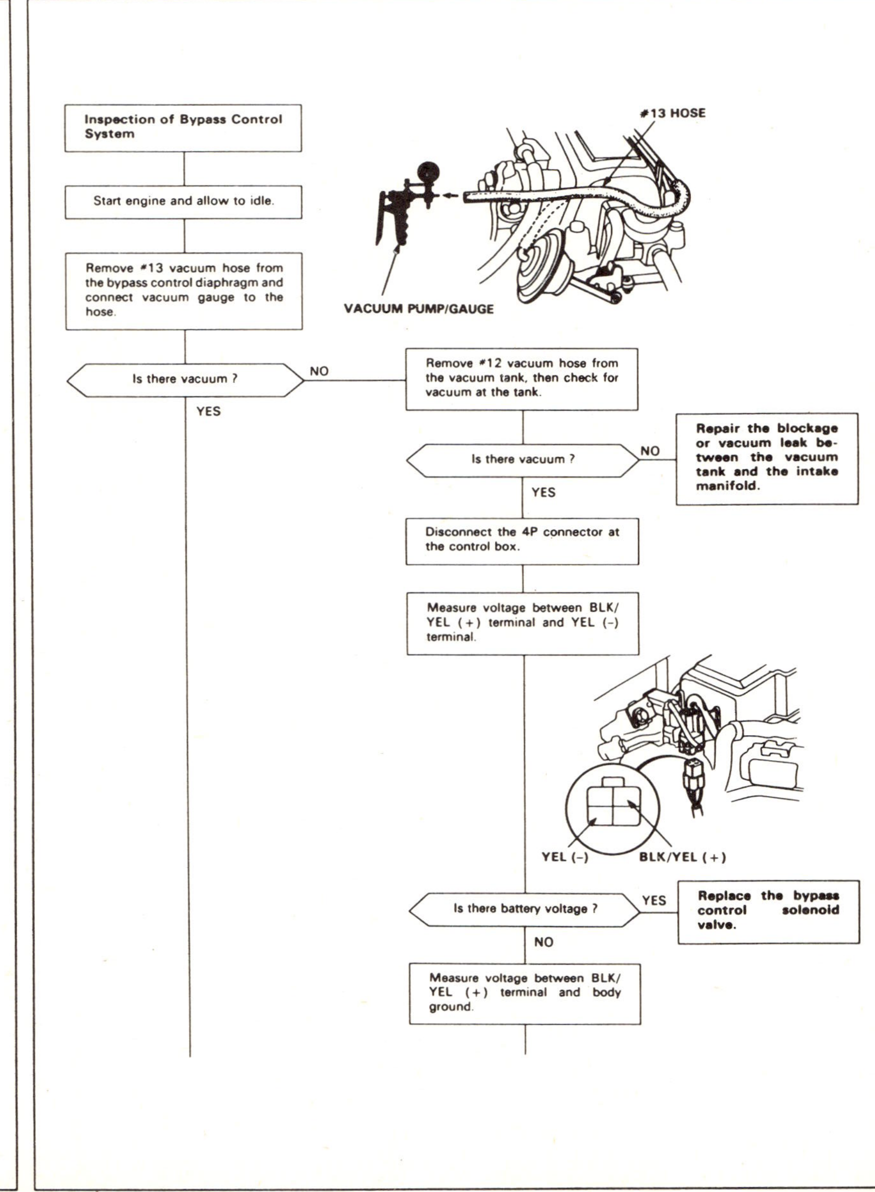

## BYPASS CONTROL SYSTEM—PRELUDE

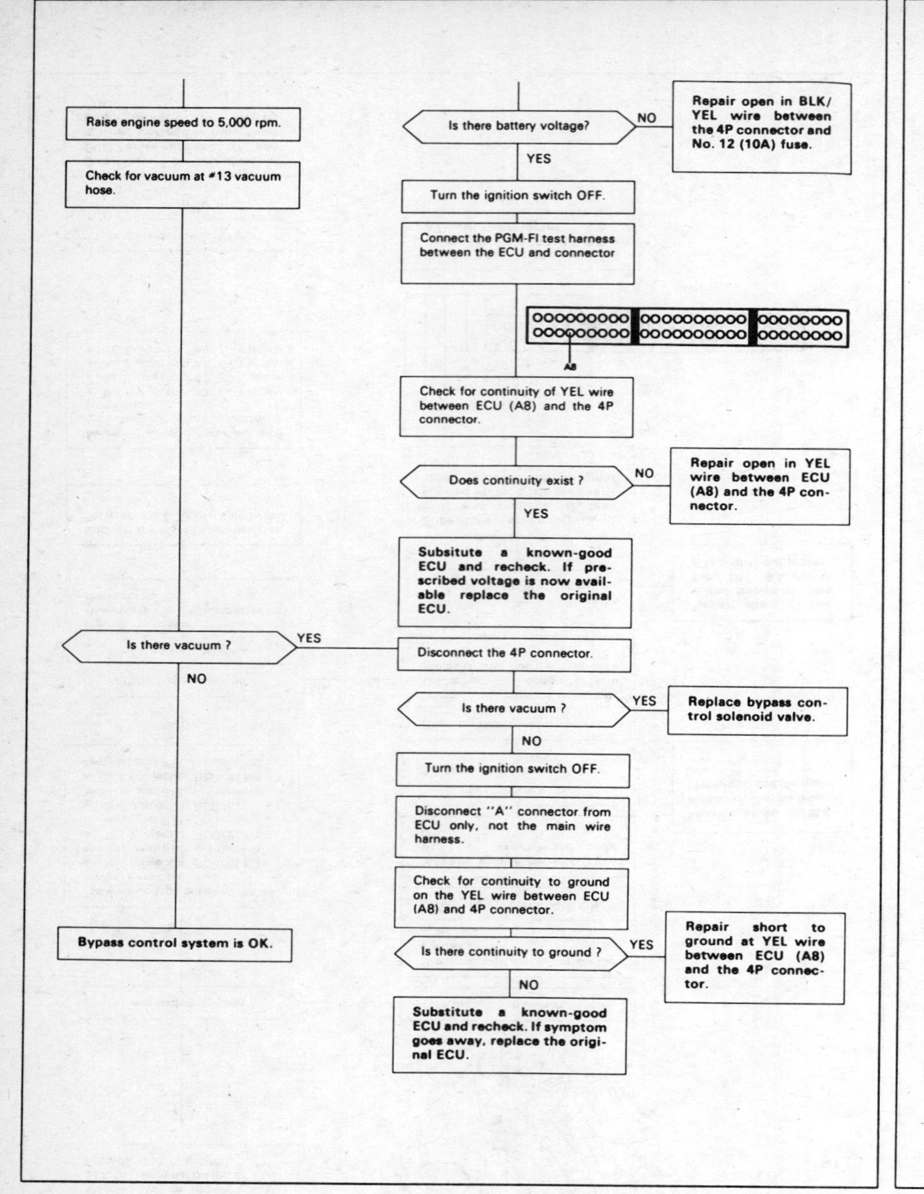


## EGR SYSTEM—CODE 12—PRELUDE

(12) Self diagnosis indicator blinks twelve times. Most likely a problem in the Exhaust Gas Recirculation (EGR) system.

The EGR System is designed to reduce oxides of nitrogen emissions (NOx) by recirculating exhaust gas through the EGR valve and the intake manifold into the combustion chambers. It is composed of the EGR valve, CVC valve, EGR control Solenoid valve, PGM-FI ECU and various sensors.

The ECU contains memories for ideal EGR valve lifts for varying operating conditions. The EGR valve lift sensor detects the amount of EGR valve lift and sends the information to the ECU. The ECU then compares it with the ideal EGR valve lift which is determined by signals sent from the other sensors. If there is any difference between the two, the ECU cuts current to the EGR control solenoid valve to reduce vacuum applied to the EGR valve.

TO No. 12 FUSE
EGR CONTROL SOLENOID VALVE
CVC VALVE
EGR VALVE
AIR CHAMBER
MAP SENSOR
TDC/CRANK SENSOR
CYL SENSOR
TW SENSOR
THROTTLE ANGLE SENSOR
ECU

(12)

- Check Engine Warning light has been reported on.
- LED indicates CODE 12.

Verify that all connectors between the EGR valve lift sensor and the PGM-FI ECU are securely connected.

Warm up engine to normal operating temperature (cooling fan comes on.)

Disconnect #16 hose from EGR valve.

Connect a vacuum pump to EGR valve and apply vacuum.

Does EGR valve hold vacuum ? Did engine stall ?
- NO → Replace EGR valve.
- YES

#16 HOSE
VACUUM PUMP/GAUGE
EGR VALVE

# EGR SYSTEM—CODE 12—PRELUDE

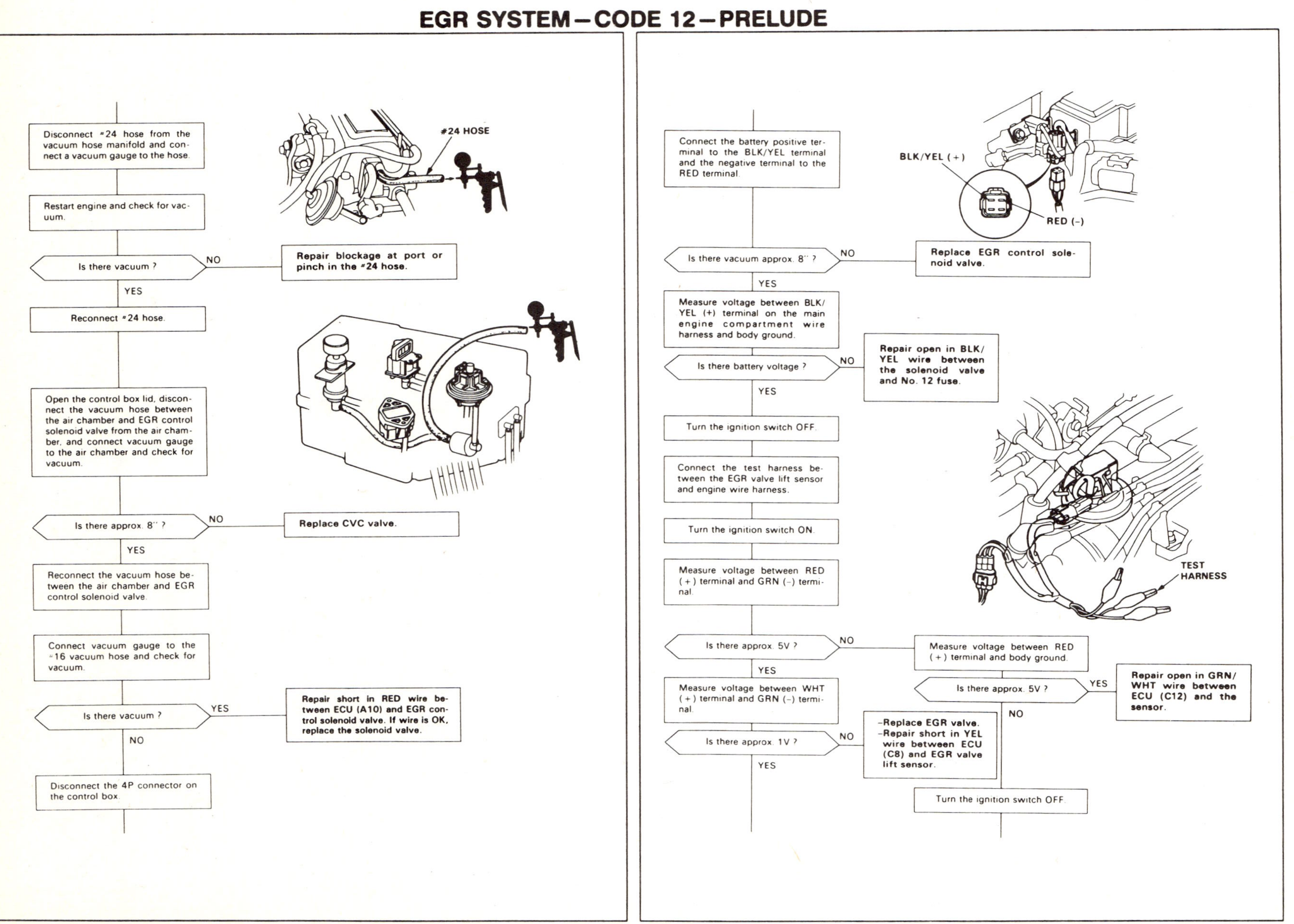

## EGR SYSTEM—CODE 12—PRELUDE

- Turn the ignition switch OFF.
- Connect the PGM-FI test harness between the ECU and connector
- Turn the ignition switch ON.
- Measure voltage between C8 (+) terminal and C12 (–) terminal.

C8 (+)
C12 (–)
1 V?

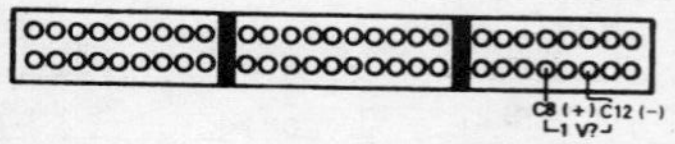

- Is there approx. 1V ?
  - NO → Repair open in YEL wire between ECU (C8) and the sensor.
  - YES ↓
- Disconnect "A" connector from the ECU only, not main wire harness.
- Measure voltage between A10 (+) terminal and A18 (–) terminal.

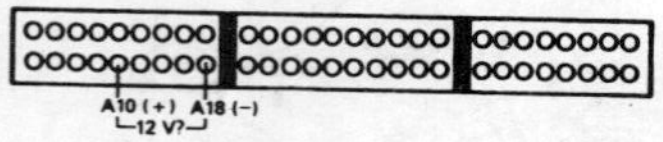

A10 (+)
A18 (–)
12 V?

- Is there battery voltage ?
  - NO → Repair open in RED wire between ECU (A10) and EGR control solenoid valve.
  - YES → Substitute a known-good ECU and recheck. If symptom/indication goes away, replace the original ECU.

- Connect the PGM-FI test harness between the ECU and connector
- Turn the ignition switch ON.
- Measure voltage between C13 (+) terminal and C12 (–) terminal.

C13 (+)
5 V?
C12 (–)

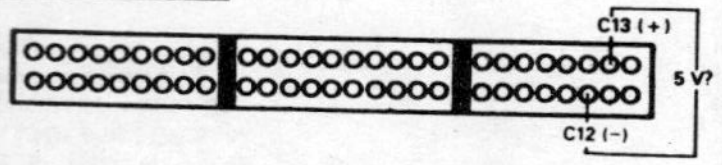

- Is there approx. 5 V ?
  - YES → Repair open in WHT/YEL wire between ECU (C13) and the sensor.
  - NO → Substitute a known-good ECU and recheck. If prescribed voltage is now available, replace the original ECU.

## PURGE CUT-OFF SOLENOID VALVE—PRELUDE

When the engine coolant temperature is above the set temperature of the TW sensor, the purge cut-off solenoid valve is activated by the control unit receiving signals from each sensor.

- **Inspection of Purge Cut-off Solenoid Valve**
- Remove #3 vacuum hose from the charcoal canister and connect a vacuum pump/gauge to the disconnected hose.
- Start the engine

  NOTE: Engine coolant temperature must be below 64°C (147°F)

- Is there manifold vacuum?
  - NO
  - YES ↓
    - Turn the ignition switch OFF.
    - Disconnect the 2P connector of the solenoid valve.
    - Start the engine.
    - Measure voltage between BLK/YEL (+) terminal on the main wire harness and body ground.
    - Is there battery voltage?
      - NO → Repair open in BLK/YEL wire between the solenoid valve and No.12 fuse.
      - YES

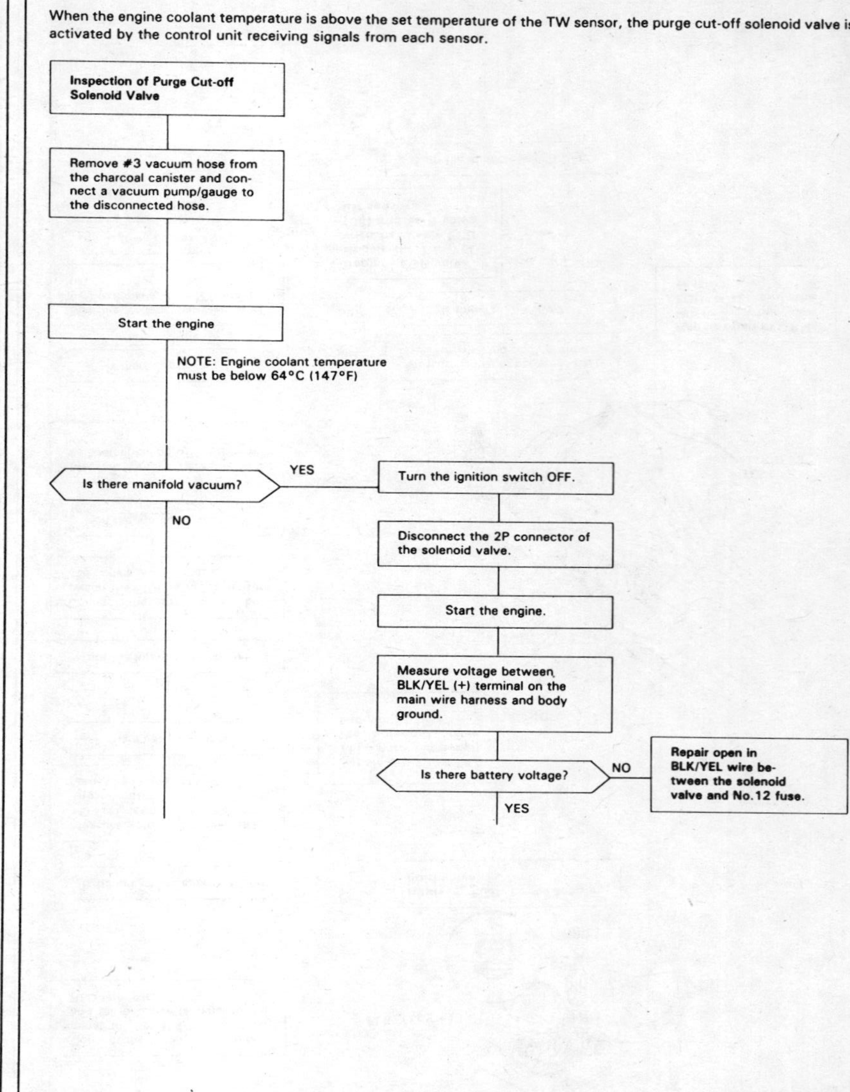

## PURGE CUT-OFF SOLENOID VALVE—PRELUDE

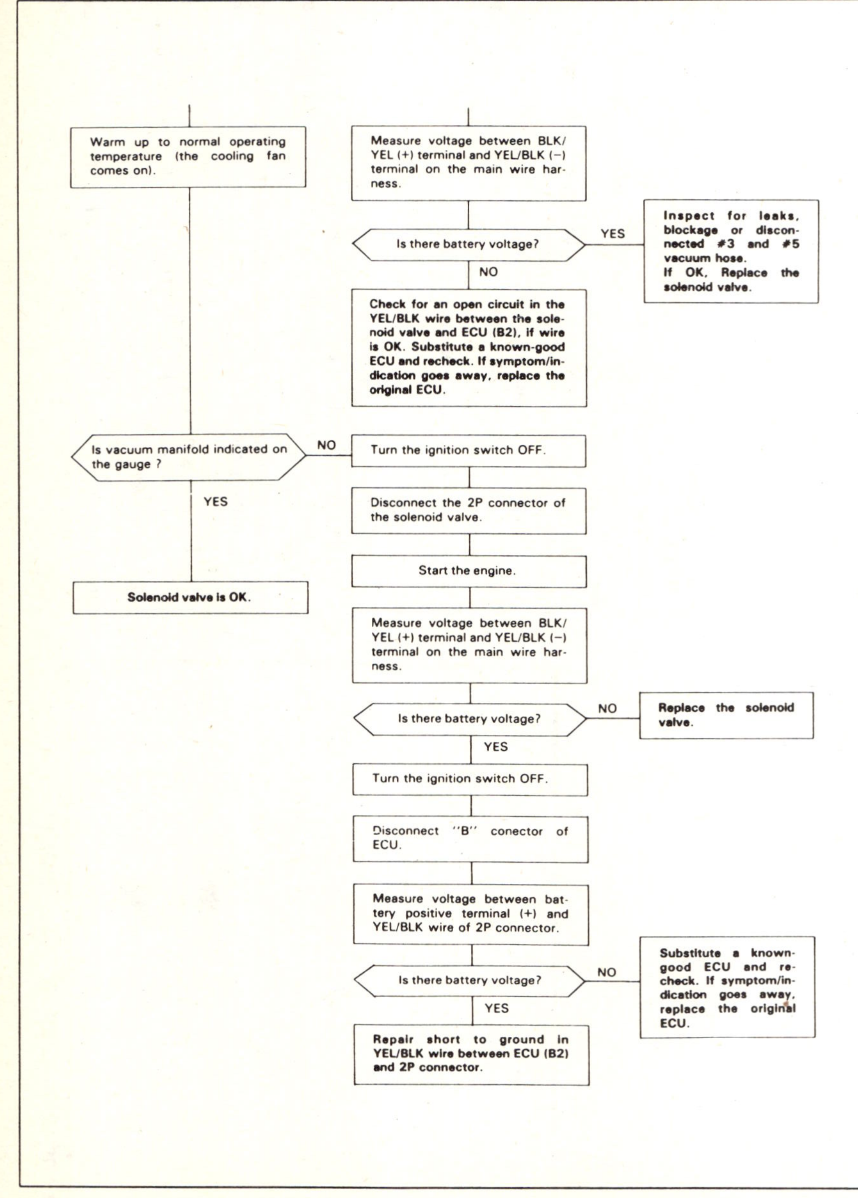


## A/C CLUTCH CUT SYSTEM—PRELUDE

To improve performance, during starting and acceleration the signal from the ECU to the compressor clutch is momentarily interrupted.

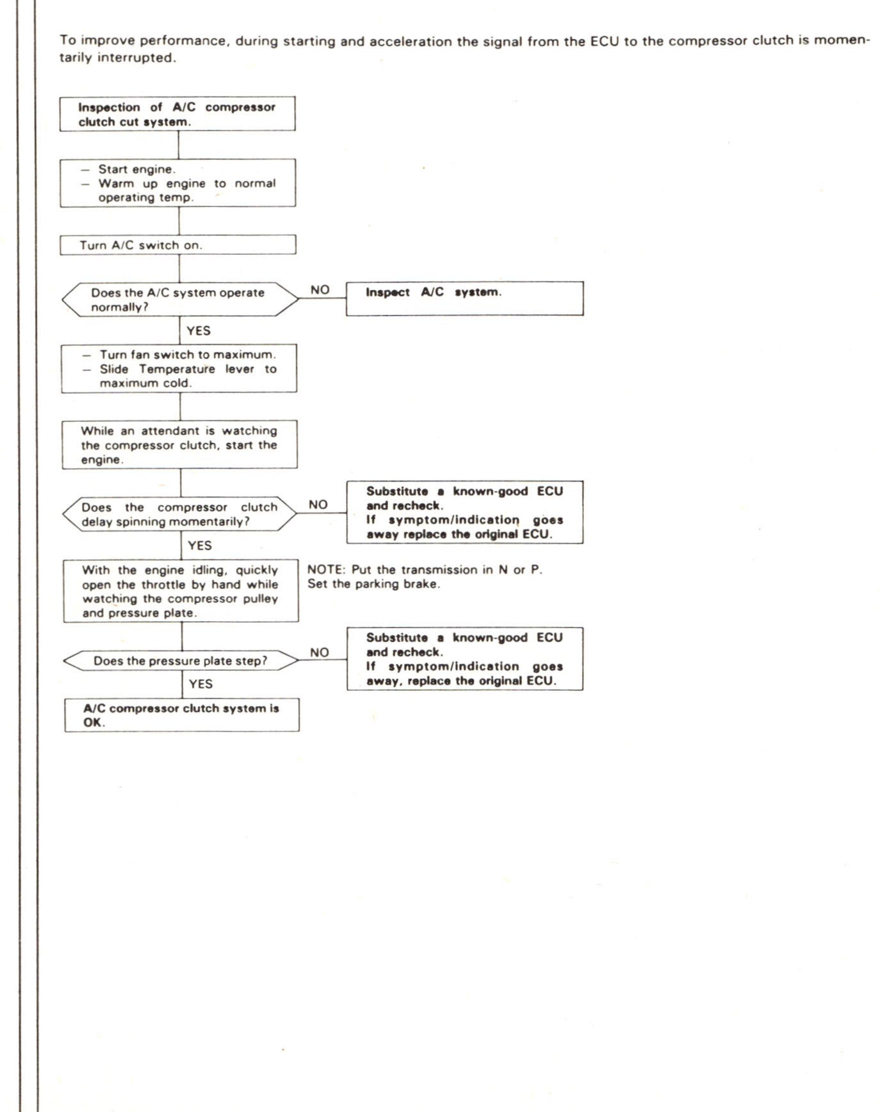

## FUEL PRESSURE

### Testing

**ALL MODELS**

1. Relieve fuel pressure.
2. Remove service bolt at the top of the fuel filter and attach fuel pressure gauge.
3. Start engine. Measure fuel pressure with engine idling and vacuum hose of pressure regulator disconnected. Pressure should be 35–41 psi (240–279 kPa).
4. If fuel pressure is not as specified, first check fuel pump. If pump is OK, check following:
   a. If pressure is higher than specified, inspect for pinched or clogged fuel return hose or piping, and faulty pressure regulator.
   b. If pressure is lower than specified, inspect for clogged fuel filter, pinched or clogged fuel hose from fuel tank to fuel pump, pressure regulator failure, leakage in fuel line, or pinched, broken or disconnected regulator vacuum hose.

## MAIN RELAY

### Testing

**ALL MODELS**

1. Remove main relay, near under-dash fuse box.
2. Connect battery positive terminal to No. 4 terminal and battery negative terminal to No. 8 terminal of main relay.
3. Check for continuity between No. 5 terminal and No. 7 terminal of main relay. If no continuity, replace main relay.
4. Connect battery positive terminal to No. 5 terminal and battery negative terminal to No. 2 terminal of main relay.
5. Check that there is continuity between No. 1 terminal and No. 3 terminal of main relay. If there is no continuity, replace main relay.
6. Connect battery positive terminal to No. 3 terminal and battery negative terminal to No. 8 terminal of main relay.
7. Check that there is continuity between No. 5 terminal and No. 7 terminal of main relay. If there is no continuity, replace main relay.

## HARNESS

### Testing

**CIVIC**

1. Keep ignition switch in **OFF** position.
2. Disconnect main relay connector.

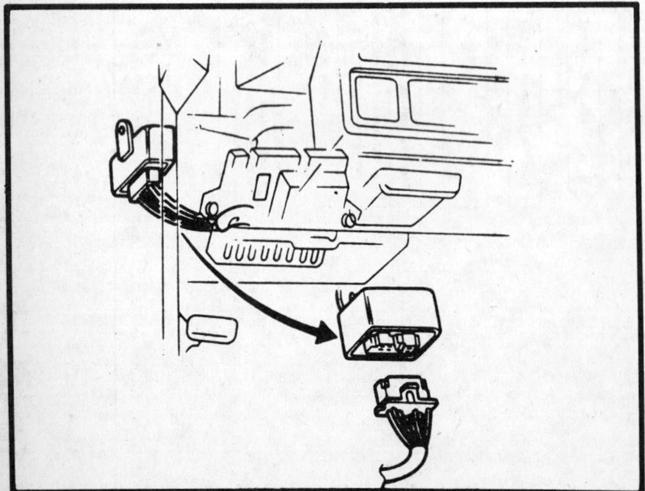

**Harness and main relay location—typical**

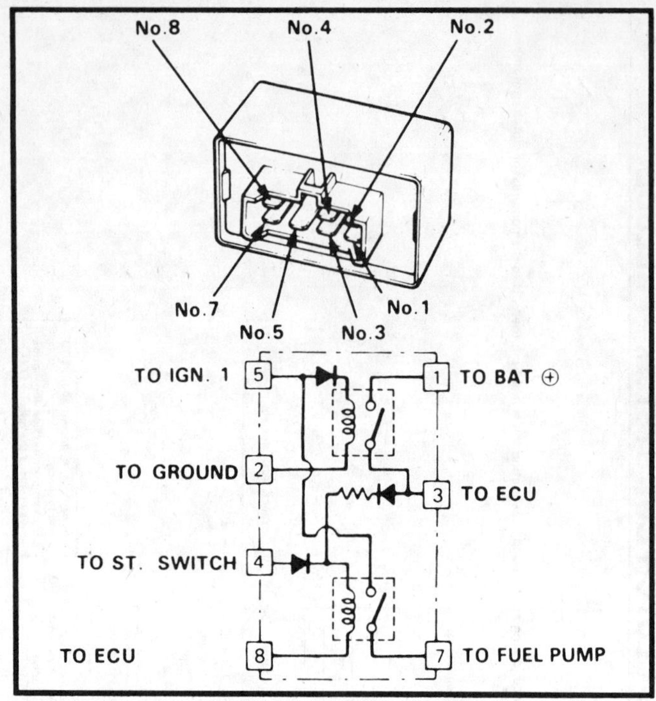


**Main relay—Connections**

3. Check continuity between Black wire in connector and body ground.
4. Connect positive probe of circuit tester to Yellow/Black wire and negative probe of tester to Black wire. Tester should read battery voltage.
5. If there is no voltage, check wiring between battery and main relay as well as ECU (15A) fuse.
6. Connect positive probe of tester to Black/Yellow wire of connector and ground negative probe of tester to Black wire.
7. Turn ignition switch **ON**. Tester should read battery voltage.
8. If no voltage, check wiring from ignition switch and main relay as well as fuse No. 14.
9. Connect positive probe of tester to Blue/White wire and negative probe to Black wire.
10. Turn ignition switch to **START** position. Tester should read 10 volts or more.
11. If no voltage, check wiring between ignition switch and main relay as well as No. 2 (10A) fuse.
12. Connect a jumper wire between 2 Black/Yellow wires in connector. Turn ignition switch **ON**. Fuel pump should run.
13. If pump does not run, check wiring between battery and fuel pump and wiring from fuel pump to ground (Black wire).

**PRELUDE**

1. Keep ignition switch in **OFF** position.
2. Disconnect main relay connector.
3. Check continuity between Black wire in connector and body ground.
4. Connect positive probe of circuit tester to Yellow/Blue wire and negative probe of tester to Black wire. Tester should read battery voltage.
5. If there is no voltage, check wiring between battery and main relay as well as EFI ECU (10A) fuse, located in the under hood relay box.
6. Connect positive probe of tester to Black/Yellow wire of connector and ground negative probe of tester to Black wire.
7. Turn ignition switch **ON**. Tester should read battery voltage.

8. If no voltage, check wiring from ignition switch and main relay as well as fuse No. 12 (10A).
9. Connect positive probe of tester to Blue/Red wire and negative probe to Black wire.
10. Turn ignition switch to **START** position. Tester should read battery voltage.
11. If no voltage, check wiring between ignition switch and main relay as well as No. 1 (7.5A) fuse.
12. Connect a jumper wire between 2 Black/Yellow wires in connector. Turn ignition switch **ON**. Fuel pump should run.
13. If pump does not run, check wiring between battery and fuel pump and wiring from fuel pump to ground (Black wire).

**1988–89 ACCORD**

1. Keep ignition switch in **OFF** position.
2. Disconnect main relay connector.
3. Check continuity between Black wire in connector and body ground.
4. Connect positive probe of circuit tester to Yellow/Blue wire and negative probe of tester to Black wire. Tester should read battery voltage.
5. If there is no voltage, check wiring between battery and main relay as well as ECU (10A) fuse, located in the under hood relay box.
6. Connect positive probe of tester to Black/Yellow wire of connector and ground negative probe of tester to Black wire.
7. Turn ignition switch **ON**. Tester should read battery voltage.
8. If no voltage, check wiring from ignition switch and main relay as well as fuse No. 1 (15A).
9. Connect positive probe of tester to Blue/Red wire and negative probe to Black wire.
10. Turn ignition switch to **START** position. Tester should read battery voltage.
11. If no voltage, check wiring between ignition switch and main relay as well as No. 10 (7.5A) fuse.
12. Connect a jumper wire between 2 Black/Yellow wires in connector. Turn ignition switch **ON**. Fuel pump should run.
13. If pump does not run, check wiring between battery and fuel pump and wiring from fuel pump to ground (Black wire).

**1990 ACCORD**

1. Keep ignition switch in **OFF** position.
2. Disconnect main relay connector.
3. Check continuity between Black wire in connector and body ground.
4. Connect positive probe of circuit tester to Yellow/Blue wire and negative probe of tester to Black wire. Tester should read battery voltage.
5. If there is no voltage, check wiring between battery and main relay as well as ECU (10A) fuse.
6. Connect positive probe of tester to Black/Yellow wire of connector and ground negative probe of tester to Black wire.
7. Turn ignition switch **ON**. Tester should read battery voltage.
8. If no voltage, check wiring from ignition switch and main relay as well as fuse No. 2 (15A).
9. Connect positive probe of tester to Blue/Red – Automatic transmission or Black/Green – Manual transmission wire and negative probe to Black wire.
10. Turn ignition switch to **START** position. Tester should read at least 10 volts.
11. If no voltage, check wiring between ignition switch and main relay as well as No. 9 (7.5A) fuse.
12. Connect a jumper wire between Black/Yellow and Yellow wires in connector. Turn ignition switch **ON**. Fuel pump should run.
13. If pump does not run, check wiring between battery and fuel pump and wiring from fuel pump to ground (Black wire).

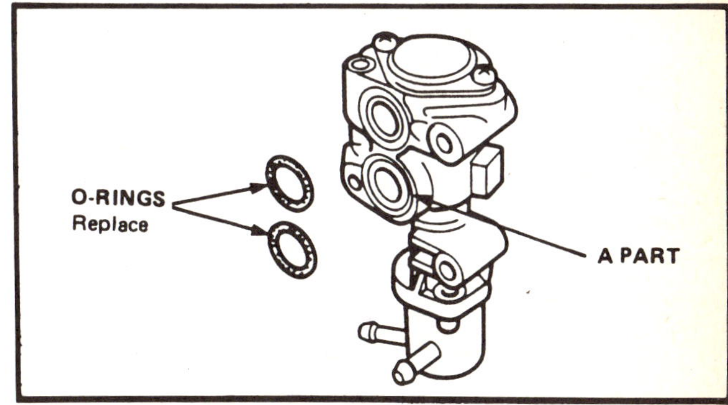

**Fast idle valve assembly – typical**

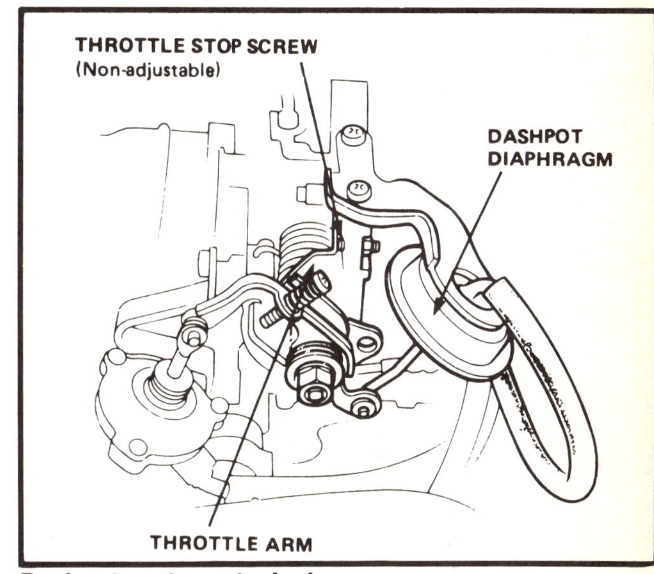

**Dashpot system – typical**

## FAST IDLE VALVE

### Testing

**1988–89 ACCORD**

**NOTE: Fast idle valve is factory adjusted, it should not be disassembled.**

1. Start engine.
2. Remove cover of fast idle valve.
3. Make sure that there is air flow with engine cold (coolant temperature below 86°F [30°C]). It can be detected by putting finger on valve seat area.
4. If no air flow, replace fast idle valve.
5. Warm up to normal operating temperature (cooling fan comes on).
6. Check that valve is completely closed. If not, air is being sucked from valve seat area. It can be detected by putting finger on valve seat area.
7. If any suction sound is heard, valve is leaking. Replace fast idle valve.

## THROTTLE CONTROL DIAPHRAGM

### Testing

**CIVIC (EXCEPT 1988 CRX SI)**

1. Start the engine and allow to reach normal operating temperature.

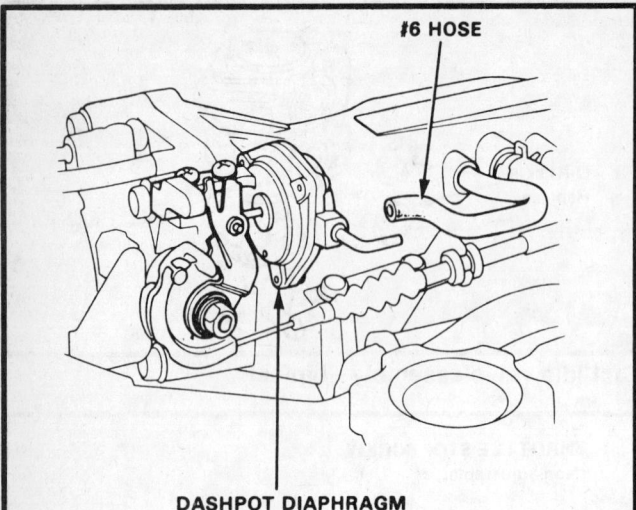

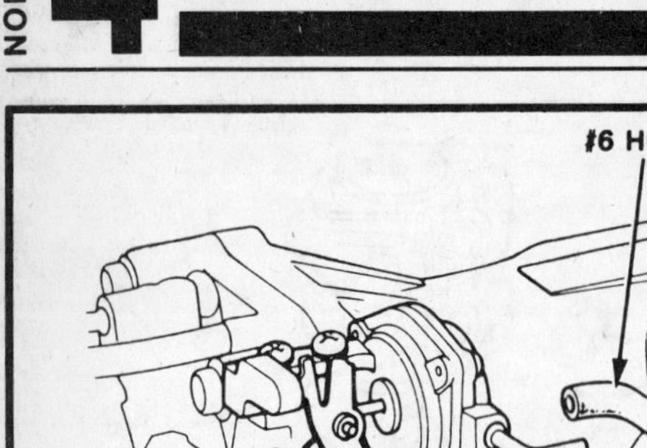


Throttle control diaphragm – all except 1988 CRX Si

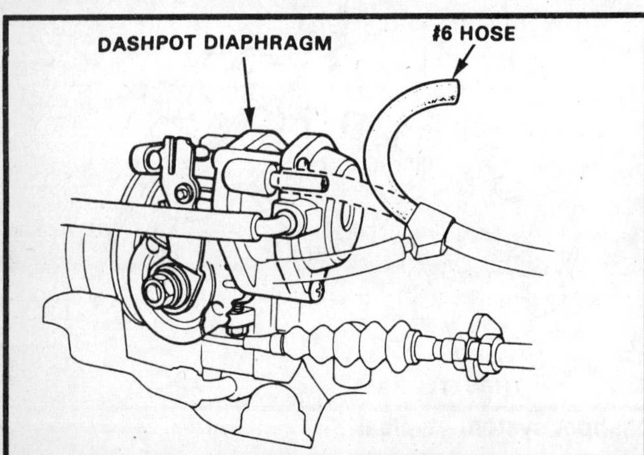


Throttle control diaphragm – 1988 CRX Si

2. Disconnect the No. 6 vacuum line from the dashpot diaphragm and check engine rpm.
3. If engine rpm is higher than 2000–3000 rpm, adjust by bending tab.
4. If the engine speed does not change, connect a vacuum gauge to the No. 6 hose and check for vacuum.
5. If there is vacuum, replace the diaphragm. If there is no vacuum, check hose for blockage or leakage.
6. Reconnect vacuum hose and adjust idle as needed.

**1988 CIVIC CRX SI**

1. Start the engine and allow to reach normal operating temperature.
2. Disconnect the No. 6 vacuum line from the dashpot diaphragm and check engine rpm.
3. If the engine speed does not change, connect a vacuum gauge to the No. 6 hose and check for vacuum.
4. If there is vacuum, replace the diaphragm. If there is no vacuum, check hose for blockage or leakage.
5. If engine rpm is out of range, inspect the throttle body.

## CANISTER TWO-WAY VALVE

### Testing

**CIVIC, PRELUDE AND 1990 ACCORD**

1. Remove the fuel tank cap.

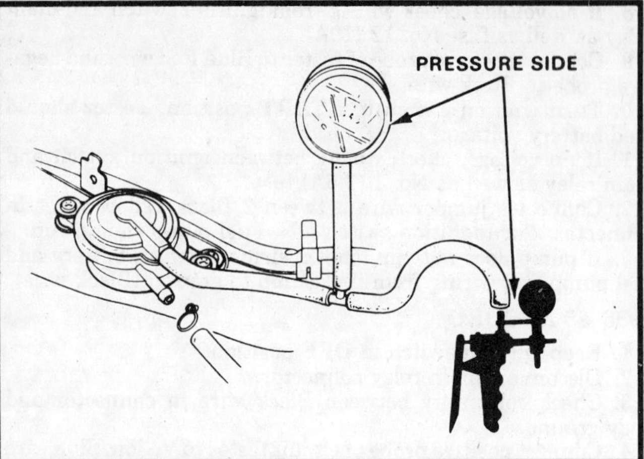


Canister 2-way valve

2. Remove the vapor line from the fuel tank and connect to T-fitting from a vacuum test gauge and vacuum pump.
3. Apply vacuum slowly and continuously. Vacuum should stabilize momentarily at 0.2–0.6 in. Hg (5–15mm Hg). If not, install a new valve and retest.
4. Move vacuum pump hose from vacuum to pressure fitting and move the gauge hose from the vacuum to pressure side.
5. Slowly pressurize the vapor line while watching the gauge. Pressure should stabilize at 0.4–1.4 in. Hg. If not, replace the valve and retest.

**1988–89 ACCORD**

1. Remove the fuel tank cap.
2. Remove the vapor line from the fuel tank and connect to T-fitting from a vacuum test gauge and vacuum pump.
3. Apply vacuum slowly and continuously. Vacuum should stabilize momentarily at 0.2–0.6 in. Hg. If not, install a new valve and retest.
4. Move vacuum pump hose from vacuum to pressure fitting and move the gauge hose from the vacuum to pressure side.
5. Slowly pressurize the vapor line while watching the gauge. Pressure should stabilize at 1.0–2.2 in. Hg. If not, replace the valve and retest.

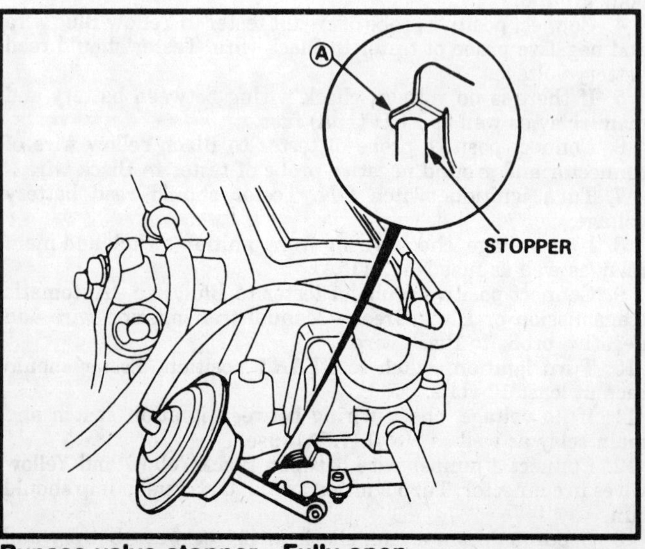


Bypass valve stopper – Fully open

B
FULL-CLOSE
SCREW

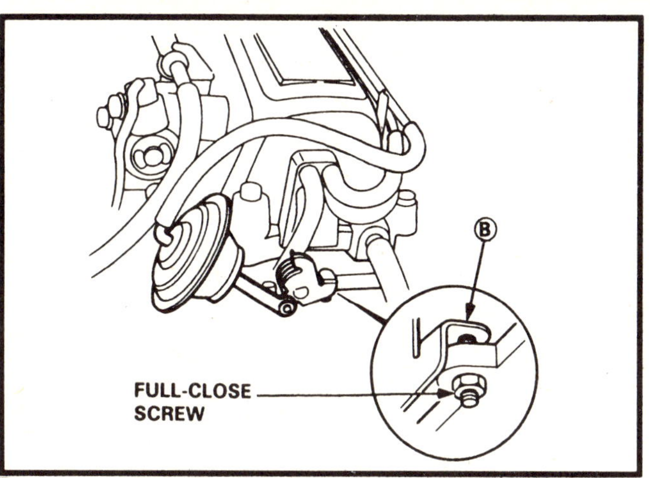

**Bypass valve fully closed screw**

## BYPASS VALVE

### Testing

**PRELUDE**

1. Check the bypass valve shaft for binding.
2. Check valve for smooth operation.
3. Check that the lever end (A), of the bypass valve is in close contact with the stopper when the valve is fully open.
4. Check the the lever end (B) is in close contact with the full-closed screw when the valve is fully closed.

**NOTE: The full-closed screw is preset at the factory and should not be adjusted.**

5. If any problems are found, clean linkage and shafts with carburetor cleaner.
6. If problem still exists, disassemble the intake manifold and examine the bypass valve for damage.

SERVICE BOLT
12 N·m (1.2 kg-m, 9 lb-ft)
SHOP TOWEL
FUEL FILTER

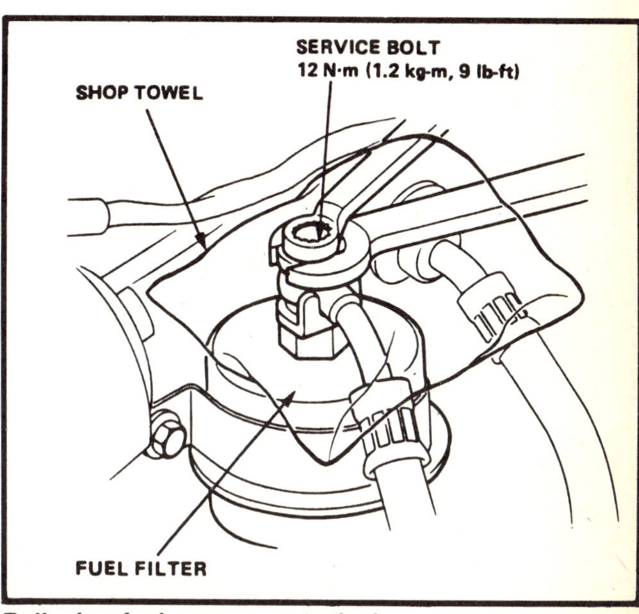

**Relieving fuel pressure—typical**

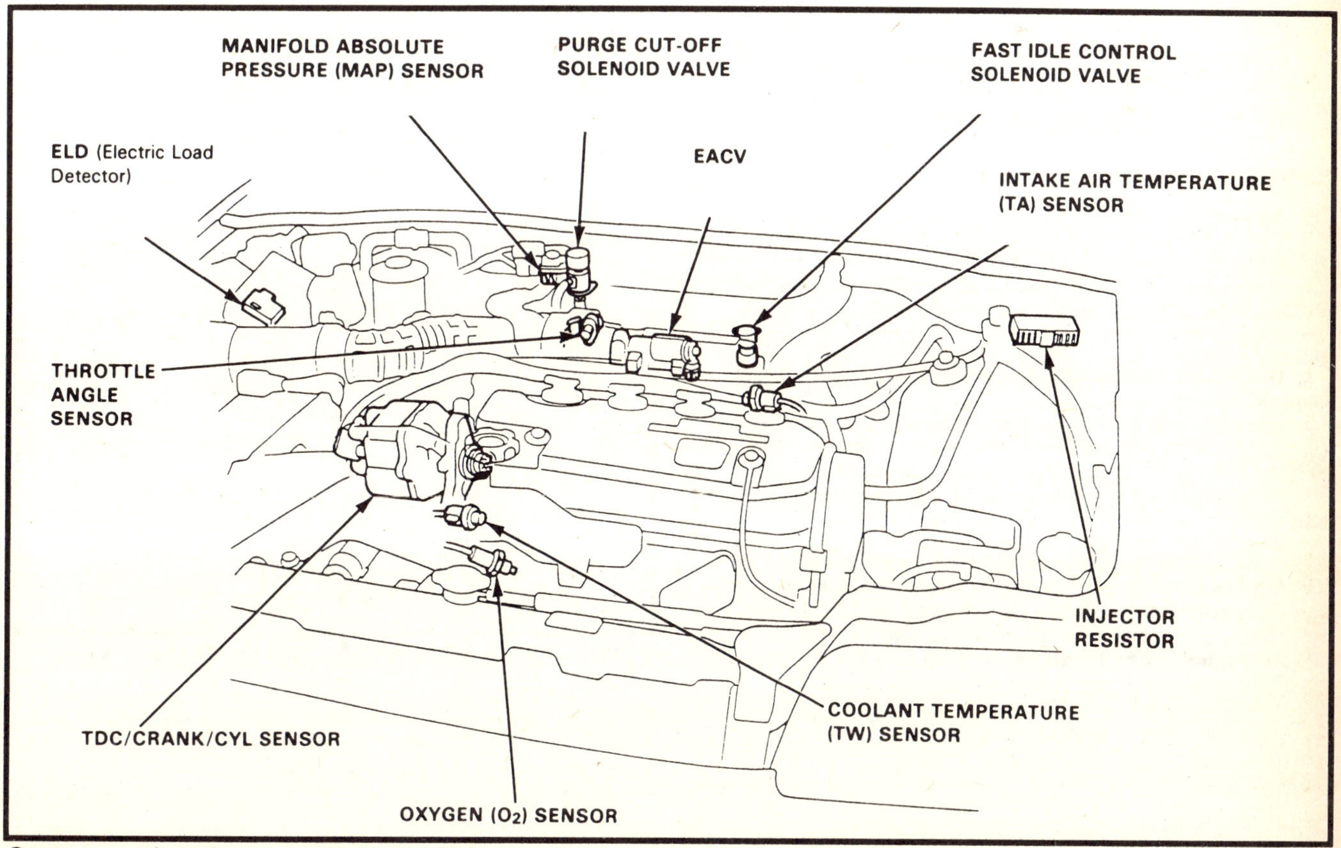

MANIFOLD ABSOLUTE
PRESSURE (MAP) SENSOR
PURGE CUT-OFF
SOLENOID VALVE
FAST IDLE CONTROL
SOLENOID VALVE
ELD (Electric Load
Detector)
EACV
INTAKE AIR TEMPERATURE
(TA) SENSOR
THROTTLE
ANGLE
SENSOR
INJECTOR
RESISTOR
TDC/CRANK/CYL SENSOR
COOLANT TEMPERATURE
(TW) SENSOR
OXYGEN (O2) SENSOR

**Component locations—typical 1.6L Civic**

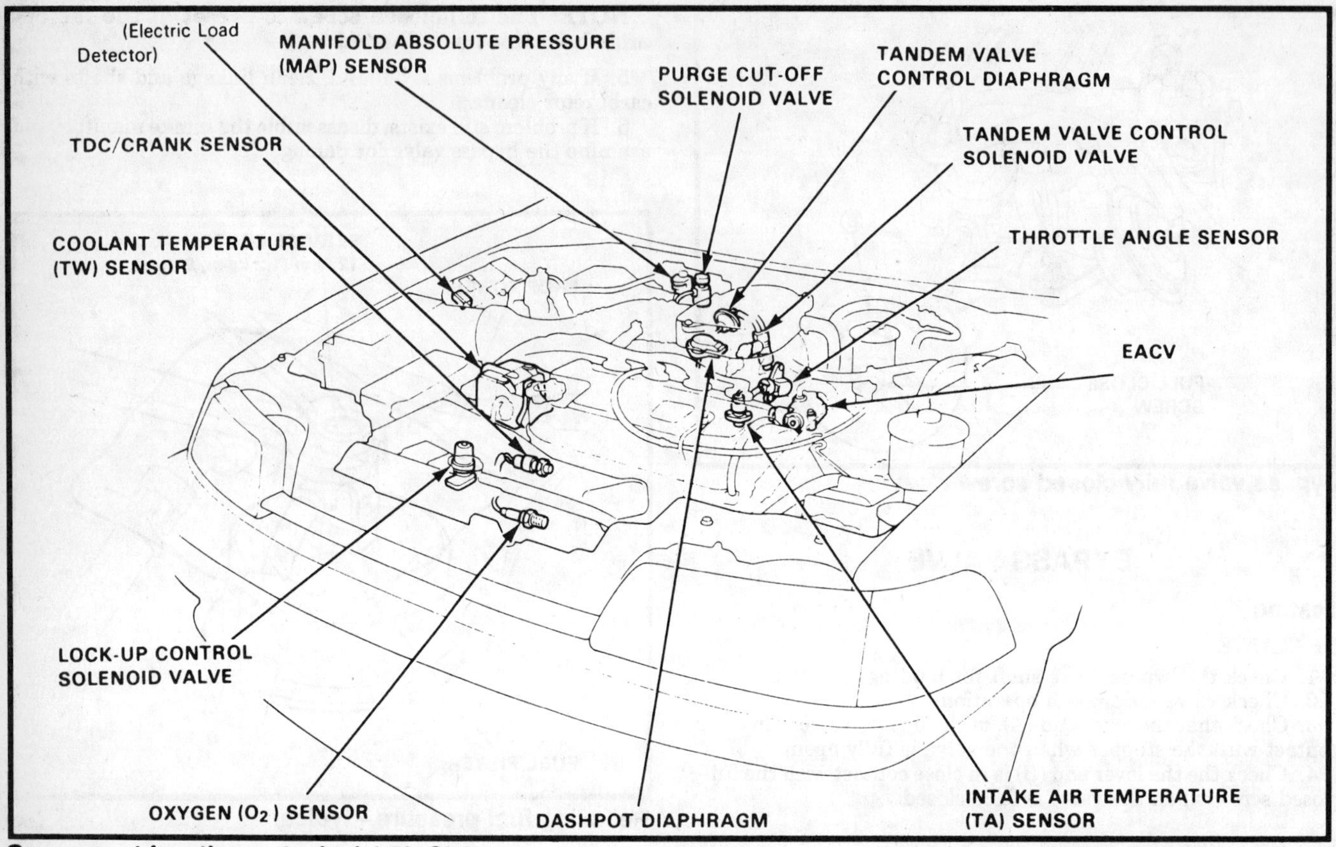

(Electric Load Detector)
MANIFOLD ABSOLUTE PRESSURE (MAP) SENSOR
PURGE CUT-OFF SOLENOID VALVE
TANDEM VALVE CONTROL DIAPHRAGM
TANDEM VALVE CONTROL SOLENOID VALVE
TDC/CRANK SENSOR
COOLANT TEMPERATURE (TW) SENSOR
THROTTLE ANGLE SENSOR
EACV
LOCK-UP CONTROL SOLENOID VALVE
OXYGEN ($O_2$) SENSOR
DASHPOT DIAPHRAGM
INTAKE AIR TEMPERATURE (TA) SENSOR

**Component locations—typical 1.5L Civic**

# Component Replacement

## FUEL SYSTEM

**Pressure Relieving**

**ALL MODELS**

**CAUTION**

***Keep open flames or sparks from work area. Do not smoke while working on fuel system. Be sure to relieve fuel pressure while engine is off.***

1. Disconnect battery negative cable from battery negative terminal.
2. Use a box end wrench on 6mm service bolt at top of fuel filter, while holding special Banjo bolt with another wrench.
3. Place a rag or a shop towel over 6mm service bolt.
4. Slowly loosen 6mm service bolt 1 complete turn. Pressurized fuel will be release from this connection.

**NOTE: A fuel pressure gauge can be attached at 6mm service bolt hole. Always replace washer between service bolt and special Banjo bolt, whenever service bolt is loosened to relieve fuel pressure. Replace all washers whenever bolts are removed to disassemble parts.**

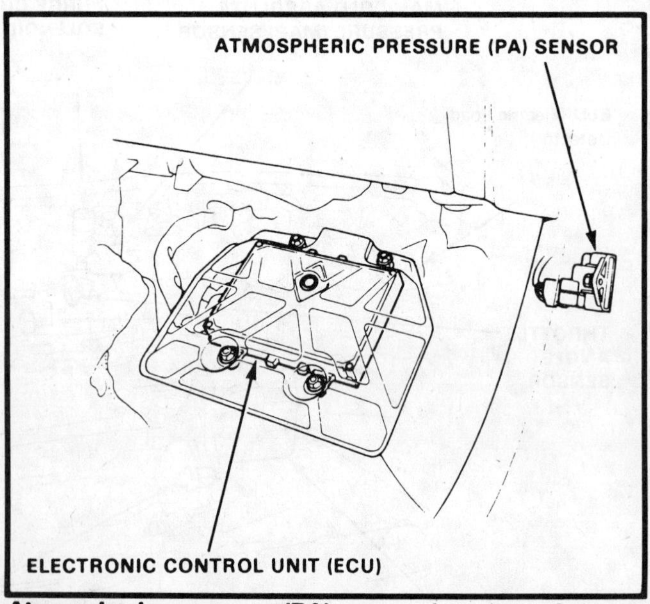

ATMOSPHERIC PRESSURE (PA) SENSOR
ELECTRONIC CONTROL UNIT (ECU)

**Atmospheric pressure (PA) sensor location—Civic shown**

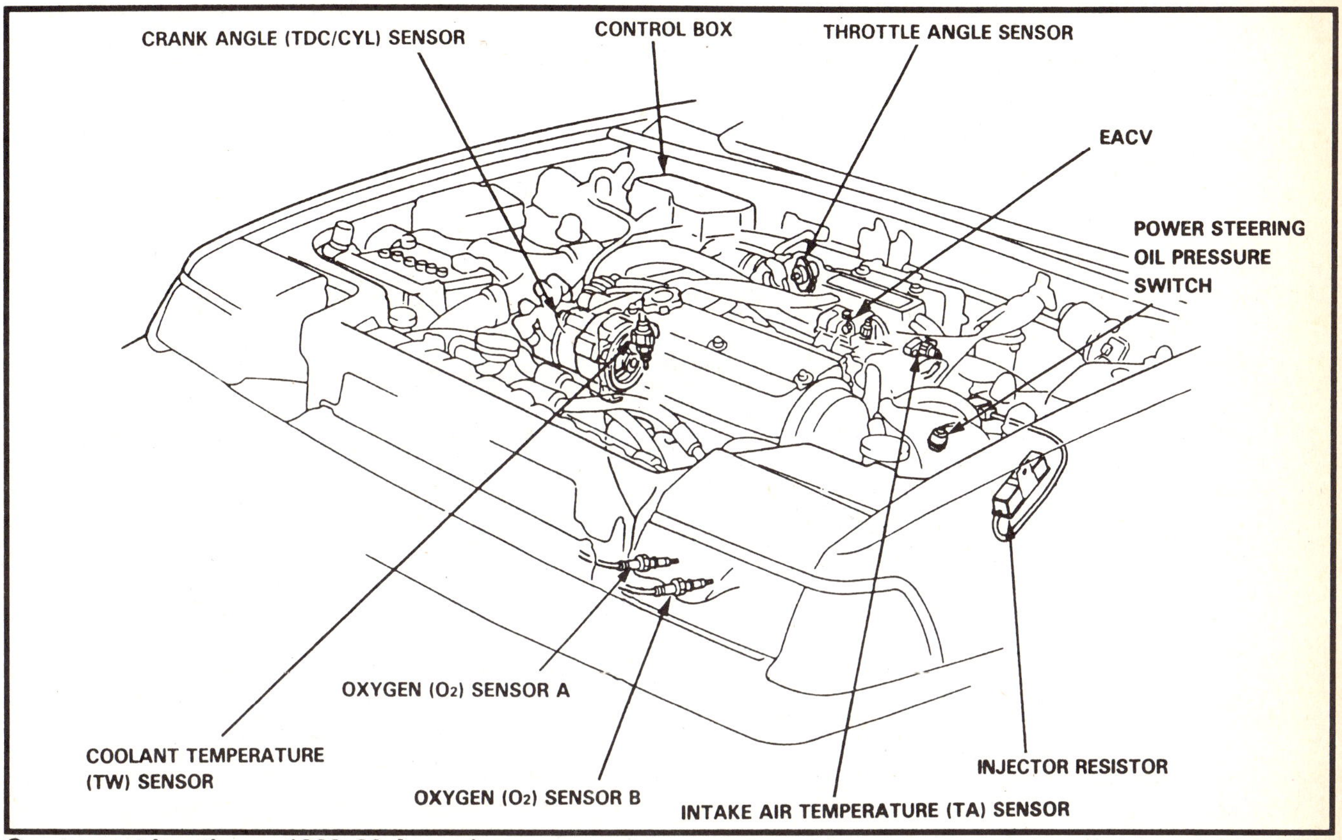

CRANK ANGLE (TDC/CYL) SENSOR
CONTROL BOX
THROTTLE ANGLE SENSOR
EACV
POWER STEERING OIL PRESSURE SWITCH
OXYGEN ($O_2$) SENSOR A
COOLANT TEMPERATURE (TW) SENSOR
OXYGEN ($O_2$) SENSOR B
INJECTOR RESISTOR
INTAKE AIR TEMPERATURE (TA) SENSOR

**Component locations – 1988–89 Accord**

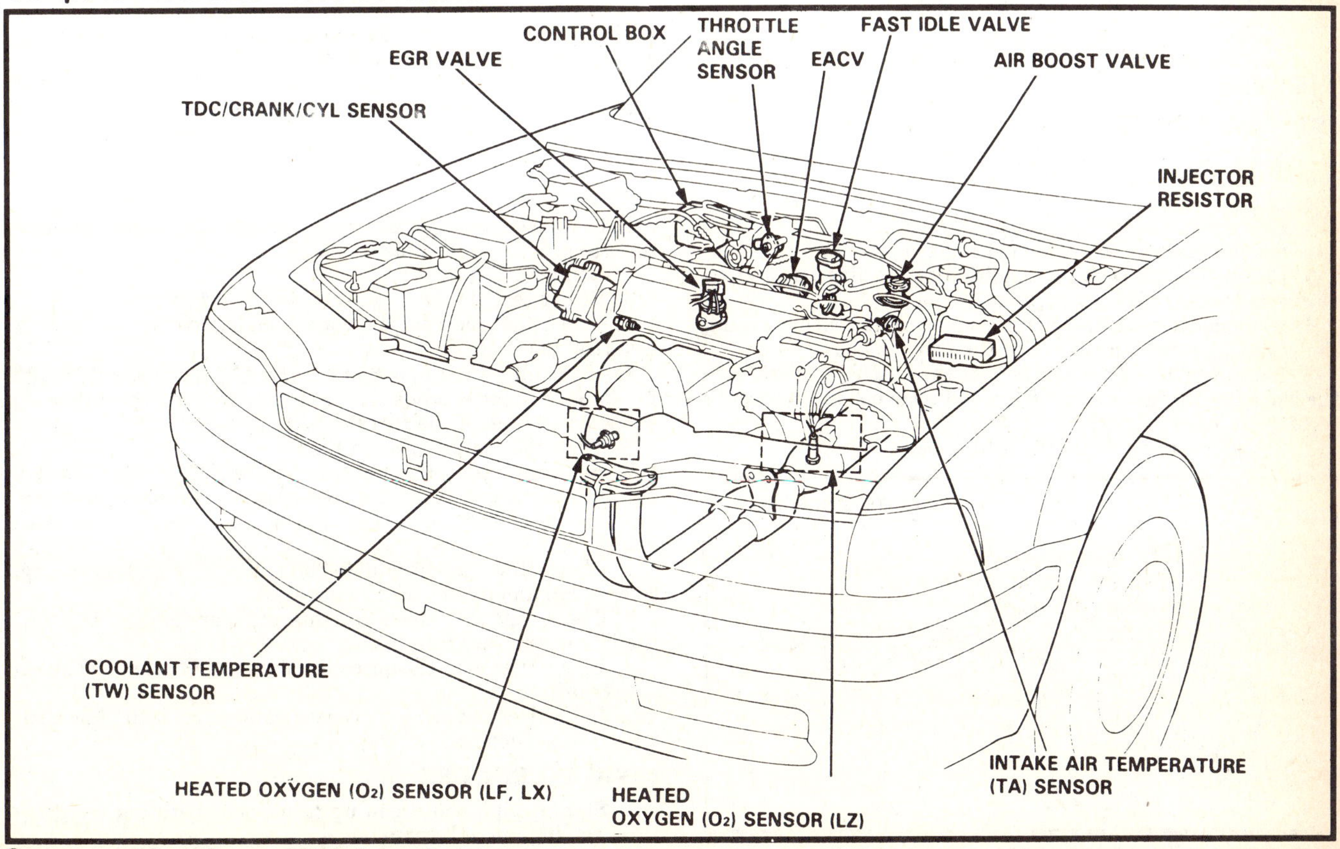

CONTROL BOX
THROTTLE ANGLE SENSOR
FAST IDLE VALVE
EGR VALVE
EACV
AIR BOOST VALVE
TDC/CRANK/CYL SENSOR
INJECTOR RESISTOR
COOLANT TEMPERATURE (TW) SENSOR
HEATED OXYGEN ($O_2$) SENSOR (LF, LX)
HEATED OXYGEN ($O_2$) SENSOR (LZ)
INTAKE AIR TEMPERATURE (TA) SENSOR

**Component locations – 1990 Accord**

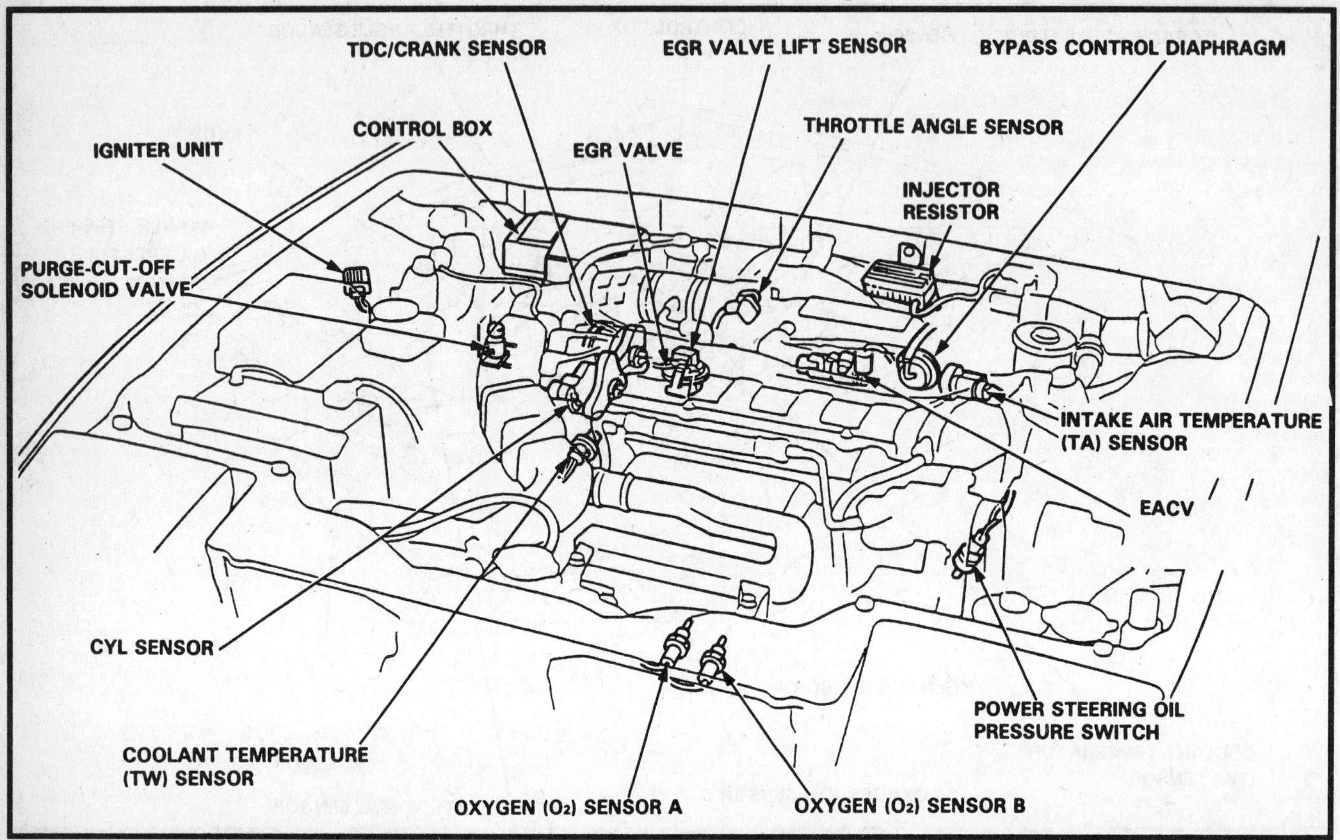

TDC/CRANK SENSOR
EGR VALVE LIFT SENSOR
BYPASS CONTROL DIAPHRAGM
CONTROL BOX
THROTTLE ANGLE SENSOR
IGNITER UNIT
EGR VALVE
INJECTOR RESISTOR
PURGE-CUT-OFF SOLENOID VALVE
INTAKE AIR TEMPERATURE (TA) SENSOR
EACV
CYL SENSOR
POWER STEERING OIL PRESSURE SWITCH
COOLANT TEMPERATURE (TW) SENSOR
OXYGEN ($O_2$) SENSOR A
OXYGEN ($O_2$) SENSOR B

**Component locations—typical Prelude**

## IDLE SPEED

### Adjustment

**1988–89 ACCORD**

1. Start engine and warm up to normal operating temperature; cooling fan will come on twice.
2. Connect tachometer.
3. Disconnect the 2P connector on the EACV.
4. Set the steering in the straight ahead position.
5. Check the idle with air conditioning, cooling fan and all electrical loads **OFF**.
6. If the idle is not 600–700 rpm, adjust as needed by turning the adjusting screw on the top of the throttle body.
7. Reconnect the 2P connector on the EACV, then remove the No. 11 (10A) fuse in the underhood relay box for at least 10 seconds, to reset the ECU.
8. Set the steering in the straight ahead position.

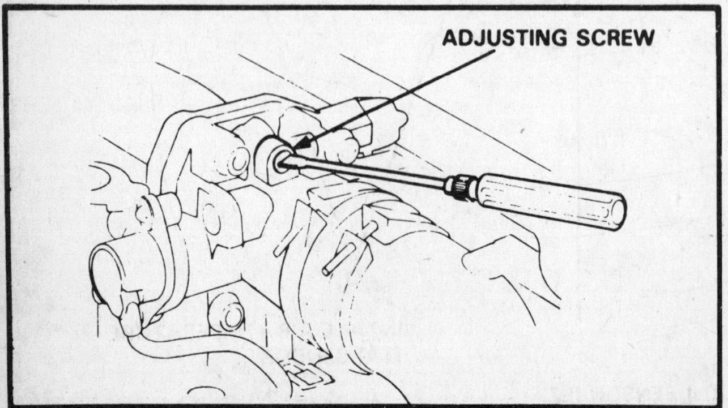


**Idle speed adjustment—Accord type shown**

9. Check that the idle is 700–800 rpm with air conditioning, cooling fan and all electrical loads **OFF**.
10. Check that with the headlights and rear defogger **ON** and while turning the steering wheel the idle remains constant.
11. If idle does not remain in specifications, refer to "Idle Control Troubleshooting" chart.

**1990 ACCORD**

1. Start engine and warm up to normal operating temperature; cooling fan will come on twice.
2. Connect tachometer.
3. Disconnect the 2P connector on the EACV.
4. Check the idle with air conditioning, cooling fan and all electrical loads **OFF**.
5. If the idle is not 550–650 rpm for U.S models or 500–600 for Canada models, adjust as needed by turning the adjusting screw on the top of the throttle body.
6. Turn the ignition switch **OFF**.
7. Reconnect the 2P connector on the EACV, then remove the back-up fuse in the underhood relay box for at least 10 seconds, to reset the ECU. This is the last Step for Canadian vehicles.
8. Check that the idle is 650–750 rpm with air conditioning, cooling fan and all electrical loads **OFF**.
9. Check that with the headlights and rear defogger **ON** that the idle is 720–820 rpm.
10. Check that with the air conditioning on **HI** that the idle is 720–820 rpm.
11. If idle does not remain in specifications, refer to "Idle Control Troubleshooting" chart.

**CIVIC EXCEPT CRX**

1. Start engine and warm up to normal operating temperature; cooling fan will come on.
2. Connect tachometer.

3. Disconnect the 2P connector on the EACV.
4. Check the idle with air conditioning, cooling fan and all electrical loads **OFF**.
5. The idle should be:
1988–90 1.5L – 575–675 rpm
1988–90 1.6L – 500–600 rpm
6. If the rpm is not within specifications, adjust as needed by turning the idle adjusting screw. If the idle speed is excessively high, check the throttle control diaphragm system.
7. Reconnect the 2P connector on the EACV, then remove the hazard fuse in the underhood relay box for at least 10 seconds, to reset the ECU.
8. With the air conditioning, cooling fan and all electrical loads **OFF** the idle should be:
1988 1.5L – 675–775 rpm
1989 1.5L – 700–800 rpm
1990 1.5L (except Canada) – 700–800 rpm
1990 1.5L (Canada) – 750–850 rpm
1988–89 1.6L – 700–800 rpm
1990 1.6L (except Canadian wagon) – 700–800 rpm
1990 1.6L (Canadian wagon) – 730–830 rpm
9. With the headlights and rear defogger **ON** the idle should be:
1988–89 1.5L – 750–850 rpm
1990 1.5L (except Canada and wagon) – 760–860 rpm
1990 1.5L (Canada and wagon) – 750–850 rpm
1989–90 1.6L – 730–830 rpm
10. With the air conditioner and blower on **HI** the idle should be:
1988 1.5L – 730–830 rpm
1989 1.5L – 750–850 rpm
1990 1.5L (except Canada and wagon) – 760–860 rpm
1990 1.5L (Canada and wagon) – 750–850 rpm
1989 1.6L (except wagon with A/T) – 750–850 rpm
1989 1.6L (wagon with A/T) – 770–870 rpm
1990 1.6L (except A/T 4WD) – 760–860 rpm
1990 1.6L (A/T 4WD) – 780–880 rpm
11. If idle speed is out of specification and LED does not blinking a Code 14, check air conditioning, alternator, transmission shift, brake, starter and power steering signals.
12. Also check EACV O-rings, fast idle control and all hose and wire connections.
13. If idle speed is still out of specification and LED does not blink a Code 14, replace the EACV and readjust the idle.
14. If idle speed is still out of specification and LED still does not blink a Code 14, replace the ECU and readjust the idle.

## CIVIC CRX

1. Start engine and warm up to normal operating temperature; cooling fan will come on.
2. Connect tachometer.
3. Disconnect the 2P connector on the EACV.
4. Check the idle with air conditioning, cooling fan and all electrical loads **OFF**.
5. The idle should be:
1988–90 1.5L Std – 575–675 rpm
1988–90 1.5L HF – 450–550 rpm
1988–90 1.5L Si – 500–600 rpm
6. If the rpm is not within specifications, adjust as needed by turning the idle adjusting screw. If the idle speed is excessively high, check the throttle control diaphragm system.
7. Reconnect the 2P connector on the EACV, then remove the Hazard fuse in the underhood relay box for at least 10 seconds, to reset the ECU.
8. With the air conditioning, cooling fan and all electrical loads **OFF** the idle should be:
1988 1.5L Std – 675–775 rpm
1988 1.5L HF (except Calif.) – 550–650 rpm
1988 1.5L HF (Calif.) – 600–700 rpm
1989 1.5L Std – 700–800 rpm
1989–90 1.5L HF (except Calif. and Hi. Alt.) – 550–650 rpm
1989–90 1.5L HF (Calif. and Hi. Alt.) – 600–700 rpm
1990 1.5L Std (except Canada) – 700–800 rpm
1990 1.5L Std (Canada) – 750–850 rpm
1988–90 1.6L Si – 700–800 rpm
9. With the headlights and rear defogger **ON** the idle should be:
1988 1.5L Std – 730–830 rpm
1988 1.5L HF – 700–800 rpm
1989–90 1.5L Std – 750–850 rpm
1989–90 1.5L HF (except Calif. and Hi. Alt.) – 550–650 rpm
1989–90 1.5L HF (Calif. and Hi. Alt.) – 700–800 rpm
1988 1.6L Si – 700–800 rpm
1989–90 1.6L Si – 730–830 rpm
10. With the air conditioner and blower on **HI** the idle should be:
1988 1.5L Std – 730–830 rpm
1989–90 1.5L Std – 750–850 rpm
1988–89 1.5L HF – 700–800 rpm
1990 1.5L HF (except Canada) – 700–800 rpm
1990 1.5L HF (Canada) – 760–860 rpm
1988 1.6L Si – 730–830 rpm
1989 1.6L Si – 750–850 rpm
1990 1.6L Si – 760–860 rpm
11. If idle speed is out of specification and LED does not blinking a Code 14, check air conditioning, alternator, transmission shift, brake, starter and power steering signals.
12. Also check EACV O-rings, fast idle control and all hose and wire connections.
13. If idle speed is still out of specification and LED does not blink a Code 14, replace the EACV and readjust the idle.
14. If idle speed is still out of specification and LED still does not blink a Code 14, replace the ECU and readjust the idle.

## PRELUDE

1. Start engine and warm up to normal operating temperature; cooling fan will come on twice.
2. Connect tachometer.
3. Disconnect the 2P connector on the EACV.
4. Set the steering in the straight ahead position.

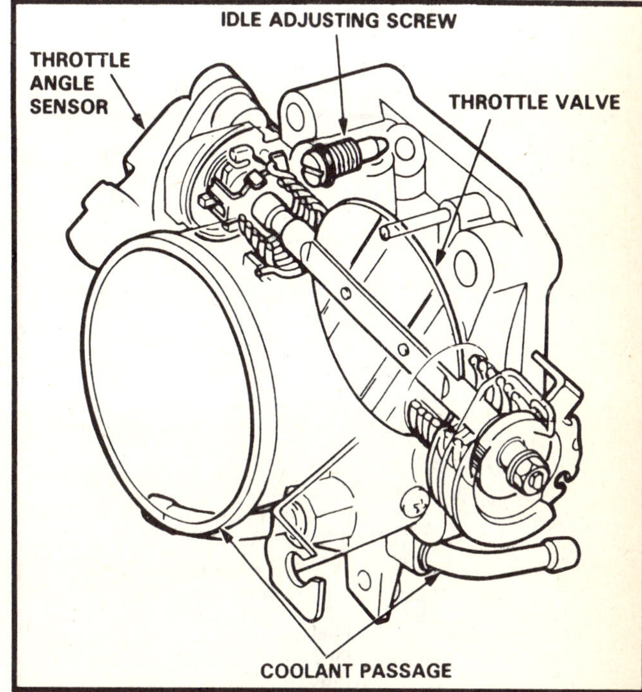

**Throttle body – 1989 Prelude shown**

5. Check the idle with air conditioning, cooling fan and all electrical loads **OFF**.
6. If the idle is not 600–700 rpm, adjust as needed by turning the adjusting screw on the top of the throttle body.
7. Turn the ignition switch **OFF**.
8. Reconnect the 2P connector on the EACV, then remove the Clock (10A) fuse in the underhood relay box for at least 10 seconds, to reset the ECU.
9. Set the steering in the straight ahead position.
10. Check that the idle is 700–800 rpm for manual transmission or 740–840 rpm for an automatic transmission, with air conditioning, cooling fan and all electrical loads **OFF**.
11. Check that with the headlights and rear defogger **ON** and while turning the steering wheel the idle remains constant at 700–800 rpm.
12. Check that with the air conditioning and blower on **HI** the idle remains constant at 700–800 rpm.
13. If idle does not remain in specifications, refer to "Idle Control Troubleshooting" chart.

## FUEL INJECTOR

### Removal and Installation

#### PORT INJECTION MODELS

1. Disconnect battery negative cable from battery negative.
2. Relieve fuel pressure.
3. Disconnect connectors from the injectors.
4. Disconnect vacuum hose and fuel return hose from pressure regulator.

**NOTE: Place a rag or shop towel over hose and tube before disconnecting.**

5. Loosen retainer nuts on fuel pipe.
6. On Prelude models, disconnect fuel pipe. On all models, remove the EACV from the intake manifold.
7. Remove injector from intake manifold.

**To install:**

8. Slide new cushion onto injector.
9. Coat new O-rings with clean engine oil and put O-rings on injectors.
10. Insert injectors into fuel pipe first.

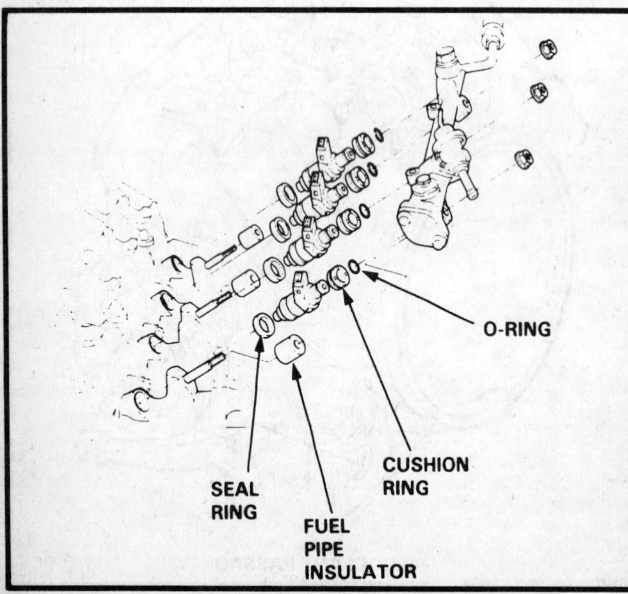


**Port fuel injectors—1990 Accord shown**

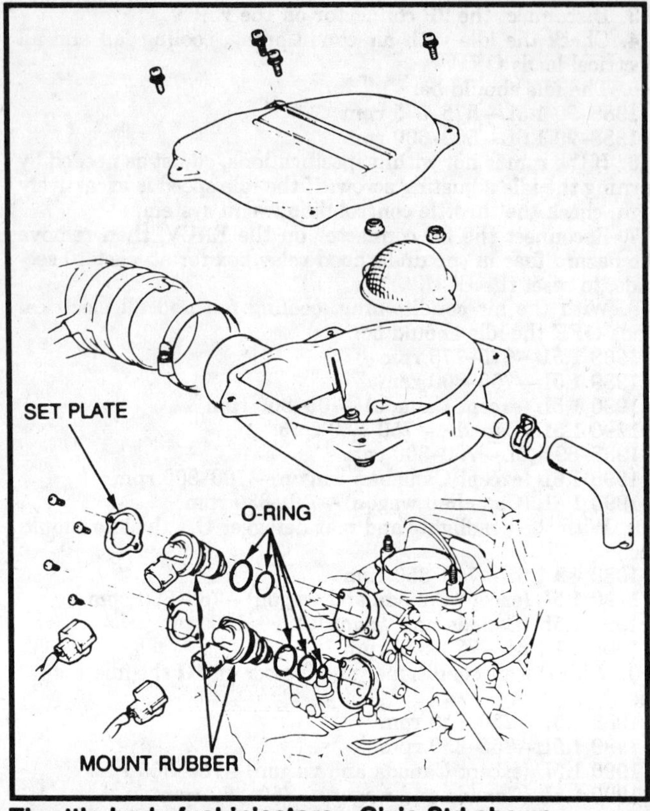


**Throttle body fuel injectors—Civic Std shown**

**NOTE: To prevent damage to O-ring, insert injector into fuel pipe squarely and carefully, then install them in intake manifold.**

11. Coat new seal rings with clean engine oil and press into intake manifold.
12. Install injector and fuel pipe assembly in manifold.

**NOTE: To prevent damage to O-ring, install injectors in fuel pipe first, then install in intake manifold.**

13. Align center line on coupler with mark on fuel pipe.
14. Install and tighten retainer nuts, install the EACV if removed.
15. Connect vacuum hose and fuel return hose to pressure regulator.
16. Install connector to the injectors.
17. Turn ignition switch **ON** but do not operate starter. After fuel pump runs for approximately 2 seconds, fuel pressure in fuel line rises. Repeat this 3 times, then check whether there is any fuel leakage.

#### THROTTLE BODY INJECTION MODELS

1. Disconnect battery negative cable from battery negative.
2. Relieve fuel pressure.
3. Remove the air intake chamber.
4. Disconnect the 2P connector from the injector.
5. Loosen the screws and remove the injector from the throttle body.
6. Cover the throttle body opening with a clean rag.

**To install:**

7. Coat new O-rings with clean engine oil and assemble them.
8. Insert the injector into the throttle body.

**NOTE: After installing the injector, make certain it can smoothly turn about 30 degrees.**

9. Turn ignition switch **ON** but do not operate starter. After

fuel pump runs for approximately 2 seconds, fuel pressure in fuel line rises. Repeat this 3 times, then check whether there is any fuel leakage.

## FUEL PRESSURE REGULATOR

### Removal and Installation

**ALL MODELS**

1. Disconnect the negative battery cable.
2. Relieve fuel pressure.
3. Remove the air intake chamber, if necessary to gain access.
4. Disconnect the vacuum hose and fuel hose.
5. Remove the 6mm retaining bolts.

**To install:**

6. Install a new O-ring.
7. Apply clean engine oil to the O-ring and mating surface of fuel pressure regulator.
8. Install the regulator.
9. Connect the vacuum and fuel hoses.
10. Install the air chamber, if removed.
11. Connect the negative battery cable.
12. Turn ignition switch **ON** but do not operate starter. After fuel pump runs for approximately 2 seconds, fuel pressure in fuel line rises. Repeat this 3 times, then check whether there is any fuel leakage.

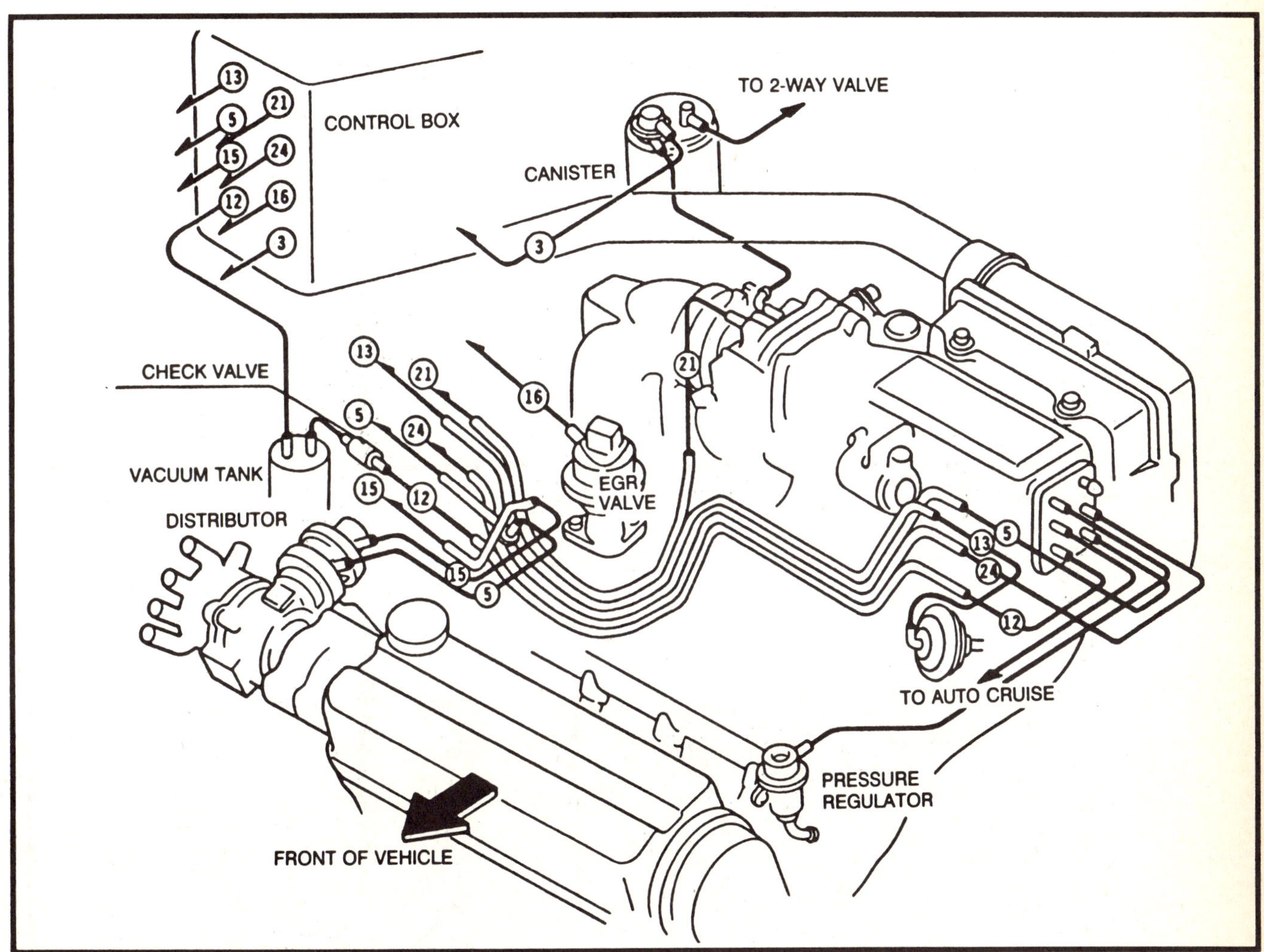


**Vacuum identification diagram—1988–89 Accord**

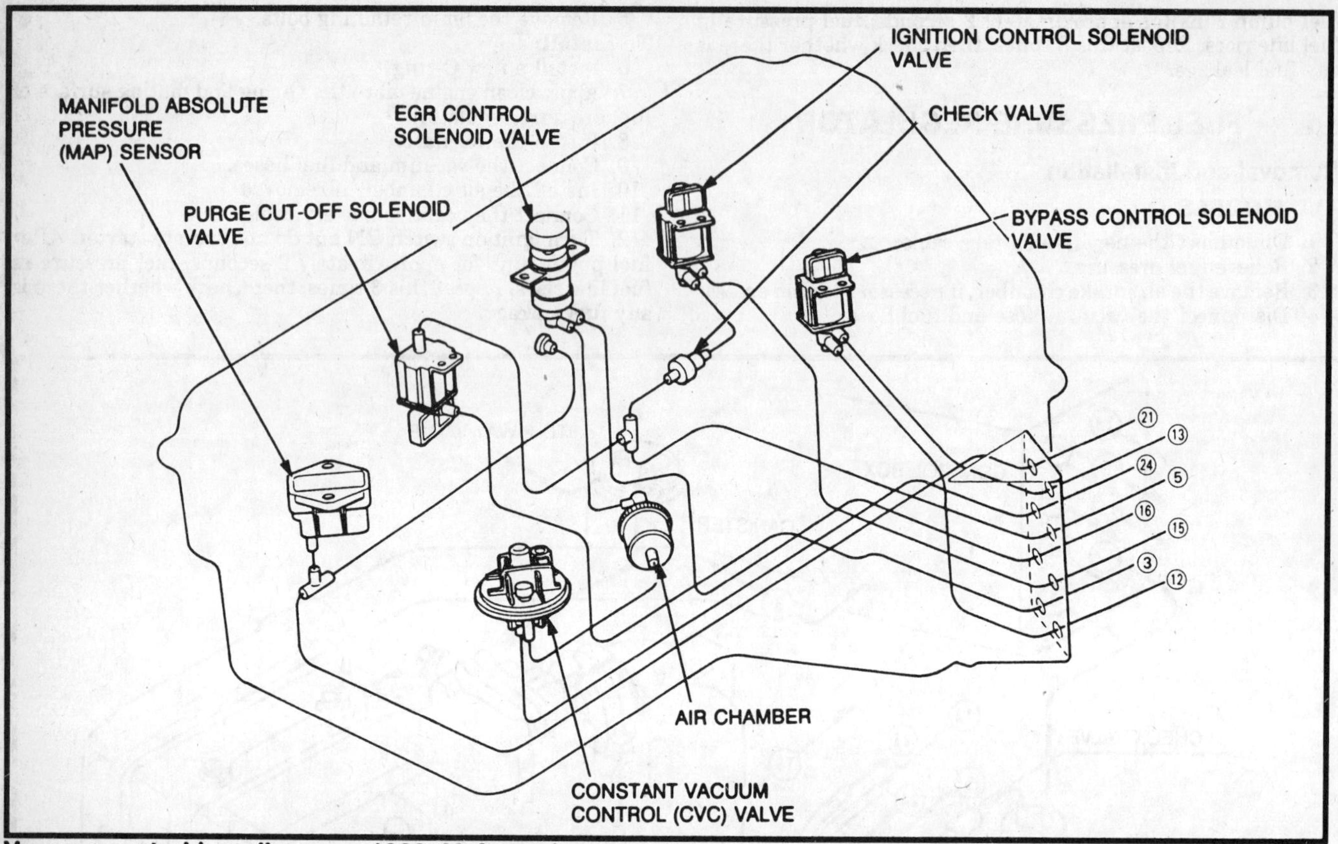

IGNITION CONTROL SOLENOID VALVE
MANIFOLD ABSOLUTE PRESSURE (MAP) SENSOR
EGR CONTROL SOLENOID VALVE
CHECK VALVE
PURGE CUT-OFF SOLENOID VALVE
BYPASS CONTROL SOLENOID VALVE
21
13
24
5
16
15
3
12
AIR CHAMBER
CONSTANT VACUUM CONTROL (CVC) VALVE

**Vacuum control box diagram – 1988–89 Accord**

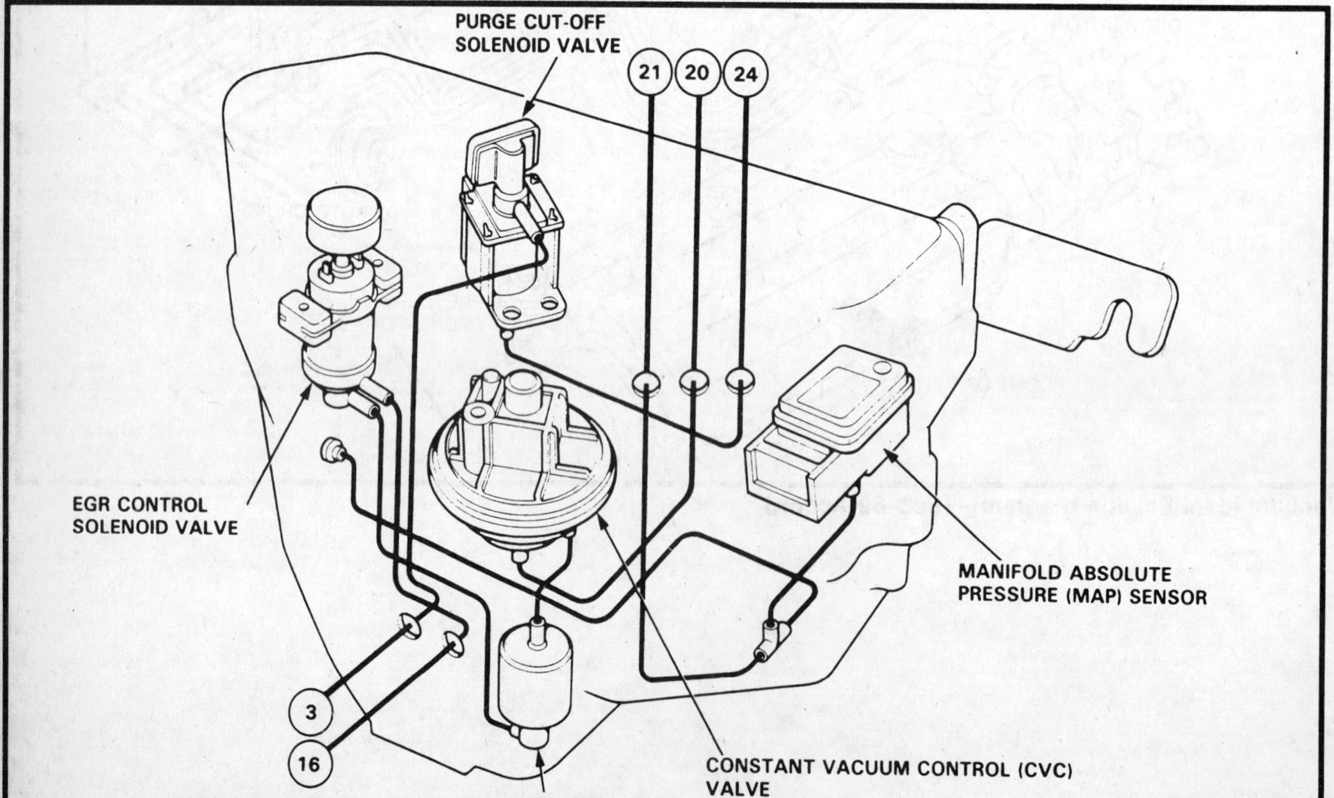

PURGE CUT-OFF SOLENOID VALVE
21
20
24
EGR CONTROL SOLENOID VALVE
MANIFOLD ABSOLUTE PRESSURE (MAP) SENSOR
3
16
CONSTANT VACUUM CONTROL (CVC) VALVE
AIR CHAMBER

**Vacuum control box diagram – 1990 Accord**

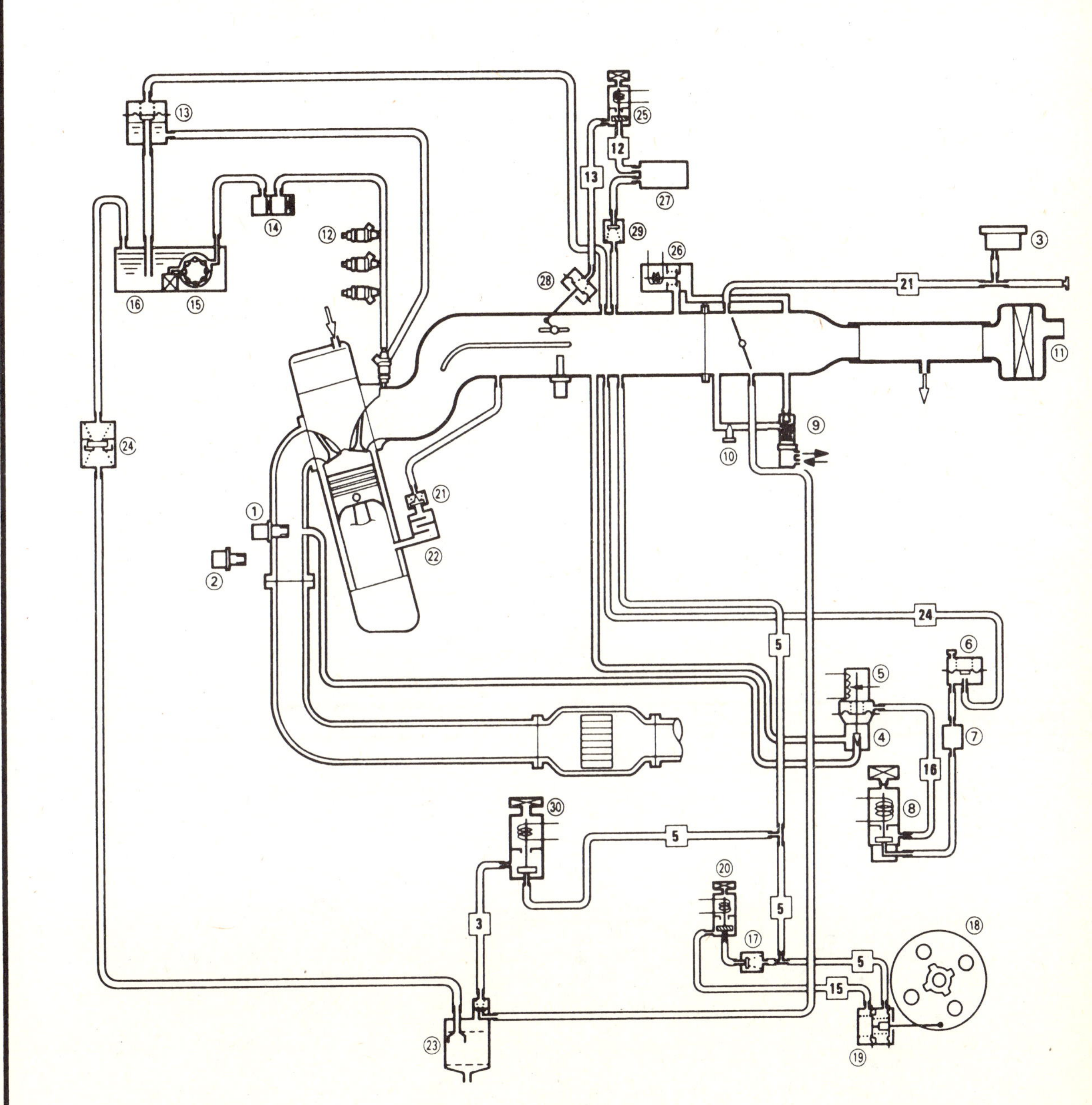

1. Oxygen sensor A
2. Oxygen sensor B
3. Manifold absolute pressure (MAP) sensor
4. EGR valve
5. EGR valve lift sensor
6. Constant Vacuum Control (CVC) valve
7. Air chamber
8. EGR control solenoid valve
9. Fast idle valve
10. Idle air adjusting screw
11. Air cleaner
12. Fuel injector
13. Pressure regulator
14. Fuel filter
15. Fuel pump
16. Fuel tank
17. Check valve
18. Distributor
19. Vacuum advance diaphragm
20. Ignition control solenoid valve
21. PCV valve
22. Breather chamber
23. Charcoal canister
24. 2-way valve
25. Bypass control solenoid valve
26. Electronic Air Control Valve (EACV)
27. Vacuum tank
28. Bypass control diaphragm
29. Check valve
30. Purge cut-off solenoid valve

**Vacuum circuit diagram – 1988–89 Accord**

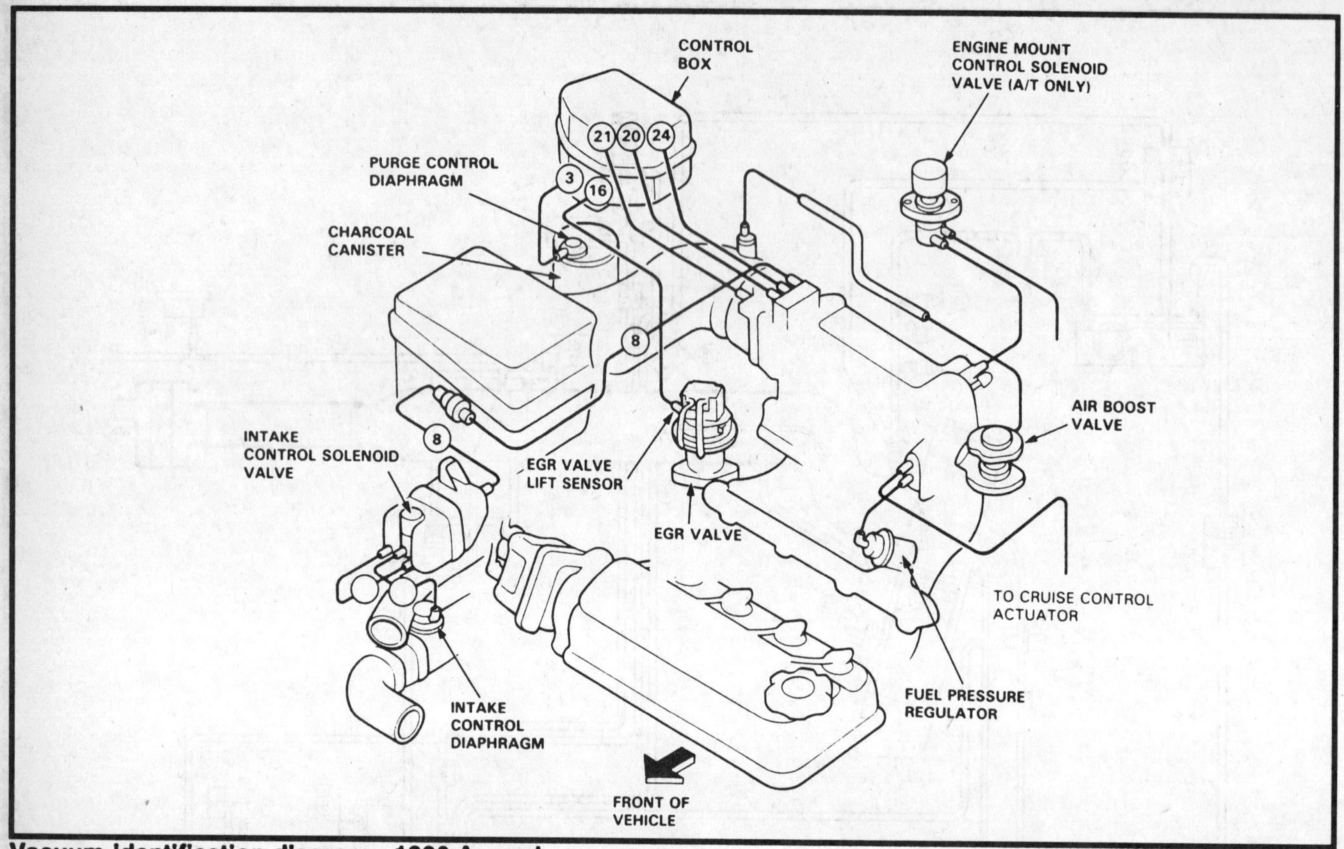

CONTROL BOX
ENGINE MOUNT CONTROL SOLENOID VALVE (A/T ONLY)
PURGE CONTROL DIAPHRAGM
CHARCOAL CANISTER
21 20 24
3 16
8
8
AIR BOOST VALVE
INTAKE CONTROL SOLENOID VALVE
EGR VALVE LIFT SENSOR
EGR VALVE
TO CRUISE CONTROL ACTUATOR
FUEL PRESSURE REGULATOR
INTAKE CONTROL DIAPHRAGM
FRONT OF VEHICLE

**Vacuum identification diagram – 1990 Accord**

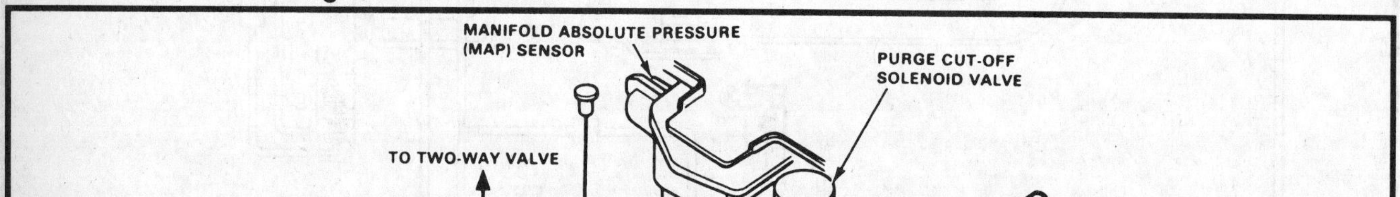

MANIFOLD ABSOLUTE PRESSURE (MAP) SENSOR
PURGE CUT-OFF SOLENOID VALVE
TO TWO-WAY VALVE
7
17
12
6
21
CHARCOAL CANISTER
FRONT OF VEHICLE

**Vacuum identification diagram – 1988 Civic 1.5L and 1989–90 1.5L Civic except Calif. auto trans and CRX HF**

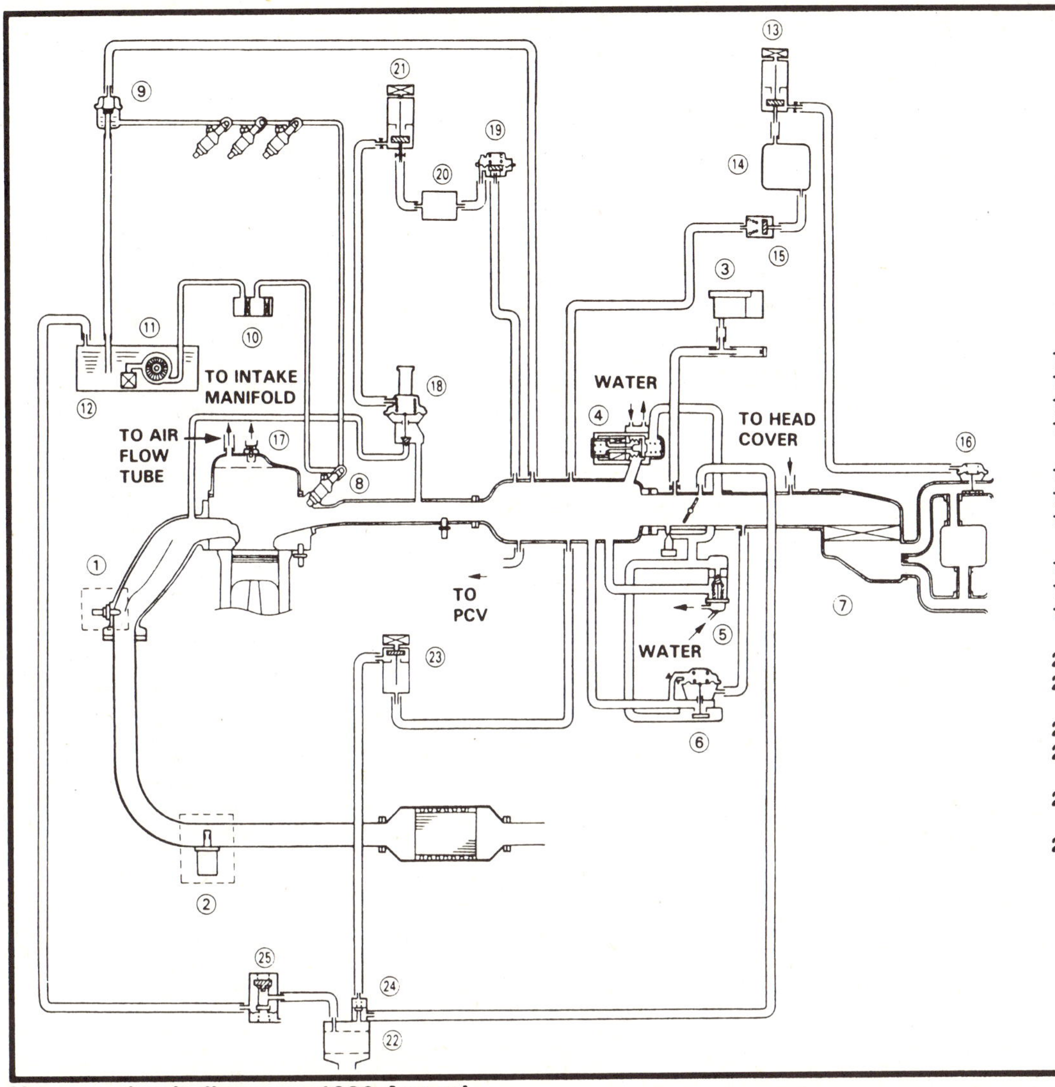

1. Oxygen sensor (DX, LX)
2. Oxygen sensor (EX)
3. Manifold Absolute Pressure (MAP) sensor
4. Electronic Air Control Valve (EACV)
5. Fast idle valve.
6. Air boost valve
7. Air cleaner
8. Fuel injector
9. Pressure regulator
10. Fuel filter
11. Fuel pump
12. Fuel tank
13. Intake control solenoid valve
14. Air chamber
15. Check valve
16. Intake control diaphragm
17. PCV valve
18. EGR valve
19. Constant Vacuum Control (CVC) valve
20. Air chamber
21. EGR control solenoid valve
22. Charcoal canister
23. Purge cut-off solenoid valve
24. Purge control diaphragm valve
25. 2-way valve

**Vacuum circuit diagram—1990 Accord**

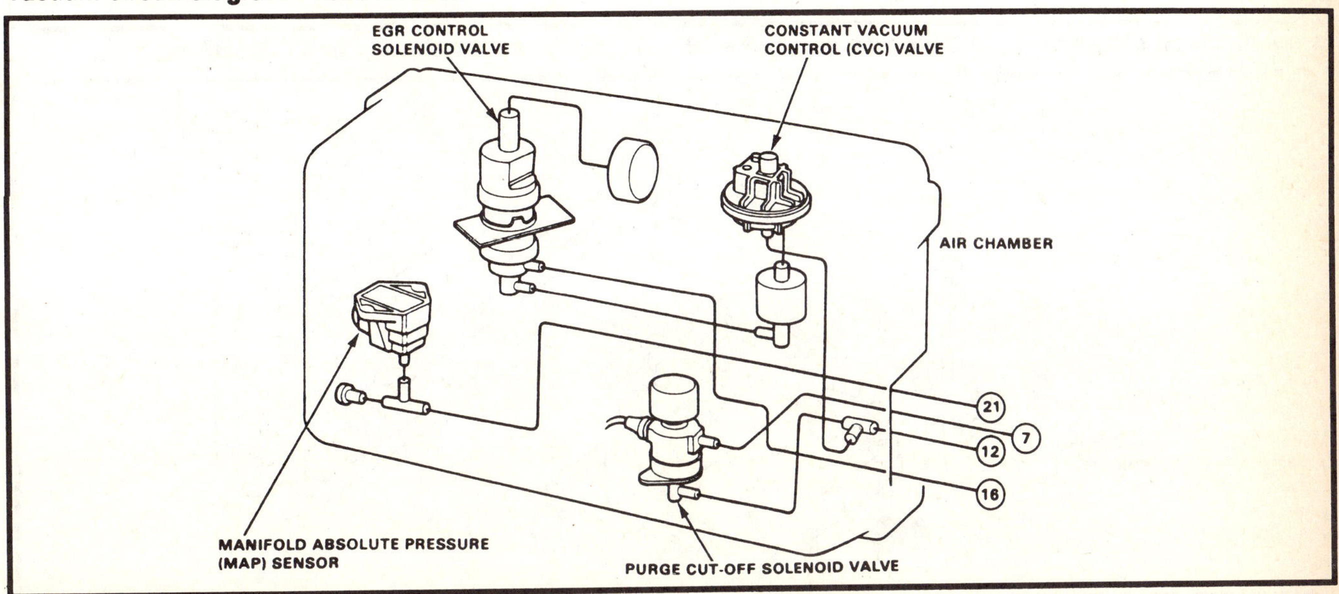


**Vacuum control box diagram1989–90 1.5L Civic Calif. A/T except CRX HF**

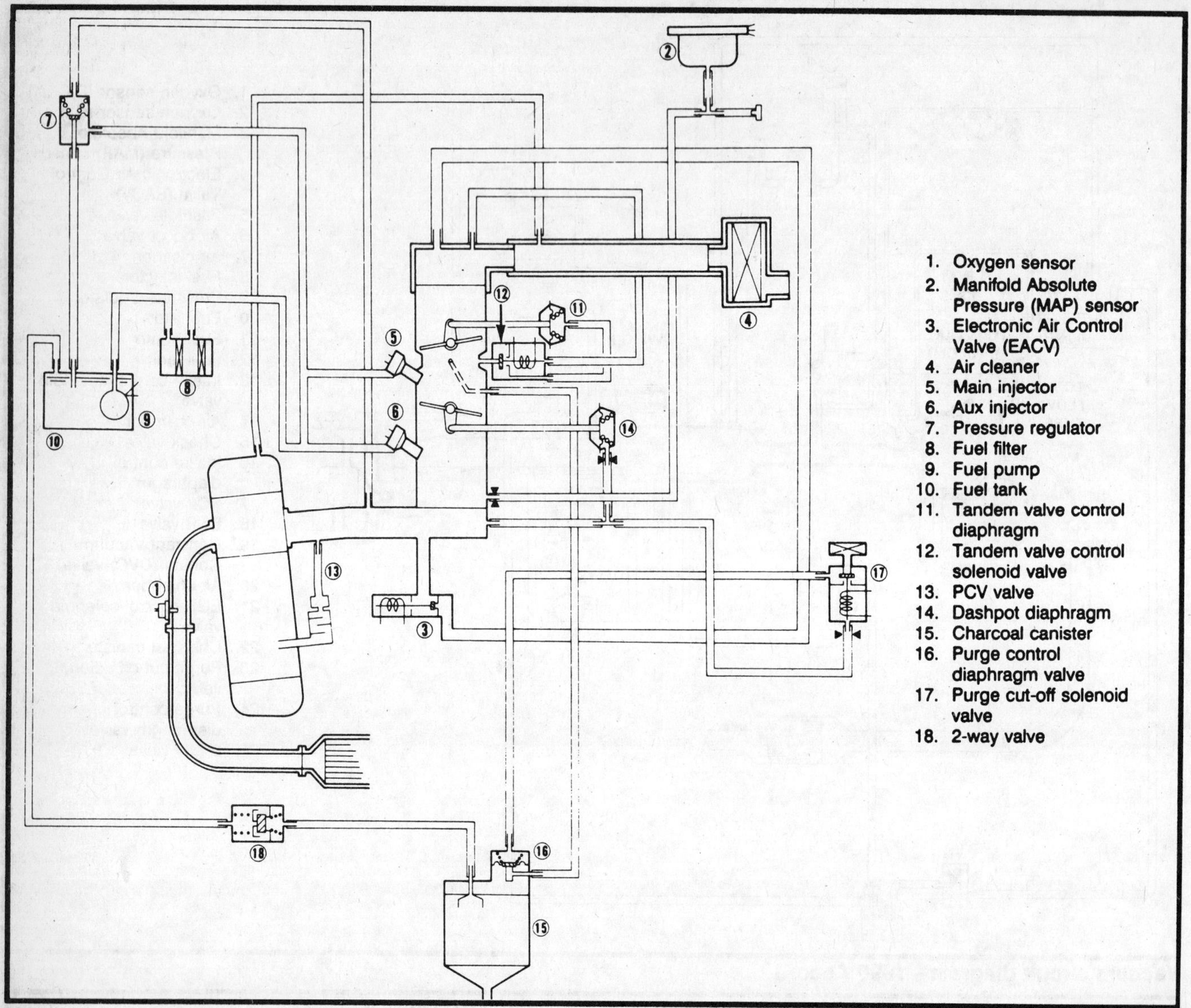

1. Oxygen sensor
2. Manifold Absolute Pressure (MAP) sensor
3. Electronic Air Control Valve (EACV)
4. Air cleaner
5. Main injector
6. Aux injector
7. Pressure regulator
8. Fuel filter
9. Fuel pump
10. Fuel tank
11. Tandem valve control diaphragm
12. Tandem valve control solenoid valve
13. PCV valve
14. Dashpot diaphragm
15. Charcoal canister
16. Purge control diaphragm valve
17. Purge cut-off solenoid valve
18. 2-way valve

**Vacuum circuit diagram – 1988 Civic 1.5L and 1989–90 1.5L Civic except Calif. A/T and CRX HF**

CONTROL BOX
TO TWO-WAY VALVE
CHARCOAL CANISTER
FRONT OF VEHICLE
16
6
21
7
12
10

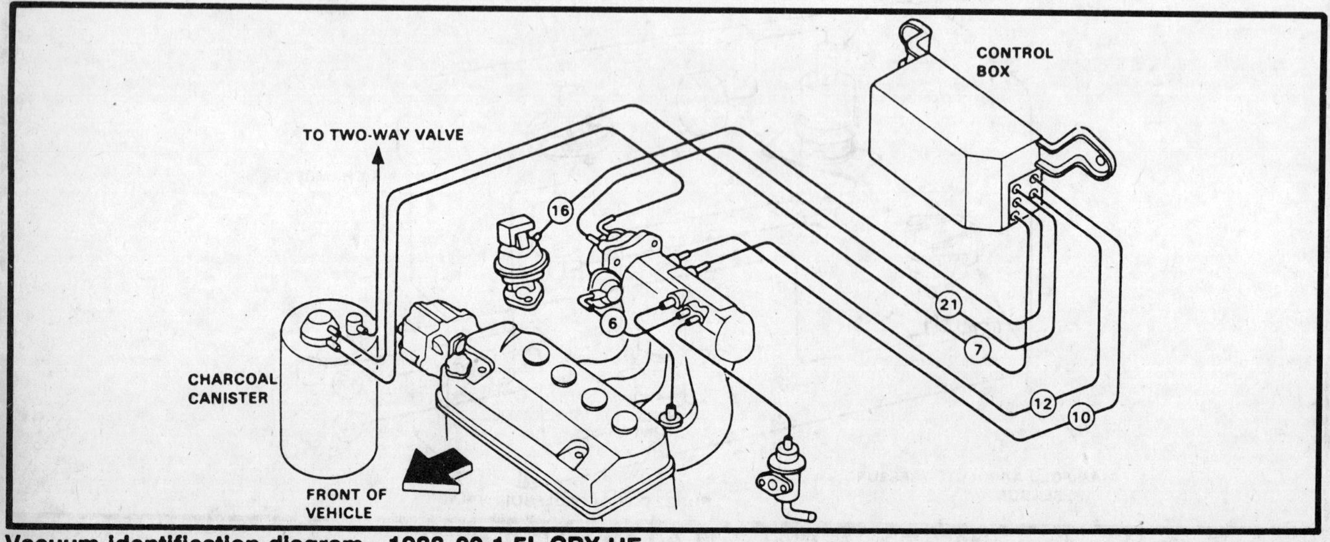

**Vacuum identification diagram – 1988–90 1.5L CRX HF**

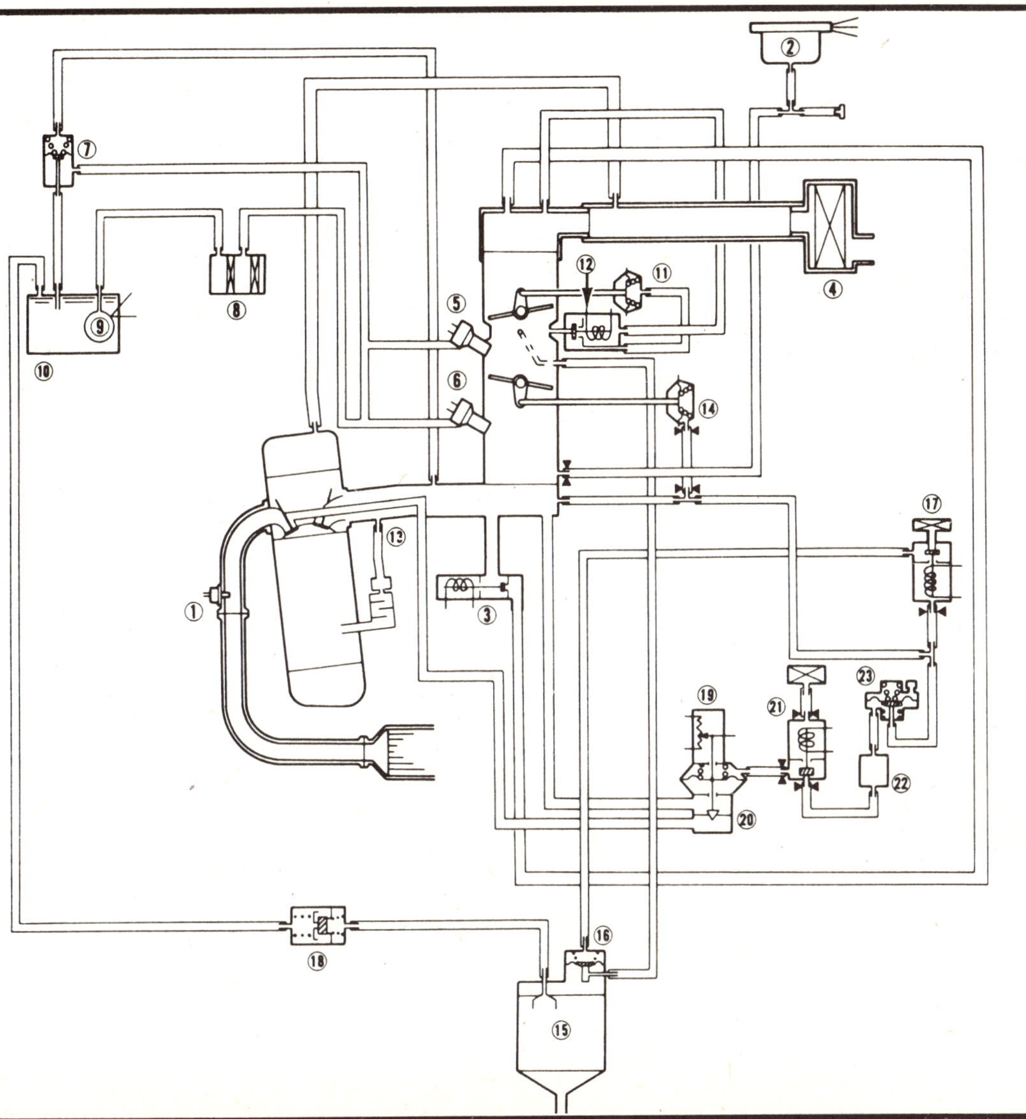

1. Oxygen sensor
2. Manifold Absolute Pressure (MAP) sensor
3. Electronic Air Control Valve (EACV)
4. Air cleaner
5. Main injector
6. Aux injector
7. Pressure regulator
8. Fuel filter
9. Fuel pump
10. Fuel tank
11. Tandem valve control diaphragm
12. Tandem valve control solenoid valve
13. PCV valve
14. Dashpot diaphragm
15. Charcoal canister
16. Purge control diaphragm valve
17. Purge cut-off solenoid valve
18. 2-way valve
19. EGR valve lift sensor
20. EGR valve
21. EGR control solenoid valve
22. Air chamber
23. Constant Vacuum Control (CVC) valve

**Vacuum circuit diagram – 1989–90 1.5L Civic Calif. auto trans except CRX HF**

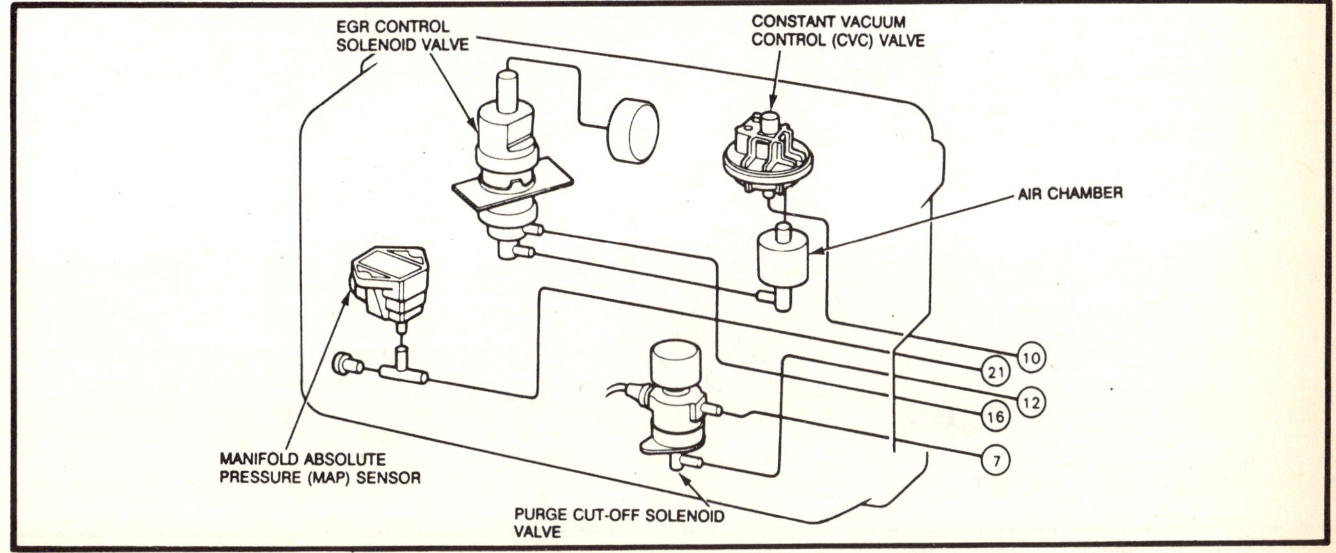


**Vacuum control box diagram1988–90 1.5L CRX HF**

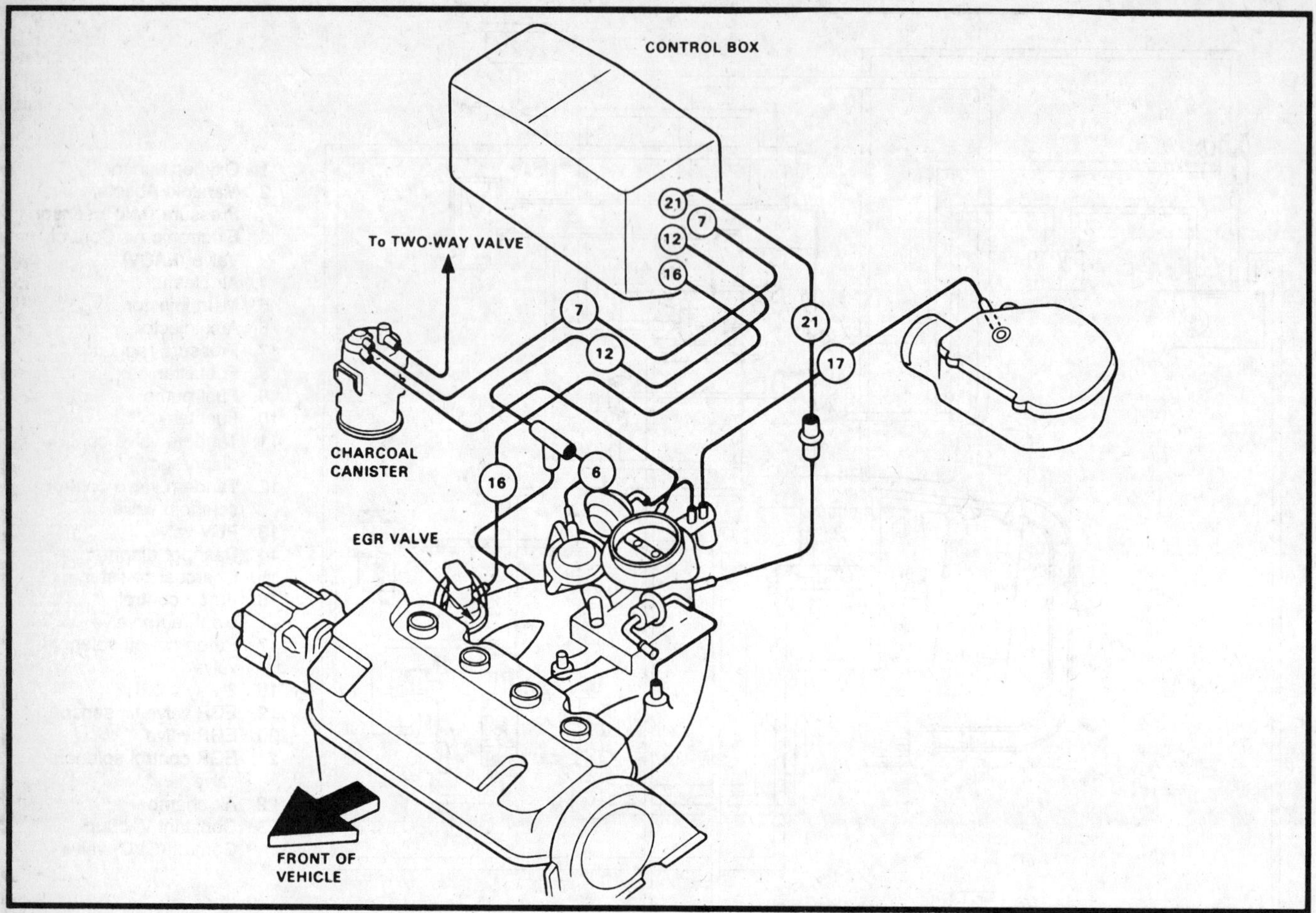

Vacuum identification diagram – 1989–90 1.5L Civic Calif. A/T except CRX HF

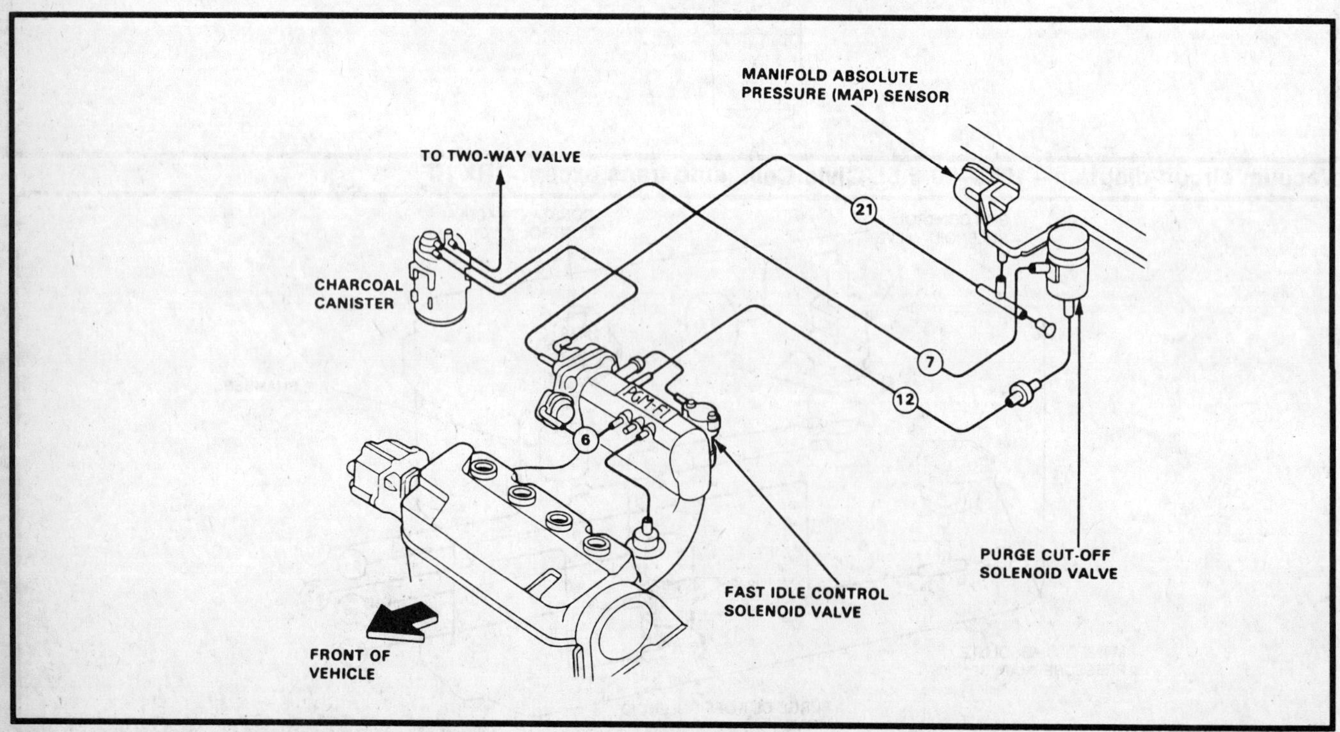


Vacuum identification diagram – 1988–90 1.6L except 1988 CRX Si

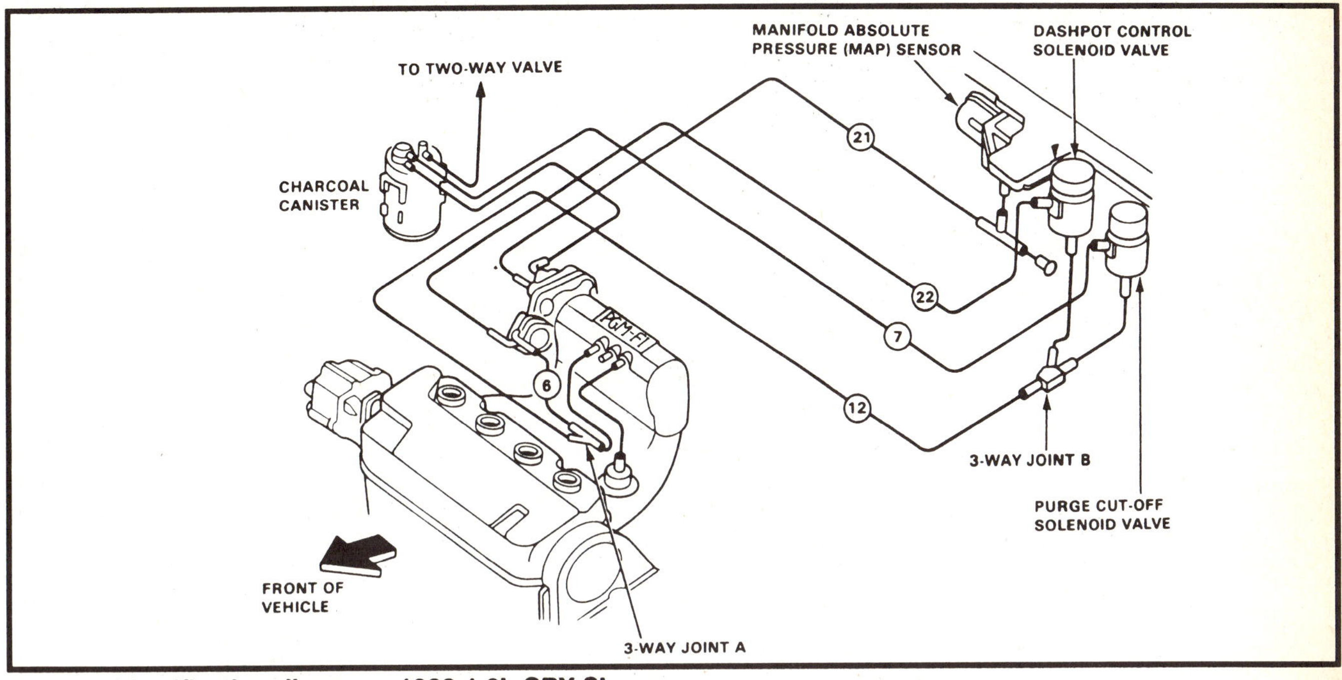


**Vacuum identification diagram—1988 1.6L CRX Si**

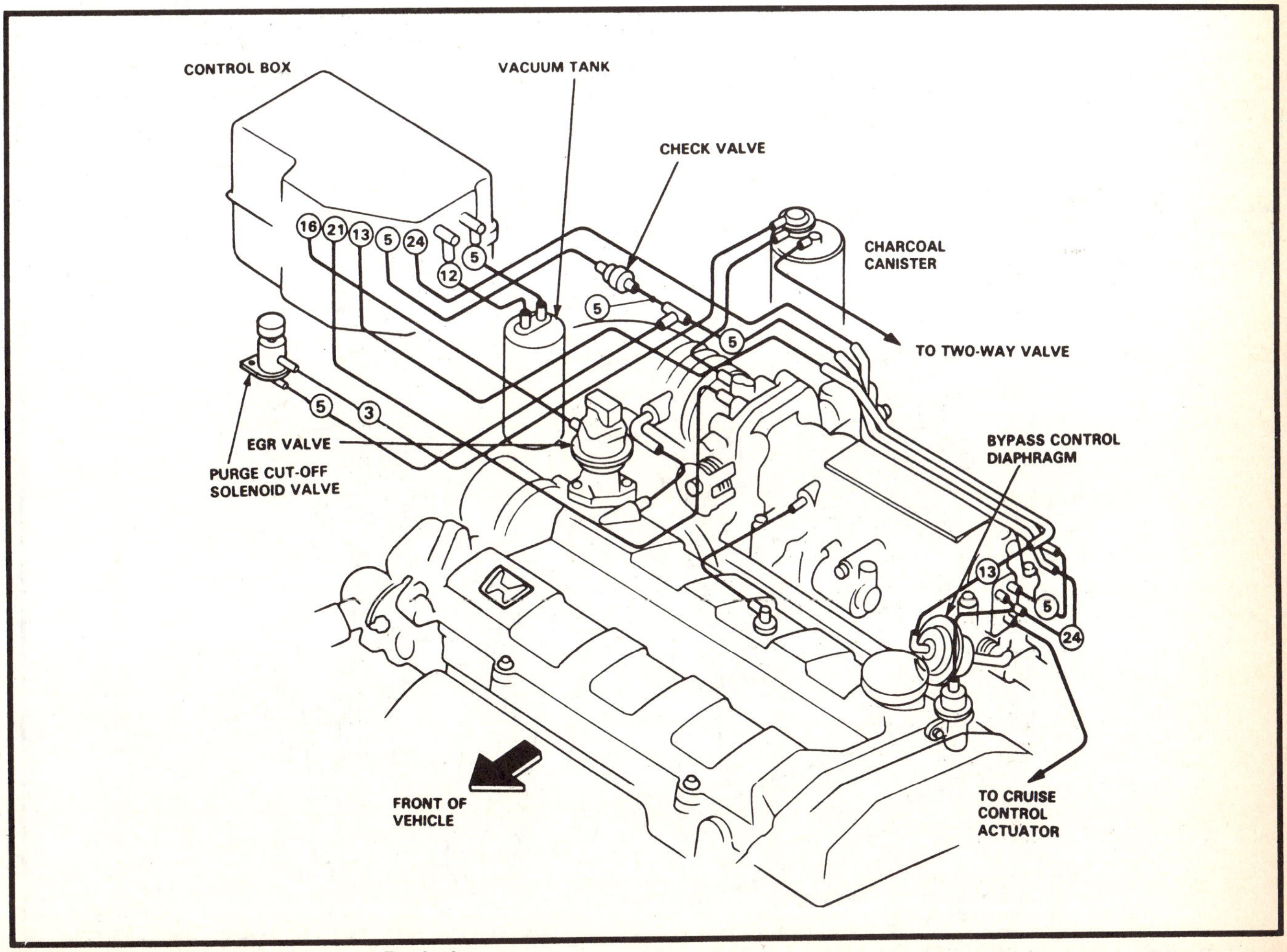

CONTROL BOX
VACUUM TANK
CHECK VALVE
CHARCOAL CANISTER
TO TWO-WAY VALVE
EGR VALVE
PURGE CUT-OFF SOLENOID VALVE
BYPASS CONTROL DIAPHRAGM
FRONT OF VEHICLE
TO CRUISE CONTROL ACTUATOR

**Vacuum identification diagram—Prelude**

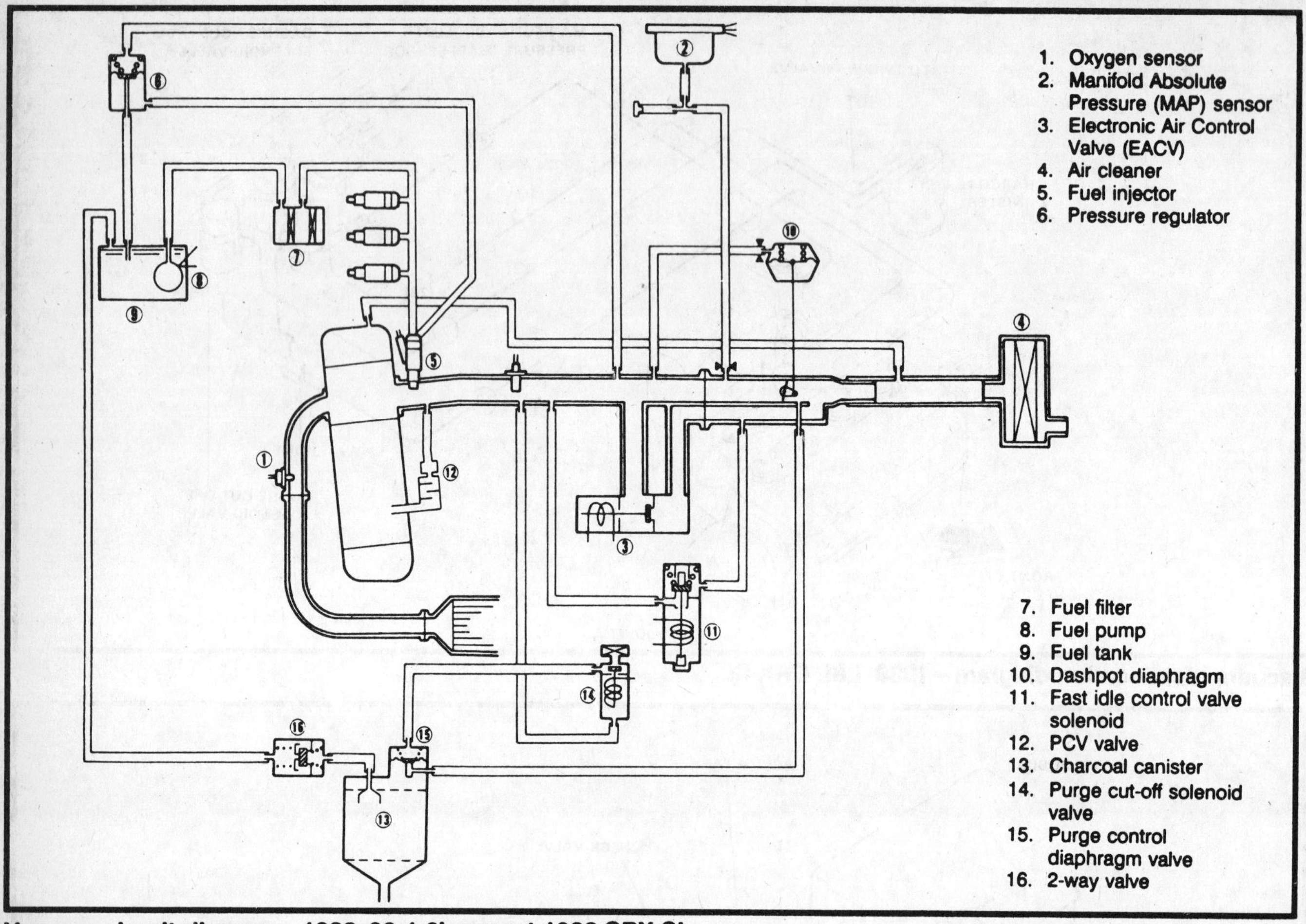

1. Oxygen sensor
2. Manifold Absolute Pressure (MAP) sensor
3. Electronic Air Control Valve (EACV)
4. Air cleaner
5. Fuel injector
6. Pressure regulator
7. Fuel filter
8. Fuel pump
9. Fuel tank
10. Dashpot diaphragm
11. Fast idle control valve solenoid
12. PCV valve
13. Charcoal canister
14. Purge cut-off solenoid valve
15. Purge control diaphragm valve
16. 2-way valve

**Vacuum circuit diagram—1988–90 1.6L except 1988 CRX SI**

EGR CONTROL SOLENOID VALVE
BYPASS CONTROL SOLENOID VALVE
CONSTANT VACUUM CONTROL (CVC) VALVE
AIR CHAMBER
MANIFOLD ABSOLUTE PRESSURE (MAP) SENSOR
16
21
13
5
24
12
5

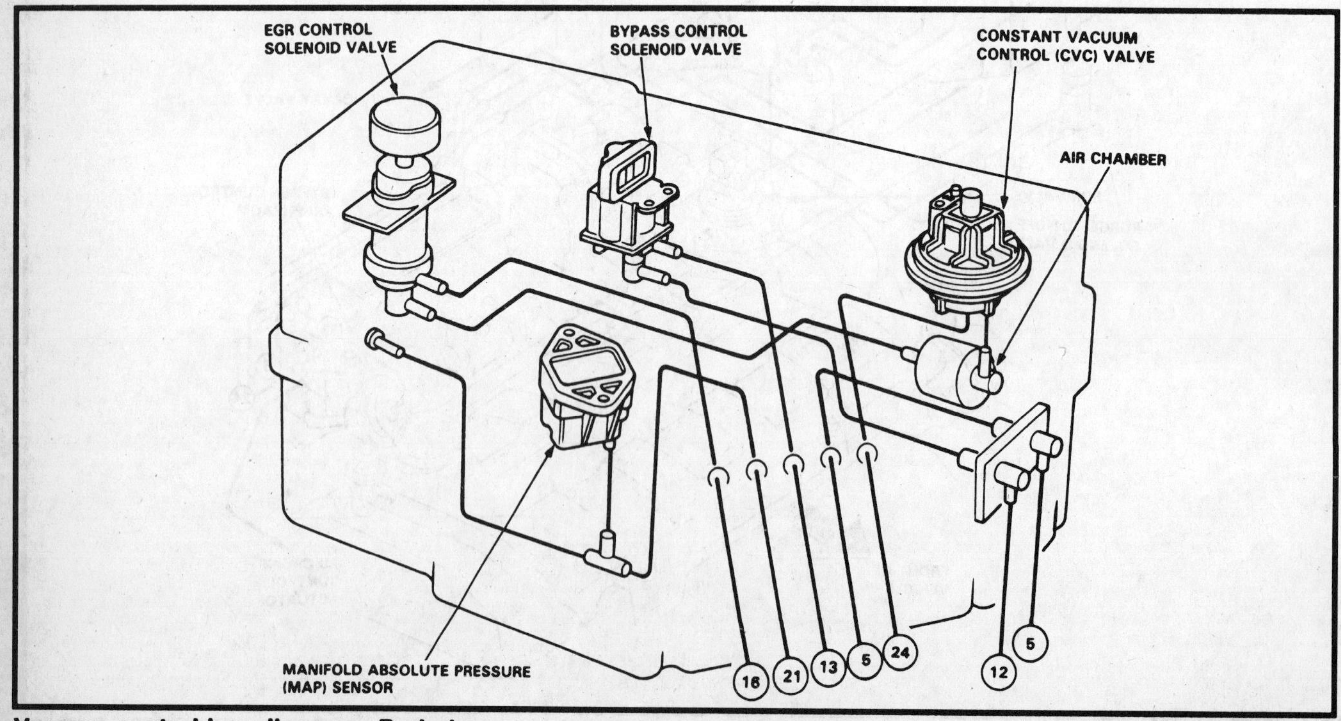

**Vacuum control box diagram—Prelude**

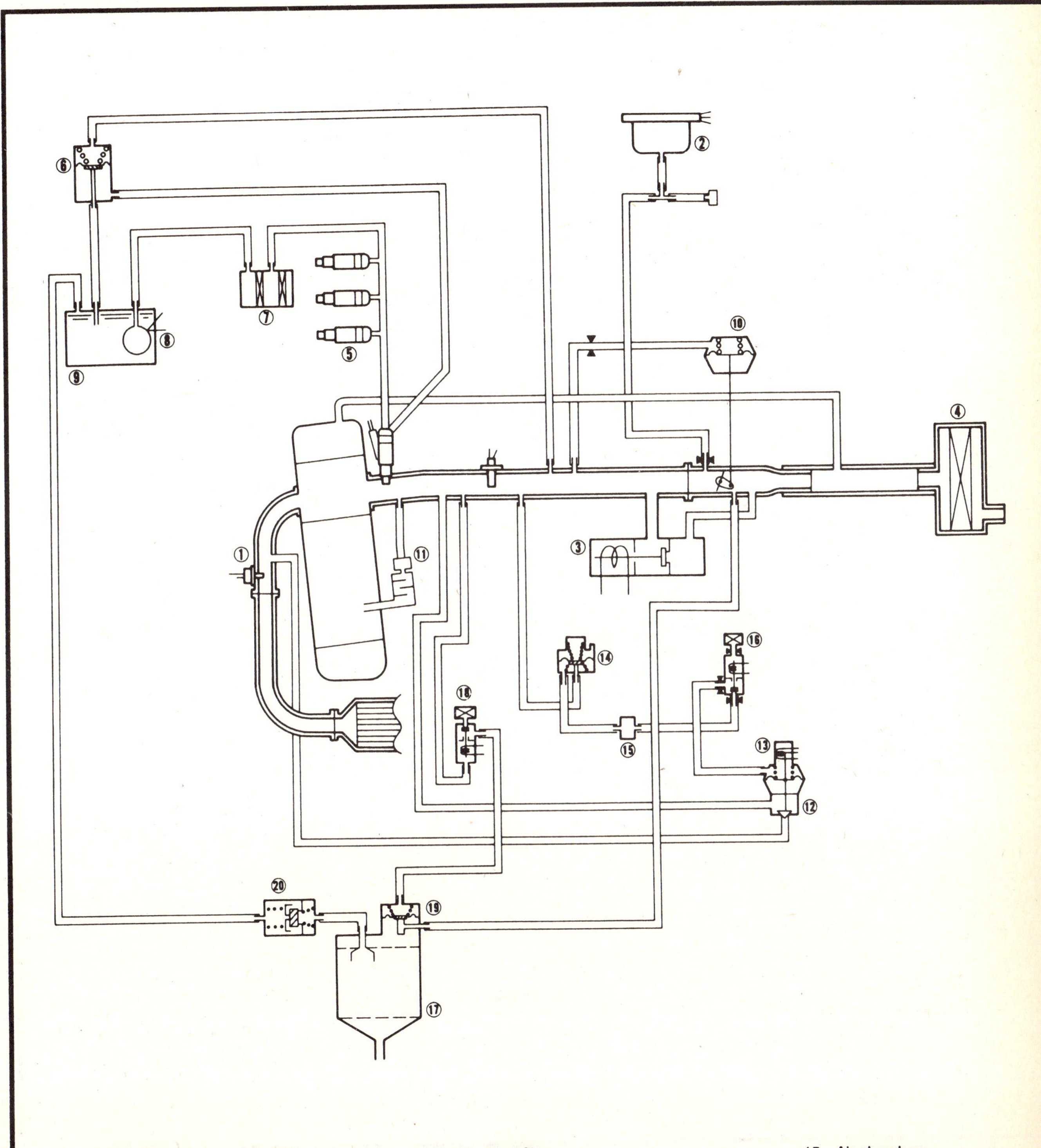

1. Oxygen sensor
2. Manifold Absolute Pressure (MAP) sensor
3. Electronic Air Control Valve (EACV)
4. Air cleaner
5. Fuel injector
6. Pressure regulator
7. Fuel filter
8. Fuel pump
9. Fuel tank
10. Dashpot diaphragm
11. PCV valve
12. EGR valve
13. EGR valve lift sensor
14. Constant Vacuum Control (CVC) valve
15. Air chamber
16. EGR control solenoid valve
17. Charcoal canister
18. Purge cut-off solenoid valve
19. Purge control diaphragm valve
20. 2-way valve

**Vacuum circuit diagram—1988–90 1.5L CRX HF**

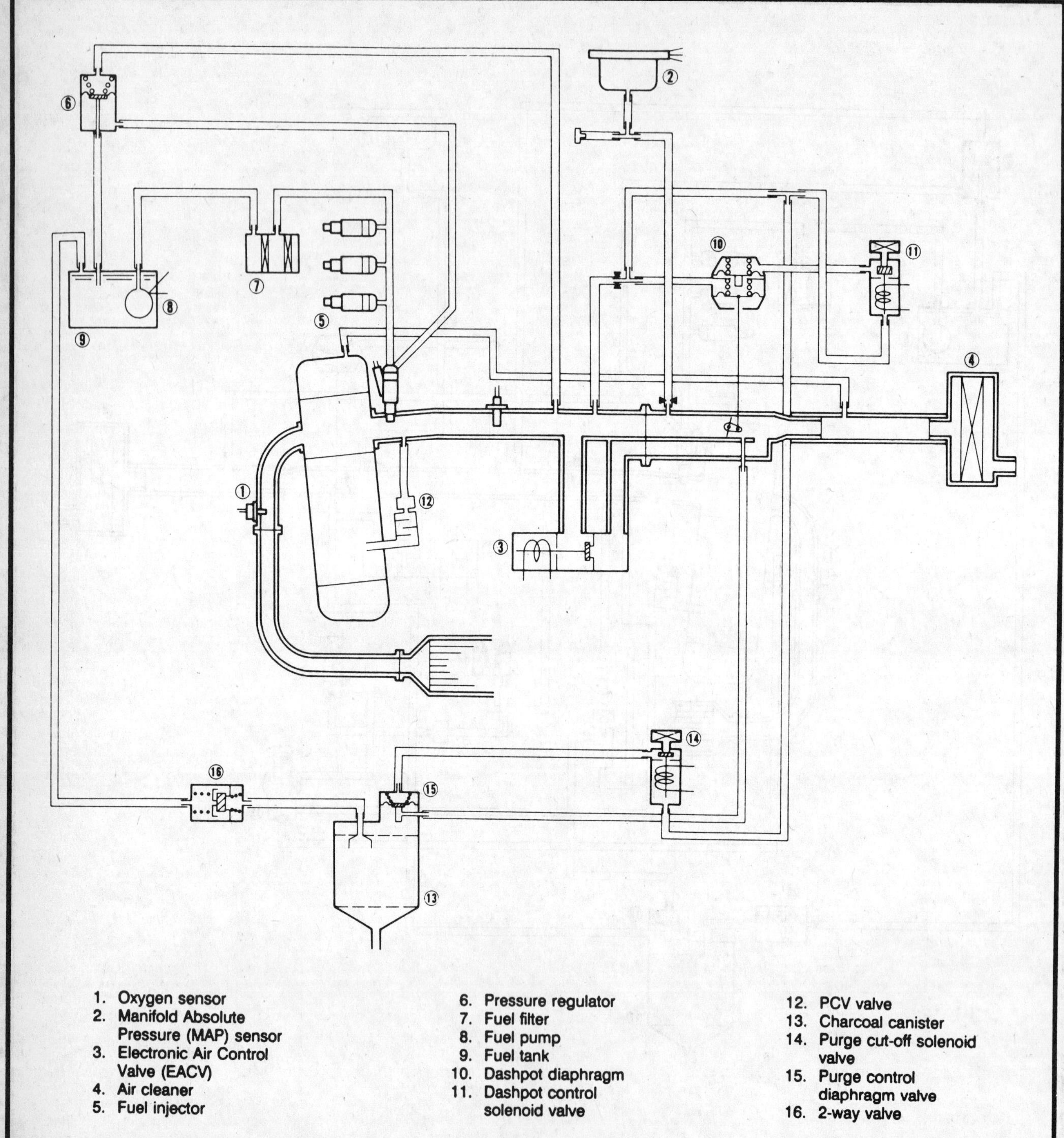
1. Oxygen sensor
2. Manifold Absolute Pressure (MAP) sensor
3. Electronic Air Control Valve (EACV)
4. Air cleaner
5. Fuel injector
6. Pressure regulator
7. Fuel filter
8. Fuel pump
9. Fuel tank
10. Dashpot diaphragm
11. Dashpot control solenoid valve
12. PCV valve
13. Charcoal canister
14. Purge cut-off solenoid valve
15. Purge control diaphragm valve
16. 2-way valve

**Vacuum circuit diagram—1988 1.6LCRX Si**

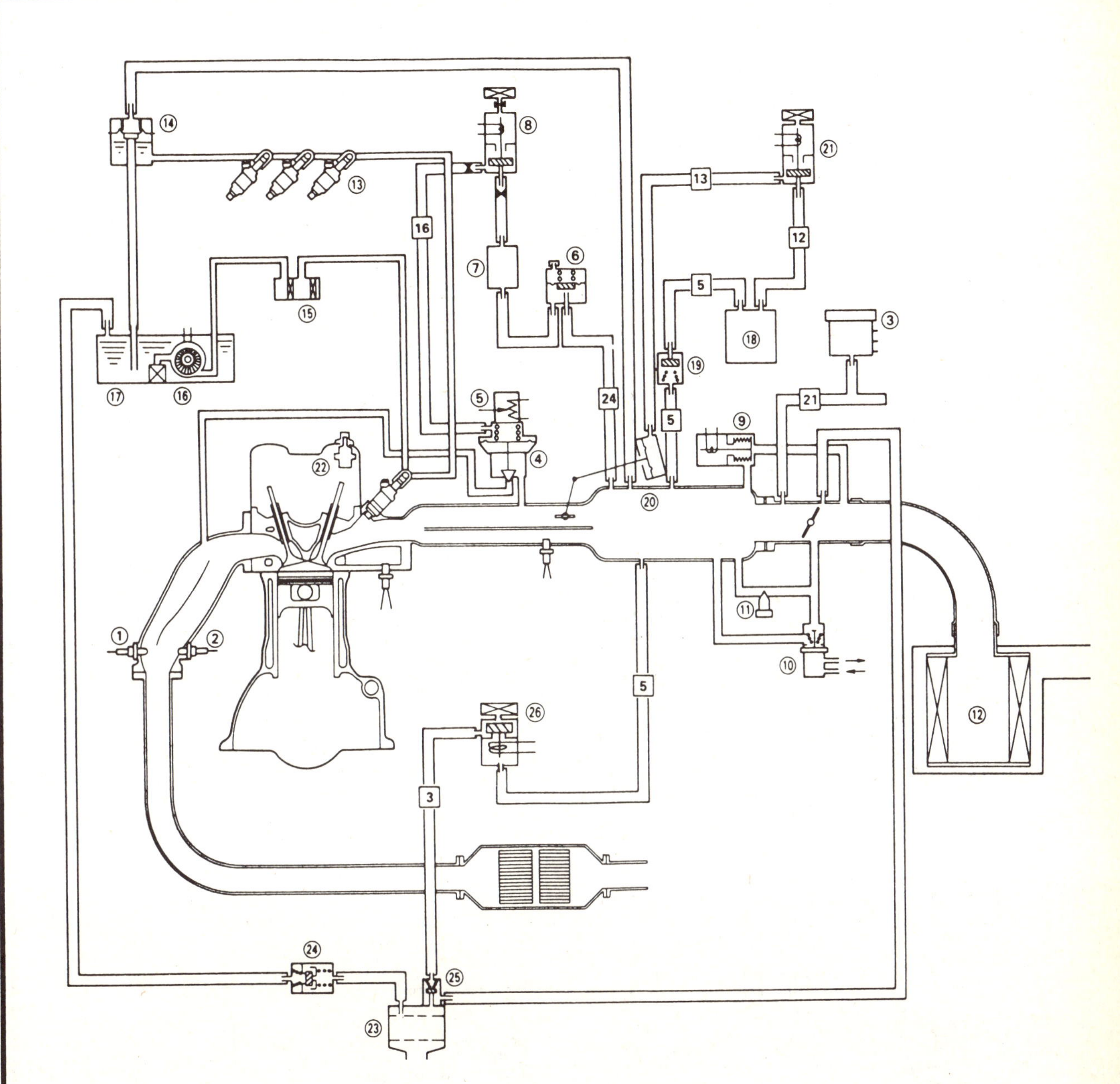

1. Oxygen sensor A
2. Oxygen sensor B
3. Manifold Absolute Pressure (MAP) sensor
4. EGR valve
5. EGR valve lift sensor
6. Constant Vacuum Control (CVC) valve
7. Air chamber
8. EGR control solenoid valve
9. Electronic Air Control Valve (EACV)
10. Fast idle valve
11. Idle air adjusting screw
12. Air cleaner
13. Fuel injector
14. Pressure regulator
15. Fuel filter
16. Fuel pump
17. Fuel tank
18. Vacuum tank
19. Check valve
20. Bypass control diaphragm
21. Bypass control solenoid valve
22. PCV valve
23. Charcoal canister
24. 2-way valve
25. Purge control diaphragm valve
26. Purge cut-off solenoid valve

**Vacuum circuit diagram—Prelude**

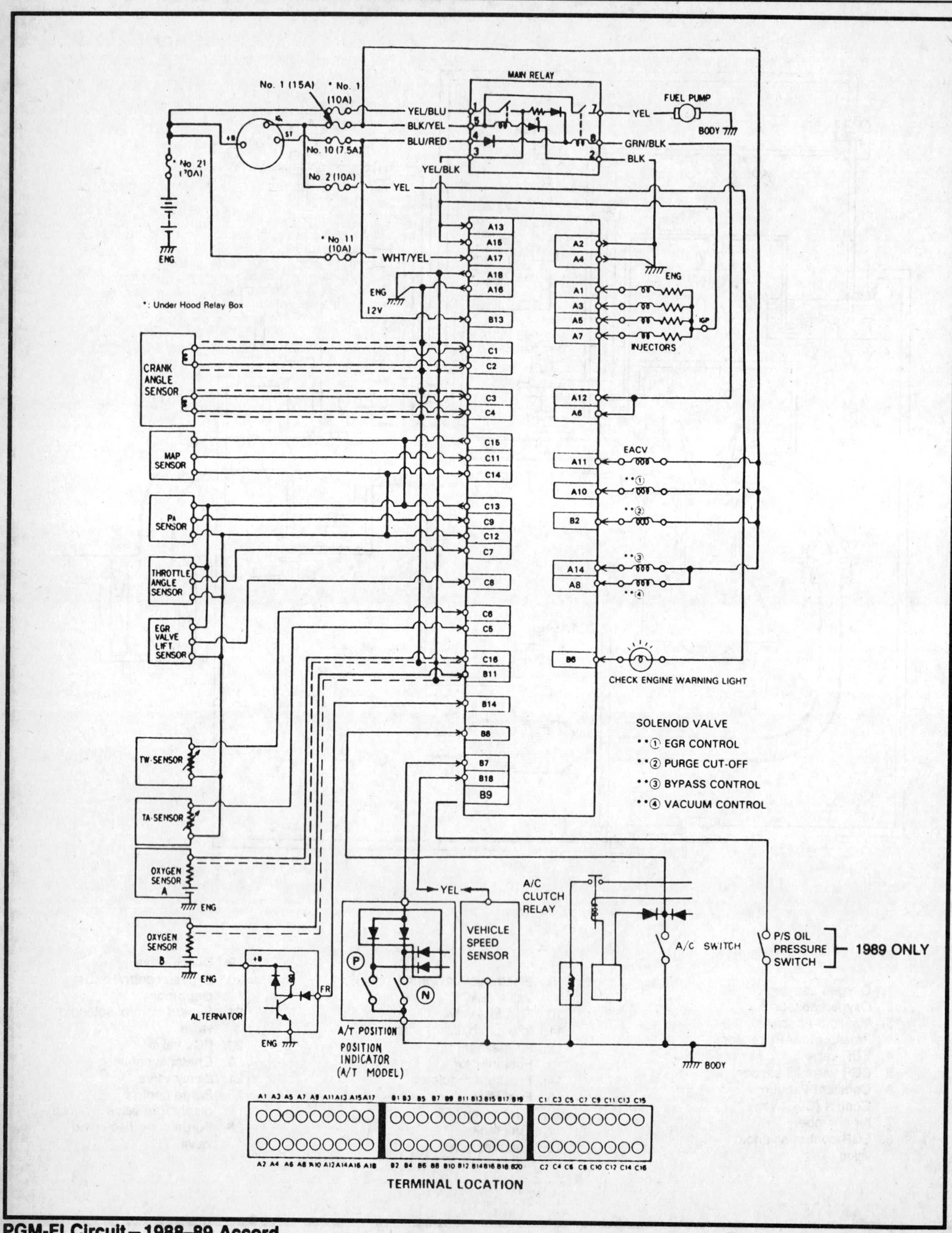


PGM-FI Circuit – 1988–89 Accord

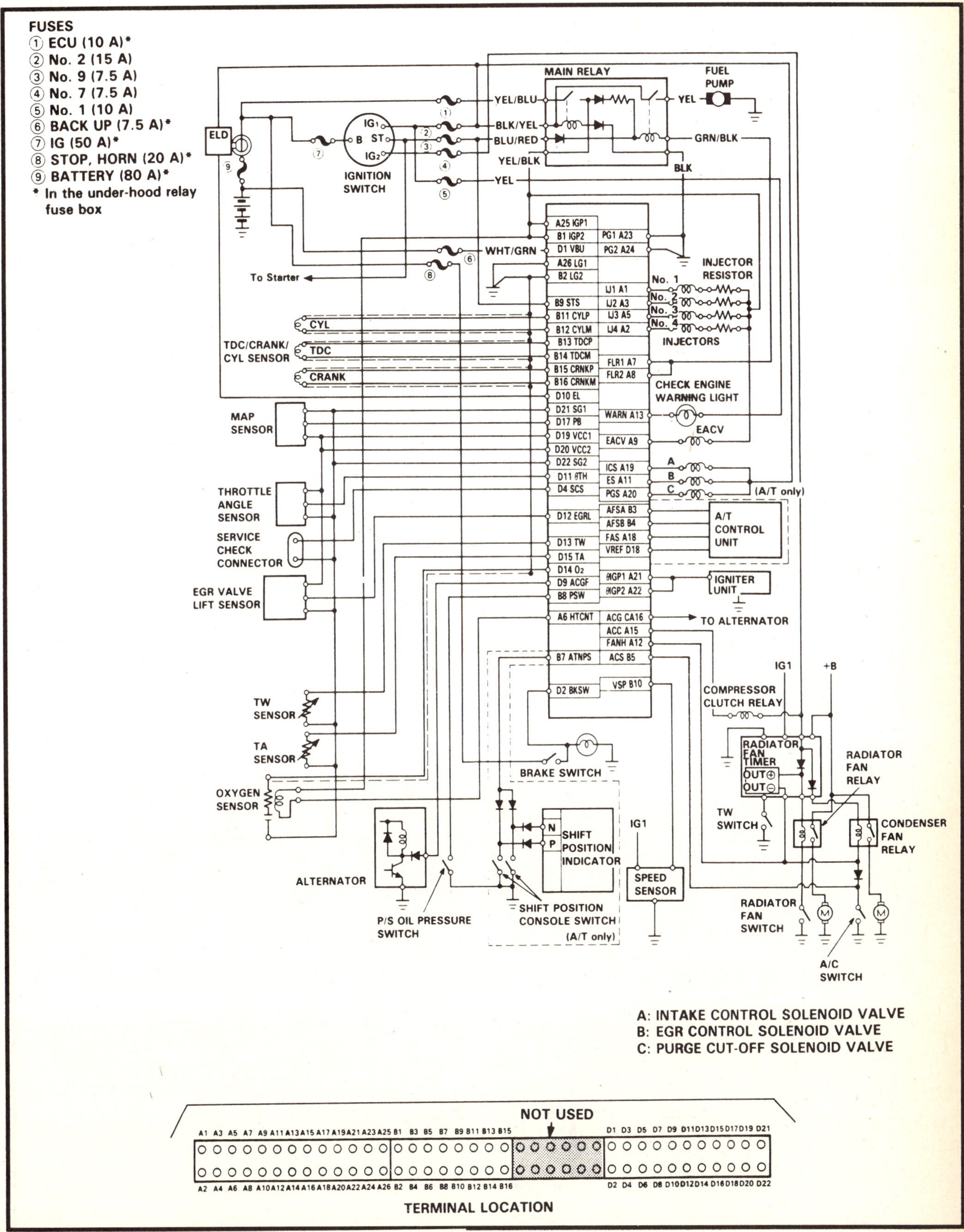


PGM-FI Circuit—1990 Accord

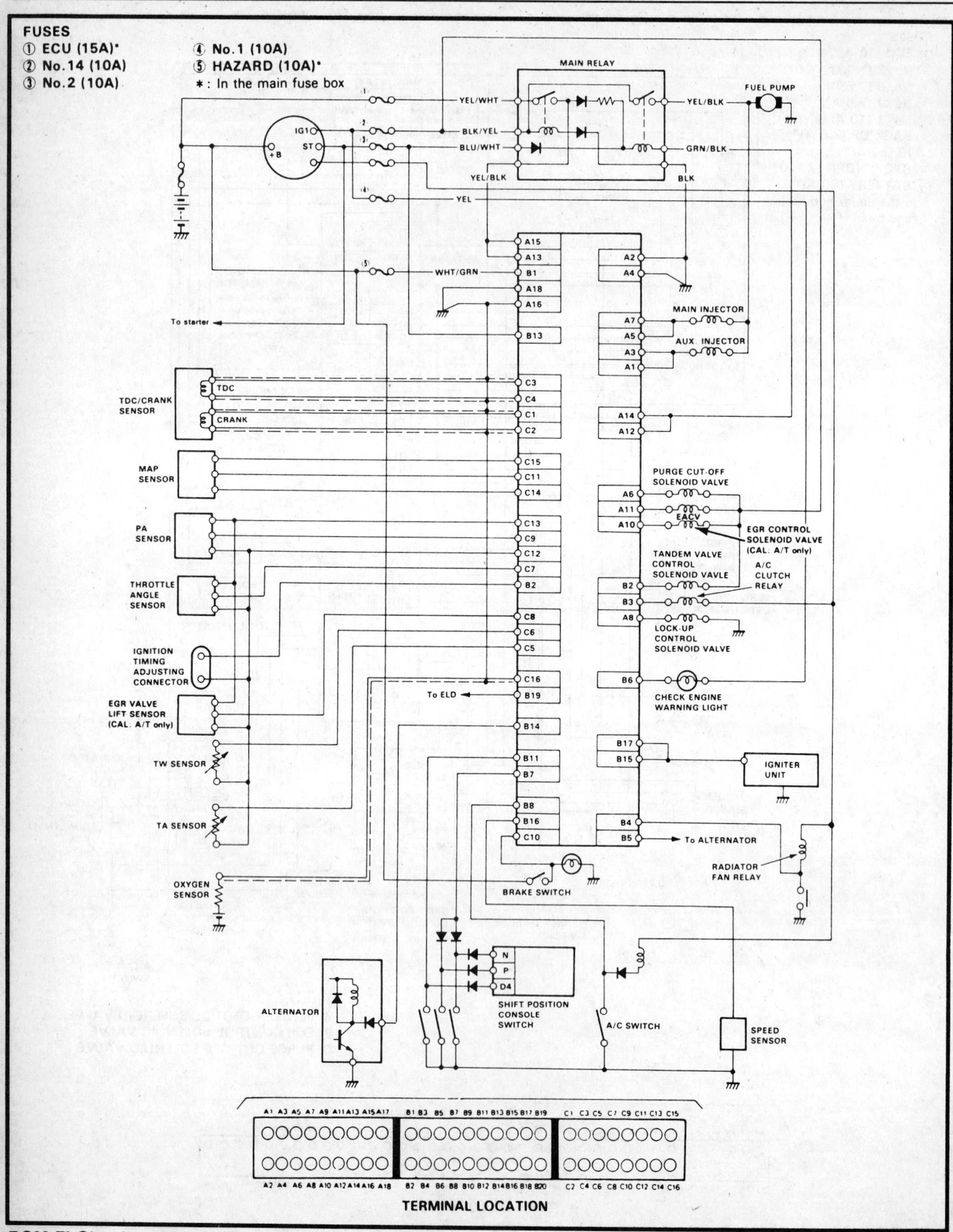


PGM-FI Circuit—1988–90 Civic 1.5L except CRX HF

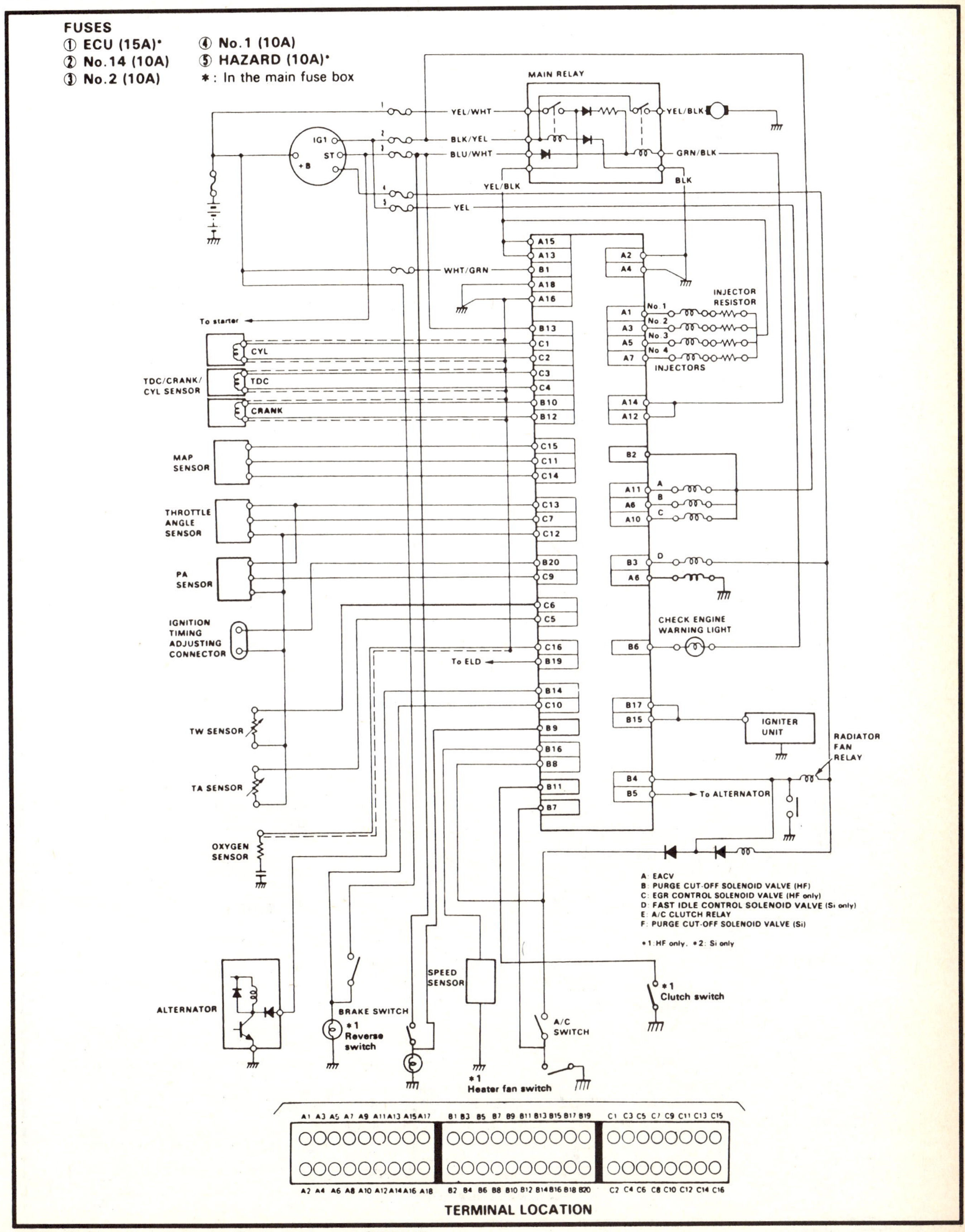


PGM-FI Circuit—1988–90 Civic 1.6L and CRX HF

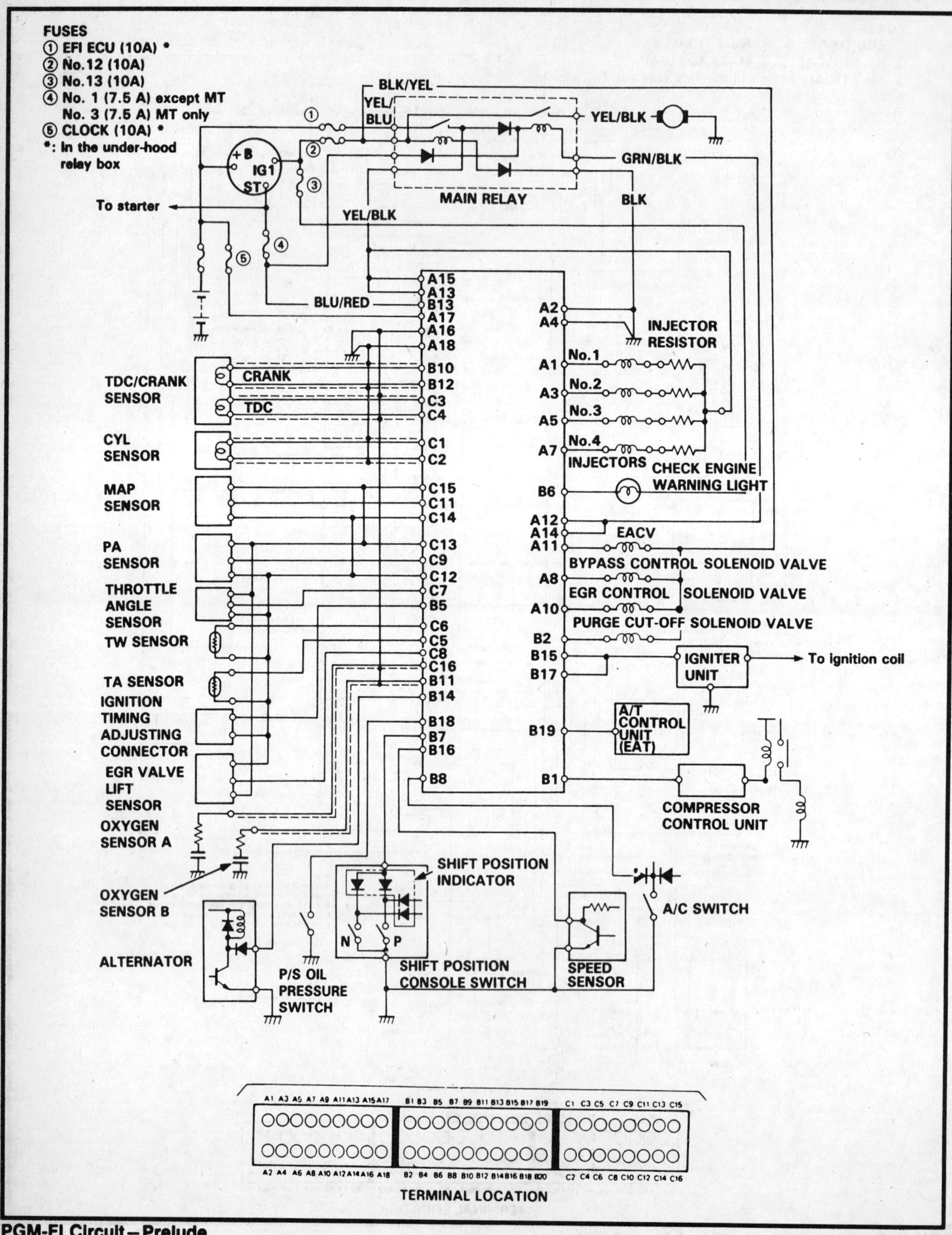


PGM-FI Circuit—Prelude

# HYUNDAI MULTI-POINT FUEL INJECTION (MPI)

## ENGINE CONTROL SYSTEM APPLICATION CHART

| Year | Model | Engine cc (liter) | Engine VIN | Fuel System | Ignition System |
|---|---|---|---|---|---|
| 1988 | Excel | 1468 (1.5) | J | FBC | Electronic |
| 1989 | Excel | 1468 (1.5) | J | FBC | Electronic |
| | Sonata | 2351 (2.4) | S | MPI | Electronic |
| | | 2972 (3.0) | T | MPI | Electronic |
| 1990 | Excel | 1468 (1.5) | J | FBC | Electronic |
| | | 1468 (1.5) | J | MPI | Electronic |
| | Sonata | 2351 (2.4) | S | MPI | Electronic |
| | | 2972 (3.0) | T | MPI | Electronic |

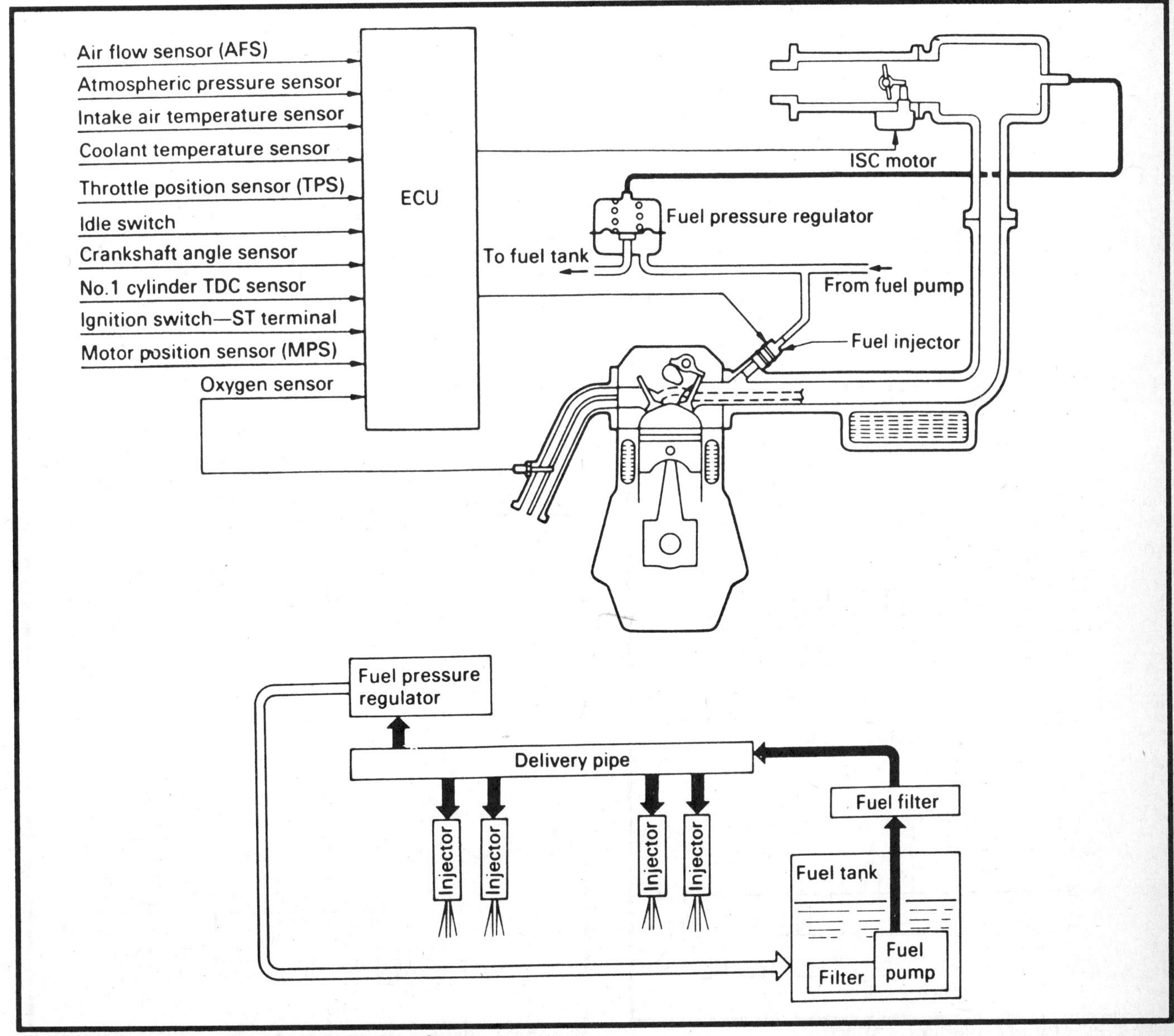


Basic MPI system – Sonata 4 cylinder shown

## General Description

The air-fuel ratio is controlled by varying the injector driving time. An electric fuel pump supplies sufficient fuel to the injection system and the pressure regulator maintains a constant pressure to the injectors. These injectors inject a metered quantity of fuel into the intake manifold in accordance with signals from the Electronic Control Unit (ECU) or engine computer. After pressure regulation excess fuel is returned to the fuel tank.

When the injector is energized, the valve inside the injector opens fully to inject the fuel. The injectors inject fuel to each manifold port, in the sequential firing order of each cylinder, as controlled by the ECU. The injectors fire simultaneous or non-synchronous injection, when during engine starting. During acceleration, additional fuel in a proportionate amount is injected for a fixed time of 10 milliseconds to 2 cylinders during intake and exhaust strokes. After starting and during normal engine operation, the ECU activates the injectors sequentially or as synchronous injection. Meaning that after each stroke the injectors are activated at the exhaust stroke of each cylinder.

Based on information from various sensors the ECU computes an optimum control for fuel mixture and ignition timing. The ECU consists of an 8-bit microprocessor, Random Access Memory (RAM), Read Only Memory (ROM) and input and output signal interface system.

## ELECTRONIC CONTROL SYSTEM

The ECU monitors the input and output signals of several sensors. Based on the sensor output the ECU controls fuel and ignition timing functions through the use of actuators.

### Air Flow Sensor (AFS)

The AFS measures the intake air volume. The ECU uses this intake air volume signal to decide the basic fuel injection duration.

### Atmosphere Pressure Sensor

The atmosphere pressure sensor signal is used by the ECU to compute the altitude of the vehicle and so the ECU can correct the ignition timing and air-fuel ratio. The atmosphere pressure sensor is contain in the AFS.

### Intake Air Temperature Sensor

The air temperature sensor is a resistor based sensor for detecting the intake air temperature. The ECU provides fuel injection control based on this information. The air temperature sensor is in located on the AFS.

### Engine Coolant Temperature Sensor

The coolant temperature sensor is located in the coolant passage of the intake manifold. The ECU uses this signal to determine the base fuel enrichment for cold and warm engine operation.

### Throttle Position Sensor (TPS)

The TPS is a rotating type variable resistor that rotates with the throttle body shaft to sense the throttle valve opening. Based on TPS voltage signals, the ECU computes the throttle valve opening and accordingly corrects fuel for engine acceleration.

### Idle Switch

The idle switch is a contact type switch. The switch is installed at the tip of the ISC. This switch provides the ECU with idle or off idle signal.

### Motor Position Sensor (MPS)

The MPS is a variable resistor type sensor and is installed in the ISC servo. The MPS senses the ISC servo plunger position and sends the signal to the ECU. THe ECU controls the valve opening, and consequently the idle speed by using the MPS signal, idle signal, engine coolant temperature signal, load signals and vehicle speed sensor.

### Cylinder TDC and Crankshaft Angle Sensor

The No. 1 TDC sensor and the crankshaft angle sensor are composed of a disc and unit assembly in the distributor. The No. 1 cylinder TDC is detected by the signal obtained through the single inner slit of the disc. The ECU, based upon this signal, determines the fuel injection cylinder. The crankshaft angle sensor signal comes from the the 4 slits at the outer circumference of the disc serve to detect the position of the crankshaft. The ECU, based on this signal, determines the fuel injection timing, and also calculates the amount of intake air, the ignition timing, etc.

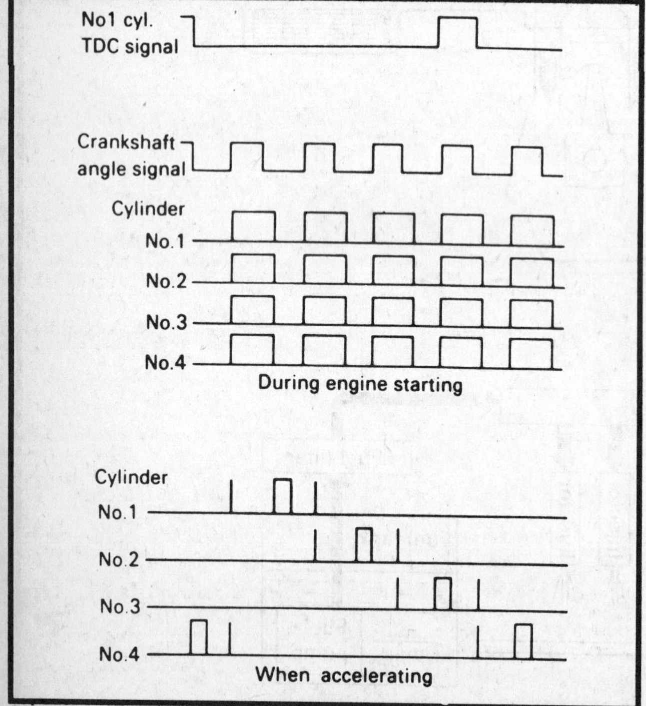


Non-synchronous injection pattern

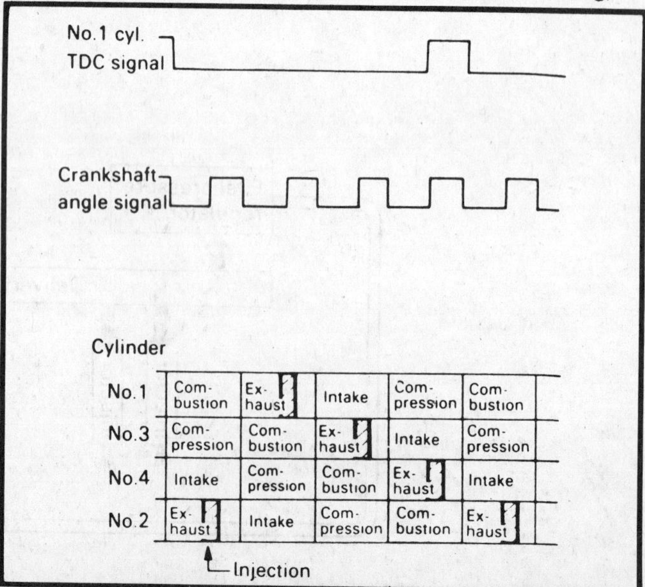


Synchronous injection pattern

for each stroke of engine.

### Oxygen Sensor

The oxygen sensor, by detecting the oxygen content in the exhaust gas, maintains the stoichiometric air/fuel ratio. In operation, the ECU receives the signals from the sensor and changes the duration during which fuel is injected. The oxygen sensor is located in the exhaust manifold.

### Vehicle Speed Sensor

The vehicle speed sensor uses a reed type switch. The speed sensor is built into the speedometer and converts the speedometer gear revolution into pulse signals, which are sent to the ECU.

### Neutral (Inhibitor) Switch

The ECU based on this signal senses when the automatic transaxle is in **NEUTRAL** or **PARK** and operates the ISC servo to keep the idle speed correct.

### Fuel Injectors

The injectors are solenoid valves. When the solenoid coil is energized, the plunger is retracted. The needle valve that is attached to the plunger is pulled to the full open position.

### Idle Speed Control (ISC) Servo

The ISC servo consists of a motor, worm gear, worm wheel and plunger. The MPS is used to detect plunger position and an idle switch to detect closed throttle position.

As the motor rotates, according to signals from the ECU, the plunger extends or retracts depending on direction of the motor rotation. This actuates the throttle valve via the ISC lever. The idle speed is controlled by changing the throttle open by the ECU through this function.

### Fuel Pump

The fuel pump is a compact impeller design and is installed inside the fuel tank. The pump assembly consists of the impeller (driven by the motor), the pump casing (which forms the pumping chamber), and cover of the pump. This pump is called a wet-type pump, because the inside is also filled with fuel. Never operate this type of pump when removed from the vehicle or explosion will result, due to fuel fumes, electric sparks and fresh air, which is not available in the fuel tank.

## SERVICE PRECAUTIONS

- Do not operate the fuel pump when the fuel lines are empty.
- Do not operate the fuel pump when removed from the fuel tank.
- Do not reuse fuel hose clamps.
- Make sure all ECU harness connectors are fastened securely. A poor connection can cause an extremely high surge voltage in the coil and condenser and result in damage to integrated circuits.
- Keep all ECU parts and harnesses dry during service.
- Before attempting to remove any parts, turn **OFF** the ignition switch and disconnect the battery ground cable.
- Do not attempt to disconnect the battery cables with the engine running.
- Do not disconnect wiring connector with the engine running, unless instructed to do so.
- Do not depress the accelerator pedal when starting.
- Do not rev up the engine immediately after starting or just prior to shutdown.
- Do not apply battery power directly to injectors.

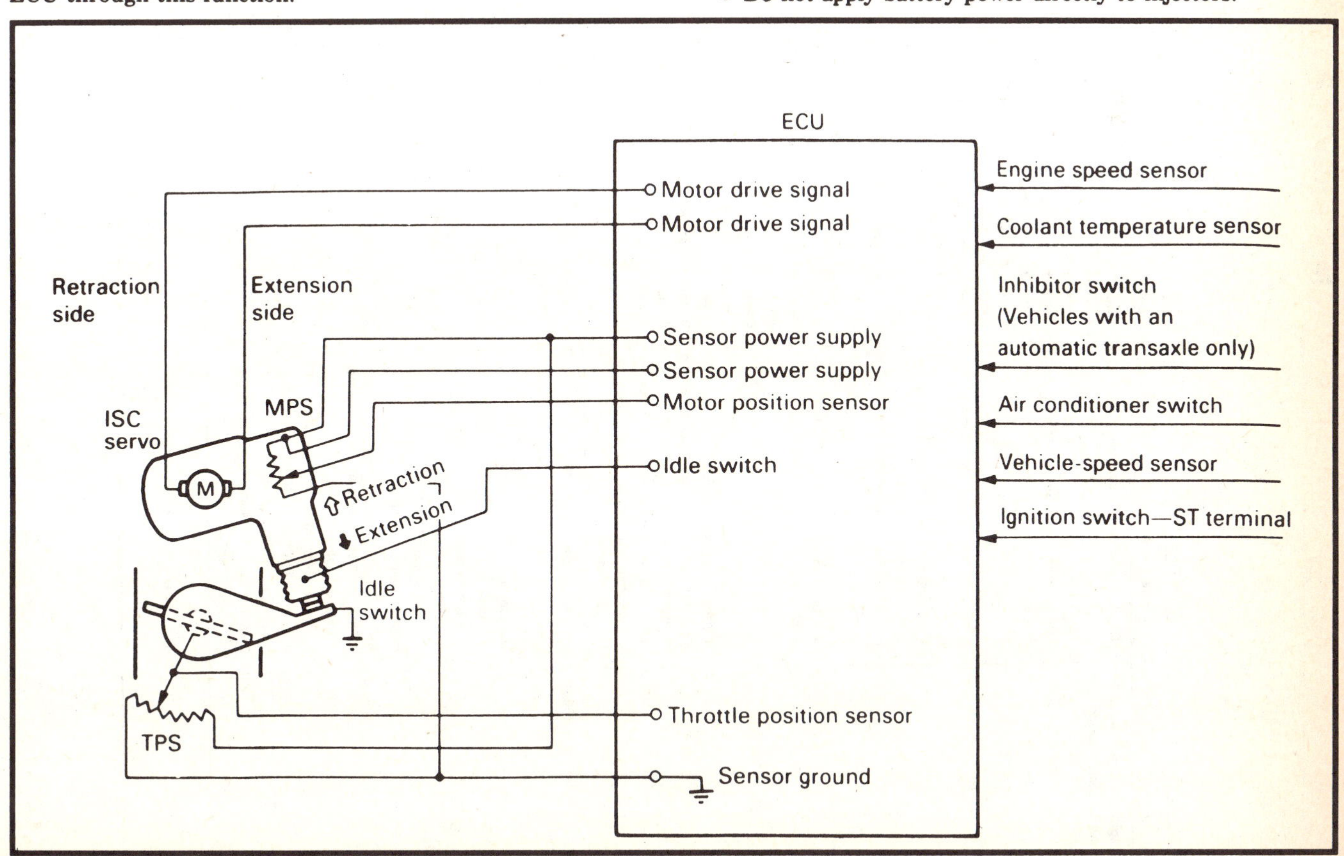


**Idle Speed Control (ISC) system—typical**

## MPI BASIC TROUBLESHOOTING

| Symptom | Probable cause | Remedy |
|---|---|---|
| Engine will not start or hard to start (cranks OK) | Trouble in the MPI system | Check for output of self-diagnosis code with a voltmeter or multi-use tester (MUT) |
| | Malfunction of the fuel pump drive control system | Perform cranking check with a multi-use tester (MUT)<br>Check the fuel pump drive control system and the fuel pump |
| | Malfunction of the ignition timing control system | Perform cranking check with a multi-use tester (MUT) |
| | Malfunction of the power transistor | Check the power transistor as a single unit |
| | Power is not supplied to ECU | Perform cranking check with a multi-use tester (MUT)<br>Check the power supply circuit |
| | Malfunction of the control relay | Replace |
| | Malfunction of the injector | Perform cranking check with a multi-use tester (MUT)<br>Check the injector drive circuit<br>Check the injector as a single unit |
| | Improper fuel pressure | Check the fuel pressure |
| | Vacuum hose disconnected or damaged | Repair or replace |
| | Malfunction of ECU | Replace |
| | Damaged or disconnected harness, or short-circuit; improper connection of the connector | Repair or replace |
| Rough idle or engine stalls | Trouble in the MPI system | Check for output of self-diagnosis code with a voltmeter or multi-use tester (MUT) |
| | Malfunction of the sensor<br>o Intake air temperature sensor<br>o Engine coolant temperature sensor<br>o Atmospheric pressure sensor<br>o Ignition switch<br>o Idle position sensor<br>o Throttle position sensor<br>o No.1 cylinder top dead center sensor, crank angle sensor<br>o Power steering oil pressure switch<br>o Air conditioner switch<br>o Inhibitor switch (A/T)<br>o Motor position sensor (MPS)<br>o Oxygen sensor | Check the sensor with a multi-use tester (MUT)<br>Check the sensor-related circuit<br>Check the sensor as a single unit |
| Rough idle or engine stalls | Malfunction of the engine control system<br>o Air-flow sensor<br>o Injector<br>o Power transistor | Check the actuator with a multi-use tester (MUT) |
| | Malfunction of the vehicle-speed reed switch | Check the vehicle speed reed switch |
| | The fuel pressure is not proper | Check the fuel pressure |
| | Vacuum hose disconnected or damaged | Repair or replace |
| | Malfunction of ECU | Replace |
| | Damaged or disconnected harness, or short-circuit; improper connection of the connector | Repair or replace |
| Engine hesitates or poor acceleration | Trouble in the MPI system | Check for output of self-diagnosis code with a voltmeter or multi-use tester (MUT) |
| | Malfunction of the sensor<br>o Intake air temperature sensor<br>o Engine coolant temperature sensor<br>o Atmospheric pressure sensor<br>o Ignition switch<br>o Idle position switch<br>o Throttle position sensor<br>o No.1 cylinder top dead center sensor, crank angle sensor<br>o Power steering oil pressure switch<br>o Air conditioner switch<br>o Inhibitor switch (A/T)<br>o Motor position sensor (MPS)<br>o Oxygen sensor | Check the sensor with a multi-use tester (MUT)<br>Check the sensor-related circuit<br>Check the sensor as a single unit |
| | Malfunction of the engine control system<br>o Air-flow sensor<br>o Injector<br>o Power transistor | Check the actuator with a multi-use tester (MUT) |
| | Malfunction of the air conditioner power relay control system | Check the system, and the components if the system is found defective |
| | The fuel pressure is not proper | Check the fuel pressure |
| | Vacuum hose disconnected or damaged | Repair or replace |
| | Malfunction of ECU | Replace |
| | Damaged or disconnected harness, or short-circuit; improper connection of the connector | Repair or replace |

## MPI BASIC TROUBLESHOOTING

| Symptom | Probable cause | Remedy |
|---|---|---|
| Poor fuel mileage | Trouble in the MPI system | Check for output of self-diagnosis code with a voltmeter or multi-use tester (MUT) |
| | Malfunction of the sensor<br>o Intake air temperature sensor<br>o Engine coolant temperature sensor<br>o Atmospheric pressure sensor<br>o Ignition switch<br>o Idle position switch<br>o Throttle position sensor<br>o No.1 cylinder top dead center sensor, crank angle sensor<br>o Power steering oil pressure switch<br>o Air conditioner switch<br>o Inhibitor switch (A/T)<br>o Motor position sensor (MPS)<br>o Oxygen sensor | Check the sensor with a multi-use tester (MUT)<br>Check the sensor-related circuit<br>Check the sensor as a single unit |
| | Malfunction of the engine control system<br>o Air-flow sensor<br>o Injector<br>o Power transistor | Check the actuator with a multi-use tester (MUT) |
| | The fuel pressure is not proper | Check the fuel pressure |

## DIAGNOSIS TROUBLE CODE CHARTS—4 CYLINDER

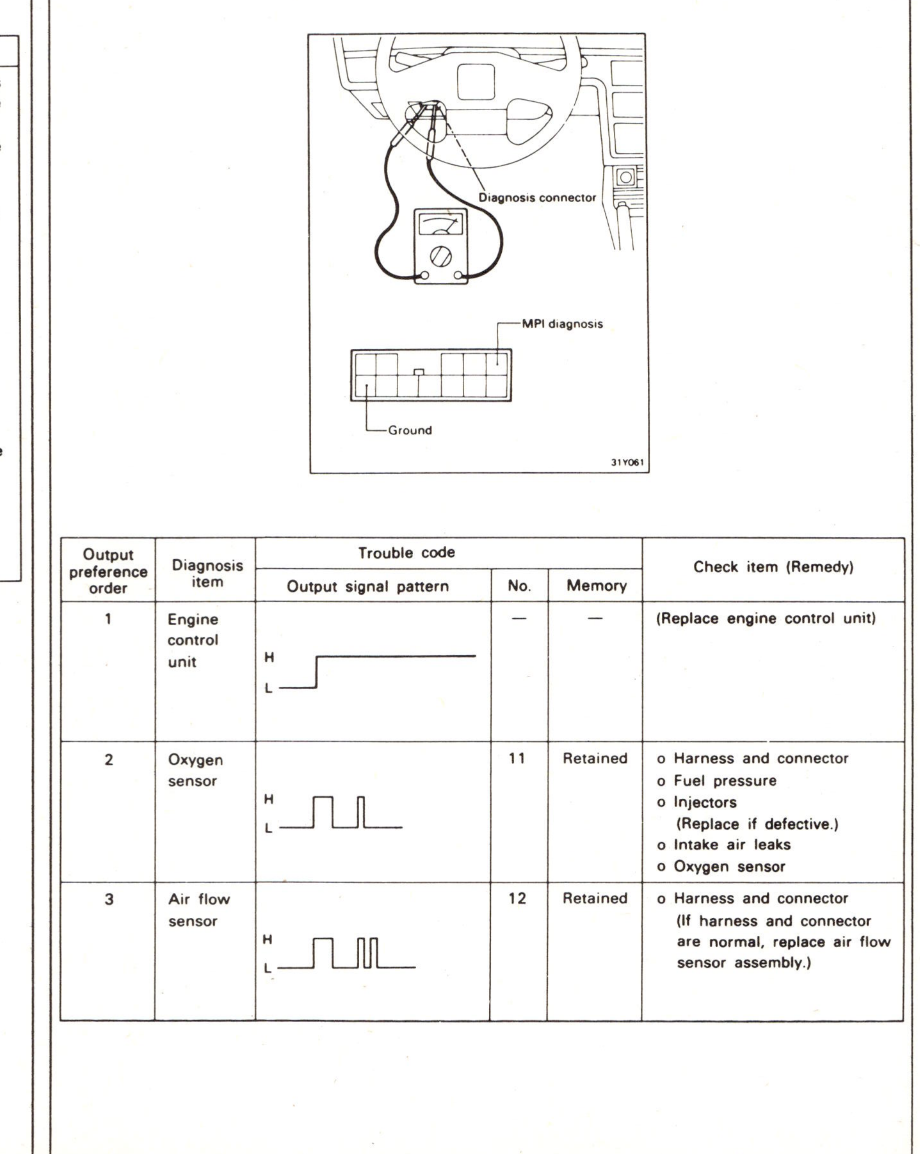


| Output preference order | Diagnosis item | Trouble code: Output signal pattern | Trouble code: No. | Trouble code: Memory | Check item (Remedy) |
|---|---|---|---|---|---|
| 1 | Engine control unit | H / L | — | — | (Replace engine control unit) |
| 2 | Oxygen sensor | H / L | 11 | Retained | o Harness and connector<br>o Fuel pressure<br>o Injectors (Replace if defective.)<br>o Intake air leaks<br>o Oxygen sensor |
| 3 | Air flow sensor | H / L | 12 | Retained | o Harness and connector (If harness and connector are normal, replace air flow sensor assembly.) |

# DIAGNOSIS TROUBLE CODE CHARTS—4 CYLINDER

| Output preference order | Diagnosis item | Trouble code: Output signal pattern | No. | Memory | Check item (Remedy) |
|---|---|---|---|---|---|
| 4 | Air temperature sensor |  H/L | 13 | Retained | o Harness and connector<br>o Air temperature sensor |
| 5 | Throttle position sensor | 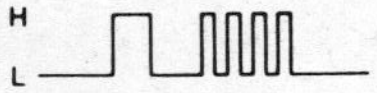 H/L | 14 | Retained | o Harness and connector<br>o Throttle position sensor<br>o Idle position switch |
| 6 | Motor position sensor | 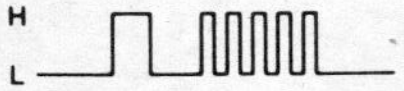 H/L | 15 | Retained | o Harness and connector<br>o Motor position sensor |
| 7 | Engine coolant temperature sensor | 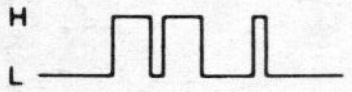 H/L | 21 | Retained | o Harness and connector<br>o Engine coolant temperature sensor |
| 8 | Crank angle sensor | 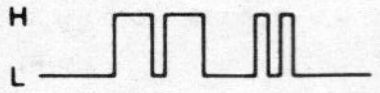 H/L | 22 | Retained | o Harness and connector (If harness and connector are normal, replace distributor assembly.) |
| 9 | No.1 cylinder top dead center sensor | 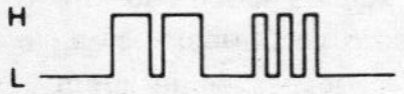 H/L | 23 | Retained | o Harness and connector (If harness and connector are normal, replace distributor assembly.) |
| 10 | Vehicle-speed sensor (reed switch) | 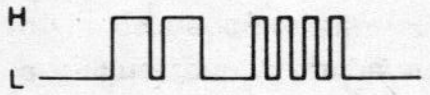 H/L | 24 | Retained | o Harness and connector<br>o Vehicle-speed sensor (reed switch) |

| Output preference order | Diagnosis item | Trouble code :ode: Output signal pattern | No. | Memory | Check item (Remedy) |
|---|---|---|---|---|---|
| 11 | Barometric pressure sensor | 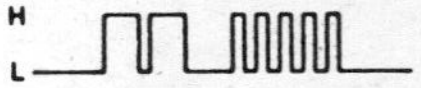 H/L | 25 | Retained | o Harness and connector (If harness and connector are normal, replace barometric pressure sensor assembly.) |
| 12 | Injector | 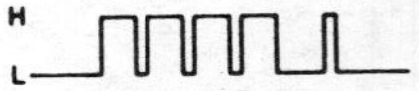 H/L | 41 | Retained | o Harness and connector<br>o Injector coil resistance |
| 13 | Fuel pump | 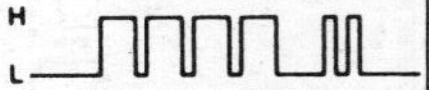 H/L | 42 | Retained | o Harness and connector<br>o Control relay |
| 14 | EGR* |  H/L | 43 | Retained | o Harness and connector<br>o EGR temperature sensor<br>o EGR valve<br>o EGR control solenoid valve<br>o EGR valve control vacuum |
| 15 | Normal state |  H/L | — | — | — |

**NOTE**

1. **Replace the engine control unit if a trouble code is read although the inspection reveals that there are no problems with the diagnosis item.**
2. **The diagnosis item marked * is applicable to California vehicles only.**

# DIAGNOSIS TROUBLE CODE CHARTS—SONATA V6

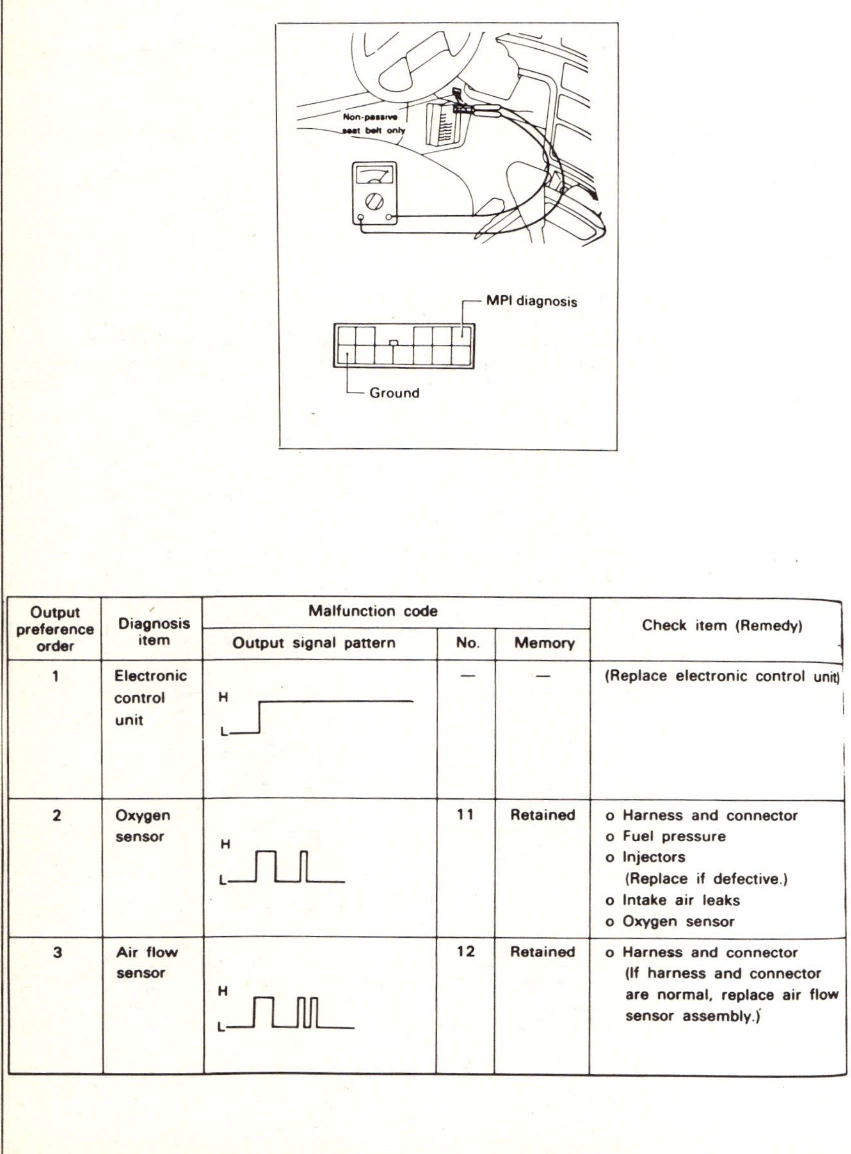


| Output preference order | Diagnosis item | Malfunction code | | | Check item (Remedy) |
|---|---|---|---|---|---|
| | | Output signal pattern | No. | Memory | |
| 1 | Electronic control unit | H L | — | — | (Replace electronic control unit) |
| 2 | Oxygen sensor | H L | 11 | Retained | o Harness and connector<br>o Fuel pressure<br>o Injectors (Replace if defective.)<br>o Intake air leaks<br>o Oxygen sensor |
| 3 | Air flow sensor | H L | 12 | Retained | o Harness and connector (If harness and connector are normal, replace air flow sensor assembly.) |
| 4 | Intake air temperature sensor | H L 31Y065 | 13 | Retained | o Harness and connector<br>o Intake air temperature sensor |
| 5 | Throttle position sensor | H L 31Y066 | 14 | Retained | o Harness and connector<br>o Throttle position sensor<br>o Idle position switch |
| 6 | Coolant temperature sensor | H L 31Y068 | 21 | Retained | o Harness and connector<br>o Coolant temperature sensor |
| 7 | Crank angle sensor | H L 31Y069 | 22 | Retained | o Harness and connector (If harness and connector are normal, replace distributor assembly.) |
| 8 | No.1 cylinder top dead center sensor | H L 31Y070 | 23 | Retained | o Harness and connector (If harness and connector are normal, replace distributor assembly.) |
| 9 | Vehicle-speed sensor (reed switch) | H L 31Y071 | 24 | Retained | o Harness and connector<br>o Vehicle-speed sensor (reed switch) |
| 10 | Barometric pressure sensor | H L 31Y072 | 25 | Retained | o Harness and connector (If harness and connector are normal, replace barometric pressure sensor assembly.) |

## DIAGNOSIS TROUBLE CODE CHARTS—4 CYLINDER

| Output preference order | Diagnosis item | Malfunction code | | | Check item (Remedy) |
|---|---|---|---|---|---|
| | | Output signal pattern | No. | Memory | |
| 11 | Injector | H / L 31Y073 | 41 | Retained | o Harness and connector<br>o Injector coil resistance |
| 12 | Fuel pump | H / L 31Y074 | 42 | Retained | o Harness and connector<br>o Control relay |
| 13 | EGR* | H / L 31Y075 | 43 | Retained | o Harness and connector<br>o EGR temperature sensor<br>o EGR valve<br>o EGR control solenoid valve<br>o EGR valve control vacuum |
| 14 | Normal state | H / L 31Y076 | — | — | — |

**NOTE**

1. **Replace the ECU only when all other possible causes for a malfunction have been explored.**
2. **The diagnostic item marked * is applicable to the vehicles for California only.**

## DIAGNOSIS FLOW CHARTS—SONATA 4 CYLINDER

Self-diagnosis check connector
Multi-use tester
Cigarette lighter socket

**Cranking Check** Item No.: Multi-use tester code number

| Check points | Check conditions | Test specification | Probable cause of malfunction (or action) |
|---|---|---|---|
| Battery voltage<br>o Service data<br>o Item No.16 | o Ignition switch: ON | 11—13 V | o Low battery voltage<br>o Power not supplied to the engine control unit<br>1) Check the power-supply circuit.<br>2) Check the ignition switch.<br>3) Check the control relay.<br>o Malfunction of the engine control unit ground circuit |

# DIAGNOSIS FLOW CHARTS—SONATA 4 CYLINDER

| Check points | Check conditions | | Test specification | Probable causes of malfunction (or action) |
|---|---|---|---|---|
| Malfunction code read out<br>o Self diagnosis | o After holding for 15 seconds or longer with the ignition switch at "ON", move the ignition switch to "LOCK" and disconnect the ISC servo connector.<br>o Crank for four seconds or longer.<br>o Ignition switch: "ON"<br>(Check for damage or disconnection of the injector, crank angle sensor circuit and other circuits.) | | Normal | o Check in accordance with the diagnosis code.<br>(Note that the diagnosis code will be erased if there is disconnection or damage of the engine control unit back-up power-supply circuit.)<br>o If various diagnosis codes are output, the most frequent cause is damage or disconnection of the power-supply or ground circuit. |
| Fuel pump<br>o Actuator test<br>o Item No.7 | Try under both conditions:<br>o Engine cranking<br>o Fuel pump forced activation | Pinch closed the return hose. | The pulsations of the fuel flow can be felt. | o Voltage is not supplied to the fuel pump.<br>1) Check the ignition switch (IG and ST)<br>2) Check the control relay.<br>3) Check the related circuits.<br>o Malfunction of the fuel pump |
| | | Listen close to the fuel tank. | The pump drive sound can be heard. | |
| Crank angle sensor<br>o Service data<br>o Item No.22 | o Engine cranking<br>o Tachometer connected (check, by using the tachometer for interruptions of the ignition coil primary current.) | Cranking speed (rpm) | Engine speed (rpm) | o If the tachometer reading is 0, there is no interruption of the ignition coil primary current.<br>1) Check the power transistor and the control circuit.<br>2) Check the ignition coil and the coil power supply circuit.<br>o If the multi-use tester's indicated rpm is abnormal<br>1) Malfunction of the crank angle sensor circuit<br>2) Malfunction of the crank angle sensor<br>3) Malfunction of the timing belt |
| | | Approx. 200 | Approx. 200 | |

| Check points | Check conditions | | Test specification | Probable causes of malfunction (or action) |
|---|---|---|---|---|
| Injectors<br>o Service data<br>o Item No.41 | o Engine cranking | Listen for operation sounds. | Operation sounds of the injectors can be heard. | o Injector malfunction<br>o Improper contact of the connector and control relay contacts |
| | | Engine coolant temperature [°C (°F)] | Injector activation time *2 (msec.) | o Malfunction of the engine coolant temperature sensor<br>o Malfunction of the ignition switch—ST. |
| | | 0 (32)*1 | Approx. 21 | |
| | | 20 (68) | Approx. 43 | |
| | | 80 (176) | Approx. 10 | |

NOTE
*1: When the engine coolant temperature is 0°C (32°F), injectors inject simultaneously at 4 cylinders.
*2: Injector activation times are indicated at a battery voltage of 11V and a cranking speed of 250 rpm or lower.

# DIAGNOSIS FLOW CHARTS—SONATA 4 CYLINDER

**Sensor Check**

| Check points | Check conditions | | Test specification | Probable cause of malfunction (or action) |
|---|---|---|---|---|
| Self-diagnosis output | o Engine: idling (2 minutes or more after engine start) | | Normal | o Check in accordance with the diagnosis code. (Note that the diagnosis code will be erased if there is disconnection or damage of the engine control unit back-up power-supply circuit.) o If various diagnosis codes are output, the most frequent cause is damage or disconnection of the power-supply or earth circuit. |
| Oxygen sensor o Service data o Item No.11 | o Engine warm-up (Make the mixture lean by engine speed reduction, and rich by racing.) | Engine condition | Voltage (mV) | o If the oxygen sensor output voltage is high during sudden deceleration 1) Check for injector leakage. 2) Check the oxygen sensor signal circuit. o If the oxygen sensor output voltage is low during engine racing 1) Check the oxygen sensor and signal circuit. |
| | | When sudden deceleration from 4,000 rpm | 250 or lower | |
| | | When engine is suddenly raced | 500—1,000 | |
| Air-flow sensor o Service data o Item No.12 | o Engine warm-up | Engine condition | Frequency (Hz) | o If the air-flow sensor output frequency suddenly changes greatly, improper contact of the air-flow sensor or connector is probable. o If the output frequency of the air-flow sensor is unusually high or low, check the air cleaner element. o If the output frequency of the air-flow sensor is high, an increase of engine resistance or leakage of compression pressure is probable. |
| | | 750 rpm (Idling) | 25—50 | |
| | | 2,000 rpm | 70—130 | |
| | | Racing | Increase caused by racing | |

| Check points | Check conditions | | Test specification | Probable cause of malfunction (or action) |
|---|---|---|---|---|
| Air temperature sensor o Service data o Item No.13 | o Ignition switch: ON, or engine running | Air temperature °C (°F) | Temperature °C (°F) | o Malfunction of air temperature or related circuitry |
| | | -20 (-4) | -20 (-4) | |
| | | 0 (32) | 0 (32) | |
| | | 20 (68) | 20 (68) | |
| | | 40 (104) | 40 (104) | |
| | | 80 (176) | 80 (176) | |
| | o Ignition switch: ON | Warm by using hair dryer or other method. | Increases | |
| Throttle-position sensor o Service data o Item No.14 | o Hold for 15 seconds or longer with the ignition switch at "ON". | Throttle valve | Voltage (mV) | o Throttle-position sensor maladjustment o Throttle-position sensor or related circuitry malfunction |
| | | Idling position | 480-520 | |
| | | Opens slowly. | Becomes higher in proportion to valve opening | |
| | | Fully open | 4,500—5,500 | |
| Motor position sensor o Service data o Item No.15 | o Engine: idling after warm-up (The idle position switch must be ON.) | Engine condition | Voltage (mV) | o If the voltage is low, check whether or not there is air intake. o If the voltage is high, the probable cause is: 1) Deposits adhered to the throttle valve 2) Increased engine resistance o If the voltage or idling speed is abnormal even though each part of the engine is normal, the probable cause is: 1) Improper adjustment of the idle speed control and/or the throttle position sensor 2) Malfunction of the motor position sensor or the related circuit |
| | | Idle (750 rpm) | 500—1,300 | |

# DIAGNOSIS FLOW CHARTS—SONATA 4 CYLINDER

| Check points | Check conditions | | Test specification | Probable cause of malfunction (or action) |
|---|---|---|---|---|
| Motor position sensor<br>o Service data<br>o Item No.15 | (The compressor clutch must be activated when the air conditioner switch is switched ON.) | Air conditioner switch ON (900 rpm) | 800—1,800 | o If the engine speed does not increase when the air conditioner switch is switched from OFF to ON, check the DC motor drive circuit<br>o Check the air conditioner system |
| | | Air conditioner switch: ON<br>Shift lever: "D" range (700 rpm) | 900—1,900 | o Check the inhibitor switch and the signal circuit |
| Crank signal<br>o Service data<br>o Item No.18 | o Ignition switch: ON | | OFF | o Ignition switch—ST signal circuit check<br>o Ignition switch check |
| Engine coolant temperature sensor<br>o Service data<br>o Item No.21 | o Ignition switch: ON, or engine running | Engine coolant temperature °C (°F) | Temperature °C (°F) | o Engine coolant temperature sensor or related circuitry malfunction |
| | | -20 (-4) | -20 (-4) | |
| | | 0 (32) | 0 (32) | |
| | | 20 (68) | 20 (68) | |
| | | 40 (104) | 40 (104) | |
| | | 80 (176) | 80 (176) | |
| Crank angle sensor<br>o Service data<br>o Item No.22 | o Engine: idling (Check with the ignition switch ON.) | Engine coolant temperature °C (°F) | Idling rpm | o If the idling speed suddenly becomes greater, a malfunction of the crank angle sensor or improper contact of the connector is probable |
| | | -20 (-4) | 1,500—1,700 | |
| | | 0 (32) | 1,350—1,550 | |
| | | 20 (68) | 1,200—1,400 | |
| | | 40 (104) | 900—1,100 | |
| | | 80 (176) | 650—850 | |

| Check points | Check conditions | | Test specification | Probable cause of malfunction (or action) |
|---|---|---|---|---|
| Barometric pressure sensor<br>o Service data<br>o Item No.25 | o Ignition switch: ON | Altitude m (ft.) | Pressure mm Hg | o Barometric pressure sensor or related circuitry malfunction.<br>(If the barometric pressure sensor pressure is low at high speed, restriction of the air cleaner element is probable.) |
| | | 0 (0) | 760 | |
| | | 600 (1,968) | 710 | |
| | | 1,200 (3,937) | 660 | |
| | | 1,800 (5,905) | 610 | |
| | o Engine: racing at 2,000 rpm | Gradually close the air-intake duct by using a hand. | Decreases. | |
| Idle switch<br>o Service data<br>o Item No.26 | o Ignition switch: ON<br>(Checking by using the accelerator pedal several times) | Throttle valve idling position | ON | o Idle position switch or related circuitry malfunction<br>o Improper adjustment of the accelerator cable or cruise control cable. |
| | | Open the throttle valve slightly. | OFF | o Malfunction of the setting of the fixed SAS. |
| Power-steering switch<br>o Service data<br>o Item No.27 | o Engine: idling | Steering wheel neutral position (wheels straight-ahead direction) | OFF | o Power-steering oil-pressure switch or signal circuit malfunction |
| | | Steering wheel half turn | ON | |
| Air-conditioner switch<br>o Service data<br>o Item No.28 | o Engine: idling (The air conditioner compressor could be activated when the air-conditioner switch is ON.) | Air-conditioner switch "OFF" | OFF | o Check the air conditioner system. |
| | | Air-conditioner switch "ON" | ON | |
| Inhibitor switch<br>o Service data<br>o Item No.29 | o Ignition switch: ON<br>(A/T models only) | Shift lever: "P" or "N" | "P", "N" | o Malfunction of the inhbitor switch or the signal circuit.<br>o Improper adjustment of the control cable between the shift lever and the inhibitor switch. |
| | | Shift lever: "D", "2", "L" or "R" | "D", "2", "L", "R" | |

# DIAGNOSIS FLOW CHARTS—SONATA 4 CYLINDER

| Check points | Check conditions | | Test specification | Probable cause of malfunction (or action) |
|---|---|---|---|---|
| EGR temperature sensor (California)<br>o Service data<br>o Item No.43 | o Engine: warmed up (Engine is maintained in a constant state for 2 minutes or more) | Engine condition | Temperature °C (°F) | o Check the EGR temperature sensor.<br>o Check the EGR control system.<br>o Check the EGR valve.<br>o Check the EGR control solenoid valve.<br>o Check the EGR control vacuum. |
| | | 750 rpm (idling) | 100 (212) or less | |
| | | o 3,500 rpm<br>o Disconnect the vacuum hose (yellow stripe) from the A port nipple of the throttle body, and pinch the hose end with fingers. | 150 (302) or more | |
| Injectors<br>o Actuator test<br>o Item No. 1—4 | o Engine: idling after warm-up (Cut off the injectors in sequence during idling after engine warm-up; check the idling condition.) | Injector No. | Engine | o If the idling condition doesn't change, check the cylinder.<br>1) Check the injector operation sound.<br>2) Check the spark plug and high-tension cable.<br>3) Check the power transistor unit and control circuit. |
| | | 1 | Idling condition changes more. (Becomes more unstable, or engine stalls.) | |
| | | 2 | | |
| | | 3 | | |
| | | 4 | | |
| Injector<br>o Service data<br>o Item No.41 | o Engine: warmed up | Engine condition | Activation time (msec.) | o If the injector activation time is unusually long or short, there is a malfunction of the air-flow sensor, engine coolant temperature sensor, intake-air temperature sensor, or barometric pressure sensor.<br>o If the injector activation time is long, increased engine resistance or leakage of compression pressure is probable. |
| | | 750 rpm (Idling) | 2.5—4.0 | |
| | | 2,000 rpm | 2.5—4.0 | |
| | | Rapid racing | Increases. | |
| Ignition advance (power transistor)<br>o Service data<br>o Item No.44 | o Engine: warmed up<br>o Timing light: set | Engine speed (rpm) | Ignition advance (°BTDC) | o If the ignition advance and actual ignition timing are different, adjust the ignition timing.<br>[The ignition timing may fluctuate during idling, but this is not a problem. The advance is greater (approx. 5°) at high altitude.] |
| | | 750 (Idling) | 5—15 | |
| | | 2,000 | 30—40 | |

| Check points | Check conditions | | Test specifications | Probable cause of malfunction (or action) |
|---|---|---|---|---|
| Air conditioner relay<br>o Service data<br>o Item No.49 | o Engine: idling after warm-up | Air conditioner switch | Air conditioner relay | o If the air conditioner relay output is abnormal, check the air conditioner signal input circuit and the air conditioner system.<br>o If the activation of the air conditioner compressor clutch is not normal, check the compressor clutch and the relay circuit. |
| | | OFF | OFF (compressor clutch non-activation) | |
| | | ON | ON (compressor clutch activation) | |
| Purge control solenoid valve<br>o Actuator test<br>o Item No.8 | o Ignition switch: ON (Engine stop) | Actuator forced actuation | Operation sound is audible | o Check the purge control solenoid valve.<br>o Check the purge control solenoid valve drive circuit. |
| EGR control solenoid valve (California)<br>o Actuator test<br>o Item No.10 | o Ignition switch: ON (Engine stop) | Actuator forced actuation | Operation sound is audible | o Check the EGR control solenoid valve.<br>o Check the EGR control solenoid valve drive circuit. |

**Cranking Check (should be performed when the engine fails to start, or before starting the engine)**

Item No.: Multi-use tester code number

| Check points | Check conditions | Test specification | Probable cause of malfunction |
|---|---|---|---|
| Battery voltage<br>o Service data<br>o Item No.16 | o Ignition switch: ON | 11—13 V | o Low battery voltage<br>o Power not supplied to the electronic control unit<br>1) Check the power-supply circuit.<br>2) Check the ignition switch.<br>3) Check the control relay.<br>o Malfunction of the electronic control unit ground circuit |

# DIAGNOSIS FLOW CHARTS—EXCEL

| Check points | Check conditions | | Test specification | Probable cause of malfunction |
|---|---|---|---|---|
| Malfunction code read out<br>o Self diagnostic | o After holding for 15 seconds or longer with the ignition switch at "ON", move the ignition switch to "LOCK" and disconnect the ISC servo connector.<br>o Crank for four seconds or longer.<br>o Ignition switch: "ON"<br>(Check for damage or disconnection of the injector or crank angle sensor circuit.) | | Normal | o Check in accordance with the diagnostic code.<br>(Note that the diagnostic code will be erased if there is disconnection or damage to the electronic control unit back-up power-supply circuit.)<br>o If various diagnostic codes are output, the most frequent cause is damage or disconnection of the power-supply or ground circuit. |
| Fuel pump<br>o Actuator test<br>o Item No.7 | Try under both conditions:<br>o Engine cranking<br>o Forced fuel pump activation | Pinch off the return hose. | The pulsations of the fuel flow can be felt. | o Voltage is not supplied to the fuel pump.<br>1) Check the ignition switch (IG and ST)<br>2) Check the control relay.<br>3) Check the related circuits.<br>o Malfunction of the fuel pump. |
| | | Listen closely to the fuel tank. | The pump drive sound can be heard. | |
| Crank angle sensor<br>o Service data<br>o Item No.22 | o Engine cranking<br>o Tachometer connected (check, by using the tachometer for interruptions of the ignition coil primary current.) | Cranking speed (rpm) | Engine speed (rpm) | o If the tachometer reading is 0, there is no interruption of the ignition coil primary current.<br>1) Check the power transistor and the control circuit.<br>2) Check the ignition coil and the coil power supply circuit.<br>o If the multi-use tester's indicated rpm is abnormal<br>1) Malfunction of the crank angle sensor circuit.<br>2) Malfunction of the crank angle sensor.<br>3) Malfunction of the timing belt. |
| | | Approx. 200 | Approx. 200 | |

| Check points | Check conditions | | Test specification | Probable cause of malfunction |
|---|---|---|---|---|
| Injectors<br>o Service data<br>o Item No.41 | o Engine cranking | Listen for operation | Injectors should be heard | o Injector malfunction.<br>o Improper contact of the connector and control relay contacts. |
| | | Engine coolant temperature [°C (°F)] | Injector activation time (msec.) | o Malfunction of the engine coolant temperature sensor.<br>o Malfunction of the ignition switch—ST. |
| | | 0 (32) | Approx. 17 | |
| | | 20 (68) | Approx. 35 | |
| | | 80 (176) | Approx. 8.5 | |

# DIAGNOSIS FLOW CHARTS—EXCEL

| Check points | Check conditions | | Test specification | Probable cause of malfunction |
|---|---|---|---|---|
| Self-diagnostic output | o Engine: idling (2 minutes or more after engine start) | | Normal | o Check in accordance with the diagnostic code. (Note that the diagnostic code will be erased if there is disconnection or damage of the engine control unit back-up power-supply circuit.)<br>o If numerous diagnostic codes are output, the most frequent cause is damage or disconnection of the power-supply or ground circuit. |
| Oxygen sensor<br>o Service data<br>o Item No.11 | o Engine warm (Make the mixture lean by engine speed reduction, and rich by racing.) | Engine condition | Voltage (mV) | o If the oxygen sensor output voltage is high during sudden deceleration<br>1) Check for injector leakage.<br>2) Check the oxygen sensor signal circuit.<br>o If the oxygen sensor output voltage is low during high engine speed<br>1) Check the oxygen sensor and signal circuit. |
| | | When sudden deceleration from 4,000 rpm | 200 or lower | |
| | | When engine is suddenly revved | 600—1,000 | |
| | o Engine warm (Using the oxygen sensor signal, check the air/fuel mixture ratio, and also check the condition of the electronic control unit.) | Engine speed (rpm) | Voltage (mV) | o If the oxygen sensor signal is normal, the electronic control unit is regulating the air/fuel mixture ratio normally.<br>o If the oxygen sensor output voltage is low at all times, check for intake air restriction.<br>o If the oxygen sensor output voltage is high at all times, check for leakage of the injector. |
| | | 700 (idle) | 400 or lower | |
| | | 2,000 | (changes)<br>600—1,000 | |
| Air-flow sensor<br>o Service data<br>o Item No.12 | o Engine warm | Engine condition | Frequency (Hz) | o If the air-flow sensor output suddenly changes greatly, improper contact of the air-flow sensor or connector is probable.<br>o If the output frequency of the air-flow sensor is unusually high or low, check the air cleaner element.<br>o If the output frequency of the air-flow sensor is high, an increase of engine resistance or leakage of compression pressure is probable. |
| | | 700 rpm (Idle) | 25—40 | |
| | | 2,000 rpm | 63—83 | |
| | | Revving | Increases caused by racing | |

| Check points | Check conditions | | Test specification | Probable cause of malfunction |
|---|---|---|---|---|
| Motor position sensor<br>o Sensor data<br>o Item No.15 | The compressor clutch must be activated when the air conditioner switch is switched ON.) | Air conditioner switch ON (800 rpm) | 800—1,800 | o If the engine speed does not increase when the air conditioner switch is switched from OFF to ON, check the DC motor drive circuit<br>o Check the air conditioner system |
| | | Air conditioner switch: ON Shift lever: "D" range (700 rpm) | 900—1,900 | o Check the inhibitor switch and the signal circuit |
| Crank signal<br>o Service data<br>o Item No.18 | o Ignition switch: ON | | OFF | o Ignition switch—ST signal circuit check<br>o Ignition switch check |
| Engine coolant temperature sensor<br>o Service data<br>o Item No.21 | o Ignition switch: ON, or engine running | Engine coolant temperature °C (°F) | Temperature °C (°F) | o Engine coolant temperature sensor or related circuit malfunction |
| | | -20 (-4) | -20 (-4) | |
| | | 0 (32) | 0 (32) | |
| | | 20 (68) | 20 (68) | |
| | | 40 (104) | 40 (104) | |
| | | 80 (176) | 80 (176) | |
| Crank angle sensor<br>o Service data<br>o Item No.22 | o Engine: idle (Check with the ignition switch ON.) | Engine coolant temperature °C (°F) | Idle rpm | o If the idle speed suddenly increases, a malfunction of the crank angle sensor or improper contact of the connector is probable. |
| | | -20 (-4) | 1,500—1,700 | |
| | | 0 (32) | 1,350—1,550 | |
| | | 20 (68) | 1,150—1,350 | |
| | | 40 (104) | 950—1,150 | |
| | | 80 (176) | 650—850 | |

# DIAGNOSIS FLOW CHARTS—EXCEL

| Check points | Check conditions | | Test specification | Probable cause of malfunction |
|---|---|---|---|---|
| Intake-air temperature sensor<br>o Service data<br>o Item No.13 | o Ignition switch: ON, or engine running | Air temperature °C (°F) | Temperature °C (°F) | o Malfunction of the intake-air temperature or related circuit |
| | | -20 (-4) | -20 (-4) | |
| | | 0 (32) | 0 (32) | |
| | | 20 (68) | 20 (68) | |
| | | 40 (104) | 40 (104) | |
| | | 80 (176) | 80 (176) | |
| | o Ignition switch: ON | Warm by using a hair dryer or other method. | Increases | |
| Throttle-position sensor<br>o Service data<br>o Item No.14 | o Hold for 15 seconds or longer with the ignition switch at "ON". | Throttle valve | Voltage (mV) | o Throttle position sensor misadjusted<br>o Throttle position sensor or related circuit malfunction |
| | | Idle position | 480—520 | |
| | | Opens slowly. | Becomes higher in proportion to valve opening | |
| | | Fully open | 4,500—5,500 | |
| Motor position sensor<br>o Service data<br>o Item No.15 | o Engine: Idle after warm-up (The idle position switch must be ON.) | Engine condition | Voltage (mV) | o If the voltage is low, check whether or not there is air intake.<br>o If the voltage is high, the probable cause is:<br>1) Deposits adhered to the throttle valve<br>2) Increased engine resistance<br>o If the voltage or idling speed is abnormal even though each part of the engine is normal, the probable cause is:<br>1) Improper adjustment of the idle speed control and/or the throttle position sensor<br>2) Malfunction of the motor position sensor or the related circuit. |
| | | Idle (700 rpm) | 500—1,300 | |

| Check points | Check conditions | | Test specification | Probable cause of malfunction |
|---|---|---|---|---|
| Barometric pressure sensor<br>o Service data<br>o Item No.25 | o Ignition switch: ON | Altitude m (ft.) | Pressure mm Hg | o Barometric pressure sensor or related circuit malfunction.<br>(If the barometric pressure sensor pressure is low at high speed, clogging of the air cleaner element is probable.) |
| | | 0 (0) | 760 | |
| | | 600 (1,968) | 710 | |
| | | 1,200 (3,937) | 660 | |
| | | 1,800 (5,905) | 610 | |
| | o Engine: 2,000 rpm | Gradually close the air-intake duct by hand. | Decreases. | |
| Idle switch<br>o Service data<br>o Item No.26 | o Ignition switch: ON (Checking by depressing the accelerator pedal several times) | Throttle valve idling position | ON | o Idle position switch or related circuit malfunction<br>o Improper adjustment of the accelerator cable or the cruise control cable.<br>o Misadjusted fixed SAS. |
| | | Open the throttle valve slightly. | OFF | |
| Power-steering switch<br>o Service data<br>o Item No.27 | o Engine: Idle | Steering wheel neutral position (wheels straight-ahead direction) | OFF | o Power steering oil-pressure switch or signal circuit malfunction |
| | | Steering wheel half turn | ON | |
| Air-conditioner switch<br>o Service data<br>o Item No.28 | o Engine: idle (The air conditioner compressor will be activated when the air-conditioner switch is ON.) | Air-conditioner switch "OFF" | OFF | o Check the air conditioner system. |
| | | Air-conditioner switch "ON" | ON | |
| Inhibitor switch<br>o Service data<br>o Item No.29 | o Ignition switch: ON (A/T models only) | Shift lever: "P" or "N" | "P", "N" | o Malfunction of the inhibitor switch or the signal circuit.<br>o Improper adjustment of the control cable between the shift lever and the inhibitor switch. |
| | | Shift lever: "D", "2", "L" or "R" | "D", "2", "L", "R" | |

# DIAGNOSIS FLOW CHARTS – EXCEL

| Check points | Check conditions | | Test specification | Probable cause of malfunction |
|---|---|---|---|---|
| EGR temperature sensor (California vehicle only)<br>o Service data<br>o Item No.43 | o Engine: warm | Engine condition | Temperature °C (°F) | o Check the EGR temperature sensor.<br>o Check the EGR control system.<br>o Check the EGR valve.<br>o Check the EGR control solenoid valve.<br>o Check the EGR control vacuum. |
| | | 700 rpm (Idle) | 70 (158) or less | |
| | | 3,500 rpm<br>o Disconnect the vacuum hose (yellow) stripe from the A port nipple of the throttle body, and pinch the hose end with your fingers. | 70 (158) or more | |
| Injectors<br>o Actuator test<br>o Item No.1—4 | o Engine: idle after warm-up (Cut off the injectors in sequence during idle after engine warm-up; check the idle condition.) | Injector No. | Engine | o If the idling condition doesn't change, check the cylinder.<br>1) Check the injector operation sound.<br>2) Check the spark plug and high-tension cable.<br>3) Check the power transistor unit and control circuit.<br>4) Check injecting condition |
| | | 1 | Unstable idle | |
| | | 2 | | |
| | | 3 | | |
| | | 4 | | |
| Injector<br>o Service data<br>o Item No.41 | o Engine: warmed up | Engine condition | Activation time (msec.) | o If the injector activation time is unusually long or short, there is a malfunction of the air-flow sensor, engine coolant temperature sensor, intake-air temperature sensor, or barometric pressure sensor.<br>o If the injector activation time is long, increased engine resistance or leakage of compression pressure is probable. |
| | | 700 rpm (Idle) | 2.2—2.9 | |
| | | 2,000 rpm | 1.8—2.6 | |
| | | Rapid racing | Increases | |
| Ignition advance (power transistor)<br>o Service data<br>o Item No.44 | o Engine: warmed up<br>o Timing light: set | Engine speed (rpm) | Ignition advance (°BTDC) | o If the ignition advance and actual ignition timing are different, adjust the ignition timing.<br>[The ignition timing may fluctuate during idling, but this is not a problem. The advance is greater (approx. 5°) at high altitude.] |
| | | 700 (Idle) | 8—12 | |
| | | 2,000 | 26—34 | |

| Check points | Check conditions | | Test specification | Probable cause of malfunction |
|---|---|---|---|---|
| Air conditioner relay<br>o Service data<br>o Item No.49 | o Engine: idle after warm-up | Air conditioner switch | Air conditioner relay | o If the air conditioner relay output is abnormal, check the air conditioner signal input circuit and the air conditioner system.<br>o If the activation of the air conditioner compressor clutch is not normal, check the compressor clutch and the relay circuit. |
| | | OFF | OFF (compressor clutch non-activation) | |
| | | ON | ON (compressor clutch activation) | |
| Purge control solenoid valve<br>o Actuator test<br>o Item No.8 | o Ignition switch: ON (Engine stop) | Actuator forced actuation | Operation sound is audible | o Check the purge control solenoid valve.<br>o Check the purge control solenoid valve drive circuit. |
| EGR control solenoid valve (California vehicle only)<br>o Actuator test<br>o Item No.10 | o Ignition switch: ON (Engine stop) | Actuator forced actuation | Operation sound is audible | o Check the EGR control solenoid valve.<br>o Check the EGR control solenoid valve drive circuit. |

# DIAGNOSIS FLOW CHARTS—SONATA V6

| Check points | Check conditions | Test specification | Probable cause of malfunction |
|---|---|---|---|
| Malfunction code read out<br>o Self diagnostic | o Crank for four seconds or longer.<br>o Ignition switch: "ON"<br>(Check for damage or disconnection of the injector or crank angle sensor circuit.) | Normal | o Check in accordance with the diagnostic code.<br>(Note that the diagnostic code will be erased if there is disconnection or damage of the electronic control unit back-up power-supply circuit.)<br>o If various diagnostic codes are output, the most frequent cause is damage or disconnection of the power-supply or ground circuit. |
| Fuel pump<br>o Actuator test<br>o Item No.7 | Try under both conditions:<br>o Engine cranking<br>o Forced fuel pump activation<br>— Pinch the return hose.<br>— Listen at the fuel tank. | The pulsations of the fuel flow can be felt.<br>The pump can be heard. | o Voltage is not supplied to the fuel pump.<br>1) Check the ignition switch (IG and ST)<br>2) Check the control relay.<br>3) Check the related circuits.<br>o Malfunction of the fuel pump. |
| Crank angle sensor<br>o Service data<br>o Item No.22 | o Engine cranking<br>o Tachometer connected (check, by using the tachometer for interruptions of the ignition coil primary current.)<br>— Cranking speed (rpm): Approx. 200 | Engine speed (rpm): Approx. 200 | o If the tachometer reading is 0, there is no interruption of the ignition coil primary current.<br>1) Check the power transistor and the control circuit.<br>2) Check the ignition coil and the coil power supply circuit.<br>o If the multi-use tester's indicated rpm is abnormal<br>1) Malfunction of the crank angle sensor circuit.<br>2) Malfunction of the crank angle sensor.<br>3) Damage to the timing belt. |
| Crank signal<br>o Service data<br>o Item No.18 | Ignition switch: ON<br>— Engine stop<br>— Cranking | OFF<br>ON | o Ignition switch-ST signal circuit check<br>o Ignition switch check |

| Check points | Check conditions | | Test specification | Probable cause of malfunction |
|---|---|---|---|---|
| Injectors<br>o Service data<br>o Item No.41 | o Engine cranking | Listen for operation | Injectors should be heard | o Injector malfunction.<br>o Improper contact of the connector and control relay contacts. |
| | | Engine coolant temperature [°C (°F)] | Injector activation time *2 (msec.) | o Malfunction of the engine coolant temperature sensor.<br>o Malfunction of the ignition switch—ST. |
| | | 0 (32)*1 | Approx. 14 | |
| | | 20 (68) | Approx. 40 | |
| | | 80 (176) | Approx. 9 | |

NOTE
*1: When the engine coolant temperature is 0°C (32°F), injectors inject simultaneously at all cylinders.
*2: Injector activation times are at a battery voltage of 11V and a cranking speed of 250 rpm or lower.

## DIAGNOSIS FLOW CHARTS—SONATA V6

| Check points | Check conditions | | Test specification | Probable cause of malfunction |
|---|---|---|---|---|
| Self-diagnostic output | o Engine: idling (2 minutes or more after engine start) | | Normal | o Check in accordance with the diagnostic code. (Note that the diagnostic code will be erased if there is disconnection or damage of the engine control unit back-up power-supply circuit.)<br>o If numerous diagnostic codes are output, the most frequent cause is damage or disconnection of the power-supply or earth circuit. |
| Oxygen sensor<br>o Service data<br>o Item No.11 | o Engine warm (Make the mixture lean by engine speed reduction, and rich by racing.) | Engine condition | Voltage (mV) | o If the oxygen sensor output voltage is high during sudden deceleration<br>1) Check for injector leakage.<br>2) Check the oxygen sensor signal circuit.<br>o If the oxygen sensor output voltage is low during high engine speed<br>1) Check the oxygen sensor and signal circuit. |
| | | When sudden deceleration from 4,000 rpm | 200 or lower | |
| | | When engine is suddenly revved | 600—1,000 | |
| | o Engine warm (Using the oxygen sensor signal, check the air/fuel mixture ratio, and also check the condition of the electronic control unit.) | Engine speed (rpm) | Voltage (mV) | o If the oxygen sensor signal is normal, the electronic control unit is regulating the air/fuel mixture ratio normally.<br>o If the oxygen sensor output voltage is low at all times, check for intake air restriction.<br>o If the oxygen sensor output voltage is high at all times, check for leakage of the injector. |
| | | 700 (idle) | 400 or lower | |
| | | 2,000 | (changes)<br>600—1,000 | |
| Air-flow sensor<br>o Service data<br>o Item No.12 | o Engine warm | Engine condition | Frequency (Hz) | o If the air-flow sensor output suddenly changes greatly, improper contact of the air-flow sensor or connector is probable.<br>o If the output frequency of the air-flow sensor is unusually high or low, check the air cleaner element.<br>o If the output frequency of the air-flow sensor is high, an increase of engine resistance or leakage of compression pressure is probable. |
| | | 700 rpm (Idle) | 30—45 | |
| | | 2,000 rpm | 85—105 | |
| | | Revving | Increases | |
| Intake-air temperature sensor<br>o Service data<br>o Item No.13 | o Ignition switch: ON, or engine running | Intake-air temperature °C (°F) | Temperature °C (°F) | o Malfunction of the intake-air temperature or related circuit |
| | | -20 (-4) | -20 (-4) | |
| | | 0 (32) | 0 (32) | |
| | | 20 (68) | 20 (68) | |
| | | 40 (104) | 40 (104) | |
| | | 80 (176) | 80 (176) | |
| | o Ignition switch: ON | Warm by using a hair dryer or other method. | Increases | |
| Throttle-position sensor<br>o Service data<br>o Item No.14 | o Hold for 15 seconds or longer with the ignition switch "ON". | Throttle valve | Voltage (mV) | o Throttle position sensor misadjusted<br>o Throttle position sensor or related circuit malfunction |
| | | Idling position | 480-520 | |
| | | Opens slowly. | Becomes higher in proportion to valve opening | |
| | | Fully open | 4,500—5,500 | |
| Battery voltage<br>o Service data<br>o Item No.16 | o Ignition switch: ON | | 11—13V | o Measure the battery voltage<br>o Check the circuit that supplies the ECU power. |
| Crank signal<br>o Service data<br>o Item No.18 | o Ignition switch: ON | | OFF | o Ignition switch—ST signal circuit check<br>o Ignition switch check |
| Coolant temperature sensor<br>o Service data<br>o Item No.21 | o Ignition switch: ON | Engine coolant temperature °C (°F) | Temperature °C (°F) | o Coolant temperature sensor or related circuit malfunction |
| | | -20 (-4) | -20 (-4) | |
| | | 0 (32) | 0 (32) | |
| | | 20 (68) | 20 (68) | |
| | | 40 (104) | 40 (104) | |
| | | 80 (176) | 80 (176) | |

# DIAGNOSIS FLOW CHARTS—SONATA V6

| Check points | Check conditions | | Test specification | Probable cause of malfunction |
|---|---|---|---|---|
| Crank angle sensor<br>o Service data<br>o Item No.22 | o Engine: idling (Check with the ignition switch ON.) | Engine coolant temperature °C (°F) | Idling rpm | o If the idle speed suddenly increases, a malfunction of the crank angle sensor or improper contact of the connector is probable. |
| | | -20 (-4) | 1,500—1,700 | |
| | | 0 (32) | 1,350—1,550 | |
| | | 20 (68) | 1,150—1,350 | |
| | | 40 (104) | 950—1,150 | |
| | | 80 (176) | 600—800 | |
| Barometric pressure sensor<br>o Service data<br>o Item No.25 | o Ignition switch: ON | Altitude m (ft.) | Pressure mm Hg | o Barometric pressure sensor or related circuit malfunction. (If the barometric pressure sensor pressure is low at high speed, clogging of the air cleaner element is probable.) |
| | | 0 (0) | 760 | |
| | | 600 (1,968) | 710 | |
| | | 1,200 (3,937) | 660 | |
| | | 1,800 (5,905) | 610 | |
| | o Engine: 2,000 rpm | Gradually close the air-intake duct by using a hand. | Decreases. | |
| Idle position switch<br>o Service data<br>o Item No.26 | o Ignition switch: ON (Check by pressing the accelerator pedal several times) | Throttle valve idling position | ON | o Idle position switch or related circuit malfunction<br>o Improper adjustment of the accelerator cable or the auto-cruise cable.<br>o Misadjusted fixed SAS. |
| | | Open the throttle valve slightly. | OFF | |
| Power steering oil-pressure switch<br>o Service data<br>o Item No.27 | o Engine: idling | Steering wheel neutral position (wheels in a straight-ahead direction) | OFF | o Power steering oil-pressure switch or signal circuit malfunction |
| | | Steering wheel half turn | ON | |

| Check points | Check conditions | | Test specification | Probable cause of malfunction |
|---|---|---|---|---|
| Air-conditioner switch<br>o Service data<br>o Item No.28 | o Engine: idling (The air conditioner compressor will be activated when the air-conditioner switch is ON.) | Air-conditioner switch "OFF" | OFF | o Check the air conditioner system. |
| | | Air-conditioner switch "ON" | ON | |
| Inhibitor switch<br>o Service data<br>o Item No.29 | o Ignition switch: ON | Shift lever: "P" or "N" | "P", "N" | o Malfunction of the inhibitor switch or the signal circuit.<br>o Improper adjustment of the control cable between the shift lever and the inhibitor switch. |
| | | Shift lever: "D", "2", "L" or "R" | "D", "2", "L", "R" | |
| EGR temperature sensor (California Only)<br>o Service data<br>o Item No.43 | o Engine: warm | Engine condition | Temperature °C (°F) | o Check the EGR temperature sensor.<br>o Check the EGR control system.<br>o Check the EGR valve.<br>o Check the EGR control solenoid valve.<br>o Check the EGR control vacuum. |
| | | 700 rpm (idling) | 100 (212) or less | |
| | | o 3,500 rpm<br>o Disconnect the vacuum hose (yellow stripe) from the A port nipple of the throttle body, and pinch the hose. | 150 (302) or more | |
| Injectors<br>o Actuator test<br>o Item No. 1—6 | o Engine: idling after warm-up (Cut off the injectors in sequence during idle after engine warm-up; check the idle condition.) | Injector No. | Engine | o If the idling condition doesn't change, check the cylinder.<br>1) Check the injector operation sound.<br>2) Check the spark plug and high-tension cable.<br>3) Check the power transistor unit and control circuit. |
| | | 1 | Unstable idle | |
| | | 2 | | |
| | | 3 | | |
| | | 4 | | |
| | | 5 | | |
| | | 6 | | |

## DIAGNOSIS FLOW CHARTS—SONATA V6

| Check points | Check conditions | | Test specification | Probable cause of malfunction |
|---|---|---|---|---|
| Injector<br>o Service data<br>o Item No.41 | o Engine: warmed up | Engine condition | Activation time (msec.) | o If the injector activation time is unusually long or short, there is a malfunction of the air-flow sensor, engine coolant temperature sensor, intake-air temperature sensor, or barometric pressure sensor.<br>o If the injector activation time is long, increased engine resistance or leakage of compression pressure is probable. |
| | | 700 rpm (Idling) | 2.7—3.2 | |
| | | 2,000 rpm | 2.4—2.9 | |
| | | Rapid racing | Increases. | |
| Ignition advance (power transistor)<br>o Service data<br>o Item No.44 | o Engine: warmed up<br>o Timing light: set (The timing light is set so as to check the actual ignition timing.) | Engine speed (rpm) | Ignition advance (°BTDC) | o If the ignition advance and actual ignition timing are different, adjust the ignition timing.<br>[The ignition timing may fluctuate during idling, but this is not a problem. The advance is greater (approx. 5°) at high altitude.] |
| | | 700 (Idling) | 5—15 | |
| | | 2,000 | 30—40 | |
| Stepper motor<br>o Service data<br>o Item No.45 | Engine: After warming up, idle the engine.<br>NOTE<br>The compressor clutch operates when the air conditioner switch is turned on. | Engine condition | Step | o If the number of steps increases to 100 or 120 or decreases to 0, a malfunction of the stepper motor or the activation circuit is probable.<br>o If the number of steps is small, check whether or not air is being sucked in.<br>o If the number of steps is large, either of the following is probable:<br>1) Deposits adhered to the throttle valve part.<br>2) Increased engine resistance<br>o If the number of steps is abnormal even though the engine is normal, adjust the basic idle speed. |
| | | 700 rpm (Idling) | 2—12 | |
| | | Air conditioner switch ON (900 rpm) | 30—70 | o Check the air conditioner system.<br>o If the engine speed does not increases when the air conditioner switch is switched from OFF to ON, check the stepper motor or ther activation circuit. |
| | | Air conditioner switch ON<br>Shift lever "D" (700 rpm) | 20—60 | o Malfunction of the inhibitor switch and signal circuit.<br>o Incorrect adjustment of the control cable between the shift lever and inhibitor switch. |

| Check points | Check conditions | | Test specification | Probable cause of malfunction |
|---|---|---|---|---|
| Air conditioner relay<br>o Service data<br>o Item No.49 | o Engine: idling after warm-up | Air conditioner switch | Air conditioner relay | o If the air conditioner relay output is abnormal, check the air conditioner signal input circuit and the air conditioner system.<br>o If the activation of the air conditioner compressor clutch is not normal, check the compressor clutch and the relay circuit. |
| | | OFF | OFF (compressor clutch not-activated) | |
| | | ON | ON (compressor clutch actived) | |
| Purge control solenoid valve<br>o Actuator test<br>o Item No.8 | o Ignition switch: ON (Engine stop) | Solenoid is turned ON | Operation is audible | o Check the purge control solenoid valve.<br>o Check the purge control solenoid valve drive circuit. |
| EGR control solenoid valve (California Only)<br>o Actuator test<br>o Item No.10 | o Ignition switch: ON (Engine stop) | Solenoid is turned ON | Operation is audible | o Check the EGR control solenoid valve.<br>o Check the EGR control solenoid valve drive circuit. |

# Diagnosis and Testing

## SELF-DIAGNOSTIC SYSTEM

**NOTE: For Self-Diagnostic System and Accessing Trouble Code Memory, see Section 3 "SELF-DIAGNOSTIC SYSTEM".**

## FUEL PRESSURE

### Relieving

1. Disconnect the fuel pump harness at the rear of the tank.
2. Start engine and allow to idle.
3. After engine stalls, fuel pressure is reduced.
4. Disconnect the negative battery cable and reconnect the fuel pump wiring.

### Testing

1. Relieve fuel pressure.
2. Turn ignition switch **OFF** and disconnect the negative battery cable.
3. Disconnect the high pressure hose at the delivery side. Using a fuel pressure gauge and adapter, install the adapter to the delivery pipe.
4. Connect the negative battery cable.
5. Apply battery voltage to the fuel pump drive terminal.
6. Fuel pressure with the pressure regulator vacuum hose plugged should be 47–50 psi (330–350 kPa).
7. Remove jumper wire from fuel pump drive terminal.
8. Reconnect vacuum hose and start engine. Fuel pressure should read 39 psi (270 kPa).
9. Relieve fuel pressure.
10. Disconnect gauge and reconnect fuel lines.
11. Connect battery cable, start engine and check for fuel leaks.

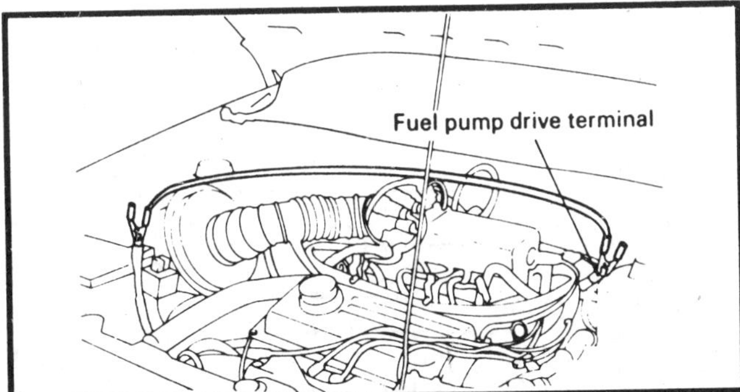


**Fuel pump drive terminal – Excel**

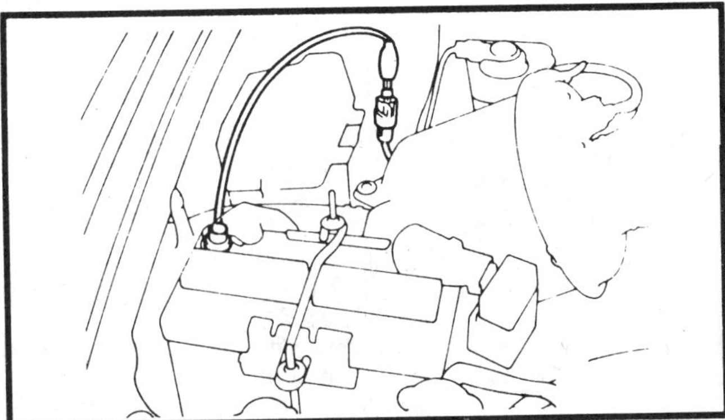

**Fuel pump drive terminal – Sonata**

## AIR FLOW SENSOR (AFS)

### Testing

1. Connect voltmeter between terminals 4 and 5 of the AFS connector.
2. Warm engine to normal operating temperature.
3. Measure the voltage at the terminals.
4. Raise the engine from idle to 3000 rpm. The voltage should vary between 2.7–3.2 volts. If not, replace the AFS.

## ATMOSPHERIC PRESSURE SENSOR

### Testing

1. Connect voltmeter between terminals 3 and 4 of the AFS connector.
2. Warm engine to normal operating temperature.
3. Measure the voltage at the terminals.
4. Slowly cover about half of the air cleaner air intake, while checking the voltage.
5. The voltage should vary from approximately 0.8 volts at 5.85 in. Hg to approximately 4 volts at 29 in. Hg. If not, replace the AFS.

## AIR TEMPERATURE SENSOR

### Testing

1. Disconnect the AFS connector.
2. Connect an ohmmeter between terminals 2 and 4. Check the resistance as follows:
   6 ohms – 32°F (0°C)
   2.7 ohms – 68°F (20°C)
   0.4 ohms – 176°F (80°C)

The higher the temperature the lower the resistance should be.

3. If the resistance is not within specification or does not change smoothly to temperature changes, replace the AFS assembly.

## ENGINE COOLANT TEMPERATURE SENSOR

### Testing

1. Remove the sensor from the intake manifold.

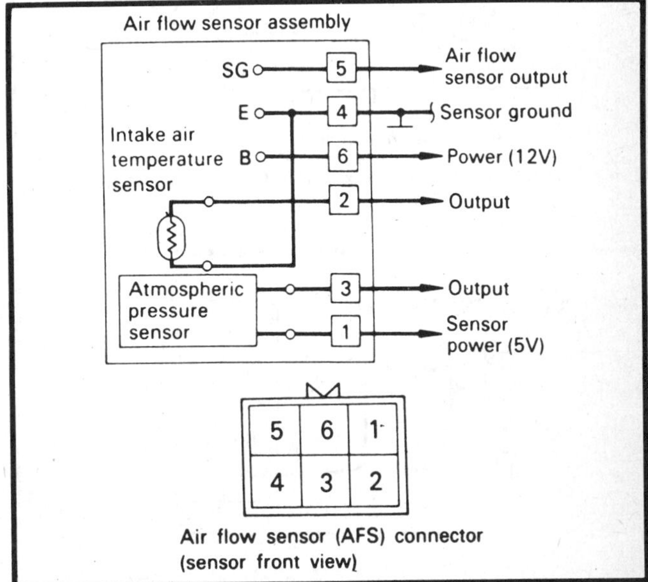


**Air flow sensor connector and assembly – Sonata 4 cylinder shown**

2. Place the sensor end in hot water, with the housing 0.12 in. (3mm) away from the water.

3. Connect an ohmmeter between the sensor terminals. Check the resistance as follows:

6 ohms – 32°F (0°C)
2.5 ohms – 68°F (20°C)
1.1 ohms – 104°F (40°C)
0.3 ohms – 176°F (80°C)

4. If the resistance is not within specification or does not change smoothly to temperature changes, replace the sensor.

## THROTTLE POSITION SENSOR (TPS)

### Testing

**4 CYLINDER**

1. Disconnect the throttle position sensor connector.
2. Measure the resistance between terminals 1 (ground) and 3 (power) of the connector.
3. The resistance should be 3.5–6.5 ohms.
4. Connect an analog type ohmmeter between terminals 1 and 2.
5. Slowly open the throttle valve from the idle position to the wide open throttle position. The resistance should change smoothly from approximately 0.5 ohms to 3.5–6.5 ohms.
6. If any test is not within specification, replace the TPS.

**6 CYLINDER**

1. Disconnect the throttle position sensor connector.
2. Measure the resistance between terminals 2 (ground) and 3 (power) of the connector.
3. The resistance should be 3.5–6.5 ohms.
4. Connect an analog type ohmmeter between terminals 2 and 4 (TPS).
5. Slowly open the throttle valve from the idle position to the wide open throttle position. The resistance should change smoothly from approximately 0.5 ohms to 3.5–6.5 ohms.
6. If any test is not within specification, replace the TPS.

## IDLE SWITCH

### Testing

**4 CYLINDER**

1. Disconnect the ISC motor.
2. Check for continuity between terminals 2 and ground.
3. Depress the accelerator and check that there is no continuity.
4. Release the pedal; there should be continuity.
5. If test fails, replace switch.

**6 CYLINDER**

1. Disconnect the ISC motor.
2. Check for continuity between terminals 1 and ground.
3. Depress the accelerator and check that there is no continuity.
4. Release the pedal; there should be continuity.
5. If test fails, replace switch.

## MOTOR POSITION SENSOR (MPS)

### Testing

1. Connect an ohmmeter between terminals 3 and 4 of the motor position sensor connector.
2. The resistance should be 4–6 ohms.
3. Connect an ohmmeter between terminals 1 and 4 of the motor position sensor connector.
4. Connect a 6 volt battery between terminals 1 and 4 of the ISC motor connector and check that the resistance changes smoothly from approximately 0.5 ohms to 4–6 ohms.
5. If any test fails, replace the ISC assembly.

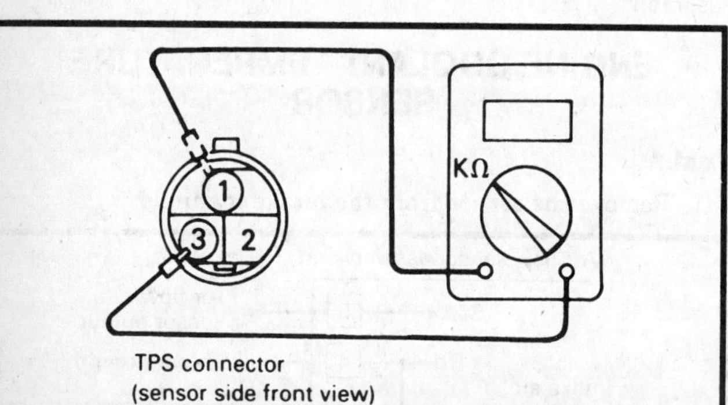

Throttle position sensor connector – 4 cylinder

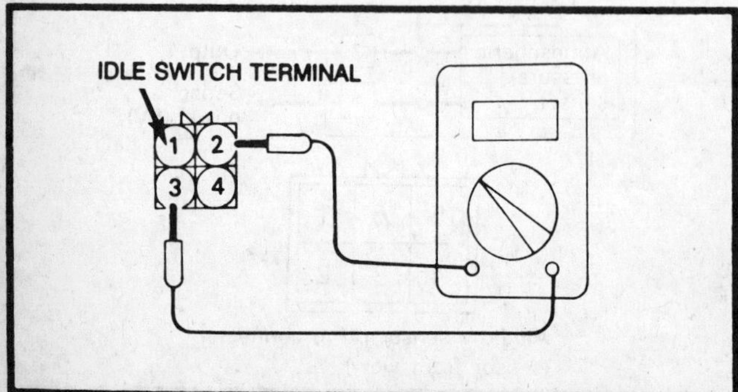

Throttle position sensor connector – 6 cylinder

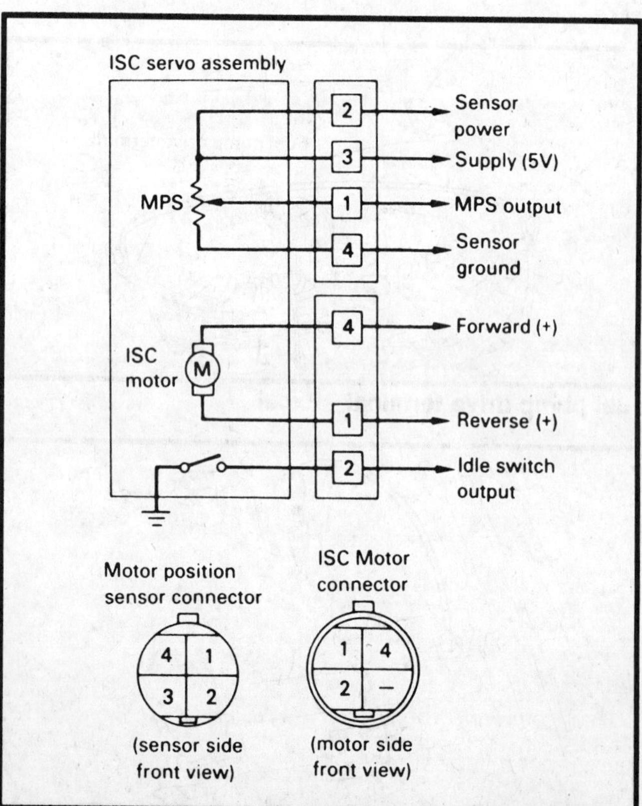

ISC and MPS connectors and circuit – 4 cylinder

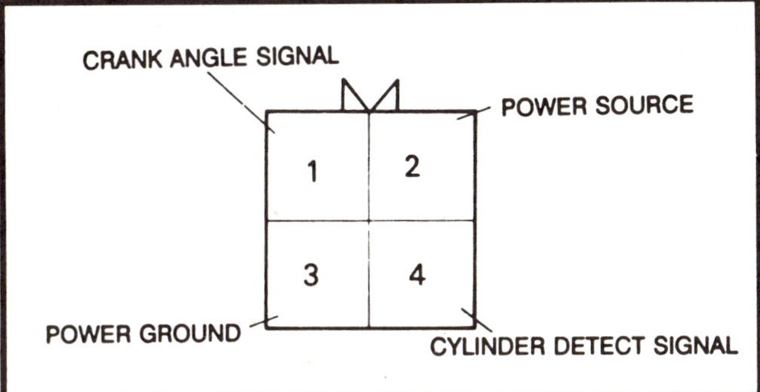


Crank angle sensor terminal connector

## TDC AND CRANK ANGLE SENSORS

### Testing

1. Connect a voltmeter across terminals 1 and 2.
2. Crank engine. Voltage should fluctuate between 0.2–1.2 volts.
3. Connect a voltmeter across terminals 1 and 3.
4. Crank engine. Voltage should be between 1.8–2.5 volts.
5. When the voltage is abnormal, check the sensor power and ground circuit.
6. If power and ground circuit check good, disassembly distributor and check sensors.

## OXYGEN SENSOR

### Testing

**NOTE: On the V6 heated type oxygen sensor, special testing harness and battery are necessary to test the oxygen sensor.**

1. Disconnect the oxygen sensor connector and connect a voltmeter to the oxygen sensor connector.
2. While increase engine rpm, measure the oxygen sensor output voltage.
3. Oxygen sensor voltage should vary between 0.2 and 0.9 volts, if not replace sensor.
4. On heated type oxygen sensors, the resistance across the heater terminals should be 30 ohms or more at 750°F (400°C).

## INJECTORS

### Testing

1. With engine running, use a stethoscope to listen for each injector operation.
2. Increase engine rpm, the ticking from the injectors should be at shorter intervals as the engine speed is increased.
3. Disconnect the connector for the injector.
4. Measure the resistance between the injector terminals.
5. The resistance should be 13–16 ohms at 68°F (20°C).
6.Replace defective injectors.

## IDLE SPEED CONTROL (ISC) SERVO

### Testing

**4 CYLINDER**

1. Disconnect the ISC motor connector.
2. Check the resistance of the ISC motor coils, terminals 1 and 4.
3. The resistance should be 5–35 ohms at 68°F (20°C).
4. Connect a 6 volt battery between terminal 1 and terminal 4 of the ISC motor connector. The servo should operate.

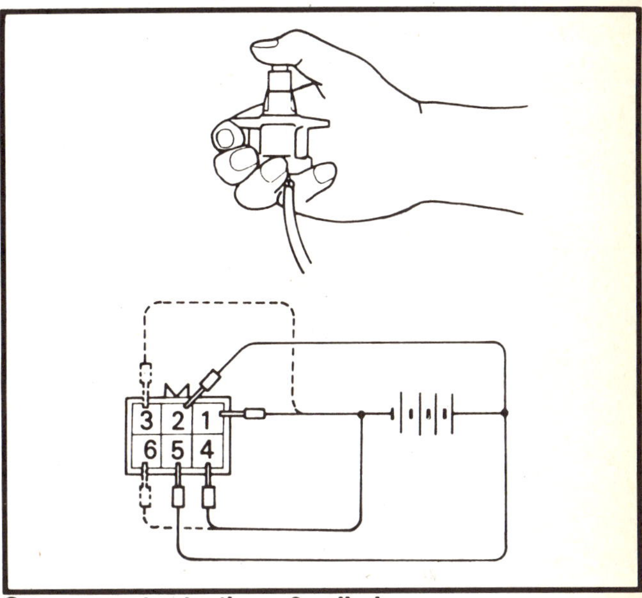


Stepper motor testing – 6 cylinder

5. If any test fails, replace the ISC assembly.

**6 CYLINDER**

1. Measure the resistance between terminals 2 and 3 and terminal 1.
2. Measure the resistance between terminals 5 and 5 and terminal 1.
3. The resistance should be 28–33 ohms at 68°F (20°C).
4. Connect the positive of a 6 volt power supply to the center terminals 2 and 5 on the connector.
5. Momentarily connect the negative of the battery to terminals 1–4, 3–4, 3–6, 1–6, and 1–4.
6. If movement occurs and resistance is with specification the stepper motor is considered to be functioning properly.

## CONTROL RELAY

### Testing

**NOTE: When applying battery voltage to the relay, make certain to apply it to the correct terminals or damage may result.**

**4 CYLINDER**

1. Using an ohmmeter, check the resistance across the terminals.
2. There should be no continuity across terminals 1–7 or 3–7.
3. There should be approximately 95 ohm across terminals 2–5 and across terminals 2–3.
4. There should be approximately 35 ohm across terminals 6–4.
5. Terminal 8 incorporates a diode; there should be no continuity between terminal 4–8, but there should be continuity between terminals 8–4.
6. Apply battery voltage across terminals 4–6; there should now be continuity between terminals 1 and 7.
7. Apply battery voltage across terminals 4–8; there should now be continuity between terminals 3 and 7.
8. If any test fails, replace the relay.

**6 CYLINDER**

1. Using an ohmmeter, check the resistance across the terminals.
2. There should be no continuity across terminals 3–4 or 1–6.
3. There should be approximately 95 ohm across terminals 3–9.

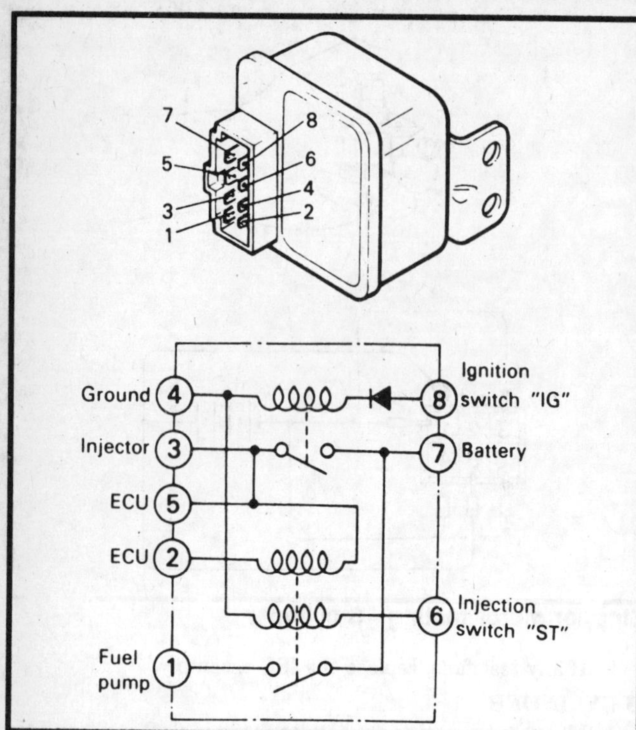


Control relay circuit–4 cylinder

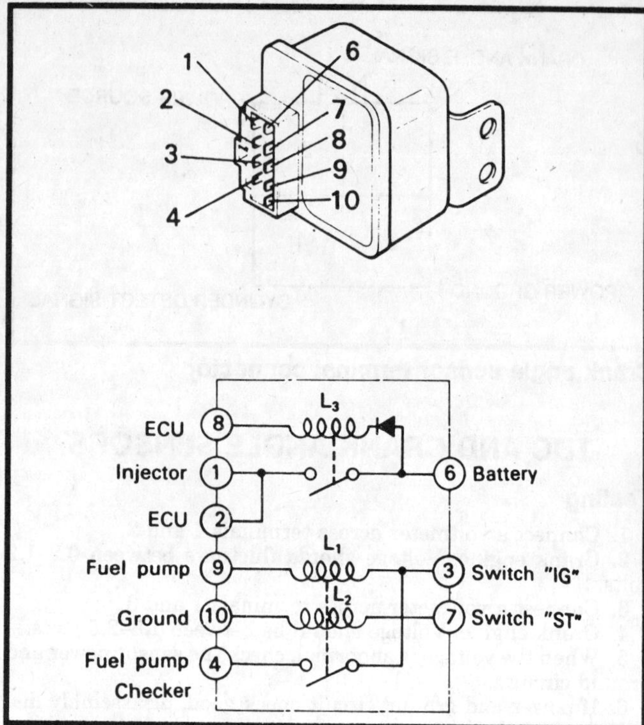


Control relay circuit–6 cylinder

4. There should be approximately 35 ohm across terminals 10–7.

5. Terminal 6 incorporates a diode; there should be no continuity between terminal 8–6, but there should be continuity between terminals 6–8.

6. Apply battery voltage across terminals 7–10; there should now be continuity between terminals 3 and 4.

7. Apply battery voltage across terminals 6–8; there should now be continuity between terminals 1 and 6.

8. If any test fails, replace the relay.

## POWER TRANSISTOR

### Testing

1. Connect an analog type ohmmeter across terminals 2 and 3 of the power transistor.
2. There should be no continuity.
3. Connect the negative terminal of a 3 volt battery to terminal 2 and the positive terminal to terminal 1 of the power transistor. The ohmmeter should show continuity.
4. Replace the power transistor, if there is a malfunction.

# Component Replacement

## THROTTLE BODY

### Removal and Installation

**NOTE: The throttle valve must not be removed from the throttle body, service as an assembly.**

1. Relieve fuel pressure.
2. Disconnect the negative battery cable
3. Drain the cooling system.
4. Remove the air intake hose.
5. Tag and remove all cables and hoses from the throttle body.
6. Remove the throttle body.

**To install:**

7. Install the throttle body with a new gasket and torque the bolts to 8 ft. lbs. (12 Nm).
8. Connect all hoses and cables to the proper locations.
9. Fill cooling system and connect the negative battery cable.

## INJECTORS

### Removal and Installation

1. Relieve fuel pressure.
2. Disconnect the negative battery cable
3. Drain the cooling system.
4. Remove the throttle body, as necessary to gain acess to the fuel delivery tube.
5. Remove the surge tank, if equipped.
6. Remove the fuel lines from delivery pipe.

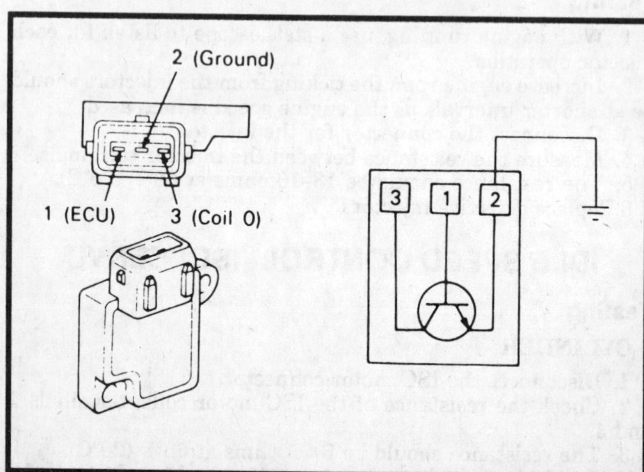


Power transistor connector

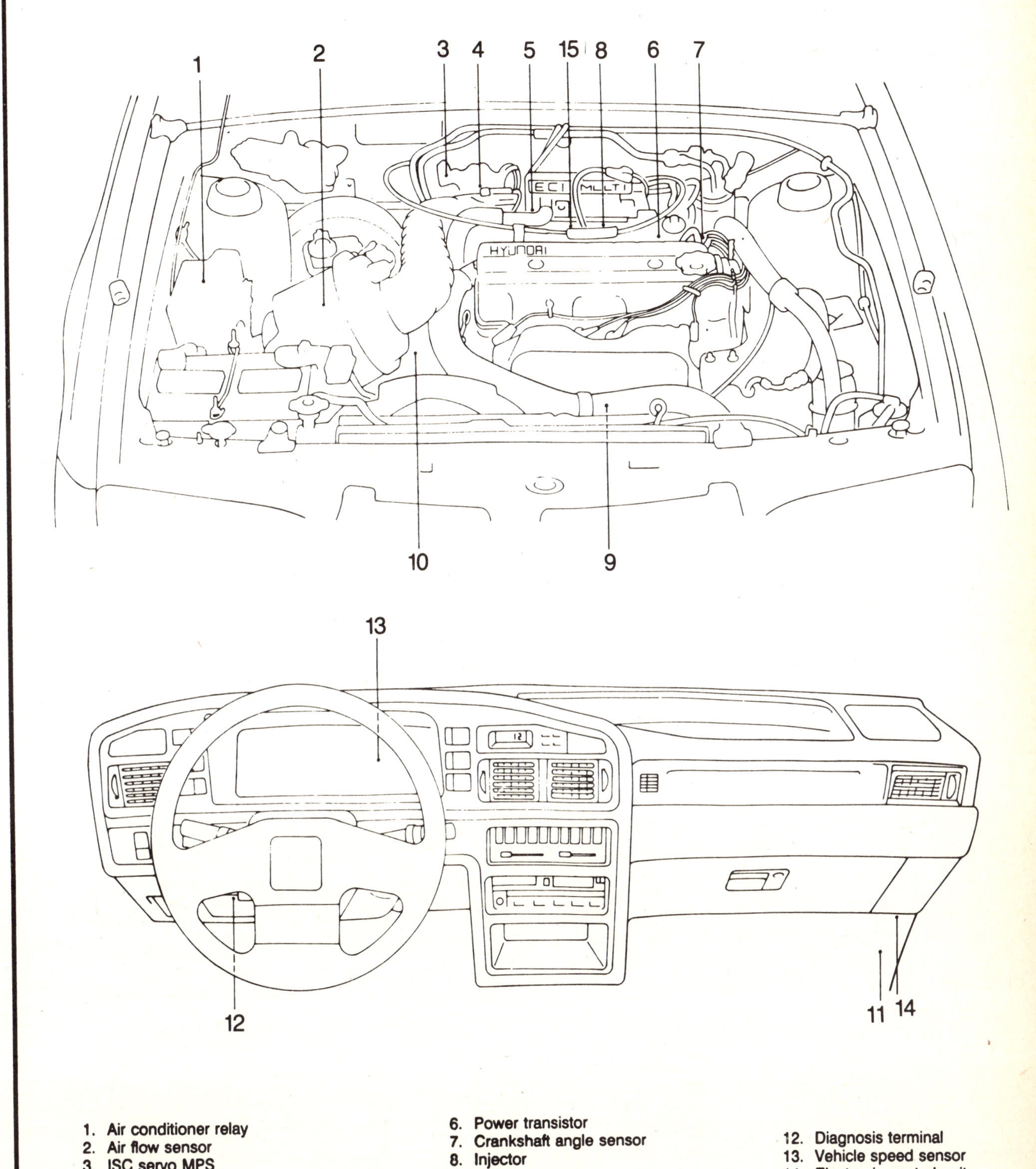


1. Air conditioner relay
2. Air flow sensor
3. ISC servo MPS
4. Throttle position sensor
5. Coolant temperature sensor
6. Power transistor
7. Crankshaft angle sensor
8. Injector
9. Oxygen sensor
10. Inhibitor switch
11. MPI control relay
12. Diagnosis terminal
13. Vehicle speed sensor
14. Electronic control unit
15. EGR temperature sensor (Calif. only)

**Component locations – Sonata 4 cylinder shown**

7. Remove the delivery pipe with the fuel injectors.

**NOTE: Take care during removal as fuel may flow out of injectors or delivery pipe.**

8. Remove the injectors from the delivery pipe.

**To install:**

9. Install a new insulator to the manifold.
10. Install a new O-ring to the injector.
11. Apply a coat of gasoline to the O-ring.
12. While turning the injector, install it onto the delivery pipe.
13. Make certain the injector turns smoothly; if not, replace the O-ring.
14. Make certain the insulators are correct when installing the delivery pipe.
15. Coat fuel line O-rings with gasoline and install.
16. Connect the negative battery cable.
17. Check all fuel lines for proper installation. Energize fuel pump and check for leaks.
18. Install surge tank and throttle body, as needed.
19. Fill the cooling system.

## IDLE AND TPS

### Adjustment

**4 CYLINDER**

1. Bring engine to normal operating temperature.
2. Make certain all accessories and cooling fan are **OFF**.
3. If equipped with power steering, place wheels in straight ahead position.
4. Make certain base timing is correct.
5. If a hand scan tool is not available, connect a digital voltmeter to the TPS ground and power supply without disconnecting the TPS and connect a tachometer.
6. Turn the ignition switch to **ON** without the engine running for 15 seconds. Check that the ISC is fully retracted.

**NOTE: When the ignition is turned ON the ISC extends to fast idle, after 15 seconds it retracts to fully closed. The motor position sensor at 0.9 volts.**

7. Turn the ignition switch **OFF**.

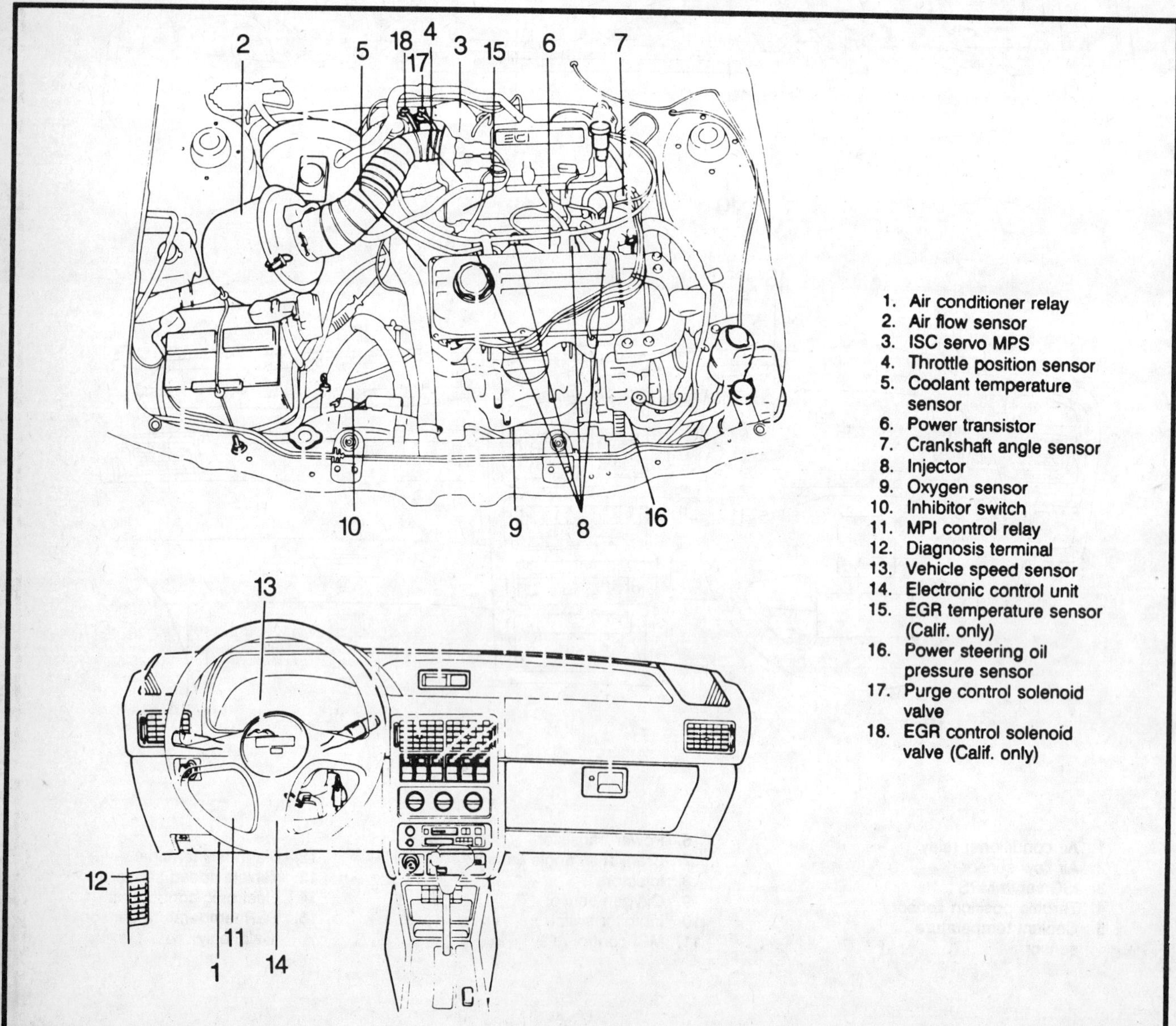


**Component locations – Excel**

8. Loosen the accelerator cable. Disconnect the ISC connector and secure the ISC motor at the fully retracted position.
9. In order to prevent the throttle valve from sticking, open it several times and then allow to click shut. Sufficiently loosen the fixed Speed Adjustment Screw (SAS).
10. Start the engine and allow to idle.
11. Check that the engine rpm is 600–800 rpm.
12. If the engine rpm is not correct, adjust the ISC adjusting screw to obtain the correct rpm.

**NOTE: If engine stalls or idle speed is low, check throttle body for carbon build up and clean as necessary.**

13. Tighten the fixed SAS until the engine speed starts to increase. Then loosen it until the engine speed just ceases to drop and loosen ½ turn.
14. Stop the engine.
15. Turn the ignition switch to **ON**, with engine not running, check that the TPS output voltage is 0.48–0.52 volts. If not correct, adjust TPS as necessary.
16. Adjust accelerator cable play.
    Excel with auto trans – 0.12–0.2 in. (3–5mm)
    Excel with manual trans – 0.04–0.08 in. (1–2mm)
    Sonata auto trans – 0.08–0.12 in. (2–3mm)
    Sonata with manual trans – 0–0.04 in. (0–1mm)
17. Connect the ISC connector.
18. Disconnect the voltmeter and connect the TPS.
19. Start the engine and make certain idle speed is correct.
20. Turn ignition switch **OFF** and disconnect the negative battery cable to remove codes stored during adjustment.

**6 CYLINDER IDLE**

1. Bring engine to normal operating temperature and connect a tachometer.
2. Make certain all accessories and cooling fan are **OFF**.

**NOTE: Inserting a paper clip into the lock hook (harness) side of the connector between the primary side of coil and noise filter connector so it can be used as a tachometer connection. Do not allow terminal to ground out and remember to remove paper clip.**

3. If equipped with power steering, place wheels in straight ahead position.
4. Make certain base timing is 3–7 degrees BTDC.
5. Turn the ignition switch **OFF** and ground the self-diagnosis check terminal.
6. **Run engine at 2500 rpm for at least 5 seconds.**
7. If idle is not 600–800 rpm, adjust SAS screw as needed.

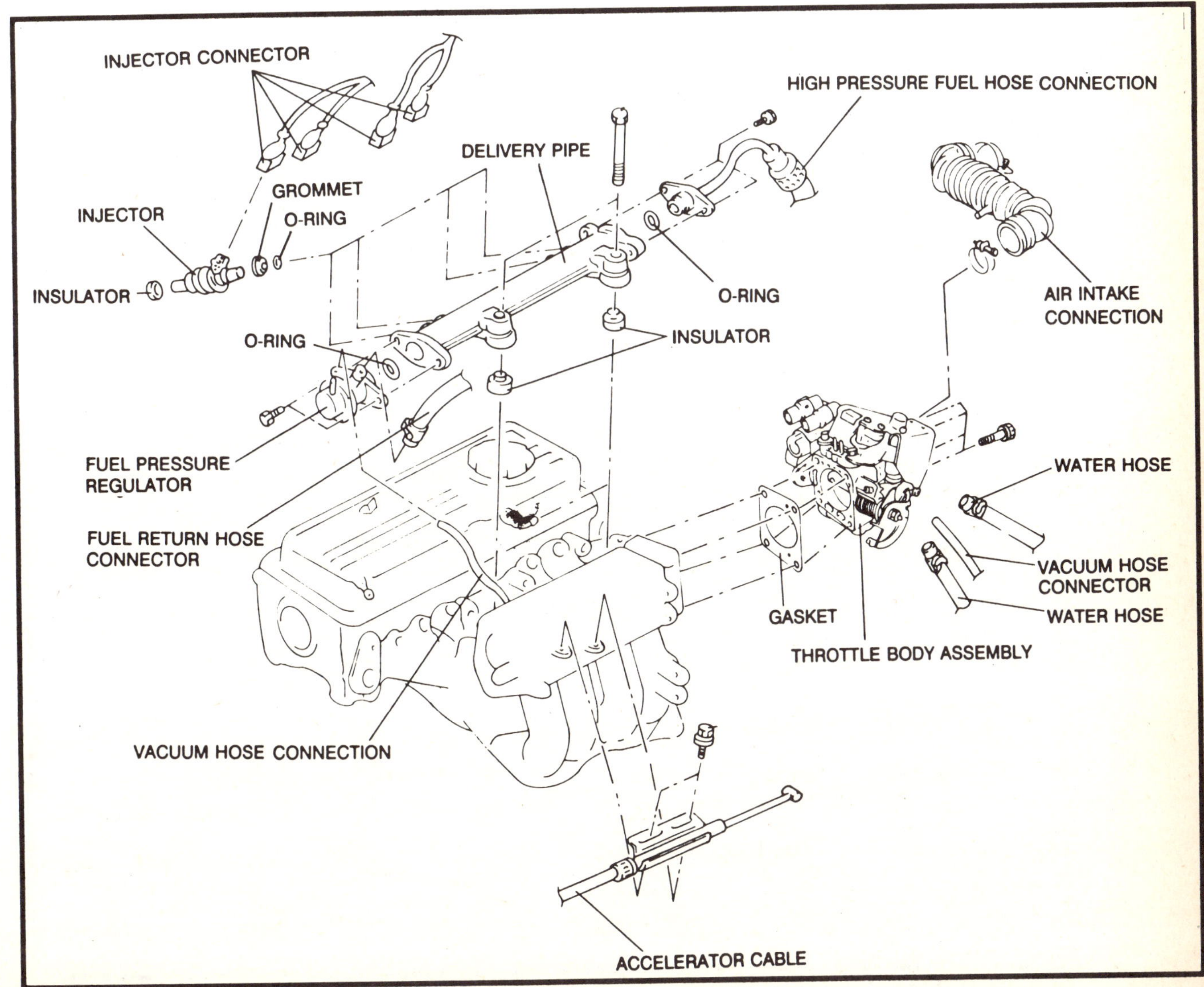


**Throttle body and injectors – 1990 Excel shown**

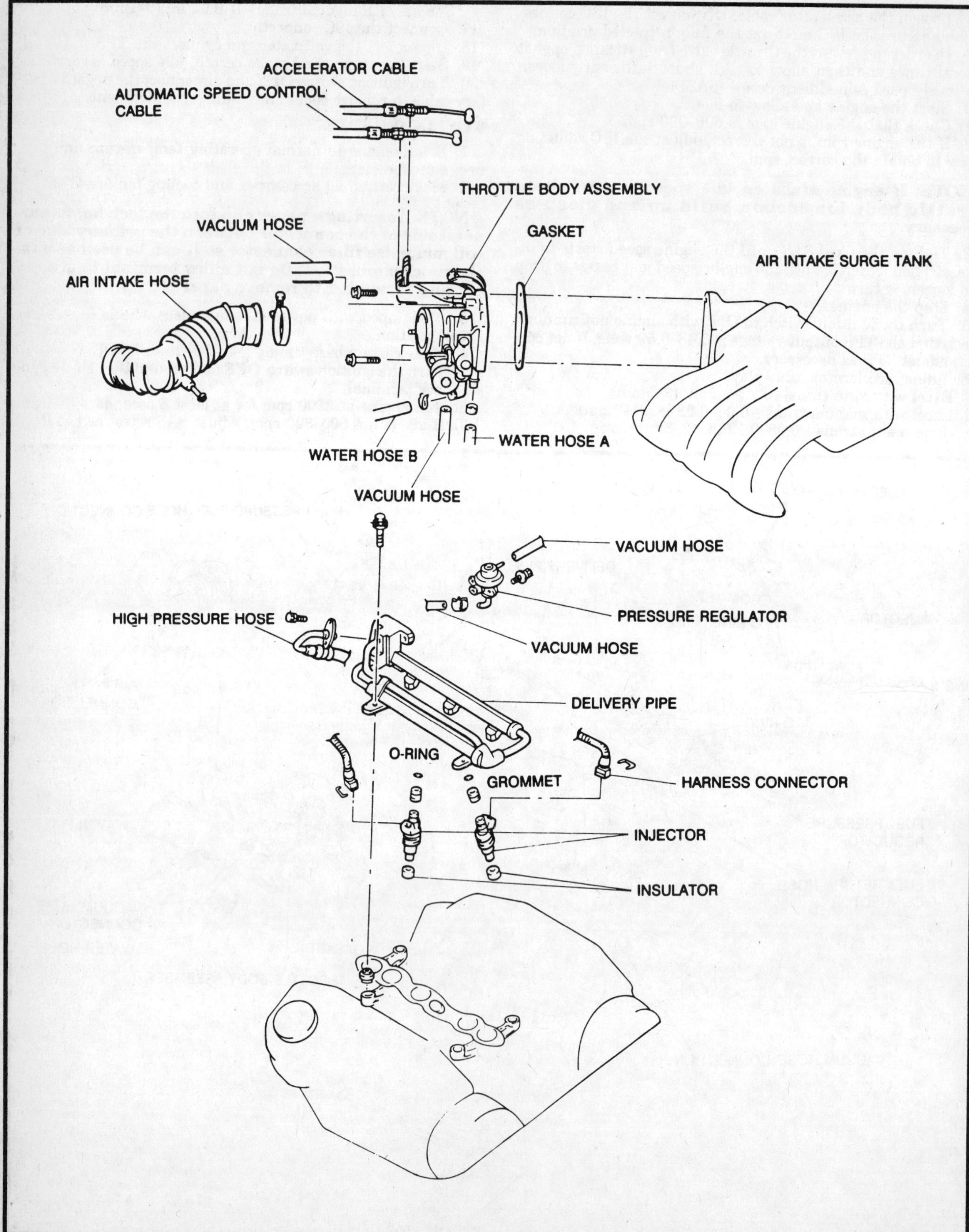


**Throttle body and injectors – 6 cylinder**

8. If the idle speed is higher than specified after adjustment, check that the idle switch (fixed SAS) has been misadjusted. If so, continue with Step 9; if the idle switch has not been misadjusted, continue with Step 16.
9. Disconnect the idle switch connector.
10. Loosen the idle switch locknut.
11. Turn the idle switch counterclockwise until the throttle valve closes.
12. Connect an ohmmeter between the terminal of the switch and the ground.
13. Screw the idle switch in until continuity is found. Screw the idle switch 1¼ turn passed that point.
14. Tighten the locknut and connect the wiring.
15. Readjust curb idle (SAS) to obtain 600–800 rpm.
16. Turn ignition switch **OFF**, remove ground from diagnosis connector, remove paper clip from connector and disconnect the negative battery cable for at least 15 seconds to clear trouble codes.

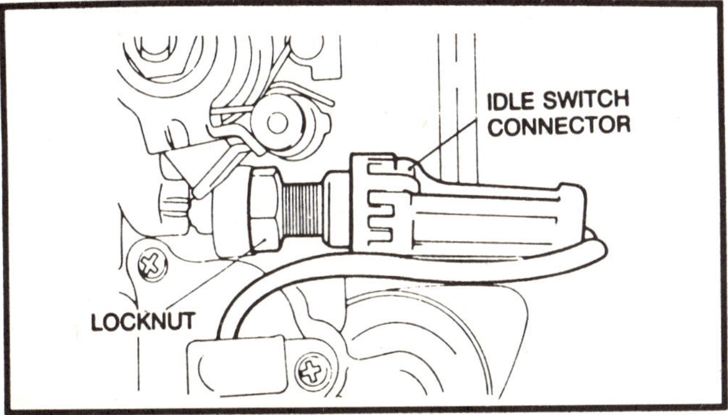


**Idle switch – V6 Sonata**

17. Check and adjust TPS as necessary.

**6 CYLINDER TPS**

1. Bring engine to normal operating temperature.
2. Make certain all accessories and cooling fan are **OFF**.
3. If equipped with power steering, place wheels in straight ahead position.
4. If a hand scan tool is not available, connect a digital voltmeter to the TPS terminals 2 and 4 without disconnecting the TPS and connect a tachometer.
5. Turn the ignition switch to **ON**, with engine not running, check that the TPS output voltage is 0.48–0.52 volts. If not correct, adjust TPS as necessary.
6. Disconnect negative battery cable for at least 5 seconds, to remove codes stored during adjustment.
7. Adjust accelerator cable tension, as needed.

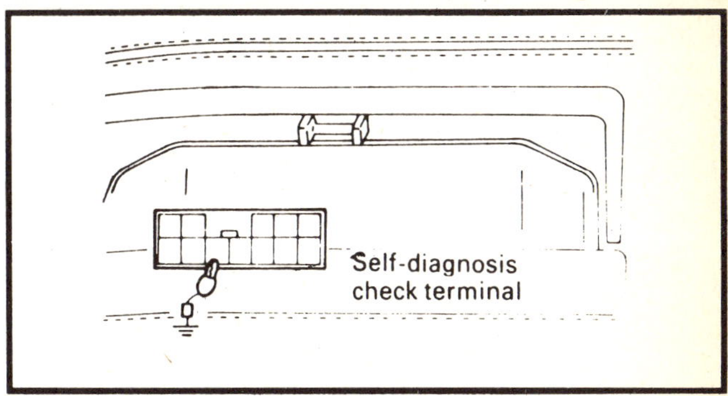


**Self-diagnosis terminal – V6 shown**

**Adjusting Speed Adjust Screw (SAS) – V6 shown**

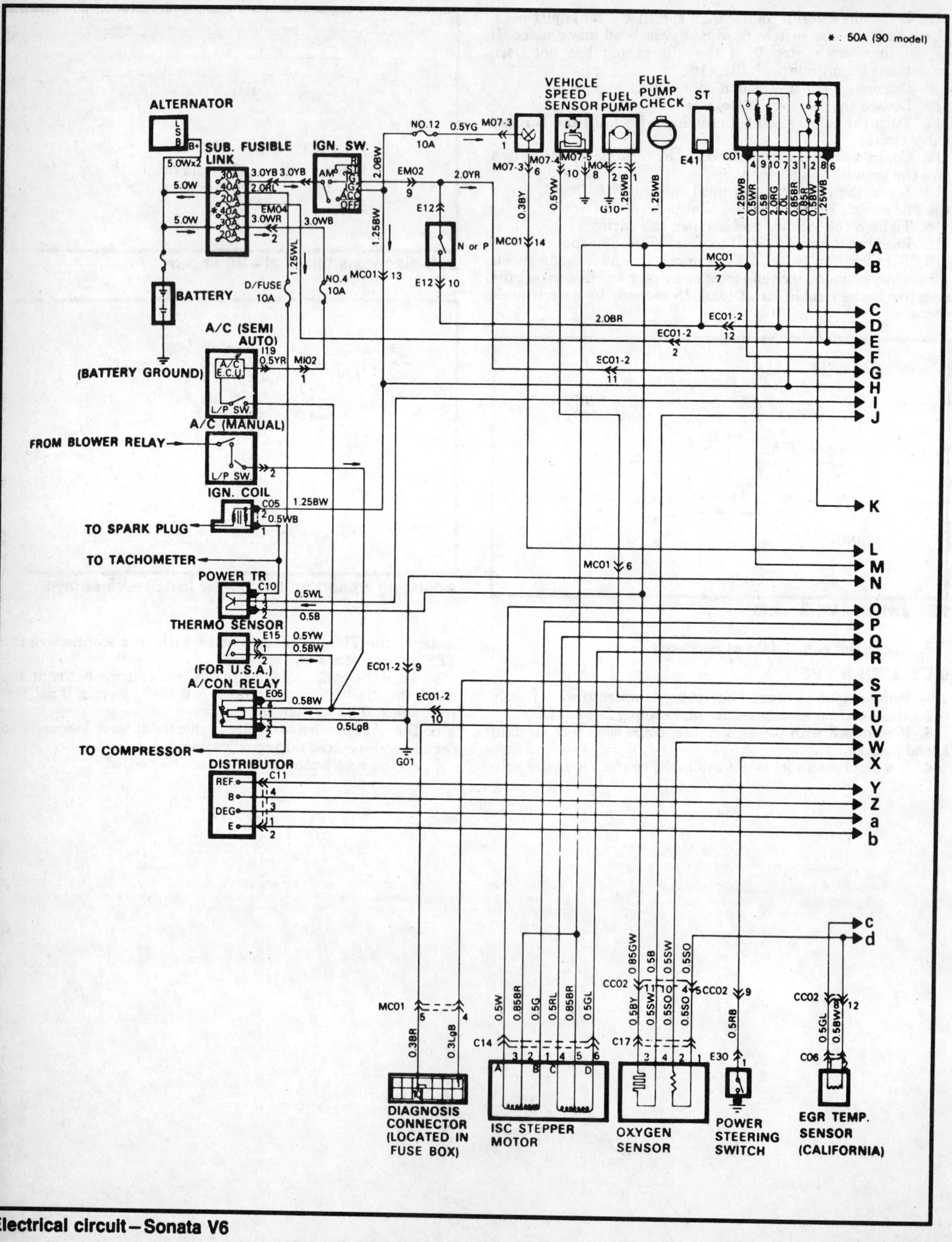

* : 50A (90 model)

ALTERNATOR
SUB. FUSIBLE LINK
IGN. SW.
VEHICLE SPEED SENSOR
FUEL PUMP
FUEL PUMP CHECK
ST
BATTERY
D/FUSE 10A
NO.4 10A
NO.12 10A
A/C (SEMI AUTO)
(BATTERY GROUND)
A/C (MANUAL)
FROM BLOWER RELAY
IGN. COIL
TO SPARK PLUG
TO TACHOMETER
POWER TR
THERMO SENSOR
(FOR U.S.A.)
A/CON RELAY
TO COMPRESSOR
DISTRIBUTOR
DIAGNOSIS CONNECTOR (LOCATED IN FUSE BOX)
ISC STEPPER MOTOR
OXYGEN SENSOR
POWER STEERING SWITCH
EGR TEMP. SENSOR (CALIFORNIA)

Electrical circuit – Sonata V6

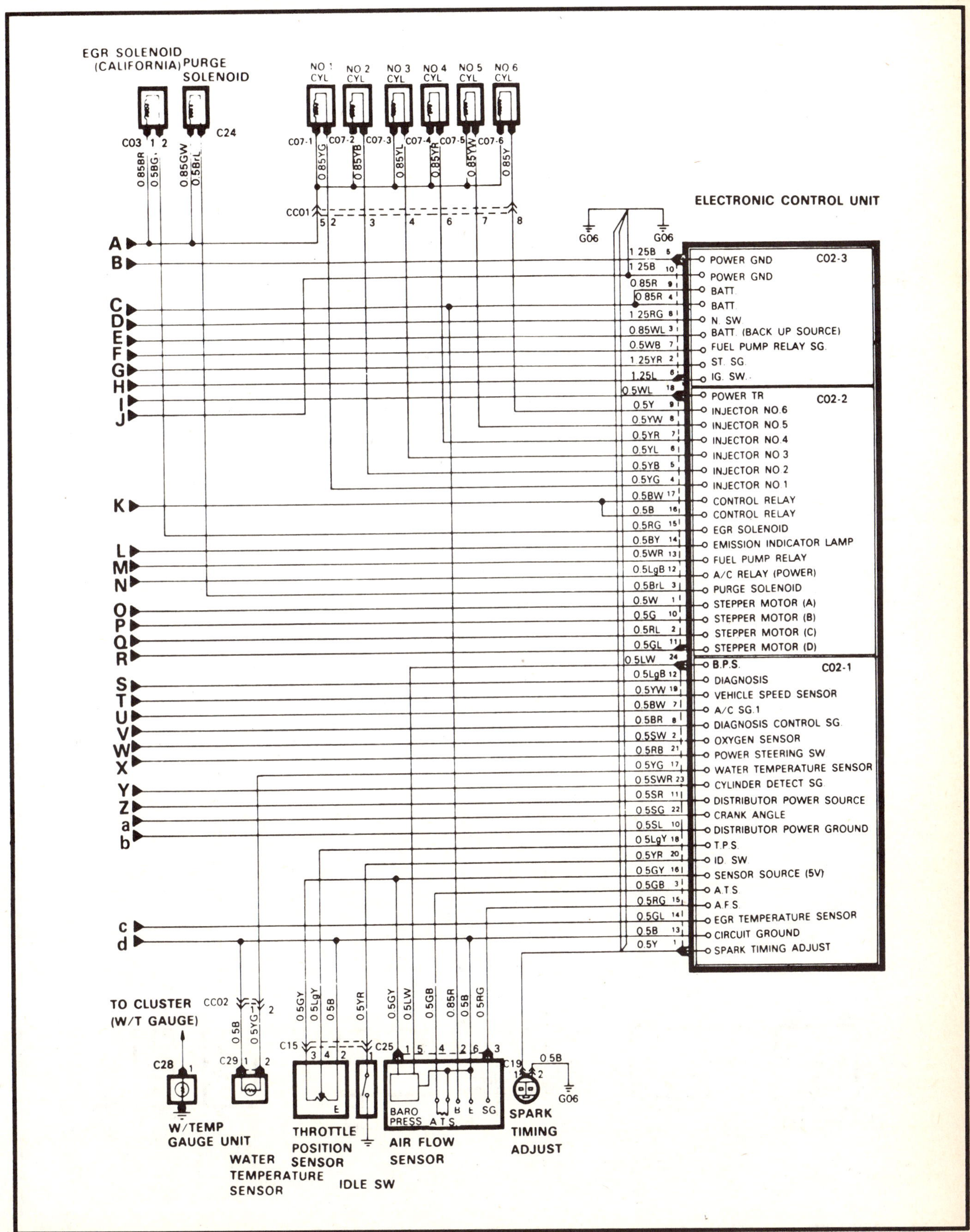


Electrical circuit—Sonata V6

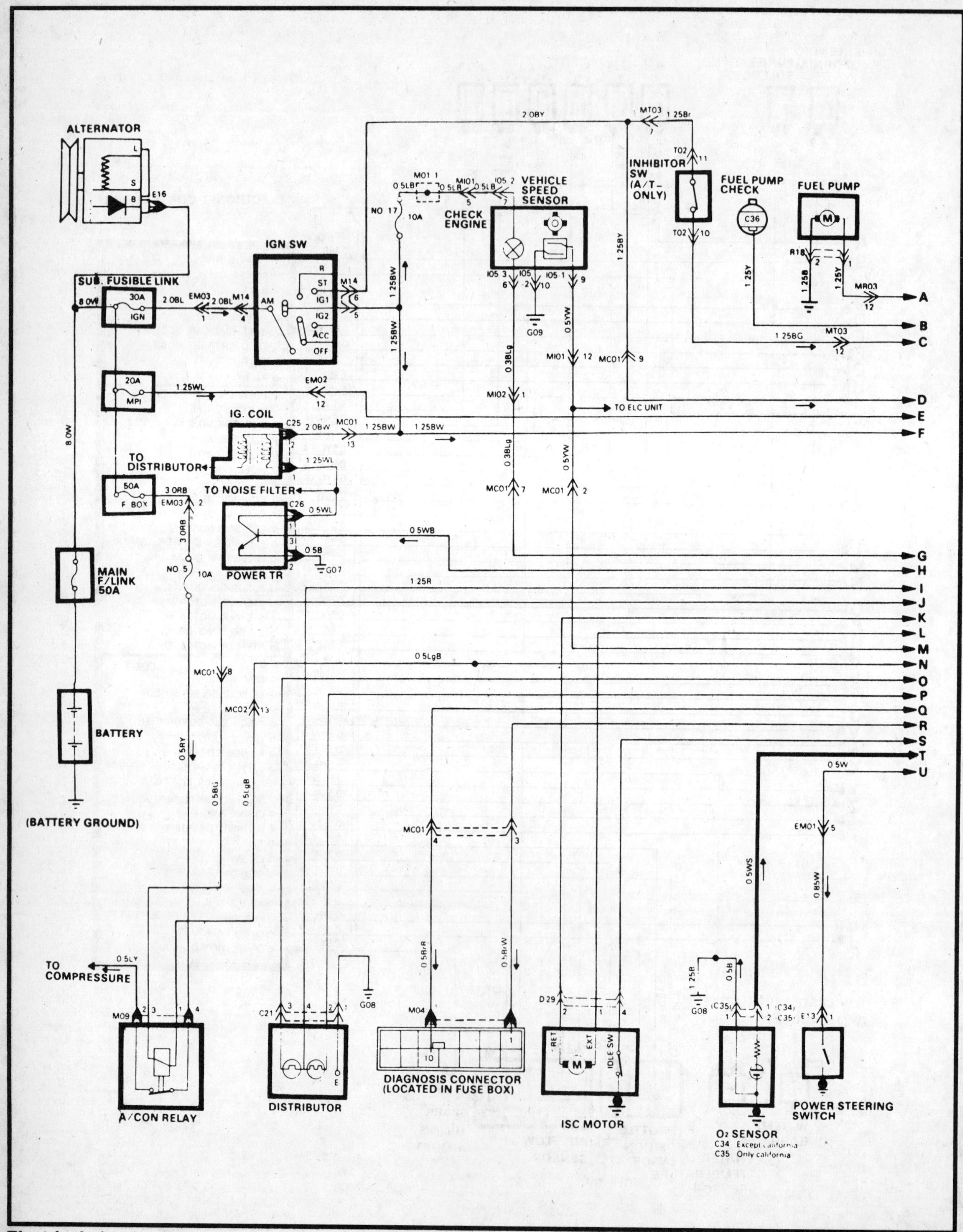


Electrical circuit—Excel

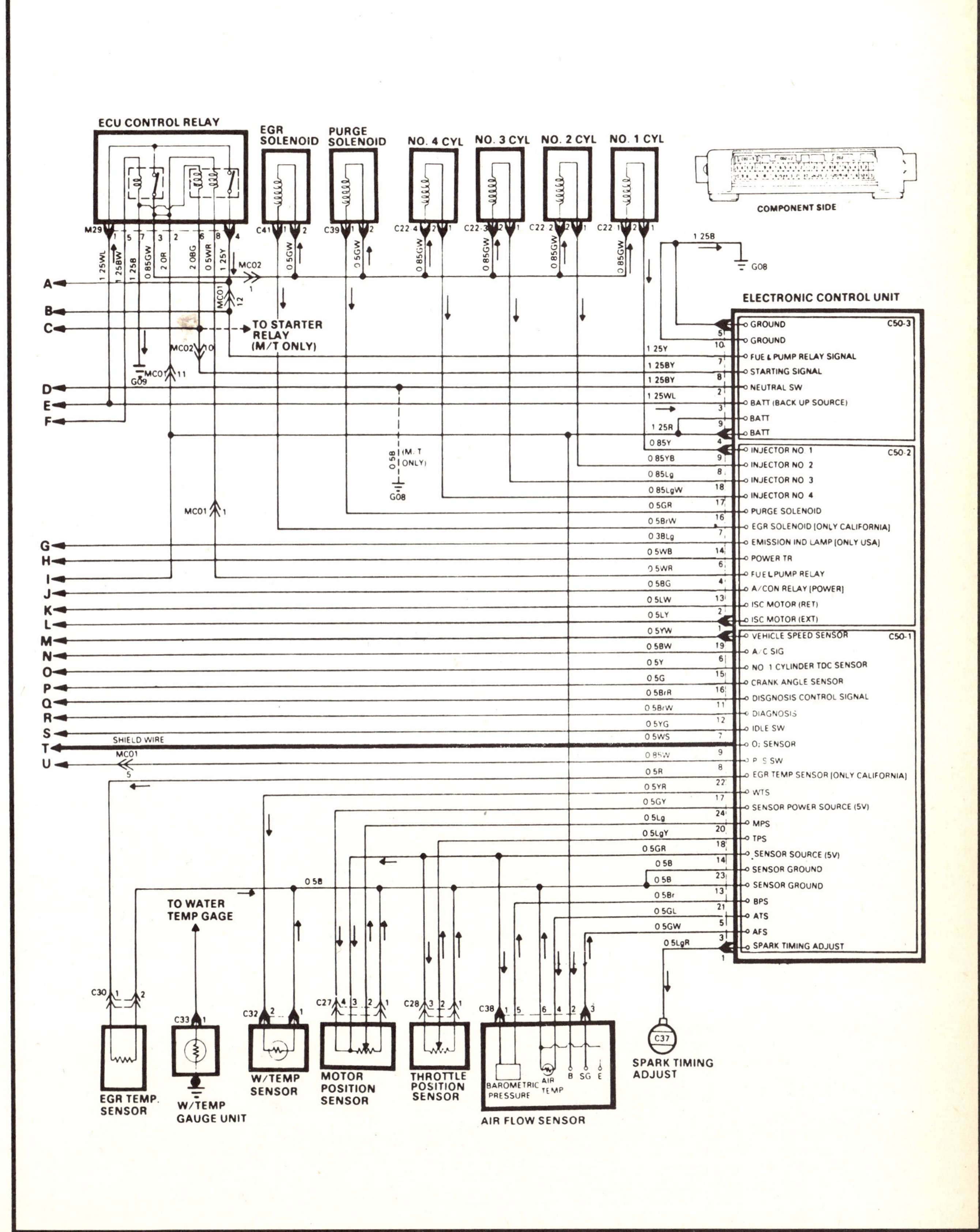


Electrical circuit—Excel

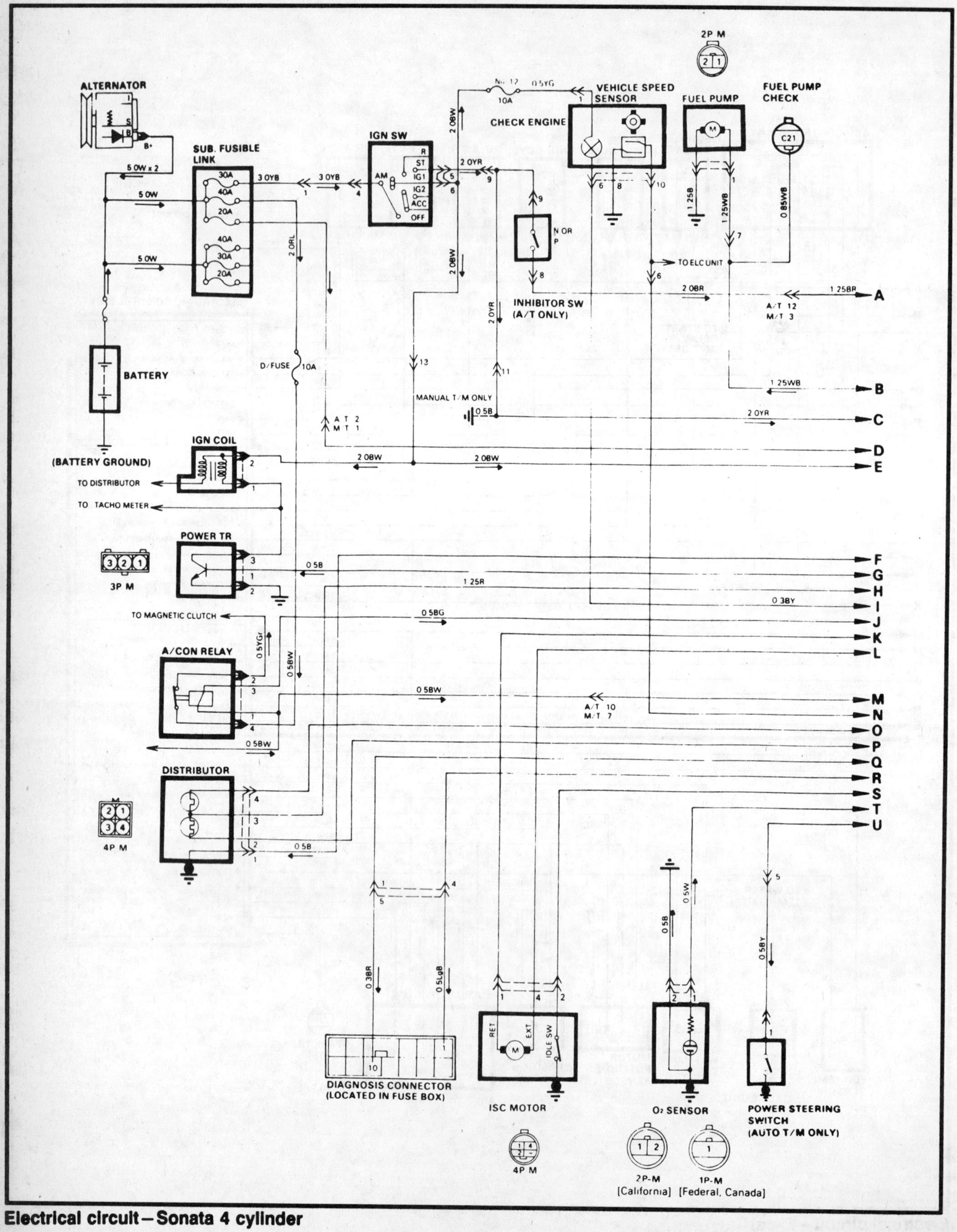

Electrical circuit – Sonata 4 cylinder

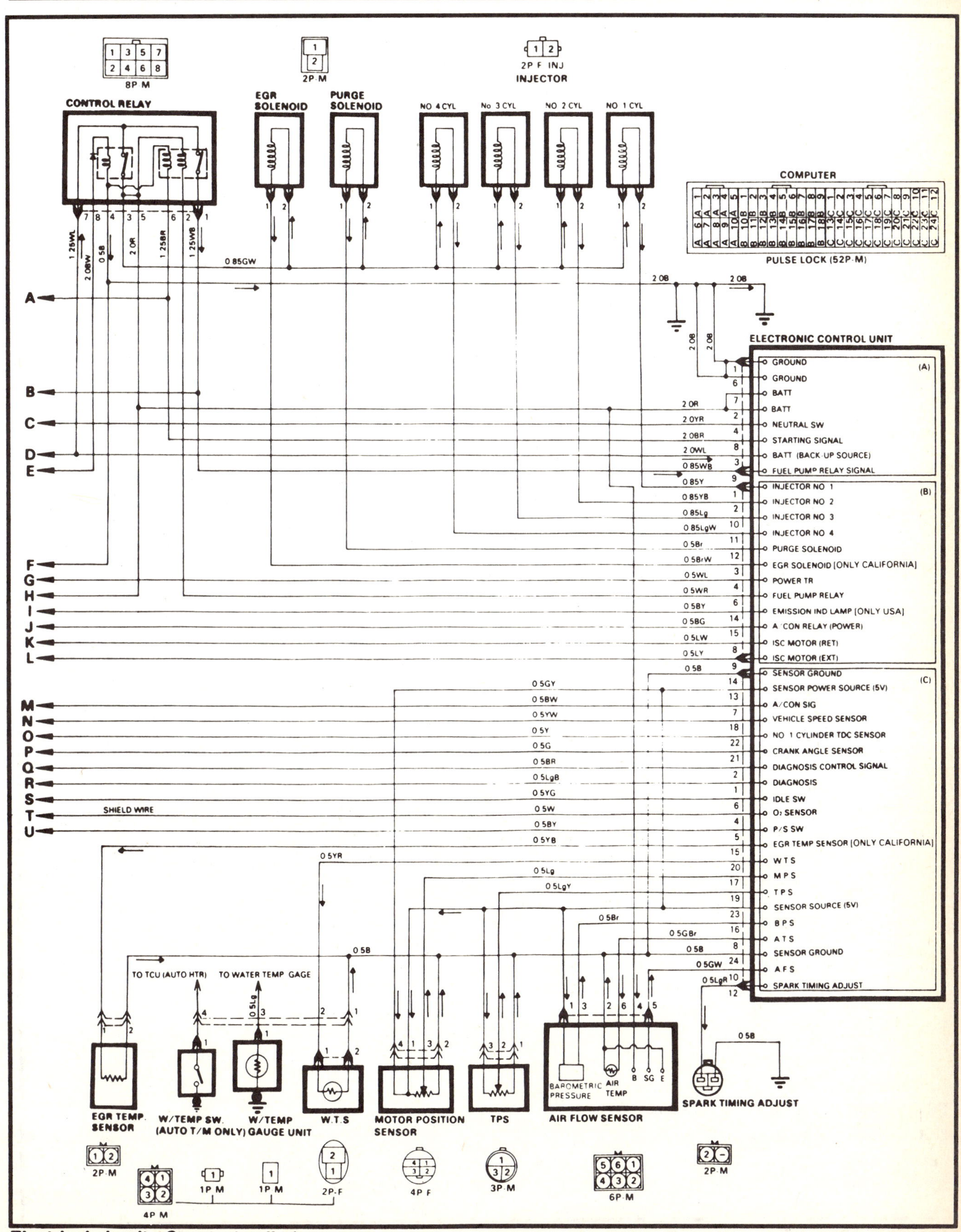


Electrical circuit—Sonata 4 cylinder

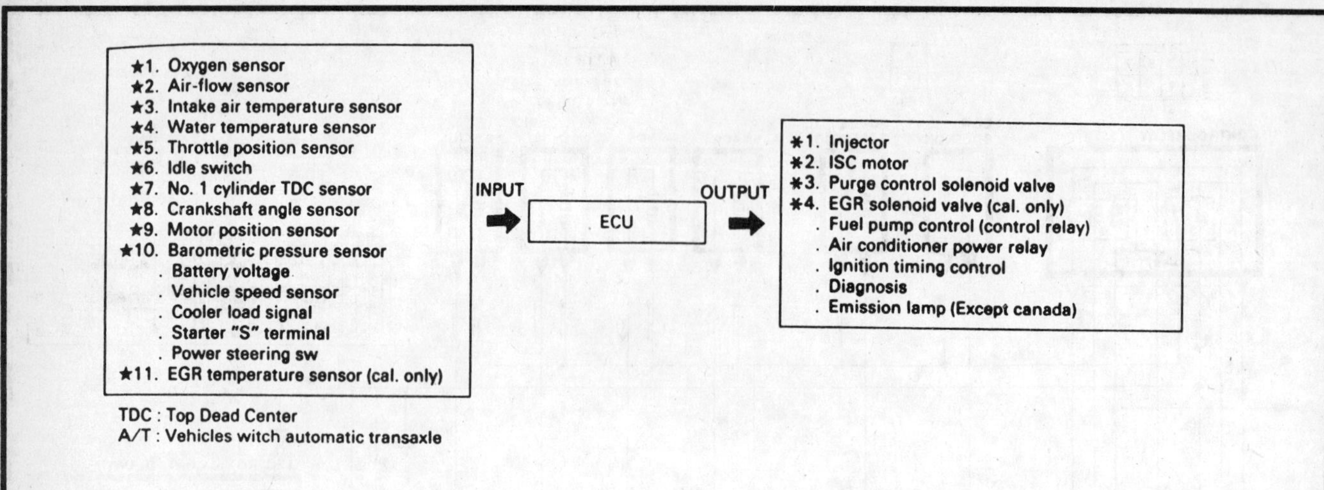


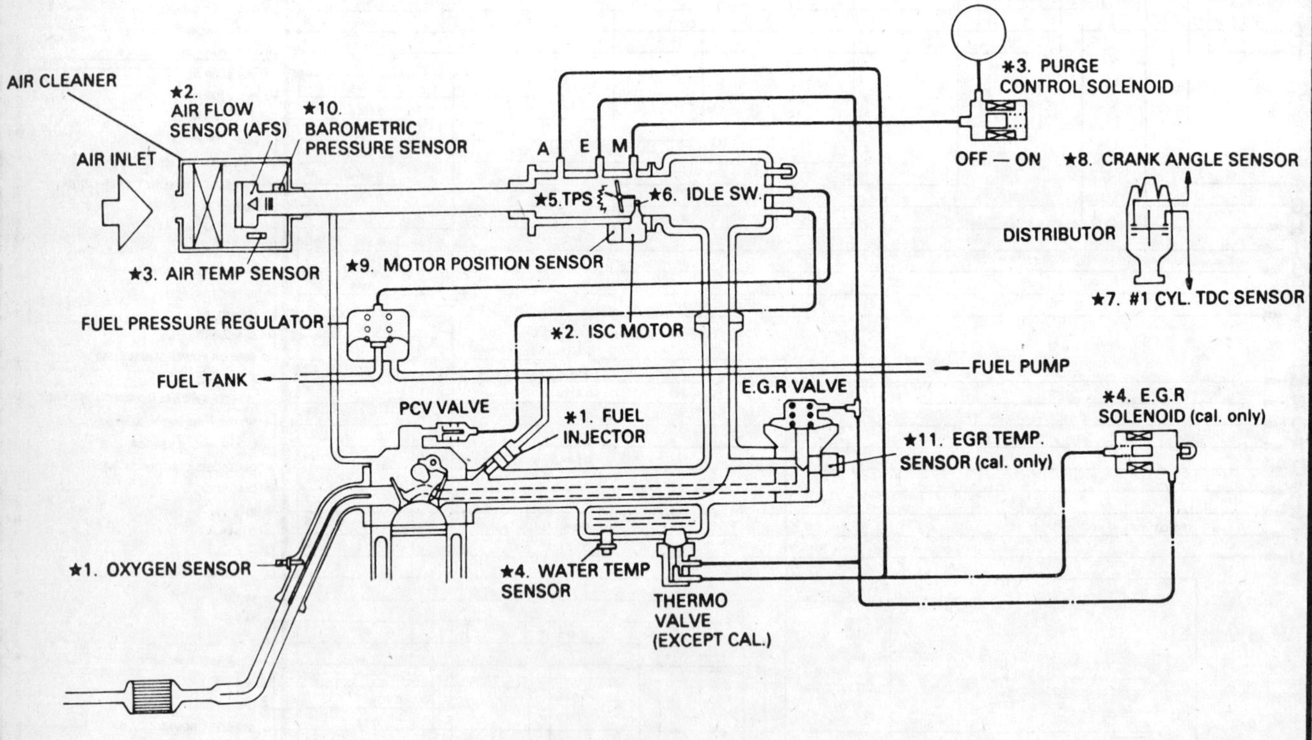


PCV : Positive Crankcase Ventilation
UCC : Underfloor Catalytic Converter
FED : Federal
CAL : California

**MPI system schematic—typical 4 cylinder**

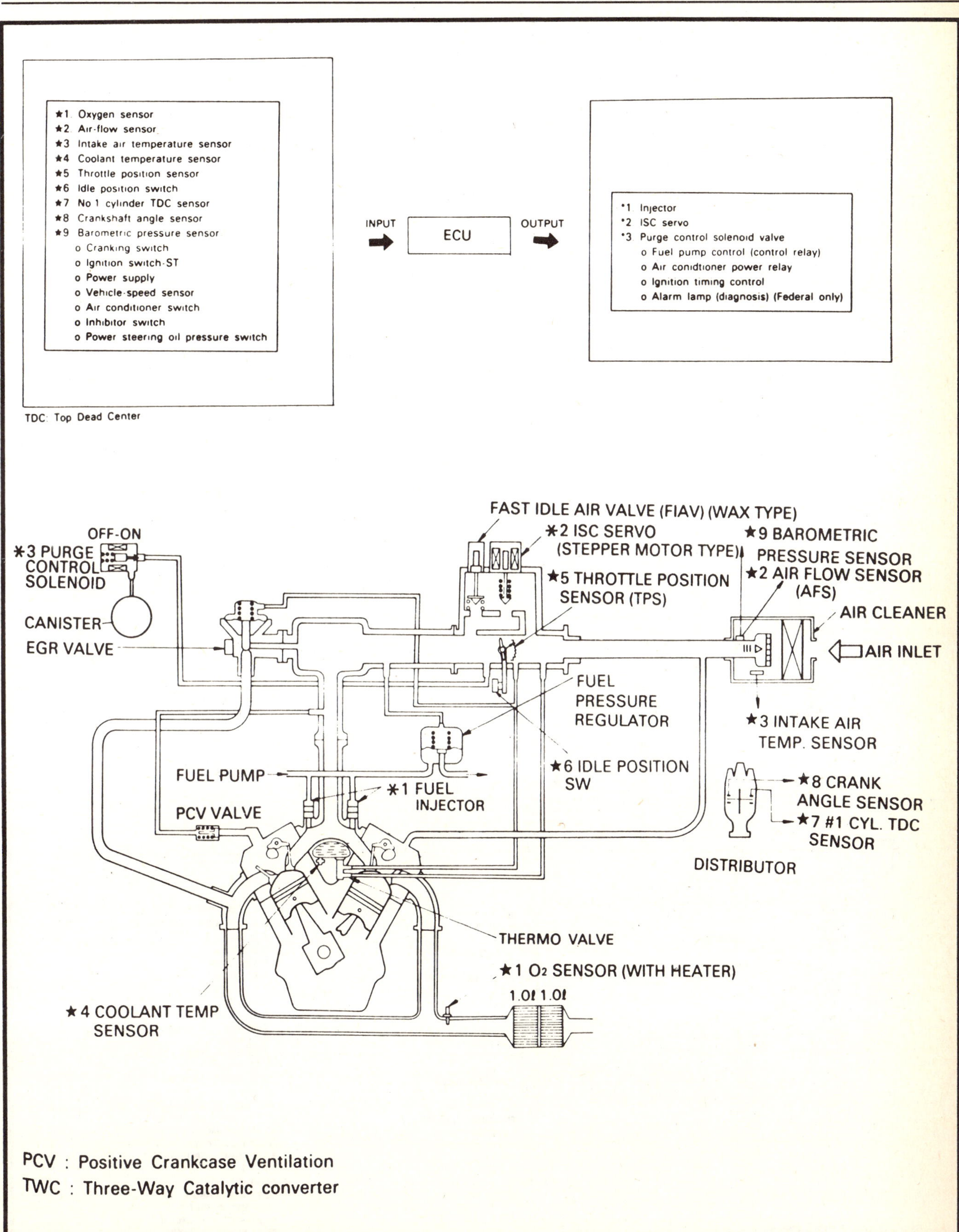


**MPI system schematic—typical 6 cylinder**

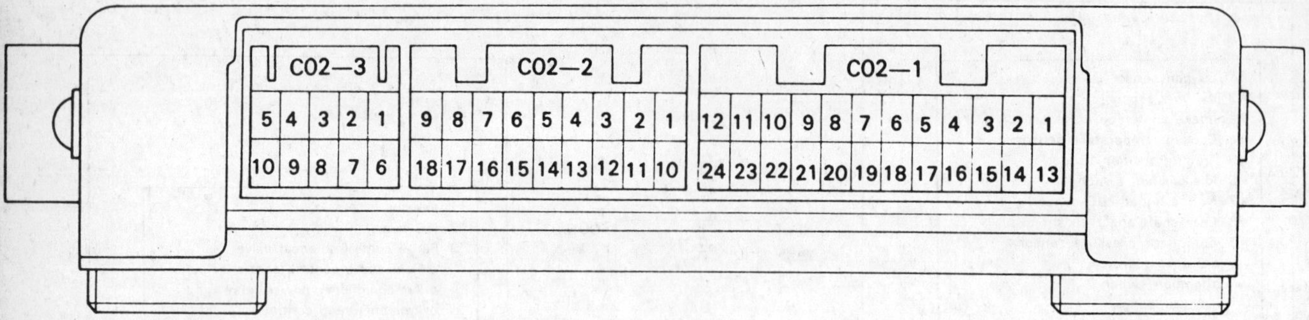

C02—3
1.
2. ST. SG
3. BATT (BACK UP SOURCE)
4. BATT
5. POWER GND
6. IG. SW
7. FUEL PUMP RELAY SG.
8. N SW
9. BATT
10 POWER GND

C02—2
1. STEPPER MOTOR (A)
2. STEPPER MOTOR (C)
3. PURGE SOLENOID
4. INJ. #1 CYL.
5. INJ. #2 CYL.
6. INJ. #3 CYL.
7. INJ. #4 CYL.
8. INJ. #5 CYL.
9. INJ. # CYL.
10 STEPPER MOTOR (B)
11 STEPPER MOTOR (D)
13 FUEL PUMP RELAY
14 EMISSION INDICATOR LAMP
15 EGR SOLENOID
16 CONTROL RELAY
17 CONTROL RELAY
18 POWER TR

C02—1
1. SPARK TIMING ADJUST
2. OXYGEN SENSOR
3. A.T.S. (AIR TEMP. SENSOR)
4.
5.
6.
7. A/C SG. 1
8. DIAGNOSIS CONTROL SG.
9.
10 DISTRIBUTOR POWER GND
11 DISTRIBUTOR POWER SOURCE
12 DIAGNOSIS
13 CIRCUIT GND
14 EGR TEMPERATURE SENSOR
15 AIR FLOW SENSOR (A.F.S.)
16 SENSOR SOURCE
17 WATER TEMPERATURE SENSOR (W.T.S.)
18 THROTTLE POSITION SENSOR (T.P.S.)
19 VEHICLE SPEED SENSOR
20 ID SW.
21 POWER STEERING SW.
22 CRANK ANGLE
23 CYLINDER DETECT SG.
24 BAROMETRIC PRESS. SENSOR (B.P.S.)

**ECU terminal identification – V6 shown**

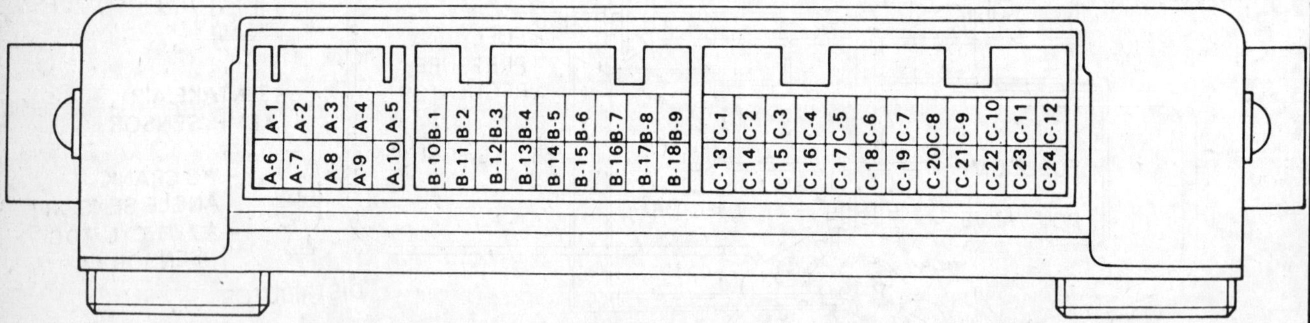

A-1 GND
A-2 BATT
A-3 BATT (BACK-UP SOURCE)
A-4 NEUTRAL SW.
A-5 N.A (NOT AVAILABLE)
*A-6 GND
*A-7 BATT
*A-8 ST. SG
*A-9 FUEL PUMP RELAY SG
*A-10 N.A

B-1 INJ. #1 CYL.
B-2 INJ. #2 CYL.
B-3 N.A
EGR SOLENOID
(ONLY CALIFORNIA)
B-4 IG. COIL (POWER TR)
B-5 N.A
B-6 FUEL PUMP RELAY
B-7 N.A
B-8 ISC MOTOR (RET.)
B-9 ISC MOTOR (EXT.)
*B-10 INJ. #3 CYL.
*B-11 INJ. #4 CYL.
*B-12 PURGE SOLENOID
*B-13 N.A
*B-14 N.A
EMISSION INDICATOR
LAMP (ONLY USA)
*B-15 A/C RELAY (POWER)
*B-16 N.A
*B-17 N.A
*B-18 N.A

C-1 DIAGNOSIS
C-2 DIAGNOSIS CONTROL SG
C-3 N.A
C-4 $O^2$ SG
C-5 P/S SW.
C-6 I.D. SW
C-7 A/C SIG
C-8 A.T.S.
C-9 N.A
C-10 AFS
C-11 N.A
C-12 SPARK TIMING ADJUST (+)
*C-13 SENSOR SOURCE (+)
*C-14 SENSOR GROUND
*C-15 N.A
EGR TEMP. SENSOR
(ONLY CALIFORNIA)
*C-16 BAROMETRIC PRESS SENSOR
*C-17 ISC MOTOR POSI. SENSOR
*C-18 VEHICLE SPEED (0 km/h)
*C-19 TPS
*C-20 W.T.S
*C-21 CRANK ANGLE
*C-22 CYL. DETECT SG.
*C-23 SENSOR SOURCE (5V)
*C-24 SENSOR GROUND

* Mark is gold plating terminal

**ECU terminal identification – 4 cylinder Sonata shown**

# FUEL INJECTION SYSTEM

## ISUZU APPLICATION CHART

| Year | Model | Engine cc (liter) | Engine VIN | Fuel System | Ignition System |
|---|---|---|---|---|---|
| 1988 | I-Mark | 1471 (1.5) | 4XC1 | 2bbl | Electronic |
| | I-Mark (Turbo) | 1471 (1.5) | 4XC1-T | EFI | Electronic |
| | Impulse | 2254 (2.3) | 4Z01 | EFI | Electronic |
| | Impulse (Turbo) | 1994 (2.0) | 4ZC1-T | EFI | Electronic |
| | Pick-up | 2254 (2.3) | 4ZD1 | 2bbl | Electronic |
| | Pick-up | 2559 (2.6) | 4ZE1 | EFI | Electronic |
| | Trooper II | 2559 (2.6) | 4ZE1 | EFI | Electronic |
| 1989-90 | Amigo | 2254 (2.3) | 4ZD1 | 2bbl | Electronic |
| | Amigo | 2559 (2.6) | 4ZE1 | EFI | Electronic |
| | I-Mark | 1471 (1.5) | 4XC1 | 2bbl | Electronic |
| | I-Mark (Turbo) | 1471 (1.5) | 4XC1-T | EFI | Electronic |
| | I-Mark | 1588 (1.6) | 4XE1 | EFI | Electronic |
| | Impulse | 2254 (2.3) | 4ZD1 | EFI | Electronic |
| | Impulse (Turbo) | 1994 (2.0) | 4ZC1-T | EFI | Electronic |
| | Pick-up | 2254 (2.3) | 4ZD1 | 2bbl | Electronic |
| | Pick-up | 2559 (2.6) | 4ZE1 | EFI | Electronic |
| | Trooper<br>Trooper II | 2559 (2.6) | 4ZE1 | EFI | Electronic |
| | Trooper<br>Trooper II | 2837 (2.8) | LL2 | EFI | Electronic |

## General Information

### SYSTEM OPERATION

**ALL EXCEPT 2.8L ENGINE**

The fuel injection control system constantly monitors and controls engine operation, which in turn helps lower emissions while maintaining the fuel economy and driveability . The Electronic Contro Unit (ECU) controls the fuel injection system, the ignition system and the turbocharger control system (if so equipped).

The ignition system consists of the following components; the crank angle sensor, the throttle valve switch, the vehicle speed sensor, the coolant temperature sensor, air flow sensor, knock sensor, transistorized ignition coil and the ECU.

In general, the fuel injection system consists of the following components; the crank angle sensor, the throttle valve switch, the vehicle speed sensor, the coolant temperature sensor, air flow sensor, oxygen sensor, fuel injectors and the ECU.

If equipped, the turbocharger system consists of the following components; the throttle valve switch, the throttle position sensor, the knock sensor, the coolant temperature sensor, air flow sensor, oxygen sensor, the stepping motor, the turbocharger controller and the ECU.

2.8L ENGINE

The electronic throttle body fuel injection system used on the 2.8L engine is a fuel metering system with the amount of fuel delivered by the throttle body injector(s) (TBI) determined by an electronic signal supplied by the Electronic Control Module (ECM). The ECM monitors various engine and vehicle conditions to calculate the fuel delivery time (pulse width) of the injector(s). The fuel pulse may be modified by the ECM to account for special operating conditions, such as cranking, cold starting, altitude, acceleration, and deceleration.

The ECM controls the exhaust emissions by modifying fuel delivery to achieve, as near as possible, an air/fuel ratio of 14.7:1. The injector on-time is determined by various inputs to the ECM. By increasing the injector pulse, more fuel is delivered, enriching the air/fuel ratio. Decreasing the injector pulse leans the air/fuel ratio. Pulses are sent to the injector in 2 different modes: synchronized and nonsynchronized.

### SYNCHRONIZED MODE

In synchronized mode operation, the injector is pulsed once for each distributor reference pulse. In dual injector throttle body systems, the injectors are pulse alternately.

### NON-SYNCHRONIZED MODE

In non-synchronized mode operation, the injector is pulsed once every 12.5 milliseconds or 6.25 milliseconds depending on calibration. This pulse time is totally independent of distributor reference pulses.

Non-synchronized mode results only under the following conditions:

1. The fuel pulse width is too small to be delivered accurately by the injector (approximately 1.5 milliseconds).
2. During the delivery of prime pulses (prime pulses charge

the intake manifold with fuel during or just prior to engine starting).

3. During acceleration enrichment.

4. During deceleration leanout.

The basic TBI unit is made up of 2 major casting assemblies: (1) a throttle body with a valve to control airflow and (2) a fuel body assembly with an integral pressure regulator and fuel injector to supply the required fuel. An electronically operated device to control the idle speed and a device to provide information regarding throttle valve position are included as part of the TBI unit.

Each fuel injector is a solenoid-operated device controlled by the ECM. The incoming fuel is directed to the lower end of the injector assembly which has a fine screen filter surrounding the injector inlet. The ECM actuates the solenoid, which lifts a normally closed ball valve off a seat. The fuel under pressure is injected in a conical spray pattern at the walls of the throttle body bore above the throttle valve. The excess fuel passes through a pressure regulator before being returned to the vehicle's fuel tank.

The pressure regulator is a diaphragm-operated relief valve with injector pressure on one side and air cleaner pressure on the other. The function of the regulator is to maintain a constant pressure drop across the injector throughout the operating load and speed range of the engine.

The throttle body portion of the TBI may contain ports located at, above or below the throttle valve. These ports generate the vacuum signals for the EGR valve, MAP sensor, and the canister purge system.

The Throttle Position Sensor (TPS) is a variable resistor used to convert the degree of throttle plate opening to an electrical signal to the ECM. The ECM uses this signal as a reference point of throttle valve position. In addition, an Idle Air Control (IAC) assembly, mounted in the throttle body is used to control idle speeds. A cone-shaped valve in the IAC assembly is located in an air passage in the throttle body that leads from the point beneath the air cleaner to below the throttle valve. The ECM monitors idle speeds and, depending on engine load, moves the IAC cone in the air passage to increase or decrease air bypassing the throttle valve to the intake manifold for control of idle speeds.

## SYSTEM COMPONENTS

### Air Conditioner Request Signal

**2.8L ENGINE**

This signal indicates to the ECM that an air conditioning mode is selected at the switch and that the A/C low pressure switch is closed. The ECM controls the A/C and adjusts the idle speed in response to this signal.

### Air Flow Sensor

**ALL EXCEPT 2.8L ENGINE**

The air flow sensor is usually located in the air cleaner housing assembly. The purpose of the air flow sensor is to measure the volume (rate) of air that is coming into the engine.

### Back-Up Control System

**ALL EXCEPT 2.8L ENGINE**

This system is used in case there is a malfunction with the microcomputer within the ECU, the back-up control system works to maintain the necessary functions of the control unit to permit continuous operation of the vehicle.

### Coolant Temperature Sensor

**ALL EXCEPT 2.8L ENGINE**

This sensor is usually located on the engine block, under the intake manifold. It sends the coolant temperature information back to the ECU. The ECU then uses this information to determine the engine temperature for calculating the required air/fuel mixture.

**2.8L ENGINE**

The coolant sensor is a thermister (a resistor which changes value based on temperature) mounted in the engine coolant stream. As the temperature of the engine coolant changes, the resistance of the coolant sensor changes. Low coolant temperature produces a high resistance (100,000 ohms at −40°F/−40°C), while high temperature causes low resistance (70 ohms at 266°F/130°C).

The ECM supplies a 5 volt signal to the coolant sensor and measures the voltage that returns. By measuring the voltage change, the ECM determines the engine coolant temperature. The voltage will be high when the engine is cold and low when the engine is hot. This information is used to control fuel management, IAC, spark timing, EGR, canister purge and other engine operating conditions.

A failure in the coolant sensor circuit should either set a Code 14 or 15. These codes indicate a failure in the coolant temperature sensor circuit.

### Crank Angle Sensor

**ALL EXCEPT 2.8L ENGINE**

The crank angle sensor is usually located inside the distributor housing, it is used to detect the engine speed and relative position of each piston in its cylinder. Using these parameters, the ECU calculates the proper ignition timing and dwell angle. The ECU then sends a signal to the transistorized ignition coil to create a spark.

### Crankshaft and Camshaft Sensor

**2.8L ENGINE**

These sensors are mounted on the engine block, near the engine crankshaft, and also near the camshaft on some engines. The sensors are used to send a signal through the Direct Ignition System (DIS) module to the ECM. The ECM uses this reference signal to calculate engine speed and crankshaft position.

The engine uses a sensor called a Hall effect switch. With the direct ignition connected to the vehicle electrical system, the system voltage is applied to the Hall effect switch located near the tip of the sensor. A small permanent magnet creates a magnetic field in the Hall effect switch circuit. As the disc with the slots rotates past the sensor tip, the magnetic field in the Hall effect switch changes and a change in the voltage occurs at the Hall effect switch output terminal.

Since this terminal is connected to the ignition module, the module senses this change in voltage and correlates the frequency of the voltage curve to determine the engine speed. The ignition module uses this voltage input to help determine when to close and open the ignition coil primary circuit and fire the spark plug.

### Direct Ignition System (DIS)

**2.8L ENGINE**

Components of the Direct Ignition System (DIS) are a coil pack, ignition module, crankshaft reluctor ring, magnetic sensor and the ECM. The coil pack consists of 2 separate, interchangeable, ignition coils. These coils operate in the same manner as previous coils. 2 coils are needed because each coil fires for 2 cylinders. The ignition module is located under the coil pack and is connected to the ECM by a 6 pin connector. The ignition module controls the primary circuits to the coils, turning them on and off and controls spark timing below 400 rpm and if the ECM bypass circuit becomes open or grounded.

The magnetic pickup sensor inserts through the engine block, just above the pan rail in proximity to the crankshaft reluctor ring. Notches in the crankshaft reluctor ring trigger the magnetic pickup sensor to provide timing information to the ECM. The magnetic pickup sensor provides a cam signal to identify correct firing sequence and crank signals to trigger each coil at the proper time.

This system uses EST and control wires from the ECM, as with the distributor systems. The ECM controls the timing using crankshaft position, engine rpm, engine temperature and manifold absolute pressure sensing.

### Electronic Control Unit

**ALL EXCEPT 2.8L ENGINE**

The ECU is usually located under the instrument panel. The ECU analyzes all electrical data signals from the sensors. It controls the fuel injection system, the ignition system and the turbocharger control system (is so equipped). The ECU has a built in back-up, diagnostic and fail safe control systems.

### Electronic Spark Timing (EST)

**2.8L ENGINE**

Electronic spark timing (EST) is used on all engines. The EST distributor contains no vacuum or centrifugal advance and uses a 7-terminal distributor module. It also has 4 wires going to a 4-terminal connector in addition to the connectors normally found on HEI distributors. A reference pulse, indicating both engine rpm and crankshaft position, is sent to the ECM. The ECM determines the proper spark advance for the engine operating conditions and sends an EST pulse to the distributor.

The EST system is designed to optimize spark timing for better control of exhaust emissions and for fuel economy improvements. The ECM monitors information from various engine sensors, computes the desired spark timing and changes the timing accordingly. A backup spark advance system is incorporated in the module in case of EST failure.

### Electronic Spark Control (ESC)

**2.8L ENGINE**

When engines are equipped with ESC in conjunction with EST, ESC is used to reduce spark advance under conditions of detonation. A knock sensor signals a separate ESC controller to retard the timing when it senses engine knock. The ESC controller signals the ECM which reduces spark advance until no more signals are received from the knock sensor.

### Fuel Injector

**ALL MODELS**

The fuel injector is controlled by the ECU and injects fuel when ever it is energized by the ECU.

### Knock Sensor

**ALL EXCEPT 2.8L ENGINE**

This sensor is usually located in the cylinder head. The purpose of this sensor is to send electrical impulses to the ECU when ever engine "knocking" occurs. The ECU uses these impulses to retard the ignition timing.

### Manifold Absolute Pressure Sensor

**2.8L ENGINE**

The Manifold Absolute Pressure (MAP) sensor measures the changes in the intake manifold pressure which result from engine load and speed changes. The pressure measured by the MAP sensor is the difference between barometric pressure (outside air) and manifold pressure (vacuum). A closed throttle engine coastdown would produce a relatively low MAP value (approximately 20–35 kPa), while wide-open throttle would produce a high value (100 kPa). This high value is produced when the pressure inside the manifold is the same as outside the manifold, and 100% of outside air (or 100 kPa) is being measured. This MAP output is the opposite of what you would measure on a vacuum gauge. The use of this sensor also allows the ECM to adjust automatically for different altitude.

The ECM sends a 5 volt reference signal to the MAP sensor. As the MAP changes, the electrical resistance of the sensor also changes. By monitoring the sensor output voltage the ECM can determine the manifold pressure. A higher pressure, lower vacuum (high voltage) requires more fuel, while a lower pressure, higher vacuum (low voltage) requires less fuel. The ECM uses the MAP sensor to control fuel delivery and ignition timing. A failure in the MAP sensor circuit should set a Code 33 or Code 34.

### Manifold Air Temperature Sensor

**2.8L ENGINE**

The Manifold Air Temperature (MAT) sensor is a thermistor mounted in the intake manifold. A thermistor is a resistor which changes resistance based on temperature. Low manifold air temperature produces a high resistance (100,000 ohms at −40°F/−40°C), while high temperature cause low resistance (70 ohms at 266°F/130°C).

The ECM supplies a 5 volt signal to the MAT sensor through a resistor in the ECM and monitors the voltage. The voltage will be high when the manifold air is cold and low when the air is hot. By monitoring the voltage, the ECM calculates the air temperature and uses this data to help determine the fuel delivery and spark advance. A failure in the MAT circuit should set either a Code 23 or Code 25.

### Oil Pressure Switch

**2.8L ENGINE**

The oil pressure switch is usually mounted on the back of the engine, just below the intake manifold. Some vehicles use the oil pressure switch as a parallel power supply (with the fuel pump relay) and will provide voltage to the fuel pump after approximately 4 psi (28 kPa) of oil pressure is reached. This switch will also help prevent engine seizure by shutting off the power to the fuel pump and causing the engine to stop when the oil pressure is lower than 4 psi.

### Oxygen Sensor

**ALL EXCEPT 2.8L ENGINE**

This sensor is usually threaded into the exhaust manifold. The oxygen sensor measures and produces an electrical signal proportional to the amount of the oxygen present in the exhaust gases.

**2.8L ENGINE**

The exhaust oxygen sensor is mounted in the exhaust system where it can monitor the oxygen content of the exhaust gas stream. The oxygen content in the exhaust reacts with the oxygen sensor to produce a voltage output. This voltage ranges from approximately 100 millivolts (high oxygen–lean mixture) to 900 millivolts (low oxygen–rich mixture).

By monitoring the voltage output of the oxygen sensor, the ECM will determine what fuel mixture command to give to the injector (lean mixture–low voltage–rich command, rich mixture–high voltage–lean command).

Remember that the oxygen sensor indicates to the ECM what is happening in the exhaust. It does not cause things to happen. It is a type of gauge: high oxygen content = lean mixture; low oxygen content = rich mixture. The ECM adjusts fuel to keep the system working.

The oxygen sensor, if open, should set a Code 13. A constant low voltage in the sensor circuit should set a Code 44 while a constant high voltage in the circuit should set a Code 45. Codes 44 and 45 could also be set as a result of fuel system problems.

### Park/Neutral Switch

**2.8L ENGINE**

**NOTE: Vehicle should not be driven with the park/neutral switch disconnected as idle quality may be affected in PARK or NEUTRAL and a Code 24 (VSS) may be set.**

This switch indicates to the ECM when the transmission is in **P** or **N**. The information is used by the ECM for control on the torque converter clutch, EGR, and the idle air control valve operation.

### Power Steering Pressure Switch

**2.8L ENGINE**

The power steering pressure switch is used so that the power steering pump load will not effect the engine idle. Turning the steering wheel increases the power steering oil pressure and pump load on the engine. The power steering pressure switch will close before the load can cause an idle problem.

### Stepper Motor

**ALL EXCEPT 2.8L ENGINE**

The stepper motor is used on turbocharged vehicles only. The stepper motor is rotated by a voltage pulse sent to it by the turbocharger control unit. The stepper motor is able to rotate 90 degrees, depending on the number of pulses it receives from the turbocharger control unit. A microswitch inside the stepper motor unit tells the turbocharger control unit when the stepper motor is at zero degree angle. 250 pulses from the turbocharger control unit are required to move the stepper motor 90 degrees.

### Throttle Position Sensor

**ALL EXCEPT 2.8L ENGINE**

The throttle position sensor (TPS) is used on turbocharged models only. The sensor controls the fuel cut system.

**2.8L ENGINE**

The Throttle Position Sensor (TPS) is connected to the throttle shaft and is controlled by the throttle mechanism. A 5 volt reference signal is sent to the TPS from the ECM. As the throttle valve angle is changed (accelerator pedal moved), the resistance of the TPS also changes. At a closed throttle position, the resistance of the TPS is high, so the output voltage to the ECM will be low (approximately 0.5 volts). As the throttle plate opens, the resistance decreases so that, at wide open throttle, the output voltage should be approximately 5 volts. At closed throttle position, the voltage at the TPS should be less than 1.25 volts.

By monitoring the output voltage from the TPS, the ECM can determine fuel delivery based on throttle valve angle (driver demand). The TPS can either be misadjusted, shorted, open or loose. Misadjustment might result in poor idle or poor wide-open throttle performance. An open TPS signals the ECM that the throttle is always closed, resulting in poor performance. This usually sets a Code 22. A shorted TPS gives the ECM a constant wide-open throttle signal and should set a Code 21. A loose TPS indicates to the ECM that the throttle is moving. This causes intermittent bursts of fuel from the injector and an unstable idle. On some vehicles, the TPS is adjustable and therefore can be adjusted to correct any complications caused by toohigh or too low a voltage signal.

### Throttle Valve Position Switch

**ALL EXCEPT 2.8L ENGINE**

The throttle position switch is usually located on the throttle body. It is used to detect the throttle valve position at engine idle, part throttle and wide open throttle.

### Vacuum Switching Valve

**ALL EXCEPT 2.8L ENGINE**

The vacuum switching valve is controlled by a signal from the ECU. This valve controls the fuel pressure according to the vacuum developed in the intake manifold.

### Vehicle Speed Sensor

**ALL EXCEPT 2.8L ENGINE**

The vehicle speed sensor is incorporated into the speedometer. The ECU receives electrical impulses from the vehicle speed sensor.

**2.8L ENGINE**

**NOTE: A vehicle equipped with a speed sensor, should not be driven without the speed sensor connected, as idle quality may be affected.**

The Vehicle Speed Sensor (VSS) is mounted behind the speedometer in the instrument cluster or on the transmission/speedometer drive gear. It provides electrical pulses to the ECM from the speedometer head. The pulses indicate the road speed. The ECM uses this information to operate the IAC, canister purge, and TCC.

Some vehicles equipped with digital instrument clusters use a Permanent Magnet (PM) generator to provide the VSS signal. The PM generator is located in the transmission and replaces the speedometer cable. The signal from the PM generator drives a stepper motor which drives the odometer. A failure in the VSS circuit should set a Code 24.

## SERVICE PRECAUTIONS

Be careful not to get water on any fuel injection system component. Pay close attention to the relay box and throttle valve switch connector. The connector is not water proofed and will be damaged by water.

When charging the battery be sure to remove it from the vehicle first. Never disconnect the battery cable from the battery when the engine is running. The generation of surge voltage may damage the control unit and other electrical parts such as the multi-drive monitor.

When replacing parts or checking the system, make sure to set the starter switch to the **OFF** position. When measuring voltage at the control unit harness connector, disconnect all the control unit harnesses first, then set the starter switch to the **ON** position.

When checking the electrical terminals of the control unit with a tester, do not apply the probe to terminal directly but insert a pin into the terminal from the harness side and perform the measurement through the pin. If the tester probe is held against the terminal directly, the terminal will be deformed, causing poor contact. Connect each harness correctly and firmly to insure a good contact.

The wiring connectors for the fuel injector, throttle valve switch, air regulator and water temperature sensor are provided with locked wires. To unlock the connector, pull and shake it gently.

System cables must be placed at least 4 in. away from the tension cables. Be careful not to apply any shock to the system components such as the air flow sensor, crank angle sensor, and control unit. Component parts of the fuel injection system are precisely set. Even a slight distortion or dent will seriously affect performance.

The fuel pump must not be operated without fuel. Since fuel lubricates the pump, noise or other serious problems such as parts seizure will result. It is also prohibited to use any fuel other than gasoline.

## I-MARK TURBO

### MAJOR COMPONENTS

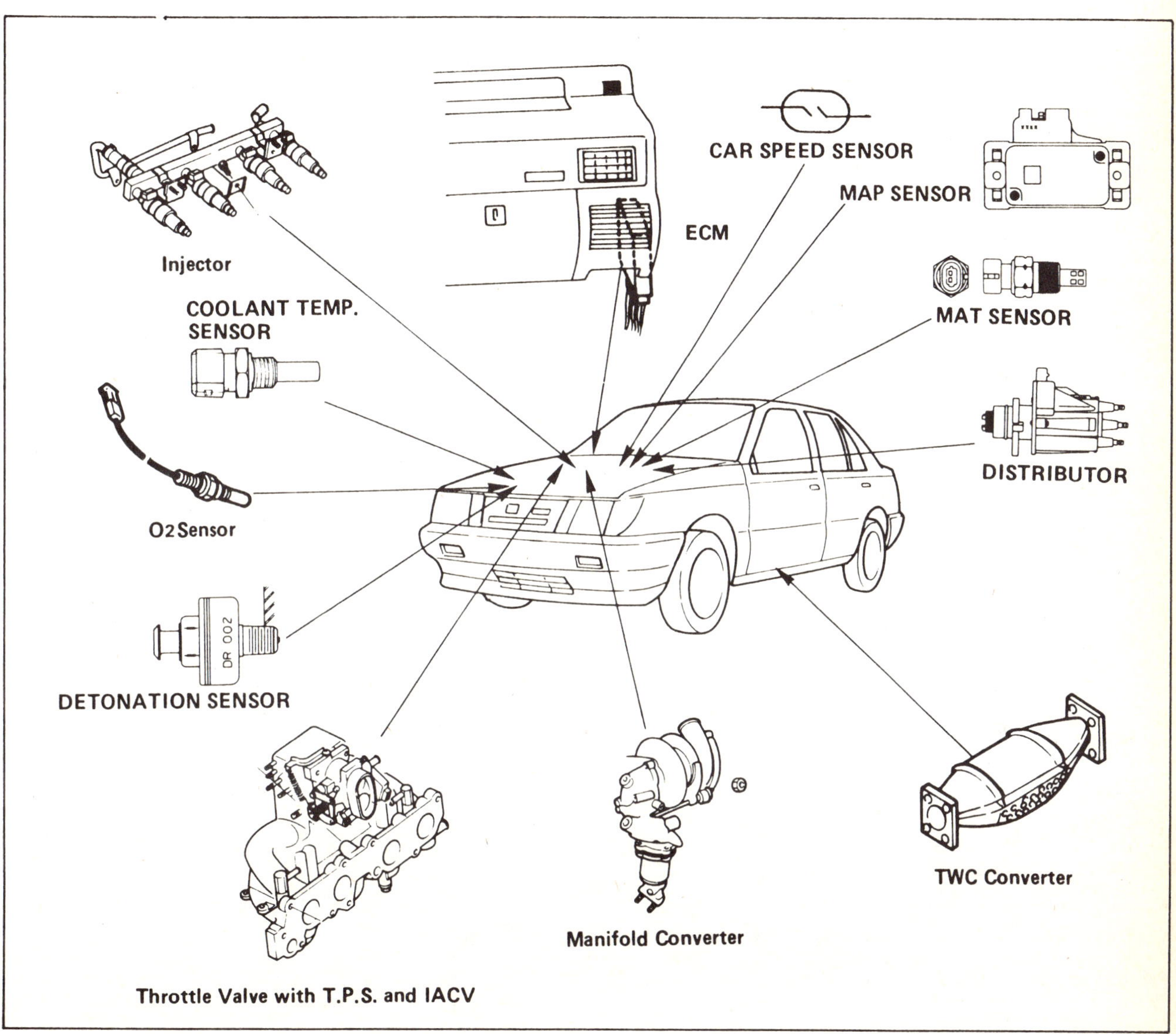


- Throttle valve with TPS
- IACV (idle air control valve)
- Detonation temp. sensor
- Coolant temp. sensor
- Car speed sensor
- MAT sensor
- Distributor
- Fuel injector
- Manifold absolute pressure sensor (MAP sensor)
- Oxygen sensor
- Electronic control module (ECM)
- Multi port fuel injectors
- Three-way catalytic converter
- Manifold converter

## I-MARK TURBO

### TROUBLE CODE IDENTIFICATION

The "CHECK ENGINE" light will only be "ON" if the malfunction exists under the conditions listed below. It takes up to five seconds minimum for the light to come on when a problem occurs. If the malfunction clears, the light will go out and a trouble code will be set in the ECM. Code 12 does not store in memory. If the light comes "on" intermittently, but no code is stored, go to the "Driver Comments" section. Any codes stored will be erased if no problem reoccurs within 50 engine starts.

The trouble codes indicate problems as follows:

| | |
|---|---|
| TROUBLE CODE 12 | No distributor reference pulses to the ECM and will only flash while the fault is present. |
| TROUBLE CODE 13 | Oxygen Sensor Circuit — The engine must run up to two minutes at part throttle, under road load, before this code will set. |
| TROUBLE CODE 14 | Coolant temperature sensor circuit (High temperature indicated) |
| TROUBLE CODE 15 | Coolant temperature sensor circuit (Low temperature indicated) |
| TROUBLE CODE 21 | Throttle position sensor circuit (Signal voltage high) |
| TROUBLE CODE 22 | Throttle position sensor circuit (Signal voltage low) |
| TROUBLE CODE 23 | Manifold air temperature (MAT) sensor (Low temperature indicated) |
| TROUBLE CODE 24 | Vehicle speed sensor (VSS) circuit |
| TROUBLE CODE 25 | Manifold air temperature (MAT) sensor circuit (High temperature indicated) |
| TROUBLE CODE 31 | Turbocharger wastgate control |
| TROUBLE CODE 32 | EGR system fault |
| TROUBLE CODE 33 | MAP sensor circuit (Signal voltage high) |
| TROUBLE CODE 34 | MAP sensor circuit (Signal voltage low) |
| TROUBLE CODE 42 | Electronic spark timing (EST) circuit |
| TROUBLE CODE 43 | Electronic spark control (ESC) (Knock control failure) |
| TROUBLE CODE 44 | Oxygen sensor circuit (Lean exhaust indication) |
| TROUBLE CODE 45 | Oxygen sensor circuit (Rich exhaust indication) |
| TROUBLE CODE 51 | MEM-CAL error (Faulty or incorrect MEM-CAL) |

# Diagnosis and Testing

## I-MARK TURBO

### CODE 13

**OXYGEN SENSOR CIRCUIT (OPEN CIRCUIT)**

Code 13 will set:

- Engine at normal operating temperature
- Allow at least 2 minutes engine running time after start.
- $O_2$ signal voltage at terminal M2-23 of ECM steadily between 0.35 and 0.55 volts for more than one minutes.
- Throttle position sensor signal above 7 % (about 0.35 volts above closed throttle voltage.)
- All conditions must be met for about 60 seconds.

The ECM supplies a voltage of about 0.45 volt between terminals "M2-23" and "M2-22". (If measured with a 10 megohm digital voltmeter, this may read as low as 0.32 volts.)
The $O_2$ sensor varies voltage within a range of about 1 volt if the exhaust is rich, down through about 0.10 volt if exhaust is lean.

The sensor is like an open circuit and produces no voltage when it is below 360°C (600°F). An open sensor circuit or cold sensor causes open loop operation.

1. Grounding the diagnostic terminal with the engine running enables the "Field Service Mode", which allows the ECM to confirm either open or closed loop operation using the "CHECK ENGINE" light.
2. This step simulates a lean exhaust. If the ECM and wiring are OK, the ECM will see the lean condition and turn the "CHECK ENGINE" light off for at least 30 seconds after engine start, and then flash "open loop". It should be considered normal if the light remains off for a longer period of time before flashing open loop.

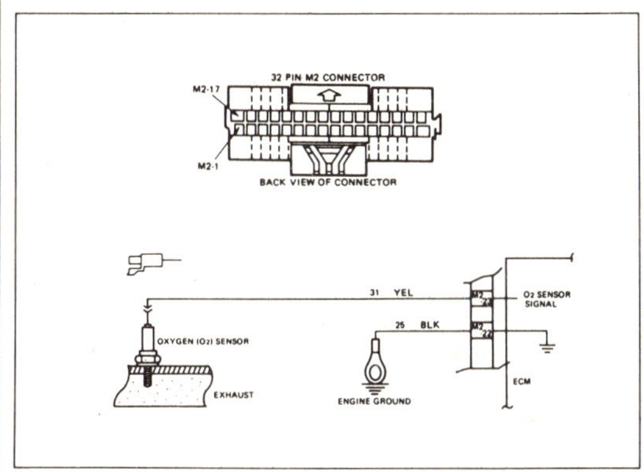


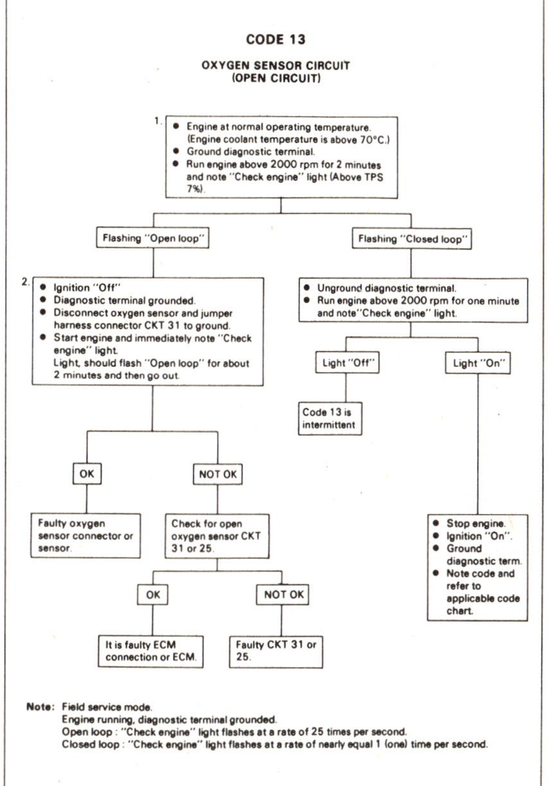


## I-MARK TURBO

### CODE 14

**COOLANT TEMPERATURE SENSOR CIRCUIT (HIGH TEMPERATURE INDICATED)**

The Coolant Temperature Sensor uses a thermistor to control the signal voltage to the ECM. The ECM applies a voltage to CKT 37 to the sensor. When the engine is cold the sensor (thermistor) resistance is high, therefore the ECM will see high signal voltage at CKT 37.

As the engine warms, the sensor resistance becomes less, and the voltage drops.

Code 14 will flash if:
Signal voltage indicates a coolant temperature above 145°C (293°F) for more than 120 seconds.

Coolant temperature is one of the inputs used to control:

- Fuel delivery
- Electronic spark timing control (EST)
- Idle air control (IAC)
- Turbocharger wastegate control
- Canister purge control
- AIR control
- EGR control
- EGR system diagnosis

1. If voltage is above 4 volts, the ECM and wiring are OK.
2. If checking resistance at the coolant sensor is difficult because of sensor location, disconnect the black ECM connectors and check resistance between harness connector terminals M2-10 and M2-18.

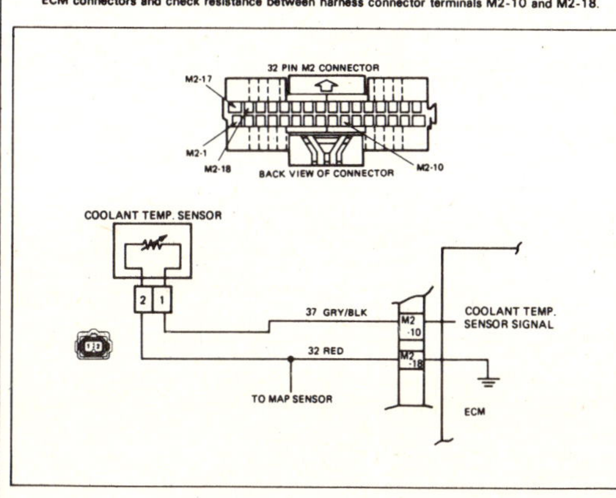


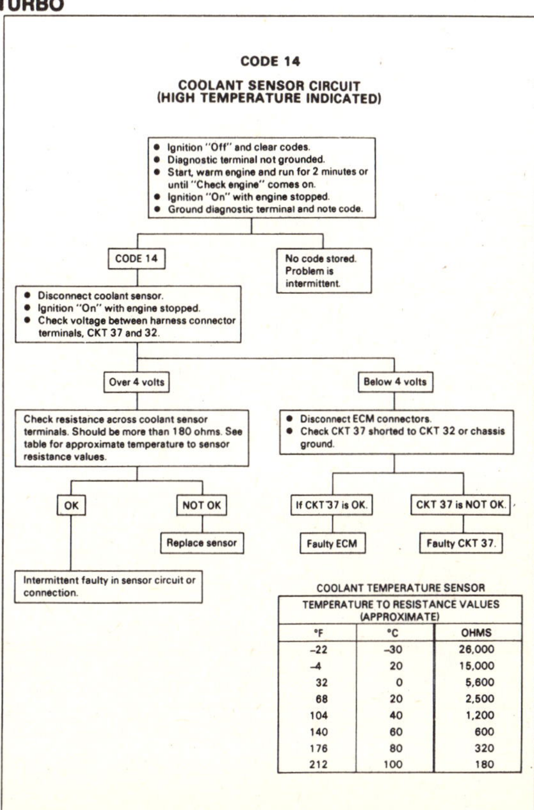


COOLANT TEMPERATURE SENSOR

TEMPERATURE TO RESISTANCE VALUES (APPROXIMATE)

| °F | °C | OHMS |
|---|---|---|
| –22 | –30 | 26,000 |
| –4 | 20 | 15,000 |
| 32 | 0 | 5,600 |
| 68 | 20 | 2,500 |
| 104 | 40 | 1,200 |
| 140 | 60 | 600 |
| 176 | 80 | 320 |
| 212 | 100 | 180 |

## I-MARK TURBO

### CODE 15

**COOLANT TEMPERATURE SENSOR CIRCUIT (LOW TEMPERATURE INDICATED)**

The Coolant Temperature Sensor uses a thermistor to control the signal voltage to the ECM. The ECM applies a voltage on CKT 37 to the sensor. When the engine is cold the sensor (thermistor) resistance is high, therefore the ECM will see high signal voltage at CKT 37.

As the engine warms, the sensor resistance becomes less, and the voltage drops. At normal engine operating temperature the voltage will measure about 1 to 1.5 volts at the ECM terminal M2-10.

Code 15 will flash if:

- Signal voltage indicates a coolant temperature less than −38.5°C (−37.3°F) for 10 seconds.
- Time since engine start is more than 1 minutes.

Coolant temperature is one of the inputs used to control:

- Fuel delivery
- Electronic spark timing (EST)
- Idle air control (IAC)
- Turbocharger wastegate control
- Canister purge control
- AIR control
- EGR control
- EGR system diagnostic

If The coolant temperature sensor CKT 37 is open with the ignition off, the ECM will see −40°C (−40°F) and deliver fuel for this temperature. If the actual temperature is above approx. −7°C (20°F) the engine will not start due to the rich mixture unless "Clear Flood" is used by fully depressing the accelerator. Engine will start using "Clear Flood" (which is Wide Open Throttle). However, "Check engine" light will not come on, and code will not be stored, until engine has run for one minute.

1. If voltage is above 4 volts, the ECM and wiring are OK.
2. If location of sensor makes it hard to check, disconnect ECM connector and check resistance between connector terminals M2-10 and M2-18.

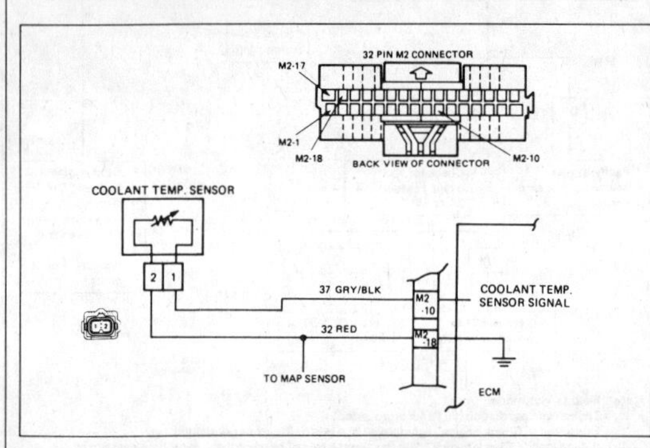


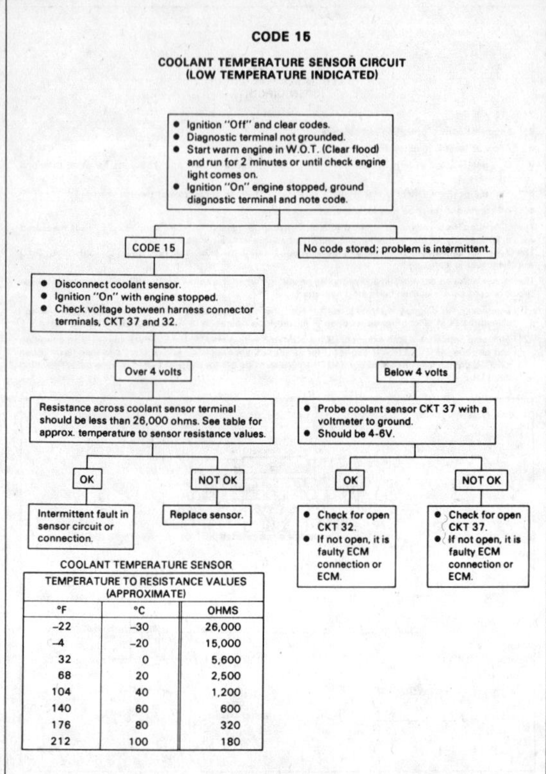


COOLANT TEMPERATURE SENSOR

| TEMPERATURE TO RESISTANCE VALUES (APPROXIMATE) | | |
|---|---|---|
| °F | °C | OHMS |
| −22 | −30 | 26,000 |
| −4 | −20 | 15,000 |
| 32 | 0 | 5,600 |
| 68 | 20 | 2,500 |
| 104 | 40 | 1,200 |
| 140 | 60 | 600 |
| 176 | 80 | 320 |
| 212 | 100 | 180 |

## I-MARK TURBO

### CODE 21

**THROTTLE POSITION SENSOR CIRCUIT (SIGNAL VOLTAGE HIGH)**

The Throttle Sensor (TPS) provides a voltage signal that changes relative to the throttle valve. Signal voltage will vary from less than 0.45 volts at idle to about 4.0 volts at wide open throttle (WOT).

The TPS signal is one of the most important inputs used by the ECM for fuel control and for many of the ECM controlled outputs.

Code 21 will flash if:

- TPS signal voltage is greater than 4.7 volts for 5 seconds.
- Engine speed less than 1200 rpm. This is the purpose for placing shift lever in drive.
- MAP is less than 50 kPa (7.5 psi) or equal to a no load condition.

1. Confirms Code 21, and that fault is present.
2. Simulates Code 22: If the ECM recognizes the low signal voltage and set Code 22, the ECM and wiring are OK.

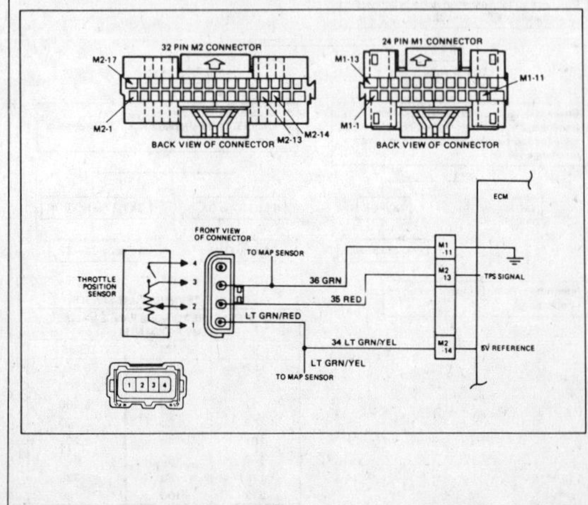


### CODE 21

**THROTTLE POSITION SENSOR CIRCUIT (SIGNAL VOLTAGE HIGH)**

- Diagnostic terminal not grounded.
- Ignition "Off".
- Clear codes.
- Start engine and idle A/C off, for 20 sec. or until "Check" engine light comes on.

**Light on**

- Ignition "On", with engine stopped.
- Ground diagnostic terminal and note codes.

- **CODE 21**
  - Diagnostic terminal not grounded.
  - Ignition "Off".
  - Clear codes.
  - Disconnect TPS sensor.
  - Start engine and idle, A/C off, for 20 sec. or until "Check engine" light comes on.
  - Ignition "On", engine stopped.
  - Ground diagnostic terminal and note code.
    - **CODE 22**: Probe TPS harness connector CKT 36 with a test light to 12 volts.
      - **Light on**: Faulty TPS connection or sensor.
      - **Light off**: Open CKT 36 on faulty ECM.
    - **CODE 21**: CKT 35 shorted to voltage or faulty ECM.
- Any other code, see applicable code chart.

**Light off**

Code 21 is intermittent.

## I-MARK TURBO

### CODE 22

**THROTTLE POSITION SENSOR CIRCUIT
(SIGNAL VOLTAGE LOW)**

The Throttle Position Sensor (TPS) provides a voltage signal that changes relative to the throttle valve. Signal voltage will vary from less than 0.45 volts at idle to about 4.0 volts at wide open throttle (WOT).

The TPS signal is one of the most important inputs used by the ECM for fuel control and for many of the ECM controlled outputs.

Code 22 will flash if:

- TPS signal voltage is less than 0.2 volts.

Possible causes of Code 22:

- Faulty TPS
- Faulty wiring or terminals
- Faulty ECM

1. Confirm code 22, and that fault is present.
2. Simulates Code 21: If the ECM recognizes the high signal voltage and sets Code 21, the ECM and wiring are OK.
3. Checks for reference voltage from the ECM. To prevent damage to ECM, be sure to disconnect connector when checking circuit wiring for open or shorts to ground.

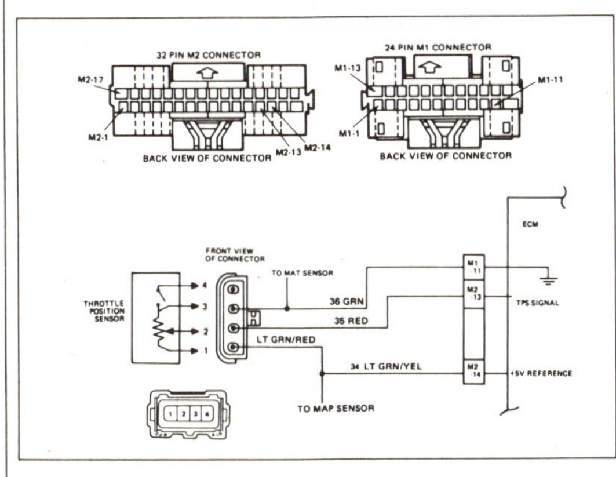


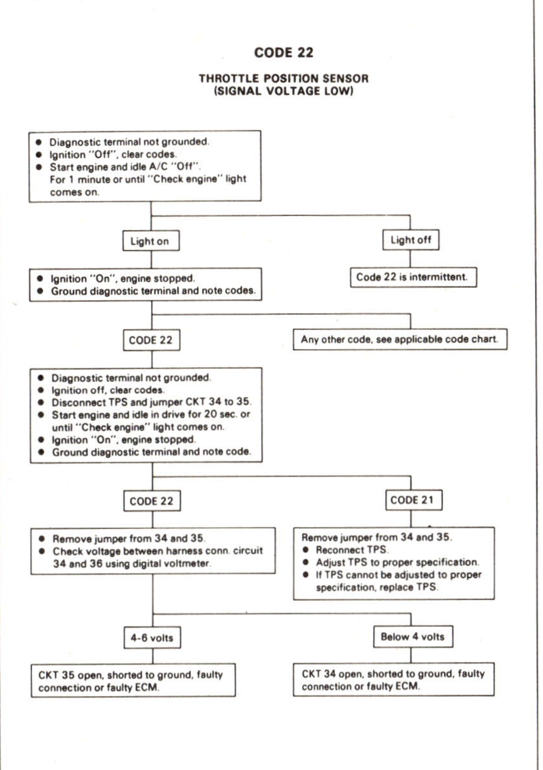


## I-MARK TURBO

### CODE 23

**MANIFOLD AIR TEMPERATURE (MAT) SENSOR
(LOW TEMPERATURE INDICATED)**

The MAT Sensor uses a thermistor to control the signal voltage to the ECM.
The ECM applies a voltage (4-6 volts) to CKT 9 to the sensor. When the air is cold, the sensor (thermistor) resistance is high, therefore the ECM will see a high signal voltage. If the air is warm, the sensor resistance is low therefore the ECM will see a low voltage.

Code 23 will flash if:

- A signal voltage indicates a manifold air temperature below −38°C (−36.4°F).
- Time since engine start is 1 minute or longer.

Due to the conditions necessary to set a Code 23, the check engine light will only stay on when both conditions are met.

1. A Code 23 will set due to an open sensor, wire, or connection. This test will determine if the wiring and ECM are OK.
2. If the resistance is greater than 25,000 ohms, replace the sensor.

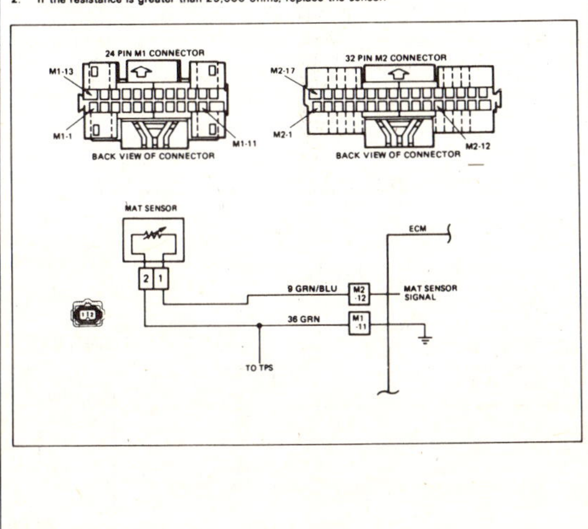


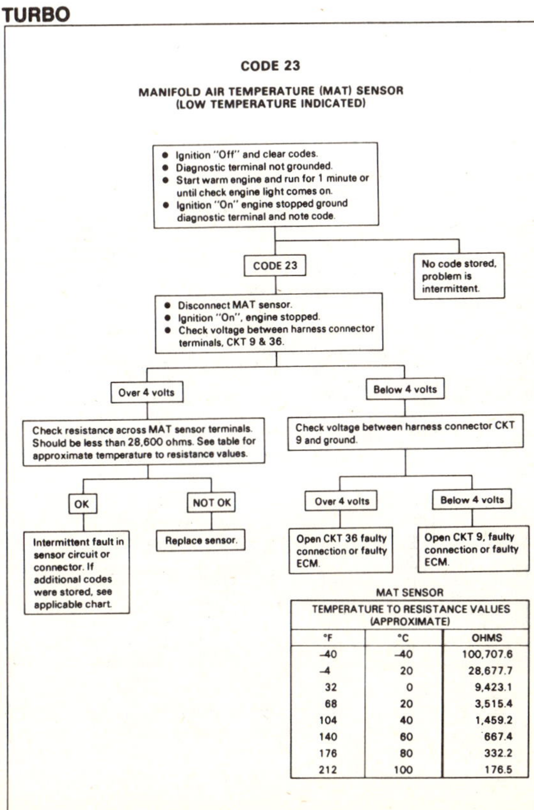


MAT SENSOR

| TEMPERATURE TO RESISTANCE VALUES (APPROXIMATE) | | |
|---|---|---|
| °F | °C | OHMS |
| −40 | −40 | 100,707.6 |
| −4 | 20 | 28,677.7 |
| 32 | 0 | 9,423.1 |
| 68 | 20 | 3,515.4 |
| 104 | 40 | 1,459.2 |
| 140 | 60 | 667.4 |
| 176 | 80 | 332.2 |
| 212 | 100 | 176.5 |

## I-MARK TURBO

### CODE 24

#### VEHICLE SPEED SENSOR (VSS) CIRCUIT

Code 24 will flash if:

- CKT 38 voltage is constant.
- Engine speed between 2400 and 4400 rpm.
- Low load condition.
- All conditions must be met for 10 seconds.

The ECM applies and monitors 12 volts on CKT 38. CKT 38 connects to the Vehicle Speed Sensor which alternately grounds CKT 38 when drive wheels are turning. This pulsing action takes place about 4096 times per mile and the ECM will calculate vehicle speeds based on the time between "pulses".

1. This test monitors the ECM voltage on CKT 38. With the wheels turning, the pulsing action will result in a varing voltage. The variation will be greater at low wheel speeds, to an average off 4-6 volts at about 20 mph (32 km/h).
2. A voltage of less than 1 volt at the ECM connector indicates that the CKT 38 wire is shorted to ground. Disconnect CKT 38 at the Vehicle Speed Sensor. If voltage now reads above 10 volts, the Vehicle Speed Sensor is faulty. If voltage remains less than 10 volts, then CKT 38 wire is grounded. If 38 is not grounded, check for a faulty ECM connector or ECM.
3. A steady 8-12 volts at the ECM connector indicates CKT 38 is open, or a faulty Vehicle Speed Sensor.

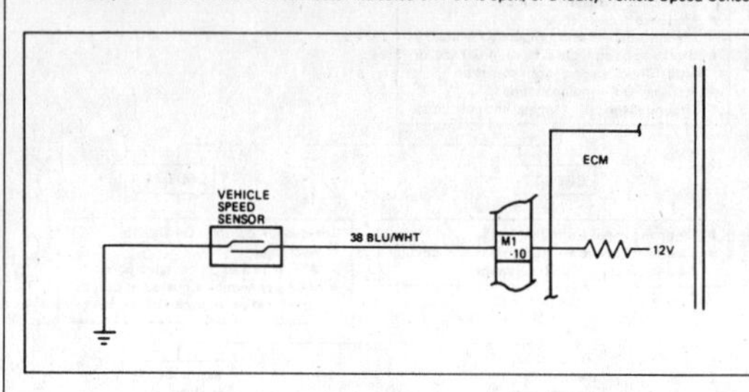


### CODE 24

#### VEHICLE SPEED SENSOR (VSS) CIRCUIT

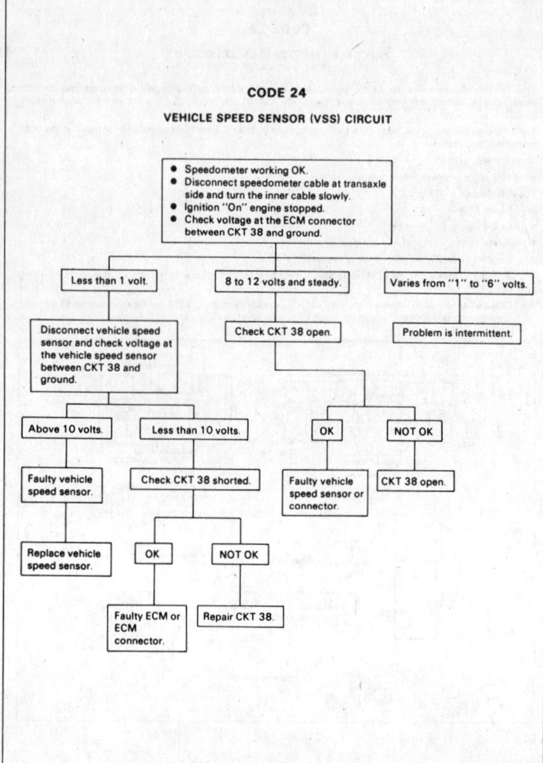


## I-MARK TURBO

### CODE 25

#### MANIFOLD AIR TEMPERATURE (MAT) SENSOR CIRCUIT (HIGH TEMPERATURE INDICATED)

The Manifold Air Temperature Sensor uses thermistor to control the signal-voltage to the ECM. The ECM applies a voltage (4-6) on CKT 9 to the sensor. When manifold air is cold, the sensor (Thermistor) resistance is high, therefore the ECM will see a high signal voltage. As the air warms, the sensor resistance becomes less, and the voltage drops.

Code 25 will set if:

- Signal voltage indicates a manifold air temperature greater than 145°C (293°F).
- Time since engine start is 2 minutes or longer.
- Turbocharger wastegate control is not operated.

Due to the conditions necessary to set a Code 25, the check engine light will only remain on while the signal is low.

1. If voltage is above 4 volts, the ECM and wiring are OK.
2. If the resistance is less than 176 ohms, replace the sensor.

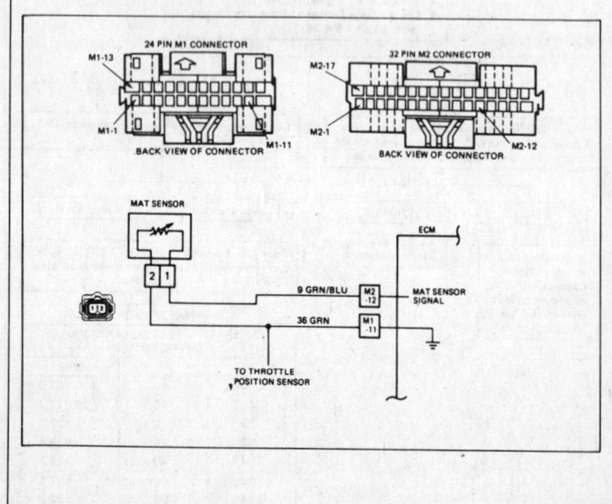


### CODE 25

#### MANIFOLD AIR TEMPERATURE (MAT) SENSOR CIRCUIT (HIGH TEMPERATURE INDICATED)

- Ignition "Off", and clear codes.
- Diagnostic terminal not grounded.
- Start warm engine and run for 1 minute or until check engine light comes on.
- Ignition "On", engine stopped, ground diagnostic terminal and note code.

CODE 25

- Disconnect MAT sensor.
- Ignition "ON", engine stopped.
- Check voltage between harness connector terminals, CKT 9 & 36.

Over 4 volts

Check resistance across MAT sensor terminals should be less than 176 ohms. See table for approximate temperature to resistance values.

OK

Intermittent fault in sensor circuit or connector. If additional codes were stored, see applicable chart.

NOT OK

Replace sensor.

Below 4 volts

CKT 9 shorted to ground or faulty ECM.

No code stored, problem is intermittent.

MAT SENSOR

| TEMPERATURE TO RESISTANCE VALUES (APPROXIMATE) | | |
|---|---|---|
| °F | °C | OHMS |
| –40 | –40 | 100707.6 |
| –4 | 20 | 28677.7 |
| 32 | 0 | 9423.1 |
| 68 | 20 | 3515.4 |
| 104 | 40 | 1459.2 |
| 140 | 60 | 667.4 |
| 176 | 80 | 332.2 |
| 212 | 100 | 176.5 |

## I-MARK TURBO

### CODE 31

#### TURBOCHARGER WASTGATE CONTROL FAILURE (OVER BOOST)

The turbo pressure is controlled by ECM. Code 31 is registered when the turbo pressure has risen to an abnormally high level.

The conditions under which Code 31 appears are:

- Neither Code 33 nor 34 (MAP sensor, high/low) is registered.
- MAP has dropped below 204 kPa within 0.2 seconds after the High Load Fuel Cut Mode (MAP is equal to or greater than 204 kPa) was reached.

(In reality, however, a relief valve opens to reduce the turbo pressure before the High Load Fuel Cut Mode is reached.)

When the turbo pressure is not controlled by ECM, the pressure that acts on the waste gate control diaphragm is equal to the turbo pressure, and this pressure is maintained constant as the turbo pressure remains in balance with the amount the waste gate valve opens.

However, when the turbo pressure is controlled by ECM, the pressure on the waste gate control diaphragm is released to the atmosphere through VSV. Therefore, this pressure drops below the actual turbo pressure and the waste gate valve closes by that amount, which in turn causes the turbo pressure to rise by the corresponding amount. Our system is designed to achieve desired turbo pressures using the duty-control of VSV, which adjusts the air flow through the VSV.

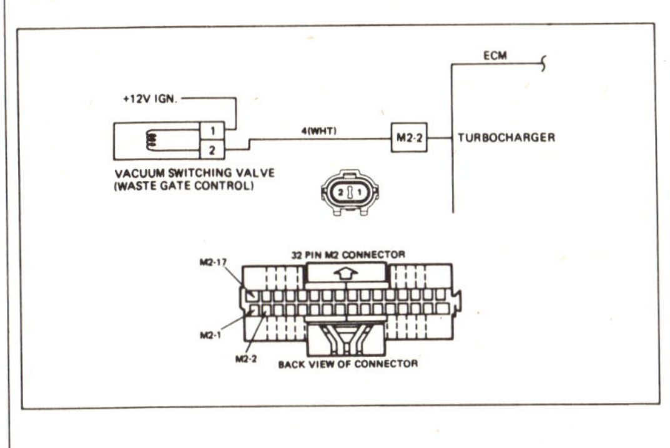


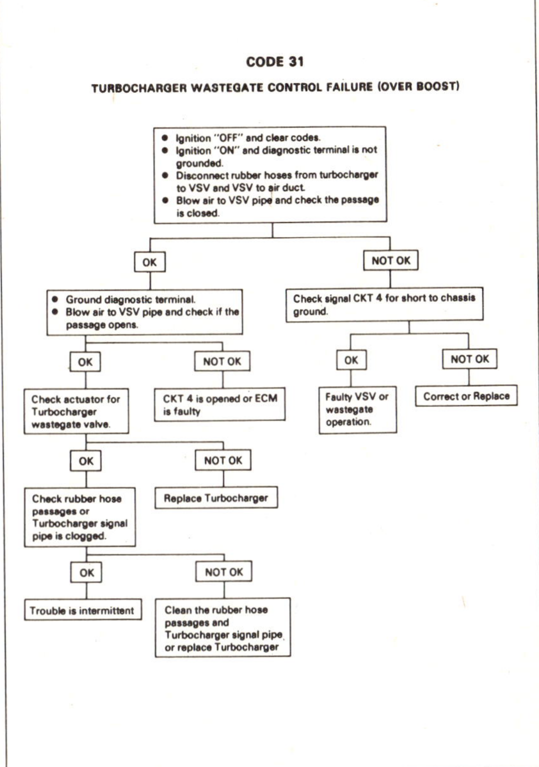


## I-MARK TURBO

### CODE 32

#### EGR SYSTEM FAILURE

The EGR temperature sensor determines the difference between the temperature when the EGR is on and when the EGR is off. If the difference remains below a certain level for a predetermined time, the system registers Code 32.

The conditions chosen as typical of the range where EGR is on are:

Engine speed: 1800 rpm — 5000 rpm
MAP: 48 kPa — 68 kPa

And the condition typical of the range where EGR is off is at idle.

The system checks the difference in temperatures when the following conditions exist:

Coolant temperature: 75°C — 100°C
Intake air temperature (MAT) : 50°C — 100°C

When idling continues at least 20 seconds under these conditions, the temperature is memorized as that of the range where EGR is off. On the other hand, if the aforementioned conditions typical of the range where EGR is on continue 8 seconds or more, the temperature at that time is read as that of the range where EGR is on. When the two temperatures are compared and if the difference were 3°C or less, then the system recognizes an EGR system malfunction. Code 32 is registered when this reading has occured three times consecutively.

If, however, the temperature difference is 6°C or more, the EGR system is recognized as normal and the "check engine" lamp turns off.

**Note:** The EGR valve is controlled by a solenoid, which in turn is controlled by ECM. This solenoid normally remains off and comes on as the CKT7 is grounded. It comes on only in predetermined ranges of coolant temperature, MAP, TPS and engine speed. It also comes on when the ignition is on and when, with the engine stopped, the diagnosis terminal has been grounded.

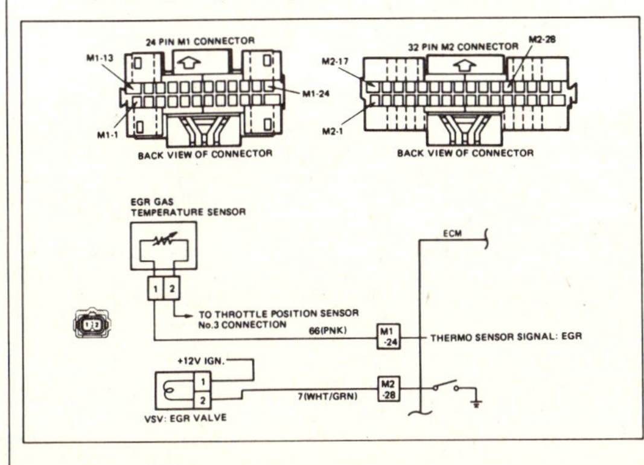


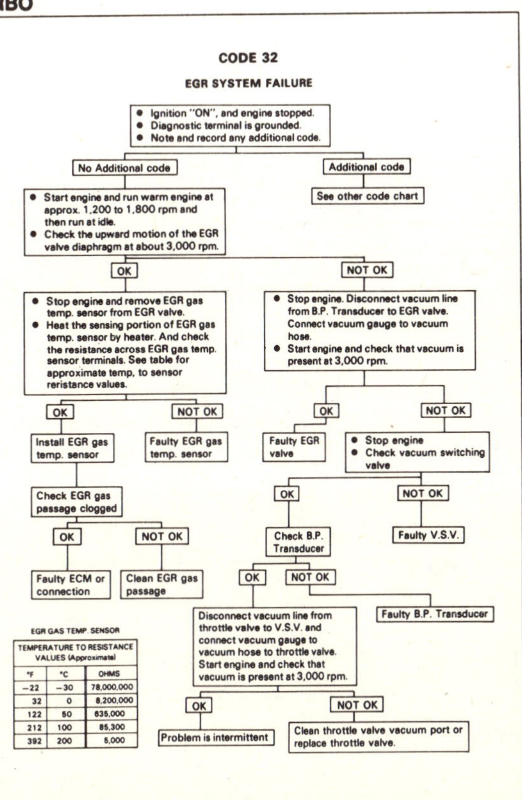


| TEMPERATURE TO RESISTANCE VALUES (Approximate) | | |
|---|---|---|
| °F | °C | OHMS |
| −22 | −30 | 78,000,000 |
| 32 | 0 | 8,200,000 |
| 122 | 50 | 635,000 |
| 212 | 100 | 85,300 |
| 392 | 200 | 5,000 |

## I-MARK TURBO

### CODE 33

**MAP SENSOR CIRCUIT
(SIGNAL VOLTAGE HIGH)**

The Manifold Absolute Pressure Sensor (MAP) responds to change in manifold pressure and vacuum. The ECM receives this information as a signal voltage that will vary from about 1 to 1.5 volts at idle to 4-4.5 volts at wide open throttle.

If the MAP sensor fails, the ECM will substitute a fixed MAP value and use the Throttle Position Sensor (TPS) to control fuel delivery.

Code 33 will set when:

- Signal is too high for a time greater than 5 seconds.
- TPS voltage indicates throttle is closed.

Engine misfire or a low unstable idle may set Code 33. Disconnect MAP sensor and system will go into backup mode.

1. Confirms Code 33 and that fault is present.
2. If the ECM recognizes and sets code 34, low MAP signal, the ECM and wiring are OK.

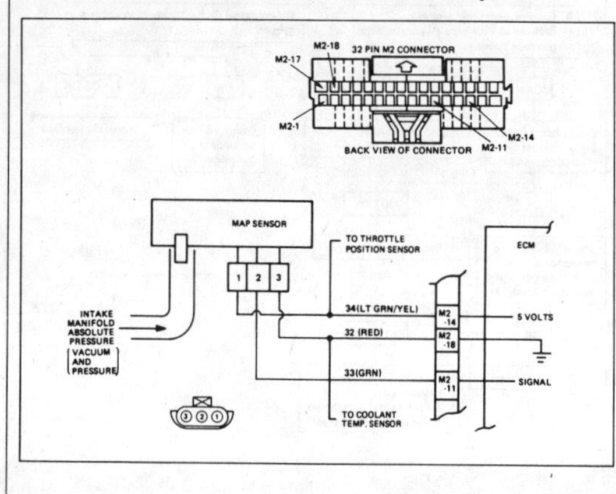


### CODE 33

**MAP SENSOR CIRCUIT
(SIGNAL VOLTAGE HIGH)**

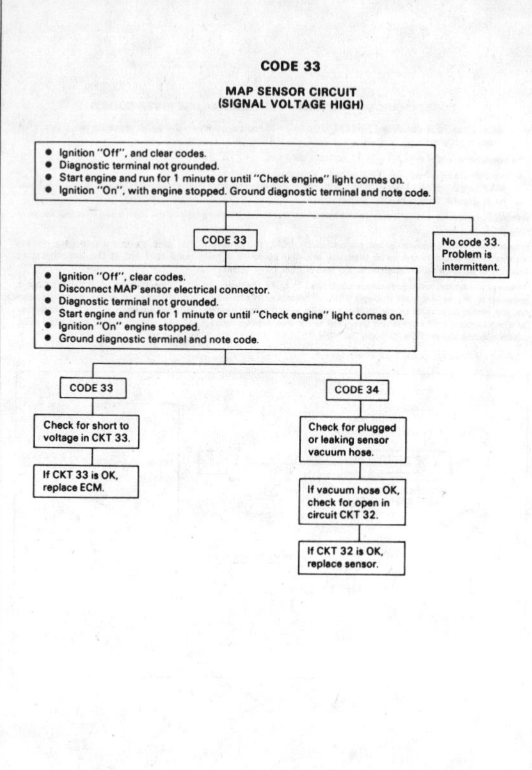


## I-MARK TURBO

### CODE 34

**MAP SENSOR CIRCUIT
(SIGNAL VOLTAGE LOW)**

The Manifold Absolute Pressure Sensor (MAP) responds to changes in manifold pressure and vacuum. The ECM receives this information as a signal voltage that will vary from about 1 to 1.5 volts at idle to 4-4.5 volts at wide open throttle.

If the MAP sensor fails, the ECM will substitute a fixed MAP value and use the Throttle Position Sensor (TPS) to control fuel delivery.

Code 34 will set when signal is too low and ignition is on.

1. Confirms Code 34 and that fault is present.
2. If the ECM recognizes and sets Code 33, high MAP signal, the ECM and wiring are OK.

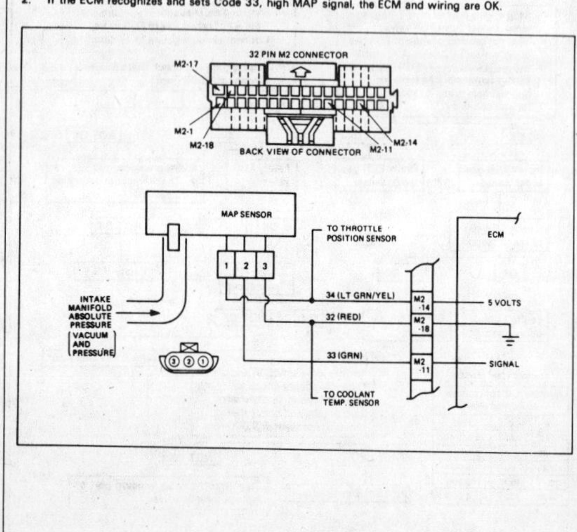


### CODE 34

**MAP SENSOR CIRCUIT
(SIGNAL VOLTAGE LOW)**

- Ignition "Off", and clear codes.
- Diagnostic terminal not grounded.
- Start engine and run for 1 minute or until check engine light comes on.
- Ignition "On", engine stopped.
- Ground diagnostic terminal and note code.

CODE 34 | No code stored; problem is intermittent.

CODE 34:

- Ignition "Off", and clear codes.
- Disconnect MAP sensor and jumper harness connector terminal "2" to "1".
- Diagnostic terminal not grounded.
- Start engine and run for 1 minute or until check engine light comes on.
- Ignition "On", engine stopped.
- Grounded diagnostic terminal and note code.

CODE 33: Replace sensor

CODE 34:

- Remove jumper from terminal "2" to "1".
- Check voltage between harness connector terminal "3" and "1" using voltmeter.

4-6 volts:

Check for open or short to ground in CKT 33.

CKT 33 OK, faulty ECM connector terminal M2-11 or ECM.

Below 4 volts:

Check for open or short to ground in CKT 34.

CKT 34 OK, faulty ECM connector terminal M2-14 or ECM.

## I-MARK TURBO

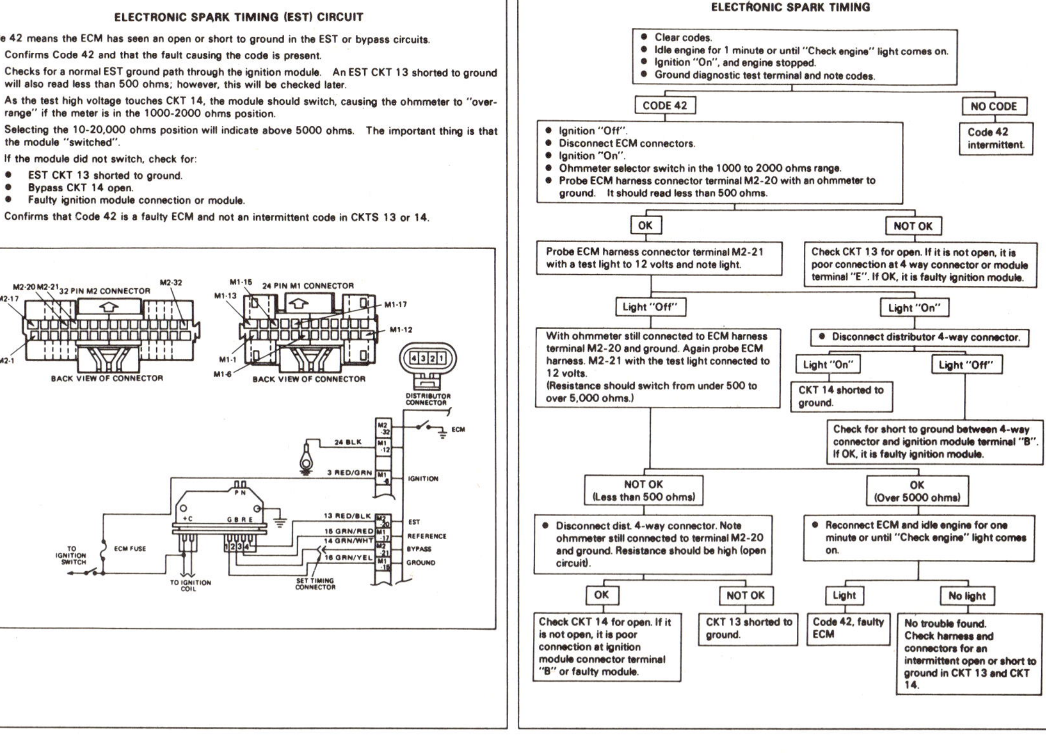

### CODE 42

#### ELECTRONIC SPARK TIMING (EST) CIRCUIT

Code 42 means the ECM has seen an open or short to ground in the EST or bypass circuits.

1. Confirms Code 42 and that the fault causing the code is present.
2. Checks for a normal EST ground path through the ignition module. An EST CKT 13 shorted to ground will also read less than 500 ohms; however, this will be checked later.
3. As the test high voltage touches CKT 14, the module should switch, causing the ohmmeter to "over-range" if the meter is in the 1000-2000 ohms position.
   Selecting the 10-20,000 ohms position will indicate above 5000 ohms. The important thing is that the module "switched".
4. If the module did not switch, check for:
   - EST CKT 13 shorted to ground.
   - Bypass CKT 14 open.
   - Faulty ignition module connection or module.
5. Confirms that Code 42 is a faulty ECM and not an intermittent code in CKTS 13 or 14.

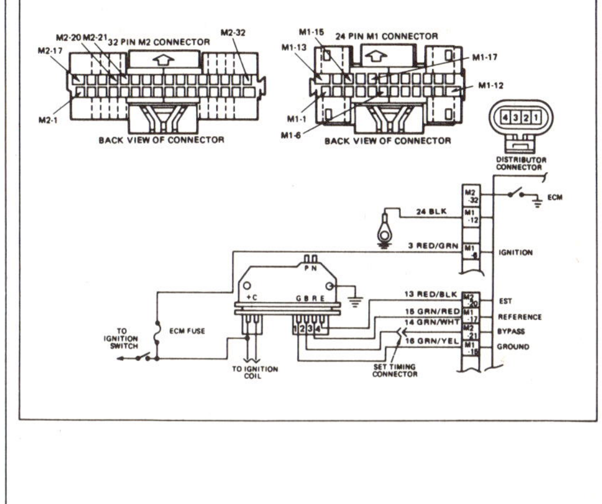


### CODE 42

#### ELECTRONIC SPARK TIMING

- Clear codes.
- Idle engine for 1 minute or until "Check engine" light comes on.
- Ignition "On", and engine stopped.
- Ground diagnostic test terminal and note codes.

**NO CODE** → Code 42 intermittent.

**CODE 42** →

- Ignition "Off".
- Disconnect ECM connectors.
- Ignition "On".
- Ohmmeter selector switch in the 1000 to 2000 ohms range.
- Probe ECM harness connector terminal M2-20 with an ohmmeter to ground. It should read less than 500 ohms.

**NOT OK** → Check CKT 13 for open. If it is not open, it is poor connection at 4 way connector or module terminal "E". If OK, it is faulty ignition module.

**OK** → Probe ECM harness connector terminal M2-21 with a test light to 12 volts and note light.

- **Light "On"** → • Disconnect distributor 4-way connector.
  - **Light "On"** → CKT 14 shorted to ground.
  - **Light "Off"** → Check for short to ground between 4-way connector and ignition module terminal "B". If OK, it is faulty ignition module.
- **Light "Off"** → With ohmmeter still connected to ECM harness terminal M2-20 and ground. Again probe ECM harness. M2-21 with the test light connected to 12 volts. (Resistance should switch from under 500 to over 5,000 ohms.)
  - **NOT OK (Less than 500 ohms)** → • Disconnect dist. 4-way connector. Note ohmmeter still connected to terminal M2-20 and ground. Resistance should be high (open circuit).
    - **OK** → Check CKT 14 for open. If it is not open, it is poor connection at ignition module connector terminal "B" or faulty module.
    - **NOT OK** → CKT 13 shorted to ground.
  - **OK (Over 5000 ohms)** → • Reconnect ECM and idle engine for one minute or until "Check engine" light comes on.
    - **Light** → Code 42, faulty ECM
    - **No light** → No trouble found. Check harness and connectors for an intermittent open or short to ground in CKT 13 and CKT 14.

## I-MARK TURBO

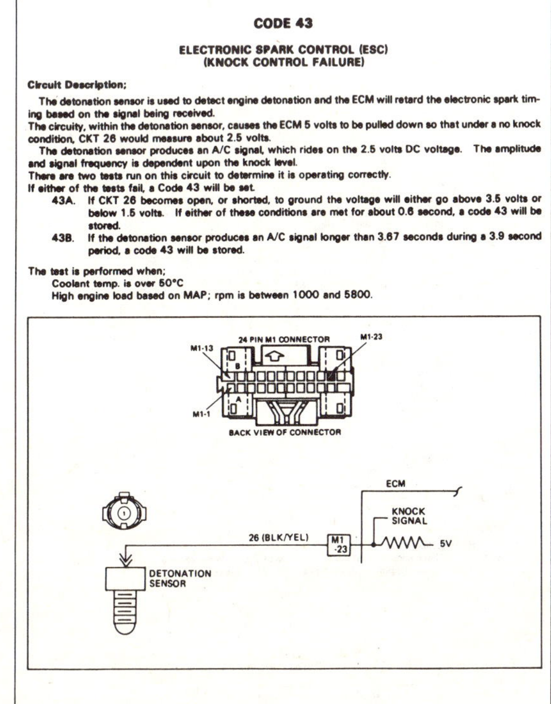

### CODE 43

#### ELECTRONIC SPARK CONTROL (ESC) (KNOCK CONTROL FAILURE)

**Circuit Description;**

The detonation sensor is used to detect engine detonation and the ECM will retard the electronic spark timing based on the signal being received.

The circuity, within the detonation sensor, causes the ECM 5 volts to be pulled down so that under a no knock condition, CKT 26 would measure about 2.5 volts.

The detonation sensor produces an A/C signal, which rides on the 2.5 volts DC voltage. The amplitude and signal frequency is dependent upon the knock level.

There are two tests run on this circuit to determine it is operating correctly.

If either of the tests fail, a Code 43 will be set.

43A. If CKT 26 becomes open, or shorted, to ground the voltage will either go above 3.5 volts or below 1.5 volts. If either of these conditions are met for about 0.6 second, a code 43 will be stored.

43B. If the detonation sensor produces an A/C signal longer than 3.67 seconds during a 3.9 second period, a code 43 will be stored.

The test is performed when;

Coolant temp. is over 50°C

High engine load based on MAP; rpm is between 1000 and 5800.



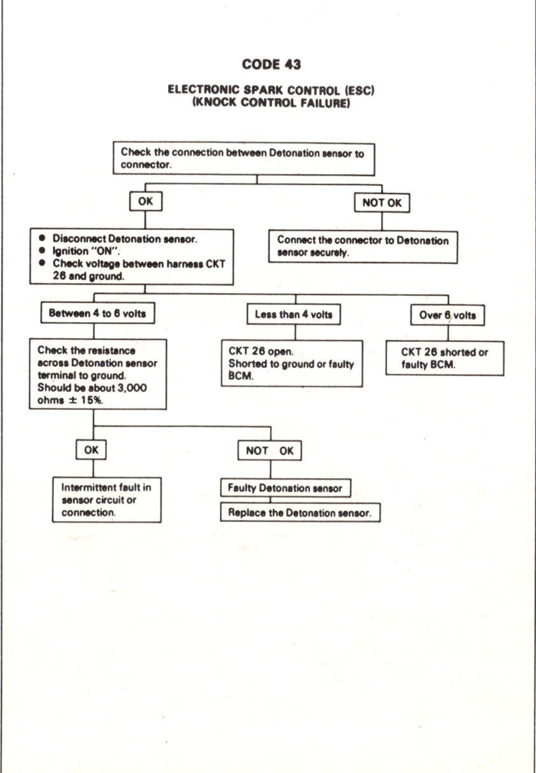

### CODE 43

#### ELECTRONIC SPARK CONTROL (ESC) (KNOCK CONTROL FAILURE)

Check the connection between Detonation sensor to connector.

**NOT OK** → Connect the connector to Detonation sensor securely.

**OK** →

- Disconnect Detonation sensor.
- Ignition "ON".
- Check voltage between harness CKT 26 and ground.

**Less than 4 volts** → CKT 26 open. Shorted to ground or faulty BCM.

**Over 6 volts** → CKT 26 shorted or faulty BCM.

**Between 4 to 6 volts** → Check the resistance across Detonation sensor terminal to ground. Should be about 3,000 ohms ± 15%.

- **OK** → Intermittent fault in sensor circuit or connection.
- **NOT OK** → Faulty Detonation sensor → Replace the Detonation sensor.

## I-MARK TURBO

### CODE 44

**OXYGEN SENSOR CIRCUIT**
**(LEAN EXHAUST INDICATION)**

The ECM supplies a voltage of about 0.45 volt between terminals M2-22 and M2-23. (If measured with a 10 megohm digitial voltmeter, this may read as low as 0.32 volts.) The $O_2$ sensor varies the voltage within a range of about 1 volt if the exhaust is rich, down through about 0.10 volt if exhaust is lean.

The sensor is like an open circuit and produces no voltage when it is below about 310°C (600°F). An open sensor circuit or cold sensor causes open loop operation.

Code 44 is set when the $O_2$ sensor signal voltage at the ECM Circuit 31:

- Remains below 0.2 volt for 8 seconds or more.

1. Grounding the diagnostic terminal with the engine running, enables "Field Service Mode" and allows the ECM to confirm either open or closed loop operation.
2. A light out or "Open Loop" indicates the fault is present. Disconnecting the $O_2$ sensor will raise the signal voltage above 0.2 volt. If the ECM and wiring are OK, the ECM should recognize the higher voltage, 0.35 to 0.55, and flash open loop when the engine is started.
3. The Code 44 or lean exhaust is most likely caused by one of the following:

- Circuit 25. If circuit 25 is open, the voltage on CKT 31 will be over one volt.
- Fuel Pressure. System will be lean if pressure is too low. It may be necessary to monitor fuel pressure while driving the car at various road speeds and/or loads to confirm.
- Fuel contamination. Water, even in small amounts, near the in-tank fuel pump inlet can be delivered to the injector. The water causes a lean exhaust and can set a Code 44.
- EGR. In normal operation, the ECM delivers less fuel and advances spark when EGR comes in. If the EGR does not open, the system will go lean and may have slight spark knock.
- MAP sensor. An output that causes the ECM to sense a lower than normal manifold pressure (high vacuum) will cause the system to go lean.
- If the above are OK, and after following the instructions at the top of the chart, a Code 44 is set, or "Check engine" light is "Off" more than "On", or flashing "Open loop", it is a faulty Oxygen Sensor.

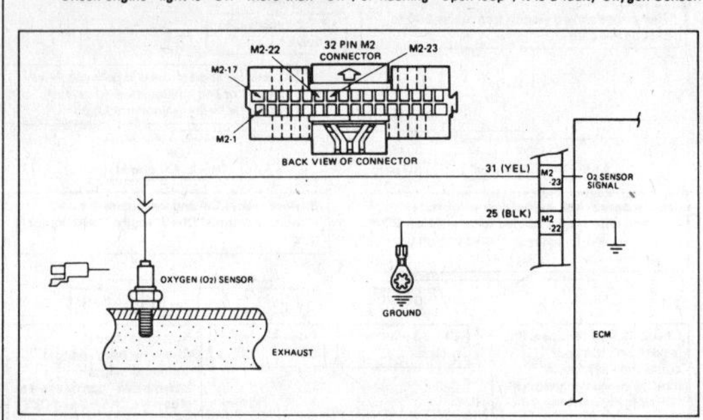


### CODE 44

**OXYGEN SENSOR CIRCUIT**
**(LEAN EXHAUST INDICATION)**

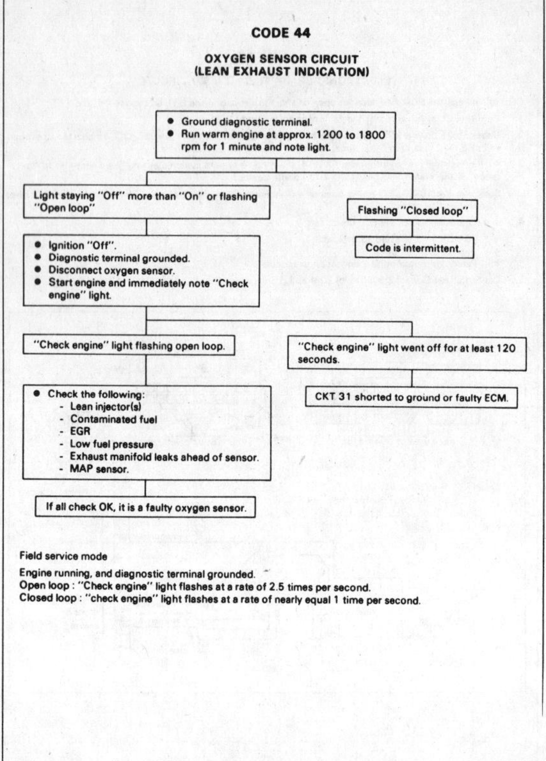


Field service mode
Engine running, and diagnostic terminal grounded.
Open loop : "Check engine" light flashes at a rate of 2.5 times per second.
Closed loop : "check engine" light flashes at a rate of nearly equal 1 time per second.

## I-MARK TURBO

### CODE 45

**OXYGEN SENSOR CIRCUIT**
**(RICH EXHAUST INDICATION)**

The ECM supplies a voltage of about 0.45 volt between terminals "M2-23 and M2-22". (If measured with a 10 megohm digital voltmeter, this may read as low as 0.32 volts.) The $O_2$ sensor varies the voltage within a range of about 1 volt if the exhaust is rich, down through about .10 volt if exhaust is lean.

The sensor is like an open circuit and produces no voltage when it is below about 360°C (600°F). An open sensor circuit or cold sensor causes open loop operation.

Code 45 is set when the $O_2$ sensor sgnal voltage at the ECM M2 connector terminal M2-22.

- Remains above 0.6 volt for 120 seconds;

1. Grounding the diagnostic terminal with the engine running, enables the "Field Service Mode" and allows the ECM to confirm either open or closed loop operation using the "Check engine" light.
2. A steady light or "Open loop" indicates the fault is present. Grounding CKT 31 causes a low $O_2$ signal voltage. If the ECM and wiring are OK, the ECM should recognize the low voltage and confirm the lean signal by turning off the "Check engine" light for at least 30 seconds.
3. Diagnostic aids:

- Fuel Pressure. System will go rich if pressure is too high. The ECM can compensate for some increase. However, if it gets too high, a Code 45 will be set. See Fuel System Diagnosis Chart.
- Leaking injector. See Fuel System Diagnosis chart.
- HEI Shielding. An open ground CKT 16 may result in EMI, or indicated electical "noise". The ECM looks at this "noise" as distributor pulses. The additional pulses result in a higher than actual engine speed signal. The ECM then delivers too much fuel, causing system to go rich. Engine tachometer will also show higher than actual engine speed, which can help in diagnosing this problem.
- Canister purge. Check for fuel saturation. If full of fuel, check canister control and hoses. See canister purge section.
- MAP sensor. An output that causes the ECM to sense a higher than normal manifold pressure (low vacuum) can cause the system to go rich. Disconnecting the MAP sensor will allow the ECM to set a fixed value for the MAP sensor. Substitute a different MAP sensor if the rich condition is gone while the sensor is disconnected.
- Check for leaking fuel pressure regulator diaphram by checking vacuum line to regulator for fuel.
- TPS. An intermittent TPS output will cause the system to go rich, due to a false indication of the engine accelerating.
- Inspect Oxygen Sensor for silicone contamination from fuel, or use of improper RTV sealant. The sensor may have a white, powdery coating and result in a high but false signal voltage (rich exhaust indication). The ECM will then reduce the amount of fuel delivered to the engine, causing a severe driveability problem.

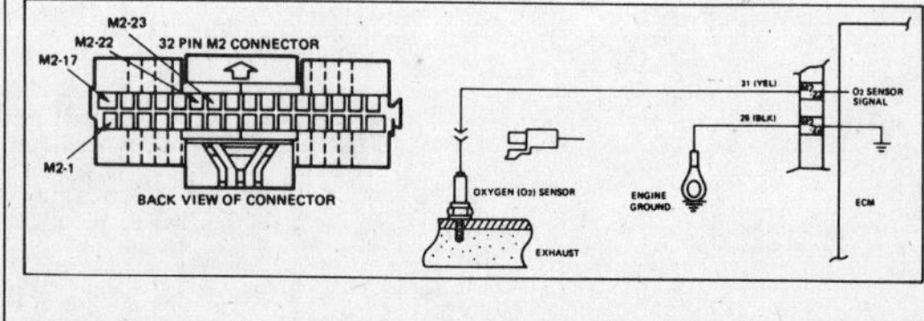


### CODE 45

**OXYGEN SENSOR CIRCUIT**
**(RICH EXHAUST INDICATION)**

- Ground diagnostic terminal.
- Run warm engine at approx. 1200 to 1800 rpm for 1 minute and note "Check engine" light.

Light staying "On" more than "Off" or flashing "Open loop".

- Ignition "Off".
- Diagnostic terminal grounded.
- Disconnect oxygen sensor and jumper harness connector signal CKT 31 to ground.
- Start engine and immediately note "Check engine" light.

"Check engine" light went off for at least 30 seconds.

System rich

Check the followings
- Rich injector(s)
- ECR at idle
- High fuel pressure.
- MAP sensor
- HEI shielding
- TPS
- Canister purge

If all check OK, it is a faulty oxygen sensor.

Steady light

Faulty ECM

Flashing "Closed loop".

Code is intermittent.

Field service mode
Engine running, diagnostic terminal grounded.
Open loop : "Check engine" light flashes at a rate of 2.5 times per second.
Closed loop : "check engine" light flashes at a rate of 1 time per second.

### CODE 51

**MEM-CAL ERROR**
**(FAULTY OR INCORECT MEM-CAL)**

Check that all pins are fully inserted in the socket.
If OK, replace MEM-CAL. clear memory, and recheck. If code 51 reappears, replace ECM.

## I-MARK TURBO

### IDLE AIR CONTROL

The ECM will control engine idle speed by moving the IAC valve to control air flow around the throttle plate. It does this by sending voltage pulses to the proper motor winding for each IAC motor. This will cause the motor shaft and valve to move "IN" or "OUT" of the motor a given distance for each pulse received. ECM pulses are referred to as "counts".

- To increase idle speed - ECM will send enough counts to retract the IAC valve and allow more air to flow through the idle air passage and bypass the throttle plate until idle speed reaches the proper RPM. This will increase the ECM counts.
- To decrease idle speed - ECM will send enough counts to extend the IAC valve and reduce air flow through the idle passage around the throttle plate. This will reduce the ECM counts.

Each time the engine is started and then the ignition is turned off the ECM will reset the IAC valve. This is done by sending enough counts to seat the valve. The fully seated valve is the ECM reference Zero. A given number of counts are then issued to open the valve, and normal ECM countrol of IAC will begin from this point. The number of counts are then added by the decrease. This is how the ECM knows what the motor position is for a given idle speed.

The ECM uses the following information to control idle speed.

- Battery voltage
- Coolant temperature
- Throttle position sensor
- Engine speed
- A/C clutch signal
- Power steering pressure switch
- Vehicle speed sensor

Don't apply battery voltage across the IAC motor terminals. It will permanently damage the IAC motor windings.

1. Be sure to disconnect the IAC valve prior to this test. The test light will confirm the ECM signals by a steady or flashing light, all circuit.
2. Before replacing an ECM, be sure to check the resistance at the IAC motor windings. Failure to do so may result in a repeat ECM failure.
3. Diagnostic aids
   Engine idle speed can be adversely affected by the following.

- Leaking injector(s) will cause fuel imbalance and poor idle quality due to excess fuel.
- Vacuum leaks can cause higher than normal idle.
- When the throttle shaft or throttle position sensor is binding or sticking in an open throttle position, the ECM does not know if the vehicle has stopped and does not control idle.
- Faulty battery cables can result in voltage variations. The ECM will try to compensate, which results in erratic idle speeds.
- The ECM will compensate for AC compressor clutch loads. Loss of this signal would be most apparent in neutral.
- Power steering pressure switch - If the ECM thinks the Pressure switch is closed the idle may be too high.

### IDLE AIR CONTROL

Clear codes and confirm "Closed loop" operation and no "Check engine light.

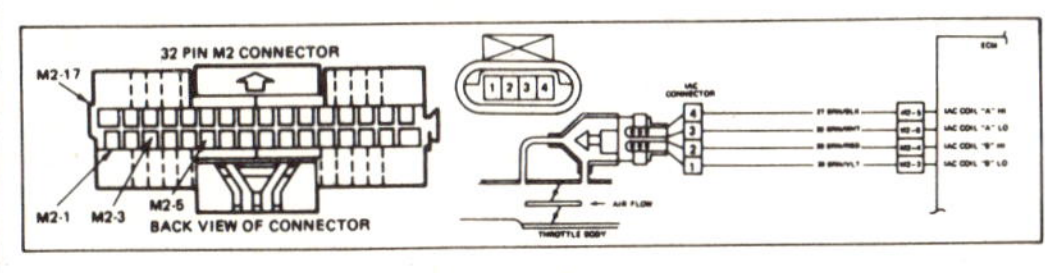

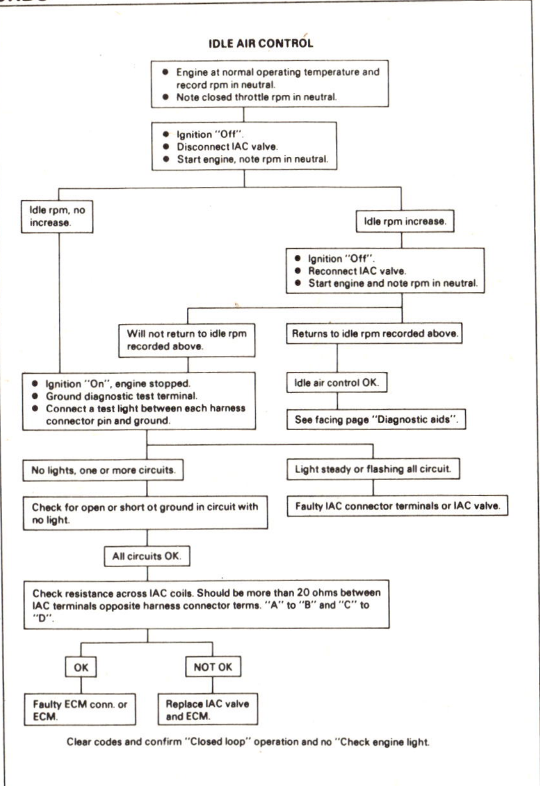

## I-MARK TURBO

### CLOSED LOOP EMISSION CONTROL SYSTEM

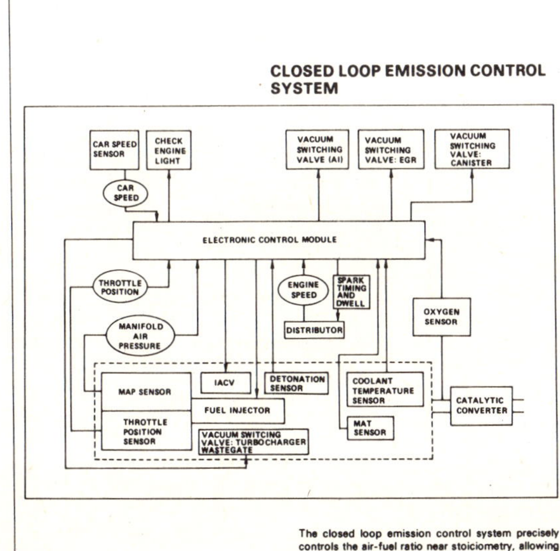

The closed loop emission control system precisely controls the air-fuel ratio near stoiciometry, allowing the use of a three-way catalyst to reduce oxides of nitrogen and oxidized hydrocarbons and carbon monoxide.

### CONTROL SYSTEM DIAGRAM

**Ignition timing control**

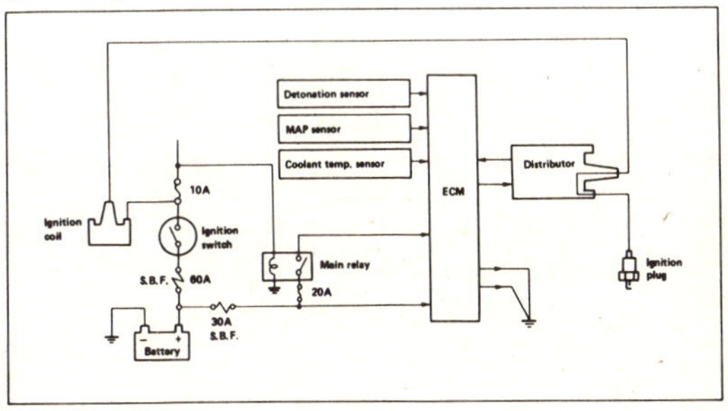

**Fuel injection control**

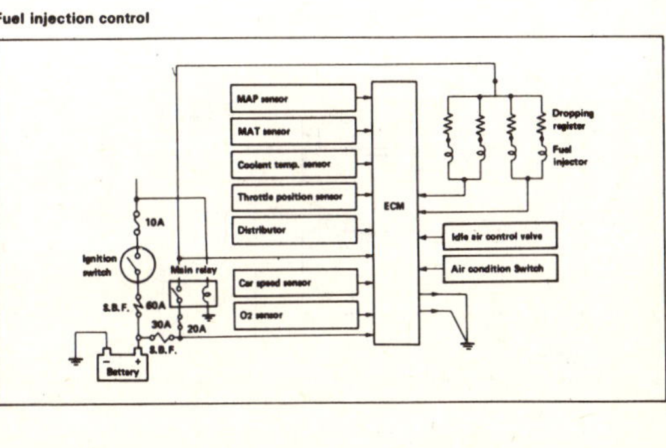

### I-MARK TURBO

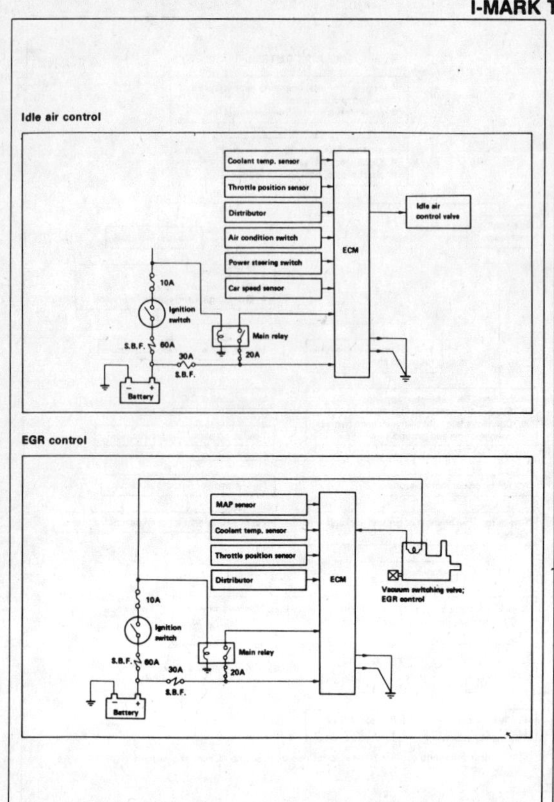


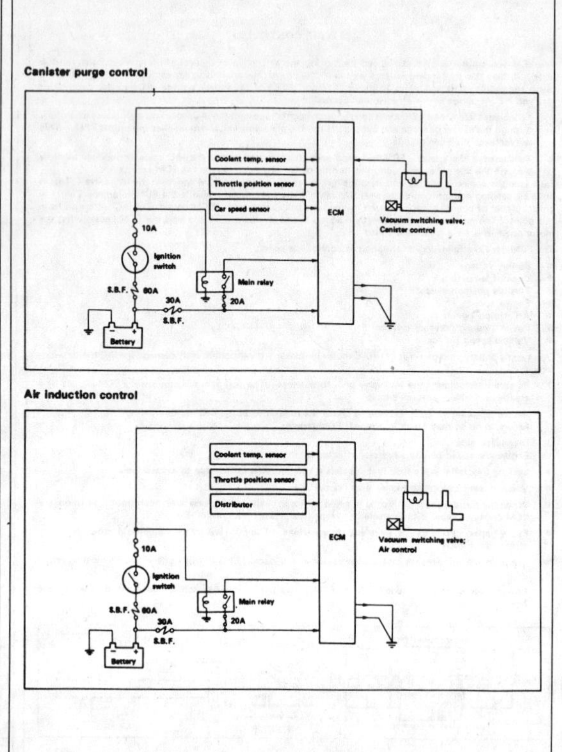


### I-MARK TURBO

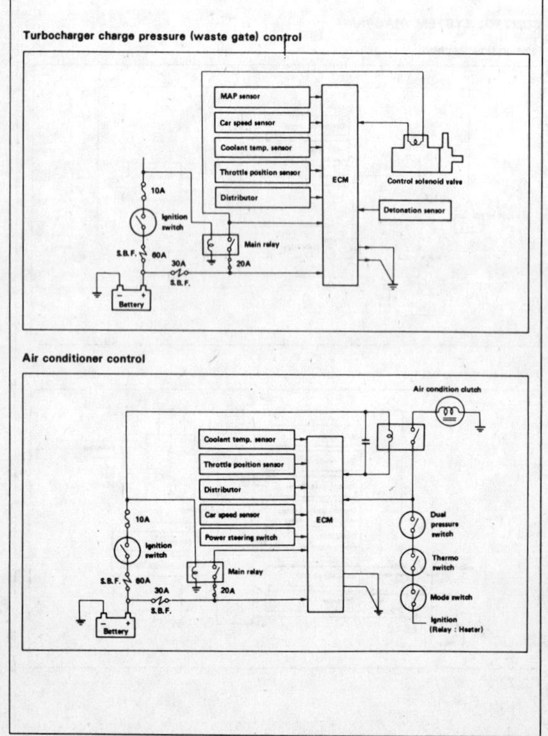


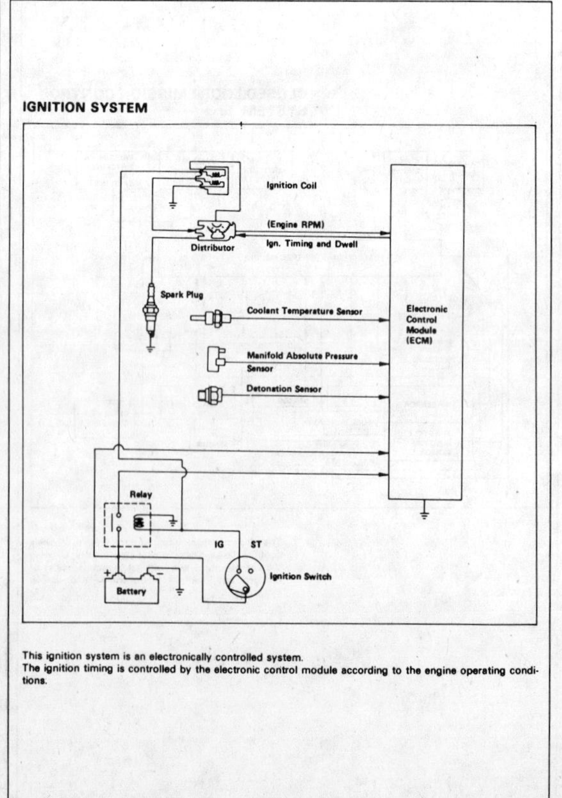


This ignition system is an electronically controlled system.
The ignition timing is controlled by the electronic control module according to the engine operating conditions.

## I-MARK TURBO

### FUEL SYSTEM

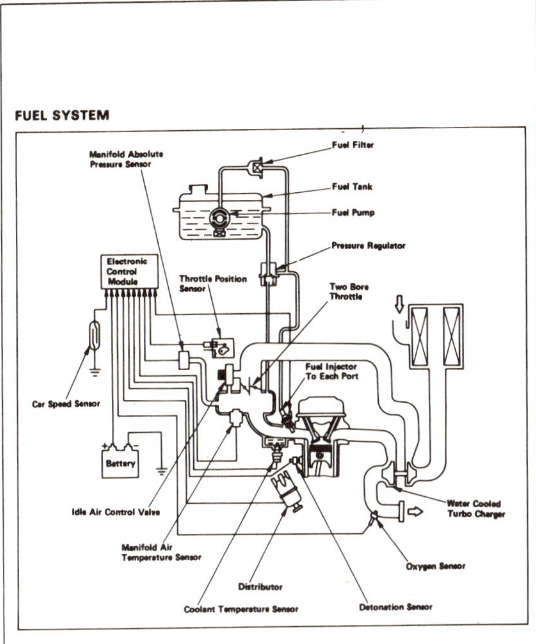


This fuel system is of electronically controlled type and consists of electronic control module (ECM), electric fuel pump, pressure regulator, fuel injectors and sensors.

### IDLE AIR CONTROL SYSTEM

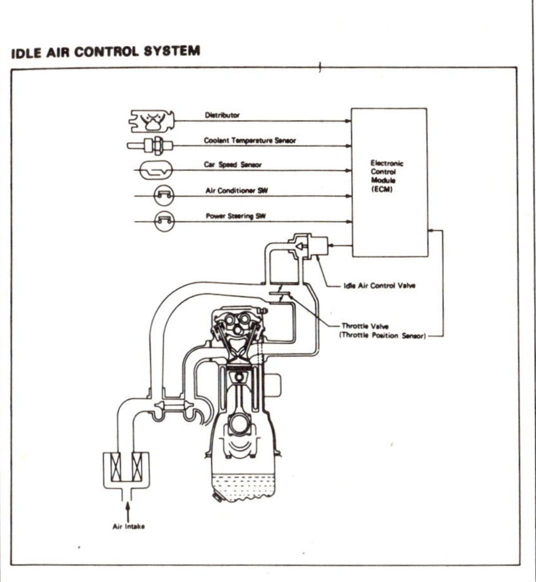


This system is an engine idle speed control system and consists of idle air control valve, electronic control module and sensors.

## I-MARK TURBO

### ELECTRONIC TURBOCHARGER CONTROL SYSTEM

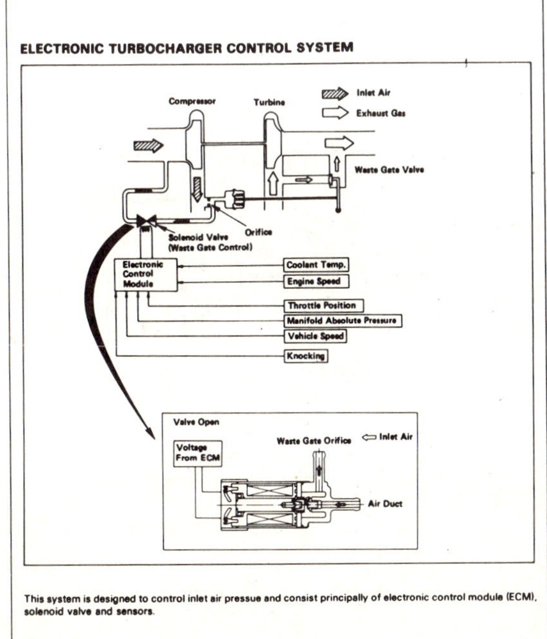


This system is designed to control inlet air pressue and consist principally of electronic control module (ECM), solenoid valve and sensors.

### COMPONENT LAYOUT AND SCHEMATIC DRAWING

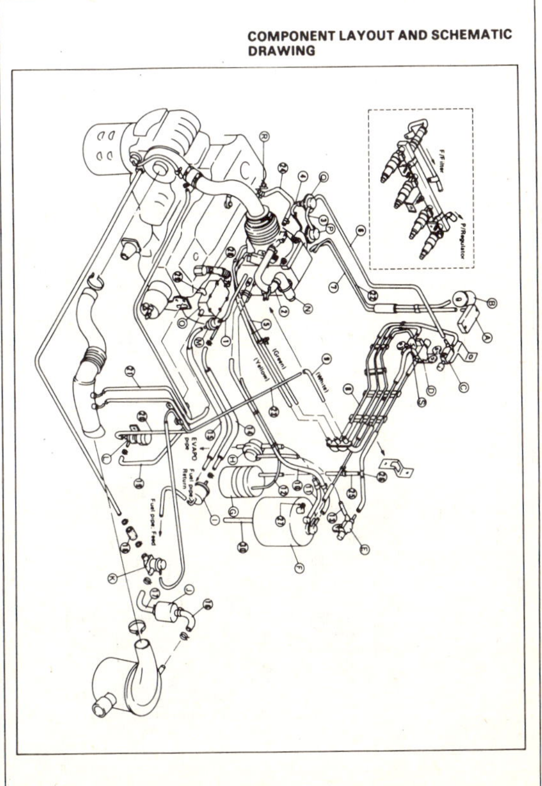

## I-MARK TURBO

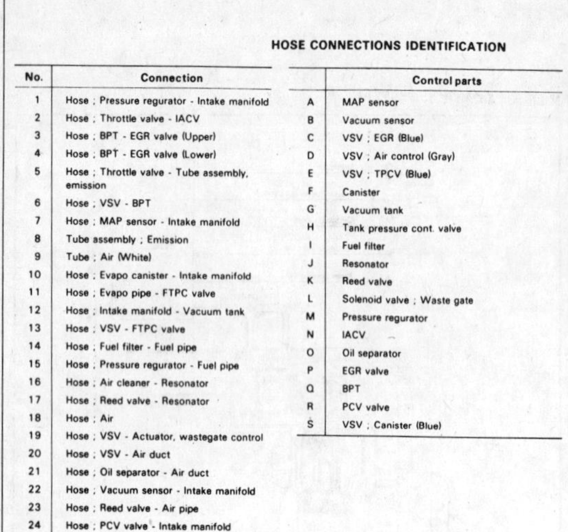

### HOSE CONNECTIONS IDENTIFICATION

| No. | Connection |
|---|---|
| 1 | Hose ; Pressure regurator - Intake manifold |
| 2 | Hose ; Throttle valve - IACV |
| 3 | Hose ; BPT - EGR valve (Upper) |
| 4 | Hose ; BPT - EGR valve (Lower) |
| 5 | Hose ; Throttle valve - Tube assembly, emission |
| 6 | Hose ; VSV - BPT |
| 7 | Hose ; MAP sensor - Intake manifold |
| 8 | Tube assembly ; Emission |
| 9 | Tube ; Air (White) |
| 10 | Hose ; Evapo canister - Intake manifold |
| 11 | Hose ; Evapo pipe - FTPC valve |
| 12 | Hose ; Intake manifold - Vacuum tank |
| 13 | Hose ; VSV - FTPC valve |
| 14 | Hose ; Fuel filter - Fuel pipe |
| 15 | Hose ; Pressure regurator - Fuel pipe |
| 16 | Hose ; Air cleaner - Resonator |
| 17 | Hose ; Reed valve - Resonator |
| 18 | Hose ; Air |
| 19 | Hose ; VSV - Actuator, wastegate control |
| 20 | Hose ; VSV - Air duct |
| 21 | Hose ; Oil separator - Air duct |
| 22 | Hose ; Vacuum sensor - Intake manifold |
| 23 | Hose ; Reed valve - Air pipe |
| 24 | Hose ; PCV valve - Intake manifold |
| 25 | Hose ; 3 way - Vacuum tank |
| 26 | 3 Way ; VSV - Vacuum tank |
| 27 | Rubber cap ; Canister |
| 28 | Hose ; Oil separator - Head cover |
| 29 | Hose ; Oil separator - cyl. head |
| 30 | Hose ; Canister drain |

| | Control parts |
|---|---|
| A | MAP sensor |
| B | Vacuum sensor |
| C | VSV ; EGR (Blue) |
| D | VSV ; Air control (Gray) |
| E | VSV ; TPCV (Blue) |
| F | Canister |
| G | Vacuum tank |
| H | Tank pressure cont. valve |
| I | Fuel filter |
| J | Resonator |
| K | Reed valve |
| L | Solenoid valve ; Waste gate |
| M | Pressure regurator |
| N | IACV |
| O | Oil separator |
| P | EGR valve |
| Q | BPT |
| R | PCV valve |
| S | VSV ; Canister (Blue) |

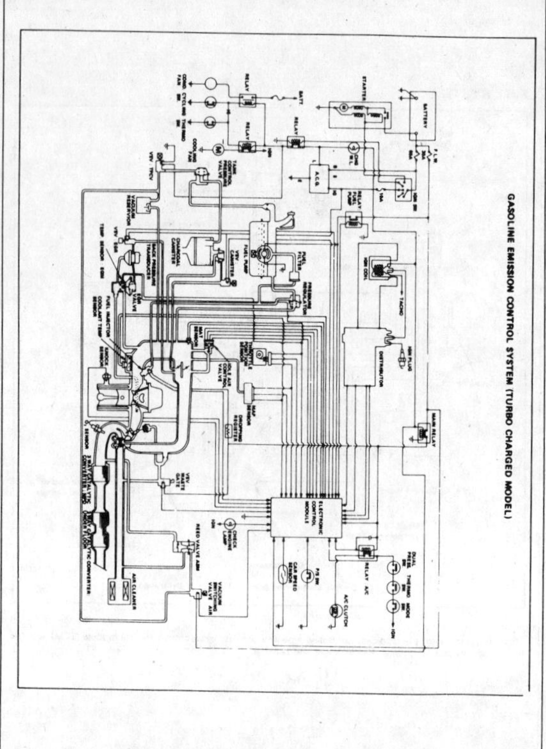


## I-MARK TURBO

### ENGINE EMISSION ELECTRICAL CIRCUIT

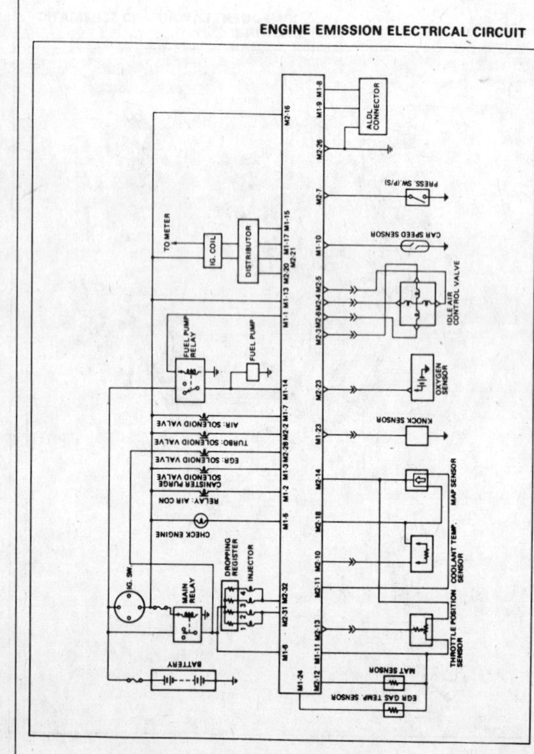


### VACUUM HOSE ROUTING DIAGRAM

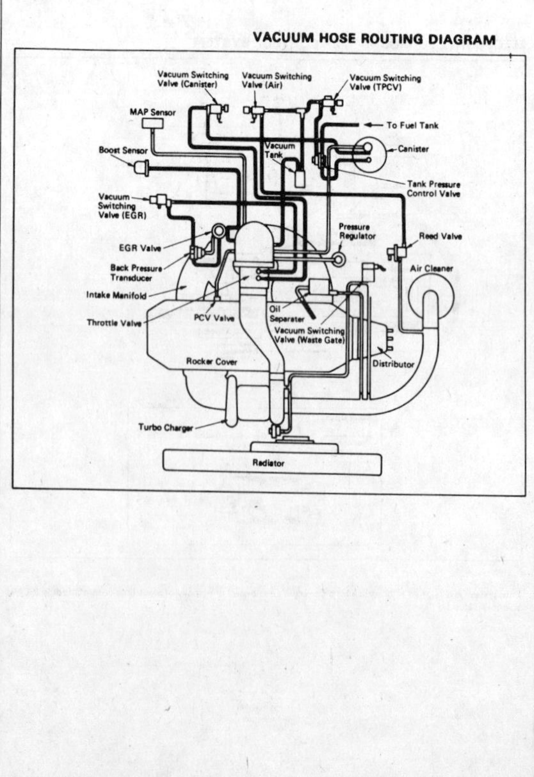

## I-MARK NON-TURBO

### COMPONENT LOCATION

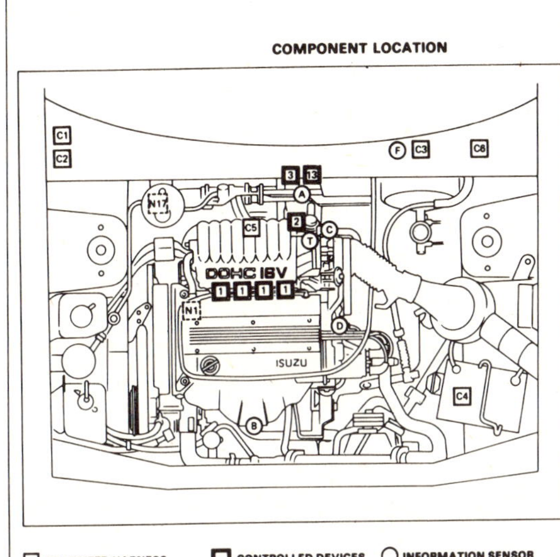


**COMPUTER HARNESS**

- **C1** Electronic Control Module (ECM)
- **C2** ALDL diagnostic connector
- **C3** "CHECK ENGINE" light
- **C4** ECM power
- **C5** ECM harness ground
- **C6** Fuse panel

**CONTROLLED DEVICES**

- **1** Fuel injector
- **2** Idle air control valve
- **3** Fuel pump relay
- **13** A/C compressor relay

**INFORMATION SENSOR**

- **A** Manifold pressure (M.A.P.)
- **B** Exhaust oxygen
- **C** Throttle position
- **D** Coolant temperature
- **F** Vehicle speed
- **T** Manifold Air Temperature

**NOT ECM CONNECTED**

- **N1** Crankcase vent valve (PCV)
- **N17** Fuel vapor canister

### "SCAN" DIAGNOSTIC CIRCUIT CHECK

The "Scan" Diagnostic Circuit Check is an organized approach to identifying a problem created by an Electronic Engine Control System malfunction. It must be the starting point for any driveability complaint diagnosis, because it directs the Service Technician to the next logical step in diagnosing the complaint. The "Scan Data" listed in the table may be used for comparison, after completing the Diagnostic Circuit Check and finding the on-board diagnostics functioning properly and no trouble codes displayed. The "Typical Values" are an average of display values recorded from normally operating vehicles and are intended to represent what a normally functioning system would typically display.

**A "SCAN" tool that displays faulty data should not be used, and the problem should be reported to the manufacturer. The use of a faulty "SCAN" can result in misdiagnosis and unnecessary parts replacement.**

Only the parameters listed below are used in this manual for diagnosis. If a "Scan" reads other parameters, the values are not recommended by ISUZU Motors for use in diagnosis. For more description on the values and use of the "Scan" to diagnosis ECM inputs, refer to the applicable diagnosis section in Section "C".

**"SCAN" DATA**

**Idle / Upper radiator hose hot / Closed throttle / Park or neutral / Acc. off**

| "SCAN" position | Units displayed | Typical data value |
|---|---|---|
| RPM | RPM | ± 150 RPM from desired RPM |
| Coolant Temp. | C° | 85° - 105° |
| MAT Temp. | C° | 10° - 80° (depends on underhood temperature) |
| MAP | Volts | 1.0 - 2.0 (depends on Vac. & Baro pressure) |
| BPW (base pulse width) | m/Sec | .8 - 3.0 |
| $O_2$ | Volts | .10 - 1.0 and varies |
| TPS | Volts | .4 - 1.25 |
| Throttle Angle | 0 - 100% | 0 |
| IAC | Counts (steps) | 1 - 30 |
| INT (Integrator) | Counts | 110 - 140 Normal (64 - 208) Limits |
| BLM (Block Learn) | Counts | 119 - 138 Normal (117- 160) Limits |
| Open/Closed Loop | Open/Closed | Closed Loop (Open with extended idle) |
| Vehicle Speed | MPH | 0 |
| TCC | On/Off | Off/(on with TCC commanded) |
| Spark Advance | # of Degrees | Varies |
| Battery VOHS | Volts | 13.5 - 14.5 |
| Fan | On/Off | Off (below 102°C) |
| A/C Request | Yes/No | No (yes, with A/C requested) |
| A/C Clutch | On/Off | Off (on, with A/C commanded on) |
| Power Steering Switch | Normal/Hi pressure | Normal |
| Fuel Pump Voltage | Volts | 13.5 - 14.5 |
| EGR Temp. Signal | Volts | 0 - 5 |
| Baro | Volts | 2.3 - 2.5 (depends on Altitude & Baro Pressure) |

**If all values are within the range illustrated, refer to symptoms in section "B".**

## I-MARK NON-TURBO

### "SCAN" DIAGNOSTIC CIRCUIT CHECK

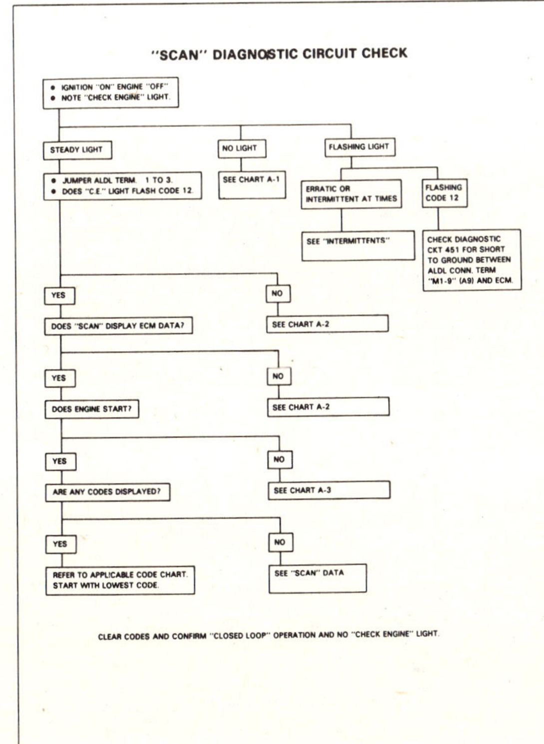


### CHART A-1

#### NO "CHECK ENGINE" LIGHT

**Circuit Description:**

There should always be a steady "Check Engine" light when the ignition is "ON" and engine stopped. Battery ignition voltage is supplied directly to the light bulb from fuse #11. The Electronic Control Module (ECM) will control the light and turn it on by providing a ground path through CKT 419 to the ECM.

**Test Description:**

Numbers below refer to circled numbers on the diagnostic chart.

1. Engine runs ok, check:
- Faulty light bulb.
- CKT 419 open.
- Fuse #11 blown. This will result in an inoperative tachometer, oil or generator lights, seat belt reminder, etc.
  Engine cranks but will not run.
- ECM fusible link open.
- #14 ignition fuse or fuel pump fusible link open.
- Battery CKT 240 to ECM open.
- Ignition CKT 439 to ECM open.
- Poor connection to ECM.
- #13 fuse open or faulty main power relay.

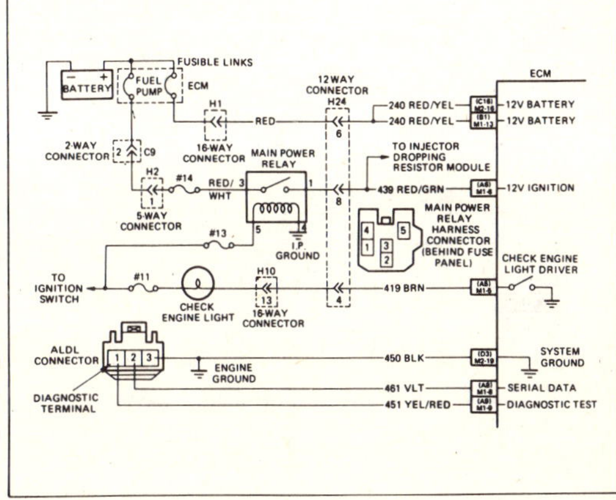

## I-MARK NON-TURBO

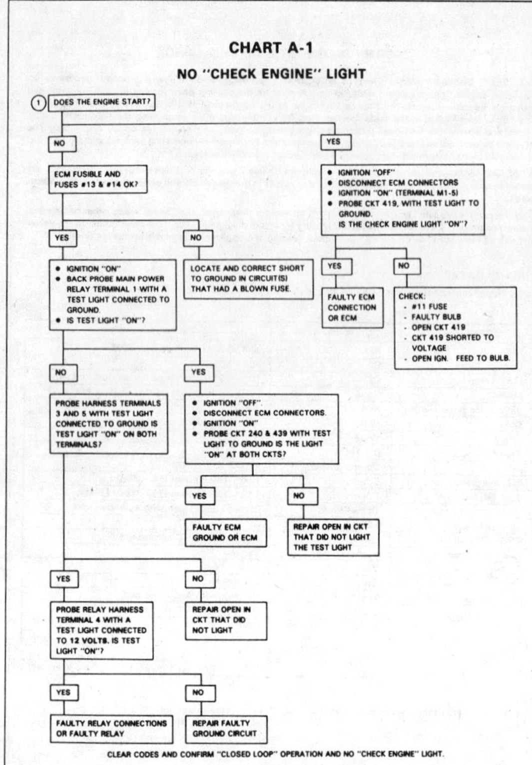


### CHART A-2

### NO ALDL DATA OR WON'T FLASH CODE 12 ("CHECK ENGINE" LIGHT ON STEADY)

**Circuit Description:**

There should always be a steady "Check Engine" light when the ignition is "ON" and engine stopped. Battery ignition voltage is supplied directly to the light. The Electronic Control Module (ECM) will turn the light on by grounding CKT 419 to the ECM.

With the diagnostic terminal grounded, the light should flash a Code 12, followed by any trouble code(s) stored in memory.

A steady light suggests a short to ground in the light control CKT 419, or an open in diagnostic CKT 451. A steady, but dim light, would indicate a failed Quad-driver. The CHART A-2 will confirm and suggest the cause.

**Test Description:**

Numbers below refer to circled numbers on the diagnostic chart.

1. If the light goes off when the ECM connector is disconnected, then CKT 419 is not shorted to ground. Take this opportunity to physically check the connector terminals for proper contact.
2. If there is a problem with the ECM that causes a "Scan" tool to not read Serial data, then the ECM should not flash a Code 12. If Code 12 does flash, be sure that the "Scan" tool is working properly on another vehicle. If the "Scan" is functioning properly and CKT 461 is OK, the Mem-Cal or ECM may be at fault for the NO ALDL symptom.
3. This step will check for an open diagnostic CKT 451.
4. At this point the "Check Engine" light wiring is okay. The problem is a faulty ECM or Mem-Cal. Replace the ECM using the Mem-Cal from the original ECM.
5. Replace the Mem-Cal using the original ECM.
6. If the new Mem-Cal or ECM have been damaged by one another in steps 4 or 5 it will be necessary to replace them both with known good parts.

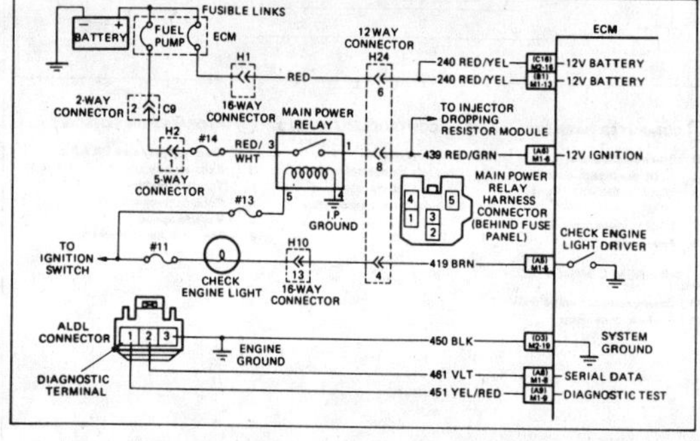


## I-MARK NON-TURBO

### CHART A-2

### NO ALDL DATA OR WON'T FLASH CODE 12 ("CHECK ENGINE" LIGHT ON STEADY)

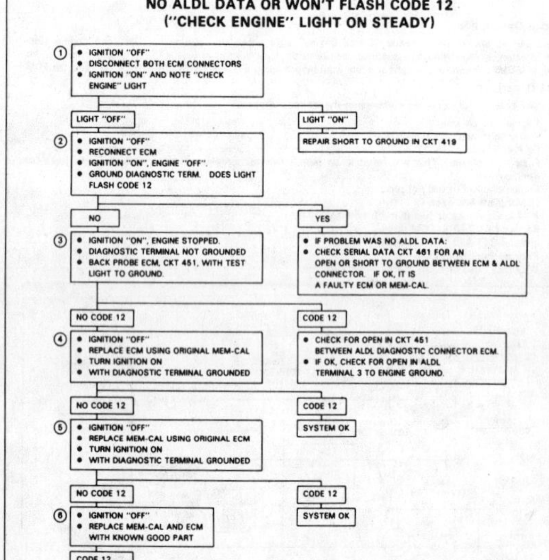


### CHART A-3

### (PAGE 1 OF 2) ENGINE CRANKS BUT WON'T RUN

**Circuit Description:**

This chart assumes that battery condition and engine cranking speed are OK, and there is adequate fuel in the tank.

**Test Description:**

Numbers below refer to circled numbers on the diagnostic chart.

1. A "Check Engine" light "ON" is basic test to determine if there is a 12 volt supply and ignition 12 volts to ECM. No ALDL may be due to an ECM problem and CHART A-2 will diagnose the ECM. If TPS is over 2.5 volts the engine may be in the clear flood mode which will cause starting problems. The engine will not start without reference pulses and therefore the "Scan" should read rpm (reference) during crank.
2. No spark may be caused by one of several components related to the Ignition System. CHART C-4 will address all problems related to the causes of a no spark condition.
3. The test light should blink, indicating the ECM is controlling the injectors ok. How bright the light blinks is not important. However, the test light should be J-34730-B or equivalent. The engine may not start if only 2 injectors are functioning and therefore one injector of each pair should be tested.
4. Use fuel pressure gage J-34730-B or equivalent. Wrap a shop towel around the fuel pressure tap to absorb any small amount of fuel leakage that may occur when installing the gage.

**Diagnostic Aids:**

- An EGR valve sticking open can cause a low air/fuel ratio during cranking. Unless engine enters "Clear Flood" at the first indication of a flooding condition, it can result in a no start.
- Check for fouled plugs.

If above are all OK, refer to Driveability Symptoms

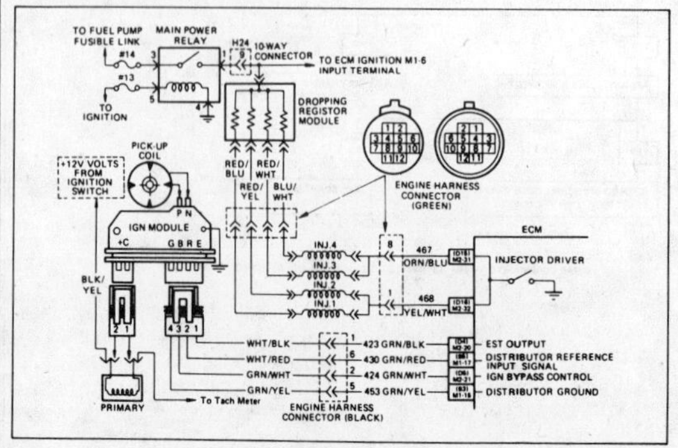

## I-MARK NON-TURBO

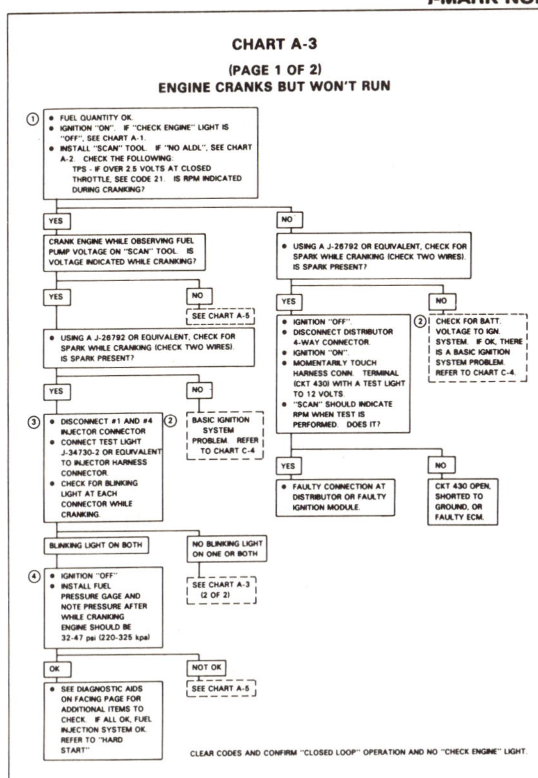


### CHART A-3

**(Page 2 of 2)**
**ENGINE CRANKS BUT WON'T RUN**

**Test Description:**

Numbers below refer to circled numbers on the diagnostic chart.

1. Checks for 12 volt supply to injectors. Due to the injectors wired in parallel there should be a light "ON" on both terminals.
2. Checks continuity of CKT 467 and CKT 468 to ECM.
3. All checks made to this point would indicate that the ECM is at fault. However, there is a possibility of CKT 467 or 468 being shorted to a voltage source either in the engine harness or in the injector harness.

To test for this condition:

- Disconnect all injectors
- Ignition "ON"
- Probe CKT's 467 and 468 on the ECM side of injector harness with a test light connected to ground. There should be no light. If light is "ON" repair short to voltage.
- If OK, check the resistance of the injectors.
- Should be 10 ohms or more.
- Check injector harness connector. Be sure terminals are not backed out of connector and contacting each other.
- If all OK, replace ECM.

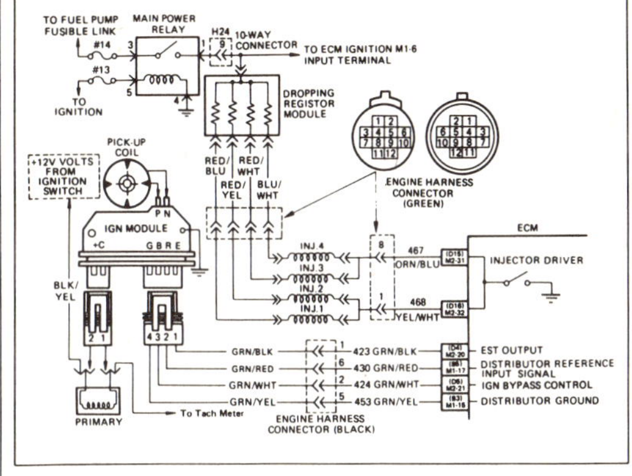


## I-MARK NON-TURBO

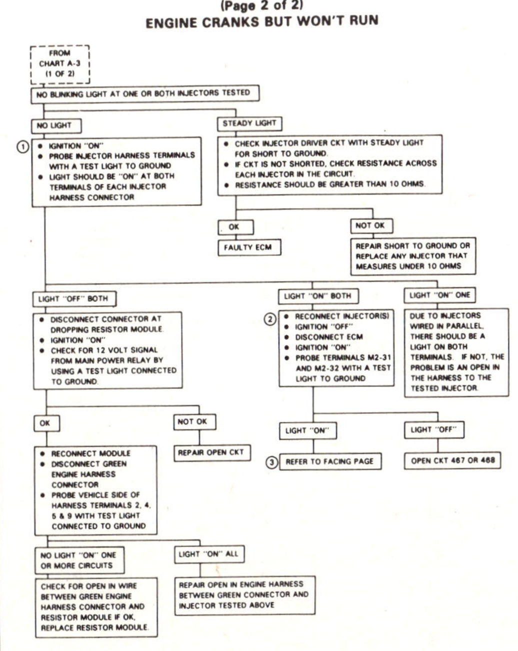


### CHART A-4

**FUEL PUMP RELAY CIRCUIT**

**Circuit Description:**

When the ignition switch is turned "ON", the Electronic Contol Module (ECM) energizes its fuel pump relay control output. It will provide +12 volts to the fuel pump relay, as long as the engine is cranking or running, and the ECM is receiving distributor reference input signal pulses.

If there are no distributor reference pulses, (key "ON", engine stopped) the ECM will shut "OFF" the fuel pump relay within 2 seconds.

**Test Description:**

Numbers below refer to circled numbers on the diagnostic chart.

1. If the fuse is blown, this test will check for short to ground on CKT 120. To prevent mis-diagnosis, be sure fuel pump is disconnected and ignition is "OFF" before performing this test.
2. Determines if the ECM and fuel pump relay circuit is operating correctly. The ECM should turn "ON" the pump relay. Since the engine is not cranking or running, the ECM will turn "OFF" the relay within 2 seconds after ignition is turned "ON".

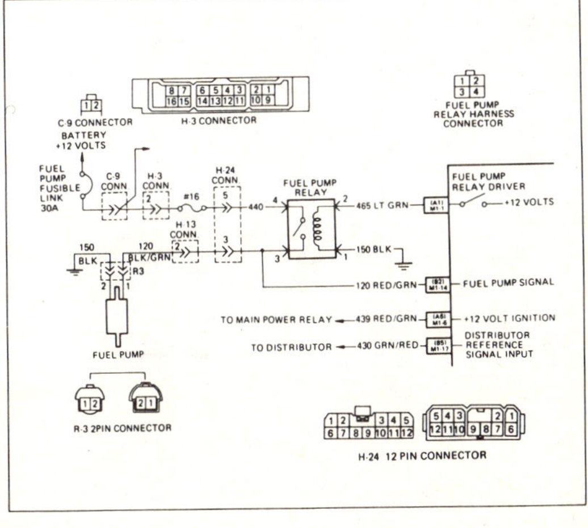

## I-MARK NON-TURBO

### CHART A-4

#### FUEL PUMP RELAY CIRCUIT

- IGNITION "OFF" FOR 10 SECONDS.
- IGNITION "ON" LISTEN FOR IN-TANK FUEL PUMP
- PUMP SHOULD RUN FOR 2 SEC. AFTER IGNITION "ON". DOES IT?

**YES:** NO TROUBLE FOUND. IF PROBLEM WAS "SCAN" TOOL DOES NOT READ FUEL PUMP VOLTAGE, CHECK FOR OPEN IN CKT 120 BETWEEN FUEL PUMP RELAY AND TERMINAL "M1-14" AT ECM.

**NO:** 15 # 16 FUSE BLOWN?

**YES:** (1)
- DISCONNECT FUEL PUMP ELECTRICAL CONNECTOR AND FUEL PUMP RELAY.
- PROBE FUEL PUMP RELAY CKT 120 WITH A TEST LIGHT CONNECTED TO 12 VOLTS.

LIGHT "ON":
- REPLACE #16 FUSE AND RECONNECT FUEL PUMP RELAY.
- IGNITION "ON" AND RECHECK FUSE

FUSE "OK":
- IGNITION "OFF" RECONNECT FUEL PUMP
- IGNITION "ON"
- RECHECK FUSE

FUSE "OK": PROBLEM WAS AN INTERMITTENT SHORT TO GROUND IN FUEL PUMP CIRCUT.

FUSE "NOT OK": SHORTED FUEL PUMP

FUSE "NOT OK": REPLACE FUEL PUMP RELAY

LIGHT "ON": REPAIR SHORT TO GROUND IN CKT 120

**NO:** (2)
- IGNITION "OFF" FOR TEN SECONDS.
- BACKPROBE RELAY HARNESS TERMINAL 3(CKT 120) WITH A TEST LIGHT CONNECTED TO GROUND.
- IGNITION "ON". NOTE TEST LIGHT WITHIN 2 SECONDS AFTER IGNITION "ON"

LIGHT "ON": SEE CHART A-5

LIGHT "OFF":
- DISCONNECT FUEL PUMP RELAY.
- IGN. "ON", ENGINE STOPPED.
- PROBE RELAY HARNESS CONNECTOR TERMINAL "4" WITH A TEST LIGHT TO GROUND.

LIGHT "OFF": REPAIR OPEN CKT 440 TO RELAY

LIGHT "ON": CONNECT A TEST LIGHT BETWEEN HARNESS CONNECTOR TERMINALS "1" AND "4". (CKT 440 & 150)

LIGHT "OFF": REPAIR OPEN GROUND CIRCUIT

LIGHT "ON":
- CONNECT TEST LIGHT BETWEEN TERMINAL "2" (CKT 465) AND GROUND.
- IGN. "OFF" FOR 10 SECONDS.
- NOTE TEST LIGHT DURING 2 SECONDS AFTER IGN. "ON".

LIGHT "ON": FAULTY RELAY

LIGHT "OFF": CHECK FOR OPEN OR SHORT TO GROUND IN CKT 465. IF OK, IT'S A FAULTY ECM.

CLEAR CODES AND CONFIRM "CLOSED LOOP" OPERATION AND NO "CHECK ENGINE" LIGHT.

### CHART A-5

#### (Page 1 of 2)
#### FUEL DELIVERY SYSTEM

**Circuit Description:**

Fuel is drawn from the tank by the electric fuel pump and is fed under pressure through the fuel damper. Pulsations caused by the fuel pump are dampened by the fuel damper. From the damper, the fuel flows through a fuel filter and continues on into the fuel rail and then is injected into the ports through the fuel injectors.

Fuel pressure in the system is governed by the pressure regulator, in such a manner that a certain pressure difference between fuel pressure and inlet manifold pressure is maintained. Excess fuel above the regulated pressure is returned to the fuel tank by the pressure regulator and the fuel return lines.

The fuel pump has a check valve in it to maintain pressure at the fuel rail after the pump stops running. The check valve plays an important part in the fuel delivery system: to keep the fuel rail "charged" with fuel after the pump shuts off. When the engine begins cranking to start, there is no delay before fuel injection begins, and quick starting is insured. The check valve is part of the fuel pump, and is not serviced separately.

**Test Description:**

Numbers below refer to circled numbers on the diagnostic chart.

1. If the engine does not start, be sure to start with CHART A-3 to prevent misdiagnosis.
2. 2 is to purge any air from the lines after installing the gauge. It also serves to cool the fuel rail for more accurate pressure testing if the engine is hot. 3 is to allow the ECM to "power down", so the next time the ignition is turned "ON", the ECM will energize its fuel pump relay control for 2 seconds. 4 & 5 indicate what the pressure should do when the ignition is turned "ON". There are two things to note: (A)-pressure reading, and (B)-that the pressure does not continue to drop after the pump stops running.
3. At this point, the regulated pressure should be within specification, and the pressure does not drop when the pump stops running. This check is to see if the pressure regulator will modulate the regulated fuel pressure when the vacuum signal to it changes. During normal engine operation, the regulated pressure can change, based on intake manifold pressure. When the manifold pressure is at its lowest (engine idling), fuel pressure will be at its lowest regulated pressure. When intake manifold pressure is at its highest (wide open throttle), regulated fuel pressure will be at its highest regulated pressure.
4. If the pressure continues to drop after the pump stops running, there is a leak somewhere. Either the check valve in the pump is leaking pressure back into the tank, the regulator has an internal leak allowing fuel to leak from the pressure side to the return side, or an injector is leaking (dripping).

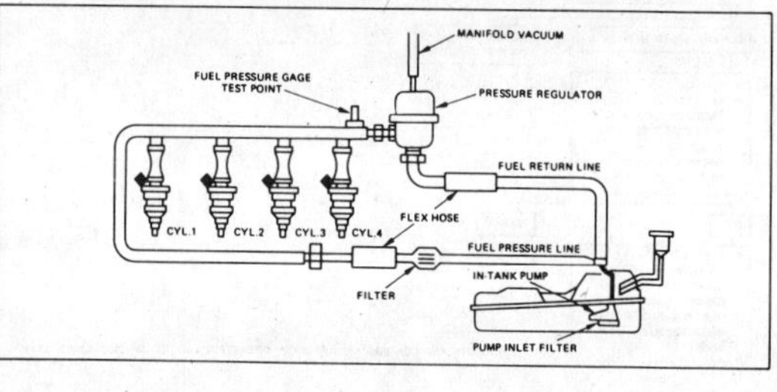


## I-MARK NON-TURBO

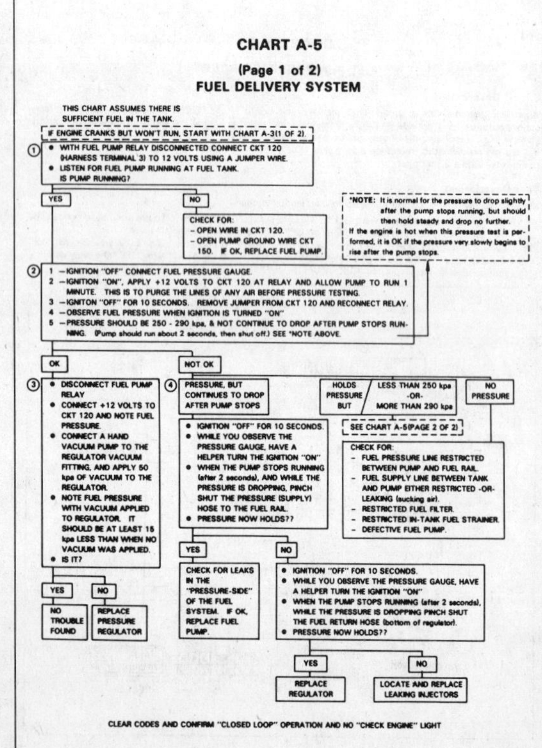


### CHART A-5

#### (Page 2 of 2)
#### FUEL DELIVERY SYSTEM

**Circuit Description:**

Fuel is drawn from the tank by the electric fuel pump and is fed under pressure through the fuel damper. Pulsations caused by the fuel pump are dampened by the fuel damper. From the damper, the fuel flows through a fuel filter and continues on into the fuel rail and then is injected into the ports through the fuel injectors.

Fuel pressure in the system is governed by the pressure regulator, in such a manner that a certain pressure difference between fuel pressure and inlet manifold pressure is maintained. Excess fuel above the regulated pressure is returned to the fuel tank through the pressure regulator and the fuel return lines.

The fuel pump has a check valve in it to maintain pressure at the fuel rail after the pump stops running. The check valve plays an important part in the fuel delivery system: to keep the fuel rail "charged" with fuel after the pump shuts off. When the engine begins cranking to start, there is no delay before fuel injection begins, and quick starting is insured. The check valve is part of the fuel pump, and is not serviced separately.

**Test Description:**

Numbers below refer to circled numbers on the diagnostic chart.

5. If any fuel lines, fittings, filter, or components are restricted, full pressure cannot be attained.
6. At this point, the pressure checks are OK, but a complete test has not been performed. Return to CHART A-5 (page 1 of 2) step 3 to complete the testing.
7. When the fuel return hose is pinched shut, there is no pressure regulator control to limit the pressure. The pressure reading will be whatever the fuel pump is capable of producing. This is the same as attaching a pressure gauge directly to the output of the fuel pump. The pressure reading would be a "maximum pressure - no flow" reading, and should be well over 290 kPa.
8. This is to determine if the cause of the high pressure is a restricted fuel return line, or a defective pressure regulator. If the pressure is normal when the regulator outlet (hose attached to bottom of regulator) is connected only to an open hose, then the problem is a restricted return line between the regulator and the tank.

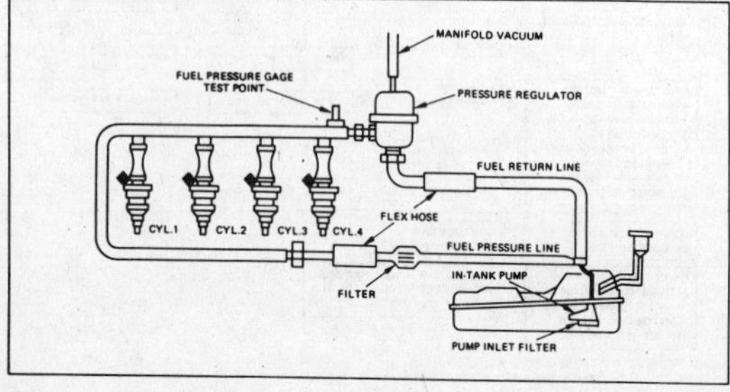

## I-MARK NON-TURBO

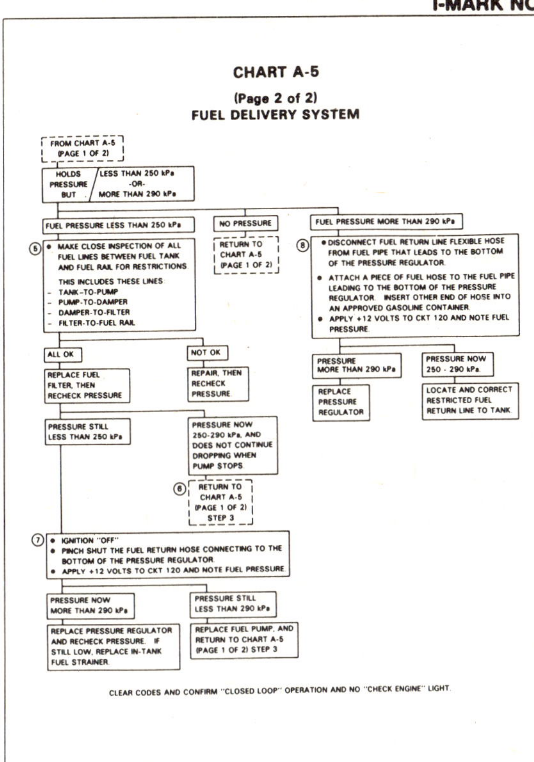


## CODE 13

### OXYGEN SENSOR CIRCUIT (OPEN CIRCUIT)

**Circuit Description:**

The ECM supplies a voltage of about .45 volt between terminals "M2-23" and "M2-22". (If measured with 10 megohm digital voltmeter, this may read as low as .32 volts).
The $O_2$ sensor varies the voltage within a range of about 1 volt, if the exhaust is rich, down through about .10 volt, if exhaust is lean.
The sensor is like an open circuit and produces no voltage, when it is below 360°C(600°F). An open sensor circuit, or cold sensor, causes "Open Loop" operation.

**Test Description:**

Numbers below refer to circled numbers on the diagnostic chart.

1. Code 13 WILL SET:

- Engine at normal operating temperature.
- At least 2 minutes engine run time after start.
- $O_2$ signal voltage steady between .42 and .55 volts.
- Throttle angle above 12%.
- All conditions must be met for about 5 seconds.

If the conditions for a Code 13 exist, the system will not go "Closed Loop".

2. This test determines if the $O_2$ sensor is the problem, or, if the ECM and wiring are at fault.
3. In doing this test, use only a high impedance digital volt ohm meter. This test checks the continuity of CKT's 412 and 413. If CKT 413 is open, the ECM voltage on CKT 412 will measure over .6 volts(600mV).

**Diagnostic Aids:**

Normal "Scan" voltage varies between 100 mv to 999 mv(.1 and 1.0 volt), while in "Closed Loop". Code 13 sets in 5 seconds, if voltage remains between .42 and .55 volts, but the system will go "Open Loop" in about 15 seconds.
Verify a clean, tight ground connection for CKT 413. Open CKT(s), 412 or 413 will result in a Code 13. If Code 13 is intermittent, refer to Section "B".

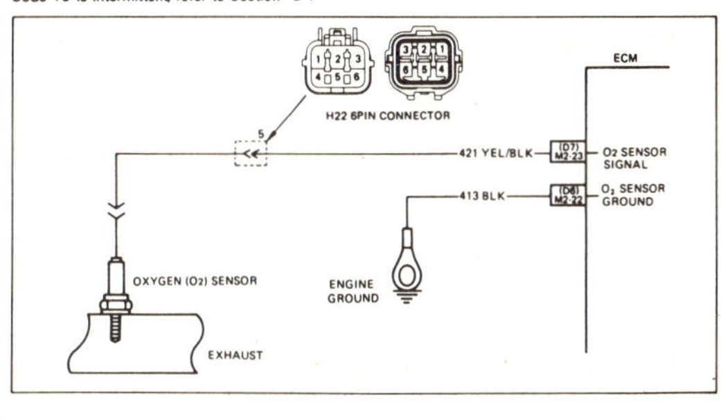


## I-MARK NON-TURBO

## CODE 13

### OXYGEN SENSOR CIRCUIT (OPEN CIRCUIT)

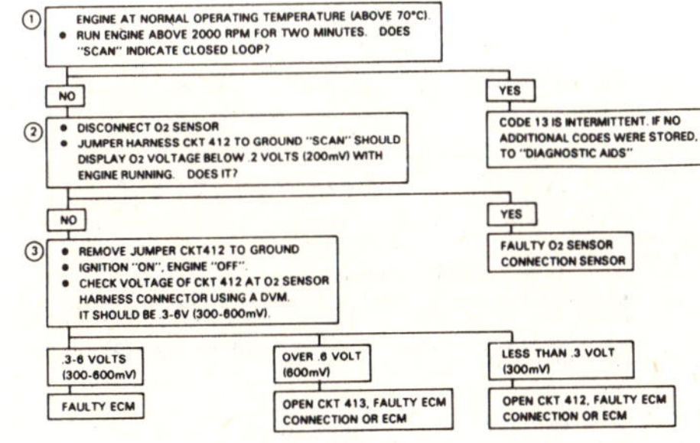


CLEAR CODES AND CONFIRM "CLOSED LOOP" OPERATION AND NO "CHECK ENGINE" LIGHT.

## CODE 14

### COOLANT TEMPERATURE SENSOR CIRCUIT (HIGH TEMPERATURE INDICATED)

**Circuit Description:**

The Coolant Temperature Sensor uses a thermistor to control the signal voltage at the ECM. The ECM applies a voltage on CKT 410 to the sensor. When the engine is cold, the sensor (thermistor) resistance is high, therefore, the ECM will see high signal voltage.

As the engine warms, the sensor resistance becomes less, and the voltage drops. At normal engine operating temperature, the voltage will measure about 1.5 to 2.0 volts at the ECM terminal "M2-10".
Coolant temperature is one of the inputs used to control:

- Fuel delivery
- Engine Spark Timing(EST)
- Idle(IAC)

**Test Description:**

Numbers below refer to circled numbers on the diagnostic chart.

1. Checks to see if code was set as result of hard failure or intermittent condition.
   Code 14 will set if:
   - Signal Voltage indicates a coolant temperature above 145°C (293°F) for 2 minutes.
2. This test simulates conditions for a Code 15. If the ECM recognizes the open circuit (high voltage), and displays a low temperature, the ECM and wiring are OK.

**Diagnostic Aids:**

A "Scan" tool reads engine temperature in degrees centigrade.
After the engine is started, the temperature should rise steadily to about 82°, then stabilize, when the thermostat opens.
A Code 14 will result if CKT 410 is shorted to ground.
If Code 14 is intermittent, refer to Section "B"

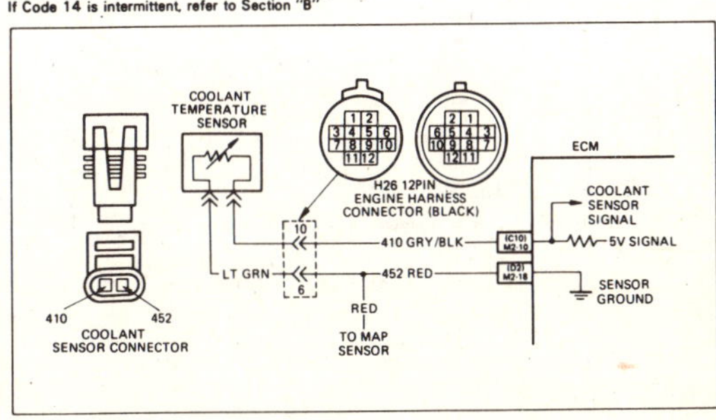

## I-MARK NON-TURBO

### CODE 14

#### COOLANT TEMPERATURE SENSOR CIRCUIT (HIGH TEMPERATURE INDICATED)

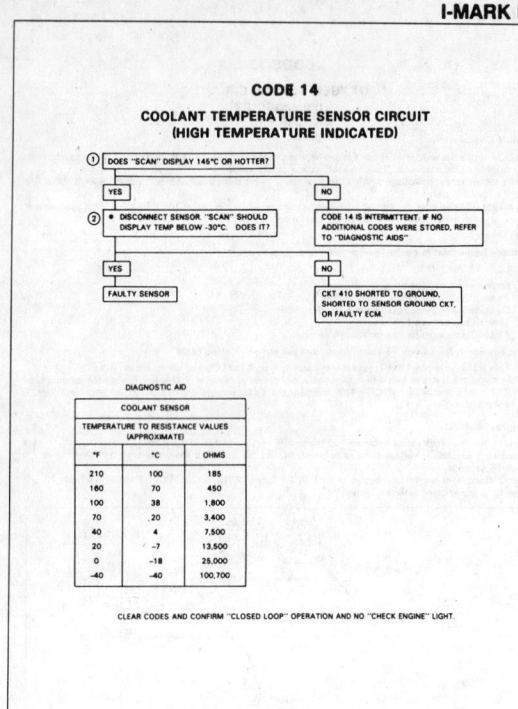

① DOES "SCAN" DISPLAY 145°C OR HOTTER?

- YES → ② • DISCONNECT SENSOR. "SCAN" SHOULD DISPLAY TEMP BELOW -30°C. DOES IT?
  - YES → FAULTY SENSOR
  - NO → CKT 410 SHORTED TO GROUND, SHORTED TO SENSOR GROUND CKT, OR FAULTY ECM.
- NO → CODE 14 IS INTERMITTENT. IF NO ADDITIONAL CODES WERE STORED, REFER TO "DIAGNOSTIC AIDS"

DIAGNOSTIC AID

| COOLANT SENSOR | | |
|---|---|---|
| TEMPERATURE TO RESISTANCE VALUES (APPROXIMATE) | | |
| °F | °C | OHMS |
| 210 | 100 | 185 |
| 160 | 70 | 450 |
| 100 | 38 | 1,800 |
| 70 | .20 | 3,400 |
| 40 | 4 | 7,500 |
| 20 | –7 | 13,500 |
| 0 | –18 | 25,000 |
| –40 | –40 | 100,700 |

CLEAR CODES AND CONFIRM "CLOSED LOOP" OPERATION AND NO "CHECK ENGINE" LIGHT.

### CODE 15

#### COOLANT TEMPERATURE SENSOR CIRCUIT (LOW TEMPERATURE INDICATED)

**Circuit Description:**

The Coolant Temperature Sensor uses a thermistor to control the signal voltage at the ECM. The ECM applies a voltage on CKT 410 to the sensor. When the engine is cold, the sensor (thermistor) resistance is high, therefore, the ECM will see high signal voltage.

As the engine warms, the sensor resistance becomes less, and the voltage drops. At normal engine operating temperature, the voltage will measure about 1.5 to 2.0 volts at the ECM terminal "M2-10" (C10). Coolant temperature is one of the inputs used to control:

- Fuel delivery
- Engine Spark Timing(EST)
- Idle(IAC)

**Test Description:**

Numbers below refer to circled numbers on the diagnostic chart.

1. Checks to see if code was set as result of hard failure or intermittent condition. Code 15 will set if:
   - The engine has been running for 2 minute.
   - Signal Voltage indicates a coolant temperature below -37°C.
2. This test simulates conditions for a Code 14. If the ECM recognizes the grounded circuit (low voltage), and displays a high temperature, the ECM and wiring are OK.
3. This test will determine if there is a wiring problem or a faulty ECM. If CKT 452 is open, there may also be a Code 33 stored. Be sure to carefully check terminals at the black engine harness connector.

**Diagnostic Aids:**

A "Scan" tool reads engine temperature in degrees centigrade.
After the engine is started, the temperature should rise steadily to about 82°, then stabilize, when the thermostat opens.
A Code 15 will result if CKT's 410 or 452 is open.
If Code 15 is intermittent, refer to Section "B".

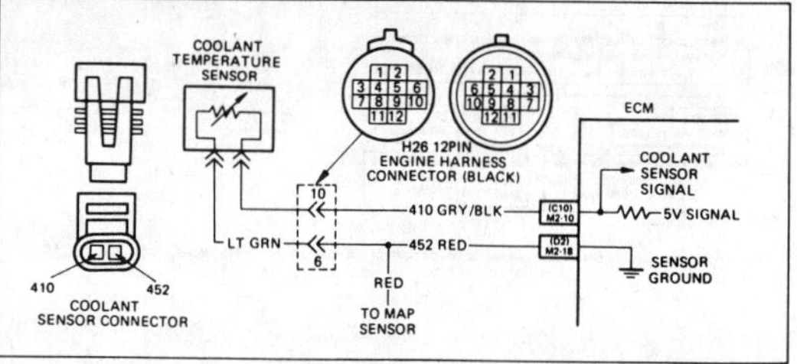


## I-MARK NON-TURBO

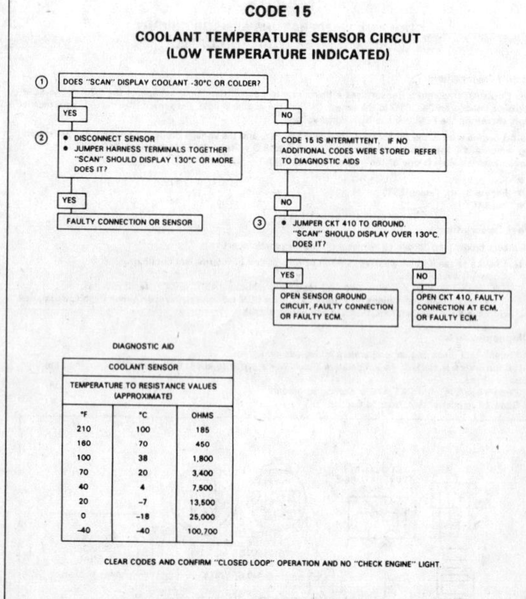


| COOLANT SENSOR | | |
|---|---|---|
| TEMPERATURE TO RESISTANCE VALUES (APPROXIMATE) | | |
| °F | °C | OHMS |
| 210 | 100 | 185 |
| 160 | 70 | 450 |
| 100 | 38 | 1,800 |
| 70 | 20 | 3,400 |
| 40 | 4 | 7,500 |
| 20 | –7 | 13,500 |
| 0 | –18 | 25,000 |
| –40 | –40 | 100,700 |

CLEAR CODES AND CONFIRM "CLOSED LOOP" OPERATION AND NO "CHECK ENGINE" LIGHT.

### CODE 21

#### THROTTLE POSITION SENSOR(TPS) CIRCUIT (SIGNAL VOLTAGE HIGH)

**Circuit Description:**

The Throttle Position Sensor (TPS) provides a voltage signal that changes relative to the throttle valve. Signal voltage will vary from less than 1.25 volts at idle to about 4.5 volts at wide open throttle(WOT).

The TPS signal is one of the most important inputs used by the ECM for fuel control and for many of the ECM controlled outputs.

**Test Description:**

Numbers below refer to circled numbers on the diagnostic chart.

1. This step checks to see if Code 21 is the result of a hard failure or an intermittent condition. A Code 21 will set if.
   - TPS reading above 4.5 volts.
   - Engine speed less than 1200 rpm.
   - MAP reading below 50 kPa.
   - All of the above conditions present for 5 seconds.
2. This step simulates conditions for a Code 22. If the ECM recognizes the change of state, the ECM and CKTs 416 and 417 are OK.
3. This step isolates a faulty sensor, ECM, or an open CKT 452. If it is determined CKT 452 is the fault be sure to check the terminals at the black engine harness connector.

**Diagnostic Aids:**

A "Scan" tool displays throttle position in volts. Closed throttle voltage should be less than 1.25 volts. TPS voltage should increase at a steady rate as throttle is moved to WOT.

A Code 21 will result if CKT 452 is open or CKT 417 is shorted to voltage. If Code 21 is intermittent, refer to Section "B".

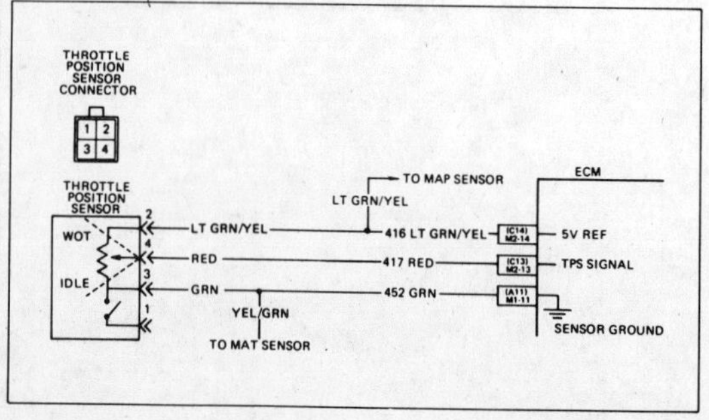

## I-MARK NON-TURBO

### CODE 21

**THROTTLE POSITION SENSOR(TPS) CIRCUIT (SIGNAL VOLTAGE HIGH)**

① • THROTTLE CLOSED. DOES "SCAN" DISPLAY TPS OVER 2.5 VOLTS?

- YES → ② • DISCONNECT SENSOR "SCAN" SHOULD DISPLAY TPS BELOW .2 VOLTS(200 mV). DOES IT?
  - YES → CKT 417 SHORTED TO VOLTAGE OR FAULTY ECM
  - NO → ③ • PROBE SENSOR GROUND CIRCUIT WITH A TEST LIGHT CONNECTED TO 12 VOLTS.
    - LIGHT "ON" → FAULTY CONNECTION OR SENSOR
    - LIGHT "OFF" → OPEN SENSOR GROUND CIRCUIT OR FAULTY ECM
- NO → CODE 21 IS INTERMITTENT. IF NO ADDITIONAL CODES WERE STORED, REFER TO DIAGNOSTIC AIDS

CLEAR CODES AND CONFIRM "CLOSED LOOP" OPERATION AND NO "CHECK ENGINE" LIGHT.

### CODE 22

**THROTTLE POSITION SENSOR(TPS) CIRCUIT (SIGNAL VOLTAGE LOW)**

**Circuit Description:**

The Throttle Position Sensor (TPS) provides a voltage signal that changes, relative to the throttle valve. Signal voltage will vary from less than 1.25 volts at idle to about 4.5 volts at wide open throttle(WOT).

The TPS signal is one of the most important inputs used by the ECM for fuel control and for many of the ECM controlled outputs.

**Test Description:**

Numbers below refer to circled numbers on the diagnostic chart.

1. This step checks to see if Code 22 is the result of a hard failure or an intermittent condition.
   A Code 22 will set if:
   - The engine is running
   - TPS voltage is below .20 volts(200mv).
2. This step simulates conditions for a Code 21. If a Code 21 is set, or the "Scan" tool displays over 4 volts, the ECM and wiring are OK.
3. The "Scan" tool may not display 12 volts. The important thing is that the ECM recognizes the voltage as over 4 volts, indicating that CKT 417 and the ECM are OK.
4. If CKT 416 is open or shorted to ground, there may also be a stored Code 34. If it is determined that the CKT is open be sure to check terminals at the black engine harness connector.

**Diagnostic Aids:**

A "Scan" tool displays throttle position in volts. Closed throttle voltage should be less than 1.25 volts. TPS voltage should increase at a steady rate as throttle is moved to WOT.

An open or grounded 416 or 417 will result in a Code 22.

If Code 22 is intermittent, refer to Section "B"

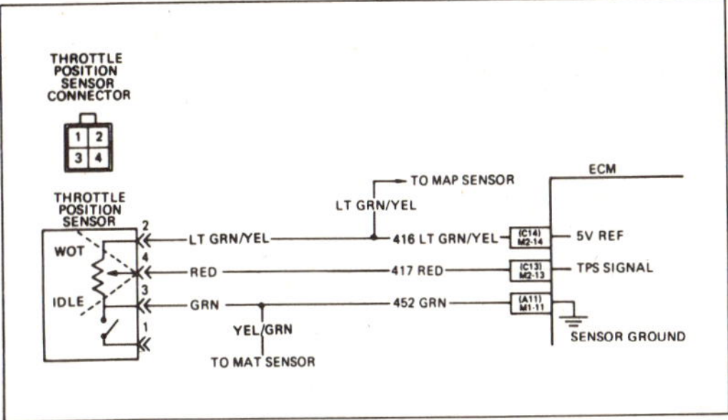


## I-MARK NON-TURBO

### CODE 22

**THROTTLE POSITION SENSOR(TPS) CIRCUIT (SIGNAL VOLTAGE LOW)**

① • THROTTLE CLOSED
• DOES "SCAN" DISPLAY TPS .2V (200mV) OR BELOW?

- YES → ② • DISCONNECT SENSOR. • JUMPER CKT'S 416 & 417 TOGETHER. • "SCAN" SHOULD DISPLAY TPS OVER 4.0V. DOES IT?
  - NO → ③ • PROBE CKT 417 WITH A TEST LIGHT CONNECTED TO 12 VOLTS. • "SCAN" TOOL SHOULD DISPLAY TPS OVER 4.0V. DOES IT?
    - YES → ④ CKT 416 OPEN, SHORTED TO GROUND, INTERNALLY SHORTED MAP SENSOR, FAULTY CONNECTION OR FAULTY ECM.
    - NO → CKT 417 OPEN OR SHORTED TO GROUND OR FAULTY CONNECTION OR ECM
  - YES → REPLACE SENSOR
- NO → CODE 22 IS INTERMITTENT. IF NO ADDITIONAL CODES WERE STORED, REFER TO DIAGNOSTIC AIDS

CLEAR CODES AND CONFIRM "CLOSED LOOP" OPERATION AND NO "CHECK ENGINE" LIGHT.

### CODE 23

**MANIFOLD AIR TEMPERATURE(MAT) SENSOR CIRCUIT (LOW TEMPERATURE INDICATED)**

**Circuit Description:**

The Manifold Air Temperature Sensor uses a thermistor to control the signal voltage to the ECM. The ECM applies a voltage (4-6 volts) on CKT 472 to the sensor. When manifold air is cold, the sensor (thermistor) resistance is high, therefore, the ECM will see a high signal voltage. As the air warms, the sensor resistance becomes less and the voltage drops.

**Test Description:**

Numbers below refer to circled numbers on the diagnostic chart.

1. This step checks to see if Code 23 is the result of a hard failure or an intermittent condition.
   A Code 23 will set if:
   - Signal voltage indicates a MAT temperature less than -30°C
   - Engine is running for longer than 3 minutes.
2. This test simulates conditions for a Code 25. If the "Scan" tool displays a high temperature, the ECM and wiring are ok.
3. This step checks continuity of CKT's 472 and 452. If CKT 452 is open there may also be a Code 21. When checking for an open CKT be sure to check terminals at the black engine harness connector.

**Diagnostic Aids:**

If the engine has been allowed to cool to an ambient temperature (overnight), coolant and MAT temperatures may be checked with a "Scan" tool and should read close to each other.

A Code 23 will result if CKT's 472 or 452 become open.

If Code 23 is intermittent, refer to Section "B".

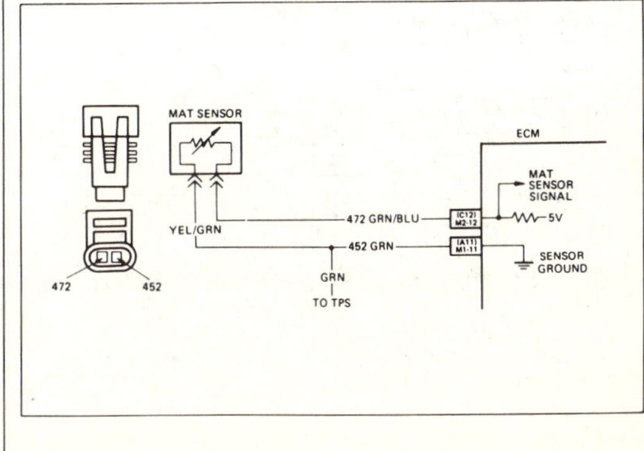

## I-MARK NON-TURBO

### CODE 23
### MANIFOLD AIR TEMPERATURE(MAT) SENSOR CIRCUIT (LOW TEMPERATURE INDICATED)

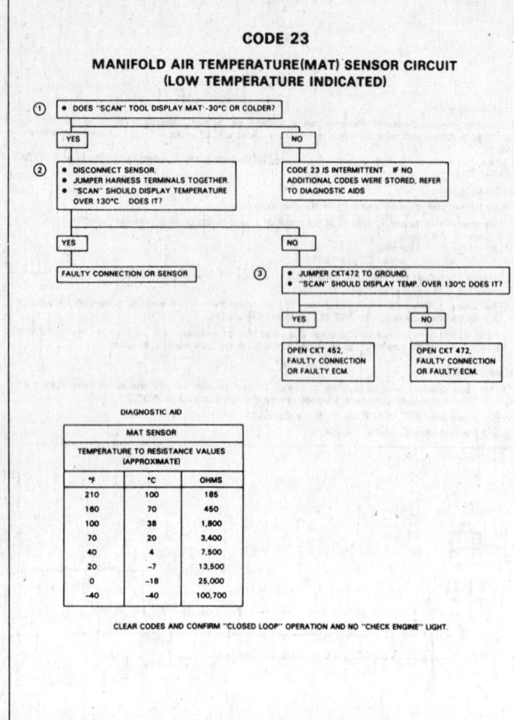


DIAGNOSTIC AID

| MAT SENSOR | | |
|---|---|---|
| TEMPERATURE TO RESISTANCE VALUES (APPROXIMATE) | | |
| °F | °C | OHMS |
| 210 | 100 | 185 |
| 160 | 70 | 450 |
| 100 | 38 | 1,800 |
| 70 | 20 | 3,400 |
| 40 | 4 | 7,500 |
| 20 | -7 | 13,500 |
| 0 | -18 | 25,000 |
| -40 | -40 | 100,700 |

CLEAR CODES AND CONFIRM "CLOSED LOOP" OPERATION AND NO "CHECK ENGINE" LIGHT.

### CODE 24
### VEHICLE SPEED SENSOR(VSS) CIRCUIT

**Circuit Description:**

The ECM applies and monitors 12 volts on CKT 437. CKT 437 connects to the Vehicle Speed Sensor which alternately grounds CKT 437 when drive wheels are turning. This pulsing action takes place 4096 times per mile, and the ECM will calculate vehicle speed based on the time between "pulses".

Code 24 will set:

- Speed sensor indicates speed less than 5 mph.
- RPM between 1500 and 4400
- Coolant greater than 85°C.
- MAP less than 49 kPa.
- Conditions present for 3 seconds.

**Test Description:**

Numbers below refer to circled numbers on the diagnostic chart.

1. This test uses the "Scan" tool to verify the VSS sensor operation.
2. This checks for the presence of the ECM-supplied 12 volt signal, which is monitored by the ECM for pulses, indicating VSS operation.
3. CKT439 is the ignition feed that supplies operating power to the VSS. If this line is open, CKT437 will be at 12 volts at all times.
4. CKT 450 supplies the ground path for VSS operation. If CKT 450 is open, VSS cannot pulse CKT 437 to ground.

**Diagnostic Aids:**

"Scan" should indicate a vehicle speed whenever the drive wheels are turning greater than 3 mph.
If Code 24 is intermittent, refer to Section "B"

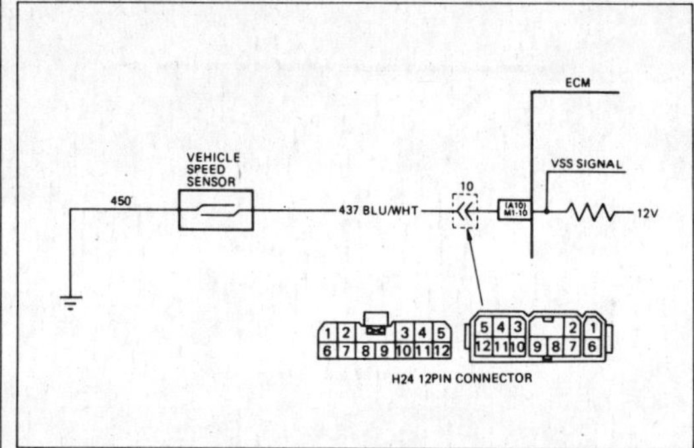


## I-MARK NON-TURBO

### CODE 24
### VEHICLE SPEED SENSOR(VSS) CIRCUIT

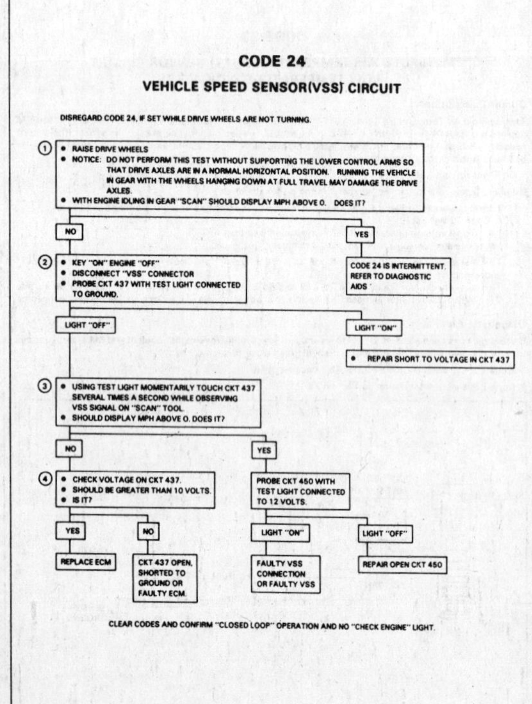


### CODE 25
### MANIFOLD AIR TEMPERATURE(MAT) SENSOR CIRCUIT (HIGH TEMPERATURE INDICATED)

**Circuit Description:**

The Manifold Air Temperature Sensor uses a thermistor to control the signal voltage to the ECM. The ECM applies a voltage (4-6 volts) on CKT 472 to the sensor. When manifold air is cold, the sensor (thermistor) resistance is high, therefore, the ECM will see a high signal voltage. As the air warms, the sensor resistance becomes less and the voltage drops.

**Test Description:**

Numbers below refer to circled numbers on the diagnostic chart.

1. This check determines if the Code 25 is the result of a hard failure or an intermittent condition.
   A Code 25 will set if:
   - A MAT temperature greater than 145°C is detected for a time longer than 2 minutes.

**Diagnostic Aids:**

If the engine has been allowed to cool to an ambient temperature (overnight), coolant and MAT temperatures may be checked with a "Scan" tool and should read close to each other.

A Code 25 will result if CKT 472 is shorted to ground.

If Code 25 is intermittent, refer to Section "B".

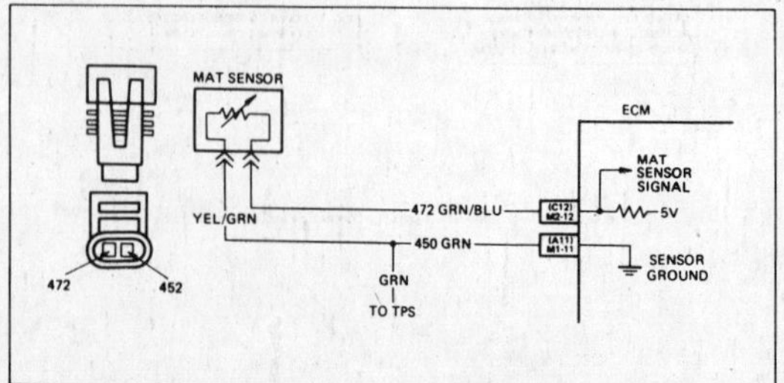

## I-MARK NON-TURBO

### CODE 25

#### MANIFOLD AIR TEMPERATURE(MAT) SENSOR CIRCUIT (HIGH TEMPERATURE INDICATED)

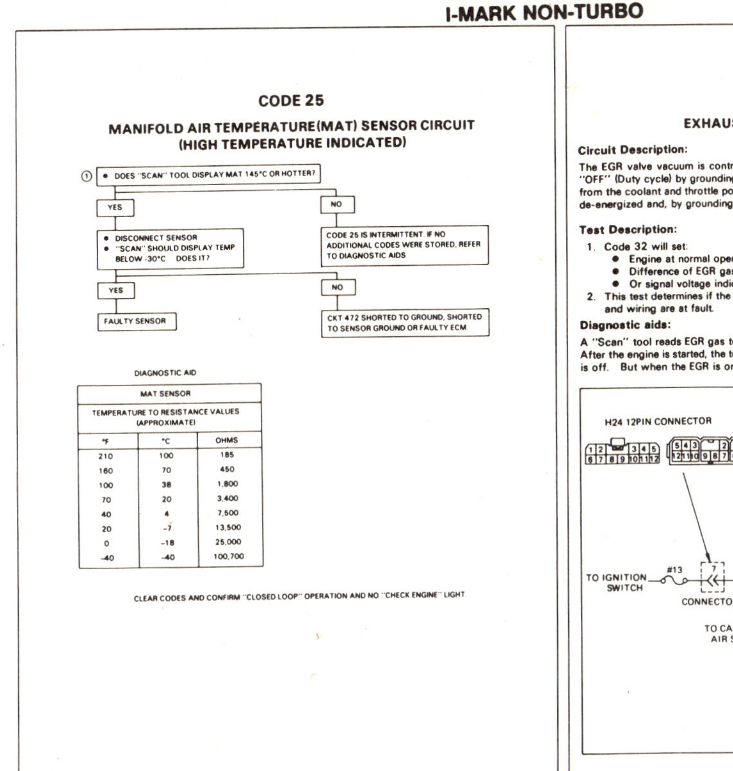


DIAGNOSTIC AID

| MAT SENSOR | | |
|---|---|---|
| TEMPERATURE TO RESISTANCE VALUES (APPROXIMATE) | | |
| °F | °C | OHMS |
| 210 | 100 | 185 |
| 160 | 70 | 450 |
| 100 | 38 | 1,800 |
| 70 | 20 | 3,400 |
| 40 | 4 | 7,500 |
| 20 | -7 | 13,500 |
| 0 | -18 | 25,000 |
| -40 | -40 | 100,700 |

CLEAR CODES AND CONFIRM "CLOSED LOOP" OPERATION AND NO "CHECK ENGINE" LIGHT.

### CODE 32

#### EXHAUST GAS RECIRCULATION(EGR) CIRCUIT

**Circuit Description:**

The EGR valve vacuum is control by an ECM controlled solenoid. The ECM will turn the EGR "ON" and "OFF" (Duty cycle) by grounding CKT435. The duty cycle is calculated by the ECM, based on information from the coolant and throttle position sensor. With the ignition "ON", engine stopped, the EGR solenoid is de-energized and, by grounding the diagnostic terminal, the solenoid is energized.

**Test Description:**

1. Code 32 will set:
   - Engine at normal operating temperature.
   - Difference of EGR gas temperature between EGR off and EGR on is less than 3°C.
   - Or signal voltage indicates a EGR gas temperature above 350°C or below –36°C.
2. This test determines if the EGR gas temperature sensor or the EGR valve is the problem, or, if the ECM and wiring are at fault.

**Diagnostic aids:**

A "Scan" tool reads EGR gas temperature in degrees centigrade.
After the engine is started, the temperature should become nearly to the manifold air temperature, when EGR is off. But when the EGR is on, the EGR gas temperature should be about between 100°C and 250°C.

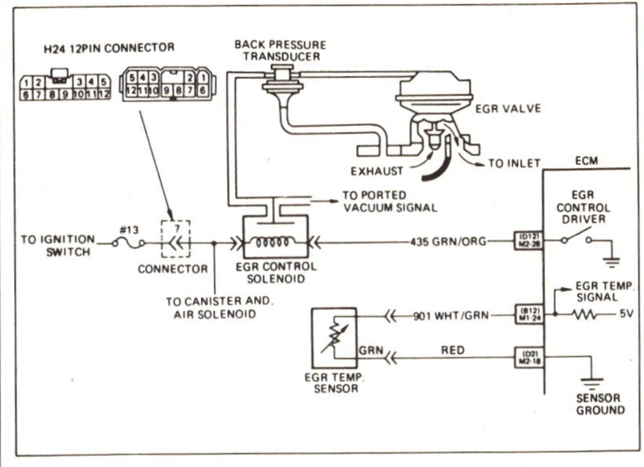


## I-MARK NON-TURBO

### CODE 32

#### EXHAUST GAS RECIRCULATION(EGR) CIRCUIT

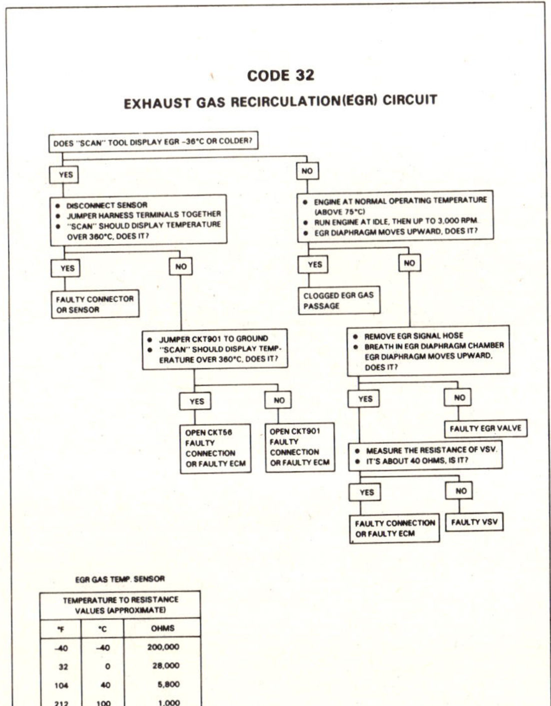


EGR GAS TEMP. SENSOR

| TEMPERATURE TO RESISTANCE VALUES (APPROXIMATE) | | |
|---|---|---|
| °F | °C | OHMS |
| -40 | -40 | 200,000 |
| 32 | 0 | 28,000 |
| 104 | 40 | 5,800 |
| 212 | 100 | 1,000 |
| 392 | 200 | 135 |

### CODE 33

#### MANIFOLD ABSOLUTE PRESSURE(MAP) SENSOR CIRCUIT (SIGNAL VOLTAGE HIGH-LOW VACUUM)

**Circuit Description:**

The Manifold Absolute Pressure (MAP) Sensor responds to changes in manifold pressure (vacuum). The ECM receives this information as a signal voltage that will vary from about 1 to 1.5 volts, at closed throttle idle, to 4-4.5 volts at wide open throttle (low vacuum).
If the MAP sensor fails, the ECM will substitute a fixed MAP value and use the Throttle Position Sensor (TPS) to control fuel delivery.

**Test Description:**

Numbers below refer to circled numbers on the diagnostic chart.

1. This step will determine if Code 33 is the result of a hard failure or an intermittent condition.

A Code 33 will set if:
- MAP signal indicates greater than 120 kPa (over 4V) (low vacuum).
- TPS less than 5%.
- These conditions are present for a time longer than 10 seconds.

2. This step simulates conditions for a Code 34. If the ECM recognizes the change, the ECM, and CKT's 416 and 432, are OK. If CKT 452 is open, there may also be a Code 15 stored.

**Diagnostic Aids:**

With the ignition "ON" and the engine stopped, the manifold pressure is equal to atmospheric pressure and the signal voltage will be high. This information is used by the ECM as an indication of vehicle altitude and is referred to as BARO. Comparison of this BARO reading with a known good vehicle with the same sensor is a good way to check accuracy of a "suspect" sensor. Readings should be the same ±4 volt.

A Code 33 will result if CKT 452 is open, or if CKT 432 is shorted to voltage or to CKT 416.

If Code 33 is intermittent, refer to Section "B".

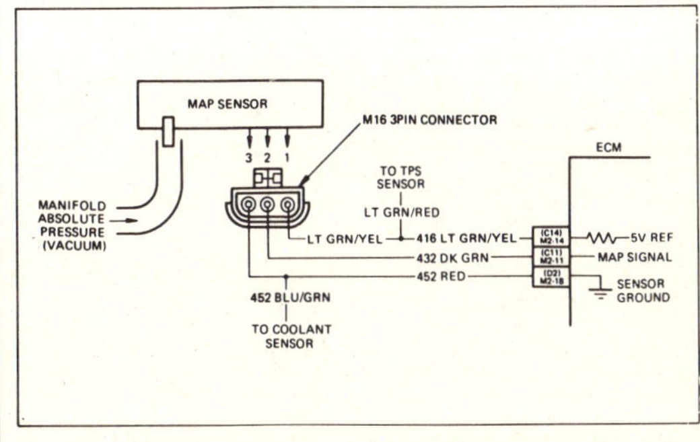

## I-MARK NON-TURBO

### CODE 33

### MANIFOLD ABSOLUTE PRESSURE (MAP) SENSOR CIRCUIT (SIGNAL VOLTAGE HIGH-LOW VACUUM)

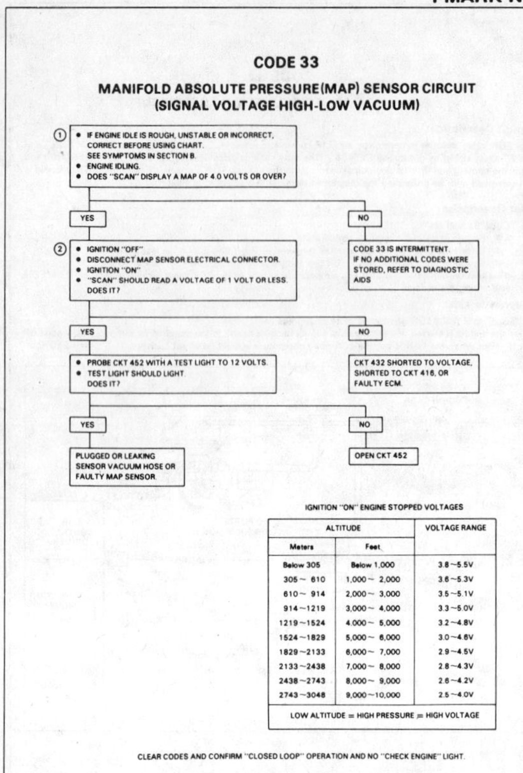


IGNITION "ON" ENGINE STOPPED VOLTAGES

| ALTITUDE | | VOLTAGE RANGE |
|---|---|---|
| Meters | Feet | |
| Below 305 | Below 1,000 | 3.8~5.5V |
| 305~ 610 | 1,000~ 2,000 | 3.6~5.3V |
| 610~ 914 | 2,000~ 3,000 | 3.5~5.1V |
| 914~1219 | 3,000~ 4,000 | 3.3~5.0V |
| 1219~1524 | 4.000~ 5,000 | 3.2~4.8V |
| 1524~1829 | 5,000~ 6,000 | 3.0~4.6V |
| 1829~2133 | 6,000~ 7,000 | 2.9~4.5V |
| 2133~2438 | 7,000~ 8,000 | 2.8~4.3V |
| 2438~2743 | 8,000~ 9,000 | 2.6~4.2V |
| 2743~3048 | 9,000~10,000 | 2.5~4.0V |
| LOW ALTITUDE = HIGH PRESSURE = HIGH VOLTAGE | | |

CLEAR CODES AND CONFIRM "CLOSED LOOP" OPERATION AND NO "CHECK ENGINE" LIGHT.

### CODE 34

### MANIFOLD ABSOLUTE PRESSURE (MAP) SENSOR CIRCUIT (SIGNAL VOLTAGE LOW-HIGH VACUUM)

**Circuit Description:**

The Manifold Absolute Pressure (MAP) Sensor responds to changes in manifold pressure (vacuum). The ECM receives this information as a signal voltage that will vary from about 1 to 1.5 volts at closed throttle idle, to 4-4.5 volts at wide open throttle.

If the MAP sensor fails, the ECM will substitute a fixed MAP value and use the Throttle Position Sensor (TPS) to control fuel delivery.

**Test Description:**

Numbers below refer to circled numbers on the diagnostic chart.

1. This step determines if Code 34 is the result of a hard failure or an intermittent condition.
   A Code 34 will set when:
   - Engine rpm is less than 1200 rpm.
   - MAP signal voltage is too low (14 kPa).

   OR
   - MAP reading less than 14 kPa.
   - Engine rpm greater than 1200 rpm.
   - TPS is less than 15.2%.
2. Jumpering harness terminals "2" to "1", 5 volt to signal, will determine if the sensor is at fault, or if there is a problem with the ECM or wiring.
3. The "Scan" tool may not display 12 volts. The important thing is that the ECM recognizes the voltage as more than 4 volts, indicating that the ECM and CKT 432 are OK.

**Diagnostic Aids:**

With the ignition "ON" and the engine stopped, the manifold pressure is equal to atmospheric pressure and the signal voltage will be high. This information is used by the ECM as an indication of vehicle altitude and is referred to as BARO. Comparison of this BARO reading with a known good vehicle with the same sensor is a good way to check accuracy of a "suspect" senser. Readings should be the same ±.4 volt.

A Code 34 will result if CKTs 416 or 432 are open or shorted to ground.

If Code 34 is intermittent, refer to Section "B".

An internally shorted TPS sensor will cause a Code 34.

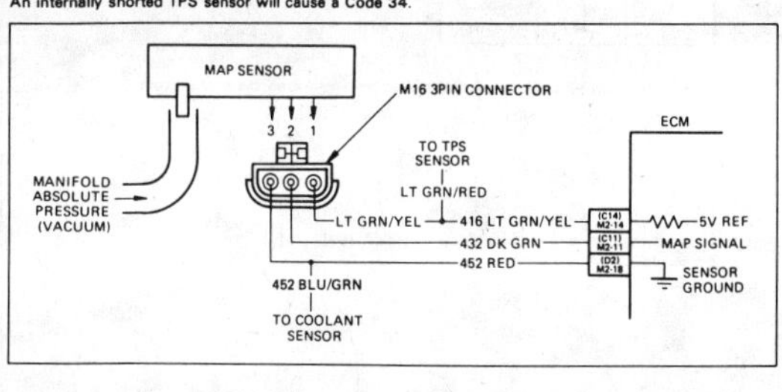


## I-MARK NON-TURBO

### CODE 34

### MANIFOLD ABSOLUTE PRESSURE (MAP) SENSOR CIRCUIT (SIGNAL VOLTAGE LOW-HIGH VACUUM)

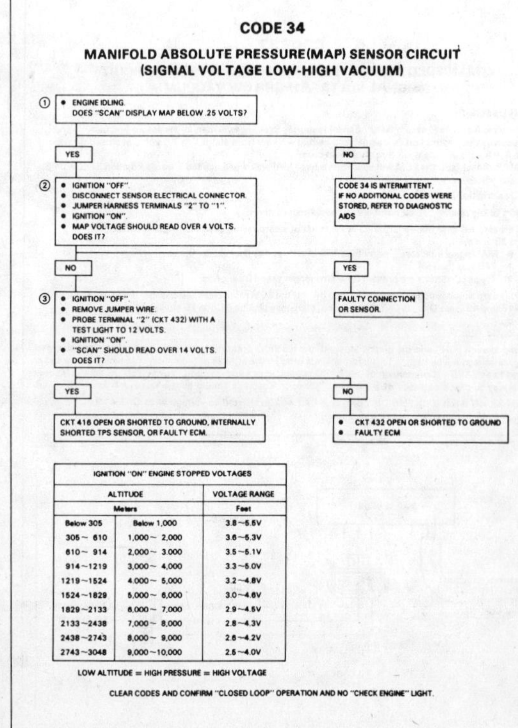


| IGNITION "ON" ENGINE STOPPED VOLTAGES | | |
|---|---|---|
| ALTITUDE | | VOLTAGE RANGE |
| Meters | | Feet |
| Below 305 | Below 1,000 | 3.8~5.5V |
| 305~ 610 | 1,000~ 2,000 | 3.6~5.3V |
| 610~ 914 | 2,000~ 3.000 | 3.5~5.1V |
| 914~1219 | 3,000~ 4,000 | 3.3~5.0V |
| 1219~1524 | 4.000~ 5,000 | 3.2~4.8V |
| 1524~1829 | 5,000~ 6,000 | 3.0~4.6V |
| 1829~2133 | 6,000~ 7,000 | 2.9~4.5V |
| 2133~2438 | 7,000~ 8,000 | 2.8~4.3V |
| 2438~2743 | 8,000~ 9,000 | 2.6~4.2V |
| 2743~3048 | 9,000~10,000 | 2.5~4.0V |

LOW ALTITUDE = HIGH PRESSURE = HIGH VOLTAGE

CLEAR CODES AND CONFIRM "CLOSED LOOP" OPERATION AND NO "CHECK ENGINE" LIGHT.

### CODE 44

### OXYGEN SENSOR CIRCUIT (LEAN EXHAUST INDICATED)

**Circuit Description:**

The ECM supplies a voltage of about .45 volt between terminals "M2-23" and "M2-22". (If measured with a 10 megohm digital voltmeter, this may read as low as .32 volts.) The $O_2$ sensor varies the voltage within a range of about 1 volt, if the exhaust is rich, down through about .10 volt, if exhaust is lean.

The sensor is like an open circuit and produces no voltage, when it is below about 360°(600°F). An open sensor circuit, or cold sensor, causes "Open Loop" operation.

**Test Description:**

Numbers below refer to circled numbers on the diagnostic chart.

1. Code 44 is set, when the $O_2$ sensor signal voltage on CKT 412:
   - Remains below 0.05 volt for 5 seconds or more;
   - And the system is operating in "Closed Loop".

**Diagnostic Aids:**

Using the "Scan", observe the block learn value at different rpms. The "Scan", also, displays the block cells, so the block learn values can be checked in each of the cells, to determine when the Code 44 may have been set. If the conditions for Code 44 exists, the block learn values will be around 150.

- $O_2$ Sensor Wire - Sensor pigtail may be mispositioned and contacting the exhaust manifold.
- Check for ground in wire between connector and sensor.
- **Fuel Contamination** — Water, even in small amounts, near the in-tank fuel pump inlet can be delivered to the injector. The water causes a lean exhaust and can set a Code 44.
- **Fuel Pressure** — System will be lean if pressure is too low. It may be necessary to monitor fuel pressure, while driving the car at various road speeds and/or loads to confirm. See Fuel System diagnosis CHART A-5 (Page 42).
- **Exhaust Leaks** — If there is an exhaust leak, the engine can cause outside air to be pulled into the exhaust and past the sensor. Vacuum or crankcase leaks can cause a lean condition.
- If Code 44 intermittent, refer to Section "B".

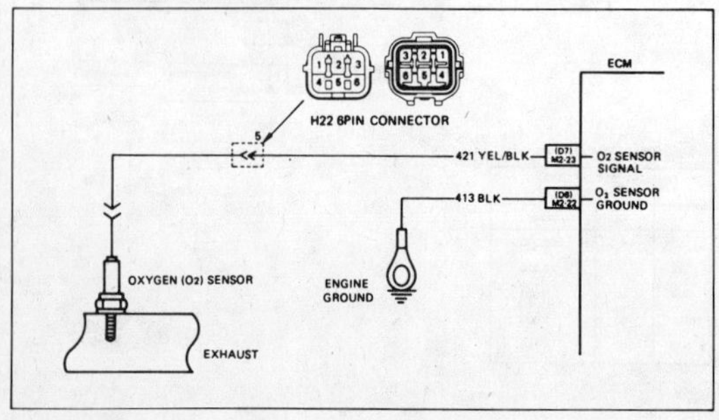

## I-MARK NON-TURBO

### CODE 44

#### OXYGEN SENSOR CIRCUIT (LEAN EXHAUST INDICATED)

① • RUN WARM ENGINE (75°C TO 95°C) AT 1200 RPM
• DOES "SCAN" INDICATE O2 VOLTAGE FIXED BELOW .30 VOLTS(300 mV)?

YES
• DISCONNECT O2 SENSOR
• WITH ENGINE IDLING "SCAN" SHOULD DISPLAY O2 BETWEEN .35 VOLTS AND .55 VOLTS (350 mV AND 550 mV). DOES IT?

YES
REFER TO DIAGNOSTIC AIDS ON FACING PAGE

NO
CKT 412 SHORTED TO GROUND OR FAULTY ECM

NO
CODE 44 IS INTERMITTENT. IF NO ADDITIONAL CODES WERE STORED, REFER TO DIAGNOSTIC AIDS

CLEAR CODES AND CONFIRM "CLOSED LOOP" OPERATION AND NO "CHECK ENGINE" LIGHT.

### CODE 45

#### OXYGEN SENSOR CIRCUIT (RICH EXHAUST INDICATED)

**Circuit Description:**

The ECM supplies a voltage of about .45 volt between terminals "M2-23" and "M2-22". (If measured with a 10 megohm digital voltmeter, this may read as low as .32 volts.) The $O_2$ sensor varies the voltage within a range of about 1 volt, if the exhaust is rich, down through about .10 volt, if exhaust is lean.

The sensor is like an open circuit and produces no voltage, when it is below about 360°C (600°F). An open sensor circuit, or cold sensor, causes "Open Loop" operation.

**Test Description:**

Numbers below refer to circled numbers on the diagnostic chart.

1. Code 45 is set, when the $O_2$ sensor signal voltage on CKT 412:
   - Remains above 0.85 volt for 5 seconds or more; and in "Closed Loop".
   - And run engine less then 3000 rpm, vehicle speed between 15 and 50 miles par hour.

**Diagnostic Aids:**

The Code 45, or rich exhaust, is most likely caused by one of the flowing:

- **Fuel Pressure** — System will go rich, if pressure is too high. The ECM can compensate for some increase. However, if it gets too high, a Code 45 will be set. See Fuel System diagnosis CHART A-5.
- **IGN Shielding** — An open ground CKT 450 may result in EMI, or induced electrical "noise". The ECM looks at this "noise" as reference pulses. The additional pulses result in a higher than actual engine speed signal. The ECM then delivers too much fuel, causing system to go rich. Engine tachometer will, also, show higher than actual engine speed, which can help in diagnosing this problem.
- **Canister Purge** — Check for fuel saturation. If full of fuel, check canister control and hoses. See Canister Purge, Section "C3".
- **MAP Sensor** — An output that causes the ECM to sense a higher than normal manifold pressure (low vacuum) can cause the system to go rich. Disconnecting the MAP sensor will allow the ECM to set a fixed value for the MAP sensor. Substitute a different MAP sensor, if the rich condition is gone, while the sensor is disconnected.
- **TPS** — An intermittent TPS output will cause the system to go rich, due to a false indication of the engine accelerating.
- **$O_2$ Sensor Contamination** — Inspect Oxygen Sensor for silicone contamination from fuel, or use of improper RTV sealant. The sensor may have a white, powdery coating and result in a high, but false signal voltage (rich exhaust indication). The ECM will then reduce the amount of fuel delivered to the engine, causing a severe surge driveability problem.
  If Code 45 is intermittent, refer to Section "B".
- If the EGR valve is stuck open it may result in a lean exhaust and could possibly set a Code 45, especially at idle. Refer to CHART A-5 to check EGR system.

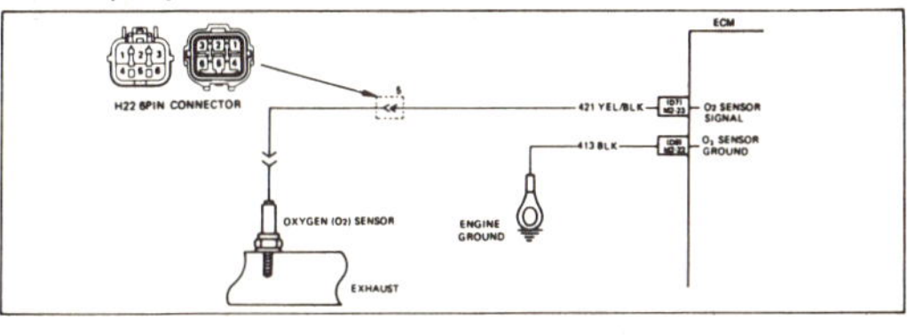


## I-MARK NON-TURBO

### CODE 45

#### OXYGEN SENSOR CIRCUIT (RICH EXHAUST INDICATED)

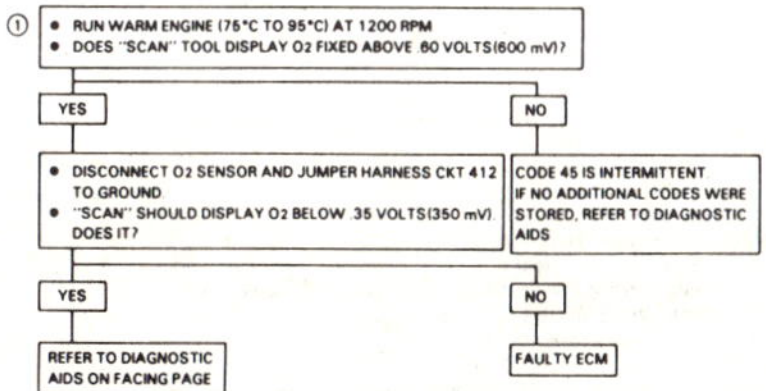


CLEAR CODES AND CONFIRM "CLOSED LOOP" OPERATION AND NO "CHECK ENGINE" LIGHT

### CODE 51

#### PROM ERROR (FAULTY OR INCORRECT MEM-CAL)

CHECK THAT MEM-CAL IS PROPERLY SEATED. IF OK, REPLACE MEM-CAL AND RECHECK. IF CODE 51 REAPPEARS, REPLACE ECM.

CLEAR CODES AND CONFIRM "CLOSED LOOP" OPERATION AND NO "CHECK ENGINE" LIGHT

**BEFORE STARTING**

Before using this section you should have performed the DIAGNOSTIC CIRCUIT CHECK.

Verify the customer complaint, and locate the correct SYMPTOM below. Check the items indicated under that symptom.

If the ENGINE CRANKS BUT WILL NOT RUN, see CHART A-3.

Several of the following symptom procedures call for a careful visual (physical) check.

**The importance of this step cannot be stressed too strongly - it can lead to correcting a problem without further checks and can save valuable time.**

This check should include:

- Vacuum hoses for splits, kinks, and proper connections, as shown on vacuum hose routing diagram label.
- Air leaks at throttle body mounting and intake manifold.
- Ignition wires for cracking, hardness, proper routing, and carbon tracking.
- Wiring for proper connections, pinches, and cuts.

**INTERMITTENTS**

Problem may or may not turn "ON" the "Check Engine" light, or store a code.

DO NOT use the Trouble Code Charts in Section "A" for intermittent problems. The fault must be present to locate the problem. If a fault is intermittent, use of Trouble Code Charts may result in replacement of good parts.

- Most intermittent problems are caused by faulty electrical connections or wiring. Perform careful check of suspect circuits for:
  - Poor mating of the connector halves, or terminals not fully seated in the connector body (backed out).
  - Improperly formed or damaged terminals. All connector terminals in problem circuit should be carefully reformed to increase contact tension.
  - Poor terminal to wire connection.
- If a visual (physical) check does not find the cause of the problem, the car can be driven with a voltmeter connected to a suspected circuit, or a "Scan" tool may be used. An abnormal voltage reading when the problem occurs indicates the problem may be in that circuit. If the wiring and connectors check OK, and a Trouble Code was stored for a circuit having a sensor, except for Codes 44 and 45, substitute a known good sensor and recheck.
- Loss of trouble code memory. To check, disconnect TPS and idle engine until "Check Engine" light comes "ON". Code 22 should be stored, and kept in memory when ignition is turned "OFF" for at least 10 seconds. If not, the ECM is faulty.
- An intermittent "Check Engine" light and No Trouble Codes may be caused by:
  - Electrical system interference caused by a defective relay, ECM driven solenoid, or switch. They can cause a sharp electrical surge. Normally, the problem will occur when the faulty component is operated.
  - Improper installation of electrical options, such as lights, 2-way radios, etc.
  - EST wires should be routed away from spark plug wires, ignition system components, and generator. Wire for CKT 453 from ECM to ignition system should be a good ground.
  - Ignition secondary shorted to ground.
  - CKTs 419 ("Check Engine" light) and 451 (Diagnostic Test) intermittently shorted to ground.
  - ECM grounds.

## I-MARK NON-TURBO

### HARD START

Definition: Engine cranks OK, but does not start for a long time. Does eventually run, or may start but immediately dies.

- CHECK:
  - For water contaminated fuel.
  - Fuel system pressure CHART A-4.
  - TPS for sticking or binding should read less than 1.25 volts on a "Scan" tool.
  - EGR operation. Valve should be fully seated closed with engine "OFF". CHART C-6.
  - For a faulty in-tank fuel pump check valve which would allow the fuel in the lines to drain back to the tank after the engine is stopped. To check for this condition, perform steps in CHART A-5.
- Check ignition system for:
  - Proper Output with J-26792.
  - Worn shaft.
  - Bare and shorted wires.
  - Pickup coil resistance and connections.
  - Loose ignition coil connections.
  - Moisture in distributor cap.
  - Spark plugs, wet plugs, cracks, wear, improper gap, burned electrodes, or heavy deposits.

### SURGES AND/OR CHUGGLE

**Definition:**

Engine power variation under steady throttle or cruise. Feels like the car speeds up and slows down with no change in the accelerator pedal.

- Use a "Scan" tool to make sure reading of VSS matches vehicle speedometer.
- CHECK:
  - For intermittent EGR at idle. See CHART C-6.
  - Ignition timing. See Emission Control Information label.
  - Inline fuel filter for dirt or restriction.
  - Fuel pressure. See CHART A-4.
  - Generator output voltage. Repair, if less than 9 or more than 16 volts.
- Inspect Oxygen sensor for silicon contamination from fuel, or use of improper RTV sealant. The sensor may have a white, powdery coating and result in a high but false signal voltage (rich exhaust indication). The ECM will then reduce the amount of fuel delivered to the engine, causing a severe driveability problem.
- Remove spark plugs. Check for cracks, wear, improper gap, burned electrodes, or heavy deposits. Also check condition of the rest of the Ignition System.

### LACK OF POWER, SLUGGISH, OR SPONGY

**Definition:**

Engine delivers less than expected power. Little or no increase in speed when accelerator pedal is pushed down part way.

- Compare customer's car to similar unit. Make sure the customer's car has an actual problem.
- Remove air cleaner and check air filter for dirt, or for being plugged. Replace as necessary.
- CHECK:
  - Ignition timing. See Emission Control Information label.
  - Restricted air filter.
  - For restricted fuel filter, contaminated fuel or improper fuel pressure. See CHART A-4.
  - ECM Grounds.
  - EGR operation for being open or partly open all the time - CHART C-6.
  - Generator output voltage. Repair if less than 9 or more than 16 volts.
  - Engine valve timing and compression.
  - Engine for proper or worn camshaft.
  - Secondary ignition voltage using a scope or J-26792.
  - Secondary air control system. See CHART C-2C.
- Check Exhaust system for restriction using CHART B-1.

### DETONATION/SPARK KNOCK

**Definition:**

A mild to severe ping, usually worse under acceleration. The engine makes sharp metallic knocks that change with throttle opening.

- CHECK for obvious overheating problems.
  - Low coolant.
  - Loose water pump belt.
  - Restricted air flow to radiator, or restricted water flow thru radiator.
  - Faulty or incorrect thermostat.
  - Inoperative electric cooling fan circuit.
  - Correct coolant solution - should be a 50/50 mix of anti-freeze coolant (or equiv.) and water.
- CHECK:
  - For poor fuel quality, proper octane rating.
  - For correct Mem-Cal.
  - Spark plugs for correct heat range.
  - Ignition timing. See Vehicle Emission Control Information label.
  - Fuel system for low pressure. See CHART A-4.
  - Check EGR valve for not opening - CHART C-6.
  - For proper transmission shift points.
  - For incorrect basic engine parts such as cam, heads, pistons, etc.
  - Excessive oil entering combustion chamber.
- Remove carbon with top engine cleaner. Follow instructions on can.

## I-MARK NON-TURBO

### HESITATION, SAG, STUMBLE

**Definition:**

Momentary lack of response as the accelerator is pushed down. Can occur at all car speeds. Usually most severe when first trying to make the car move, as from a stop sign. May cause the engine to stall, if severe enough.

- Perform careful visual (physical) check
- CHECK:
  - Fuel pressure. See CHART A-5.
  - Water contaminated fuel.
  - TPS for binding or sticking.
  - Ignition timing. See Emission Control Information label.
  - MAP Sensor - CHART C-1A.
  - Generator output voltage. Repair if less than 9 or more than 16 volts.
  - For open Ignition System ground, CKT 453.
  - Canister purge system for proper operation.
  - EGR valve operation CHART C-6.
  - Perform Injector Balance test, CHART C-2A.

### CUTS OUT, MISSES

**Definition:**

Steady pulsation or jerking that follows engine speed, usually more pronounced as engine load increases. The exhaust has a steady spitting sound at idle or low speed.

- Perform careful visual (physical) check
- If Ignition System is suspected of causing a miss at idle or cutting out under load:
- Check for missing cylinder by:
  1. Start engine. Disconnect IAC valve. Remove one spark plug wire at a time using insulated pliers.
  2. If there is an rpm drop on all cylinders, (equal to within 50 rpm), go to ROUGH, UNSTABLE, OR INCORRECT IDLE, OR STALLING symptom. Stop Engine. Reconnect IAC valve.
  3. If there is no rpm drop on one or more cylinders, or excessive variation in drop, check for spark on the suspected cylinder(s) with J-26792 Spark tester or equivalent. If no spark, see Section "B" for Intermittent Operation or Miss. If there is spark, remove spark plug(s) in these cylinders and check for:
     - Cracks
     - Wear
     - Improper Gap
     - Burned Electrodes
     - Heavy Deposits
- Check spark plug wire resistance (should not exceed 30,000 ohms).
- Use a test light and check each injector control circuit. If OK, perform injector balance test, CHART C-2A.
- If the previous checks did not find the problem:
  - Visually inspect ignition system for moisture, dust, cracks, burns, etc. Spray plug wires with fine water mist to check for shorts.
  - Fuel System - Plugged fuel filter, water, low pressure. See CHART A-5.
  - Use "Scan" tool and check for erratic TPS voltage.
  - Check Injector harness connectors for intermittent connections.
  - Perform compression check.
  - Valve Timing.
  - Remove head covers. Check for broken or weak valve springs, worn camshaft lobes. Repair as necessary. See Section.

### POOR FUEL ECONOMY

**Definition:**

Fuel economy, as measured by an actual road test, is noticeably lower than expected. Also, economy is noticeably lower than it was on this car at one time, as previously shown by an actual road test.

- CHECK:
  - Engine thermostat for faulty part (always open) or for wrong heat range.
  - Fuel Pressure. See CHART A-5.
- Check owner's driving habits.
  - Is A/C "ON" full time?
  - Are tires at correct pressure?
  - Are excessively heavy loads being carried?
  - Is acceleration too much, too often?
  - Suggest driver read "Driving For Best Fuel Economy" in Owner's Manual.
- Check air cleaner element (filter) for dirt or being plugged.
- Check for proper calibration of speedometer.
- Visually (physically) Check:
  - Vacuum hoses for splits, kinks and proper connections as shown on vacuum hose routing diagram information label.
  - Ignition wires for cracking, hardness and proper connections.
- Check Ignition timing. See Emission Control Information label.
- Removes spark plugs. Check for cracks, wear, improper gap, burned electrodes or heavy deposits. Repair or replace, as necessary.
- Check compression.
- Check for dragging brakes.
- Suggest owner fill fuel tank and recheck fuel economy.
- Check for exhaust system restriction. See CHART B-1.

### ROUGH, UNSTABLE, OR INCORRECT IDLE, STALLING

**Definition:**

The engine runs unevenly at idle. If bad enough, the car may shake. Also, the idle may vary in RPM (called "hunting"). Either condition may be severe enough to cause stalling. Engine idles at incorrect speed.

- CHECK:
  - Ignition timing. See Emission Control Information label.
  - For injector balance, CHART C-2A.
  - IAC - See CHART C-2B.
  - If a sticking throttle shaft or binding linkage causes a high TPS voltage (open throttle indication) the ECM will not control idle. Monitor TPS voltage. "Scan" and/or Voltmeter should read less than 1.2 volts with throttle closed.
  - EGR "ON" while idling will cause roughness, stalling, and hard starting. CHART C-6.
  - Battery cables and ground straps should be clean and secure. Erratic voltage will cause IAC to change its position resulting in poor idle quality.
  - IAC valve will not move if system voltage is below 9 or greater than 17.8 volts.
  - Power Steering - CHART C-1B. ECM should compensate for Power Steering loads. Loss of this signal would be most noticable when parking and steering loads are high.
  - MAP Sensor - Ignition "ON", engine stopped. Compare MAP voltage with known good vehicle. Voltage should be the same ± 400 mv (.4 volts).

## I-MARK NON-TURBO

OR

Start and idle engine. Disconnect MAP sensor electrical connector. If idle improves, substitute a known good sensor and recheck.

- A/C compressor or relay. If inoperative, refer to CHART C-7.
- A/C Refrigerant Pressure too high. Check for overcharge or faulty pressure switch.
- Cooling fan inoperative.
- Canister Purge System
- PCV valve for proper operation by placing finger over inlet hole in valve end several times. Valve should snap back. If not replace valve.
- Run a cylinder compression check
- Inspect Oxygen sensor for silicon contamination from fuel, or use of improper RTV sealant. The sensor will have a white, powdery coating, and will result in a high but false signal voltage (rich exhaust indication). The ECM will then reduce the amount of fuel delivered to the engine, causing a severe driveability problem.
- Check for fuel in pressure regulator vacuum hose. If present, replace pressure regulator assembly.

### ABOVE NORMAL EMISSIONS(ODORS)

- If test shows higher than normal CO and HC, (also has excessive odors), check items that will cause engine to run rich:
- CHECK:
  - For high fuel pressure. See CHART A-5.
  - For incorrect timing. See Vehicle Emission Control Information Label.
  - Injector Balance, CHART C-2A.
  - Canister for fuel loading.
  - For stuck PCV valve or blocked PCV hose.
  - Condition of ignition system.
  - For lead contamination of catalytic converter (look for removal of fuel filler neck restrictor).

### DIESELING, RUN-ON

**Definition:**

Engine continues to run after key is turned "OFF", but runs very roughly. If engine runs smoothly, check ignition switch and adjustment.

- Check injector for leaking. Use CHART A-5 or perform Injector Balance test CHART C-2A.

### BACKFIRE

**Definition:**

Fuel ignites in intake manifold, or in exhaust system, making a loud popping noise.

- CHECK:
  - Ignition Timing, see Emission Control Information label on car.
  - EGR operation for being open all the time. See CHART C-6.
  - Output voltage of ignition coil.
  - For crossfire between spark plugs (distributor cap, spark plug wires, and proper routing of plug wires).
  - Engine timing - See Emission Control Information label.
  - For faulty spark plugs and/or plug wires or boots.
- Perform a compression check - look for sticking or leaking valves.
  - For proper valve timing.
  - Broken or worn valve train parts.

### CHART B-1

### RESTRICTED EXHAUST SYSTEM CHECK

Proper diagnosis for a restricted exhaust system is essential before any components are replaced. Either of the following procedures may be used for diagnosis, depending upon engine or tool used:

**Check at A.I.R. Pipe:**

1. Remove the rubber hose at the exhaust manifold A.I.R. pipe check valve. Remove check valve.
2. Connect a fuel pump pressure gauge to a hose and nipple from a Propane Enrichment Device (J26911) (See illustration).
3. Insert the nipple into the exhaust manifold A.I.R. pipe.

**OR**

**CHECK AT $O_2$ SENSOR:**

1. Carefully remove $O_2$ sensor.
2. Install Borroughs Exhaust Backpressure Tester (BT 8515 or BT 8603) or equivalent in place of $O_2$ sensor.
3. After completing test described below, be sure to coat threads of $O_2$ sensor with antiseize compound AC MS-8572 or equivalent prior to re-installation.

**DIAGNOSIS:**

1. With the engine idling at normal operating temperature, observe the exhaust system backpressure reading on the gauge. Reading should not exceed 1 1/4 psi (8.6 kPa).
2. Accelerate engine to 2000 rpm and observe gauge. Reading should not exceed 3 psi(20.7 kPa).
3. If the backpressure, at either rpm, exceeds specification, a restricted exhaust system is indicated.
4. Inspect the entire exhaust system for a collapsed pipe, heat distress, or possible internal muffler failure.
5. If there are no obvious reasons for the excessive backpressure, a restricted catalytic converter should be suspected and replaced using current recommended procedures.

## I-MARK NON-TURBO

### CHART C-1A

### MAP OUTPUT CHECK

**Circuit Description:**

The Manifold Absolute Pressure Sensor (MAP) measures manifold pressure (vacuum) and sends that signal to the ECM. The MAP Sensor is mainly used to calculate engine load, which is a fundamental input for spark and fuel calculations. The MAP Sensor is also used to determine the barometric pressure.

**Test Description:**

Numbers below refer to circled numbers on the diagnostic chart.

1. Checks MAP sensor output voltage to the ECM. This voltage, without engine running, represents a barometer reading to the ECM. Comparison of this BARO reading with a known good vehicle, with the same sensor, is a good way to check accuracy of a "suspect" sensor. Readings should be the same ± .4 volt.
2. Applying 34 kPa (10" Hg) vacuum to the MAP sensor should cause the voltage to be 1.2 volts less than the voltage at Step 1. Upon applying vacuum to the sensor, the change in voltage should be instantaneous. A slow voltage change indicates a faulty sensor.
3. Check vacuum hose to sensor for leaking or restriction. Be sure no other vacuum devices are connected to the MAP hose.

**NOTE:**

**The engine must be running in this step or the "Scanner" will not indicate a change in voltage. It is normal for the "Check Engine" light to come on and for the system to set a Code 33 during this step. Make sure the code is cleared when this test is completed.**

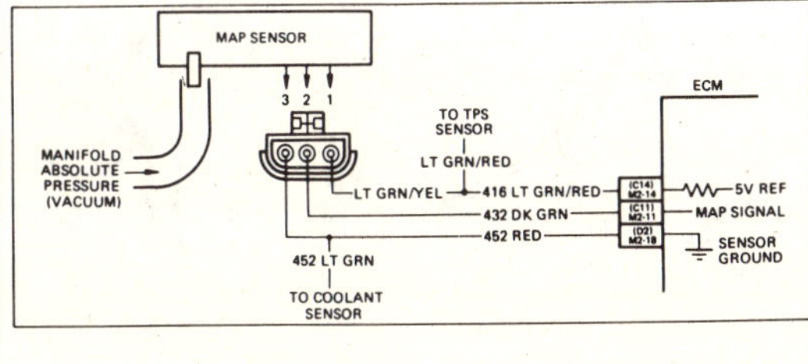


### CHART C-1A

### MAP OUTPUT CHECK

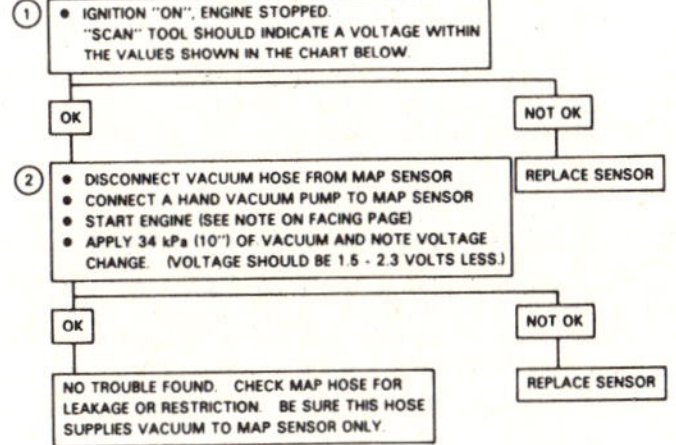


| ALTITUDE | | VOLTAGE RANGE |
|---|---|---|
| Meters | Feet | |
| Below 305 | Below 1,000 | 3.8 ~ 5.5V |
| 305 ~ 610 | 1,000 ~ 2,000 | 3.6 ~ 5.3V |
| 610 ~ 914 | 2,000 ~ 3,000 | 3.5 ~ 5.1V |
| 914 ~ 1219 | 3,000 ~ 4,000 | 3.3 ~ 5.0V |
| 1219 ~ 1524 | 4,000 ~ 5,000 | 3.2 ~ 4.8V |
| 1524 ~ 1829 | 5,000 ~ 6,000 | 3.0 ~ 4.6V |
| 1829 ~ 2133 | 6,000 ~ 7,000 | 2.9 ~ 4.5V |
| 2133 ~ 2438 | 7,000 ~ 8,000 | 2.8 ~ 4.3V |
| 2438 ~ 2743 | 8,000 ~ 9,000 | 2.6 ~ 4.2V |
| 2743 ~ 3048 | 9,000 ~ 10,000 | 2.5 ~ 4.0V |
| LOW ALTITUDE=HIGH PRESSURE=HIGH VOLTAGE | | |

CLEAR CODES AND CONFIRM "CLOSED LOOP" OPERATION AND NO "CHECK ENGINE" LIGHT.

## I-MARK NON-TURBO

### CHART C-1B

#### POWER STEERING PRESSURE SWITCH (PSPS) DIAGNOSIS

**Circuit Description:**

The power steering pressure switch is normally open to ground, and CKT 901 will be near the battery voltage. Turning the steering wheel increases power steering oil pressure and its load on an idling engine. The pressure switch will close before the load can cause an idle problem.

Closing the switch causes CKT 901 to read less than 1 volt. The ECM will increase the idle air rate and disengage the A/C relay.

- A pressure switch that will not close, or an open CKT 901 or 450, may cause the engine to stop when power steering loads are high.
- A switch that will not open, or a CKT 901 shorted to ground, may affect idle quality and will cause the A/C relay to be de-energized.

**Test Description:**

Numbers below refer to circled numbers on the diagnostic chart.

1. Different makes of "Scan" tools may display the state of this switch in different ways. Refer to "Scan" tool operator's manual to determine how this input is indicated.
2. Checks to determine if CKT 901 is shorted to ground.
3. This should simulate a closed switch.

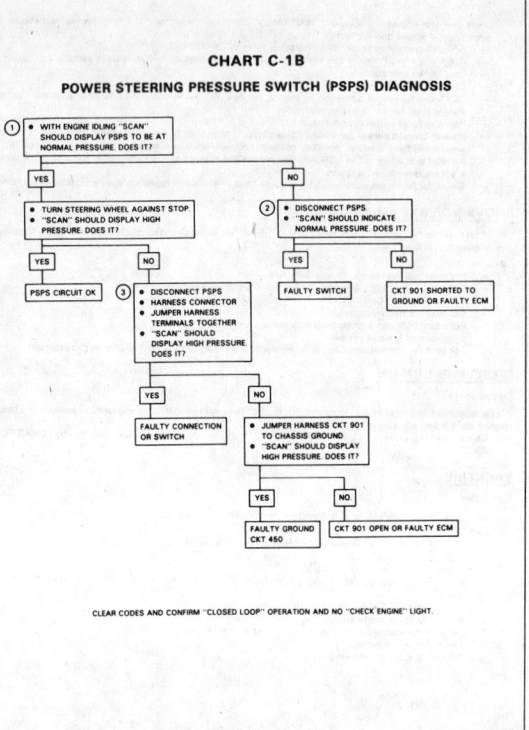

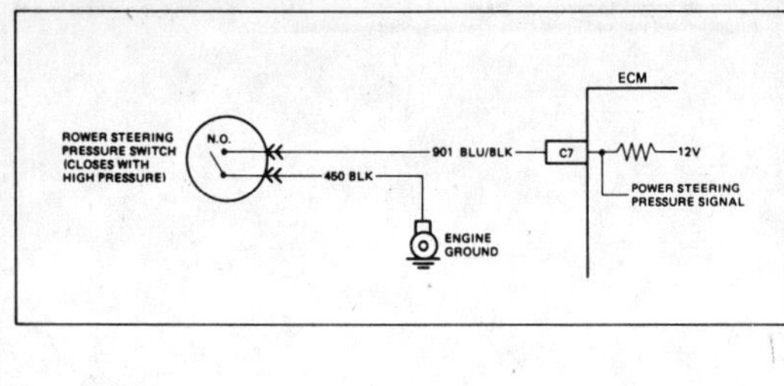


### CHART C-1B

#### POWER STEERING PRESSURE SWITCH (PSPS) DIAGNOSIS

(1)
- WITH ENGINE IDLING "SCAN" SHOULD DISPLAY PSPS TO BE AT NORMAL PRESSURE. DOES IT?

YES
- TURN STEERING WHEEL AGAINST STOP.
- "SCAN" SHOULD DISPLAY HIGH PRESSURE. DOES IT?

YES → PSPS CIRCUIT OK

NO → (3)
- DISCONNECT PSPS
- HARNESS CONNECTOR
- JUMPER HARNESS TERMINALS TOGETHER
- "SCAN" SHOULD DISPLAY HIGH PRESSURE. DOES IT?

YES → FAULTY CONNECTION OR SWITCH

NO
- JUMPER HARNESS CKT 901 TO CHASSIS GROUND
- "SCAN" SHOULD DISPLAY HIGH PRESSURE. DOES IT?

YES → FAULTY GROUND CKT 450

NO. → CKT 901 OPEN OR FAULTY ECM

NO → (2)
- DISCONNECT PSPS.
- "SCAN" SHOULD INDICATE NORMAL PRESSURE. DOES IT?

YES → FAULTY SWITCH

NO → CKT 901 SHORTED TO GROUND OR FAULTY ECM

CLEAR CODES AND CONFIRM "CLOSED LOOP" OPERATION AND NO "CHECK ENGINE" LIGHT.

## I-MARK NON-TURBO

### CHART C-2A

#### INJECTOR BALANCE TEST

The injector tester is a timer used to turn each injector on for a precise amount of time. This time allows a measured amount of fuel to be sprayed into the intake manifold thereby reducing the pressure in the fuel rail. All injectors in the engine should measure about the same pressure drop (± 10kPa).

**STEP 1**

Connect fuel gage J-347301 or equivalent to fuel pressure tap. Wrap a shop towel around fitting while connecting gage to avoid fuel spillage.

Disconnect harness connectors at all injectors, and connect injector tester J-34730-3 or equivalent to one injector. Ignition must be off at least 10 seconds to complete ECM shutdown cycle. Fuel pump should run about 2 seconds after ignition is turned "ON". At this point, insert clear tubing attached to vent valve into a suitable container and bleed air from gage and hose to insure accurate gage operation.

**STEP 2**

Turn ignition "OFF" for 10 seconds and then on again to get fuel pressure to its maximum. This insures that fuel pressure is precisely the same for each injector tested. Energize tester one time and note pressure drop the instant the gage needle stops. The pressure may increase for a few seconds after the initial pressure drop. This increase should not be considered in the test, because it may vary depending on temperature.

**Note:** The entire test should not be repeated more than once to prevent flooding.

**STEP 3**

This example shows how faulty injectors would appear, as compared to good ones. Usually, good injectors will have virtually the same drop. Retest any injector that has a pressure difference of 10 kPa, either more or less than the average of the other injectors on the engine. Replace any injector that also fails the retest. If the pressure drop of all injectors is within 10 kPa of this average, the injectors appear to be flowing properly.

### CHART C-2A

#### INJECTOR BALANCE TEST

Before performing test, perform the fuel pressure test in Section A.

Step 1. Connect fuel pressure gage and injector tester.

1. Ignition "OFF".
2. Connect fuel pressure gage and injector tester.
3. Ignition "ON".
4. Bleed off air in gage.

Step 2. Connect fuel pressure gage and injector tester.

1. Ignition "OFF" for 10 sec.
2. Ignition "ON".
3. Turn injector on by depressing button on injector tester and note pressure at the instant the gage needle stops.

Step 3.

1. Repeat step 2 on all injectors and record pressure drop on each.
   Retest injectors that appear faulty. Replace any injectors that have a 10 kPa difference (either more or less) in pressure.

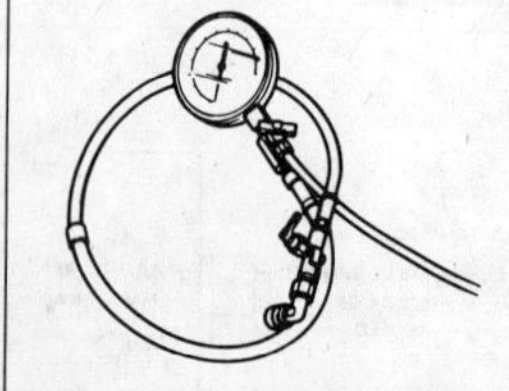

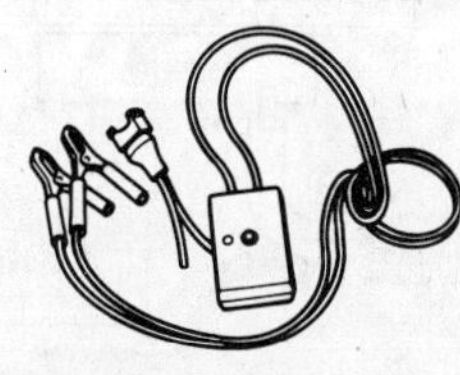

## I-MARK NON-TURBO

### CHART C-2B

#### IDLE AIR CONTROL (IAC) VALVE CHECK

**Circuit Description:**

The ECM controls idle rpm with the IAC valve. To increase idle rpm, the ECM moves the IAC valve out, allowing more air to pass by the throttle plate. To decrease rpm, it moves the IAC valve in, reducing air flow by the throttle plate. A "Scan" tool will read the ECM commands to the IAC valve in counts. The higher the counts, the more air allowed (higher idle). The lower the counts, the less air allowed (lower idle).

**Test Description:**

Numbers below refer to circled numbers on the diagnostic chart.

1. Continue with test, even if engine will not idle. If idle is too low, "Scan" will display 80 or more counts, or steps. If idle is high, it will display "0" counts. Occasionally, an erratic or unstable idle may occur. Engine speed may vary 200 rpm, or more, up and down. Disconnect IAC. If the condition is unchanged, the IAC is not at fault.
2. When the engine was stopped, the IAC Valve retracted (more air) to a fixed "Part" position for increased air flow and idle speed during the next engine start. A "Scan" will display 80 or more counts. When performing this test, immediately note rpm on start up, because, on a warm engine, the rpm will decrease rapidly.
3. Be sure to disconnect the IAC valve prior to this test. The test light will confirm the ECM signals by a steady or flashing light on all circuits.
4. There is a remote possibility that one of the circuits is shorted to voltage, which would have been indicated by a steady light. Disconnect ECM and turn the ignition "ON" and probe terminals to check for this condition.

**Diagnostic Aids:**

A slow unstable idle may be caused by a system problem that cannot be overcome by the IAC. "Scan" counts will be above 60 counts, if too low, and "0" counts, if too high.
If idle is too high, stop engine. Remove hose from IAC and plug IAC valve side. Start engine. If idle speed is above 450 rpm in drive, locate and correct vacuum leak. If rpm is less than 450 rpm, adjust minimum idle speed, or correct other conditions, which may affect idle. Refer to "Rough, Unstable, Incorrect Idle or Stalling

- System too lean (High Air/Fuel Ratio)
  Idle speed may be too high or too low. Engine speed may vary up and down, disconnecting IAC does not help. May set Code 44.
  "Scan" and/or Voltmeter will read an oxygen sensor output less than 300 mv (.3 volts). Check for low regulated fuel pressure or water in fuel. A lean exhaust, with an oxygen sensor output fixed above 800 mv (.8 volts), will be a contaminated sensor, usually silicone. This may also set a Code 45.
- System too rich (Low Air/Fuel Ratio)
  Idle speed too low. "Scan" counts usually above 80. System obviously rich and may exhibit black smoke exhaust.
  "Scan" tool and/or Voltmeter will read an oxygen sensor signal fixed above 800 mv (.8 volts).
  Check:
  - High fuel pressure
  - Injector leaking or sticking
- Throttle Body - Remove IAC and inspect bore for foreign material or evidence of IAC valve dragging in the bore.
- If above are all OK, refer to "Rough, Unstable, Incorrect Idle or Stalling"

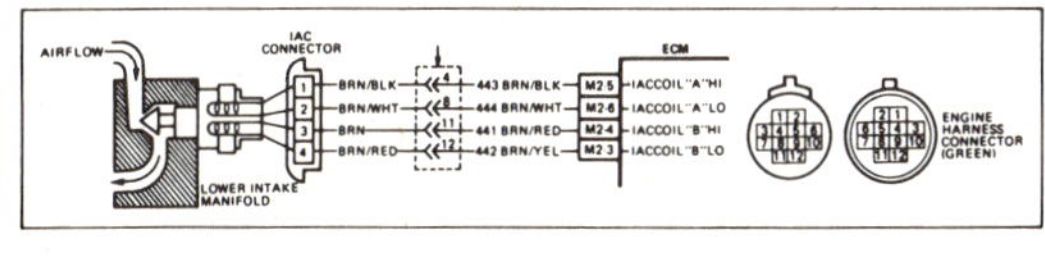


### CHART C-2B

#### IDLE AIR CONTROL (IAC) VALVE CHECK

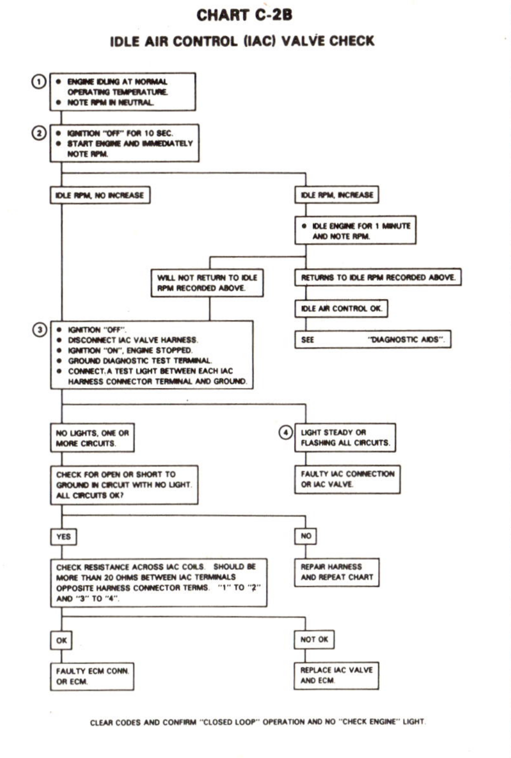


## I-MARK NON-TURBO

### CHART C-2C

#### SECONDARY AIR SYSTEM CHECK

**Circuit Description:**

The secondary air control system is used to control air to the secondary intake valves when engine RPM is above about 3500. Using this system allows for reduced intake air at idle to improve idle quality and then opens at higher RPM to improve engine performance and increase horsepower.

This system consists of the secondary port intake valves mounted in the lower intake manifold and a vacuum operated diaphragm which moves the secondary valve linkage. The vacuum to the diaphragm is turned "ON" or "OFF" by a vacuum switching solenoid which is electrically operated by the electronic control module (ECM).

**Test Description:**

Numbers below refer to circled numbers on the diagnostic chart.

1. Without the proper vacuum being available at the solenoid the system will not function properly.
2. When the ignition is "ON" and the engine not running the solenoid should not be energized.
   Since the solenoid is normally closed the vacuum should not be able to pass to the actuator.
3. With the diagnostic terminal grounded the ECM should ground circuit 476 which should energize the solenoid and allow vacuum to pass to the actuator.
4. This step checks the ability of the vacuum actuator to be able to move the secondary intake valve linkage.
5. This step as well as the following steps checks for circuit continuity.

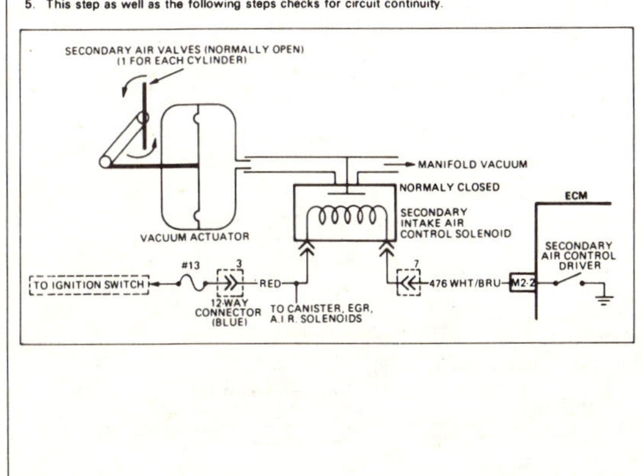


### CHART C-2C

#### SECONDARY AIR SYSTEM CHECK

① • DISCONNECT MANIFOLD VACUUM HOSE FROM SECONDARY INTAKE AIR CONTROL SOLENOID.
• CONNECT VACUUM GAGE TO HOSE.
• WITH ENGINE IDLING, VACUUM SHOULD BE AT LEAST 10" VACUUM.

OK / NOT OK

NOT OK → REPAIR VACUUM SOURCE

OK → ② • CONNECT VACUUM PUMP WITH GAGE TO ENGINE MANIFOLD SIDE OF SOLENOID.
• IGNITION "ON", ENGINE "OFF".
• APPLY VACUUM AND OBSERVE VACUUM ACTUATOR LINKAGE.
• LINKAGE SHOULD NOT MOVE. DOES IT?

NO / YES

YES → • DISCONNECT SOLENOID ELECTRICAL CONNECTOR.
• REPEAT TEST.
• LINKAGE SHOULD NOT MOVE. DOES IT?

YES → CIRCUIT 476 SHORTED TO GROUND OR FAULTY ECM.

NO → REPLACE SOLENOID

NO → ③ • GROUND DIAGNOSTIC TERMINAL (JUMPER ALDL CONNECTOR 1 TO 3).
• APPLY VACUUM AGAIN WHILE OBSERVING LINKAGE.
• LINKAGE SHOULD MOVE. DOES IT?

NO / YES

YES → NO PROBLEM FOUND. SECONDARY INTAKE AIR SYSTEM OK.

NO → ④ • CONNECT VACUUM PUMP TO ACTUATOR HOSE OR DIRECTLY TO ACTUATOR.
• WITH VACUUM APPLIED THE ACTUATOR LINKAGE SHOULD MOVE. DOES IT?

NO / YES

NO → HOSE LEAKING, LINKAGE BINDING FAULTY ACTUATOR.

YES → ⑤ • DISCONNECT SOLENOID ELECTRICAL CONNECTOR.
• IGNITION "ON", ENGINE "OFF".
• DIAGNOSTIC TERMINAL GROUNDED.
• CONNECT A TEST LIGHT ACROSS HARNESS TERMINALS.

LIGHT "ON" / LIGHT "OFF"

LIGHT "ON" → FAULTY SOLENOID

LIGHT "OFF" → PROBE IGNITION FEED CIRCUIT WTH A TEST LIGHT CONNECTED TO GROUND.

LIGHT "ON" → CIRCUIT 476 IS OPEN, HAS A FAULTY CONNECTION AT ECM OR ECM IS FAULTY.

LIGHT "OFF" → REPAIR OPEN IN 12 VOLT IGNITION CIRCUIT.

CLEAR CODES AND CONFIRM "CLOSED LOOP" OPERATION AND NO "CHECK ENGINE" LIGHT.

## I-MARK NON-TURBO

### CHART C-3

#### CANISTER PURGE VALVE CHECK

**Circuit Description:**

Canister purge is controlled by a solenoid that allows manifold vacuum to purge the canister when energized. The ECM supplies a ground to energize the solenoid (purge "ON").

If the diagnostic test terminal is ungrounded with the engine stopped, or when the following conditions are met, the purge solenoid is energized (purge "ON").

- Engine run time after start more than 1 minute.
- Coolant temperature above 75°C.
- Vehicle speed above 2 mph.
- Throttle off idle.

**Test Description:**

Numbers below refer to circled numbers on the diagnostic chart.

1. Checks to see if the solenoid is opened or closed. The solenoid is normally closed in this step, so the valve should hold vacuum.
2. Checks for a complete circuit and checks the ECM for the ability to energize the solenoid. Normally, there is ignition voltage on CKT 39 and the ECM provides a ground on CKT 428.
3. Completes functional check by grounding test terminal. This should normally energize the solenoid and allow the vacuum to drop (purge "ON").

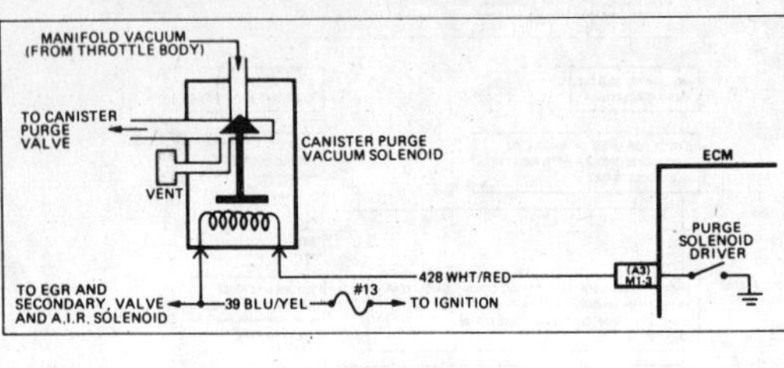


### CHART C-3

#### CANISTER PURGE VALVE CHECK

NOTE: THIS CHART ONLY COVERS THE SOLENOID PORTION OF THE CANISTER CONTROL PURGE SYSTEM. TO TEST THE CONTROL VALVE(S) SEE DIAGNOSIS UNDER GENERAL DESCRIPTION. ALSO BE SURE ALL HOSES ARE OK.

① 
- IGNITION "ON", ENGINE STOPPED.
- DISCONNECT VACUUM HOSES FROM PURGE SOLENOID.
- AT THE SOLENOID, APPLY VACUUM(10" Hg OR 34 kPa) TO THROTTLE BODY SIDE OF PURGE SOLENOID.

ABLE TO GET 10" VACUUM

② 
- GROUND DIAGNOSTIC TERMINAL
- NOTE VACUUM

NO DROP → DISCONNECT SOLENOID

③ 
- PROBE EACH HARNESS CONNECTOR TERMINAL WITH A TEST LIGHT TO GROUND

LIGHT ON ONE TERMIVAL → CONNECT A TEST LIGHT BETWEEN HARNESS TERMINALS "A" AND "B"

NO LIGHT → OPEN CKT 428, FAULTY CONNECTION AT ECM OR FAULTY ECM

LIGHT → FAULTY SOLENOID

NO LIGHT → REPAIR OPEN IGNITION CIRCUIT

LIGHT BOTH TERMINALS → REPAIR SHORT TO VOLTAGE IN CKT 428 → REPEAT TEST USING ORIGINAL ECM

DROPS → NO TROUBLE FOUND

UNABLE TO GET 10" VACUUM

- DISCONNECT ELECTRICAL CONNECTOR FROM PURGE SOLENOID
- CONNECT TEST LIGHT BETWEEN HARNESS CONNECTOR TERMENAL

NO LIGHT → FAULTY SOLENOID

LIGHT → CHECK CKT 428 IS NOT GROUND. IF CKT 428 IS NOT SHORTED TO GROUND, IT IS A FAULTY ECM.

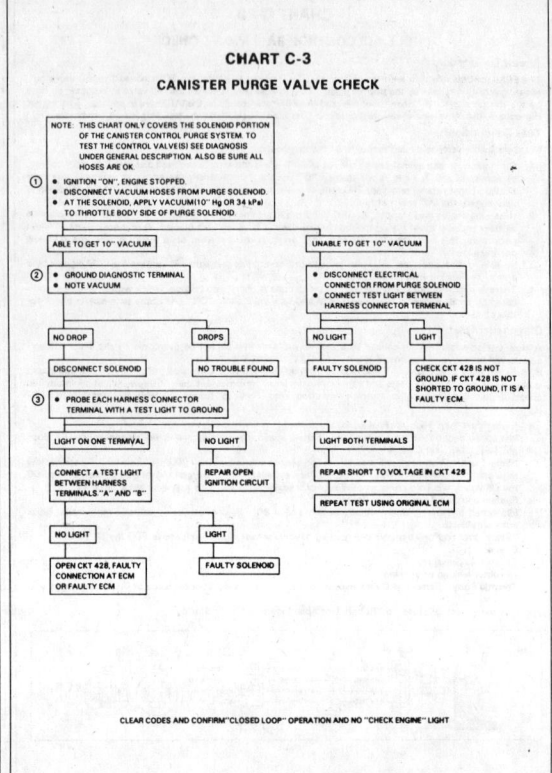

CLEAR CODES AND CONFIRM "CLOSED LOOP" OPERATION AND NO "CHECK ENGINE" LIGHT

## I-MARK NON-TURBO

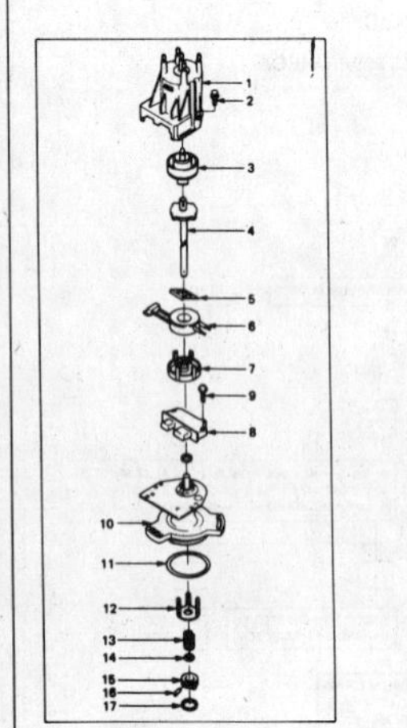

| DISTRIBUTOR PARTS LIST | |
|---|---|
| 1 | DISTRIBUTOR CAP |
| 2 | SCREW ASM-DISTRIBUTOR CAP RETAINING |
| 3 | ROTOR |
| 4 | SHAFT ASSEMBLY |
| 5 | RETAINER-PICK UP COIL |
| 6 | COIL-PICK UP |
| 7 | POLE PIECE-STATIONARY |
| 8 | MODULE |
| 9 | SCREW ASM-MODULE RETAINING |
| 10 | HOUSING |
| 10a | PIN-POLE PIECE LOCATING (not shown separately) |
| 10b | SEAL-LUBRICATION (shown but not numbered seperately) |
| 11 | "O" RING (on housing) |
| 12 | WASHER-TANG (between housing and thrust spring) |
| 13 | SPRING-THRUST |
| 14 | WASHER-FLAT (between thrust spring and drive coupling) |
| 15 | COUPLING-DRIVE |
| 16 | PIN |
| 17 | SPRING-PIN RETAINING (around coupling) |

### CHART C-4A

#### IGNITION SYSTEM CHECK

**Test Description:**

Numbers below refer to circled numbers on the diagnostic chart.

1. Two wires are checked, to ensure that an open is not present in a spark plug wire.

1A. If spark occurs with 4 terminal distributor connector disconnected, pick-up coil output is too low for EST operation.

2. A spark indicates the problem must be the distributor cap or rotor.
3. Normally, there should be battery voltage at the "C" and "+" terminals. Low voltage would indicate an open or a high resistance circuit from the distributor to the coil or ignition switch. If "C" term. voltage was low, but "+" term. voltage is 10 volts or more, circuit from "C" term. to Ign. coil or ignition coil primary winding is open.
4. Checks for a shorted module or grounded circuit from the ignition coil to the module. The dist. module should be turned "OFF", so normal voltage should be about 12 volts.

   If the module is turned "ON", the voltage would be low, but above 1 volt. This could cause the ign, coil to fail from excessive heat.

   With an open ignition coil primary winding, a small amount of voltage will leak through the module from the "BAT." to the tach terminal.
5. Applying a voltage (1.5 to 8V) to module terminal "P" should turn the module "ON" and the tach. term. voltage should drop to about 7-9 volts. This test will determine whether the module or coil is faulty or if the pick-up coil is not generating the proper signal to turn the module "ON". This test can be performed by using a DC battery with a rating of 1.5 to 8 volts. The use of the test light is mainly to allow the "P" terminal to be probed more easily.

   Some digital multi-meters can also be used to trigger the module by selecting ohms, usually the diode position. In this position the meter may have a voltage across its terminals which can be used to trigger the module. The voltage in the ohms position can be checked by using a second meter or by checking the manufactures specification of the tool being used.
6. This should turn "OFF" the module and cause a spark. If no spark occurs, the fault is most likely in the ignition coil because most module problems would have been found before this point in the procedure. A module tester (J24642) could determine which is at fault.

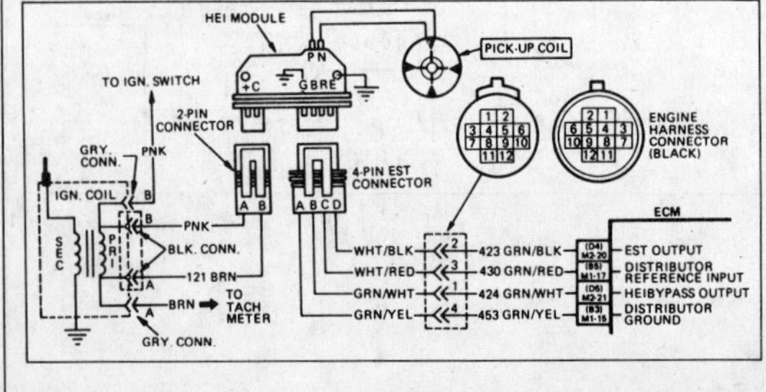

## I-MARK NON-TURBO

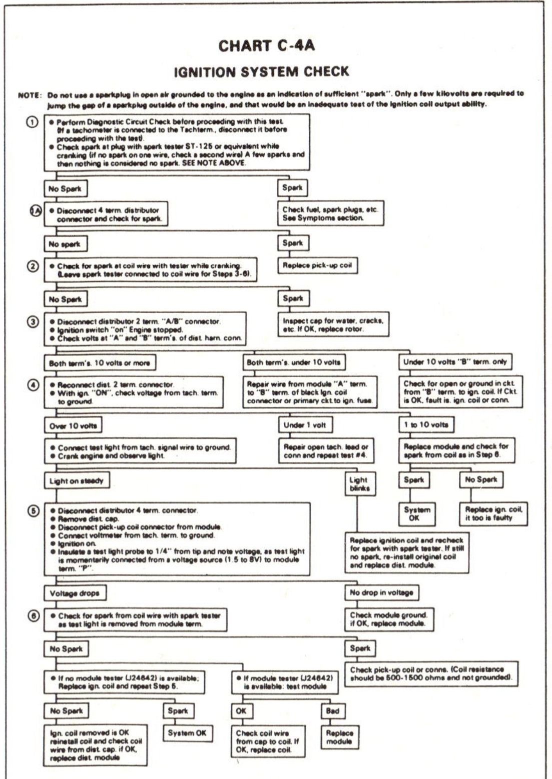


### CHART C-5A

### AIR INJECTION REACTON (AIR) SYSTEM CHECK

**Circuit Description:**

When the engine is first started and coolant temperature is below about 50°C (122°F) the ECM will energize the puls air solenoid valve by grounding CKT 436. With the solenoid energized, vacuum is allowed to be applied to the reed valve diaphragm. When the vacuum is applied to the reed valve diaphragm the valve opens which allows the exhaust pulses to pull inlet air through the reed valve and into the exhaust stream. The check valve restricts the positive pressure exhaust pulses from getting into the inlet duct.

**Test Description:**

Numbers below refer to circled numbers on the diagnostic chart.

1. The reed valve should close off the path for air to flow in this step.
2. This step checks to make sure that the solenoid valve is not leaking. Also during this test the solenoid should not be energized and vacuum should not be being applied to the reed valve.
3. If air can be blown through the hose during this test it may be due to a faulty reed valve, faulty solenoid, or a solenoid that is being energized. The solenoid should not be energized during this step.
4. With the ignition "ON" and the diagnostic terminal grounded, the ECM should energize the puls air solenoid valve and allow vacuum to be applied to the reed valve diaphragm.

**Diagnostic Aids:**

When the puls air system is active, noise is generally heard at the system resonator, but will quiet down when de-activated. The amount of time that the system is activated is dependent on the coolant temperature and the ECM command can be monitored with a "Scan" tool. On a warmed up engine the AIR will be directed to the exhaust ports for a minimum of 5 seconds. The noise from the resonator can be monitored at the same time as the "Scan" tool. If exhaust deposits are present in the inlet duct side of the system it is an indication that the check valve in the reed valve is leaking and should be replaced. This condition may result in an unstable idle condition.

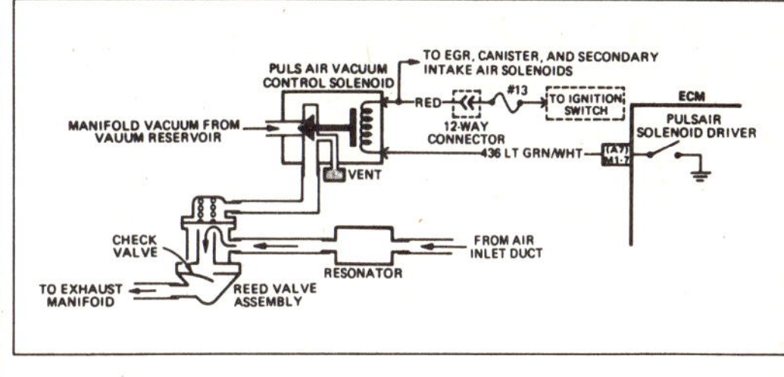


## I-MARK NON-TURBO

### CHART C-5A

### AIR INJECTION REACTON (AIR) SYSTEM CHECK

(1) VISUALLY INSPECT THE SYSTEM HOSES, LINES AND CONNECTONS FOR LOOSENESS, CRACKS, KINKS OR ANY OTHER DAMAGE.
- IGNITION "OFF"
- REMOVE PULS AIR HOSE FROM AIR CLEANER ASSEMBLY
- SHOULD NOT BE ABLE TO BLOW THROUGH HOSE

OK → (2)

NOT OK → CHECK HOSES FOR LEAKS. IF OK, REPLACE REED VALVE.

(2)
- IGNITION "ON", ENGINE "OFF"
- REMOVE VACUUM HOSE FROM MANIFOLD SIDE OF THE PULS AIR SOLENOID.
- USING A HAND HELD VACUUM PUMP, APPLY 10" VACUUM TO SOLENOID.
- VACUUM SHOULD HOLD AND NOT LEAK DOWN. DOES SOLENOID LEAK?

NO → (3)

YES → REPLACE PULS AIR SOLENOID

(3)
- WITH VACUUM STILL APPLIED AGAIN TRY BLOWING THROUGH HOSE
- SHOULD NOT BE ABLE TO BLOW THROUGH HOSE

OK → (4)

NOT OK →
- DISCONNECT SOLENOID ELECTRICAL CONNECTOR
- IGNITION "ON", ENGINE "OFF".
- PROBE CKT436 WITH A TEST LIGHT CONNECTED TO 12 VOLTS.
- LIGHT SHOULD NOT BE "ON". IS IT?

  - YES → CKT436 SHORTED TO GROUND OR FAULTY ECM
  - NO → FAULTY VACUUM CONTROL SOLENOID

(4)
- IGNITION "ON", ENGINE "OFF"
- GROUND DIAGNOSTIC TERMINAL
- APPLY 10" VACUUM TO PULS AIR SOLENOID
- SHOULD NOW BE ABLE TO BLOW THROUGH HOSE

OK → SHOULD NOT BE ABLE TO SUCK THROUGH HOSE
- OK → CHECK FOR ENGINE VACUUM PRESENT TO SOLENOID WITH ENGINE RUNNING. IF OK: AIR SYSTEM IS OK.
- NOT OK → FAULTY CHECK VALVE, REPLACE REED VALVE.

NOT OK →
- DISCONNECT SOLENOID ELECTRICAL CONNECTOR
- CONNECT TEST LIGHT BETWEEN HARNESS TERMINALS

LIGHT "ON" →
- APPLY 10" VACUUM DIRECTLY TO REED VALVE.
- CAN AIR NOW BE BLOWN THROUGH PULS AIR HOSE THAT WAS CONNECTED TO AIR CLEANER ASSEMBLY?

  - YES → RESTRICTION IN VACUUM HOSE OR FAULTY PULS AIR SOLENOID
  - NO → RESTRICTION IN AIR HOSE OR FAULTY REED VALVE

LIGHT "OFF" → CONNECT TEST LIGHT FROM EACH HARNESS TERMINAL AND GROUND AND NOTE LIGHT
- LIGHT "ON" BOTH → REPAIR SHORT TO VOLTAGE ON CKT436
- LIGHT "ON" ONE → OPEN IN CKT436 OR FAULTY ECM
- NOT LIGHT → REPAIR OPEN IN CKT BETWEEN #13 FUSE AND PULS AIR SOLENOID

### CHART C-6

### EGR SYSTEM CHECK

**Circuit Description:**

The EGR valve is controlled by a normally closed solenoid (allows a vacuum to pass when energized). The ECM energizes the solenoid to turn the EGR on, and de-energizes the solenoid to turn EGR off. EGR is commanded "ON" by the ECM when the engine temperature is above 30°C, TPS signal not indicating idle, RPM greater than about 1000, and MAP indicating the engine is under a load.

**Test Description:**

Numbers below refer to circled numbers on the diagnostic chart.

1. With the ignition "ON" engine stopped, the solenoid should not be energized and vacuum should not pass to the EGR valve.
2. Grounding the diagnostic terminal will energize the solenoid and allow vacuum to pass to valve.
3. When the EGR solenoid is turned off the vacuum in the EGR valve should bleed off through the vent.
4. Checks for plugged EGR passages. If passages are plugged, the engine may have severe detonation on acceleration.

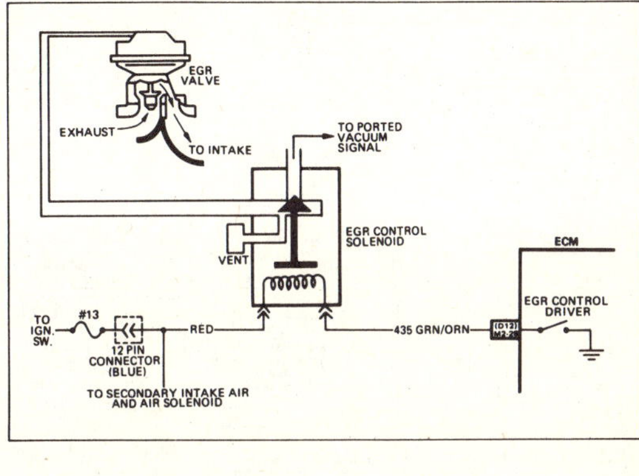

## I-MARK NON-TURBO

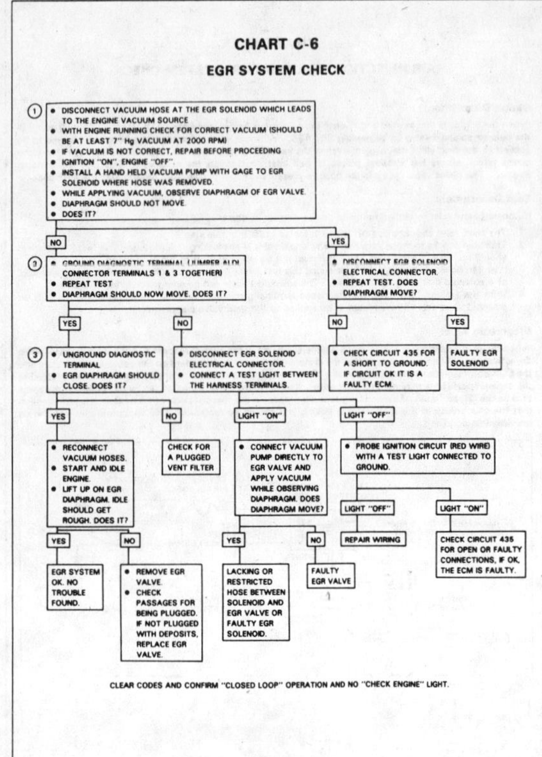


### CHART C-7
### A/C CLUTCH CONTROL

**Circuit Description:**

ECM control of the A/C clutch improves idle quality and performance by:

- delaying clutch apply until the idle air rate is increased.
- releasing clutch when idle speed is too low or during high power steering loads.
- releasing clutch at wide open throttle.
- smooths cycling of the compressor by providing additional fuel at the instant clutch is applied.

Voltage is supplied to the A/C Clutch Control relay on CKT 67 by the A/C Control Switch through a low pressure evaporator switch and the pressure cycling switch. This same voltage is suppplied as a signal to ECM pin B8. After a time delay of about 1/2 second the ECM will ground terminal A2, CKT 458, and close the A/C relay contacts.

When relay is energized, battery voltage from CKT 67 is supplied to the A/C clutch through the relay and CKT 59.

**Test Description:**

Numbers below refer to circled numbers on the diagnostic chart.

1. The ECM will only energize the A/C relay, when the engine is running. This test will determine if the relay, or CKT 458, is faulty.
2. In order for the clutch to properly be engaged, the pressure cycling switch must be closed to provide 12 volts to the relay, and the evaporator switch must be closed, so the A/C request (12 volts) will be present at the ECM.
3. Determines if the signal is reaching the ECM on CKT 459 from the A/C control panel. Signal should only be present when the A/C mode or defrost mode has been selected.
4. A short to ground in any part of the A/C request circuit, CKT 67 to the relay CKT59, CKT 459 to the A/C clutch, or the A/C clutch, could be the cause of the blown fuse.
5. With the ignition "ON" and the diagnostic terminal grounded, the ECM should be grounding CKT 458 which should cause the test light to be "ON".

**Diagnostic Aids:**

If complaint was insufficient cooling, the problem may be caused by an inoperative cooling fan. The engine cooling fan should turn "ON", when A/C is "ON".

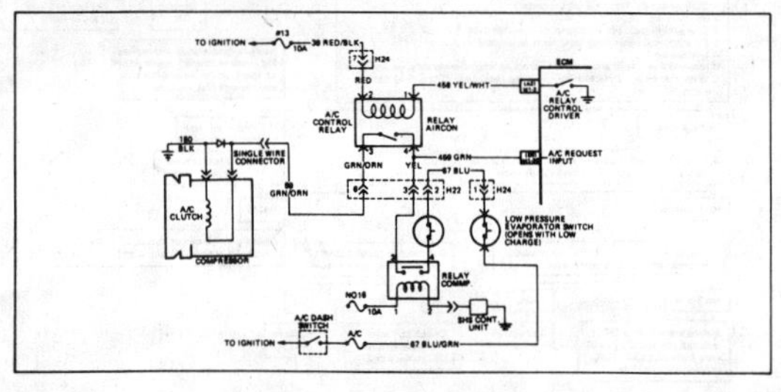

## I-MARK NON-TURBO

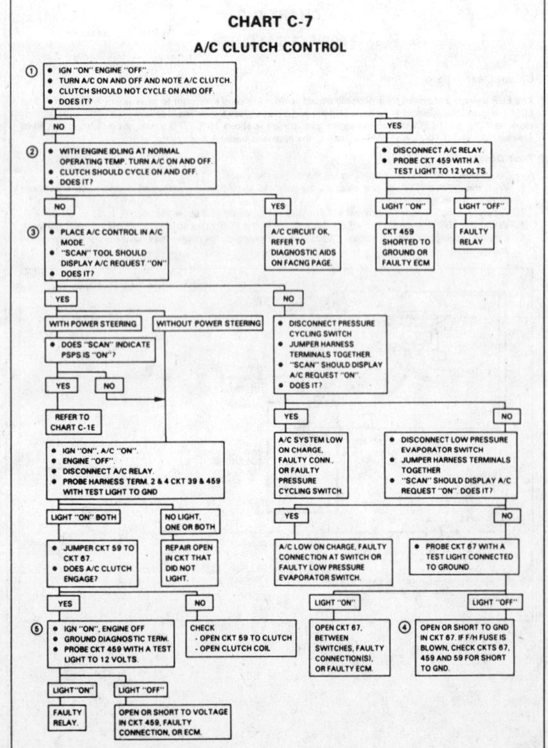


### "NON-SCAN" DIAGNOSTIC CIRCUIT CHECK

**Circuit Description:**

The Diagnostic Circuit Check is an organized approach for identifying a problem caused by the Fuel Injection System.

Driver comments normally fall into one of the following areas:

- Steady "Check Engine" light
- Driveability Problem
- Engine "Cranks But Will Not Run"

Understanding the chart and using it correctly will reduce diagnosis time and prevent the unnecessary replacement of parts.

**Test Description:**

Numbers below refer to circled numbers on the diagnostic chart.

1. A steady "Check Engine" light with the ignition "ON" and engine stopped confirms battery and ignition voltage to the Electronic Control Module (ECM).
2. Ground diagnosis terminal by jumpering terminal "1" to "3" in the ALDL connector located behind the passenger kick pad.
   The ECM will cause the "Check Engine" light to flash Code 12, indicating that the ECM diagnostics are working. Code 12 will flash three (3) times, followed by any other trouble codes stored in the memory. Each additional code will flash three (3) times, starting with the lowest code, and then start over again with Code 12. If there are no other codes, Code 12 will flash until the diagnostic "test" terminal jumper is disconnected or the engine is started.
3. Record all stored codes except for Code 12. If the problem is "Engine Cranks But Will Not Run", go to CHART A-3.
4. If no additional codes were recorded, see Section B for driveability symptoms and recommended service procedures. Depending on the severity of the problem, the "Field Service Mode" may be helpful in diagnosis.
   With the engine running and the diagnostic terminal grounded, the ECM will respond to the oxygen sensor signal voltage and use the "Check Engine" light to display this information as follows:
   A. "Closed loop" confirms that the oxygen sensor signal is being used by the ECM to control fuel delivery and that the system is working normally. Signal voltage will swing quickly from below .35 to above .55 volts. Light flashes at a rate of 1 per second, confirming "closed loop" operation.
   B. "Open loop" indicates that oxygen sensor voltage signal is not usable to the ECM. Signal voltage is at a constant value between .35 and .55 volts. Light flashes at a rate of 2.5 per second indicating "open loop" operation. System will flash "open loop" from 30 seconds to 2 minutes after engine starts or until sensor reaches normal operating temperature. If system fails to go "closed loop", see Code 13 chart.
   C. "Check Engine" light "OFF" indicates that exhaust is lean. $O_2$ sensor signal voltage will be less than .35 volts and steady. See Code 44 chart.
   D. "Check Engine" light "ON" steady indicates that exhaust is rich. Sensor signal voltage will be above .55 volts and steady. See Code 45 chart.
5. Road test of the system using the "Field Service Mode" should be done only at steady road speeds. Because the vehicle operates differently in the "Field Service Mode", the following conditions may be observed and should be considered normal.

- Acceleration - Light may be "ON" too long due to acceleration enrichment.
- Deceleration - Light may be "OFF" too long due to decel enleanment or fuel cut-off.
- Idle - Light may be "ON" too long with idle below 1200 rpm.

6. Clearing codes. Ignition "OFF". Disconnect battery for ten seconds.

## I-MARK NON-TURBO

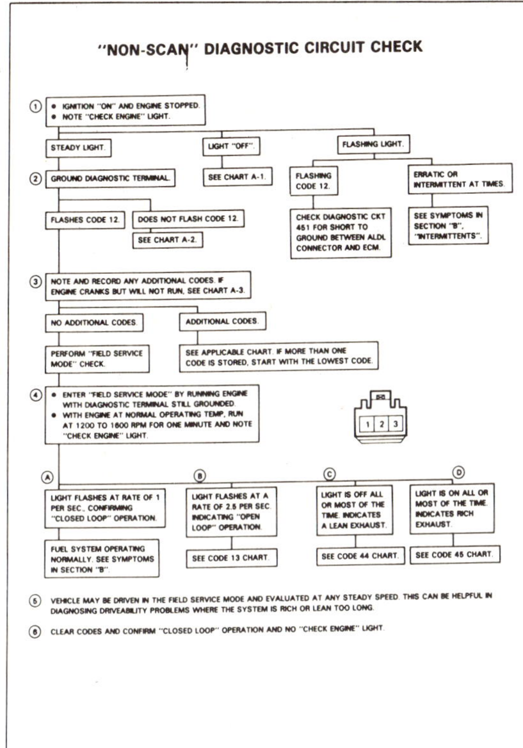

"NON-SCAN" DIAGNOSTIC CIRCUIT CHECK
1
• IGNITION "ON" AND ENGINE STOPPED.
• NOTE "CHECK ENGINE" LIGHT.
STEADY LIGHT.
LIGHT "OFF".
FLASHING LIGHT.
2
GROUND DIAGNOSTIC TERMINAL.
SEE CHART A-1.
FLASHING CODE 12.
ERRATIC OR INTERMITTENT AT TIMES.
FLASHES CODE 12.
DOES NOT FLASH CODE 12.
SEE CHART A-2.
CHECK DIAGNOSTIC CKT 451 FOR SHORT TO GROUND BETWEEN ALDL CONNECTOR AND ECM.
SEE SYMPTOMS IN SECTION "B", "INTERMITTENTS".
3
NOTE AND RECORD ANY ADDITIONAL CODES. IF ENGINE CRANKS BUT WILL NOT RUN, SEE CHART A-3.
NO ADDITIONAL CODES.
ADDITIONAL CODES.
PERFORM "FIELD SERVICE MODE" CHECK.
SEE APPLICABLE CHART. IF MORE THAN ONE CODE IS STORED, START WITH THE LOWEST CODE.
4
• ENTER "FIELD SERVICE MODE" BY RUNNING ENGINE WITH DIAGNOSTIC TERMINAL STILL GROUNDED.
• WITH ENGINE AT NORMAL OPERATING TEMP, RUN AT 1200 TO 1600 RPM FOR ONE MINUTE AND NOTE "CHECK ENGINE" LIGHT.
1 2 3
A
B
C
D
LIGHT FLASHES AT RATE OF 1 PER SEC., CONFIRMING "CLOSED LOOP" OPERATION.
LIGHT FLASHES AT A RATE OF 2.5 PER SEC. INDICATING "OPEN LOOP" OPERATION.
LIGHT IS OFF ALL OR MOST OF THE TIME. INDICATES A LEAN EXHAUST.
LIGHT IS ON ALL OR MOST OF THE TIME. INDICATES RICH EXHAUST.
FUEL SYSTEM OPERATING NORMALLY. SEE SYMPTOMS IN SECTION "B".
SEE CODE 13 CHART.
SEE CODE 44 CHART.
SEE CODE 45 CHART.
5
VEHICLE MAY BE DRIVEN IN THE FIELD SERVICE MODE AND EVALUATED AT ANY STEADY SPEED. THIS CAN BE HELPFUL IN DIAGNOSING DRIVEABILITY PROBLEMS WHERE THE SYSTEM IS RICH OR LEAN TOO LONG.
6
CLEAR CODES AND CONFIRM "CLOSED LOOP" OPERATION AND NO "CHECK ENGINE" LIGHT.


### CODE 13

### OXYGEN SENSOR CIRCUIT (OPEN CIRCUIT)

**Circuit Description:**

The ECM supplies a voltage of about .45 volt between terminals "M2-23" and "M2-22". (If measured with 10 megohm digital voltmeter, this may read as low as .32 volts).
The $O_2$ sensor varies the voltage within a range of about 1 volt, if the exhaust is rich, down through about 10 volt, if exhaust is lean.
The sensor is like an open circuit and produces no voltage, when it is below 360°C (600°F). An open sensor circuit, or cold sensor, causes "Open Loop" operation.

**Test Description:**

Numbers below refer to circled numbers on the diagnostic chart.

1. Code 13 will set:
   - Engine at normal operating temperature.
   - At least 2 minutes engine run time after start.
   - $O_2$ signal voltage steady between .35 and .55 volts.
   - Throttle angle above 7%.
   - All conditions must be met for about 60 seconds.

   If the conditions for a Code 13 exist, the system will not go "Closed Loop".
2. This test determines if the $O_2$ sensor is the problem, or, if the ECM and wiring are at fault.
3. This step simulates a lean exhaust. If the ECM and wiring are OK the ECM will see the lean condition and turn the "Check Engine" light "OFF" for at least 30 seconds after engine start, and then flash "open loop". It should be considered normal if the light remains "OFF" for a longer period of time before flashing open loop.
4. In doing this test, use only a high impedence digital volt ohm meter. This test checks the continuity of CKT's 412 and 413. If CKT 413 is open, the ECM voltage on CKT 412 will measure over .6 volts (600 mv).

**Diagnostic Aids:**

Normal voltage varies between 100 mv to 999 mv (.1 and 1.0 volt), while in "Closed Loop". Code 13 sets in one minute, if voltage remains between .35 and .55 volts, but the system will go "Open Loop" in about 15 seconds.
Verify a clean, tight ground connection for CKT 413. Open CKT(s), 412 or 413 will result in a Code 13.
If Code 13 is intermittent, refer to Section "B"

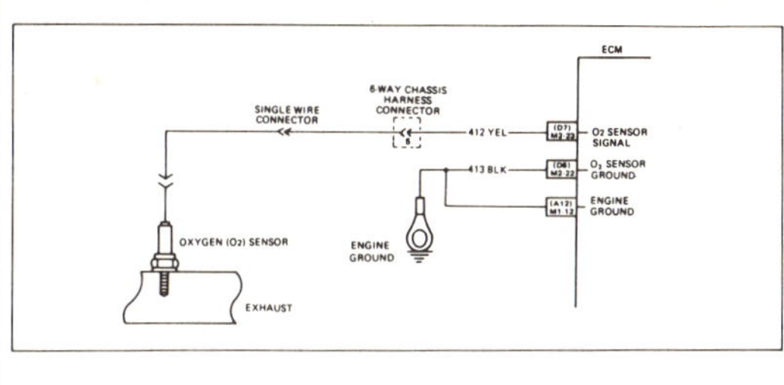


## I-MARK NON-TURBO

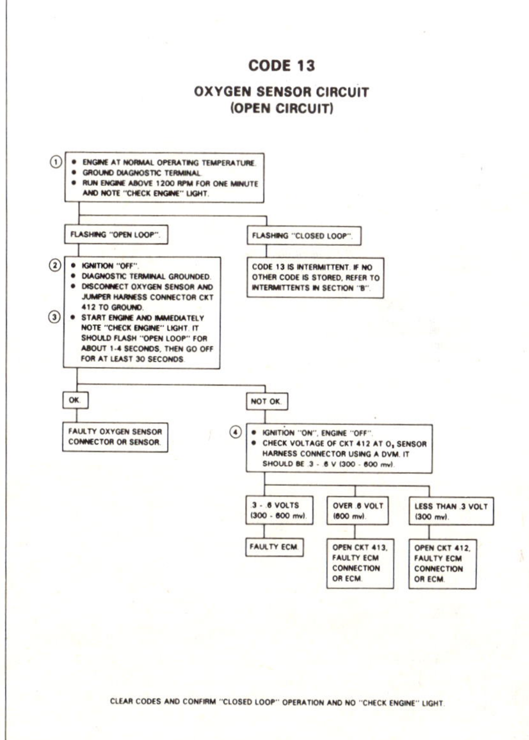

CODE 13
OXYGEN SENSOR CIRCUIT
(OPEN CIRCUIT)
1
• ENGINE AT NORMAL OPERATING TEMPERATURE.
• GROUND DIAGNOSTIC TERMINAL.
• RUN ENGINE ABOVE 1200 RPM FOR ONE MINUTE AND NOTE "CHECK ENGINE" LIGHT.
FLASHING "OPEN LOOP".
FLASHING "CLOSED LOOP".
2
• IGNITION "OFF".
• DIAGNOSTIC TERMINAL GROUNDED.
• DISCONNECT OXYGEN SENSOR AND JUMPER HARNESS CONNECTOR CKT 412 TO GROUND.
3
• START ENGINE AND IMMEDIATELY NOTE "CHECK ENGINE" LIGHT. IT SHOULD FLASH "OPEN LOOP" FOR ABOUT 1-4 SECONDS, THEN GO OFF FOR AT LEAST 30 SECONDS.
CODE 13 IS INTERMITTENT. IF NO OTHER CODE IS STORED, REFER TO INTERMITTENTS IN SECTION "B".
OK.
NOT OK.
FAULTY OXYGEN SENSOR CONNECTOR OR SENSOR.
4
• IGNITION "ON", ENGINE "OFF".
• CHECK VOLTAGE OF CKT 412 AT O2 SENSOR HARNESS CONNECTOR USING A DVM. IT SHOULD BE .3 - .6 V (300 - 600 mv).
.3 - .6 VOLTS (300 - 600 mv).
OVER .6 VOLT (600 mv).
LESS THAN .3 VOLT (300 mv).
FAULTY ECM.
OPEN CKT 413, FAULTY ECM CONNECTION OR ECM.
OPEN CKT 412, FAULTY ECM CONNECTION OR ECM.
CLEAR CODES AND CONFIRM "CLOSED LOOP" OPERATION AND NO "CHECK ENGINE" LIGHT.


### CODE 14

### COOLANT TEMPERATURE SENSOR CIRCUIT (SIGNAL VOLTAGE LOW-HIGH TEMPERATURE INDICATED)

**Circuit Description:**

The Coolant Temperature Sensor uses a thermistor to control the signal voltage at the ECM. The ECM applies a voltage on CKT 410 to the sensor. When the engine is cold, the sensor (thermistor) resistance is high, therefore, the ECM will see high signal voltage.
As the engine warms, the sensor resistance becomes less, and the voltage drops. At normal engine operating temperature, the voltage will measure about 1.5 to 2.0 volts at the ECM terminal "M2-10".
Coolant temperature is one of the inputs used to control:

- Fuel delivery
- Engine Spark Timing (EST)
- Idle (IAC)

**Test Description:**

Numbers below refer to circled numbers on the diagnostic chart.

1. Checks to see if code was set as result of hard failure or intermittent condition.
   Code 14 will set if:
   - Signal Voltage indicates a coolant temperature above 135°C (275°F) for 2 minutes.
2. This test simulates conditions for a Code 15. If the ECM recognizes the open circuit (high voltage), and displays a low temperature, the ECM and wiring are OK.

**Diagnostic Aids:**

When a Code 14 is set, the ECM will turn "ON" the Engine Cooling Fan.
A Code 14 will result if CKT 410 is shorted to ground.
If Code 14 is intermittent, refer to Section "B".

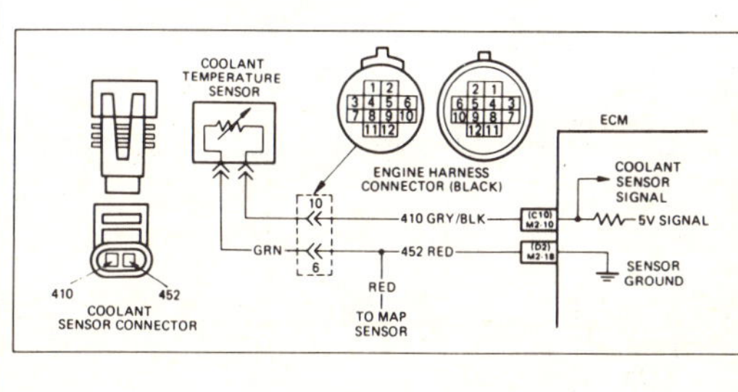

## I-MARK NON-TURBO

### CODE 14

**COOLANT TEMPERATURE SENSOR CIRCUIT
(SIGNAL VOLTAGE LOW-HIGH TEMPERATURE INDICATED)**

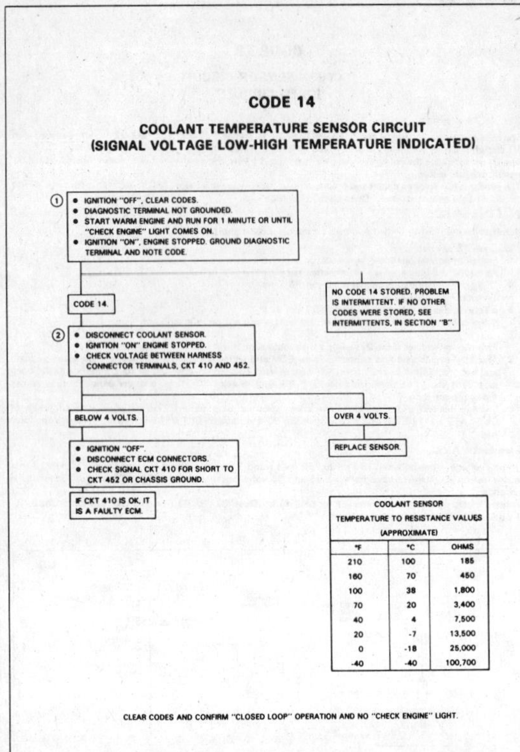


| COOLANT SENSOR TEMPERATURE TO RESISTANCE VALUES (APPROXIMATE) | | |
|---|---|---|
| °F | °C | OHMS |
| 210 | 100 | 185 |
| 160 | 70 | 450 |
| 100 | 38 | 1,800 |
| 70 | 20 | 3,400 |
| 40 | 4 | 7,500 |
| 20 | -7 | 13,500 |
| 0 | -18 | 25,000 |
| -40 | -40 | 100,700 |

CLEAR CODES AND CONFIRM "CLOSED LOOP" OPERATION AND NO "CHECK ENGINE" LIGHT.

### CODE 15

**COOLANT TEMPERATURE SENSOR CIRCUIT
(SIGNAL VOLTAGE HIGH-LOW TEMPERATURE INDICATED)**

**Circuit Description:**

The Coolant Temperature Sensor uses a thermistor to control the signal voltage at the ECM. The ECM applies a voltage on CKT 410 to the sensor. When the engine is cold, the sensor (thermistor) resistance is high, therefore, the ECM will see high signal voltage.

As the engine warms, the sensor resistance becomes less, and the voltage drops. At normal engine operating temperature, the voltage will measure about 1.5 to 2.0 volts at the ECM terminal "M2-10".

Coolant temperature is one of the inputs used to control:

- Fuel delivery
- Engine Spark Timing (EST)
- Idle (IAC)

**Test Description:**

Numbers below refer to circled numbers on the diagnostic chart.

1. Checks to see if code was set as result of hard failure or intermittent condition.
   Code 15 will set if:
   - The engine has been running for 1 minute.
   - Signal Voltage indicates a coolant temperature below -37°C.
2. This test simulates conditions for a Code 14. If the ECM recognizes the ground circuit (low voltage), and displays a high temperature, the ECM and wiring are OK.
3. This test will determine if there is a wiring problem or a faulty ECM. If CKT 452 is open, there may also be a Code 33 stored. Be sure to carefully check terminals at the engine harness connectors.

**Diagnostic Aids:**

When a Code 15 is set, the ECM will turn on the Engine Cooling Fan.
A Code 15 will result if CKT's 410 or 452 are open.
If Code 15 is intermittent, refer to Section "B".

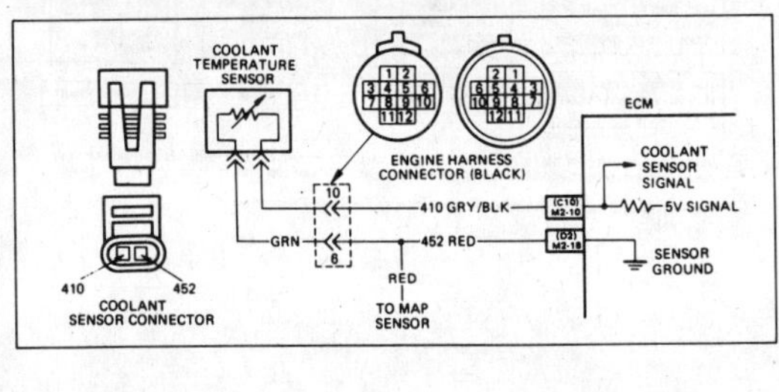


## I-MARK NON-TURBO

### CODE 15

**COOLANT TEMPERATURE SENSOR CIRCUIT
(SIGNAL VOLTAGE HIGH-LOW TEMPERATURE INDICATED)**

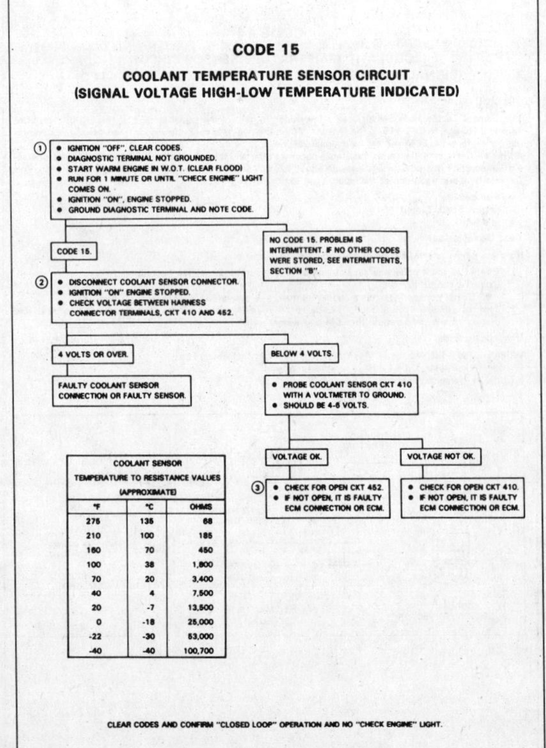


| COOLANT SENSOR TEMPERATURE TO RESISTANCE VALUES (APPROXIMATE) | | |
|---|---|---|
| °F | °C | OHMS |
| 275 | 135 | 68 |
| 210 | 100 | 185 |
| 160 | 70 | 450 |
| 100 | 38 | 1,800 |
| 70 | 20 | 3,400 |
| 40 | 4 | 7,500 |
| 20 | -7 | 13,500 |
| 0 | -18 | 25,000 |
| -22 | -30 | 53,000 |
| -40 | -40 | 100,700 |

CLEAR CODES AND CONFIRM "CLOSED LOOP" OPERATION AND NO "CHECK ENGINE" LIGHT.

### CODE 21

**THROTTLE POSITION SENSOR (TPS) CIRCUIT
(SIGNAL VOLTAGE HIGH)**

**Circuit Description:**

The Throttle Position Sensor (TPS) provides a voltage signal that changes relative to the throttle valve. Signal voltage will vary from less than 1.25 volts at idle to about 4.5 volts at wide open throttle (WOT).

The TPS signal is one of the most important inputs used by the ECM for fuel control and for many of the ECM controlled outputs.

**Test Description:**

Numbers below refer to circled numbers on the diagnostic chart.

1. This step checks to see if Code 21 is the result of a hard failure or an intermittent condition.
   A Code 21 will set if:
   - TPS reading above 4.5 volts.
   - Engine speed less than 1200 rpm.
   - MAP reading below 50 kPa.
   - All of the above conditions present for 5 seconds.
2. This step simulates conditions for a Code 22. If the ECM recognizes the change of state, the ECM and CKTs 416 and 417 are OK.
3. This step isolates a faulty sensor, ECM, or an open CKT 452. If it is determined CKT 452 is the fault be sure to check the terminals at the green engine harness connector.

**Diagnostic Aids:**

Closed throttle voltage should be less than 1.25 volts. TPS voltage should increase at a steady rate as throttle is moved to WOT.
A Code 21 will result if CKT 452 is open or CKT 417 is shorted to voltage. If Code 21 is intermittent, refer to Section "B". (page 03C-73)

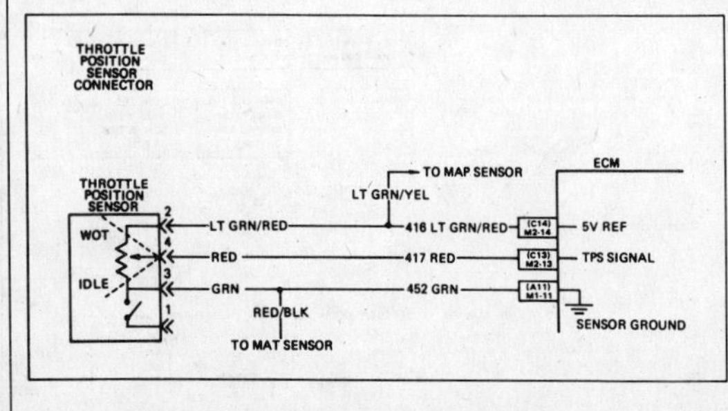

## I-MARK NON-TURBO

### CODE 21

**THROTTLE POSITION SENSOR (TPS) CIRCUIT (SIGNAL VOLTAGE HIGH)**

(1)
- ENGINE AT NORMAL OPERATING TEMPERATURE.
- DIAGNOSTIC TERM. NOT GROUNDED.
- IGNITION "OFF". CLEAR CODES.
- START ENGINE AND IDLE IN NEUTRAL, A/C OFF, FOR 1 MINUTE OR UNTIL "CHECK ENGINE" LIGHT COMES ON.
- IGNITION "ON", ENGINE STOPPED.
- GROUND DIAGNOSTIC TERMINAL AND NOTE CODE.

NO CODE 21 STORED. PROBLEM IS INTERMITTENT. IF NO OTHER CODE IS STORED, SEE INTERMITTENTS IN SECTION B.

CODE 21.

(2)
- DIAGNOSTIC TERMINAL NOT GROUNDED.
- IGNITION "OFF". CLEAR CODES.
- DISCONNECT SENSOR.
- START ENGINE AND IDLE IN NEUTRAL, A/C OFF, FOR 1 MINUTE OR UNTIL "CHECK ENGINE" LIGHT COMES ON.
- IGNITION "ON", ENGINE STOPPED.
- GROUND DIAGNOSTIC TERMINAL AND NOTE CODE.

(3) CODE 22.
- PROBE TPS HARNESS CONNECTOR CKT 452 WITH TEST LIGHT TO 12 VOLTS.

LIGHT "ON" — FAULTY TPS CONNECTION OR SENSOR.

LIGHT "OFF" — REPAIR OPEN CKT 452.

CODE 21.
CHECK:
- CKT 417 FOR SHORT TO VOLTAGE.
- CKT 417 FOR SHORT TO CKT 416.
- FOR FAULTY ECM.

CLEAR CODES AND CONFIRM "CLOSED LOOP" OPERATION AND NO "CHECK ENGINE" LIGHT.

### CODE 22

**THROTTLE POSITION SENSOR (TPS) CIRCUIT (SIGNAL VOLTAGE LOW)**

**Circuit Description:**

The Throttle Position Sensor (TPS) provides a voltage signal that changes, relative to the throttle valve. Signal voltage will vary from less than 1.25 volts at idle to about 4.5 volts at wide open throttle (WOT).

The TPS signal is one of the most important inputs used by the ECM for fuel control and for many of the ECM controlled outputs.

**Test Description:**

Numbers below refer to circled numbers on the diagnostic chart.

1. This step checks to see if Code 22 is the result of a hard failure or an intermittent condition.
   A Code 22 will set if:
   - The engine is running.
   - TPS voltage is below .20 volts (200 mv).
2. This step simulates conditions for a Code 21. If the ECM recognizes the high signal voltage and sets code 21, the ECM and wiring are OK.
3. Checks for reference voltage from the ECM. The importanting is that the ECM recognizes the voltage as over 4 volts, indicating that CKT417 and the ECM are OK.
4. If CKT 416 is open or shorted to ground, there may also be a stored Code 34. If it is determined that the CKT is open be sure to check terminals at the engine harness connecotor.

**Diagnostic Aids:**

Closed throttle voltage should be less than 1.25 volts. TPS voltage should increase at a steady rate as throttle is moved to WOT.

An open or grounded 416 or 417 will result in a Code 22.

If Code 22 is intermittent, refer to Section "B".

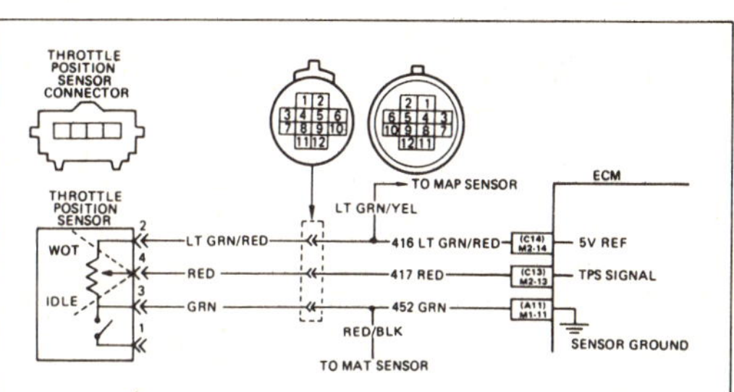


## I-MARK NON-TURBO

### CODE 22

**THROTTLE POSITION SENSOR (TPS) CIRCUIT (SIGNAL VOLTAGE LOW)**

(1)
- DIAGNOSTIC TERMINAL NOT GROUNDED.
- IGNITION "OFF".
- CLEAR CODES.
- START ENGINE AND IDLE IN NEUTRAL, A/C OFF, FOR 1 MINUTE OR UNTIL "CHECK ENGINE" LIGHT COMES ON.
- IGNITION "ON", ENGINE STOPPED.
- GROUND DIAGNOSTIC TERMINAL AND NOTE CODE.

NO CODE 22. PROBLEM IS INTERMITTENT. IF NO OTHER CODE STORED, SEE INTERMITTENTS IN SECTION B.

CODE 22.

(2)
- DIAGNOSTIC TERMINAL NOT GROUNDED.
- IGNITION "OFF". CLEAR CODES.
- DISCONNECT TPS AND JUMPER CKTS 416 TO 417.
- START ENGINE AND IDLE IN NEUTRAL, A/C OFF, FOR 1 MINUTE OR UNTIL "CHECK ENGINE" LIGHT COMES ON.
- IGNITION "ON", ENGINE STOPPED.
- GROUND DIAGNOSTIC TERMINAL AND NOTE CODE.

CODE 22.

(3)
- REMOVE JUMPER FROM 416 AND 417.
- CHECK VOLTAGE BETWEEN CKT 452 AND 416 USING DIGITAL VOLTMETER.

4-6 VOLTS. — DISCONNECT ECM CONNECTOR AND CHECK FOR OPEN OR SHORT TO GROUND IN CKT 417. CHECK FOR INTERNALLY SHORTED MAP SENSOR. IF OK, IT IS FAULTY ECM CONNECTOR TERMINAL OR ECM.

BELOW 4 VOLTS. — (4) DISCONNECT ECM CONNECTOR. CHECK FOR OPEN OR SHORT TO GROUND IN CKT 416. IF OK, IT IS FAULTY ECM CONNECTOR TERMINAL OR ECM.

CODE 21. — REPLACE TPS.

CLEAR CODES AND CONFIRM "CLOSED LOOP" OPERATION AND NO "CHECK ENGINE" LIGHT.

### CODE 23

**MANIFOLD AIR TEMPERATURE (MAT) SENSOR CIRCUIT (SIGNAL VOLTAGE HIGH)**

**Circuit Description:**

The Manifold Air Temperature Sensor uses a thermistor to control the signal voltage to the ECM. The ECM applies a voltage (4-6 volts) on CKT 472 to the sensor. When manifold air is cold, the sensor (thermistor) resistance is high, therefore, the ECM will see a high signal voltage. As the air warms, the sensor resistance becomes less and the voltage drops.

**Test Description:**

Numbers below refer to circled numbers on the diagnostic chart.

1. A Code 23 will set due to an open sensor, wire, or connection. This test will determine if the wiring and ECM are OK.
2. If the resistance is greater than 25,000 ohms, replace the sensor.
   If CKT 452 is open there may also be a Code 21.
   If Code 23 is intermittent, refer to Section "B".

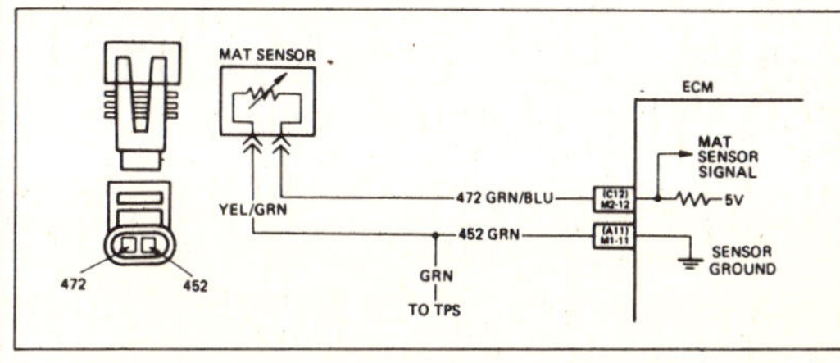

## I-MARK NON-TURBO

### CODE 23
### MANIFOLD AIR TEMPERATURE (MAT) SENSOR CIRCUIT (SIGNAL VOLTAGE HIGH)

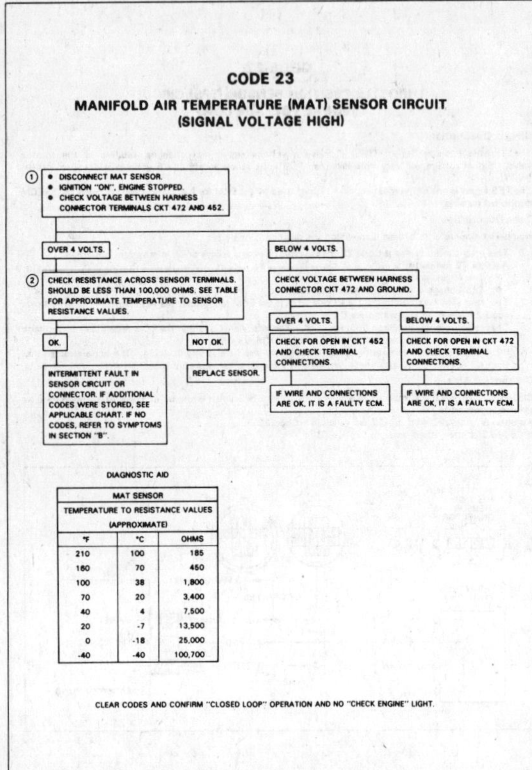


DIAGNOSTIC AID

| MAT SENSOR | | |
|---|---|---|
| TEMPERATURE TO RESISTANCE VALUES (APPROXIMATE) | | |
| °F | °C | OHMS |
| 210 | 100 | 185 |
| 160 | 70 | 450 |
| 100 | 38 | 1,800 |
| 70 | 20 | 3,400 |
| 40 | 4 | 7,500 |
| 20 | -7 | 13,500 |
| 0 | -18 | 25,000 |
| -40 | -40 | 100,700 |

CLEAR CODES AND CONFIRM "CLOSED LOOP" OPERATION AND NO "CHECK ENGINE" LIGHT.

### CODE 24
### VEHICLE SPEED SENSOR (VSS) CIRCUIT

**Circuit Description:**

The ECM applies and monitors 12 volts on CKT 437. CKT 437 connects to the Vehicle Speed Sensor which alternately grounds CKT 437 when drive wheels are turning. This pulsing action takes place 2500 times per kilometer, and the ECM will calculate vehicle speed based on the time between "pulses".

Code 24 will set:

- Speed sensor indicates speed less than 8 kPh.
- RPM between 2400 and 4400.
- Coolant greater than 85°C.
- MAP less than 20 kPa.
- Conditions present for 10 seconds.

**Test Description:**

Numbers below refer to circled numbers on the diagnostic chart.

1. This test uses the volt meter to verify the VSS sensor operation, by checking for change in voltage.
2. This checks for the presence of the ECM-supplied 12 volt signal, which is monitored by the ECM for pulses, indicating VSS operation.
3. CKT 439 is the ignition feed that supplies operating power to the VSS. If this line is open, CKT 437 will be at 12 volts at all times.
4. CKT 450 supplies the ground path for VSS operation. If CKT 450 is open, VSS cannot pulse CKT 437 to ground.

**Diagnostic Aids:**

Volt meter should indicate a changing voltage at vehicle speed sensor whenever the drive wheels are turning. If Code 24 is intermittent, refer to Section "B".

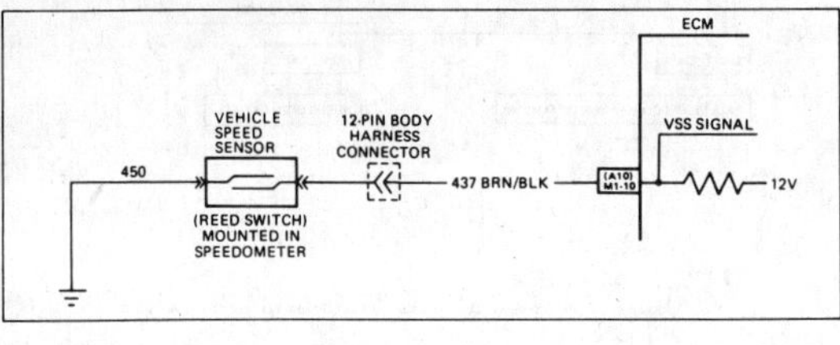


## I-MARK NON-TURBO

### CODE 24
### VEHICLE SPEED SENSOR (VSS) CIRCUIT

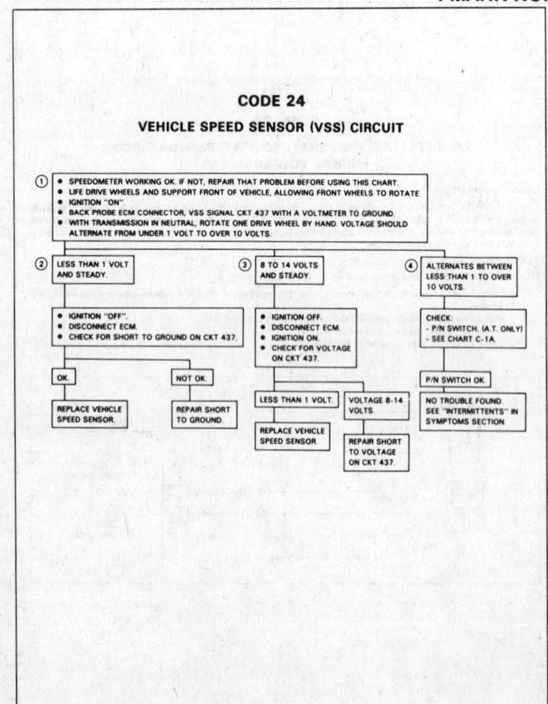


### CODE 25
### MANIFOLD AIR TEMPERATURE (MAT) SENSOR CIRCUIT (SIGNAL VOLTAGE LOW)

**Circuit Description:**

The Manifold Air Temperature Sensor uses a thermistor to control the signal voltage to the ECM. The ECM applies a voltage (4-6 volts) on CKT 472 to the sensor. When manifold air is cold, the sensor (thermistor) resistance is high, therefore, the ECM will see a high signal voltage. As the air warms, the sensor resistance becomes less and the voltage drops.

**Test Description:**

Numbers below refer to circled numbers on the diagnostic chart.

1. If voltage is above 4 volts, the ECM and wiring are OK.
2. If the resistance is less than 185 ohms, replace the sensor.

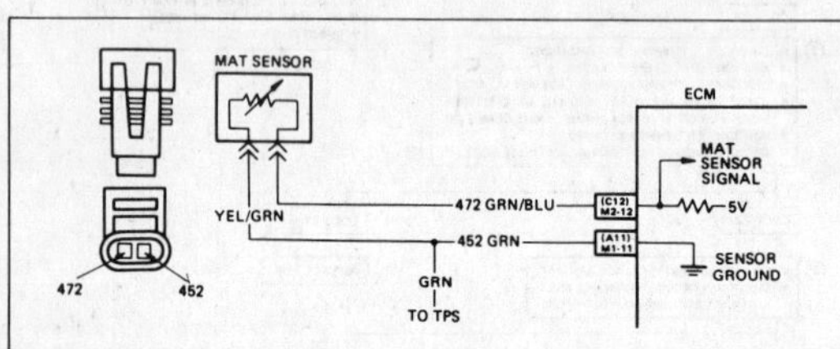

## I-MARK NON-TURBO

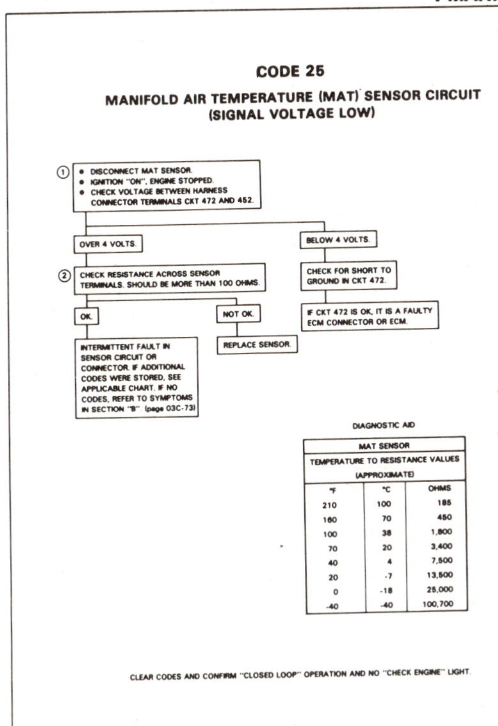


MAT SENSOR
TEMPERATURE TO RESISTANCE VALUES (APPROXIMATE)

| °F | °C | OHMS |
|---|---|---|
| 210 | 100 | 185 |
| 160 | 70 | 450 |
| 100 | 38 | 1,800 |
| 70 | 20 | 3,400 |
| 40 | 4 | 7,500 |
| 20 | -7 | 13,500 |
| 0 | -18 | 25,000 |
| -40 | -40 | 100,700 |

### CODE 32

#### EGR SYSTEM FAILURE

The EGR temperature sensor determines the difference between the temperature when the EGR is on and that when the EGR is off. If the difference remains below a certain level for predetermined time, the system registers Code 32.

The conditions chosen as typical of the range where EGR is on are:

Engine speed: 1800 rpm — 5000 rpm
MAP: 48 kPa — 68 kPa

And the condition typical of the range where EGR is off is idle.

The system checks the difference in temperatures when the following conditions exist:

Coolant temperature: 75°C — 100°C
Intake air temperature (MAT) : 50°C — 100°C

When idling continues at least 20 seconds under these conditions, the temperature is memorized as that of the range where EGR is off. On the other hand, if the aforementioned conditions typical of the range where EGR is on continue 8 seconds or more, the temperature at that time is read in as that of the range where EGR is on. When the two temperatures are compared and if the difference were 3°C or less, then the system recognizes an EGR system malfunction. Code 32 is registered when this reading has occured three times consecutively.

If, however, the temperature difference were 6°C or more, the EGR system is recognized as normal and the "check engine" lamp turns off.

**Note:** The EGR valve is controlled by a solenoid, which in turn is controlled by ECM. This solenoid normally remains off and comes on as the CKT7 is grounded. It comes on only in predetermined ranges of coolant temperature, MAP, TPS and engine speed. It also comes on when the ignition is on and when, with the engine stopped, the diagnosis terminal has been grounded.

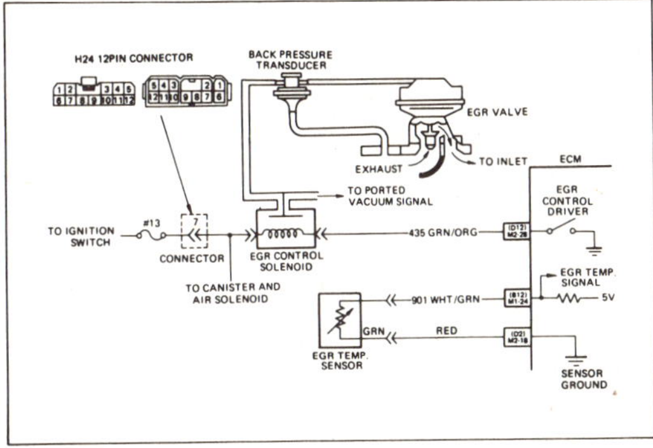


## I-MARK NON-TURBO

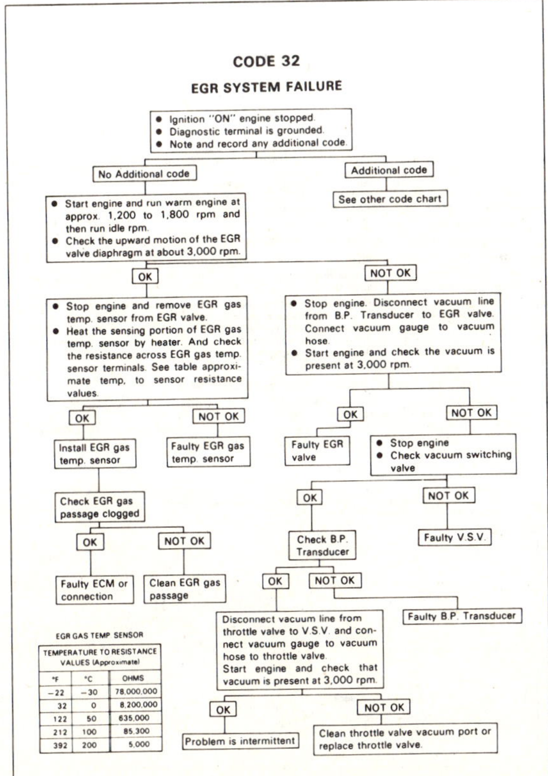


EGR GAS TEMP. SENSOR
TEMPERATURE TO RESISTANCE VALUES (Approximate)

| °F | °C | OHMS |
|---|---|---|
| −22 | −30 | 78,000,000 |
| 32 | 0 | 8,200,000 |
| 122 | 50 | 635,000 |
| 212 | 100 | 85,300 |
| 392 | 200 | 5,000 |

### CODE 33

#### MANIFOLD ABSOLUTE PRESSURE (MAP) SENSOR CIRCUIT (SIGNAL VOLTAGE HIGH-LOW VACUUM)

**Circuit Description:**

The Manifold Absolute Pressure (MAP) Sensor responds to changes in manifold pressure (vacuum). The ECM receives this information as a signal voltage that will vary from about 1 to 1.5 volts, at closed throttle idle, to 4-4.5 volts at wide open throttle (low vacuum).
If the MAP sensor fails, the ECM will substitute a fixed MAP value and use the Throttle Position Sensor (TPS) to control fuel delivery.

**Test Description:**

Numbers below refer to circled numbers on the diagnostic chart.

1. This step will determine if Code 33 is the result of a hard failure or an intermittent condition.
   A Code 33 will set if:
   - MAP signal indicates greater than 80 kPa (over 4V)(low vacuum).
   - TPS less than 5%.
   - These conditions are present for a time longer than 10 seconds.
2. This step simulates conditions for a Code 34. If the ECM recognizes the change, the ECM, and CKT's 416 and 432, are OK. If CKT 452 is open, there may also be a Code 15 stored.

**Diagnostic Aids:**

With the ignition "ON" and the engine stopped, the manifold pressure is equal to atmospheric pressure and the signal voltage will be high. This information is used by the ECM as an indication of vehicle altitude and is referred to as BARO. Comparison of this BARO reading with a known good vehicle with the same sensor is a good way to check accuracy of a "suspect" sensor. Readings should be the same ± .4 volt.
A Code 33 will result if CKT 452 is open, or if CKT 432 is shorted to voltage or to CKT 416.
If Code 33 is intermittent, refer to Section "B".

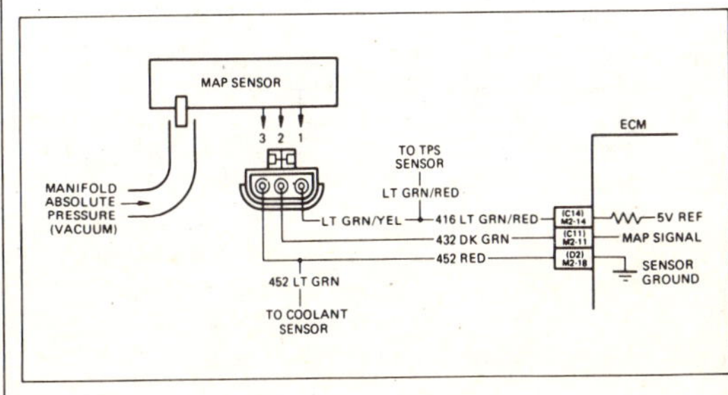

## I-MARK NON-TURBO

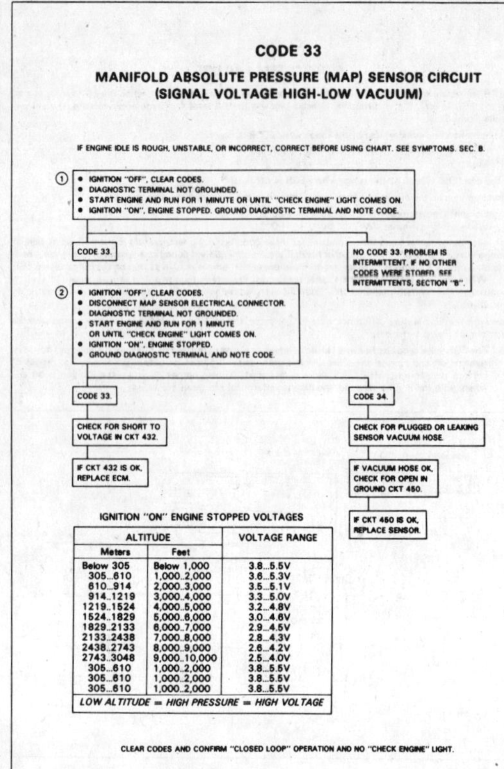


IGNITION "ON" ENGINE STOPPED VOLTAGES

| ALTITUDE | | VOLTAGE RANGE |
|---|---|---|
| Meters | Feet | |
| Below 305 | Below 1,000 | 3.8...5.5V |
| 305...610 | 1,000..2,000 | 3.6...5.3V |
| 610...914 | 2,000..3,000 | 3.5...5.1V |
| 914..1219 | 3,000..4,000 | 3.3...5.0V |
| 1219..1524 | 4,000..5,000 | 3.2...4.8V |
| 1524..1829 | 5,000..6,000 | 3.0...4.6V |
| 1829..2133 | 6,000..7,000 | 2.9...4.5V |
| 2133..2438 | 7,000..8,000 | 2.8...4.3V |
| 2438..2743 | 8,000..9,000 | 2.6...4.2V |
| 2743..3048 | 9,000..10,000 | 2.5...4.0V |
| 305...610 | 1,000..2,000 | 3.8...5.5V |
| 305...610 | 1,000..2,000 | 3.8...5.5V |
| 305...610 | 1,000..2,000 | 3.8...5.5V |
| *LOW ALTITUDE = HIGH PRESSURE = HIGH VOLTAGE* | | |

CLEAR CODES AND CONFIRM "CLOSED LOOP" OPERATION AND NO "CHECK ENGINE" LIGHT.

### CODE 34

### MANIFOLD ABSOLUTE PRESSURE (MAP) SENSOR CIRCUIT (SIGNAL VOLTAGE LOW-HIGH VACUUM)

**Circuit Description:**

The Manifold Absolute Pressure (MAP) Sensor responds to changes in manifold pressure (vacuum). The ECM receives this information as a signal voltage that will vary from about 1 to 1.5 volts at closed throttle idle, to 4-4.5 volts at wide open throttle.
If the MAP sensor fails, the ECM will substitute a fixed MAP value and use the Throttle Position Sensor (TPS) to control fuel delivery.

**Test Description:**

Numbers below refer to circled numbers on the diagnostic chart.

1. This step determines if Code 34 is the result of a hard failure or an intermittent condition.
   A Code 34 will set when:
   - Engine rpm is less than 1200 rpm.
   - MAP signal voltage is too low (14 kPa).
     OR
   - MAP reading less than 14 kPa.
   - Engine rpm greater than 1200 rpm.
   - TPS is less than 3.5%.
2. Jumpering harness terminals "2" to "1", 5 volt to signal, will determine if the sensor is at fault, or if there is a problem with the ECM or wiring.

**Diagnostic Aids:**

With the ignition "ON" and the engine stopped, the manifold pressure is equal to atmospheric pressure and the signal voltage will be high. This information is used by the ECM as an indication of vehicle altitude and is referred to as BARO. Comparison of this BARO reading with a known good vehicle with the same sensor is a good way to check accuracy of a "suspect" sensor. Readings should be the same ± .4 volt.
A Code 34 will result if CKTs 416 or 432 are open or shorted to ground.
If Code 34 is intermittent, refer to Section "B".
An internally shorted TPS sensor will cause a Code 34.

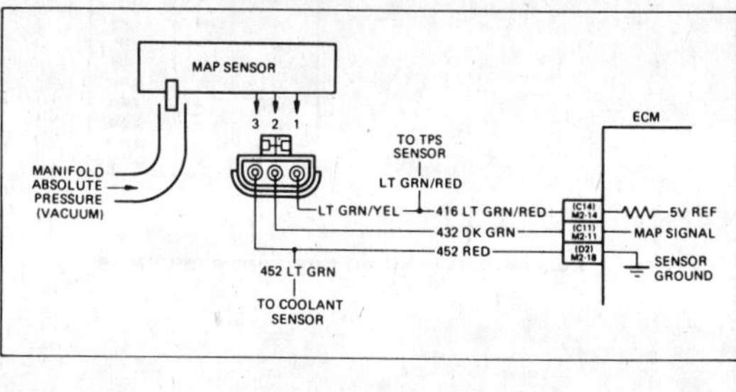


## I-MARK NON-TURBO

### CODE 34

### MANIFOLD ABSOLUTE PRESSURE (MAP) SENSOR CIRCUIT (SIGNAL VOLTAGE LOW-HIGH VACUUM)

① 
- IGNITION "OFF", CLEAR CODES.
- DIAGNOSTIC TERMINAL NOT GROUNDED.
- START ENGINE AND RUN FOR 1 MINUTE OR UNTIL "CHECK ENGINE" LIGHT COMES ON.
- IGNITION "ON", ENGINE STOPPED.
- GROUND DIAGNOSTIC TERMINAL AND NOTE CODE.

CODE 34. | NO CODE 34. PROBLEM IS INTERMITTENT. IF NO OTHER CODES WERE STORED, SEE "INTERMITTENTS", SECTION "B".

② 
- IGNITION "OFF", CLEAR CODES.
- DISCONNECT MAP SENSOR AND JUMPER HARNESS CONNECTOR TERMINAL "2" TO "1".
- DIAGNOSTIC TERMINAL NOT GROUNDED.
- START ENGINE AND RUN FOR 1 MINUTE.
- OR UNTIL "CHECK ENGINE" LIGHT COMES ON. IGNITION "ON", ENGINE STOPPED.
- GROUND DIAGNOSTIC TERMINAL AND NOTE CODE.

CODE 34.
- REMOVE JUMPER FROM TERMINAL "B" TO "C".
- CHECK VOLTAGE BETWEEN HARNESS CONNECTOR TERMINAL "A" AND "C" USING VOLTMETER.

4 TO 6 VOLTS.

CHECK FOR:
- OPEN OR SHORT TO GROUND IN CKT 432.
- INTERNALLY SHORTED TPS SENSOR.

CKT 432 OK, FAULTY ECM CONNECTOR TERMINAL OR ECM.

BELOW 4 TO 6 VOLTS.

OPEN OR SHORT TO GROUND IN CKT 416.

CKT 416 OK, FAULTY ECM CONNECTOR TERMINAL OR ECM.

CODE 33.

REPLACE SENSOR.

CLEAR CODES AND CONFIRM "CLOSED LOOP" OPERATION AND NO "CHECK ENGINE" LIGHT.

### CODE 44

### OXYGEN SENSOR CIRCUIT (LEAN EXHAUST INDICATED) 1.6L DOHC (PFI)

**Circuit Description:**

The ECM supplies a voltage of about .45 volt between terminals "D7" and "D6". (If measured with a 10 megohm digital voltmeter, this may read as low as .32 volts.) The $O_2$ sensor varies the voltage within a range of about 1 volt, if the exhaust is rich, down through about .10 volt, if exhaust is lean.
The sensor is like an open circuit and produces no voltage, when it is below about 360°C (600°F). An open sensor circuit, or cold sensor, causes "Open Loop" operation.

**Test Description:**

Numbers below refer to circled numbers on the diagnostic chart.

Code 44 is set, when the $O_2$ sensor signal voltage on CKT 412:
- Remains below .2 volt for 2 minutes or more;
- And the system is operating in "Closed Loop".

1. Grounding the diagnostic terminal with the engine running enables the "Field Service Mode" and allows the ECM to confirm either open or closed loop operation.
2. A light out or "Open Loop" indicates the fault is present. Disconnecting the $O_2$ sensor will raise the signal voltage above .2 volt. If the ECM and wiring are OK, the ECM should recognize the higher voltage (.35 - .55 V) and flash open loop when the engine is started.

**Diagnostic Aids:**

The Code 44 or lean exhaust is most likely caused by one of the following:
- **$O_2$ Sensor Wire** - Sensor pigtail may be mispositioned and contacting the exhaust manifold.
- **Check for ground in wire between connector and sensor.**
- **Fuel Contamination** - Water, even in small amounts, near the in-talk fuel pump inlet can be delivered to the injector. The water causes a lean exhaust and can set a Code 44.
- **Fuel Pressure** - System will be lean if pressure is too low. It may be necessary to monitor fuel pressure, while driving the car at various road speeds and/or loads to confirm. See Fuel Delivery system CHART A-5. (page 03C-42)
- **Exhaust Leaks** - If there is an exhaust leak, the engine can cause outside air to be pulled into the exhaust and past the sensor. Vacuum or crankcase leaks can cause a lean condition.
- **If Code 44 is intermittent, refer to Section "B".**

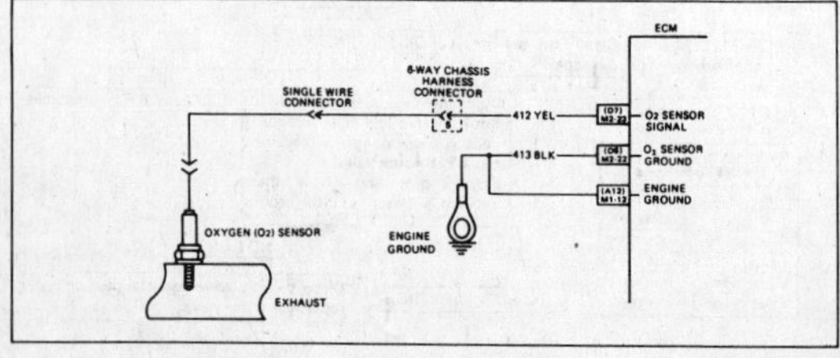

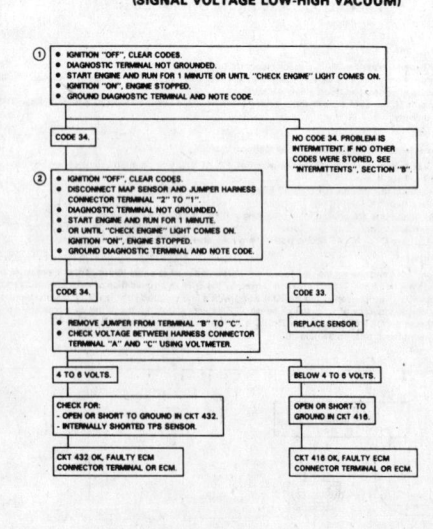

## I-MARK NON-TURBO

### CODE 44

**OXYGEN SENSOR CIRCUIT (LEAN EXHAUST INDICATED)**

① • GROUND DIAGNOSTIC TERMINAL.
• RUN WARM ENGINE AT APPROX. 1200 TO 1800 RPM FOR 1 MINUTE AND NOTE "CHECK ENGINE" LIGHT.

LIGHT STAYING "OFF" MORE THAN "ON" OR FLASHING "OPEN LOOP".

FLASHING "CLOSED LOOP".

CODE IS INTERMITTENT. IF NO ADDITIONAL CODES WERE STORED, REFER TO INTERMITTENTS SECTION B.

② • IGNITION "OFF".
• DIAGNOSTIC TERMINAL GROUNDED.
• DISCONNECT OXYGEN SENSOR.
• START ENGINE AND IMMEDIATELY NOTE "CHECK ENGINE" LIGHT.

LIGHT FLASHING OPEN LOOP.

• SEE THE FACING PAGE DIAGNOSTIC AIDS TO CHECK THE FOLLOWING:
- SENSOR(S).
- LEAN INJECTOR.
- CONTAMINATED FUEL.
- EGR.
- FOR LOW FUEL PRESSURE.
- EXHAUST MANIFOLD LEAKS AHEAD OF SENSOR.

ALL CHECKS OK.

FAULTY OXYGEN SENSOR.

LIGHT WENT OFF FOR AT LEAST 30 SECONDS.

CHECK SIGNAL CKT 412 FOR SHORT TO GROUND.

CKT 412 OK.

IT IS A FAULTY ECM.

FIELD SERVICE MODE:
ENGINE RUNNING, DIAGNOSTIC TERMINAL GROUNDED.
OPEN-LOOP, "CHECK ENGINE" LIGHT FLASHES AT A RATE OF 2 TIMES PER SECOND.
CLOSED-LOOP, "CHECK ENGINE" LIGHT FLASHES AT A RATE OF 1 TIME PER SECOND.

CLEAR CODES AND CONFIRM "CLOSED LOOP" OPERATION AND NO "CHECK ENGINE" LIGHT.

### CODE 45

**OXYGEN SENSOR CIRCUIT (RICH EXHAUST INDICATED)**

**Circuit Description:**

The ECM supplies a voltage of about .45 volt between terminals M2-23 (D7) and M2-22 (D6). (If measured with a 10 megohm digital voltmeter, this may read as low as .32 volts.) The $O_2$ sensor varies the voltage within a range of about 1 volt, if the exhaust is rich, down through about .10 volt, if exhaust is lean.
The sensor is like an open circuit and produces no voltage, when it is below about 360°C (600°F). An open sensor circuit, or cold sensor, causes "Open Loop" operation.

**Test Description:**

Numbers below refer to circled numbers on the diagnostic chart.

Code 45 is set, when the $O_2$ sensor signal voltage on CKT 412:

- Remains above .6 volt for 2 minutes or more; and in "Closed Loop".

1. Grounding the diagnostic terminal with the engine running enables the "Field Service Mode" and allows the ECM to confirm either open or closed loop operation using the "Check Engine" light.
2. A steady light or "Open Loop" indicates the fault is present. Grounding CKT 412 causes a low $O_2$ signal voltage. If the ECM and wiring are OK, the ECM should recognize the low voltage and confirm the lean signal by turning off the "Check Engine" light for at least 30 seconds.

**Diagnostic Aids:**

The Code 45, or rich exhaust, is most likely caused by one of the following:

- **Fuel Pressure** - System will go rich, if pressure is too high. The ECM can compensate for some increase. However, if it gets too high, a Code 45 will be set. See Fuel System Diagnosis CHART A-7.
- **HEI Shielding** - An open ground CKT 450 (distributor ground) may result in EMI, or induced electrical "noise". The ECM looks at this "noise" as reference pulses. The additional pulses result in a higher than actual engine speed signal. The ECM then delivers too much fuel, causing system to go rich. Engine tachometer will, also, show higher than actual engine speed, which can help in diagnosing this problem.
- **Canister Purge** - Check for fuel saturation. If full of fuel, check canister control and hoses. See Canister Purge, Section "C3".
- **MAP Sensor** - An output that causes the ECM to sense a higher than normal manifold pressure (low vacuum) can cause the system to go rich. Disconnecting the MAP sensor will allow the ECM to set a fixed value for the MAP sensor. Substitute a different MAP sensor, if the rich condition is gone, while the sensor is disconnected.
- **TPS** - An intermittent TPS output will cause the system to go rich, due to a false indication of the engine accelerating.
- **$O_2$ Sensor Contamination** - Inspect Oxygen Sensor for silicone contamination from fuel, or use of improper RTV sealant. The sensor may have a white, powdery coating and result in a high, but false signal voltage (rich exhaust indication). The ECM will then reduce the amount of fuel delivered to the engine, causing a severe surge driveability problem.
- If Code 45 is intermittent, refer to Section "B".

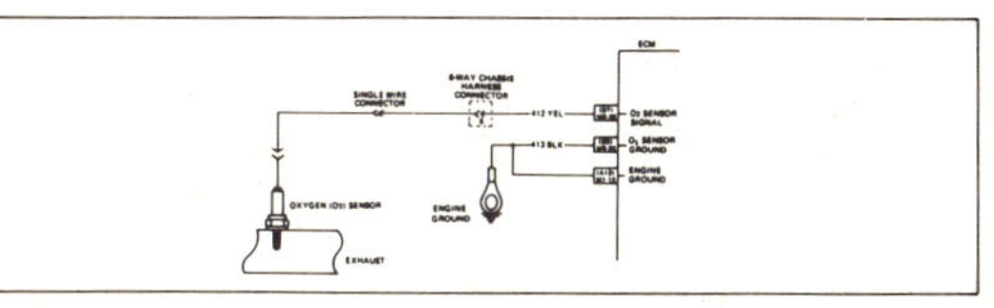

## I-MARK NON-TURBO

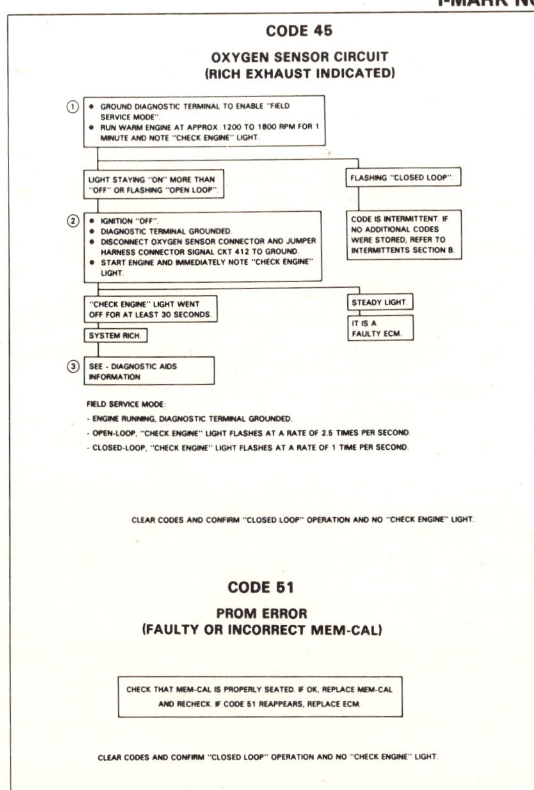


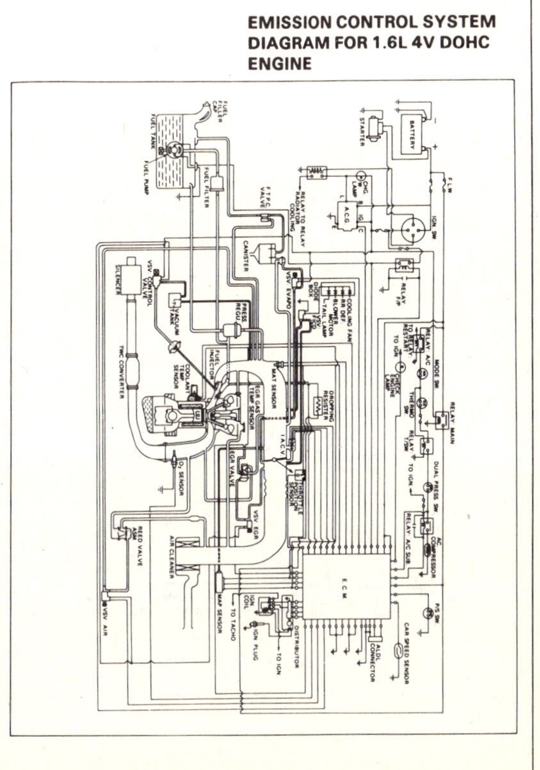

**EMISSION CONTROL SYSTEM DIAGRAM FOR 1.6L 4V DOHC ENGINE**

## I-MARK NON-TURBO

### COMPONENT LAYOUT AND SCHEMATIC DROWING

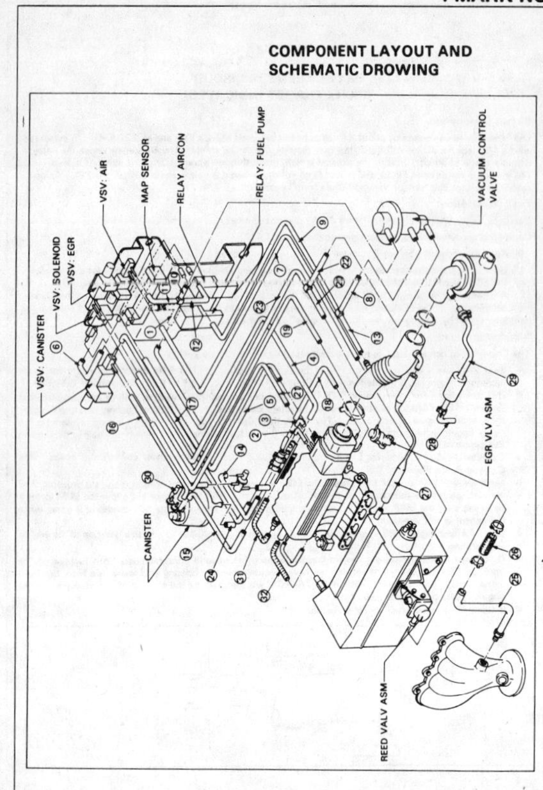


### HOSE CONNECTIONS IDENTIFICATION

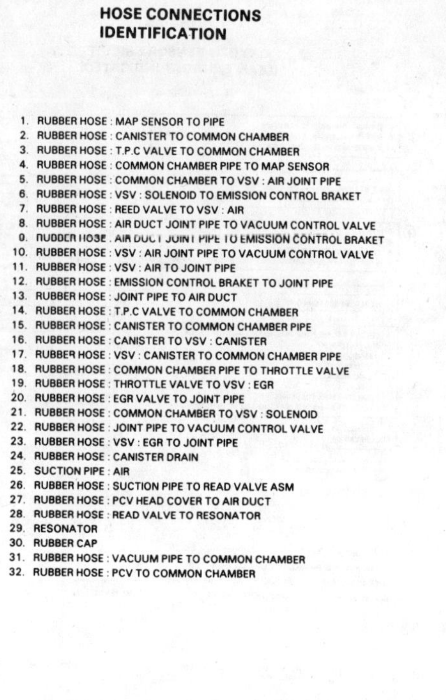

1. RUBBER HOSE : MAP SENSOR TO PIPE
2. RUBBER HOSE : CANISTER TO COMMON CHAMBER
3. RUBBER HOSE : T.P.C VALVE TO COMMON CHAMBER
4. RUBBER HOSE : COMMON CHAMBER PIPE TO MAP SENSOR
5. RUBBER HOSE : COMMON CHAMBER TO VSV : AIR JOINT PIPE
6. RUBBER HOSE : VSV : SOLENOID TO EMISSION CONTROL BRAKET
7. RUBBER HOSE : REED VALVE TO VSV : AIR
8. RUBBER HOSE : AIR DUCT JOINT PIPE TO VACUUM CONTROL VALVE
9. RUBBER HOSE : AIR DUCT JOINT PIPE TO EMISSION CONTROL BRAKET
10. RUBBER HOSE : VSV : AIR JOINT PIPE TO VACUUM CONTROL VALVE
11. RUBBER HOSE : VSV : AIR TO JOINT PIPE
12. RUBBER HOSE : EMISSION CONTROL BRAKET TO JOINT PIPE
13. RUBBER HOSE : JOINT PIPE TO AIR DUCT
14. RUBBER HOSE : T.P.C VALVE TO COMMON CHAMBER
15. RUBBER HOSE : CANISTER TO COMMON CHAMBER PIPE
16. RUBBER HOSE : CANISTER TO VSV : CANISTER
17. RUBBER HOSE : VSV : CANISTER TO COMMON CHAMBER PIPE
18. RUBBER HOSE : COMMON CHAMBER PIPE TO THROTTLE VALVE
19. RUBBER HOSE : THROTTLE VALVE TO VSV : EGR
20. RUBBER HOSE : EGR VALVE TO JOINT PIPE
21. RUBBER HOSE : COMMON CHAMBER TO VSV : SOLENOID
22. RUBBER HOSE : JOINT PIPE TO VACUUM CONTROL VALVE
23. RUBBER HOSE : VSV : EGR TO JOINT PIPE
24. RUBBER HOSE : CANISTER DRAIN
25. SUCTION PIPE : AIR
26. RUBBER HOSE : SUCTION PIPE TO READ VALVE ASM
27. RUBBER HOSE : PCV HEAD COVER TO AIR DUCT
28. RUBBER HOSE : READ VALVE TO RESONATOR
29. RESONATOR
30. RUBBER CAP
31. RUBBER HOSE : VACUUM PIPE TO COMMON CHAMBER
32. RUBBER HOSE : PCV TO COMMON CHAMBER

## I-MARK NON-TURBO

### ECM WIRING DIAGRAM (1)

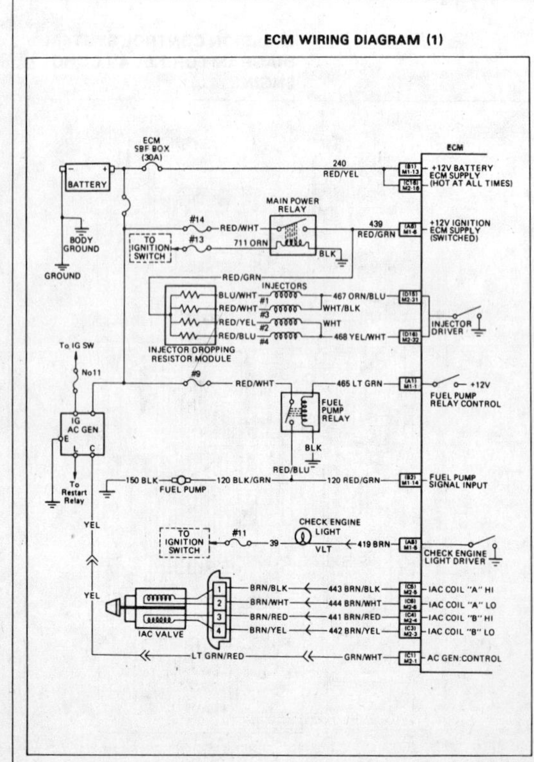


### ECM WIRING DIAGRAM (2)

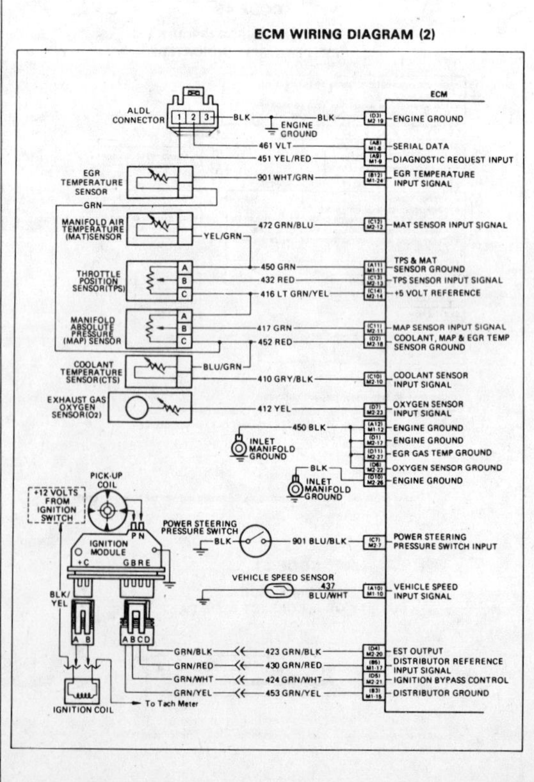

## I-MARK NON-TURBO

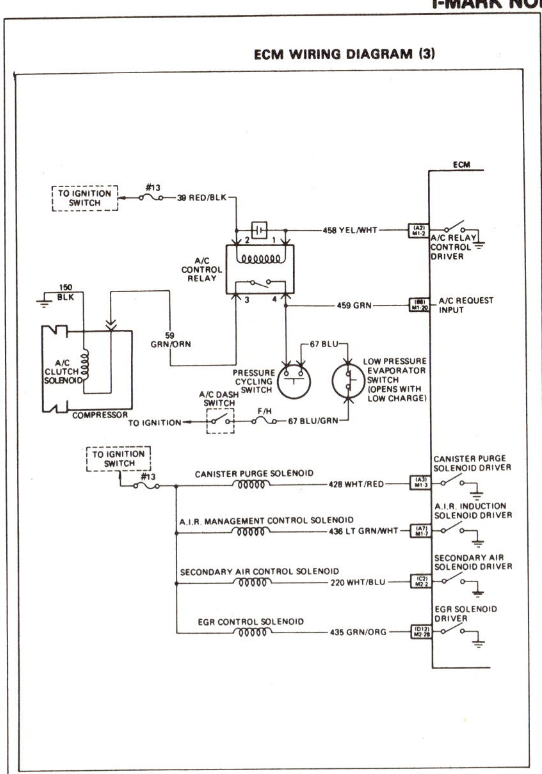


### ECM CONNECTOR IDENTIFICATION

This ECM voltage chart is for use with a digital voltmeter to further aid in diagnosis. The voltages you get may vary due to low battery charge or other reasons, but they should be very close to specification.

**The following conditions must be met before testing:**

- Engine at operating temperature
- Engine idling in close loop (for "Engine Run" column)
- ALDL Test terminal not grounded
- ALDL tool not installed
- Transaxle in "Park" or "Neutral"
- All voltages showing "B+" indicates battery or charging system voltage
- All voltages showing "O" are ground circuits, and should read less than 0.10 volts

| VOLTAGE DC KEY "ON" | VOLTAGE DC ENG. RUN | CIRCUIT | PIN | WIRE COLOR |
|---|---|---|---|---|
| 01 | B+ | FUEL PUMP RELAY | (A1) M1-1 | LT GRN |
| B+ | B+ | AIR CON RELAY | (A2) M1-2 | YEL/WHT |
| B+ | B+ | CANISTER PURGE SOLENOID | (A3) M1-3 | WHT/RED |
| | | | (A4) M1-4 | |
| 15 | B+ | CHECK ENGINE LAMP | (A5) M1-5 | BRN |
| B+ | B+ | MAIN RELAY | (A6) M1-6 | RED/GRN |
| B+ | B+ | AIR MANAGEMENT | (A7) M1-7 | LT GRN/WHT |
| 3-5 | 3-5 | SERIAL DATA OUTPUT | (A8) M1-8 | VLT |
| 5.0 | 5.0 | DIAGNOSTIC REQUEST | (A9) M1-9 | YEL/RED |
| .1 TO 10 | .1 TO 10 | VSS SIGNAL INPUT | (A10) M1-10 | BLU/WHT |
| 0 | 0 | TPS, MAT SENSOR GROUND | (A11) M1-11 | GRN |
| 0 | 0 | ENGINE GROUND | (A12) M1-12 | BLK |

| WIRE COLOR | PIN | CIRCUIT | VOLTAGE DC KEY "ON" | VOLTAGE DC ENG RUN |
|---|---|---|---|---|
| RED/YEL | (B1) M1-13 | SBF BOX (30A) | B+ | B+ |
| RED/GRN | (B2) M1-14 | FUEL PUMP SIGNAL | 01 | B+ |
| GRN/YEL | (B3) M1-15 | DISTRIBUTOR REFERENCE GROUND | 0 | 0 |
| | (B4) M1-16 | | | |
| GRN/RED | (B5) M1-17 | DISTRIBUTOR REFERENCE INPUT SIGNAL | 15 | 13 |
| | (B6) M1-18 | | | |
| | (B7) M1-19 | | | |
| GRN | (B8) M1-20 | AIR CONDITIONING REQUEST INPUT | 1 | 1 |
| | (B9) M1-21 | | | |
| | (B10) M1-22 | | | |
| | (B11) M1-23 | | | |
| WHT/GRN | (B12) M1-24 | EGR TEMPERATURE SENSOR INPUT | | |

BACK VIEW OF CONNECTOR — 24 PIN M1 CONNECTOR

| VOLTAGE DC KEY "ON" | VOLTAGE DC ENG. RUN | CIRCUIT | PIN | WIRE COLOR |
|---|---|---|---|---|
| | | ACG CONTROL | (C1) M2-1 | GRN/WHT |
| B+ | B+ | A.I.R. INDUCTION | (C2) M2-2 | WHT/BLU |
| SAME AS C5 | SEE TEXT | IAC (COIL B LOW) | (C3) M2-3 | BRN/YEL |
| SAME AS C6 | SEE TEXT | IAC (COIL B HIGH) | (C4) M2-4 | BRN/RED |
| SAME AS C3 | SEE TEXT | IAC (COIL A HIGH) | (C5) M2-5 | BRN/BLK |
| SAME AS C4 | SEE TEXT | IAC (COIL A LOW) | (C6) M2-6 | BRN/WHT |
| B+ | B+ | POWER STEERING SWITCH INPUT | (C7) M2-7 | BLU/BLK |
| | | | (C8) M2-8 | |
| | | | (C9) M2-9 | |
| 1.2 | 1.5 | COOLANT TEMP SENSOR SIGNAL INPUT | (C10) M2-10 | GRY/BLK |
| 4.6 | 1.2 | MAP SENSOR SIGNAL INPUT | (C11) M2-11 | GRN |
| 1-2 | 2-3 | MAT SENSOR SIGNAL INPUT | (C12) M2-12 | GRN/BLU |
| .9 | .9 | TPS SIGNAL INPUT | (C13) M2-13 | RED |
| 5.0 | 5.0 | +5V REFERENCE VOLTAGE OUTPUT | (C14) M2-14 | LT GRN/YEL |
| | | | (C15) M2-15 | |
| B+ | B+ | SBF BOX (30A) | (C16) M2-16 | RED/YEL |

| WIRE COLOR | PIN | CIRCUIT | VOLTAGE DC KEY "ON" | VOLTAGE DC ENG RUN |
|---|---|---|---|---|
| BLK | (D1) M2-17 | ENGINE GROUND | 0 | 0 |
| RED | (D2) M2-18 | MAP, COOLANT, EGR TEMP SENSOR GROUND | 0 | 0 |
| BLK | (D3) M2-19 | ENGINE GROUND | 0 | 0 |
| GRN/BLK | (D4) M2-20 | EST OUTPUT | 15 | 12 |
| GRN/WHT | (D5) M2-21 | IGNITION MODULE BYPASS CIRCUIT | 15 | 45 |
| BLK | (D6) M2-22 | OXYGEN SENSOR ENGINE GROUND CIRCUIT | 0 | 0 |
| YEL/BLK | (D7) M2-23 | OXYGEN SENSOR INPUT SIGNAL | | |
| | (D8) M2-24 | | | |
| | (D9) M2-25 | | | |
| BLK | (D10) M2-26 | ENGINE GROUND | 0 | 0 |
| BLK | (D11) M2-27 | ENGINE GROUND | 0 | 0 |
| GRN/ORG | (D12) M2-28 | EGR CONTROL | B+ | B+ |
| | (D13) M2-29 | | | |
| | (D14) M2-30 | | | |
| ORN/BLU | (D15) M2-31 | INJECTOR 3 & 4 DRIVE | B+ | B+ |
| YEL/WHT | (D16) M2-32 | INJECTOR 1 & 2 DRIVE | B+ | B+ |

BACK VIEW OF CONNECTOR — 32 PIN M2 CONNECTOR

- All voltages showing "B+" indicates battery or charging system voltage
- All voltages showing "O" are ground circuits, and should read less than 0.10 volts

## I-TEC DIAGNOSTIC CODES AND TROUBLESHOOTING CHART—1988 IMPULSE

| DIAGNOSED ITEM | DIAGNOSED CONTENT: FAULT MODE | DIAGNOSED CONTENT: MICRO COMPUTER INPUT | CODE NO | CODE MEMORY | FAIL-SAFE FUNCTION OF MICRO-COMPUTER | ENGINE CONDITION | VEHICLE CONDITION |
|---|---|---|---|---|---|---|---|
| Engine is not started | – | Engine speed less than 200 rpm | 12 | No | None | Normal | Normal |
| $O_2$ sensor system | Harness open Sensor deterioration | Intermediate voltage | 13 | Yes | Fuel is not compensated by $O_2$ signal | Exhaust is worsened | No noticeable abnormal operation |
| | Incorrect signal (Lean) | Lean signal (Low voltage) | 44 | Yes | | | |
| | Incorrect signal (Rich) | Rich signal (High voltage) | 45 | Yes | | | |
| Water temperature sensor system | Shorted with ground | Insufficient signal | 14 | Yes | Coolant temperature is assumed to be 85°C (185°F) | · The engine does not operate normally when temperature is low and signal is insufficient, or when temperature is high and signal is excessive. Or the engine cannot be started<br>· The engine does not operate smoothly when temperature is low after the fail-safe function is actuated | |
| | Incorrect signal | Signal is less than 0°C (32°F) even after the engine is warmed up sufficiently | 15 | Yes | | | |
| | Harness open | Excessive signal | 16 | Yes | | | |
| Throttle valve switch system — Both idle contact and full contact | Both idle contact and full make contact simultaneously | Both contacts make simultaneously | 21 | Yes | Both signals are assumed to be OFF | Fuel is not cut after the fail-safe function is actuated, and lean fuel is resulted when the throttle valve is fully opened | Fuel consumption rate is worsened after the fail-safe function is actuated. Fuel tends to be lean when the throttle valve is fully opened. |
| Throttle valve switch system — Idle contact | Normally make contact | Signal is sent continuously (Not diagnosed when the air flow sensor system is defective) | 43 | Yes | Assumed to be OFF | · Fuel cut range appears during running<br>· Fuel is not cut after the fail-safe function is actuated | · The vehicle does not run smoothly<br>· Fuel consumption rate is worsened after the fail-safe function is actuated |
| Throttle valve switch system — Full contact | Normally make contact | Signal is sent continuously | 65 | Yes | Assumed to be OFF | · Air-fuel ratio is high when partially loaded<br>· Fuel tends to be lean when the throttle valve is fully opened after the fail-safe function is actuated | · Fuel consumption rate and exhaust are worsened. The spark plugs are carboned or the engine stalls depending on the condition.<br>· Fuel tends to be lean when the throttle valve is fully opened after the fail-safe function is actuated |

## I-TEC DIAGNOSTIC CODES AND TROUBLESHOOTING CHART—1988 IMPULSE

| DIAGNOSED ITEM | DIAGNOSED CONTENT | | CODE NO. | CODE MEMORY | FAIL-SAFE FUNCTION OF MICRO-COMPUTER | ENGINE CONDITION | VEHICLE CONDITION |
|---|---|---|---|---|---|---|---|
| | FAULT MODE | MICRO-COMPUTER INPUT | | | | | |
| Starter signal system | Normally open | Signal is not input | 22 | Yes | None | Normal | No abnormality is felt |
| Crank angle sensor system | · No signal arrives.<br>· Faulty signal | · No angle sensor signal is input.<br>· Idling speed is lower than the actual speed. | 41 | Yes | None | · The engine stalls. The engine cannot be started.<br>· Air-fuel ratio tends to be high. | · Stalls or cannot be started<br>· Both fuel consumption rate and exhaust are worsened. |
| Air flow sensor system | Harness open, shorting with ground, or broken hot wire. | Insufficient signal | 61 | Yes | Injection pulse is changed over at throttle valve position. | · The engine stalls depending on the condition.<br>· The engine may be slightly unstable after the fail-safe function is actuated. | · The engine stalls depending on the condition.<br>· The accelerator response is worsened after the fail-safe function is actuated. The engine may not operate smoothly. |
| | Broken cold wire | Excessive signal | 62 | Yes | | | |
| Car speed sensor system | No signal arrives | No signal is input (Not diagnosed when the air flow sensor system is defective) | 63 | Yes | None | The fuel cut operates even when running under a low speed. | Does not run smoothly and shakes. |
| Knocking sensor system | Harness open or shorting with ground. | Excessive or insufficient signal | 66 | Yes | Ignition timing is delayed. | · Engine knocking.<br>· Output drops after the fail-safe function is actuated. | · Engine knocking.<br>· Output drops after the fail-safe function is actuated. |
| Micro-computer unit | Abnormal LSI (1) | – | 51 | Yes | · Injection pulse is changed over at throttle valve position<br>· Fixed ignition timing. | · In the worst case, the engine stalls, or the engine does not operate smoothly at a certain time.<br>· The engine does not stall but it does not satisfy the specification | · In the worst case, the engine stalls, or the engine does not operate smoothly at a certain time.<br>· The exhaust and running performance deviate from the specification. |
| | Abnormal LSI (2) | – | 52 | Yes | | | |
| | Abnormal LSI (3) | – | 55 | Yes | | | |

## I-TEC DIAGNOSTIC CODES AND TROUBLESHOOTING CHART—1988 IMPULSE

| DIAGNOSED ITEM | DIAGNOSED CONTENT | | CODE NO. | CODE MEMORY | FAIL-SAFE FUNCTION OF MICRO-COMPUTER | ENGINE CONDITION | VEHICLE CONDITION |
|---|---|---|---|---|---|---|---|
| | FAULT MODE | MICRO-COMPUTER INPUT | | | | | |
| Power transistor system for ignition | Output terminal shorted with ground. | – | 23 | Yes | None | · The engine stalls.<br>· Cannot be started. | · The engine stalls.<br>· Cannot be started. |
| | Harness open | – | 35 | Yes | | | |
| | Defective transistor or grounding system | – | 54 | Yes | | | |
| Vacuum switching valve system | Output terminal shorted with ground or harness open | – | 25 | Yes | None | Output terminal shorted with ground.<br>· Fuel pressure rises continuously when the engine is cold, causing fuel to be rich.<br>· When the engine is warmed up, automatically corrected as the $O_2$ sensor compensates<br>Same as below for harness open. | Output terminal shorted with ground.<br>· Fuel consumption rate and exhaust gas level are worsened when the engine is cold or when accelerated rapidly.<br>Same as below for harness open. |
| | Defective transistor or grounding system. | – | 53 | Yes | None | When the engine is overheated, air-fuel ratio tends to be lean.<br>No problem when the engine is warmed up normally. | The engine cannot be restarted smoothly when coolant temperature is high. |
| Fuel injector system. | Output terminal shorted with ground or harness open | – | 33 | Yes | None | · The engine stalls.<br>· Cannot be started. | · The engine stalls.<br>· Cannot be started. |
| | Defective transistor or grounding system. | – | 64 | Yes | | | |
| Throttle position sensor system TURBO CONTROL SYSTEM | Abnormal signal | Insufficient signal | 71 | Yes | None | The engine stalls depending on the condition | The engine stalls depending on the condition |
| EGR vacuum switching valve system | Output terminal shorted with ground or harness open | – | 72 | Yes | None | Exhaust is worsened | No noticeable abnormal operation |
| | Defective transistor or grounding system | – | 73 | | | | |

## SIGNAL CODE NUMBER 12—1988 IMPULSE

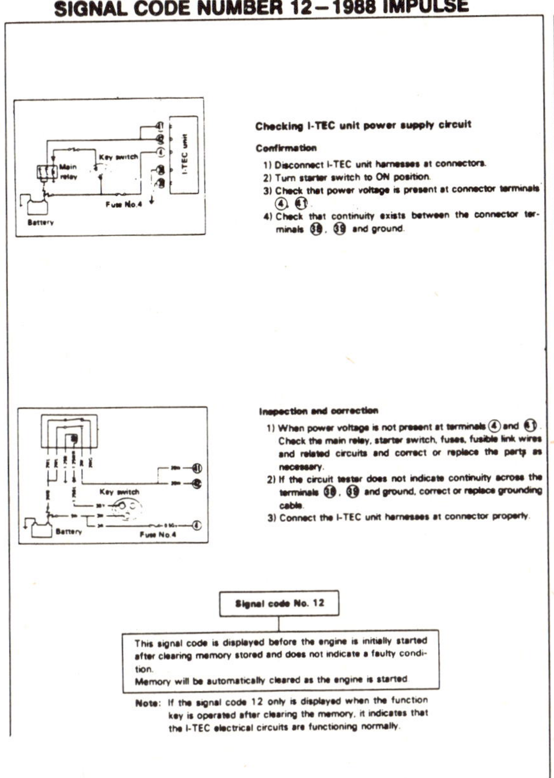

**Checking I-TEC unit power supply circuit**

**Confirmation**

1) Disconnect I-TEC unit harnesses at connectors.
2) Turn starter switch to ON position.
3) Check that power voltage is present at connector terminals ④, ㊶.
4) Check that continuity exists between the connector terminals ㊳, ㊴ and ground.

**Inspection and correction**

1) When power voltage is not present at terminals ④ and ㊶. Check the main relay, starter switch, fuses, fusible link wires and related circuits and correct or replace the parts as necessary.
2) If the circuit tester does not indicate continuity across the terminals ㊳, ㊴ and ground, correct or replace grounding cable.
3) Connect the I-TEC unit harnesses at connector properly.

**Signal code No. 12**

This signal code is displayed before the engine is initially started after clearing memory stored and does not indicate a faulty condition.
Memory will be automatically cleared as the engine is started.

**Note:** If the signal code 12 only is displayed when the function key is operated after clearing the memory, it indicates that the I-TEC electrical circuits are functioning normally.

## SIGNAL CODES NUMBER 13, 44 AND 45 NO SIGNALS OR ABNORMAL SIGNALS FROM THE OXYGEN SENSOR—1988 IMPULSE

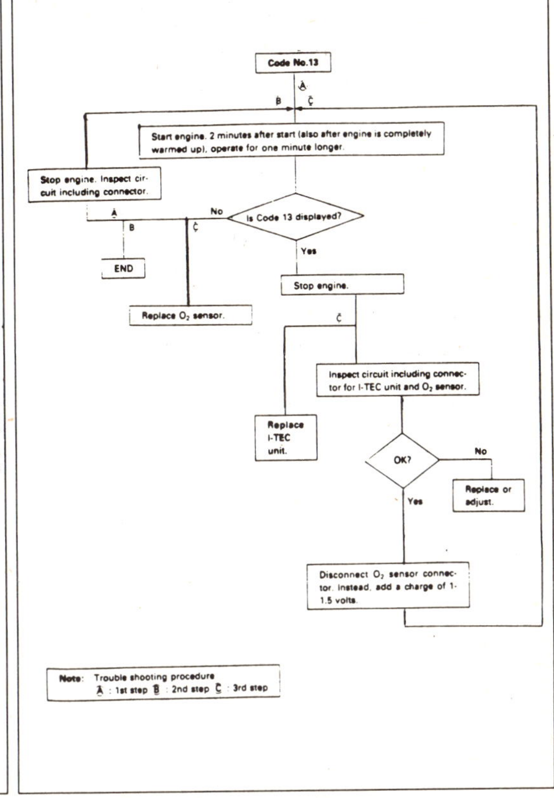


## SIGNAL CODES NUMBER 44 AND 45—1988 IMPULSE

Code No.44 and 45

Warm up engine (to 70°C or more.) Operate engine at 1/4 throttle level, under 2500 r.p.m., run steadily for more than 2 minutes.

Inspect circuit including connector.

Is code 44 or code 45 displayed? — No → END

Yes

Read output of $O_2$ sensor at I-TEC unit connector section (while engine is racing lightly, and with no load).

Reading is steady, or with slight fluctuation, and is under 0.3V or above 0.6V. → Inspect circuit including connector.

Reading moves back and forth from less than 0.35V to above 0.50V. → Stop engine. Check I-TEC unit connector.

Is I-TEC connector OK? — No → Correct

Yes → Replace I-TEC unit.

## INSPECTION FOR SIGNAL CODES NUMBER 13, 44 AND 45—1988 IMPULSE

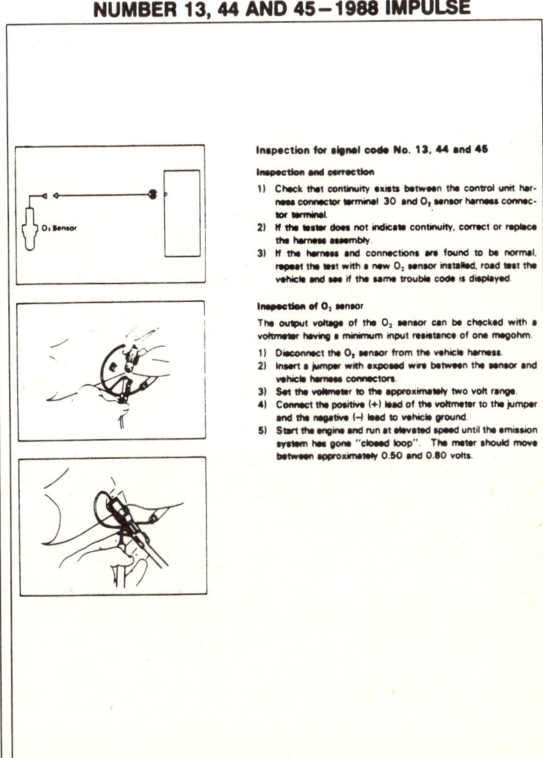

**Inspection for signal code No. 13, 44 and 45**

**Inspection and correction**

1) Check that continuity exists between the control unit harness connector terminal 30 and $O_2$ sensor harness connector terminal.
2) If the tester does not indicate continuity, correct or replace the harness assembly.
3) If the harness and connections are found to be normal, repeat the test with a new $O_2$ sensor installed, road test the vehicle and see if the same trouble code is displayed.

**Inspection of $O_2$ sensor**

The output voltage of the $O_2$ sensor can be checked with a voltmeter having a minimum input resistance of one megohm.

1) Disconnect the $O_2$ sensor from the vehicle harness.
2) Insert a jumper with exposed wire between the sensor and vehicle harness connectors.
3) Set the voltmeter to the approximately two volt range.
4) Connect the positive (+) lead of the voltmeter to the jumper and the negative (−) lead to vehicle ground.
5) Start the engine and run at elevated speed until the emission system has gone "closed loop". The meter should move between approximately 0.50 and 0.80 volts.

## SIGNAL CODE NUMBER 14—1988 IMPULSE

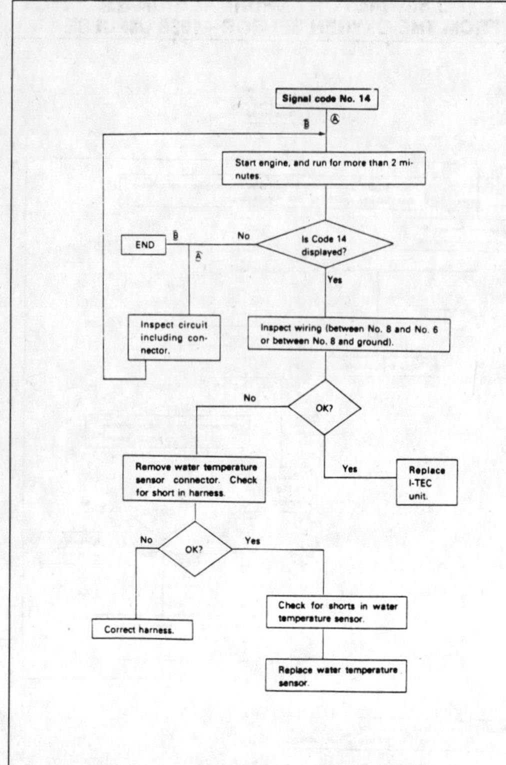


## SIGNAL CODES NUMBER 15 AND 16—1988 IMPULSE

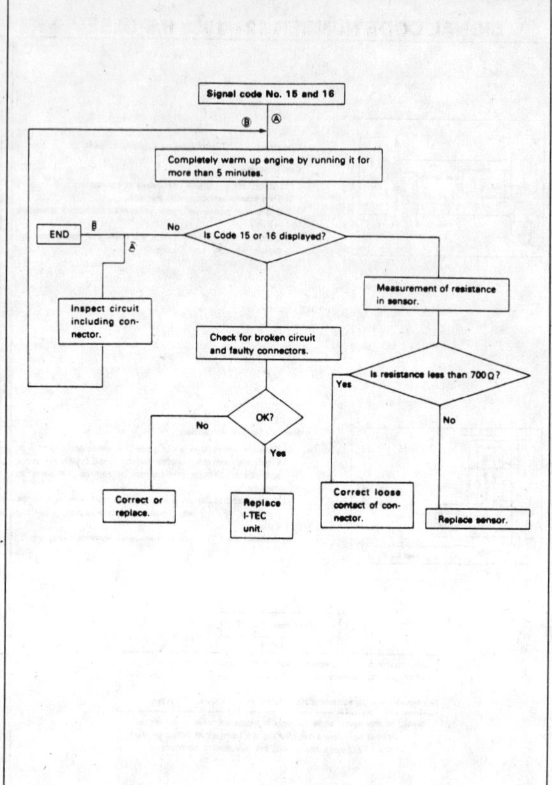


## INSPECTION FOR SIGNAL CODES NUMBER 14, 15 AND 16—1988 IMPULSE

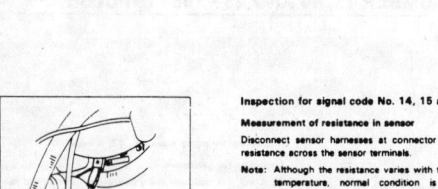

**Inspection for signal code No. 14, 15 and 16**

**Measurement of resistance in sensor**

Disconnect sensor harnesses at connector and measure the resistance across the sensor terminals.

**Note:** Although the resistance varies with the engine coolant temperature, normal condition is indicated when measured resistance falls within the range shown in the following table:

| Coolant temperatue °C (°F) | Normal resistance KΩ |
|---|---|
| -10 ( 14) | 7— 12 |
| 10 ( 50) | 3 — 5 |
| 20 ( 68) | 2 — 3 |
| 50 (122) | 0.7 — 1 |
| 80 (176) | 0.2 — 0.4 |

If measured resistance deviates from the specified range, replace water temperature sensor.

**Wiring**

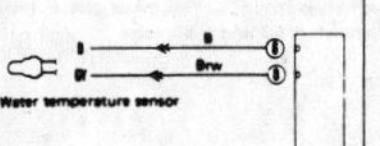


| Signal code No. | Wiring | Ω |
|---|---|---|
| 14 | (8)——B | ∞ |
| | (8)——Ground | ∞ |
| 15, 16 | (8)——GY | 0 |
| | (6)——B | 0 |
| | (6) Ground | ∞ |

When abnormal condition is indicated, correct or replace applicable harnesses.

## SIGNAL CODE NUMBER 21—1988 IMPULSE

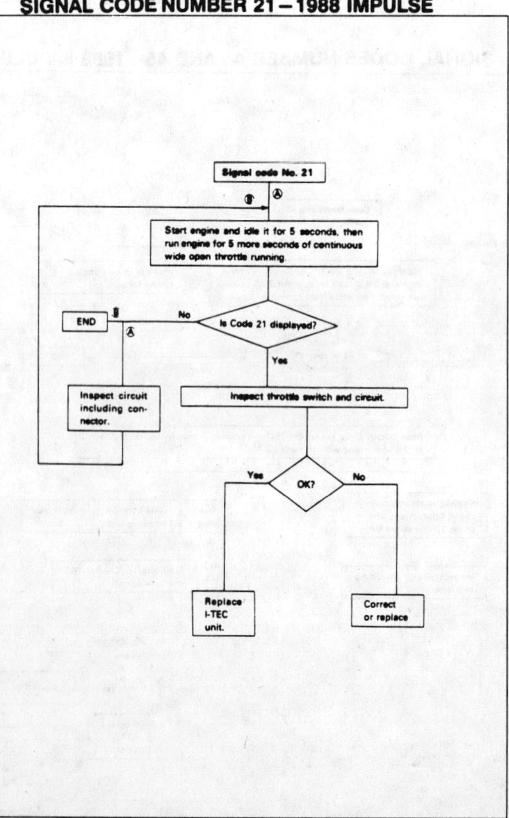

## SIGNAL CODES NUMBER 43 AND 65—1988 IMPULSE

## THROTTLE SWITCH INSPECTION—1988 IMPULSE

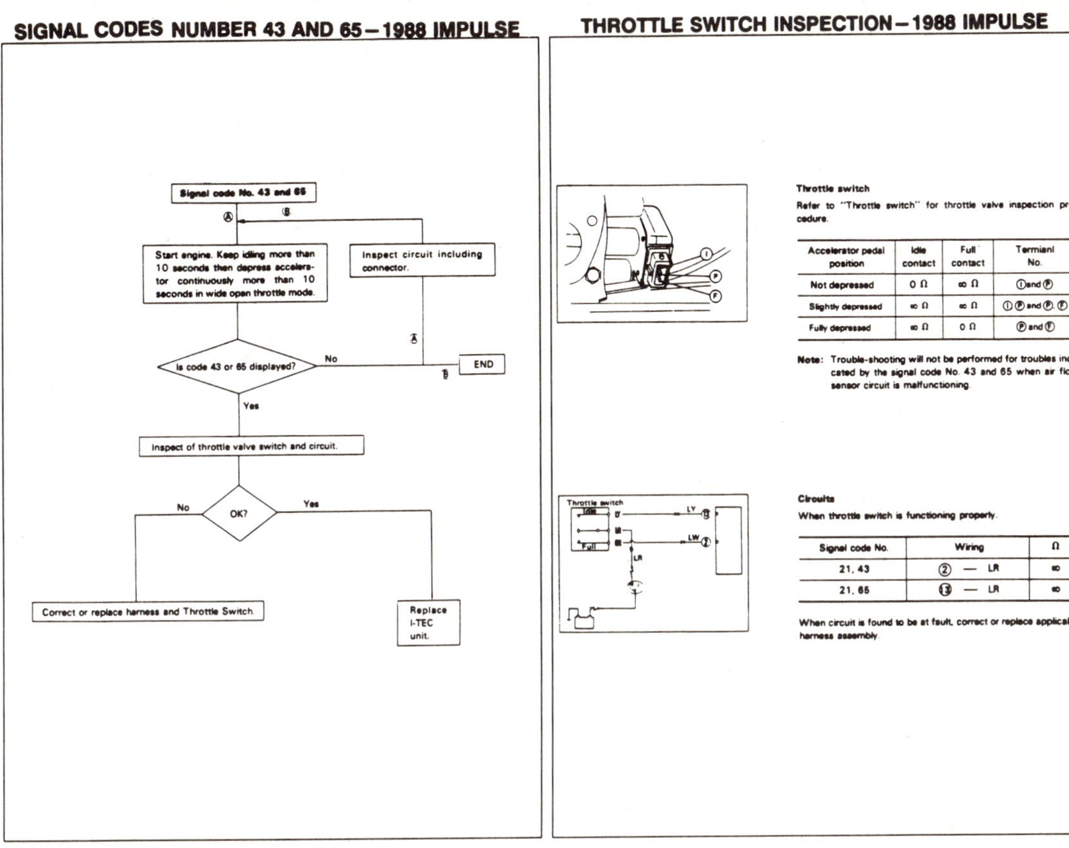


**Throttle switch**

Refer to "Throttle switch" for throttle valve inspection procedure.

| Accelerator pedal position | Idle contact | Full contact | Termianl No. |
|---|---|---|---|
| Not depressed | 0 Ω | ∞ Ω | Ⓘ and Ⓟ |
| Slightly depressed | ∞ Ω | ∞ Ω | Ⓘ Ⓟ and Ⓟ, Ⓕ |
| Fully depressed | ∞ Ω | 0 Ω | Ⓟ and Ⓕ |

**Note:** Trouble-shooting will not be performed for troubles indicated by the signal code No. 43 and 65 when air flow sensor circuit is malfunctioning.

**Circuits**

When throttle switch is functioning properly.

| Signal code No. | Wiring | Ω |
|---|---|---|
| 21, 43 | ② — LR | ∞ |
| 21, 65 | ⑬ — LR | ∞ |

When circuit is found to be at fault, correct or replace applicable harness assembly.

## SIGNAL CODE NUMBER 22—1988 IMPULSE

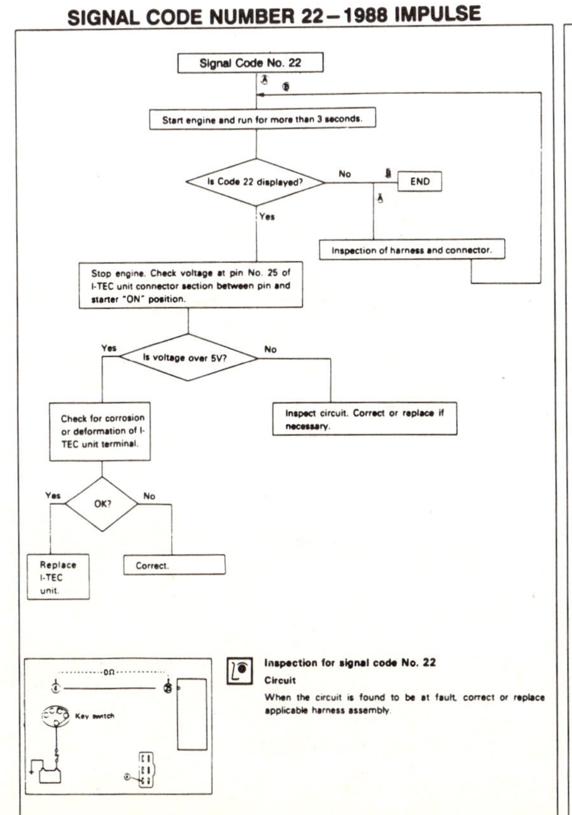


## SIGNAL CODE NUMBER 41—1988 IMPULSE

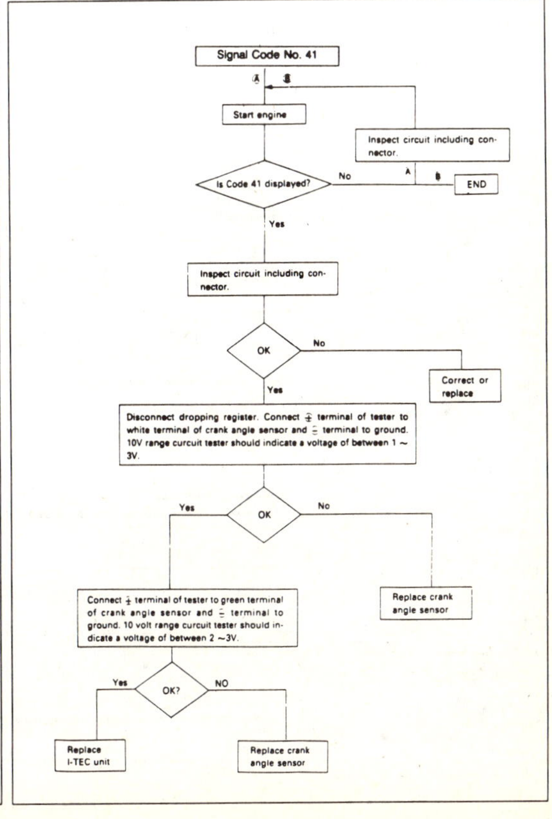

## INSPECTION FOR SIGNAL CODES NUMBER 22 AND 41—1988 IMPULSE

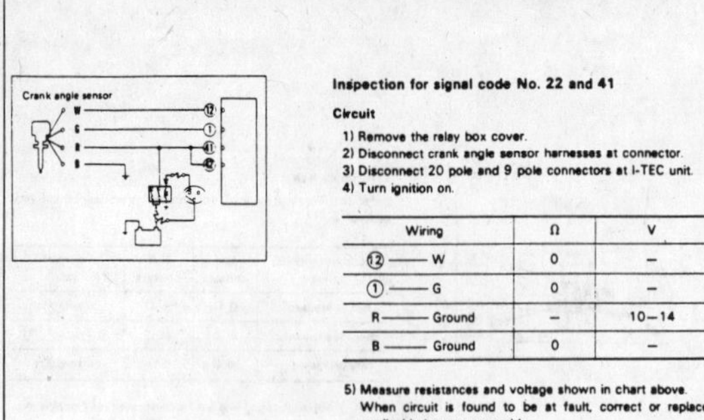


**Inspection for signal code No. 22 and 41**

**Circuit**

1) Remove the relay box cover.
2) Disconnect crank angle sensor harnesses at connector.
3) Disconnect 20 pole and 9 pole connectors at I-TEC unit.
4) Turn ignition on.

| Wiring | Ω | V |
|---|---|---|
| ⑫ —— W | 0 | – |
| ① —— G | 0 | – |
| R —— Ground | – | 10–14 |
| B —— Ground | 0 | – |

5) Measure resistances and voltage shown in chart above. When circuit is found to be at fault, correct or replace applicable harness assembly.
6) Install the relay box cover and connect harness at connector securely.

**When circuit is found to be in normal condition, replace distributor assembly.**

## SIGNAL CODES NUMBER 61 AND 62—1988 IMPULSE

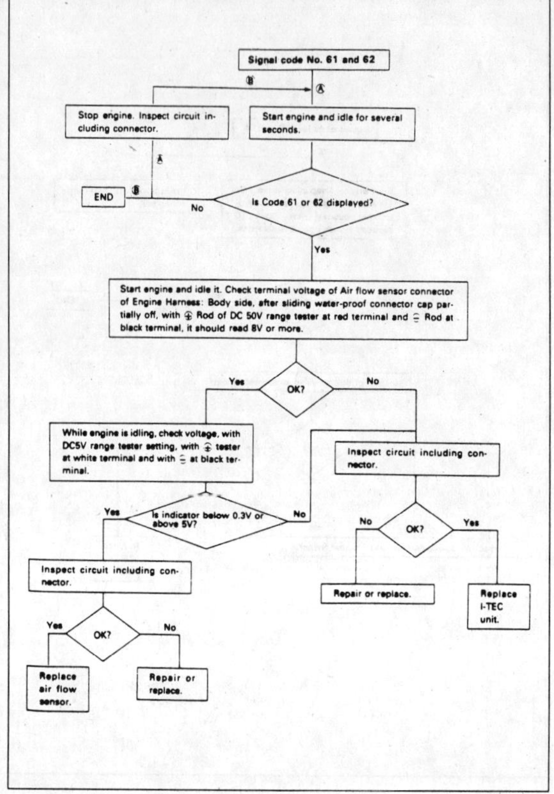


## INSPECTION FOR SIGNAL CODES NUMBER 61 AND 62—1988 IMPULSE

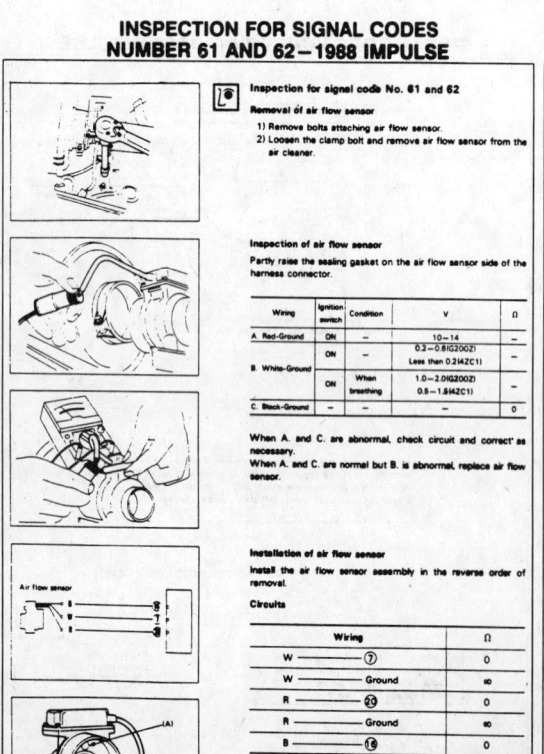

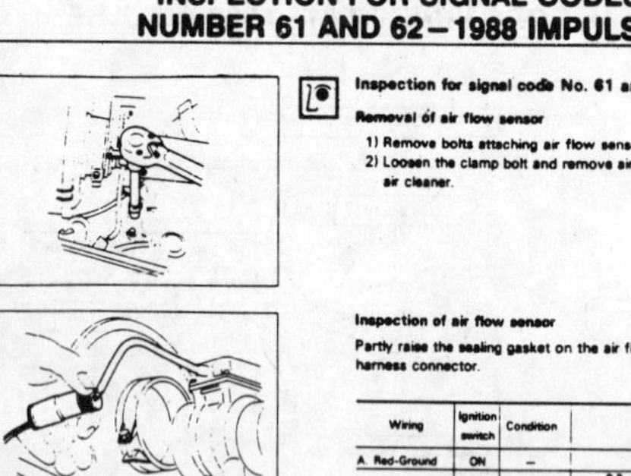

**Inspection for signal code No. 61 and 62**

**Removal of air flow sensor**

1) Remove bolts attaching air flow sensor.
2) Loosen the clamp bolt and remove air flow sensor from the air cleaner.

**Inspection of air flow sensor**

Partly raise the sealing gasket on the air flow sensor side of the harness connector.

| Wiring | Ignition switch | Condition | V | Ω |
|---|---|---|---|---|
| A. Red-Ground | ON | – | 10–14 | – |
| B. White-Ground | ON | – | 0.2–0.6(G200Z)<br>Less than 0.2(4ZC1) | – |
| | ON | When breathing | 1.0–2.0(G200Z)<br>0.5–1.5(4ZC1) | – |
| C. Black-Ground | – | – | – | 0 |

When A. and C. are abnormal, check circuit and correct as necessary.
When A. and C. are normal but B. is abnormal, replace air flow sensor.

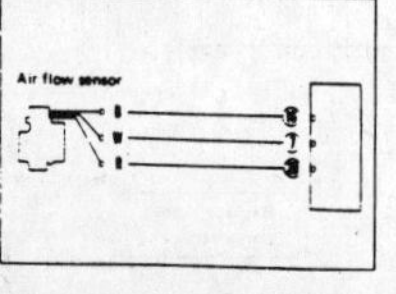


**Installation of air flow sensor**

Install the air flow sensor assembly in the reverse order of removal.

**Circuits**

| Wiring | Ω |
|---|---|
| W —— ⑦ | 0 |
| W —— Ground | ∞ |
| R —— ⑳ | 0 |
| R —— Ground | ∞ |
| B —— ⑯ | 0 |

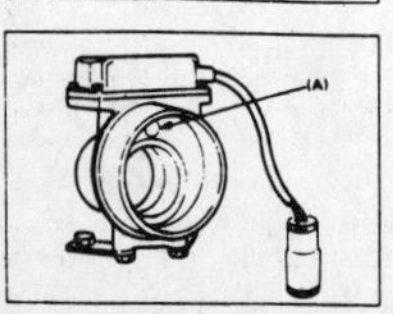


When circuit is found to be at fault, correct or replace applicable harness assembly.

**Note:** If the air flow sensor has been removed, check by-pass circuit (A) for contamination and clean with carburetor cleaner as necessary.

## SIGNAL CODE NUMBER 63—1988 IMPULSE

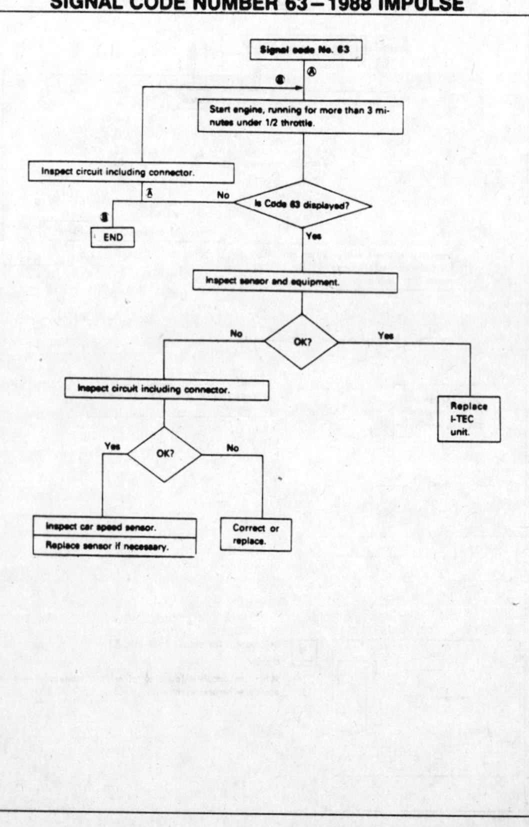

## INSPECTION FOR SIGNAL CODE NUMBER 63–1988 IMPULSE

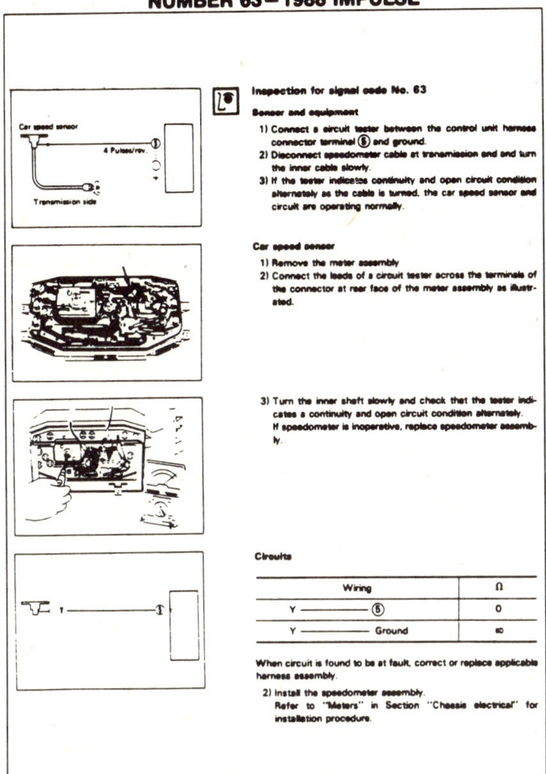


**Inspection for signal code No. 63**

**Sensor and equipment**

1) Connect a circuit tester between the control unit harness connector terminal ⑤ and ground.
2) Disconnect speedometer cable at transmission end and turn the inner cable slowly.
3) If the tester indicates continuity and open circuit condition alternately as the cable is turned, the car speed sensor and circuit are operating normally.

**Car speed sensor**

1) Remove the meter assembly
2) Connect the leads of a circuit tester across the terminals of the connector at rear face of the meter assembly as illustrated.
3) Turn the inner shaft slowly and check that the tester indicates a continuity and open circuit condition alternately.
If speedometer is inoperative, replace speedometer assembly.

**Circuits**

| Wiring | Ω |
|---|---|
| Y ——— ⑤ | 0 |
| Y ——— Ground | ∞ |

When circuit is found to be at fault, correct or replace applicable harness assembly.

2) Install the speedometer assembly.
Refer to "Meters" in Section "Chassis electrical" for installation procedure.

## SIGNAL CODE NUMBER 66–1988 IMPULSE

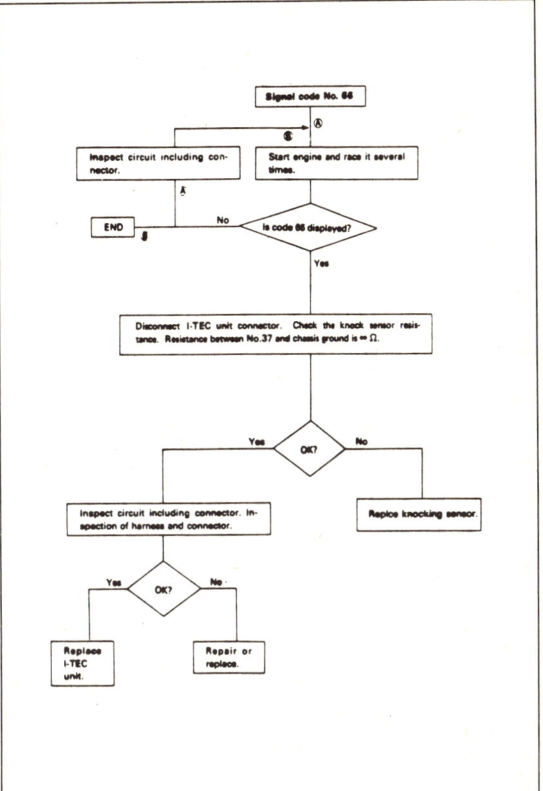


## INSPECTION FOR SIGNAL CODE NUMBER 66 1988 IMPULSE

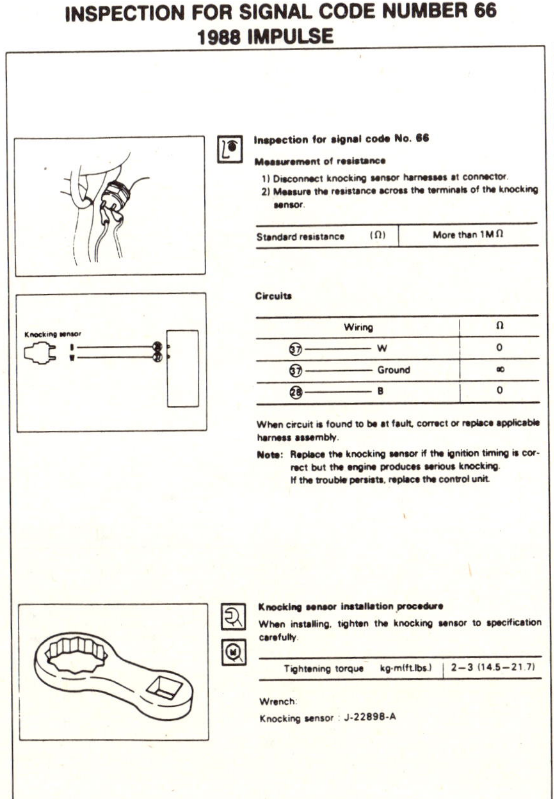


**Inspection for signal code No. 66**

**Measurement of resistance**

1) Disconnect knocking sensor harnesses at connector.
2) Measure the resistance across the terminals of the knocking sensor.

| Standard resistance (Ω) | More than 1M Ω |
|---|---|

**Circuits**

| Wiring | Ω |
|---|---|
| ㊲ ——— W | 0 |
| ㊲ ——— Ground | ∞ |
| ㉘ ——— B | 0 |

When circuit is found to be at fault, correct or replace applicable harness assembly.

**Note:** Replace the knocking sensor if the ignition timing is correct but the engine produces serious knocking.
If the trouble persists, replace the control unit.

**Knocking sensor installation procedure**

When installing, tighten the knocking sensor to specification carefully.

| Tightening torque kg-m(ft.lbs.) | 2–3 (14.5–21.7) |
|---|---|

Wrench:
Knocking sensor : J-22898-A

## SIGNAL CODES NUMBER 51, 52 AND 55 1988 IMPULSE

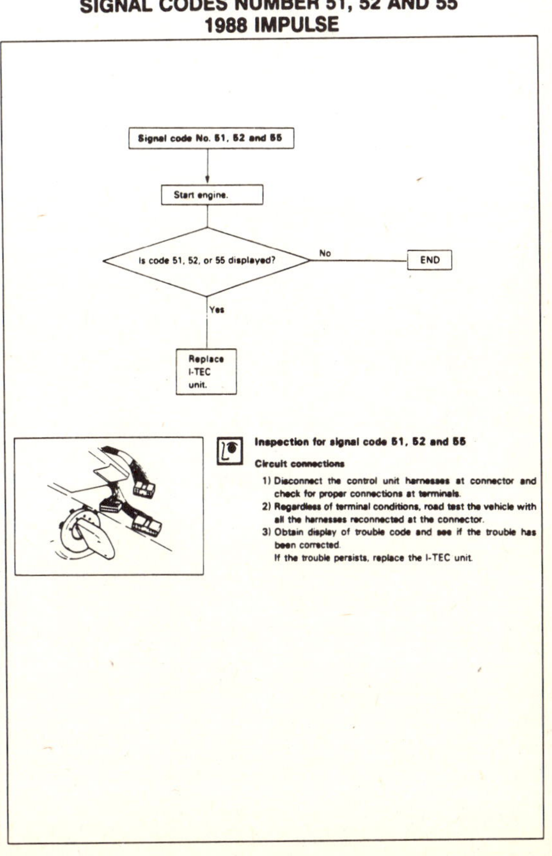


**Inspection for signal code 51, 52 and 55**

**Circuit connections**

1) Disconnect the control unit harnesses at connector and check for proper connections at terminals.
2) Regardless of terminal conditions, road test the vehicle with all the harnesses reconnected at the connector.
3) Obtain display of trouble code and see if the trouble has been corrected.
If the trouble persists, replace the I-TEC unit.

## SIGNAL CODES NUMBER 23, 35 AND 54 1988 IMPULSE

Signal code No. 23, 35 and 54

Start engine.

Is Code 23, 35 and 54 displayed?

No → END

Code 54 → Replace I-TEC unit.

Code 23, 35 → Inspect circuit including connector.

Stop engine. Check and adjust irregularities by pulling and pushing related connectors.

OK?

No → Correct or replace.

Yes → Inspect power transistor.

OK?

Yes → Replace I-TEC unit.

No → Correct or replace.

## INSPECTING THE POWER TRANSISTOR 1988 IMPULSE

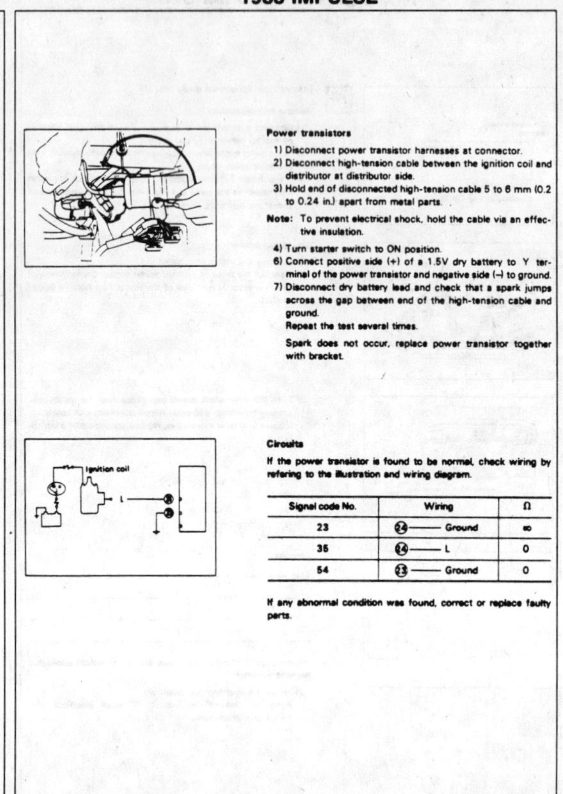


**Power transistors**

1) Disconnect power transistor harnesses at connector.
2) Disconnect high-tension cable between the ignition coil and distributor at distributor side.
3) Hold end of disconnected high-tension cable 5 to 6 mm (0.2 to 0.24 in.) apart from metal parts.

**Note:** To prevent electrical shock, hold the cable via an effective insulation.

4) Turn starter switch to ON position.
6) Connect positive side (+) of a 1.5V dry battery to Y terminal of the power transistor and negative side (−) to ground.
7) Disconnect dry battery lead and check that a spark jumps across the gap between end of the high-tension cable and ground.
Repeat the test several times.
Spark does not occur, replace power transistor together with bracket.

**Circuits**

If the power transistor is found to be normal, check wiring by refering to the illustration and wiring diagram.

| Signal code No. | Wiring | Ω |
|---|---|---|
| 23 | (24) —— Ground | ∞ |
| 35 | (24) —— L | 0 |
| 54 | (23) —— Ground | 0 |

If any abnormal condition was found, correct or replace faulty parts.

## SIGNAL CODES NUMBER 25 AND 53 1988 IMPULSE

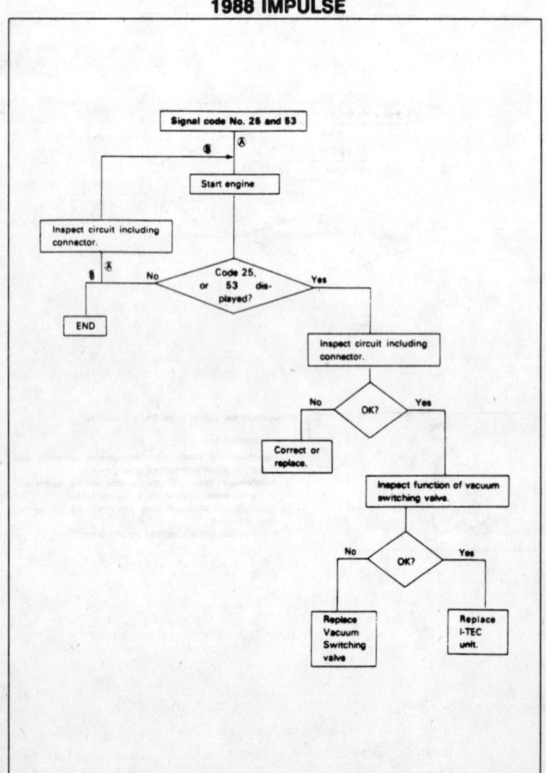


## INSPECTION FOR SIGNAL CODES NUMBER 25 AND 53 1988 IMPULSE

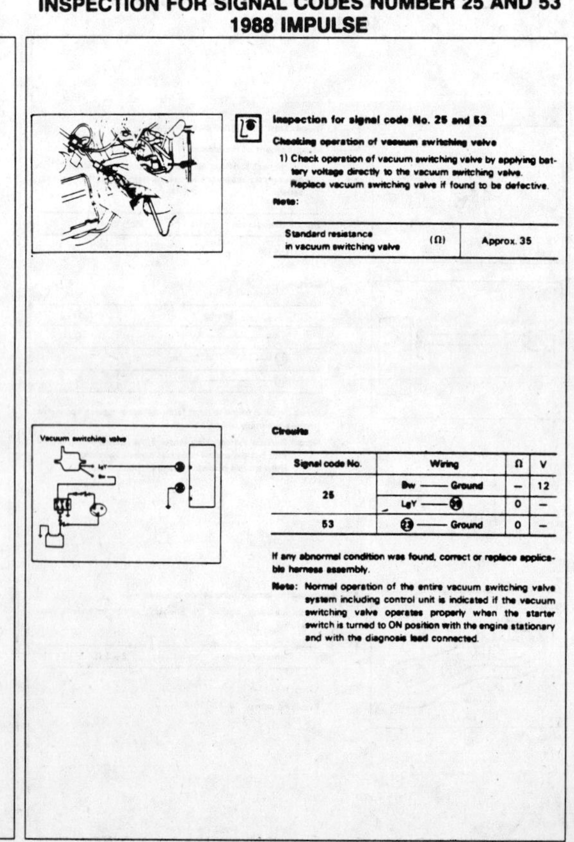


**Inspection for signal code No. 25 and 53**

**Checking operation of vacuum switching valve**

1) Check operation of vacuum switching valve by applying battery voltage directly to the vacuum switching valve.
Replace vacuum switching valve if found to be defective.

**Note:**

| Standard resistance in vacuum switching valve | (Ω) | Approx. 35 |
|---|---|---|

**Circuits**

| Signal code No. | Wiring | Ω | V |
|---|---|---|---|
| 25 | Bw —— Ground | − | 12 |
| | LgY —— (24) | 0 | − |
| 53 | (23) —— Ground | 0 | − |

If any abnormal condition was found, correct or replace applicable harness assembly.

**Note:** Normal operation of the entire vacuum switching valve system including control unit is indicated if the vacuum switching valve operates properly when the starter switch is turned to ON position with the engine stationary and with the diagnosis lead connected.

## SIGNAL CODE NUMBER 33—1988 IMPULSE

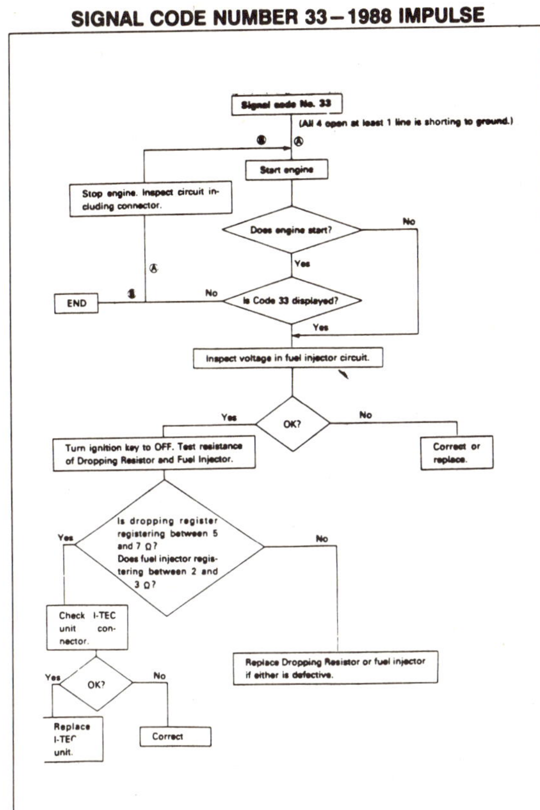


## INSPECTION FOR SIGNAL CODE NUMBER 33 1988 IMPULSE

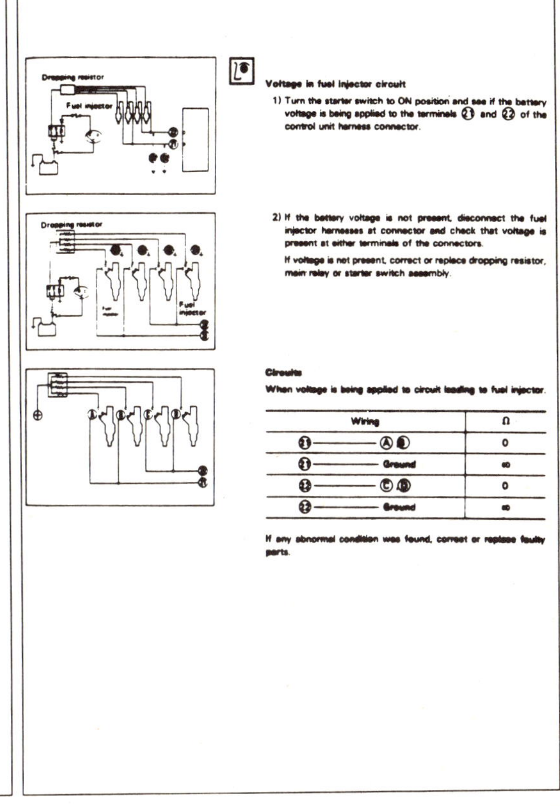


**Voltage in fuel injector circuit**

1) Turn the starter switch to ON position and see if the battery voltage is being applied to the terminals ㉑ and ㉒ of the control unit harness connector.

2) If the battery voltage is not present, disconnect the fuel injector harnesses at connector and check that voltage is present at either terminals of the connectors.

If voltage is not present, correct or replace dropping resistor, main relay or starter switch assembly.

**Circuits**

When voltage is being applied to circuit leading to fuel injector.

| Wiring | Ω |
|---|---|
| ㉑ ——— Ⓐ Ⓑ | 0 |
| ㉑ ——— Ground | ∞ |
| ㉒ ——— Ⓒ Ⓓ | 0 |
| ㉒ ——— Ground | ∞ |

If any abnormal condition was found, correct or replace faulty parts.

## SIGNAL CODE NUMBER 64—1988 IMPULSE

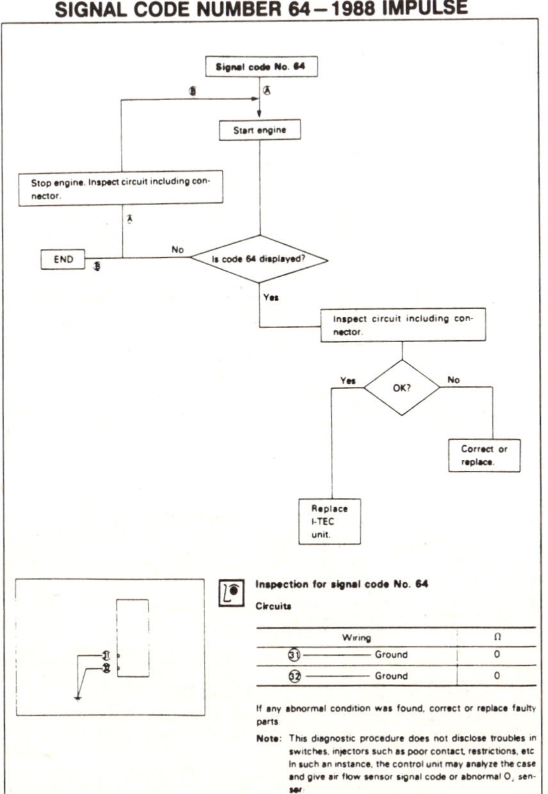


**Inspection for signal code No. 64**

**Circuits**

| Wiring | Ω |
|---|---|
| ㉛ ——— Ground | 0 |
| ㉜ ——— Ground | 0 |

If any abnormal condition was found, correct or replace faulty parts.

**Note:** This diagnostic procedure does not disclose troubles in switches, injectors such as poor contact, restrictions, etc. In such an instance, the control unit may analyze the case and give air flow sensor signal code or abnormal $O_2$ sensor.

## SIGNAL CODE NUMBER 71—1988 IMPULSE

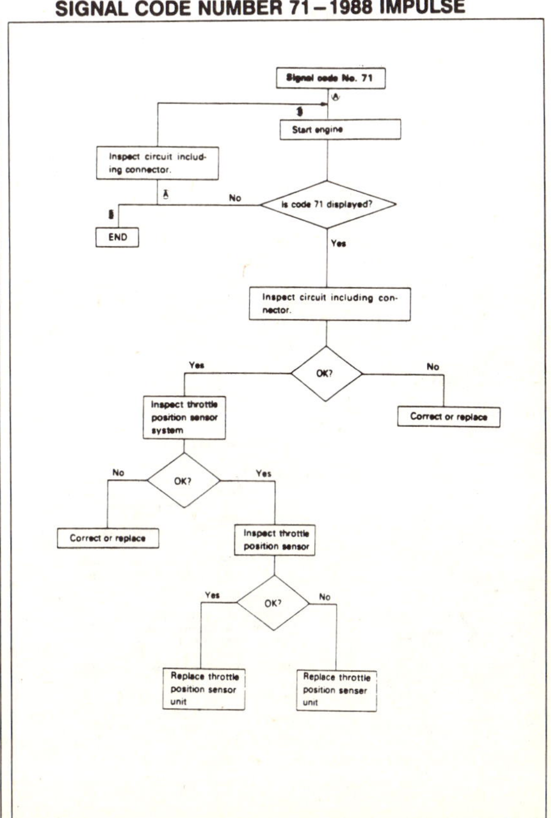

### SIGNAL CODES NUMBER 72 AND 73—1988 IMPULSE

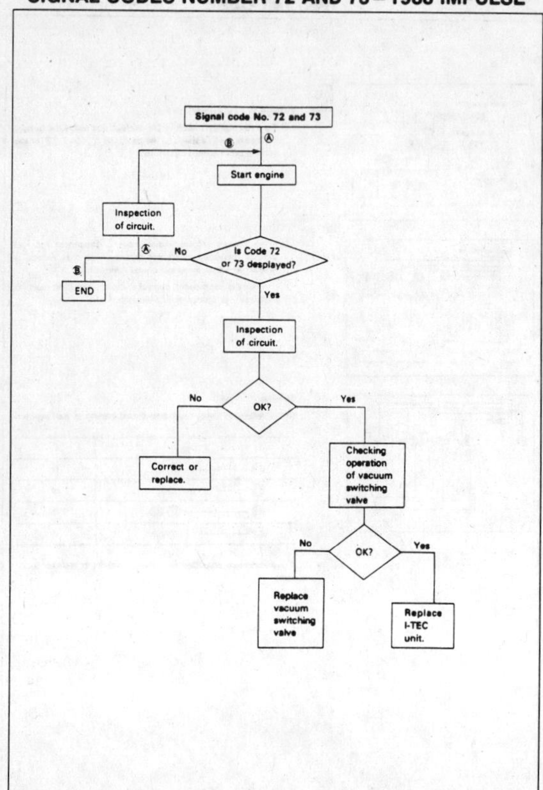


### INSPECTION FOR SIGNAL CODES 71, 72 AND 73 1988 IMPULSE

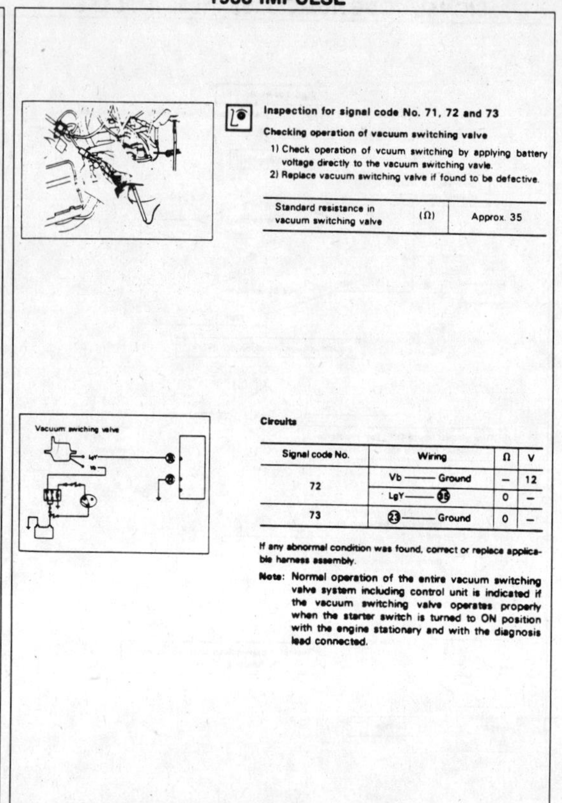

**Inspection for signal code No. 71, 72 and 73**

**Checking operation of vacuum switching valve**

1) Check operation of vcuum switching by applying battery voltage directly to the vacuum switching vavle.
2) Replace vacuum switching valve if found to be defective.

| Standard resistance in vacuum switching valve | (Ω) | Approx. 35 |
|---|---|---|

**Circuits**

| Signal code No. | Wiring | Ω | V |
|---|---|---|---|
| 72 | Vb —— Ground | – | 12 |
| | LgY —— 35 | 0 | – |
| 73 | 23 —— Ground | 0 | – |

If any abnormal condition was found, correct or replace applicable harness assembly.

**Note:** Normal operation of the entire vacuum switching valve system including control unit is indicated if the vacuum switching valve operates properly when the starter switch is turned to ON position with the engine stationary and with the diagnosis lead connected.

### TURBOCHARGER CONTROL SYSTEM SELF-DIAGNOSIS AND TROUBLESHOOTING PROCEDURES 1988 IMPULSE

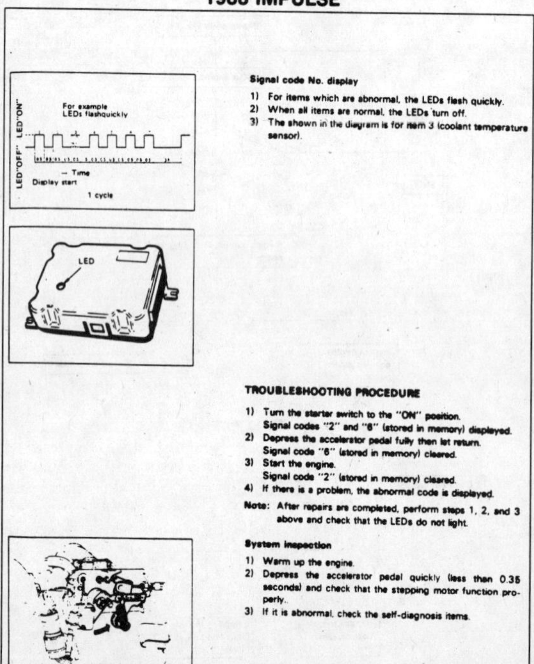


**Signal code No. display**

1) For items which are abnormal, the LEDs flash quickly.
2) When all items are normal, the LEDs turn off.
3) The shown in the diagram is for item 3 (coolant temperature sensor).

**TROUBLESHOOTING PROCEDURE**

1) Turn the starter switch to the "ON" position.
Signal codes "2" and "6" (stored in memory) displayed.
2) Depress the accelerator pedal fully then let return.
Signal code "6" (stored in memory) cleared.
3) Start the engine.
Signal code "2" (stored in memory) cleared.
4) If there is a problem, the abnormal code is displayed.

**Note:** After repairs are completed, perform steps 1, 2, and 3 above and check that the LEDs do not light.

**System inspection**

1) Warm up the engine.
2) Depress the accelerator pedal quickly (less than 0.35 seconds) and check that the stepping motor function properly..
3) If it is abnormal, check the self-diagnosis items.

### TURBOCHARGER CONTROL SYSTEM DIAGNOSTIC CODES TROUBLESHOOTING CHART—1988 IMPULSE

| Diagnosis item | Trouble assessing conditions | System operation | LED display |
|---|---|---|---|
| Knock control circuit | Abnormal knock control signal | Stopped | 1 |
| Ignition signal circuit | No input of ignition signal for about 0.3 sec. or more. | Stopped | 2 |
| Water temperature sensor circuit | Short or open circuit between Terminal ⑩ and ㉑ (water temperature) sensor. | Water temperature correction; fixed to θw = 250 | 3 |
| Stepping motor circuit (micro switch circuit) | 0V of voltage between Terminals ⑧ and ④ when the control unit is not driving the stepping motor. | Stopped | 4 |
| | More than 0V voltage between Terminals ⑧ and ④ when the operating stepping motor. | During assessment; Operated After assessment; Stopped | |
| Throttle sensor circuit | Shortcircuiting or disconnection between terminal ⑤ and ④ (throttle sensor output terminal). | Stopped | 5 |
| Throttle sensor circuit | No voltage variance or abnormal voltage between terminals ⑤ (throttle sensor output) and ④. | Stopped | 6 |
| | No wide open throttle operation since engine was started. | Operated | |
| Power supply voltage | Less than 10V or more than 16V power supply voltage. | Stopped | 7 |

**Note:** The numbers in the circles refer to the controller's pin numbers. The controller's pin numbers are as follows.

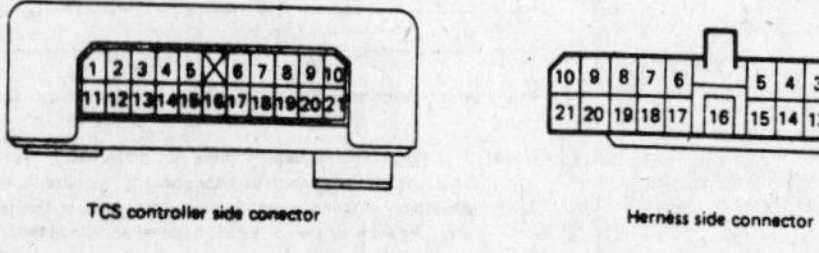

## TROUBLE CODE NUMBER 1—1988 TURBO IMPULSE

## SIGNAL CODE 66—1988 IMPULSE

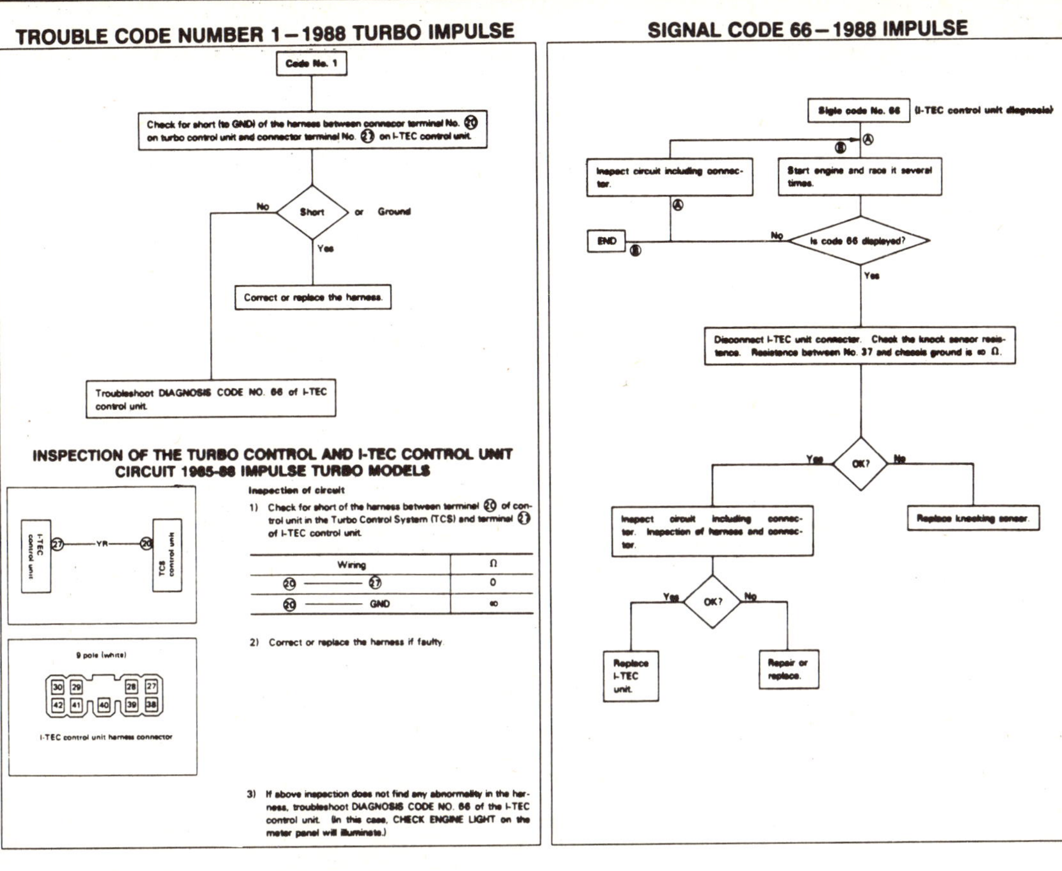


### INSPECTION OF THE TURBO CONTROL AND I-TEC CONTROL UNIT CIRCUIT 1985-88 IMPULSE TURBO MODELS

**Inspection of circuit**

1) Check for short of the harness between terminal 20 of control unit in the Turbo Control System (TCS) and terminal 27 of I-TEC control unit.

| Wiring | Ω |
|---|---|
| 20 ——— 27 | 0 |
| 20 ——— GND | ∞ |

2) Correct or replace the harness if faulty.

3) If above inspection does not find any abnormality in the harness, troubleshoot DIAGNOSIS CODE NO. 66 of the I-TEC control unit. (In this case, CHECK ENGINE LIGHT on the meter panel will illuminate.)

## INSPECTING THE KNOCK SENSOR CIRCUIT 1988 TURBO IMPULSE

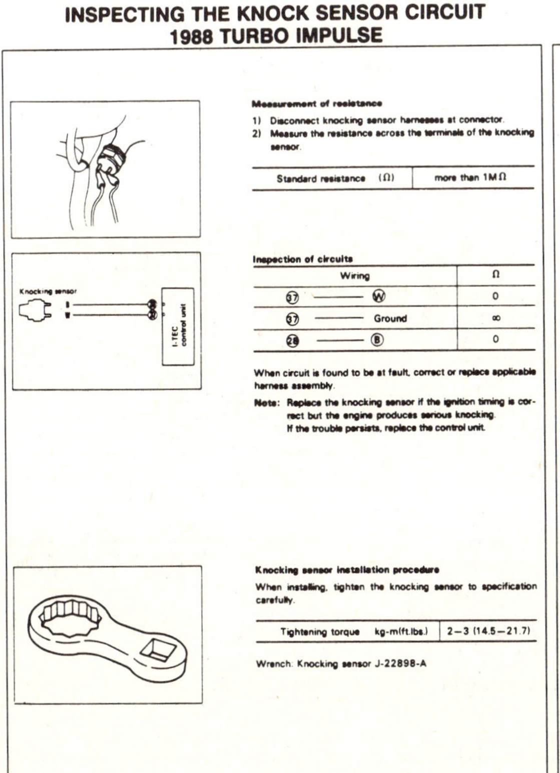


**Measurement of resistance**

1) Disconnect knocking sensor harnesses at connector.
2) Measure the resistance across the terminals of the knocking sensor.

| Standard resistance (Ω) | more than 1M Ω |
|---|---|

**Inspection of circuits**

| Wiring | Ω |
|---|---|
| 37 ——— W | 0 |
| 37 ——— Ground | ∞ |
| 28 ——— B | 0 |

When circuit is found to be at fault, correct or replace applicable harness assembly.

**Note:** Replace the knocking sensor if the ignition timing is correct but the engine produces serious knocking. If the trouble persists, replace the control unit.

**Knocking sensor installation procedure**

When installing, tighten the knocking sensor to specification carefully.

| Tightening torque kg-m(ft.lbs.) | 2–3 (14.5–21.7) |
|---|---|

Wrench: Knocking sensor J-22898-A

## TROUBLE CODE NUMBER 2—1988 TURBO IMPULSE

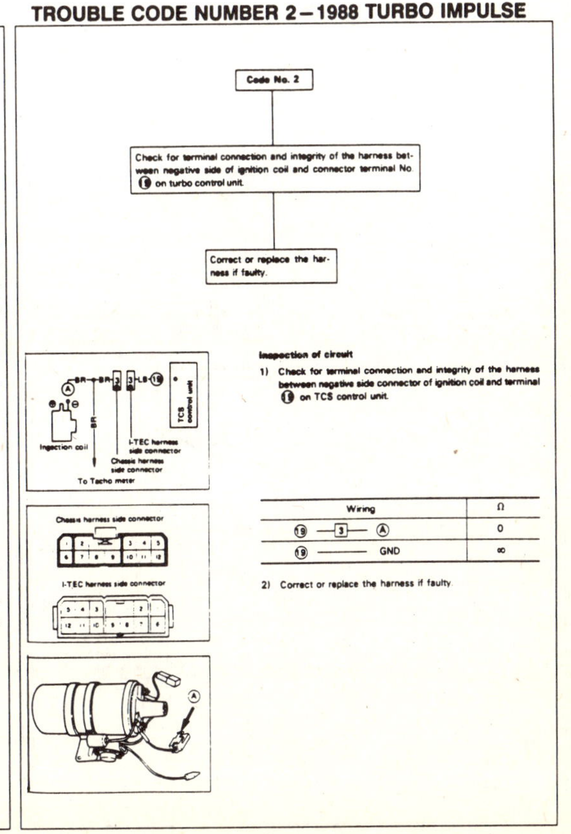

Code No. 2
Check for terminal connection and integrity of the harness between negative side of ignition coil and connector terminal No. 1 on turbo control unit.
Correct or replace the harness if faulty.
BR
LB
A
Ingnection coil
I-TEC harness side connector
Chassis harness side connector
To Tacho meter
TCS control unit
Chassis harness side connector
I-TEC harness side connector
A


**Inspection of circuit**

1) Check for terminal connection and integrity of the harness between negative side connector of ignition coil and terminal 1 on TCS control unit.

| Wiring | Ω |
|---|---|
| 19 —3— A | 0 |
| 19 ——— GND | ∞ |

2) Correct or replace the harness if faulty.

## INSPECTION FOR TROUBLE CODE NUMBER 3 1988 TURBO IMPULSE

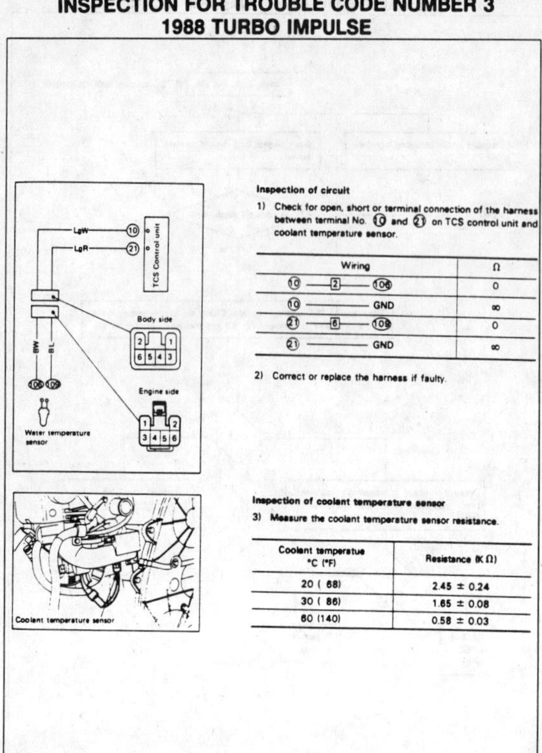

**Inspection of circuit**

1) Check for open, short or terminal connection of the harness between terminal No. ⑩ and ㉑ on TCS control unit and coolant temperature sensor.

| Wiring | Ω |
|---|---|
| ⑩ —[2]— 106 | 0 |
| ⑩ —— GND | ∞ |
| ㉑ —[6]— 109 | 0 |
| ㉑ —— GND | ∞ |

2) Correct or replace the harness if faulty.

**Inspection of coolant temperature sensor**

3) Measure the coolant temperature sensor resistance.

| Coolant temperatue °C (°F) | Resistance (K Ω) |
|---|---|
| 20 ( 68) | 2.45 ± 0.24 |
| 30 ( 86) | 1.65 ± 0.08 |
| 60 (140) | 0.58 ± 0.03 |

Coolant temperature sensor

## TROUBLE CODE NUMBER 3—1988 TURBO IMPULSE

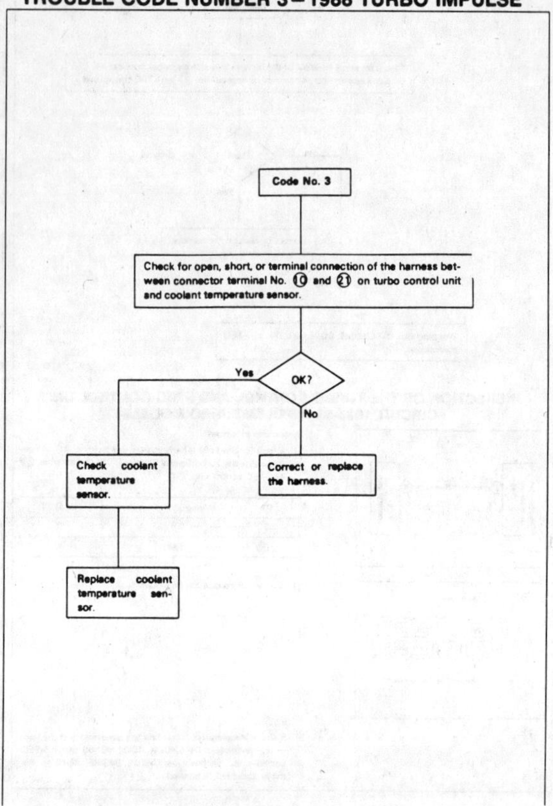


## TROUBLE CODE NUMBER 4—1988 TURBO IMPULSE

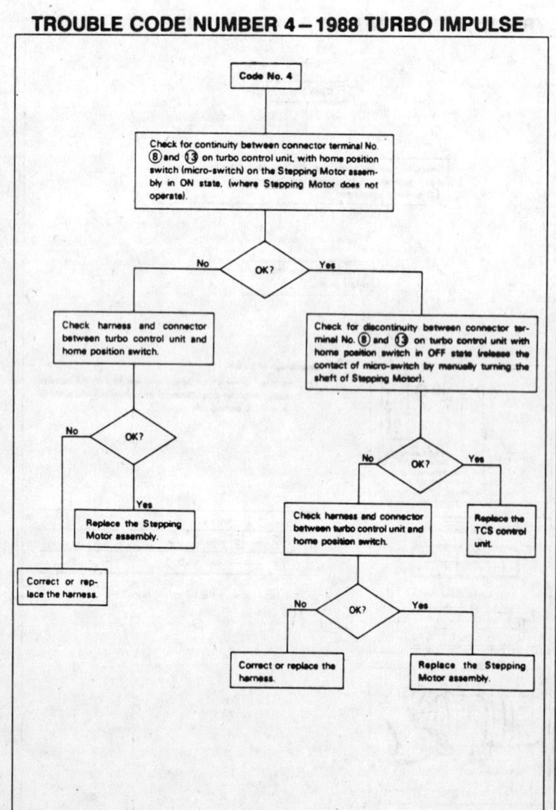


## INSPECTION FOR TROUBLE CODE NUMBER 4 1988 TURBO IMPULSE

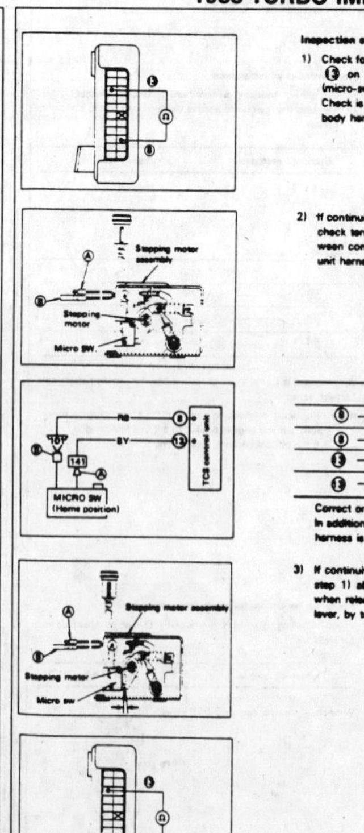


**Inspection of circuit**

1) Check for continuity between connector terminal No. ⑧ and ⑬ on the TCS control unit with home position switch (micro-switch) on the Stepping Motor assembly in ON state. Check is to be carried out, with TCS control unit and Engine body harness being connected.

2) If continuity is not confirmed between terminals ⑧ and ⑬, check terminal connection and integrity of the harness between connector terminals ⑧ and ⑬ on the TCS control unit harness side and micro-switch terminals.

| Wiring | Ω |
|---|---|
| ⑧ —— 101 | 0 |
| ⑧ —— GND | ∞ |
| ⑬ —— [141] | 0 |
| ⑬ —— GND | ∞ |

Correct or replace the harness if faulty.
In addition, replace the Stepping Motor assembly unless the harness is faulty.

3) If continuity is confirmed between terminals ⑧ and ⑬ in step 1) above, make sure that continuity is not obtained when releasing the contact between the micro-switch and lever, by turning the Stepping Motor manually.

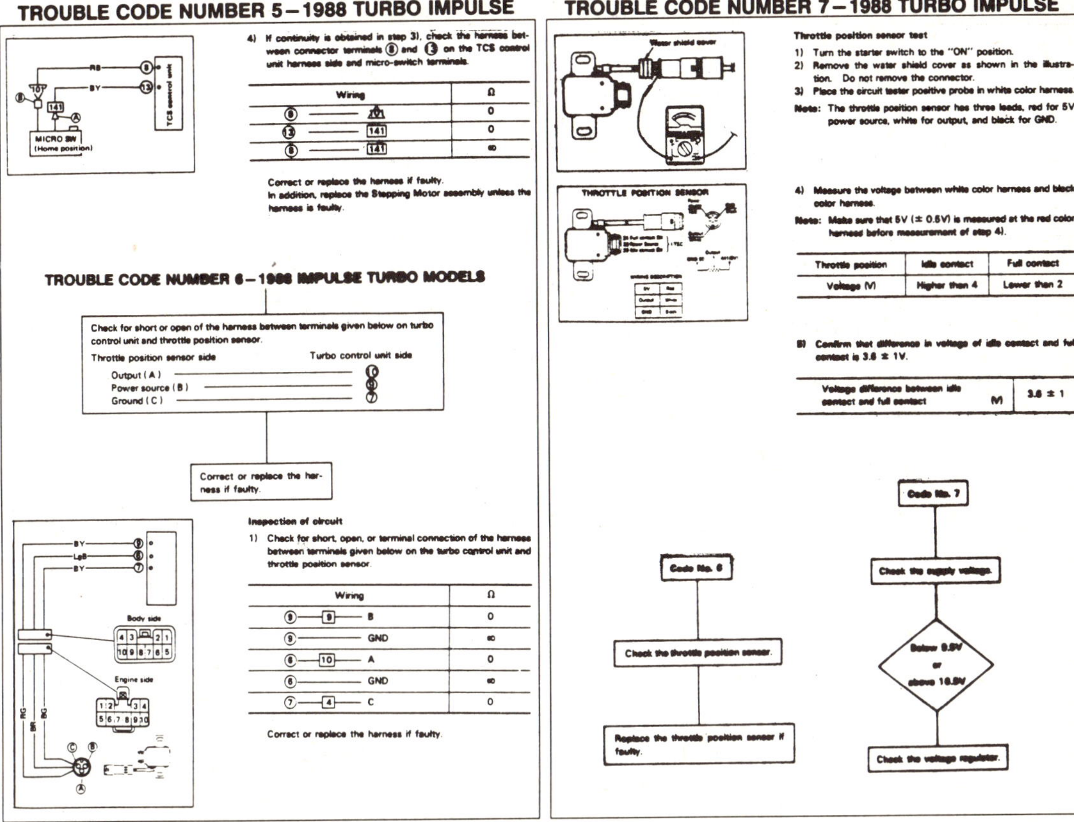

## TROUBLE CODE NUMBER 5—1988 TURBO IMPULSE

4) If continuity is obtained in step 3), check the harness between connector terminals ⑧ and ⑬ on the TCS control unit harness side and micro-switch terminals.

| Wiring | Ω |
|---|---|
| ⑧ —— 101 | 0 |
| ⑬ —— 141 | 0 |
| ⑧ —— 141 | ∞ |

Correct or replace the harness if faulty.
In addition, replace the Stepping Motor assembly unless the harness is faulty.

## TROUBLE CODE NUMBER 6—1988 IMPULSE TURBO MODELS

Check for short or open of the harness between terminals given below on turbo control unit and throttle position sensor.

Throttle position sensor side — Turbo control unit side
Output (A) — ⑩
Power source (B) — ⑨
Ground (C) — ⑦

Correct or replace the harness if faulty.

**Inspection of circuit**

1) Check for short, open, or terminal connection of the harness between terminals given below on the turbo control unit and throttle position sensor.

| Wiring | Ω |
|---|---|
| ⑨ — 9 — B | 0 |
| ⑨ —— GND | ∞ |
| ⑥ — 10 — A | 0 |
| ⑥ —— GND | ∞ |
| ⑦ — 4 — C | 0 |

Correct or replace the harness if faulty.

## TROUBLE CODE NUMBER 7—1988 TURBO IMPULSE

**Throttle position sensor test**

1) Turn the starter switch to the "ON" position.
2) Remove the water shield cover as shown in the illustration. Do not remove the connector.
3) Place the circuit tester positive probe in white color harness.

Note: The throttle position sensor has three leads, red for 5V power source, white for output, and black for GND.

4) Measure the voltage between white color harness and black color harness.

Note: Make sure that 5V (± 0.5V) is measured at the red color harness before measurement of step 4).

| Throttle position | Idle contact | Full contact |
|---|---|---|
| Voltage (V) | Higher than 4 | Lower than 2 |

5) Confirm that difference in voltage of idle contact and full contact is 3.6 ± 1V.

| Voltage difference between idle contact and full contact (V) | 3.6 ± 1 |
|---|---|

Code No. 6
Check the throttle position sensor.
Replace the throttle position sensor if faulty.

Code No. 7
Check the supply voltage.
Below 9.5V or above 16.5V
Check the voltage regulator.

## 1989-90 IMPULSE TURBO AND NON-TURBO

### COMPONENTS LAYOUT

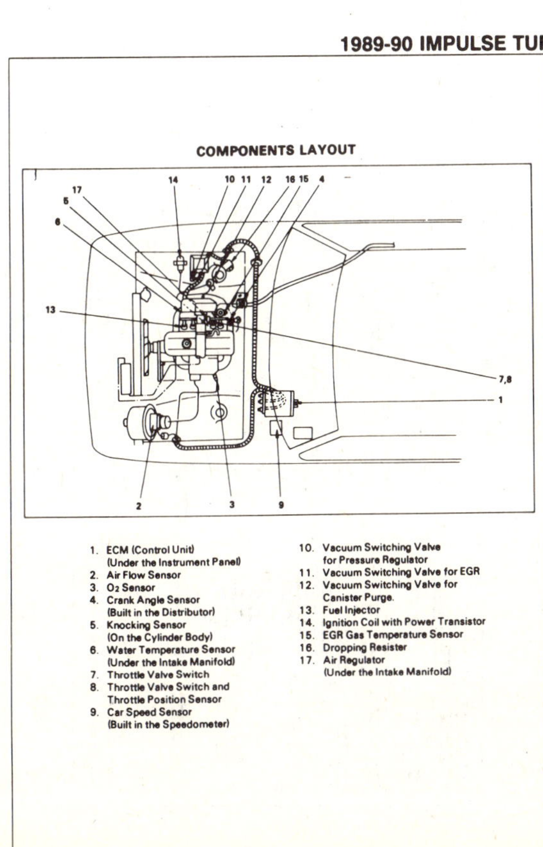

1. ECM (Control Unit) (Under the Instrument Panel)
2. Air Flow Sensor
3. $O_2$ Sensor
4. Crank Angle Sensor (Built in the Distributor)
5. Knocking Sensor (On the Cylinder Body)
6. Water Temperature Sensor (Under the Intake Manifold)
7. Throttle Valve Switch
8. Throttle Valve Switch and Throttle Position Sensor
9. Car Speed Sensor (Built in the Speedometer)
10. Vacuum Switching Valve for Pressure Regulator
11. Vacuum Switching Valve for EGR
12. Vacuum Switching Valve for Canister Purge.
13. Fuel Injector
14. Ignition Coil with Power Transistor
15. EGR Gas Temperature Sensor
16. Dropping Resister
17. Air Regulator (Under the Intake Manifold)

### INSPECTION AND MAINTENANCE

NORMAL CONDITION; • SIGNAL CODE NO. 12 WILL DISPLAY –
• THE MEMORY CLEARED BY No. 4 FUSE

I. DISCONNECT THE SENSOR OR ACTUATOR WIRING CONNECTOR (WHEN CHECKING THE PART/COMPONENT INDEPENDENTLY)

| Code No. | Item | | Measure Terminal: Tester + Side | Measure Terminal: Tester – Side | Key SW Posit. | Condition to Check | Standard Value |
|---|---|---|---|---|---|---|---|
| 21 43 65 | Throttle valve SW. | | Center Terminal (Terminal P) | Upper Terminal (Terminal I) | OFF | Ac. Pedal Not Depressed | 0 Ω |
| | | | | | | Ac. Pedal Slightly Depressed | ∞ Ω |
| | | | | | | Ac. Pedal Full Depressed | ∞ Ω |
| | | | Center Terminal (Terminal P) | Lower Terminal (Terminal F) | OFF | Ac. Pedal Not Depressed | ∞ Ω |
| | | | | | | Ac. Pedal Slightly Depressed | ∞ Ω |
| | | | | | | Ac. Pedal Full Depressed | 0 Ω |
| 33 64 | Fuel Injector | #1 | Terminal | Terminal | OFF | | 2~3 Ω |
| | | #2 | Terminal | Terminal | OFF | | 2~3 Ω |
| | | #3 | Terminal | Terminal | OFF | | 2~3 Ω |
| | | #4 | Terminal | Terminal | OFF | | 2~3 Ω |
| 23 35 54 | Ignition Coil | | + Terminal | – Terminal | OFF | Primary | 0.81~0.99 Ω |
| | | | + Terminal | Center Terminal | OFF | Secondary | 7.52~11.28K Ω |
| 25 53 | Vacuum Switching Valve for pressure regulator | | Terminal | Terminal | OFF | | Approx. 36 Ω |
| 26 27 | Vacuum Switching Valve for Canister Purge | | Terminal | Terminal | OFF | | Approx. 36 Ω |
| 34 72 73 | Vacuum Switching Valve for EGR (4ZC1-T only) | | Terminal | Terminal | OFF | | Approx. 36 Ω |

### 1989-90 IMPULSE TURBO AND NON-TURBO

| Code No. | Item | Measure Terminal: Tester + Side | Measure Terminal: Tester − Side | Key SW Posit. | Condition to Check | | Standard Value |
|---|---|---|---|---|---|---|---|
| 14<br>15 | Water Temperature Sensor | Terminal | Terminal | OFF | Water Temp. °C(°F) | −10 (14)<br>10 (50)<br>20 (68)<br>50 (122)<br>80 (176) | 7~12K Ω<br>3~ 5K Ω<br>2~ 3K Ω<br>0.7~1K Ω<br>0.2~0.4K Ω |
| 33 | Dropping Resistor | Center Terminal | Another Terminal | OFF | | | 5~7 Ω/Pc. |
| 63 | Car Speed Sensor | 41 | Ground | OFF | Turn the Inner Cable Slowly | | 4 Pulse/rev. |
| 66 | Knock Sensor | Terminal | Terminal | OFF | | | More than 1 M Ω |
| | Air Regulator | Terminal | Terminal | OFF | | | Approx. 40 Ω |
| | Main Relay | Terminal (WR) | Terminal (B) | OFF | | | Approx. 80 Ω |
| | Air Pump Relay | Terminal (WB) | Terminal (BY) | OFF | | | Approx. 80 Ω |
| | Fuel Pump Relay | Terminal (WR) | Terminal (B) | OFF | | | Approx. 80 Ω |

**INSPECTION AND MAINTENANCE**

II. DISCONNECT THE ECM UNIT WIRING CONNECTORS

| Code No. | Item | Measure Terminal: Tester + Side | Measure Terminal: Tester − Side | Key SW Posit. | Condition to Check | Standard Value |
|---|---|---|---|---|---|---|
| | Power Supply Circuit | 38 | Ground (51 or 52) | ON | | 10~14V |
| | | 39 (40) | Ground (51 or 52) | ON | | 10~14V |
| | | 9 & 10 | Ground (51 or 52) | OFF | | 0 Ω |
| | | 6 & 7 | Ground (51 or 52) | OFF | | 0 Ω |
| | | 51 or 52 | Ground (Body) | OFF | | 0 Ω |
| 33<br>64 | Fuel Injector #1<br>#2<br>#3<br>#4 | 1<br>2<br>3<br>8 | Ground (6 or 7)<br>Ground (6 or 7)<br>Ground (6 or 7)<br>Ground (6 or 7) | ON<br>ON<br>ON<br>ON | | 10~14V<br>10~14V<br>10~14V<br>10~14V |
| 22 | Starter Signal | 16 | Ground (51 or 52) | ST | Disconnect Starter Relay | 10~14V |
| 25<br>53 | P/R V.S.V. (Vacuum Switching Valve) | 13 | Ground (9 or 10) | ON | | 10~14V |
| 21<br>43<br>65 | Throttle Valve SW. | 18 | Ground (51 or 52) | ON | Ac. Pedal Not Depressed | 10~14V |
| | | | | | Ac. Pedal Slightly Depressed | 0V |
| | | | | | Ac. Pedal Full Depressed | 0V |
| | | 27 | Ground (51 or 52) | ON | Ac. Pedal Not Depressed | 0V |
| | | | | | Ac. Pedal Slightly Depressed | 0V |
| | | | | | Ac. Pedal Full Depressed | 10~14V |
| 26<br>27 | Canister V.S.V. | 12 | Ground (9 or 10) | ON | | 10~14V |
| 23<br>35<br>54 | Ignition-Terminal (Power Transistor for Ignition) | Ground (Dry Batt. + Side) | P-Transistor Yellow Harness (Dry Batt. − Side) | ON | Connected ECM Wiring Connector. Keep 5~6 mm Gap in Between the High Tension Cable and Ground. and Then, Disconnect either the Plus or Minus Probe and See if There are Sparks. If Sparks occur, This System is Normal. | |
| 13<br>44<br>45 | $O_2$ sensor | 30 | Ground (51 or 52) | ON | The Engine Speed About 2000 RPM For Approx. 10 Min. | 0.3~0.8V |
| 34<br>72<br>73 | EGR V.S.V (4ZC1-T only) | 20 | Ground (9 or 10) | ON | | 10~14V |

### 1989-90 IMPULSE TURBO AND NON-TURBO

**MALFUNCTION CODES VS FAULTY COMPONENTS**

| Code No. | | 12 | 13 | 14 | 15 | 21 | 22 | 23 | 25 | 26 | 27 | 32 | 33 | 34 | 35 | 41 | 43 | 44 | 45 | 51 | 52 | 53 | 54 | 61 | 62 | 63 | 64 | 65 | 66 | 71 | 72 | 73 |
|---|---|---|---|---|---|---|---|---|---|---|---|---|---|---|---|---|---|---|---|---|---|---|---|---|---|---|---|---|---|---|---|---|
| Input Sensors | Air Flow Sensor | | | | | | | | | | | | | | | | | | | | | | | ⊗ | ⊗ | x | | x | | | | |
| | Coolant Temp. Sensor | | x | ⊗ | ⊗ | | | | | | | x | | x | | | | x | x | | | | | | | | | | | | | |
| | Oxygen Sensor | | ⊗ | | | | | | | | | | | | | | | ⊗ | ⊗ | | | | | | | | | | | | | |
| | Crank Angle Sensor | x | | | | | | | | | | x | | | | ⊗ | | | | | | | | | | | | | x | | | |
| | Throttle Valve Switch | | | | | ⊗ | | | | | | x | | x | | | ⊗ | x | x | | | | | | | | | x | | x | | |
| | Starter Switch | | | | | | ⊗ | | | | | | | | | x | | | | | | | | | | | | | | | | |
| | EGR Gas Temp. Sensor | | | | | | | | | | | ⊗ | | ⊗ | | | | | | | | | | | | | | | | | | |
| | Ignition Coil | x | | | | | x | | | | | x | | | | x | | x | x | | | | | | | | | | | | | |
| | Computer | | | | | | | | | | | | | | | | | | | ⊗ | ⊗ | | | x | x | | | | | | | |
| | Car Speed Sensor | | | | | | x | | | | | x | | x | | | | | | | | | | | | ⊗ | | | | | | |
| | Knock Sensor | | | | | | | | | | | | | | | | | | | | | | | | | | | | ⊗ | | | |
| | Throttle Position Sensor | | | | | | | | | | | | | | | | | | | | | | | | | | | | | ⊗ | | |
| Output Actuators | Fuel Injector | | | | | | | | | | | | ⊗ | | | | | | | | | | | | | | | | | | | |
| | Power Switch (Ignition Coil Driver) | | | | | | | x | | | | | | x | | | | | | | | ⊗ | | | | x | | | | | | |
| | V.S.V.; Pressure regulator | | | | | | | | ⊗ | | | | | | | | | | | | x | | | | | | | | | | | |
| | V.S.V; Canister | | | | | | | | | ⊗ | x | | | | | | | | | | | | | | | | | | | | | |
| | V.S.V; EGR | | | | | | | | | | | | | | | | | | | | | | | | | | | | | | ⊗ | x |

**NOTE:**

**Code 12 does not indicate malfunction, but checks whether ECM is functioning normally. Listed malfunction codes are only those which show malfunction of sensed parameters.**

**ECM determines whether a malfunction of that code exists by evaluating the parameters marked "⊗". Parameters marked "X" are used to set the zones in which existence or nonexistence of a malfunction is to be determined.**

**DIAGNOSTIC CODES**

| DIAGNOSED ITEM | | DIAGNOSED CONTENT: FAULT MODE | DIAGNOSED CONTENT: MICRO-COMPUTER INPUT | CODE NO. | REF. PAGE | CODE MEMORY | FAIL-SAFE FUNCTION OF MICRO-COMPUTER | ENGINE CONDITION | VEHICLE CONDITION |
|---|---|---|---|---|---|---|---|---|---|
| Engine is not started | | — | Engine speed less than 200 rpm | 12 | 03-55 | No | None | Normal | Normal |
| $O_2$ sensor system | | Harness open. Sensor deterioration | Intermediate voltage | 13 | 03-56 | Yes | Fuel is not compensated by $O_2$ signal | Exhaust is worsened | No noticeable abnormal operation |
| | | Incorrect signal (Lean) | Lean signal (Low voltage) | 44 | 03-57 | Yes | | | |
| | | Incorrect signal (Rich) | Rich signal (High voltage) | 45 | 03-57 | Yes | | | |
| Water temperature sensor system | | Shorted with ground | Insufficient signal | 14 | 03-63 | Yes | Coolant temperature is assumed to be 85°C (185°F) | · The engine does not operate normally when temperature is low and signal is insufficient, or when temperature is high and signal is excessive. Or the engine cannot be started.<br>· The engine does not operate smoothly when temperature is low after the fail-safe function is actuated. | |
| | | Harness open | Excessive signal | 15 | 03-64 | Yes | | | |
| Throttle valve switch system | Both idle contact and full contact | Both idle contact and full make contact simultaneously | Both contacts make simultaneously | 21 | 03-66 | Yes | Both signals are assumed to be OFF. | Fuel is not cut after the fail-safe function is actuated, and lean fuel is resulted when the throttle valve is fully opened. | Fuel consumption rate is worsened after the fail-safe function is actuated. Fuel tends to be lean when the throttle valve is fully opened. |
| | Idle contact | Normally make contact | Signal is sent continuously<br>Not diagnosed when the air flow sensor system is defective | 43 | 03-67 | Yes | Assumed to be OFF. | · Fuel cut range appears during running.<br>· Fuel is not cut after the fail-safe function is actuated. | · The vehicle does not run smoothly.<br>· Fuel consumption rate is worsened after the fail-safe function is actuated. |
| | Full contact | Normally make contact | Signal is sent continuously | 65 | 03-67 | Yes | | · Air-fuel ratio is high when partially loaded.<br>· Fuel tends to be lean when the throttle valve is fully opened after the fail-safe function is actuated. | · Fuel consumption rate and exhaust are worsened. The spark plugs are carboned or the engine stalls depending on the condition.<br>· Fuel tends to be lean when the throttle valve is fully opened after the fail-safe function is actuated. |

## 1989-90 IMPULSE TURBO AND NON-TURBO

| DIAGNOSED ITEM | DIAGNOSED CONTENT | | CODE NO. | REF. PAGE | CODE MEMORY | FAIL-SAFE FUNCTION OF MICRO-COMPUTER | ENGINE CONDITION | VEHICLE CONDITION |
|---|---|---|---|---|---|---|---|---|
| | FAULT MODE | MICRO-COMPUTER INPUT | | | | | | |
| Starter signal system | Normally open | Signal is not input | 22 | 03-69 | Yes | None | Normal | No abnormality is felt. |
| Crank angle sensor system | · No signal arrives. Faulty signal | · No angle sensor signal is input. · Idling speed is lower than the actual speed. | 41 | 03-70 | Yes | None | · The engine stalls. The engine cannot be started. · Air-fuel ratio tends to be high. | · Stalls or cannot be started. · Both fuel consumption rate and exhaust are worsened. |
| Air flow sensor system | Harness open, shorting with ground, or broken hot wire. | Insufficient signal | 61 | 03-71 | Yes | Injection pulse is changed over at throttle valve position. | · The engine stalls depending on the condition. · The engine may be slightly unstable after the fail-safe function is actuated. | · The engine stalls depending on the condition. · The accelerator response is worsened after the fail-safe function is actuated. The engine may not operate smoothly. |
| | Broken cold wire. | Excessive signal | 62 | 03-71 | Yes | | | |
| Car speed sensor system | No signal arrives | No signal is input (Not diagnosed if the air flow sensor system is defective) | 63 | 03-74 | Yes | None | The fuel cut operates even when running under a low speed. | Does not run smoothly and shakes. |
| Knocking sensor system | Harness open or shorting with ground. | Excessive or insufficient signal | 66 | 03-76 | Yes | Ignition timing is delayed. | · Engine knocking · Output drops after the fail-safe function is actuated. | · Engine knocking. · Output drops after the fail-safe function is actuated. |
| Micro-computer unit | Abnormal LSI (1) | – | 51 | 03-78 | Yes | · Injection pulse is changed over at throttle valve position · Fixed ignition timing. | · In the worst case, the engine stalls, or the engine does not operate smoothly at a certain time. · The engine does not stall but it does not satisfy the specification. | · In the worst case, the engine stalls, or the engine does not operate smoothly at a certain time. · The exhaust and running performance deviate from the specification. |
| | Abnormal LSI (2) | – | 52 | 03-78 | Yes | | | |
| Power transistor system for ignition | Output terminal shorted with ground. | – | 23 | 03-79 | Yes | None | · The engine stalls. · Cannot be started. | · The engine stalls. · Cannot be started. |
| | Harness open. | – | 35 | 03-79 | Yes | | | |
| | Defective transistor or grounding system. | – | 54 | 03-79 | Yes | | | |

| DIAGNOSED ITEM | DIAGNOSED CONTENT | | CODE NO. | REF. PAGE | CODE MEMORY | FAIL-SAFE FUNCTION OF MICRO-COMPUTER | ENGINE CONDITION | VEHICLE CONDITION |
|---|---|---|---|---|---|---|---|---|
| | FAULT MODE | MICRO-COMPUTER INPUT | | | | | | |
| Vacuum switching valve system for canister purge | Output terminal shorted with ground or harness open | – | 26 | 03-81 | Yes | None | · Output terminal shortened with ground. · The canister isn't purged (always) · Same as below for harness open. | · The canister may be filled with gasoline · Same as below for harness open. |
| | Defective transistor or grounding system. | – | 27 | 03-81 | Yes | None | · Opening the throttle valve slightly causes high air-fuel ratio, depending on the condition. | · Closing the throttle valve Slightly may cause surge. |
| Vacuum switching valve system for pressure regulator | Output terminal shorted with ground or harness open | – | 25 | 03-84 | Yes | None | Output terminal shorted with ground. Fuel pressure rises continuously when the engine is cold, causing fuel to be rich. When the engine is warmed up, automatically corrected as the $O_2$ sensor compensates Same as below for harness open. | Output terminal shorted with ground. · Fuel consumption rate and exhaust gas level are worsened when the engine is cold or when accelerated rapidly. Same as below for harness open. |
| | Defective transistor or grounding system. | – | 53 | 03-84 | Yes | None | When the engine is overheated, air-fuel ratio tends to be lean. No problem when the engine is warmed up normally. | The engine cannot be restarted smoothly when coolant temperature is high. |
| Fuel injector system. | Output terminal shorted with ground or harness open | – | 33 | 03-87 | Yes | None | · The engine stalls. · Cannot be started. | · The engine stalls. · Cannot be started. |
| | Defective transistor or grounding system. | – | 64 | 03-89 | Yes | | | |
| EGR temperature sensor system | Malfunction | – | 32 | 03-92 | Yes | None | Exhaust is worsened | No noticeable abnormal operation |
| | EGR temperature sensor or harness NG | – | 34 | 03-97 | Yes | None | Normal | No abnormality is felt. |
| Throttle position sensor system (TURBO CONTROL SYSTEM) | Abnormal signal | Insufficient signal | 71 | 03-100 | Yes | None | The engine stalls depending on the condition | The engine stalls depending on the condition |

## 1989-90 IMPULSE TURBO AND NON-TURBO

| DIAGNOSED ITEM | DIAGNOSED CONTENT | | CODE NO. | REF. PAGE | CODE MEMORY | FAIL-SAFE FUNCTION OF MICRO-COMPUTER | ENGINE CONDITION | VEHICLE CONDITION |
|---|---|---|---|---|---|---|---|---|
| | FAULT MODE | MICRO-COMPUTER INPUT | | | | | | |
| EGR vacuum switching valve system | Output terminal shorted with ground or harness open. | – | 72 34 | 03-102 | Yes | None | Exhaust is worsened | No noticeable abnormal operation |
| | Defective transistor grounding system. | – | 73 | 03-102 | | | | |

**BEFORE STARTING**

1. Verify the customer complaint and locate the correct symptoms. If the fault relates fundamental engine problems, such as engine lacks power or, fuel consumption excessive etc., see "DIAGNOSIS" in GASOLINE ENGINE SECTION 01, first.
2. Several of the symptom procedures below call for a careful visual check. This check should include:
   - Vacuum hoses for splits, kinks, and proper connections, as shown on Emission Control Information label.
   - Air leaks at throttle valve mounting, common chamber mounting and intake manifold.
   - Ignition wires for cracking, harness, proper routing, and carbon tracking.
   - Wiring for proper connections, pinches, and cuts.
     The importance of this step cannot be stressed too strongly — it can lead to correcting a problem without further checks and can save valuable time.
3. Be careful to check the system components (Closed Loop Emission Control System), especially handling the electrical parts, ECM etc..
   Refer to "CHASSIS ELECTRICAL" SECTION 04B and "Handling precautions" in this section.
   - When checking the ECM harness connector terminals from the connector wire side with the tester probe, be careful not to touch the other terminal by mistake.
     Do not insert the circuit tester test probes into the connector open side to test the continuity.
   - Before starting with a voltmeter connected to a suspected circuit, check to see if the battery is in full charged condition.
   - Before disconnecting the ECM connectors, check the ignition switch "OFF" condition, and check and record the stored memory in ECM.
     When disconnecting the ECM connectors, all three connectors (10P, 18P, 24P) should be done, and note that all stored codes also are cleared at the same time.
   - To avoid the connector terminals being damaged or touching the wrong terminal, short jumper cables which have staked terminals are useful.
   - When connecting a suspected circuit to the Battery + terminal or Engine ground, it will be advisable to install a test light (3.2W, 12V) in between, to protect electrical parts from being overcharged or damaged.

## 1989-90 IMPULSE TURBO AND NON-TURBO

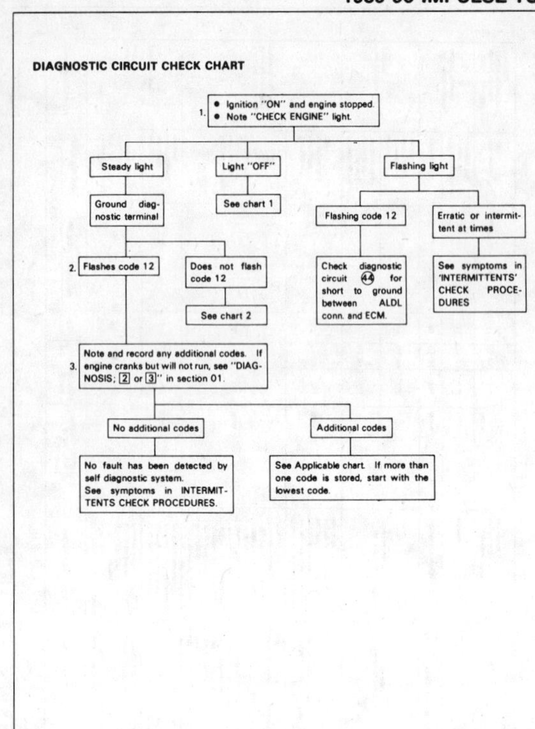


### CHART 1

### NO "CHECK ENGINE" LIGHT

There should always be a steady "CHECK ENGINE" Light when the ignition is "ON" and engine stopped. Battery is supplied directly to the light bulb. The Electronic Control Module (ECM) will control the light and turn it on by providing a ground path through circuit 21 to the ECM.

Engine runs ok, check:

- Faulty light bulb.
- Circuit 21 open.
- Gage fuse blown. This will result in no meter lights, turn signal lights or reminder buzzer etc.

Engine cranks but will not run.

- Continuous battery - fuse or FLW open.
- ECM ignition fuse open.
- Battery circuit 39, 40 to ECM open.
- Ignition circuit 5 to ECM open.
- Poor connection to ECM.

1. Solenoids and relays are turned "ON" and "OFF" by the ECM, using internal electronic switches called "drivers". Solenoid and relay coil resistance must measure more than 20 ohms. Less resistance will cause early failure of the ECM "driver".

   Failure of outer load such as shorted mode can damage the ECM driver.

   Before replacing ECM, be sure to check the coil resistance of all solenoids and relays controlled by the ECM. See ECM wiring diagram for the solenoid(s) and relay(s) and the coil terminal identification.

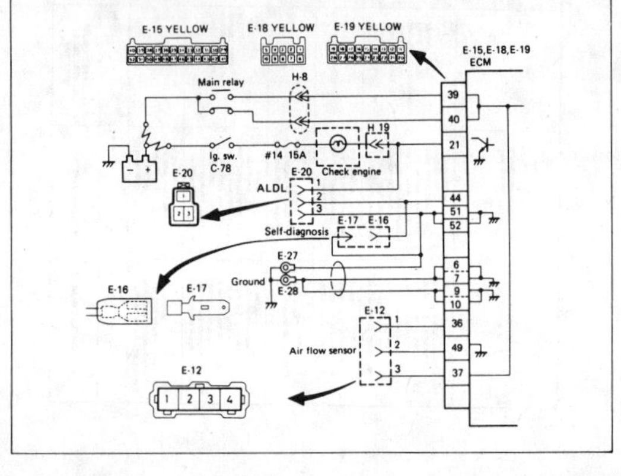


## 1989-90 IMPULSE TURBO AND NON-TURBO

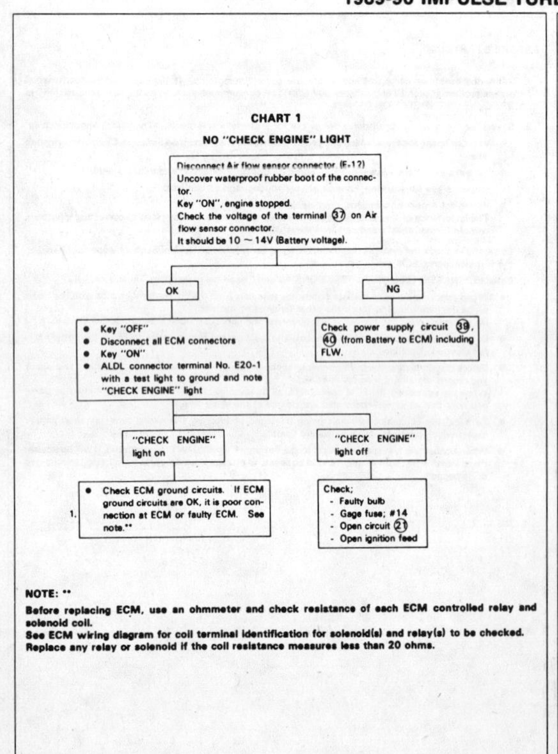


**NOTE: ****

**Before replacing ECM, use an ohmmeter and check resistance of each ECM controlled relay and solenoid coil.**
**See ECM wiring diagram for coil terminal identification for solenoid(s) and relay(s) to be checked.**
**Replace any relay or solenoid if the coil resistance measures less than 20 ohms.**

### CHART 2

### WON'T FLASH CODE 12 AND "CHECK ENGINE" LIGHT STAYS ON

There should always be a steady "CHECK ENGINE" Light when the ignition is "on" and engine stopped. Battery ignition voltage is supplied to the light bulb. The Electronic Control Module (ECM) will turn the light on by grounding circuit E20-1 or 21 at the ECM.
With the diagnostic terminal grounded, the light should flash a Code 12, followed by any trouble code(s) stored in memory.
A steady light suggests a short to ground in the light control circuit E20-1 or 21 or an open in diagnostic circuit 44 or E16.

1. If the light goes off when the ECM connector is disconnected, then circuit E20-1 or 21 is not shorted to ground. Also, check the connector terminals physically for proper contact at this time.
2. This step will check for an open diagnostic circuit 44 or E16.
3. At this point the "CHECK ENGINE" light wiring is okay. The problem is a faulty ECM connector or ECM.

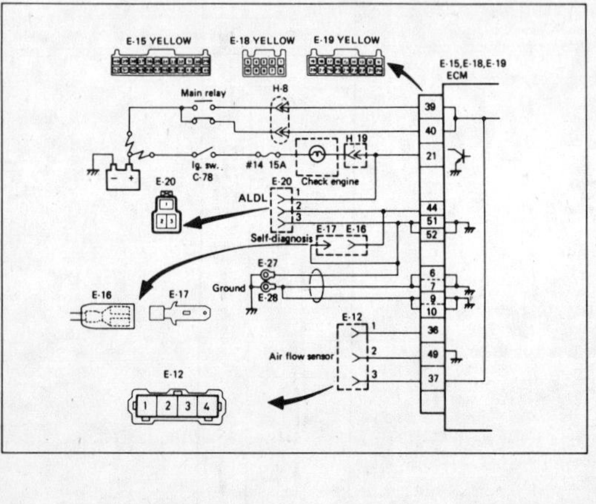

## 1989-90 IMPULSE TURBO AND NON-TURBO

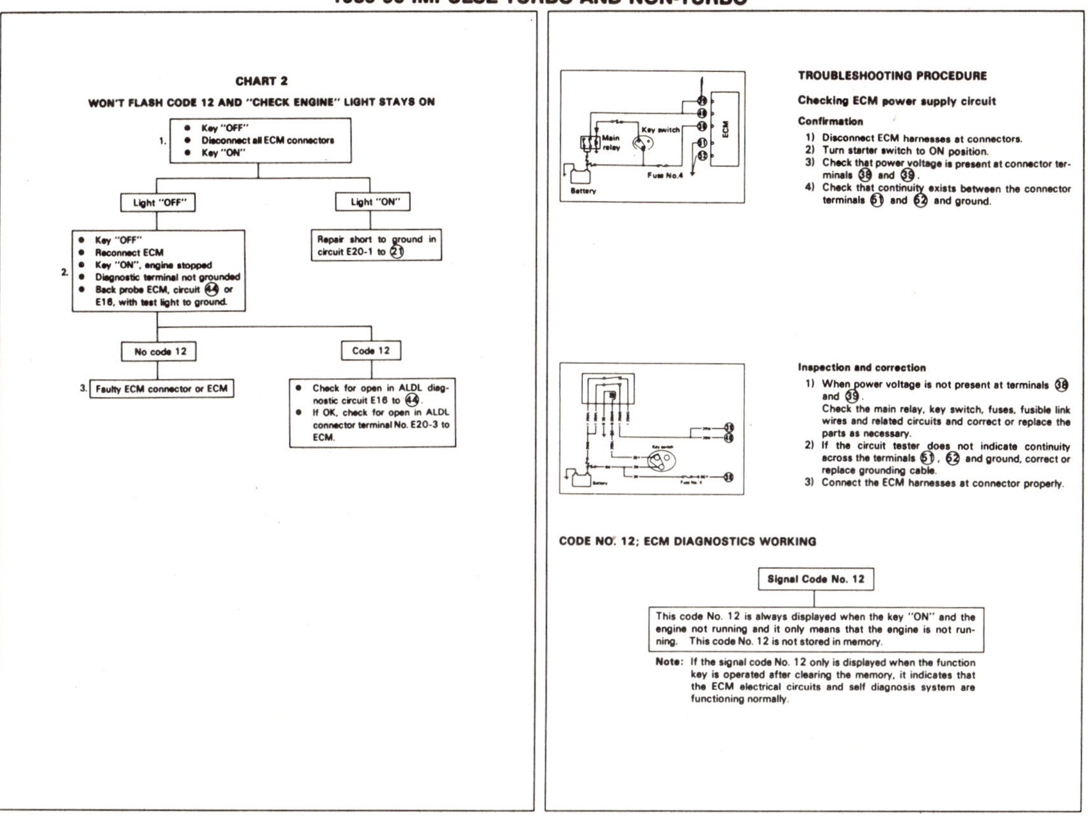


## 1989-90 IMPULSE TURBO AND NON-TURBO

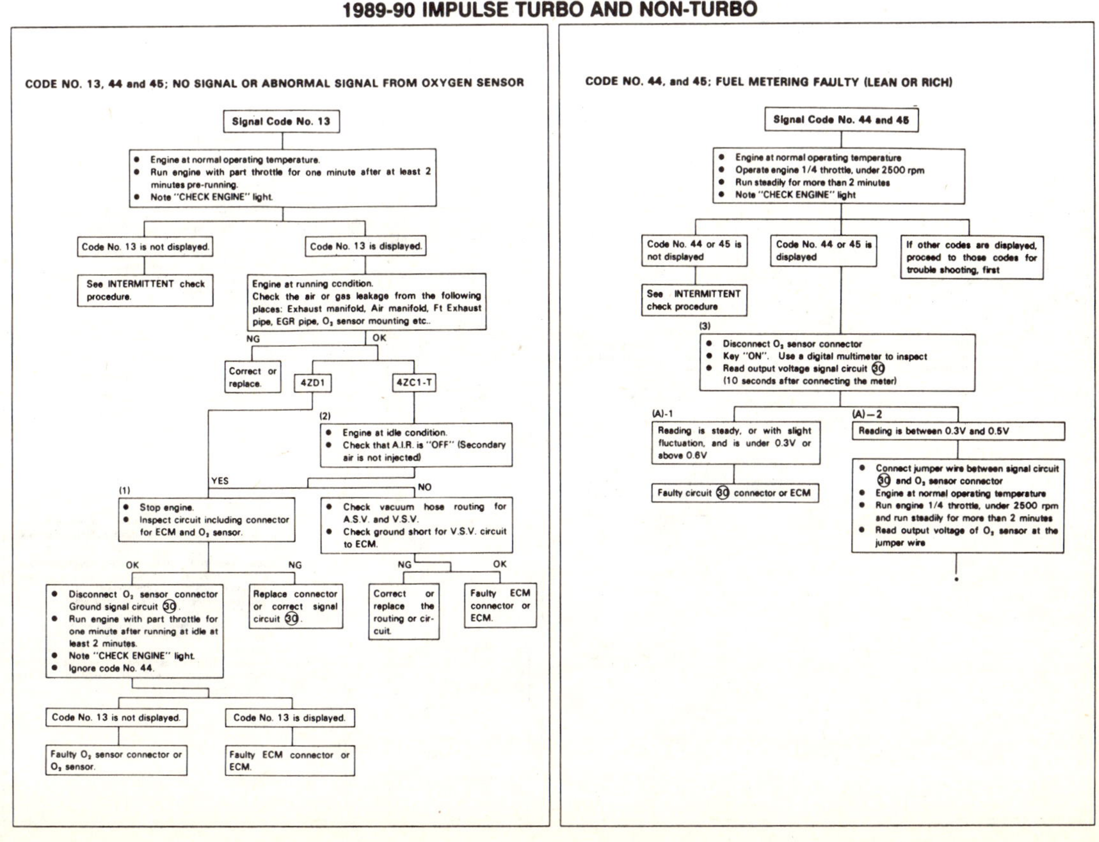

## 1989-90 IMPULSE TURBO AND NON-TURBO

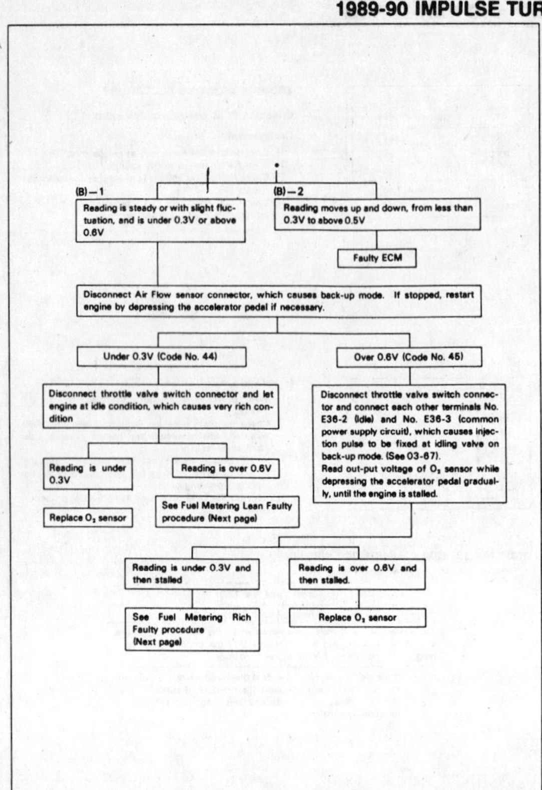


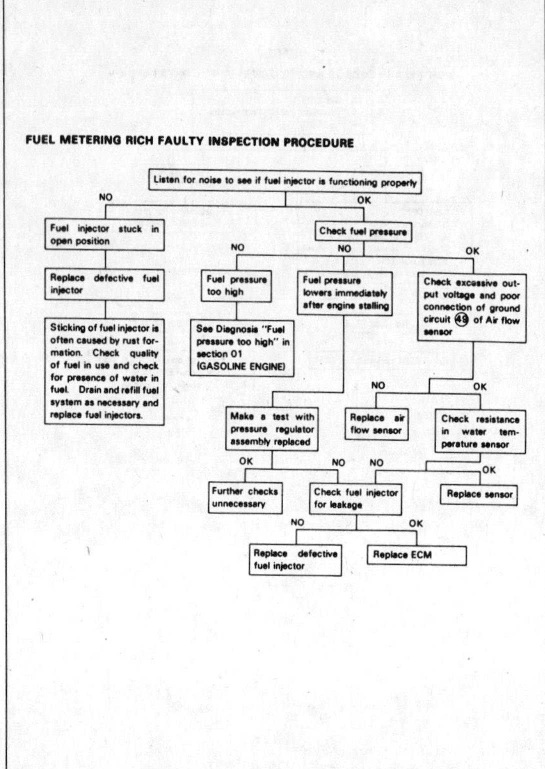


## 1989-90 IMPULSE TURBO AND NON-TURBO

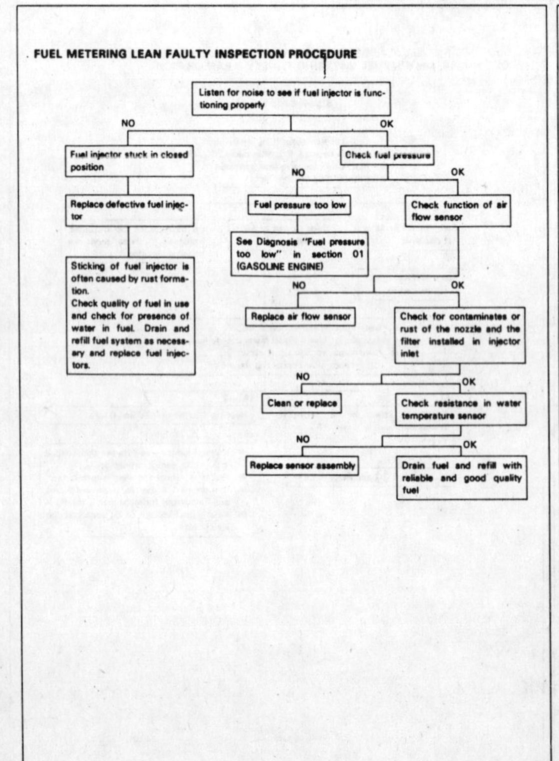


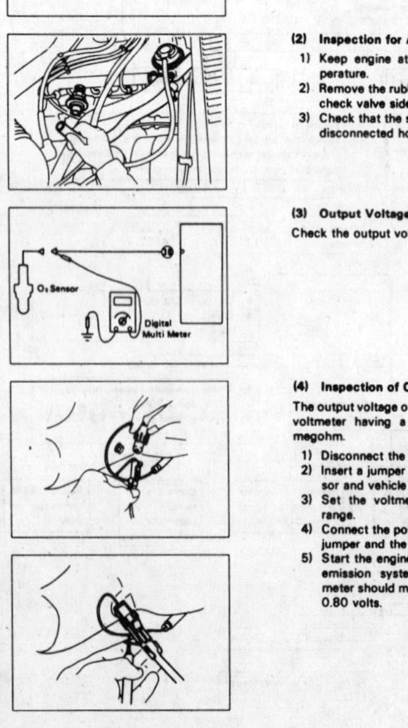


**Inspection for Signal Code No. 13, 44 and 45**

**(1) Circuit Continuity Check**

1) Check that continuity exists between the ECM harness connector terminal (30) and $O_2$ sensor harness connector terminal.
2) If the tester does not indicate continuity, correct or replace the harness assembly.

**(2) Inspection for AIR Control (4ZC1-T only)**

1) Keep engine at idle, under normal operating temperature.
2) Remove the rubber hose from ASV to check valve at check valve side, and plug the check valve.
3) Check that the secondary air is not injected from the disconnected hose.

**(3) Output Voltage Check**

Check the output voltage of the signal circuit (30).

**(4) Inspection of $O_2$ Sensor**

The output voltage of the $O_2$ sensor can be checked with a voltmeter having a minimum input resistance of one megohm.

1) Disconnect the $O_2$ sensor from the vehicle harness.
2) Insert a jumper with exposed wire between the sensor and vehicle harness connectors.
3) Set the voltmeter to the approximately two volt range.
4) Connect the positive (+) lead of the voltmeter to the jumper and the negative (–) lead to vehicle ground.
5) Start the engine and run at elevated speed until the emission system has gone "closed loop". The meter should move between approximately 0.50 and 0.80 volts.

## 1989-90 IMPULSE TURBO AND NON-TURBO

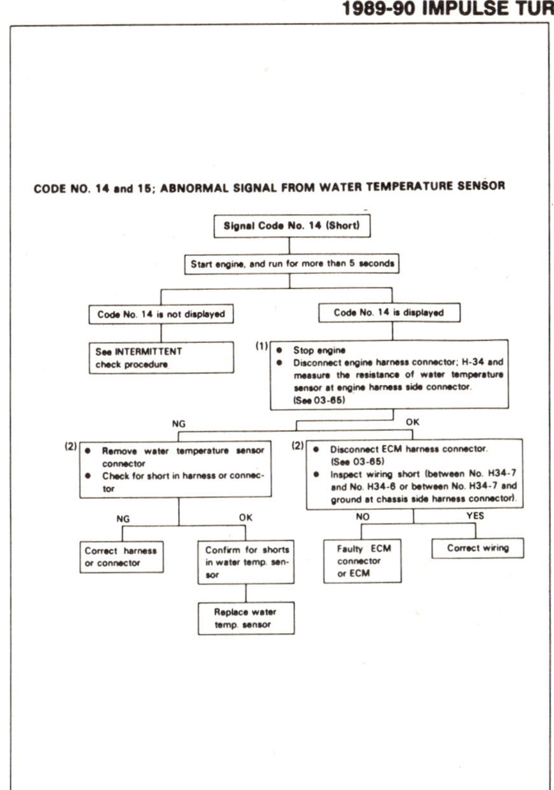


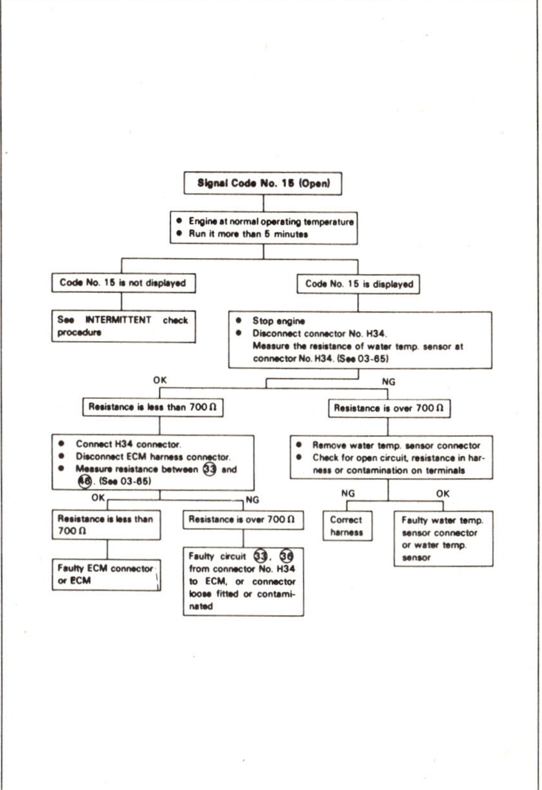


## 1989-90 IMPULSE TURBO AND NON-TURBO

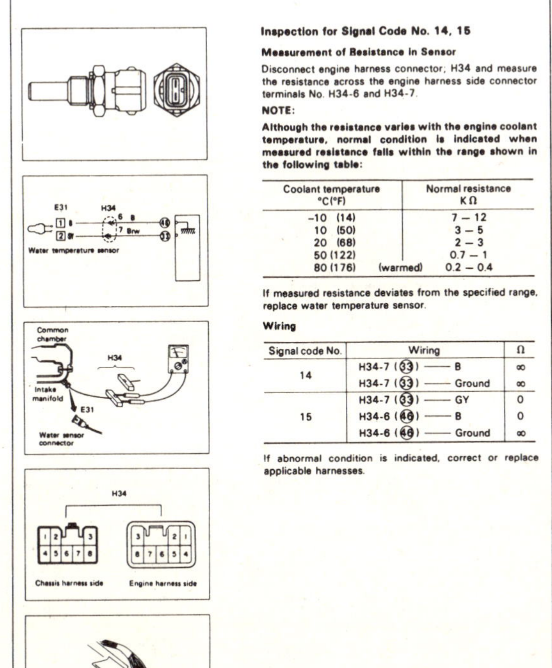

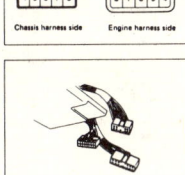

**Inspection for Signal Code No. 14, 15**

**Measurement of Resistance in Sensor**

Disconnect engine harness connector; H34 and measure the resistance across the engine harness side connector terminals No. H34-6 and H34-7.

**NOTE:**

**Although the resistance varies with the engine coolant temperature, normal condition is indicated when measured resistance falls within the range shown in the following table:**

| Coolant temperature °C(°F) | Normal resistance KΩ |
|---|---|
| −10 (14) | 7 — 12 |
| 10 (50) | 3 — 5 |
| 20 (68) | 2 — 3 |
| 50 (122) | 0.7 — 1 |
| 80 (176) (warmed) | 0.2 — 0.4 |

If measured resistance deviates from the specified range, replace water temperature sensor.

**Wiring**

| Signal code No. | Wiring | Ω |
|---|---|---|
| 14 | H34-7 (33) —— B | ∞ |
| | H34-7 (33) —— Ground | ∞ |
| 15 | H34-7 (33) —— GY | 0 |
| | H34-6 (46) —— B | 0 |
| | H34-6 (46) —— Ground | ∞ |

If abnormal condition is indicated, correct or replace applicable harnesses.

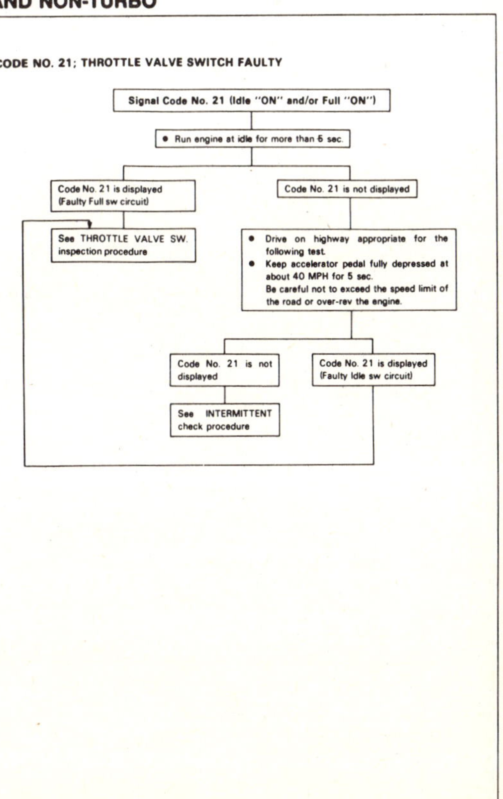

## 1989-90 IMPULSE TURBO AND NON-TURBO

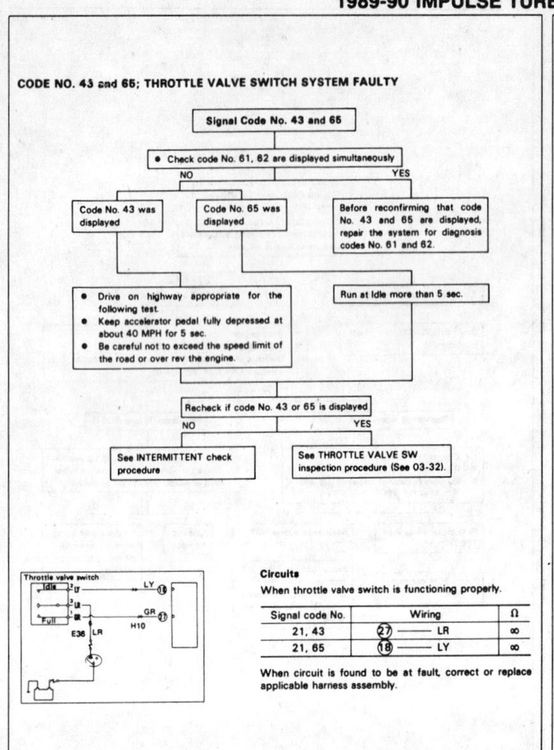


**Circuits**

When throttle valve switch is functioning properly.

| Signal code No. | Wiring | Ω |
|---|---|---|
| 21, 43 | 27 —— LR | ∞ |
| 21, 65 | 18 —— LY | ∞ |

When circuit is found to be at fault, correct or replace applicable harness assembly.

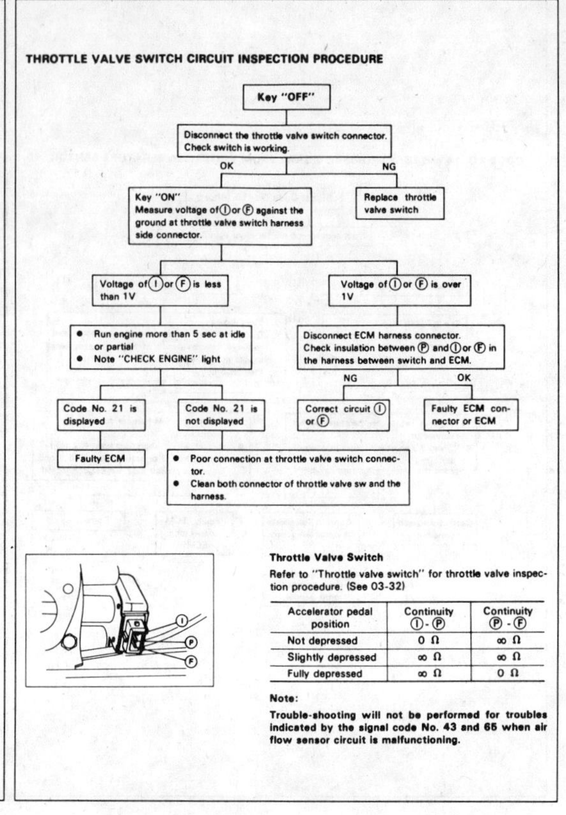

THROTTLE VALVE SWITCH CIRCUIT INSPECTION PROCEDURE
Key "OFF"
Disconnect the throttle valve switch connector. Check switch is working.
OK
NG
Key "ON" Measure voltage of Ⓘ or Ⓕ against the ground at throttle valve switch harness side connector.
Replace throttle valve switch
Voltage of Ⓘ or Ⓕ is less than 1V
Voltage of Ⓘ or Ⓕ is over 1V
• Run engine more than 5 sec at idle or partial
• Note "CHECK ENGINE" light
Disconnect ECM harness connector. Check insulation between Ⓟ and Ⓘ or Ⓕ in the harness between switch and ECM.
NG
OK
Code No. 21 is displayed
Code No. 21 is not displayed
Correct circuit Ⓘ or Ⓕ
Faulty ECM connector or ECM
Faulty ECM
• Poor connection at throttle valve switch connector.
• Clean both connector of throttle valve sw and the harness.
I
P
F


**Throttle Valve Switch**

Refer to "Throttle valve switch" for throttle valve inspection procedure. (See 03-32)

| Accelerator pedal position | Continuity Ⓘ - Ⓟ | Continuity Ⓟ - Ⓕ |
|---|---|---|
| Not depressed | 0 Ω | ∞ Ω |
| Slightly depressed | ∞ Ω | ∞ Ω |
| Fully depressed | ∞ Ω | 0 Ω |

**Note:**

**Trouble-shooting will not be performed for troubles indicated by the signal code No. 43 and 65 when air flow sensor circuit is malfunctioning.**

## 1989-90 IMPULSE TURBO AND NON-TURBO

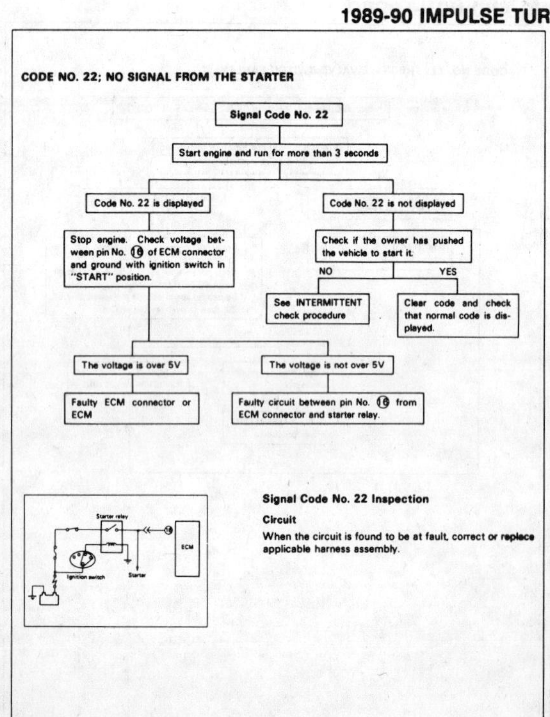


**Signal Code No. 22 Inspection**

**Circuit**

When the circuit is found to be at fault, correct or replace applicable harness assembly.

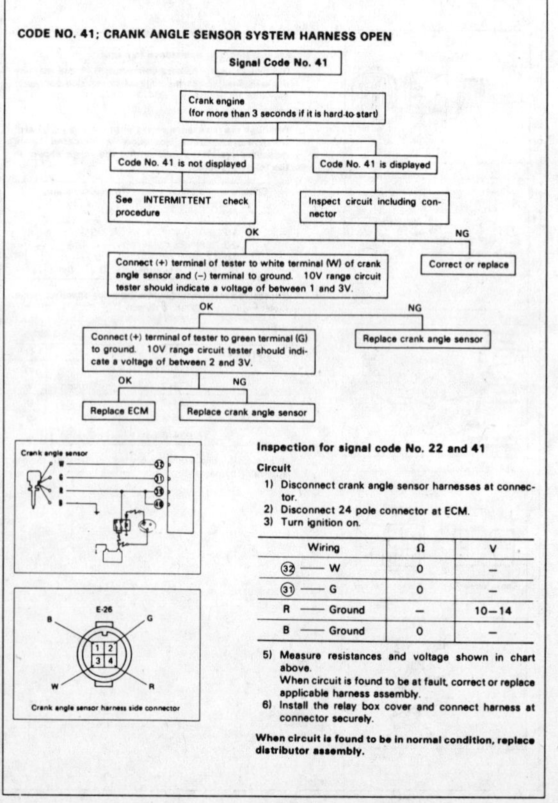


**Inspection for signal code No. 22 and 41**

**Circuit**

1) Disconnect crank angle sensor harnesses at connector.
2) Disconnect 24 pole connector at ECM.
3) Turn ignition on.

| Wiring | Ω | V |
|---|---|---|
| 32 —— W | 0 | — |
| 31 —— G | 0 | — |
| R —— Ground | — | 10–14 |
| B —— Ground | 0 | — |

5) Measure resistances and voltage shown in chart above.
When circuit is found to be at fault, correct or replace applicable harness assembly.
6) Install the relay box cover and connect harness at connector securely.

**When circuit is found to be in normal condition, replace distributor assembly.**

## 1989-90 IMPULSE TURBO AND NON-TURBO

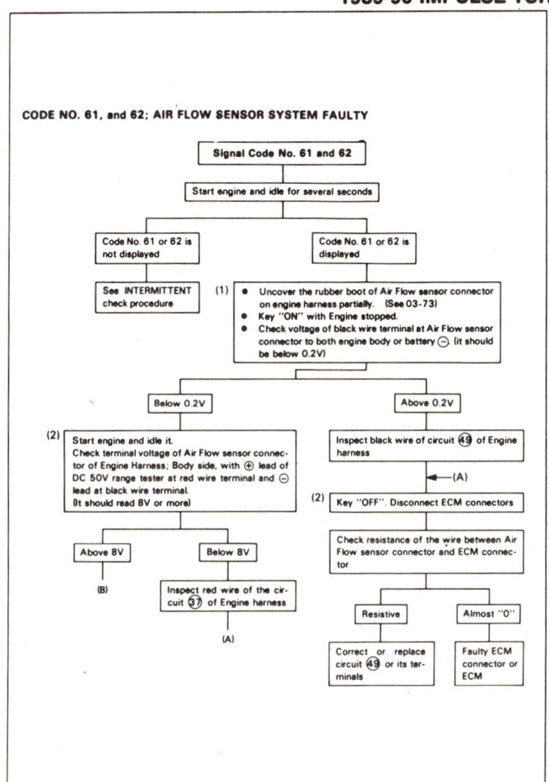


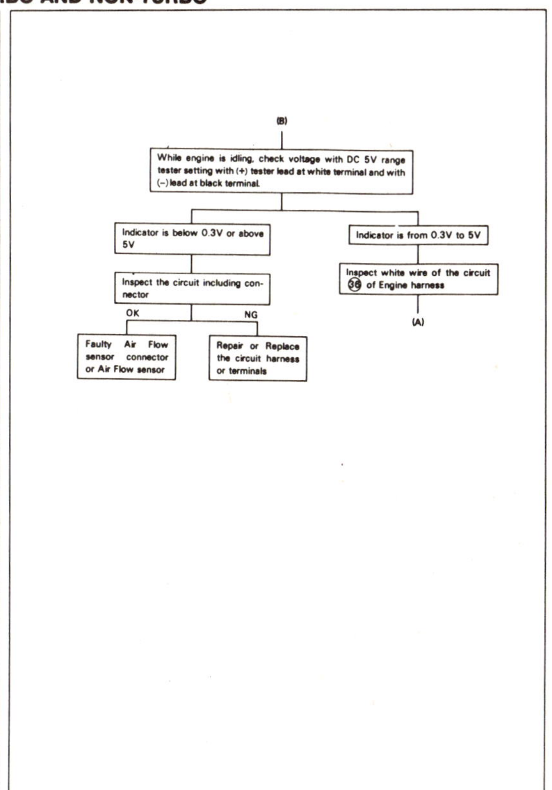


## 1989-90 IMPULSE TURBO AND NON-TURBO

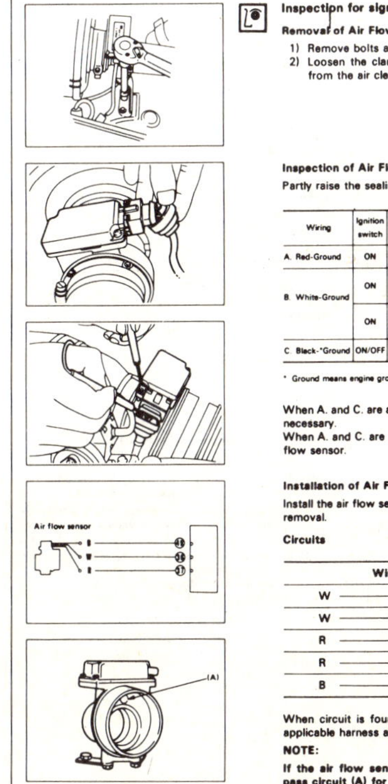


### Inspection for signal code No. 61 and 62

**Removal of Air Flow Sensor**

1) Remove bolts attaching air flow sensor.
2) Loosen the clamp bolt and remove air flow sensor from the air cleaner.

**Inspection of Air Flow Sensor**

Partly raise the sealing cover on the harness connector.

| Wiring | Ignition switch | Condition | V | Ω |
|---|---|---|---|---|
| A. Red-Ground | ON | – | 10–14 | – |
| B. White-Ground | ON | Engine stop | Less than 0.2 | – |
| | ON | Engine idling | 0.5–1.5 | – |
| C. Black-*Ground | ON/OFF | – | Less than 0.2 | 0 |

* Ground means engine ground.

When A. and C. are abnormal, check circuit and correct as necessary.
When A. and C. are normal but B. is abnormal, replace air flow sensor.

**Installation of Air Flow Sensor**

Install the air flow sensor assembly in the reverse order of removal.

**Circuits**

| Wiring | | Ω |
|---|---|---|
| W | 36 | 0 |
| W | Ground | ∞ |
| R | 37 | 0 |
| R | Ground | ∞ |
| B | 49 | 0 |

When circuit is found to be at fault, correct or replace applicable harness assembly.

**NOTE:**

**If the air flow sensor has been removed, check by-pass circuit (A) for contamination and clean with carburetor cleaner as necessary.**

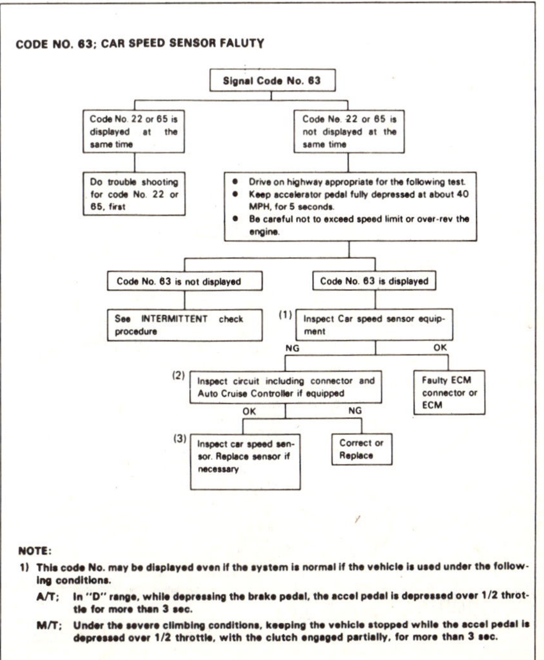


**NOTE:**

**1) This code No. may be displayed even if the system is normal if the vehicle is used under the following conditions.**

**A/T; In "D" range, while depressing the brake pedal, the accel pedal is depressed over 1/2 throttle for more than 3 sec.**

**M/T; Under the severe climbing conditions, keeping the vehicle stopped while the accel pedal is depressed over 1/2 throttle, with the clutch engaged partially, for more than 3 sec.**

## 1989-90 IMPULSE TURBO AND NON-TURBO

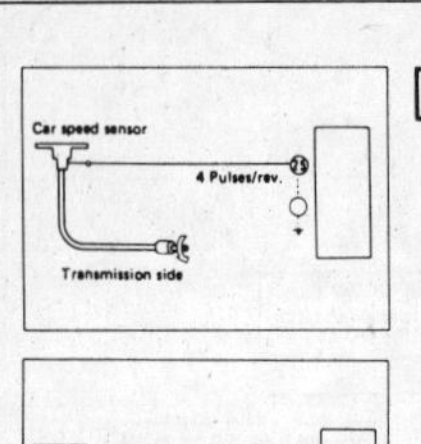


**Inspection for signal code No. 63**

**(1) Sensor and Equipment**

1) Connect a circuit tester between the control unit harness connector terminal 25 and ground.
2) Disconnect speedometer cable at transmission end and turn the inner cable slowly.
3) If the tester indicates continuity and open circuit condition alternately as the cable is turned, the car speed sensor and circuit are operating normally.

**(2) Circuits**

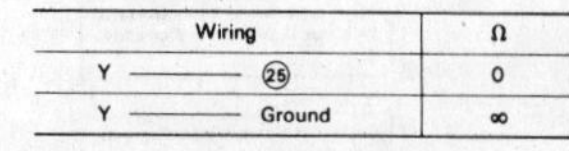

| Wiring | Ω |
|---|---|
| Y ——— ㉕ | 0 |
| Y ——— Ground | ∞ |

When circuit is found to be at fault, correct or replace applicable harness assembly.

2) Install the speedometer assembly. Refer to "Meters" in Section "Chassis electrical" for installation procedure.

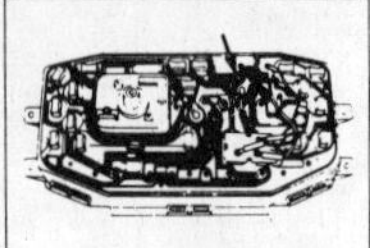

**(3) Car Speed Sensor**

1) Remove the meter assembly
2) Connect the leads of a circuit tester across the terminals of the connector at rear face of the meter assembly as illustrated.

3) Turn the inner shaft slowly and check that the tester indicates a continuity and open circuit condition alternately. If speedometer is inoperative, replace speedometer assembly.

**CODE NO. 66; KNOCK SENSOR FAULTY**

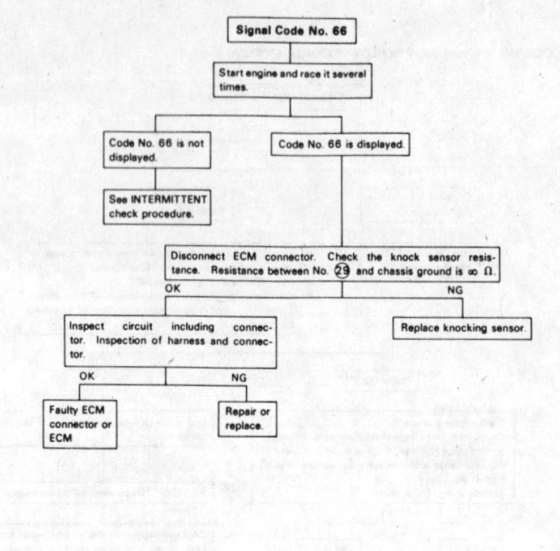


## 1989-90 IMPULSE TURBO AND NON-TURBO

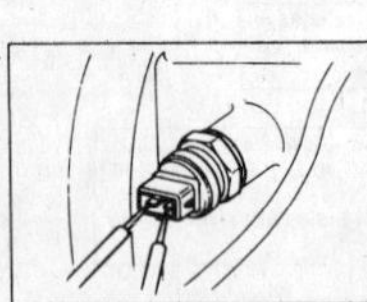

**Inspection for signal code No. 66**

**Measurement of Resistance**

1) Disconnect knocking sensor harnesses at connector.
2) Measure the resistance across the terminals of the knocking sensor.

| Standard resistance (Ω) | More than 1M Ω |
|---|---|

**Circuits**

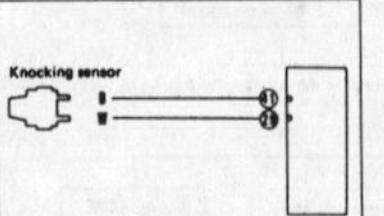


| Wiring | Ω |
|---|---|
| ㉘ ——— W | 0 |
| ㉙ ——— Ground | ∞ |
| ㊶ ——— B | 0 |

When circuit is found to be at fault, correct or replace applicable harness assembly.

**NOTE:**

**Replace the knocking sensor if the ignition timing is correct but the engine produces serious knocking. If the trouble persists, replace the control unit.**

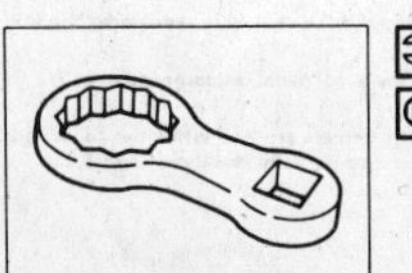

**Knocking Sensor Installation Procedure**

When installing, carefully tighten the knocking sensor to specification.

| Tightening torque kg-m(ft.lbs.) | 2–3 (14.5–21.7) |
|---|---|

Wrench, knocking sensor : J-22898-A

**CODE NO. 51, 52; ROM (CODE NO. 51) NG, RAM (CODE NO. 52) NG**

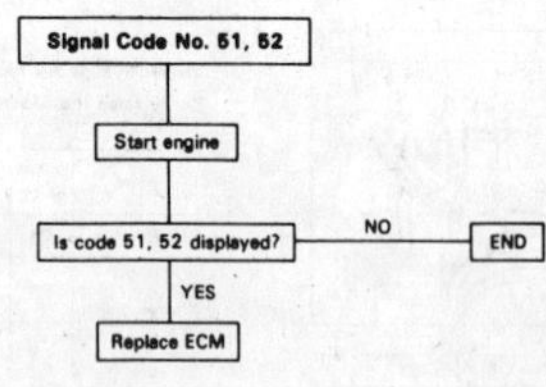

**Inspection for Signal Code No. 51 and 52**

**Circuit Connections**

1. Disconnect the ECM harnesses at connector and check for proper connections at terminals.
2. Regardless of terminal conditions, road test the vehicle with all the harnesses reconnected at the connector.
3. Obtain display of trouble code and see if the trouble has been corrected. If the trouble persists, replace the ECM.

## 1989-90 IMPULSE TURBO AND NON-TURBO

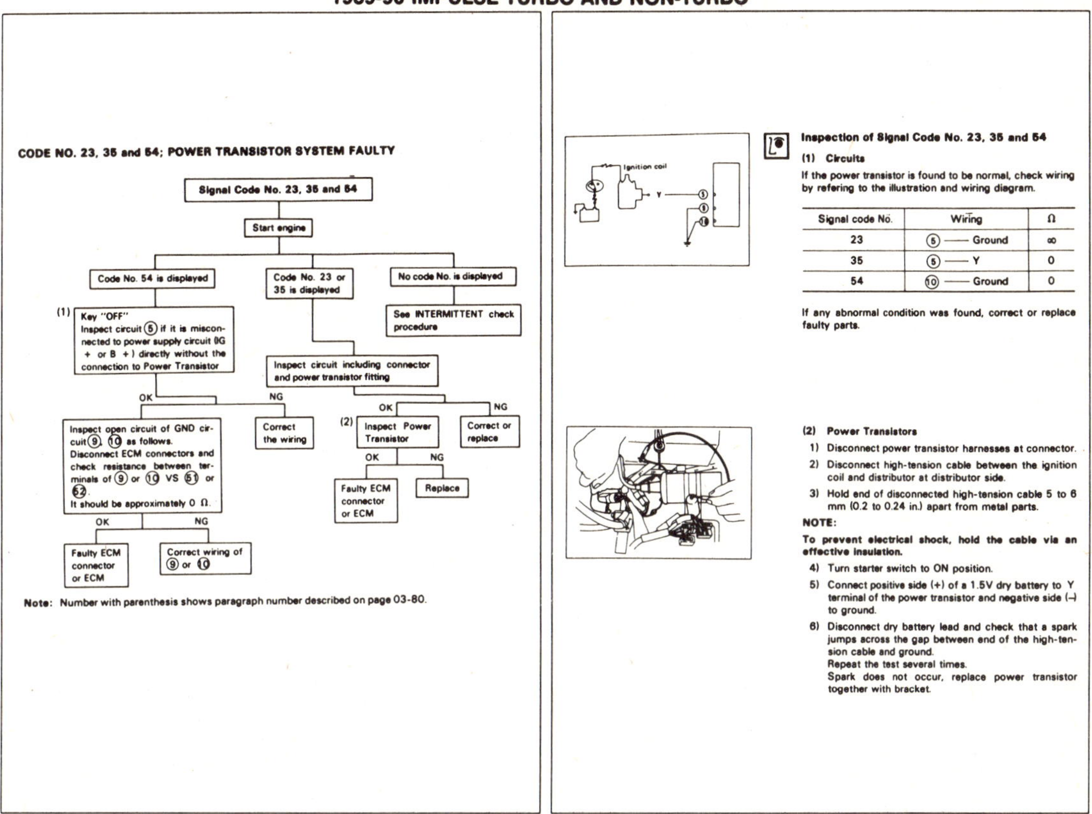

**CODE NO. 23, 35 and 54; POWER TRANSISTOR SYSTEM FAULTY**

Signal Code No. 23, 35 and 54

Start engine

- Code No. 54 is displayed
  - (1) Key "OFF" Inspect circuit ⑤ if it is misconnected to power supply circuit (IG + or B +) directly without the connection to Power Transistor
    - OK: Inspect open circuit of GND circuit ⑨, ⑩ as follows. Disconnect ECM connectors and check resistance between terminals of ⑨ or ⑩ VS 51 or 52. It should be approximately 0 Ω.
      - OK: Faulty ECM connector or ECM
      - NG: Correct wiring of ⑨ or ⑩
    - NG: Correct the wiring
- Code No. 23 or 35 is displayed
  - Inspect circuit including connector and power transistor fitting
    - OK: (2) Inspect Power Transistor
      - OK: Faulty ECM connector or ECM
      - NG: Replace
    - NG: Correct or replace
- No code No. is displayed
  - See INTERMITTENT check procedure

**Note:** Number with parenthesis shows paragraph number described on page 03-80.

Ignition coil; Y; ⑤; ⑨; ⑩

**Inspection of Signal Code No. 23, 35 and 54**

**(1) Circuits**

If the power transistor is found to be normal, check wiring by refering to the illustration and wiring diagram.

| Signal code No. | Wiring | Ω |
|---|---|---|
| 23 | ⑤ —— Ground | ∞ |
| 35 | ⑤ —— Y | 0 |
| 54 | ⑩ —— Ground | 0 |

If any abnormal condition was found, correct or replace faulty parts.

**(2) Power Transistors**

1) Disconnect power transistor harnesses at connector.
2) Disconnect high-tension cable between the ignition coil and distributor at distributor side.
3) Hold end of disconnected high-tension cable 5 to 6 mm (0.2 to 0.24 in.) apart from metal parts.

**NOTE:**

**To prevent electrical shock, hold the cable via an effective insulation.**

4) Turn starter switch to ON position.
5) Connect positive side (+) of a 1.5V dry battery to Y terminal of the power transistor and negative side (–) to ground.
6) Disconnect dry battery lead and check that a spark jumps across the gap between end of the high-tension cable and ground.
   Repeat the test several times.
   Spark does not occur, replace power transistor together with bracket.

## 1989-90 IMPULSE TURBO AND NON-TURBO

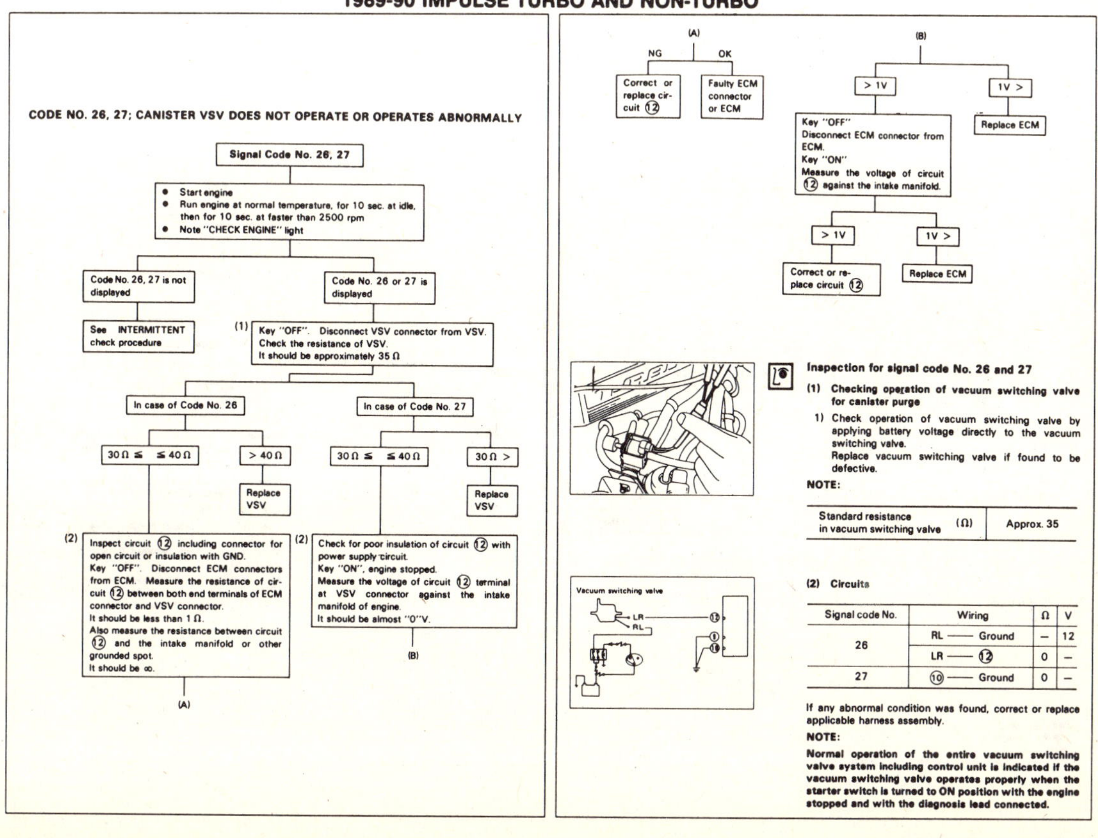

**CODE NO. 26, 27; CANISTER VSV DOES NOT OPERATE OR OPERATES ABNORMALLY**

Signal Code No. 26, 27

- Start engine
- Run engine at normal temperature, for 10 sec. at idle, then for 10 sec. at faster than 2500 rpm
- Note "CHECK ENGINE" light

- Code No. 26, 27 is not displayed
  - See INTERMITTENT check procedure
- Code No. 26 or 27 is displayed
  - (1) Key "OFF". Disconnect VSV connector from VSV. Check the resistance of VSV. It should be approximately 35 Ω
    - In case of Code No. 26
      - 30 Ω ≤ ≤ 40 Ω
        - (2) Inspect circuit ⑫ including connector for open circuit or insulation with GND. Key "OFF". Disconnect ECM connectors from ECM. Measure the resistance of circuit ⑫ between both end terminals of ECM connector and VSV connector. It should be less than 1 Ω. Also measure the resistance between circuit ⑫ and the intake manifold or other grounded spot. It should be ∞.
          - (A)
      - > 40 Ω: Replace VSV
    - In case of Code No. 27
      - 30 Ω ≤ ≤ 40 Ω
        - (2) Check for poor insulation of circuit ⑫ with power supply circuit. Key "ON", engine stopped. Measure the voltage of circuit ⑫ terminal at VSV connector against the intake manifold of engine. It should be almost "0"V.
          - (B)
      - 30 Ω >: Replace VSV

(A)
- NG: Correct or replace circuit ⑫
- OK: Faulty ECM connector or ECM

(B)
- > 1V
  - Key "OFF" Disconnect ECM connector from ECM. Key "ON" Measure the voltage of circuit ⑫ against the intake manifold.
    - > 1V: Correct or replace circuit ⑫
    - 1V >: Replace ECM
- 1V >: Replace ECM

**Inspection for signal code No. 26 and 27**

**(1) Checking operation of vacuum switching valve for canister purge**

1) Check operation of vacuum switching valve by applying battery voltage directly to the vacuum switching valve.
   Replace vacuum switching valve if found to be defective.

**NOTE:**

| Standard resistance in vacuum switching valve | (Ω) | Approx. 35 |
|---|---|---|

Vacuum switching valve; LR; RL; ⑫; ⑨; ⑩

**(2) Circuits**

| Signal code No. | Wiring | Ω | V |
|---|---|---|---|
| 26 | RL —— Ground | – | 12 |
| | LR —— ⑫ | 0 | – |
| 27 | ⑩ —— Ground | 0 | – |

If any abnormal condition was found, correct or replace applicable harness assembly.

**NOTE:**

**Normal operation of the entire vacuum switching valve system including control unit is indicated if the vacuum switching valve operates properly when the starter switch is turned to ON position with the engine stopped and with the diagnosis lead connected.**

## 1989-90 IMPULSE TURBO AND NON-TURBO

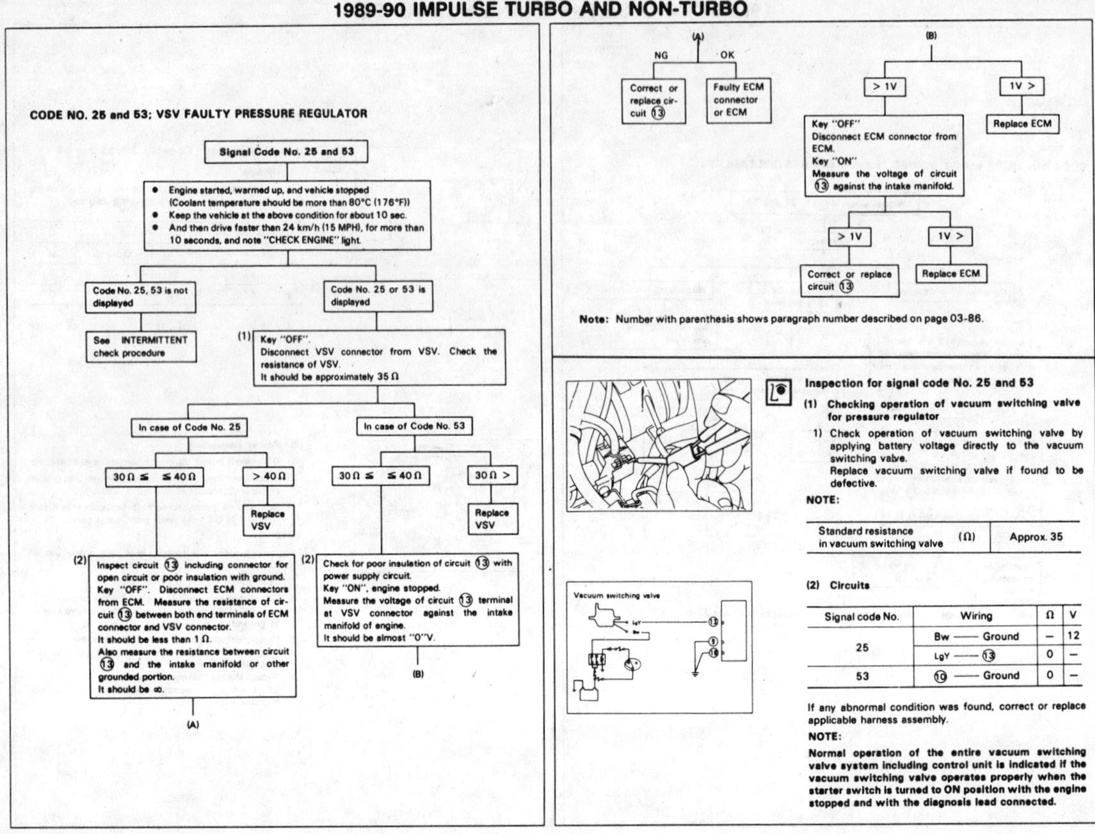


## 1989-90 IMPULSE TURBO AND NON-TURBO

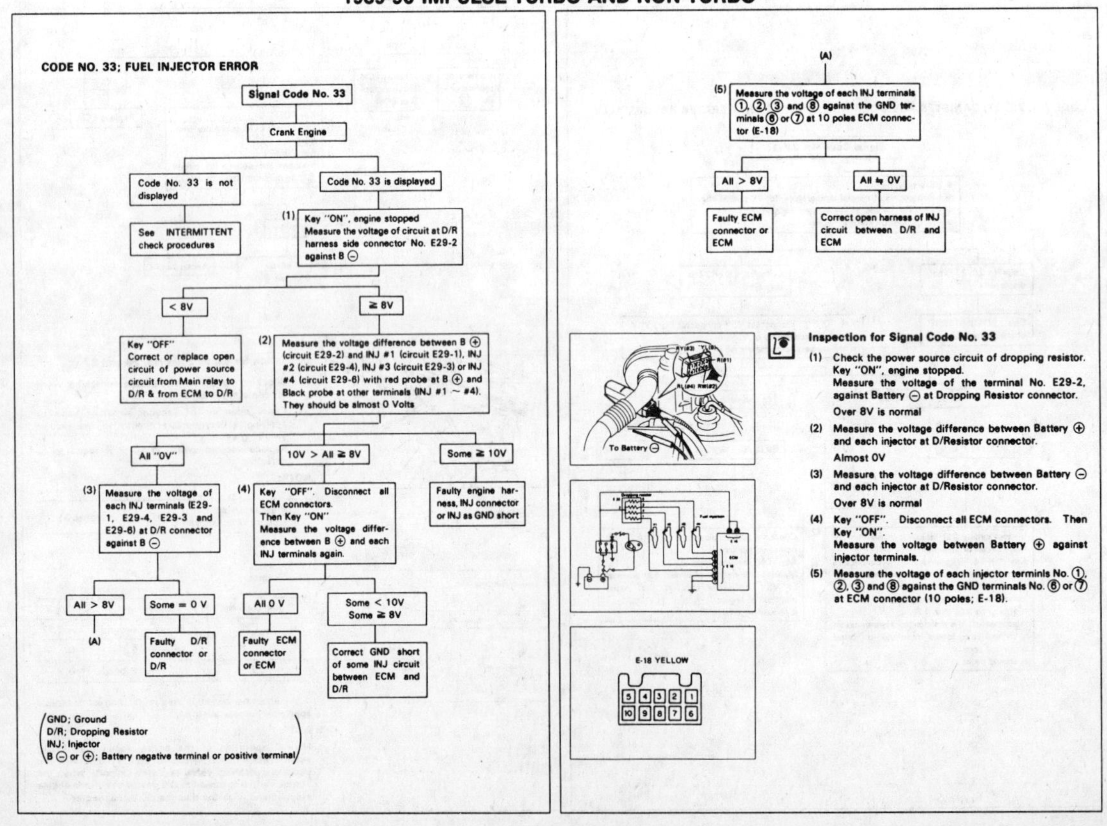

CODE NO. 33; FUEL INJECTOR ERROR
Signal Code No. 33
Crank Engine
Code No. 33 is not displayed
See INTERMITTENT check procedures
Code No. 33 is displayed
(1) Key "ON", engine stopped. Measure the voltage of circuit at D/R harness side connector No. E29-2 against B ⊖
< 8V
Key "OFF" Correct or replace open circuit of power source circuit from Main relay to D/R & from ECM to D/R
≧ 8V
(2) Measure the voltage difference between B ⊕ (circuit E29-2) and INJ #1 (circuit E29-1), INJ #2 (circuit E29-4), INJ #3 (circuit E29-3) or INJ #4 (circuit E29-6) with red probe at B ⊕ and Black probe at other terminals (INJ #1 ~ #4). They should be almost 0 Volts
All "0V"
10V > All ≧ 8V
Some ≧ 10V
(3) Measure the voltage of each INJ terminals (E29-1, E29-4, E29-3 and E29-6) at D/R connector against B ⊖
(4) Key "OFF". Disconnect all ECM connectors. Then Key "ON" Measure the voltage difference between B ⊕ and each INJ terminals again.
Faulty engine harness, INJ connector or INJ as GND short
All > 8V
(A)
Some = 0 V
Faulty D/R connector or D/R
All 0 V
Faulty ECM connector or ECM
Some < 10V Some ≧ 8V
Correct GND short of some INJ circuit between ECM and D/R
GND; Ground
D/R; Dropping Resistor
INJ; Injector
B ⊖ or ⊕; Battery negative terminal or positive terminal
(A)
(5) Measure the voltage of each INJ terminals ①, ②, ③ and ⑧ against the GND terminals ⑥ or ⑦ at 10 poles ECM connector (E-18)
All > 8V
Faulty ECM connector or ECM
All ≒ 0V
Correct open harness of INJ circuit between D/R and ECM
Inspection for Signal Code No. 33
(1) Check the power source circuit of dropping resistor. Key "ON", engine stopped. Measure the voltage of the terminal No. E29-2, against Battery ⊖ at Dropping Resistor connector. Over 8V is normal
(2) Measure the voltage difference between Battery ⊕ and each injector at D/Resistor connector. Almost 0V
(3) Measure the voltage difference between Battery ⊖ and each injector at D/Resistor connector. Over 8V is normal
(4) Key "OFF". Disconnect all ECM connectors. Then Key "ON". Measure the voltage between Battery ⊕ against injector terminals.
(5) Measure the voltage of each injector terminals No. ①, ②, ③ and ⑧ against the GND terminals No. ⑥ or ⑦ at ECM connector (10 poles; E-18).
To Battery ⊖
E-18 YELLOW
5 4 3 2 1
10 9 8 7 6

## 1989-90 IMPULSE TURBO AND NON-TURBO

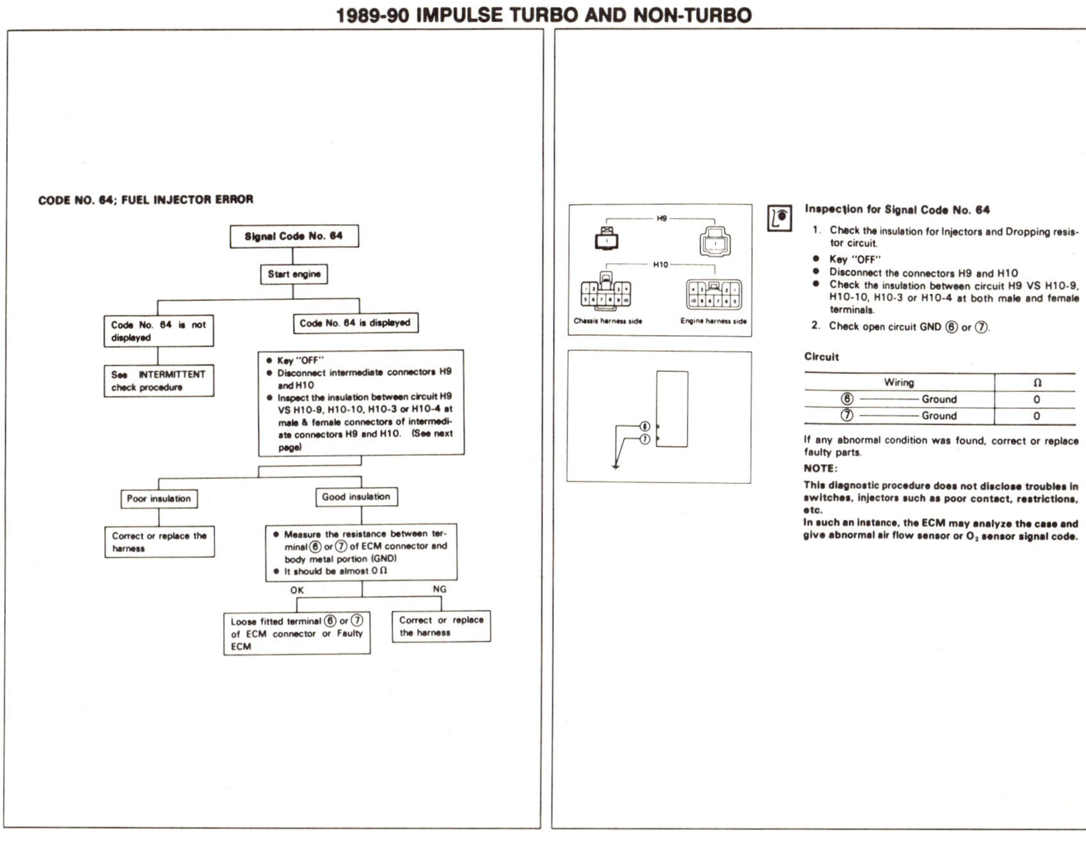

CODE NO. 64; FUEL INJECTOR ERROR
Signal Code No. 64
Start engine
Code No. 64 is not displayed
Code No. 64 is displayed
See INTERMITTENT check procedure
● Key "OFF"
● Disconnect intermediate connectors H9 and H10
● Inspect the insulation between circuit H9 VS H10-9, H10-10, H10-3 or H10-4 at male & female connectors of intermediate connectors H9 and H10. (See next page)
Poor insulation
Good insulation
Correct or replace the harness
● Measure the resistance between terminal ⑥ or ⑦ of ECM connector and body metal portion (GND)
● It should be almost 0 Ω
OK
NG
Loose fitted terminal ⑥ or ⑦ of ECM connector or Faulty ECM
Correct or replace the harness
H9
H10
Chassis harness side
Engine harness side
⑥
⑦


**Inspection for Signal Code No. 64**

1. Check the insulation for Injectors and Dropping resistor circuit.
   - Key "OFF"
   - Disconnect the connectors H9 and H10
   - Check the insulation between circuit H9 VS H10-9, H10-10, H10-3 or H10-4 at both male and female terminals.
2. Check open circuit GND ⑥ or ⑦.

**Circuit**

| Wiring | Ω |
|---|---|
| ⑥ ———— Ground | 0 |
| ⑦ ———— Ground | 0 |

If any abnormal condition was found, correct or replace faulty parts.

**NOTE:**

**This diagnostic procedure does not disclose troubles in switches, injectors such as poor contact, restrictions, etc.**
**In such an instance, the ECM may analyze the case and give abnormal air flow sensor or $O_2$ sensor signal code.**

## 1989-90 IMPULSE TURBO AND NON-TURBO

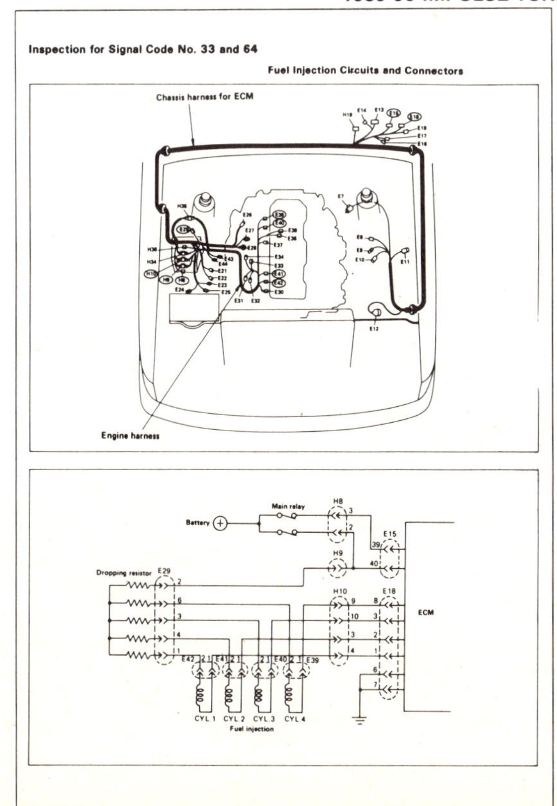


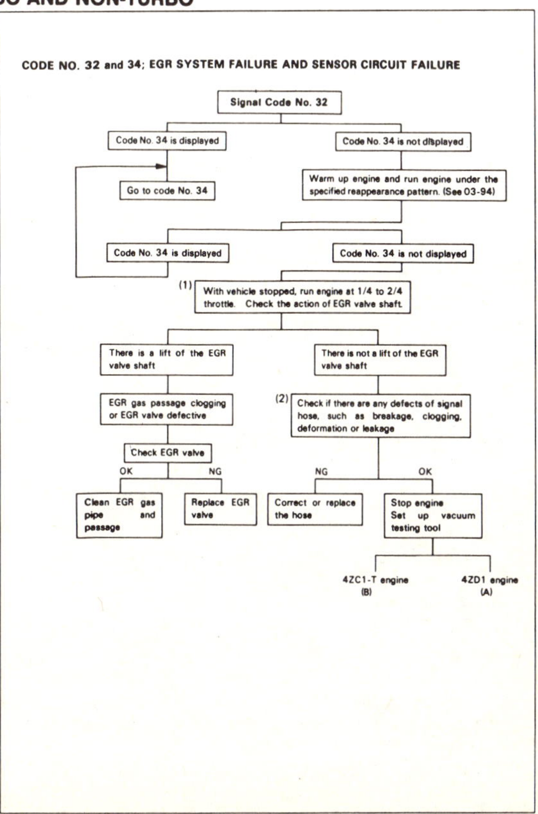

## 1989-90 IMPULSE TURBO AND NON-TURBO

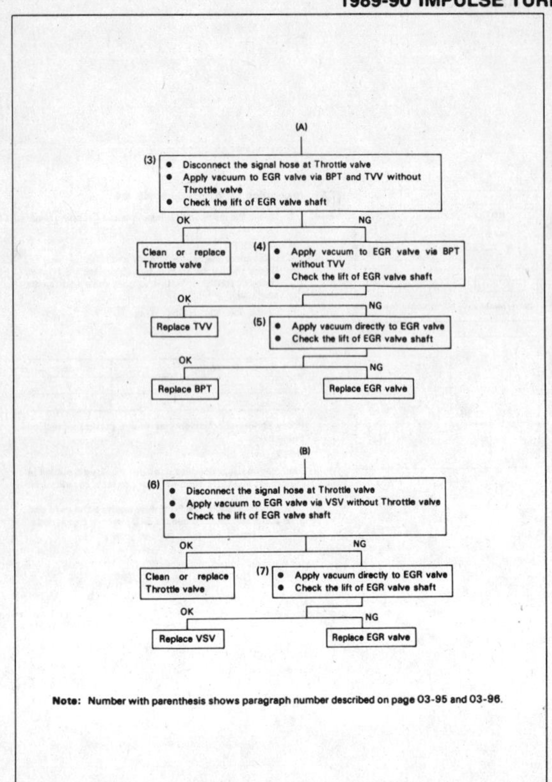


**Note:** Number with parenthesis shows paragraph number described on page 03-95 and 03-96.

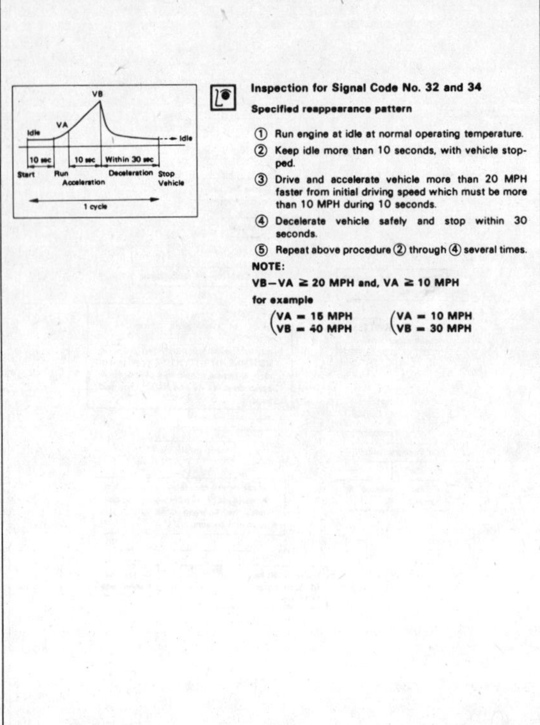


**Inspection for Signal Code No. 32 and 34**

**Specified reappearance pattern**

① Run engine at idle at normal operating temperature.
② Keep idle more than 10 seconds, with vehicle stopped.
③ Drive and accelerate vehicle more than 20 MPH faster from initial driving speed which must be more than 10 MPH during 10 seconds.
④ Decelerate vehicle safely and stop within 30 seconds.
⑤ Repeat above procedure ② through ④ several times.

**NOTE:**

**VB−VA ≧ 20 MPH and, VA ≧ 10 MPH**

**for example**

(VA = 15 MPH, VB = 40 MPH) (VA = 10 MPH, VB = 30 MPH)

## 1989-90 IMPULSE TURBO AND NON-TURBO

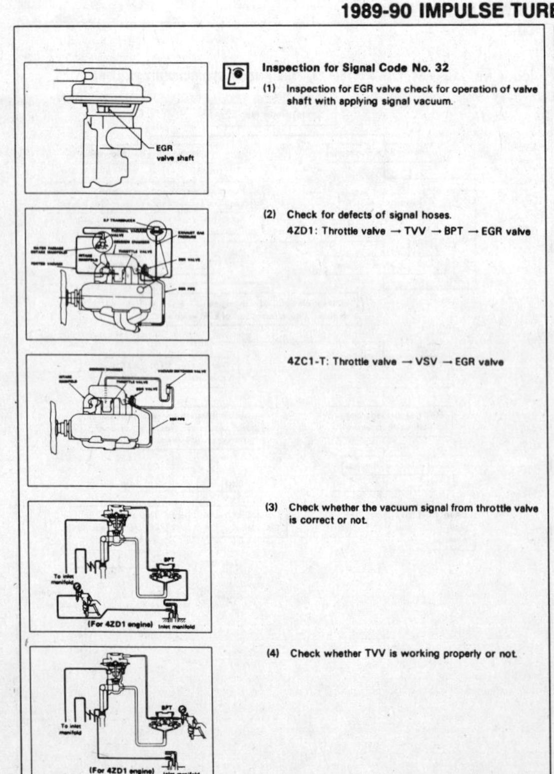


**Inspection for Signal Code No. 32**

(1) Inspection for EGR valve check for operation of valve shaft with applying signal vacuum.

(2) Check for defects of signal hoses.
4ZD1: Throttle valve → TVV → BPT → EGR valve

4ZC1-T: Throttle valve → VSV → EGR valve

(3) Check whether the vacuum signal from throttle valve is correct or not.

(4) Check whether TVV is working properly or not.

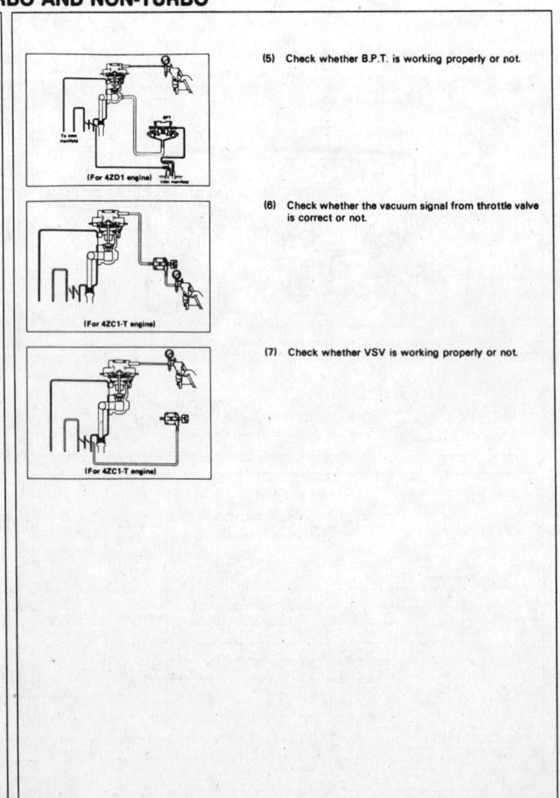


(5) Check whether B.P.T. is working properly or not.

(6) Check whether the vacuum signal from throttle valve is correct or not.

(7) Check whether VSV is working properly or not.

## 1989-90 IMPULSE TURBO AND NON-TURBO

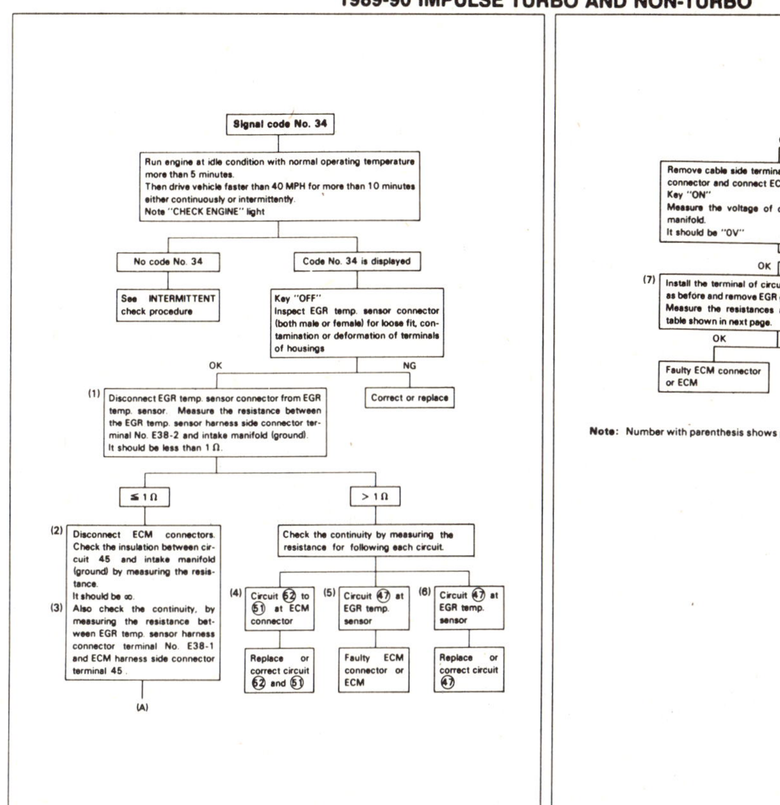


(A)

OK / NG

Remove cable side terminal of circuit 45 from ECM connector and connect ECM connectors to ECM.
Key "ON"
Measure the voltage of circuit 45 against intake manifold.
It should be "OV"

OK / NG

(7) Install the terminal of circuit 45 to ECM connector, as before and remove EGR gas temp. sensor.
Measure the resistances and compare it with the table shown in next page.

OK: Faulty ECM connector or ECM

NG: Replace EGR temp. sensor

NG: Correct or replace circuit 45

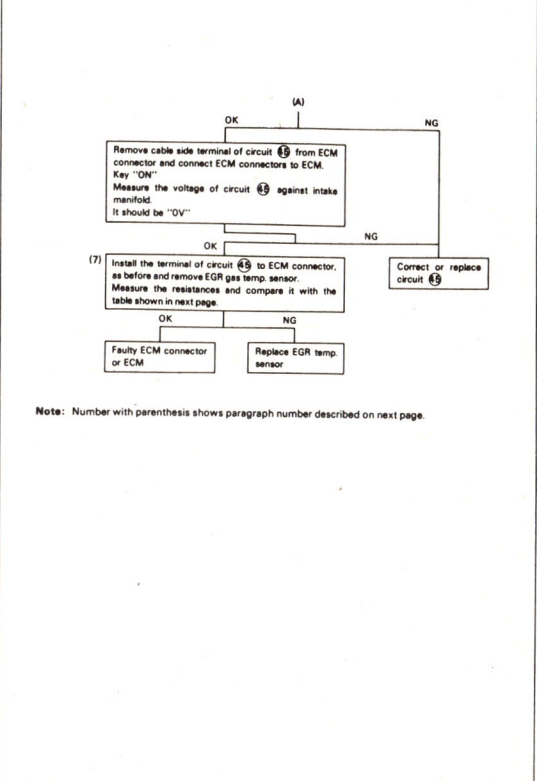

**Note:** Number with parenthesis shows paragraph number described on next page.

## 1989-90 IMPULSE TURBO AND NON-TURBO

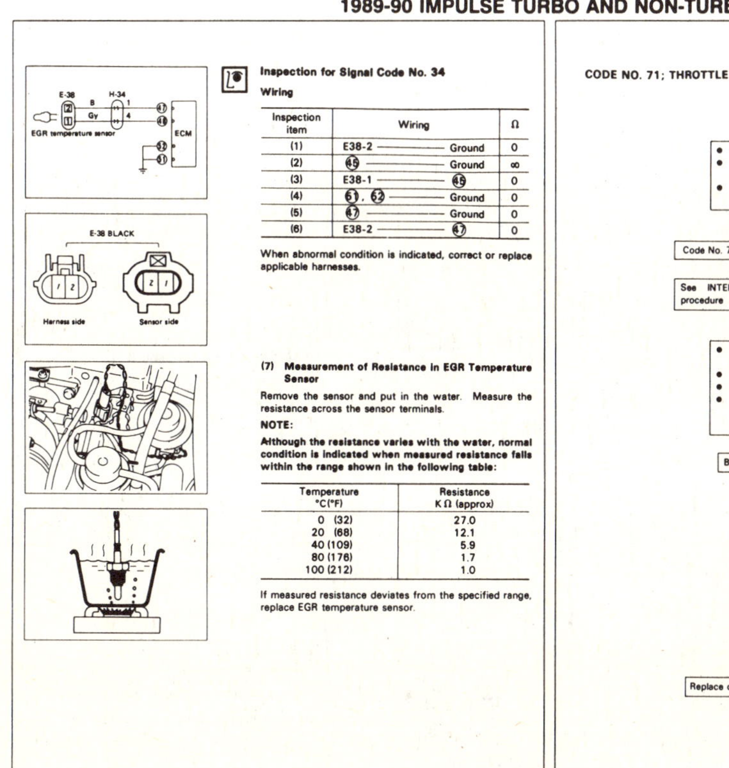


**Inspection for Signal Code No. 34**

**Wiring**

| Inspection item | Wiring | Ω |
|---|---|---|
| (1) | E38-2 ——— Ground | 0 |
| (2) | 45 ——— Ground | ∞ |
| (3) | E38-1 ——— 45 | 0 |
| (4) | 61, 62 ——— Ground | 0 |
| (5) | 47 ——— Ground | 0 |
| (6) | E38-2 ——— 47 | 0 |

When abnormal condition is indicated, correct or replace applicable harnesses.

**(7) Measurement of Resistance in EGR Temperature Sensor**

Remove the sensor and put in the water. Measure the resistance across the sensor terminals.

**NOTE:**

**Although the resistance varies with the water, normal condition is indicated when measured resistance falls within the range shown in the following table:**

| Temperature °C(°F) | Resistance K Ω (approx) |
|---|---|
| 0 (32) | 27.0 |
| 20 (68) | 12.1 |
| 40 (109) | 5.9 |
| 80 (176) | 1.7 |
| 100 (212) | 1.0 |

If measured resistance deviates from the specified range, replace EGR temperature sensor.

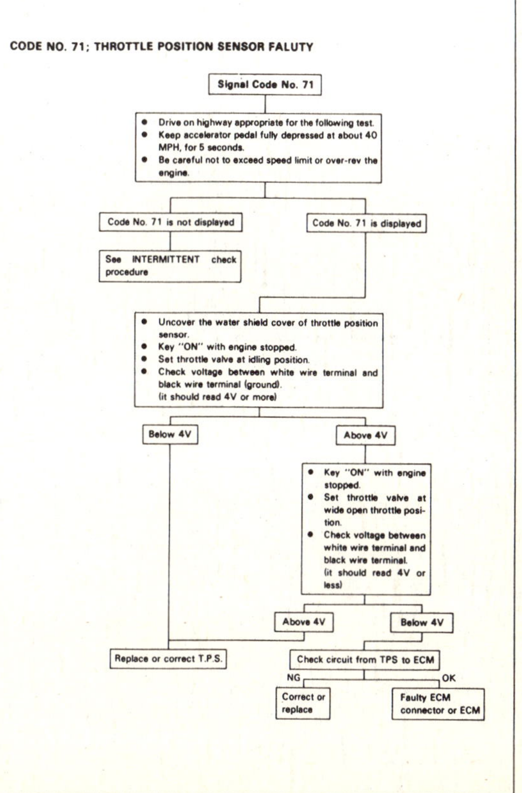

## 1989-90 IMPULSE TURBO AND NON-TURBO

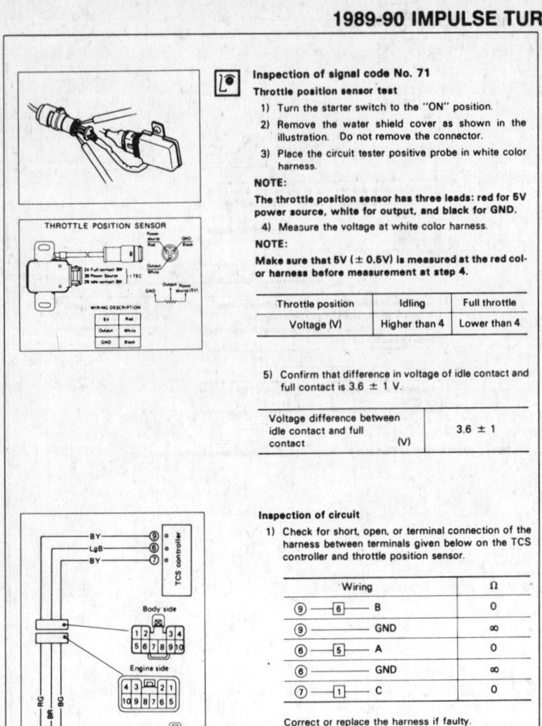


**Inspection of signal code No. 71**

**Throttle position sensor test**

1) Turn the starter switch to the "ON" position.
2) Remove the water shield cover as shown in the illustration. Do not remove the connector.
3) Place the circuit tester positive probe in white color harness.

**NOTE:**

**The throttle position sensor has three leads: red for 5V power source, white for output, and black for GND.**

4) Measure the voltage at white color harness.

**NOTE:**

**Make sure that 5V (± 0.5V) is measured at the red color harness before measurement at step 4.**

| Throttle position | Idling | Full throttle |
|---|---|---|
| Voltage (V) | Higher than 4 | Lower than 4 |

5) Confirm that difference in voltage of idle contact and full contact is 3.6 ± 1 V.

| Voltage difference between idle contact and full contact (V) | 3.6 ± 1 |
|---|---|

**Inspection of circuit**

1) Check for short, open, or terminal connection of the harness between terminals given below on the TCS controller and throttle position sensor.

| Wiring | Ω |
|---|---|
| ⑨ — [6] — B | 0 |
| ⑨ — GND | ∞ |
| ⑥ — [5] — A | 0 |
| ⑥ — GND | ∞ |
| ⑦ — [1] — C | 0 |

Correct or replace the harness if faulty.

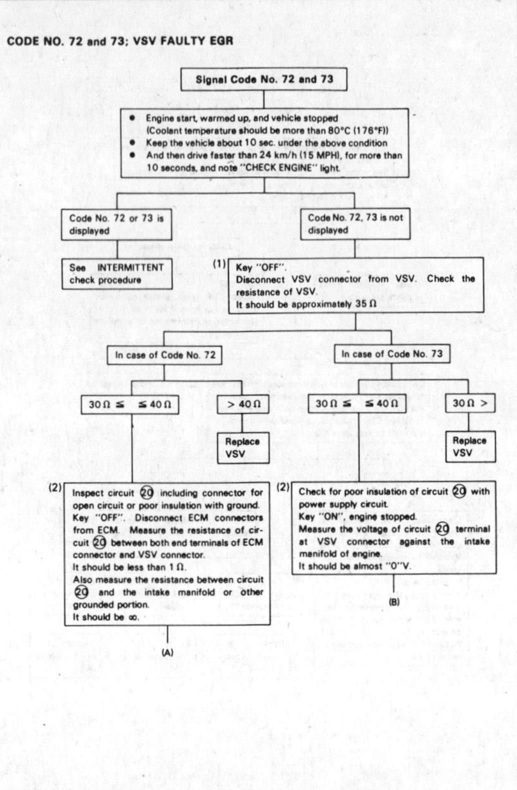

**CODE NO. 72 and 73; VSV FAULTY EGR**

**Signal Code No. 72 and 73**

- Engine start, warmed up, and vehicle stopped (Coolant temperature should be more than 80°C (176°F))
- Keep the vehicle about 10 sec. under the above condition
- And then drive faster than 24 km/h (15 MPH), for more than 10 seconds, and note "CHECK ENGINE" light.

Code No. 72 or 73 is displayed → See INTERMITTENT check procedure

Code No. 72, 73 is not displayed →

(1) Key "OFF".
Disconnect VSV connector from VSV. Check the resistance of VSV.
It should be approximately 35 Ω

In case of Code No. 72:
- > 40 Ω → Replace VSV
- 30 Ω ≤ ≤ 40 Ω → (2) Inspect circuit ⑳ including connector for open circuit or poor insulation with ground.
Key "OFF". Disconnect ECM connectors from ECM. Measure the resistance of circuit ⑳ between both end terminals of ECM connector and VSV connector.
It should be less than 1 Ω.
Also measure the resistance between circuit ⑳ and the intake manifold or other grounded portion.
It should be ∞.
→ (A)

In case of Code No. 73:
- 30 Ω > → Replace VSV
- 30 Ω ≤ ≤ 40 Ω → (2) Check for poor insulation of circuit ⑳ with power supply circuit.
Key "ON", engine stopped.
Measure the voltage of circuit ⑳ terminal at VSV connector against the intake manifold of engine.
It should be almost "0"V.
→ (B)

## 1989-90 IMPULSE TURBO AND NON-TURBO

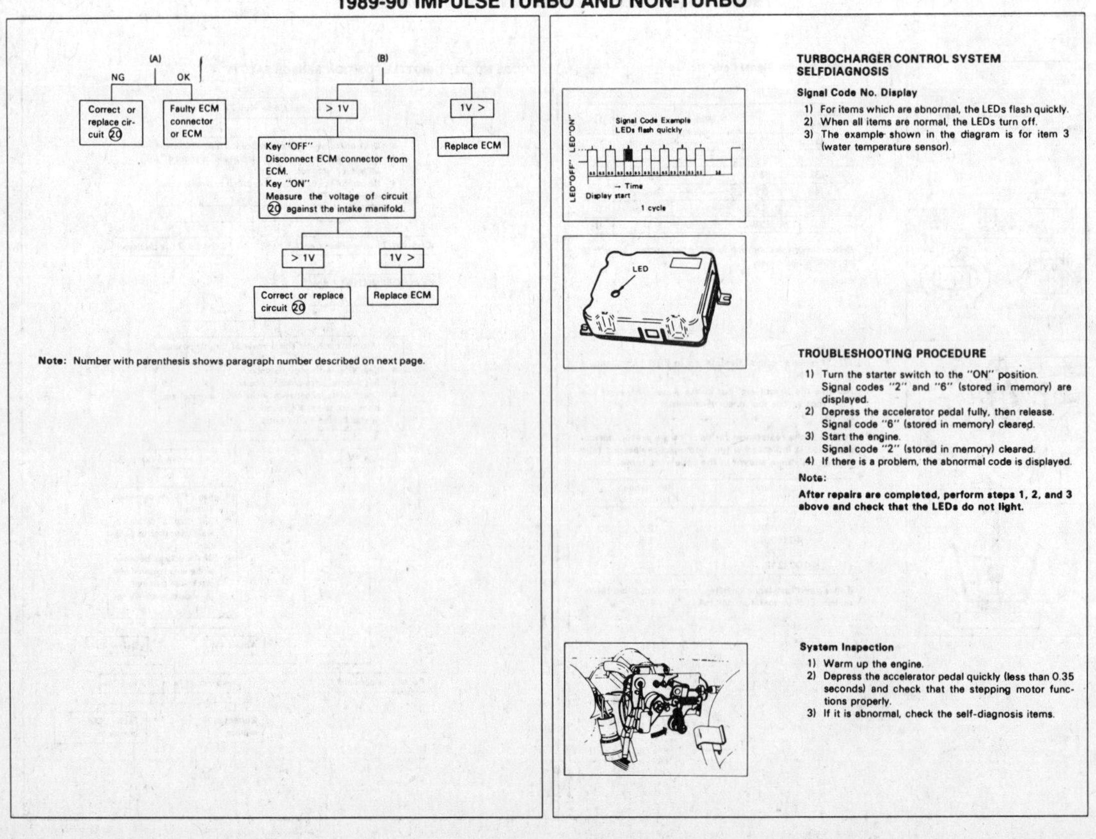


(A)
- NG → Correct or replace circuit ⑳
- OK → Faulty ECM connector or ECM

(B)
- 1V > → Replace ECM
- > 1V → Key "OFF"
Disconnect ECM connector from ECM.
Key "ON"
Measure the voltage of circuit ⑳ against the intake manifold.
  - > 1V → Correct or replace circuit ⑳
  - 1V > → Replace ECM

**Note:** Number with parenthesis shows paragraph number described on next page.

**TURBOCHARGER CONTROL SYSTEM SELFDIAGNOSIS**

**Signal Code No. Display**

1) For items which are abnormal, the LEDs flash quickly.
2) When all items are normal, the LEDs turn off.
3) The example shown in the diagram is for item 3 (water temperature sensor).

**TROUBLESHOOTING PROCEDURE**

1) Turn the starter switch to the "ON" position.
Signal codes "2" and "6" (stored in memory) are displayed.
2) Depress the accelerator pedal fully, then release.
Signal code "6" (stored in memory) cleared.
3) Start the engine.
Signal code "2" (stored in memory) cleared.
4) If there is a problem, the abnormal code is displayed.

**Note:**

**After repairs are completed, perform steps 1, 2, and 3 above and check that the LEDs do not light.**

**System Inspection**

1) Warm up the engine.
2) Depress the accelerator pedal quickly (less than 0.35 seconds) and check that the stepping motor functions properly.
3) If it is abnormal, check the self-diagnosis items.

## 1989-90 IMPULSE TURBO AND NON-TURBO

### SIGNAL CODE NUMBER DIAGNOSIS CONTENT

| Diagnosis item | Trouble assessing conditions | System operation | LED display |
|---|---|---|---|
| Knock control circuit | Abnormal knock control signal | Stopped | 1 |
| Ignition signal circuit | No input of ignition signal for about 0.3 sec. or more. | Stopped | 2 |
| Water temperature sensor circuit | Short or open circuit between Terminal ⑩ and ㉑ (water temperature) sensor. | Water temperature correction; fixed to θw = 250 | 3 |
| Stepping motor circuit (micro switch circuit) | 0 volts between Terminals ⑧ and ④ when the control unit is not driving the stepping motor. | Stopped | 4 |
| | More than 0 volts between Terminals ⑧ and ④ during operation of stepping motor. | During assessment; Operated After assessment; Stopped | |
| Throttle sensor circuit | Shortcircuiting or disconnection between terminal ⑥ and ④ (throttle sensor output terminal). | Stopped | 5 |
| Throttle sensor circuit | No voltage variance or abnormal voltage between terminals ⑥ (throttle sensor output) and ④ | Stopped | 6 |
| | No wide open throttle operation since engine was started. | Operated | |
| Power supply voltage | Less than 10V or more than 16V power supply voltage. | Stopped | 7 |

**Note:**
**The numbers in the circles refer to the controller's pin numbers. The controller's pin numbers are as follows.**

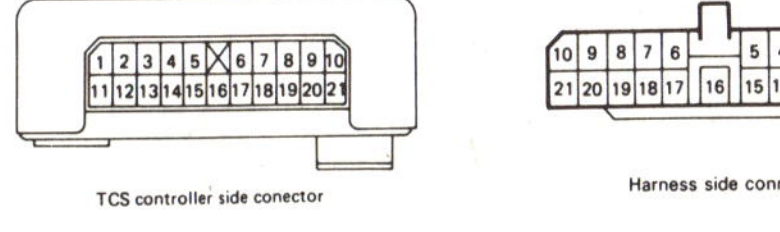


TCS controller side conector

Harness side connector

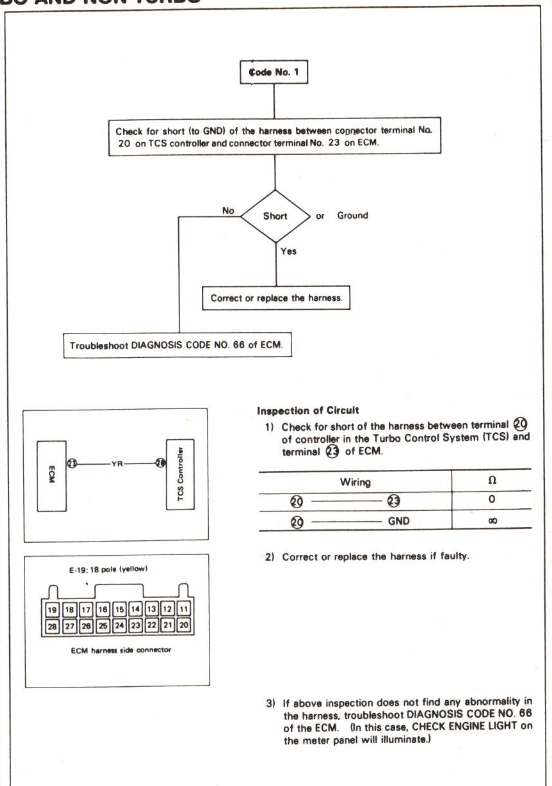


**Inspection of Circuit**

1) Check for short of the harness between terminal ⑳ of controller in the Turbo Control System (TCS) and terminal ㉓ of ECM.

| Wiring | Ω |
|---|---|
| ⑳ —— ㉓ | 0 |
| ⑳ —— GND | ∞ |

2) Correct or replace the harness if faulty.

3) If above inspection does not find any abnormality in the harness, troubleshoot DIAGNOSIS CODE NO. 66 of the ECM. (In this case, CHECK ENGINE LIGHT on the meter panel will illuminate.)

## 1989-90 IMPULSE TURBO AND NON-TURBO

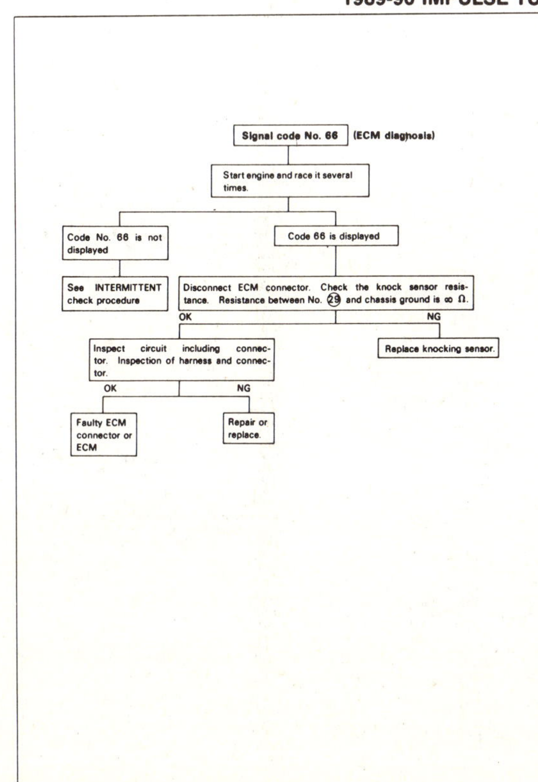


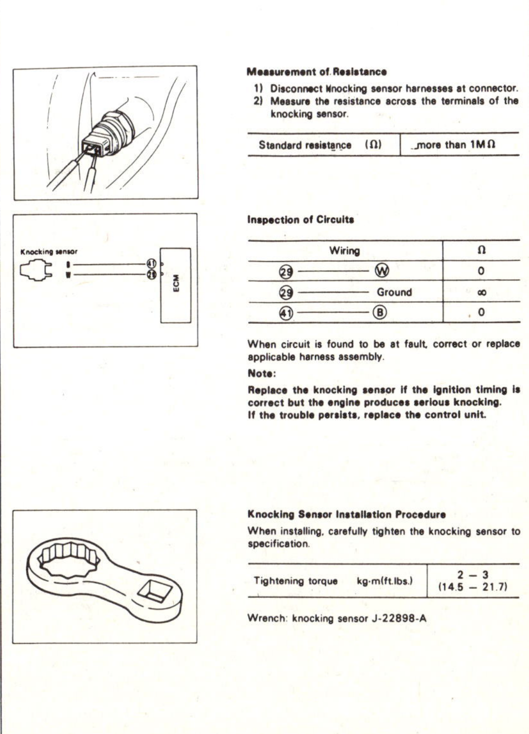


**Measurement of Resistance**

1) Disconnect knocking sensor harnesses at connector.
2) Measure the resistance across the terminals of the knocking sensor.

| Standard resistance (Ω) | more than 1M Ω |
|---|---|

**Inspection of Circuits**

| Wiring | Ω |
|---|---|
| ㉙ —— Ⓦ | 0 |
| ㉙ —— Ground | ∞ |
| ㊶ —— Ⓑ | 0 |

When circuit is found to be at fault, correct or replace applicable harness assembly.

**Note:**
**Replace the knocking sensor if the ignition timing is correct but the engine produces serious knocking. If the trouble persists, replace the control unit.**

**Knocking Sensor Installation Procedure**

When installing, carefully tighten the knocking sensor to specification.

| Tightening torque kg-m(ft.lbs.) | 2 – 3 (14.5 – 21.7) |
|---|---|

Wrench: knocking sensor J-22898-A

## 1989-90 IMPULSE TURBO AND NON-TURBO

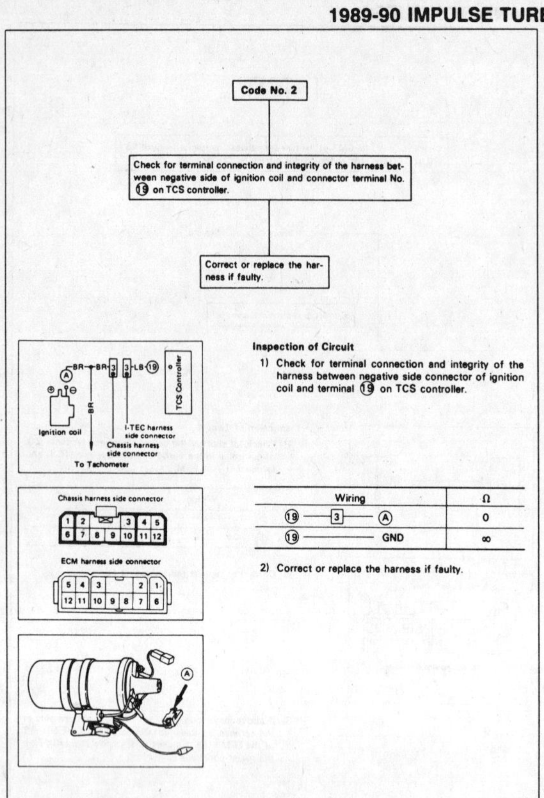

Code No. 2
Check for terminal connection and integrity of the harness between negative side of ignition coil and connector terminal No. ⑲ on TCS controller.
Correct or replace the harness if faulty.
Ignition coil
To Tachometer
I-TEC harness side connector
Chassis harness side connector
TCS Controller
Chassis harness side connector
ECM harness side connector


**Inspection of Circuit**

1) Check for terminal connection and integrity of the harness between negative side connector of ignition coil and terminal ⑲ on TCS controller.

| Wiring | Ω |
|---|---|
| ⑲ —[3]— Ⓐ | 0 |
| ⑲ ——— GND | ∞ |

2) Correct or replace the harness if faulty.

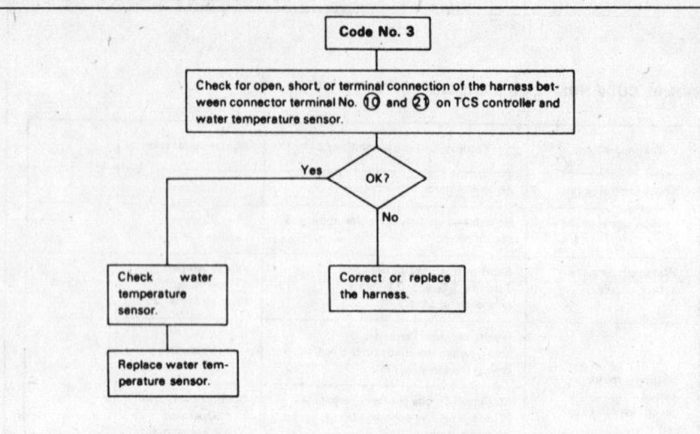


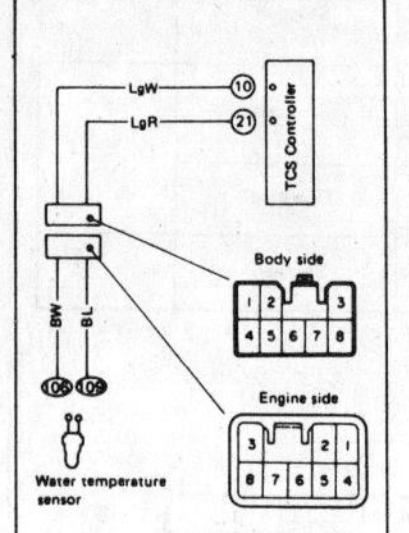


**Inspection of Circuit**

1) Check for open, short or terminal connection of the harness between terminal No. ⑩ and ㉑ on TCS controller and water temperature sensor.

| Wiring | Ω |
|---|---|
| ⑩ —[2]— ⑩⑥ | 0 |
| ⑩ ——— GND | ∞ |
| ㉑ —[5]— ⑩⑨ | 0 |
| ㉑ ——— GND | ∞ |

2) Correct or replace the harness if faulty.

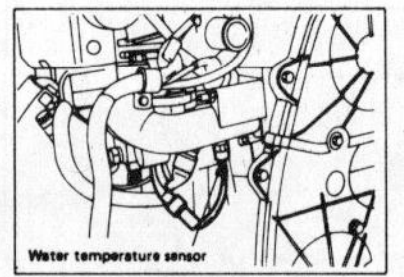

Water temperature sensor

**Inspection of Water Temperature Sensor**

3) Measure the water temperature sensor resistance.

| Coolant temperatue °C (°F) | Resistance (K Ω) |
|---|---|
| 20 ( 68) | 2.45 ± 0.24 |
| 30 ( 86) | 1.65 ± 0.08 |
| 60 (140) | 0.58 ± 0.03 |

## 1989-90 IMPULSE TURBO AND NON-TURBO

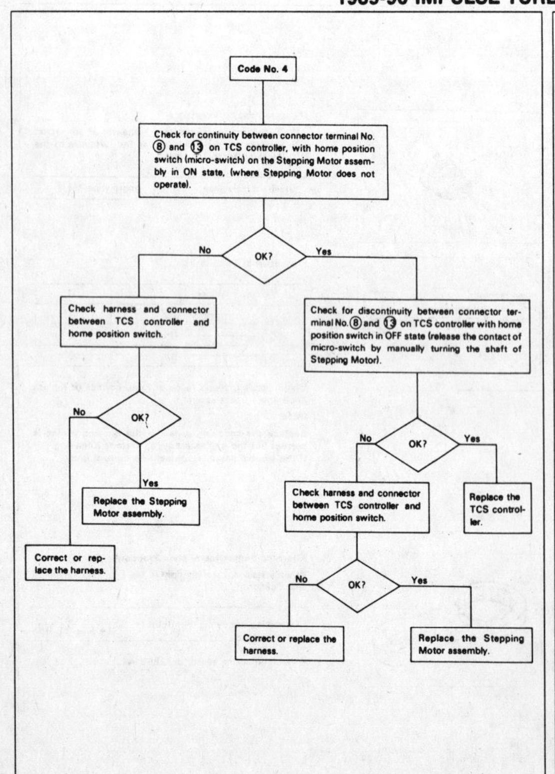


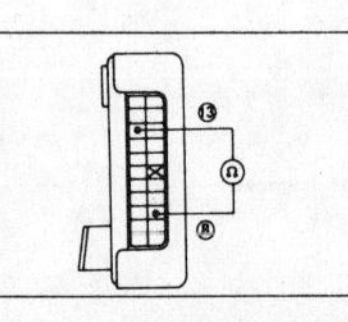

**Inspection of Circuit**

1) Check for continuity between connector terminal No. ⑧ and ⑬ on the TCS controller with home position switch (micro-switch) on the Stepping Motor assembly in ON state.
Check is to be carried out, with TCS controller and Engine body harness being connected.

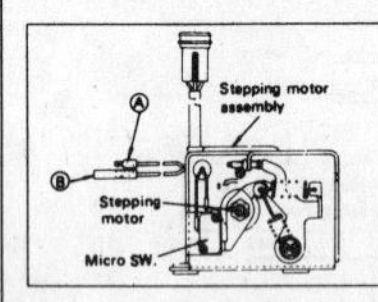


2) If continuity is not confirmed between terminals ⑧ and ⑬, check terminal connection and integrity of the harness between connector terminals ⑧ and ⑬ on the TCS controller harness side and micro-switch terminals.

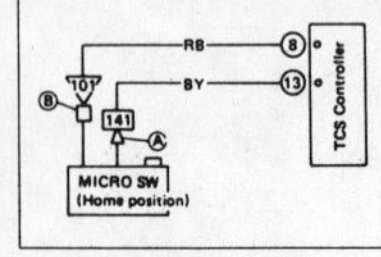


| Wiring | Ω |
|---|---|
| ⑧ ——— 101 | 0 |
| ⑧ ——— GND | ∞ |
| ⑬ ——— [141] | 0 |
| ⑬ ——— GND | ∞ |

Correct or replace the harness if faulty.
Replace the stepping motor assembly if the harness is OK.

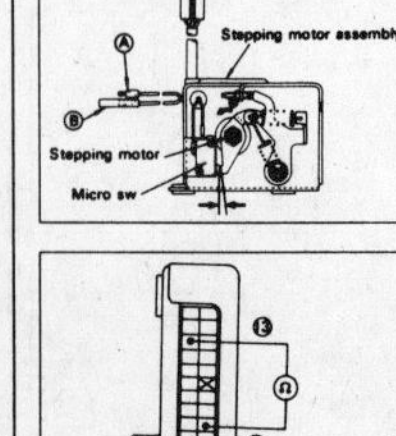


3) If continuity is confirmed between terminals 8 and 13 in step 1) above, make sure that continuity is not obtained when releasing the contact between the micro-switch and lever, by turning the Stepping Motor manually.

## 1989-90 IMPULSE TURBO AND NON-TURBO

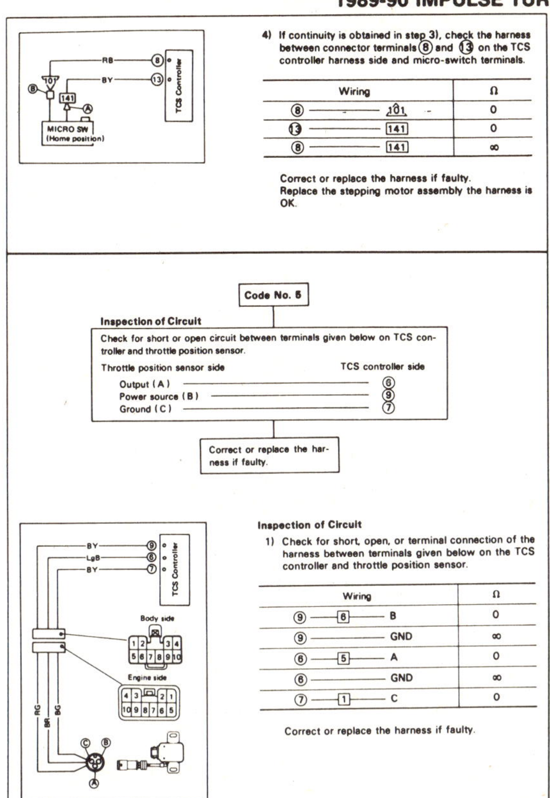

4) If continuity is obtained in step 3), check the harness between connector terminals ⑧ and ⑬ on the TCS controller harness side and micro-switch terminals.

| Wiring | Ω |
|---|---|
| ⑧ — 101 | 0 |
| ⑬ — 141 | 0 |
| ⑧ — 141 | ∞ |

Correct or replace the harness if faulty.
Replace the stepping motor assembly the harness is OK.

Code No. 5

Inspection of Circuit

Check for short or open circuit between terminals given below on TCS controller and throttle position sensor.

| Throttle position sensor side | TCS controller side |
|---|---|
| Output (A) | ⑥ |
| Power source (B) | ⑨ |
| Ground (C) | ⑦ |

Correct or replace the harness if faulty.

Inspection of Circuit

1) Check for short, open, or terminal connection of the harness between terminals given below on the TCS controller and throttle position sensor.

| Wiring | Ω |
|---|---|
| ⑨ — 6 — B | 0 |
| ⑨ — GND | ∞ |
| ⑥ — 5 — A | 0 |
| ⑥ — GND | ∞ |
| ⑦ — 1 — C | 0 |

Correct or replace the harness if faulty.

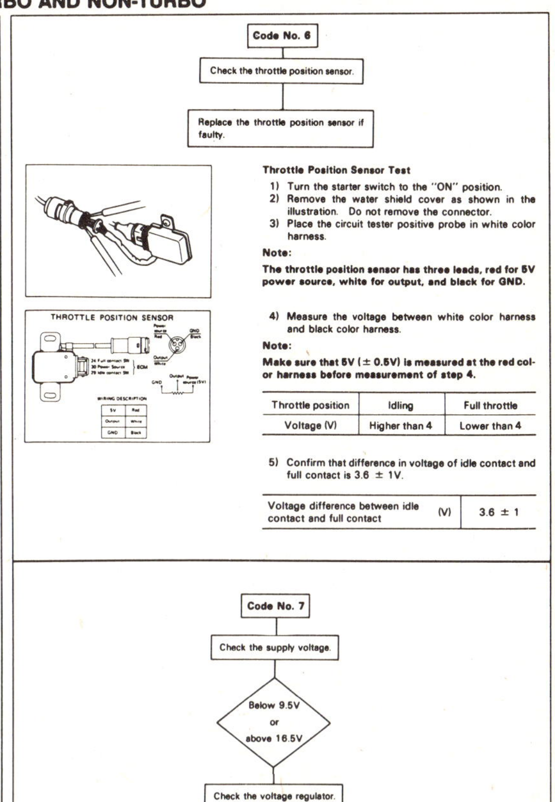

Code No. 6

Check the throttle position sensor.

Replace the throttle position sensor if faulty.

Throttle Position Sensor Test

1) Turn the starter switch to the "ON" position.
2) Remove the water shield cover as shown in the illustration. Do not remove the connector.
3) Place the circuit tester positive probe in white color harness.

**Note:**

**The throttle position sensor has three leads, red for 5V power source, white for output, and black for GND.**

4) Measure the voltage between white color harness and black color harness.

**Note:**

**Make sure that 5V (± 0.5V) is measured at the red color harness before measurement of step 4.**

| Throttle position | Idling | Full throttle |
|---|---|---|
| Voltage (V) | Higher than 4 | Lower than 4 |

5) Confirm that difference in voltage of idle contact and full contact is 3.6 ± 1V.

| Voltage difference between idle contact and full contact | (V) | 3.6 ± 1 |
|---|---|---|

Code No. 7

Check the supply voltage.

Below 9.5V or above 16.5V

Check the voltage regulator.

## 1989-90 IMPULSE TURBO AND NON-TURBO

WIRING DIAGRAM

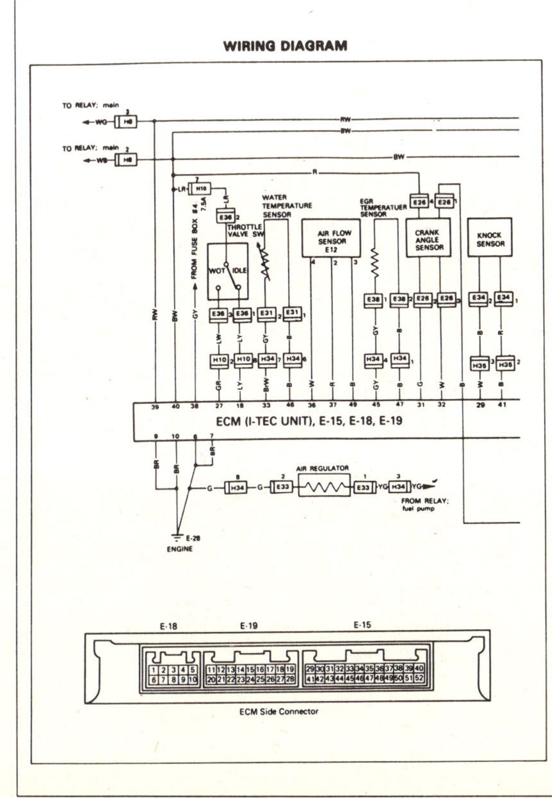


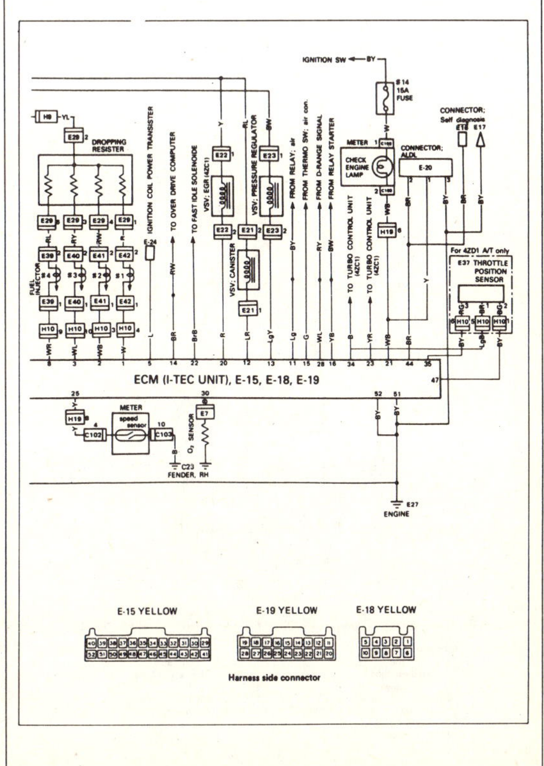

## 1989-90 IMPULSE TURBO

### GENERAL DESCRIPTION

### SYSTEM DIAGRAM

For model with 4ZC1-T engine.

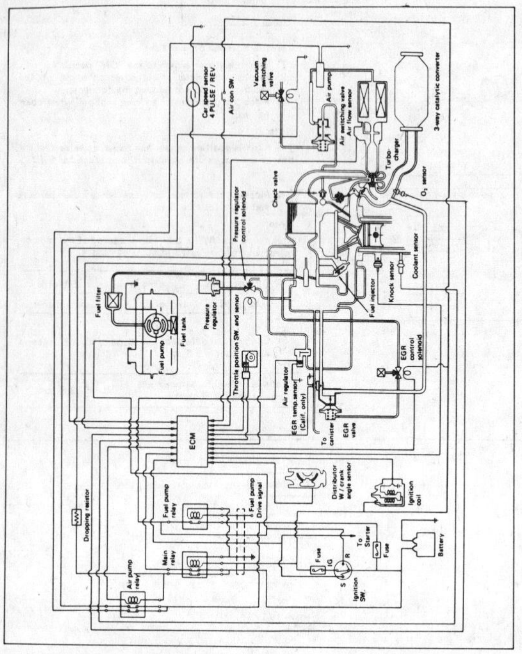

## 1989-90 IMPULSE NON-TURBO

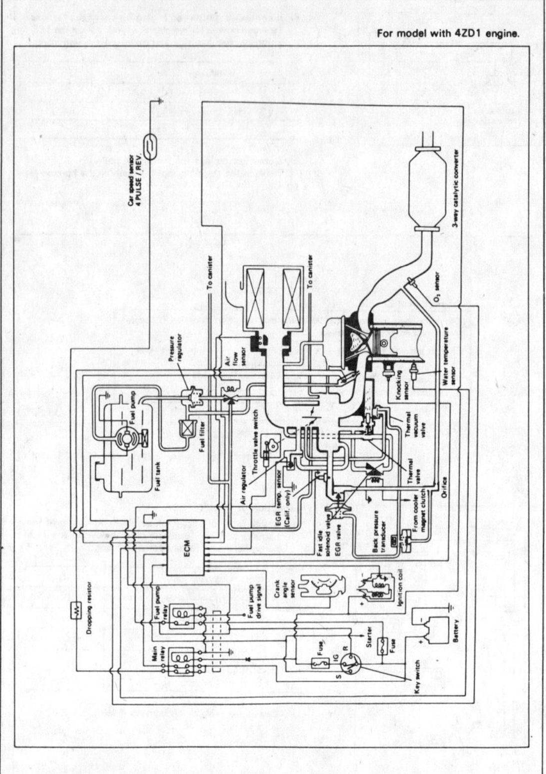

For model with 4ZD1 engine.

## 1989-90 IMPULSE TURBO AND NON-TURBO

### VACUUM HOSE ROUTING DIAGRAM

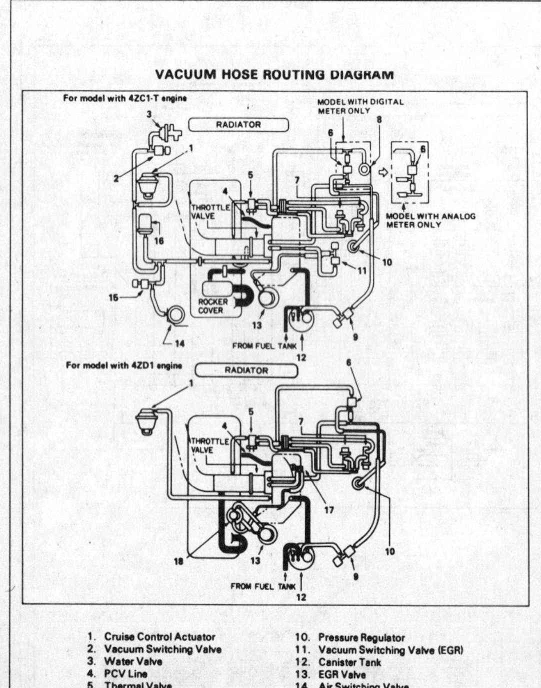

1. Cruise Control Actuator
2. Vacuum Switching Valve
3. Water Valve
4. PCV Line
5. Thermal Valve
6. Vacuum Switching Valve (P/REG.)
7. Fast Idle Solenoid
8. Boost Pressure Switch
9. Vacuum Switching Valve (CANISTER)
10. Pressure Regulator
11. Vacuum Switching Valve (EGR)
12. Canister Tank
13. EGR Valve
14. Air Switching Valve
15. Vacuum Switching Valve (AIR)
16. Vacuum Tank
17. Thermal Vacuum Valve
18. EGR B/P Transducer

## TROOPER/TROOPER II (EXCEPT 2.8L ENGINE) AMIGO/PICK-UP

### COMPONENTS LAYOUT

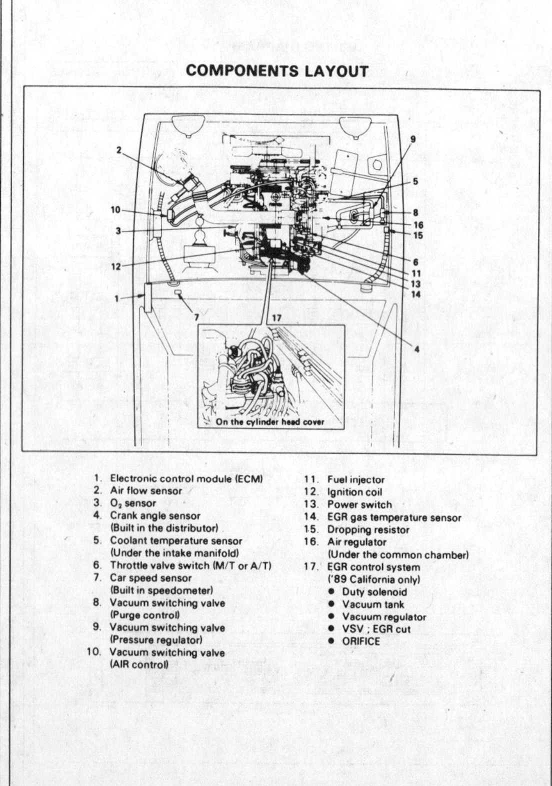

1. Electronic control module (ECM)
2. Air flow sensor
3. $O_2$ sensor
4. Crank angle sensor (Built in the distributor)
5. Coolant temperature sensor (Under the intake manifold)
6. Throttle valve switch (M/T or A/T)
7. Car speed sensor (Built in speedometer)
8. Vacuum switching valve (Purge control)
9. Vacuum switching valve (Pressure regulator)
10. Vacuum switching valve (AIR control)
11. Fuel injector
12. Ignition coil
13. Power switch
14. EGR gas temperature sensor
15. Dropping resistor
16. Air regulator (Under the common chamber)
17. EGR control system ('89 California only)
   - Duty solenoid
   - Vacuum tank
   - Vacuum regulator
   - VSV ; EGR cut
   - ORIFICE

## TROOPER/TROOPER II (EXCEPT 2.8L ENGINE), AMIGO/PICK-UP

### INSPECTION AND MAINTENANCE

NORMAL CONDITION; • SIGNAL CODE NO. 12 WILL DISPLAY
• THE MEMORY CLEARED BY MAIN FUSE

I. DISCONNECT THE SENSOR OR ACTUATOR WIRING CONNECTOR (WHEN CHECKING THE PART/COMPONENT INDEPENDENTLY)

| Code No. | Item | | Measure Terminal: Tester + Side | Measure Terminal: Tester − Side | Key SW Posit. | Condition to Check | | Standard Value |
|---|---|---|---|---|---|---|---|---|
| 21<br>43<br>65 | Throttle valve SW | | Center Terminal (Terminal P) | Upper Terminal (Terminal I) | OFF | Ac. Pedal Not Depressed | | 0 Ω |
| | | | | | | Ac. Pedal Slightly Depressed | | ∞ Ω |
| | | | | | | Ac. Pedal Full Depressed | | ∞ Ω |
| | | | Center Terminal (Terminal P) | Lower Terminal (Terminal F) | OFF | Ac. Pedal Not Depressed | | ∞ Ω |
| | | | | | | Ac. Pedal Slightly Depressed | | ∞ Ω |
| | | | | | | Ac. Pedal Full Depressed | | 0 Ω |
| 33<br>64 | Fuel Injector | #1 | Terminal | Terminal | OFF | | | 2~3 Ω |
| | | #2 | Terminal | Terminal | OFF | | | 2~3 Ω |
| | | #3 | Terminal | Terminal | OFF | | | 2~3 Ω |
| | | #4 | Terminal | Terminal | OFF | | | 2~3 Ω |
| 23<br>35<br>54 | Ignition Coil | | + Terminal | − Terminal | OFF | Primary | | 0.81~0.99 Ω |
| | | | + Terminal | Center Terminal | OFF | Secondary | | 7.52~11.28K Ω |
| 25<br>53 | Vacuum Switching Valve | | Terminal | Terminal | OFF | | | Approx. 36 Ω |
| 14<br>15 | Water Tempera-ture Sensor | | Terminal | Terminal | OFF | Water Temp. °C(°F) | −10 (14)<br>10 (50)<br>20 (68)<br>50 (122)<br>80 (176) | 7~12K Ω<br>3~ 5K Ω<br>2~ 3K Ω<br>0.7~1K Ω<br>0.2~0.4K Ω |
| 33 | Dropping Resistor | | Center Terminal | Another Terminal | OFF | | | 5~7 Ω/Pc. |
| 63 | Car Speed Sensor | | 41 | Ground | OFF | Turn the Inner Cable Slowly | | 4 Pulse/rev. |
| | Air Regulator | | Terminal | Terminal | OFF | | | Approx. 40 Ω |
| | Main Relay | | Terminal (WR) | Terminal (B) | OFF | | | Approx. 80 Ω |
| | Charge Relay | | Terminal (WG, B) | Terminal (B) | OFF | | | Approx. 80 Ω |
| | Fuel Pump Relay | | Terminal (WR) | Terminal (B) | OFF | | | Approx. 80 Ω |

### INSPECTION AND MAINTENANCE

II. DISCONNECT THE ECM UNIT WIRING CONNECTORS

| Code No. | Item | | Measure Terminal: Tester + Side | Measure Terminal: Tester − Side | Key SW Posit. | Condition to Check | Standard Value |
|---|---|---|---|---|---|---|---|
| | Power Supply Circuit | | 33 | Ground (11 or 20) | ON | | 10~14V |
| | | | 12 (21) | Ground (11 or 20) | ON | | 10~14V |
| | | | 9 & 10 | Ground (11 or 20) | OFF | | 0 Ω |
| | | | 1 & 2 | Ground (11 or 20) | OFF | | 0 Ω |
| | | | 11 or 20 | Ground (Body) | OFF | | 0 Ω |
| 33<br>64 | Fuel Injector | #1, #4 | 6 | Ground (1 or 2) | ON | | 10~14V |
| | | #2, #3 | 7 | Ground (1 or 2) | ON | | 10~14V |
| 22 | Starter Signal | | 16 | Ground (11 or 20) | ST | Disconnect Starter Relay | 10~14V |
| 25<br>53 | P/R V.S.V (Vacuum Switch-ing Valve) | | 4 | Ground (9 or 10) | ON | | 10~14V |
| 21<br>43<br>65 | Throttle Valve SW | | 17 | Ground (11 or 20) | ON | Ac. Pedal Not Depressed | 10~14V |
| | | | | | | Ac. Pedal Slightly Depressed | 0V |
| | | | | | | Ac. Pedal Full Depressed | 0V |
| | | | 14 | Ground (11 or 20) | ON | Ac. Pedal Not Depressed | 0V |
| | | | | | | Ac. Pedal Slightly Depressed | 0V |
| | | | | | | Ac. Pedal Full Depressed | 10~14V |
| 26<br>27 | Canister V.S.V. | | 8 | Ground (9 or 10) | ON | | 10~14V |
| 23<br>35<br>54 | Ignition-Terminal (Power Transistor for Ignition) | | Ground (Dry Batt. + Side) | P-Transistor Yellow Harness (Dry Batt. − Side) | ON | Connected ECM Wiring Connector. Keep 5~6 mm Gap in Between the High Tension Cable and Ground, and Then, Disconnect either the Plus or Minus Probe and See if There are Sparks. If Sparks occur, This System is Normal. | |
| 13<br>44<br>45 | $O_2$ sensor | | 27 | Ground (11 or 20) | ON | The Engine Speed About 2000 RPM For Approx. 10 Min. | 0.3~0.8V |

## TROOPER/TROOPER II (EXCEPT 2.8L ENGINE), AMIGO/PICK-UP

### MALFUNCTION CODES VS FAULTY COMPONENTS (2.6L)

| | Code No | 12 | 13 | 14 | 15 | 21 | 22 | 23 | 25 | 26 | 29 | 32 | 33 | 34 | 35 | 41 | 43 | 44 | 45 | 51 | 52 | 53 | 54 | 61 | 62 | 63 | 64 | 65 |
|---|---|---|---|---|---|---|---|---|---|---|---|---|---|---|---|---|---|---|---|---|---|---|---|---|---|---|---|---|
| Input Sensors | Air Flow Sensor | | X | | | | | | | | | | | | | | X | X | X | | | | | ⓧ | ⓧ | X | | X |
| | Coolant Temp. Sensor | | X | ⓧ | ⓧ | | | | | | | X | | X | | | | X | X | | | | | | | | | |
| | Oxygen Sensor | | ⓧ | | | | | | | | | | | | | | | ⓧ | ⓧ | | | | | | | | | |
| | Crank Angle Sensor | X | | | | | X | | | | | X | | | | ⓧ | X | X | X | | | | | X | X | X | | X |
| | Idle/WOT Position | | X | | | ⓧ | | | | | | X | | X | | | ⓧ | X | X | | | | | | | | | ⓧ |
| | Starter Switch | | | | | | ⓧ | | | | | | | | | X | | | | | | | | | | | | |
| | EGR Gas Temp. Sensor | | | | | | | | | | | ⓧ | | ⓧ | | | | | | | | | | | | | | |
| | Ignition Coil | X | | | | | ⓧ | | | | | X | | | | X | | X | X | | | | | | | | | |
| | Computer | | | | | | | | | | | | | | | | | | | ⓧ | ⓧ | | | X | X | | | |
| | Car Speed Sensor | | X | | | | X | | | | | X | | X | | | | | X | | | | | | | ⓧ | | |
| Output Actuaters | Fuel Injector | | | | | | | | | | | | ⓧ | | | | | | | | | | | | | | | |
| | Power Switch (Ignition Coil Driver) | | | | | | | X | | | | | | | X | | | | | | | | ⓧ | | | | X | |
| | V.S.V.; Pressure regulator | | | | | | | | ⓧ | | | | | | | | | | | | | X | | | | | | |

**Note:**

**Code 12 does not indicate malfunction, but checks whether ECM is functioning normally.**
**Listed malfunction codes are only those which show malfunction of sensed parameters.**

**ECM determines whether a malfunction of that code exists by evaluating the parameters marked "ⓧ". Parameters marked "X" are used to set the zones in which existence or nonexistence of a malfunction is to be determined.**

**Parameter marked "X" can not be main factors, but are used to judge the existance of these relative codes, together with other parameters.**

### DIAGNOSTIC CODES

| DIAGNOSED ITEM | | DIAGNOSED CONTENT: FAULT MODE | DIAGNOSED CONTENT: MICRO-COMPUTER INPUT | CODE NO. | PAGE | CODE MEMORY | FAIL-SAFE FUNCTION OF MICRO-COMPUTER | ENGINE CONDITION | VEHICLE CONDITION |
|---|---|---|---|---|---|---|---|---|---|
| Engine is not started | | — | Engine speed less than 200 rpm | 12 | 59 | No | None | Normal | Normal |
| $O_2$ sensor system | | Harness open Sensor deterior-ation | Intermediate voltage | 13 | 60-66 | Yes | Fuel is not compensated by $O_2$ signal | Exhaust emission is worsened | No noticeable abnormal operation |
| Fuel metering system | | Incorrect signal (Lean) | Lean signal (Low voltage) | 44 | 61-66 | Yes | | | |
| | | Incorrect signal (Rich) | Rich signal (High voltage) | 45 | 61-66 | Yes | | | |
| Water temperature sensor system | | Shorted with ground | Insufficient signal | 14 | 67-69 | Yes | Coolant temperature is assumed to be 85°C (185°F) | · The engine does not operate normally when temperature is low and signal is insufficient, or when temperature is high and signal is excessive. Or the engine cannot be started.<br>· The engine does not operate smoothly when temperature is low after the fail-safe function is actuated. | |
| | | Harness open | Excessive signal | 15 | 68-69 | Yes | | | |
| Throttle valve switch system | Both idle contact and full contact | Both idle contact and full make contact simultaneously | Both contacts make simultaneously | 21 | 70-72 | Yes | Both signals are assumed to be OFF | Coasting Fuel-cut is not made, after the fail-safe function is actuated, and lean fuel is resulted when the throttle valve is fully opened. | Fuel consumption rate is worsened after the fail-safe function is actuated. Fuel tends to be lean when the throttle valve is fully opened. |
| | Idle contact | Continuously make contact | Signal is sent continuously<br>Not diagnosed when the air flow sensor system is defective | 43 | 71-72 | Yes | Assumed to be OFF | · Fuel cut range appears during running before fail-safe function is actuated.<br>· Coasting Fuel-cut is not made, after the fail-safe function is actuated | · The vehicle does not run smoothly<br>· Fuel consumption rate is worsened after the fail-safe function is actuated |
| | Full contact | Continuously make contact | Signal is sent continuously | 65 | 71-72 | Yes | Assumed to be OFF | · Air-fuel ratio is high when partially loaded.<br>· Fuel tends to be lean when the throttle valve is fully opened after the fail-safe function is actuated. | · Fuel consumption rate and exhaust are worsened. The spark plugs are carboned or the engine stalls depending on the condition.<br>· Fuel tends to be lean when the throttle valve is fully opened after the fail-safe function is actuated |

## TROOPER/TROOPER II (EXCEPT 2.8L ENGINE), AMIGO/PICK-UP

| DIAGNOSED ITEM | DIAGNOSED CONTENT: FAULT MODE | DIAGNOSED CONTENT: MICRO-COMPUTER INPUT | CODE NO. | PAGE | CODE MEMORY | FAIL-SAFE FUNCTION OF MICRO-COMPUTER | ENGINE CONDITION | VEHICLE CONDITION |
|---|---|---|---|---|---|---|---|---|
| Starter signal system | Harness open or ground shorted | Signal is not input | 22 | 73 | Yes | None | Normal | No abnormality is felt. |
| Crank angle sensor system | · Harness open<br>· Faulty signal | · No angle sensor signal is input.<br>· Idling speed is lower than the actual speed. | 41 | 74 | Yes | None | · The engine stalls. The engine cannot be started.<br>· Ignition timing may be changed. | · Stalls or cannot be started.<br>· The driveability is effected. |
| Air flow sensor system | Harness open, shorting with ground, or broken hot wire. | Insufficient signal | 61 | 75-77 | Yes | Injection pulse is changed over at throttle valve position. | · The engine stalls depending on the condition.<br>· The engine may be slightly unstable after the fail-safe function is actuated. | · The engine stalls depending on the condition.<br>· The accelerator response is worsened after the fail-safe function is actuated. The engine may not operate smoothly. |
| | Broken cold wire. | Excessive signal | 62 | 75-77 | Yes | | | |
| Car speed sensor system | · Harness open or ground shorted.<br>· Car speed sensor broken. | No signal is input (Not diagnosed with the air flow sensor system is defective) | 63 | 78-79 | Yes | None | The fuel cut operates even when running under a low speed. | Does not run smoothly and shakes. |
| Micro-computer unit | Abnormal LSI (1) | – | 51 | 80 | Yes | · Injection pulse is changed over at throttle valve position<br>· Fixed ignition timing. | · In the worst case, the engine stalls, or the engine does not operate smoothly at a certain time.<br>· The engine does not stall but it does not satisfy the specification. | · In the worst case, the engine stalls, or the engine does not operate smoothly at a certain time.<br>· The exhaust and running performance deviate from the specification. |
| | Abnormal LSI (2) | – | 52 | 80 | Yes | | | |
| Power transistor system for ignition | Output terminal shorted with ground. | – | 23 | 81-82 | Yes | None | · The engine stalls.<br>· Cannot be started. | · The engine stalls<br>· Cannot be started |
| | Harness open. | – | 35 | I | Yes | | | |
| | Defective transistor or grounding system. | – | 54 | I | Yes | | | |
| Vacuum switching valve system for canister purge | · Open circuit harness connector VSV or inside of ECM | – | 26 | 86-88 | Yes | None | Air/Fuel ratio tends to be rich under the idle or partial conditions before warmed up | Fuel consumption rate is worsened when engine is cold. |
| | · Ground shorted harness connector VSV or inside of ECM | – | | | | | Air/Fuel ratio tends to be lean at hot restart | The engine can not be restarted smoothly when coolant temp. is very high. |
| | · Driver transister in ECM open | – | 27 | I | Yes | | | |
| Vacuum switching valve system for pressure regulator | · Open circuit harness connector VSV or inside of ECM | | 25 | 83-85 | Yes | None | Air/Fuel ratio tends to be rich under the idle or partial conditions before warmed up | Fuel consumption rate is worsened when engine is cold. |
| | · Ground shorted harness connector VSV or inside of ECM | | | | | | Air/Fuel ratio tends to be lean at hot restart | The engine can not be restarted smoothly when coolant temp. is very high. |
| | · Driver transister in ECM open | – | 53 | I | Yes | | | |
| Fuel injector system. | · Open circuit harness connector injector, dropping resister, or inside of ECM | – | 33 | 89-90 | Yes | None | Air/Fuel ratio tends to be rich under the idle or partial conditions before warmed up | Fuel consumption rate is worsened when engine is cold. |
| | · Ground shorted harness connector injector, dropping register, or inside of ECM | – | | | | | Air/Fuel ratio tends to be lean at hot restart | The engine can not be restarted smoothly when coolant temp. is very high. |
| | · Driver transister in ECM open | – | 64 | 91-94 | Yes | | | |
| EGR temperature sensor system | Mulfunction | – | 32 | 95-99 | Yes | None | Exhaust is worsened | No noticeable abnormal operation |
| | EGR temperature sensor or harness NG | – | 34 | 100-102 | Yes | None | Normal | No abnormality is felt. |

## TROOPER/TROOPER II (EXCEPT 2.8L ENGINE), AMIGO/PICK-UP

### BEFORE STARTING

1. Verify the customer complaint and locate the correct symptoms. If the fault relates fundamental engine problems, such as engine lacks power or, fuel consumption excessive etc., see "DIAGNOSIS" in ENGINE SECTION 6, first.
2. Several of the symptom procedures below call for a careful visual check. This check should include:
   - Vacuum hoses for splits, kinks, and proper connections, as shown on Emission Control Information label.
   - Air leaks at throttle valve mounting, common chamber mounting and intake manifold.
   - Ignition wires for cracking, harness, proper routing, and carbon tracking.
   - Wiring for proper connections, pinches, and cuts.
     The importance of this step cannot be stressed too strongly — it can lead to correcting a problem without further checks and can save valuable time.
3. Be careful to check the system components (Closed Loop Emission Control System), especially handling the electrical parts, ECM etc..

   Refer to "ELECTRICAL—BODY AND CHASSIS" SECTION 8 and "Handling precautions"
   - When checking the ECM harness connector terminals from the connector wire side with the tester probe, be careful not to touch the other terminal by mistake.
     Do not insert the circuit tester test probes into the connector open side to test the continuity.
   - Before starting with a voltmeter connected to a suspected circuit, check to see if the battery is in full charged condition.
   - Before disconnecting the ECM connectors, check the ignition switch "OFF" condition, and check and record the stored memory in ECM.
     When disconnecting the ECM connectors, all three connectors (10P, 18P, 14P) should be done, and note that all stored codes also be cleared at the same time.
   - To avoid the connector terminals being damaged or touching the wrong terminal, short jumper cables which are staked terminals are useful.
   - When connecting a suspected circuit to the Battery + terminal or Engine ground, it will be advisable to install a test light (3.2W, 12V) in between, to protect electrical parts overcharged or damaged.

### INTERMITTENTS CHECK PROCEDURES

Problem may or may not turn "on" the "CHECK ENGINE" light, or store a code.

"INTERMITTENTS" means that "CHECK ENGINE" light comes on at times, but does not stay on.

If the trouble is found "INTERMITTENTS" in course of "Trouble Code Chart" refer to this column to locate the problem.

If a fault is intermittent, use of Trouble Code Charts may result in replacement of good parts.

- Most intermittent problems are caused by faulty electrical connections or wiring. Perform careful check as described below.
- Poor mating of the connector halves, or terminals not fully seated in the connector body (backed out).
- Improperly formed or damaged terminals. All connector terminals in problem circuit should be carefully reformed to increase contact tension.
- Poor terminal to wire connection. This requires removing the terminal from the connector body to check.
- Poor ground connections.
- If a visual check does not find the cause of the problem, the car can be driven with a voltmeter connected to a suspected circuit. An abnormal voltage reading when the problem occurs indicates the problem may be in that circuit.
  An intermittent "CHECK ENGINE" light with no stored code may be caused by;
- Ignition coil to ground and arcing at spark plug wires or plugs.
- Main relay which is in power supply circuit is faulty.
- Poor matings of power supply circuits connections.
- "CHECK ENGINE" light wire to ECM shorted to ground (circuit ㉒).
- Diagnostic "Test" Terminal wire to ECM, shorted to ground (circuit ㉔).
- ECM power grounds are poor connected to Intake manifold. See ECM wiring diagrams.
- Check for an electrical system interference caused by a defective relay, ECM driven solenoid, or switch. They can cause a sharp electrical surge. Normally, the problem will occur when the faulty component is operated.
- Check for improper installation of electrical options, such as lights, wireless control, etc.

## TROOPER/TROOPER II (EXCEPT 2.8L ENGINE), AMIGO/PICK-UP

### DIAGNOSTIC CIRCUIT CHECK

The Diagnostic Circuit Check is an organized approach for identifying a problem caused by the Fuel Injection System.

Driver comments normally fall into one of the following areas:

- Steady "CHECK ENGINE" light with engine running
- Driveability Problem
- Engine "Cranks But Will Not Run"

Understanding the chart and using it correctly will reduce diagnosis time and prevent the unneccessary replacement of parts.

1. A steady "CHECK ENGINE" light with the ignition "ON" and engine stopped confirms battery and ignition voltage to the Electronic Control Module (ECM).
2. Connect "Trouble Code Test" terminals with each other, near ECM connector.
   The ECM will cause the "CHECK ENGINE" light to flash Code 12, indicating that the ECM diagnostics are working. Code 12 will flash three (3) times, followed by any other trouble codes stored in the memory. Each additional code will flash three (3) times, starting with the lowest code, and then start over again with Code 12. If there are no other codes, Code 12 will flash until the "Trouble Code Test" terminals are disconnected or the engines is started.
3. Record all stored codes except for Code 12. If the problem is "Engine Cranks But Will Not Run", go to "HARD STARTING" DIAGNOSIS (4ZE1) in Engine section 6.
4. If no additional codes were recorded, see driveability symptoms and recommended service procedures.
5. Road test of the system should be done, under the safety conditions and traffic rules.
   If available, it will be better to use a chassis dynamometer.
6. Clearing codes. Ignition off; remove fuse No. 3 (10A) located in the lower-left part of the Instrument panel or Main fuse (60A) in the fuse and relay box of Engine room or remove Battery – terminal, for at least 10 seconds.

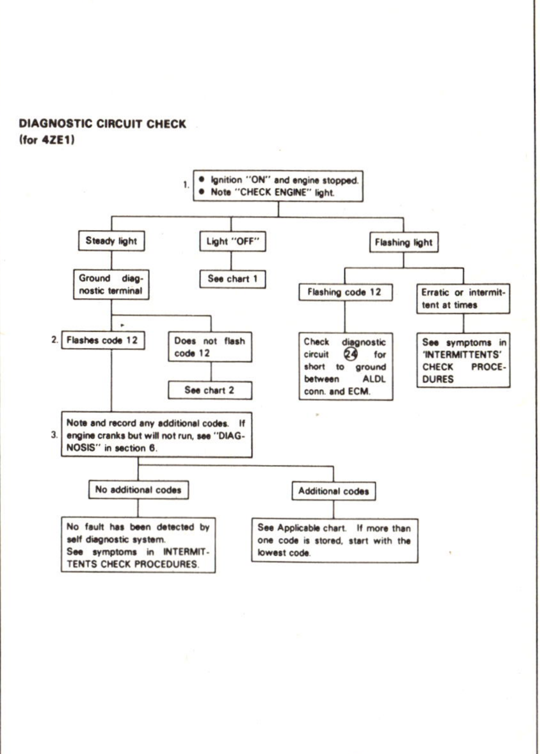


## TROOPER/TROOPER II (EXCEPT 2.8L ENGINE), AMIGO/PICK-UP

### CHART 1

#### NO "CHECK ENGINE" LIGHT

#### 4ZE1 Fuel Injection

There should always be a steady "CHECK ENGINE" Light when the ignition is "on" and engine stopped. Battery is supplied directly to the light bulb. The Electronic Control Module (ECM) will control the light and turn it on by providing a ground path through circuit (22) to the ECM.

Engine runs ok, check:

- Faulty light bulb.
- Circuit (22) open.
- Gage fuse blown. This will result in no meter lights, turn signal lights or reminder buzzer etc.

Engine cranks but will not run.

- Continuous battery - fuse or main fuse open.
- ECM ignition fuse open.
- Battery circuit (12), (21) to ECM open.
- Ignition circuit (5) to ECM open.
- Poor connection to ECM.

1. Solenoids and relays are turned "ON" and "OFF" by the ECM, using internal electronic switches called "drivers". Solenoid and relay coil resistance must measure more than 20 ohms. Less resistance will cause early failure of the ECM "driver".

   Failure of outer load such as shorted mode can damage the ECM driver.

   Before replacing ECM, be sure to check the coil resistance of all solenoids and relays controlled by the ECM. See ECM wiring diagram for the solenoid(s) and relay(s) and the coil terminal identification.

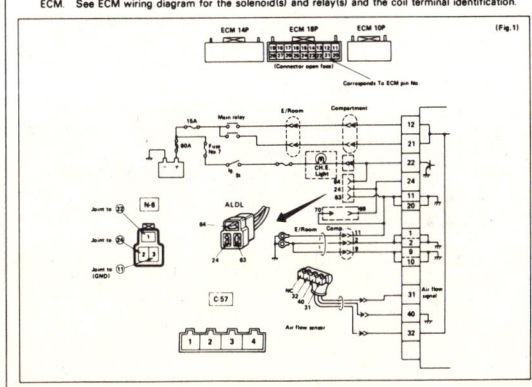

(Fig. 1)

### CHART 1

#### NO "CHECK ENGINE" LIGHT

#### 4ZE1 Fuel Injection

Disconnect Air flow sensor connector.
Uncover waterproof rubber boot of the connector.
Key "ON", engine stopped.
Check the voltage of the terminal (32) on Air flow sensor connector.
It should be 10 ~ 14V (Battery voltage).

- **OK:**
  - Key "OFF"
  - Disconnect all ECM connectors
  - Key "ON"
  - ALDL Terminal of circuit (84) with a test light to ground and note "CHECK ENGINE" light
    - **"CHECK ENGINE" light on:** 1. Check ECM ground circuits. If ECM ground circuits are OK, it is poor connection at ECM or faulty ECM. See note.**
    - **"CHECK ENGINE" light off:** Check;
      - Faulty bulb
      - Gage fuse
      - Open circuit (22)
      - Open ignition feed
- **Not OK:** Check power supply circuit (12), (21) (from Battery to ECM) including 15A fuse.

**Note: ****

**Before replacing ECM, use an ohmmeter and check resistance of each ECM controlled relay and solenoid coil.**
**See ECM wiring diagram for coil terminal identification for solenoid(s) and relay(s) to be checked.**
**Replace any relay or solenoid if the coil resistance measures less than 20 ohms.**

## TROOPER/TROOPER II (EXCEPT 2.8L ENGINE), AMIGO/PICK-UP

### CHART 2

**WON'T FLASH CODE 12/"CHECK ENGINE" LIGHT ON STEADY**

**4ZE1 Fuel Injection**

There should always be a steady "CHECK ENGINE" Light when the ignition is "on" and engine stopped. Battery ignition voltage is supplied to the light bulb. The Electronic Control Module (ECM) will turn the light on by grounding circuit (64) or (22) at the ECM.
With the diagnostic terminal grounded, the light should flash a Code 12, followed by any trouble code(s) stored in memory.
A steady light suggests a short to ground in the light control circuit (64) or (22) or an open in diagnostic circuit (24) or (69).

1. If the light goes off when the ECM connector is disconnected, then circuit (64) or (22) is not shorted to ground. Also, check the connector terminals physically for proper contact at this time.
2. This step will check for an open diagnostic circuit (24) or (69).
3. At this point the "CHECK ENGINE" light wiring is okay. The problem is a faulty ECM connector or ECM.

(Fig.1)

ECM 14P ECM 18P ECM 10P (Connector open face)

15A Main relay 60A Fuse No.1 E/Room Compartment CH.E. Light ALDL 64 24 63 70 69 N-7 N-8 Air flow sensor Air flow signal Trouble Code Test Terminal

12 21 22 24 11 20 1 2 9 10 31 40 32

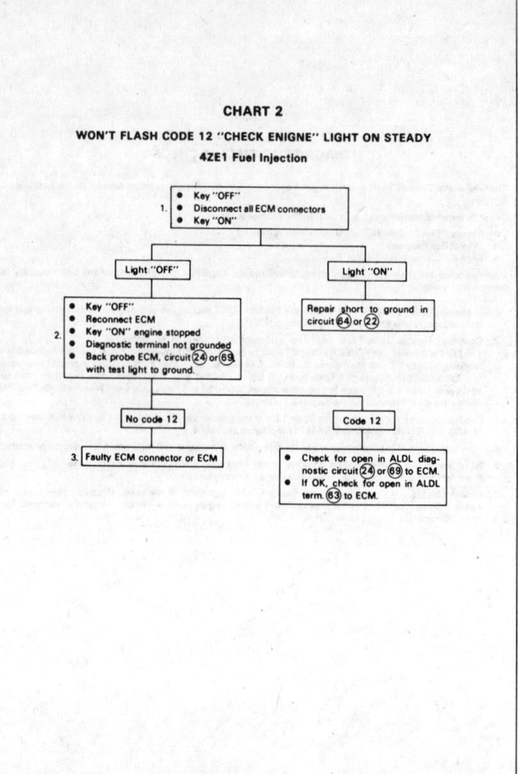


## TROOPER/TROOPER II (EXCEPT 2.8L ENGINE), AMIGO/PICK-UP

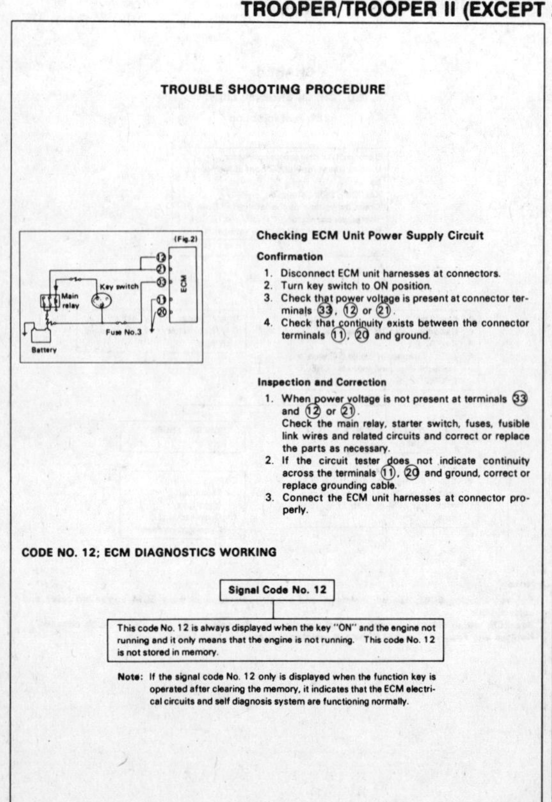

### TROUBLE SHOOTING PROCEDURE

(Fig.2)



**Checking ECM Unit Power Supply Circuit**

**Confirmation**

1. Disconnect ECM unit harnesses at connectors.
2. Turn key switch to ON position.
3. Check that power voltage is present at connector terminals (33), (12) or (21).
4. Check that continuity exists between the connector terminals (11), (20) and ground.

**Inspection and Correction**

1. When power voltage is not present at terminals (33) and (12) or (21).
   Check the main relay, starter switch, fuses, fusible link wires and related circuits and correct or replace the parts as necessary.
2. If the circuit tester does not indicate continuity across the terminals (11), (20) and ground, correct or replace grounding cable.
3. Connect the ECM unit harnesses at connector properly.

**CODE NO. 12; ECM DIAGNOSTICS WORKING**

Signal Code No. 12

This code No. 12 is always displayed when the key "ON" and the engine not running and it only means that the engine is not running. This code No. 12 is not stored in memory.

**Note:** If the signal code No. 12 only is displayed when the function key is operated after clearing the memory, it indicates that the ECM electrical circuits and self diagnosis system are functioning normally.

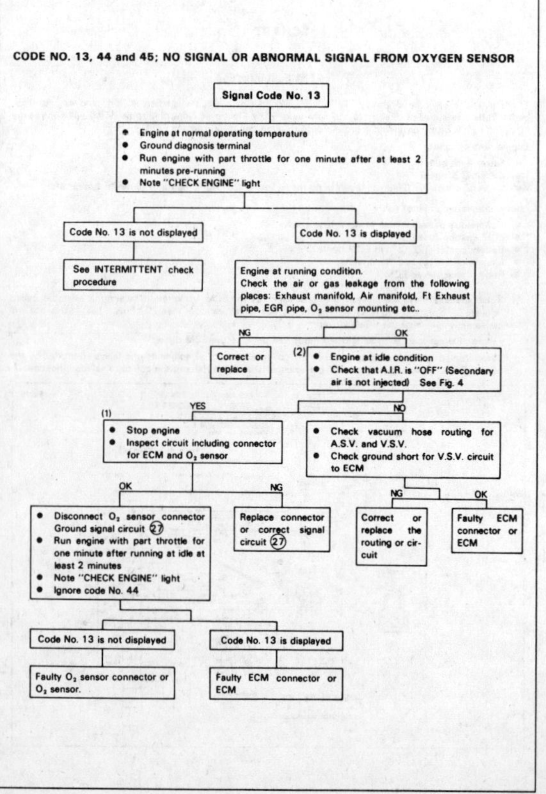

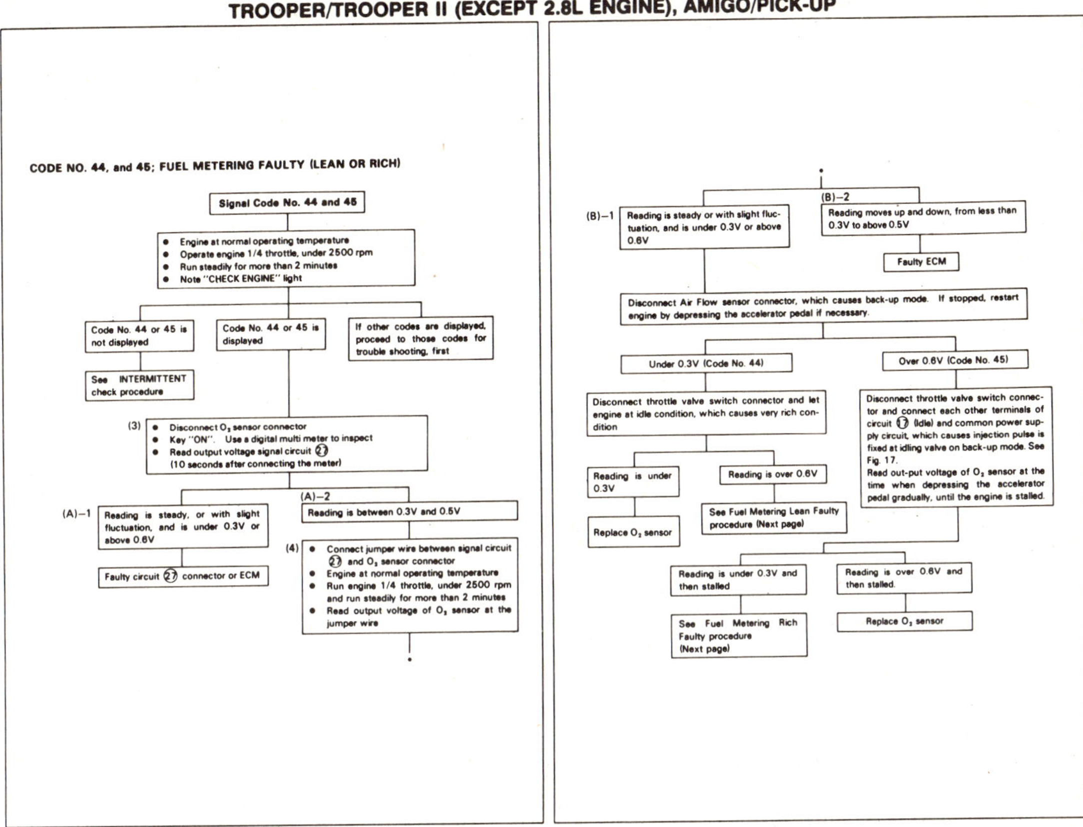


TROOPER/TROOPER II (EXCEPT 2.8L ENGINE), AMIGO/PICK-UP

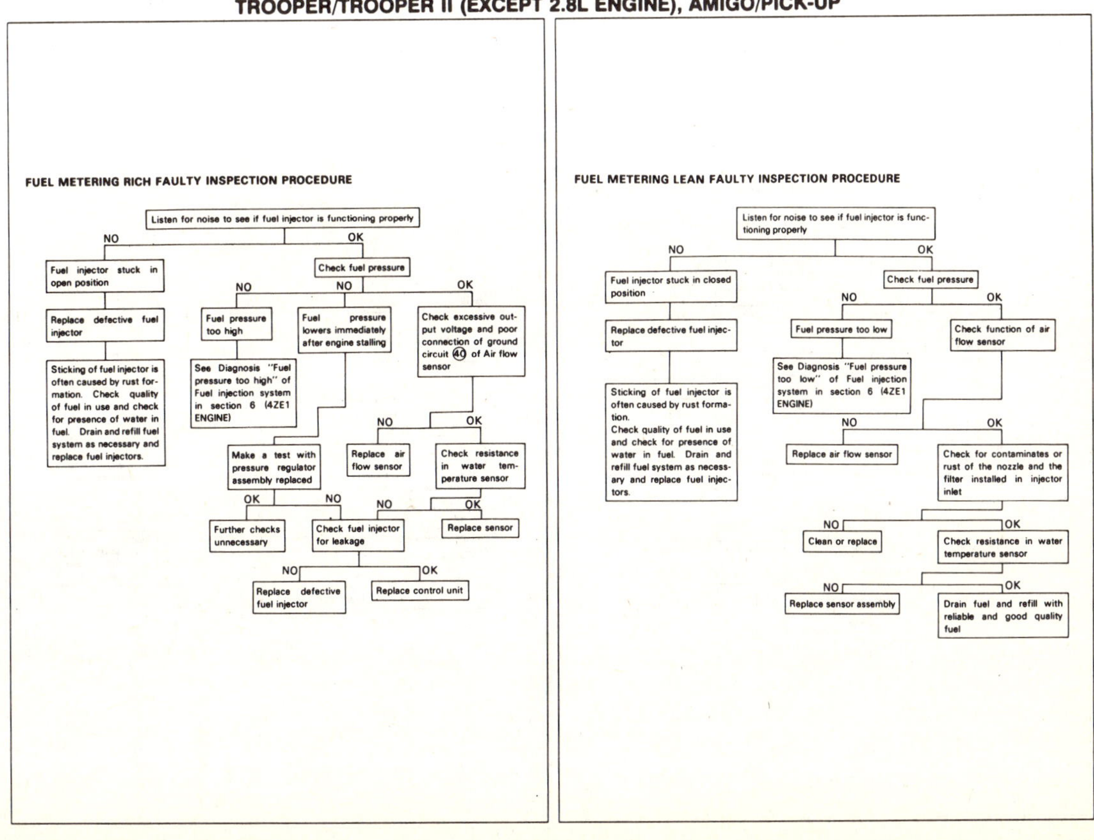

## TROOPER/TROOPER II (EXCEPT 2.8L ENGINE), AMIGO/PICK-UP

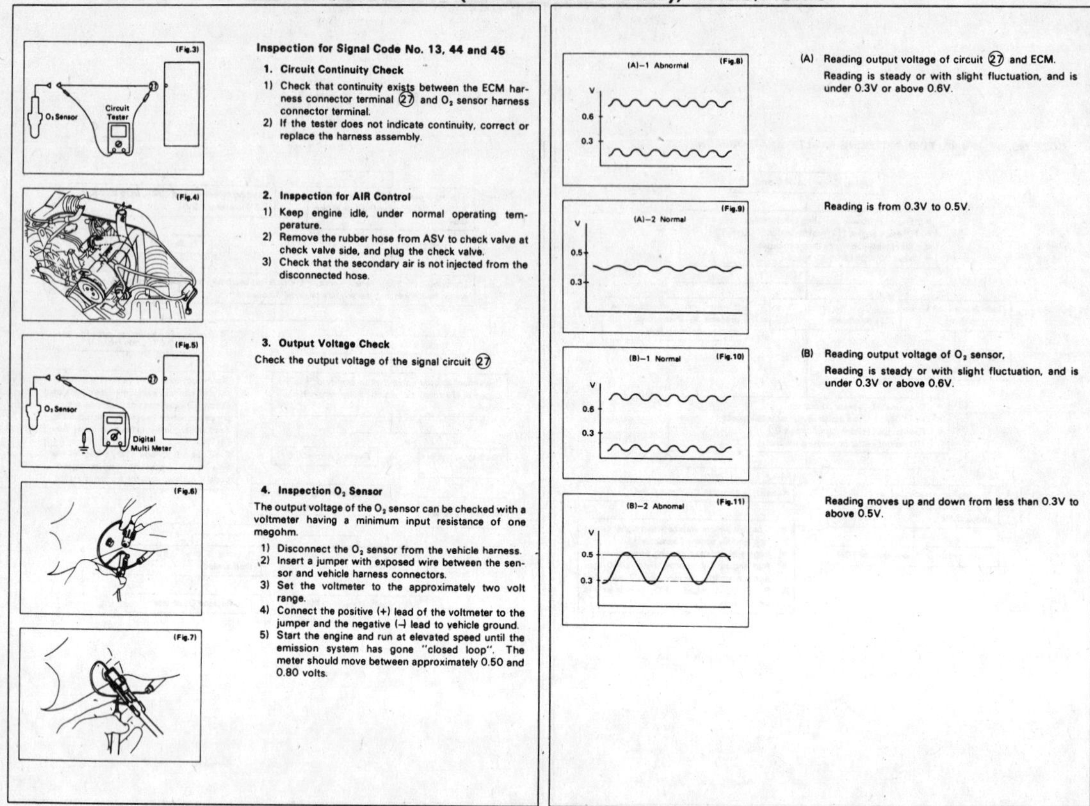

**Inspection for Signal Code No. 13, 44 and 45**

**1. Circuit Continuity Check**

1) Check that continuity exists between the ECM harness connector terminal ㉗ and $O_2$ sensor harness connector terminal.
2) If the tester does not indicate continuity, correct or replace the harness assembly.

**2. Inspection for AIR Control**

1) Keep engine idle, under normal operating temperature.
2) Remove the rubber hose from ASV to check valve at check valve side, and plug the check valve.
3) Check that the secondary air is not injected from the disconnected hose.

**3. Output Voltage Check**

Check the output voltage of the signal circuit ㉗

**4. Inspection $O_2$ Sensor**

The output voltage of the $O_2$ sensor can be checked with a voltmeter having a minimum input resistance of one megohm.

1) Disconnect the $O_2$ sensor from the vehicle harness.
2) Insert a jumper with exposed wire between the sensor and vehicle harness connectors.
3) Set the voltmeter to the approximately two volt range.
4) Connect the positive (+) lead of the voltmeter to the jumper and the negative (−) lead to vehicle ground.
5) Start the engine and run at elevated speed until the emission system has gone "closed loop". The meter should move between approximately 0.50 and 0.80 volts.

(A) Reading output voltage of circuit ㉗ and ECM.
Reading is steady or with slight fluctuation, and is under 0.3V or above 0.6V.

Reading is from 0.3V to 0.5V.

(B) Reading output voltage of $O_2$ sensor.
Reading is steady or with slight fluctuation, and is under 0.3V or above 0.6V.

Reading moves up and down from less than 0.3V to above 0.5V.



## TROOPER/TROOPER II (EXCEPT 2.8L ENGINE), AMIGO/PICK-UP

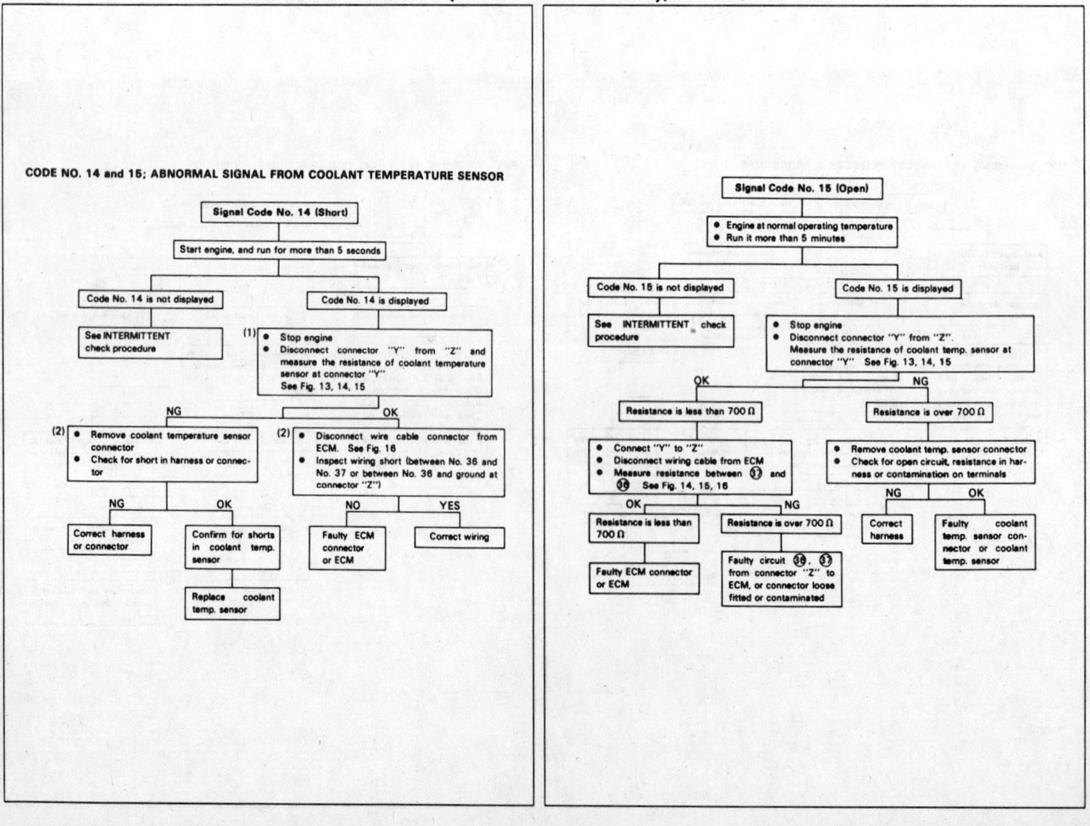

**CODE NO. 14 and 15; ABNORMAL SIGNAL FROM COOLANT TEMPERATURE SENSOR**

Signal Code No. 14 (Short)

Start engine, and run for more than 5 seconds

- Code No. 14 is not displayed → See INTERMITTENT check procedure
- Code No. 14 is displayed → (1) • Stop engine • Disconnect connector "Y" from "Z" and measure the resistance of coolant temperature sensor at connector "Y" See Fig. 13, 14, 15
  - NG → (2) • Remove coolant temperature sensor connector • Check for short in harness or connector
    - NG → Correct harness or connector
    - OK → Confirm for shorts in coolant temp. sensor → Replace coolant temp. sensor
  - OK → (2) • Disconnect wire cable connector from ECM. See Fig. 16 • Inspect wiring short (between No. 36 and No. 37 or between No. 36 and ground at connector "Z")
    - NO → Faulty ECM connector or ECM
    - YES → Correct wiring

Signal Code No. 15 (Open)

• Engine at normal operating temperature • Run it more than 5 minutes

- Code No. 15 is not displayed → See INTERMITTENT check procedure
- Code No. 15 is displayed → • Stop engine • Disconnect connector "Y" from "Z". Measure the resistance of coolant temp. sensor at connector "Y" See Fig. 13, 14, 15
  - OK → Resistance is less than 700 Ω → • Connect "Y" to "Z" • Disconnect wiring cable from ECM • Measure resistance between ㊲ and ㊱ See Fig. 14, 15, 16
    - OK → Resistance is less than 700 Ω → Faulty ECM connector or ECM
    - NG → Resistance is over 700 Ω → Faulty circuit ㊱, ㊲ from connector "Z" to ECM, or connector loose fitted or contaminated
  - NG → Resistance is over 700 Ω → • Remove coolant temp. sensor connector • Check for open circuit, resistance in harness or contamination on terminals
    - NG → Correct harness
    - OK → Faulty coolant temp. sensor connector or coolant temp. sensor

## TROOPER/TROOPER II (EXCEPT 2.8L ENGINE), AMIGO/PICK-UP

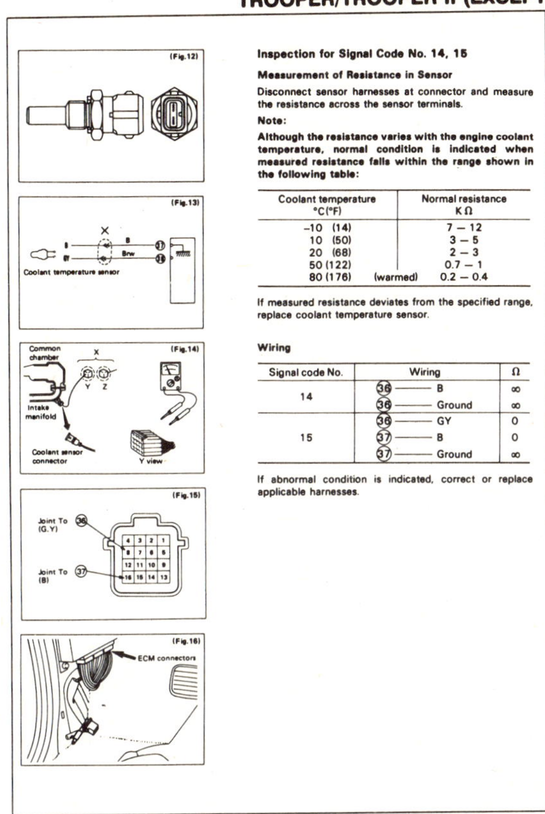



**Inspection for Signal Code No. 14, 15**

**Measurement of Resistance in Sensor**

Disconnect sensor harnesses at connector and measure the resistance across the sensor terminals.

**Note:**

**Although the resistance varies with the engine coolant temperature, normal condition is indicated when measured resistance falls within the range shown in the following table:**

| Coolant temperature °C(°F) | Normal resistance KΩ |
|---|---|
| −10 (14) | 7 — 12 |
| 10 (50) | 3 — 5 |
| 20 (68) | 2 — 3 |
| 50 (122) | 0.7 — 1 |
| 80 (176) (warmed) | 0.2 — 0.4 |

If measured resistance deviates from the specified range, replace coolant temperature sensor.

**Wiring**

| Signal code No. | Wiring | Ω |
|---|---|---|
| 14 | 36 —— B | ∞ |
| | 36 —— Ground | ∞ |
| 15 | 36 —— GY | 0 |
| | 37 —— B | 0 |
| | 37 —— Ground | ∞ |

If abnormal condition is indicated, correct or replace applicable harnesses.

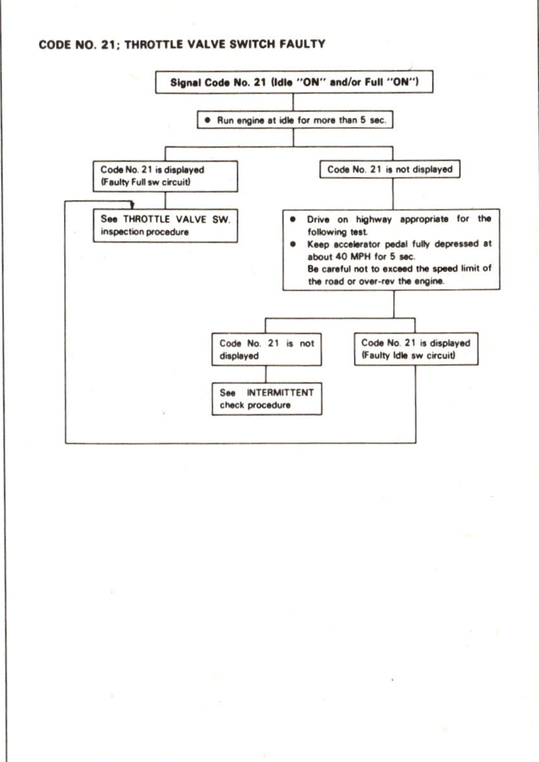

**CODE NO. 21; THROTTLE VALVE SWITCH FAULTY**

Signal Code No. 21 (Idle "ON" and/or Full "ON")

- Run engine at idle for more than 5 sec.

Code No. 21 is displayed (Faulty Full sw circuit) → See THROTTLE VALVE SW. inspection procedure

Code No. 21 is not displayed →
- Drive on highway appropriate for the following test.
- Keep accelerator pedal fully depressed at about 40 MPH for 5 sec.
  Be careful not to exceed the speed limit of the road or over-rev the engine.

→ Code No. 21 is not displayed → See INTERMITTENT check procedure

→ Code No. 21 is displayed (Faulty Idle sw circuit) → See THROTTLE VALVE SW. inspection procedure

## TROOPER/TROOPER II (EXCEPT 2.8L ENGINE), AMIGO/PICK-UP

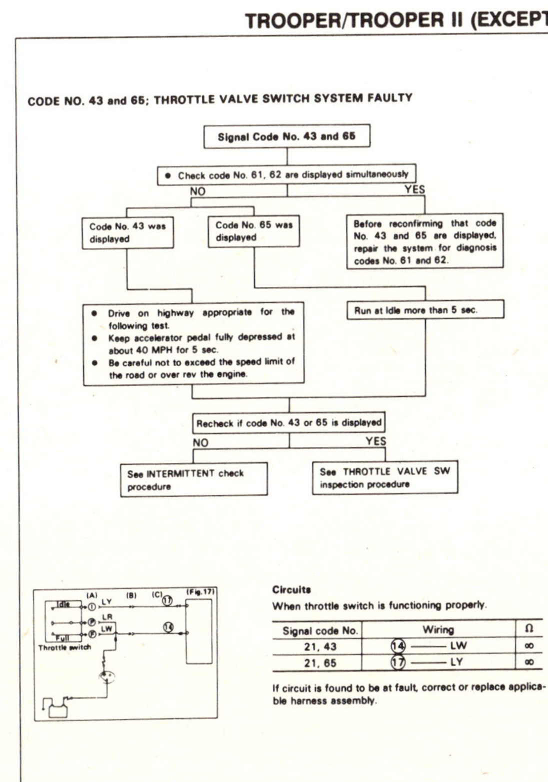

**CODE NO. 43 and 65; THROTTLE VALVE SWITCH SYSTEM FAULTY**

Signal Code No. 43 and 65

- Check code No. 61, 62 are displayed simultaneously

NO:
- Code No. 43 was displayed →
  - Drive on highway appropriate for the following test.
  - Keep accelerator pedal fully depressed at about 40 MPH for 5 sec.
  - Be careful not to exceed the speed limit of the road or over rev the engine.
- Code No. 65 was displayed → Run at Idle more than 5 sec.

YES: Before reconfirming that code No. 43 and 65 are displayed, repair the system for diagnosis codes No. 61 and 62.

Recheck if code No. 43 or 65 is displayed
- NO → See INTERMITTENT check procedure
- YES → See THROTTLE VALVE SW inspection procedure



**Circuits**

When throttle switch is functioning properly.

| Signal code No. | Wiring | Ω |
|---|---|---|
| 21, 43 | 14 —— LW | ∞ |
| 21, 65 | 17 —— LY | ∞ |

If circuit is found to be at fault, correct or replace applicable harness assembly.

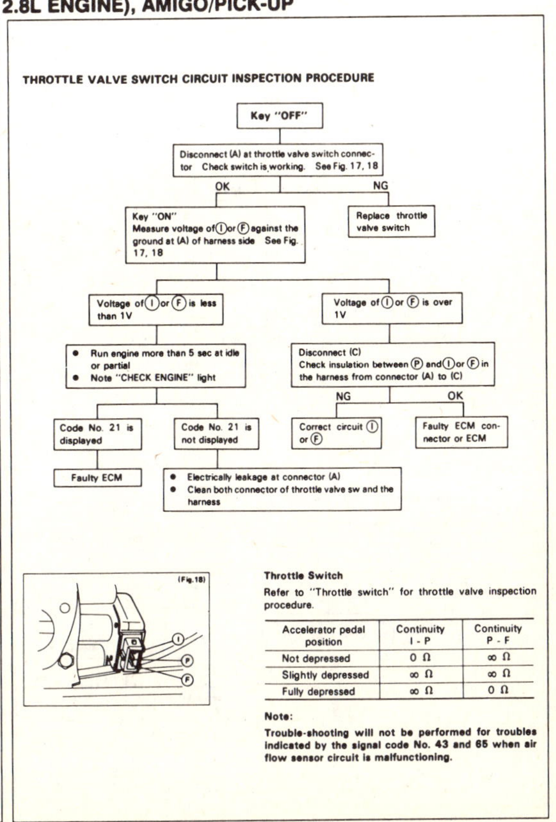

**THROTTLE VALVE SWITCH CIRCUIT INSPECTION PROCEDURE**

Key "OFF"

Disconnect (A) at throttle valve switch connector. Check switch is working. See Fig. 17, 18

- NG → Replace throttle valve switch
- OK → Key "ON" Measure voltage of Ⓘ or Ⓕ against the ground at (A) of harness side. See Fig. 17, 18
  - Voltage of Ⓘ or Ⓕ is less than 1V →
    - Run engine more than 5 sec at idle or partial
    - Note "CHECK ENGINE" light
    - Code No. 21 is displayed → Faulty ECM
    - Code No. 21 is not displayed →
      - Electrically leakage at connector (A)
      - Clean both connector of throttle valve sw and the harness
  - Voltage of Ⓘ or Ⓕ is over 1V → Disconnect (C) Check insulation between Ⓟ and Ⓘ or Ⓕ in the harness from connector (A) to (C)
    - NG → Correct circuit Ⓘ or Ⓕ
    - OK → Faulty ECM connector or ECM



**Throttle Switch**

Refer to "Throttle switch" for throttle valve inspection procedure.

| Accelerator pedal position | Continuity I - P | Continuity P - F |
|---|---|---|
| Not depressed | 0 Ω | ∞ Ω |
| Slightly depressed | ∞ Ω | ∞ Ω |
| Fully depressed | ∞ Ω | 0 Ω |

**Note:**

**Trouble-shooting will not be performed for troubles indicated by the signal code No. 43 and 65 when air flow sensor circuit is malfunctioning.**

## TROOPER/TROOPER II (EXCEPT 2.8L ENGINE), AMIGO/PICK-UP

### CODE NO. 22; NO SIGNAL FROM THE STARTER

Signal Code No. 22

Start engine and run for more than 3 seconds

- Code No. 22 is displayed
  - Stop engine. Check voltage at pin No. 16 of ECM connector between pin and starter "ON" position.
    - The voltage is over 5V
      - Faulty ECM connector or ECM
    - The voltage is not over 5V
      - Faulty the circuit of pin No. 16 from ECM connector and starter "ON" position
- Code No. 22 is not displayed
  - Check if the owner has pushed the vehicle to start it.
    - NO: See INTERMITTENT check procedure
    - YES: Clear code and check that normal code is displayed.

(Fig.19) 0Ω, 1, 16, Key Switch

(Fig.20) Starter SW; B, 1 2 3, 4 5 6

**Signal Code No. 22 Inspection**

**Circuit**

When the circuit is found to be at fault, correct or replace applicable harness assembly.

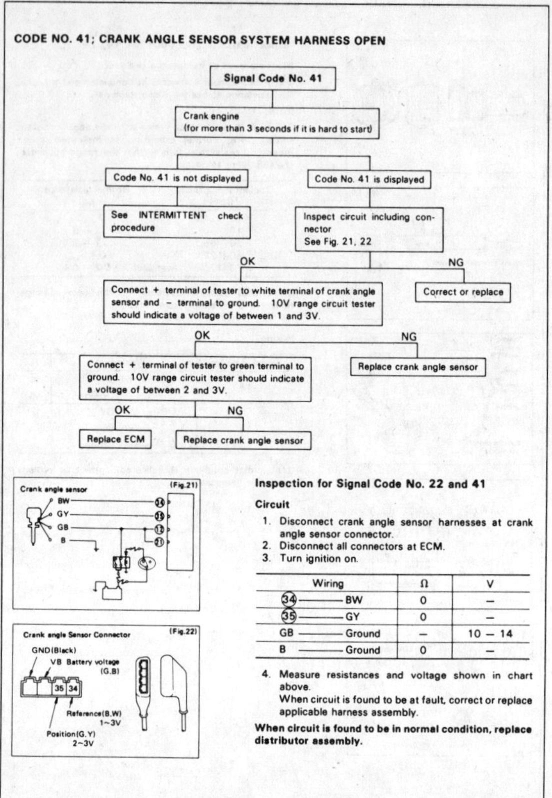


**Inspection for Signal Code No. 22 and 41**

**Circuit**

1. Disconnect crank angle sensor harnesses at crank angle sensor connector.
2. Disconnect all connectors at ECM.
3. Turn ignition on.

| Wiring | Ω | V |
|---|---|---|
| 34 —— BW | 0 | – |
| 35 —— GY | 0 | – |
| GB —— Ground | – | 10 – 14 |
| B —— Ground | 0 | – |

4. Measure resistances and voltage shown in chart above.
   When circuit is found to be at fault, correct or replace applicable harness assembly.

**When circuit is found to be in normal condition, replace distributor assembly.**

## TROOPER/TROOPER II (EXCEPT 2.8L ENGINE), AMIGO/PICK-UP

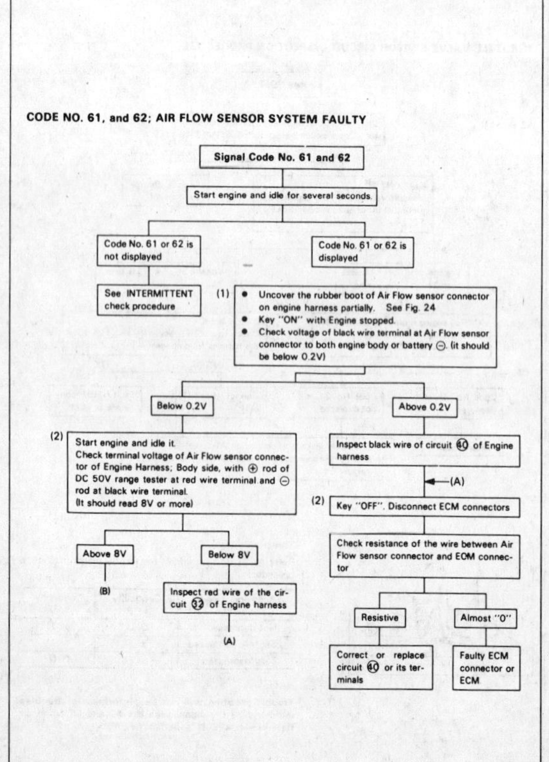

CODE NO. 61, and 62; AIR FLOW SENSOR SYSTEM FAULTY
Signal Code No. 61 and 62
Start engine and idle for several seconds.
Code No. 61 or 62 is not displayed
Code No. 61 or 62 is displayed
See INTERMITTENT check procedure
(1) • Uncover the rubber boot of Air Flow sensor connector on engine harness partially. See Fig. 24
• Key "ON" with Engine stopped.
• Check voltage of black wire terminal at Air Flow sensor connector to both engine body or battery ⊖. (it should be below 0.2V)
Below 0.2V
Above 0.2V
(2) Start engine and idle it. Check terminal voltage of Air Flow sensor connector of Engine Harness; Body side, with ⊕ rod of DC 50V range tester at red wire terminal and ⊖ rod at black wire terminal. (It should read 8V or more)
Inspect black wire of circuit 40 of Engine harness
(A)
(2) Key "OFF". Disconnect ECM connectors
Check resistance of the wire between Air Flow sensor connector and EOM connector
Above 8V
Below 8V
(B)
Inspect red wire of the circuit 32 of Engine harness
(A)
Resistive
Almost "0"
Correct or replace circuit 40 or its terminals
Faulty ECM connector or ECM


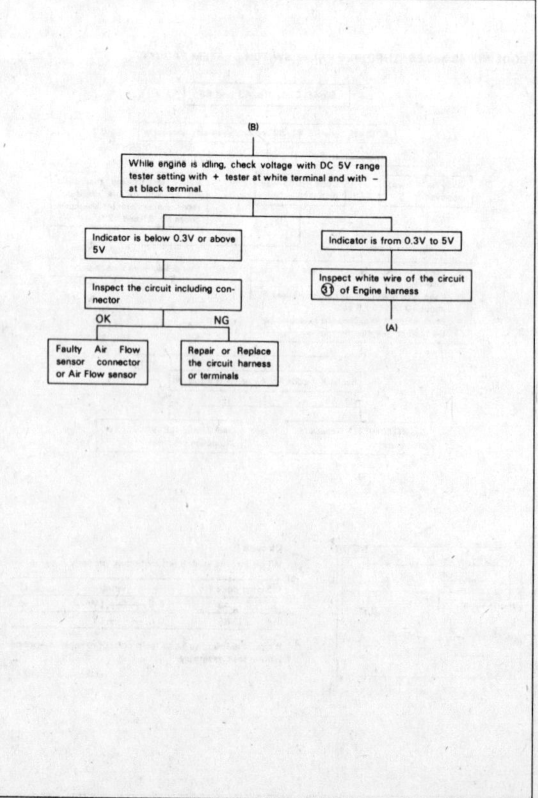

## TROOPER/TROOPER II (EXCEPT 2.8L ENGINE), AMIGO/PICK-UP

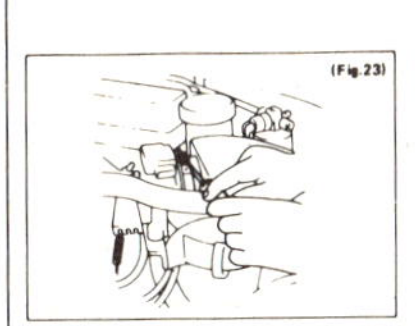

**Inspection for Signal Code No. 61 and 62**

**Removal of Air Flow Sensor**

1. Remove bolts attaching air flow sensor.
2. Loosen the clamp bolt and remove air flow sensor from the air cleaner.

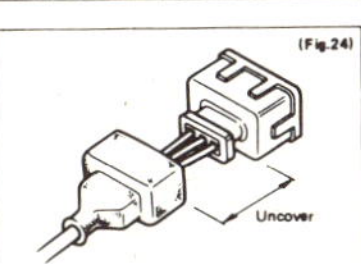


**1. Inspection of Air Flow Sensor**

Uncover the rubber boot on the air flow sensor side of the harness connector.

| Wiring | Ignition switch | Condition | V | Ω |
|---|---|---|---|---|
| A. Red-Ground | ON | – | 10–14 | – |
| | ON | – | Less than 0.2 | – |
| B. White-Ground | ON | When breathing (See Fig-27) | 0.5–1.5 | – |
| C. Black-Ground* | ON/OFF | – | Less than 0.2 | 0 |

* Ground means engine ground.

When A. and C. are abnormal, check circuit and correct as necessary.
When A. and C. are normal but B. is abnormal, replace air flow sensor.

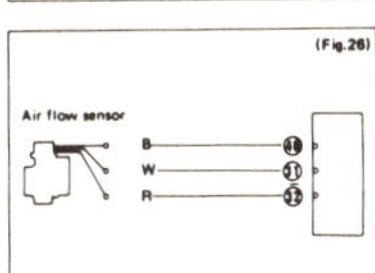


**Inspection of Air Flow Sensor**

Install the air flow sensor assembly in the reverse order of removal.

**2. Circuits**

| Wiring | | Ω |
|---|---|---|
| W | ㉛ | 0 |
| W | Ground | ∞ |
| R | ㉜ | 0 |
| R | Ground | ∞ |
| B | ㊵ | 0 |

When circuit is found to be at fault, correct or replace applicable harness assembly.

**Note:**

**If the air flow sensor has been removed, check by-pass circuit Ⓐ for contamination and clean with carburetor cleaner as necessary.**

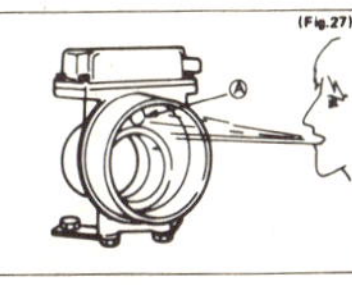


**CODE NO. 63; CAR SPEED SENSOR FALUTY**

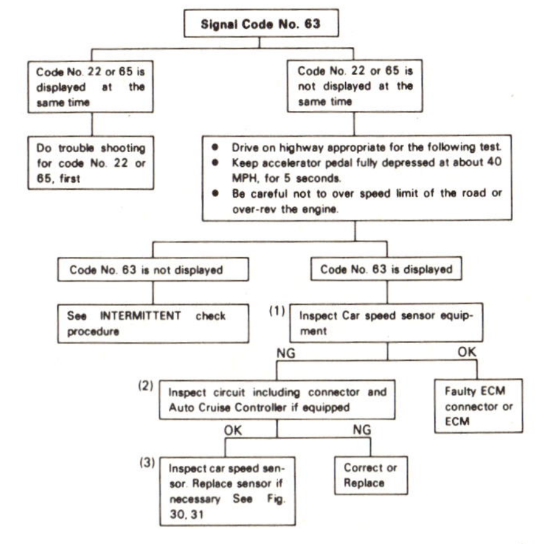


**Note:**

**This code No. may be displayed even the system is normal if the vehicle is used under the following conditions.**

**A/T: In "D" range, while depressing the brake pedal, the accel pedal is depressed over 1/2 throttle for more than 3 sec.**

**M/T: Under the severe climbing conditions, keep the vehicle stopped while the accel pedal depressing over 1/2 throttle, with the clutch engaged partially, for more than 3 sec.**

## TROOPER/TROOPER II (EXCEPT 2.8L ENGINE), AMIGO/PICK-UP

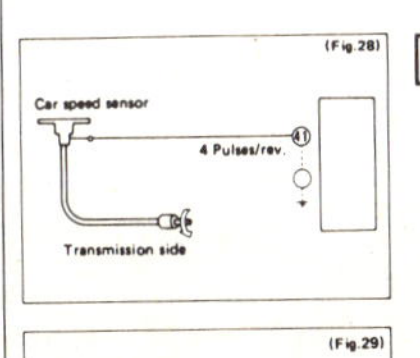


**Inspection for Signal Code No. 63**

**1. Sensor and Equipment**

1) Connect a circuit tester between the control unit harness connector terminal ㊶ and ground.
2) Disconnect speedometer cable at transmission end and turn the inner cable slowly.
3) If the tester indicates continuity and open circuit condition alternately as the cable is turned, the car speed sensor and circuit are operating normally.

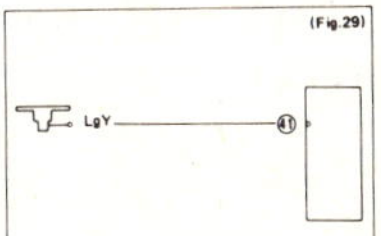


**2. Circuits**

| Wiring | | Ω |
|---|---|---|
| LgY | ㊶ | 0 |
| B | Ground | ∞ |

When circuit is found to be at fault, correct or replace applicable harness assembly.

Install the speedometer assembly.
Refer to "Meters" in Section 8 "ELECTRICAL-BODY AND CHASSIS" for installation procedure.

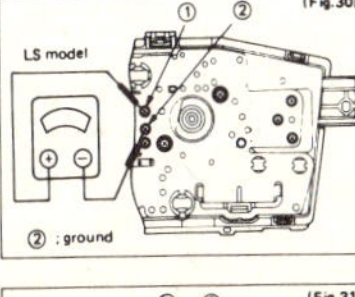


**3. Car Speed Sensor (Incorporated in the Speedometer)**

1) Remove the meter assembly.
2) Connect the leads of a circuit tester across the terminals (①, ②) at rear face of the meter assembly as illustrated.

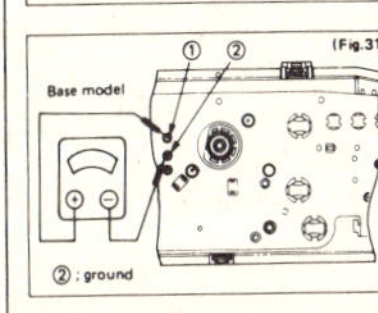


3) Turn the inner shaft slowly and check that the tester indicates a continuity and open circuit condition alternately.
If car speed sensor is inoperative, replace speedometer assembly.

**CODE NO. 51, 52; ROM (CODE NO. 51) NG, RAM (CODE NO. 52) NG**

Signal Code No. 51, 52
→ Start engine
→ Is code 51, 52 displayed?
- NO → END
- YES → Replace ECM

(Fig.32)

**Inspection for Signal Code No. 51 and 52**

**Circuit Connections**

1. Disconnect the control unit harnesses at connector and check for proper connections at terminals.
2. Regardless of terminal conditions, road test the vehicle with all the harnesses reconnected at the connector.
3. Obtain display of trouble code and see if the trouble has been corrected.
If the trouble persists, replace the ECM.

## TROOPER/TROOPER II (EXCEPT 2.8L ENGINE), AMIGO/PICK-UP

**CODE NO. 23, 35 and 54; POWER TRANSISTOR SYSTEM FAULTY**

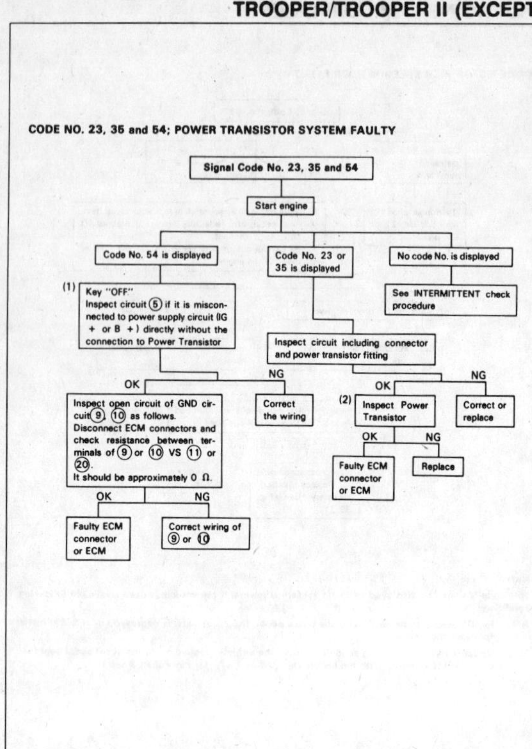


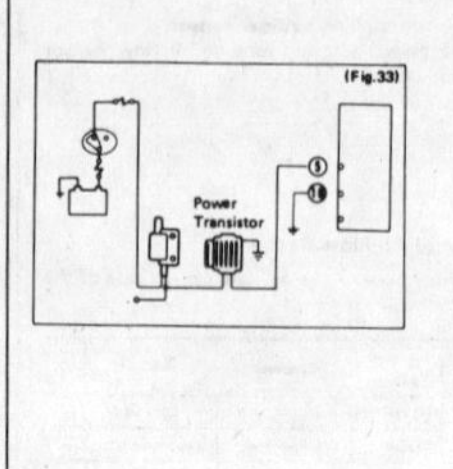


**Inspection for Signal Code No. 23, 35 and 54**

**1. Circuits**

Check wiring by refering to the illustration and wiring diagram.

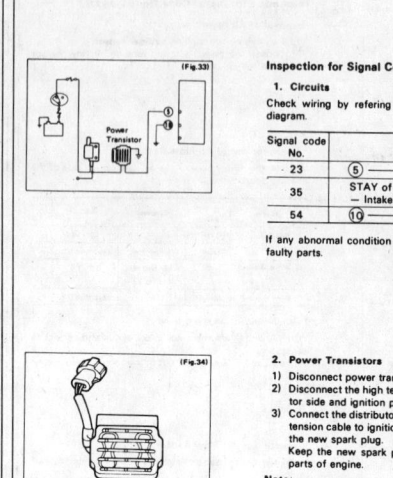

| Signal code No. | Wiring | Ω |
|---|---|---|
| 23 | ⑤ —— Ground | ∞ |
| 35 | STAY of power Transister — Intake manifold | 0 |
| 54 | ⑩ —— Ground | 0 |

If any abnormal condition was found, correct or replace faulty parts.

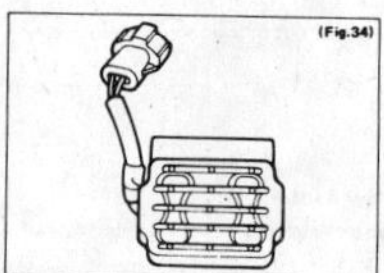


**2. Power Transistors**

1) Disconnect power transistor harnesses at connector.
2) Disconnect the high tension cable from both distributor side and ignition plug side.
3) Connect the distributor side of the disconnected high tension cable to ignition coil and the plug side of it to the new spark plug.
Keep the new spark plug in contact with the metal parts of engine.

**Note:**

**To prevent electrical shock, hold the cable via an effective insulation.**

4) Turn starter switch to ON position.
5) Connect positive side (+) of a 1.5V dry battery to Y terminal of the power transistor and negative side (–) to ground.
6) Disconnect dry battery lead and check that a spark jumps across the plug gap.
Repeat the test several times.
If spark does not occur, replace power transistor together with bracket.

## TROOPER/TROOPER II (EXCEPT 2.8L ENGINE), AMIGO/PICK-UP

**CODE NO. 25 and 53; VSV FAULTY PRESSURE REGULATOR**

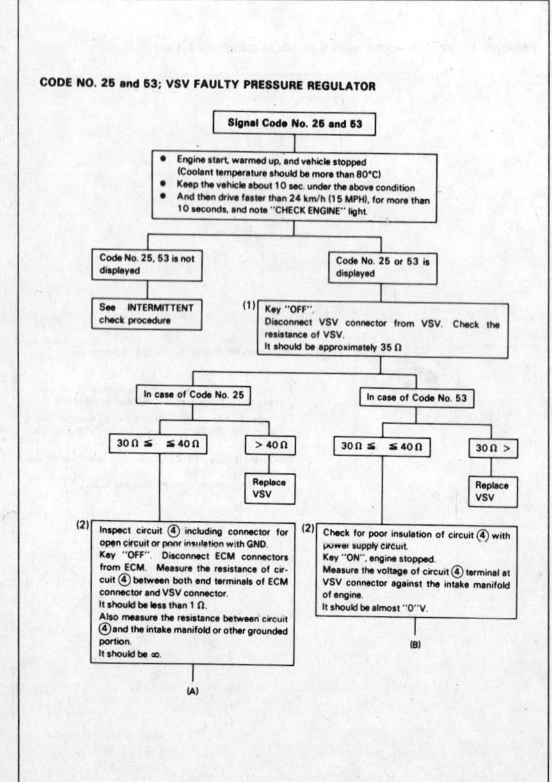


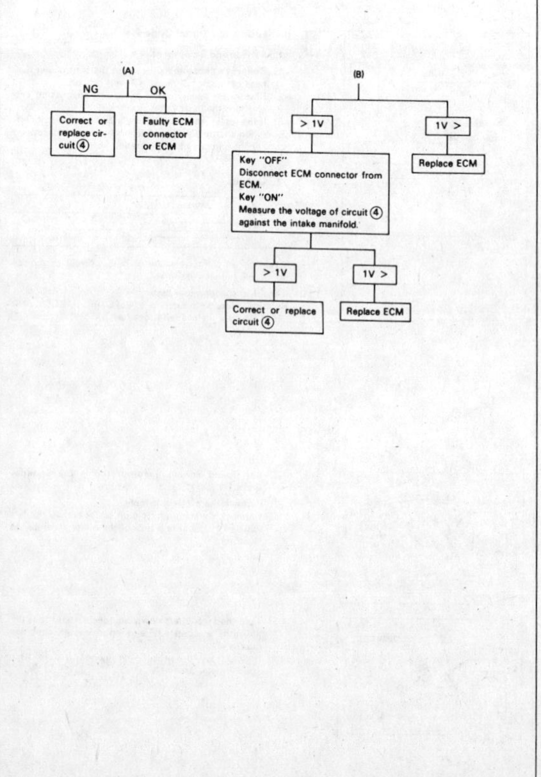

## TROOPER/TROOPER II (EXCEPT 2.8L ENGINE), AMIGO/PICK-UP

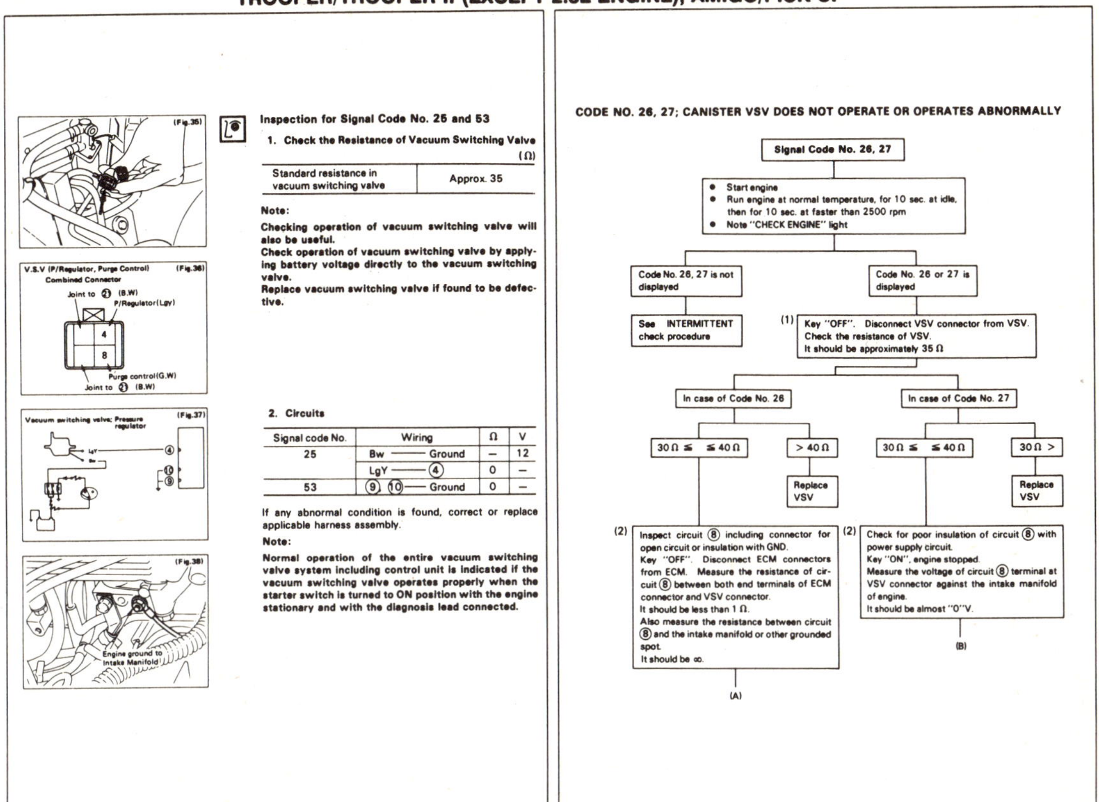



**Inspection for Signal Code No. 25 and 53**

1. **Check the Resistance of Vacuum Switching Valve** (Ω)

| Standard resistance in vacuum switching valve | Approx. 35 |
|---|---|

**Note:**

**Checking operation of vacuum switching valve will also be useful.**
**Check operation of vacuum switching valve by applying battery voltage directly to the vacuum switching valve.**
**Replace vacuum switching valve if found to be defective.**

2. **Circuits**

| Signal code No. | Wiring | Ω | V |
|---|---|---|---|
| 25 | Bw —— Ground | — | 12 |
| | LgY —— ④ | 0 | — |
| 53 | ⑨ ⑩ —— Ground | 0 | — |

If any abnormal condition is found, correct or replace applicable harness assembly.

**Note:**

**Normal operation of the entire vacuum switching valve system including control unit is indicated if the vacuum switching valve operates properly when the starter switch is turned to ON position with the engine stationary and with the diagnosis lead connected.**

**CODE NO. 26, 27; CANISTER VSV DOES NOT OPERATE OR OPERATES ABNORMALLY**

Signal Code No. 26, 27

- Start engine
- Run engine at normal temperature, for 10 sec. at idle, then for 10 sec. at faster than 2500 rpm
- Note "CHECK ENGINE" light

Code No. 26, 27 is not displayed → See INTERMITTENT check procedure

Code No. 26 or 27 is displayed → (1) Key "OFF". Disconnect VSV connector from VSV. Check the resistance of VSV. It should be approximately 35 Ω

In case of Code No. 26:
- 30 Ω ≦ ≦ 40 Ω → (2)
- > 40 Ω → Replace VSV

In case of Code No. 27:
- 30 Ω ≦ ≦ 40 Ω → (2)
- 30 Ω > → Replace VSV

(2) Inspect circuit ⑧ including connector for open circuit or insulation with GND. Key "OFF". Disconnect ECM connectors from ECM. Measure the resistance of circuit ⑧ between both end terminals of ECM connector and VSV connector. It should be less than 1 Ω. Also measure the resistance between circuit ⑧ and the intake manifold or other grounded spot. It should be ∞. → (A)

(2) Check for poor insulation of circuit ⑧ with power supply circuit. Key "ON", engine stopped. Measure the voltage of circuit ⑧ terminal at VSV connector against the intake manifold of engine. It should be almost "0" V. → (B)

## TROOPER/TROOPER II (EXCEPT 2.8L ENGINE), AMIGO/PICK-UP

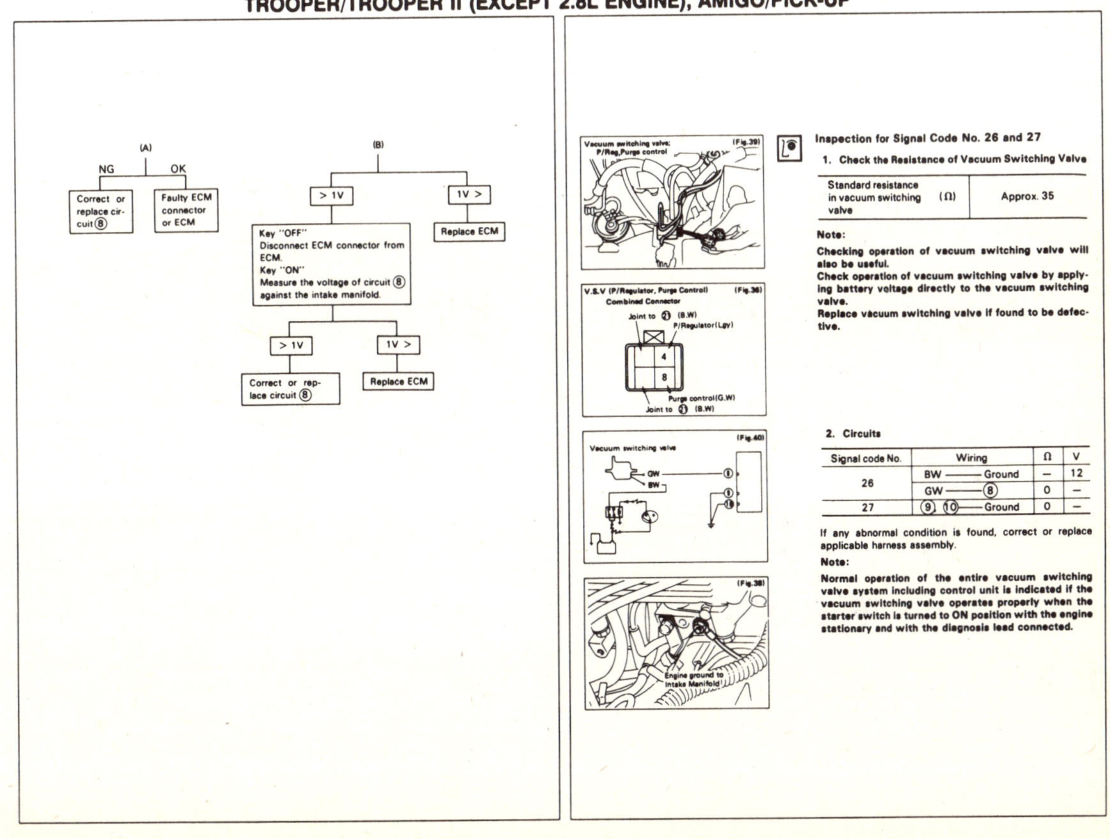

(A)
- NG → Correct or replace circuit ⑧
- OK → Faulty ECM connector or ECM

(B)
- > 1V → Key "OFF" Disconnect ECM connector from ECM. Key "ON" Measure the voltage of circuit ⑧ against the intake manifold.
  - > 1V → Correct or replace circuit ⑧
  - 1V > → Replace ECM
- 1V > → Replace ECM



**Inspection for Signal Code No. 26 and 27**

1. **Check the Resistance of Vacuum Switching Valve**

| Standard resistance in vacuum switching valve (Ω) | Approx. 35 |
|---|---|

**Note:**

**Checking operation of vacuum switching valve will also be useful.**
**Check operation of vacuum switching valve by applying battery voltage directly to the vacuum switching valve.**
**Replace vacuum switching valve if found to be defective.**

2. **Circuits**

| Signal code No. | Wiring | Ω | V |
|---|---|---|---|
| 26 | BW —— Ground | — | 12 |
| | GW —— ⑧ | 0 | — |
| 27 | ⑨ ⑩ —— Ground | 0 | — |

If any abnormal condition is found, correct or replace applicable harness assembly.

**Note:**

**Normal operation of the entire vacuum switching valve system including control unit is indicated if the vacuum switching valve operates properly when the starter switch is turned to ON position with the engine stationary and with the diagnosis lead connected.**

## TROOPER/TROOPER II (EXCEPT 2.8L ENGINE), AMIGO/PICK-UP

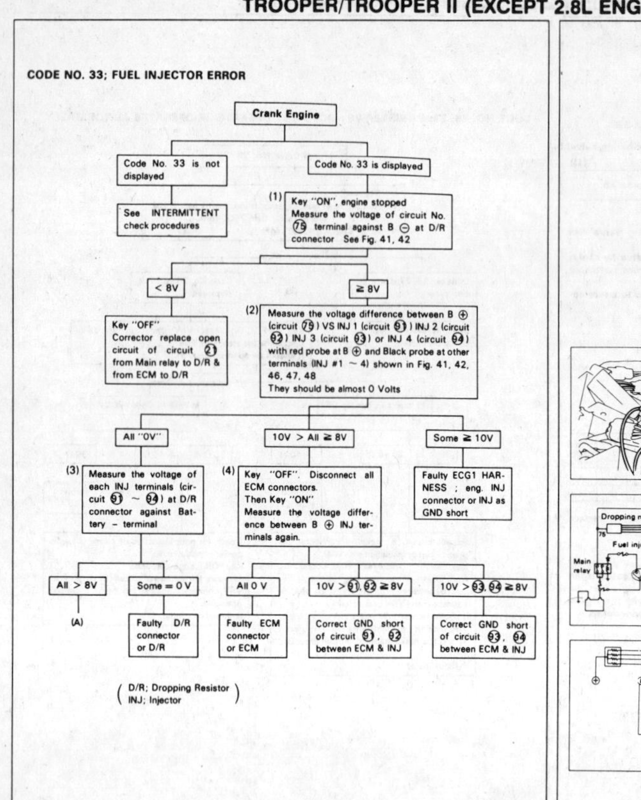


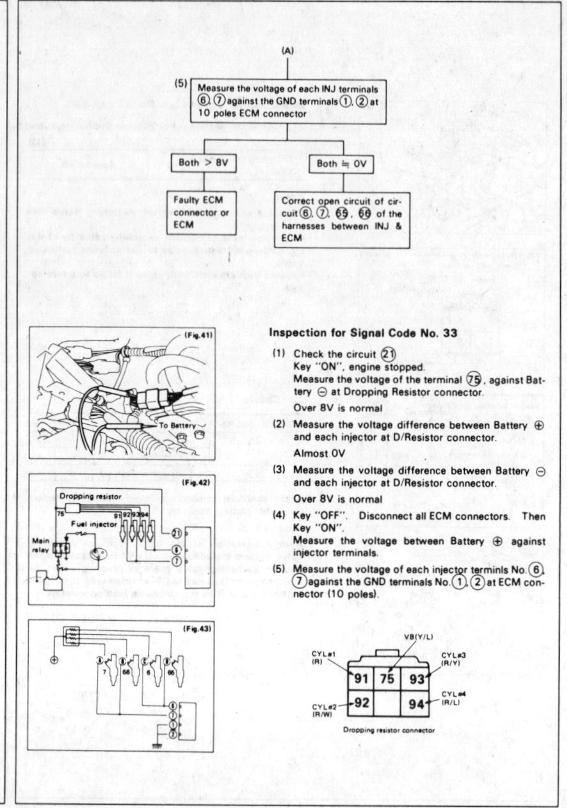


**Inspection for Signal Code No. 33**

(1) Check the circuit 21
Key "ON", engine stopped.
Measure the voltage of the terminal 75, against Battery ⊖ at Dropping Resistor connector.
Over 8V is normal

(2) Measure the voltage difference between Battery ⊕ and each injector at D/Resistor connector.
Almost 0V

(3) Measure the voltage difference between Battery ⊖ and each injector at D/Resistor connector.
Over 8V is normal

(4) Key "OFF". Disconnect all ECM connectors. Then Key "ON".
Measure the voltage between Battery ⊕ against injector terminals.

(5) Measure the voltage of each injector terminls No. 6, 7 against the GND terminals No. 1, 2 at ECM connector (10 poles).

## TROOPER/TROOPER II (EXCEPT 2.8L ENGINE), AMIGO/PICK-UP

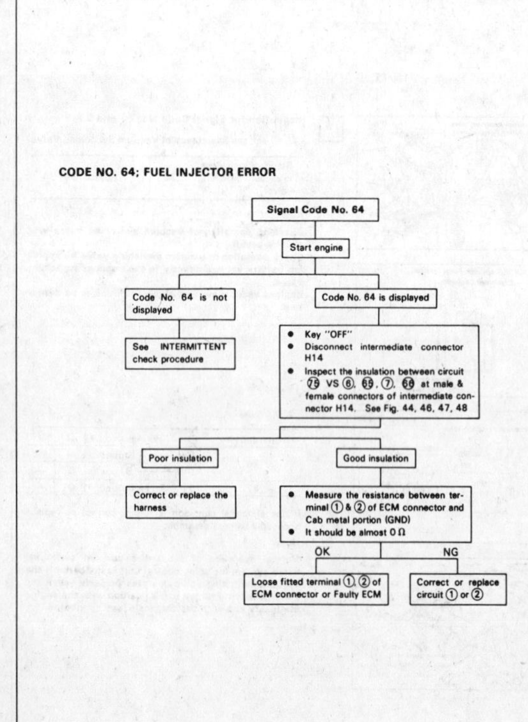


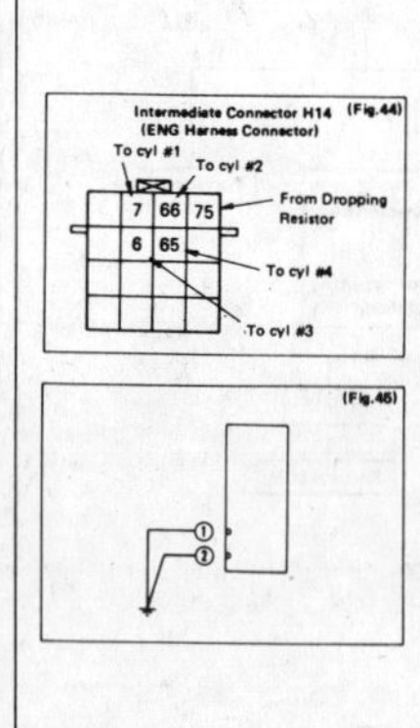


**Inspection for Signal Code No. 64**

1. Check the insulation for Injectors and Dropping resistor circuit.
   - Key "OFF"
   - Disconnect the connector H14
   - Check the insulation between circuit 75 VS 6, 65, 7, 66 at both male and female terminals.
2. Check open for the GND circuit 1, 2.

**Circuit**

| Wiring | Ω |
|---|---|
| 1 — Ground | 0 |
| 2 — Ground | 0 |

If any abnormal condition was found, correct or replace faulty parts.

**Note:**

**This diagnostic procedure does not disclose troubles in switches, injectors such as poor contact, restrictions, etc.**
**In such an instance, the control unit may analyze the case and give abnormal air flow sensor or $O_2$ sensor signal code.**

## TROOPER/TROOPER II (EXCEPT 2.8L ENGINE), AMIGO/PICK-UP

**Inspection for Signal Code No. 33 and 64**

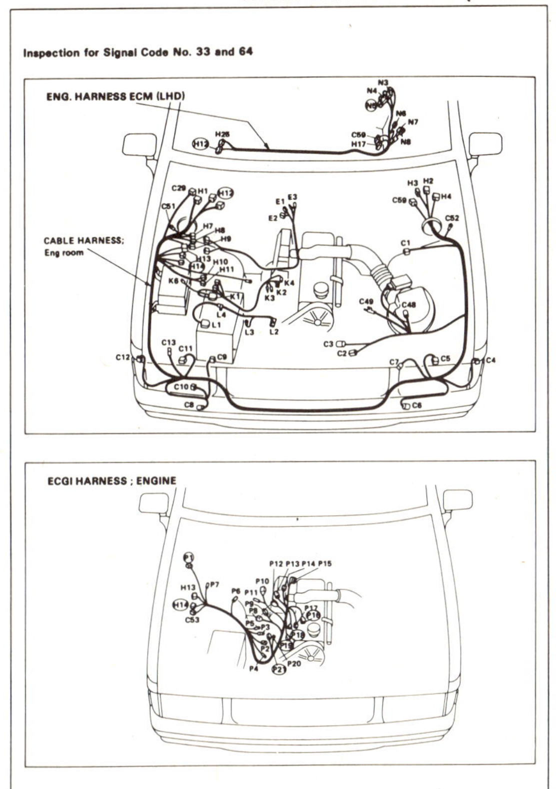


**FUEL INJECTION CIRCUITS AND CONNECTORS**

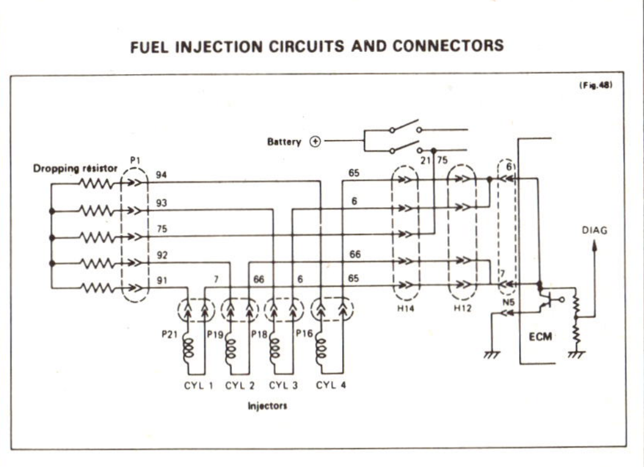


| Harness Name | Connector No. | Parts Name |
|---|---|---|
| CABLE HARNESS; Eng. room | H12 | Engine room harness—Engine harness ECM (RH) |
| | H14 | Engine room harness—closed loop harness |
| ENGINE HARNESS ECM (LHD) | N5 | ECM controller 10 pin |
| ECGI HARNESS; Engine | P1 | Dropping resistor |
| | P16 | Resistor ; fuel injector (No. 4) |
| | P18 | Resistor ; fuel injector (No. 3) |
| | P19 | Resistor ; fuel injector (No. 2) |
| | P21 | Resistor ; fuel injector (No. 1) |

## TROOPER/TROOPER II (EXCEPT 2.8L ENGINE), AMIGO/PICK-UP

**CODE NO. 32 and 34; EGR SYSTEM FAILURE AND SENSOR CIRCUIT FAILURE**

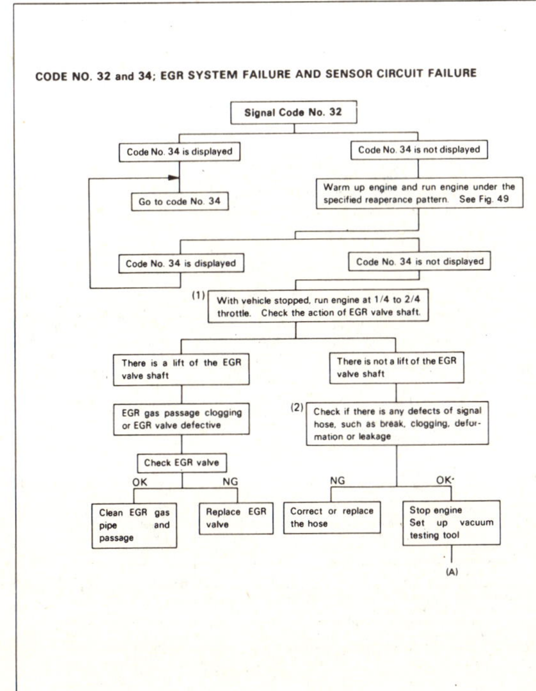


**'88 Model & '89 Federal**

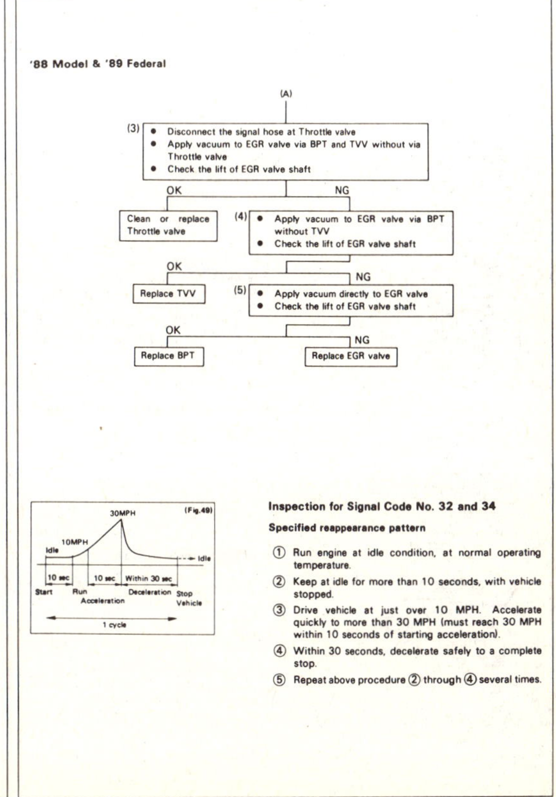


**Inspection for Signal Code No. 32 and 34**

**Specified reappearance pattern**

① Run engine at idle condition, at normal operating temperature.

② Keep at idle for more than 10 seconds, with vehicle stopped.

③ Drive vehicle at just over 10 MPH. Accelerate quickly to more than 30 MPH (must reach 30 MPH within 10 seconds of starting acceleration).

④ Within 30 seconds, decelerate safely to a complete stop.

⑤ Repeat above procedure ② through ④ several times.

## TROOPER/TROOPER II (EXCEPT 2.8L ENGINE), AMIGO/PICK-UP

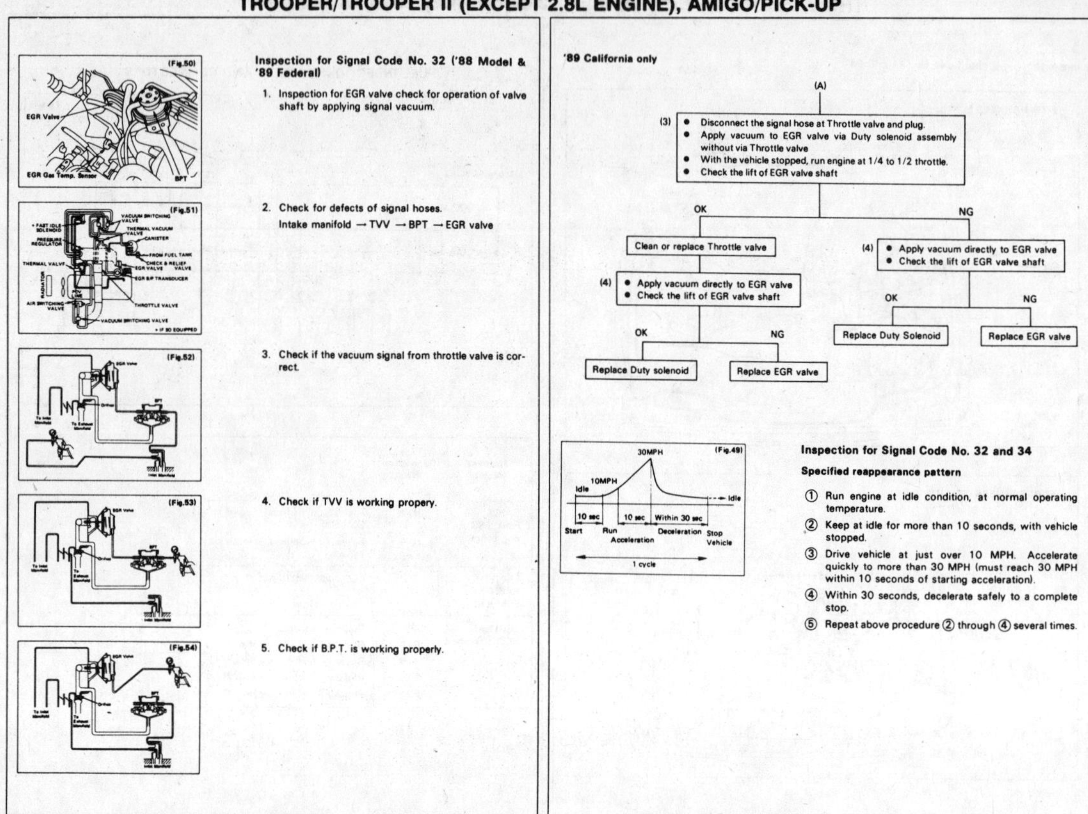

**Inspection for Signal Code No. 32 ('88 Model & '89 Federal)**

1. Inspection for EGR valve check for operation of valve shaft by applying signal vacuum.
2. Check for defects of signal hoses.
   Intake manifold → TVV → BPT → EGR valve
3. Check if the vacuum signal from throttle valve is correct.
4. Check if TVV is working propery.
5. Check if B.P.T. is working properly.

**'89 California only**

(A)

(3)
- Disconnect the signal hose at Throttle valve and plug.
- Apply vacuum to EGR valve via Duty solenoid assembly without via Throttle valve
- With the vehicle stopped, run engine at 1/4 to 1/2 throttle.
- Check the lift of EGR valve shaft

OK → Clean or replace Throttle valve → (4)
- Apply vacuum directly to EGR valve
- Check the lift of EGR valve shaft

  OK → Replace Duty solenoid
  NG → Replace EGR valve

NG → (4)
- Apply vacuum directly to EGR valve
- Check the lift of EGR valve shaft

  OK → Replace Duty Solenoid
  NG → Replace EGR valve

**Inspection for Signal Code No. 32 and 34**

**Specified reappearance pattern**

① Run engine at idle condition, at normal operating temperature.
② Keep at idle for more than 10 seconds, with vehicle stopped.
③ Drive vehicle at just over 10 MPH. Accelerate quickly to more than 30 MPH (must reach 30 MPH within 10 seconds of starting acceleration).
④ Within 30 seconds, decelerate safely to a complete stop.
⑤ Repeat above procedure ② through ④ several times.

## TROOPER/TROOPER II (EXCEPT 2.8L ENGINE), AMIGO/PICK-UP

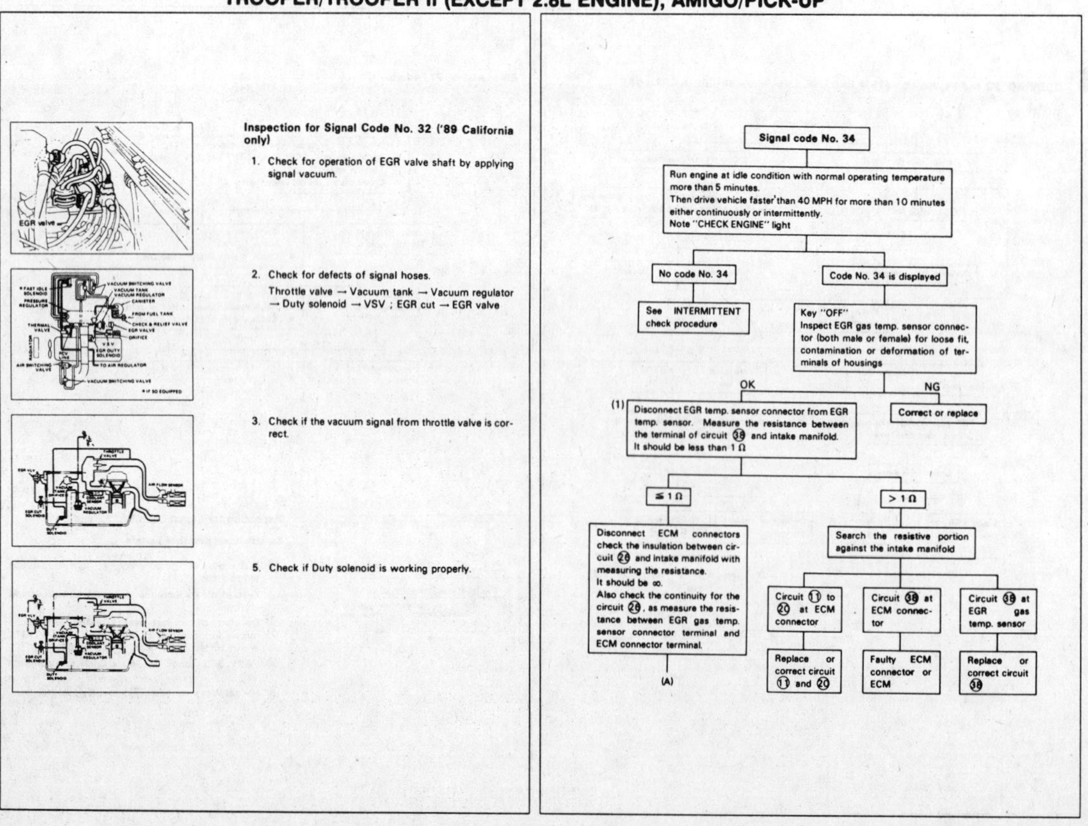

**Inspection for Signal Code No. 32 ('89 California only)**

1. Check for operation of EGR valve shaft by applying signal vacuum.
2. Check for defects of signal hoses.
   Throttle valve → Vacuum tank → Vacuum regulator → Duty solenoid → VSV ; EGR cut → EGR valve
3. Check if the vacuum signal from throttle valve is correct.
5. Check if Duty solenoid is working properly.

**Signal code No. 34**

Run engine at idle condition with normal operating temperature more than 5 minutes.
Then drive vehicle faster than 40 MPH for more than 10 minutes either continuously or intermittently.
Note "CHECK ENGINE" light

- No code No. 34 → See INTERMITTENT check procedure
- Code No. 34 is displayed → Key "OFF"
  Inspect EGR gas temp. sensor connector (both male or female) for loose fit, contamination or deformation of terminals of housings
  - NG → Correct or replace
  - OK → (1) Disconnect EGR temp. sensor connector from EGR temp. sensor. Measure the resistance between the terminal of circuit (38) and intake manifold.
    It should be less than 1 Ω
    - ≦ 1 Ω → Disconnect ECM connectors check the insulation between circuit (26) and intake manifold with measuring the resistance.
      It should be ∞.
      Also check the continuity for the circuit (26), as measure the resistance between EGR gas temp. sensor connector terminal and ECM connector terminal. → (A)
    - > 1 Ω → Search the resistive portion against the intake manifold
      - Circuit (11) to (20) at ECM connector → Replace or correct circuit (11) and (20)
      - Circuit (38) at ECM connector → Faulty ECM connector or ECM
      - Circuit (38) at EGR gas temp. sensor → Replace or correct circuit (38)

## TROOPER/TROOPER II (EXCEPT 2.8L ENGINE), AMIGO/PICK-UP

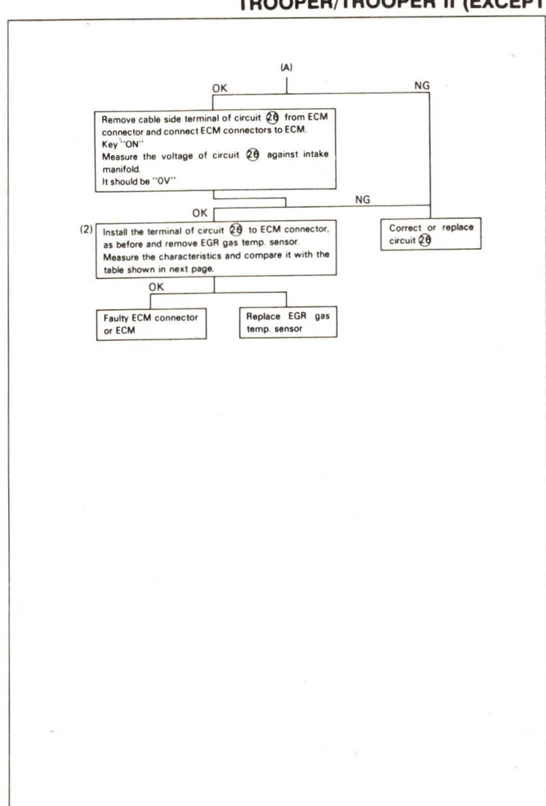


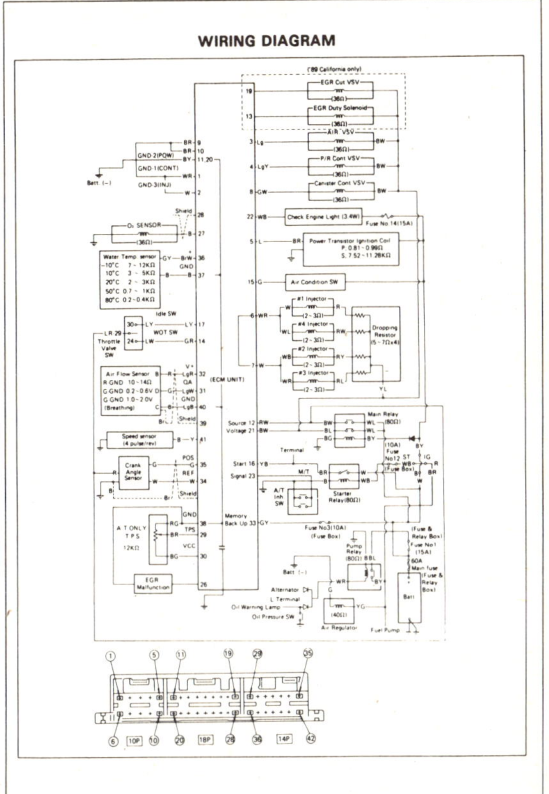


## TROOPER/TROOPER II (EXCEPT 2.8L ENGINE) AMIGO/PICK-UP

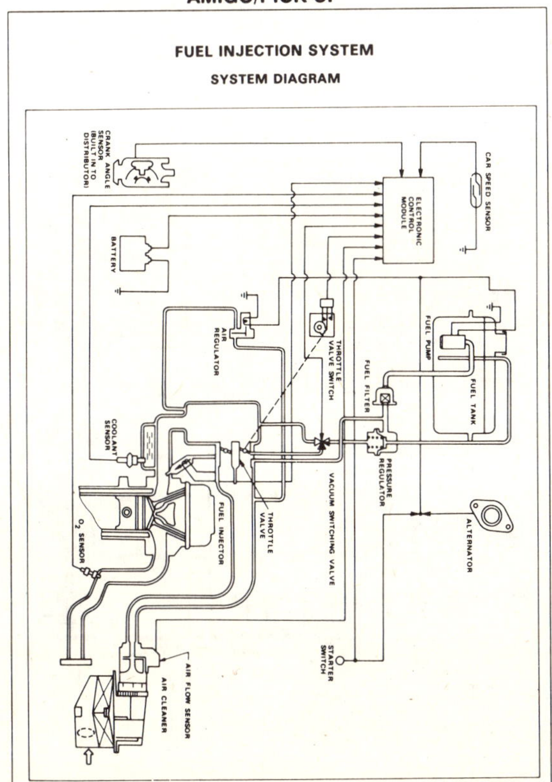


## 1988-89 FEDERAL TROOPER/TROOPER II (EXCEPT 2.8L ENGINE), AMIGO/PICK-UP

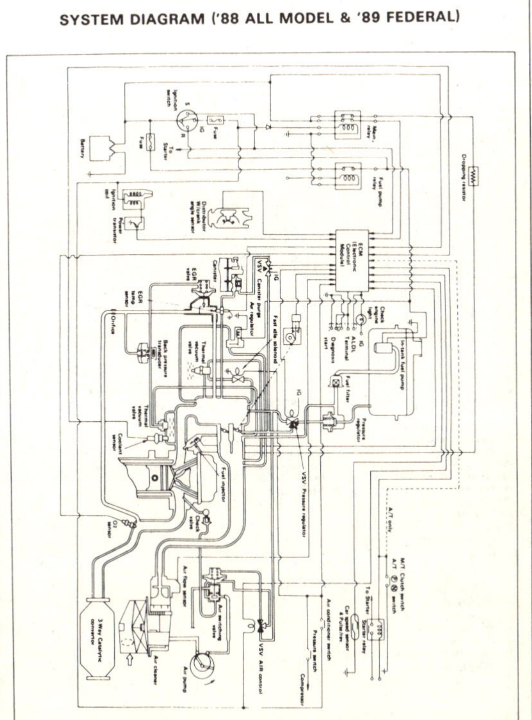

## 1989 CALIFORNIA TROOPER/TROOPER II (EXCEPT 2.8L ENGINE), AMIGO/PICK-UP

### GENERAL DESCRIPTION

SYSTEM DIAGRAM ('89 CALIFORNIA)

## 1988-89 FEDERAL TROOPER/TROOPER II (EXCEPT 2.8L ENGINE), AMIGO/PICK-UP

VACUUM HOSE ROUTING DIAGRAM ('88 ALL MODEL & '89 FEDERAL)

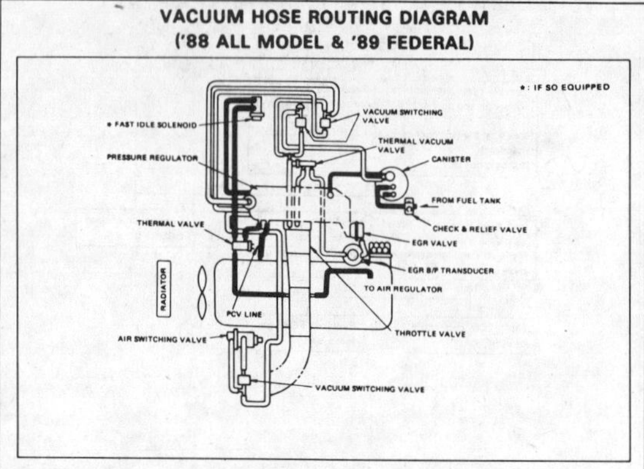


## 1989 CALIFORNIA TROOPER/TROOPER II (EXCEPT 2.8L ENGINE), AMIGO/PICK-UP

VACUUM HOSE ROUTING DIAGRAM ('89 CALIFORNIA)

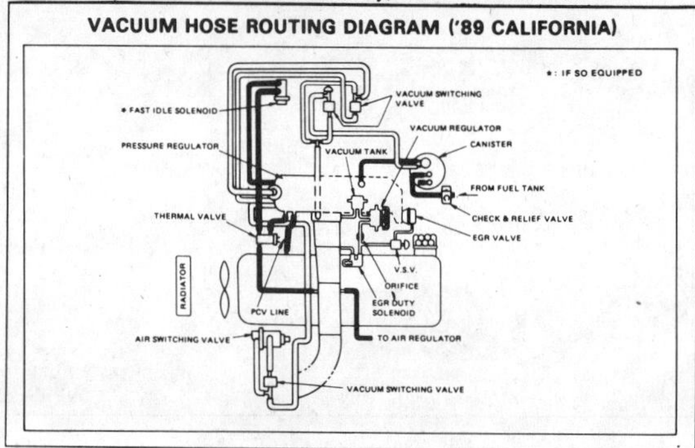


## TROOPER/TROOPER II WITH 2.8L ENGINE

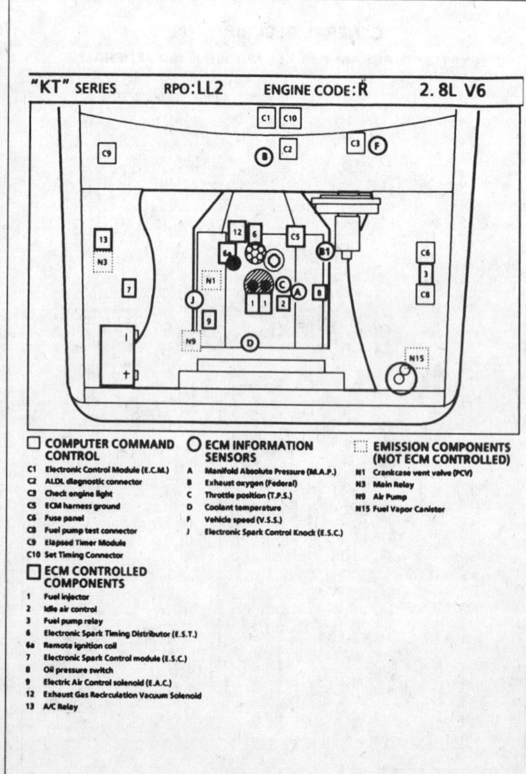


## TROOPER/TROOPER II WITH 2.8L ENGINE

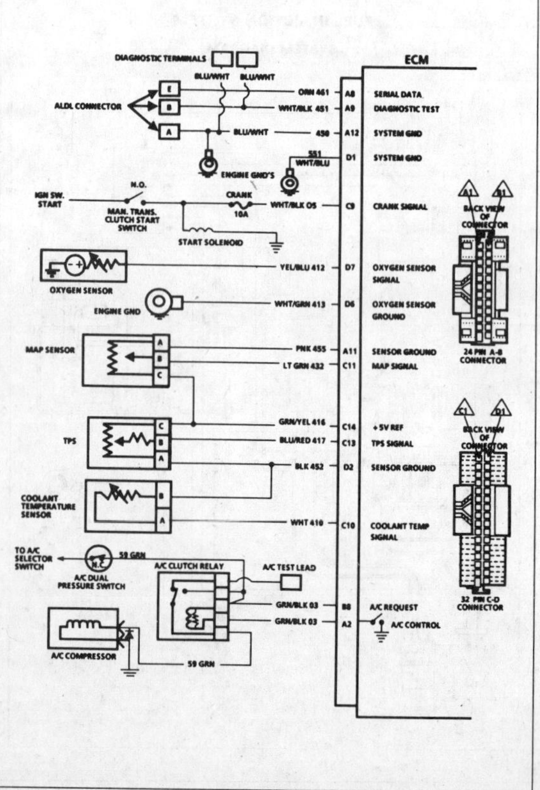

## TROOPER/TROOPER II WITH 2.8L ENGINE

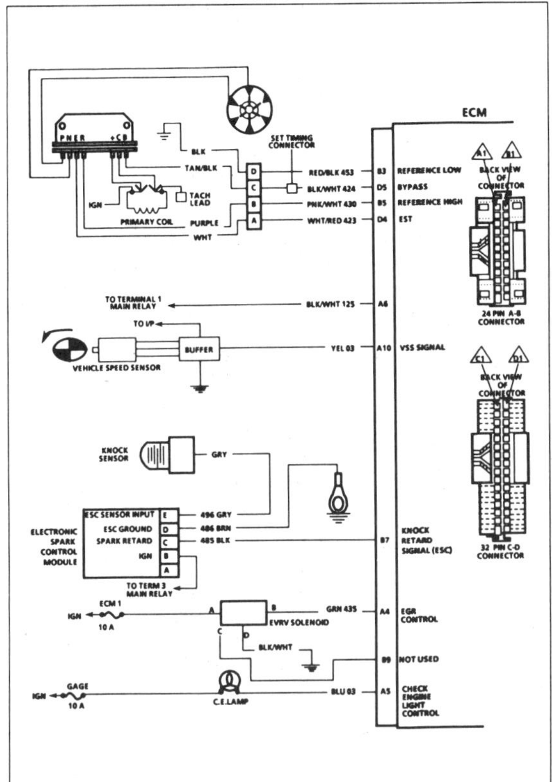


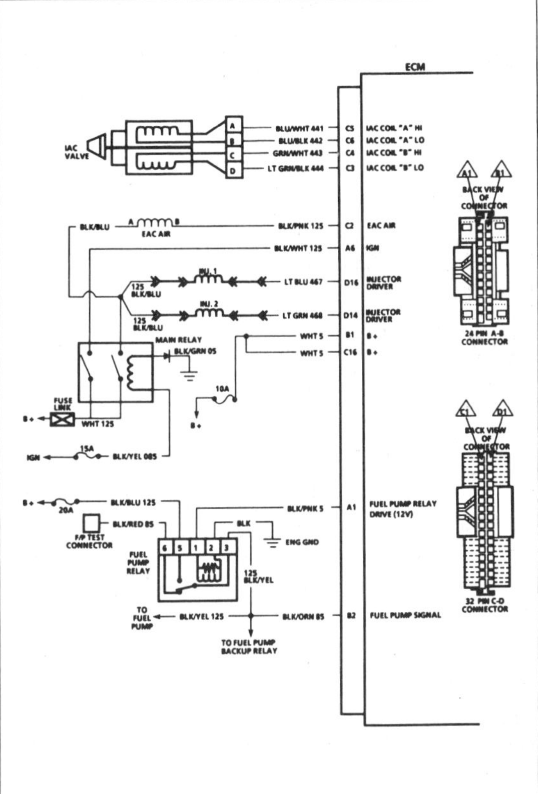


## TROOPER/TROOPER II WITH 2.8L ENGINE

### TBI FUEL INJECTION ECM CONNECTOR IDENTIFICATION

This ECM voltage chart is for use with a digital voltmeter to further aid in diagnosis. The voltages you get may vary due to low battery charge or other reasons, but they should be very close.

THE FOLLOWING CONDITIONS MUST BE MET BEFORE TESTING:

● Engine at operating temperature ● Engine idling in closed loop (for "Engine Run" column)
● Test terminal not grounded ● ALCL tool not installed

| Note | VOLTAGE KEY "ON" | VOLTAGE ENG. RUN | CIRCUIT | PIN | WIRE COLOR |
|---|---|---|---|---|---|
| ④ | 0 | 14 | FUEL PUMP RELAY | A1 | BLK/PNK |
| ⑦ | 12 | 14 | A/C CLUTCH CONTROL | A2 | GRN/BLK |
| | | | NOT USED | A3 | |
| | 12 | 14 | EGR CONTROL | A4 | GRN |
| | 0 | 14 | "CHECK ENGINE" LIGHT | A5 | BLU |
| | 12 | 14 | IGN - ECM FUSE | A6 | BLK/WHT |
| | | | | A7 | |
| | 2-5 | 2.5 | SERIAL DATA | A8 | ORN |
| | 5 | 5 | DIAG. TERM. | A9 | WHT/BLU |
| | ① | ① | SPEED SENSOR SIGNAL | A10 | YELLOW |
| | 0 | 0 | MAP GROUND | A11 | PNK |
| | 0 | 0 | SYSTEM GROUND | A12 | WHT/BLK |
| | | | NOT USED | C1 | |
| ⑨ | 12 | 14 | EAC SOL. | C2 | BLK/PNK |
| | NOT USEABLE | | IAC "B" LO | C3 | LT GRN/BLK |
| | NOT USEABLE | | IAC "B" HI | C4 | GRN/WHT |
| | NOT USEABLE | | IAC "A" HI | C5 | BLU/WHT |
| | NOT USEABLE | | IAC "A" LO | C6 | BLU/BLK |
| | | | NOT USED | C7 | |
| | | | NOT USED | C8 | |
| | 0 | 0 | CRANK SIGNAL | C9 | WHT/BLK |
| ② | 1.9 | 1.7 | COOLANT TEMP. SIGNAL | C10 | WHT |
| | 4.9 | 2.0 | MAP SIGNAL | C11 | LT GRN |
| | | | NOT USED | C12 | |
| ⑤ | .48 | .48 | TPS SIGNAL | C13 | BLU/RED |
| | 5 | 5 | 5 VOLT REFERENCE | C14 | GRN/YEL |
| | | | NOT USED | C15 | |
| | 12 | 14 | BATTERY | C16 | WHT |

24 PIN A-B CONNECTOR — BACK VIEW OF CONNECTOR (A1, B1)

32 PIN C-D CONNECTOR — BACK VIEW OF CONNECTOR (C1, D1)

| WIRE COLOR | PIN | CIRCUIT | VOLTAGE KEY "ON" | VOLTAGE ENG. RUN | Note |
|---|---|---|---|---|---|
| WHT | B1 | BATTERY | 12 | 14 | |
| BLK/ORN | B2 | FUEL PUMP SIGNAL | 0 | 14 | ④ |
| RED/BLK | B3 | EST REF LOW | 0 | 0 | |
| | B4 | NOT USED | | | |
| PNK/WHT | B5 | EST REFERENCE HI | 0 | 1.6 | |
| | B6 | NOT USED | | | |
| BLK | B7 | ESC SIGNAL | 9 | 9 | |
| GRN/BLU | B8 | A/C SIGNAL | 0 | 0/12 | ⑧ |
| BRN | B9 | NOT USED | | | |
| | B10 | NOT USED | | | |
| | B11 | NOT USED | | | |
| | B12 | NOT USED | | | |
| WHT/BLU | D1 | SYSTEM GROUND | 0 | 0 | |
| BLK | D2 | SV RETURN | 0 | 0 | |
| | D3 | NOT USED | | | |
| WHT/RED | D4 | EST CONTROL | 0 | 1.0 | |
| BLK/WHT | D5 | EST BYPASS | 0 | 4.75 | |
| WHT/GRN | D6 | SENSOR GRD. ($O_2$) | 0 | 0 | |
| YEL/BLU | D7 | $O^2$ SENSOR SIGNAL | 0 | ③ | |
| | D8 | NOT USED | | | |
| | D9 | NOT USED | | | |
| | D10 | NOT USED | | | |
| | D11 | NOT USED | | | |
| | D12 | NOT USED | | | |
| | D13 | NOT USED | | | |
| GRN | D14 | INJECTOR B | 12 | 14 | |
| | D15 | NOT USED | | | |
| BLU | D16 | INJECTOR A | 12 | 14 | |

ENGINE 2.8L LL2

1 Varies from .60 to battery voltage depending on position of drive wheels.
2 Varies with temperature.
3 Varies. (toggles)
4 12V First two seconds.
5 Measured between terminals C13 and D2 (± .06V).
6 12 V in gear
7 0 V with A/C "on"
8 12 V with A/C "on"
9 12 V when cold

### DIAGNOSTIC CIRCUIT CHECK

The Diagnostic Circuit Check is an organized approach to identifying a problem created by an electronic engine control system malfunction. It must be the starting point for any driveability complaint diagnosis, because it directs the service technician to the next logical step in diagnosing the complaint.

The "Scan Data" listed in the table may be used for comparison, after completing the diagnostic circuit check and finding the on-board diagnostics functioning properly and no trouble codes displayed. The "Typical Values" are an average of display values recorded from normally operating vehicles and are intended to represent what a normally functioning system would typically display.

**A "SCAN" TOOL THAT DISPLAYS FAULTY DATA SHOULD NOT BE USED, AND THE PROBLEM SHOULD BE REPORTED TO THE MANUFACTURER. THE USE OF A FAULTY "SCAN" CAN RESULT IN MISDIAGNOSIS AND UNNECESSARY PARTS REPLACEMENT.**

Only the parameters listed below are used in this manual for diagnosis. If a "Scan" tool reads other parameters, the values are not recommended by Isuzu for use in diagnosis

### "SCAN" DATA

Idle / Upper Radiator Hose Hot / Closed Throttle / Park or Neutral / Closed Loop / Acc. off

| "SCAN" Position | Units Displayed | Typical Data Value |
|---|---|---|
| Coolant Temp. | C° | 85° - 105° |
| TPS | Volts | .48 + - .06 |
| MAP | Volts | 1 - 2 (depends on Vac. & Baro pressure) |
| INT (Integrator) | Counts | Varies |
| BLM (Block Learn) | Counts | 118 - 138 |
| IAC | Counts (steps) | 1 - 50 |
| RPM | RPM | 650 ± 100 RPM |
| $O_2$ | Volts | .001 - 999 and varies |
| Open/Closed Loop | Open/Closed | Closed Loop (will go open at idle) |
| A/C Request | Yes/No | No (yes, with A/C requested) |
| VSS | MPH | 0 |
| Battery | Volts | 13.5 - 14.5 |
| Air Divert | Normal/Divert | Normal |
| Air Switch | Converter/Port | Converter |
| Knock Signal | Yes/No | No |

### TROOPER/TROOPER II WITH 2.8L ENGINE

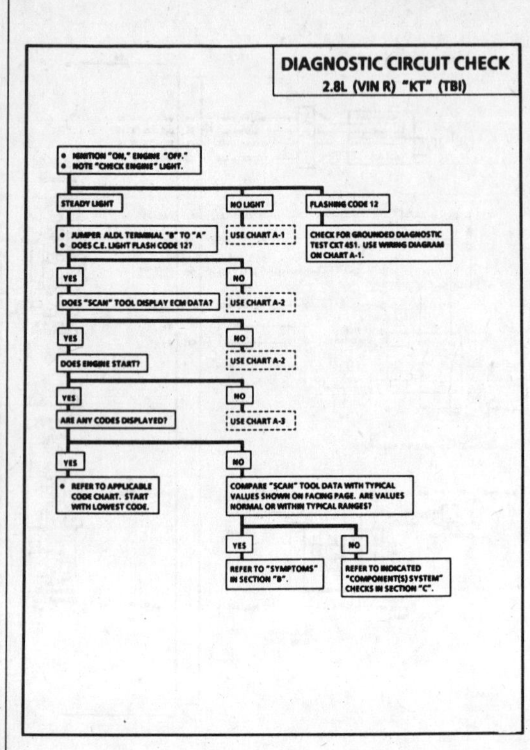


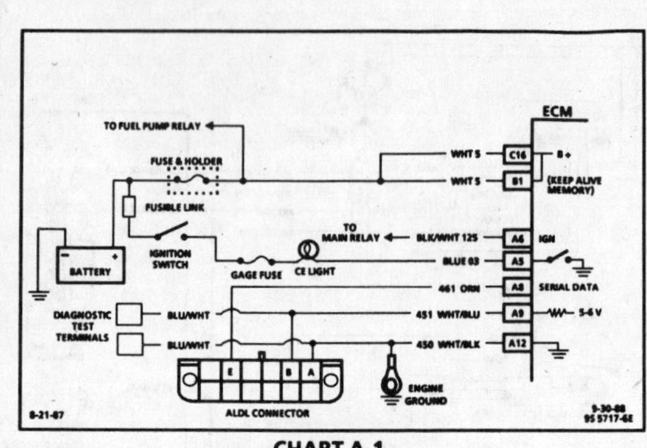


### CHART A-1
#### NO "CHECK ENGINE" LIGHT
#### 2.8L (VIN R) "KT" (TBI)

**Circuit Description:**

There should always be a steady "Check Engine" light, when the ignition is "ON" and engine stopped. Ignition voltage is supplied directly to the light bulb. The electronic control module (ECM) will control the light and turn it "ON" by providing a ground path through CKT 03 to the ECM.

**Test Description:** Numbers below refer to circled numbers on the diagnostic chart.

1. Battery feed CKT 5 is protected by a 10A in-line fuse. If this fuse was blown, refer to wiring diagram on the facing page of Code 54.
2. Using a test light connected to 12 volts, probe each of the system ground circuits to be sure a good ground is present. See ECM terminal end view in front of this section for ECM pin locations of ground circuits.

**Diagnostic Aids:**

Engine runs OK, check:

- Faulty light bulb
- CKT 03 open
- Gage fuse blown. This will result in no oil, or generator lights, seat belt reminder, etc.

Engine cranks, but will not run.

- Continuous battery - fuse or fusible link open
- ECM ignition fuse open
- Battery CKT 5 to ECM open
- Ignition CKT 125 to ECM open
- Poor connection to ECM

### TROOPER/TROOPER II WITH 2.8L ENGINE

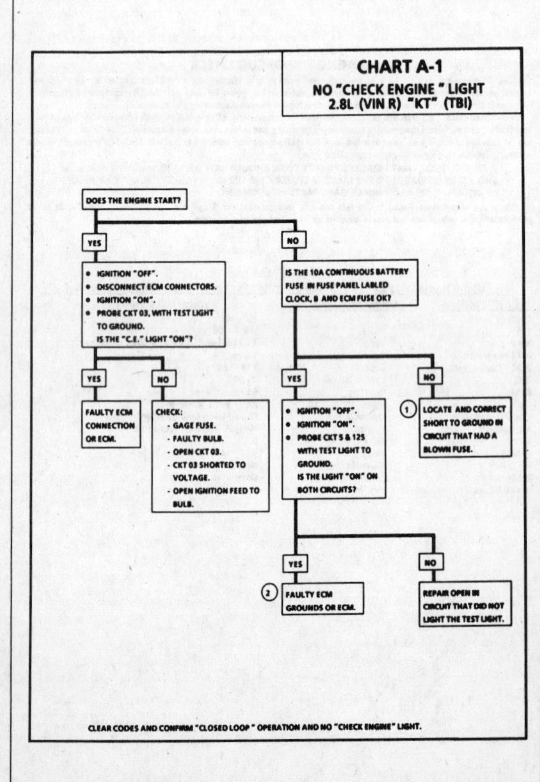


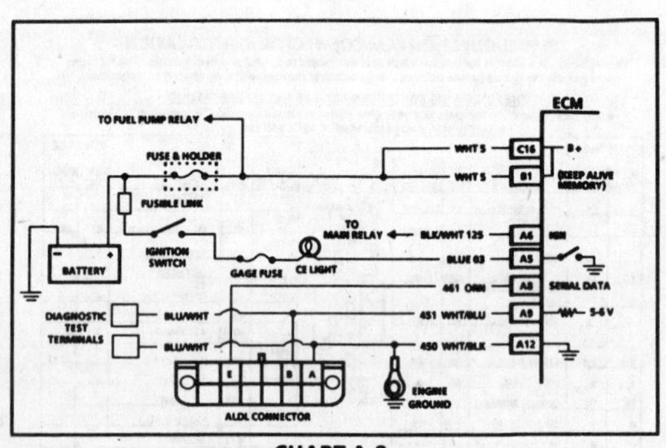


### CHART A-2
#### NO ALDL DATA OR WON'T FLASH CODE 12
#### "CHECK ENGINE" LIGHT "ON" STEADY
#### 2.8L (VIN R) "KT" (TBI)

**Circuit Description:**

There should always be a steady "Check Engine" light, when the ignition is "ON" and engine stopped. Ignition voltage is supplied directly to the light bulb. The electronic control module (ECM) will turn the light "ON" by grounding CKT 03 at the ECM.

With the diagnostic terminal grounded, the light should flash a Code 12, followed by any trouble code(s) stored in memory.

A steady light suggests a short to ground in the light control CKT 03, or an open in diagnostic CKT 451.

**Test Description:** Numbers below refer to circled numbers on the diagnostic chart.

1. If there is a problem with the ECM that causes a "Scan" tool to not read serial data, then the ECM should not flash a Code 12. If Code 12 does flash, be sure that the "Scan" tool is working properly on another vehicle. If the "Scan" is functioning properly and CKT 461 is OK, the Mem-Cal or ECM may be at fault for the NO ALDL symptom.
2. If the light goes "OFF," when the ECM connector is disconnected, then CKT 03 is not shorted to ground.
3. This step will check for an open diagnostic CKT 451.
4. At this point, the "Check Engine" light wiring is OK. The problem is a faulty ECM or PROM. If Code 12 does not flash, the ECM should be replaced using the original PROM. Replace the PROM only after trying an ECM, as a defective PROM is an unlikely cause of the problem.

## TROOPER/TROOPER II WITH 2.8L ENGINE

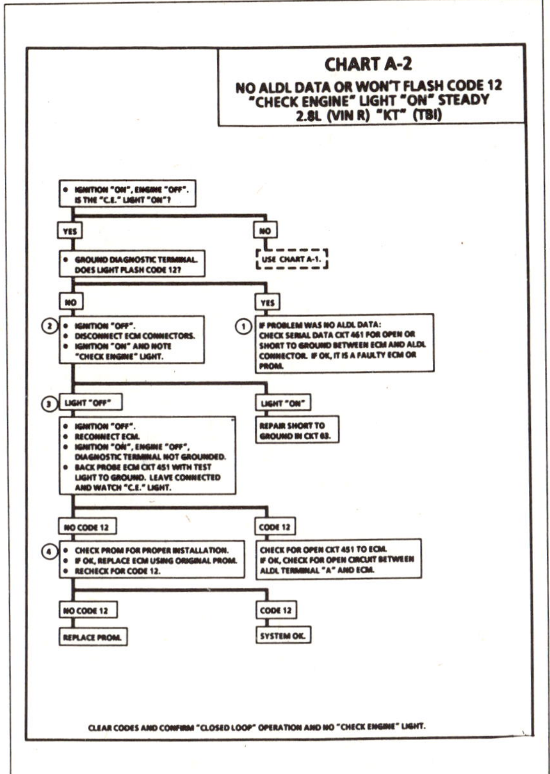


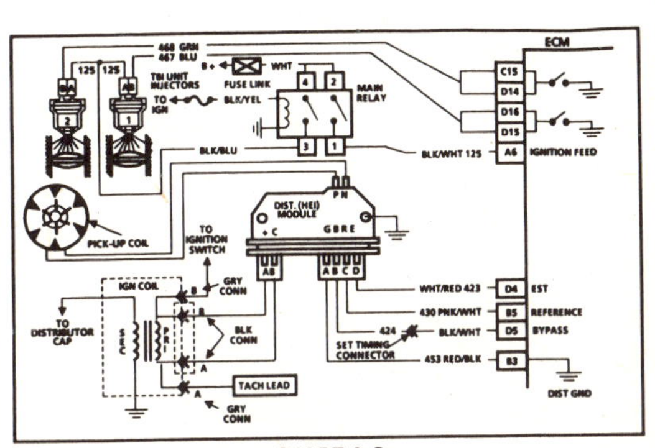


### CHART A-3

(Page 2 of 2)
ENGINE CRANKS BUT WILL NOT RUN
2.8L (VIN R) "KT" (TBI)

**Circuit Description:**

This chart assumes that battery condition and engine cranking speed are OK, and there is adequate fuel in the tank.

**Test Description:** Numbers below refer to circled numbers on the diagnostic chart.

1. No fuel spray from one injector indicates a faulty fuel injector or no ECM control of injector. If the test light "blinks" while cranking, then ECM control should be considered OK. Be sure test light makes good contact between connector terminals during test. The light may be a little dim when "blinking." This is due to current draw of the test light. How bright it "blinks" is not important. The test light bulb should be a J 36400-1 or equivalent.
2. CKTs BLK/BLU 125 supply ignition voltage to the injectors. Probe each connector terminal with a test light to ground. There should be a light "ON" at one terminal. If the test light confirms ignition voltage at the connector, the ECM injector control CKT 467 or 468 may be open. Reconnect the injector and using a test light connected to ground, check for a light at the applicable ECM connector terminal "D14" or "D16". A light at this point indicates that the injector drive circuit involved is OK.
If an ECM repeat failure has occurred, the injector is shorted. Replace the injector and ECM.

## TROOPER/TROOPER II WITH 2.8L ENGINE

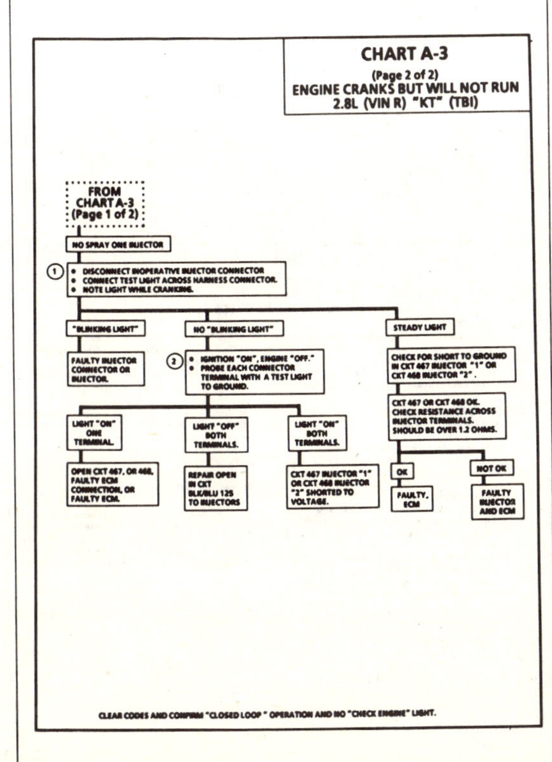


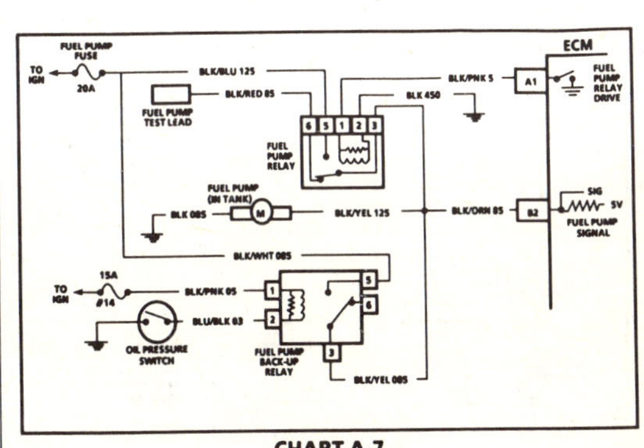


### CHART A-7

(Page 1 of 2)
FUEL SYSTEM DIAGNOSIS
2.8L (VIN R) "KT" (TBI)

**Circuit Description:**

When the ignition switch is turned "ON," the electronic control module (ECM) will turn "ON" the in-tank fuel pump. It will remain "ON" as long as the engine is cranking or running, and the ECM is receiving ignition reference pulses.

If there are no reference pulses, the ECM will shut "OFF" the fuel pump within 2 seconds after key "ON."

The pump will deliver fuel to the TBI unit, where the system pressure is controlled to 62 to 90 kPa (9 to 13 psi). Excess fuel is then returned to the fuel tank.

The fuel pump test terminal is located in the left side of the engine compartment. When the engine is stopped, the pump can be turned "ON" by applying battery voltage to the test terminal.

**Test Description:** Numbers below refer to circled numbers on the diagnostic chart.

1. Fuel pressure should be noted while fuel pump is running. Fuel pressure will drop immediately after fuel pump stops running due to a controlled bleed in the fuel system.

**Diagnostic Aids:**

Improper fuel system pressure can result in one of the following symptoms:

- Cranks, but won't run.
- Code 44.
- Code 45.
- Cuts out, may feel like ignition problem.
- Poor fuel economy, loss of power.
- Hesitation.

### TROOPER/TROOPER II WITH 2.8L ENGINE

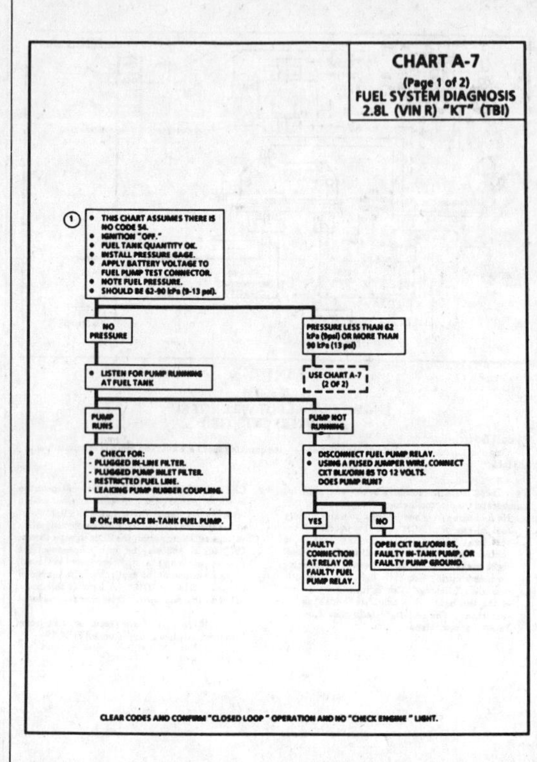


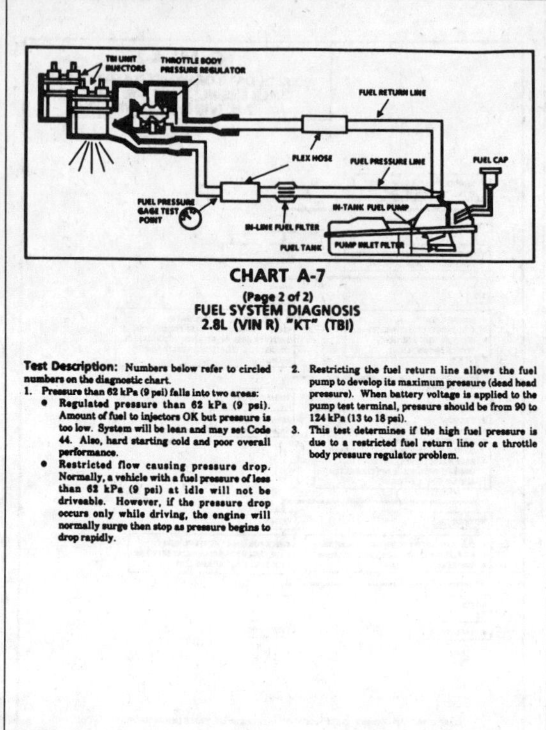


### CHART A-7

(Page 2 of 2)
FUEL SYSTEM DIAGNOSIS
2.8L (VIN R) "KT" (TBI)

**Test Description:** Numbers below refer to circled numbers on the diagnostic chart.

1. Pressure than 62 kPa (9 psi) falls into two areas:
   - Regulated pressure than 62 kPa (9 psi). Amount of fuel to injectors OK but pressure is too low. System will be lean and may set Code 44. Also, hard starting cold and poor overall performance.
   - Restricted flow causing pressure drop. Normally, a vehicle with a fuel pressure of less than 62 kPa (9 psi) at idle will not be driveable. However, if the pressure drop occurs only while driving, the engine will normally surge then stop as pressure begins to drop rapidly.
2. Restricting the fuel return line allows the fuel pump to develop its maximum pressure (dead head pressure). When battery voltage is applied to the pump test terminal, pressure should be from 90 to 124 kPa (13 to 18 psi).
3. This test determines if the high fuel pressure is due to a restricted fuel return line or a throttle body pressure regulator problem.

### TROOPER/TROOPER II WITH 2.8L ENGINE

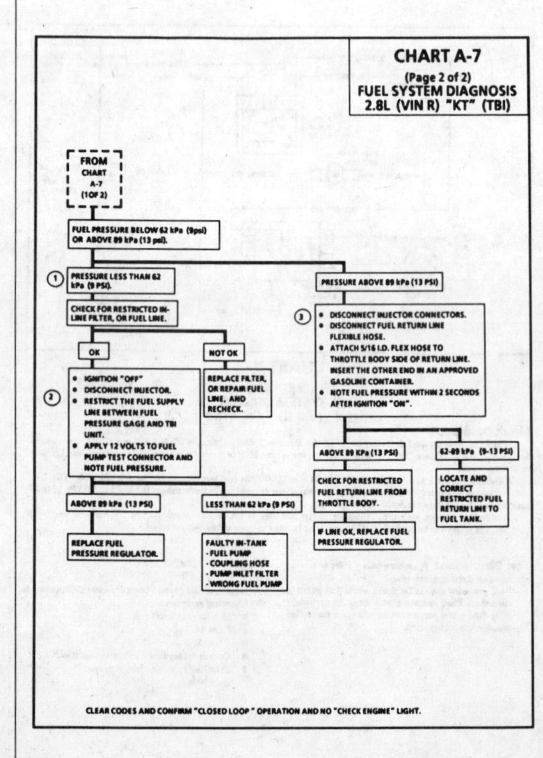


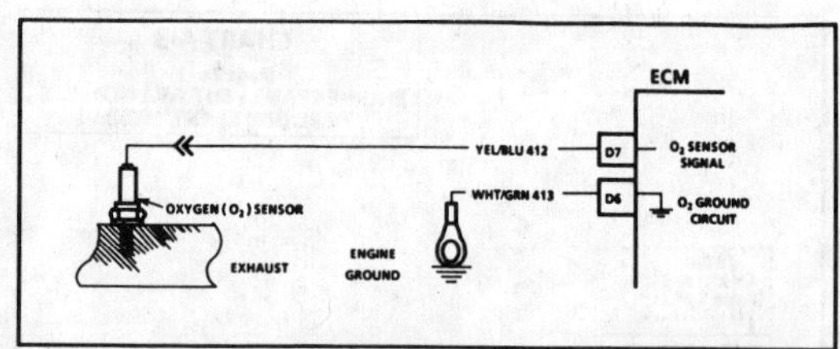


### CODE 13

OXYGEN SENSOR CIRCUIT
(OPEN CIRCUIT)
2.8L (VIN R) "KT" (TBI)

**Circuit Description:**

The ECM supplies a voltage of about .45 volt between terminals "D7" and "D6". (If measured with a 10 megohm digital voltmeter, this may read as low as .32 volt). The $O_2$ sensor varies the voltage within a range of about 1 volt if the exhaust is rich, down through about .10 volt if exhaust is lean.

The sensor is like an open circuit and produces no voltage when it is below 315°C (600°F). An open sensor circuit or cold sensor causes "Open Loop" operation.

**Test Description:** Numbers below refer to circled numbers on the diagnostic chart.

1. Code 13 will set:
   - Engine at normal operating temperature.
   - At least 2 minutes engine time after start.
   - $O_2$ signal voltage steady between .35 and .55 volt.
   - rpm above 1600.
   - Throttle position sensor signal above 5% (about .3 volt above closed throttle voltage).
   - All conditions must be met for about 60 seconds.

   If the conditions for a Code 13 exist, the system will not go "Closed Loop."
2. This will determine if the sensor is at fault or the wiring or ECM is the cause of the Code 13.
3. In doing this test, use only a high impedance digital volt ohmmeter. This test checks the continuity of CKT(s) 412 and 413. If CKT 413 is open, the ECM voltage on CKT 412 will be over .6 volt (600 mV).

**Diagnostic Aids:**

Normal "Scan" voltage varies between 100 mv to 999 mV (.1 and 1.0 volt), while in "Closed Loop." Code 13 sets in one minute, if voltage remains between .35 and .55 volt, but the system will go "Open Loop" in about 15 seconds.

## TROOPER/TROOPER II WITH 2.8L ENGINE

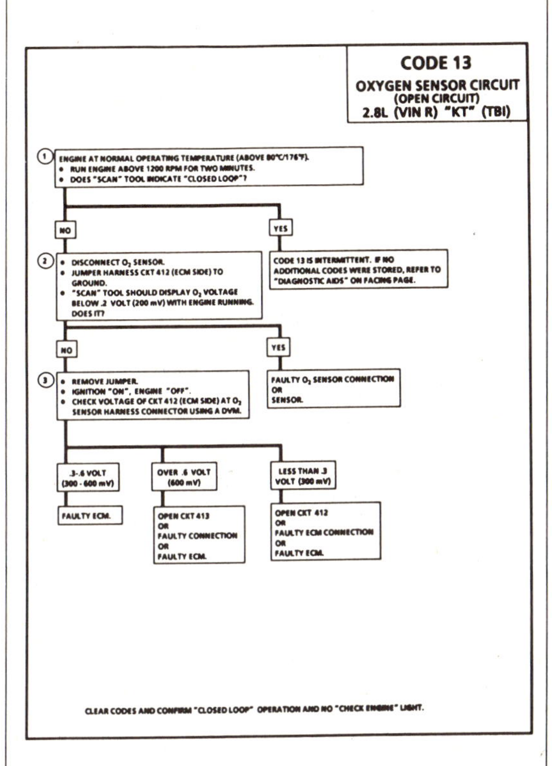


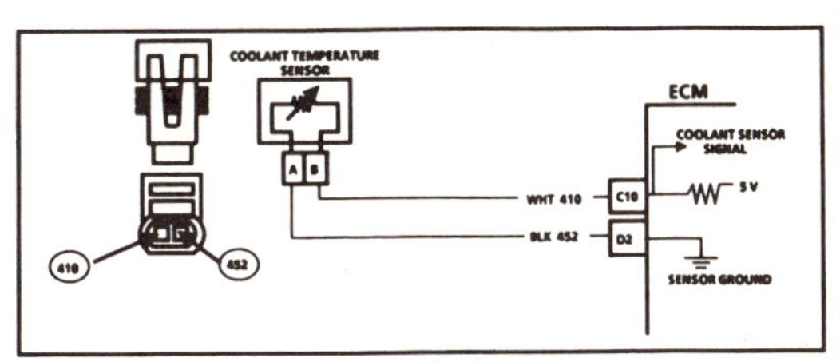


### CODE 14

### COOLANT TEMPERATURE SENSOR CIRCUIT (HIGH TEMPERATURE INDICATED) 2.8L (VIN R) "KT" (TBI)

**Circuit Description:**

The coolant temperature sensor uses a thermistor to control the signal voltage to the ECM. The ECM applies a voltage on CKT 410 to the sensor. When the engine is cold the sensor (thermistor) resistance is high, therefore the ECM will see high signal voltage.

As the engine warms, the sensor resistance becomes less, and the voltage drops. At normal engine operating temperature (85°C to 95°C) the voltage will measure about 1.5 to 2.0 volts.

**Test Description:** Numbers below refer to circled numbers on the diagnostic chart.

1. Code 14 will set if:
   - Signal voltage indicates a coolant temperature above 135°C (275°F) for 2 seconds.
2. This test will determine if CKT 410 is shorted to ground which will cause the conditions for Code 14.

**Diagnostic Aids:**

Check harness routing for a potential short to ground in CKT 410.

"Scan" tool displays engine temp. in degrees centigrade. After engine is started, the temperature should rise steadily to about 90°C then stabilize when thermostat opens.

## TROOPER/TROOPER II WITH 2.8L ENGINE

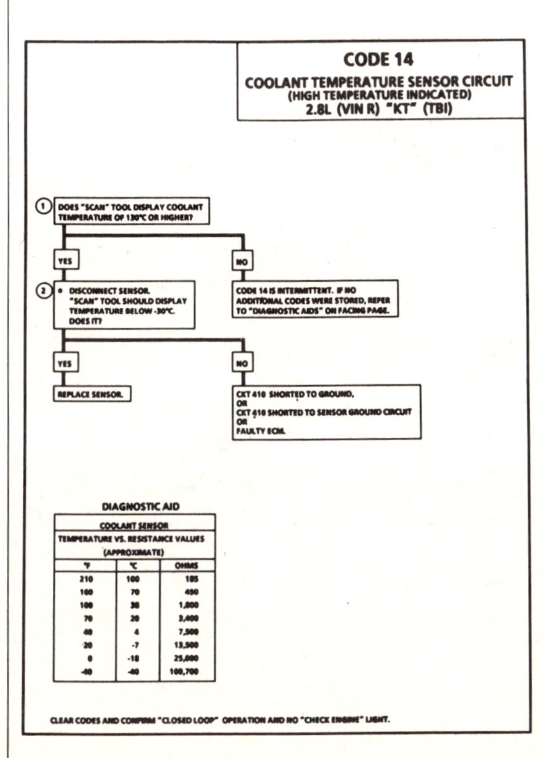


COOLANT SENSOR
TEMPERATURE VS. RESISTANCE VALUES
(APPROXIMATE)

| °F | °C | OHMS |
|---|---|---|
| 210 | 100 | 185 |
| 160 | 70 | 450 |
| 100 | 38 | 1,800 |
| 70 | 20 | 3,400 |
| 40 | 4 | 7,500 |
| 20 | -7 | 13,500 |
| 0 | -18 | 25,000 |
| -40 | -40 | 100,700 |

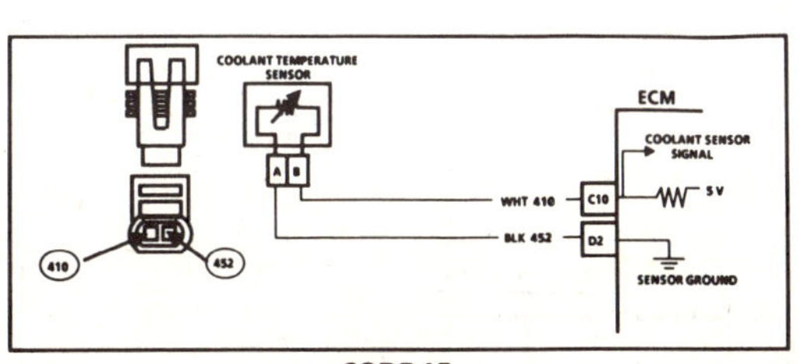


### CODE 15

### COOLANT TEMPERATURE SENSOR CIRCUIT (LOW TEMPERATURE INDICATED) 2.8L (VIN R) "KT" (TBI)

**Circuit Description:**

The coolant temperature sensor uses a thermistor to control the signal voltage to the ECM. The ECM applies a voltage on CKT 410 to the sensor. When the engine is cold the sensor (thermistor) resistance is high, therefore the ECM will see high signal voltage.

As the engine warms, the sensor resistance becomes less, and the voltage drops. At normal engine operating temperature (85°C to 95°C) the voltage will measure about 1.5 to 2.0 volts at the ECM.

**Test Description:** Numbers below refer to circled numbers on the diagnostic chart.

1. Code 15 will set if:
   - Engine running longer than 30 seconds.
   - Coolant temp. less than -30°C (-22°F), for 3 seconds.
2. This test simulates a Code 14. If the ECM recognizes the low signal voltage, (high temperature) and "Scan" reads 130°C or above, the ECM and wiring are OK.
3. This test will determine if CKT 410 is open. There should be 5 volts present at sensor connector if measured with a DVM.

**Diagnostic Aids:**

A "Scan" tool reads engine temperature in degrees centigrade. After engine is started the temperature should rise steadily to about 90°C then stabilize when thermostat opens.

If Code 21 is also set, check CKT 452 for faulty wiring or connections. Check terminals at sensor for good contact.

## TROOPER/TROOPER II WITH 2.8L ENGINE

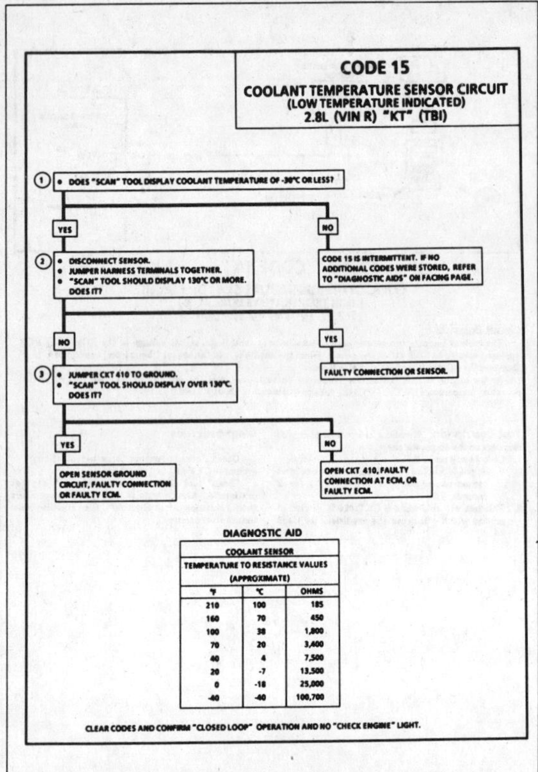


| COOLANT SENSOR | | |
|---|---|---|
| TEMPERATURE TO RESISTANCE VALUES (APPROXIMATE) | | |
| °F | °C | OHMS |
| 210 | 100 | 185 |
| 160 | 70 | 450 |
| 100 | 38 | 1,800 |
| 70 | 20 | 3,400 |
| 40 | 4 | 7,500 |
| 20 | -7 | 13,500 |
| 0 | -18 | 25,000 |
| -40 | -40 | 100,700 |

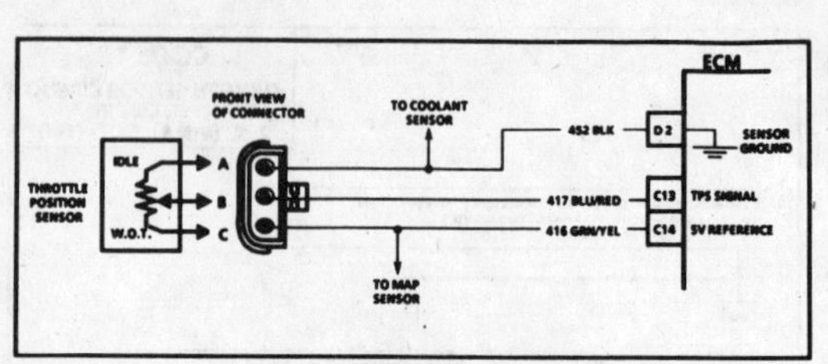


### CODE 21
### THROTTLE POSITION SENSOR (TPS) CIRCUIT
### (SIGNAL VOLTAGE HIGH)
### 2.8L (VIN R) "KT" (TBI)

**Circuit Description:**

The throttle position sensor (TPS) provides a voltage signal that changes relative to the throttle blade. Signal voltage will vary from about .5 at idle to about 5 volts at wide open throttle.

The TPS signal is one of the most important inputs used by the ECM for fuel control and for most of the ECM control outputs.

**Test Description:** Numbers below refer to circled numbers on the diagnostic chart.

1. Code 21 will set if:
   - TPS signal voltage is greater than 2.5 volts.
   - All conditions met for 8 seconds.
   - MAP less than 52 kPa.
2. With the TPS sensor disconnected, the TPS voltage should go low if the ECM and wiring are OK.
3. Probing CKT 452 with a test light to 12 volts checks the sensor ground CKT. A faulty sensor ground will cause a Code 21.

**Diagnostic Aids:**

A "Scan" tool reads throttle position in volts. Should read less than 1.25 volts with throttle closed and ignition "ON" or at idle. Voltage should increase at a steady rate as throttle is moved toward WOT.

An open in CKT 452 will result in a Code 21.

## TROOPER/TROOPER II WITH 2.8L ENGINE

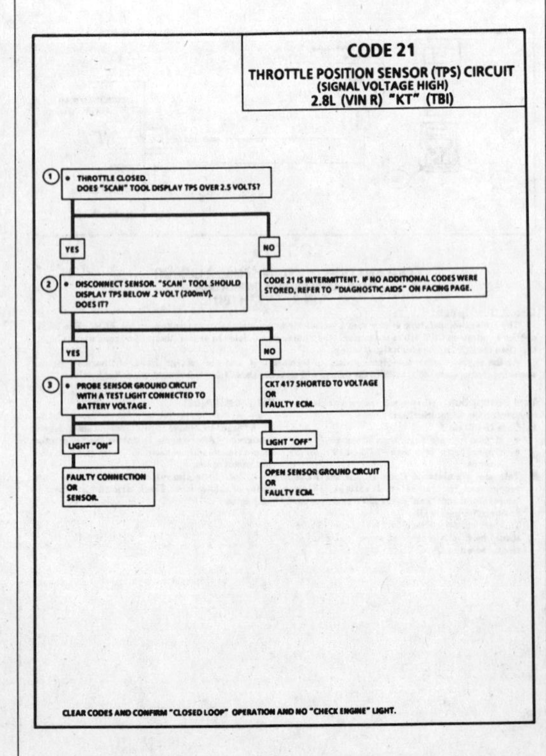


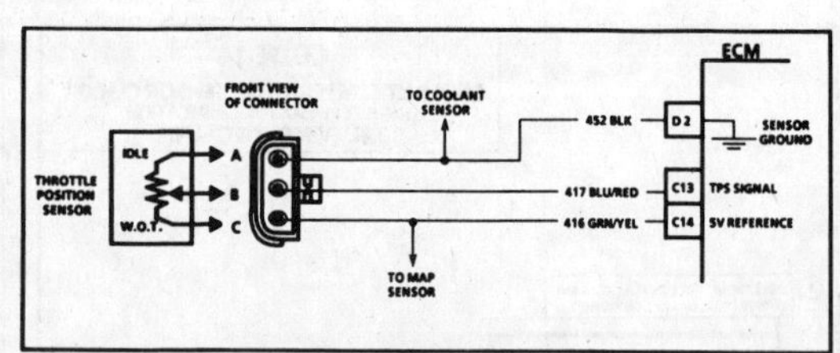


### CODE 22
### THROTTLE POSITION SENSOR (TPS) CIRCUIT
### (SIGNAL VOLTAGE LOW)
### 2.8L (VIN R) "KT" (TBI)

**Circuit Description:**

The throttle position sensor (TPS) provides a voltage signal that changes relative to the throttle blade. Signal voltage will vary from about .5 at idle to about 5 volts at wide open throttle.

The TPS signal is one of the most important inputs used by the ECM for fuel control and for most of the ECM control outputs.

**Test Description:** Numbers below refer to circled numbers on the diagnostic chart.

1. Code 22 will set if:
   - Engine is running
   - TPS signal voltage is less than about .2 volt for 2 seconds.
2. Simulates Code 21: (high voltage) If the ECM recognizes the high signal voltage then the ECM and wiring are OK.
3. This simulates a high signal voltage to check for an open in CKT 417. The "Scan" tool will not read up to 12 volts, but what is important is that the ECM recognizes the signal on CKT 417.
4. There should be 5 volts at terminal "C" if measured with a DVM when ignition is "ON."

**Diagnostic Aids:**

A "Scan" tool reads throttle position in volts. Should read less than 1.25 volts with throttle closed and ignition "ON" or at idle. Voltage should increase at a steady rate as throttle is moved toward WOT.

An open or short to ground in CKTs 416 or 417 will result in a Code 22.

If a Code 22 is also set check CKT 416 carefully for open or short to ground.

## TROOPER/TROOPER II WITH 2.8L ENGINE

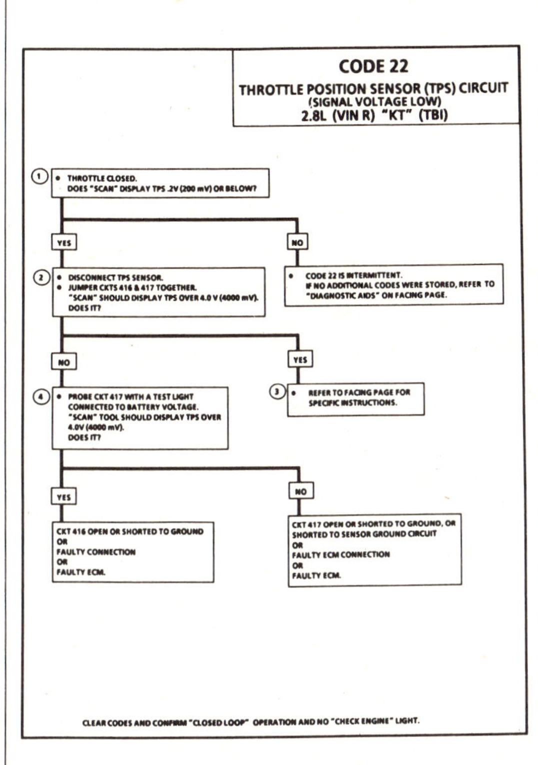


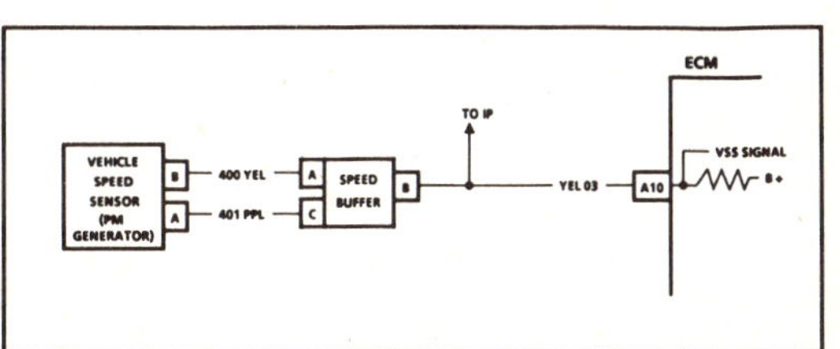


### CODE 24

#### VEHICLE SPEED SENSOR (VSS) CIRCUIT
#### 2.8L (VIN R) "KT" (TBI)

**Circuit Description:**

The ECM applies and monitors 12 volts on CKT yellow 03. CKT yellow 03 connects to the vehicle speed sensor which alternately grounds CKT yellow 03 when drive wheels are turning. This pulsing action takes place about 2000 times per mile and the ECM will calculate vehicle speed based on the time between "pulses."

"Scan" reading should closely match with speedometer reading with drive wheels turning. Disregard a Code 24 set when drive wheels are not turning.

**Test Description:** Numbers below refer to circled numbers on the diagnostic chart.

1. Code 24 will set if vehicle speed equals 0 mph when:
   - Engine speed is between 1200 and 4400 rpm
   - TPS is less than 2% (closed throttle)
   - Low load condition (high vacuum). Less than 25 kPa.
   - All conditions met for 5 seconds.

   These conditions are met during a road load deceleration. Disregard Code 24 that sets when drive wheels are not turning.
2. 8-12 volts, at the I/P connector, indicates CKT yellow 03 is open between the I/P connector and the VSS, or there is a faulty vehicle speed sensor. A voltage of less than 1 volt, at the I/P connector, indicates that CKT yellow 03 wire is shorted to ground. If, after disconnecting CKT yellow 03 at the vehicle speed sensor, the voltage reads above 10 volts, the vehicle speed sensor is faulty. If voltage remains less than 8 volts, then CKT yellow 03 wire is grounded. If yellow 03 is not grounded, there is a faulty connection at the ECM, or a faulty ECM.

**Diagnostic Aids:**

"Scan" should indicate a vehicle speed whenever the drive wheels are turning.

## TROOPER/TROOPER II WITH 2.8L ENGINE

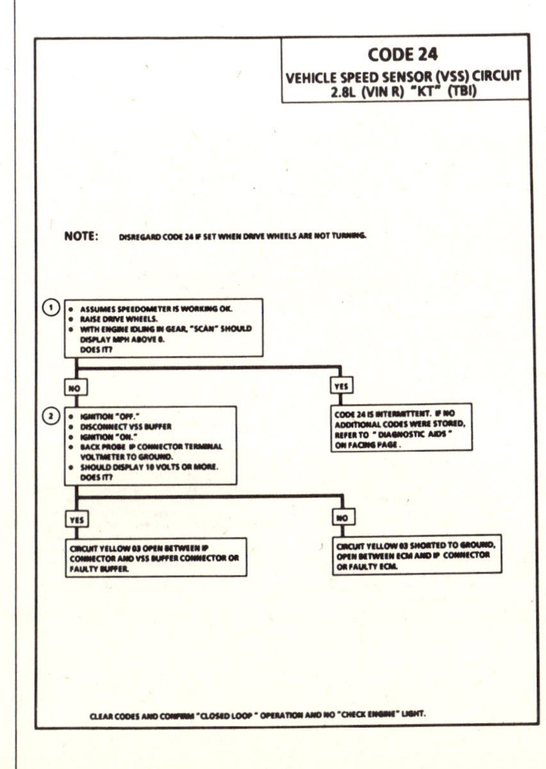


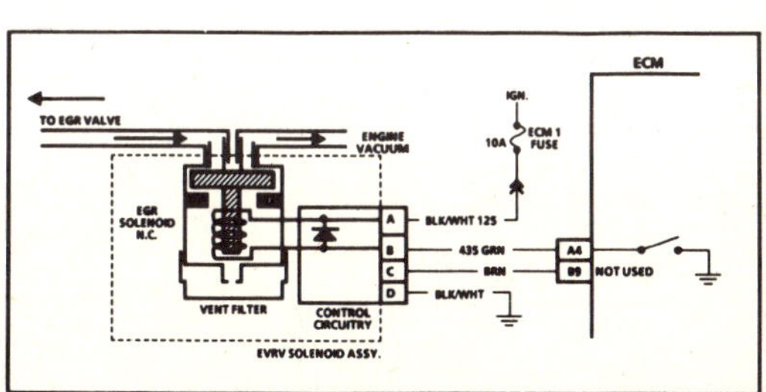


### CODE 32

#### EGR SYSTEM FAILURE
#### 2.8L (VIN R) "KT" (TBI)

**Circuit Description:**

The ECM operates a solenoid to control the exhaust gas recirculation (EGR) valve. This solenoid is normally closed. By providing a ground path, the ECM energizes the solenoid which then allows vacuum to pass to the EGR valve.

The ECM monitors EGR effectiveness by de-energizing the EGR control solenoid thereby shutting "OFF" vacuum to the EGR valve diaphragm. With the EGR valve closed, fuel integrator counts will be greater than they were during normal EGR operation. If the change is not within the calibrated window, a Code 32 will be set.

The ECM will check EGR operation when:

- Vehicle speed is above 50 mph.
- Engine vacuum is between 40 and 51 kPa.
- No change in throttle position while test is being run.

**Test Description:** Numbers below refer to circled numbers on the diagnostic chart.

1. With the ignition "ON," engine stopped, the solenoid should not be energized and vacuum should not pass to the EGR valve. Grounding the diagnostic terminal will energize the solenoid and allow vacuum to pass to valve.
2. Checks for plugged EGR passages. If passages are plugged, the engine may have severe detonation on acceleration.
3. The vehicle must be driven during this test in order to produce sufficient engine load to operate the EGR. Lightly accelerating (approximately 1/4 throttle) will produce a large and stable enough reading to determine if the ECM is commanding the system "ON."

**Diagnostic Aids:**

- Before replacing ECM, use an ohmmeter and check the resistance of each ECM controlled relay and solenoid coil.

  See ECM wiring diagram for coil terminal identification of solenoid(s) and relay(s) to be checked.

  Replace any solenoid where resistance measures less than 20 ohms.

### TROOPER/TROOPER II WITH 2.8L ENGINE

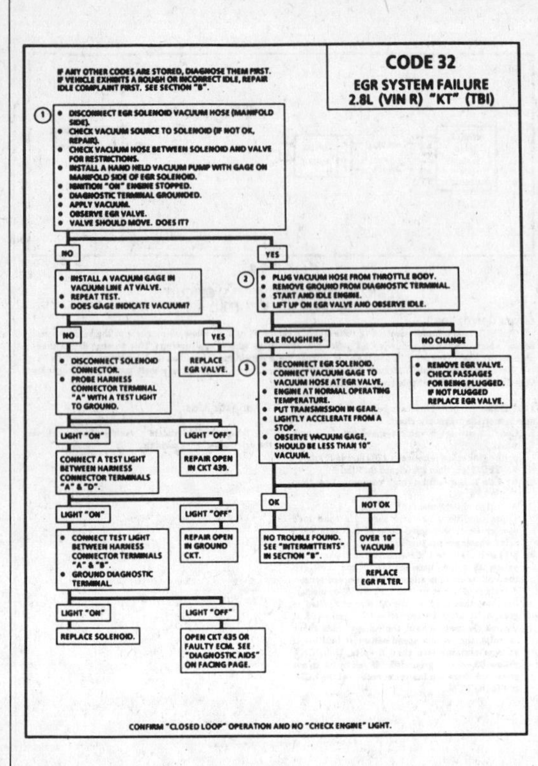


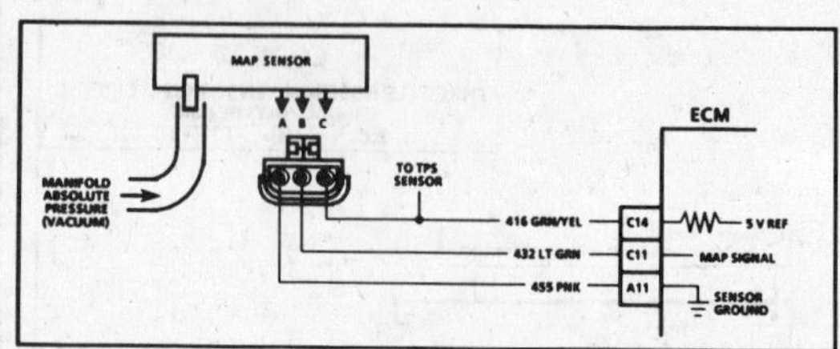


## CODE 33

### MANIFOLD ABSOLUTE PRESSURE (MAP) SENSOR CIRCUIT (SIGNAL VOLTAGE HIGH - LOW VACUUM) 2.8L (VIN R) "KT" (TBI)

**Circuit Description:**

The manifold absolute pressure sensor (MAP) responds to changes in manifold pressure (vacuum). The ECM receives this information as a signal voltage that will vary from about 1-1.5 volts at idle to 4-4.5 volts at wide open throttle.

A "Scan" displays manifold pressure in volts. Low pressure (high vacuum) reads a low voltage while a high pressure (low vacuum) reads a high voltage.

If the MAP sensor fails the ECM will substitute a fixed MAP value and use the throttle position sensor (TPS) to control fuel delivery.

**Test Description:** Numbers below refer to circled numbers on the diagnostic chart.

1. Code 33 will set when:
   - Signal is too high, (kPa greater than 68 kPa), for a time greater than 5 seconds.
   - TPS less than 4%.

   Engine misfire or a low unstable idle may set Code 33. Disconnect MAP sensor and system will go into backup mode. If the misfire or idle condition remains, see "Symptoms" in Section "B".
2. If the ECM recognizes the low MAP signal, the ECM and wiring are OK.

**Diagnostic Aids:**

If idle is rough or unstable refer to "Symptoms" in Section "B" for items which can cause an unstable idle.

An open in CKT 455 or the connection will result in a Code 33.

With the ignition "ON" and the engine stopped, the manifold pressure is equal to atmospheric pressure and the signal voltage will be high. This information is used by the ECM as an indication of vehicle altitude and is referred to as BARO. Comparison of this BARO reading with a known good vehicle with the same sensor is a good way to check accuracy of a "suspect" sensor. Reading should be the same ± .4 volt.

### TROOPER/TROOPER II WITH 2.8L ENGINE

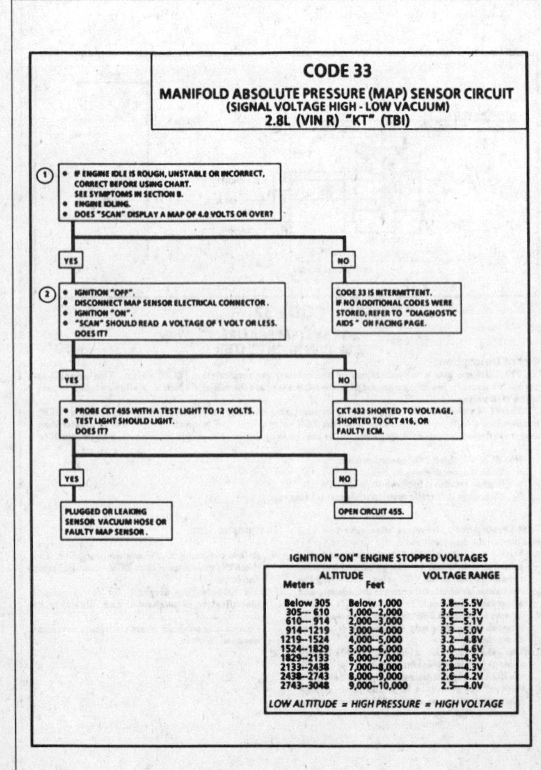


IGNITION "ON" ENGINE STOPPED VOLTAGES

| ALTITUDE Meters | ALTITUDE Feet | VOLTAGE RANGE |
|---|---|---|
| Below 305 | Below 1,000 | 3.8—5.5V |
| 305—610 | 1,000—2,000 | 3.6—5.3V |
| 610—914 | 2,000—3,000 | 3.5—5.1V |
| 914—1219 | 3,000—4,000 | 3.3—5.0V |
| 1219—1524 | 4,000—5,000 | 3.2—4.8V |
| 1524—1829 | 5,000—6,000 | 3.0—4.6V |
| 1829—2133 | 6,000—7,000 | 2.9—4.5V |
| 2133—2438 | 7,000—8,000 | 2.8—4.3V |
| 2438—2743 | 8,000—9,000 | 2.6—4.2V |
| 2743—3048 | 9,000—10,000 | 2.5—4.0V |

*LOW ALTITUDE = HIGH PRESSURE = HIGH VOLTAGE*

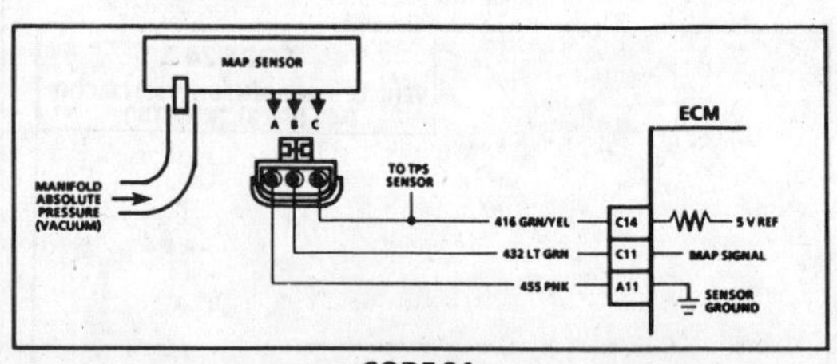


## CODE 34

### MANIFOLD ABSOLUTE PRESSURE (MAP) SENSOR CIRCUIT (SIGNAL VOLTAGE LOW - HIGH VACUUM) 2.8L (VIN R) "KT" (TBI)

**Circuit Description:**

The manifold absolute pressure sensor (MAP) responds to changes in manifold pressure (vacuum). The ECM receives this information as a signal voltage that will vary from about 1-1.5 volts at idle to 4-4.5 volts at wide open throttle.

If the MAP sensor fails the ECM will substitute a fixed MAP value and use the throttle position sensor (TPS) to control fuel delivery.

**Test Description:** Numbers below refer to circled numbers on the diagnostic chart.

1. Code 34 will set when:
   - Signal is too low (kPa less than 14) and engine running less than 1200 rpm.

     OR
   - Engine running greater than 1200 rpm.
   - Throttle position greater than 21% (over 1.5 volts).
2. If the ECM recognizes the high MAP signal, the ECM and wiring are OK.
3. The "Scan" tool may not display 12 volts. The important thing is that the ECM recognises the voltage as more than 4 volts, indicating that the ECM and CKT 432 are OK.

**Diagnostic Aids:**

An intermittent open in CKT's 432 or 416 will result in a Code 34.

With the ignition "ON" and the engine stopped, the manifold pressure is equal to atmospheric pressure and the signal voltage will be high. This information is used by the ECM as an indication of vehicle altitude and is referred to as BARO. Comparison of this BARO reading with a known good vehicle with the same sensor is a good way to check accuracy of a "suspect" sensor. Reading should be the same ± .4 volt.

Also CHART C-1D can be used to test the MAP sensor.

## TROOPER/TROOPER II WITH 2.8L ENGINE

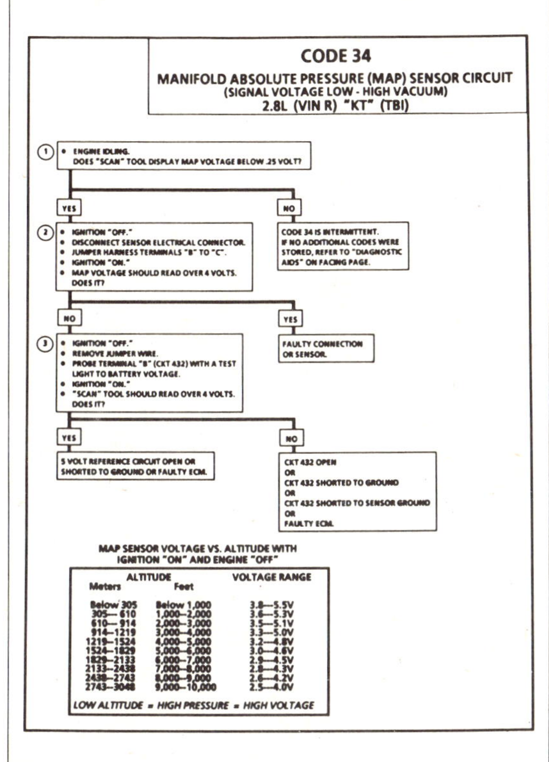


MAP SENSOR VOLTAGE VS. ALTITUDE WITH IGNITION "ON" AND ENGINE "OFF"

| ALTITUDE Meters | Feet | VOLTAGE RANGE |
|---|---|---|
| Below 305 | Below 1,000 | 3.8—5.5V |
| 305—610 | 1,000—2,000 | 3.6—5.3V |
| 610—914 | 2,000—3,000 | 3.5—5.1V |
| 914—1219 | 3,000—4,000 | 3.3—5.0V |
| 1219—1524 | 4,000—5,000 | 3.2—4.8V |
| 1524—1829 | 5,000—6,000 | 3.0—4.6V |
| 1829—2133 | 6,000—7,000 | 2.9—4.5V |
| 2133—2438 | 7,000—8,000 | 2.8—4.3V |
| 2438—2743 | 8,000—9,000 | 2.6—4.2V |
| 2743—3048 | 9,000—10,000 | 2.5—4.0V |

*LOW ALTITUDE = HIGH PRESSURE = HIGH VOLTAGE*

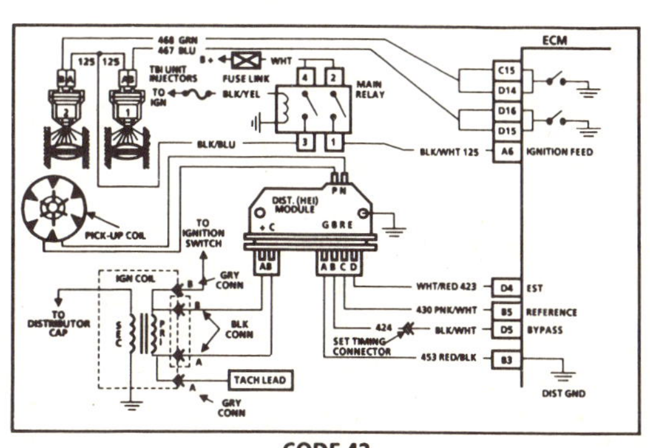


## CODE 42

### ELECTRONIC SPARK TIMING (EST) CIRCUIT 2.8L (VIN R) "KT" (TBI)

**Circuit Description:**

When the system is running on the ignition module, that is, no voltage on the bypass line, the ignition module grounds the EST signal. The ECM expects to see no voltage on the EST line during this condition. If it sees a voltage, it sets Code 42 and will not go into the EST mode.

When the rpm for EST is reached (about 400 rpm), and bypass voltage applied, the EST should no longer be grounded in the ignition module so the EST voltage should be varying.

If the bypass line is open or grounded, the ignition module will not switch to EST mode so the EST voltage will be low and Code 42 will be set.

If the EST line is grounded, the ignition module will switch to EST, but because the line is grounded there will be no EST signal. A Code 42 will be set.

**Test Description:** Numbers below refer to circled numbers on the diagnostic chart.

1. Code 42 means the ECM has seen an open or short to ground in the EST or bypass circuits. This test confirms Code 42 and that the fault causing the code is present.
2. Checks for a normal EST ground path through the ignition module. An EST CKT 423 shorted to ground will also read less than 500 ohms; however, this will be checked later.
3. As the test light voltage touches CKT 424, the module should switch causing the ohmmeter to "overrange" if the meter is in the 1000-2000 ohms position. Selecting the 10-20,000 ohms position will indicate above 5000 ohms. The important thing is that the module "switched".
4. The module did not switch and this step checks for:
   - EST CKT 423 shorted to ground.
   - Bypass CKT 424 open.
   - Faulty ignition module connection or module.
5. Confirms that Code 42 is a faulty ECM and not an intermittent in CKTs 423 or 424.

**Diagnostic Aids:**

If a Code 42 was stored and the customer complains of a "hard start," the problem is most likely a grounded EST line (CKT 423).

The "Scan" tool does not have any ability to help diagnose a Code 42 problem.

A PROM not fully seated in the ECM can result in a Code 42.

## TROOPER/TROOPER II WITH 2.8L ENGINE

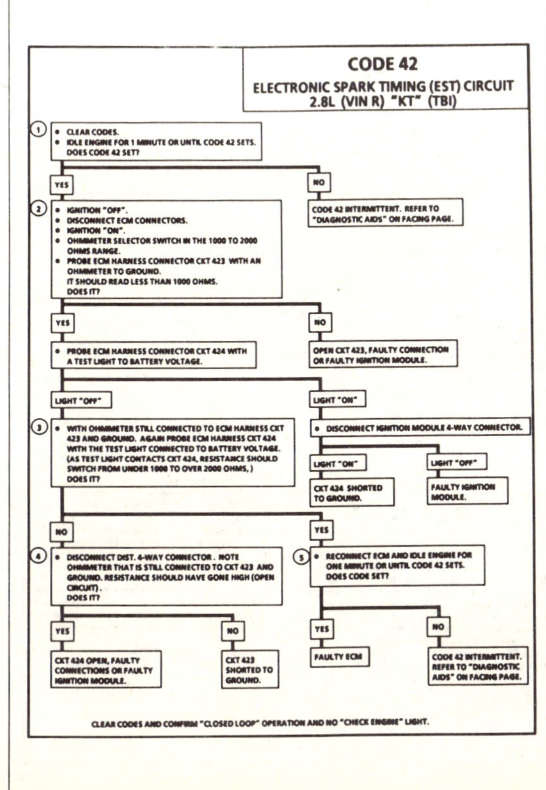


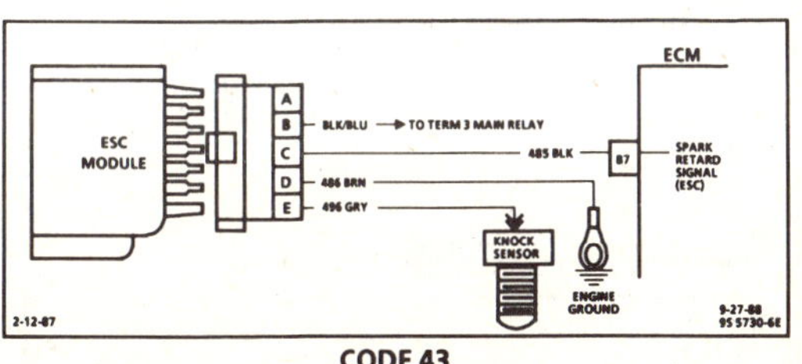


## CODE 43

### ELECTRONIC SPARK CONTROL (ESC) CIRCUIT 2.8L (VIN R) "KT" (TBI)

**Circuit Description:**

Electronic spark control is accomplished with a module that sends a voltage signal to the ECM. As the knock sensor detects engine knock, the voltage from the ESC module to the ECM drops, and this signals the ECM to retard timing. The ECM will retard the timing when knock is detected and rpm is above about 900 rpm.

Code 43 means the ECM has been low voltage at CKT 485 terminal "B7" for longer than 5 seconds with the engine running or the system has failed the functional check.

This system performs a functional check once per start up to check the ESC system. To perform this test the ECM will advance the spark when coolant is above 95°C and at a high load condition (near W.O.T.). The ECM then checks the signal at "B7" to see if a knock is detected. The functional check is performed once per start up and if knock is detected when coolant is below 95°C (194°F) the test has passed and the functional check will not be run. If the functional check fails, the "Service Engine Soon" light will remain "ON" until ignition is turned "OFF" or until a knock signal is detected.

**Test Description:** Numbers below refer to circled numbers on the diagnostic chart.

1. If the conditions for a Code 43 are present the "Scan" will always display "yes." There should not be a knock at idle unless an internal engine problem, or a system problem exists.
2. This test will determine if the system is functioning at this time. Usually a knock signal can be generated by tapping on the right exhaust manifold. If no knock signal is generated try tapping on block close to the area of the sensor.
3. Because Code 43 sets when the signal voltage on CKT 485 remains low this test should cause the signal on CKT 485 to go high. The 12 volts signal should be seen by the ECM as "no knock" if the ECM and wiring are OK.
4. This test will determine if the knock signal is being detected on CKT 496 or if the ESC module is at fault.
5. If CKT 496 is routed to close to secondary ignition wires the ESC module may see the interference as a knock signal.
6. This checks the ground circuit to the module. An open ground will cause the voltage on CKT 485 to be about 12 volts which would cause the Code 43 functional test to fail.
7. Contacting CKT 496 with a test light to 12 volts should generate a knock signal. This will determine if the ESC module is operating correctly.

**Diagnostic Aids:**

Code 43 can be caused by a faulty connection at the knock sensor at the ESC module or at the ECM. Also check CKT 485 for possible open or short to ground.

## TROOPER/TROOPER II WITH 2.8L ENGINE

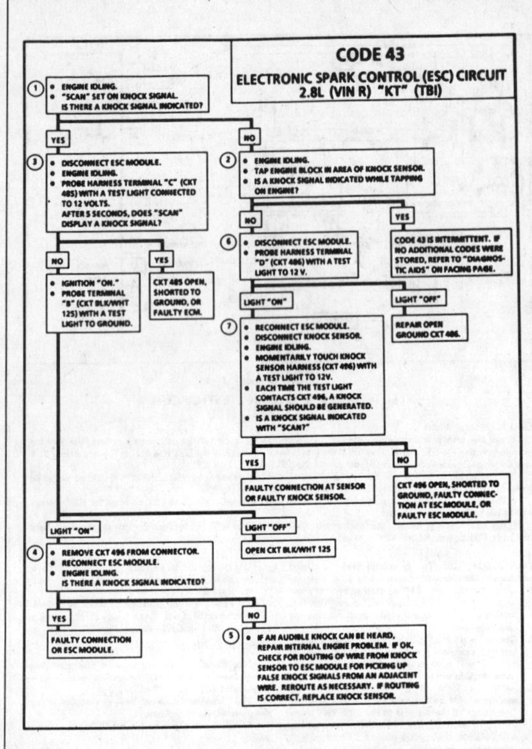


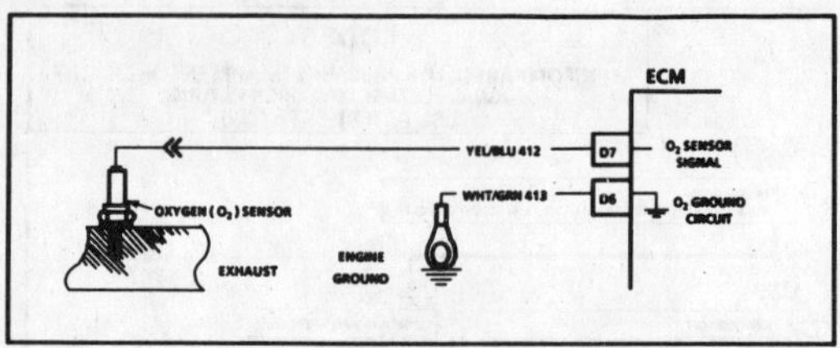


### CODE 44

**OXYGEN SENSOR CIRCUIT**
**(LEAN EXHAUST INDICATED)**
**2.8L (VIN R) "KT" (TBI)**

**Circuit Description:**

The ECM supplies a voltage of about .45 volt between terminals "D6" and "D7". (If measured with a 10 megohm digital voltmeter, this may read as low as .32 volt.) The $O_2$ sensor varies the voltage within a range of about 1 volt if the exhaust is rich, down through about .10 volt if exhaust is lean.

The sensor is like an open circuit and produces no voltage when it is below about 360°C (600°F) An open sensor circuit or cold sensor causes "Open Loop" operation.

**Test Description:** Numbers below refer to circled numbers on the diagnostic chart.

1. Code 44 is set when the $O_2$ sensor signal voltage on CKT 412.
   - Remains below .2 volt for 50 seconds.
   - And the system is operating in "Closed Loop."

**Diagnostic Aids:**

Using the "Scan," observe the block learn values at different rpm and air flow conditions to determine when the Code 44 may have been set. If the conditions for Code 44 exists the block learn values will be around 150.

- $O_2$ Sensor Wire. Sensor pigtail may be mispositioned and contacting the exhaust manifold.
- Check for intermittent ground in wire between connector and sensor.
- MAP Sensor. A (MAP) sensor output that causes the ECM to sense a higher than normal vacuum will cause the system to go lean. Disconnect the MAP sensor and if the lean condition is gone, replace the sensor.
- Lean Injector(s).
- Fuel Contamination. Water, even in small amounts, near the in-tank fuel pump inlet can be delivered to the injectors. The water causes a lean exhaust and can set a Code 44.
- Fuel Pressure. System will be lean if pressure is too low. It may be necessary to monitor fuel pressure while driving the car at various road speeds and/or loads to confirm. See "Fuel System Diagnosis" CHART A-7.
- Exhaust Leaks. If there is an exhaust leak, the engine can pull outside air into the exhaust and past the sensor. Vacuum or crankcase leaks can cause a lean condition.
- If the above are OK, it is a faulty oxygen sensor.

## TROOPER/TROOPER II WITH 2.8L ENGINE

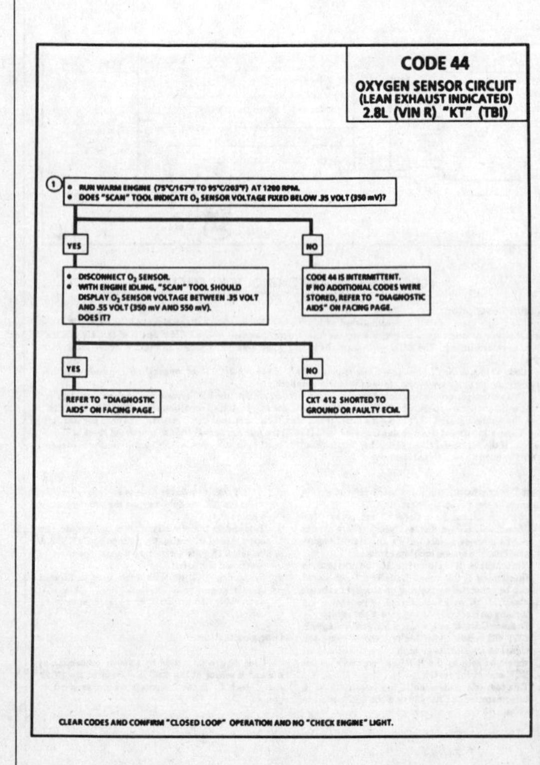


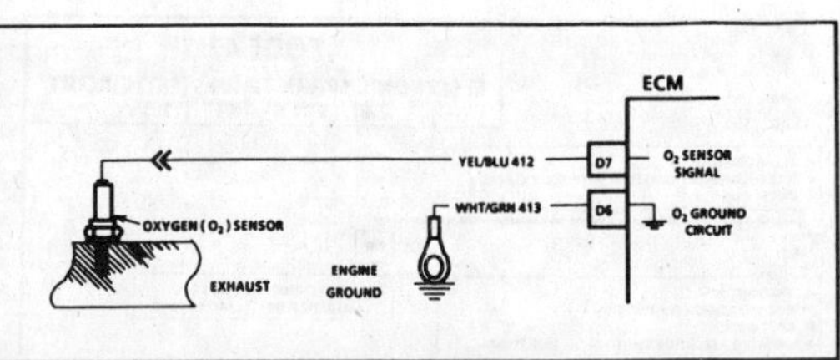


### CODE 45

**OXYGEN SENSOR CIRCUIT**
**(RICH EXHAUST INDICATED)**
**2.8L (VIN R) "KT" (TBI)**

**Circuit Description:**

The ECM supplies a voltage of about .45 volt between terminals "D6" and "D7". (If measured with a 10 megohm digital voltmeter, this may read as low as .32 volt.) The $O_2$ sensor varies the voltage within a range of about 1 volt if the exhaust is rich, down through about .10 volt if exhaust is lean.

The sensor is like an open circuit and produces no voltage when it is below about 360°C (600°F) An open sensor circuit or cold sensor causes "Open Loop" operation.

**Test Description:** Numbers below refer to circled numbers on the diagnostic chart.

1. Code 45 is set when the $O_2$ sensor signal voltage or CKT 412.
   - Remains above .7 volt for 50 seconds; and in "Closed Loop."
   - Engine time after start is 1 minute or more.
   - Throttle angle greater than 2% (about .2 volt above idle voltage) but less than 25%.

**Diagnostic Aids:**

Using the "Scan," observe the block learn values at different rpm conditions to determine when the Code 45 may have been set. If the conditions for Code 45 exists, The block learn values will be around 115.

- Fuel Pressure. System will go rich if pressure is too high. The ECM can compensate for some increase. However, if it gets too high, a Code 45 may be set. See "Fuel System Diagnosis" CHART A-7.
- Leaking injector. See CHART A-7.
- Check for fuel contaminated oil.
- HEI Shielding. An open ground CKT 453 (ignition system ref. low) may result in EMI, or induced electrical "noise." The ECM looks at this "noise" as reference pulses. The additional pulses result in a higher than actual engine speed signal. The ECM then delivers too much fuel, causing system to go rich. Engine tachometer will also show higher than actual engine speed, which can help in diagnosing this problem.
- Canister purge. Check for fuel saturation. If full of fuel, check canister control and hoses. See "Canister Purge" Section "C3".
- MAP sensor. An output that causes the ECM to sense a lower than normal vacuum can cause the system to go rich. Disconnecting the MAP sensor will allow the ECM to set a fixed value for the sensor. Substitute a different MAP sensor if the rich condition is gone while the sensor is disconnected.
- TPS. An intermittent TPS output will cause the system to go rich, due to a false indication of the engine accelerating.

## TROOPER/TROOPER II WITH 2.8L ENGINE

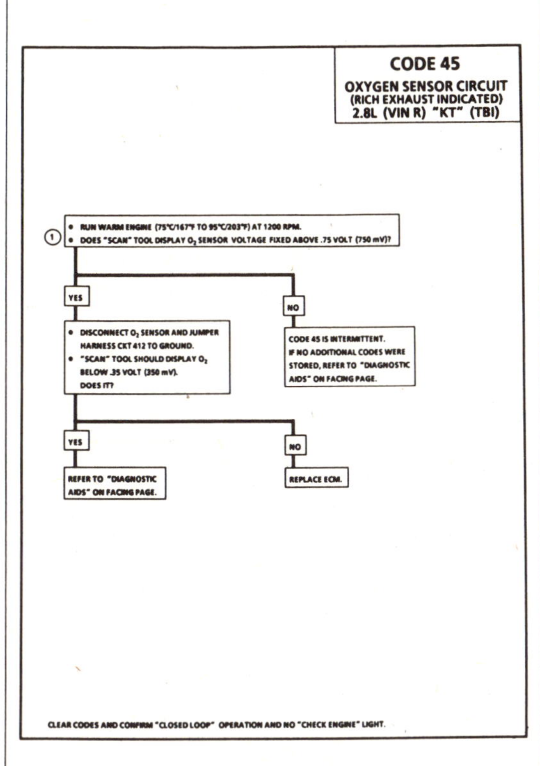


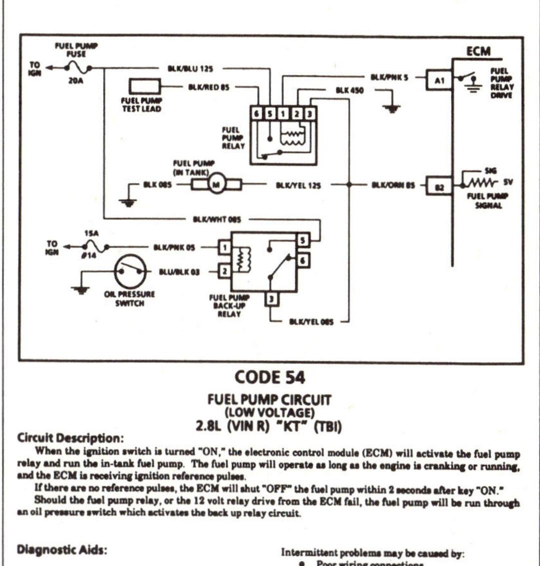


### CODE 54
**FUEL PUMP CIRCUIT**
**(LOW VOLTAGE)**
**2.8L (VIN R) "KT" (TBI)**

**Circuit Description:**

When the ignition switch is turned "ON," the electronic control module (ECM) will activate the fuel pump relay and run the in-tank fuel pump. The fuel pump will operate as long as the engine is cranking or running, and the ECM is receiving ignition reference pulses.

If there are no reference pulses, the ECM will shut "OFF" the fuel pump within 2 seconds after key "ON."

Should the fuel pump relay, or the 12 volt relay drive from the ECM fail, the fuel pump will be run through an oil pressure switch which activates the back up relay circuit.

**Diagnostic Aids:**

An inoperative fuel pump relay can result in long cranking times, particularly if the engine is cold or engine oil pressure is low. The extended crank period is caused by the time necessary for oil pressure to build enough to close the oil pressure switch and turn "ON" the fuel pump.

Intermittent problems may be caused by:
- Poor wiring connections
- Corrosion
- All connections should be checked visually

## TROOPER/TROOPER II WITH 2.8L ENGINE

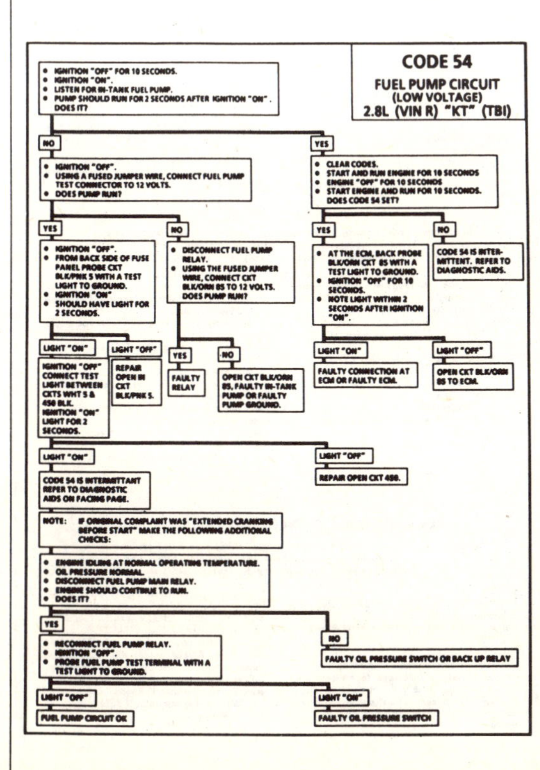


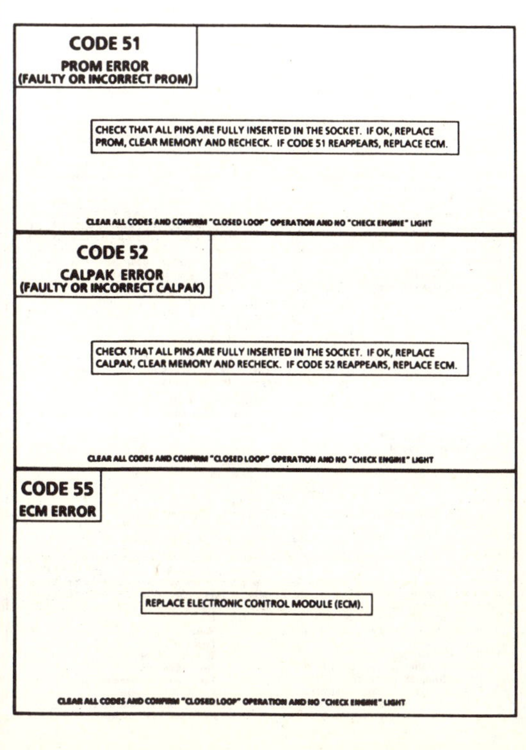

## TROOPER/TROOPER II WITH 2.8L ENGINE

### INTERMITTENTS

Problem may or may not turn "ON" the "Check Engine" light, or store a code.

DO NOT use the Trouble Code Charts in Section "A" for intermittent problems. The fault must be present to locate the problem. If a fault is intermittent, use of Trouble Code Charts may result in replacement of good parts.

- Most intermittent problems are caused by faulty electrical connections or wiring. Perform careful check of suspect circuits for:
  - Poor mating of the connector halves, or terminals, not fully seated in the connector body (backed out).
  - Improperly formed or damaged terminals. All connector terminals in problem circuit should be carefully reformed or replaced to insure proper contact tension.
  - Poor terminal to wire connection. This requires removing the terminal from the connector body to check as outlined in the Introduction to Section "6E".
- If a visual (physical) check does not find the cause of the problem, the car can be driven with a voltmeter connected to a suspected circuit or a "Scan" tool may be used. An abnormal voltage reading, when the problem occurs, indicates the problem may be in that circuit. If the wiring and connectors check OK, and a Trouble Code was stored for a circuit having a sensor, except for Codes 44 and 45, substitute a known good sensor and recheck.
- Loss of trouble code memory. To check, disconnect TPS and idle engine until "Check Engine" light comes "ON." Code 22 should be stored, and kept in memory, when ignition is turned "OFF" for at least 10 seconds. If not, the ECM is faulty.
- An intermittent "Check Engine" light, and No Trouble Codes, may be caused by:
  - Electrical system interference caused by a defective relay, ECM driven solenoid, or switch. They can cause a sharp electrical surge. Normally, the problem will occur when the faulty component is operated.
  - Improper installation of electrical options, such as lights, 2-way radios, etc.
  - EST wires should be routed away from spark plug wires, ignition system components, and generator. Wire for CKT 453 from ECM to ignition system should be a good ground.
  - Ignition secondary shorted to ground.
  - CKTs BLU/PNK 03 ("Check Engine" light) or 451 (Diagnostic Test) intermittently shorted to ground.
  - ECM power grounds.

### HARD START

**Definition:** Engine cranks OK, but does not start for a long time. Does eventually run, or may start but immediately dies.

- CHECK:
  - For water contaminated fuel.
  - Fuel system pressure CHART A-7.
  - TPS for sticking or binding should read less than 1.25 volts on a "Scan" tool.
  - No crank signal; see CHART C-1B.
  - EGR operation; CHART C-7.
  - Fuel pump relay - Connect test light between pump test terminal and ground. Light should be "ON" for 2 seconds following ignition "ON." If not, refer to CODE 54 CHART.
  - For a faulty in-tank fuel pump check valve, which would allow the fuel in the lines to drain back to the tank after the engine is stopped. To check for this condition:
    1. Ignition "OFF."
    2. Disconnect fuel line at the filter.
    3. Remove the tank filler cap.
    4. Connect a radiator test pump to the line and apply 103 kPa (15 psi) pressure. If the pressure will hold for 60 seconds, the check valve is OK.
- Check ignition system for:
  - Proper output with ST-125.
  - Worn shaft.
  - Bare and shorted wires.
  - Pickup coil resistance and connections.
  - Loose ignition coil connections.
  - Moisture in distributor cap.
  - Spark plugs, wet plugs, cracks, wear, improper gap, burned electrodes, or heavy deposits.
- If engine starts but then, immediately stalls, open distributor bypass line. If engine then starts, and runs OK, replace distributor pickup coil.
- Check CKT 423 (EST) for short to ground.

### SURGES AND/OR CHUGGLE

**Definition:** Engine power variation, under steady throttle or cruise. Feels like the car speeds up and slows down, with no change in the accelerator pedal.

- Use a "Scan" tool to make sure reading of VSS matches vehicle speedometer. See "Special Information," Section "6E".
- CHECK:
  - For intermittent EGR at idle. See appropriate CHART C-7.
  - Ignition timing. See Emission Control Information label.
  - Inline fuel filter for dirt or restriction.
  - Fuel pressure. See CHART A-7.
  - Generator output voltage. Repair, if less than 9, or more than 16 volts.
- Inspect oxygen sensor for silicon contamination from fuel, or use of improper RTV sealant. The sensor may have a white, powdery coating and result in a high but false signal voltage (rich exhaust indication). The ECM will then reduce the amount of fuel delivered to the engine, causing a severe driveability problem.
- Remove spark plugs. Check for cracks, wear, improper gap, burned electrodes, or heavy deposits. Also, check condition of the rest of the Ignition System.

### LACK OF POWER, SLUGGISH, OR SPONGY

**Definition:** Engine delivers less than expected power. Little or no increase in speed, when accelerator pedal is pushed down part way.

- Compare customer's car to similar unit. Make sure the customer's car has an actual problem.
- Remove air cleaner and check air filter for dirt, or for being plugged. Replace as necessary.
- If there is spray from only one injector, then there is a malfunction in the injector assembly, or in the signal to the injector assembly. The malfunction can be isolated, by switching the injector connectors. If the problem remains with the original injector, after switching the connector, then, the injector is defective. Replace the injector. If the problem moves with the injector connector, then, the problem is an improper signal in the injector circuits, see CHART A-3.
- CHECK:
  - Ignition timing. See Emission Control Information label.
  - For restricted fuel filter, contaminated fuel or improper fuel pressure. See CHART A-7.
  - ECM grounds.
  - EGR operation for being open, or partly open, all the time - CHART C-7.
  - Generator output voltage. Repair, if less than 9, or more than 16 volts.
  - Engine valve timing and compression.
  - Engine, for proper or worn camshaft. See Section "6A".
  - Secondary ignition voltage, using a scope or ST-125.
  - Proper operation of EST. See Section "C4".
- Check exhaust system for restriction, See CHART B-1:
  1. With engine at normal operating temperature, connect a vacuum gage to any convenient vacuum port on intake manifold.
  2. Run engine at 1000 rpm and record vacuum reading.
  3. Increase rpm slowly to 2500 rpm. Note vacuum reading at steady 2500 rpm.
  4. If vacuum at 2500 rpm decreases more than 3" from reading at 1000 rpm, the exhaust system should be inspected for restrictions.
  5. Disconnect exhaust pipe from engine and repeat steps 3 & 4. If vacuum still drops more than 3" with exhaust disconnected, check valve timing.

## TROOPER/TROOPER II WITH 2.8L ENGINE

### DETONATION / SPARK KNOCK

**Definition:** A mild to severe ping, usually worse under acceleration. The engine makes sharp metallic knocks that change with throttle opening.

- CHECK for obvious overheating problems.
  - Low coolant.
  - Loose water pump belt.
  - Restricted air flow to radiator, or restricted water flow through radiator.
  - Faulty or incorrect thermostat.
  - Coolant sensor, which has shifted in value.
  - Correct coolant solution - should be a 50/50 mix of anti-freeze coolant and water.
- CHECK:
  - For poor fuel quality, proper octane rating.
  - For correct PROM.
  - Spark plugs for correct heat range.
  - Ignition timing. See Vehicle Emission Control Information label.
  - Proper operation of Thermac.
  - Fuel system for low pressure. See CHART A-7.
  - Check EGR system for not opening - CHART C-7.
  - For incorrect basic engine parts such as cam, heads, pistons, etc.
  - Excessive oil entering combustion chamber.
- Remove carbon with top engine cleaner. Follow instructions on can.
- If there is spray from only one injector, then, there is a malfunction in the injector assembly, or in the signal to the injector assembly. The malfunction can be isolated, by switching the injector connectors. If the problem remains with the original injector, after switching the connector, then, the injector is defective. Replace the injector. If the problem moves with the injector connector, then, the problem is an improper signal in the injector circuits, see CHART A-3.

### HESITATION, SAG, STUMBLE

**Definition:** Momentary lack of response as the accelerator is pushed down. Can occur at all car speeds. Usually most severe when first trying to make the car move, as from a stop sign. May cause the engine to stall if severe enough.

- Perform careful visual (physical) check, as described at start of Section "B".
- CHECK:
  - Fuel pressure. See CHART A-7.
  - Water contaminated fuel.
  - TPS for binding or sticking.
  - Ignition timing. See Emission Control Information label.
  - Generator output voltage. Repair if less than 9 or more than 16 volts.
  - For open Ignition System ground, CKT 453.
  - Canister purge system for proper operation. See Section "C3".
  - EGR valve operation CHART C-7.

### CUTS OUT, MISSES

**Definition:** Steady pulsation or jerking that follows engine speed, usually more pronounced as engine load increases. The exhaust has a steady spitting sound at idle or low speed.

- Perform careful visual (physical) check, as described at start of Section "B".
- If ignition system is suspected of causing a miss at idle or cutting, out under load:
- Check for missing cylinder by:
  1. Disconnect IAC motor. Start engine. Remove one spark plug wire at a time, using insulated pliers.
  2. If there is an rpm drop, on all cylinders, (equal to within 50 rpm), go to ROUGH, UNSTABLE, OR INCORRECT IDLE, OR STALLING symptom. Reconnect IAC motor.
  3. If there is no rpm drop on one or more cylinders, or excessive variation in drop, check for spark, on the suspected cylinder(s) with J 26792 (ST-125) spark tester or equivalent. If no spark, see Section "6D" for Intermittent Operation or Miss. If there is spark, remove spark plug(s) in these cylinders and check for:
     - Cracks
     - Wear
     - Improper gap
     - Burned electrodes
     - Heavy deposits
     - Perform compression check on questionable cylinder.
- Check wire resistance (should not exceed 30,000 ohms), also, check rotor and distributor cap.
- If the previous checks did not find the problem:
  - Visually inspect ignition system for moisture, dust, cracks, burns, etc. Spray plug wires with fine water mist to check for shorts.
  - Fuel System - Plugged fuel filter, water, low pressure. See CHART A-7.
  - Perform compression check.
  - Valve Timing.
  - Remove rocker covers. Check for bent pushrods, worn rocker arms, broken or weak valve springs, worn camshaft lobes. Repair as necessary. See Section "6A".
- If there is spray from only one injector, then, there is a malfunction in the injector assembly, or in the signal to the injector assembly. The malfunction can be isolated, by switching the injector connectors. If the problem remains with the original injector, after switching the connector, then, the injector is defective. Replace the injector. If the problem moves with the injector connector, then, the problem is an improper signal in the injector circuits, see CHART A-3.

### POOR FUEL ECONOMY

**Definition:** Fuel economy, as measured by an actual road test, is noticeably lower than expected. Also, economy is noticeably lower than it was on this car at one time, as previously shown by an actual road test.

- CHECK:
  - Engine thermostat for faulty part (always open) or for wrong heat range.
  - Fuel Pressure. See CHART A-7.
- Check owner's driving habits.
  - Is A/C "ON" full time (Defroster mode "ON")?
  - Are tires at correct pressure?
  - Are excessively heavy loads being carried?
  - Is acceleration too much, too often?
  - Suggest driver read "Important Facts on Fuel Economy" in Owner's Manual.
- Perform "Diagnostic Circuit Check."
- Check air cleaner element (filter) for dirt or being plugged.
- Check for proper calibration of speedometer.
- Visually (physically) Check:
  - Vacuum hoses for splits, kinks, and proper connections, as shown on Vehicle Emission Control Information label.
  - Ignition wires for cracking, hardness, and proper connections.
- Check Ignition timing. See Emission Control Information label.
- Remove spark plugs. Check for cracks, wear, improper gap, burned electrodes or heavy deposits. Repair or replace, as necessary.
- Check compression. See Section "6A".
- Check for dragging brakes.
- Suggest owner fill fuel tank and recheck fuel economy.
- Check for exhaust system restriction. See CHART B-1.

### ROUGH, UNSTABLE, OR INCORRECT IDLE, STALLING

**Definition:** The engine runs unevenly at idle. If bad enough, the car may shake. Also, the idle may vary in rpm (called "hunting"). Either condition may be severe enough to cause stalling. Engine idles at incorrect speed.

- CHECK:
  - Ignition timing. See Emission Control Information label.
  - For injector(s) leaking. Check Fuel Pressure CHART A-7.
  - IAC - See CHART C-2C.
  - If a sticking throttle shaft or binding linkage causes a high TPS voltage (open throttle indication), the ECM will not control idle. Monitor TPS voltage. "Scan" and/or voltmeter should read less than 1.25 volts with throttle closed.
  - Vacuum leaks can cause higher than normal idle.
  - EGR "ON," while idling, will cause roughness, stalling, and hard starting. CHART C-7.
  - Battery cables and ground straps should be clean and secure. Erratic voltage will cause IAC to change its position, resulting in poor idle quality.
  - IAC valve will not move, if system voltage is below 9, or greater than 17.8 volts.
  - Use "Scan" tool to determine if ECM is receiving A/C request signal.

## TROOPER/TROOPER II WITH 2.8L ENGINE

- MAP Sensor - Ignition "ON," engine stopped. Compare MAP voltage with known good vehicle. Voltage should be the same ± 400 mV (.4 volt).

  or

  Start and idle engine. Disconnect sensor electrical connector. If idle improves, substitute a known good sensor and recheck.
- A/C Refrigerant Pressure too high. Check for overcharge or faulty pressure switch.
- PCV valve for proper operation by placing finger over inlet hole in valve end several times. Valve should snap back. If not, replace valve.
- Run a cylinder compression check. See Section "6".
- Inspect oxygen sensor for silicon contamination from fuel, or use of improper RTV sealant. The sensor will have a white, powdery coating, and will result in a high but false signal voltage (rich exhaust indication). The ECM will then reduce the amount of fuel delivered to the engine, causing a severe driveability problem.

### EXCESSIVE EXHAUST EMISSIONS OR ODORS

**Definition:** Vehicle fails an emission test. Vehicle has excessive "rotten egg" smell. Excessive odors do not necessarily indicate excessive emissions.

- Perform "Diagnostic Circuit Check."
- IF TEST SHOWS EXCESSIVE CO AND HC, (or also has excessive odors):
  - Check items which cause car to run RICH.
    - Make sure engine is at normal operating temperature.
- CHECK:
  - Fuel pressure. See CHART A-7.
  - Incorrect timing. See Vehicle Emission Control Information label.
  - Canister for fuel loading. See CHART C-3.
  - PCV valve for being plugged, stuck, or blocked PCV hose, or fuel in the crankcase.
  - Spark plugs, plug wires, and ignition components. See Section "6D".
  - Check for lead contamination of catalytic converter (look for removal of fuel filler neck restrictor).
  - Check for properly installed fuel cap.
  - If the system is running rich, (block learn less than 118), refer to "Diagnostic Aids" on facing page of Code 45.
- IF TEST SHOWS EXCESSIVE NOx:
  - Check items which cause car to run LEAN, or to run too hot.
    - EGR valve for not opening. See CHART C-7.
    - Vacuum leaks.
    - Coolant system and coolant fan for proper operation. See CHART C-12.
    - Remove carbon with top engine cleaner. Follow instructions on can.
    - Check ignition timing for excessive base advance. See Emission Control Information label.
- If the system is running lean, (block learn greater than 138), refer to "Diagnostic Aids"

### DIESELING, RUN-ON

**Definition:** Engine continues to run after key is turned "OFF," but runs very roughly. If engine runs smoothly, check ignition switch and adjustment.

- Check injector for leaking. Apply 12 volts to fuel pump test terminal to turn "ON" fuel pump and pressurize fuel system. Visually check injector and TBI assembly for fuel leakage.

### BACKFIRE

**Definition:** Fuel ignites in intake manifold, or in exhaust system, making a loud popping noise.

- CHECK:
  - EGR operation for being open all the time. See CHART C-7.
  - Output voltage of ignition coil(s).
  - For crossfire between spark plugs (distributor cap, spark plug wires, and proper routing of plug wires).
  - Engine timing - See Emission Control Information label.
  - For faulty spark plugs and/or plug wires or boots.
- Perform a compression check - look for sticking or leaking valves.
  - For proper valve timing.
  - Broken or worn valve train parts.

## CHART B-1

### RESTRICTED EXHAUST SYSTEM CHECK

### ALL ENGINES

Proper diagnosis for a restricted exhaust system is essential before any components are replaced. Either of the following procedures may be used for diagnosis, depending upon engine or tool used:

**CHECK AT A. I. R. PIPE:**

1. Remove the rubber hose at the exhaust manifold A.I.R. pipe check valve. Remove check valve.
2. Connect a fuel pump pressure gauge to a hose and nipple from a Propane Enrichment Device (J 26911) (see illustration).
3. Insert the nipple into the exhaust manifold A.I.R. pipe.

**OR**

**CHECK AT $O_2$ SENSOR:**

1. Carefully remove $O_2$ sensor.
2. Install Borroughs exhaust backpressure tester (BT 8515 or BT 8603) or equivalent in place of $O_2$ sensor (see illustration).
3. After completing test described below, be sure to coat threads of $O_2$ sensor with anti-seize compound prior to re-installation.

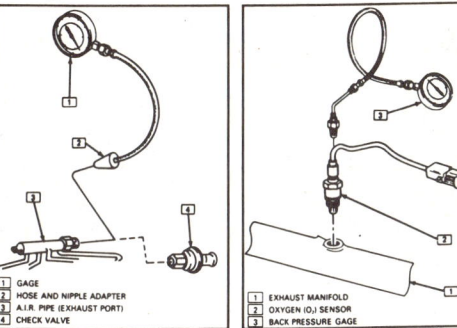

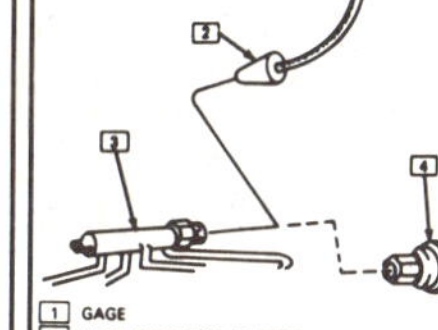

1 GAGE
2 HOSE AND NIPPLE ADAPTER
3 A.I.R. PIPE (EXHAUST PORT)
4 CHECK VALVE

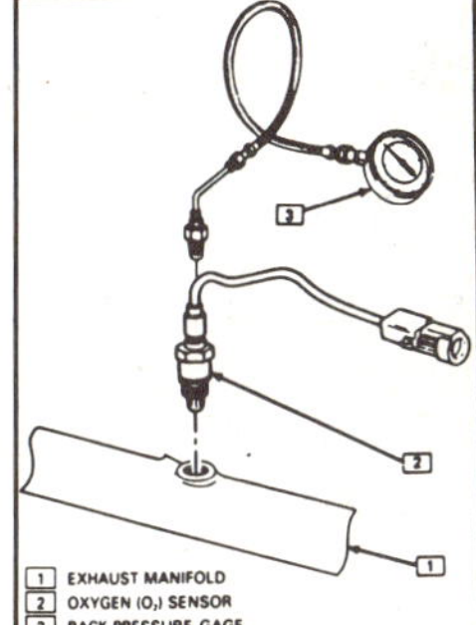

1 EXHAUST MANIFOLD
2 OXYGEN ($O_2$) SENSOR
3 BACK PRESSURE GAGE

**DIAGNOSIS:**

1. With the engine idling at normal operating temperature, observe the exhaust system backpressure reading on the gage. Reading should not exceed 8.6 kPa (1.25 psi).
2. Increase engine speed to 2000 rpm and observe gage. Reading should not exceed 20.7 kPa (3 psi).
3. If the backpressure at either speed exceeds specification, a restricted exhaust system is indicated.
4. Inspect the entire exhaust system for a collapsed pipe, heat distress, or possible internal muffler failure.
5. If there are no obvious reasons for the excessive backpressure, the catalytic converter is suspected to be restricted and should be replaced using current recommended procedures.

## TROOPER/TROOPER II WITH 2.8L ENGINE

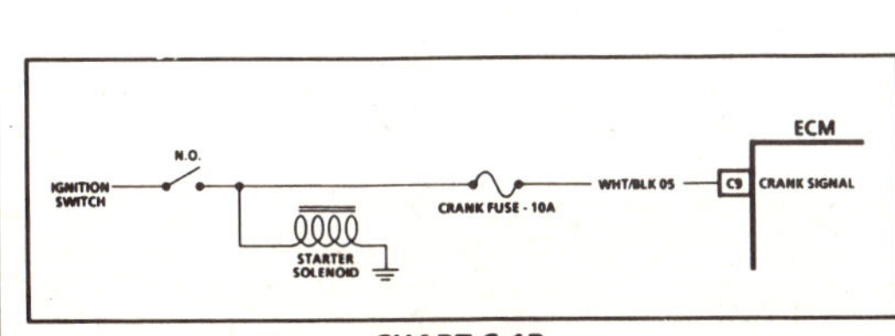


## CHART C-1B

### CRANK SIGNAL
### 2.8L (VIN R) "KT" (TBI)

**Circuit Description:**

Crank signal is a 12 volt signal to the ECM during cranking to allow enrichment and cancel diagnostics until engine is running 12 volts is no longer on circuit.

**Test Description:** Numbers below refer to circled numbers on the diagnostic chart.

1. Checks for normal (cranking) voltage to terminal "C9" of ECM. Test light should be "ON" during cranking.
2. Checks to determine if source of blown fuse was a faulty ECM.

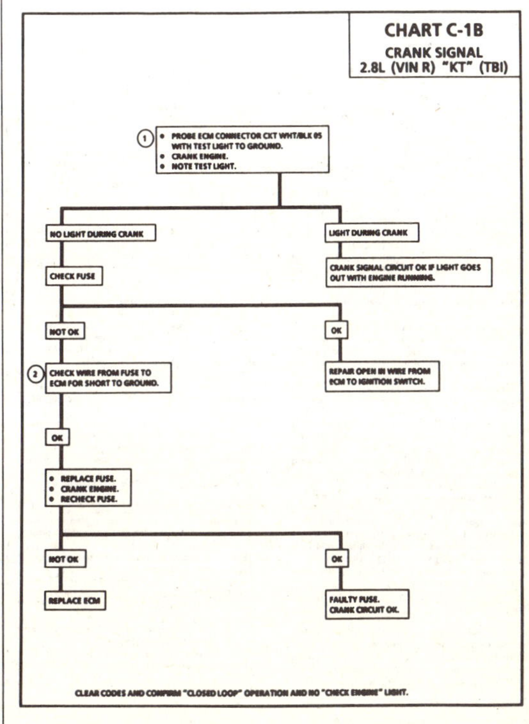

### TROOPER/TROOPER II WITH 2.8L ENGINE

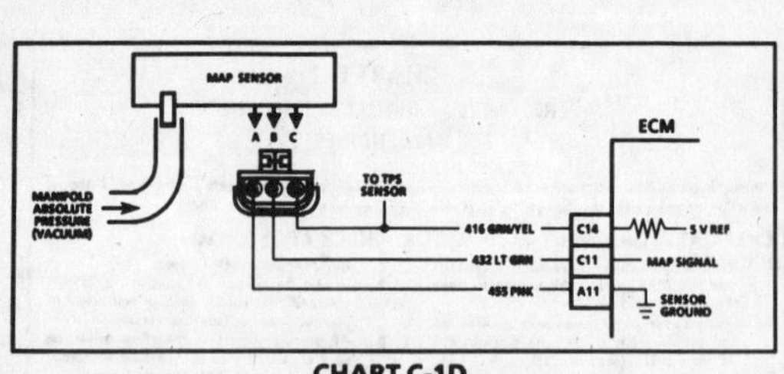


## CHART C-1D
### MAP OUTPUT CHECK
### 2.8L (VIN R) "KT" (TBI)

**Circuit Description:**

The Manifold Absolute Pressure (MAP) sensor measures manifold pressure (vacuum) and sends that signal to the ECM. The ECM uses this information for fuel and spark control.

**Test Description:** Numbers below refer to circled numbers on the diagnostic chart.

1. Checks MAP sensor output voltage to the ECM. This voltage, without engine running, represents a barometer reading to the ECM.
2. Applying 34 kPa (10 inches Hg) vacuum to the MAP sensor should cause the voltage to be 1.2 volts less than the voltage at Step 1. Upon applying vacuum to the sensor, the change in voltage should be instantaneous. A slow voltage change indicates a faulty sensor.
3. Check vacuum hose to sensor for leaking or restriction. Be sure no other vacuum devices are connected to the MAP hose.

   With the ignition "ON" and the engine stopped, the manifold pressure is equal to atmospheric pressure and the signal voltage will be high. This information is used by the ECM as an indication of vehicle altitude and is referred to as BARO. Comparison of this BARO reading with a known good vehicle with the same sensor is a good way to check accuracy of a "suspect" sensor. Reading should be the same ± .4 volt.

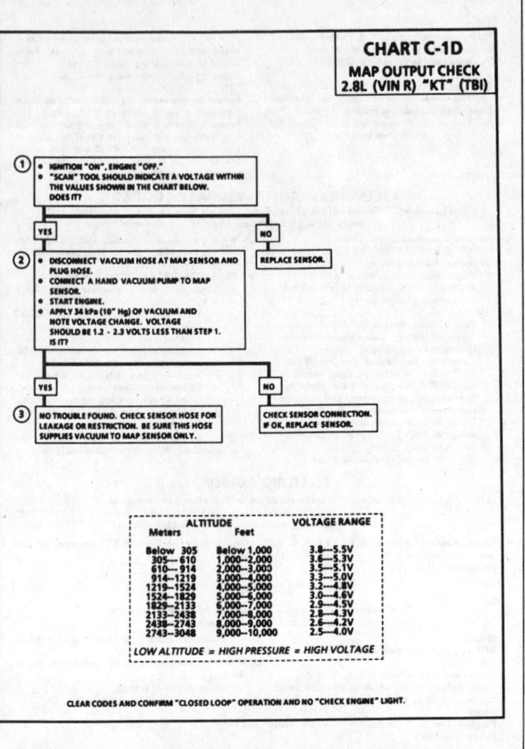


| ALTITUDE Meters | Feet | VOLTAGE RANGE |
|---|---|---|
| Below 305 | Below 1,000 | 3.8—5.5V |
| 305—610 | 1,000—2,000 | 3.6—5.3V |
| 610—914 | 2,000—3,000 | 3.5—5.1V |
| 914—1219 | 3,000—4,000 | 3.3—5.0V |
| 1219—1524 | 4,000—5,000 | 3.2—4.8V |
| 1524—1829 | 5,000—6,000 | 3.0—4.6V |
| 1829—2133 | 6,000—7,000 | 2.9—4.5V |
| 2133—2438 | 7,000—8,000 | 2.8—4.3V |
| 2438—2743 | 8,000—9,000 | 2.6—4.2V |
| 2743—3048 | 9,000—10,000 | 2.5—4.0V |

*LOW ALTITUDE = HIGH PRESSURE = HIGH VOLTAGE*

CLEAR CODES AND CONFIRM "CLOSED LOOP" OPERATION AND NO "CHECK ENGINE" LIGHT.

### TROOPER/TROOPER II WITH 2.8L ENGINE

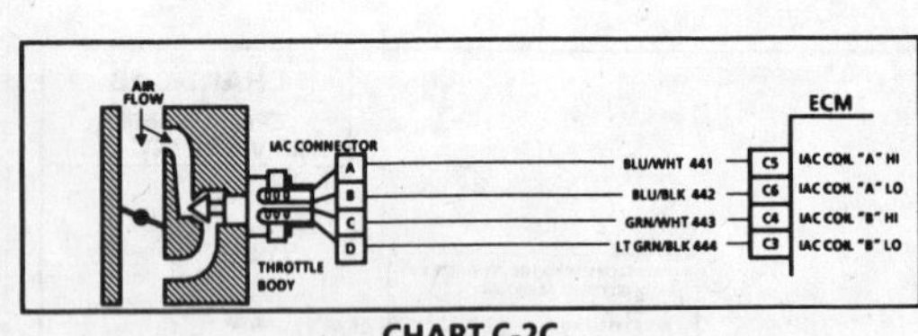


## CHART C-2C
### IDLE AIR CONTROL (IAC) VALVE CHECK
### 2.8L (VIN R) "KT" (TBI)

**Circuit Description:**

The ECM controls idle rpm with the IAC valve. To increase idle rpm, the ECM moves the IAC valve out, allowing more air to pass by the throttle plate. To decrease rpm, it moves the IAC valve in, reducing air flow by the throttle plate. A "Scan" tool will read the ECM commands to the IAC valve in counts. The higher the counts, the more air allowed (higher idle). The lower the counts, the less air allowed (lower idle).

**Test Description:** Numbers below refer to circled numbers on the diagnostic chart.

1. Continue with test, even if engine will not idle. If idle is to low, "Scan" will display 80 or more counts, or steps. If idle is high, it will display "0" counts. Occasionally, an erratic or unstable idle may occur. Engine speed may vary 200 rpm, or more, up and down. Disconnect IAC. If the condition is unchanged, the IAC is not at fault.
2. When the engine was stopped, the IAC valve retracted (more air) to a fixed "park" position for increased air flow and idle speed during the next engine start. A "Scan" will display 100 or more counts. When performing this test, immediately note rpm on start up, because, on a warm engine, the rpm will decrease rapidly.
3. Be sure to disconnect the IAC valve prior to this test. The test light will confirm the ECM signals by a steady or flashing light on all circuits.
4. There is a remote possibility that one of the circuits is shorted to voltage, which would have been indicated by a steady light. Disconnect ECM and turn the ignition "ON" and probe terminals to check for this condition.

**Diagnostic Aids:**

A slow unstable idle may be caused by a system problem that cannot be overcome by the IAC. "Scan" counts will be above 80 counts, if engine speed is too low, and "0" counts, if engine speed is too high.

If idle is too high, stop engine. Ignition "ON." Ground diagnostic terminal. Wait 30 seconds for IAC to seat, then, disconnect IAC. Unground diagnostic terminal and start engine. If idle speed is above 450 rpm in drive, locate and correct vacuum leak. If rpm is less than 450 rpm, refer to "Rough Unstable or Incorrect Idle," in "Symptoms," Section "B".

- <u>System too lean (High Air/Fuel Ratio)</u>
  Idle speed may be too high or too low. Engine speed may vary up and down, disconnecting IAC does not help. May set Code 44.
  "Scan" and/or voltmeter will read an oxygen sensor output less than 300 mV (.3 volt). Check for low regulated fuel pressure or water in fuel. A lean exhaust, with an oxygen sensor output fixed above 800 mV (.8 volt), will be a contaminated sensor, usually silicone. This may also set a Code 45.
- <u>System too rich (Low Air/Fuel Ratio)</u>
  Idle speed too low. "Scan" counts usually above 80. System obviously rich and may exhibit black smoke exhaust.
  "Scan" tool and/or voltmeter will read an oxygen sensor signal fixed above 800 mV (.8 volt).
  Check:
  - High fuel pressure
  - Injector leaking or sticking
- <u>Throttle Body</u>
  Remove IAC and inspect bore for foreign material or evidence of IAC valve dragging the bore.
- If above are all OK, refer to "Rough, Unstable, Incorrect Idle or Stalling "

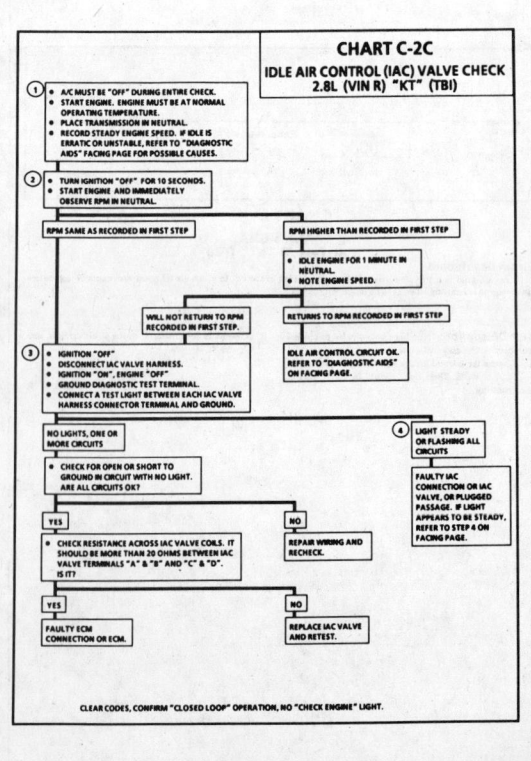

## CHART C-2C
### IDLE AIR CONTROL (IAC) VALVE CHECK
### 2.8L (VIN R) "KT" (TBI)

(1)
- A/C MUST BE "OFF" DURING ENTIRE CHECK.
- START ENGINE. ENGINE MUST BE AT NORMAL OPERATING TEMPERATURE.
- PLACE TRANSMISSION IN NEUTRAL.
- RECORD STEADY ENGINE SPEED. IF IDLE IS ERRATIC OR UNSTABLE, REFER TO "DIAGNOSTIC AIDS" FACING PAGE FOR POSSIBLE CAUSES.

(2)
- TURN IGNITION "OFF" FOR 10 SECONDS.
- START ENGINE AND IMMEDIATELY OBSERVE RPM IN NEUTRAL.

RPM SAME AS RECORDED IN FIRST STEP

RPM HIGHER THAN RECORDED IN FIRST STEP

- IDLE ENGINE FOR 1 MINUTE IN NEUTRAL.
- NOTE ENGINE SPEED.

WILL NOT RETURN TO RPM RECORDED IN FIRST STEP.

RETURNS TO RPM RECORDED IN FIRST STEP

IDLE AIR CONTROL CIRCUIT OK. REFER TO "DIAGNOSTIC AIDS" ON FACING PAGE.

(3)
- IGNITION "OFF"
- DISCONNECT IAC VALVE HARNESS.
- IGNITION "ON", ENGINE "OFF"
- GROUND DIAGNOSTIC TEST TERMINAL.
- CONNECT A TEST LIGHT BETWEEN EACH IAC VALVE HARNESS CONNECTOR TERMINAL AND GROUND.

NO LIGHTS, ONE OR MORE CIRCUITS

- CHECK FOR OPEN OR SHORT TO GROUND IN CIRCUIT WITH NO LIGHT. ARE ALL CIRCUITS OK?

YES

- CHECK RESISTANCE ACROSS IAC VALVE COILS. IT SHOULD BE MORE THAN 20 OHMS BETWEEN IAC VALVE TERMINALS "A" & "B" AND "C" & "D". IS IT?

YES

FAULTY ECM CONNECTION OR ECM.

NO

REPLACE IAC VALVE AND RETEST.

NO

REPAIR WIRING AND RECHECK.

(4) LIGHT STEADY OR FLASHING ALL CIRCUITS

FAULTY IAC CONNECTION OR IAC VALVE, OR PLUGGED PASSAGE. IF LIGHT APPEARS TO BE STEADY, REFER TO STEP 4 ON FACING PAGE.

CLEAR CODES, CONFIRM "CLOSED LOOP" OPERATION, NO "CHECK ENGINE" LIGHT.

## TROOPER/TROOPER II WITH 2.8L ENGINE

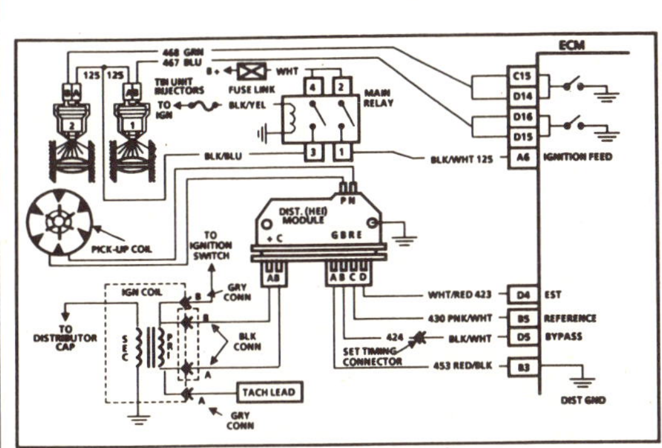


### CHART C-4B

**IGNITION SYSTEM CHECK**
**(REMOTE COIL / SEALED MODULE CONNECTOR DISTRIBUTOR)**
**2.8L (VIN R) "KT" (TBI)**

**Test Description:** Numbers below refer to circled numbers on the diagnostic chart.

1. Two wires are checked, to ensure that an open is not present in a spark plug wire.

1A. If spark occurs with EST connector disconnected, pick-up coil output is too low for EST operation.

2. A spark indicates the problem must be the distributor cap or rotor.
3. Normally, there should be battery voltage at the "C" and "+" terminals. Low voltage would indicate an open or a high resistance circuit from the distributor to the coil or ignition switch. If "C" term. voltage was low, but "+" term. voltage is 10 volts or more, circuit from "C" term. to ignition coil or ignition coil primary winding is open.
4. Checks for a shorted module or grounded circuit from the ignition coil to the module. The distributor module should be turned "OFF," so normal voltage should be about 12 volts.

   If the module is turned "ON", the voltage would be low, but above 1 volt. This could cause the ignition coil to fail from excessive heat.

   With an open ignition coil primary winding, a small amount of voltage will leak through the module from the "Bat." to the tach terminal.
5. Applying a voltage (1.5 to 8 volts) to module terminal "P" should turn the module "ON" and the tach. term. voltage should drop to about 7-9 volts. This test will determine whether the module or coil is faulty or if the pick-up coil is not generating the proper signal to turn the module "ON." This test can be performed by using a DC battery with a rating of 1.5 to 8 volts. The use of the test light is mainly to allow the "P" terminal to be probed more easily. Some digital multi-meters can also be used to trigger the module by selecting ohms, usually the diode position. In this position the meter may have a voltage across it's terminals which can be used to trigger the module. The voltage in the ohm's position can be checked by using a second meter or by checking the manufacturers specification of the tool being used.
6. This should turn "OFF" the module and cause a spark. If no spark occurs, the fault is most likely in the ignition coil because most module problems would have been found before this point in the procedure. A module tester could determine which is at fault.

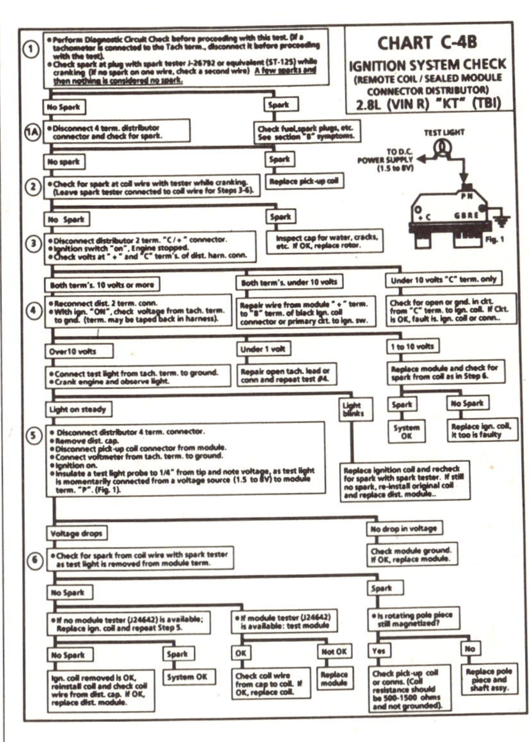


## TROOPER/TROOPER II WITH 2.8L ENGINE

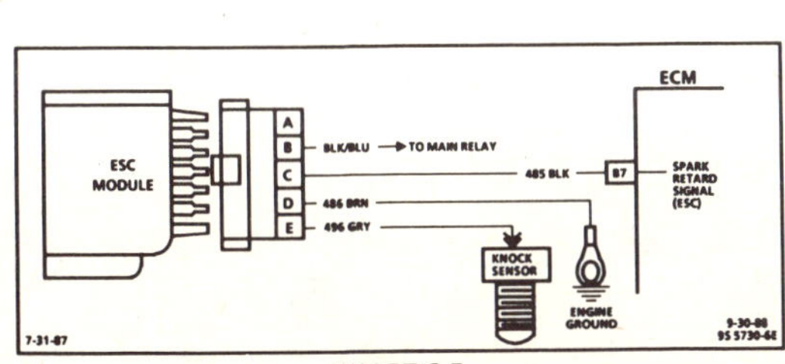


### CHART C-5

**ELECTRONIC SPARK CONTROL (ESC) SYSTEM CHECK**
**(ENGINE KNOCK, POOR PERFORMANCE, OR POOR ECONOMY)**
**2.8L (VIN R) "KT" (TBI)**

**Circuit Description:**

Electronic spark control is accomplished with a module that sends a voltage signal to the ECM. As the knock sensor detects engine knock, the voltage from the ESC module to the ECM is shut "OFF" and this signals the ECM to retard timing, if engine rpm is over about 900.

**Test Description:** Numbers below refer to circled numbers on the diagnostic chart.

1. If a Code 43 is not set, but a knock signal is indicated while running at 1500 rpm, listen for an internal engine noise. Under a no load condition, there should not be any detonation, and if knock is indicated, an internal engine problem may exist.
2. Usually a knock signal can be generated by tapping on the right exhaust manifold. This test can also be performed at idle. Test number 1 was run at 1500 rpm, to determine if a constant knock signal was present, which would affect engine performance.
3. This tests whether the knock signal is due to the sensor, a basic engine problem, or the ESC module.
4. If the module ground circuit is faulty, the ESC module will not function correctly. The test light should light indicating the ground circuit is OK.
5. Contacting CKT 496, with a test light to 12 volts, should generate a knock signal to determine whether the knock sensor is faulty, or the ESC module can't recognize a knock signal.

**Diagnostic Aids:**

If the ESC system checks OK, but detonation is the complaint, refer to "Detonation/Spark Knock" symptom

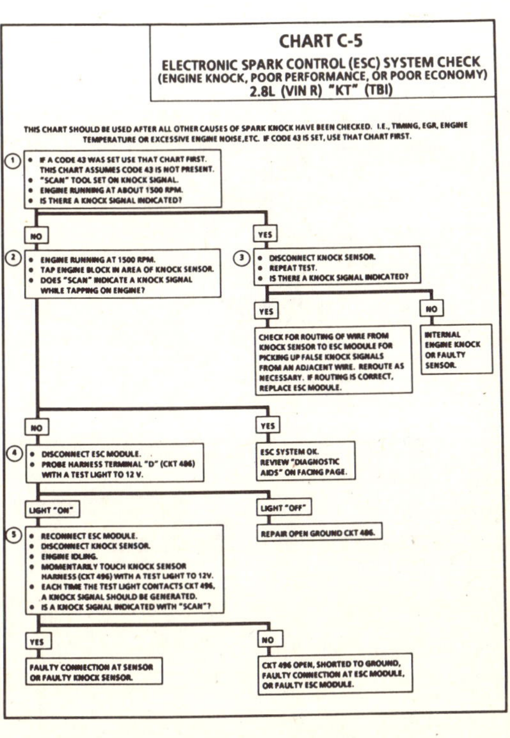

## TROOPER/TROOPER II WITH 2.8L ENGINE

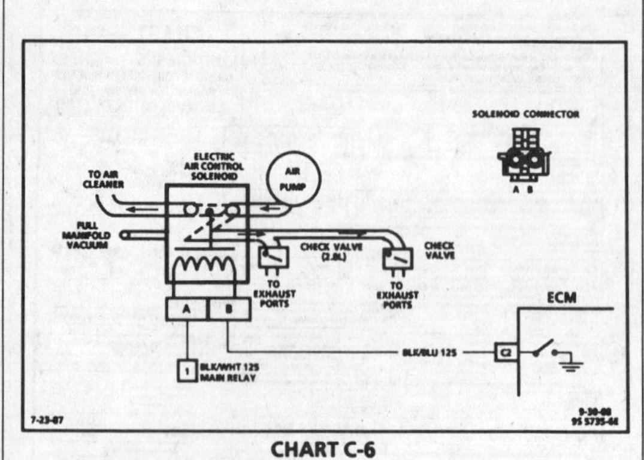


### CHART C-6

**AIR MANAGEMENT CHECK**
**(ELECTRONIC AIR CONTROL VALVE)**
**2.8L (VIN R) "KT" (TBI)**

**Circuit Description:**

An electric air control valve solenoid directs air into the exhaust ports or the air cleaner. During cold start the ECM completes the ground circuit, the EAC solenoid is energized, and air is directed to the exhaust ports. As "coolant" temperature increases, or system goes to "Closed Loop," the ECM opens the ground circuit, the EAC solenoid is de-energized, and air goes to the air cleaner. If the system is not operating properly, check manifold vacuum signal (10"Hg/34kPa) at the valve and check the electrical circuit from the solenoid to the ECM.

**Test Description:** Numbers below refer to circled numbers on the diagnostic chart.

1. This is a system performance test. When vehicle goes to "Closed Loop," air will switch from the ports and divert to the air cleaner.
2. Tests for a grounded electric divert circuit. Normal system light will be "OFF."
3. Checks for an open control circuit. Grounding diagnostic terminal will energise the solenoid, if ECM and circuits are normal. In this step, if test light is "ON," circuits are normal and fault is in valve connections or valve.
4. Checks for voltage from battery through a fuse to the solenoid.

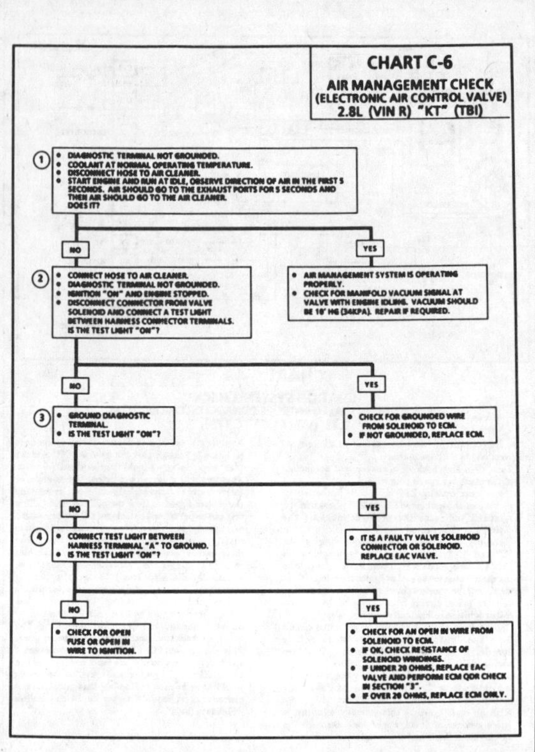


## TROOPER/TROOPER II WITH 2.8L ENGINE

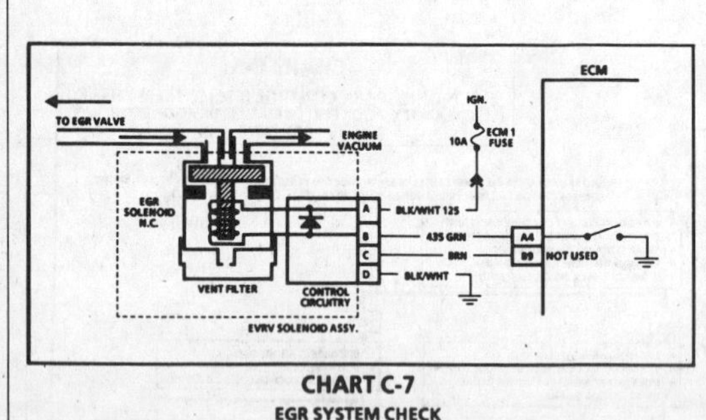


### CHART C-7

**EGR SYSTEM CHECK**
**2.8L (VIN R) "KT" (TBI)**

**Circuit Description:**

The EGR valve is controlled by a normally closed solenoid (allow a vacuum to pass when energized). The ECM pulses the solenoid to turn "ON" and regulate the EGR. The ECM diagnoses the system using an internal EGR test procedure.

The ECM control of the EGR is based on the following inputs:

- Engine coolant temperature - above 25°C.
- TPS - "OFF" idle
- MAP

If Code 24 is stored, use that chart first.

Code 32 will detect a faulty solenoid, vacuum supply, EGR valve or plugged passage. This chart checks for plugged EGR passages, a sticking EGR valve, or a stuck open or inoperative solenoid.

**Test Description:** Numbers below refer to circled numbers on the diagnostic chart.

1. With the ignition "ON," engine stopped, the solenoid should not be energized and vacuum should not pass to the EGR valve.
2. Grounding the diagnostic terminal will energize the solenoid and allow vacuum to pass to valve.
3. Checks for plugged EGR passages. If passages are plugged, the engine may have severe detonation on acceleration.
4. The EGR solenoid will not be energized in park or neutral. This test will determine if the switch input is being received by the ECM.

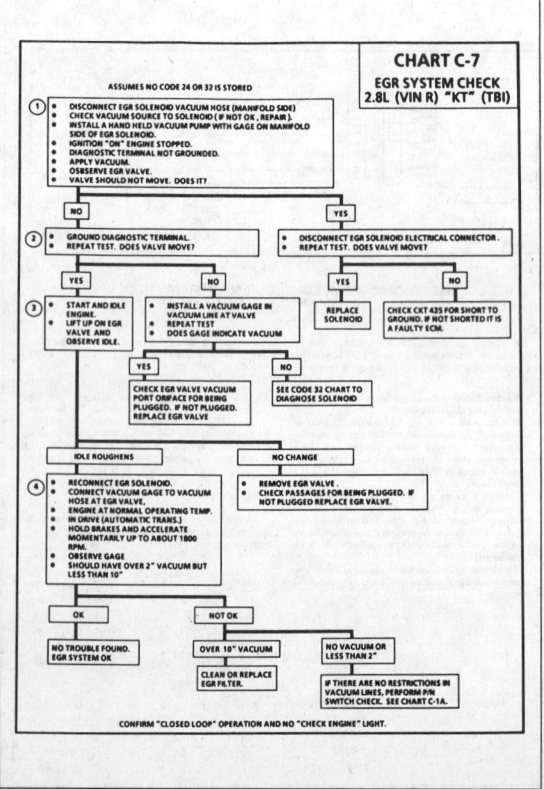

# TROOPER/TROOPER II WITH 2.8L ENGINE

A/C DUAL PRESSURE SWITCH
NC
A/C CLUTCH CONTROL RELAY
59 GRN
A/C COMPRESSOR TEST LEAD RIGHT HAND RELAY CTR.
ECM
A/C REQUEST
GRN/BLU 05
GRN/BLK 03
B8
A2
A/C CONTROL
TO COMPRESSOR
GRN/WHT
GRN/BLK
EVAPO THERMO SW.
A/C CONTROL RELAY
GRN/WHT
10A
TO FUSE BOX
MODE SW.
A/C OFF
NO
BLOWER SW.
A/C COMPRESSOR CUTOUT RELAY
GRN/ORN
GRN/BLK

## CHART C-10

### A/C CLUTCH CONTROL CIRCUIT DIAGNOSIS 2.8L (VIN R) "KT" (TBI)

**Circuit Description:**

The A/C clutch control relay is ECM controlled to delay A/C clutch engagement about .4 second after A/C is turned "ON." This allows the IAC to adjust engine rpm before the A/C clutch engages. The ECM also causes the relay to disengage the A/C clutch during WOT, when high power steering pressure is present, or if engine is overheating. The A/C clutch control relay is energized when the ECM provides a ground path for CKT grn/blk 03.

The dual pressure switch (located in the receiver dryer) will open if A/C pressure is less than 40 psi (276 kPa) or if A/C pressure exceeds 440 psi (3034 kPa).

ECM control of the A/C clutch improves idle quality and performance by:

- Delaying clutch apply until the idle air rate is increased.
- Releasing clutch when idle speed is too low.
- Releasing clutch at wide open throttle.
- Smooths cycling of the compressor by providing additional fuel at the instant clutch is applied.

Turning "ON" air conditioning supplies CKT grn/blu 05 battery voltage to the clutch control relay and terminal "B8" of the ECM connector. After a time delay of about 1/2 second the ECM will ground terminal "A2" of the ECM connector, CKT grn/blk 03, and close the control relay. A/C compressor clutch will engage.

**Test Description:** Numbers refer to circled numbers on the diagnostic chart.

1. Checks for low refrigerant or dual pressure switch as cause for no A/C.
2. This and following tests check for faulty A/C control relay.
3. This and following tests check for open ECM wiring or faulty ECM.

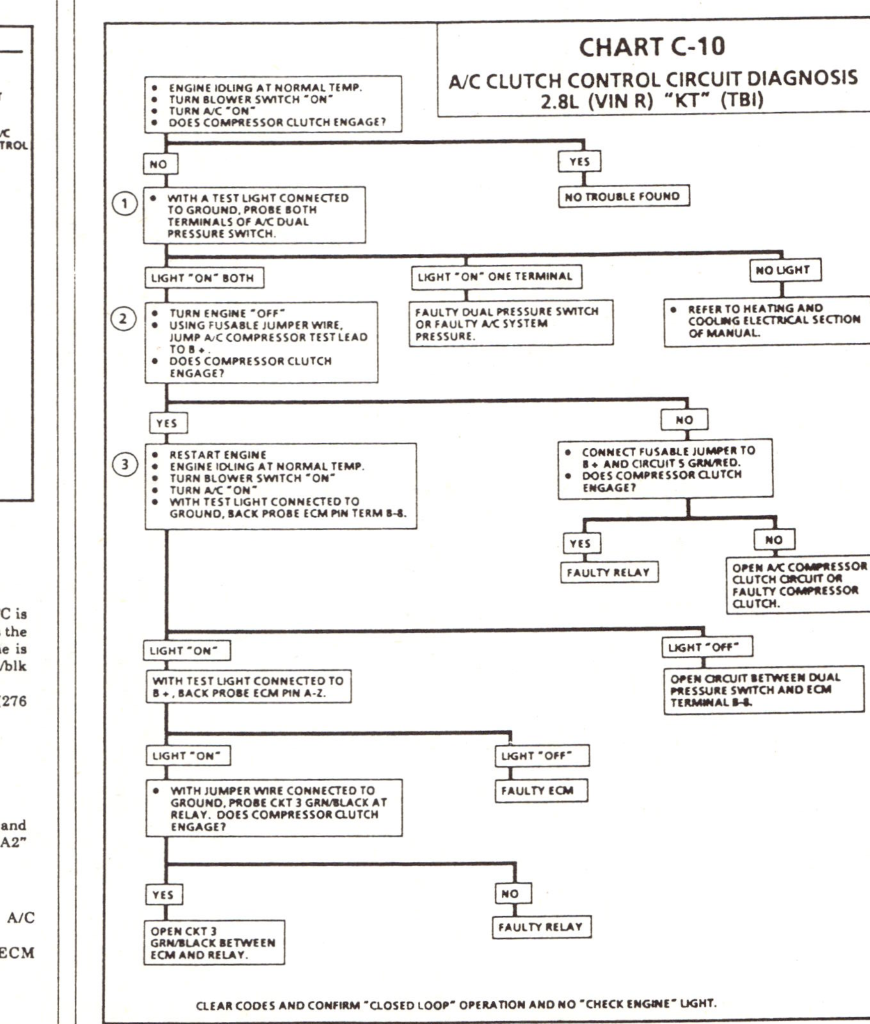

## SELF-DIAGNOSTIC SYSTEMS

### I-MARK TURBO

#### Entering the System

The diagnostic circuit check makes sure that the self-diagnostic system works, determines that the trouble codes will display and guides diagnosis to other problem areas. When the engine is running and a problem develops in the system which the self-diagnosis can evaluate, the **CHECK ENGINE** light will come on and a trouble code will be stored in the ECM trouble code memory. The light will remain on with the engine running as long as there is a problem. If the problem is intermittent the **CHECK ENGINE** light will go out, but the trouble codes will be stored in the ECM trouble code memory.

With the ignition turned **ON** and the engine stopped, the **CHECK ENGINE** lamp should be on. This is a bulb check to indicate the light is working properly. The trouble code test lead is a 3 terminal connector, located near the ECM connector. This connector is used to actuate the trouble code system in the ECM. This connector is also known as the ALDL (assembly line diagnostic link) or the ALDL (assembly line communications link). 2 terminals of this connector (A & C) are used to activate the trouble code system in the ECM.

The trouble code memory is activated by placing the ignition switch in the **ON** position and running a jumper wire between terminals **A** and C of the ALDL connector.

The **CHECK ENGINE** light will begin to flash a trouble Code 12. Code 12 consists of one flash , a short pause and then 2 more flashes. There will be a longer pause and a Code 12 will repeat 2 more times. The check indicates that the self-diagnostic system is working. This cycle will repeat itself until the engine is started or the ignition switch is turned **OFF**. If more than 1 fault code is stored in the memory, the lowest number code will flash 3 times followed by the next highest code number until all the codes have been flashed. The faults will then repeat in the same order. In most cases, codes will be checked with the engine running since no codes other than Code 12 or 51 will be present on the initial "KEY ON". Remove the ground from the test terminal before starting the engine.

A trouble code indicates a problem in a given circuit. For example, a trouble Code 14 indicates a problem in the coolant sensor circuit. This includes the coolant sensor, connector harness and ECM. The procedure for finding the problem can be found by using the apppropriate troubleshooting chart(s). When the engine is started the **CHECK ENGINE** light will go off. If the **CHECK ENGINE** light remains on, the self-diagnostic system has detected a fault.

If a trouble code can be obtained when the **CHECK ENGINE**light is off with the engine running, the trouble code must be evaluated. A determination must be made to see if the fault is intermittent or if the engine must be at certain operating conditions to turn the **CHECK ENGINE** light on. Faults indicated by trouble Codes 13, 31, 44 and 45 require engine operation at part throttle for up to 5 minutes after the engine has reached normal operating temperature before the **CHECK ENGINE** light will come on and store a trouble code.

The fault indicated by trouble Code 15 takes 5 minutes of engine operation before it will display. The diagnostic charts for trouble Codes 13, 31, 44 and 45 should be used if any of these trouble codes can be obtained. Be sure to remove the ground from the test terminal before starting the engine.

#### Clearing the Trouble Code Memory

The trouble code memory is fed a continuous 12 volts even with the key in the **OFF** position. After a fault has been corrected, it will be necessary to remove the voltage for 10 seconds to clear any stored codes. The quickest way to remove the voltage is to remove the "ECM" fuse from the fuse block for 10 seconds. The voltage can also be removed by disconnecting the negative battery cable, but this will mean if the vehicle is equipped with electronic instrumentation, such as a clock and radio, they would have to be reset.

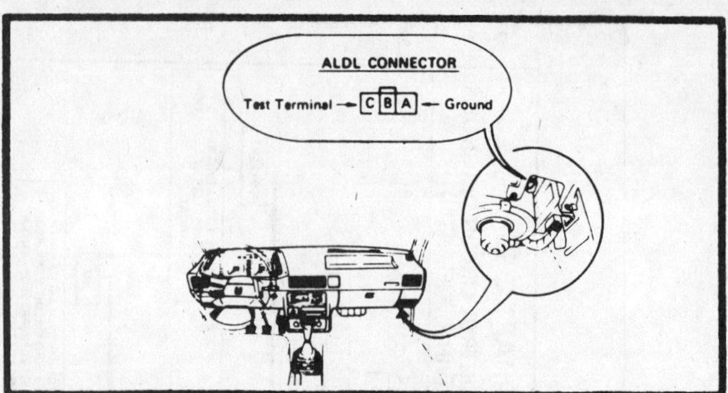


**ALDL connector location – I-Mark turbo and non-turbo**

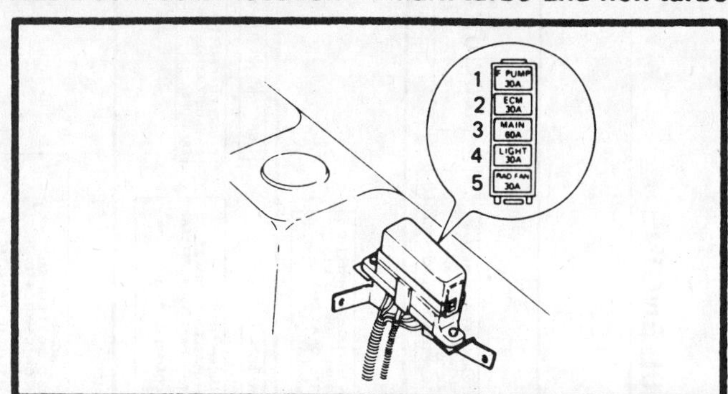


**On I-Mark turbo, remove voltage from the ECM by pulling the "ECM 30A" fuse**

#### Driver Comments

After performing the Diagnostic Circuit Check, and there is no **CHECK ENGINE** light with a warm running engine, then refer to the Driver Comments for an emission non-compliance problem or an engine performance problem (odor, surge, fuel economy, etc.). This ultimately will lead to the System Performance Check which is also used after a repair had been made.

#### System Performance Check

The System Performance Check is used to check the operation of the mixture control (M/C) solenoid and the main metering circuit using a tachometer, and to check the operation of the Closed Loop Emission Control System by the use of a dwell meter.

### I-MARK NON-TURBO, TROOPER AND TROOPER II WITH 2.8L ENGINE

The Electronic Control Module (ECM) is equipped with a self-diagnostic system which detects component and system failures and helps the technician to identify the faulty circuit by displaying a trouble code.

#### Check Engine Light

The **CHECK ENGINE** light is located on the instrument panel and has 2 functions:

1. It alerts the driver that a problem has occured in the system and that the car should be taken to the service station for evalution as soon as possible.
2. Once in the shop, the light allows the technician to read Trouble Codes to help diagnose system problems. Normally, the light will go on with the key **ON** and the engine not running. If the ECM has detected a problem within the system, the light will remain on. If the problem goes away, the light will go out in most cases but a trouble code will remain stored in the ECM.

### Intermittent Check Engine Light

Fault codes are generally divided into 2 catagories: intermittent and hard.

An intermittent code is defined as one that does not re-set itself, but the fault condition is not active when the techinician is working on the car. The most probable cause of intermittent faults are loose wiring connections.

A hard code is defined as one that is present when the vehicle is brought in for service and one that the ECM recognizes and displays to the technician.

### Trouble Codes

The ECM uses a network of sensors throughout the engine to measure a number of engine operating parameters. It compares these readings to pre-programmed reference values stored in the ECM's memory. If a sensor reading is not what the ECM thinks it should be, the ECM will illuminate the **CHECK ENGINE** light on the instrument panel, assign a trouble code to the fault and store the code in the ECM's memory. The trouble code does not always identify the faulty component, but alerts the technician to what circuit the fault is in. A engine circuit consists of the sensor, attendant wiring, connectors and the ECM itself.

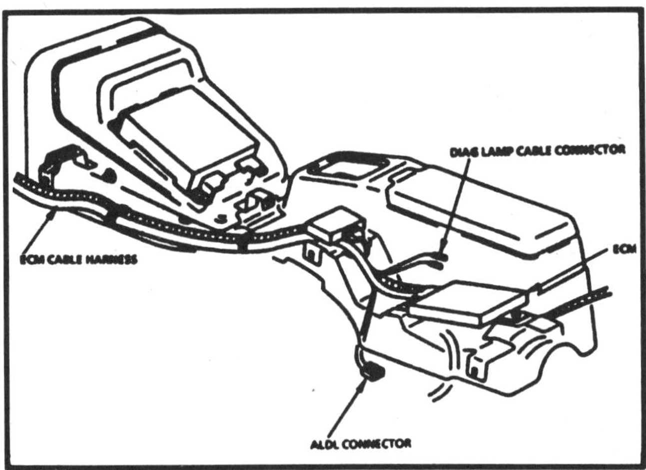


**ALDL and diagnostic lead connector location—Trooper and Trooper II (except 2.8L engine)**

### ALDL Connector

The Assembly Line Diagnostic Link (ALDL) connector is a diagnostic connector located in the passenger compartment. The connector has terminals used at the point of assembly to make sure that the engine is operating to design specifications before leaving the plant. Terminal **1** of the connector is the diagnostic terminal and can be connected to terminal **3** to enter the DIAGNOSTIC mode or ground to enter the FIELD SERVICE mode.

### Diagnostic Mode

**NOTE: In the Diagnostic mode, codes can be read only with the engine stopped. Grounding the diagnostic terminal with the engine running sends the ECM into the FIELD SERVICE MODE described below.**

If the diagnostic terminal of the ALDL connector is grounded with the ignition switch in the **ON** position and the engine not running, the ECM will enter the Diagnostic mode. In this mode the ECM will perform the following functions:

1. Display a Code 12 by flashing the **CHECK ENGINE** light which indicates that the system is functioning properly. A Code 12 consists of 1 flash, followed by a short pause, then 2 quick flashes. The code will be repeated 3 times. If no other codes are stored in the system, code will continue to flash until the diagnostic terminal is ungrounded.

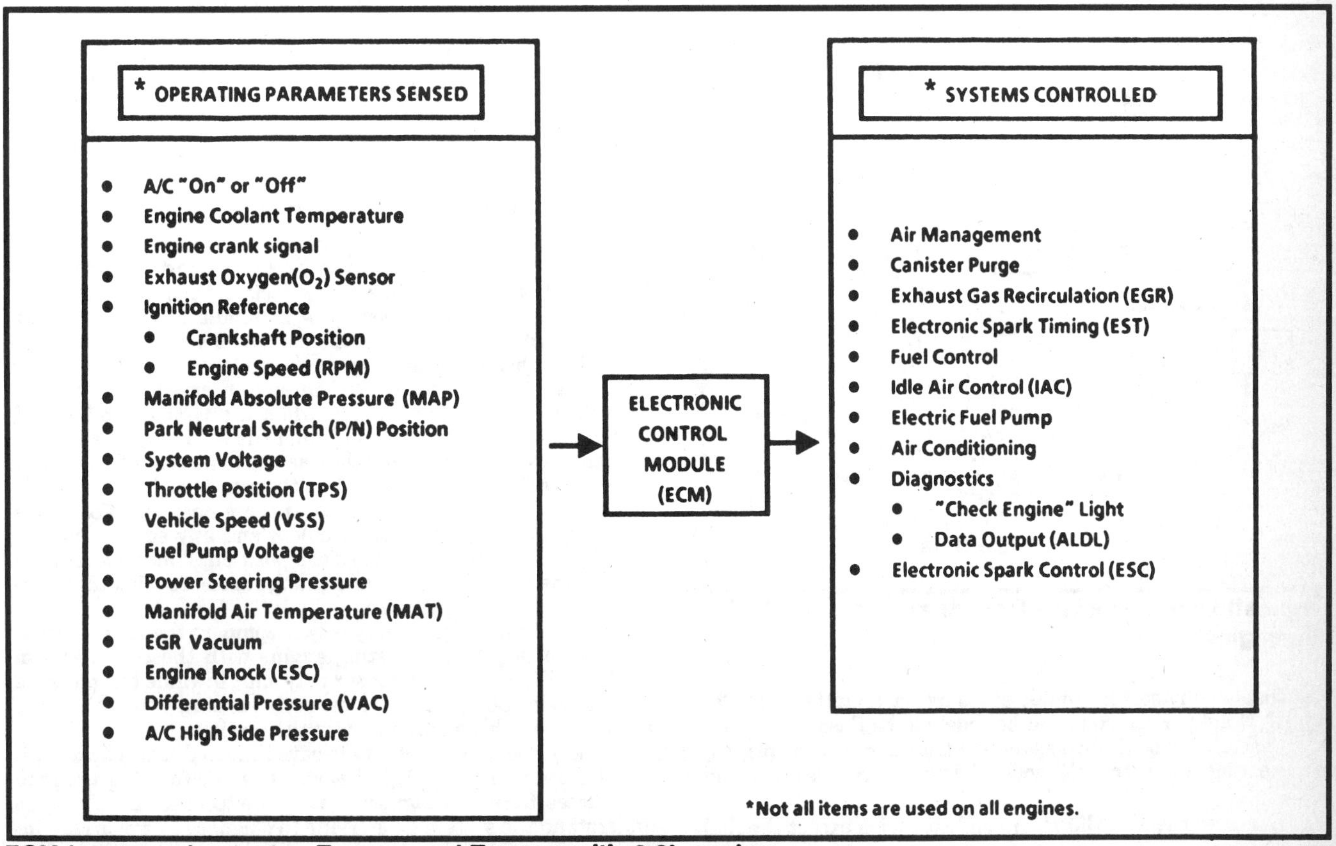

*** OPERATING PARAMETERS SENSED**

- A/C "On" or "Off"
- Engine Coolant Temperature
- Engine crank signal
- Exhaust Oxygen($O_2$) Sensor
- Ignition Reference
  - Crankshaft Position
  - Engine Speed (RPM)
- Manifold Absolute Pressure (MAP)
- Park Neutral Switch (P/N) Position
- System Voltage
- Throttle Position (TPS)
- Vehicle Speed (VSS)
- Fuel Pump Voltage
- Power Steering Pressure
- Manifold Air Temperature (MAT)
- EGR Vacuum
- Engine Knock (ESC)
- Differential Pressure (VAC)
- A/C High Side Pressure

→ **ELECTRONIC CONTROL MODULE (ECM)** →

*** SYSTEMS CONTROLLED**

- Air Management
- Canister Purge
- Exhaust Gas Recirculation (EGR)
- Electronic Spark Timing (EST)
- Fuel Control
- Idle Air Control (IAC)
- Electric Fuel Pump
- Air Conditioning
- Diagnostics
  - "Check Engine" Light
  - Data Output (ALDL)
- Electronic Spark Control (ESC)

*Not all items are used on all engines.

**ECM inputs and outputs—Trooper and Trooper with 2.8L engine**

F E D C B A
G M

TERMINAL IDENTIFICATION

| | | | |
|---|---|---|---|
| A | GROUND | E | SERIAL DATA |
| B | DIAGNOSTIC TERMINAL | F | TCC (IF USED) |
| C | A.I.R. (IF USED) | G | FUEL PUMP (IF USED) |
| D | CHECK ENGINE LIGHT (IF USED) | M | SERIAL DATA (IF USED) |

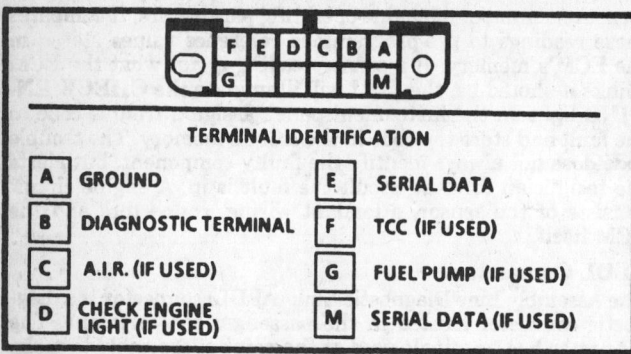

**ALDL connector terminal identification – Trooper and Trooper II with 2.8L engine**

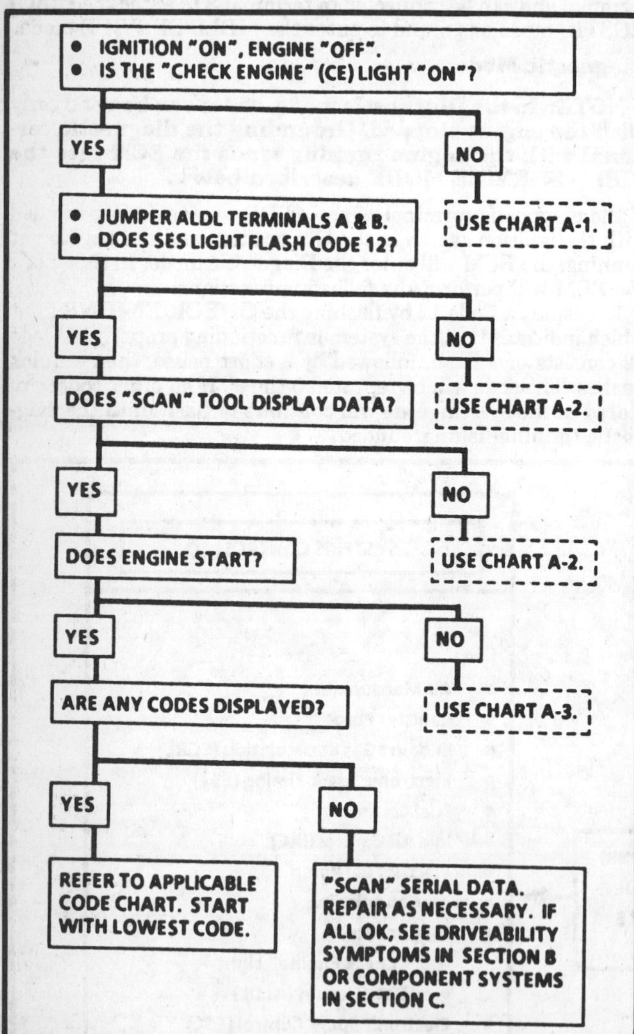


**Diagnostic circuit check – Trooper and Trooper II with 2.8L engine**

2. Display any stored trouble codes by flashing the **CHECK ENGINE** light. Each code will be code will be flashed 3 times, then Code 12 will be flashed again. If a trouble code is displayed, use the diagnostic trouble code charts to troubleshoot the system.

3. Energize the all ECM relays and solenoids except the fuel pump relay.

4. Move the AIC valve to the fully extended position.

### Field Service Mode

The Field Service Mode is entered by grounding the diagnostic terminal with the engine running and will indicate to the technician whether the ECM is in "Open" or "Closed" loop.

In Open Loop, the **CHECK ENGINE** light flashes 2½ times a second.

In Closed Loop, the **CHECK ENGINE** light flashes once per second.

In Closed Loop, the light will stay out most of the time if the system is too lean. The light will stay lit most of the time if the system is too rich.

When the system is in the Field Service Mode:

1. New trouble codes cannot be stored in the ECM.
2. The closed loop timer is by-passed.

### Clearing Trouble Codes

When the ECM sets a trouble code, the **CHECK ENGINE** light will illuminate and the code will be stored in the computer's memory. If the problem is intermittent, the light will go out in 10 seconds when the fault goes away. However, the code will remain in the ECM's memory until battery voltage to the ECM is removed. Removing battery voltage for 30 seconds will clear all stored trouble codes. Trouble codes will be cleared from the ECM after the fault is repaired.

**CAUTION**

***To prevent damage to the ECM, the ignition key must be turned to the OFF position when disconnecting voltage to the ECM (battery cable, ECM pigtail, ECM fuse, jumper cables).***

### ECM Learning Ability

The ECM has an inherent "learning" ability which allows the computer to make corrections for minor variations in the fuel system to improve driveability. If the battery cable was disconnected to clear a diagnostic code, or if a repair has been performed, the "learning" process has to be re-initiated. After a repair, a change may be noticed in the vehicle's normal performance. To "teach" the vehicle, warm up the engine to normal operating temperature and drive at part throttle with moderate accelerator and idle conditions until normal performance resumes.

### Scan Tools Used with Intermittents

In some scan tool applications, such as when trying to detect an intermittent problem which lasts for a very short time, the data update rate makes the tool less effective than a voltmeter, . However, the scan tool does allow manipulation of the wiring harness or components under the hood with the engine not running while observing the scan tool's readout.

The scan tool can be plugged in and observed while driving the vehicle under the condition when the **CHECK ENGINE** light turns on momentarily or when the engine driveability is momentarily poor. If the problem seems to be related to certain parameters that can be checked on the scan tool, they should be checked while driving the vehicle. If there does not seem to be any correlation between the problem and any specific circuit, the scan tool can be checked on each position. Watching for a period of time to see if there is any change in the reading that indicates intermittent operation.

The scan tool is also an easy way to compare the operating parameters of a poorly operating engine with those of a known good one. For example, a sensor may shift in value but not set a trouble code. Comparing the sensor's reading with those of a known good vehicle may uncover the problem.

The scan tool has the ability to save time in diagnosis and prevent the replacement of good parts. The key to using the scan tool successfully for diagnosis lies in the technician's ability to understand the system he is trying to diagnose as well as an understanding of the scan tool's operation and limitations. The technician should read the tool manufacturer's operating manual to become familiar with the tool's operation.

### PULSAR, AMIGO, PICK-UP, TROOPER AND TROOPER II – EXCEPT 2.8L ENGINE

The self-diagnosis system is so designed that the circuits handling the input signals from the sensors and output signals for the driving actuator are continuously monitored by the control unit. In the event of a failure, the control unit stores it in memory and operates the **CHECK ENGINE** light on the instrument panel when the nature of trouble is important, to warn the operator of failure.

**NOTE: The self-diagnosis system is capable of troubleshooting the electrical circuits in the Closed Loop Emission Control system only and does not cover the trouble in the sensors, actuators or the engine itself.**

When a failure has developed in the following systems while driving, the **CHECK ENGINE** light within the instrument panel is operated to warn the driver of a system failure.

1. Air flow sensor system
2. Coolant temperature sensor
3. Fuel injector system
4. Micro computer
5. Oxygen sensor
6. Vehicle speed sensor
7. Knock sensor system (turbo only)

#### Trouble Codes

When the diagnosis lead in the vicinity of the control unit is connected with the ignition switch in the **ON** position (engine not running), the trouble code stored in the memory is displayed by the **CHECK ENGINE** light.

With the engine not running, first a Code 12 is flashed 3 times, which indicates that the ECM is functioning normally. If there are other trouble codes in the ECM, it displays each one 3 times in ascending order of code numbers.

The malfunction codes are also displayed when the engine is running, but a Code 12 will not be indicated.

The actual trouble code is indicated by intermittent flashing of the **CHECK ENGINE** light which represent the digits of the code.

The flashing consists of patterns and individual patterns will repeat to indicate the fault. When there is more than 1 fault in the ECM's memory, related patterns will be repeated 3 times each. When all the codes are displayed, the computer will go through the entire code display again.

When the engine is not running, a Code 12 is indicated first followed by the code with the smallest number. In this mode, each code is indicated three times at 3.2 second intervals. When all the codes are displyed, Code 12 is flashed again then the entire code cycle is repeated.

#### Clearing Trouble Codes

After completion of the service operation , clear the trouble codes stored in memory by disconnecting the 60A main fuse in the fuse junction block, then check that only Code 12 is displayed.

**NOTE: All the codes stored in memory will be cleared automatically when the 13-pole connector in the control unit is disconnected. Since all the memory will be cleared when number 4 fuse is disconnected, it will be necessary to reset the clock and other electrical equipment.**

#### Tools Needed For Inspection

1. Ohmmeter
2. Voltmeter
3. Test light
4. 1.5V dry cell battery

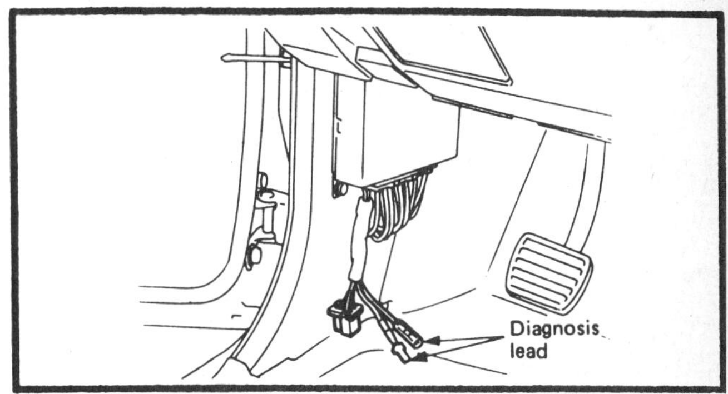


**Diagnostic lead location – Amigo and Pick-Up**

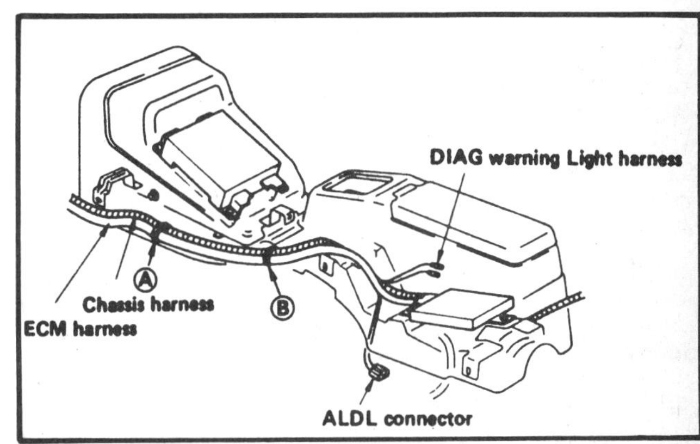


**ALDL and diagnostic lead connector location – Trooper and Trooper II (except 2.8L engine)**

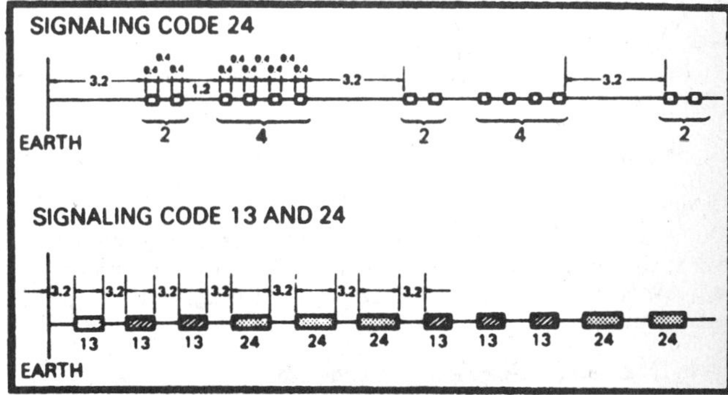


**Trouble code indication – Amigo, Pick-Up, Trooper and Trooper II (except 2.8L engine)**

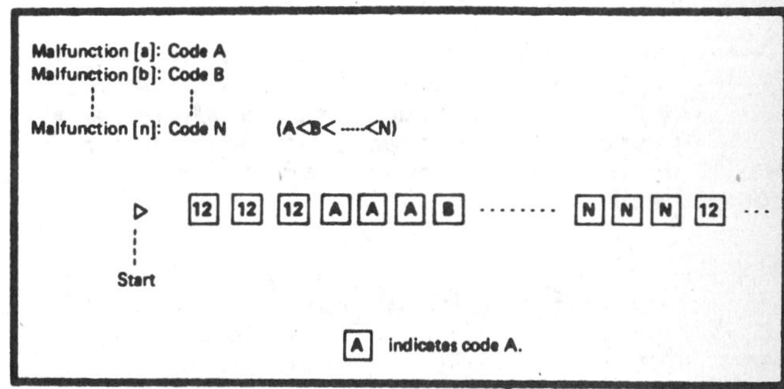


**Engine not running trouble code indication – Amigo, Pick-Up, Trooper and Trooper II (except 2.8L engine)**

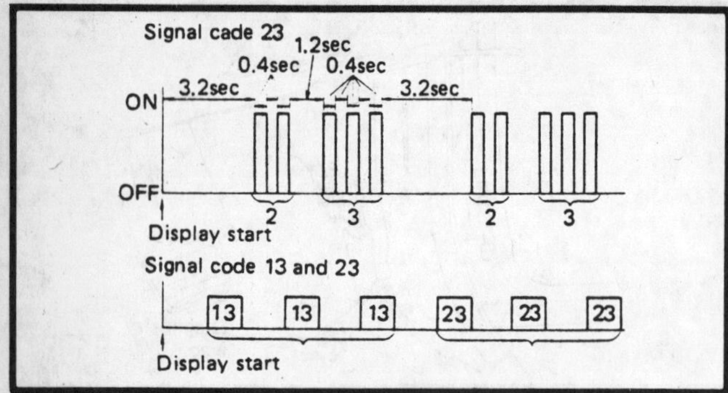

Trouble code display signals

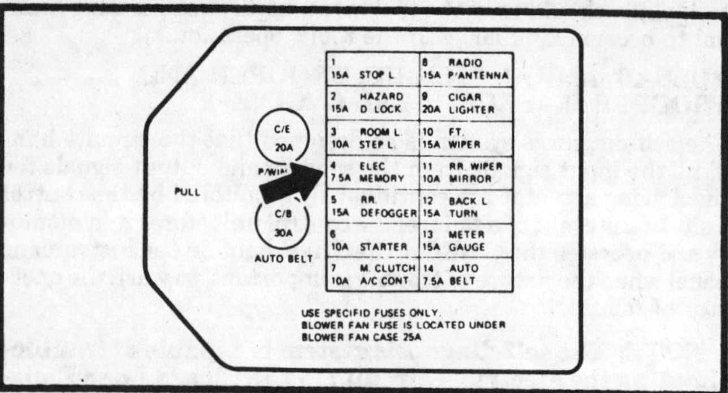

Number 4 fuse location – Impulse

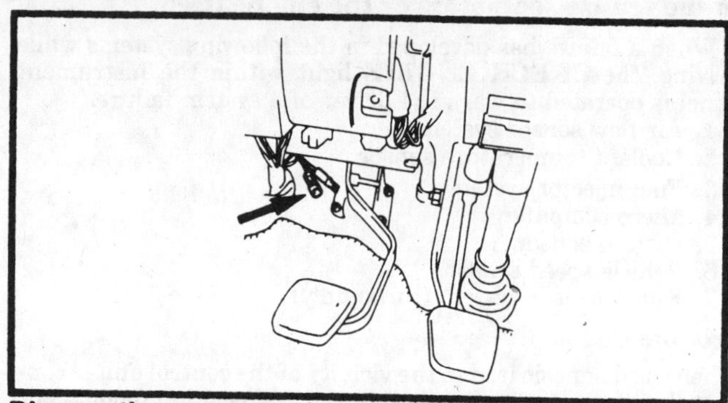
Diagnostic connector location – Impluse

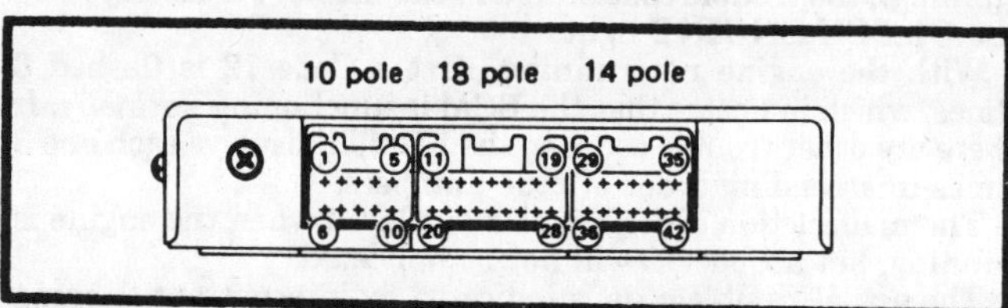

Control unit harness connector terminal identification – Amigo, Pick-Up, Trooper and Trooper II (except 2.8L engine)

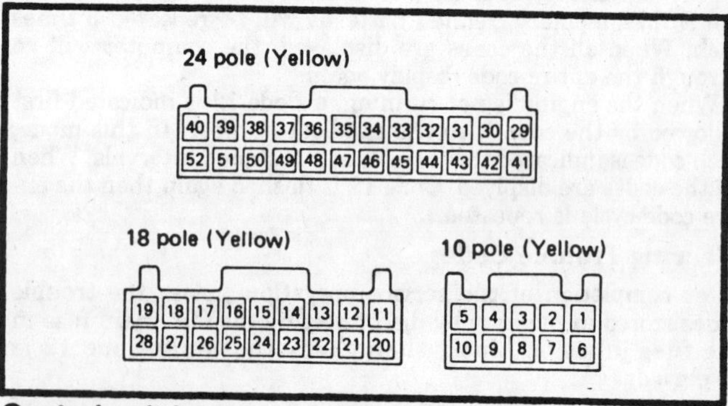

Control unit harness connector terminal identification – Impulse

5. A 3- meter wire with pin and alligator clip
6. 4 jumper wires

**How To Read Code Numbers On The Control Unit Harness Connector Terminals**

The control unit harnesses are connected with 3 types of connectors, each of which has specific numbers. In the following, inspection procedures will be described with reference made to the specific numbers.

**NOTE: When checking the system, note the numbers carefully to avoid the wrong connection since battery power is applied to some terminals only when the ignition switch is in the ON position.**

**Inspection Procedure**

The inspection based on the monitor codes should be performed in the following steps:

1. When trouble is not found through checks in Steps 1 and 2, proceed as follows:
   a. Clear the memory.
   b. Reconnect the circuit properly.
   c. Road test the vehicle.
   d. Obtain the code display to see if the trouble has been corrected.
   e. If the trouble persists, replace the control unit.
2. When making a continuity test or a short–out test on the circuit, disconnect the wiring at the control unit, sensor or actuator.
3. When checking the control unit harness connector terminals, make the connection by inserting the pin at the end of the lead. Avoid connecting the tester probe directly to the control unit terminal.

**Tools Needed to Service the System**

The system requires a scan tool, tachometer, test light, ohmmeter, digital voltmeter with a 10 megohms impedance, vacuum gauge and jumper wires for diagnosis. A test light or voltmeter must be used when specified in the procedures.

**NOTE: Some vehicles will use more sensors than others. Also, a complete general diagnostic section is outlined. The steps and procedures can be altered as required (if necessary) by the technician according to the specific model being diagnosed and the sensors it is equipped with.**

## FUEL PRESSURE

### Test

**I-MARK**

1. Loosen the clip on the fuel hose between the pressure regulator and the fuel distribution pipe. Disconnect the fuel hose carefully.

**NOTE: The fuel line is still under pressure. When pulling out the fuel hose, cover up the joint with a shop towel to prevent the gasoline from spraying.**

2. Connect a suitable fuel pressure gauge (J–33945 for turbo and J–34730–B for non-turbo) across the fuel pressure regulator and fuel distributor pipe correctly.

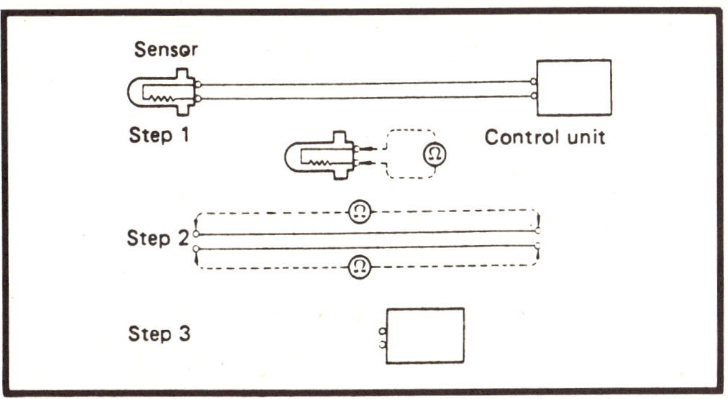


**Inspecting the control unit**

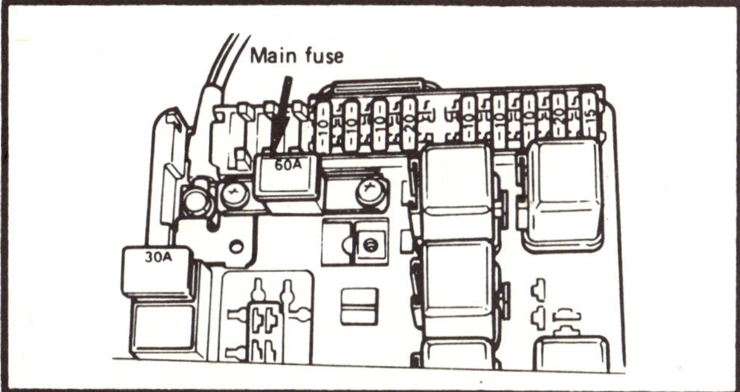


**Main fuse location—Amigo and Pick-Up**

3. Start the engine and measure the fuel pressure under the two different conditions listed below:

a. With the vacuum hose of the pressure regulator disconnected (intake manifold end of the hose plugged), the pressure should be 35.6 psi.

b. With the vacuum hose of the pressure regulator connected (and an idle speed of 900 rpm), the pressure should be 28.4 psi.

**NOTE: The following method can be used to operate the fuel pump without running the engine. Operate the fuel pump by applying battery voltage directly to the RED/BLU (turbo) or RED/GRN (non-turbo) wires connected to the fuel pump relay within the right side of the engine compartment.The measurement as described in Step 3, cannot be obtained when this method of measurement is employed.**

4. Remove the fuel pressure gauage and reconnect the fuel lines. Start the engine and check for fuel leaks.

5. If the fuel pressure was too low, use the following procedure. Be sure to check the fuel pressure with battery voltage applied directly to the fuel pump using suitable jumper wires:

a. If the fuel pressure does not reach normal level, check the fuel pump circuit and correct as necessary. The fuel pump relay may be defective.

b. If the fuel pressure remains unchanged, check for a restriction in the hose on the intake side of the fuel pump. The pressure regulator may be defective or the fuel pump may be malfunctioning. Replace parts as necessarys.

6. If the fuel pressure is too high, the pressure regulator may be defective; replace as necessary. There could be a restriction in the fuel return circuit; clean it as necessary. The fuel pump may be defective, replace as necessary.

7. If the fuel pump pressure lowers immediately after the fuel pump stops, the fuel pump may be defective. The pressure regulator may be defective; replace as necessary. The fuel injectors may be leaking.

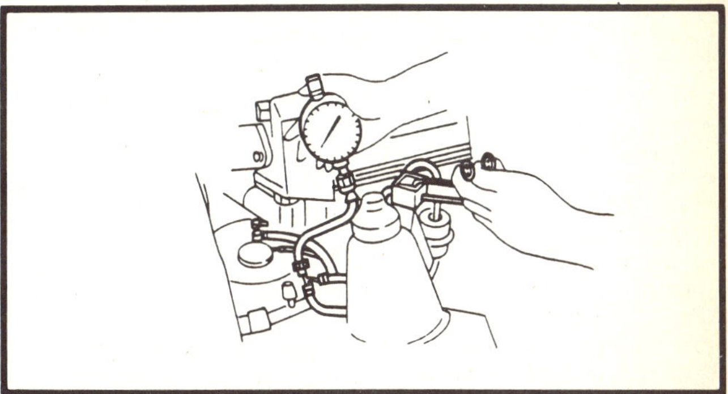

**Connecting the fuel pressure gauge—I-Mark, Impulse and 1988 Trooper II**

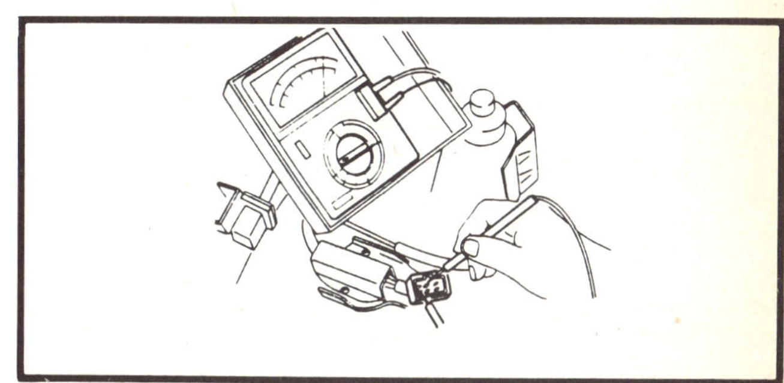

**Measuring the reisistance at the dropping resistor**

**IMPULSE AND 1988 TROOPER II**

1. Loosen the clip on the fuel hose between the pressure regulator and the fuel distribution pipe. Disconnect the fuel hose carefully.

**NOTE: The fuel line is still under pressure. When pulling out the fuel hose, cover up the joint with a shop towel to prevent the gasoline from spraying.**

2. Connect a suitable fuel pressure gauge (J–33945 or equivalent) across the fuel pressure regulator and fuel distributor pipe correctly. Disconnect the vacuum switching valve at the harness connector.

3. Start the engine and measure the fuel pressure under the 4 different conditions as listed below:

a. With the vacuum hose of the pressure regulator disconnected (intake manifold end of the hose plugged), the pressure should be 42.6 psi.

b. With the vacuum hose of the pressure regulator connected (and an idle speed of 900 rpm), the pressure should be 35.6 psi.

c. Engine stopped and vacuum hose disconnected, the pressure should be 35.6 psi.

d. Battery voltage applied directly to the vacuum switching valve under condition 3b above. The pressure should be 42.6 psi.

If the fuel pressure remains unchanged through the check in Step 3d, replace the vacuum switching valve.

**NOTE: The following method can be used to operate the fuel pump without running the engine. Remove the fuel pump relay from the fuse box and remove the fuse box attaching bolts. Remove the right hand wire harness clip and turn the fuse box over. Operate the fuel pump by applying battery voltage directly to cable 1.25BY connected to the fuel pump relay within the relay box. The measurement as described in Step 3b can not be obtained when this method of measurement is employed.**

4. Remove the fuel pressure gauge and reconnect the fuel lines. Start the engine and check for fuel leaks.

5. If the fuel pressure was too low, use the following procedure. Be sure to check the fuel pressure with battery voltage applied directly to the fuel pump using suitable jumper wires:

a. If the fuel pressure does not reach normal level, check the fuel pump circuit and correct as necessary. The fuel pump relay may be defective.

b. If the fuel pressure remains unchanged, check for a restriction in the hose on the intake side of the fuel pump. The pressure regulator may be defective or the fuel pump is malfunctioning. Replace parts as necessary.

6. If the fuel pressure is too high, the pressure regulator may be defective; replace as necessary. There could be a restriction in the fuel return circuit. The fuel pump may be defective; check and replace as necessary.

7. If the fuel pump pressure lowers immediately after the fuel pump stops, the fuel pump may be defective. The pressure regulator may be defective or a fuel injector nmay be leaking.

### 1989-90 TROOPER – TROOPER II (EXCEPT 2.8L ENGINE), AMIGO AND PICK-UP

1. Loosen the clip on the fuel hose betewen the pressure regulator and the fuel distribution pipe. Disconnect the fuel hose carefully.

**NOTE: The fuel line is still under pressure. When pulling out the fuel hose, cover up the joint with a shop towel to prevent the gasoline from spraying.**

2. Connect a suitable fuel pressure gauge (J-35957-1 and 2 or equivalent) across the fuel pressure regulator and fuel distributor pipe correctly. Disconnect the vacuum switching valve at the harness connector.

3. Start the engine and measure the fuel pressure under the 4 different conditions as listed below:

a. With the vacuum hose of the pressure regulator disconnected (intake manifold side end of the hose plugged), the pressure should be approximately 42 psi.

b. With the vacuum hose of the pressure regulator connected (and an idle speed of 900 rpm), the pressure should be approximately 35 psi.

c. Battery voltage applied directly to the vacuum switching valve under condition B above. The pressure should be approximately 42.6 psi.

d. Vacuum hose disconnected and plugged from the pressure regulator. Stop engine and check for fuel leakage. Fuel pressure must remain constant at or above 35 psi for at least 4 minutes after the engine is turned off.

**NOTE: The following method can be used to operate the fuel pump without running the engine. Remove the fuel pump relay from the fuse and relay box and short terminal 1 (1.25BY) and terminal 3 (1.25BL) of the relay female terminals with a jumper wire as shown.**

4. Remove the fuel pressure gauage and reconnect the fuel lines. Start the engine and check for fuel leaks.

5. If the fuel pressure was to low use the following procedure (be sure to check the fuel pressure with battery voltage applied directly to the fuel pump using suitable jumper wires:

a. If the fuel pressure does not reach normal level, check the fuel pump circuit and correct as necessary or the fuel pump relay may be defective.

b. If the fuel pressure remains unchanged, check for a restriction in the hose on the intake side of the fuel pump. The pressure regulator may be defective or the fuel pump is malfunctioning, replace parts as necessary.

### 1989-90 TROOPER – TROOPER II WITH 2.8L ENGINE

A fuel system pressure test is part of several diagnostic charts and symptom checks. To perform this test, proceed as follows:

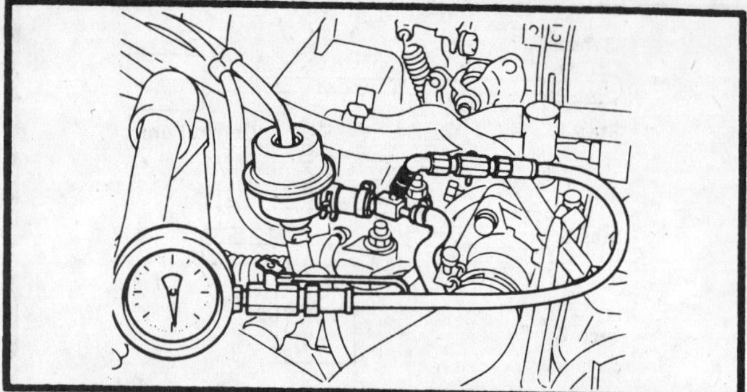

**Fuel pressure gauge hook-up – 1989–90 Trooper and Trooper II (except 2.8L engine), Amigo and Pick-Up**

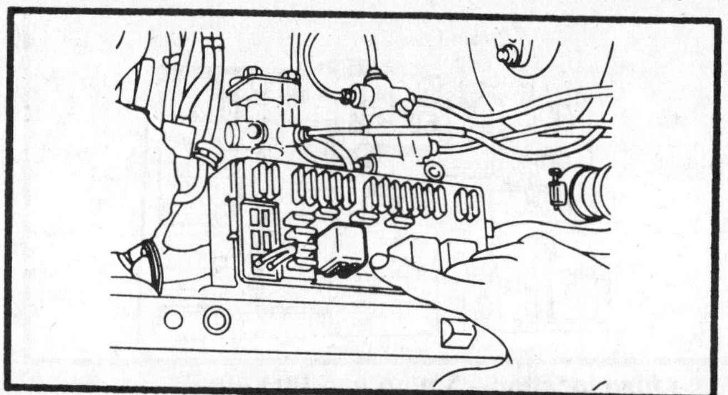

**Fuel pump relay location – 1989–90 Trooper and Trooper II (except 2.8L engine), Amigo and Pick-Up**

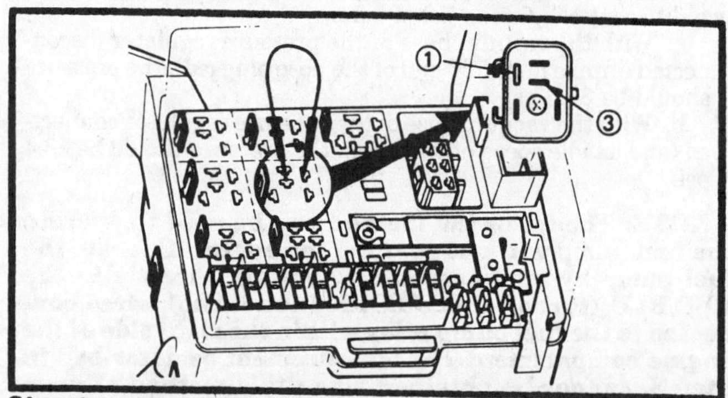


**Shorting the fuel pump relay terminals – 1989–90 Trooper and Trooper II (except 2.8L engine), Amigo and Pick-Up**

1. Disconnect the negative battery cable.

2. Loosen the fuel filler cap to relieve the vapor pressure from the fuel tank.

3. Relieve the fuel pressure from the fuel system. Turn the ignition **OFF**. At this point all the fuel presure will be relieved from the system because the internal constant bleed feature of the TBI unit automatically relieves the pressure as soon as the ignition switch is turned to the **OFF** position.

4. Uncouple the fuel supply flexible hose in the engine compartment and install fuel pressure gauge J-29658-B and adpater J-29658-85 or equivalents between the steel line and the flex hose.

5. Be sure to tighten the fuel line to the gauge to ensure that there are no leaks during testing.

6. Connect the negative battery cable.

7. Start the engine and observe the fuel pressure reading. The fuel pressure should be 9–13 psi (62–90 kPa).

8. Relieve the fuel pressure and loosen the fuel filler cap to relieve the fuel tank vapor pressure. Remove the fuel pressure gauge and reinstall the fuel line. Be sure to install a new O-ring on the fuel feed line.
9. Start the engine and check for fuel leaks.

## FUEL INJECTOR

### Inspection

#### I-MARK

1. To check the operation of the injectors with the engine running:
   a. Using a stethoscope or suitable metal bar, check for a regular clicking sound which should alternately increase and decrease in frequency with increases and decreases in engine idle speed.
   b. If a constant clicking sound is not heard, the injector(s) is not functioning properly and must be replaced.
2. To test for leakage:
   a. Remove the injector inlet pipe and throttle valve assembly.
   b. Remove all the injectors together with fuel hoses.
   c. Apply battery voltage directly to the fuel pump relay at the 0.85 **BG** terminal for 1988 models and **BLK/GRN** terminal for 1989-90 models. Check the leakage rate from the injector tip after applying voltage to the fuel pump relay terminal.
   c. The leakage rate should be 2 drops per minute or less.
   d. If the leak rate is greater than 2 drops per minute, replace the injector.
3. To test injector resistance:
   a. Disconnect the fuel injector harness connector and measure the resistance across the injector teminals.
   b. The resistance should be 2-3 ohms.
   c. If the resistance is not as specified, replace the injector(s).

#### EXCEPT I-MARK

1. Disconnect the fuel injector harness at the connector. Using a suitable ohmmeter, measure the resistance across the terminals.
2. The standard resistance should be 2–3 ohms.
3. If the injector resistance deviates from the specified range, replace the injector.
4. With the engine running, check the fuel injector operating noise using a metal bar or stethoscope.
   a. Normal operation of the injector is indicated when a regular click is heard which varies with engine speed.
   b. If a regular click is not heard, the injector is malfunctioning and must be replaced.
5. Test for leakage as follows:
   a. Remove the common chamber assembly. Remove all the injectors with the fuel hoses still connected.
   b. Check for fuel leakage by operating the fuel pump with the battery voltage applied directly to the fuel pump relay terminal.
   c. The leakage should be less than 2 drops per minute.
   d. If the amount of leakage is beyond the set limit, replace the injector.
   e. Install the parts in the reverse order of the removal. Start the engine and check for fuel leaks.

## FUEL CUT-OFF SYSTEM

### Inspection

#### IMPULSE, AMIGO, PICK-UP, TROOPER AND TROOPER II

1. Start the engine and let it run to reach normal operating temperature. Turn the ignition switch to the **OFF** position.
2. Disconnect the throttle valve switch harness at the connector, insert a fine wire into the idle terminal (upper), then make the connections at the connector with the wire pulled part way out.

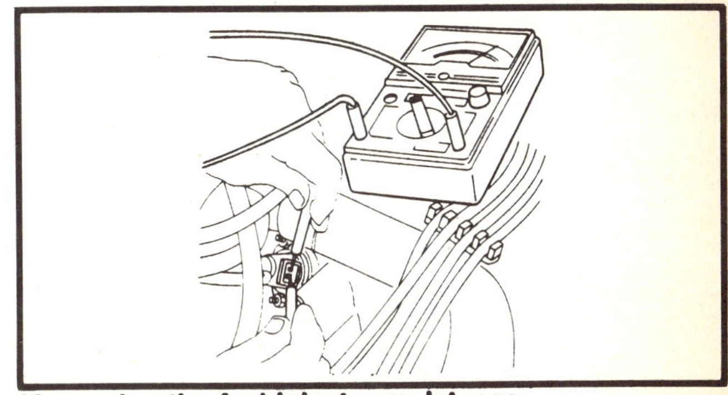

Measuring the fuel injector resistance

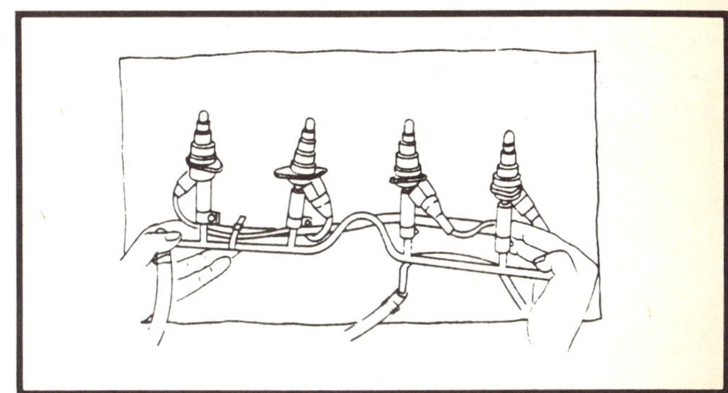

Checking for a fuel injector leak down

3. Disconnect the throttle position sensor at the connector. Use a suitable jumper wire to connect the harness side connector terminals.

**NOTE: Do not connect the harness side connector and throttle position sensor with the piece of wire in place. Shorting will result.**

4. Start the engine and keep the engine running at 2500–3000 rpm. Check to make sure the engine speed drops to approximately 1250 rpm when battery voltage is applied to the wire extending from the idle terminal.

**NOTE: Engine hunting may occur at approximately 1250 rpm, but does not indicate abnormal conditions.**

## DROPPING RESISTOR

### Inspection

#### ALL MODELS

1. Disconnect the wiring connector from the dropping resistor.
2. Using a suitable ohmmeter, measure the resistance across the center terminal of the dropping resistor side connectors and other terminals.
3. The standard resistance should be 5–7 ohms.
4. If the measured resistance deviates from the specified range, replace the dropping resistor assembly.

## AIR REGULATOR FAST IDLE

### Inspection

#### ALL MODELS EXCEPT I-MARK

The engine must be cold to perform this inspection.

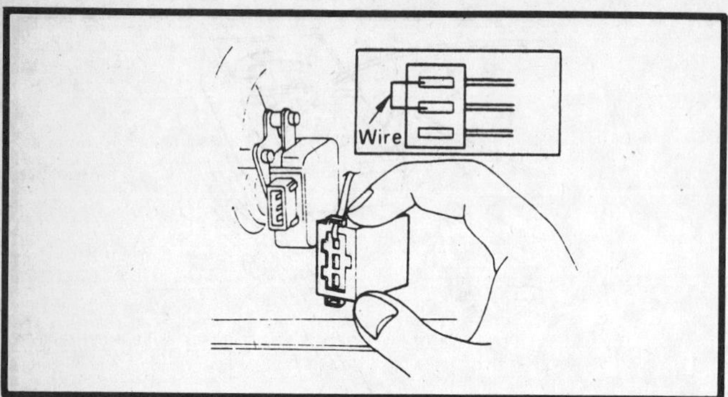

Placing the jumper wire into the throttle valve switch harness

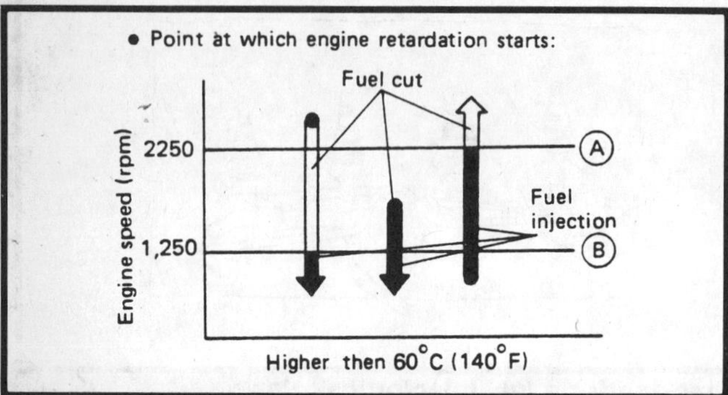

Fuel cut off system graph chart

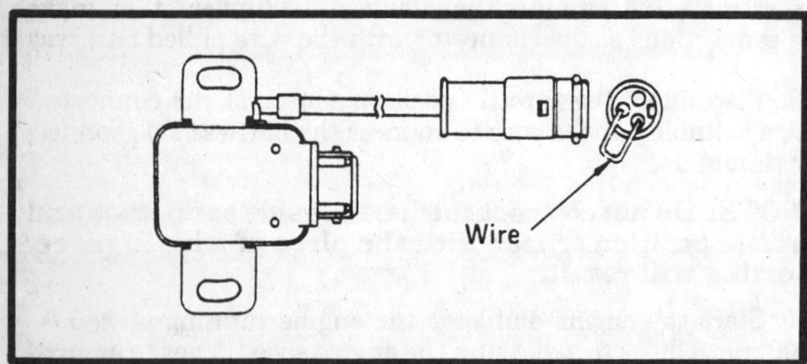

Placing the jumper wire into the throttle position sensor

**NOTE: This test does not apply to the I-Mark.**

1. The engine idling speed must be slightly higher than normal immediately after starting the engine.
2. The engine idle speed should normalize as the engine temperature increases.
3. When fast idle speed is not obtained, use the following procedure:
   a. Disconnect the air regulator harness at the connector.
   b. Using a suitable ohmmeter, measure the resistance across the air regulator terminals.
   c. The standard resistance should be 38–42 ohms.
   d. If the measured resistance deviates from the specified range, replace the air regulator assembly.
4. Turn the ignition switch to the **ON** position and, using a suitable voltmeter, check for voltage at the air regulator connectors.
   a. When voltage is present, check the wiring leading to the air regulator and correct as necessary. The fuel pump may be defective; replace it as necessary.
   b. When voltage is present, the air regulator is defective and should be replaced.

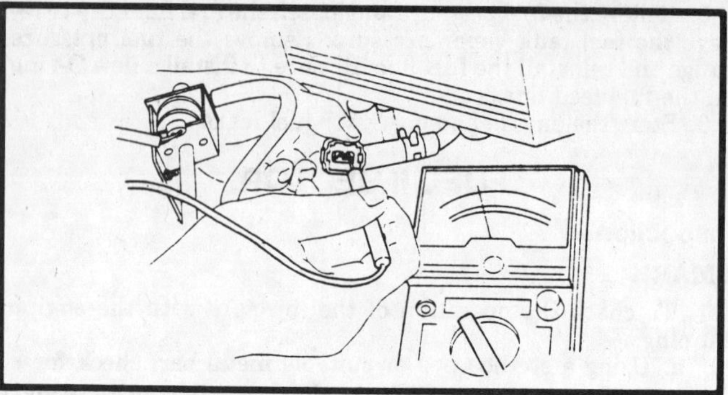
Measuring the air regulator resistance

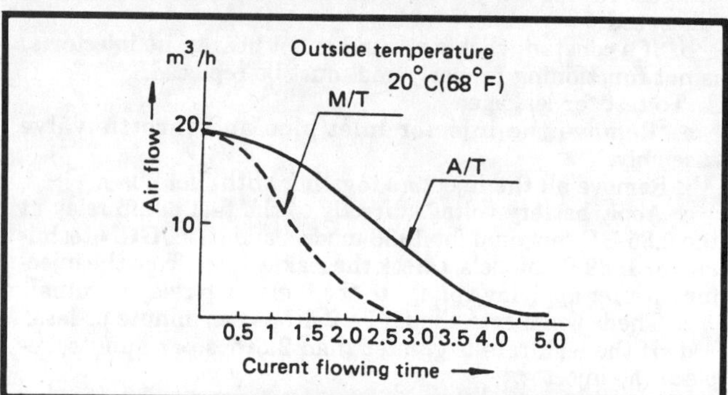

Air regulator fast idle air flow graph chart

## AIR REGULATOR

### Leakage Inspection

#### ALL MODELS EXCEPT I-MARK

**NOTE: This test does not apply to the I-Mark.**

Check for leakage after allowing the engine to warm up thoroughly.

1. Disconnect the hose between the air regulator and throttle valve body at the throttle valve body side.
2. Close the hole in the throttle valve body. Start the engine.
3. Check for variation in the engine speed by closing and opening the end of the air regulator hose (this test is made to determine the function of the air regulator).
4. The variation of the engine speed should be less than 50 rpm. If the variation in engine speed is more than specified, replace the air regulator assembly.

## THROTTLE VALVE SWITCH

### Inspection and Test

#### ALL MODELS EXCEPT I-MARK

**NOTE: This inspection and test does not apply to the I-Mark.**

1. Check that the operation of the accelerator pedal and control cable is smooth.
2. Check that the valve stopper returns and makes a firm contact with the throttle valve stopper screw when the throttle valve is released.

**NOTE: The throttle valve stopper screw is factory set and sealed with paint. Setting of the screw should not be disturbed unless absolutely necessary.**

3. Test the throttle switch as follows:
   a. Disconnect the throttle valve switch harness connector.

b. Using a suitable ohmmeter, make a continuity test across the terminals with the accelerator pedal depressed in steps.

c. With the accelerator pedal NOT depressed, the resistance should be 0 ohms at idle contact and infinity ohms at full contact when testing terminals **I** and **P**.

d. With the accelerator pedal slightly depressed, the resistance should be infinity ohms at idle contact and infinity ohms at full contact when testing terminals **I and P** and **P and F**.

e. With the accelerator pedal fully depressed, the resistance should be infinity ohms at idle contact and 0 ohms at full contact when testing terminals **I and P**.

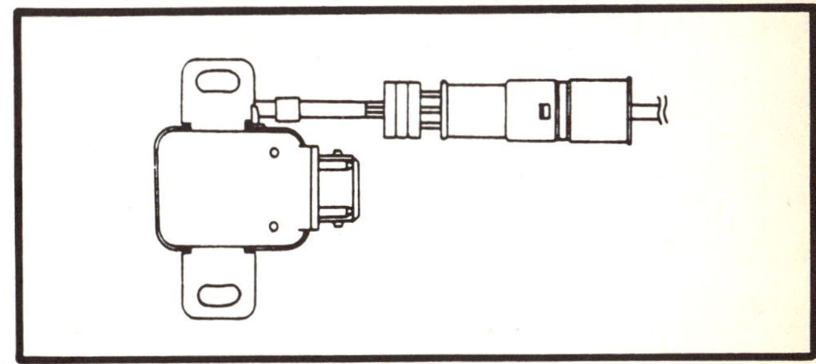
Testing the throttle position sensor

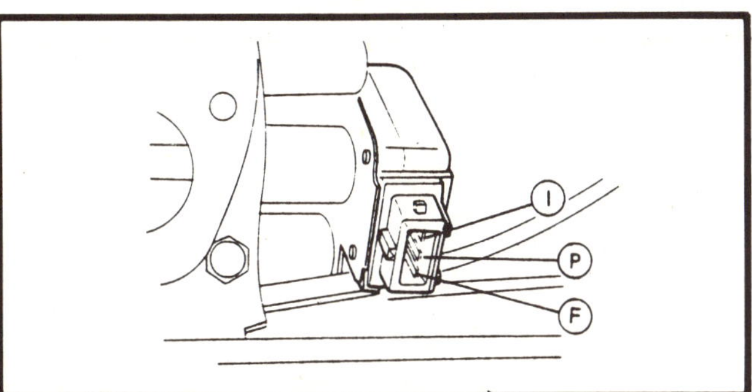

Making the throttle valve switch test

## THROTTLE POSITION SENSOR

### Inspection

**ALL MODELS EXCEPT I-MARK**

1. Turn the ignition switch to the **ON** position. Remove the water shield cover.
2. Using a suitable voltmeter, place the positive probe into the white color wire harness.
3. Measure the voltage between the white color harness of the throttle position sensor.

**NOTE: The throttle position sensor has 3 leads, red for 5 volt power source, white for output and black for ground.**

4. The voltage should be higher then 4 volts at idle contact and lower than 2 volts at full contact.

**NOTE: Make sure that 5 volts (± 0.5 volts) is measured at the red colored harness before making the voltage check in Step 4.**

5. Confirm the difference in the voltage of the idle contact and full contact is 3.6 ± 1 volt.
6. If the throttle position sensor fails any part of this inspection, replace it with a new one.

**I-MARK**

Before performing the adjustment, make sure that the acclerator pedal and control cable operate smoothly. Make sure the stopper valve returns and makes a firm contact with the throtle valve stopper screw when the throttle valve is released.

**NOTE: The throttle valve stopper screw is set and sealed at the factory. The screw setting should not be disturbed unless absolutely necessary.**

To adjust the throttle position sensor perform the following:

1. Make sure that the throttle valve is closed all the way.

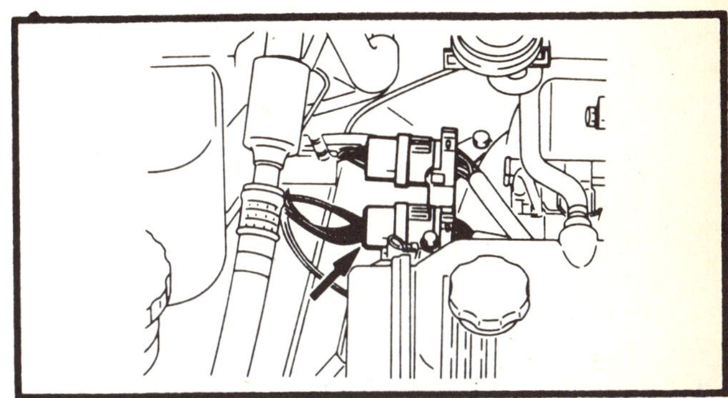
Location of H25 connector for TPS adjustment—I-Mark

2. Remove the water shield cover, but do not remove the connector (H25).
3. Place the positive (+) probe of a voltmeter in the red color wire (H25 pin No. 6).
4. Turn the ignition switch to the **ON** postion.
5. Measure the voltage between the red color wire and ground. The voltage should be between 0.31-0.39 volts.
6. If the voltage is not as specified, loosen the TPS attaching screw and move the TPS until the voltage is within specifications.

# Component Replacement

## FUEL INJECTOR

### Removal and Installation

1. Relieve the fuel system pressure and disconnect the negative battery cable.
2. Disconnect the air cleaner duct hose.
3. Disconnect the accelerator cable from the throttle body.
4. Remove the throttle valve assembly, as required.
5. Remove the suppport bracket from the common chamber and the cylinder head.
6. Move the charcoal canister out of the way, as required.
7. Remove the fuel line clips from the common chamber.
8. Remove the valve cover breather hose.
9. Disconnect the presure regulator vacuum line.
10. Disconnect the fuel return line from the fuel rail. Cover the line joint with a shop towel to absorb any excess fuel.
11. Disconnect the fuel injector harness connectors.
12. Remove the fuel rail attaching bolts.
13. Disconnect the fuel feed pipe from the fuel rail. Cover the line joint with a shop towel to absorb any excess fuel.
14. Withdraw the fuel rail and injector assembly from the cylinder head.
15. Installation of the fuel rail/injector assembly is performed in reverse of removal. Use new O-ring seals.

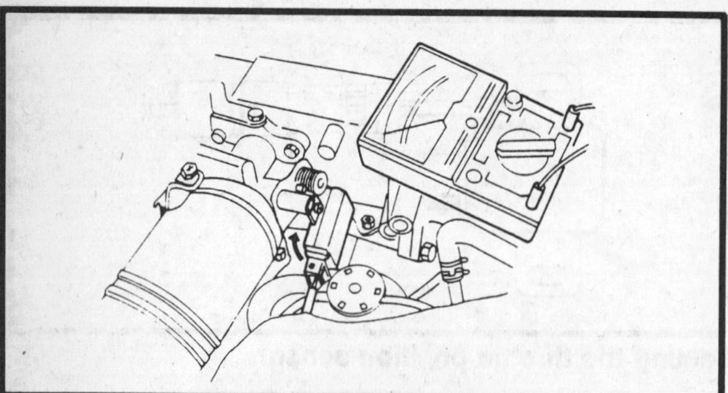

Making the throttle valve switch adjustment

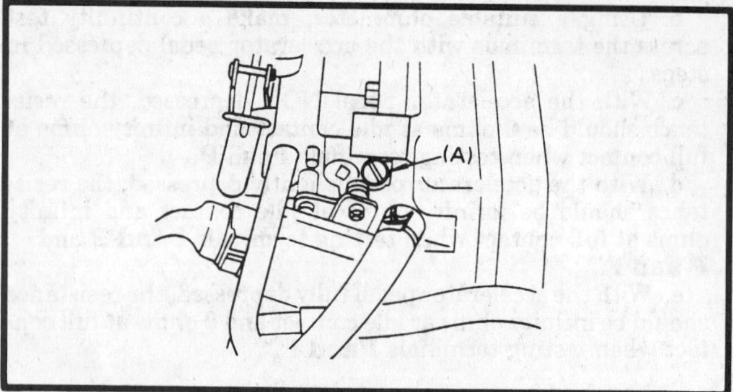


Location of the idle speed screw

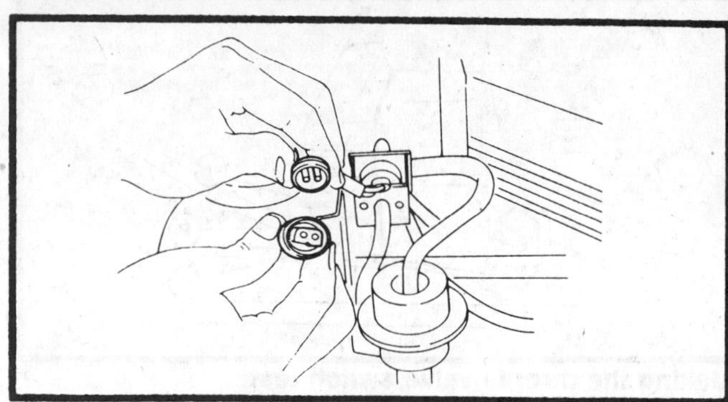

VSV connector location – I-Mark

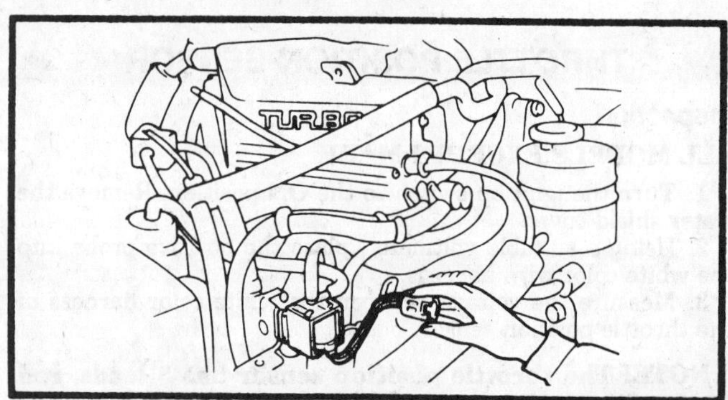

VSV connector location – Impulse

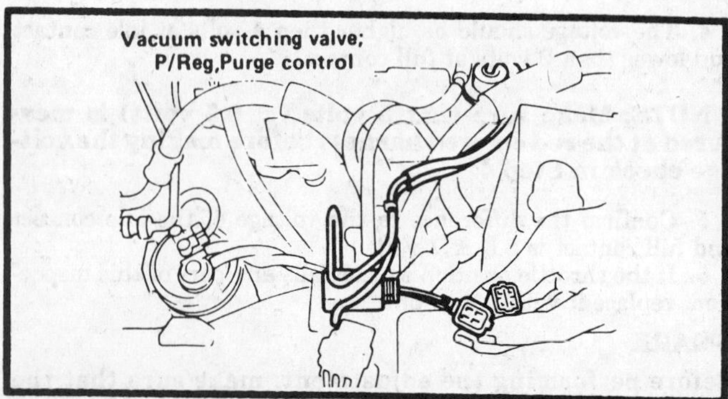


Pressure regulator VSV location – Amigo, Pick-Up, Trooper and Trooper II

## THROTTLE SWITCH

### Adjustment

#### ALL MODELS EXCEPT I-MARK, 1990 AMIGO AND PICK-UP

**NOTE: This adjustment does not apply to the I-Mark. For throttle switch adjustment on 1990 Amigo and Pick-Up, refer to separate procdure.**

1. Check that the throttle valve is completely closed. Loosen the throttle valve switch mounting screws slightly.
2. Using an ohmmeter and while checking the continuity across the terminals, turn the switch body in the clockwise direction until a continuity reading is obtained.
3. When a continuity is obtained, further turn the switch 1 degree in the clockwise direction and lock in that position.

**NOTE: Turning of the switch body 1 degree is equivalent to approximately 0.6 mm of a stroke of the throttle valve stopper bolt.**

#### 1990 AMIGO AND PICK-UP

1. Manually open the throttle valve by moving the throttle linkage away from the stop screws.
2. Place a 0.024 in. feeler gauge on top of the stop screw and allow the linkage to close on the feeler gauge.
3. Remove the connector from the throttle switch.
4. Connect the negative (–) lead of a suitable ohmmeter to the pin **P** of the switch and the positive (+) lead to pin **I**. The ohmmeter should read continuity across the switch.
5. If the ohmmeter does not read continuity, loosen the throttle valve switch mounting screws slightly and rotate the switch just to the point where continuity is read. At this point, tighten the mounting screws.
6. Depress the gas pedal and look at the ohmmeter. There should be no continuity when the gas pedal is depressed.
7. Move the throttle linkage to wide open position. Move the positive (+) lead from the **I** pin to the **F** pin and make sure that continuity exists.

## IDLE SPEED

### Adjustment

#### ALL MODELS

#### Conditions

Perform the idle speed adjustment under the following conditions:

- Front wheels blocked and parking brake firmly applied.
- Transmission in **N** (Manual trans.) or **P** position (automatic trans.)
- Engine warmed up to normal operating temperature.
- Throttle valve completely closed and idle contact on.
- All electrical accessories turned to the **OFF** position.
- A/C (if equipped) turned the **OFF** position.
- Harness for pressure regulator VSV disconected – all models.

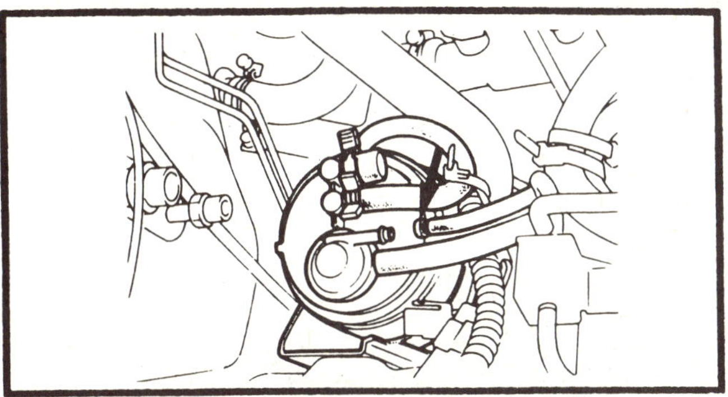
Canister purge line location – Amigo, Pick-Up, Trooper and Trooper II

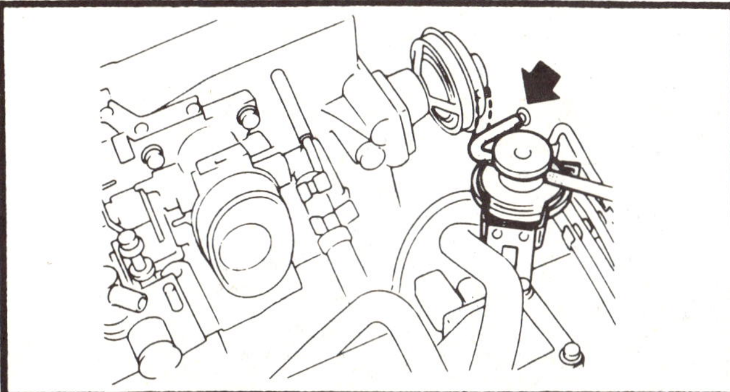
EGR vacuum signal line location – Amigo, Pick-Up, Trooper and Trooper II

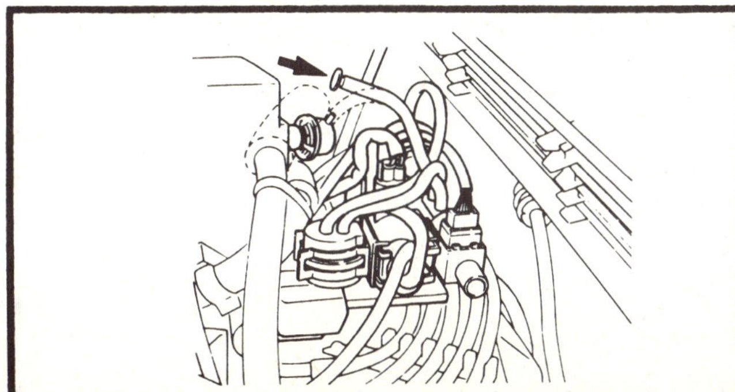
EGR cut VSV line location – 1989–90 California Amigo and Pick-Up

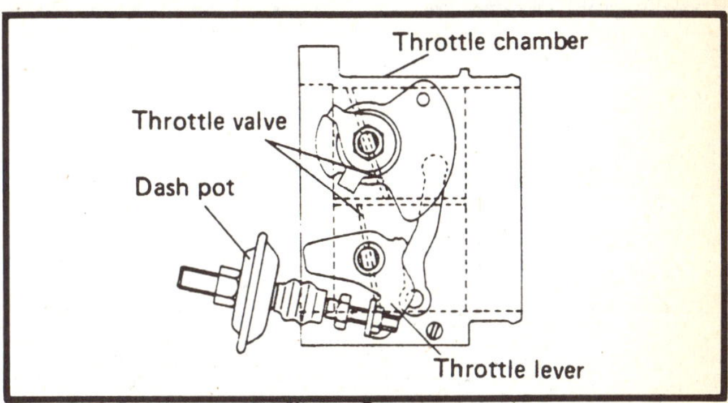

Making the dash pot adjustment

- Canister purge line disconnected and plugged – Amigo, Pick-Up, Trooper and Trooper II
- EGR vacuum signal line disconnected and plugged – Amigo, Pick-Up, Trooper and Trooper II
- EGR cut VSV disconnected and plugged – 1989-90 Amigo and Pick-Up (California only)

### Adjustment Procedure

1. Connect a tachometer to the engine.
2. Adjust the idle speed to the following specifications by turning the idle adjust screw **A** on the throttle body:
   a. I-Mark (turbo) – 950 rpm
   b. I-Mark (non-Turbo) – 900 rpm
   c. Impulse – 850-950 rpm
   d. Amigo and Pick-Up – 850-950
   e. Trooper and Trooper II – 850-950
3. After the idle speed adjustment, make sure to connect the VSV harness securely.

**NOTE: Check and clean the idle port(s) as necessary; restriction in the port(s) will cause fluctuation in the idle speed.**

## DASHPOT

### Adjustment

#### IMPULSE ONLY

1. Set the engine speed to 2000 rpm with the throttle lever.
2. Tighten the adjusting screw until it just makes contact with the dash pot shaft head.
3. Lock the adjusting screw in position with the lock nut.

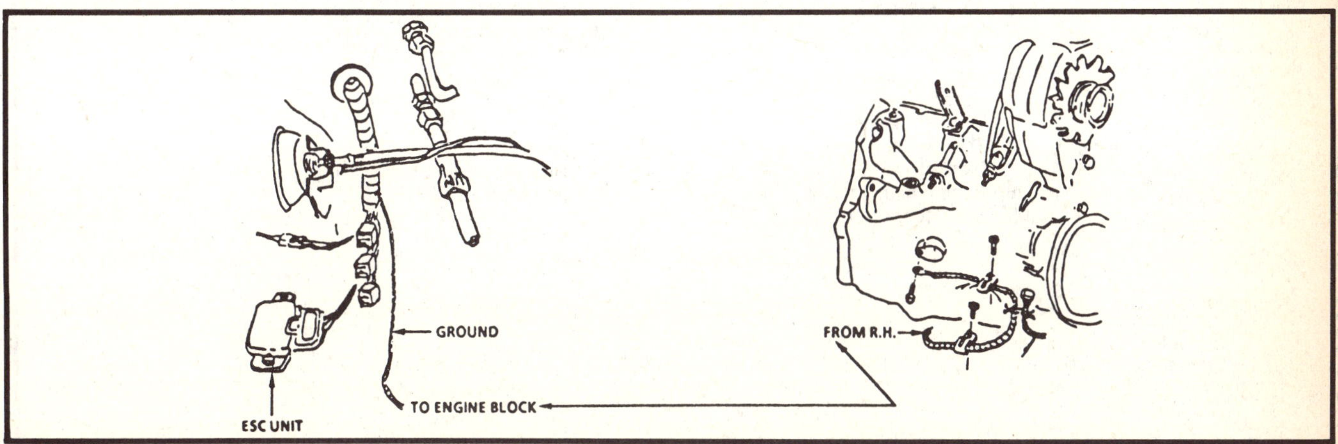

Engine ground strap location – V6 engine

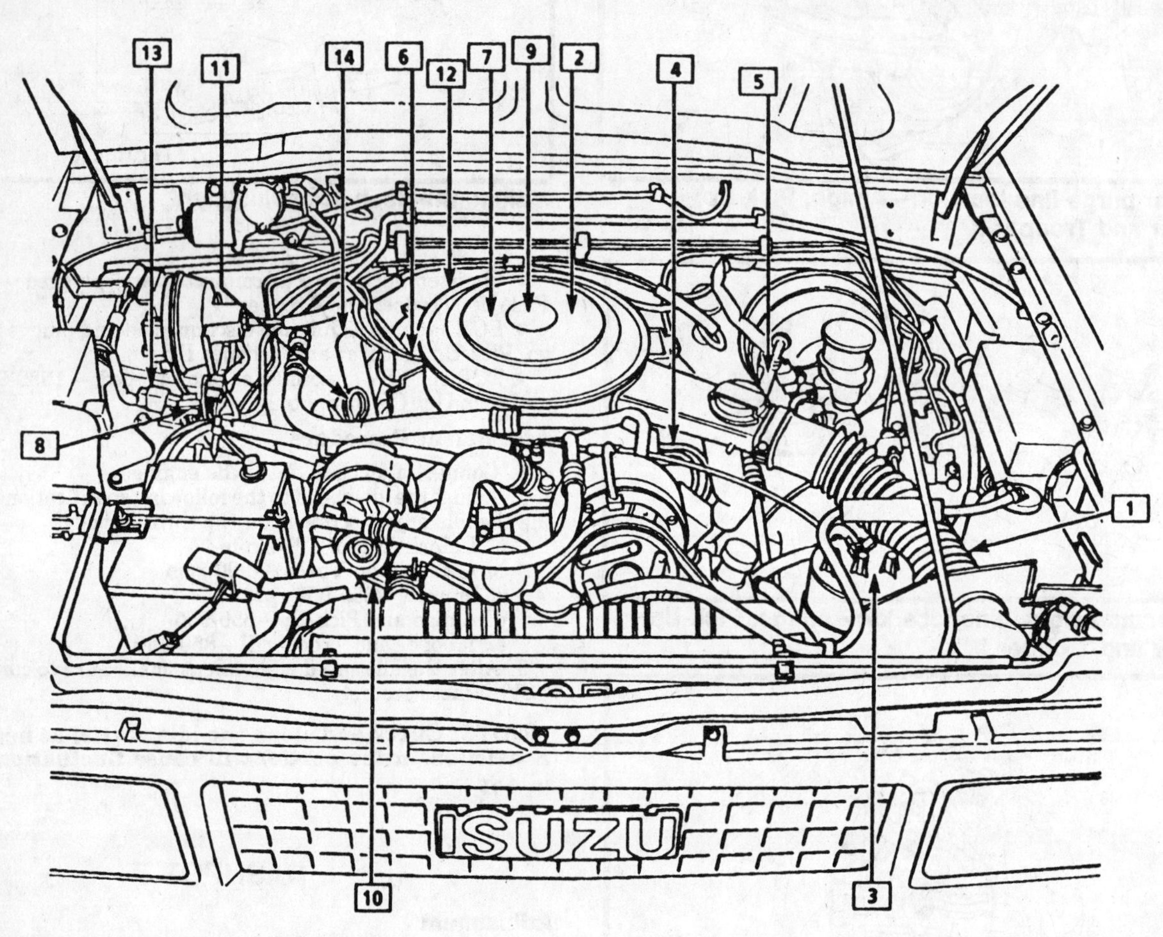


1. Air cleaner duct hose
2. Air cleaner
3. Canister
4. Fuel lines
5. Master vac vacuum hose
6. Vacuum hoses
7. Throttle cable and bracket
8. Engine harness bulk head connectors
9. Throttle body injector connectors
10. Ground strap
11. Alternator harness
12. Main harness and connector
13. Vacuum switching valves
14. Heater hoses, brackets and clips

Engine component locations – V6 engine

# MAZDA FUEL INJECTION SYSTEMS

**NOTE: On vehicles equipped with Audio Anti-Theft system, before performing any repairs, the Personal Code Number (P.C.N.) must be obtained from the individual owner of the vehicle, or an authorized Mazda dealer.**

## General Information

### FUEL SYSTEM

The fuel system, supplies the necessary fuel to the injectors to achieve combustion at a constant pressure. Fuel is metered and injected into the intake manifold according to the signals from the engine control unit.

The fuel system consists of the fuel pump, fuel filter, distribution pipe, pressure regulator, injectors, transfer pump (4WD models) and a circuit opening relay (1990 models).

### ENGINE CONTROL UNIT (ECU)

The ECU, through various input signals, monitors battery voltage, engine rpm, amount of air intake, cranking signal, intake temperature, coolant temperature, oxygen concentration in the exhaust gases, throttle opening, atmospheric pressure gearshift position, clutch engagement, braking, power steering operation and A/C compressor operation.

The ECU controls the operation of the fuel injection system, idle-up system, fuel evaporation system and ignition timing. The ECU has a built in fail-safe mechanism. If a fault occurs while driving, The ECU will substitute pre-programmed values. Driving performance will be affected, but the vehicle will still be driveable.

### PRESSURE REGULATOR CONTROL (PRC) SYSTEM

This system, prevents fuel lose during idle after the engine is hot re-started, by cutting the vacuum to the pressure regulator causing the fuel pressure to increase.

### EVAPORATIVE EMISSION CONTROL (EEC) SYSTEM

This system is controlled by signals sent from the water thermo sensor, intake air thermo sensor, air flow sensor and the engine speed sensor (ignition coil). The ECU calculates the engine operating conditions from the signals, and controls the EEC system by the operation of the solenoid valves for the No. 1 purge control valve and the vacuum switching valve.

| Terminal | Resistance (Ω) | |
|---|---|---|
| | Fully closed | Fully open |
| E2↔Vs | 20—400 | 20—1,000 |
| E2↔Vc | 100—300 | |
| E2↔VB | 200—400 | |
| E2↔THAA (Intake air temp. sensor) | -20°C ( -4°F) 13,600—18,400<br>20°C ( 68°F) 2,210— 2,690<br>60°C (140°F) 493— 667 | |
| E1↔Fc | ∞ | 0 |

**Air flow meter terminal specifications—All models except 1989 323 and 1990 Miata and 929 with DOHC engine**

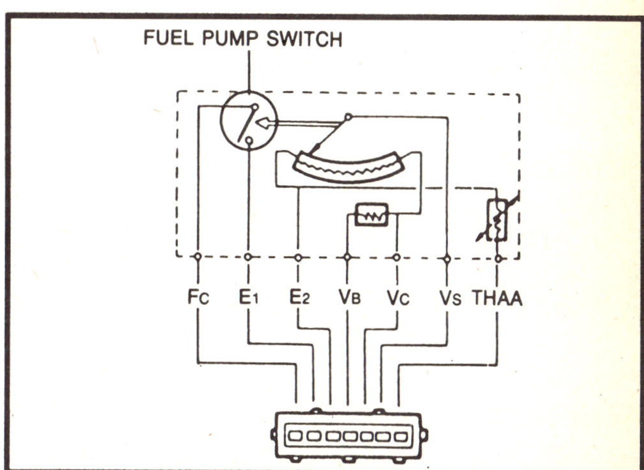


**Air flow meter terminal identification—All models except 1990 Miata and 929 with DOHC engine**

### IDLE SPEED CONTROL (ISC) SYSTEM

The ISC system controls the amount of air detected by the air flow meter, by regulating the amount of bypass air that passes through the throttle body, which helps in maintaining a steady idle speed.

## Diagnosis and Testing

**NOTE: For further diagnosis and testing of the fuel injection system and its related components, refer to the appropriate troubleshooting charts.**

### AIR FLOW METER

#### Inspection

**EXCEPT B2600 AND 1990 MPV WITH 4 CYLINDER ENGINE**

1. Inspect the air flow meter body for cracks. Using a suitable ohmmeter, check the resistance between the terminals.
2. Using a suitable tool, open the measuring plate. Measure the resistance between E1 and FC (fuel pump switch) and between E2 and VS.
3. The resistance between E1 and FC should be infinity when the measuring plate is fully closed. When the measuring plate is fully opened, the ohmmeter should read 0 ohms.
4. The resistance between E2 and VS should be 20–400 ohms when the measuring plate is fully closed and 20–1000 ohms when fully opened.
5. On all models except 1990 Miata, the resistance between E2 and VC should be 100–300 ohms.

| Terminal | Resistance (Ω) |
|---|---|
| E2 ↔ Vs | 20 — 400 |
| E2 ↔ Vc | 100 — 300 |
| E2 ↔ VB | 200 — 400 |
| E2 ↔ THA (Intake air thermosensor) | -20°C ( -4°F) 14,600—17,800<br>20°C ( 68°F) 2,210— 2,690<br>60°C (140°F) 493— 667 |
| E1 ↔ Fc | ∞ |

**Air flow meter terminal specifications—1989 323**

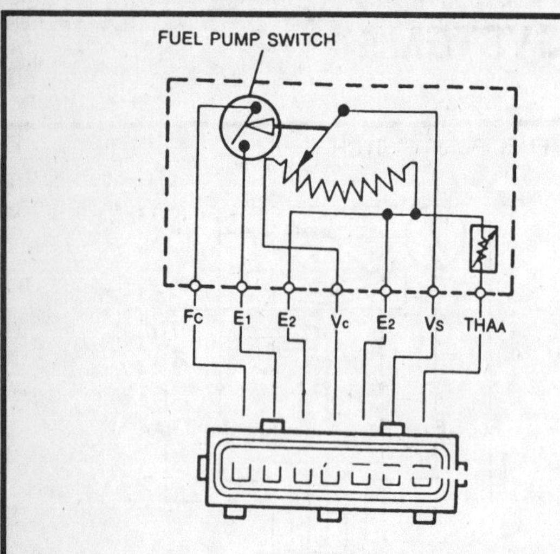


| Terminal | Resistance (Ω) | |
|---|---|---|
| | Fully closed | Fully open |
| E2↔Vs | 200—600 | 20—1,000 |
| E2↔Vc | 200—400 | |
| E2↔THAA (Intake air thermosensor) | −20°C (−4°F) 13,600—18,400<br>20°C (68°F) 2,210—2,690<br>60°C (140°F) 493—667 | |
| E1↔Fc | ∞ | 0 |

**Air flow meter terminal identification and resistance specifications – 1990 Miata and 929 with DOHC engine**

5. On 1990 Miata and 929 with DOHC engine, the resistance between E2 and VC should be 200–400 ohms.

6. On all models except 1990 Miata and 929 with DOHC engine, the resistance between E2 and VB should be 200–400 ohms when the measuring plate is fully opened or fully closed.

7. On all models, if specifications are not as indicated, replace the air flow meter.

## AIR FLOW SENSOR

### Inspection

**B2600 AND 1990 MPV WITH 4 CYLINDER ENGINE**

1. Slide back the rubber boot from the air flow sensor connector.

2. Using a suitable voltmeter, check the terminal voltages.

3. If specifications are not as indicated, check the wiring harness for an open circuit.

4. If no problem is found in the wiring harness, check the burn-off operation as follows:

a. Disconnect the battery ground cable, then reconnect it.

b. Start the engine allow it to reach normal operating temperature. Slide back the rubber boot from the air flow sensor connector.

c. With the transmission in **NEUTRAL**, operate the engine for approximately 3 minutes at 2000 rpm.

d. Turn the ignition switch to the **OFF** position.

e. On 1989 B2600 and 1990 MPV with 4 cylinder engine, using a suitable voltmeter, check the voltage of the air flow sensor wire (G/O) and terminal (2K) on the ECU. Voltmeter should read 0 volts when the ignition switch is turned **OFF** and 8–12 volts momentarily, 2–5 seconds after the ignition switch is turned **OFF**.

f. On 1990 B2600, using a suitable voltmeter, check the voltage of the air flow sensor wire (G/O) and terminal (2H) on the ECU. Voltmeter should read 0 volts when the ignition switch is turned Off and 8–12 volts momentarily, 2–5 seconds after the ignition switch is turned Off.

g. If specifications are not as indicated, check the voltage of the ECU terminals 2E, 2I and 2J on 1989 B2600 and 1990 MPV with 4 cylinder engine, then the wiring harness. On 1990 B2600, check the voltage of the ECU terminals 2P, 2Q and 1l, then the wiring harness. Replace the air flow sensor if necessary.

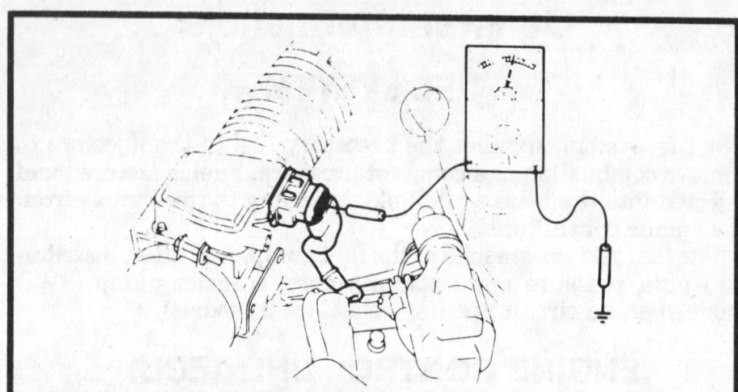

**Air flow sensor testing – 1989–90 B2600 and 1990 MPV with 4 cylinder engine**

| Terminal wire \ Condition | Ignition switch ON | Engine running |
|---|---|---|
| B/W (Power supply) | Approx. 12V | |
| G/O (Burn-off) | 0V | |
| G/B (Airflow mass) | 1.0—2.0V | 1.9—5V |
| B/W (Ground) | Approx. 0V | |
| B/O (Ground) | Approx. 0V | |

**Air flow sensor terminal specifications – 1990 MPV with 4 cylinder engine**

| Terminal wire \ Condition | Ignition switch ON | Engine running |
|---|---|---|
| B/Y (Power supply) | Approx. 12V | |
| G/O (Burn-off) | 0V | |
| G/B (Airflow mass) | 1.0—2.0V | 1.9—5V |
| G/Y (Ground) | Approx. 0V | |
| B/O (Ground) | Approx. 0V | |

**Air flow sensor terminal specifications – 1989–90 B2600**

## THROTTLE BODY

### Inspection

1. Check the throttle valve operation. Check the free play of the accelerator cable. Free play should be 0.04–0.12 in. (1–3mm).

2. If freeplay is not as specified, depress the accelerator pedal to the floor and check that the throttle valve is fully opened. Adjust by using bolt (B), if necessary.

## THROTTLE POSITION SENSOR

### Inspection

**1988–89 323 AND 1990 MIATA**

1. Disconnect the electrical connector from the throttle position sensor.

2. Connect throttle position sensor tester (49–9200–165) and

adapter (49-9200-166) or equivalent to the throttle position sensor. If testers are not available, use a suitable ohmmeter.

3. On 1988-89 323, insert a feeler gauge of 0.020 in. (0.5mm), then 0.027 in. (0.7mm), between the throttle stop screw and the stop lever.

4. On 1990 Miata, insert a feeler gauge of 0.016in (0.4mm), then 0.027 in. (0.7mm), between the throttle stop screw and the stop lever.

5. On all models, note the operation of the buzzer (on tester) or the continuity between the terminals.

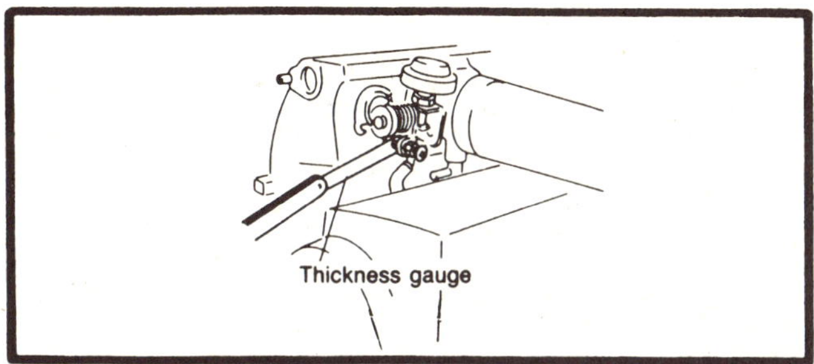


**Using a feeler gauge to test throttle sensor – 1988-89 323**

### 1990 323 WITH MANUAL TRANSAXLE

1. Disconnect the electrical connector from the throttle sensor.

2. Connect a suitable ohmmeter to the throttle sensor terminals.

3. Insert a 0.004 in. (0.1mm) feeler gauge between the throttle stop screw and the stop lever. Check the continuity between the terminals.

4. Insert a 0.039 in. (1.0mm) feeler gauge between the throttle stop screw and the stop lever. Check the continuity between the terminals.

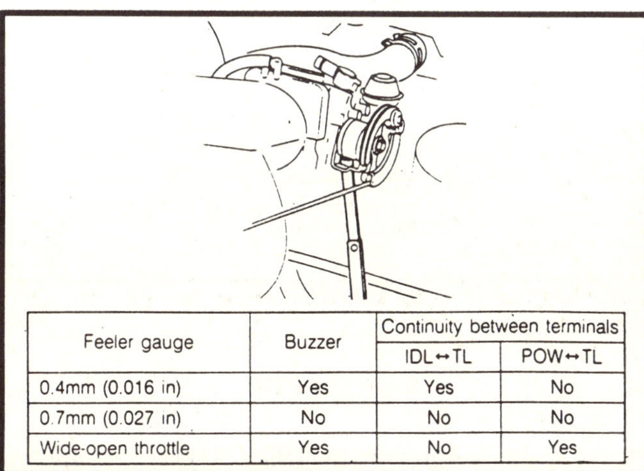

| Feeler gauge | Buzzer | Continuity between terminals | |
|---|---|---|---|
| | | IDL↔TL | POW↔TL |
| 0.4mm (0.016 in) | Yes | Yes | No |
| 0.7mm (0.027 in) | No | No | No |
| Wide-open throttle | Yes | No | Yes |

**Using a feeler gauge to test throttle sensor – 1990 Miata**

### 1990 323 WITH AUTO TRANSAXLE

1. Disconnect the electrical connector from the throttle sensor.

2. Connect a suitable ohmmeter between terminals E and IDL on the throttle sensor.

3. Insert a 0.004 in. (0.1mm) feeler gauge between the throttle stop screw and the stop lever. Continuity should exists between the terminals.

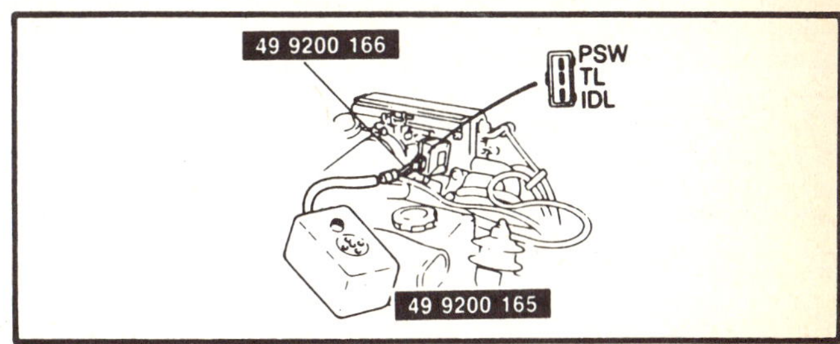


**Connecting throttle sensor testing tools – 1988-89 323**

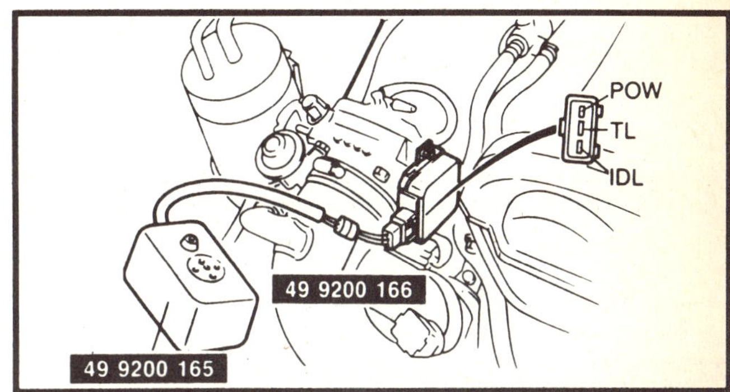


**Connecting throttle sensor testing tools – 1990 Miata**

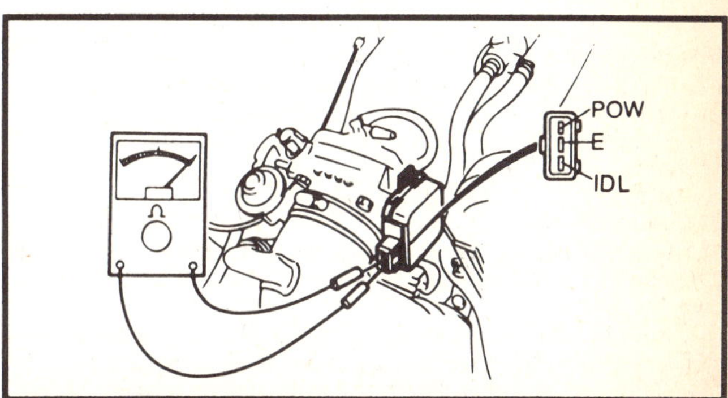


**Testing the throttle sensor using an ohmmeter – 1990 323 with M/T**

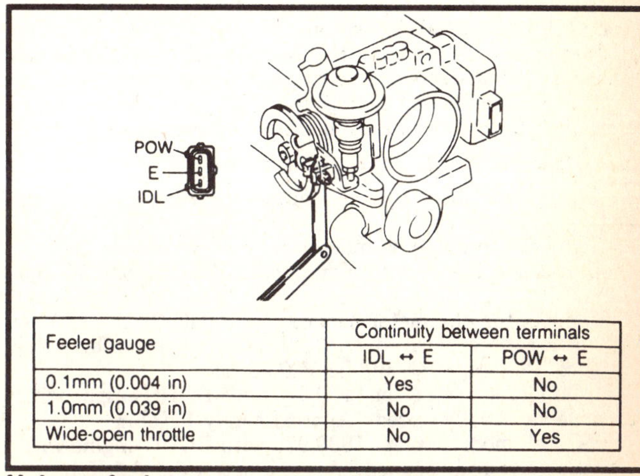


| Feeler gauge | Continuity between terminals | |
|---|---|---|
| | IDL ↔ E | POW ↔ E |
| 0.1mm (0.004 in) | Yes | No |
| 1.0mm (0.039 in) | No | No |
| Wide-open throttle | No | Yes |

**Using a feeler gauge to test throttle sensor – 1990 323 with M/T**

4. Insert a 0.024 in. (0.6mm) feeler gauge between the throttle stop screw and the stop lever. No continuity should exists between the terminals.

5. Connect a suitable ohmmeter between terminals VT and E on the throttle sensor. Ohmmeter should read as follows:

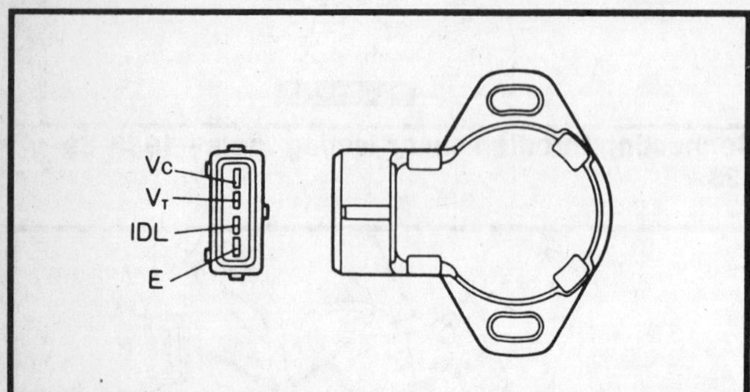

**Throttle sensor terminal identification – 1990 323 with A/T**

a. Throttle valve fully closed – Below 1 Kohm
b. Throttle valve fully open – Approximately 5 Kohms

**626 AND MX-6, B2600 AND 1990 MPV WITH 4 CYLINDER ENGINE**

1. Disconnect the air hose from the throttle body assembly, then disconnect the throttle sensor electrical connector (3-pin).
2. Connect Adapter Harness (49-G018-901) between the throttle sensor and the wiring harness.
3. Turn the ignition switch to the **ON** position. Check that the throttle valve is in the fully closed position.
4. Using a suitable voltmeter, measure the Black and Red wire voltages. Voltage should be as follows:
   Black wire – Approximately 0 volts
   Red wire – 4.5–5.5 volts
5. If specifications are not as indicated, check the battery voltage and the wiring harness. If problems are indicated, replace the ECU.
6. Note the Red wire voltage. Compare the Blue wire voltage to the recorded Red wire voltage as specified in the chart.
7. Hold the throttle valve in the fully opened position. Compare the Blue wire voltage to the recorded Red wire voltage as specified in the chart.
8. Check that the Blue wire voltage increases smoothly, when opening the throttle valve from the fully closed to the fully opened positions. If not, replace the throttle sensor. Turn the ignition switch to the **OFF** position.
9. Disconnect the Adapter Harness (49-G018-901) and reconnect the throttle sensor electrical connector. Disconnect the battery ground cable and depress the brake pedal for at least 5 seconds, to eliminate the ECU malfunction memory.

**929 AND MPV WITH 6 CYLINDER ENGINE**

1. Disconnect the electrical connector from the throttle sensor.
2. Connect a suitable ohmmeter between terminals (A) and (D) on the throttle sensor. Ohmmeter should read 3.5–6.5 Kohms.
3. Connect the ohmmeter between terminals (B) and (D) on the throttle sensor. Slowly open the throttle valve and check the resistance. Resistance should be as follows:
   a. On all models except 1990 929 with DOHC engine with throttle valve fully closed – Approximately 1 Kohm.
   b. On all models except 1990 929 with DOHC engine with throttle valve fully opened – Approximately 3.5–6.5 Kohm.
   c. 1990 929 with DOHC engine with throttle valve fully closed – Approximately 0.2–0.6 Kohm.
   d. 1990 929 with DOHC engine with throttle valve fully opened – Approximately 3.3–7.0 Kohm.
4. On all models, if specifications are not as indicated, check the throttle sensor adjustment.

## FUEL INJECTORS

### On-Vehicle Inspection

1. Start the engine and allow it to reach normal operating temperature, and let run at idle.
2. Apply a sound scope or equivalent to the injector and check for operating sound.
3. If no sound is detected, check the wiring harness, injector resistance and the following ECU terminal voltages:
   1988 323 – Terminals 3C, 3E, 3B and 3D
   1988 323 with turbo and 1989 323 – Terminals 3C and 3E
   1990 323 – Terminals 2A, 2U and 2V
4. On 1989 B2600 and 1990 MPV with 4 cylinder engine, check the resistance of the injector harness connector.

### Resistance Test

1. Disconnect the electrical connector from the fuel injector.
2. Using a suitable ohmmeter, measure the resistance across the fuel injector connector terminals. Ohmmeter should read as follows:
   1988 323 non-turbo – 11–15 ohms
   1989 – 90 323, 1990 Miata and 1988–90 929 – 12–16 ohms
   1989–90 B2600 – 12–16 ohms
   1988–89 626 and MX-6 non-turbo and 1990 626 and MX-6 – 12–16 ohms
   1988–89 626 and MX-6 with turbo–11–15 ohms
3. If specifications are not as indicated, replace the fuel injector.
4. Repeat Step 1 through 3 for the remaining fuel injectors.

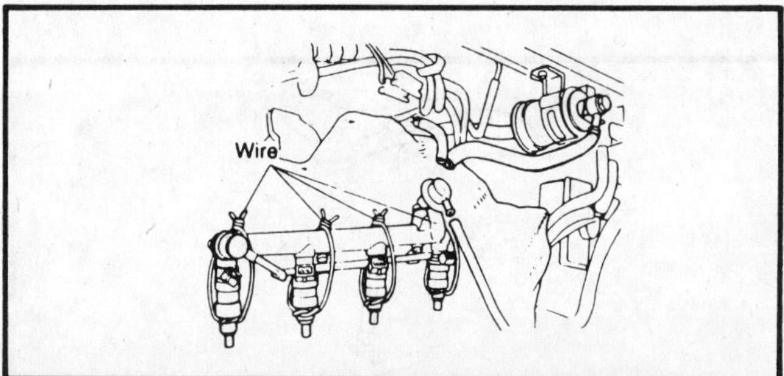

**Attaching fuel injectors to delivery pipe with wire – 6 cylinder similar**

## Leak Test

**1988–89 323**

1. Relieve the fuel pressure from the fuel system, the disconnect the battery ground cable.
2. Remove the delivery pipe, injectors and pressure regulator as an assembly.
3. Firmly attach each injector to the distribution pipe with a piece of wire. Ensure that the injectors are not allowed to move or rotate on the distribution pipe.
4. Connect the distribution pipe assembly between the fuel filter and return pipe, then connect the return hose to the pressure regulator.
5. Connect the battery ground cable. Install a jumper to the test connector (yellow wire).
6. Turn the ignition switch to the **ON** position, and check the injectors for fuel leakage.

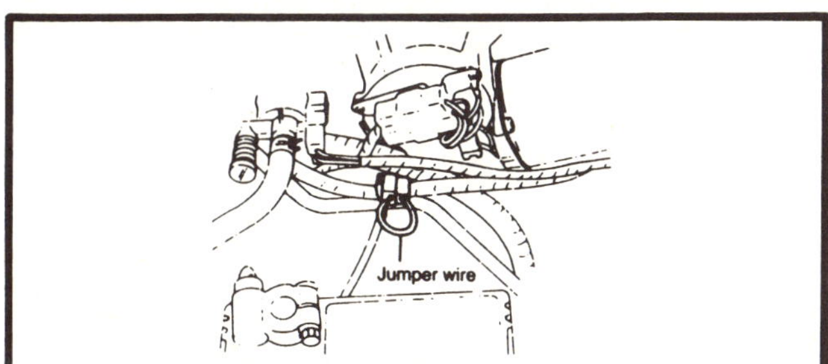


**Installing jumper wire to test connector for fuel leakage test—1988–89 323**

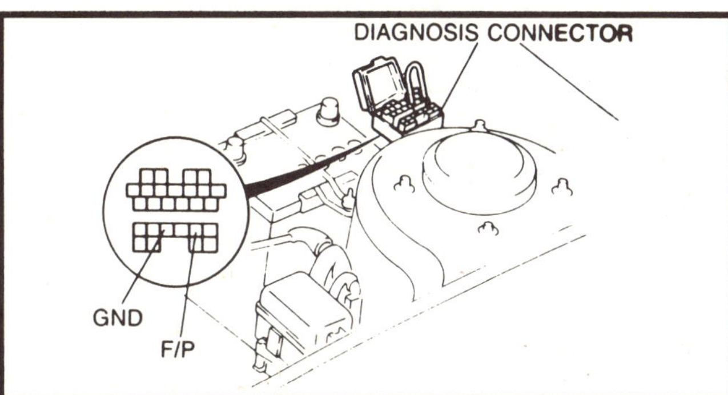


**Installing jumper wire across diagnosis connector terminals F/P and GND for fuel leakage test—1990 323 and Miata**

**NOTE: After approximately 5 minutes, a slight amount of fuel leakage is acceptable.**

7. If fuel leaks from any injector, replace it.

**1989–90 B2600**

1. Relieve the fuel pressure from the fuel system, then disconnect the battery ground cable.
2. Remove the injectors and delivery pipe as an assembly.
3. Firmly attach each injector to the distribution pipe with a piece of wire. Ensure that the injectors are not allowed to move or rotate on the distribution pipe.
5. Connect the battery ground cable. Install a jumper to the test connector (yellow wire).
6. Turn the ignition switch to the **ON** position. Tilt injectors approximately 60 degrees and check the injectors for fuel leakage.

**NOTE: After approximately 1 minute, a slight amount of fuel leakage is acceptable.**

7. If fuel leaks from any injector, replace it.

**1990 323 AND MIATA**

1. Relieve the fuel pressure from the fuel system.
2. On Miata, remove the air valve.
3. On all models, disconnect the electrical connector from each fuel injector.
4. Remove the delivery pipe attaching bolts, then affix the injectors to the to the delivery pipe with some suitable wire.

**NOTE: Affix the injectors firmly to the distribution pipe so that no movement of the injectors is possible.**

5. Install a jumper wire between terminals F/P and GND on the diagnosis connector. Turn the ignition switch to the **ON** position.
6. Tilt the injectors approximately 60 degrees, and check that no fuel leaks from the injector nozzles. If fuel leaks from any injector, replace that injector.

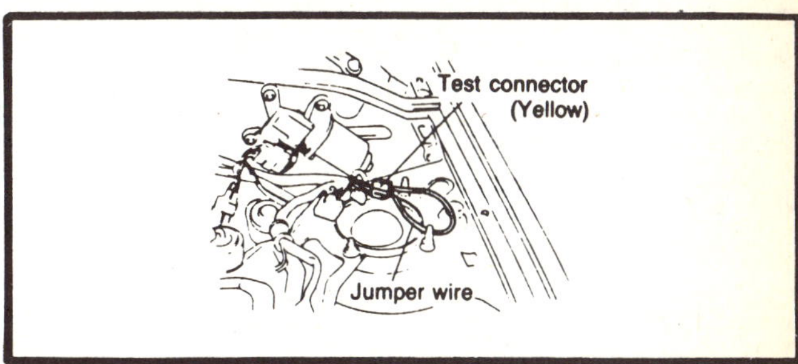


**Installing jumper wire across test connector for fuel leakage test—1988–90 626/MX-6**

**NOTE: After approximately 1 minute, a slight amount of fuel leakage is acceptable.**

7. Turn the ignition switch to the **OFF** position and remove the jumper wire.

**1988–90 929**
**1988–90 626 AND MX-6**

1. Relieve the fuel pressure from the fuel system.
2. On 626 and MX-6, lift the dynamic chamber upward.
3. On all models, disconnect the electrical connector from each injector, then remove the delivery pipe with the injectors attached.
4. Affix the injectors to the to the delivery pipe with some suitable wire.

**NOTE: Affix the injectors firmly to the distribution pipe so that no movement of the injectors is possible.**

5. Connect jumper wire between the terminals of the fuel pump test connector (yellow). Turn the ignition switch to the **ON** position.
6. On 1990 626 and MX-6, ground the test connector (green: 1-pin) using a suitable jumper wire. Open the throttle valve to release the air in the injectors
7. On all models, Check that no fuel leaks from the injector nozzles. If fuel leaks, replace the injector.

**NOTE: After 1 minute, a slight amount of fuel leakage is acceptable.**

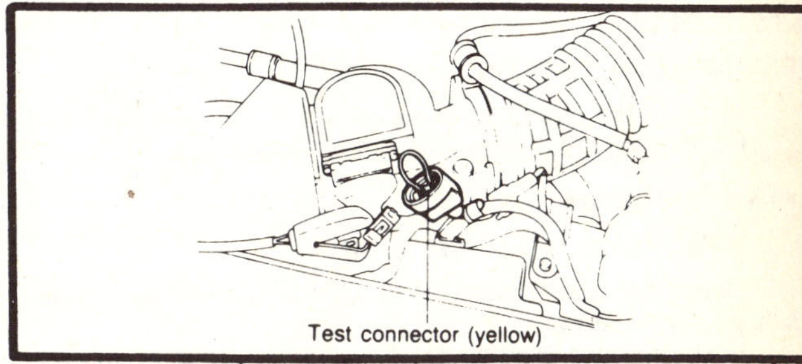


**Installing jumper wire across test connector for fuel leakage test—1988–89 929 (1990 929 similar)**

## Volume Test

**1988–89 323 AND 929**
**1988-90 626 AND MX-6**

1. Connect a suitable hose to the of each injector. Place the end of the hose into a suitable container or graduated container.
2. Connect a jumper wire across both terminals of check connector.
3. Connect Injector Tester (49–B092–953) between the battery and the injector.

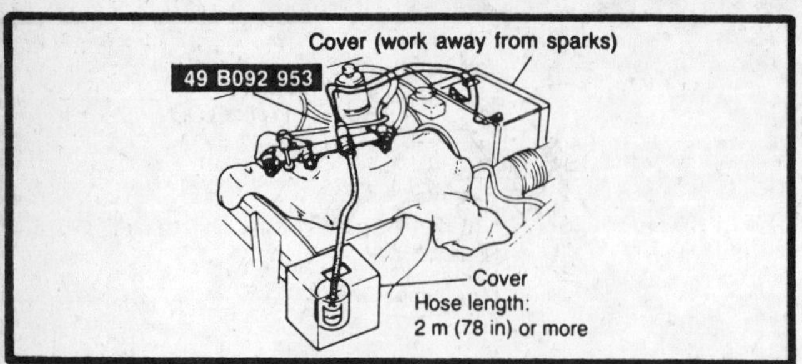


**Fuel volume test – 1988–89 323 and 1988–90 626/MX-6**

4. Turn the ignition switch to the **ON** position and apply battery voltage to the injector for approximately 15 seconds. The volume discharge from the injector should be as follows:

323 – Within 1.95–2.50 cu. in. (32–41cc)
929 – Within 2.68–3.97 cu. in. (44–65cc)
1988–89 626 and MX-6 non-turbo – 2.68–3.72 cu. in. (44–61cc)
1988–89 626 and MX-6 turbo – 4.45–5.49 cu. in. (73–90cc)

5. If specifications are not as specified, replace the injector.
6. Repeat Steps 1 through 5 for the remaining injectors.
7. Turn the ignition switch to the **OFF** position and disconnect the Injector Tester. Remove the jumper wire and the hose.

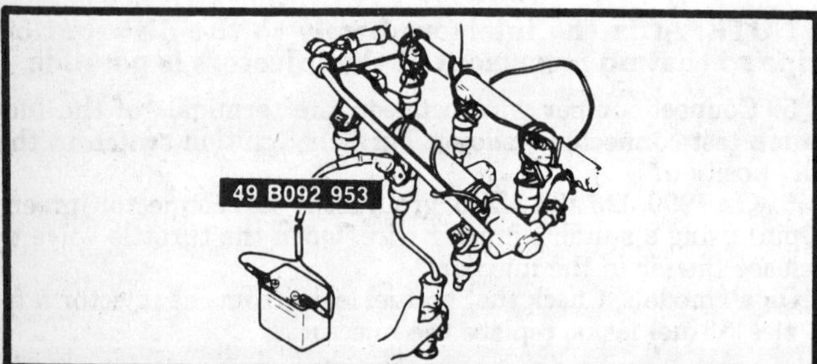


**Fuel volume test – 1988–89 929**

## CIRCUIT OPENING RELAY

### Circuit Test

#### 1988–90 626 AND MX-6

1. Remove the circuit opening relay.
2. Check the circuit terminals as indicated.

### Terminal Voltage Test

#### EXCEPT 1988–90 626 AND MX-6

1. Using a suitable voltmeter, measure the voltage between each terminal and ground.
2. If the voltage reading at B terminal is not as specified, check the fuse or wiring harness from the ignition switch.

| Terminal \ Condition | Ignition switch ON (V) | Ignition switch START (V) | Idling (V) |
|---|---|---|---|
| Fp (B/P) | 0 | Approx. 12 | Approx. 12 |
| Fc (LG) | Approx. 12 | 0 | 0 |
| B (W/R) | Approx. 12 | Approx. 12 | Approx. 12 |
| STA (V) | 0 | Approx. 12 | 0 |
| E1 (B) | 0 | 0 | 0 |

**Circuit opening relay terminal voltage specification chart – All models except 1988–90 626**

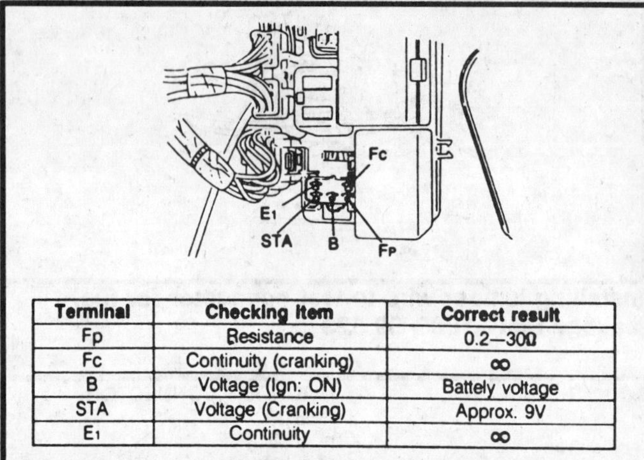


| Terminal | Checking item | Correct result |
|---|---|---|
| Fp | Resistance | 0.2—30Ω |
| Fc | Continuity (cranking) | ∞ |
| B | Voltage (Ign: ON) | Battely voltage |
| STA | Voltage (Cranking) | Approx. 9V |
| E1 | Continuity | ∞ |

**Testing the circuit opening relay circuit – 1988–90 626/MX-6**

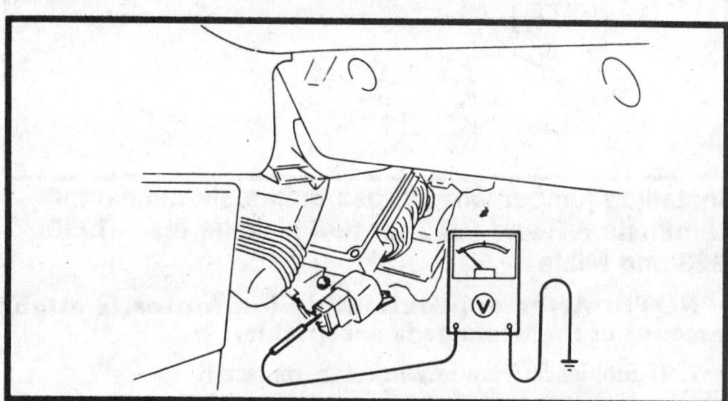


**Testing circuit opening relay using a voltmeter**

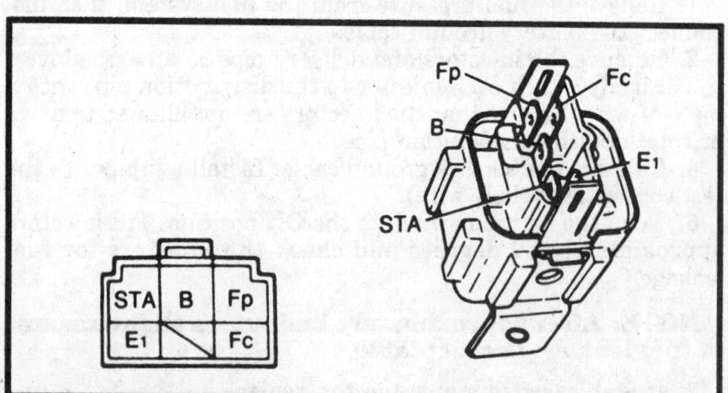


**Circuit opening relay**

3. If the voltage reading at the STA terminal is not as specified, check the wiring harness at the ignition switch.
4. If the voltage reading at the $E_1$ terminal is not as specified, check the ground harness.
5. If the voltage reading at the FC terminal is not as specified, check the air flow meter or test on the circuit opening relay resistance.
6. If the voltage reading at the FP terminal is not as specified, check the circuit opening relay resistance.

### Resistance Test

1. Using a suitable ohmmeter, measure the resistance across the indicated terminals.
2. If the specifications are not as specified, replace the circuit opening relay.

| Between terminals | Resistance (Ω) |
|---|---|
| STA ↔ E1 | 21—43 |
| B ↔ Fc | 109—226 |
| B ↔ Fp | ∞ |

**Circuit opening relay terminal resistance chart–All models except 1988–89 323, 929 and 626/MX-6**

| Between terminals | Resistance (Ω) |
|---|---|
| STA ↔ E1 | 15—30 |
| B ↔ Fc | 80—150 |
| B ↔ Fp | ∞ |

**Circuit opening relay terminal resistance chart–1988–89 323, 929 and 626/MX-6**

## TROUBLESHOOTING WITH THE SELF-DIAGNOSIS CHECKER

The self diagnosis checker (49–H018–9A1) and System Selector (49–B019–9A0, used on 1990 Miata) are used to retrieve code numbers of malfunctions which have happened and were memorized or are continuing. The malfunction is indicated by the code number and a buzzer.

If there is more than 1 malfunction, the code numbers will display on the self diagnosis checker 1 by 1 in numerical order. In the case of malfunctions, 09, 13, and 01, the code numbers are displayed in a order of 01, 09 and then 13.

The memory of malfunctions is canceled by disconnecting the negative battery cable for at least 5 seconds.

The ECU has a built in fail-safe mechanism for the main input sensors. If a malfunction occurs, the emission control unit will substitute values. This will slightly effect the driving performance, but the vehicle may still be driven.

The ECU continuously checks for malfunctions of the input devices. But, the ECU checks checks for malfunctions of the output devices within 3 seconds after turning the ignition switch to the **ON** position and the test connector is grounded.

The malfunction indicator light indicates a pattern the same as the buzzer of the self-diagnosis checker when the self-diagnosis check connector is grounded. When the self-diagnosis check connector is not grounded, the lamp illuminates steady while malfunction of the main input sensor occurs and goes out if the malfunction recovers. However, the malfunction code is memorized in the emission control unit.

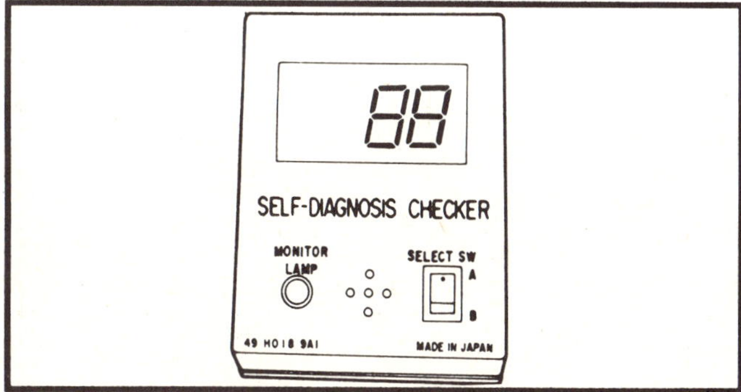

**Self-diagnosis checker**

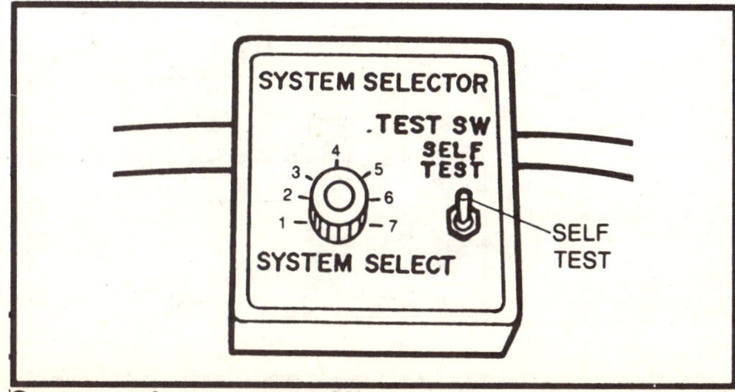

**System selector–1990 Miata**

### Inspection Procedure

1. On all models except 1990 Miata, Connect the Self Diagnosis Checker (49–H018–9A1) or equivalent to the check connector. The check connector is usually located above the right side wheel housing.
2. On 1990 Miata, connect Self Diagnosis Checker (49–H018–9A1) and System Selector (49–B019–9A0) to the diagnosis connector and ground.

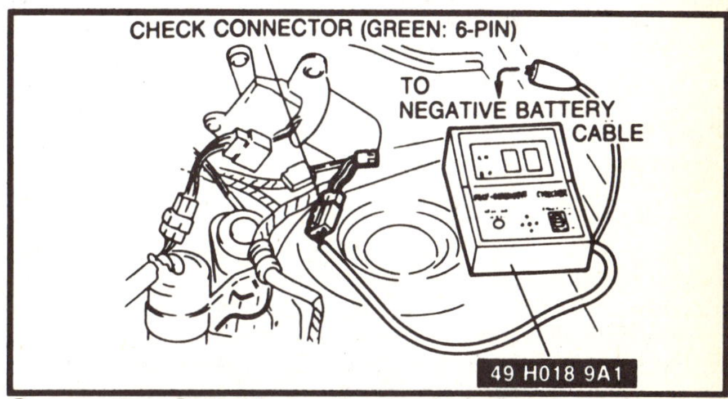

**Connecting Self-diagnosis checker to the check connector–All models except 1990 Miata**

**Connecting Self-diagnosis checker to the test connector–All models except 1990 Miata**

3. On all models, set the select switch on the Self Diagnosis Checker to the "A" position.
4. On 1990 Miata, set the System Selector (49–B019–9A0) to position 1 and the SELF TEST switch to the up position.
5. On all models except 1990 Miata, ground the test connector using a suitable jumper wire.
6. On all models, turn the ignition switch to the **ON** position. Check that the number "88" flashes on the digital display and the buzzer sounds for 3 seconds after turning the ignition switch **ON**.
7. If the number "88" does not flash, check the main relay, power supply circuit and the check the check connector wiring.

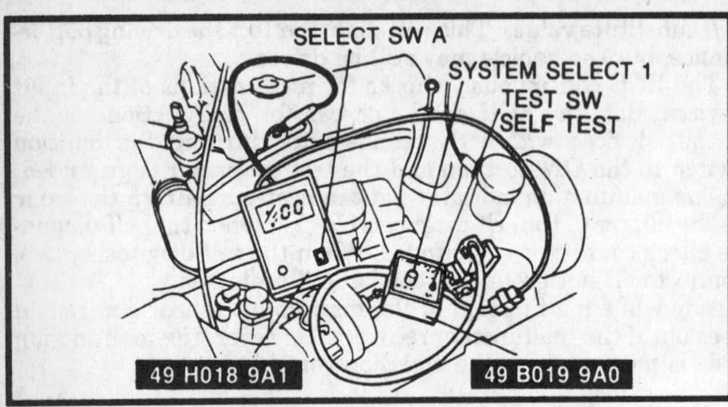


**Connecting Self-diagnosis checker and System selector – 1990 Miata**

8. On all models except 1990 Miata, if the number "88" flashes and the buzzer sounds continuously for more than 20 seconds, replace the ECU and perform Steps 3 and 4 again.

9. On 1990 Miata, if the number "88" flashes and the buzzer sounds continuously for more than 20 seconds, check for a short circuit between the ECU and terminal 1F on the diagnosis connector. Replace the ECU if necessary.

10. On all models, note the code numbers and check the causes, repair as necessary. Be sure to recheck the code numbers by performing the "After Repair Procedure," after repairing.

### After Repair Procedure

1. Cancel the memory of malfunctions by disconnecting the battery negative cable and depressing the brake pedal for at least 5 seconds, then reconnect the battery ground cable.

2. On all models except 1990 Miata, connect the Self-Diagnosis Checker 49–H018–9A1 to the check connector. Ground the test connector (green: 1 pin) using a suitable jumper wire.

3. On 1990 Miata, connect Self-Diagnosis Checker (49–H018–9A1) and System Selector (49–B019–9A0) to the diagnosis connector.

4. On all models, turn the ignition switch to the **ON** position, but do not start the engine for approximately 6 seconds.

5. Start the engine and allow it to reach normal operating temperature, the run the engine at 2000 rpm for 2 minutes. Check that no code numbers are displayed.

## TROUBLESHOOTING WITH THE ENGINE SIGNAL MONITOR

The Engine Signal Monitor (49–9200–162), Adapter Harness (49–9200–163 or 49–G018–903 on 1990 B2600, 929 and 626/MX-6) and Engine Digital Monitor Sheet (49–G018–904 used on 1990 B2600, 929 and 626/MX-6) are used to check the control unit terminal voltages. This monitor checks the individual terminal voltages through selection by the monitor switch.

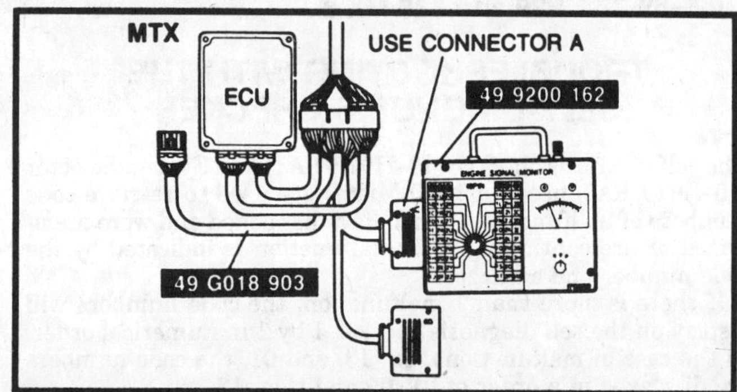


**Connecting engine signal monitor and adapter harness – 1990 626/MX-6 with M/T and 1990 929 and B2600**

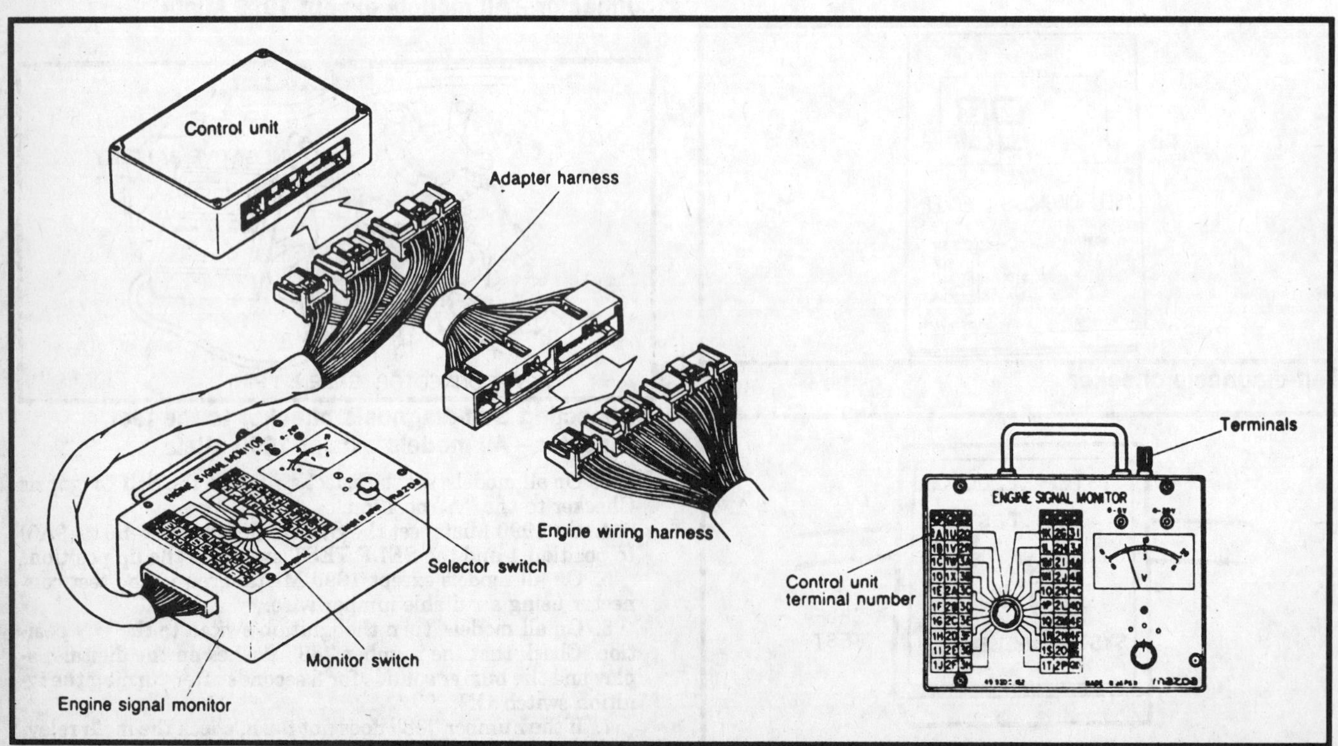


**Connecting engine signal monitor – All models except 1990 B2600, 929 and 626/MX-6**

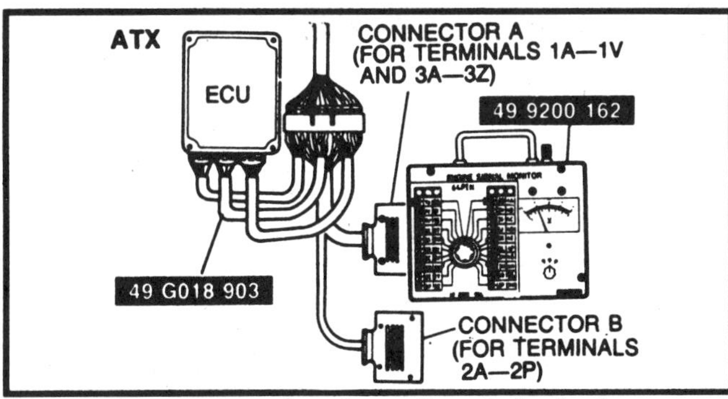


**Connecting engine signal monitor and adapter harness—1990 626/MX-6 with A/T**

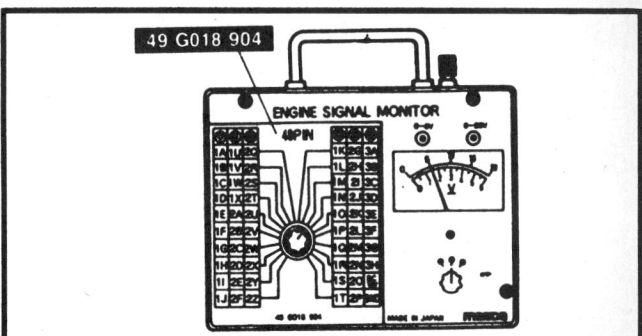

**Installing engine signal monitor sheet—1990 626/MX-6, 929 and B2600**

### Inspection

1. Start the engine and allow it to reach normal operating temperature. Stop the engine.
2. Disconnect the ECU connector.
2. On 1989 B2600 and 1990 MPV, connect Engine Signal Monitor (49-9200-162) and Adapter Harness (49-9200-167) between the ECU and The ECU wiring harness.
3. On all 1988-89 models, except 1989 B2600, connect Engine Signal Monitor (49-9200-162) and the Adapter Harness (49-9200-163) between the ECU and the ECU wiring harness.
4. On 1990 models, connect Engine Signal Monitor (49-9200-162), and Adapter Harness (49-G018-903) between the ECU and the ECU wiring harness, then install Engine Signal Monitor Sheet (49-G018-904) on the Engine Signal Monitor.

**NOTE: On 1990 626/MX-6 equipped with M/T, use connector (A) of the Adapter. On models equipped with A/T, use connector (A) of the Adapter to check voltages of terminals 1A through 1V and 3A through 3Z, and use connector (B) to check voltages of terminals 2A through 2P.**

5. On all models, position the select switch and the monitor switch on the Engine Signal Monitor to the terminal number to be tested. Check the voltage of each terminal.
6. If any ECU terminal volatge is incorrect, check the input or output component and related wiring. If components and wiring are normal, replace the ECU.

## SYSTEM SCHEMATIC – 1988 323 NON-TURBO

## COMPONENT LOCATION – 1988 323 NON-TURBO

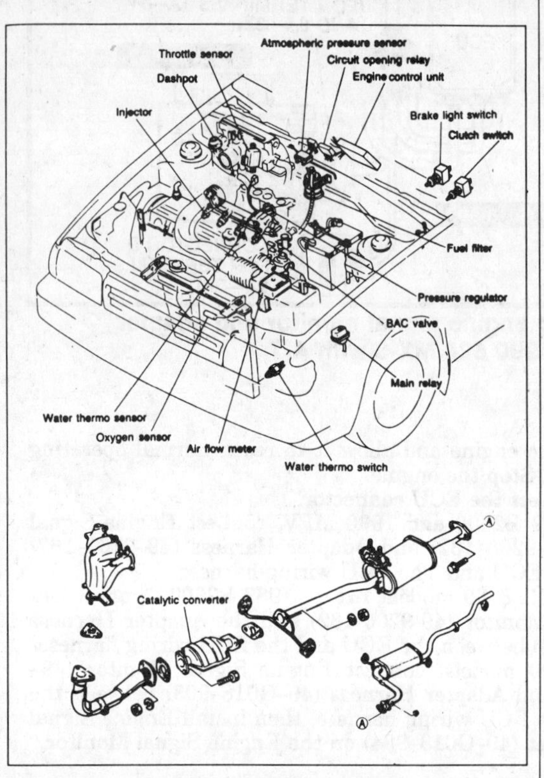


## COMPONENT DESCRIPTION – 1988 323 NON-TURBO

| No. | COMPONENT | FUNCTION | REMARKS |
|---|---|---|---|
| 1 | Air cleaner | Filters air into the combustion chamber | — |
| 2 | Air flow meter | Detects intake air amount; sends signal to the engine control unit. (for determination of fuel injection amount) | Intake air thermo sensor and fuel pump switch are integrated. |
| 3 | Atmospheric pressure sensor | Detects atmospheric pressure to prevent over rich mixture; sends signal to engine control unit. | |
| 4 | Air valve | When engine is cold, supplies bypass air into dynamic chamber for quick warm-up and smooth idle. | • Engine speed is increased to shorten warm-up period.<br>• Thermo wax type<br>• Installed into BAC valve |
| 5 | Brake light switch | Detects brake operation (deceleration); sends signal to engine control unit. | |
| 6 | Catalytic converter | Reduce HC and CO by oxidation. Reduce NOx. | Honeycomb construction |
| 7 | Charcoal canister | Stores fuel tank fumes while engine is stopped. | |
| 8 | Check connector | For Self-diagnosis checker | 6 pin connector (Green) |
| 9 | Circuit opening relay | Supplies voltage for fuel pump while engine running. | |
| 10 | Clutch switch | Detects in-gear condition; sends signal to engine control unit. | Switch is closed when clutch pedal is released. |
| 11 | Engine control unit | Detects the following;<br>1. Engine speed<br>2. Intake air amount<br>3. Engine coolant temperature<br>4. Engine load condition<br>5. Oxygen concentration in exhaust gas<br>6. In-gear condition<br>7. Intake air temperature<br>8. Atmospheric pressure<br>9. A/C operation<br>10. P/S operation<br>11. E/L (Electrical load) operation<br>12. Starting signal<br>13. Initial set signal<br>Controls operation of the following;<br>1. Fuel injection amount<br>2. Idle speed control system<br>3. Fail-safe system<br>4. Monitor switch function | 1. Ignition coil (–) terminal<br>2. Air flow meter<br>3. Water thermo sensor<br>4. Throttle sensor (Point type)<br>5. Oxygen sensor<br>6. Clutch switch and neutral switch<br>7. Intake air thermo sensor (in air flow meter)<br>8. Atmospheric pressure sensor<br>9. A/C switch<br>10. P/S switch<br>11. E/L switch<br>12. Starter switch (Ignition switch)<br>13. Test terminal<br>1. Injector<br>2. BAC valve (ISC solenoid valve)<br>3. Self-diagnosis checker and MIL<br>4. Monitor lamp (Self-diagnosis checker) |
| 12 | Dashpot | Gradually allows throttle valve closing during deceleration. | Adjustment speed<br>MTX....2800 ± 150 rpm<br>ATX....2800 ± 300 rpm (in neutral) |
| 13 | Fuel filter | Filters particles from fuel | |
| 14 | Fuel pump | Provides fuel to injectors | • Operates while engine is running<br>• Installed in fuel tank |
| 15 | Intake air thermo sensor | Detects intake air temperature; compensates fuel injection amount through engine control unit. | Thermistor |
| 16 | Injector | Injects fuel to intake port | Controlled by signals from engine control unit. |
| 17 | Intank Filter | Filters particles from fuel | Installed in low-pressure side |
| 18 | ISC valve | Supplies bypass air to dynamic chamber for smooth idle | Insalled into BAC valve |
| 19 | Neutral switch | Detects transaxle condition; sends signal to control unit | |
| 20 | Oxygen Sensor | Detects oxygen concentration in exhaust gas; sends signal to control unit; compensates fuel injection amount. | Ziconia ceramic with platinum coating |
| 21 | Pressure Regulator | Regulates fuel pressure to injectors | |
| 22 | No.1 Purge Control Valve | Opens and closes evaporative vapor passage from canister to intake manifold | During open throttle |
| 23 | No.2 Purge Control Valve | Positive pressure and negative pressure valves operate in accordance with fuel tank pressure. | Prevents canister from flooding. |
| 24 | Throttle Sensor (Point type) | Detects throttle opening angle; sends signal to engine control unit; compensates fuel injection amount. | |
| 25 | Solenoid Valve (for No.1 purge control valve) | Opens and closes vacuum passage to No.1 purge control valve. | Controlled by signal from engine control unit |
| | Solenoid Valve (for vacuum switch valve) | Opens and closes vacuum passage to vacuum switch valve. | Controlled by signal from engine control unit |
| 26 | Vacuum Switch Valve | Opens passage of vacuum line when vacuum applied. | Vacuum from three-way solenoid valve |
| 27 | Water Thermo Sensor | Detects coolant temperature; sends signal to engine control unit; compensates fuel injection amount. | Thermistor |
| 28 | Water Thermo Switch | Detects radiator coolant temperature; sends signal to control unit; increases fuel injection amount. | Above 17°C (63°F): ON |

## SPECIFICATION CHART—1988 323 NON-TURBO

**SPECIFICATIONS**

| Item | | Transaxle type | Manual transaxle | Automatic transaxle |
|---|---|---|---|---|
| Idle speed | | rpm | 850 ± 50 in Neutral | 850 ± 50 in P range |
| **Throttle body** | | | | |
| Type | | | Horizontal draft (1-barrel) | |
| Throat diameter | | mm (in) | 45 (1.77) | |
| **Air flow meter** | | | | |
| Resistor | Ω | E2—Vs | Fully closed: 20—400 Fully open: 20—1,000 | |
| | | E2—Vc | 100—300 | |
| | | E2—VB | 200—400 | |
| | | E2—THA | −20°C (−4°F) 10,000—20,000<br>20°C (68°F) 2,000—3,000<br>60°C (140°F) 400—700 | |
| **Fuel pump** | | | | |
| Type | | | Impeller (in tank) | |
| Output pressure | | kPa (kg/cm², psi) | 441—588 (4.5—6.0, 64.0—85.3) | |
| Feeding capacity | | cc (cu-in)/10 sec | 220—380 (13.4—23.2) when fuel pressure at 250 kPa (2.55 kg/cm², 36.3 psi) | |
| **Fuel filter** | | | | |
| Type | Low pressure side | | Nylon 6 (250 mesh) element | |
| | High pressure side | | Paper element | |
| **Pressure regulator** | | | | |
| Type | | | Diaphragm | |
| Regulating pressure | | kPa (kg/cm², psi) | 240—279 (2.45—2.85, 34.8—40.5) (Vacuum hose disconnected) | |
| **Injector** | | | | |
| Type | | | High-ohmic | |
| Type of drive | | | Voltage | |
| Resistance | | Ω | 11—15 | |
| Injection amount | | cc (cu in)/15 sec | 32—41 (1.95—2.50) | |
| **Idle speed control valve** | | | | |
| Solenoid resistance | | Ω | 5—20 | |
| **Fuel tank** | | | | |
| Capacity | | liters (US gal, Imp gal) | 48 (12.7, 10.6) | |
| **Air cleaner** | | | | |
| Element type | | | Wet | |
| **Accelerator cable** | | | | |
| Free play | | mm (in) | 1—3 (0.039—0.118) | |
| **Fuel** | | | | |
| Specification | | | Unleaded gasoline | |

## TROUBLESHOOTING AND RELATIONSHIP CHART 1988 323 NON-TURBO

**RELATIONSHIP CHART**
**Output Devices and Input Devices**

| OUTPUT DEVICE / INPUT DEVICE | INJECTOR: FUEL INJECTION AMOUNT | INJECTOR: FUEL INJECTION TIMING | SOLENOID (PRESSURE REGULATOR) | BAC VALVE: AIR VALVE | BAC VALVE: ISC VALVE | PURGE SOLENOID: No.1 | PURGE SOLENOID: No.2 |
|---|---|---|---|---|---|---|---|
| IGNITION COIL | O | O | X | X | O | X | O |
| AIRFLOW METER | O | X | X | X | X | X | O |
| IDLE SWITCH | O | X | O | X | O | X | X |
| PSW SWITCH | O | X | X | X | X | X | X |
| WATER THERMO SENSOR | O | X | O | X | O | O | X |
| INTAKE AIR THERMO SENSOR | O | X | O | X | O | O | X |
| ATMOSPHERIC PRESSURE SENSOR | O | X | X | X | O | X | X |
| OXYGEN SENSOR | O | X | X | X | O | O | X |
| BRAKE LIGHT SWITCH | O | X | X | X | X | X | X |
| WATER THERMO SWITCH | O | X | X | X | O | O | X |
| NEUTRAL AND CLUTCH SWITCH | O | X | O | X | O | O | X |
| STARTER SWITCH | O | X | O | X | X | X | X |
| E/L SWITCH | X | X | X | X | O | X | X |
| A/C SWITCH | X | X | X | X | O | X | X |
| P/S SWITCH | X | X | X | X | O | X | X |
| TEST CONNECTOR | X | X | X | X | O | X | X |

O Related
X Not related

## OUTPUT COMPONENTS AND ENGINE CONDITION 1988 323 NON-TURBO

| OUTPUT DEVICES / ENGINE CONDITION | | CRANKING (COLD ENGINE) | WARMING UP (DURING IDLE) | MEDIUM LOAD: COLD | MEDIUM LOAD: WARM | ACCELERATION | HEAVY LOAD | DECELERATION | IDLE (THROTTLE VALVE FULLY CLOSED) | IGN: ON (ENGINE NOT RUNNING) | REMARKS |
|---|---|---|---|---|---|---|---|---|---|---|---|
| INJECTOR | INJECTION (Air Fuel Mixture) | Rich | Rich | Rich | Rich and Lean | Rich | Rich | Fuel cut off | Rich | Does not inject | |
| | INJECTION TIMING | 1 Group | 1 Group | 1 Group | 1 Group | 1 Group | 1 Group | Fuel cut off | 1 Group | Does not inject | Above 6,400 rpm fuel cut off |
| BAC VALVE | AIR VALVE | * Open | * Open | * Open | Closed | Closed | Closed | Closed | Closed | Closed | * Coolant temp: below 60°C (140°F) |
| | ISC VALVE | Large amount of bypass air | Large amount of bypass air | Small amount of bypass air | Small amount of bypass air | Small amount of bypass air | Small amount of bypass air | Small amount of bypass air | * Large and small amount of bypass air | Does not operate | * Test connector grounded: small amount of air |
| PURGE CONTROL SOLENOID | No.1 | OFF (Vacuum cut off) | OFF (Vacuum cut off) | * ON (Vacuum to No.1 purge control valve) | * ON (Vacuum to No.1 purge control valve) | * ON (Vacuum to No.1 purge control valve) | * ON (Vacuum to No.1 purge control valve) | * ON (Vacuum to No.1 purge control valve) | OFF (Vacuum cut off) | OFF (Vacuum cut off) | * Engine speed: Above 1,500 rpm |
| | No.2 | OFF (Vacuum cut off) | OFF (Vacuum cut off) | * ON (Vacuum to vacuum switch valve) | * ON (Vacuum to vacuum switch valve) | * ON (Vacuum to vacuum switch valve) | * ON (Vacuum to vacuum switch valve) | OFF (Vacuum cut off) | OFF (Vacuum cut off) | OFF (Vacuum cut off) | * Engine speed: Above 1,500 rpm |

## TROUBLESHOOTING CHART—1988 323 NON-TURBO

| SYMPTOM / POSSIBLE CAUSE | INPUT DEVICES: Ignition coil | Air flow meter | Water thermo sensor | Intake air thermo sensor (In Air flow meter) | Atmospheric pressure sensor | Oxygen sensor | Feedback system | | OUTPUT DEVICES: Solenoid valve (No.1 purge control valve) | Solenoid valve (Vacuum switch valve) | BAC Valve (idle speed control) | |
|---|---|---|---|---|---|---|---|---|---|---|---|---|
| 1 Fault indicated by SST Code No. | 01 | 08 | 09 | 10 | 14 | 15 | 17 | | 26 | 27 | 34 | |

| No. | Symptom | |
|---|---|---|
| 2 | Hard start or won't start (Crank OK) | |
| 3 | Engine stall | While warming up |
| | | After warming up |
| 4 | Rough idle | While warming up |
| | | After warming up |
| 5 | High idle speed after warming up | |
| 6 | Poor acceleration, hesitation, or lack of power | |
| 7 | Runs rough on deceleration | |
| 8 | Afterburn in exhaust system | |
| 9 | Poor fuel consumption | |
| 10 | Fail emission test | |

**TROUBLESHOOTING PROCEDURE:**

**Note**
**Step 1 under symptom is to quickly determine what system or parts may be at fault by use of the SST. (Self-Diagnosis checker 49 H018 9A1)**

**1st** Check input sensors and output solenoid valves with **SST**

**2nd** Check other switches with **SST**

**3rd** Check the following items:

**Electrical system**
1) Battery condition
2) Fuses

**Ignition system**
1) Spark plugs
2) Ignition timing

**Fuel system**
1) Fuel level
2) Fuel leakage
3) Fuel filter
4) Idle speed (with test connector grounded)

**Intake air system**
1) Air cleaner element
2) Vacuum or air leakage
3) Vacuum hose routing
4) Accelerator cable

**Engine**
1) Compression
2) Overheating

**Others**
1) Clutch slippage
2) Brake dragging

**4th** Check the Fuel and Emission Control Systems

### TROUBLESHOOTING CHART—1988 323 NON-TURBO

| SYMPTOM 10 | 9 | 8 | 7 | 6 | 5 | 4 | 4 | 3 | 3 | 2 | | POSSIBLE CAUSE |
|---|---|---|---|---|---|---|---|---|---|---|---|---|
| 5 | | 3 | | 2 | 2 | 5 | 4 | 4 | 3 | 2 | | Intake air system (Poor connection of components, throttle body) |
| 6 | 2 | 4 | 3 | 3 | | 4 | 3 | 3 | 2 | 1 | | Fuel system (Fuel injection, Fuel pressure) |
| 3 | | 1 | 2 | | 1 | 2 | 1 | 2 | 1 | | | ISC (Idle speed control) system (Air valve or Idle speed control malfunction) |
| | | | | | | 1 | 2 | 1 | | | | PCV (Positive crank case ventilation) system (System clogged) |
| 2 | 1 | 2 | 1 | | | | | | | | | Deceleration control system (Fuel cut operation malfunction) |
| 4 | | | | 1 | | 3 | | | | | | Evaporative emission control system |
| 1 | 3 | | | 4 | | | | | | | | Exhaust system (System clogged) |

**Note**
**The number of the list such shown a priorities of inspection from the most possible to that with the lowest possibility.**
**These were determined on following basis:**
- **Ease of inspection**
- **Most possible system**
- **Most possible point in system**

### MALFUNCTION INDICATOR LIGHT (MIL) OPERATION 1988 3423 NON-TURBO

Malfunction codes are determined as below

**1. Malfunction code cycle break**
The time between malfunction code cycles is 4.0 sec. (the time the light is off).

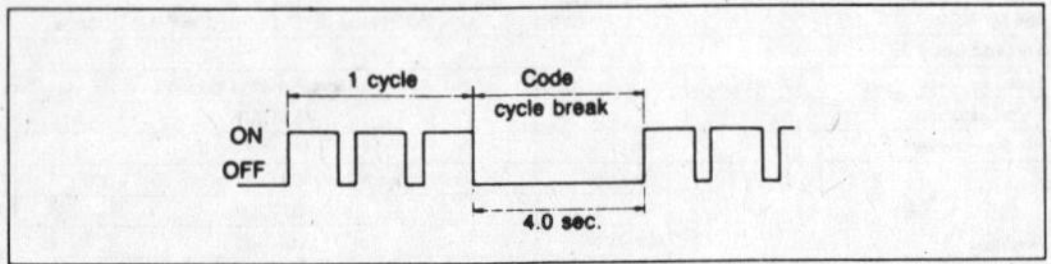


**2. Second digit of malfunction code (ones position)**
The digit in the ones position of the malfunction code represents the number of times the buzzer is or MIL on 0.4 sec. during one cycle.

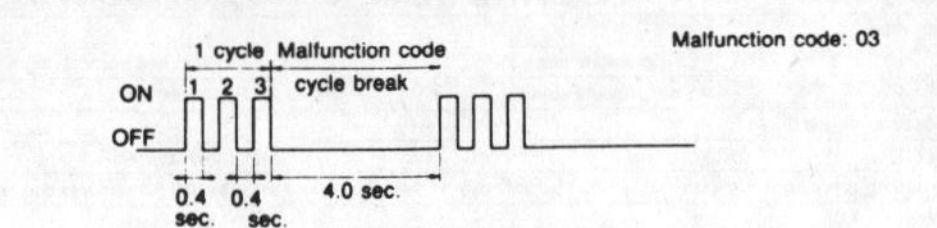


**3. First digit of malfunction code (tens position)**
The digit in the tens position of the warning code represents the number of times the buzzer or MIL is on 1.2 sec during one cycle.

The buzzer and MIL are off for 1.6 sec. between the long and short pulses.

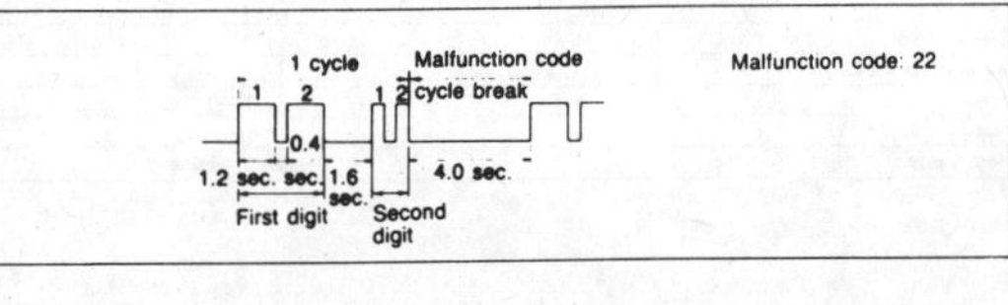


### TROUBLE CODE IDENTIFICATION CHART—1988 323 NON-TURBO

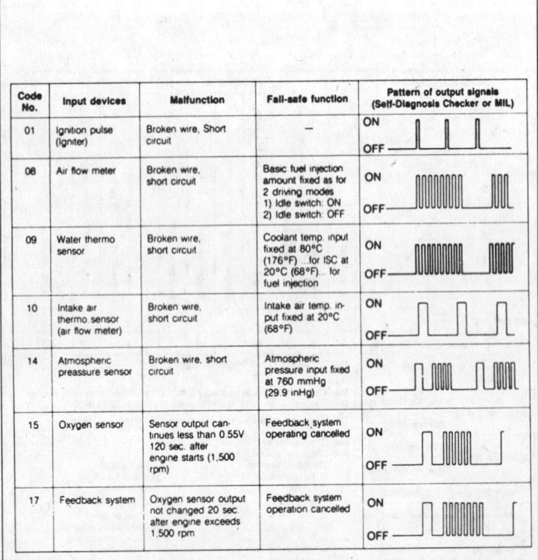

| Code No. | Input devices | Malfunction | Fail-safe function | Pattern of output signals (Self-Diagnosis Checker or MIL) |
|---|---|---|---|---|
| 01 | Ignition pulse (Igniter) | Broken wire, Short circuit | — | ON / OFF |
| 08 | Air flow meter | Broken wire, short circuit | Basic fuel injection amount fixed as for 2 driving modes 1) Idle switch: ON 2) Idle switch: OFF | ON / OFF |
| 09 | Water thermo sensor | Broken wire, short circuit | Coolant temp. input fixed at 80°C (176°F) ...for ISC at 20°C (68°F)... for fuel injection | ON / OFF |
| 10 | Intake air thermo sensor (air flow meter) | Broken wire, short circuit | Intake air temp. input fixed at 20°C (68°F) | ON / OFF |
| 14 | Atmospheric preassure sensor | Broken wire, short circuit | Atmospheric pressure input fixed at 760 mmHg (29.9 inHg) | ON / OFF |
| 15 | Oxygen sensor | Sensor output continues less than 0.55V 120 sec. after engine starts (1,500 rpm) | Feedback system operating cancelled | ON / OFF |
| 17 | Feedback system | Oxygen sensor output not changed 20 sec. after engine exceeds 1,500 rpm | Feedback system operation cancelled | ON / OFF |

The warning codes indicate a broken wire or short circuit in an output device.

**NOTE**
**When inspecting for output device malfunctions, the test connector must be grounded before the ignition switch is turned ON.**

| Code No. | Output devices | Pattern of output signals (Self-Diagnosis Checker or MIL) |
|---|---|---|
| 25 | Solenoid valve (for pressure regulator control) (If equipped) | ON / OFF |
| 26 | Solenoid valve (for vacuum switch valve) | ON / OFF |
| 27 | Solenoid valve (for No.1 purge control) | ON / OFF |
| 34 | Solenoid valve (for idle speed control valve) | ON / OFF |

**Caution**
**a) If there is more than one failure present, the lowest number warning code is displayed first, the remaining codes are displayed in order.**
**b) After repairing all failures, turn off the ignition switch, disconnect the negative battery cable, and depress the brake pedal for at least 5 seconds to erase the warning code memory.**

## TROUBLE CODE DIAGNOSTIC CHART—1988 323 NON-TURBO

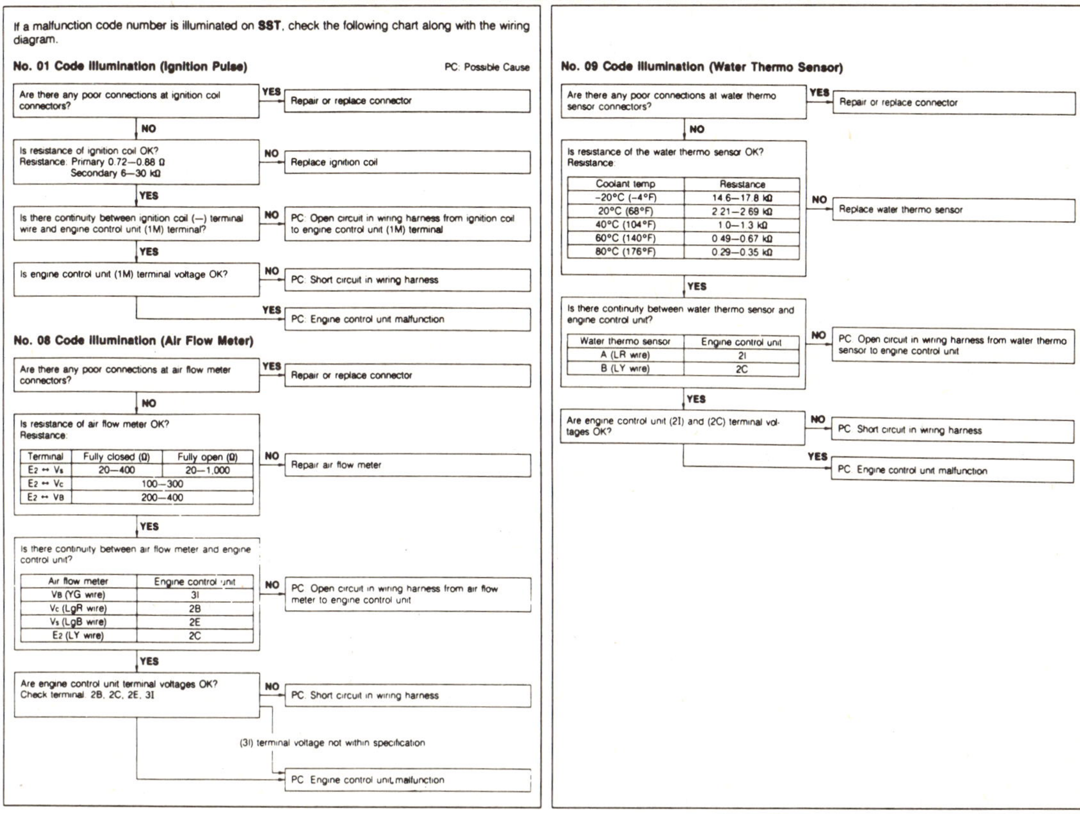


## TROUBLE CODE DIAGNOSTIC CHART—1988 323 NON-TURBO

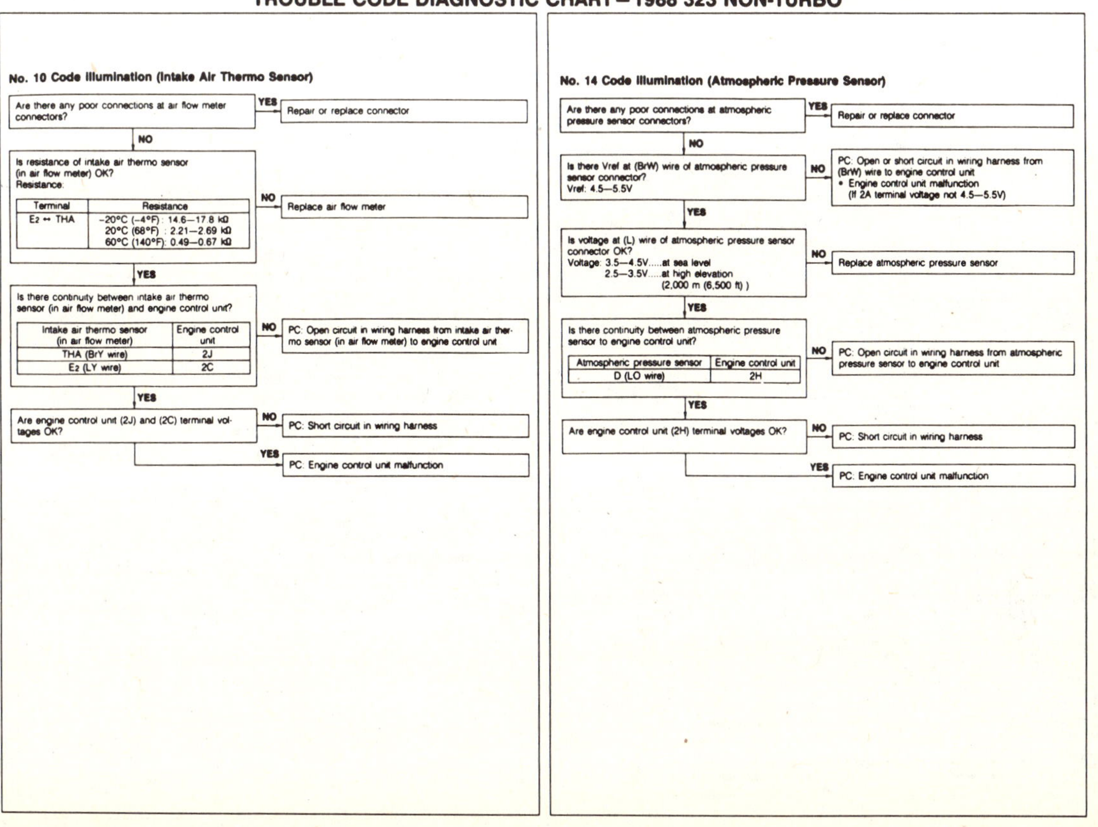

### TROUBLE CODE DIAGNOSTIC CHART—1988 323 NON-TURBO

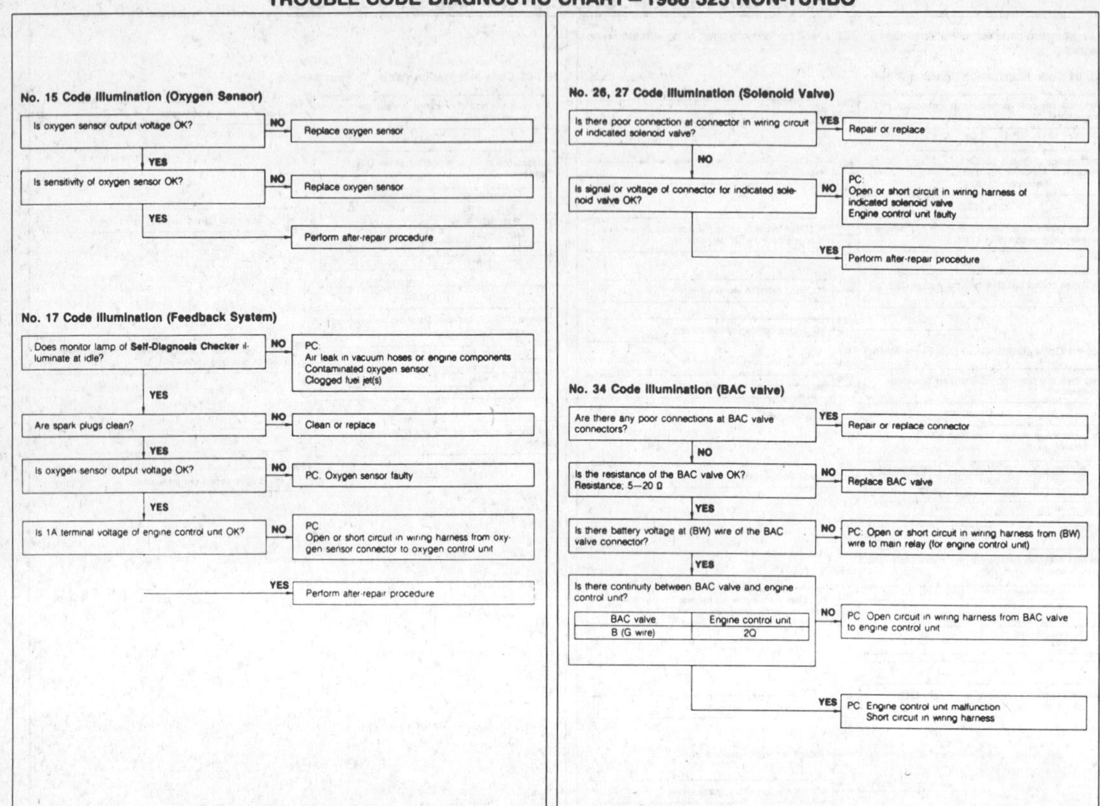


### CONTROL UNIT PIN LOCATIONS AND TEST VOLTAGES—1988 323 NON-TURBO

| Terminal | Connected to | Voltage | Condition | Remark |
|---|---|---|---|---|
| 1A (Output) | MIL | Below 2.5V | Ignition switch OFF → ON for 3 sec. | Test connector grounded |
| | | Approx. 12V | After 3 sec. | |
| 1B (Output) | Self-Diagnosis Checker (for Code No.) | Below 2.5V | Ignition switch OFF → ON for 3 sec. | • Test connector grounded • Checker connected |
| | | Approx. 12V | After 3 sec. | |
| 1C | — | — | — | — |
| 1D (Output) | Self-Diagnosis Checker (for Monitor lamp) | Approx. 5V | Ignition switch OFF → ON for 3 sec. | • Test connector grounded • Checker connected |
| | | Approx. 10V | After 3 sec. | |
| 1E (Input) | Throttle sensor (IDL switch) | Approx. 12V | Accelerator pedal depressed | |
| | | Below 1.5V | Accelerator pedal released | |
| 1F (Output) | A/C control relay | Approx. 12V | Ignition switch ON | |
| | | Below 1.5V | A/C switch ON (at idle) | |
| 1G (Input) | Neutral/clutch switch | Approx. 12V | Clutch pedal depressed | In-gear condition (Neutral: constant 12 V) |
| | | Below 1.5V | Clutch pedal released | |
| 1H (Input) | Water thermo switch (Radiator) | Approx. 12V | Below 17°C (63°F) | |
| | | Below 1.5V | Above 17°C (63°F) | |
| 1I (Input) | Electrical load (E/L) switch | Approx. 2.5V | E/L switch ON | |
| | | Approx. 10V | E/L switch OFF | |
| 1J (Input) | Brake light switch | Approx. 12V | Brake pedal depressed | |
| | | Below 1.5V | Brake pedal released | |
| 1K (Input) | Power steering switch | Approx. 12V | Power steering switch OFF | |
| | | Below 1.5V | Power steering switch ON | |
| 1L (Input) | A/C switch | Approx. 12V | A/C switch OFF | Blower motor ON |
| | | Below 2.5V | A/C switch ON | |
| 1M (Input) | Ignition coil | Approx. 12V | Ignition switch ON | (When engine running) Engine Signal Monitor: Green and red light flash |
| | | Approx. 12V | At idle | |
| 1N | — | — | — | — |
| 1O | — | — | — | — |
| 1P | — | — | — | — |
| 1Q | — | — | — | — |
| 1R | — | — | — | — |
| 1S | — | — | — | — |
| 1T | — | — | — | — |
| 1U (Output) | Igniter | Below 1.5V | Ignition switch ON | |
| | | Approx. 12V | At idle | |
| 1V (Input) | MT switch (ground) | Below 1.5V | — | ATX: constant 12V |
| 1W (Input) | Test connector | Below 1.5V | Test connector grounded | |
| | | Approx. 12V | Test connector not grounded | |
| 1X | — | — | — | — |
| 2A (Output) | Vref | 4.5—5.5V | — | — |
| 2B (Input) | Air flow meter (Vc) | 7—9V | — | — |
| 2C | Ground (E2) | Below 1.5V | — | — |
| 2D (Input) | Oxygen sensor | 0.3—0.7V | At idle | |
| | | More than 0.45V | During acceleration | |
| | | Less than 0.45V | During deceleration | |
| 2E (Input) | Air flow meter (Vs) | Approx. 2V | Ignition switch ON | |
| | | 4—5V | At idle | |
| 2F | — | — | — | — |
| 2G (Input) | Throttle sensor (PSW switch) | Approx. 12V | Accelerator pedal released | |
| | | Below 1.5V | Accelerator pedal depressed (fully open throttle) | |
| 2H (Input) | Atmospheric pressure sensor | Approx. 4V | — | At sea level |
| 2I (Input) | Water thermo sensor | Approx. 0.5V | Normal operating temperature | |
| 2J (Input) | Intake air thermo sensor (Air flow meter) | 2—3V | Intake air temperature: 20°C (68°F) | |

| Terminal | Connected to | Voltage | Condition | Remark |
|---|---|---|---|---|
| 2K (Output) | Pressure regulator control valve (PRCV) solenoid | Below 1.5V | Intake air temp. more than 58°C (136°F) Water temp. more than 90°C (194°F) | If PRCV solenoid is equipped. |
| | | Approx. 12V | Other | |
| 2L | — | — | — | — |
| 2M | — | — | — | — |
| 2N | — | — | — | — |
| 2O | No.2 purge control solenoid | Approx. 12V | Less than 1,500 rpm | |
| | | Below 1.5V | More than 1,500 rpm | |
| 2P | No.1 purge control valve solenoid | Below 1.5V | Intake air temp. more than 50°C (122°F) Water temp. more than 50°C (122°F) | In-gear condition. • Jumper wire connect to the Neutral switch (MTX) • Disconnect the inhibitor switch connector (ATX) |
| | | Approx. 12V | Other | |
| 2Q | Idle speed control (ISC) valve | 1.5—11.6V | At idle | Engine Signal Monitor: Green and red light flash |
| 2R | Ground | Below 1.5V | — | — |
| 3A | Ground | Below 1.5V | — | — |
| 3B | Starter switch | Below 2.5V | Ignition switch ON | |
| | | 7—9V | While cranking | |
| 3C | Injector No.2, No.4 | Approx. 12V | At idle | Engine Signal Monitor: Green and red light flash |
| 3D | Inhibitor switch | Below 1.5V | "N" or "P" range | MTX constant 0V |
| | | Approx. 12V | Other range | |
| 3E | Injector No.1 and No.3 | Approx. 12V | At idle | Engine Signal Monitor: Green and red light flash |
| 3F | — | — | — | — |
| 3G | Ground | Below 1.5V | — | — |
| 3H | — | — | — | — |
| 3I | Main relay | Approx. 12V | Ignition switch ON | |
| 3J | Battery | Approx. 12V | — | — |

**Engine control unit connector**

| 3I | 3G | 3E | 3C | 3A | 2Q | 2O | 2M | 2K | 2I | 2G | 2E | 2C | 2A | 1W | 1U | 1S | 1Q | 1O | 1M | 1K | 1I | 1G | 1E | 1C | 1A |
|---|---|---|---|---|---|---|---|---|---|---|---|---|---|---|---|---|---|---|---|---|---|---|---|---|---|
| 3J | 3H | 3F | 3D | 3B | 2R | 2P | 2N | 2L | 2J | 2H | 2F | 2D | 2B | 1X | 1V | 1T | 1R | 1P | 1N | 1L | 1J | 1H | 1F | 1D | 1B |

## CONTROL UNIT CONNECTOR TEST 1988 323 NON-TURBO

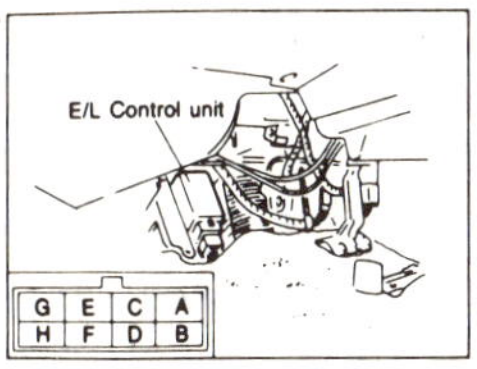

**E/L CONTROL UNIT**

**Inspection**

1. Connect a voltmeter between the E/L control unit and ground
2. Start the engine and check the terminal voltages as described below.

| Terminal | Input | Output | Connection to | Voltage (after warm-up) Ignition switch: ON | Condition |
|---|---|---|---|---|---|
| A (YG) | — | — | Main relay | Approx. 12V | |
| B (YG) | ○ | | Electrical fan relay | Approx. 12V | Coolant temp.: below 97°C (206.6°F) |
| | | | | Below 1.5V | Coolant temp.: above 97°C (206.6°F) |
| C (B) | — | — | Ground | 0V | |
| D | — | — | — | — | — |
| E (L) | | ○ | Control unit (1H) | Below 1.5V | E/L: ON |
| | | | | Approx. 12V | E/L: OFF |
| F (RB) | ○ | | Combination switch | Approx. 12V | Combination switch: ON |
| | | | | Below 1.5V | Combination switch: OFF |
| G (LG) | ○ | | Blower motor switch | Below 1.5V | Blower motor switch: ON (2nd, 3rd or 4th position) |
| | | | | Approx. 12V | Others |
| H (BY) | ○ | | Rear defroster switch | Below 1.5V | Rear defroster switch: ON |
| | | | | Approx. 12V | Rear defroster switch: OFF |

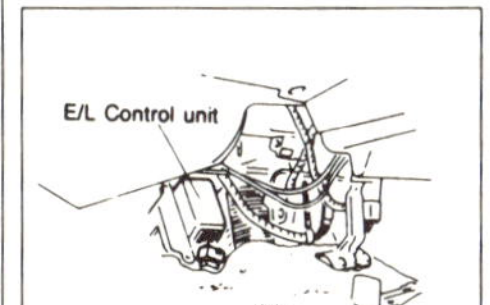


**Replacement**

1. Disconnect the connector from the E/L control unit.
2. Replace the E/L control unit.
3. Install in the reverse order of removal.

## SYSTEM SCHEMATIC—1988 323 TURBO

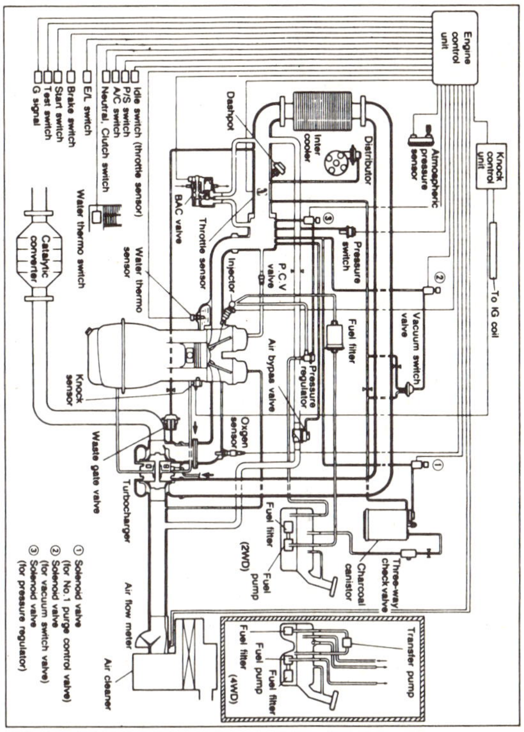

## COMPONENTS LOCATION—1988 323 TURBO

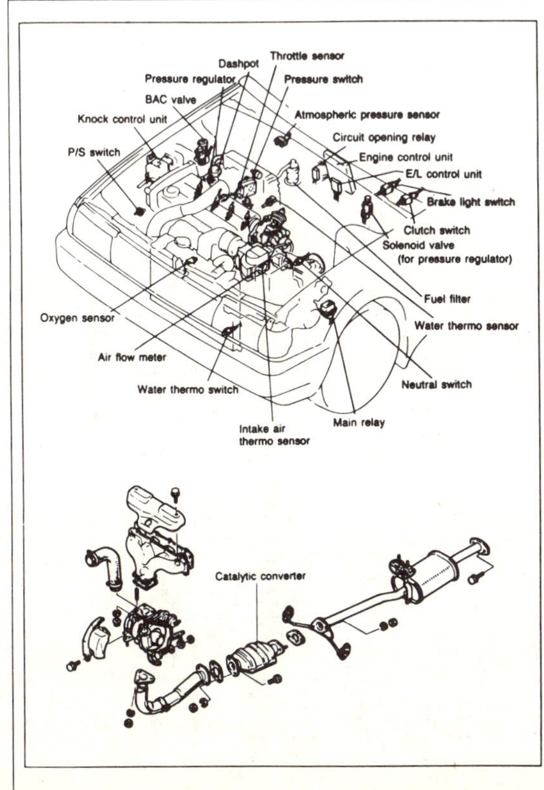


## COMPONENTS DESCRIPTION—1988 323 TURBO

| No. | COMPONENT | FUNCTION | REMARKS |
|---|---|---|---|
| 1 | Air cleaner | Filters air into the combustion chamber | |
| 2 | Air flow meter | Detects intake air amount; sends signal to the engine control unit. (for determination of fuel injection amount) | Intake air thermo sensor and fuel pump switch are integrated. |
| 3 | Atmospheric pressure sensor | Detects atmospheric pressure to prevent over rich mixture; sends signal to engine control unit. | |
| 4 | Air valve | When engine is cold, supplies bypass air into dynamic chamber for quick warm-up and smooth idle. | • Thermo wax type<br>• Installed into BAC valve |
| 5 | Brake light switch | Detects brake operation (deceleration); sends signal to control unit. | |
| 6 | Catalytic converter | Reduce HC and CO by oxidation. Reduce NOx. | Honeycomb construction |
| 7 | Charcoal canister | Stores fuel tank fumes while engine is stopped for evaporative emission. | |
| 8 | Check connector | For Self-diagnosis checker | 6 pin connector (Green) |
| 9 | Circuit opening relay | Supplies voltage for fuel pump while engine running. | |
| 10 | Clutch switch | Detects in-gear condition; sends signal to control unit. | Switch closed when clutch pedal is released. |
| 11 | Engine control unit | Detects the following;<br>1. Engine speed<br>2. Intake air amount<br>3. Engine coolant temperature<br>4. Engine load condition<br>5. Oxygen concentration<br>6. In-gear condition<br>7. Intake air temperature<br>8. Atmospheric pressure<br>9. A/C operation<br>10. P/S operation<br>11. E/L operation<br>12. Starting signal<br>13. Initial set signal<br>Controls operation of the following;<br>1. Fuel injection amount<br>2. Idle speed control system<br>3. Pressure regulator control system<br>4. Fail-safe system<br>5. Monitor switch function | 1. Ignition coil (–) terminal<br>2. Air flow meter<br>3. Water thermo sensor<br>4. Throttle sensor (Point type)<br>5. Oxygen sensor<br>6. Clutch switch and neutral switch<br>7. Intake air thermo sensor (in air flow meter)<br>8. Atmospheric pressure sensor<br>9. A/C switch<br>10. P/S switch<br>11. E/L switch<br>12. Starter switch (Ignition switch)<br>13. Test terminal<br>1. Injector<br>2. BAC valve<br>3. Solenoid valve (for pressure regulator)<br>4. Self-diagnosis checker and MIL<br>5. Monitor lamp (Self-diagnosis checker) |
| 12 | Dashpot | Gradually allows throttle valve closing during deceleration. | Adjustment speed<br>MTX....2,000 ± 150 rpm |
| 13 | Fuel filter | Filters particles from fuel | |
| 14 | Fuel pump | Provides fuel to injectors | • Operates while engine is running<br>• Installed in fuel tank |
| 15 | Injector | Injects fuel to intake port | Controlled by signals from engine control unit. |
| 16 | Intake Air Thermo Sensor | Detects intake air temperature; compensates fuel injection amount through engine control unit. | Thermistor |
| 17 | Intercooler | Cools intake air temperature after turbocharger | Air cooled |

## COMPONENTS DESCRIPTION—1988 323 TURBO

| No. | COMPONENT | FUNCTION | REMARKS |
|---|---|---|---|
| 18 | Intank Filter | Filters particles from fuel | Installed in low-pressure side |
| 19 | ISC valve | Supplies bypass air to intake manifold assembly for smooth idle | Installed into BAC valve |
| 20 | Neutral switch | Detects transaxle condition; sends signal to control unit | |
| 21 | Oxygen Sensor | Detects oxygen concentration in exhaust gas; sends signal to engine control unit; compensates fuel injection amount | Zirconia ceramic with platinum coating |
| 22 | Pressure Regulator | Regulates fuel pressure to injectors | |
| 23 | Pressure Switch (For Overboost Detection) | Detects overboost condition; sends signal to engine control unit | |
| 24 | No.1 Purge Control Valve | Open and closes evaporative vapor passage from canister to intake manifold | During open throttle |
| 25 | No.2 Purge Control Valve | Positive pressure and negative pressure valves operate in accordance with fuel tank pressure | Prevents canister from flooding |
| 26 | Throttle Sensor (Variable resistor type) | Detects throttle opening angle; sends signal to control unit; compensates fuel injection amount | |
| 27 | Solenoid Valve (for No.1 purge control valve) | Opens and closes vacuum passage to No.1 purge control valve | Controlled by signal from engine control unit |
| | Solenoid Valve (for vacuum switch valve) | Opens and closes vacuum passage to vacuum switch valve | Controlled by signal from engine control unit |
| | Solenoid valve (for pressure regulator) | Closes vacuum passage between dynamic chamber and pressure regulator | Only during hot condition |
| 28 | Transfer Pump | Pumps fuel from one side of tank to other to maintain balance | |
| 29 | Turbocharger | Pressurizes intake air utilizing exhaust gas flow | Water cooled |
| 30 | Vacuum Switch Valve | Opens passage of vacuum line when vacuum applied | Vacuum from three-way solenoid valve |
| 31 | Water Thermo Sensor | Detects coolant temperature; sends signal to control unit; compensates fuel injection amount | Thermistor |
| 32 | Water Thermo Switch | Detects radiator coolant temperature; sends signal to control unit; increases fuel injection amount | Above 17°C (63°F): ON |
| 33 | Waste Gate Valve | Allows bypassing of exhaust gas to control turbocharger boost pressure | |

## SPECIFICATIONS CHART—1988 323 TURBO

| Item \ Engine model | | | Turbo |
|---|---|---|---|
| Idle-speed rpm | | | 850 ± 50 in Neutral |
| **Throttle body** | | | |
| Type | | | Horizontal draft (1-barrel) |
| Throat diameter | | mm (in) | 50 (1.968) |
| **Air flow meter** | | | |
| Resistance Ω | E2—Vs | | Fully closed: 20—400 Fully open: 20—1,000 |
| | E2—Vc | | 100—300 |
| | E2—VB | | 200—400 |
| | E2—THA | | −20°C (−4°F) 10,000—20,000<br>20°C (68°F) 2,000—3,000<br>60°C (140°F) 400—700 |
| **Fuel pump** | | | |
| Type | | | Impeller (in tank) |
| Output pressure | | kPa (kg/cm², psi) | 441—588 (4.5—6.0, 64—85) |
| Feeding capacity | | cc (cu in)/10 sec. | 220—380 (13.4—23.2) when fuel pressure is at 250 kPa |
| **Transfer pump** | | | |
| Feeding capacity | | cc (cu in)/10 sec. | 278—388 (16.95—23.7) |
| **Pressure regulator** | | | |
| Type | | | Diaphragm |
| Regulating pressure | | kPa (kg/cm², psi) | 240—279 (2.45—2.85, 34.8—40.5) |
| **Fuel filter** | | | |
| Type | Low-pressure side | | Nylon 6 (250 mesh) element |
| | High-pressure side | | Paper element |
| **Injector** | | | |
| Type | | | High-ohmic |
| Type of drive | | | Voltage |
| Resistance | | Ω | 12—16 |
| Injection amount | | cc (cu in)/15 sec | 66—82 (4.0—5.0) |
| **Turbocharger** | | | |
| Type | | | Water cooled |
| Lubrication | | | Engine oil |
| Boost pressure (Max) | | kPa (kg/cm², psi) | 55—64 (0.56—0.65, 8.0—9.2) |
| **Waste-gate valve** | | | |
| Operating pressure | | kPa (kg/cm², psi) | 48.1—58.9 (0.49—0.60, 7.0—8.6) |
| **Idle-speed control valve** | | | |
| Solenoid resistance | | Ω | 5—20 |
| **Fuel tank** | | | |
| Capacity | | liters (US gal, Imp gal) | 50 (13, 11) |
| **Air cleaner** | | | |
| Element type | | | Wet |
| **Accelerator cable** | | | |
| Free play | | mm (in) | 1—3 (0.039—0.118) |
| **Fuel** | | | |
| Specification | | | Unleaded gasoline |

## TROUBLESHOOTING AND RELATIONSHIP CHART 1988 323 TURBO

| INPUT DEVICE \ OUTPUT DEVICE | INJECTOR | | PRCV SOLENOID | BAC VALVE | | PURGE SOLENOID | |
|---|---|---|---|---|---|---|---|
| | FUEL INJECTION AMOUNT | FUEL INJECTION TIMING | | AIR VALVE | ISC VALVE | No.1 | No.2 |
| IGNITION COIL | ○ | ○ | x | x | ○ | x | ○ |
| AIR FLOW METER | ○ | x | x | x | x | x | ○ |
| IDLE SWITCH | ○ | x | ○ | x | ○ | x | x |
| THROTTLE SENSOR | ○ | x | x | x | x | x | x |
| WATER THERMO SENSOR | ○ | x | ○ | x | ○ | ○ | x |
| INTAKE AIR THERMO SENSOR | ○ | x | ○ | x | ○ | ○ | x |
| ATMOSPHERIC PRESSURE SENSOR | ○ | x | x | x | ○ | x | x |
| OXYGEN SENSOR | ○ | x | x | x | ○ | ○ | x |
| PRESSURE SWITCH | ○ | x | x | x | x | x | x |
| BRAKE LIGHT SWITCH | ○ | x | x | x | x | x | x |
| WATER THERMO SWITCH | ○ | x | x | x | ○ | ○ | x |
| NEUTRAL AND CLUTCH SWITCH | ○ | x | ○ | x | ○ | ○ | x |
| START SWITCH | ○ | ○ | ○ | x | x | x | x |
| FF SWITCH | ○ | x | x | x | x | x | x |
| A/C SWITCH | x | x | x | x | ○ | x | x |
| P/S SWITCH | x | x | x | x | ○ | x | x |
| G SENSOR | x | ○ | x | x | x | x | x |
| TEST CONNECTOR | x | x | x | x | ○ | x | x |

## OUTPUT COMPONENTS AND ENGINE CONDITION 1988 323 TURBO

**Output Devices and Engine Conditions (Turbocharged Engine)**

| OUTPUT DEVICES \ ENGINE CONDITION | | CRANKING (COLD ENGINE) | WARMING UP (DURING IDLE) | MEDIUM LOAD | | ACCELERATION | HEAVY LOAD | DECELERATION | IDLE (THROTTLE VALVE FULLY CLOSED) | IGN: ON (ENGINE NOT RUNNING) | REMARKS |
|---|---|---|---|---|---|---|---|---|---|---|---|
| | | | | COLD | WARM | | | | | | |
| INJECTOR | INJECTION | Rich | | | Rich and Lean | Rich | | Fuel Cut | Rich | Does not inject | |
| | INJECTION TIMING | 1 Group | 2 Group | | | | | | 2 Group | | Above 6,800 rpm fuel cut |
| PRCV SOLENOID | | ON (Vacuum cut) | OFF (Vacuum to pressure regulator) | | | | | | * After start ON (Vacuum cut) | Does not operate | * During hot starting |
| BAC VALVE | AIR VALVE | * Open | | | Close | | | | | | * Coolant temp below 60°C (140°F) |
| | ISC VALVE | Large amount of bypass air | | Small amount of bypass air | | | | | * Large and small amount of bypass air | Does not operate | * Test connector grounded: small amount |
| PURGE SOLENOID | No.1 | OFF (Vacuum cut) | | | * ON (Vacuum to No.1 purge control valve) | | | | OFF (Vacuum cut) | | * Positive pressure OFF |
| | No.2 | OFF (Vacuum cut) | | * ON (Vacuum to vacuum switch valve) | | | | OFF (Vacuum cut) | | | * Engine speed above 1,500 rpm |

## TROUBLESHOOTING CHART—1988 323 TURBO

| SYMPTOM \ POSSIBLE CAUSE | Ignition coil | Group sensor (Distributor) | Air flow meter | Water thermo sensor | Intake air thermo sensor (in Air flow meter) | Throttle sensor (Variable resistor type) | Atmospheric pressure sensor | Oxygen sensor | Feedback system | | Solenoid valve (Pressure regulator) | Solenoid valve (No.1 purge control valve) | Solenoid valve (Vacuum switch valve) | BAC Valve (Idle speed control valve) |
|---|---|---|---|---|---|---|---|---|---|---|---|---|---|---|
| | INPUT DEVICES | | | | | | | | | | OUTPUT DEVICES | | | |
| 1 Fault indicated by SST Code NO. | 01 | 03 | 08 | 09 | 10 | 12 | 14 | 15 | 17 | | 25 | 26 | 27 | 34 |

Symptoms:

1. Fault indicated by SST Code NO.
2. Hard start or won't start (Crank: OK)
3. Engine stall — Only while warming up / Only after warming up
4. Rough idle — Only while warming up / Only after warming up
5. High idle speed after warming up
6. Poor acceleration, hesitation, or lack of power
7. Runs rough on deceleration
8. Knocking
9. Excessive fuel consumption
10. Abnormal noise
11. Vibration
12. White smoke
13. Excessive oil consumption
14. Afterburn in exhaust system
15. Engine stalls or rough after hot starting
16. Fail emission test

**TROUBLESHOOTING PROCEDURE:**

**Note**
**Step 1 under symptom is to quickly determine what system or parts may be at fault using the self-diagnosis checker (49 H018 9A1)**

**1st** Check input sensors and switches and output solenoid valves self-diagnosed with **Self-diagnosis checker**

**2nd** Check other switches with **Self-diagnosis checker**

**3rd** Check the following items

| **Electric system** | **Ignition system** |
|---|---|
| 1) Battery condition | 1) Spark plugs |
| 2) Fuses | 2) Ignition timing |

| **Fuel system** | **Intake air system** |
|---|---|
| 1) Fuel amount | 1) Air cleaner element |
| 2) Fuel leakage | 2) Vacuum or air leakage |
| 3) Fuel filter | 3) Vacuum hose routing |
| 4) Idle speed | 4) Accelerator cable |

**4th** Check the Fuel and Emission Control Systems

| POSSIBLE CAUSE | Symptom 16 | 15 | 14 | 13 | 12 | 11 | 10 | 9 | 8 | 7 | 6 | 5 | 4 (a) | 4 (b) | 3 (a) | 3 (b) | 2 |
|---|---|---|---|---|---|---|---|---|---|---|---|---|---|---|---|---|---|
| Intake air system (Poor connection of components, throttle body) | 5 | | 3 | | | | | | | | 2 | 2 | 5 | 4 | 4 | 3 | 2 |
| Fuel system (Fuel injection, fuel pressure) | 6 | 2 | 4 | | | | | 2 | | 3 | 3 | | 4 | 3 | 3 | 2 | 1 |
| ISC (Idle speed control) system (Air valve, ISC solenoid valve) | 3 | | 1 | | | | | | | 2 | | 1 | 2 | 1 | 2 | 1 | |
| PRC (Pressure regulator control) system | | 1 | | | | | | | | | | | | | | | |
| Turbocharging system (Oil and water passage, turbine, and compressor wheels malfunction) | | | | 1 | 1 | 1 | 1 | | 2 | | 5 | | | | | | |
| PCV (Positive crank case ventilation) system | | | | | | | | | | | | | 1 | 2 | 1 | | |
| Knock control system | | | | | | | | | 1 | | | | | | | | |
| Evaporative emission control system (Vacuum switch valve, No.1, No.2 purge valve malfunction) | 4 | | | | | | | | | | 1 | | 3 | | | | |
| Deceleration system (Fuel cut operation malfunction) | 2 | | 2 | | | | | 1 | | 1 | | | | | | | |
| Exhaust system (System clogged) | 1 | | | | | | | 3 | | | 4 | | | | | | |

**The number of the list show the priorities of inspections from the most possible to that with the lowest possibility.**
**These were determined on the following basis:**
**• Ease of inspection • Most possible system • Most possible point in the system**

## MALFUNCTION INDICATOR LIGHT (MIL) OPERATION 1988 323 TURBO

Malfunction codes are determined as below

**1. Malfunction code cycle break**
The time between malfunction code cycles is 4.0 sec (the time the light is off).

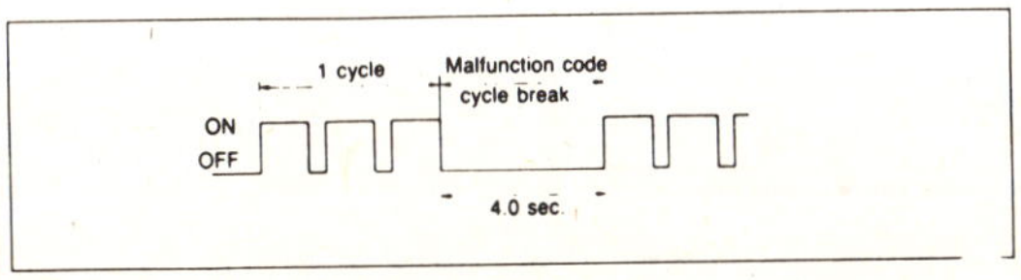

1 cycle
Malfunction code cycle break
ON
OFF
4.0 sec.

**2. Second digit of malfunction code (ones position)**
The digit in the ones position of the malfunction code represents the number of times the buzzer is on 0.4 sec during one cycle.

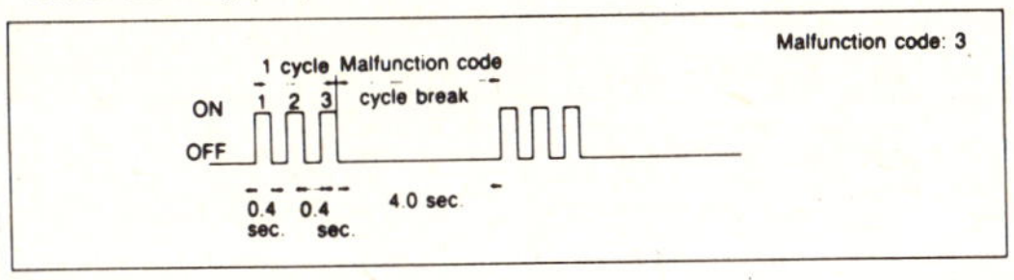

Malfunction code: 3
1 cycle
Malfunction code cycle break
ON
OFF
1 2 3
0.4 sec.
0.4 sec.
4.0 sec.

**3. First digit of malfunction code (tens position)**
The digit in the tens position of the malfunction code represents the number of times the buzzer is on 1.2 sec during one cycle.

It should also be noted that, the light goes off for 1.6 sec. between the long and short pulses of buzzer.

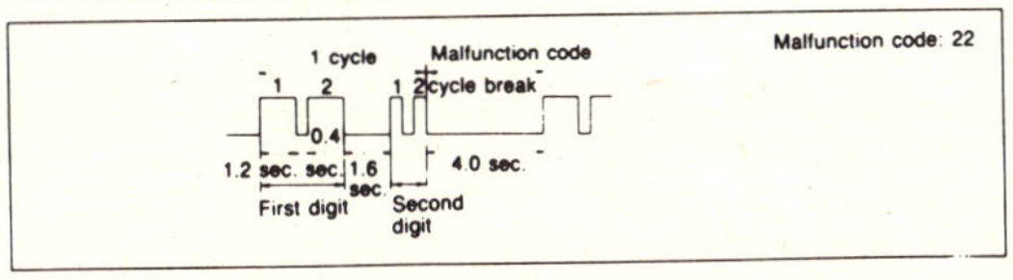

Malfunction code: 22
1 cycle
Malfunction code cycle break
1 2
1 2
0.4
1.2 sec.
sec.
1.6 sec.
4.0 sec.
First digit
Second digit

## TROUBLE CODE IDENTIFICATION CHART 1988 323 TURBO

**Input Devices**

| Code No. | Input devices | Malfunction | Fail-safe function | Pattern of output signals (Self-Diagnosis Checker or MIL) |
|---|---|---|---|---|
| 01 | Ignition pulse (Igniter) | Broken wire, Short circuit | — | ON / OFF |
| 03 | Distributor (G signal) | Broken wire, short circuit | — | ON / OFF |
| 08 | Air flow meter | Broken wire, short circuit | Basic fuel injection amount fixed as for 2 driving modes 1) Idle switch: ON 2) Idle switch: OFF | ON / OFF |
| 09 | Water thermo sensor | Broken wire, short circuit | Coolant temp input fixed at 80°C (176°F) ...for ISC at 60°C (14°F) ...for fuel injection | ON / OFF |
| 10 | Intake air thermo sensor (air flow meter) | Broken wire, short circuit | Intake air temp input fixed at 20°C (68°F) | ON / OFF |
| 12 | Throttle sensor | Broken wire, short circuit | Throttle valve opening angle signal input fixed at full open | ON / OFF |
| 14 | Atmospheric preassure sensor | Broken wire, short circuit | Atmospheric pressure input fixed at 760 mmHg (29.9 inHg) | ON / OFF |
| 15 | Oxygen sensor | Oxygen sensor output below 0.55V 120 sec after engine at 1,500 rpm | Cancels EGI feedback operation | ON / OFF |
| 17 | Feedback system | $O_2$ sensor output below 0.45V 20 sec after engine exceeds 1,500 rpm | Cancels EGI feedback operation | ON / OFF |

## TROUBLE CODE IDENTIFICATION CHART 1988 323 TURBO

**Output Devices**
The warning codes indicate a broken wire or short circuit in an output device.

**NOTE**
**When inspecting for malfunctions of output device, the test connector must be grounded before the ignition switch ON.**

| Code No. | Output devices | Pattern of output signals (Self-Diagnosis Checker or MIL) |
|---|---|---|
| 25 | Solenoid valve (for pressure regulator) | ON / OFF |
| 26 | Solenoid valve (for vacuum switch valve) | ON / OFF |
| 27 | Solenoid valve (for No.1 purge control valve) | ON / OFF |
| 34 | Solenoid valve (for idle speed) | ON / OFF |

**Caution**
**a) If there is more than one failure present, the lowest number warning code is displayed first, the subsequent warning codes light up in order.**
**b) After repairing all failures, to erase thr memory of a warning code, turn off the ignition switch and disconnect the negative battery cable for at least 5 seconds.**

## TROUBLE CODE DIAGNOSTIC CHART 1988 323 TURBO

If a malfunction code number is illuminated on **SST**, check the following chart along with the wiring diagram.

**No. 01 code Illumination (Ignition Pulse)**

PC: Possible Cause

- Are there any poor connections at ignition coil connectors?
  - YES → Repair or replace connector
  - NO ↓
- Is resistance of ignition coil OK? Resistance: Primary 0.72—0.88 Ω, Secondary 6—30 kΩ
  - NO → Replace ignition coil
  - YES ↓
- Is there continuity between ignition coil (—) terminal wire and engine control unit (1M) terminal?
  - NO → PC: Open circuit in wiring harness from ignition coil to engine control unit (1M) terminal
  - YES ↓
- Is engine control unit (1M) terminal voltage OK?
  - NO → PC: Short circuit in wiring harness
  - YES → PC: Engine control unit malfunction

**No. 03 Code Illumination (G Signal)**

- Are there any poor connection at the distributor connectors?
  - YES → Repair or replace connector
  - NO ↓
- Measure (1N) terminal voltage of engine control unit at idle.
  - 0V → PC: Open circuit in the wiring harness from (1N) terminal of engine control unit to distributor. Malfunction of G signal generator
  - Approx. 3.0V → PC: Engine control unit defective

## TROUBLE CODE DIAGNOSTIC CHART—1988 323 TURBO

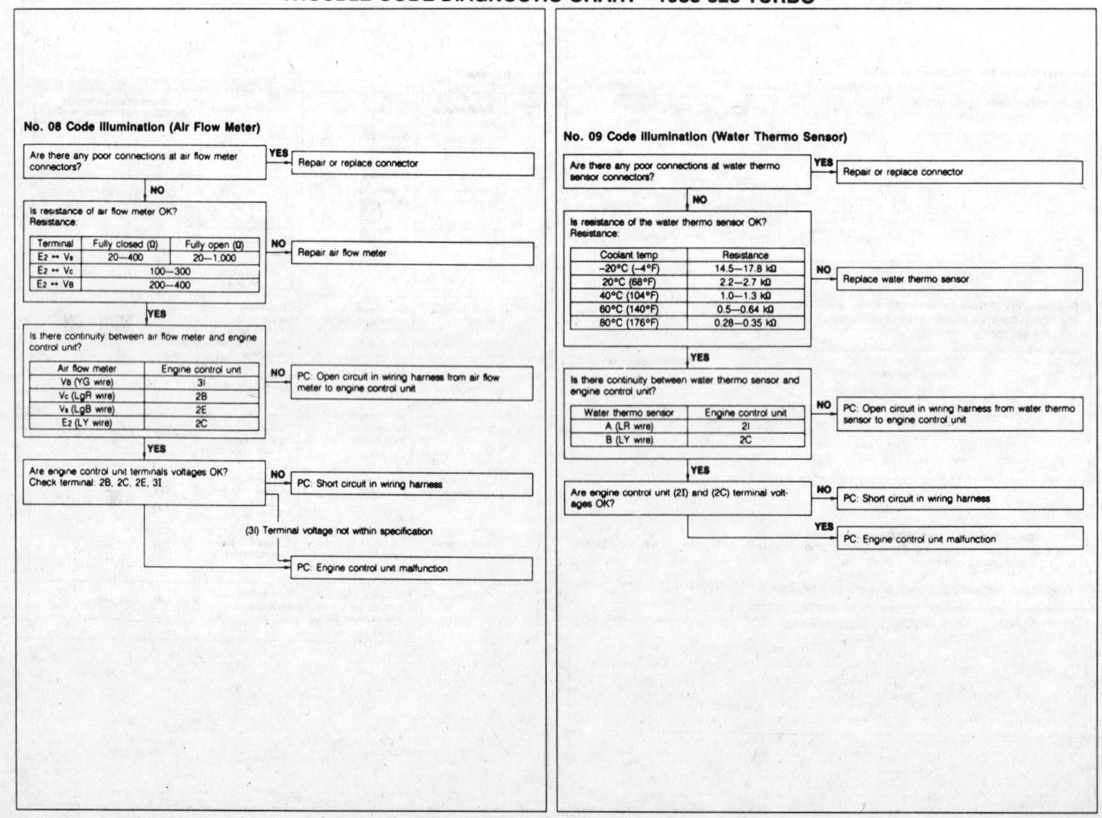

## TROUBLE CODE DIAGNOSTIC CHART—1988 323 TURBO

**No. 10 Code Illumination (Intake Air Thermo Sensor)**

Are there any poor connections at air flow meter connectors?
YES → Repair or replace connector
NO ↓

Is resistance of intake air thermo sensor (in air flow meter) OK?
Resistance

| Terminal | Resistance |
|---|---|
| $E_2$ ↔ THA | −20°C (−4°F) 13.6—18.4 kΩ<br>20°C (68°F) 2.21—2.69 kΩ<br>60°C (140°F) 0.493—0.667 kΩ |

NO → Replace air flow meter
YES ↓

Is there continuity between intake air thermo sensor (in air flow meter) and engine control unit?

| Intake air thermo sensor (in air flow meter) | Engine control unit |
|---|---|
| THA (BrY wire) | 2J |
| $E_2$ (LY wire) | 2C |

NO → PC: Open circuit in wiring harness from intake air thermo sensor (in air flow meter) to engine control unit
YES ↓

Are engine control unit (2J) and (2C) terminal voltages OK?
NO → PC: Short circuit in wiring harness
YES → PC: Engine control unit malfunction

**No. 12 Code Illumination (Throttle Sensor)**

Are there any poor connections at throttle sensor connectors?
YES → Repair or replace connector the terminal
NO ↓

Is the resistance of the throttle sensor OK?
NO → Replace the throttle sensor
YES ↓

Is there 4.5—5.5V at the (1C) terminal of the throttle sensor connector?
NO → PC:
Open or short circuit in the wiring harness from the (1C) terminal to the throttle sensor to engine control unit.
Engine control unit defective
YES ↓

Measure (2G) terminal voltage of the engine control unit
More than 4.5V → PC:
Open circuit in the wiring harness from (1A) terminal of throttle sensor to engine control unit
Open circuit in the wiring harness of ground
Less than 0.25V → PC: Short circuit in the wiring harness from (1A) terminal of throttle sensor to engine control unit.
0.25—4.5V → PC: Engine control unit defective

## TROUBLE CODE DIAGNOSTIC CHART—1988 323 TURBO

**No. 14 Code Illumination (Atmospheric Pressure Sensor)**

Are there any poor connections at atmospheric pressure sensor connectors?
YES → Repair or replace connector
NO ↓

Is there Vref at (YG) wire of atmospheric pressure sensor connector?
Vref: 4.5—5.5V
NO → PC: Open or short circuit in wiring harness from (YG) wire to control unit
- Engine control unit malfunction (If 2A terminal voltage not 4.5—5.5V)

YES ↓

Is voltage at (LO) wire of atmospheric pressure sensor connector OK?
Voltage: 3.5—4.5V.....at sea level
2.5—3.5V.....at high elevation (2,000 m (6,500 ft) )
NO → Replace atmospheric pressure sensor
YES ↓

Is there continuity between atmospheric pressure sensor to engine control unit?

| Atmospheric pressure sensor | Engine control unit |
|---|---|
| D (Lg W wire) | 2H |

NO → PC: Open circuit in wiring harness from atmospheric pressure sensor to engine control unit
YES ↓

Are engine control unit (2H) terminal voltages OK?
NO → PC: Short circuit in wiring harness
YES → PC: Engine control unit malfunction

**No. 15 Code display Illumination (Oxygen Sensor)**

Is oxygen sensor output voltage OK?
NO → Replace oxygen sensor.
YES ↓

Is sensitivity of oxygen sensor OK?
NO → Replace oxygen sensor
YES → Perform after-repair procedure

**No. 17 Code display Illumination (Feedback System)**

Does monitor lamp of **Self-Diagnosis Checker** illuminate at idle?
NO → PC: Air leak in vacuum hoses or emission components. Contaminated oxygen sensor
Clogged fuel jet(s)
YES ↓

Are spark plugs clean?
NO → Clean or replace spark plugs
YES ↓

Is oxygen sensor output voltage OK?
NO → PC: oxygen sensor faulty
YES ↓

Is 1A terminal voltage of engine control unit OK?
NO → PC: Open or short circuit in wiring harness from oxygen sensor connector to engine control unit
YES → Perform after-repair procedure

### TROUBLE CODE DIAGNOSTIC CHART 1988 323 TURBO

**No. 25, 26, 27 Code Illumination (Solenoid Valve)**

- Is there poor connection at connector in wiring circuit of indicated solenoid valve?
  - YES → Repair or replace connector
  - NO → Is signal or voltage of connector for indicated solenoid valve OK?
    - NO → PC: Open or short circuit in wiring harness of indicated solenoid valve. Emission control unit faulty
    - YES → Perform after-repair procedure

**No. 34 Code Illumination (BAC Valve)**

- Are there any poor connections at BAC valve connectors?
  - YES → Repair or replace connector
  - NO → Is the resistance of the BAC valve OK? Resistance: 5–20 Ω
    - NO → Replace BAC valve
    - YES → Is there battery voltage at (YG) wire of the BAC valve connector?
      - NO → PC: Open or short circuit in wiring harness from (YG) wire to main relay (for engine control unit)
      - YES → Is there continuity between BAC valve and engine control unit?

| BAC valve | Control unit |
|---|---|
| B (G wire) | 2Q |

- NO → PC: Open circuit in wiring harness from BAC valve to engine control unit
- YES → PC • Engine control unit malfunction • Short circuit in wiring harness

### CONTROL UNIT PIN LOCATIONS AND TEST VOLTAGES 1988 323 TURBO

| Terminal | Connected to | Voltage | Condition | Remark |
|---|---|---|---|---|
| 1A (Output) | MIL | Below 2.5V | Ignition switch OFF → ON for 3 sec. | Test connector grounded |
| | | Approx. 12V | After 3 sec. | |
| 1B (Output) | Self-Diagnosis Checker (for Code No.) | Below 2.5V | Ignition switch OFF → ON for 3 sec. | Test connector grounded Checker connected |
| | | Approx. 12V | After 3 sec. | |
| 1C | — | — | — | — |
| 1D (Output) | Self-Diagnosis Checker (for Monitor lamp) | Approx. 5V | Ignition switch OFF → ON for 3 sec. | Test connector grounded Checker connected |
| | | Approx. 12V | After 3 sec. | |
| 1E (Input) | Throttle sensor (IDL switch) | Approx. 12V | Accelerator pedal depressed | |
| | | Below 1.5V | Accelerator pedal released | |
| 1F (Output) | A/C control relay | Approx. 12V | Ignition switch ON | |
| | | Below 1.5V | A/C switch ON (at idle) | |
| 1G (Input) | Neutral/clutch switch | Approx. 12V | Clutch pedal depressed | In-gear condition (Neutral: Constant 12V) |
| | | Below 1.5V | Clutch pedal released | |
| 1H (Input) | Water thermo switch (Radiator) | Approx. 12V | Below 17°C (63°F) | |
| | | Below 1.5V | Above 17°C (63°F) | |
| 1I (Input) | Electrical load (E/L) switch | Approx. 2.5V | E/L switch ON | |
| | | Approx. 12V | E/L switch OFF | |
| 1J (Input) | Brake light switch | Approx. 12V | Brake pedal depressed | |
| | | Below 1.5V | Brake pedal released | |
| 1K (Input) | Power steering switch | Approx. 12V | Power steering switch OFF | |
| | | Below 1.5V | Power steering switch ON | |
| 1L (Input) | A/C switch | Approx. 12V | A/C switch OFF | Blower motor ON |
| | | Below 2.5V | A/C switch ON | |
| 1M (Input) | Ignition coil | Approx. 12V | Ignition switch ON | (When engine running) Engine Signal Monitor: Green and red light flash |
| | | Approx. 12V | At idle | |
| 1N | G sensor (Distributor) | Below 1.5V | Ignition switch ON | |
| | | Approx. 3V | At idle | |
| 1O | — | — | — | — |
| 1P | — | — | — | — |
| 1Q | — | — | — | — |
| 1R | — | — | — | — |
| 1S | — | — | — | — |
| 1T | — | — | — | — |
| 1U (Output) | Knock control unit (I terminal) | Below 1.5V | Ignition switch ON | |
| | | Approx. 12V | At idle | |
| 1V (Input) | FF switch | Below 1.5V | 4x4 | |
| | | Approx. 12V | FF | |
| 1W (Input) | Test connector | Below 1.5V | Test connector grounded | |
| | | Approx. 12V | Test connector not grounded | |
| 1X | — | — | — | — |
| 2A (Output) | Vref | 4.5—5.5V | — | — |
| 2B (Input) | Air flow meter (Vc) | 7—9V | — | — |
| 2C | Ground (E2) | Below 1.5V | — | — |
| 2D (Input) | Oxygen sensor | 0.3—0.7V | At idle | |
| | | More than 0.45V | During acceleration | |
| | | Less than 0.45V | During deceleration | |

### CONTROL UNIT PIN LOCATIONS AND TEST VOLTAGES 1988 323 TURBO

| Terminal | Connected to | Voltage | Condition | Remark |
|---|---|---|---|---|
| 2E (Input) | Air flow meter (Vs) | Approx. 2V | Ignition switch ON | |
| | | 4—5V | At idle | |
| 2F | — | — | — | — |
| 2G (Input) | Throttle sensor | Approx. 0.5V | Accelerator pedal released | |
| | | Approx. 4V | Accelerator pedal depressed | |
| 2H (Input) | Atmospheric pressure sensor | Approx. 4V | — | At sea level |
| 2I (Input) | Water thermo sensor | Approx. 0.5V | Normal operating temperature | |
| 2J (Input) | Intake air thermo sensor (Air flow meter) | 2—3V | Intake air temperature: 20°C (68°F) | |
| 2K (Output) | Pressure regulator control valve (PRCV) solenoid | Below 2.5V | Intake air temp. more than 58°C (136°F) Water temp. more than 90°C (194°F) | |
| | | Approx. 12V | Other | |
| 2L (Output) | Pressure switch | Approx. 12V | At idle | Air pressure 71.8—79.8 kPa (0.73—0.81 kg/cm², 10.4—11.6 psi) |
| | | Below 1.5V | At overboost | |
| 2M (Output) | Knock control unit (f terminal) | Below 1.5V | At idle | Coolant temp: More than 80°C (176°F) Intake air temp: More than 0°C (32°F) |
| | | Approx. 12V | Engine speed 1,000 rpm (Positive pressure) | |
| 2N (Output) | Indicator light | Approx. 12V | At idle | 71.8—79.8 kPa (0.73—0.81 kg/cm², 10.4—11.6 psi) |
| | | Below 1.5V | At overboost | |
| 2O | No.2 purge control solenoid | Approx. 12V | Less than 1,500 rpm | |
| | | Below 1.5V | More than 1,500 rpm | |
| 2P | No.1 purge control valve solenoid | Below 1.5V | Intake air temp. more than 50°C (122°F) Water temp. more than 50°C (122°F) | In-gear condition. Jumper wire connect to the Neutral switch |
| | | Approx. 12V | Other | |
| 2Q | Idle speed control (ISC) valve | 1.5—11.6V | At idle | Engine Signal Monitor: Green and red light flash |
| 2R | Ground | Below 1.5V | — | — |
| 3A | Ground | Below 1.5V | — | — |
| 3B | Starter switch | Below 2.5V | Ignition switch ON | |
| | | 7—9V | While cranking | |
| 3C | Injector No.2, No.4 | Approx. 12V | At idle | Engine Signal Monitor: Green and red light flash |
| 3D | — | — | — | — |
| 3E | Injector No.1, No.3 | Approx. 12V | At idle | Engine signal Monitor: Green and red light flash |
| 3F | — | — | — | — |
| 3G | Ground | Below 1.5V | — | — |
| 3H | — | — | — | — |
| 3I | Main relay | Approx. 12V | Ignition switch ON | |
| 3J | Battery | Approx. 12V | — | — |

**Engine control unit connector**

| 3I | 3G | 3E | 3C | 3A | 2Q | 2O | 2M | 2K | 2I | 2G | 2E | 2C | 2A | 1W | 1U | 1S | 1Q | 1O | 1M | 1K | 1I | 1G | 1E | 1C | 1A |
|---|---|---|---|---|---|---|---|---|---|---|---|---|---|---|---|---|---|---|---|---|---|---|---|---|---|
| 3J | 3H | 3F | 3D | 3B | 2R | 2P | 2N | 2L | 2J | 2H | 2F | 2D | 2B | 1X | 1V | 1T | 1R | 1P | 1N | 1L | 1J | 1H | 1F | 1D | 1B |

### CONTROL UNIT CONNECTOR TEST—1988 323 TURBO

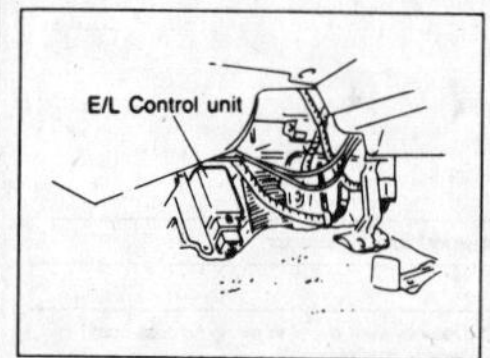


**ELECTRICAL LOAD (E/L) CONTROL UNIT Inspection**

1. Connect a voltmeter between the E/L control unit and ground.
2. Start the engine and check the terminal voltages as described below.

| Terminal | Input | Output | Connection to | Voltage (after warm-up) Ignition switch: ON | Voltage (after warm-up) Idle | Remarks |
|---|---|---|---|---|---|---|
| A (YG) | — | — | Ignition switch | Approx. 12V | | |
| B (YG) | ○ | | Electrical fan relay | Approx. 12V | | Coolant temp. below 97°C (206.6°F) |
| | | | | Below 1.5V | | Coolant temp. above 97°C (206.6°F) |
| C (B) | — | — | Ground | 0V | | |
| D | — | — | — | — | — | — |
| E (L) | | ○ | Control unit (1H) | | Below 1.5V | E/L: ON |
| | | | | | Approx. 12V | E/L: OFF |
| F (RB) | ○ | | Combination switch | | Approx. 12V | Combination switch: ON |
| | | | | | Below 1.5V | Combination switch: OFF |
| G (LG) | ○ | | Blower motor switch | | Below 1.5V | Blower motor switch: ON (2nd, 3rd or 4th position) |
| | | | | | Approx. 12V | Others |
| H (BY) | ○ | | Rear defroster switch | | Below 1.5V | Rear defroster switch: ON |
| | | | | | Approx. 12V | Rear defroster switch: OFF |

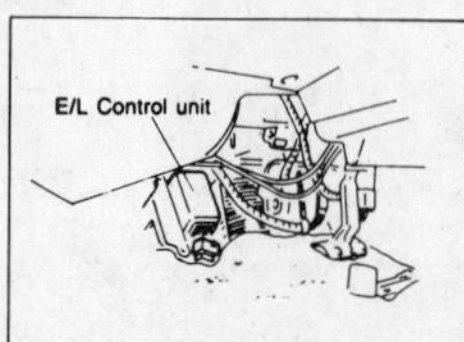


**Replacement**

1. Disconnect the connector from the E/L control unit.
2. Replace the E/L control unit.
3. Install in the reverse order of removal.

## SYSTEM SCHEMATIC—1989 323 NON-TURBO

## COMPONENTS LOCATION—1989 323 NON-TURBO

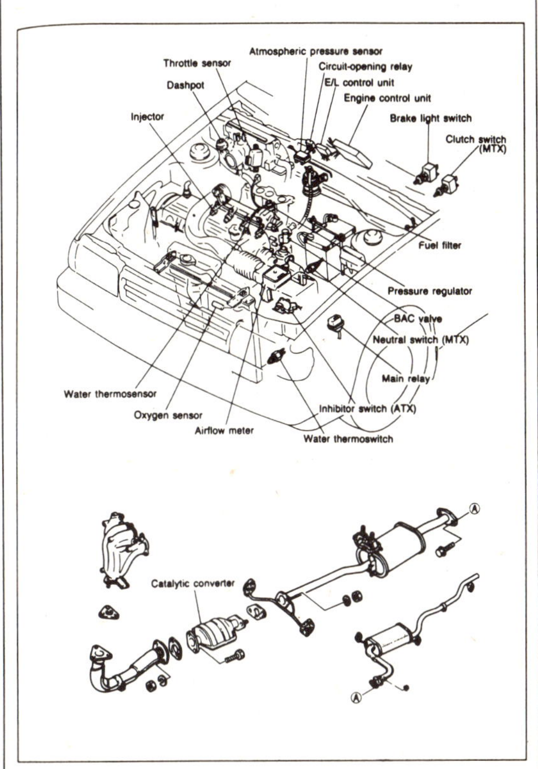


## COMPONENTS DESCRIPTION—1989 323 NON-TURBO

| COMPONENT | FUNCTION | REMARKS |
|---|---|---|
| Air cleaner | Filters air entering combustion chamber | |
| Airflow meter | Detects intake air amount; sends signal to the engine control unit (for determination of fuel injection amount) | Includes intake air thermosensor and fuel pump switch |
| Atmospheric pressure sensor | Detects atmospheric pressure; sends signal to engine control unit | |
| Air valve | When engine is cold, supplies bypass air into dynamic chamber for quick warm-up and smooth idle | • Engine speed increased to shorten warm-up period<br>• Thermo wax type<br>• Installed in BAC valve |
| BAC valve | Supplies bypass air into dynamic chamber | Consists of air valve and ISC valve |
| Brake light switch | Detects brake operation (deceleration), sends signal to engine control unit | |
| Catalytic converter | Reduces HC and CO by oxidization; reduces NOx by reduction | Honeycomb construction |
| Charcoal canister | Stores fuel tank fumes while engine is stopped | |
| Check connector | For Self-Diagnosis Checker | 6-pin connector (Green) |
| Circuit-opening relay | Supplies voltage for fuel pump while engine is running | |
| Clutch switch | Detects in-gear condition; sends signal to engine control unit | Switch closed when clutch pedal released |
| Engine control unit | Detects the following:<br>1. Engine speed<br>2. Intake air amount<br>3. Engine coolant temperature<br>4. Engine load condition<br>5. Oxygen concentration in exhaust gas<br>6. In-gear condition<br>7. Intake air temperature<br>8. Atmospheric pressure<br>9. A/C operation<br>10. P/S operation<br>11. E/L (Electrical load) operation<br>12. Starting signal<br>13. Initial set signal<br>Controls operation of the following:<br>1. Fuel injection amount<br>2. Idle-speed control system<br>3. Purge control system<br>4. Fail-safe system<br>5. Monitor switch function | 1. Ignition coil (–) terminal<br>2. Airflow meter<br>3. Water thermosensor<br>4. Throttle sensor (Point type)<br>5. Oxygen sensor<br>6. Clutch switch and neutral switch<br>7. Intake air thermosensor (in airflow meter)<br>8. Atmospheric pressure sensor<br>9. A/C switch<br>10. P/S pressure switch<br>11. E/L control unit<br>12. Ignition switch<br>13. Test connector<br>1. Injector<br>2. BAC valve<br>3. Solenoid valve (No.1 purge control, No.2 purge control)<br>4. Self-Diagnosis Checker and MIL<br>5. Monitor lamp (Self-Diagnosis Checker) |
| Dashpot | Causes gradual throttle valve closing during deceleration | Adjustment speed<br>MTX....2,800 ± 150 rpm<br>ATX....2,800 ± 300 rpm |
| Fuel filter | Filters particles from fuel | |
| Fuel pump | Provides fuel to injectors | • Operates while engine is running<br>• Installed in fuel tank |
| Inhibitor switch | Detects in-gear condition; sends signal to engine control unit | Switch closed in N or P range (ATX) |
| Intake air thermosensor | Detects intake air temperature; sends signal to engine control unit | Thermistor |
| Injector | Injects fuel to intake port | Controlled by signals from engine control unit |
| In-tank fuel filter | Filters particles from fuel | |
| ISC valve | Supplies bypass air to dynamic chamber for smooth idle | Installed in BAC valve |
| Neutral switch | Detects transaxle condition; sends signal to engine control unit | |
| Oxygen sensor | Detects oxygen concentration in exhaust gas; sends signal to engine control unit | Zirconia ceramic with platinum coating |
| Pressure regulator | Regulates fuel pressure to injectors | |
| P/S pressure switch | Detects P/S operation; sends signal to engine control unit | When steering wheel turned right or left |
| No.1 Purge control valve | Opens and closes evaporative vapor passage from canister to intake manifold | |
| No.2 Purge control valve | Positive pressure and negative pressure valves operate in accordance with fuel tank pressure | Prevents canister from flooding |
| Throttle sensor (point type) | Detects throttle opening angle; sends signal to engine control unit | |
| Solenoid valve (No.1 purge control) | Opens and closes vacuum passage to No.1 purge control valve | Controlled by signal from engine control unit |
| Solenoid valve (No.2 purge control) | Opens and closes vacuum passage to vacuum switch valve | Controlled by signal from engine control unit |
| Vacuum switch valve | Opens passage of vacuum line when vacuum applied | |
| Water thermosensor | Detects coolant temperature; sends signal to engine control unit | Thermistor |
| Water thermoswitch | Detects coolant temperature; sends signal to engine control unit | Above 17°C (63°F), ON |

## SPECIFICATIONS CHART—1989 323 NON-TURBO

| Item | | Transaxle type | Manual transaxle | Automatic transaxle |
|---|---|---|---|---|
| **Idle speed** (with test connector grounded) | | rpm | 850 ± 50 (Neutral) | 850 ± 50 (P range) |
| **Throttle body** | | | | |
| Type | | | Horizontal draft (1-barrel) | |
| Throat diameter | | mm (in) | 45 (1.77) | |
| **Airflow meter** | | | | |
| Resistor Ω | E2—Vs | | Fully closed: 20—400 Fully open: 20—1,000 | |
| | E2—Vc | | 100—300 | |
| | E2—VB | | 200—400 | |
| | E2—THA | | −20°C (−4°F) 14,600—17,800<br>20°C (68°F) 2,210—2,690<br>60°C (140°F) 493—667 | |
| **Fuel pump** | | | | |
| Type | | | Impeller (in tank) | |
| Output pressure | | kPa (kg/cm², psi) | 441—588 (4.5—6.0, 64—85) | |
| Feeding capacity | | cc (cu in)/10 sec | 220 (13.4) min. when fuel pressure at 250 kPa (2.55 kg/cm², 36.3 psi) | |
| **Fuel filter** | | | | |
| Type | Low-pressure side | | Nylon element | |
| | High-pressure side | | Paper element | |
| **Pressure regulator** | | | | |
| Type | | | Diaphragm | |
| Regulating pressure | | kPa (kg/cm², psi) | 235—275 (2.4—2.8, 34—40) (Vacuum hose disconnected) | |
| **Injector** | | | | |
| Type | | | High-ohmic | |
| Type of drive | | | Voltage | |
| Resistance | | Ω | 12—16 | |
| Injection amount | | cc (cu in)/15 sec | 32—41 (1.95—2.50) | |
| **Idle-speed control valve** | | | | |
| Solenoid resistance | | Ω | 5—20 | |
| **Fuel tank** | | | | |
| Capacity | | liters (US gal, imp gal) | 48 (12.7, 10.6) | |
| **Air cleaner** | | | | |
| Element type | | | Oil permeated | |
| **Accelerator cable** | | | | |
| Free play | | mm (in) | 1—3 (0.039—0.118) | |
| **Fuel** | | | | |
| Specification | | | Unleaded regular | |

## TROUBLESHOOTING CHART—1989 323 NON-TURBO

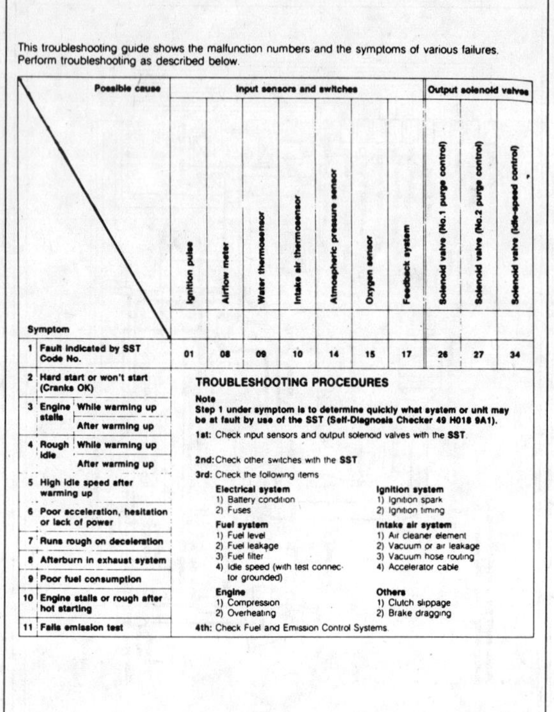

This troubleshooting guide shows the malfunction numbers and the symptoms of various failures. Perform troubleshooting as described below.

| Symptom \ Possible cause | Input sensors and switches: Ignition pulse | Airflow meter | Water thermosensor | Intake air thermosensor | Atmospheric pressure sensor | Oxygen sensor | Feedback system | Output solenoid valves: Solenoid valve (No.1 purge control) | Solenoid valve (No.2 purge control) | Solenoid valve (Idle-speed control) |
|---|---|---|---|---|---|---|---|---|---|---|
| 1 Fault indicated by SST Code No. | 01 | 08 | 09 | 10 | 14 | 15 | 17 | 26 | 27 | 34 |
| 2 Hard start or won't start (Cranks OK) | | | | | | | | | | |
| 3 Engine stalls — While warming up | | | | | | | | | | |
| 3 Engine stalls — After warming up | | | | | | | | | | |
| 4 Rough idle — While warming up | | | | | | | | | | |
| 4 Rough idle — After warming up | | | | | | | | | | |
| 5 High idle speed after warming up | | | | | | | | | | |
| 6 Poor acceleration, hesitation or lack of power | | | | | | | | | | |
| 7 Runs rough on deceleration | | | | | | | | | | |
| 8 Afterburn in exhaust system | | | | | | | | | | |
| 9 Poor fuel consumption | | | | | | | | | | |
| 10 Engine stalls or rough after hot starting | | | | | | | | | | |
| 11 Fails emission test | | | | | | | | | | |

### TROUBLESHOOTING PROCEDURES

**Note**
**Step 1 under symptom is to determine quickly what system or unit may be at fault by use of the SST (Self-Diagnosis Checker 49 H018 9A1).**

**1st:** Check input sensors and output solenoid valves with the **SST**.

**2nd:** Check other switches with the **SST**.

**3rd:** Check the following items

**Electrical system**
1) Battery condition
2) Fuses

**Ignition system**
1) Ignition spark
2) Ignition timing

**Fuel system**
1) Fuel level
2) Fuel leakage
3) Fuel filter
4) Idle speed (with test connector grounded)

**Intake air system**
1) Air cleaner element
2) Vacuum or air leakage
3) Vacuum hose routing
4) Accelerator cable

**Engine**
1) Compression
2) Overheating

**Others**
1) Clutch slippage
2) Brake dragging

**4th:** Check Fuel and Emission Control Systems.

## TROUBLESHOOTING CHART (CONT.) 1989 NON-TURBO

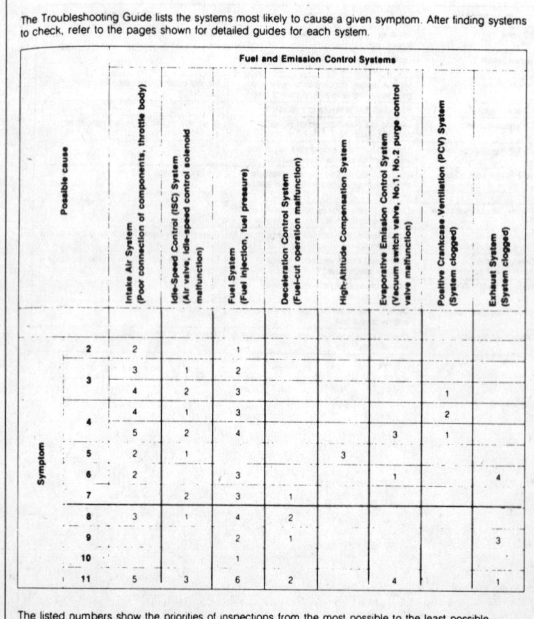

The Troubleshooting Guide lists the systems most likely to cause a given symptom. After finding systems to check, refer to the pages shown for detailed guides for each system.

| Symptom \ Possible cause (Fuel and Emission Control Systems) | Intake Air System (Poor connection of components, throttle body) | Idle-Speed Control (ISC) System (Air valve, idle-speed control solenoid malfunction) | Fuel System (Fuel injection, fuel pressure) | Deceleration Control System (Fuel-cut operation malfunction) | High-Altitude Compensation System | Evaporative Emission Control System (Vacuum switch valve, No.1, No.2 purge control valve malfunction) | Positive Crankcase Ventilation (PCV) System (System clogged) | Exhaust System (System clogged) |
|---|---|---|---|---|---|---|---|---|
| 2 | 2 | | 1 | | | | | |
| 3 | 3 | 1 | 2 | | | | | |
| 3 | 4 | 2 | 3 | | | | 1 | |
| 4 | 4 | 1 | 3 | | | | 2 | |
| 4 | 5 | 2 | 4 | | | 3 | 1 | |
| 5 | 2 | 1 | | | 3 | | | |
| 6 | 2 | | 3 | | | 1 | | 4 |
| 7 | | 2 | 3 | 1 | | | | |
| 8 | 3 | 1 | 4 | 2 | | | | |
| 9 | | | 2 | 1 | | | | 3 |
| 10 | | | 1 | | | | | |
| 11 | 5 | 3 | 6 | 2 | | 4 | | 1 |

The listed numbers show the priorities of inspections from the most possible to the least possible. These were determined on the following basis:
- Ease of inspection
- Most possible system
- Most possible point in system

## MALFUNCTION INDICATOR LIGHT (MIL) OPERATION 1989 323 NON-TURBO

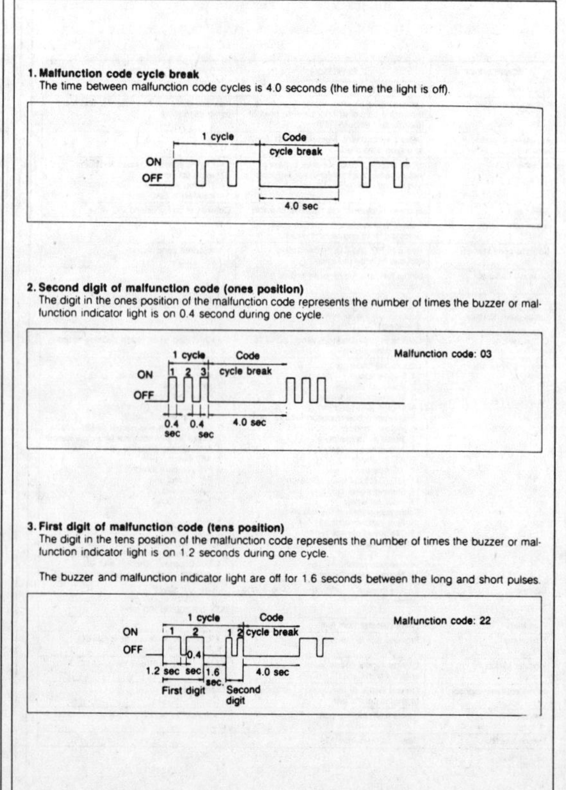


**1. Malfunction code cycle break**
The time between malfunction code cycles is 4.0 seconds (the time the light is off).

**2. Second digit of malfunction code (ones position)**
The digit in the ones position of the malfunction code represents the number of times the buzzer or malfunction indicator light is on 0.4 second during one cycle.

**3. First digit of malfunction code (tens position)**
The digit in the tens position of the malfunction code represents the number of times the buzzer or malfunction indicator light is on 1.2 seconds during one cycle.

The buzzer and malfunction indicator light are off for 1.6 seconds between the long and short pulses.

## TROUBLE CODE DIAGNOSTIC CHART 1989 323 NON-TURBO

| Code No. | Sensor or subsystem | Malfunction | Fail-safe function | Pattern of output signals (Self-Diagnosis Checker or MIL) |
|---|---|---|---|---|
| 01 | Ignition pulse | No ignition signal | – | ON / OFF |
| 08 | Airflow meter | Broken wire, short circuit | Basic fuel injection amount fixed as for 2 driving modes 1) Idle switch ON 2) Idle switch OFF | ON / OFF |
| 09 | Water thermosensor | Broken wire, short circuit | Coolant temp input fixed at 80°C (176°F) for ISC, at 20°C (68°F) for fuel injection | ON / OFF |
| 10 | Intake air thermosensor (airflow meter) | Broken wire, short circuit | Intake air temp input fixed at 20°C (68°F) | ON / OFF |
| 14 | Atmospheric pressure sensor | Broken wire, short circuit | Atmospheric pressure input fixed at 760 mmHg (29.9 inHg) | ON / OFF |
| 15 | Oxygen sensor | Sensor output continues less than 0.55V 120 sec after engine at above 1,500 rpm | Feedback system operation canceled | ON / OFF |
| 17 | Feedback system | Oxygen sensor output not changed 30 sec after engine exceeds 1,500 rpm | Feedback system operation canceled | ON / OFF |
| 26 | Solenoid valve (No 1 purge control) | Broken wire, short circuit | – | ON / OFF |
| 27 | Solenoid valve (No 2 purge control) | Broken wire short circuit | – | ON / OFF |
| 34 | Solenoid valve (idle-speed control) | Broken wire short circuit | – | ON / OFF |

**Caution**

**a) If more than one failure is present, the lowest number malfunction code is displayed first and the remaining codes are displayed sequentially.**

**b) After repairing all failures, turn off the ignition switch, disconnect the negative battery cable, and depress the brake pedal for at least 5 seconds to erase the malfunction code memory.**

If a malfunction code number is shown on the **SST**, check the following chart along with the wiring diagram

**Code No. 01 (Ignition pulse)**

**PC: Possible Cause**

Are there poor connections in ignition coil circuit? — **YES** → Repair or replace connector

**NO** ↓

Is resistance of ignition coil OK?
Resistance Primary 0.72—0.88 Ω
Secondary 6—30 kΩ
— **NO** → Replace ignition coil

**YES** ↓

Is there continuity between ignition coil (–) terminal wire and engine control unit (1M) terminal? — **NO** → PC Open circuit in wiring harness from ignition coil to engine control unit (1M) terminal

**YES** ↓

Is same Code No. present after performing after-repair procedure? — **NO** → Ignition pulse and circuit OK

**YES** ↓

Is engine control unit (1M) terminal voltage OK? — **NO** → PC • No power supply to ignition coil • Short circuit in wiring harness

**YES** → PC Engine control unit malfunction

## TROUBLE CODE DIAGNOSTIC CHART—1989 323 NON-TURBO

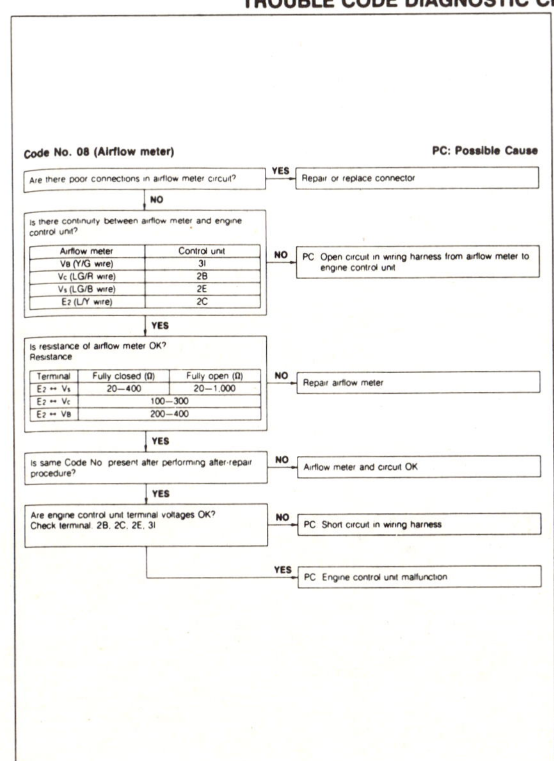

**Code No. 08 (Airflow meter)**

**PC: Possible Cause**

Are there poor connections in airflow meter circuit? — **YES** → Repair or replace connector

**NO** ↓

Is there continuity between airflow meter and engine control unit?

| Airflow meter | Control unit |
|---|---|
| $V_B$ (Y/G wire) | 3I |
| $V_C$ (LG/R wire) | 2B |
| $V_S$ (LG/B wire) | 2E |
| $E_2$ (L/Y wire) | 2C |

— **NO** → PC Open circuit in wiring harness from airflow meter to engine control unit

**YES** ↓

Is resistance of airflow meter OK?
Resistance

| Terminal | Fully closed (Ω) | Fully open (Ω) |
|---|---|---|
| $E_2 \leftrightarrow V_S$ | 20—400 | 20—1,000 |
| $E_2 \leftrightarrow V_C$ | 100—300 | |
| $E_2 \leftrightarrow V_B$ | 200—400 | |

— **NO** → Repair airflow meter

**YES** ↓

Is same Code No. present after performing after-repair procedure? — **NO** → Airflow meter and circuit OK

**YES** ↓

Are engine control unit terminal voltages OK?
Check terminal. 2B, 2C, 2E, 3I
— **NO** → PC Short circuit in wiring harness

**YES** → PC Engine control unit malfunction

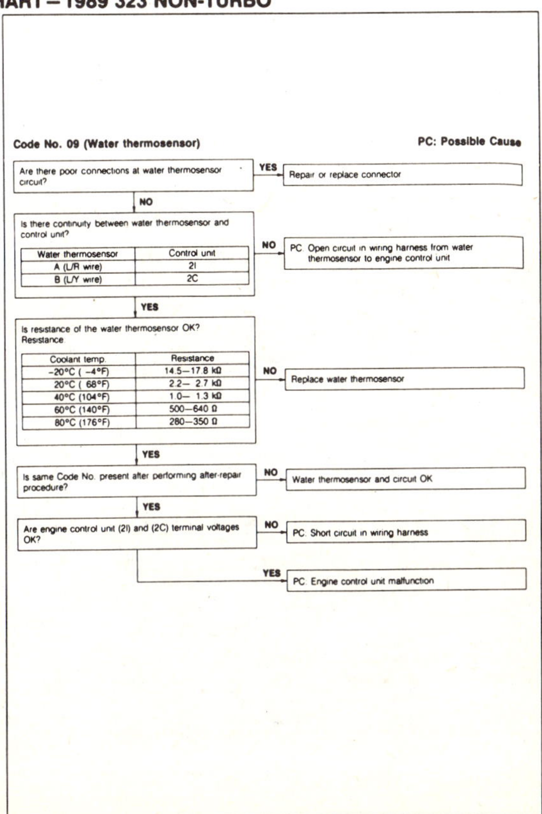

**Code No. 09 (Water thermosensor)**

**PC: Possible Cause**

Are there poor connections at water thermosensor circuit? — **YES** → Repair or replace connector

**NO** ↓

Is there continuity between water thermosensor and control unit?

| Water thermosensor | Control unit |
|---|---|
| A (L/R wire) | 2I |
| B (L/Y wire) | 2C |

— **NO** → PC. Open circuit in wiring harness from water thermosensor to engine control unit

**YES** ↓

Is resistance of the water thermosensor OK?
Resistance

| Coolant temp. | Resistance |
|---|---|
| −20°C (−4°F) | 14.5—17.8 kΩ |
| 20°C (68°F) | 2.2— 2.7 kΩ |
| 40°C (104°F) | 1.0— 1.3 kΩ |
| 60°C (140°F) | 500—640 Ω |
| 80°C (176°F) | 280—350 Ω |

— **NO** → Replace water thermosensor

**YES** ↓

Is same Code No. present after performing after-repair procedure? — **NO** → Water thermosensor and circuit OK

**YES** ↓

Are engine control unit (2I) and (2C) terminal voltages OK? — **NO** → PC. Short circuit in wiring harness

**YES** → PC. Engine control unit malfunction

## TROUBLE CODE DIAGNOSTIC CHART—1989 323 NON-TURBO

**Code No. 10 (Intake air thermosensor)** **PC: Possible Cause**

Are there poor connections in airflow meter circuit?
- YES → Repair or replace connector
- NO ↓

Is there continuity between intake air thermosensor (in airflow meter) and engine control unit?

| Intake air thermosensor (in airflow meter) | Control unit |
|---|---|
| THA (BR/Y wire) | 2J |
| $E_2$ (L/Y wire) | 2C |

- NO → PC: Open circuit in wiring harness from intake air thermosensor (in airflow meter) to engine control unit
- YES ↓

Is resistance of intake air thermosensor (in airflow meter) OK?
Resistance:

| Terminal | Resistance |
|---|---|
| $E_2$ ↔ THA | -20°C ( -4°F) : 14.6—17.8 kΩ<br>20°C ( 68°F) : 2.21—2.69 kΩ<br>60°C (140°F) : 493—667 Ω |

- NO → Replace airflow meter
- YES ↓

Is same Code No. present after performing after-repair procedure?
- NO → Intake air thermosensor and circuit OK
- YES ↓

Are engine control unit (2J) and (2C) terminal voltages OK?
- NO → PC: Short circuit in wiring harness
- YES → PC: Engine control unit malfunction

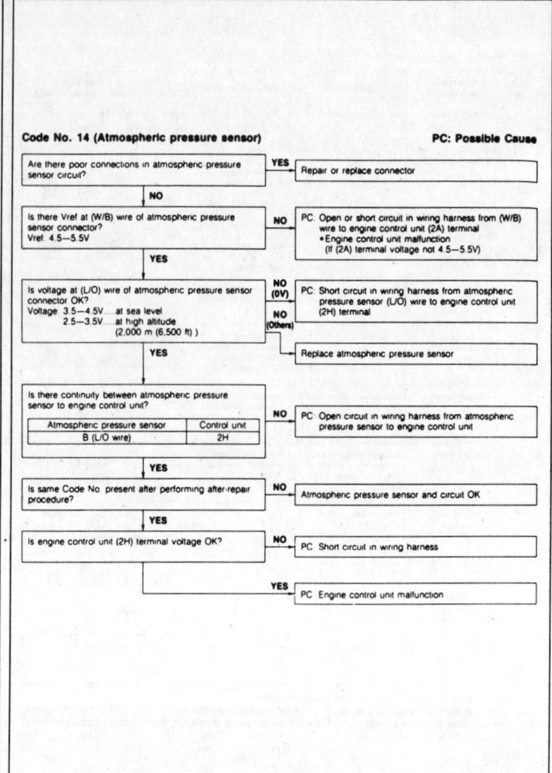


## TROUBLE CODE DIAGNOSTIC CHART—1989 323 NON-TURBO

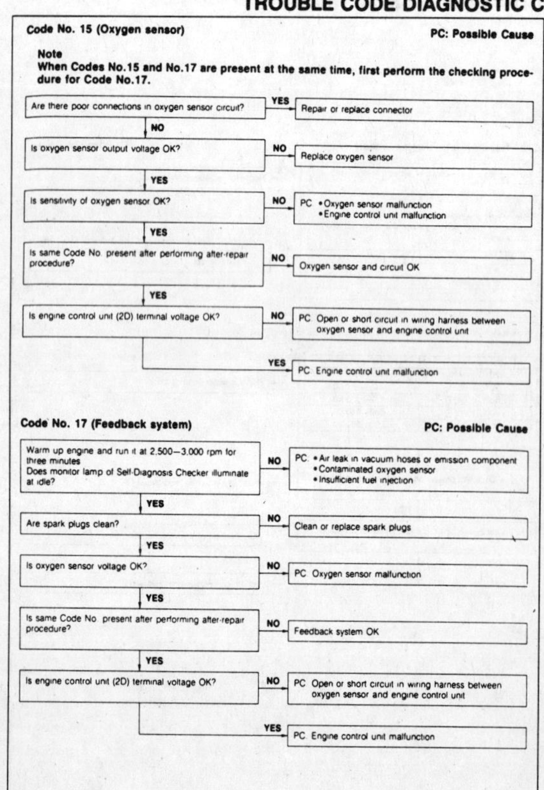


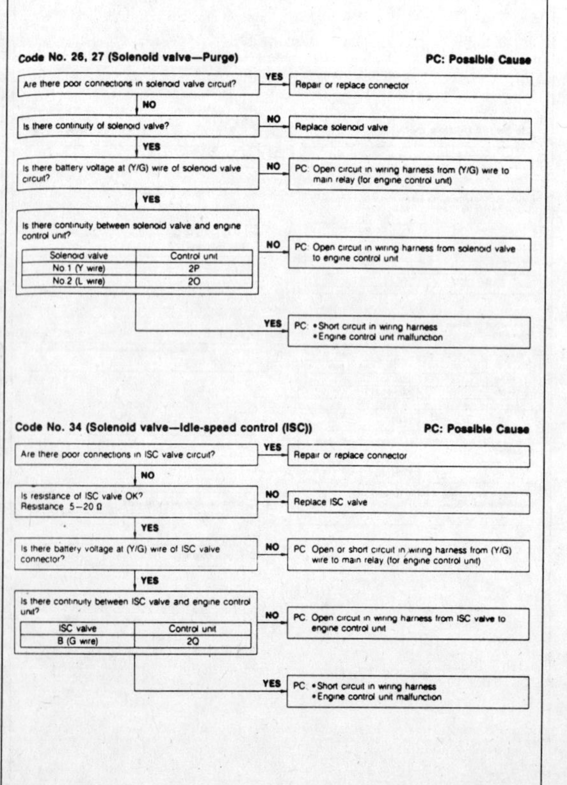

Code No. 26, 27 (Solenoid valve—Purge)
PC: Possible Cause
Are there poor connections in solenoid valve circuit?
YES
Repair or replace connector
NO
Is there continuity of solenoid valve?
NO
Replace solenoid valve
YES
Is there battery voltage at (Y/G) wire of solenoid valve circuit?
NO
PC: Open circuit in wiring harness from (Y/G) wire to main relay (for engine control unit)
YES
Is there continuity between solenoid valve and engine control unit?
Solenoid valve
Control unit
No.1 (Y wire)
2P
No.2 (L wire)
2O
NO
PC: Open circuit in wiring harness from solenoid valve to engine control unit
YES
PC: •Short circuit in wiring harness
•Engine control unit malfunction
Code No. 34 (Solenoid valve—Idle-speed control (ISC))
PC: Possible Cause
Are there poor connections in ISC valve circuit?
YES
Repair or replace connector
NO
Is resistance of ISC valve OK?
Resistance 5—20 Ω
NO
Replace ISC valve
YES
Is there battery voltage at (Y/G) wire of ISC valve connector?
NO
PC: Open or short circuit in wiring harness from (Y/G) wire to main relay (for engine control unit)
YES
Is there continuity between ISC valve and engine control unit?
ISC valve
Control unit
B (G wire)
2Q
NO
PC: Open circuit in wiring harness from ISC valve to engine control unit
YES
PC: •Short circuit in wiring harness
•Engine control unit malfunction

## CONTROL UNIT PIN LOCATIONS AND TEST VOLTAGES—1989 323 NON-TURBO

| Terminal | Connected to | Voltage | Condition | Remarks |
|---|---|---|---|---|
| 1A (Output) | MIL | Below 2.5V | Ignition switch OFF → ON for 3 sec | Test connector grounded |
| | | Approx. 12V | After 3 sec | |
| 1B (Output) | Self-Diagnosis Checker (for Code No.) | Below 2.5V | Ignition switch OFF → ON for 3 sec | • Test connector grounded • Checker connected |
| | | Approx. 12V | After 3 sec | |
| 1C | — | — | — | — |
| 1D (Output) | Self-Diagnosis Checker (for monitor light) | Approx. 5V | Ignition switch OFF → ON for 3 sec | • Test connector grounded • Checker connected |
| | | Approx. 10V | After 3 sec | |
| 1E (Input) | Throttle sensor (IDL switch) | Approx. 12V | Accelerator pedal depressed | |
| | | Below 1.5V | Accelerator pedal released | |
| 1F (Output) | A/C control relay | Approx. 12V | Ignition switch ON | |
| | | Below 1.5V | A/C switch ON (at idle) | |
| 1G (Input) | Neutral/clutch switch | Approx. 12V | Clutch pedal depressed | In-gear condition (Neutral: constant 12 V) |
| | | Below 1.5V | Clutch pedal released | |
| 1H (Input) | Water thermoswitch (Radiator) | Approx. 12V | Below 17°C (63°F) | |
| | | Below 1.5V | Above 17°C (63°F) | |
| 1I (Input) | Electrical load (E/L) control unit | Approx. 2.5V | E/L switch ON | |
| | | Approx. 10V | E/L switch OFF | |
| 1J (Input) | Brake light switch | Approx. 12V | Brake pedal depressed | |
| | | Below 1.5V | Brake pedal released | |
| 1K (Input) | Power steering pressure switch | Approx. 12V | Power steering switch OFF | |
| | | Below 1.5V | Power steering switch ON | |
| 1L (Input) | A/C switch | Approx. 12V | A/C switch OFF | Blower motor ON |
| | | Below 2.5V | A/C switch ON | |
| 1M (Input) | Ignition coil | Approx. 12V | Ignition switch ON | (When engine running) Engine Signal Monitor Green and red lights flash |
| | | Approx. 12V | At idle | |
| 1N | — | — | — | — |
| 1O | — | — | — | — |
| 1P | — | — | — | — |
| 1Q | — | — | — | — |
| 1R | — | — | — | — |
| 1S | — | — | — | — |
| 1T | — | — | — | — |
| 1U (Output) | Igniter | Below 1.5V | Ignition switch ON | |
| | | Approx. 12V | At idle | |
| 1V (Input) | M/T switch (ground) | Below 1.5V | | ATX, constant 12V |
| 1W (Input) | Test connector | Below 1.5V | Test connector grounded | |
| | | Approx. 12V | Test connector not grounded | |
| 1X | — | — | — | — |
| 2A (Output) | Vref | 4.5—5.5V | | |
| 2B (Input) | Airflow meter (Vc) | 7—9V | | |
| 2C | Ground ($E_2$) | Below 1.5V | | |
| 2D (Input) | Oxygen sensor | 0.3—0.7V | At idle | |
| | | More than 0.55V | During acceleration | |
| | | Less than 0.55V | During deceleration | |
| 2E (Input) | Airflow meter (Vs) | Approx. 2V | Ignition switch ON | |
| | | 4—5V | At idle | |
| 2F | — | — | — | — |
| 2G (Input) | Throttle sensor (PSW switch) | Approx. 12V | Accelerator pedal released | |
| | | Below 1.5V | Accelerator pedal depressed (fully open throttle) | |
| 2H (Input) | Atmospheric pressure sensor | Approx. 4V | | At sea level |
| 2I (Input) | Water thermosensor | Approx. 0.5V | Normal operating temperature | |
| 2J (Input) | Intake air thermosensor (Airflow meter) | 2—3V | Intake air temperature: 20°C (68°F) | |

| Terminal | Connected to | Voltage | Condition | Remarks |
|---|---|---|---|---|
| 2K (Output) | | Below 1.5V | Intake air temp. more than 58°C (136°F) Coolant temp. more than 90°C (194°F) After starting for 180 sec. | If solenoid valve (PRC) attached |
| | | Approx. 12V | Other | |
| 2L | — | — | — | — |
| 2M | — | — | — | — |
| 2N | — | — | — | — |
| 2O | Solenoid valve (No.2 purge control) | Approx. 12V | Less than 1,500 rpm | |
| | | Below 1.5V | More than 1,500 rpm | |
| 2P | Solenoid valve (No.1 purge control) | Below 1.5V | Intake air temp. more than 50°C (122°F) Coolant temp. more than 50°C (122°F) | In-gear condition • Connect jumper wire to neutral switch (MTX) • Disconnect inhibitor switch (ATX) |
| | | Approx. 12V | Other | |
| 2Q | Idle-speed control (ISC) valve | 1.5—11.6V | At idle | Engine Signal Monitor Green and red lights flash |
| 2R | Ground | Below 1.5V | | |
| 3A | Ground | Below 1.5V | | |
| 3B | Ignition switch | Below 2.5V | Ignition switch ON | |
| | | 7—9V | While cranking | |
| 3C | Injectors No.2 and No.4 | Approx. 12V | At idle | Engine Signal Monitor Green and red lights flash |
| 3D | Inhibitor switch | Below 1.5V | "N" or "P" range | MTX constant 0V |
| | | Approx. 12V | Other range | |
| 3E | Injectors No.1 and No.3 | Approx. 12V | At idle | Engine Signal Monitor Green and red lights flash |
| 3F | — | — | — | — |
| 3G | Ground | Below 1.5V | | |
| 3H | — | — | — | — |
| 3I | Main relay | Approx. 12V | Ignition switch ON | |
| 3J | Battery | Approx. 12V | | |

**Engine control unit connector**

| 3I | 3G | 3E | 3C | 3A | 2Q | 2O | 2M | 2K | 2I | 2G | 2E | 2C | 2A | 1W | 1U | 1S | 1Q | 1O | 1M | 1K | 1I | 1G | 1E | 1C | 1A |
|---|---|---|---|---|---|---|---|---|---|---|---|---|---|---|---|---|---|---|---|---|---|---|---|---|---|
| 3J | 3H | 3F | 3D | 3B | 2R | 2P | 2N | 2L | 2J | 2H | 2F | 2D | 2B | 1X | 1V | 1T | 1R | 1P | 1N | 1L | 1J | 1H | 1F | 1D | 1B |

## OUTPUT COMPONENTS AND ENGINE CONDITION 1989 323 NON-TURBO

**Output Devices and Engine Condition**

| OUTPUT DEVICES | | CRANKING (COLD ENGINE) | WARMING UP (DURING IDLE) | MEDIUM LOAD COLD | MEDIUM LOAD WARM | ACCELERATION | HEAVY LOAD | DECELERATION | IDLE (THROTTLE VALVE FULLY CLOSED) | IGN: ON (ENGINE NOT RUNNING) | REMARKS |
|---|---|---|---|---|---|---|---|---|---|---|---|
| INJECTOR | INJECTION (Air Fuel Mixture) | Rich | | | Rich and Lean | Rich | | Fuel cutoff | Rich | Does not inject | |
| | INJECTION TIMING | 1 Group | | | | | | | 1 Group | | Above 6,300 rpm fuel cutoff |
| BAC VALVE | AIR VALVE | * Open | | | Closed | | | | | | * Coolant temp: below 60°C (140°F) |
| | ISC VALVE | Large amount of bypass air | | Small amount of bypass air | | | | | * Large and small amount of bypass air | Does not operate | * Test connector grounded: small amount of air |
| SOLENOID VALVE | No.1 PURGE CONTROL | OFF (Vacuum cutoff) | | | *1 ON (Vacuum to No.1 purge control valve) | | | | OFF (Vacuum cutoff) | | *1 Engine speed: above 2,000 rpm |
| | No.2 PURGE CONTROL | OFF (Vacuum cutoff) | | *2 ON (Vacuum to vacuum switch valve) | | | | OFF (Vacuum cutoff) | | | *2 Engine speed: above 1,500 rpm |

## TROUBLESHOOTING AND RELATIONSHIP CHART 1989 323 NON-TURBO

**Input Devices and Output Devices**

| INPUT DEVICE \ OUTPUT DEVICE | INJECTOR: FUEL INJECTION AMOUNT | INJECTOR: FUEL INJECTION TIMING | BAC VALVE: AIR VALVE | BAC VALVE: ISC VALVE | SOLENOID VALVE: No.1 PURGE CONTROL | SOLENOID VALVE: No.2 PURGE CONTROL |
|---|---|---|---|---|---|---|
| IGNITION COIL | O | O | X | O | O | O |
| AIRFLOW METER | O | X | X | X | O | O |
| IDLE SWITCH | O | X | X | O | O | X |
| PSW SWITCH | O | X | X | X | X | X |
| WATER THERMO-SENSOR | O | X | X | O | O | X |
| INTAKE AIR THERMO-SENSOR | O | X | X | O | O | X |
| ATMOSPHERIC PRESSURE SENSOR | O | X | X | O | X | X |
| OXYGEN SENSOR | O | X | X | X | O | X |
| BRAKE LIGHT SWITCH | O | X | X | X | X | X |
| WATER THERMO-SWITCH | O | X | X | O | O | X |
| NEUTRAL AND CLUTCH SWITCH | O | X | X | O | O | X |
| INHIBITOR SWITCH | O | X | X | O | O | X |
| IGNITION SWITCH | O | O | X | X | X | X |
| E/L CONTROL UNIT | X | X | X | O | X | X |
| A/C SWITCH | X | X | X | O | X | X |
| P/S PRESSURE SWITCH | X | X | X | O | X | X |
| TEST CONNECTOR | X | X | X | O | X | X |

O: Related
X: Not related

## SYSTEM SCHEMATIC – 1989 323 TURBO

## COMPONENT LOCATIONS – 1989 323 TURBO

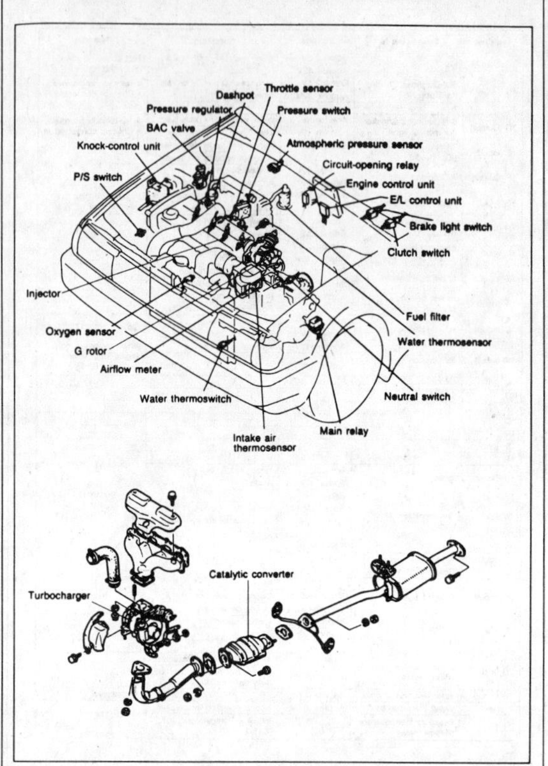


## COMPONENT DESCRIPTIONS – 1989 323 TURBO

| COMPONENT | FUNCTION | REMARKS |
|---|---|---|
| Air cleaner | Filters air entering combustion chamber | |
| Airflow meter | Detects intake air amount; sends signal to the engine control unit (for determination of fuel injection amount) | Includes intake air thermosensor and fuel pump switch |
| Atmospheric pressure sensor | Detects atmospheric pressure; sends signal to engine control unit | |
| Air valve | When engine is cold, supplies bypass air into dynamic chamber for quick warm-up and smooth idle | • Engine speed increased to shorten warm-up period<br>• Thermo wax type<br>• Installed in BAC valve |
| BAC valve | Supplies bypass air into dynamic chamber | Consists of air valve and ISC valve |
| Brake light switch | Detects brake operation (deceleration); sends signal to engine control unit | |
| Catalytic converter | Reduces HC and CO by oxidization; reduces NOx by reduction | Honeycomb construction |
| Charcoal canister | Stores fuel tank fumes while engine is stopped | |
| Check connector | For Self-Diagnosis Checker | 6-pin connector (Green) |
| Circuit-opening relay | Supplies voltage for fuel pump while engine is running | |
| Clutch switch | Detects in-gear condition; sends signal to engine control unit | Switch closed when clutch pedal released |
| Engine control unit | Detects the following.<br>1. Engine speed<br>2. Intake air amount<br>3. Engine coolant temperature<br>4. Engine load condition<br>5. Oxygen concentration in exhaust gas<br>6. In-gear condition<br>7. Intake air temperature<br><br>8. Atmospheric pressure<br>9. A/C operation<br>10. P/S operation<br>11. E/L (Electrical load) operation<br>12. Starting signal<br>13. Initial set signal<br>Controls operation of the following:<br>1. Fuel injection amount<br>2. Idle-speed control system<br>3. Pressure regulator control system<br><br>4. Purge control system<br><br>5. Fail-safe system<br>6. Monitor switch function | <br>1. Ignition coil (–) terminal<br>2. Airflow meter<br>3. Water thermosensor<br>4. Throttle sensor (variable resister type)<br>5. Oxygen sensor<br>6. Clutch switch and neutral switch<br>7. Intake air thermosensor (in airflow meter)<br>8. Atmospheric pressure sensor<br>9. A/C switch<br>10. P/S pressure switch<br>11. E/L control unit<br>12. Ignition switch<br>13. Test connector<br><br>1. Injector<br>2. BAC valve<br>3. Solenoid valve (Pressure regulator control)<br>4. Solenoid valves (No.1 purge control, No.2 purge control)<br>5. Self-Diagnosis Checker and MIL<br>6. Monitor lamp (Self-Diagnosis Checker) |
| Dashpot | Causes gradual throttle valve closing during deceleration | Adjustment speed 2,000 ± 150 rpm |
| Fuel filter | Filters particles from fuel | |
| Fuel pump | Provides fuel to injectors | • Operates while engine is running<br>• Installed in fuel tank |
| Injector | Injects fuel to intake port | Controlled by signals from engine control unit |
| Intake air thermosensor | Detects intake air temperature; sends signal to engine control unit | Thermistor |
| Intercooler | Cools intake air temperature after turbocharger | Air-cooled type |
| In-tank fuel filter | Filters particles from fuel | |
| ISC valve | Supplies bypass air to dynamic chamber for smooth idle | Installed in BAC valve |
| Neutral switch | Detects transaxle condition; sends signal to engine control unit | |
| Oxygen sensor | Detects oxygen concentration in exhaust gas; sends signal to engine control unit | Zirconia ceramic with platinum coating |
| Pressure regulator | Regulates fuel pressure to injectors | |
| Pressure switch (overboost detection) | Detects overboost condition; sends signal to engine control unit | |
| P/S pressure switch | Detects P/S operation; sends signal to engine control unit | When steering wheel turned right or left |
| No.1 Purge control valve | Opens and closes evaporative vapor passage from canister to intake manifold | |
| No.2 Purge control valve | Positive pressure and negative pressure valves operate in accordance with fuel tank pressure | Prevents canister from flooding |
| Throttle sensor (variable resister type) | Detects throttle opening angle; sends signal to engine control unit; compensates fuel injection amount | |
| Solenoid valve (No.1 purge control) | Opens and closes vacuum passage to No.1 purge control valve | Controlled by signal from engine control unit |
| Solenoid valve (No.2 purge control) | Opens and closes vacuum passage to vacuum switch valve | Controlled by signal from engine control unit |
| Solenoid valve (pressure regulator control) | Closes vacuum passage between dynamic chamber and pressure regulator | Only during hot condition |
| Transfer pump | Pumps fuel from one side of tank to other to maintain balance | Only 4WD |
| Turbocharger | Pressurizes intake air utilizing exhaust gas flow | Water cooled |
| Vacuum switch valve | Opens passage of vacuum line when vacuum applied | |
| Water thermosensor | Detects coolant temperature; sends signal to engine control unit | Thermistor |
| Water thermoswitch | Detects coolant temperature; sends signal to engine control unit | Above 17°C (63°F): ON |
| Waste gate valve | Allows bypassing of exhaust gas to control turbocharger boost pressure | |

## SPECIFICATIONS CHART—1989 323 TURBO

| Item | | Engine model | Turbo 2WD | Turbo 4WD |
|---|---|---|---|---|
| **Idle speed** (with test connector grounded) | | rpm | 850 ± 50 (Neutral) | |
| **Throttle body** | | | | |
| Type | | | Horizontal draft (1-barrel) | |
| Throat diameter | | mm (in) | 50 (1.97) | |
| **Airflow meter** | | | | |
| Resistor | E2—Vs | Ω | Fully closed: 20—400 Fully open: 20—1,000 | |
| | E2—Vc | | 100—300 | |
| | E2—VB | | 200—400 | |
| | E2—THA | | −20°C (−4°F) 14,600—17,800<br>20°C (68°F) 2,210— 2,690<br>60°C (140°F) 493— 667 | |
| **Fuel pump** | | | | |
| Type | | | Impeller (in tank) | |
| Output pressure | | kPa (kg/cm², psi) | 441—588 (4.5—6.0, 64—85) | |
| Feeding capacity | | cc (cu-in)/10 sec | 220 (13.4) min. when fuel pressure at 250 kPa (2.55 kg/cm², 36.3 psi) | |
| **Transfer pump** | | | | |
| Feeding capacity | | cc (cu-in)/10 sec | 200 (12.2) min. | |
| **Pressure regulator** | | | | |
| Type | | | Diaphragm | |
| Regulating pressure | | kPa (kg/cm², psi) | 235—275 (2.4—2.8, 34—40) (Vacuum hose disconnected) | |
| **Fuel filter** | | | | |
| Type | Low-pressure side | | Nylon element | |
| | High-pressure side | | Paper element | |
| **Injector** | | | | |
| Type | | | High-ohmic | |
| Type of drive | | | Voltage | |
| Resistance | | Ω | 12—16 | |
| Injection amount | | cc (cu-in)/15 sec | 66—82 (4.0—5.0) | |
| **Turbocharger** | | | | |
| Type | | | Water cooled | |
| Lubrication | | | Engine oil | |
| Boost pressure (Max) | | kPa (kg/cm², psi) | 55—64 (0.56—0.65, 8.0—9.2) | |
| **Waste gate valve** | | | | |
| Operating pressure | | kPa (kg/cm², psi) | 48.1—58.9 (0.49—0.60, 7.0—8.5) | |
| **Idle-speed control valve** | | | | |
| Solenoid resistance | | Ω | 5—20 | |
| **Fuel tank** | | | | |
| Capacity | | liters (US gal, Imp gal) | 48 (12.7, 10.6) | 50 (13.2, 11.0) |
| **Air cleaner** | | | | |
| Element type | | | Oil permeated | |
| **Accelerator cable** | | | | |
| Free play | | mm (in) | 1—3 (0.039—0.118) | |
| **Fuel** | | | | |
| Specification | | | Unleaded regular | |

## TROUBLESHOOTING CHART 1989 323 TURBO

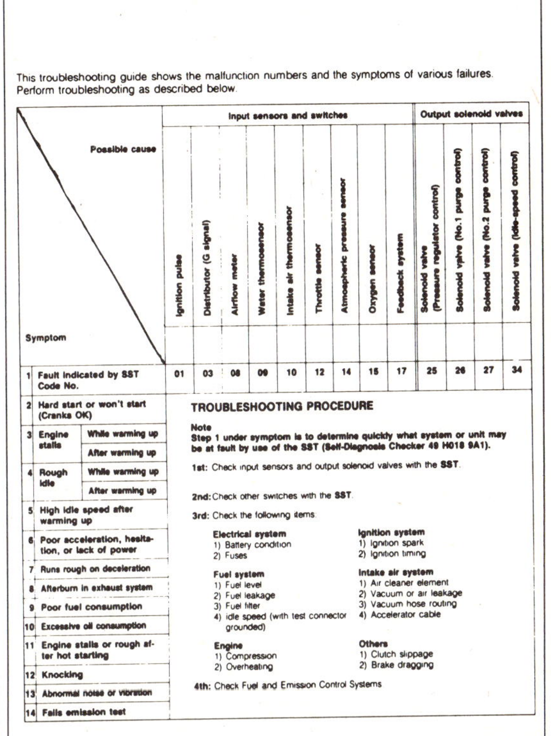

This troubleshooting guide shows the malfunction numbers and the symptoms of various failures. Perform troubleshooting as described below.

| Symptom \ Possible cause | Ignition pulse | Distributor (G signal) | Airflow meter | Water thermosensor | Intake air thermosensor | Throttle sensor | Atmospheric pressure sensor | Oxygen sensor | Feedback system | Solenoid valve (Pressure regulator control) | Solenoid valve (No.1 purge control) | Solenoid valve (No.2 purge control) | Solenoid valve (Idle-speed control) |
|---|---|---|---|---|---|---|---|---|---|---|---|---|---|
| | Input sensors and switches | | | | | | | | | Output solenoid valves | | | |
| 1 Fault indicated by SST Code No. | 01 | 03 | 08 | 09 | 10 | 12 | 14 | 15 | 17 | 25 | 26 | 27 | 34 |

Symptoms:

2 Hard start or won't start (Cranks OK)
3 Engine stalls — While warming up / After warming up
4 Rough idle — While warming up / After warming up
5 High idle speed after warming up
6 Poor acceleration, hesitation, or lack of power
7 Runs rough on deceleration
8 Afterburn in exhaust system
9 Poor fuel consumption
10 Excessive oil consumption
11 Engine stalls or rough after hot starting
12 Knocking
13 Abnormal noise or vibration
14 Fails emission test

**TROUBLESHOOTING PROCEDURE**

**Note**
**Step 1 under symptom is to determine quickly what system or unit may be at fault by use of the SST (Self-Diagnosis Checker 49 H018 9A1).**

**1st:** Check input sensors and output solenoid valves with the **SST**.

**2nd:** Check other switches with the **SST**.

**3rd:** Check the following items:

**Electrical system**
1) Battery condition
2) Fuses

**Ignition system**
1) Ignition spark
2) Ignition timing

**Fuel system**
1) Fuel level
2) Fuel leakage
3) Fuel filter
4) idle speed (with test connector grounded)

**Intake air system**
1) Air cleaner element
2) Vacuum or air leakage
3) Vacuum hose routing
4) Accelerator cable

**Engine**
1) Compression
2) Overheating

**Others**
1) Clutch slippage
2) Brake dragging

**4th:** Check Fuel and Emission Control Systems

## TROUBLESHOOTING CHART (CONT.) 1989 323 TURBO

The Troubleshooting Guide lists the systems most likely to cause a given symptom. After finding systems to check, refer to the pages shown for detailed guides for each system.

| Symptom \ Possible cause | Intake Air System (Poor connection of components, throttle body) | Idle-Speed Control (ISC) System (Air valve, idle-speed control solenoid malfunction) | Fuel System (Fuel injection, fuel pressure) | Pressure Regulator Control (PRC) System | Turbocharging System (Oil & water passage, turbine and compressor wheels malfunction) | Knock-Control System | Deceleration Control System (Fuel-cut operation malfunction) | High-Altitude Compensation System | Evaporative Emission Control System (Vacuum switch valve, No.1 and No.2 purge control valve malfunction) | Positive Crankcase Ventilation (PCV) System (System clogged) | Exhaust System (System clogged) |
|---|---|---|---|---|---|---|---|---|---|---|---|
| | Fuel and Emission Control Systems | | | | | | | | | | |
| | | | | | | | | | | | |
| 2 | 3 | | 2 | | | 1 | | | | | |
| 3 | 3 | 1 | 2 | | | | | | | | |
| 3 | 4 | 2 | 3 | | | | | | | 1 | |
| 4 | 4 | 1 | 3 | | | | | | | 2 | |
| 4 | 5 | 2 | 4 | | | | | | 3 | 1 | |
| 5 | 2 | 1 | | | | | | 3 | | | |
| 6 | 2 | | 3 | | 5 | | | 6 | 1 | | 4 |
| 7 | | 2 | 3 | | | | 1 | | | | |
| 8 | 3 | 1 | 4 | | | | 2 | | | | |
| 9 | | | 2 | | | | 1 | | | | 3 |
| 10 | | | | | 1 | | | | | | |
| 11 | | | 2 | 1 | | | | | | | |
| 12 | | | | | 2 | 1 | | 3 | | | |
| 13 | | | | | 1 | | | | | | |
| 14 | 5 | 3 | 6 | | | | 2 | | 4 | | 1 |

The listed numbers show the priorities of inspections from the most possible to the least possible. These were determined on the following basis:
- Ease of inspection
- Most possible system
- Most possible point in system

## MALFUNCTION INDICATOR LIGHT (MIL) OPERATION 1989 323 TURBO

**1. Malfunction code cycle break**
The time between malfunction code cycles is 4.0 seconds (the time the light is off).

1 cycle
Code cycle break
ON
OFF
4.0 sec

**2. Second digit of malfunction code (ones position)**
The digit in the ones position of the malfunction code represents the number of times the buzzer or malfunction indicator light is on 0.4 second during one cycle.

Malfunction code: 03
1 cycle
Code
1 2 3
cycle break
ON
OFF
0.4 sec
0.4 sec
4.0 sec

**3. First digit of malfunction code (tens position)**
The digit in the tens position of the malfunction code represents the number of times the buzzer or malfunction indicator light is on 1.2 seconds during one cycle.

The buzzer and malfunction indicator light are off for 1.6 seconds between the long and short pulses.

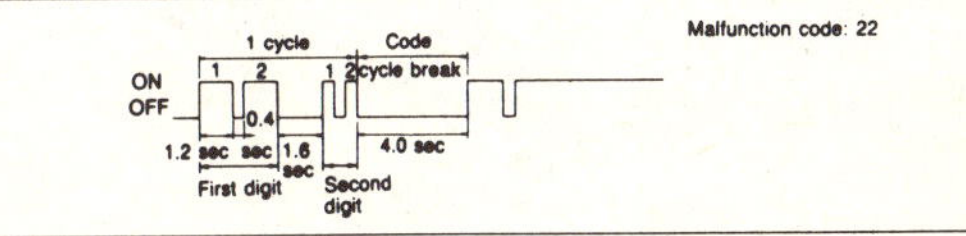

## TROUBLE CODE IDENTIFICATION CHART 1989 323 TURBO

| Code No. | Sensor or subsystem | Malfunction | Fail-safe function | Pattern of output signals (Self-Diagnosis Checker or MIL) |
|---|---|---|---|---|
| 01 | Ignition pulse | No ignition signal | – | ON / OFF |
| 03 | Distributor (G signal) | No G signal | – | ON / OFF |
| 08 | Airflow meter | Broken wire, short circuit | Basic fuel injection amount fixed as for 2 driving modes 1) Idle switch: ON 2) Idle switch: OFF | ON / OFF |
| 09 | Water thermosensor | Broken wire, short circuit | Coolant temp. input fixed at 80°C (176°F) ...for ISC, at 60°C (140°F) ...for fuel injection | ON / OFF |
| 10 | Intake air thermosensor (airflow meter) | Broken wire, short circuit | Intake air temp. input fixed at 20°C (68°F) | ON / OFF |
| 12 | Throttle sensor | Broken wire, short circuit | Throttle valve opening angle signal input fixed at full open | ON / OFF |
| 14 | Atmospheric pressure sensor | Broken wire, short circuit | Atmospheric pressure input fixed at 760 mmHg (29.9 inHg) | ON / OFF |
| 15 | Oxygen sensor | Sensor output continues less than 0.55V 120 sec after engine at above 1,500 rpm | Feedback system operation canceled | ON / OFF |
| 17 | Feedback system | Oxygen sensor output not changed 30 sec after engine exceeds 1,500 rpm | Feedback system operation canceled | ON / OFF |
| 25 | Solenoid valve (pressure regulator control) | Broken wire, short circuit | – | ON / OFF |
| 26 | Solenoid valve (No.1 purge control) | Broken wire, short circuit | – | ON / OFF |
| 27 | Solenoid valve (No.2 purge control) | Broken wire, short circuit | – | ON / OFF |
| 34 | Solenoid valve (idle-speed control) | Broken wire, short circuit | – | ON / OFF |

**Caution**

**a) If more than one failure is present, the lowest number malfunction code is displayed first and the remaining codes are displayed sequentially.**

**b) After repairing all failures, turn off the ignition switch, disconnect the negative battery cable, and depress the brake pedal for at least 5 seconds to erase the memory of a malfunction code.**

## TROUBLE CODE DIAGNOSTIC CHART 1989 323 TURBO

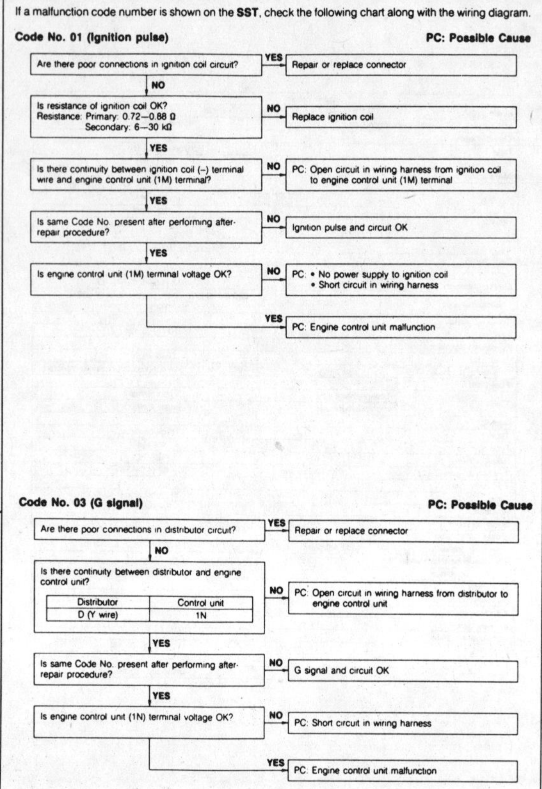


## TROUBLE CODE DIAGNOSTIC CHART—1989 323 TURBO

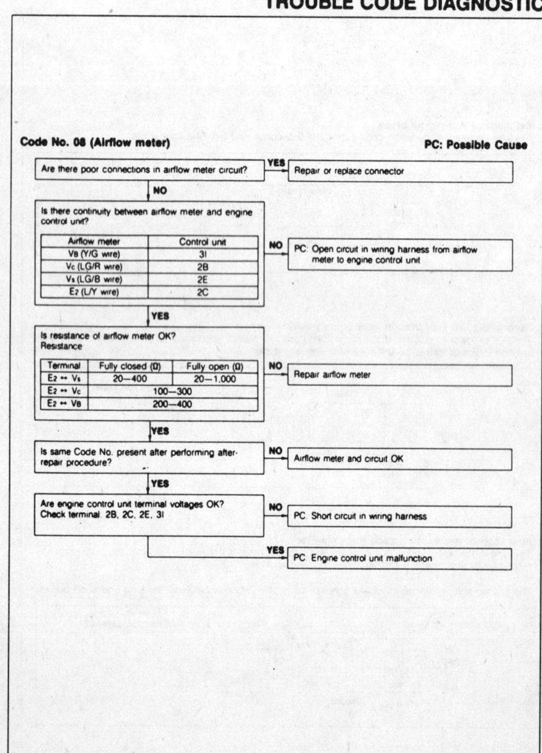


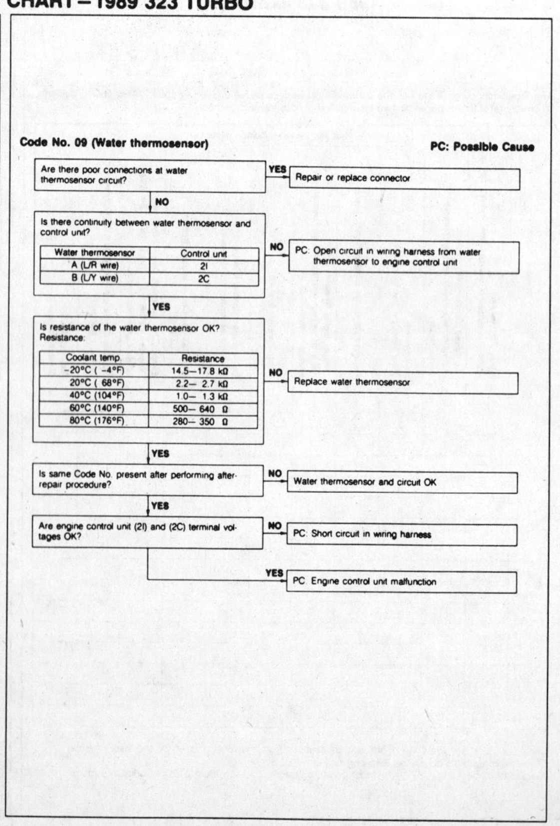

## TROUBLE CODE DIAGNOSTIC CHART—1989 323 TURBO

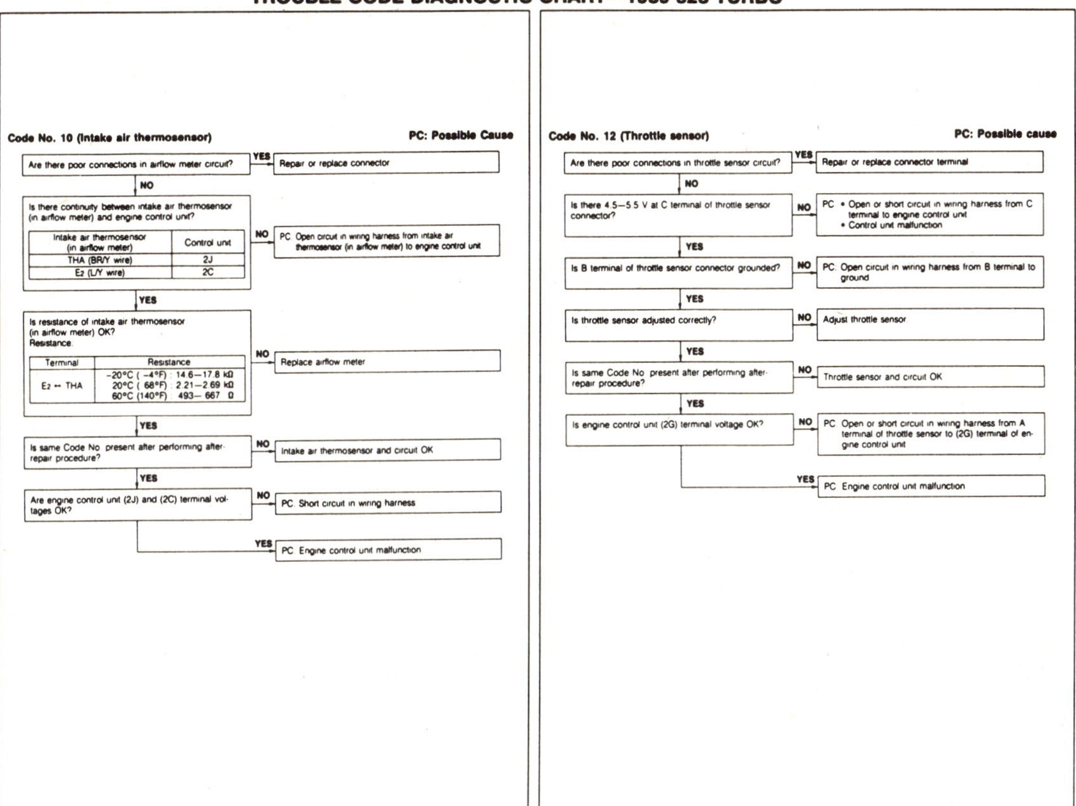


## TROUBLE CODE DIAGNOSTIC CHART—1989 323 TURBO

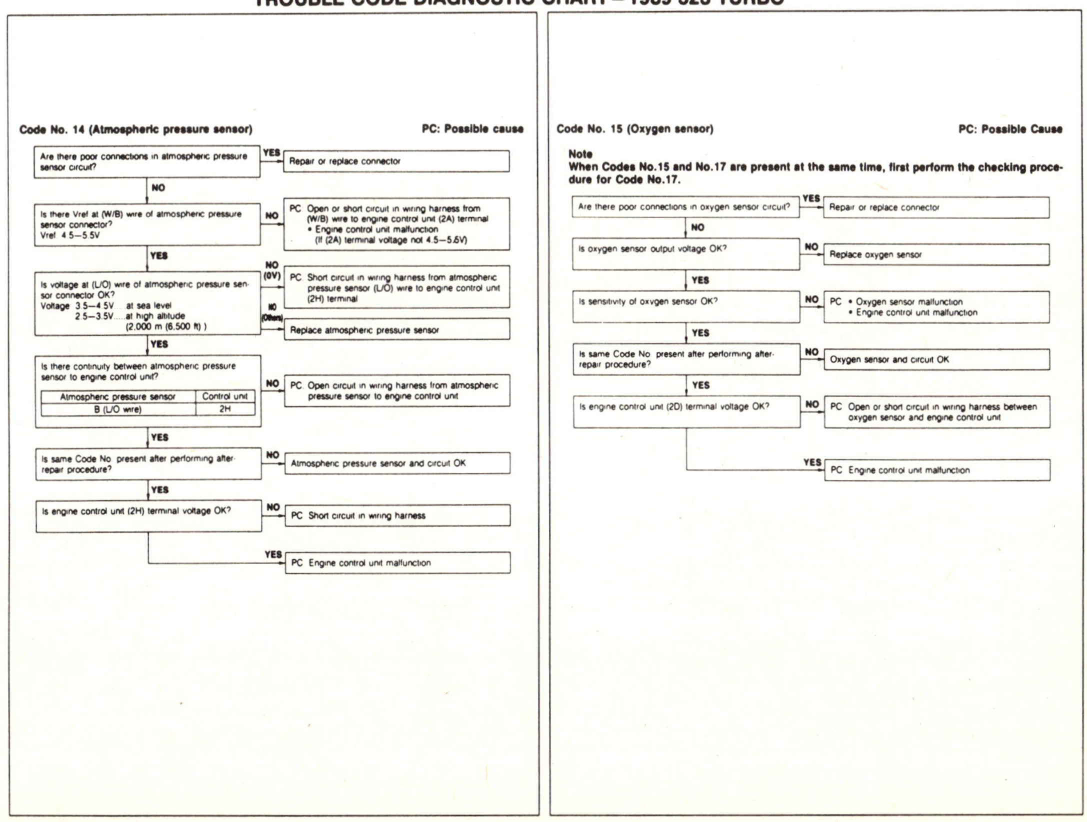

## TROUBLE CODE DIAGNOSTIC CHART—1989 323 TURBO

**Code No. 17 (Feedback system)** **PC: Possible Cause**

Warm up engine and run it at 2,500–3,000 rpm for three minutes. Does monitor lamp of Self-Diagnosis Checker illuminate at idle?
NO → PC • Air leak in vacuum hoses or emission component • Contaminated oxygen sensor • Insufficient fuel injection
YES ↓

Are spark plugs clean?
NO → Clean or replace spark plugs
YES ↓

Is oxygen sensor voltage OK?
NO → PC Oxygen sensor malfunction
YES ↓

Is same Code No. present after performing after-repair procedure?
NO → Feedback system OK
YES ↓

Is engine control unit (2D) terminal voltage OK?
NO → PC Open or short circuit in wiring harness between oxygen sensor and engine control unit
YES → PC Engine control unit malfunction

**Code No. 25 (Solenoid valve-Pressure regulator control)** **PC: Possible Cause**

Are there poor connections in solenoid valve circuit?
YES → Repair or replace connector
NO ↓

Is there continuity of solenoid valve?
NO → Replace solenoid valve
YES ↓

Is there battery voltage at (Y/G) wire of solenoid valve connector?
NO → PC Open circuit in wiring harness from (Y/G) wire to main relay (for engine control unit)
YES ↓

Is there continuity between solenoid valve and engine control unit?

| Solenoid valve | Control unit |
|---|---|
| B (Y/G wire) | 2K |

NO → PC Open circuit in wiring harness from solenoid valve to engine control unit
YES → PC • Short circuit in wiring harness • Engine control unit malfunction

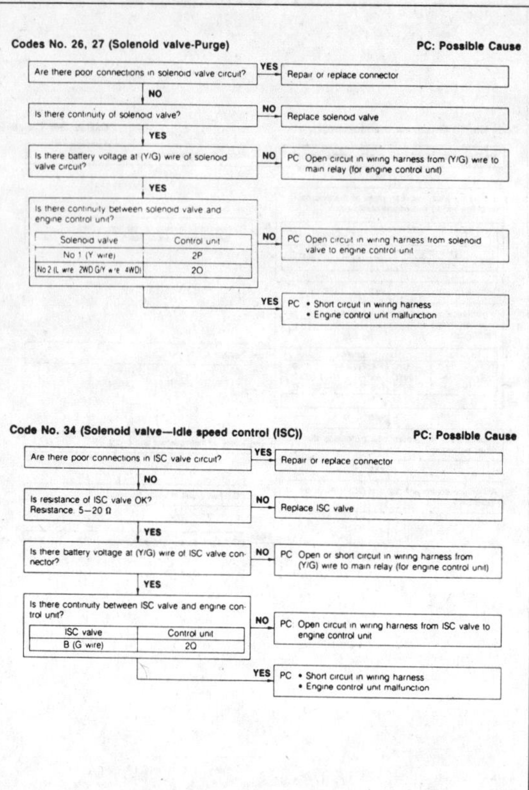


## CONTROL UNIT PIN LOCATIONS AND TEST VOLTAGES—1989 323 TURBO

| Terminal | Connected to | Voltage | Condition | Remarks |
|---|---|---|---|---|
| 1A (Output) | MIL | Below 2.5V | Ignition switch OFF → ON for 3 sec | Test connector grounded |
| | | Approx. 12V | After 3 sec | |
| 1B (Output) | Self-Diagnosis Checker (for Code No.) | Below 2.5V | Ignition switch OFF → ON for 3 sec | Test connector grounded Checker connected |
| | | Approx. 12V | After 3 sec | |
| 1C | — | — | — | — |
| 1D (Output) | Self-Diagnosis Checker (for monitor light) | Approx. 5V | Ignition switch OFF → ON for 3 sec | Test connector grounded Checker connected |
| | | Approx. 12V | After 3 sec | |
| 1E (Input) | Throttle sensor (IDL switch) | Approx. 12V | Accelerator pedal depressed | |
| | | Below 1.5V | Accelerator pedal released | |
| 1F (Output) | A/C control relay | Approx. 12V | Ignition switch ON | |
| | | Below 1.5V | A/C switch ON (at idle) | |
| 1G (Input) | Neutral/clutch switch | Approx. 12V | Clutch pedal depressed | In-gear condition (Neutral: Constant 12V) |
| | | Below 1.5V | Clutch pedal released | |
| 1H (Input) | Water thermoswitch (Radiator) | Approx. 12V | Below 17°C (63°F) | |
| | | Below 1.5V | Above 17°C (63°F) | |
| 1I (Input) | Electrical load (E/L) control unit | Approx. 2.5V | E/L switch ON | |
| | | Approx. 12V | E/L switch OFF | |
| 1J (Input) | Brake light switch | Approx. 12V | Brake pedal depressed | |
| | | Below 1.5V | Brake pedal released | |
| 1K (Input) | Power steering pressure switch | Approx. 12V | Power steering switch OFF | |
| | | Below 1.5V | Power steering switch ON | |
| 1L (Input) | A/C switch | Approx. 12V | A/C switch OFF | Blower motor ON |
| | | Below 2.5V | A/C switch ON | |
| 1M (Input) | Ignition coil | Approx. 12V | Ignition switch ON | (When engine running) Engine Signal Monitor: Green and red lights flash |
| | | Approx. 12V | At idle | |
| 1N | G sensor (Distributor) | Below 1.5V | Ignition switch ON | |
| | | Approx. 3V | At idle | |
| 1O | — | — | — | — |
| 1P | — | — | — | — |
| 1Q | — | — | — | — |
| 1R | — | — | — | — |
| 1S | — | — | — | — |
| 1T | — | — | — | — |
| 1U (Output) | Knock-control unit (I terminal) | Below 1.5V | Ignition switch ON | |
| | | Approx. 12V | At idle | |
| 1V (Input) | FF switch | Below 1.5V | 4x4 | |
| | | Approx. 12V | FF | |
| 1W (Input) | Test connector | Below 1.5V | Test connector grounded | |
| | | Approx. 12V | Test connector not grounded | |
| 1X | — | — | — | — |
| 2A (Output) | Vref | 4.5–5.5V | | |
| 2B (Input) | Airflow meter (Vc) | 7–9V | | |
| 2C | Ground (E2) | Below 1.5V | | |
| 2D (Input) | Oxygen sensor | 0.3–0.7V | At idle | |
| | | More than 0.55V | During acceleration | |
| | | Less than 0.55V | During deceleration | |
| 2E (Input) | Airflow meter (Vs) | Approx. 2V | Ignition switch ON | |
| | | 4–5V | At idle | |
| 2F | — | — | — | — |
| 2G (Input) | Throttle sensor | Approx. 0.5V | Accelerator pedal released | |
| | | Approx. 4V | Accelerator pedal depressed | |
| 2H (Input) | Atmospheric pressure sensor | Approx. 4V | | At sea level |
| 2I (Input) | Water thermosensor | Approx. 0.5V | Normal operating temperature | |
| 2J (Input) | Intake air thermosensor (Airflow meter) | 2–3V | Intake air temperature: 20°C (68°F) | |
| 2K (Output) | Solenoid valve (PRC) | Below 2.5V | Intake air temp. more than 58°C (136°F) Coolant temp. more than 90°C (194°F) After starting for 180 sec. | |
| | | Approx. 12V | Other | |
| 2L (Output) | Pressure switch | Approx. 12V | At idle | Air pressure 71.6–79.5 kPa (0.73–0.81 kg/cm², 10.4–11.5 psi) |
| | | Below 1.5V | At overboost | |
| 2M (Output) | Knock control unit (f terminal) | Below 1.5V | At idle | Coolant temp. More than 80°C (176°F) Intake air temp. More than 0°C (32°F) |
| | | Approx. 12V | Engine speed 1,000 rpm (Positive pressure) | |
| 2N (Output) | Indicator light | Approx. 12V | At idle | 71.6–79.5 kPa (0.73–0.81 kg/cm², 10.4–11.5 psi) |
| | | Below 1.5V | At overboost | |
| 2O | Solenoid valve (No.2 purge control) | Approx. 12V | Less than 1,500 rpm | |
| | | Below 1.5V | More than 1,500 rpm | |
| 2P | Solenoid valve (No.1 purge control) | Below 1.5V | Intake air temp. more than 50°C (122°F) Coolant temp. more than 50°C (122°F) | In-gear condition Connect jumper wire to neutral switch |
| | | Approx. 12V | Other | |
| 2Q | Idle-speed control (ISC) valve | 1.5–11.6V | At idle | Engine Signal Monitor Green and red lights flash |
| 2R | Ground | Below 1.5V | | |
| 3A | Ground | Below 1.5V | | |
| 3B | Ignition switch | Below 2.5V | Ignition switch ON | |
| | | 7–9V | While cranking | |
| 3C | Injectors No.2 and No.4 | Approx. 12V | At idle | Engine Signal Monitor Green and red lights flash |
| 3D | — | — | — | — |
| 3E | Injectors No.1 and No.3 | Approx. 12V | At idle | Engine signal Monitor Green and red lights flash |
| 3F | — | — | — | — |
| 3G | Ground | Below 1.5V | | |
| 3H | — | — | — | — |
| 3I | Main relay | Approx. 12V | Ignition switch ON | |
| 3J | Battery | Approx. 12V | | |

**Engine control unit connector**

| 3I | 3G | 3E | 3C | 3A | 2Q | 2O | 2M | 2K | 2I | 2G | 2E | 2C | 2A | 1W | 1U | 1S | 1Q | 1O | 1M | 1K | 1I | 1G | 1E | 1C | 1A |
|---|---|---|---|---|---|---|---|---|---|---|---|---|---|---|---|---|---|---|---|---|---|---|---|---|---|
| 3J | 3H | 3F | 3D | 3B | 2R | 2P | 2N | 2L | 2J | 2H | 2F | 2D | 2B | 1X | 1V | 1T | 1R | 1P | 1N | 1L | 1J | 1H | 1F | 1D | 1B |

## OUTPUT COMPONENTS AND ENGINE CONDITIONS—1989 323 TURBO

**Output Devices and Engine Conditions (Turbocharged Engine)**

| OUTPUT DEVICES \ ENGINE CONDITION | | CRANKING (COLD ENGINE) | WARMING UP (DURING IDLE) | MEDIUM LOAD COLD | MEDIUM LOAD WARM | ACCELERATION | HEAVY LOAD | DECELERATION | IDLE (THROTTLE VALVE FULLY CLOSED) | IGN: ON (ENGINE NOT RUNNING) | REMARKS |
|---|---|---|---|---|---|---|---|---|---|---|---|
| INJECTOR | INJECTION | Rich | | | Rich and Lean | Rich | | Fuel cut | Rich | Does not inject | |
| | INJECTION TIMING | 1 Group | 2 Group | | | | | | 2 Group | | Above 6,800 rpm fuel cut Intake manifold pressure Above 77 kPa (0.78 kg/cm², 11 psi) fuel cut |
| SOLENOID VALVE (PRC) | | ON (Vacuum cut) | OFF (Vacuum to pressure regulator) | | | | | | * After start ON (Vacuum cut) | Does not operate | * During hot starting |
| BAC VALVE | AIR VALVE | * Open | | | Close | | | | | | * Coolant temp below 60°C (140°F) |
| | ISC VALVE | Large amount of bypass air | | Small amount of bypass air | | | | | * Large and small amount of bypass air | Does not operate | * Test connector grounded small amount |
| SOLENOID VALVE | No.1 PURGE CONTROL | OFF (Vacuum cut) | | | * ON (Vacuum to No.1 purge control valve) | | | | OFF (Vacuum cut) | | * Positive pressure OFF |
| | No.2 PURGE CONTROL | OFF (Vacuum cut) | | * ON (Vacuum to vacuum switch valve) | | | | OFF (Vacuum cut) | | | * Engine speed above 1,500 rpm |

## TROUBLESHOOTING AND RELATIONSHIP CHART—1989 323 TURBO

**Input Devices and Output Devices**

| INPUT DEVICE \ OUTPUT DEVICE | INJECTOR: FUEL INJECTION AMOUNT | INJECTOR: FUEL INJECTION TIMING | SOLENOID (PRC) | BAC VALVE: AIR VALVE | BAC VALVE: ISC VALVE | SOLENOID VALVE: No.1 PURGE CONTROL | SOLENOID VALVE: No.2 PURGE CONTROL |
|---|---|---|---|---|---|---|---|
| IGNITION COIL | ○ | ○ | x | x | ○ | ○ | ○ |
| AIRFLOW METER | ○ | x | x | x | x | ○ | ○ |
| IDLE SWITCH | ○ | x | ○ | x | ○ | ○ | x |
| THROTTLE SENSOR | ○ | x | x | x | x | x | x |
| WATER THERMO-SENSOR | ○ | x | ○ | x | ○ | ○ | x |
| INTAKE AIR THERMO-SENSOR | ○ | x | ○ | x | ○ | ○ | x |
| ATMOSPHERIC PRESSURE SENSOR | ○ | x | x | x | ○ | x | x |
| OXYGEN SENSOR | ○ | x | x | x | x | ○ | x |
| PRESSURE SWITCH | ○ | x | x | x | x | x | x |
| BRAKE LIGHT SWITCH | ○ | x | x | x | x | x | x |
| WATER THERMO-SWITCH | ○ | x | x | x | ○ | ○ | x |
| NEUTRAL AND CLUTCH SWITCH | ○ | x | ○ | x | ○ | ○ | x |
| IGNITION SWITCH | ○ | ○ | ○ | x | x | x | x |
| E/L CONTROL UNIT | x | x | x | x | ○ | x | x |
| FF SWITCH | ○ | x | x | x | x | x | x |
| A/C SWITCH | x | x | x | x | ○ | x | x |
| P/S PRESSURE SWITCH | x | x | x | x | ○ | x | x |
| G SENSOR | x | ○ | x | x | x | x | x |
| TEST CONNECTOR | x | x | x | x | ○ | x | x |

## EXPLODED VIEW OF THE AIR INTAKE SYSTEM—1990 323 WITH B6 ENGINE

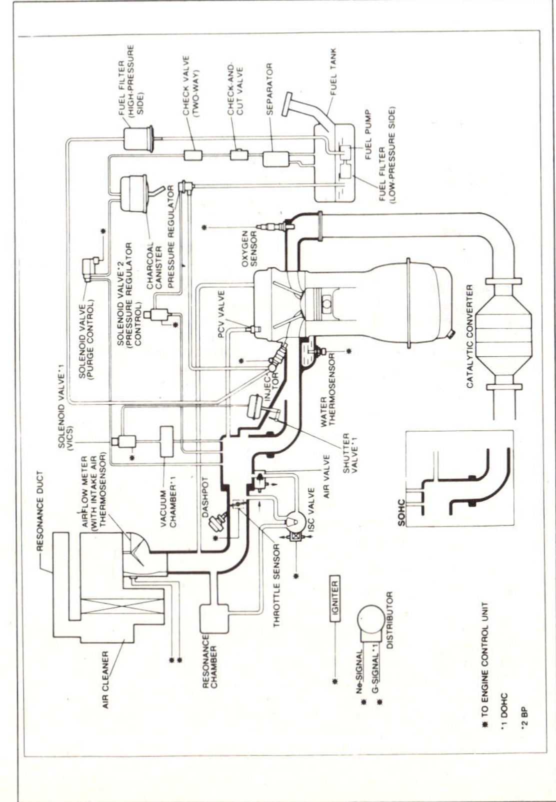


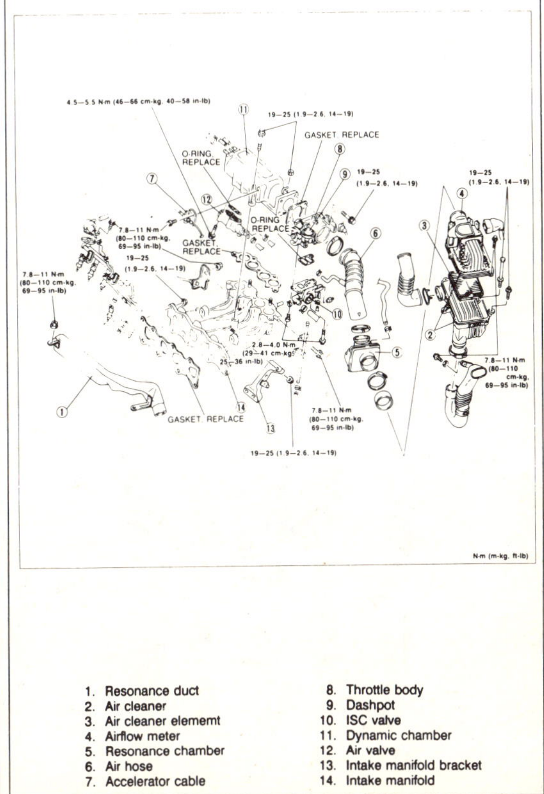


1. Resonance duct
2. Air cleaner
3. Air cleaner elememt
4. Airflow meter
5. Resonance chamber
6. Air hose
7. Accelerator cable
8. Throttle body
9. Dashpot
10. ISC valve
11. Dynamic chamber
12. Air valve
13. Intake manifold bracket
14. Intake manifold

### EXPLODED VIEW OF THE AIR INTAKE SYSTEM – 1990 323 WITH DOHC ENGINE

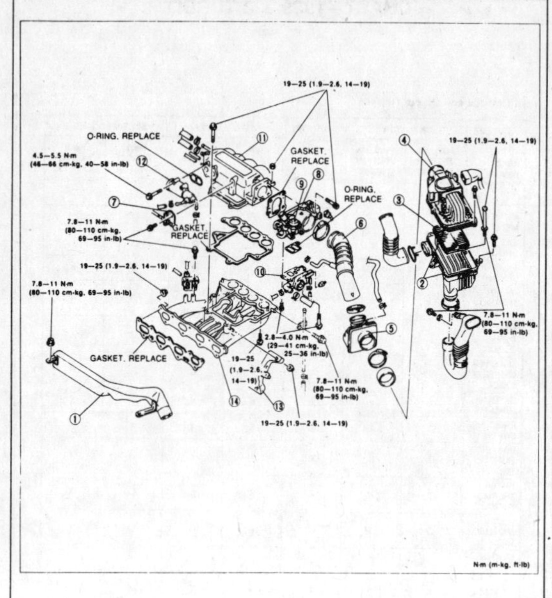


1. Resonance duct
2. Air cleaner
3. Air cleaner element
4. Airflow meter
5. Resonance chamber
6. Air hose
7. Accelerator cable
8. Throttle body
9. Dashpot
10. ISC valve
11. Dynamic chamber
12. Air valve
13. Intake manifold bracket
14. Intake manifold

**COMPONENT PARTS**
**Removal / Inspection / Installation**
1. Remove in the order shown in the figure, referring to **Removal Note**.
2. Inspect the intake air system components visually and repair or replace if necessary.
3. Install in the reverse order of removal, referring to **Installation Note**.

**BP DOHC**

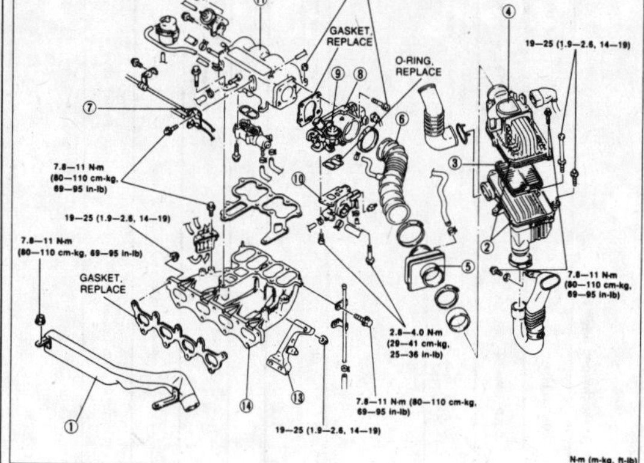


1. Resonance duct
2. Air cleaner
3. Air cleaner element
4. Airflow meter
5. Resonance chamber
6. Air hose
7. Accelerator cable
8. Throttle body
9. Dashpot
10. ISC valve
11. Dynamic chamber
12. Air valve
13. Intake manifold bracket
14. Intake manifold

### COMPONENT LOCATION – 1990 323

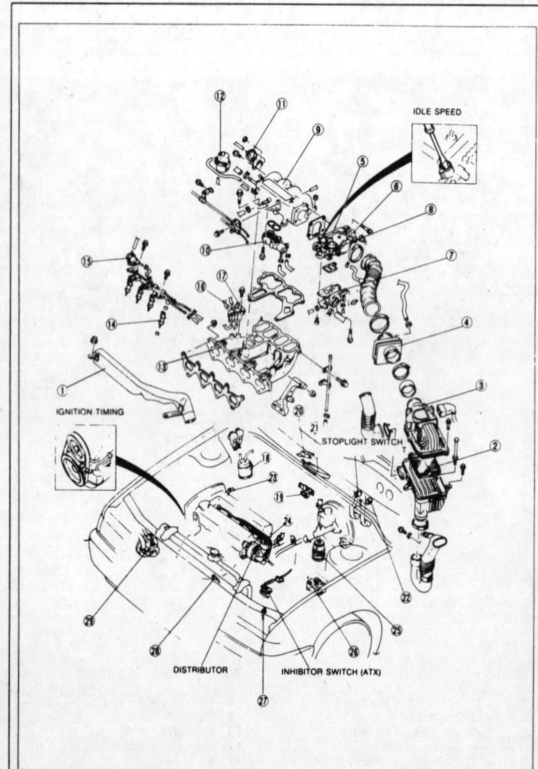


1. Resonance duct
2. Air cleaner element
3. Airflow meter
4. Resonance chamber
5. Throttle body
6. Throttle sensor
7. ISC valve
8. Dashpot
9. Dynamic chamber
10. Air valve
11. Shutter valve actuator (VICS), DOHC
12. Vacuum chamber DOHC
13. Intake manifold
14. Injector
15. Pressure regulator
16. Solenoid valve (pressure regulator control, BP engine)
17. Solenoid valve (VICS, DOHC engine
18. Charcoal canister
19. Solenoid valve (purge control)
20. Circuit opening relay
21. Engine control unit (ECU)
22. Clutch switch (M/T)
23. PCV valve
24. Water thermosensor
25. Fuel filter (high-pressure side)
26. Main relay (fuel inject. relay)
27. Neutral switch (M/T)
28. Oxygen sensor
29. P/S pressure switch

## SPECIFICATION CHART—1990 323

| Item | | | Engine | B6 SOHC | BP SOHC | BP DOHC |
|---|---|---|---|---|---|---|
| Idle speed*1 *2 | | | rpm | 750 ± 50 | | |
| Ignition timing*2 | | | BTDC | 7 ± 1° | 5 ± 1° | 10 ± 1° |
| **Fuel pump** | | | | | | |
| Maximum output pressure | | | kPa (kg/cm², psi) | 441—589 (4.5—6.0, 64—85) | | |
| **Fuel filter** | | | | | | |
| Type | Low-pressure side | | | Nylon element (in fuel pump) | | |
| | High-pressure side | | | Paper element | | |
| **Pressure regulator** | | | | | | |
| Regulating pressure | | | kPa (kg/cm², psi) | 265—314 (2.7—3.2, 38—46) | | |
| **Injector** | | | | | | |
| Type | | | | High-ohmic | | |
| Type of drive | | | | Electromechanical | | |
| Resistance | | | Ω | 12—16 | | |
| **Idle speed control (ISC) valve** | | | | | | |
| Type | | | | Rotary | | |
| Resistance | | | Ω | 11—13 | | |
| **Purge control solenoid valve** | | | | | | |
| Resistance | | | Ω | 23—27 | | |
| **Water thermosensor** | | | | | | |
| Resistance kΩ | | -20°C (-4°F) | | 14.6—17.8 | | |
| | | 20°C (68°F) | | 2.21—2.69 | | |
| | | 40°C (104°F) | | 1.0—1.3 | | |
| | | 80°C (176°F) | | 0.29—0.35 | | |
| **Airflow meter** | | | | | | |
| Resistance Ω | E2↔Vs | Fully closed | | 200—600 | | |
| | | Fully open | | 20—1,200 | | |
| | E2↔Vc | | | 200—400 | | |
| | E2↔THAA (Intake air thermosensor) | -20°C (-4°F) | | 13,600—18,400 | | |
| | | 20°C (68°F) | | 2,210—26,900 | | |
| | | 60°C (140°F) | | 493—667 | | |
| | E1↔Fc | Fully closed | | ∞ | | |
| | | Fully open | | 0 | | |
| **Fuel tank** | | | | | | |
| Capacity | | | liters (US gal, Imp gal) | Hatchback...50 (13.2, 11.0), Sedan...55 (14.5, 12.1) | | |
| **Air cleaner** | | | | | | |
| Element type | | | | Oil permeated | | |
| **Fuel** | | | | | | |
| Specification | | | | Unleaded regular (RON 91 or higher) | | |

***1 With parking brake applied (Canada)**
***2 TEN terminal of diagnosis connector grounded**

## COMPONENT DESCRIPTIONS—1990 323

| Component | Function | Remark |
|---|---|---|
| **Air cleaner** | Fillers air entering throttle body | |
| **Airflow meter** | Detects amount of intake air, sends signal to ECU | Intake air thermosensor and fuel pump switch included |
| **Air valve** | Supplies bypass air into dynamic chamber when engine cold | • Engine speed increased to shorten warm-up period<br>• Thermowax type<br>• Installed in dynamic chamber |
| **Atmospheric pressure sensor** | Detects atmospheric pressure, sends signal to ECU | Built in ECU |
| **Catalytic converter** | Reduces HC, CO, and NOx by chemical reaction | Monolith type |
| **Charcoal canister** | Stores fuel tank fumes while engine stopped | |
| **Check valve** | Controls pressure in fuel tank | Two-way type |
| **Circuit opening relay** | Voltage for fuel pump while engine running | |
| **Clutch switch (MTX)** | Detects clutch condition, sends signal to ECU | Switch OFF when clutch pedal released |
| **Diagnosis connector** | Concentrated service connector<br>Concentrated terminals<br>1. EGI self-diagnostic terminal<br>2. EC-AT self-diagnostic terminal<br>3. Test terminal<br>4. Fuel pump check terminal<br>5. Cooling fan check terminal<br>6. Engine rpm output terminal | 21-pin (located near left suspension mounting block) |
| **Dynamic chamber** | Interconnects all cylinders | |
| **Engine control unit (ECU)** | **Detects the following:**<br>1. A/C operation<br>2. Air/fuel ratio (Oxygen concentration)<br>3. Atmospheric pressure<br>4. Braking signal<br>5. Cranking signal<br>6. DRL (Daytime Running light) operation<br>7. E/L operation<br>8. Engine coolant temperature<br>9. Engine speed<br>10. Ignition ON signal<br>11. In-gear condition<br>12. Intake air amount<br>13. Intake air temperature<br>14. No.1 piston TDC of compression<br>15. P/S operation<br>16. Test signal (Ignition timing, idle speed, Malfunction Code No.)<br>17. Throttle valve fully closed/fully open condition<br>18. Throttle valve opening angle | 1. A/C switch<br>2. Oxygen sensor<br>3. Atmospheric pressure sensor<br>4. Stoplight switch (MTX)<br>5. Ignition switch (START Position)<br>6. DRL relay (Canada)<br>7. Blower motor switch, cooling fan relay, head light switch, and rear window defroster switch<br>8. Water thermosensor<br>9. Distributor (Ne-SIGNAL)<br>10. Ignition switch<br>11. Neutral and clutch switches (MTX) Inhibitor switch (ATX)<br>12. Airflow meter<br>13. Intake air thermosensor (In airflow meter)<br>14. Distributor (G-signal) (DOHC)<br>15. P/S pressure switch<br>16. Diagnosis connector (TEN terminal)<br>17. Throttle sensor (MTX)<br>18. Throttle sensor (ATX) |

## COMPONENT DESCRIPTIONS (CONT.)—1990 323

| Component | Function | Remark |
|---|---|---|
| **Engine control unit (ECU) (Cont'd)** | **Controls operation of the following:**<br>1. A/C (Cut-off)<br>2. Fail-safe function<br>3. Fuel injection system<br>4. Idle speed control<br>5. Ignition timing control system<br>6. Lockup control system<br>7. Monitor function<br>8. Pressure regulator control system<br>9. Purge control system<br>10. VICS | 1. A/C relay<br>2. Self-Diagnosis Checker and MIL<br>3. Injector<br>4. Idle speed control (ISC) valve<br>5. Igniter<br>6. EC-AT control unit (ATX)<br>7. Monitor lamp (Self-Diagnosis Checker)<br>8. Solenoid valve (Pressure regulator control) (BP)<br>9. Purge control solenoid valve<br>10. Solenoid valve (VICS) (DOHC) |
| **Fuel filter** | Filters particles from fuel | |
| **Fuel pump** | Provides fuel to injectors | • Operates while engine running<br>• Installed in fuel tank |
| **Igniter** | Receives spark signal from ECU and generates high voltage in ignition coil | |
| **Ignition switch (START position)** | Engine cranking signal sent to ECU | |
| **Inhibitor switch (ATX)** | Detects in-gear condition, sends signal to ECU | Switch ON in N or P range |
| **Injector** | Injects fuel into intake port | • Controlled by signals from ECU<br>• High-ohmic injector |
| **Intake air thermosensor** | Detects intake air temperature, sends signal to ECU | Installed on dynamic chamber |
| **Idle speed control (ISC) valve** | Controls bypass air amount | Controlled by duty signal from ECU |
| **Main relay (FUEL INJ relay)** | Supplies electric current to injectors, ECU, etc | |
| **MIL (Malfunction indicator lamp)** | Lamp flashes to indicate malfunction code number of input and output devices | TEN terminal grounded |
| **Neutral switch** | Detects in gear condition, sends signal to ECU | Switch ON in neutral |
| **Oxygen sensor** | Detects oxygen concentration, sends signal to ECU | Zirconia ceramic and platinum coating |
| **PCV valve** | Controls blowby gas introduced into engine | |
| **P/S pressure switch** | Detects P/S operation, sends signal to ECU | P/S ON when steering wheel turned right or left |
| **Pressure regulator** | Adjusts fuel pressure supplied to injectors | |
| **Resonance chamber** | Reduces intake air noise | |
| **Resonance duct** | Reduces intake air noise | |
| **Separator** | Prevents fuel from flowing into charcoal canister | |

| Component | | Function | Remark |
|---|---|---|---|
| **Stoplight switch** | | Detects braking operation (deceleration), sends signal to ECU | MTX |
| **Solenoid valve (Purge control)** | | Controls evaporative fumes from canister to intake manifold | Controlled by duty signal from ECU |
| **Shutter valve actuator (DOHC)** | | Closes/opens shutter valve to improve torque characteristics | For Variable inertia charging system (VICS*) |
| **Throttle body** | | Controls intake air quantity | Integrated throttle sensor, dashpot, and ISC valve |
| **Throttle sensor** | | • Detects throttle valve fully closed/fully opened condition (MTX)<br>• Detects throttle valve opening angle (ATX)<br>• Sends signals to ECU | |
| **Three-way solenoid valve** | **VICS*** | Controls vacuum to shutter valve actuator | Cuts vacuum when engine speed above 5,000 rpm |
| | **Pressure regulator control** | Controls vacuum to pressure regulator | Cuts vacuum just after starting when engine hot starting |
| **Vacuum chamber (DOHC)** | | Stores vacuum for use during wide open throttle | For VICS* |
| **Water thermosensor** | | Detects coolant temperature, sends signal to ECU | |

* VICS: Variable Inertia Charging System

## TROUBLESHOOTING CHART—1990 323

**ENGINE CONTROL OPERATION CHART**
**Input Device and Engine Conditions**

**Note**
• **The data in this chart is for reference only.**

| INPUT DEVICES / ENGINE CONDITIONS | APPROXIMATE TIME (BASED ON 10—16°C, 50—60°F AMBIENT) | SENSORS: DISTRIBUTOR (G-SIGNAL)*1 (Ne-SIGNAL) | SENSORS: WATER THERMOSENSOR | SENSORS: OXYGEN SENSOR | SENSORS: AIRFLOW METER | SENSORS: INTAKE AIR THERMOSENSOR | THROTTLE SENSOR: Vт TERMINAL (ATX) | THROTTLE SENSOR: POW TERMINAL (MTX) | THROTTLE SENSOR: IDL TERMINAL |
|---|---|---|---|---|---|---|---|---|---|
| CRANKING —COLD ENGINE • COLD AIR • COLD COOLANT | Zero | | | Signal has no effect on ECU | Signal has no effect on ECU | Signal has no effect on ECU | Signal has no effect on ECU | Signal has no effect on ECU | Signal has no effect on ECU |
| COLD START —FAST IDLE • COLD AIR • COLD COOLANT | One minute | | Cool to warm Medium voltage (3.5V and dropping) | Sensor cold Low to high voltage (0—0.9V) | Low volume airflow High voltage (Above 3.0V) | Cool (Below 20°C (68°F)) Above 2.3V | Closed throttle Low voltage (0.2—0.8V) | Closed throttle Low voltage (Approx 5.0V) | Closed throttle Low voltage (Approx 0V) |
| COLD DRIVEAWAY —PART THROTTLE • COLD AIR • COLD COOLANT | Two minutes | | | | | | | | |
| WARM DRIVEAWAY —PART THROTTLE • WARM AIR • WARM COOLANT | Three minutes | Ne-SIGNAL Sends all cylinders TDC signal to ECU | Warm Medium voltage (Approx 0.7V and dropping) | Sensor hot switching from high voltage (0.9V) | Moderate volume air flow Low to medium voltage (1.0—3.5V) | | Part throttle Medium voltage (0.8—3.0V) | Part throttle Medium voltage Approx 5.0V) | |
| HOT CRUISE • WARM AIR • WARM COOLANT | | G SIGNAL*1 Sends No. 1 cylinder TDC (compression) signal to ECU | | to low voltage (0.1V) | | | | | Open throttle High voltage (battery voltage) |
| HOT ACCELERATION —60% THROTTLE | | | | High voltage (0.9V) | Moderate to strong volume of airflow Low voltage (0.5—1.5V) | Cool to warm (−20—40°C (−4—104°F)) 1.5—4.3V | | | |
| HOT ACCELERATION —WIDE OPEN THROTTLE | More than four minutes | | Hot Low voltage (Below 0.5V) | | | | Wide open throttle High voltage (Above 3.0V) | Wide open throttle High voltage (Approx 0V) | |
| DECELERATION —CLOSED THROTTLE | | | | Low voltage (0V) | Low volume of airflow (Above 3.0V) | | Closed throttle Low voltage (0.2—0.8V) | Closed throttle Low voltage (Approx 5.0V) | Closed throttle Low voltage (Approx 0V) |
| HOT CURB IDLE —EXTENDED | | | | Switching from high to low voltage (0.75—0.25V) | | | | | |
| HOT ENGINE SHUTDOWN | — | OFF | OFF | Sensor hot Low voltage (0.1V) until sensor cools | OFF | OFF | OFF | OFF | OFF |

*1 DOHC

| ENGINE CONDITIONS | SENSORS: STOPLIGHT SWITCH (MTX) | SENSORS: NEUTRAL AND CLUTCH SWITCHES (MTX) | SENSORS: INHIBITOR SWITCH (ATX) | SENSORS: A/C SWITCH | SENSORS: P/S PRESSURE SWITCH | SENSORS: E/L SIGNAL* | IGNITION SWITCH: START POSITION | IGNITION SWITCH: ON POSITION | DIAGNOSIS CONNECTOR TEN TERMINAL (TEST TERMINAL) | DRL (Daytime Running Light) CONTROL UNIT (CANADA) | EC-AT CONTROL UNIT (IGNITION TIMING RETARD SIGNAL) |
|---|---|---|---|---|---|---|---|---|---|---|---|
| CRANKING | Signal has no effect on ECU | Signal has no effect on ECU | Signal has no effect on ECU | Signal has no effect on ECU | Signal has no effect on ECU | Signal has no effect on ECU | Sends signal to ECU (Battery voltage) | Signal has no effect on ECU | Signal has no effect on ECU | OFF (Battery voltage) | |
| COLD START | Brake pedal depressed Sends signal to ECU (Battery voltage) | In neutral Low voltage signal to ECU (Approx 0V) | In N or P range No signal to ECU (Approx 0V) | | | | | | | | OFF (Approx 0V) |
| COLD DRIVEAWAY | | | | | | Headlight or rear window defroster switch ON High voltage signal to ECU (Battery voltage) OFF Low voltage signal to ECU (Below 1.5V) Blower switch*2 or Electrical fan ON Low voltage signal to ECU (Below 1.5V) OFF High voltage signal to ECU (Battery voltage) | | | | | |
| WARM DRIVEAWAY | No signal sent to ECU (Below 1.5V) | | | A/C switch ON Sends signal to ECU (Below 1.5V) A/C switch OFF No signal to ECU (Battery voltage) | Steering wheel turned Low voltage signal to ECU (Below 1.5V) Steering wheel straight ahead High voltage signal to ECU (Battery voltage) | | | | Terminal not grounded High voltage Signal to ECU (Battery voltage) Terminal grounded Low voltage signal to ECU (Below 1.5V) | | |
| HOT CRUISE | | Driving in any gear High voltage signal to ECU (Battery voltage) | In D range No signal to ECU (Battery voltage) | | | | No signal to ECU (Below 1.5V) | Sends signal to ECU (Battery voltage) | | DRL ON (Approx 0V) DRL OFF (Battery voltage) | Down-shift ON (Battery voltage) |
| HOT ACCELERATION —60% THROTTLE | | | | | | | | | | | |
| HOT ACCELERATION —WIDE OPEN THROTTLE | | | | | | | | | | | |
| DECELERATION | Brake pedal depressed Sends signal to ECU (Battery voltage) | | | | | | | | | | OFF |
| HOT CURB IDLE | | In neutral Low voltage signal to ECU (Approx 0V) | In N or P range Sends signal to ECU (Approx 0V) | | | | | | | | |
| HOT ENGINE SHUTDOWN | OFF | OFF | OFF | OFF | OFF | OFF | OFF | OFF | OFF | OFF | |

*2 **Blower switch: 2nd position or more**

## OUTPUT COMPONENTS AND ENGINE CONDITIONS 1990 323

**Output Devices and Engine Conditions**

**Note**
• **The data in this chart is for reference only.**

| OUTPUT DEVICES / ENGINE CONDITIONS | APPROXIMATE TIME (BASED ON 10—16°C, 50—60°F AMBIENT) | INJECTOR: INJECTION | INJECTOR: INJECTION TIMING | AIR VALVE | ISC VALVE | SOLENOID VALVE (PURGE CONTROL) | THREE-WAY SOLENOID VALVE (PRESSURE REGULATOR CONTROL)*1 | A/C RELAY | EC-AT CONTROL UNIT (ATX) |
|---|---|---|---|---|---|---|---|---|---|
| CRANKING —COLD ENGINE • COLD AIR • COLD COOLANT | Zero | | All cylinders each ignition pulse | | | | | OFF (A/C OFF) | |
| COLD START —FAST IDLE • COLD AIR • COLD COOLANT | One minute | Rich | | Open coolant temperature Below 40°C (104°F) | Large amount of bypass air | OFF (Purge cut) | | OFF (A/C OFF Approx 5 sec [DOHC] 10 sec [SOHC]) | Coolant temperature Below 72°C (162°F) ON (Lock-up inhibit) |
| COLD DRIVEAWAY —PART THROTTLE • COLD AIR • COLD COOLANT | Two minutes | | | | | | | | |
| WARM DRIVEAWAY —PART THROTTLE • WARM AIR • WARM COOLANT | Three minutes | Rich and lean | 2-group*1 1-group*2 | | | Operates (Duty values [purge gas amount] change) | OFF (Vacuum to pressure regulator) | ON (A/C ON) | |
| HOT CRUISE • WARM AIR • WARM COOLANT | | | | | | | | | |
| HOT ACCELERATION —60% THROTTLE | | Rich | | | Small amount of bypass air | | | | OFF (Lockup) |
| HOT ACCELERATION —WIDE OPEN THROTTLE | More than four minutes | | | | | | | OFF (A/C cut) | |
| DECELERATION —CLOSED THROTTLE | | Fuel cut | | Closed | | OFF (Purge cut) | | ON (A/C ON) | |
| HOT CURB IDLE —EXTENDED | | Rich and lean | 2-group*1 1-group*2 | | | | After starting ON during hot start only (Vacuum cut) | | |
| HOT ENGINE SHUTDOWN | — | Does not inject | | | OFF | OFF | OFF | OFF | OFF |

*1 **DOHC**, *2 **SOHC**, *3 **BP**

## TROUBLESHOOTING AND RELATIONSHIP CHART 1990 323

| OUTPUT DEVICE / INPUT DEVICE | INJECTOR: FUEL INJECTION AMOUNT | INJECTOR: FUEL INJECTION TIMING | ISC VALVE | SOLENOID VALVE (PURGE CONTROL) | A/C RELAY (A/C CUT-OFF) | EC-AT CONTROL UNIT (LOCK-UP CONTROL) | IGNITER (IGNITION TIMING CONTROL) | THREE-WAY SOLENOID VALVE: PRESSURE REGULATOR*1 | THREE-WAY SOLENOID VALVE: VICS*2 | SELF-DIAGNOSIS CHECKER/ MIL*3 (MALFUNCTION CODE) | SELF-DIAGNOSIS CHECKER (MONITOR LAMP) |
|---|---|---|---|---|---|---|---|---|---|---|---|
| EC-AT CONTROL UNIT | | | | | | | ○ | | | | |
| DIAGNOSIS CONNECTOR (TEN TERMINAL) | | | | | | | ○ | | | ○ | ○ |
| IGNITION SWITCH (START POSITION) | ○ | ○ | ○ | | ○ | | ○ | ○ | | | |
| E/L SIGNAL*4 | | | ○ | | | | | | | | ○ |
| P/S PRESSURE SWITCH | | | ○ | | ○ | | | | | | |
| A/C SWITCH | | | ○ | | ○ | | | | | | ○ |
| INHIBITOR SWITCH (ATX) | ○ | | ○ | ○ | ○ | | ○ | ○ | | | ○ |
| NEUTRAL AND CLUTCH SWITCHES (MTX) | ○ | | ○ | ○ | ○ | | ○ | ○ | | | ○ |
| STOPLIGHT SWITCH (MTX) | ○ | | | | | | | | | | ○ |
| ATMOSPHERIC PRESSURE SENSOR (IN ECU) | ○ | | ○ | ○ | ○ | | | | | ○ | |
| THROTTLE SENSOR | ○ | ○ | ○ | ○ | ○ | | ○ | ○ | | ○ (ATX) | ○ |
| INTAKE AIR THERMOSENSOR | ○ | | ○ | ○ | | | | ○ | | ○ | |
| AIRFLOW METER | ○ | ○ | | ○ | | | ○ | | | ○ | |
| OXYGEN SENSOR | ○ | | | ○ | | | | | | ○ | ○ |
| WATER THERMOSENSOR | ○ | | ○ | ○ | | ○ | ○ | ○ | | ○ | |
| DISTRIBUTOR: G-SIGNAL*2 | | ○ | | | | | | | | ○ | |
| DISTRIBUTOR: Ne-SIGNAL | ○ | ○ | ○ | ○ | ○ | | ○ | | ○ | ○ | |

*1 **BP**
*2 **DOHC**
*3 **MIL: Malfunction Indicator Lamp**
*4 **E/L SIGNAL: Blower fan control switch second position or higher, cooling fan operating, headlights ON, or rear window defroster switch ON**

## TROUBLESHOOTING CHART—1990 323

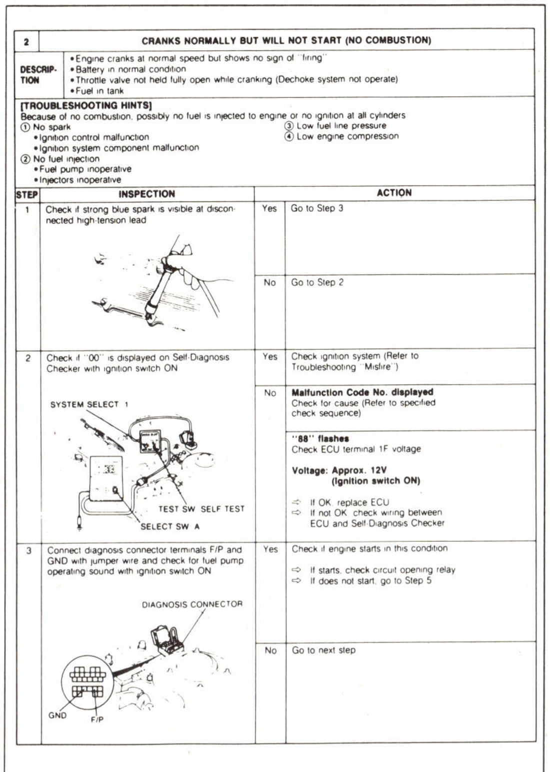

| 2 | CRANKS NORMALLY BUT WILL NOT START (NO COMBUSTION) |
|---|---|
| DESCRIPTION | •Engine cranks at normal speed but shows no sign of "firing"<br>•Battery in normal condition<br>•Throttle valve not held fully open while cranking (Dechoke system not operate)<br>•Fuel in tank |

**[TROUBLESHOOTING HINTS]**
Because of no combustion, possibly no fuel is injected to engine or no ignition at all cylinders
① No spark
•Ignition control malfunction
•Ignition system component malfunction
② No fuel injection
•Fuel pump inoperative
•Injectors inoperative
③ Low fuel line pressure
④ Low engine compression

| STEP | INSPECTION | | ACTION |
|---|---|---|---|
| 1 | Check if strong blue spark is visible at disconnected high-tension lead | Yes | Go to Step 3 |
| | | No | Go to Step 2 |
| 2 | Check if "00" is displayed on Self-Diagnosis Checker with ignition switch ON<br>SYSTEM SELECT 1<br>TEST SW SELF TEST<br>SELECT SW A | Yes | Check ignition system (Refer to Troubleshooting "Misfire") |
| | | No | **Malfunction Code No. displayed**<br>Check for cause (Refer to specified check sequence)<br>**"88" flashes**<br>Check ECU terminal 1F voltage<br>**Voltage: Approx. 12V (Ignition switch ON)**<br>⇨ If OK, replace ECU<br>⇨ If not OK, check wiring between ECU and Self-Diagnosis Checker |
| 3 | Connect diagnosis connector terminals F/P and GND with jumper wire and check for fuel pump operating sound with ignition switch ON<br>DIAGNOSIS CONNECTOR<br>GND F/P | Yes | Check if engine starts in this condition<br>⇨ If starts, check circuit opening relay<br>⇨ If does not start, go to Step 5 |
| | | No | Go to next step |

| STEP | INSPECTION | | ACTION |
|---|---|---|---|
| 4 | Check if approx. 12V exists at fuel pump connector B/P wire with ignition switch ON<br>B/P<br>B | Yes | Check continuity of fuel pump (Between terminals B/P and B) |
| | | No | Check circuit opening relay |
| 5 | Check for injector operating sound while cranking engine | Yes | Go to Step 7 |
| | | No | Go to next step |
| 6 | Check if approx. 12V exists at injector connector (W/R) wire with ignition switch ON | Yes | Check ECU terminals N, 2A, 2U and 2V voltages |
| | | No | Check for open circuit in wiring between main relay (FUEL INJ relay) and injector |
| 7 | Connect diagnosis connector terminals F/P and GND with jumper wire and check for correct fuel line pressure with ignition switch ON<br>**Fuel Line pressure:**<br>**265—314 kPa (2.7—3.2 kg/cm², 38—46 psi)**<br>INSTALL CLAMPS<br>GND F/P | Yes | Go to next step |
| | | No | **Low pressure**<br>Check fuel line pressure while pinching fuel return hose<br>⇨ If fuel line pressure quickly increases, check pressure regulator<br>⇨ If fuel line pressure gradually increases, check for clogging between fuel pump and pressure regulator<br>If not clogged, check fuel pump maximum pressure<br>**High pressure**<br>Check if fuel return hose is clogged or restricted<br>⇨ If OK, replace pressure regulator<br>⇨ If not OK, repair or replace |
| 8 | Check for correct engine compression<br>**Engine compression:**<br>**• BP SOHC**<br>**834 kPa (8.5 kg/cm², 121 psi)-300 rpm**<br>**• BP DOHC**<br>**883 kPa (9.0 kg/cm², 128 psi)-300 rpm**<br>**• B6**<br>**932 kPa (9.5 kg/cm², 135 psi)-300 rpm** | Yes | Go to next step |
| | | No | Check engine condition<br>•Worn piston, piston rings or cylinder wall<br>•Defective cylinder head gasket<br>•Distorted cylinder head<br>•Improper valve seating<br>•Valve sticking in guide |
| 9 | Check if all spark plugs are OK<br>WEAR AND CARBON BUILDUP<br>PLUG GAP 1.0—1.1mm (0.039—0.043 in)<br>BURNS<br>DAMAGE AND DETERIORATION<br>DAMAGE | Yes | Go to next step |
| | | No | Clean or replace |
| 10 | Try known good ECU and check if condition improves | | |

## TROUBLESHOOTING CHART—1990 323

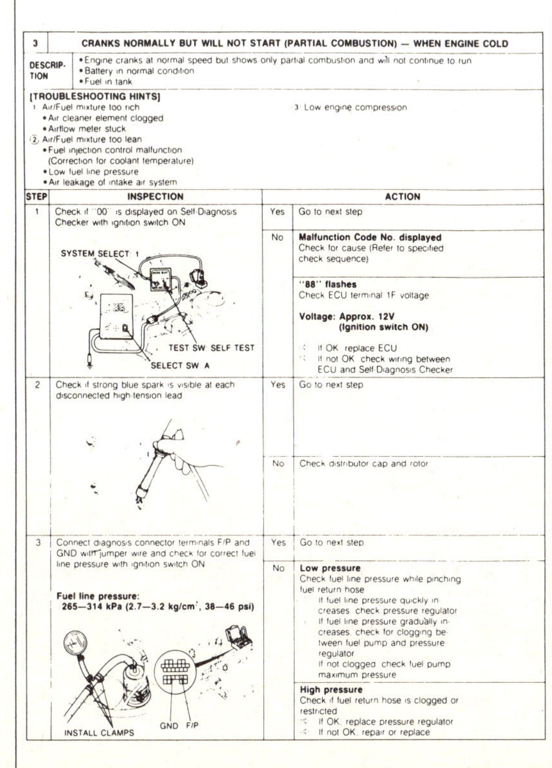

| 3 | CRANKS NORMALLY BUT WILL NOT START (PARTIAL COMBUSTION) — WHEN ENGINE COLD |
|---|---|
| DESCRIPTION | •Engine cranks at normal speed but shows only partial combustion and will not continue to run<br>•Battery in normal condition<br>•Fuel in tank |

**[TROUBLESHOOTING HINTS]**
① Air/Fuel mixture too rich
•Air cleaner element clogged
•Airflow meter stuck
② Air/Fuel mixture too lean
•Fuel injection control malfunction (Correction for coolant temperature)
•Low fuel line pressure
•Air leakage of intake air system
③ Low engine compression

| STEP | INSPECTION | | ACTION |
|---|---|---|---|
| 1 | Check if "00" is displayed on Self-Diagnosis Checker with ignition switch ON<br>SYSTEM SELECT 1<br>TEST SW SELF TEST<br>SELECT SW A | Yes | Go to next step |
| | | No | **Malfunction Code No. displayed**<br>Check for cause (Refer to specified check sequence)<br>**"88" flashes**<br>Check ECU terminal 1F voltage<br>**Voltage: Approx. 12V (Ignition switch ON)**<br>⇨ If OK, replace ECU<br>⇨ If not OK, check wiring between ECU and Self-Diagnosis Checker |
| 2 | Check if strong blue spark is visible at each disconnected high-tension lead | Yes | Go to next step |
| | | No | Check distributor cap and rotor |
| 3 | Connect diagnosis connector terminals F/P and GND with jumper wire and check for correct fuel line pressure with ignition switch ON<br>**Fuel line pressure:**<br>**265—314 kPa (2.7—3.2 kg/cm², 38—46 psi)**<br>INSTALL CLAMPS<br>GND F/P | Yes | Go to next step |
| | | No | **Low pressure**<br>Check fuel line pressure while pinching fuel return hose<br>⇨ If fuel line pressure quickly increases, check pressure regulator<br>⇨ If fuel line pressure gradually increases, check for clogging between fuel pump and pressure regulator<br>If not clogged, check fuel pump maximum pressure<br>**High pressure**<br>Check if fuel return hose is clogged or restricted<br>⇨ If OK, replace pressure regulator<br>⇨ If not OK, repair or replace |

| STEP | INSPECTION | | ACTION |
|---|---|---|---|
| 4 | Check if ECU terminal voltages are OK (Especially 1C, 2D and 2Q) | Yes | Go to next step |
| | | No | Check for cause |
| 5 | Check for air leakage of intake air system components | Yes | Repair or replace |
| | | No | Go to next step |
| 6 | Check if engine starts when water thermosensor is disconnected | Yes | Check water thermosensor |
| | | No | Go to next step |
| 7 | Manually check if airflow meter moves smoothly from fully closed to fully open | Yes | Go to next step |
| | | No | Repair or replace |
| 8 | Check for correct engine compression<br>⇨ page B1-10 B2-10<br>**Engine compression:**<br>**• BP SOHC**<br>**834 kPa (8.5 kg/cm², 121 psi)-300 rpm**<br>**• BP DOHC**<br>**883 kPa (9.0 kg/cm², 128 psi)-300 rpm**<br>**• B6**<br>**932 kPa (9.5 kg/cm², 135 psi)-300 rpm** | Yes | Go to next step |
| | | No | Check engine condition<br>•Worn piston, piston rings or cylinder wall<br>•Defective cylinder head gasket<br>•Distorted cylinder head<br>•Improper valve seating<br>•Valve sticking in guide |
| 9 | Check if spark plugs are OK<br>WEAR AND CARBON BUILDUP<br>PLUG GAP 1.0—1.1mm (0.039—0.043 in)<br>BURNS<br>DAMAGE AND DETERIORATION<br>DAMAGE | Yes | Go to next step |
| | | No | Clean, or replace |
| 10 | Try known good ECU and check if condition improves | | |

## TROUBLESHOOTING CHART—1990 323

| 4 | CRANKS NORMALLY BUT WILL NOT START (PARTIAL COMBUSTION) — AFTER WARM-UP | | |
|---|---|---|---|
| DESCRIPTION | • Engine cranks at normal speed and shows partial combustion but will not continue to run after running and hot soaked<br>• Battery in normal condition<br>• Engine starts normally when cold | | |

[TROUBLESHOOTING HINTS]
1. Air/Fuel mixture too rich
   - Insufficient fuel injection control (Correction for coolant temperature)
   - Injector fuel leakage
2. Vapor lock
   - Fuel pressure not held in fuel line after engine stopped
   - High RVP (winter) fuel used in warm weather
3. Pressure regulator control system malfunction [BP]

| STEP | INSPECTION | | ACTION |
|---|---|---|---|
| 1 | Warm-up engine to normal operating temperature and stop it<br>Connect diagnosis connector terminals F/P and GND with jumper wire for 3 minutes with ignition switch ON, then check if engine starts<br>GND F/P | Yes | Change fuel to another brand |
| | | No | **BP engine**<br>Go to next step |
| | | | **B6 engine**<br>Go to Step 3 |
| 2 | Remove vacuum hose from pressure regulator and plug it<br>Check if engine starts<br>PLUG | Yes | Check pressure regulator control system |
| | | No | Go to next step |
| 3 | Connect diagnosis connector terminals F/P and GND with jumper wire and check for correct fuel line pressure with ignition switch ON<br>**Fuel line pressure: 265—314 kPa (2.7—3.2 kg/cm², 38—46 psi)**<br>INSTALL CLAMPS GND F/P | Yes | Go to next step |
| | | No | **Low pressure**<br>Check fuel line pressure while pinching fuel return hose<br>⇨ If fuel line pressure quickly increases, check pressure regulator<br>⇨ If fuel line pressure gradually increases, check fuel line and filter for clogging<br>If not clogged, check fuel pump maximum pressure |
| | | | **High pressure**<br>Check if fuel return hose is clogged or restricted<br>⇨ If OK, replace pressure regulator<br>⇨ If not OK, repair or replace |
| 4 | In same condition as Step 2, check if fuel line pressure is held after ignition switch is turned OFF<br>**Fuel line pressure: More than 147 kPa (1.5 kg/cm², 21 psi) for 5 min.** | Yes | Go to Step 6 |
| | | No | Go to next step |

| STEP | INSPECTION | | ACTION |
|---|---|---|---|
| 5 | Plug outlet of pressure regulator and check if fuel line pressure is held after ignition switch is turned OFF<br>**Fuel line pressure: More than 147 kPa (1.5 kg/cm², 21 psi) for 5 min.** | Yes | Replace pressure regulator |
| | | No | Check fuel pump hold pressure<br>⇨ If OK, check injector for fuel leakage<br>⇨ If not OK, replace fuel pump |
| 6 | Disconnect water thermosensor connector and check if engine starts | Yes | Check water thermosensor<br>⇨ If OK, check connections of water thermosensor and ECU 2Q terminal<br>⇨ If not OK, replace water thermosensor |
| | | No | Go to next step |
| 7 | Check if "00" is displayed on Self-Diagnosis Checker with ignition switch ON<br>SYSTEM SELECT: 1<br>TEST SW: SELF TEST<br>SELECT SW: A | Yes | Go to next step |
| | | No | **Malfunction Code No. displayed**<br>Check for cause (Refer to specified check sequence) |
| | | | **"88" flashes**<br>Check ECU terminal 1F voltage<br>**Voltage: Approx. 12V (Ignition switch ON)**<br>⇨ If OK, replace ECU<br>⇨ If not OK, check wiring between ECU and Self-Diagnosis Checker |
| 8 | Check if ECU terminal voltages are OK (Especially 1C, 2D and 2Q) | Yes | Go to next step |
| | | No | Check for cause |
| 9 | Try known good ECU and check if condition improves | Yes | Replace ECU |
| | | No | Change fuel to another brand |

## TROUBLESHOOTING CHART—1990 323

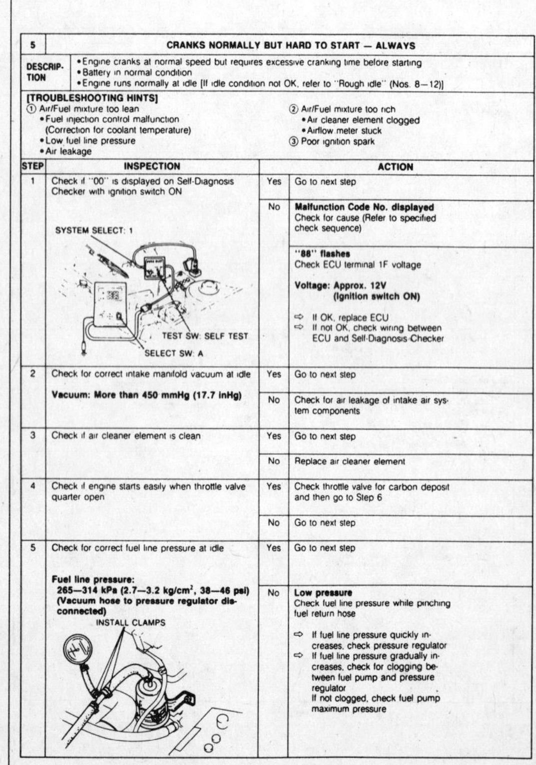

| 5 | CRANKS NORMALLY BUT HARD TO START — ALWAYS | | |
|---|---|---|---|
| DESCRIPTION | • Engine cranks at normal speed but requires excessive cranking time before starting<br>• Battery in normal condition<br>• Engine runs normally at idle [If idle condition not OK, refer to "Rough idle" (Nos. 8—12)] | | |

[TROUBLESHOOTING HINTS]
1. Air/Fuel mixture too lean
   - Fuel injection control malfunction (Correction for coolant temperature)
   - Low fuel line pressure
   - Air leakage
2. Air/Fuel mixture too rich
   - Air cleaner element clogged
   - Airflow meter stuck
3. Poor ignition spark

| STEP | INSPECTION | | ACTION |
|---|---|---|---|
| 1 | Check if "00" is displayed on Self-Diagnosis Checker with ignition switch ON<br>SYSTEM SELECT: 1<br>TEST SW: SELF TEST<br>SELECT SW: A | Yes | Go to next step |
| | | No | **Malfunction Code No. displayed**<br>Check for cause (Refer to specified check sequence) |
| | | | **"88" flashes**<br>Check ECU terminal 1F voltage<br>**Voltage: Approx. 12V (Ignition switch ON)**<br>⇨ If OK, replace ECU<br>⇨ If not OK, check wiring between ECU and Self-Diagnosis Checker |
| 2 | Check for correct intake manifold vacuum at idle<br>**Vacuum: More than 450 mmHg (17.7 inHg)** | Yes | Go to next step |
| | | No | Check for air leakage of intake air system components |
| 3 | Check if air cleaner element is clean | Yes | Go to next step |
| | | No | Replace air cleaner element |
| 4 | Check if engine starts easily when throttle valve quarter open | Yes | Check throttle valve for carbon deposit and then go to Step 6 |
| | | No | Go to next step |
| 5 | Check for correct fuel line pressure at idle<br>**Fuel line pressure: 265—314 kPa (2.7—3.2 kg/cm², 38—46 psi) (Vacuum hose to pressure regulator disconnected)**<br>INSTALL CLAMPS | Yes | Go to next step |
| | | No | **Low pressure**<br>Check fuel line pressure while pinching fuel return hose<br>⇨ If fuel line pressure quickly increases, check pressure regulator<br>⇨ If fuel line pressure gradually increases, check for clogging between fuel pump and pressure regulator<br>If not clogged, check fuel pump maximum pressure |

| STEP | INSPECTION | | ACTION |
|---|---|---|---|
| 6 | Disconnect ISC valve connector at idle and check if engine speed increases<br>ISC VALVE | Yes | Go to next step |
| | | No | Replace ISC valve |
| 7 | Check if ECU terminal voltages are OK (Especislly 1C, 2D and 2Q) | Yes | Go to next step |
| | | No | Check for cause |
| 8 | Connect diagnosis connector terminals F/P and GND with jumper wire and check if engine starts normally<br>DIAGNOSIS CONNECTOR<br>GND F/P | Yes | Check circuit opening relay<br>⇨ If relay OK, repair or replace wire harness<br>⇨ If relay not OK, replace relay |
| | | No | Go to next step |
| 9 | Connect System Selector to diagnosis connector and set Test Switch to "SELF TEST" and check for correct ignition timing at idle after warm-up<br>**Ignition timing (BTDC)**<br>**• BP DOHC : 10 ± 1°**<br>**• BP SOHC : 5 ± 1°**<br>**• B6 : 7 ± 1°**<br>TEST SWITCH | Yes | Go to next step |
| | | No | Adjust ignition timing |
| 10 | Check for correct engine compression<br>**Engine compression (Minimum):**<br>**• BP DOHC 883 kPa (9.0 kg/cm², 128 psi)-300 rpm**<br>**• BP SOHC 834 kPa (8.5 kg/cm², 121 psi)-300 rpm**<br>**• B6 932 kPa (9.5 kg/cm², 135 psi)-300 rpm** | Yes | Go to next step |
| | | No | Check engine condition<br>• Worn piston, piston rings or cylinder wall<br>• Defective cylinder head gasket<br>• Distorted cylinder head<br>• Improper valve seating<br>• Valve sticking in guide |
| 11 | Try known good ECU and check if condition improves | | |

## TROUBLESHOOTING CHART—1990 323

| 6 | CRANKS NORMALLY BUT HARD TO START — WHEN ENGINE COLD |
|---|---|
| DESCRIP-TION | • Engine cranks at normal speed but requires excessive cranking time before starting<br>• Battery in normal condition<br>• Restarts OK after warm-up<br>• Engine runs normally at idle [If idle condition is not OK, refer to "Rough idle" (Nos 8—12)] |

**[TROUBLESHOOTING HINTS]**

① Air/Fuel mixture too rich
- Airflow meter sticking
- Air cleaner element clogged
- Idle speed control malfunction

② Air/Fuel mixture too lean
- Fuel injection control malfunction (Correction for coolant temperature)

③ Poor atomization
- Low RVP (summer) fuel is used in cold weather

| STEP | INSPECTION | | ACTION |
|---|---|---|---|
| 1 | Check if "00" is displayed on Self-Diagnosis Checker with ignition switch ON<br>SYSTEM SELECT: 1<br>TEST SW: SELF TEST<br>SELECT SW: A | Yes | Go to next step |
| | | No | **Malfunction Code No. displayed**<br>Check for cause (Refer to specified check sequence) |
| | | | **"88" flashes**<br>Check ECU terminal 1F voltage<br>**Voltage: Approx. 12V (Ignition switch ON)**<br>⇨ If OK, replace ECU<br>⇨ If not OK, check wiring between ECU and Self-Diagnosis Checker |
| 2 | Check if ECU terminal voltages are OK (Especially 1C, 2D, and 2Q) | Yes | Go to next step |
| | | No | Check for cause |
| 3 | Check if engine starts easily when throttle valve quarter open | Yes | Check if ISC valve is OK<br>⇨ If OK, check air valve<br>⇨ If not OK, replace ISC valve |
| | | No | Go to next step |
| 4 | Check for correct intake manifold vacuum at idle<br>**Vacuum: More than 450 mmHg (17.7 inHg)** | Yes | Go to next step |
| | | No | Check for air leakage of intake air system components |

| STEP | INSPECTION | | ACTION |
|---|---|---|---|
| 5 | Check if air cleaner element is clean | Yes | Go to next step |
| | | No | Replace air cleaner element |
| 6 | Disconnect ISC valve connector at idle and check if engine speed increases<br>ISC VALVE | Yes | Go to next step |
| | | No | Replace ISC valve |
| 7 | Connect diagnosis connector terminals F/P and GND with jumper wire and check for correct fuel line pressure with ignition switch ON<br>**Fuel line pressure: 265—314 kPa (2.7—3.2 kg/cm², 38—46 psi)**<br>INSTALL CLAMPS<br>GND F/P | Yes | Go to next step |
| | | No | **Low pressure**<br>Check fuel line pressure while pinching fuel return hose<br>⇨ If fuel line pressure quickly increases, check pressure regulator<br>⇨ If fuel line pressure gradually increases, check fuel line for clogging between fuel pump and pressure regulator<br>If not clogged, check fuel pump maximum pressure |
| | | | **High pressure**<br>Check if fuel return hose is clogged or restricted<br>⇨ If OK, replace pressure regulator<br>⇨ If not OK, repair or replace hose |
| 8 | Try known good ECU and check if condition improves | Yes | Replace ECU |
| | | No | Change fuel to another brand |

## TROUBLESHOOTING CHART—1990 323

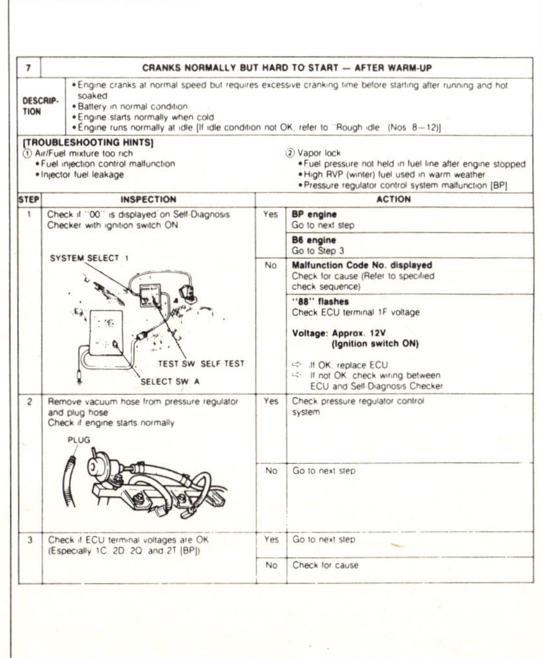

| 7 | CRANKS NORMALLY BUT HARD TO START — AFTER WARM-UP |
|---|---|
| DESCRIP-TION | • Engine cranks at normal speed but requires excessive cranking time before starting after running and hot soaked<br>• Battery in normal condition<br>• Engine starts normally when cold<br>• Engine runs normally at idle [If idle condition not OK, refer to "Rough idle" (Nos 8—12)] |

**[TROUBLESHOOTING HINTS]**

① Air/Fuel mixture too rich
- Fuel injection control malfunction
- Injector fuel leakage

② Vapor lock
- Fuel pressure not held in fuel line after engine stopped
- High RVP (winter) fuel used in warm weather
- Pressure regulator control system malfunction [BP]

| STEP | INSPECTION | | ACTION |
|---|---|---|---|
| 1 | Check if "00" is displayed on Self-Diagnosis Checker with ignition switch ON<br>SYSTEM SELECT: 1<br>TEST SW: SELF TEST<br>SELECT SW: A | Yes | **BP engine**<br>Go to next step |
| | | | **B6 engine**<br>Go to Step 3 |
| | | No | **Malfunction Code No. displayed**<br>Check for cause (Refer to specified check sequence) |
| | | | **"88" flashes**<br>Check ECU terminal 1F voltage<br>**Voltage: Approx. 12V (Ignition switch ON)**<br>⇨ If OK, replace ECU<br>⇨ If not OK, check wiring between ECU and Self-Diagnosis Checker |
| 2 | Remove vacuum hose from pressure regulator and plug hose<br>Check if engine starts normally<br>PLUG | Yes | Check pressure regulator control system |
| | | No | Go to next step |
| 3 | Check if ECU terminal voltages are OK (Especially 1C, 2D, 2Q and 2T [BP]) | Yes | Go to next step |
| | | No | Check for cause |

| STEP | INSPECTION | | ACTION |
|---|---|---|---|
| 4 | Run engine at idle and check if fuel line pressure is held after ignition switch turned OFF<br>**Fuel line pressure: More than 147 kPa (1.5 kg/cm², 21 psi) for 5 min.**<br>INSTALL CLAMPS | Yes | Go to next step |
| | | No | Plug outlet of pressure regulator and check if fuel line pressure is held after ignition switch turned OFF<br>⇨ If OK, replace pressure regulator<br>⇨ If not OK, check fuel pump hold pressure<br>If fuel pump is OK, check injectors for fuel leakage |
| 5 | Warm-up engine to normal operating temperature and stop it<br>Connect diagnosis connector terminals F/P and GND with jumper wire for 3 minutes with ignition switch ON, then check if engine starts easily<br>DIAGNOSIS CONNECTOR<br>GND F/P | Yes | Change fuel to another brand |
| | | No | Go to next step |
| 6 | Try known good ECU and check if condition improves | Yes | Replace ECU |
| | | No | Change fuel to another brand |

## TROUBLESHOOTING CHART—1990 323

**8 — ROUGH IDLE/ENGINE STALLS AT IDLE — ALWAYS**

**DESCRIPTION** • Engine starts normally but stalls or vibrates excessively at idle in every condition

**[TROUBLESHOOTING HINTS]**

① Air/Fuel mixture too lean
- Air leakage
- Fuel injection control malfunction
- Low fuel line pressure

② One or more injectors clogged or not operating
③ Low intake air amount
④ Incorrect idle speed
- Idle speed misadjustment
- Idle speed control malfunction

⑤ One or more spark plugs not firing
⑥ Low engine compression
⑦ Airflow meter stuck

| STEP | INSPECTION | | ACTION |
|---|---|---|---|
| 1 | Check if "00" is displayed on Self-Diagnosis Checker with ignition switch ON<br>SYSTEM SELECT: 1<br>TEST SW: SELF TEST<br>SELECT SW: A | Yes | Go to next step |
| | | No | **Malfunction Code No. displayed**<br>Check for cause (Refer to specified check sequence) |
| | | | **"88" flashes**<br>Check ECU terminal 1F voltage<br>**Voltage: Approx. 12V (Ignition switch ON)**<br>⇨ If OK, replace ECU<br>⇨ If not OK, check wiring between ECU and Self-Diagnosis Checker |
| 2 | Check if ECU terminal voltages are OK (Especially 2D, 2O and 2Q) | Yes | Go to next step |
| | | No | Check for cause |
| 3 | Disconnect high-tension lead at idle and check if engine speed decreases equally at each cylinder | Yes | Go to Step 8 |
| | | No | Go to next step |
| 4 | Check for injector operating sound at idle | Yes | Go to Step 6 |
| | | No | Go to Step 5 |
| 5 | Check if approx. 12V exists at injector connector (W/R) wire | Yes | Check if injector resistance is OK<br>**Resistance: Approx. 12—16Ω**<br>⇨ If OK, check wiring between ECU and injector<br>⇨ If not OK, replace injector |
| | | No | Check wiring between ECU and injector |
| 6 | Check if strong blue spark visible at disconnected high-tension lead | Yes | Go to next step |
| | | No | Check high-tension lead<br>⇨ If OK, check distributor cap and rotor<br>⇨ If not OK, replace high-tension lead |
| 7 | Check if spark plugs are OK<br>WEAR AND CARBON BUILDUP<br>PLUG GAP 1.0—1.1mm (0.039—0.043 in)<br>BURNS<br>DAMAGE AND DETERIORATION<br>DAMAGE | Yes | Check for correct engine compression<br>⇨ If OK, replace injector<br>⇨ If not OK, check for cause |
| | | No | Clean or replace |
| 8 | Check for correct fuel line pressure at idle<br>**Fuel line pressure: 265—314 kPa (2.7—3.2 kg/cm², 38—48 psi) (Vacuum hose to pressure regulator disconnected)**<br>INSTALL CLAMPS | Yes | Go to next step |
| | | No | **Low pressure**<br>Check fuel line pressure while pinching fuel return hose<br>⇨ If fuel line pressure quickly increases, check pressure regulator<br>⇨ If fuel line pressure gradually increases, check for clogging between fuel pump and pressure regulator<br>If not clogged, check fuel pump maximum pressure |
| 9 | Manually check if airflow meter moves smoothly from fully closed to fully open | Yes | Go to next step |
| | | No | Repair or replace |
| 10 | Check for air leakage at intake air system components | Yes | Repair or replace |
| | | No | Go to next step |
| 11 | Check for correct engine compression<br>**Engine compression (Minimum):**<br>**• BP SOHC 834 kPa (8.5 kg/cm², 121 psi)-300 rpm**<br>**• BP DOHC 883 kPa (9.0 kg/cm², 128 psi)-300 rpm**<br>**• B6 932 kPa (9.5 kg/cm², 135 psi)-300 rpm** | Yes | Go to next step |
| | | No | Check for cause |
| 12 | Connect System Selector to diagnosis connector and set Test Switch to "SELF TEST" and check for correct ignition timing at idle after warm-up<br>**Ignition timing (BTDC):**<br>**BP DOHC : 10 ± 1°**<br>**BP SOHC : 5 ± 1°**<br>**B6 : 7 ± 1°** | Yes | Try known good ECU and check if condition improves |
| | | No | Adjust ignition timing |

## TROUBLESHOOTING CHART—1990 323

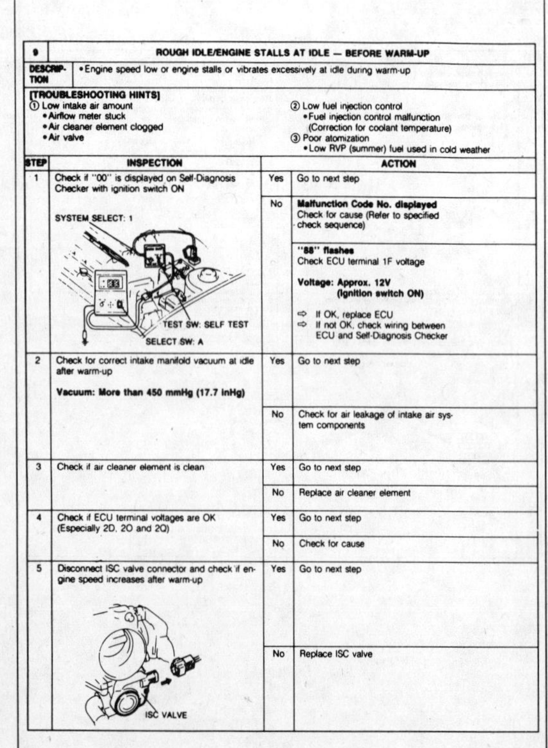

**9 — ROUGH IDLE/ENGINE STALLS AT IDLE — BEFORE WARM-UP**

**DESCRIPTION** • Engine speed low or engine stalls or vibrates excessively at idle during warm-up

**[TROUBLESHOOTING HINTS]**

① Low intake air amount
- Airflow meter stuck
- Air cleaner element clogged
- Air valve

② Low fuel injection control
- Fuel injection control malfunction (Correction for coolant temperature)

③ Poor atomization
- Low RVP (summer) fuel used in cold weather

| STEP | INSPECTION | | ACTION |
|---|---|---|---|
| 1 | Check if "00" is displayed on Self-Diagnosis Checker with ignition switch ON<br>SYSTEM SELECT: 1<br>TEST SW: SELF TEST<br>SELECT SW: A | Yes | Go to next step |
| | | No | **Malfunction Code No. displayed**<br>Check for cause (Refer to specified check sequence) |
| | | | **"88" flashes**<br>Check ECU terminal 1F voltage<br>**Voltage: Approx. 12V (Ignition switch ON)**<br>⇨ If OK, replace ECU<br>⇨ If not OK, check wiring between ECU and Self-Diagnosis Checker |
| 2 | Check for correct intake manifold vacuum at idle after warm-up<br>**Vacuum: More than 450 mmHg (17.7 inHg)** | Yes | Go to next step |
| | | No | Check for air leakage of intake air system components |
| 3 | Check if air cleaner element is clean | Yes | Go to next step |
| | | No | Replace air cleaner element |
| 4 | Check if ECU terminal voltages are OK (Especially 2D, 2O and 2Q) | Yes | Go to next step |
| | | No | Check for cause |
| 5 | Disconnect ISC valve connector and check if engine speed increases after warm-up<br>ISC VALVE | Yes | Go to next step |
| | | No | Replace ISC valve |

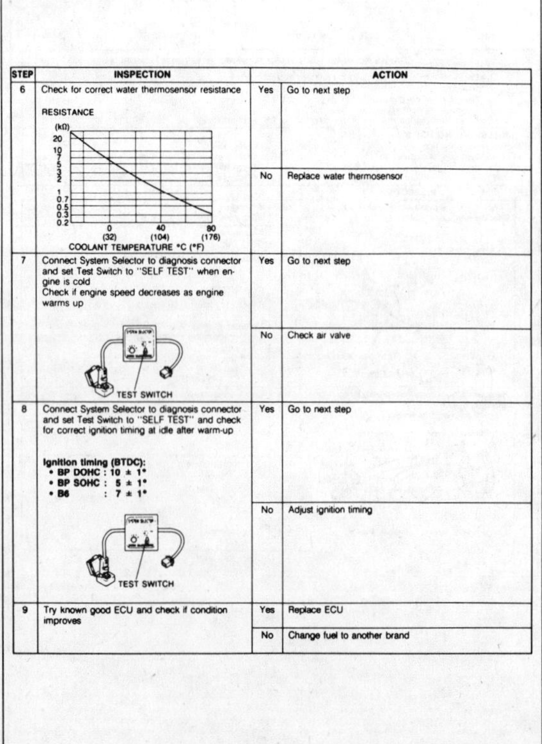

| STEP | INSPECTION | | ACTION |
|---|---|---|---|
| 6 | Check for correct water thermosensor resistance<br>RESISTANCE (kΩ): 20, 10, 7, 5, 3, 2, 1, 0.7, 0.5, 0.3, 0.2<br>0 (32), 40 (104), 80 (176)<br>COOLANT TEMPERATURE °C (°F) | Yes | Go to next step |
| | | No | Replace water thermosensor |
| 7 | Connect System Selector to diagnosis connector and set Test Switch to "SELF TEST" when engine is cold<br>Check if engine speed decreases as engine warms up<br>TEST SWITCH | Yes | Go to next step |
| | | No | Check air valve |
| 8 | Connect System Selector to diagnosis connector and set Test Switch to "SELF TEST" and check for correct ignition timing at idle after warm-up<br>**Ignition timing (BTDC):**<br>**• BP DOHC : 10 ± 1°**<br>**• BP SOHC : 5 ± 1°**<br>**• B6 : 7 ± 1°**<br>TEST SWITCH | Yes | Go to next step |
| | | No | Adjust ignition timing |
| 9 | Try known good ECU and check if condition improves | Yes | Replace ECU |
| | | No | Change fuel to another brand |

## TROUBLESHOOTING CHART—1990 323

| 10 | ROUGH IDLE/ENGINE STALLS AT IDLE — AFTER WARM-UP | | |
|---|---|---|---|
| DESCRIPTION | • Engine runs normally at idle during warm-up but engine stalls or vibrates excessively after warm-up | | |
| [TROUBLESHOOTING HINTS] ① Idle speed control system malfunction ② Air/Fuel mixture too lean • Air leakage • Low fuel line pressure ③ Air/Fuel mixture too rich • Fuel injection control malfunction (Correction for coolant temperature) ④ Poor ignition ⑤ Low engine compression | | | |
| STEP | INSPECTION | | ACTION |
| 1 | Check if "00" is displayed on Self-Diagnosis Checker with ignition switch ON<br>SYSTEM SELECT: 1 / TEST SW: SELF TEST / SELECT SW: A | Yes | Go to next step |
| | | No | **Malfunction Code No. displayed**<br>Check for cause (Refer to specified check sequence)<br>**"88" flashes**<br>Check ECU terminal 1F voltage<br>**Voltage: Approx. 12V (Ignition switch ON)**<br>⇨ If OK, replace ECU<br>⇨ If not OK, check wiring between ECU and Self-Diagnosis Checker |
| 2 | Check if ECU terminal voltages are OK (Especially 2D, 2O and 2Q) | Yes | Go to next step |
| | | No | Check for cause |
| 3 | Disconnect high-tension lead at idle and check if engine speed decreases equally at all cylinders | Yes | Go to next step |
| | | No | Go to Step 10 |
| 4 | Check for correct intake manifold vacuum at idle<br>**Vacuum: More than 450 mmHg (17.7 inHg)** | Yes | Go to next step |
| | | No | Check for air leakage of intake air system components |
| 5 | Check if air cleaner element is clean | Yes | Go to next step |
| | | No | Replace air cleaner element |
| 6 | Disconnect ISC valve connector at idle and check if engine speed increases<br>ISC VALVE | Yes | Go to next step |
| | | No | Replace ISC valve |

| STEP | INSPECTION | | ACTION |
|---|---|---|---|
| 7 | Check for correct fuel line pressure at idle<br>**Fuel line pressure: 265—314 kPa (2.7—3.2 kg/cm², 38—46 psi) (Vacuum hose to pressure regulator disconnected)**<br>INSTALL CLAMPS | Yes | Go to next step |
| | | No | **Low pressure**<br>Check fuel line pressure while pinching fuel return hose<br>⇨ If fuel line pressure quickly increases, check pressure regulator<br>⇨ If fuel line pressure gradually increases, check for clogging between fuel pump and pressure regulator<br>If not clogged, check fuel pump maximum pressure |
| 8 | Connect System Selector to diagnosis connector and set Test Switch to "SELF TEST" and check for correct ignition timing at idle after warm-up<br>**Ignition timing (BTDC):**<br>• **BP DOHC : 10 ± 1°**<br>• **BP SOHC : 5 ± 1°**<br>• **B6 : 7 ± 1°** | Yes | Go to next step |
| | | No | Adjust ignition system |
| 9 | Disconnect water thermosensor connector and check if engine condition improves | Yes | Replace water thermosensor |
| | | No | Try known good ECU and check if condition improves |
| 10 | Check for injector operating sound at idle | Yes | Go to next step |
| | | No | Check if injector resistance is OK<br>**Resistance: 12—16Ω**<br>⇨ If OK, check wiring between ECU and injector<br>⇨ If not OK, replace injector |
| 11 | Check for correct engine compression<br>**Engine compression (Minimum):**<br>• **BP SOHC : 834 kPa (8.5 kg/cm², 121 psi)-300 rpm**<br>• **BP DOHC : 883 kPa (9.0 kg/cm², 128 psi)-300 rpm**<br>• **B6 : 932 kPa (9.5 kg/cm², 135 psi)-300 rpm** | Yes | Go to next step |
| | | No | Check engine |
| 12 | Check if strong blue spark is visible at disconnected high-tension lead | Yes | Go to next step |
| | | No | Check high-tension lead<br>⇨ If OK, check distributor cap and rotor<br>⇨ If not OK, replace high-tension lead |
| 13 | Check if spark plugs are OK | Yes | Try known good ECU and check if condition imprives |
| | | No | Repair or replace |

## TROUBLESHOOTING CHART—1990 323

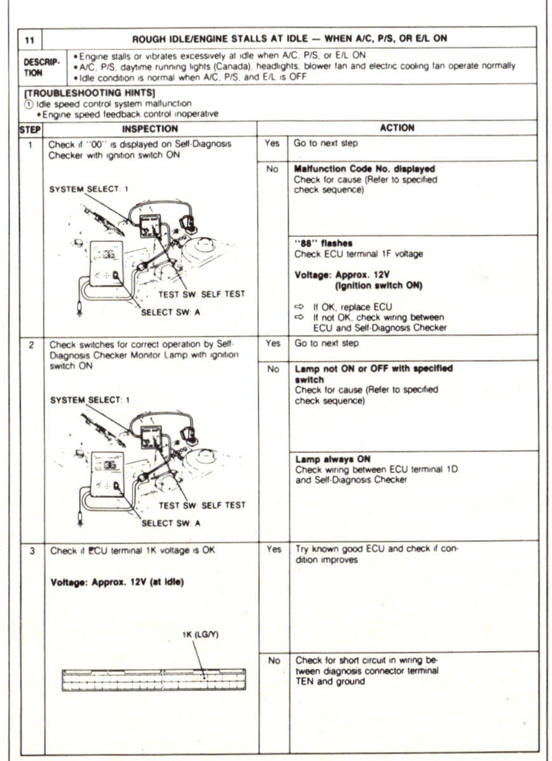

| 11 | ROUGH IDLE/ENGINE STALLS AT IDLE — WHEN A/C, P/S, OR E/L ON | | |
|---|---|---|---|
| DESCRIPTION | • Engine stalls or vibrates excessively at idle when A/C, P/S, or E/L ON<br>• A/C, P/S, daytime running lights (Canada), headlights, blower fan and electric cooling fan operate normally<br>• Idle condition is normal when A/C, P/S, and E/L is OFF | | |
| [TROUBLESHOOTING HINTS] ① Idle speed control system malfunction • Engine speed feedback control inoperative | | | |
| STEP | INSPECTION | | ACTION |
| 1 | Check if "00" is displayed on Self-Diagnosis Checker with ignition switch ON<br>SYSTEM SELECT: 1 / TEST SW: SELF TEST / SELECT SW: A | Yes | Go to next step |
| | | No | **Malfunction Code No. displayed**<br>Check for cause (Refer to specified check sequence)<br>**"88" flashes**<br>Check ECU terminal 1F voltage<br>**Voltage: Approx. 12V (Ignition switch ON)**<br>⇨ If OK, replace ECU<br>⇨ If not OK, check wiring between ECU and Self-Diagnosis Checker |
| 2 | Check switches for correct operation by Self-Diagnosis Checker Monitor Lamp with ignition switch ON<br>SYSTEM SELECT: 1 / TEST SW: SELF TEST / SELECT SW: A | Yes | Go to next step |
| | | No | **Lamp not ON or OFF with specified switch**<br>Check for cause (Refer to specified check sequence)<br>**Lamp always ON**<br>Check wiring between ECU terminal 1D and Self-Diagnosis Checker |
| 3 | Check if ECU terminal 1K voltage is OK<br>**Voltage: Approx. 12V (at idle)**<br>1K (LG/Y) | Yes | Try known good ECU and check if condition improves |
| | | No | Check for short circuit in wiring between diagnosis connector terminal TEN and ground |

| 12 | ROUGH IDLE/ENGINE STALLS JUST AFTER STARTING | | |
|---|---|---|---|
| DESCRIPTION | • Engine starts normally but vibrates excessively or stalls just after starting (acceleration from idle)<br>• Idle conditions are normal in the other conditions | | |
| [TROUBLESHOOTING HINTS] ① Fuel injection control system or idle speed control system malfunction • Start signal not input to ECU ② Idle speed misadjustment ③ Ignition timing misadjustment | | | |
| STEP | INSPECTION | | ACTION |
| 1 | Check if "00" is displayed on Self-Diagnosis Checker with ignition switch ON<br>SYSTEM SELECT: 1 / TEST SW: SELF TEST / SELECT SW: A | Yes | Go to next step |
| | | No | **Malfunction Code No. displayed**<br>Check for cause (Refer to specified check sequence)<br>**"88" flashes**<br>Check ECU terminal 1F voltage<br>**Voltage: Approx. 12V (Ignition switch ON)**<br>⇨ If OK, replace ECU<br>⇨ If not OK, check wiring between ECU and Self-Diagnosis Checker |
| 2 | Check if ECU terminal voltages are OK (Especially 1N) | Yes | Go to next step |
| | | No | Check for cause |
| 3 | Connect System Selector to diagnosis connector and set Test Switch to "SELF TEST" and check for correct ignition timing at idle after warm-up<br>**Ignition timing (BTDC):**<br>• **BP DOHC : 10 ± 1°**<br>• **BP SOHC : 5 ± 1°**<br>• **B6 : 7 ± 1°**<br>TEST SWITCH | Yes | In same condition as Step 3 inspection and apply parking brake<br>Check for correct idle speed<br>**Idle speed: 750 ± 50 rpm**<br>⇨ If OK, go to next step<br>⇨ If not OK, adjust idle speed |
| | | No | Adjust |
| 4 | Check continuity between throttle sensor terminals IDL and E<br>MTX / ATX / E / IDL | Yes | Go to next step |
| | | No | Adjust |
| 5 | Try known good ECU and check if condition improves | | |

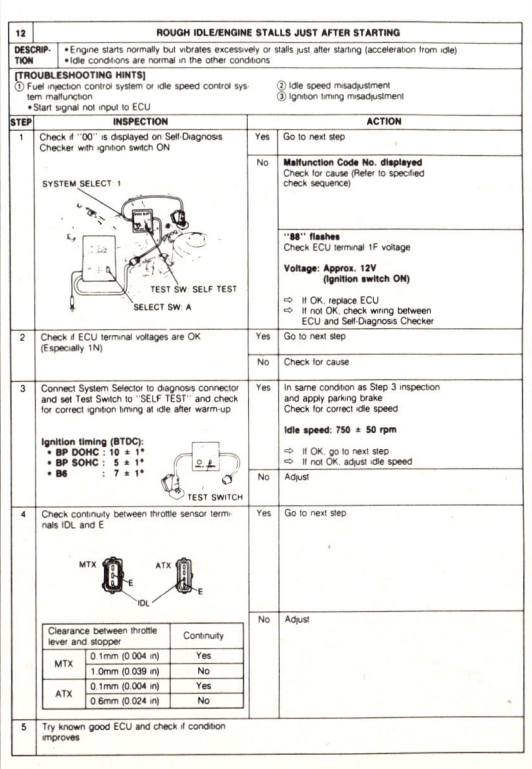

| Clearance between throttle lever and stopper | | Continuity |
|---|---|---|
| MTX | 0.1mm (0.004 in) | Yes |
| | 1.0mm (0.039 in) | No |
| ATX | 0.1mm (0.004 in) | Yes |
| | 0.6mm (0.024 in) | No |

## TROUBLESHOOTING CHART—1990 323

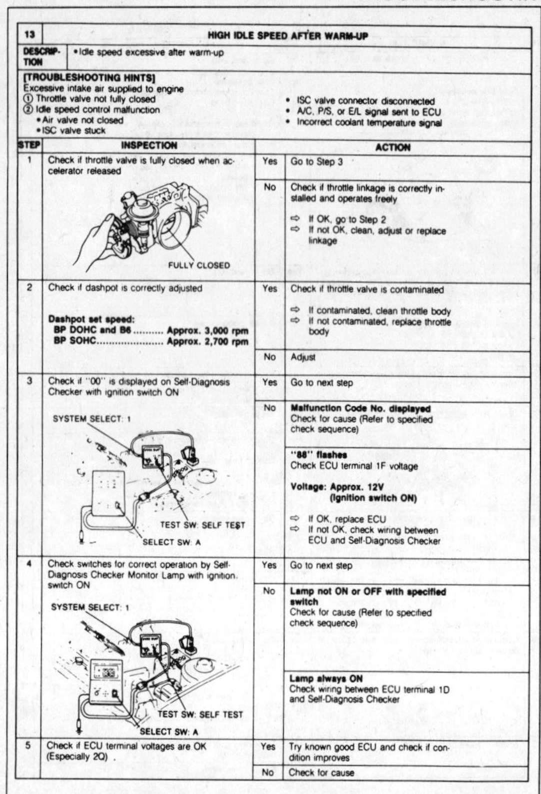

| 13 | HIGH IDLE SPEED AFTER WARM-UP | | |
|---|---|---|---|
| DESCRIPTION | • Idle speed excessive after warm-up | | |
| [TROUBLESHOOTING HINTS]<br>Excessive intake air supplied to engine<br>① Throttle valve not fully closed<br>② Idle speed control malfunction<br>• Air valve not closed<br>• ISC valve stuck<br>• ISC valve connector disconnected<br>• A/C, P/S, or E/L signal sent to ECU<br>• Incorrect coolant temperature signal | | | |
| STEP | INSPECTION | ACTION | |
| 1 | Check if throttle valve is fully closed when accelerator released<br>FULLY CLOSED | Yes | Go to Step 3 |
| | | No | Check if throttle linkage is correctly installed and operates freely<br>⇨ If OK, go to Step 2<br>⇨ If not OK, clean, adjust or replace linkage |
| 2 | Check if dashpot is correctly adjusted<br>**Dashpot set speed:**<br>**BP DOHC and B6 .......... Approx. 3,000 rpm**<br>**BP SOHC...................... Approx. 2,700 rpm** | Yes | Check if throttle valve is contaminated<br>⇨ If contaminated, clean throttle body<br>⇨ If not contaminated, replace throttle body |
| | | No | Adjust |
| 3 | Check if "00" is displayed on Self-Diagnosis Checker with ignition switch ON<br>SYSTEM SELECT: 1<br>TEST SW: SELF TEST<br>SELECT SW: A | Yes | Go to next step |
| | | No | **Malfunction Code No. displayed**<br>Check for cause (Refer to specified check sequence) |
| | | | **"88" flashes**<br>Check ECU terminal 1F voltage<br>**Voltage: Approx. 12V (Ignition switch ON)**<br>⇨ If OK, replace ECU<br>⇨ If not OK, check wiring between ECU and Self-Diagnosis Checker |
| 4 | Check switches for correct operation by Self-Diagnosis Checker Monitor Lamp with ignition switch ON<br>SYSTEM SELECT: 1<br>TEST SW: SELF TEST<br>SELECT SW: A | Yes | Go to next step |
| | | No | **Lamp not ON or OFF with specified switch**<br>Check for cause (Refer to specified check sequence) |
| | | | **Lamp always ON**<br>Check wiring between ECU terminal 1D and Self-Diagnosis Checker |
| 5 | Check if ECU terminal voltages are OK (Especially 2Q) | Yes | Try known good ECU and check if condition improves |
| | | No | Check for cause |

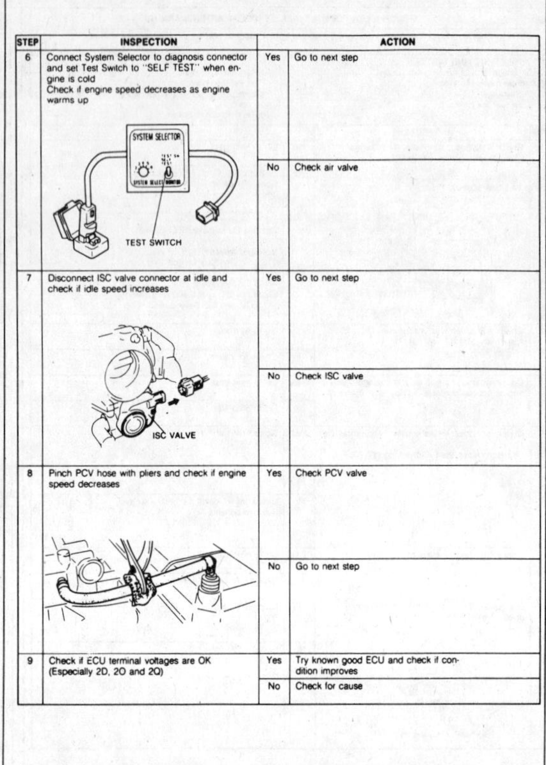

| STEP | INSPECTION | ACTION | |
|---|---|---|---|
| 6 | Connect System Selector to diagnosis connector and set Test Switch to "SELF TEST" when engine is cold<br>Check if engine speed decreases as engine warms up<br>SYSTEM SELECTOR<br>TEST SWITCH | Yes | Go to next step |
| | | No | Check air valve |
| 7 | Disconnect ISC valve connector at idle and check if idle speed increases<br>ISC VALVE | Yes | Go to next step |
| | | No | Check ISC valve |
| 8 | Pinch PCV hose with pliers and check if engine speed decreases | Yes | Check PCV valve |
| | | No | Go to next step |
| 9 | Check if ECU terminal voltages are OK (Especially 2D, 2O and 2Q) | Yes | Try known good ECU and check if condition improves |
| | | No | Check for cause |

## TROUBLESHOOTING CHART—1990 323

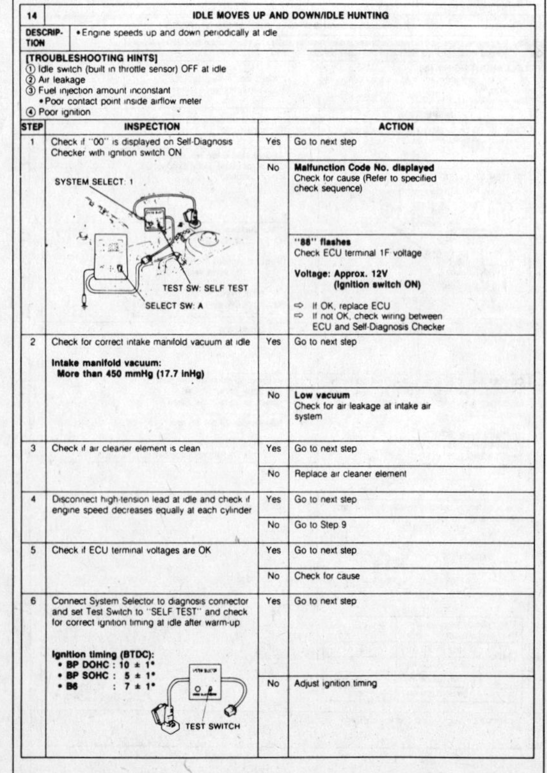

| 14 | IDLE MOVES UP AND DOWN/IDLE HUNTING | | |
|---|---|---|---|
| DESCRIPTION | • Engine speeds up and down periodically at idle | | |
| [TROUBLESHOOTING HINTS]<br>① Idle switch (built in throttle sensor) OFF at idle<br>② Air leakage<br>③ Fuel injection amount inconstant<br>• Poor contact point inside airflow meter<br>④ Poor ignition | | | |
| STEP | INSPECTION | ACTION | |
| 1 | Check if "00" is displayed on Self-Diagnosis Checker with ignition switch ON<br>SYSTEM SELECT: 1<br>TEST SW: SELF TEST<br>SELECT SW: A | Yes | Go to next step |
| | | No | **Malfunction Code No. displayed**<br>Check for cause (Refer to specified check sequence) |
| | | | **"88" flashes**<br>Check ECU terminal 1F voltage<br>**Voltage: Approx. 12V (Ignition switch ON)**<br>⇨ If OK, replace ECU<br>⇨ If not OK, check wiring between ECU and Self-Diagnosis Checker |
| 2 | Check for correct intake manifold vacuum at idle<br>**Intake manifold vacuum:**<br>**More than 450 mmHg (17.7 inHg)** | Yes | Go to next step |
| | | No | **Low vacuum**<br>Check for air leakage at intake air system |
| 3 | Check if air cleaner element is clean | Yes | Go to next step |
| | | No | Replace air cleaner element |
| 4 | Disconnect high-tension lead at idle and check if engine speed decreases equally at each cylinder | Yes | Go to next step |
| | | No | Go to Step 9 |
| 5 | Check if ECU terminal voltages are OK | Yes | Go to next step |
| | | No | Check for cause |
| 6 | Connect System Selector to diagnosis connector and set Test Switch to "SELF TEST" and check for correct ignition timing at idle after warm-up<br>**Ignition timing (BTDC):**<br>**• BP DOHC : 10 ± 1°**<br>**• BP SOHC : 5 ± 1°**<br>**• B6 : 7 ± 1°**<br>TEST SWITCH | Yes | Go to next step |
| | | No | Adjust ignition timing |

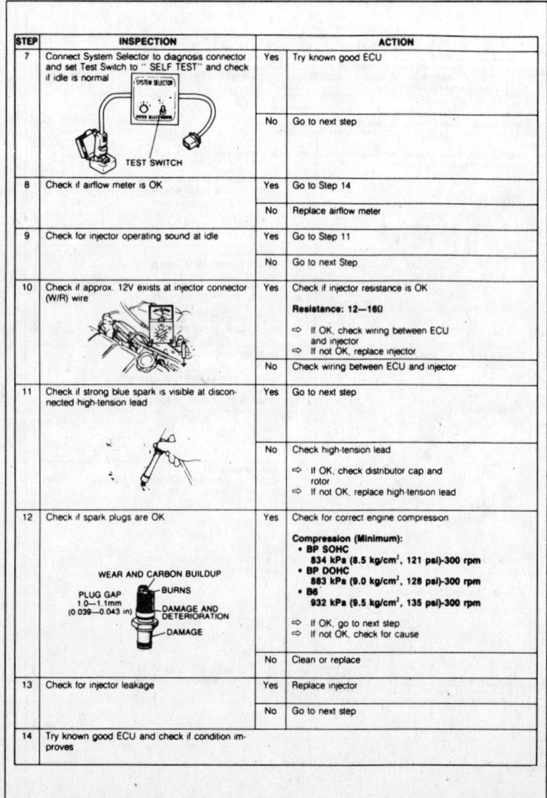

| STEP | INSPECTION | ACTION | |
|---|---|---|---|
| 7 | Connect System Selector to diagnosis connector and set Test Switch to "SELF TEST" and check if idle is normal<br>SYSTEM SELECTOR<br>TEST SWITCH | Yes | Try known good ECU |
| | | No | Go to next step |
| 8 | Check if airflow meter is OK | Yes | Go to Step 14 |
| | | No | Replace airflow meter |
| 9 | Check for injector operating sound at idle | Yes | Go to Step 11 |
| | | No | Go to next Step |
| 10 | Check if approx. 12V exists at injector connector (W/R) wire | Yes | Check if injector resistance is OK<br>**Resistance: 12—16Ω**<br>⇨ If OK, check wiring between ECU and injector<br>⇨ If not OK, replace injector |
| | | No | Check wiring between ECU and injector |
| 11 | Check if strong blue spark is visible at disconnected high-tension lead | Yes | Go to next step |
| | | No | Check high-tension lead<br>⇨ If OK, check distributor cap and rotor<br>⇨ If not OK, replace high-tension lead |
| 12 | Check if spark plugs are OK<br>WEAR AND CARBON BUILDUP<br>BURNS<br>PLUG GAP 1.0—1.1mm (0.039—0.043 in)<br>DAMAGE AND DETERIORATION<br>DAMAGE | Yes | Check for correct engine compression<br>**Compression (Minimum):**<br>**• BP SOHC 834 kPa (8.5 kg/cm², 121 psi)-300 rpm**<br>**• BP DOHC 883 kPa (9.0 kg/cm², 128 psi)-300 rpm**<br>**• B6 932 kPa (9.5 kg/cm², 135 psi)-300 rpm**<br>⇨ If OK, go to next step<br>⇨ If not OK, check for cause |
| | | No | Clean or replace |
| 13 | Check for injector leakage | Yes | Replace injector |
| | | No | Go to next step |
| 14 | Try known good ECU and check if condition improves | | |

## TROUBLESHOOTING CHART—1990 323

| 15 | ENGINE STALLS ON DECELERATION | | |
|---|---|---|---|
| DESCRIPTION | • Engine unexpectedly stops running while decelerating or after deceleration<br>• Idle condition normal | | |
| [TROUBLESHOOTING HINTS]<br>Engine speed drops too much when releasing accelerator or poor connector connection disconnected by deceleration force<br>① Idle speed control malfunction<br>② Fuel cut control malfunction<br>③ Idle speed misadjustment | | | |
| **STEP** | **INSPECTION** | **ACTION** | |
| 1 | Check if idle condition is normal | Yes | Go to next step |
| | | No | Adjust or perform troubleshooting Nos. 8—11 "ROUGH IDLE" |
| 2 | Check if "00" is displayed on Self-Diagnosis Checker with ignition switch ON<br>SYSTEM SELECT: 1<br>TEST SW: SELF TEST<br>SELECT SW: A | Yes | Go to next step |
| | | No | **Malfunction Code No. displayed**<br>Check for cause (Refer to specified check sequence) |
| | | | **"88" flashes**<br>Check ECU terminal 1F voltage<br>**Voltage: Approx. 12V (Ignition switch ON)**<br>⇨ If OK, replace ECU<br>⇨ If not OK, check wiring between ECU and Self-Diagnosis Checker |
| 3 | Check switches for correct operation with Self-Diagnosis Checker Monitor Lamp with ignition switch ON<br>SYSTEM SELECT: 1<br>TEST SW: SELF TEST<br>SELECT SW: A | Yes | Go to next step |
| | | No | **Lamp not ON or OFF with specified switch**<br>Check for cause (Refer to specified check sequence) |
| | | | **Lamp always ON**<br>Check wiring between ECU terminal 1D and Self-Diagnosis Checker |
| 4 | Check if ECU terminal voltages are OK (Especially 1V, 2D, 2O, 2U, 2V and 2Q) | Yes | Go to next step |
| | | No | Check for cause |
| 5 | Check for poor connection of following parts<br>• Ignition coil<br>• Igniter<br>• Distributor<br>• High-tension lead<br>• Injector<br>• Circuit opening relay<br>• ECU | Yes | Repair or replace |
| | | No | Go to next step |
| 6 | Check if dashpot is correctly adjusted<br>**Dashpot set speed:**<br>**BP DOHC and B6...Approx. 3,000 rpm**<br>**BP SOHC..............Approx. 2,700 rpm** | Yes | Go to next step |
| | | No | Adjust dashpot |
| 7 | Disconnect ISC valve connector at idle and check if engine speed increases<br>ISC VALVE | Yes | Go to next step |
| | | No | Replace ISC valve |
| 8 | Try known good ECU and check if condition improves | | |

## TROUBLESHOOTING CHART—1990 323

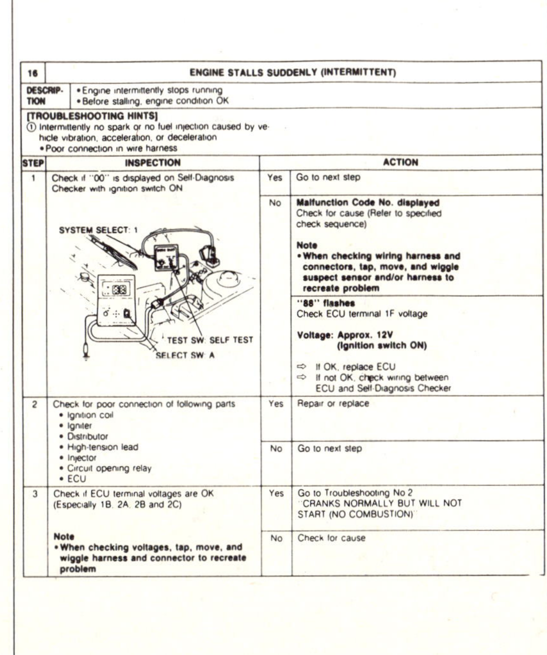

| 16 | ENGINE STALLS SUDDENLY (INTERMITTENT) | | |
|---|---|---|---|
| DESCRIPTION | • Engine intermittently stops running<br>• Before stalling, engine condition OK | | |
| [TROUBLESHOOTING HINTS]<br>① Intermittently no spark or no fuel injection caused by vehicle vibration, acceleration, or deceleration<br>• Poor connection in wire harness | | | |
| **STEP** | **INSPECTION** | **ACTION** | |
| 1 | Check if "00" is displayed on Self-Diagnosis Checker with ignition switch ON<br>SYSTEM SELECT: 1<br>TEST SW: SELF TEST<br>SELECT SW: A | Yes | Go to next step |
| | | No | **Malfunction Code No. displayed**<br>Check for cause (Refer to specified check sequence)<br>**Note**<br>**• When checking wiring harness and connectors, tap, move, and wiggle suspect sensor and/or harness to recreate problem** |
| | | | **"88" flashes**<br>Check ECU terminal 1F voltage<br>**Voltage: Approx. 12V (Ignition switch ON)**<br>⇨ If OK, replace ECU<br>⇨ If not OK, check wiring between ECU and Self-Diagnosis Checker |
| 2 | Check for poor connection of following parts<br>• Ignition coil<br>• Igniter<br>• Distributor<br>• High-tension lead<br>• Injector<br>• Circuit opening relay<br>• ECU | Yes | Repair or replace |
| | | No | Go to next step |
| 3 | Check if ECU terminal voltages are OK (Especially 1B, 2A, 2B and 2C)<br>**Note**<br>**• When checking voltages, tap, move, and wiggle harness and connector to recreate problem** | Yes | Go to Troubleshooting No 2 "CRANKS NORMALLY BUT WILL NOT START (NO COMBUSTION)" |
| | | No | Check for cause |

| 17 | HESITATES/STUMBLES ON ACCELERATION | | |
|---|---|---|---|
| DESCRIPTION | • Flat spot occurs just after accelerator depressed or mild jerking occurs during acceleration | | |
| [TROUBLESHOOTING HINTS]<br>① Air/Fuel mixture becomes lean when depressing accelerator<br>• Fuel injection control malfunction (Correction for accelerating condition)<br>• Air leakage<br>• Fuel line pressure decreases<br>• Spark advance control malfunction | | | |
| **STEP** | **INSPECTION** | **ACTION** | |
| 1 | Check if "00" is displayed on Self-Diagnosis Checker with ignition switch ON<br>SYSTEM SELECT: 1<br>TEST SW: SELF TEST<br>SELECT SW: A | Yes | Go to next step |
| | | No | **Malfunction Code No. displayed**<br>Check for cause (Refer to specified check sequence) |
| | | | **"88" flashes**<br>Check ECU terminal 1F voltage<br>**Voltage: Approx. 12V (Ignition switch ON)**<br>⇨ If OK, replace ECU<br>⇨ If not OK, check wiring between ECU and Self-Diagnosis Checker |
| 2 | Check for correct intake manifold vacuum at idle<br>**Vacuum: More than 450 mmHg (17.7 inHg)** | Yes | Go to next step |
| | | No | Check for air leakage at intake air system components |
| 3 | Check if air cleaner element is clean | Yes | Go to next step |
| | | No | Replace air cleaner element |
| 4 | Check if ECU terminal voltages are OK [Especially 1N, 2K, 2M (ATX), 2L (MTX)] | Yes | Go to next step |
| | | No | Check for cause |
| 5 | Check if throttle linkage is correctly installed and operates freely | Yes | Go to next step |
| | | No | Correct, clean, or replace as required any binding or damaged linkage and adjust cable deflection at throttle body |
| 6 | Check continuity between throttle sensor terminals IDL and E<br>MTX ATX E IDL | Yes | Go to next step |
| | | No | Adjust |

| Clearance between throttle lever and stopper | | Continuity |
|---|---|---|
| MTX | 0.1mm (0.004 in) | Yes |
| | 1.0mm (0.039 in) | No |
| ATX | 0.1mm (0.004 in) | Yes |
| | 0.6mm (0.024 in) | No |

## TROUBLESHOOTING CHART—1990 323

| STEP | INSPECTION | | ACTION |
|---|---|---|---|
| 7 | Check for correct fuel line pressure at idle<br><br>**Fuel line pressure:**<br>**265—314 kPa (2.7—3.2 kg/cm², 38—46 psi)**<br>**(Vacuum hose to pressure regulator disconnected)**<br>INSTALL CLAMPS | Yes | Check if fuel line pressure decreases when accelerating quickly<br>⇨ If decreases, check fuel line and filter for clogging<br>⇨ If no decrease, go to next step |
| | | No | **Low pressure**<br>Check fuel line pressure while pinching fuel return hose<br>⇨ If fuel line pressure quickly increases, check pressure regulator<br>⇨ If fuel line pressure gradually increases, check for clogging between fuel pump and pressure regulator<br>If not clogged, check fuel pump maximum pressure |
| 8 | Connect System Selector to diagnosis connector and set Test Switch to "SELF TEST" and check for correct ignition timing at idle after warm-up<br><br>**Ignition timing (BTDC):**<br>**• BP DOHC : 10 ± 1°**<br>**• BP SOHC : 5 ± 1°**<br>**• B6 : 7 ± 1°**<br>TEST SWITCH | Yes | Check if ignition timing advances when accelerating<br>⇨ If advances, go to next step<br>⇨ If not advance, replace ECU |
| | | No | Adjust |
| 9 | Check if air duct and air hoses are correctly installed | Yes | Go to next step |
| | | No | Repair |
| 10 | Check if exhaust system is restricted | Yes | Repair or replace |
| | | No | **SOHC**<br>Go to Step 12<br>**DOHC**<br>Go to next step |
| 11 | Check if variable inertia control system (VICS) is OK | Yes | Go to next step |
| | | No | Repair or replace |
| 12 | Try known good ECU and check if condition improves | | |

| 18 | SURGES WHILE CRUISING | | |
|---|---|---|---|
| DESCRIPTION | • Unexpected change in engine speed which is usually repetitive | | |
| [TROUBLESHOOTING HINTS]<br>① Air/Fuel mixture too lean<br>② Misfire<br>③ Poor connection in wiring harness | | | |
| **STEP** | **INSPECTION** | | **ACTION** |
| 1 | Check if "00" is displayed on Self-Diagnosis Checker with ignition switch ON<br>SYSTEM SELECT: 1<br>TEST SW: SELF TEST<br>SELECT SW: A | Yes | Go to next step |
| | | No | **Malfunction Code No. displayed**<br>Check for cause (Refer to specified check sequence)<br><br>**"88" flashes**<br>Check ECU terminal 1F voltage<br>**Voltage: Approx. 12V**<br>**(Ignition switch ON)**<br>⇨ If OK, replace ECU<br>⇨ If not OK, check wiring between ECU and Self-Diagnosis Checker |
| 2 | Check switches for correct operation by Self-Diagnosis Checker Monitor Lamp with ignition switch ON<br>SYSTEM SELECT: 1<br>TEST SW: SELF TEST<br>SELECT SW: A | Yes | Go to next step |
| | | No | **Lamp not ON or OFF with specified input switch**<br>Check for cause (Refer to specified check sequence)<br><br>**Lamp always ON**<br>Check ECU terminal 1D voltage |
| 3 | Check if throttle sensor is OK | Yes | Go to next step |
| | | No | Adjust |
| 4 | Disconnect oxygen sensor connector and check if condition improves<br>OXYGEN SENSOR CONNECTOR | Yes | Check oxygen sensor |
| | | No | Go to next step |
| 5 | Check if ECU terminal voltages are OK | Yes | Go to next step |
| | | No | Check for cause |

## TROUBLESHOOTING CHART—1990 323

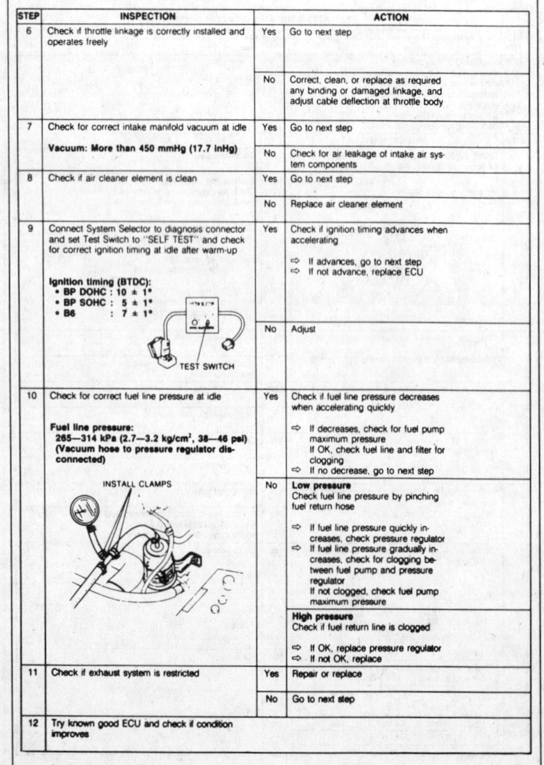

| STEP | INSPECTION | | ACTION |
|---|---|---|---|
| 6 | Check if throttle linkage is correctly installed and operates freely | Yes | Go to next step |
| | | No | Correct, clean, or replace as required any binding or damaged linkage, and adjust cable deflection at throttle body |
| 7 | Check for correct intake manifold vacuum at idle<br><br>**Vacuum: More than 450 mmHg (17.7 inHg)** | Yes | Go to next step |
| | | No | Check for air leakage of intake air system components |
| 8 | Check if air cleaner element is clean | Yes | Go to next step |
| | | No | Replace air cleaner element |
| 9 | Connect System Selector to diagnosis connector and set Test Switch to "SELF TEST" and check for correct ignition timing at idle after warm-up<br><br>**Ignition timing (BTDC):**<br>**• BP DOHC : 10 ± 1°**<br>**• BP SOHC : 5 ± 1°**<br>**• B6 : 7 ± 1°**<br>TEST SWITCH | Yes | Check if ignition timing advances when accelerating<br>⇨ If advances, go to next step<br>⇨ If not advance, replace ECU |
| | | No | Adjust |
| 10 | Check for correct fuel line pressure at idle<br><br>**Fuel line pressure:**<br>**265—314 kPa (2.7—3.2 kg/cm², 38—46 psi)**<br>**(Vacuum hose to pressure regulator disconnected)**<br>INSTALL CLAMPS | Yes | Check if fuel line pressure decreases when accelerating quickly<br>⇨ If decreases, check for fuel pump maximum pressure<br>If OK, check fuel line and filter for clogging<br>⇨ If no decrease, go to next step |
| | | No | **Low pressure**<br>Check fuel line pressure by pinching fuel return hose<br>⇨ If fuel line pressure quickly increases, check pressure regulator<br>⇨ If fuel line pressure gradually increases, check for clogging between fuel pump and pressure regulator<br>If not clogged, check fuel pump maximum pressure<br><br>**High pressure**<br>Check if fuel return line is clogged<br>⇨ If OK, replace pressure regulator<br>⇨ If not OK, replace |
| 11 | Check if exhaust system is restricted | Yes | Repair or replace |
| | | No | Go to next step |
| 12 | Try known good ECU and check if condition improves | | |

| 19 | LACK OF POWER | | |
|---|---|---|---|
| DESCRIPTION | • Performance poor under load when throttle valve wide open<br>• Reduced maximum speed<br>• Idle condition normal | | |
| [TROUBLESHOOTING HINTS]<br>① Factors other than engine malfunction<br>• Clutch slipping<br>• ATX slipping<br>• Brake dragging<br>• Low tire pressure<br>• Incorrect tire size<br>• Overloaded<br>② Low intake air amount<br>• Throttle valve not open fully<br>• Clogged intake air system<br>③ Air/Fuel mixture too lean<br>• Fuel line pressure decreases<br>• Fuel injection malfunction<br>④ Poor ignition<br>⑤ Low engine compression<br>⑥ Alcohol blended fuel used | | | |
| **STEP** | **INSPECTION** | | **ACTION** |
| 1 | Check factors other than engine<br>• Clutch slipping<br>• ATX slipping<br>• Brake dragging<br>• Low tire pressure<br>• Incorrect tire size | Yes | Go to next step |
| | | No | Repair |
| 2 | Check if throttle valve fully opened when accelerator depressed fully | Yes | Go to next step |
| | | No | Check if accelerator cable is correctly installed<br>⇨ If OK, check throttle body<br>⇨ If not OK, install accelerator cable correctly |
| 3 | Check if "00" is displayed on Self-Diagnosis Checker with ignition switch ON<br>SYSTEM SELECT: 1<br>TEST SW: SELF TEST<br>SELECT SW: A | Yes | Go to next step |
| | | No | **Malfunction Code No. displayed**<br>Check for cause (Refer to specified check sequence)<br><br>**"88" flashes**<br>Check ECU terminal 1F voltage<br>**Voltage: Approx. 12V**<br>**(Ignition switch ON)**<br>⇨ If OK, replace ECU<br>⇨ If no OK, check wiring between ECU and Self-Diagnosis Checker |
| 4 | Connect System Selector to diagnosis connector and set Test Switch to "SELF TEST" and check for correct ignition timing at idle after warm-up<br><br>**Ignition timing (BTDC):**<br>**• BP DOHC : 10 ± 1°**<br>**• BP SOHC : 5 ± 1°**<br>**• B6 : 7 ± 1°**<br>TEST SWITCH | Yes | Check if ignition timing advances when accelerating<br>⇨ If advances, go to next step<br>⇨ If not advance, check ECU terminal voltages |
| | | No | Adjust |
| 5 | Check if ECU terminal voltages are OK [Especially 1K, 1N, 2L (MTX), 2M (ATX), 2K, and 2S] | Yes | Go to next step |
| | | No | Check for cause |

## TROUBLESHOOTING CHART—1990 323

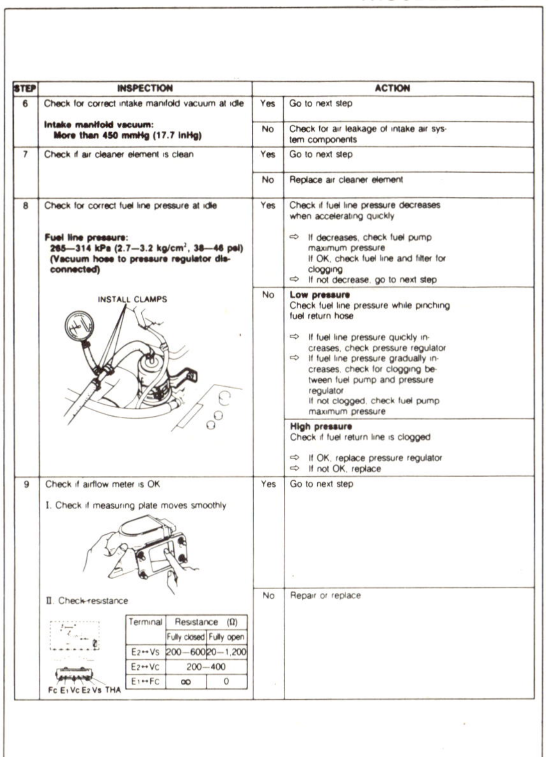

| STEP | INSPECTION | | ACTION |
|---|---|---|---|
| 6 | Check for correct intake manifold vacuum at idle<br>**Intake manifold vacuum:**<br>**More than 450 mmHg (17.7 inHg)** | Yes | Go to next step |
| | | No | Check for air leakage of intake air system components |
| 7 | Check if air cleaner element is clean | Yes | Go to next step |
| | | No | Replace air cleaner element |
| 8 | Check for correct fuel line pressure at idle<br>**Fuel line pressure:**<br>**265—314 kPa (2.7—3.2 kg/cm², 38—46 psi) (Vacuum hose to pressure regulator disconnected)**<br>INSTALL CLAMPS | Yes | Check if fuel line pressure decreases when accelerating quickly<br>⇨ If decreases, check fuel pump maximum pressure<br>If OK, check fuel line and filter for clogging<br>⇨ If not decrease, go to next step |
| | | No | **Low pressure**<br>Check fuel line pressure while pinching fuel return hose<br>⇨ If fuel line pressure quickly increases, check pressure regulator<br>⇨ If fuel line pressure gradually increases, check for clogging between fuel pump and pressure regulator<br>If not clogged, check fuel pump maximum pressure<br>**High pressure**<br>Check if fuel return line is clogged<br>⇨ If OK, replace pressure regulator<br>⇨ If not OK, replace |
| 9 | Check if airflow meter is OK<br>I. Check if measuring plate moves smoothly<br>II. Check resistance | Yes | Go to next step |
| | | No | Repair or replace |

| Terminal | Resistance (Ω) Fully closed | Resistance (Ω) Fully open |
|---|---|---|
| E2↔Vs | 200—600 | 20—1,200 |
| E2↔Vc | 200—400 | |
| E1↔Fc | ∞ | 0 |

Fc E1 Vc E2 Vs THA

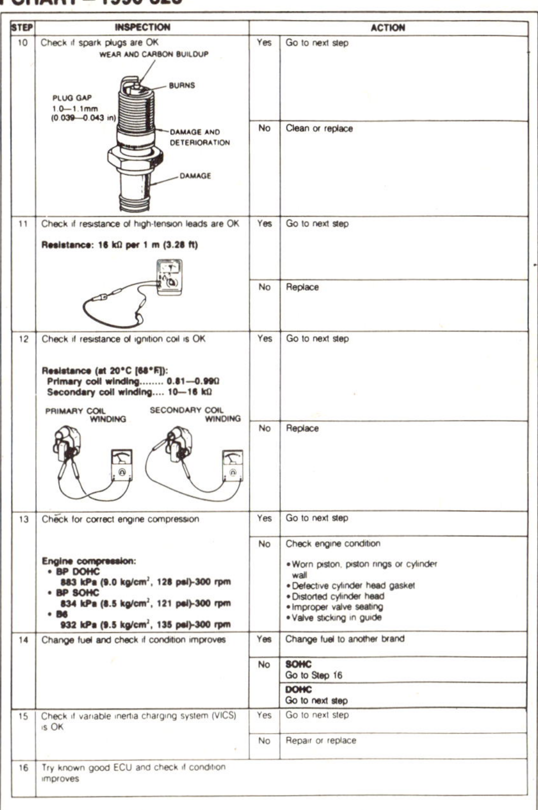

| STEP | INSPECTION | | ACTION |
|---|---|---|---|
| 10 | Check if spark plugs are OK<br>WEAR AND CARBON BUILDUP<br>BURNS<br>PLUG GAP 1.0—1.1mm (0.039—0.043 in)<br>DAMAGE AND DETERIORATION<br>DAMAGE | Yes | Go to next step |
| | | No | Clean or replace |
| 11 | Check if resistance of high-tension leads are OK<br>**Resistance: 16 kΩ per 1 m (3.28 ft)** | Yes | Go to next step |
| | | No | Replace |
| 12 | Check if resistance of ignition coil is OK<br>**Resistance (at 20°C [68°F]):**<br>**Primary coil winding........ 0.81—0.99Ω**<br>**Secondary coil winding.... 10—16 kΩ**<br>PRIMARY COIL WINDING<br>SECONDARY COIL WINDING | Yes | Go to next step |
| | | No | Replace |
| 13 | Check for correct engine compression<br>**Engine compression:**<br>**• BP DOHC**<br>**883 kPa (9.0 kg/cm², 128 psi)-300 rpm**<br>**• BP SOHC**<br>**834 kPa (8.5 kg/cm², 121 psi)-300 rpm**<br>**• B6**<br>**932 kPa (9.5 kg/cm², 135 psi)-300 rpm** | Yes | Go to next step |
| | | No | Check engine condition<br>• Worn piston, piston rings or cylinder wall<br>• Defective cylinder head gasket<br>• Distorted cylinder head<br>• Improper valve seating<br>• Valve sticking in guide |
| 14 | Change fuel and check if condition improves | Yes | Change fuel to another brand |
| | | No | **SOHC**<br>Go to Step 16<br>**DOHC**<br>Go to next step |
| 15 | Check if variable inertia charging system (VICS) is OK | Yes | Go to next step |
| | | No | Repair or replace |
| 16 | Try known good ECU and check if condition improves | | |

## TROUBLESHOOTING CHART—1990 323

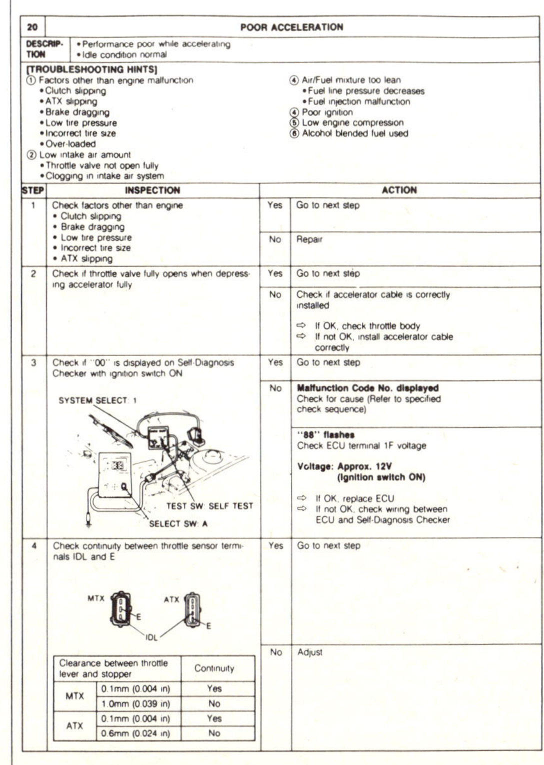

| 20 | POOR ACCELERATION |
|---|---|
| DESCRIPTION | • Performance poor while accelerating<br>• Idle condition normal |

**[TROUBLESHOOTING HINTS]**

① Factors other than engine malfunction
- • Clutch slipping
- • ATX slipping
- • Brake dragging
- • Low tire pressure
- • Incorrect tire size
- • Over-loaded

② Low intake air amount
- • Throttle valve not open fully
- • Clogging in intake air system

④ Air/Fuel mixture too lean
- • Fuel line pressure decreases
- • Fuel injection malfunction

④ Poor ignition

⑤ Low engine compression

⑥ Alcohol blended fuel used

| STEP | INSPECTION | | ACTION |
|---|---|---|---|
| 1 | Check factors other than engine<br>• Clutch slipping<br>• Brake dragging<br>• Low tire pressure<br>• Incorrect tire size<br>• ATX slipping | Yes | Go to next step |
| | | No | Repair |
| 2 | Check if throttle valve fully opens when depressing accelerator fully | Yes | Go to next step |
| | | No | Check if accelerator cable is correctly installed<br>⇨ If OK, check throttle body<br>⇨ If not OK, install accelerator cable correctly |
| 3 | Check if "00" is displayed on Self-Diagnosis Checker with ignition switch ON<br>SYSTEM SELECT: 1<br>TEST SW: SELF TEST<br>SELECT SW: A | Yes | Go to next step |
| | | No | **Malfunction Code No. displayed**<br>Check for cause (Refer to specified check sequence)<br>**"88" flashes**<br>Check ECU terminal 1F voltage<br>**Voltage: Approx. 12V**<br>**(Ignition switch ON)**<br>⇨ If OK, replace ECU<br>⇨ If not OK, check wiring between ECU and Self-Diagnosis Checker |
| 4 | Check continuity between throttle sensor terminals IDL and E<br>MTX ATX IDL E | Yes | Go to next step |
| | | No | Adjust |

| | Clearance between throttle lever and stopper | Continuity |
|---|---|---|
| MTX | 0.1mm (0.004 in) | Yes |
| | 1.0mm (0.039 in) | No |
| ATX | 0.1mm (0.004 in) | Yes |
| | 0.6mm (0.024 in) | No |

| STEP | INSPECTION | | ACTION |
|---|---|---|---|
| 5 | Connect System Selector to diagnosis connector and set Test Switch to "SELF TEST" and check for correct ignition timing at idle after warm-up<br>**Ignition timing (BTDC):**<br>**• BP DOHC : 10 ± 1°**<br>**• BP SOHC : 5 ± 1°**<br>**• B6 : 7 ± 1°**<br>TEST SWITCH | Yes | Check if ignition timing advances when accelerating<br>⇨ If advances, go to next step<br>⇨ If not advance, check ECU terminal voltages |
| | | No | Adjust |
| 6 | Check if ECU terminal voltages are OK | Yes | Go to next step |
| | | No | Check for cause |
| 7 | Check for correct intake manifold vacuum at idle<br>**Intake manifold vacuum:**<br>**More than 450 mmHg (17.7 inHg)** | Yes | Go to next step |
| | | No | Check for air leakage of intake air system components |
| 8 | Check for correct fuel line pressure at idle<br>**Fuel line pressure:**<br>**265—314 kPa (2.7—3.2 kg/cm², 38—46 psi) (Vacuum hose to pressure regulator disconnected)**<br>INSTALL CLAMPS | Yes | Check if fuel line pressure decreases when accelerating quickly<br>⇨ If decrease, check fuel pump maximum pressure<br>If OK, check fuel line and filter for clogging<br>⇨ If not decrease, go to next step |
| | | No | **Low pressure**<br>Check for fuel line pressure while pinching fuel return hose<br>⇨ If fuel line pressure quickly increases, check pressure regulator<br>⇨ If fuel line pressure gradually increases, check for clogging between fuel pump and pressure regulator<br>If not clogged, check fuel pump maximum pressure<br>**High pressure**<br>Check if fuel line is clogged<br>⇨ If OK, replace pressure regulator<br>⇨ If not OK, replace |

## TROUBLESHOOTING CHART – 1990 323

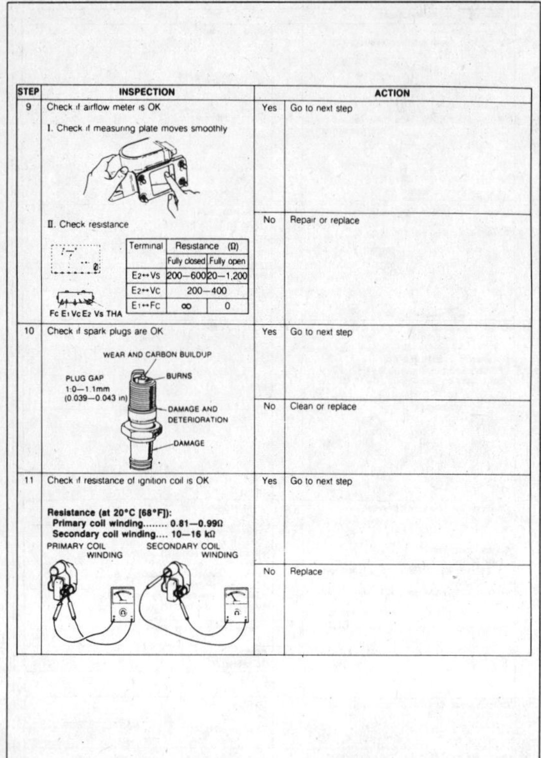

| STEP | INSPECTION | | ACTION |
|---|---|---|---|
| 9 | Check if airflow meter is OK<br>I. Check if measuring plate moves smoothly | Yes | Go to next step |
| | II. Check resistance | No | Repair or replace |
| 10 | Check if spark plugs are OK | Yes | Go to next step |
| | | No | Clean or replace |
| 11 | Check if resistance of ignition coil is OK<br>**Resistance (at 20°C [68°F]):**<br>**Primary coil winding........ 0.81—0.99Ω**<br>**Secondary coil winding.... 10—16 kΩ** | Yes | Go to next step |
| | | No | Replace |

| Terminal | Resistance (Ω) Fully closed | Resistance (Ω) Fully open |
|---|---|---|
| E2↔Vs | 200—600 | 20—1,200 |
| E2↔Vc | 200—400 | |
| E1↔Fc | ∞ | 0 |



| STEP | INSPECTION | | ACTION |
|---|---|---|---|
| 12 | Check for correct engine compression<br>**Engine compression (Minimum):**<br>**• BP DOHC**<br>**883 kPa (9.0 kg/cm², 128 psi)-300 rpm**<br>**• BP SOHC**<br>**834 kPa (8.5 kg/cm², 121 psi)-300 rpm**<br>**• B6**<br>**932 kPa (9.5 kg/cm², 135 psi)-300 rpm** | Yes | Go to next step |
| | | No | Check engine condition<br>• Worn piston, piston rings or cylinder wall<br>• Defective cylinder head gasket<br>• Distorted cylinder head<br>• Improper valve seating<br>• Valve sticking in guide |
| 13 | Change fuel and check if acceleration improves | Yes | Change fuel to another brand |
| | | No | Go to next step |
| 14 | Check if A/C cut-off control system is OK | Yes | **SOHC**<br>Go to Step 16 |
| | | | **DOHC**<br>Go to next step |
| | | No | Repair or replace |
| 15 | Check if variable inertia charging system (VICS) is OK | Yes | Go to next step |
| | | No | Repair or replace |
| 16 | Try known good ECU and check if condition improves | | |

## TROUBLESHOOTING CHART – 1990 323

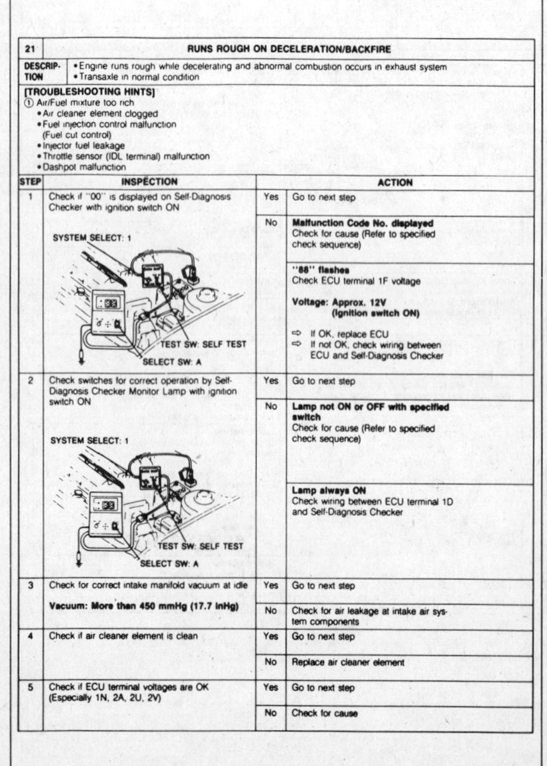

| 21 | RUNS ROUGH ON DECELERATION/BACKFIRE |
|---|---|
| DESCRIPTION | • Engine runs rough while decelerating and abnormal combustion occurs in exhaust system<br>• Transaxle in normal condition |

**[TROUBLESHOOTING HINTS]**
① Air/Fuel mixture too rich
- Air cleaner element clogged
- Fuel injection control malfunction (Fuel cut control)
- Injector fuel leakage
- Throttle sensor (IDL terminal) malfunction
- Dashpot malfunction

| STEP | INSPECTION | | ACTION |
|---|---|---|---|
| 1 | Check if "00" is displayed on Self-Diagnosis Checker with ignition switch ON | Yes | Go to next step |
| | | No | **Malfunction Code No. displayed**<br>Check for cause (Refer to specified check sequence) |
| | | | **"88" flashes**<br>Check ECU terminal 1F voltage<br>**Voltage: Approx. 12V (Ignition switch ON)**<br>⇨ If OK, replace ECU<br>⇨ If not OK, check wiring between ECU and Self-Diagnosis Checker |
| 2 | Check switches for correct operation by Self-Diagnosis Checker Monitor Lamp with ignition switch ON | Yes | Go to next step |
| | | No | **Lamp not ON or OFF with specified switch**<br>Check for cause (Refer to specified check sequence) |
| | | | **Lamp always ON**<br>Check wiring between ECU terminal 1D and Self-Diagnosis Checker |
| 3 | Check for correct intake manifold vacuum at idle<br>**Vacuum: More than 450 mmHg (17.7 inHg)** | Yes | Go to next step |
| | | No | Check for air leakage at intake air system components |
| 4 | Check if air cleaner element is clean | Yes | Go to next step |
| | | No | Replace air cleaner element |
| 5 | Check if ECU terminal voltages are OK (Especially 1N, 2A, 2U, 2V) | Yes | Go to next step |
| | | No | Check for cause |



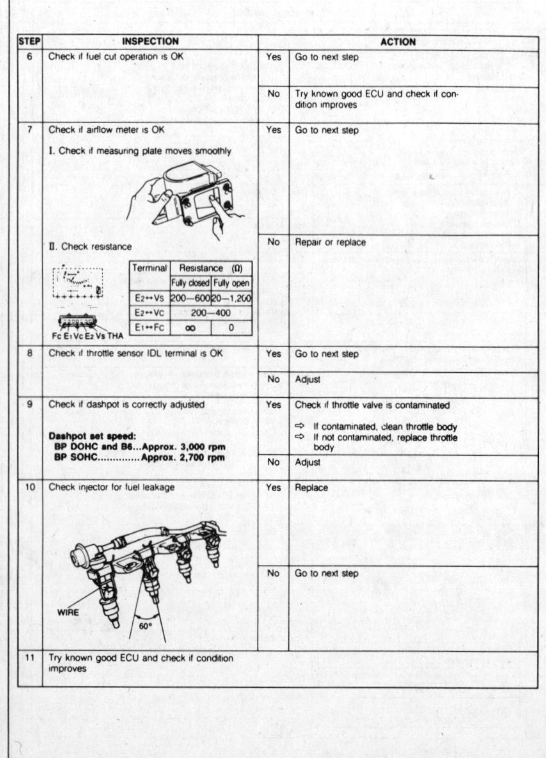

| STEP | INSPECTION | | ACTION |
|---|---|---|---|
| 6 | Check if fuel cut operation is OK | Yes | Go to next step |
| | | No | Try known good ECU and check if condition improves |
| 7 | Check if airflow meter is OK<br>I. Check if measuring plate moves smoothly | Yes | Go to next step |
| | II. Check resistance | No | Repair or replace |
| 8 | Check if throttle sensor IDL terminal is OK | Yes | Go to next step |
| | | No | Adjust |
| 9 | Check if dashpot is correctly adjusted<br>**Dashpot set speed:**<br>**BP DOHC and B6...Approx. 3,000 rpm**<br>**BP SOHC.............Approx. 2,700 rpm** | Yes | Check if throttle valve is contaminated<br>⇨ If contaminated, clean throttle body<br>⇨ If not contaminated, replace throttle body |
| | | No | Adjust |
| 10 | Check injector for fuel leakage | Yes | Replace |
| | | No | Go to next step |
| 11 | Try known good ECU and check if condition improves | | |

| Terminal | Resistance (Ω) Fully closed | Resistance (Ω) Fully open |
|---|---|---|
| E2↔Vs | 200—600 | 20—1,200 |
| E2↔Vc | 200—400 | |
| E1↔Fc | ∞ | 0 |

## TROUBLESHOOTING CHART—1990 323

**22 — KNOCKING**

**DESCRIPTION:** • Abnormal combustion accompanied by audible "pinging" noise

**[TROUBLESHOOTING HINTS]**
① Incorrect ignition timing (Too advanced)
② Carbon deposits in cylinder
③ Overheating
④ Air/Fuel mixture too lean
• Fuel injection amount not correct
• Fuel line pressure decreases while accelerating

| STEP | INSPECTION | | ACTION |
|---|---|---|---|
| 1 | Connect System Selector to diagnosis connector and set Test Switch to "SELF TEST" and check for correct ignition timing at idle after warm-up<br>**Ignition timing (BTDC):**<br>• **BP DOHC : 10 ± 1°**<br>• **BP SOHC : 5 ± 1°**<br>• **B6 : 7 ± 1°** 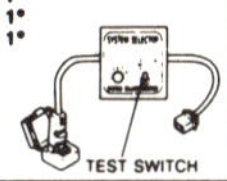<br>TEST SWITCH | Yes | Go to next step |
| | | No | Adjust |
| 2 | Check if "00" is displayed on Self-Diagnosis Checker with ignition switch ON 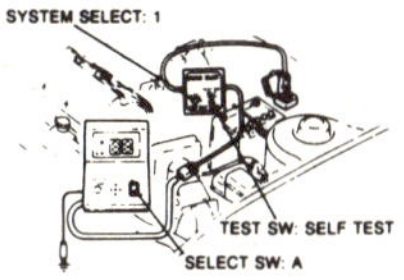<br>SYSTEM SELECT: 1<br>TEST SW: SELF TEST<br>SELECT SW: A | Yes | Go to next step |
| | | No | **Malfunction Code No. displayed**<br>Check for cause (Refer to specified check sequence) |
| | | | **"88" flashes**<br>Check ECU terminal 1F voltage<br>**Voltage: Approx. 12V (Ignition switch ON)**<br>⇨ If OK, replace ECU<br>⇨ If not OK, check wiring between ECU and Self-Diagnosis Checker |
| 3 | Check for correct intake manifold vacuum at idle<br>**Vacuum: More than 450 mmHg (17.7 inHg)** | Yes | Go to next step |
| | | No | Check for air leakage of intake air system components |
| 4 | Check for correct engine compression<br>**Engine compression (Minimum):**<br>• **BP DOHC**<br>**883 kPa (9.0 kg/cm², 128 psi)-300 rpm**<br>• **BP SOHC**<br>**834 kPa (8.5 kg/cm², 121 psi)-300 rpm**<br>• **B6**<br>**932 kPa (9.5 kg/cm², 135 psi)-300 rpm** | Yes | Go to next step |
| | | No | **High compression**<br>Check engine condition<br>• Carbon deposits |
| 5 | Check for correct fuel line pressure at idle<br>**Fuel line pressure:**<br>**265—314 kPa (2.7—3.2 kg/cm², 38—46 psi)**<br>**(Vacuum hose to pressure regulator disconnected)** 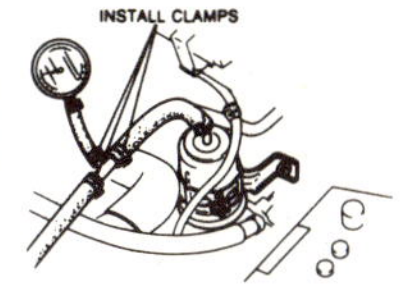<br>INSTALL CLAMPS | Yes | Check if fuel line pressure decreases when accelerating quickly<br>⇨ If decreases, check for clogging between fuel pump and pressure regulator<br>⇨ If not decrease, go to next step |
| | | No | **Low pressure**<br>Check fuel line pressure while pinching fuel return hose<br>⇨ If fuel line pressure quickly increases, check pressure regulator<br>⇨ If fuel line pressure gradually increases, check for clogging between fuel pump and pressure regulator<br>If not clogged, check fuel pump maximum pressure |
| 6 | Check if cooling system is OK | Yes | Go to next step |
| | | No | Repair or replace<br>• Thermostat<br>• Electric cooling fan<br>• Radiator |
| 7 | Try known good ECU and check if condition improves | Yes | Replace ECU |
| | | No | Change fuel to another brand or use higher octane fuel |

## TROUBLESHOOTING CHART—1990 323

**23 — FUEL ODOR**

**DESCRIPTION:** • Gasoline odor in cabin

**[TROUBLESHOOTING HINTS]**
① Poor connection or damaged at fuel system or evaporative emission control system
② Charcoal canister overflow due to evaporative emission control system malfunction

| STEP | INSPECTION | | ACTION |
|---|---|---|---|
| 1 | Check if fuel leakage or damage of fuel system and evaporative emission control system | Yes | Repair or replace |
| | | No | Go to next step |
| 2 | Check if "00" is displayed on Self-Diagnosis Checker with ignition switch ON 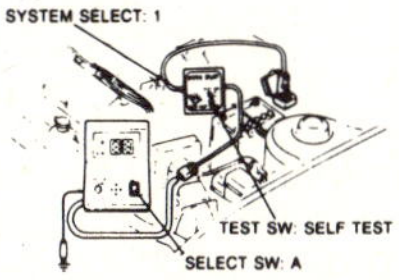<br>SYSTEM SELECT: 1<br>TEST SW: SELF TEST<br>SELECT SW: A | Yes | Go to next step |
| | | No | **Malfunction Code No. displayed**<br>Check for cause (Refer to specified check sequence) |
| | | | **"88" flashes**<br>Check ECU terminal 1F voltage<br>**Voltage: Approx. 12V (Ignition switch ON)**<br>⇨ If OK, replace ECU<br>⇨ If not OK, check wiring between ECU and Self-Diagnosis Checker |
| 3 | Check if vacuum is felt at solenoid valve (purge control) with engine running and throttle valve opened (Neutral switch connector disconnected) 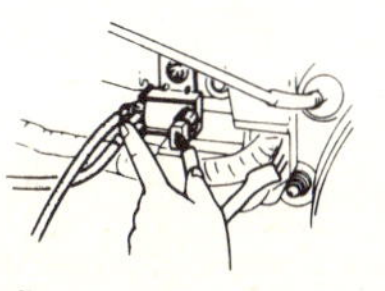 | Yes | Go to Step 5 |
| | | No | Check for solenoid valve operating sound<br>⇨ If OK, check vacuum hoses for clogging<br>⇨ If not OK, go to next step |
| 4 | Apply 12V and ground to solenoid valve (purge control) and check if vacuum is felt at solenoid valve at idle 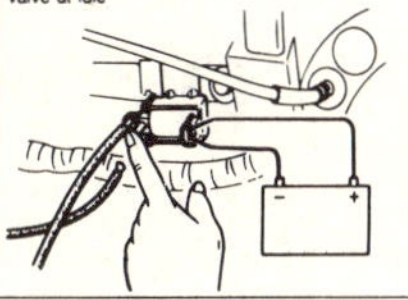 | Yes | Check ECU terminal voltages |
| | | No | Replace solenoid valve |
| 5 | Try known good ECU and check if condition improves | | |

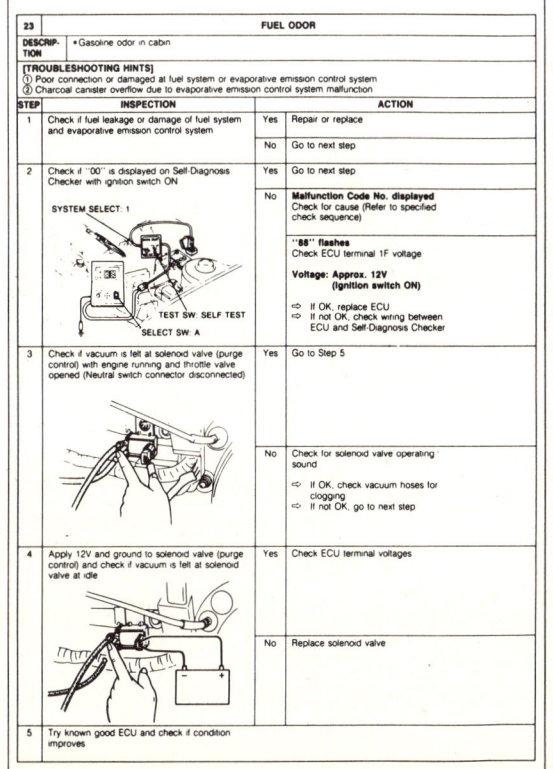

**24 — EXHAUST SULFER SMELL**

**DESCRIPTION:** • Exhaust gas smells abnormally

**[TROUBLESHOOTING HINTS]**
High sulfer content fuel used

| STEP | INSPECTION | ACTION |
|---|---|---|
| 1 | Change fuel to another brand | |

**25 — HIGH OIL CONSUMPTION**

**DESCRIPTION:** • Oil consumption excessive

**[TROUBLESHOOTING HINTS]**
① PCV system malfunction
② Engine malfunction (Oil working up, working down, or leakage)

| STEP | INSPECTION | | ACTION |
|---|---|---|---|
| 1 | Check if PCV hose, ventilation hose or their attaching nipples are separated, damaged, clogged, or restricted | Yes | Repair or replace |
| | | No | Go to next step |
| 2 | Check if air pressure or oil is present at ventilation hose 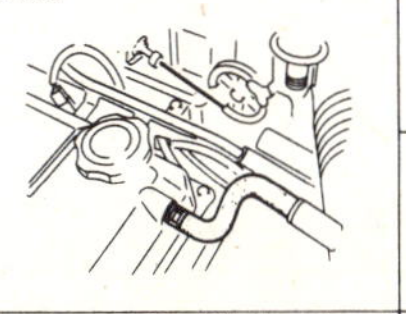 | Yes | Go to next step |
| | | No | Check engine condition<br>• Oil leakage<br>• Worn valve seal<br>• Worn valve stem<br>• Worn valve guide |
| 3 | Check if vacuum is felt at PCV valve at idle 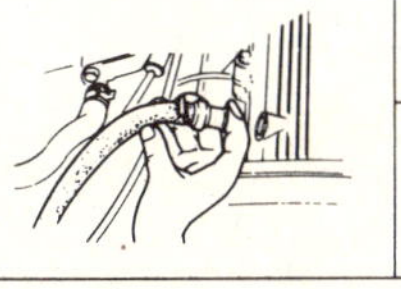 | Yes | Check engine condition<br>• Worn piston ring groove<br>• Stuck piston rings<br>• Worn piston or cylinder |
| | | No | Replace PCV valve |

## TROUBLESHOOTING CHART—1990 323

| 26 | POOR FUEL ECONOMY | | |
|---|---|---|---|
| DESCRIPTION | •Fuel economy unsatisfactory | | |

**[TROUBLESHOOTING HINTS]**
While fuel consumption is drastically increased by city driving, short-run operation, stop and go driving, extended winter warm-up periods, etc., as opposed to "trip" mileage, an attempt should be made to determine these factors when confronted with "poor mileage" conditions. However, since the operator is not always at fault, the following is offered.

① Operator depressing accelerator more than usual due to low engine power
- •Poor ignition
- •Low intake air amount
- •Electronic spark advance control system malfunction
- •Clutch slipping/ATX slipping
- •Exhaust component restricted

② Air/Fuel mixture too rich
- •High fuel line pressure

③ Alcohol blended fuel used

④ High vehicle load
- •Low tire pressure
- •Incorrect tire size
- •Brake dragging

| STEP | INSPECTION | | ACTION |
|---|---|---|---|
| 1 | Check factors other than engine<br>• Low tire pressure<br>• Unrecommended tire size<br>• Clutch slipping<br>• Brake dragging<br>• Exhaust component restricted | Yes | Go to next step |
| | | No | Repair |
| 2 | Check if air hoses are connected correctly | Yes | Go to next step |
| | | No | Repair |
| 3 | Check if "00" is displayed on Self-Diagnosis Checker with ignition switch ON<br>SYSTEM SELECT: 1<br>TEST SW: SELF TEST<br>SELECT SW: A | Yes | Go to next step |
| | | No | **Malfunction Code No. displayed**<br>Check for cause (Refer to specified check sequence)<br>**"88" flashes**<br>Check ECU terminal 1F voltage<br>**Voltage: Approx. 12V (Ignition switch ON)**<br>⇨ If OK, replace ECU<br>⇨ If not OK, check wiring between ECU and Self-Diagnosis Checker |
| 4 | Check for correct intake manifold vacuum at idle<br>**Vacuum: More than 450 mmHg (17.7 inHg)** | Yes | Go to next step |
| | | No | Check for air leakage at intake air system components |
| 5 | Check if air cleaner element is clean | Yes | Go to next step |
| | | No | Replace air cleaner element |

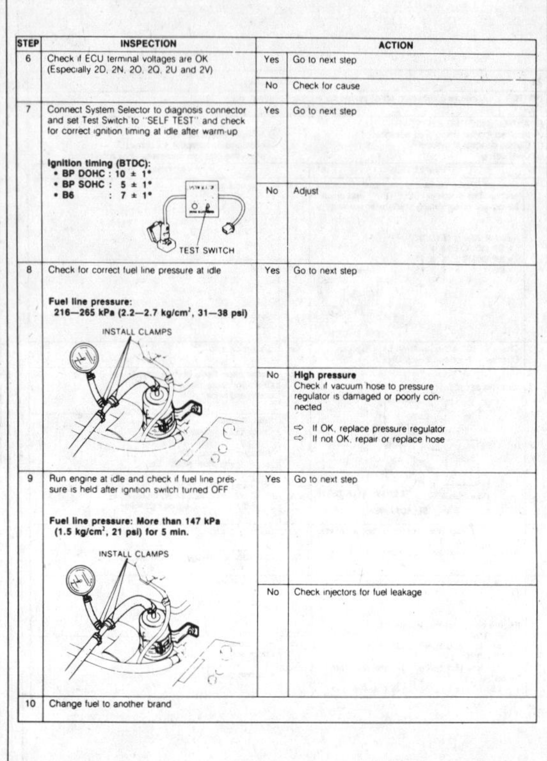

| STEP | INSPECTION | | ACTION |
|---|---|---|---|
| 6 | Check if ECU terminal voltages are OK (Especially 2D, 2N, 2O, 2Q, 2U and 2V) | Yes | Go to next step |
| | | No | Check for cause |
| 7 | Connect System Selector to diagnosis connector and set Test Switch to "SELF TEST" and check for correct ignition timing at idle after warm-up<br>**Ignition timing (BTDC):**<br>**• BP DOHC : 10 ± 1°**<br>**• BP SOHC : 5 ± 1°**<br>**• B6 : 7 ± 1°**<br>TEST SWITCH | Yes | Go to next step |
| | | No | Adjust |
| 8 | Check for correct fuel line pressure at idle<br>**Fuel line pressure:**<br>**216—265 kPa (2.2—2.7 kg/cm², 31—38 psi)**<br>INSTALL CLAMPS | Yes | Go to next step |
| | | No | **High pressure**<br>Check if vacuum hose to pressure regulator is damaged or poorly connected<br>⇨ If OK, replace pressure regulator<br>⇨ If not OK, repair or replace hose |
| 9 | Run engine at idle and check if fuel line pressure is held after ignition switch turned OFF<br>**Fuel line pressure: More than 147 kPa (1.5 kg/cm², 21 psi) for 5 min.**<br>INSTALL CLAMPS | Yes | Go to next step |
| | | No | Check injectors for fuel leakage |
| 10 | Change fuel to another brand | | |

## TROUBLESHOOTING CHART—1990 323

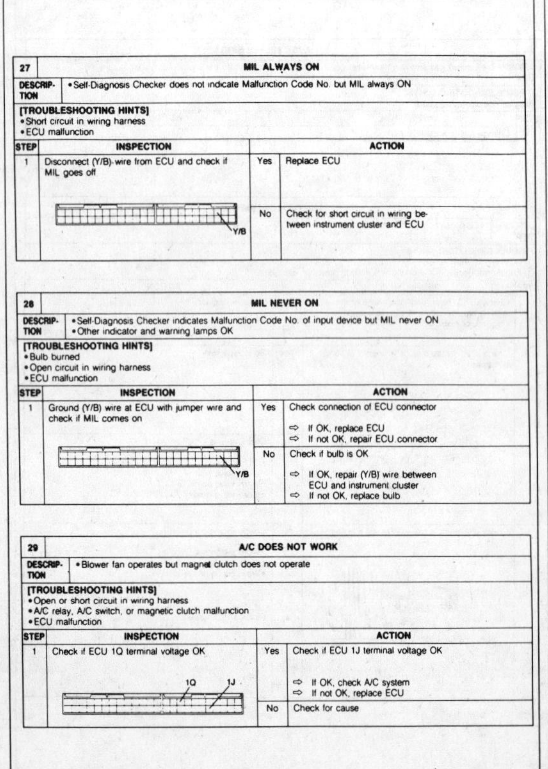

| 27 | MIL ALWAYS ON | | |
|---|---|---|---|
| DESCRIPTION | •Self-Diagnosis Checker does not indicate Malfunction Code No. but MIL always ON | | |

**[TROUBLESHOOTING HINTS]**
- •Short circuit in wiring harness
- •ECU malfunction

| STEP | INSPECTION | | ACTION |
|---|---|---|---|
| 1 | Disconnect (Y/B) wire from ECU and check if MIL goes off<br>Y/B | Yes | Replace ECU |
| | | No | Check for short circuit in wiring between instrument cluster and ECU |

| 28 | MIL NEVER ON | | |
|---|---|---|---|
| DESCRIPTION | •Self-Diagnosis Checker indicates Malfunction Code No. of input device but MIL never ON<br>•Other indicator and warning lamps OK | | |

**[TROUBLESHOOTING HINTS]**
- •Bulb burned
- •Open circuit in wiring harness
- •ECU malfunction

| STEP | INSPECTION | | ACTION |
|---|---|---|---|
| 1 | Ground (Y/B) wire at ECU with jumper wire and check if MIL comes on<br>Y/B | Yes | Check connection of ECU connector<br>⇨ If OK, replace ECU<br>⇨ If not OK, repair ECU connector |
| | | No | Check if bulb is OK<br>⇨ If OK, repair (Y/B) wire between ECU and instrument cluster<br>⇨ If not OK, replace bulb |

| 29 | A/C DOES NOT WORK | | |
|---|---|---|---|
| DESCRIPTION | •Blower fan operates but magnet clutch does not operate | | |

**[TROUBLESHOOTING HINTS]**
- •Open or short circuit in wiring harness
- •A/C relay, A/C switch, or magnetic clutch malfunction
- •ECU malfunction

| STEP | INSPECTION | | ACTION |
|---|---|---|---|
| 1 | Check if ECU 1Q terminal voltage OK<br>1Q 1J | Yes | Check if ECU 1J terminal voltage OK<br>⇨ If OK, check A/C system<br>⇨ If not OK, replace ECU |
| | | No | Check for cause |

**Troubleshooting**
If a malfunction code number is shown on the **SST**, check for the cause by using the chart related to the code number shown.

| CODE No. | 02 (DISTRIBUTOR Ne-SIGNAL) | | |
|---|---|---|---|
| STEP | INSPECTION | | ACTION |
| 1 | Check distributor circuit for poor connection | Yes | Repair or replace connector |
| | | No | **SOHC**<br>Go to Step 3<br>**DOHC**<br>Go to next step |
| 2 | Check if Code No.03 is also present | Yes | Go to next step |
| | | No | Go to Step 5 |
| 3 | Check terminal-wire (B/LG) for continuity | Yes | Go to next step |
| | | No | Repair or replace |
| 4 | Check if battery voltage exists at distributor terminal-wire (W/R) | Yes | Go to next step |
| | | No | Check for open circuit in wiring from distributor to main relay (FUEL INJ relay) |
| 5 | Check terminal-wire (W) between distributor and ECU terminal 2E for continuity | Yes | Go to next step |
| | | No | Repair or replace |
| 6 | Check if ECU terminal 2E voltage is OK | Yes | Replace ECU |
| | | No | Go to next step |
| 7 | Check if approx. 0V or approx. 5V exists at distributor terminal-wire (W) | Yes | Replace distributor |
| | | No | Go to next step |
| 8 | Check if approx. 5V exists at ECU terminal 2E (With distributor connector disconnected) | Yes | Check for short circuit in wiring from distributor to ECU |
| | | No | Replace ECU |

**Circuit Diagram**

## TROUBLE CODE IDENTIFICATIONS CHART—1990 323

**Code Numbers**

| Sensor or subsystem | Condition | Fail-safe | Malfunction display | |
|---|---|---|---|---|
| | | | Malfunction Code No. | MIL output signal pattern |
| Ne-signal | No Ne-signal | – | 02 | ON / OFF |
| G-signal [DOHC] | No G-signal | Cancels 2-group injection | 03 | ON / OFF |
| Airflow meter | Open or short circuit | Basic fuel injection amount fixed as for 2 driving modes (1) Idle switch: ON (2) Idle switch: OFF | 08 | ON / OFF |
| Water thermosensor | Open or short circuit | Maintains constant 20°C (68°F) [BP], 40°C (104°F) [B6] command | 09 | ON / OFF |
| Intake air thermosensor | Open or short circuit | Maintains constant 20°C (68°F) command | 10 | ON / OFF |
| Throttle sensor [ATX] | Open or short circuit | Maintains constant command of throttle valve fully open | 12 | ON / OFF |
| Atmospheric pressure sensor (In ECU) | Open or short circuit | Maintains constant command of sea level pressure | 14 | ON / OFF |
| Oxygen sensor | Sensor output continues less than 0.55 V 95 sec. after engine starts (1,500 rpm) | Cancels engine feedback operation | 15 | ON / OFF |
| Feedback system | Sensor output continues unchanged 50 sec. after engine exceeds 1,500 rpm | Cancels engine feedback operation | 17 | ON / OFF |
| Solenoid valve (Pressure regulator) [BP ENGINE] | Open or short circuit | – | 25 | ON / OFF |
| Solenoid valve (Purge control) | | – | 26 | ON / OFF |
| ISC valve | | – | 34 | ON / OFF |
| Solenoid valve (VICS) [DOHC] | | – | 41 | ON / OFF |

**Caution**
- **If there is more than one failure present, the code numbers will be indicated in numerical order, lowest number first.**
- **After repairing a failure, turn off the ignition switch, disconnect the negative battery cable, and depress the brake pedal for at least five (5) seconds to erase the malfunction code(s) from the ECU memory.**

## TROUBLE CODE DIAGNOSTIC CHART—1990 323

| CODE No. | 03 (DISTRIBUTOR G-SIGNAL) — DOHC | | |
|---|---|---|---|
| STEP | INSPECTION | ACTION | |
| 1 | Check distributor circuit for poor connection | Yes | Repair or replace connector |
| | | No | Go to next step |
| 2 | Check if Code No.02 is also present | Yes | Go to next step |
| | | No | Go to Step 5 |
| 3 | Check terminal-wire (B/LG) for continuity | Yes | Go to next step |
| | | No | Repair or replace |
| 4 | Check if battery voltage exists at distributor terminal-wire (W/R) | Yes | Go to next step |
| | | No | Check for open circuit in wiring from distributor to main relay |
| 5 | Check terminal-wire (Y/L) between distributor and ECU terminal 2G for continuity | Yes | Go to next step |
| | | No | Repair or replace |
| 6 | Check if ECU terminal 2G voltage is OK | Yes | Replace ECU |
| | | No | Go to next step |
| 7 | Check if approx. 0V or approx. 5V exists at distributor terminal-wire (Y/L) | Yes | Replace distributor |
| | | No | Go to next step |
| 8 | Check if approx. 5V exists at ECU terminal 2G (With distributor connector disconnected) | Yes | Check for short circuit in wiring from distributor to ECU |
| | | No | Replace ECU |

**Circuit Diagram**

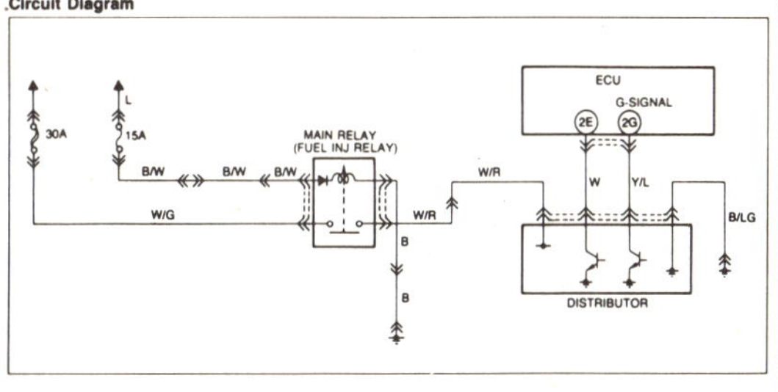


## TROUBLE CODE DIAGNOSTIC CHART—1990 323

| CODE No. | 08 (AIRFLOW METER) | | |
|---|---|---|---|
| STEP | INSPECTION | ACTION | |
| 1 | Check airflow meter circuit for poor connection | Yes | Repair or replace connector |
| | | No | Go to next step |
| 2 | Check if Code No.10 is also present | Yes | Check for open circuit in wiring from airflow meter terminal-wire (B/LG) to ground |
| | | No | Go to next step |
| 3 | Check if resistance of airflow meter is OK | Yes | Go to next step |
| | (see table below) | No | Replace airflow meter |
| 4 | Check wire harness between airflow meter and ECU for continuity | Yes | Go to next step |
| | (see table below) | No | Repair or replace wire harness |
| 5 | Check if ECU terminals 2D, 2K, and 2O voltages are OK | Yes | Replace ECU |
| | | No | Check for short circuit in wiring from airflow meter to ECU |

Step 3:

| Airflow meter | Fully closed (Ω) | Fully open (Ω) |
|---|---|---|
| D(LG/R)—F(R) | 200—600 | 20—1,200 |
| D(LG/R)—C(B/LG) | 200—400 | |

Step 4:

| Airflow meter | ECU |
|---|---|
| D (LG/R) | 2K (LG/R) |
| F (R) | 2O (R) |

**Circuit Diagram**

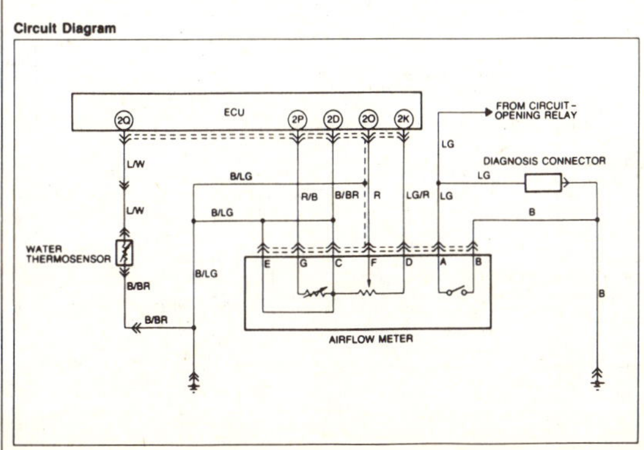


| CODE No. | 09 (WATER THERMOSENSOR) | | |
|---|---|---|---|
| STEP | INSPECTION | ACTION | |
| 1 | Check water thermosensor circuit for poor connection | Yes | Repair or replace connector |
| | | No | Go to next step |
| 2 | Check wire harness between water thermosensor and ECU for continuity | Yes | Go to next step |
| | (see table below) | No | Repair or replace |
| 3 | Check if resistance of water thermosensor is OK | Yes | Go to next step |
| | (see table below) | No | Replace water thermosensor |
| 4 | Check if same Code No. is present following after-repair procedure | Yes | Go to next step |
| | | No | Water thermosensor and circuit OK |
| 5 | Check if ECU terminals 2D and 2Q voltages are OK | Yes | Replace ECU |
| | | No | Check for short circuit in wiring from water thermosensor to ECU |

Step 2:

| Water thermosensor | ECU |
|---|---|
| A (L/W) | 2Q (L/W) |
| B (B/BR) | 2D (B/BR) |

Step 3:

| Coolant temp. | Resistance (kΩ) |
|---|---|
| -20°C ( -4°F) | 14.6—17.8 |
| 20°C ( 68°F) | 2.21—2.69 |
| 80°C (176°F) | 0.29—0.35 |

**Circuit Diagram**

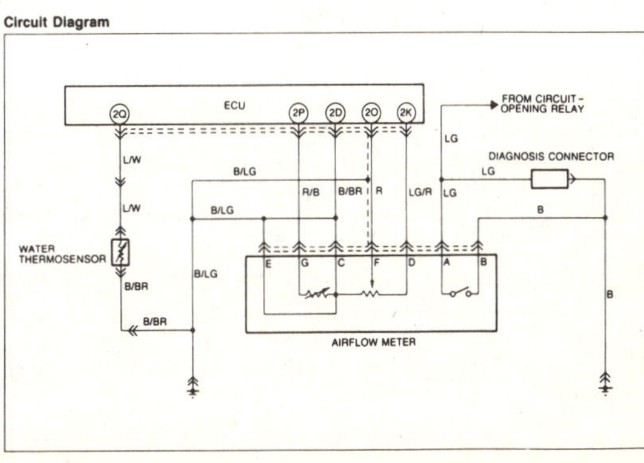

## TROUBLE CODE DIAGNOSTIC CHART—1990 323

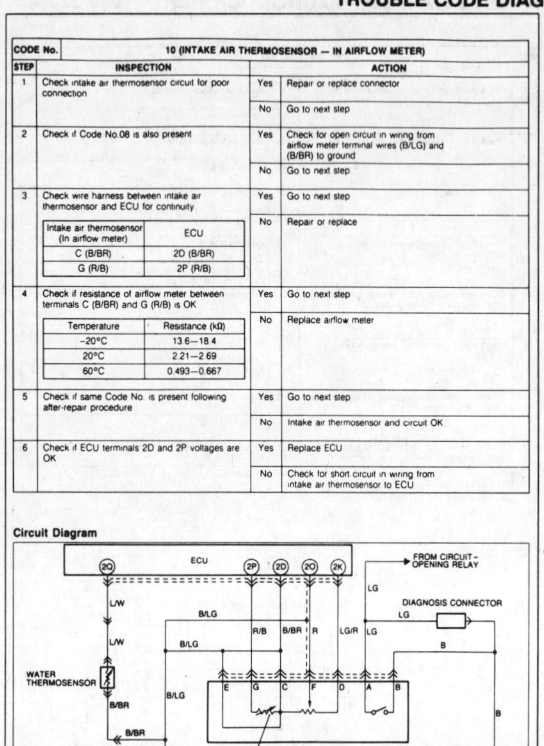

| CODE No. | 10 (INTAKE AIR THERMOSENSOR — IN AIRFLOW METER) | | |
|---|---|---|---|
| STEP | INSPECTION | ACTION | |
| 1 | Check intake air thermosensor circuit for poor connection | Yes | Repair or replace connector |
| | | No | Go to next step |
| 2 | Check if Code No.08 is also present | Yes | Check for open circuit in wiring from airflow meter terminal wires (B/LG) and (B/BR) to ground |
| | | No | Go to next step |
| 3 | Check wire harness between intake air thermosensor and ECU for continuity<br>Intake air thermosensor (In airflow meter) / ECU:<br>C (B/BR) — 2D (B/BR)<br>G (R/B) — 2P (R/B) | Yes | Go to next step |
| | | No | Repair or replace |
| 4 | Check if resistance of airflow meter between terminals C (B/BR) and G (R/B) is OK<br>Temperature / Resistance (kΩ):<br>−20°C — 13.6—18.4<br>20°C — 2.21—2.69<br>60°C — 0.493—0.667 | Yes | Go to next step |
| | | No | Replace airflow meter |
| 5 | Check if same Code No. is present following after-repair procedure | Yes | Go to next step |
| | | No | Intake air thermosensor and circuit OK |
| 6 | Check if ECU terminals 2D and 2P voltages are OK | Yes | Replace ECU |
| | | No | Check for short circuit in wiring from intake air thermosensor to ECU |

**Circuit Diagram**



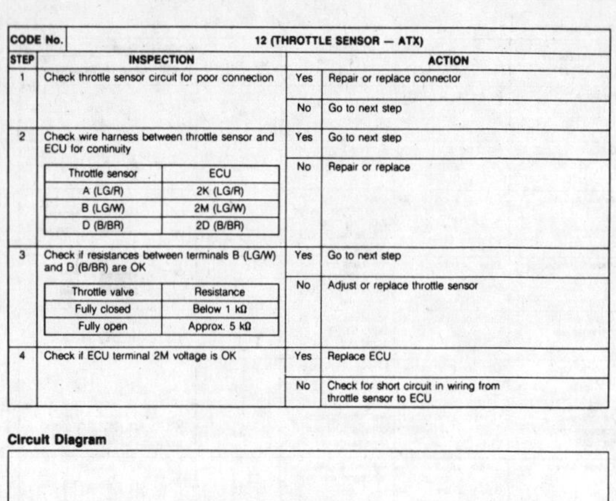

| CODE No. | 12 (THROTTLE SENSOR — ATX) | | |
|---|---|---|---|
| STEP | INSPECTION | ACTION | |
| 1 | Check throttle sensor circuit for poor connection | Yes | Repair or replace connector |
| | | No | Go to next step |
| 2 | Check wire harness between throttle sensor and ECU for continuity<br>Throttle sensor / ECU:<br>A (LG/R) — 2K (LG/R)<br>B (LG/W) — 2M (LG/W)<br>D (B/BR) — 2D (B/BR) | Yes | Go to next step |
| | | No | Repair or replace |
| 3 | Check if resistances between terminals B (LG/W) and D (B/BR) are OK<br>Throttle valve / Resistance:<br>Fully closed — Below 1 kΩ<br>Fully open — Approx. 5 kΩ | Yes | Go to next step |
| | | No | Adjust or replace throttle sensor |
| 4 | Check if ECU terminal 2M voltage is OK | Yes | Replace ECU |
| | | No | Check for short circuit in wiring from throttle sensor to ECU |

**Circuit Diagram**

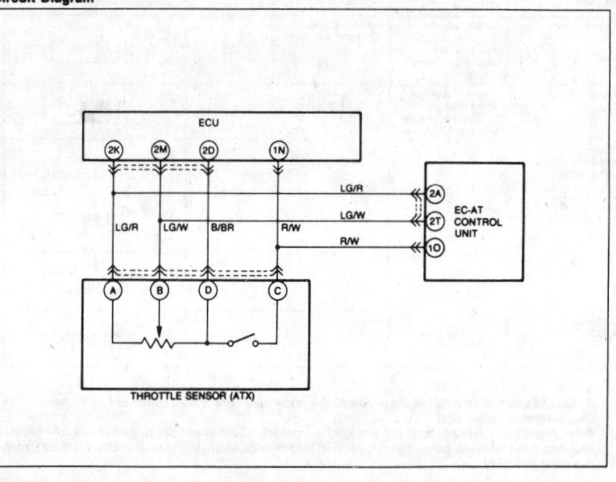



## TROUBLE CODE DIAGNOSTIC CHART—1990 323

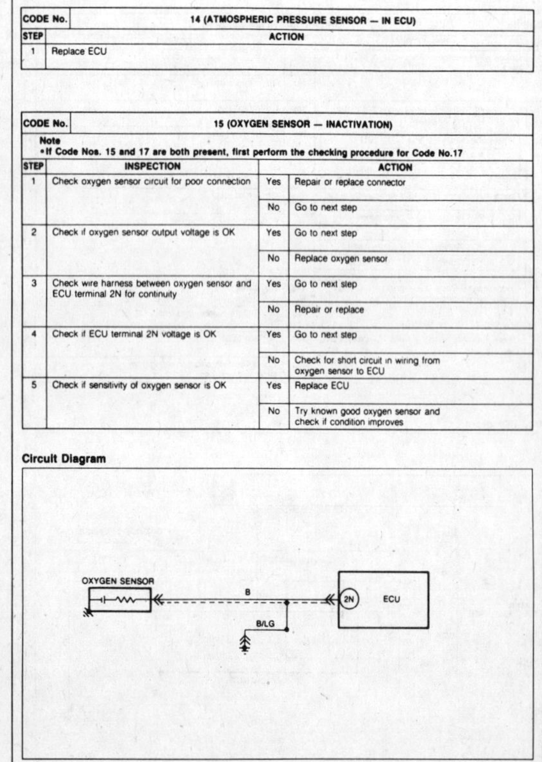

| CODE No. | 14 (ATMOSPHERIC PRESSURE SENSOR — IN ECU) |
|---|---|
| STEP | ACTION |
| 1 | Replace ECU |

| CODE No. | 15 (OXYGEN SENSOR — INACTIVATION) | | |
|---|---|---|---|
| **Note** •**If Code Nos. 15 and 17 are both present, first perform the checking procedure for Code No.17** | | | |
| STEP | INSPECTION | ACTION | |
| 1 | Check oxygen sensor circuit for poor connection | Yes | Repair or replace connector |
| | | No | Go to next step |
| 2 | Check if oxygen sensor output voltage is OK | Yes | Go to next step |
| | | No | Replace oxygen sensor |
| 3 | Check wire harness between oxygen sensor and ECU terminal 2N for continuity | Yes | Go to next step |
| | | No | Repair or replace |
| 4 | Check if ECU terminal 2N voltage is OK | Yes | Go to next step |
| | | No | Check for short circuit in wiring from oxygen sensor to ECU |
| 5 | Check if sensitivity of oxygen sensor is OK | Yes | Replace ECU |
| | | No | Try known good oxygen sensor and check if condition improves |

**Circuit Diagram**



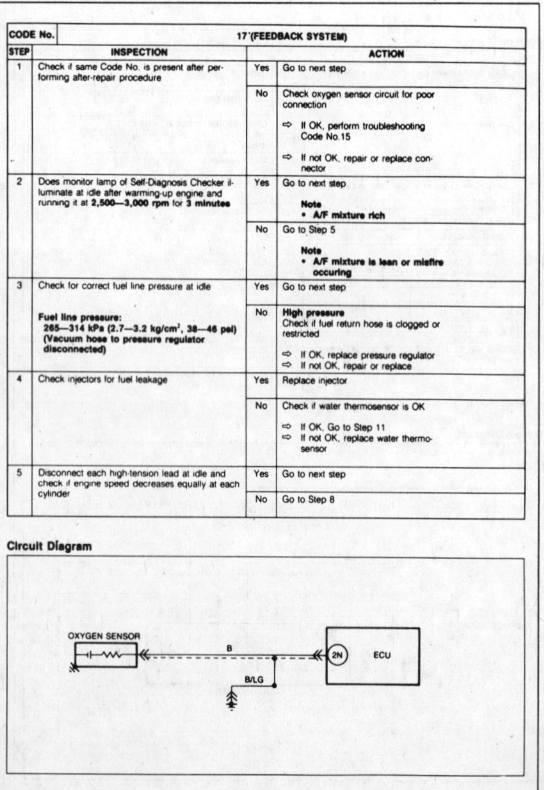

| CODE No. | 17 (FEEDBACK SYSTEM) | | |
|---|---|---|---|
| STEP | INSPECTION | ACTION | |
| 1 | Check if same Code No. is present after performing after-repair procedure | Yes | Go to next step |
| | | No | Check oxygen sensor circuit for poor connection<br>⇨ If OK, perform troubleshooting Code No.15<br>⇨ If not OK, repair or replace connector |
| 2 | Does monitor lamp of Self-Diagnosis Checker illuminate at idle after warming-up engine and running it at **2,500—3,000 rpm** for **3 minutes** | Yes | Go to next step<br>**Note**<br>• **A/F mixture rich** |
| | | No | Go to Step 5<br>**Note**<br>• **A/F mixture is lean or misfire occuring** |
| 3 | Check for correct fuel line pressure at idle<br>**Fuel line pressure:**<br>**265—314 kPa (2.7—3.2 kg/cm², 38—46 psi)**<br>**(Vacuum hose to pressure regulator disconnected)** | Yes | Go to next step |
| | | No | **High pressure**<br>Check if fuel return hose is clogged or restricted<br>⇨ If OK, replace pressure regulator<br>⇨ If not OK, repair or replace |
| 4 | Check injectors for fuel leakage | Yes | Replace injector |
| | | No | Check if water thermosensor is OK<br>⇨ If OK, Go to Step 11<br>⇨ If not OK, replace water thermosensor |
| 5 | Disconnect each high-tension lead at idle and check if engine speed decreases equally at each cylinder | Yes | Go to next step |
| | | No | Go to Step 8 |

**Circuit Diagram**

## TROUBLE CODE DIAGNOSTIC CHART—1990 323

| STEP | INSPECTION | | ACTION |
|---|---|---|---|
| 6 | Check for correct fuel line pressure at idle<br><br>**Fuel line pressure: 265—314 kPa (2.7—3.2 kg/cm², 38—46 psi) (Vacuum hose to pressure regulator disconnected)** | Yes | Go to next step |
| | | No | **Low pressure**<br>Check fuel line pressure while pinching fuel return hose<br>⇨ If fuel line pressure quickly increases, check pressure regulator<br>⇨ If fuel line pressure gradually increases, check for clogging between fuel pump and pressure regulator<br>If not clogged, check fuel pump maximum pressure |
| 7 | Check intake air system components for air leakage | Yes | Go to Step 11 |
| | | No | Replace |
| 8 | Check for misfire of dead cylinder from Step 5 inspection | Yes | Repair or replace ignition system |
| | | No | Go to next step |
| 9 | Check for injector operating sound at idle of dead cylinder from Step 5 inspection | Yes | Go to next step |
| | | No | Check for approx. 12V at injector connector terminal-wire (W/R)<br>⇨ If Yes, replace injector<br>⇨ If No, check wire harness for short or open |
| 10 | Replace injector at dead cylinder from Step 5 inspection<br><br>Check if same Code No. is present following performing after-repair procedure | Yes | Go to next step |
| | | No | Removed injector at fault |
| 11 | Try known good ECU and check if condition improves | | |

| CODE No. | 25 (SOLENOID VALVE — PRESSURE REGULATOR CONTROL)—BP ENGINE | | |
|---|---|---|---|
| STEP | INSPECTION | | ACTION |
| 1 | Disconnect connector from ECU and check if battery voltage exists at ECU terminal 2T wire (G/O) with ignition switch ON | Yes | Check ECU terminal connector for poor connection<br>⇨ If OK, replace ECU<br>⇨ If not OK, repair or replace connector |
| | | No | Go to next step |
| 2 | In same condition as Step 1, check if battery voltage exists at solenoid valve connector terminal-wire (G/O) | Yes | Repair or replace wire (G/O) |
| | | No | Go to next step |
| 3 | Check if solenoid valve is OK | Yes | Check for short or open circuit in wiring from main relay (FUEL INJ relay) to solenoid valve |
| | | No | Replace solenoid valve |

| CODE No. | 26 (SOLENOID VALVE — PURGE CONTROL) | | |
|---|---|---|---|
| STEP | INSPECTION | | ACTION |
| 1 | Disconnect connector from ECU and check if battery voltage exists at ECU terminal 2X wire harness (W/L) with ignition switch ON | Yes | Check ECU terminal connector for poor connection<br>⇨ If OK, replace ECU<br>⇨ If not OK, repair or replace connector |
| | | No | Go to next step |
| 2 | In same condition as Step 1, check if battery voltage exists at solenoid valve connector terminal-wire (W/L) | Yes | Repair or replace wire harness (W/L) |
| | | No | Go to next step |
| 3 | Check if solenoid valve is OK | Yes | Check for short or open circuit in wiring from main relay (FUEL INJ relay) to solenoid valve |
| | | No | Replace solenoid valve |

**Circuit Diagram**

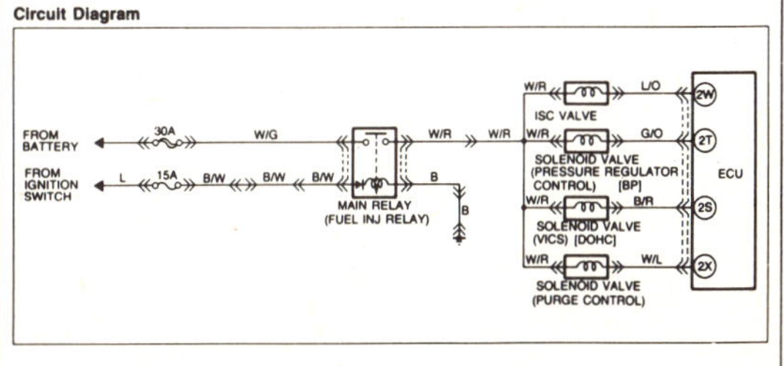


## TROUBLE CODE DIAGNOSTIC CHART—1990 323

| CODE No. | 34 (ISC VALVE) | | |
|---|---|---|---|
| STEP | INSPECTION | | ACTION |
| 1 | Disconnect connector from ECU and check if battery voltage exists at ECU terminal 2W wire (L/O) with ignition switch ON | Yes | Check ECU terminal connector for poor connection<br>⇨ If OK, replace ECU<br>⇨ If not OK, repair or replace connector |
| | | No | Go to next step |
| 2 | In same condition as Step 1, check if battery voltage exists at ISC valve connector terminal-wire (L/O) | Yes | Repair or replace wire (L/O) |
| | | No | Go to next step |
| 3 | Check ISC valve for correct resistance<br><br>**Resistance: 11—13Ω [at 20°C(68°F)]** | Yes | Check for short or open circuit in wiring from main relay (FUEL INJ relay) to ISC valve |
| | | No | Replace ISC valve |

| CODE No. | 41 (SOLENOID VALVE — VICS)—DOHC | | |
|---|---|---|---|
| STEP | INSPECTION | | ACTION |
| 1 | Disconnect connector from ECU and check if battery voltage exists at ECU terminal 2S wire (B/R) with ignition switch ON | Yes | Check ECU terminal connector for poor connection<br>⇨ If OK, replace ECU<br>⇨ If not OK, repair or replace connector |
| | | No | Go to next step |
| 2 | In same condition as Step 1, check if battery voltage exists at solenoid valve connector terminal-wire (B/R) | Yes | Repair or replace wire (B/R) |
| | | No | Go to next step |
| 3 | Check if solenoid valve is OK | Yes | Check for short or open circuit in wiring from main relay to solenoid valve |
| | | No | Replace solenoid valve |

**Circuit Diagram**

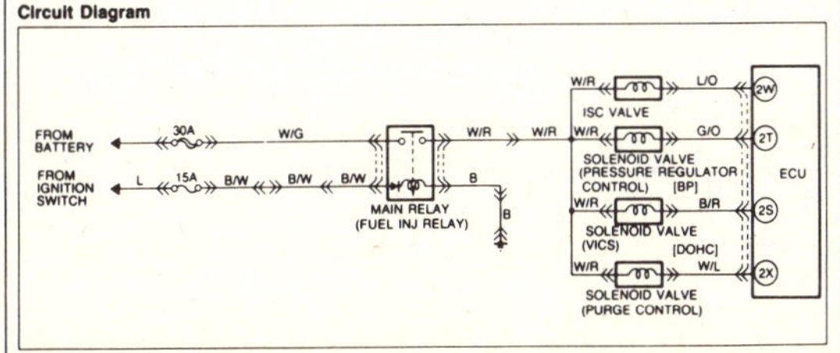


## CONTROL UNIT PIN LOCATIONS AND TEST VOLTAGES 1990 323

**Tmerminal voltage**

| Terminal | Input | Output | Connection to | Test condition | | Correct voltage | Remark |
|---|---|---|---|---|---|---|---|
| 1A | — | — | Battery | Constant | | Approx. 12V | For backup |
| 1B | ○ | | Main relay (FUEL INJ relay) | Ignition switch | OFF | Approx. 0V | — |
| | | | | | ON | Approx. 12V | |
| 1C | ○ | | Ignition switch (START) | While cranking | | Approx. 10V | — |
| | | | | Ignition switch ON | | Approx. 0V | |
| 1D | | ○ | Self-Diagnosis Checker (Monitor lamp) | Test switch at "SELF-TEST" Lamp illuminated for 3 sec. after ignition switch OFF→ON | | Approx. 5V | With Self-Diagnosis Checker and System Selector |
| | | | | Lamp not illuminated after 3 sec. | | Approx. 12V | |
| | | | | Test switch at "$O_2$ MONITOR" at idle Monitor lamp illuminated | | Approx. 5V | |
| | | | | Test switch at "$O_2$ MONITOR" at idle Monitor lamp not illuminated | | Approx. 12V | |
| 1E | | ○ | Malfunction indicator lamp (MIL) | Lamp illuminated for 3 sec. after ignition switch OFF→ON | | Below 2.5V | With System Selector test switch at "SELF-TEST" |
| | | | | Lamp not illuminated after 3 sec. | | Approx. 12V | |
| | | | | Lamp illuminated | | Below 2.5V | |
| | | | | Lamp not illuminated | | Approx. 12V | |
| 1F | | ○ | Self-Diagnosis Checker (Code Number) | Buzzer sounded for 3 sec. after ignition switch OFF→ON | | Below 2.5V | • With Self-Diagnosis Checker and System Selector<br>• With System Selector test switch at "SELF-TEST" |
| | | | | Buzzer not sounded after 3 sec. | | Approx. 12V | |
| | | | | Buzzer sounded | | Below 2.5V | |
| | | | | Buzzer not sounded | | Approx. 12V | |
| 1G | | ○ | Igniter | Ignition switch ON | | Approx. 0V | — |
| | | | | Idle | | Approx. 0.2V | |
| 1H | — | — | — | — | | — | — |
| 1I | — | — | — | — | | — | — |
| 1J | | ○ | A/C relay | Ignition switch ON | | Approx. 12V | — |
| | | | | A/C switch ON at idle | | Below 2.5V | |
| | | | | A/C switch OFF at idle | | Approx. 12V | |
| 1K | ○ | | Diagnosis connector (TEN terminal) | System Selector test switch at "$O_2$ MONITOR" | | Approx. 12V | — |
| | | | | System Selector test switch at "SELF-TEST" | | Below 1.0V | |
| 1L | ○ | | DRL relay (Canada) | Parking brake pulled with ignition switch ON (DRL OFF) | | Approx. 12V | • DRL: Daytime Running Lights |
| | | | | Idle (DRL ON) | | Below 2.5V | |

| 2Y | 2W | 2U | 2S | 2Q | 2O | 2M | 2K | 2I | 2G | 2E | 2C | 2A | U | S | Q | O | M | K | I | G | E | C | A |
|---|---|---|---|---|---|---|---|---|---|---|---|---|---|---|---|---|---|---|---|---|---|---|---|
| 2Z | 2X | 2V | 2T | 2R | 2P | 2N | 2L | 2J | 2H | 2F | 2D | 2B | V | T | R | P | N | L | J | H | F | D | B |

### CONTROL UNIT PIN LOCATIONS AND TEST VOLTAGES – 1990 323

| Terminal | Input | Output | Connection to | Test condition | Correct voltage | Remark |
|---|---|---|---|---|---|---|
| 1M | – | – | – | – | – | – |
| 1N | ○ | | Throttle sensor (Idle switch) (MTX/ATX) EC-AT control unit (ATX) | Accelerator pedal released | Below 1.0V | Ignition switch ON |
| | | | | Accelerator pedal depressed | Approx. 12V | |
| 1O | ○ | | Stoplight switch (MTX) | Brake pedal released | Below 1.0V | – |
| | | | | Brake pedal depressed | Approx. 12V | |
| | | | EC-AT CONTROL (ATX) | – | Below 1.0V | – |
| 1P | ○ | | P/S pressure switch | Ignition switch ON | Approx. 12V | – |
| | | | | P/S ON at idle | Below 1.0V | |
| | | | | P/S OFF at idle | Approx. 12V | |
| 1Q | ○ | | A/C switch | A/C switch ON | Below 2.5V | Ignition switch ON and blower motor ON |
| | | | | A/C switch OFF | Approx. 12V | |
| 1R | ○ | | Fan switch | Fan operating (Engine coolant temperature over 97°C (207°F) or diagnosis connector terminal TFA grounded) | Below 1.0V | – |
| | | | | Fan not operating (Idle) | App.ox. 12V | |
| 1S | ○ | | Blower control switch | Blower control switch OFF or 1st position | Approx. 12V | Ignition switch ON |
| | | | | Blower control switch 2nd or higher position | Below 1.0V | |
| 1T | ○ | | Rear window defroster switch | Rear window defroster switch OFF | Below 1.0V | Ignition switch ON |
| | | | | Rear window defroster switch ON | Approx. 12V | |
| 1U | ○ | | Headlight switch | Headlights ON | Approx. 12V | – |
| | | | | Headlights OFF | Below 1.0V | |
| 1V | ○ | | Neutral/Clutch switches (MTX) | Neutral position or clutch pedal depressed | Below 1.0V | – |
| | | | | Others | Approx. 12V | |
| | | | Inhibitor switch (ATX) | N or P range | Below 1.0V | |
| | | | | Others | Approx. 12V | |

| 2Y | 2W | 2U | 2S | 2O | 2O | 2M | 2K | 2I | 2G | 2E | 2C | 2A | U | S | Q | O | M | K | I | G | E | C | A |
|---|---|---|---|---|---|---|---|---|---|---|---|---|---|---|---|---|---|---|---|---|---|---|---|
| 2Z | 2X | 2V | 2T | 2R | 2P | 2N | 2L | 2J | 2H | 2F | 2D | 2B | V | T | R | P | N | L | J | H | F | D | B |

| Terminal | Input | Output | Connection to | Test condition | Correct voltage | Remark |
|---|---|---|---|---|---|---|
| 2A | – | – | Ground (Injector) | Constant | 0V | – |
| 2B | – | – | Ground (Output) | Constant | 0V | – |
| 2C | – | – | Ground (CPU) | Constant | 0V | – |
| 2D | – | – | Ground (Input) | Constant | 0V | – |
| 2E | ○ | | Distributor (Ne-signal) | Ignition switch ON | Approx. 0V or 5V | – |
| | | | | Idle | Approx. 2V | |
| 2F | – | – | – | – | – | – |
| 2G | ○ | | Distributor (G-signal) [DOHC] | Ignition switch ON | Approx. 0V or 5V | |
| | | | | Idle | Approx. 1.5V | |
| 2H | ○ | | Ground (California) | Constant | 0V | – |
| | | | Open (Federal) | Constant | Approx. 1–2.5V | |
| | | | Main relay (Canada) | Ignition switch ON | Approx. 12V | |
| 2I | – | – | – | – | – | – |
| 2J | ○ | | Open (MTX) | Constant | Approx. 1–2.5V | – |
| | | | Main relay (FUEL INJ relay) (ATX) | Ignition switch ON | Approx. 12V | |
| 2K | ○ | | Throttle sensor (ATX)/EC-AT control unit (ATX)/Airflow meter | Constant | 4.5–5.5V | – |
| 2L | ○ | | Throttle sensor (Power switch) (MTX) | Accelerator pedal released | Approx. 5V | – |
| | | | | Accelerator pedal fully depressed | Below 1.0V | |
| 2M | ○ | | Throttle sensor (ATX)/EC-AT control unit (ATX) | Accelerator pedal released | Approx. 0.5V | – |
| | | | | Accelerator pedal fully depressed | Approx. 4.0V | |
| 2N | ○ | | Oxygen sensor | Ignition switch ON | 0V | – |
| | | | | Idle (Cold engine) | 0V | |
| | | | | Idle (After warm-up) | 0–1.0V | |
| | | | | Increasing engine speed (After warm-up) | 0.5–1.0V | |
| | | | | Deceleration | 0–0.4V | |
| 2O | ○ | | Airflow meter | Ignition switch ON | Approx. 3.8V | – |
| | | | | Idle | Approx. 3.3V | |
| 2P | ○ | | Intake air thermosensor | Ambient air temperature 20°C (68°F) | Approx. 2.5V | Built in airflow meter |
| 2Q | ○ | | Water thermosensor | Engine coolant temperature 20°C (68°F) | Approx. 2.5V | – |
| | | | | After warm-up | Below 0.5V | |
| 2R | – | – | – | – | – | – |

| 2Y | 2W | 2U | 2S | 2O | 2O | 2M | 2K | 2I | 2G | 2E | 2C | 2A | U | S | Q | O | M | K | I | G | E | C | A |
|---|---|---|---|---|---|---|---|---|---|---|---|---|---|---|---|---|---|---|---|---|---|---|---|
| 2Z | 2X | 2V | 2T | 2R | 2P | 2N | 2L | 2J | 2H | 2F | 2D | 2B | V | T | R | P | N | L | J | H | F | D | B |

### CONTROL UNIT PIN LOCATIONS AND TEST VOLTAGES 1990 323

| Terminal | Input | Output | Connection to | Test condition | Correct voltage | Remark |
|---|---|---|---|---|---|---|
| 2S | | ○ | Solenoid valve (VICS) | Engine speed below 5,000 rpm | Below 1.5V | VICS: Variable Inertia Charging System [DOHC] |
| | | | | Engine speed above 5,000 rpm | Approx. 12V | |
| 2T | | ○ | Solenoid valve (Pressure regulator) [BP] | 60 [DOHC]/120 [SOHC] seconds after engine started when engine coolant temperature above 90°C (194°F) and intake air temperature above 58°C (136°F) [DOHC]/50°C (122°F) [SOHC] | Below 1.5V | – |
| | | | | Other condition at idle | Approx. 12V | |
| 2U | | ○ | Injector (Nos. 1, 3) | Ignition switch ON | Approx. 12V | * Engine Signal Monitor: Green and red lamps flash |
| | | | | Idle | Approx. 12V* | |
| | | | | Engine speed above 2,000 rpm on deceleration (After warm-up) | Approx. 12V | |
| 2V | | ○ | Injector (Nos. 2, 4) | Ignition switch at idle | Approx. 12V | |
| | | | | Idle | Approx. 12V* | |
| | | | | Engine speed above 2,000 rpm on deceleration (After warm-up) | Approx. 12V | |
| 2W | | ○ | ISC valve | Ignition switch ON | Approx. 7V | – |
| | | | | Idle | Approx. 9V | |
| 2X | | ○ | Solenoid valve (Purge control) | Ignition switch ON | Approx. 12V | – |
| | | | | Idle | Approx. 12V | |
| 2Y | – | – | – | – | – | – |
| 2Z | | ○ | EC-AT control unit (ATX) | Engine coolant temperature below 72°C (162°F) at idle | Below 2.5V | |
| | | | | Engine coolant temperature above 72°C (162°F) at idle | Approx. 12V | |

| 2Y | 2W | 2U | 2S | 2Q | 2O | 2M | 2K | 2I | 2G | 2E | 2C | 2A | U | S | Q | O | M | K | I | G | E | C | A |
|---|---|---|---|---|---|---|---|---|---|---|---|---|---|---|---|---|---|---|---|---|---|---|---|
| 2Z | 2X | 2V | 2T | 2R | 2P | 2N | 2L | 2J | 2H | 2F | 2D | 2B | V | T | R | P | N | L | J | H | F | D | B |

### CONTROL UNIT PIN LOCATIONS AND TEST VOLTAGES TROUBLESHOOTING CHART – 1990 323

| Incorrect voltage | Possible cause |
|---|---|
| Above 0V | • Poor contact at ground terminal<br>• Open circuit in wiring from ECU to ground |
| Always approx. 0V or approx. 2V | • Refer to Code No. 02 Troubleshooting |
| – | – |
| Always approx. 0V or approx. 1.5V | • Refer to Code No. 03 Troubleshooting |
| Approx. 5V (California) | • Open circuit in wiring from ECU terminal 2H to ground |
| Always 0V (Federal) | • Short circuit in wiring from ECU terminal 2H to ground |
| Always 0V (Canada) | • Short circuit in wiring from ECU terminal 2H to main relay (FUEL INJ relay) |
| – | – |
| Approx. 5V (MTX) | • ECU malfunction |
| Always 0V (ATX) | • Open circuit in wiring from ECU terminal 2J to ground |
| Always 0V | • Short circuit in wiring from ECU terminal 2K to throttle sensor (ATX), EC-AT control unit (ATX), or airflow meter<br>• Poor connection at ECU connector<br>• ECU malfunction |
| Below 4.5V or above 5.5V | • ECU malfunction |
| Always 0V | • Throttle sensor malfunction<br>• Short circuit in wiring from ECU terminal 2L to throttle sensor<br>• Poor connection at ECU connector<br>• ECU malfunction |
| Always approx. 5V | • Throttle sensor misadjustment<br>• Open circuit in wiring from ECU terminal 2L to throttle sensor<br>• Open circuit in wiring from ECU terminal 2L to ECU terminal 2D |
| Always constant | • Open circuit in wiring from ECU terminal 2M to throttle sensor<br>• Open circuit in wiring from ECU terminal 2K to throttle sensor<br>• Open circuit in wiring from ECU terminal 2D to throttle sensor |
| Always above 1V | • Throttle sensor misadjustment |
| 0V after warm-up | • Refer to Code No. 15 Troubleshooting |
| Always approx. 1V after warm-up | • Refer to Code No. 17 Troubleshooting |
| Always approx. 0V or approx. 5V | • Refer to Code No. 08 Troubleshooting |
| Always approx. 0V or approx. 5V | • Refer to Code No. 10 Troubleshooting |
| Always approx. 0V or approx. 5V | • Refer to Code No. 09 Troubleshooting |
| – | – |

## CONTROL UNIT PIN LOCATIONS AND TEST VOLTAGES TROUBLESHOOTING CHART—1990 323

| Incorrect voltage | | Possible cause |
|---|---|---|
| — | | — |
| Always below 1.0V | | • Throttle sensor misadjustment<br>• Short circuit in wiring from throttle sensor to ECU terminal 1N<br>• ECU malfunction |
| Always approx. 12V | | • Throttle sensor misadjustment<br>• Open circuit in wiring from throttle sensor to ECU terminal 1N<br>• Open circuit in wiring from throttle sensor to ECU terminal 2D |
| Always below 1.0V (Stoplight OK) | | • Open circuit in wiring from stoplight switch to ECU terminal 1O |
| Always below 1.0V | | • P/S pressure switch malfunction<br>• Short circuit in wiring from P/S pressure switch to ECU terminal 1P<br>• ECU malfunction |
| Always approx. 12V | | • P/S pressure switch malfunction<br>• Open circuit in wiring from P/S pressure switch to ECU terminal 1P<br>• Open circuit in wiring from P/S pressure switch to ground |
| Always below 2.5V (Blower fan OK) | | • A/C switch malfunction<br>• Short circuit in wiring from A/C switch to ECU terminal 1Q<br>• Poor connection at ECU connector<br>• ECU malfunction |
| Always approx. 12V (Blower fan OK) | | • A/C switch malfunction<br>• Open circuit in wiring from A/C switch to ECU terminal 1Q<br>• Open circuit in wiring from A/C switch to blower control switch |
| Always below 1.0V (Electrical cooling fan OK) | | • Open or short circuit in wiring from fan relay to ECU terminal 1R<br>• ECU malfunction |
| Always below 1.0V (Blower fan OK) | | • Short circuit in wiring from blower control switch to ECU terminal 1S<br>• Poor connection at ECU connector<br>• ECU malfunction |
| Always approx. 12V (Blower fan OK) | | • Open circuit in wiring from blower control switch to ECU terminal 1S |
| Always below 1.0V | Illumination lamp ON when rear window defroster switch ON | • Open or short circuit in wiring from rear window defroster switch to ECU terminal 1T |
| | Illumination lamp never ON | • Open circuit in wiring from ignition switch to rear window defroster switch<br>• Rear window defroster switch malfunction |
| Always below 1.0V (Headlights OK) | | • Open or short circuit in wiring from headlight relay to ECU terminal 1U |
| Always below 1.0V (MTX) | | • Neutral switch malfunction<br>• Clutch switch malfunction<br>• Short circuit in wiring from ECU terminal 1V to neutral or clutch switch |
| Always approx. 12V (MTX) | | • Neutral switch malfunction<br>• Clutch switch malfunction<br>• Open circuit in wiring from ECU terminal 1V to neutral or clutch switch<br>• Poor connection at ECU connector |
| Always below 1.0V (ATX) | | • Inhibitor switch malfunction<br>• Short circuit in wiring from inhibitor switch to ECU terminal 1V |
| Always approx. 12V (ATX) | | • Inhibitor switch malfunction<br>• Open circuit in wiring from inhibitor switch to ECU terminal 1V |

| Incorrect voltage | | Possible cause |
|---|---|---|
| Always 0V | | • ROOM 10A fuse burned<br>• Open circuit in wiring from ROOM 10A fuse to ECU terminal 1A |
| Always 0V | | • Main relay malfunction<br>• Open or short circuit in wiring from main relay to ECU terminal 1B |
| Always 0V (Starter turns) | | • Open or short circuit in wiring from starter interlock switch (USA MTX), ignition switch (CANADA MTX), or inhibitor switch (ATX) to ECU terminal 1C |
| Always 0V | | • Main relay (FUEL INJ relay) malfunction<br>• Open circuit in wiring from main relay to diagnosis connector terminal +B<br>• Open or short circuit in wiring from diagnosis connector terminal MEN to ECU terminal 1D |
| Always approx. 12V | | • Poor connection at ECU connector<br>• ECU malfunction |
| Always approx. 5V | | • ECU malfunction |
| Always below 2.5V | MIL always ON | • Short circuit in wiring from combination meter to ECU terminal 1E<br>• ECU malfunction |
| | MIL never ON | • Open circuit in wiring from combination meter to ECU terminal 1E |
| Always approx. 12V | | • Poor connection at ECU connector<br>• ECU malfunction |
| Always below 2.5V | No display on Self-Diagnosis Checker | • Main relay (FUEL INJ relay) malfunction<br>• Open circuit in wiring from main relay to diagnosis connector terminal +B |
| | "88" displayed and buzzer sounds continuously | • Open or short circuit in wiring from diagnosis connector terminal FEN to ECU terminal 1F |
| Always approx. 12V | | • Poor connection at ECU connector<br>• ECU malfunction |
| Always 0V | | • Short circuit in wiring from igniter to ECU terminal 1G |
| — | | — |
| — | | — |
| Always below 2.5V | A/C does not operate | • A/C relay malfunction<br>• Open circuit in wiring from ignition switch to A/C relay<br>• Open circuit in wiring from A/C relay to ECU terminal 1J |
| | A/C switch OFF but A/C operates | • Short circuit in wiring from A/C relay to ECU terminal 1J<br>• ECU malfunction |
| Always approx. 12V | | • Poor connection at ECU connector<br>• ECU malfunction |
| Always below 1.0V | | • Short circuit in wiring from diagnosis connector terminal TEN to ECU terminal 1K |
| Always approx. 12V | | • Open circuit in wiring from diagnosis connector terminal TEN to ECU terminal 1K<br>• Open circuit in wiring from diagnosis connector terminal GND to ground |
| Always below 2.5V | DRL ON when ignition switch ON | • Short circuit in wiring from DRL relay to ECU terminal 1L<br>• Short circuit in wiring from DRL relay to DRL unit |
| | DRL never ON | • Open circuit in wiring from DRL relay to ignition switch |
| Always approx. 12V | | • Parking brake switch always ON<br>• DRL unit malfunction |

## CONTROL UNIT PIN LOCATIONS AND TEST VOLTAGES TROUBLESHOOTING CHART—1990 323

| Incorrect voltage | Possible cause |
|---|---|
| Always approx. 0V or approx. 12V | • Refer to Code No. 41 Troubleshooting |
| Always approx. 0V or approx. 12V | • Refer to Code No. 25 Troubleshooting |
| Always approx. 0V | • Main relay (FUEL INJ relay) malfunction<br>• Open or short circuit in wiring from injector to ECU terminal 2U or 2V |
| Always approx. 12V | • ECU malfunction |
| Always approx. 0V or approx. 12V | • Refer to Code No. 34 Troubleshooting |
| Always approx. 0V or approx. 12V | • Refer to Code No. 26 Troubleshooting |
| — | — |
| Always approx. 0V or approx. 12V | • Open or short circuit in wiring from ECU terminal 2Z to EC-AT control unit terminal 1N |

## CONTROL UNIT WIRING SCHEMATIC—1990 323

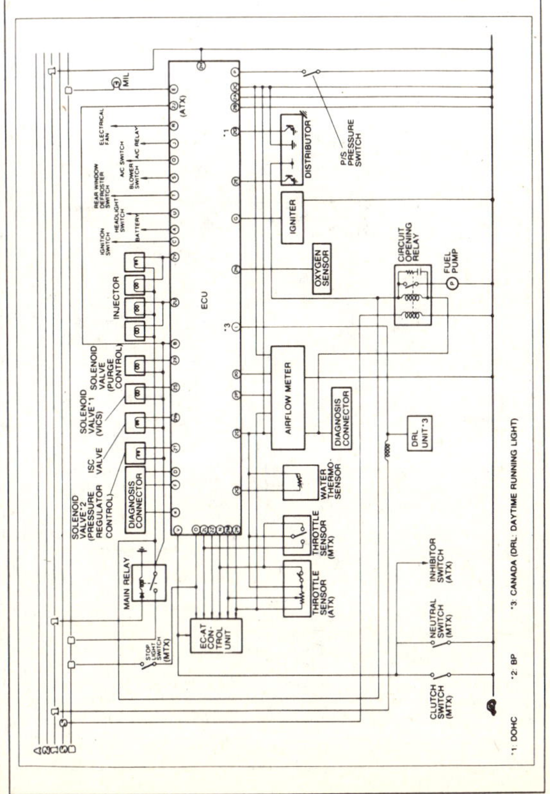

## SYSTEM SCHEMATIC
## 1988–89 626/MX-6 NON-TURBO

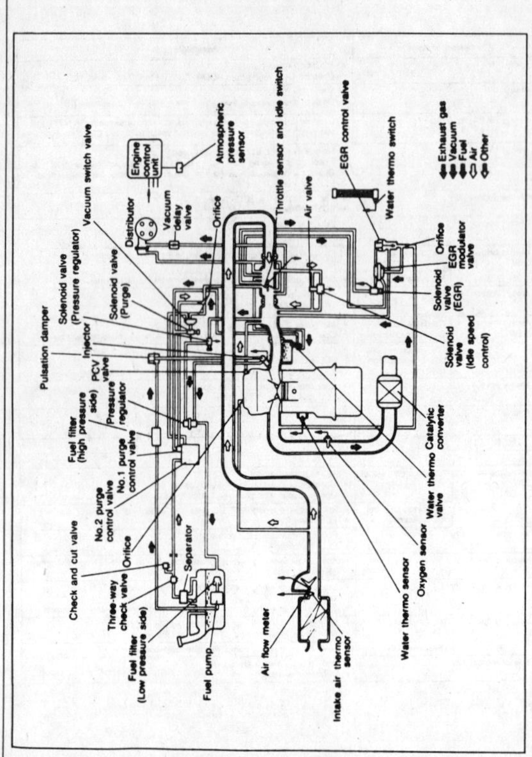


## COMPONENT LOCATION (INPUT)
## 1988 626/MX-6 NON-TURBO

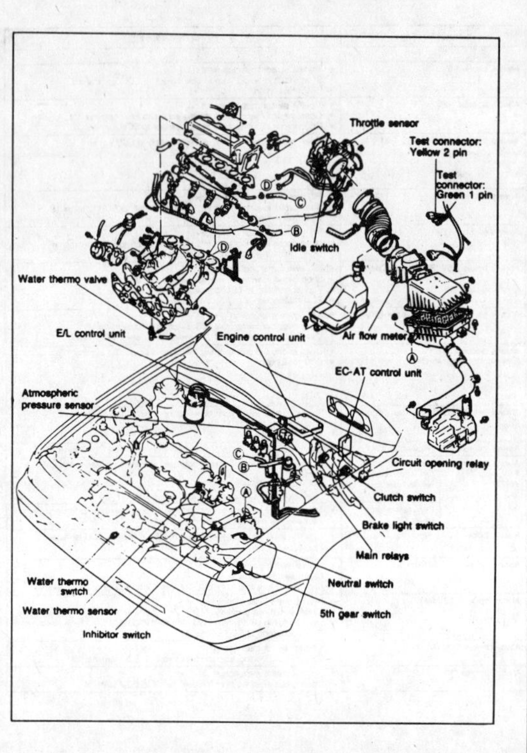


## COMPONENT LOCATION (INPUT)
## 1989 626/MX-6 NON-TURBO

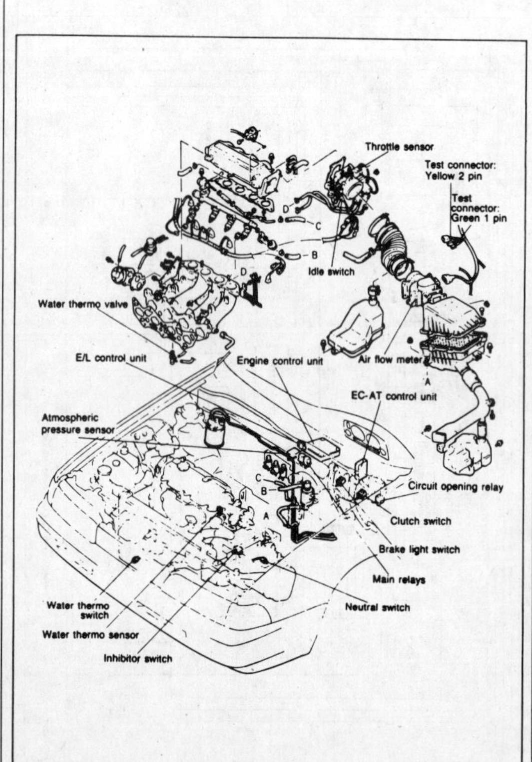


## EXPLODED VIEW OF INTAKE AIR SYSTEM
## 1988–89 626/MX-6 NON-TURBO

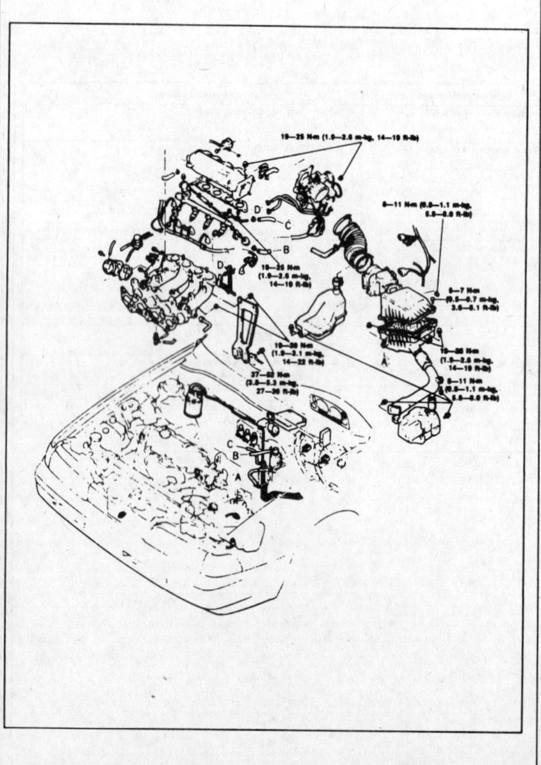

## FUEL COMPONENTS LOCATION (OUTPUT) 1988–89 626/MX-6 NON-TURBO

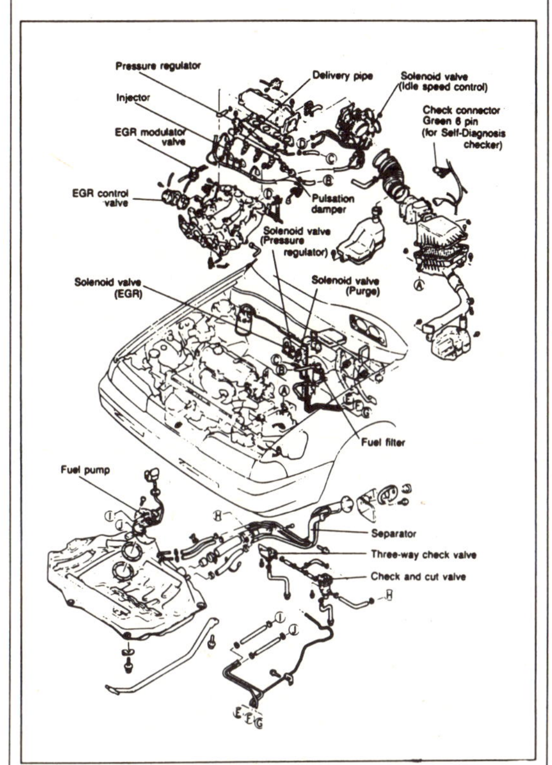


## VACUUM HOSE ROUTING 1988–89 626/MX-6 NON-TURBO

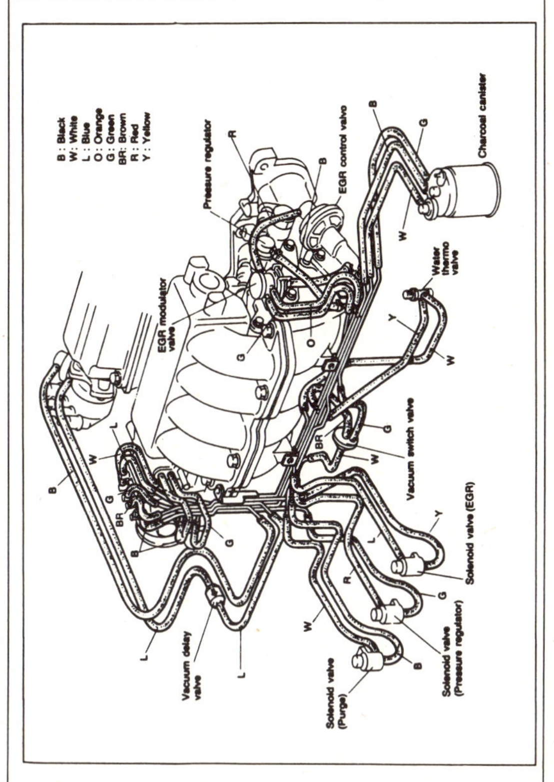


## SPECIFICATIONS CHART 1988–89 626/MX-6 NON-TURBO

| Item | | Engine type: Non-Turbo Engine |
|---|---|---|
| Idle speed rpm | | 750 ± 25 (ATX: P range)* |
| **Throttle body** | | |
| Type | | Horizontal draft (2-barrel) |
| Throat diameter mm (in) | No. 1 | MTX: 40 (1.6), ATX: 46 (1.8) |
| | No. 2 | MTX: 46 (1.8), ATX: 40 (1.6) |
| **Air flow meter** | | |
| Resistor Ω | E2—Vs | Fully closed: 20—400 Fully open: 20—1,000 |
| | E2—Vc | 100—400 |
| | E2—VB | 200—400 |
| | E2—THA | -20°C (-4°F) 13,600—18,400<br>20°C (68°F) 2,210— 2,690<br>60°C (140°F) 493— 667 |
| **Fuel pump** | | |
| Type | | Impeller (in tank) |
| Output pressure kPa (kg/cm², psi) | | 441—588 (4.5—6.0, 64—85) |
| Feeding capacity cc (cu in)/10 sec. | | 220 (13.4) min |
| **Fuel filter** | | |
| Type | Low pressure side | Nylon element |
| | High pressure side | Paper element |
| **Pressure regulator** | | |
| Type | | Diaphragm |
| Regulating pressure kPa (kg/cm², psi) | | 235—275 (2.4—2.8, 34—40) |
| **Injector** | | |
| Type | | High-ohmic |
| Type of drive | | Voltage |
| Resistance Ω | | 12—16 |
| Injection amount cc (cu in)/15 seconds | | 44—61 (2.68—3.72) |
| **Idle speed control valve** | | |
| Solenoid resistance Ω | | 6.3—9.9 |
| **Fuel tank** | | |
| Capacity liters (US gal, Imp gal) | | 60 (15.9, 13.2) |
| **Air cleaner** | | |
| Element type | | Oil permeated |
| **Fuel** | | |
| Specification | | Unleaded regular |

* With test connector grounded

## TROUBLESHOOTING CHART 1988–89 626/MX-6 NON-TURBO

This troubleshooting guide shows the malfunction numbers and the symptoms of various failures. Perform troubleshooting as described below.

| Symptom \ Possible cause | Input sensors and switches: Ignition pulse | Air flow meter | Water thermo sensor | Intake air thermo sensor | Throttle sensor | Atmospheric pressure sensor | Oxygen sensor | Feedback system | Output solenoid valves: Solenoid valve (Pressure regulator) | Solenoid valve (Purge) | Solenoid valve (EGR) | Solenoid valve (Idle speed control) |
|---|---|---|---|---|---|---|---|---|---|---|---|---|
| 1 Fault indicated by SST Code No. | 01 | 08 | 09 | 10 | 12 | 14 | 15 | 17 | 25 | 26 | 28 | 34 |
| 2 Hard start or won't start (Cranks OK) | | | | | | | | | | | | |
| 3 Engine stalls — While warming up | | | | | | | | | | | | |
| 3 Engine stalls — After warming up | | | | | | | | | | | | |
| 4 Rough idle — While warming up | | | | | | | | | | | | |
| 4 Rough idle — After warming up | | | | | | | | | | | | |
| 5 High idle speed after warming up | | | | | | | | | | | | |
| 6 Poor acceleration, hesitation or lack of power | | | | | | | | | | | | |
| 7 Runs rough on deceleration | | | | | | | | | | | | |
| 8 Afterburn in exhaust system | | | | | | | | | | | | |
| 9 Poor fuel consumption | | | | | | | | | | | | |
| 10 Engine stalls or rough after hot starting | | | | | | | | | | | | |
| 11 Fails emission test | | | | | | | | | | | | |

### TROUBLESHOOTING PROCEDURE

**Note**
**Step 1 under symptom is to quickly determine what system or unit may be at fault by use of the SST. (Self-Diagnosis Checker 49 H018 9A1)**

**1st:** Check input sensors and output solenoid valves with the **SST**.

**2nd:** Check other switches with the **SST**.

**3rd:** Check the following items:

**Electrical system**
1) Battery condition
2) Fuses

**Ignition system**
1) Ignition spark
2) Ignition timing

**Fuel system**
1) Fuel level
2) Fuel leakage
3) Fuel filter
4) Idle speed (with test connector grounded)

**Intake air system**
1) Air cleaner element
2) Vacuum or air leakage
3) Vacuum hose routing
4) Accelerator cable

**Engine**
1) Compression
2) Overheating

**Others**
1) Clutch slippage
2) Brake dragging

4th. Check Fuel and Emission Control Systems.

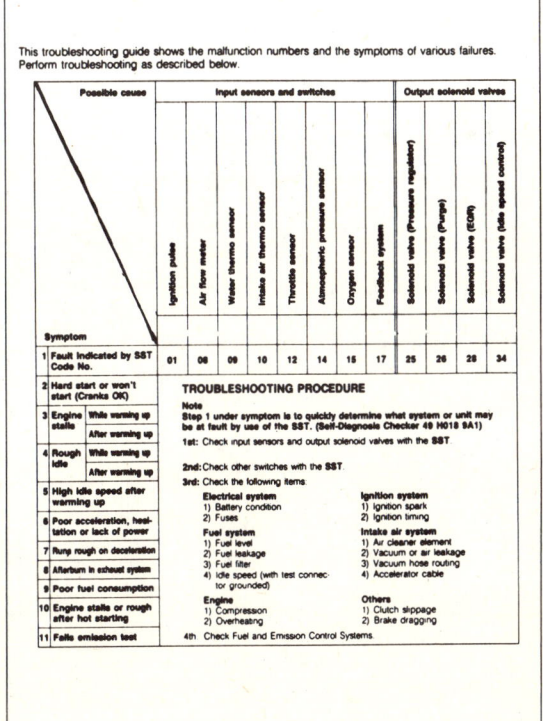

## TROUBLESHOOTING CHART (CONT.) 1988-89 626/MX-6 NON-TURBO

The Troubleshooting Guide lists the systems most likely to cause a given symptom. After finding systems to check, refer to the pages shown for detailed guides for each system.

| Symptom \ Possible cause | Fuel and Emission Control Systems: Intake Air System (Poor connection of components, throttle body) | Fuel System (Fuel injection, Fuel pressure) | Pressure Regulator Control System | Idle Speed Control (ISC) System (Air valve, Idle speed control solenoid malfunction) | EGR System (EGR control valve stuck and open) | EEC system (Vacuum switch valve, No.1 purge valve malfunction) | PCV System (System clogged) | Deceleration System (Fuel cut operation malfunction) | Exhaust system (System clogged) |
|---|---|---|---|---|---|---|---|---|---|
| 2 | 2 | 1 | | | | | | | |
| 3 | 4 | 3 | | 1 | 2 | | | | |
| 3 | 5 | 4 | | 2 | 3 | | 1 | | |
| 4 | 5 | 4 | | 1 | 3 | | 2 | | |
| 4 | 6 | 5 | | 2 | 3 | 4 | 1 | | |
| 5 | 2 | | | 1 | | | | | |
| 6 | 3 | 4 | | | 1 | 2 | | | 5 |
| 7 | | 3 | | 2 | | | | 1 | |
| 8 | 3 | 4 | | 1 | | | | 2 | |
| 9 | | 2 | | | 3 | | | 1 | 4 |
| 10 | | 2 | 1 | | | | | | |
| 11 | 6 | 7 | | 4 | 2 | 5 | | 3 | 1 |

The numbers of the list show the priorities of inspections from the most possible to that with the lowest possibility. These were determined on the following basis:

- Ease of inspection
- Most possible system
- Most possible point in system

## TROUBLE CODE IDENTIFICATION CHART 1988–89 626/MX6 NON-TURBO

| Malfunction display: Code No. | MIL output signal pattern | Sensor or subsystem | Self-diagnosis | Fail-safe |
|---|---|---|---|---|
| 01 | ON / OFF | Ignition pulse | No ignition signal | — |
| 08 | ON / OFF | Air flow meter | Open or short circuit | Maintains basic signal at preset value |
| 09 | ON / OFF | Water thermo sensor | Open or short circuit | Maintains constant command • 35°C (95°F) for EGI • 50°C (122°F) for ISC control use |
| 10 | ON / OFF | Intake air thermo sensor (air flow meter) | Open or short circuit | Maintains constant 20°C (68°F) command |
| 12 | ON / OFF | Throttle sensor | Open or short circuit | Maintains constant command of throttle valve fully open |
| 14 | ON / OFF | Atmospheric pressure sensor | Open or short circuit | Maintains constant command of sea level pressure |
| 15 | ON / OFF | Oxygen sensor | Sensor output continues less than 0.55V 120 sec. after engine starts (1,500 rpm) | Cancels EGI feedback operation |
| 17 | ON / OFF | Feedback system | Sensor output not changed 20 sec. after engine exceeds 1,500 rpm | Cancels EGI feedback operation |
| 25 | ON / OFF | Solenoid valve (pressure regulator) | Open or short circuit | – |
| 26 | ON / OFF | Solenoid valve (purge control) | | – |
| 28 | ON / OFF | Solenoid valve (EGR) | | – |
| 34 | ON / OFF | Solenoid valve (Idle speed control) | | – |

**Caution**

**a) If there is more than one failure present, the lowest number malfunction code is displayed first, the remaining codes are displayed sequentially.**

**b) After repairing a failure, turn off the ignition switch and disconnect the negative battery cable and depress the brake pedal for at least 5 seconds to erase the memory of a malfunction code.**

## TROUBLE CODE DIAGNOSTIC CHART—1988–89 626/MX6 NON-TURBO

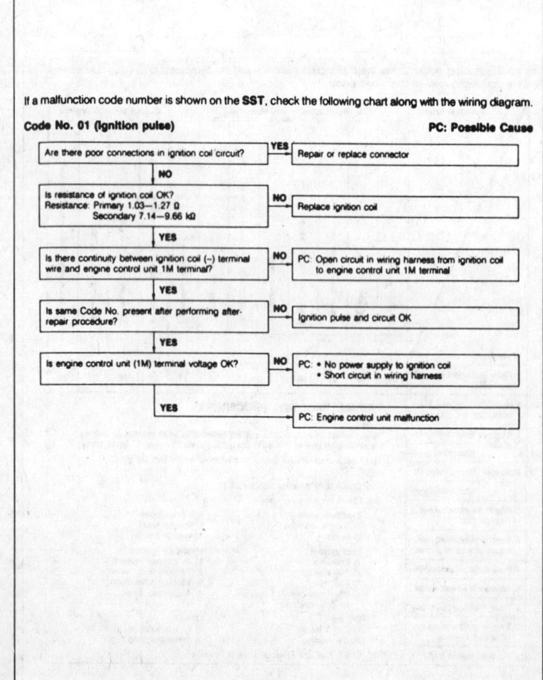


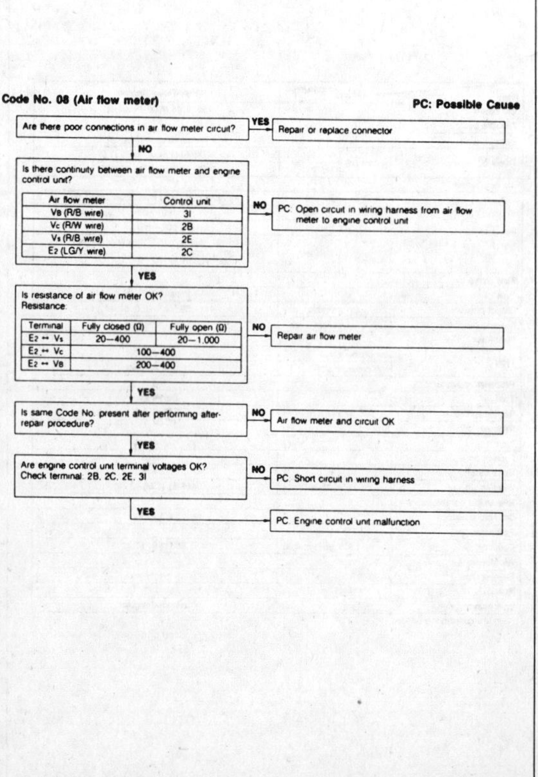

## TROUBLE CODE DIAGNOSTIC CHART—1988–89 626/MX6 NON-TURBO

### Code No. 09 (Water thermo sensor)

**PC: Possible Cause**

Are there poor connections at water thermo sensor circuit?
- **YES** → Repair or replace connector
- **NO** ↓

Is there continuity between water thermo sensor and control unit?

| Water thermo sensor | Control unit |
|---|---|
| A (Y/B wire) | 2I |
| B (LG/Y wire) | 2C |

- **NO** → PC: Open circuit in wiring harness from water thermo sensor to engine control unit
- **YES** ↓

Is resistance of the water thermo sensor OK?
Resistance

| Coolant temp | Resistance |
|---|---|
| −20°C (−4°F) | 14.5—17.8 kΩ |
| 20°C (68°F) | 2.2—2.7 kΩ |
| 40°C (104°F) | 1.0—1.3 kΩ |
| 60°C (140°F) | 500—640 Ω |
| 80°C (176°F) | 280—350 Ω |

- **NO** → Replace water thermo sensor
- **YES** ↓

Is same Code No. present after performing after-repair procedure?
- **NO** → Water thermo sensor and circuit OK
- **YES** ↓

Are engine control unit 2I and 2C terminal voltages OK?
- **NO** → PC: Engine short circuit in wiring harness
- **YES** → PC: Engine control unit malfunction

### Code No. 10 (Intake air thermo sensor)

**PC: Possible Cause**

Are there poor connections in air flow meter circuit?
- **YES** → Repair or replace connector
- **NO** ↓

Is there continuity between intake air thermo sensor (in air flow meter) and engine control unit?

| Intake air temperature sensor (in air flow meter) | Control unit |
|---|---|
| THA (R wire) | 2J |
| $E_2$ (LG/Y wire) | 2C |

- **NO** → PC: Open circuit in wiring harness from intake air thermo sensor (in air flow meter) to engine control unit
- **YES** ↓

Is resistance of intake air thermo sensor (in air flow meter) OK?
Resistance

| Terminal | Resistance |
|---|---|
| $E_2$ ↔ THA | −20°C (−4°F): 13.6—18.4 kΩ<br>20°C (68°F): 2.21—2.69 kΩ<br>60°C (140°F): 493—667 Ω |

- **NO** → Replace air flow meter
- **YES** ↓

Is same Code No. present after performing after-repair procedure?
- **NO** → Intake air thermo sensor and circuit OK
- **YES** ↓

Are engine control unit 2J and 2C terminal voltages OK?
- **NO** → PC: Short circuit in wiring harness
- **YES** → PC: Engine control unit malfunction

## TROUBLE CODE DIAGNOSTIC CHART—1988–89 626/MX6 NON-TURBO

### Code No. 12 (Throttle sensor)

**PC: Possible cause**

Are there poor connections in throttle sensor circuit?
- **YES** → Repair or replace connector terminal
- **NO** ↓

Is there 4.5—5.5 V at A terminal of throttle sensor connector?
- **NO** → PC: •Open or short circuit in wiring harness from A terminal to 2A terminal of control unit •Control unit malfunction
- **YES** ↓

Is C terminal of throttle sensor connector grounded?
- **NO** → PC: Open circuit in wiring harness from C terminal to ground
- **YES** ↓

Is throttle sensor adjusted correctly?
- **NO** → Adjust throttle sensor
- **YES** ↓

Is same Code No. present after performing after-repair procedure?
- **NO** → Throttle sensor and circuit OK
- **YES** ↓

Is engine control unit 2G terminal voltage OK?
- **NO** → PC: Open or short circuit in wiring harness from B terminal of throttle sensor to 2G terminal of engine control unit
- **YES** → PC: Engine control unit malfunction

### Code No. 14 (Atmospheric pressure sensor)

**PC: Possible cause**

Are there poor connections in atmospheric pressure sensor circuit?
- **YES** → Repair or replace connector
- **NO** ↓

Is there Vref at (LG/R) wire of atmospheric pressure sensor connector?
Vref: 4.5—5.5V
- **NO** → PC: Open or short circuit in wiring harness from (LG/R) wire to engine control unit • Engine control unit malfunction (If 2A terminal voltage not 4.5—5.5V)
- **YES** ↓

Is voltage at (Y) wire of atmospheric pressure sensor connector OK?
Voltage: 3.5—4.5V.....at sea level
2.5—3.5V.....at high altitude (2,000 m (6,500 ft))
- **NO (0V)** → PC: Short circuit in wiring harness from atmospheric pressure sensor (Y) wire to engine control unit 2H terminal
- **NO (Other)** → Replace atmospheric pressure sensor
- **YES** ↓

Is there continuity between atmospheric pressure sensor to engine control unit?

| Atmospheric pressure sensor | Control unit |
|---|---|
| A (LG/Y wire) | 2C |
| D (Y wire) | 2H |

- **NO** → PC: Open circuit in wiring harness from atmospheric pressure sensor to engine control unit
- **YES** ↓

Is same Code No. present after performing after-repair procedure?
- **NO** → Atmospheric pressure sensor and circuit OK
- **YES** ↓

Are engine control unit 2C and 2H terminal voltages OK?
- **NO** → PC: Short circuit in wiring harness
- **YES** → PC: Engine control unit malfunction

## TROUBLE CODE DIAGNOSTIC CHART—1988–89 626/MX6 NON-TURBO

### Code No. 15 (Oxygen sensor)

**PC: Possible Cause**

**Note**
**When Codes No.15 and 17 are present at the same time, first perform the checking procedure for Code No.17.**

- Are there poor connections in oxygen sensor circuit?
  - YES → Repair or replace connector
  - NO ↓
- Is oxygen sensor output voltage OK?
  - NO → Replace oxygen sensor
  - YES ↓
- Is sensitivity of oxygen sensor OK?
  - NO → PC: •Oxygen sensor malfunction •Engine control unit malfunction
  - YES ↓
- Is same Code No. present after performing after-repair procedure?
  - NO → Oxygen sensor and circuit OK
  - YES ↓
- Is engine control unit 2D terminal voltage OK?
  - NO → PC: Open or short circuit in wiring harness between oxygen sensor and engine control unit 2D terminal
  - YES → PC: Engine control unit malfunction

### Code No. 17 (Feedback system)

**PC: Possible Cause**

- Warm up engine and run it at 2,500—3,000 rpm for three minutes. Does monitor lamp of Self-Diagnosis Checker illuminate at idle?
  - NO → PC: • Air leak in vacuum hoses or emission component • Contaminated oxygen sensor • Insufficient fuel injection
  - YES ↓
- Are spark plugs clean?
  - NO → Clean or replace spark plugs
  - YES ↓
- Is oxygen sensor voltage OK?
  - NO → PC: Oxygen sensor malfunction
  - YES ↓
- Is same Code No. present after performing after-repair procedure?
  - NO → Feedback system OK
  - YES ↓
- Is engine control unit 2D terminal voltage OK?
  - NO → PC: Open or short circuit in wiring harness between oxygen sensor and engine control unit 2D terminal
  - YES → PC: Engine control unit malfunction

## TROUBLE CODE DIAGNOSTIC CHART—1988–89 626/MX6 NON-TURBO

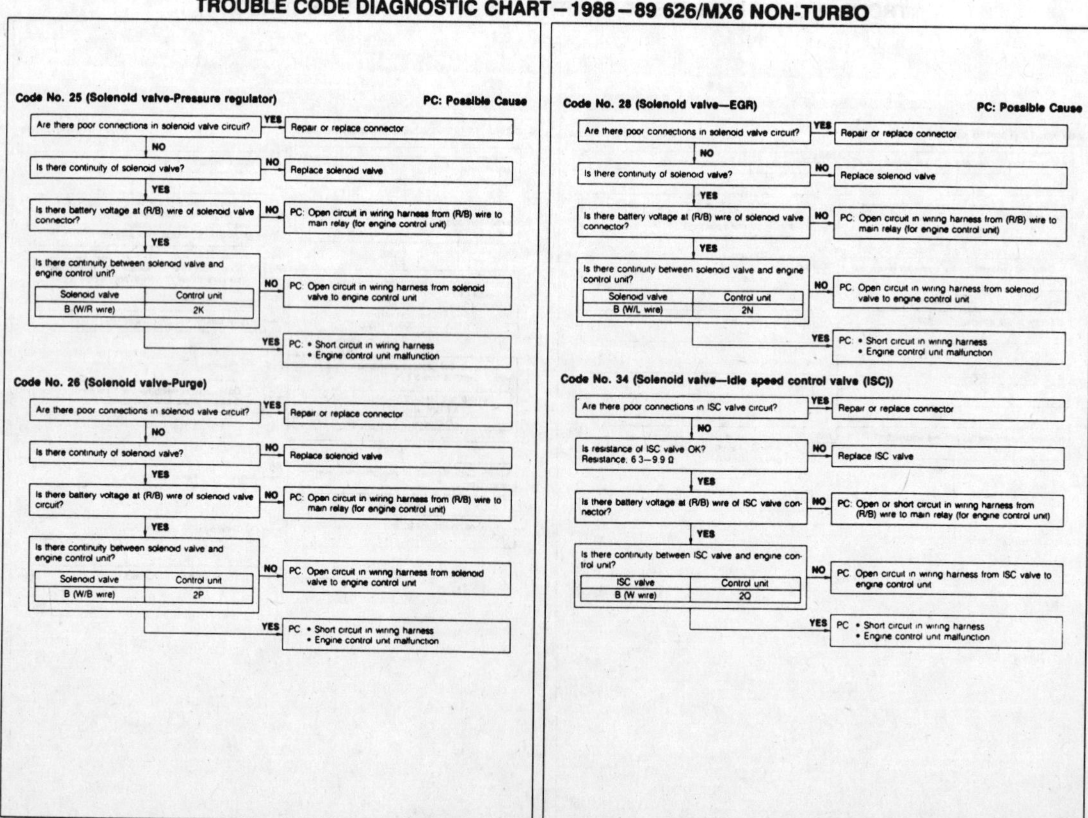

## CONTROL PIN LOCATIONS AND TEST VOLTAGES 1988–89 626/MX-6 NON-TURBO

**Terminal Voltage**
If the input and output devices and related wiring are normal, but the engine control unit terminal voltage is incorrect, replace the engine control unit.

| Terminal | Input | Output | Connection to | Voltage (After warming-up) IGN: ON | Voltage (After warming-up) Idle | Remarks |
|---|---|---|---|---|---|---|
| 1A | | O | Malfunction indicator light | For 3sec. after ignition switch OFF → ON: below 4.8V (light illuminates) After 3sec.: Battery voltage (light does not illuminate) | | • Test connector grounded • Light illuminates: below 4.8V • Light does not illuminate: Battery voltage |
| 1B | | O | Self-Diagnosis Checker (Code number) | For 3sec. after ignition switch OFF → ON: below 6.2V (Buzzer sounds) After 3sec.: Battery voltage (Buzzer does not sound) | | • Using Self-Diagnosis Checker and test connector grounded • Buzzer sounds: below 6.2V • Buzzer does not sound: Battery voltage |
| 1C | — | — | — | — | | — |
| 1D | | O | Self-Diagnosis Checker (Monitor lamp) | Test connector grounded For 3sec. after ignition switch OFF → ON: below 6.2V (light illuminates) After 3sec.: Battery voltage (light does not illuminate) | (Test connector grounded) approx. 5V (Test connector not grounded) Monitor lamp ON: below 6.2V Monitor lamp OFF: Battery voltage | With Self-Diagnosis Checker |
| 1E | O | | Idle switch | Accelerator pedal released: below 0.5V Accelerator pedal depressed: above 7.7V | | |
| 1F | | O | A/C relay | Battery voltage | A/C switch ON: below 2.5V A/C switch OFF: Battery voltage | Blower motor ON |
| 1G | — | — | — | — | | — |
| 1H | O | | Water thermo switch | Above 7.3V | | Radiator temp.: below 17°C (63°F) |
| | | | | Below 1.5V | | Radiator temp.: above 17°C (63°F) |
| 1I | O | | Electrical load control unit | Electrical load ON: below 1.5V Electrical load OFF: above 7.3V | | Electrical load: Rear defroster Headlight Blower motor (3rd & 4th position) Electrical fan |
| 1J | O | | Brake light switch | Brake pedal released: below 3.6V Brake pedal depressed: above 10.0V | | |
| 1K | O | | P/S pressure switch | Constant above 10.5V | P/S ON: below 1.5V P/S OFF: above 10.5V | |
| 1L | O | | A/C switch | A/C switch ON: below 1.5V A/C switch OFF: above 10.0V | | Blower motor: ON |
| 1M | O | | Ignition coil ⊖ terminal | Battery voltage | *1 Battery voltage | *1 Engine Signal Monitor: green and red lights flash |
| 1N | — | — | — | — | | — |

## CONTROL PIN LOCATIONS AND TEST VOLTAGES 1988 626/MX-6 NON-TURBO

| Terminal | Input | Output | Connection to | Voltage (After warming-up) IGN: ON | Voltage (After warming-up) Idle | Remarks |
|---|---|---|---|---|---|---|
| 1O | — | — | — | — | | — |
| 1P | — | — | — | — | | — |
| 1Q | — | — | — | — | | — |
| 1R | — | — | — | — | | — |
| 1S | | O | Shift indicator light (MTX) | Battery voltage | | Shift indicator light illuminates: below 4.8V |
| 1T | — | — | — | — | | — |
| 1U | — | — | — | — | | — |
| 1V | O | | — | Constant below 1.5V | | MTX |
| | | | | Constant above 10.5V | | ATX |
| 1W | O | | Test connector | Test connector grounded: below 0.5V Test connector not grounded: above 10.5V | | Green connector, 1-pin |
| 1X | — | — | — | — | | — |
| 2A | | O | V ref | 4.5—5.5V | | |
| 2B | O | | Air flow meter (Vc) | 7—9V | | |
| 2C | — | — | Ground (E2) | 0V | | |
| 2D | O | | Oxygen sensor | 0V | 0—1.0V | • Cold engine: 0V at idle • After warming-up: Increase engine speed: 0.5—1.0V Deceleration: 0—0.4V |
| 2E | O | | Air flow meter (Vs) | Approx. 1.7V | Approx. 3—5V | Increase engine speed: voltage increases |
| 2F | — | — | — | — | | — |
| 2G | O | | Throttle sensor | Accelerator pedal released: approx. 0.5V (depends on 2A terminal voltage) | | Max. voltage (Throttle valve fully open): approx. 4.3V |
| 2H | O | | Atmospheric pressure sensor | At sea level: approx. 4.0V | | |
| 2I | O | | Water thermo sensor | 0.3—0.6 V | | Engine coolant temp. 20°C (68°F): approx. 2.5V |
| 2J | O | | Air flow meter (Intake air thermo sensor) | Approx. 2.5V at 20°C (68°F) | | |
| 2K | | O | Solenoid valve (Pressure regulator control) | For 120 sec. after ignition switch OFF → ON: below 3.5V | For 120 sec. after starting: below 3.5V | Hot condition: Coolant temp. above 70°C (158°F) Intake air temp. above 20°C (63°F) |
| | | | | Battery voltage | | Other conditions |
| 2L | O | | Speedometer | Battery voltage | | • Above 113 mph (180 km/h): below 1.0V |
| 2M | — | — | — | — | | — |

## CONTROL PIN LOCATIONS AND TEST VOLTAGES 1989 626/MX-6 NON-TURBO

| Terminal | Input | Output | Connection to | Voltage (After warming-up) IGN: ON | Voltage (After warming-up) Idle | Remarks |
|---|---|---|---|---|---|---|
| 1O | — | — | — | — | | — |
| 1P | — | — | — | — | | — |
| 1Q | — | — | — | — | | — |
| 1R | — | — | — | — | | — |
| 1S | — | — | — | — | | — |
| 1T | — | — | — | — | | — |
| 1U | — | — | — | — | | — |
| 1V | O | | — | Constant below 1.5V | | MTX |
| | | | | Constant above 10.5V | | ATX |
| 1W | O | | Test connector | Test connector grounded: below 0.5V Test connector not grounded: above 10.5V | | Green connector, 1-pin |
| 1X | — | — | — | — | | — |
| 2A | | O | V ref | 4.5—5.5V | | |
| 2B | O | | Air flow meter (Vc) | 7—9V | | |
| 2C | — | — | Ground (E2) | 0V | | |
| 2D | O | | Oxygen sensor | 0V | 0—1.0V | • Cold engine: 0V at idle • After warming-up: Increase engine speed: 0.5—1.0V Deceleration: 0—0.4V |
| 2E | O | | Air flow meter (Vs) | Approx. 1.7V | Approx. 3—5V | Increase engine speed: voltage increases |
| 2F | — | — | — | — | | — |
| 2G | O | | Throttle sensor | Accelerator pedal released: approx. 0.5V (depends on 2A terminal voltage) | | Max. voltage (Throttle valve fully open): approx. 4.3V |
| 2H | O | | Atmospheric pressure sensor | At sea level: approx. 4.0V | | |
| 2I | O | | Water thermo sensor | 0.3—0.6 V | | Engine coolant temp. 20°C (68°F): approx. 2.5V |
| 2J | O | | Air flow meter (Intake air thermo sensor) | Approx. 2.5V at 20°C (68°F) | | |
| 2K | | O | Solenoid valve (Pressure regulator control) | For 120 sec. after ignition switch OFF → ON: below 3.5V | For 120 sec. after starting: below 3.5V | Hot condition: Coolant temp. above 70°C (158°F) Intake air temp. above 20°C (63°F) |
| | | | | Battery voltage | | Other conditions |
| 2L | O | | Speedometer | Battery voltage | | • Above 113 mph (180 km/h): below 1.0V |
| 2M | — | — | — | — | | — |

## CONTROL PIN LOCATIONS AND TEST VOLTAGES 1988–89 626/MX-6 NON-TURBO

| Terminal | Input | Output | Connection to | Voltage (After warming-up) IGN: ON | Voltage (After warming-up) Idle | Remarks |
|---|---|---|---|---|---|---|
| 2N | | O | Solenoid valve (EGR) | Below 3.5V | | • Cold engine: below 3.5V Radiator coolant temp.—below 17°C (63°F) or Engine coolant temp.—below 70°C (158°F) • Engine above approx. 1,500 rpm: Battery voltage |
| 2O | — | — | — | — | | — |
| 2P | | O | Solenoid valve (Purge control valve) | Battery voltage | | • Medium and high load: below 3.5V |
| 2Q | | O | Solenoid valve (Idle speed control valve) | Approx. 1.7—11V | | |
| 2R | — | — | Ground ($E_{02}$) | 0V | | |
| 3A | — | — | Ground ($E_{01}$) | 0V | | |
| 3B | O | | Ignition switch (Start position) | Below 2.5V | | While cranking: battery voltage |
| 3C | | O | Injector (No. 4 and No. 2) | Battery voltage | *1 Battery voltage | *1 Engine Signal Monitor green and red lights flash |
| 3D | O | | Inhibitor switch through EC-AT unit | "N" or "P" range: below 2.5V Other ranges: battery voltage | | ATX |
| | | | Neutral and clutch switch | In-gear condition Clutch pedal depressed: battery voltage Clutch pedal released: below 0.5V | | MTX (Neutral: constant battery voltage) |
| 3E | | O | Injector (No. 1 and No. 3) | Battery voltage | *1 Battery voltage | *1 Engine Signal Monitor: green and red lights flash |
| 3F | — | — | — | — | | — |
| 3G | — | — | Ground ($E_1$) | 0V | | |
| 3H | — | — | — | — | | — |
| 3I | O | — | Main relay | Battery voltage | | |
| 3J | — | — | Battery | Battery voltage | | For back-up |

## COMPONENT DESCRIPTIONS 1988–89 626/MX-6 NON-TURBO

| Component | Function | Remarks | Application: New model | Application: Previous model |
|---|---|---|---|---|
| Air flow meter | Detects amount of intake air; sends signal to control unit | Intake air temp sensor and fuel pump switch are integrated | ○ | ○ |
| Atmospheric pressure sensor | Detects atmospheric pressure; sends signal to control unit | | ○ | ○ |
| Circuit opening relay | Voltage for fuel pump while engine running | | ○ | ○ |
| Clutch switch | Detects in-gear condition; sends signal to control unit | Switch ON when clutch pedal released | ○ | ○ |
| EC-AT control unit | Detects N or P range; sends signal to control unit | | ○ | X |
| Engine control unit | Detects signals from input sensors and switches; controls injector operation | | ○ | ○ |
| Fuel filter | Filters particles from fuel | | ○ | ○ |
| Fuel pump | Provides fuel to injectors | • Operates while engine running<br>• Installed in fuel tank | ○ | ○ |
| Idle switch | Detects when throttle valve fully closed; sends signal to control unit | Installed on throttle body | ○ | ○ |
| Ignition coil (−) terminal | Detects engine speed; sends signal to control unit | | ○ | ○ |
| Ignition switch (ST position) | Sends engine cranking signal to control unit | | ○ | ○ |
| Inhibitor switch | Detects in-gear condition; sends signal to EC-AT control unit | Switch ON in "N" or "P" range | ○ | ○ |
| Injector | Injects fuel into intake port | • Controlled by signals from control unit<br>• High-ohmic injector | ○ | ○ |
| Intake air thermo sensor | Detects intake air temperature; sends signal to control unit | Installed in air flow meter | ○ | ○ |
| Main relay | Supplies electric current to injectors and control unit | | ○ | ○ |
| Neutral switch | Detects in-gear condition; sends signal to control unit | Switch ON when in-gear | ○ | ○ |
| Oxygen sensor | Detects Oxygen concentration; sends signal to control unit | Zirconia ceramic and platinum coating | ○ | ○ |
| Pressure regulator | Adjusts fuel pressure supplied to injectors | | ○ | ○ |
| Pulsation damper | Absorbs fuel pulsation | | ○ | ○ |
| Speedometer | Detects vehicle speed; sends signal to control unit | ON: Above 113 mph (180 km/h) | ○ | X |
| Throttle sensor | Detects throttle valve opening angle; sends signal to control unit | Installed on throttle body | ○ | ○ |
| Water thermo sensor | Detects coolant temperature; sends signal to control unit | | ○ | ○ |
| Water thermo switch | Detects radiator coolant temperature; sends signal to control unit | ON above 17°C (63°F) | ○ | ○ |

## COMPONENT RELATIONSHIP CHART 1988 626/MX-6 NON-TURBO

| INPUT DEVICES \ OUTPUT DEVICES | INJECTOR: FUEL INJECTION AMOUNT | INJECTOR: FUEL INJECTION TIMING | BAC VALVE: AIR VALVE | BAC VALVE: ISC VALVE | SOLENOID VALVE (EGR) | SOLENOID VALVE (PURGE) | SOLENOID VALVE (PRESSURE REGULATOR) | SHIFT INDICATOR LIGHT |
|---|---|---|---|---|---|---|---|---|
| TEST CONNECTOR | × | × | × | ○ | × | × | × | × |
| 5TH GEAR SWITCH | × | × | × | × | × | × | × | ○ |
| BRAKE LIGHT SWITCH | ○ | × | × | × | × | × | × | × |
| VEHICLE SPEED SWITCH | ○ | × | × | × | × | × | × | × |
| ELECTRICAL LOAD CONTROL UNIT | × | × | × | ○ | × | × | × | × |
| P/S PRESSURE SWITCH | × | × | × | ○ | × | × | × | × |
| A/C SWITCH | ○ | × | × | ○ | × | ○ | × | × |
| IGNITION SWITCH (STA POSITION) | ○ | ○ | × | × | × | × | ○ | ○ |
| INHIBITOR SWITCH and EC—AT CONTROL UNIT | ○ | × | × | ○ | ○ | × | × | × |
| NEUTRAL AND CLUTCH SWITCH | ○ | × | × | ○ | × | × | × | ○ |
| OXYGEN SENSOR | ○ | × | × | × | × | × | × | × |
| WATER THERMO SWITCH (RADIATOR) | ○ | × | × | × | ○ | × | × | × |
| ATMOSPHERIC PRESSURE SENSOR | ○ | × | × | ○ | × | × | × | × |
| INTAKE AIR THERMO SENSOR | ○ | × | × | ○ | × | × | ○ | × |
| WATER THERMO SENSOR | ○ | × | × | ○ | ○ | ○ | ○ | × |
| IDLE SWITCH | ○ | ○ | × | ○ | × | × | × | ○ |
| THROTTLE SENSOR | ○ | ○ | × | ○ | ○ | × | ○ | ○ |
| AIR FLOW METER | ○ | × | × | × | × | ○ | × | ○ |
| IGNITION COIL | ○ | ○ | × | ○ | ○ | ○ | ○ | ○ |

○: Related
X: Not related

## OUTPUT COMPONENTS AND ENGINE CONDITIONS 1988 626/MX-6 NON-TURBO

| OUTPUT DEVICES \ ENGINE CONDITIONS | | CRANKING (COLD ENGINE) | WARMING UP (DURING IDLE) | MEDIUM LOAD: COLD | MEDIUM LOAD: WARM | ACCELERATION | HEAVY LOAD | DECELERATION | IDLE (THROTTLE VALVE FULLY CLOSED) | IGN: ON (ENGINE NOT RUNNING) | REMARKS |
|---|---|---|---|---|---|---|---|---|---|---|---|
| INJECTOR | INJECTION | Rich | | | Rich and lean | Rich | | Fuel cut | Rich and lean | No injection | |
| | INJECTION TIMING | 1 group (once per revolution) | | | | 1 group (once per revolution) (once per two revolutions)* | | | 1 group (once per revolution) | | Above 6,300 rpm: fuel cut *Above 4,500 rpm |
| BAC VALVE | AIR VALVE | Open* | | | Close | | | | | | *Coolant temp: below 50°C (122°F) |
| | ISC VALVE | Large amount of bypass air | Large amount of bypass air* | Small amount of bypass air | | | | | | No bypass | *In extreme cold condition |
| SOLENOID VALVE (EGR) | | ON (EGR cut) | | | OFF (EGR) | ON (EGR cut) | OFF (EGR) | ON (EGR cut) | ON (EGR cut)* | ON | *ATX D range: OFF (NO EGR due to no vacuum to system) |
| SOLENOID VALVE (PURGE) | | OFF (2nd stage not operated) | | | ON (2nd stage operates) | | | | OFF (2nd stage not operated) | OFF | 1st stage: controlled by water thermo valve |
| SOLENOID VALVE (PRESSURE REGULATOR CONTROL) | | OFF (Vacuum to pressure regulator) | | | | | | | After starting* ON (Vacuum cut) | OFF | *During hot start only |
| SHIFT INDICATOR LIGHT | | Not illuminated | | Illuminates (Except in 5th gear)* | | | | Not illuminated | | | *Intake manifold vacuum less than specified value (acceleration or load) |

## COMPONENT RELATIONSHIP CHART 1989 626/MX-6 NON-TURBO

| INPUT DEVICES \ OUTPUT DEVICES | INJECTOR: FUEL INJECTION AMOUNT | INJECTOR: FUEL INJECTION TIMING | BAC VALVE: AIR VALVE | BAC VALVE: ISC VALVE | SOLENOID VALVE (EGR) | SOLENOID VALVE (PURGE) | SOLENOID VALVE (PRESSURE REGULATOR) |
|---|---|---|---|---|---|---|---|
| TEST CONNECTOR | × | × | × | ○ | × | × | × |
| BRAKE LIGHT SWITCH | ○ | × | × | × | × | × | × |
| VEHICLE SPEED SWITCH | ○ | × | × | × | × | × | × |
| ELECTRICAL LOAD CONTROL UNIT | × | × | × | ○ | × | × | × |
| P/S PRESSURE SWITCH | × | × | × | ○ | × | × | × |
| A/C SWITCH | ○ | × | × | ○ | × | ○ | × |
| IGNITION SWITCH (STA POSITION) | ○ | ○ | × | × | × | × | ○ |
| INHIBITION SWITCH and EC-AT CONTROL UNIT | ○ | × | × | ○ | ○ | × | × |
| NEUTRAL AND CLUTCH SWITCH | ○ | × | × | ○ | × | × | × |
| OXYGEN SENSOR | ○ | × | × | × | × | × | × |
| WATER THERMO SWITCH (RADIATOR) | ○ | × | × | × | ○ | × | × |
| ATMOSPHERIC PRESSURE SENSOR | ○ | × | × | ○ | × | × | × |
| INTAKE AIR THERMO SENSOR | ○ | × | × | ○ | × | × | ○ |
| WATER THERMO SENSOR | ○ | × | × | ○ | ○ | ○ | ○ |
| IDLE SWITCH | ○ | ○ | × | ○ | × | × | × |
| THROTTLE SENSOR | ○ | ○ | × | ○ | ○ | × | ○ |
| AIR FLOW METER | ○ | × | × | × | × | ○ | × |
| IGNITION COIL | ○ | ○ | × | ○ | ○ | ○ | ○ |

○: Related
X: Not related

## OUTPUT COMPONENTS AND ENGINE CONDITIONS 1989 626/MX-6 NON-TURBO

| OUTPUT DEVICES | | CRANKING (COLD ENGINE) | WARMING UP (DURING IDLE) | MEDIUM LOAD COLD | MEDIUM LOAD WARM | ACCEL-ERATION | HEAVY LOAD | DECEL-ERATION | IDLE (THROT-TLE VALVE FULLY CLOSED) | IGN: ON (ENGINE NOT RUNNING) | REMARKS |
|---|---|---|---|---|---|---|---|---|---|---|---|
| INJEC-TOR | INJECTION | Rich | | | Rich and lean | Rich | | Fuel cut | Rich and lean | No injection | |
| | INJECTION TIMING | 1 group (once per revolution) | | | | 1 group (once per revolution) (once per two revolutions)* | | | 1 group (once per revolution) | | Above 6,300 rpm fuel cut *Above 4,500 rpm |
| BAC VALVE | AIR VALVE | Open* | | | Close | | | | | | *Coolant temp below 50°C (122°F) |
| | ISC VALVE | Large amount of bypass air | Large amount of bypass air* | Small amount of bypass air | | | | | | No bypass | *In extreme cold condition |
| SOLENOID VALVE (EGR) | | ON (EGR cut) | | | OFF (EGR) | ON (EGR cut) | OFF (EGR) | ON (EGR cut) | ON (EGR cut)* | ON | *ATX D range OFF (NO EGR due to no vacuum to system) |
| SOLENOID VALVE (PURGE) | | OFF (2nd stage not operated) | | | ON (2nd stage operates) | | | | OFF (2nd stage not operated) | OFF | 1st stage controlled by water thermo valve |
| SOLENOID VALVE (PRESSURE REGULA-TOR CONTROL) | | OFF (Vacuum to pressure regulator) | | | | | | | After starting* ON (Vacuum cut) | OFF | *During hot start only |

## WIRING SCHEMATIC—1988 626/MX-6 NON-TURBO

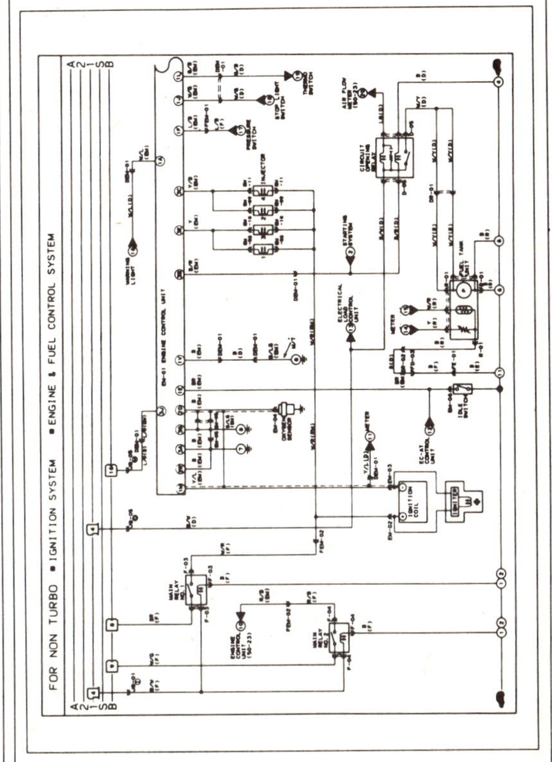


## WIRING SCHEMATIC—1989 626/MX-6 NON-TURBO

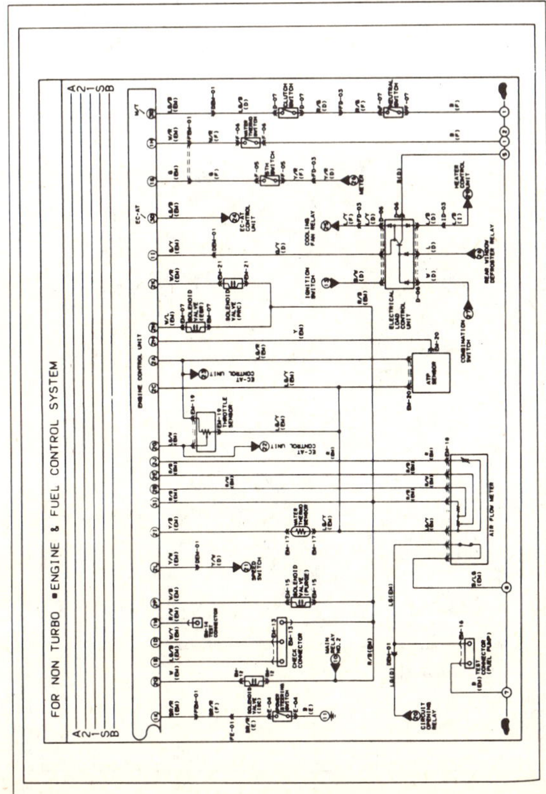


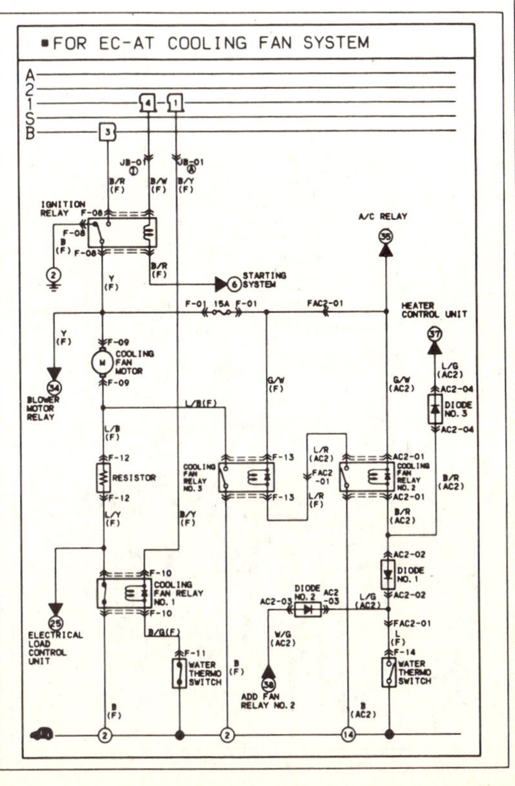

## WIRING SCHEMATIC – 1989 626/MX-6 NON-TURBO

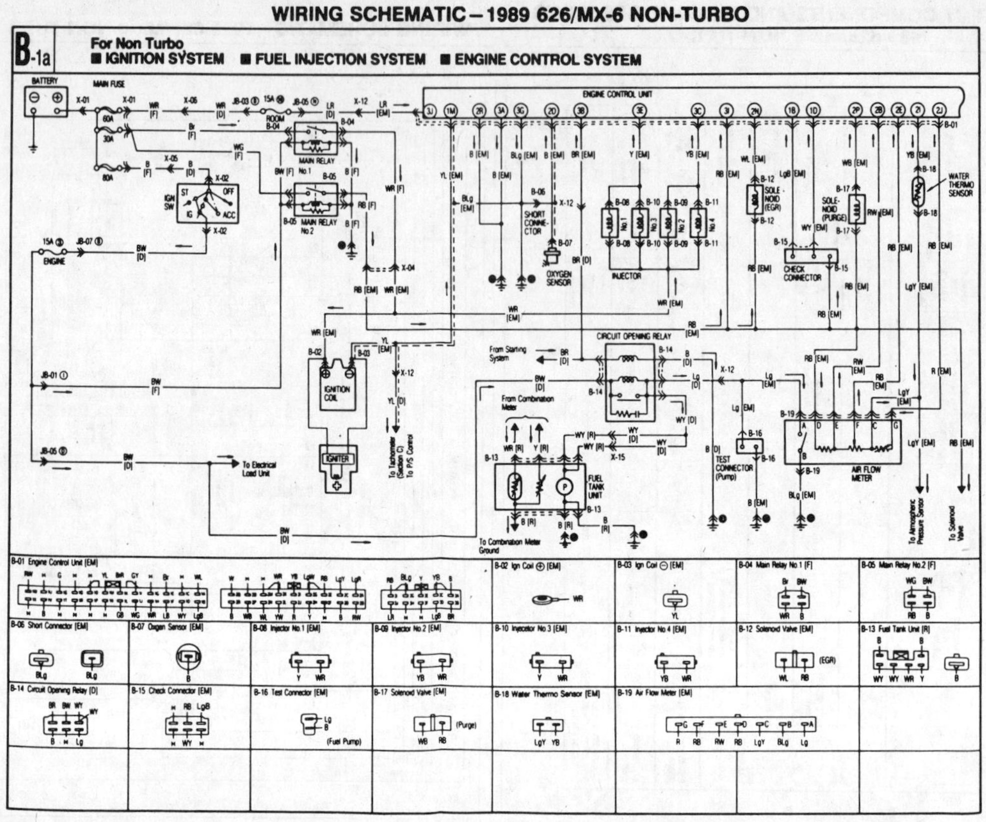


## SYSTEM SCHEMATIC – 1988–89 626/MX-6 WITH TURBO

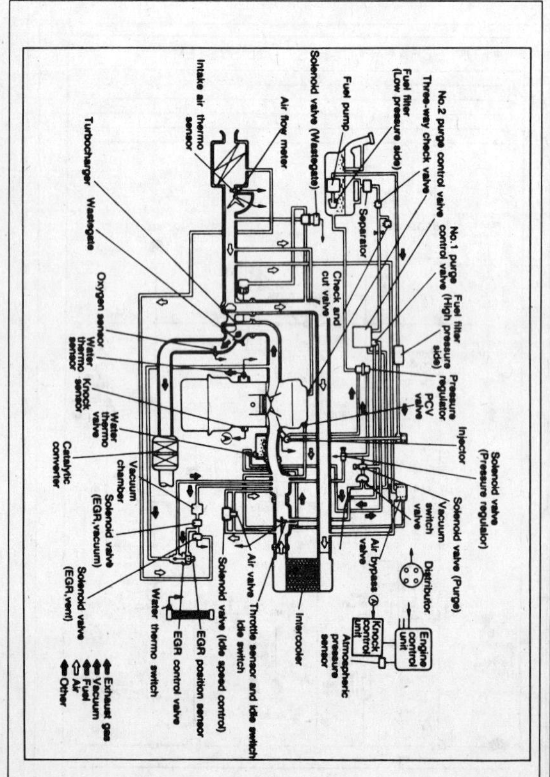


## COMPONENT LOCATION (INPUT) 1988–89 626/MX-6 WITH TURBO

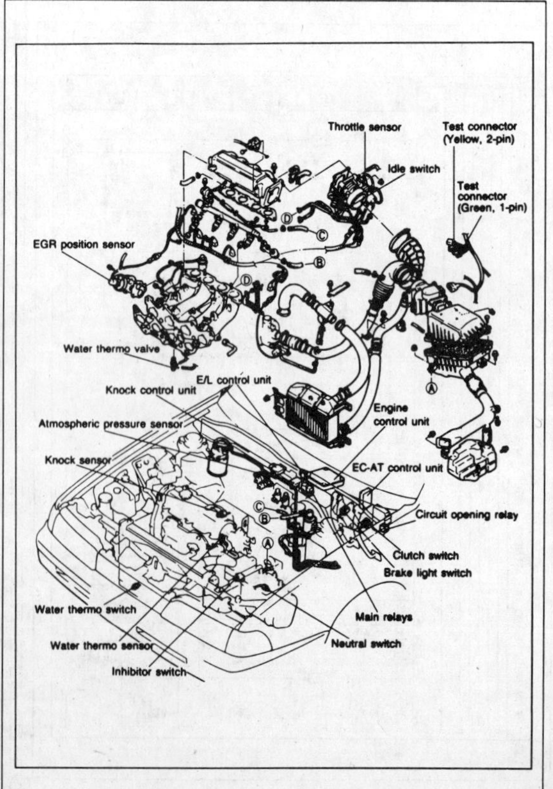

## EXPLODED VIEW OF FUEL SYSTEM COMPONENTS (OUTPUT)—1988–89 626/MX-6 WITH TURBO

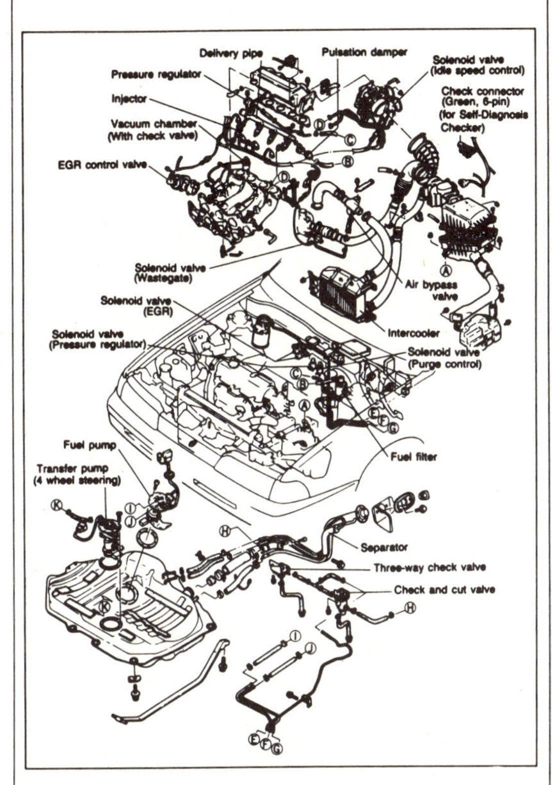


## EXPLODED VIEW OF THE INTAKE AIR SYSTEM 1988–89 626/MX-6 WITH TURBO

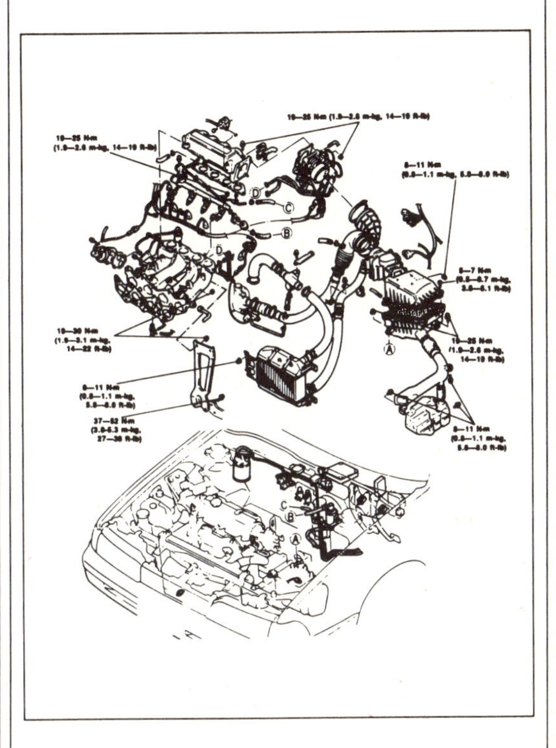

## VACUUM HOSE ROUTING 1988–89 626/MX-6 WITH TURBO

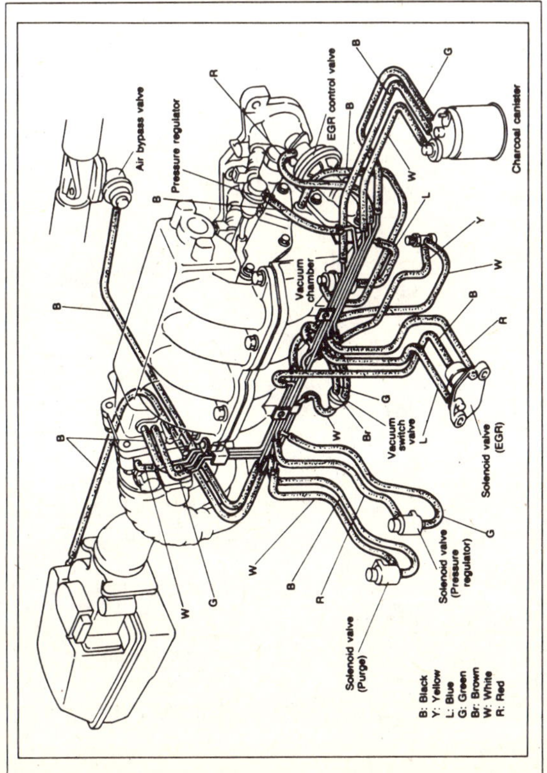


## SPECIFICATIONS CHART 1988–89 626/MX-6 WITH TURBO

| Item | Engine type | | Turbo Engine |
|---|---|---|---|
| Idle speed | rpm | | 750 ± 25 (ATX: P range)* |
| **Throttle body** | | | |
| Type | | | Horizontal draft (2-barrel) |
| Throat diameter | mm (in) | No. 1 | MTX: 40 (1.6), ATX: 46 (1.8) |
| | | No. 2 | MTX: 46 (1.8), ATX: 40 (1.6) |
| **Air flow meter** | | | |
| Resistor | Ω | $E_2$—Vs | Fully closed: 20—400 Fully open: 20—1,000 |
| | | $E_2$—Vc | 100—400 |
| | | $E_2$—VB | 200—400 |
| | | $E_2$—THA | −20°C (−4°F) 13,600—18,400<br>20°C (68°F) 2,210— 2,690<br>60°C (140°F) 493— 667 |
| **Fuel pump** | | | |
| Type | | | Impeller (in tank) |
| Output pressure | kPa (kg/cm², psi) | | Main pump: 441—588 (4.5—6.0, 64—85)<br>Transfer pump: 39 (0.4, 5.7) max. |
| Feeding capacity | cc (cu in)/10 seconds | | Main pump: 220 (13.4) min.<br>Transfer pump: 190 (11.6) min. |
| **Fuel filter** | | | |
| Type | Low pressure side | | Nylon element |
| | High pressure side | | Paper element |
| **Pressure regulator** | | | |
| Type | | | Diaphragm |
| Regulating pressure | kPa (kg/cm², psi) | | 235—275 (2.4—2.8, 34—40) |
| **Injector** | | | |
| Type | | | High-ohmic |
| Type of drive | | | Voltage |
| Resistance | Ω | | 11—15 |
| Injection amount | cc (cu in)/15 seconds | | 73—90 (4.45—5.49) |
| **Idle speed control valve** | | | |
| Solenoid resistance | Ω | | 6.3—9.9 |
| **Turbocharger** | | | |
| Cooling method | | | Engine coolant |
| Lubrication method | | | Engine oil |
| Boost pressure (Maximum) | kPa (kg/cm², psi) | | 60 (0.61, 8.7): Solenoid duty value 100%<br>45 (0.46, 6.5): Solenoid duty value 0% |
| **Fuel tank** | | | |
| Capacity | liters (US gal, Imp gal) | | 60 (15.9, 13.2), 57 (15.0, 12.5): 4-wheel steering vehicle |
| **Air cleaner** | | | |
| Element type | | | Oil permeated |
| **Fuel** | | | |
| Specification | | | Unleaded premium (Unleaded regular) |

* With test connector grounded

## TROUBLESHOOTING CHART – 1988–89 626/MX-6 with TURBO

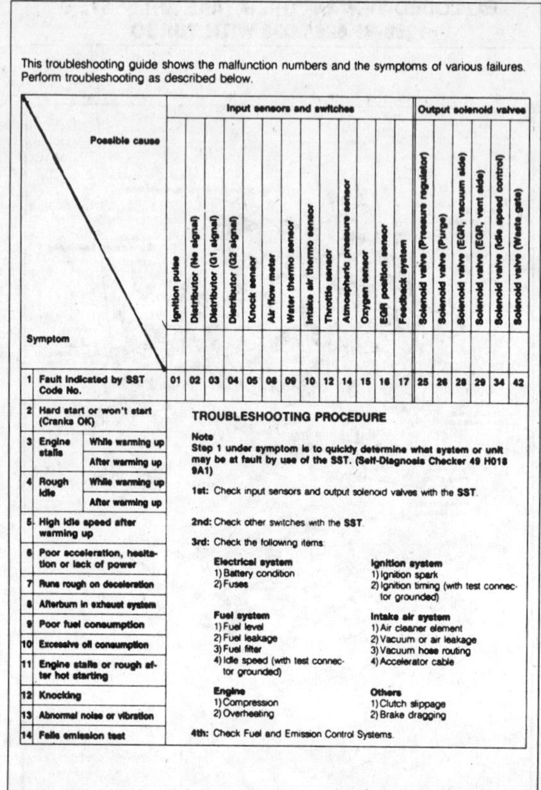

This troubleshooting guide shows the malfunction numbers and the symptoms of various failures. Perform troubleshooting as described below.

| Symptom \ Possible cause | | Ignition pulse | Distributor (Ne signal) | Distributor (G1 signal) | Distributor (G2 signal) | Knock sensor | Air flow meter | Water thermo sensor | Intake air thermo sensor | Throttle sensor | Atmospheric pressure sensor | Oxygen sensor | EGR position sensor | Feedback system | Solenoid valve (Pressure regulator) | Solenoid valve (Purge) | Solenoid valve (EGR, vacuum side) | Solenoid valve (EGR, vent side) | Solenoid valve (Idle speed control) | Solenoid valve (Waste gate) |
|---|---|---|---|---|---|---|---|---|---|---|---|---|---|---|---|---|---|---|---|---|
| | | Input sensors and switches | | | | | | | | | | | | | Output solenoid valves | | | | | |
| 1 Fault indicated by SST Code No. | | 01 | 02 | 03 | 04 | 05 | 08 | 09 | 10 | 12 | 14 | 15 | 16 | 17 | 25 | 26 | 28 | 29 | 34 | 42 |
| 2 Hard start or won't start (Cranks OK) | | | | | | | | | | | | | | | | | | | | |
| 3 Engine stalls | While warming up | | | | | | | | | | | | | | | | | | | |
| | After warming up | | | | | | | | | | | | | | | | | | | |
| 4 Rough idle | While warming up | | | | | | | | | | | | | | | | | | | |
| | After warming up | | | | | | | | | | | | | | | | | | | |
| 5 High idle speed after warming up | | | | | | | | | | | | | | | | | | | | |
| 6 Poor acceleration, hesitation or lack of power | | | | | | | | | | | | | | | | | | | | |
| 7 Runs rough on deceleration | | | | | | | | | | | | | | | | | | | | |
| 8 Afterburn in exhaust system | | | | | | | | | | | | | | | | | | | | |
| 9 Poor fuel consumption | | | | | | | | | | | | | | | | | | | | |
| 10 Excessive oil consumption | | | | | | | | | | | | | | | | | | | | |
| 11 Engine stalls or rough after hot starting | | | | | | | | | | | | | | | | | | | | |
| 12 Knocking | | | | | | | | | | | | | | | | | | | | |
| 13 Abnormal noise or vibration | | | | | | | | | | | | | | | | | | | | |
| 14 Fails emission test | | | | | | | | | | | | | | | | | | | | |

### TROUBLESHOOTING PROCEDURE

**Note**
**Step 1 under symptom is to quickly determine what system or unit may be at fault by use of the SST. (Self-Diagnosis Checker 49 H018 9A1)**

**1st:** Check input sensors and output solenoid valves with the **SST**.

**2nd:** Check other switches with the **SST**.

**3rd:** Check the following items:

**Electrical system**
1) Battery condition
2) Fuses

**Ignition system**
1) Ignition spark
2) Ignition timing (with test connector grounded)

**Fuel system**
1) Fuel level
2) Fuel leakage
3) Fuel filter
4) Idle speed (with test connector grounded)

**Intake air system**
1) Air cleaner element
2) Vacuum or air leakage
3) Vacuum hose routing
4) Accelerator cable

**Engine**
1) Compression
2) Overheating

**Others**
1) Clutch slippage
2) Brake dragging

**4th:** Check Fuel and Emission Control Systems.

The Troubleshooting Guide lists the systems most likely to cause a given symptom. After finding systems to check, refer to the pages shown for detailed guides for each system.

| Symptom | | Intake Air System (Poor connection of components, throttle body) | Fuel System (Fuel injection, Fuel pressure) | Pressure Regulator Control System | Idle Speed Control (ISC) System (Air valve, idle speed control solenoid malfunction) | Turbocharging System (Oil & water passage, Turbine and compressor wheels malfunction) | Electronic Spark Advance (ESA) System (Knock control system) | EGR System (EGR control valve stuck and open) | EEC System (Vacuum switch valve, No.1 purge valve malfunction) | PCV System (System clogged) | Deceleration System (Fuel cut operation malfunction) | Exhaust System (System clogged) |
|---|---|---|---|---|---|---|---|---|---|---|---|---|
| | | Fuel and Emission Control Systems | | | | | | | | | | |
| 2 | | 3 | 2 | | | | 1 | | | | | |
| 3 | | 4 | 3 | | 1 | | | 2 | | | | |
| | | 5 | 4 | | 2 | | | 3 | | 1 | | |
| 4 | | 5 | 4 | | 1 | | | 3 | | 2 | | |
| | | 6 | 5 | | 2 | | | 3 | 4 | 1 | | |
| 5 | | 2 | | | 1 | | | | | | | |
| 6 | | 3 | 4 | | | 6 | | 1 | 2 | | | 5 |
| 7 | | | 3 | | 2 | | | | | | 1 | |
| 8 | | 3 | 4 | | 1 | | | | | | 2 | |
| 9 | | | 2 | | | | | 3 | | | 1 | 4 |
| 10 | | | | | | 1 | | | | | | |
| 11 | | | 2 | 1 | | | | | | | | |
| 12 | | | | | | 2 | 1 | | | | | |
| 13 | | | | | | 1 | | | | | | |
| 14 | | 6 | 7 | | 4 | | | 2 | 5 | | 3 | 1 |

The numbers of the list show the priorities of inspections from the most possible to that with the lowest possibility.
These were determined on the following basis:
• Ease of inspection • Most possible system • Most possible point in system

## MALFUNCTION INDICATOR LIGHT (MIL) OPERATION 1988–89 626/MX-6 WITH TURBO

**PRINCIPLE OF CODE CYCLE**

Malfunction codes are determined as shown below

**1. Code cycle break**
The time between malfunction code cycles is 4.0 sec (the time the light is off).

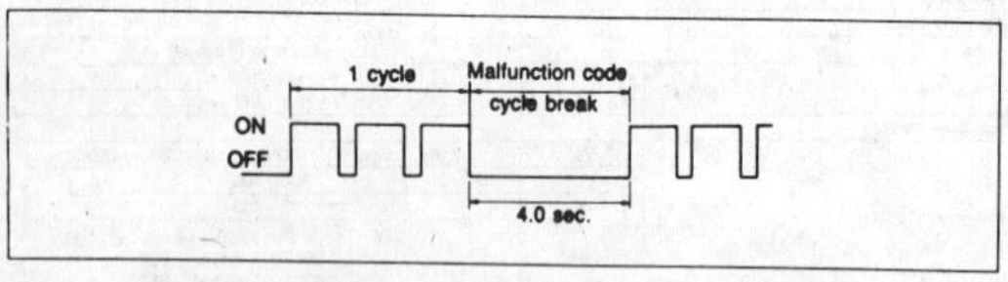

1 cycle
Malfunction code cycle break
ON
OFF
4.0 sec.

**2. Second digit of malfunction code (ones position)**
The digit in the ones position of the malfunction code represents the number of times the buzzer is on 0.4 sec during one cycle.

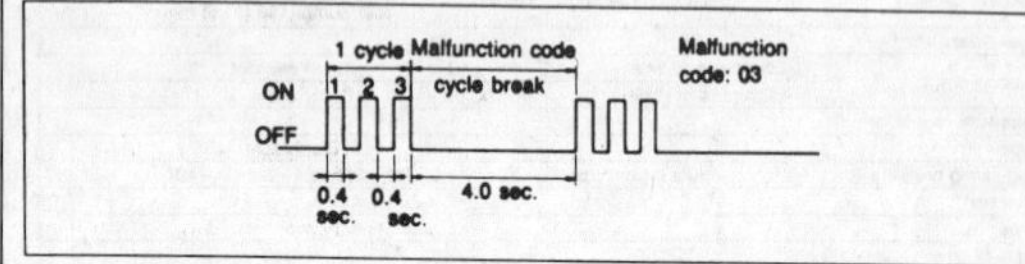


**3. First digit of malfunction code (tens position)**
The digit in the tens position of the malfunction code represents the number of times the buzzer is on 1.2 sec during one cycle.

It should also be noted that the light goes off for 1.6 sec. between the long and short pulses of the buzzer.

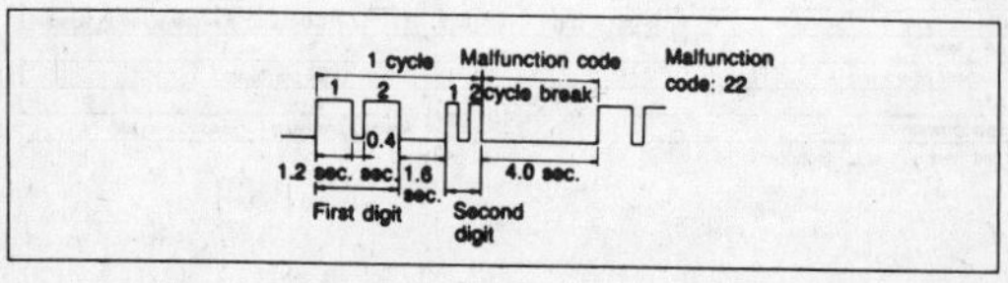

1 cycle
Malfunction code cycle break
Malfunction code: 22
1 2
1 2
0.4
1.2 sec.
sec.
1.6 sec.
4.0 sec.
First digit
Second digit

## TROUBLE CODE IDENTIFICATION CHART 1988–89 626/MX-6 WITH TURBO

**CODE NUMBER**

| Malfunction display: Malfunction code no. | Malfunction display: MIL output signal pattern | Sensor or subsystem | Self-diagnosis | Fail-safe |
|---|---|---|---|---|
| 01 | ON / OFF | Ignition pulse | No ignition signal | — |
| 02 | ON / OFF | Ne signal | No Ne signal from crank angle sensor | — |
| 03 | ON / OFF | $G_1$ signal | No $G_1$ signal | Neither $G_1$ nor $G_2$ signal: Engine stopped |
| 04 | ON / OFF | $G_2$ signal | No $G_2$ signal | |
| 05 | ON / OFF | Knock sensor and knock control unit | Open or short circuit | • Retards ignition timing 6° in heavy-load condition<br>• Waste gate opens earlier |
| 08 | ON / OFF | Air flow meter | Open or short circuit | Maintains basic signal at preset value |
| 09 | ON / OFF | Water thermo sensor | Open or short circuit | Maintains constant command<br>• 35°C (95°F) for EGI<br>• 50°C (122°F) for ISC control use |
| 10 | ON / OFF | Intake air thermo sensor (air flow meter) | Open or short circuit | Maintains constant 20°C (68°F) command |
| 12 | ON / OFF | Throttle sensor | Open or short circuit | Maintains constant command of throttle valve fully open |
| 14 | ON / OFF | Atmospheric pressure sensor | Open or short circuit | Maintains constant command of sea level pressure |
| 15 | ON / OFF | Oxygen sensor | Sensor output continues less than 0.55V 120 sec. after engine starts (1,500 rpm) | Cancels EGI feedback operation |
| 16 | ON / OFF | EGR position sensor | Open or short circuit | Cuts off EGR |
| | | | Sensor output does not match target value (incorrect output) | — |
| 17 | ON / OFF | Feedback system | Sensor output not changed 20 sec. after engine exceeds 1,500 rpm | Cancels EGI feedback operation |

## TROUBLE CODE IDENTIFICATION CHART (CONT.) 1988–89 626/MX-6 WITH TURBO

| Malfunction display | | Sensor or subsystem | Self-diagnosis | Fail-safe |
|---|---|---|---|---|
| Malfunction code no. | MIL output signal pattern | | | |
| 25 | ON / OFF | Solenoid valve (pressure regulator) | Open or short circuit | — |
| 26 | ON / OFF | Solenoid valve (purge control) | | — |
| 28 | ON / OFF | Solenoid valve (EGR-vacuum) | | — |
| 29 | ON / OFF | Solenoid valve (EGR-vent) | | — |
| 34 | ON / OFF | Solenoid valve (idle speed control) | | — |
| 42 | ON / OFF | Solenoid valve (waste gate) | | — |

**Caution**

**a) If there is more than one failure present, the lowest number malfunction code is displayed first, the remaining codes are displayed sequentially.**

**b) After repairing a failure, turn off the ignition switch and disconnect the negative battery cable and depress the brake pedal for at least 5 seconds to erase the memory of a malfunction code.**

## TROUBLE CODE DIAGNOSTIC CHART 1988–89 626/MX-6 WITH TURBO

**Code No. 01 (Ignition pulse)** — **PC: Possible Cause**

- Are there poor connections in igniter circuit? — YES → Repair or replace connector
- NO ↓ Is there battery voltage at igniter terminal WR? — NO → PC: Open circuit in wiring harness from main relay to igniter
- YES ↓ Is igniter OK? (Refer to section 5) — NO → Replace igniter
- YES ↓ Is engine control unit 1X terminal OK? — NO → PC: Malfunction of IGt signal or engine control unit. Open circuit in related wiring harness
- YES ↓ Is same Code No. present after performing after-repair procedure? — NO → Ignition pulse and circuit OK
- YES ↓ Is engine control unit 1M terminal OK? — NO → PC: Open circuit in wiring harness between engine control unit 1M terminal and igniter
- YES → PC: Engine control unit malfunction

## TROUBLE CODE DIAGNOSTIC CHART—1988–89 626/MX-6 WITH TURBO

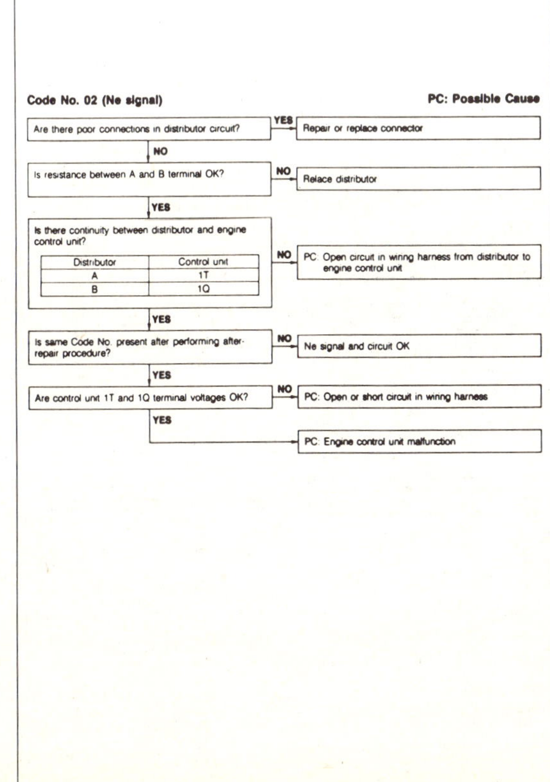


**Code No. 03 (G1 signal)** — **PC: Possible Cause**

- Are there poor connections in distributor circuit? — YES → Repair or replace connector
- NO ↓ Is resistance between E and F terminal OK? — NO → Replace distributor
- YES ↓ Is there continuity between distributor and engine control unit? — NO → PC: Open circuit in wiring harness from distributor to engine control unit

| Distributor | Control unit |
|---|---|
| E | 1N |
| F | 1P |

- YES ↓ Is same Code No. present after performing after-repair procedure? — NO → G1 signal and circuit OK
- YES ↓ Are engine control unit 1N and 1P terminal voltages OK? — NO → PC: Short circuit in wiring harness
- YES → PC: Engine control unit malfunction

## TROUBLE CODE DIAGNOSTIC CHART – 1988–89 626/MX-6 WITH TURBO

**Code No. 04 (G2 signal)**

**PC: Possible Cause**

Are there poor connections in distributor circuit? — **YES** → Repair or replace connector

**NO**

Is resistance between C and D terminal OK? — **NO** → Replace distributor

**YES**

Is there continuity between distributor and engine control unit?

| Distributor | Control unit |
|---|---|
| C | 1O |
| D | 1P |

— **NO** → PC: Open circuit in wiring harness from distributor to engine control unit

**YES**

Is same Code No. present after performing after-repair procedure? — **No** → G2 signal and circuit OK

**YES**

Are engine control unit 1O and 1P terminal voltages OK? — **NO** → PC: Short circuit in wiring harness

**YES** → PC: Engine control unit malfunction

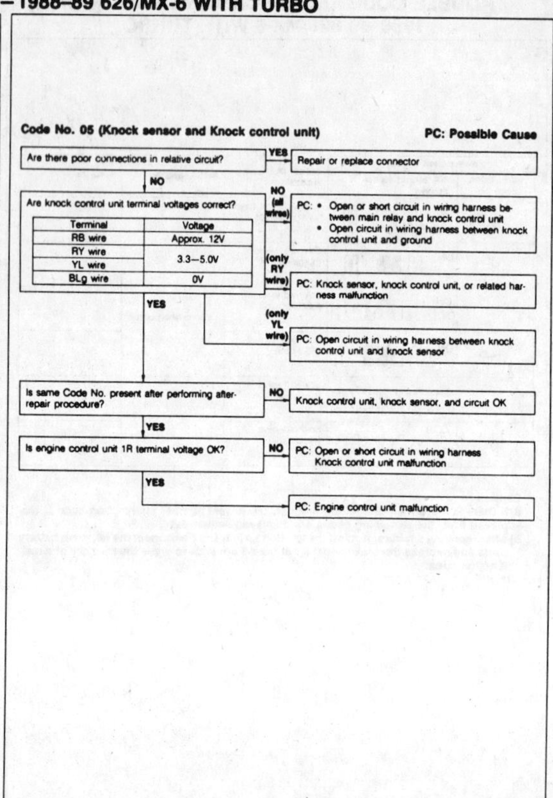


## TROUBLE CODE DIAGNOSTIC CHART – 1988–89 626/MX-6 WITH TURBO

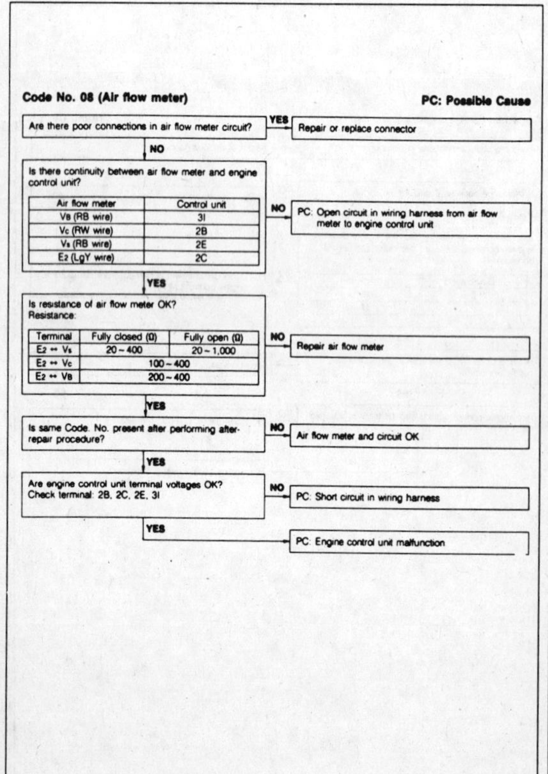


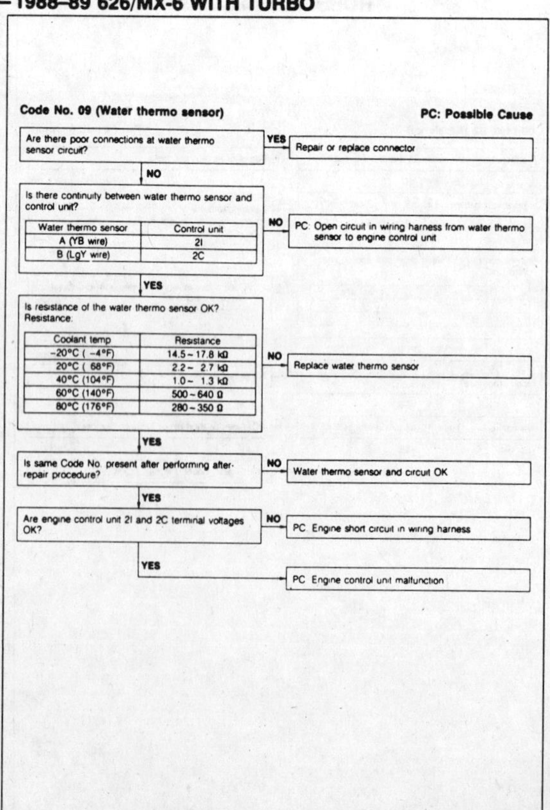

## TROUBLE CODE DIAGNOSTIC CHART—1988–89 626/MX-6 WITH TURBO

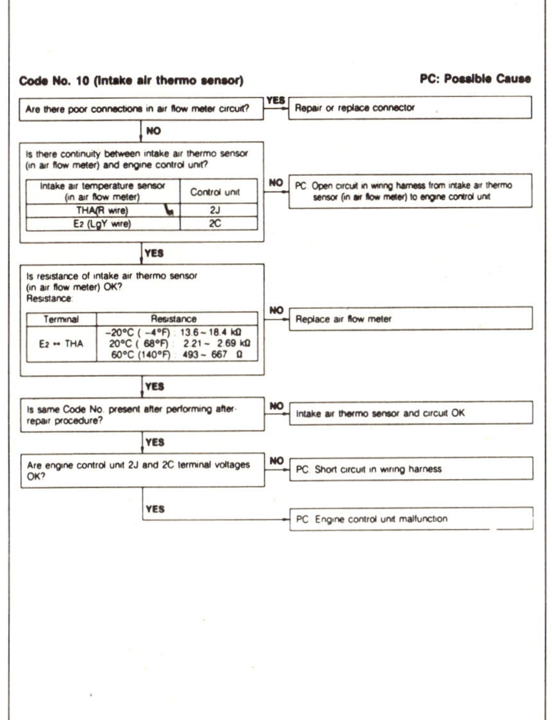


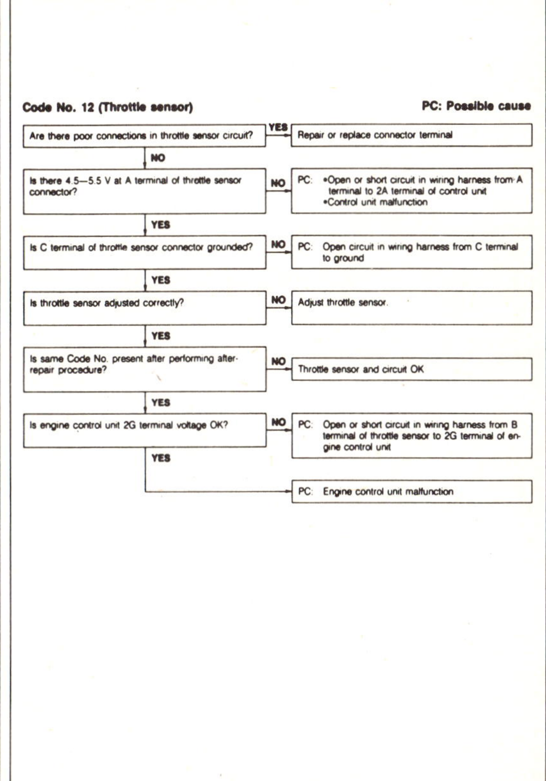


## TROUBLE CODE DIAGNOSTIC CHART—1988–89 626/MX-6 WITH TURBO

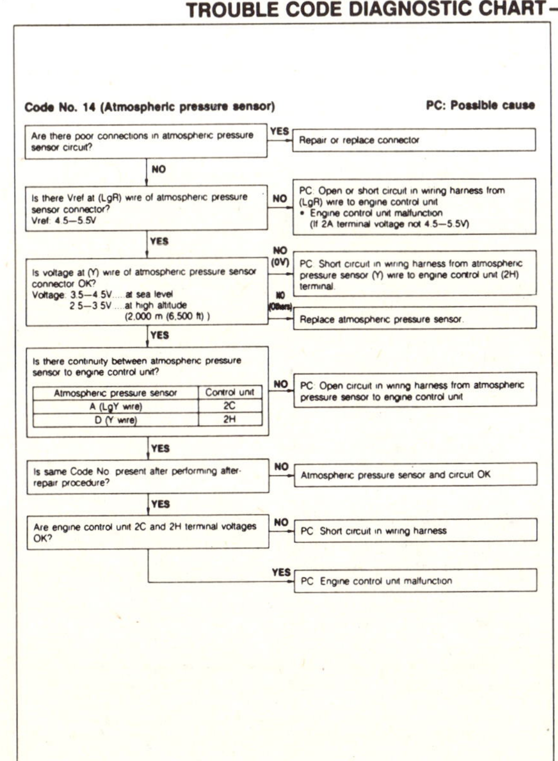


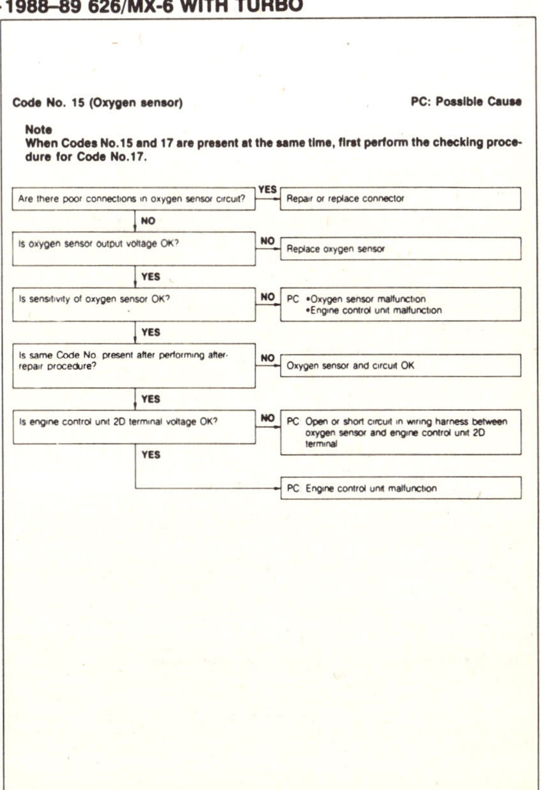

## TROUBLE CODE DIAGNOSTIC CHART—1988–89 626/MX-6 WITH TURBO

**Code No. 16 (EGR position sensor)** — **PC: Possible Cause**

- Are there poor connections in sensor circuit? — YES → Repair or replace connector
- NO ↓ Is EGR control valve OK? (Refer to page 4B—94) — NO → Replace EGR control valve
- YES ↓ Is resistance of EGR position sensor OK? — NO → Replace EGR control valve
- YES ↓ Is there continuity between EGR position sensor and engine control unit? — NO → PC: Open circuit in wiring harness from EGR position sensor to engine control unit

| EGR position sensor | Engine control unit |
|---|---|
| (C) (YL wire) | 2F |
| (B) (LgY wire) | 2C |
| (A) (LgR wire) | 2A |

- YES ↓ Is there 4.5—5.5V at A terminal of EGR position sensor connector? — NO → PC: Short circuit in wiring harness from EGR position sensor A terminal to engine control unit 2A terminal • Engine control unit malfunction
- YES ↓ Is same Code No. present after performing after-repair procedure? — NO → EGR position sensor and circuit OK
- YES ↓ Is engine control unit 2F terminal voltage OK? — NO → PC: Short circuit in wiring harness from EGR position sensor C terminal to engine control unit 2F terminal
- YES → Replace engine control unit

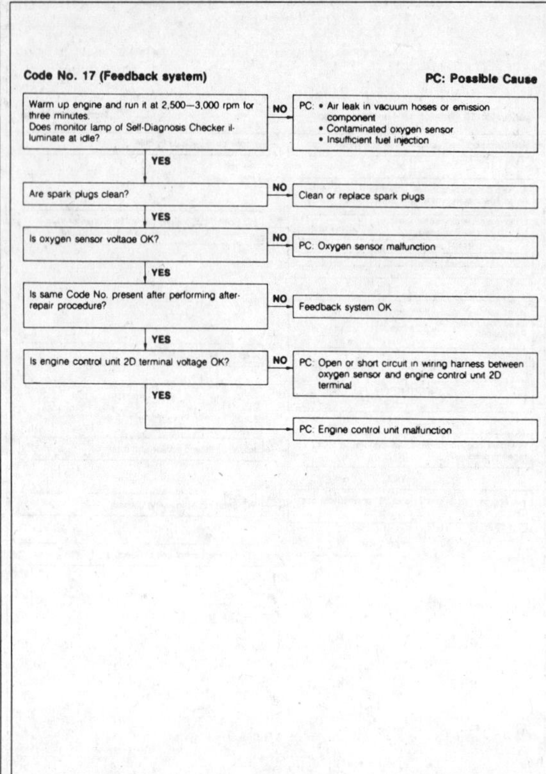


## TROUBLE CODE DIAGNOSTIC CHART—1988–89 626/MX-6 WITH TURBO

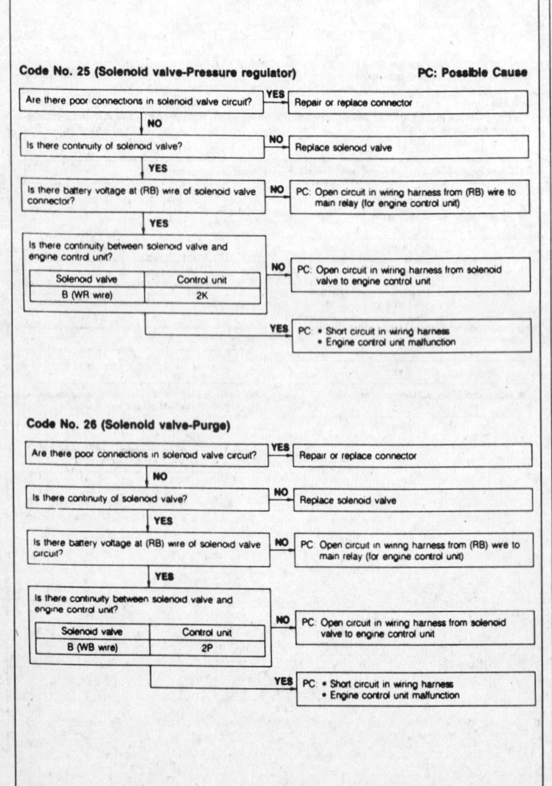


**Code No. 28 (Solenoid valve—EGR, vacuum side)** — **PC: Possible Cause**

- Are there poor connections in solenoid valve circuit? — YES → Repair or replace connector
- NO ↓ Is there continuity of solenoid valve? — NO → Replace solenoid valve
- YES ↓ Is there battery voltage at RB wire of solenoid valve connector? — NO → PC: Open circuit in wiring harness from RB wire to main relay (for engine control unit)
- YES ↓ Is there continuity between solenoid valve and engine control unit? — NO → PC: Open circuit in wiring harness from solenoid valve to engine control unit

| Solenoid valve | Control unit |
|---|---|
| B (WL wire) | 2N |

- YES → PC: • Short circuit in wiring harness • Engine control unit malfunction

**Code No. 29 (Solenoid valve—EGR, vent side)**

- Are there poor connections in solenoid valve circuit? — YES → Repair or replace connector
- NO ↓ Is there continuity of solenoid valve? — NO → Replace solenoid valve
- YES ↓ Is there battery voltage of RB wire of solenoid valve connector? — NO → PC: Open circuit in wiring harness from RB wire to main relay (for engine control unit)
- YES ↓ Is there continuity between solenoid valve and engine control unit? — NO → PC: Open circuit in wiring harness from solenoid valve to engine control unit

| Solenoid valve | Control unit |
|---|---|
| B (LW wire) | 2M |

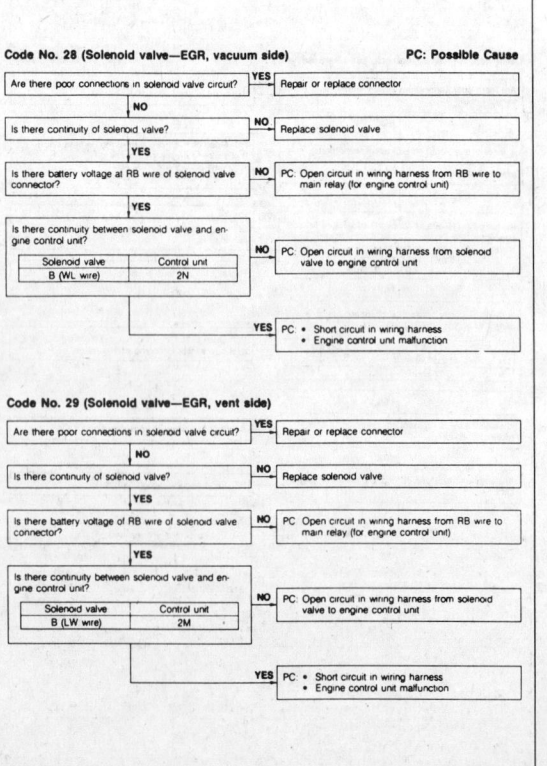

- YES → PC: • Short circuit in wiring harness • Engine control unit malfunction

## TROUBLE CODE DIAGNOSTIC CHART 1988–89 626/MX-6 WITH TURBO

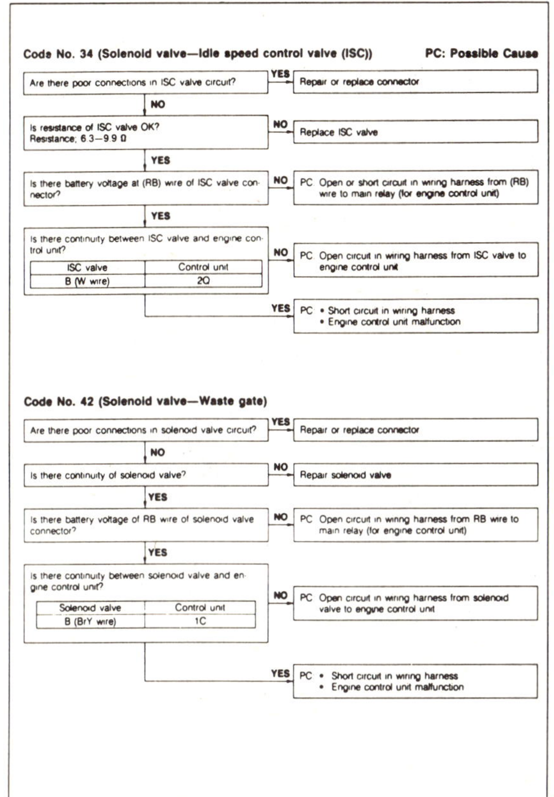


## CONTROL UNIT CONNECTOR TESTING 1988–89 626/MX-6 WITH TURBO

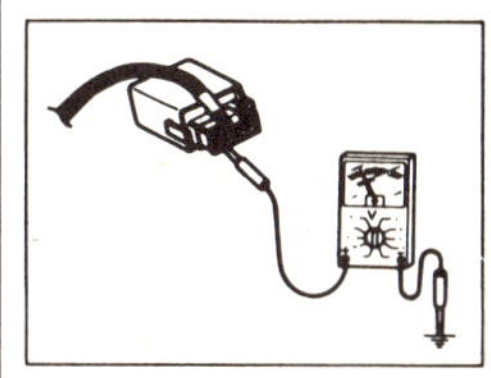

**E/L CONTROL UNIT**

**Inspection**

1. Connect a voltmeter between the E/L control unit and ground.
2. Start the engine and check the terminal voltages as described below.

| Terminal | Input | Output | Connection to | Voltage (after warm-up) | | Remarks |
|---|---|---|---|---|---|---|
| | | | | Ignition switch: ON | Idle | |
| A (BW) | — | — | Ignition switch | Battery voltage | | |
| B | — | — | — | — | — | — |
| C (B) | — | — | Ground | 0V | | |
| D (LY) | ○ | | Electrical fan relay | Battary voltage | | Coolant temp.: below 97°C (207°F) |
| | | | | Below 1.5V | | Coolant temp. above 97°C (207°F) |
| E (GY) | | ○ | Control unit (1I) | 0V | | E/L ON |
| | | | | Battery voltage | | E/L OFF |
| F (W) | ○ | | Headlight switch | Battery voltage | | Headlight switch ON |
| | | | | Below 1.5V | | Headlight switch OFF |
| G (LB) | ○ | | Blower motor switch | Below 1.5V | | Blower motor switch ON (3rd or 4th position) |
| | | | | Approx. 5V | | Others |
| H (BL) | ○ | | Rear defroster switch | Below 1.5V | | Rear defroster switch ON |
| | | | | Battery voltage | | Rear defroster switch: OFF |

Center console

E/L control unit

**Replacement**

1. Remove the engine control unit.
2. Replace the E/L control unit.
3. Install in the reverse order of removal.

## CONTROL UNIT PIN LOCATIONS AND TEST VOLTAGES—1988–89 626/MX-6 WITH TURBO

**Terminal Voltage**

If the input and output devices and related wiring are normal, but the engine control unit terminal voltage is incorrect, replace the engine control unit.

| Terminal | Input | Output | Connection to | Voltage (After warming-up) | | Remarks |
|---|---|---|---|---|---|---|
| | | | | IGN: ON | Idle | |
| 1A | | ○ | Malfunction indicator light | For 3 sec. after ignition switch OFF → ON: below 4.8V (Light illuminates)<br>After 3 sec.: Battery voltage (Light does not illuminate) | | • Test connector grounded<br>• Light illuminates: below 4.8V<br>• Light does not illuminate: Battery voltage |
| 1B | | ○ | Self-Diagnosis Checker (Code number) | For 3 sec. after ignition switch OFF → ON: below 6.2V (Buzzer sounds)<br>After 3 sec.: Battery voltage (Buzzer does not sound) | | • Using Self-Diagnosis Checker and test connector grounded<br>• Buzzer sounds: below 6.2V<br>• Buzzer does not sound: Battery voltage |
| 1C | | ○ | Solenoid valve (Waste gate) | Battery voltage | | • Suddenly increase engine speed to above 4,500 rpm: below 3.5V |
| 1D | | ○ | Self-Diagnosis Checker (Monitor lamp) | Test connector grounded<br>For 3 sec. after ignition switch OFF → ON:<br>below 6.2V (Light illuminates)<br>After 3 sec.: Battery voltage (light does not illuminate) | (Test connector grounded) approx. 5V<br>(Test connector not grounded) Monitor lamp ON: below 6.2V<br>Monitor lamp OFF: Battery voltage | With Self-Diagnosis Checker |
| 1E | ○ | | Idle switch | Accelerator pedal released: below 0.5V<br>Accelerator pedal depressed: above 7.7V | | |
| 1F | | ○ | A/C relay | Battry voltage | A/C switch ON: below 2.5V<br>A/C switch OFF: battery voltage | Blower motor: ON |
| 1G | — | — | — | — | | — |
| 1H | ○ | | Water thermo switch | Above 7.3V | | Radiator temp.: below 17°C (63°F) |
| | | | | Below 1.5V | | Radiator temp.: above 17°C (63°F) |
| 1I | ○ | | Electrical load control unit | Electrical load ON: below 1.5V<br>Electrical load OFF: above 7.3V | | Electrical load:<br>Rear defroster<br>Headlight<br>Blower motor (3rd & 4th position)<br>Electrical fan |
| 1J | ○ | | Brake light switch | Brake pedal released: below 3.6V<br>Brake pedal depressed: above 10.0V | | |
| 1K | ○ | | P/S pressure switch | Constant above 10.5V | P/S ON: below 1.5V<br>P/S OFF: above 10.5V | |
| 1L | ○ | | A/C switch | A/C switch ON: below 1.5V<br>A/C switch OFF: above 10.0V | | Blower motor: ON |
| 1M | ○ | | Igniter (IGf signal) | Below 1.0V | 0.1—1.8V | |
| 1N | ○ | | Distributor (G1 ⊕ signal) | Approx. 0.6—0.8V | | |

| Terminal | Input | Output | Connection to | Voltage (After warming-up) | | Remarks |
|---|---|---|---|---|---|---|
| | | | | IGN: ON | Idle | |
| 1O | ○ | | Distributor (G2 ⊕ signal) | Approx. 0.6—0.8V | | |
| 1P | ○ | | Distributor (G1, G2 ⊖ signal) | Approx. 0.6—0.8V | | |
| 1Q | ○ | | Distributor (Ne ⊖ signal) | Approx. 0.6—0.8V | | |
| 1R | ○ | | Knock control unit | 3.3—5.0V | | Knocking: 1.3—2.6V |
| 1S | | ○ | Warning buzzer Overboost | TypeA[*1]: Approx. 0.08V<br>Type B[*2]: Approx. 12V | | Buzzer sounds:<br>Type A: 0V<br>Type B: Blow 1V |
| 1T | ○ | | Distributor (Ne ⊕ signal) | Approx. 0.6—0.8V | | |
| 1U | — | — | — | — | | — |
| 1V | ○ | | — | Constant below 1.5V | | MTX |
| | | | | Constant above 10.5V | | ATX |
| 1W | ○ | | Test connector | Test connector grounded: below 0.5V<br>Test connector not grounded: above 10.5V | | Green connector, 1-pin |
| 1X | | ○ | Igniter (IGt signal) | Approx. 0V | Approx. 0.6—0.8V | |
| 2A | | ○ | V ref | 4.5—5.5V | | |
| 2B | ○ | | Air flow meter (Vc) | 7—9V | | |
| 2C | — | — | Ground (E2) | 0V | | |
| 2D | ○ | | Oxygen sensor | 0V | 0—1.0V | • Cold engine: 0V at idle<br>• After warming-up: Increase engine speed: 0.5—1.0V<br>Deceleration: 0—0.4V |
| 2E | ○ | | Air flow meter (Vs) | Approx. 1.7V | Approx. 4—6V | Increase engine speed:<br>voltage increases |
| 2F | ○ | | EGR position sensor | 0.25—0.95V | | |
| 2G | ○ | | Throttle sensor | Accelerator pedal released: Approx. 0.5V (depends on 2A terminal voltage) | | Max. voltage (Throttle valve fully opened): approx. 4.3V |
| 2H | ○ | | Atmospheric pressure sensor | At sea level: approx. 4.0V | | |
| 2I | ○ | | Water thermo sensor | 0.3—0.6V | | Engine coolant temp. 20°C (68°F) approx. 2.5V |
| 2J | ○ | | Air flow meter (Intake air thermo sensor) | Approx. 2.5V at 20°C (68°F) | | |
| 2K | | ○ | Solenoid valve (Pressure regulator control) | For 120 sec. after ignition switch OFF → ON: below 3.5V | For 120 sec. after starting: below 3.5V | Hot condition: Coolant temp. above 70°C (158°F) and Intake air temp. above 20°C (68°F) |
| | | | | Battery voltage | | Other conditions |
| 2L | — | — | — | — | | — |

[*1] Type A: Manufacturer for CPU of body electrical system is "NEC".
[*2] Type B: Manufacturer for CPU of body electrical system is "YAZAKI or U-shin"

## CONTROL UNIT PIN LOCATIONS AND TEST VOLTAGES 1988–89 626/MX-6 WITH TURBO

| Terminal | Input | Output | Connection to | Voltage (After warming-up) IGN: ON | Idle | Remarks |
|---|---|---|---|---|---|---|
| 2M | | O | Solenoid valve (EGR-Vent side) | Battery voltage | | • Voltages change depending on driving condition (EGR amount) • Cold engine: battery voltage Radiator coolant temp-below 17°C (63°F) or Engine coolant temp-below 40°C (104°F) |
| 2N | | O | Solenoid valve (EGR-vacuum side) | Battery voltage | | |
| 2O | | O | Circuit opening relay | Battery voltage | Below 3.5V | |
| 2P | | O | Solenoid valve (Purge control valve) | Battery voltage | | Medium and high load: below 3.5V |
| 2Q | | O | Solenoid valve (Idle speed control valve) | Approx. 1.7—11V | | |
| 2R | — | — | Ground (E02) | 0V | | |
| 3A | — | — | Ground (E01) | 0V | | |
| 3B | O | | Ignition switch (Start position) | Below 2.5V | | While cranking: battery voltage |
| 3C | | O | Injector (No. 4 and No. 2) | Battery voltage | *1 Battery voltage | *1 Engine Signal Monitor green and red lights flash |
| 3D | O | | Inhibitor switch through EC-AT unit | "N" or "P" range: below 2.5V Other ranges: Battery voltage | | ATX |
| | | | Neutral and clutch switch | In-gear condition Clutch pedal depressed: Battery voltage Clutch pedal released: below 0.5V | | MTX (Neutral: constant Battery voltage) |
| 3E | | O | Injector (No. 1 and No. 3) | Battery voltage | *1 Battery voltage | *1 Engine Signal Monitor: green and red lights flash |
| 3F | — | — | — | — | | — |
| 3G | — | — | Ground (E1) | 0V | | |
| 3H | — | — | — | — | | — |
| 3I | O | — | Main relay | Battery voltage | | |
| 3J | — | — | Battery | Battery voltage | | For back-up |

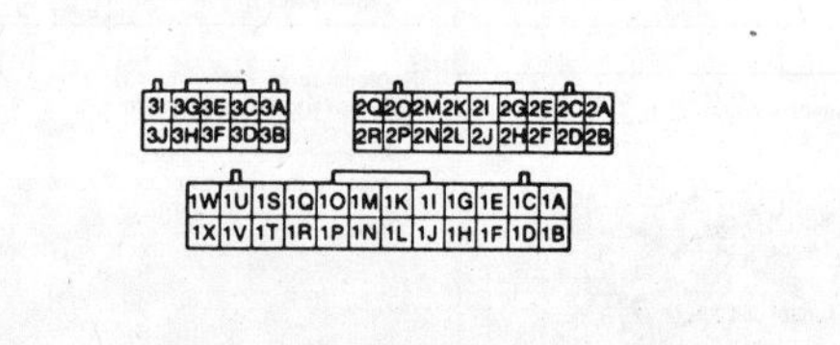


## COMPONENT DESCRIPTIOS – 1988–89 626/MX6 WITH TURBO

### COMPONENT DESCRIPTIONS

| Component | Function | Remarks | Application New model | Application Previous model |
|---|---|---|---|---|
| Air flow meter | Detects amount of intake air; sends signal to control unit | Intake air temp sensor and fuel pump switch are integrated | O | O |
| Atmospheric pressure sensor | Detects atmospheric pressure; sends signal to control unit | | O | O |
| Circuit opening relay | Voltage for fuel pump while engine running | | O | O |
| Clutch switch | Detects in-gear condition; sends signal to control unit | Switch ON when clutch pedal released | O | O |
| EC-AT control unit | Detects N or P range; sends signal to control unit | | O | X |
| Engine control unit | Detects signals from input sensors and switches; controls injector operation | | O | O |
| Fuel filter | Filters particles from fuel | | O | O |
| Fuel pump | Provides fuel to injectors | • Operates while engine running • Installed in fuel tank | O | O |
| G rotor and pick-up | Detects No.1 and No.4 cylinders TDC; sends signal to control unit | For determining fuel injection timing and ignition timing | O | X |
| Idle switch | Detects when throttle valve fully closed; sends signal to control unit | Installed on throttle body | O | O |
| Ignition switch (ST position) | Sends engine cranking signal to control unit | | O | O |
| Inhibitor switch | Detects in-gear condition; sends signal to EC-AT control unit | Switch ON in "N" or "P" range | O | O |
| Injector | Injects fuel into intake port | • Controlled by signals from control unit • High-ohmic injector | O | O |
| Intake air thermo sensor | Detects intake air temperature; sends signal to control unit | Installed in air flow meter | O | O |
| Main relay | Supplies electric current to injectors and control unit | | O | O |
| Ne rotor and pick-up | Detects crank angle at 30° intervals; sends signal to control unit | Engine speed signal | O | X |
| Neutral switch | Detects in-gear condition; sends signal to control unit | Switch ON when in-gear | O | O |
| Oxygen sensor | Detects Oxygen concentration; sends signal to control unit | Zirconia ceramic and platinum coating | O | O |
| Pressure regulator | Adjusts fuel pressure supplied to injectors | | O | O |
| Pulsation damper | Absorbs fuel pulsation | | O | O |
| Throttle sensor | Detects throttle valve opening angle; sends signal to control unit | Installed on throttle body | O | O |
| Water thermo sensor | Detects coolant temperature; sends signal to control unit | | O | O |
| Water thermo switch | Detects radiator coolant temperature; sends signal to control unit | ON: above 17°C (63°F) | O | O |

## OUTPUT COMPONENTS AND ENGINE CONDITIONS 1988–89 626/MX-6 WITH TURBO

| Output devices \ Engine conditions | | Cranking (cold engine) | Warming up (during idle) | Medium load Cold | Medium load Warm | Acceleration | Heavy load | Deceleration | Idle (throttle valve fully closed) | IGN: ON (engine not running) | Remarks |
|---|---|---|---|---|---|---|---|---|---|---|---|
| Injector | Injection | Rich | | | Rich and lean | Rich | | Fuel cut | Rich and lean | No injection | |
| | Injection timing | 1 group (once per revolution) | 2 group (once per two revolutions) | | | | | | 2 group (once per two revolutions) | | Above 6,300 rpm: fuel cut |
| BAC valve | Air valve | Open* | | | Close | | | | | | *Coolant temp: below 50°C (122°F) |
| | ISC valve | Large amount of bypass air | Large amount of bypass air* | Small amount of bypass air | | | | | | No bypass | *In extremly cold condition |
| Solenoid valve (EGR Vent side) | | OFF (Duty value 0%) (Atmospheric pressure to EGR valve) — EGR cut — | | | Operate [Duty values (EGR amount) change*] | | OFF (Duty value 0%) (Atmospheric pressure to EGR valve) — EGR cut — | | | Does not operate | *Depends on engine condition |
| Solenoid valve (EGR Vacuum side) | | OFF (No vacuum to EGR valve) | | | | | OFF (No vacuum to EGR valve) | | | | |
| Solenoid valve (Purge) | | OFF (2nd stage not operated) | | | ON (2nd stage operates) | | | | OFF (2nd stage not operated) | OFF | 1st stage: controlled by water thermo valve |
| Solenoid valve (Pressure regulator control) | | OFF (Vacuum to pressure regulator) | | | | | | | After starting: ON* (Vacuum cut) | OFF | *During hot start only |
| Solenoid valve (Wastegate) | | OFF (Boost pressure not released) | | ON (Duty value 100%) | | ON (Duty value changes)* (Boost pressure controlled) | | OFF (Boost pressure not released) | | OFF | *When knocking occurs |
| Igniter (Ignition timing) | | Fixed at BTDC 6° | Advanced: depends on engine condition | | | Retarded: depends on intensity of knocking | | Advanced: depends on engine speed | | — | |

## COMPONENT RELATIONSHIP CHART 1988–89 626/MX-6 WITH TURBO

| Input devices \ Output devices | Injector: Fuel injection amount | Injector: Fuel injection timing | BAC valve: Air valve | BAC valve: ISC valve | Solenoid valve (EGR Vent side) | Solenoid valve (EGR Vacuum side) | Solenoid valve (Purge) | Solenoid valve (Pressure regulator) | Solenoid valve (Wastegate) | Igniter |
|---|---|---|---|---|---|---|---|---|---|---|
| TEST CONNECTOR | X | X | X | O | X | X | X | X | X | O |
| KNOCK SENSOR | X | X | X | X | X | X | X | X | O | O |
| BRAKE LIGHT SWITCH | O | X | X | X | X | X | X | X | X | X |
| EGR POSITION SENSOR | X | X | X | X | O | O | X | X | X | O |
| ELECTRICAL LOAD CONTROL UNIT | X | X | X | O | X | X | X | X | X | X |
| P/S PRESSURE SWITCH | X | X | X | O | X | X | X | X | X | X |
| A/C SWITCH | O | X | X | O | X | X | O | X | X | X |
| IGNITION SWITCH (STA POSITION) | O | O | X | X | X | X | X | O | X | O |
| INHIBITOR SWITCH and EC—AT CONTROL UNIT | O | X | X | O | X | X | X | X | X | O |
| NEUTRAL AND CLUTCH SWITCH | O | X | X | O | X | X | X | X | X | O |
| OXYGEN SENSOR | O | X | X | X | X | X | X | X | X | X |
| WATER THERMO SWITCH (RADIATOR) | O | X | X | X | O | O | X | X | X | O |
| ATMOSPHERIC PRESSURE SENSOR | O | X | X | O | X | X | X | X | X | X |
| INTAKE AIR THERMO SENSOR | O | X | X | O | X | X | X | O | O | X |
| WATER THERMO SENSOR | O | O | X | O | O | O | O | O | O | O |
| IDLE SWITCH | O | O | X | O | O | O | X | X | X | O |
| THROTTLE SENSOR | O | O | X | O | O | O | X | O | X | O |
| AIR FLOW METER | O | X | X | X | O | O | O | X | O | O |
| Ne SIGNAL | O | O | X | O | O | O | O | O | O | O |
| G2 SIGNAL | X | O | X | X | X | X | X | X | X | O |
| G1 SIGNAL | X | O | X | X | X | X | X | X | X | O |

O Related X Not related

## WIRING SCHEMATIC—1988 626/MX-6 WITH TRUBO

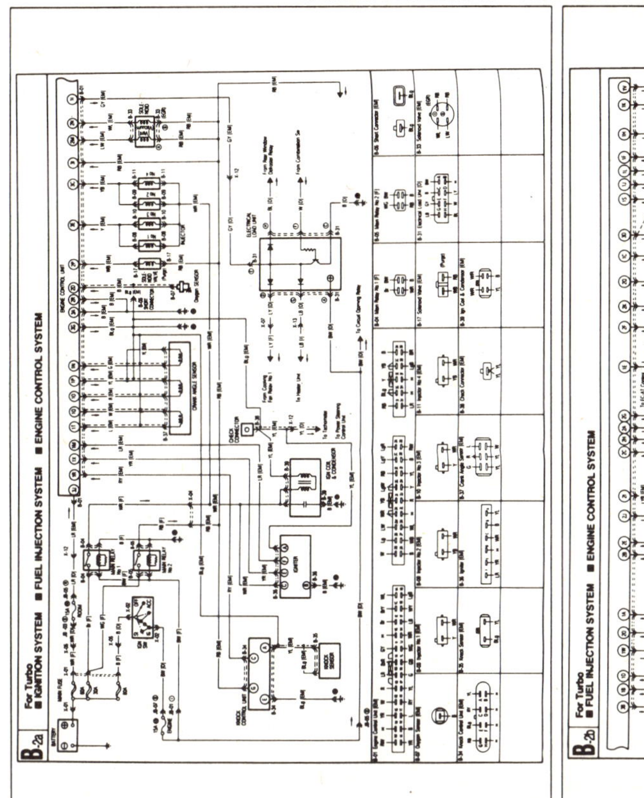

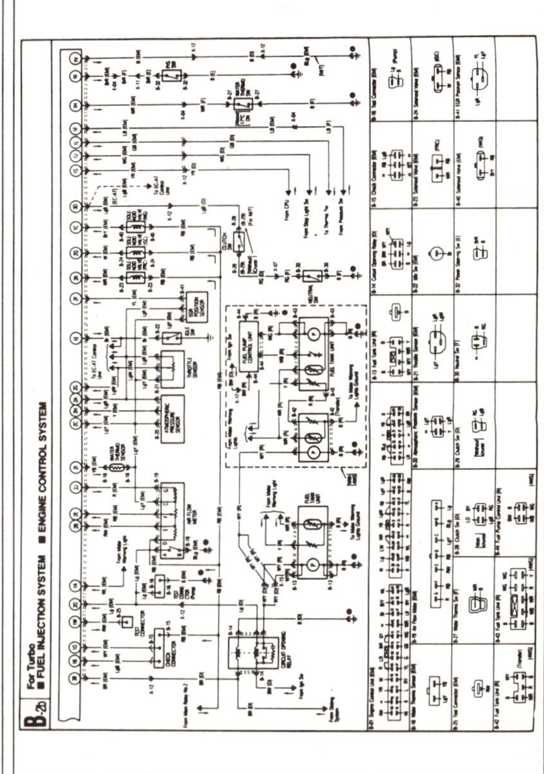

## WIRING SCHEMATIC—1989 626/MX-6 WITH TURBO

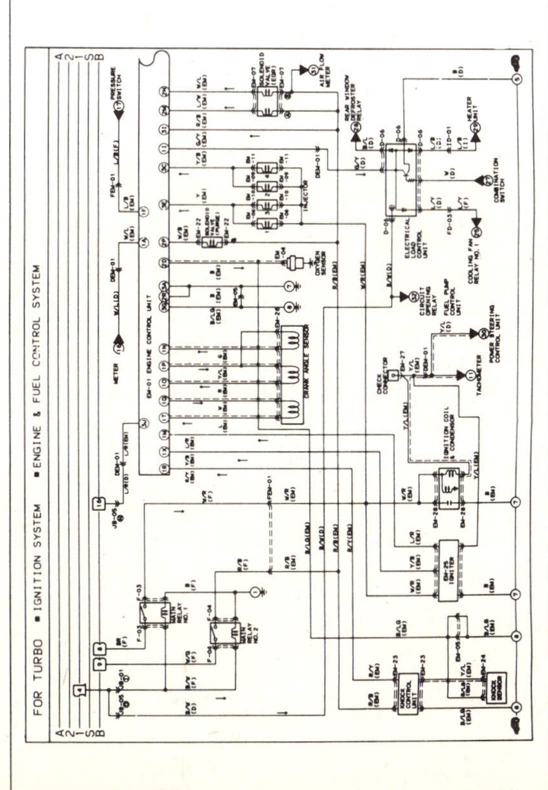

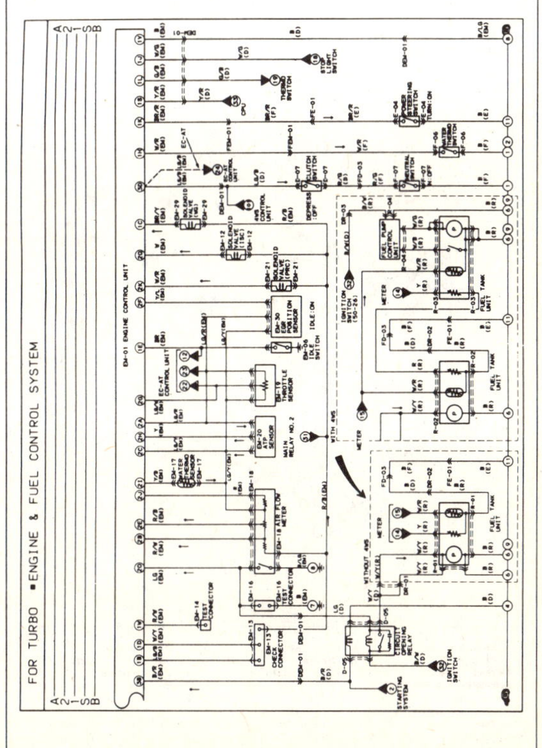

## SYSTEM SCHEMATIC – 1990 626/MX-6 NON-TURBO

## VACUUM HOSE ROUTING – 1990 626/MX-6 NON-TURBO

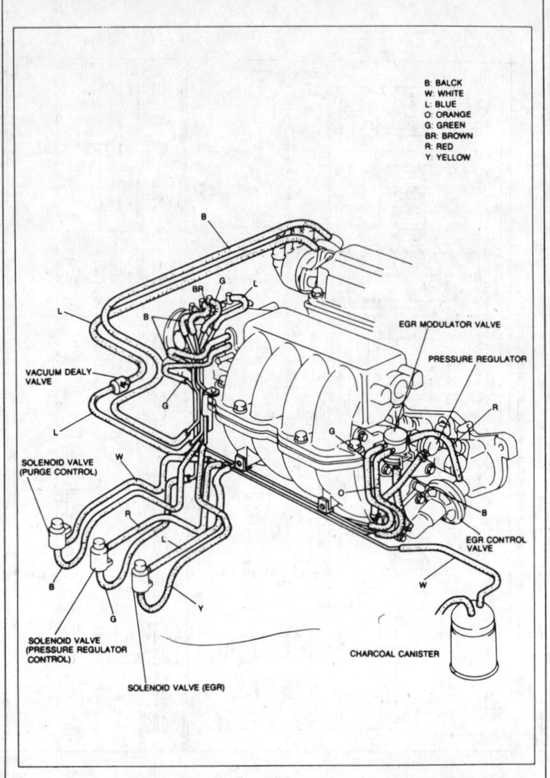


## EXPLODED VIEW OF INTAKE AIR SYSTEM 1990 626/MX-6 NON-TURBO

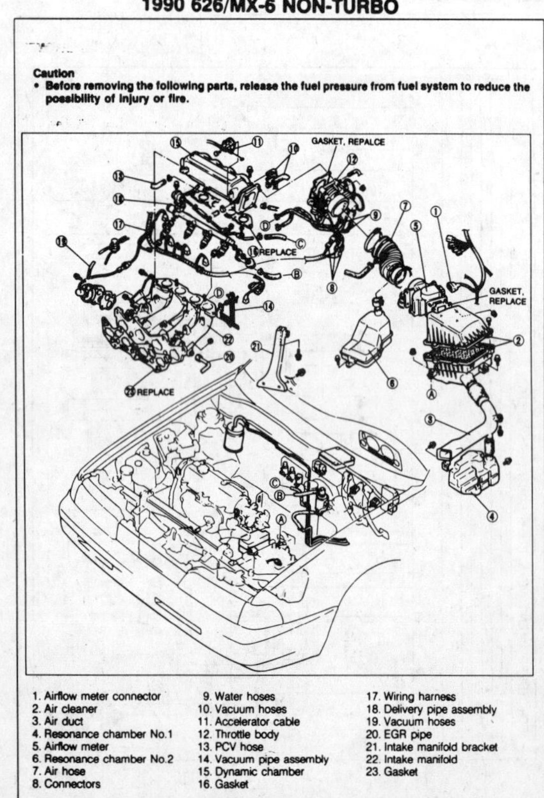


**Caution**

- **Before removing the following parts, release the fuel pressure from fuel system to reduce the possibility of injury or fire.**

1. Airflow meter connector
2. Air cleaner
3. Air duct
4. Resonance chamber No.1
5. Airflow meter
6. Resonance chamber No.2
7. Air hose
8. Connectors
9. Water hoses
10. Vacuum hoses
11. Accelerator cable
12. Throttle body
13. PCV hose
14. Vacuum pipe assembly
15. Dynamic chamber
16. Gasket
17. Wiring harness
18. Delivery pipe assembly
19. Vacuum hoses
20. EGR pipe
21. Intake manifold bracket
22. Intake manifold
23. Gasket

## EXPLODED VIEW OF CONTROL SYSTEM COMPONENTS 1990 626/MX-6 NON-TURBO

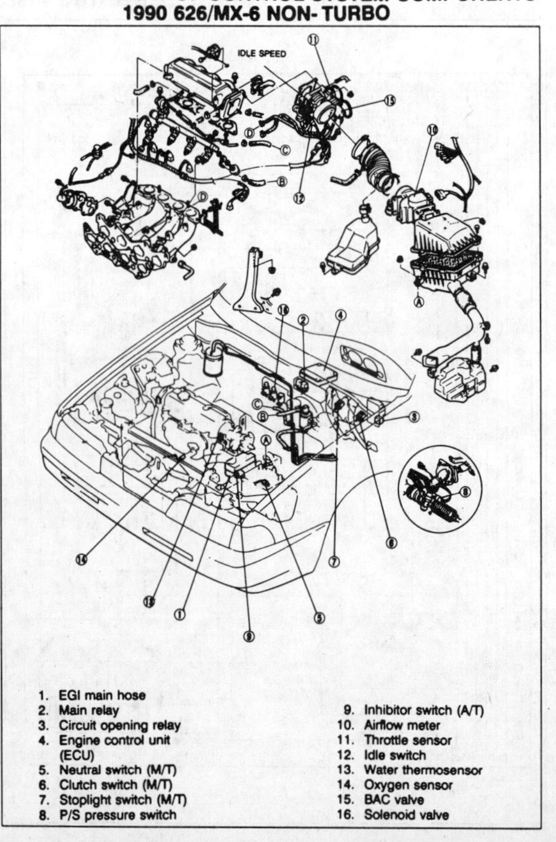


1. EGI main hose
2. Main relay
3. Circuit opening relay
4. Engine control unit (ECU)
5. Neutral switch (M/T)
6. Clutch switch (M/T)
7. Stoplight switch (M/T)
8. P/S pressure switch
9. Inhibitor switch (A/T)
10. Airflow meter
11. Throttle sensor
12. Idle switch
13. Water thermosensor
14. Oxygen sensor
15. BAC valve
16. Solenoid valve

## EXPLODED VIEW OF FUEL AND EMISSION SYSTEM COMPONENTS – 1990 626/MX-6 NON-TURBO

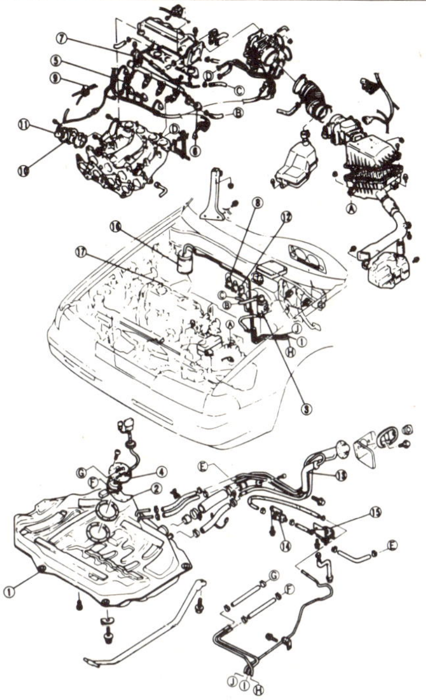

1. Fuel tank
2. Fuel filter (low pressure side)
3. Fuel filter (high pressure side)
4. Fuel pump
5. Injector
6. Pulsation damper
7. Pressure regulator
8. EGR solenoid valve
9. EGR modulator valve
10. EGR control valve
11. EGR position sensor (California models)
12. Purge control solenoid valve
13. Separator
14. 2-way check valve
15. Check and cur-valve
16. Charcoal canister
17. PCV valve

## SPECIFICATIONS CHART – 1990 626/MX-6 NON-TURBO

| Item | | Engine type | Non-Turbo Engine |
|---|---|---|---|
| Idle speed | | rpm | 750 ± 25 (ATX: P range)* |
| **Throttle body** | | | |
| Type | | | Horizontal draft (2-barrel) |
| Throat diameter | mm (in) | No.1 | MTX: 40 (1.6), ATX: 46 (1.8) |
| | | No.2 | MTX: 46 (1.8), ATX: 40 (1.6) |
| **Airflow meter** | | | |
| Resistor | Ω | $E_2$—Vs | Fully closed: 20—400 Fully open: 20—1,000 |
| | | $E_2$—Vc | 100—400 |
| | | $E_2$—VB | 200—400 |
| | | $E_2$—THA | −20°C (−4°F) 13,600—18,400<br>20°C (68°F) 2,210— 2,690<br>60°C (140°F) 493— 667 |
| **Fuel pump** | | | |
| Type | | | Impeller (in tank) |
| Output pressure | | kPa (kg/cm², psi) | 441—588 (4.5—6.0, 64—85) |
| Feeding capacity | | cc (cu in)/10 sec. | 220 (13.4) min |
| **Fuel filter** | | | |
| Type | Low-pressure side | | Nylon element |
| | High-pressure side | | Paper element |
| **Pressure regulator** | | | |
| Type | | | Diaphragm |
| Regulating pressure | | kPa (kg/cm², psi) | 235—275 (2.4—2.8, 34—40) |
| **Injector** | | | |
| Type | | | High-ohmic |
| Type of drive | | | Voltage |
| Resistance | | Ω | 12—16 |
| Injection amount | | cc (cu in)/15 sec. | 44—61 (2.68—3.72) |
| **Idle speed control valve** | | | |
| Solenoid resistance | | Ω | 6.3—9.9 |
| **Fuel tank** | | | |
| Capacity | | liters (US gal, Imp gal) | 60 (15.9, 13.2) |
| **Air cleaner** | | | |
| Element type | | | Oil permeated |
| **Fuel** | | | |
| Specification | | | Unleaded regular (RON 87 or higher) |

* With test connector grounded.

## TROUBLESHOOTING CHART – 1990 626/MX-6 NON-TURBO

This troubleshooting guide shows the malfunction numbers and the symptoms of various failures. Perform troubleshooting as described below.

| Symptom \ Possible cause | Ignition pulse | Airflow meter | Water thermosensor | Intake air thermosensor | Throttle sensor | Atmospheric pressure sensor | Oxygen sensor | EGR position sensor (California only) | Feedback system | Solenoid valve (Pressure regulator) | Solenoid valve (Purge) | Solenoid valve (EGR) | Solenoid valve (Idle speed control) |
|---|---|---|---|---|---|---|---|---|---|---|---|---|---|
| | **Input sensors and switches** | | | | | | | | | **Output solenoid valves** | | | |
| 1 Fault indicated by SST Code No. | 01 | 08 | 09 | 10 | 12 | 14 | 15 | 16 | 17 | 25 | 26 | 28 | 34 |

Symptoms:

2. Hard start or won't start (Cranks OK)
3. Engine stalls — While warming up / After warming up
4. Rough idle — While warming up / After warming up
5. High idle speed after warming up
6. Poor acceleration, hesitation or lack of power
7. Runs rough on deceleration
8. Afterburn in exhaust system
9. Poor fuel consumption
10. Engine stalls or rough after hot starting

### TROUBLESHOOTING PROCEDURE

**Note**

- **Step 1 under symptom is to quickly determine what system or unit may be at fault by use of the SST. (Self-Diagnosis Checker 49 H018 9A1)**

**1st:** Check input sensors and output solenoid valves with the **SST**.

**2nd:** Check other switches with the **SST**.

**3rd:** Check the following items:

**Electrical system**
1) Battery condition
2) Fuses

**Fuel system**
1) Fuel level
2) Fuel leakage
3) Fuel filter
4) Idle speed (with test connector grounded)

**Engine**
1) Compression
2) Overheating

**Ignition system**
1) Ignition spark
2) Ignition timing

**Intake air system**
1) Air cleaner element
2) Vacuum or air leakage
3) Vacuum hose routing
4) Accelerator cable

**Others**
1) Clutch slippage
2) Brake dragging

**4th:** Check Fuel and Emission Control Systems.

The Troubleshooting Guide lists the systems most likely to cause a given symptom. After finding systems to check, refer to the pages shown for detailed guides for each system.

| Symptom (Possible cause →) | Intake Air System (Poor connection of components, throttle body) | Fuel System (Fuel injection, Fuel pressure) | Pressure Regulator Control System | Idle Speed Control (ISC) System (Air valve, idle speed control solenoid malfunction) | EGR System (EGR control valve stuck and open) | Evaporative Emission Control system (Solenoid valve [Purge control] malfunction) | PCV System (System clogged) | Deceleration System (Fuel cut operation malfunction) | Exhaust system (System clogged) |
|---|---|---|---|---|---|---|---|---|---|
| | **Fuel and Emission Control Systems** | | | | | | | | |
| 2 | 2 | 1 | | | | | | | |
| 3 | 4 | 3 | | 1 | 2 | | | | |
| 3 | 5 | 4 | | 2 | 3 | | 1 | | |
| 4 | 5 | 4 | | 1 | 3 | | 2 | | |
| 4 | 6 | 5 | | 2 | 3 | 4 | 1 | | |
| 5 | 2 | | | 1 | | | | | |
| 6 | 3 | 4 | | | 1 | 2 | | | 5 |
| 7 | | 3 | | 2 | | | | 1 | |
| 8 | 3 | 4 | | 1 | | | | 2 | |
| 9 | | 2 | | | 3 | | | 1 | 4 |
| 10 | | 2 | 1 | | | | | | |

The numbers of the list show the priorities of inspections from the most possible to that with the lowest possibility. These were determined on the following basis:

- Ease of inspection
- Most possible system
- Most possible point in system

## MALFUNCTION INDICATOR LIGHT (MIL) OPERATION 1990 626/MX-6 NON-TURBO

**PRINCIPLE OF CODE CYCLE**

Malfunction codes are determined as shown below

**1. Code cycle break**
The time between malfunction code cycles is 4.0 sec (the time the light is off).

1 CYCLE MALFUNCTION CODE
CYCLE BREAK
ON
OFF
4.0 SEC.

**2. Second digit of malfunction code (ones position)**
The digit in the ones position of the malfunction code represents the number of times the buzzer is on 0.4 sec during one cycle.

1 CYCLE MALFUNCTION CODE
1 2 3 CYCLE BREAK
MALFUNCTION CODE: 03
ON
OFF
0.4 SEC. 0.4 SEC. 4.0 SEC.

**3. First digit of warning code (tens position)**
The digit in the tens position of the malfunction code represents the number of times the buzzer is on 1.2 sec during one cycle.

It should also be noted that the light goes off for 1.6 sec. between the long and short pulses of the buzzer.

1 CYCLE MALFUNCTION CODE
1 2 1 2 CYCLE BREAK
MALFUNCTION CODE: 22
0.4
1.2 SEC. SEC. 1.6 SEC 4.0 SEC.
FIRST DIGIT SECOND DIGIT

## TROUBLE CODE IDENTIFICATION CHART 1990 626/MX-6 NON-TURBO

**CODE NUMBER**

| Malfunction display | | Sensor or subsystem | Self-diagnosis | Fail-safe |
|---|---|---|---|---|
| Code No. | MIL output signal pattern | | | |
| 01 | ON OFF | Ignition pulse | No ignition signal | — |
| 08 | ON OFF | Airflow meter | Open or short circuit | Maintains basic signal at preset value |
| 09 | ON OFF | Water thermosensor | Open or short circuit | Maintains constant command •40°C (104°F) for EGI •50°C (122°F) for ISC control use |
| 10 | ON OFF | Intake air thermosensor (airflow meter) | Open or short circuit | Maintains constant 20°C (68°F) command |
| 12 | ON OFF | Throttle sensor | Open or short circuit | Maintains constant command of throttle valve fully open |
| 14 | ON OFF | Atmospheric pressure sensor | Open or short circuit | Maintains constant command of sea level pressure |
| 15 | ON OFF | Oxygen sensor | Sensor output continues less than 0.55V 120 sec. after engine starts (1,500 rpm) | Cancels EGI feedback operation |
| 16 | ON OFF | EGR position sensor (California only) | Open short circuit | Cuts off EGR |
| 17 | ON OFF | Feedback system | Sensor output not changed 20 sec. after engine exceeds 1,500 rpm | Cancels EGI feedback operation |
| 25 | ON OFF | Solenoid valve (pressure regulator) | Open or short circuit | — |
| 26 | ON OFF | Solenoid valve (purge control) | | — |
| 28 | ON OFF | Solenoid valve (EGR) | | — |
| 34 | ON OFF | ISC valve | | — |

**Caution**
- **If there is more than one failure present, the lowest number malfunction code is displayed first, the remaining codes are displayed sequentially.**
- **After repairing a failure, turn off the ignition switch and disconnect the negative battery cable and depress the brake pedal for at least 5 seconds to erase the memory of a malfunction code.**

## TROUBLE CODE DIAGNOSTIC CHART—1990 626/MX-6 NON-TURBO

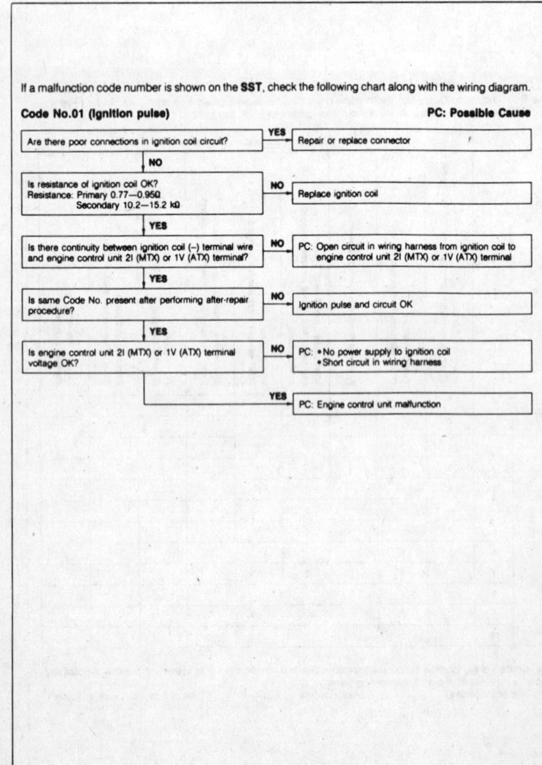


**Code No.08 (Airflow meter)** — **PC: Possible Cause**

Are there poor connections in airflow meter circuit? — YES → Repair or replace connector

NO ↓

Is there continuity between airflow meter and engine control unit?

| Airflow meter | Control unit |
|---|---|
| VB (R/B wire) | 1B |
| Vc (R/W wire) | 2J (MTX) 2A (ATX) |
| Vs (R/B wire) | 2O (MTX) 2B (ATX) |
| E2 (LG/Y wire) | 2D (MTX) 3D (ATX) |

NO → PC: Open circuit in wiring harness from airflow meter to engine control unit

YES ↓

Is resistance of airflow meter OK?
Resistance:

| Terminal | Fully closed (Ω) | Fully open (Ω) |
|---|---|---|
| E2 ↔ Vs | 20—400 | 20—1,000 |
| E2 ↔ Vc | 100—400 | |
| E2 ↔ VB | 200—400 | |

NO → Replace airflow meter

YES ↓

Is same Code No. present after performing after-repair procedure? — NO → Airflow meter and circuit OK

YES ↓

Are engine control unit terminal voltages OK?
Check terminal: 1B,2D,2J,2O (MTX) 1B,2A,2B,3D (ATX)
NO → PC: Short circuit in wiring harness

YES → PC: Engine control unit malfunction

## TROUBLE CODE DIAGNOSTIC CHART—1990 626/MX-6 NON-TURBO

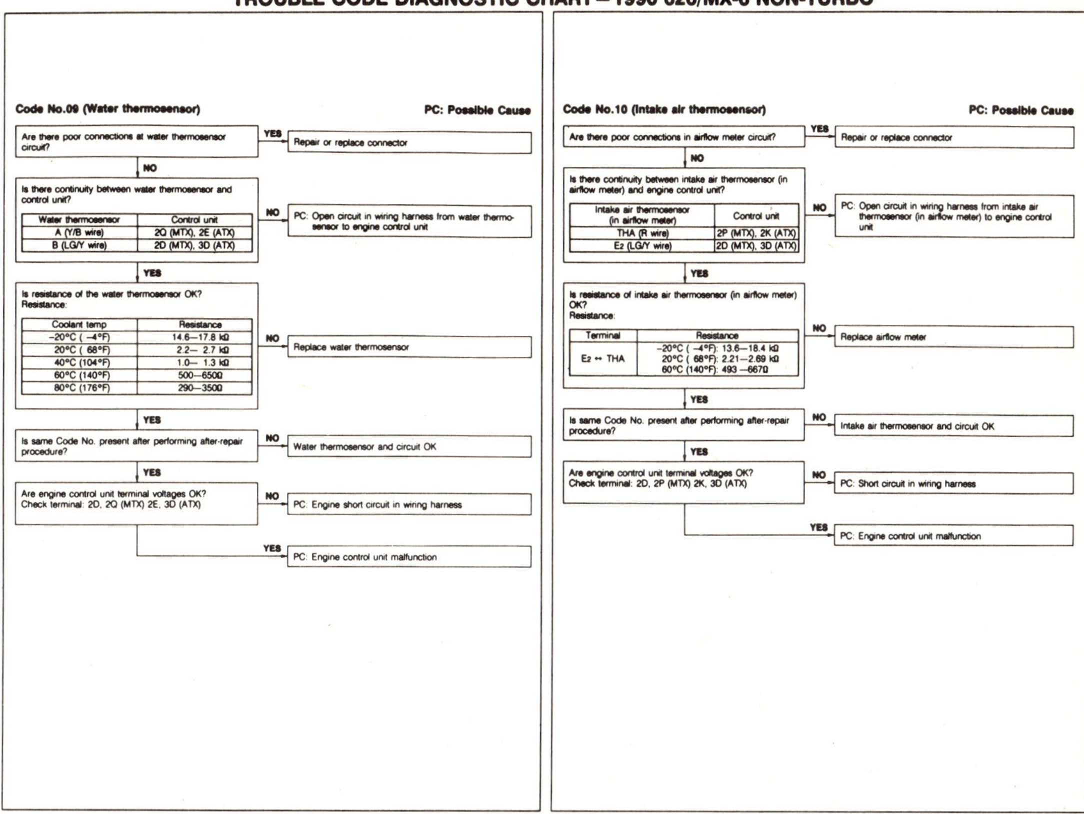


## TROUBLE CODE DIAGNOSTIC CHART—1990 626/MX-6 NON-TURBO

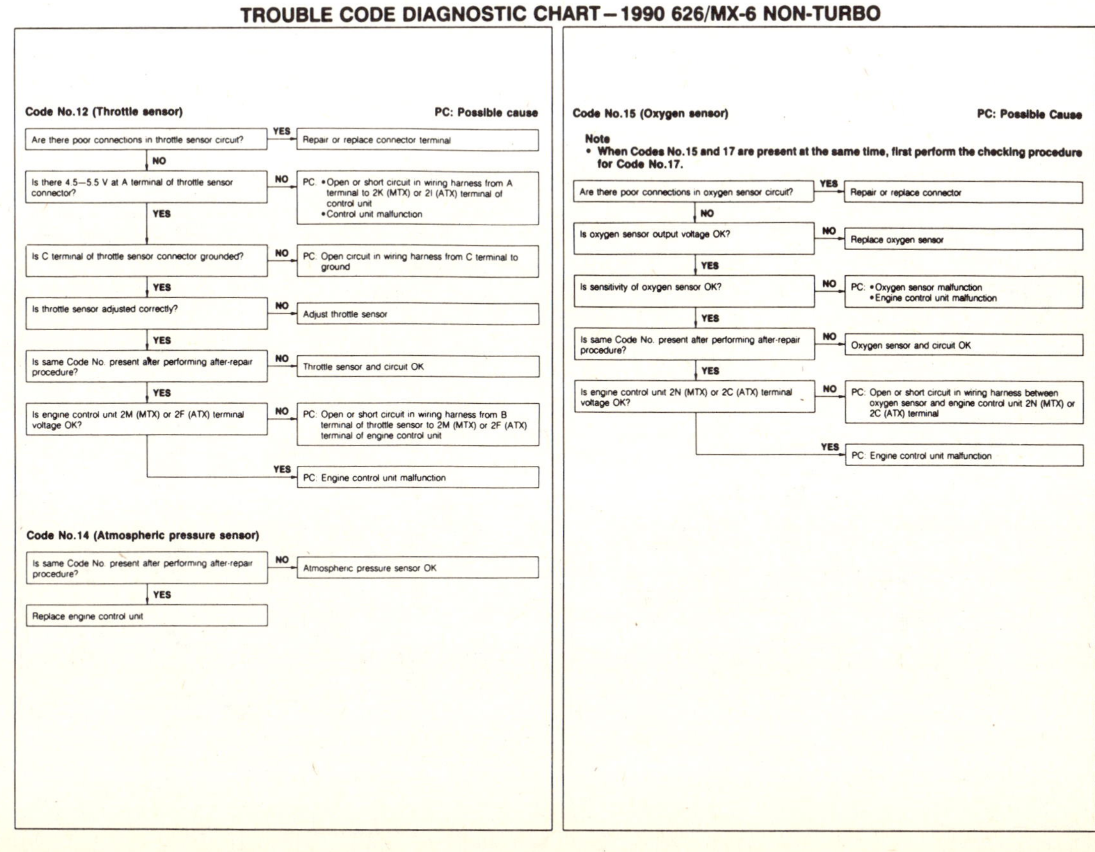

## TROUBLE CODE DIAGNOSTIC CHART—1990 626/MX-6 NON-TURBO

**Code No.16 (EGR position sensor) (California only)** — **PC: Possible Cause**

**Note**
- **Inspect the vacuum hose to the EGR control valve for air leakage, blockage and damage if the MIL illuminates only during cruising.**

Are there poor connections in sensor circuit? — YES → Repair or replace connector
NO ↓

Is EGR control valve OK? — NO → Replace EGR control valve
YES ↓

Is resistance of EGR position sensor OK? — NO → Replace EGR control valve
YES ↓

Is there continuity between EGR position sensor and engine control unit?

| EGR position sensor | Engine control unit |
|---|---|
| (C) (Y/L wire) | 2L (MTX), 2J (ATX) |
| (B) (LG/Y wire) | 2D (MTX), 3D (ATX) |
| (A) (LG/R wire) | 2K (MTX), 2I (ATX) |

NO → PC: Open circuit in wiring harness from EGR position sensor to engine control unit
YES ↓

Is there 4.5—5.5V at A terminal of EGR position sensor connector? — NO → PC: Short circuit in wiring harness from EGR position sensor A terminal to engine control unit 2K (MTX) or 2I (ATX) terminal
•Engine control unit malfunction
YES ↓

Is same Code No. present after performing after-repair procedure? — NO → EGR position sensor and circuit OK
YES ↓

Is engine control unit 2L (MTX) or 2J (ATX) terminal voltage OK? — NO → PC: Short circuit in wiring harness from EGR position sensor C terminal to engine control unit 2L (MTX) or 2J (ATX) terminal
YES → Replace engine control unit

**Code No.17 (Feedback system)** — **PC: Possible Cause**

Warm up engine and run it at 2,500—3,000 rpm for three minutes.
Does monitor lamp of Self-Diagnosis Checker illuminate at idle? — NO → PC: •Air leak in vacuum hoses or emission component
•Contaminated oxygen sensor
•Insufficient fuel injection
YES ↓

Are spark plugs clean? — NO → Clean or replace spark plugs
YES ↓

Is oxygen sensor voltage OK? — NO → PC: Oxygen sensor malfunction
YES ↓

Is same Code No. present after performing after-repair procedure? — NO → Feedback system OK
YES ↓

Is engine control unit 2N (MTX) or 2C (ATX) terminal voltage OK? — NO → PC: Open or short circuit in wiring harness between oxygen sensor and engine control unit 2N (MTX) or 2C (ATX) terminal
YES → PC: Engine control unit malfunction

## TROUBLE CODE DIAGNOSTIC CHART—1990 626/MX-6 NON-TURBO

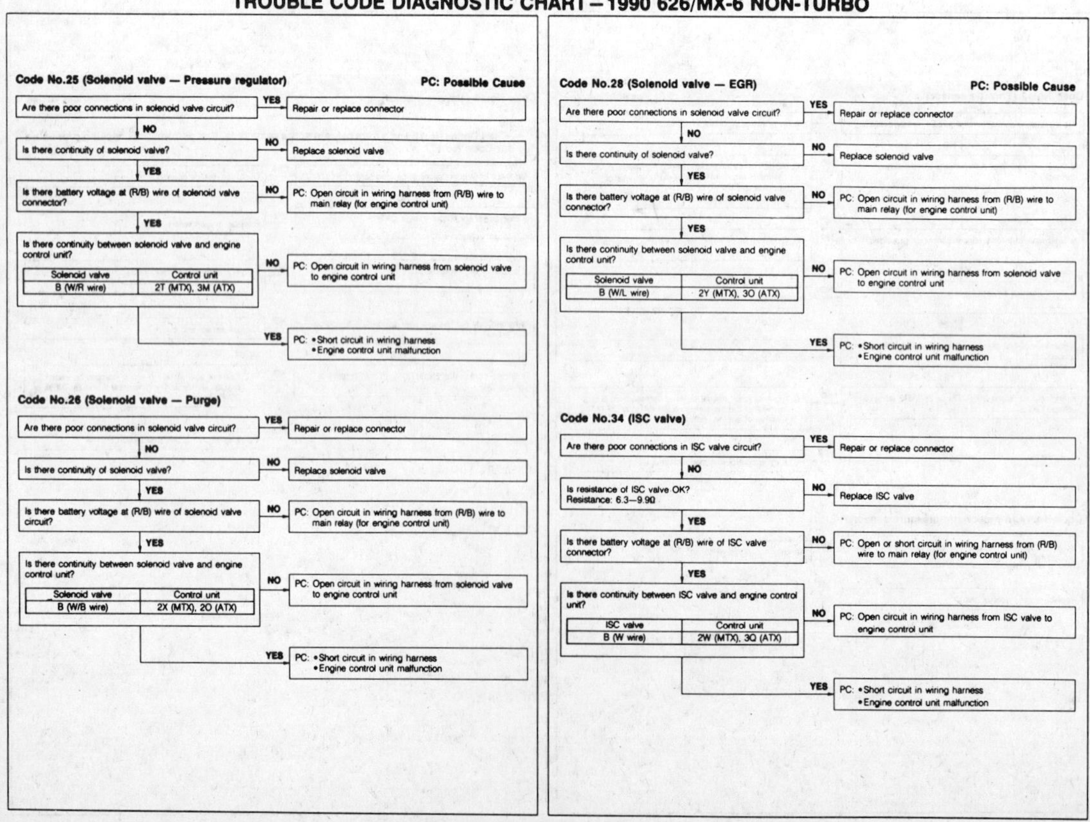

## CONTROL UNIT PIN LOCATIONS AND VOLTAGE TEST—1990 626/MX-6 NON-TURBO WITH M/T

**Terminal voltage**
**MTX**

| Terminal | Input | Output | Connected to | Voltage (After warming-up) | | Remark |
|---|---|---|---|---|---|---|
| | | | | IGN: ON | Idle | |
| 1A | — | — | Battery | Battery voltage | | For back-up |
| 1B | — | — | Main relay | Battery voltage | | — |
| 1C | ○ | | Ign. switch (START) | Below 2.5V | | While cranking: Battery voltage |
| 1D | | ○ | Self-Diagnosis Checker (Monitor lamp) | Test connector grounded<br>• For 3 sec. after ign. switch OFF→ON Below 6.2V (lamp illuminates)<br>• After 3 sec.: Battery voltage (lamp not illuminate) | Test connector not grounded<br>• Lamp illuminates: Below 6.2V<br>• Lamp not illuminate: Battery voltage<br>Test connector grounded: Approx. 5V | Using Self-Diagnosis Checker |
| 1E | | ○ | Malfunction indicator lamp (MIL) | • For 3 sec. after ign. switch OFF→ON: Below 4.8V (lamp illuminates)<br>• After 3 sec.: Battery voltage (lamp not illuminate) | | • Test connector grounded<br>• Lamp illuminates: Below 4.8V<br>• Lamp not illuminate: Battery voltage |
| 1F | | ○ | Self-Diagnosis Checker (Code number) | • For 3 sec. after ign. switch OFF→ON: Below 6.2V (Buzzer sounds)<br>• After 3 sec.: Battery voltage (Buzzer not sound) | | • Using Self-Diagnosis Checker and test connector grounded<br>• Buzzer sounds: Below 6.2V<br>• Buzzer not sound: Battery voltage |
| 1G | — | — | — | — | | — |
| 1H | — | — | — | — | | — |
| 1I | — | — | — | — | | — |
| 1J | | ○ | A/C relay | Battery voltage | • A/C switch ON: Below 2.5V<br>• A/C switch OFF: Battery voltage | Blower motor ON |
| 1K | ○ | | Test connector | • Test connector grounded: Below 1.5V<br>• Test connector not grounded: Above 10.5V | | Test connector: 1-pin, Green connector |
| 1L | — | — | — | — | | — |
| 1M | ○ | | Daytime running light control unit (Canada only) | • Parking brake lever pulled up: Battery voltage<br>• Parking brake lever released: Below 1.5V | | |
| 1N | ○ | | Idle switch | • Accelerator pedal released: Below 0.5V<br>• Accelerator pedal depressed: Above 7.7V | | — |
| 1O | ○ | | Stoplight switch | • Brake pedal released: Below 3.6V<br>• Brake pedal depressed: Above 10.0V | | |
| 1P | ○ | | Power steering pressure switch | Above 10.5V | • P/S ON: Below 1.5V<br>• P/S OFF: Above 10.5V | — |
| 1Q | | | A/C switch | • A/C switch ON: Below 1.5V<br>• A/C switch OFF: Above 10.0V | | Blower motor ON |
| 1R | ○ | | Electrical fan (Water thermoswitch) | Battery voltage | | Coolant temp.: Below 97C° (207°F) |
| | | | | Below 1.5V | | Coolant temp.: Above 97C° (207°F) |

| Terminal | Input | Output | Connected to | Voltage (After warming-up) | | Remark |
|---|---|---|---|---|---|---|
| | | | | IGN: ON | Idle | |
| 1S | ○ | | Blower fan switch | • Switch less than 2nd position: Battery voltage<br>• Switch 3rd or 4th position: Below 1.5V | | |
| 1T | ○ | | Rear window defroster switch | • Switch OFF: Battery voltage<br>• Switch ON: Below 1.5V | | — |
| 1U | ○ | | Headlight switch | • Headlight OFF: Below 1.5V<br>• Headlight ON: Battery voltage | | — |
| 1V | ○ | | Neutral and clutch switch | In-gear condition<br>• Clutch pedal depressed: Below 0.5V<br>• Clutch pedal released: Battery voltage | | Neutral: Below 0.5V |
| 2A | — | — | Ground (EO1) | 0V | | — |
| 2B | — | — | Ground (EO2) | 0V | | — |
| 2C | — | — | Ground (E1) | 0V | | — |
| 2D | — | — | Ground (E2) | 0V | | — |
| 2E | — | — | — | — | | — |
| 2F | — | — | — | — | | — |
| 2G | ○ | | Speedometer | Battery voltage | | Above 113 mph (180 km/h): Below 1.0V |
| 2H | — | — | — | — | | — |
| 2I | ○ | | ignition coil-terminal | Battery voltage | *Battery voltage | *Engine signal monitor green and red lamp flash |
| 2J | ○ | | Airflow meter (Vc) | 7—9V | | — |
| 2K | — | — | Vref | 4.5—5.5V | | — |
| 2L | ○ | | EGR position sensor (California only) | 0.25—0.95V | | |
| 2M | ○ | | Throttle sensor | Accelerator pedal released: Approx. 0.5V (depends on terminal voltage) | | Throttle valve fully open: 4.3V |
| 2N | ○ | | Oxygen sensor | 0V | 0—1.0V | • Cold engine at idle: 0V<br>• After warming-up Acceleration: 0.5—1.0V Deceleration: 0—0.4V |
| 2O | ○ | | Airflow meter (Vs) | Approx. 1.7V | Approx. 3—5V | Increase engine speed: voltage increase |
| 2P | ○ | | Air flow meter (Intake air thermosensor) | Approx. 2.5V at 20°C (68°F) | | — |
| 2Q | ○ | | Water thermosensor | 0.3—0.6V | | Coolant temp. 20°C (68°F): Approx. 2.5V |
| 2R | — | — | — | — | | — |
| 2S | — | — | — | — | | — |
| 2T | | ○ | Solenoid valve (Pressure regulator control) | For 120 sec. after ign. switch OFF ON: Below 3.5V | For 120 sec. after starting: Below 3.5V | Coolant temp. above 70°C (158°F) and intake air temp. above 20°C (68°F) |
| 2U | | ○ | Injector (No.1 and No.3) | Battery voltage | *Battery voltage | *Engine signal monitor green and red lamps flash |
| 2V | | ○ | Injector (No.2 and No.4) | Battery voltage | *Battery voltage | *Engine signal monitor green and red lamps flash |

## CONTROL UNIT PIN LOCATIONS AND VOLTAGE TEST—1990 626/MX-6 NON-TURBO WITH M/T

**ATX**

| Terminal | Input | Output | Connected to | Voltage (After warming-up) | | Remark |
|---|---|---|---|---|---|---|
| | | | | IGN: ON | Idle | |
| 1A | — | — | Battery | Battery voltage | | For back-up |
| 1B | — | — | Main relay | Battery voltage | | — |
| 1C | ○ | | Inhibitor switch (ATX) | Below 2.5V | | While cranking: Battery voltage |
| 1D | | ○ | Self-Diagnosis Cheker (Monitor lamp) | Test connect grounded<br>• For 3 sec. after ign. switch OFF→ON. Below 6.2V (lamp illuminates)<br>• After 3 sec.: Battery voltage (lamp not illuminate) | Test connector not grounded<br>• Lamp illuminates: Below 6.2V<br>• Lamp not illuminate: Battery voltage<br>Test connector grounded: Approx. 5V | Using Self-Diagnosis Checker |
| 1E | | ○ | Malfunction indicator lamp (MIL) | • For 3 sec. after ign. switch OFF→ON: Below 4.8V (lamp illuminates)<br>• After 3 sec.: Battery voltage (lamp not illuminate) | | • Test connector grounded<br>• Lamp illuminates: Below 4.8V<br>• Lamp not illuminate: Battery voltage |
| 1F | | ○ | Self-Diagnosis Checker (Code number) | • For 3 sec. after ign. switch OFF→ON: Below 6.2V (Buzzer sounds)<br>• After 3 sec.: Battery voltage (Buzzer not sound) | | • Using Self-Diagnosis Checker and test connector grounded<br>• Buzzer sounds: Below 6.2V<br>• Buzzer not sound: Battery voltage |
| 1G | — | — | — | — | | — |
| 1H | ○ | | Headlight switch | • Headlight OFF: Below 1.5V<br>• Headlight ON: Battery voltage | | — |
| 1I | ○ | | Test connector | • Test connector grounded: Below 1.5V<br>• Test connector not grounded: Above 10.5V | | Test connector: 1-pin, Green connector |
| 1J | ○ | | Rear window defroster switch | • Switch OFF: Battery voltage<br>• Switch ON: Below 1.5V | | — |
| 1K | — | — | — | — | | — |
| 1L | | ○ | A/C relay | Battery voltage | • A/C switch ON: Below 2.5V<br>• A/C switch OFF: Battery voltage | Blower motor ON |
| 1M | ○ | | Vehicle speed sensor | Approx. 4.5V or below 1.5V | | During driving: Approx. 4.5V |
| 1N | ○ | | Power steering pressure switch | Above 10.5V | • P/S ON: Below 1.5V<br>• P/S OFF: Above 10.5V | — |
| 1O | | | A/C switch | • A/C switch ON: Below 1.5V<br>• A/C switch OFF: above 10.0V | | Blower motor ON |
| 1P | ○ | | Blower fan switch | • Switch less than 2nd position: Battery voltage<br>• Switch 3rd or 4th position: Below 1.5V | | |
| 1Q | ○ | | Stoplight switch | • Brake pedal released: Below 3.6V<br>• Brake pedal depressed: Above 10.0V | | |
| 1R | ○ | | Inhibitor switch (N and P range) | • N or P range: Below 1.5V<br>• Others: Battery voltage | | |

| Terminal | Input | Output | Connected to | Voltage (After warming-up) | | Remark |
|---|---|---|---|---|---|---|
| | | | | IGN: ON | Idle | |
| 2W | | ○ | ISC valve | Engine signal monitor green and red lamps flash | | — |
| 2X | | ○ | Solenoid valve (Purge control) | Battery voltage | | — |
| 2Y | | ○ | Solenoid valve (EGR) | Below 3.5V | | • Engine coolant temp. —below 50°C Below 3.5V<br>• Engine speed above approx. 1,500 rpm: Battery voltage |
| 2Z | — | — | — | — | | — |

**Terminal locations**

| 2Y | 2W | 2U | 2S | 2Q | 2O | 2M | 2K | 2I | 2G | 2E | 2C | 2A | | 1U | 1S | 1Q | 1O | 1M | 1K | 1I | 1G | 1E | 1C | 1A |
|---|---|---|---|---|---|---|---|---|---|---|---|---|---|---|---|---|---|---|---|---|---|---|---|---|
| 2Z | 2X | 2V | 2T | 2R | 2P | 2N | 2L | 2J | 2H | 2F | 2D | 2B | | 1V | 1T | 1R | 1P | 1N | 1L | 1J | 1H | 1F | 1D | 1B |

## CONTROL UNIT PIN LOCATIONS AND VOLTAGE TEST—1990 626/MX-6 NON-TURBO WITH A/T.

| Terminal | Input | Output | Connected to | Voltage (After warming-up) | | Remark |
|---|---|---|---|---|---|---|
| | | | | IGN: ON | Idle | |
| 1T | ○ | | Idle switch | • Accelerator pedal released: Below 0.5V<br>• Accelerator pedal depressed: Above 7.7V | | — |
| 1U | — | — | Ignition switch (IG1) | Battery voltage | | For EC-AT shift-solenoid valves |
| 1V | ○ | | Ignition coil ⊖ terminal | Battery voltage | *Battery voltage | *Engine signal monitor: green and red lamp flash |
| 2A | ○ | | Airflow meter (Vc) | 7—9V | | — |
| 2B | ○ | | Airflow meter (Vs) | Approx. 1.7V | Approx. 3—5V | Increase engine speed: voltage increase |
| 2C | ○ | | Oxygen sensor | 0V | 0—10V | • Cold engine at idle: 0V<br>• After warming-up Acceleration: 0.5—1.0V Deceleration: 0—0.4V |
| 2D | ○ | | Electrical fan [Low] (No.1 water thermoswitch) | Battery voltage | | Coolant temp.: Below 97°C (207°F) |
| | | | | Below 1.5V | | Coolant temp.: Above 97°C (207°F) |
| 2E | ○ | | Water thermosensor | 0.3—0.6V | | Coolant temp. 20°C (68°F): Approx. 2.5V |
| 2F | ○ | | Throttle sensor | Accelerator pedal released: Approx. 0.5V (depends on 2I terminal voltage) | | Throttle valve fully open: 4.3V |
| 2G | ○ | | Electrical fan [High] (No.2 thermoswitch) | Battery voltage | | Coolant temp.: Below 108°C (226°F) |
| | | | | Below 1.5V | | Coolant temp.: Above 108°C (226F°) |
| 2H | ○ | | Hold switch | • Switch depressed: Battery voltage<br>• Switch released: Below 1.5V | | — |
| 2I | — | — | Vref | 4.5—5.5V | | — |
| 2J | ○ | | EGR position sensor (California only) | 0.25—0.95V | | |
| 2K | ○ | | Airflow meter (Intake air thermosensor) | Approx. 2.5V at 20°C (68°F) | | — |
| 2L | ○ | | Mode switch (Power side) | • POWER mode: Below 1.5V<br>• ECONOMY mode or HOLD mode: Battery voltage | | |
| 2M | ○ | | Pulse generator | Below 1.5V | *Battery voltage | *P or N range |
| 2N | — | — | Pulse generator | Below 1.5V | | Ground |
| 2O | | ○ | Solenoid valve (Purge control) | Battery voltage | | — |
| 2P | | ○ | Hold indicator | • Hold mode: Below 1.5V<br>• Other modes: Battery voltage | | — |
| 3A | — | — | Ground (EO1) | 0V | | — |
| 3B | — | — | Ground (EO2) | 0V | | — |
| 3C | — | — | Ground (E1) | 0V | | — |
| 3D | — | — | Ground (E2) | 0V | | — |
| 3E | ○ | | Inhibitor switch (D range) | • D range: Battery voltage<br>• Other range: Below 1.5V | | — |
| 3F | ○ | | Daytime running light control unit (Canada only) | • Parking brake lever pulled up: Battery voltage<br>• Parking brake lever released: Below 1.5V | | |
| 3G | ○ | | Inhibitor switch (L range) | • L range: Battery to voltage<br>• Other range: Below 1.5V | | — |
| 3H | ○ | | Inhibitor switch (S range) | • S range: Battery voltage<br>• Other range: Below 1.5V | | — |
| 3I | — | — | — | — | | — |
| 3J | — | — | — | — | | — |
| 3L | | ○ | Mode indicator | • HOLD mode: Battery voltage<br>• POWER or ECONOMY mode: Below 1.5V | | |
| 3M | | ○ | Solenoid valve (Pressure regulator control) | For 120 sec. after ign. Switch OFF→ON: Below 3.5V | For 120 sec. after starting: Below 3.5V | Coolant temp. above 70°C (158°F) and intake air temp. above 20°C (63°F) |
| 3N | ○ | | Fluid thermoswitch | • Fluid temp. below 143°C (389°F): Approx.10—12V<br>• Fluid temp. above 150°C (302°F): Below 1.5V | | — |
| 3O | | ○ | Solenoid valve (EGR) | Below 3.5V | | • Engine coolant temp. —below 50°C Below 3.5V<br>• Engine speed above approx. 1,500 rpm: Battery voltage |
| 3P | — | — | — | — | | — |
| 3Q | | ○ | ISC valve | Engine signal monitor green and red lamps flash | | — |
| 3R | — | — | — | — | | — |
| 3S | — | — | — | — | | — |
| 3T | — | — | — | — | | — |
| 3U | | ○ | Injector (No.1 and No.3) | Battery voltage | *Battery voltage | *Engine signal monitor green and red lamps flash |
| 3V | | ○ | Injector (No.2 and No.4) | Battery voltage | *Battery voltage | *Engine signal monitor green and red lamps flash |
| 3W | | ○ | 1—2 shift solenoid valve | • Solenoid valve ON: Battery voltage<br>• Solenoid valve OFF: Below 1.5V | | Refer to next page |
| 3X | | ○ | 2—3 shift solenoid valve | • Solenoid valve ON: Battery voltage<br>• Solenoid valve OFF: Below 1.5V | | Refer to next page |
| 3Y | | ○ | 3—4 shift solenoid valve | • Solenoid valve ON: Battery voltage<br>• Solenoid valve OFF: Below 1.5V | | Refer to next page |
| 3Z | | ○ | Lock-up solenoid valve | • Lock-up: Battery voltage<br>• Not lock-up: Below voltage | | Refer to next page |

**Terminal locations**

| 3Y | 3W | 3U | 3S | 3Q | 3O | 3M | 3K | 3I | 3G | 3E | 3C | 3A | 2O | 2M | 2K | 2I | 2G | 2E | 2C | 2A | 1U | 1S | 1Q | 1O | 1M | 1K | 1I | 1G | 1E | 1C | 1A |
|---|---|---|---|---|---|---|---|---|---|---|---|---|---|---|---|---|---|---|---|---|---|---|---|---|---|---|---|---|---|---|---|
| 3Z | 3X | 3V | 3T | 3R | 3P | 3N | 3L | 3J | 3H | 3F | 3D | 3B | 2P | 2N | 2L | 2J | 2H | 2F | 2D | 2B | 1V | 1T | 1R | 1P | 1N | 1L | 1J | 1H | 1F | 1D | 1B |

## COMPONENT DESCRIPTIONS 1990 626/MX-6 NON-TURBO

**COMPONENT DESCRIPTIONS**

| Component | Function | Remarks |
|---|---|---|
| **Airflow meter** | Detects amount of intake air; sends signal to control unit | Intake air thermosensor and fuel pump switch are integrated |
| **Atmospheric pressure sensor** | Detects atmospheric pressure; sends signal to control unit | |
| **Circuit opening relay** | Voltage for fuel pump while engine running | |
| **Clutch switch (MTX)** | Detects in-gear condition; sends signal to control unit | Switch ON when clutch pedal released |
| **Engine control unit** | Detects signals from input sensors and switches; controls injector operation | |
| **Fuel filter** | Filters particles from fuel | |
| **Fuel pump** | Provides fuel to injectors | • Operates while engine running<br>• Installed in fuel tank |
| **Idle switch** | Detects when throttle valve fully closed; sends signal to control unit | Installed on throttle body |
| **Ignition coil (–) terminal** | Detects engine speed; sends signal to control unit | |
| **Ignition switch (ST position)** | Sends engine cranking signal to control unit | |
| **Inhibitor switch (ATX)** | Detects in-gear condition; sends signal to engine control unit | Switch ON in "N" or "P" range |
| **Injector** | Injects fuel into intake port | • Controlled by signals from control unit<br>• High-ohmic injector |
| **Intake air thermosensor** | Detects intake air temperature; sends signal to control unit | Installed in airflow meter |
| **Main relay** | Supplies electric current to injectors and control unit | |
| **Neutral switch (MTX)** | Detects in-gear condition; sends signal to control unit | Switch ON when in-gear |
| **Oxygen sensor** | Detects Oxygen concentration; sends signal to control unit | Zirconia ceramic and platinum coating |
| **Pressure regulator** | Adjusts fuel pressure supplied to injectors | |
| **Pulsation damper** | Absorbs fuel pulsation | |
| **Speedometer** | Detects vehicle speed; sends signal to control unit | ON: Above 113 mph (180 km/h) |
| **Throttle sensor** | Detects throttle valve opening angle; sends signal to control unit | Installed on throttle body |
| **Water thermosensor** | Detects coolant temperature; sends signal to control unit | |

## COMPONENT RELATIONSHIP CHART 1990 626/MX-6 NON-TURBO

**Output Devices and Input Devices**

○: Related ×: Not related

| INPUT DEVICES / OUTPUT DEVICES | FUEL INJECTION AMOUNT (INJECTOR) | FUEL INJECTION TIMING (INJECTOR) | AIR VALVE (BAC VALVE) | ISC VALVE (BAC VALVE) | SOLENOID VALVE (EGR) | SOLENOID VALVE (PURGE) | SOLENOID VALVE (PRESSURE REGULATOR) |
|---|---|---|---|---|---|---|---|
| TEST CONNECTOR | × | × | × | ○ | × | × | × |
| STOPLIGHT SWITCH | ○ | × | × | × | × | × | × |
| VEHICLE SPEED SWITCH | ○ | × | × | × | × | × | × |
| ELECTRICAL LOAD | × | × | × | ○ | × | × | × |
| P/S PRESSURE SWITCH | × | × | × | ○ | × | × | × |
| A/C SWITCH | ○ | × | × | ○ | × | × | × |
| IGNITION SWITCH (STA POSITION) | ○ | ○ | × | × | × | × | ○ |
| INHIBITIOR SWITCH | ○ | × | × | ○ | ○ | ○ | × |
| NEUTRAL AND CLUTCH SWITCH | ○ | × | × | ○ | × | ○ | × |
| OXYGEN SENSOR | ○ | × | × | × | × | ○ | × |
| ATMOSPHERIC PRESSURE SENSOR | ○ | × | × | ○ | × | ○ | × |
| INTAKE AIR THERMOSENSOR | ○ | × | × | ○ | × | ○ | ○ |
| WATER THERMOSENSOR | ○ | × | × | ○ | ○ | ○ | ○ |
| IDLE SWITCH | ○ | ○ | × | ○ | × | × | × |
| THROTTLE SENSOR | ○ | ○ | × | ○ | ○ | ○ | ○ |
| AIRFLOW METER | ○ | × | × | × | × | ○ | × |
| IGNITION COIL | ○ | ○ | × | ○ | ○ | ○ | ○ |

## OUTPUT COMPONENTS AND ENGINE CONDITIONS 1990 626/MX-6 NON-TURBO

**Output Devices and Engine Conditions**

| OUTPUT DEVICES \ ENGINE CONDITIONS | | CRANKING (COLD ENGINE) | WARMING UP (DURING IDLE) | MEDIUM LOAD | | ACCEL-ERATION | HEAVY LOAD | DECEL-ERATION | IDLE (THROT-TLE VALVE FULLY CLOSED) | IGN: ON (ENGINE NOT RUNNING) | REMARKS |
|---|---|---|---|---|---|---|---|---|---|---|---|
| | | | | COLD | WARM | | | | | | |
| INJECTOR | INJECTION | Rich | | | Rich and lean | Rich | | Fuel cut | Rich and lean | No injection | |
| | INJECTION TIMING | 1 group (once per revolution) | | | | 1 group (once per revolution) (once per two revolutions)* | | Fuel cut | 1 group (once per revolution) | No injection | Above 6,300 rpm: fuel cut *Above 4,500 rpm |
| BAC VALVE | AIR VALVE | Open* | | | Close | | | | | | *Coolant temp: below 50°C (122°F) |
| | ISC VALVE | Large amount of bypass air | Large amount of bypass air* | Small amount of bypass air | | | | | | No bypass | *In extreme cold condition |
| SOLENOID VALVE (EGR) | | ON (EGR cut) | | | OFF (EGR) | ON (EGR cut) | OFF (EGR) | ON (EGR cut) | ON (EGR cut)* | ON | *ATX D range: OFF (NO EGR due to no vacuum to system) |
| SOLENOID VALVE (PURGE) | | OFF (Purge cut) | | | Operates (Duty valves [purge gas amount] change*) | | | OFF (Purge cut) | | | *Depends on engine condition |
| SOLENOID VALVE (PRESSURE REGULATOR CONTROL) | | OFF (Vacuum to pressure regulator) | | | | | | | After starting*: ON (Vacuum cut) | OFF | *During hot start only |

## WIRING SCHEMATIC – 1990 626/MX-6 NON-TURBO

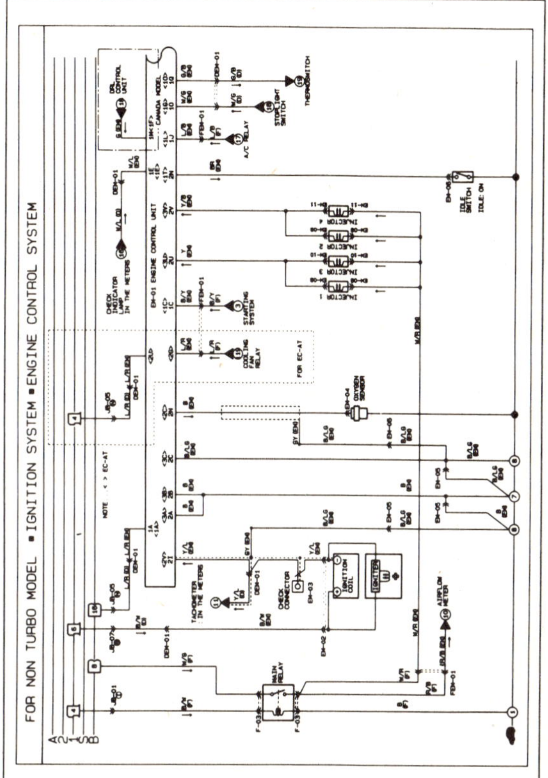


## WIRING SCHEMATIC (CONT.) – 1990 626/MX-6 NON-TURBO

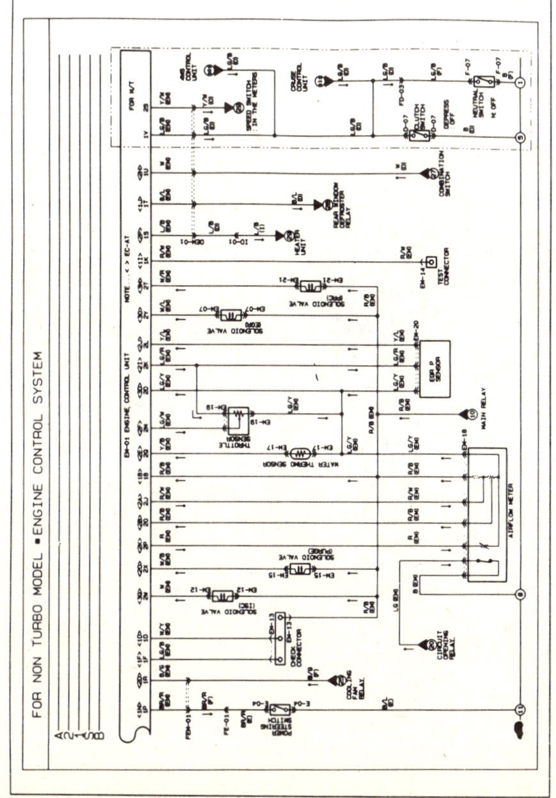


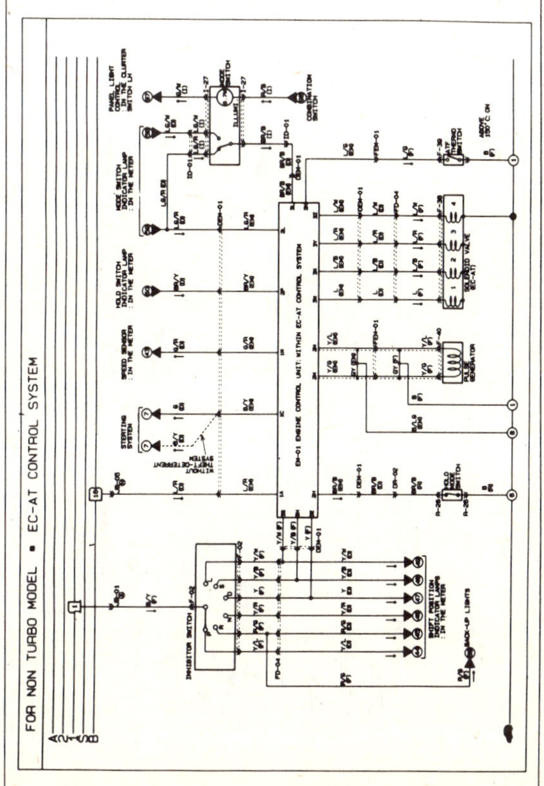

## WIRING SCHEMATIC (CONT.)—1990 626/MX-6 NON-TURBO

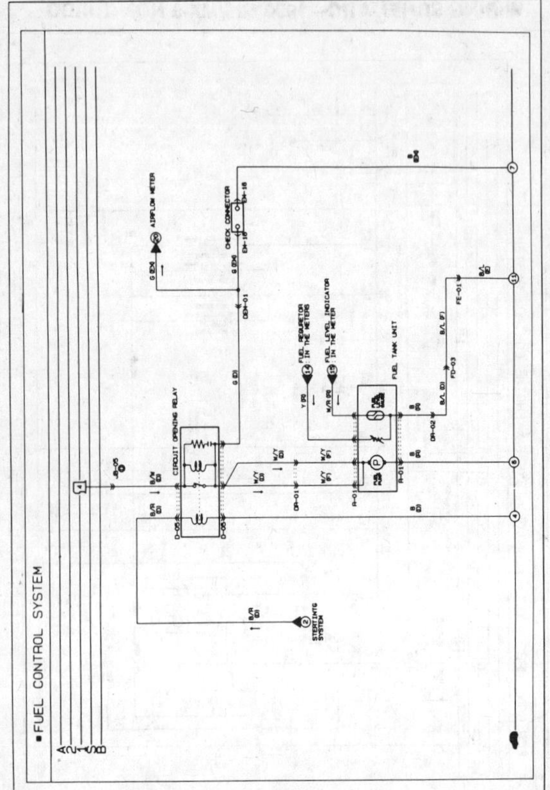

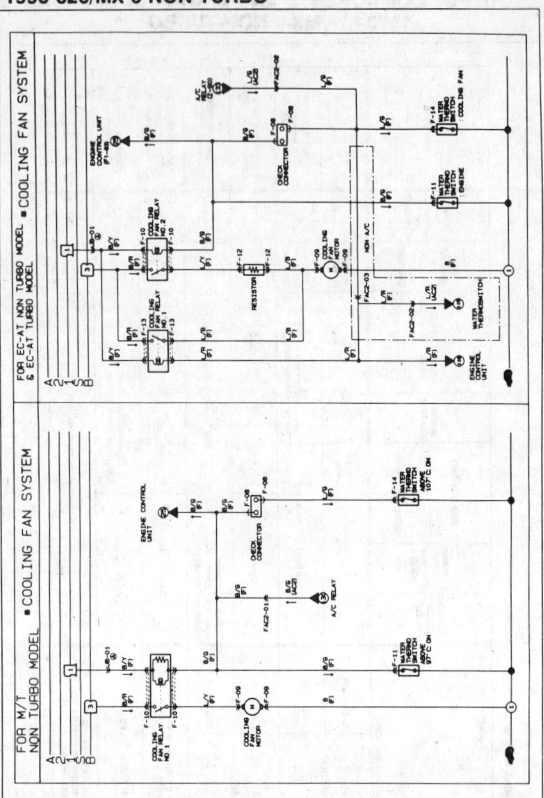

## SYSTEM SCHEMATIC—1990 626/MX-6 TURBO

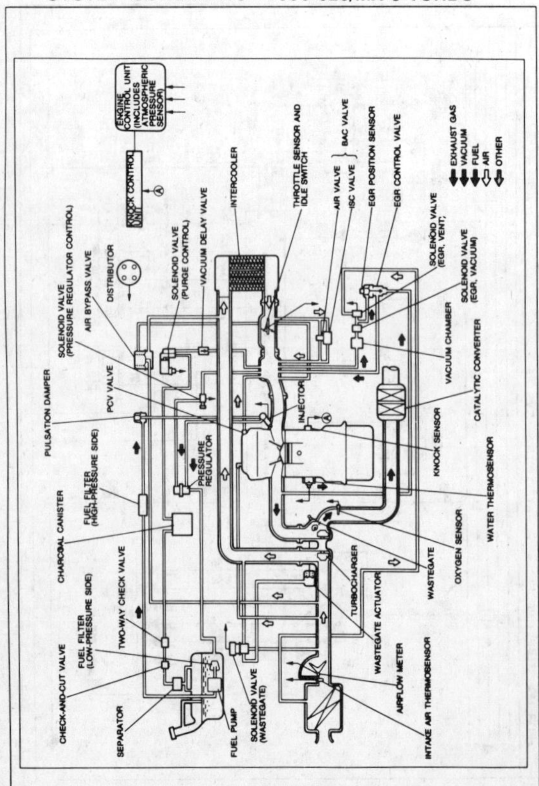

## VACUUM HOSE ROUTING—1990 626/MX-6 TURBO

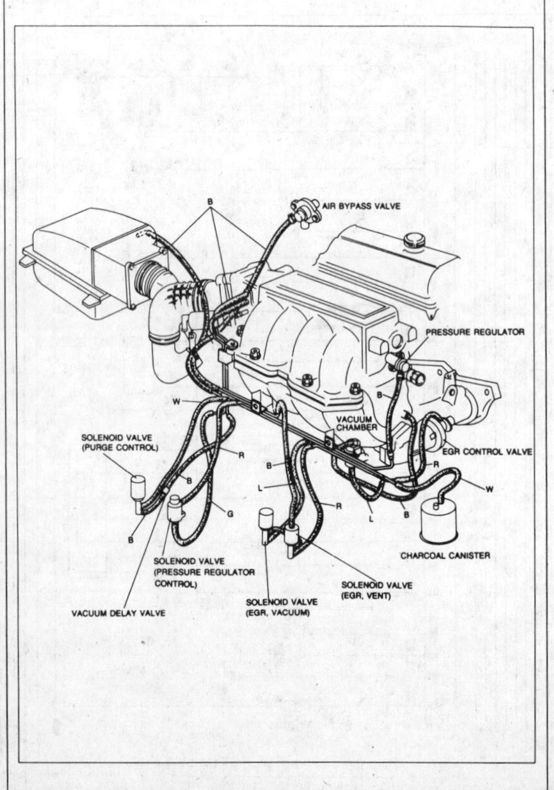

## EXPLODED VIEW OF AIR INTAKE AIR SYSTEM 1990 626/MX-6 TURBO

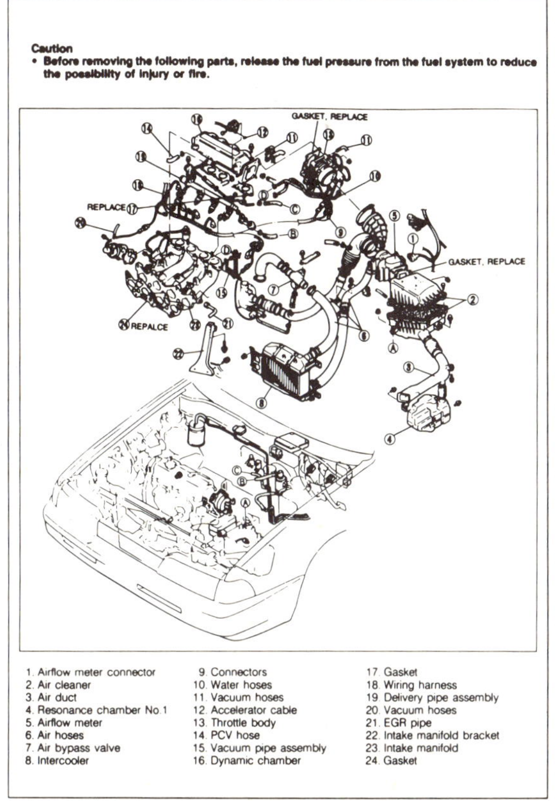


1. Airflow meter connector
2. Air cleaner
3. Air duct
4. Resonance chamber No.1
5. Airflow meter
6. Air hoses
7. Air bypass valve
8. Intercooler
9. Connectors
10. Water hoses
11. Vacuum hoses
12. Accelerator cable
13. Throttle body
14. PCV hose
15. Vacuum pipe assembly
16. Dynamic chamber
17. Gasket
18. Wiring harness
19. Delivery pipe assembly
20. Vacuum hoses
21. EGR pipe
22. Intake manifold bracket
23. Intake manifold
24. Gasket

## EXPLODED VIEW OF CONTROL SYSTEM COMPONENTS 1990 626/MX-6 TURBO

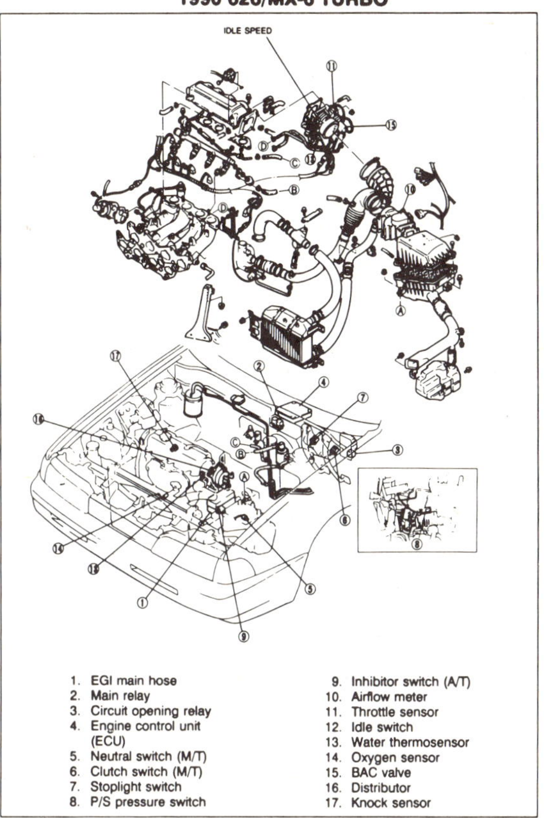


1. EGI main hose
2. Main relay
3. Circuit opening relay
4. Engine control unit (ECU)
5. Neutral switch (M/T)
6. Clutch switch (M/T)
7. Stoplight switch
8. P/S pressure switch
9. Inhibitor switch (A/T)
10. Airflow meter
11. Throttle sensor
12. Idle switch
13. Water thermosensor
14. Oxygen sensor
15. BAC valve
16. Distributor
17. Knock sensor

## EXPLODED VIEW FUEL AND EMISSION COMPONENTS 1990 626/MX-6 TURBO

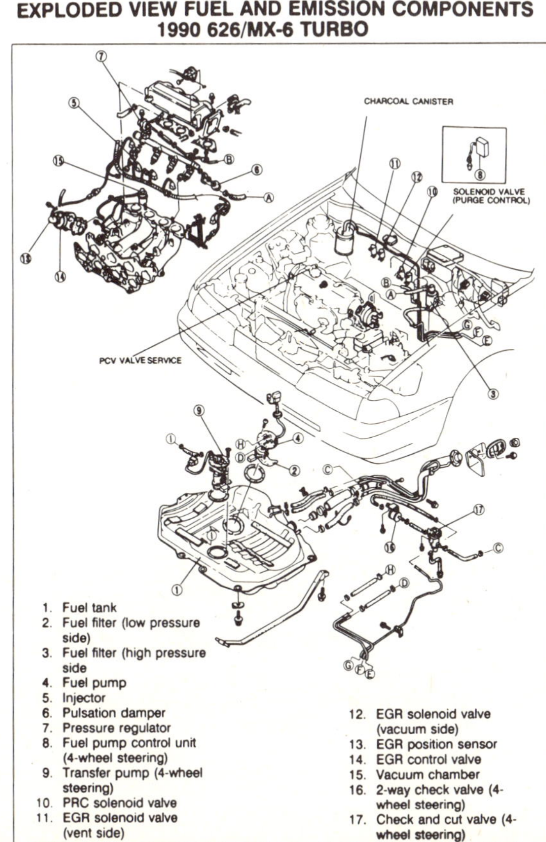


1. Fuel tank
2. Fuel filter (low pressure side)
3. Fuel filter (high pressure side
4. Fuel pump
5. Injector
6. Pulsation damper
7. Pressure regulator
8. Fuel pump control unit (4-wheel steering)
9. Transfer pump (4-wheel steering)
10. PRC solenoid valve
11. EGR solenoid valve (vent side)
12. EGR solenoid valve (vacuum side)
13. EGR position sensor
14. EGR control valve
15. Vacuum chamber
16. 2-way check valve (4-wheel steering)
17. Check and cut valve (4-wheel steering)

## SPECIFICATIONS CHART—1990 626/MX-6 TURBO

**SPECIFICATIONS**

| Item | | Engine type | Turbo Engine |
|---|---|---|---|
| Idle speed | rpm | | 750 ± 25 (ATX: P range)* |
| **Throttle body** | | | |
| Type | | | Horizontal draft (2-barrel) |
| Throat diameter | mm (in) | No.1 | MTX: 40 (1.6), ATX: 46 (1.8) |
| | | No.2 | MTX: 46 (1.8), ATX: 40 (1.6) |
| **Airflow meter** | | | |
| Resistor | Ω | E2—Vs | Fully closed: 20—400 Fully open: 20—1,000 |
| | | E2—Vc | 100—400 |
| | | E2—VB | 200—400 |
| | | E2—THA | −20°C (−4°F) 13,600—18,400<br>20°C (68°F) 2,210—2,690<br>60°C (140°F) 493—667 |
| **Fuel pump** | | | |
| Type | | | Impeller (in tank) |
| Output pressure | kPa (kg/cm², psi) | | Main pump: 441—588 (4.5—6.0, 64—85)<br>Transfer pump: 39 (0.4, 5.7) max. |
| Feeding capacity | cc (cu in)/10 seconds | | Main pump: 220 (13.4) min.<br>Transfer pump: 190 (11.6) min. |
| **Fuel filter** | | | |
| Type | | Low-pressure side | Nylon element |
| | | High-pressure side | Paper element |
| **Pressure regulator** | | | |
| Type | | | Diaphragm |
| Regulating pressure | kPa (kg/cm², psi) | | 235—275 (2.4—2.8, 34—40) |
| **Injector** | | | |
| Type | | | High-ohmic |
| Type of drive | | | Voltage |
| Resistance | Ω | | 12—16 |
| Injection amount | cc (cu in)/15 seconds | | 73—90 (4.45—5.49) |
| **Idle speed control valve** | | | |
| Solenoid resistance | Ω | | 6.3—9.9 |
| **Turbocharger** | | | |
| Cooling method | | | Engine coolant |
| Lubrication method | | | Engine oil |
| Boost pressure (Maximum) | kPa (kg/cm², psi) | | 60 (0.61, 8.7): Solenoid duty value 100%<br>45 (0.46, 6.5): Solenoid duty value 0% |
| **Fuel tank** | | | |
| Capacity | liters (US gal, Imp gal) | | 60 (15.9, 13.2), 57 (15.0, 12.5): 4-wheel steering vehicle |
| **Air cleaner** | | | |
| Element type | | | Oil permeated |
| **Fuel** | | | |
| Specification | | | Unleaded premium (Unleaded regular) |

* With test connector grounded.

## COMPONENTS DESCRIPTION – 1990 626/MX-6 TURBO

### COMPONENT DESCRIPTIONS

| Component | Function | Remarks |
|---|---|---|
| **Airflow meter** | Detects amount of intake air; sends signal to control unit | Intake air thermosensor and fuel pump switch are integrated |
| **Atmospheric pressure sensor** | Detects atmospheric pressure; sends signal to control unit | |
| **Circuit opening relay** | Voltage for fuel pump while engine running | |
| **Clutch switch (MTX)** | Detects in-gear condition; sends signal to control unit | Switch ON when clutch pedal released |
| **EC-AT control unit (ATX)** | Detects N or P range; sends signal to control unit | |
| **Engine control unit** | Detects signals from input sensors and switches; controls injector operation | |
| **Fuel filter** | Filters particles from fuel | |
| **Fuel pump** | Provides fuel to injectors | •Operates while engine running<br>•Installed in fuel tank |
| **Fuel pump control unit** | Detects signal from transfer pump switch; controls operation of transfer pupm | |
| **G rotor and pick-up** | Detects No.1 and No.4 cylinders TDC; sends signal to control unit | For determining fuel injection timing and ignition timing |
| **Idle switch** | Detects when throttle valve fully closed; sends signal to control unit | Installed on throttle body |
| **Ignition switch (ST position)** | Sends engine cranking signal to control unit | |
| **Inhibitor switch (ATX)** | Detects in-gear condition; sends signal to EC-AT control unit | Switch ON in "N" or "P" range |
| **Injector** | Injects fuel into intake port | •Controlled by signals from control unit<br>•High-ohmic injector |
| **Intake air thermosensor** | Detects intake air temperature; sends signal to control unit | Installed in airflow meter |
| **Main relay** | Supplies electric current to injectors and control unit | |
| **Ne rotor and pick-up** | Detects crank angle at 30° intervals; sends signal to control unit | Engine speed signal |
| **Neutral switch (MTX)** | Detects in-gear condition; sends signal to control unit | Switch ON when in-gear |
| **Oxygen sensor** | Detects Oxygen concentration; sends signal to control unit | Zirconia ceramic and platinum coating |
| **Pressure regulator** | Adjusts fuel pressure supplied to injectors | |
| **Pulsation damper** | Absorbs fuel pulsation | |
| **Throttle sensor** | Detects throttle valve opening angle; sends signal to control unit | Installed on throttle body |
| **Transfer pump (4-wheel steering)** | Pumps fuel from one side of the tank to other to maintain balance | |
| **Water thermosensor** | Detects coolant temperature; sends signal to control unit | |

## TROUBLESHOOTING CHART – 1990 626/MX-6 TURBO

### TROUBLESHOOTING GUIDE

This troubleshooting guide shows the malfunction numbers and the symptoms of various failures. Perform troubleshooting as described below.

| Symptom \ Possible cause | Input sensors and switches | | | | | | | | | | | | | Output solenoid valves | | | | | |
|---|---|---|---|---|---|---|---|---|---|---|---|---|---|---|---|---|---|---|---|
| | Ignition pulse | Distributor (Ne signal) | Distributor (G1 signal) | Distributor (G2 signal) | Knock sensor | Airflow meter | Water thermosensor | Intake air thermosensor | Throttle sensor | Atmospheric pressure sensor | Oxygen sensor | EGR position sensor | Feedback system | Solenoid valve (Pressure regulator) | Solenoid valve (Purge) | Solenoid valve (EGR, vacuum side) | Solenoid valve (EGR, vent side) | ISC valve | Solenoid valve (Waste gate) |
| 1 Fault indicated by SST Code No. | 01 | 02 | 03 | 04 | 05 | 08 | 09 | 10 | 12 | 14 | 15 | 16 | 17 | 25 | 26 | 28 | 29 | 34 | 42 |

| No. | Symptom | |
|---|---|---|
| 2 | Hard start or won't start (Cranks OK) | |
| 3 | Engine stalls | While warming up |
| | | After warming up |
| 4 | Rough idle | While warming up |
| | | After warming up |
| 5 | High idle speed after warming up | |
| 6 | Poor acceleration, hesitation or lack of power | |
| 7 | Runs rough on deceleration | |
| 8 | Afterburn in exhaust system | |
| 9 | Poor fuel consumption | |
| 10 | Excessive oil consumption | |
| 11 | Engine stalls or rough after hot starting | |
| 12 | Knocking | |
| 13 | Abnormal noise or vibration | |

#### TROUBLESHOOTING PROCEDURE

**Note**
- **Step 1 under symptom is to quickly determine what system or unit may be at fault by use of the SST. (Self-Diagnosis Checker 49 H018 9A1)**

**1st:** Check input sensors and output solenoid valves with the **SST**.

**2nd:** Check other switches with the **SST**.

**3rd:** Check the following items:

**Electrical system**
1) Battery condition
2) Fuses

**Ignition system**
1) Ignition spark
2) Ignition timing (with test connector grounded)

**Fuel system**
1) Fuel level
2) Fuel leakage
3) Fuel filter
4) Idle speed (with test connector grounded)

**Intake air system**
1) Air cleaner element
2) Vacuum or air leakage
3) Vacuum hose routing
4) Accelerator cable

**Engine**
1) Compression
2) Overheating

**Others**
1) Clutch slippage
2) Brake dragging

**4th:** Check Fuel and Emission Control Systems.

## TROUBLESHOOTING CHART (CONT.) 1990 626/MX-6 TURBO

The Troubleshooting Guide lists the systems most likely to cause a given symptom. After finding systems to check, refer to the pages shown for detailed guides for each system.

| Symptom \ Possible cause | Intake Air System (Poor connection of components, throttle body) | Fuel System (Fuel Injection, Fuel pressure) | Pressure Regulator Control System | Idle Speed Control (ISC) System (Air valve, ISC valve malfunction) | Turbocharging System (Oil & water passage, Turbine and compressor wheels malfunction) | Electronic Spark Advance (ESA) System (Knock control system) | EGR System (EGR control valve stuck and open) | Evaporative Emission Control System (Solenoid valve [Purge control] malfunction) | PCV System (System clogged) | Deceleration System (Fuel cut operation malfunction) | Exhaust System (System clogged) |
|---|---|---|---|---|---|---|---|---|---|---|---|
| 2 | 3 | 2 | | | | 1 | | | | | |
| 3 | 4 | 3 | | 1 | | | 2 | | | | |
| | 5 | 4 | | 2 | | | 3 | | 1 | | |
| 4 | 5 | 4 | | 1 | | | 3 | | 2 | | |
| | 6 | 5 | | 2 | | | 3 | 4 | 1 | | |
| 5 | 2 | | | 1 | | | | | | | |
| 6 | 3 | 4 | | | 6 | | 1 | 2 | | | 5 |
| 7 | | 3 | | 2 | | | | | | 1 | |
| 8 | 3 | 4 | | 1 | | | | | | 2 | |
| 9 | | 2 | | | | | 3 | | | 1 | 4 |
| 10 | | | | | 1 | | | | | | |
| 11 | | 2 | 1 | | | | | | | | |
| 12 | | | | | 2 | 1 | | | | | |
| 13 | | | | | 1 | | | | | | |

The numbers of the list show the priorities of inspections from the most possible to that with the lowest possibility. These were determined on the following basis:
- Ease of inspection
- Most possible system
- Most possible point in system

## MALFUNCTION INDICATOR LIGHT (MIL) OPERATION 1990 626/MX-6 TURBO

**PRINCIPLE OF CODE CYCLE**

Malfunction codes are determined as shown below

**1. Code cycle break**
The time between malfunction code cycles is 4.0 sec (the time the light is off).

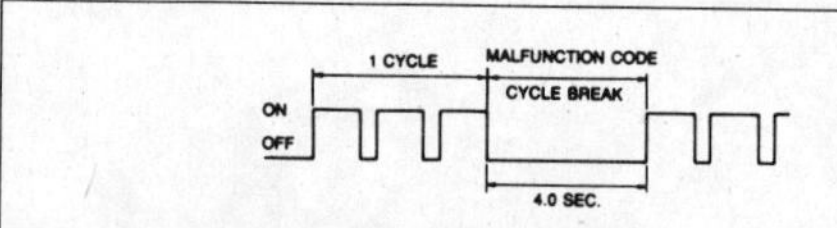


**2. Second digit of malfunction code (ones position)**
The digit in the ones position of the malfunction code represents the number of times the buzzer is on 0.4 sec during one cycle.

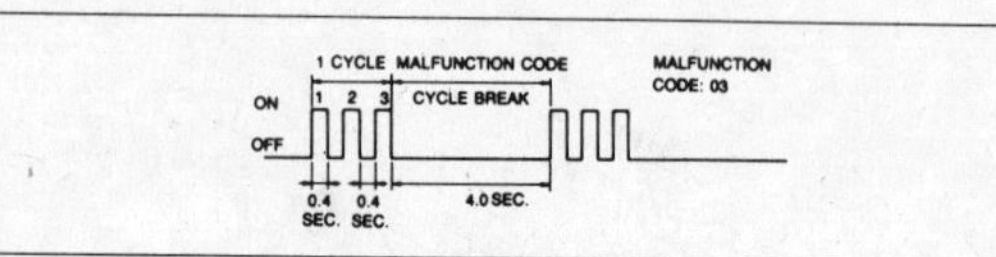


**3. First digit of warning code (tens position)**
The digit in the tens position of the malfunction code represents the number of times the buzzer is on 1.2 sec during one cycle.

It should also be noted that the light goes off for 1.6 sec. between the long and short pulses of the buzzer.

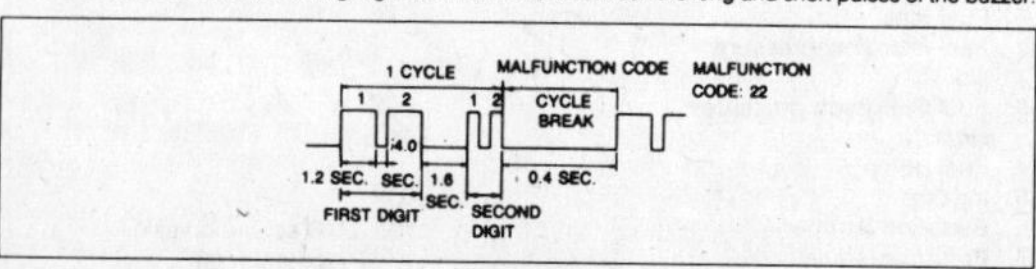

## TROUBLE CODE IDENTIFICATION CHART—1990 626/MX-6 TURBO

**CODE NUMBER**

| Malfunction display | | Sensor or subsystem | Self-diagnosis | Fail-safe |
|---|---|---|---|---|
| Code No. | MIL output signal pattern | | | |
| 01 | ON OFF | Ignition pulse | No ignition signal | — |
| 02 | ON OFF | Ne signal | No Ne signal from crank angle sensor | — |
| 03 | ON OFF | G1 signal | No G1 signal | Neither G1 nor G2 signal: Engine stopped |
| 04 | ON OFF | G2 signal | No G2 signal | |
| 05 | ON OFF | Knock sensor and knock control unit | Open or short circuit | • Retards ignition timing 6° in heavy-load condition<br>• Waste gate opens earlier |
| 08 | ON OFF | Airflow meter | Open or short circuit | Maintains basic signal at preset value |
| 09 | ON OFF | Water thermosensor | Open or short circuit | Maintains constant command<br>• 40°C (104°F) for EGI<br>• 50°C (122°F) for ISC control use |
| 10 | ON OFF | Intake air thermosensor (airflow meter) | Open or short circuit | Maintains constant 20°C (68°F) command |
| 12 | ON OFF | Throttle sensor | Open or short circuit | Maintains constant command of throttle valve fully open |
| 14 | ON OFF | Atmospheric pressure sensor | Open or short circuit | Maintains constant command of sea level pressure |
| 15 | ON OFF | Oxygen sensor | Sensor output continues less than 0.55V 120 sec. after engine starts (1,500 rpm) | Cancels EGI feedback operation |
| 16 | ON OFF | EGR position sensor | Open or short circuit | Cuts off EGR |
| | | | Sensor output does not match target value (incorrect output) | — |
| 17 | ON OFF | Feedback system | Sensor output not changed 20 sec. after engine exceeds 1,500 rpm | Cancels EGI feedback operation |

| Malfunction display | | Sensor or subsystem | Self-diagnosis | Fail-safe |
|---|---|---|---|---|
| Code No. | MIL output signal pattern | | | |
| 25 | ON OFF | Solenoid valve (pressure regulator) | Open or short circuit | — |
| 26 | ON OFF | Solenoid valve (purge control) | | — |
| 28 | ON OFF | Solenoid valve (EGR-vacuum) | | — |
| 29 | ON OFF | Solenoid valve (EGR-vent) | | — |
| 34 | ON OFF | ISC valve | | — |
| 42 | ON OFF | Solenoid valve (waste gate) | | — |

**Caution**

- **If there is more than one failure present, the lowest number malfunction code is displayed first, the remaining codes are displayed sequentially.**
- **After repairing a failure, turn off the ignition switch and disconnect the negative battery cable and depress the brake pedal for at least 5 seconds to erase the memory of a malfunction code.**

## TROUBLE CODE DIAGNOSTIC CHART—1990 626/MX-6 TURBO

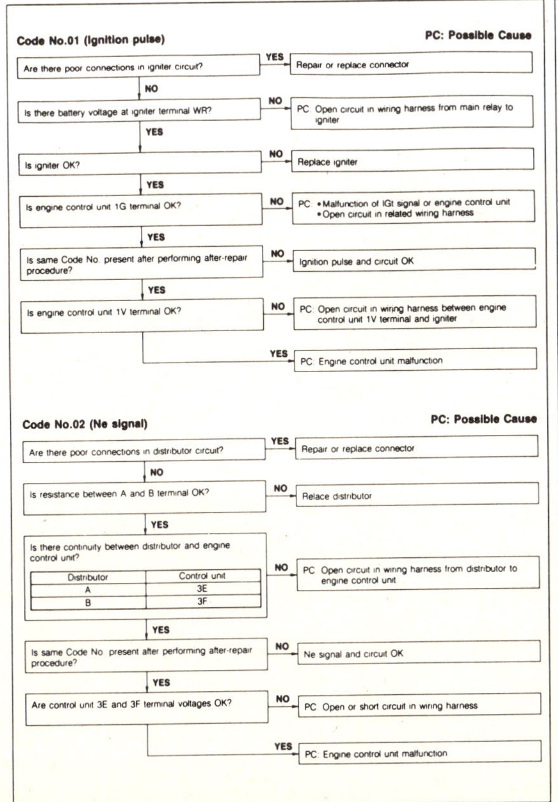


**Code No.03 (G1 signal)** — PC: Possible Cause

- Are there poor connections in distributor circuit?
  - YES → Repair or replace connector
  - NO ↓
- Is resistance between B and D terminal OK? (Refer to page F2–80)
  - NO → Replace distributor
  - YES ↓
- Is there continuity between distributor and engine control unit?

| Distributor | Control unit |
|---|---|
| B | 3F |
| D | 3G |

  - NO → PC: Open circuit in wiring harness from distributor to engine control unit
  - YES ↓
- Is same Code No. present after performing after-repair procedure?
  - NO → G1 signal and circuit OK
  - YES ↓
- Are engine control unit 3F and 3G terminal voltages OK?
  - NO → PC: Short circuit in wiring harness
  - YES → PC: Engine control unit malfunction

**Code No.04 (G2 signal)** — PC: Possible Cause

- Are there poor connections in distributor circuit?
  - YES → Repair or replace connector
  - NO ↓
- Is resistance between B and C terminal OK?
  - NO → Replace distributor
  - YES ↓
- Is there continuity between distributor and engine control unit?

| Distributor | Control unit |
|---|---|
| B | 3F |
| C | 3H |

  - NO → PC: Open circuit in wiring harness from distributor to engine control unit
  - YES ↓
- Is same Code No. present after performing after-repair procedure?
  - NO → G2 signal and circuit OK
  - YES ↓
- Are engine control unit 3F and 3H terminal voltages OK?
  - NO → PC: Short circuit in wiring harness
  - YES → PC: Engine control unit malfunction

## TROUBLE CODE DIAGNOSTIC CHART—1990 626/MX-6 TURBO

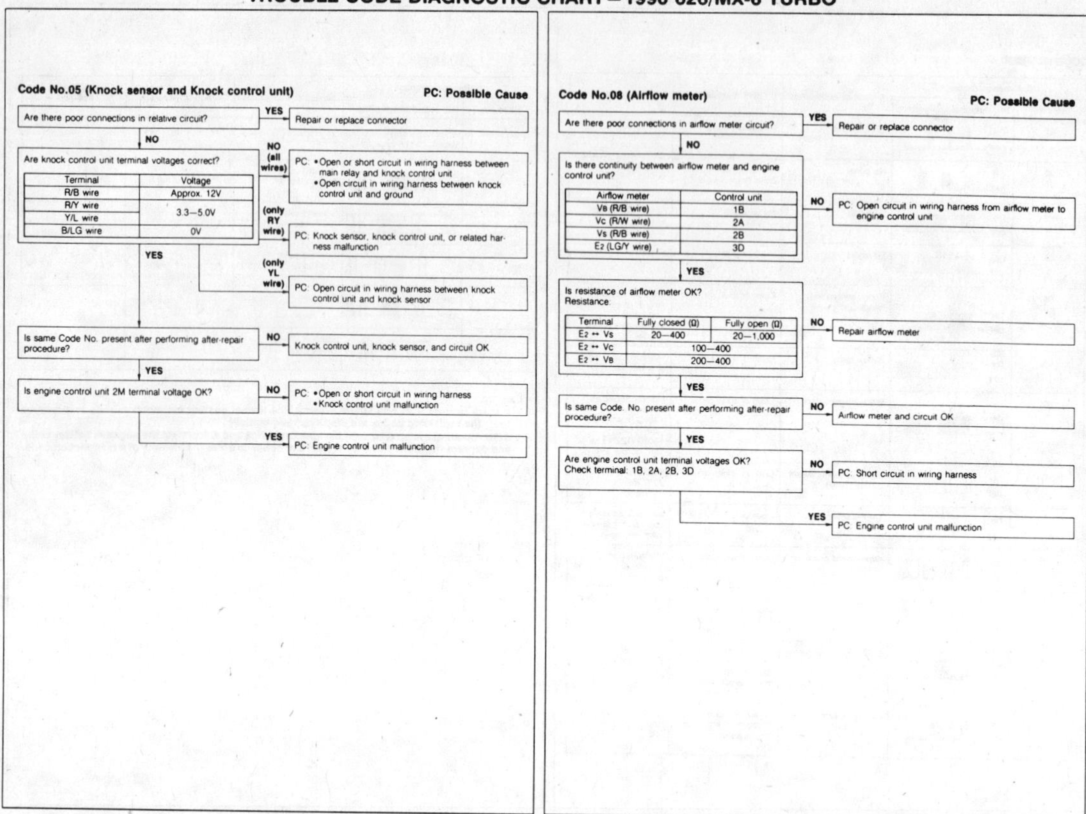


## TROUBLE CODE DIAGNOSTIC CHART—1990 626/MX-6 TURBO

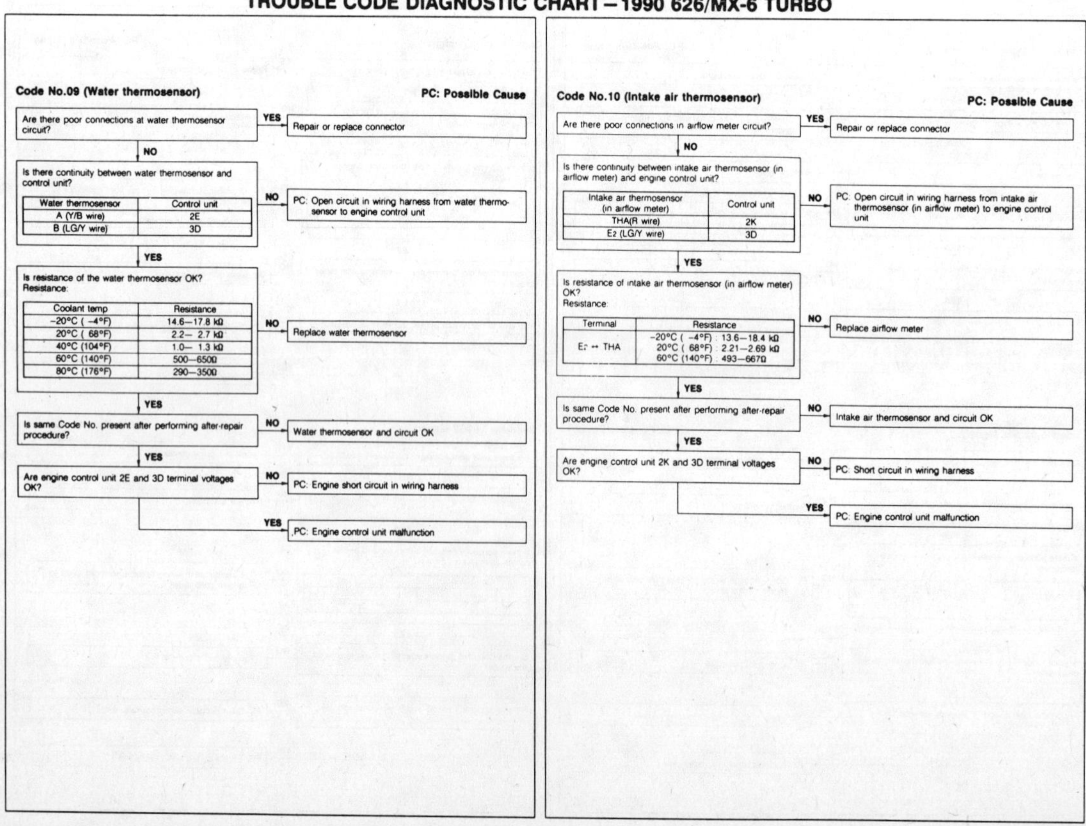

## TROUBLE CODE DIAGNOSTIC CHART—1990 626/MX-6 TURBO

### Code No.12 (Throttle sensor)

**PC: Possible cause**

- Are there poor connections in throttle sensor circuit? — YES → Repair or replace connector terminal
- NO ↓
- Is there 4.5—5.5 V at A terminal of throttle sensor connector? — NO → PC • Open or short circuit in wiring harness from A terminal to 2I terminal of control unit • Control unit malfunction
- YES ↓
- Is C terminal of throttle sensor connector grounded? — NO → PC Open circuit in wiring harness from C terminal to ground
- YES ↓
- Is throttle sensor adjusted correctly? — NO → Adjust throttle sensor
- YES ↓
- Is same Code No. present after performing after-repair procedure? — NO → Throttle sensor and circuit OK
- YES ↓
- Is engine control unit 2F terminal voltage OK? — NO → PC Open or short circuit in wiring harness from B terminal of throttle sensor to 2F terminal of engine control unit
- YES → PC Engine control unit malfunction

### Code No.14 (Atmospheric pressure sensor)

- Is same Code No. present after performing after-repair procedure? — NO → Atmospheric pressure sensor OK
- YES ↓
- Replace engine control unit

### Code No.15 (Oxygen sensor)

**PC: Possible Cause**

**Note**
- **When Codes No.15 and 17 are present at the same time, first perform the checking procedure for Code No.17.**

- Are there poor connections in oxygen sensor circuit? — YES → Repair or replace connector
- NO ↓
- Is oxygen sensor output voltage OK? — NO → Replace oxygen sensor
- YES ↓
- Is sensitivity of oxygen sensor OK? — NO → PC • Oxygen sensor malfunction • Engine control unit malfunction
- YES ↓
- Is same Code No. present after performing after-repair procedure? — NO → Oxygen sensor and circuit OK
- YES ↓
- Is engine control unit 2C terminal voltage OK? — NO → PC Open or short circuit in wiring harness between oxygen sensor and engine control unit 2C terminal
- YES → PC Engine control unit malfunction

## TROUBLE CODE DIAGNOSTIC CHART—1990 626/MX-6 TURBO

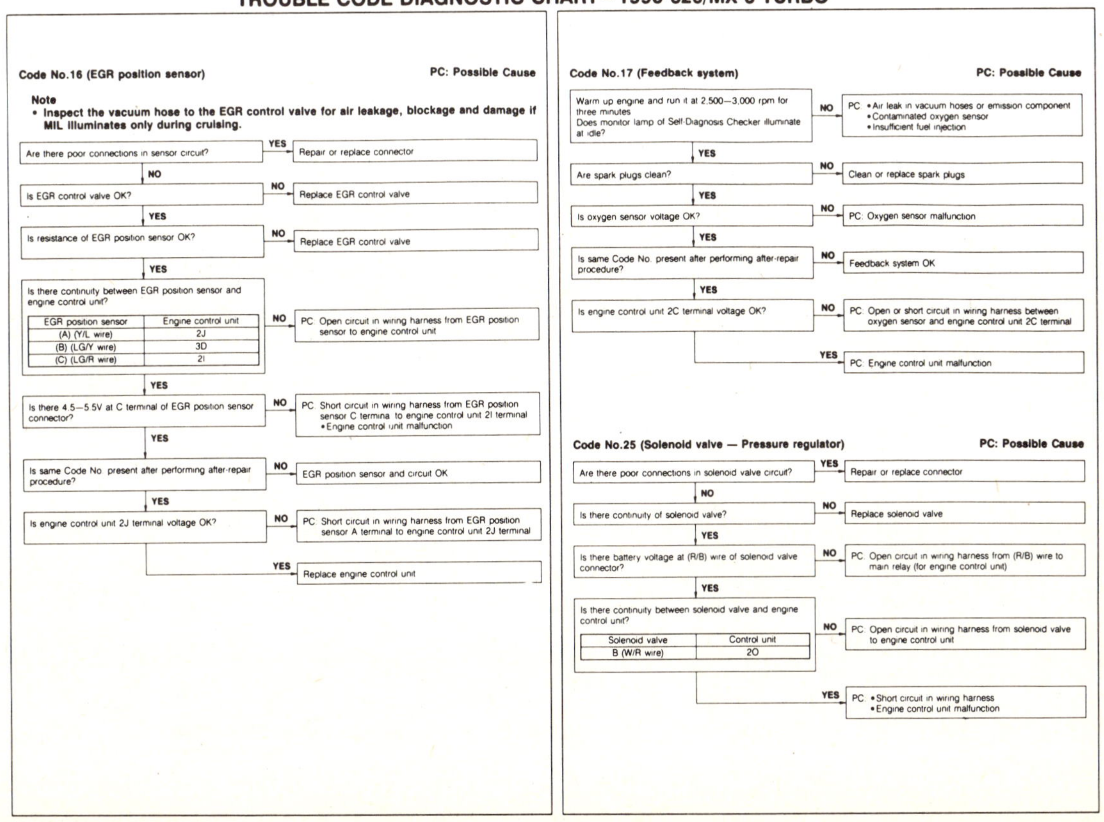

### TROUBLE CODE DIAGNOSTIC CHART—1990 626/MX-6 TURBO

**Code No.26 (Solenoid valve — Purge)**

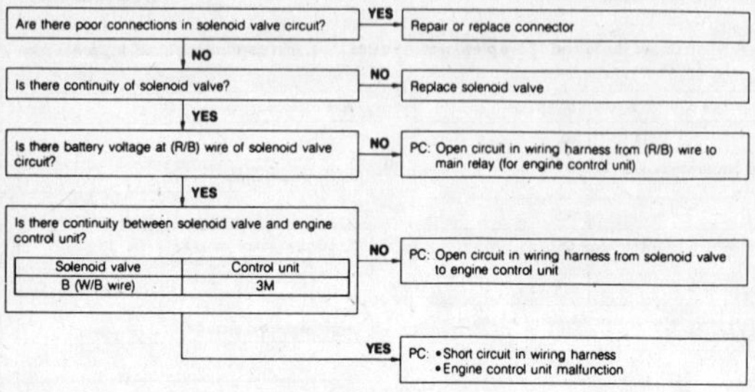

- Are there poor connections in solenoid valve circuit?
  - YES → Repair or replace connector
  - NO ↓
- Is there continuity of solenoid valve?
  - NO → Replace solenoid valve
  - YES ↓
- Is there battery voltage at (R/B) wire of solenoid valve circuit?
  - NO → PC: Open circuit in wiring harness from (R/B) wire to main relay (for engine control unit)
  - YES ↓
- Is there continuity between solenoid valve and engine control unit?

| Solenoid valve | Control unit |
|---|---|
| B (W/B wire) | 3M |

  - NO → PC: Open circuit in wiring harness from solenoid valve to engine control unit
  - YES → PC: • Short circuit in wiring harness • Engine control unit malfunction

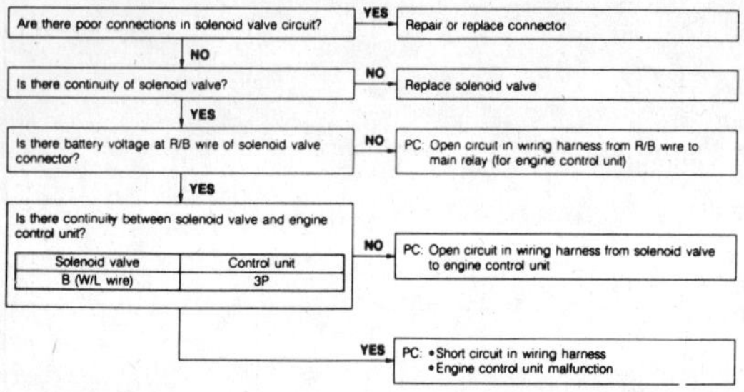

**Code No.29 (Solenoid valve — EGR, vent side)**

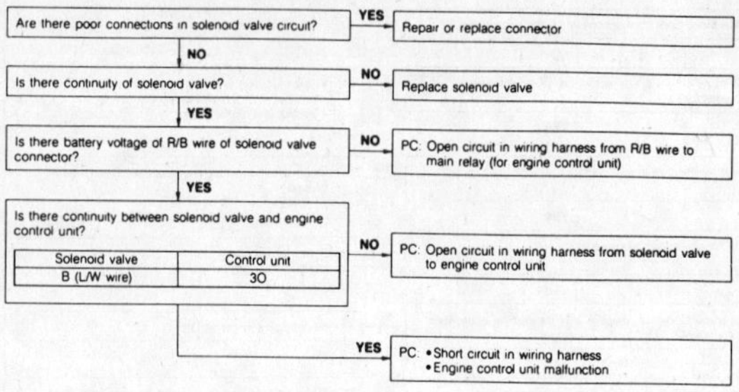

- Are there poor connections in solenoid valve circuit?
  - YES → Repair or replace connector
  - NO ↓
- Is there continuity of solenoid valve?
  - NO → Replace solenoid valve
  - YES ↓
- Is there battery voltage of R/B wire of solenoid valve connector?
  - NO → PC: Open circuit in wiring harness from R/B wire to main relay (for engine control unit)
  - YES ↓
- Is there continuity between solenoid valve and engine control unit?

| Solenoid valve | Control unit |
|---|---|
| B (L/W wire) | 3O |

  - NO → PC: Open circuit in wiring harness from solenoid valve to engine control unit
  - YES → PC: • Short circuit in wiring harness • Engine control unit malfunction

**Code No.28 (Solenoid valve — EGR, vacuum side)**

PC: Possible Cause

- Are there poor connections in solenoid valve circuit?
  - YES → Repair or replace connector
  - NO ↓
- Is there continuity of solenoid valve?
  - NO → Replace solenoid valve
  - YES ↓
- Is there battery voltage at R/B wire of solenoid valve connector?
  - NO → PC: Open circuit in wiring harness from R/B wire to main relay (for engine control unit)
  - YES ↓
- Is there continuity between solenoid valve and engine control unit?

| Solenoid valve | Control unit |
|---|---|
| B (W/L wire) | 3P |

  - NO → PC: Open circuit in wiring harness from solenoid valve to engine control unit
  - YES → PC: • Short circuit in wiring harness • Engine control unit malfunction

**Code No.34 (ISC valve)**

PC: Possible Cause

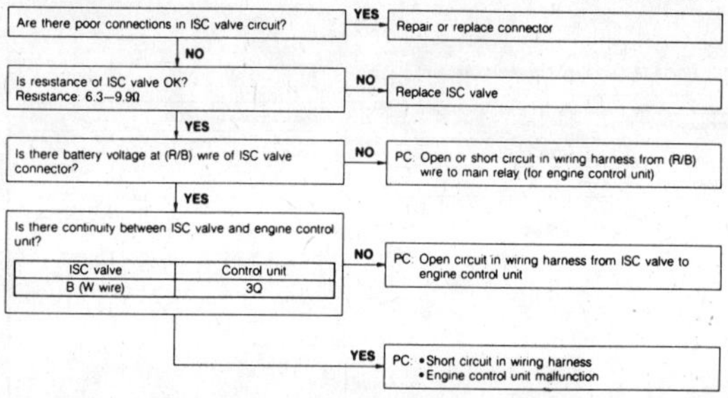

- Are there poor connections in ISC valve circuit?
  - YES → Repair or replace connector
  - NO ↓
- Is resistance of ISC valve OK? Resistance: 6.3—9.9Ω
  - NO → Replace ISC valve
  - YES ↓
- Is there battery voltage at (R/B) wire of ISC valve connector?
  - NO → PC: Open or short circuit in wiring harness from (R/B) wire to main relay (for engine control unit)
  - YES ↓
- Is there continuity between ISC valve and engine control unit?

| ISC valve | Control unit |
|---|---|
| B (W wire) | 3Q |

  - NO → PC: Open circuit in wiring harness from ISC valve to engine control unit
  - YES → PC: • Short circuit in wiring harness • Engine control unit malfunction

### TROUBLE CODE DIAGNOSTIC CHART 1990 626/MX-6 TURBO

**Code No.42 (Solenoid valve — Waste gate)**

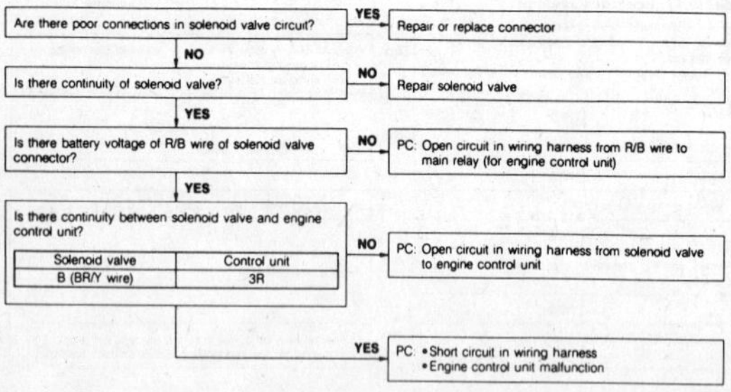

- Are there poor connections in solenoid valve circuit?
  - YES → Repair or replace connector
  - NO ↓
- Is there continuity of solenoid valve?
  - NO → Repair solenoid valve
  - YES ↓
- Is there battery voltage of R/B wire of solenoid valve connector?
  - NO → PC: Open circuit in wiring harness from R/B wire to main relay (for engine control unit)
  - YES ↓
- Is there continuity between solenoid valve and engine control unit?

| Solenoid valve | Control unit |
|---|---|
| B (BR/Y wire) | 3R |

  - NO → PC: Open circuit in wiring harness from solenoid valve to engine control unit
  - YES → PC: • Short circuit in wiring harness • Engine control unit malfunction

### CONTROL UNIT PIN LOCATIONS AND VOLTAGE TEST 1990 626/MX-6 TURBO

**Terminal voltage**

| Terminal | Input | Output | Connected to | Voltage (After warming-up) IGN: ON | Voltage (After warming-up) Idle | Remark |
|---|---|---|---|---|---|---|
| 1A | — | — | Battery | Battery voltage | | For back-up |
| 1B | — | — | Main relay | Battery voltage | | — |
| 1C | O | | Ign. switch [START] (MTX) / Inhibitor switch (ATX) | Below 2.5V | | While cranking: Battery voltage |
| 1D | | O | Self-Diagnosis Checker (Monitor lamp) | Test connect grounded<br>• For 3 sec. after ign. switch OFF→ON: Below 6.2V (lamp illuminates)<br>• After 3 sec.: Battery voltage (lamp not illuminate) | Test connector not grounded<br>• Lamp illuminates: Below 6.2V<br>• Lamp not illuminate: Battery voltage<br>Test connector grounded: Approx. 5V | Using Self-Diagnosis Checker |
| 1E | | O | Malfunction indicator lamp (MIL) | • For 3 sec. after ign. switch OFF→ON: Below 4.8V (lamp illuminates)<br>• After 3 sec.: Battery voltage (lamp not illuminate) | | • Test connector grounded<br>• Lamp illuminates: Below 4.8V<br>• Lamp not illuminate: Battery voltage |
| 1F | | O | Self-Diagnosis Checker (Code number) | • For 3 sec. after ign. switch OFF→ON: Below 6.2V (Buzzer sounds)<br>• After 3 sec.: Battery voltage (Buzzer not sound) | | • Using Self-Diagnosis Checker and test connector grounded<br>• Buzzer sounds: Below 6.2V<br>• Buzzer not sound: Battery voltage |
| 1G | | O | Igniter | 0V | Approx. 0.6—0.8V | — |
| 1H | O | | Headlight switch | • Headlight OFF: Below 1.5V<br>• Headlight ON : Battery voltage | | — |
| 1I | O | | Test connector | • Test connector grounded: Below 1.5V<br>• Test connector not grounded: Above 10.5V | | Test connector: 1-pin, Green connector |
| 1J | O | | Rear window defroster switch | • Switch OFF: Battery voltage<br>• Switch ON : Below 1.5V | | — |
| 1K | — | — | — | — | | — |
| 1L | | O | A/C relay | Battery voltage | • A/C switch ON: Below 2.5V<br>• A/C switch OFF: Battery voltage | Blower motor ON |
| 1M | O | | Vehicle speed switch | Batter voltage | | Above 130 mph (210 km/h ): Below 1.0V |
| 1N | O | | Power steering pressure switch | Above 10.5V | • P/S ON: Below 1.5V<br>• P/S OFF: Above 10.5V | — |
| 1O | | | A/C switch | • A/C switch ON : Below 1.5V<br>• A/C switch OFF: above 10.0V | | Blower motor ON |
| 1P | O | | Blower fan switch | • Switch less than 2nd position: Battery voltage<br>• Switch 3rd or 4th position: Below 1.5V | | |

## CONTROL UNIT PIN LOCATIONS AND VOLTAGE TEST (CONT.)—1990 626/MX-6 TURBO

| Terminal | Input | Output | Connected to | Voltage (After warming-up) IGN: ON | Voltage (After warming-up) Idle | Remark |
|---|---|---|---|---|---|---|
| 1Q | ○ | | Stoplight switch (MTX) | • Brake pedal released: Below 3.6V • Brake pedal depressed: Above 10.0V | | |
| | | | EC-AT control unit (ATX) | Below 1.5V | | 3-2, 2-1 shifting and throttle valve more than 5/8 open: Battery voltage |
| 1R | ○ | | Inhibitor switch through EC-AT unit | "N" or "P" range: below 2.5V Other ranges: Battery voltage | | ATX |
| | | | Neutral and clutch switch | In-gear condition Clutch pedal depressed: below 0.5V Clutch pedal released: Battery voltage | | MTX (Neutral: Below 0.5V) |
| 1S | ○ | | Daytime running light control unit (Canada only) | • Parking brake lever pulled up: Battery voltage • Parking brake lever released: Below 1.5V | | — |
| 1T | ○ | | Idle switch | • Accelerator pedal released: Below 0.5V • Accelerator pedal depressed: Above 7.7V | | — |
| 1U | ○ | | — | Constant below 1.5V | | MTX |
| | | | | Constant above 10.5V | | ATX |
| 1V | ○ | | Igniter | Below 1.0V | 0.1—1.8V | |
| 2A | ○ | | Airflow meter (Vc) | 7—9V | | — |
| 2B | ○ | | Airflow meter (Vs) | Approx. 1.7V | Approx. 3—5V | Increase engine speed: voltage increase |
| 2C | ○ | | Oxygen sensor | 0V | 0—1.0V | • Cold engine at idle: 0V • After warming-up Acceleration: 0.5V—1.0V Deceleration: 0—0.4V |
| 2D | ○ | | Electrical fan [Low] (No.1 water thermoswitch) | Battery voltage | | Coolant temp.: Below 97°C (207°F) |
| | | | | Below 1.5V | | Coolant temp.: Above 97°C (207°F) |
| 2E | ○ | | Water thermosensor | 0.3—0.6V | | Coolant temp.: 20°C (68°F): Approx. 2.5V |
| 2F | ○ | | Throttle sensor | Accelerator pedal released: Approx. 0.5V (depends on 21 terminal voltage) | | Throttle valve fully open: 4.3V |
| 2G | ○ | | Electrical fan [High] (No.2 water thermoswitch) (ATX) | Battery voltage | | Coolant temp.: Below 108°C (226°F) |
| | | | | Below 1.5V | | Coolant temp.: Above 108°C (226°F) |
| 2H | — | — | — | — | | — |
| 2I | — | — | Vref | 4.5—5.5V | | — |
| 2J | ○ | | ECR position sensor (California only) | 0.25—0.95V | | |
| 2K | ○ | | Airflow meter (Intake air thermosensor) | Approx. 2.5V at 20C° (68°F) | | — |
| 2L | — | — | — | — | | — |
| 2M | ○ | | Knock control unit | 3.3—5.0V | | Knocking: 1.3—2.6V |
| 2N | — | — | — | — | | — |
| 2O | | ○ | Solenoid valve (Purge control) | Battery voltage | | — |
| 2P | — | — | — | — | | — |
| 3A | — | — | Ground (E01) | 0V | | — |

| Terminal | Input | Output | Connected to | Voltage (After warming-up) IGN: ON | Voltage (After warming-up) Idle | Remark |
|---|---|---|---|---|---|---|
| 3B | — | — | Ground (E02) | 0V | | — |
| 3C | — | — | Ground (E1) | 0V | | — |
| 3D | — | — | Ground (E2) | 0V | | — |
| 3G | ○ | | Distributor (G1 ⊕signal) | Approx. 0.6—0.8V | | |
| 3H | ○ | | Distributor (G2 ⊕signal) | Approx. 0.6—0.8V | | |
| 3F | ○ | | Distributor (Ne, G1, G2, ⊖ signal) | Approx. 0.6—0.8V | | |
| 3E | ○ | | Distributor (Ne ⊕signal) | Approx. 0.6—0.8V | | |
| 3I | — | — | — | — | | — |
| 3J | — | — | — | — | | — |
| 3K | — | — | — | — | | — |
| 3L | | ○ | Warning buzzer Overboost | Type A*1: Approx. 0.08V Type B*2: Approx. 12V | | Buzzer sounds: Type A: 0V Type B: Below 1V |
| 3M | | ○ | Solenoid valve (Pressure regulator control) | For 120 sec. after ign. switch OFF → ON: Below 3.5V | For 120 sec. after staring: Below 3.5V | Coolant temp. above 70°C (158°F) and intake air temp. above 20°C (63°F) |
| 3N | — | — | — | — | | — |
| 3O | | ○ | Solenoid valve (EGR-vent side) | Battery voltage | | |
| 3P | | ○ | Solenoid valve (EGR-vacuum side) | Battery voltage | | • Voltages change depending on driving condition (EGR amount) • Cold engine: battery voltage Engine coolant temp-below 40°C (104°F) |
| 3Q | | ○ | ISC valve | Engine signal monitor green and red lamps flash | | |
| 3R | | ○ | Solenoid valve (Waste gate) | Battery voltage | | • Suddenly increase engine speed to above 4,500 rpm: Below 3.5V |
| 3S | – | — | — | — | | — |
| 3T | | ○ | Circuit opening relay | Battery voltage | Below 3.5V | |
| 3U | | ○ | Injector (No.1 and No.3) | Battery voltage | Battery voltage | * Engine signal monitor green and red lamps flash |
| 3V | | ○ | Injector (No.2 and No.4) | | | |
| 3W | — | — | — | — | | — |
| 3X | — | — | — | — | | — |
| 3Y | — | — | — | — | | — |
| 3Z | — | — | — | — | | — |

*1 Type A: Manufacturer for CPU of body electrical system is "NEC"
*2 Type B: Manufacturer for CPU of body electrical system is "YAZAKI or U-shin"

## COMPONENT RELATIONSHIP CHART 1990 626/MX-6 TURBO

Output Devices and Input Devices

○: Related ×: Not related

| INPUT DEVICES \ OUTPUT DEVICES | INJECTOR: FUEL INJECTION AMOUNT | INJECTOR: FUEL INJECTION TIMING | BAC VALVE: AIR VALVE | BAC VALVE: ISC VALVE | SOLENOID VALVE (EGR Vent side) | SOLENOID VALVE (EGR Vacuum side) | SOLENOID VALVE (PURGE) | SOLENOID VALVE (PRESSURE REGULATOR) | SOLENOID VALVE (WASTEGATE) | IGNITER |
|---|---|---|---|---|---|---|---|---|---|---|
| TEST CONNECTOR | × | × | × | ○ | × | × | × | × | × | ○ |
| KNOCK SENSOR | × | × | × | × | × | × | × | × | ○ | ○ |
| STOPLIGHT SWITCH | ○ | × | × | × | × | × | × | × | × | × |
| EGR POSITION SENSOR | × | × | × | × | ○ | ○ | × | × | × | ○ |
| ELECTRICAL LOAD | × | × | × | ○ | × | × | × | × | × | × |
| P/S PRESSURE SWITCH | × | × | × | ○ | × | × | × | × | × | × |
| A/C SWITCH | ○ | × | × | ○ | × | × | × | × | × | × |
| IGNITION SWITCH (STA POSITION) | ○ | ○ | × | × | × | × | × | ○ | × | ○ |
| INHIBITOR SWITCH and EC-AT CONTROL UNIT | ○ | × | × | ○ | × | × | ○ | × | × | ○ |
| NEUTRAL AND CLUTCH SWITCH | ○ | × | × | ○ | × | × | ○ | × | × | ○ |
| OXYGEN SENSOR | ○ | × | × | × | × | × | ○ | × | × | × |
| ATMOSPHERIC PRESSURE SENSOR | ○ | × | × | ○ | × | × | ○ | × | × | × |
| INTAKE AIR THERMOSENSOR | ○ | × | × | ○ | × | × | ○ | ○ | ○ | × |
| WATER THERMOSENSOR | ○ | ○ | × | ○ | ○ | ○ | ○ | ○ | ○ | ○ |
| IDLE SWITCH | ○ | ○ | × | ○ | ○ | ○ | × | × | × | ○ |
| THROTTLE SENSOR | ○ | ○ | × | ○ | ○ | ○ | ○ | ○ | × | ○ |
| AIRFLOW METER | ○ | × | × | × | ○ | ○ | ○ | × | ○ | ○ |
| Ne SIGNAL | ○ | ○ | × | ○ | ○ | ○ | ○ | ○ | ○ | ○ |
| G2 SIGNAL | × | ○ | × | × | × | × | × | × | × | ○ |
| G1 SIGNAL | × | ○ | × | × | × | × | × | × | × | ○ |

## OUTPUT COMPONENTS AND ENGINE CONDITION 1990 626/MX-6 TURBO

| OUTPUT DEVICES \ ENGINE CONDITIONS | | CRANKING (COLD ENGINE) | WARMING UP (DURING IDLE) | MEDIUM LOAD: COLD | MEDIUM LOAD: WARM | ACCELE-RATION | HEAVY LOAD | DECELE-RATION | IDLE (THROTTLE VALVE FULLY CLOSED) | IGN: ON (ENGINE NOT RUNNING) | REMARKS |
|---|---|---|---|---|---|---|---|---|---|---|---|
| INJECTOR | INJECTION | Rich | | | Rich and lean | Rich | | Fuel cut | Rich and lean | No injection | |
| | INJECTION TIMING | 1 group (once per revolution) | 2 group (once per two revolutions) | | | | | | 2 group (once per two revolutions) | | Above 6,300 rpm fuel cut |
| BAC VALVE | AIR VALVE | Open* | | | Close | | | | | | * Coolant temp below 50°C (122°F) |
| | ISC VALVE | Large amount of bypass air | Large amount of bypass air* | Small amount of bypass air | | | | | | No bypass | * In extremly cold condition |
| SOLENOID VALVE (EGR Vent side) | | OFF (Duty value 0%) (Atmospheric pressure to EGR valve) — EGR cut — | | | Operate [Duty values (EGR amount) change*] | | OFF (Duty value 0%) (Atmospheric pressure to EGR valve) — EGR cut — | | | Does not operate | * Depends on engine condition |
| SOLENOID VALVE (EGR Vacuum side) | | OFF (No vacuum to EGR valve) | | | | | OFF (No vacuum to EGR valve) | | | | |
| SOLENOID VALVE (PURGE) | | OFF (Purge cut) | | | Operate [Duty values (purge amount) change*] | | | OFF (Purge cut) | | | * Depends on engine condition |
| SOLENOID VALVE (PRESSURE REGULATOR CONTROL) | | OFF (Vacuum to pressure regulator) | | | | | | | After starting: ON* (Vacuum cut) | OFF | * During hot start only |
| SOLENOID VALVE (WASTEGATE) | | OFF (Boost pressure not released) | | ON (Duty value 100%) | | ON (Duty value changes)* (Boost pressure controlled) | | OFF (Boost pressure not released) | | OFF | * When knocking occurs |
| IGNITER (Ignition timing) | | Fixed at BTDC 6° | Advanced: depends on engine condition | | | Retarded: depends on intensity of knocking | | Advanced: depends on engine speed | | — | |

## WIRING SCHEMATIC – 1990 626/MX-6 TURBO

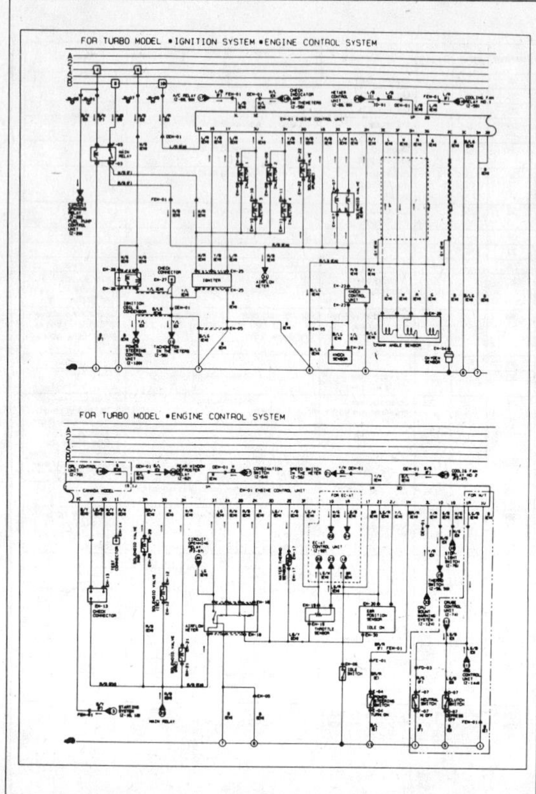


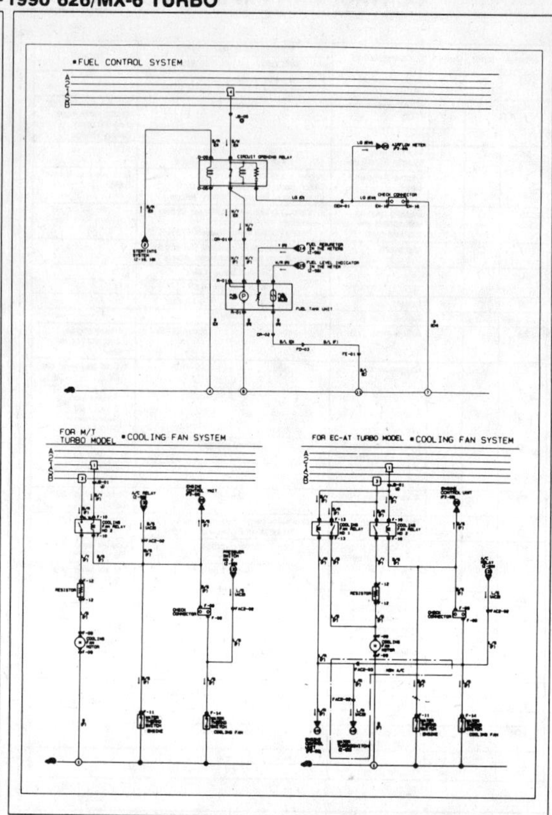


## SYSTEM SCHEMATIC – 1990 MIATA

SYSTEM DIAGRAM

AIR CLEANER
AIRFLOW METER
SOLENOID VALVE (PURGE CONTROL)
FUEL FILTER (HIGH-PRESSURE SIDE)
CHARCOAL CANISTER
OXYGEN SENSOR
DASHPOT
PRESSURE REGULATOR
PCV VALVE
TWO-WAY CHECK VALVE
DYNAMIC CHAMBER
THROTTLE SENSOR
FUEL VAPOR VALVE
AIR VALVE
ISC VALVE
INJECTOR
WATER THERMOSENSOR
FUEL TANK
FUEL PUMP
FUEL FILTER (LOW-PRESSURE SIDE)
CATALYTIC CONVERTER

## EXPLODED VIEW OF CONTROL SYSTEM COMPONENTS 1990 MIATA

STRUCTURAL VIEW

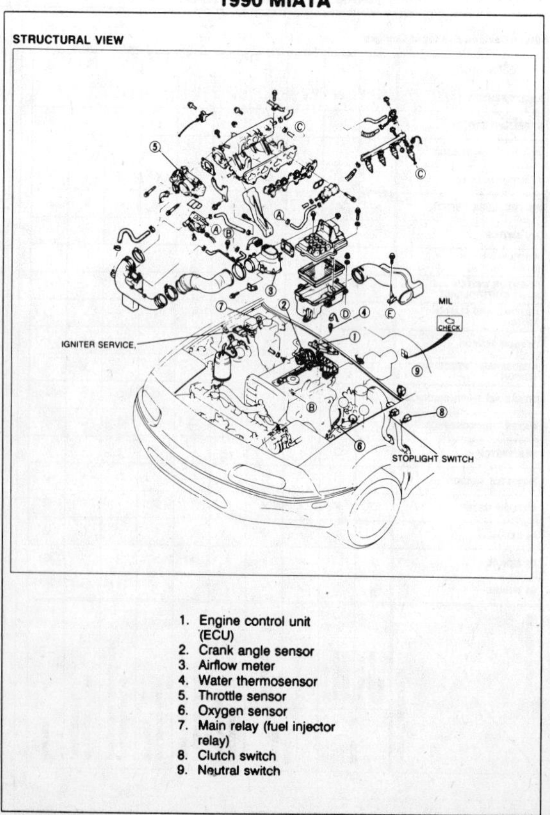


1. Engine control unit (ECU)
2. Crank angle sensor
3. Airflow meter
4. Water thermosensor
5. Throttle sensor
6. Oxygen sensor
7. Main relay (fuel injector relay)
8. Clutch switch
9. Neutral switch

## EXPLODED VIEW OF FUEL SYSTEM COMPONENTS 1990 MIATA

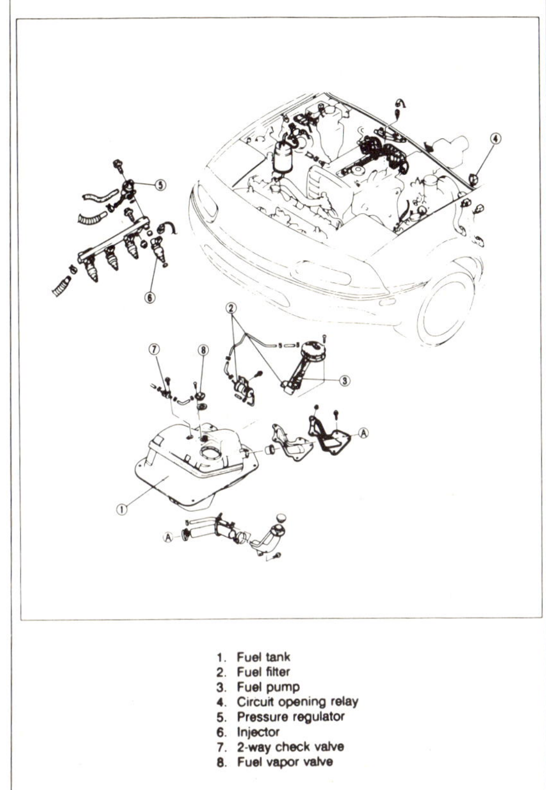

1. Fuel tank
2. Fuel filter
3. Fuel pump
4. Circuit opening relay
5. Pressure regulator
6. Injector
7. 2-way check valve
8. Fuel vapor valve

## EXPLODED VIEW OF INTAKE AIR SYSTEM 1990 MIATA

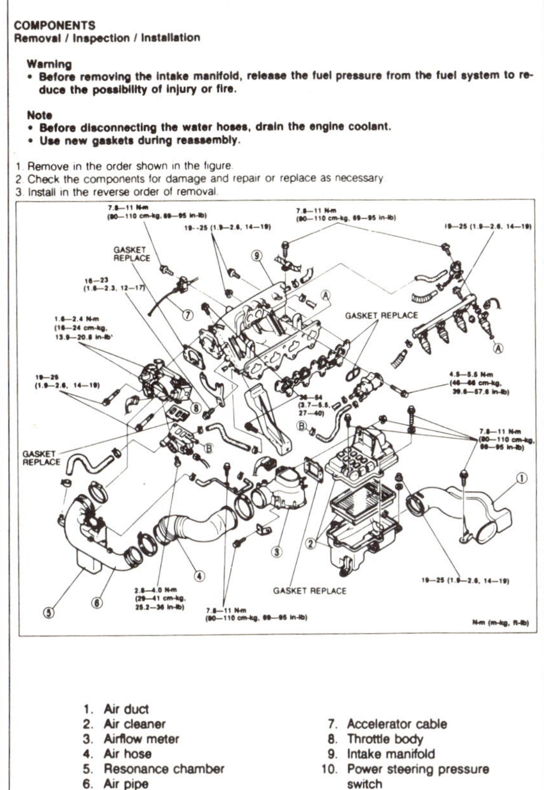

COMPONENTS
Removal / Inspection / Installation
Warning
• Before removing the intake manifold, release the fuel pressure from the fuel system to reduce the possibility of injury or fire.
Note
• Before disconnecting the water hoses, drain the engine coolant.
• Use new gaskets during reassembly.
1. Remove in the order shown in the figure.
2. Check the components for damage and repair or replace as necessary.
3. Install in the reverse order of removal.
GASKET REPLACE
GASKET REPLACE
GASKET REPLACE
GASKET REPLACE
N·m (m-kg, ft-lb)


1. Air duct
2. Air cleaner
3. Airflow meter
4. Air hose
5. Resonance chamber
6. Air pipe
7. Accelerator cable
8. Throttle body
9. Intake manifold
10. Power steering pressure switch

## VACUUM HOSE ROUTING – 1990 MIATA

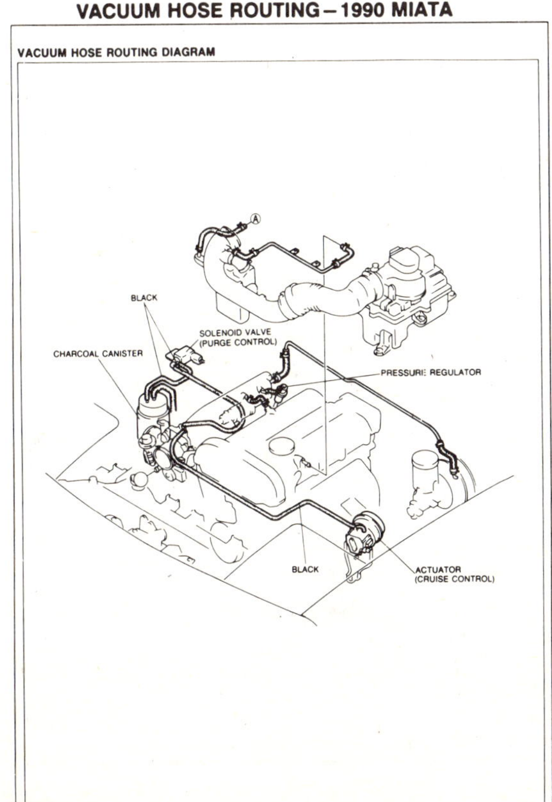


## SPECIFICATIONS CHART – 1990 MIATA

**SPECIFICATIONS**

| Item | | | Specification |
|---|---|---|---|
| Idle speed rpm | | | 850 ± 50* |
| Ignition timing BTDC | | | 10° ± 1°* |
| **Throttle body** | | | |
| Type | | | Horizontal draft |
| Throat diameter mm (in) | | | 55 (2.2) |
| **Fuel pump** | | | |
| Type | | | Impeller (in-tank) |
| Output pressure kPa (kg/cm², psi) | | | 441–589 (4.5–6.0, 64–85) |
| **Fuel filter** | | | |
| Type | Low-pressure side | | Nylon element |
| | High-pressure side | | Paper element |
| **Pressure regulator** | | | |
| Type | | | Diaphragm |
| Regulating pressure kPa (kg/cm², psi) | | | 265–314 (2.7–3.2, 38–46) |
| **Injector** | | | |
| Type | | | High-ohmic |
| Type of drive | | | Voltage |
| Resistance Ω | | | 12–16 (at 20°C, 68°F) |
| **ISC valve** | | | |
| Solenoid resistance Ω | | | 11–13 (at 20°C, 68°F) |
| **Air valve** | | | |
| Opening temperature | | | Below 40°C (104°F) |
| **Solenoid valve (Purge control)** | | | |
| Solenoid resistance Ω | | | 23–27 (at 20°C, 68°F) |
| **Crank angle sensor** | | | |
| Type | | | Optical pickup |
| **Airflow meter** | | | |
| Resistance Ω | E2↔Vs | Fully closed | 200–600 |
| | | Fully open | 20–1,000 |
| | E2↔Vc | | 200–400 |
| | E2↔THAA (Intake air thermosensor) | -20°C (-4°F) | 13,600–18,400 |
| | | 20°C (68°F) | 2,210–26,900 |
| | | 60°C (140°F) | 493–667 |
| | E1↔Fc | Fully closed | ∞ |
| | | Fully open | 0 |
| **Water thermosensor** | | | |
| Resistance kΩ | | -20°C (-4°F) | 14.6–17.8 |
| | | 20°C (68°F) | 2.2–2.7 |
| | | 80°C (179°F) | 0.29–0.35 |
| **Circuit opening relay** | | | |
| Resistance Ω | | STA–E1 | 21–43 |
| | | B–Fc | 109–226 |
| | | B–Fp | ∞ |
| **Fuel tank** | | | |
| Capacity liters (US gal, Imp gal) | | | 45 (11.9, 9.9) |
| **Air cleaner** | | | |
| Element type | | | Oil permeated |
| **Accelerator cable** | | | |
| Free play mm (in) | | | 1–3 (0.039–0.118) |
| **Fuel** | | | |
| Specification | | | Unleaded regular (RON 87 or higher) |

*... With System Selector (49 B019 9A0) test switch at SELF TEST

## COMPONENT DESCRIPTION—1990 MIATA

### COMPONENT DESCRIPTIONS

| Component | Function | Remark |
|---|---|---|
| **Air cleaner** | Filters air entering throttle body | |
| **Airflow meter** | Detects amount of intake air; sends signal to engine control unit | •Intake air thermosensor and fuel pump switch included<br>•Use Vref (5 volt) as power source |
| **Air valve** | When cold, supplies bypass air into dynamic chamber | •Engine speed increased to shorten warm-up period<br>•Thermowax type |
| **Atmospheric pressure sensor** | Detects atmospheric pressure; sends signal to engine control unit | Built-in ECU |
| **Catalytic converter** | Reduces HC, CO, and NOx by chemical reaction | Monolith type |
| **Charcoal canister** | Stores gas tank fumes while engine stopped | |
| **Circuit opening relay** | Voltage for fuel pump while engine running | |
| **Clutch switch (M/T)** | Detects in-gear condition; sends signal to engine control unit | Switch OFF when clutch pedal released |
| **Diagnosis connector** | Concentrated service connector<br>Concentrated terminals are:<br>1. EGI self-diagnosis terminal<br>2. Initial set terminal<br>3. Fuel pump check terminal<br>4. Cooling fan check terminal | 21-pin (Black) |
| **Crank angle sensor**<br>**G-signal**<br>**Ne-signal** | Detects No.1 cylinder TDC; sends signal to engine control unit<br>Detects crank angle at 180° intervals; sends signal to control unit | |
| **Dynamic chamber** | Interconnects all cylinders | |
| **Engine control unit (ECU)** | **Detects the following:**<br>1. A/C operation<br>2. Air/fuel ratio (Oxygen concentration)<br>3. Atmospheric pressure<br>4. Braking signal<br>5. Cranking signal<br>6. E/L operation<br>7. Engine coolant temperature<br>8. Engine speed<br>9. In-gear condition<br>10. Intake air amount<br>11. Intake air temperature<br>12. No.1 piston TDC<br>13. P/S operation<br>14. Initial set signal<br>15. Throttle valve fully closed<br>16. Throttle valve opening amount<br>**Controls operation of the following:**<br>1. A/C (Cut off)<br>2. Fail-safe function<br>3. Fuel injection system<br>4. Idle speed control<br>5. Ignition timing control system<br>6. Monitor function<br>7. Purge control system | 1. A/C switch<br>2. Oxygen sensor<br>3. Atmospheric pressure sensor<br>4. Stoplight switch<br>5. Ignition switch (START position)<br>6. Cooling fan relay<br>Light and turn relay<br>Blower motor switch<br>7. Water thermosensor<br>8. Crank angle sensor (Ne-signal)<br>9. Neutral and clutch switches<br>10. Airflow meter<br>11. Intake air thermosensor<br>12. Crank angle sensor (G-signal)<br>13. P/S pressure switch<br>14. Diagnosis connector<br>15. Throttle sensor (IDL signal)<br>16. Throttle sensor (POW signal)<br><br>1. A/C relay<br>2. Self-diagnosis checker and MIL<br>3. Injector<br>4. ISC valve<br>5. Igniter<br>6. Monitor lamp (Self-diagnosis checker)<br>7. Solenoid valve (Purge control) |
| **Fuel vapor valve** | Prevents fuel from entering canister during vehicle roll over | |
| **Fuel filter** | Filters particles from fuel | |
| **Fuel pump** | Provides fuel to injectors | •Operates while engine running<br>•Installed in fuel tank |
| **Igniter** | Receives spark signal from engine control unit and generates high voltage in ignition coil<br>Detects high voltage ignition; sends substitute ignition signal to engine control unit | |
| **Ignition switch (START position)** | Sends engine cranking signal to engine control unit | |
| **Injector** | Injects fuel into intake port | •Controlled by signals from engine control unit<br>•High-ohmic injector<br>•Single port injector nozzle |
| **Intake air thermosensor** | Detects intake air temperature; sends signal to engine control unit | Installed in airflow meter |
| **ISC valve** | Controls bypass air amount | •Controlled by duty signal from engine control unit<br>•Controls idle-up |
| **Solenoid valve (Purge control)** | Controls evaporative fumes from canister to intake manifold | •Controlled by duty signal from engine control unit |
| **Main relay** | Supplies electric current to injectors, engine control unit, etc. | |
| **MIL (Malfunction indicator lamp)** | Lamp illuminates when input device malfunctions | (TEN) terminal of diagnosis connector |
| | Lamp flashes to indicate malfunction code number of input and output devices | (TEN) terminal of diagnosis connector |
| **Neutral switch** | Detects in-gear condition; sends signal to engine control unit | Switch ON when in neutral |
| **Oxygen sensor** | Detects oxygen concentration; sends signal to engine control unit | Zirconia ceramic and platinum coating |
| **PCV valve** | Controls blowby gas amount introduced into engine | |
| **P/S pressure switch** | Detects P/S operation; sends signal to engine control unit | P/S ON when steering wheel turned |
| **Pressure regulator** | Adjusts fuel pressure supplied to injectors | |
| **Resonance chamber** | Improves mid range torque characteristics | |
| **Stoplight switch** | Detects braking operation (deceleration); sends signal to engine control unit | |
| **Throttle body** | Controls intake air quantity | Integrated throttle sensor, ISC valve, and dashpot |
| **Throttle sensor IDL signal** | Detects position of the throttle valve fully closed; sends signal to engine control unit | |
| **Throttle sensor POW signal** | Detects the throttle valve opening at a preset angle; sends signal to engine control unit | |
| **Two-way check valve** | Controls pressure in fuel tank | |
| **Water thermosensor** | Detects coolant temperature; sends signal to engine control unit | |

## TROUBLESHOOTING CHART—1990 MIATA

### ENGINE CONTROL OPERATION CHART
**Input Devices and Engine Conditions**

| INPUT DEVICES / ENGINE CONDITIONS | APPROXIMATE TIME (BASED ON 10—16°C, or 50—60°F AMBIENT) | SENSORS: CRANK ANGLE SENSOR G-SIGNAL | CRANK ANGLE SENSOR NE-SIGNAL | AIRFLOW METER INTAKE AIR THERMOSENSOR | AIRFLOW METER POTENTIOMETER | WATER THERMOSENSOR | OXYGEN SENSOR | IGF SIGNAL | ATMOSPHERIC PRESSURE SENSOR |
|---|---|---|---|---|---|---|---|---|---|
| CRANKING —COLD ENGINE • COLD AIR • COLD COOLANT | ZERO | | | SIGNAL HAS NO EFFECT ON ECU | SIGNAL HAS NO EFFECT ON ECU | | SIGNAL HAS NO EFFECT ON ECU | | |
| COLD START —FAST IDLE • COLD AIR • COLD COOLANT | ONE MINUTE | | | | LOW VOLUME AIRFLOW (3.0V) | COOL TO WARM: MEDIUM VOLTAGE (2.3V and DROPPING) | SENSOR COLD: LOW TO HIGH VOLTAGE (0—0.8V) | | |
| COLD DRIVEAWAY —PART THROTTLE • WARM AIR • WARM COOLANT | TWO MINUTES | | | | | | | | |
| WARM DRIVEAWAY —PART THROTTLE • WARM AIR • WARM COOLANT | THREE MINUTES | | | | MODERATE VOLUME AIRFLOW: LOW TO MEDIUM VOLTAGE (1—3.5V) | WARM: MEDIUM VOLTAGE (APPROX 0.7V AND DROPPING) | SENSOR WARM: HIGH VOLTAGE (ABOVE 0.6V) | | |
| HOT CRUISE • WARM AIR • WARM COOLANT | | | | | | | | | |
| HOT ACCELERATION —60% THROTTLE | | SENDS No 1 CYLINDER TDC SIGNAL TO ECU | SENDS ENGINE SPEED SIGNAL TO ECU | COOL TO WARM: MEDIUM VOLTAGE (4.3—1.5V) | MODERATE TO STRONG VOLUME AIRFLOW (0.5—1.5V) | | SENSOR HOT: SWITCHING FROM HIGH VOLTAGE (ABOVE 0.6V RICH) TO LOW VOLTAGE (BELOW 0.4V LEAN) | SENDS IGNITION SPARK SIGNAL TO ECU | SENDS VOLTAGE SIGNAL TO ECU THAT VARIES WITH ALTITUDE: VOLTAGE (APPROX 4V) |
| HOT ACCELERATION —WIDE OPEN THROTTLE | MORE THAN FOUR MINUTES | | | | | HOT: LOW VOLTAGE (APPROX 0.4V) | HIGH VOLTAGE (ABOVE 0.6V RICH) | | |
| DECELERATION —CLOSED THROTTLE | | | | | | | LOW VOLTAGE (BELOW 0.4V LEAN) | | |
| HOT CURB IDLE —EXTENDED | | | | | LOW VOLUME AIRFLOW (3.0V) | | SWITCHING FROM HIGH TO LOW VOLTAGE (0.4—0.6V) | | |
| HOT ENGINE SHUTDOWN | | OFF | OFF | OFF | OFF | OFF | SENSOR HOT: LOW VOLTAGE (0.1V LEAN) UNTIL SENSOR COOLS DOWN | OFF | OFF |

| ENGINE CONDITIONS | SWITCHES: THROTTLE SENSOR POWER TERMINAL | THROTTLE SENSOR IDLE TERMINAL | A/C SWITCH | NEUTRAL AND CLUTCH SWITCHES | STOPLIGHT SWITCH | HEADLIGHT SWITCH | BLOWER CONTROL SWITCH | COOLING FAN SWITCH | P/S PRESSURE SWITCH | IGNITION SWITCH START POSITION | TEST TERMINAL |
|---|---|---|---|---|---|---|---|---|---|---|---|
| CRANKING —COLD ENGINE • COLD AIR • COLD COOLANT | SEND SIGNAL TO ECU WHEN PEDAL DEPRESSED | SIGNAL HAS NO EFFECT ON ECU | SIGNAL HAS NO EFFECT ON ECU | SIGNAL HAS NO EFFECT ON ECU | SIGNAL HAS NO EFFECT ON ECU | SIGNAL HAS NO EFFECT ON ECU | SIGNAL HAS NO EFFECT ON ECU | SIGNAL HAS NO EFFECT ON ECU | SIGNAL HAS NO EFFECT ON ECU | SENDS SIGNAL TO ECU | SIGNAL HAS NO EFFECT ON ECU |
| COLD START —FAST IDLE • COLD AIR • COLD COOLANT | | LOW VOLTAGE SIGNAL TO ECU (BELOW 1.5V) | | IN NEUTRAL: LOW VOLTAGE SIGNAL TO ECU (APPROX. 0V) | | | | | | | |
| COLD DRIVEAWAY —PART THROTTLE • WARM AIR • WARM COOLANT | | | | | | | | | | | |
| WARM DRIVEAWAY —PART THROTTLE • WARM AIR • WARM COOLANT | HIGH VOLTAGE SIGNAL TO ECU (APPROX. 5V) | | | | | | | | | | |
| HOT CRUISE • WARM AIR • WARM COOLANT | | | | | NO SIGNAL TO ECU (BELOW 1.5V) | | | | | | |
| HOT ACCELERATION —60% THROTTLE | | HIGH VOLTAGE SIGNAL TO ECU (APPROX. 12V) | A/C SWITCH ON: SENDS SIGNAL TO ECU (APPROX. 0V)<br>A/C SWITCH OFF: NO SIGNAL TO ECU (APPROX. 12V) | DRIVING IN ANY GEAR: HIGH VOLTAGE SIGNAL TO ECU (APPROX. 12V) | | HEADLIGHT SWITCH ON: HIGH VOLTAGE SIGNAL TO ECU (APPROX. 12V)<br>HEADLIGHT SWITCH OFF: LOW VOLTAGE SIGNAL TO ECU (BELOW 1.5V) | BLOWER CONTROL SWITCH ON: LOW VOLTAGE SIGNAL TO ECU (APPROX. 0V)<br>BLOWER CONTROL SWITCH OFF: HIGH VOLTAGE SIGNAL TO ECU (APPROX. 12V) | COOLING FAN SWITCH ON: LOW VOLTAGE SIGNAL TO ECU (APPROX. 0V)<br>COOLING FAN SWITCH ON: HIGH VOLTAGE SIGNAL TO ECU (APPROX. 12V) | STEERING WHEEL TURNED: LOW VOLTAGE SIGNAL TO ECU (APPROX. 0V)<br>STEERING WHEEL STRAIGHT AHEAD: HIGH VOLTAGE TO ECU (APPROX. 12V) | NO SIGNAL TO ECU (BELOW 1.5V) | TERMINAL NOT GROUNDED: HIGH VOLTAGE TO ECU (APPROX. 12V) |
| HOT ACCELERATION —WIDE OPEN THROTTLE | LOW VOLTAGE SIGNAL TO ECU (APPROX. 0V) | | | | | | | | | | |
| DECELERATION —CLOSED THROTTLE | | | | | | | | | | | |
| HOT CURB IDLE —EXTENDED | HIGH VOLTAGE SIGNAL TO ECU (APPROX. 5V) | LOW VOLTAGE SIGNAL TO ECU (APPROX. 0V) | | IN NEUTRAL: LOW VOLTAGE SIGNAL TO ECU (APPROX. 0V) | BRAKE PEDAL DEPRESSED: SENDS SIGNAL TO ECU (APPROX. 12V) | | | | | | LOW VOLTAGE SIGNAL TO ECU WHEN CONNECTOR GROUNDED (BELOW 1.5V) |
| HOT ENGINE SHUTDOWN | OFF | OFF | OFF | OFF | OFF | OFF | OFF | OFF | OFF | OFF | OFF |

## TROUBLESHOOTING CHART—1990 MIATA

### Output Devices and Engine Conditions

| ENGINE CONDITIONS \ OUTPUT DEVICES | APPROXIMATE TIME (BASED ON 10—16°C or 50—60°F AMBIENT) | INJECTOR: INJECTION | INJECTOR: INJECTION TIMING | ISC VALVE | SOLENOID VALVE (PURGE CONTROL) | A/C RELAY (A/C CUT-OFF) | MAIN RELAY | IGNITER |
|---|---|---|---|---|---|---|---|---|
| CRANKING —COLD ENGINE • COLD AIR • COLD COOLANT | ZERO | | ALL CYLINDERS EACH Ne-SIGNAL | | | OFF (A/C CUT) | | |
| COLD START —FAST IDLE • COLD AIR • COLD COOLANT | ONE MINUTE | RICH | | LARGE AMOUNT OF BYPASS AIR | OFF (PURGE CUT) | | | |
| COLD DRIVEAWAY —PART THROTTLE • COLD AIR • COLD COOLANT | TWO MINUTES | | | | | ON (A/C ON) | | |
| WARM DRIVEAWAY —PART THROTTLE • WARM AIR • WARM COOLANT | THREE MINUTES | | 2-GROUP | | | | | |
| HOT CRUISE • WARM AIR • WARM COOLANT | | RICH AND LEAN | | SMALL AMOUNT OF BYPASS AIR | OPERATES (DUTY VALUES [PURGE GAS AMOUNT] CHANGE) | | ON | IGNITION SPARK ADVANCE SIGNAL |
| HOT ACCELERATION —60% THROTTLE | | RICH | | | | | | |
| HOT ACCELERATION —WIDE OPEN THROTTLE | MORE THAN FOUR MINUTES | | | | | OFF (A/C CUT) | | |
| DECELERATION —CLOSED THROTTLE | | FUEL CUT | | LARGE AND SMALL AMOUNT OF BYPASS AIR | OFF (PURGE CUT) | ON (A/C ON) | | |
| HOT CURB IDLE —EXTENDED | | RICH AND LEAN | 2-GROUP | SMALL AMOUT OF BYPASS AIR | | | | |
| HOT ENGINE SHUT DOWN | — | DOES NOT INJECT | | OFF | OFF | OFF | OFF | OFF |

| Input Devices \ Output Devices | | Injector: Fuel Injection Amount | Injector: Fuel Injection Timing | ISC Valve | Solenoid Valve (Purge Control) | A/C Relay (A/C Cut-Off) | Igniter (Ignition Timing Control) |
|---|---|---|---|---|---|---|---|
| Test Terminal | | × | × | ○ | × | × | ○ |
| Ignition Switch (Start Position) | | ○ | ○ | ○ | ○ | ○ | ○ |
| P/S Pressure Switch | | × | × | ○ | × | × | × |
| Cooling Fan Switch | | × | × | ○ | × | × | × |
| Blower Control Switch | | × | × | ○ | × | × | × |
| Headlight Switch | | × | × | ○ | × | × | × |
| Stoplight Switch | | ○ | × | × | × | × | × |
| Neutral and Clutch Switches | | ○ | × | ○ | ○ | × | ○ |
| A/C Switch | | × | × | ○ | × | ○ | × |
| Throttle Sensor | Idle Terminal (IDL) | ○ | × | ○ | ○ | × | ○ |
| | Power Terminal (POW) | ○ | × | × | × | ○ | × |
| Atmospheric Pressure Sensor | | ○ | × | ○ | × | × | × |
| IGF Signal | | ○ | × | × | × | × | × |
| Oxygen Sensor | | ○ | × | × | ○ | × | × |
| Water Thermosensor | | ○ | × | ○ | ○ | × | ○ |
| Airflow Meter | Potentiometer | ○ | × | × | ○ | × | ○ |
| | Intake Air Thermosensor | ○ | × | ○ | ○ | × | × |
| Crank Angle Sensor | Ne-Signal | ○ | ○ | ○ | ○ | × | ○ |
| | G-Signal | × | ○ | × | × | × | ○ |

## TROUBLESHOOTING CHART DESCRIPTION 1990 MIATA

| No. | TROUBLESHOOTING ITEM | DESCRIPTION |
|---|---|---|
| | | |
| 2 | Cranks normally but will not start (No combustion) | Engine cranks at normal speed but shows no sign of "firing" |
| 3 | Cranks normally but will not start (Partial combustion) — When engine is cold | Engine cranks at normal speed but shows partial combustion and will not continue to run |
| 4 | Cranks normally but will not start (Partial combustion) — After warm-up | Engine cranks at normal speed but shows partial combustion and will not continue to run after running and hot soaked |
| 5 | Cranks normally but hard to start — Always | Engine cranks at normal speed but requires excessive cranking time (more than 5 sec.) before starting |
| 6 | Cranks normally but hard to start — When engine is cold | Same condition as No.5 when engine is cold Restarts OK after warm-up |
| 7 | Cranks normally but hard to start — After warm-up | Same condition as No.5 after running and hot soaked Starts normally when cold |
| 8 | Rough idle — Always | Engine vibrates excessively at idle in every condition |
| 9 | Low idle speed/Rough idle — Before warm-up | Engine speed low or engine vibrates excessively at idle during warm-up |
| 10 | Low idle speed/Rough idle — After warm-up | Engine runs normally at idle during warm-up but vibrates excessively after warm-up |
| 11 | High idle speed — After warm-up | Engine idle excessive for operation mode |
| 12 | Low idle speed — When A/C, P/S, or E/L ON | Engine speed decreases at idle when A/C, P/S, or E/L is ON |
| 13 | Rough idle just after starting | Engine starts normally but vibrates excessively only just after starting |
| 14 | Idle moves up and down | Engine speed up and down periodically at idle |
| 15 | Engine stalls at idle — Always | Engine starts normally but vibrates excessively and stalls at idle in every condition |
| 16 | Engine stalls at idle — Before warm-up | Engine starts normally but vibrates excessively and stalls at idle before warm-up |
| 17 | Engine stalls at idle — After warm-up | Engine runs normally at idle during warm-up but becomes rough and stalls after warm-up |
| 18 | Engine stalls during start-up | Engine unexpectedly stops running while starting |
| 19 | Engine stalls on deceleration | Engine unexpectedly stops running while decelerating or after deceleration |
| 20 | Engine stalls at idle — When A/C, P/S, or E/L ON | Engine unexpectedly stops running at idle when A/C, P/S, or E/L is ON |
| 21 | Engine stalls suddenly (Intermittent) | Engine intermittently stops running |
| 22 | Hesitates/Stumbles on acceleration | Flat spot occurs just after accelerator is depressed or mild jerking occurs during acceleration |
| 23 | Surges while cruising | Unexpected, usually repetitive change in engine speed |
| 24 | Lack of power | Performance poor under load Maximum speed reduced |
| 25 | Poor acceleration | Performance poor while accelerating |
| 26 | Runs rough on deceleration/Backfire | Engine runs rough while decelerating and abnormal combustion in exhaust system |
| 27 | Knocking | Abnormal combustion accompanied by audible "pinging" noise |
| 28 | Fuel odor | Gasoline odor in cabin |
| 29 | Exhaust sulfur smell | Exhaust gas smells abnormal (rotten egg smell) |
| 30 | High oil consumption | Oil consumption excessive |
| 31 | Poor fuel economy | Fuel economy unsatisfactory |
| 32 | MIL always ON | Self-Diagnosis Checker does not indicate Malfunction Code No. but MIL comes on |
| 33 | MIL never ON | Self-Diagnosis Checker indicates Malfunction Code No. of input device but MIL never ON |
| 34 | A/C does not work | Blower fan operates but no cool air discharged |

## TROUBLESHOOTING CHART—1990 MIATA

**2 — CRANKS NORMALLY BUT WILL NOT START (NO COMBUSTION)**

**DESCRIPTION**
- Engine cranks at normal speed but shows no sign of "firing"
- Battery in normal condition
- Throttle valve not held fully open while cranking
- Fuel in tank

**[TROUBLESHOOTING HINTS]**
Because of no combustion, possibly no fuel is injected to engine or no ignition at all cylinders

① No spark
- Ignition control malfunction
- Ignition system component malfunction

② No fuel injection
- Fuel pump does not operate
- Injector does not operate

③ Low fuel line pressure

④ Low engine compression

| STEP | INSPECTION | | ACTION |
|---|---|---|---|
| 1 | Check if strong blue spark is visible at disconnected high-tension lead while cranking engine 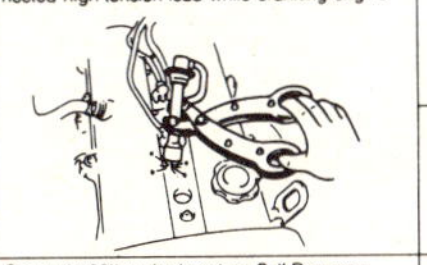 | Yes | Go to Step 3 |
| | | No | Go to Step 2 |
| 2 | Check if "00" is displayed on Self-Diagnosis Checker with ignition switch ON 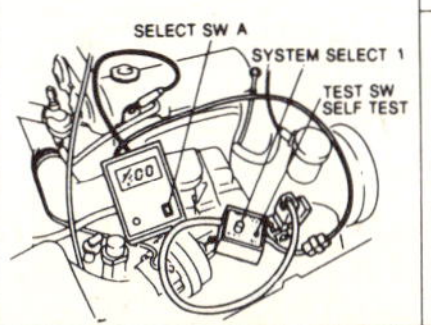<br>SELECT SW A<br>SYSTEM SELECT 1<br>TEST SW SELF TEST | Yes | Check ignition system (Refer to Troubleshooting "Misfire") |
| | | No | **Malfunction Code No. displayed**<br>Check for cause (Refer to specified check sequence) |
| | | | **"88" flashes**<br>Check ECU terminal 1F voltage<br>**Voltage: Approx. 12V (Ignition switch ON)**<br>⇨ If OK, replace ECU<br>⇨ If not OK, check wiring between ECU and Self-Diagnosis Checker |
| 3 | Connect diagnosis connector terminals F/P and GND with jumper wire and check for fuel pump operating sound with ignition switch ON 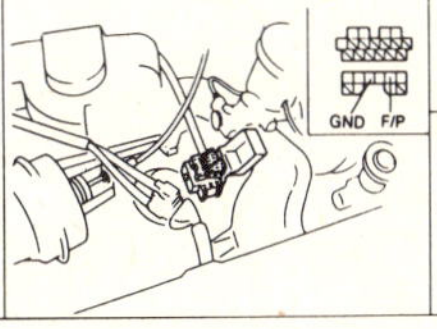<br>GND F/P | Yes | Check if engine starts in this condition<br>⇨ If starts, check circuit opening relay<br>⇨ If does not start, go to Step 5 |
| | | No | Go to Step 4 |

## TROUBLESHOOTING CHART—1990 MIATA

| STEP | INSPECTION | | ACTION |
|---|---|---|---|
| 4 | Check if approx. 12V exists at fuel pump connector (L/R) wire with jumper wire connected (Step 3) | Yes | Check continuity of fuel pump |
| | | No | Check circuit opening relay |
| 5 | Check for injector operating sound while cranking engine | Yes | Go to Step 7 |
| | | No | Go to Step 6 |
| 6 | Check if approx. 12V exists at injector connector (W/R) wire with ignition switch ON | Yes | Check ECU terminals 2A, 2U and 2V voltages |
| | | No | Check for open circuit in wiring between main relay and injector |
| 7 | Connect diagnosis connector terminals F/P and GND with jumper wire and check for correct fuel line pressure with ignition switch ON<br>**Fuel Line pressure:**<br>**265—314 kPa (2.7—3.2 $kg/cm^2$, 38—46 psi)**<br>GND F/P INSTALL CLAMPS | Yes | Go to next step |
| | | No | **Low pressure**<br>Check fuel line pressure while pinching fuel return hose<br>⇨ If fuel line pressure quickly increases, check pressure regulator<br>⇨ If fuel line pressure gradually increases, check for clogging between fuel pump and pressure regulator<br>If not clogged, check fuel pump maximum pressure |
| 8 | Check for correct engine compression<br>**Engine compression: 1,324—932 kPa (13.5—9.5 $kg/cm^2$, 192—135 psi) - 300 rpm** | Yes | Go to next step |
| | | No | Check engine condition<br>• Worn piston, piston rings or cylinder wall<br>• Defective cylinder head gasket<br>• Distorted cylinder head<br>• Improper valve seating<br>• Valve sticking in guide |
| 9 | Check if spark plugs are OK<br>WEAR AND CARBON BUILDUP<br>BURNS<br>PLUG GAP 1.0—1.1mm (0.039—0.043 in)<br>DAMAGE AND DETERIORATION<br>DAMAGE | Yes | Go to next step |
| | | No | Repair, clean, or replace |
| 10 | Try known good ECU and check if condition improves | | |

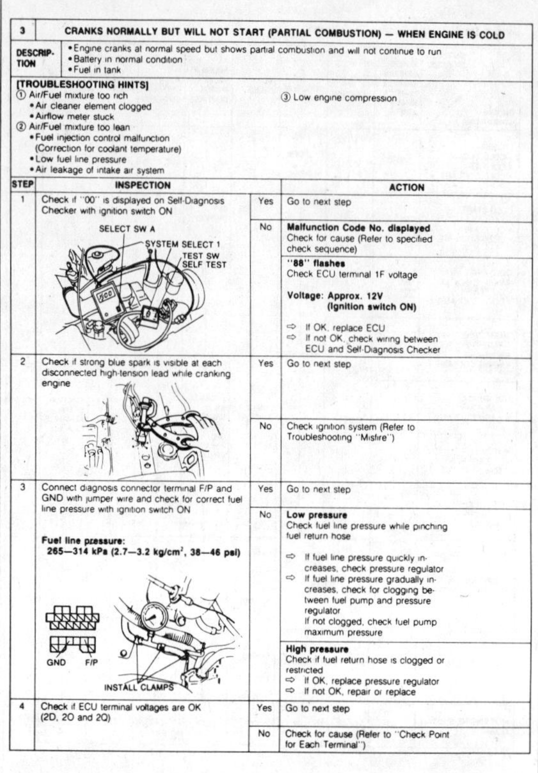

| 3 | CRANKS NORMALLY BUT WILL NOT START (PARTIAL COMBUSTION) — WHEN ENGINE IS COLD | | |
|---|---|---|---|
| DESCRIPTION | • Engine cranks at normal speed but shows partial combustion and will not continue to run<br>• Battery in normal condition<br>• Fuel in tank | | |
| [TROUBLESHOOTING HINTS] | ① Air/Fuel mixture too rich<br>• Air cleaner element clogged<br>• Airflow meter stuck<br>② Air/Fuel mixture too lean<br>• Fuel injection control malfunction (Correction for coolant temperature)<br>• Low fuel line pressure<br>• Air leakage of intake air system<br>③ Low engine compression | | |
| STEP | INSPECTION | | ACTION |
| 1 | Check if "00" is displayed on Self-Diagnosis Checker with ignition switch ON<br>SELECT SW A<br>SYSTEM SELECT 1<br>TEST SW<br>SELF TEST | Yes | Go to next step |
| | | No | **Malfunction Code No. displayed**<br>Check for cause (Refer to specified check sequence)<br>**"88" flashes**<br>Check ECU terminal 1F voltage<br>**Voltage: Approx. 12V (Ignition switch ON)**<br>⇨ If OK, replace ECU<br>⇨ If not OK, check wiring between ECU and Self-Diagnosis Checker |
| 2 | Check if strong blue spark is visible at each disconnected high-tension lead while cranking engine | Yes | Go to next step |
| | | No | Check ignition system (Refer to Troubleshooting "Misfire") |
| 3 | Connect diagnosis connector terminal F/P and GND with jumper wire and check for correct fuel line pressure with ignition switch ON<br>**Fuel line pressure:**<br>**265—314 kPa (2.7—3.2 $kg/cm^2$, 38—46 psi)**<br>GND F/P INSTALL CLAMPS | Yes | Go to next step |
| | | No | **Low pressure**<br>Check fuel line pressure while pinching fuel return hose<br>⇨ If fuel line pressure quickly increases, check pressure regulator<br>⇨ If fuel line pressure gradually increases, check for clogging between fuel pump and pressure regulator<br>If not clogged, check fuel pump maximum pressure<br>**High pressure**<br>Check if fuel return hose is clogged or restricted<br>⇨ If OK, replace pressure regulator<br>⇨ If not OK, repair or replace |
| 4 | Check if ECU terminal voltages are OK (2D, 2O and 2Q) | Yes | Go to next step |
| | | No | Check for cause (Refer to "Check Point for Each Terminal") |

## TROUBLESHOOTING CHART—1990 MIATA

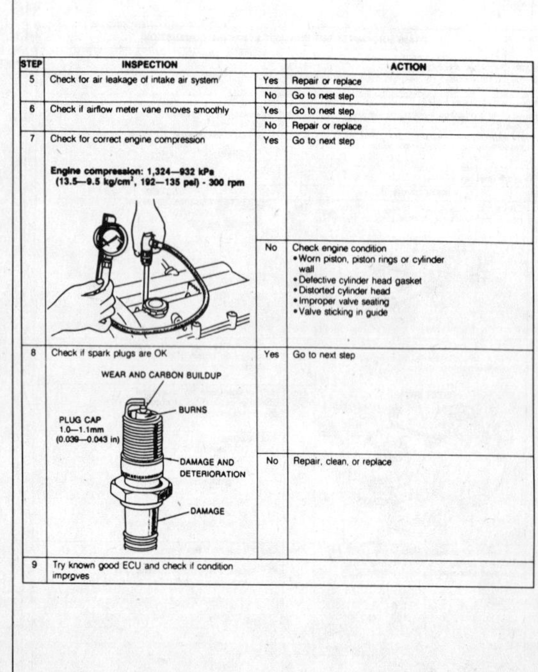

| STEP | INSPECTION | | ACTION |
|---|---|---|---|
| 5 | Check for air leakage of intake air system | Yes | Repair or replace |
| | | No | Go to next step |
| 6 | Check if airflow meter vane moves smoothly | Yes | Go to next step |
| | | No | Repair or replace |
| 7 | Check for correct engine compression<br>**Engine compression: 1,324—932 kPa (13.5—9.5 $kg/cm^2$, 192—135 psi) - 300 rpm** | Yes | Go to next step |
| | | No | Check engine condition<br>• Worn piston, piston rings or cylinder wall<br>• Defective cylinder head gasket<br>• Distorted cylinder head<br>• Improper valve seating<br>• Valve sticking in guide |
| 8 | Check if spark plugs are OK<br>WEAR AND CARBON BUILDUP<br>BURNS<br>PLUG CAP 1.0—1.1mm (0.039—0.043 in)<br>DAMAGE AND DETERIORATION<br>DAMAGE | Yes | Go to next step |
| | | No | Repair, clean, or replace |
| 9 | Try known good ECU and check if condition imprgves | | |

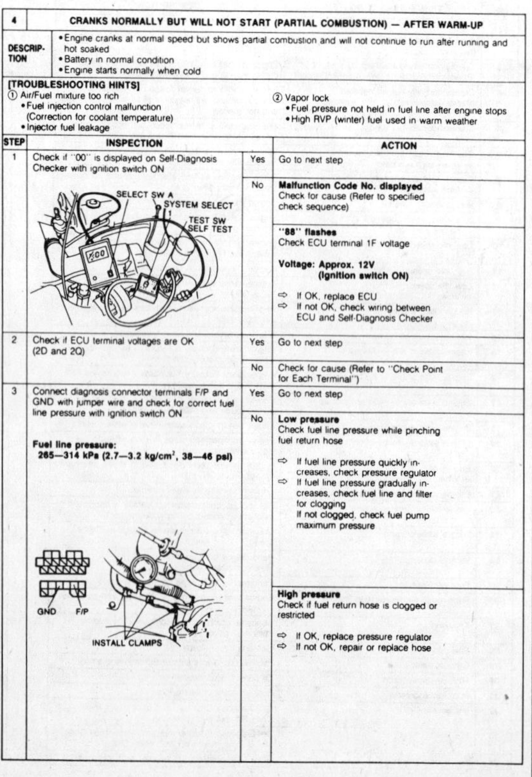

| 4 | CRANKS NORMALLY BUT WILL NOT START (PARTIAL COMBUSTION) — AFTER WARM-UP | | |
|---|---|---|---|
| DESCRIPTION | • Engine cranks at normal speed but shows partial combustion and will not continue to run after running and hot soaked<br>• Battery in normal condition<br>• Engine starts normally when cold | | |
| [TROUBLESHOOTING HINTS] | ① Air/Fuel mixture too rich<br>• Fuel injection control malfunction (Correction for coolant temperature)<br>• Injector fuel leakage<br>② Vapor lock<br>• Fuel pressure not held in fuel line after engine stops<br>• High RVP (winter) fuel used in warm weather | | |
| STEP | INSPECTION | | ACTION |
| 1 | Check if "00" is displayed on Self-Diagnosis Checker with ignition switch ON<br>SELECT SW A<br>SYSTEM SELECT<br>TEST SW<br>SELF TEST | Yes | Go to next step |
| | | No | **Malfunction Code No. displayed**<br>Check for cause (Refer to specified check sequence)<br>**"88" flashes**<br>Check ECU terminal 1F voltage<br>**Voltage: Approx. 12V (Ignition switch ON)**<br>⇨ If OK, replace ECU<br>⇨ If not OK, check wiring between ECU and Self-Diagnosis Checker |
| 2 | Check if ECU terminal voltages are OK (2D and 2Q) | Yes | Go to next step |
| | | No | Check for cause (Refer to "Check Point for Each Terminal") |
| 3 | Connect diagnosis connector terminals F/P and GND with jumper wire and check for correct fuel line pressure with ignition switch ON<br>**Fuel line pressure:**<br>**265—314 kPa (2.7—3.2 $kg/cm^2$, 38—46 psi)**<br>GND F/P INSTALL CLAMPS | Yes | Go to next step |
| | | No | **Low pressure**<br>Check fuel line pressure while pinching fuel return hose<br>⇨ If fuel line pressure quickly increases, check pressure regulator<br>⇨ If fuel line pressure gradually increases, check fuel line and filter for clogging<br>If not clogged, check fuel pump maximum pressure<br>**High pressure**<br>Check if fuel return hose is clogged or restricted<br>⇨ If OK, replace pressure regulator<br>⇨ If not OK, repair or replace hose |

## TROUBLESHOOTING CHART—1990 MIATA

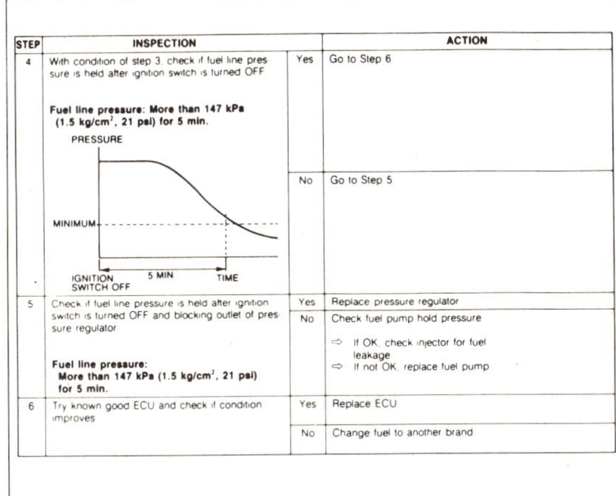

| STEP | INSPECTION | | ACTION |
|---|---|---|---|
| 4 | With condition of step 3, check if fuel line pressure is held after ignition switch is turned OFF<br>**Fuel line pressure: More than 147 kPa (1.5 kg/cm², 21 psi) for 5 min.**<br>PRESSURE / MINIMUM / IGNITION SWITCH OFF / 5 MIN / TIME | Yes | Go to Step 6 |
| | | No | Go to Step 5 |
| 5 | Check if fuel line pressure is held after ignition switch is turned OFF and blocking outlet of pressure regulator<br>**Fuel line pressure: More than 147 kPa (1.5 kg/cm², 21 psi) for 5 min.** | Yes | Replace pressure regulator |
| | | No | Check fuel pump hold pressure<br>⇨ If OK, check injector for fuel leakage<br>⇨ If not OK, replace fuel pump |
| 6 | Try known good ECU and check if condition improves | Yes | Replace ECU |
| | | No | Change fuel to another brand |

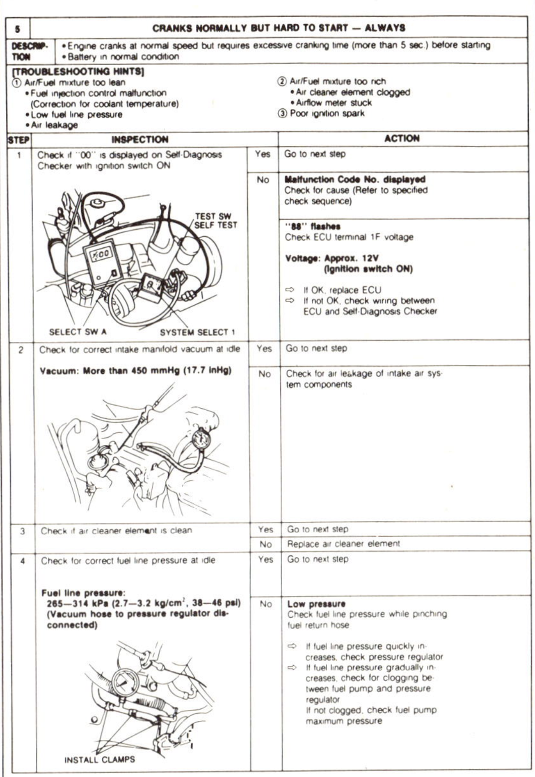

| 5 | CRANKS NORMALLY BUT HARD TO START — ALWAYS | | |
|---|---|---|---|
| DESCRIPTION | • Engine cranks at normal speed but requires excessive cranking time (more than 5 sec.) before starting<br>• Battery in normal condition | | |
| [TROUBLESHOOTING HINTS] | ① Air/Fuel mixture too lean<br>• Fuel injection control malfunction (Correction for coolant temperature)<br>• Low fuel line pressure<br>• Air leakage<br>② Air/Fuel mixture too rich<br>• Air cleaner element clogged<br>• Airflow meter stuck<br>③ Poor ignition spark | | |
| **STEP** | **INSPECTION** | | **ACTION** |
| 1 | Check if "00" is displayed on Self-Diagnosis Checker with ignition switch ON<br>TEST SW SELF TEST / SELECT SW A / SYSTEM SELECT 1 | Yes | Go to next step |
| | | No | **Malfunction Code No. displayed**<br>Check for cause (Refer to specified check sequence)<br>**"88" flashes**<br>Check ECU terminal 1F voltage<br>**Voltage: Approx. 12V (Ignition switch ON)**<br>⇨ If OK, replace ECU<br>⇨ If not OK, check wiring between ECU and Self-Diagnosis Checker |
| 2 | Check for correct intake manifold vacuum at idle<br>**Vacuum: More than 450 mmHg (17.7 inHg)** | Yes | Go to next step |
| | | No | Check for air leakage of intake air system components |
| 3 | Check if air cleaner element is clean | Yes | Go to next step |
| | | No | Replace air cleaner element |
| 4 | Check for correct fuel line pressure at idle<br>**Fuel line pressure: 265—314 kPa (2.7—3.2 kg/cm², 38—46 psi) (Vacuum hose to pressure regulator disconnected)**<br>INSTALL CLAMPS | Yes | Go to next step |
| | | No | **Low pressure**<br>Check fuel line pressure while pinching fuel return hose<br>⇨ If fuel line pressure quickly increases, check pressure regulator<br>⇨ If fuel line pressure gradually increases, check for clogging between fuel pump and pressure regulator<br>If not clogged, check fuel pump maximum pressure |

## TROUBLESHOOTING CHART—1990 MIATA

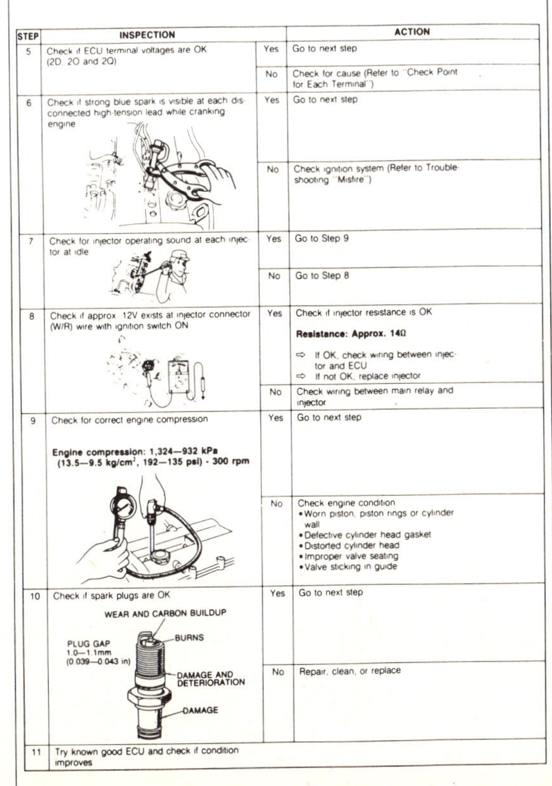

| STEP | INSPECTION | | ACTION |
|---|---|---|---|
| 5 | Check if ECU terminal voltages are OK (2D, 2O and 2Q) | Yes | Go to next step |
| | | No | Check for cause (Refer to "Check Point for Each Terminal") |
| 6 | Check if strong blue spark is visible at each disconnected high-tension lead while cranking engine | Yes | Go to next step |
| | | No | Check ignition system (Refer to Troubleshooting "Misfire") |
| 7 | Check for injector operating sound at each injector at idle | Yes | Go to Step 9 |
| | | No | Go to Step 8 |
| 8 | Check if approx. 12V exists at injector connector (W/R) wire with ignition switch ON | Yes | Check if injector resistance is OK<br>**Resistance: Approx. 14Ω**<br>⇨ If OK, check wiring between injector and ECU<br>⇨ If not OK, replace injector |
| | | No | Check wiring between main relay and injector |
| 9 | Check for correct engine compression<br>**Engine compression: 1,324—932 kPa (13.5—9.5 kg/cm², 192—135 psi) - 300 rpm** | Yes | Go to next step |
| | | No | Check engine condition<br>• Worn piston, piston rings or cylinder wall<br>• Defective cylinder head gasket<br>• Distorted cylinder head<br>• Improper valve seating<br>• Valve sticking in guide |
| 10 | Check if spark plugs are OK<br>WEAR AND CARBON BUILDUP / BURNS / PLUG GAP 1.0—1.1mm (0.039—0.043 in) / DAMAGE AND DETERIORATION / DAMAGE | Yes | Go to next step |
| | | No | Repair, clean, or replace |
| 11 | Try known good ECU and check if condition improves | | |

| 6 | CRANKS NORMALLY BUT HARD TO START — WHEN ENGINE IS COLD | | |
|---|---|---|---|
| DESCRIPTION | • Engine cranks at normal speed but requires excessive cranking time before starting<br>• Battery in normal condition<br>• Restarts OK after warm-up | | |
| [TROUBLESHOOTING HINTS] | ① Air/Fuel mixture too rich<br>• Airflow meter stuck<br>• Air cleaner element clogged<br>• Idle speed control malfunction<br>② Air/Fuel mixture too lean<br>• Fuel injection control malfunction (Correction for coolant temperature)<br>③ Poor atomization of fuel<br>• Low RVP (summer) fuel used in cold weather | | |
| **STEP** | **INSPECTION** | | **ACTION** |
| 1 | Check if "00" is displayed on Self-Diagnosis Checker with ignition switch ON<br>TEST SW SELF TEST / SELECT SW A / SYSTEM SELECT 1 | Yes | Go to next step |
| | | No | **Malfunction Code No. displayed**<br>Check for cause (Refer to specified check sequence)<br>**"88" flashes**<br>Check ECU terminal 1F voltage<br>**Voltage: Approx. 12V (Ignition switch ON)**<br>⇨ If OK, replace ECU<br>⇨ If not OK, check wiring between ECU and Self-Diagnosis Checker |
| 2 | Check if ECU terminal voltages are OK (1C, 2D, 2O and 2Q) | Yes | Go to next step |
| | | No | Check for cause (Refer to "Check Point for Each Terminal") |
| 3 | Check if engine starts easily when depressing accelerator while cranking | Yes | Check if ISC valve is OK<br>⇨ If OK, check air valve<br>⇨ If not OK, replace ISC valve |
| | | No | Go to next step |
| 4 | Check for correct intake manifold vacuum at idle<br>**Vacuum: More than 450 mmHg (17.7 inHg)** | Yes | Go to next step |
| | | No | Check for air leakage of intake air system components |
| 5 | Check if air cleanerr element is clean | Yes | Go to next step |
| | | No | Replace air cleaner element |
| 6 | Try known good ECU and check if condition improves | Yes | Replace ECU |
| | | No | Change fuel to another brand |

## TROUBLESHOOTING CHART—1990 MIATA

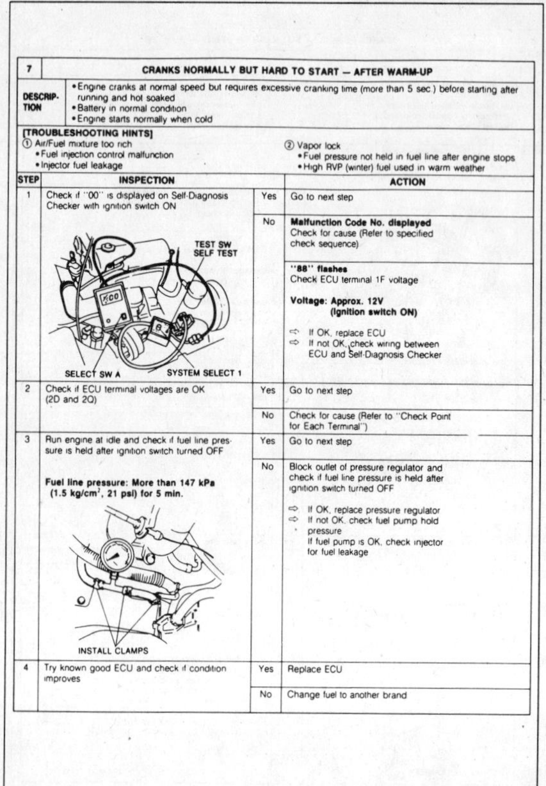

| 7 | CRANKS NORMALLY BUT HARD TO START — AFTER WARM-UP | | |
|---|---|---|---|
| DESCRIPTION | • Engine cranks at normal speed but requires excessive cranking time (more than 5 sec.) before starting after running and hot soaked<br>• Battery in normal condition<br>• Engine starts normally when cold | | |
| [TROUBLESHOOTING HINTS] | ① Air/Fuel mixture too rich<br>• Fuel injection control malfunction<br>• Injector fuel leakage<br>② Vapor lock<br>• Fuel pressure not held in fuel line after engine stops<br>• High RVP (winter) fuel used in warm weather | | |
| STEP | INSPECTION | | ACTION |
| 1 | Check if "00" is displayed on Self-Diagnosis Checker with ignition switch ON | Yes | Go to next step |
| | | No | **Malfunction Code No. displayed**<br>Check for cause (Refer to specified check sequence) |
| | | | **"88" flashes**<br>Check ECU terminal 1F voltage<br>**Voltage: Approx. 12V (Ignition switch ON)**<br>⇨ If OK, replace ECU<br>⇨ If not OK, check wiring between ECU and Self-Diagnosis Checker |
| 2 | Check if ECU terminal voltages are OK (2D and 2Q) | Yes | Go to next step |
| | | No | Check for cause (Refer to "Check Point for Each Terminal") |
| 3 | Run engine at idle and check if fuel line pressure is held after ignition switch turned OFF<br>**Fuel line pressure: More than 147 kPa (1.5 kg/cm², 21 psi) for 5 min.** | Yes | Go to next step |
| | | No | Block outlet of pressure regulator and check if fuel line pressure is held after ignition switch turned OFF<br>⇨ If OK, replace pressure regulator<br>⇨ If not OK, check fuel pump hold pressure<br>If fuel pump is OK, check injector for fuel leakage |
| 4 | Try known good ECU and check if condition improves | Yes | Replace ECU |
| | | No | Change fuel to another brand |



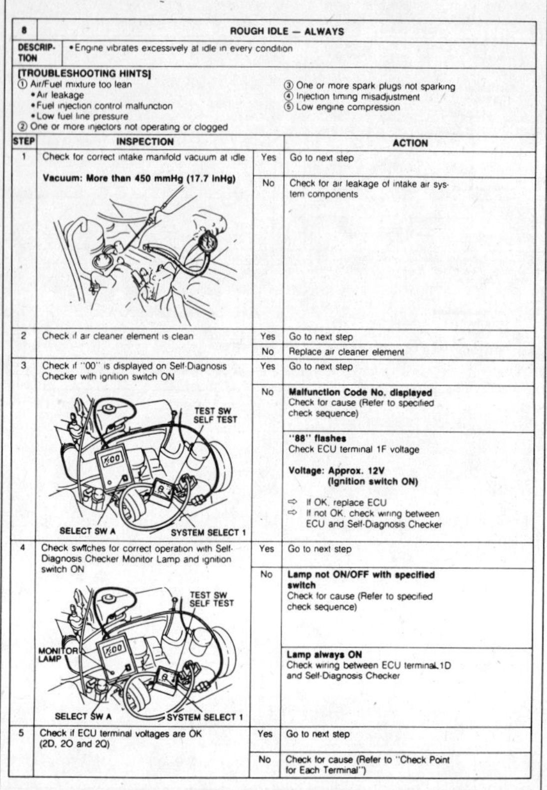

| 8 | ROUGH IDLE — ALWAYS | | |
|---|---|---|---|
| DESCRIPTION | • Engine vibrates excessively at idle in every condition | | |
| [TROUBLESHOOTING HINTS] | ① Air/Fuel mixture too lean<br>• Air leakage<br>• Fuel injection control malfunction<br>• Low fuel line pressure<br>② One or more injectors not operating or clogged<br>③ One or more spark plugs not sparking<br>④ Injection timing misadjustment<br>⑤ Low engine compression | | |
| STEP | INSPECTION | | ACTION |
| 1 | Check for correct intake manifold vacuum at idle<br>**Vacuum: More than 450 mmHg (17.7 inHg)** | Yes | Go to next step |
| | | No | Check for air leakage of intake air system components |
| 2 | Check if air cleaner element is clean | Yes | Go to next step |
| | | No | Replace air cleaner element |
| 3 | Check if "00" is displayed on Self-Diagnosis Checker with ignition switch ON | Yes | Go to next step |
| | | No | **Malfunction Code No. displayed**<br>Check for cause (Refer to specified check sequence) |
| | | | **"88" flashes**<br>Check ECU terminal 1F voltage<br>**Voltage: Approx. 12V (Ignition switch ON)**<br>⇨ If OK, replace ECU<br>⇨ If not OK, check wiring between ECU and Self-Diagnosis Checker |
| 4 | Check switches for correct operation with Self-Diagnosis Checker Monitor Lamp and ignition switch ON | Yes | Go to next step |
| | | No | **Lamp not ON/OFF with specified switch**<br>Check for cause (Refer to specified check sequence) |
| | | | **Lamp always ON**<br>Check wiring between ECU terminal 1D and Self-Diagnosis Checker |
| 5 | Check if ECU terminal voltages are OK (2D, 2O and 2Q) | Yes | Go to next step |
| | | No | Check for cause (Refer to "Check Point for Each Terminal") |



## TROUBLESHOOTING CHART—1990 MIATA

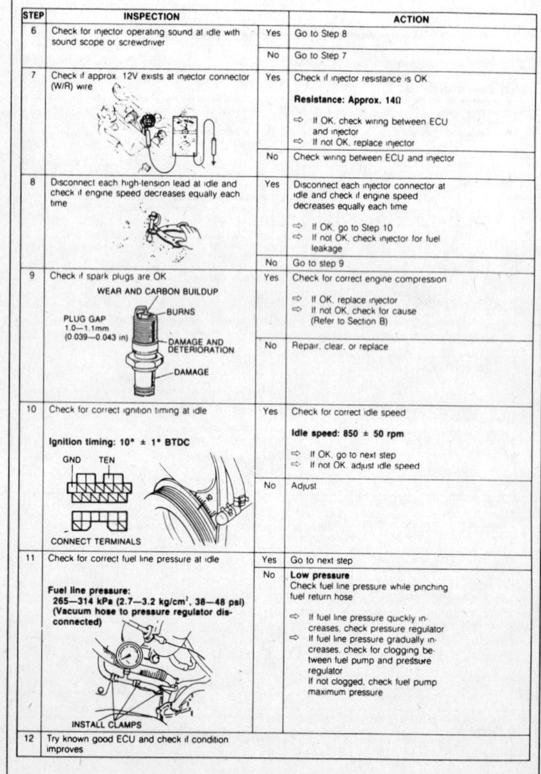

| STEP | INSPECTION | | ACTION |
|---|---|---|---|
| 6 | Check for injector operating sound at idle with sound scope or screwdriver | Yes | Go to Step 8 |
| | | No | Go to Step 7 |
| 7 | Check if approx. 12V exists at injector connector (W/R) wire | Yes | Check if injector resistance is OK<br>**Resistance: Approx. 14Ω**<br>⇨ If OK, check wiring between ECU and injector<br>⇨ If not OK, replace injector |
| | | No | Check wiring between ECU and injector |
| 8 | Disconnect each high-tension lead at idle and check if engine speed decreases equally each time | Yes | Disconnect each injector connector at idle and check if engine speed decreases equally each time<br>⇨ If OK, go to Step 10<br>⇨ If not OK, check injector for fuel leakage |
| | | No | Go to step 9 |
| 9 | Check if spark plugs are OK | Yes | Check for correct engine compression<br>⇨ If OK, replace injector<br>⇨ If not OK, check for cause (Refer to Section B) |
| | | No | Repair, clear, or replace |
| 10 | Check for correct ignition timing at idle<br>**Ignition timing: 10° ± 1° BTDC** | Yes | Check for correct idle speed<br>**Idle speed: 850 ± 50 rpm**<br>⇨ If OK, go to next step<br>⇨ If not OK, adjust idle speed |
| | | No | Adjust |
| 11 | Check for correct fuel line pressure at idle<br>**Fuel line pressure: 265—314 kPa (2.7—3.2 kg/cm², 38—48 psi) (Vacuum hose to pressure regulator disconnected)** | Yes | Go to next step |
| | | No | **Low pressure**<br>Check fuel line pressure while pinching fuel return hose<br>⇨ If fuel line pressure quickly increases, check pressure regulator<br>⇨ If fuel line pressure gradually increases, check for clogging between fuel pump and pressure regulator<br>If not clogged, check fuel pump maximum pressure |
| 12 | Try known good ECU and check if condition improves | | |



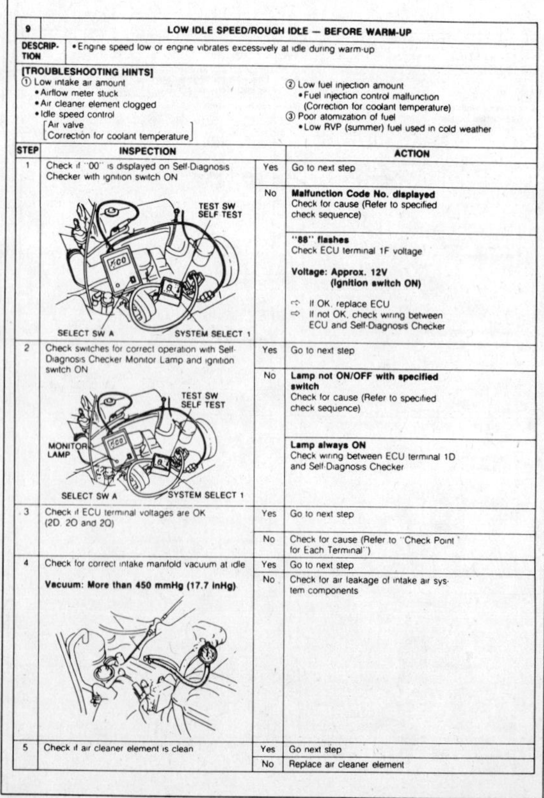

| 9 | LOW IDLE SPEED/ROUGH IDLE — BEFORE WARM-UP | | |
|---|---|---|---|
| DESCRIPTION | • Engine speed low or engine vibrates excessively at idle during warm-up | | |
| [TROUBLESHOOTING HINTS] | ① Low intake air amount<br>• Airflow meter stuck<br>• Air cleaner element clogged<br>• Idle speed control [Air valve, Correction for coolant temperature]<br>② Low fuel injection amount<br>• Fuel injection control malfunction (Correction for coolant temperature)<br>③ Poor atomization of fuel<br>• Low RVP (summer) fuel used in cold weather | | |
| STEP | INSPECTION | | ACTION |
| 1 | Check if "00" is displayed on Self-Diagnosis Checker with ignition switch ON | Yes | Go to next step |
| | | No | **Malfunction Code No. displayed**<br>Check for cause (Refer to specified check sequence) |
| | | | **"88" flashes**<br>Check ECU terminal 1F voltage<br>**Voltage: Approx. 12V (Ignition switch ON)**<br>⇨ If OK, replace ECU<br>⇨ If not OK, check wiring between ECU and Self-Diagnosis Checker |
| 2 | Check switches for correct operation with Self-Diagnosis Checker Monitor Lamp and ignition switch ON | Yes | Go to next step |
| | | No | **Lamp not ON/OFF with specified switch**<br>Check for cause (Refer to specified check sequence) |
| | | | **Lamp always ON**<br>Check wiring between ECU terminal 1D and Self-Diagnosis Checker |
| 3 | Check if ECU terminal voltages are OK (2D, 2O and 2Q) | Yes | Go to next step |
| | | No | Check for cause (Refer to "Check Point for Each Terminal") |
| 4 | Check for correct intake manifold vacuum at idle<br>**Vacuum: More than 450 mmHg (17.7 inHg)** | Yes | Go to next step |
| | | No | Check for air leakage of intake air system components |
| 5 | Check if air cleaner element is clean | Yes | Go next step |
| | | No | Replace air cleaner element |

## TROUBLESHOOTING CHART—1990 MIATA

| STEP | INSPECTION | | ACTION |
|---|---|---|---|
| 6 | Connect System Selector to diagnosis connector and set Test Switch to "SELF TEST" when engine is cold<br>Check if engine speed decreases as engine warms up<br>(GND, TEN, TEST SW.) | Yes | Go to next step |
| | | No | Check air valve |
| 7 | With condition of Step 5 check for correct ignition timing at idle after warm-up<br>**Ignition timing: 10° ± 1° BTDC** | Yes | Go to next step |
| | | No | Adjust |
| 8 | Try known good ECU and check if condition improves | Yes | Replace ECU |
| | | No | Change fuel to another brand |

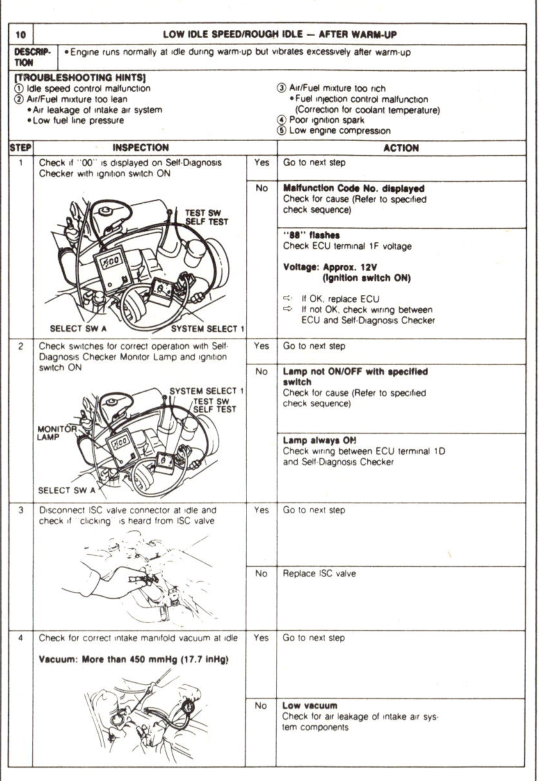

| 10 | LOW IDLE SPEED/ROUGH IDLE — AFTER WARM-UP | | |
|---|---|---|---|
| DESCRIPTION | • Engine runs normally at idle during warm-up but vibrates excessively after warm-up | | |

**[TROUBLESHOOTING HINTS]**
① Idle speed control malfunction
② Air/Fuel mixture too lean
• Air leakage of intake air system
• Low fuel line pressure
③ Air/Fuel mixture too rich
• Fuel injection control malfunction (Correction for coolant temperature)
④ Poor ignition spark
⑤ Low engine compression

| STEP | INSPECTION | | ACTION |
|---|---|---|---|
| 1 | Check if "00" is displayed on Self-Diagnosis Checker with ignition switch ON<br>(TEST SW SELF TEST, SELECT SW A, SYSTEM SELECT 1) | Yes | Go to next step |
| | | No | **Malfunction Code No. displayed**<br>Check for cause (Refer to specified check sequence) |
| | | | **"88" flashes**<br>Check ECU terminal 1F voltage<br>**Voltage: Approx. 12V (Ignition switch ON)**<br>⇨ If OK, replace ECU<br>⇨ If not OK, check wiring between ECU and Self-Diagnosis Checker |
| 2 | Check switches for correct operation with Self-Diagnosis Checker Monitor Lamp and ignition switch ON<br>(SYSTEM SELECT 1, TEST SW SELF TEST, MONITOR LAMP, SELECT SW A) | Yes | Go to next step |
| | | No | **Lamp not ON/OFF with specified switch**<br>Check for cause (Refer to specified check sequence) |
| | | | **Lamp always ON**<br>Check wiring between ECU terminal 1D and Self-Diagnosis Checker |
| 3 | Disconnect ISC valve connector at idle and check if "clicking" is heard from ISC valve | Yes | Go to next step |
| | | No | Replace ISC valve |
| 4 | Check for correct intake manifold vacuum at idle<br>**Vacuum: More than 450 mmHg (17.7 inHg)** | Yes | Go to next step |
| | | No | **Low vacuum**<br>Check for air leakage of intake air system components |

## TROUBLESHOOTING CHART—1990 MIATA

| STEP | INSPECTION | | ACTION |
|---|---|---|---|
| 5 | Check if ECU terminal voltages are OK (2D, 2O and 2Q) | Yes | Go to next step |
| | | No | Check for cause (Refer to "Check Point for Each Terminal") |
| 6 | Check for correct ignition timing at idle<br>**Ignition timing: 10° ± 1° BTDC**<br>(GND, TEN, CONNECT TERMINALS) | Yes | Check for correct idle speed<br>**Idle speed: 850 ± 50 rpm**<br>⇨ If OK, go to next step<br>⇨ If not OK, adjust idle speed |
| | | No | Adjust |
| 7 | Check for correct fuel line pressure at idle<br>**Fuel line pressure: 265—314 kPa (2.7—3.2 kg/cm², 38—46 psi) (Vacuum hose to pressure regulator disconnected)**<br>(INSTALL CLAMPS) | Yes | Go to next step |
| | | No | **Low pressure**<br>Check fuel line pressure while pinching fuel return hose<br>⇨ If fuel line pressure quickly increases, check pressure regulator<br>⇨ If fuel line pressure gradually increases, check for clogging between fuel pump and pressure regulator<br>If not clogged, check fuel pump maximum pressure |
| 8 | Check if strong blue spark is visible at each disconnected high-tension lead while cranking engine | Yes | Go to next step |
| | | No | Check ignition system (Refer to Troubleshooting "Misfire") |
| 9 | Check for correct engine compression<br>**Engine compression: 1,324—932 kPa (13.5—9.5 kg/cm², 192—135) - 300 rpm** | Yes | Go to next step |
| | | No | Check engine condition<br>• Worn piston, piston rings or cylinder wall<br>• Defective cylinder head gasket<br>• Distorted cylinder head<br>• Improper valve seating<br>• Valve sticking in guide |
| 10 | Check if spark plugs are OK<br>(WEAR AND CARBON BUILDUP, BURNS, PLUG GAP 1.0—1.1mm (0.039—0.043 in), DAMAGE AND DETERIORATION, DAMAGE) | Yes | Go to next step |
| | | No | Repair, clear, or replace |
| 11 | Try known good ECU and check if condition improves | | |

| 11 | HIGH IDLE SPEED — AFTER WARM-UP | | |
|---|---|---|---|

**[TROUBLESHOOTING HINTS]**
Excessive intake air supplied to engine
① Throttle valve not fully closed
② Idle speed control malfunction
• Air valve not closing
• ISC valve stuck
• Incorrect coolant temperature signal

| STEP | INSPECTION | | ACTION |
|---|---|---|---|
| 1 | Check if throttle valve is fully closed when accelerator released | Yes | Go to Step 3 |
| | | No | Check if throttle linkage is correctly installed and operates freely<br>⇨ If OK, go to Step 2<br>⇨ If not OK, clean, adjust or replace linkage |
| 2 | Check if dashpot is correctly adjusted<br>**Dashpot set speed: 2,500 ± 150 rpm** | Yes | Check if throttle valve is contaminated<br>⇨ If contaminated, clean throttle body<br>⇨ If not contaminated, replace throttle body |
| | | No | Adjust |
| 3 | Check if "00" is displayed on Self-Diagnosis Checker with ignition switch ON<br>(SELECT SW A, SYSTEM SELECT 1, TEST SW SELF TEST) | Yes | Go to next step |
| | | No | **Malfunction Code No. displayed**<br>Check for cause (Refer to specified check sequence) |
| | | | **"88" flashes**<br>Check ECU terminal 1F voltage<br>**Voltage: Approx. 12V (Ignition switch ON)**<br>⇨ If OK, replace ECU<br>⇨ If not OK, check wiring between ECU and Self-Diagnosis Checker |
| 4 | Check switches for correct operation with Self-Diagnosis Checker Monitor Lamp and ignition switch ON<br>(SELECT SW A, SYSTEM SELECT 1, TEST SW SELF TEST, MONITOR LAMP) | Yes | Go to next step |
| | | No | **Lamp not ON/OFF with specified switch**<br>Check for cause (Refer to specified check sequence) |
| | | | **Lamp always ON**<br>Check wiring between ECU terminal 1D and Self-Diagnosis Checker |

## TROUBLESHOOTING CHART—1990 MIATA

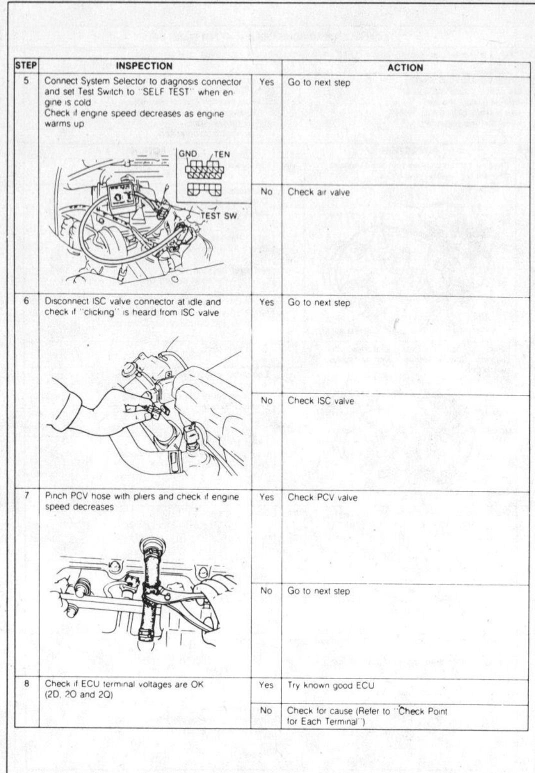

| STEP | INSPECTION | | ACTION |
|---|---|---|---|
| 5 | Connect System Selector to diagnosis connector and set Test Switch to "SELF TEST" when engine is cold<br>Check if engine speed decreases as engine warms up<br>GND TEN<br>TEST SW | Yes | Go to next step |
| | | No | Check air valve |
| 6 | Disconnect ISC valve connector at idle and check if "clicking" is heard from ISC valve | Yes | Go to next step |
| | | No | Check ISC valve |
| 7 | Pinch PCV hose with pliers and check if engine speed decreases | Yes | Check PCV valve |
| | | No | Go to next step |
| 8 | Check if ECU terminal voltages are OK (2D, 2O and 2Q) | Yes | Try known good ECU |
| | | No | Check for cause (Refer to "Check Point for Each Terminal") |

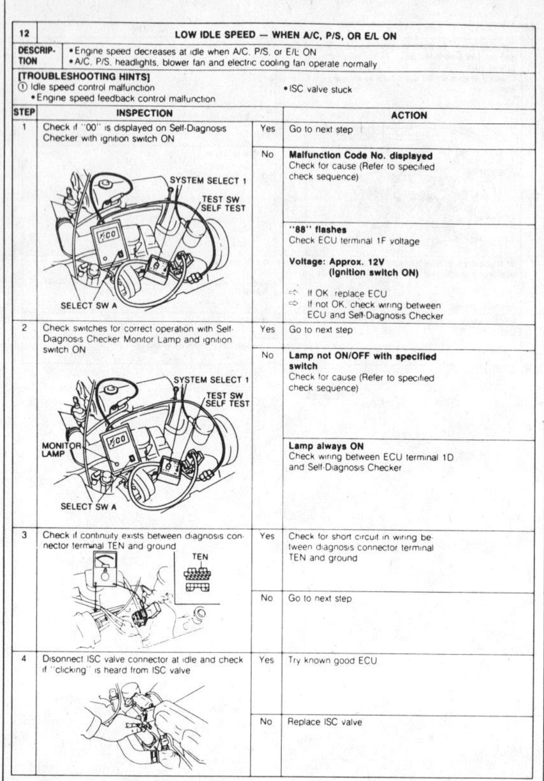

| 12 | LOW IDLE SPEED — WHEN A/C, P/S, OR E/L ON | | |
|---|---|---|---|
| DESCRIPTION | • Engine speed decreases at idle when A/C, P/S, or E/L ON<br>• A/C, P/S, headlights, blower fan and electric cooling fan operate normally | | |
| [TROUBLESHOOTING HINTS] | ① Idle speed control malfunction<br>• Engine speed feedback control malfunction<br>• ISC valve stuck | | |
| **STEP** | **INSPECTION** | | **ACTION** |
| 1 | Check if "00" is displayed on Self-Diagnosis Checker with ignition switch ON<br>SYSTEM SELECT 1<br>TEST SW SELF TEST<br>SELECT SW A | Yes | Go to next step |
| | | No | **Malfunction Code No. displayed**<br>Check for cause (Refer to specified check sequence) |
| | | | **"88" flashes**<br>Check ECU terminal 1F voltage<br>**Voltage: Approx. 12V (Ignition switch ON)**<br>⇨ If OK, replace ECU<br>⇨ If not OK, check wiring between ECU and Self-Diagnosis Checker |
| 2 | Check switches for correct operation with Self-Diagnosis Checker Monitor Lamp and ignition switch ON<br>SYSTEM SELECT 1<br>TEST SW SELF TEST<br>MONITOR LAMP<br>SELECT SW A | Yes | Go to next step |
| | | No | **Lamp not ON/OFF with specified switch**<br>Check for cause (Refer to specified check sequence) |
| | | | **Lamp always ON**<br>Check wiring between ECU terminal 1D and Self-Diagnosis Checker |
| 3 | Check if continuity exists between diagnosis connector terminal TEN and ground<br>TEN | Yes | Check for short circuit in wiring between diagnosis connector terminal TEN and ground |
| | | No | Go to next step |
| 4 | Disonnect ISC valve connector at idle and check if "clicking" is heard from ISC valve | Yes | Try known good ECU |
| | | No | Replace ISC valve |

## TROUBLESHOOTING CHART—1990 MIATA

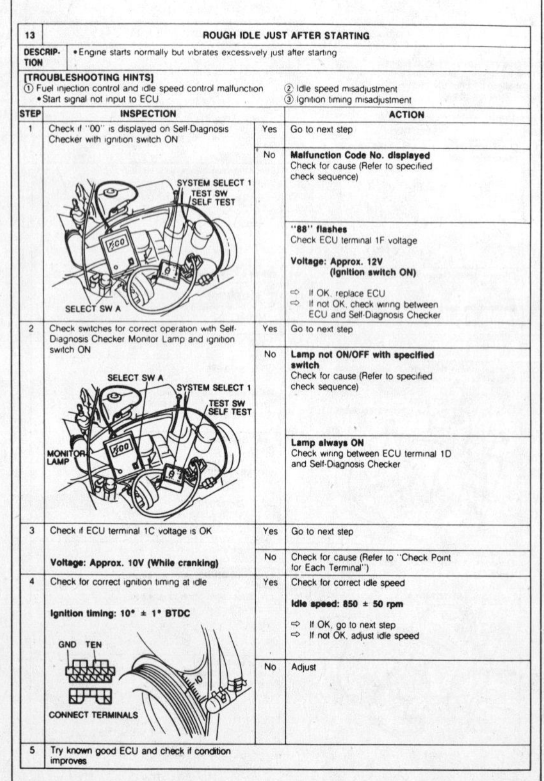

| 13 | ROUGH IDLE JUST AFTER STARTING | | |
|---|---|---|---|
| DESCRIPTION | • Engine starts normally but vibrates excessively just after starting | | |
| [TROUBLESHOOTING HINTS] | ① Fuel injection control and idle speed control malfunction<br>• Start signal not input to ECU<br>② Idle speed misadjustment<br>③ Ignition timing misadjustment | | |
| **STEP** | **INSPECTION** | | **ACTION** |
| 1 | Check if "00" is displayed on Self-Diagnosis Checker with ignition switch ON<br>SYSTEM SELECT 1<br>TEST SW SELF TEST<br>SELECT SW A | Yes | Go to next step |
| | | No | **Malfunction Code No. displayed**<br>Check for cause (Refer to specified check sequence) |
| | | | **"88" flashes**<br>Check ECU terminal 1F voltage<br>**Voltage: Approx. 12V (Ignition switch ON)**<br>⇨ If OK, replace ECU<br>⇨ If not OK, check wiring between ECU and Self-Diagnosis Checker |
| 2 | Check switches for correct operation with Self-Diagnosis Checker Monitor Lamp and ignition switch ON<br>SELECT SW A<br>SYSTEM SELECT 1<br>TEST SW SELF TEST<br>MONITOR LAMP | Yes | Go to next step |
| | | No | **Lamp not ON/OFF with specified switch**<br>Check for cause (Refer to specified check sequence) |
| | | | **Lamp always ON**<br>Check wiring between ECU terminal 1D and Self-Diagnosis Checker |
| 3 | Check if ECU terminal 1C voltage is OK<br>**Voltage: Approx. 10V (While cranking)** | Yes | Go to next step |
| | | No | Check for cause (Refer to "Check Point for Each Terminal") |
| 4 | Check for correct ignition timing at idle<br>**Ignition timing: 10° ± 1° BTDC**<br>GND TEN<br>CONNECT TERMINALS | Yes | Check for correct idle speed<br>**Idle speed: 850 ± 50 rpm**<br>⇨ If OK, go to next step<br>⇨ If not OK, adjust idle speed |
| | | No | Adjust |
| 5 | Try known good ECU and check if condition improves | | |

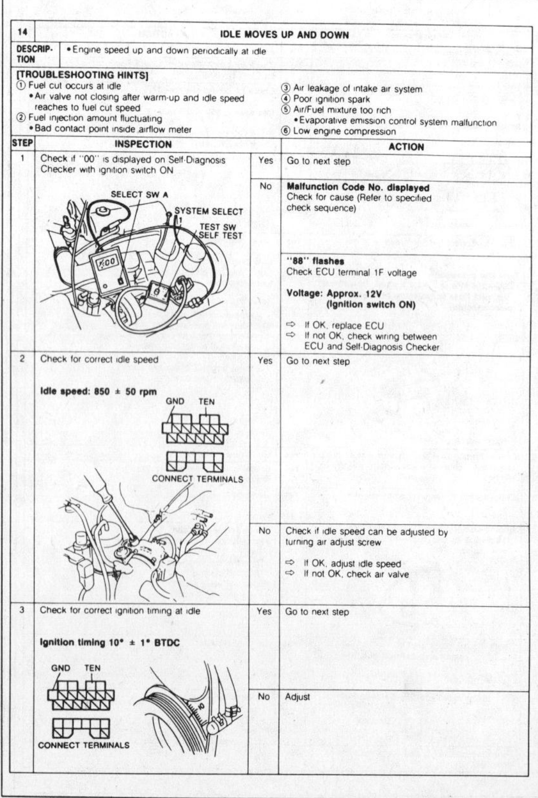

| 14 | IDLE MOVES UP AND DOWN | | |
|---|---|---|---|
| DESCRIPTION | • Engine speed up and down periodically at idle | | |
| [TROUBLESHOOTING HINTS] | ① Fuel cut occurs at idle<br>• Air valve not closing after warm-up and idle speed reaches to fuel cut speed<br>② Fuel injection amount fluctuating<br>• Bad contact point inside airflow meter<br>③ Air leakage of intake air system<br>④ Poor ignition spark<br>⑤ Air/Fuel mixture too rich<br>• Evaporative emission control system malfunction<br>⑥ Low engine compression | | |
| **STEP** | **INSPECTION** | | **ACTION** |
| 1 | Check if "00" is displayed on Self-Diagnosis Checker with ignition switch ON<br>SELECT SW A<br>SYSTEM SELECT 1<br>TEST SW SELF TEST | Yes | Go to next step |
| | | No | **Malfunction Code No. displayed**<br>Check for cause (Refer to specified check sequence) |
| | | | **"88" flashes**<br>Check ECU terminal 1F voltage<br>**Voltage: Approx. 12V (Ignition switch ON)**<br>⇨ If OK, replace ECU<br>⇨ If not OK, check wiring between ECU and Self-Diagnosis Checker |
| 2 | Check for correct idle speed<br>**Idle speed: 850 ± 50 rpm**<br>GND TEN<br>CONNECT TERMINALS | Yes | Go to next step |
| | | No | Check if idle speed can be adjusted by turning air adjust screw<br>⇨ If OK, adjust idle speed<br>⇨ If not OK, check air valve |
| 3 | Check for correct ignition timing at idle<br>**Ignition timing 10° ± 1° BTDC**<br>GND TEN<br>CONNECT TERMINALS | Yes | Go to next step |
| | | No | Adjust |

## TROUBLESHOOTING CHART—1990 MIATA

| STEP | INSPECTION | | ACTION |
|---|---|---|---|
| 4 | Check for correct intake manifold vacuum at idle<br>**Intake manifold vacuum: More than 450 mmHg (17.7 inHg)** | Yes | Go to next step |
| | | No | **Low vacuum**<br>Check for air leakage of intake air system |
| 5 | Check for injector operating sound at idle with sound scope or screwdriver | Yes | Go to Step 7 |
| | | No | Go to Step 6 |
| 6 | Check if approx. 12V exists at injector connector (W/R) wire | Yes | Check if injector resistance is OK<br>**Resistance: Approx. 14Ω**<br>⇨ If OK, check wiring between ECU and injector<br>⇨ If not OK, replace injector |
| | | No | Check wiring between ECU and injector |
| 7 | Disconnect each high-tension lead at idle and check if engine speed decreases equally each time | Yes | Disconnect each injector connector at idle and check if engine speed decreases equally each time<br>⇨ If OK, go to Step 9<br>⇨ If not OK, check injector for fuel leakage |
| | | No | Go to Step 8 |
| 8 | Check if spark plugs are OK<br>WEAR AND CARBON BUILDUP; PLUG GAP 1.0—1.1mm (0.039—0.043 in); BURNS; DAMAGE AND DETERIORATION; DAMAGE | Yes | Check for correct engine compression<br>⇨ If OK, replace injector<br>⇨ If not OK, check for cause (Refer to Section B) |
| | | No | Repair, clean, or replace |
| 9 | Check if ECU terminal voltages are OK (2D, 2Q and 2X) | Yes | Go to next step |
| | | No | Check for cause (Refer to "Check Point for Each Terminal") |
| 10 | Check if vacuum is felt at solenoid valve (purge control) at idle | Yes | Check solenoid valve |
| | | No | Go to next step |
| 11 | Try known good ECU and check if condition improves | | |

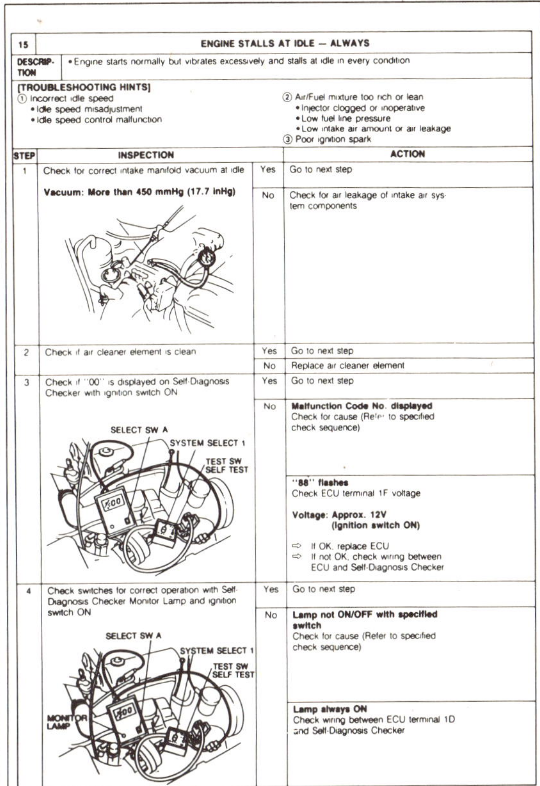

| 15 | ENGINE STALLS AT IDLE — ALWAYS | | |
|---|---|---|---|
| DESCRIPTION | • Engine starts normally but vibrates excessively and stalls at idle in every condition | | |
| [TROUBLESHOOTING HINTS] | ① Incorrect idle speed<br>• Idle speed misadjustment<br>• Idle speed control malfunction<br>② Air/Fuel mixture too rich or lean<br>• Injector clogged or inoperative<br>• Low fuel line pressure<br>• Low intake air amount or air leakage<br>③ Poor ignition spark | | |
| STEP | INSPECTION | | ACTION |
| 1 | Check for correct intake manifold vacuum at idle<br>**Vacuum: More than 450 mmHg (17.7 inHg)** | Yes | Go to next step |
| | | No | Check for air leakage of intake air system components |
| 2 | Check if air cleaner element is clean | Yes | Go to next step |
| | | No | Replace air cleaner element |
| 3 | Check if "00" is displayed on Self-Diagnosis Checker with ignition switch ON<br>SELECT SW A; SYSTEM SELECT 1; TEST SW SELF TEST | Yes | Go to next step |
| | | No | **Malfunction Code No. displayed**<br>Check for cause (Refer to specified check sequence)<br>**"88" flashes**<br>Check ECU terminal 1F voltage<br>**Voltage: Approx. 12V (Ignition switch ON)**<br>⇨ If OK, replace ECU<br>⇨ If not OK, check wiring between ECU and Self-Diagnosis Checker |
| 4 | Check switches for correct operation with Self-Diagnosis Checker Monitor Lamp and ignition switch ON<br>SELECT SW A; SYSTEM SELECT 1; TEST SW SELF TEST; MONITOR LAMP | Yes | Go to next step |
| | | No | **Lamp not ON/OFF with specified switch**<br>Check for cause (Refer to specified check sequence)<br>**Lamp always ON**<br>Check wiring between ECU terminal 1D and Self-Diagnosis Checker |

## TROUBLESHOOTING CHART—1990 MIATA

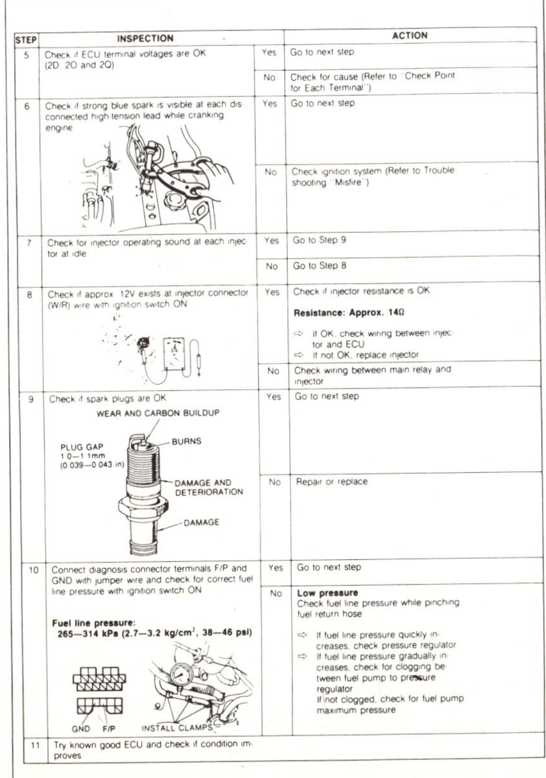

| STEP | INSPECTION | | ACTION |
|---|---|---|---|
| 5 | Check if ECU terminal voltages are OK (2D, 2O and 2Q) | Yes | Go to next step |
| | | No | Check for cause (Refer to "Check Point for Each Terminal") |
| 6 | Check if strong blue spark is visible at each disconnected high-tension lead while cranking engine | Yes | Go to next step |
| | | No | Check ignition system (Refer to Troubleshooting "Misfire") |
| 7 | Check for injector operating sound at each injector at idle | Yes | Go to Step 9 |
| | | No | Go to Step 8 |
| 8 | Check if approx. 12V exists at injector connector (W/R) wire with ignition switch ON | Yes | Check if injector resistance is OK<br>**Resistance: Approx. 14Ω**<br>⇨ If OK, check wiring between injector and ECU<br>⇨ If not OK, replace injector |
| | | No | Check wiring between main relay and injector |
| 9 | Check if spark plugs are OK<br>WEAR AND CARBON BUILDUP; PLUG GAP 1.0—1.1mm (0.039—0.043 in); BURNS; DAMAGE AND DETERIORATION; DAMAGE | Yes | Go to next step |
| | | No | Repair or replace |
| 10 | Connect diagnosis connector terminals F/P and GND with jumper wire and check for correct fuel line pressure with ignition switch ON<br>**Fuel line pressure: 265—314 kPa (2.7—3.2 kg/cm², 38—46 psi)**<br>GND; F/P; INSTALL CLAMPS | Yes | Go to next step |
| | | No | **Low pressure**<br>Check fuel line pressure while pinching fuel return hose<br>⇨ If fuel line pressure quickly increases, check pressure regulator<br>⇨ If fuel line pressure gradually increases, check for clogging between fuel pump to pressure regulator<br>If not clogged, check for fuel pump maximum pressure |
| 11 | Try known good ECU and check if condition improves | | |

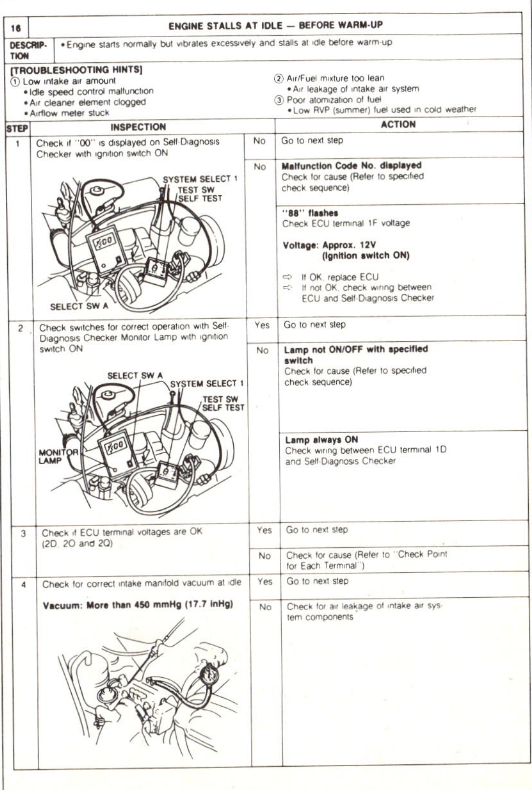

| 16 | ENGINE STALLS AT IDLE — BEFORE WARM-UP | | |
|---|---|---|---|
| DESCRIPTION | • Engine starts normally but vibrates excessively and stalls at idle before warm-up | | |
| [TROUBLESHOOTING HINTS] | ① Low intake air amount<br>• Idle speed control malfunction<br>• Air cleaner element clogged<br>• Airflow meter stuck<br>② Air/Fuel mixture too lean<br>• Air leakage of intake air system<br>③ Poor atomization of fuel<br>• Low RVP (summer) fuel used in cold weather | | |
| STEP | INSPECTION | | ACTION |
| 1 | Check if "00" is displayed on Self-Diagnosis Checker with ignition switch ON<br>SYSTEM SELECT 1; TEST SW SELF TEST; SELECT SW A | No | Go to next step |
| | | No | **Malfunction Code No. displayed**<br>Check for cause (Refer to specified check sequence)<br>**"88" flashes**<br>Check ECU terminal 1F voltage<br>**Voltage: Approx. 12V (Ignition switch ON)**<br>⇨ If OK, replace ECU<br>⇨ If not OK, check wiring between ECU and Self-Diagnosis Checker |
| 2 | Check switches for correct operation with Self-Diagnosis Checker Monitor Lamp with ignition switch ON<br>SELECT SW A; SYSTEM SELECT 1; TEST SW SELF TEST; MONITOR LAMP | Yes | Go to next step |
| | | No | **Lamp not ON/OFF with specified switch**<br>Check for cause (Refer to specified check sequence)<br>**Lamp always ON**<br>Check wiring between ECU terminal 1D and Self-Diagnosis Checker |
| 3 | Check if ECU terminal voltages are OK (2D, 2O and 2Q) | Yes | Go to next step |
| | | No | Check for cause (Refer to "Check Point for Each Terminal") |
| 4 | Check for correct intake manifold vacuum at idle<br>**Vacuum: More than 450 mmHg (17.7 inHg)** | Yes | Go to next step |
| | | No | Check for air leakage of intake air system components |

## TROUBLESHOOTING CHART—1990 MIATA

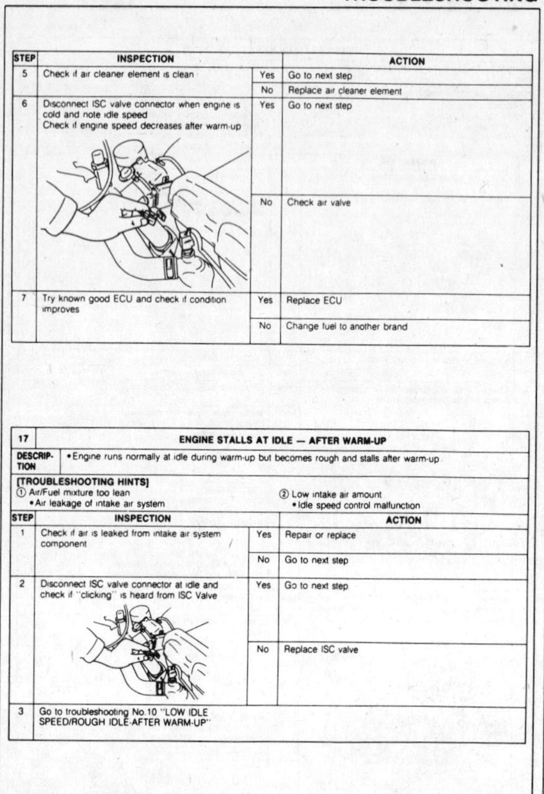

| STEP | INSPECTION | | ACTION |
|---|---|---|---|
| 5 | Check if air cleaner element is clean | Yes | Go to next step |
| | | No | Replace air cleaner element |
| 6 | Disconnect ISC valve connector when engine is cold and note idle speed<br>Check if engine speed decreases after warm-up | Yes | Go to next step |
| | | No | Check air valve |
| 7 | Try known good ECU and check if condition improves | Yes | Replace ECU |
| | | No | Change fuel to another brand |

| 17 | ENGINE STALLS AT IDLE — AFTER WARM-UP | | |
|---|---|---|---|
| DESCRIPTION | • Engine runs normally at idle during warm-up but becomes rough and stalls after warm-up | | |
| [TROUBLESHOOTING HINTS] | ① Air/Fuel mixture too lean<br>• Air leakage of intake air system<br>② Low intake air amount<br>• Idle speed control malfunction | | |
| **STEP** | **INSPECTION** | | **ACTION** |
| 1 | Check if air is leaked from intake air system component | Yes | Repair or replace |
| | | No | Go to next step |
| 2 | Disconnect ISC valve connector at idle and check if "clicking" is heard from ISC Valve | Yes | Go to next step |
| | | No | Replace ISC valve |
| 3 | Go to troubleshooting No.10 "LOW IDLE SPEED/ROUGH IDLE-AFTER WARM-UP" | | |

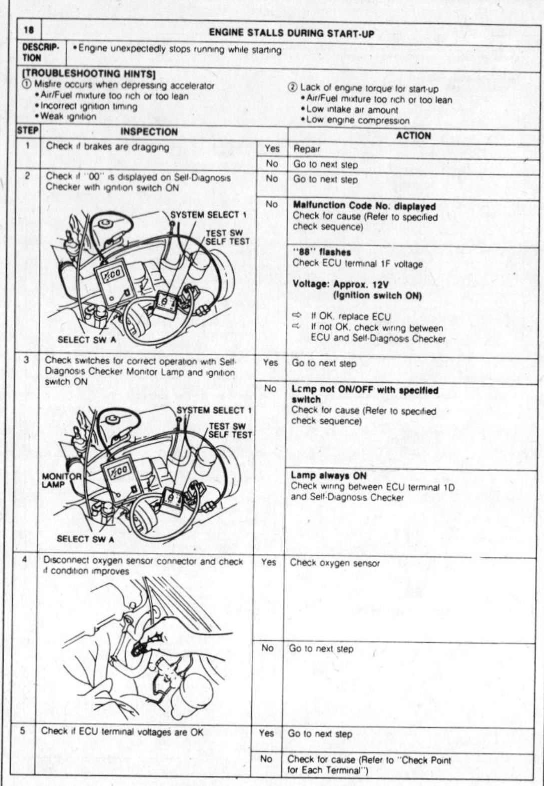

| 18 | ENGINE STALLS DURING START-UP | | |
|---|---|---|---|
| DESCRIPTION | • Engine unexpectedly stops running while starting | | |
| [TROUBLESHOOTING HINTS] | ① Misfire occurs when depressing accelerator<br>• Air/Fuel mixture too rich or too lean<br>• Incorrect ignition timing<br>• Weak ignition<br>② Lack of engine torque for start-up<br>• Air/Fuel mixture too rich or too lean<br>• Low intake air amount<br>• Low engine compression | | |
| **STEP** | **INSPECTION** | | **ACTION** |
| 1 | Check if brakes are dragging | Yes | Repair |
| | | No | Go to next step |
| 2 | Check if "00" is displayed on Self-Diagnosis Checker with ignition switch ON<br>SYSTEM SELECT 1<br>TEST SW SELF TEST<br>SELECT SW A | No | Go to next step |
| | | No | **Malfunction Code No. displayed**<br>Check for cause (Refer to specified check sequence)<br>**"88" flashes**<br>Check ECU terminal 1F voltage<br>**Voltage: Approx. 12V (Ignition switch ON)**<br>⇨ If OK, replace ECU<br>⇨ If not OK, check wiring between ECU and Self-Diagnosis Checker |
| 3 | Check switches for correct operation with Self-Diagnosis Checker Monitor Lamp and ignition switch ON<br>SYSTEM SELECT 1<br>TEST SW SELF TEST<br>MONITOR LAMP<br>SELECT SW A | Yes | Go to next step |
| | | No | **Lamp not ON/OFF with specified switch**<br>Check for cause (Refer to specified check sequence)<br>**Lamp always ON**<br>Check wiring between ECU terminal 1D and Self-Diagnosis Checker |
| 4 | Disconnect oxygen sensor connector and check if condition improves | Yes | Check oxygen sensor |
| | | No | Go to next step |
| 5 | Check if ECU terminal voltages are OK | Yes | Go to next step |
| | | No | Check for cause (Refer to "Check Point for Each Terminal") |

## TROUBLESHOOTING CHART—1990 MIATA

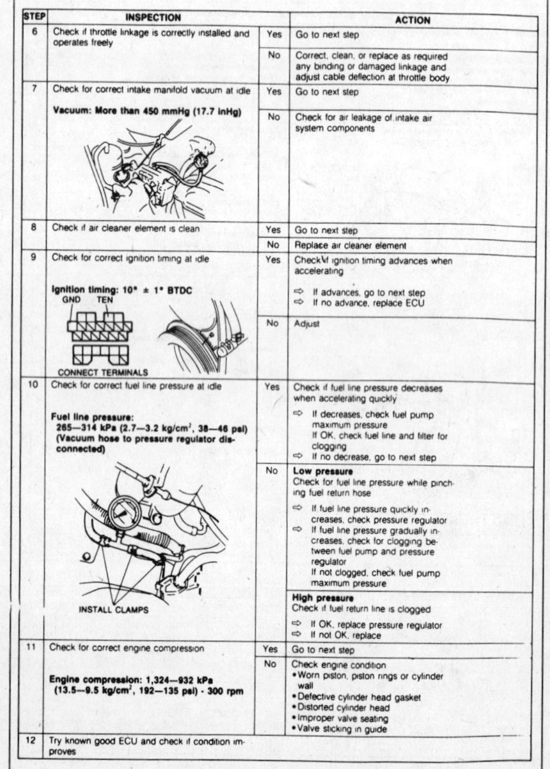

| STEP | INSPECTION | | ACTION |
|---|---|---|---|
| 6 | Check if throttle linkage is correctly installed and operates freely | Yes | Go to next step |
| | | No | Correct, clean, or replace as required any binding or damaged linkage and adjust cable deflection at throttle body |
| 7 | Check for correct intake manifold vacuum at idle<br>**Vacuum: More than 450 mmHg (17.7 inHg)** | Yes | Go to next step |
| | | No | Check for air leakage of intake air system components |
| 8 | Check if air cleaner element is clean | Yes | Go to next step |
| | | No | Replace air cleaner element |
| 9 | Check for correct ignition timing at idle<br>**Ignition timing: 10° ± 1° BTDC**<br>GND TEN<br>CONNECT TERMINALS | Yes | Check if ignition timing advances when accelerating<br>⇨ If advances, go to next step<br>⇨ If no advance, replace ECU |
| | | No | Adjust |
| 10 | Check for correct fuel line pressure at idle<br>**Fuel line pressure: 265—314 kPa (2.7—3.2 kg/cm², 38—46 psi) (Vacuum hose to pressure regulator disconnected)**<br>INSTALL CLAMPS | Yes | Check if fuel line pressure decreases when accelerating quickly<br>⇨ If decreases, check fuel pump maximum pressure<br>If OK, check fuel line and filter for clogging<br>⇨ If no decrease, go to next step |
| | | No | **Low pressure**<br>Check for fuel line pressure while pinching fuel return hose<br>⇨ If fuel line pressure quickly increases, check pressure regulator<br>⇨ If fuel line pressure gradually increases, check for clogging between fuel pump and pressure regulator<br>If not clogged, check fuel pump maximum pressure<br>**High pressure**<br>Check if fuel return line is clogged<br>⇨ If OK, replace pressure regulator<br>⇨ If not OK, replace |
| 11 | Check for correct engine compression<br>**Engine compression: 1,324—932 kPa (13.5—9.5 kg/cm², 192—135 psi) - 300 rpm** | Yes | Go to next step |
| | | No | Check engine condition<br>• Worn piston, piston rings or cylinder wall<br>• Defective cylinder head gasket<br>• Distorted cylinder head<br>• Improper valve seating<br>• Valve sticking in guide |
| 12 | Try known good ECU and check if condition improves | | |

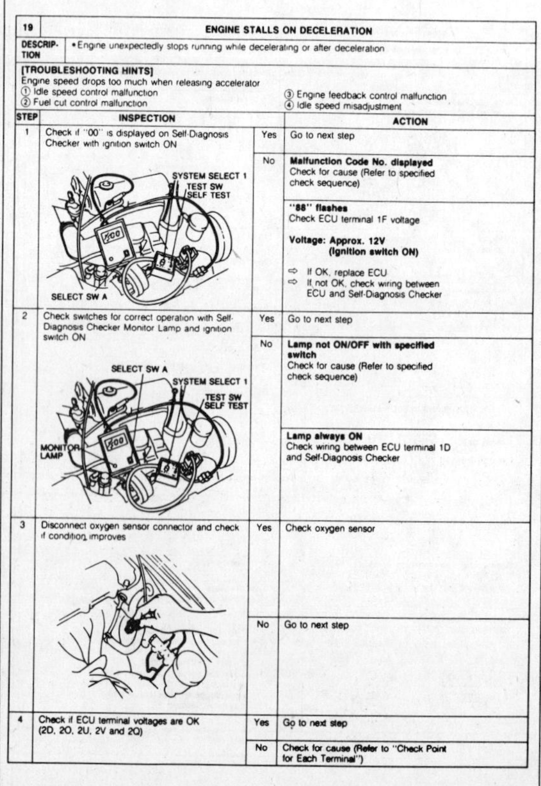

| 19 | ENGINE STALLS ON DECELERATION | | |
|---|---|---|---|
| DESCRIPTION | • Engine unexpectedly stops running while decelerating or after deceleration | | |
| [TROUBLESHOOTING HINTS] | Engine speed drops too much when releasing accelerator<br>① Idle speed control malfunction<br>② Fuel cut control malfunction<br>③ Engine feedback control malfunction<br>④ Idle speed misadjustment | | |
| **STEP** | **INSPECTION** | | **ACTION** |
| 1 | Check if "00" is displayed on Self-Diagnosis Checker with ignition switch ON<br>SYSTEM SELECT 1<br>TEST SW SELF TEST<br>SELECT SW A | Yes | Go to next step |
| | | No | **Malfunction Code No. displayed**<br>Check for cause (Refer to specified check sequence)<br>**"88" flashes**<br>Check ECU terminal 1F voltage<br>**Voltage: Approx. 12V (Ignition switch ON)**<br>⇨ If OK, replace ECU<br>⇨ If not OK, check wiring between ECU and Self-Diagnosis Checker |
| 2 | Check switches for correct operation with Self-Diagnosis Checker Monitor Lamp and ignition switch ON<br>SELECT SW A<br>SYSTEM SELECT 1<br>TEST SW SELF TEST<br>MONITOR LAMP | Yes | Go to next step |
| | | No | **Lamp not ON/OFF with specified switch**<br>Check for cause (Refer to specified check sequence)<br>**Lamp always ON**<br>Check wiring between ECU terminal 1D and Self-Diagnosis Checker |
| 3 | Disconnect oxygen sensor connector and check if condition improves | Yes | Check oxygen sensor |
| | | No | Go to next step |
| 4 | Check if ECU terminal voltages are OK (2D, 2O, 2U, 2V and 2Q) | Yes | Go to next step |
| | | No | Check for cause (Refer to "Check Point for Each Terminal") |

## TROUBLESHOOTING CHART—1990 MIATA

| STEP | INSPECTION | | ACTION |
|---|---|---|---|
| 5 | Disconnect ISC valve connector at idle and check if "clicking" is heard from ISC valve | Yes | Go to next step |
| | | No | Replace ISC valve |
| 6 | Check for correct idle speed<br>**Idle speed: 850 ± 50 rpm**<br>GND TEN<br>CONNECT TERMINALS | Yes | Go to next step |
| | | No | Adjust |
| 7 | Try known good ECU and check if condition improves | | |

| 20 | ENGINE STALLS AT IDLE — WHEN A/C, P/S, OR E/L ON | | |
|---|---|---|---|
| DESCRIPTION | • Engine unexpectedly stops running at idle when A/C, P/S, or E/L ON<br>• A/C, P/S, headlight, blower fan and electric cooling fan OK | | |
| [TROUBLESHOOTING HINTS]<br>① Idle speed control malfunction<br>• No input signal from switch<br>• Idle speed misadjustment<br>• ISC valve stuck | | | |
| **STEP** | **INSPECTION** | | **ACTION** |
| 1 | Check if "00" is displayed on Self-Diagnosis Checker with ignition switch ON<br>SYSTEM SELECT 1<br>TEST SW<br>SELF TEST<br>SELECT SW A | Yes | Go to next step |
| | | No | **Malfunction Code No. displayed**<br>Check for cause (Refer to specified check sequence)<br>**"88" flashes**<br>Check ECU terminal 1F voltage<br>**Voltage: Approx. 12V (Ignition switch ON)**<br>⇨ If OK, replace ECU<br>⇨ If not OK, check wiring between ECU and Self-Diagnosis Checker |
| 2 | Check switches for correct operation with Self-Diagnosis Checker Monitor Lamp and ignition switch ON<br>SELECT SW A<br>SYSTEM SELECT 1<br>TEST SW<br>SELF TEST<br>MONITOR LAMP | Yes | Go to next step |
| | | No | **Lamp not ON/OFF with specified switch**<br>Check for cause (Refer to specified check sequence)<br>**Lamp always ON**<br>Check wiring between ECU terminal 1D and Self-Diagnosis Checker |
| 3 | Check if ECU terminal voltages are OK (1G, 1P, 1U, 2D, 2Q and 2W) | Yes | Go to next step |
| | | No | Check for cause (Refer to "Check Point for Each Terminal") |
| 4 | Check for correct idle speed<br>**Idle speed: 850 ± 50 rpm**<br>GND TEN<br>CONNECT TERMINALS | Yes | Go to next step |
| | | No | Adjust |
| 5 | Disconnect ISC valve connector at idle and check if "clicking" is heard from ISC valve | Yes | Go to next step |
| | | No | Replace ISC valve |
| 6 | Try known good ECU and check if condition improves | | |

## TROUBLESHOOTING CHART—1990 MIATA

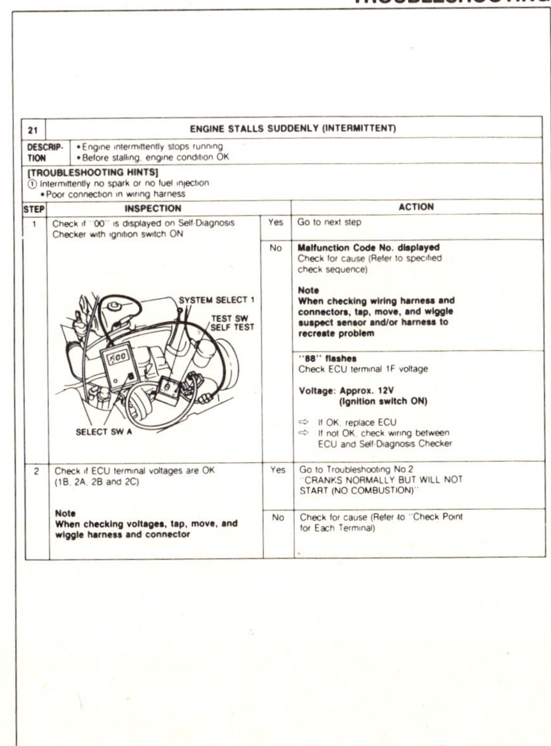

| 21 | ENGINE STALLS SUDDENLY (INTERMITTENT) | | |
|---|---|---|---|
| DESCRIPTION | • Engine intermittently stops running<br>• Before stalling, engine condition OK | | |
| [TROUBLESHOOTING HINTS]<br>① Intermittently no spark or no fuel injection<br>• Poor connection in wiring harness | | | |
| **STEP** | **INSPECTION** | | **ACTION** |
| 1 | Check if "00" is displayed on Self-Diagnosis Checker with ignition switch ON<br>SYSTEM SELECT 1<br>TEST SW<br>SELF TEST<br>SELECT SW A | Yes | Go to next step |
| | | No | **Malfunction Code No. displayed**<br>Check for cause (Refer to specified check sequence)<br>**Note**<br>**When checking wiring harness and connectors, tap, move, and wiggle suspect sensor and/or harness to recreate problem**<br>**"88" flashes**<br>Check ECU terminal 1F voltage<br>**Voltage: Approx. 12V (Ignition switch ON)**<br>⇨ If OK, replace ECU<br>⇨ If not OK, check wiring between ECU and Self-Diagnosis Checker |
| 2 | Check if ECU terminal voltages are OK (1B, 2A, 2B and 2C)<br>**Note**<br>**When checking voltages, tap, move, and wiggle harness and connector** | Yes | Go to Troubleshooting No.2 "CRANKS NORMALLY BUT WILL NOT START (NO COMBUSTION)" |
| | | No | Check for cause (Refer to "Check Point for Each Terminal) |

| 22 | HESITATES/STUMBLES ON ACCELERATION | | |
|---|---|---|---|
| DESCRIPTION | • Flat spot occurs just after accelerator depressed or mild jerking occurs during acceleration | | |
| [TROUBLESHOOTING HINTS]<br>① Air/Fuel mixture leaning when depressing accelerator<br>• Fuel injection control malfunction<br>(Correction for accelerating condition)<br>• Air leakage of intake air system<br>• Fuel line pressure low<br>• Spark advance control malfunction | | | |
| **STEP** | **INSPECTION** | | **ACTION** |
| 1 | Check if "00" is displayed on Self-Diagnosis Checker with ignition switch ON<br>SYSTEM SELECT 1<br>TEST SW<br>SELF TEST<br>SELECT SW A | Yes | Go to next step |
| | | No | **Malfunction Code No. displayed**<br>Check for cause (Refer to specified check sequence)<br>**"88" flashes**<br>Check ECU terminal 1F voltage<br>**Voltage: Approx. 12V (Ignition switch ON)**<br>⇨ If OK, replace ECU<br>⇨ If not OK, check wiring between ECU and Self-Diagnosis Checker |
| 2 | Check switches for correct operation with Self-Diagnosis Checker Monitor Lamp and ignition switch ON<br>SELECT SW A<br>SYSTEM SELECT 1<br>TEST SW<br>SELF TEST<br>MONITOR LAMP | Yes | Go to next step |
| | | No | **Lamp not ON/OFF with specified switch**<br>Check for cause (Refer to specified check sequence)<br>**Lamp always ON**<br>Check wiring between ECU terminal 1D and Self-Diagnosis Checker |
| 3 | Disconnect oxygen sensor connector and check if condition improves | Yes | Check oxygen sensor |
| | | No | Go to next step |
| 4 | Check if ECU terminal voltages are OK | Yes | Go to next step |
| | | No | Check for cause (Refer to "Check Point for Each Terminal") |
| 5 | Check if throttle linkage is correctly installed and operates freely | Yes | Go to next step |
| | | No | Correct, clean, or replace as required any binding or damaged linkage and adjust cable deflection at throttle body |

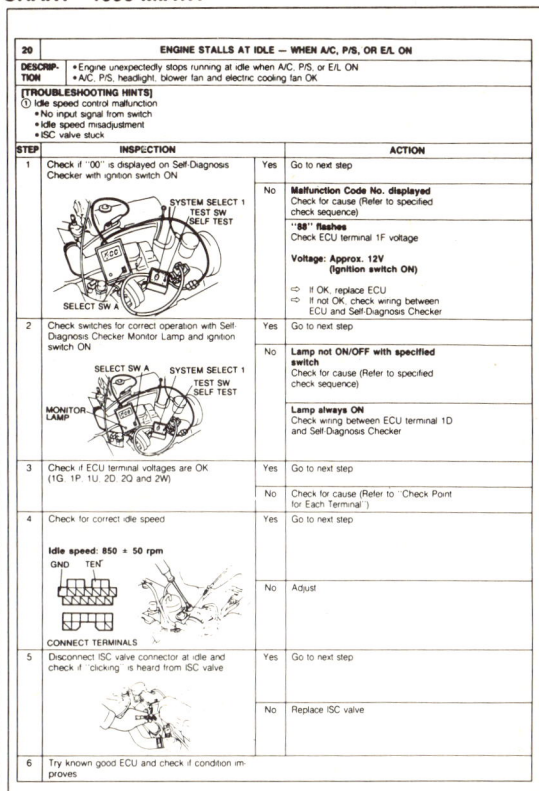

## TROUBLESHOOTING CHART—1990 MIATA

| STEP | INSPECTION | | ACTION |
|---|---|---|---|
| 6 | Check if air duct and air hoses are correctly installed | Yes | Go to next step |
| | | No | Repair |
| 7 | Check for correct intake manifold vacuum at idle<br>**Vacuum: More than 450 mmHg (17.7 inHg)** | Yes | Go to next step |
| | | No | Check for air leakage of intake air system components |
| 8 | Check if air cleaner element is clean | Yes | Go to next step |
| | | No | Replace air cleaner element |
| 9 | Check for correct ignition timing at idle<br>**Ignition timing: 10° ± 1° BTDC**<br>GND TEN<br>CONNECT TERMINALS | Yes | Check if ignition timing advances when accelerating<br>⇨ If advances, go to next step<br>⇨ If no advance, replace ECU |
| | | No | Adjust |
| 10 | Check for correct fuel line pressure at idle<br>**Fuel line pressure: 265—314 kPa (2.7—3.2 kg/cm², 38—46 psi) (Vacuum hose to pressure regulator disconnected)**<br>INSTALL CLAMPS | Yes | Check if fuel line pressure decreases when accelerating quickly<br>⇨ If decreases, check fuel line and filter for clogging<br>⇨ If no decrease, go to next step |
| | | No | **Low pressure**<br>Check for fuel line pressure while pinching fuel return hose<br>⇨ If fuel line pressure quickly increases, check pressure regulator<br>⇨ If fuel line pressure gradually increases, check for clogging between fuel pump and pressure regulator<br>If not clogged, check fuel pump maximum pressure |
| 11 | Check if exhaust system is restricted | Yes | Repair or replace |
| | | No | Go to next step |
| 12 | Try known good ECU and check if condition improves | | |

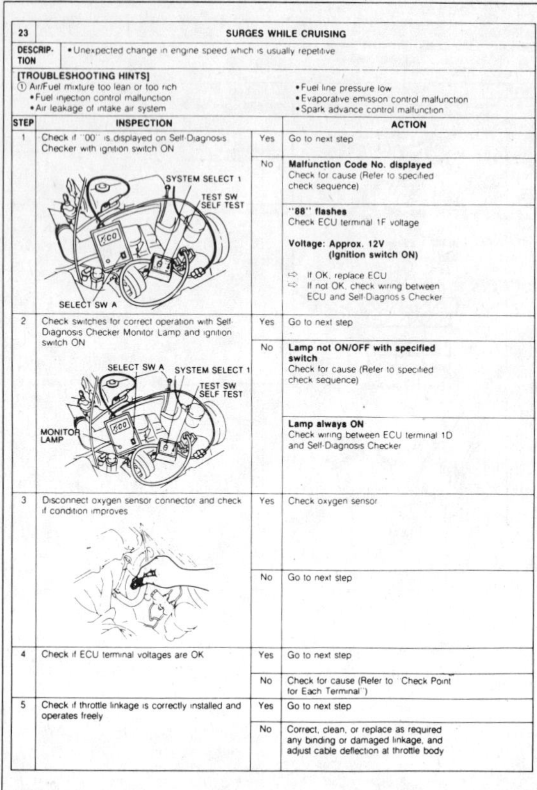

| 23 | SURGES WHILE CRUISING | | |
|---|---|---|---|
| DESCRIPTION | • Unexpected change in engine speed which is usually repetitive | | |
| [TROUBLESHOOTING HINTS]<br>① Air/Fuel mixture too lean or too rich<br>• Fuel injection control malfunction<br>• Air leakage of intake air system<br>• Fuel line pressure low<br>• Evaporative emission control malfunction<br>• Spark advance control malfunction | | | |
| **STEP** | **INSPECTION** | | **ACTION** |
| 1 | Check if "00" is displayed on Self-Diagnosis Checker with ignition switch ON<br>SYSTEM SELECT 1<br>TEST SW SELF TEST<br>SELECT SW A | Yes | Go to next step |
| | | No | **Malfunction Code No. displayed**<br>Check for cause (Refer to specified check sequence)<br>**"88" flashes**<br>Check ECU terminal 1F voltage<br>**Voltage: Approx. 12V (Ignition switch ON)**<br>⇨ If OK, replace ECU<br>⇨ If not OK, check wiring between ECU and Self-Diagnosis Checker |
| 2 | Check switches for correct operation with Self-Diagnosis Checker Monitor Lamp and ignition switch ON<br>SELECT SW A<br>SYSTEM SELECT 1<br>TEST SW SELF TEST<br>MONITOR LAMP | Yes | Go to next step |
| | | No | **Lamp not ON/OFF with specified switch**<br>Check for cause (Refer to specified check sequence)<br>**Lamp always ON**<br>Check wiring between ECU terminal 1D and Self-Diagnosis Checker |
| 3 | Disconnect oxygen sensor connector and check if condition improves | Yes | Check oxygen sensor |
| | | No | Go to next step |
| 4 | Check if ECU terminal voltages are OK | Yes | Go to next step |
| | | No | Check for cause (Refer to "Check Point for Each Terminal") |
| 5 | Check if throttle linkage is correctly installed and operates freely | Yes | Go to next step |
| | | No | Correct, clean, or replace as required any binding or damaged linkage, and adjust cable deflection at throttle body |

## TROUBLESHOOTING CHART—1990 MIATA

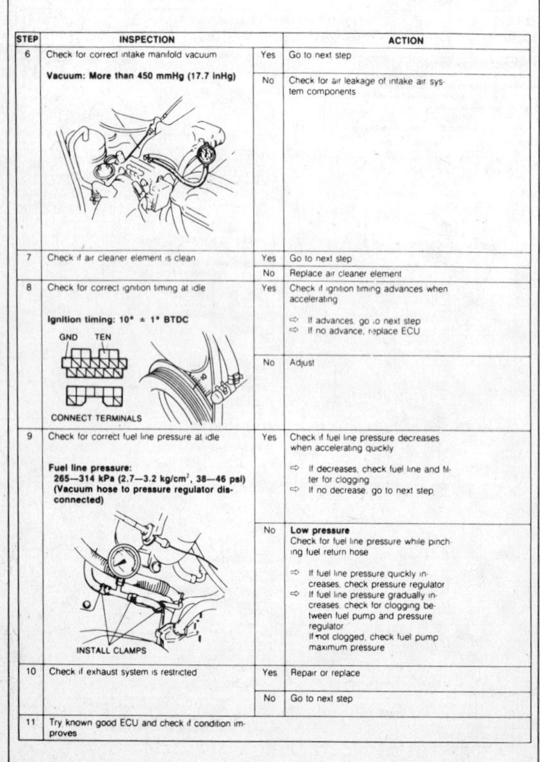

| STEP | INSPECTION | | ACTION |
|---|---|---|---|
| 6 | Check for correct intake manifold vacuum<br>**Vacuum: More than 450 mmHg (17.7 inHg)** | Yes | Go to next step |
| | | No | Check for air leakage of intake air system components |
| 7 | Check if air cleaner element is clean | Yes | Go to next step |
| | | No | Replace air cleaner element |
| 8 | Check for correct ignition timing at idle<br>**Ignition timing: 10° ± 1° BTDC**<br>GND TEN<br>CONNECT TERMINALS | Yes | Check if ignition timing advances when accelerating<br>⇨ If advances, go to next step<br>⇨ If no advance, replace ECU |
| | | No | Adjust |
| 9 | Check for correct fuel line pressure at idle<br>**Fuel line pressure: 265—314 kPa (2.7—3.2 kg/cm², 38—46 psi) (Vacuum hose to pressure regulator disconnected)**<br>INSTALL CLAMPS | Yes | Check if fuel line pressure decreases when accelerating quickly<br>⇨ If decreases, check fuel line and filter for clogging<br>⇨ If no decrease, go to next step |
| | | No | **Low pressure**<br>Check for fuel line pressure while pinching fuel return hose<br>⇨ If fuel line pressure quickly increases, check pressure regulator<br>⇨ If fuel line pressure gradually increases, check for clogging between fuel pump and pressure regulator<br>If not clogged, check fuel pump maximum pressure |
| 10 | Check if exhaust system is restricted | Yes | Repair or replace |
| | | No | Go to next step |
| 11 | Try known good ECU and check if condition improves | | |

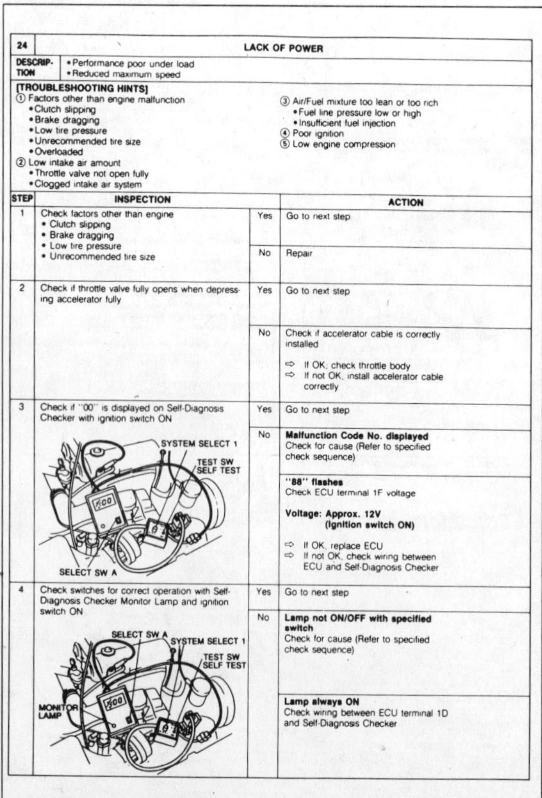

| 24 | LACK OF POWER | | |
|---|---|---|---|
| DESCRIPTION | • Performance poor under load<br>• Reduced maximum speed | | |
| [TROUBLESHOOTING HINTS]<br>① Factors other than engine malfunction<br>• Clutch slipping<br>• Brake dragging<br>• Low tire pressure<br>• Unrecommended tire size<br>• Overloaded<br>② Low intake air amount<br>• Throttle valve not open fully<br>• Clogged intake air system<br>③ Air/Fuel mixture too lean or too rich<br>• Fuel line pressure low or high<br>• Insufficient fuel injection<br>④ Poor ignition<br>⑤ Low engine compression | | | |
| **STEP** | **INSPECTION** | | **ACTION** |
| 1 | Check factors other than engine<br>• Clutch slipping<br>• Brake dragging<br>• Low tire pressure<br>• Unrecommended tire size | Yes | Go to next step |
| | | No | Repair |
| 2 | Check if throttle valve fully opens when depressing accelerator fully | Yes | Go to next step |
| | | No | Check if accelerator cable is correctly installed<br>⇨ If OK, check throttle body<br>⇨ If not OK, install accelerator cable correctly |
| 3 | Check if "00" is displayed on Self-Diagnosis Checker with ignition switch ON<br>SYSTEM SELECT 1<br>TEST SW SELF TEST<br>SELECT SW A | Yes | Go to next step |
| | | No | **Malfunction Code No. displayed**<br>Check for cause (Refer to specified check sequence)<br>**"88" flashes**<br>Check ECU terminal 1F voltage<br>**Voltage: Approx. 12V (Ignition switch ON)**<br>⇨ If OK, replace ECU<br>⇨ If not OK, check wiring between ECU and Self-Diagnosis Checker |
| 4 | Check switches for correct operation with Self-Diagnosis Checker Monitor Lamp and ignition switch ON<br>SELECT SW A<br>SYSTEM SELECT 1<br>TEST SW SELF TEST<br>MONITOR LAMP | Yes | Go to next step |
| | | No | **Lamp not ON/OFF with specified switch**<br>Check for cause (Refer to specified check sequence)<br>**Lamp always ON**<br>Check wiring between ECU terminal 1D and Self-Diagnosis Checker |

## TROUBLESHOOTING CHART—1990 MIATA

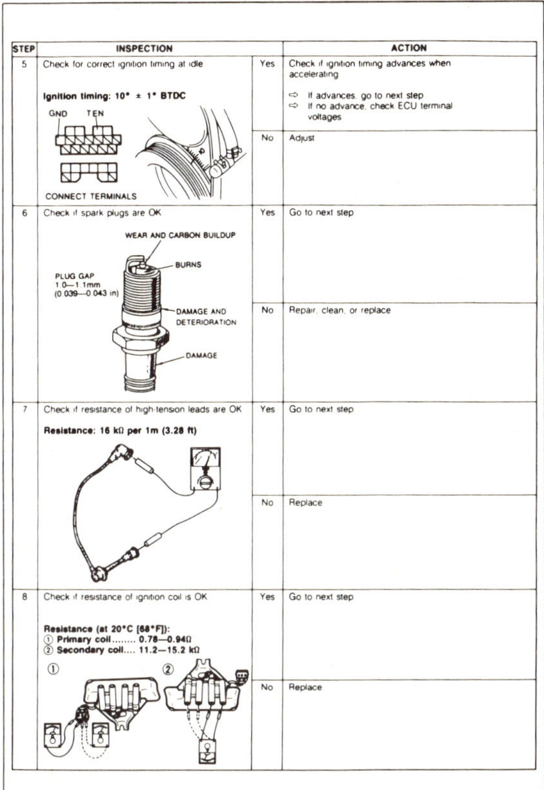

| STEP | INSPECTION | | ACTION |
|---|---|---|---|
| 5 | Check for correct ignition timing at idle<br>**Ignition timing: 10° ± 1° BTDC**<br>GND TEN<br>CONNECT TERMINALS | Yes | Check if ignition timing advances when accelerating<br>⇨ If advances, go to next step<br>⇨ If no advance, check ECU terminal voltages |
| | | No | Adjust |
| 6 | Check if spark plugs are OK<br>WEAR AND CARBON BUILDUP<br>BURNS<br>PLUG GAP 1.0—1.1mm (0.039—0.043 in)<br>DAMAGE AND DETERIORATION<br>DAMAGE | Yes | Go to next step |
| | | No | Repair, clean, or replace |
| 7 | Check if resistance of high-tension leads are OK<br>**Resistance: 16 kΩ per 1m (3.28 ft)** | Yes | Go to next step |
| | | No | Replace |
| 8 | Check if resistance of ignition coil is OK<br>**Resistance (at 20°C [68°F]):**<br>**① Primary coil........ 0.78—0.94Ω**<br>**② Secondary coil.... 11.2—15.2 kΩ** | Yes | Go to next step |
| | | No | Replace |

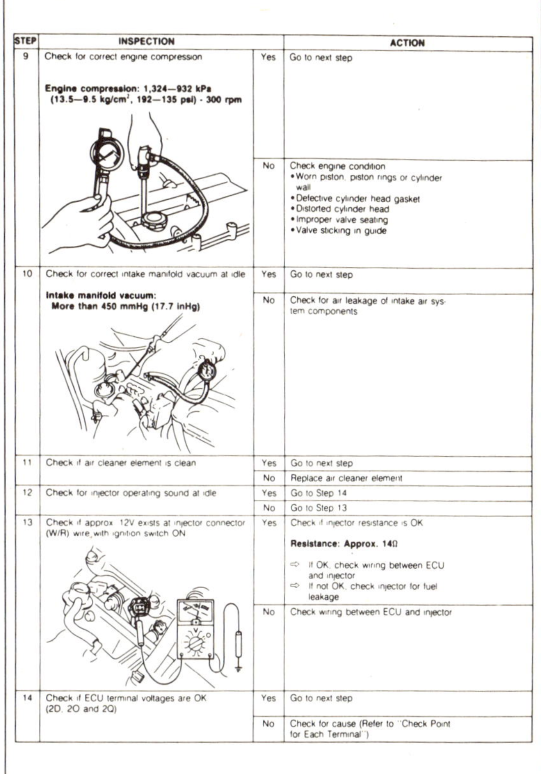

| STEP | INSPECTION | | ACTION |
|---|---|---|---|
| 9 | Check for correct engine compression<br>**Engine compression: 1,324—932 kPa (13.5—9.5 kg/cm², 192—135 psi) - 300 rpm** | Yes | Go to next step |
| | | No | Check engine condition<br>• Worn piston, piston rings or cylinder wall<br>• Defective cylinder head gasket<br>• Distorted cylinder head<br>• Improper valve seating<br>• Valve sticking in guide |
| 10 | Check for correct intake manifold vacuum at idle<br>**Intake manifold vacuum: More than 450 mmHg (17.7 inHg)** | Yes | Go to next step |
| | | No | Check for air leakage of intake air system components |
| 11 | Check if air cleaner element is clean | Yes | Go to next step |
| | | No | Replace air cleaner element |
| 12 | Check for injector operating sound at idle | Yes | Go to Step 14 |
| | | No | Go to Step 13 |
| 13 | Check if approx. 12V exists at injector connector (W/R) wire with ignition switch ON | Yes | Check if injector resistance is OK<br>**Resistance: Approx. 14Ω**<br>⇨ If OK, check wiring between ECU and injector<br>⇨ If not OK, check injector for fuel leakage |
| | | No | Check wiring between ECU and injector |
| 14 | Check if ECU terminal voltages are OK (2D, 2O and 2Q) | Yes | Go to next step |
| | | No | Check for cause (Refer to "Check Point for Each Terminal") |

## TROUBLESHOOTING CHART—1990 MIATA

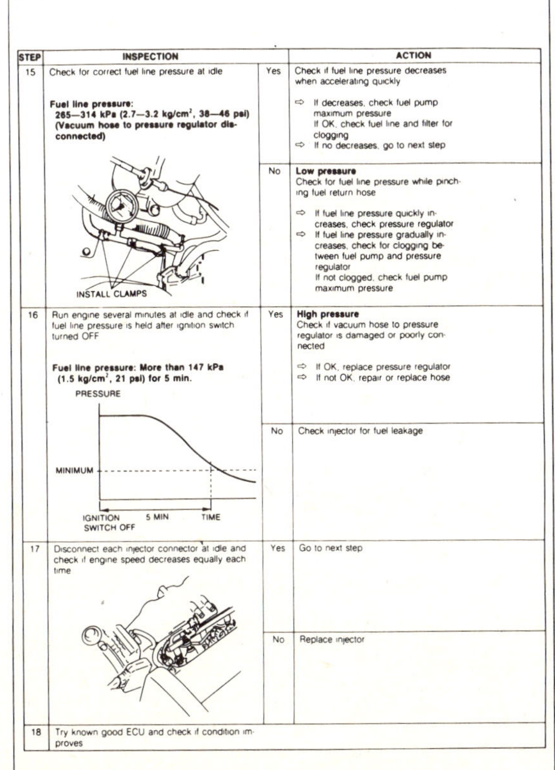

| STEP | INSPECTION | | ACTION |
|---|---|---|---|
| 15 | Check for correct fuel line pressure at idle<br>**Fuel line pressure: 265—314 kPa (2.7—3.2 kg/cm², 38—46 psi) (Vacuum hose to pressure regulator disconnected)**<br>INSTALL CLAMPS | Yes | Check if fuel line pressure decreases when accelerating quickly<br>⇨ If decreases, check fuel pump maximum pressure<br>If OK, check fuel line and filter for clogging<br>⇨ If no decreases, go to next step |
| | | No | **Low pressure**<br>Check for fuel line pressure while pinching fuel return hose<br>⇨ If fuel line pressure quickly increases, check pressure regulator<br>⇨ If fuel line pressure gradually increases, check for clogging between fuel pump and pressure regulator<br>If not clogged, check fuel pump maximum pressure |
| 16 | Run engine several minutes at idle and check if fuel line pressure is held after ignition switch turned OFF<br>**Fuel line pressure: More than 147 kPa (1.5 kg/cm², 21 psi) for 5 min.**<br>PRESSURE<br>MINIMUM<br>IGNITION SWITCH OFF<br>5 MIN<br>TIME | Yes | **High pressure**<br>Check if vacuum hose to pressure regulator is damaged or poorly connected<br>⇨ If OK, replace pressure regulator<br>⇨ If not OK, repair or replace hose |
| | | No | Check injector for fuel leakage |
| 17 | Disconnect each injector connector at idle and check if engine speed decreases equally each time | Yes | Go to next step |
| | | No | Replace injector |
| 18 | Try known good ECU and check if condition improves | | |

| 25 | POOR ACCELERATION |
|---|---|
| DESCRIPTION | • Performance poor while accelerating |

**[TROUBLESHOOTING HINTS]**

① Factors other than engine malfunction
• Clutch slipping
• Brake dragging
• Low tire pressure
• Unrecommended tire size
• Overloaded

② Low intake air amount
• Throttle valve not open fully
• Clogged intake air system

③ Air/Fuel mixture too lean or too rich
• Fuel line pressure low or high
• Insufficient fuel injection

④ Poor ignition

⑤ Low engine compression

| STEP | INSPECTION | | ACTION |
|---|---|---|---|
| 1 | Check factors other than engine<br>• Clutch slipping<br>• Brake dragging<br>• Low tire pressure<br>• Unrecommended tire size | Yes | Go to next step |
| | | No | Repair |
| 2 | Check if throttle valve fully opens when depressing accelerator fully | Yes | Go to next step |
| | | No | Check if accelerator cable is correctly installed<br>⇨ If OK, check throttle body<br>⇨ If not OK, install accelerator cable correctly |
| 3 | Check if "00" is displayed on Self-Diagnosis Checker with ignition switch ON<br>SYSTEM SELECT 1<br>TEST SW SELF TEST<br>SELECT SW A | Yes | Go to next step |
| | | No | **Malfunction Code No. displayed**<br>Check for cause (Refer to specified check sequence)<br>**"88" flashes**<br>Check ECU terminal 1F voltage<br>**Voltage: Approx. 12V (Ignition switch ON)**<br>⇨ If OK, replace ECU<br>⇨ If not OK, check wiring between ECU and Self-Diagnosis Checker |
| 4 | Check switches for correct operation with Self-Diagnosis Checker Monitor Lamp and ignition switch ON<br>SELECT SW A<br>SYSTEM SELECT 1<br>TEST SW SELF TEST<br>MONITOR LAMP | Yes | Go to next step |
| | | No | **Lamp not ON/OFF with specified switch**<br>Check for cause (Refer to specified check sequence)<br>**Lamp always ON**<br>Check wiring between ECU terminal 1D and Self-Diagnosis Checker |

## TROUBLESHOOTING CHART—1990 MIATA

| STEP | INSPECTION | | ACTION |
|---|---|---|---|
| 5 | Check for correct ignition timing at idle<br>**Ignition timing: 10° ± 1° BTDC** | Yes | Check if ignition timing advances when accelerating<br>⇨ If advances, go to next step<br>⇨ If no advance, check ECU terminal voltages |
| | | No | Adjust |
| 6 | Check if spark plugs are OK | Yes | Go to next step |
| | | No | Repair, clean, or replace |
| 7 | Check if resistance of high-tension leads are OK<br>**Resistance: 16 kΩ per 1m (3.28 ft)** | Yes | Go to next step |
| | | No | Replace |
| 8 | Check if resistance of ignition coil is OK<br>**Resistance (at 20°C [68°F]):**<br>**① Primary coil........ 0.78—0.94Ω**<br>**② Secondary coil.... 11.2—15.2 kΩ** | Yes | Go to next step |
| | | No | Replace |

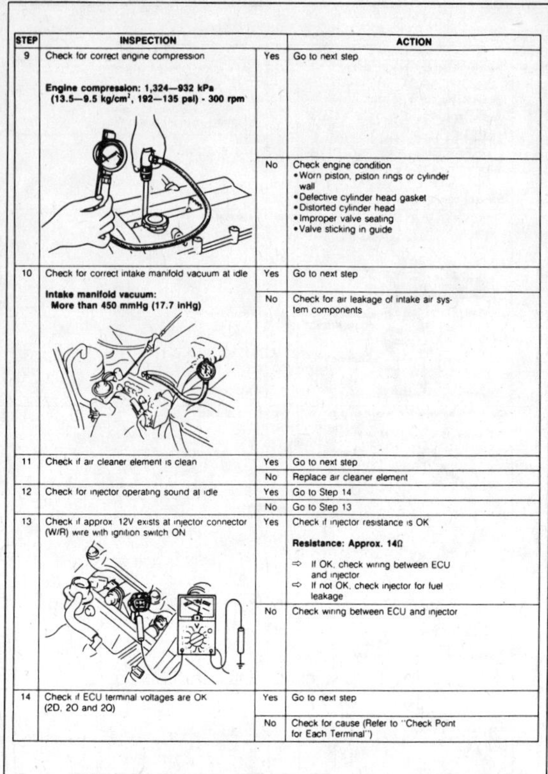

| STEP | INSPECTION | | ACTION |
|---|---|---|---|
| 9 | Check for correct engine compression<br>**Engine compression: 1,324—932 kPa (13.5—9.5 kg/cm², 192—135 psi) - 300 rpm** | Yes | Go to next step |
| | | No | Check engine condition<br>• Worn piston, piston rings or cylinder wall<br>• Defective cylinder head gasket<br>• Distorted cylinder head<br>• Improper valve seating<br>• Valve sticking in guide |
| 10 | Check for correct intake manifold vacuum at idle<br>**Intake manifold vacuum: More than 450 mmHg (17.7 inHg)** | Yes | Go to next step |
| | | No | Check for air leakage of intake air system components |
| 11 | Check if air cleaner element is clean | Yes | Go to next step |
| | | No | Replace air cleaner element |
| 12 | Check for injector operating sound at idle | Yes | Go to Step 14 |
| | | No | Go to Step 13 |
| 13 | Check if approx. 12V exists at injector connector (W/R) wire with ignition switch ON | Yes | Check if injector resistance is OK<br>**Resistance: Approx. 14Ω**<br>⇨ If OK, check wiring between ECU and injector<br>⇨ If not OK, check injector for fuel leakage |
| | | No | Check wiring between ECU and injector |
| 14 | Check if ECU terminal voltages are OK (2D, 2O and 2Q) | Yes | Go to next step |
| | | No | Check for cause (Refer to "Check Point for Each Terminal") |

## TROUBLESHOOTING CHART—1990 MIATA

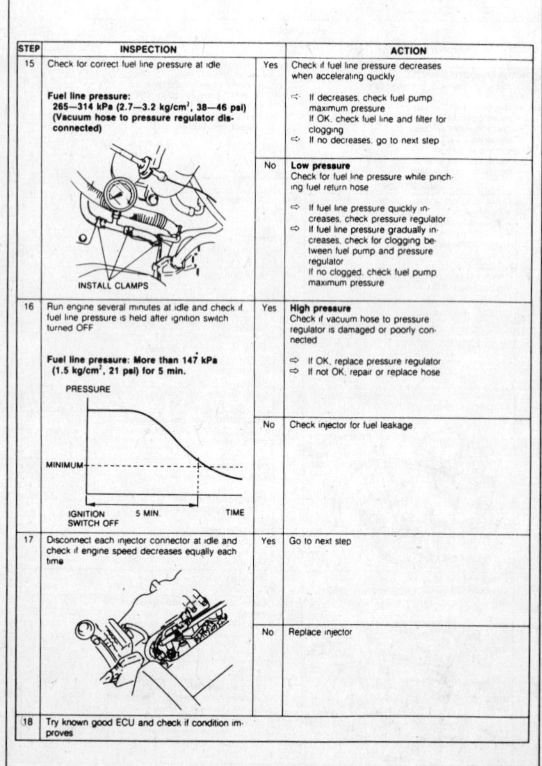

| STEP | INSPECTION | | ACTION |
|---|---|---|---|
| 15 | Check for correct fuel line pressure at idle<br>**Fuel line pressure: 265—314 kPa (2.7—3.2 kg/cm², 38—46 psi) (Vacuum hose to pressure regulator disconnected)** | Yes | Check if fuel line pressure decreases when accelerating quickly<br>⇨ If decreases, check fuel pump maximum pressure<br>If OK, check fuel line and filter for clogging<br>⇨ If no decreases, go to next step |
| | | No | **Low pressure**<br>Check for fuel line pressure while pinching fuel return hose<br>⇨ If fuel line pressure quickly increases, check pressure regulator<br>⇨ If fuel line pressure gradually increases, check for clogging between fuel pump and pressure regulator<br>If no clogged, check fuel pump maximum pressure |
| 16 | Run engine several minutes at idle and check if fuel line pressure is held after ignition switch turned OFF<br>**Fuel line pressure: More than 147 kPa (1.5 kg/cm², 21 psi) for 5 min.** | Yes | **High pressure**<br>Check if vacuum hose to pressure regulator is damaged or poorly connected<br>⇨ If OK, replace pressure regulator<br>⇨ If not OK, repair or replace hose |
| | | No | Check injector for fuel leakage |
| 17 | Disconnect each injector connector at idle and check if engine speed decreases equally each time | Yes | Go to next step |
| | | No | Replace injector |
| 18 | Try known good ECU and check if condition improves | | |

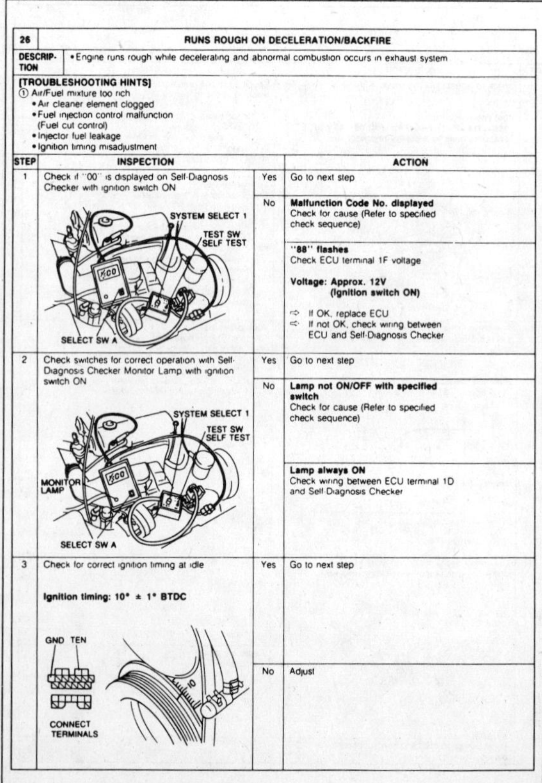

| 26 | RUNS ROUGH ON DECELERATION/BACKFIRE | | |
|---|---|---|---|
| DESCRIPTION | • Engine runs rough while decelerating and abnormal combustion occurs in exhaust system | | |
| **[TROUBLESHOOTING HINTS]**<br>① Air/Fuel mixture too rich<br>• Air cleaner element clogged<br>• Fuel injection control malfunction (Fuel cut control)<br>• Injector fuel leakage<br>• Ignition timing misadjustment | | | |
| **STEP** | **INSPECTION** | | **ACTION** |
| 1 | Check if "00" is displayed on Self-Diagnosis Checker with ignition switch ON | Yes | Go to next step |
| | | No | **Malfunction Code No. displayed**<br>Check for cause (Refer to specified check sequence) |
| | | | **"88" flashes**<br>Check ECU terminal 1F voltage<br>**Voltage: Approx. 12V (Ignition switch ON)**<br>⇨ If OK, replace ECU<br>⇨ If not OK, check wiring between ECU and Self-Diagnosis Checker |
| 2 | Check switches for correct operation with Self-Diagnosis Checker Monitor Lamp with ignition switch ON | Yes | Go to next step |
| | | No | **Lamp not ON/OFF with specified switch**<br>Check for cause (Refer to specified check sequence) |
| | | | **Lamp always ON**<br>Check wiring between ECU terminal 1D and Self-Diagnosis Checker |
| 3 | Check for correct ignition timing at idle<br>**Ignition timing: 10° ± 1° BTDC** | Yes | Go to next step |
| | | No | Adjust |

## TROUBLESHOOTING CHART—1990 MIATA

| STEP | INSPECTION | | ACTION |
|---|---|---|---|
| 4 | Check if fuel cut operation is OK during deceleration<br>**Fuel cut: Above 1,900 rpm after warm-up** | Yes | Go to next step |
| | | No | Try known good ECU |
| 5 | Run engine several minutes at idle and check if fuel line pressure is held after ignition switch turned OFF<br>**Fuel line pressure: More than 147 kPa (1.5 kg/cm², 21 psi) for 5 min.**<br>INSTALL CLAMPS | Yes | Go to next step |
| | | No | Check injector for fuel leakage |
| 6 | Check if air cleaner element is clean | Yes | Go to next step |
| | | No | Replace |
| 7 | Try known good ECU and check if condition improves | | |

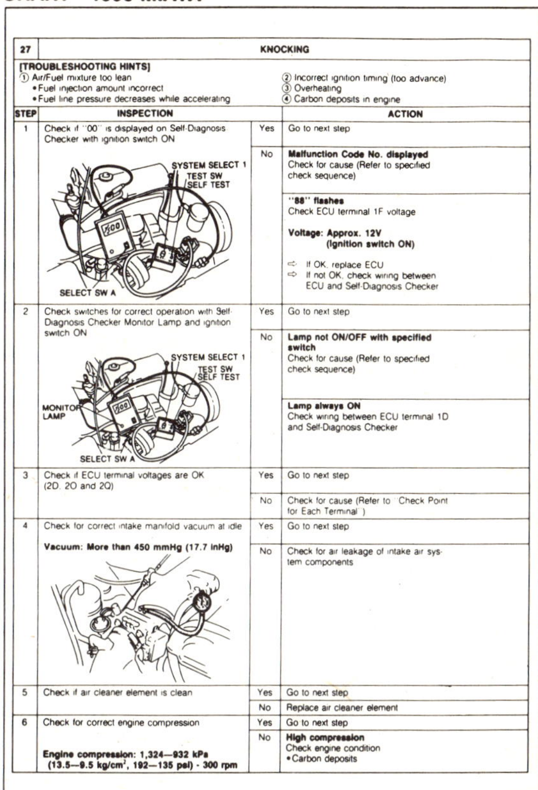

| 27 | KNOCKING | | |
|---|---|---|---|
| [TROUBLESHOOTING HINTS]<br>① Air/Fuel mixture too lean<br>• Fuel injection amount incorrect<br>• Fuel line pressure decreases while accelerating<br>② Incorrect ignition timing (too advance)<br>③ Overheating<br>④ Carbon deposits in engine | | | |
| **STEP** | **INSPECTION** | | **ACTION** |
| 1 | Check if "00" is displayed on Self-Diagnosis Checker with ignition switch ON<br>SYSTEM SELECT 1<br>TEST SW<br>SELF TEST<br>SELECT SW A | Yes | Go to next step |
| | | No | **Malfunction Code No. displayed**<br>Check for cause (Refer to specified check sequence) |
| | | | **"88" flashes**<br>Check ECU terminal 1F voltage<br>**Voltage: Approx. 12V (Ignition switch ON)**<br>⇨ If OK, replace ECU<br>⇨ If not OK, check wiring between ECU and Self-Diagnosis Checker |
| 2 | Check switches for correct operation with Self-Diagnosis Checker Monitor Lamp and ignition switch ON<br>SYSTEM SELECT 1<br>TEST SW<br>SELF TEST<br>MONITOR LAMP<br>SELECT SW A | Yes | Go to next step |
| | | No | **Lamp not ON/OFF with specified switch**<br>Check for cause (Refer to specified check sequence) |
| | | | **Lamp always ON**<br>Check wiring between ECU terminal 1D and Self-Diagnosis Checker |
| 3 | Check if ECU terminal voltages are OK (2D, 2O and 2Q) | Yes | Go to next step |
| | | No | Check for cause (Refer to "Check Point for Each Terminal") |
| 4 | Check for correct intake manifold vacuum at idle<br>**Vacuum: More than 450 mmHg (17.7 inHg)** | Yes | Go to next step |
| | | No | Check for air leakage of intake air system components |
| 5 | Check if air cleaner element is clean | Yes | Go to next step |
| | | No | Replace air cleaner element |
| 6 | Check for correct engine compression<br>**Engine compression: 1,324—932 kPa (13.5—9.5 kg/cm², 192—135 psi) - 300 rpm** | Yes | Go to next step |
| | | No | **High compression**<br>Check engine condition<br>• Carbon deposits |

## TROUBLESHOOTING CHART—1990 MIATA

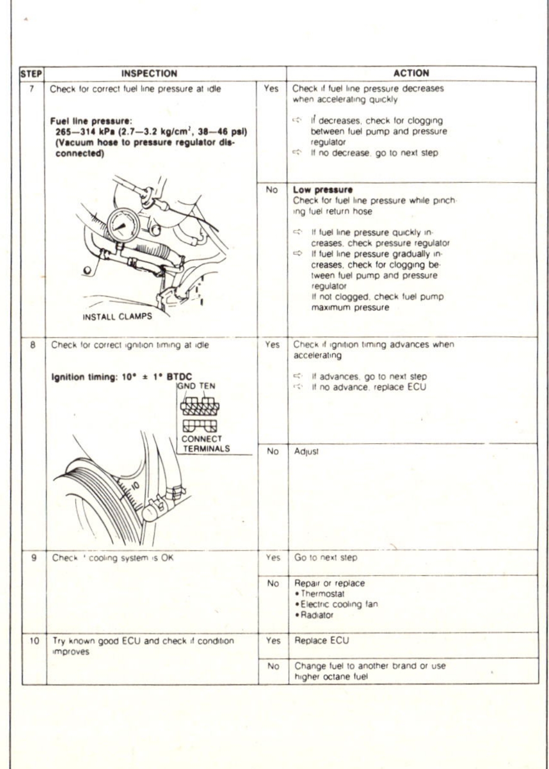

| STEP | INSPECTION | | ACTION |
|---|---|---|---|
| 7 | Check for correct fuel line pressure at idle<br>**Fuel line pressure: 265—314 kPa (2.7—3.2 kg/cm², 38—46 psi) (Vacuum hose to pressure regulator disconnected)**<br>INSTALL CLAMPS | Yes | Check if fuel line pressure decreases when accelerating quickly<br>⇨ If decreases, check for clogging between fuel pump and pressure regulator<br>⇨ If no decrease, go to next step |
| | | No | **Low pressure**<br>Check for fuel line pressure while pinching fuel return hose<br>⇨ If fuel line pressure quickly increases, check pressure regulator<br>⇨ If fuel line pressure gradually increases, check for clogging between fuel pump and pressure regulator<br>If not clogged, check fuel pump maximum pressure |
| 8 | Check for correct ignition timing at idle<br>**Ignition timing: 10° ± 1° BTDC**<br>GND TEN<br>CONNECT TERMINALS | Yes | Check if ignition timing advances when accelerating<br>⇨ If advances, go to next step<br>⇨ If no advance, replace ECU |
| | | No | Adjust |
| 9 | Check if cooling system is OK | Yes | Go to next step |
| | | No | Repair or replace<br>• Thermostat<br>• Electric cooling fan<br>• Radiator |
| 10 | Try known good ECU and check if condition improves | Yes | Replace ECU |
| | | No | Change fuel to another brand or use higher octane fuel |

| 28 | FUEL ODOR | | |
|---|---|---|---|
| DESCRIPTION | • Gasoline odor in cabin | | |
| [TROUBLESHOOTING HINTS]<br>① Poor connection or damaged fuel system or evaporative emission control system<br>② Charcoal canister overflow due to evaporative emission control system malfunction | | | |
| **STEP** | **INSPECTION** | | **ACTION** |
| 1 | Check if fuel leak or damage are visible at fuel system and evaporative emission control system | Yes | Repair or replace |
| | | No | Go to next step |
| 2 | Check if "00" is displayed on Self-Diagnosis Checker with ignition switch ON<br>SELECT SW A<br>SYSTEM SELECT 1<br>TEST SW<br>SELF TEST<br>49 H018 9A1<br>49 B019 9A0 | Yes | Go to next step |
| | | No | **Malfunction Code No. displayed**<br>Check for cause (Refer to specified check sequence) |
| | | | **"88" flashes**<br>Check ECU terminal 1F voltage<br>**Voltage: Approx. 12V (Ignition switch ON)**<br>⇨ If OK, replace ECU<br>⇨ If not OK, check wiring between ECU and Self-Diagnosis Checker |
| 3 | Check if vacuum is felt at solenoid valve (purge control) with engine running and throttle valve opened (Neutral switch connector disconnected) | Yes | Go to Step 5 |
| | | No | Check for solenoid valve operating sound in this condition<br>⇨ If OK, check vacuum hoses for clogging<br>⇨ If not OK, go to next step |
| 4 | Apply 12V and ground to solenoid valve (purge control) and check if air flows through valve | Yes | Check ECU terminal 2X voltage |
| | | No | Replace solenoid valve |
| 5 | Try known good ECU | | |

## TROUBLESHOOTING CHART—1990 MIATA

### 29 EXHAUST SULFER SMELL

**DESCRIPTION**
- Exhaust gas smells abnormal (Rotten egg smell)

**[TROUBLESHOOTING HINTS]**
High sulfer content fuel used

| STEP | INSPECTION | ACTION |
|---|---|---|
| 1 | Change fuel to another brand | |

### 30 HIGH OIL CONSUMPTION

**[TROUBLESHOOTING HINTS]**
① PCV system malfunction
② Engine malfunction (Oil working up, working down, or leakage)

| STEP | INSPECTION | | ACTION |
|---|---|---|---|
| 1 | Check if PCV hose, ventilation hose or their attaching nipples are separated, damaged, clogged, or restricted | Yes | Repair or replace |
| | | No | Go to next step |
| 2 | Check if air pressure or oil is present at ventilation hose<br>AIR PRESSURE OR OIL | Yes | Go to next step |
| | | No | Check engine condition<br>• Oil leakage<br>• Worn valve seal<br>• Worn valve stem<br>• Worn valve guide |
| 3 | Check if vacuum is felt at PCV valve at idle | Yes | Check engine condition<br>• Worn piston ring groove<br>• Stuck piston ring<br>• Worn piston or cylinder |
| | | No | Replace PCV valve |

### 31 POOR FUEL ECONOMY

**[TROUBLESHOOTING HINTS]**
While fuel consumption is drastically increased during city driving, short-run operation, stop and go driving, extended winter warm-up periods, etc., as opposed to "trip" mileage, an attempt should be made to determine these factors when confronted with "poor mileage" conditions. However, since the operator is not always at fault, the following is offered.

① Operator depressing accelerator more than usual due to low engine power
- Poor ignition
- Low intake air amount
- Electric spark advance control malfunction
- Clutch slipping
- Exhaust component restricted

② Air/Fuel mixture too rich
- High fuel line pressure

③ Alcohol blended fuel used
④ High vehicle load
- Low tire pressure
- Unrecommended tire used
- Brake dragging

⑤ Fuel cut control malfunction
⑥ High idle speed
(Refer to page F-36)

| STEP | INSPECTION | | ACTION |
|---|---|---|---|
| 1 | Check factors other than engine<br>• Low tire pressure<br>• Unrecommended tire used<br>• Clutch slipping<br>• Brake dragging<br>• Exhaust component resticted | Yes | Go to next step |
| | | No | Repair |
| 2 | Check if air hoses are connected correctly | Yes | Go to next step |
| | | No | Repair |
| 3 | Check if air cleaner element is clean | Yes | Go to next step |
| | | No | Replace |
| 4 | Check if "00" is displayed on Self-Diagnosis Checker with ignition switch ON<br>SYSTEM SELECT 1<br>TEST SW SELF TEST<br>SELECT SW A | Yes | Go to next step |
| | | No | **Malfunction Code No. displayed**<br>Check for cause (Refer to specified check sequence)<br><br>**"88" flashes**<br>Check ECU terminal 1F voltage<br>**Voltage: Approx. 12V (Ignition switch ON)**<br>⇨ If OK, replace ECU<br>⇨ If not OK, check wiring between ECU and Self-Diagnosis Checker |
| 5 | Check switches for correct operation with Self-Diagnosis Checker Monitor Lamp and ignition switch ON<br>SYSTEM SELECT 1<br>MONITOR LAMP<br>TEST SW SELF TEST<br>SELECT SW A | Yes | Go to next step |
| | | No | **Lamp not ON/OFF with specified switch**<br>Check for cause (Refer to specified check sequence)<br><br>**Lamp always ON**<br>Check wiring between ECU terminal 1F and Self-Diagnosis Checker |

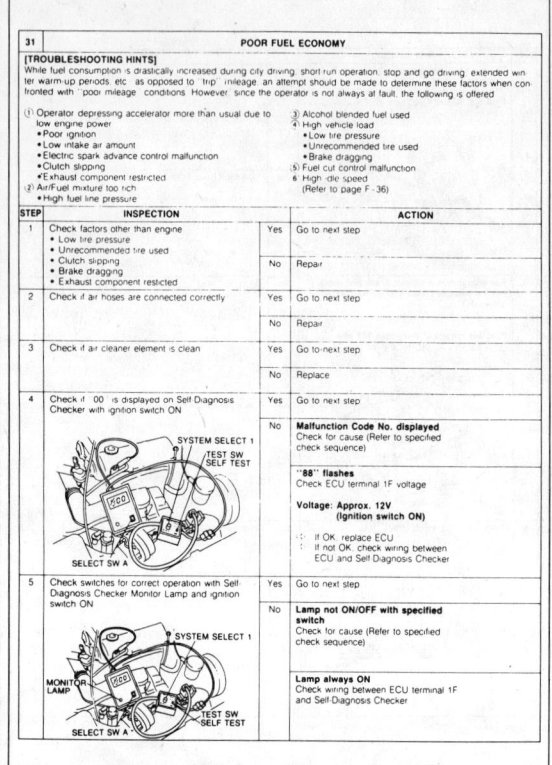

## TROUBLESHOOTING CHART—1990 MIATA

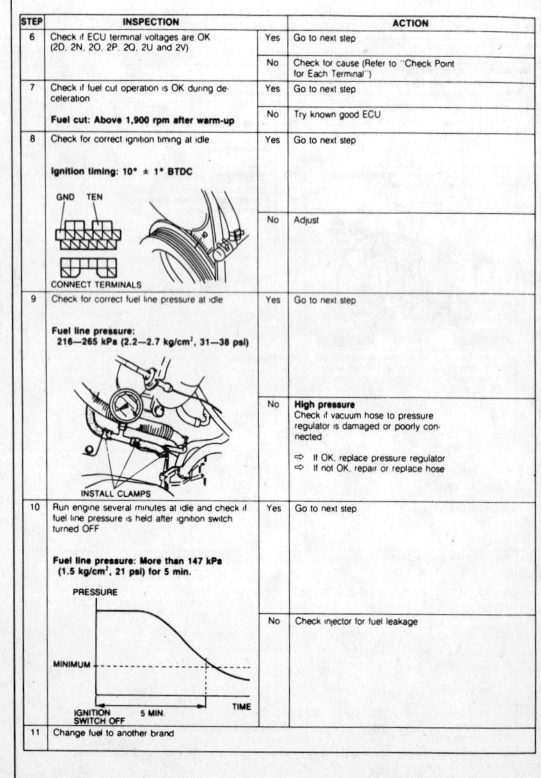

| STEP | INSPECTION | | ACTION |
|---|---|---|---|
| 6 | Check if ECU terminal voltages are OK (2D, 2N, 2O, 2P, 2Q, 2U and 2V) | Yes | Go to next step |
| | | No | Check for cause (Refer to "Check Point for Each Terminal") |
| 7 | Check if fuel cut operation is OK during deceleration<br>**Fuel cut: Above 1,900 rpm after warm-up** | Yes | Go to next step |
| | | No | Try known good ECU |
| 8 | Check for correct ignition timing at idle<br>**Ignition timing: 10° ± 1° BTDC**<br>GND TEN<br>CONNECT TERMINALS | Yes | Go to next step |
| | | No | Adjust |
| 9 | Check for correct fuel line pressure at idle<br>**Fuel line pressure: 216—265 kPa (2.2—2.7 kg/cm², 31—38 psi)**<br>INSTALL CLAMPS | Yes | Go to next step |
| | | No | **High pressure**<br>Check if vacuum hose to pressure regulator is damaged or poorly connected<br>⇨ If OK, replace pressure regulator<br>⇨ If not OK, repair or replace hose |
| 10 | Run engine several minutes at idle and check if fuel line pressure is held after ignition switch turned OFF<br>**Fuel line pressure: More than 147 kPa (1.5 kg/cm², 21 psi) for 5 min.**<br>PRESSURE<br>MINIMUM<br>TIME<br>IGNITION SWITCH OFF 5 MIN. | Yes | Go to next step |
| | | No | Check injector for fuel leakage |
| 11 | Change fuel to another brand | | |

### 32 MIL ALWAYS ON

**DESCRIPTION**
- Self-Diagnosis Checker does not indicate Malfunction Code No. but MIL always ON

**[TROUBLESHOOTING HINTS]**
- Short circuit in wiring harness
- ECU malfunction

| STEP | INSPECTION | | ACTION |
|---|---|---|---|
| 1 | Disconnect (Y/B) wire from ECU and check if MIL goes off<br>Y/B | Yes | Replace ECU |
| | | No | Check for short circuit in wiring between combination meter and ECU |

### 33 MIL NEVER ON

**DESCRIPTION**
- Self-Diagnosis Checker indicates Malfunction code No. of input device but MIL never ON
- Other indicator and warning lamps OK

**[TROUBLESHOOTING HINTS]**
- Bulb burned out
- Open circuit in wiring harness
- ECU malfunction

| STEP | INSPECTION | | ACTION |
|---|---|---|---|
| 1 | Ground (Y/B) wire at ECU with jumper wire and check if MIL comes on<br>Y/B | Yes | Check connection condition of ECU connector<br>⇨ If OK, replace ECU<br>⇨ If not OK, repair ECU connector |
| | | No | Check if bulb is OK<br>⇨ If OK, repair (Y/B) wire between ECU and combination meter<br>⇨ If not OK, replace bulb |

## TROUBLESHOOTING CHART—1990 MIATA

| 34 | A/C DOES NOT WORK | | |
|---|---|---|---|
| DESCRIPTION | • Blower fan operates but cool air not expelled | | |
| [TROUBLESHOOTING HINTS]<br>• Open or short circuit in wiring harness<br>• A/C relay malfunction<br>• A/C switch malfunction<br>• Magnetic clutch malfunction<br>• ECU malfunction | | | |
| STEP | INSPECTION | ACTION | |
| 1 | Ground (L/B) wire at A/C relay with jumper wire and check if condenser fan operates when ignition switch turned ON  | Yes | Go to next step |
| | | No | Go to A/C systen Troubleshooting |
| 2 | Ground (L/B) wire at ECU with jumper wire and check if condenser fan operates when ignition switch turned ON 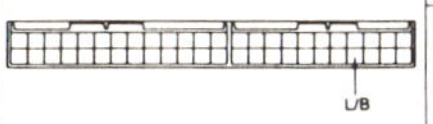 | Yes | Go to next step |
| | | No | Repair (L/B) wire between ECU and A/C relay |
| 3 | Check if A/C switch is OK | Yes | Try known good ECU |
| | | No | Check A/C switch and wiring |

## MALFUNCTION INDICATOR LIGHT (MIL) OPERATION 1990 MIATA

**Principle of Code Cycle**

Malfunction codes are determined as shown below.

**1. Code cycle break**

The time between malfunction code cycles is 4.0 seconds (the time the lamp is off).

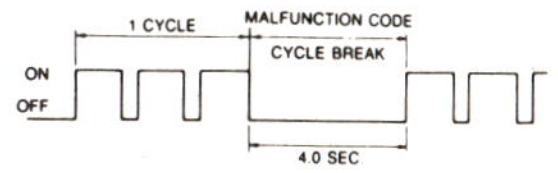

**2. Second digit of malfunction code (ones position)**

The digit in the ones position of the malfunction code represents the number of times the buzzer sounds 0.4 second during one cycle.

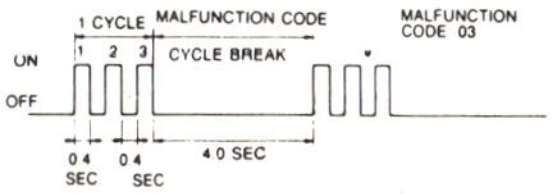

**3. First digit of malfunction code (tens position)**

The digit in the tens position of the malfunction code represents the number of times the buzzer is on 1.2 seconds during one cycle.

It should also be noted that the light goes off for 1.6 seconds between the long and short pulses of the buzzer.

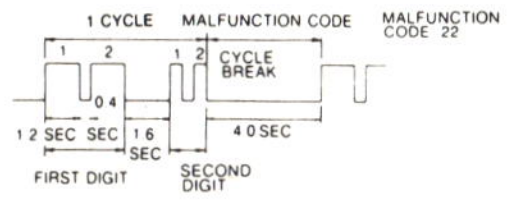

## TROUBLE CODE IDENTIFICATION CHART—1990 MIATA

**Code Numbers**

| Malfunction display: Code No. | Malfunction display: Pattern of output signal (Self-Diagnosis Checker) | Sensor or subsystem | Self-diagnosis | Fail-safe |
|---|---|---|---|---|
| 01 | | Ignition pulse | No IGf-signal | — |
| 02 | | Ne-signal | No Ne-signal | — |
| 03 | | G-signal | No G-signal | — |
| 08 | | Airflow meter | Open or short circuit | Basic fuel injection amount fixed as for two driving modes<br>(1) Idle switch: ON<br>(2) Idle switch: OFF |
| 09 | | Water thermosensor | Open or short circuit | Maintains constant 20°(68°F) command |
| 10 | | Intake air thermosensor (Airflow meter) | Open or short circuit | Maintains constant 20°C (68°F) command |
| 14 | | Atmospheric pressure sensor | Open or short circuit | Maintains constant command of sea level pressure |
| 15 | | Oxygen sensor | Sensor output continues less than 0.45V 180 sec. after engine exceeds 1,500 rpm | Cancels engine feedback operation |
| 17 | | Feedback system | Sensor output continues unchanged 20 sec. after engine exceeds 1,500 rpm | Cancels engine feedback operation |
| 26 | | Solenoid valve (Purge control) | — | — |
| 34 | | ISC valve | | — |

**Caution**

- **If there is more than one failure present, the lowest number malfunction code is displayed first, the remaining codes are displayed in order.**
- **After repairing all failures, turn off the ignition switch, disconnect the negative battery cable, and depress the brake pedal for 5 seconds to erase the memory of a malfunction code from the engine control unit.**

## TROUBLE CODE DIAGNOSTIC CHART—1990 MIATA

**Troubleshooting**

If a malfunction code number is shown on the **SST**, check for the cause by using the chart related to the code number shown.

| CODE NO.01 IGf-SIGNAL | | | |
|---|---|---|---|
| STEP | INSPECTION | ACTION | |
| 1 | Are there any poor connections at ignition coil connectors and igniter connectors? | Yes | Repair or replace connector |
| | | No | Go to next step |
| 2 | Does tachometer operates? | Yes | Go to next step |
| | | No | Check for open circuit in wiring from igniter to ECU terminal 2I |
| 3 | Is resistance of ignition coil OK?<br>Resistance Primary 0.78—0.94Ω<br>Secondary 11.2—15.2 kΩ | Yes | Go to next step |
| | | No | Replace ignition coil |
| 4 | Is there continuity between ignition coil and igniter?<br>Ignition coil A (W) – Igniter A (W)<br>Ignition coil B (Y) – Igniter H (Y) | Yes | Go to next step |
| | | No | Check for open circuit in wiring from ignition coil to igniter |
| 5 | Is ignition coil terminal wire (L) voltage OK? | Yes | Go to next step |
| | | No | Check for open circuit in wiring from ignition coil to ignition switch |
| 6 | Is igniter terminal wire (L) voltage OK? | Yes | Go to next step |
| | | No | Check for open circuit in wiring from igniter to ignition switch |
| 7 | Is there continuity between igniter and ground? | Yes | Go to next step |
| | | No | Check for open circuit in wiring from igniter to ground |
| 8 | Check igniter | Yes | Replace ECU |
| | | No | Replace igniter |

**Circuit Diagram**

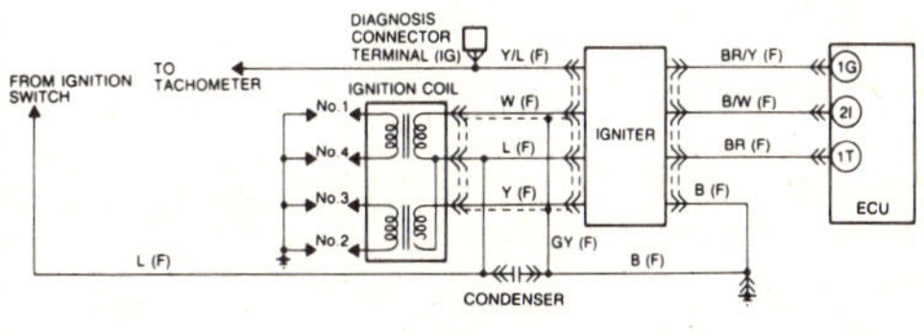

### TROUBLE CODE DIAGNOSTIC CHART—1990 MIATA

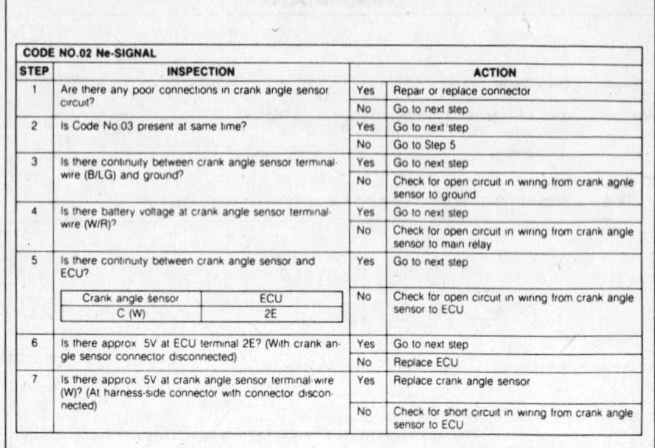

**CODE NO.02 Ne-SIGNAL**

| STEP | INSPECTION | ACTION | |
|---|---|---|---|
| 1 | Are there any poor connections in crank angle sensor circuit? | Yes | Repair or replace connector |
| | | No | Go to next step |
| 2 | Is Code No.03 present at same time? | Yes | Go to next step |
| | | No | Go to Step 5 |
| 3 | Is there continuity between crank angle sensor terminal-wire (B/LG) and ground? | Yes | Go to next step |
| | | No | Check for open circuit in wiring from crank agnle sensor to ground |
| 4 | Is there battery voltage at crank angle sensor terminal-wire (W/R)? | Yes | Go to next step |
| | | No | Check for open circuit in wiring from crank angle sensor to main relay |
| 5 | Is there continuity between crank angle sensor and ECU?<br>Crank angle sensor \| ECU<br>C (W) \| 2E | Yes | Go to next step |
| | | No | Check for open circuit in wiring from crank angle sensor to ECU |
| 6 | Is there approx. 5V at ECU terminal 2E? (With crank angle sensor connector disconnected) | Yes | Go to next step |
| | | No | Replace ECU |
| 7 | Is there approx. 5V at crank angle sensor terminal-wire (W)? (At harness-side connector with connector disconnected) | Yes | Replace crank angle sensor |
| | | No | Check for short circuit in wiring from crank angle sensor to ECU |

**CODE NO.03 G-SIGNAL**

| STEP | INSPECTION | ACTION | |
|---|---|---|---|
| 1 | Are there any poor connections in crank angle sensor circuit? | Yes | Repair or replace connector |
| | | No | Go to next step |
| 2 | Is Code No 02 also present? | Yes | Go to next step |
| | | No | Go to Step 5 |
| 3 | Is there continuity between crank angle sensor terminal-wire (B/LG) and ground? | Yes | Go to next step |
| | | No | Check for open circuit in wiring from crank angle sensor to ground |
| 4 | Is there battery voltage at crank angle sensor terminal-wire (W/R)? | Yes | Go to next step |
| | | No | Check for open circuit in wiring from crank angle sensor to main relay |
| 5 | Is there continuity between crank angle sensor and ECU?<br>Crank angle sensor \| ECU<br>D (Y/L) \| 2G | Yes | Go to next step |
| | | No | Check for open circuit in wiring from crank angle sensor to ECU |
| 6 | Is there approx. 5V at ECU terminal 2E? (With crank angle sensor connector disconnected) | Yes | Go to next step |
| | | No | Replace ECU |
| 7 | Is there approx. 5V at crank angle sensor terminal-wire (Y/L)? (At harness-side connector with connector disconnected) | Yes | Replace crank angle sensor |
| | | No | Check for short circuit in wiring from crank angle sensor to ECU |

**CODE NO.08 AIRFLOW METER**

| STEP | INSPECTION | ACTION | |
|---|---|---|---|
| 1 | Are there any poor connections in airflow meter circuit? | Yes | Repair or replace connector |
| | | No | Go to next step |
| 2 | Is Code No.10 present at same time? | Yes | Check for open circuit in wiring from airflow meter terminal-wire (B/LG) to ground |
| | | No | Go to next step |
| 3 | Is resistance of airflow meter OK?<br>Airflow meter \| Fully closed Ω \| Fully open Ω<br>D (LG/R)—F (G/Y) \| 200—600 \| 20—1,000<br>D (LG/R)—C (B/LG) \| 200—400 | Yes | Go to next step |
| | | No | Replace airflow meter |
| 4 | Is there continuity between airflow meter connector and ECU?<br>Airflow meter \| ECU<br>D (LG/R) \| 2K<br>F (G/Y) \| 2O | Yes | Go to next step |
| | | No | Check for open circuit in wiring from airflow meter to ECU |
| 5 | Are ECU terminals 2D and 2K voltages OK? | Yes | Replace ECU |
| | | No | Check for short circuit in wiring from airflow meter to ECU |

**Circuit Diagram**

### TROUBLE CODE DIAGNOSTIC CHART—1990 MIATA

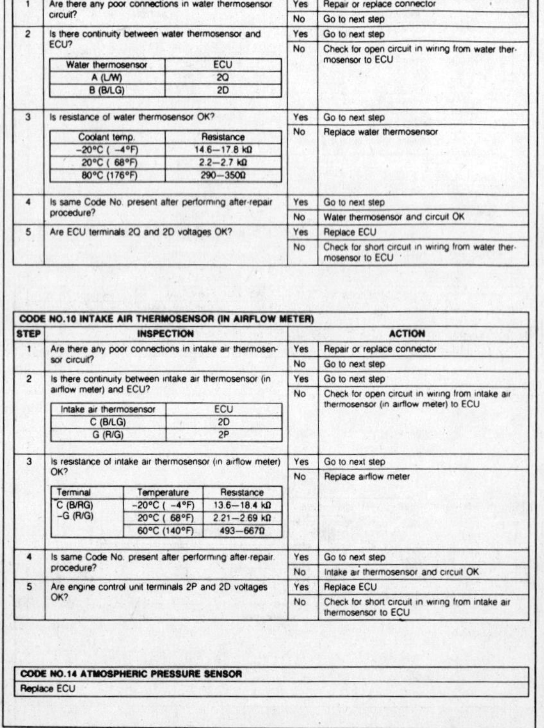

**CODE NO.09 WATER THERMOSENSOR**

| STEP | INSPECTION | ACTION | |
|---|---|---|---|
| 1 | Are there any poor connections in water thermosensor circuit? | Yes | Repair or replace connector |
| | | No | Go to next step |
| 2 | Is there continuity between water thermosensor and ECU?<br>Water thermosensor \| ECU<br>A (L/W) \| 2Q<br>B (B/LG) \| 2D | Yes | Go to next step |
| | | No | Check for open circuit in wiring from water thermosensor to ECU |
| 3 | Is resistance of water thermosensor OK?<br>Coolant temp. \| Resistance<br>−20°C (−4°F) \| 14.6—17.8 kΩ<br>20°C (68°F) \| 2.2—2.7 kΩ<br>80°C (176°F) \| 290—350Ω | Yes | Go to next step |
| | | No | Replace water thermosensor |
| 4 | Is same Code No. present after performing after-repair procedure? | Yes | Go to next step |
| | | No | Water thermosensor and circuit OK |
| 5 | Are ECU terminals 2Q and 2D voltages OK? | Yes | Replace ECU |
| | | No | Check for short circuit in wiring from water thermosensor to ECU |

**CODE NO.10 INTAKE AIR THERMOSENSOR (IN AIRFLOW METER)**

| STEP | INSPECTION | ACTION | |
|---|---|---|---|
| 1 | Are there any poor connections in intake air thermosensor circuit? | Yes | Repair or replace connector |
| | | No | Go to next step |
| 2 | Is there continuity between intake air thermosensor (in airflow meter) and ECU?<br>Intake air thermosensor \| ECU<br>C (B/LG) \| 2D<br>G (R/G) \| 2P | Yes | Go to next step |
| | | No | Check for open circuit in wiring from intake air thermosensor (in airflow meter) to ECU |
| 3 | Is resistance of intake air thermosensor (in airflow meter) OK?<br>Terminal \| Temperature \| Resistance<br>C (B/RG)—G (R/G) \| −20°C (−4°F) \| 13.6—18.4 kΩ<br>\| 20°C (68°F) \| 2.21—2.69 kΩ<br>\| 60°C (140°F) \| 493—667Ω | Yes | Go to next step |
| | | No | Replace airflow meter |
| 4 | Is same Code No. present after performing after-repair procedure? | Yes | Go to next step |
| | | No | Intake air thermosensor and circuit OK |
| 5 | Are engine control unit terminals 2P and 2D voltages OK? | Yes | Replace ECU |
| | | No | Check for short circuit in wiring from intake air thermosensor to ECU |

**CODE NO.14 ATMOSPHERIC PRESSURE SENSOR**

Replace ECU

**CODE NO.15 OXYGEN SENSOR**

**Note**
- **If Code Nos.15 and 17 are both present, first perform the checking procedure for Code No.17**

| STEP | INSPECTION | ACTION | |
|---|---|---|---|
| 1 | Are there any poor connections in oxygen sensor circuit? | Yes | Repair or replace connector |
| | | No | Go to next step |
| 2 | Is oxygen sensor output voltage OK? | Yes | Go to next step |
| | | No | Replace oxygen sensor |
| 3 | Is there continuity between oxygen sensor and ECU terminal 2N? | Yes | Go to next step |
| | | No | Check for open circuit in wiring from oxygen sensor to ECU |
| 4 | Is ECU terminal 2N voltage OK? | Yes | Go to next step |
| | | No | Check for short circuit in wiring from oxygen sensor to ECU |
| 5 | Is sensitivity of oxygen sensor OK? | Yes | Replace ECU |
| | | No | Replace oxygen sensor |

**CODE NO.17 FEEDBACK SYSTEM**

| STEP | INSPECTION | ACTION | |
|---|---|---|---|
| 1 | Warm up engine and run it at 2,500—3,000 rpm for 3 min. | | |
| 2 | Does monitor lamp of Self-Diagnosis Checker illuminate at idle? | Yes | Go to next step |
| | | No | Check for air leak in vacuum hoses or emission components<br>Check for contaminated oxygen sensor<br>Check for insufficient fuel injection |
| 3 | Are spark plugs clean? | Yes | Go to next step |
| | | No | Clean or replace spark plugs |
| 4 | Is oxygen sensor voltage OK? | Yes | Go to next step |
| | | No | Replace oxygen sensor |
| 5 | Is same Code No. present after performing after-repair procedure? | Yes | Go to next step |
| | | No | Check for short circuit in wiring from oxygen sensor to ECU terminal 2N |
| 6 | Is there continuity between oxygen sensor and ECU terminal 2N? | Yes | Go to next step |
| | | No | Check for open circuit in wiring from oxygen sensor to ECU |
| 7 | Is ECU terminal 2N voltage OK? | Yes | Replace ECU |
| | | No | Check for short circuit in wiring from oxygen sensor to ECU |

**Circuit Diagram**

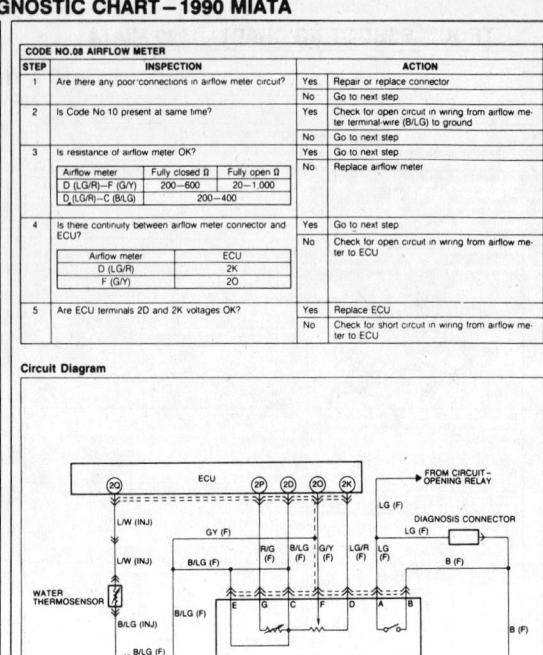

## TROUBLE CODE DIAGNOSTIC CHART—1990 MIATA

**CODE NO.26 SOLENOID VALVE (PURGE CONTROL)**

| STEP | INSPECTION | | ACTION |
|---|---|---|---|
| 1 | Are there any poor connections in solenoid valve circuit? | Yes | Repair or replace connector |
| | | No | Go to next step |
| 2 | Is resistance of solenoid valve OK?<br>**Resistance: 25 ± 2Ω** | Yes | Go to next step |
| | | No | Replace solenoid valve |
| 3 | Is there battery voltage at terminal wire (W/R) of solenoid valve circuit? | Yes | Go to next step |
| | | No | Check for open circuit in wiring from solenoid valve to main relay |
| 4 | Is there continuity between solenoid valve and ECU?<br>Solenoid valve: B (Y/R) — ECU: 2X | Yes | Go to next step |
| | | No | Check for open circuit in wiring from solenoid valve to ECU |
| 5 | Is ECU terminal (2X) voltage OK? | Yes | Replace ECU |
| | | No | Check for short circuit in wiring from solenoid valve to ECU |

**CODE NO.34 ISC VALVE**

| STEP | INSPECTION | | ACTION |
|---|---|---|---|
| 1 | Are there any poor connections in ISC valve circuit? | Yes | Repair or replace connector |
| | | No | Go to next step |
| 2 | Is resistance of ISC valve OK?<br>**Resistance: 12 ± 1Ω** | Yes | Go to next step |
| | | No | Replace ISC valve |
| 3 | Is there battery voltage at terminal-wire (W/R) of ISC valve circuit? | Yes | Go to next step |
| | | No | Check for open circuit in wiring from ISC valve to main relay |
| 4 | Is there continuity between ISC valve and ECU?<br>ISC valve: B (L/O) — ECU: 2W | Yes | Go to next step |
| | | No | Check for open circuit in wiring from ISC valve to ECU |
| 5 | Is ECU terminal 2W voltage OK? | Yes | Replace ECU |
| | | No | Check for short circuit in wiring from ISC valve to ECU |

**Circuit Diagram**

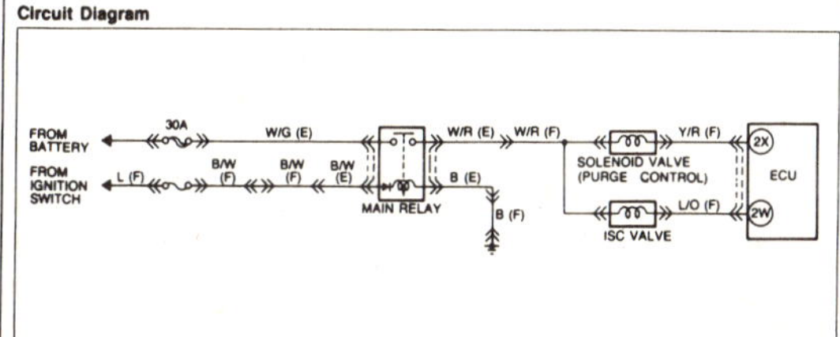


## CONTROL UNIT PIN LOCATION AND VOLTAGE TEST 1990 MIATA

| Terminal | Input | Output | Connection to | Test condition | Voltage | Remark |
|---|---|---|---|---|---|---|
| 1A | — | — | Battery | Constant | Approx. 12V | For backup |
| 1B | ○ | | Main relay | Ignition switch OFF | Approx. 0V | |
| | | | | Ignition switch ON | Approx. 12V | |
| 1C | ○ | | Ignition switch (Start position) | While cranking | Approx. 10V | |
| | | | | Ignition switch ON | Approx. 0V | |
| 1D | | ○ | Self-Diagnosis Checker (Monitor lamp) | Test switch at "SELF-TEST" Lamp illuminated for 3 sec. after ignition switch OFF→ON | Approx. 5V | With Self-Diagnosis Checker and System Selector |
| | | | | Lamp not illuminated after 3 sec. | Approx. 12V | |
| | | | | Test switch at "$O_2$ MONITOR" at idle Monitor lamp illuminated | Approx. 5V | |
| | | | | Test switch at "$O_2$ MONITOR" at idle Monitor lamp not illuminated | Approx. 12V | |
| 1E | | ○ | Malfunction indicator lamp | Lamp illuminated for 3 sec. after ignition switch OFF→ON | Below 2.5V | With System Selector test switch at "SELF-TEST" |
| | | | | Lamp not illuminated after 3 sec. | Approx. 12V | |
| | | | | Lamp illuminated | Below 2.5V | |
| | | | | Lamp not illuminated | Approx. 12V | |
| 1F | | ○ | Self-Diagnosis Checker (Code number) | Buzzer sound for 3 sec. after ignition switch OFF→ON | Below 2.5V | • With Self-Diagnosis Checker and System Selector • With System Selector test switch at "SELF-TEST" |
| | | | | Buzzer not sounded after 3 sec. | Approx. 12V | |
| | | | | Buzzer sounded | Below 2.5V | |
| | | | | Buzzer not sounded | Approx. 12V | |
| 1G | | ○ | Igniter | Ignition switch ON | Approx. 0V | |
| | | | | Idle | Approx. 0.2V | |
| 1H | | ○ | Igniter | Ignition switch ON | Approx. 0V | |
| | | | | Idle | Approx. 0.2V | |
| 1I | — | — | — | — | — | — |
| 1J | | ○ | A/C relay | Ignition switch ON | Approx. 12V | |
| | | | | A/C switch ON at idle | Below 2.5V | |
| | | | | A/C switch OFF at idle | Approx. 12V | |
| 1K | ○ | | Diagnosis connector | System Selector test switch at "$O_2$ MONITOR" | Approx. 12V | |
| | | | | System Selector test switch at "SELF-TEST" | Approx. 0V | |
| 1L | — | — | — | — | — | — |
| 1M | — | — | — | — | — | — |
| 1N | ○ | | Throttle sensor (Idle point) | Accelerator pedal released | Approx. 0V | Ignition switch ON |
| | | | | Accelerator pedal depressed | Approx. 12V | |
| 1O | ○ | | Stoplight switch | Brake pedal released | 0V | |
| | | | | Brake pedal depressed | Approx. 12V | |
| 1P | ○ | | P/S pressure switch | Ignition switch ON | Approx. 12V | |
| | | | | P/S ON (at idle) | 0V | |
| | | | | P/S OFF (at idle) | Approx. 12V | |
| 1Q | ○ | | A/C switch | A/C switch ON (Ignition switch ON) | Below 2.5V | Blower motor ON |
| | | | | A/C switch OFF (Ignition switch ON) | Approx. 12V | |

## CONTROL UNIT PIN LOCATION AND VOLTAGE TEST (CONT.)—1990 MIATA

| Terminal | Input | Output | Connection to | Test condition | Voltage | Remark |
|---|---|---|---|---|---|---|
| 1R | ○ | | Fan switch | Fan operating (Engine coolant temperature over 97°C (207°F) or diagnosis connector terminal TFA grounded) | Approx. 0V | |
| | | | | Fan not operating (Idle) | Approx. 12V | |
| 1S | ○ | | Blower control switch | Blower control switch at mid, high or super high position | Approx. 0V | Ignition switch ON |
| | | | | Blower control switch OFF or low | Approx. 12V | |
| 1T | — | — | — | — | — | |
| 1U | ○ | | Headlight switch | Headlights ON (Tail, parking, low beam/high beam) | Approx. 12V | |
| | | | | Headlights OFF | 0V | |
| 1V | ○ | | Neutral or clutch switch | Neutral position or clutch pedal depressed | Approx. 0V | |
| | | | | Other conditions | Approx. 12V | |
| 2A | — | — | Ground (Injector) | Constant | 0V | |
| 2B | — | — | Ground (Output) | Constant | 0V | |
| 2C | — | — | Ground (CPU) | Constant | 0V | |
| 2D | — | — | Ground (Input) | Constant | 0V | |
| 2E | ○ | | Crank angle sensor (Ne-signal) | Ignition switch ON | Approx. 0V or 5V | |
| | | | | Idle | Approx. 2V | |
| 2F | — | — | — | — | — | — |
| 2G | ○ | | Crank angle sensor (G-signal) | Ignition switch ON | Approx. 0V or 5V | |
| | | | | Idle | Approx. 1.5V | |
| 2H | ○ | | Ground | California spec. | 0V | |
| | | | Open | Federal and Canada spec. | Approx. 2V | |
| 2I | ○ | | Igniter | Ignition switch ON | Below 0.5V | |
| | | | | Idle | Approx. 1V | |
| 2J | ○ | | Ground | Constant | 0V | |
| 2K | | ○ | Airflow meter | Constant | 4.5—5.5V | |
| 2L | ○ | | Throttle sensor (Power terminal) | Accelerator pedal released | Approx. 5V | |
| | | | | Accelerator pedal fully depressed | Approx. 0V | |
| 2M | — | — | — | — | — | — |
| 2N | ○ | | Oxygen sensor | Ignition switch ON | 0V | |
| | | | | Idle (Cold engine) | 0V | |
| | | | | Idle (After warm-up) | 0—1V | |
| | | | | Increase engine speed (After warm-up) | 0.5—1V | |
| | | | | Deceleration | 0—0.4V | |
| 2O | ○ | | Airflow meter | Ignition switch ON | Approx. 3.8V | |
| | | | | Idle | Approx. 3.3V | |
| 2P | ○ | | Airflow sensor (Intake air thermosensor) | At 20°C (68°F) | Approx. 2.5V | |
| 2Q | ○ | | Water thermosensor | Engine coolant temperature 20°C (68°F) | Approx. 2.5V | |
| | | | | After warm up | Approx. 0.4V | |
| 2R | — | — | — | — | — | — |

| Terminal | Input | Output | Connection to | Test condition | Voltage | Remark |
|---|---|---|---|---|---|---|
| 2S | — | — | — | — | — | — |
| 2T | — | — | — | — | — | — |
| 2U | | ○ | Injector (Nos 1, 3) (Nos 2, 4) | Ignition switch ON | Approx. 12V | * Engine Signal Monitor Green and red lights flash |
| | | | | Idle | Approx. 12V* | |
| 2V | | ○ | | Deceleration from 3,000 rpm to 1,900 rpm (After warm up) | Approx. 12V | |
| 2W | | ○ | ISC valve | Ignition switch ON | Approx. 7V | |
| | | | | Idle | Approx. 9V | |
| 2X | | ○ | Solenoid valve (Purge control) | Ignition switch ON | Approx. 12V | |
| | | | | Idle | Approx. 12V | |
| 2Y | — | — | — | — | — | — |
| 2Z | — | — | — | — | — | — |

**Terminal location**

| 2Y | 2W | 2U | 2S | 2Q | 2O | 2M | 2K | 2I | 2G | 2E | 2C | 2A | 1U | 1S | 1Q | 1O | 1M | 1K | 1I | 1G | 1E | 1C | 1A |
|---|---|---|---|---|---|---|---|---|---|---|---|---|---|---|---|---|---|---|---|---|---|---|---|
| 2Z | 2X | 2V | 2T | 2R | 2P | 2N | 2L | 2J | 2H | 2F | 2D | 2B | 1V | 1T | 1R | 1P | 1N | 1L | 1J | 1H | 1F | 1D | 1B |

### CONTROL UNIT PIN LOCATION AND VOLTAGE TEST TROUBLESHOOTING CHART – 1990 MIATA

**Check Point for Each Terminal**

| Terminal | Connection to | Abnormal voltage | Possible cause |
|---|---|---|---|
| 1A | Battery | Always approx. 0V (Battery OK) | • ROOM 10A fuse burned<br>• Open circuit in wiring from ROOM 10A fuse to ECU terminal 1A |
| 1B | Main relay | Always approx. 0V | • Main relay malfunction<br>• Open circuit in wiring from main relay to ECU terminal 1B |
| 1C | Ignition switch (Start position) | Always approx. 0V (Starter turns) | • Open circuit in wiring from starter interlock switch to ECU terminal 1C |
| 1D | Self-Diagnosis Checker (Monitor lamp) | Always approx. 0V | • Main relay malfunction<br>• Open circuit in wiring from main relay to diagnosis connector terminal +B<br>• Open or short circuit in wiring from diagnosis connector terminal MEN to ECU terminal 1D |
| | | Always approx. 12V | • Poor connection at ECU connector<br>• ECU malfunction |
| | | Always approx. 5V | • ECU malfunction |
| 1E | Malfunction indicator lamp (MIL) | Always below 2.5V (MIL always ON) | • Short circuit in wiring from combination meter to ECU terminal 1E<br>• ECU malfunction |
| | | Always below 2.5V (MIL never ON) | • Open circuit in wiring from combination meter to ECU terminal 1E |
| | | Always approx. 12V | • Poor connection at ECU connector<br>• ECU malfunction |
| 1F | Self-Diagnosis Checker (Code No.) | Always below 2.5V (No display on Self-Diagnosis Checker) | • Main relay malfunction<br>• Open circuit in wiring from main relay to diagnosis connector terminal +B |
| | | Always below 2.5V ("88" is displayed and buzzer sounds continuously) | • Open or short circuit in wiring from diagnosis connector terminal FEN to ECU terminal 1F |
| | | Always approx. 12V | • Poor connection at ECU connector<br>• ECU malfunction |
| 1G<br>1H | Igniter | Always approx. 0V | Refer to Code No.01 troubleshooting |
| 1J | A/C relay | Always below 2.5V (A/C does not operate) | • A/C relay malfunction<br>• Open circuit in wiring from main relay to A/C relay<br>• Open circuit in wiring from A/C relay to ECU terminal 1J |
| | | Always below 2.5V (A/C switch OFF but A/C operates) | • Short circuit in wiring from A/C relay to ECU terminal 1J<br>• ECU malfuction |
| | | Always approx. 12V | • A/C switch malfunction<br>• Poor connection at ECU connector<br>• ECU malfunction |
| 1K | Diagnosis connector (Terminal TEN) | Always approx. 0V | • Short circuit in wiring from ECU terminal 1K to diagnosis connector terminal TEN |
| | | Always approx. 12V | • Open circuit in wiring from ECU terminal 1K to diagnosis connector terminal TEN<br>• Open circuit in wiring from diagnosis connector terminal GND to ground |
| 1N | Throttle sensor (Idle terminal) | Always approx. 0V | • Throttle sensor misadjustment<br>• Short circuit in wiring from ECU terminal 1N to throttle sensor<br>• ECU malfunction |
| | | Always approx. 12V | • Throttle sensor misadjustment<br>• Open circuit in wiring from ECU terminal 1N to throttle sensor<br>• Open circuit in wiring from throttle sensor to ground |
| 1O | Stoplight switch | Always approx. 0V (Stoplights OK) | • Open circuit in wiring from stoplight switch to ECU terminal 1O |
| 1P | P/S pressure switch | Always approx. 0V | • P/S pressure switch malfunction<br>• Short circuit in wiring from ECU terminal 1P to P/S pressure switch<br>• ECU malfunction |
| | | Always approx. 12V | • P/S pressure switch malfunction<br>• Open circuit in wiring from ECU terminal 1P to P/S pressure switch<br>• Open circuit in wiring from P/S pressure switch to ground |
| 1Q | A/C switch | Always approx. 0V (with blower switch ON) | • A/C switch malfunction<br>• Short circuit in wiring from ECU terminal 1Q to A/C switch<br>• Poor connection at ECU connector<br>• ECU malfunction |
| | | Always approx. 12V (with blower switch ON) (Blower fan OK) | • A/C switch malfunction<br>• Open circuit in wiring from ECU terminal 1Q to A/C switch<br>• Open circuit in wiring from A/C switch to blower control switch |
| 1R | Fan switch | Always approx. 0V (Cooling fan OK) | • Open or short circuit in wiring from electric cooling fan relay to ECU terminal 1R<br>• ECU malfunction |
| 1S | Blower control switch | Always approx. 0V (Blower fan OK) | • Short circuit in wiring from blower control switch to ECU terminal 1S<br>• Poor connection at ECU connector<br>• ECU malfunction |
| | | Always approx. 12V (Blower fan OK) | • Open circuit in wiring from blower control switch to ECU terminal 1S |
| 1U | Headlight switch | Always approx. 0V (Headlights OK) | • Open or short circuit in wiring from headlight relay to ECU terminal 1U |
| 1V | Neutral switch<br>Clutch switch | Always approx. 0V | • Neutral switch malfunction<br>• Clutch switch malfunction<br>• Short circuit in wiring from ECU terminal 1V to neutral or clutch switch<br>• Poor connection at ECU connector<br>• ECU malfunction<br>• Open circuit in wiring from ECU terminal 1V to neutral and clutch switches |
| 2A<br>2B<br>2C<br>2D | Ground | More than 0V | • Poor contact at ground terminal<br>• Open circuit in wiring from ECU to ground |
| 2E | Crank angle sensor (Ne-signal) | Always approx. 0V or approx. 5V | • Refer to Code No.02 troubleshooting |
| 2G | Crank angle sensor (G-signal) | Always approx. 0V or approx. 5V | • Refer to Code No.03 troubleshooting |
| 2H | Ground (California) | Approx. 5V | • Open circuit in wiring from ECU terminal 2H to ground |
| | —<br>(Federal and Canada) | Approx. 0V | • Short circuit in wiring from ECU terminal 2H to ground |
| 2I | Igniter | Always Approx. 0V | • Refer to Code No.01 troubleshooting |
| 2J | Ground | Approx. 5V | • Open circuit in wiring from ECU terminal 2J to ground |

### CONTROL UNIT PIN LOCATION AND VOLTAGE TEST TROUBLESHOOTING CHART (CONT.) – 1990 MIATA

| Terminal | Connection to | Abnormal voltage | Possible cause |
|---|---|---|---|
| 2K | Airflow meter | Always approx. 0V | • Short circuit in wiring from ECU terminal 2K to airflow meter<br>• Poor connection at ECU connector<br>• ECU malfunction |
| | | Below 4.5V or above 5.5V | • ECU malfunction |
| 2L | Throttle sensor (Power terminal) | Always approx. 0V | • Throttle sensor malfunction<br>• Short circuit in wiring from ECU terminal 2L to throttle sensor<br>• Poor connection at ECU connector<br>• ECU malfunction |
| | | Always approx. 5V | • Throttle sensor misadjustment<br>• Open circuit in wiring from ECU terminal 2L to throttle sensor<br>• Open circuit in wiring from throttle sensor to ground |
| 2N | Oxygen sensor | 0V after warm-up | • Refer to Code No.15 troubleshooting |
| | | Always approx. 1V after warm-up | • Refer to Code No 17 troubleshooting |
| 2O | Airflow meter | Always approx. 0V or approx. 5V | • Refer to Code No.08 troubleshooting |
| 2P | Airflow meter (Intake air thermosensor) | Always approx. 0V or approx. 5V | • Refer to Code No.10 troubleshooting |
| | | | • Intake air thermosensor malfunction |
| 2Q | Water thermosensor | Always approx. 0V or approx. 5V | • Refer to Code No.09 troubleshooting |
| | | | • Water thermosensor malfunction |
| 2U<br>2V | Injector | Always approx. 0V | • Main relay malfunction<br>Open or short circuit in wiring from injector to ECU terminal 2U or 2V |
| | | Always approx. 12V | • ECU malfnction |
| 2W | ISC valve | Always approx. 0V or approx. 12V | • Refer to Code No.34 troubleshooting |
| | | | • ISC valve malfunction |
| 2X | Solenoid valve (Purge control) | Always approx. 0V or approx. 12V | • Refer to Code No 26 troubleshooting |
| | | | • Solenoid valve (Purge control) malfunction |

### WIRING SCHEMATIC – 1990 MIATA

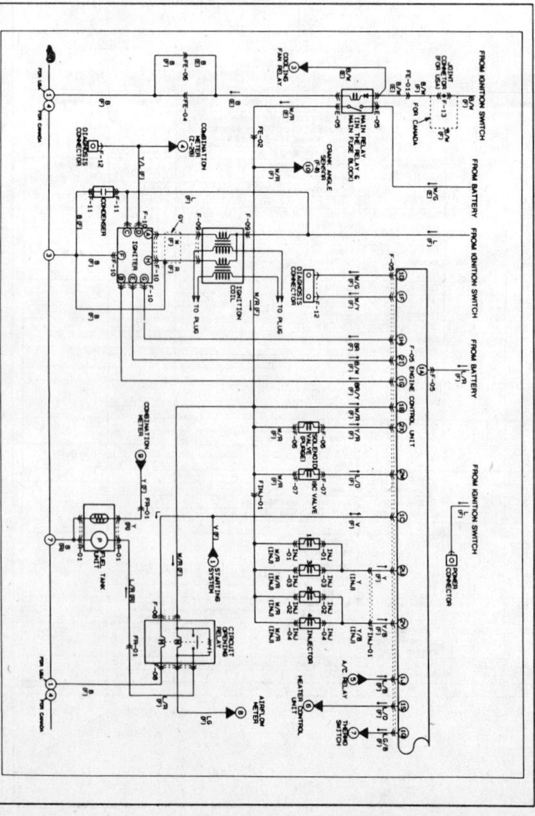

## WIRING SCHEMATIC—1990 MIATA

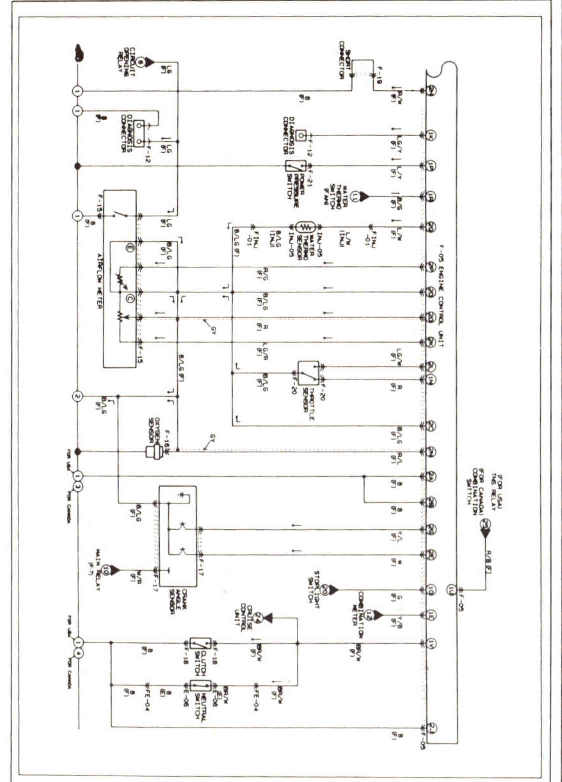

## SYSTEM SCHEMATIC—1988–89 929

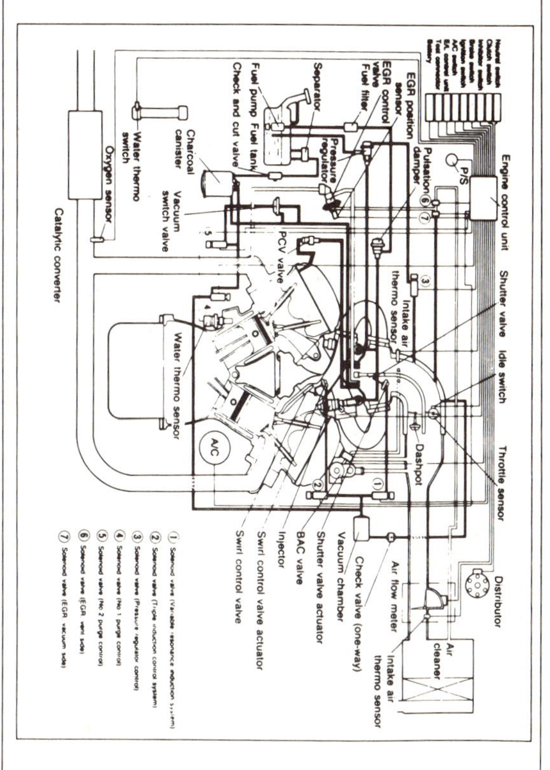


## VACUUM HOSE ROUTING—1988–89 929

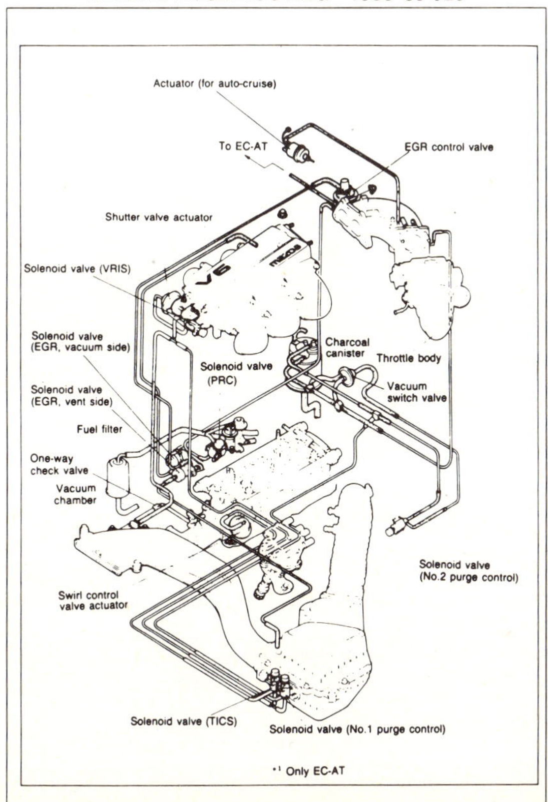


## EXPLODED VIEW OF FUEL AND EMISSION COMPONENTS—1988–89 929

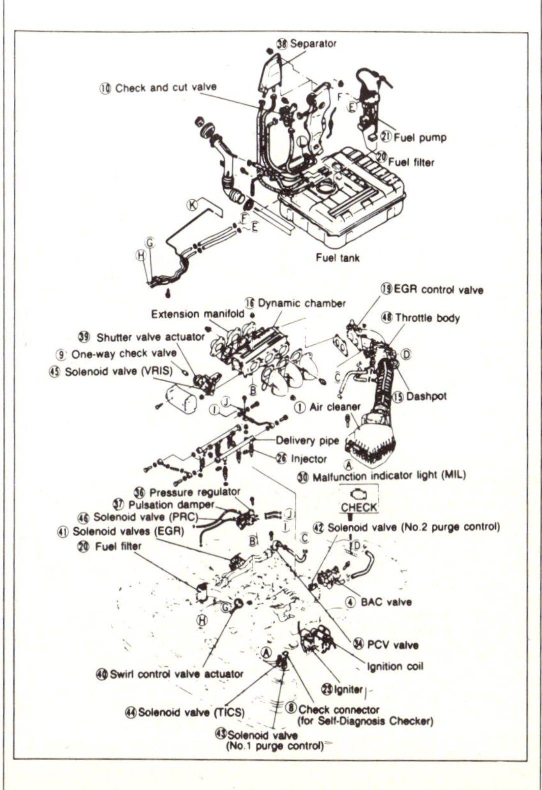

## EXPLODED VIEW OF INTAKE AIR SYSTEM COMPONENTS—1988–89 929

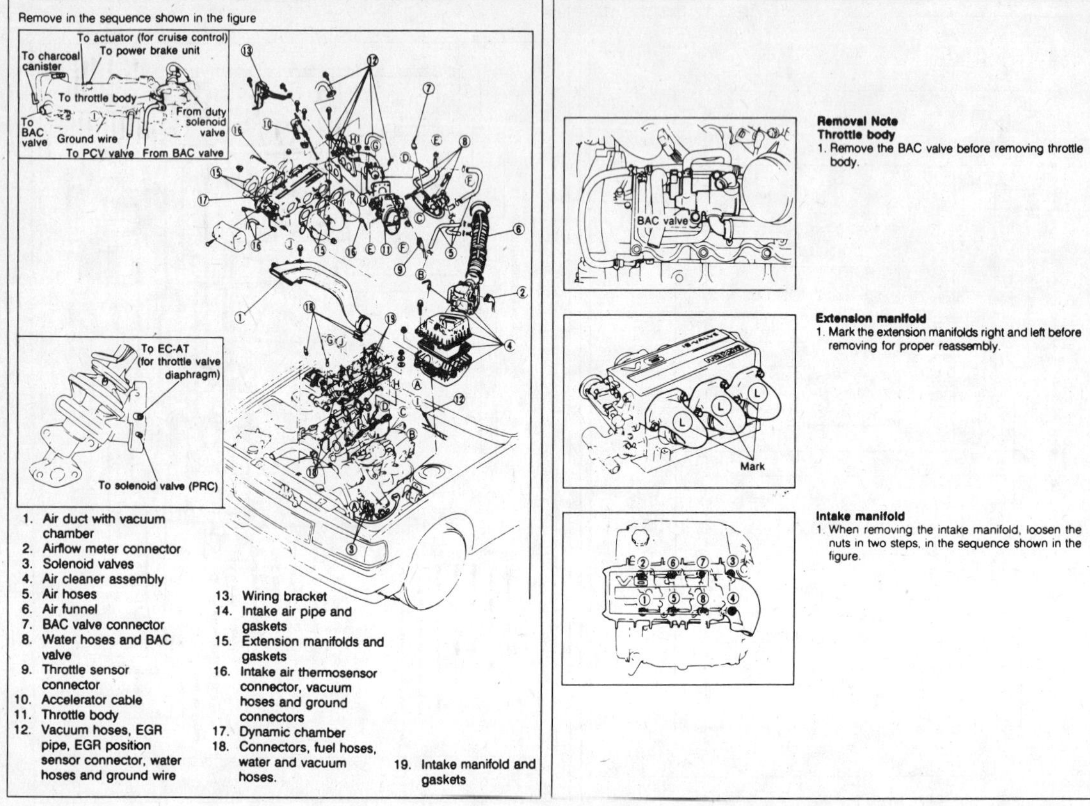


## SPECIFICATION CHART—1988–89 929

| Item | | | |
|---|---|---|---|
| Idle speed [rpm] | | | 650 ± 20 (A/T in P range) |
| Air cleaner | Element type | | Wet |
| Throttle body | Throat diameter [mm (in)] | | 45 (1.772) x 2 |
| Dashpot | Adjustment speed [rpm] | | 3,500 ± 300 |
| Throttle sensor | Idle switch (adjustment clearance) [mm (in)] | | 0.5 (0.020) (Continuity)<br>0.7 (0.028) (No continuity) |
| | Resistance [kΩ] | | Ⓐ – Ⓓ 3.5–6.5<br>Ⓑ – Ⓓ below 1 ↔ 3.5–6.5 |
| Air flow meter | Resistance [Ω] | $E_2$ ↔ Vs | 20–400 (Measuring plate fully closed), 20–1,000 (fully open) |
| | | $E_2$ ↔ Vc | 100–300 |
| | | $E_2$ ↔ $V_B$ | 200–400 |
| | | $E_2$ ↔ THA | −20°C (−4°F) 13,600–18,400<br>20°C (68°F) 2,210–2,690<br>60°C (140°F) 493–667 |
| | | $E_1$ ↔ Fc | Fully closed: ∞ Fully open: 0 |
| Fuel tank | Capacity [liters (US gal, Imp gal)] | | 70 (18.5, 15.4) |
| Fuel filter | Low pressure | | Nylon 6 (250 mesh) |
| | High pressure | | Filter paper |
| Fuel pump | Type | | Impeller (in tank) |
| | Output pressure [kPa (kg/cm², psi)] | | 441–588 (4.5–6.0, 64.0–85.3) |
| | Feeding capacity [liters (US gal, Imp gal)/min] | | 1.3–2.3 (0.34–0.61, 0.29–0.51) |
| Pressure regulator | Type | | Diaphragm |
| | Regulated pressure [kPa (kg/cm², psi)] | | 264–314 (2.7–3.2, 38.3–45.5) |
| Injector | Drive | | Voltage |
| | Type | | High-ohmic injector |
| | Resistance [Ω] | | 12–16 |
| | Injection amount [cc (cu in)/15 sec] | | 44–65 (2.68–3.97) |
| Ignition timing | | | BTDC 15 ± 1° (A/T in P range) |
| BAC valve | Resistance [Ω] | | 10.8–12.3 |
| Fuel specification | | | Unleaded gasoline |

## COMPONENTS DESCRIPTION—1988–89 929

| No. | Component | Function | Remark |
|---|---|---|---|
| 1 | **Air cleaner** | Filters air entering throttle body | |
| 2 | **Air flow meter** | Detects amount of intake air; sends signal to engine control unit | Intake air thermo sensor and fuel pump switch included |
| 3 | **Atmospheric pressure sensor** | Detects atmospheric pressure; sends signal to engine control unit | |
| 4 | **Air valve** | Supplies bypass air into dynamic chamber (engine cold) | • Engine speed increased to shorten warm-up period<br>• Thermo wax type<br>• Installed in BAC valve |
| | **BAC valve** | Supplies bypass air into dynamic chamber | Consists of air valve and ISC valve |
| | **ISC valve** | Supplies bypass air into dynamic chamber | • Controlled by duty signal from engine control unit<br>• Works independent of idle-up<br>• Installed in BAC valve |
| 5 | **Brake light switch** | Detects braking (deceleration); sends signal to engine control unit | |
| 6 | **Catalytic converter** | Reduce HC, CO and NOx by chemical reaction | Monolith type |
| 7 | **Charcoal canister** | Stores fuel tank fumes (engine stopped) | |
| 8 | **Check connectors** | For Self-Diagnosis Checker and fuel pump | • 6-pin connector (Green); Self-Diagnosis Checker<br>• 2-pin connector (Yellow); Fuel pump |
| 9 | **Check valve** | Prevents malfunction of shutter valve or swirl control valve during heavy load operation | |
| 10 | **Check and cut valve** | Controls pressure in fuel tank | |
| 11 | **Circuit opening relay** | Voltage for fuel pump (engine running) | |
| 12 | **Clutch switch** | Detects in-gear condition; sends signal to engine control unit | Switch closed when clutch pedal released |
| 13 | **Control unit (Engine control unit)** | Detects the following:<br>1. Engine speed<br>2. No.1 and No.4 piston TDC<br>3. Intake air amount<br>4. Engine coolant temperature<br>5. Radiator temperature<br>6. Throttle valve opening angle<br>7. Throttle valve fully closed<br>8. Air/fuel ratio (oxygen concentration)<br>9. In-gear condition<br>10. Intake air temperature<br>11. Engine air temperature<br>12. A/C operation<br>13. P/S operation<br>14. E/L operation<br>15. Cranking signal<br>16. Initial set signal<br>17. Braking signal<br>18. EGR operation signal<br>Controls operation of the following:<br>1. Fuel injection system<br>2. Idle speed control system<br>3. Pressure regulator control system<br>4. Variable resonance induction system<br>5. Triple induction control system<br>6. Ignition system (Electronic spark advance)<br>7. Purge control system<br>8. EGR control system<br>9. Fail-safe system<br>10. Switch monitor function | 1. Ne rotor (in distributor)<br>2. G rotors (in distributor)<br>3. Air flow meter<br>4. Water thermo sensor<br>5. Water thermo switch<br>6. Throttle sensor<br>7. Idle switch (in throttle sensor)<br>8. Oxygen sensor<br>9. Neutral, clutch, and inhibitor switches<br>10. Intake air thermo sensor (in air flow meter)<br>11. Intake air thermo sensor (on dynamic chamber)<br>12. A/C switch<br>13. P/S switch<br>14. E/L switch, E/L control unit<br>15. Ignition switch<br>16. Test connector<br>17. Brake light switch<br>18. EGR position sensor<br>1. Injector<br>2. BAC valve<br>3. Solenoid valve (Pressure regulator control)<br>4. Solenoid valve (VRIS)<br>5. Solenoid valve (TICS)<br>6. Igniter<br>7. Solenoid valve (Purge control)<br>8. Solenoid valves (EGR)<br>9. Self-Diagnosis Checker and MIL<br>10. Monitor lamp (Self-Diagnosis Checker) |

## COMPONENTS DESCRIPTION (CONT.)—1988–89 929

| No. | Component | Function | Remark |
|---|---|---|---|
| 14 | Distributor | Detects crank angle at 30° intervals, No.1 cylinder TDC and No.4 cylinder TDC; sends signal to engine control unit | • Ne (Number of engine speed) signal<br>• G1 (Group 1) signal<br>• G2 (Group 2) signal |
| | G1 coil | Detects No.4 cylinder TDC; sends signal to engine control unit | For determining fuel injection timing |
| | G2 coil | Detects No.1 cylinder TDC; sends signal to engine control unit | For determining fuel injection timing |
| 15 | Dashpot | Allows gradual throttle valve closing during deceleration | Adjustment speed: 3,500 ± 300 rpm |
| 16 | Dynamic chamber | Interconnects all cylinders | Contains shutter valve |
| 17 | Electrical load control unit | Detects E/L operation; sends signal to engine control unit | Electrical load<br>• Rear defroster<br>• Blower motor<br>• Headlight |
| 18 | Exhaust gas recirculation (EGR) position sensor | Detects EGR control valve opening; sends signal to engine control unit | Installed on EGR control valve |
| 19 | Exhaust gas recirculation control valve | Recirculates portion of exhaust gas | Contains EGR position sensor |
| 20 | Fuel filter | Filters fuel | |
| 21 | Fuel pump | Provides fuel to injectors | • Operates only while engine running<br>• Installed in fuel tank |
| 22 | Idle switch | Detects when throttle valve fully closed; sends signal to engine control unit | Installed in throttle sensor<br>Switch closed when throttle valve fully closed |
| | Throttle sensor | Detects throttle valve opening angle; sends signal to engine control unit | • Idle switch included<br>• Installed on throttle body |
| 23 | Igniter | Receives spark signal from engine control unit; generates high voltage in ignition coil | High voltage distributed by rotor in distributor |
| 24 | Ignition switch | Detects engine cranking; sends signal to engine control unit | |
| 25 | Inhibitor switch | Detects in-gear condition; sends signal to EC-AT control unit | Switch closed in N or P range (A/T) |
| 26 | Injector | Injects fuel into intake port | • Controlled by signals from engine control unit<br>• High-ohmic injector |
| 27 | Intake manifold | Supplies intake air to all cylinders | Contains swirl control valve |
| 2 | Intake air thermo sensor | Detects intake air temperature; sends signal to engine control unit | Installed in air flow meter |
| 28 | | Detects engine air temperature; sends signal to engine control unit | Installed on dynamic chamber |
| 29 | Main relay | Supplies current to injectors and engine control unit | |
| 30 | MIL (Malfunction indicator light) | Light illuminates when input device malfunctions | Installed in instrument cluster |
| 31 | Neutral switch | Detects in-gear condition; sends signal to engine control unit | Switch closed when in-gear |
| 32 | Oxygen sensor | Detects oxygen concentration; sends signal to engine control unit | Zirconia ceramic and platinum coating |
| 33 | Oxygen sensor relay | Supplies current to oxygen sensor heater | |

| No. | Component | | Function | Remark |
|---|---|---|---|---|
| 34 | PCV valve | | Controls blowby gas amount pulled into engine | |
| 35 | P/S switch | | Detects P/S operation; sends signal to engine control unit | ON when steering wheel turned right or left |
| 36 | Pressure regulator | | Adjusts fuel pressure supplied to injectors | |
| 37 | Pulsation damper | | Absorbs fuel pulsation | |
| 38 | Separator | | Prevents fuel from flowing into charcoal canister | |
| 39 | Shutter valve actuator | | Controls shutter valve operation | For variable resonance induction system (VRIS) |
| 40 | Swirl control valve actuator | | Controls swirl control valve operation | For triple induction control system (TICS) |
| 41 | Solenoid valve | EGR (Vent side, Vacuum side) | Controls EGR control valve operation | • Exhaust gas recirculation<br>• Duty solenoid valve |
| 42 | | No.2 purge control | Controls vacuum switch valve | |
| 43 | | No.1 purge control | Controls vacuum to No.1 purge control valve on charcoal canister | |
| 44 | | TICS | Controls vacuum to swirl control valve actuator | Triple induction control system (TICS) |
| 45 | | VRIS | Controls vacuum to shutter valve actuator | • Variable resonance induction system (VRIS) |
| 46 | | Pressure regulator control | Blocks vacuum from intake manifold to pressure regulator | Hot engine only |
| 47 | Test connector | | For Self-Diagnosis Checker and idle speed / ignition timing adjustment | 1-pin connector (Green) |
| 48 | Throttle body | | Controls intake air amount | Integrated throttle sensor and dashpot incorporated |
| 49 | Water thermo sensor | | Detects coolant temperature; sends signal to engine control unit | |
| 50 | Water thermo switch | | Detects radiator coolant temperature; sends signal to engine control unit | ON: above 17°C (62.6°F) |

## TROUBLESHOOTING CHART—1988–89 929

This troubleshooting guide shows the malfunction code numbers and the symptoms of various failures. Perform troubleshooting as described below.

| | Symptoms \ Possible cause | Input sensors | | | | | | | | | | | | | Output solenoid valves | |
|---|---|---|---|---|---|---|---|---|---|---|---|---|---|---|---|---|
| | | Ignition pulse | Distributor (Ne signal) | Distributor (G1 signal) | Distributor (G2 signal) | Air flow meter | Water thermo sensor | Intake air thermo sensor (Air flow meter) | Intake air thermo sensor (Dynamic chamber) | Throttle sensor | Atmospheric pressure sensor | Oxygen sensor | EGR position sensor | Fuel feedback system | Solenoid valve (Pressure regulator control) | Solenoid valve (No. 2 purge control) |
| 1 | Fault indicated by SST malfunction code No. | 01 | 02 | 03 | 04 | 08 | 09 | 10 | 11 | 12 | 14 | 15 | 16 | 17 | 25 | 26 |
| 2 | Hard starts or won't start (cranks OK) | | | | | | | | | | | | | | | |
| 3 | Engine stalls — While warming up | | | | | | | | | | | | | | | |
| | Engine stalls — After warming up | | | | | | | | | | | | | | | |
| 4 | Rough idle — While warming up | | | | | | | | | | | | | | | |
| | Rough idle — After warming up | | | | | | | | | | | | | | | |
| 5 | High speed idle after warming up | | | | | | | | | | | | | | | |
| 6 | Poor acceleration, hesitation or lack of power | | | | | | | | | | | | | | | |
| 7 | Runs rough on deceleration | | | | | | | | | | | | | | | |
| 8 | Afterburn in exhaust system | | | | | | | | | | | | | | | |
| 9 | Poor fuel consumption | | | | | | | | | | | | | | | |
| 10 | Excessive oil consumption | | | | | | | | | | | | | | | |
| 11 | Engine stalls or runs rough after hot starting | | | | | | | | | | | | | | | |
| 12 | Fails emission test | | | | | | | | | | | | | | | |

**TROUBLESHOOTING PROCEDURE**

**Note**

**Symptom 1 shows the code No. to quickly determine what system or unit may be at fault by use of the SST.**
**(SST: Self-Diagnosis Checker 49 H018 9A1)**

1st Check the electrical system's parts
1) Battery condition
2) Main relays
3) Main fuses
4) Fuses

2nd Check input sensors and output solenoid valves self-diagnosed with **SST**

3rd Check input switches with **SST**

4th Check the following items.

Ignition system
1) Ignition spark
2) Ignition timing (with test connector grounded)

Intake air system
1) Air cleaner element
2) Vacuum of air leakage
3) Vacuum hose and air hose routing
4) Accelerator cable

Fuel system
1) Fuel amount
2) Fuel leakage
3) Fuel filter
4) Idle speed (with test connector grounded)

Engine
1) Compression
2) Overheating

Others
1) Clutch slippage
2) Brake dragging

5th Check Fuel and Emission Control Systems

The troubleshooting guide lists the systems most likely to cause a given symptom. After finding systems to check, refer to the pages shown for detailed guides for each system.

| Output solenoid valves | | | | | |
|---|---|---|---|---|---|
| Solenoid valve (No. 1 purge control) | Solenoid valve (EGR, vacuum side) | Solenoid valve (EGR, vent side) | Idle speed control valve | Solenoid valve (TICS) or Oxygen sensor relay | Solenoid valve (VRIS) |
| 27 | 28 | 29 | 34 | 40 | 41 |

| Possible cause / Symptom | Intake air system (Poor connection of components, throttle body) | Idle speed control system (Air valve, Idle speed control valve malfunction) | VRIS (Variable resonance induction system) | TICS (Triple induction control system) | Fuel system (Fuel injection, fuel pressure) | Pressure regulator control system | Electronic spark advance control system | Deceleration control system (Fuel cut operation malfunction) | EGR system (EGR control valve struck and open) | Evaporative emission control system (Vacuum switch valve, No. 1 purge control valve malfunction) | PCV system (System clogged) | Exhaust system (System clogged) |
|---|---|---|---|---|---|---|---|---|---|---|---|---|
| 2 | 3 | | | | 2 | | 1 | | 4 | | | |
| 3 | 4 | 1 | | | 3 | | | | 2 | | | |
| | 5 | 2 | | | 4 | | | | 3 | | 1 | |
| 4 | 6 | 1 | | | 5 | | | | 3 | 4 | 2 | |
| | 6 | 2 | | | 5 | | | | 3 | 4 | 1 | |
| 5 | 3 | 2 | | | 4 | | | 1 | | | | |
| 6 | 3 | | 5 | 6 | 4 | | | | 1 | 2 | | 7 |
| 7 | | 2 | | | 3 | | | 1 | | | | |
| 8 | 3 | 2 | | | 4 | | | 1 | | | | |
| 9 | | | | 2 | 4 | 5 | | 1 | 3 | | | 6 |
| 10 | | | | | | | | | | | 1 | |
| 11 | | 2 | | | 3 | 1 | | | | | | |
| 12 | 6 | 5 | | | 7 | 8 | | 3 | 2 | 4 | | 1 |

The numbers of the list show the priorities of inspections from the most possible to that with the lowest possibility.
These were determined on the following basis:
- Ease of inspection
- Most possible system
- Most possible point in system

### MALFUNCTION INDICATOR LIGHT (MIL) OPERATION 1988–89 929

**MALFUNCTION CODE FUNCTION**

Malfunction codes are determined as below

**1. Malfunction code cycle break**
The time between malfunction code cycles is 4.0 sec (the time the MIL and buzzer are is off)

1 cycle
Malfunction code cycle break
ON
OFF
4.0 sec.

**2. Second digit of malfunction code (ones position)**
The digit in the ones position of the malfunction code represents the number of times the MIL and buzzer are on 0.4 sec during one cycle

1 cycle Malfunction code
1 2 3 cycle break
ON
OFF
0.4 sec. 0.4 sec.
4.0 sec.

**3. First digit of malfunction code (tens position)**
The digit in the tens position of the malfunction code represents the number of times the MIL and buzzer are on 1.2 sec during one cycle

The MIL and buzzre are off for 1.6 sec. between the long and short pulses

1 cycle Malfunction code
1 2 1 2 cycle break
0.4
1.2 sec. sec. 1.6 sec.
4.0 sec.
First digit
Second digit

### TROUBLE CODE IDENTIFICATION CHART—1988–89 929

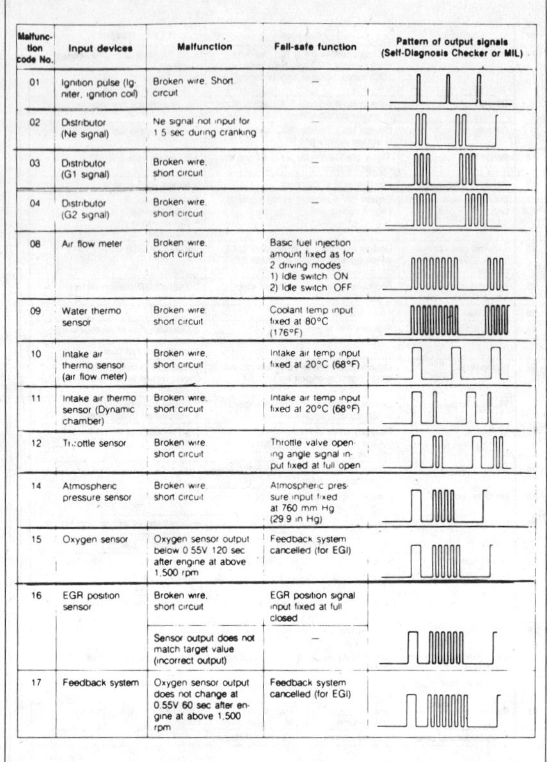

| Malfunction code No. | Input devices | Malfunction | Fail-safe function | Pattern of output signals (Self-Diagnosis Checker or MIL) |
|---|---|---|---|---|
| 01 | Ignition pulse (Igniter, ignition coil) | Broken wire, Short circuit | — | |
| 02 | Distributor (Ne signal) | Ne signal not input for 1.5 sec during cranking | — | |
| 03 | Distributor (G1 signal) | Broken wire, short circuit | — | |
| 04 | Distributor (G2 signal) | Broken wire, short circuit | — | |
| 08 | Air flow meter | Broken wire, short circuit | Basic fuel injection amount fixed as for 2 driving modes 1) Idle switch: ON 2) Idle switch: OFF | |
| 09 | Water thermo sensor | Broken wire, short circuit | Coolant temp input fixed at 80°C (176°F) | |
| 10 | Intake air thermo sensor (air flow meter) | Broken wire, short circuit | Intake air temp input fixed at 20°C (68°F) | |
| 11 | Intake air thermo sensor (Dynamic chamber) | Broken wire, short circuit | Intake air temp input fixed at 20°C (68°F) | |
| 12 | Throttle sensor | Broken wire, short circuit | Throttle valve opening angle signal input fixed at full open | |
| 14 | Atmospheric pressure sensor | Broken wire, short circuit | Atmospheric pressure input fixed at 760 mm Hg (29.9 in Hg) | |
| 15 | Oxygen sensor | Oxygen sensor output below 0.55V 120 sec after engine at above 1,500 rpm | Feedback system cancelled (for EGI) | |
| 16 | EGR position sensor | Broken wire, short circuit | EGR position signal input fixed at full closed | |
| | | Sensor output does not match target value (incorrect output) | — | |
| 17 | Feedback system | Oxygen sensor output does not change at 0.55V 60 sec after engine at above 1,500 rpm | Feedback system cancelled (for EGI) | |

### TROUBLE CODE IDENTIFICATION CHART (CONT.) 1988–89 929

**Output Devices**

| Malfunction code | Output devices | Pattern of output signals (Self-Diagnosis Checker or MIL) |
|---|---|---|
| 25 | Solenoid valve(Pressure regulator control) | |
| 26 | Solenoid valve (No. 2 purge control) | |
| 27 | Solenoid valve (No. 1 purge control) | |
| 28 | Solenoid valve (EGR, vacuum side) | |
| 29 | Solenoid valve (EGR, vent side) | |
| 34 | Idle speed control valve (ISC valve) | |
| 40 | Solenoid valve (Triple induction control system) and oxygen sensor relay | |
| 41 | Solenoid valve (Variable resonance induction system) | |

**Caution**
**a) If there is more than one failure present, the lowest number malfunction code is displayed first, the subsequent malfunction codes light up in order.**
**b) After repairing all failures, do to erase the memory of malfunction codes.**

**c) Inspect the open circuit of solenoid valve (TICS) or oxygen sensor relay according to the following procedure.**

**Procedure**
1. Warm up the engine to normal operating temperature and stop it.
2. Connect the **SST** to the check connector (Green, 6-pin) and the negative battery terminal.
3. Connect a jumper wire between the test connector (Green, 1-pin) and a ground.
4. Remove the oxygen sensor relay and turn the ignition switch ON. Check that malfunction code No. 40 is not illuminated. If malfunction code No. 40 is illuminated, check the solenoid valve (TICS)
5. Turn the ignition switch OFF. Install the oxygen sensor relay, and disconnect the solenoid valve (TICS) connector.
6. Turn the ignition switch ON, and check that malfunction code No. 40 is not illuminated. If the malfunction code No. 40 is illuminated, check the oxygen sensor relay

Oxygen sensor relay
Solenoid valve (TICS)

### TROUBLE CODE DIAGNOSTIC CHART—1988–89 929

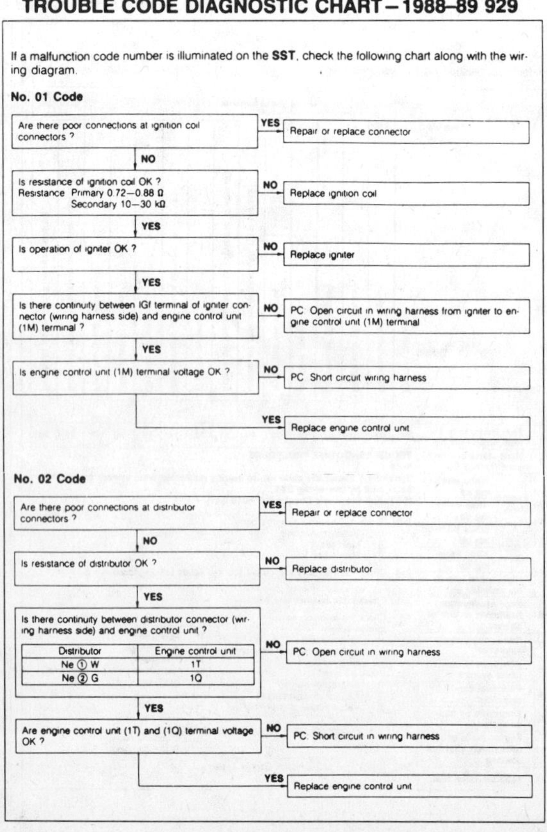

If a malfunction code number is illuminated on the **SST**, check the following chart along with the wiring diagram.

**No. 01 Code**

- Are there poor connections at ignition coil connectors? — YES → Repair or replace connector
- NO ↓
- Is resistance of ignition coil OK? Resistance: Primary 0.72–0.88 Ω, Secondary 10–30 kΩ — NO → Replace ignition coil
- YES ↓
- Is operation of igniter OK? — NO → Replace igniter
- YES ↓
- Is there continuity between IGf terminal of igniter connector (wiring harness side) and engine control unit (1M) terminal? — NO → PC: Open circuit in wiring harness from igniter to engine control unit (1M) terminal
- YES ↓
- Is engine control unit (1M) terminal voltage OK? — NO → PC: Short circuit wiring harness
- YES → Replace engine control unit

**No. 02 Code**

- Are there poor connections at distributor connectors? — YES → Repair or replace connector
- NO ↓
- Is resistance of distributor OK? — NO → Replace distributor
- YES ↓
- Is there continuity between distributor connector (wiring harness side) and engine control unit? — NO → PC: Open circuit in wiring harness

| Distributor | Engine control unit |
|---|---|
| Ne ① W | 1T |
| Ne ② G | 1Q |

- YES ↓
- Are engine control unit (1T) and (1Q) terminal voltage OK? — NO → PC: Short circuit in wiring harness
- YES → Replace engine control unit

## TROUBLE CODE DIAGNOSTIC CHART—1988–89 929

### No. 03 Code

Are there poor connections at distributor connectors ? — YES → Repair or replace connector

NO ↓

Is resistance of distributor OK ? — NO → Replace distributor

YES ↓

Is there continuity between distributor connector (wiring harness side) and engine control unit ? — NO → PC Open circuit in wiring harness

| Distributor | Engine control unit |
|---|---|
| G1 ① L | 1N |
| G1 ② YG | 1P |

YES ↓

Are engine control unit (1N) and (1P) terminal voltage OK ? — NO → PC Short circuit in wiring harness

YES → Replace engine control unit

### No. 04 Code

Are there poor connections at distributor connectors ? — YES → Repair or replace connector

NO ↓

Is resistance of distributor OK ? — NO → Replace distributor

YES ↓

Is there continuity between distributor connector (wiring harness side) and engine control unit ? — NO → PC Open circuit in wiring harness

| Distributor | Engine control unit |
|---|---|
| G2 ① R | 1O |
| G2 ② YL | 1P |

YES ↓

Are engine control unit (1O) and (1P) terminal voltage OK ? — NO → PC Short circuit in wiring harness

YES → Replace engine control unit

### No. 08 Code

Are there poor connections at air flow meter connectors — YES → Repair or replace connector

NO ↓

Is resistance of air flow meter OK ?
Resistance — NO → Repair air flow meter

| Terminal | Fully closed (Ω) | Fully open (Ω) |
|---|---|---|
| E2 ↔ Vs | 20—400 | 20—1,000 |
| E2 ↔ Vc | 100—300 | |
| E2 ↔ VB | 200—400 | |

YES ↓

Is there continuity between air flow meter connector (wiring harness side) and engine control unit ? — NO → PC Open circuit in wiring harness from air flow meter to engine control unit

| Air flow meter | Control unit |
|---|---|
| VB (BW wire) | 3I |
| Vc (BrY wire) | 2B |
| Vs (L wire) | 2E |
| E2 (RB wire) | 2C |

YES ↓

Are engine control unit terminal voltages OK ?
Check terminal 2B, 2C, 2E, 3I — NO → PC Short circuit in wiring harness

YES → Replace engine control unit

## TROUBLE CODE DIAGNOSTIC CHART—1988–89 929

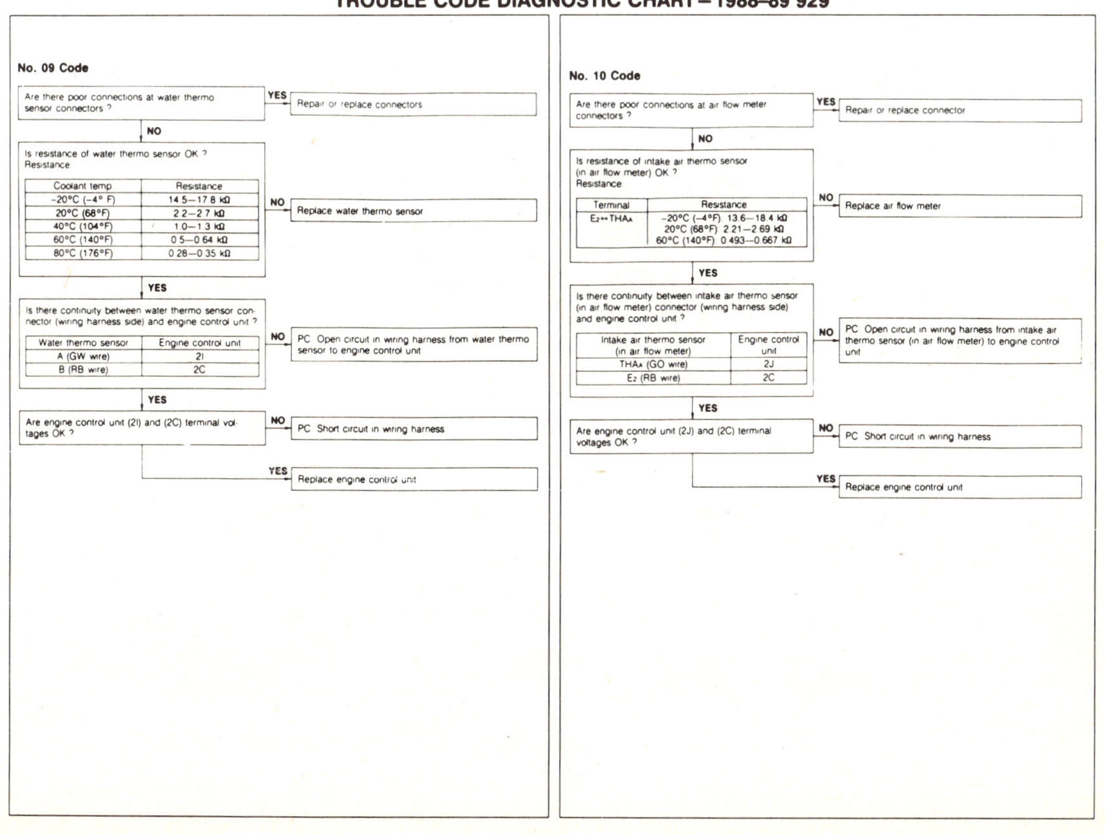

## TROUBLE CODE DIAGNOSTIC CHART—1988–89 929

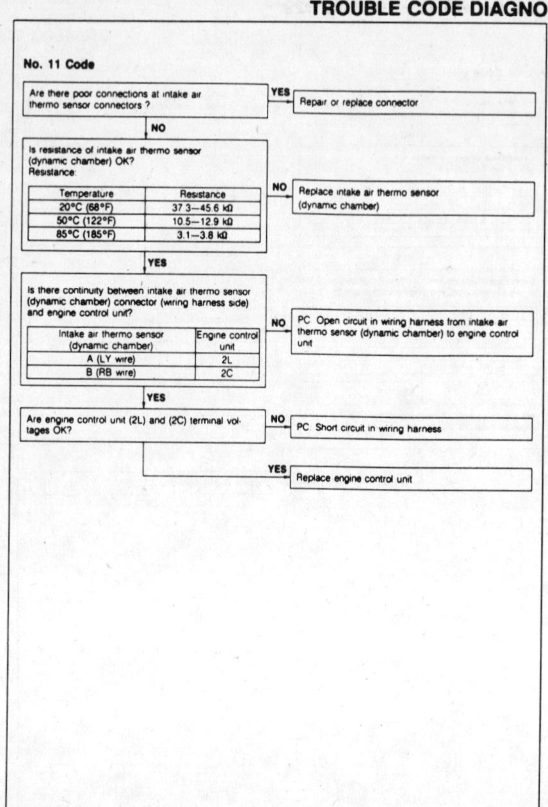


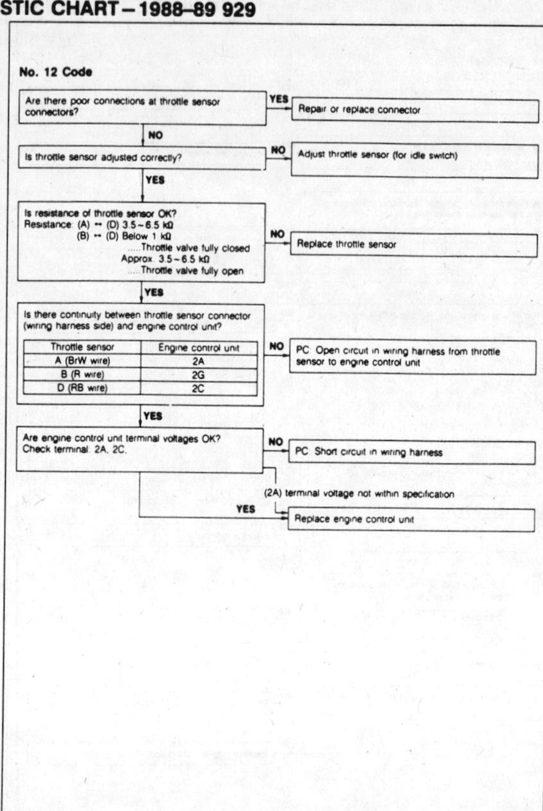


## TROUBLE CODE DIAGNOSTIC CHART—1988–89 929

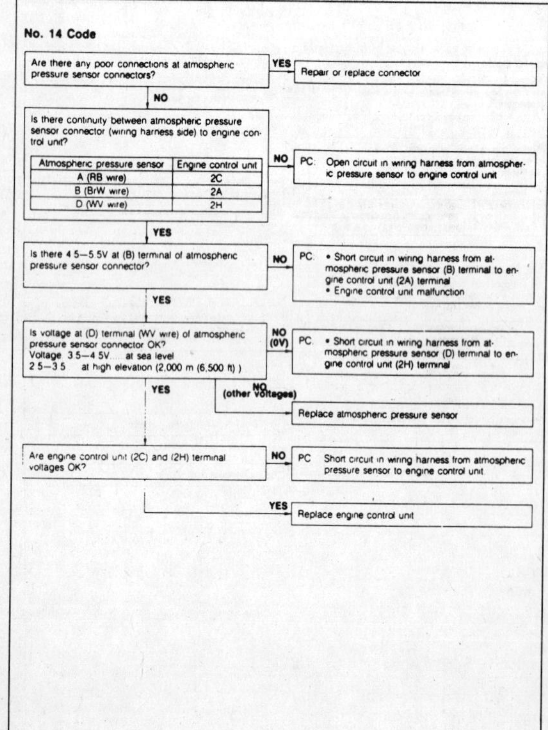


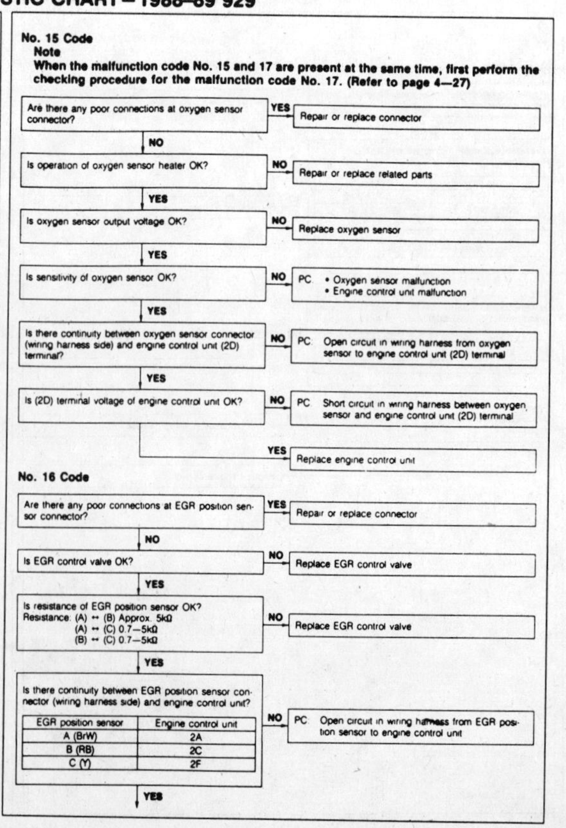

## TROUBLE CODE DIAGNOSTIC CHART—1988–89 929

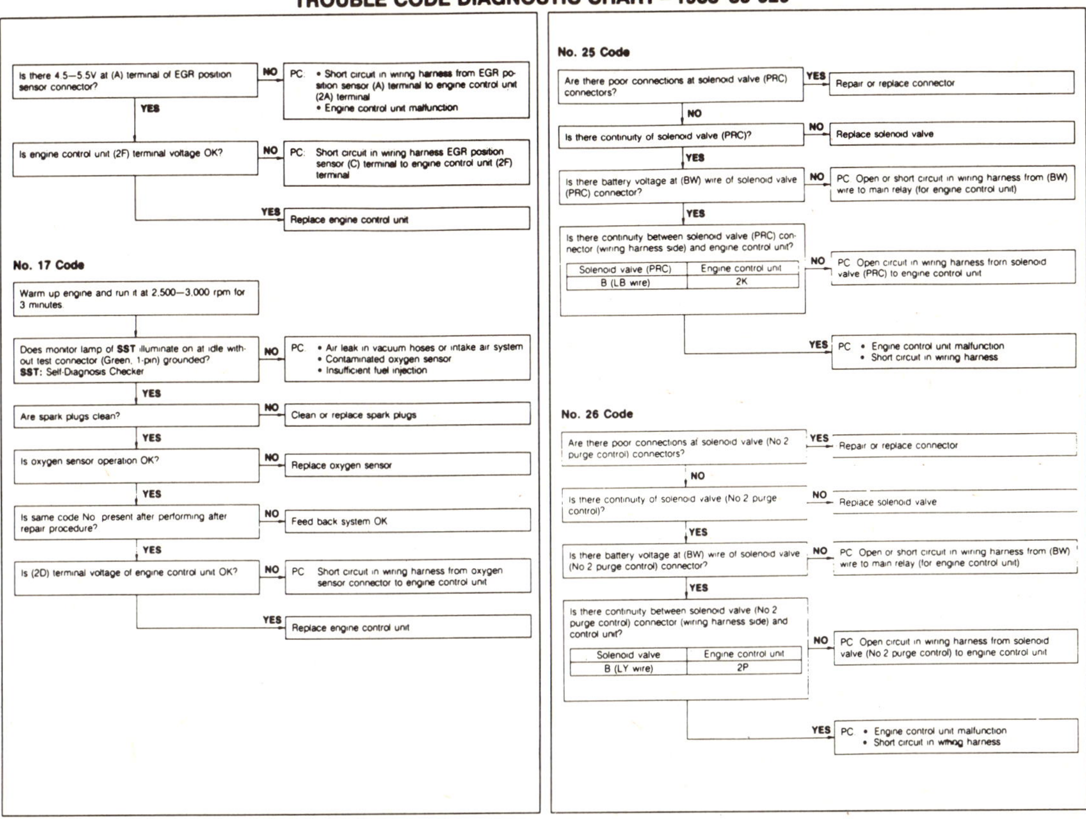


## TROUBLE CODE DIAGNOSTIC CHART—1988–89 929

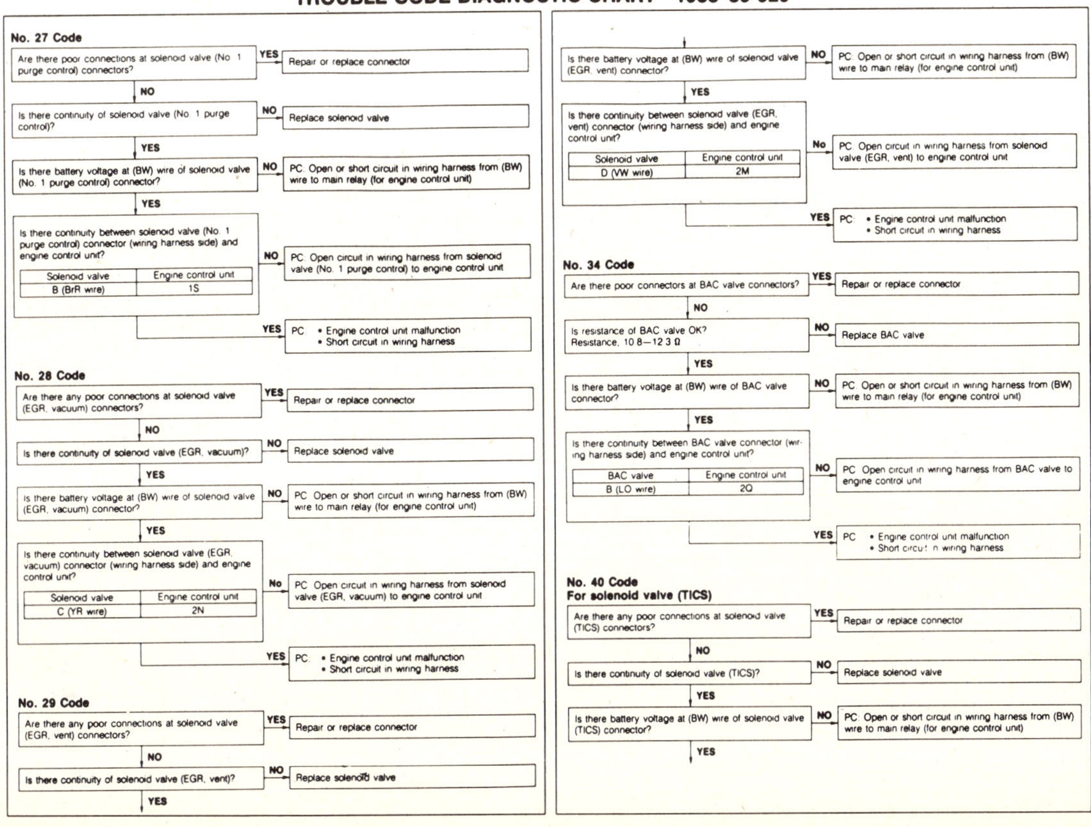

## TROUBLE CODE DIAGNOSTIC CHART—1988–89 929

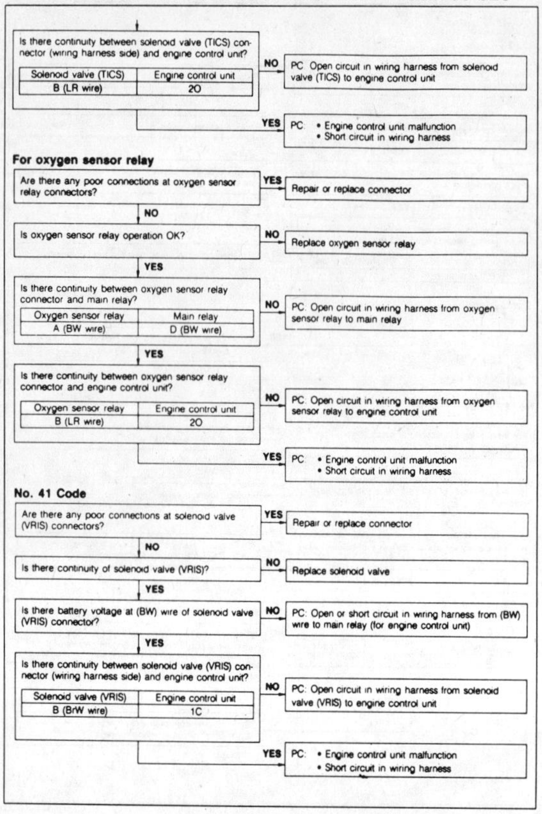

Is there continuity between solenoid valve (TICS) connector (wiring harness side) and engine control unit?

| Solenoid valve (TICS) | Engine control unit |
|---|---|
| B (LR wire) | 2O |

- NO → PC: Open circuit in wiring harness from solenoid valve (TICS) to engine control unit
- YES → PC: • Engine control unit malfunction • Short circuit in wiring harness

### For oxygen sensor relay

Are there any poor connections at oxygen sensor relay connectors?
- YES → Repair or replace connector
- NO ↓

Is oxygen sensor relay operation OK?
- NO → Replace oxygen sensor relay
- YES ↓

Is there continuity between oxygen sensor relay connector and main relay?

| Oxygen sensor relay | Main relay |
|---|---|
| A (BW wire) | D (BW wire) |

- NO → PC: Open circuit in wiring harness from oxygen sensor relay to main relay
- YES ↓

Is there continuity between oxygen sensor relay connector and engine control unit?

| Oxygen sensor relay | Engine control unit |
|---|---|
| B (LR wire) | 2O |

- NO → PC: Open circuit in wiring harness from oxygen sensor relay to engine control unit
- YES → PC: • Engine control unit malfunction • Short circuit in wiring harness

### No. 41 Code

Are there any poor connections at solenoid valve (VRIS) connectors?
- YES → Repair or replace connector
- NO ↓

Is there continuity of solenoid valve (VRIS)?
- NO → Replace solenoid valve
- YES ↓

Is there battery voltage at (BW) wire of solenoid valve (VRIS) connector?
- NO → PC: Open or short circuit in wiring harness from (BW) wire to main relay (for engine control unit)
- YES ↓

Is there continuity between solenoid valve (VRIS) connector (wiring harness side) and engine control unit?

| Solenoid valve (VRIS) | Engine control unit |
|---|---|
| B (BrW wire) | 1C |

- NO → PC: Open circuit in wiring harness from solenoid valve (VRIS) to engine control unit
- YES → PC: • Engine control unit malfunction • Short circuit in wiring harness

## CONTROL UNIT PIN LOCATION AND VOLTAGE TEST 1988–89 929

**Terminal voltage**

| Terminal | Input | Output | Connection to | Voltage (after warm-up) | | Remarks |
|---|---|---|---|---|---|---|
| | | | | Ignition switch: ON | Idle | |
| 1A | | O | MIL (Malfunction Indicator Light) | For 3 sec after ignition switch OFF → ON: below 2.5V (light illuminates) After 3 sec: approx. 12V | | • Test connector grounded • Light illuminates: below 2.5V • Light does not illuminate: approx. 12V |
| 1B | | O | Self-Diagnosis checker (for Code No) | For 3 sec after ignition switch OFF → ON: below 2.5V (Buzzer sounds) After 3 sec: approx. 12V (Buzzer does not sound) | | • With Self-Diagnosis Checker and Test connector grounded • Buzzer sounds: below 2.5V • Buzzer does not sound: approx. 12V |
| 1C | | O | Solenoid valve (Variable resonance induction system (VRIS) ) | approx. 12V | | • 3,700—5,750 rpm: below 2.0V • Other conditions: approx. 12V |
| 1D | | O | Self-Diagnosis Checker (Monitor lamp) | <Test connector grounded> For 3 sec after ignition switch OFF → ON: approx. 5V (lamp illuminates) After 3 sec: approx. 12V | <Test connector grounded> approx. 5V <Test connector is not grounded> Monitor lamp ON; Below 6V Monitor lamp OFF approx. 12V | With Self-Diagnosis Checker |
| 1E | O | | Idle switch | Pedal depressed: approx. 10V Pedal released: 0V | | |
| 1F | | O | A/C relay | A/C switch ON; below 2.5V A/C switch OFF; approx. 12V | | Blower motor: ON |
| 1G | O | | Neutral/clutch switch | Clutch pedal depressed: approx. 12V Clutch pedal released: 0V | | In-gear condition (Neutral constant 12V) |
| 1H | O | | Water thermo switch | 0V | | Radiator temp below 17°C (62.6°F): approx. 12V |
| 1I | O | | Electrical load (E/L) control unit | E/L switch ON: below 2.5V E/L switch OFF: approx. 10—12V | | Electrical load switch: Blow motor switch Rear defroster switch Headlight switch |
| 1J | O | | Brake light switch | Brake pedal depressed: approx. 12V Brake pedal released: 0V | | |
| 1K | O | | Power steering (P/S) switch | Constant approx. 12V | P/S: ON: 0V P/S: OFF: approx. 12V | |
| 1L | O | | A/C switch | A/C switch ON: below 2.5V A/C switch OFF: approx. 12V | | Blower motor: ON |
| 1M | O | | Igniter | 0V | Approx. 0.6—0.8V | Engine Signal Monitor green and red lights flash |
| 1N | O | | Distributor (G1 signal) | Approx. 0.6—0.8V | | G1: Group 1 |

## CONTROL UNIT PIN LOCATION AND VOLTAGE TEST (CONT.)—1988–89 929

| Terminal | Input | Output | Connection to | Voltage (after warm-up) | | Remarks |
|---|---|---|---|---|---|---|
| | | | | Ignition switch: ON | Idle | |
| 1O | O | | Distributor (G2 signal) | Approx. 0.6—0.8V | | G2: Group 2 |
| 1P | O | | Distributor (G1, G2 ground) | Approx. 0.6—0.8V | | |
| 1Q | O | | Distributor (Ne ground) | Approx. 0.6—0.8V | | Ne: Number of engine speed |
| 1R | — | — | — | — | | — |
| 1S | | O | Solenoid valve (No. 1 purge control) | Approx. 12V | Approx. 12V | No-load condition |
| | | | | | Initial of acceleration or Above 900 rpm: below 2.5V Other condition: Approx. 12V | <Load condition> M/T: Disconnect neutral switch connector and connect jumper wire to connector EC-AT: Disconnect EC-AT control unit connector (20-pin) |
| 1T | O | | Distributor (Ne signal) | Approx. 0.6—0.8V | | Ne: Number of engine speed |
| 1U | — | — | — | For 180 sec after ignition switch OFF → ON: Approx. 12V | For 180 sec after starting: Approx. 12V | During hot condition: Coolant temp.: above 90°C (194°F) and engine air temp.: above 85°C (185°F) |
| | | | | Below 2.5V | | Other condition |
| 1V | O | | — | Below 1.5V | | M/T |
| | | | | Approx. 12V | | EC-AT |
| 1W | O | | Test Connector | Test connector: grounded 0v Test connector: Not grounded approx. 12V | | Green connector, 1-pin |
| 1X | | O | Igniter | 0V | Approx. 0.8V | |
| 2A | | O | Vref | 4.5—5.5V | | |
| 2B | O | | Air flow meter (Vc) | 7—9V | | |
| 2C | — | — | Ground (E2) | 0V | | |
| 2D | O | | Oxygen sensor | 0V | 0—1V | • Cold engine: 0V at idle • After warming-up: Acceleration: 0.7—1.0V Deceleration: 0—0.2V |
| 2E | O | | Air flow meter (Vs) | Approx. 2.0V | Approx. 3.5—5V | |
| 2F | O | | EGR position sensor | Approx. 0.25—0.9V | | |
| 2G | O | | Throttle sensor | Throttle valve fully closed: approx. 0.3V Throttle valve fully open: approx. 4V | | |
| 2H | O | | Atmospheric pressure sensor | Approx. 3.5—4.5V at sea level | | |

| Terminal | Input | Output | Connection to | Voltage (after warm-up) | | Remarks |
|---|---|---|---|---|---|---|
| | | | | Ignition switch: ON | Idle | |
| 2I | O | | Water thermo sensor | Approx. 0.3—0.6V | | |
| 2J | O | | Intake air thermo sensor (air flow meter) | Approx. 2—3V at 20°C (68°F) | | |
| 2K | | O | Solenoid valve (Pressure regulator control) | For 240 sec after ignition switch OFF → ON: below 2.5V | For 240 sec after starting: below 2.5V | During hot condition: coolant temp.: above 80°C (176°F) and engine air temp.: above 75°C (167°F) |
| | | | | Approx. 12V | | Other conditions |
| 2L | O | | Intake air thermo sensor (Dynamic chamber) | Approx. 1—2V at 80°C (176°F) | | |
| 2M | | O | Solenoid valve (EGR, vent side) | Approx. 12V | | Initial of acceleration; Engine Signal Monitor green and red light flash |
| 2N | | O | Solenoid valve (EGR, vacuum side) | Approx. 12V | | Initial of acceleration; Engine Signal Monitor green and red light flash |
| 2O | | O | Solenoid valve (Triple induction control system (TICS) ) and oxygen sensor relay | Below 2.5V | | Above approx. 4,000 rpm: approx. 12V |
| 2P | | O | Solenoid valve (No.2 purge control) | Approx. 12V | | Acceleration over 1,500 rpm: below 2.5V |
| 2Q | | O | Idle speed control (ISC) valve | Approx. 2—5V | Approx. 9—11V | Engine Signal Monitor green and red lights flash |
| 2R | — | — | Ground ($E_{02}$) | 0V | | |
| 3A | — | — | Ground ($E_{01}$) | 0V | | |
| 3B | O | | Ignition switch (START position) | Below 2.5V | | While cranking: 6—11V |
| 3C | | O | Injector (No.3 and No.4) | Approx. 12V | | Engine Signal Monitor green and red lights flash |
| 3D | O | | Inhibitor switch | "N" or "P" range: 0V Other ranges: approx. 12V | | M/T: constant 0V |
| 3E | | O | Injector (No.1 and No.2) | Approx. 12V | | Engine Signal Monitor green and red lights flash |
| 3F | | O | Injector (No.5 and No.6) | Approx. 12V | | Engine Signal Monitor green and red lights flash |
| 3G | — | — | Ground ($E_1$) | 0V | | — |
| 3H | — | — | — | — | | — |
| 3I | O | | Main relay | Approx. 12V | | |
| 3J | — | — | Battery | Approx. 12V | | For back-up |

## COMPONENT RELATIONSHIP CHART—1988–89 929

| OUTPUT DEVICE \ INPUT DEVICE | | IG COIL | DISTRIBUTOR: Ne ROTOR (1Q, 1T) | DISTRIBUTOR: G ROTOR (1R, 1O, 1P) | AIR FLOW METER (2E) | THROTTLE SENSOR: IDLE SWITCH (1E) | THROTTLE SENSOR: THROTTLE SENSOR (2G) | WATER THERMO SENSOR (2I) | WATER THERMO SWITCH (1H) | INTAKE AIR THERMO SENSOR: AIR FLOW METER (2J) | INTAKE AIR THERMO SENSOR: DYNAMIC CHAMBER (2L) | OXYGEN SENSOR (2D) | Atmospheric pressure sensor (2H) | EGR position sensor (2F) | TEST CONNECTOR (1W) | NEUTRAL CLUTCH INHIBITOR SWITCH (1G, 3D) | A/C SWITCH (1L) | P/S SWITCH (1K) |
|---|---|---|---|---|---|---|---|---|---|---|---|---|---|---|---|---|---|---|
| INJECTOR (3C, 3E, 3F) | FUEL INJECTION AMOUNT | X | O | X | O | O | O | O | O | O | X | O | O | X | X | O | O | X |
| | FUEL INJECTION TIMING | X | O | O | X | O | O | X | X | X | X | X | X | X | X | X | X | X |
| PRESSURE REGULATOR (2K) | | X | O | X | X | X | O | O | X | X | O | X | X | X | X | X | X | X |
| BAC VALVE | AIR VALVE | X | X | X | X | X | X | X | X | X | X | X | X | X | X | X | X | X |
| | ISC VALVE (3Q) | X | O | X | X | O | O | O | X | O | O | X | X | X | O | O | O | O |
| SOLENOID VALVE (VRIS) (1C) | | X | O | X | O | X | X | X | X | X | X | X | X | X | X | X | X | X |
| SOLENOID VALVE (TICS) (2O) | | X | O | X | O | X | X | X | X | X | X | X | X | X | X | X | X | X |
| IGNITER (1X) | | X | O | O | O | O | O | O | X | X | X | X | X | O | O | O | O | X |
| EVAPORATIVE EMISSION CONTROL SYSTEM | SOLENOID VALVE (No.1 PURGE CONTROL)(1S) | X | O | X | X | O | X | O | X | X | X | X | X | X | X | O | X | X |
| | SOLENOID VALVE (No.2 PURGE CONTROL)(2P) | X | O | X | O | X | X | O | X | X | X | X | O | X | X | X | X | X |
| EGR | Vent side (2M) | X | O | X | O | O | O | O | O | X | X | X | X | O | X | X | X | X |
| | Vacuum side (2N) | X | O | X | O | O | O | O | O | X | X | X | X | O | X | X | X | X |

O: Related X: Not related

## OUTPUT COMPONENTS AND ENGINE CONDITION 1988–89 929

| OUTPUT DEVICE \ ENGINE CONDITION | | CRANKING (COLD ENGINE) | WARM-UP (DURING IDLE) | MEDIUM LOAD: COLD | MEDIUM LOAD: WARM | ACCELERATION | HEAVY LOAD | DECELERATION | IDLE (THROTTLE VALVE FULLY CLOSED) | IG: ON (ENGINE NOT RUNNING) | REMARKS |
|---|---|---|---|---|---|---|---|---|---|---|---|
| INJECTOR | No.1 No.2 / No.3 No.4 / No.5 No.6 | Rich | | | *1 Rich and lean (Fuel feedback operation) | Rich | | Fuel cut | *1 Rich and lean (Fuel feedback operation) | Does not inject | *1 Operated by oxygen sensor |
| | INJECTION TIMING | 1 group | 3 group | | | | | | 3 group | | Fuel cut: above 6,500 rpm |
| SOLENOID VALVE (PRC) | | OFF (Vacuum to pressure regulator) | | | | | | | *1 After start ON (Vacuum cut) | Does not operate | *1 During hot condition only |
| BAC VALVE | AIR VALVE | *1 Open | | | Close | | | | | | *1 Coolant temp.: below 60°C (140°F) |
| | ISC VALVE | *1 Large amount of bypass air | | Large and small amount of bypass air | | | | | *1 Large and small amount of bypass air | Does not operate | *1 Engine speed feedback system operates when the test connector open |
| SOLENOID VALVE (VRIS) | | OFF (Shutter valve closed) | | *1 ON (Shutter valve open) | | | | OFF (Shutter valve closed) | | Does not operate | *1 ON: approx. 3,700—5,750 rpm |
| SOLENOID VALVE (TICS) | | ON (Swirl control valve closed) | | OFF (Swirl control valve open) | | | | ON (Swirl control valve closed) | | Does not operate | |
| IGNITER | | BTDC 7° | Max.; BTDC 42° Min.; BTDC 11° | Advance | Max.; BTDC 42° Min.; BTDC −6° | | | Gradually returned to initial timing | Max.; BTDC 32° Min.; BTDC 11° | Does not operate | |
| EVAPORATIVE EMISSION CONTROL | SOLENOID VALVE (No.1 PURGE CONTROL) | OFF (Vacuum cut) | | *1 *2 ON | *2 ON (Vacuum to No.1 purge control valve) | | | ON (Above 900 rpm) | OFF (Vacuum cut) | Does not operate | *1: Coolant temp.: above 55°C (131°F) *2: In-gear |
| | SOLENOID VALVE (No.2 PURGE CONTROL) | OFF (Vacuum cut) | | | ON (Vacuum to vacuum switch valve) | | | OFF (Vacuum cut) | | Does not operate | |
| EGR CONTROL | SOLENOID VALVE (VENT SIDE) | OFF (No vacuum to EGR valve) — EGR cut — | | | *1 System operates | | | OFF (No vacuum to EGR valve) — EGR cut — | | Does not operate | *1 Depends on engine condition |
| | SOLENOID VALVE (VACUUM SIDE) | OFF (Duty value 100%) (Atmospheric pressure to EGR valve) | | | (Amount of EGR changes) | | | OFF (Duty valve 100 %) (Atmospheric pressure to EGR valve) | | | |

## WIRING SCHEMATIC—1988 929

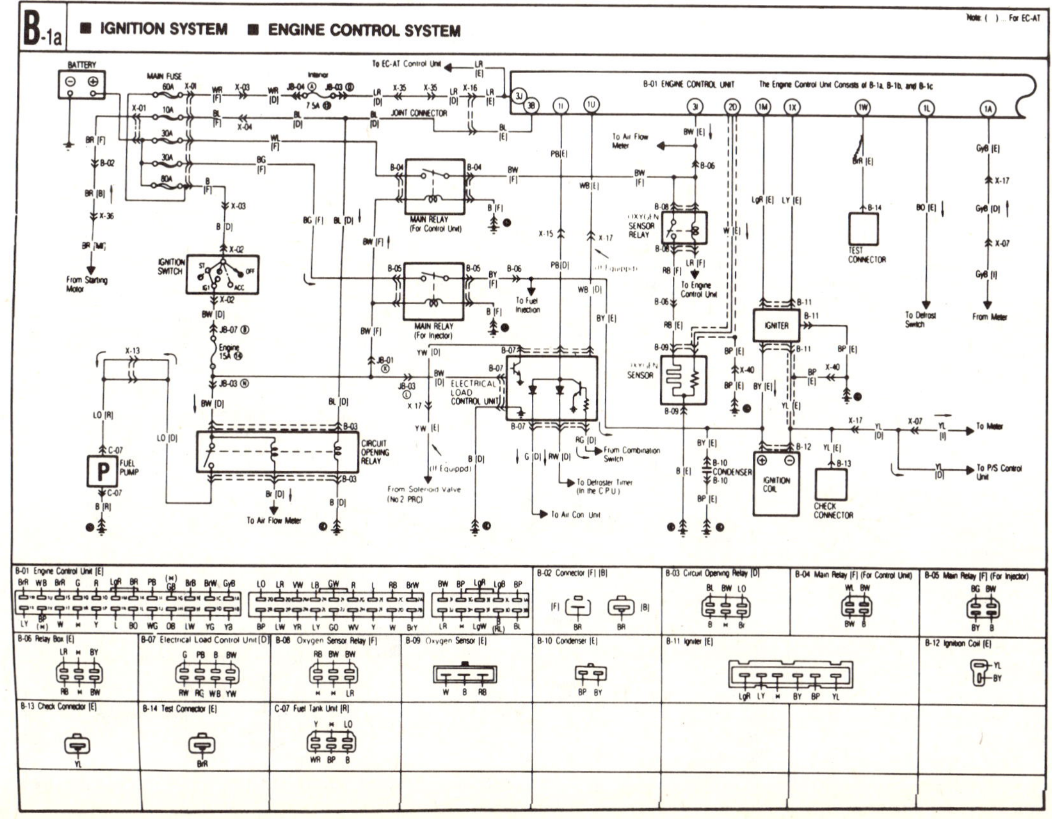

## WIRING SCHEMATIC (CONT.)—1988 929

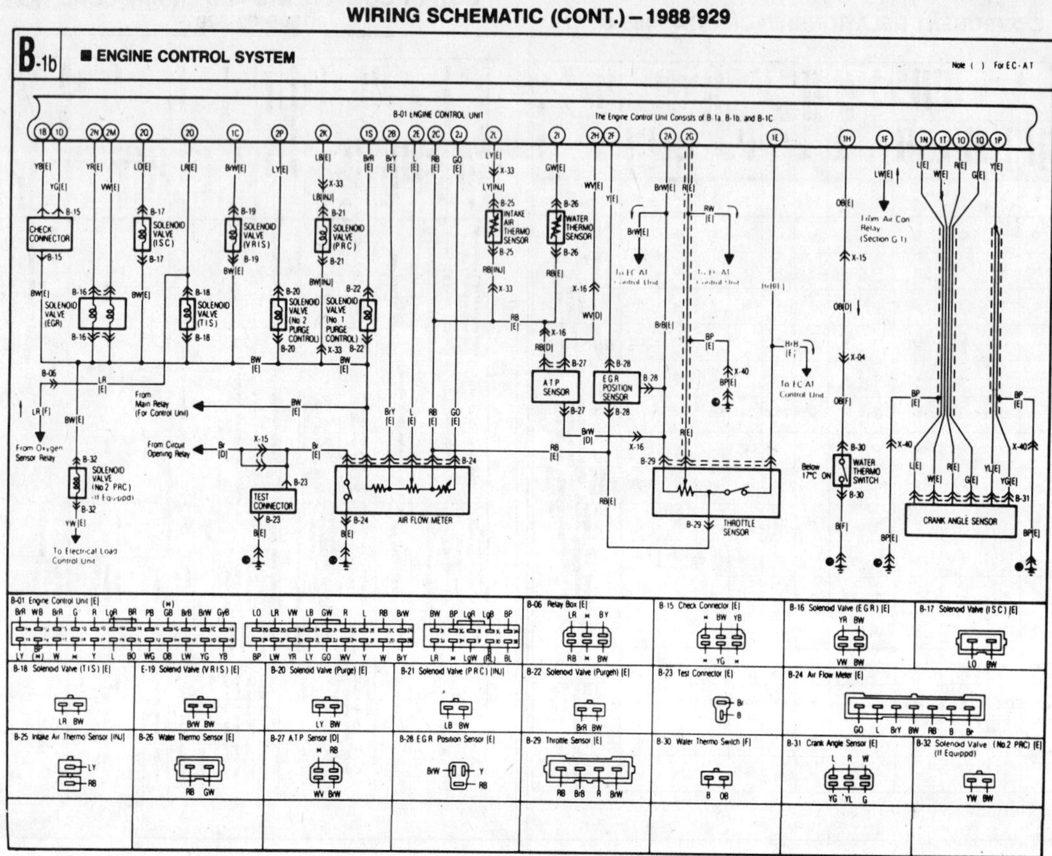


## WIRING SCHEMATIC—1989 929

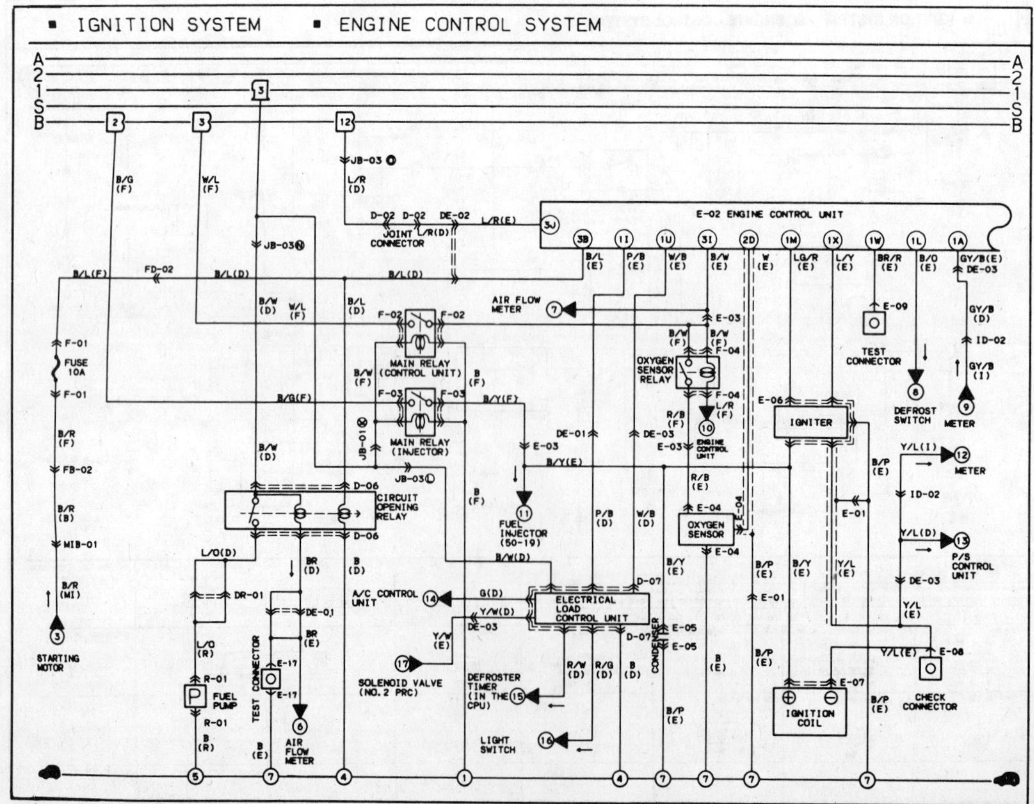

## WIRING SCHEMATIC (CONT.)—1989 929

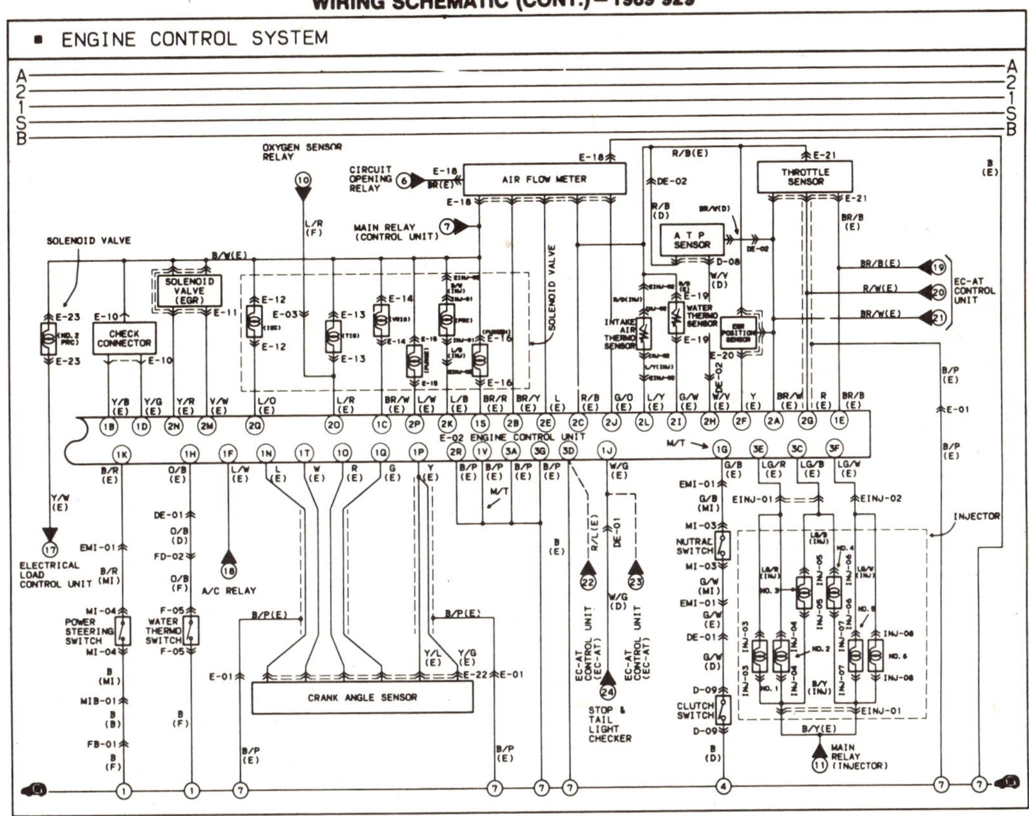


## SYSTEM SCHEMATIC—1990 929 WITH SOHC ENGINE

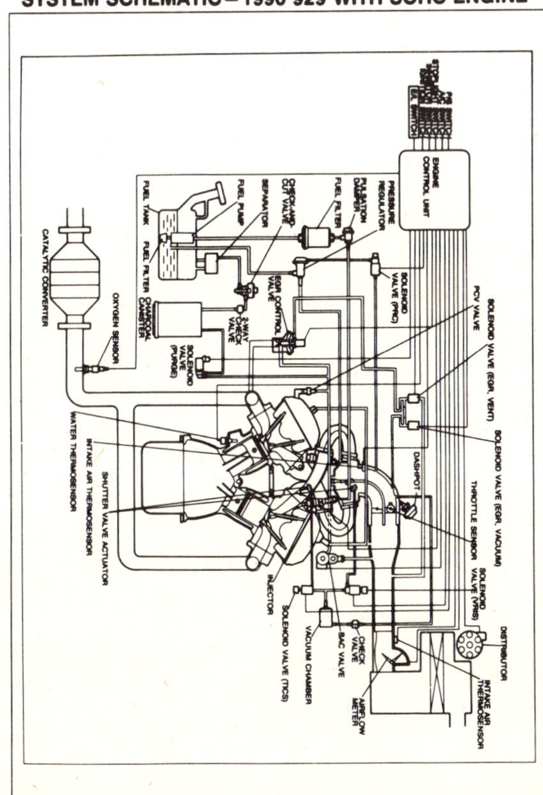


## EXPLODED VIEW OF FUEL AND OUTPUT SYSTEM COMPONENTS—1990 929 WITH SOHC ENGINE

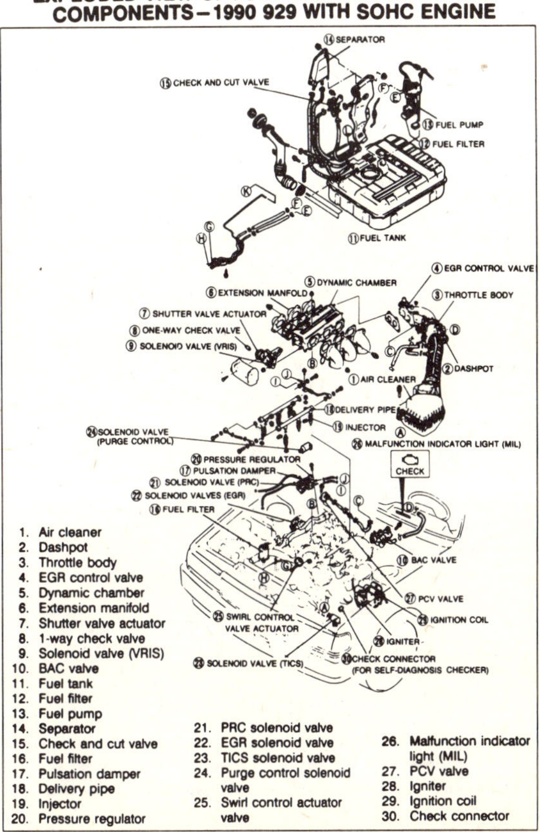


1. Air cleaner
2. Dashpot
3. Throttle body
4. EGR control valve
5. Dynamic chamber
6. Extension manifold
7. Shutter valve actuator
8. 1-way check valve
9. Solenoid valve (VRIS)
10. BAC valve
11. Fuel tank
12. Fuel filter
13. Fuel pump
14. Separator
15. Check and cut valve
16. Fuel filter
17. Pulsation damper
18. Delivery pipe
19. Injector
20. Pressure regulator
21. PRC solenoid valve
22. EGR solenoid valve
23. TICS solenoid valve
24. Purge control solenoid valve
25. Swirl control actuator valve
26. Malfunction indicator light (MIL)
27. PCV valve
28. Igniter
29. Ignition coil
30. Check connector

## VACUUM HOSE ROUTING 1990 929 WITH SOHC ENGINE

VACUUM HOSE ROUTING DIAGRAM

ACTUATOR (FOR AUTO-CRUISE)
TO EC-AT
EGR CONTROL VALVE
SOLENOID VALVE (PURGE CONTROL)
SHUTTER VALVE ACTUATOR
SOLENOID VALVE (VRIS)
SOLENOID VALVE (EGR, VACUUM)
SOLENOID VALVE (PRC)
THROTTLE BODY
CHARCOAL CANISTER
SOLENOID VALVE (EGR, VENT)
ONE-WAY CHECK VALVE
VACUUM CHAMBER
SWIRL CONTROL VALVE ACTUATOR
SOLENOID VALVE (TICS)

## EXPLODED VIEW OF INTAKE AIR SYSTEM 1990 929 WITH SOHC ENGINE

Remove in the sequence shown in the figure

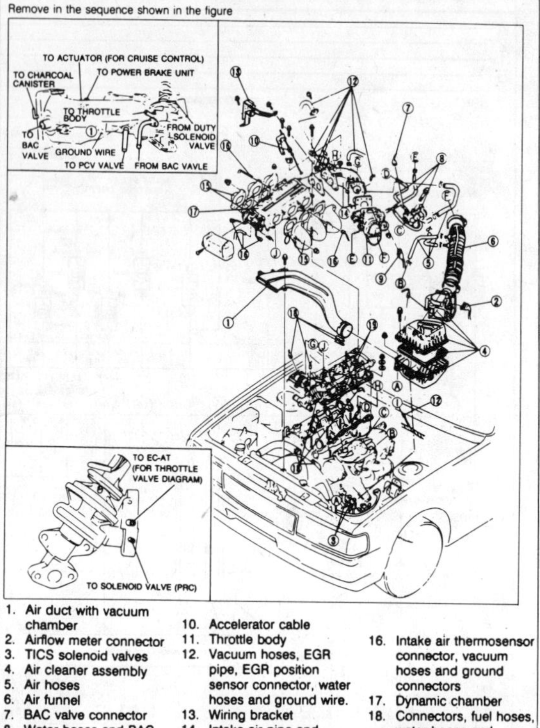


1. Air duct with vacuum chamber
2. Airflow meter connector
3. TICS solenoid valves
4. Air cleaner assembly
5. Air hoses
6. Air funnel
7. BAC valve connector
8. Water hoses and BAC valve
9. Throttle sensor connector
10. Accelerator cable
11. Throttle body
12. Vacuum hoses, EGR pipe, EGR position sensor connector, water hoses and ground wire.
13. Wiring bracket
14. Intake air pipe and gaskets
15. Extension manifolds and gaskets
16. Intake air thermosensor connector, vacuum hoses and ground connectors
17. Dynamic chamber
18. Connectors, fuel hoses, water hoses and vacuum hose
19. Intake manifold and gaskets

## EXPLODED VIEW OF CONTROL SYSTEM COMPONENTS 1990 929 WITH SOHC ENGINE

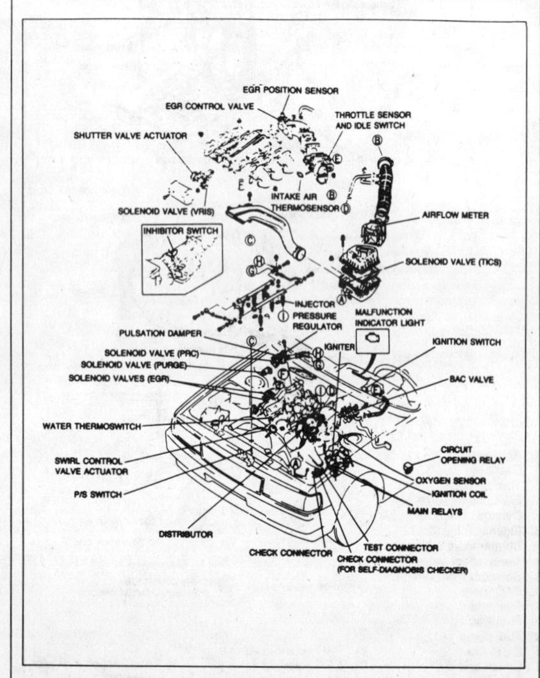


The control system consists of input devices and the control unit, the control unit controls the fuel injection amount (EGI), fuel injection pressure, bypass air amount, ignition timing, shutter valve operation (VRIS), swirl control valve operation (TICS), emission devices operation, monitor switch function, and fail-safe function.

## SPECIFICATIONS CHART 1990 929 WITH SOHC ENGINE

| Item | | | |
|---|---|---|---|
| Idle speed (with test connector grounded) rpm | | | 650 ± 20 (A/T: P range) |
| Air cleaner | Element type | | Wet |
| Throttle body | Throat diameter mm (in) | | 45 (1.772) x 2 |
| Dashpot | Adjustment speed rpm | | 3,500 ± 300 |
| Throttle sensor | Idle switch (adjustment clearance) mm (in) | | 0.5 (0.020) (Continuity)<br>0.7 (0.028) (No continuity) |
| | Resistance kΩ | | Ⓐ — Ⓓ 3.5—6.5<br>Ⓑ — Ⓓ below 1 ↔ 3.5—6.5 |
| Airflow meter | Resistance Ω | E2 ↔ Vs | 20—400 (Measuring plate fully closed), 20—1,000 (fully open) |
| | | E2 ↔ Vc | 100—300 |
| | | E2 ↔ VB | 200—400 |
| | | E2 ↔ THA | −20°C (−4°F) 13,600—18,400<br>20°C (68°F) 2,210— 2,690<br>60°C (140°F) 493— 667 |
| | | E1 ↔ Fc | Fully closed; ∞ Fully open: 0 |
| Fuel tank | Capacity liters (US gal, Imp gal) | | 70 (18.5, 15.4) |
| Fuel filter | Low pressure | | Nylon 6 (250 mesh) |
| | High pressure | | Filter paper |
| Fuel pump | Type | | Impeller (in tank) |
| | Fuel pressure kPa (kg/cm², psi) | | 441—588 (4.5—6.0, 64.0—85.3) |
| | Feeding capacity liters (US gal, Imp gal)/min | | 1.3—2.3 (0.34—0.61, 0.29—0.51) |
| Pressure regulator | Type | | Diaphragm |
| | Regulated pressure kPa (kg/cm², psi) | | 264—314 (2.7—3.2, 38.4—45.5)...Vacuum cut |
| Injector | Drive | | Voltage |
| | Type | | High-ohmic injector |
| | Resistance Ω | | 12—16 |
| | Injection amount cc (cu in)/15 sec | | 44—65 (2.68—3.97) |
| Ignition timing (with test connector grounded) | | | BTDC 15 ± 1° |
| BAC valve | Resistance Ω | | 10.8—12.3 |
| Fuel specification | | | Unleaded gasoline |

## COMPONENT DESCRIPTION—1990 929 WITH SOHC ENGINE

| No. | Component | Function | Remark |
|---|---|---|---|
| 1 | **Air cleaner** | Filters air entering throttle body | |
| 2 | **Airflow meter** | Detects amount of intake air; sends signal to engine control unit | Intake air thermosensor and fuel pump switch included |
| 3 | **Atmospheric pressure sensor** | Detects atmospheric pressure; sends signal to engine control unit | Built-in engine control unit |
| 4 | **Air valve** | Supplies bypass air into dynamic chamber (engine cold) | • Engine speed increased to shorten warm-up period<br>• Thermowax type<br>• Installed in BAC valve |
| | **BAC valve** | Supplies bypass air into dynamic chamber | Consists of air valve and ISC valve |
| | **ISC valve** | Supplies bypass air into dynamic chamber | • Controlled by duty signal from engine control unit<br>• Works independent of idle-up<br>• Installed in BAC valve |
| 5 | **Brake light switch** | Detects braking (deceleration); sends signal to engine control unit | |
| 6 | **Catalytic converter** | Reduces HC, CO, and NOx by chemical reaction | Monolith type |
| 7 | **Charcoal canister** | Stores fuel tank fumes (engine stopped) | |
| 8 | **Check connectors** | For Self-Diagnosis Checker and fuel pump | • 6-pin connector (Green); Self-Diagnosis Checker<br>• 2-pin connector (Yellow); Fuel pump |
| 9 | **Check valve** | Prevents misoperation of shutter valve or swirl control valve during heavy load operation | |
| 10 | **Check-and-cut valve** | Controls pressure in fuel tank | |
| 11 | **Circuit opening relay** | Voltage for fuel pump (engine running) | |
| 12 | **Control unit (Engine control unit)** | Detects the following:<br>1. Engine speed<br>2. No.1 and No.4 piston TDC<br>3. Intake air amount<br>4. Engine coolant temperature<br>5. Atmospheric pressure<br>6. Throttle valve opening angle<br>7. Throttle valve fully closed<br>8. Air/fuel ratio (oxygen concentration)<br>9. In-gear condition<br>10. Intake air temperature<br>11. Engine air temperature<br>12. A/C operation<br>13. P/S operation<br>14. E/L operation<br>15. Cranking signal<br>16. Initial set signal<br>17. EGR operation signal<br><br>Controls operation of the following:<br>1. Fuel injection system<br>2. Idle speed control system<br>3. Pressure regulator control system<br>4. Variable resonance induction system<br>5. Triple induction control system<br>6. Ignition system (Electronic spark advance)<br>7. Purge control system<br>8. EGR control system<br>9. Fail-safe system<br>10. Switch monitor function | 1. Ne rotor (in distributor)<br>2. G rotors (in distributor)<br>3. Airflow meter<br>4. Water thermosensor<br>5. Atomospheric pressure sensor<br>6. Throttle sensor<br>7. Idle switch (in throttle sensor)<br>8. Oxygen sensor<br>9. Inhibitor switch<br>10. Intake air thermosensor (in airflow meter)<br>11. Intake air thermosensor (on dynamic chamber)<br>12. A/C switch<br>13. P/S switch<br>14. E/L switches<br>15. Ignition switch<br>16. Test connector<br>17. EGR position sensor<br><br>1. Injector<br>2. BAC valve<br>3. Solenoid valve (Pressure regulator control)<br>4. Solenoid valve (VRIS)<br>5. Solenoid valve (TICS)<br>6. Igniter<br>7. Solenoid valve (Purge control)<br>8. Solenoid valves (EGR)<br>9. Self-Diagnosis Checker and MIL<br>10. Monitor lamp (Self-Diagnosis Checker) |

| No. | Component | Function | Remark |
|---|---|---|---|
| 13 | **Distributor** | Detects crank angle at 30° intervals, No.1 cylinder TDC and No.4 cylinder TDC; sends signal to engine control unit | • Ne (Number of engine speed) signal<br>• G1 (Group 1) signal<br>• G2 (Group 2) signal |
| | **G1 coil** | Detects No.4 cylinder TDC; sends signal to engine control unit | For determining fuel injection timing |
| | **G2 coil** | Detects No.1 cylinder TDC; sends signal to engine control unit | For determining fuel injection timing |
| 14 | **Dashpot** | Causes gradual throttle valve closing during deceleration | Adjustment speed; 3,500 ± 300 rpm |
| 15 | **Dynamic chamber** | Interconnects all cylinders | Contains shutter valve |
| 16 | **EC-AT control unit** | Detect N ro P range | |
| 17 | **Exhaust gas recirculation (EGR) position sensor** | Detects EGR control valve opening; sends signal to engine control unit | Installed on EGR control valve |
| 18 | **Exhaust gas recirculation control valve** | Recirculates portion of exhaust gas | Contains EGR position sensor |
| 19 | **Fuel filter** | Filters fuel | |
| 20 | **Fuel pump** | Provides fuel to injectors | • Operates only while engine running<br>• Installed in fuel tank |
| 21 | **Idle switch** | Detects when throttle valve fully closed; sends signal to engine control unit | Installed in throttle sensor<br>Switch closed when throttle valve fully closed |
| | **Throttle sensor** | Detects throttle valve opening angle; sends signal to engine control unit | • Idle switch included<br>• Installed on throttle body |
| 22 | **Igniter** | Receives spark signal from engine control unit; generates high voltage in ignition coil | High voltage distributed by rotor in distributor |
| 23 | **Ignition switch** | Detects engine cranking; sends signal to engine control unit | |
| 24 | **Inhibitor switch** | Detects in-gear condition; sends signal through EC-AT control unit to engine control unit | Switch closed in N or P range (A/T) |
| 25 | **Injector** | Injects fuel into intake port | • Controlled by signals from engine control unit<br>• High-ohmic injector |
| 26 | **Intake manifold** | Supplies intake air to all cylinders | Contains swirl control valve |
| 27 | **Intake air thermosensor** | Detects intake air temperature; sends signal to engine control unit | Installed in airflow meter |
| | | Detects engine air temperature; sends signal to engine control unit | Installed on dynamic chamber |
| 28 | **Main relay** | Supplies current to injectors and engine control unit | |
| 29 | **MIL (Malfunction indicator light)** | Light illuminates when input device malfunctions | Installed in instrument cluster |
| 30 | **Oxygen sensor** | Detects oxygen concentration; sends signal to engine control unit | Zirconia ceramic and platinum coating |
| 31 | **PCV valve** | Controls blowby gas amount pulled into engine | |
| 32 | **P/S switch** | Detects P/S operation; sends signal to engine control unit | ON when steering wheel turned right or left |
| 33 | **Pressure regulator** | Adjusts fuel pressure supplied to injectors | |
| 34 | **Pulsation damper** | Absorbs fuel pulsation | |

## COMPONENT DESCRIPTION (CONT.) 1990 929 WITH SOHC ENGINE

| No. | Component | | Function | Remark |
|---|---|---|---|---|
| 35 | **Separator** | | Prevents fuel from flowing into charcoal canister | |
| 36 | **Shutter valve actuator** | | Controls shutter valve operation | For variable resonance induction system (VRIS) |
| 37 | **Swirl control valve actuator** | | Controls swirl control valve operation | For triple induction control system (TICS) |
| 38 | Solenoid valve | **EGR (Vent side, Vacuum side)** | Controls EGR control valve operation | • Exhaust gas recirculation<br>• Duty solenoid valve |
| 39 | | **Purge control** | Controls evaporative fumes from canister | Controlled by duty signal from ECU |
| 40 | | **TICS** | Controls vacuum to swirl control valve actuator | Triple induction control system (TICS) |
| 41 | | **VRIS** | Controls vacuum to shutter valve actuator | Variable resonance induction system (VRIS) |
| 42 | | **Pressure regulator control** | Blocks vacuum from intake manifold to pressure regulator | Hot engine start only |
| 43 | **Test connector** | | For Self-Diagnosis Checker and idle speed/ignition timing adjustment | 1-pin connector (Green) |
| 44 | **Throttle body** | | Controls intake air amount | Integrated throttle sensor and dashpot incorporated |
| 45 | **Water thermosensor** | | Detects engine coolant temperature; sends signal to engine control unit | |

## TROUBLESHOOTING CHART 1990 929 WITH SOHC ENGINE

This troubleshooting guide shows the malfunction code numbers and the symptoms of various failures. Perform troubleshooting as described below.

| | Symptoms \ Possible cause | | Input sensors | | | | | | | | | | | | | Output solenoid valves | |
|---|---|---|---|---|---|---|---|---|---|---|---|---|---|---|---|---|---|
| | | | Ignition pulse | Distributor (Ne signal) | Distributor (G1 signal) | Distributor (G2 signal) | Airflow meter | Water thermosensor | Intake air thermosensor (Airflow meter) | Intake air thermosensor (Dynamic chamber) | Throttle sensor | Atmospheric pressure sensor | Oxygen sensor | EGR position sensor | Fuel feedback system | Solenoid valve (Pressure regulator control) | Solenoid valve (Purge control) |
| 1 | **Fault indicated by SST malfunction Code No.** | | 01 | 02 | 03 | 04 | 08 | 09 | 10 | 11 | 12 | 14 | 15 | 16 | 17 | 25 | 26 |
| 2 | **Hard starts or won't start (cranks OK)** | | | | | | | | | | | | | | | | |
| 3 | **Engine stalls** | **While warming up** | | | | | | | | | | | | | | | |
| | | **After warming up** | | | | | | | | | | | | | | | |
| 4 | **Rough idle** | **While warming up** | | | | | | | | | | | | | | | |
| | | **After warming up** | | | | | | | | | | | | | | | |
| 5 | **High speed idle after warming up** | | | | | | | | | | | | | | | | |
| 6 | **Poor acceleration, hesitation, or lack of power** | | | | | | | | | | | | | | | | |
| 7 | **Runs rough on deceleration** | | | | | | | | | | | | | | | | |
| 8 | **Afterburn in exhaust system** | | | | | | | | | | | | | | | | |
| 9 | **Poor fuel consumption** | | | | | | | | | | | | | | | | |
| 10 | **Excessive oil consumption** | | | | | | | | | | | | | | | | |
| 11 | **Engine stalls or runs rough after hot starting** | | | | | | | | | | | | | | | | |
| 12 | **Fails emission test** | | | | | | | | | | | | | | | | |

**TROUBLESHOOTING PROCEDURE**

**Note**
- **Symptom 1 shows the Code No. to quickly determine what system or unit may be at fault by use of the SST. (SST: Self-Diagnosis Checker 49 H018 9A1)**

1st. Check the electrical system.
1) Battery condition
2) Main relays
3) Main fuses
4) Fuses

2nd. Check input sensors and output solenoid valves diagnosed with the **SST**.

3rd. Check input switches with **SST**.

4th. Check the following items:
Ignition system
1) Ignition spark
2) Ignition timing (with test connector grounded)
Intake air system
1) Air cleaner element
2) Vacuum or air leakage
3) Vacuum hose and air hose routing
4) Accelerator cable
Fuel system
1) Fuel amount
2) Fuel leakage
3) Fuel filter
4) Idle speed (with test connector grounded)
Engine
1) Compression
2) Overheating
Others
1) Clutch slippage
2) Brake dragging

5th. Check Fuel and Emission Control Systems.

## TROUBLESHOOTING CHART 1990 929 WITH SOHC ENGINE

The troubleshooting guide lists the systems most likely to cause a given symptom. After finding systems to check, refer to the pages shown for detailed guides for each system.

| Symptom \ Possible cause | Output solenoid valves: Solenoid valve (EGR, vacuum side) 28 | Solenoid valve (EGR, vent side) 29 | Idle speed control valve 34 | Solenoid valve (TICS) or Oxygen sensor heater 40 | Solenoid valve (VRIS) 41 | Fuel and Emission Control Systems: Intake air system (Poor connection of components, Throttle body) | Idle speed control system (Air valve, Idle speed control valve malfunction) | VRIS (Variable resonance induction system) | TICS (Triple induction control system) | Fuel system (Fuel injection, Fuel pressure) | Pressure regulator control system | Electronic spark advance control system | Deceleration control system (Fuel cut operation malfunction) | EGR system (EGR control valve open) | Evaporative emission control system | PCV system (System clogged) | Exhaust system (System clogged) |
|---|---|---|---|---|---|---|---|---|---|---|---|---|---|---|---|---|---|
| 2 | | | | | | 3 | | | | 2 | | 1 | | 4 | | | |
| 3 | | | | | | 4 | 1 | | | 3 | | | | 2 | | | |
| | | | | | | 5 | 2 | | | 4 | | | | 3 | | 1 | |
| 4 | | | | | | 6 | 1 | | | 5 | | | | 3 | 4 | 2 | |
| | | | | | | 6 | 2 | | | 5 | | | | 3 | 4 | 1 | |
| 5 | | | | | | 3 | 2 | | | 4 | | | 1 | | | | |
| 6 | | | | | | 3 | | 5 | 6 | 4 | | | | 1 | 2 | | 7 |
| 7 | | | | | | | 2 | | | 3 | | | 1 | | | | |
| 8 | | | | | | 3 | 2 | | | 4 | | | 1 | | | | |
| 9 | | | | | | | | | 2 | 4 | 5 | | 1 | 3 | | | 6 |
| 10 | | | | | | | | | | | | | | | | 1 | |
| 11 | | | | | | | 2 | | | 3 | 1 | | | | | | |
| 12 | | | | | | 6 | 5 | | | 7 | 8 | | 3 | 2 | 4 | | 1 |

The numbers of the list show the priorities of inspections from the most possible to that with the lowest possibility. These were determined on the following basis:

- Ease of inspection
- Most possible system
- Most possible point in system

## OUTPUT COMPONENTS AND ENGINE CONDITION 1990 929 WITH SOHC ENGINE

| OUTPUT DEVICE \ ENGINE CONDITION | | CRANKING (COLD ENGINE) | WARM-UP (DURING IDLE) | MEDIUM LOAD COLD | MEDIUM LOAD WARM | ACCELE-RATION | HEAVY LOAD | DECELE-RATION | IDLE (THROTTLE VALVE FULLY CLOSED) | IG: ON (ENGINE NOT RUNNING) | REMARKS |
|---|---|---|---|---|---|---|---|---|---|---|---|
| INJECTOR | No.1, No.2 / No.3, No.4 / No.5, No.6 | Rich | | | *1 Rich and lean (Fuel feedback operation) | Rich | | Fuel cut | *1 Rich and lean (Fuel feedback operation) | Does not inject | *1 Operated by oxygen sensor |
| | INJECTION TIMING | 1 group | 3 group | | | | | | 3 group | | Fuel cut: above 6,500 rpm |
| SOLENOID VALVE (PRC) | | OFF (Vacuum to pressure regulator) | | | | | | | *1 After start ON (Vacuum cut) | Does not operate | *1 During hot condition only |
| BAC VALVE | AIR VALVE | *1 Open | | | Close | | | | | | *1 Coolant temp.: below 60°C (140°F) |
| | ISC VALVE | *1 Large amount of bypass air | | Large and small amount of bypass air | | | | | *1 Large and small amount of bypass air | Does not operate | *1 Engine speed feed-back system operates when the test connector open |
| SOLENOID VALVE (VRIS) | | OFF (Shutter valve closed) | | *1 ON (Shutter valve open) | | | | OFF (Shutter valve closed) | | Does not operate | *1 ON: approx. 3,700—5,750 rpm |
| SOLENOID VALVE (TICS) | | ON (Swirl control valve closed) | | OFF (Swirl control valve open) | | | | ON (Swirl control valve closed) | | Does not operate | |
| IGNITER | | BTDC 7° | Max.; BTDC 42° Min.; BTDC 11° | Advance | Max.; BTDC 42° Min.; BTDC −6° | | | Gradually returned to initial timing | Max.; BTDC 32° Min.; BTDC 11° | Does not operate | |
| SOLENOID VALVE (PURGE CONTROL) | | OFF (Purge cut) | | | *1 Operate [Duty values (purge amount) change] | | | OFF (Purge cut) | | | *1 Depends on engine condition |
| EGR CONTROL | SOLENOID VALVE (VENT SIDE) | OFF (No vacuum to EGR valve) | | | *1 System operates | | | OFF (No vacuum to EGR valve) | | Does not operate | *1 Depends on engine condition |
| | | EGR cut | | | | | | EGR cut | | | |
| | SOLENOID VALVE (VACUUM SIDE) | OFF (Duty value 100%) (Atmospheric pressure to EGR valve) | | | (Amount of EGR changes) | | | OFF (Duty valve 100%) (Atmospheric pressure to EGR valve) | | | |

## COMPONENT REALTIONSHIP CHART 1990 929 WITH SOHC ENGINE

○: Related
X: Not related

| INPUT DEVICE \ OUTPUT DEVICE | | INJECTOR: FUEL INJECTION AMOUNT | INJECTOR: FUEL INJECTION TIMING | PRESSURE REGULATOR | BAC VALVE: AIR VALVE | BAC VALVE: ISC VALVE | SOLENOID VALVE (VRIS) | SOLENOID VALVE (TICS) | IGNITER | SOLENOID VALVE (PURGE CONTROL) | EGR: VENT SIDE | EGR: VACUUM SIDE |
|---|---|---|---|---|---|---|---|---|---|---|---|---|
| P/S SWITCH | | × | × | × | × | ○ | × | × | × | × | × | × |
| A/C SWITCH | | ○ | × | × | × | ○ | × | × | ○ | ○ | × | × |
| INHIBITOR SWITCH | | ○ | × | × | × | ○ | × | × | ○ | ○ | × | × |
| TEST CONNECTOR | | × | × | × | × | ○ | × | × | ○ | × | × | × |
| EGR POSITION SENSOR | | × | × | × | × | × | × | × | ○ | × | ○ | ○ |
| ATMOSPHERIC PRESSURE SENSOR | | ○ | × | × | × | × | × | × | × | ○ | × | × |
| OXYGEN SENSOR | | ○ | × | × | × | × | × | × | × | ○ | × | × |
| INTAKE AIR THERMOSENSOR | DYNAMIC CHAMBER | × | × | ○ | × | ○ | × | × | × | × | × | × |
| | AIRFLOW METER | ○ | × | × | × | ○ | × | × | × | × | × | × |
| WATER THERMOSENSOR | | ○ | × | ○ | × | ○ | × | × | ○ | ○ | ○ | ○ |
| THROTTLE SENSOR | THROTTLE SENSOR | ○ | ○ | ○ | × | ○ | × | × | ○ | × | ○ | ○ |
| | IDLE SWITCH | ○ | ○ | × | × | ○ | × | × | ○ | ○ | ○ | ○ |
| AIRFLOW METER | | ○ | × | × | × | × | ○ | ○ | ○ | ○ | ○ | ○ |
| DISTRIBUTOR | G ROTOR | × | ○ | × | × | × | × | × | ○ | × | × | × |
| | Ne ROTOR | ○ | ○ | ○ | × | ○ | ○ | ○ | ○ | ○ | ○ | ○ |
| IG COIL | | × | × | × | × | × | × | × | × | ○ | × | × |

## MALFUNCTION INDICATOR LIGHT (MIL) OPERATION 1990 929 WITH SOHC ENGINE

**MALFUNCTION CODE EXPLANATION**
Malfunction codes are determined as below.

**1. Malfunction code cycle break**
The time between malfunction code cycles is 4.0 seconds (the time the MIL and buzzer are off).

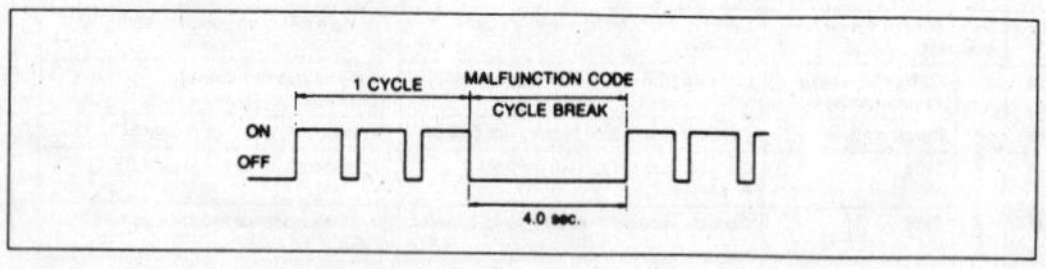


**2. Second digit of malfunction code (ones position)**
The digit in the ones position of the malfunction code represents the number of times the MIL and buzzer are on for 0.4 second and during one cycle.

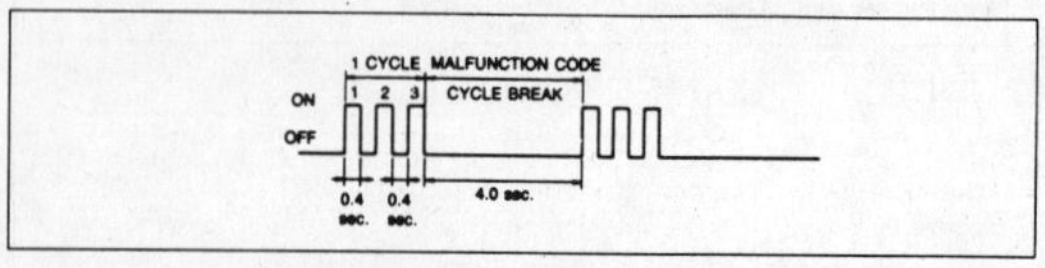


**3. First digit of malfunction code (tens position)**
The digit in the tens position of the malfunction code represents the number of times the MIL and buzzer are on 1.2 seconds during one cycle.
The MIL and buzzer are off for 1.6 seconds and between the long and short pulses.

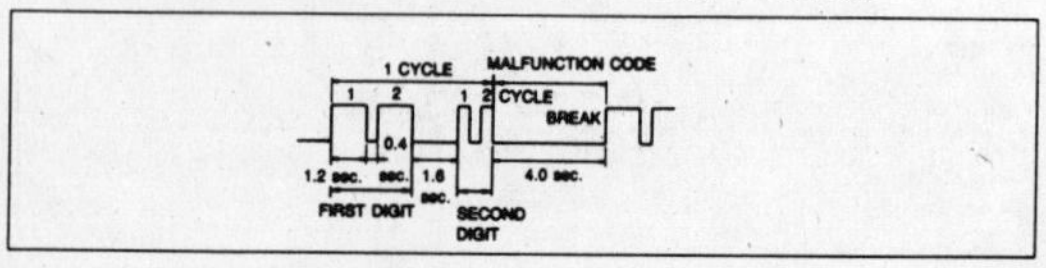

## TROUBLE CODE IDENTIFICATION CHART—1990 929 WITH SOHC ENGINE

**Input Devices**

| Malfunction Code No. | Input devices | Malfunction | Fail-safe function | Pattern of output signals (Self-Diagnosis Checker or MIL) |
|---|---|---|---|---|
| 01 | Ignition pulse (Igniter, Ignition coil) | Broken wire, short circuit | – | |
| 02 | Distributor (Ne signal) | Ne signal not input for 1.5 sec. during cranking | – | |
| 03 | Distributor (G1 signal) | Broken wire, short circuit | – | |
| 04 | Distributor (G2 signal) | Broken wire, short circuit | – | |
| 08 | Airflow meter | Broken wire, short circuit | Basic fuel injection amount fixed as for 2 driving modes 1) Idle switch: ON 2) Idle switch: OFF | |
| 09 | Water thermosensor | Broken wire, short circuit | Coolant temp. input fixed at 80°C (176°F) | |
| 10 | Intake air thermosensor (Airflow meter) | Broken wire, short circuit | Intake air temp. input fixed at 20°C (68°F) | |
| 11 | Intake air thermosensor (Dynamic chamber) | Broken wire, short circuit | Intake air temp. input fixed at 20°C (68°F) | |
| 12 | Throttle sensor | Broken wire, short circuit | Throttle valve opening angle signal input fixed at fully open | |
| 14 | Atmospheric pressure sensor | ECU | Atmospheric pressure input fixed at 760 mmHg (29.9 in Hg) | |
| 15 | Oxygen sensor | Oxygen sensor output below 0.55V 120 sec. after engine at 1,500 rpm | Feedback system canceled (for EGI) | |
| 16 | EGR position sensor | Broken wire, short circuit | EGR position signal input fixed at fully closed | |
| | | Sensor output does not match target value (incorrect output) | – | |
| 17 | Feedback system | Oxygen sensor output does not change from 0.55V 60 sec. after engine at 1,500 rpm | Feedback system canceled (for EGI) | |

**Output Devices**

| Malfunction code | Output devices | Pattern of output signals (Self-Diagnosis Checker or MIL) |
|---|---|---|
| 25 | Solenoid valve (Pressure regulator control) | |
| 26 | Solenoid valve (Purge control) | |
| 28 | Solenoid valve (EGR, vacuum side) | |
| 29 | Solenoid valve (EGR, vent side) | |
| 34 | Idle speed control valve (ISC valve) | |
| 40 | Solenoid valve (Triple induction control system) and oxygen sensor heater | |
| 41 | Solenoid valve (Variable resonance induction system) | |

**Caution**

- **If there is more than one failure present, the lowest number malfunction code is displayed first, the subsequent malfunction codes light up in order.**
- **After repairing all failures, erase the memory of malfunction codes.**
- **Inspect for an open circuit of the solenoid valve (TICS) or oxygen sensor heater according to the following procedure.**

**Procedure**

1. Warm up the engine to normal operating temperature and stop it.
2. Connect the **SST** to the check connector (Green: 6-pin) and the negative battery terminal.
3. Connect a jumper wire between the test connector (Green: 1-pin) and a ground.
4. Disconnect the oxygen sensor connector and turn the ignition switch ON. Check that malfunction Code No. 40 is not illuminated. If malfunction Code No. 40 is illuminated, check the solenoid valve (TICS)
5. Turn the ignition switch OFF. Connect the oxygen sensor connector, and disconnect the solenoid valve (TICS) connector.
6. Turn the ignition switch ON, and check that malfunction Code No. 40 is not illuminated. If malfunction Code No. 40 is illuminated, check the oxygen sensor heater.

Ⓒ
Ⓑ
Ⓐ
OXYGEN SENSOR
SOLENOID VALVE (TICS)

## TROUBLE CODE DIAGNOSTIC CHART—1990 929 WITH SOHC ENGINE

If a malfunction code number is illuminated on the **SST**, check the following chart along with the wiring diagram.

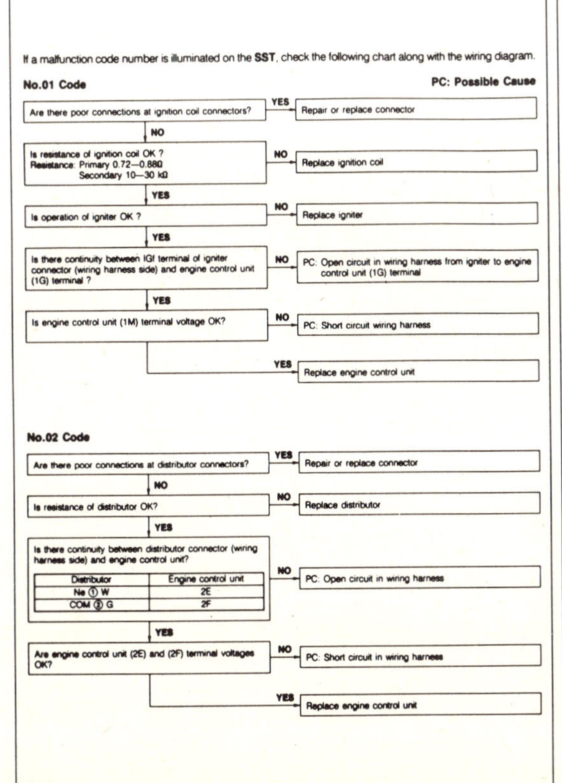


**No.03 Code** — **PC: Possible Cause**

- Are there poor connections at distributor connectors?
  - YES → Repair or replace connector
  - NO ↓
- Is resistance of distributor OK?
  - NO → Replace distributor
  - YES ↓
- Is there continuity between distributor connector (wiring harness side) and engine control unit?

| Distributor | Engine control unit |
|---|---|
| G1 ① L | 2G |

  - NO → PC: Open circuit in wiring harness
  - YES ↓
- Are engine control unit (2G) terminal voltages OK?
  - NO → PC: Short circuit in wiring harness
  - YES → Replace engine control unit

**No.04 Code**

- Are there poor connections at distributor connectors?
  - YES → Repair or replace connector
  - NO ↓
- Is resistance of distributor OK?
  - NO → Replace distributor
  - YES ↓
- Is there continuity between distributor connector (wiring harness side) and engine control unit?

| Distributor | Engine control unit |
|---|---|
| G2 ① R | 2H |

  - NO → PC: Open circuit in wiring harness
  - YES ↓
- Are engine control unit (2H) terminal voltages OK?
  - NO → PC: Short circuit in wiring harness
  - YES → Replace engine control unit

### TROUBLE CODE DIAGNOSTIC CHART – 1990 929 WITH SOHC ENGINE

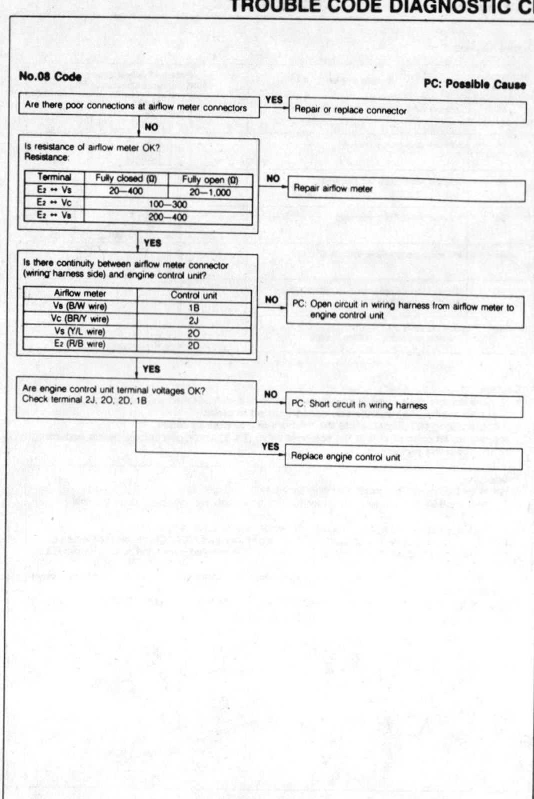


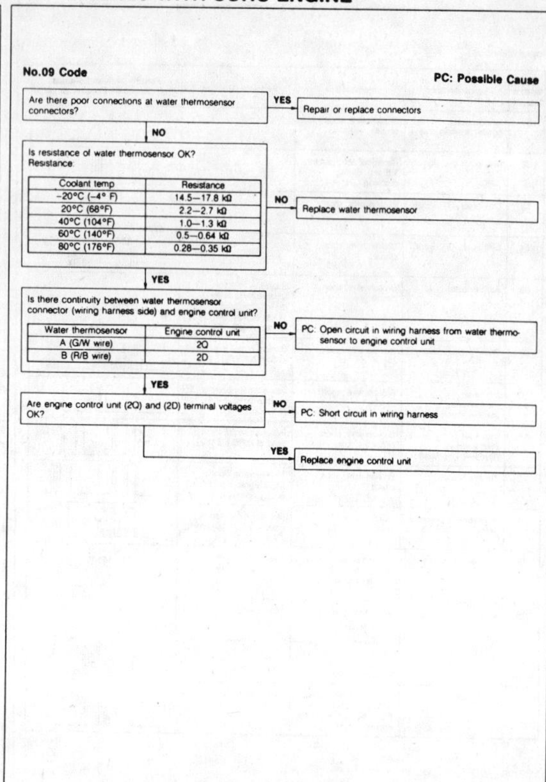


### TROUBLE CODE DIAGNOSTIC CHART – 1990 929 WITH SOHC ENGINE

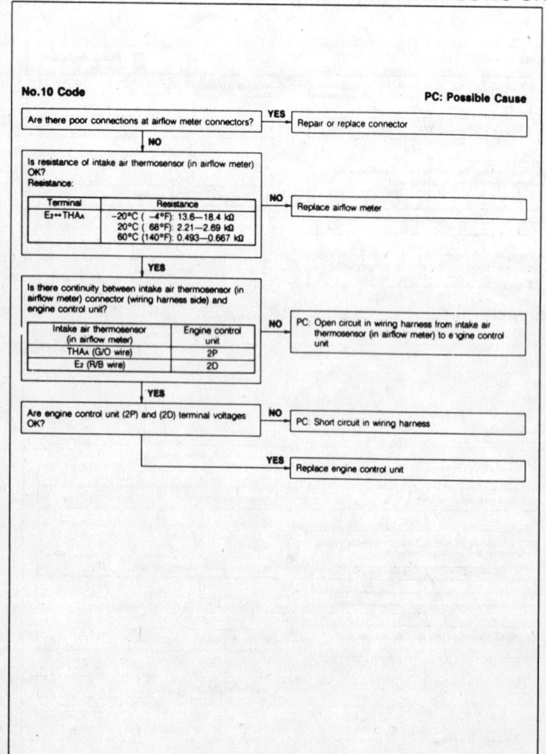


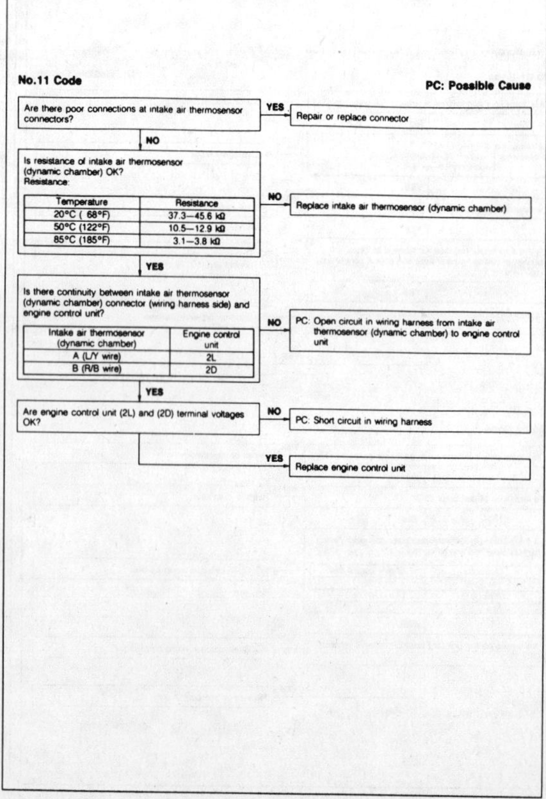

## TROUBLE CODE DIAGNOSTIC CHART—1990 929 WITH SOHC ENGINE

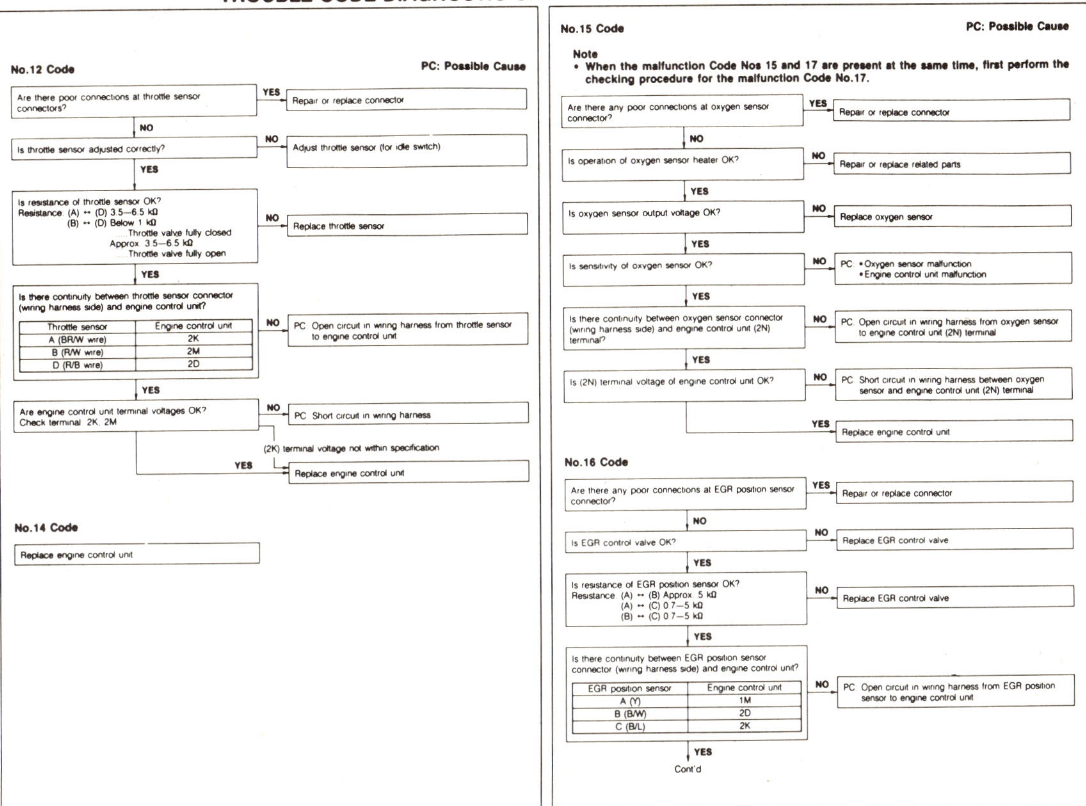


## TROUBLE CODE DIAGNOSTIC CHART—1990 929 WITH SOHC ENGINE

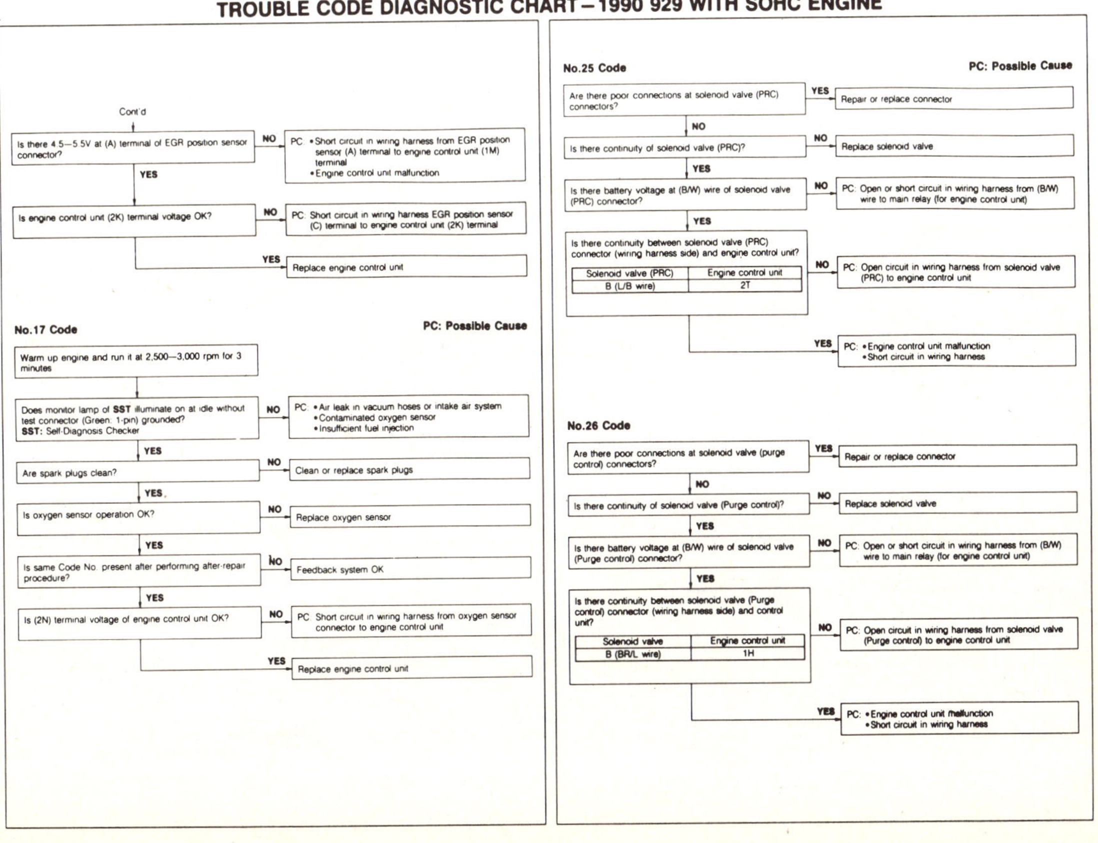

## TROUBLE CODE DIAGNOSTIC CHART—1990 929 WITH SOHC ENGINE

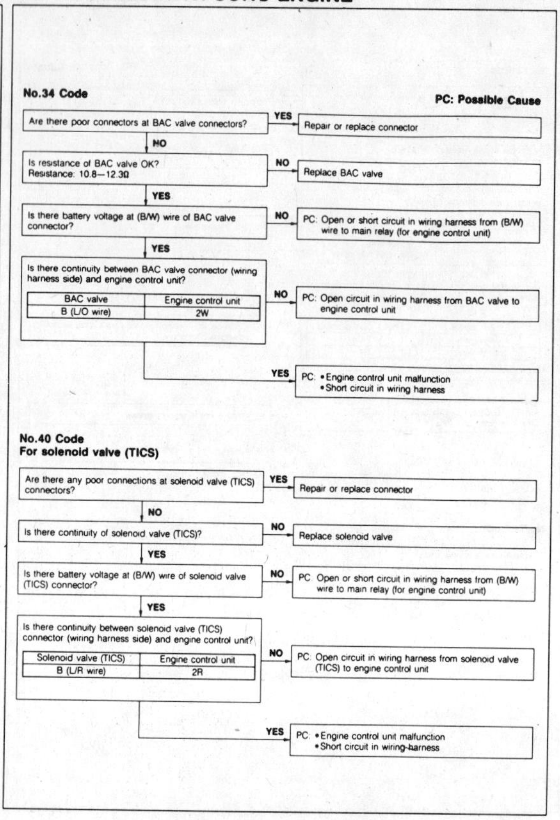

PC: Possible Cause

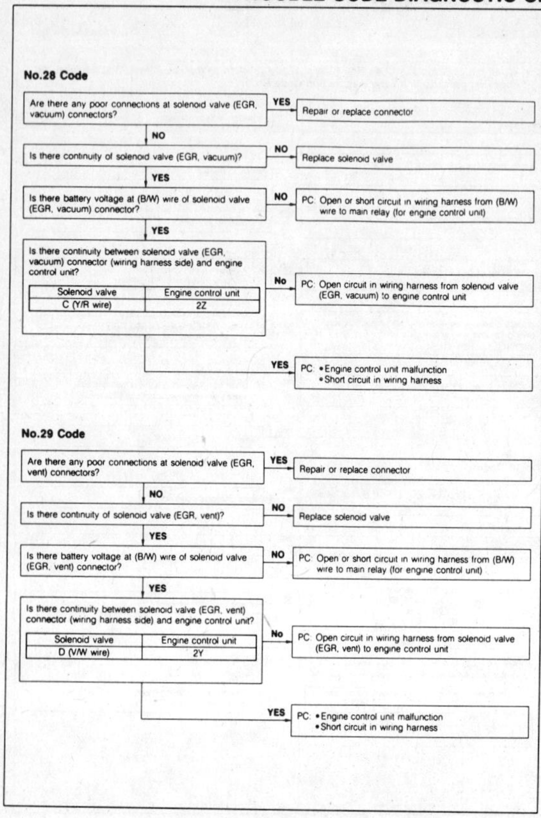

**No.28 Code**

Are there any poor connections at solenoid valve (EGR, vacuum) connectors? — YES → Repair or replace connector

NO ↓

Is there continuity of solenoid valve (EGR, vacuum)? — NO → Replace solenoid valve

YES ↓

Is there battery voltage at (B/W) wire of solenoid valve (EGR, vacuum) connector? — NO → PC: Open or short circuit in wiring harness from (B/W) wire to main relay (for engine control unit)

YES ↓

Is there continuity between solenoid valve (EGR, vacuum) connector (wiring harness side) and engine control unit?

| Solenoid valve | Engine control unit |
|---|---|
| C (Y/R wire) | 2Z |

— No → PC: Open circuit in wiring harness from solenoid valve (EGR, vacuum) to engine control unit

YES → PC: • Engine control unit malfunction • Short circuit in wiring harness

**No.29 Code**

Are there any poor connections at solenoid valve (EGR, vent) connectors? — YES → Repair or replace connector

NO ↓

Is there continuity of solenoid valve (EGR, vent)? — NO → Replace solenoid valve

YES ↓

Is there battery voltage at (B/W) wire of solenoid valve (EGR, vent) connector? — NO → PC: Open or short circuit in wiring harness from (B/W) wire to main relay (for engine control unit)

YES ↓

Is there continuity between solenoid valve (EGR, vent) connector (wiring harness side) and engine control unit?

| Solenoid valve | Engine control unit |
|---|---|
| D (V/W wire) | 2Y |

— No → PC: Open circuit in wiring harness from solenoid valve (EGR, vent) to engine control unit

YES → PC: • Engine control unit malfunction • Short circuit in wiring harness

**No.34 Code**

Are there poor connectors at BAC valve connectors? — YES → Repair or replace connector

NO ↓

Is resistance of BAC valve OK? Resistance: 10.8—12.3Ω — NO → Replace BAC valve

YES ↓

Is there battery voltage at (B/W) wire of BAC valve connector? — NO → PC: Open or short circuit in wiring harness from (B/W) wire to main relay (for engine control unit)

YES ↓

Is there continuity between BAC valve connector (wiring harness side) and engine control unit?

| BAC valve | Engine control unit |
|---|---|
| B (L/O wire) | 2W |

— NO → PC: Open circuit in wiring harness from BAC valve to engine control unit

YES → PC: • Engine control unit malfunction • Short circuit in wiring harness

**No.40 Code**
**For solenoid valve (TICS)**

Are there any poor connections at solenoid valve (TICS) connectors? — YES → Repair or replace connector

NO ↓

Is there continuity of solenoid valve (TICS)? — NO → Replace solenoid valve

YES ↓

Is there battery voltage at (B/W) wire of solenoid valve (TICS) connector? — NO → PC: Open or short circuit in wiring harness from (B/W) wire to main relay (for engine control unit)

YES ↓

Is there continuity between solenoid valve (TICS) connector (wiring harness side) and engine control unit?

| Solenoid valve (TICS) | Engine control unit |
|---|---|
| B (L/R wire) | 2R |

— NO → PC: Open circuit in wiring harness from solenoid valve (TICS) to engine control unit

YES → PC: • Engine control unit malfunction • Short circuit in wiring harness

## TROUBLE CODE DIAGNOSTIC CHART 1990 929 WITH SOHC ENGINE

PC: Possible Cause

**For oxygen sensor heater**

Are there any poor connections at oxygen sensor heater connectors? — YES → Repair or replace connector

NO ↓

Is oxygen sensor heater operation OK? — NO → Replace oxygen sensor

YES ↓

Is there continuity between oxygen sensor heater connector and main relay?

| Oxygen sensor | Main relay |
|---|---|
| C (B/W wire) | D (B/W wire) |

— NO → PC: Open circuit in wiring harness from oxygen sensor to main relay

YES ↓

Is there continuity between oxygen sensor heater connector and engine control unit?

| Oxygen sensor | Engine control unit |
|---|---|
| B (L/R wire) | 2R |

— NO → PC: Open circuit in wiring harness from oxygen sensor to engine control unit

YES → PC: • Engine control unit malfunction • Short circuit in wiring harness

**No.41 Code**

Are there any poor connections at solenoid valve (VRIS) connectors? — YES → Repair or replace connector

NO ↓

Is there continuity of solenoid valve (VRIS)? — NO → Replace solenoid valve

YES ↓

Is there battery voltage at (B/W) wire of solenoid valve (VRIS) connector? — NO → PC: Open or short circuit in wiring harness from (B/W) wire to main relay (for engine control unit)

YES ↓

Is there continuity between solenoid valve (VRIS) connector (wiring harness side) and engine control unit?

| Solenoid valve (VRIS) | Engine control unit |
|---|---|
| B (V wire) | 2S |

— NO → PC: Open circuit in wiring harness from solenoid valve (VRIS) to engine control unit

YES → PC: • Engine control unit malfunction • Short circuit in wiring harness

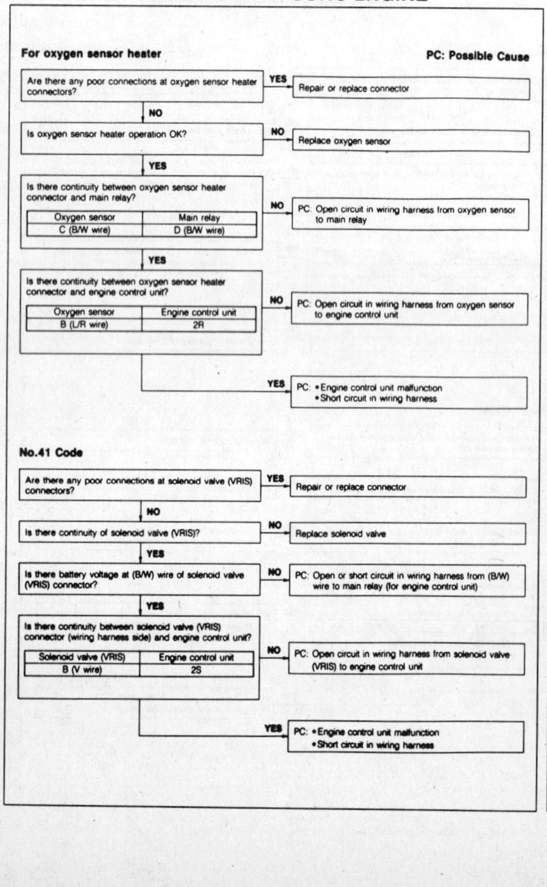

## CONTROL UNIT PIN LOCATION AND VOLTAGE TEST 1990 929 WITH SOHC ENGINE

**Terminal voltage**

| Terminal | Input | Output | Connection to | Voltage (after warm-up) Ignition switch: ON | Voltage (after warm-up) Idle | Remarks |
|---|---|---|---|---|---|---|
| 1A | — | — | Battery | Approx. 12V | | For back-up |
| 1B | O | | Main relay | Approx. 12V | | |
| 1C | O | | Ignition switch (START position) | Below 2.5V | | While cranking 6—11V |
| 1D | | O | Self-Diagnosis Checker (Monitor lamp) | <Test connector grounded> For 3 sec after ignition switch OFF→ON: Approx. 5V (lamp illuminates) After 3 sec: Approx. 12V | <Test connector grounded> Approx. 5V <Test connector is not grounded> Monitor lamp ON: Below 6V Monitor lamp OFF: Approx. 12V | With Self-Diagnosis Checker |
| 1E | | O | MIL (Malfunctin Indicator Light) | For 3 sec. after ignition switch OFF→ON: Below 2.5V (light illuminates) After 3 sec: Approx. 12V | | • Test connector grounded • Light illuminates: Below 2.5V • Light does not illuminates: Approx. 12V |
| 1F | | O | Self-Diagnosis Checker (for Code No.) | For 3 sec. after ignition switch OFF→OFF: Below 2.5V (Buzzer sounds) After 3 sec: Approx. 12V (Buzzer does not sound) | | • With Self-Diagnosis Checker and Test connector grounded • Buzzer sounds: Below 2.5V • Buzzer does not sound: Approx. 12V |
| 1G | | O | Igniter | 0V | Approx. 0.8V | |
| 1H | | O | Solenoid valve (Purge control) | Approx. 12V | | Driving in-gear 5—15V (Engine Signal Monitor: green and red lights flash) |
| 1I | — | — | — | For 180 sec after ignition switch OFF →ON: Approx. 12V | For 180 sec after starting: Approx. 12V | During hot condition: Coolant temp.: above 90°C (194°F) and engine air temp.: above 85°C (185°F) |
| | | | | Below 2.5V | | Other condition |
| 1J | | O | A/C relay | A/C switch ON: Below 2.5V A/C switch OFF: Approx. 12V | | Blower motor: ON |
| 1K | O | | Test connector | Test connector: Grounded 0V Test connector: Not grounded approx. 12V | | Green connector: 1-pin |
| 1L | O | | — | Below 1.5V | | M/T |
| | | | | Approx. 12V | | EC-AT |
| 1M | O | | EGR position sensor | Approx. 0.25—0.9V | | |
| 1N | O | | Idle switch | Pedal depressed: Approx. 10V Pedal released: 0V | | |
| 1P | O | | Power steering (P/S) switch | Constant approx. 12V | P/S: ON: 0V P/S: OFF: Approx. 12V | |
| 1Q | O | | A/C switch | A/C switch ON: Below 2.5V A/C switch OFF: Approx. 12V | | Blower motor: ON |
| 1S | O | | Blow motor switch | Blow motor switch ON: Below 2.5V Blow motor switch OFF: Approx. 10—12V | | |
| 1T | O | | Rear defroster switch | Rear defroster switch ON: Below 2.5V Rear defroster switch OFF: Approx. 10—12V | | |
| 1U | O | | Headlight switch | Headlight switch ON: below 2.5V Headlight switch OFF: Approx. 10—12V | | |

## CONTROL UNIT PIN LOCATION AND VOLTAGE TEST (CONT.)—929 WITH SOHC

| Terminal | Input | Output | Connection to | Voltage (after warm-up) | | Remark |
|---|---|---|---|---|---|---|
| | | | | Ignition switch: ON | Idle | |
| 1V | O | | Neutral/clutch switch | Clutch pedal depressed: Approx. 12V<br>Clutch pedal released: 0V | | Neutral condition |
| 2A | — | — | Ground ($E_{01}$) | 0V | | |
| 2B | — | — | Ground ($E_{02}$) | 0V | | |
| 2C | — | — | Ground ($E_1$) | 0V | | — |
| 2D | — | — | Ground ($E_2$) | 0V | | |
| 2E | O | | Distributor (Ne signal) | Approx. 0.6—0.8V | | Ne: Number of engine speed |
| 2F | O | | Distributor (COM) | Approx. 0.6—0.8V | | |
| 2G | O | | Distributor ($G_1$ signal) | Approx. 0.6—0.8V | | $G_1$: Group 1 |
| 2H | O | | Distributor ($G_2$ signal) | Approx. 0.6—0.8V | | $G_2$: Group 2 |
| 2I | O | | Igniter | 0V | Approx. 0.6—0.8V | Engine Signal Monitor green and red lights flash |
| 2J | O | | Airflow meter (Vc) | 7—9V | | |
| 2K | | O | Vref | 4.5—5.5V | | |
| 2L | O | | Intake air thermosensor (Dynamic chamber) | Approx. 1—2V at 80°C (176°F) | | |
| 2M | O | | Throttle sensor | Throttle valve fully closed: Approx. 0.3V<br>Throttle valve fully open: Approx. 4V | | |
| 2N | O | | Oxygen sensor | 0V | 0—1V | • Cold engine 0V at idle<br>• After warming-up:<br>Acceleration: 0.7—1.0V<br>Deceleration: 0—0.2V |
| 2O | O | | Airflow meter (Vs) | Approx. 2.0V | Approx. 3.5—5V | |
| 2P | O | | Intake air thermosensor (airflow meter) | Approx. 2—3V at 20°C (68°F) | | |
| 2Q | O | | Water thermosensor | Approx. 0.3—0.6V | | |
| 2R | | O | Solenoid valve (Trip induction control system (TICS)) and oxygen sensor heater | Below 2.5V | | Above approx. 4,000 rpm: Approx. 12V |
| 2S | | O | Solenoid valve (Variable resonance induction system (VRIS)) | Approx. 12V | | • 3,700—6,500 rpm: Below 2.0V<br>• Other conditions: Approx. 12V |
| 2T | | O | Solenoid valve (Pressure regulator control) | For 240 sec after ignition switch OFF→ON Below 2.5V | For 240 sec after starting: Below 2.5V | During hot condition: coolant temp.: Above 80°C (176°F) and engine air temp.: above 75°C (167°F) |
| | | | | Approx. 12V | | Other conditions |
| 2U | | O | Injector (No.1 and No.2) | Approx. 12V | | Engine Signal Monitor green and red lights flash |
| 2V | | O | Injector (No.3 and No.4) | Approx. 12V | | Engine Signal Monitor green and red lights flash |
| 2W | | O | Idle speed control (ISC) valve | Approx. 2—5V | Approx. 9—11V | Engine Signal Monitor green and red lights flash |
| 2X | | O | Injector (No.5 and No.6) | Approx. 12V | | Engine Signal Monitor green and red lights flash |
| 2Y | | O | Solenoid valve (EGR, vent side) | Approx. 12V | | Initial acceleration: Engine Signal Monitor green and red lights flash |
| 2Z | | O | Solenoid valve (EGR, vacuum side) | Approx. 12V | | Initial acceleration: Engine Signal Monitor green and red lights flash |

**Control unit connector (Control unit side)**

| 2Y | 2W | 2U | 2S | 2Q | 2O | 2M | 2K | 2I | 2G | 2E | 2C | 2A | 1U | 1S | 1Q | 1O | 1M | 1K | 1I | 1G | 1E | 1C | 1A |
|---|---|---|---|---|---|---|---|---|---|---|---|---|---|---|---|---|---|---|---|---|---|---|---|
| 2Z | 2X | 2V | 2T | 2R | 2P | 2N | 2L | 2J | 2H | 2F | 2D | 2B | 1V | 1T | 1R | 1P | 1N | 1L | 1J | 1H | 1F | 1D | 1B |

## WIRING SCHEMATIC—929 WITH SOHC ENGINE

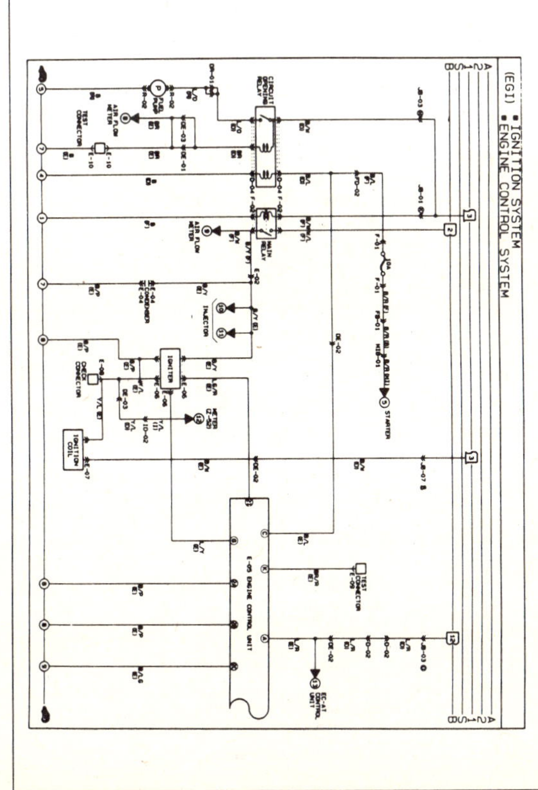


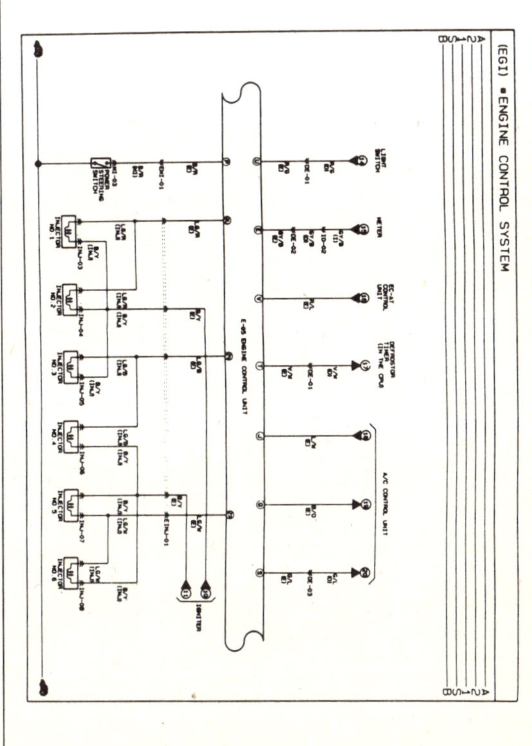

**WIRING SCHEMATIC – 929 WITH SOHC ENGINE**

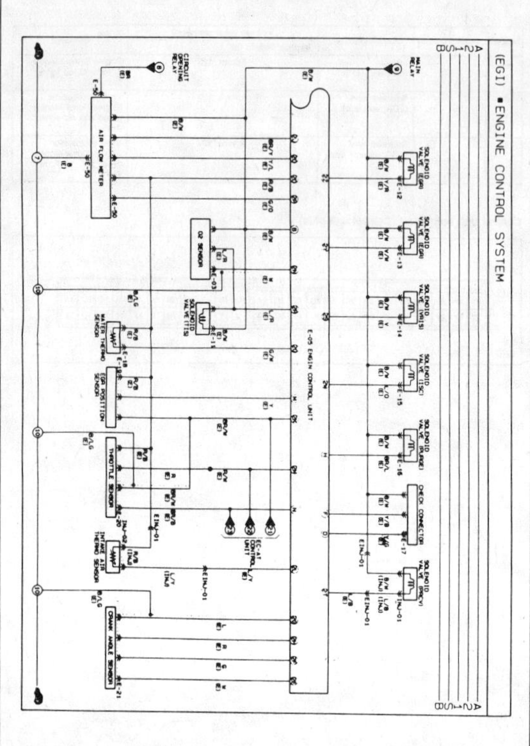


**SYSTEM SCHEMATIC – 929 WITH DOHC ENGINE**

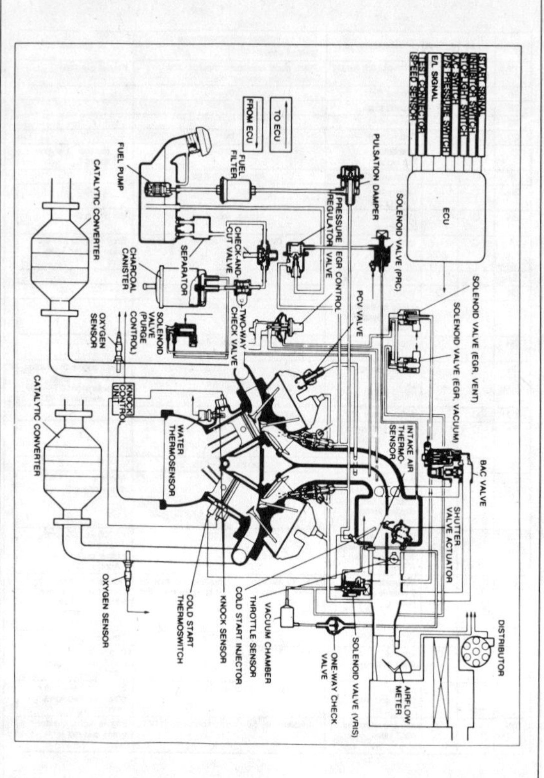


**VACUUM HOSE ROUTING – 929 WITH DOHC ENGINE**

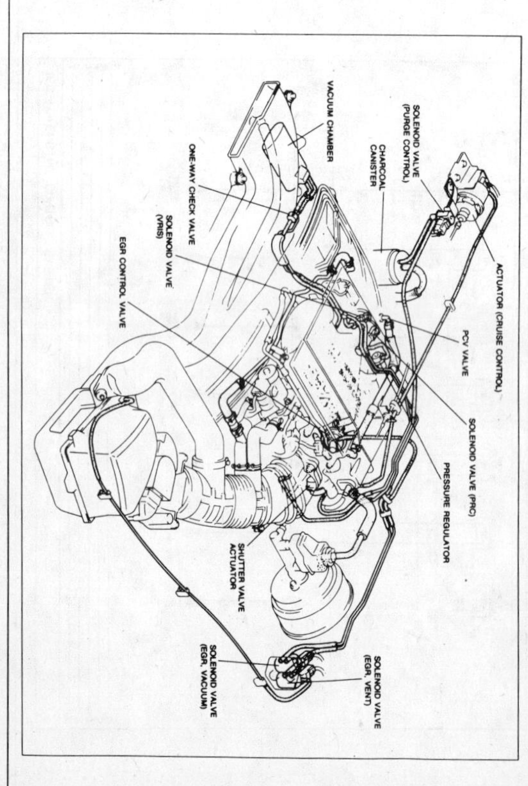


**EXPLODED VIEW OF FUEL SYSTEM COMPONENTS 1990 929 WITH DOHC ENGINE**

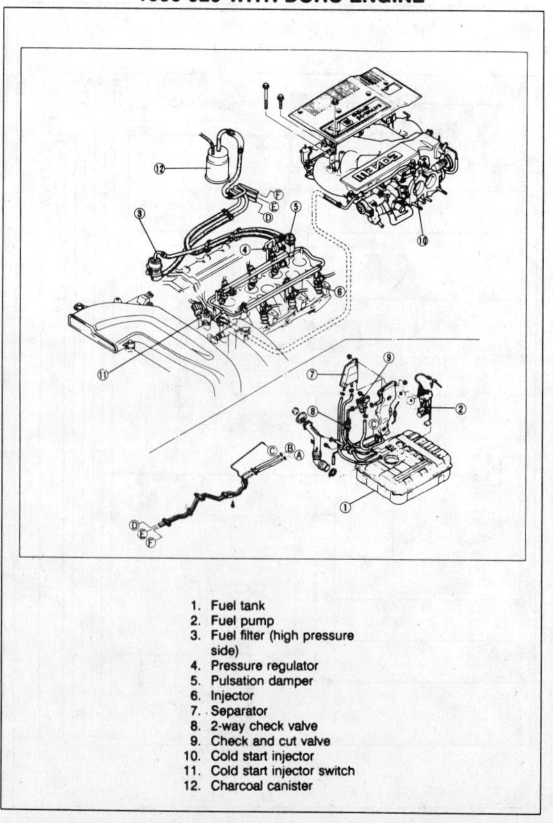

## EXPLODED VIEW OF CONTROL SYSTEM OUTPUT COMPONENTS—1990 929 WITH DOHC ENGINE

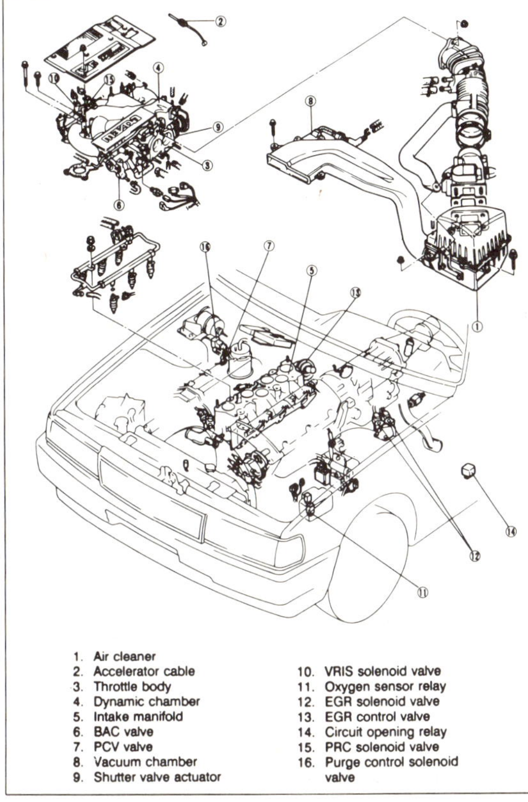

1. Air cleaner
2. Accelerator cable
3. Throttle body
4. Dynamic chamber
5. Intake manifold
6. BAC valve
7. PCV valve
8. Vacuum chamber
9. Shutter valve actuator
10. VRIS solenoid valve
11. Oxygen sensor relay
12. EGR solenoid valve
13. EGR control valve
14. Circuit opening relay
15. PRC solenoid valve
16. Purge control solenoid valve

## EXPLODED VIEW OF CONTROL SYSTEM INPUT COMPONENTS—1990 929 WITH DOHC ENGINE

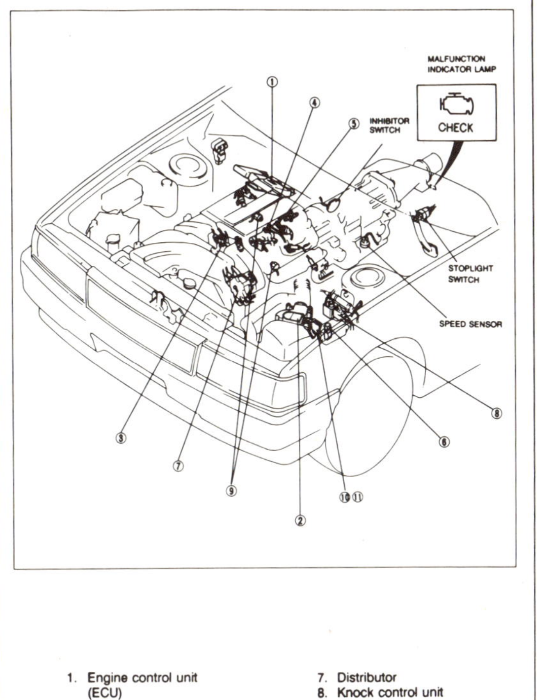


1. Engine control unit (ECU)
2. Airflow meter
3. Water thermosensor
4. Intake air thermosensor
5. Throttle sensor
6. EGI main relay
7. Distributor
8. Knock control unit
9. Knock sensor
10. Check connector (green: 6-pin)
11. Test connector (green: 1-pin)

## SPECIFICATIONS CHART 1990 929 WITH DOHC ENGINE

| Item | | | JE DOHC |
|---|---|---|---|
| Idle speed (Test connector grounded) rpm | | | 700 ± 20 (P range) |
| Ignition timing (Test connector grounded) | | | BTDC 8 ± 1° |
| Air cleaner | Element type | | Oil permeated |
| Throttle body | Throat diameter mm (in) | | 46 (1.811) x 2 |
| Throttle sensor | Idle switch (adjustment clearance) mm (in) | | 0.1 (0.004) (Continuity)<br>0.3 (0.012) (No continuity) |
| | Resistance kΩ | Vref ↔ GND | 3—7 |
| | | TVO ↔ GND | 0.2—6.0 (Fully closed)<br>3.3—7.0 (Fully open) |
| Airflow meter | Resistance Ω | $E_2$ ↔ Vs | 20—600 (Measuring plate fully closed)<br>20—1,000 (Fully open) |
| | | $E_2$ ↔ Vc | 200—400 |
| | | $E_1$ ↔ Fc | 0 (Fully open)<br>∞ (Fully closed) |
| | | $E_2$ ↔ THA | -20°C (-4°F) 13,600—18,400<br>20°C (68°F) 2,210—2,690<br>60°C (140°F) 493—667 |
| Fuel tank | Capacity liters (US gal, Imp gal) | | 70 (18.5) |
| Fuel filter | Low-pressure | | Nylon 6 (250 mesh) |
| | High-pressure | | Filter paper |
| Fuel pump | Type | | Impeller (in tank) |
| | Fuel pressure kPa ($kg/cm^2$, psi) | | 441—588 (4.5—6.0, 64.0—85.3) |
| | Feeding capacity liters (US gal, Imp gal)/min | | 1.3—2.3 (0.34—0.61, 0.29—0.51) |
| Pressure regulator | Type | | Diaphragm |
| | Fuel pressure kPa ($kg/cm^2$, psi) | | 264—314 (2.7—3.2, 38.4—45.5)... Vacuum cut |
| Injector | Drive | | Voltage |
| | Type | | High-ohmic injector |
| | Resistance Ω | | 12—16 |
| | Injection amount cc (cu in)/15 sec | | 60—65 (3.66—3.97) |
| Cold start injector | Resistance Ω | | 3—5 |
| | Injection amount cc (cu in)/15 sec | | 32—40 (1.95—2.44) |
| BAC valve | Resistance Ω | | 10—13 |
| Solenoid valve | Purge control | Resistance Ω | 29—35 |
| | EGR vacuum side | Resistance Ω | 29—35 |
| | EGR vent side | Resistance Ω | 29—35 |
| Fuel specification | | | Unleaded gasoline (RON 91 or higher) |

## COMPONENT DESCRIPTIONS 1990 929 WITH DOHC ENGINE

| Component | Function | Remark |
|---|---|---|
| **Air cleaner** | Filters air entering throttle body | |
| **Airflow meter** | Detects amount of intake air, sends signal to engine control unit | Intake air thermosensor and fuel pump switch included |
| **Air valve** | Supplies bypass air into dynamic chamber (engine cold) | • Engine speed increased to shorten warm-up period<br>• Thermowax type<br>• Installed in BAC valve |
| **Bypass aie control (BAC) valve** | Supplies bypass air into dynamic chamber | Consists of air valve and ISC valve |
| **Idle speed control (ISC) valve** | Supplies bypass air into dynamic chamber | • Controlled by duty signal from engine control unit<br>• Works independent of idle-up<br>• Installed in BAC valve |
| **Brake light switch** | Detects braking, sends signal to engine control unit | |
| **Catalytic converter** | Reduces HC, CO, and NOx by chemical reaction | Monolith type |
| **Charcoal canister** | Stores fuel tank fumes (engine stopped) | |
| **Check connectors** | For Self-Diagnosis Checker and fuel pump | • 6-pin connector (Green); Self-Diagnosis Checker<br>• 2-pin connector (Yellow); Fuel pump |
| **Check-and-cut valve** | Controls pressure in fuel tank | |
| **Circuit opening relay** | Voltage for fuel pump (engine running) | |
| **Engine control unit (ECU)** | Detects the following<br>1. Engine speed<br>2. No.1 and No.4 piston TDC<br>3. Intake air amount<br>4. Engine coolant temperature<br>5. Atmospheric pressure<br>6. Throttle valve opening angle<br>7. Throttle valve fully closed<br>8. Air/fuel ratio (oxygen concentration)<br>9. In-gear condition<br>10. Intake air temperature<br>11. Engine air temperature<br><br>12. A/C operation<br>13. P/S operation<br>14. E/L operation<br>15. Cranking signal<br>16. Initial set signal<br>17. Braking signal<br>18. EGR operation signal<br>19. Knocking signal<br>Controls operation of the following<br>1. Fuel injection system<br>2. Idle speed control system<br>3. Pressure regulator control system<br>4. Variable resonance induction system<br>5. Ignition system (Electronic spark advance)<br>6. Purge control system<br>7. EGR control system<br>8. Fail-safe system<br>9. Switch monitor function | 1. Ne rotor (in distributor)<br>2. G rotor (in distributor)<br>3. Airflow meter<br>4. Water thermosensor<br>5. Atmospheric pressure sensor (in ECU)<br>6. Throttle sensor<br>7. Idle switch (in throttle sensor)<br>8. Oxygen sensor<br>9. EC-AT control unit<br>10. Intake air thermosensor (in airflow meter)<br>11. Intake air thermosensor (on dynamic chamber)<br>12. A/C switch<br>13. P/S switch<br>14. E/L switch<br>15. Ignition switch<br>16. Test connector<br>17. Brake light switch<br>18. EGR position sensor<br>19. Knock control unit<br><br>1. Injector<br>2. ISC valve<br>3. Solenoid valve (PRC)<br>4. Solenoid valve (VRIS)<br>5. Igniter<br>6. Solenoid valve (Purge control)<br>7. Solenoid valves (EGR)<br>8. Self-Diagnosis Checker and MIL<br>9. Monitor lamp (Self-Diagnosis Checker) |
| **Cold start injector** | Injects fuel into throttle body (engine cold) | |
| **Cold start thermoswitch** | Controls cold start injector | |

## COMPONENT DESCRIPTIONS (CONT.) – 1990 929 WITH DOHC ENGINE

| Component | Function | Remark |
|---|---|---|
| **Distributor** | Detects crank angle at 30° intervals, No.1 cylinder TDC and No.4 cylinder TDC, sends signal to ECU | • Ne (Engine speed) signal<br>• G1 signal<br>• G2 signal |
| **G1 coil** | Detects No.4 cylinder TDC, sends signal to ECU | For determining fuel injection timing |
| **G2 coil** | Detects No.1 cylinder TDC, sends signal to ECU | For determining fuel injection timing |
| **Dynamic chamber** | Interconnects all cylinders | Contains shutter valve |
| **EC-AT control unit** | Detects N or P range, sends signal to ECU | |
| **Exhaust gas recirculation (EGR) position sensor** | Detects EGR control valve opening, sends signal to ECU | Installed on EGR control valve |
| **Exhaust gas recirculation (EGR) control valve** | Recirculates portion of exhaust gas | Contains EGR position sensor |
| **Fuel filter** | Filters fuel | |
| **Fuel pump** | Provides fuel to injectors | • Operates only while engine running<br>• Installed in fuel tank |
| **Idle switch** | Detects when throttle valve fully closed; sends signal to ECU | Installed in throttle sensor<br>Switch closed when throttle valve fully closed |
| **Throttle sensor** | Detects throttle valve opening angle; sends signal to ECU | • Idle switch included<br>• Installed on throttle body |
| **Igniter** | Receives spark signal from ECU; generates high voltage in ignition coil | High voltage distributed by rotor in distributor |
| **Ignition switch** | Detects engine cranking; sends signal to ECU | |
| **Inhibitor switch** | Detects in-gear condition; sends signal to EC-AT control unit | Switch closed in N or P range |
| **Injector** | Injects fuel into intake port | • Controlled by signals from ECU<br>• High-ohmic injector |
| **Intake manifold** | Supplies intake air to all cylinders | |
| **Intake air thermosensor** | Detects intake air temperature; sends signal to ECU | Installed in airflow meter |
| | Detects engine air temperature; sends signal to ECU | Installed on dynamic chamber |
| **Knock control unit** | Receives knock signal from knock sensor; sends signal to ECU | |
| **Knock sensor** | Detects engine knocking; sends signal to knock control unit | |
| **Main relay** | Supplies current to injectors and ECU | |
| **MIL (Malfunction indicator lamp)** | Light illuminates if input device malfunctions | Installed in instrument cluster |
| **One-way check valve** | Prevents misoperation of shutter valve during heavy-load operations | |
| **Oxygen sensor** | Detects oxygen concentration; sends signal to ECU | Zirconia ceramic and platinum coating |
| **Oxygen sensor relay** | Supplies current to oxygen sensor heater | |

| Component | | Function | Remark |
|---|---|---|---|
| **PCV valve** | | Controls blowby gas amount pulled into engine | |
| **Power steering (P/S) switch** | | Detects P/S operation; sends signal to ECU | ON when steering wheel turned right or left |
| **Pressure regulator** | | Adjusts fuel pressure supplied to injectors | |
| **Pulsation damper** | | Absorbs fuel pulsations | |
| **Separator** | | Prevents fuel from flowing into charcoal canister | |
| **Shutter valve actuator** | | Controls shutter valve operation | For VRIS |
| **Solenoid valve** | **EGR (Vent side, Vacuum side)** | Controls EGR control valve operation | • Exhaust gas recirculation<br>• Duty solenoid valve |
| | **Purge control** | Controls evaporative fumes from canister to intake manifold | Controlled by duty signal from ECU |
| | **Variable resonance induction system (VRIS)** | Controls vacuum to shutter valve actuator | |
| | **Pressure regulator control (PRC)** | Blocks vacuum from intake manifold to pressure regulator | Hot engine restart only |
| **Test connector** | | For Self-Diagnosis Checker and idle speed/ignition timing adjustment | Green: 1-pin connector |
| **Throttle body** | | Controls intake air amount | Integrated throttle sensor and cold start injector incorporated |
| **Water thermosensor** | | Detects engine coolant temperature; sends signal to ECU | |

## COMPONENT RELATIONSHIP CHART 1990 929 WITH DOHC ENGINE

| OUTPUT DEVICE \ INPUT DEVICE | | DISTRIBUTOR: Ne ROTOR | DISTRIBUTOR: G ROTOR | AIR-FLOW METER | THROTTLE SENSOR: IDLE SWITCH | THROTTLE SENSOR: THROTTLE SENSOR | WATER THERMO-SENSOR | INTAKE AIR THERMO-SENSOR: AIR-FLOW METER | INTAKE AIR THERMO-SENSOR: DYNAMIC CHAMBER | OXYGEN SENSOR | ATMOSPHERIC PRESSURE SENSOR | EGR POSITION SENSOR | EC-AT CONTROL UNIT | KNOCK CONTROL UNIT | A/C SWITCH | P/S SWITCH | IGNITION SWITCH (START SIGNAL) | BRAKE LIGHT SWITCH | E/L SWITCH | TEST CONNECTOR |
|---|---|---|---|---|---|---|---|---|---|---|---|---|---|---|---|---|---|---|---|---|
| INJECTOR | FUEL INJECTION AMOUNT | O | X | O | O | O | O | O | O | O | O | X | O | X | O | X | O | X | X | X |
| | FUEL INJECTION TIMING | O | O | X | X | O | X | X | X | X | X | X | X | X | X | X | O | X | X | X |
| IGNITER | | O | O | O | O | O | O | X | X | X | O | O | O | O | O | X | O | X | X | O |
| BAC VALVE (ISC VALVE) | | O | X | X | O | O | O | O | O | X | X | X | O | X | O | O | X | X | O | O |
| SOLENOID VALVE (PRC) | | O | X | X | X | O | O | X | O | X | X | X | X | X | X | X | O | X | X | X |
| SOLENOID VALVE (VRIS) | | O | X | X | X | O | X | X | X | X | X | X | X | X | X | X | O | X | X | X |
| SOLENOID VALVE (PURGE CONTROL) | | O | X | O | O | O | O | O | X | O | O | X | O | X | X | X | X | X | X | X |
| SOLENOID VALVE (EGR) | VENT SIDE | O | X | O | O | O | O | X | X | X | O | X | X | X | X | X | X | X | X | X |
| | VACUUM SIDE | O | X | O | O | O | O | X | X | X | O | X | X | X | X | X | X | X | X | X |
| EC-AT CONTROL UNIT (LOCKUP CONTROL) | | O | X | X | X | X | O | X | X | X | X | X | X | X | X | X | O | X | X | X |
| OXYGEN SENSOR HEATER RELAY | | O | X | O | X | X | O | X | X | X | X | X | X | X | X | X | O | X | X | X |
| A/C RELAY | | O | X | X | X | O | X | X | X | X | X | X | O | X | O | X | O | X | X | X |

O: Related
X: Not related

## OUTPUT COMPONENT AND ENGINE OPERATION 1990 929 WITH DOHC ENGINE

| OUTPUT DEVICE \ ENGINE CONDITION | | CRANKING (COLD ENGINE) | WARM-UP (DURING IDLE) | MEDIUM LOAD: COLD | MEDIUM LOAD: WARM | ACCELERATION | HEAVY LOAD | DECELERATION | IDLE (THROTTLE VALVE FULLY CLOSED) | IGN: ON (ENGINE NOT RUNNING) | REMARK |
|---|---|---|---|---|---|---|---|---|---|---|---|
| INJECTOR | No.1 / No.2 / No.3 / No.4 / No.5 / No.6 | Rich | | | *1 Rich and lean (Fuel feedback operation) | Rich | | Fuel cut | *1 Rich and lean (Fuel feedback operation) | Does not inject | *1 Operated by oxygen sensor |
| | INJECTION TIMING | Sequential injection | | | | | | | Sequential injection | | Fuel cut: Above 6,500 rpm |
| IGNITER | | Fixed at BTDC 7° | Advanced: Depends on engine condition | | | Retarded: Depends on intensity of knocking | | Advanced: Depends on engine condition | | – | |
| BAC VALVE | AIR VALVE | *2 Open | | | Closed | | | | | | *2 Collant temp: Below 60°C (140°F) |
| | ISC VALVE | Large amount of bypass air | | | Small amount of bypass air | | | Large and small amount of bypass air | *3 Large and small amount of bypass air | Does not operate | *3 Engine speed feedback system operates when test connector open |
| SOLENOID VALVE (PRC) | | OFF (Vacuum to pressure regulator) | | | | | | | *4 After start ON (Vacuum cut) | Does not operate | *4 During hot condition only |
| SOLENOID VALVE (VRIS) | | OFF (Shutter valve closed) | | *5 ON (Shutter valve open) | | | | OFF (Shutter valve closed) | | Does not operate | *5 ON: Below 3,800 rpm |
| SOLENOID VALVE (PURGE CONTROL) | | OFF (Purge cut) | | | Operates (Duty values [purge gas amount] change) | | | | OFF (Purge cut) | Does not operate | |
| SOLENOID VALVE (EGR) | VENT SIDE | OFF (No vacuum to EGR valve) — EGR cut — | | | *6 System operates | | | OFF (No vacuum to EGR valve) — EGR cut — | | Does not operate | *6 Depends on engine condition |
| | VACUUM SIDE | OFF (Duty value 100%) (Atmospheric pressure to EGR valve) | | | (Amount of EGR changes) | | | OFF (Duty value 100%) (Atmosheric pressure to EGR valve) | | | |
| OXYGEN SENSOR HEATER RELAY | | OFF | ON | | | | OFF | ON | | Does not operate | |
| A/C RELAY | | OFF (A/C OFF) | ON (A/C ON) | | | | OFF (A/C OFF) | ON (A/C ON) | | ON (A/C OFF) | |

## TROUBLESHOOTING CHART
## 1990 929 WITH DOHC ENGINE

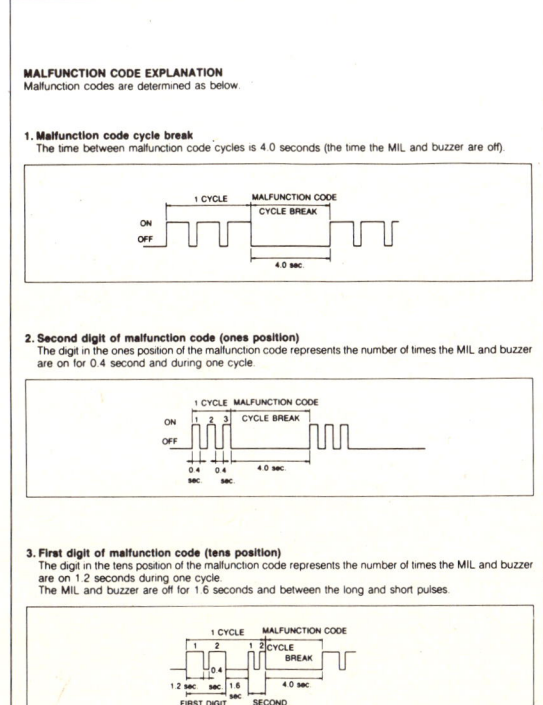

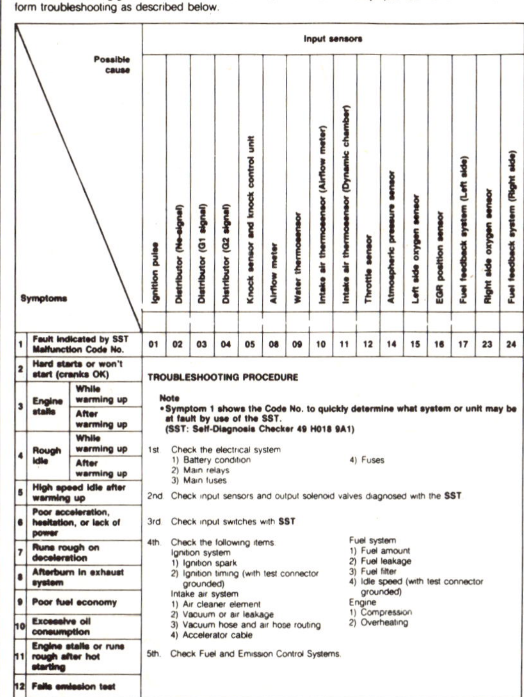

This troubleshooting guide shows the malfunction code numbers and the symptoms of various failures. Perform troubleshooting as described below.

| Symptoms \ Possible cause | | Input sensors | | | | | | | | | | | | | | | |
|---|---|---|---|---|---|---|---|---|---|---|---|---|---|---|---|---|---|
| | | Ignition pulse | Distributor (Ne-signal) | Distributor (G1 signal) | Distributor (G2 signal) | Knock sensor and knock control unit | Airflow meter | Water thermosensor | Intake air thermosensor (Airflow meter) | Intake air thermosensor (Dynamic chamber) | Throttle sensor | Atmospheric pressure sensor | Left side oxygen sensor | EGR position sensor | Fuel feedback system (Left side) | Right side oxygen sensor | Fuel feedback system (Right side) |
| 1 | Fault indicated by SST Malfunction Code No. | 01 | 02 | 03 | 04 | 05 | 08 | 09 | 10 | 11 | 12 | 14 | 15 | 16 | 17 | 23 | 24 |
| 2 | Hard starts or won't start (cranks OK) | | | | | | | | | | | | | | | | |
| 3 | Engine stalls — While warming up | | | | | | | | | | | | | | | | |
| | Engine stalls — After warming up | | | | | | | | | | | | | | | | |
| 4 | Rough idle — While warming up | | | | | | | | | | | | | | | | |
| | Rough idle — After warming up | | | | | | | | | | | | | | | | |
| 5 | High speed idle after warming up | | | | | | | | | | | | | | | | |
| 6 | Poor acceleration, hesitation, or lack of power | | | | | | | | | | | | | | | | |
| 7 | Runs rough on deceleration | | | | | | | | | | | | | | | | |
| 8 | Afterburn in exhaust system | | | | | | | | | | | | | | | | |
| 9 | Poor fuel economy | | | | | | | | | | | | | | | | |
| 10 | Excessive oil consumption | | | | | | | | | | | | | | | | |
| 11 | Engine stalls or runs rough after hot starting | | | | | | | | | | | | | | | | |
| 12 | Fails emission test | | | | | | | | | | | | | | | | |

**TROUBLESHOOTING PROCEDURE**

**Note**
- **Symptom 1 shows the Code No. to quickly determine what system or unit may be at fault by use of the SST. (SST: Self-Diagnosis Checker 49 H018 9A1)**

1st. Check the electrical system
1) Battery condition
2) Main relays
3) Main fuses
4) Fuses

2nd. Check input sensors and output solenoid valves diagnosed with the **SST**

3rd. Check input switches with **SST**

4th. Check the following items.
Ignition system
1) Ignition spark
2) Ignition timing (with test connector grounded)
Intake air system
1) Air cleaner element
2) Vacuum or air leakage
3) Vacuum hose and air hose routing
4) Accelerator cable
Fuel system
1) Fuel amount
2) Fuel leakage
3) Fuel filter
4) Idle speed (with test connector grounded)
Engine
1) Compression
2) Overheating

5th. Check Fuel and Emission Control Systems.

## TROUBLESHOOTING CHART (CONT.)
## 1990 929 WITH DOHC ENGINE

The troubleshooting guide lists the systems most likely to cause a given symptom. After finding systems to check, refer to the pages shown for detailed guides for each system.

| Output solenoid valves | | | | | | | | Fuel and Emission Control Systems | | | | | | | | | | |
|---|---|---|---|---|---|---|---|---|---|---|---|---|---|---|---|---|---|---|
| Solenoid valve (PRC) | Solenoid valve (Purge control) | Solenoid valve (EGR, vacuum side) | Solenoid valve (EGR, vent side) | Idle speed control valve | Oxygen sensor heater relay | Solenoid valve (VRIS) | Possible cause | Intake air system (Poor connection of components, Throttle body) | Idle speed control system (Air valve, Idle speed control valve malfunction) | Variable resonance induction system (VRIS) | Fuel system (Fuel injection, Fuel pressure) | Pressure regulator control (PRC) system | Electronic spark advance (ESA) control system | Deceleration control system (Fuel cut operation malfunction) | EGR system (EGR control valve open) | Evaporative emission control system | PCV system (System clogged) | Exhaust system (System clogged) |
| 25 | 26 | 28 | 29 | 34 | 36 | 41 | Symptom | | | | | | | | | | | |
| | | | | | | | 2 | 3 | | | 2 | | 1 | | 4 | | | |
| | | | | | | | 3 | 4 | 1 | | 3 | | | | 2 | | | |
| | | | | | | | | 5 | 2 | | 4 | | | | 3 | | 1 | |
| | | | | | | | 4 | 6 | 1 | | 5 | | | | 3 | 4 | 2 | |
| | | | | | | | | 6 | 2 | | 5 | | | | 3 | 4 | 1 | |
| | | | | | | | 5 | 2 | 1 | | 3 | | | | | | | |
| | | | | | | | 6 | 3 | | 5 | 4 | | | | 1 | 2 | | 6 |
| | | | | | | | 7 | | 2 | | 3 | | | 1 | | | | |
| | | | | | | | 8 | 3 | 2 | | 4 | | | 1 | | | | |
| | | | | | | | 9 | | | | 3 | 4 | | 1 | 2 | | | 5 |
| | | | | | | | 10 | | | | | | | | | | 1 | |
| | | | | | | | 11 | | 2 | | 3 | 1 | | | | | | |
| | | | | | | | 12 | 6 | 5 | | 7 | 8 | | 3 | 2 | 4 | | 1 |

The numbers of the list show the priorities of inspections from the most possible to that with the lowest possibility. These were determined on the following basis:
- Ease of inspection
- Most possible system
- Most possible point in system

## MALFUNCTION INDICATOR LIGHT (MIL) OPERATION
## 1990 929 WITH DOHC ENGINE

**MALFUNCTION CODE EXPLANATION**
Malfunction codes are determined as below.

**1. Malfunction code cycle break**
The time between malfunction code cycles is 4.0 seconds (the time the MIL and buzzer are off).

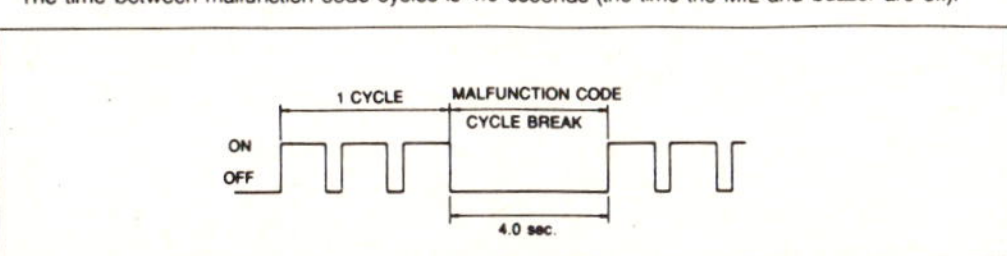


**2. Second digit of malfunction code (ones position)**
The digit in the ones position of the malfunction code represents the number of times the MIL and buzzer are on for 0.4 second and during one cycle.

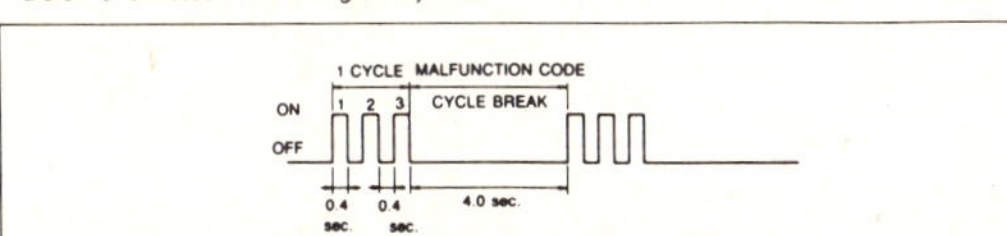


**3. First digit of malfunction code (tens position)**
The digit in the tens position of the malfunction code represents the number of times the MIL and buzzer are on 1.2 seconds during one cycle.
The MIL and buzzer are off for 1.6 seconds and between the long and short pulses.

## TROUBLE CODE IDENTIFICATION CHART
## 1990 929 WITH DOHC ENGINE

**Input Devices**

| Malfunction Code No. | Input devices | Malfunction | Fail-safe function | Pattern of output signals (Self-Diagnosis Checker or MIL) |
|---|---|---|---|---|
| 01 | Ignition pulse (Igniter, Ignition coil) | Broken wire, short circuit | — | |
| 02 | Distributor (Ne-signal) | Ne-signal not input for 1.5 sec. during cranking | — | |
| 03 | Distributor (G1 signal) | Broken wire, short circuit | — | |
| 04 | Distributor (G2 signal) | Broken wire, short circuit | — | |
| 05 | Knock sensor and knock control unit | Broken wire, short circuit | — | |
| 08 | Airflow meter | Broken wire, short circuit | Basic fuel injection amount fixed as for 2 driving modes 1) Idle switch: ON 2) Idle switch: OFF | |
| 09 | Water thermosensor | Broken wire, short circuit | Coolant temp. input fixed at 80°C (176°F) | |
| 10 | Intake air thermosensor (Airflow meter) | Broken wire, short circuit | Intake air temp. input fixed at 20°C (68°F) | |
| 11 | Intake air thermosensor (Dynamic chamber) | Broken wire, short circuit | Intake air temp. input fixed at 20°C (68°F) | |
| 12 | Throttle sensor | Broken wire, short circuit | Throttle valve opening angle signal input fixed at fully open | |
| 14 | Atmospheric pressure sensor | Short circuit | Atmospheric pressure input fixed at 760 mmHg (29.9 in Hg) | |
| 15 | Left side oxygen sensor | Oxygen sensor output below 0.55V 120 sec. after engine at 1,500 rpm | Feedback system canceled (for EGI) | |
| 16 | EGR position sensor | Broken wire, short circuit | EGR position signal input fixed at fully closed | |
| | | Sensor output does not match target value (incorrect output) | — | |
| 17 | Feedback system (Left side) | Oxygen sensor output does not change from 0.55V 60 sec. after engine at 1,500 rpm | Feedback system canceled (for EGI) | |

## TROUBLE CODE IDENTIFICATION CHART (CONT.) 929 WITH DOHC ENGINE

| Malfunction Code No. | Input devices | Malfunction | Fail-safe function | Pattern of output signals (Self-Diagnosis Checker or MIL) |
|---|---|---|---|---|
| 23 | Right side oxygen sensor | Oxygen sensor output below 0.55V 120 sec after engine at 1,500 rpm | Feedback system canceled (for EGI) | |
| 24 | Feedback system (Right side) | Oxygen sensor output does not change from 0.55V 60 sec. after engine at 1,500 rpm | Feedback system canceled (for EGI) | |

**Output Devices**

| Malfunction code | Output devices | Pattern of output signals (Self-Diagnosis Checker or MIL) |
|---|---|---|
| 25 | Solenoid valve (PRC) | |
| 26 | Solenoid valve (Purge control) | |
| 28 | Solenoid valve (EGR, vacuum side) | |
| 29 | Solenoid valve (EGR, vent side) | |
| 34 | ISC valve | |
| 36 | Oxygen sensor heater relay | |
| 41 | Solenoid valve (VRIS) | |

**Caution**

- **If there is more than one failure present, the lowest number malfunction code is displayed first, any subsequent malfunction codes light up in order.**
- **After repairing all failures, turn off the ignition switch and disconnect the negative battery cable and depress the brake pedal for at least five seconds to erase the memory of a malfunction code.**

## TROUBLE CODE DIAGNOSTIC CHART 1990 929 WITH DOHC ENGINE

If a malfunction code number is illuminated on the **SST**, check the following chart along with the wiring diagram.

**No.01 Code**

Are there poor connections at ignition coil connectors? — YES → Repair or replace connector

NO ↓

Is resistance of ignition coil OK ?
Resistance: Primary 0.72—0.88Ω
Secondary 10—30 kΩ
— NO → Replace ignition coil

YES ↓

Is operation of igniter OK ? — NO → Replace igniter

YES ↓

Is there continuity between IGf terminal of igniter connector (wiring harness side) and engine control unit (1V) terminal? — NO → Open circuit in wiring harness from igniter to engine control unit (1V) terminal

YES ↓

Is engine control unit (1V) terminal voltage OK? — NO → Short circuit in wiring harness

YES → Replace engine control unit

**No.02 Code**

Are there poor connections at distributor connectors? — YES → Repair or replace connector

NO ↓

Is resistance of distributor OK? — NO → Replace distributor

YES ↓

Is there continuity between distributor connector (wiring harness side) and engine control unit?

| Distributor | Engine control unit |
|---|---|
| Ne ① W | 3E |
| Ne ② G | 3F |

— NO → Open circuit in wiring harness

YES ↓

Are engine control unit (3E) and (3F) terminal voltages OK? — NO → Short circuit in wiring harness

YES → Replace engine control unit

## TROUBLE CODE DIAGNOSTIC CHART—1990 929 WITH DOHC ENGINE

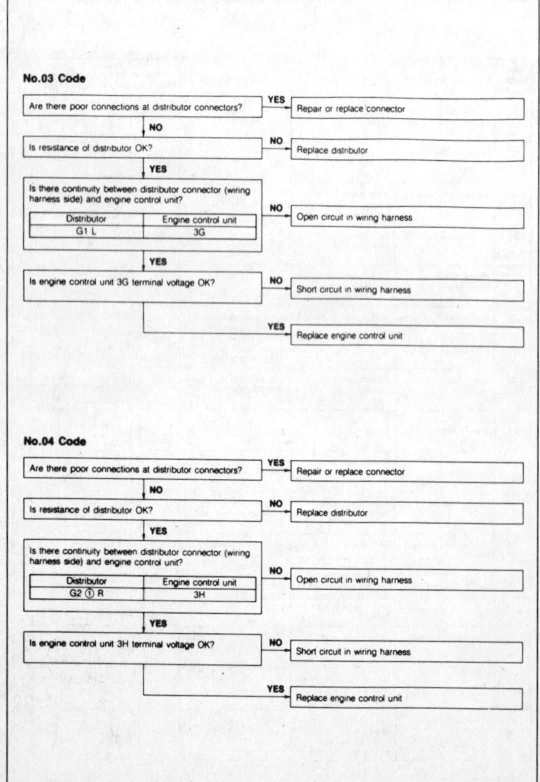


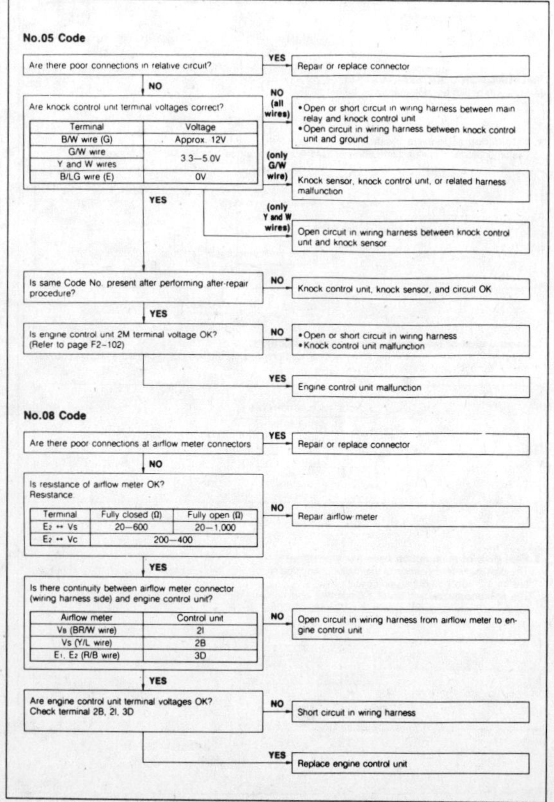

## TROUBLE CODE DIAGNOSTIC CHART—1990 929 WITH DOHC ENGINE

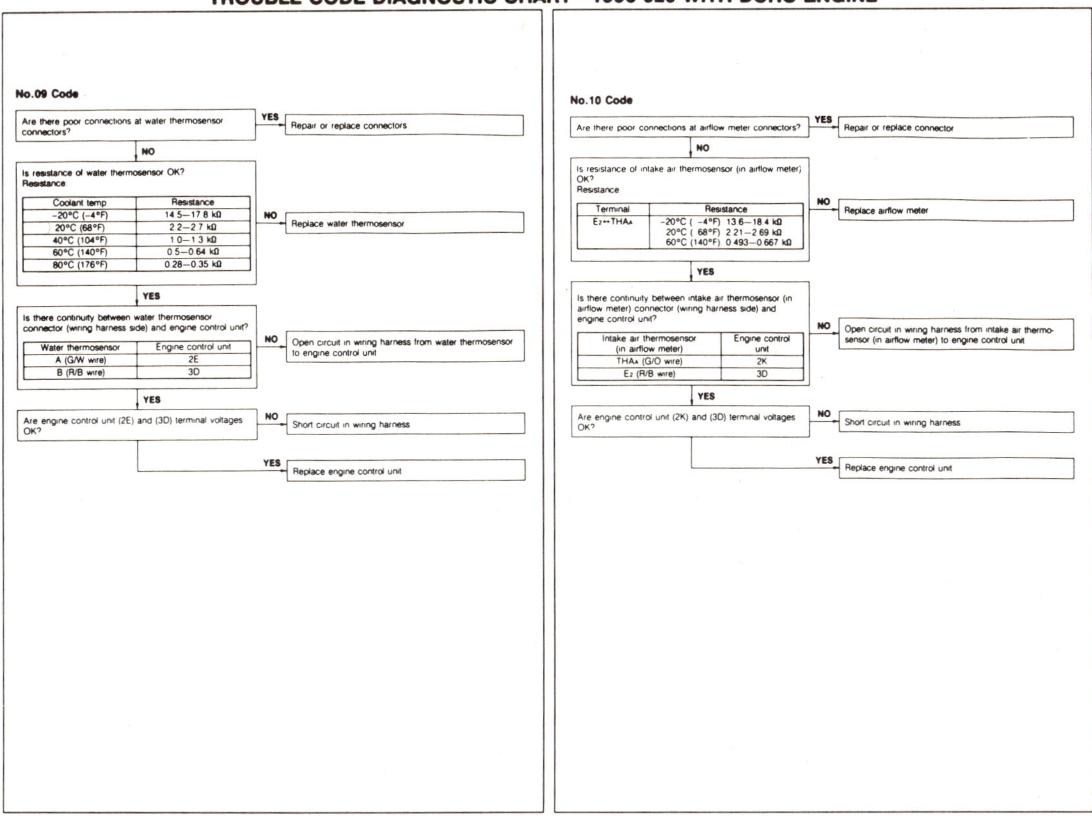


## TROUBLE CODE DIAGNOSTIC CHART—1990 929 WITH DOHC ENGINE

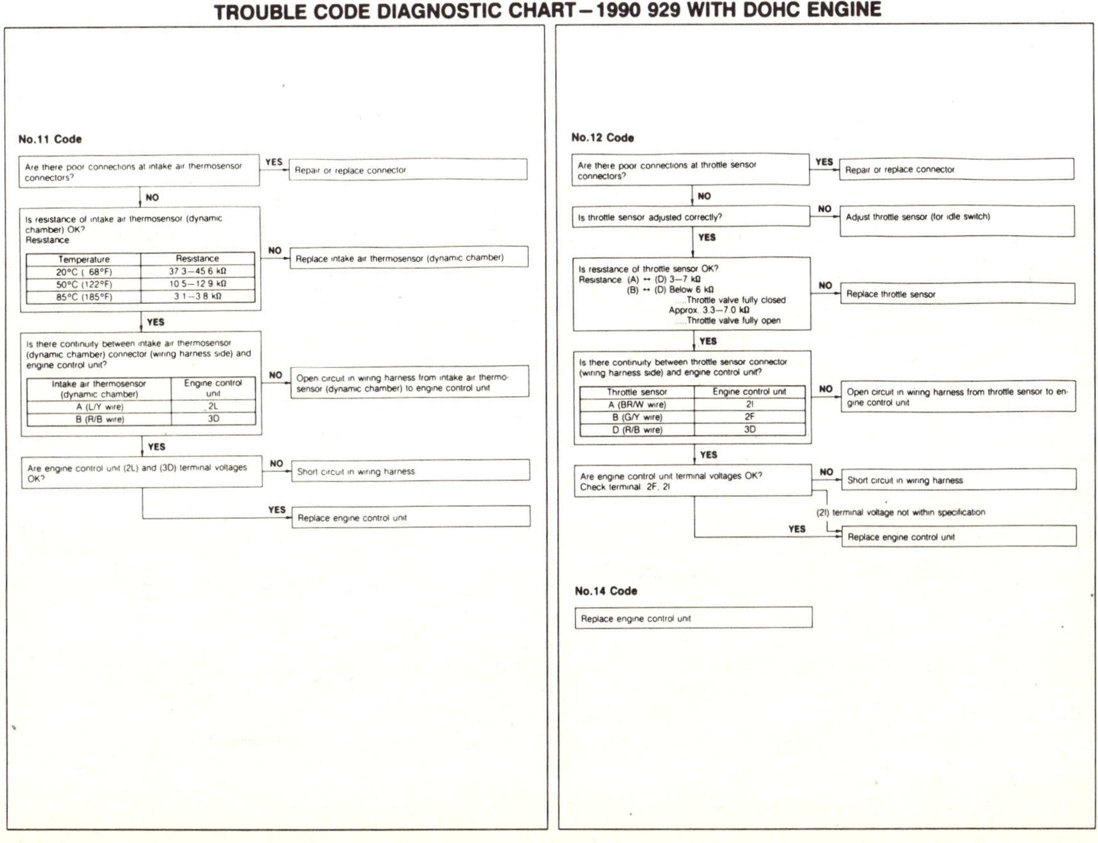

## TROUBLE CODE DIAGNOSTIC CHART – 1990 929 WITH DOHC ENGINE

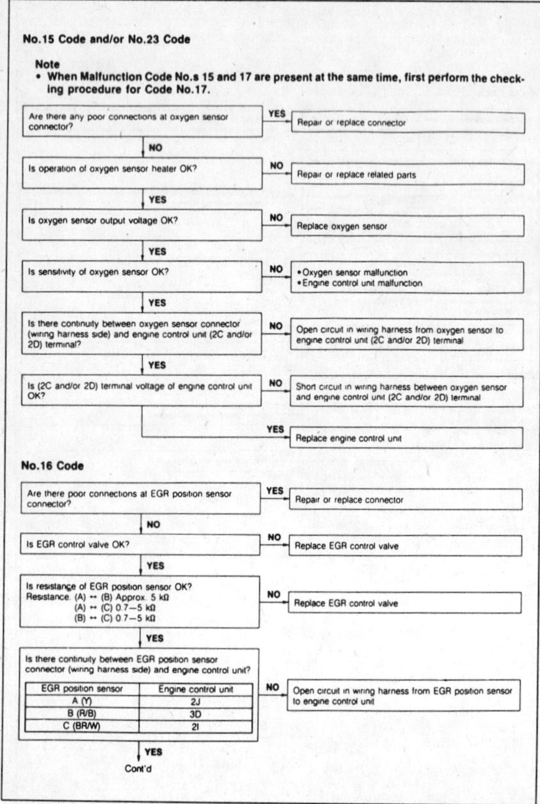


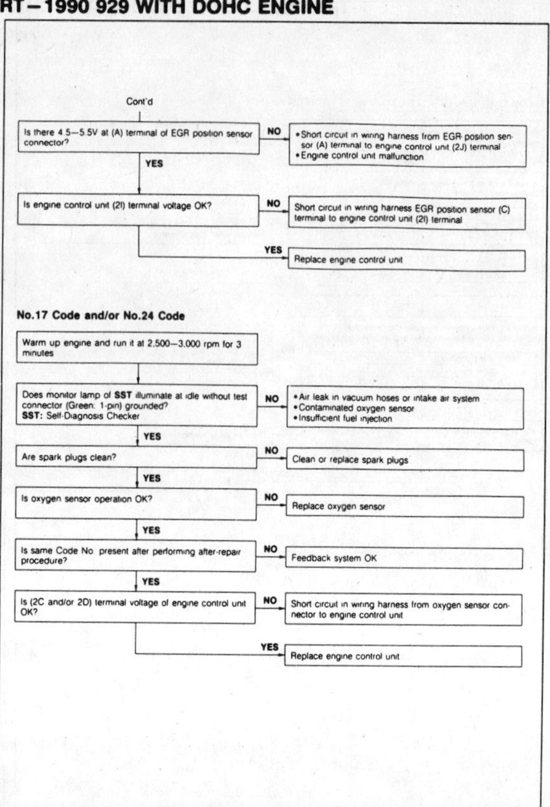


## TROUBLE CODE DIAGNOSTIC CHART – 1990 929 WITH DOHC ENGINE

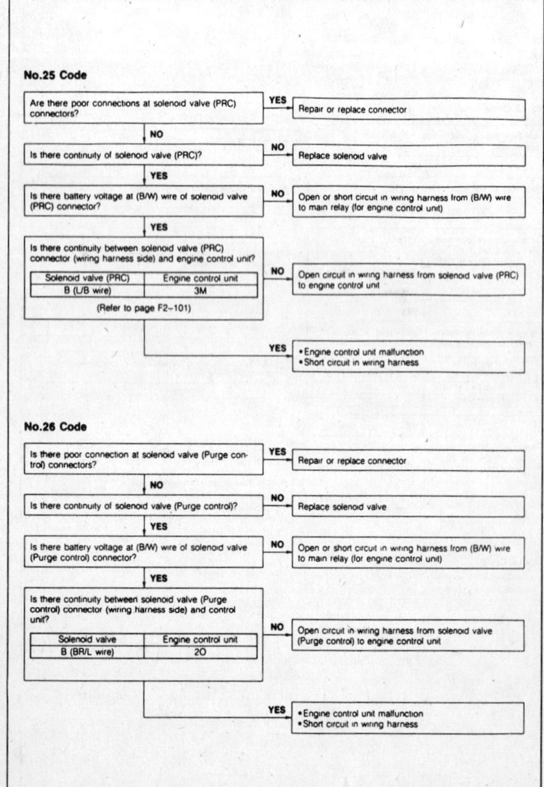


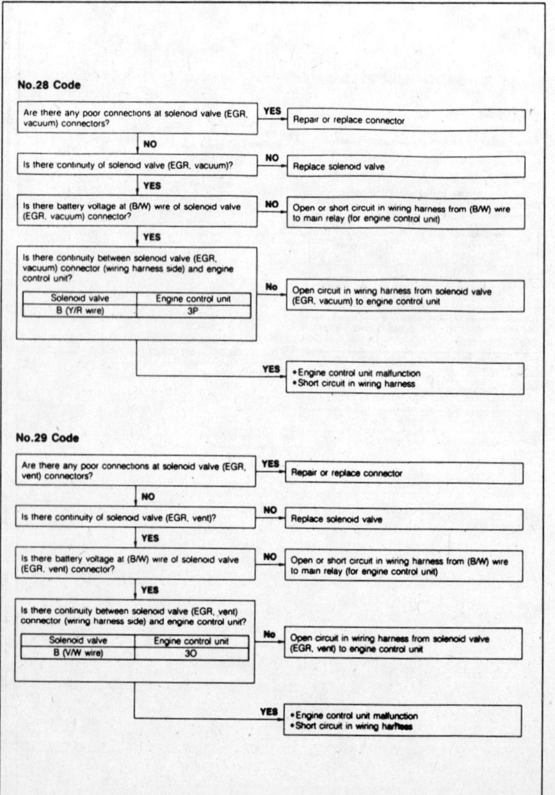

## TROUBLE CODE DIAGNOSTIC CHART—1990 929 WITH DOHC ENGINE

### No.34 Code

1. Are there poor connectors at BAC valve connectors?
   - YES → Repair or replace connector
   - NO → go to 2
2. Is resistance of BAC valve OK? Resistance: 10—13Ω
   - NO → Replace BAC valve
   - YES → go to 3
3. Is there battery voltage at (B/W) wire of BAC valve connector?
   - NO → Open or short circuit in wiring harness from (B/W) wire to main relay (for engine control unit)
   - YES → go to 4
4. Is there continuity between BAC valve connector (wiring harness side) and engine control unit?

| BAC valve | Engine control unit |
|---|---|
| B (L/O wire) | 3Q |

   - NO → Open circuit in wiring harness from BAC valve to engine control unit
   - YES → • Engine control unit malfunction • Short circuit in wiring harness

### No.36 Code

1. Are there poor connections at oxygen sensor heater relay connectors?
   - YES → Repair or replace connector
   - NO → go to 2
2. Is oxygen sensor heater relay operation OK?
   - NO → Replace oxygen sensor relay
   - YES → go to 3
3. Is there continuity between oxygen sensor heater relay connector and main relay?

| Oxygen sensor relay | Main relay |
|---|---|
| A (B/W wire) | D (B/W wire) |

   - NO → Open circuit in wiring harness from oxygen sensor heater relay to main relay
   - YES → go to 4
4. Is there continuity between oxygen sensor heater relay connector and engine control unit?

| Oxygen sensor | Engine control unit |
|---|---|
| B (R/Y wire) | 3K |

   - NO → Open circuit in wiring harness from oxygen sensor heater relay to engine control unit
   - YES → • Engine control unit malfunction • Short circuit in wiring harness

### No.41 Code

1. Are there poor connections at solenoid valve (VRIS) connectors?
   - YES → Repair or replace connector
   - NO → go to 2
2. Is there continuity of solenoid valve (VRIS)?
   - NO → Replace solenoid valve
   - YES → go to 3
3. Is there battery voltage at (B/W) wire of solenoid valve (VRIS) connector?
   - NO → Open or short circuit in wiring harness from (B/W) wire to main relay (for engine control unit)
   - YES → go to 4
4. Is there continuity between solenoid valve (VRIS) connector (wiring harness side) and engine control unit?

| Solenoid valve (VRIS) | Engine control unit |
|---|---|
| B (V wire) | 3J |

   - NO → Open circuit in wiring harness from solenoid valve (VRIS) to engine control unit
   - YES → • Engine control unit malfunction • Short circuit in wiring harness

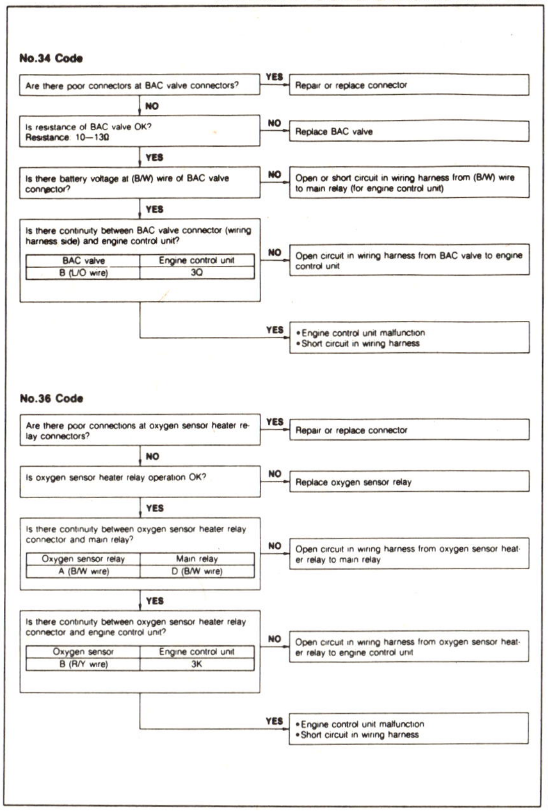

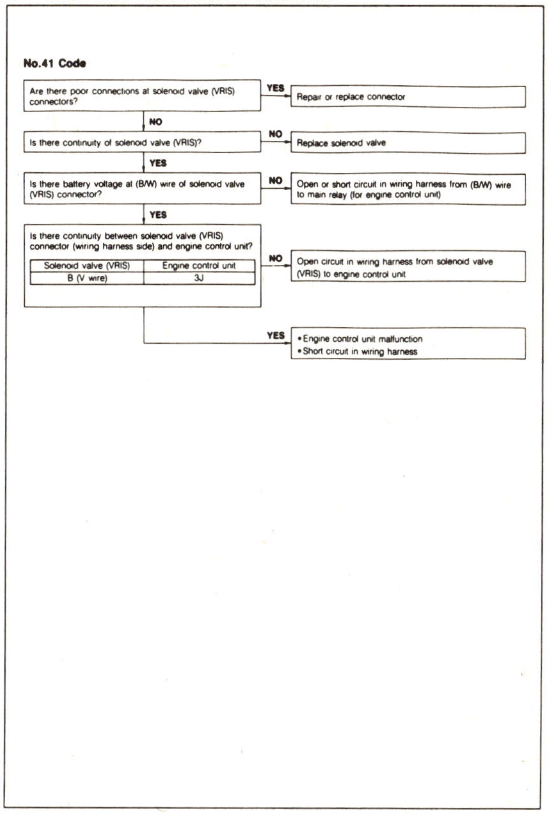

## CONTROL UNIT PIN LOCATION AND VOLTAGE TEST—1990 929 WITH DOHC ENGINE

| Terminal | Input | Output | Connected to | Voltage (after warm-up) | | Remark |
|---|---|---|---|---|---|---|
| | | | | Ignition switch: ON | Idle | |
| 1A | ○ | | Battery | Approx. 12V | | For backup |
| 1B | ○ | | Main relay | Approx. 12V | | |
| 1C | ○ | | Ignition switch (START position) | Below 2.5V ("P" or "N" range) | | While cranking: 6—11V |
| 1D | | ○ | Self-Diagnosis Checker (Monitor lamp) | <Test connector grounded> For 3 sec. after ignition switch OFF → ON: Approx. 5V (lamp illuminates) After 3 sec. approx. 12V | <Test connector grounded> Approx. 5V <Test connector not grounded> Monitor lamp ON, Below 6V Monitor lamp OFF, Approx. 12V | With Self-Diagnosis Checker |
| 1E | | ○ | MIL (Malfunction Indicator Lamp) | For 3 sec. after ignition switch OFF → ON: Below 2.5V (light illumination) After 3 sec.: approx. 12V | | • Test connector grounded • Lamp illuminates: Below 2.5V • Lamp does not illuminate: Approx. 12V |
| 1F | | ○ | Self-Diagnosis Checker (For Code No.) | For 3 sec. after ignition switch OFF → ON Below 2.5V (Buzzer sounds) After 3 sec.: approx. 12V (Buzzer does not sound) | | • With Self-Diagnosis Checker and test connector grounded • Buzzer sounds: Below 2.5V • Buzzer does not sound: Approx. 12V |
| 1G | | ○ | Igniter | 0V | Approx. 0.8V | |
| 1H | ○ | | Headlight switch | Headlight switch ON: Approx. 10—12V Headlight switch OFF: Below 2.5V | | |
| 1I | ○ | | Test connector | Test connector grounded: 0V Test connector not grounded: Approx. 12V | | Green connector: 1-pin |
| 1J | ○ | | Rear defroster switch | Rear defroster switch ON: Below 2.5V Rear defroster switch OFF: Approx. 12V | | |
| 1K | | ○ | EC-AT control unit | Approx. 12V | | |
| 1L | | ○ | A/C relay | A/C switch ON: Below 2.5V A/C switch OFF: Approx. 12V | | Blower motor: ON |
| 1M | ○ | | Speed sensor | 0V or approx. 5V | | |
| 1N | ○ | | Power steering (P/S) switch | Constant approx. 12V | P/S ON: 0V P/S OFF Approx. 12V | |
| 1O | ○ | | A/C switch | A/C switch ON: Below 2.5V A/C switch OFF: Approx. 12V | | Blower motor: ON |
| 1P | ○ | | Blower motor switch | Fan speed control ON (5th—8th position) Below 1.5V Others: Approx. 12V | | |
| 1Q | ○ | | Brake light switch | Brake pedal depressed: Approx. 12V Brake pedal released: 0V | | |
| 1R | ○ | | EC-AT control unit | P or N range: Below 2.5V Other ranges: Approx. 12V | | |
| 1S | — | — | — | — | | — |
| 1T | ○ | | Idle switch | Accelerator pedal depressed: Approx. 12V Accelerator peal released: 0V | | |
| 1U | — | — | — | — | | — |
| 1V | ○ | | Igniter | 0V | Approx. 0.8—1.2V | Engine Signal Monitor: Green and red lights flash |
| 2A | — | — | — | — | — | |
| 2B | ○ | | Airflow meter (Vs) | Approx. 4V | Approx. 3.5—5V | |
| 2C | ○ | | Oxygen sensor (Right side) | 0V | 0 ↔ 1V | • Cold engine: 0V at idle • After warming-up: Acceleration: 0.7—1.0V Deceleration: 0—0.2V |
| 2D | ○ | | Oxygen sensor (Left side) | | | |
| 2E | ○ | | Water thermosensor | Approx. 0.3—0.6V | | |
| 2F | ○ | | Throttle sensor | Throttle valve fully closed: Approx. 0.3V Throttle valve fully open: Approx. 4V | | |
| 2G | — | — | — | — | | |
| 2H | — | — | — | — | | |
| 2I | | ○ | Vref | 4.5—5.5V | | |
| 2J | ○ | | EGR position sensor | Approx. 0.25—0.9V | | |
| 2K | ○ | | Intake air thermosensor (Airflow meter) | Approx. 2—3V at 20°C (68°F) | | |
| 2L | ○ | | Intake air thermosensor (Dynamic chamber) | Approx. 1—2V at 80°C (176°F) | | |
| 2M | ○ | | Knock control unit | Approx. 4—5V | | |
| 2N | — | — | — | — | | |
| 2O | | ○ | Solenoid valve (Purge control) | Approx. 12V | | Engine Signal Monitor: Green and red lights flash |
| 2P | — | — | — | — | | |
| 3A | — | — | Ground | 0V | | |
| 3B | — | — | Ground | 0V | | |
| 3C | — | — | Ground | 0V | | |
| 3D | — | — | Ground | 0V | | |
| 3E | ○ | | Distributor (Ne-signal) | 0V | Approx. 0.6—0.8V | Ne: Number of engine speed (RPM) |
| 3F | — | — | Distributor (Ground) | 0V | | |
| 3G | ○ | | Distributor (G1 signal) | 0V | Approx. 0.6—0.8V | |
| 3H | ○ | | Distributor (G2 signal) | 0V | Approx. 0.6—0.8V | |
| 3I | — | — | — | — | | |
| 3J | | ○ | Solenoid valve (VRIS) | Approx. 12V | Approx. 13—14V | |
| 3K | | ○ | Oxygen sensor heater relay | Below 2.5V | | Above approx. 4,000 rpm: approx. 12V |
| 3L | — | — | — | — | | — |
| 3M | | ○ | Solenoid valve (PRC) | For 240 sec. after ignition switch OFF → ON: Below 2.5V | For 240 sec. after starting: Below 2.5V | During hot condition: Coolant temp.: Above 80°C (176°F) and Engine air temp.: Above 75°C (167°F) |
| | | | | Approx. 12V | | Other conditions |
| 3N | — | — | — | — | | — |
| 3O | | ○ | Solenoid valve (EGR, Vent side) | Approx. 12V | | Initial acceleration: Engine Signal Monitor; green and red lights flash |

### CONTROL UNIT PIN LOCATION AND VOLTAGE TEST (CONT.) – 1990 929 WITH DOHC ENGINE

| Terminal | Input | Output | Connection to | Voltage (after warm-up) | | Remark |
|---|---|---|---|---|---|---|
| | | | | Ignition switch: ON | Idle | |
| 3P | | ○ | Solenoid valve (EGR, Vacuum side) | Approx. 12V | | Initial acceleration: Engine Signal Monitor; green and red lights flash |
| 3Q | | ○ | Idle Speed Control (ISC) valve | Approx. 12V | Approx. 9–11V | Engine Signal Monitor: Green and red lights flash |
| 3R | — | — | — | — | | |
| 3S | — | — | — | — | | |
| 3T | — | — | — | — | | |
| 3U | | ○ | Injector No.1 | Approx. 12V | | Engine Signal Monitor: Green and red lights flash |
| 3V | | ○ | Injector No.2 | | | |
| 3W | | ○ | Injector No.3 | | | |
| 3X | | ○ | Injector No.4 | | | |
| 3Y | | ○ | Injector No.5 | | | |
| 3Z | | ○ | Injector No.6 | | | |

**Control Unit Connector (Control Unit Side)**

| 3Y | 3W | 3U | 3S | 3Q | 3O | 3M | 3K | 3I | 3G | 3E | 3C | 3A | 2O | 2M | 2K | 2I | 2G | 2E | 2C | 2A | 1U | 1S | 1Q | 1O | 1M | 1K | 1I | 1G | 1E | 1C | 1A |
|---|---|---|---|---|---|---|---|---|---|---|---|---|---|---|---|---|---|---|---|---|---|---|---|---|---|---|---|---|---|---|---|
| 3Z | 3X | 3V | 3T | 3R | 3P | 3N | 3L | 3J | 3H | 3F | 3D | 3B | 2P | 2N | 2L | 2J | 2H | 2F | 2D | 2B | 1V | 1T | 1R | 1P | 1N | 1L | 1J | 1H | 1F | 1D | 1B |

### WIRING SCHEMATIC – 1990 929 WITH DOHC ENGINE

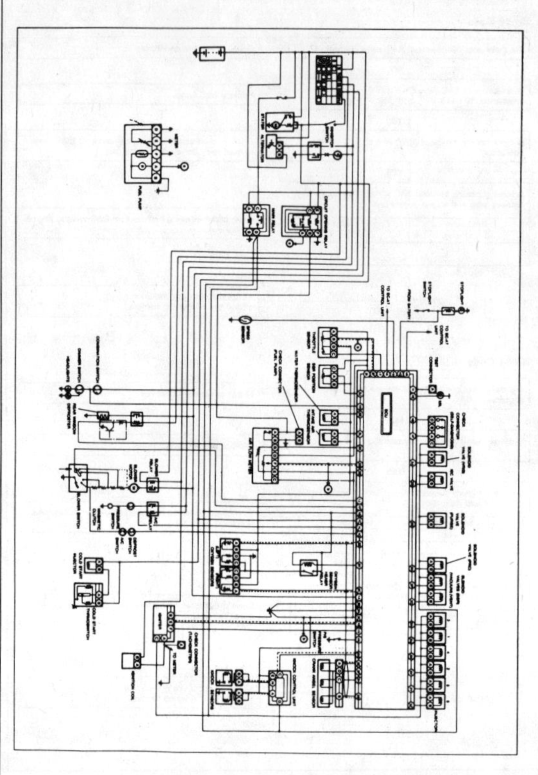

### SYSTEM SCHEMATIC – 1989 B2600

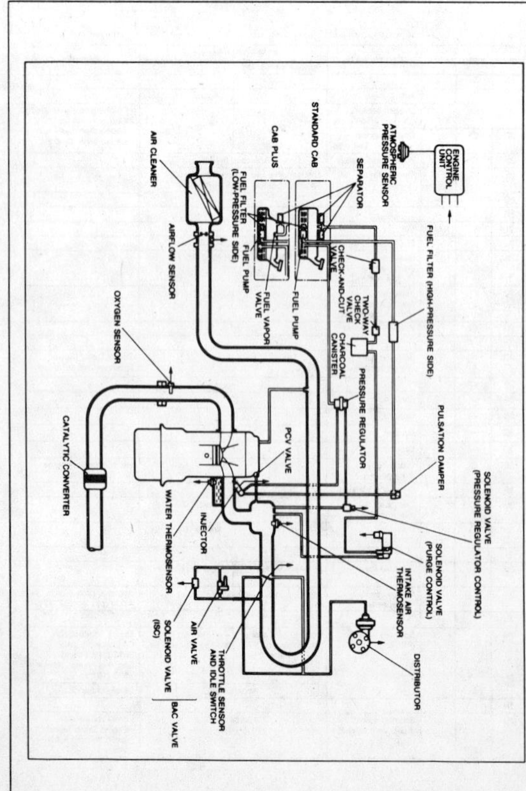


### VACUUM HOSE ROUTING – 1989 B2600

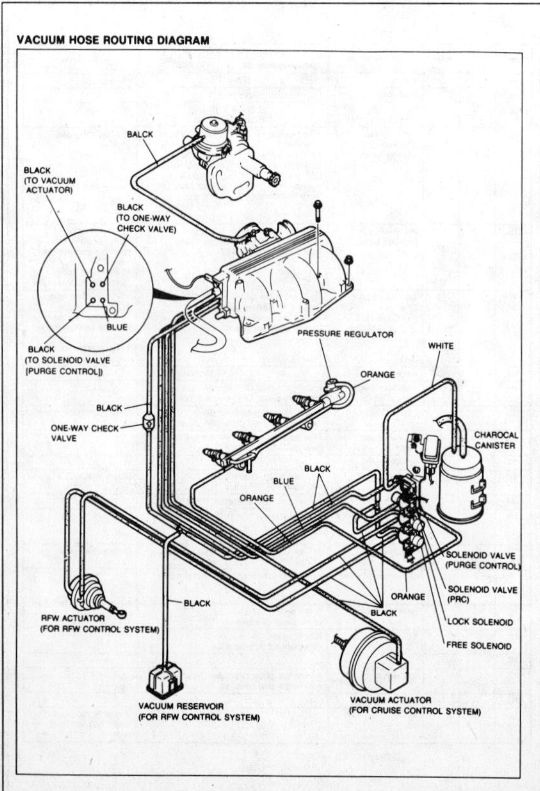

## EXPLODED VIEW OF CONTROL UNIT COMPONENTS 1989 B2600

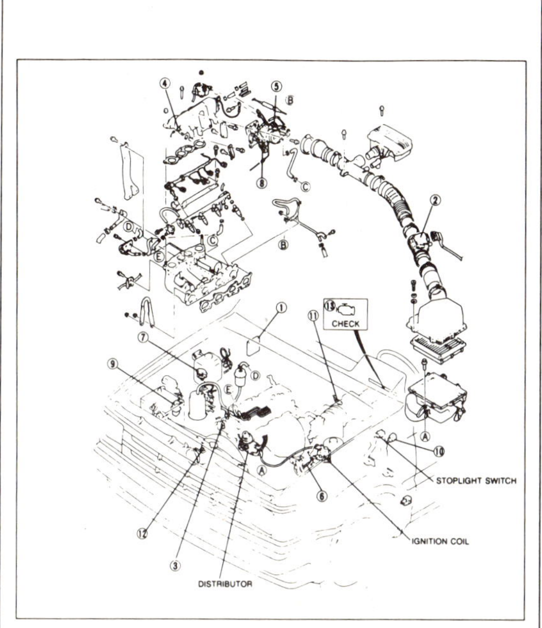


1. Engine control unit (ECU)
2. Airflow sensor
3. Water thermosensor
4. Intake air thermosensor
5. Throttle sensor
6. Oxygen sensor
7. Atmospheric pressure sensor
8. Idle switch
9. Main relay
10. Clutch switch
11. Neutral switch
12. P/S pressure switch
13. Malfunction indicator lamp (MIL)

## EXPLODED VIEW OF FUEL SYSTEM COMPONENTS 1989 B2600

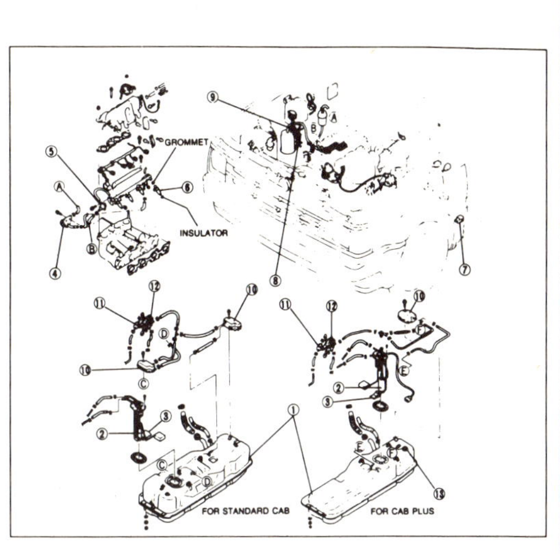


1. Fuel tank
2. Fuel pump
3. Fuel filter
4. Pulsation damper
5. Pressure regulator
6. Injector
7. Circuit opening relay
8. PRC solenoid valve
9. Purge control solenoid valve
10. Separator
11. 2-way check valve
12. Check and cut valve
13. Fuel vapor valve

## EXPLODED VIEW OF INTAKE AIR SYSTEM COMPONENTS—1989 B2600

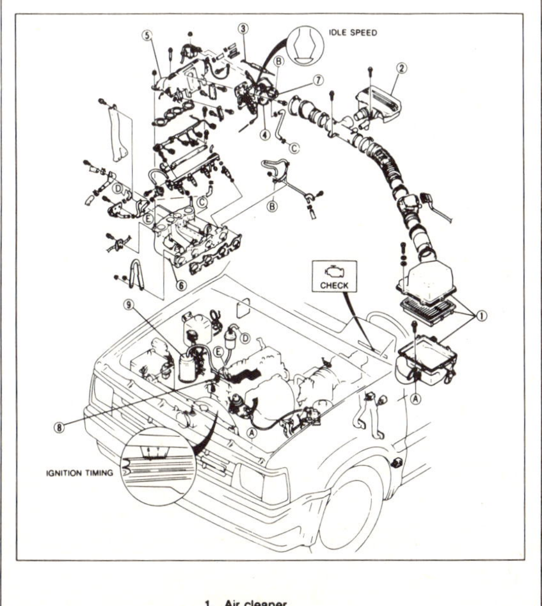


1. Air cleaner
2. Resonance chamber
3. Accelerator cable
4. Throttle body
5. Dynamic chamber
6. Intake manifold
7. BAC valve
8. PCV valve
9. Charcoal canister

## SPECIFICATIONS CHART—1989 B2600

| Item | | | Specification |
|---|---|---|---|
| Idle speed | rpm | | 750 ± 20*1 |
| Ignition timing | BTDC | | 5° ± 1°*1 |
| **Throttle body** | | | |
| Type | | | Horizontal draft (2-barrel) |
| Throat diameter | mm (in) | No.1 | 40 (1.6) |
| | | No.2 | 46 (1.8) |
| **Fuel pump** | | | |
| Type | | | Impeller (in-tank) |
| Output pressure | kPa (kg/cm², psi) | | 441—589 (4.5—6.0, 64—85) |
| **Fuel filter** | | | |
| Type | | Low-pressure side | Nylon element |
| | | High-pressure side | Paper element |
| **Pressure regulator** | | | |
| Type | | | Diaphragm |
| Regulating pressure | kPa (kg/cm², psi) | | 265—314 (2.7—3.2, 38—46) |
| **Injector** | | | |
| Type | | | High-ohmic |
| Type of drive | | | Voltage |
| Resistance | Ω | | 12—16 (at 20°C, 68°F) |
| **BAC valve (solenoid valve [idle speed control])** | | | |
| Solenoid resistance | Ω | | 7.7—9.3 (at 23°C, 73°F) |
| **Solenoid valve (Purge control)** | | | |
| Solenoid resistance | Ω | | 30—34 (at 20°C, 68°F) |
| **Water thermosensor** | | | |
| Resistance | kΩ | −20°C (−4°F) | 14.5—17.8 |
| | | 20°C (68°F) | 2.2—2.7 |
| | | 80°C (176°F) | 0.28—0.35 |
| **Intake air thermosensor** | | | |
| Resistance | kΩ | 25°C (77°F) | 29.7—36.3 |
| | | 85°C (185°F) | 3.3—3.7 |
| **Circuit opening relay** | | | |
| Resistance | Ω | STA—E1 | 21—43 |
| | | B—Fc | 109—226 |
| | | B—Fp | ∞ |
| **Fuel tank** | | | |
| Capacity | liters (US gal, Imp gal) | | 56 (14.8, 12.3) |
| **Air cleaner** | | | |
| Element type | | | Dry |
| **Accelerator cable** | | | |
| Free play | mm (in) | | 1—3 (0.039—0.118) |
| **Fuel** | | | |
| Specification | | | Unleaded regular (RON 87 or higher) |

*1 ... Test connector grounded

### COMPONENT DESCRIPTIONS—1989 B2600

**COMPONENT DESCRIPTIONS**

| Component | Function | Remarks |
|---|---|---|
| **Air cleaner** | Filters air entering throttle body | |
| **Airflow sensor** | Detects amount of intake air; sends signal to engine control unit | |
| **Air valve** | When cold, supplies bypass air into dynamic chamber | • Engine speed increased to shorten warm-up period<br>• Thermowax type<br>• Installed in BAC valve |
| **Atmospheric pressure sensor** | Detects atmospheric pressure; sends signal to engine control unit | |
| **BAC valve** | Supplies bypass air into dynamic chamber | Consists of air valve and ISC valve |
| **Catalytic converter** | Reduces HC, CO, and NOx by chemical reaction | Monolith type |
| **Charcoal canister** | Stores gas tank fumes when engine stopped | |
| **Check connector** | For Self-Diagnosis Checker | 6-pin connector (Green) |
| **Check-and-cut valve** | Releases excessive pressure or vacuum in fuel tank to atmosphere | |
| **Circuit opening relay** | Voltage for fuel pump while engine running | |
| **Clutch switch** | Detects in-gear condition; sends signal to engine control unit | Switch ON when clutch pedal depressed |
| **Dynamic chamber** | Interconnects all cylinders | |
| **Engine control unit** | Detects following:<br>1. Engine speed<br>2. No.1 piston TDC<br>3. Intake air amount<br>4. Engine coolant temperature<br>5. Ignition ON signal<br>6. Throttle valve opening angle<br>7. Throttle valve fully closed<br>8. Air/fuel ratio (Oxygen concentration)<br>9. In-gear condition<br>10. Intake air temperature<br>11. Atmospheric pressure<br>12. A/C operation<br>13. P/S operation<br>14. E/L operation<br>15. Cranking signal<br>16. Test signal (idle speed, malfunction code No.)<br>17. Braking signal<br>Controls operation of the following<br>1. Fuel injection system<br>2. Idle speed control<br>3. Pressure regulator control system<br>4. Purge control system<br>5. Fail-safe function<br>6. Monitor function<br>7. Burn-off system<br>8. Ignition timing control system<br>9. Fuel pump<br>10. A/C (cut off)<br>11. Main relay control | 1. Ignition coil<br>2. G-signal<br>3. Airflow sensor<br>4. Water thermosensor<br>5. Ignition switch<br>6. Throttle sensor<br>7. Idle switch<br>8. Oxygen sensor<br>9. Neutral and clutch switches<br>10. Intake air thermosensor (on dynamic chamber)<br>11. Atmospheric pressure sensor<br>12. A/C switch<br>13. P/S pressure switch<br>14. Headlight and blower switches<br>15. Ignition switch (START position)<br>16. Test connector<br>17. Stoplight switch<br>1. Injector<br>2. Solenoid valve (Idle speed control)<br>3. Solenoid valve (Pressure regulator control)<br>4. Solenoid valve (Purge control)<br>5. Self-Diagnosis Checker and MIL<br>6. Monitor lamp (Self-Diagnosis Checker)<br>7. Airflow sensor<br>8. Igniter<br>9. Circuit opening relay<br>10. A/C relay<br>11. Main relay |

| Component | Function | Remarks |
|---|---|---|
| **Fuel filter** | Filters particles from fuel | |
| **Fuel pump** | Provides fuel to injectors | • Operates while engine running<br>• Installed in fuel tank |
| **Fuel vapor valve** | Prevents fuel from flowing into charcoal canister | |
| **G rotor and pick-up** | Detects No.1 cylinder TDC; sends signal to engine control unit | For determining fuel injection timing |
| **Idle switch** | Detects when throttle valve fully closed; sends signal to engine control unit | Installed on throttle body |
| **Igniter** | Receives spark signal from signal rotor and generates high voltage to ignition coil | |
| **Ignition coil (−) terminal** | Detects engine speed; sends signal to engine control unit | |
| **Ignition switch (START position)** | Sends engine cranking signal to engine control unit | |
| **Injector** | Injects fuel into intake port | • Controlled by signals from engine control unit<br>• High-ohmic injector<br>• Two port injector nozzle |
| **Intake air thermosensor** | Detects intake air temperature; sends signal to engine control unit | Installed on dynamic chamber |
| **Main relay** | Supplies electric current to injectors and engine control unit | |
| **MIL (Malfunction indicator lamp)** | (For Federal)<br>Lamp illuminates to indicate the maintenance schedule for the emission control system | Every 60,000 and 80,000 miles |
| | (For California)<br>Lamp illuminates when input device malfunctions | Test connector not grounded |
| | (For California)<br>Lamp flashes to indicate malfunction code No. of input and output devices | Test connector grounded |
| **Neutral switch** | Detects in-gear condition; sends signal to engine control unit | Switch ON when neutral |
| **Oxygen sensor** | Detects oxygen concentration; sends signal to engine control unit | Zirconia ceramic and platinum coating |
| **PCV valve** | Controls amount of blowby gas introduced into engine | |
| **P/S pressure switch** | Detects P/S operation; sends signal to engine control unit | P/S ON when steering wheel turned right or left |
| **Pressure regulator** | Adjusts fuel pressure supplied to injectors | |
| **Resonance chamber** | Improves mid-range torque characteristics | |
| **Separator** | Prevents fuel from flowing into charcoal canister | |
| **Solenoid valve — Idle speed control** | Controls bypass air amount | • Controlled by duty signal from engine control unit<br>• With integrated air valve<br>• Controls idle-up |
| **Solenoid valve — Pressure regulator control** | Controls vacuum to pressure regulator | Cuts vacuum passage when hot |
| **Solenoid valve — Purge control** | Controls evaporative fumes from canister to intake manifold | |
| **Stoplight switch** | Detects braking operation (deceleration); sends signal to engine control unit | |

### COMPONENT DESCRIPTION (CONT.)—1989 B2600

| Component | Function | Remarks |
|---|---|---|
| **Test connector** | For Self-Diagnosis Checker and idle speed adjustment | 1-pin connector (Green) |
| **Throttle body** | Controls intake air quantity | Integrated throttle sensor and idle switch |
| **Throttle sensor** | Detects throttle valve opening angle; sends signal to engine control unit | Installed on throttle body |
| **Two-way check valve** | Controls pressure in fuel tank | |
| **Vacuum delay valve (for distributor)** | Retards ignition timing during transition from deceleration to acceleration | |
| **Water thermosensor** | Detects coolant temperature; sends signal to engine control unit | |

### COMPONENT RELATIONSHIP CHART—1989 B2600

| INPUT DEVICES / OUTPUT DEVICES | INJECTOR: FUEL INJECTION AMOUNT | INJECTOR: FUEL INJECTION TIMING | BAC VALVE: AIR VALVE | BAC VALVE: ISC VALVE | SOLENOID VALVE (PURGE CONTROL) | SOLENOID VALVE (PRESSURE REGULATOR CONTROL) | A/C RELAY (A/C CUT-OFF) | AIRFLOW SENSOR (BURN-OFF) | CIRCUIT OPENING RELAY (FUEL PUMP CONTROL) | IGNITER (IGNITION TIMING CONTROL) |
|---|---|---|---|---|---|---|---|---|---|---|
| TEST CONNECTOR | × | × | × | ○ | × | × | × | × | × | |
| IGNITION SWITCH (ON POSITION) | × | × | × | × | × | × | × | ○ | × | × |
| IGNITION SWITCH (START POSITION) | ○ | ○ | × | ○ | × | ○ | ○ | × | × | ○ |
| HEADLIGHT AND BLOWER SWITCH | × | × | × | ○ | × | × | × | × | × | × |
| P/S PRESSURE SWITCH | ○ | × | × | ○ | × | × | × | × | × | × |
| A/C SWITCH | ○ | × | × | ○ | × | × | ○ | × | × | ○ |
| NEUTRAL AND CLUTCH SWITCH | ○ | × | × | ○ | ○ | × | ○ | × | × | ○ |
| STOPLIGHT SWITCH | ○ | × | × | × | × | × | × | × | × | × |
| IDLE SWITCH | ○ | × | × | ○ | ○ | ○ | × | × | × | ○ |
| ATMOSPHERIC PRESSURE SENSOR | ○ | × | × | ○ | × | × | × | × | × | ○ |
| THROTTLE SENSOR | ○ | × | × | × | × | ○ | ○ | × | × | ○ |
| INTAKE AIR THERMOSENSOR | ○ | × | × | × | × | ○ | × | × | × | × |
| AIRFLOW SENSOR | ○ | × | × | ○ | ○ | × | × | ○ | × | × |
| OXYGEN SENSOR | ○ | × | × | × | ○ | × | × | × | × | × |
| WATER THERMOSENSOR | ○ | × | × | ○ | ○ | ○ | × | ○ | × | × |
| IGNITION COIL | ○ | ○ | × | ○ | ○ | ○ | × | ○ | ○ | × |
| DISTRIBUTOR (G-SIGNAL) | × | ○ | × | × | × | × | × | × | × | × |

## INPUT COMPONENTS AND ENGINE CONDITIONS—1989 B2600

**Input Devices and Engine Conditions**

| ENGINE CONDITIONS \ INPUT DEVICES | APPROXIMATE TIME (BASED ON 10—16°C or 50—60°F AMBIENT) | SENSORS: DISTRIBUTOR (G-SIGNAL) | IGNITION COIL | WATER THERMOSENSOR | OXYGEN SENSOR | AIRFLOW SENSOR | INTAKE AIR THERMOSENSOR | THROTTLE SENSOR | ATMOSPHERIC PRESSURE SENSOR |
|---|---|---|---|---|---|---|---|---|---|
| CRANKING —COLD ENGINE • COLD AIR • COLD COOLANT | Zero | | | | Signal has no effect on ECU | Signal has no effect on ECU | Signal has no effect on ECU | Signal has no effect on ECU | |
| COLD START —FAST IDLE • COLD AIR • COLD COOLANT | One minutes | | | Cool to warm medium voltage (3.5V and dropping) | Sensor cold low to high voltage (0—0.9V) | Low volume airflow low to high voltage (2.4—2.6V) | | Closed throttle low voltage (0.3—0.7V) | |
| COLD DRIVEAWAY —PART THROTTLE • COLD AIR • COLD COOLANT | Two minutes | | | | | | | | |
| WARM DRIVEAWAY —PART THROTTLE • WARM AIR • WARM COOLANT | Three minutes | | | Warm medium voltage (Approx 0.7V and dropping) | Sensor warm high voltage (0.9V) | Moderate volume airflow low to medium voltage (3.0V) | | Part throttle medium voltage (1—3.5V) | |
| HOT CRUISE • WARM AIR • WARM COOLANT | | Sends No.1 cylinder TDC signal to ECU | Sends engine speed signal to ECU | | Sensor hot switching from high voltage (0.9V) to low voltage (0.1V) | | Cool to warm medium voltage (1.4—3.4V) | | Sends voltage signal to ECU that varies with altitude voltage (approx 4V at sea level) |
| HOT ACCELERATION —60% THROTTLE | | | | | | Moderate to strong volume of airflow (3.8V) | | | |
| HOT ACCELERATION —WIDE OPEN THROTTLE | More than four minutes | | | Hot low voltage (Approx 0.4V) | High voltage (0.9V) | Strong volume of airflow (4.0V) | | Wide open throttle high voltage (Approx 4.0V) | |
| DECELERATION —CLOSED THROTTLE | | | | | Low voltage (0V) | Low volume of airflow (2.4V) | | Closed throttle low voltage (0.3—0.7V) | |
| HOT CURB IDLE —EXTENDED | | | | | Switching from high to low voltage (0.75—0.25V) | | | | |
| HOT ENGINE SHUTDOWN | — | OFF | OFF | OFF | Sensor hot low voltage (0.1V) until sensor cools | OFF | OFF | OFF | OFF |

| ENGINE CONDITIONS | SWITCHES: IDLE SWITCH | STOP-LIGHT SWITCH | NEUTRAL AND CLUTCH SWITCHES | A/C SWITCH | P/S PRESSURE SWITCH | HEAD-LIGHT SWITCH | BLOWER SWITCH | IGNITION SWITCH: START POSITION | IGNITION SWITCH: ON POSITION | TEST CONNECTOR |
|---|---|---|---|---|---|---|---|---|---|---|
| CRANKING | Signal has no effect on ECU | Signal has no effect on ECU | Signal has no effect on ECU | Signal has no effect on ECU | Signal has no effect on ECU | Signal has no effect on ECU | Signal has no effect on ECU | Sends signal to ECU (approx 12V) | Signal has no effect on ECU | Signal has no effect on ECU |
| COLD START | Low voltage signal to ECU (below 1.5V) | Brake pedal depressed: sends signal to ECU (approx. 12V) | In neutral: low voltage signal to ECU (approx 0V) | | | | | | | |
| COLD DRIVEAWAY | | | | | | | | | | |
| WARM DRIVEAWAY | | | | | | | | | | |
| HOT CRUISE | High voltage signal to ECU (approx. 12V) | No signal send to ECU (below 1.5V) | Driving in any gear high voltage signal to ECU (approx 12V) | A/C switch ON sends signal to ECU (approx 12V) A/C switch OFF no signal to ECU (below 1.5V) | Steering wheel turned low voltage signal to ECU (below 1.5V) Steering wheel straight ahead high voltage signal to ECU (approx 12V) | Headlight switch ON: low voltage signal to ECU (below 1.5V) Headlight switch OFF high voltage signal to ECU (approx 12V) | Blower switch ON low voltage signal to ECU (below 1.5V) Blower switch OFF high voltage signal to ECU (approx. 12V) | No signal to ECU (below 1.5V) | Sends signal to ECU (approx 12V) | Connector not grounded: high voltage signal to ECU (approx. 12V) |
| HOT ACCELERATION —60% THROTTLE | | | | | | | | | | |
| HOT ACCELERATION —WIDE OPEN THROTTLE | | | | | | | | | | |
| DECELERATION | Low voltage signal to ECU (below 1.5V) | Brake pedal depressed sends signal to ECU (approx 12V) | | | | | | | | |
| HOT CURB IDLE | | | In neutral low voltage signal to ECU (approx. 0V) | | | | | | | Low voltage signal to ECU when connector grounded (below 1.5V) |
| HOT ENGINE SHUTDOWN | OFF | OFF | OFF | OFF | OFF | OFF | OFF | OFF | OFF | OFF |

## OUTPUT COMPONENTS AND ENGINE CONDITIONS—1989 B2600

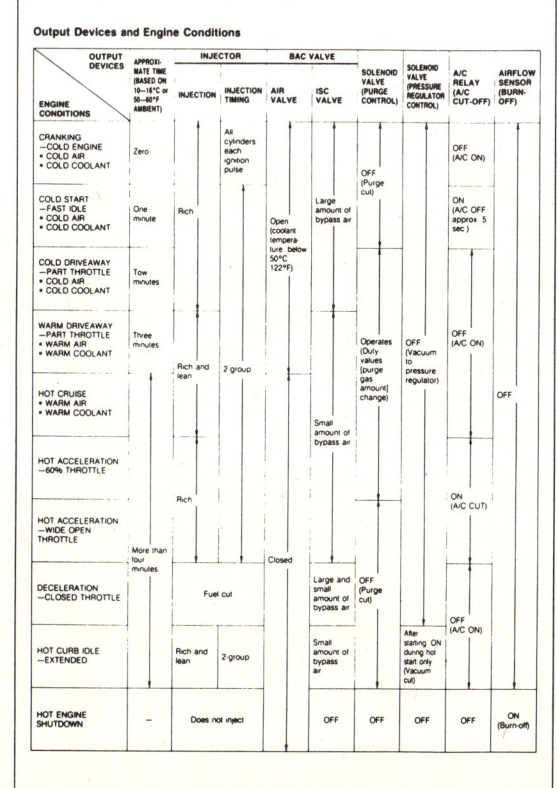

**Output Devices and Engine Conditions**

| ENGINE CONDITIONS \ OUTPUT DEVICES | APPROXIMATE TIME (BASED ON 10—16°C or 50—60°F AMBIENT) | INJECTOR: INJECTION | INJECTOR: INJECTION TIMING | BAC VALVE: AIR VALVE | BAC VALVE: ISC VALVE | SOLENOID VALVE (PURGE CONTROL) | SOLENOID VALVE (PRESSURE REGULATOR CONTROL) | A/C RELAY (A/C CUT-OFF) | AIRFLOW SENSOR (BURN-OFF) |
|---|---|---|---|---|---|---|---|---|---|
| CRANKING —COLD ENGINE • COLD AIR • COLD COOLANT | Zero | | All cylinders each ignition pulse | | | | | OFF (A/C ON) | |
| COLD START —FAST IDLE • COLD AIR • COLD COOLANT | One minute | Rich | | Open (coolant temperature below 50°C 122°F) | Large amount of bypass air | OFF (Purge cut) | | ON (A/C OFF approx. 5 sec.) | |
| COLD DRIVEAWAY —PART THROTTLE • COLD AIR • COLD COOLANT | Tow minutes | | | | | | | | |
| WARM DRIVEAWAY —PART THROTTLE • WARM AIR • WARM COOLANT | Three minutes | Rich and lean | 2-group | | | Operates (Duty values [purge gas amount] change) | OFF (Vacuum to pressure regulator) | OFF (A/C ON) | |
| HOT CRUISE • WARM AIR • WARM COOLANT | | | | | Small amount of bypass air | | | | OFF |
| HOT ACCELERATION —60% THROTTLE | | Rich | | | | | | ON (A/C CUT) | |
| HOT ACCELERATION —WIDE OPEN THROTTLE | More than four minutes | | | Closed | | | | | |
| DECELERATION —CLOSED THROTTLE | | Fuel cut | | | Large and small amount of bypass air | OFF (Purge cut) | | | |
| HOT CURB IDLE —EXTENDED | | Rich and lean | 2-group | | Small amount of bypass air | | After starting ON during hot start only (Vacuum cut) | OFF (A/C ON) | |
| HOT ENGINE SHUTDOWN | — | Does not inject | | | OFF | OFF | OFF | OFF | ON (Burn-off) |

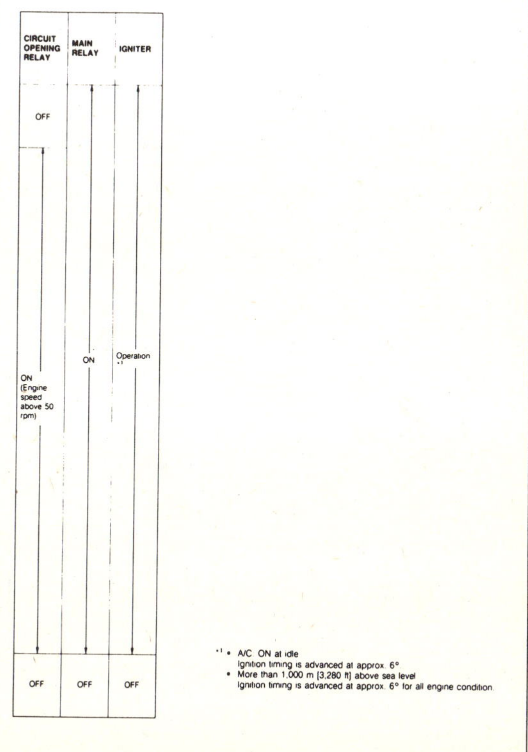

| ENGINE CONDITIONS | CIRCUIT OPENING RELAY | MAIN RELAY | IGNITER |
|---|---|---|---|
| CRANKING | OFF | | |
| COLD START | | | |
| COLD DRIVEAWAY | | | |
| WARM DRIVEAWAY | | | |
| HOT CRUISE | ON (Engine speed above 50 rpm) | ON | Operation [*1] |
| HOT ACCELERATION —60% THROTTLE | | | |
| HOT ACCELERATION —WIDE OPEN THROTTLE | | | |
| DECELERATION | | | |
| HOT CURB IDLE | | | |
| HOT ENGINE SHUTDOWN | OFF | OFF | OFF |

*1 • A/C ON at idle
Ignition timing is advanced at approx. 6°
• More than 1,000 m (3,280 ft) above sea level
Ignition timing is advanced at approx. 6° for all engine condition

### TROUBLESHOOTING CHART – 1989 B2600

**Cranks normally but won't start (No combustion)**

| STEP | QUICK INSPECTION | | ACTION | | POSSIBLE CAUSE |
|---|---|---|---|---|---|
| 1 | Check for malfunction code (01) with SST [IGN ON, Test connector (Green 1-pin) grounded] | Yes | Check for cause by referring to check sequence | | |
| | | No | Go to Next Step | | |
| 2 | Check for spark by disconnecting high-tension lead while cranking | Yes | Go to Next Step | | |
| | | No | Check ignition system | | |
| 3 | Check for fuel pump operating sound from fuel filler port [IGN ON, Test connector (Yellow: 2-pin) connected] | Yes | Check if engine starts in this condition | Yes | Check circuit opening relay switching operation |
| | | | | | Check circuit opening relay circuit (IGN: START) |
| | | | | No | Go to Next Step |
| | | No | Check circuit opening relay switching operation | Yes | Check circuit opening relay circuit |
| | | | | | Check fuel pump operation |
| | | | | No | Replace circuit opening relay |
| 4 | Check fuel line pressure [IGN ON, Test connector (Yellow: 2-pin) connected] **Fuel line pressure: 265—314 kPa (2.7—3.2 kg/cm$^2$, 38—46 psi)** | Yes | Go to Next Step | | |
| | | No | Check fuel pump maximum pressure **Fuel pump maximum pressure: 441—588 kPa (4.5—6.0 kg/cm$^2$, 64—85 psi)** | Yes | Replace pressure regulator |
| | | | | No | Replace fuel pump |
| 5 | Check for injector operating sound while cranking | Yes | Go to Next Step | | |
| | | No | Check voltage at ECU (3C) and (3E) terminals with SST **Voltage: Approx. 12V (IGN ON)** | Yes | Check throttle sensor |
| | | | | | Replace ECU |
| | | | | No | Check wiring for short or open |
| | | | | | Poor ground circuit from ECU (3A) terminal (Check terminal voltage with SST) |
| 6 | Substitute a well-known ECU Check if the condition improves | Yes | | | ECU malfunction |
| | | No | Check ground circuit from ECU (2R) terminal with SST **Voltage: 0V (IGN ON)** | Yes | Go to Next Step |
| | | | | No | Poor ground circuit |
| 7 | | | | | Low compression |

**Cranks normally but hard to start (Always)**

| STEP | QUICK INSPECTION | | ACTION | | POSSIBLE CAUSE |
|---|---|---|---|---|---|
| 1 | Check if vacuum hoses and the air hoses are connected correctly | Yes | Go to Next Step | | |
| | | No | Connect correctly | | |
| 2 | Check air cleaner element for clogging | Yes | Go to Next Step | | |
| | | No | Clean air cleaner element | | |
| 3 | Check ignition timing at idle after warm up **Ignition timing: 5° ± 1° BTDC** [Test connector (Green: 1-pin) grounded] | Yes | Go to Next Step | | |
| | | No | Adjust ignition timing | | |
| | Note: At high—altitude area **[1,000 m (3,280 ft)]**, ignition timing advances to **11° ± 2° BTDC** when test connector (Green: 1-pin) is not grounded | | | | |
| 4 | Disconnect high-tension lead of each cylinder at idle Check if engine condition changes | Yes | Go to Next Step | | |
| | | No | Check ignition system | Yes | Replace injector (If step 3 OK) |
| | | | | No | Check spark plug |
| | | | | | Check high-tension lead |
| | | | | | Check distributor cap |
| 5 | Check for injector operating sound at idle | Yes | Go to Next Step | | |
| | | No | Check resistance at injector harness connector (EMINJ-01) | Yes | Check wiring short or open |
| | | | Terminals: (B/Y)—(LG/B), (B/Y)—(LG/R); Resistance: 6—8Ω | No | Check injector resistance |
| | | | | | Check wiring short or open |
| 6 | Check fuel line pressure [IGN ON, Test connector (Yellow: 2-pin) connected] **Fuel line pressure: 265—314 kPa (2.7—3.2 kg/cm$^2$, 38—46 psi)** | Yes | Go to Next Step | | |
| | | No | Check if fuel filter has been replaced according to maintenance schedule | Yes | Check fuel line for clogging |
| | | | | No | Replace fuel filter |
| | | | Check fuel pump maximum pressure **Fuel pump maximum pressure: 441—588 kPa (4.5—6.0 kg/cm$^2$, 64—85 psi)** | Yes | Replace pressure regulator |
| | | | | No | Replace fuel pump |

**Cranks normally but hard to start (Always) (Cont'd)**

| STEP | QUICK INSPECTION | | ACTION | | POSSIBLE CAUSE |
|---|---|---|---|---|---|
| 7 | Operate fuel pump [IGN ON, Test connector (Yellow: 2-pin) connected] Turn ignition switch OFF and observe fuel pressure **for 5 minutes** **Fuel pressure: More than 147 kPa (1.5 kg/cm2, 21 psi)** | Yes | Go to Next Step | | |
| | | No | Check fuel pump pressure drop | No | Replace fuel pump |
| | | | Check pressure regulator pressure drop | Yes | Check injector fuel leakage |
| | | | | No | Replace pressure regulator |
| 8 | | | | | Check compression |

### TROUBLESHOOTING CHART – 1989 B2600

**Cranks normally but hard to start (Only when engine is cold)**

| STEP | QUICK INSPECTION | | ACTION | | POSSIBLE CAUSE |
|---|---|---|---|---|---|
| 1 | Check specific gravity of battery using a hydrometer **Specific gravity: Above 1.200** | Yes | Go to Next Step | | |
| | | No | Recharge battery | | |
| 2 | Check for malfunction code (09) (26) with SST [IGN ON, Test connector (Green: 1-pin) grounded] | Yes | Check for cause by referring to check sequence | | |
| | | No | Go to Next Step | | |
| 3 | Disconnect high-tension lead of each cylinder at idle Check if engine condition changes | Yes | Go to Next Step | | |
| | | | Check ignition system | | Check spark plug |
| | | | | | Check high-tension lead |
| | | | | | Check distributor cap |
| 4 | Check fuel line pressure [IGN ON, Test connector (Yellow: 2-pin) connected] **Fuel line pressure: 265—314 kPa (2.7—3.2 kg/cm$^2$, 38—46 psi)** | Yes | Go to Next Step | | |
| | | No | Check for fuel leaks | | |
| | | | Check if fuel filter has been replaced according to maintenance schedule | Yes | Check fuel line for clogging |
| | | | | No | Replace fuel filter |
| | | | Check fuel pump maximum pressure **Fuel pump maximum pressure: 441—588 kPa (4.5—6.0 kg/cm$^2$, 64—85 psi)** | Yes | Replace pressure regulator |
| | | | | No | Replace fuel pump |
| 5 | Disconnect ISC valve connector when engine is cold Check if idle speed decreases during warm up | Yes | Go to Next Step | | |
| | | No | | | Check if BAC valve (air valve) opens when cold |
| 6 | Check voltage at ECU (3B) terminal with SST **Voltage: Approx. 10V (while cranking)** | Yes | Go to Next Step | | |
| | | No | Check starter interlock switch | Yes | Check related wiring |
| | | | | No | Replace switch |
| 7 | Check voltage at ECU (2I) terminal with SST **Voltage: Approx. 2.5V (IGN ON, Engine coolant temperature 20°C [68°F])** | Yes | Go to Next Step | | |
| | | No | | | Check water thermosensor |
| 8 | | | | | Check compression |

**Cranks normally but hard to start (Only when engine is warm)**

| STEP | QUICK INSPECTION | | ACTION | | POSSIBLE CAUSE AND DETAILED INSPECTION |
|---|---|---|---|---|---|
| 1 | Check for malfunction code (26) with SST [IGN ON, Test connector (Green: 1-pin) grounded] | Yes | Check for cause by referring to check sequence | | |
| | | No | Go to Next Step | | |
| 2 | Disconnect water thermosensor connector Check if condition improves | Yes | Check water thermosensor connector condition as follows: 1. Shake connector and check if condition changes 2. Check condition of terminal (burned or damaged) 3. Connect a good terminal to harness side connector and check for looseness | Yes | Check water thermosensor |
| | | | | No | Poor contact of water thermosensor connector |
| | | No | Go to Next Step | | |
| 3 | Operate fuel pump [IGN ON, Test connector (Yellow: 2-pin) connected] Turn ignition switch OFF and observe fuel pressure **for 5 minutes** **Fuel pressure: More than 147 kPa (1.5 kg/cm$^2$, 21 psi)** | Yes | Go to Next Step | | |
| | | No | Check fuel pump pressure drop | No | Replace fuel pump |
| | | | Check pressure regulator pressure drop | Yes | Check injector fuel leakage |
| | | | | No | Replace pressure regulator |
| 4 | | | | | ECU malfunction |

## TROUBLESHOOTING CHART—1989 B2600

**Cranks normally but hard to start (Only after heat soak)**

| STEP | QUICK INSPECTION | | ACTION | | POSSIBLE CAUSE |
|---|---|---|---|---|---|
| 1 | Check for malfunction code (09), (11), (25), (26) with SST [IGN ON. Test connector (Green 1-pin) grounded] | Yes | Check for cause by referring to check sequence | | |
| | | No | Go to Next Step | | |
| 2 | Circulate fuel by operating fuel pump **for 20 secondes** [IGN ON, Test connector (Yellow 2-pin) connected] Check if condition improves | Yes | Go to Step 3 | | |
| | | No | Go to Step 4 | | |
| 3 | Disconnect vacuum hose from pressure regulator Check if condition improves | Yes | Check the components related to pressure regulator control system | | Check water thermo-sensor |
| | | | | | Check intake air thermosensor |
| | | | | | Check solenoid valve (PRC) |
| | | | | | ECU malfunction (Check (2N) terminal voltage) |
| | | No | Go to Next Step | | |
| 4 | Operate fuel pump [IGN ON. Test connector (Yellow 2-pin) connected] Turn ignition switch OFF and observe fuel pressure **for 5 minutes** **Fuel pressure: More than 147 kPa (1.5 kg/cm², 21 psi)** | Yes | Go to Next Step | | |
| | | No | Check fuel pump pressure drop | No | Replace fuel pump |
| | | | Check pressure regulator pressure drop | Yes | Check injector fuel leakage |
| | | | | No | Replace pressure regulator |
| 5 | Change fuel with specified one Check if condition improves | Yes | | | Poor fuel quality |
| | | No | Go to Next Step | | |
| 6 | | | | | ECU malfunction |

**Cranks normally but won't start (Intermittent)**

| STEP | QUICK INSPECTION | | ACTION | POSSIBLE CAUSE |
|---|---|---|---|---|
| 1 | Shake connector of ignition coil, main relay and ECU while cranking Check if the engine starts | Yes | There may be a poor contact of the connector. Repair or replace the wiring | |
| | | No | Go to troubleshooting "Cranks normally but hard to start (Always)" | |

**Rough idle (Always)**

| STEP | QUICK INSPECTION | | ACTION | | POSSIBLE CAUSE |
|---|---|---|---|---|---|
| 1 | Check for malfunction code (08), (09), (14), (17), (26), (34) with SST [IGN ON. Test connector (Green 1-pin) grounded] | Yes | Check for cause by referring to the check sequence | | |
| | | No | **"88" flashing** Check voltage at ECU (3G) terminal with SST **Voltage: 0V (IG ON)** | Yes | Replace ECU |
| | | | | No | Poor ground circuit |
| | | | **"00"** Go to Next Step | | |
| 2 | Check ignition at idle after warm up **Ignition timing: 5° ± 1° BTDC** [Test connector (Green: 1-pin) grounded] | Yes | Go to Next Step | | |
| | | No | Adjust ignition timing (If possible) | | |
| 3 | Disconnect high-tension lead of each cylinder at idle Check if engine condition changes | Yes | Go to Next Step | | |
| | | No | Check ignition svstem | Yes | Replace injector (If Step 3 OK) |
| | | | | No | Check spark plug |
| | | | | | Check high-tension lead |
| | | | | | Check distributor cap |
| 4 | Check idle speed after warm up **Idle speed: 750 ± 20 rpm** [Test connector (Green: 1-pin) grounded] | Yes | Go to Next Step | | |
| | | No | Adjust idle speed (If possible) | | |
| 5 | Check for injector operating sound at idle | Yes | Go to Next Step | | |
| | | No | Check resistance at injector harness connector (EMINJ-01) **Terminals / Resistanc: (B/Y)—(LG/B), (BY)—(LG/R) 6—8Ω** | Yes | Check wiring short or open |
| | | | | No | Check injector resistance |
| | | | | | Check wiring short or open |
| 6 | Check fuel line pressure [IGN ON. Test connector (Yellow 2-pin) connected] **Fuel line pressure: 265—314 kPa (2.7—3.2 kg/cm², 38—46 psi)** | Yes | Go to Next Step | | |
| | | No | Check for fuel leakage | | |
| | | | Substitute a good fuel filter and retest | Yes | Replace fuel filter |
| | | | Check fuel pump maximum pressure **Fuel pump maximum pressure: 441—588 kPa (4.5—6.0 kg/cm², 64—85 psi)** | Yes | Replace pressure regulator |
| | | | | No | Replace fuel pump |
| 7 | Check intake manifold vacuum at idle **Vacuum: 520 ± 20 mmHg (20.5 ± 0.79 inHg)** | Yes | Go to Next Step | | |
| | | No | Check for air leaks | Yes | Intake air system components damaged |
| | | | | | Vacuum and intake air hoses loose or damaged |
| | | | | | Bolts or nuts loose |
| | | | | | Gaskets damaged |
| | | | | No | Check throttle valve closing condition |
| 8 | Substitute a well-known ECU Check if condition improves | Yes | | | ECU malfunction |
| | | No | Check voltage at ECU (3G) terminal with SST **Voltage: 0V (IGN ON)** | Yes | Go to Next Step |
| | | | | No | Poor ground circuit |
| 9 | | | | | Check compression |

## TROUBLESHOOTING CHART—1989 B2600

**Rough idle (Only when engine is cold)**

| STEP | QUICK INSPECTION | | ACTION | | POSSIBLE CAUSE |
|---|---|---|---|---|---|
| 1 | Check for malfunction code (08), (09), (26) with SST [IGN ON, Test connector (Green 1-pin) grounded] | Yes | Check for cause by referring to check sequence | | |
| | | No | Go to Next Step | | |
| 2 | Check ignition timing advance (Centrifugal advance) | Yes | Go to Next Step | | |
| | | No | | | Distributor malfunction |
| 3 | Disconnect high-tension lead of each cylinder at idle Check if engine condition changes | Yes | Go to Next Step | | |
| | | No | Check ignition system | Yes | Replace injector (If step 4 OK) |
| | | | | No | Check spark plug |
| | | | | | Check high-tension lead |
| | | | | | Check distributor cap |
| 4 | Check for injector operating sound at idle | Yes | Go to Next Step | | |
| | | No | Check resistance at injector harness connector (EMINJ-01) **Terminals / Resistance: (B/Y)—(LG/B), (B/Y)—(LG/R) 6—8Ω** | Yes | Check wiring short or open |
| | | | | No | Injector malfunction (Check resistance) |
| | | | | | Wiring short or open |
| 5 | Disconnect ISC valve connector at idle when engine is cold Check if idle speed decreases during warm up | Yes | Go to Next Step | | |
| | | No | | | Check if BAC valve (air valve) opens when cold |
| 6 | Check if specified engine oil is used | Yes | Go to Next Step | | |
| | | No | Change engine oil to specified oil | | |
| 7 | Subsitute a well-known ECU Check if condition improves | Yes | | | ECU malfunction |
| | | No | | | Check airflow sensor |

**Rough idle (Only when engine is warm)**

| STEP | QUICK INSPECTION | | ACTION | | POSSIBLE CAUSE |
|---|---|---|---|---|---|
| 1 | Run engine at **2,000 rpm** for more than **20 seconds** Check for malfunction code (08), (09), (15), (17), (26), (34) with SST [IGN ON, Test connector (Green: 1-pin) grounded] | Yes | Check for cause by referring to check sequence | | |
| | | No | Go to Next Step | | |
| 2 | Check idle speed after warm up **Idle speed: 750 ± 20 rpm** [Test connector (Green: 1-pin) grounded] | Yes | Go to Next Step | | |
| | | No | Adjust idle speed (If possible) | | |
| 3 | Check for flashing of SST monitor lamp after warm up **Monitor lamp: Flashes more than 8 times/10 seconds at 2,000—3,000 rpm** [Test connector (Green: 1-pin) not grounded] | Yes | Go to Next Step | | |
| | | No | | | Replace oxygen sensor |
| 4 | Disconnect ISC valve connector after warm up Check if engine speed drops | Yes | Go to Next Step | | |
| | | No | | | Check ISC valve |
| 5 | Disconnect water thermosensor connector Check if condition improves | Yes | Check water thermosensor connector condition as follows. 1. Shake connector and check if condition changes 2. Check condition of terminal (burned or damaged) 3. Connect a good terminal to harness side connector and check for looseness | Yes | Check water thermosensor |
| | | | | No | Poor contact of water thermosensor connector |
| | | No | Go to Next Step | | |
| 6 | Disconnect high-tension lead of each cylinder at idle Check if engine condition changes | Yes | Go to Next Step | | |
| | | No | Check ignition system | Yes | Replace injector (If step 7 OK) |
| | | | | No | Check spark plug |
| | | | | | Check high-tension lead |
| | | | | | Check distributor cap |
| | | | Note: If spark plug is wet, injector may be leaking | | |

### TROUBLESHOOTING CHART—1989 B2600

**Rough idle (Only when engine is warm) (Cont'd)**

| STEP | QUICK INSPECTION | | ACTION | | POSSIBLE CAUSE |
|---|---|---|---|---|---|
| 7 | Check for injector operating sound at idle | Yes | Go to Next Step | | |
| | | No | Check resistance at injector harness connector (EMINJ-01)<br>**Terminals**: (B/Y)—(LG/B), (B/Y)—(LG/R); **Resistance**: 6—8Ω | Yes | Check wiring short or open |
| | | | | No | Check injector resistance |
| | | | | | Check wiring short or open |
| 8 | Check for air leaks by listening for sucking noise | Yes | Go to Next Step | | |
| | | No | | | Intake air system components damaged |
| | | | | | Vacuum and intake air hoses loose or damaged |
| | | | | | Bolts or nuts loose |
| | | | | | Gaskets damaged |
| 9 | | | | | Check compression |

**Rough idle (Only after heat soak)**

| STEP | QUICK INSPECTION | | ACTION | | POSSIBLE CAUSE |
|---|---|---|---|---|---|
| 1 | Run engine at **2,000 rpm** for more than **20 seconds**<br>Check for malfunction code (09), (11), (15), (17), (25), (26) with SST<br>[IGN ON, Test connector (Green: 1-pin) grounded] | Yes | Check for cause by referring to check sequence | | |
| | | No | Go to Next Step | | |
| 2 | Check switches with SST<br>Neutral switch<br>Clutch switch<br>[IGN ON, Test connector (Green: 1-pin) grounded] | Yes | Go to Next Step | | |
| | | No | Check for cause by referring to check sequence | | |
| 3 | Check for flashing of SST monitor lamp after warm up<br>**Monitor lamp: Flashes more than 8 times 10 seconds at 2,000—3,000 rpm**<br>[Test connector (Green: 1-pin) not grounded] | Yes | Go to Next Step | | |
| | | No | | | Replace oxygen sensor |
| 4 | Disconnect vacuum hose from pressure regulator<br>Check if condition improve | Yes | Check components related to pressure regulator control system | | Check water thermosensor |
| | | | | | Check intake air thermosensor |
| | | | | | Check solenoid valve (PRC) |
| | | | | | ECU malfunction (Check (2N) terminal voltage) |
| | | No | Go to Next Step | | |
| 5 | Run engine at idle and stop it<br>Observe fuel pressure **for 5 minutes**<br>**Fuel pressure: More than 147 kPa (1.5 kg/cm², 21 psi)** | Yes | Go to Next Step | | |
| | | No | Check fuel pump pressure drop | No | Replace fuel pump malfunction |
| | | | Check pressure regulator pressure drop | Yes | Check injector fuel leakage |
| | | | | No | Replace pressure regulator |
| 6 | Disconnect high-tension lead of each cylinder at idle<br>Check if engine condition changes | Yes | Go to Next Step | | |
| | | No | Check ignition system | Yes | Replace injector (If step 3 OK) |
| | | | | No | Check spark plug |
| | | | | | Check high-tension lead |
| | | | | | Check distributor cap |
| 7 | Check for injector operating sound at idle | Yes | Go to Next Step | | |
| | | No | Check resistance at injector harness connector (EMINJ-01)<br>**Terminals**: (B/Y)—(LG/B), (B/Y)—(LG/R); **Resistance**: 6—8Ω | Yes | Check wiring short or open |
| | | | | No | Check injector resistance |
| | | | | | Check wiring short or open |
| 8 | Change fuel to specified grade<br>Check if condition improves | Yes | | | Poor fuel quality |
| | | No | Go to Next Step | | |
| 9 | | | | | ECU malfunction |

### TROUBLESHOOTING CHART—1989 B2600

**Rough idle just after starting**

| STEP | QUICK INSPECTION | | ACTION | | POSSIBLE CAUSE |
|---|---|---|---|---|---|
| 1 | Check for malfunction code (09), (34) with SST<br>[IGN ON, Test connector (Green 1-pin) grounded] | Yes | Check for cause by referring to check sequence | | |
| | | No | Go to Next Step | | |
| 2 | Check idle switch with SST<br>[IGN ON, Test connector (Green: 1-pin) grounded] | Yes | Go to Next Step | | |
| | | No | Check for cause by referring to check sequence | | |
| 3 | Check ignition timing at idle after warm up<br>**Ignition timing: 5° ± 1° BTDC at sea level; 11° ± 2° BTDC at high altitude [1,000 m (3,280 ft)]**<br>[Test connector (Green: 1-pin) not grounded] | Yes | Go to Next Step | | |
| | | No | Check ignition timing after warm up<br>**Ignition timing: 5° ± 1 °BTDC**<br>[Test connector (Green: 1-pin) grounded] | Yes | Check atmospheric pressure sensor |
| | | | | No | Adjust ignition timing |
| 4 | Check idle speed after warm up<br>**Idle speed: 750 ± 20 rpm**<br>[Test connector (Green: 1-pin) grounded] | Yes | Go to Next Step | | |
| | | No | Try to adjust idle speed | Yes | Idle-speed misadjustment |
| | | | | No | Check accelerator cable free play |
| | | | | | Check ISC valve (Stuck closed) |
| | | | | | Check throttle body |
| 5 | Substitute a well-known ECU<br>Check if condition improves | Yes | | | ECU malfunction |
| | | No | Check voltage at ECU (3B) terminal with SST<br>**Voltage: Approx. 10V (While cranking)** | Yes | Go to Next Step |
| | | | | No | Check starter interlock switch |
| | | | | | Check related wiring |
| 6 | | | | | Poor quality engine oil |

**Low idle speed (When A/C, P/S, E/L ON)**

| STEP | QUICK INSPECTION | | ACTION | POSSIBLE CAUSE |
|---|---|---|---|---|
| 1 | Check for malfunction code (34) with SST<br>[IGN ON, Test connector (Green: 1-pin) grounded] | Yes | Check for cause by referring to check sequence | |
| | | No | Go to Next Step | |
| 2 | Disconnect ISC valve connector at idle<br>Check if the condition does not change | Yes | Go to Next Step | |
| | | No | | Check coolant level |
| | | | | Check engine oil |
| 3 | Check switches with SST<br>Idle switch<br>Neutral switch<br>Clutch switch<br>[IGN ON, Test connector (Green: 1-pin) grounded] | Yes | Go to Next Step | |
| | | No | Check for cause by referring to check sequence | |
| 4 | Check continuity between test connector (Green: 1-pin) and ground | | | Wiring short to ground |

**Note:**
**In case of low idle speed with A/C ON, if the problem cannot be solved by the above steps, it may be an A/C compressor malfunction.**

**High idle speed after warm up**

| STEP | QUICK INSPECTION | | ACTION | | POSSIBLE CAUSE |
|---|---|---|---|---|---|
| 1 | Check for malfunction code (09), (26), (34) with SST<br>[IGN ON, Test connector (Green 1-pin) grounded] | Yes | Check for cause by referring to check sequence | | |
| | | No | Go to Next Step | | |
| 2 | Check ignition timing at idle after warm up<br>**Ignition timing: 5° ± 1° BTDC**<br>[Test connector (Green: 1-pin) grounded] | Yes | Go to Next Step | | |
| | | No | Adjust ignition timing | | |
| 3 | Check intake manifold vacuum at idle | Yes | Go to Next Step | | |
| | | No | Check for air leaks | Yes | Intake air system components damaged |
| | | | | | Vacuum and intake air hoses loose or damaged |
| | | | | | Bolts or nuts loose |
| | | | | | Gaskets damaged |
| | | | | No | Check vacuum hose routing |
| | | | | | Check throttle body |
| 4 | Check idle speed after warm up<br>**Idle speed: 750 ± 20 rpm**<br>[Test connector (Green: 1-pin) grounded] | Yes | | | Check ISC valve |
| | | No | Try to adjust idle speed | Yes | Idle speed misadjustment |
| | | | | No | Gc to Next Step |
| 5 | Disconnect ISC valve connector at idle when engine is cold<br>Check if idle speed decreases during warm up | Yes | Go to Next Step | | |
| | | No | | | Check air valve |
| 6 | Disconnect water thermosensor connector and check if condition improves | Yes | Check water thermosensor connector condition as follows:<br>1. Shake connector and check if condition changes<br>2. Check condition of terminal (burned or damaged)<br>3. Connect a good terminal to harness side connector and check for looseness | Yes | Check water thermosensor |
| | | | | No | Poor contact of water thermosensor connector |
| | | No | Go to Next Step | | |
| 7 | | | | | ECU malfunction |

## TROUBLESHOOTING CHART—1989 B2600

**Idle hunting or surging**

| STEP | QUICK INSPECTION | | ACTION | | POSSIBLE CAUSE |
|---|---|---|---|---|---|
| 1 | (If trouble occurs only at warm condition)<br>Run engine at **2,000 rpm** for more than **20 seconds**<br>Check for malfunction code (09), (15), (17), (26), (34) with SST<br>[IGN ON, Test connector (Green: 1-pin) grounded] | Yes | Check for cause by referring to check sequence | | |
| | | No | Go to Next Step | | |
| 2 | (If trouble occurs only at warm condition)<br>Check for flashing of SST monitor lamp after warm up<br>**Monitor lamp:**<br>**Flashes more than 8 times 10 seconds at 2,000—3,000 rpm**<br>[Test connector (Green: 1-pin) not grounded] | Yes | Go to Next Step | | |
| | | No | | | Replace oxygen sensor |
| 3 | Check intake manifold vacuum at idle<br>**Vacuum: 520 ± 20 mmHg (20.5 ± 0.79 inHg)** | Yes | Go to Next Step | | |
| | | No | Check for air leaks | Yes | Intake air system components damaged |
| | | | | | Vacuum and air intake hoses loose or damaged |
| | | | | | Bolts or nuts loose |
| | | | | | Gaskets damaged |
| | | | | No | Check throttle body |
| 4 | Pinch PCV hose<br>Check if condition improves | Yes | | | Check PCV valve |
| | | No | Go to Next Step | | |
| 5 | Check fuel line pressure [IGN ON, Test connector (Yellow: 2-pin) connected]<br>**Fuel line pressure:**<br>**265—314 kPa (2.7—3.2 kg/cm², 38—46 psi)** | Yes | Go to Next Step | | |
| | | No | Check for fuel leaks | | |
| | | | Substitute a good fuel filter and retest | Yes | Replace fuel filter |
| | | | Check fuel pump maximum pressure<br>**Fuel pump maximum pressure:**<br>**441—588 kPa (4.5—6.0 kg/cm², 64—85 psi)** | Yes | Replace pressure regulator |
| | | | | No | Replace fuel pump |
| 6 | | | | | ECU malfunction |

**Engine stall at idle (Always)**

| STEP | QUICK INSPECTION | | ACTION | | POSSIBLE CAUSE |
|---|---|---|---|---|---|
| 1 | Check for malfunction code (08), (26) with SST<br>[IGN ON, Test connector (Green: 1-pin) grounded] | Yes | Check for cause by referring to the check sequence | | |
| | | No | Go to Next Step | | |
| 2 | Check if strong blue spark is visible at spark plug while cranking | Yes | Go to Next Step | | |
| | | No | Check ignition system | | Check spark plug |
| | | | | | Check high-tension lead |
| | | | | | Check distributor cap |
| 3 | Check fuel line pressure<br>[IGN ON, Test connector (Yellow: 2-pin) connected]<br>**Fuel line pressure:**<br>**265—314 kPa (2.7—3.2 kg/cm², 38—46 psi)** | Yes | Go to Next Step | | |
| | | No | Check for fuel leaks | | |
| | | | Check if fuel filter has been replaced according to maintenance schedule | Yes | Check fuel line for clogging |
| | | | | No | Replace fuel filter |
| | | | Check fuel pump maximum pressure<br>**Fuel pump maximum pressure:**<br>**441—588 kPa (4.5—6.0 kg/cm², 64—85 psi)** | Yes | Replace pressure regulator |
| | | | | No | Replace fuel pump |
| 4 | Check if vacuum hoses and the air hoses are connected correctly | Yes | Go to Next Step | | |
| | | No | Connect correctly | | |
| 5 | | | | | Airflow sensor |
| 6 | | | | | ECU malfunction |

**Engine stall at idle (Only when engine is cold)**

| STEP | QUICK INSPECTION | | ACTION | POSSIBLE CAUSE |
|---|---|---|---|---|
| 1 | Check for malfunction code (09), (26) with SST<br>[IGN ON, Test connector (Green: 1-pin) grounded] | Yes | Check for cause by referring to check sequence | |
| | | No | Go to Next Step | |
| 2 | | | | Check BAC valve (air valve) |
| 3 | | | | ECU malfunction |

## TROUBLESHOOTING CHART—1989 B2600

**Engine stall at idle (Only when engine is warm)**

| STEP | QUICK INSPECTION | | ACTION | | POSSIBLE CAUSE |
|---|---|---|---|---|---|
| 1 | Check for malfunction code (09), (26) with SST<br>[IGN ON, Test connector (Green 1-pin) grounded] | Yes | Check for cause by referring to check sequence | | |
| | | No | Go to Next Step | | |
| 2 | Disconnect water thermosensor connector<br>Check if condition improves | Yes | Check water thermosensor connector as follows:<br>1. Shake connector and check if condition changes<br>2. Check condition of terminal (burned or damaged)<br>3. Connect a good terminal to harness side connector and check for looseness | Yes | Check water thermosensor |
| | | | | No | Poor contact of water thermosensor connector |
| | | No | Go to Next Step | | |
| 3 | | | | | ECU malfunction |

**Engine stall at idle (When A/C, P/S, E/L O/N)**

| STEP | QUICK INSPECTION | | ACTION | POSSIBLE CAUSE |
|---|---|---|---|---|
| 1 | Check for malfunction code (34) with SST<br>[IGN ON, Test connector (Green 1-pin) grounded] | Yes | Check for cause by referring to check sequence | |
| | | No | Go to Next Step | |
| 2 | Check switches with SST<br>• Headlight switch<br>• Blower switch<br>[IGN ON, Test connector (Green 1-pin) grounded] | Yes | Go to Next Step | |
| | | No | Check for cause by referring to check sequence | |
| 3 | Disconnect ISC valve connector at idle<br>[Test connector (Green 1-pin) grounded]<br>Check if the condition does not change | Yes | Go to Next Step | |
| | | No | | Check ISC valve |
| | | | | Check engine oil |
| 4 | Check idle speed after warm up<br>**Idle speed: 750 ± 20 rpm**<br>[Test connector (Green 1-pin) grounded] | Yes | Go to Next Step | |
| | | No | Adjust idle speed | |
| 5 | | | | ECU malfunction |

**Note:**
**Engine stalls at idle with A/C ON, if the trouble can not be fixed after checking above steps, it may be A/C compressior malfunction**

**Engine stall during start up**

| STEP | QUICK INSPECTION | | ACTION | | POSSIBLE CAUSE |
|---|---|---|---|---|---|
| 1 | Check for malfunction code (08) with SST<br>[IGN ON, Test connector (Green 1-pin) grounded] | Yes | Check for cause by referring to check sequence | | |
| | | No | Go to Next Step | | |
| 2 | Check idle speed after warm up<br>**Idle speed: 750 ± 20 rpm**<br>[Test connector (Green: 1-pin) grounded] | Yes | Go to Next Step | | |
| | | No | Adjust idle speed | | |
| 3 | Check for injector operating sound at idle | Yes | Go to Next Step | | |
| | | No | Check resistance at injector harness connector (EMINJ-01)<br>**Terminal / Resistance:**<br>**(B/Y)—(LG/B), (B/Y)—(LG/R): 6—8Ω** | Yes | Check wiring short or open |
| | | | | No | Cheek injector resistance |
| | | | | | Check wiring |
| 4 | Check ignition timing advance | Yes | Go to Next Step | | |
| | | No | | | Insufficient centrifugal advance: Distributor malfunction |
| | | | Insufficient vacuum advance: Check vacuum hose to distributor for damage or misrouting | Yes | Distributor malfunction |
| | | | | No | Vacuum hose damaged or misrouted |
| 5 | | | | | ECU malfunction |

**Engine stall on deceleration**

| STEP | QUICK INSPECTION | | ACTION | | POSSIBLE CAUSE |
|---|---|---|---|---|---|
| 1 | Check for malfunction code (08), (09), (15), (17) with SST<br>[IG ON, Test connector (Green 1-pin) grounded] | Yes | Check for cause by referring to check sequence | | |
| | | No | Go to Next Step | | |
| 2 | Check idle switch and stoplight switch with SST<br>[IGN ON, Test connector (Green 1-pin) grounded] | Yes | Go to Next Step | | |
| | | No | Check for cause by referring to check sequence | | |
| 3 | Check for flashing of monitor lamp after warm up<br>**Monitor lamp: Flashes more than 8 times/10 seconds at 2,000—3,000 rpm**<br>[The connector (Green: 1-pin) not grounded] | Yes | Go to Next Step | | |
| | | No | | | Replace oxygen sensor |
| 4 | Check for fuel cut operation during deceleration<br>**Fuel cut:**<br>**Above 1,600 rpm after warm up** | Yes | Go to Next Step | | |
| | | No | Check water thermosensor | Yes | Replace ECU |
| | | | | No | Replace water thermosensor |

## TROUBLESHOOTING CHART—1989 B2600

**Engine stall on deceleration (Cont'd)**

| STEP | QUICK INSPECTION | | ACTION | | POSSIBLE CAUSE |
|---|---|---|---|---|---|
| 5 | Check ignition timing advance | Yes | Go to Next Step | | |
| | | No | | | Insufficient centrifugal advance: Distributor malfunction |
| | | | Insufficient vacuum advance:<br>Check for vacuum routing | Yes | Distributor |
| | | | | No | Vacuum hose |
| 6 | | | | | Check ISC valve |

**Engine stall at idle (Intermittent)**

| STEP | QUICK INSPECTION | | ACTION | POSSIBLE CAUSE |
|---|---|---|---|---|
| 1 | Shake connector of ignition coil, main relay and ECU while cranking<br>Check if the engine starts | Yes | There may be a poor contact at the connector. Repair or replace the wiring | |
| | | No | Go to troubleshooting "Engine stall at idle (Always)" | |

**Hesitates/Stumbles on acceleration**

| QUICK | INSPECTION | | ACTION | | POSSIBLE CAUSE |
|---|---|---|---|---|---|
| 1 | Run engine at **2,000 rpm** for **20 seconds** and stop it<br>Check for malfunction code (09), (15), (17) with SST<br>[IGN ON, Test connector (Green: 1-pin) grounded] | Yes | Check for cause by referring to check sequence | | |
| | | No | Go to Next Step | | |
| 2 | Check idle switch with SST<br>[IGN ON, Test connector (Green: 1-pin) grounded] | Yes | Go to Next Step | | |
| | | No | Check for cause by referring to check sequence | | |
| 3 | Disconnect oxygen sensor connector<br>Check if condition improves | Yes | | | Check oxygen Sensor |
| | | No | Go to Next Step | | |
| 4 | Check fuel line pressure while accelerating<br>(Vacuum hose to pressure regulator disconnected)<br>**Fuel line pressure: Keeps 265—314 kPa 2.7—3.2 kg/cm$^2$, 38—46 psi)** | Yes | Go to Next Step | | |
| | | No | Check if fuel filter has been replaced according to maintenance schedule | Yes | Check fuel line for clogging |
| | | | | No | Replace fuel filter |
| | | | | | Pressure regulator Replace |
| 5 | Check for air leaks with throttle valve open by listening for sucking noise | Yes | | | Intake air system components damaged |
| | | | | | Vacuum and intake air hoses loose or damaged |
| | | | | | Bolts or nuts loose |
| | | | | | Gaskets damaged |
| | | No | Go to Next Step | | |
| 6 | Substitute a well-known ECU<br>Check if condition improves | Yes | | | ECU malfunction |
| | | No | | | Check airflow sensor |
| | | | | | Check throttle body |
| | | | | | Check spark plug |
| 7 | Check other systems | | | | Clutch slipping |

**Hesitates at steady speed**

| STEP | QUICK INSPECTION | | ACTION | | POSSIBLE CAUSE |
|---|---|---|---|---|---|
| 1 | Run engine at **2,000 rpm** for **20 seconds** and stop it<br>Check for malfunction code (08), (09), (12), (14), (15), (17) with SST<br>[IGN ON, Test connector (Green: 1-pin) grounded] | Yes | Check for cause by referring to check sequence | | |
| | | No | Go to Next Step | | |
| 2 | Disconnect oxygen sensor connector<br>Check if condition improves | Yes | | | Check oxygen sensor |
| | | No | Go to Next Step | | |
| 3 | Check for air leaks with throttle valve open by listening for sucking noise | Yes | Go to Next Step | | |
| | | No | | | Intake air system components damaged |
| | | | | | Vacuum and intake air hoses loose or damaged |
| | | | | | Nuts or bolts loose |
| | | | | | Gasket damaged |
| 4 | Check fuel line pressure while accelerating<br>(Vacuum hose to pressure regulator disconnected)<br>**Fuel line pressure: Keeps 265—314 kPa (2.7—3.2 kg/cm$^2$, 38—46 psi)** | Yes | Go to Next Step | | |
| | | No | Check if fuel filter has been replaced according to maintenance schedule | Yes | Check fuel line for clogging |
| | | | | No | Replace fuel filter |
| | | | | | Replace pressure regulator |
| 5 | Check condition of ignition coil and airflow meter connectors<br>(Burned or damaged) | Yes | | | Poor contact |
| | | No | Go to Next Step | | |
| 6 | Gradually open throttle valve<br>Check if engine speed increases smoothly | Yes | Go to Next Step | | |
| | | No | | | Check airflow sensor |
| | | | | | Check throttle valve |
| | | | | | Check throttle sensor |
| 7 | | | | | Check spark plug |
| 8 | Change fuel to specified grade<br>Check if condition improves | Yes | | | Poor fuel quality |
| | | No | Go to Next Step | | |
| 9 | | | | | ECU malfunction |

## TROUBLESHOOTING CHART—1989 B2600

**Jerking on acceleration**

| STEP | QUICK INSPECTION | | ACTION | | POSSIBLE CAUSE |
|---|---|---|---|---|---|
| 1 | Run engine at **2,000 rpm** for **20 seconds** and stop it<br>Check for malfunction code (09), (15), (17) with SST<br>[IGN ON, Test connector (Green: 1-pin) grounded] | Yes | Check for cause by referring to check sequence | | |
| | | No | Go to Next Step | | |
| 2 | Check idle switch with SST<br>[IGN ON, Test connector (Green: 1-pin) grounded] | Yes | Go to Next Step | | |
| | | No | Check for cause by referring to check sequence | | |
| 3 | Disconnect oxygen sensor connector<br>Check if condition improves | Yes | | | Check oxygen Sensor |
| | | No | Go to Next Step | | |
| 4 | Check fuel line pressure while accelerating<br>(Vacuum hose to pressure regulator disconnected)<br>**Fuel line pressure: Keeps 265—314 kPa 2.7—3.2 kg/cm$^2$, 38—46 psi)** | Yes | Go to Next Step | | |
| | | No | Check if fuel filter has been replaced according to maintenance schedule | Yes | Check fuel line for clogging |
| | | | | No | Replace fuel filter |
| | | | | | Pressure regulator Replace |
| 5 | Check for air leaks with throttle valve open by listening for sucking noise | Yes | | | Intake air system components damaged |
| | | | | | Vacuum and intake air hoses loose or damaged |
| | | | | | Bolts or nuts loose |
| | | | | | Gaskets damaged |
| | | No | Go to Next Step | | |
| 6 | Substitute a well-known ECU<br>Check if condition improves | Yes | | | ECU malfunction |
| | | No | | | Check airflow sensor |
| | | | | | Check throttle body |
| | | | | | Check spark plug |
| 7 | Check other systems | | | | Clutch slipping |

**Knocking**

| STEP | QUICK INSPECTION | | ACTION | | POSSIBLE CAUSE |
|---|---|---|---|---|---|
| 1 | Check malfunction code (09) with SST<br>[IGN ON, Test connector (Green 1-pin) grounded] | Yes | Check for cause by referring to the check sequence | | |
| | | No | Go to Step 2 (at sea level)<br>Go to Step 3 [at high altitude, **1,000 m, (3,280 ft)**] | | |
| 2 | Check ignition timing after warm up<br>**Ignition timing: 5° ± 1° BTDC**<br>[Test connector (Green: 1-pin) not grounded] | Yes | Go to Next Step | | |
| | | No | Check atmospheric pressure sensor output voltage<br>**Voltage: 3.5—4.5 V at sea level (IG ON)** | Yes | Ignition timing misadjust |
| | | | | No | Replace atmospheric pressure sensor |
| 3 | Disconnect water thermosensor connector<br>Check if condition improves | Yes | | | Check water thermosensor |
| | | No | Go to Next Step | | |
| 4 | Check vacuum routing | Yes | Go to Next Step | | |
| | | No | | | Vacuum hose |
| 5 | Observe fuel line pressure while accelerating from idle<br>**Fuel line pressure: Keeps 265—314 kPa (2.7—3.2 kg/cm$^2$, 38—46 psi)**<br>(Vacuum hose to pressure regulator disconnected) | Yes | Go to Next Step | | |
| | | No | Check fuel pump maximum pressure<br>**Fuel pump maximum pressure: 441—588 kPa (4.5—6.0 kg/cm$^2$, 64—85 psi)** | Yes | Replace fuel filter |
| | | | | No | Replace fuel pump |
| 6 | | | | | Check airflow sensor |
| 7 | | | | | Check spark plug |
| 8 | Change fuel to specified grade<br>Check if condition improves | Yes | | | Poor fuel quality |
| | | No | Go to Next Step | | |
| 9 | Check cooling system | | | | Thermostat |
| | | | | | Radiator |
| 10 | | | | | ECU malfunction |

## TROUBLESHOOTING CHART—1989 B2600

**Poor acceleration**

| STEP | QUICK INSPECTION | | ACTION | | POSSIBLE CAUSE |
|---|---|---|---|---|---|
| 1 | Check for malfunction code (08), (09), (12), (14), (26) with SST [IGN ON, Test connector (Green 1-pin) grounded] | Yes | Check for cause by referring to check sequence | | |
| | | No | Go to Next Step | | |
| 2 | Check idle switch with SST [IGN ON, Test connector (Green 1-pin) grounded] | Yes | Go to Next Step | | |
| | | No | Check for cause by referring to check sequence | | |
| 3 | Disconnect high—tension lead of each cylinder at idle. Check if engine condition changes [ISC valve connector disconnected] | Yes | Go to Next Step | | |
| | | No | Check ignition system | Yes | Replace injector |
| | | | | No | Check spark plug |
| | | | | | Check high-tension |
| | | | | | Check distributor cup |
| 4 | Check ignition timing advance | Yes | Go to Next Step | | |
| | | No | | | Insufficient centrifugal advance: Distributor malfunction |
| | | | Insufficient vacuum advance. Check vacuum hose routing | Yes | Distributor |
| | | | | No | Vacuum hose |
| 5 | Check for air leaks by listening for sucking noise | Yes | | | Intake air system components damaged |
| | | | | | Vacuum and air intake hoses loose or damaged |
| | | | | | Nuts or bolts loose |
| | | | | | Gasket damaged |
| | | No | Go to Next Step | | |
| 6 | Observe fuel line pressure while accelerating from idle<br>**Fuel line pressure: Keeps 265—314 kPa (2.7—3.2 kg/cm², 38—46 psi)**<br>[Vacuum hose to pressure regulator disconnected] | Yes | Go to Next Step | | |
| | | No | Check if fuel filter has been replaced according to maintenance schedule | No | Replace pressure regulator |
| | | | | Yes | Replace fuel filter |
| 7 | Gradually depress accelerator from idle. Check if engine speed increases smoothly | Yes | Go to Next Step | | |
| | | No | Check accelerator cable free play | Yes | Check airflow sensor |
| | | | | | Check throttle body |
| | | | | No | Adjust |
| 8 | Check fuel to specified grade. Check if condition improves | Yes | | | Poor fuel quality |
| | | No | Go to Next Step | | |
| 9 | Substitute a well-known ECU. Check if condition improves | Yes | | | ECU malfunction |
| | | No | Go to Next Step | | |
| 10 | Check other systems | | | | Clutch slipping |
| | | | | | Transmission (M/T) |
| | | | | | Brake dragging |
| | | | | | Belt tension |

**Lack of power**

| STEP | QUICK INSPECTION | | ACTION | | POSSIBLE CAUSE |
|---|---|---|---|---|---|
| 1 | Check for malfunction code (08) and (14) (only high-altitude) with SST [IGN ON, Test connector (Green 1-pin) grounded] | Yes | Check for cause by referring to check sequence | | |
| | | No | Go to Step 2 (High-altitude)<br>Go to Step 3 (Others) | | |
| 2 | (Only at high-altitude) Check ignition timing at idle<br>**Ignition timing: 11° ± 2° BTDC**<br>[Test connector (Green 1-pin) not grounded] | Yes | Go to Next Step | | |
| | | No | Check ignition timing<br>**Ignition timing: 5° ± 1° BTDC**<br>[Test connector (Green 1-pin) grounded] | Yes | Check atmospheric pressure sensor |
| | | | | No | Adjust ignition timing |
| 3 | Check ignition timing advance | Yes | Go to Next Step | | |
| | | No | | | Insufficient centrifugal advance: Distributor malfunction |
| | | | Insufficient vacuum advance. Check for vacuum routing | Yes | Distributor malfunction |
| | | | | No | Vacuum hose |
| 4 | Disconnect ISC valve connector and the high-tension lead of each cylinder. Check if condition changes | Yes | Go to Next Step | | |
| | | No | Check ignition system | Yes | Replace injector (If step 5 OK) |
| | | | | No | Check high-tension lead |
| | | | | | Check distributor cap |
| | | | | | Check spark plug |
| 5 | Check for injector operating sound at idle | Yes | Go to Next Step | | |
| | | No | Chcek resistance at injector harness connector (EMINJ-01)<br>**Terminals** (B/Y)—(LG/B), (B/Y)—(LG/R); **Resistance** 6—8Ω | Yes | Check wiring short or open |
| | | | | No | Check injector resistance |
| | | | | | Check wiring short or open |
| 6 | Check air cleaner element for clogging | Yes | Go to Next Step | | |
| | | No | Clean air cleaner element | | |
| 7 | Check for air leaks by listening for sucking noises<br>• At idle<br>• When throttle valve is open | Yes | | | Intake air system |
| | | | | | Components damaged |
| | | | | | Vacuum and air intake hoses loose or damaged |
| | | | | | Nuts or bolts loose |
| | | | | | Gasket damaged |

## TROUBLESHOOTING CHART—1989 B2600

**Lack of power (Cont'd)**

| STEP | QUICK INSPECTION | | ACTION | | POSSIBLE CAUSE |
|---|---|---|---|---|---|
| 8 | Check fuel line pressure [IGN ON, Test connector (Yellow 2-pin) connected]<br>**Fuel line pressure: 265—314 kPa (2.7—3.2 kg/cm², 38—46 psi)** | Yes | Go to Next Step | | |
| | | No | Check for fuel leakage | | |
| | | | Substitute a good fuel filter and retest | Yes | Replace fuel filter |
| | | | Check fuel pump maximum pressure<br>**Fuel pump maximum pressure: 441—588 kPa (4.5—6.0 kg/cm², 64—85 psi)** | Yes | Replace pressure regulator |
| | | | | No | Replace fuel pump |
| 9 | Check fuel line pressure at idle<br>**Fuel line pressure: 216—264 kPa (2.2—2.7 kg/cm², 31—38 psi)** | Yes | Go to Next Step | | |
| | | No | | | Replace pressure regulator |
| 10 | Check if fuel line pressure drops while accelerating (Vacuum hose disconnected) | Yes | Check if fuel filter has been replaced according to maintenance schedule | Yes | Check fuel line for clogging |
| | | | | No | Replace fuel filter |
| | | No | Go to Next Step | | |
| 11 | Check exhaust system for damage | Yes | Go to Next Step | | |
| | | No | Repair or replace | | |
| 12 | Check A/C, P/S and alternator belts tensions | Yes | Go to Next Step | | |
| | | No | Adjust belt tension | | |
| 13 | Check if accelerator can be depressed fully | Yes | Go to Next Step | | |
| | | No | Check accelerator cable | Yes | Throttle body |
| | | | | No | Accelerator cable |
| 14 | Substitute a well-known ECU. Check if condition improves | Yes | | | ECU malfunction |
| | | No | | | Check air flow sensor |
| | | | | | Check throttle sensor |
| | | | | | Go to Next Step |
| 15 | Substitute a specified fuel. Check if condition improves | Yes | | | Poor fuel quality |
| | | No | Go to Next Step | | |
| 16 | Check other systems | | | | Brake |
| | | | | | Clutch |
| | | | | | Engine |

**Bucking at high speed**

| STEP | QUICK INSPECTION | | ACTION | | POSSIBLE CAUSE |
|---|---|---|---|---|---|
| 1 | Run engine at **2,000 rpm** for more than **20 seconds**. Check for malfunction code (08), (09), (12), (14), (15), (17) with SST [IGN ON, Test connector (Green 1-pin) grounded] | Yes | Check for cause by referring to check sequence | | |
| | | No | Go to Next Step | | |
| 2 | Disconnect oxygen sensor connector. Check if condition improves | Yes | | | Check oxygen sensor |
| | | No | Go to Next Step | | |
| 3 | Observe fuel line pressure while accelerating from idle<br>**Fuel line pressure: Keeps 265—314 kPa (2.7—3.2 kg/cm², 38—46 psi)**<br>[Vacuum hose to pressure regulator disconnected] | Yes | Go to Next Step | | |
| | | No | Check if fuel filter has been replaced according to maintenance schedule | Yes | Check fuel line for clogging |
| | | | | No | Replace fuel filter |
| | | | | | Replace pressure regulator |
| 4 | Check for air leaks by listening sucking noise | Yes | Go to Next Step | | |
| | | No | | | Intake air system components damaged |
| | | | | | Vacuum and air intake hoses loose or damaged |
| | | | | | Nuts or bolts loose |
| | | | | | Gasket damaged |
| 5 | Check ignition timing advance | Yes | Go to Next Step | | |
| | | No | | | Insufficient centrifugal advance: Distributor malfunction |
| | | | Insufficient vacuum advance. Check vacuum hose routing | Yes | Distributor |
| | | | | No | Vacuum hose |
| 6 | Gradually open throttle valve from idle check if engine speed increases smoothly | Yes | Go to Next Step | | |
| | | No | | | Check airflow sensor |
| 7 | | | | | Check spark plug |
| 8 | | | | | ECU malfunction |

## TROUBLESHOOTING CHART—1989 B2600

**Bucking on deceleration**

| STEP | QUICK INSPECTION | | ACTION | POSSIBLE CAUSE |
|---|---|---|---|---|
| 1 | Check for malfunction code (08), (09), (12), (26), (34) with SST [IGN ON. Test connector (Green 1-pin) grounded] | Yes | Check for cause by referring to the check sequence | |
| | | No | Go to Next Step | |
| 2 | Check switches with SST [IGN ON, Test connector (Green: 1-pin) grounded] • Idle switch • Stoplight switch | Yes | Go to Next Step | |
| | | No | Check for cause by referring to the check sequence | |
| 3 | Substitute a well-known ECU Check if condition improves | Yes | | ECU malfunction |
| | | No | | Check throttle sensor |
| | | | | Go to Next Step |
| 4 | | | | Check spark plug |
| 5 | | | | Check clutch slipping |
| 6 | | | | Check compression between cylinders |

**Rotten egg smell**

| STEP | QUICK INSPECTION | ACTION | POSSIBLE CAUSE |
|---|---|---|---|
| 1 | Change fuel to specified grade Check if condition improves | | Poor fuel quality |

**Poor fuel economy**

| STEP | QUICK INSPECTION | | ACTION | | POSSIBLE CAUSE |
|---|---|---|---|---|---|
| 1 | Run the engine at **2,000 rpm** for more than **20 seconds** after warm up and stop it Check for malfunction code (15) (17) with SST [IGN ON, Test connector (Green: 1-pin ) grounded] | Yes | Check for cause by referring to check sequence | | |
| | | No | Go to Next Step | | |
| 2 | Check idle switch with SST [IGN ON, Test connector (Green: 1-pin) grounded] | Yes | Go to Next Step | | |
| | | No | Check for cause by referring to check sequence | | |
| 3 | Check for flashing of monitor lamp after warm up<br>**Monitor lamp: Flashes more than 8 times /10 seconds at 2,000—3,000 rpm**<br>[Test connector (Green: 1-pin) not grounded] | Yes | Go to Next Step | | |
| | | No | | | Replace oxygen sensor |
| 4 | Check fuel line pressure at idle<br>**Fuel line pressure: 216—264 kPa (2.2—2.7 kg/cm², 31—38 psi)** | Yes | Go to Next Step | | |
| | | No | Check vacuum line to pressure regulator for clogging or air leakage | Yes | Vacuum line clogging or damaged |
| | | | | No | Check solenoid valve (PRC) |
| | | | | | ECU malfunction (Check (2N) terminal voltage) |
| | | | | | Replace pressure regulator |
| 5 | Check for fuel cut operation during deceleration<br>**Fuel cut: Above 1,600 rpm after warm up** | Yes | Go to Next Step | | |
| | | No | Check water thermosensor | Yes | Replace ECU |
| | | | | No | Replace water thermosensor |
| 6 | Check ignition timing advance | Yes | Go to Next Step | | |
| | | No | | | Insufficient centrifugal advance: Distributor malfunction |
| | | | Insufficient vacuum advance Check for vacuum routing | Yes | Distributor |
| | | | | No | Vacuum hose |
| 7 | Check other systems | | | | Clutch slipping |
| | | | | | Brake |
| | | | | | Tire air pressure |

## TROUBLESHOOTING CHART—1989 B2600

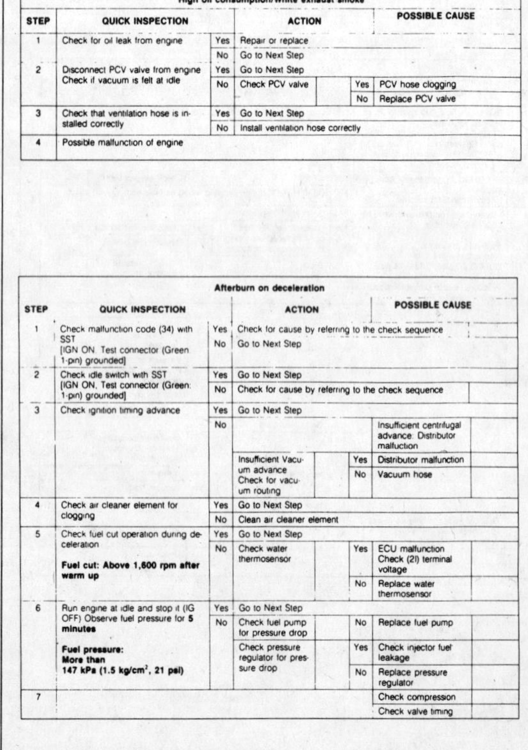

**High oil consumption/White exhaust smoke**

| STEP | QUICK INSPECTION | | ACTION | | POSSIBLE CAUSE |
|---|---|---|---|---|---|
| 1 | Check for oil leak from engine | Yes | Repair or replace | | |
| | | No | Go to Next Step | | |
| 2 | Disconnect PCV valve from engine Check if vacuum is felt at idle | Yes | Go to Next Step | | |
| | | No | Check PCV valve | Yes | PCV hose clogging |
| | | | | No | Replace PCV valve |
| 3 | Check that ventilation hose is installed correctly | Yes | Go to Next Step | | |
| | | No | Install ventilation hose correctly | | |
| 4 | Possible malfunction of engine | | | | |

**Afterburn on deceleration**

| STEP | QUICK INSPECTION | | ACTION | | POSSIBLE CAUSE |
|---|---|---|---|---|---|
| 1 | Check malfunction code (34) with SST [IGN ON. Test connector (Green 1-pin) grounded] | Yes | Check for cause by referring to the check sequence | | |
| | | No | Go to Next Step | | |
| 2 | Check idle switch with SST [IGN ON, Test connector (Green: 1-pin) grounded] | Yes | Go to Next Step | | |
| | | No | Check for cause by referring to the check sequence | | |
| 3 | Check ignition timing advance | Yes | Go to Next Step | | |
| | | No | | | Insufficient centrifugal advance: Distributor malfuction |
| | | | Insufficient Vacuum advance Check for vacuum routing | Yes | Distributor malfunction |
| | | | | No | Vacuum hose |
| 4 | Check air cleaner element for clogging | Yes | Go to Next Step | | |
| | | No | Clean air cleaner element | | |
| 5 | Check fuel cut operation during deceleration<br>**Fuel cut: Above 1,600 rpm after warm up** | Yes | Go to Next Step | | |
| | | No | Check water thermosensor | Yes | ECU malfunction Check (2I) terminal voltage |
| | | | | No | Replace water thermosensor |
| 6 | Run engine at idle and stop it (IG OFF) Observe fuel pressure for **5 minutes**<br>**Fuel pressure: More than 147 kPa (1.5 kg/cm², 21 psi)** | Yes | Go to Next Step | | |
| | | No | Check fuel pump for pressure drop | No | Replace fuel pump |
| | | | Check pressure regulator for pressure drop | Yes | Check injector fuel leakage |
| | | | | No | Replace pressure regulator |
| 7 | | | | | Check compression |
| | | | | | Check valve timing |

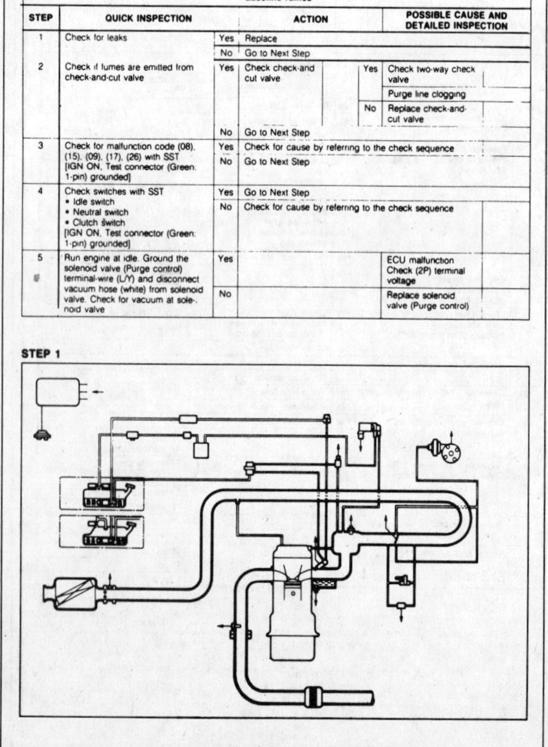

**Gasoline fumes**

| STEP | QUICK INSPECTION | | ACTION | | POSSIBLE CAUSE AND DETAILED INSPECTION |
|---|---|---|---|---|---|
| 1 | Check for leaks | Yes | Replace | | |
| | | No | Go to Next Step | | |
| 2 | Check if fumes are emitted from check-and-cut valve | Yes | Check check-and cut valve | Yes | Check two-way check valve |
| | | | | | Purge line clogging |
| | | | | No | Replace check-and-cut valve |
| | | No | Go to Next Step | | |
| 3 | Check for malfunction code (08), (15), (09), (17), (26) with SST [IGN ON, Test connector (Green: 1-pin) grounded] | Yes | Check for cause by referring to the check sequence | | |
| | | No | Go to Next Step | | |
| 4 | Check switches with SST • Idle switch • Neutral switch • Clutch switch [IGN ON, Test connector (Green: 1-pin) grounded] | Yes | Go to Next Step | | |
| | | No | Check for cause by referring to the check sequence | | |
| 5 | Run engine at idle. Ground the solenoid valve (Purge control) terminal-wire (L/Y) and disconnect vacuum hose (white) from solenoid valve. Check for vacuum at solenoid valve | Yes | | | ECU malfunction Check (2P) terminal voltage |
| | | No | | | Replace solenoid valve (Purge control) |

STEP 1

## TROUBLESHOOTING CHART—1989 B2600

**MIL always ON**

| STEP | QUICK INSPECTION | | ACTION | | POSSIBLE CAUSE |
|---|---|---|---|---|---|
| 1 | (California) Check for malfunction code with SST [IGN ON, Test connector (Green: 1-pin) grounded] | | **"88"** Replace ECU | | |
| | | | **"00"** Wiring between ECU (1A) terminal and MIL short to ground | | |
| 2 | (Federal) Check if emission system parts replacement time has come | Yes | Check if MIL has been reset by exchanging MIL set screw position | Yes | Replace mileage sensor |
| | | | | No | Reset the MIL |
| | **Emission system parts replacement schedule: Every 60,000 and 80,000 miles** | No | | | Replace mileage sensor |

**MIL never ON**

| STEP | QUICK INSPECTION | | ACTION | | POSSIBLE CAUSE |
|---|---|---|---|---|---|
| 1 | Check if other indicator lamps illuminate | Yes | Go to Next Step | | |
| | | No | Check power supply circuit to combination meter | | |
| 2 | Check bulb of the MIL | Yes | (California only) Ground ECU (1A) terminal Check if MIL illuminates | Yes | Replace ECU |
| | | | | No | Wiring between ECU and MIL open |
| | | | | | (Federal) MIL set connector loose or disconnected |
| | | | | | (Federal) Replace mileage sensor |
| | | No | Replace | | |

**A/C does not work**

| STEP | QUICK INSPECTION | | ACTION | | POSSIBLE CAUSE |
|---|---|---|---|---|---|
| 1 | Check if condenser fan operates when grounding A/C relay terminal-wire (L/W) (IGN ON) | Yes | Check voltage at ECU (1L) terminal with SST Voltage at idle after warm up: 0V (A/C and blower switches ON) | Yes | ECU malfunction (Check (1C) terminal voltage) |
| | | | | | Wiring between ECU (1C) and A/C relay open |
| | | | | No | A/C system malfunction |
| | | No | Check A/C system | | |

## MALFUNCTION INDICATOR LIGHT (MIL) OPERATION 1989 B2600

**PRINCIPLE OF CODE CYCLE**
Malfunction codes are determined as shown below

**1. Code cycle break**
The time between malfunction code cycles is 4.0 sec (the time the MIL (California only) and the buzzer are off).

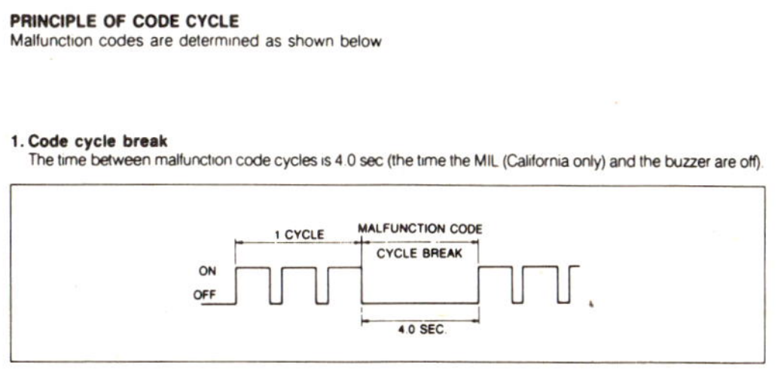


**2. Second digit of malfunction code (ones position)**
The digit in the ones position of the malfunction code represents the number of times the MIL (California only) and the buzzer are on 0.4 sec during one cycle.

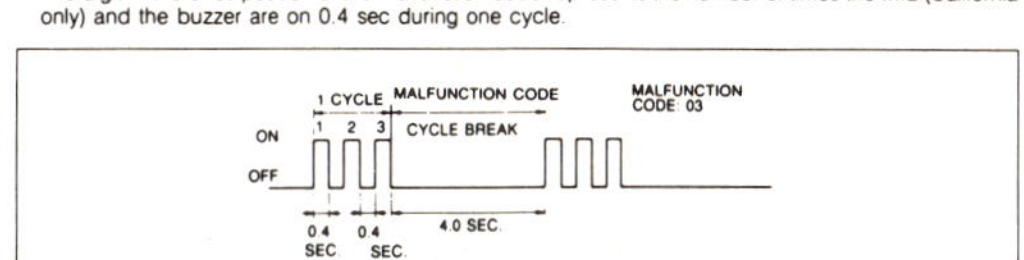


**3. First digit of malfunction code (tens position)**
The digit in the tens position of the malfunction code represents the number of times the MIL (California only) and the buzzer are on 1.2 sec during one cycle.

It should also be noted that the light goes off for 1.6 sec. between the long and short pulses of the MIL (California only) and the buzzer.

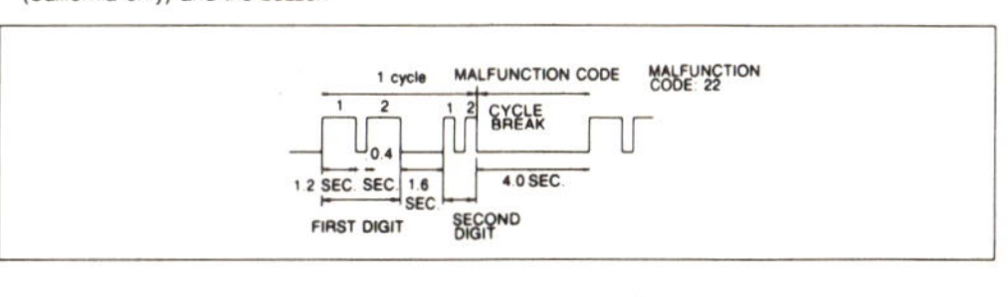


## TROUBLE CODE IDENTIFICATION CHART 1989 B2600

**CODE NUMBERS**

| Code No. | Malfunction display: Pattern of output signal (Self-Diagnosis Checker or MIL (California only)) | Sensor or subsystem | Self-diagnosis | Fail-safe |
|---|---|---|---|---|
| 01 | ON / OFF | Ignition pulse | No ignition signal | |
| 03 | ON / OFF | G signal | No G signal | Cancels 2-group injection |
| 08 | ON / OFF | Airflow sensor | Open or short circuit | Basic fuel injection amount fixed as for two driving modes (1) Idle switch: ON (2) Idle switch: OFF |
| 09 | ON / OFF | Water thermosensor | Open or short circuit | Maintains constant 20°(68°F) command |
| 11 | ON / OFF | Intake air thermosensor (dynamic chamber) | Open or short circuit | Maintains constant 20°C (68°F) command |
| 12 | ON / OFF | Throttle sensor | Open or short circuit | Maintains constant command of throttle valve fully open |
| 14 | ON / OFF | Atmospheric pressure sensor | Open or short circuit | Maintains constant command of sea level pressure |
| 15 | ON / OFF | Oxygen sensor | Sensor output continues less than 0.45V 180 sec after engine exceeds 1,500 rpm | Cancels engine feedback operation |
| 17 | ON / OFF | Feedback system | Sensor output not changed 20 sec. after engine exceeds 1,500 rpm | Cancels engine feedback operation |
| 25 | ON / OFF | Solenoid valve (pressure regulator control) | Open or short circuit | — |
| 26 | ON / OFF | Solenoid valve (purge control) | | — |
| 34 | ON / OFF | Solenoid valve (idle speed control) | | — |

**Caution**
**a) If there is more than one failure present, the lowest number malfunction code is displayed first, the remaining codes are displayed in order.**
**b) After repairing all failures, turn off the ignition switch, disconnect the negative battery cable, and depress the brake pedal for 5 sec to erase the memory of a malfunction code from the engine control unit.**

## TROUBLE CODE DIAGNOSTIC CHART—1989 B2600

If a malfunction code number is shown on the **SST**, check the following chart along with the wiring diagram.

**Code No.01 (Ignition pulse)** **PC: Possible Cause**

| | | |
|---|---|---|
| Are there poor connections in ignition coil circuit? | YES | Repair or replace connector |
| | NO | ↓ |
| Is resistance of ignition coil OK? Resistance: Primary 0.77—0.95Ω Secondary 6—30 kΩ | NO | Replace ignition coil |
| | YES | ↓ |
| Is there continuity between ignition coil (−) terminal wire and engine control unit 1M terminal? | NO | PC Open circuit in wiring harness from ignition coil to engine control unit 1M terminal |
| | YES | ↓ |
| Is same Code No. present after performing after-repair procedure? | NO | Ignition pulse and circuit OK |
| | YES | ↓ |
| Is engine control unit (1M) terminal voltage OK? | NO | PC • No power supply to ignition coil • Short circuit in wiring harness |
| | YES | PC Engine control unit malfunction |

**Code No.03 (G signal)** **PC: Possible Cause**

| | | |
|---|---|---|
| Are there any poor connections at distributor connectors? | YES | Repair or replace connector |
| | NO | ↓ |
| Is G rotor OK? (visual inspection) | NO | Replace G rotor |
| | YES | ↓ |
| Is there continuity between distributor G signal (R/L wire) and engine control unit (1N) terminal? | NO | PC Open circuit in wiring harness from G signal to engine control unit (1N) terminal |
| | YES | ↓ |
| Is engine control unit (1N) terminal voltage OK? | NO | PC • Short circuit in wiring harness • Replace pickup coil |
| | YES | PC: Engine control unit malfunction |

## TROUBLE CODE DIAGNOSTIC CHART—1989 B2600

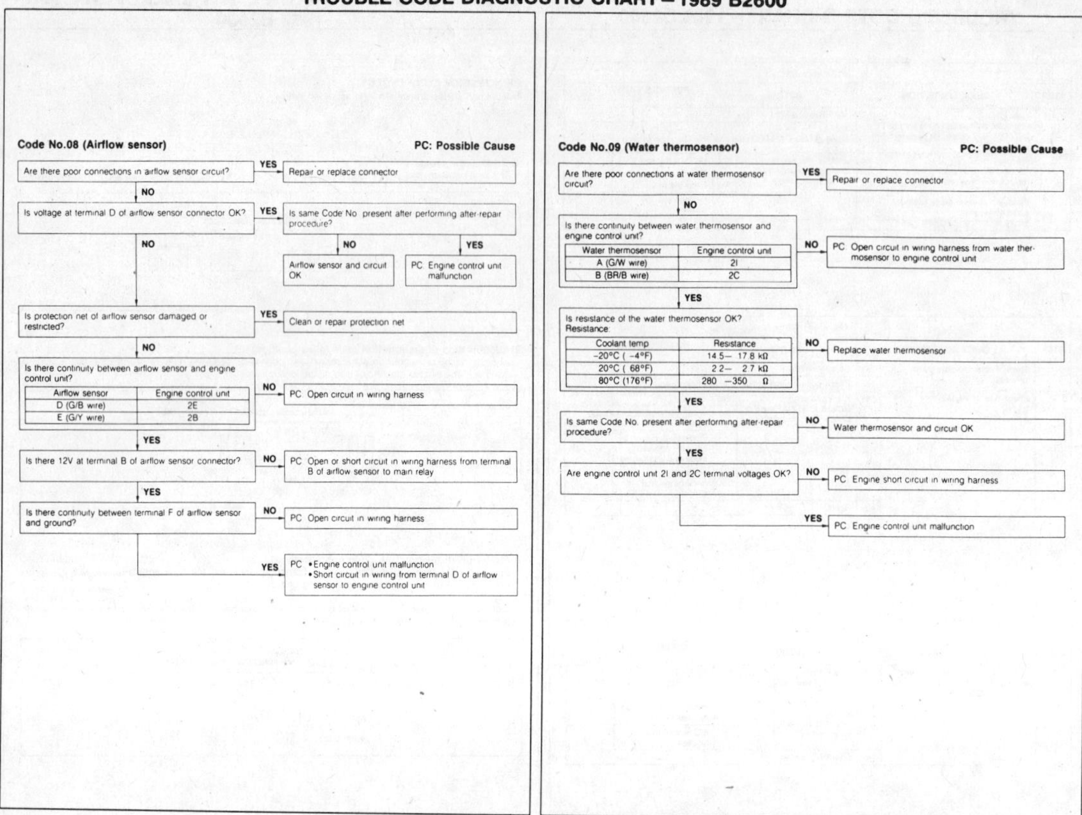


## TROUBLE CODE DIAGNOSTIC CHART—1989 B2600

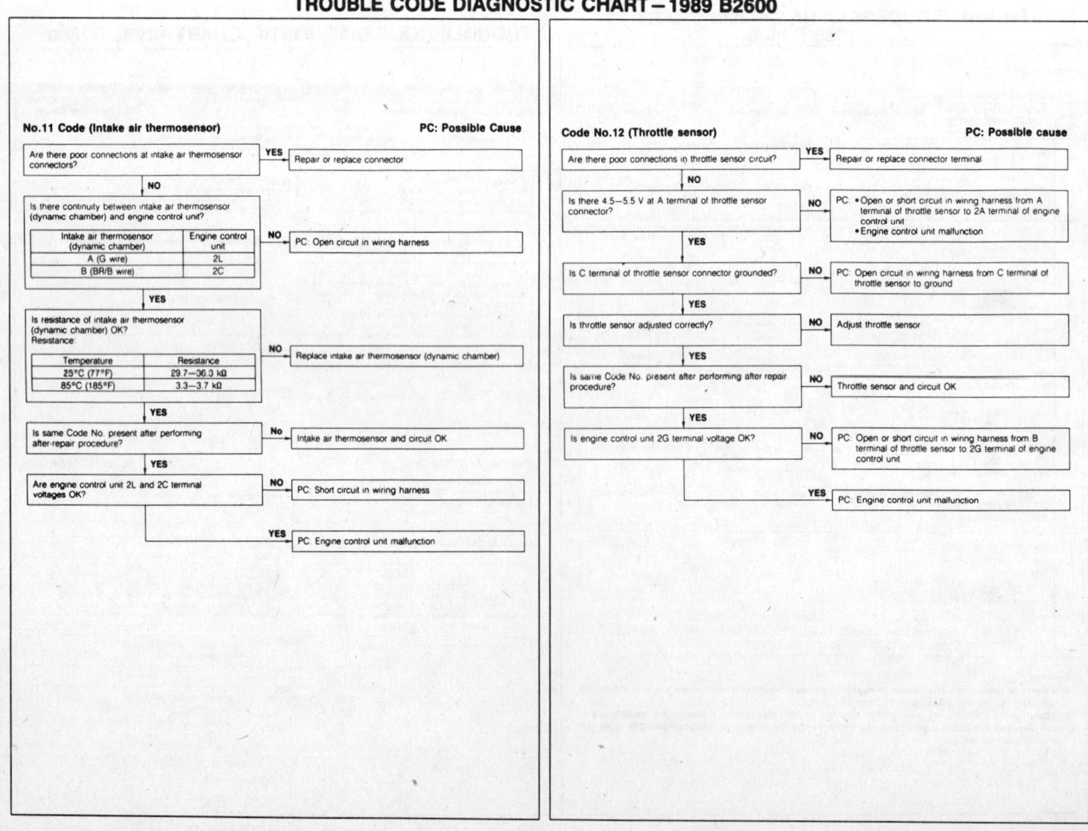

## TROUBLE CODE DIAGNOSTIC CHART—1989 B2600

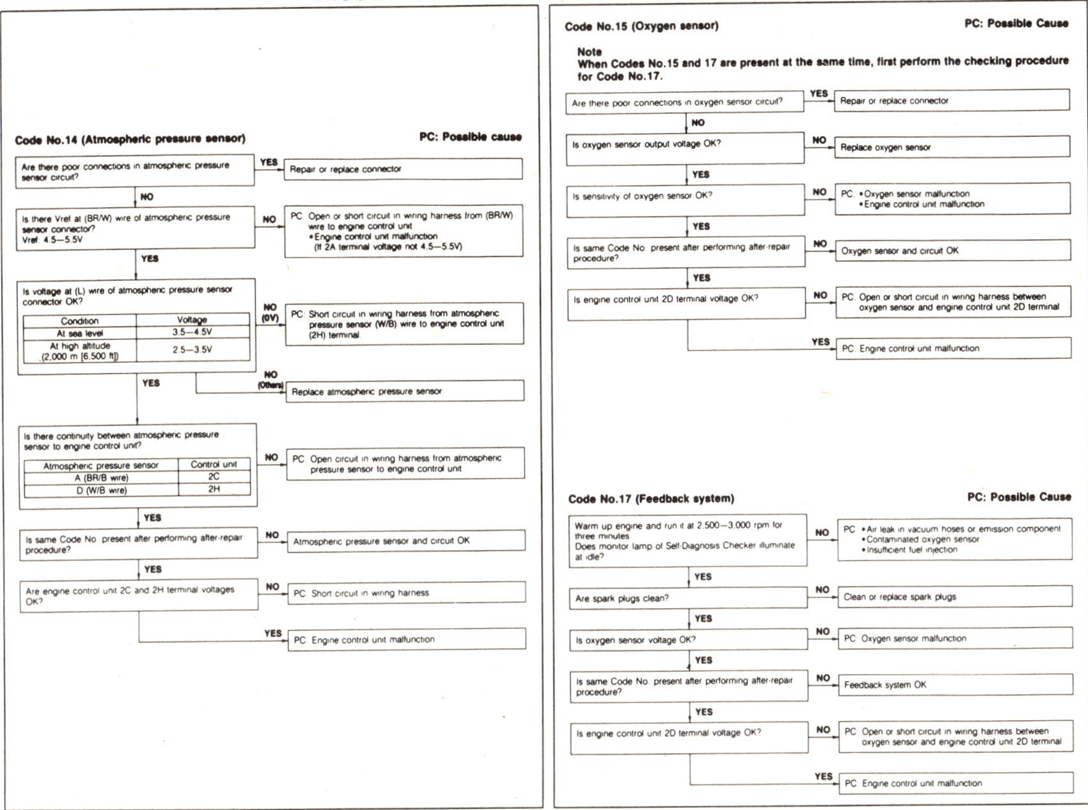


## TROUBLE CODE DIAGNOSTIC CHART—1989 B2600

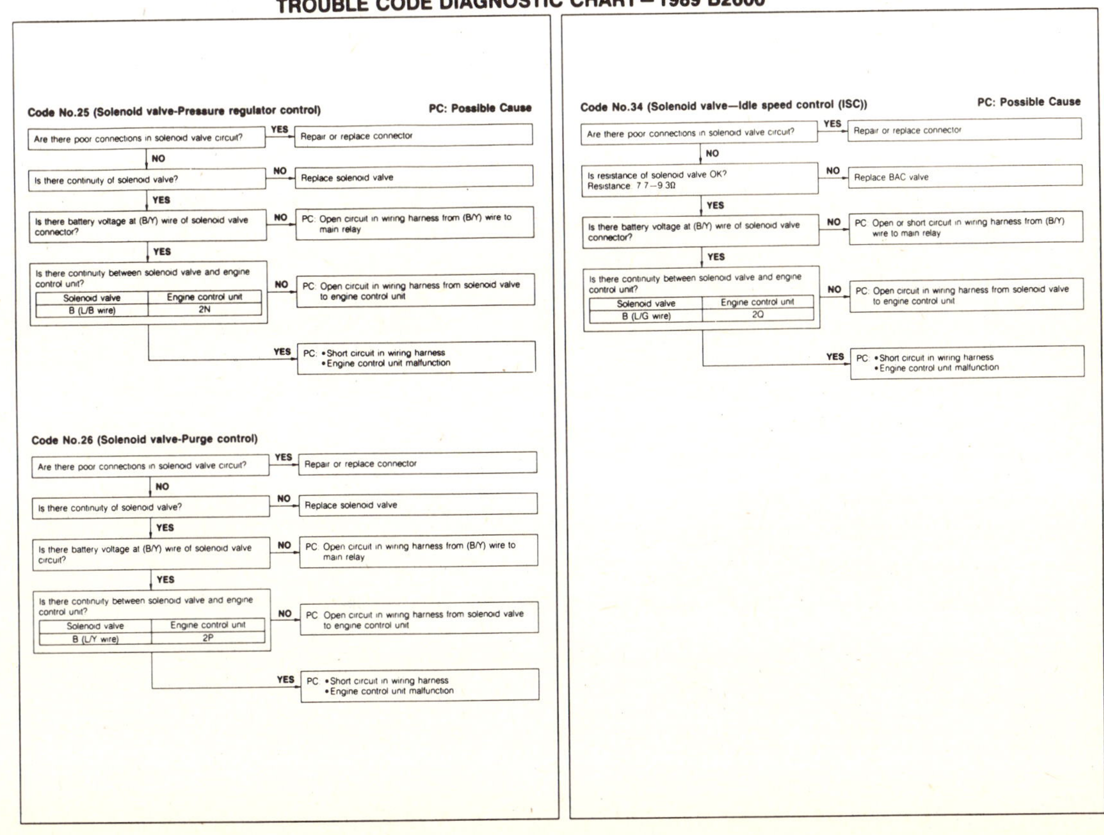

### CONTROL UNIT PIN LOCATION AND VOLTAGE TEST—1989 B2600

**Terminal voltage**

| Terminal | Input | Output | Connection to | Test condition | Voltage | Remarks |
|---|---|---|---|---|---|---|
| 1A | | ○ | Malfunction indicator lamp (California only) | For 3 seconds after ignition switch OFF→ON (Lamp illuminates) | Below 2.5V | Test connector (Green 1-pin) grounded |
| | | | | After 3 seconds (Lamp dose not illuminates) | Approx. 12V | |
| | | | | Lamp illuminates | Below 2.5V | |
| | | | | Lamp not illuminate | Approx. 12V | |
| 1B | | ○ | Self-Diagnosis checker (Code number) | For 3 seconds after ignition switch OFF→ON (Buzzer sounds) | Below 2.5V | • With Self-Diagnosis Checker • Test connector (Green 1-pin) grounded |
| | | | | After 3 seconds (Buzzer does not sounded) | Approx. 12V | |
| | | | | Buzzer sounds | Below 2.5V | |
| | | | | Buzzer not sounded | Approx. 12V | |
| 1C | | ○ | A/C relay | Ignition switch ON | Approx. 12V | Blower motor ON |
| | | | | For 10 seconds After fully depressing accelerator pedal with A/C switch ON (A/C does not operate) (in-gear, ignition switch ON) | Approx. 12V | |
| | | | | After 10 seconds | Below 2.5V | |
| | | | | For 5 seconds after cranking with A/C switch ON (A/C does not operate) | Approx. 12V | |
| | | | | After 5 seconds (A/C operates) | Below 2.5V | |
| | | | | A/C switch ON at idle | Below 2.5V | |
| | | | | A/C switch OFF at idle | Approx. 12V | |
| 1D | | ○ | Self-Diagnosis Checker (Monitor lamp) | Test connector (Green 1-pin) grounded For 3 seconds after ignition switch OFF→ON (Lamp illuminates) | Approx. 5V | With Self-Diagnosis Checker |
| | | | | After 3 seconds (Lamp does not illuminated) | Approx. 12V | |
| | | | | Test connector (Green 1-pin) not grounded at idle. Monitor lamp ON | Approx. 5V | |
| | | | | Test connector (Green 1-pin) not grounded at idle. Monitor lamp OFF | Approx. 12V | |
| 1E | ○ | | Idle switch | Accelerator pedal released | 0V | Ignition switch ON |
| | | | | Accelerator pedal depressed | Approx. 12V | |
| 1F | — | — | — | — | — | — |
| 1G | ○ | | Headlight switch | Headlight ON | Below 1.5V | Ignition switch ON |
| | | | | Headlight OFF | Approx. 12V | |
| 1H | ○ | | Blower switch | Blower ON | Below 1.5V | |
| | | | | Blower OFF | Approx. 12V | |
| 1I | | ○ | Circuit opening relay | Ignition switch ON | Approx. 12V | |
| | | | | During cranking or at idle | Below 2.5V | |
| 1J | ○ | | Stoplight switch | Brake pedal released | 0V | Ignition switch ON |
| | | | | Brake pedal depressed | Approx. 12V | |
| 1K | ○ | | P/S pressure switch | Ignition switch ON | Approx. 12V | |
| | | | | P/S ON (at idle) | 0V | |
| | | | | P/S OFF (at idle) | Approx. 12V | |
| 1L | ○ | | A/C switch | A/C switch ON (Ignition switch ON) | Below 2.5V | Blower motor: ON |
| | | | | A/C switch OFF (Ignition switch ON) | Approx. 12V | |
| 1M | ○ | | Ignition coil ⊖ terminal | Ignition switch ON | Approx. 12V | * Engine Signal Monitor: Green and red lights flash |
| | | | | Idle | Approx. 12V* | |
| 1N | ○ | | Distributor (G-signal) | Ignition switch ON | 0 or 5V | |
| | | | | Idle | Approx. 1.2V | |

| Terminal | Input | Output | Connection to | Test condition | Voltage | Remarks |
|---|---|---|---|---|---|---|
| 2A | | ○ | Vref | Ignition switch ON | 4.5—5.5V | |
| 2B | ○ | | Airflow sensor (Ground) | Constant | 0V | |
| 2C | — | — | Ground (E2) | Constant | 0V | |
| 2D | ○ | | Oxygen sensor | Ignition switch ON | 0V | |
| | | | | Idle (Cold engine) | 0V | |
| | | | | Idle (After warm up) | 0—1.0V | Needle moves from 0V to 1V |
| | | | | Increase engine speed (After warm up) | 0.5—1.0V | |
| | | | | Deceleration | 0—0.4V | |
| 2E | ○ | | Airflow sensor (Intake air mass) | Ignition switch ON | 1.0—2.0V | |
| | | | | Idle (After warm up) | 1.9—2.6V | |
| | | | | Increase engine speed (After warm up) | 2—5V | |
| 2F | ○ | | Test connector | Test connector (Green: 1-pin) not grounded | Approx. 12V | Ignition switch ON |
| | | | | Test connector (Green: 1-pin) grounded | 0V | |
| 2G | ○ | | Throttle sensor | Accelerator pedal released | Approx. 0.5V | Ignition switch ON |
| | | | | Accelerator pedal fully depressed | Approx. 4.3V | |
| 2H | ○ | | Atmospheric pressure sensor | At sea level | 3.5—4.5V | |
| | | | | At high altitude (2,000 [6,500 ft]) | 2.5—3.5V | |
| 2I | ○ | | Water thermosensor | Engine coolant temperature 20°C (68°F) | Approx. 2.5V | Ignition switch ON |
| | | | | After warm up | Approx. 0.4V | |
| 2J | ○ | | Ignition switch (ON position) | Ignition switch OFF | 0V | |
| | | | | Ignition switch ON | Approx. 12V | |
| 2K | | ○ | Airflow sensor (Burn-off) | Just after ignition switch OFF | 0V | Burn-off functions momentarily |
| | | | | Burn off (2-5 seconds after ignition switch OFF) | 8—12V | |
| 2L | ○ | | Intake air thermosensor (Dynamic chamber) | At 20°C (68°F) | Approx. 2.5V | |
| 2M | | ○ | Igniter | At sea level | Approx. 12V | Ignition switch ON |
| | | | | At high altitude (1,500 m [4,920 ft]) | 0V | |
| 2N | | ○ | Solenoid valve (PRC) | For 120 seconds after ignition switch OFF→ON | Below 2.5V | During hot condition. Coolant temp. above 90°C (194°F) Intake air temp. above 75°C (167°F) |
| | | | | For 120 seconds after starting | Below 2.5V | |
| | | | | Ignition switch ON | Approx. 12V | Other conditions |
| 2P | | ○ | Solenoid valve (Purge control) | Ignition switch ON | Approx. 12V | |
| | | | | Idle | Approx. 12V | * Engine signal monitor: Green and red lights flash |
| | | | | Driving in gear | 5—15V* | |
| 2Q | | ○ | Solenoid valve (Idle speed control) | Ignition switch ON | Approx. 11V | Engine signal monitor: Green and red lights flash |
| | | | | Idle | Approx. 10V | |
| 2R | — | — | Ground (E02) | Constant | 0V | |

### CONTROL UNIT PIN LOCATION AND VOLTAGE TEST (CONT.)—1989 B2600

| Terminal | Input | Output | Connection to | Test condition | Voltage | Remarks |
|---|---|---|---|---|---|---|
| 3A | — | — | Ground (E01) | Constant | 0V | |
| 3B | ○ | | Ignition switch (Start position) | While cranking | Approx. 10V | |
| | | | | Ignition switch ON | Approx. 0V | |
| 3C | | ○ | Injector (No.1, 2) | Ignition switch ON | Approx. 12V | * Engine Signal Monitor: Green and red lights flash |
| | | | | Idle | Approx. 12V* | |
| 3D | ○ | | Neutral or clutch switch | Neutral or clutch pedal depressed | 0V | Ignition switch ON |
| | | | | Other condition | Approx. 12V | |
| 3E | | ○ | Injector (No.3, 4) | Ignition switch ON | Approx. 12V | * Engine Signal Monitor: Green and red lights flash |
| | | | | Idle | Approx. 12V* | |
| 3F | | ○ | Main relay | Ignition switch OFF | Approx. 12V | |
| | | | | Ignition switch ON | Approx. 0V | |
| 3G | — | — | Ground (E1) | Constant | 0V | |
| 3H | — | — | — | — | — | — |
| 3I | ○ | | Main relay | Ignition switch OFF | Approx. 0V | |
| | | | | Ignition switch ON | Approx. 12V | |
| | | | | During burn-off (airflow sensor) | | |
| 3J | — | — | Battery | Constant | Approx. 12V | For backup |

**Terminal location**

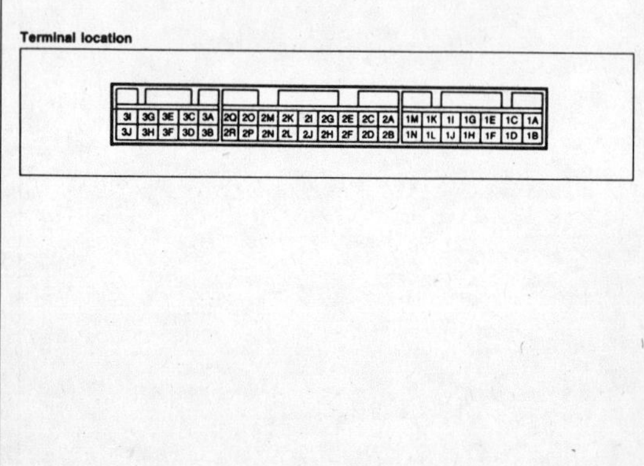


### WIRING SCHEMATIC—1989 B2600

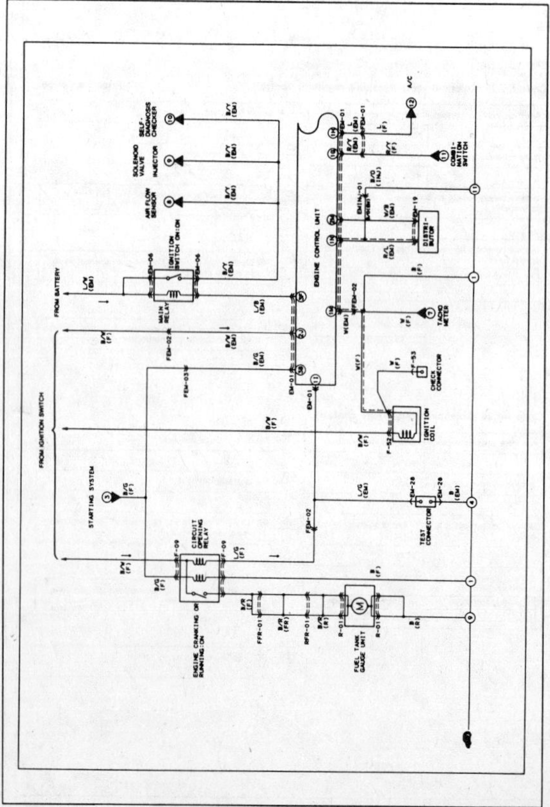

WIRING SCHEMATIC – 1989 B2600

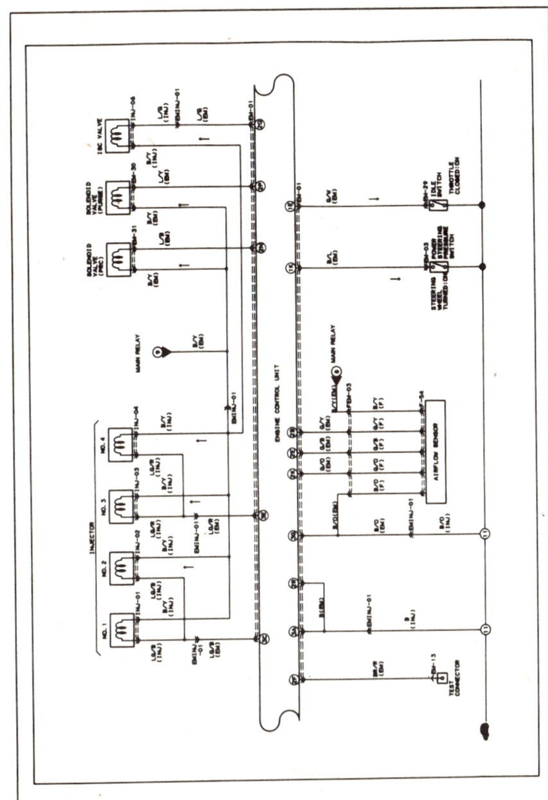

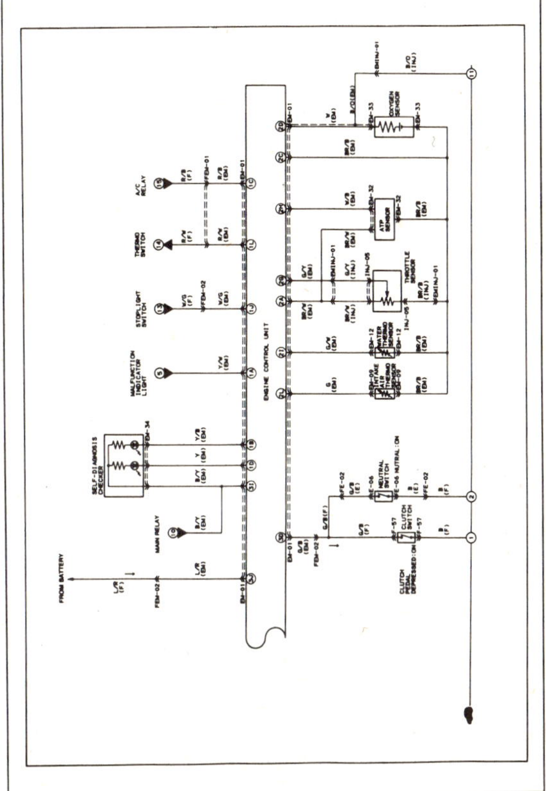

SYSTEM SCHEMATIC – 1990 B2600

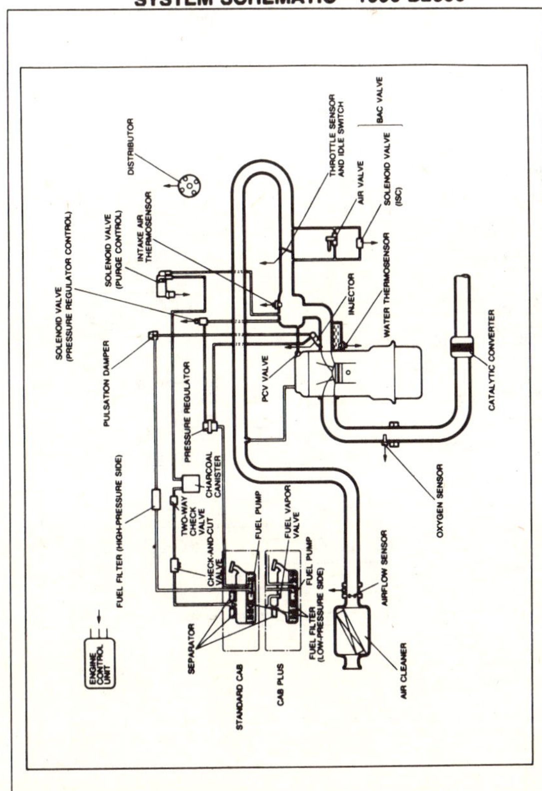


VACUUM HOSE ROUTING – 1990 B2600

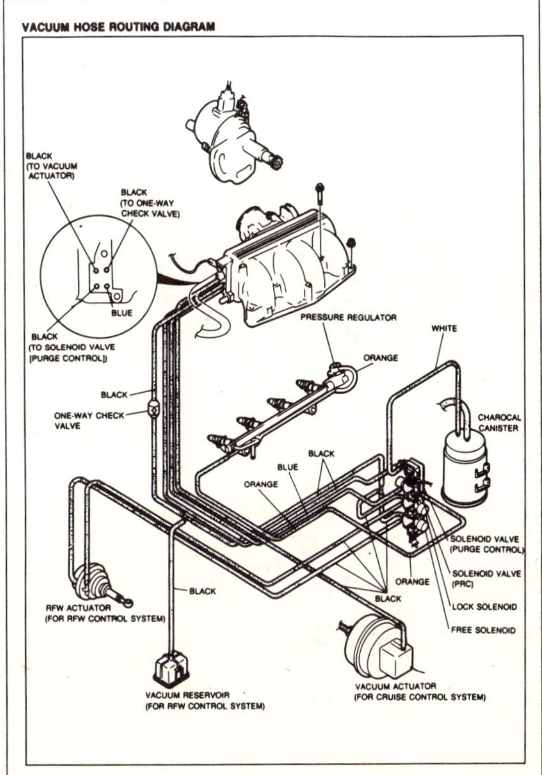

## EXPLODED VIEW OF AIR INTAKE SYSTEM – 1990 B2600

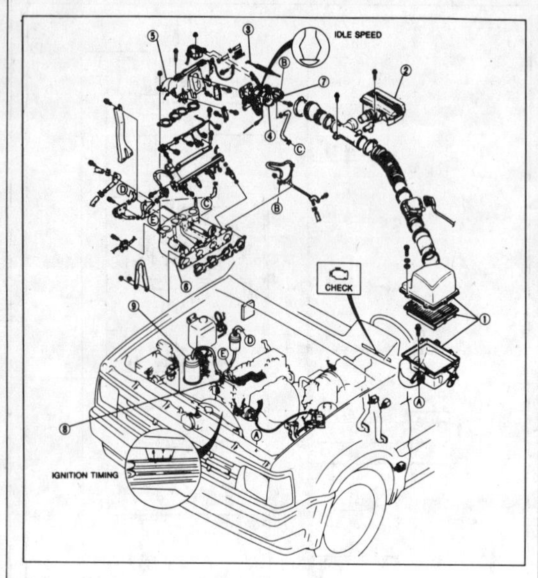


1. Air cleaner
2. Resonance chamber (G6 engine)
3. Accelerator cable
4. Throttle body
5. Dynamic chamber
6. Intake manifold
7. BAC valve
8. PCV valve
9. Charcoal canister

## EXPLODED VIEW FUEL AND EMISSION SYSTEM COMPONENTS – 1990 B2600

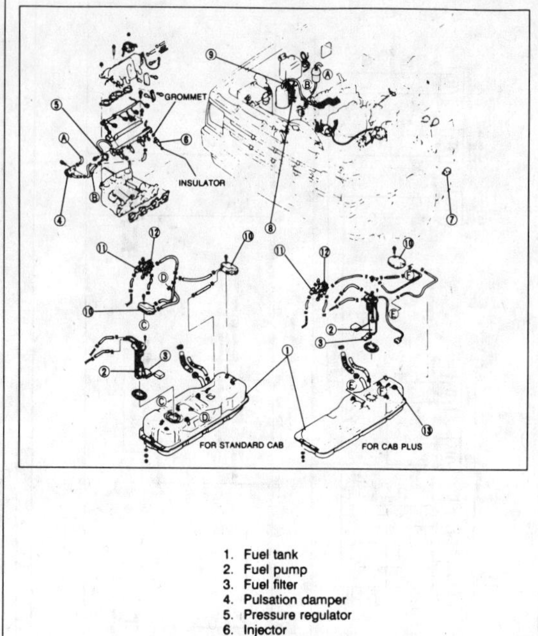


1. Fuel tank
2. Fuel pump
3. Fuel filter
4. Pulsation damper
5. Pressure regulator
6. Injector
7. Circuit opening relay
8. PRC solenoid valve
9. Purge control solenoid valve
10. Separator
11. 2-way check valve
12. Check and cut valve
13. Fuel vapor valve

## EXPLODED VIEW OF CONTROL SYSTEM COMPONENTS 1990 B2600

STRUCTURAL VIEW

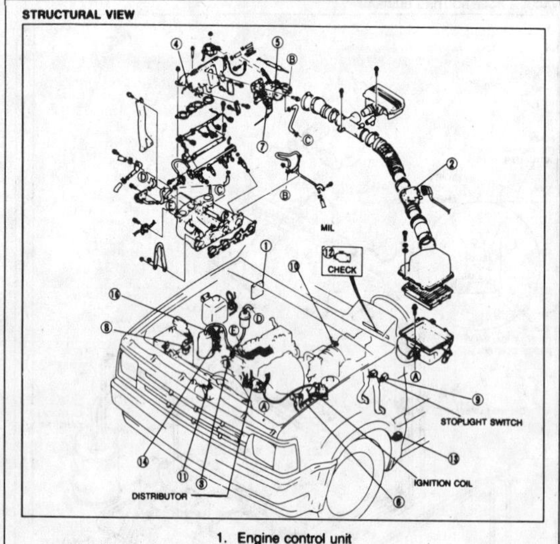


1. Engine control unit (ECU)
2. Airflow sensor
3. Water thermosensor
4. Intake air thermosensor
5. Throttle sensor
6. Oxygen sensor
7. Idle switch
8. Main relay
9. Clutch switch
10. Neutral switch
11. P/S pressure switch
12. Malfunction indicator lamp (MIL)
13. Circuit opening relay
14. PRC solenoid valve
15. Purge control solenoid valve

## SPECIFICATION CHART – 1990 B2600

| Item | | | Specification |
|---|---|---|---|
| Idle speed*1 | | rpm | M/T: 750 ± 20, A/T: 770 ± 20 (P range) |
| Ignition timing*1 | | BTDC | **G6**: 5° ± 1°, **F2**: 6° ± 1° |
| **Throttle body** | | | |
| Type | | | Horizontal draft (G6: 2 barrel, F2: 1 barrel) |
| Throat diameter | No.1 | mm (in) | **G6** 40 (1.6); **F2**: 50 (2.0) |
| | No.2 | | **G6** 46 (1.8) |
| **Fuel pump** | | | |
| Type | | | Impeller (in-tank) |
| Output pressure | | kPa (kg/cm², psi) | 441–589 (4.5–6.0, 64–85) |
| **Fuel filter** | | | |
| Type | Low-pressure side | | Nylon element |
| | High-pressure side | | Paper element |
| **Pressure regulator** | | | |
| Type | | | Diaphragm |
| Regulating pressure | | kPa (kg/cm², psi) | 265–314 (2.7–3.2, 38–46) |
| **Injector** | | | |
| Type | | | High-ohmic |
| Type of drive | | | Voltage |
| Resistance | | Ω | 12–16 (at 20°C, 68°F) |
| Volume | | | **G6**: 74–89 cc (4.51–5.43 cu in)/15 sec. |
| | | | **F2**: 50–62 cc (3.05–3.78 cu in)/15 sec. |
| **BAC valve (solenoid valve [idle speed control])** | | | |
| Solenoid resistance | | Ω | 7.7–9.3 (at 23°C, 73°F) |
| **Solenoid valve (Purge control)** | | | |
| Solenoid resistance | | Ω | 30–34 (at 20°C, 68°F) |
| **Water thermosensor** | | | |
| Resistance | −20°C (−4°F) | kΩ | 14.5–17.8 |
| | 20°C (68°F) | | 2.2–2.7 |
| | 80°C (176°F) | | 0.28–0.35 |
| **Intake air thermosensor** | | | |
| Resistance | 25°C (77°F) | kΩ | 29.7–36.3 |
| | 85°C (185°F) | | 3.3–3.7 |
| **Circuit opening relay** | | | |
| Resistance | STA—$E_1$ | Ω | 21–43 |
| | B—Fc | | 109–226 |
| | B—Fp | | ∞ |
| **Fuel tank** | | | |
| Capacity | | liters (US gal, Imp gal) | 56 (14.8, 12.3) |
| **Air cleaner** | | | |
| Element type | | | Dry |
| **Accelerator cable** | | | |
| Free play | | mm (in) | 1–3 (0.039–0.118) |
| **Fuel** | | | |
| Specification | | | Unleaded regular (RON 87 or higher) |

*1..... Test connector grounded

## COMPONENT DESCRIPTIONS—1990 B2600

### COMPONENT DESCRIPTIONS

| Component | Function | Remarks |
|---|---|---|
| **Air cleaner** | Filters air entering throttle body | |
| **Airflow sensor** | Detects amount of intake air; sends signal to engine control unit | |
| **Air valve** | When cold, supplies bypass air into dynamic chamber | • Engine speed increased to shorten warm-up period<br>• Thermowax type<br>• Installed in BAC valve |
| **Atmospheric pressure sensor** | Detects atmospheric pressure | In ECU |
| **BAC valve** | Supplies bypass air into dynamic chamber | Consists of air valve and ISC valve |
| **Catalytic converter** | Reduces HC, CO, and NOx by chemical reaction | Monolith type |
| **Charcoal canister** | Stores gas tank fumes when engine stopped | |
| **Check connector** | For Self-Diagnosis Checker | 6-pin connector (Green) |
| **Check-and-cut valve** | Releases excessive pressure or vacuum in fuel tank to atmosphere | |
| **Circuit opening relay** | Voltage for fuel pump while engine running | |
| **Clutch switch** | Detects in-gear condition; sends signal to engine control unit | Switch ON when clutch pedal depressed |
| **Crank angle sensor (In distributor)** | 1. Detects No.1 cylinder TDC; sends sig nal to engine control unit<br>2. Detects engine speed; sends signal to engine | For determining fuel injection timing |
| **Dynamic chamber** | Interconnects all cylinders | |
| **Engine control unit** | Detects following:<br>1. Engine speed<br>2. No.1 piston TDC<br>3. Intake air amount<br>4. Engine coolant temperature<br>5. Ignition ON signal<br>6. Throttle valve opening angle<br>7. Throttle valve fully closed<br>8. Air/fuel ratio (Oxygen concentration)<br>9. In-gear condition<br>10. Intake air temperature<br>11. Atmospheric pressure<br>12. A/C operation<br>13. P/S operation<br>14. E/L operation<br>15. Cranking signal<br>16. Test signal (idle speed, malfunction code No.)<br>17. Braking signal<br>Controls operation of the following:<br>1. Fuel injection system<br>2. Idle speed control<br>3. Pressure regulator control system<br>4. Purge control system<br>5. Fail-safe function<br>6. Monitor function<br>7. Burn-off system<br>8. Ignition timing control system<br>9. Fuel pump<br>10. A/C (cut off)<br>11. Main relay control | 1. Ne-Signal<br>2. G-signal<br>3. Airflow sensor<br>4. Water thermosensor<br>5. Ignition switch<br>6. Throttle sensor<br>7. Idle switch<br>8. Oxygen sensor<br>9. Neutral and clutch switches<br>10. Intake air thermosensor (on dynamic chamber)<br>11. Atmospheric pressure sensor (In ECU)<br>12. A/C switch<br>13. P/S pressure switch<br>14. Headlight and blower switches<br>15. Ignition switch (START position)<br>16. Test connector<br>17. Stoplight switch<br>1. Injector<br>2. Solenoid valve (Idle speed control)<br>3. Solenoid valve (Pressure regulator control)<br>4. Solenoid valve (Purge control)<br>5. Self-Diagnosis Checker and MIL<br>6. Monitor lamp (Self-Diagnosis Checker)<br>7. Airflow sensor<br>8. Igniter<br>9. Circuit opening relay<br>10. A/C relay<br>11. Main relay |
| **Fuel filter** | Filters particles from fuel | |
| **Fuel pump** | Provides fuel to injectors | • Operates while engine running<br>• Installed in fuel tank |
| **Fuel vapor valve** | Prevents fuel from flowing into charcoal canister | |
| **Idle switch** | Detects when throttle valve fully closed; sends signal to engine control unit | Installed on throttle body |
| **Igniter** | Receives spark signal from signal ECU and generates high voltage to ignition coil | |
| **Ignition switch (START position)** | Sends engine cranking signal to engine control unit | |
| **Injector** | Injects fuel into intake port | • Controlled by signals from engine control unit<br>• High-ohmic injector<br>• Two port injector nozzle **(G6)** |
| **Intake air thermosensor** | Detects intake air temperature; sends signal to engine control unit | Installed on dynamic chamber |
| **Main relay** | Supplies electric current to injectors and engine control unit. | |
| **MIL (Malfunction indicator lamp)** | (For Federal and Canada)<br>Lamp illuminates to indicate the maintenance schedule for the emission control system | Every 60,000 and 80,000 miles (Federal) or 90,000 and 130,000 km (Canada) |
| | (For California)<br>Lamp illuminates when input device malfunctions | Test connector not grounded |
| | (For California)<br>Lamp flashers to indicate malfunction code No. of input and output devices | Test connector grounded |
| **Neutral switch** | Detects in-gear condition; sends signal to engine control unit | Switch ON when neutral |
| **Oxygen sensor** | Detects oxygen concentration; sends signal to engine control unit | Zirconia ceramic and platinum coating |
| **PCV valve** | Controls amount of blowby gas introduced into engine | |
| **P/S pressure switch** | Detects P/S operation; sends signal to engine control unit | P/S: ON when steering wheel turned right or left |
| **Pressure regulator** | Adjusts fuel pressure supplied to injectors | |
| **Resonance chamber** | Improves mid-range torque characteristics | |
| **Separator** | Prevents fuel from flowing into charcoal canister | |
| **Solenoid valve — Idle speed control** | Controls bypass air amount | • Controlled by duty signal from engine control unit<br>• With integrated air valve<br>• Controls idle-up |
| **Solenoid valve — Pressure regulator control** | Controls vacuum to pressure regulator | Cuts vacuum passage when hot |
| **Solenoid valve — Purge control** | Controls evaporative fumes from canister to intake manifold | |
| **Stoplight switch** | Detects braking operation (deceleration); sends signal to engine control unit | |
| **Test connector** | For Self-Diagnosis Checker and idle speed ignition timing adjustment | 1-pin connector (Green) |
| **Throttle body** | Controls intake air quantity | Integrated throttle sensor and idle switch |
| **Throttle sensor** | Detects throttle valve opening angle; sends signal to engine control unit | Installed on throttle body |
| **Two-way check valve** | Controls pressure in fuel tank | |
| **Water thermosensor** | Detects coolant temperature; sends signal to engine control unit | |

## COMPONENT RELATIONSHIP CHART—1990 B2600

### RELATIONSHIP CHART

| INPUT DEVICES \ OUTPUT DEVICES | INJECTOR: FUEL INJECTION AMOUNT | INJECTOR: FUEL INJECTION TIMING | BAC VALVE: AIR VALVE | BAC VALVE: ISC VALVE | SOLENOID VALVE (PURGE CONTROL) | SOLENOID VALVE (PRESSURE REGULATOR CONTROL) | A/C RELAY (A/C CUT-OFF) | AIRFLOW SENSOR (BURN-OFF) | CIRCUIT OPENING RELAY (FUEL PUMP CONTROL) | IGNITER (IGNITION TIMING CONTROL) |
|---|---|---|---|---|---|---|---|---|---|---|
| TEST CONNECTOR | × | × | × | ○ | × | × | × | × | × | ○ |
| IGNITION SWITCH (ON POSITION) | × | × | × | × | × | × | × | ○ | × | × |
| IGNITION SWITCH (START POSITION) | ○ | ○ | × | ○ | × | ○ | ○ | × | × | ○ |
| HEADLIGHT AND BLOWER SWITCH | × | × | × | ○ | × | × | × | × | × | × |
| P/S PRESSURE SWITCH | × | × | × | ○ | × | × | × | × | × | × |
| A/C SWITCH | × | × | × | ○ | × | × | ○ | × | × | × |
| NEUTRAL AND CLUTCH SWITCH | ○ | × | × | ○ | ○ | × | ○ | × | × | × |
| STOPLIGHT SWITCH | ○ | × | × | × | × | × | × | × | × | × |
| IDLE SWITCH | ○ | × | × | ○ | ○ | ○ | × | × | × | ○ |
| ATMOSPHERIC PRESSURE SENSOR | ○ | × | × | F2 ○ G6 × | × | × | × | × | × | × |
| THROTTLE SENSOR | ○ | × | × | × | × | ○ | ○ | × | × | × |
| INTAKE AIR THERMOSENSOR | ○ | × | × | × | × | ○ | × | × | × | × |
| AIRFLOW SENSOR | ○ | × | × | ○ | ○ | × | × | ○ | × | ○ |
| OXYGEN SENSOR | ○ | × | × | × | ○ | × | × | × | × | × |
| WATER THERMOSENSOR | ○ | × | × | ○ | ○ | ○ | × | ○ | × | ○ |
| DISTRIBUTOR (Ne-SIGNAL) | ○ | ○ | × | ○ | ○ | ○ | × | ○ | ○ | ○ |
| DISTRIBUTOR (G-SIGNAL) | × | ○ | × | × | × | × | × | × | × | × |

## INPUT COMPONENTS AND ENGINE CONDITIONS 1990 B2600

### Input Devices and Engine Conditions

| ENGINE CONDITIONS \ INPUT DEVICES | APPROXIMATE TIME (BASED ON 10—16°C or 50—60°F AMBIENT) | SENSORS: DISTRIBUTOR (G-SIGNAL) | SENSORS: DISTRIBUTOR (Ne-SIGNAL) | SENSORS: WATER THERMOSENSOR | SENSORS: OXYGEN SENSOR | SENSORS: AIRFLOW SENSOR | SENSORS: INTAKE AIR THERMOSENSOR | SENSORS: THROTTLE SENSOR | SENSORS: ATMOSPHERIC PRESSURE SENSOR (IN ECU) |
|---|---|---|---|---|---|---|---|---|---|
| CRANKING —COLD ENGINE • COLD AIR • COLD COOLANT | Zero | | | | Signal has no effect on ECU | Signal has no effect on ECU | Signal has no effect on ECU | Signal has no effect on ECU | |
| COLD START —FAST IDLE • COLD AIR • COLD COOLANT | One minutes | | | Cool to warm: medium voltage (3.5V and dropping) | Sensor cold: low to high voltage (0—0.9V) | Low volume airflow: low to high voltage (2.4—2.6V) | | Closed throttle: low voltage (0.3—0.7V) | |
| COLD DRIVEAWAY —PART THROTTLE • COLD AIR • COLD COOLANT | Two minutes | | | | | | | | |
| WARM DRIVEAWAY —PART THROTTLE • WARM AIR • WARM COOLANT | Three minutes | | | Warm: medium voltage (Approx. 0.7V and dropping) | Sensor warm: high voltage (0.9V) | Moderate volume airflow: low to medium voltage (3.0V) | | Part throttle: medium voltage (1—3.5V) | Sends voltage signal to ECU that varies with altitude: voltage (approx 4V at sea level) |
| HOT CRUISE • WARM AIR • WARM COOLANT | | Sends No.1 cylinder TDC signal to ECU | Sends engine speed signal to ECU | | Sensor hot: switching from high voltage (0.9V) ↕ to low voltage (0.1V) | | Cool to warm: medium voltage (1.4—3.4V) | | |
| HOT ACCELERATION —60% THROTTLE | | | | | | Moderate to strong volume of airflow (3.8V) | | | |
| HOT ACCELERATION —WIDE OPEN THROTTLE | More than four minutes | | | Hot: low voltage (Approx. 0.4V) | High voltage (0.9V) | Strong volume of airflow (4.0V) | | Wide open throttle: high voltage (Approx. 4.0V) | |
| DECELERATION —CLOSED THROTTLE | | | | | Low voltage (0V) | Low volume of airflow (2.4V) | | Closed throttle: low voltage (0.3—0.7V) | |
| HOT CURB IDLE —EXTENDED | | | | | Switching from high to low voltage (0.75—0.25V) | | | | |
| HOT ENGINE SHUTDOWN | — | OFF | OFF | OFF | Sensor hot: low voltage (0.1V) until sensor cools | OFF | OFF | OFF | OFF |

## INPUT COMPONENTS AND ENGINE CONDITIONS (CONT.)1990 B2600

| SWITCHES | | | | | | | | | |
|---|---|---|---|---|---|---|---|---|---|
| | | | | | | | IGNITION SWITCH | | |
| IDLE SWITCH | STOP-LIGHT SWITCH | NEUTRAL AND CLUTCH SWITCHES | A/C SWITCH | P/S PRESSURE SWITCH | HEAD-LIGHT SWITCH | BLOWER SWITCH | START POSITION | ON POSITION | TEST CONNECTOR |
| Signal has no effect on ECU | Signal has no effect on ECU | Signal has no effect on ECU | Signal has no effect on ECU | Signal has no effect on ECU | Signal has no effect on ECU | Signal has no effect on ECU | Sends signal to ECU (approx. 12V) | Signal has no effect on ECU | Signal has no effect on ECU |
| Low voltage signal to ECU (below 1.5V) | Brake pedal depressed: sends signal to ECU (approx. 12V) | In neutral: low voltage signal to ECU (approx. 0V) | | | | | | | |
| High voltage signal to ECU (approx 12V) | No signal send to ECU (below 1.5V) | Driving in any gear: high voltage signal to ECU (approx. 12V) | A/C switch ON: sends signal to ECU (approx. 12V) A/C switch OFF: no signal to ECU (below 1.5V) | Steering wheel turned: low voltage signal to ECU (below 1.5V) Steering wheel straight ahead: high voltage signal to ECU (approx. 12V) | Headlight switch ON: low voltage signal to ECU (below 1.5V) Headlight switch OFF: high voltage signal to ECU (approx. 12V) | Blower switch ON: low voltage signal to ECU (below 1.5V) Blower switch OFF: high voltage signal to ECU (approx. 12V) | No signal to ECU (below 1.5V) | Sends signal to ECU (approx. 12V) | Connector not grounded: high voltage signal to ECU (approx. 12V) |
| Low voltage signal to ECU (below 1.5V) | Brake pedal depressed: sends signal to ECU (approx 12V) | In neutral: low voltage signal to ECU (approx. 0V) | | | | | | | Low voltage signal to ECU when connector grounded (below 1.5V) |
| OFF | OFF | OFF | OFF | OFF | OFF | OFF | OFF | OFF | OFF |

## OUTPUT COMPONENTS AND ENGINE CONDITIONS 1990 B2600

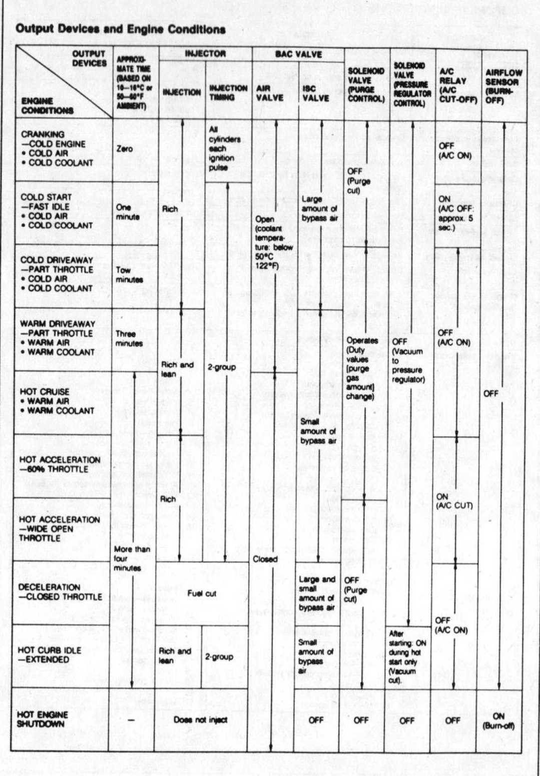

**Output Devices and Engine Conditions**

| OUTPUT DEVICES / ENGINE CONDITIONS | APPROXIMATE TIME (BASED ON 10—16°C or 50—60°F AMBIENT) | INJECTOR: INJECTION | INJECTOR: INJECTION TIMING | BAC VALVE: AIR VALVE | BAC VALVE: ISC VALVE | SOLENOID VALVE (PURGE CONTROL) | SOLENOID VALVE (PRESSURE REGULATOR CONTROL) | A/C RELAY (A/C CUT-OFF) | AIRFLOW SENSOR (BURN-OFF) |
|---|---|---|---|---|---|---|---|---|---|
| CRANKING —COLD ENGINE • COLD AIR • COLD COOLANT | Zero | | All cylinders each ignition pulse | | | OFF (Purge cut) | | OFF (A/C ON) | |
| COLD START —FAST IDLE • COLD AIR • COLD COOLANT | One minute | Rich | | Open (coolant temperature: below 50°C 122°F) | Large amount of bypass air | | | ON (A/C OFF: approx. 5 sec.) | |
| COLD DRIVEAWAY —PART THROTTLE • COLD AIR • COLD COOLANT | Tow minutes | | | | | | | | |
| WARM DRIVEAWAY —PART THROTTLE • WARM AIR • WARM COOLANT | Three minutes | Rich and lean | 2-group | | | Operates (Duty values [purge gas amount] change) | OFF (Vacuum to pressure regulator) | OFF (A/C ON) | |
| HOT CRUISE • WARM AIR • WARM COOLANT | | | | | Small amount of bypass air | | | | OFF |
| HOT ACCELERATION —60% THROTTLE | | | | | | | | | |
| HOT ACCELERATION —WIDE OPEN THROTTLE | | Rich | | | | | | ON (A/C CUT) | |
| DECELERATION —CLOSED THROTTLE | More than four minutes | Fuel cut | | Closed | Large and small amount of bypass air | OFF (Purge cut) | | | |
| HOT CURB IDLE —EXTENDED | | Rich and lean | 2-group | | Small amount of bypass air | | After starting: ON during hot start only (Vacuum cut). | OFF (A/C ON) | |
| HOT ENGINE SHUTDOWN | — | Does not inject | | | OFF | OFF | OFF | OFF | ON (Burn-off) |

## OUTPUT COMPONENTS AND ENGINE CONDITIONS 1990 B2600

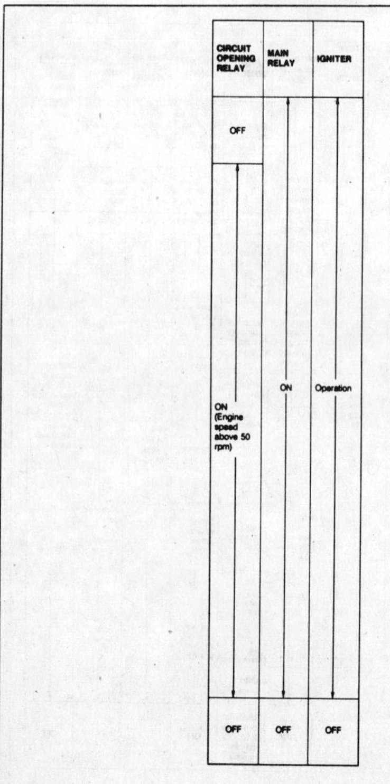

| CIRCUIT OPENING RELAY | MAIN RELAY | IGNITER |
|---|---|---|
| OFF | | |
| ON (Engine speed above 50 rpm) | ON | Operation |
| OFF | OFF | OFF |

## TROUBLESHOOTING CHARTS—1990 B2600

**Cranks normally but won't start (No combustion)**

| STEP | QUICK INSPECTION | | ACTION | | POSSIBLE CAUSE |
|---|---|---|---|---|---|
| 1 | Check for malfunction code (02) with SST [IGN ON, Test connector (Green: 1-pin) grounded] | Yes | Check for cause by referring to check sequence | | |
| | | No | Go to Next Step | | |
| 2 | Check for spark by disconnecting high-tension lead while cranking | Yes | Go to Next Step | | |
| | | No | Check ignition system | | |
| 3 | Check for fuel pump operating sound from fuel filler port [IGN ON, Test connector (Yellow: 2-pin) connected] | Yes | Check if engine starts in this condition | Yes | Check circuit opening relay switching operation |
| | | | | | Check circuit opening relay circuit (IGN: START) |
| | | | | No | Go to Next Step |
| | | No | Check circuit opening relay switching operation | Yes | Check circuit opening relay circuit |
| | | | | | Check fuel pump operation |
| | | | | No | Replace circuit opening relay |
| 4 | Check fuel line pressure [IGN ON, Test connector (Yellow: 2-pin) connected] **Fuel line pressure: 265—314 kPa (2.7—3.2 kg/cm², 38—46 psi)** | Yes | Go to Next Step | | |
| | | No | Check fuel pump maximum pressure **Fuel pump maximum pressure: 441—588 kPa (4.5—6.0 kg/cm², 64—85 psi)** | Yes | Replace pressure regulator |
| | | | | No | Replace fuel pump |
| 5 | Check for injector operating sound while cranking | Yes | Go to Next Step | | |
| | | No | Check voltage at ECU (2U) and (2V) terminals with SST **Voltage: Approx. 12V (IGN ON)** | Yes | Check throttle sensor |
| | | | | | Replace ECU |
| | | | | No | Check wiring for short or open |
| | | | | | Poor ground circuit from ECU (2A) terminal (Check terminal voltage with SST) |
| 6 | Substitute a well-known ECU Check if the condition improves | Yes | | | ECU malfunction |
| | | No | Check ground circuit from ECU (2B) terminal with SST **Voltage: 0V (IGN ON)** | Yes | Go to Next Step |
| | | | | No | Poor ground circuit |
| 7 | | | | | Low compression |

## TROUBLESHOOTING CHARTS – 1990 B2600

**Cranks normally but hard to start (Always)**

| STEP | QUICK INSPECTION | | ACTION | | POSSIBLE CAUSE |
|---|---|---|---|---|---|
| 1 | Check if vacuum hoses and the air hoses are connected correctly | Yes | Go to Next Step | | |
| | | No | Connect correctly | | |
| 2 | Check air cleaner element for clogging | Yes | Go to Next Step | | |
| | | No | Clean air cleaner element | | |
| 3 | Check ignition timing at idle after warm up<br>**Ignition timing:**<br>**BTDC 5° ± 1° (G6)**<br>**6° ± 1° (F2)**<br>[Test connector (Green: 1-pin) grounded] | Yes | Go to Next Step | | |
| | | No | Adjust ignition timing | | |
| 4 | Disconnect high-tension lead of each cylinder at idle<br>Check if engine condition changes | Yes | Go to Next Step | | |
| | | No | Check ignition system | Yes | Replace injector (If step 3 OK) |
| | | | | No | Check spark plug |
| | | | | | Check high-tension lead |
| | | | | | Check distributor cap |
| 5 | Check for injector operating sound at idle | Yes | Go to Next Step | | |
| | | No | Check resistance at injector harness connector (EMINJ-01)<br>Terminals: (B/Y)—(LG/R), (B/Y)—(LG/R)<br>Resistance: 6—9Ω | Yes | Check wiring short or open |
| | | | | No | Check injector resistance |
| | | | | | Check wiring short or open |
| 6 | Check fuel line pressure<br>[IGN ON, Test connector (Yellow: 2-pin) connected]<br>**Fuel line pressure:**<br>**265—314 kPa (2.7—3.2 kg/cm², 38—46 psi)** | Yes | Go to Next Step | | |
| | | No | Check if fuel filter has been replaced according to maintenance schedule | Yes | Check fuel line for clogging |
| | | | | No | Replace fuel filter |
| | | | Check fuel pump maximum pressure<br>**Fuel pump maximum pressure:**<br>**441—588 kPa (4.5—6.0 kg/cm², 64—85 psi)** | Yes | Replace pressure regulator |
| | | | | No | Replace fuel pump |
| 7 | Operate fuel pump<br>[IGN ON, Test connector (Yellow: 2-pin) connected]<br>Turn ignition switch OFF and observe fuel pressure **for 5 minutes**<br>**Fuel pressure:**<br>**More than**<br>**147 kPa (1.5 kg/cm², 21 psi)** | Yes | Go to Next Step | | |
| | | No | Check fuel pump pressure drop | No | Replace fuel pump |
| | | | Check pressure regulator pressure drop | Yes | Check injector fuel leakage |
| | | | | No | Replace pressure regulator |
| 8 | | | | | Check compression |

**Cranks normally but hard to start (Only when engine is cold)**

| STEP | QUICK INSPECTION | | ACTION | | POSSIBLE CAUSE |
|---|---|---|---|---|---|
| 1 | Check specific gravity of battery using a hydrometer<br>**Specific gravity: Above 1.200** | Yes | Go to Next Step | | |
| | | No | Recharge battery | | |
| 2 | Check for malfunction code (09) (26) with SST<br>[IGN ON, Test connector (Green: 1-pin) grounded] | Yes | Check for cause by referring to check sequence | | |
| | | No | Go to Next Step | | |
| 3 | Disconnect high-tension lead of each cylinder at idle<br>Check if engine condition changes | Yes | Go to Next Step | | |
| | | | Check ignition system | | Check spark plug |
| | | | | | Check high-tension lead |
| | | | | | Check distributor cap |
| 4 | Check fuel line pressure<br>[IGN ON, Test connector (Yellow: 2-pin) connected]<br>**Fuel line pressure:**<br>**265—314 kPa (2.7—3.2 kg/cm², 38—46 psi)** | Yes | Go to Next Step | | |
| | | No | Check for fuel leaks | | |
| | | | Check if fuel filter has been replaced according to maintenance schedule | Yes | Check fuel line for clogging |
| | | | | No | Replace fuel filter |
| | | | Check fuel pump maximum pressure<br>**Fuel pump maximum pressure:**<br>**441—588 kPa (4.5—6.0 kg/cm², 64—85 psi)** | Yes | Replace pressure regulator |
| | | | | No | Replace fuel pump |
| 5 | Disconnect ISC valve connector when engine is cold<br>Check if idle speed decreases during warm up | Yes | Go to Next Step | | |
| | | No | | | Check if BAC valve (air valve) opens when cold |
| 6 | Check voltage at ECU (1C) terminal with SST<br>**Voltage: Approx. 10V (while cranking)** | Yes | Go to Next Step | | |
| | | No | Check starter interlock switch | Yes | Check related wiring |
| | | | | No | Replace switch |
| 7 | Check voltage at ECU (2Q) terminal with SST<br>**Voltage: Approx. 2.5V (IGN ON, Engine coolant temperature 20°C [68°F])** | Yes | Go to Next Step | | |
| | | No | | | Check water thermosensor |
| 8 | | | | | Check compression |

## TROUBLESHOOTING CHARTS – 1990 B2600

**Cranks normally but hard to start (Only when engine is warm)**

| STEP | QUICK INSPECTION | | ACTION | | POSSIBLE CAUSE |
|---|---|---|---|---|---|
| 1 | Check for malfunction code with SST<br>[IGN ON, Test connector (Green: 1-pin) grounded] | Yes | Check for cause by referring to check sequence | | |
| | | No | Go to Next Step | | |
| 2 | Disconnect water thermosensor connector<br>Check if condition improves | Yes | Check water thermosensor connector condition as follows:<br>1. Shake connector and check if condition changes<br>2. Check condition of terminal (burned or damaged)<br>3. Connect a good terminal to harness side connector and check for looseness | Yes | Check water thermosensor |
| | | | | No | Poor contact of water thermosensor connector |
| | | No | Go to Next Step | | |
| 3 | Operate fuel pump<br>[IGN ON, Test connector (Yellow: 2-pin) connected]<br>Turn ignition switch OFF and observe fuel pressure **for 5 minutes**<br>**Fuel pressure:**<br>**More than**<br>**147 kPa (1.5 kg/cm², 21 psi)** | Yes | Go to Next Step | | |
| | | No | Check fuel pump pressure drop | No | Replace fuel pump |
| | | | Check pressure regulator pressure drop | Yes | Check injector fuel leakage |
| | | | | No | Replace pressure regulator |
| 4 | | | | | ECU malfunction |

**Cranks normally but hard to start (Only after heat soak)**

| STEP | QUICK INSPECTION | | ACTION | | | POSSIBLE CAUSE |
|---|---|---|---|---|---|---|
| 1 | Check for malfunction code with SST<br>[IGN ON, Test connector (Green: 1-pin) grounded] | Yes | Check for cause by referring to check sequence | | | |
| | | No | Go to Next Step | | | |
| 2 | Circulate fuel by operating fuel pump **for 20 seconds**<br>[IGN ON, Test connector (Yellow: 2-pin) connected]<br>Check if condition improves | Yes | Go to Step 3 | | | |
| | | No | Go to Step 4 | | | |
| 3 | Disconnect vacuum hose from pressure regulator<br>Check if condition improves | Yes | Check the components related to pressure regulator control system | | | Check water thermosensor |
| | | | | | | Check intake air thermosensor |
| | | | | | | Check solenoid valve (PRC) |
| | | | | | | ECU malfunction (Check (2T) terminal voltage) |
| | | No | Go to Next Step | | | |
| 4 | Operate fuel pump<br>[IGN ON, Test connector (Yellow: 2-pin) connected]<br>Turn ignition switch OFF and observe fuel pressure **for 5 minutes**<br>**Fuel pressure:**<br>**More than**<br>**147 kPa (1.5 kg/cm², 21 psi)** | Yes | Go to Next Step | | | |
| | | No | Check fuel pump pressure drop | **F2–148** | No | Replace fuel pump |
| | | | Check pressure regulator pressure drop | **F2–152** | Yes | Check injector fuel leakage |
| | | | | | No | Replace pressure regulator |
| 5 | Change fuel with specified one<br>Check if condition improves | Yes | | | | Poor fuel quality |
| | | No | Go to Next Step | | | |
| 6 | | | | | | ECU malfunction |

**Cranks normally but won't start (Intermittent)**

| STEP | QUICK INSPECTION | | ACTION | POSSIBLE CAUSE |
|---|---|---|---|---|
| 1 | Shake connector of ignition coil, main relay and ECU while cranking<br>Check if the engine starts | Yes | There may be a poor contact of the connector. Repair or replace the wiring | |
| | | No | Go to troubleshooting "Cranks normally but hard to start (Always)" | |

## TROUBLESHOOTING CHARTS—1990 B2600

**Rough idle (Always)**

| STEP | QUICK INSPECTION | | ACTION | | POSSIBLE CAUSE |
|---|---|---|---|---|---|
| 1 | Check for malfunction code with SST<br>[IGN ON, Test connector (Green: 1-pin) grounded] | Yes | Check for cause by referring to the check sequence | | |
| | | No | **"88" flashing**<br>Check voltage at ECU (2C) terminal with SST<br>**Voltage: 0V (IG ON)** | Yes | Replace ECU |
| | | | | No | Poor ground circuit |
| | | | **"00"** Go to Next Step | | |
| 2 | Check ignition at idle after warm up<br>**Ignition timing: BTDC 5° ± 1° (G6) 6° ± 1° (F2)**<br>[Test connector (Green: 1-pin) grounded] | Yes | Go to Next Step | | |
| | | No | Adjust ignition timing (If possible) | | |
| 3 | Disconnect high-tension lead of each cylinder at idle<br>Check if engine condition changes | Yes | Go to Next Step | | |
| | | No | Check ignition system | Yes | Replace injector (If Step 3 OK) |
| | | | | No | Check spark plug |
| | | | | | Check high-tension lead |
| | | | | | Check distributor cap |
| 4 | Check idle speed after warm up<br>**Idle speed: 750 ± 20 rpm (M/T) 770 ± 20 rpm (A/T, P range)**<br>[Test connector (Green: 1-pin) grounded] | Yes | Go to Next Step | | |
| | | No | Adjust idle speed (If possible) | | |
| 5 | Check for injector operating sound at idle | Yes | Go to Next Step | | |
| | | No | Check resistance at injector harness connector (EMINJ-01)<br>**Terminals / Resistance:** (B/Y)—(LG/B), (B/Y)—(LG/R): 6—8Ω | Yes | Check wiring short or open |
| | | | | No | Check injector resistance |
| | | | | | Check wiring short or open |
| 6 | Check fuel line pressure<br>[IGN ON, Test connector (Yellow: 2-pin) connected]<br>**Fuel line pressure: 265—314 kPa (2.7—3.2 kg/cm², 38—46 psi)** | Yes | Go to Next Step | | |
| | | No | Check for fuel leakage | | |
| | | | Substitute a good fuel filter and retest | Yes | Replace fuel filter |
| | | | Check fuel pump maximum pressure<br>**Fuel pump maximum pressure: 441—588 kPa (4.5—6.0 kg/cm², 64—85 psi)** | Yes | Replace pressure regulator |
| | | | | No | Replace fuel pump |
| 7 | Check intake manifold vacuum at idle<br>**Vacuum: 520 ± 20 mmHg (20.5 ± 0.79 inHg)** | Yes | Go to Next Step | | |
| | | No | Check for air leaks | Yes | Intake air system components damaged |
| | | | | | Vacuum and intake air hoses loose or damaged |
| | | | | | Bolts or nuts loose |
| | | | | | Gaskets damaged |
| | | | | No | Check throttle valve closing condition |

**Rough idle (Always) (Cont'd)**

| STEP | QUICK INSPECTION | | ACTION | | POSSIBLE CAUSE |
|---|---|---|---|---|---|
| 8 | Substitute a well-known ECU<br>Check if condition improves | Yes | | | ECU malfunction |
| | | No | Check voltage at ECU (2C) terminal with SST<br>**Voltage: 0V (IGN ON)** | Yes | Go to Next Step |
| | | | | No | Poor ground circuit |
| 9 | | | | | Check compression |

**Rough idle (Only when engine is cold)**

| STEP | QUICK INSPECTION | | ACTION | | POSSIBLE CAUSE |
|---|---|---|---|---|---|
| 1 | Check for malfunction code with SST<br>[IGN ON, Test connector (Green: 1-pin) grounded] | Yes | Check for cause by referring to check sequence | | |
| | | No | Go to Next Step | | |
| 2 | Check ignition at idle after warm up<br>**Ignition timing: BTDC 5° ± 1° (G6) 6° ± 1° (F2)**<br>[Test connector (Green: 1-pin) grounded] | Yes | Go to Next Step | | |
| | | No | Adjust ignition timing (If possible) | | |
| 3 | Disconnect high-tension lead of each cylinder at idle<br>Check if engine condition changes | Yes | Go to Next Step | | |
| | | No | Check ignition system | Yes | Replace injector (If step 4 OK) |
| | | | | No | Check spark plug |
| | | | | | Check high-tension lead |
| | | | | | Check distributor cap |
| 4 | Check for injector operating sound at idle | Yes | Go to Next Step | | |
| | | No | Check resistance at injector harness connector (EMINJ-01)<br>**Terminals / Resistance:** (B/Y)—(LG/B), (B/Y)—(LG/R): 6—8Ω | Yes | Check wiring short or open |
| | | | | No | Injector malfunction (Check resistance) |
| | | | | | Wiring short or open |
| 5 | Disconnect ISC valve connector at idle when engine is cold<br>Check if idle speed decreases during warm up | Yes | Go to Next Step | | |
| | | No | | | Check if BAC valve (air valve) opens when cold |
| 6 | Check if specified engine oil is used | Yes | Go to Next Step | | |
| | | No | Change engine oil to specified oil | | |
| 7 | Subsitute a well-known ECU<br>Check if condition improves | Yes | | | ECU malfunction |
| | | No | | | Check airflow sensor |

## TROUBLESHOOTING CHARTS—1990 B2600

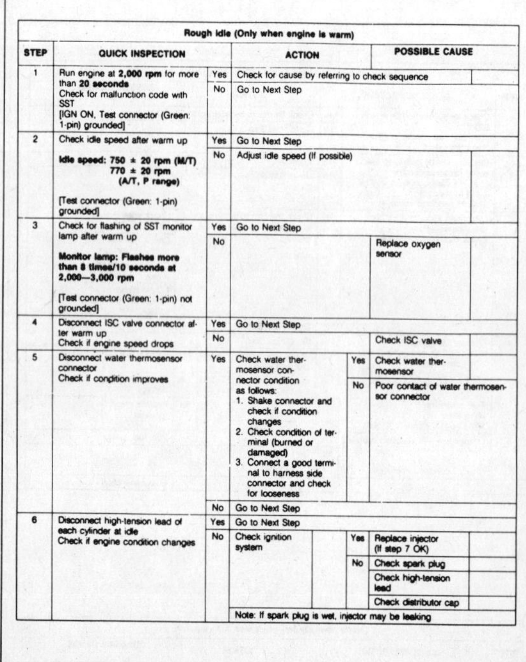

**Rough idle (Only when engine is warm)**

| STEP | QUICK INSPECTION | | ACTION | | POSSIBLE CAUSE |
|---|---|---|---|---|---|
| 1 | Run engine at **2,000 rpm** for more than **20 seconds**<br>Check for malfunction code with SST<br>[IGN ON, Test connector (Green: 1-pin) grounded] | Yes | Check for cause by referring to check sequence | | |
| | | No | Go to Next Step | | |
| 2 | Check idle speed after warm up<br>**Idle speed: 750 ± 20 rpm (M/T) 770 ± 20 rpm (A/T, P range)**<br>[Test connector (Green: 1-pin) grounded] | Yes | Go to Next Step | | |
| | | No | Adjust idle speed (If possible) | | |
| 3 | Check for flashing of SST monitor lamp after warm up<br>**Monitor lamp: Flashes more than 8 times/10 seconds at 2,000—3,000 rpm**<br>[Test connector (Green: 1-pin) not grounded] | Yes | Go to Next Step | | |
| | | No | | | Replace oxygen sensor |
| 4 | Disconnect ISC valve connector after warm up<br>Check if engine speed drops | Yes | Go to Next Step | | |
| | | No | | | Check ISC valve |
| 5 | Disconnect water thermosensor connector<br>Check if condition improves | Yes | Check water thermosensor connector condition as follows:<br>1. Shake connector and check if condition changes<br>2. Check condition of terminal (burned or damaged)<br>3. Connect a good terminal to harness side connector and check for looseness | Yes | Check water thermosensor |
| | | | | No | Poor contact of water thermosensor connector |
| | | No | Go to Next Step | | |
| 6 | Disconnect high-tension lead of each cylinder at idle<br>Check if engine condition changes | Yes | Go to Next Step | | |
| | | No | Check ignition system | Yes | Replace injector (If step 7 OK) |
| | | | | No | Check spark plug |
| | | | | | Check high-tension lead |
| | | | | | Check distributor cap |
| | | | Note: If spark plug is wet, injector may be leaking | | |

**Rough idle (Only when engine is warm) (Cont'd)**

| STEP | QUICK INSPECTION | | ACTION | | POSSIBLE CAUSE |
|---|---|---|---|---|---|
| 7 | Check for injector operating sound at idle | Yes | Go to Next Step | | |
| | | No | Check resistance at injector harness connector (EMINJ-01)<br>**Terminals / Resistance:** (B/Y)—(LG/B), (B/Y)—(LG/R): 6—8Ω | Yes | Check wiring short or open |
| | | | | No | Check injector resistance |
| | | | | | Check wiring short or open |
| 8 | Check for air leaks by listening for sucking noise | Yes | Go to Next Step | | |
| | | No | | | Intake air system components damaged |
| | | | | | Vacuum and intake air hoses loose or damaged |
| | | | | | Bolts or nuts loose |
| | | | | | Gaskets damaged |
| 9 | | | | | Check compression |

## TROUBLESHOOTING CHARTS—1990 B2600

**Rough idle (Only after heat soak)**

| STEP | QUICK INSPECTION | | ACTION | | POSSIBLE CAUSE |
|---|---|---|---|---|---|
| 1 | Run engine at **2,000 rpm** for more than **20 seconds**<br>Check for malfunction code with SST<br>[IGN ON, Test connector (Green: 1-pin) grounded] | Yes | Check for cause by referring to check sequence | | |
| | | No | Go to Next Step | | |
| 2 | Check switches with SST<br>Neutral switch<br>Clutch switch<br>[IGN ON, Test connector (Green: 1-pin) grounded] | Yes | Go to Next Step | | |
| | | No | Check for cause by referring to check sequence | | |
| 3 | Check for flashing of SST monitor lamp after warm up<br>**Monitor lamp: Flashes more than 8 times 10 seconds at 2,000—3,000 rpm**<br>[Test connector (Green: 1-pin) not grounded] | Yes | Go to Next Step | | |
| | | No | | | Replace oxygen sensor |
| 4 | Disconnect vacuum hose from pressure regulator<br>Check if condition improve | Yes | Check components related to pressure regulator control system | | Check water thermosensor |
| | | | | | Check intake air thermosensor |
| | | | | | Check solenoid valve (PRC) |
| | | | | | ECU malfunction (Check (2T) terminal voltage) |
| | | No | Go to Next Step | | |
| 5 | Run engine at idle and stop it<br>Observe fuel pressure **for 5 minutes**<br>**Fuel pressure: More than 147 kPa (1.5 kg/cm², 21 psi)** | Yes | Go to Next Step | | |
| | | No | Check fuel pump pressure drop | No | Replace fuel pump malfunction |
| | | | Check pressure regulator pressure drop | Yes | Check injector fuel leakage |
| | | | | No | Replace pressure regulator |
| 6 | Disconnect high-tension lead of each cylinder at idle<br>Check if engine condition changes | Yes | Go to Next Step | | |
| | | No | Check ignition system [Refer to ignition system troubleshooting (Misfire)] | Yes | Replace injector (If step 3 OK) |
| | | | | No | Check spark plug |
| | | | | | Check high-tension lead |
| | | | | | Check distributor cap |
| 7 | Check for injector operating sound at idle | Yes | Go to Next Step | | |
| | | No | Check resistance at injector harness connector (EMINJ-01)<br>**Terminals:** (B/Y)—(LG/B), (B/Y)—(LG/R) **Resistance:** 6—8Ω | Yes | Check wiring short or open |
| | | | | No | Check injector resistance |
| | | | | | Check wiring short or open |
| 8 | Change fuel to specified grade<br>Check if condition improves | Yes | | | Poor fuel quality |
| | | No | Go to Next Step | | |
| 9 | | | | | ECU malfunction |

**Rough idle just after starting**

| STEP | QUICK INSPECTION | | ACTION | | POSSIBLE CAUSE |
|---|---|---|---|---|---|
| 1 | Check for malfunction code with SST<br>[IGN ON, Test connector (Green: 1-pin) grounded] | Yes | Check for cause by referring to check sequence | | |
| | | No | Go to Next Step | | |
| 2 | Check idle switch with SST<br>[IGN ON, Test connector (Green: 1-pin) grounded] | Yes | Go to Next Step | | |
| | | No | Check for cause by referring to check sequence | | |
| 3 | Check ignition timing at idle after warm up<br>**Ignition timing: BTDC 5° ± 1° (G6), 6° ± 1° (F2)**<br>[Test connector (Green: 1-pin) not grounded] | Yes | Go to Next Step | | |
| | | No | Adjust ignition timing | | |
| 4 | Check idle speed after warm up<br>**Idle speed: 750 ± 20 rpm (M/T) 770 ± 20 rpm (A/T, P range)**<br>[Test connector (Green: 1-pin) grounded] | Yes | Go to Next Step | | |
| | | No | Try to adjust idle speed | Yes | Idle-speed misadjustment |
| | | | | No | Check accelerator cable free play |
| | | | | | Check ISC valve (Stuck closed) |
| | | | | | Check throttle body |
| 5 | Substitute a well-known ECU<br>Check if condition improves | Yes | | | ECU malfunction |
| | | No | Check voltage at ECU (1C) terminal with SST<br>**Voltage: Approx. 10V (While cranking)** | Yes | Go to Next Step |
| | | | | No | Check starter interlock switch |
| | | | | | Check related wiring |
| 6 | | | | | Poor quality engine oil |

**Low idle speed (When A/C, P/S, E/L ON)**

| STEP | QUICK INSPECTION | | ACTION | POSSIBLE CAUSE |
|---|---|---|---|---|
| 1 | Check for malfunction code with SST<br>[IGN ON, Test connector (Green: 1-pin) grounded] | Yes | Check for cause by referring to check sequence | |
| | | No | Go to Next Step | |
| 2 | Disconnect ISC valve connector at idle<br>Check if the condition does not change | Yes | Go to Next Step | |
| | | No | | Check coolant level |
| | | | | Check engine oil |
| 3 | Check switches with SST<br>Idle switch<br>Neutral switch<br>Clutch switch<br>[IGN ON, Test connector (Green: 1-pin) grounded] | Yes | Go to Next Step | |
| | | No | Check for cause by referring to check sequence | |
| 4 | Check continuity between test connector (Green: 1-pin) and ground | | | Wiring short to ground |

**Note:**
**In case of low idle speed with A/C ON, if the problem cannot be solved by the above steps, it may be an A/C compressor malfunction.**

## TROUBLESHOOTING CHARTS—1990 B2600

**High idle speed after warm up**

| STEP | QUICK INSPECTION | | ACTION | | POSSIBLE CAUSE |
|---|---|---|---|---|---|
| 1 | Check for malfunction code with SST<br>[IGN ON, Test connector (Green: 1-pin) grounded] | Yes | Check for cause by referring to check sequence | | |
| | | No | Go to Next Step | | |
| 2 | Check ignition timing at idle after warm up<br>**Ignition timing: BTDC 5° ± 1° (G6) 6° ± 1° (F2)**<br>[Test connector (Green: 1-pin) grounded] | Yes | Go to Next Step | | |
| | | No | Adjust ignition timing | | |
| 3 | Check if throttle valve is fully closed when accelerator released | Yes | Go to Next Step | | |
| | | No | Check if throttle linkage is correctly in stalled and operates freely | | |
| 4 | Check idle speed after warm up<br>**Idle speed: 750 ± 20 rpm (M/T)**<br>**Idle speed: 770 ± 20 rpm (A/T, P range)**<br>[Test connector (Green: 1-pin) grounded] | Yes | | | Check ISC valve |
| | | No | Try to adjust idle speed | Yes | Idle speed misadjustment |
| | | | | No | Go to Next Step |
| 5 | Disconnect ISC valve connector at idle when engine is cold<br>Check if idle speed decreases during warm up | Yes | Go to Next Step | | |
| | | No | | | Check air valve |
| 6 | Disconnect water thermosensor connector and check if condition improves | Yes | Check water thermosensor connector condition as follows:<br>1. Shake connector and check if condition changes<br>2. Check condition of terminal (burned or damaged)<br>3. Connect a good terminal to harness side connector and check for looseness | Yes | Check water thermosensor |
| | | | | No | Poor contact of water thermosensor connector |
| | | No | Go to Next Step | | |
| 7 | | | | | ECU malfunction |

**Idle hunting or surging**

| STEP | QUICK INSPECTION | | ACTION | | | POSSIBLE CAUSE |
|---|---|---|---|---|---|---|
| 1 | (If trouble occurs only at warm condition)<br>Run engine at **2,000 rpm** for more than **20 seconds**<br>Check for malfunction code with SST<br>[IGN ON, Test connector (Green: 1-pin) grounded] | Yes | Check for cause by referring to check sequence | | | |
| | | No | Go to Next Step | | | |
| 2 | (If trouble occurs only at warm condition)<br>Check for flashing of SST monitor lamp after warm up<br>**Monitor lamp: Flashes more than 8 times 10 seconds at 2,000—3,000 rpm**<br>[Test connector (Green: 1-pin) not grounded] | Yes | Go to Next Step | | | |
| | | No | | | | Replace oxygen sensor |
| 3 | Check intake manifold vacuum at idle<br>**Vacuum: G6 520 ± 20 mmHg (20.5 ± 0.79 inHg) F2 530 ± 20 mmHg (20.9 ± 0.79 inHg)** | Yes | Go to Next Step | | | |
| | | No | Check for air leaks | F2–133 | Yes | Intake air system components damaged |
| | | | | | | Vacuum and air intake hoses loose or damaged |
| | | | | | | Bolts or nuts loose |
| | | | | | | Gaskets damaged |
| | | | | | No | Check throttle body |
| 4 | Pinch PCV hose<br>Check if condition improves | Yes | | | | Check PCV valve |
| | | No | Go to Next Step | | | |
| 5 | Check fuel line pressure [IGN ON, Test connector (Yellow: 2-pin) connected]<br>**Fuel line pressure: 265—314 kPa (2.7—3.2 kg/cm², 38—46 psi)** | Yes | Go to Next Step | | | |
| | | No | Check for fuel leaks | | | |
| | | | Substitute a good fuel filter and retest | | Yes | Replace fuel filter |
| | | | Check fuel pump maximum pressure<br>**Fuel pump maximum pressure: 441—588 kPa (4.5—6.0 kg/cm², 64—85 psi)** | F2–142 | Yes | Replace pressure regulator |
| | | | | | No | Replace fuel pump |
| 6 | | | | | | ECU malfunction |

## TROUBLESHOOTING CHARTS—1990 B2600

**Engine stall at idle (Always)**

| STEP | QUICK INSPECTION | | ACTION | | POSSIBLE CAUSE |
|---|---|---|---|---|---|
| 1 | Check for malfunction code with SST [IGN ON, Test connector (Green: 1-pin) grounded] | Yes | Check for cause by referring to the check sequence | | |
| | | No | Go to Next Step | | |
| 2 | Check if strong blue spark is visible at spark plug while cranking | Yes | Go to Next Step | | |
| | | No | Check ignition system | | Check spark plug |
| | | | | | Check high-tension lead |
| | | | | | Check distributor cap |
| 3 | Check fuel line pressure [IGN ON, Test connector (Yellow: 2-pin) connected] **Fuel line pressure: 265—314 kPa (2.7—3.2 kg/cm², 38—46 psi)** | Yes | Go to Next Step | | |
| | | No | Check for fuel leaks | | |
| | | | Check if fuel filter has been replaced according to maintenance schedule | Yes | Check fuel line for clogging |
| | | | | No | Replace fuel filter |
| | | | Check fuel pump maximum pressure **Fuel pump maximum pressure: 441—588 kPa (4.5—6.0 kg/cm², 64—85 psi)** | Yes | Replace pressure regulator |
| | | | | No | Replace fuel pump |
| 4 | Check if vacuum hoses and the air hoses are connected correctly | Yes | Go to Next Step | | |
| | | No | Connect correctly | | |
| 5 | | | | | Airflow sensor |
| 6 | | | | | ECU malfunction |

**Engine stall at idle (Only when engine is cold)**

| STEP | QUICK INSPECTION | | ACTION | POSSIBLE CAUSE |
|---|---|---|---|---|
| 1 | Check for malfunction code with SST [IGN ON, Test connector (Green: 1-pin) grounded] | Yes | Check for cause by referring to check sequence | |
| | | No | Go to Next Step | |
| 2 | | | | Check BAC valve (air valve) |
| 3 | | | | ECU malfunction |

**Engine stall at idle (Only when engine is warm)**

| STEP | QUICK INSPECTION | | ACTION | | POSSIBLE CAUSE |
|---|---|---|---|---|---|
| 1 | Check for malfunction code with SST [IGN ON, Test connector (Green: 1-pin) grounded] | Yes | Check for cause by referring to check sequence | | |
| | | No | Go to Next Step | | |
| 2 | Disconnect water thermosensor connector Check if condition improves | Yes | Check water thermosensor connector as follows: 1. Shake connector and check if condition changes 2. Check condition of terminal (burned or damaged) 3. Connect a good terminal to harness side connector and check for looseness | Yes | Check water thermosensor |
| | | | | No | Poor contact of water thermosensor connector |
| | | No | Go to Next Step | | |
| 3 | | | | | ECU malfunction |

**Engine stall at idle (When A/C, P/S, È/L O/N)**

| STEP | QUICK INSPECTION | | ACTION | POSSIBLE CAUSE |
|---|---|---|---|---|
| 1 | Check for malfunction code with SST [IGN ON, Test connector (Green: 1-pin) grounded] | Yes | Check for cause by referring to check sequence | |
| | | No | Go to Next Step | |
| 2 | Check switches with SST • Headlight switch • Blower switch [IGN ON, Test connector (Green: 1-pin) grounded] | Yes | Go to Next Step | |
| | | No | Check for cause by referring to check sequence | |
| 3 | Disconnect ISC valve connector at idle [Test connector (Green: 1-pin) grounded] Check if the condition does not change | Yes | Go to Next Step | |
| | | No | | Check ISC valve |
| | | | | Check engine oil |
| 4 | Check idle speed after warm up **Idle speed: 750 ± 20 rpm (M/T) 770 ± 20 rpm (A/T, P range)** [Test connector (Green: 1-pin) grounded] | Yes | Go to Next Step | |
| | | No | Adjust idle speed | |
| 5 | | | | ECU malfunction |

**Note:**
**Engine stalls at idle with A/C ON, If the trouble can not be fixed after checking above steps, It may be A/C compressor malfunction**

## TROUBLESHOOTING CHARTS—1990 B2600

**Engine stall during start up**

| STEP | QUICK INSPECTION | | ACTION | | POSSIBLE CAUSE |
|---|---|---|---|---|---|
| 1 | Check for malfunction code with SST [IGN ON, Test connector (Green: 1-pin) grounded] | Yes | Check for cause by referring to check sequence | | |
| | | No | Go to Next Step | | |
| 2 | Check idle speed after warm up **Idle speed: 750 ± 20 rpm (M/T) 770 ± 20 rpm (A/T, P range)** [Test connector (Green: 1-pin) grounded] | Yes | Go to Next Step | | |
| | | No | Adjust idle speed | | |
| 3 | Check for injector operating sound at idle | Yes | Go to Next Step | | |
| | | No | Check resistance at injector harness connector (EMINJ-01) **Terminal / Resistance:** (B/Y)—(LG/B), (B/Y)—(LG/R): 6—8Ω | Yes | Check wiring short or open |
| | | | | No | Check injector resistance |
| | | | | | Check wiring |
| 4 | Check ignition timing at idle after warm up **Ignition timing: BTDC 5° ± 1° (G6) 6° ± 1° (F2)** [Test connector (Green: 1-pin) grounded] | Yes | Go to Next Step | | |
| | | No | Adjust ignition timing | | |
| 5 | | | | | ECU malfunction |

**Engine stall on deceleration**

| STEP | QUICK INSPECTION | | ACTION | | POSSIBLE CAUSE |
|---|---|---|---|---|---|
| 1 | Check for malfunction code with SST [IG ON, Test connector (Green: 1-pin) grounded] | Yes | Check for cause by referring to check sequence | | |
| | | No | Go to Next Step | | |
| 2 | Check idle switch and stoplight switch with SST [IGN ON, Test connector (Green: 1-pin) grounded] | Yes | Go to Next Step | | |
| | | No | Check for cause by referring to check sequence | | |
| 3 | Check for flashing of monitor lamp after warm up **Monitor lamp: Flashes more than 8 times/10 seconds at 2,000—3,000 rpm** [The connector (Green: 1-pin) not grounded] | Yes | Go to Next Step | | |
| | | No | | | Replace oxygen sensor |
| 4 | Check for fuel cut operation during deceleration **Fuel cut: after warm up Above 1,600 rpm (G6) Above 1,900 rpm (F2)** | Yes | Go to Next Step | | |
| | | No | Check water thermosensor | Yes | Replace ECU |
| | | | | No | Replace water thermosensor |

**Engine stall on deceleration (Cont'd)**

| STEP | QUICK INSPECTION | | ACTION | POSSIBLE CAUSE |
|---|---|---|---|---|
| 5 | Check idle speed after warm up **Idle speed: 750 ± 20 rpm (M/T) 770 ± 20 rpm (A/T, P range)** [Test connector (Green: 1-pin) grounded] | Yes | Go to Next Step | |
| | | No | Adjust idle speed (If possible) | |
| 6 | Check ignition timing at idle after warm up **Ignition timing: BTDC 5° ± 1° (G6) 6° ± 1° (F2)** [Test connector (Green: 1-pin) not grounded] | Yes | Go to Next Step | |
| | | No | Adjust ignition timing | |
| 7 | | | | Check ISC valve |

**Hesitates/Stumbles on acceleration**

| QUICK | INSPECTION | | ACTION | | POSSIBLE CAUSE |
|---|---|---|---|---|---|
| 1 | Run engine at **2,000 rpm** for **20 seconds** and stop it Check for malfunction code with SST [IGN ON, Test connector (Green: 1-pin) grounded] | Yes | Check for cause by referring to check sequence | | |
| | | No | Go to Next Step | | |
| 2 | Check idle switch with SST [IGN ON, Test connector (Green: 1-pin) grounded] | Yes | Go to Next Step | | |
| | | No | Check for cause by referring to check sequence | | |
| 3 | Disconnect oxygen sensor connector Check if condition improves | Yes | | | Check oxygen Sensor |
| | | No | Go to Next Step | | |
| 4 | Check fuel line pressure while accelerating (Vacuum hose to pressure regulator disconnected) **Fuel line pressure: Keeps 265—314 kPa 2.7—3.2 kg/cm², 38—46 psi)** | Yes | Go to Next Step | | |
| | | No | Check if fuel filter has been replaced according to maintenance schedule | Yes | Check fuel line for clogging |
| | | | | No | Replace fuel filter |
| | | | | | Pressure regulator Replace |
| 5 | Check for air leaks with throttle valve open by listening for sucking noise | Yes | | | Intake air system components damaged |
| | | | | | Vacuum and intake air hoses loose or damaged |
| | | | | | Bolts or nuts loose |
| | | | | | Gaskets damaged |
| | | No | Go to Next Step | | |
| 6 | Substitute a well-known ECU Check if condition improves | Yes | | | ECU malfunction |
| | | No | | | Check airflow sensor |
| | | | | | Check throttle body |
| | | | | | Check spark plug |
| 7 | Check other systems | | | | Clutch slipping |

## TROUBLESHOOTING CHARTS—1990 B2600

**Engine stall at idle (Intermittent)**

| STEP | QUICK INSPECTION | | ACTION | POSSIBLE CAUSE |
|---|---|---|---|---|
| 1 | Shake connector of ignition coil, main relay and ECU while cranking<br>Check if the engine starts | Yes | There may be a poor contact at the connector. Repair or replace the wiring | |
| | | No | Go to troubleshooting "Engine stall at idle (Always)" | |

**Hesitates at steady speed**

| STEP | QUICK INSPECTION | | ACTION | | POSSIBLE CAUSE |
|---|---|---|---|---|---|
| 1 | Run engine at **2,000 rpm** for **20 seconds** and stop it<br>Check for malfunction code with SST<br>[IGN ON, Test connector (Green: 1-pin) grounded] | Yes | Check for cause by referring to check sequence | | |
| | | No | Go to Next Step | | |
| 2 | Disconnect oxygen sensor connector<br>Check if condition improves | Yes | | | Check oxygen sensor |
| | | No | Go to Next Step | | |
| 3 | Check for air leaks with throttle valve open by listening for sucking noise | Yes | Go to Next Step | | |
| | | No | | | Intake air system components damaged |
| | | | | | Vacuum and intake air hoses loose or damaged |
| | | | | | Nuts or bolts loose |
| | | | | | Gasket damaged |
| 4 | Check fuel line pressure while accelerating<br>(Vacuum hose to pressure regulator disconnected)<br>**Fuel line pressure: Keeps 265—314 kPa (2.7—3.2 kg/cm², 38—46 psi)** | Yes | Go to Next Step | | |
| | | No | Check if fuel filter has been replaced according to maintenance schedule | Yes | Check fuel line for clogging |
| | | | | No | Replace fuel filter |
| | | | | | Replace pressure regulator |
| 5 | Check condition of ignition coil and airflow meter connectors<br>(Burned or damaged) | Yes | | | Poor contact |
| | | No | Go to Next Step | | |
| 6 | Gradually open throttle valve<br>Check if engine speed increases smoothly | Yes | Go to Next Step | | |
| | | No | | | Check airflow sensor |
| | | | | | Check throttle valve |
| | | | | | Check throttle sensor |
| 7 | | | | | Check spark plug |
| 8 | Change fuel to specified grade<br>Check if condition improves | Yes | | | Poor fuel quality |
| | | No | Go to Next Step | | |
| 9 | | | | | ECU malfunction |

**Jerking on acceleration**

| STEP | QUICK INSPECTION | | ACTION | | POSSIBLE CAUSE |
|---|---|---|---|---|---|
| 1 | Run engine at **2,000 rpm** for **20 seconds** and stop it<br>Check for malfunction code with SST<br>[IGN ON, Test connector (Green: 1-pin) grounded] | Yes | Check for cause by referring to check sequence | | |
| | | No | Go to Next Step | | |
| 2 | Check idle switch with SST<br>[IGN ON, Test connector (Green: 1-pin) grounded] | Yes | Go to Next Step | | |
| | | No | Check for cause by referring to check sequence | | |
| 3 | Disconnect oxygen sensor connector<br>Check if condition improves | Yes | | | Check oxygen Sensor |
| | | No | Go to Next Step | | |
| 4 | Check fuel line pressure while accelerating<br>(Vacuum hose to pressure regulator disconnected)<br>**Fuel line pressure: Keeps 265—314 kPa 2.7—3.2 kg/cm², 38—46 psi)** | Yes | Go to Next Step | | |
| | | No | Check if fuel filter has been replaced according to maintenance schedule | Yes | Check fuel line for clogging |
| | | | | No | Replace fuel filter |
| | | | | | Pressure regulator Replace |
| 5 | Check for air leaks with throttle valve open by listening for sucking noise | Yes | | | Intake air system components damaged |
| | | | | | Vacuum and intake air hoses loose or damaged |
| | | | | | Bolts or nuts loose |
| | | | | | Gaskets damaged |
| | | No | Go to Next Step | | |
| 6 | Substitute a well-known ECU<br>Check if condition improves | Yes | | | ECU malfunction |
| | | No | | | Check airflow sensor |
| | | | | | Check throttle body |
| | | | | | Check spark plug |
| 7 | Check other systems | | | | Clutch slipping |

## TROUBLESHOOTING CHARTS—1990 B2600

**Knocking**

| STEP | QUICK INSPECTION | | ACTION | | POSSIBLE CAUSE |
|---|---|---|---|---|---|
| 1 | Check malfunction code with SST<br>[IGN ON, Test connector (Green: 1-pin) grounded] | Yes | Check for cause by referring to the check sequence | | |
| | | No | Go to Step 2 | | |
| 2 | Check ignition timing at idle after warm up<br>**Ignition timing: BTDC 5° ± 1° (M/T) 6° ± 1° (A/T, P range)**<br>[Test connector (Green: 1-pin) not grounded] | Yes | Go to Next Step | | |
| | | No | Adjust ignition timing | | |
| 3 | Disconnect water thermosensor connector<br>Check if condition improves | Yes | | | Check water thermosensor |
| | | No | Go to Next Step | | |
| 4 | Check vacuum routing | Yes | Go to Next Step | | |
| | | No | | | Vacuum hose |
| 5 | Observe fuel line pressure while accelerating from idle<br>**Fuel line pressure: Keeps 265—314 kPa (2.7—3.2 kg/cm², 38—46 psi)**<br>(Vacuum hose to pressure regulator disconnected) | Yes | Go to Next Step | | |
| | | No | Check fuel pump maximum pressure<br>**Fuel pump maximum pressure: 441—588 kPa (4.5—6.0 kg/cm², 64—85 psi)** | Yes | Replace fuel filter |
| | | | | No | Replace fuel pump |
| 6 | | | | | Check airflow sensor |
| 7 | | | | | Check spark plug |
| 8 | Change fuel to specified grade<br>Check if condition improves | Yes | | | Poor fuel quality |
| | | No | Go to Next Step | | |
| 9 | Check cooling system | | | | Thermostat |
| | | | | | Radiator |
| 10 | | | | | ECU malfunction |

**Poor acceleration**

| STEP | QUICK INSPECTION | | ACTION | | POSSIBLE CAUSE |
|---|---|---|---|---|---|
| 1 | Check for malfunction code with SST<br>[IGN ON, Test connector (Green: 1-pin) grounded] | Yes | Check for cause by referring to check sequence | | |
| | | No | Go to Next Step | | |
| 2 | Check idle switch with SST<br>[IGN ON, Test connector (Green: 1-pin) grounded] | Yes | Go to Next Step | | |
| | | No | Check for cause by referring to check sequence | | |
| 3 | Disconnect high—tension lead of each cylinder at idle. Check if engine condition changes<br>[ISC valve connector disconnected] | Yes | Go to Next Step | | |
| | | No | Check ignition system | Yes | Replace injector |
| | | | | No | Check spark plug |
| | | | | | Check high-tension |
| | | | | | Check distributor cup |
| 4 | Check ignition at idle after warm up<br>**Ignition timing: BTDC 5° ± 1° (G6) 6° ± 1° (F2)**<br>[Test connector (Green: 1-pin) grounded] | Yes | Go to Next Step | | |
| | | No | Adjust ignition timing | | |
| 5 | Check for air leaks by listening for sucking noise | Yes | | | Intake air system components damaged |
| | | | | | Vacuum and air intake hoses loose or damaged |
| | | | | | Nuts or bolts loose |
| | | | | | Gasket damaged |
| | | No | Go to Next Step | | |
| 6 | Observe fuel line pressure while accelerating from idle<br>**Fuel line pressure: Keeps 265—314 kPa (2.7—3.2 kg/cm², 38—46 psi)**<br>[Vacuum hose to pressure regulator disconnected] | Yes | Go to Next Step | | |
| | | No | Check if fuel filter has been replaced according to maintenance schedule | No | Replace pressure regulator |
| | | | | Yes | Replace fuel filter |
| 7 | Gradually depress accelerator from idle Check if engine speed increases smoothly | Yes | Go to Next Step | | |
| | | No | Check accelerator cable free play | Yes | Check airflow sensor |
| | | | | | Check throttle body |
| | | | | No | Adjust |
| 8 | Check fuel to specified grade<br>Check if condition improves | Yes | | | Poor fuel quality |
| | | No | Go to Next Step | | |
| 9 | Substitute a well-known ECU<br>Check if condition improves | Yes | | | ECU malfunction |
| | | No | Go to Next Step | | |
| 10 | Check other systems | | | | Clutch slipping |
| | | | | | Transmission (M/T) |
| | | | | | Brake dragging |
| | | | | | Belt tension |

### TROUBLESHOOTING CHARTS—1990 B2600

**Lack of power**

| STEP | QUICK INSPECTION | | ACTION | | POSSIBLE CAUSE |
|---|---|---|---|---|---|
| 1 | Check for malfunction code and (only high-altitude) with SST [IGN ON, Test connector (Green:1-pin) grounded] | Yes | Check for cause by referring to check sequence | | |
| | | No | Go to Step 2 (High-altitude)<br>Go to Step 3 (Others) | | |
| 2 | Check ignition timing at idle after warm up<br>**Ignition timing:**<br>**BTDC 5° ± 1° (G6)**<br>**6° ± 1° (F2)**<br>[Test connector (Green: 1-pin) grounded] | Yes | Go to Next Step | | |
| | | No | Adjust ignition timing | | |
| 3 | Disconnect ISC valve connector and the high-tension lead of each cylinder<br>Check if condition changes | Yes | Go to Next Step | | |
| | | No | Check ignition system | Yes | Replace injector (If step 4 OK) |
| | | | | No | Check high-tension lead |
| | | | | | Check distributor cap |
| | | | | | Check spark plug |
| 4 | Check for injector operating sound at idle | Yes | Go to Next Step | | |
| | | No | Chcek resistance at injector harness connector (EMINJ-01)<br>Terminals: (B/Y)—(LG/B), (B/Y)—(LG/R); Resistance: 6—8Ω | Yes | Check wiring short or open |
| | | | | No | Check injector resistance |
| | | | | | Check wiring short or open |
| 5 | Check air cleaner element for clogging | Yes | Go to Next Step | | |
| | | No | Clean air cleaner element | | |
| 6 | Check for air leaks by listening for sucking noises<br>• At idle<br>• When throttle valve is open | Yes | | | Intake air system |
| | | | | | Components damaged |
| | | | | | Vacuum and air intake hoses loose or damaged |
| | | | | | Nuts or bolts loose |
| | | | | | Gasket damaged |

**Lack of power (Cont'd)**

| STEP | QUICK INSPECTION | | ACTION | | POSSIBLE CAUSE |
|---|---|---|---|---|---|
| 7 | Check fuel line pressure [IGN ON, Test connector (Yellow: 2-pin) connected]<br>**Fuel line pressure:**<br>**265—314 kPa (2.7—3.2 kg/cm², 38—46 psi)** | Yes | Go to Next Step | | |
| | | No | Check for fuel leakage | | |
| | | | Substitute a good fuel filter and retest | Yes | Replace fuel filter |
| | | | Check fuel pump maximum pressure<br>**Fuel pump maximum pressure:**<br>**441—588 kPa (4.5—6.0 kg/cm², 64—85 psi)** | Yes | Replace pressure regulator |
| | | | | No | Replace fuel pump |
| 8 | Check fuel line pressure at idle<br>**Fuel line pressure:**<br>**216—264 kPa (2.2—2.7 kg/cm², 31—38 psi)** | Yes | Go to Next Step | | |
| | | No | | | Replace pressure regulator |
| 9 | Check if fuel line pressure drops while accelerating (Vacuum hose disconnected) | Yes | Check if fuel filter has been replaced according to maintenance schedule | Yes | Check fuel line for clogging |
| | | | | No | Replace fuel filter |
| | | No | Go to Next Step | | |
| 10 | Check exhaust system for damage | Yes | Go to Next Step | | |
| | | No | Repair or replace | | |
| 11 | Check A/C, P/S and alternator belts tensions | Yes | Go to Next Step | | |
| | | No | Adjust belt tension | | |
| 12 | Check if accelerator can be depressed fully | Yes | Go to Next Step | | |
| | | No | Check accelerator cable | Yes | Throttle body |
| | | | | No | Accelerator cable |
| 13 | Substitute a well-known ECU<br>Check if condition improves | Yes | | | ECU malfunction |
| | | No | | | Check air flow sensor |
| | | | | | Check throttle sensor |
| | | | | | Go to Next Step |
| 14 | Substitute a specified fuel<br>Check if condition improves | Yes | | | Poor fuel quality |
| | | No | Go to Next Step | | |
| 15 | Check other systems | | | | Brake |
| | | | | | Clutch |
| | | | | | Engine |

### TROUBLESHOOTING CHARTS—1990 B2600

**Bucking at high speed**

| STEP | QUICK INSPECTION | | ACTION | | POSSIBLE CAUSE |
|---|---|---|---|---|---|
| 1 | Run engine at **2,000 rpm** for more than **20 seconds**<br>Check for malfunction code with SST<br>[IGN ON, Test connector (Green: 1-pin) grounded] | Yes | Check for cause by referring to check sequence | | |
| | | No | Go to Next Step | | |
| 2 | Disconnect oxygen sensor connector<br>Check if condition improves | Yes | | | Check oxygen sensor |
| | | No | Go to Next Step | | |
| 3 | Observe fuel line pressure while accelerating from idle<br>**Fuel line pressure:**<br>**Keeps 265—314 kPa (2.7—3.2 kg/cm², 38—46 psi)**<br>[Vacuum hose to pressure regulator disconnected] | Yes | Go to Next Step | | |
| | | No | Check if fuel filter has been replaced according to maintenance schedule | Yes | Check fuel line for clogging |
| | | | | No | Replace fuel filter |
| | | | | | Replace pressure regulator |
| 4 | Check for air leaks by listening sucking noise | Yes | Go to Next Step | | |
| | | No | | | Intake air system components damaged |
| | | | | | Vacuum and air intake hoses loose or damaged |
| | | | | | Nuts or bolts loose |
| | | | | | Gasket damaged |
| 5 | Check ignition timing at idle after warm up<br>**Ignition timing:**<br>**BTDC 5° ± 1° (G6)**<br>**6° ± 1° (F2)**<br>[Test connector (Green: 1-pin) grounded] | Yes | Go to Next Step | | |
| | | No | Adjust ignition timing | | |
| 6 | Gradually open throttle valve from idle check if engine speed increases smoothly | Yes | Go to Next Step | | |
| | | No | | | Check airflow sensor |
| 7 | | | | | Check spark plug |
| 8 | | | | | ECU malfunction |

**Bucking on deceleration**

| STEP | QUICK INSPECTION | | ACTION | POSSIBLE CAUSE |
|---|---|---|---|---|
| 1 | Check for malfunction code with SST [IGN ON, Test connector (Green: 1-pin) grounded] | Yes | Check for cause by referring to the check sequence | |
| | | No | Go to Next Step | |
| 2 | Check switches with SST [IGN ON, Test connector (Green: 1-pin) grounded]<br>• Idle switch<br>• Stoplight switch | Yes | Go to Next Step | |
| | | No | Check for cause by referring to the check sequence | |
| 3 | Substitute a well-known ECU<br>Check if condition improves | Yes | | ECU malfunction |
| | | No | | Check throttle sensor |
| | | | | Go to Next Step |
| 4 | | | | Check spark plug |
| 5 | | | | Check clutch slipping |
| 6 | | | | Check compression between cylinders |

## TROUBLESHOOTING CHARTS—1990 B2600

**Poor fuel economy**

| STEP | QUICK INSPECTION | ACTION | | | POSSIBLE CAUSE |
|---|---|---|---|---|---|
| 1 | Run the engine at **2,000 rpm** for more than **20 seconds** after warm up and stop it Check for malfunction code with SST [IGN ON, Test connector (Green: 1-pin ) grounded] | Yes | Check for cause by referring to check sequence | | |
| | | No | Go to Next Step | | |
| 2 | Check idle switch with SST [IGN ON, Test connector (Green: 1-pin) grounded] | Yes | Go to Next Step | | |
| | | No | Check for cause by referring to check sequence | | |
| 3 | Check for flashing of monitor lamp after warm up<br>**Monitor lamp: Flashes more than 8 times /10 seconds at 2,000—3,000 rpm**<br>[Test connector (Green: 1-pin) not grounded] | Yes | Go to Next Step | | |
| | | No | | | Replace oxygen sensor |
| 4 | Check fuel line pressure at idle<br>**Fuel line pressure: 196—255 kPa (2.0—2.6 kg/cm², 28—37 psi)** | Yes | Go to Next Step | | |
| | | No | Check vacuum line to pressure regulator for clogging or air leakage | Yes | Vacuum line clogging or damaged |
| | | | | No | Check solenoid valve (PRC) |
| | | | | | ECU malfunction (Check (2T) terminal voltage) |
| | | | | | Replace pressure regulator |
| 5 | Check for fuel cut operation during deceleration<br>**Fuel cut: after warm up Above 1,600 rpm (G6) Above 1,900 rpm (F2)** | Yes | Go to Next Step | | |
| | | No | Check water thermosensor | Yes | Replace ECU |
| | | | | No | Replace water thermosensor |
| 6 | Check ignition timing at idle after warm up<br>**Ignition timing: BTDC 5° ± 1° (G6) 6° ± 1° (F2)**<br>[Test connector (Green: 1-pin) grounded] | Yes | Go to Next Step | | |
| | | No | Adjust ignition timing | | |
| 7 | Check other systems | | | | Clutch slipping |
| | | | | | Brake |
| | | | | | Tire air pressure |

**High oil consumption/White exhaust smoke**

| STEP | QUICK INSPECTION | ACTION | | | POSSIBLE CAUSE |
|---|---|---|---|---|---|
| 1 | Check for oil leak from engine | Yes | Repair or replace | | |
| | | No | Go to Next Step | | |
| 2 | Disconnect PCV valve from engine Check if vacuum is felt at idle | Yes | Go to Next Step | | |
| | | No | Check PCV valve | Yes | PCV hose clogging |
| | | | | No | Replace PCV valve |
| 3 | Check that ventilation hose is installed correctly | Yes | Go to Next Step | | |
| | | No | Install ventilation hose correctly | | |
| 4 | Possible malfunction of engine Check for cause by referring to the check sequence of Section B2 | | | | |

**Afterburn on deceleration**

| STEP | QUICK INSPECTION | ACTION | | | POSSIBLE CAUSE |
|---|---|---|---|---|---|
| 1 | Check malfunction code with SST [IGN ON, Test connector (Green: 1-pin) grounded] | Yes | Check for cause by referring to the check sequence | | |
| | | No | Go to Next Step | | |
| 2 | Check idle switch with SST [IGN ON, Test connector (Green: 1-pin) grounded] | Yes | Go to Next Step | | |
| | | No | Check for cause by referring to the check sequence | | |
| 3 | Check ignition timing at idle after warm up<br>**Ignition timing: BTDC 5° ± 1° (G6) 6° ± 1° (F2)**<br>[Test connector (Green: 1-pin) grounded] | Yes | Go to Next Step | | |
| | | No | Adjust ignition timing | | |
| 4 | Check air cleaner element for clogging | Yes | Go to Next Step | | |
| | | No | Clean air cleaner element | | |
| 5 | Check fuel cut operation during deceleration<br>**Fuel cut: after warm up Above 1,600 rpm (G6) Above 1,900 rpm (F2)** | Yes | Go to Next Step | | |
| | | No | Check water thermosensor | Yes | ECU malfunction Check (2Q) terminal voltage |
| | | | | No | Replace water thermosensor |
| 6 | Run engine at idle and stop it (IG OFF) Observe fuel pressure for **5 minutes**<br>**Fuel pressure: More than 147 kPa (1.5 kg/cm², 21 psi)** | Yes | Go to Next Step | | |
| | | No | Check fuel pump for pressure drop | No | Replace fuel pump |
| | | | Check pressure regulator for pressure drop | Yes | Check injector fuel leakage |
| | | | | No | Replace pressure regulator |
| 7 | | | | | Check compression |
| | | | | | Check valve timing |

**Rotten egg smell**

| STEP | QUICK INSPECTION | ACTION | POSSIBLE CAUSE |
|---|---|---|---|
| 1 | Change fuel to specified grade Check if condition improves | | Poor fuel quality |

## TROUBLESHOOTING CHARTS—1990 B2600

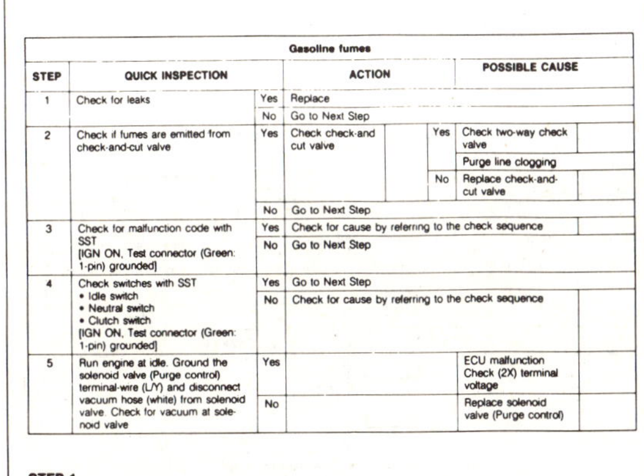

**Gasoline fumes**

| STEP | QUICK INSPECTION | ACTION | | | POSSIBLE CAUSE |
|---|---|---|---|---|---|
| 1 | Check for leaks | Yes | Replace | | |
| | | No | Go to Next Step | | |
| 2 | Check if fumes are emitted from check-and-cut valve | Yes | Check check-and cut valve | Yes | Check two-way check valve |
| | | | | | Purge line clogging |
| | | | | No | Replace check-and-cut valve |
| | | No | Go to Next Step | | |
| 3 | Check for malfunction code with SST [IGN ON, Test connector (Green: 1-pin) grounded] | Yes | Check for cause by referring to the check sequence | | |
| | | No | Go to Next Step | | |
| 4 | Check switches with SST<br>• Idle switch<br>• Neutral switch<br>• Clutch switch<br>[IGN ON, Test connector (Green: 1-pin) grounded] | Yes | Go to Next Step | | |
| | | No | Check for cause by referring to the check sequence | | |
| 5 | Run engine at idle. Ground the solenoid valve (Purge control) terminal-wire (L/Y) and disconnect vacuum hose (white) from solenoid valve. Check for vacuum at solenoid valve | Yes | | | ECU malfunction Check (2X) terminal voltage |
| | | No | | | Replace solenoid valve (Purge control) |

**STEP 1**

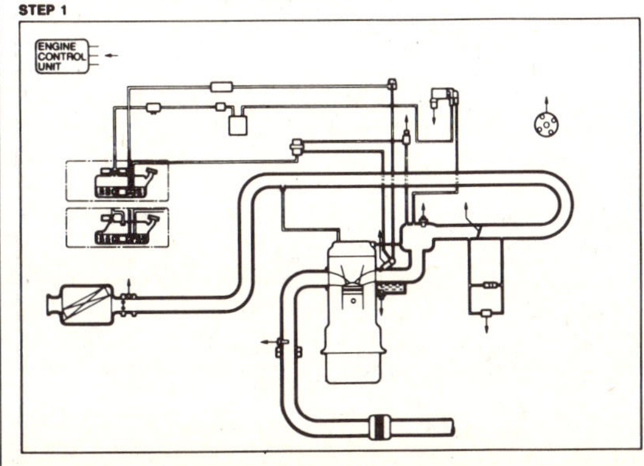


**MIL always ON**

| STEP | QUICK INSPECTION | ACTION | | | POSSIBLE CAUSE |
|---|---|---|---|---|---|
| 1 | (California) Check for malfunction code with SST [IGN ON, Test connector (Green: 1-pin) grounded] | "88" Replace ECU | | | |
| | | "00" Wiring between ECU (1E) terminal and MIL short to ground | | | |
| 2 | (Federal and Canada) Check if emission system parts replacement time has come<br>**Emission system parts replacement schedule: Every 60,000 and 80,000 miles (Federal) or 90,000 and 130,000 km (Canada)** | Yes | Check if MIL has been reset by exchanging MIL set connector | Yes | Replace mileage sensor |
| | | | | No | Reset the MIL |
| | | No | | | Replace mileage sensor |

**MIL never ON**

| STEP | QUICK INSPECTION | ACTION | | | POSSIBLE CAUSE |
|---|---|---|---|---|---|
| 1 | Check if other indicator lamps illuminate | Yes | Go to Next Step | | |
| | | No | Check power supply circuit to combination meter | | |
| 2 | Check bulb of the MIL | Yes | (California only) Ground ECU (1E) terminal Check if MIL illuminates | Yes | Replace ECU |
| | | | | No | Wiring between ECU and MIL open |
| | | | | | (Federal and Canada) MIL set connector loose or disconnected |
| | | | | | (Federal and Canada) Replace mileage sensor |
| | | No | Replace | | |

**A/C does not work**

| STEP | QUICK INSPECTION | ACTION | | | POSSIBLE CAUSE |
|---|---|---|---|---|---|
| 1 | Check if condenser fan operates when grounding A/C relay terminal-wire (L/W) (IGN ON) | Yes | Check voltage at ECU (1Q) terminal with SST<br>Voltage at idle after warm up: 0V (A/C and blower switches ON) | Yes | ECU malfunction (Check (1J) terminal voltage) |
| | | | | | Wiring between ECU (1J) and A/C relay open |
| | | | | No | A/C system malfunction |
| | | No | Check A/C system | | |

## MALFUNCTION INDICATOR LIGHT (MIL) OPERATION 1990 B2600

**PRINCIPLE OF CODE CYCLE**
Malfunction codes are determined as shown below

**1. Code cycle break**
The time between malfunction code cycles is 4.0 sec (the time the MIL (California only) and the buzzer are off).

1 CYCLE
MALFUNCTION CODE
CYCLE BREAK
ON
OFF
4.0 SEC.

**2. Second digit of malfunction code (ones position)**
The digit in the ones position of the malfunction code represents the number of times the MIL (California only) and the buzzer are on 0.4 sec during one cycle.

1 CYCLE MALFUNCTION CODE
MALFUNCTION CODE: 03
1 2 3 CYCLE BREAK
ON
OFF
0.4 SEC. 0.4 SEC. 4.0 SEC.

**3. First digit of malfunction code (tens position)**
The digit in the tens position of the malfunction code represents the number of times the MIL (California only) and the buzzer are on 1.2 sec during one cycle.

It should also be noted that the light goes off for 1.6 sec. between the long and short pulses of the MIL (California only) and the buzzer.

1 cycle MALFUNCTION CODE
MALFUNCTION CODE: 22
1 2 1 2 CYCLE BREAK
0.4
1.2 SEC. SEC. 1.6 SEC. 4.0 SEC.
FIRST DIGIT SECOND DIGIT

## TROUBLE CODE IDENTIFICATION CHART 1990 B2600

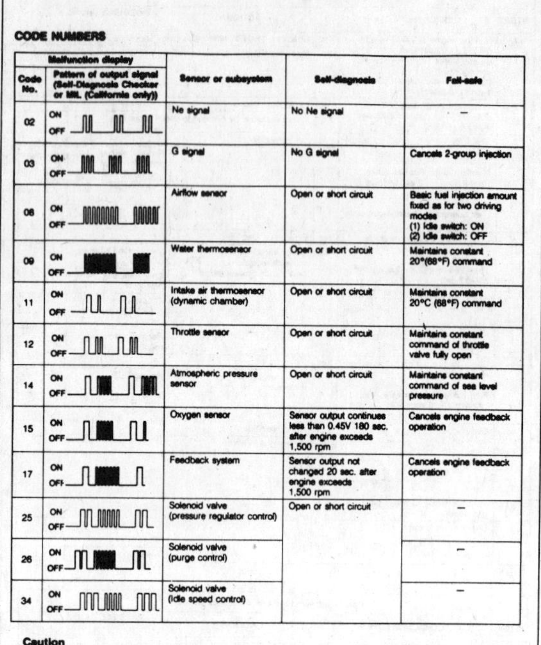

**CODE NUMBERS**

| Malfunction display: Code No. | Malfunction display: Pattern of output signal (Self-Diagnosis Checker or MIL (California only)) | Sensor or subsystem | Self-diagnosis | Fail-safe |
|---|---|---|---|---|
| 02 | ON OFF | Ne signal | No Ne signal | — |
| 03 | ON OFF | G signal | No G signal | Cancels 2-group injection |
| 08 | ON OFF | Airflow sensor | Open or short circuit | Basic fuel injection amount fixed as for two driving modes (1) Idle switch: ON (2) Idle switch: OFF |
| 09 | ON OFF | Water thermosensor | Open or short circuit | Maintains constant 20°(68°F) command |
| 11 | ON OFF | Intake air thermosensor (dynamic chamber) | Open or short circuit | Maintains constant 20°C (68°F) command |
| 12 | ON OFF | Throttle sensor | Open or short circuit | Maintains constant command of throttle valve fully open |
| 14 | ON OFF | Atmospheric pressure sensor | Open or short circuit | Maintains constant command of sea level pressure |
| 15 | ON OFF | Oxygen sensor | Sensor output continues less than 0.45V 180 sec. after engine exceeds 1,500 rpm | Cancels engine feedback operation |
| 17 | ON OFF | Feedback system | Sensor output not changed 20 sec. after engine exceeds 1,500 rpm | Cancels engine feedback operation |
| 25 | ON OFF | Solenoid valve (pressure regulator control) | Open or short circuit | — |
| 26 | ON OFF | Solenoid valve (purge control) | | — |
| 34 | ON OFF | Solenoid valve (idle speed control) | | — |

**Caution**
**a) If there is more than one failure present, the lowest number malfunction code is displayed first, the remaining codes are displayed in order.**
**b) After repairing all failures, turn off the ignition switch, disconnect the negative battery cable, and depress the brake pedal for 5 sec to erase the memory of a malfunction code from the engine control unit.**

## TROUBLE CODE DIAGNOSTIC CHART—1990 B2600

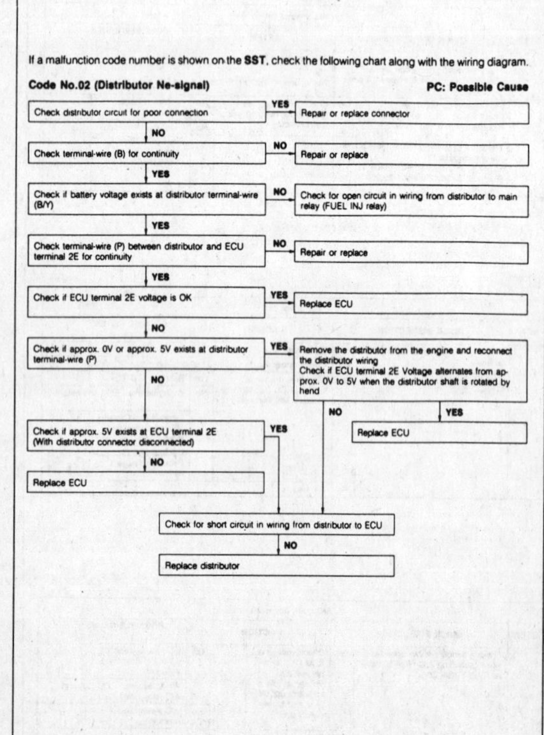


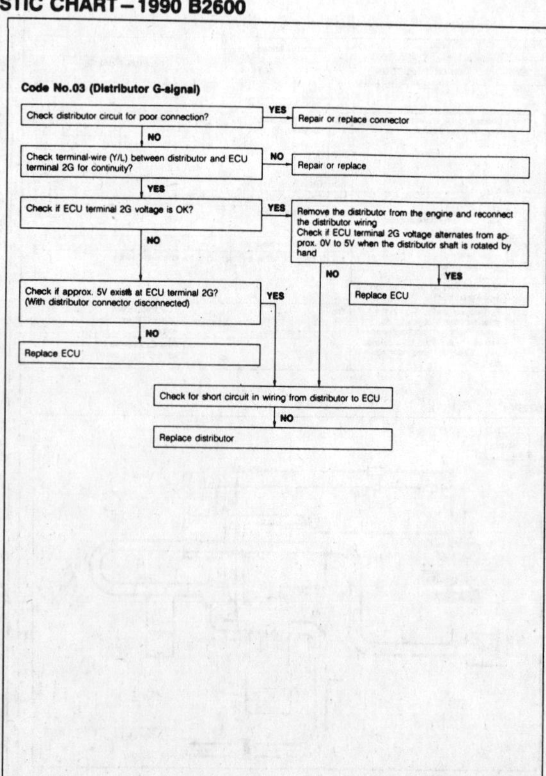

## TROUBLE CODE DIAGNOSTIC CHART—1990 B2600

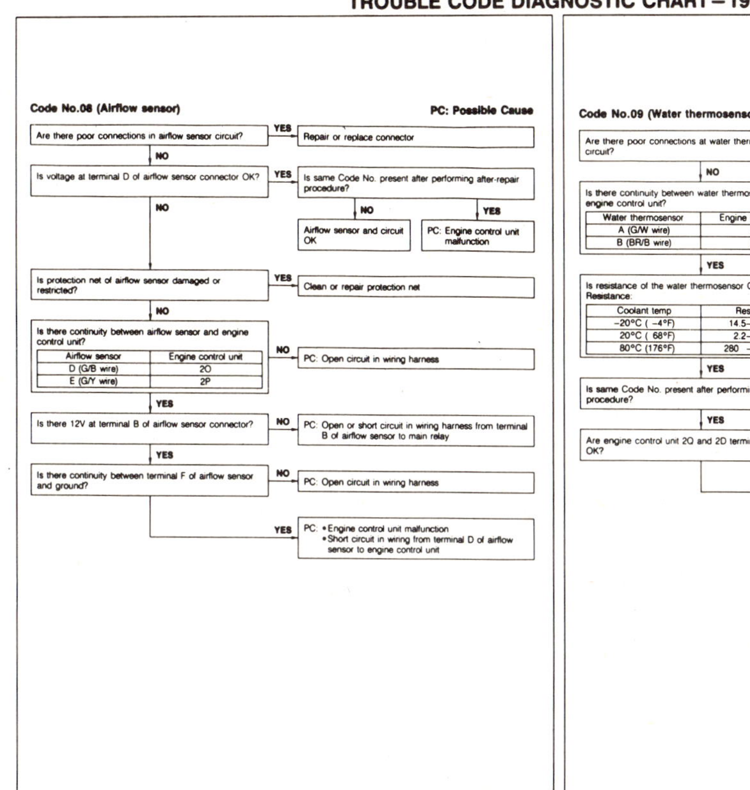


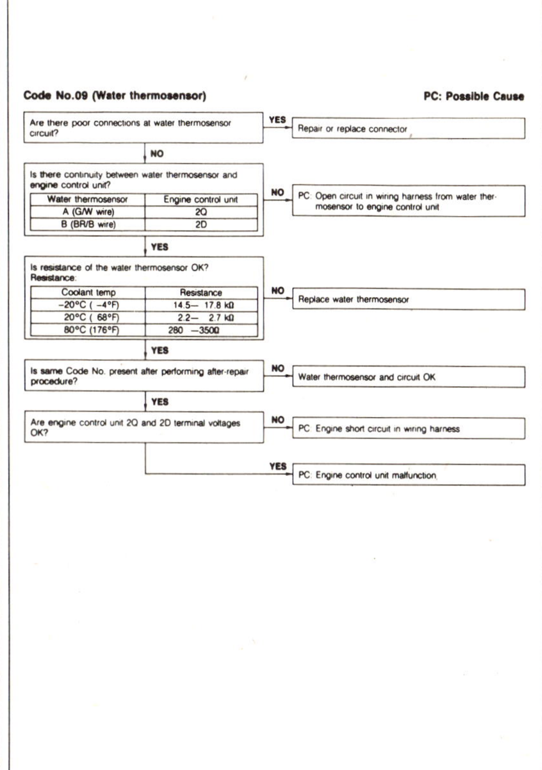


## TROUBLE CODE DIAGNOSTIC CHART—1990 B2600

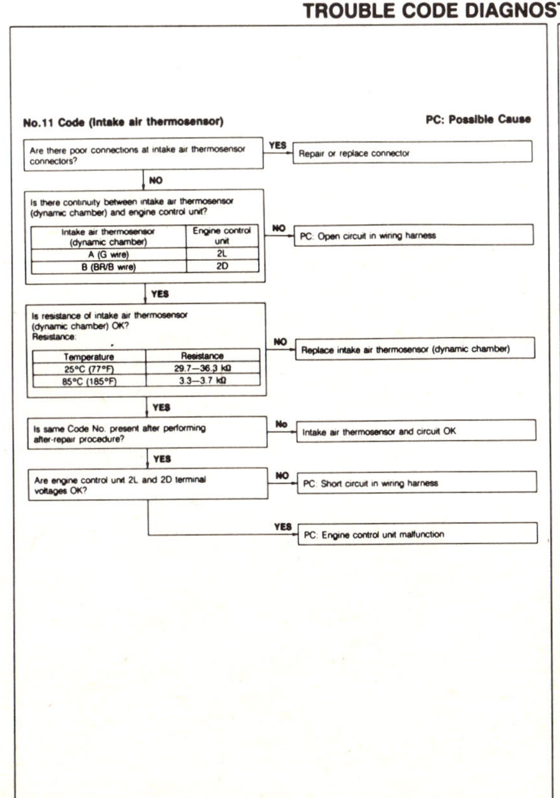


**Code No.12 (Throttle sensor)** — **PC: Possible cause**

| | | |
|---|---|---|
| Are there poor connections in throttle sensor circuit? | YES | Repair or replace connector terminal |
| | NO | (continue) |
| Is there 4.5—5.5 V at A terminal of throttle sensor connector? | NO | PC: •Open or short circuit in wiring harness from A terminal of throttle sensor to 2K terminal of engine control unit<br>•Engine control unit malfunction |
| | YES | (continue) |
| Is C terminal of throttle sensor connector grounded? | NO | PC: Open circuit in wiring harness from C terminal of throttle sensor to ground |
| | YES | (continue) |
| Is throttle sensor adjusted correctly? | NO | Adjust throttle sensor |
| | YES | (continue) |
| Is same Code No. present after performing after-repair procedure? | NO | Throttle sensor and circuit OK |
| | YES | (continue) |
| Is engine control unit 2M terminal voltage OK? | NO | PC: Open or short circuit in wiring harness from B terminal of throttle sensor to 2M terminal of engine control unit |
| | YES | PC: Engine control unit malfunction |

**Code No.15 (Oxygen sensor)** — **PC: Possible Cause**

**Note**
**When Codes No.15 and 17 are present at the same time, first perform the checking procedure for Code No.17.**

| | | |
|---|---|---|
| Are there poor connections in oxygen sensor circuit? | YES | Repair or replace connector |
| | NO | (continue) |
| Is oxygen sensor output voltage OK? | NO | Replace oxygen sensor |
| | YES | (continue) |
| Is sensitivity of oxygen sensor OK? | NO | PC: •Oxygen sensor malfunction<br>•Engine control unit malfunction |
| | YES | (continue) |
| Is same Code No. present after performing after-repair procedure? | NO | Oxygen sensor and circuit OK |
| | YES | (continue) |
| Is engine control unit 2N terminal voltage OK? | NO | PC: Open or short circuit in wiring harness between oxygen sensor and engine control unit 2N terminal |
| | YES | PC: Engine control unit malfunction |

## TROUBLE CODE DIAGNOSTIC CHART—1990 B2600

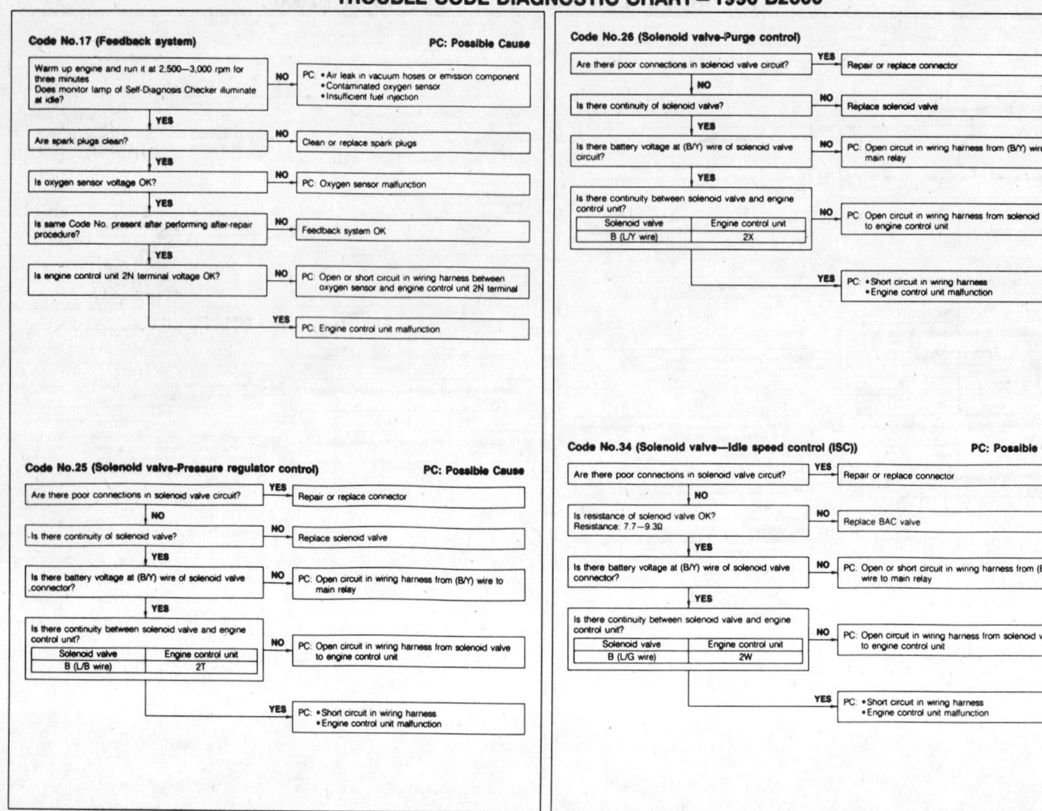


## CONTROL UNIT PIN LOCATION AND VOLTAGE TEST—1990 B2600

**Terminal voltage**

| Terminal | Input | Output | Connection to | Test condition | Voltage | Remarks |
|---|---|---|---|---|---|---|
| 1A | — | — | Battery | Constant | Approx. 12V | For backup |
| 1B | O | | Main relay | Ignition switch OFF | Approx. 0V | |
| | | | | Ignition switch ON | Approx. 12V | |
| | | | | During burn-off (airflow sensor) | | |
| 1C | O | | Ignition switch (Start position) | While cranking | Approx. 10V | |
| | | | | Ignition switch ON | Approx. 0V | |
| 1D | | O | Self-Diagnosis Checker (Monitor lamp) | Test connector (Green: 1-pin) grounded For 3 seconds after ignition switch OFF→ON (Lamp illuminates) | Approx. 5V | With Self-Diagnosis Checker |
| | | | | After 3 seconds (Lamp does not illuminated) | Approx. 12V | |
| | | | | Test connector (Green: 1-pin) not grounded at idle. Monitor lamp ON | Approx. 5V | |
| | | | | Test connector (Green: 1-pin) not grounded at idle. Monitor lamp OFF | Approx. 12V | |
| 1E | | O | Malfunction indicator lamp (California only) | For 3 seconds after ignition switch OFF→ON (Lamp illuminates) | Below 2.5V | Test connector (Green: 1-pin) grounded |
| | | | | After 3 seconds (Lamp dose not illuminates) | Approx. 12V | |
| | | | | Lamp illuminates | Below 2.5V | |
| | | | | Lamp not illuminate | Approx. 12V | |
| 1F | | O | Self-Diagnosis checker (Code number) | For 3 seconds after ignition switch OFF→ON (Buzzer sounds) | Below 2.5V | • With Self-Diagnosis Checker • Test connector (Green: 1-pin) grounded |
| | | | | After 3 seconds (Buzzer does not sounded) | Approx. 12V | |
| | | | | Buzzer sounds | Below 2.5V | |
| | | | | Buzzer not sounded | Approx. 12V | |
| 1G | | O | Main relay | Ignition switch OFF | Approx. 12V | |
| | | | | During burn-off (airflow sensor) Ignition switch ON | Approx. 0V | |
| 1H | | O | Circuit opening relay | Ignition switch ON | Approx. 12V | |
| | | | | During cranking or at idle | Below 2.5V | |
| 1I | O | | Ignition switch (ON position) | Ignition switch OFF | 0V | |
| | | | | Ignition switch ON | Approx. 12V | |
| 1J | | O | A/C relay | Ignition switch ON | Approx. 12V | Blower motor: ON |
| | | | | For 10 seconds After fully depressing accelerator pedal with A/C switch ON (A/C does not operate) (in-gear, ignition switch ON) | Approx. 12V | |
| | | | | After 10 seconds | Below 2.5V | |
| | | | | For 5 seconds after cranking with A/C switch ON (A/C does not operate) | Approx. 12V | |
| | | | | After 5 seconds (A/C operates) | Below 2.5V | |
| | | | | A/C switch ON at idle | Below 2.5V | |
| | | | | A/C switch OFF at idle | Approx. 12V | |
| 1K | O | | Test connector | Test connector (Green: 1-pin) not grounded | Approx. 12V | Ignition switch ON |
| | | | | Test connector (Green: 1-pin) grounded | 0V | |
| 1L | O | | Ground (M/T) | Ignition switch ON | 0V | |
| | | | Open (A/T) | Ignition switch ON | Approx. 12V | |
| 1M | O | | Speed sensor (A/T) | Ignition switch ON | 0 or 4.5V | |
| | | | | Idle | Approx 4.5V | |
| 1N | O | | Idle switch | Accelerator pedal released | 0V | Ignition switch ON |
| | | | | Accelerator pedal depressed | Approx. 12V | |
| 1O | O | | Stoplight switch | Brake pedal released | 0V | Ignition switch ON |
| | | | | Brake pedal depressed | Approx. 12V | |
| 1P | O | | P/S pressure switch | Ignition switch ON | Approx. 12V | |
| | | | | P/S ON (at idle) | 0V | |
| | | | | P/S OFF (at idle) | Approx. 12V | |
| 1Q | O | | A/C switch | A/C switch ON (Ignition switch ON) | Below 2.5V | Blower motor: ON |
| | | | | A/C switch OFF (Ignition switch ON) | Approx. 12V | |

**Terminal voltage**

| Terminal | Input | Output | Connection to | Test condition | Voltage | Remarks |
|---|---|---|---|---|---|---|
| 1R | O | | Ground (EC-AT) | Ignition switch ON | 0V | For G6 |
| | | | Open (M/T, HAT) | Ignition switch ON | Approx. 12V | |
| 1S | O | | Blower switch | Blower OFF | Approx 12V | Ignition switch ON |
| | | | | Blower ON | Below 1.5V | |
| 1T | — | — | — | — | — | — |
| 1U | O | | Headlight switch | Headlight ON | Approx. 12V | |
| | | | | Headlight OFF | Below 1.5V | |
| 1V | O | | Neutral or clutch switch (Inhibitor switch) | Neutral or clutch pedal depressed (P or N rahges) | 0V | Ignition switch ON |
| | | | | Other condition | Approx. 12V | |
| 2A | — | — | Ground (E01) | Constant | 0V | |
| 2B | — | — | Ground (E02) | Constant | 0V | |
| 2C | — | — | Ground (E1) | Constant | 0V | |
| 2D | — | — | Ground (E2) | Constant | 0V | |
| 2E | | O | Distributor | Ignition switch ON | 0 or 5V | Ne-Signal |
| | | | | Idle | 2V | |
| 2F | | O | Igniter | Ignition switch ON | 0 or 5V | Ignition-timing signal |
| | | | | Idle | Approx. 0.5V | |
| 2G | O | | Distributor | Ignition switch ON | 0 or 5V | G-Signal |
| | | | | Idle | Approx. 1.2V | |
| 2H | | O | Airflow sensor (Burn-off) | Just after ignition switch OFF | 0V | Burn-off functions momentarily |
| | | | | Burn off (2-5 seconds after ignition switch OFF) | 8—12V | |
| 2I | — | — | — | — | — | — |
| 2J | — | — | — | — | — | — |
| 2K | | O | Vref | Ignition switch ON | 4.5—5.5V | |
| 2L | O | | Intake air thermosensor (Dynamic chamber) | At 20°C (68°F) | Approx. 2.5V | |
| 2M | O | | Throttle sensor | Accelerator pedal released | Approx. 0.5V | Ignition switch ON |
| | | | | Accelerator pedal fully depressed | Approx. 4.3V | |
| 2N | O | | Oxygen sensor | Ignition switch ON | 0V | |
| | | | | Idle (Cold engine) | 0V | |
| | | | | Idle (After warm up) | 0—1.0V | Needle moves from 0V to 1V |
| | | | | Increase engine speed (After warm up) | ·0.5—1.0V | |
| | | | | Deceleration | 0—0.4V | |
| 2O | O | | Airflow sensor (Intake air mass) | Ignition switch ON | 1.0—2.0V | |
| | | | | Idle (After warm up) | 1.9—2.6V | |
| | | | | Increase engine speed (After warm up) | 2—5V | |
| 2P | O | | Airflow sensor (Ground) | Constant | 0V | |
| 2Q | O | | Water thermosensor | Engine coolant termperautre 20°C (68°F) | Approx. 2.5V | Ignition switch ON |
| | | | | After warm up | Approx. 0.4V | |
| 2R | — | — | — | — | — | — |
| 2S | — | — | — | — | — | — |
| 2T | | O | Solenoid valve (PRC) | For 120 seconds after ignition switch OFF→ON | Below 2.5V | During hot condition. Coolant temp. above 90°C (194°F) Intake air temp. above 75°C (167°F) |
| | | | | For 120 seconds after starting | Below 2.5V | |
| | | | | Ignition switch ON | Approx. 12V | Other conditions |
| 2U | | O | Injector G6 (No.3, 4) F2 (No.1, 3) | Ignition switch ON | Approx. 12V | * Engine Signal Monitor: Green and red lights flash |
| | | | | Idle | Approx. 12V* | |

## CONTROL UNIT PIN LOCATION AND VOLTAGE TEST (CONT.) – 1990 B2600

**Terminal voltage**

| Terminal | Input | Output | Connection to | Test condition | Voltage | Remarks |
|---|---|---|---|---|---|---|
| 2V | | ○ | Injector G6 (No.1, 2) F2 (No.2, 4) | Ignition switch ON | Approx. 12V | * Engine Signal Monitor: Green and red lights flash |
| | | | | Idle | Approx. 12V* | |
| 2W | | ○ | Solenoid valve (Idle speed control) | Ignition switch ON | Approx. 11V | Engine signal monitor: Green and red lights flash |
| | | | | Idle | Approx. 10V | |
| 2X | | ○ | Solenoid valve (Purge control) | Ignition switch ON | Approx. 12V | |
| | | | | Idle | Approx. 12V | * Engine signal monitor: Green and red lights flash |
| | | | | Driving in gear | 5—1.5V* | |
| 2Y | | ○ | HAT control unit | Ignition switch ON | Approx. 12V | For G6 HAT |
| | | | | Accelera for pedal fully depessed | 0 | |
| 2Z | — | — | — | — | — | — |

**Terminal location**

| 2Y | 2W | 2U | 2S | 2Q | 2O | 2M | 2K | 2I | 2G | 2E | 2C | 2A | U | S | Q | O | M | K | I | G | E | C | A |
|---|---|---|---|---|---|---|---|---|---|---|---|---|---|---|---|---|---|---|---|---|---|---|---|
| 2Z | 2X | 2V | 2T | 2R | 2P | 2N | 2L | 2J | 2H | 2F | 2D | 2B | V | T | R | P | N | L | J | H | F | D | B |

## WIRING SCHEMATIC – 1990 B2600

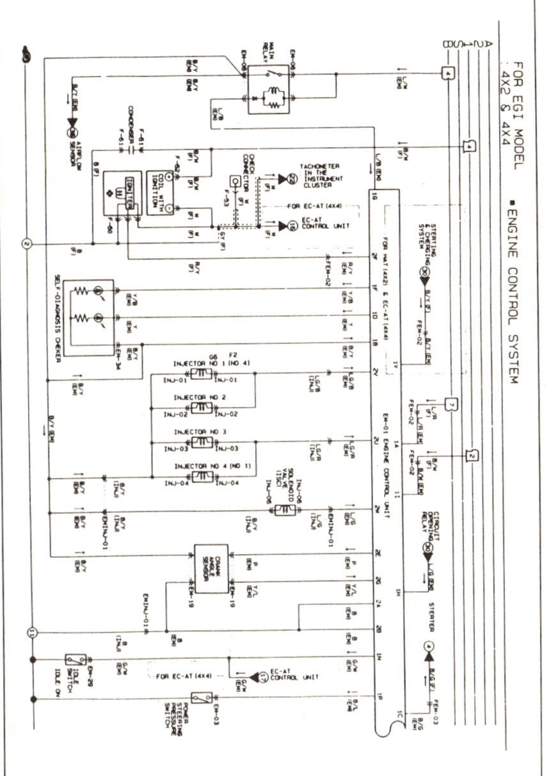


## WIRING SCHEMATIC (CONT.) – 1990 B2600

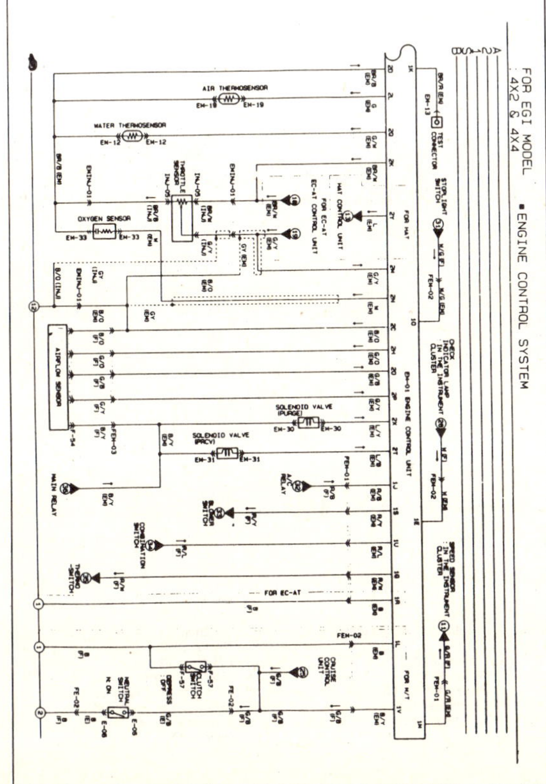


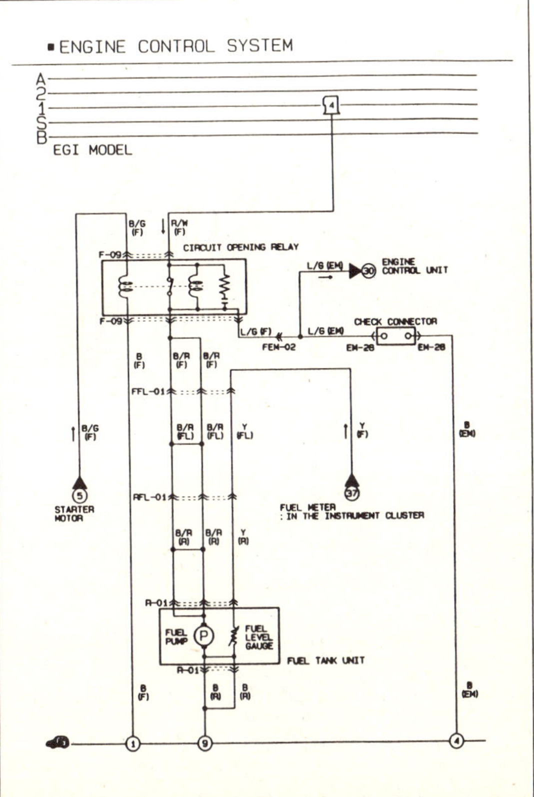

## SYSTEM SCHEMATIC 1990 MPV WITH 4 CYLINDER ENGINE

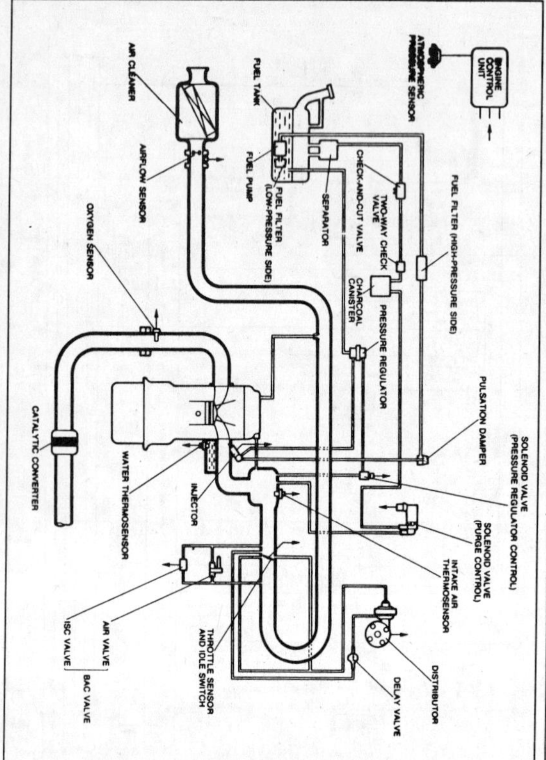


## VACUUM HOSE ROUTING 1990 MPV WITH 4 CYLINDER ENGINE

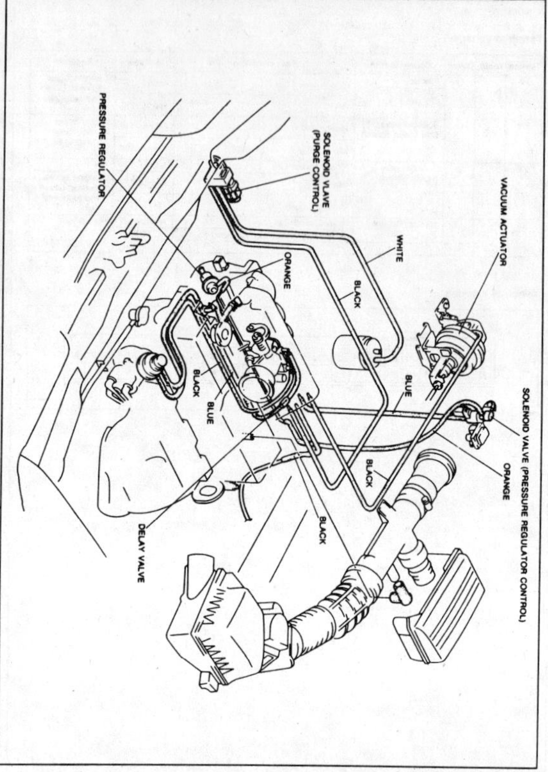


## EXPLODED VIEW OF INTAKE AIR SYSTEM COMPONENTS–1990 MPV WITH 4 CYLINDER ENGINE

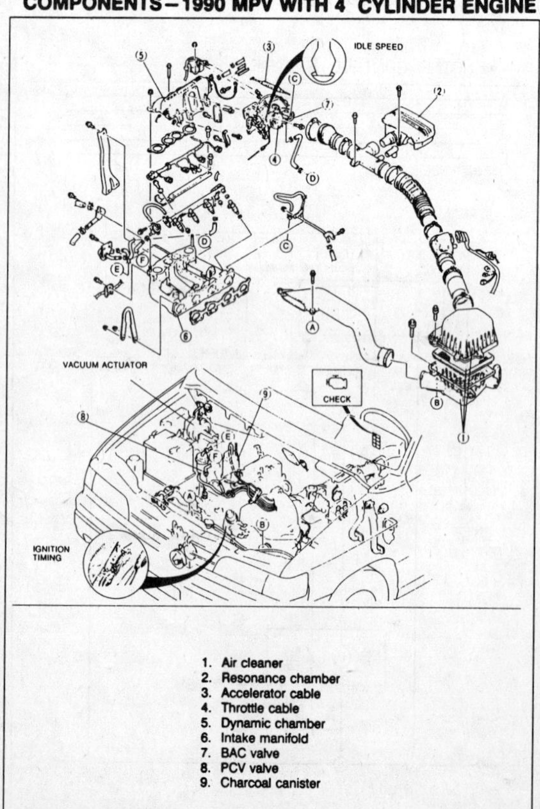


1. Air cleaner
2. Resonance chamber
3. Accelerator cable
4. Throttle cable
5. Dynamic chamber
6. Intake manifold
7. BAC valve
8. PCV valve
9. Charcoal canister

## EXPLODED VIEW OF CONTROL UNIT COMPONENTS 1990 MPV WITH 4 CYLINDER ENGINE

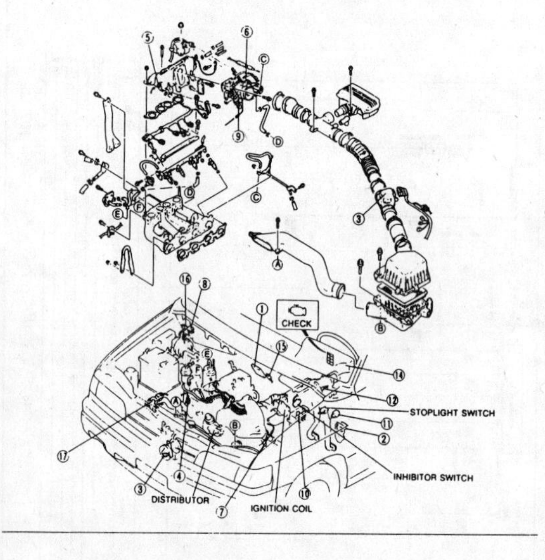


1. Engine control unit (ECU)
2. Central processing unit (CPU)
3. Airflow sensor
4. Water thermosensor
5. Intake air thermosensor
6. Throttle sensor
7. Oxygen sensor
8. Atmospheric pressure sensor
9. Idle switch
10. Main relay
11. Clutch switch
12. Neutral switch
13. P/S pressure switch
14. Mileage sensor
15. Circuit opening relay
16. PRC solenoid valve
17. Purge control solenoid valve

## EXPLODED VIEW OF FUEL AND EMISSION SYSTEM COMPONENTS—1990 MPV WITH 4 CYLINDER ENGINE

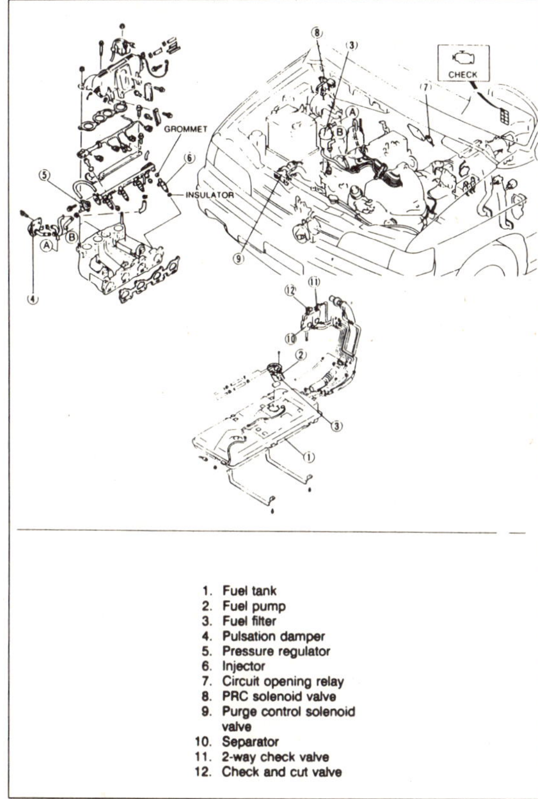


1. Fuel tank
2. Fuel pump
3. Fuel filter
4. Pulsation damper
5. Pressure regulator
6. Injector
7. Circuit opening relay
8. PRC solenoid valve
9. Purge control solenoid valve
10. Separator
11. 2-way check valve
12. Check and cut valve

## SPECIFICATIONS CHART 1990 MPV WITH 4 CYLINDER ENGINE

| Item | | | Specification |
|---|---|---|---|
| Idle speed | | rpm | M/T 750 ± 20 (Neutral), A/T 770 ± 20 (P range) *1 |
| Ignition timing | | BTDC | 5 ± 1° (A/T P range) *2 |
| **Throttle body** | | | |
| Type | | | Horizontal draft (2 barrel) |
| Throat diameter | mm (in) | No.1 | M/T 40 (1.6), A/T 46 (1.8) |
| | | No.2 | M/T 46 (1.8), A/T 40 (1.6) |
| **Fuel pump** | | | |
| Type | | | Impeller (in-tank) |
| Output pressure | | kPa (kg/cm², psi) | 441–589 (4.5–6.0, 64–85) |
| **Fuel filter** | | | |
| Type | Low-pressure side | | Nylon element |
| | High-pressure side | | Paper element |
| **Pressure regulator** | | | |
| Type | | | Diaphragm |
| Regulating pressure | | kPa (kg/cm², psi) | 265–314 (2.7–3.2, 38–46) |
| **Injector** | | | |
| Type | | | High-ohmic |
| Type of drive | | | Voltage |
| Resistance | | Ω | 12–16 (at 20°C, 68°F) |
| **BAC valve (solenoid valve [idle speed control])** | | | |
| Solenoid resistance | | Ω | 7.7–9.3 (at 23°C, 73°F) |
| **Solenoid valve (Purge control)** | | | |
| Solenoid resistance | | Ω | 23–27 (at 20°C, 68°F) |
| **Water thermosensor** | | | |
| Resistance | kΩ | −20°C (−4°F) | 14.5–17.8 |
| | | 20°C (68°F) | 2.2–2.7 |
| | | 80°C (176°F) | 0.28–0.35 |
| **Intake air thermosensor** | | | |
| Resistance | kΩ | 25°C (77°F) | 29.7–36.3 |
| | | 85°C (185°F) | 3.3–3.7 |
| **Circuit opening relay** | | | |
| Resistance | Ω | STA–E1 | 21–43 |
| | | B–Fc | 109–226 |
| | | B–Fp | ∞ |
| **Fuel tank** | | | |
| Capacity | | liters (US gal, Imp gal) | 60 (15.9, 13.2) |
| **Air cleaner** | | | |
| Element type | | | Dry |
| **Accelerator cable** | | | |
| Free play | | mm (in) | 1–3 (0.039–0.118) |
| **Fuel** | | | |
| Specification | | | Unleaded regular (RON 87 or higher) |

*1 Test connector grounded
*2 Vacuum hoses disconnected and test connector grounded

## COMPONENTS DECSRIPTION CHART—1990 MPV WITH 4 CYLINDER ENGINE

### COMPONENT DESCRIPTIONS

| Component | Function | Remarks |
|---|---|---|
| **Air cleaner** | Filters air entering throttle body | |
| **Airflow sensor** | Detects amount of intake air; sends signal to engine control unit | |
| **Air valve** | When cold, supplies bypass air into dynamic chamber | • Engine speed increased to shorten warm-up period<br>• Thermowax type<br>• Installed in BAC valve |
| **Atmospheric pressure sensor** | Detects atmospheric pressure; sends signal to engine control unit | |
| **BAC valve** | Supplies bypass air into dynamic chamber | Consists of air valve and ISC valve |
| **Catalytic converter** | Reduces HC, CO, and NOx by chemical reaction | Monolith type |
| **Central processing unit (CPU)** | Detects that E/L is being applied; sends singal to engine control unit | |
| **Charcoal canister** | Stores gas tank fumes when engine stopped | |
| **Check connector** | For Self-Diagnosis Checker | 6-pin connector (Green) |
| **Check-and-cut valve** | Releases excessive pressure or vacuum in fuel tank to atmosphere | |
| **Circuit opening relay** | Voltage for fuel pump while engine running | |
| **Clutch switch (M/T)** | Detects in-gear condition; sends signal to engine control unit | Switch ON when clutch pedal depressed |
| **Dynamic chamber** | Interconnects all cylinders | |
| **Engine control unit** | Detects following:<br>1. Engine speed<br>2. No.1 piston TDC<br>3. Intake air amount<br>4. Engine coolant temperature<br>5. Ignition ON signal<br>6. Throttle valve opening angle<br>7. Throttle valve fully closed<br>8. Air/fuel ratio (Oxygen concentration)<br>9. In-gear condition<br>10. Intake air temperature<br>11. Atmospheric pressure<br>12. A/C operation<br>13. P/S operation<br>14. E/L operation<br>15. Cranking signal<br>16. Test signal (idle speed, malfunction code No.)<br>17. Braking signal<br>Controls operation of the following:<br>1. Fuel injection system<br>2. Idle speed control<br>3. Pressure regulator control system<br>4. Purge control system<br>5. Fail-safe function<br>6. Monitor function<br>7. Burn-off system<br>8. Lockup inhibition control<br>9. Ignition timing control system<br>10. Fuel pump<br>11. A/C (cut off)<br>12. Main relay control | 1. Ignition coil<br>2. G-signal<br>3. Airflow sensor<br>4. Water thermosensor<br>5. Ignition switch<br>6. Throttle sensor<br>7. Idle switch<br>8. Oxygen sensor<br>9. Neutral and clutch switches (M/T) Inhibitor switch (A/T)<br>10. Intake air thermosensor (on dynamic chamber)<br>11. Atmospheric pressure sensor<br>12. A/C switch<br>13. P/S pressure switch<br>14. Central processing unit (CPU)<br>15. Ignition switch (START position)<br>16. Test connector<br>17. Stoplight switch<br>1. Injector<br>2. Solenoid valve (Idle speed control)<br>3. Solenoid valve (Pressure regulator control)<br>4. Solenoid valve (Purge control)<br>5. Self-Diagnosis Checker and MIL<br>6. Monitor lamp (Self-Diagnosis Checker)<br>7. Airflow sensor<br>8. 4 A/T control unit<br>9. Igniter<br>10. Circuit opening relay<br>11. A/C relay<br>12. Main relay |

| Component | | Function | Remarks |
|---|---|---|---|
| **Fuel filter** | | Filters particles from fuel | |
| **Fuel pump** | | Provides fuel to injectors | • Operates while engine running<br>• Installed in fuel tank |
| **G rotor and pick-up** | | Detects No.1 cylinder TDC; sends signal to engine control unit | For determining fuel injection timing |
| **Idle switch** | | Detects when throttle valve fully closed; sends signal to engine control unit | Installed on throttle body |
| **Igniter** | | Receives spark signal from signal rotor and generates high voltage to ignition coil | |
| **Ignition coil (−) terminal** | | Detects engine speed; sends signal to engine control unit | |
| **Ignition switch (START position)** | | Sends engine cranking signal to engine control unit | |
| **Inhibitor switch (A/T)** | | Detects in-gear condition; sends signal to engine control unit | Switch ON in "N" or "P" range |
| **Injector** | | Injects fuel into intake port | • Controlled by signals from engine control unit<br>• High-ohmic injector<br>• Two port injector nozzle |
| **Intake air thermosensor** | | Detects intake air temperature; sends signal to engine control unit | Installed on dynamic chamber |
| **Main relay** | | Supplies electric current to injectors and engine control unit. | |
| **MIL (Malfunction indicator lamp)** | | Light illuminates when input device malfunctions and oxygen sensor replacement time comes | Test connector not grounded |
| | | Light flashes to indicate malfunction code No. of input and output devices | Test connector grounded |
| | | Lamp stays ON when oxygen sensor replacement time comes | |
| **Neutral switch** | | Detects in-gear condition; sends signal to engine control unit | Switch ON when neutral |
| **Oxygen sensor** | | Detects oxygen concentration; sends signal to engine control unit | Zirconia ceramic and platinum coating |
| **PCV valve** | | Controls amount of blowby gas introduced into engine | |
| **P/S pressure switch** | | Detects P/S operation; sends signal to engine control unit | P/S: ON when steering wheel turned right or left |
| **Pressure regulator** | | Adjusts fuel pressure supplied to injectors | |
| **Resonance chamber** | | Improves mid-range torque characteristics | |
| **Separator** | | Prevents fuel from flowing into charcoal canister | |
| **Solenoid valve** | **Idle speed control** | Controls bypass air amount | • Controlled by duty signal from engine control unit<br>• With integrated air valve<br>• Controls idle-up |
| | **Pressure regulator control** | Controls vacuum to pressure regulator | Cuts vacuum passage when hot |
| | **Purge control** | Controls evaporative fumes from canister to intake manifold | |
| **Stoplight switch** | | Detects braking operation (deceleration); sends signal to engine control unit | |

## COMPONENTS DECSRIPTION CHART 1990 MPV WITH 4 CYLINDER ENGINE

| Component | Function | Remarks |
|---|---|---|
| **Test connector** | For Self-Diagnosis Checker and idle speed adjustment | 1-pin connector (Green) |
| **Throttle body** | Controls intake air quantity | Integrated throttle sensor and idle switch |
| **Throttle sensor** | Detects throttle valve opening angle; sends signal to engine control unit | Installed on throttle body |
| **Two-way check valve** | Controls pressure in fuel tank | |
| **Vacuum delay valve (for distributor)** | Retards ignition timing during transition from deceleration to acceleration | |
| **Water thermosensor** | Detects coolant temperature; sends signal to engine control unit | |

## COMPONENTS RELATIONSHIP CHART 1990 MPV WITH 4 CYLINDER ENGINE

**RELATIONSHIP CHART**

| INPUT DEVICES \ OUTPUT DEVICES | INJECTOR: FUEL INJECTION AMOUNT | INJECTOR: FUEL INJECTION TIMING | BAC VALVE: AIR VALVE | BAC VALVE: ISC VALVE | SOLENOID VALVE (PURGE CONTROL) | SOLENOID VALVE (PRESSURE REGULATOR CONTROL) | A/C RELAY (A/C CUT-OFF) | 4A/T CONTROL UNIT (LOCKUP INHIBITION CONTROL) | AIRFLOW SENSOR (BURN-OFF) | CIRCUIT OPENING RELAY (FUEL PUMP CONTROL) | IGNITER (IGNITION TIMING CONTROL) |
|---|---|---|---|---|---|---|---|---|---|---|---|
| TEST CONNECTOR | × | × | × | | × | × | × | × | × | × | - |
| IGNITION SWITCH (ON POSITION) | × | × | × | × | × | × | × | × | ○ | × | × |
| IGNITION SWITCH (START POSITION) | ○ | ○ | × | ○ | × | ○ | ○ | × | × | × | ○ |
| CENTRAL PROCESSING UNIT (CPU) | × | × | × | ○ | × | × | × | × | × | × | × |
| P/S PRESSURE SWITCH | ○ | × | × | ○ | × | × | × | × | × | × | × |
| A/C SWITCH | ○ | × | × | ○ | × | × | ○ | × | × | × | ○ |
| INHIBITOR SWITCH (A/T) | ○ | × | × | ○ | ○ | × | ○ | × | × | × | ○ |
| NEUTRAL AND CLUTCH SWITCH (M/T) | ○ | × | × | ○ | ○ | × | ○ | ○ | × | × | ○ |
| STOPLIGHT SWITCH | ○ | × | × | × | × | × | × | × | × | × | × |
| IDLE SWITCH | ○ | × | × | ○ | ○ | ○ | × | × | × | × | ○ |
| ATMOSPHERIC PRESSURE SENSOR | ○ | × | × | ○ | × | × | × | × | × | × | ○ |
| THROTTLE SENSOR | ○ | × | × | × | × | ○ | ○ | × | × | × | ○ |
| INTAKE AIR THERMOSENSOR | ○ | × | × | × | × | ○ | × | × | × | × | × |
| AIRFLOW SENSOR | ○ | × | × | ○ | ○ | × | × | ○ | ○ | × | × |
| OXYGEN SENSOR | ○ | × | × | × | ○ | × | × | × | × | × | × |
| WATER THERMOSENSOR | ○ | × | × | ○ | ○ | ○ | × | × | ○ | × | × |
| IGNITION COIL | ○ | ○ | × | ○ | ○ | ○ | × | × | ○ | ○ | × |
| DISTRIBUTOR (G-SIGNAL) | × | ○ | × | × | × | × | × | × | × | × | × |

## INPUT COMPONENTS AND ENGINE CONDITIONS 1990 MPV WITH 4 CYLINDER ENGINE

**Input Devices and Engine Conditions**

| ENGINE CONDITIONS \ INPUT DEVICES | APPROXIMATE TIME (BASED ON 10–16°C or 50–60°F AMBIENT) | SENSORS: DISTRIBUTOR (G-SIGNAL) | IGNITION COIL | WATER THERMOSENSOR | OXYGEN SENSOR | AIRFLOW SENSOR | INTAKE AIR THERMOSENSOR | THROTTLE SENSOR | ATMOSPHERIC PRESSURE SENSOR |
|---|---|---|---|---|---|---|---|---|---|
| CRANKING —COLD ENGINE • COLD AIR • COLD COOLANT | Zero | | | | Signal has no effect on ECU | Signal has no effect on ECU | Signal has no effect on ECU | Signal has no effect on ECU | |
| COLD START —FAST IDLE • COLD AIR • COLD COOLANT | One minutes | | | Cool to warm: medium voltage (3.5V and dropping) | Sensor cold: low to high voltage (0—0.9V) | Low volume airflow: low to high voltage (2.4—2.6V) | | Closed throttle: low voltage (0.3—0.7V) | |
| COLD DRIVEAWAY —PART THROTTLE • COLD AIR • COLD COOLANT | Two minutes | | | | | | | | |
| WARM DRIVEAWAY —PART THROTTLE • WARM AIR • WARM COOLANT | Three minutes | | | Warm: medium voltage (Approx 0.7V and dropping) | Sensor warm: high voltage (0.9V) | Moderate volume airflow: low to medium voltage (3.0V) | | Part throttle: medium voltage (1—3.5V) | |
| HOT CRUISE • WARM AIR • WARM COOLANT | | Sends No.1 cylinder TDC signal to ECU | Sends engine speed signal to ECU | | Sensor hot: switching from high voltage (0.9V) ↕ to low voltage (0.1V) | | | | Sends voltage signal to ECU that varies with altitude: voltage (approx. 4V at sea level) |
| HOT ACCELERATION —60% THROTTLE | | | | | | Moderate to strong volume of airflow: (3.8V) | Cool to warm: medium voltage (1.4—3.4V) | | |
| HOT ACCELERATION —WIDE OPEN THROTTLE | More than four minutes | | | Hot: low voltage (Approx. 0.4V) | High voltage (0.9V) | Strong volume of airflow: (4.0V) | | Wide open throttle: high voltage (Approx. 4.0V) | |
| DECELERATION —CLOSED THROTTLE | | | | | Low voltage (0V) | | | | |
| HOT CURB IDLE —EXTENDED | | | | | Switching from high to low voltage (0.75—0.25V) | Low volume of airflow: (2.4V) | | Closed throttle low voltage (0.3—0.7V) | |
| HOT ENGINE SHUTDOWN | — | OFF | OFF | OFF | Sensor hot: low voltage (0.1V) until sensor cools | OFF | OFF | OFF | OFF |

## INPUT COMPONENTS AND ENGINE CONDITIONS (CONT.)—1990 MPV WITH 4 CYLINDER ENGINE

| SWITCHES: IDLE SWITCH | STOP-LIGHT SWITCH | NEUTRAL AND CLUTCH SWITCHES (M/T) | INHIBITOR SWITCH (A/T) | A/C SWITCH | P/S PRESSURE SWITCH | CPU (E/L SIGNAL) | IGNITION SWITCH: START POSITION | IGNITION SWITCH: ON POSITION | TEST CONNECTOR |
|---|---|---|---|---|---|---|---|---|---|
| Signal has no effect on ECU | Signal has no effect on ECU | Signal has no effect on ECU | Signal has no effect on ECU | Signal has no effect on ECU | Signal has no effect on ECU | Signal has no effect on ECU | Sends signal to ECU (approx. 12V) | Signal has no effect on ECU | Signal has no effect on ECU |
| Low voltage signal to ECU (below 1.5V) | Brake pedal depressed: sends signal to ECU (approx. 12V) | In neutral: low voltage signal to ECU (approx. 0V) | In Nor P range: sends signal to ECV (approx. 12V) | | | | | | |
| | | | | | | | | | |
| | | | | | | | | | |
| High voltage signal to ECU (approx. 12V) | No signal send to ECU (below 1.5V) | Driving in any gear: high voltage signal to ECU (approx. 12V) | In D range: no voltage signal to ECU (below 1.5V) | A/C switch ON: sends signal to ECU (approx. 12V) | Steering wheel turned: low voltage signal to ECU (below 1.5V) | Headlight or blower or rear window defroster switch ON: low voltage signal to ECU (below 1.5V) | No signal to ECU (below 1.5V) | Sends signal to ECU (approx. 12V) | Connector not grounded: high voltage signal to ECU (approx. 12V) |
| | | | | A/C switch OFF: no signal to ECU (below 1.5V) | Steering wheel straight ahead: high voltage signal to ECU (approx. 12V) | Headlight and blower and rear window defroster switches OFF: high voltage signal to ECU (approx. 12V) | | | |
| | | | | | | | | | |
| Low voltage signal to ECU (below 1.5V) | Brake pedal depressed: sends signal to ECU (approx. 12V) | | | | | | | | |
| | | In neutral: low voltage signal to ECU (approx. 0V) | In Nor P range: sends signal to ECU (approx. 12V) | | | | | | Low voltage signal to ECU when connector grounded (below 1.5V) |
| OFF | OFF | OFF | OFF | OFF | OFF | OFF | OFF | OFF | OFF |

## OUTPUT COMPONENTS AND ENGINE CONDITIONS—1990 MPV WITH 4 CYLINDER ENGINE

**Output Devices and Engine Conditions**

| OUTPUT DEVICES / ENGINE CONDITIONS | APPROXIMATE TIME (BASED ON 10—16°C or 50—60°F AMBIENT) | INJECTOR: INJECTION | INJECTOR: INJECTION TIMING | BAC VALVE: AIR VALVE | BAC VALVE: ISC VALVE | SOLENOID VALVE (PURGE CONTROL) | SOLENOID VALVE (PRESSURE REGULATOR CONTROL) | A/C RELAY (A/C CUT-OFF) | 4 A/T CONTROL UNIT (LOCK UP INHIBITOR CONTROL) | AIRFLOW SENSOR (BURN-OFF) | CIRCUIT OPENING RELAY | MAIN RELAY | IGNITER |
|---|---|---|---|---|---|---|---|---|---|---|---|---|---|
| CRANKING —COLD ENGINE • COLD AIR • COLD COOLANT | Zero | | All cylinders each ignition pulse | | | OFF (Purge cut) | | OFF (A/C ON) | | | OFF | | |
| COLD START —FAST IDLE • COLD AIR • COLD COOLANT | One minute | Rich | | Open (coolant temperature below 50°C 122°F) | Large amount of bypass air | | | ON (A/C OFF approx. 5 sec.) | | | | | |
| COLD DRIVEAWAY —PART THROTTLE • COLD AIR • COLD COOLANT | Tow minutes | | | | | | | | OFF (Lock up) | | | | |
| WARM DRIVEAWAY —PART THROTTLE • WARM AIR • WARM COOLANT | Three minutes | Rich and lean | 2-group | | | Operates (Duty values [purge gas amount] change) | OFF (Vacuum to pressure regulator) | OFF (A/C ON) | | | | | |
| HOT CRUISE • WARM AIR • WARM COOLANT | | | | | Small amount of bypass air | | | | | OFF | ON (Engine speed above 50 rpm) | ON | Operation *1 |
| HOT ACCELERATION —60% THROTTLE | | | | | | | | | | | | | |
| HOT ACCELERATION —WIDE OPEN THROTTLE | More than four minutes | Rich | | Closed | | | | ON (A/C CUT) | ON (Lock up inhibit) | | | | |
| DECELERATION —CLOSED THROTTLE | | Fuel cut | | | Large and small amount of bypass air | OFF (Purge cut) | | OFF (A/C ON) | OFF (Lock up) | | | | |
| HOT CURB IDLE —EXTENDED | | Rich and lean | 2-group | | Small amount of bypass air | | After starting ON during hot start only (Vacuum cut) | | | | | | |
| HOT ENGINE SHUTDOWN | — | Does not inject | | | OFF | OFF | OFF | OFF | OFF | ON (Burn-off) | OFF | OFF | OFF |

*1 • A/C: ON at idle
Ignition timing is advanced at approx. 6°.
• More than 1,000 m [3,280 ft] above sea level
Ignition timing is advanced at approx. 6° for all engine condition.

## TROUBLESHOOTING CHART—1990 MPV WITH 4 CYLINDER ENGINE

**Cranks normally but won't start (No combustion)**

| STEP | QUICK INSPECTION | | ACTION | | POSSIBLE CAUSE |
|---|---|---|---|---|---|
| 1 | Check for malfunction code (01) with SST [IGN ON, Test connector (Green: 1-pin) grounded] | Yes | Check for cause by referring to check sequence | | |
| | | No | Go to Next Step | | |
| 2 | Check for spark by disconnecting high-tension lead while cranking | Yes | Go to Next Step | | |
| | | No | Check ignition system | | |
| 3 | Check for fuel pump operating sound from fuel filler port [IGN ON, Test connector (Yellow: 2-pin) connected] | Yes | Check if engine starts in this condition | Yes | Check circuit opening relay switching operation |
| | | | | | Check circuit opening relay circuit (IGN: START) |
| | | | | No | Go to Next Step |
| | | No | Check circuit opening relay switching operation | Yes | Check circuit opening relay circuit |
| | | | | | Check fuel pump circuit |
| | | | | | Check fuel pump operation |
| | | | | No | Replace circuit opening relay |
| 4 | Check fuel line pressure [IGN ON, Test connector (Yellow: 2-pin) connected] **Fuel line pressure: 265—314 kPa (2.7—3.2 kg/cm$^2$, 38—46 psi)** | Yes | Go to Next Step | | |
| | | No | Check fuel pump maximum pressure **Fuel pump maximum pressure: 441—588 kPa (4.5—6.0 kg/cm$^2$, 64—85 psi)** | Yes | Replace pressure regulator |
| | | | | No | Replace fuel pump |
| 5 | Check for injector operating sound while cranking | Yes | Go to Next Step | | |
| | | No | Check voltage at ECU (3C) and (3E) terminals with SST **Voltage: Approx. 12V (IGN ON)** | Yes | Check throttle sensor |
| | | | | | Replace ECU |
| | | | | No | Check wiring for short or open |
| | | | | | Poor ground circuit from ECU (3A) terminal (Check terminal voltage with SST) |
| 6 | Check continuity between check connector [for tachometer (White: 1-pin)] and ground (IGN ON) | Yes | | | Wiring short to ground |
| | | No | Go to Next Step | | |
| 7 | Substitute a well-known ECU. Check if the condition improves | Yes | | | ECU malfunction |
| | | No | Check ground circuit from ECU (2R) terminal with SST **Voltage: 0V (IGN ON)** | Yes | Go to Next Step |
| | | | | No | Poor ground circuit |
| 8 | | | | | Low compression |

**Cranks normally but hard to start (Always)**

| STEP | QUICK INSPECTION | | ACTION | | POSSIBLE CAUSE |
|---|---|---|---|---|---|
| 1 | Check ignition timing at idle after warm up **Ignition timing: 5 ± 1° BTDC** [Test connector (Green: 1-pin) grounded Vacuum hoses disconnected] | Yes | Go to Next Step | | |
| | | No | Adjust ignition timing | | |
| | Note: At high—altitude area **[1,000 m (3,280 ft)]**, ignition timing advances to **11 ± 2° BTDC** when test connector (Green: 1-pin) is not grounded | | | | |
| 2 | Disconnect high-tension lead of each cylinder at idle. Check if engine condition changes | Yes | Go to Next Step | | |
| | | No | Check ignition system | Yes | Replace injector (If step 3 OK) |
| | | | | No | Check spark plug |
| | | | | | Check high-tension lead |
| | | | | | Check distributor cap |
| 3 | Check for injector operating sound at idle | Yes | Go to Next Step | | |
| | | No | Check resistance at injector harness connector (EMINJ-01)<br>**Terminals / Resistance:** (B/Y)—(LG/B), (B/Y)—(LG/R): **6—8Ω** | Yes | Check wiring short or open |
| | | | | No | Check injector resistance |
| | | | | | Check wiring short or open |
| 4 | Check fuel line pressure [IGN ON, Test connector (Yellow: 2-pin) connected] **Fuel line pressure: 265—314 kPa (2.7—3.2 kg/cm$^2$, 38—46 psi)** | Yes | Go to Next Step | | |
| | | No | Substitute a good fuel filter and retest | Yes | Replace fuel filter |
| | | | Check fuel pump maximum pressure **Fuel pump maximum pressure: 441—588 kPa (4.5—6.0 kg/cm$^2$, 64—85 psi)** | Yes | Replace pressure regulator |
| | | | | No | Replace fuel pump |
| 5 | Operate fuel pump [IGN ON, Test connector (Yellow: 2-pin) connected] Turn ignition switch OFF and observe fuel pressure **for 5 minutes** **Fuel pressure: More than 147 kPa (1.5 kg/cm$^2$, 21 psi)** | Yes | Go to Next Step | | |
| | | No | Check fuel pump pressure drop | No | Replace fuel pump |
| | | | Check pressure regulator pressure drop | Yes | Check injector fuel leakage |
| | | | | No | Replace pressure regulator |
| 6 | | | | | Check compression |
| 7 | | | | | ECU malfunction |

## TROUBLESHOOTING CHART—1990 MPV WITH 4 CYLINDER ENGINE

**Cranks normally but hard to start (Only when engine is cold)**

| STEP | QUICK INSPECTION | | ACTION | | POSSIBLE CAUSE |
|---|---|---|---|---|---|
| 1 | Check specific gravity of battery using a hydrometer<br>**Specific gravity: Above 1.200** | Yes | Go to Next Step | | |
| | | No | Recharge battery | | |
| 2 | Check for malfunction code (09) (26) with SST<br>[IGN ON, Test connector (Green: 1-pin) grounded] | Yes | Check for cause by referring to check sequence | | |
| | | No | Go to Next Step | | |
| 3 | Disconnect high-tension lead of each cylinder at idle<br>Check if engine condition changes | Yes | Go to Next Step | | |
| | | | Check ignition system | | Check spark plug |
| | | | | | Check high-tension lead |
| | | | | | Check distributor cap |
| 4 | Check fuel line pressure<br>[IGN ON, Test connector (Yellow: 2-pin) connected]<br>**Fuel line pressure: 265—314 kPa (2.7—3.2 kg/cm², 38—46 psi)** | Yes | Go to Next Step | | |
| | | No | Check for fuel leaks | | |
| | | | Substitute a good fuel filter and retest | Yes | Replace fuel filter |
| | | | Check fuel pump maximum pressure<br>**Fuel pump maximum pressure: 441—588 kPa (4.5—6.0 kg/cm², 64—85 psi)** | Yes | Replace pressure regulator |
| | | | | No | Replace fuel pump |
| 5 | Disconnect ISC valve connector when engine is cold<br>Check if idle speed decreases during warm up | Yes | Go to Next Step | | |
| | | No | | | Check if BAC valve (air valve) opens when cold |
| 6 | | | | | ECU malfunction |

**Cranks normally but hard to start (Only when engine is warm)**

| STEP | QUICK INSPECTION | | ACTION | | POSSIBLE CAUSE |
|---|---|---|---|---|---|
| 1 | Check for malfunction code (26) with SST<br>[IGN ON, Test connector (Green: 1-pin) grounded] | Yes | Check for cause by referring to check sequence | | |
| | | No | Go to Next Step | | |
| 2 | Disconnect water thermosensor connector<br>Check if condition improves | Yes | Check water thermosensor connector condition as follows:<br>1. Shake connector and check if condition changes<br>2. Check condition of terminal (burned or damaged)<br>3. Connect a good terminal to harness side connector and check for looseness | Yes | Check water thermosensor |
| | | | | No | Poor contact of water thermosensor connector |
| | | No | Go to Next Step | | |
| 3 | Operate fuel pump<br>[IGN ON, Test connector (Yellow: 2-pin) connected]<br>Turn ignition switch OFF and observe fuel pressure **for 5 minutes**<br>**Fuel pressure: More than 147 kPa (1.5 kg/cm², 21 psi)** | Yes | Go to Next Step | | |
| | | No | Check fuel pump pressure drop | No | Replace fuel pump |
| | | | Check pressure regulator pressure drop | Yes | Check injector fuel leakage |
| | | | | No | Replace pressure regulator |
| 4 | | | | | ECU malfunction |

## TROUBLESHOOTING CHART—1990 MPV WITH 4 CYLINDER ENGINE

**Cranks normally but hard to start (Only after heat soak)**

| STEP | QUICK INSPECTION | | ACTION | | POSSIBLE CAUSE |
|---|---|---|---|---|---|
| 1 | Check for malfunction code (09), (11), (25), (26) with SST<br>[IGN ON, Test connector (Green: 1-pin) grounded] | Yes | Check for cause by referring to check sequence | | |
| | | No | Go to Next Step | | |
| 2 | Circulate fuel by operating fuel pump **for 20 secondes**<br>[IGN ON, Test connector (Yellow: 2-pin) connected]<br>Check if condition improves | Yes | Go to Step 3 | | |
| | | No | Go to Step 4 | | |
| 3 | Disconnect vacuum hose from pressure regulator<br>Check if condition improves | Yes | Check the components related to pressure regulator control system | | Check water thermosensor |
| | | | | | Check intake air thermosensor |
| | | | | | Check solenoid valve (PRC) |
| | | | | | ECU malfunction (Check (2N) terminal voltage) |
| | | No | Go to Next Step | | |
| 4 | Operate fuel pump<br>[IGN ON, Test connector (Yellow: 2-pin) connected]<br>Turn ignition switch OFF and observe fuel pressure **for 5 minutes**<br>**Fuel pressure: More than 147 kPa (1.5 kg/cm², 21 psi)** | Yes | Go to Next Step | | |
| | | No | Check fuel pump pressure drop | No | Replace fuel pump |
| | | | Check pressure regulator pressure drop | Yes | Check injector fuel leakage |
| | | | | No | Replace pressure regulator |
| 5 | Change fuel with specified one<br>Check if condition improves | Yes | | | Poor fuel quality |
| | | No | Go to Next Step | | |
| 6 | | | | | ECU malfunction |

**Cranks normally but won't start (Intermittent)**

| STEP | QUICK INSPECTION | | ACTION | POSSIBLE CAUSE |
|---|---|---|---|---|
| 1 | Shake connector of ignition coil, main relay and ECU while cranking<br>Check if the engine starts | Yes | There may be a poor contact of the connector. Repair or replace the wiring | |
| | | No | Go to troubleshooting "Cranks normally but hard to start (Always)" | |

**Rough idle (Always)**

| STEP | QUICK INSPECTION | | ACTION | | | POSSIBLE CAUSE |
|---|---|---|---|---|---|---|
| 1 | Check for malfunction code (08), (09), (14), (17), (26), (34) with SST<br>[IGN ON, Test connector (Green: 1-pin) grounded] | Yes | Check for cause by referring to the check sequence | | | |
| | | No | **"88" flashing**<br>Check voltage at ECU (3G) terminal with SST<br>**Voltage: 0V (IG ON)** | F2–181 | Yes | Replace ECU |
| | | | | | No | Poor ground circuit |
| | | | **"00"** Go to Next Step | | | |
| 2 | Check ignition at idle after warm up<br>**Ignition timing: 5 ± 1° BTDC**<br>[Test connector (Green: 1-pin) grounded<br>Vacuum hoses disconnected] | Yes | Go to Next Step | | | |
| | | No | Adjust ignition timing (If possible) | | | |
| 3 | Disconnect high-tension lead of each cylinder at idle<br>Check if engine condition changes | Yes | Go to Next Step | | | |
| | | No | Check ignition system | | Yes | Replace injector (If Step 3 OK) |
| | | | | | No | Check spark plug |
| | | | | | | Check high-tension lead |
| | | | | | | Check distributor cap |
| 4 | Check idle speed after warm up<br>**Idle speed: 750 ± 20 rpm (M/T) 770 ± 20 rpm (A/T)**<br>[Test connector (Green: 1-pin) grounded] | Yes | Go to Next Step | | | |
| | | No | Adjust idle speed (If possible) | | | |
| 5 | Check for injector operating sound at idle | Yes | Go to Next Step | | | |
| | | No | Check resistance at injector harness connector (EMINJ-01)<br>Terminals: (B/Y)—(LG/B), (BY)—(LG/R); Resistanc: 6—8Ω | | Yes | Check wiring short or open |
| | | | | | No | Check injector resistance |
| | | | | | | Check wiring short or open |
| 6 | Check fuel line pressure<br>[IGN ON, Test connector (Yellow: 2-pin) connected]<br>**Fuel line pressure: 265—314 kPa (2.7—3.2 kg/cm², 38—46 psi)** | Yes | Go to Next Step | | | |
| | | No | Check for fuel leakage | | | |
| | | | Substitute a good fuel filter and retest | | Yes | Replace fuel filter |
| | | | Check fuel pump maximum pressure<br>**Fuel pump maximum pressure: 441—588 kPa (4.5—6.0 kg/cm², 64—85 psi)** | | Yes | Replace pressure regulator |
| | | | | | No | Replace fuel pump |
| 7 | Check intake manifold vacuum at idle<br>**Vacuum: 480 ± 20 mmHg (18.9 ± 0.79 inHg)** | Yes | Go to Next Step | | | |
| | | No | Check for air leaks | | Yes | Intake air system components damaged |
| | | | | | | Vacuum and intake air hoses loose or damaged |
| | | | | | | Bolts or nuts loose |
| | | | | | | Gaskets damaged |
| | | | | | No | Check throttle valve closing condition |

## TROUBLESHOOTING CHART—1990 MPV WITH 4 CYLINDER ENGINE

**Rough idle (Always) (Cont'd)**

| STEP | QUICK INSPECTION | | ACTION | | | POSSIBLE CAUSE |
|---|---|---|---|---|---|---|
| 8 | Substitute a well-known ECU<br>Check if condition improves | Yes | | | | ECU malfunction |
| | | No | Check voltage at ECU (3G) terminal with SST<br>**Voltage: 0V (IGN ON)** | | Yes | Go to Next Step |
| | | | | | No | Poor ground circuit |
| 9 | | | | | | Check compression |

**Rough idle (Only when engine is cold)**

| STEP | QUICK INSPECTION | | ACTION | | | POSSIBLE CAUSE |
|---|---|---|---|---|---|---|
| 1 | Check for malfunction code (08), (09), (26) with SST<br>[IGN ON, Test connector (Green: 1-pin) grounded] | Yes | Check for cause by referring to check sequence | | | |
| | | No | Go to Next Step | | | |
| 2 | Check ignition timing advance (Centrifugal advance) | Yes | Go to Next Step | | | |
| | | No | | | | Distributor malfunction |
| 3 | Disconnect high-tension lead of each cylinder at idle<br>Check if engine condition changes | Yes | Go to Next Step | | | |
| | | No | Check ignition system | **G–20** | Yes | Replace injector (If step 4 OK) |
| | | | | | No | Check spark plug |
| | | | | | | Check high-tension lead |
| | | | | | | Check distributor cap |
| 4 | Check for injector operating sound at idle | Yes | Go to Next Step | | | |
| | | No | Check resistance at injector harness connector (EMINJ-01)<br>**Terminals:** (B/Y)—(LG/B), (B/Y)—(LG/R); **Resistance:** 6—8Ω | **F2–163** | Yes | Check wiring short or open |
| | | | | | No | Injector malfunction (Check resistance) |
| | | | | | | Wiring short or open |
| 5 | Disconnect ISC valve connector at idle when engine is cold<br>Check if idle speed decreases during warm up | Yes | Go to Next Step | | | |
| | | No | | | | Check if BAC valve (air valve) opens when cold |
| 6 | Check if specified engine oil is used | Yes | Go to Next Step | | | |
| | | No | Change engine oil to specified oil | | | |
| 7 | Subsitute a well-known ECU<br>Check if condition improves | Yes | | | | ECU malfunction |
| | | No | | | | Check airflow sensor |

**Rough idle (Only when engine is warm)**

| STEP | QUICK INSPECTION | | ACTION | | | POSSIBLE CAUSE |
|---|---|---|---|---|---|---|
| 1 | Run engine at **2,000 rpm** for more than **20 seconds**<br>Check for malfunction code (08), (09), (15), (17), (26), (34) with SST<br>[IGN ON, Test connector (Green: 1-pin) grounded] | Yes | Check for cause by referring to check sequence | | | |
| | | No | Go to Next Step | | | |
| 2 | Check idle speed after warm up<br>**Idle speed: 750 ± 20 rpm (M/T) 770 ± 20 rpm (A/T)**<br>[Test connector (Green: 1-pin) grounded] | Yes | Go to Next Step | | | |
| | | No | Adjust idle speed (If possible) | | | |
| 3 | Check for flashing of SST monitor lamp after warm up<br>**Monitor lamp: Flashes more than 8 times/10 seconds at 2,000—3,000 rpm**<br>[Test connector (Green: 1-pin) not grounded] | Yes | Go to Next Step | | | |
| | | No | | | | Replace oxygen sensor |
| 4 | Disconnect ISC valve connector after warm up<br>Check if engine speed drops | Yes | Go to Next Step | | | |
| | | No | | | | Check ISC valve |
| 5 | Disconnect water thermosensor connector<br>Check if condition improves | Yes | Check water thermosensor connector condition as follows:<br>1. Shake connector and check if condition changes<br>2. Check condition of terminal (burned or damaged)<br>3. Connect a good terminal to harness side connector and check for looseness | | Yes | Check water thermosensor |
| | | | | | No | Poor contact of water thermosensor connector |
| | | No | Go to Next Step | | | |
| 6 | Disconnect high-tension lead of each cylinder at idle<br>Check if engine condition changes | Yes | Go to Next Step | | | |
| | | No | Check ignition system | **G–20** | Yes | Replace injector (If step 7 OK) |
| | | | | | No | Check spark plug |
| | | | | | | Check high-tension lead |
| | | | | | | Check distributor cap |
| | | | Note: If spark plug is wet, injector may be leaking | | | |

## TROUBLESHOOTING CHART—1990 MPV WITH 4 CYLINDER ENGINE

**Rough idle (Only when engine is warm) (Cont'd)**

| STEP | QUICK INSPECTION | | ACTION | | | POSSIBLE CAUSE |
|---|---|---|---|---|---|---|
| 7 | Check for injector operating sound at idle | Yes | Go to Next Step | | | |
| | | No | Check resistance at injector harness connector (EMINJ-01)<br>**Terminals:** (B/Y)—(LG/B), (B/Y)—(LG/R); **Resistance:** 6—8Ω | **F2–163** | Yes | Check wiring short or open |
| | | | | | No | Check injector resistance |
| | | | | | | Check wiring short or open |
| 8 | Check for air leaks by listening for sucking noise | Yes | Go to Next Step | | | |
| | | No | | | | Intake air system components damaged |
| | | | | | | Vacuum and intake air hoses loose or damaged |
| | | | | | | Bolts or nuts loose |
| | | | | | | Gaskets damaged |
| 9 | | | | | | ECU malfunction |

**Rough idle (Only after heat soak)**

| STEP | QUICK INSPECTION | | ACTION | | | POSSIBLE CAUSE |
|---|---|---|---|---|---|---|
| 1 | Run engine at, **2,000 rpm** for more than **20 seconds**<br>Check for malfunction code (09), (11), (15), (17), (25), (26) with SST<br>[IGN ON, Test connector (Green: 1-pin) grounded] | Yes | Check for cause by referring to check sequence | | | |
| | | No | Go to Next Step | | | |
| 2 | Check switches with SST<br>Neutral switch (M/T)<br>Clutch switch (M/T)<br>Inhibitor switch (A/T)<br>[IGN ON, Test connector (Green: 1-pin) grounded] | Yes | Go to Next Step | | | |
| | | No | Check for cause by referring to check sequence | | | |
| 3 | Check for flashing of SST monitor lamp after warm up<br>**Monitor lamp: Flashes more than 8 times 10 seconds at 2,000—3,000 rpm**<br>[Test connector (Green: 1-pin) not grounded] | Yes | Go to Next Step | | | |
| | | No | | | | Replace oxygen sensor |
| 4 | Disconnect vacuum hose from pressure regulator<br>Check if condition improve | Yes | Check components related to pressure regulator control system | | | Check water thermosensor |
| | | | | | | Check intake air thermosensor |
| | | | | | | Check solenoid valve (PRC) |
| | | | | | | ECU malfunction (Check (2N) terminal voltage) |
| | | No | Go to Next Step | | | |
| 5 | Run engine at idle and stop it<br>Observe fuel pressure **for 5 minutes**<br>**Fuel pressure: More than 147 kPa (1.5 kg/cm$^2$, 21 psi)** | Yes | Go to Next Step | | | |
| | | No | Check fuel pump pressure drop | | No | Replace fuel pump malfunction |
| | | | Check pressure regulator pressure drop | | Yes | Check injector fuel leakage |
| | | | | | No | Replace pressure regulator |
| 6 | Disconnect high-tension lead of each cylinder at idle<br>Check if engine condition changes | Yes | Go to Next Step | | | |
| | | No | Check ignition system | | Yes | Replace injector (If step 3 OK) |
| | | | | | No | Check spark plug |
| | | | | | | Check high-tension lead |
| | | | | | | Check distributor cap |

## TROUBLESHOOTING CHART—1990 MPV WITH 4 CYLINDER ENGINE

**Rough idle (Only after heat soak) (Cont'd)**

| STEP | QUICK INSPECTION | | ACTION | | POSSIBLE CAUSE |
|---|---|---|---|---|---|
| 7 | Check for injector operating sound at idle | Yes | Go to Next Step | | |
| | | No | Check resistance at injector harness connector (EMINJ-01)<br>Terminals: (B/Y)—(LG/B), (B/Y)—(LG/R); Resistance: 6—8Ω | Yes | Check wiring short or open |
| | | | | No | Check injector resistance |
| | | | | | Check wiring short or open |
| 8 | Change fuel to specified grade<br>Check if condition improves | Yes | | | Poor fuel quality |
| | | No | Go to Next Step | | |
| 9 | | | | | ECU malfunction |

**Rough idle just after starting**

| STEP | QUICK INSPECTION | | ACTION | | POSSIBLE CAUSE |
|---|---|---|---|---|---|
| 1 | Check for malfunction code (09), (34) with SST<br>[IGN ON, Test connector (Green: 1-pin) grounded] | Yes | Check for cause by referring to check sequence | | |
| | | No | Go to Next Step | | |
| 2 | Check idle switch with SST<br>[IGN ON, Test connector (Green 1-pin) grounded] | Yes | Go to Next Step | | |
| | | No | Check for cause by referring to check sequence | | |
| 3 | Check ignition timing at idle after warm up<br>**Ignition timing:**<br>**5 ± 1° BTDC at sea level**<br>**11 ± 2° BTDC at high altitude [1,000 m (3,280 ft)]**<br>[Test connector (Green 1-pin) not grounded<br>Vacuum hoses disconnected] | Yes | Go to Next Step | | |
| | | No | Check ignition timing after warm up<br>**Ignition timing: 5 ± 1 °BTDC**<br>[Test connector (Green 1-pin) grounded Vacuum hoses disconnected] | Yes | Check atmospheric pressure sensor |
| | | | | No | Adjust ignition timing |
| 4 | Check idle speed after warm up<br>**Idle speed: 750 ± 20 rpm (M/T)**<br>**770 ± 20 rpm (A/T)**<br>[Test connector (Green 1-pin) grounded] | Yes | Go to Next Step | | |
| | | No | Try to adjust idle speed | Yes | Idle-speed misadjustment |
| | | | | No | Check accelerator cable free play |
| | | | | | Check ISC valve (Stuck closed) |
| | | | | | Check throttle body |
| 5 | Substitute a well-known ECU<br>Check if condition improves | Yes | | | ECU malfunction |
| | | No | Check voltage at ECU (3B) terminal with SST<br>**Voltage: Approx. 10V (While cranking)** | Yes | Go to Next Step |
| | | | | No | Check starter interlock switch (M/T) |
| | | | | | Check inhibitor switch (A/T) |
| | | | | | Check related wiring |
| 6 | | | | | Poor quality engine oil |

**Low idle speed (When A/C, P/S, E/L ON)**

| STEP | QUICK INSPECTION | | ACTION | POSSIBLE CAUSE |
|---|---|---|---|---|
| 1 | Check for malfunction code (34) with SST<br>[IGN ON, Test connector (Green: 1-pin) grounded] | Yes | Check for cause by referring to check sequence | |
| | | No | Go to Next Step | |
| 2 | Disconnect ISC valve connector at idle<br>Check if the condition does not change | Yes | Go to Next Step | |
| | | No | | Check coolant level |
| | | | | Check engine oil |
| 3 | Check switches with SST<br>Idle switch<br>Neutral switch (M/T)<br>Clutch switch (M/T)<br>Inhibitor switch (A/T)<br>[IGN ON, Test connector (Green: 1-pin) grounded] | Yes | Go to Next Step | |
| | | No | Check for cause by referring to check sequence | |
| 4 | Check continuity between test connector (Green: 1-pin) and ground | | | Wiring short to ground |

**Note**

- **In case of low idle speed with A/C ON, if the problem cannot be solved by the above steps, it may be an A/C compressor malfunction.**

## TROUBLESHOOTING CHART—1990 MPV WITH 4 CYLINDER ENGINE

**High idle speed after warm up**

| STEP | QUICK INSPECTION | | ACTION | | POSSIBLE CAUSE |
|---|---|---|---|---|---|
| 1 | Check for malfunction code (09), (26), (34) with SST<br>[IGN ON, Test connector (Green: 1-pin) grounded] | Yes | Check for cause by referring to check sequence | | |
| | | No | Go to Next Step | | |
| 2 | Check ignition timing at idle after warm up<br>**Ignition timing: 5 ± 1° BTDC**<br>[Test connector (Green: 1-pin) grounded<br>Vacuum hoses disconnected] | Yes | Go to Next Step | | |
| | | No | Adjust ignition timing | | |
| 3 | Check intake manifold vacuum at idle | Yes | Go to Next Step | | |
| | | No | Check for air leaks | Yes | Intake air system components damaged |
| | | | | | Vacuum and intake air hoses loose or damaged |
| | | | | | Bolts or nuts loose |
| | | | | | Gaskets damaged |
| | | | | No | Check vacuum hose routing |
| | | | | | Check throttle body |
| 4 | Check idle speed after warm up<br>**Idle speed: 750 ± 20 rpm**<br>**770 ± 20 rpm**<br>[Test connector (Green: 1-pin) grounded] | Yes | | | Check ISC valve |
| | | No | Try to adjust idle speed | Yes | Idle speed misadjustment |
| | | | | No | Go to Next Step |
| 5 | Disconnect ISC valve connector at idle when engine is cold<br>Check if idle speed decreases during warm up | Yes | Go to Next Step | | |
| | | No | | | Check air valve |
| 6 | Disconnect water thermosensor connector and check if condition improves | Yes | Check water thermosensor connector condition as follows:<br>1. Shake connector and check if condition changes<br>2. Check condition of terminal (burned or damaged)<br>3. Connect a good terminal to harness side connector and check for looseness | Yes | Check water thermosensor |
| | | | | No | Poor contact of water thermosensor connector |
| | | No | Go to Next Step | | |
| 7 | | | | | ECU malfunction |

**Idle hunting or surging**

| STEP | QUICK INSPECTION | | ACTION | | POSSIBLE CAUSE |
|---|---|---|---|---|---|
| 1 | (If trouble occurs only at warm condition)<br>Run engine at **2,000 rpm** for more than **20 seconds**<br>Check for malfunction code (09), (15), (17), (26), (34) with SST<br>[IGN ON, Test connector (Green: 1-pin) grounded] | Yes | Check for cause by referring to check sequence | | |
| | | No | Go to Next Step | | |
| 2 | (If trouble occurs only at warm condition)<br>Check for flashing of SST monitor lamp after warm up<br>**Monitor lamp:**<br>**Flashes more than 8 times 10 seconds at 2,000—3,000 rpm**<br>[Test connector (Green: 1-pin) not grounded] | Yes | Go to Next Step | | |
| | | No | | | Replace oxygen sensor |
| 3 | Check intake manifold vacuum at idle<br>**Vacuum: 480 ± 20 mmHg (18.9 ± 0.79 inHg)** | Yes | Go to Next Step | | |
| | | No | Check for air leaks | Yes | Intake air system components damaged |
| | | | | | Vacuum and air intake hoses loose or damaged |
| | | | | | Bolts or nuts loose |
| | | | | | Gaskets damaged |
| | | | | No | Check throttle body |
| 4 | Pinch PCV hose<br>Check if condition improves | Yes | | | Check PCV valve |
| | | No | Go to Next Step | | |
| 5 | Check fuel line pressure [IGN ON, Test connector (Yellow: 2-pin) connected]<br>**Fuel line pressure: 265—314 kPa (2.7—3.2 kg/cm², 38—46 psi)** | Yes | Go to Next Step | | |
| | | No | Check for fuel leaks | | |
| | | | Substitute a good fuel filter and retest | Yes | Replace fuel filter |
| | | | Check fuel pump maximum pressure<br>**Fuel pump maximum pressure: 441—588 kPa (4.5—6.0 kg/cm², 64—85 psi)** | Yes | Replace pressure regulator |
| | | | | No | Replace fuel pump |
| 6 | | | | | ECU malfunction |

## TROUBLESHOOTING CHART—1990 MPV WITH 4 CYLINDER ENGINE

**Engine stall at idle (Always)**

| STEP | QUICK INSPECTION | | ACTION | | POSSIBLE CAUSE |
|---|---|---|---|---|---|
| 1 | Check for malfunction code (08), (26) with SST [IGN ON, Test connector (Green 1-pin) grounded] | Yes | Check for cause by referring to the check sequence | | |
| | | No | Go to Next Step | | |
| 2 | Slightly open throttle valve and disconnect high-tension lead of each cylinder Check if engine condition changes | Yes | Go to Next Step | | |
| | | No | Check ignition system | | Check spark plug |
| | | | | | Check high-tension lead |
| | | | | | Check distributor cap |
| 3 | Check fuel line pressure [IGN ON, Test connector (Yellow 2-pin) connected] **Fuel line pressure: 265—314 kPa (2.7—3.2 kg/cm², 38—46 psi)** | Yes | Go to Next Step | | |
| | | No | Check for fuel leaks | | |
| | | | Substitute a good fuel filter and retest | Yes | Replace fuel filter |
| | | | Check fuel pump maximum pressure **Fuel pump maximum pressure: 441—588 kPa (4.5—6.0 kg/cm², 64—85 psi)** | Yes | Replace pressure regulator |
| | | | | No | Replace fuel pump |
| 4 | Check if vacuum hoses and the air hoses are connected correctly | Yes | Go to Next Step | | |
| | | No | Connect correctly | | |
| 5 | | | | | Airflow sensor |
| 6 | | | | | ECU malfunction |

**Engine stall at idle (Only when engine is cold)**

| STEP | QUICK INSPECTION | | ACTION | POSSIBLE CAUSE |
|---|---|---|---|---|
| 1 | Check for malfunction code (09), (26) with SST [IGN ON, Test connector (Green 1-pin) grounded] | Yes | Check for cause by referring to check sequence | |
| | | No | Go to Next Step | |
| 2 | | | | Check BAC valve (air valve) |
| 3 | | | | ECU malfunction |

**Engine stall at idle (Only when engine is warm)**

| STEP | QUICK INSPECTION | | ACTION | | POSSIBLE CAUSE |
|---|---|---|---|---|---|
| 1 | Check for malfunction code (09), (26) with SST [IGN ON, Test connector (Green: 1-pin) grounded] | Yes | Check for cause by referring to check sequence | | |
| | | No | Go to Next Step | | |
| 2 | Disconnect water thermosensor connector Check if condition improves | Yes | Check water thermosensor connector as follows 1 Shake connector and check if condition changes 2 Check condition of terminal (burned or damaged) 3 Connect a good terminal to harness side connector and check for looseness | Yes | Check water thermosensor |
| | | | | No | Poor contact of water thermosensor connector |
| | | No | Go to Next Step | | |
| 3 | | | | | ECU malfunction |

**Engine stall at idle (Only after heat soak)**

| STEP | QUICK INSPECTION | ACTION | POSSIBLE CAUSE |
|---|---|---|---|
| 1 | | | ECU malfunction |

## TROUBLESHOOTING CHART—1990 MPV WITH 4 CYLINDER ENGINE

**Engine stall at idle (When A/C, P/S, E/L is ON)**

| STEP | QUICK INSPECTION | | ACTION | POSSIBLE CAUSE |
|---|---|---|---|---|
| 1 | Check for malfunction code (34) with SST [IGN ON, Test connector (Green: 1-pin) grounded] | Yes | Check for cause by referring to check sequence | |
| | | No | Go to Next Step | |
| 2 | Check switches with SST • Headlight switch • Rear defroster switch • Blower switch [IGN ON, Test connector (Green 1-pin) grounded] | Yes | Go to Next Step | |
| | | No | Check for cause by referring to check sequence | |
| 3 | Disconnect ISC valve connector at idle [Test connector (Green 1-pin) grounded] Check if the condition does not change | Yes | Go to Next Step | |
| | | No | | Check ISC valve |
| | | | | Check engine oil |
| 4 | Check idle speed after warm up **Idle speed: 750 ± 20 rpm (M/T) 770 ± 20 rpm (A/T)** [Test connector (Green: 1-pin) grounded] | Yes | Go to Next Step | |
| | | No | Adjust idle speed | |
| 5 | | | | ECU malfunction |

**Note**
- **Engine stalls at idle with A/C ON, if the trouble can not be fixed after checking above steps, it may be A/C compressior malfunction**

**Engine stall at idle (In "D" range)**

| STEP | QUICK INSPECTION | | ACTION | POSSIBLE CAUSE |
|---|---|---|---|---|
| 1 | Check for malfunction code (34) with SST [IGN ON, Test connector (Green 1-pin) grounded] | Yes | Check for cause by referring to check sequence | |
| | | No | Go to Next Step | |
| 2 | Check inhibitor switch with SST [IGN ON, Test connector (Green 1-pin) grounded] | Yes | Go to Next Step | |
| | | No | Check for cause by referring to check sequence | |
| 3 | | | | ECU malfunction |

**Engine stall during start up**

| STEP | QUICK INSPECTION | | ACTION | | POSSIBLE CAUSE |
|---|---|---|---|---|---|
| 1 | Check for malfunction code (08) with SST [IGN ON, Test connector (Green: 1-pin) grounded] | Yes | Check for cause by referring to check sequence | | |
| | | No | Go to Next Step | | |
| 2 | Check idle speed after warm up **Idle speed: 750 ± 20 rpm (M/T) 770 ± 20 rpm (A/T)** [Test connector (Green: 1-pin) grounded] | Yes | Gc to Next Step | | |
| | | No | Adjust idle speed | | |
| 3 | Check for injector operating sound at idle | Yes | Go to Next Step | | |
| | | No | Check resistance at injector harness connector (EMINJ-01) **Terminal: (B/Y)—(LG/B), (B/Y)—(LG/R); Resistance: 6—8Ω** | Yes | Check wiring short or open |
| | | | | No | Cheek injector resistance |
| | | | | | Check wiring |
| 4 | Check ignition timing advance | Yes | Go to Next Step | | |
| | | No | | | Insufficient centrifugal advance: Distributor malfunction |
| | | | Insufficient vacuum advance: Check vacuum hose to distributor for damage or misrouting and delay valve direction | Yes | Distributor malfunction |
| | | | | No | Vacuum hose damaged or misrouted |
| | | | | | Delay valve |
| 5 | | | | | ECU malfunction |

**Engine stall on deceleration (Only when engine is cold)**

| STEP | QUICK INSPECTION | | ACTION | POSSIBLE CAUSE |
|---|---|---|---|---|
| 1 | Check for malfunction code (08), (09) with SST [IG ON, Test connector (Green: 1-pin) grounded] | Yes | Check for cause by referring to check sequence | |
| | | No | Go to Next Step | |
| 2 | | | | ECU malfunction |

## TROUBLESHOOTING CHART—1990 MPV WITH 4 CYLINDER ENGINE

**Engine stall on deceleration (Only after warm up)**

| STEP | QUICK INSPECTION | | ACTION | POSSIBLE CAUSE |
|---|---|---|---|---|
| 1 | Check for malfunction code (08), (09) with SST [IGN ON, Test connector (Green: 1-pin) grounded] | Yes | Check for cause by referring to check sequence | |
| | | No | Go to Next Step | |
| 2 | Check stoplight switch with SST [IGN ON, Test connector (Green: 1-pin) grounded] | Yes | Go to Next Step | |
| | | No | Check for cause by referring to check sequence | |
| 3 | | | | ECU malfunction |

**Engine stall at idle (Intermittent)**

| STEP | QUICK INSPECTION | | ACTION | POSSIBLE CAUSE |
|---|---|---|---|---|
| 1 | Shake connector of ignition coil, main relay and ECU while cranking Check if the engine starts | Yes | There may be a poor contact at the connector. Repair or replace the wiring | |
| | | No | Go to troubleshooting "Engine stall at idle (Always)" | |

**Hesitates/Stumbles on acceleration**

| QUICK | INSPECTION | | ACTION | | POSSIBLE CAUSE |
|---|---|---|---|---|---|
| 1 | Run engine at **2,000 rpm** for **20 seconds** and stop it Check for malfunction code (09), (15), (17) with SST [IGN ON, Test connector (Green: 1-pin) grounded] | Yes | Check for cause by referring to check sequence | | |
| | | No | Go to Next Step | | |
| 2 | Check idle switch with SST [IGN ON, Test connector (Green: 1-pin) grounded] | Yes | Go to Next Step | | |
| | | No | Check for cause by referring to check sequence | | |
| 3 | Disconnect oxygen sensor connector Check if condition improves | Yes | | | Check oxygen Sensor |
| | | No | Go to Next Step | | |
| 4 | Check fuel line pressure while accelerating (Vacuum hose to pressure regulator disconnected) **Fuel line pressure: Keeps 265—314 kPa 2.7—3.2 kg/cm², 38—46 psi)** | Yes | Go to Next Step | | |
| | | No | Substitute a good fuel filter and retest | Yes | Replace fuel filter |
| | | | | | Pressure regulator Replace |
| 5 | Check for air leaks with throttle valve open by listening for sucking noise | Yes | | | Intake air system components damaged |
| | | | | | Vacuum and intake air hoses loose or damaged |
| | | | | | Bolts or nuts loose |
| | | | | | Gaskets damaged |
| | | No | Go to Next Step | | |
| 6 | Substitute a well-known ECU Check if condition improves | Yes | | | ECU malfunction |
| | | No | | | Check airflow sensor |
| | | | | | Check throttle body |
| | | | | | Check vacuum delay valve |
| | | | | | Check spark plug |
| 7 | Check other systems | | | | Clutch slipping |
| | | | | | AT kick-down control malfunction |

## TROUBLESHOOTING CHART—1990 MPV WITH 4 CYLINDER ENGINE

**Hesitates at steady speed**

| STEP | QUICK INSPECTION | | ACTION | | POSSIBLE CAUSE |
|---|---|---|---|---|---|
| 1 | Run engine at **2,000 rpm** for **20 seconds** and stop it Check for malfunction code (08), (09), (12), (14), (15), (17) with SST [IGN ON, Test connector (Green: 1-pin) grounded] | Yes | Check for cause by referring to check sequence | | |
| | | No | Go to Next Step | | |
| 2 | Disconnect oxygen sensor connector Check if condition improves | Yes | | | Check oxygen sensor |
| | | No | Go to Next Step | | |
| 3 | Check for air leaks with throttle valve open by listening for sucking noise | Yes | Go to Next Step | | |
| | | No | | | Intake air system components damaged |
| | | | | | Vacuum and intake air hoses loose or damaged |
| | | | | | Nuts or bolts loose |
| | | | | | Gasket damaged |
| 4 | Check fuel line pressure while accelerating (Vacuum hose to pressure regulator disconnected) **Fuel line pressure: Keeps 265—314 kPa (2.7—3.2 kg/cm², 38—46 psi)** | Yes | Go to Next Step | | |
| | | No | Substitute a good fuel filter and retest | Yes | Replace fuel filter |
| | | | | | Replace pressure regulator |
| 5 | Check condition of ignition coil and airflow meter connectors (Burned or damaged) | Yes | | | Poor contact |
| | | No | Go to Next Step | | |
| 6 | Gradually open throttle valve Check if engine speed increases smoothly | Yes | Go to Next Step | | |
| | | No | | | Check airflow sensor |
| | | | | | Check throttle valve |
| | | | | | Check throttle sensor |
| 7 | | | | | Check spark plug |
| 8 | Change fuel to specified grade Check if condition improves | Yes | | | Poor fuel quality |
| | | No | Go to Next Step | | |
| 9 | | | | | ECU malfunction |

**Jerking on acceleration**

| STEP | QUICK INSPECTION | | ACTION | | POSSIBLE CAUSE |
|---|---|---|---|---|---|
| 1 | Run engine at **2,000 rpm** for **20 seconds** and stop it Check for malfunction code (09), (15), (17) with SST [IGN ON, Test connector (Green: 1-pin) grounded] | Yes | Check for cause by referring to check sequence | | |
| | | No | Go to Next Step | | |
| 2 | Check idle switch with SST [IGN ON, Test connector (Green: 1-pin) grounded] | Yes | Go to Next Step | | |
| | | No | Check for cause by referring to check sequence | | |
| 3 | Disconnect oxygen sensor connector Check if condition improves | Yes | | | Check oxygen Sensor |
| | | No | Go to Next Step | | |
| 4 | Check fuel line pressure while accelerating (Vacuum hose to pressure regulator disconnected) **Fuel line pressure: Keeps 265—314 kPa 2.7—3.2 kg/cm², 38—46 psi)** | Yes | Go to Next Step | | |
| | | No | Substitute a good fuel filter and retest | Yes | Replace fuel filter |
| | | | | | Pressure regulator Replace |
| 5 | Check for air leaks with throttle valve open by listening for sucking noise | Yes | | | Intake air system components damaged |
| | | | | | Vacuum and intake air hoses loose or damaged |
| | | | | | Bolts or nuts loose |
| | | | | | Gaskets damaged |
| | | No | Go to Next Step | | |
| 6 | Substitute a well-known ECU Check if condition improves | Yes | | | ECU malfunction |
| | | No | | | Check airflow sensor |
| | | | | | Check throttle body |
| | | | | | Check vacuum delay valve |
| | | | | | Check spark plug |
| 7 | Check other systems | | | | Clutch slipping |
| | | | | | AT kick-down control malfunction |

## TROUBLESHOOTING CHART—1990 MPV WITH 4 CYLINDER ENGINE

**Knocking**

| STEP | QUICK INSPECTION | | ACTION | | POSSIBLE CAUSE |
|---|---|---|---|---|---|
| 1 | Check malfunction code (09) with SST [IGN ON, Test connector (Green: 1-pin) grounded] | Yes | Check for cause by referring to the check sequence | | |
| | | No | Go to Step 2 (at sea level) Go to Step 3 [at high altitude; **1,000 m, (3,280 ft)**] | | |
| 2 | Check ignition timing after warm up **Ignition timing: 5 ± 1° BTDC** [Test connector (Green: 1-pin) not grounded. Vacuum hoses disconnected] | Yes | Go to Next Step | | |
| | | No | Check atmospheric pressure sensor output voltage **Voltage: 3.5—4.5 V at sea level (IG ON)** | Yes | Ignition timing misadjust |
| | | | | No | Replace atmospheric pressure sensor |
| 3 | Disconnect water thermosensor connector Check if condition improves | Yes | | | Check water thermo-sensor |
| | | No | Go to Next Step | | |
| 4 | Check vacuum routing and vacuum delay valve direction | Yes | Go to Next Step | | |
| | | No | | | Vacuum hose |
| | | | | | Vacuum delay valve |
| 5 | Observe fuel line pressure while accelerating from idle **Fuel line pressure: Keeps 265—314 kPa (2.7—3.2 kg/cm², 38—46 psi)** (Vacuum hose to pressure regulator disconnected) | Yes | Go to Next Step | | |
| | | No | Check fuel pump maximum pressure **Fuel pump maximum pressure: 441—588 kPa (4.5—6.0 kg/cm², 64—85 psi)** | Yes | Replace fuel filter |
| | | | | No | Replace fuel pump |
| 6 | | | | | Check airflow sensor |
| 7 | | | | | Check spark plug |
| 8 | Change fuel to specified grade Check if condition improves | Yes | | | Poor fuel quality |
| | | No | Go to Next Step | | |
| 9 | Check cooling system | | | | Thermostat |
| | | | | | Radiator |
| 10 | | | | | ECU malfunction |

**Poor acceleration**

| STEP | QUICK INSPECTION | | ACTION | | POSSIBLE CAUSE |
|---|---|---|---|---|---|
| 1 | Check for malfunction code (08), (09), (12), (14), (26) with SST [IGN ON, Test connector (Green 1-pin) grounded] | Yes | Check for cause by referring to check sequence | | |
| | | No | Go to Next Step | | |
| 2 | Check idle switch with SST [IGN ON, Test connector (Green 1-pin) grounded] | Yes | Go to Next Step | | |
| | | No | Check for cause by referring to check sequence | | |
| 3 | Disconnect high—tension lead of each cylinder at idle. Check if engine condition changes [ISC valve connector disconnected] | Yes | Go to Next Step | | |
| | | No | Check ignition system | Yes | Replace injector |
| | | | | No | Check spark plug |
| | | | | | Check high-tension |
| | | | | | Check distributor cup |
| 4 | Check ignition timing advance | Yes | Go to Next Step | | |
| | | No | | | Insufficient centrifugal advance. Distributor malfunction |
| | | | Insufficient vacuum advance Check vacuum hose routing and delay valve direction | Yes | Distributor |
| | | | | No | Vacuum hose |
| | | | | | Delay valve |
| 5 | Check for air leaks by listening for sucking noise | Yes | | | Intake air system components damaged |
| | | | | | Vacuum and air intake hoses loose or damaged |
| | | | | | Nuts or bolts loose |
| | | | | | Gasket damaged |
| | | No | Go to Next Step | | |
| 6 | Observe fuel line pressure while accelerating from idle **Fuel line pressure: Keeps 265—314 kPa (2.7—3.2 kg/cm², 38—46 psi)** [Vacuum hose to pressure regulator disconnected] | Yes | Go to Next Step | | |
| | | No | Substitute a good fuel filter and retest | Yes | Replace fuel filter |
| | | | | No | Replace pressure regulator |
| 7 | Gradually depress accelerator from idle Check if engine speed increases smoothly | Yes | Go to Next Step | | |
| | | No | Check accelerator cable free play | Yes | Check airflow sensor |
| | | | | | Check throttle body |
| | | | | No | Adjust |
| 8 | Check fuel to specified grade Check if condition improves | Yes | | | Poor fuel quality |
| | | No | Go to Next Step | | |
| 9 | Substitute a well-known ECU Check if condition improves | Yes | | | ECU malfunction |
| | | No | Go to Next Step | | |
| 10 | Check other systems | | | | Clutch slipping |
| | | | | | Transmission (M/T) |
| | | | | | Transmission (A/T) |
| | | | | | Brake |
| | | | | | Belt tension |

## TROUBLESHOOTING CHART—1990 MPV WITH 4 CYLINDER ENGINE

**Lack of power**

| STEP | QUICK INSPECTION | | ACTION | | POSSIBLE CAUSE |
|---|---|---|---|---|---|
| 1 | Check for malfunction code (08) and (14) (only high-altitude) with SST [IGN ON, Test connector (Green: 1-pin) grounded] | Yes | Check for cause by referring to check sequence | | |
| | | No | Go to Step 2 (High-altitude) Go to Step 3 (Others) | | |
| 2 | (Only at high-altitude) Check ignition timing at idle **Ignition timing: 11 ± 2° BTDC** [Test connector (Green: 1-pin) not grounded Vacuum hoses disconnected] | Yes | Go to Next Step | | |
| | | No | Check ignition timing **Ignition timing: 5 ± 1° BTDC** [Test connector (Green: 1-pin) grounded Vacuum hoses disconnected] | Yes | Check atmospheric pressure sensor |
| | | | | No | Adjust ignition timing |
| 3 | Check ignition timing advance | Yes | Go to Next Step | | |
| | | No | | | Insufficient centrifugal advance: Distributor malfunction |
| | | | Insufficient vacuum advance: Check for vacuum routing and delay valve direction | Yes | Distributor malfunction |
| | | | | No | Vacuum hose |
| | | | | | Delay valve |
| 4 | Disconnect ISC valve connector and the high-tension lead of each cylinder Check if condition changes | Yes | Go to Next Step | | |
| | | No | Check ignition system [Refer to ignition system troubleshooting (Misfire)] | Yes | Replace injector (If step 5 OK) |
| | | | | No | Check high-tension lead |
| | | | | | Check distributor cap |
| | | | | | Check spark plug |
| 5 | Check for injector operating sound at idle | Yes | Go to Next Step | | |
| | | No | Chcek resistance at injector harness connector (EMINJ-01) Terminals: [illegible] Resistance: 6—8Ω | Yes | Check wiring short or open |
| | | | | No | Check injector resistance |
| | | | | | Check wiring short or open |
| 6 | Check air cleaner element for clogging | Yes | Go to Next Step | | |
| | | No | Clean air cleaner element | | |
| 7 | Check for air leaks by listening for sucking noises • At idle • When throttle valve is open | Yes | | | Intake air system |
| | | | | | Components damaged |
| | | | | | Vacuum and air intake hoses loose or damaged |
| | | | | | Nuts or bolts loose |
| | | | | | Gasket damaged |

**Lack of power (Cont'd)**

| STEP | QUICK INSPECTION | | ACTION | | POSSIBLE CAUSE |
|---|---|---|---|---|---|
| 8 | Check fuel line pressure [IGN ON, Test connector (Yellow: 2-pin) connected] **Fuel line pressure: 265—314 kPa (2.7—3.2 kg/cm², 38—46 psi)** | Yes | Go to Next Step | | |
| | | No | Check for fuel leakage | | |
| | | | Substitute a good fuel filter and retest | Yes | Replace fuel filter |
| | | | Check fuel pump maximum pressure **Fuel pump maximum pressure: 441—588 kPa (4.5—6.0 kg/cm², 64—85 psi)** F2–156 | Yes | Replace pressure regulator |
| | | | | No | Replace fuel pump |
| 9 | Check fuel line pressure at idle **Fuel line pressure: 206—255 kPa (2.1—2.6 kg/cm², 30—37 psi)** | Yes | Go to Next Step | | |
| | | No | | | Replace pressure regulator |
| 10 | Check if fuel line pressure drops while accelerating (Vacuum hose disconnected) | Yes | Substitute a good fuel filter and retest | Yes | Replace fuel filter |
| | | No | Go to Next Step | | |
| 11 | Check exhaust system for damage | Yes | Go to Next Step | | |
| | | No | Repair or replace | | |
| 12 | Check A/C, P/S and alternator belts tensions | Yes | Go to Next Step | | |
| | | No | Adjust belt tension | | |
| 13 | Check if accelerator can be depressed fully | Yes | Go to Next Step | | |
| | | No | Check accelerator cable | Yes | Throttle body |
| | | | | No | Accelerator cable |
| 14 | Substitute a well-known ECU Check if condition improves | Yes | | | ECU malfunction |
| | | No | | | Check air flow sensor |
| | | | | | Check throttle sensor |
| | | | | | Go to Next Step |
| 15 | Substitute a specified fuel Check if condition improves | Yes | | | Poor fuel quality |
| | | No | Go to Next Step | | |
| 16 | Check other systems | | | | Brake |
| | | | | | Clutch |
| | | | | | A/T |
| | | | | | Engine |

## TROUBLESHOOTING CHART—1990 MPV WITH 4 CYLINDER ENGINE

**Bucking at high speed**

| STEP | QUICK INSPECTION | | ACTION | | POSSIBLE CAUSE |
|---|---|---|---|---|---|
| 1 | Run engine at **2,000 rpm** for more than **20 seconds** Check for malfunction code (08), (09), (12), (14), (15), (17) with SST [IGN ON, Test connector (Green 1-pin) grounded] | Yes | Check for cause by referring to check sequence | | |
| | | No | Go to Next Step | | |
| 2 | Disconnect oxygen sensor connector Check if condition improves | Yes | | | Check oxygen sensor |
| | | No | Go to Next Step | | |
| 3 | Observe fuel line pressure while accelerating from idle **Fuel line pressure: Keeps 265—314 kPa (2.7—3.2 kg/cm², 38—46 psi)** [Vacuum hose to pressure regulator disconnected] | Yes | Go to Next Step | | |
| | | No | Substitute a good fuel filter and retest | Yes | Replace fuel filter |
| | | | | | Replace pressure regulator |
| 4 | Check for air leaks by listening sucking noise | Yes | Go to Next Step | | |
| | | No | | | Intake air system components damaged |
| | | | | | Vacuum and air intake hoses loose or damaged |
| | | | | | Nuts or bolts loose |
| | | | | | Gasket damaged |
| 5 | Check ignition timing advance | Yes | Go to Next Step | | |
| | | No | | | Insufficient centrifugal advance: Distributor malfunction |
| | | | Insufficient vacuum advance: Check vacuum hose routing and delay valve direction | Yes | Distributor |
| | | | | No | Vacuum hose |
| | | | | | Delay valve |
| 6 | Gradually open throttle valve from idle check if engine speed increases smoothly | Yes | Go to Next Step | | |
| | | No | | | Check airflow sensor |
| 7 | | | | | Check spark plug |
| 8 | | | | | ECU malfunction |

**Bucking on deceleration**

| STEP | QUICK INSPECTION | | ACTION | POSSIBLE CAUSE |
|---|---|---|---|---|
| 1 | Check for malfunction code (08), (09), (12), (26), (34) with SST [IGN ON, Test connector (Green 1-pin) grounded] | Yes | Check for cause by referring to the check sequence | |
| | | No | Go to Next Step | |
| 2 | Check switches with SST [IGN ON, Test connector (Green 1-pin) grounded] • Idle switch • Stoplight switch | Yes | Go to Next Step | |
| | | No | Check for cause by referring to the check sequence | |
| 3 | Substitute a well-known ECU Check if condition improves | Yes | | ECU malfunction |
| | | No | | Check throttle sensor |
| | | | | Go to Next Step |
| 4 | | | | Check spark plug |
| 5 | | | | Check clutch slipping |
| 6 | | | | Check compression between cylinders |

## TROUBLESHOOTING CHART—1990 MPV WITH 4 CYLINDER ENGINE

**Poor fuel economy**

| STEP | QUICK INSPECTION | | ACTION | | POSSIBLE CAUSE |
|---|---|---|---|---|---|
| 1 | Run the engine at **2,000 rpm** for more than **20 seconds** after warm up and stop it Check for malfunction code (15) (17) with SST [IGN ON, Test connector (Green 1-pin ) grounded] | Yes | Check for cause by referring to check sequence | | |
| | | No | Go to Next Step | | |
| 2 | Check idle switch with SST [IGN ON, Test connector (Green 1-pin) grounded] | Yes | Go to Next Step | | |
| | | No | Check for cause by referring to check sequence | | |
| 3 | Check for flashing of monitor lamp after warm up **Monitor lamp: Flashes more than 8 times /10 seconds at 2,000—3,000 rpm** [Test connector (Green 1-pin) not grounded] | Yes | Go to Next Step | | |
| | | No | | | Replace oxygen sensor |
| 4 | Check fuel line pressure at idle **Fuel line pressure: 206—255 kPa (2.1—2.6 kg/cm², 30—37 psi)** | Yes | Go to Next Step | | |
| | | No | Check vacuum line to pressure regulator for clogging or air leakage | Yes | Vacuum line clogging or damaged |
| | | | | No | Check solenoid valve (PRC) |
| | | | | | ECU malfunction (Check (2N) terminal voltage) |
| | | | | | Replace pressure regulator |
| 5 | Check for fuel cut operation during deceleration **Fuel cut: Above 1,600 rpm after warm up** | Yes | Go to Next Step | | |
| | | No | Check water thermosensor | Yes | Replace ECU |
| | | | | No | Replace water thermosensor |
| 6 | Check ignition timing advance | Yes | Go to Next Step | | |
| | | No | | | Insufficient centrifugal advance: Distributor malfunction |
| | | | Insufficient vacuum advance: Check for vacuum routing and delay valve direction | Yes | Distributor |
| | | | | No | Vacuum hose |
| | | | | | Delay valve |
| 7 | Check other systems | | | | Clutch slipping |
| | | | | | A/T malfunction |
| | | | | | Brake |
| | | | | | Tire air pressure |

**High oil consumption/White exhaust smoke**

| STEP | QUICK INSPECTION | | ACTION | | POSSIBLE CAUSE |
|---|---|---|---|---|---|
| 1 | Check for oil leak from engine | Yes | Repair or replace | | |
| | | No | Go to Next Step | | |
| 2 | Disconnect PCV valve from engine Check if vacuum is felt at idle | Yes | Go to Next Step | | |
| | | No | Check PCV valve | Yes | PCV hose clogging |
| | | | | No | Replace PCV valve |
| 3 | Check that ventilation hose is installed correctly | Yes | Go to Next Step | | |
| | | No | Install ventilation hose correctly | | |
| 4 | Possible malfunction of engine | | | | |

**Afterburn on deceleration**

| STEP | QUICK INSPECTION | | ACTION | | POSSIBLE CAUSE |
|---|---|---|---|---|---|
| 1 | Check malfunction code (34) with SST [IGN ON, Test connector (Green 1-pin) grounded] | Yes | Check for cause by referring to the check sequence | | |
| | | No | Go to Next Step | | |
| 2 | Check idle switch with SST [IGN ON, Test connector (Green 1-pin) grounded] | Yes | Go to Next Step | | |
| | | No | Check for cause by referring to the check sequence | | |
| 3 | Check ignition timing advance | Yes | Go to Next Step | | |
| | | No | | | Insufficient centrifugal advance: Distributor malfuction |
| | | | Insufficient Vacuum advance. Check for vacuum routing and delay valve direction | Yes | Distributor malfunction |
| | | | | No | Vacuum hose |
| | | | | | Delay valve |
| 4 | Check air cleaner element for clogging | Yes | Go to Next Step | | |
| | | No | Clean air cleaner element | | |
| 5 | Check fuel cut operation during deceleration **Fuel cut: Above 1,600 rpm after warm up** | Yes | Go to Next Step | | |
| | | No | Check water thermosensor | Yes | ECU malfunction Check (2I) terminal voltage |
| | | | | No | Replace water thermosensor |
| 6 | Run engine at idle and stop it (IG OFF) Observe fuel pressure for **5 minutes** **Fuel pressure: More than 147 kPa (1.5 kg/cm², 21 psi)** | Yes | Go to Next Step | | |
| | | No | Check fuel pump for pressure drop | No | Replace fuel pump |
| | | | Check pressure regulator for pressure drop | Yes | Check injector fuel leakage |
| | | | | No | Replace pressure regulator |
| 7 | | | | | Check compression |
| | | | | | Check valve timing |

## TROUBLESHOOTING CHART—1990 MPV WITH 4 CYLINDER ENGINE

**Rotten egg smell**

| STEP | QUICK INSPECTION | ACTION | POSSIBLE CAUSE |
|---|---|---|---|
| 1 | Change fuel to specified grade<br>Check if condition improves | | Poor fuel quality |

**Black exhaust smoke**

| STEP | QUICK INSPECTION | | ACTION | | POSSIBLE CAUSE |
|---|---|---|---|---|---|
| 1 | Run engine at **2,000 rpm** for more then **20 seconds**<br>Check for malfunction code (17) with SST<br>[IGN ON, Test connector (Green 1-pin grounded] | Yes | Check for cause by referring to the check sequence | | |
| | | No | Go to Next Step | | |
| 2 | Check fuel line pressure<br>[IGN ON, Test connector (Yellow 2-pin) connected]<br>**Fuel line pressure:**<br>**265—314 kPa (2.7—3.2 kg/cm², 38—46 psi)** | Yes | Go to Next Step | | |
| | | No (high) | | | Replace pressure regulator |
| 3 | Disconnect water thermosensor connector<br>Check if condition improves | Yes | Check water thermosensor connector condition as follows<br>1 Shake connector and check if condition changes<br>2 Check condition of terminal (burned or damaged)<br>3 Connect a good terminal to harness side connector and check for looseness | Yes | Check water thermosensor |
| | | | | No | Poor contact of water thermosensor connector |
| | | No | Go to Step 4 [at high-altitude (**1,000 m, 3,280 ft**)]<br>Go to Step 5 (at sea level) | | |
| 4 | | | | | Check atmospheric pressure sensor |
| 5 | | | | | Check airflow sensor |
| 6 | | | | | ECU malfunction |

**Fails emission test (Engine condition OK)**

| STEP | QUICK INSPECTION | | ACTION | | | POSSIBLE CAUSE |
|---|---|---|---|---|---|---|
| 1 | Run engine at **2,000 rpm** for **20 seconds** and stop it<br>Check for malfunction code (08), (09), (14), (15), (17), (26) with SST<br>[IGN ON, Test connector (Green: 1-pin) grounded] | Yes | Check for cause by referring to the check sequence | | | |
| | | No | Go to Next Step | | | |
| 2 | Check idle switch with SST<br>[IGN ON, Test connector (Green: 1-pin) grounded] | Yes | Go to Next Step | | | |
| | | No | Check for cause by referring to the check sequence | | | |
| 3 | Check for air leaks by listening for sucking noise<br>• At idle<br>• Throttle valve open | Yes | Go to Next Step | | | |
| | | No | | | | |
| 4 | Check for flashing of SST monitor lamp after warm up<br>**Monitor lamp:**<br>**Flashes more than 8 times/10 seconds at 2,000—3,000 rpm**<br>[Test connector (Green: 1-pin) not grounded] | Yes | Go to Next Step | | | |
| | | No | | | | Replace oxygen sensor |
| 5 | Check fuel line pressure<br>[IG ON, Test connector (Yellow: 2-pin) connected]<br>**Fuel line pressure:**<br>**265—314 kPa (2.7—3.2 kg/cm², 38—46 psi)** | Yes | Go to Next Step | | | |
| | | No (low) | Check for fuel leakage | | | |
| | | | Substitute a good fuel filter and retest | | Yes | Replace fuel filter |
| | | | Check fuel pump maximum pressure<br>**Fuel pump maximum pressure:**<br>**441—588 kPa (4.5—6.0 kg/cm², 64—85 psi)** | | Yes | Replace pressure regulator |
| | | | | | No | Replace fuel pump |
| | | No (high) | | | | Replace pressure regulator |
| 6 | Check fuel line pressure at idle<br>**Fuel line pressure:**<br>**206—255 kPa (2.1—2.6 kg/cm², 30—37 psi)** | Yes | Go to Next Step | | | |
| | | No | Check vacuum line to pressure regulator for clogging or air leakage | | Yes | Vacuum line |
| | | | | | No | Check solenoid valve (PRC) |
| | | | | | | ECU malfunction (2N) |
| | | | | | | Replace pressure regulator |
| 7 | Check fuel cut operation while decelerating<br>**Fuel cut:**<br>**Above 1,600 rpm after warm up** | Yes | Go to Next Step | | | |
| | | No | Check water thermosensor | | Yes | Replace ECU |
| | | | | | No | Replace water thermosensor |
| 8 | Check for fuel fumes | Yes | Check for cause by referring to troubleshooting "Gasoline fumes" | | | |
| | | No | Go to Next Step | | | |
| 9 | | | | | | Check PCV valve |
| 10 | | | | | | Check Spark plugs |
| 11 | | | | | | Poor fuel quality |

## TROUBLESHOOTING CHART—1990 MPV WITH 4 CYLINDER ENGINE

**Gasoline fumes**

| STEP | QUICK INSPECTION | | ACTION | | POSSIBLE CAUSE |
|---|---|---|---|---|---|
| 1 | Check for leaks | Yes | Replace | | |
| | | No | Go to Next Step | | |
| 2 | Check if fumes are emitted from check-and-cut valve | Yes | Check check-and cut valve | Yes | Check two-way check valve |
| | | | | | Purge line clogging |
| | | | | No | Replace check-and-cut valve |
| | | No | Go to Next Step | | |
| 3 | Check for malfunction code (08), (15), (09), (17), (26) with SST<br>[IGN ON, Test connector (Green: 1-pin) grounded] | Yes | Check for cause by referring to the check sequence | | |
| | | No | Go to Next Step | | |
| 4 | Check switches with SST<br>• Idle switch<br>• Neutral switch (M/T)<br>• Clutch switch (M/T)<br>• Inhibitor switch (A/T)<br>[IGN ON, Test connector (Green: 1-pin) grounded] | Yes | Go to Next Step | | |
| | | No | Check for cause by referring to the check sequence | | |
| 5 | Run engine at idle. Ground the solenoid valve (Purge control) terminal-wire (L/Y) and disconnect vacuum hose (white) from solenoid valve. Check for vacuum at solenoid valve | Yes | | | ECU malfunction<br>Check (2P) terminal voltage |
| | | No | | | Replace solenoid valve (Purge control) |

**STEP 1**

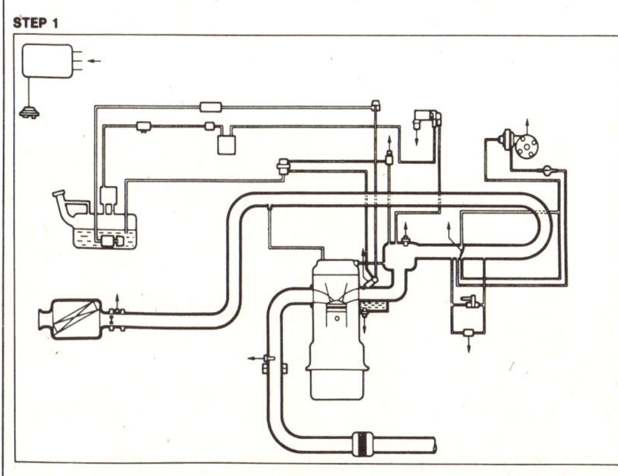

**MIL always ON (Engine condition is OK)**

| STEP | QUICK INSPECTION | | ACTION | | POSSIBLE CAUSE |
|---|---|---|---|---|---|
| 1 | Check for malfunction code with SST<br>[IGN ON, Test connector (Green: 1-pin) grounded] | | "**88**" Replace ECU | | |
| | | | "**00**" (Except California)<br>Go to Step 2 | | |
| | | | "**00**" (California only)<br>Wiring between ECU (1A) terminal and MIL short to ground | | |
| 2 | (Except California)<br>Check if oxygen sensor replacement time has come<br>**Oxygen sensor replacement schedule:**<br>**Every 80,000 miles (Federal)**<br>**Every 120,000 km (Canada)** | Yes | Check if MIL has been reset by exchanging MIL set screw position | Yes | Go to next action |
| | | | | No | Reset the MIL |
| | | | Check continuity between MIL and ground | Yes | Wiring between MIL and ECU is short to ground |
| | | | | No | Replace mileage sensor |
| | | No | Check continuity between MIL and ground | Yes | Wiring between MIL and ECU shorted |
| | | | | No | Replace mileage sensor |

**MIL never ON (Engine condition is not OK)**

| STEP | QUICK INSPECTION | | ACTION | | POSSIBLE CAUSE |
|---|---|---|---|---|---|
| 1 | Check if other indicator lamps illuminate | Yes | Go to Next Step | | |
| | | No | Check power supply circuit to combination meter | | |
| 2 | Check bulb of the MIL | Yes | Ground ECU (1A) terminal<br>Check if MIL illuminates | Yes | Replace ECU |
| | | | | No | Wiring between ECU and MIL open |
| | | No | Replace | | |
| 3 | (Except California)<br>Check if the MIL illuminated at oxygen sensor replacement time<br>**Oxygen sensor replacement time:**<br>**Every 80,000 miles (Federal)**<br>**Every 120,000 km (Canada)** | | | | (Except California)<br>MIL set screw loose or left out |
| | | | | | (Except California)<br>Replace mileage sensor |

**A/C does not work**

| STEP | QUICK INSPECTION | | ACTION | | POSSIBLE CAUSE |
|---|---|---|---|---|---|
| 1 | Check if condenser fan operates when grounding A/C relay terminal-wire (R/W) (IGN ON) | Yes | Check voltage at ECU (1L) terminal with SST<br>Voltage at idle after warm up: 0V<br>(A/C and blower switches ON) | Yes | ECU malfunction<br>(Check (1C) terminal voltage) |
| | | | | | Wiring between ECU (1C) and A/C relay open |
| | | | | No | A/C system malfunction |
| | | No | Check A/C system | | |

## MALFUNCTION INDICATOR LIGHT (MIL) OPERATION 1990 MPV WITH 4 CYLINDER ENGINE

**PRINCIPLE OF CODE CYCLE**

Malfunction codes are determined as shown below

**1. Code cycle break**
The time between malfunction code cycles is 4.0 sec (the time the light is off).

1 CYCLE
MALFUNCTION CODE
CYCLE BREAK
ON
OFF
4.0 SEC.

**2. Second digit of malfunction code (ones position)**
The digit in the ones position of the malfunction code represents the number of times the buzzer is on 0.4 sec during one cycle.

1 CYCLE
MALFUNCTION CODE
MALFUNCTION CODE: 03
1 2 3
CYCLE BREAK
ON
OFF
0.4 SEC.
0.4 SEC.
4.0 SEC.

**3. First digit of malfunction code (tens position)**
The digit in the tens position of the malfunction code represents the number of times the buzzer is on 1.2 sec during one cycle.

It should also be noted that the light goes off for 1.6 sec. between the long and short pulses of the buzzer.

1 cycle
MALFUNCTION CODE
MALFUNCTION CODE: 22
1 2
1 2
CYCLE BREAK
0.4
1.2 SEC.
SEC.
1.6 SEC.
4.0 SEC.
FIRST DIGIT
SECOND DIGIT

## TROUBLE CODE IDENTIFICATION CHART 1990 MPV WITH 4 CYLINDER ENGINE

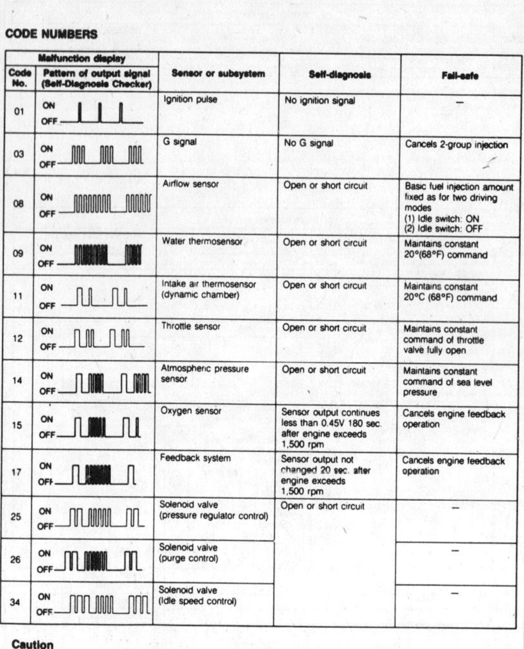

**CODE NUMBERS**

| Malfunction display | | Sensor or subsystem | Self-diagnosis | Fail-safe |
|---|---|---|---|---|
| Code No. | Pattern of output signal (Self-Diagnosis Checker) | | | |
| 01 | ON / OFF | Ignition pulse | No ignition signal | — |
| 03 | ON / OFF | G signal | No G signal | Cancels 2-group injection |
| 08 | ON / OFF | Airflow sensor | Open or short circuit | Basic fuel injection amount fixed as for two driving modes (1) Idle switch: ON (2) Idle switch: OFF |
| 09 | ON / OFF | Water thermosensor | Open or short circuit | Maintains constant 20°(68°F) command |
| 11 | ON / OFF | Intake air thermosensor (dynamic chamber) | Open or short circuit | Maintains constant 20°C (68°F) command |
| 12 | ON / OFF | Throttle sensor | Open or short circuit | Maintains constant command of throttle valve fully open |
| 14 | ON / OFF | Atmospheric pressure sensor | Open or short circuit | Maintains constant command of sea level pressure |
| 15 | ON / OFF | Oxygen sensor | Sensor output continues less than 0.45V 180 sec. after engine exceeds 1,500 rpm | Cancels engine feedback operation |
| 17 | ON / OFF | Feedback system | Sensor output not changed 20 sec. after engine exceeds 1,500 rpm | Cancels engine feedback operation |
| 25 | ON / OFF | Solenoid valve (pressure regulator control) | Open or short circuit | — |
| 26 | ON / OFF | Solenoid valve (purge control) | | — |
| 34 | ON / OFF | Solenoid valve (Idle speed control) | | — |

**Caution**

- **If there is more than one failure present, the lowest number malfunction code is displayed first, the remaining codes are displayed in order.**
- **After repairing all failures, turn off the ignition switch, disconnect the negative battery cable, and depress the brake pedal for 5 sec to erase the memory of a malfunction code from the engine control unit.**

## TROUBLE CODE DIAGNOSTIC CHART—1990 MPV WITH 4 CYLINDER ENGINE

If a malfunction code number is shown on the **SST**, check the following chart along with the wiring diagram.

**Code No.01 (Ignition pulse)** — **PC: Possible Cause**

- Are there poor connections in ignition coil circuit?
  - YES → Repair or replace connector
  - NO ↓
- Is resistance of ignition coil OK? Resistance: Primary 0.77—0.95Ω Secondary 6—30 kΩ
  - NO → Replace ignition coil
  - YES ↓
- Is there continuity between ignition coil (−) terminal wire and engine control unit 1M terminal?
  - NO → PC: Open circuit in wiring harness from ignition coil to engine control unit 1M terminal
  - YES ↓
- Is same Code No. present after performing after-repair procedure?
  - NO → Ignition pulse and circuit OK
  - YES ↓
- Is engine control unit (1M) terminal voltage OK?
  - NO → PC: • No power supply to ignition coil • Short circuit in wiring harness
  - YES → PC: Engine control unit malfunction

**Code No.03 (G signal)** — **PC: Possible Cause**

- Are there any poor connections at distributor connectors?
  - YES → Repair or replace connector
  - NO ↓
- Is G rotor OK? (visual inspection)
  - NO → Replace G rotor
  - YES ↓
- Is there continuity between distributor G signal (R/L wire) and engine control unit (1N) terminal?
  - NO → PC: Open circuit in wiring harness from G signal to engine control unit (1N) terminal
  - YES ↓
- Is engine control unit (1N) terminal voltage OK?
  - NO → PC: • Short circuit in wiring harness • Replace pickup coil
  - YES → PC: Engine control unit malfunction

**Code No.08 (Airflow sensor)** — **PC: Possible Cause**

- Are there poor connections in airflow sensor circuit?
  - YES → Repair or replace connector
  - NO ↓
- Is voltage at terminal D of airflow sensor connector OK?
  - YES → Is same Code No. present after performing after-repair procedure?
    - NO → Airflow sensor and circuit OK
    - YES → PC: Engine control unit malfunction
  - NO ↓
- Is protection net of airflow sensor damaged or restricted?
  - YES → Clean or repair protection net
  - NO ↓
- Is there continuity between airflow sensor and engine control unit?

| Airflow sensor | Engine control unit |
|---|---|
| D (G/B wire) | 2E |
| E (G/O wire) | 2B |

  - NO → PC: Open circuit in wiring harness
  - YES ↓
- Is there 12V at terminal B of airflow sensor connector?
  - NO → PC: Open or short circuit in wiring harness from terminal B of airflow sensor to main relay
  - YES ↓
- Is there continuity between terminal F of airflow sensor and ground?
  - NO → PC: Open circuit in wiring harness
  - YES → PC: • Engine control unit malfunction • Short circuit in wiring from terminal D of airflow sensor to engine control unit

## TROUBLE CODE DIAGNOSTIC CHART—1990 MPV WITH 4 CYLINDER ENGINE

**Code No.09 (Water thermosensor)** **PC: Possible Cause**

Are there poor connections at water thermosensor circuit? — YES → Repair or replace connector

NO ↓

Is there continuity between water thermosensor and engine control unit?

| Water thermosensor | Engine control unit |
|---|---|
| A (G/W wire) | 2I |
| B (BR/B wire) | 2C |

NO → PC: Open circuit in wiring harness from water thermosensor to engine control unit

YES ↓

Is resistance of the water thermosensor OK?
Resistance:

| Coolant temp | Resistance |
|---|---|
| −20°C (−4°F) | 14.5 —17.8 kΩ |
| 20°C (68°F) | 2.2 — 2.7 kΩ |
| 80°C (176°F) | 0.28— 0.35 kΩ |

NO → Replace water thermosensor

YES ↓

Is same Code No. present after performing after-repair procedure? — NO → Water thermosensor and circuit OK

YES ↓

Are engine control unit 2I and 2C terminal voltages OK? — NO → PC: Engine short circuit in wiring harness

YES → PC: Engine control unit malfunction

**No.11 Code (Intake air thermosensor)** **PC: Possible Cause**

Are there poor connections at intake air thermosensor connectors? — YES → Repair or replace connector

NO ↓

Is there continuity between intake air thermosensor (dynamic chamber) and engine control unit?

| Intake air thermosensor (dynamic chamber) | Engine control unit |
|---|---|
| A (G wire) | 2L |
| B (BR/B wire) | 2C |

NO → PC: Open circuit in wiring harness

YES ↓

Is resistance of intake air thermosensor (dynamic chamber) OK?
Resistance:

| Temperature | Resistance |
|---|---|
| 25°C (77°F) | 29.7—36.3 kΩ |
| 85°C (185°F) | 3.3— 3.7 kΩ |

NO → Replace intake air thermosensor (dynamic chamber)

YES ↓

Is same Code No. present after performing after-repair procedure? — NO → Intake air thermosensor and circuit OK

YES ↓

Are engine control unit 2L and 2C terminal voltages OK? — NO → PC: Short circuit in wiring harness

YES → PC: Engine control unit malfunction

## TROUBLE CODE DIAGNOSTIC CHART—1990 MPV WITH 4 CYLINDER ENGINE

**Code No.12 (Throttle sensor)** **PC: Possible cause**

Are there poor connections in throttle sensor circuit? — YES → Repair or replace connector terminal

NO ↓

Is there 4.5—5.5V at A terminal of throttle sensor connector? — NO → PC: • Open or short circuit in wiring harness from A terminal of throttle sensor to 2A terminal of engine control unit • Engine control unit malfunction

YES ↓

Is C terminal of throttle sensor connector grounded? — NO → PC: Open circuit in wiring harness from C terminal of throttle sensor to ground

YES ↓

Is throttle sensor adjusted correctly? — NO → Adjust throttle sensor

YES ↓

Is same Code No. present after performing after-repair procedure? — NO → Throttle sensor and circuit OK

YES ↓

Is engine control unit 2G terminal voltage OK? — NO → PC: Open or short circuit in wiring harness from B terminal of throttle sensor to 2G terminal of engine control unit

YES → PC: Engine control unit malfunction

**Code No.14 (Atmospheric pressure sensor)** **PC: Possible cause**

Are there poor connections in atmospheric pressure sensor circuit? — YES → Repair or replace connector

NO ↓

Is there Vref at (BR/W) wire of atmospheric pressure sensor connector?
Vref: 4.5—5.5V — NO → PC: Open or short circuit in wiring harness from (BR/W) wire to engine control unit • Engine control unit malfunction (If 2A terminal voltage not 4.5—5.5V)

YES ↓

Is voltage at (L) wire of atmospheric pressure sensor connector OK?

| Condition | Voltage |
|---|---|
| At sea level | 3.5—4.5V |
| At high altitude (2,000 m (6,500 ft)) | 2.5—3.5V |

NO (0V) → PC: Short circuit in wiring harness from atmospheric pressure sensor (L) wire to engine control unit (2H) terminal

NO (Others) → Replace atmospheric pressure sensor

YES ↓

Is there continuity between atmospheric pressure sensor to engine control unit?

| Atmospheric pressure sensor | Control unit |
|---|---|
| A (BR/B wire) | 2C |
| D (L wire) | 2H |

NO → PC: Open circuit in wiring harness from atmospheric pressure sensor to engine control unit

YES ↓

Is same Code No. present after performing after-repair procedure? — NO → Atmospheric pressure sensor and circuit OK

YES ↓

Are engine control unit 2C and 2H terminal voltages OK? — NO → PC: Short circuit in wiring harness

YES → PC: Engine control unit malfunction

## TROUBLE CODE DIAGNOSTIC CHART—1990 MPV WITH 4 CYLINDER ENGINE

### Code No.15 (Oxygen sensor)

**PC: Possible Cause**

**Note**
- **When Codes No.15 and 17 are present at the same time, first perform the checking procedure for Code No.17.**

- Are there poor connections in oxygen sensor circuit? — **YES** → Repair or replace connector
- **NO** ↓ Is oxygen sensor output voltage OK? — **NO** → Replace oxygen sensor
- **YES** ↓ Is sensitivity of oxygen sensor OK? — **NO** → PC: • Oxygen sensor malfunction • Engine control unit malfunction
- **YES** ↓ Is same Code No. present after performing after-repair procedure? — **NO** → Oxygen sensor and circuit OK
- **YES** ↓ Is engine control unit 2D terminal voltage OK? — **NO** → PC: Open or short circuit in wiring harness between oxygen sensor and engine control unit 2D terminal
- **YES** → PC: Engine control unit malfunction

### Code No.17 (Feedback system)

**PC: Possible Cause**

- Warm up engine and run it at 2,500–3,000 rpm for three minutes. Does monitor lamp of Self-Diagnosis Checker illuminate at idle? — **NO** → PC: • Air leak in vacuum hoses or emission component • Contaminated oxygen sensor • Insufficient fuel injection
- **YES** ↓ Are spark plugs clean? — **NO** → Clean or replace spark plugs
- **YES** ↓ Is oxygen sensor voltage OK? — **NO** → PC: Oxygen sensor malfunction
- **YES** ↓ Is same Code No. present after performing after-repair procedure? — **NO** → Feedback system OK
- **YES** ↓ Is engine control unit 2D terminal voltage OK? — **NO** → PC: Open or short circuit in wiring harness between oxygen sensor and engine control unit 2D terminal
- **YES** → PC: Engine control unit malfunction

### Code No.25 (Solenoid valve-Pressure regulator control)

**PC: Possible Cause**

- Are there poor connections in solenoid valve circuit? — **YES** → Repair or replace connector
- **NO** ↓ Is there continuity of solenoid valve? — **NO** → Replace solenoid valve
- **YES** ↓ Is there battery voltage at (B/W) wire of solenoid valve connector? — **NO** → PC: Open circuit in wiring harness from (B/W) wire to main relay
- **YES** ↓ Is there continuity between solenoid valve and engine control unit? — **NO** → PC: Open circuit in wiring harness from solenoid valve to engine control unit

| Solenoid valve | Engine control unit |
|---|---|
| B (L/B wire) | 2N |

- **YES** → PC: • Short circuit in wiring harness • Engine control unit malfunction

### Code No.26 (Solenoid valve-Purge control)

- Are there poor connections in solenoid valve circuit? — **YES** → Repair or replace connector
- **NO** ↓ Is there continuity of solenoid valve? — **NO** → Replace solenoid valve
- **YES** ↓ Is there battery voltage at (B/W) wire of solenoid valve circuit? — **NO** → PC: Open circuit in wiring harness from (B/W) wire to main relay
- **YES** ↓ Is there continuity between solenoid valve and engine control unit? — **NO** → PC: Open circuit in wiring harness from solenoid valve to engine control unit

| Solenoid valve | Engine control unit |
|---|---|
| B (L/Y wire) | 2P |

- **YES** → PC: • Short circuit in wiring harness • Engine control unit malfunction

## TROUBLE CODE DIAGNOSTIC CHART 1990 MPV WITH 4 CYLINDER ENGINE

### Code No.34 (Solenoid valve—Idle speed control (ISC))

**PC: Possible Cause**

- Are there poor connections in solenoid valve circuit? — **YES** → Repair or replace connector
- **NO** ↓ Is resistance of solenoid valve OK? Resistance: 7.7–9.3Ω — **NO** → Replace BAC valve
- **YES** ↓ Is there battery voltage at (B/W) wire of solenoid valve connector? — **NO** → PC: Open or short circuit in wiring harness from (B/W) wire to main relay
- **YES** ↓ Is there continuity between solenoid valve and engine control unit? — **NO** → PC: Open circuit in wiring harness from solenoid valve to engine control unit

| Solenoid valve | Engine control unit |
|---|---|
| B (L/G wire) | 2Q |

- **YES** → PC: • Short circuit in wiring harness • Engine control unit malfunction

## RESETTING MIL—1990 MPV WITH 4 CYLINDER ENGINE

**How To Reset the MIL**

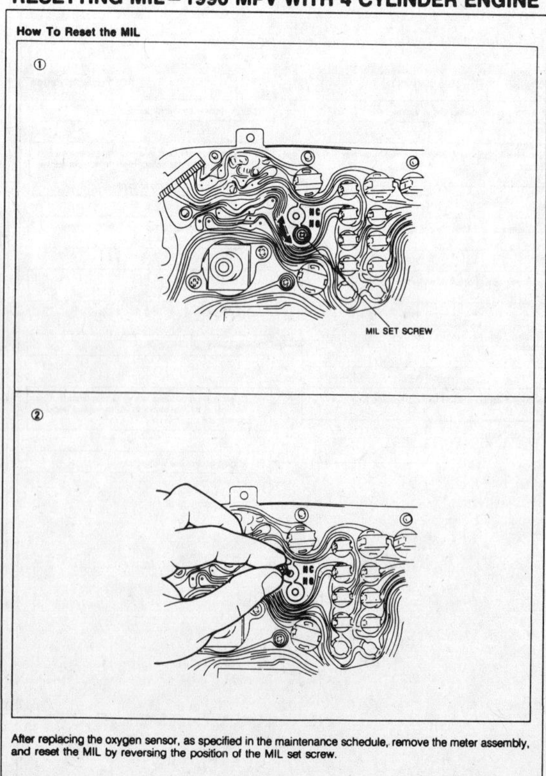


After replacing the oxygen sensor, as specified in the maintenance schedule, remove the meter assembly, and reset the MIL by reversing the position of the MIL set screw.

## CONTROL UNIT PIN LOCATION AND VOLTAGE TEST—1990 MPV WITH 4 CYLINDER ENGINE

**Terminal voltage**

| Terminal | Input | Output | Connection to | Test condition | Voltage | Remarks |
|---|---|---|---|---|---|---|
| 1A | | | Malfunction indicator lamp | For 3 seconds after ignition switch OFF→ON (Lamp illuminates) | Below 2.5V | Test connector (Green 1-pin) grounded |
| | | | | After 3 seconds (Lamp dose not illuminates) | Approx 12V | |
| | | | | Lamp illuminates | Below 2.5V | |
| | | | | Lamp not illuminate | Approx 12V | |
| 1B | | ○ | Self-Diagnosis checker (Code number) | For 3 seconds after ignition switch OFF→ON (Buzzer sounds) | Below 2.5V | • With Self-Diagnosis Checker • Test connector (Green 1-pin) grounded |
| | | | | After 3 seconds (Buzzer does not sounded) | Approx 12V | |
| | | | | Buzzer sounds | Below 2.5V | |
| | | | | Buzzer not sounded | Approx 12V | |
| 1C | | ○ | A/C relay | Ignition switch ON | Approx 12V | Blower motor ON |
| | | | | For 10 seconds After fully depressing accelerator pedal with A/C switch ON (A/C does not operate) (in-gear, ignition switch ON) | Approx 12V | |
| | | | | After 10 seconds | Below 2.5V | |
| | | | | For 5 seconds after cranking with A/C switch ON (A/C does not operate) | Approx 12V | |
| | | | | After 5 seconds (A/C operates) | Below 2.5V | |
| | | | | A/C switch ON at idle | Below 2.5V | |
| | | | | A/C switch OFF at idle | Approx 12V | |
| 1D | | ○ | Self-Diagnosis Checker (Monitor lamp) | Test connector (Green 1-pin) grounded For 3 seconds after ignition switch OFF→ON (Lamp illuminates) | Approx 5V | With Self-Diagnosis Checker |
| | | | | After 3 seconds (Lamp does not illuminated) | Approx 12V | |
| | | | | Test connector (Green 1-pin) not grounded at idle Monitor lamp ON | Approx 5V | |
| | | | | Test connector (Green 1-pin) not grounded at idle Monitor lamp OFF | Approx 12V | |
| 1E | ○ | | Idle switch | Accelerator pedal released | 0V | Ignition switch ON |
| | | | | Accelerator pedal depressed | Approx 12V | |
| 1F | — | — | — | — | — | — |
| 1G | ○ | | Ground | M/T | 0V | Ignition switch ON |
| | | | Open | A/T | Approx 12V | |
| 1H | ○ | | Central processing unit | Electrical load ON (Ignition switch ON) | Below 1.5V | Electrical load Rear defroster Headlight Blower motor (High & Super high position) |
| | | | | Electrical load OFF (Ignition switch ON) | Approx 12V | |
| 1I | | ○ | Circuit opening relay | Ignition switch ON | Approx 12V | |
| | | | | During cranking or at idle | Below 2.5V | |
| 1J | ○ | | Stoplight switch | Brake pedal released | 0V | Ignition switch ON |
| | | | | Brake pedal depressed | Approx 12V | |
| 1K | ○ | | P/S pressure switch | Ignition switch ON | Approx 12V | |
| | | | | P/S ON (at idle) | 0V | |
| | | | | P/S OFF (at idle) | Approx 12V | |
| 1L | ○ | | A/C switch | A/C switch ON (Ignition switch ON) | Below 2.5V | Blower motor ON |
| | | | | A/C switch OFF (Ignition switch ON) | Approx 12V | |
| 1M | ○ | | Ignition coil ⊖ terminal | Ignition switch ON | Approx 12V | * Engine Signal Monitor Green and red lights flash |
| | | | | Idle | Approx 12V* | |
| 1N | ○ | | Distributor (G-signal) | Ignition switch ON | 0 or 5V | |
| | | | | Idle | Approx 1.2V | |

| Terminal | Input | Output | Connection to | Test condition | Voltage | Remarks |
|---|---|---|---|---|---|---|
| 2A | | ○ | Vref | Ignition switch ON | 4.5—5.5V | |
| 2B | ○ | | Airflow sensor (Ground) | Constant | 0V | |
| 2C | — | — | Ground (E2) | Constant | 0V | |
| 2D | ○ | | Oxygen sensor | Ignition switch ON | 0V | |
| | | | | Idle (Cold engine) | 0V | |
| | | | | Idle (After warm up) | 0—1.0V | Needle moves from 0V to 1V |
| | | | | Increase engine speed (After warm up) | 0.5—1.0V | |
| | | | | Deceleration | 0—0.4V | |
| 2E | ○ | | Airflow sensor (Intake air mass) | Ignition switch ON | 1.0—2.0V | |
| | | | | Idle (After warm up) | 1.9—2.6V | |
| | | | | Increase engine speed (After warm up) | 2—5V | |
| 2F | ○ | | Test connector | Test connector (Green: 1-pin) not grounded | Approx. 12V | Ignition switch ON |
| | | | | Test connector (Green: 1-pin) grounded | 0V | |
| 2G | ○ | | Throttle sensor | Accelerator pedal released | Approx. 0.5V | Ignition switch ON |
| | | | | Accelerator pedal fully depressed | Approx. 4.3V | |
| 2H | ○ | | Atmospheric pressure sensor | At sea level | 3.5—4.5V | |
| | | | | At high altitude (2,000 [6,500 ft]) | 2.5—3.5V | |
| 2I | ○ | | Water thermosensor | Engine coolant temperature 20°C (68°F) | Approx. 2.5V | Ignition switch ON |
| | | | | After warm up | Approx. 0.4V | |
| 2J | ○ | | Ignition switch (ON position) | Ignition switch OFF | 0V | |
| | | | | Ignition switch ON | Approx. 12V | |
| 2K | | ○ | Airflow sensor (Burn-off) | Just after ignition switch OFF | 0V | Burn-off functions momentarily |
| | | | | Burn off (2-5 seconds after ignition switch OFF) | 8—12V | |
| 2L | ○ | | Intake air thermosensor (Dynamic chamber) | At 20°C (68°F) | Approx. 2.5V | |
| 2M | | ○ | Igniter | At sea level | Approx. 12V | Ignition switch ON |
| | | | | At high altitude (1.500 m [4.920 ft]) | 0V | |
| 2N | | ○ | Solenoid valve (PRC) | For 120 seconds after ignition switch OFF→ON | Below 2.5V | During hot condition. Coolant temp. above 90°C (194°F) Intake air temp. above 75°C (167°F) |
| | | | | For 120 seconds after starting | Below 2.5V | |
| | | | | Ignition switch ON | Approx. 12V | Other conditions |
| 2O | | ○ | 4 A/T control unit | Ignition switch ON | Approx. 12V | |
| | | | | Accelerator pedal fully depressed (Engine running) | 0V | |
| 2P | | ○ | Solenoid valve (Purge control) | Ignition switch ON | Approx. 12V | |
| | | | | Idle | Approx. 12V | * Engine signal monitor: Green and red lights flash |
| | | | | Driving in gear | 5—15V* | |
| 2Q | | ○ | Solenoid valve (Idle speed control) | Ignition switch ON | Approx. 11V | Engine signal monitor: Green and red lights flash |
| | | | | Idle | Approx. 10V | |
| 2R | — | — | Ground (E02) | Constant | 0V | |

## CONTROL UNIT PIN LOCATION AND VOLTAGE TEST (CONT.)—1990 MPV WITH 4 CYLINDER ENGINE

| Terminal | Input | Output | Connection to | Test condition | Voltage | Remarks |
|---|---|---|---|---|---|---|
| 3A | — | — | Ground (E01) | Constant | 0V | |
| 3B | ○ | | Ignition switch (Start position) | While cranking | Approx. 10V | |
| | | | | Ignition switch ON | Approx. 0V | |
| 3C | | ○ | Injector (No.1, 2) | Ignition switch ON | Approx. 12V | * Engine Signal Monitor: Green and red lights flash |
| | | | | Idle | Approx. 12V* | |
| 3D | ○ | | Neutral or clutch switch (M/T) | Neutral or clutch pedal depressed | 0V | Ignition switch ON |
| | | | | Other condition | Approx. 12V | |
| | | | Inhibitor switch (A/T) | N and P ranges | 0V | |
| | | | | 1, 2, D, and R ranges | Approx. 12V | |
| 3E | | ○ | Injector (No.3, 4) | Ignition switch ON | Approx. 12V | * Engine Signal Monitor: Green and red lights flash |
| | | | | Idle | Approx. 12V* | |
| 3F | | ○ | Main relay | Ignition switch OFF | Approx. 12V | |
| | | | | Ignition switch ON | Approx. 0V | |
| 3G | — | — | Ground (E1) | Constant | 0V | |
| 3H | — | — | — | — | — | — |
| 3I | ○ | | Main relay | Ignition switch OFF | Approx. 0V | |
| | | | | Ignition switch ON | Approx. 12V | |
| | | | | During burn-off (airflow sensor) | | |
| 3J | — | — | Battery | Constant | Approx. 12V | For backup |

**Terminal location**

| | | | | | | | | | | | | | | | | | | | | |
|---|---|---|---|---|---|---|---|---|---|---|---|---|---|---|---|---|---|---|---|---|
| 3I | 3G | 3E | 3C | 3A | 2Q | 2O | 2M | 2K | 2I | 2G | 2E | 2C | 2A | 1M | 1K | 1I | 1G | 1E | 1C | 1A |
| 3J | 3H | 3F | 3D | 3B | 2R | 2P | 2N | 2L | 2J | 2H | 2F | 2D | 2B | 1N | 1L | 1J | 1H | 1F | 1D | 1B |

## WIRING SCHEMATIC 1990 MPV WITH 4 CYLINDWER ENGINE

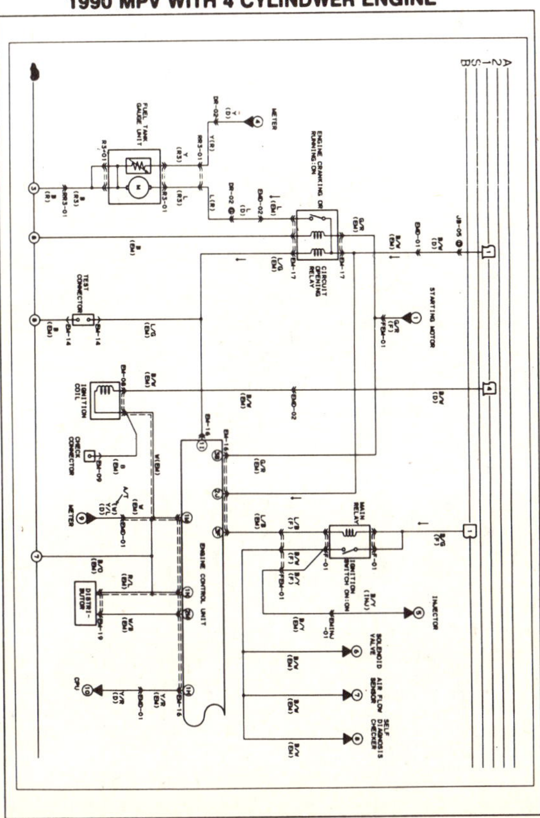

## WIRING SCHEMATIC 1990 MPV WITH 4 CYLINDER ENGINE

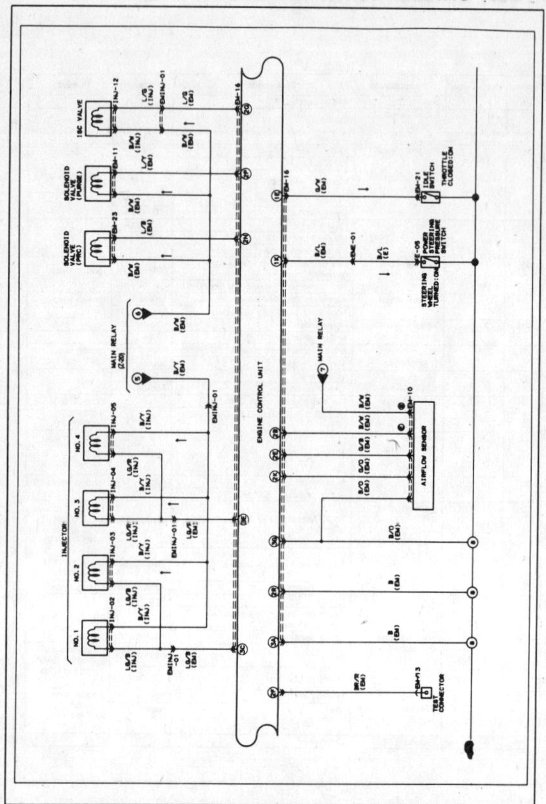

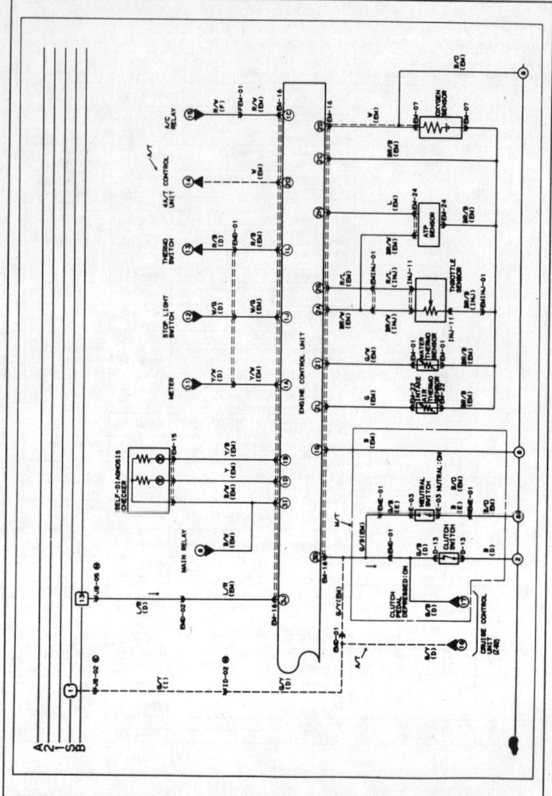

## SYSTEM SCHEMATIC 1990 MPV WITH 6 CYLINDER ENGINE

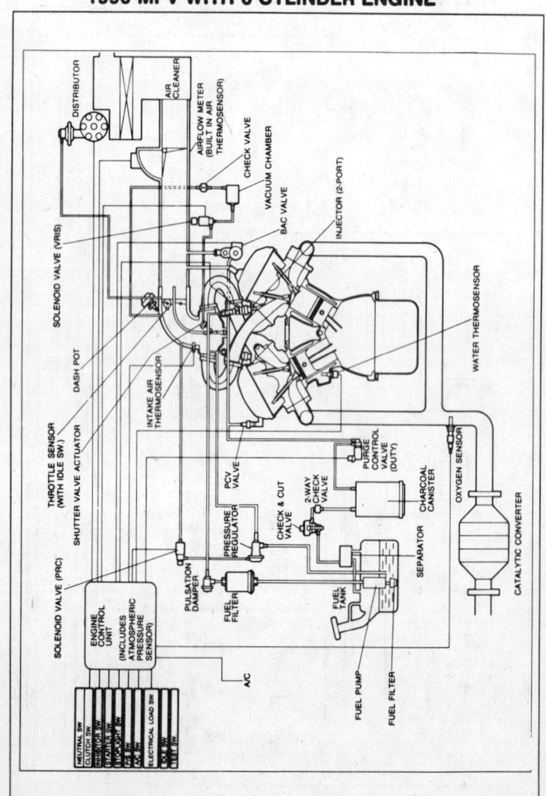


## VACUUM HOSE ROUTING 1990 MPV WITH 6 CYLINDER ENGINE

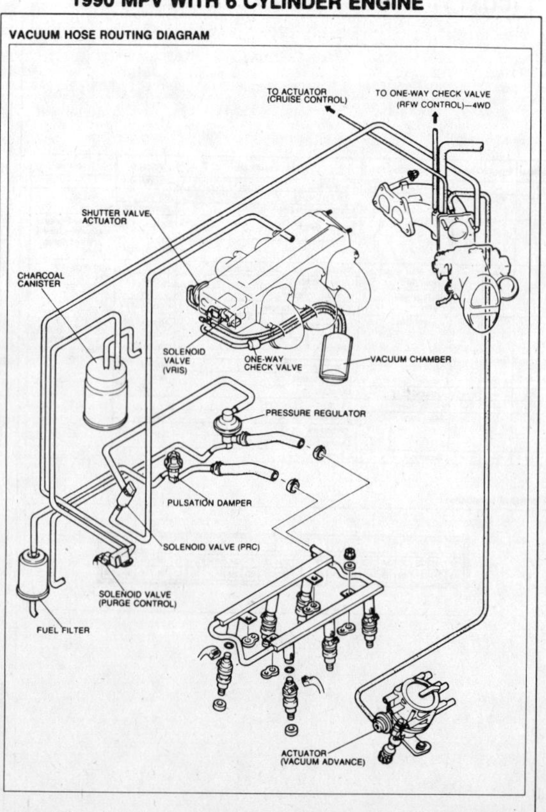

## EXPLODED VIEW OF CONTROL SYSTEM COMPONENTS (INPUT) CYLINDER ENGINE 1990 MPV WITH 6

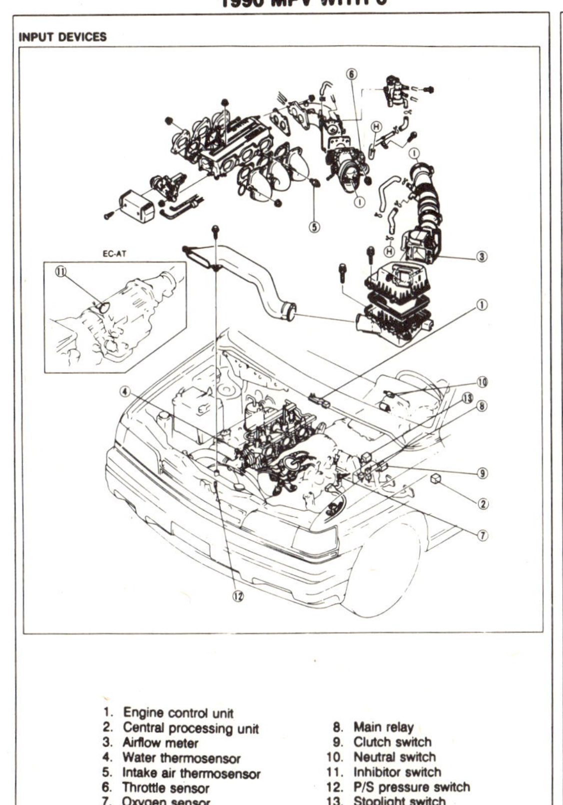


1. Engine control unit
2. Central processing unit
3. Airflow meter
4. Water thermosensor
5. Intake air thermosensor
6. Throttle sensor
7. Oxygen sensor
8. Main relay
9. Clutch switch
10. Neutral switch
11. Inhibitor switch
12. P/S pressure switch
13. Stoplight switch

## EXPLODED VIEW OF FUEL SYSTEM COMPONENTS 1990 MPV WITH 6 CYLINDER ENGINE

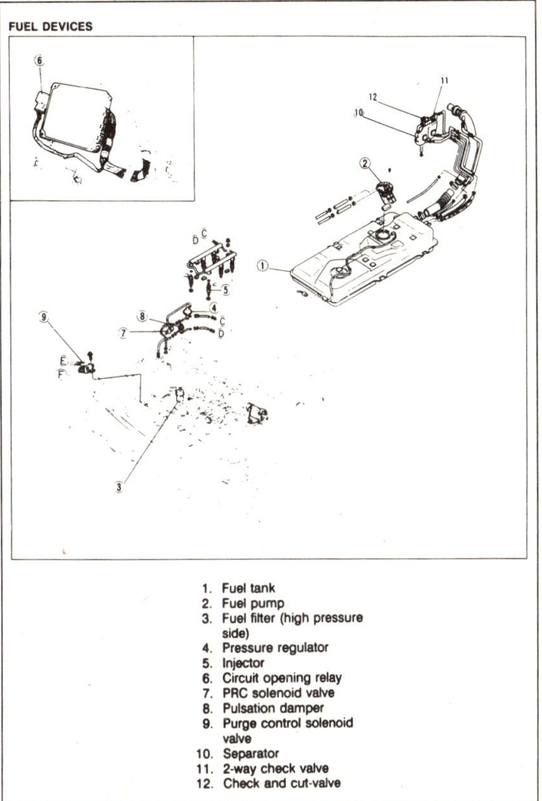


1. Fuel tank
2. Fuel pump
3. Fuel filter (high pressure side)
4. Pressure regulator
5. Injector
6. Circuit opening relay
7. PRC solenoid valve
8. Pulsation damper
9. Purge control solenoid valve
10. Separator
11. 2-way check valve
12. Check and cut-valve

## EXPLODED VIEW OF CONTROL SYSTEM COMPONENTS (OUTPUT)—1990 MPV WITH 6 CYLINDER ENGINE

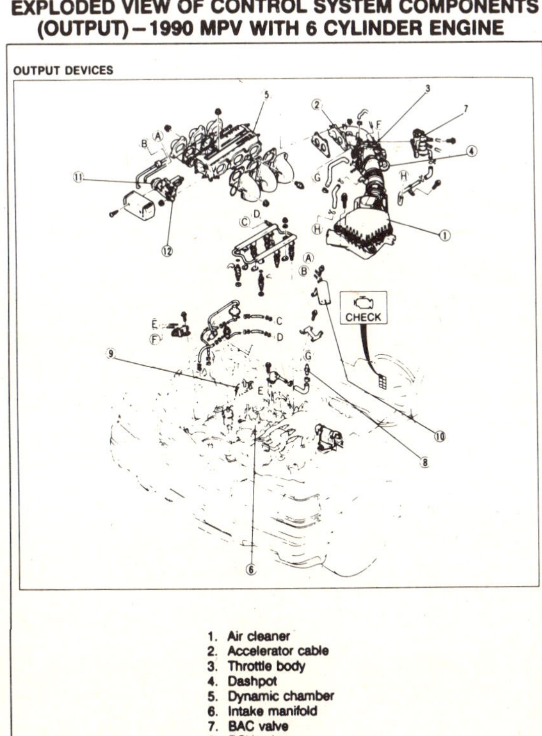


1. Air cleaner
2. Accelerator cable
3. Throttle body
4. Dashpot
5. Dynamic chamber
6. Intake manifold
7. BAC valve
8. PCV valve
9. Charcoal canister
10. Vacuum chamber
11. Shutter actuator valve
12. VRIS solenoid valve

## EXPLODED VIEW OF INTAKE AIR SYSTEM COMPONENTS—1990 MPV WITH 6 CYLINDER ENGINE

STRUCTURAL VIEW

This system controls the air required by the engine for operation.
The system consists of the air duct, air cleaner, airflow meter, throttle body, dynamic chamber, extension manifolds, and intake manifold. It also contains the VRIS control system and ISC system for improved engine power and idle smoothness.

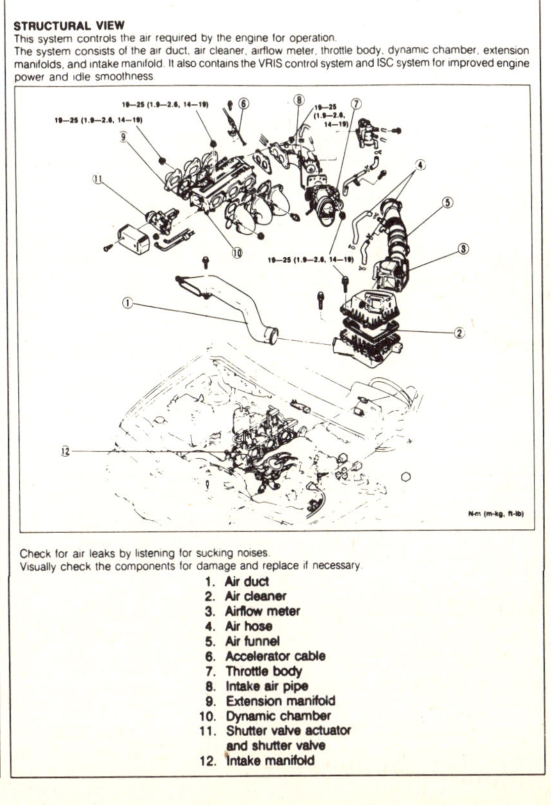


Check for air leaks by listening for sucking noises.
Visually check the components for damage and replace if necessary.

1. Air duct
2. Air cleaner
3. Airflow meter
4. Air hose
5. Air funnel
6. Accelerator cable
7. Throttle body
8. Intake air pipe
9. Extension manifold
10. Dynamic chamber
11. Shutter valve actuator and shutter valve
12. Intake manifold

## SPECIFICATIONS CHART 1990 MPV WITH 6 CYLINDER ENGINE

| Item | | | Specification |
|---|---|---|---|
| Idle speed (Test connector grounded) rpm | | | 800 ± 20 (A/T: in P range) |
| Air cleaner | Element type | | Oil permeated |
| Throttle body | Throat diameter mm (in) | | 45 (1.772) x 2 |
| Dashpot | Adjustment speed rpm | | 3,500 ± 300 |
| Throttle sensor | Idle switch (adjustment clearance) mm (in) | | 0.5 (0.020) (Continuity)<br>0.7 (0.028) (No continuity) |
| | Resistance kΩ | | Ⓐ– Ⓓ 3.5–6.5<br>Ⓑ– Ⓓ below 1 ↔ 3.5–6.5 |
| Airflow meter | Resistance Ω | E2 ↔ Vs | 20–400 (Measuring plate fully closed), 20–1,000 (fully open) |
| | | E2 ↔ Vc | 100–300 |
| | | E2 ↔ VB | 200–400 |
| | | E2 ↔ THA | -20°C (-4°F) 13,600–18,400<br>20°C (68°F) 2,210–2,690<br>60°C (140°F) 493–667 |
| | | E1 ↔ Fc | Fully closed; ∞ Fully open: 0 |
| Fuel tank | Capacity liters (US gal, Imp gal) | | 74 (19.6, 16.3) |
| Fuel filter | Low pressure | | Nylon 6 (250 mesh) |
| | High pressure | | Filter paper |
| Fuel pump | Type | | Impeller (in tank) |
| | Output pressure kPa ($kg/cm^2$, psi) | | 441–589 (4.5–6.0, 64–85) |
| Pressure regulator | Type | | Diaphragm |
| | Regulated pressure kPa ($kg/cm^2$, psi) | | 265–314 (2.7–3.2, 38–46) |
| Injector | Drive | | Voltage |
| | Type | | High-ohmic injector |
| | Resistance Ω | | 12–16 |
| Ignition timing (Test connector grounded) | | | 11 ± 1° BTDC (A/T: P range) |
| BAC valve | Resistance Ω | | 10.7–12.3 |
| Fuel specification | | | Unleaded gasoline (RON 87 or higher) |

## COMPONENTS DESCRIPTION 1990 MPV WITH 6 CYLINDER ENGINE

### COMPONENT DESCRIPTIONS

| Component | Function | Remark |
|---|---|---|
| **Air cleaner** | Filters air entering throttle body | |
| **Airflow meter** | Detects amount of intake air; sends signal to engine control unit | Intake air thermosensor and fuel pump switch included |
| **Air valve** | Supplies bypass air into dynamic chamber (engine cold) | • Engine speed increased to shorten warm-up period<br>• Thermo wax type<br>• Installed in BAC valve |
| **Atmospheric pressure sensor** | Detects atmospheric pressure; sends signal to engine control unit | Built in ECU |
| **BAC valve** | Supplies bypass air into dynamic chamber | Consists of air valve and ISC valve |
| **Catalytic converter** | Reduce HC, CO and NOx by chemical reaction | Monolith type |
| **Central processing unit (CPU) (E/L control function)** | Detects electrical load operation; sends signal to engine control unit | Electrical load<br>• Rear defroster<br>• Blower motor<br>• Headlights |
| **Charcoal canister** | Store fuel tank fumes | |
| **Check connectors** | For Self-Diagnosis Checker and fuel pump | • 6-pin connector (Green): Self-Diagnosis Checker<br>• 2-pin connector (Yellow): Fuel pump |
| **Check valve** | Prevents misoperation of shutter valve during heavy-load operation | |
| **Check-and-cut valve** | Controls pressure in fuel tank | |
| **Circuit opening relay** | Voltage for fuel pump (engine running) | |
| **Clutch switch (M/T)** | Detects clutch condition; sends signal to engine control unit | Switch opened when clutch pedal released |
| **Engine control unit (ECU)** | Detects the following.<br>1. A/C operation<br>2. Air/fuel ratio (oxygen concentration)<br>3. Atmospheric pressure<br>4. Braking signal<br>5. Cranking signal<br>6. E/L operation<br>7. Engine air temperature<br>8. Engine coolant temperature<br>9. Engine speed<br>10. In-gear condition<br>11. Initial set signal<br>12. Intake air amount<br>13. Intake air temperature<br>14. No.1 piston TDC<br>15. P/S operation<br>16. Throttle valve fully closed<br>17. Throttle valve opening angle<br>Controls operation of the following<br>1. Fail-safe system<br>2. Fuel injection system<br>3. Idle speed control system<br>4. Ignition system (Atmospheric pressure sensor)<br>5. Pressure regulator control system<br>6. Purge control system<br>7. Switch monitor function<br>8. Variable resonance induction system | 1. A/C switch<br>2. Oxygen sensor<br>3. Atmospheric pressure sensor<br>4. Stoplight switch<br>5. Ignition switch<br>6. E/L switch, central processing unit (CPU)<br>7. Intake air thermosensor (on dynamic chamber)<br>8. Water thermosensor<br>9. Ignition coil<br>10. Neutral, clutch, and inhibitor switches<br>11. Test connector<br>12. Airflow meter<br>13. Intake air thermosensor (in airflow meter)<br>14. G rotors (in distributor)<br>15. P/S pressure switch<br>16. Idle switch (in throttle sensor)<br>17. Throttle sensor<br><br>1. Self-Diagnosis Checker and MIL<br>2. Injector<br>3. BAC valve<br>4. Igniter<br>5. Solenoid valve (Pressure regulator control)<br>6. Solenoid valve (Purge control)<br>7. Monitor lamp (Self-Diagnosis Checker)<br>8. Solenoid valve (VRIS) |
| **Distributor** | Detects No.1 cylinder TDC; sends signal to engine control unit | G-signal |
| **Dashpot** | Causes gradual throttle valve closing during deceleration | Adjustment speed: 3,500 ± 300 rpm |
| **Dynamic chamber** | Interconnects all cylinders | Includes shutter valve |
| **Fuel filter** | Filters fuel | |
| **Fuel pump** | Provides fuel to injectors | • Operates only while engine running<br>• Installed in fuel tank |

## COMPONENTS DESCRIPTION (CONT.) 1990 MPV WITH 6 CYLINDER ENGINE

| Component | | Function | Remark |
|---|---|---|---|
| **G-coil** | | Detects No.1 cylinder TDC; sends signal to engine control unit | For determining fuel injection timing |
| **Idle switch** | | Detects when throttle valve fully closed; sends signal to engine control unit | Installed in throttle sensor. Switch closed when throttle valve fully closed |
| **Igniter** | | Receives spark signal of signal rotor and controls primary current flow in ignition coil | Controls spark advance by signal of ATP sensor. |
| **Ignition switch** | | Detects engine cranking; sends signal to engine control unit | |
| **Inhibitor switch (A/T)** | | Detects in-gear condition; sends signal to engine control unit | Switch closed in N or P range (EC-AT) |
| **Injector** | | Injects fuel into intake port | • Controlled by signals from engine control unit<br>• High-ohmic injector |
| **Intake manifold** | | Supplies intake air to all cylinders | |
| **Intake air thermosensor** | | Detects intake air temperature; sends signal to engine control unit | Installed in airflow meter |
| | | Detects engine air temperature; sends signal to engine control unit | Installed on dynamic chamber |
| **Main relay** | | Supplies current to injectors and engine control unit | |
| **MIL (Malfunction indicator lamp)** | | Illuminates when input device malfunctions | Installed in instrument cluster |
| **Neutral switch (M/T)** | | Detects in-gear condition, sends signal to engine control unit | Switch open when in-gear |
| **Oxygen sensor** | | Detects oxygen concentration; sends signal to engine control unit | Zirconia ceramic and platinum coating |
| **PCV valve** | | Controls blowby gas amount pulled into engine | |
| **P/S pressure switch** | | Detects P/S operation, sends signal to engine control unit | ON when steering wheel turned right or left |
| **Pressure regulator** | | Adjusts fuel pressure supplied to injectors | |
| **Pulsation damper** | | Absorbs fuel pulsations | |
| **Separator** | | Prevents fuel from flowing into charcoal canister | |
| **Shutter valve actuator** | | Controls shutter valve operation | For variable resonance induction system (VRIS) |
| **Solenoid valve** | **Idle speed control** | Supplies bypass air into dynamic chamber | • Controlled by duty signal from engine control unit<br>• Works independent of idle-up<br>• Installed in BAC valve |
| | **Pressure regulator control** | Blocks vacuum from intake manifold to pressure regulator | Hot engine only |
| | **Purge control** | Controls evaporative fumes from canister to intake manifold | |
| | **VRIS** | Cuts vacuum to shutter valve actuator | Variable resonance induction system (VRIS) |
| **Stoplight switch (M/T)** | | Detects braking (deceleration); sends signal to engine control unit | |
| **Test connector** | | For Self-Diagnosis Checker and idle speed adjustment | 1-pin connector (Green) |
| **Throttle body** | | Controls intake air amount | Integrated throttle sensor and dashpot incorporated |
| **Throttle sensor** | | Detects throttle valve opening angle; sends signal to engine control unit | • Idle switch included<br>• Installed on throttle body |
| **Water thermosensor** | | Detects coolant temperature; sends signal to engine control unit | |

## COMPONENT REALTIONSHIP CHART 1990 MPV WITH 6 CYLINDER ENGINE

| INPUT DEVICE \ OUTPUT DEVICE | INJECTOR: FUEL INJECTION AMOUNT | INJECTOR: FUEL INJECTION TIMING | PRESSURE REGULATOR | BAC VALVE: AIR VALVE | BAC VALVE: ISC VALVE | SOLENOID VALVE (VRIS) | IGNITER (ATMOSPHERIC CONTROL) | SOLENOID VALVE (PURGE) | OXYGEN SENSOR HEATER | A/C RELAY |
|---|---|---|---|---|---|---|---|---|---|---|
| E/L SIGNAL | × | × | × | × | ○ | × | × | × | × | × |
| P/S PRESSURE SW | × | × | × | × | ○ | × | × | × | × | × |
| A/C SW | × | × | × | × | ○ | × | × | × | × | ○ |
| NEUTRAL CLUTCH (M/T) INHIBITOR (A/T) SW | ○ | × | ○ | × | ○ | × | × | ○ | × | ○ |
| TEST CONNECTOR | ○ | × | × | × | ○ | × | ○ | × | × | × |
| STOPLIGHT SW (M/T) | ○ | × | × | × | ○ | × | × | × | × | × |
| ATP SENSOR | ○ | × | × | × | ○ | × | ○ | ○ | × | × |
| OXYGEN SENSOR | ○ | × | × | × | × | × | × | ○ | × | × |
| INTAKE AIR THERMOSENSOR: DYNAMIC CHAMBER | ○ | × | ○ | × | ○ | × | × | × | × | × |
| INTAKE AIR THERMOSENSOR: AIRFLOW METER | ○ | × | × | × | ○ | × | × | ○ | × | × |
| IGN SW (START) | ○ | ○ | ○ | × | ○ | × | × | × | ○ | ○ |
| WATER THERMOSENSOR | ○ | × | ○ | × | ○ | × | × | ○ | ○ | × |
| THROTTLE SENSOR: THROTTLE SENSOR | ○ | × | ○ | × | ○ | × | × | ○ | × | ○ |
| THROTTLE SENSOR: IDLE SWITCH | ○ | × | × | × | ○ | × | × | ○ | × | × |
| AIRFLOW METER | ○ | × | × | × | × | × | × | ○ | ○ | × |
| DISTRIBUTOR (G-SIGNAL) | × | ○ | × | × | × | × | × | × | × | × |
| IGN COIL (Ignition pulse) | ○ | ○ | ○ | × | ○ | ○ | × | ○ | ○ | × |

○: Related
×: Not related

## INPUT COMPONENTS AND ENGINE CONDITIONS—1990 MPV WITH 6 CYLINDER ENGINE

**Input Devices and Engine Conditions**

| ENGINE CONDITIONS / INPUT DEVICES | APPROXIMATE TIME (BASED ON 10—16°C, 50—60°F AMBIENT TEMPERATURE) | INPUT DEVICE | THROTTLE SENSOR | | | | AIRFLOW METER | | | INPUT DEVICES | | | | | | | IGNITION SWITCH | | |
|---|---|---|---|---|---|---|---|---|---|---|---|---|---|---|---|---|---|---|---|
| | | IGNITION COIL | IDLE SWITCH | THROTTLE POSITION SENSOR | WATER THERMO-SENSOR | OXYGEN SENSOR | AIRFLOW SENSOR | INTAKE AIR THERMO-SENSOR | ATMOSPHERIC PRESSURE SENSOR | DISTRIBUTOR (G-SIGNAL) | NEUTRAL AND CLUTCH SWITCH (M/T) | INHIBITOR SWITCH (A/T) | STOP-LIGHT SWITCH (M/T) | A/C SWITCH | P/S PRESSURE SWITCH | CPU (E/L SIGNAL) | START POSITION | ON POSITION | TEST CONNECTOR |
| CRANKING COLD ENGINE · COLD AIR · COLD COOLANT | ZERO | | SIGNAL HAS NO EFFECT ON ECU | SIGNAL HAS NO EFFECT ON ECU | | SIGNAL HAS NO EFFECT ON ECU | SIGNAL HAS NO EFFECT ON ECU | SIGNAL HAS NO EFFECT ON ECU | | | SIGNAL HAS NO EFFECT ON ECU | SIGNAL HAS NO EFFECT ON ECU | SIGNAL HAS NO EFFECT ON ECU | SIGNAL HAS NO EFFECT ON ECU | SIGNAL HAS NO EFFECT ON ECU | SIGNAL HAS NO EFFECT ON ECU | SENDS SIGNAL TO ECU | SIGNAL HAS NO EFFECT ON ECU | SIGNAL HAS NO EFFECT ON ECU |
| COLD START FAST IDLE · COLD AIR · COLD COOLANT | ONE MINUTE | | LOW VOLTAGE SIGNAL TO ECU (BELOW 1.5V) | CLOSED THROTTLE LOW VOLTAGE (0.3—0.7V) | COOL TO WARM MEDIUM VOLTAGE (3V AND DROPPING) | SENSOR COLD LOW TO HIGH VOLTAGE (0—0.8V) | LOW VOLUME AIR-FLOW LOW-TO MEDIUM VOLTAGE (4.0—7.0V) | | | | IN NEUTRAL SENDS SIGNAL TO ECU (APPROX 12V) | IN N or P SENDS SIGNAL TO ECU (APPROX 12V) | BRAKE PEDAL DEPRESSED. SENDS SIGNAL TO ECU (APPROX 12V) | | | | | | |
| COLD DRIVEAWAY PART THROTTLE · COLD AIR · COLD COOLANT | TWO MINUTES | | | | | | | | | | | | | | | | | | |
| WARM DRIVEAWAY PART THROTTLE (25%) · WARM INTAKE AIR · WARM COOLANT | THREE MINUTES | | | | WARM MEDIUM VOLTAGE (APPROX 0.7V AND DROPPING) | SENSOR WARM HIGH VOLTAGE (ABOVE 0.6V) | MODERATE VOLUME AIR FLOW LOW-TO-MEDIUM VOLTAGE (6.1—8.5V) | | | | | | | | | | | | |
| HOT CRUISE · WARM INTAKE AIR · HOT COOLANT | | | | PART THROTTLE MEDIUM VOLTAGE (1—2.5V) | | SENSOR HOT SWITCHING FROM HIGH VOLTAGE (ABOVE 0.6V RICH) TO LOW VOLTAGE (BELOW 0.4V LEAN) | | | SENDS VOLTAGE SIGNAL TO ECU THAT VARIES WITH ALTITUDE VOLTAGE (APPROX 4V) | | | | | | | | | | |
| HOT ACCELERATION · 60% THROTTLE | | SENDS ENGINE RPM SIGNAL TO ECU | HIGH VOLTAGE SIGNAL TO ECU (APPROX 12V) | | | | MODERATE-TO-HIGH VOLUME OF AIR-FLOW (8.1—8.7V) | COOL TO WARM MEDIUM VOLTAGE (1.4—3V) | | SENDS NO 1 CYLINDER TDC SIGNAL TO ECU | DRIVING IN ANY GEAR NO VOLTAGE SEND SIGNAL TO ECU (BELOW 1.5V) | IN D,R SECOND AND LOW NO VOLTAGE SIGNAL TO ECU (BELOW 1.5V) | NO SIGNAL TO ECU (BELOW 1.5V) | A/C SWITCH ON SENDS SIGNAL TO ECU (APPROX 12V) A/C SWITCH OFF NO SIGNAL TO ECU (BELOW 1.5V) | STEERING WHEEL TURNED LOW VOLTAGE SIGNAL TO ECU (BELOW 1.5V) STEERING WHEEL STRAIGHT AHEAD HIGH VOLTAGE SIGNAL TO ECU (APPROX 12V) | HEADLIGHTS, BLOWER OR REAR DEFOGER ON LOW VOLTAGE SIGNAL TO ECU (BELOW 2.5V) HEADLIGHTS, BLOWER, AND REAR DEFOGER OFF HIGH VOLTAGE SIGNAL TO ECU (APPROX 12V) | NO SIGNAL TO ECU (BELOW 1.5V) | SENDS SIGNAL TO ECU (APPROX 12V) | WHEN CONNECTOR NOT GROUNDED HIGH VOLTAGE SIGNAL TO ECU (APPROX 12V) |
| HOT ACCELERATION · WIDE OPEN THROTTLE | MORE THAN FOUR MINUTES | | | WIDE OPEN THROTTLE (APPROX 4.0V) | HOT LOW VOLTAGE (APPROX 0.5V) | HIGH VOLTAGE (ABOVE 0.6V RICH) | HIGH VOLUME OF AIR-FLOW (8.4—9.0V) | | | | | | | | | | | | |
| DECELERATION · CLOSED THROTTLE | | | LOW VOLTAGE SIGNAL TO ECU (BELOW 1.5V) | CLOSED THROTTLE LOW VOLTAGE (0.3—0.7V) | | LOW VOLTAGE (BELOW 0.4V LEAN) | LOW VOLUME AIR-FLOW (4.0—5.0V) | | | | | | BRAKE PEDAL DEPRESSED SENDS SIGNAL TO ECU (APPROX 12V) | | | | | | |
| HOT CURB IDLE (EXTENDED) | | | | | | SWITCHING FROM HIGH TO LOW VOLTAGE (0.4—0.6V) | | | | | IN NEUTRAL SENDS SIGNAL TO ECU (APPROX 12V) | IN N OR P RANGE. SENDS SIGNAL TO ECU (BELOW 1.5V) | | | | | | | WHEN CONNECTOR GROUNDED LOW VOLTAGE SIGNAL TO ECU (BELOW 1.5V) |
| HOT ENGINE SHUTDOWN | | OFF | OFF | OFF | OFF | SENSOR HOT LOW VOLTAGE (0.1V LEAN) UNTIL SENSOR COOLS DOWN | OFF | OFF | OFF | OFF | OFF | OFF | OFF | OFF | OFF | OFF | OFF | OFF | OFF |

## OUTPUT COMPONENTS AND ENGINE CONDITIONS—1990 MPV WITH 6 CYLINDER ENGINE

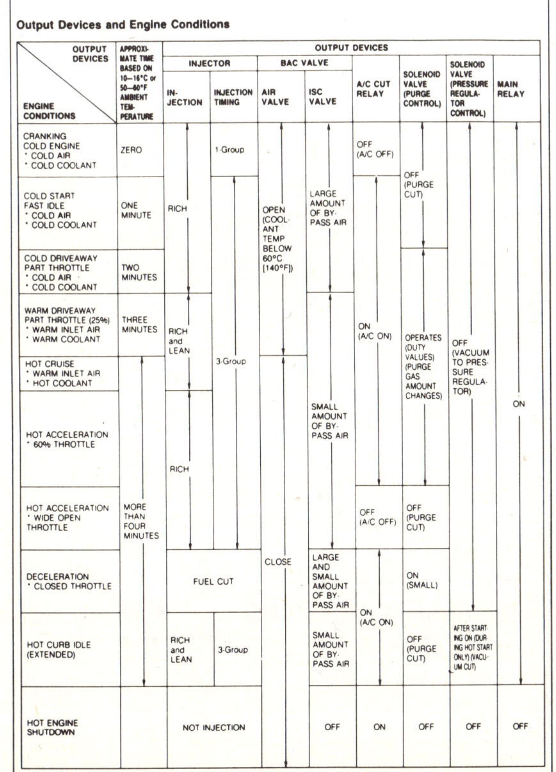

**Output Devices and Engine Conditions**

| ENGINE CONDITIONS / OUTPUT DEVICES | APPROXIMATE TIME BASED ON 10—16°C or 50—60°F AMBIENT TEMPERATURE | OUTPUT DEVICES | | | | | | | | OUTPUT DEVICES | | |
|---|---|---|---|---|---|---|---|---|---|---|---|---|
| | | INJECTOR | | BAC VALVE | | | | | | | | |
| | | INJECTION | INJECTION TIMING | AIR VALVE | ISC VALVE | A/C CUT RELAY | SOLENOID VALVE (PURGE CONTROL) | SOLENOID VALVE (PRESSURE REGULATOR CONTROL) | MAIN RELAY | IGNITER | SOLENOID VALVE (V.R.I.S) | OXYGEN SENSOR HEATER |
| CRANKING COLD ENGINE · COLD AIR · COLD COOLANT | ZERO | | 1-Group | | | OFF (A/C OFF) | | | | | OFF (V.R.I.S OFF) | OFF |
| COLD START FAST IDLE · COLD AIR · COLD COOLANT | ONE MINUTE | RICH | | OPEN (COOLANT TEMP BELOW 60°C (140°F)) | LARGE AMOUNT OF BY-PASS AIR | | OFF (PURGE CUT) | | | | | ON (BELOW 2,250 rpm) |
| COLD DRIVEAWAY PART THROTTLE · COLD AIR · COLD COOLANT | TWO MINUTES | | | | | | | | | | ON (V.R.I.S ON 3,600—5,650 rpm ONLY) | |
| WARM DRIVEAWAY PART THROTTLE (25%) · WARM INLET AIR · WARM COOLANT | THREE MINUTES | RICH and LEAN | | | | ON (A/C ON) | | | | | | |
| HOT CRUISE · WARM INLET AIR · HOT COOLANT | | | 3-Group | | | | OPERATES (DUTY VALUES) (PURGE GAS AMOUNT CHANGES) | OFF (VACUUM TO PRESSURE REGULATOR) | | ON (MORE THAN 1,000 m (3,280 ft) ABOVE SEA LEVEL) | | |
| HOT ACCELERATION · 60% THROTTLE | | RICH | | | SMALL AMOUNT OF BY-PASS AIR | | | | ON | | OFF (V.R.I.S OFF) | OFF |
| HOT ACCELERATION · WIDE OPEN THROTTLE | MORE THAN FOUR MINUTES | | | | | OFF (A/C OFF) | OFF (PURGE CUT) | | | | | |
| DECELERATION · CLOSED THROTTLE | | FUEL CUT | | CLOSE | LARGE AND SMALL AMOUNT OF BY-PASS AIR | | ON (SMALL) | | | | ON (V.R.I.S ON 3,600—5,650 rpm ONLY) | |
| HOT CURB IDLE (EXTENDED) | | RICH and LEAN | 3-Group | | SMALL AMOUNT OF BY-PASS AIR | ON (A/C ON) | OFF (PURGE CUT) | AFTER STARTING ON (DURING HOT START ONLY) (VACUUM CUT) | | | OFF (V.R.I.S OFF) | |
| HOT ENGINE SHUTDOWN | | NOT INJECTION | | | OFF | ON | OFF | OFF | OFF | OFF | OFF | OFF |

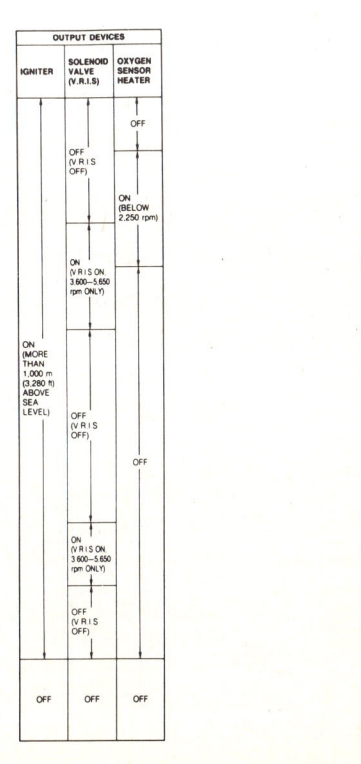

## TROUBLESHOOTING CHART—1990 MPV WITH 6 CYLINDER ENGINE

### Cranks normally but won't start (No combustion)

| STEP | QUICK INSPECTION | | ACTION | | POSSIBLE CAUSE |
|---|---|---|---|---|---|
| 1 | Check malfunction code (01) with SST [IG ON, Test connector (Green: 1-pin) grounded] | Yes | Go to Next Step | | |
| | | No | Check for the cause by referring to the check sequence | | |
| 2 | Check for spark by disconnecting high-tension leads while cranking | Yes | Go to Next Step | | |
| | | No | Check ignition system | | |
| 3 | Check for fuel pump operating sound from fuel filler port [IG ON, Test connector (Yellow: 2-pin) connected] | Yes | Check if the engine starts with this condition | Yes | Circuit opening relay and wiring (Starting circuit) |
| | | | | | Go to Next Step |
| | | No | Check circuit opening relay switching operation | Yes | Wiring open (Check relay circuit) |
| | | | | | Fuel pump malfunction |
| | | | | | Wiring open (Check fuel pump circuit) |
| | | | | No | Circuit opening relay malfunction (Replace) |
| 4 | Check fuel line pressure [IG ON, Test connector (Yellow: 2-pin) connected] **Fuel line pressure: 285—314 kPa (2.7—3.2 kg/cm², 38.3—45.5 psi)** | Yes | Go to Next Step | | |
| | | No | Check fuel pump maximum pressure **Fuel pump max. pressure: 441—589 kPa (4.5—6.0 kg/cm², 64—85 psi)** | Yes | Pressure regulator malfunction (Replace) |
| | | | | No | Fuel pump malfunction (Replace) |
| 5 | Check for injector operating sound while cranking | Yes | Go to Next Step | | |
| | | No | Check voltage at ECU 3C, 3E and 3F terminals with SST **Voltage: Approx. 12V (IG ON)** | Yes | Throttle sensor malfunction |
| | | | | | ECU malfunction (Replace) |
| | | | | No | Wiring open or short |
| | | | | | Poor ground circuit from ECU 3A terminal |
| 6 | Check continuity between check connector [for tachometer (Black: 1-pin)] and ground (IG ON) | Yes | | | Wiring short to ground |
| | | No | Go to Next Step | | |
| 7 | Substitute a well-known ECU Check if the condition becomes better | Yes | | | ECU malfunction |
| | | No | Check ground circuit from ECU 2R terminal with SST **Voltage: 0V (IG ON)** | Yes | Go to Next Step |
| | | | | No | Poor ground circuit |
| 8 | | | | | Low compression |

### Cranks normally but hard to start (Always)

| STEP | QUICK INSPECTION | | ACTION | | POSSIBLE CAUSE |
|---|---|---|---|---|---|
| 1 | Check ignition timing at idle after warm up **Ignition timing: 11 ± 1° BTDC** [Test connector (Green: 1-pin) grounded] | Yes | Go to Next Step | | |
| | | No | Adjust ignition timing | | |
| | Note: At high-altitude area **[1,000 m (3,280 ft)]**, ignition timing advances to **approx. 20° BTDC** when test connector (Green: 1-pin) is not grounded | | | | |
| 2 | Disconnect high-tension lead of each cylinder at idle Check if the engine condition changes | Yes | Go to Next Step | | |
| | | No | Check ignition system | Yes | Injector clogging (Replace) |
| | | | | No | Spark plug |
| | | | | | High-tension lead |
| | | | | | Distributor cap |
| 3 | Check injector operating sound at idle | Yes | Go to Next Step | | |
| | | No | Check injector resistance | Yes | Wiring short or open |
| | | | | No | Injector malfunction (Replace) |
| 4 | Check fuel line pressure [IG ON, Test connector (Yellow: 2-pin) connected] **Fuel line pressure: 265—314 kPa (2.7—3.2 kg/cm², 38.3—45.5 psi)** | Yes | Go to Next Step | | |
| | | No | Check fuel pump max. pressure **Fuel pump max. pressure: 441—589 kPa (4.5—6.0 kg/cm², 64—85 psi)** | Yes | Pressure regulator malfunction (Replace) |
| | | | | No | Fuel pump malfunction (Replace) |
| 5 | Operate the fuel pump [IG ON, Test connector (Yellow: 2-pin) connected] Turn the ignition switch OFF and observe fuel pressure for 5 minutes **Fuel pressure: More than 147 kPa (1.5 kg/cm², 21 psi)** | Yes | Go to Next Step | | |
| | | No | Check fuel pump pressure drop | No | Fuel pump malfunction (Replace) |
| | | | Check pressure regulator pressure drop | Yes | Injector fuel leaks |
| | | | | No | Pressure regulator malfunction (Replace) |
| 6 | | | | | Low compression |
| 7 | Substitute a well-known ECU | | | | |

## TROUBLESHOOTING CHART—1990 MPV WITH 6 CYLINDER ENGINE

### Cranks normally but hard to start (only when engine cold)

| STEP | QUICK INSPECTION | | ACTION | | POSSIBLE CAUSE |
|---|---|---|---|---|---|
| 1 | Check gravity of battery by using a hydrometer **Specific gravity: Above 1.200** | Yes | Go to Next Step | | |
| | | No | Recharge the battery | | |
| 2 | Check malfunction codes (09) (26) with SST [IG, ON, Test connector (Green: 1-pin) grounded] | Yes | Go to Next Step | | |
| | | No | Check for the cause by referring to the check sequence | | |
| 3 | Disconnect high-tension lead of each cylinder at idle Check if the engine condition changes | Yes | Go to Next Step | | |
| | | | Check ignition system | | Spark plug |
| | | | | | High-tension lead |
| | | | | | Distributor cap |
| 4 | Check fuel line pressure [IG ON, Test connector (Yellow: 2-pin) connected] **Fuel line pressure: 265—314 kPa (2.7-3.2 kg/cm², 38.3—45.5 psi)** | Yes | Go to Next Step | | |
| | | No | Check fuel pump max. pressure **Fuel pump max. pressure: 441—589 kPa (4.5—6.0 kg/cm², 64—85 psi)** | Yes | Pressure regulator malfunction (Replace) |
| | | | | No | Fuel pump malfunction (Replace) |
| 5 | Disconnect the ISC valve connector Check idle speed during warm up | Yes | Go to Next Step | | |
| | | No | | | BAC valve stuck close |
| 6 | Substitute a well-known ECU | | | | |

### Cranks normally but hard to start (only when engine warm)

| STEP | QUICK INSPECTION | | ACTION | | POSSIBLE CAUSE |
|---|---|---|---|---|---|
| 1 | Check malfunction code (26) with SST [IG ON, Test connector (Green. 1-pin) grounded] | Yes | Go to Next Step | | |
| | | No | Check for the cause by referring to the check sequence | | |
| 2 | Disconnect water thermosensor connector Check if the condition become better | Yes | Check water thermosensor connector for connecting condition as follows 1. Shake connector and check if the condition changes 2. Check condition of terminal (burning or damaged) 3. Connect a male terminal to the connector and check for connection condition | Yes | Water thermosensor malfunction |
| | | | | No | Poor contact at water thermosensor connector |
| | | No | Go to Next Step | | |
| 3 | Operate the fuel pump [IG ON, Test connector (Yellow: 2-pin) connected] Turn the ignition switch OFF and observe fuel pressure for 5 minutes **Fuel pressure: More than 147 kPa (1.5 kg/cm², 21 psi)** | Yes | Go to Next Step | | |
| | | No | Check fuel pump pressure drop | No | Fuel pump malfunction (Replace) |
| | | | Check pressure regulator pressure drop | Yes | Injector fuel leaks |
| | | | | No | Pressure regulator malfunction (Replace) |
| 4 | Substitute a well-known ECU | | | | |

### Cranks normally but hard to start (only after heat soak)

| STEP | QUICK INSPECTION | | ACTION | | POSSIBLE CAUSE |
|---|---|---|---|---|---|
| 1 | Check malfunction code (09) (11) (25) (26) with SST [IG ON, Test connector (Green. 1-pin) grounded] | Yes | Go to Next Step | | |
| | | No | Check for the cause by referring to the check sequence | | |
| 2 | Circulate the fuel by operating fuel pump **for 20 sec.** [IG ON, Test connector (Yellow: 2-pin) connected] Check if the condition becomes better | Yes | Go to Step 4 | | |
| | | No | Go to Next Step | | |
| 3 | Disconnect vacuum hose from pressure regulator Check if the condition becomes better | Yes | Check the components related to pressure regulator control system | | Water thermosensor malfunction |
| | | | | | Intake air thermosensor malfunction |
| | | | | | Solenoid valve (PRC) malfunction |
| | | | | | ECU malfunction (Check 2N terminal voltage) |
| | | No | Go to Next Step | | |
| 4 | Operate the fuel pump [IG ON. Test connector (Yellow 2-pin) connected] Turn the ignition switch OFF and observe fuel pressure **for 5 min.** **Fuel pressure: More than 147 kPa (1.5 kg/cm², 21 psi)** | Yes | Go to Next Step | | |
| | | No | Check fuel pump pressure drop | No | Fuel pump malfunction (Replace) |
| | | | Check pressure regulator pressure drop | Yes | Injector fuel leaks |
| | | | | No | Pressure regulator malfunction (Replace) |
| 5 | Change fuel according to specifications Check if the condition becomes better | Yes | | | Poor fuel quality |
| | | No | | | |
| 6 | Substitute a well-known ECU | | | | |

### Cranks normally but won't start (Intermittent)

| STEP | QUICK INSPECTION | | ACTION | | POSSIBLE CAUSE |
|---|---|---|---|---|---|
| 1 | Shake connector of ignition coil, main relay and ECU while cranking Check if the engine starts | Yes | There may be a poor contact at the connector Repair or replace the wiring | | |
| | | No | Go to troubleshooting "Cranks normally but hard to start (always)" | | |

## TROUBLESHOOTING CHART—1990 MPV WITH 6 CYLINDER ENGINE

**Rough idle (Always)**

| STEP | QUICK INSPECTION | | ACTION | | POSSIBLE CAUSE |
|---|---|---|---|---|---|
| 1 | Check malfunction code (08) (09) (14) (17) (26) (34) with SST [IG ON, Test connector (Green 1-pin) grounded] | Yes | Check for the cause by referring to the check sequence | | |
| | | No | "88" flashing Check voltage at ECU 3G terminal with SST **Voltage: 0V (IG ON)** | Yes | ECU malfunction (Replace) |
| | | | | No | Poor ground circuit |
| | | | "00" Go to Next Step | | |
| 2 | Check ignition timing at idle after warm up **Ignition timing: 11 ± 1° BTDC** [Test connector (Green: 1-pin) grounded] | Yes | Go to Next Step | | |
| | | No | Adjust ignition timing (If possible) | | |
| 3 | Disconnect high-tension lead of each cylinder at idle Check if the condition changes | Yes | Go to Next Step | | |
| | | No | Check ignition system | Yes | Injector clogging (Replace) |
| | | | | No | Spark plug |
| | | | | | High-tension lead |
| | | | | | Distributor cap |
| 4 | Check idle speed after warm up **Idle speed: 800 ± 20 rpm** [Test connector (Green: 1-pin) grounded] | Yes | Go to Next Step | | |
| | | No | Adjust idle speed (If possible) | | |
| 5 | Check injector operating sound at idle | Yes | Go to Next Step | | |
| | | No | Check injector resistance | Yes | Wiring short or open |
| | | | | No | Injector malfunction (Check resistance) |
| | | | | | Wiring short or open |
| 6 | Check fuel line pressure [IG ON, Test connector (Yellow 2-pin) connected] **Fuel line pressure: 265—314 kPa (2.7—3.2 kg/cm², 38.3—45.5 psi)** | Yes | Go to Next Step | | |
| | | No | | | Fuel leaks |
| | | | | | Fuel filter clogging |
| | | | Check fuel pump max. pressure **Fuel pump max. pressure: 441—589 kPa (4.5—6.0 kg/cm², 64—85 psi)** | Yes | Pressure regulator malfunction (Replace) |
| | | | | No | Fuel pump malfunction (Replace) |
| 7 | Check intake manifold vacuum at idle **Vacuum: 480 ± 20 mmHg (18.9 ± 0.8 inHg)** | Yes | Go to Next Step | | |
| | | No | Check for air leaks | Yes | Throttle valve |
| | | | | No | Vacuum hoses |
| | | | | | Gaskets |
| | | | | | Bolts or nuts |
| | | | | | Component broken |
| 8 | Substitute a well-known ECU Check if the condition becomes better | Yes | | | ECU malfunction |
| | | No | Check voltage at ECU 3G terminal with SST **Voltage: 0V (IG ON)** | Yes | Go to Section N |
| | | | | No | Poor ground circuit |

**Rough idle (only when engine is cold)**

| STEP | QUICK INSPECTION | | ACTION | | POSSIBLE CAUSE |
|---|---|---|---|---|---|
| 1 | Check malfunction code (08) (09) (26) with SST [IG ON, Test connector (Green 1-pin) grounded] | Yes | Go to Next Step | | |
| | | No | Check for the cause by referring to the check sequence | | |
| 2 | Check ignition timing advance (Centrifugal advance) | Yes | Go to Next Step | | |
| | | No | | | Distributor malfunction |
| 3 | Disconnect high-tension lead of each cylinder at idle Check if the engine condition changes | Yes | Go to Next Step | | |
| | | No | Check ignition system | Yes | Injector clogging |
| | | | | No | Spark plug |
| | | | | | High-tension lead |
| | | | | | Distributor cap |
| 4 | Check injector operating sound at idle | Yes | Go to Next Step | | |
| | | No | Check injector resistance | Yes | Wiring short or open |
| | | | | No | Injector malfunction (Check resistance) |
| | | | | | Wiring short or open |
| 5 | Disconnect ISC valve connector and check idle speed during warm up | Yes | Go to Next Step | | |
| | | No | | | BAC valve stuck close |
| 6 | Disconnect water thermosensor connector Check if the condition become better | Yes | Check water thermosensor connector for connecting condition as follows 1. Shake connector and check if the condition changes 2. Check condition of terminal (burning or damaged) 3. Connect a male terminal to the connector and check for connection condition | Yes | Wather thermosensor malfunction |
| | | | | No | Poor contact at water thermosensor connector |
| | | No | Go to Next Step | | |
| 7 | Check if the specified engine oil is used | Yes | Go to Next Step | | |
| | | No | Change engine oil with specified one | | |
| 8 | Substitute a well-known ECU Check if the condition becomes better | Yes | | | ECU malfunction |
| | | No | | | Airflow meter malfunction |

## TROUBLESHOOTING CHART—1990 MPV WITH 6 CYLINDER ENGINE

**Rough idle (only when engine is warm)**

| STEP | QUICK INSPECTION | | ACTION | | POSSIBLE CAUSE |
|---|---|---|---|---|---|
| 1 | Run the engine at **2,000 rpm for more than 20 sec.** Check malfunction code (08) (09) (15) (17) (26) with SST [IG ON, Test connector (Green: 1-pin) grounded] | Yes | Go to Next Step | | |
| | | No | Check for the cause by referring to the check sequence | | |
| 2 | Check idle speed after warm up **Idle speed: 800 ± 20 rpm** [Test connector (Green: 1-pin) grounded] | Yes | Go to Next Step | | |
| | | No | Adjust idle speed (if possible) | | |
| 3 | Check monitor lamp flashing with SST after warm up **Monitor lamp: Flashes ON and OFF more than 8 times/10 seconds at 2,000—3,000 rpm** [Test connector (Green: 1-pin) not grounded] | Yes | Go to Next Step | | |
| | | No | | | Oxygen sensor deterioration |
| 4 | Disconnect ISC valve connector Check if the engine speed drops | Yes | Go to Next Step | | |
| | | No | | | ISC valve malfunction |
| 5 | Disconnect water thermosensor connector Check if the condition becomes better | Yes | Check water thermosensor connector for connecting condition as follows 1. Shake connector and check if the condition changes 2. Check condition of terminal (burning or damaged) 3. Connect a male terminal to the connector and check for connection condition | Yes | Water thermosensor malfunction |
| | | | | No | Poor contact at water thermosensor connector |
| | | No | Go to Next Step | | |
| 6 | Disconnect high-tension lead of each cylinder at idle Check if the engine condition changes | Yes | Go to Next Step | | |
| | | No | Check ignition system | Yes | Injector clogging |
| | | | | No | Spark plug |
| | | | | | High-tension lead |
| | | | | | Distributor cap |
| | | | Note: If spark plug is wet,it may be from injector fuel leaks | | |
| 7 | Check for injector operating sound at idle | Yes | Go to Next Step | | |
| | | No | Check injector resistance | Yes | Wiring short or open |
| | | | | No | Injector malfunction (Check resistance) |
| | | | | | Wiring short or open |
| 8 | Check intake manifold vacuum at idle **Vacuum: 480 ± 20 mmHg (18.9 ± 0.8 inHg)** | Yes | Go to Next Step | | |
| | | No | Check for air leaks | Yes | Throttle valve |
| | | | | No | Vacuum hoses |
| | | | | | Gaskets |
| | | | | | Bolts or nuts loosen |
| | | | | | Components damaged |
| 9 | Substitute a well-known ECU | | | | ECU malfunction |

**Rough idle (only after heat soak)**

| STEP | QUICK INSPECTION | | ACTION | | POSSIBLE CAUSE |
|---|---|---|---|---|---|
| 1 | Run the engine at **2,000 rpm for more than 20 sec.** Check malfunction code (09) (11) (15) (17) (25) (26) with SST [IG ON, Test connector (Green: 1-pin) grounded] | Yes | Go to Next Step | | |
| | | No | Check for the cause by referring to the check sequence | | |
| 2 | Check switches with SST Neutral switch (M/T) Clutch switch (M/T) Inhibitor switch (A/T) [IG ON, Test connector (Green: 1-pin) grounded] | Yes | Go to Next Step | | |
| | | No | Check for the cause by referring to the check sequence | | |
| 3 | Check monitor lamp flashing with SST after warm up **Monitor lamp: Flashes ON and OFF more than 8 times/10 seconds at 2,000—3,000 rpm** [Test connector (Green: 1-pin) not grounded] | Yes | Go to Next Step | | |
| | | No | | | Oxygen sensor deterioration |
| 4 | Disconnect vacuum hose from pressure regulator Check if the condition become better | Yes | Check components related to pressure regulator control system | | Water thermo sensor malfunction |
| | | | | | Intake air thermosensor malfunction |
| | | | | | Solenoid valve (PRC) malfunction |
| | | | | | ECU malfunction (Check 2N terminal voltage) |
| | | No | Go to Next Step | | |
| 5 | Run the engine at idle and stop it Observe fuel pressure **for 5 min.** **Fuel pressure: More than 147 kPa (1.5 kg/cm², 21 psi)** | Yes | Go to Next Step | | |
| | | No | Check fuel pump pressure drop | No | Fuel pump malfunction (Replace) |
| | | | Check pressure regulator pressure drop | Yes | Injector fuel leaks |
| | | | | No | Pressure regulator malfunction |
| 6 | Disconnect high-tension lead of each cylinder at idle Check if the condition changes | Yes | Go to Next Step | | |
| | | No | Check ignition system | Yes | Injector clogging/Replace (If step 3 OK) |
| | | | | No | Spark plug |
| | | | | | High-tension lead |
| | | | | | Distributor cap |
| 7 | Check for injector operating sound at idle | Yes | Go to Next Step | | |
| | | No | Check injector resistance | Yes | Wiring short or open |
| | | | | No | Injector malfunction (Check resistance) |
| 8 | Change fuel to specified one | | | | Poor fuel quality |

## TROUBLESHOOTING CHART—1990 MPV WITH 6 CYLINDER ENGINE

**Rough idle just after starting**

| STEP | QUICK INSPECTION | | ACTION | | POSSIBLE CAUSE |
|---|---|---|---|---|---|
| 1 | Check malfunction code (09) (34) with SST<br>[IG ON, Test connector (Green: 1-pin) grounded] | Yes | Go to Next Step | | |
| | | No | Check for the cause by referring to the check sequence | | |
| 2 | Check idle switch with SST<br>[IG ON, Test connector (Green: 1-pin) grounded] | Yes | Go to Next Step | | |
| | | No | Check for the cause by referring to the check sequence | | |
| 3 | Check ignition timing at idle after warm up<br>**Ignition timing: 11 ± 1° BTDC at sea level Approx. 20° BTDC at high altitude [1,000 m (3,280 ft)]**<br>[Test connector (Green: 1-pin) grounded] | Yes | Go to Next Step | | |
| | | No | Check ignition timing after warm up<br>**Ignition timing: 11 ± 1° BTDC**<br>[Test connector (Green: 1-pin) grounded] | Yes | Atmospheric pressure sensor (ECU Replace) |
| | | | | No | Adjust ignition timing |
| 4 | Check idle speed after warm up<br>**Idle speed: 800 ± 20 rpm**<br>[Test connector (Green: 1-pin) grounded] | Yes | Go to Next Step | | |
| | | No | Try to adjust idle speed | Yes | Idle speed misadjustment |
| | | | | No | Accelerator cable free play |
| | | | | | ISC valve stuck close |
| | | | | | Throttle valve |
| 5 | Substitute a well-known ECU<br>Check if the condition become better | Yes | | | ECU malfunction |
| | | No | Check voltage at ECU 3B terminal with SST<br>**Voltage: Approx. 6—11V (While cranking)** | Yes | Go to Section H |
| | | | | No | Inhibitor switch (A/T) |
| | | | | | Interlock switch |
| | | | | | Replated wiring open |
| 6 | | | | | Poor quality engine oil |

**Low idle speed (when A/C, P/S, E/L is ON)**

| STEP | QUICK INSPECTION | | ACTION | POSSIBLE CAUSE |
|---|---|---|---|---|
| 1 | Check malfunction code (34) with SST<br>[IG ON, Test connector (Green: 1-pin) grounded] | Yes | Go to Next Step | |
| | | No | Check for the cause by referring to the check sequence | |
| 2 | Disconnect ISC valve connected at idle<br>Check if the condition does not change | Yes | Go to Next Step | |
| | | No | | Low coolant level |
| | | | | Engine oil poor quality |
| 3 | Check switches with SST<br>Idle switch<br>Neutral switch (M/T)<br>Clutch switch (M/T)<br>Inhibitor switch (A/T)<br>[IG ON, Test connector (Green: 1-pin) grounded] | Yes | Go to Next Step | |
| | | No | Check for the cause by referring to the check sequence | |
| 4 | Check the continuity between test connector (Green: 1-pin) and ground | | | Wiring short to ground |

**Note**

- **In case of low idle speed with A/C ON, if the problem can not be solved by the above steps, it may be an A/C compressor malfunction.**

## TROUBLESHOOTING CHART—1990 MPV WITH 6 CYLINDER ENGINE

**High idle speed after warm up**

| STEP | QUICK INSPECTION | | ACTION | | POSSIBLE CAUSE |
|---|---|---|---|---|---|
| 1 | Check malfunction code (09) (26) (34) with SST<br>[IG ON, Test connector (Green: 1-pin) grounded] | Yes | Go to Next Step | | |
| | | No | Check for the cause by referring to the check sequence | | |
| 2 | Check ignition timing at idle after warm up<br>**Ignition timing: 11 ± 1° BTDC**<br>[Test connector (Green: 1-pin) grounded] | Yes | Go to Next Step | | |
| | | No | Adjust ignition timing | | |
| 3 | Check intake manifold vacuum at idle | Yes | Go to Next Step | | |
| | | No | Check for air leak | Yes | Vacuum hose mis-routing |
| | | | | | Throttle valve open |
| | | | | No | Vacuum hoses |
| | | | | | Gaskets |
| | | | | | Bolts or nuts loose |
| | | | | | Intake air system components |
| 4 | Check idle speed<br>**Idle speed: 800 ± 20 rpm**<br>[Test connector (Green: 1-pin) grounded] | Yes | | | ISC valve malfunction |
| | | No | Adjust idle speed (If possible) | | |
| 5 | Disconnect ISC valve connector and check idle speed during warm up | Yes | Go to Next Step | | |
| | | No | | | Air valve stuck open |
| 6 | Disconnect water thermosensor connector and check if the condition becomes better | Yes | Check for water thermosensor connector condition as follows<br>1. Shake the connector and check if the engine condition become better<br>2. Check pin condition (burning or broken)<br>3. Connect the male pin to the connector and check for connector condition | Yes | Water thermosensor malfunction |
| | | | | No | Poor contact at water thermosensor connector |
| | | No | Go to Next Step | | |
| 7 | Substitute a well-known ECU | | | | |

**Idle hunting or surging**

| STEP | QUICK INSPECTION | | ACTION | | POSSIBLE CAUSE |
|---|---|---|---|---|---|
| 1 | (If trouble occurs at warm condition)<br>Check malfunction code (15) (34) with SST<br>[IG ON, Test connector (Green: 1-pin) grounded] | Yes | Check for the cause by referring to the check sequence | | |
| | | No | Go to Next Step | | |
| 2 | (If trouble occurs at warm condition)<br>Check monitor lamp flashing with SST after warm up<br>**Monitor lamp: Flashes ON and OFF more than 8 times/10 seconds at 2,000—3,000 rpm**<br>[Test connector (Green: 1-pin) not grounded] | Yes | Go to Next Step | | |
| | | No | | | Oxygen sensor deterioration<br>(Replace) |
| 3 | Check for intake manifold vacuum at idle<br>**Vacuum: 480 ± 20 mmHg (18.9 ± 0.8 inHg)** | Yes | Go to Next Step | | |
| | | No | | | Air leakage |
| 4 | Pinch PCV hose check if the engine condition changes | Yes | | | PCV valve malfunction |
| | | No | Go to Next Step | | |
| 5 | Disconnect the water thermosensor connector<br>Check if the engine condition changes | Yes | | | Water thermosensor malfunction |
| | | No | Go to Next Step | | |
| 6 | Connect the test connector (Yellow: 2-pin) with a jumper wire and disconnect the airflow meter connector<br>Check if the hunting stops | Yes | | | Airflow meter malfunction |
| | | No | Go to Next Step | | |
| 7 | Check fuel line pressure at idle (Vacuum hose to pressure regulator disconnected)<br>**Fuel line pressure: 265—314 kPa (2.7—3.2 kg/cm², 38—46 psi)** | Yes | Go to Next Step | | |
| | | No | Fuel line pressure does not drop when connecting vacuum hose to pressure regulator:<br>Check for vacuum at vacuum hose | Yes | Pressure regulator malfunction. (Replace) |
| | | | | No | Vacuum hose clogging (Replace) |
| | | | Fuel line pressure drops but not with in specification when connecting vacuum hose to pressure regulator<br>**Fuel line pressure: 265—314 kPa (2.7—3.2 kg/cm², 38—46 psi)** | | Pressure regulator malfunction<br>(Replace) |
| 8 | Substitute a well-known ECU | | | | ECU malfunction |
| 9 | Substitute a well-known pulsation damper | | | | Pulsation damper malfunction |

## TROUBLESHOOTING CHART—1990 MPV WITH 6 CYLINDER ENGINE

**Engine stall at idle (Always)**

| STEP | QUICK INSPECTION | | ACTION | | POSSIBLE CAUSE |
|---|---|---|---|---|---|
| 1 | Check malfunction code with SST [IG: ON, Test connector (Green: 1-pin) grounded] | Yes | Check the cause by referring to the check sequence | | |
| | | No | Go to Next Step | | |
| 2 | Disconnect high-tension lead of each cylinder (Engine speed: 1,000 rpm) Is engine condition changed? | Yes | Go to Next Step | | |
| | | No | Check ignition system | Yes | Injector malfunction |
| | | | | No | Ignition system malfunction |
| 3 | Check fuel line pressure [IG ON, test connector (Yellow: 2-pin) connected] **Fuel line pressure: 265—314 kPa (2.7—3.2 kg/cm², 38—46 psi)** | Yes | Go to Next Step | | |
| | | No | | | Fuel leaks |
| | | | | | Fuel filter clogging |
| | | | Check fuel pump maximum pressure **Fuel pump maximum pressure: 441—589 kPa (4.5—6.0 kg/cm², 64—85 psi)** | Yes | Pressure regulator malfunction (Replace) |
| | | | | No | Fuel pump malfunction (Replace) |
| 4 | Check the vacuum hoses and the air funnel for sucking air | Yes | Shut the air leak part | | |
| | | No | Go to Next Step | | |
| 5 | Connect terminals of test connector (Yellow: 2-pin) Check if engine stalls | Yes | Go to Next Step | | |
| | | No | Replace airflow meter | | |
| 6 | | | | | ECU malfunction |

**Engine stall at idle (Only when engine is cold)**

| STEP | QUICK INSPECTION | | ACTION | POSSIBLE CAUSE |
|---|---|---|---|---|
| 1 | Check 2I terminal of ECU | Yes | Go to Next Step | |
| | | No | Replace water thermosensor | Water thermosensor malfunction |
| 2 | Blow the air valve (cold condition) Check the airflow | Yes | Go to Next Step | |
| | | No | Replace the BAC valve | BAC valve malfunction (Replace) |
| 3 | Substitute a well-known ECU | | | ECU malfunction (Replace) |

**Engine stall at idle (only when engine is warm)**

| STEP | QUICK INSPECTION | | ACTION | POSSIBLE CAUSE |
|---|---|---|---|---|
| 1 | Check 2I terminal of ECU | Yes | Go to Next Step | |
| | | No | Replace water thermosensor | Water thermosensor malfunction |
| 2 | Substitute a well-known ECU | | | ECU malfunction |

**Engine stall at idle (Only after heat soak)**

| STEP | QUICK INSPECTION | | ACTION | | POSSIBLE CAUSE |
|---|---|---|---|---|---|
| 1 | Run the engine at **2,000 rpm for more than 20 sec.** Check malfunction code (09) (11) (15) (17) (25) (26) with SST [IG ON, Test connector (Green: 1-pin) grounded] | Yes | Go to Next Step | | |
| | | No | Check for the cause by referring to the check sequence | | |
| 2 | Check switches with SST Neutral switch (M/T) Clutch switch (M/T) Inhibitor switch (A/T) [IG ON, Test connector (Green: 1-pin) grounded] | Yes | Go to Next Step | | |
| | | No | Check for the cause by referring to the check sequence | | |
| 3 | Check monitor lamp flashing with SST after warm up **Monitor lamp: Flashes ON and OFF more than 8 times/10 seconds at 2,000—3,000 rpm** [Test connector (Green: 1-pin) not grounded] | Yes | Go to Next Step | | |
| | | No | | | Oxygen sensor deterioration |
| 4 | Disconnect vacuum hose from pressure regulator Check if the condition becomes better | Yes | Check components related to pressure regulator control system | | Water thermosensor malfunction |
| | | | | | Intake air thermosensor malfunction |
| | | | | | Solenoid valve (PRC) malfunction |
| | | | | | ECU malfunction (Check 2N terminal voltage) |
| | | No | Go to Next Step | | |
| 5 | Run the engine at idle and stop it Observe fuel pressure **for 5 minutes** **Fuel pressure: More than 147 kPa (1.5 kg/cm², 21 psi)** | Yes | Go to Next Step | | |
| | | No | Check fuel pump pressure drop | No | Fuel pump malfunction (Replace) |
| | | | Check pressure regulator pressure drop | Yes | Injector fuel leaks |
| | | | | No | Pressure regulator malfunction |
| 6 | Disconnect high-tension lead of each cylinder at idle Check if the condition changes | Yes | Go to Next Step | | |
| | | No | Check ignition system | Yes | Injector clogging Replace (If step 3 OK) |
| | | | | No | Spark plug |
| | | | | | High-tension lead |
| | | | | | Distributor cap |

## TROUBLESHOOTING CHART—1990 MPV WITH 6 CYLINDER ENGINE

**Engine stall at idle (When A/C, P/S, E/L (CPU) is ON)**

| STEP | QUICK INSPECTION | | ACTION | POSSIBLE CAUSE |
|---|---|---|---|---|
| 1 | Check malfunction code (34) with SST [IG ON, Test connector (Green: 1-pin) grounded] | Yes | Go to Next Step | |
| | | No | Check for the cause by referring to the check sequence | |
| 2 | Disconnect ISC valve connected at idle Check if the condition does not change | Yes | Go to Next Step | |
| | | No | | Low coolant level |
| | | | | Engine oil poor quality |
| 3 | Check switches with SST Idle switch Neutral switch (M/T) Clutch switch (M/T) Inhibitor switch (A/T) | Yes | Go to Next Step | |
| | | No | Check for the cause by referring to the check sequence | |
| 4 | Check the continuity between test connector (Green: 1-pin) and ground | | | Wiring short to ground |

**Note**

- **Engine stall at idle with A/C ON, if the trouble can not be fixed after checking above steps, it may be A/C compressor malfunction.**

**Engine stall at idle (in "D" range)**

| STEP | QUICK INSPECTION | | ACTION | POSSIBLE CAUSE |
|---|---|---|---|---|
| 1 | Check malfunction code with SST [IG: ON, Test connector (Green: 1-pin) grounded] | Yes | Check the cause by referring to the check sequence | |
| | | No | Go to Next Step | |
| 2 | Check switches with SST Inhibitor switch [IG, ON, Test connector (Green: 1-pin) grounded] | Yes | Go to Next Step | |
| | | No | Adjustment or replacement | Inhibitor switch malfunction |
| 3 | Substitute a well-known ECU | | | ECU malfunction (Replace) |

**Engine stall during start up**

| STEP | QUICK INSPECTION | | ACTION | POSSIBLE CAUSE |
|---|---|---|---|---|
| 1 | Check malfunction code (08) with SST [IG: ON, Test connector (Green: 1-pin) grounded] | Yes | Check the cause by referring to the check sequence | |
| | | No | Go to Next Step | |
| 2 | Check idle switch with SST [IG ON, Test connector (Green: 1-pin) grounded] | Yes | Go to Next Step | |
| | | No | Check for the cause by referring to the check sequence | |
| 3 | Check the idle speed **Idle speed: 800 ± 20 rpm** [Test connector (Green: 1-pin) grounded] | Yes | Go to Next Step | |
| | | No | Adjust the idle speed | |
| 4 | Check for injector operating sound at idle | Yes | Go to Next Step | |
| 5 | Check for correct ignition timing advance | Yes | Go to Next Step | |
| | | No | Adjust or replace | Distributor malfunction |
| 6 | Substitute a well-known ECU | | | ECU malfunction |

**Engine stall on deceleration (only when engine is cold)**

| STEP | QUICK INSPECTION | | ACTION | POSSIBLE CAUSE |
|---|---|---|---|---|
| 1 | Check malfunction code with SST [IG: ON, Test connector (Green: 1-pin) grounded] | Yes | Check the cause by referring to the check sequence | |
| | | No | Go to Next Step | |
| 2 | Check 2I terminal of ECU | Yes | Go to Next Step | |
| | | No | Replace the water thermosensor | Water thermosensor malfunction |
| 3 | Blow the air valve (cold condition) check the air flow | Yes | Go to Next Step | |
| | | No | Replace the BAC valve | BAC valve malfunction (Replace) |
| 4 | Check the dashpot touch rpm and the operation | Yes | Go to Next Step | |
| | | No | Adjust or replace the dashpot | Dashpot malfunction |
| 5 | Connect terminals of test connector (Yellow: 2-pin) Check the engine stalls | Yes | Go to Next Step | |
| | | No | Replace the circuit opening relay | Circuit opening relay malfunction |
| 6 | Substitute a well-known ECU | | | ECU malfunction |

## TROUBLESHOOTING CHART—1990 MPV WITH 6 CYLINDER ENGINE

**Engine stall on deceleration (Only after warm up)**

| STEP | QUICK INSPECTION | | ACTION | POSSIBLE CAUSE |
|---|---|---|---|---|
| 1 | Check malfunction code (08) with SST [IG ON, Test connector (Green: 1-pin) grounded | Yes | Check the cause by referring to the check sequence | |
| | | No | Go to Next Step | |
| 2 | Substitute a well-known ECU | | | ECU malfunction (Replace) |

**Engine stall at idle (Intermittent)**

| STEP | QUICK INSPECTION | | ACTION | POSSIBLE CAUSE |
|---|---|---|---|---|
| 1 | Shake connector of ignition coil, main relay and ECU while cranking Check if the engine starts | Yes | There may be a poor contact at the connector. Repair or replace the wiring | |
| | | No | Go to troubleshooting "Engine stall at idle (always)" | |

**Hesitates/Stumbles on acceleration**

| STEP | QUICK INSPECTION | | ACTION | POSSIBLE CAUSE |
|---|---|---|---|---|
| 1 | Run the engine at **2,000 rpm for 20 sec.** and stop it<br>Check malfunction code (08) (09) (15) (17) with SST [IG ON, Test connector (Green: 1-pin) grounded] | Yes | Go to Next Step | |
| | | No | Check for the cause by referring to the check sequence | |
| 2 | Check idle switch with SST<br>[IG ON, Test connector (Green: 1-pin) grounded] | Yes | Go to Next Step | |
| | | No | Check for the cause by referring to the check sequence | |
| 3 | Disconnect oxygen sensor connector Check if the condition becomes better | Yes | | Oxygen sensor |
| | | No | Go to Next Step | |
| 4 | Check fuel line pressure while accelerating (Vacuum hose to pressure regulator disconnected)<br>**Fuel line pressure: Keeps 264—314 kPa ($2.7—3.2\ kg/cm^2$, 38.3—45.5 psi)** | Yes | Go to Next Step | |
| | | No | | Fuel line clogging<br>Fuel filter clogging<br>Pressure regulator malfunction/(Replace) |
| 5 | Check for air leaks with throttle valve open by listening for sucking noise | Yes | Go to Next step | |
| | | No | | Vacuum hoses<br>Gaskets<br>Nuts or bolts loose<br>Damaged components |
| 6 | Disconnect water thermosensor connector Check if the condition improves | Yes | Check water thermosensor connector for connecting condition as follows<br>1. Shake connector and check if the condition changes<br>2. Check condition of terminal (burning or damaged)<br>3. Connect a male terminal to the connector and check for connection condition | Yes: Water thermosensor malfunction<br>No: Poor contact at water thermosensor connector |
| | | No | Go to Next Step | |
| 7 | Substitute a well-known ECU Check if the condition become better | Yes | | ECU malfunction |
| | | No | | Airflow meter<br>Throttle valve<br>Spark plug |
| 8 | Check other systems | | | Clutch slipping |

## TROUBLESHOOTING CHART—1990 MPV WITH 6 CYLINDER ENGINE

**Hesitates at steady speed**

| STEP | QUICK INSPECTION | | ACTION | POSSIBLE CAUSE |
|---|---|---|---|---|
| 1 | Run the engine at **2,000 rpm for 20 sec.** and stop it<br>Check malfunction code (08) (09) (12) (14) (15) (17) with SST [IG ON, Test connector (Green: 1-pin) grounded] | Yes | Go to Next Step | |
| | | No | Check for the cause by referring to the check sequence | |
| 2 | Disconnect oxygen sensor connector Check if the condition becomes better | Yes | | Oxygen sensor deterioration |
| | | No | Go to Next Step | |
| 3 | Check for air leaks with throttle valve open by listening for sucking noise | Yes | Go to Next Step | |
| | | No | | Vacuum hoses<br>Gaskets<br>Nuts or bolts loosen<br>Damaged components |
| 4 | Check fuel line pressure while accelerating (Vacuum hose to pressure regulator disconnected)<br>**Fuel line pressure: Keeps 264—314 kPa ($2.7—3.2\ kg/cm^2$, 38.3—45.5 psi)** | Yes | Go to Next Step | |
| | | No | | Pressure regulator malfunction (Replace) |
| 5 | Check condition ignition coil and airflow meter connectors terminal (Burning or damaged) | Yes | Go to Next Step | |
| | | No | | Poor contact |
| 6 | Gradually open the throttle valve Check if engine speed increases smoothly | Yes | Go to Next Step | |
| | | No | | Airflow meter<br>Throttle valve<br>Throttle sensor |
| 7 | | | | Spark plug<br>• Gap<br>• Damage |
| 8 | Change fuel with specified one<br>Check if the condition become better | Yes | | Poor fuel quality |
| | | No | Go to Next Step | |
| 9 | Substitute a well-known ECU | | | |

**Hesitates/Stumbles on acceleration**

| STEP | QUICK INSPECTION | | ACTION | POSSIBLE CAUSE |
|---|---|---|---|---|
| 1 | Run the engine at **2,000 rpm for 20 sec.** and stop it<br>Check malfunction code (08) (09) (15) (17) with SST [IG ON, Test connector (Green: 1-pin) grounded] | Yes | Go to Next Step | |
| | | No | Check for the cause by referring to the check sequence | |
| 2 | Check idle switch with SST<br>[IG ON, Test connector (Green: 1-pin) grounded] | Yes | Go to Next Step | |
| | | No | Check for the cause by referring to the check sequence | |
| 3 | Disconnect oxygen sensor connector Check if the condition becomes better | Yes | | Oxygen sensor |
| | | No | Go to Next Step | |
| 4 | Check fuel line pressure while accelerating (Vacuum hose to pressure regulator disconnected)<br>**Fuel line pressure: Keeps 264—314 kPa ($2.7—3.2\ kg/cm^2$, 38.3—45.5 psi)** | Yes | Go to Next Step | |
| | | No | | Fuel line clogging<br>Fuel filter clogging<br>Pressure regulator malfunction/(Replace) |
| 5 | Check for air leaks with throttle valve open by listening for sucking noise | Yes | Go to Next step | |
| | | No | | Vacuum hoses<br>Gaskets<br>Nuts or bolts loosen<br>Damaged components |
| 6 | Disconnect water thermosensor connector Check if the condition improves | Yes | Check water thermosensor connector for connecting condition as follows<br>1. Shake connector and check if the condition changes<br>2. Check condition of terminal (burning or damaged)<br>3. Connect a male terminal to the connector and check for connection condition | Yes: Water thermosensor malfunction<br>No: Poor contact at water thermosensor connector |
| | | No | Go to Next Step | |
| 7 | Substitute a well-known ECU Check if the condition become better | Yes | | ECU malfunction |
| | | No | | Airflow meter<br>Throttle valve<br>Spark plug |
| 8 | Check other systems | | | Clutch slipping |

## TROUBLESHOOTING CHART—1990 MPV WITH 6 CYLINDER ENGINE

**Knocking**

| STEP | QUICK INSPECTION | | ACTION | | POSSIBLE CAUSE |
|---|---|---|---|---|---|
| 1 | Check malfunction code (09) with SST [IGN ON. Test connector (Green: 1-pin) grounded] | Yes | Check for cause by referring to check sequence | | |
| | | No | Go to Step 2 (at sea level)<br>Go to Step 3 [at high altitude (1,000 m, 3,280 ft)] | | |
| 2 | Check ignition timing after warm up<br>**Ignition timing 11 ± 1° BTDC**<br>[Test connector (Green: 1-pin) grounded] | Yes | Go to Next Step | | |
| | | No | Adjust ignition timing | | |
| 3 | Check fuel line pressure at idle (Disconnecting vacuum hose to pressure regulator)<br>**Fuel line pressure: Keeps 265—314 kPa (2.7—3.2 kg/cm², 38—46 psi)** | Yes | Go to Next Step | | |
| | | No | Fuel pressure does not drop when connecting vacuum hose to pressure regulator:<br>**Fuel line pressure 216—265 kPa (2.2—2.7 kg/cm², 31—38 psi)**<br>Check for vacuum at vacuum hose | Yes | Pressure regulator malfunction (Replace) |
| | | | | No | Vacuum hose clogging (Replace) |
| | | | Fuel pressure drops but not within specification when connecting vacuum hose to pressure regulator<br>**Fuel line pressure 216—265 kPa (2.2—2.7 kg/cm², 31—38 psi)** | | Pressure regulator malfunction (Replace) |
| 4 | Check the airflow meter | Yes | Go to Next Step | | |
| | | No | Replace the airflow meter | | Airflow meter malfunction |
| 5 | Change the fuel with specified one<br>Check if the condition become better<br>**Fuel spec: RON87 or higher** | Yes | | | Poor fuel quality |
| | | No | Go to Next Step | | |
| 6 | Substitute a well-known ECU | | | | ECU malfunction |

**Poor acceleration**

| STEP | QUICK INSPECTION | | ACTION | | POSSIBLE CAUSE |
|---|---|---|---|---|---|
| 1 | Check malfunction code with SST [IG ON, Test connector (Green: 1-pin) grounded | Yes | Check the cause by referring to the check sequence | | |
| | | No | Go to Next Step | | |
| 2 | Check switches with SST<br>Idle switch | Yes | Check the cause by referring to the check sequence | | Idle switch malfunction (Adjust or replace) |
| | | No | Go to Next Step | | |
| 3 | Disconnect high-tension lead of each cylinder at idle<br>IS engine condition changed? | Yes | Go to Next Step | | |
| | | No | Check ignition system<br>Refer to G section | Yes | Injector malfunction |
| | | | | No | Ignition system malfunction |
| 4 | Check air cleaner element for clogging | Yes | Go to Next Step | | |
| | | No | Clean air cleaner element | | |
| 5 | Check ignition timing at idle after warm up<br>**Ignition timing: 11 ± 1° BTDC**<br>[Test connector (Green: 1-pin) grounded] | Yes | Go to Next Step | | |
| | | No | Adjust ignition timing | | |
| | Note: At high-altitude area **[1,000 m (3,280 ft)]**, ignition timing advances to **approx. 20° BTDC** when test connector (Green: 1-pin) is not grounded) | | | | |
| 6 | Check the vacuum hoses and the air funnel for sucking air | Yes | Shut the air leak part | | |
| | | No | Go to Next Step | | |
| 7 | Check fuel line pressure at idle (vacuum hose to pressure regulator disconnected)<br>**Fuel line pressure: 265—314 kPa (2.7—3.2 kg/cm², 38—46 psi)** | Yes | Go to Next Step | | |
| | | No | Fuel line pressure does not drop when connecting vacuum hose to pressure regulator<br>Check for vacuum at vacuum hose | Yes | Pressure regulator malfunction (Replace) |
| | | | | No | Vacuum hose clogging (Replace) |
| | | | Fuel line pressure drops but not within specification when connecting vacuum hose to pressure regulator | | Pressure regulator malfunction (Replace) |
| | | | | | Fuel pump malfunction (Replace) |
| 8 | When the accelerator pedal is depressed, engine speed increases smoothly | Yes | Go to Next Step | | |
| | | No | Airflow meter malfunction | | |
| 9 | Change the fuel with specified one<br>Check if the condition become better<br>**Fuel spec: RON87 or higher** | Yes | | | Poor fuel quality |
| | | No | Go to Next Step | | |
| 10 | Substitute a well-known ECU | | | | ECU malfunction |

## TROUBLESHOOTING CHART—1990 MPV WITH 6 CYLINDER ENGINE

**Bucking at high speed**

| STEP | QUICK INSPECTION | | ACTION | | POSSIBLE CAUSE |
|---|---|---|---|---|---|
| 1 | Check malfunction code with SST [IG ON, Test connector (Green: 1-pin) grounded] | Yes | Check for the cause by referring to the check sequence | | |
| | | No | Go to Next Step | | |
| 2 | Disconnect the oxygen sensor connector, check the engine condition better | Yes | Replace the oxygen sensor | | |
| | | No | Go to Next Step | | |
| 3 | Check fuel line pressure at idle (Vacuum hose to pressure regulator disconnected)<br>**Fuel line pressure: 265—314 kPa (2.7—3.2 kg/cm², 38—46 psi)** | Yes | Go to Next Step | | |
| | | No | Fuel line pressure does not drop when connecting vacuum hose to pressure regulator:<br>**Fuel line pressure: 206—255 kPa (2.1—2.6 kg/cm², 30—37 psi)**<br>Check for vacuum at vacuum hose | Yes | Pressure regulator malfunction Replace |
| | | | | No | Vacuum hose clogging Replace |
| | | | Fuel lien pressure drops but not within specification when connecting vacuum hose to pressure regulator | | Pressure regulator malfunction Replace |
| 4 | Check the vacuum hoses and the air funnel for sucking air | Yes | Shut the air leak part | | |
| | | No | Go to Next Step | | |
| 5 | Check ignition timing at idle after warm up<br>**Ignition timing: 11 ± 1° BTDC**<br>[Test connector (Green: 1-pin) grounded] | Yes | Go to Next Step | | |
| | | No | Adjust ignition timing | | |
| | Note: At high-altitude area **[1,000 m (3,280 ft)]**, ignition timing advances to **approx. 20° BTDC** when test connector (Green: 1-pin) is not grounded | | | | |
| 6 | When the accelerator pedal is depressed, engine speed increases smoothly | Yes | Go to Next Step | | |
| | | No | Airflow meter malfunction | | |
| 7 | Substitute a well-known ECU | | | | ECU malfunction |

**Bucking on deceleration**

| STEP | QUICK INSPECTION | | ACTION | POSSIBLE CAUSE |
|---|---|---|---|---|
| 1 | Check malfunction code (08) (09) (12) (26) (34) with SST [IG ON, Test connector (Green: 1-pin) grounded] | Yes | Go to Next Step | |
| | | No | Check for the cause by referring to the check sequence | |
| 2 | Check switches with SST [IG ON, Test connector (Green: 1-pin) grounded]<br>• Idle switch<br>• Stoplight switch | Yes | Go to Next Step | |
| | | No | Check for the cause by referring to the check sequence | |
| 3 | Check the dashpot operation | Yes | Go to Next Step | |
| | | No | Adjust or replace the dashpot | Dashpot malfunction |
| 4 | Substitute a well-known ECU<br>Check if the condition become better | Yes | | ECU malfunction |
| | | No | | Throttle sensor |
| | | | | Go to Next Step |
| 5 | | | | Spark plug gap |
| 6 | | | | Clutch slipping |
| 7 | | | | Compression differs between cylinders |

## TROUBLESHOOTING CHART—1990 MPV WITH 6 CYLINDER ENGINE

**Poor fuel economy**

| STEP | QUICK INSPECTION | | ACTION | | POSSIBLE CAUSE |
|---|---|---|---|---|---|
| 1 | Run the engine at **2,000—3,000 rpm for more than 20 sec.** after warm up and stop it. Check malfunction code (15) (17) with SST [IG ON, Test connector (Green: 1-pin) grounded] | Yes | Go to Next Step | | |
| | | No | Check for the cause by referring to the check sequence | | |
| 2 | Check idle switch with SST [IG ON, Test connector (Green: 1-pin) grounded] | Yes | Go to Next Step | | |
| | | No | Check for the cause by referring to the check sequence | | |
| 3 | Check monitor lamp flashing with SST after warm up. **Monitor lamp: Flashes ON and OFF more than 8 times/10 seconds at 2,000—3,000 rpm** [Test connector (Green: 1-pin) not grounded] | Yes | Go to Next Step | | |
| | | No | | | Oxygen sensor deterioration |
| 4 | Check fuel line pressure at idle. **Fuel line pressure: 206—255 kPa (2.1—2.6 kg/cm², 30—37 psi)** | Yes | Go to Next Step | | |
| | | No | Check vacuum line to pressure regulator for clogging or air leak | Yes | Vacuum line |
| | | | | No | Solenoid valve (PRC) malfunction |
| | | | | | ECU malfunction (Check 2N terminal voltage) |
| | | | | | Pressure regulator malfunction/Replace |
| 5 | Check for fuel cut operation while deceleration. **Fuel cut: Until approx. 1,800 rpm (In-gear and braking condition-M/T Only) Until approx. 2,600 rpm (During A/C operation) Until approx. 2,300—4,600 rpm (Other conditions) After warm up** | Yes | Go to Next Step | | |
| | | No | Check water thermosensor | Yes | ECU malfunction |
| | | | | No | Water thermosensor malfunction/Replace |
| | | | Check stoplight switch | | |
| | | | Check neutral switch | | |
| 6 | Check ignition timing advance | Yes | Go to Next Step | | |
| | | No | | | Centrifugal advance bad: Distributor malfunction |
| | | | Vacuum advance bad. Check for vacuum routing and delay valve direction | | Distributor malfunction |
| 7 | Check other systems | | | | Clutch plate slipping |
| | | | | | A/T malfunction |
| | | | | | Brake wear |
| | | | | | Tire air pressure |

**High oil consumption/White exhaust smoke**

| STEP | QUICK INSPECTION | | ACTION | | POSSIBLE CAUSE |
|---|---|---|---|---|---|
| 1 | Check for oil leak from engine | Yes | Repair or replace | | |
| | | No | Go to Next Step | | |
| 2 | Disconnect PCV valve from engine. Check if the vacuum is felt at idle | Yes | Go to Next Step | | |
| | | No | Check PCV valve | Yes | PCV hose clogging |
| | | | | No | PCV valve malfunction |
| 3 | Check that the ventilation hose is installed correctly | Yes | Go to Next Step | | |
| | | No | Install ventilation hose correctly | | |
| 4 | It may be a malfunction of the engine | | | | |

**Afterburn on deceleration**

| STEP | QUICK INSPECTION | | ACTION | | POSSIBLE CAUSE |
|---|---|---|---|---|---|
| 1 | Check malfunction code (34) with SST [IG ON, Test connector (Green: 1-pin) grounded] | Yes | Go to Next Step | | |
| | | No | Check for the cause by referring to the check sequence | | |
| 2 | Check idle switch with SST [Test connector (Green: 1-pin) grounded] | Yes | Go to Next Step | | |
| | | No | Check for the cause by referring to the check sequence | | |
| 3 | Check ignition timing at idle after warm up. **Ignition timing: 11 ± 1° BTDC** [Test connector (Green: 1-pin) grounded] | Yes | Go to Next Step | | |
| | | No | | | Centrifugal advance bad: Distributor malfunction |
| | | | Vacuum advance bad. Check for vacuum routing | | Distributor malfunction |
| 4 | Check air cleaner element for clogging | Yes | Go to Next Step | | |
| | | No | Clean air cleaner element | | |
| 5 | Check fuel cut operation while decelerating. **Fuel cut: Above approx. 1,800 rpm after warm up** | Yes | Go to Next Step | | |
| | | No | Check water thermosensor | Yes | ECU malfunction 2I terminal |
| | | | | No | Water thermosensor (Replace) |
| 6 | Run the engine at idle and stop the engine (IG OFF). Check fuel pressure drop | Yes | Go to Next Step | | |
| | | No | Check fuel pump for pressure drop | Yes | Check pressure regulator for pressure drop |
| | | | | No | Fuel pump malfunction (Replace) |
| | | | Check pressure regulator for pressure drop | Yes | Injector leak |
| | | | | No | Pressure regulator malfunction (Replace) |
| 7 | | | | | Low compression |
| | | | | | Valve timing |

## TROUBLESHOOTING CHART—1990 MPV WITH 6 CYLINDER ENGINE

**Rotten egg smell**

| STEP | QUICK INSPECTION | ACTION | POSSIBLE CAUSE |
|---|---|---|---|
| 1 | Change fuel with specified one | | Poor fuel quality |

**Black exhaust smoke**

| STEP | QUICK INSPECTION | | ACTION | | POSSIBLE CAUSE |
|---|---|---|---|---|---|
| 1 | Run the engine at **2,000 rpm for more than 20 sec.** check for malfunction code (17) with SST [IGN ON, Test connector (Green: 1-pin) grounded] | Yes | Go to Next Step | | |
| | | No | Check for the cause by referring to the check sequence | | |
| 2 | Check fuel line pressure [IGN ON, Test connector (Yellow: 2-pin) connected] **Fuel line pressure: 265—314 kPa (2.7—3.2 kg/cm², 38—46 psi)** | Yes | Go to Next Step | | |
| | | No (high) | | | Pressure regulator malfunction (Replace) |
| 3 | Disconnect water thermosensor connector. Check if the condition improves | Yes | Check water thermosensor connector for connecting condition as follows 1. Shake connector and check if the condition changes 2. Check condition of terminal (burning or damaged) 3. Connect a male terminal to the connector and check for connection condition | Yes | Water thermosensor malfunction |
| | | | | No | Poor contact at water thermosensor connector |
| | | No | Go to step 4 **[at high-altitude (1,000 m, 3,280 ft)]** Go to step 5 (at sea level) | | |
| 4 | | | | | Atmospheric pressure sensor malfunction (Replace the ECU) |
| 5 | | | | | Airflow sensor malfunction |
| 6 | Substitute a well-known ECU | | | | ECU malfunction |

**Fails emission test (Engine condition OK)**

| STEP | QUICK INSPECTION | | ACTION | | POSSIBLE CAUSE |
|---|---|---|---|---|---|
| 1 | Run the engine at **2,000 rpm for 20 sec.** and stop it. Check malfunction code (08) (09) (14) (15) (17) (26) with SST [IG ON, Test connector (Green: 1-pin) grounded] | Yes | Go to Next Step | | |
| | | No | Check for the cause by referring to the check sequence | | |
| 2 | Check idle switch with SST [IG ON, Test connector (Green: 1-pin) grounded] | Yes | Go to Next Step | | |
| | | No | Check for the cause by referring to the check sequence | | |
| 3 | Check for air leaks by hearing sucking noise • At idle • Throttle valve open | Yes | Go to Next Step | | |
| | | No | | | |
| 4 | Check monitor lamp flashing with SST after warm up. **Monitor lamp: Flashes ON and OFF more than 8 times/10 seconds at 2,000—3,000 rpm** [Test connector (Green: 1-pin) not grounded] | Yes | Go to Next Step | | |
| | | No | | | Oxygen sensor deterioration (Replace) |
| 5 | Check fuel line pressure [IG ON, Test connector (Yellow: 2-pin) connected] **Fuel line pressure: 265—314 kPa (2.7—3.2 kg/cm², 38—46 psi)** | Yes | Go to Next Step | | |
| | | No (Low) | | | Fuel leaks |
| | | | | | Fuel filter clogging |
| | | | Check fuel pump maximum pressure **F1-146**. **Fuel pump maximum pressure: 441—589 kPa (4.5—6.0 kg/cm², 64—85 psi)** | Yes | Pressure regulator malfunction (Replace) |
| | | | | No | Fuel pump malfunction (Replace) |
| | | No (High) | | | Pressure regulator malfunction (Replace) |
| 6 | Check fuel line pressure at idle. **Fuel line pressure: 206—255 kPa (2.1—2.6 kg/cm², 30—37 psi)** | Yes | Go to Next Step | | |
| | | No | Check vacuum line to pressure regulator for clogging or air leak | Yes | Vacuum line |
| | | | | No | Solenoid valve (PRC) malfunction |
| | | | | | ECU malfunction (2N) |
| | | | | | Pressure regulator malfunction (Replace) |

## TROUBLESHOOTING CHART—1990 MPV WITH 6 CYLINDER ENGINE

**Fails emission test (Engine condition OK) (Cont'd)**

| STEP | QUICK INSPECTION | | ACTION | | POSSIBLE CAUSE |
|---|---|---|---|---|---|
| 7 | Check fuel cut operation while decelerating<br>**Fuel cut: Above approx. 1,800 rpm after warm up** | Yes | Go to Next Step | | |
| | | No | Check water thermosensor | Yes | ECU malfunction (Replace) |
| | | | | No | Water thermosensor malfunction/Replace |
| 8 | Check for fuel fumes | Yes | Check for the cause by referring to troubleshooting "Gasoline fumes" | | |
| | | No | Go to Next Step | | |
| 9 | | | | | PCV valve |
| | | | | | PCV hose |
| 10 | | | | | Spark plug<br>• Gap<br>• Damage |
| 11 | | | | | Poor fuel quality |

**MIL always ON (Engine condition is OK)**

| STEP | QUICK INSPECTION | | ACTION | | POSSIBLE CAUSE |
|---|---|---|---|---|---|
| 1 | Check malfunction code with SST<br>[IG ON. Test connector (Green 1-pin) grounded] | | "88" ECU malfunction (Replace) | | |
| | | | "00" (Except california)<br>Go to Next Step | | |
| | | | "00" (California only)<br>Wiring between ECU 1A terminal and MIL short to ground | | |
| 2 | (Except California)<br>Check if the oxygen sensor replacement time has come<br>**Oxygen sensor replacement time: Every 80,000 miles (Federal) Every 120,000 km (Canada)** | Yes | Check if MIL was reset by exchanging MIL set screw position | Yes | Go to next action |
| | | | | No | Reset the MIL |
| | | | Check continuity between MIL and ground | Yes | Wiring between MIL and ECU is short to ground |
| | | | | No | Mileage sensor malfunction |
| | | No | Check continuity between MIL and ground | Yes | Wiring between MIL and ECU is short to ground |
| | | | | No | Mileage sensor malfunction |

**Gasoline fumes**

| STEP | QUICK INSPECTION | | ACTION | | POSSIBLE CAUSE |
|---|---|---|---|---|---|
| 1 | Check for fuel leaks | Yes | Go to Next Step | | |
| | | No | Replace | | |
| 2 | Check if the evaporative fume is emitted from the check-and-cut valve | Yes | Check check-and-cut valve | Yes | Two-way check valve |
| | | | | | Purge line clogging |
| | | | | No | Check-and-cut valve malfunction (Replace) |
| | | No | Go to Next Step | | |
| 3 | Check malfunction code (08) (15) (09) (17) (26) with SST [IGN ON, Test connector (Green: 1-pin) grounded] | Yes | Go to Next Step | | |
| | | No | Check for the cause by referring to the check sequence | | |
| 4 | Check switches with SST<br>• Idle switch<br>• Neutral switch (M/T)<br>• Clutch switch (M/T)<br>• Inhibitor switch (A/T)<br>[IGN ON, Test connector (Green: 1-pin) grounded] | Yes | Go to Next Step | | |
| | | No | Check for the cause by referring to the check sequence | | |
| 5 | Ground the solenoid valve (Purge control) connector (L/Y) terminal and disconnect vacuum hose from chacoal canister and check for vacuum | Yes | | | ECU malfunction (Check 2K) terminal voltage |
| | | No | | | Solenoid valve (Purge control) malfunction (Replace) |

**STEP 1**

PRESSURE REGULATOR
SOLENOID (PRC)

## TROUBLESHOOTING CHART 1990 MPV WITH 6 CYLINDER ENGINE

**MIL never ON (Engine condition is not OK)**

| STEP | QUICK INSPECTION | | ACTION | | POSSIBLE CAUSE |
|---|---|---|---|---|---|
| 1 | Check if other indicator lamp illuminate | Yes | Go to Next Step | | |
| | | No | Check power supply circuit to combination meter | | |
| 2 | Check bulb of the MIL | Yes | Ground the ECU 1A terminal check if the MIL illuminates | Yes | ECU malfunction |
| | | | | No | Wiring between ECU and MIL open |
| | | No | Replace | | |
| 3 | (Except California)<br>Check if the MIL illuminated at oxygen sensor replacement time<br>**Oxygen sensor replacement time: Every 80,000 miles (Federal) Every 120,000 km (Canada)** | | | | MIL set screw loose or left |
| | | | | | Mileage sensor malfunction |

**A/C does not work**

| STEP | QUICK INSPECTION | | ACTION | | POSSIBLE CAUSE |
|---|---|---|---|---|---|
| 1 | Check if the condenser fan operates when grounding A/C relay connector (R/W) terminal (IG ON) | Yes | Check the voltage at ECU 1L terminal with SST<br>**Voltage at idle after warm up: 0V (A/C and blower switches ON)** | Yes | ECU malfunction (Check 1C terminal voltage) |
| | | | | | Wiring between ECU 1C and A/C relay open |
| | | | | No | A/C system malfunction |
| | | No | Check A/C system | | |

## MALFUNCTION INDICATOR LIGHT (MIL) OPERATION 1990 MPV WITH 6 CYLINDER ENGINE

**PRINCIPLE OF CODE CYCLE**

Malfunction codes are determined as below using the MIL and Self-Diagnosis Checker.

**1. Malfunction code cycle break**
The time between malfunction code cycles is 4.0 sec. (the time the MIL and checker buzzer are off).

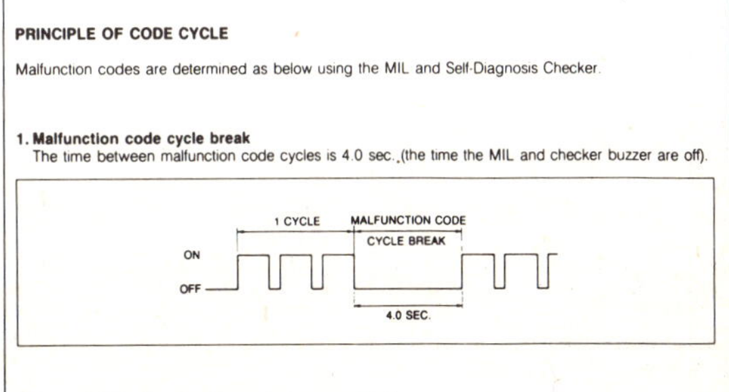


**2. Second digit of malfunction code (ones position)**
The digit in the ones position of the malfunction code represents the number of times the MIL and buzzer are on 0.4 sec. during one cycle.

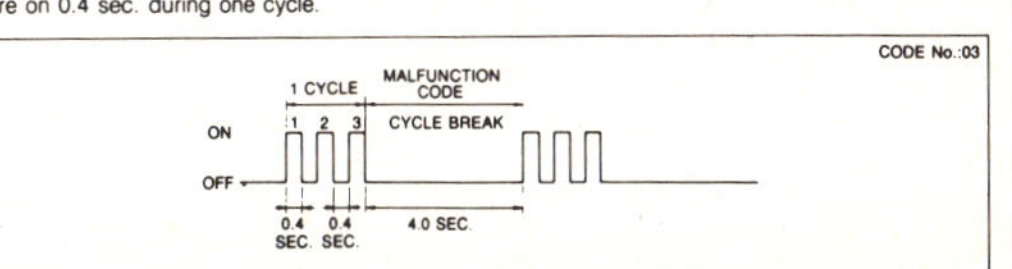


**3. First digit of malfunction code (tens position)**
The digit in the tens position of the malfunction code represents the number of times the MIL and buzzer are on 1.2 sec. during one cycle.

The MIL and buzzer are off for 1.6 sec. between the long and short pulses.

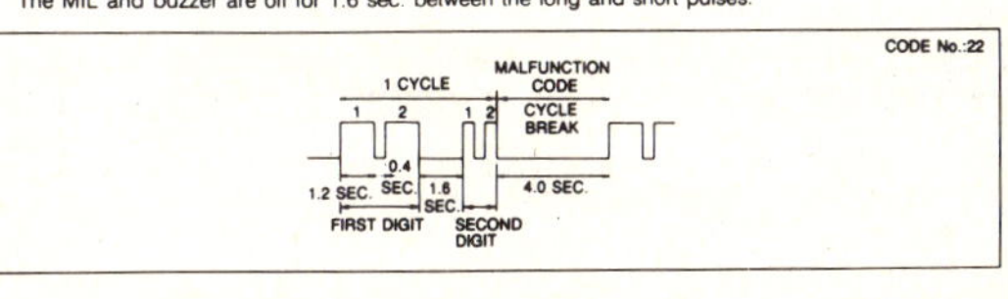

### TROUBLE CODE IDENTIFICATION CHART
### 1990 MPV WITH 6 CYLINDER ENGINE

**CODE NUMBER**
**Input Devices**

| Code No. | Input devices | Malfunction | Fail-safe function | Output signal pattern (Self-Diagnosis Checker or MIL) |
|---|---|---|---|---|
| 01 | Ignition pulse (Igniter, ignition coil) | Broken wire, short circuit | — | |
| 03 | Distributor (G-signal) | Broken wire, short circuit | — | |
| 08 | Airflow meter | Broken wire, short circuit | Basic fuel injection amount fixed as for 2 driving modes: 1) Idle switch: ON 2) Idle switch: OFF | |
| 09 | Water thermosensor | Broken wire, short circuit | Coolant temp. input fixed at 80°C (176°F) | |
| 10 | Intake air thermosensor (Airflow meter) | Broken wire, short circuit | Intake air temp. input fixed at 20°C (68°F) | |
| 11 | Intake air thermo-sensor (Dynamic chamber) | Broken wire, short circuit | Intake air temp. input fixed at 20°C (68°F) | |
| 12 | Throttle sensor | Broken wire, short circuit | Throttle valve opening angle input signal fixed at full open | |
| 14 | Atmospheric pressure sensor (within ECU) | Malfunction ECU | Atmospheric pressure input signal fixed at 760 mmHg (29.9 inHg) | |
| 15 | Oxygen sensor | Oxygen sensor output remains below 0.55V 120 sec after engine at 1,500 rpm | Feedback system canceled (for EGI) | |
| 17 | Feedback system | Oxygen sensor output remains 0.55V 60 sec after engine above 1,500 rpm | Feedback system canceled (for EGI) | |

### TROUBLE CODE IDENTIFICATION CHART (CONT.)
### 1990 MPV WITH 6 CYLINDER ENGINE

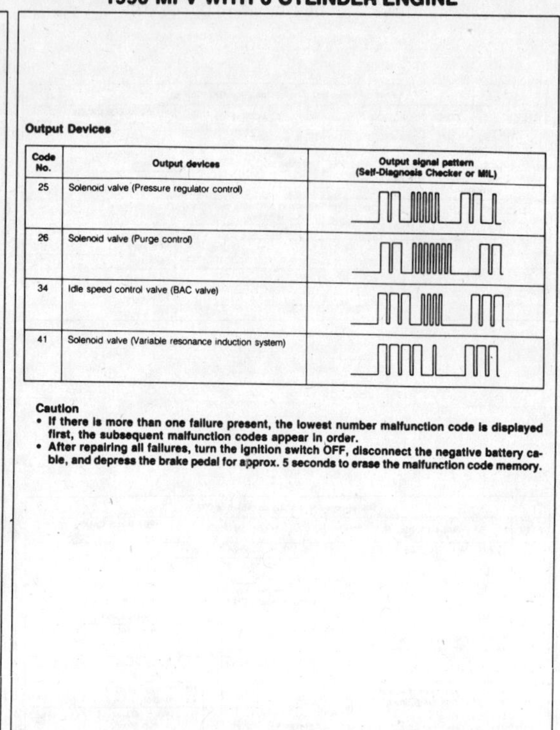

**Output Devices**

| Code No. | Output devices | Output signal pattern (Self-Diagnosis Checker or MIL) |
|---|---|---|
| 25 | Solenoid valve (Pressure regulator control) | |
| 26 | Solenoid valve (Purge control) | |
| 34 | Idle speed control valve (BAC valve) | |
| 41 | Solenoid valve (Variable resonance induction system) | |

**Caution**
- **If there is more than one failure present, the lowest number malfunction code is displayed first, the subsequent malfunction codes appear in order.**
- **After repairing all failures, turn the ignition switch OFF, disconnect the negative battery cable, and depress the brake pedal for approx. 5 seconds to erase the malfunction code memory.**

### TROUBLE CODE DFIAGNOSTIC CHART—1990 MPV WITH 6 CYLINDER ENGINE

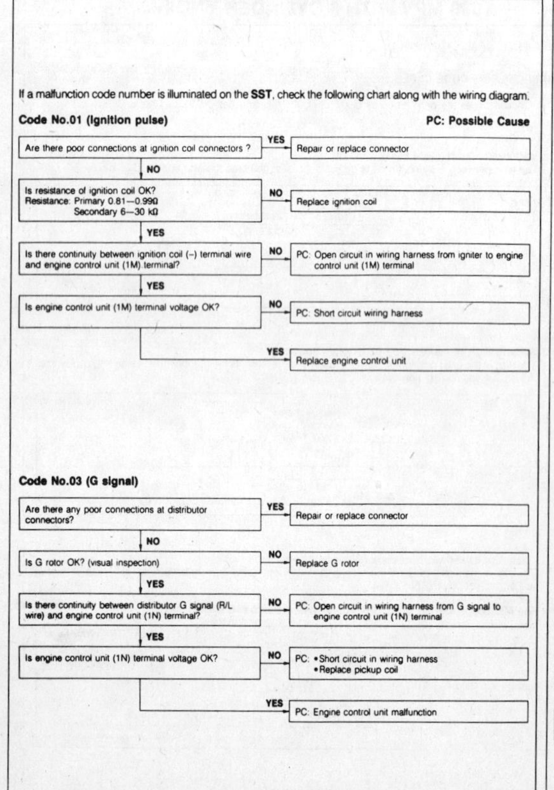


**Code No.08 (Airflow meter)** **PC: Possible Cause**

Are there poor connections at airflow meter connectors? — YES → Repair or replace connector

NO

Is resistance of airflow meter OK?
Resistance:

| Terminal | Fully closed (Ω) | Fully open (Ω) |
|---|---|---|
| $E_2 \leftrightarrow V_S$ | 20—400 | 20—1,000 |
| $E_2 \leftrightarrow V_C$ | 100—300 | |
| $E_2 \leftrightarrow V_B$ | 200—400 | |

NO → Repair airflow meter

YES

Is there continuity between airflow meter connector (wiring harness side) and engine control unit?

| Airflow meter | Control unit |
|---|---|
| $V_B$ (B/W wire) | 3I |
| $V_C$ (B/W wire) | 2B |
| $V_S$ (G/B wire) | 2E |
| $E_2$ (BR/B wire) | 2C |

NO → PC: Open circuit in wiring harness from airflow meter to engine control unit

YES

Are engine control unit terminal voltages OK? Check terminals 2B, 2C, 2E, 3I — NO → PC: Short circuit in wiring harness

YES → Replace engine control unit

**Code No.09 (Water thermosensor)**

Are there poor connections at water thermosensor connectors? — YES → Repair or replace connectors

NO

Is resistance of water thermosensor OK?
Resistance:

| Coolant temp | Resistance |
|---|---|
| −20°C ( −4°F) | 14.5—17.8 kΩ |
| 20°C ( 68°F) | 2.2— 2.7 kΩ |
| 40°C (104°F) | 1.0— 1.3 kΩ |
| 60°C (140°F) | 0.5— 0.7 kΩ |
| 80°C (176°F) | 0.28— 0.35 kΩ |

NO → Replace water thermosensor

YES

Is there continuity between water thermosensor connector (wiring harness side) and engine control unit?

| Water thermosensor | Engine control unit |
|---|---|
| (G/W wire) | 2I |

NO → PC: Open circuit in wiring harness from water thermosensor to engine control unit

YES

Are engine control unit (2I) terminal voltages OK? — NO → PC: Short circuit in wiring harness

YES → Replace engine control unit

## TROUBLE CODE DFIAGNOSTIC CHART—1990 MPV WITH 6 CYLINDER ENGINE

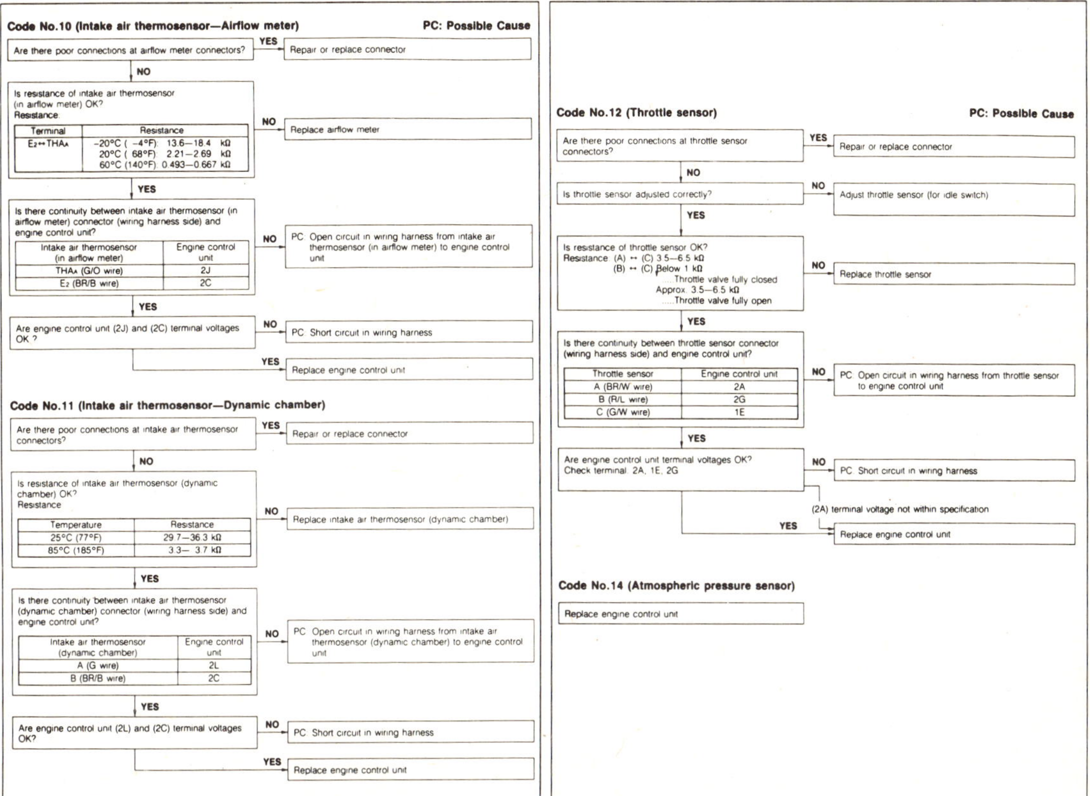


## TROUBLE CODE DFIAGNOSTIC CHART—1990 MPV WITH 6 CYLINDER ENGINE

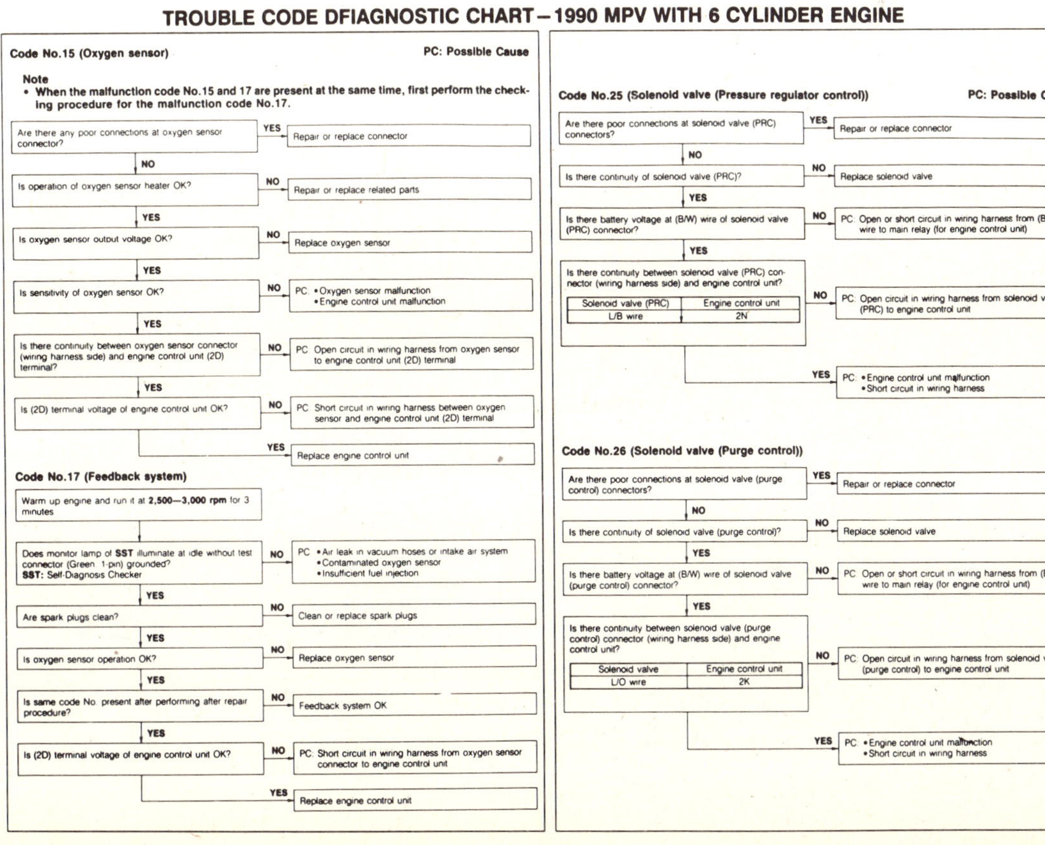

## TROUBLE CODE DFIAGNOSTIC CHART 1990 MPV WITH 6 CYLINDER ENGINE

**Code No.34 (Idle-speed control valve (BAC valve))** — **PC: Possible Cause**

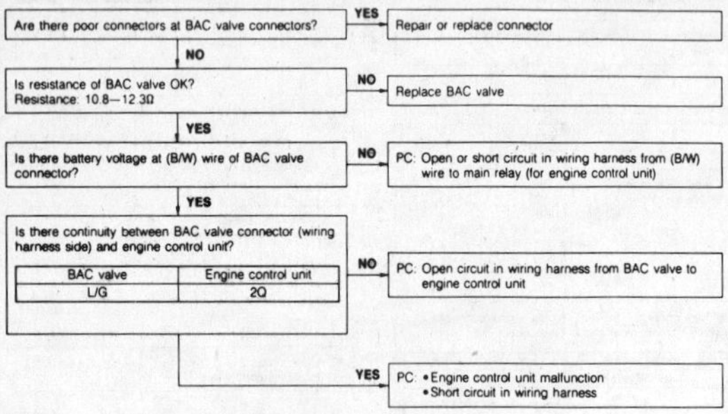

- Are there poor connectors at BAC valve connectors?
  - YES → Repair or replace connector
  - NO ↓
- Is resistance of BAC valve OK? Resistance: 10.8—12.3Ω
  - NO → Replace BAC valve
  - YES ↓
- Is there battery voltage at (B/W) wire of BAC valve connector?
  - NO → PC: Open or short circuit in wiring harness from (B/W) wire to main relay (for engine control unit)
  - YES ↓
- Is there continuity between BAC valve connector (wiring harness side) and engine control unit?

| BAC valve | Engine control unit |
|---|---|
| L/G | 2Q |

  - NO → PC: Open circuit in wiring harness from BAC valve to engine control unit
  - YES → PC: • Engine control unit malfunction • Short circuit in wiring harness

**Code No.41 (Solenoid valve (Variable resonance induction system))**

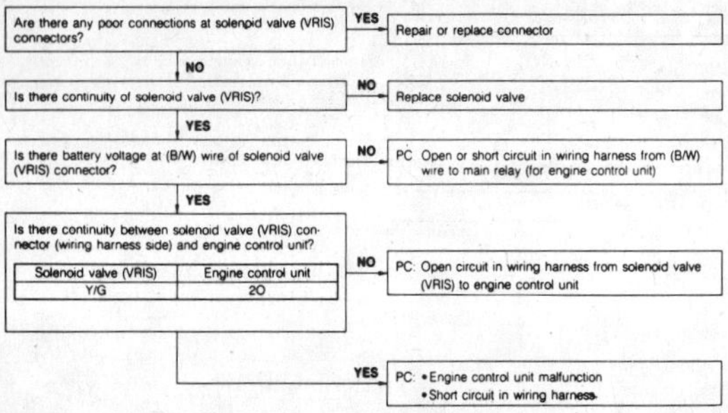

- Are there any poor connections at solenpid valve (VRIS) connectors?
  - YES → Repair or replace connector
  - NO ↓
- Is there continuity of solenoid valve (VRIS)?
  - NO → Replace solenoid valve
  - YES ↓
- Is there battery voltage at (B/W) wire of solenoid valve (VRIS) connector?
  - NO → PC Open or short circuit in wiring harness from (B/W) wire to main relay (for engine control unit)
  - YES ↓
- Is there continuity between solenoid valve (VRIS) connector (wiring harness side) and engine control unit?

| Solenoid valve (VRIS) | Engine control unit |
|---|---|
| Y/G | 2O |

  - NO → PC: Open circuit in wiring harness from solenoid valve (VRIS) to engine control unit
  - YES → PC: • Engine control unit malfunction • Short circuit in wiring harness

## CONTROL UNIT PIN LOCATION AND VOLTAGE TEST 1990 MPV WITH 6 CYLINDER ENGINE

**Terminal Voltage**

| Terminal | Input | Output | Connection to | Test condition | Voltage | Remarks |
|---|---|---|---|---|---|---|
| 1A | | ○ | Malfunction indicator light | For 3 seconds after ignition switch OFF→ON (Light illuminates) | Below 4.8V | • Test connector grounded |
| | | | | After 3 seconds (Light does not illuminate) | Approx. 12V | |
| | | | | Light illuminates | Below 4.8V | |
| | | | | Light does not illuminate | Approx. 12V | |
| 1B | | ○ | Self-Diagnosis Checker (Code number) | For 3 seconds after ignition switch OFF→ON (Buzzer sounds) | Below 6.2V | • Using Self-Diagnosis Checker and test connector grounded |
| | | | | After 3 seconds (Buzzer does not sound) | Approx. 12V | |
| | | | | Buzzer sounds | Below 6.2V | |
| | | | | Buzzer does not sound | Approx. 12V | |
| 1C | | ○ | A/C relay | Ignition switch ON | Approx. 12V | Blower motor: ON |
| | | | | For 10 seconds after cranking with A/C switch ON (A/C does not operate) | Approx. 12V | |
| | | | | After 10 seconds (A/C operates) | Below 2.5V | |
| | | | | A/C switch ON at idle | Below 2.5V | |
| | | | | A/C switch OFF at idle | Approx. 12V | |
| 1D | | ○ | Self-Diagnosis Checker (Monitor lamp) | Test connector grounded For 3 seconds after ignition switch OFF→ON (Light illuminates) | Below 6.2V | With Self-Diagnosis Checker |
| | | | | After 3 seconds (Light does not illuminate) | Approx. 12V | |
| | | | | Test connector not grounded at idle Monitor lamp ON | Below 6.2V | |
| | | | | Test connector not grounded at idle Monitor lamp OFF | Approx. 12V | |
| 1E | ○ | | Idle switch | Accelerator pedal released | Approx. 0V | Ignition switch ON |
| | | | | Accelerator pedal depressed | Approx. 12V | |
| 1F | — | — | — | — | — | — |
| 1G | ○ | | M/T Switch | M/T | Approx. 0V | Ignition switch ON |
| | | | | A/T | Approx. 12V | |
| 1H | ○ | | Central processing unit | Electrical load ON (IGN ON) | Below 3V | Electrical load Rear defroster Headlight Blower motor (3rd & 4th positions) |
| | | | | Electrical load OFF (IGN ON) | Above 7.3V | |
| 1I | | | — | — | — | — |
| 1J | ○ | | Stoplight switch (M/T) | Brake pedal released | Approx. 0V | Ignition switch ON |
| | | | | Brake pedal depressed | Approx. 12V | |
| 1K | ○ | | P/S pressure switch | P/S ON (at idle) | Approx. 0V | |
| | | | | P/S OFF (at idle) | Approx. 12V | |
| 1L | ○ | | A/C switch | A/C switch ON (Ignition switch ON) | Approx. 0V | Blower motor: ON |
| | | | | A/C switch OFF (Ignition switch ON) | Approx. 12V | |
| 1M | ○ | | Ignition coil ⊖ terminal | Ignition switch ON | Approx. 12V | * Engine Signal Monitor: green and red lights flash |
| | | | | Idle | Approx. 12V* | |
| 1N | ○ | | Distributor (G signal) | Ignition swtich ON | Approx. 0V | * Engine Signal Monitor: green and red lights flash |
| | | | | Idle | Approx. 1V* | |

## CONTROL UNIT PIN LOCATION AND VOLTAGE TEST (CONT.)—1990 MPV WITH 6 CYLINDER ENGINE

| Terminal | Input | Output | Connection to | Test condition | Voltage | Remarks |
|---|---|---|---|---|---|---|
| 2A | | ○ | Vref | Ignition switch ON | 4.5—5.5V | |
| 2B | ○ | | Airflow meter (Vc) (Ground) | Constant | 7—9V | |
| 2C | — | — | Ground (E2) | Constant | 0V | |
| 2D | ○ | | Oxygen sensor | Ignition switch ON | 0V | |
| | | | | Idle (Cold engine) | 0V | |
| | | | | Idle (After warm up) | 0—1.0V | |
| | | | | Increase engine speed (After warm up) | 0.5—1.0V | |
| | | | | Deceleration | 0—0.4V | |
| 2E | ○ | | Airflow meter (Vs) (Intake air mass) | Ignition switch ON | Approx. 2V | |
| | | | | Idle (After warm up) | Approx. 3.5—5V | |
| | | | | Increase engine speed (After warm up) | 5—9V | |
| 2F | ○ | | Test connector | Test connector (Green: 1-pin) not grounded | Approx. 12V | Ignition switch ON |
| | | | | Test connector (Green: 1pin) grounded | 0V | |
| 2G | ○ | | Throttle sensor | Accelerator pedal released | Approx. 0.3V | Ignition switch ON |
| | | | | Accelerator pedal fully depressed | Approx. 4V | |
| 2H | | | — | — | — | — |
| 2I | ○ | | Water thermo-sensor | Engine coolant temperature 20°C (68°F) | Approx. 2.5V | Ignition switch ON |
| | | | | After warm up | Approx. 0.4V | |
| 2J | ○ | | Intake air thermosensor (Airflow meter) | At 20°C (68°F) | Approx. 2—3V | |
| 2K | | ○ | Solenoid valve (Purge control) | Ignition switch ON | Approx. 12V | * Engine Signal Monitor: green and red lights flash |
| | | | | Idle | Approx. 12V | |
| | | | | Driving in-gear | 5—15V* | |
| 2L | ○ | | Intake air thermo-sensor (Dynamic chamber) | At 20°C (68°F) | Approx. 2.5V | |
| 2M | | ○ | Igniter | At sea level | Approx. 12V | Ignition switch ON |
| | | | | At high altitude (1000 m [3280 ft]) | Approx. 0V | |
| 2N | | ○ | Solenoid valve (PRC) | For 120 seconds after ignition switch OFF→ON | Below 2.5V | During hot condition. Coolant temp. above 80°C (194°F) Intake air temp. above 75°C (167°F) |
| | | | | For 120 seconds after starting | Below 2.5V | |
| | | | | Ignition switch ON | Approx. 12V | Other conditions |
| 2O | | ○ | Solenoid valve (VRIS) | 3,500—5,600 rpm | Below 2.5V | |
| | | | | Other conditions | Approx. 12V | |
| 2P | | ○ | Solenoid valve (Equip only kit) | — | — | — |
| 2Q | | ○ | Solenoid valve (Idle speed control) | Ignition switch ON | Above 8V | Engine Signal Monitor: green and red light flash |
| | | | | Idle | Approx. 9V | |
| 2R | — | — | Ground (Eo2) | Constant | 0V | |

| Terminal | Input | Output | Connection to | Test condition | Voltage | Remarks |
|---|---|---|---|---|---|---|
| 3A | — | — | Ground (Eo1) | Constant | 0V | |
| 3B | ○ | | Ignition switch (Start position) (M/T) | While cranking | 6—11V | |
| | | | | Ignition switch ON | Below 2.5V | |
| 3C | | ○ | Injector (No.3 and No.4) | Ignition switch ON | Approx. 12V | * Engine Signal Monitor: green and red lights flash |
| | | | | Idle | Approx. 12V* | |
| 3D | ○ | | Neutral or clutch switch (M/T) | Neutral or clutch pedal depressed | Approx. 0V | Ignition switch ON |
| | | | | Other conditions | Approx. 12V | |
| | | | Inhibitor switch | "N" or "P" range | Approx. 0V | |
| | | | | Other ranges | Approx. 12V | |
| 3E | | ○ | Injector (No.1 and No.2) | Ignition switch ON | Approx. 12V | * Engine Signal Monitor: green and red lights flash |
| | | | | Idle | Approx. 12V* | |
| 3F | | ○ | Injector (No.5 and No.6) | Ignition switch ON | Approx. 12V | * Engine Signal Monitor: green and red lights flash |
| | | | | Idle | Approx. 12V* | |
| 3G | — | — | Ground (E1) | Constant | 0V | |
| 3H | | ○ | Heater (Oxygen) | Above 2,250 rpm | Approx. 12V | |
| | | | | Other condition | Below 2.5V | |
| 3I | ○ | | Main relay | Ignition switch OFF | Approx. 0V | |
| | | | | Ignition switch ON | Approx. 12V | |
| 3J | — | — | Battery | Constant | Approx. 12V | For backup |

**Control Unit Connector (Control unit side)**

Control unit connector

| 3I | 3G | 3E | 3C | 3A | 2Q | 2O | 2M | 2K | 2I | 2G | 2E | 2C | 2A | 1M | 1K | 1I | 1G | 1E | 1C | 1A |
|---|---|---|---|---|---|---|---|---|---|---|---|---|---|---|---|---|---|---|---|---|
| 3J | 3H | 3F | 3D | 3B | 2R | 2P | 2N | 2L | 2J | 2H | 2F | 2D | 2B | 1N | 1L | 1J | 1H | 1F | 1D | 1B |

# CONTROL UNIT CONNECTOR VOLTAGE TEST—1990 MPV WITH 6 CYLINDER ENGINE

9MU0F1 192

**Inspection**

1. Remove the CPU and check the voltage between terminals of joint box and a body ground.
2. If the terminal inspection results are incorrect, refer to the action column and check the indicated system.
3. If the terminal inspection results are correct but system does not operate, replace the CPU.

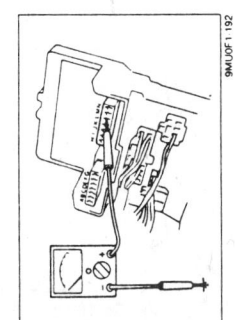

| Terminal | Connected to | Test condition | Specification | To correct |
|---|---|---|---|---|
| b | Battery (through ROOM 10A fuse) | Constant | Approx. 12V | Check ROOM 10A fuse and wiring harness |
| c | Engine control unit | Ignition switch ON | Approx. 12V | Check engine control unit and wiring harness |
| d | Ground | Constant | 0V | Check wiring harness |
| f | Headlight switch | Headlight switch ON | Approx. 12V | Check headlight switch and wiring harness |
| g | Blower fan switch | Blower fan switch High or Super-high position | 0V | Check blower fan switch and wiring harness |
| h | Rear window defroster switch | Rear window defroster switch ON | 0V | Check rear window defroster switch and wiring harness |
| i | Ignition switch | Ignition switch ON | Approx. 12V | Check ignition switch and wiring harness |

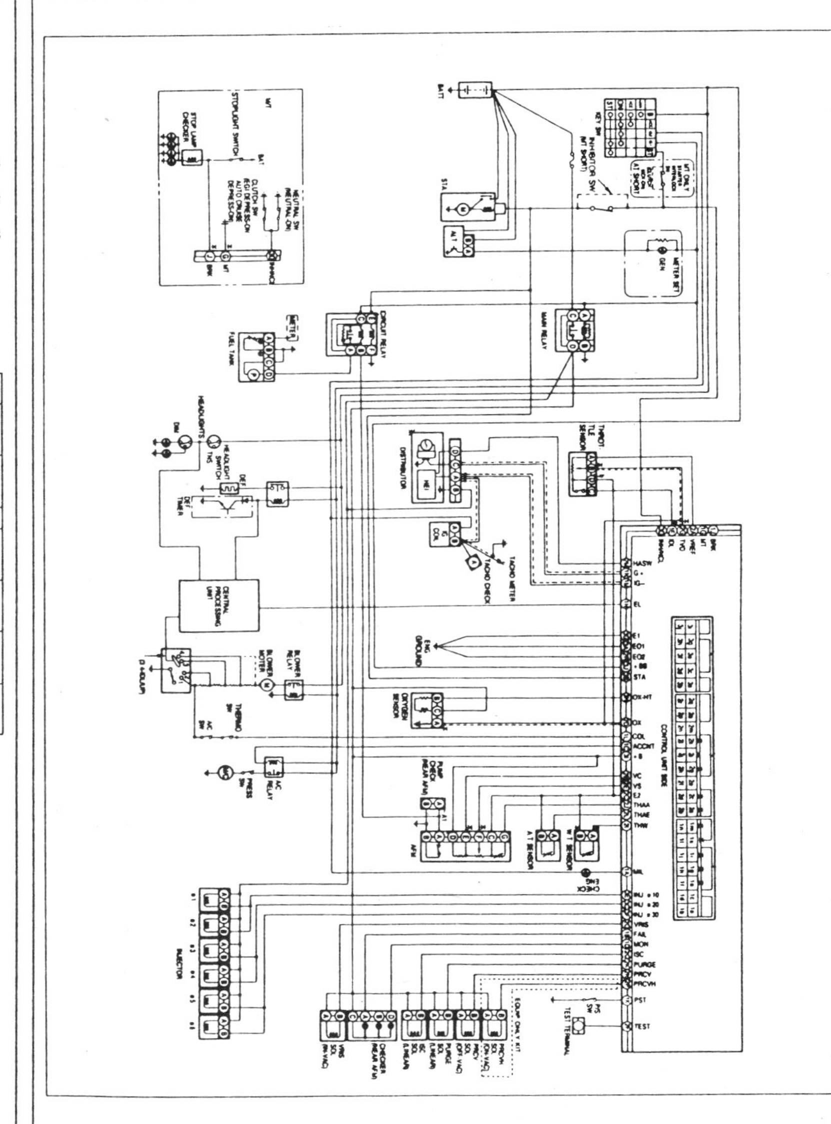

# Component Replacement

## RELIEVING FUEL PRESSURE

The fuel in the fuel system remains under high pressure even when the engine is not running. So before disconnecting any fuel lines, release the fuel pressure to reduce the risk of injury or fire. Relieve the fuel pressure as follows:

### Except 1990 323

1. Start the engine.
2. On all models except 1990 MPV, disconnect the circuit opening relay connector, with the engine running.
3. On 1990 MPV, disconnect the air flow meter connector.
4. On all models, after the engine stalls, turn the ignition switch to the **OFF** position and reconnect the circuit opening relay connector and/or the air flow meter connector.
5. Use a shop rag to cover the fuel lines when disconnecting. Plug all fuel lines after they have been disconnected.

### 1990 323

1. Start the engine.
2. Remove the rear cushion, then disconnect the fuel pump electrical connector.
3. After the engine stalls, turn the ignition switch to the **OFF** position and reconnect the fuel pump electrical connector and install the rear seat cushion.
4. Use a shop rag to cover the fuel lines when disconnecting. Plug all fuel lines after they have been disconnected.

## AIR FLOW METER

### Removal and Installation

#### EXCEPT B2600 AND 1990 MPV WITH 4 CYLINDER ENGINE

1. Disconnect the battery ground cable, then the high tension leads and connectors.
2. Loosen the hose band, then remove the intake hose.
3. Remove the air flow meter attaching bolts.
4. Turn the air cleaner cover upside down and remove the attaching nuts, then remove the air flow meter.
5. Reverse procedure to install.

## AIR FLOW SENSOR

### Removal and Installation

#### B2600 AND 1990 MPV WITH 4 CYLINDER ENGINE

1. Disconnect the battery ground cable.
2. Disconnect the air flow sensor electrical connector.
3. Loosen the air hose clamps, then remove the mounting bolts.
4. Remove the air flow sensor. Reverse procedure to install.

## THROTTLE BODY

### Removal and Installation

1. Disconnect the battery ground cable.
2. Disconnect the accelerator cable from the throttle linkage.
3. Disconnect the air funnel, then the air hoses and lines from the throttle body. Remove the BAC valve, if equipped.
4. Disconnect all electrical connectors from the throttle body assembly. Remove the throttle body attaching screws, then the throttle body assembly.
5. Reverse procedure to install.

## THROTTLE POSITION SENSOR

### Adjustment

#### 1988–89 323 AND 1990 MIATA

1. Disconnect the electrical connector from the throttle position sensor. Connect throttle position tester (49–9200–165) and adapter (49–9200–166) or equivalent to the throttle position.
2. On 1988–89 323 , insert a 0.020 in. (0.5mm) feeler gauge between the throttle stop screw and the stop lever.
3. On 1990 Miata, insert a 0.016 in. (0.4mm) feeler gauge between the throttle stop screw and the stop lever.
4. On all models, loosen the throttle position sensor attaching screws. Rotate the sensor clockwise approximately 30 degrees, then rotate it back counterclockwise until the buzzer sounds.
5. Remove the 0.020 in. (0.5mm) or 0.016 in. (0.4mm) feeler gauge and insert a 0.027 in. (0.7mm) feeler gauge. Check that the buzzer does not sound. If the buzzer sounds, repeats Steps 3 and 4.
6. Tighten the throttle position sensor attaching screw. Be sure not to move the throttle position sensor from the set position when tightening the attaching screws.
6. Operate the throttle valve, then recheck the adjustment of the throttle position sensor by performing the inspection over again.

#### 1990 323 WITH MANUAL TRANSAXLE

1. Disconnect the electrical connector from the throttle sensor.
2. Connect a suitable ohmmeter to terminals E and IDL on the throttle sensor.
3. Insert a 0.016 in. (0.4mm) feeler gauge between the throttle stop screw and the stop lever.
4. Loosen the throttle sensor attaching screws and rotate the throttle sensor clockwise approximately 30 degrees, then rotate it back counterclockwise until continuity exists.

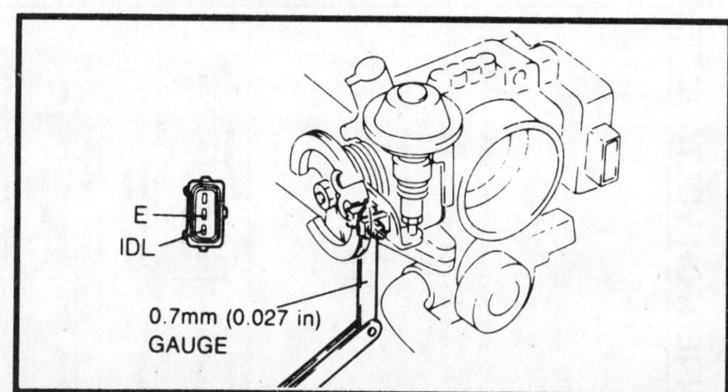


**Using a feeler gauge to adjust throttle sensor – 1990 323 with M/T**

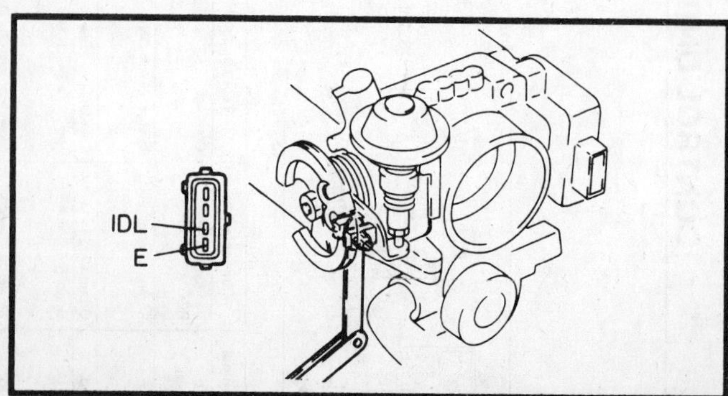


**Using a feeler gauge to adjust throttle sensor – 1990 323 with A/T**

5. Replace the feeler gauge with a 0.027 in. (0.7mm) feeler gauge. Check that continuity does not exists. If continuity exists, repeat Steps 3 through 5.

6. Tighten the throttle sensor attaching screws.

**NOTE: Do not move the throttle sensor while tighten the screws.**

7. Operate the throttle valve and recheck the adjustment of the throttle sensor.

### 1990 323 WITH AUTO TRANSAXLE

1. Disconnect the electrical connector from the throttle sensor.

2. Connect a suitable ohmmeter between terminals E and IDL on the throttle sensor.

3. Insert a 0.01 in. (0.25mm) feeler gauge between the throttle stop screw and the stop lever.

4. Loosen the throttle sensor attaching screws and rotate the sensor approximately 30 degrees, then rotate it back counterclockwise until continuity exists.

5. Replace the feeler gauge with a 0.016 in. (0.4mm) feeler gauge and check that continuity does not exists. If continuity exists, repeat Steps 3 through 5.

6. Tighten the throttle sensor attaching screws.

**NOTE: Do not move the throttle sensor while tighten the screws.**

7. Open the throttle valve and check that the resistance between the throttle sensor terminals E and VT using a suitable ohmmeter. Ohmmeter should read approximately 5 Kohms.

### 626 AND MX-6, B2600 AND 1990 MPV WITH 4 CYLINDER ENGINE

1. Disconnect the air hose from throttle body assembly, then disconnect the throttle sensor electrical connector (3-pin).

2. Connect Adapter Harness (49–G018–901) between the throttle sensor and the wiring harness.

3. Turn the ignition switch to the **ON** position. Check that the throttle valve is in the fully closed position.

4. Using a suitable voltmeter, measure the Red wire voltage and record it. Change the voltmeter connection to the Blue wire, and loosen the throttle sensor mounting bolts.

5. Turn the throttle sensor to adjust the Blue wire voltage to within the specified range for the recorded Red wire voltage. Tighten the throttle sensor mounting bolts. Recheck the Blue wire voltage.

6. Hold the throttle valve in the fully opened position, and check that the Blue wire voltage is within thew specified range.

7. Check that the Blue wire voltage increases smoothly when opening the throttle valve from its closed to fully opened position. If not, replace the throttle sensor.

8. Turn the ignition switch to the **OFF** position. Disconnect the Adapter Harness (49–G018–901) and reconnect the throttle sensor electrical connector.

9. Disconnect the battery ground cable and depress the brake pedal for at least 5 seconds to eliminate the ECU malfunction memory.

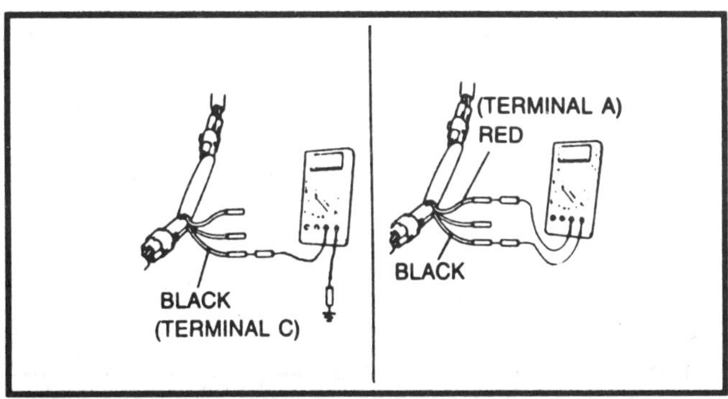


**Measuring the throttle sensor black and red wire voltages—1989–90 B2600, 1990 323 and MPV with 4 cylinder engine**

**Specification:**

| RED wire voltage (V) | BLUE wire voltage (V) | RED wire voltage (V) | BLUE wire voltage (V) |
|---|---|---|---|
| 4.50—4.59 | 0.37—0.54 | 5.10—5.19 | 0.42—0.61 |
| 4.60—4.69 | 0.38—0.55 | 5.20—5.29 | 0.43—0.62 |
| 4.70—4.79 | 0.39—0.56 | 5.30—5.39 | 0.44—0.63 |
| 4.80—4.89 | 0.40—0.57 | 5.40—5.49 | 0.44—0.64 |
| 4.90—4.99 | 0.40—0.58 | 5.50 | 0.44—0.66 |
| 5.00—5.09 | 0.41—0.60 | | |

**Throttle sensor specification chart with the throttle valve closed—1989–90 B2600, 1990 323 and MPV with 4 cylinder engine**

**Specification:**

| RED wire voltage (V) | BLUE wire voltage (V) | RED wire voltage (V) | BLUE wire voltage (V) |
|---|---|---|---|
| 4.50—4.59 | 3.58—4.23 | 5.10—5.19 | 4.05—4.79 |
| 4.60—4.69 | 3.66—4.32 | 5.20—5.29 | 4.13—4.88 |
| 4.70—4.79 | 3.74—4.41 | 5.30—5.39 | 4.21—4.98 |
| 4.80—4.89 | 3.82—4.51 | 5.40—5.49 | 4.29—5.07 |
| 4.90—4.99 | 3.90—4.60 | 5.50 | 4.29—5.17 |
| 5.00—5.09 | 3.97—4.70 | | |

**Throttle sensor specification chart with the throttle valve opened—1989–90 B2600, 1990 323 and MPV with 4 cylinder engine**

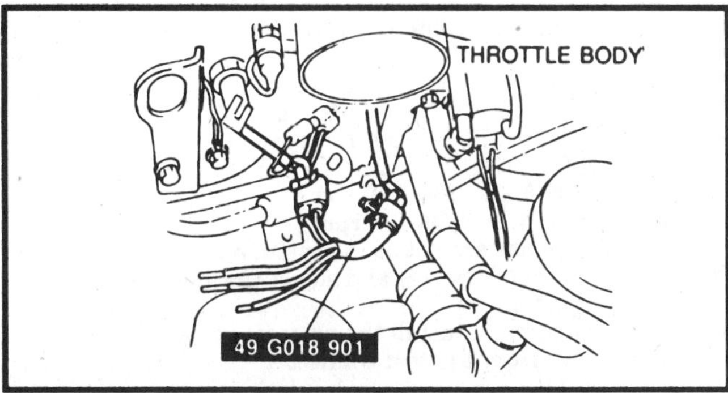


**Connecting throttle sensor adapter harness—1989–90 B2600, 1990 323 and MPV with 4 cylinder**

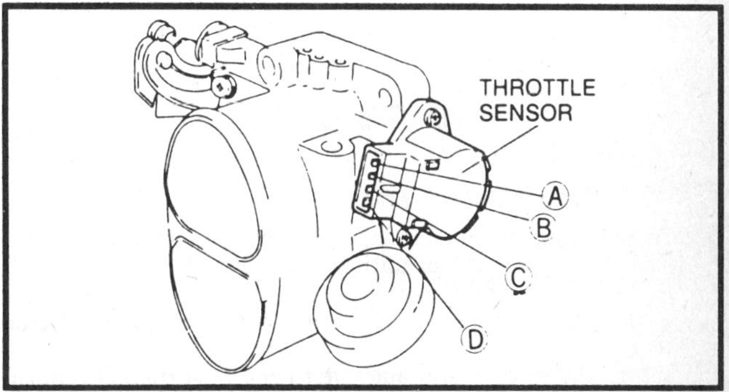


**Throttle sensor terminal identification—929 and 1990 MPV with 6 cylinder engine**

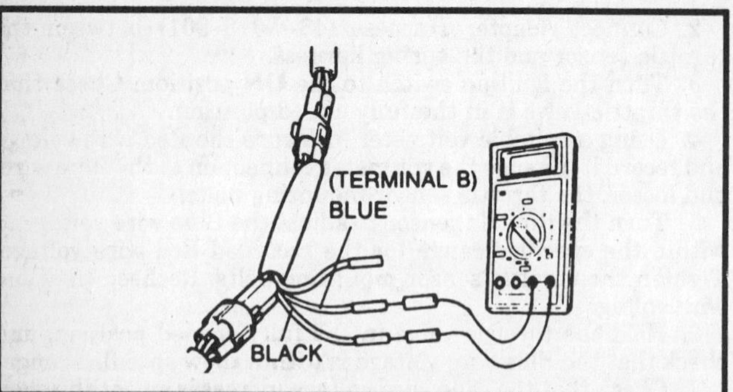


**Measuring the throttle sensor blue wire voltage – 1989–90 B2600, 1990 323 and MPV with 4 cylinder engine**

### 929 AND MPV WITH 6 CYLINDER ENGINE

1. Disconnect the throttle sensor electrical connector from the throttle sensor.
2. Connect a suitable ohmmeter between terminals (C) and (D) on the throttle sensor.
3. On all models except 1990 929 with DOHC engine, insert a 0.020 in. (0.5mm) feeler gauge between the throttle stopper screw and the throttle lever. On 1990 929 with DOHC engine, insert a 0.004 in. (0.1mm) feeler gauge between the throttle stopper screw and the throttle lever. On all models, continuity should exists between terminals (C) and (D).

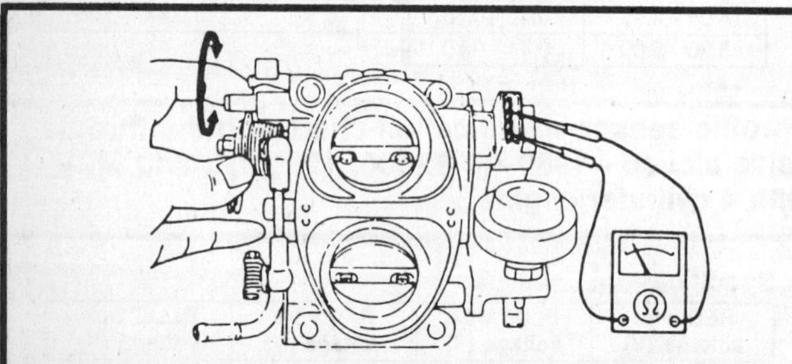

**Throttle sensor testing using an ohmmeter – 929 and 1990 MPV with 6 cylinder engine**

4. On all models except 1990 929 with DOHC engine, insert a 0.028 in. (0.7mm) feeler gauge between the throttle stopper screw and the throttle lever. On 1990 929 with DOHC engine, insert a 0.012 in. (0.3mm) feeler gauge between the Throttler stopper screw and the throttle lever. On all models, no continuity should exists between terminals (C) and (D).
5. If specifications in Steps 3 and 4 are not as indicated, loosen the throttle sensor attaching screws and turn the throttle sensor until proper specifications are obtained.
6. Reconnect the throttle sensor electrical connector.

## IDLE SPEED

### Adjustment

#### EXCEPT 1990 323 AND MIATA

Before adjusting the idle speed, check that the ignition timing, spark plugs, etc., are in normal operating condition. Turn off all lights and other unnecessary loads.

1. Connect a suitable tachometer to the check connector (white 1 pin).
2. Start the engine and allow it to reach normal operating temperature. Connect a jumper wire between the test connector (green: 1 pin) and ground.

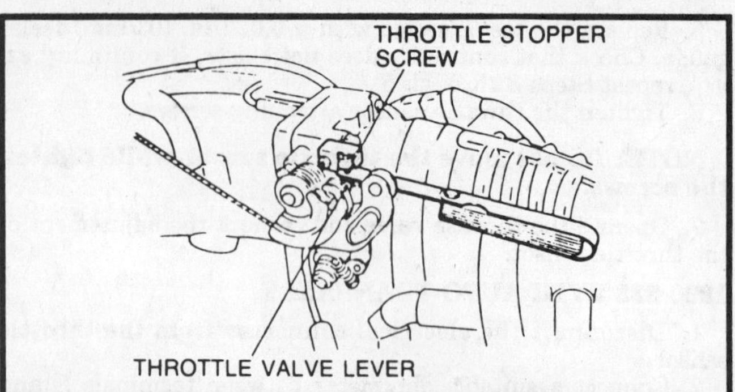


**Using a feeler gauge to test throttle sensor – 929 and 1990 MPV with 6 cylinder engine**

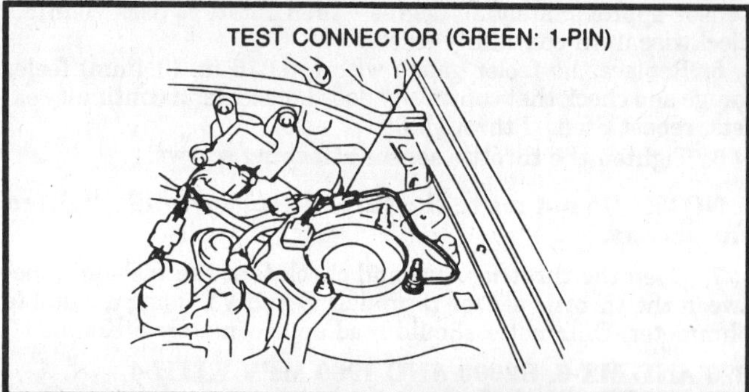


**Grounding test connector for idle speed adjustment – Except 1990 323 and Miata**

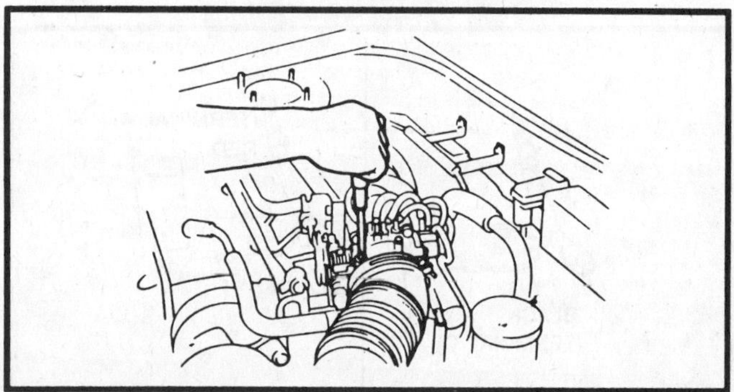

**Idle speed adjustment – Except 1990 323 and Miata**

3. Check the idle speed. Idle speed should be as follows:
   1988–89 323 – 800–900 rpm
   1988–90 626 and MX-6 – 725–775 rpm
   1988–89 929 and 1990 929 with SOHC engine – 630–670 rpm
   1990 929 with DOHC engine – 680–720 rpm
   1990 MPV with A/T – 730–770 rpm
   1990 MPV with M/T – 750–790 rpm
   1989 B2600 – 730–770 rpm
   1990 B2600 M/T – 730–770 rpm
   1990 B2600 A/T – 750–790 rpm
4. If specifications are not as indicated, remove the protector cap from the air adjust screw, and adjust the idle speed by turning the air adjust screw.
5. After adjustment, install the protector cap and disconnect the jumper wire from the test connector.

**NOTE: After adjusting the idle speed, the dashpot should be adjusted.**

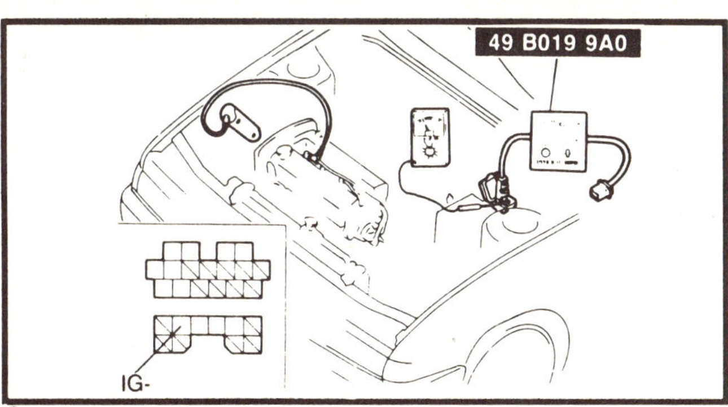

**Connecting system selector for idle speed adjustment – 1990 323 and Miata**

#### 1990 323 AND MIATA

**NOTE: When checking the idle speed, ensure that the electric cooling fan is not operating**

1. Start the engine and allow it to reach normal operating temperature. Turn off all lights and other unnecessary loads.
2. Connect System Selector 49–B019–9AO to the diagnosis connector. Connect a suitable tachometer to the IG terminal on the diagnosis connector.
3. Set switch (A) on the System Selector to position 1, then set the TEST SW to SELF-TEST.

**NOTE: If System Selector 49–B019–9A0 is not available, connect a jumper wire between the TEN terminal and GND terminal on the diagnosis connector.**

4. Apply the parking brake and check the idle speed:
   323 – 700–800 rpm
   Miata – 800–900 rpm

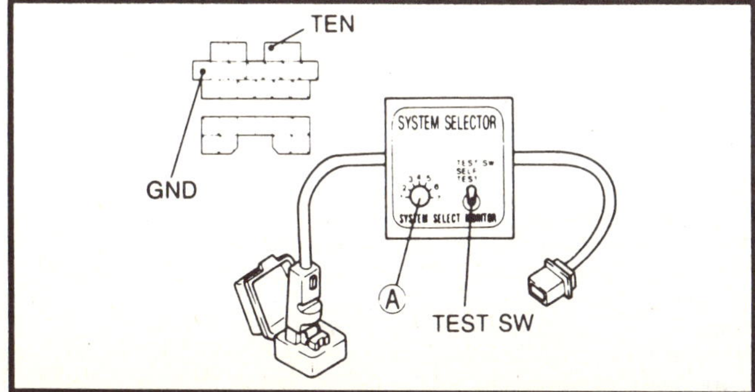

**Installing jumper wire across terminals GND and TEN on the diagnosis connector when system selector is not available**

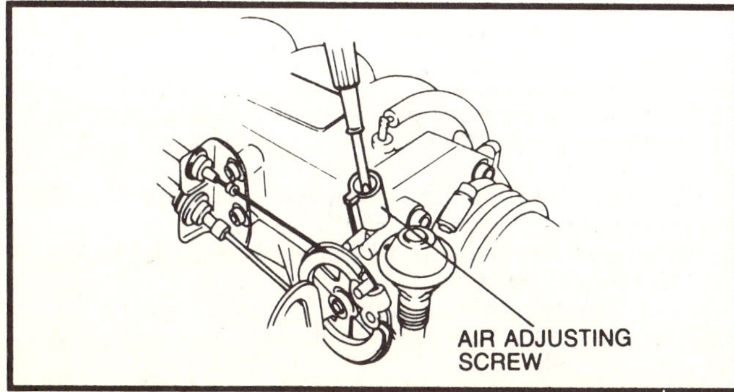

**Idle speed adjustment – 1990 323 and Miata**

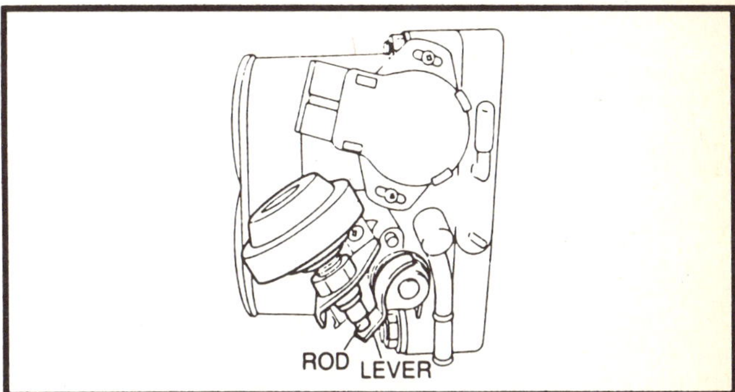

**Dashpot adjustment – 1988–89 323, 929 and 1990 MPV with 6 cylinder**

**NOTE: On 323 Canada models equipped with A/T, when the parking brake is not applied, the idle speed should be approximately 800 rpm.**

5. If specifications are not as specified, adjust the idle by turning the air adjusting screw. Disconnect the System Selector.

## DASHPOT

### Adjustment

#### 1988–89 323
#### 1990 MIATA AND MPV WITH 6 CYLINDER ENGINE

1. Push lightly on the dashpot rod and check that the rod goes into the dashpot slowly. Release the dashpot rod, and check that the rod comes out quickly.
2. Start the engine and allow it to reach normal operating temperature. Stop the engine and connect a suitable tachometer to the engine.
3. On 323 models, start the engine, and slowly increase engine speed to 3500 rpm.
4. On Miata and MPV, start the engine, and slowly increase engine speed to 4000 rpm.
5. On all models, slowly decrease engine rpm, and check that the dashpot rod touches the lever at the specified rpm.
   323 less turbo and M/T – 2650–2950 rpm
   323 less turbo and A/T – 2500–3100 rpm
   323 with turbo – 1850–2150 rpm
   Miata – 2350–2650 rpm
   MPV – 3200–3800 rpm
6. If specifications are not as specified, loosen the lock nut and adjust by turning the lock nut.

#### 1988–90 929, 1990 323

1. Open the throttle valve fully. Push lightly dashpot rod and check that the rod goes into the dashpot slowly. Release the dashpot rod, and check that the rod comes out quickly. If not, replace the dashpot.
2. Start the engine and allow it to reach normal operating temperature. Stop the engine. On 1990 323, connect a suitable tachometer to the IG terminal on the diagnosis connector. On 1988–90 929, connect a suitable tachometer to the check connector.
3. On all models, start the engine, and slowly increase engine speed to 4000 rpm.
4. Slowly decrease engine rpm, and check that the dashpot rod touches the lever at the specified rpm.
   On 323 models equipped with DOHC engine – Approximately 3000 rpm
   On 323 models equipped with SOHC engine – Approximately 2700 rpm
   On 1988–89 929 and 1990 929 with SOHC engine – 3200–3800 rpm

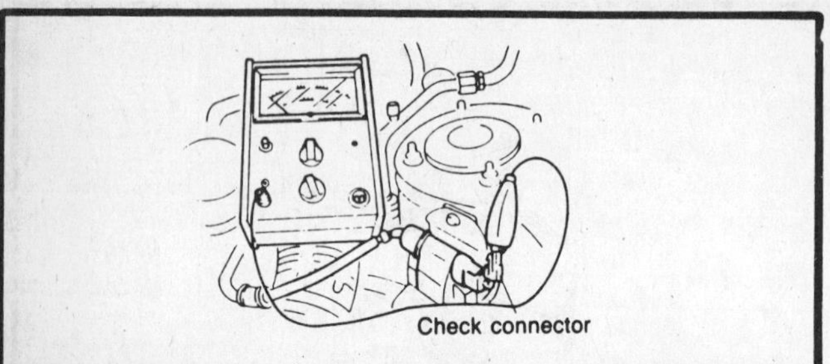


**Connecting tachometer to check connector for dashpot adjustment – 1988–89 929**

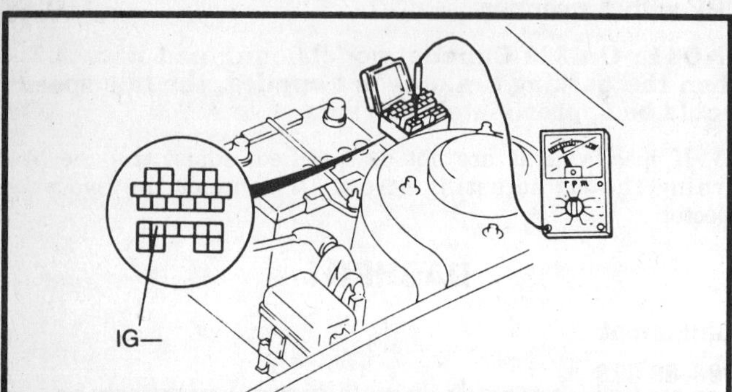


**Connecting tachometer to diagnosis connector terminal IG for dashpot adjustment – 1990 323**

5. If specifications are not as specified, loosen the locknut and turn the dashpot to adjust.

## FUEL INJECTORS

### Removal and Installation

#### 1988–89 323 and 1988–90 929

1. Relieve the fuel pressure from the fuel system, then disconnect the battery ground cable. On 929 models, remove the dynamic chamber cover.
2. On all models, remove the dynamic chamber, then disconnect the fuel hoses and vacuum hoses, then the electrical connectors from each injector.
3. Remove the delivery pipe with the pressure regulator, then the injectors.
4. Reverse procedure to install, noting the following:
   a. Use new O-rings.
   b. Coat O-ring with gasoline before installing injector.
   c. Torque delivery pipe attaching bolts to 14–19 ft. lbs. (19–25 Nm).

#### 1990 323

1. Relieve the fuel pressure from the fuel system, then disconnect the battery ground cable.
2. Disconnect the electrical connector from each injector.
3. Remove the injector harness from the delivery pipe, then the delivery pipe attaching bolts from the intake manifold.
4. Remove the delivery pipe assembly, with the injectors and pressure regulator attached.
5. Remove the injectors, then the grommets and insulators. Reverse procedure to install, noting the following:
   a. Use new O-rings and insulators.
   b. Coat O-rings with gasoline before installing.

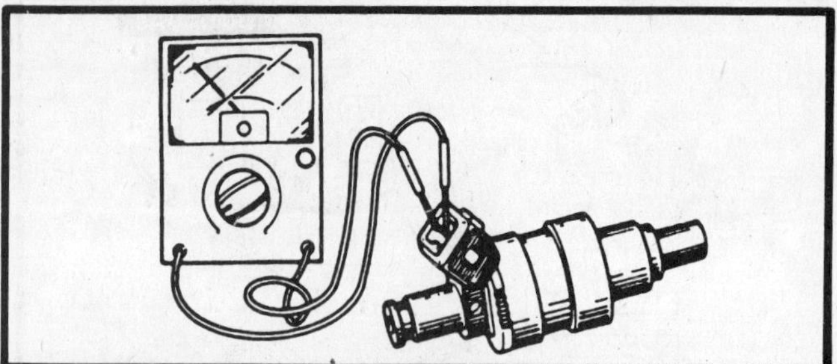

**Checking injector resistance**

   c. Torque delivery pipe attaching bolts to 14–19 ft. lbs. (19–25 Nm).

#### 1990 MIATA

1. Relieve the fuel pressure from the fuel system, then disconnect the battery ground cable.
2. Remove the air valve, then disconnect the vacuum hoses.
3. Disconnect the electrical connector from each injector.
4. Remove the delivery pipe attaching bolts, then the delivery pipe with the pressure regulator attached.
5. Remove the injectors, grommets and insulators. Reverse procedure to install, noting the following:
   a. Use new O-rings and insulators.
   b. Coat O-rings with gasoline before installing.
   c. Torque the delivery pipe attaching bolts to 14–19 ft. lbs. (19–25 Nm).

#### 1988–90 626 AND MX-6

1. Relieve the fuel pressure from the fuel system, then disconnect the battery ground cable.
2. Remove the wiring harness bracket attaching bolts, then the wiring harness bracket.
3. On non-turbocharged models, remove the EGR modulator valve, then the EGR modulator valve mounting bracket.
3.On all models, disconnect the air hose from the throttle body. Remove the engine hanger attaching bolts, then the engine hanger.
4.Remove the dynamic chamber mounting bolts and nuts, then lift the dynamic chamber assembly.
5.Disconnect the fuel return line bracket from the intake manifold, then disconnect the injector connectors.
6.Remove the delivery pipe attaching bolts, then the delivery pipe with the injectors, pressure regulator and pulsation damper.
7.Remove the injectors from the delivery pipe.
8.Reverse procedure to install, noting the following:
   a. Use new O-rings.
   b. Coat O-rings with gasoline before installing.
   c. Torque delivery pip dynamic chamber and engine hanger nuts and bolts to 14–19 ft. lbs. (19–25 Nm).

#### 1989–90 B2600 AND 1990 MPV

1. Relieve the fuel pressure from the fuel system.
2. Remove the dynamic chamber attaching nuts and bolts, then the dynamic chamber.
3. Disconnect the vacuum and fuel hoses. Remove the delivery pipe attaching bolts, then remove the delivery pipe with the pressure regulator attached.
4. Disconnect the electrical connectors from the injectors. Remove the grommets, injectors and insulators.
5. Reverse procedure to install, noting the following:
   a. Use new O-rings.
   b. Coat O-rings with gasoline before installing.
   c. Torque the delivery pipe and dynamic chamber attaching nuts and bolts to 14–19 ft. lbs. (19–25 Nm).

# MAZDA RX-7 ELECTRONIC GASOLINE INJECTION (EGI) SYSTEM

## General Information

This system is broken down into three major systems. The Fuel System, Air Induction System and Electronic Control System.

### FUEL SYSTEM

The fuel injection system supplies the necessary fuel to the injectors to achieve combustion at constant pressure. Fuel is metered and injected into the intake manifold according to the signals from the EGI control unit. The injection system consists of the fuel pump, fuel filter, distribution pipe, pulsation dampener, pressure regulator, injectors, air flow meter, air valve, and solenoid resistor.

### AIR INDUCTION SYSTEM

The air induction system provides sufficient air for the engine operation.

### ELECTRONIC CONTROL SYSTEM

The EGI control unit (ECU), through various input sensors, monitors battery voltage, engine rpm, amount of air intake, cranking signal, intake temperature, coolant temperature, oxygen concentration in the exhaust gases, throttle opening, atmospheric pressure, gearshift position, clutch engagement, braking, power steering operation and A/C compressor operation.

The ECU controls operation of the fuel injection system, idle-up system , fuel evaporation system and ignition timing. The ECU has a built in fail-safe mechanism. If a fault occurs while driving, the ECU will substitute pre-programmed values. Driving performance will be affected, but the vehicle will still be driveable.

## Diagnosis and Testing

### SERVICE PRECAUTIONS

- Do not operate the fuel pump when the fuel lines are empty.

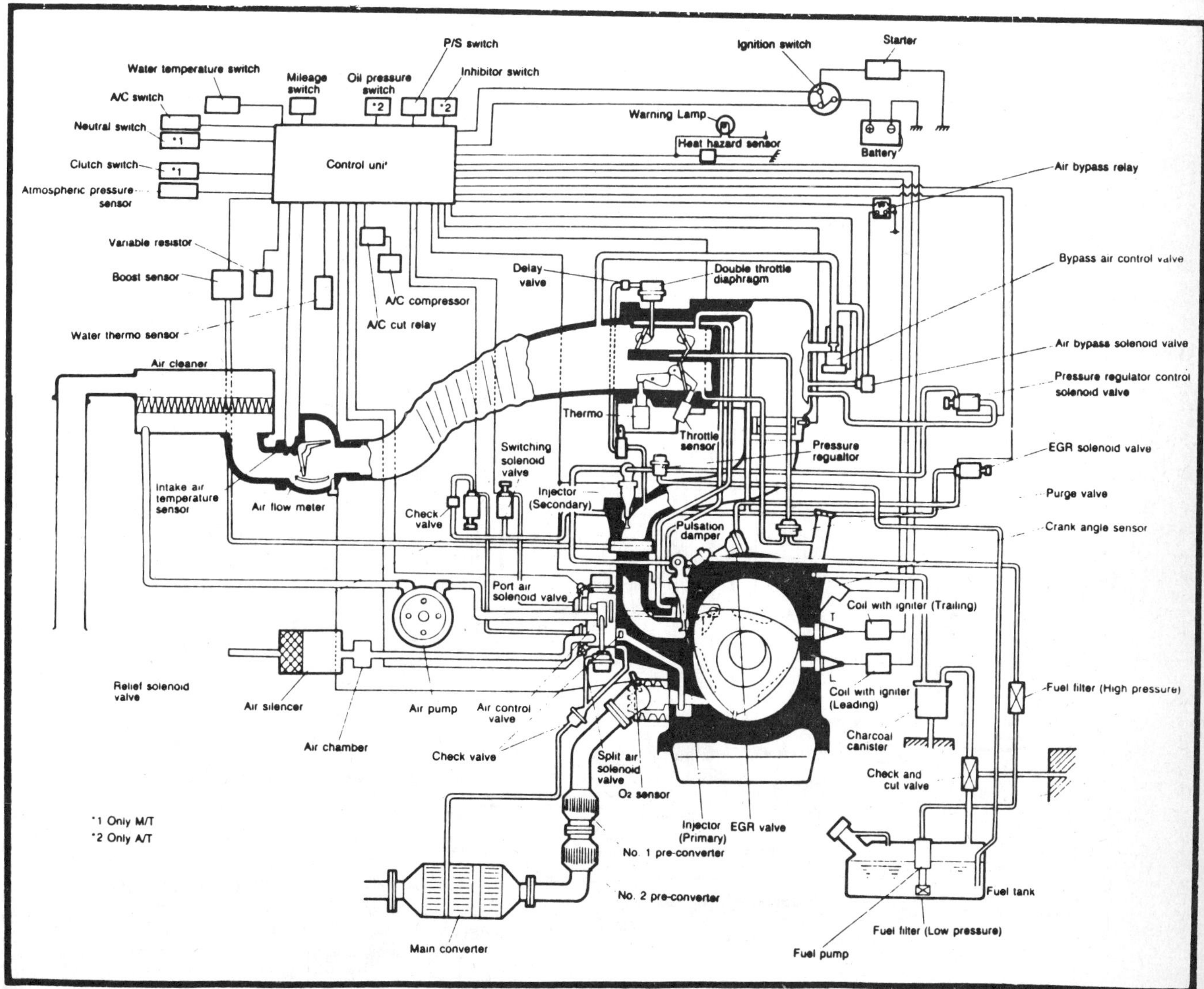


Exploded view of fuel system—1988 RX-7 except turbo

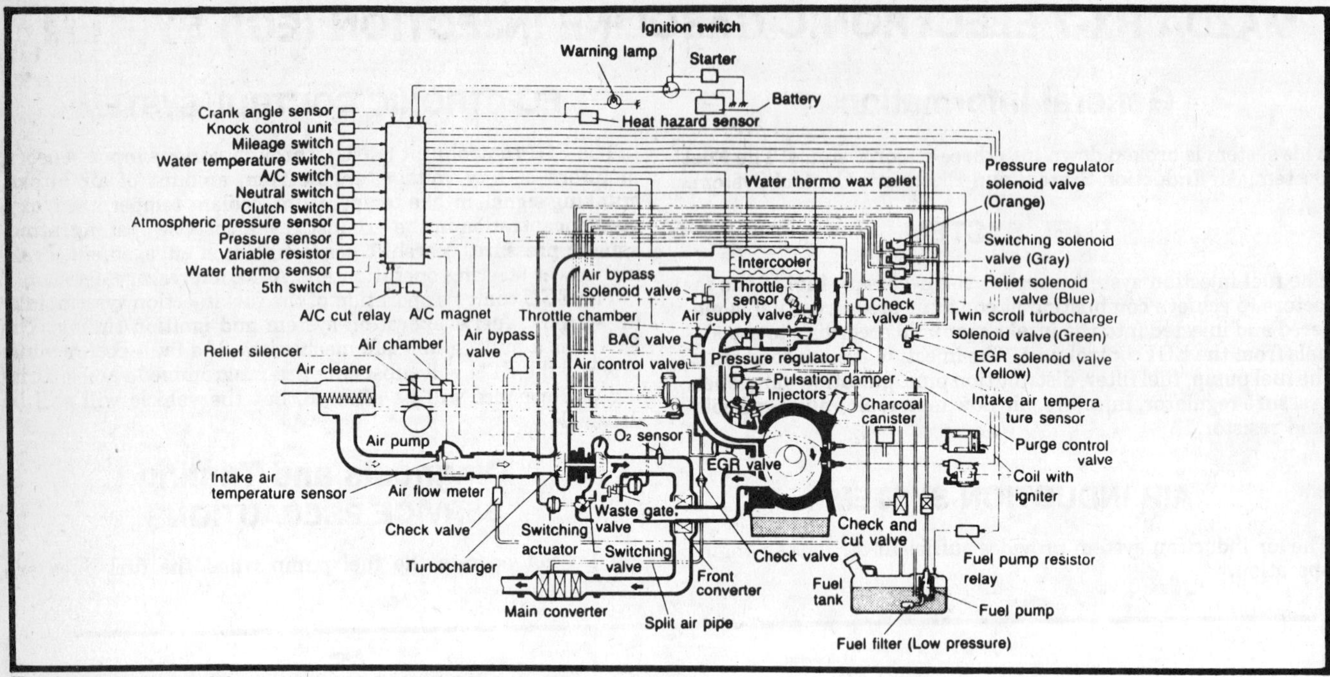

Ignition switch
Warning lamp
Starter
Battery
Heat hazard sensor
Crank angle sensor
Knock control unit
Mileage switch
Water temperature switch
A/C switch
Neutral switch
Clutch switch
Atmospheric pressure sensor
Pressure sensor
Variable resistor
Water thermo sensor
5th switch
A/C cut relay
A/C magnet
Water thermo wax pellet
Pressure regulator solenoid valve (Orange)
Switching solenoid valve (Gray)
Relief solenoid valve (Blue)
Twin scroll turbocharger solenoid valve (Green)
EGR solenoid valve (Yellow)
Intercooler
Air bypass solenoid valve
Throttle sensor
Air supply valve
Check valve
Throttle chamber
Air bypass valve
Air chamber
BAC valve
Relief silencer
Air cleaner
Air control valve
Pressure regulator
Pulsation damper
Injectors
Charcoal canister
Intake air temperature sensor
Purge control valve
Coil with igniter
Air pump
O2 sensor
EGR valve
Intake air temperature sensor
Air flow meter
Waste gate
Check valve
Switching actuator
valve
Check and cut valve
Check valve
Turbocharger
Switching valve
Front converter
Fuel tank
Fuel pump resistor relay
Main converter
Split air pipe
Fuel pump
Fuel filter (Low pressure)

**Exploded view of fuel system – 1988 RX-7 turbo**

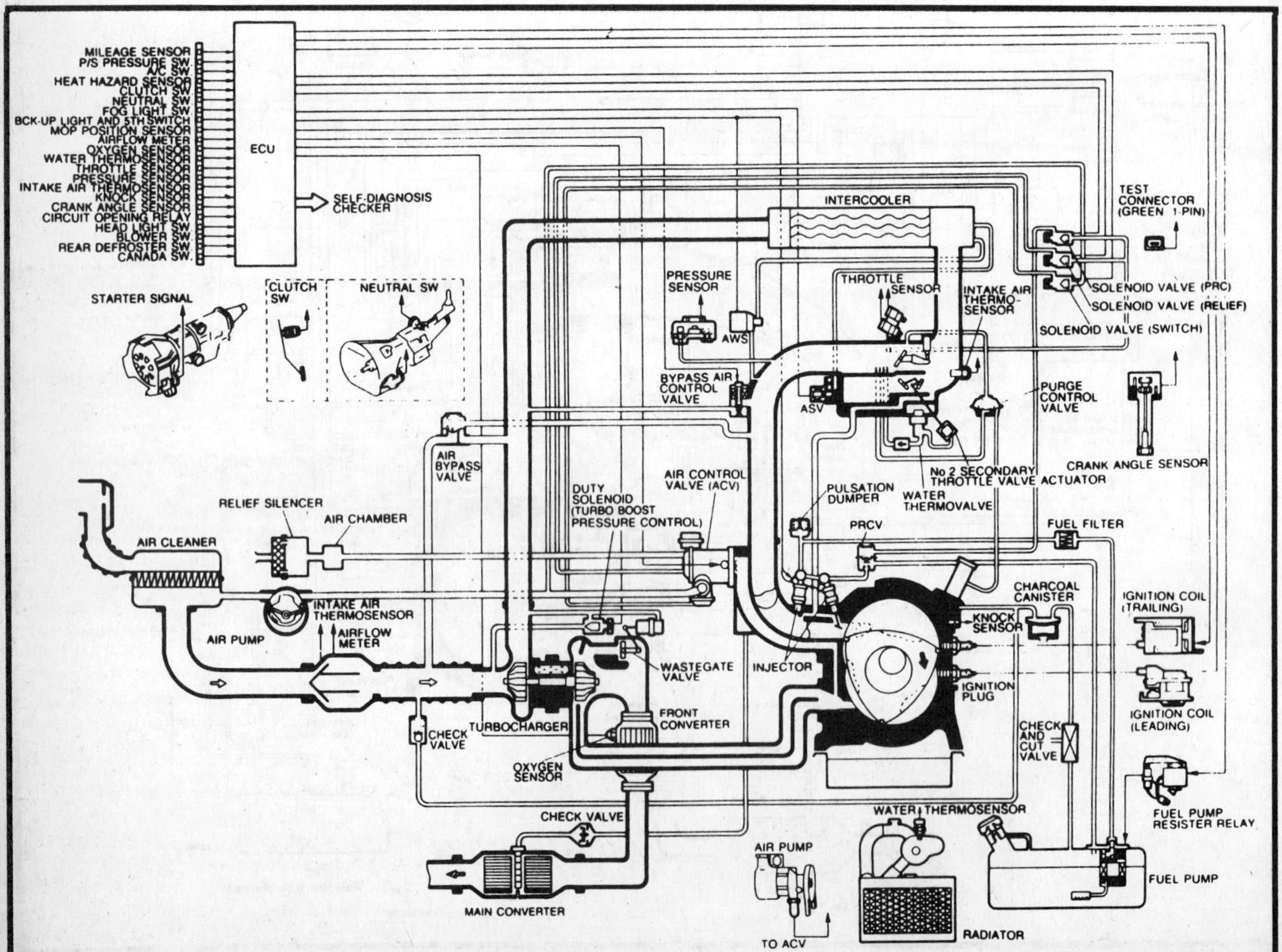

MILEAGE SENSOR
P/S PRESSURE SW.
A/C SW.
HEAT HAZARD SENSOR
CLUTCH SW.
NEUTRAL SW.
FOG LIGHT SW.
BCK-UP LIGHT AND 5TH SWITCH
MOP POSITION SENSOR
AIRFLOW METER
OXYGEN SENSOR
WATER THERMOSENSOR
THROTTLE SENSOR
PRESSURE SENSOR
INTAKE AIR THERMOSENSOR
KNOCK SENSOR
CRANK ANGLE SENSOR
CIRCUIT OPENING RELAY
HEAD LIGHT SW.
BLOWER SW.
REAR DEFROSTER SW.
CANADA SW.
ECU
SELF-DIAGNOSIS CHECKER
INTERCOOLER
TEST CONNECTOR (GREEN 1-PIN)
STARTER SIGNAL
CLUTCH SW
NEUTRAL SW
PRESSURE SENSOR
AWS
THROTTLE SENSOR
INTAKE AIR THERMO-SENSOR
SOLENOID VALVE (PRC)
SOLENOID VALVE (RELIEF)
SOLENOID VALVE (SWITCH)
BYPASS AIR CONTROL VALVE
ASV
PURGE CONTROL VALVE
No 2 SECONDARY THROTTLE VALVE ACTUATOR
CRANK ANGLE SENSOR
AIR BYPASS VALVE
AIR CONTROL VALVE (ACV)
PULSATION DUMPER
WATER THERMOVALVE
RELIEF SILENCER
AIR CHAMBER
DUTY SOLENOID (TURBO BOOST PRESSURE CONTROL)
PRCV
FUEL FILTER
AIR CLEANER
CHARCOAL CANISTER
IGNITION COIL (TRAILING)
INTAKE AIR THERMOSENSOR
AIRFLOW METER
KNOCK SENSOR
AIR PUMP
WASTEGATE VALVE
INJECTOR
IGNITION PLUG
IGNITION COIL (LEADING)
FRONT CONVERTER
CHECK VALVE
TURBOCHARGER
CHECK AND CUT VALVE
OXYGEN SENSOR
WATER THERMOSENSOR
FUEL PUMP RESISTER RELAY
CHECK VALVE
AIR PUMP
FUEL PUMP
MAIN CONVERTER
RADIATOR
TO ACV

**Exploded view of fuel system – 1989–90 RX-7 turbo**

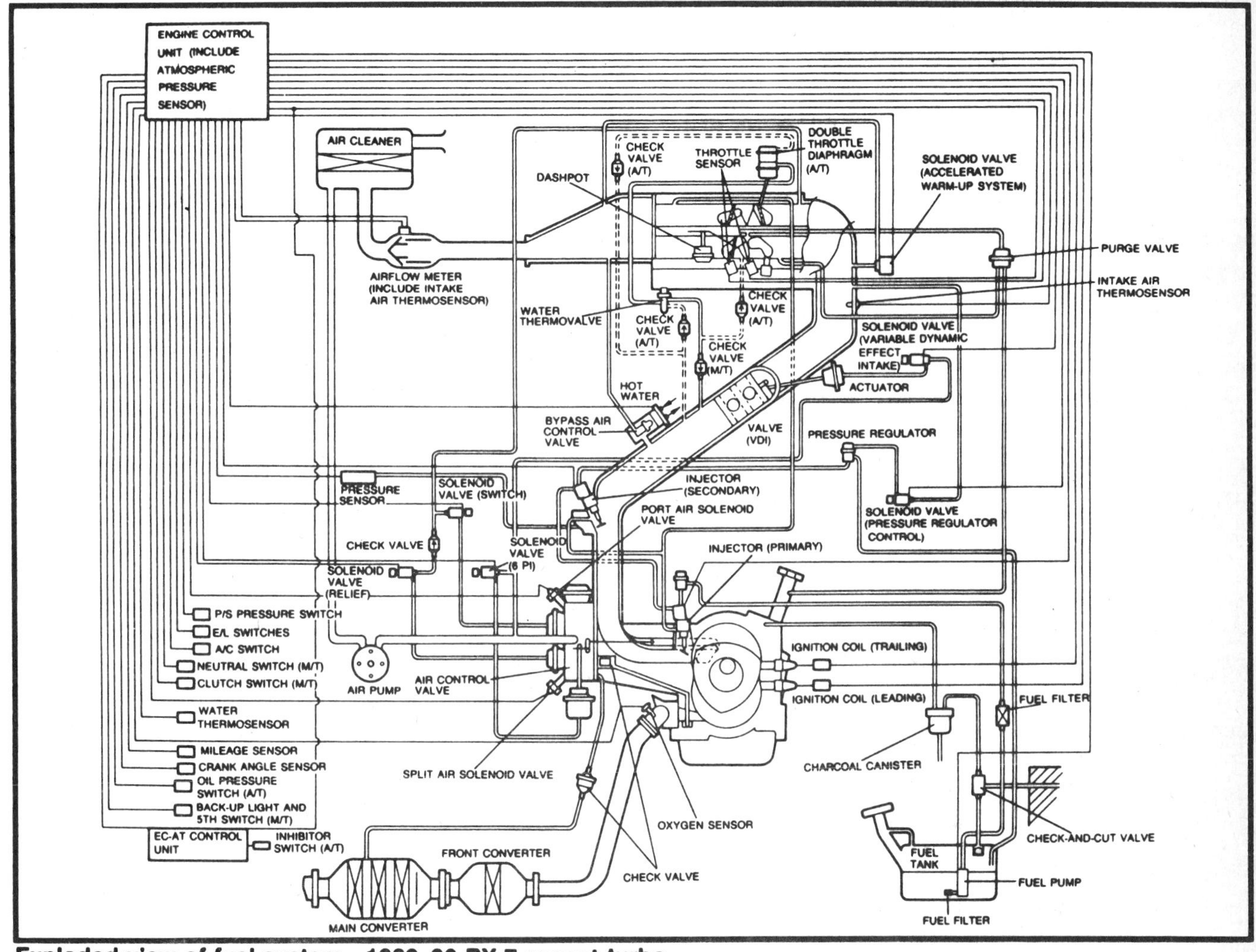


**Exploded view of fuel system – 1989–90 RX-7 except turbo**

- Do not reuse fuel hose clamps.
- Make sure all EGI harness connectors are fastened securely. A poor connection can cause an extremely high surge voltage in the coil and condenser and result in damage to integrated circuits.
- Keep the EGI harness at least 4 in. away from adjacent harnesses to prevent an EGI system malfunction due to external electronic "noise."
- Keep EGI all parts and harnesses dry during service.
- Before attempting to remove any parts, turn off the ignition switch and disconnect the battery ground cable.
- Always use a 12 volt battery as a power source.
- Do not attempt to disconnect the battery cables with the engine running.

up the engine immediately after starting or just prior to shutdown.

- Do not attempt to disassemble the EGI control unit under any circumstances.
- If installing a two-way or CB radio, keep the antenna as far as possible away from the electronic control unit. Keep the antenna feeder line at least 8 in. away from the EGI harness and do not let the two run parallel for a long distance. Be sure to ground the radio to the vehicle body.
- Do not apply battery power directly to injectors.
- Handle air flow meter carefully to avoid damage.
- Do not disassemble air flow meter or clean meter with any type of detergent.

## TESTING PRECAUTIONS

1. Before connecting or disconnecting control unit ECU harness connectors, make sure the ignition switch is OFF and the negative battery cable is disconnected to avoid the possibility of damage to the control unit.
2. When performing ECU input/output signal diagnosis, remove the pin terminal retainer from the connectors to make it easier to insert tester probes into the connector.
3. When connecting or disconnecting pin connectors from the ECU, take care not to bend or break any pin terminals. Check that there are no bends or breaks on ECU pin terminals before attempting any connections.
4. Before replacing any ECU, perform the ECU input/output signal diagnosis to make sure the ECU is functioning properly or not.
5. After checking through EGI troubleshooting, perform the EFI self-diagnosis and driving test.
6. When measuring supply voltage of ECU controlled components with a circuit tester, separate one tester probe from another. If the two tester probes accidentally make contact with each other during measurement, a short circuit will result and damage the power transistor in the ECU.

## Diagnosis and Testing

### AIR FLOW METER

#### Inspection

1. Inspect the air flow meter body for cracks.
2. Using a suitable ohmmeter, check the resistance between the terminals and check the resistance on the charts below.

#### Operational Check

**1988 MODELS ONLY**

1. Using a suitable tool press open the measuring plate. Measure the resistance between the air flow meter terminals.
2. The resistance should be within the specifications shown in the following charts.
3. If the air flow meter fails to meet these specifications, replace it.

| Condition / Terminal | Measuring plate: Fully closed | Measuring plate: Fully open |
|---|---|---|
| $E_1 \leftrightarrow F_c$ | ∞ | 0 |
| $E_2 \leftrightarrow V_s$ | 50 ~ 500 | 50 ~ 500 |

**Air flow meter operational resistance values—1988 RX-7 except turbo**

| Condition / Terminal | Measuring plate: Fully closed | Measuring plate: Fully open |
|---|---|---|
| $E_1 \leftrightarrow F_c$ | ∞ | 0 |
| $E_2 \leftrightarrow V_s$ | 200 ~ 600 | 20 ~ 1,000 |

**Air flow meter operational resistance values—1988 RX-7 turbo**

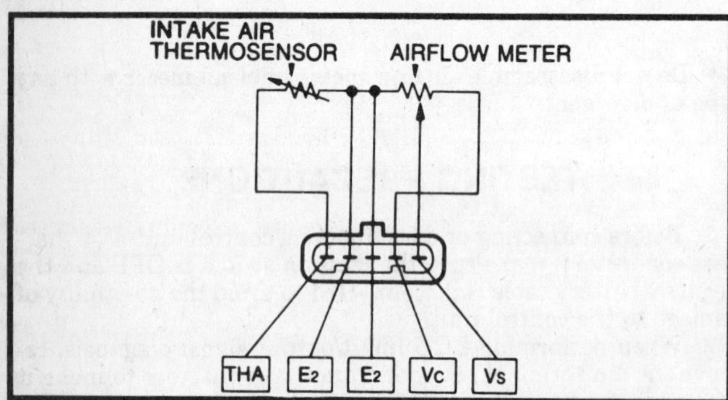


**Air flow meter terminal locations—1989–90 RX-7**

| Terminal | Resistance (Ω) |
|---|---|
| $E_2 \leftrightarrow V_s$ | 200—1,000 (Closed; 20°C (68°F))<br>20—800 (Open; 20°C (68°F)) |
| $E_2 \leftrightarrow V_c$ | 200—400 (Closed ↔ Open; 20°C (68°F)) |
| $E_2 \leftrightarrow$ THA (Intake air thermosensor) | −20°C ( −4°F) 10,000—20,000<br>0°C ( 32°F) 4,000— 7,000<br>20°C ( 68°F) 2,000— 3,000<br>40°C (104°C) 900— 1,300<br>60°C (140°F) 400— 700 |

**Air flow meter resistance values—1989–90 RX-7**

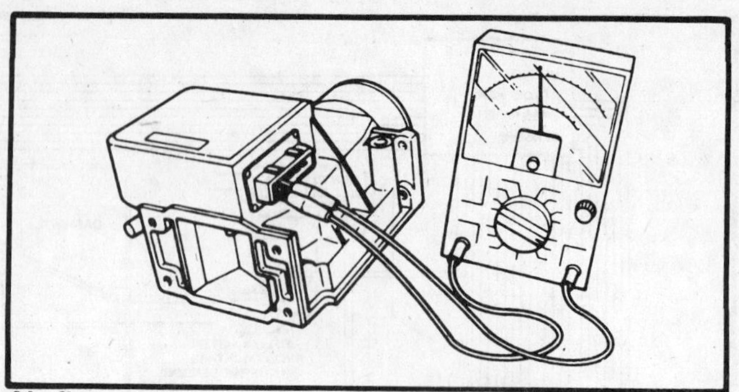

**Air flow meter resistance check—typical**

| Terminal | Resistance (Ω) |
|---|---|
| $E_2 \leftrightarrow V_s$ | 200 ~ 600 |
| $E_2 \leftrightarrow$ Vref | 200 ~ 400 |
| $E_2 \leftrightarrow$ THA (Intake air temperature sensor) | −20°C ( −4°F) 10,000 ~ 20,000<br>0°C ( 32°F) 4,000 ~ 7,000<br>20°C ( 68°F) 2,000 ~ 3,000<br>40°C (104°F) 900 ~ 1,300<br>60°C (140°F) 400 ~ 700 |
| $E_1 \leftrightarrow F_c$ | ∞ |

**Air flow meter resistance values—1988 RX-7 turbo**

| Terminal | Resistance (Ω) |
|---|---|
| $E_2 \leftrightarrow V_s$ | 50 ~ 500 |
| $E_2 \leftrightarrow$ Vref | 200 ~ 500 |
| $E_2 \leftrightarrow$ THA (Intake air temperature sensor) | −20°C ( −4°F) 10,000 ~ 20,000<br>0°C ( 32°F) 4,000 ~ 7,000<br>20°C ( 68°F) 2,000 ~ 3,000<br>40°C (104°F) 900 ~ 1,300<br>60°C (140°F) 400 ~ 700 |
| $E_1 \leftrightarrow F_c$ | ∞ |

**Air flow meter resistance values—1988 RX-7 except turbo**

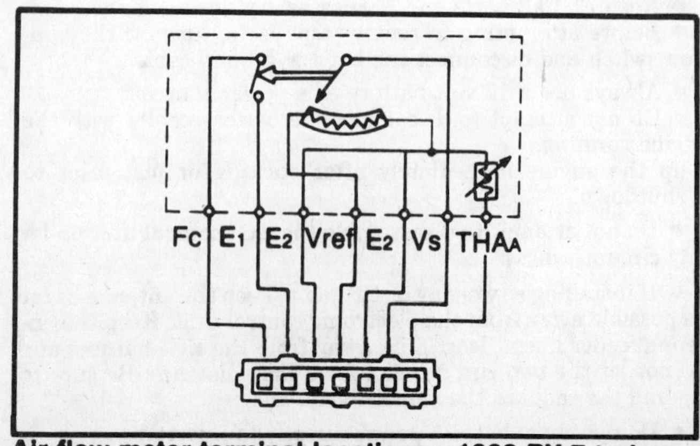


**Air flow meter terminal locations—1988 RX-7 turbo**

## THROTTLE BODY

### Checking the No. 1 Secondary Throttle Valve

1. The No. 1 secondary throttle valve starts to open when the primary throttle valve opens to (15 degrees 1988–90 Turbo), (12 degrees M/T, 8 degrees A/T 1988–90 non-Turbo) and completely opens at the same time when the primary throttle valve fully opens.
2. Check the clearance between between the primary throttle valve and the wall of the throttle bore when the No. 1 secondary throttle valve starts to open.
3. If the clearance is not within the specification shown below, bend the tab until the proper clearance is obtained.

**Standard Clearance:**

1988–90 RX-7 except Turbo – M/T 0.02–0.03 in. (0.5–0.7mm), A/T 0.03–0.06 in. (0.8–1.4mm)
1988–90 RX-7 Turbo – 0.04–0.07 in. (1.1–1.7mm)

### Checking fast idle operation

In order to perform this operation properly the vehicle and the throttle chamber must be at 77°F (25°C).
1.For proper fast idle operation the matching mark on the fast idle cam must be aligned with the center of the cam roller.
2.If the matching mark and the center of the cam roller do not align, turn the cam roller do not align, turn the cam adjusting screw until the proper alignment is obtained.

**NOTE: Fast idle adjustment is unnecessary unless it has been tampered with.**

3. Once the correct matching mark aligns with the center of the cam roller check the clearance (throttle chamber – primary throttle valve), turn the fast idle adjusting screw to the correct specification.

**Standard clearance** – 0.016–0.02 in. (0.4–0.5mm)

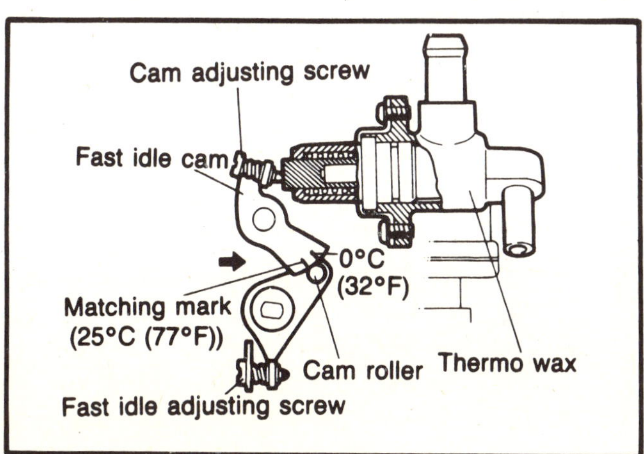

**Fast idle adjustment – 1988 RX-7**

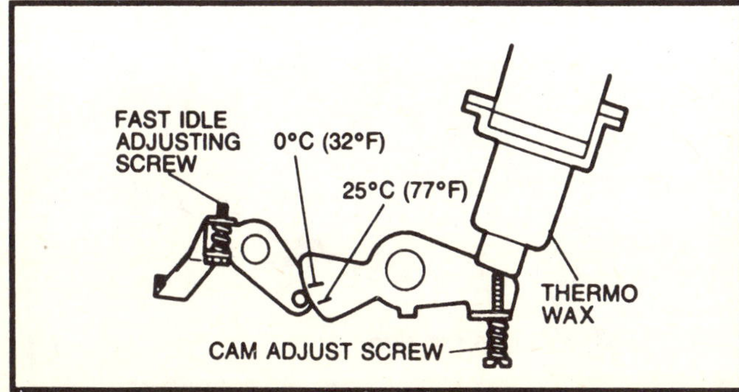

**Fast idle adjustment – 1989–90 RX-7 except turbo**

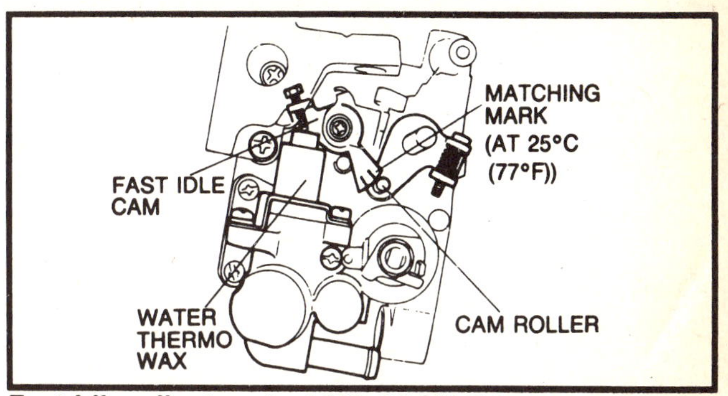

**Fast idle adjustment – 1989–90 RX-7 turbo**

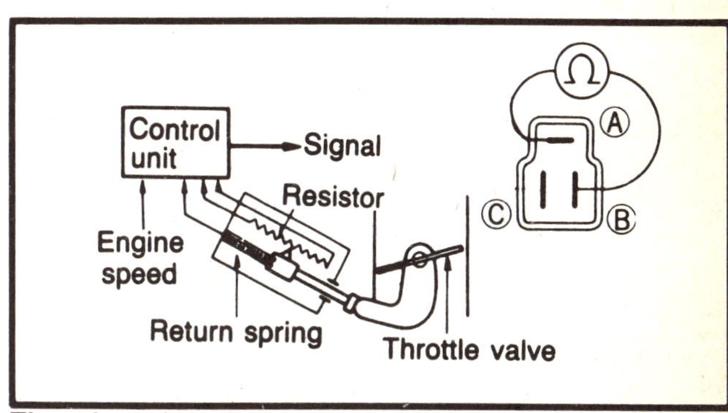

**Throttle sensor inspection – 1988 RX-7**

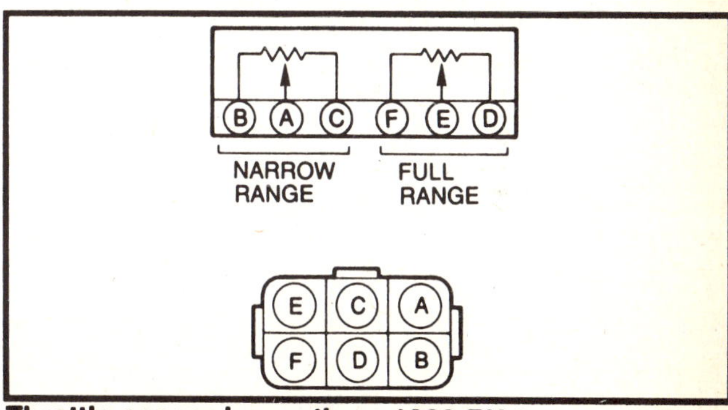

**Throttle sensor inspection – 1989 RX-7**

## THROTTLE POSITION SENSOR

### Inspection

**1988–89 MODELS**

1. Disconnect the throttle sensor connector.
2. Connect a suitable ohmmeter to the sensor as shown in the illustration.
3. On 1988 models, open the throttle valve and observe the resistance readings:

Throttle opening

**A to B**

- A–B Idle position: approximately 1 kΩ
- Full open position: approximately 4–6 kΩ

**A to C**

- A–B Idle position: approximately 4–6 kΩ
- Full open position: approximately 4–6 kΩ

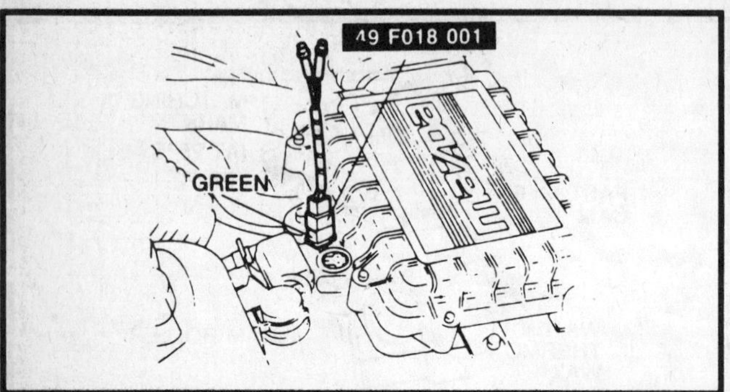

Throttle sensor adjustment

4. On 1989 Models, open the throttle valve and observe the resistance readings:

**A to B**
- A–B Idle position: 0.8–1.2 kΩ
- Full open position: 4–6 kΩ

**E to D**
- E–D Idle position: 0.6–0.9 kΩ
- E–D Full open position: 3.4–5.1 kΩ

5. Reconnect the connector.

**1990 MODELS**

1. Start the engine and allow it to reach normal operating temperature. Stop the engine.
2. Connect Engine Signal Monitor (49–9200–162) and Adapter Harness (49–G018–903) between the ECU and the ECU wiring harness.

**NOTE: When testing the throttle sensor, ensure to use connector B of the Adapter Harness (49–G018–903).**

3. Place Engine Signal Monitor Sheet (49–G018–904) on the Engine Signal Monitor. Check the voltage between terminals 2F and 2G of the ECU. Specifications should be as follows:

**Terminal 2F**
- Throttle closed: 0.75–1.25 volts
- Throttle fully opened: Approximately 5 volts.

**Terminal 2G**
- Throttle closed: 0.25–1.25 volts
- Throttle fully opened: 4.1–4.44 volts

4. If specifications are not as indicated, adjust or replace the throttle sensor as necessary.

## FUEL PUMP

### Inspection

1. Relieve the fuel system pressure.

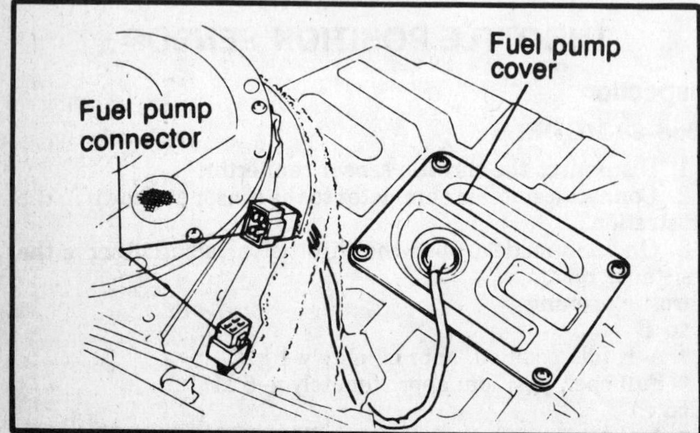

Fuel pump connector location

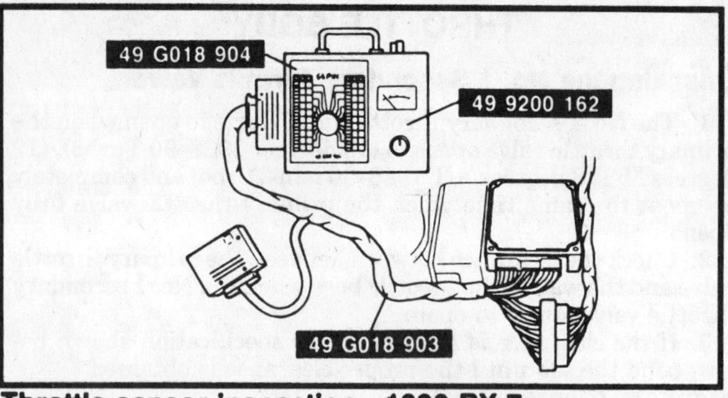

Throttle sensor inspection – 1990 RX-7

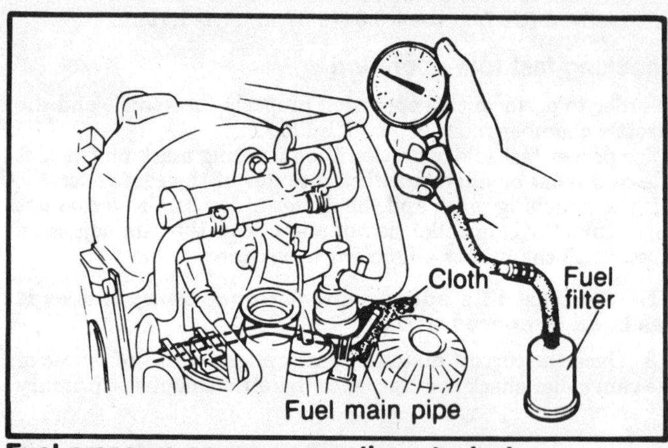

Fuel pressure gauge connection – typical

2. Disconnect the negative battery cable.
3. Disconnect the main fuel hose from the main fuel line.
4. Connect a suitable fuel pressure gauge.
5. Connect the negative battery cable.
6. Connect the terminals of the check connector with a jumper wire. Turn on the ignition switch to operate the fuel pump.
7. Observe that the fuel pressure is within the following specifications:

**Fuel Pump Pressure:**
- 1988–90 RX-7 except Turbo – 64–85.3 psi (441–588 kPa)
- 1988–90 RX-7 Turbo – 71.1–92.4 psi (490–637 kPa)

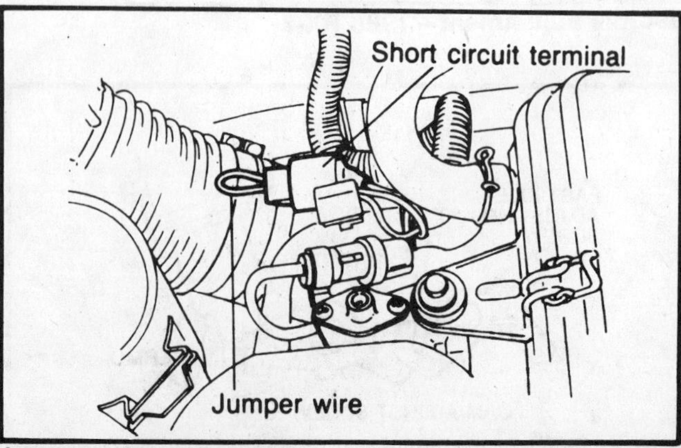

Jumper wire connection

**Fuel system pressure tester No. 49 9200 750**

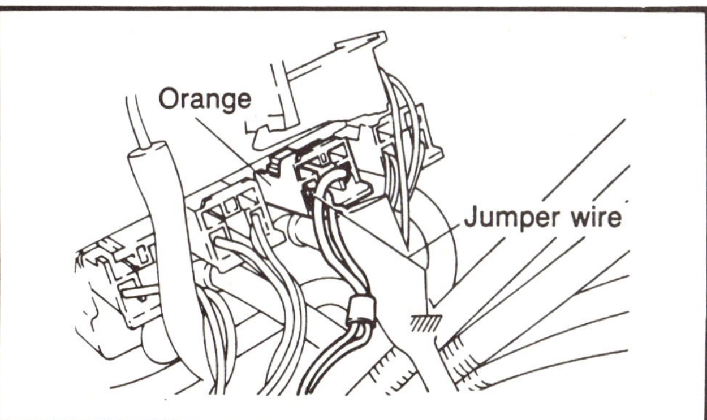


**Pressure wire solenoid valve—turbo models**

## PRESSURE REGULATOR

### Fuel Line Pressure Inspection

**1988 MODELS**

1. Relieve the fuel pressure from the fuel system.
2. Disconnect the negative battery cable. Connect the multi-pressure tester (49–9200–750) or equivalent to the fuel line.
3. Reconnect the negative battery cable. Start the engine.
4. Disconnect the vacuum hose from the pressure regulator (non-turbo models), or connect a jumper wire to the pressure regulator solenoid valve (LO) terminal and ground (turbo models).
5. Measure the fuel pressure at idle. The fuel pressure (regulating pressure) should be 35.6–37.0 psi.

**NOTE: The fuel pressure is indicated on the lower LED line (fuel pressure line) at any ranges.**

6. Reconnect the vacuum hose connected to the pressure regulator (non-turbo models), or connect a jumper wire to the pressure regulator solenoid valve (LO) terminal and ground (turbo models). The fuel pressure (regulating pressure) should be 28.4 psi.
7. If not within specifications, replace the pressure regulator.

**1989–90 MODELS**

1. Relieve the fuel pressure from the fuel system.
2. Disconnect the battery ground cable. Connect a suitable pressure gauge between the fuel filter and the pulsation damper.
3. Connect the battery ground cable. Start the engine. Pressure gauge should read 27–33 psi.
4. Install a jumper wire to the terminals of the check connector (yellow: 2-pin).
5. Turn the ignition to the **ON** position for approximately 10 seconds. Pinch the outlet hose of the pressure regulator, then turn the ignition switch to the **OFF** position.
6. After approximately 5 minutes, the pressure gauge should read 21 psi. If pressure gauge readings are not as specified, replace the pressure regulator.

## FUEL INJECTORS

### Injection Fuel Pressure Inspection

1. Relieve the fuel pressure from the fuel system.
2. Disconnect the negative battery cable. Connect the multi-pressure tester (49-9200-750) or equivalent to the fuel line.
3. Reconnect the negative battery cable. Start the engine and measure the fuel pressure at the "III" range of the multi pressure tester. The fuel pressure (injection pressure) should be approximately35.6–37.0 psi.

**NOTE: The fuel pressure is indicated on the lower LED line (fuel pressure line) at any ranges.**

4. If the injection pressure is lower than specified, check the following:
   a. Fuel pump outlet pressure.
   b. Fuel line pressure.
   c. Fuel filter clog.
5. If the injection pressure is higher than specifications, check the following:
   a. Fuel return pipe clogged.
   b. Fuel line pressure.

### Injector Inspection

Check the operating sound of the injector, using a sound scope, check that operating sounds are produced from each injector at idle and at acceleration.

1. If the injectors do not operate, check the following:
   a. Check the fuel injection system main fuse (40 amp).
   b. Turn the ignition switch **ON** and **OFF**, verify that the main relay clicks.
2. If the clicking sound is not heard at the main relay, when the ignition switch is turned **ON**, use a voltmeter and check for 12 volts at the main relay connector (BW) terminal wire.

### Injector Fuel Leak Test

**1988–90 MODELS**

1. Relieve the fuel pressure from the fuel system.
2. Remove the throttle body unit (Dynamic Chamber).
3. Loosen the distribution pipe attaching bolts. Affix the injectors to the distribution pipe with some suitable wire.

**NOTE: Affix the injectors firmly to the distribution pipe so no movement of the injectors is possible.**

4. Turn the ignition switch to the **ON** position. Use a jumper wire and connect the shot circuit terminals of the fuel pump check connector. Make sure that the fuel does not leak from the injector nozzles.
5. After approximately five minutes, a very small (slight) amount of fuel leakage from the injectors is acceptable.
6. If the injector leaks fuel at a fast rate, replace the injector. If the injectors do not leak, remove all test equipment and install the removed components in the reverse order of the removal procedure.

### Injector Volume Test

**1988 MODELS ONLY**

1. Relieve the fuel pressure from the fuel system.

2. Connect a suitable hose to the injector. Connect injector checker (49–B092–953) or equivalent to the injector.
3. Turn the ignition switch to the **ON** position. Use a jumper wire and connect the terminals of the fuel pump check connector. Make sure that the fuel does not leak from the injector nozzles.
4. Apply battery voltage to the injector checker and measure the injector volume with a graduated cylinder (measuring container). Injector volume should be as follows:
Non-Turbo – 111–118cc (6.8–7.2 cu. in.)
Turbocharged – 133–142cc (8.1–8.7 cu. in.)

**NOTE: Do not apply battery voltage directly to the injector. The injector checker must be used or the injector will be damaged.**

## TROUBLESHOOTING WITH THE SELF-DIAGNOSIS CHECKER

The self diagnosis checker (49–H018–9A1) is used to retrieve code numbers of malfunctions which have happened and were memorized or are continuing. The malfunction is indicated by the code number and a buzzer.

If there is more than 1 malfunction, the code numbers will display on the self diagnosis checker 1 by 1 in numerical order. In the case of malfunctions, 08, 13, and 01, the code numbers are displayed in a order of 01, 08 and then 13.

The memory of malfunctions is canceled by disconnecting the negative battery cable for at least 5 seconds.

The ECU has a built in fail-safe mechanism for the main input sensors. If a malfunction occurs, the emission control unit will substitute values. This will slightly effect the driving performance, but the vehicle may still be driven.

The ECU continuously checks for malfunctions of the input devices. But, the ECU checks checks for malfunctions of the output devices within 3 seconds after turning the ignition switch to the **ON** position and the test connector is grounded.

The malfunction indicator light indicates a pattern the same as the buzzer of the self-diagnosis checker when the self-diagnosis check connector is grounded. When the self-diagnosis check connector is not grounded, the lamp illuminates steady while malfunction of the main input sensor occurs and goes out if the malfunction recovers. However, the malfunction code is memorized in the emission control unit.

### Inspection Procedure

1. Start and allow the engine to reach normal operating temperature. Stop the engine.
2. Connect the Self Diagnosis Checker (49–H018–9A1) or equivalent to the check connector and the battery ground cable.
3. Set the select switch on the Self Diagnosis Checker to the "B" position on 1988 models, and the "A" position on 1989–90 models.

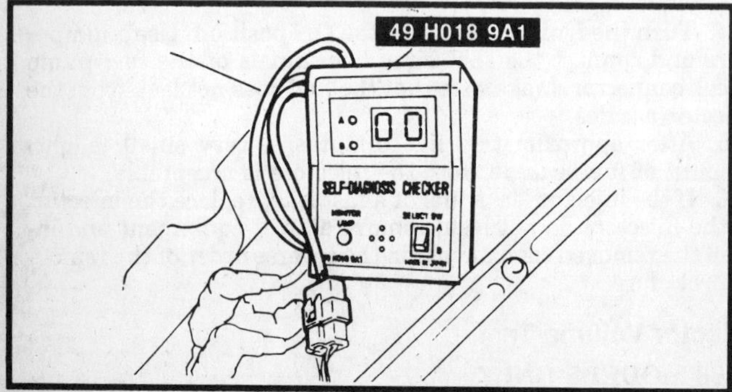


**Self-diagnosis checker 49–H018–9A1**

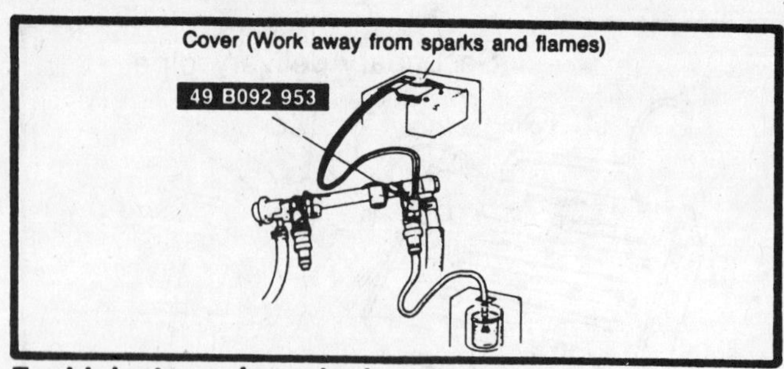


**Fuel injector volume test**

4. On 1988 models, check the Self-Diagnosis checker for trouble codes, and proceed to appropriate trouble code diagnostic chart.
5. On 1989–90 models, connect a jumper wire between the test connector and ground.
6. Turn the ignition switch to the **ON** position. Check that the number "88" flashes on the digital display and the buzzer sounds for 3 seconds after turning the ignition switch **ON**.
7. If the number "88" does not flash, check the check connector wiring.
9. If the number "88" flashes and the buzzer sounds continuously for more than 20 seconds, check for a short circuit between terminal 1F on the ECU and the check connector. Check ECU terminals 3X and 3Z voltages. Replace the ECU if necessary.
10. Note the code numbers and check the causes, repair as necessary. Be sure to recheck the code numbers by performing the "After Repair Procedure," after repairing.

### After Repair Procedure

1. Cancel the memory of malfunctions by disconnecting the battery negative cable and depressing the brake pedal for at least 5 seconds, then reconnect the battery ground cable.
2. Connect the Self-Diagnosis Checker 49–H018–9A1 to the check connector. Ground the test connector (green: 1 pin) using a suitable jumper wire.

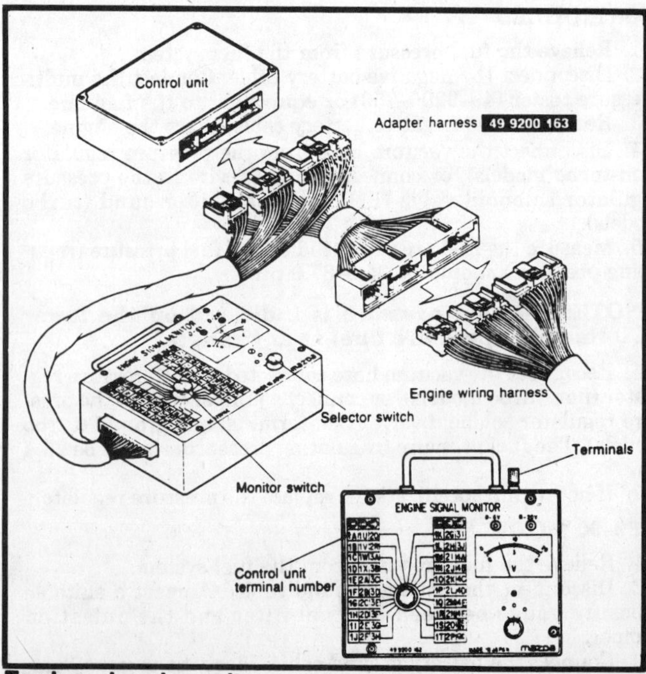


**Engine signal monitor**

4. Turn the ignition switch to the **ON** position, but do not start the engine for approximately 6 seconds.

5. Start the engine and allow it to reach normal operating temperature, the run the engine at 2000 rpm for 2 minutes. Check that no code numbers are displayed.

## TROUBLESHOOTING WITH THE ENGINE SIGNAL MONITOR

The Engine Signal Monitor (49–9200–162), Adapter Harness (49–9200–163 on 1988 models and 49–G018–903 on 1990 models) and Engine Digital Monitor Sheet (49–G018–903 used on 1990 models) are used to check the control unit terminal voltages. On 1989 models, a suitable volt meter is used to test the ECU terminal voltages. This monitor checks the individual terminal voltages through selection by the monitor switch.

### Inspection

#### EXCEPT 1989 MODELS

1. Start the engine and allow it to reach normal operating temperature. Stop the engine.
2. Lift up the floor mat on the passenger's side, and remove the ECU protective cover. Disconnect the ECU connector.
3. On all 1988 models, connect Engine Signal Monitor (49–9200–162) and the Adapter Harness (49–9200–163) between the ECU and the ECU wiring harness.
4. On 1990 models, connect Engine Signal Monitor (49–9200–162), and Adapter Harness (49–G018–903) between the ECU and the ECU wiring harness, then install Engine Signal Monitor Sheet (49–G018–904) on the Engine Signal Monitor.

**NOTE: On 1990 models, use connector (A) of the Adapter Harness to check voltages of terminals 1A through 1V and 3A through 3Z, and use connector (B) to check voltages of terminals 2A through 2P.**

5. On all models, position the select switch and the monitor switch on the Engine Signal Monitor to the terminal number to be tested. Check the voltage of each terminal.
6. If any ECU terminal voltage is incorrect, check the input or output component and related wiring. If components and wiring are normal, replace the ECU.

#### 1989 MODELS

1. Start and allow the engine to reach normal operating temperature. Stop the engine.
2. Lift up the floor mat on the passenger's side and remove the ECU protective cover.
3. Turn the ignition switch to the **ON** position.
4. Using a suitable voltmeter, check the ECU terminal voltages as shown.
5. If proper voltage is not indicated on the voltmeter, check all wiring, connections and the related component.

**CONTROL UNIT PIN LOCATIONS AND TEST VOLTAGES—1988 MODELS EXCEPT TURBO**

| Terminal | Input | Output | Connection to | Voltage (after warming up) | | Remark |
|---|---|---|---|---|---|---|
| | | | | Ignition switch: ON | Idle | |
| 1A | | O | Digital code checker | Ignition switch OFF → ON for 3 sec. below 5V, after 3sec. approx. 12V | | with digital code checker |
| 1B | | O | Digital code checker | Ignition switch OFF → ON for 3 sec. below 5V, after 3 sec. approx. 12V | | with digital code checker |
| 1C | | O | Air bypass relay | approx. 12V | | |
| 1D | | O | Digital code checker (Green lamp) | Ignition switch OFF → ON for 3 sec. below 5V, after 3 sec. approx. 12V | | with digital code checker |
| 1E | O | | A/C switch | below 2.5V (A/C: ON), approx. 12V (A/C: OFF) | | Blower motor ON |
| 1F | | O | A/C main relay | approx. 12V (A/C: OFF) | | Blower moter ON |
| 1G | O | | Neutral switch | below 1.5V (in neutral,A/T) approx. 12V (others) | | |
| 1H | O | | Water temperature switch | below 1.5V (Radiator coolant temperature above 17°C (62.6°F)) | | |
| 1I | O | | 5th switch | approx. 12V (M/T; 5th gear, A/T; others) below 1.5V (M/T others, A/T over drive) | | |
| 1J | O | | Initial set coupler | approx. 4 ~ 7V (Initial set coupler: OFF), below 1.5V (Initial set coupler: ON) | | |
| 1K | | O | Shift indicator light | below 1.5V | approx. 12V | |
| 1L | O | | Clutch switch | below 1.5V (clutch pedal; released) approx. 12V (clutch pedal; depressed, A/T) | | |
| 1M | | O | Coil with igniter (Trailing) IGf-T | below 2V | | |
| 1N | O | | Crank angle sensor G ① | below 1.0V | | G |
| 1O | O | | Mileage switch | approx. 12V (below 20,000 miles) below 1.5V (above 20,000 miles) | | |
| 1P | O | | Crank angle sensor G ② | below 1.0V | | B |
| 1Q | O | | Crank angle sensor Ne ② | below 1.0V | | W |
| 1R | O | | P/S switch | 10 ~ 12V | below 1.5V (Steering wheel turned) approx. 12V (Straight ahead) | |

**CONTROL UNIT PIN LOCATIONS AND TEST VOLTAGES—1988 MODELS EXCEPT TURBO**

| Terminal | Input | Output | Connection to | Voltage (after warming up) | | Remark |
|---|---|---|---|---|---|---|
| | | | | Ignition switch: ON | Idle | |
| 1S | | O | Port air solenoid valve | below 2.5V | | |
| 1T | O | | Crank angle sensor Ne ① | below 1.0V | | |
| 1U | | O | Coil with igniter (Trailing) IGs-T (Select signal) | approx. 4.4V | approx. 2.2V | |
| 1V | | O | Coil with igniter (Leading) IGT-L (Ignition timing signal) | 0V | approx. 0.8V | |
| 1W | O | | Heat hazard sensor | below 1.5V | approx. 12V | Floor Temp. below 110°C (230°F) |
| 1X | | O | Coil with igniter (Trailing) IGT-T (Ignition timing signal) | 0V | approx. 0.8V | |
| 2A | | O | V ref | 4.5 ~ 5.5V | | |
| 2B | O | | Boost sensor | 3.5 ~ 4.0V | | Disconnect the vacuum hose |
| 2C | — | — | Ground | 0V | | |
| 2D | O | | $O_2$ sensor | below 1.0V | | |
| 2E | O | | Air flow meter (Vs) | approx. 4V | 2.5 ~ 3.5V | |
| 2F | O | | Variable resistor | 1 ~ 4V (Varies according to the variable resistor adjustment) | | |
| 2G | O | | Throttle sensor (TVO) | approx. 1V (Throttle sensor adjusted properly) | | |
| 2H | O | | Atmospheric pressure sensor | 3.5 ~ 4.5V (at sea level) 2.5 ~ 3.5V (at 2,000 m (6,500ft)) | | |
| 2I | O | | Water thermo sensor | approx. 0.4 ~ 1.8V | | |
| 2J | O | | Air flow meter (Intake air temperature sensor) | 2 ~ 3V at 20°C (68°F) | | |
| 2K | | O | Split air solenoid valve | approx. 12V (M/T: Neutral, A/T: N,P range) | | |
| 2L | O | | Intake air temperature sensor (dynamic chamber) | 1 ~ 2V at 80°C (176°F) | | |
| 2M | | O | Pressure regulator control solenoid valve | below 2.0V | approx. 12V | |
| 2N | | O | EGR solenoid valve | approx. 12V | | |
| 2O | | O | Switching solenoid valve | approx. 12V (Throttle sensor is adjusted properly) | approx. 12V | |
| 2P | | O | Relief solenoid valve | below 2.0V (Throttle sensor is adjusted properly) | below 2.0V | |
| 2Q | | O | Bypass air control valve | 8 ~ 12V | | |
| 2R | — | — | Ground | 0V | | |
| 3A | — | — | Ground | 0V | | |
| 3B | O | | Starter switch | below 1.5V approx 12V (at cranking) | | |
| 3C | | O | Injector (rear primary) | approx. 12V | | |
| 3D | | O | Inhibitor switch | below 1.5V (A/T; N.P range, M/T) approx. 12V (A/T; others) | | |

## CONTROL UNIT PIN LOCATIONS AND TEST VOLTAGES – 1988 MODELS EXCEPT TURBO

| Terminal | Input | Output | Connection to | Voltage (after warming up) Ignition switch: ON | Idle | Remark |
|---|---|---|---|---|---|---|
| 3E | | O | Injector (front primary) | approx. 12V | | |
| 3F | | O | Injector (rear secondary) | approx. 12V | | |
| 3G | — | — | Ground | 0V | | |
| 3H | | O | Injector (front secondary) | approx. 12V | | |
| 3I | — | — | Main relay | approx. 12V | | |
| 3J | O | | Battery | approx. 12V | | |

## CONTROL UNIT PIN LOCATIONS AND TEST VOLTAGES – 1988 TURBO MODELS

| Terminal | Input | Output | Connection to | Voltage (after warming up) Ignition switch: ON | Idle | Remark |
|---|---|---|---|---|---|---|
| 1A | | O | Digital code checker | Ignition switch OFF → ON for 3 sec below 5V, after 3sec approx. 12V | | with digital code checker |
| 1B | | O | Digital code checker | Ignition switch OFF → ON for 3 sec below 5V after 3 sec approx. 12V | | with digital code checker |
| 1C | | O | Air bypass solenoid valve | Approx. 12V | | |
| 1D | | O | Digital code checker (Green lamp) | Ignition switch OFF → ON for 3 sec below 5V after 3 sec approx. 12V | | with digital code checker |
| 1E | O | | A/C switch | below 2.5V (A/C: ON), approx. 12V (A/C: OFF) | | Blower motor ON |
| 1F | | O | A/C main relay | approx. 12V (A/C: OFF) | | |
| 1G | O | | Neutral switch | below 1.5V (in neutral), approx. 12V (others) | | |
| 1H | O | | Water temperature switch | below 1.5V (water temperature, above 17°C (62.6°F)) | | |
| 1I | O | | 5th switch | below 1.5V, (5th gear), approx. 12V (others) | | |
| 1J | O | | Initial set coupler | approx. 4 ~ 7V (Initial set coupler: OFF), below 1.5V (Initial set coupler: ON) | | |
| 1K | | O | Shift indicator light | below 1.5V | approx. 12V | |
| 1L | O | | Clutch switch | below 1.5V (clutch pedal, released) approx. 12V (clutch pedal, depressed) | | |
| 1M | O | | Coil with igniter (Trailing) IGI-T | below 2V | | |
| 1N | O | | Crank angle sensor G ① | below 1.0V | | G |
| 1O | O | | Mileage switch | approx. 12V (below 20,000 miles), below 1.5V (above 20,000 miles) | | |
| 1P | O | | Crank angle sensor G ② | below 1.0V | | B |
| 1Q | O | | Crank angle sensor Ne ② | below 1.0V | | W |
| 1R | O | | Knock control unit | 3 ~ 5V | | |
| 1S | | O | Port air solenoid valve | approx. 12V | | Mileage switch ON below 1.5V |
| 1T | O | | Crank angle sensor Ne ① | below 1.0V | | R |

## CONTROL UNIT PIN LOCATIONS AND TEST VOLTAGES – 1988 TURBO MODELS

| Terminal | Input | Output | Connection to | Voltage (after warming up) Ignition switch: ON | Idle | Remark |
|---|---|---|---|---|---|---|
| 3G | — | — | Ground | 0V | | |
| 3H | | O | Injector (Front secondary) | approx. 12V | | |
| 3I | — | — | Main relay | approx. 12V | | |
| 3J | O | | Battery | approx. 12V | | |

## DIGITAL CODE CHECKER TROUBLE CODES

| Code No. | Location problem | Fail safe function |
|---|---|---|
| 01 | Crank angle sensor | — |
| 02 | Air flow meter | Maintains the basic signal at a preset value. |
| 03 | Water thermo sensor | Maintains a constant 80°C (176°F) command. |
| 04 | Intake air temperature sensor (Air flow meter) | Maintains a constant 20°C (68°F) command. |
| 05 | Oxygen ($O_2$) sensor | Stop the feedback correction |
| 06 | Throttle sensor | Maintains a constant 100% (approx. 18°) command. |
| 07 | Boost sensor | Maintains a constant −96 mmHg (3.78 inHg) command. |
| 09 | Atmospheric pressure sensor | Maintains a constant command of the sea-level pressure |
| 12 | Coil with igniter (Trailing side) | Stop the operation of ignition system (only trailing side) |
| 15 | Intake air temperature sensor (Dynamic chamber) | Maintains a constant 20°C (68°F) command |

## CONTROL UNIT PIN LOCATIONS AND TEST VOLTAGES – 1988 TURBO MODELS

| Terminal | Input | Output | Connection to | Voltage (after warming up) Ignition switch: ON | Idle | Remark |
|---|---|---|---|---|---|---|
| 1U | | O | Coil with igniter (Trailing) IGs-T (Select signal) | approx. 4.4V | approx. 2.2V | |
| 1V | | O | Coil with igniter (Leading) IGT-L (Ignition timing signal) | 0V | approx. 0.8V | |
| 1W | O | | Heat hazard sensor | below 1.5V | approx. 12V | Floor Temp below 110°C (230°F) |
| 1X | | O | Coil with igniter (Trailing) IGT-T (Ignition timing signal) | 0V | approx. 0.8V | |
| 2A | | O | V ref | 4.5 ~ 5.5V | | |
| 2B | O | | Pressure sensor | 2.3 ~ 2.7V | | Disconnect the vacuum hose |
| 2C | — | — | Ground | 0V | | |
| 2D | O | | $O_2$ sensor | below 1.0V | | |
| 2E | O | | Air flow meter (Vs) | approx. 4V | 2.5 ~ 3.5V | |
| 2F | O | | Variable resistor | 1 ~ 4V (Varies according to the variable resistor adjustment) | | |
| 2G | O | | Throttle sensor (TVO) | approx. 1V (Throttle sensor adjusted properly) | | |
| 2H | O | | Atmospheric pressure sensor | 3.5 ~ 4.5V (at sea level) 2.5 ~ 3.5V (at 2,000 m (6,500 ft)) | | |
| 2I | O | | Water thermo sensor | approx. 0.4 ~ 1.8V | | Warm engine |
| 2J | O | | Air flow meter (Intake air temperature sensor) | 2 ~ 3V at 20°C (68°F) | | |
| 2K | | O | Twin scroll turbocharger solenoid valve | below 2.0V | | |
| 2L | O | | Intake air temperature sensor (inlet air pipe) | 1 ~ 2V at 80°C (176°F) | | |
| 2M | | O | Pressure regulator control solenoid valve | below 2.0V | approx. 12V | Cranking below 2.0V |
| 2N | | O | EGR solenoid valve | approx. 12V | | |
| 2O | | O | Switching solenoid valve | approx. 12V (Throttle sensor is adjusted properly) | approx. 12V | |
| 2P | | O | Relief solenoid valve | below 2V (Throttle sensor is adjusted properly) | below 2.0V | |
| 2Q | | O | Bypass air control (BAC) valve | 8 ~ 12V | | |
| 2R | — | — | Ground | 0V | | |
| 3A | — | — | Ground | 0V | | |
| 3B | O | | Starter switch | below 1.5V approx. 12V (at cranking) | | |
| 3C | | O | Injector (Rear primary) | approx. 12V | | |
| 3D | | O | Fuel pump resistor relay | approx. 12V | below 2.0V | |
| 3E | | O | Injector (Front primary) | approx. 12V | | |
| 3F | | O | Injector (Rear secondary) | approx. 12V | | |

## TROUBLE CODE TROUBLESHOOTING CHARTS

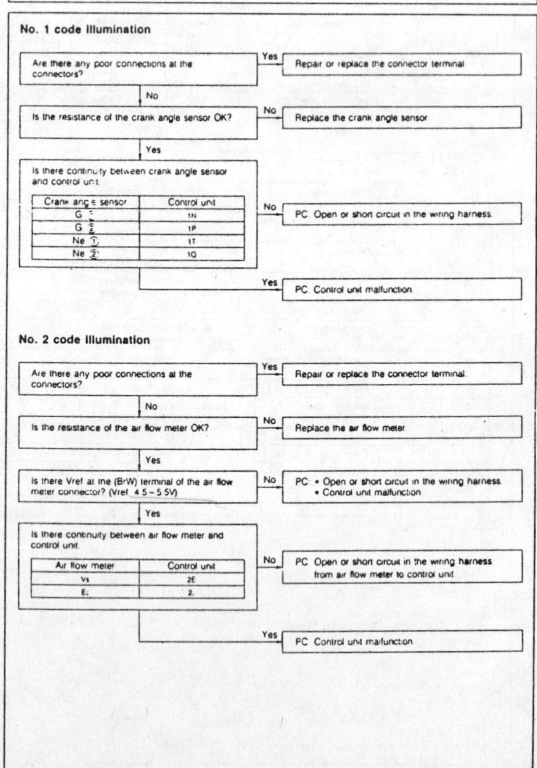

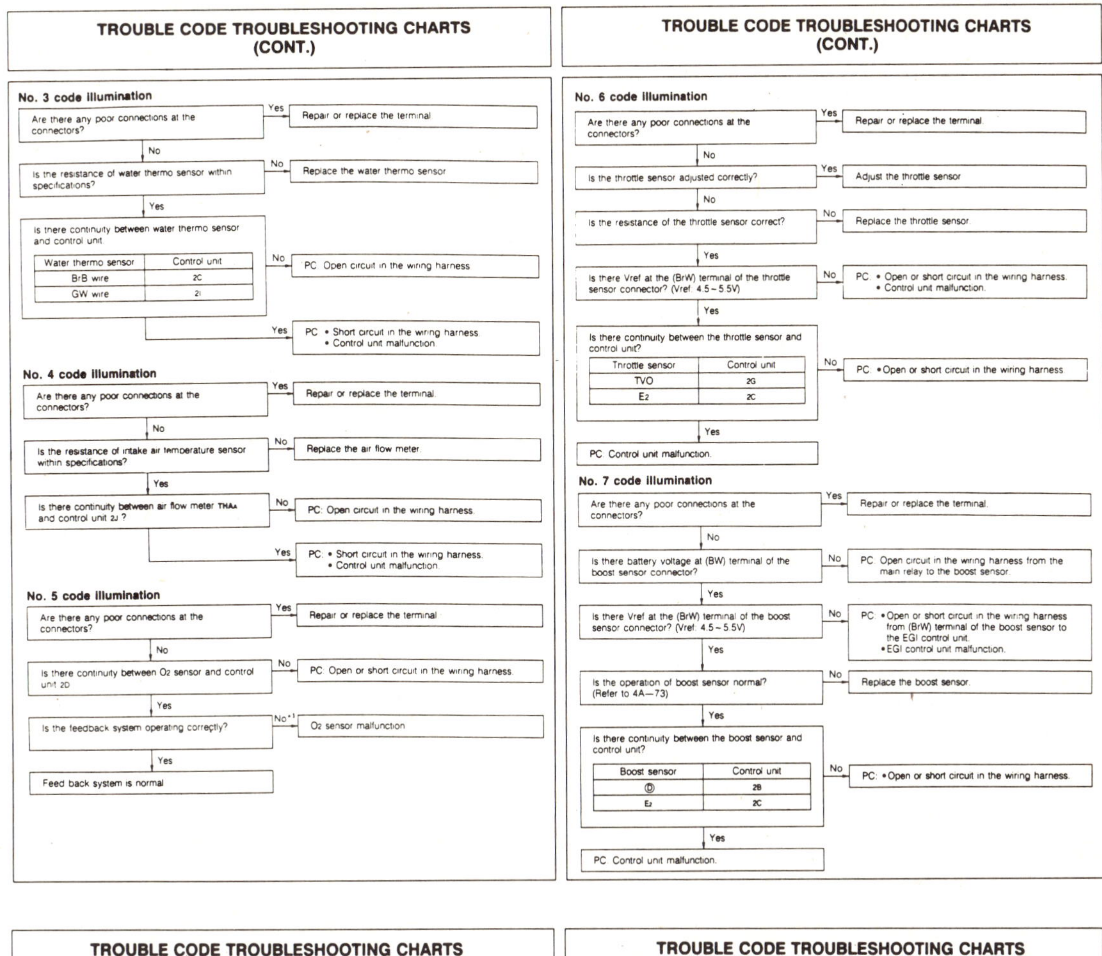
TROUBLE CODE TROUBLESHOOTING CHARTS (CONT.)
No. 3 code illumination
Are there any poor connections at the connectors?
Yes
Repair or replace the terminal
No
Is the resistance of water thermo sensor within specifications?
No
Replace the water thermo sensor
Yes
Is there continuity between water thermo sensor and control unit.
Water thermo sensor
Control unit
BrB wire
2C
GW wire
2I
No
PC: Open circuit in the wiring harness
Yes
PC: • Short circuit in the wiring harness. • Control unit malfunction.
No. 4 code illumination
Are there any poor connections at the connectors?
Yes
Repair or replace the terminal.
No
Is the resistance of intake air temperature sensor within specifications?
No
Replace the air flow meter.
Yes
Is there continuity between air flow meter THAA and control unit 2J ?
No
PC: Open circuit in the wiring harness.
Yes
PC: • Short circuit in the wiring harness. • Control unit malfunction.
No. 5 code illumination
Are there any poor connections at the connectors?
Yes
Repair or replace the terminal.
No
Is there continuity between O2 sensor and control unit 2D
No
PC: Open or short circuit in the wiring harness.
Yes
Is the feedback system operating correctly?
No*1
O2 sensor malfunction
Yes
Feed back system is normal
TROUBLE CODE TROUBLESHOOTING CHARTS (CONT.)
No. 6 code illumination
Are there any poor connections at the connectors?
Yes
Repair or replace the terminal
No
Is the throttle sensor adjusted correctly?
Yes
Adjust the throttle sensor
No
Is the resistance of the throttle sensor correct?
No
Replace the throttle sensor.
Yes
Is there Vref at the (BrW) terminal of the throttle sensor connector? (Vref: 4.5~5.5V)
No
PC: • Open or short circuit in the wiring harness. • Control unit malfunction.
Yes
Is there continuity between the throttle sensor and control unit?
Throttle sensor
Control unit
TVO
2G
E2
2C
No
PC: • Open or short circuit in the wiring harness.
Yes
PC: Control unit malfunction.
No. 7 code illumination
Are there any poor connections at the connectors?
Yes
Repair or replace the terminal.
No
Is there battery voltage at (BW) terminal of the boost sensor connector?
No
PC: Open circuit in the wiring harness from the main relay to the boost sensor.
Yes
Is there Vref at the (BrW) terminal of the boost sensor connector? (Vref: 4.5~5.5V)
No
PC: • Open or short circuit in the wiring harness from (BrW) terminal of the boost sensor to the EGI control unit. • EGI control unit malfunction.
Yes
Is the operation of boost sensor normal? (Refer to 4A—73)
No
Replace the boost sensor.
Yes
Is there continuity between the boost sensor and control unit?
Boost sensor
Control unit
Ⓓ
2B
E2
2C
No
PC: • Open or short circuit in the wiring harness.
Yes
PC: Control unit malfunction.

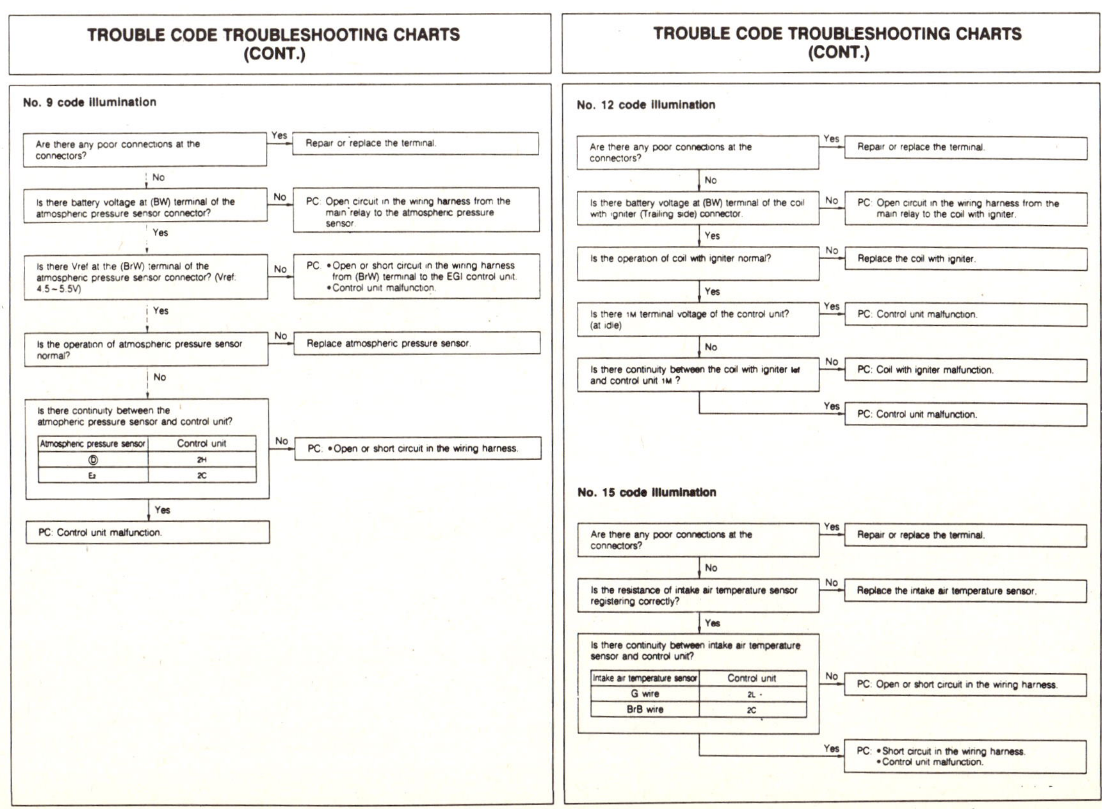
TROUBLE CODE TROUBLESHOOTING CHARTS (CONT.)
No. 9 code illumination
Are there any poor connections at the connectors?
Yes
Repair or replace the terminal.
No
Is there battery voltage at (BW) terminal of the atmospheric pressure sensor connector?
No
PC: Open circuit in the wiring harness from the main relay to the atmospheric pressure sensor.
Yes
Is there Vref at the (BrW) terminal of the atmospheric pressure sensor connector? (Vref: 4.5~5.5V)
No
PC: • Open or short circuit in the wiring harness from (BrW) terminal to the EGI control unit. • Control unit malfunction.
Yes
Is the operation of atmospheric pressure sensor normal?
No
Replace atmospheric pressure sensor.
No
Is there continuity between the atmospheric pressure sensor and control unit?
Atmospheric pressure sensor
Control unit
Ⓓ
2H
E2
2C
No
PC: • Open or short circuit in the wiring harness.
Yes
PC: Control unit malfunction.
TROUBLE CODE TROUBLESHOOTING CHARTS (CONT.)
No. 12 code illumination
Are there any poor connections at the connectors?
Yes
Repair or replace the terminal.
No
Is there battery voltage at (BW) terminal of the coil with igniter (Trailing side) connector.
No
PC: Open circuit in the wiring harness from the main relay to the coil with igniter.
Yes
Is the operation of coil with igniter normal?
No
Replace the coil with igniter.
Yes
Is there 1M terminal voltage of the control unit? (at idle)
Yes
PC: Control unit malfunction.
No
Is there continuity between the coil with igniter IGt and control unit 1M ?
No
PC: Coil with igniter malfunction.
Yes
PC: Control unit malfunction.
No. 15 code illumination
Are there any poor connections at the connectors?
Yes
Repair or replace the terminal.
No
Is the resistance of intake air temperature sensor registering correctly?
No
Replace the intake air temperature sensor.
Yes
Is there continuity between intake air temperature sensor and control unit?
Intake air temperature sensor
Control unit
G wire
2L
BrB wire
2C
No
PC: Open or short circuit in the wiring harness.
Yes
PC: • Short circuit in the wiring harness. • Control unit malfunction.

### VACUUM HOSE ROUTING – 1989–90 RX-7 TURBO

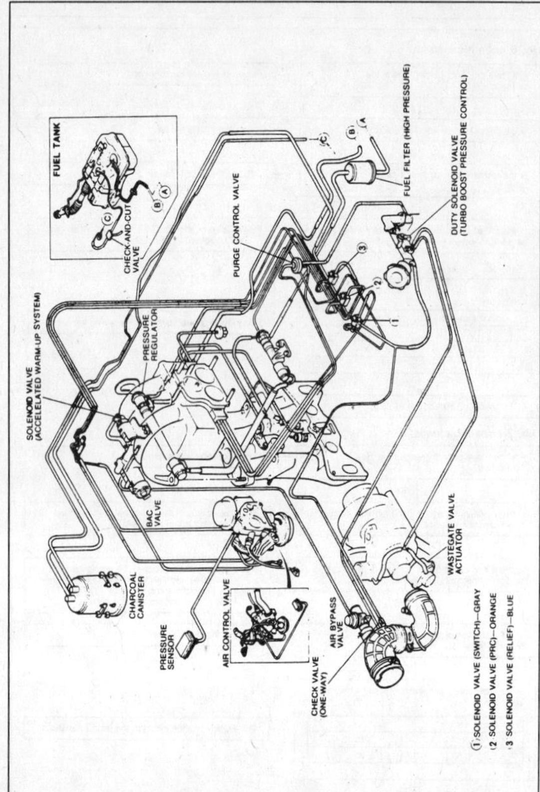


### EXPLODED VIEW OF FUEL COMPONENTS 1989–90 RX-7 TURBO

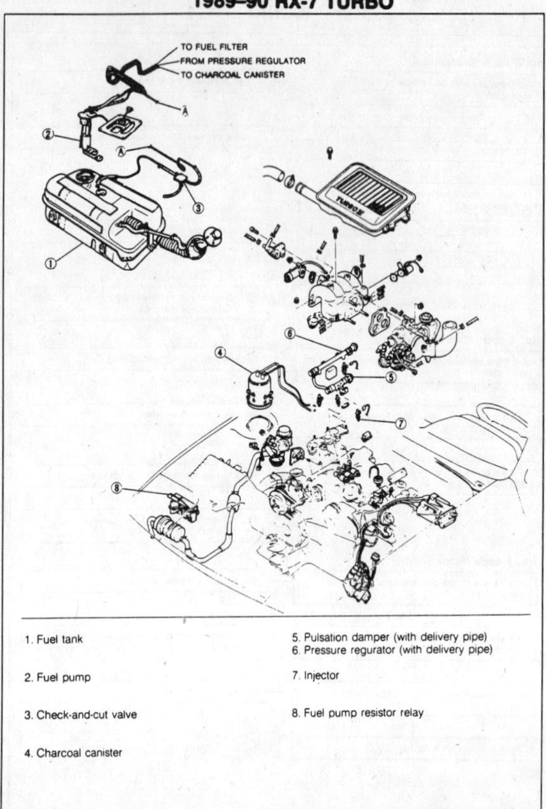


### EXPLODED VIEW OF CONTROL UNIT OUTPUT COMPONENTS – 1989–90 RX-7 TURBO

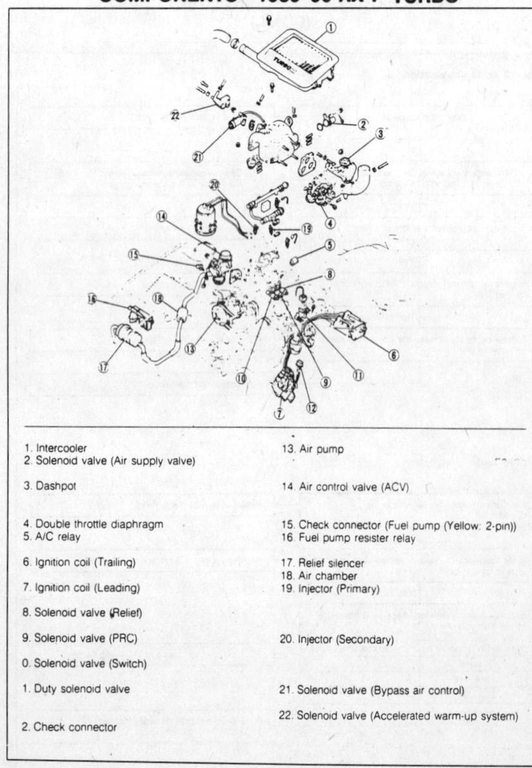


### EXPLODED VIEW OF CONTROL UNIT INPUT COMPONENTS – 1989–90 RX-7 TURBO

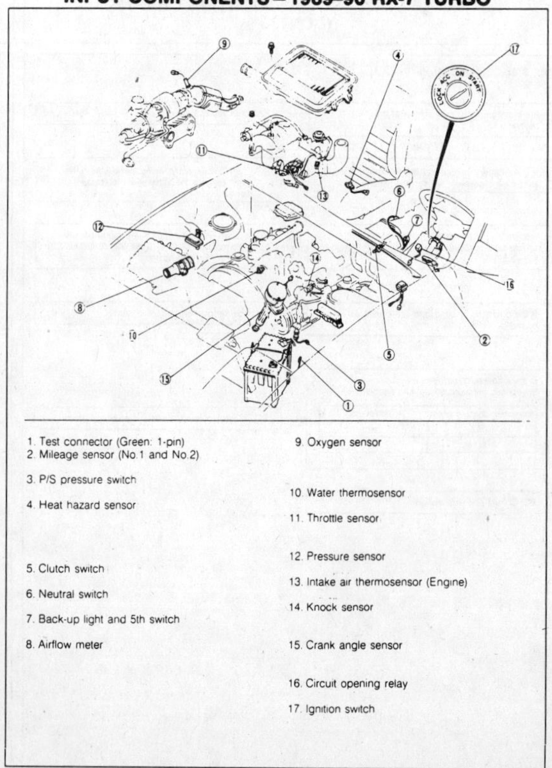

## COMPONENT RELATIONSHIP CHART 1989–90 R7 TURBO

| INPUT DEVICES \ OUTPUT DEVICES | FUEL INJECTION AMOUNT | FUEL INJECTION TIMING | FUEL PUMP RESISTOR RELAY | A/C RELAY | PORT AIR SOLENOID VALVE | SPLIT AIR SOLENOID VALVE | SOLENOID VALVE (SWITCH) | SOLENOID VALVE (RELIEF) | SOLENOID VALVE (AWS AND ASV) | SOLENOID VALVE (PRESSURE REGULATOR) | SOLENOID VALVE (BYPASS AIR CONTROL) | DUTY SOLENOID VALVE (TURBO BOOST PRESSURE) | IGNITER (TRAILING IGNITION TIMING) | IGNITER (LEADING IGNITION TIMING) |
|---|---|---|---|---|---|---|---|---|---|---|---|---|---|---|
| TEST CONNECTOR | ○ | × | × | × | × | × | × | × | × | × | ○ | × | ○ | ○ |
| IGNITION SWITCH | ○ | × | ○ | × | × | × | ○ | ○ | ○ | ○ | ○ | × | ○ | ○ |
| POWER STEERING PRESSURE SWITCH | × | × | × | × | × | × | × | × | × | × | ○ | × | × | × |
| A/C SWITCH | ○ | × | × | ○ | × | × | × | × | × | × | ○ | × | ○ | ○ |
| FOG LIGHT SWITCH (IF EQUIPPED) | ○ | × | × | × | × | × | × | × | × | × | ○ | × | ○ | ○ |
| HEAD LIGHT SWITCH | ○ | × | × | × | × | × | × | × | × | × | ○ | × | ○ | ○ |
| BLOWER SWITCH | ○ | × | × | × | × | × | × | × | × | × | ○ | × | ○ | ○ |
| REAR DEFROSTER SWITCH | ○ | × | × | × | × | × | × | × | × | × | ○ | × | ○ | ○ |
| CLUTCH SWITCH | ○ | × | ○ | ○ | × | × | ○ | ○ | ○ | ○ | ○ | × | ○ | ○ |
| NEUTRAL SWITCH | ○ | × | ○ | ○ | × | × | ○ | ○ | ○ | ○ | ○ | × | ○ | ○ |
| BACK-UP LIGHT AND 5TH SWITCH | ○ | × | × | × | × | ○ | ○ | ○ | × | × | ○ | × | ○ | ○ |
| MILEAGE SENSOR No.1 | ○ | × | × | × | ○ | × | × | × | × | × | × | × | ○ | ○ |
| MILEAGE SENSOR No.2 | ○ | × | × | × | × | × | × | × | × | × | × | ○ | ○ | ○ |
| HEAT HAZARD SENSOR | × | × | × | × | × | × | × | ○ | × | × | × | × | × | × |
| OXYGEN SENSOR | ○ | × | × | × | × | × | × | × | × | × | × | × | × | × |
| THROTTLE SENSOR (NARROW RANGE) | ○ | × | ○ | × | × | × | ○ | ○ | × | ○ | ○ | × | ○ | ○ |
| THROTTLE SENSOR (FULL RANGE) | ○ | × | × | ○ | × | × | × | ○ | ○ | × | × | ○ | ○ | ○ |
| PRESSURE SENSOR | ○ | × | × | × | ○ | × | ○ | ○ | × | × | × | ○ | ○ | ○ |
| KNOCK SENSOR | × | × | × | × | × | × | × | × | × | × | × | × | ○ | ○ |
| CRANK ANGLE SENSOR (NE) | ○ | ○ | ○ | ○ | ○ | × | ○ | ○ | ○ | × | ○ | ○ | ○ | ○ |
| CRANK ANGLE SENSOR (G+, G−) | ○ | ○ | ○ | × | × | × | × | × | × | ○ | × | × | ○ | ○ |
| WATER THERMOSENSOR | ○ | × | ○ | × | × | × | ○ | ○ | ○ | ○ | ○ | × | ○ | ○ |
| INTAKE AIR THERMOSENSOR (AFM) | ○ | × | ○ | × | × | × | × | × | × | × | × | × | × | × |
| INTAKE AIR THERMOSENSOR (ENGINE) | ○ | × | ○ | × | × | × | × | × | × | ○ | ○ | ○ | × | × |
| AIRFLOW METER (AFM) | ○ | × | ○ | × | × | × | × | × | × | × | × | × | ○ | ○ |

## OUTPUT COMPONENTS AND ENGINE CONDITION CHART – 1989–90 R7 TURBO

| OUPUT DEVICES \ ENGINE CONITION | | CRANKING (WHEN ENGINE IS COLD) | IDLE (DURING WARM-UP) | MEDIUM LOAD COLD | MEDIUM LOAD WARM | ACCELERATION | HEAVY LOAD | DECELERATION (THROTTLE VALVE FULLY CLOSED) | IDLE (WHEN ENGINE IS WARM) | IGNITION: ON (NOT ENGINE RUN) | REMARKS |
|---|---|---|---|---|---|---|---|---|---|---|---|
| FUEL INJECTION AMOUNT | PRIMARY INJECTOR | ← Rich → | | | Rich and Lean | ← Rich → | | Fuel cut* | Rich | OFF | Rear side; Above 8,500 rpm; Fuel cut |
| | SECONDARY INJECTOR | ← OFF → | | | | | | ← OFF → | | | |
| FUEL PUMP RESISTOR RELAY | | 12V drive | ← 9–10V drive → | | | ← 12V drive → | | ← 9–10V drive → | | OFF | Hot start (at idle) 12V drive |
| A/C RELAY | | ← ON → | | | | *1OFF | ← ON → | | | | *1Only 7 sec |
| PORT AIR SOLENOID VALVE | | ← Above 20,000 mile (34,000 km), ON Below 20,000 mile (34,000 km), OFF → | | | | | OFF | ← OFF (Above 2,000 mile (34,000 km) ON) → | | | |
| SPLIT AIR SOLENOID VALVE | | ← Back-up light and 5th switch ON or high altitude, ON → | | | | | | | | | |
| SOLENOID VALVE (SWITCH) | | ← OFF (Port air) → | | | ← ON (Split air) → | | | ← OFF (Port air) → | | | |
| SOLENOID VALVE (RELIEF) | | ← ON (Air injection) → | | OFF (Relief) | ON (Air injection) | ← OFF (Relief) → | | ← ON (Air injection) → | | ON | |
| SOLENOID VALVE (AWS AND ASV) | | ON | For 17 sec after start, ON | ← OFF → | | | | | | | |
| SOLENOID VALVE (PRESSURE REGULATOR) | | ← OFF → | | | | | | | After start; *2ON | ON | *2For 50 sec –90 sec. |
| SOLENOID VALVE (BYPASS AIR CONTROL) | | Large quantity of air | Feed back duty | ← Fixed duty → | | | | | Feed back duty | Fixed duty | Test connector (Green 1-pin) ground, Stop the feedback control |
| DUTY SOLENOID VALVE (TURBO BOOST PRESSURE) | | ← Feed back duty → | | | | | | | | | |
| IGNITER (TRAILING IGNITION TIMING) | | BTDC 5° | Retard | Advanced (If occuring knocking, retards ignition timing max 5°) | | | | Retard | ATDC 20° | OFF | Test connector (Green 1-pin) ON (Only ignition ON (engine does not run)) inspect the knock control system |
| IGNITER (LEADING IGNITION TIMING) | | | | | | | | | ATDC 5° | | |

## SPECIFICATIONS CHART – 1989–90 R7 TURBO

| Item | | | Model: 13B Turbocharged engine |
|---|---|---|---|
| Idle speed (Test connector grounded) | | rpm | 750 ± 25 |
| Air cleaner | Element type | | Long life wet |
| Throttle body | Type | | Horizontal draft (2 stage-3 barrel) |
| | Throat diameter | Primary mm (in) | 45 (1.772) |
| | | Secondary mm (in) | 45 (1.772) x 2 |
| | Water thermovalve | Operation temp. °C (°F) | 55–65 (131–149) or more |
| Dashpot | Adjustment (Throttle sensor (narrow range) resistance Ⓐ–Ⓑ) | kΩ | 1.8–3.8 |
| Turbocharger | Type | | Water cooled |
| | Boost pressure | kPa (kg/cm², psi) | 57.0 (0.58, 8.25) |
| | Lubrication | | Engine oil |
| Waste gate valve | | | Incorporated with turbocharger |
| Fuel tank | Capacity | liters (US gal, Imp gal) | 70 (18.5, 15.4) |
| Fuel filter | Type | Low pressure | Nylon 6 (164 & 45 mesh) |
| | | High pressure | Filter paper |
| Pressure regulator | Type | | Diaphragm |
| | Regulated pressure | kPa (kg/cm², psi) | 235–275 (2.4–2.8, 34.1–39.8) |
| Fuel pump | Type | | Impeller (in-tank) |
| | Output pressure | kPa (kg/cm², psi) | 490–637 (5.0–6.5, 71.1–92.4) |
| Injector (Primary and Secondary) | Drive | | Voltage drive |
| | Injection volume | cc (cu in)/15 sec | 133–142 (8.1–8.7) |
| Heat hazard sensor | Operation temperature | °C (°F) | 105–115 (221–239) |
| Main silencer | Capacity | cc (cu in) | 12,000 (732) x 2 |
| Ignition timing (Test connector grounded) | | | Leading 5° ± 1° ATDC<br>Trailing 20° ± 2° ATDC |
| Distribution | Type | | Control unit |
| Spark advance | Type | | Control unit |
| Idle up system | A/C | rpm | 800 (875, 1990 models) |
| Anti-afterburn valve | Operating time | sec | 1.60–2.20 |
| Intercooler | Type | | Air cooled |

## COMPONENT DESCRIPTION CHART 1989–90 R7 TURBO

| Component | Function | Remarks |
|---|---|---|
| **Accelerated Warm-up System (AWS)** | Supplies bypass air into dynamic chamber | Controlled by duty signal from control unit |
| **Anti-afterburn Valve** | Supplies fresh air into rear port during deceleration | Included in air control valve |
| **Air Bleed Socket** | Supplies fresh air into primary injector hole | |
| **Air Bypass Valve** | Reduces sound of intake air from turbocharger relieved through air flow meter during deceleration | |
| **Air Cleaner** | Filters air into throttle chamber | |
| **Air Control Valve** | Directs air to one of three locations: exhaust port, main converter, or relief air silencer | Consists of 3 valves;<br>Relief valve<br>Switching valve<br>Anti-afterburn valve |
| **Airflow Meter** | Detects amount of intake air; sends signal to control unit | |
| **Atmospheric Pressure Sensor** | Detects atmospheric pressure; sends signal to control unit | Built in ECU |
| **Air Pump** | Supplies secondary air to air control valve | |
| **Catalytic Converter** | Reduces HC, CO, and NOx | |
| **Check Valve** | Supplies the blowby gas and evaporative emission gas into the turbocharger when the intake manifold vacuum becomes positive pressure | For evaporative emission control system |
| **Charcoal Canister** | Stores fuel tank fumes when engine stops | Vented to atmosphere through charcoal and filter |
| **Check-and-cut Valve** | Controls pressure in fuel tank | |
| **Coil with Igniter** | Generates high voltage | Leading: ignite simultaneously<br>Trailing: ignite individually |
| **Crank Angle Sensor** | Detects eccentric shaft angle at 30° intervals and front rotor position; sends signal to control unit | |
| **Dashpot** | Gradually closes throttle valve during deceleration | |
| **Double Throttle System** | Gradually opens the No.2 secondary throttle valve when No.1 secondary throttle valve suddenly opens | |
| **Duty Solenoid Valve (Turbo boost pressure control)** | Controls turbo boost pressure | |
| **Dynamic Chamber** | Connects front and rear ports | Primary and secondary separated |
| **Engine Control Unit** | Detects the following:<br>1. Engine speed<br>2. Amount of intake air<br>3. Engine coolant temperature<br>4. Throttle opening<br>5. Intake manifold pressure<br>6. $O_2$ concentration<br>7. In-gear condition<br>8. Intake air temperature<br>9. Floor temperature<br>10. A/C operation<br>11. Cranking signal<br>12. Atmospheric pressure<br>13. Knocking signal<br>14. Initial set signal<br>15. Position of transmission gear<br>16. Metering oil pump (MOP) position signal<br>17. Electric load (E/L) condition<br>18. Mileage<br>Controls operation of the following:<br>1. Fuel injection system<br>2. Ignition control system | Crank angle sensor<br>Airflow meter<br>Water thermosensor<br>Throttle sensors<br>Pressure sensor<br>Oxygen ($O_2$) sensor<br>Neutral switch and clutch switch<br>Intake air thermosensor<br>Heat hazard sensor<br>A/C switch<br>Starter switch<br>Atmospheric pressure sensor<br>Knock sensor and knock control unit<br>Test connector (Green: 1-pin)<br>Back-up light and 5th switch<br>MOP position sensor<br>Headlight switch, Blower switch, Rear defroster switch, Fog light switch<br>Mileage sensor (No.1 and No.2) |

## COMPONENT DESCRIPTION CHART (CONT.)
## 1989–90 R7 TURBO

| Component | Function | Remarks |
|---|---|---|
| **Engine Control Unit (Cont'd)** | 3. ISC system<br>4. Pressure regulator control system<br>5. Secondary air injection control system<br>6. Tubo boost pressure control<br>7. Fuel pump control system | |
| **Fast Idle System** | Opens primary throttle valve slightly at idle | Only when engine is cold |
| **Fuel Filter** | Filters fuel | |
| **Fuel Pump** | Provides fuel to injectors | Operates while engine is running<br>Installed in fuel tank |
| **Fuel Pump Resistor Relay** | Controls voltage for fuel pump | |
| **Heat Hazard Sensor** | Detects floor temperature; sends signal to control unit | Heat hazard sensor turned ON: relieves secondary air |
| **Injector** | Injects fuel into intermediate housing and secondary intake manifold | Controlled by signals from control unit |
| **Intake Air Thermosensor** | Detects intake air temperature and temperature into the engine; sends signal to control unit | Located on the airflow meter and intake air pipe<br>Thermistor |
| **Intercooler** | Prevents increase of intake air temperature | Air cooled type |
| **Knock Sensor** | Detects engine knock; sends signal to engine control unit | |
| **Mileage Sensor (No.1 and No.2)** | Detects vehicle mileage; sends signal to control unit | Above 20,000 miles (34,000 km): mileage sensor No.1 is ON<br>Below 600 miles (1,000 km): mileage sensor No.2 is ON |
| **Oxygen Sensor** | Detects $O_2$ concentration; sends signal to control unit | Zirconia ceramic and platinum coating |
| **Pressure Regulator** | Adjusts fuel pressure supplied to injectors | |
| **Pulsation Damper** | Absorbs fuel pulsations | |
| **Pressure sensor** | Detects intake manifold pressure, sends signal to control unit | |
| **Purge Control Valve** | Regulates evaporative fumes from fuel tank and canister to intake manifold | |
| **Solenoid Valve (Bypass air control)** | Supplies bypass air into dynamic chamber | Controlled by duty signal from control unit |
| **Solenoid Valve (PRC)** | Shuts vacuum passage between dynamic chamber and pressure regulator | Only when engine hot<br>Orange |
| **Solenoid Valve (Switch)** | Controls switching valve of the air control valve | Gray |
| **Solenoid Valve (Relief)** | Controls relief valve | Blue |
| **Solenoid Valve (AWS)** | Controls accelerated warm-up system (AWS) | |
| **Solenoid Valve (Air supply valve)** | Supplies bypass air into dynamic chamber | During AWS operation |
| **Split Air Solenoid Valve** | Control split air | |
| **Test Connector (Green: 1-pin)** | Sends initial set signal to control unit | During adjustment of idle speed, ignition timing and knock system; connector grounded |
| **Throttle Body** | Controls intake air quantity | |
| **Throttle Sensors** | Detects primary throttle valve opening angle, sends signal to control unit | |
| **Turbocharger** | Pressurizes intake air utilizing exhaust gas flow | Twin-scroll turbocharger |
| **Wastegate Valve** | Controls amount of exhaust gas bypassing exhaust turbine to control intake air boost pressure | |
| **Water Thermosensor** | Detects engine coolant temperature; sends signal to control unit | Thermistor |

## TRPOUBLESHOOTING DIAGNOSTIC CHART
## 1989–90 R7 TURBO

This troubleshooting guide shows the malfunction code numbers retrieved by the **SST** and symptoms of various failures. Perform troubleshooting as described below.

**TROUBLESHOOTING PROCEDURE**
**Troubleshooting With SST**
Troubleshooting with the **SST** (Self-Diagnosis Checker 49 H018 9A1) is done to quickly determine what system or unit may be at fault.

**1st:** Check input sensors and output devices with the **SST**.
**2nd:** Check other switches with the **SST**.
**3rd:** Check the following items:
- Electrical system : Battery conditions, fuses
- Ignition system : Ignition spark, ignition timing (with test connector grounded)
- Fuel system : Fuel level, fuel leakage, fuel filter, idle speed (with test connector grounded)
- Intake air system : Air cleaner element, vacuum or air leakage, vacuum hose routing, accelerator cable
- Engine : Compression, overheating
- Others : Clutch slippage, brake dragging

**4th:** Check fuel and emission control systems.

**Malfunction Code No.**

| Code No. | Input device |
|---|---|
| 01 | Ignition coil (Trailing side) |
| 02 | Crank angle sensor (Ne-signal) |
| 03 | Crank angle sensor (G-signal) |
| 05 | Knock sensor |
| 08 | Airflow meter (AFM) |
| 09 | Water thermosensor |
| 10 | Intake air thermosensor (AFM) |
| 11 | Intake air thermosensor (Engine) |
| 12 | Throttle sensor (Full range) |
| 13 | Pressure sensor |
| 14 | Atmospheric pressure sensor (Built in ECU) |
| 15 | Oxygen sensor |
| 17 | Feedback system |
| 18 | Throttle sensor (Narrow range) |
| 20 | Metering oil pump position sensor |
| 27 | Metering oil pump |
| 37 | Metering oil pump |

| Code No. | Output device |
|---|---|
| 25 | Solenoid valve (Pressure regulator control (PRC)) |
| 26 | Step motor (Metering oil pump) |
| 30 | Split air solenoid valve |
| 31 | Solenoid valve (Relief) |
| 32 | Solenoid valve (Switching) |
| 33 | Port air solenoid valve |
| 34 | Solenoid valve (Bypass air control (BAC)) |
| 38 | Solenoid valve (Accelerated warm-up system (AWS) and Air supply valve (ASV)) |
| 42 | Solenoid valve (Waste gate control) |
| 51 | Fuel pump resistor relay |
| 71 | Injector (Front secondary) |
| 73 | Injector (Rear secondary) |

## SYSTEM TROUBLESHOOTING CHART
## 1989–90 R7 TURBO

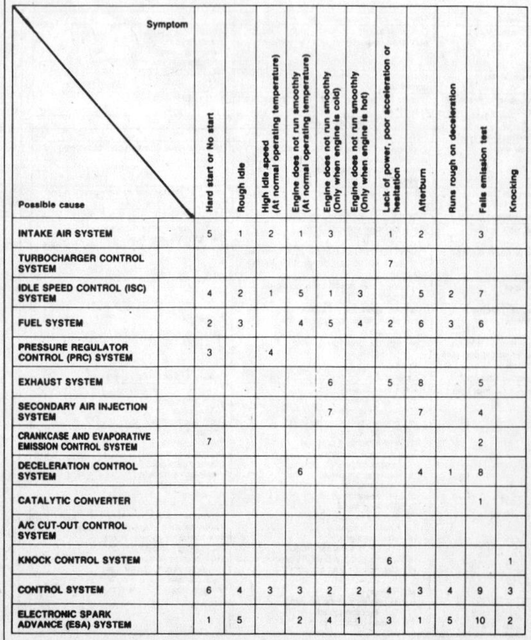

**Troubleshooting of Each System**
The troubleshooting guide lists the most likely causes to a given symptom. After finding the systems to check, refer to the pages shown for detailed guides.
The numbers of the list show the priorities of inspections from the most probable to that with the lowest probability.
These were determined on the following basis:
- Ease of inspection
- Most possible system
- Most possible point in system

| Possible cause \ Symptom | Hard start or No start | Rough idle | High idle speed (At normal operating temperature) | Engine does not run smoothly (At normal operating temperature) | Engine does not run smoothly (Only when engine is cold) | Engine does not run smoothly (Only when engine is hot) | Lack of power, poor acceleration or hesitation | Afterburn | Runs rough on deceleration | Fails emission test | Knocking |
|---|---|---|---|---|---|---|---|---|---|---|---|
| INTAKE AIR SYSTEM | 5 | 1 | 2 | 1 | 3 | | 1 | 2 | | 3 | |
| TURBOCHARGER CONTROL SYSTEM | | | | | | | 7 | | | | |
| IDLE SPEED CONTROL (ISC) SYSTEM | 4 | 2 | 1 | 5 | 1 | 3 | | 5 | 2 | 7 | |
| FUEL SYSTEM | 2 | 3 | | 4 | 5 | 4 | 2 | 6 | 3 | 6 | |
| PRESSURE REGULATOR CONTROL (PRC) SYSTEM | 3 | | 4 | | | | | | | | |
| EXHAUST SYSTEM | | | | | 6 | | 5 | 8 | | 5 | |
| SECONDARY AIR INJECTION SYSTEM | | | | | 7 | | | 7 | | 4 | |
| CRANKCASE AND EVAPORATIVE EMISSION CONTROL SYSTEM | 7 | | | | | | | | | 2 | |
| DECELERATION CONTROL SYSTEM | | | | 6 | | | | 4 | 1 | 8 | |
| CATALYTIC CONVERTER | | | | | | | | | | 1 | |
| A/C CUT-OUT CONTROL SYSTEM | | | | | | | | | | | |
| KNOCK CONTROL SYSTEM | | | | | | | 6 | | | | 1 |
| CONTROL SYSTEM | 6 | 4 | 3 | 3 | 2 | 2 | 4 | 3 | 4 | 9 | 3 |
| ELECTRONIC SPARK ADVANCE (ESA) SYSTEM | 1 | 5 | | 2 | 4 | 1 | 3 | 1 | 5 | 10 | 2 |

## MALFUNCTION INDIACTOR LAMP (MIL) OPERATION
## 1989–90 R7 TURBO

**Troubleshooting**
**Principle of code cycle**
Malfunction codes are determined as below by use of the MIL and Self-Diagnosis Checker.

**1. Malfunction code cycle break**
The time between malfunction code cycles is 4.0 sec. (the time the MIL and checker buzzer are off).

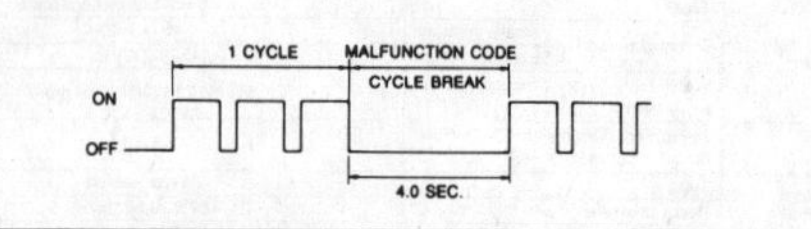


**2. Second digit of malfunction code (ones position)**
The digit in the ones position of the malfunction code represents the number of times the MIL and buzzer are on 0.4 sec. during one cycle.

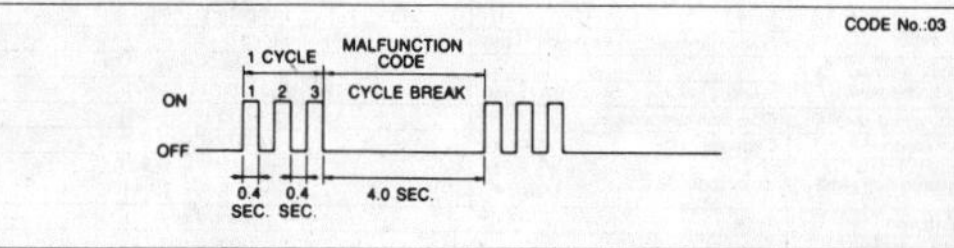


**3. First digit of malfunction code (tens position)**
The digit in the tens position of the malfunction code represents the number of times the MIL and buzzer are on 1.2 sec. during one cycle.

The MIL and buzzer are off for 1.6 sec. between the long and short pulses.

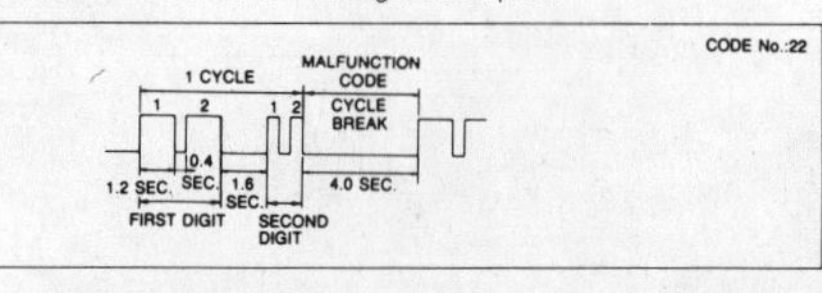

## TROUBLE CODE CHART—1989-90 R7 TURBO

**Code number**

**Caution**
**a) If there is more than one failure present, the lowest number malfunction code is displayed first, the subsequent malfunction codes appear in order.**
**b) After repairing all failures, turn the ignition switch OFF, disconnect the negative battery cable for at least 5 seconds to erase the malfunction code memory.**

**Input devices**

| Code No. | Input devices | Malfunction | Fail-safe function | Output signal pattern (Self-Diagnosis Checker or MIL) |
|---|---|---|---|---|
| 01 | Ignition coil (Trailing side) | Malfunction of spark plug, broken wire, short circuit | Trailing-side ignition pulse cut | |
| 02 | Crank angle sensor (Ne signal) | Broken wire, short circuit | Fuel injection and ignition cut | |
| 03 | Crank angle sensor (G signal) | Broken wire, short circuit | Fuel injection and ignition cut | |
| 05 | Knock sensor | Broken wire, short circuit | Ignition timing fixed | |
| 08 | Airflow meter (AFM) | Broken wire, short circuit | Basic fuel injection amount and ignition timing fixed | |
| 09 | Water thermosensor | Broken wire, short circuit | Coolant temp. input fixed at 80°C (176°F) | |
| 10 | Intake air thermosensor (AFM) | Broken wire, short circuit | Intake air temp. input fixed at 20°C (68°F) | |
| 11 | Intake air thermosensor (Engine) | Broken wire, short circuit | Intake air temp. input fixed at 20°C (68°F) | |
| 12 | Throttle sensor (Full range) | Broken wire, short circuit | Throttle valve opening angle input signal fixed at 20% open | |
| 13 | Pressure sensor (Intake manifold pressure) | Broken wire, short circuit | Intake manifold pressure input signal fixed at 760 mmHg (29.9 inHg) | |
| 14 | Atmospheric pressure sensor (ATP) | Malfunctioning ECU | Atmospheric pressure input signal fixed at 760 mmHg (29.9 inHg) | |
| 15 | Oxygen sensor | Oxygen sensor output remains below 0.55V 80 sec. after F/B system operation begining | Feedback system canceled (For EGI) | |
| 17 | Feedback system | Oxygen sensor output remains 0.55V 10 sec. after F/B system operation begining | Feedback system canceled (For EGI) | |
| 18 | Throttle sensor (Narrow range) | Broken wire, short circuit | Throttle valve opening angle input signal fixed at full open | |
| 20 | Metering oil pump position sensor | Broken wire, short circuit | MOP fixed smallest open Basic fuel injection amount and ignition timing fixed | |
| 27 | Metering oil pump (MOP) | Malfunctioning MOP, step motors, broken wire, short circuit or malfunctioning ECU | MOP fixed smallest open Basic fuel injection amount and ignition timing fixed | |
| 37 | Metering oil pump (MOP) | Malfunction MOP, step motors, broken wire, short circuit, malfunctioning ECU, alternator or battery | Basic fuel injection amount and ignition timing fixed | |

**Output devices**

| Code No. | Output devices | Output signal pattern (Self-Diagnosis Checker or MIL) |
|---|---|---|
| 25 | Solenoid valve (Pressure regulator control (PRC)) | |
| 26 | Step motor (Metering oil pump) | |
| 30 | Split air solenoid valve | |
| 31 | Solenid valve (Relief) | |
| 32 | Solenoid valve (Switch) | |
| 33 | Port air solenoid valve | |
| 34 | Solenoid valve (Bypass air control (BAC)) | |
| 38 | Solenoid valve (Accelerated warm-up system (AWS) and air supply valve (ASV)) | |
| 42 | Duty solenoid (Turbo boost pressure control) | |
| 51 | Fuel pump resistor relay | |
| 71 | Injector (Front secondary) | |
| 73 | Injector (Rear secondary) | |

## TROUBLE CODE DIAGNOSTIC CHART—1989-90 R7 TURBO

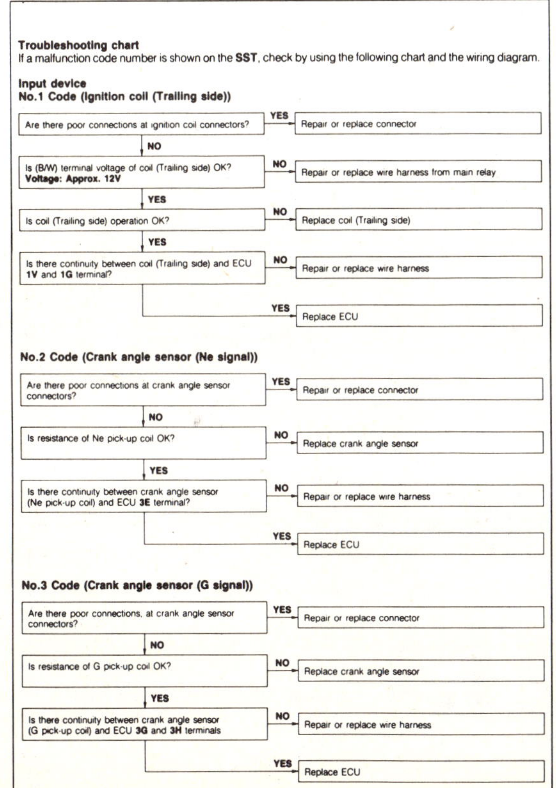

Troubleshooting chart
If a malfunction code number is shown on the SST, check by using the following chart and the wiring diagram.
Input device
No.1 Code (Ignition coil (Trailing side))
Are there poor connections at ignition coil connectors?
YES
Repair or replace connector
NO
Is (B/W) terminal voltage of coil (Trailing side) OK?
Voltage: Approx. 12V
NO
Repair or replace wire harness from main relay
YES
Is coil (Trailing side) operation OK?
NO
Replace coil (Trailing side)
YES
Is there continuity between coil (Trailing side) and ECU 1V and 1G terminal?
NO
Repair or replace wire harness
YES
Replace ECU
No.2 Code (Crank angle sensor (Ne signal))
Are there poor connections at crank angle sensor connectors?
YES
Repair or replace connector
NO
Is resistance of Ne pick-up coil OK?
NO
Replace crank angle sensor
YES
Is there continuity between crank angle sensor (Ne pick-up coil) and ECU 3E terminal?
NO
Repair or replace wire harness
YES
Replace ECU
No.3 Code (Crank angle sensor (G signal))
Are there poor connections, at crank angle sensor connectors?
YES
Repair or replace connector
NO
Is resistance of G pick-up coil OK?
NO
Replace crank angle sensor
YES
Is there continuity between crank angle sensor (G pick-up coil) and ECU 3G and 3H terminals
NO
Repair or replace wire harness
YES
Replace ECU


**No.5 Code (Knock sensor)**

Are there poor connections at knock sensor connectors? — YES → Repair or replace connector
NO ↓
Check continuity between knock sensor and ECU 2M terminal — NO → Repair or replace wire harness
YES → Replace knock sensor or ECU

**No.8 Code (Airflow meter)**

Are there poor connections at airflow meter connectors? — YES → Repair or replace connector
NO ↓
Is resistance of airflow meter OK? — NO → Replace airflow meter
YES ↓
Is there continuity between airflow meter and ECU? — NO → Repair or replace wire harness

| Airflow meter | ECU |
|---|---|
| Vs (GY) | 2B |
| $E_1$ (BR/B) | 3C and 3D |
| VREF (BR/W) | 2I |

YES → Replace ECU

**No.10 Code (Intake air thermosensor (AFM))**

Are there poor connections at thermosensor connectors? — YES → Repair or replace connector
NO ↓
Is resistance of intake air thermosensor OK? — NO → Replace airflow meter
YES ↓
Is there continuity between intake air thermosensor and ECU? — NO → Repair or replace wire harness

| Intake air thermosensor | ECU |
|---|---|
| Signal side (G/O) | 2K |
| Ground (BR/B) | 3C |

YES → Replace ECU

## TROUBLE CODE DIAGNOSTIC CHART—1990 R7 TURBO

**No.11 Code (Intake air thermosensor (Engine))**

- Are there poor connections at thermosensor connectors?
  - YES → Repair or replace connector
  - NO ↓
- Is resistance of intake air thermosensor OK?
  - NO → Replace intake air thermosensor
  - YES ↓
- Is there continuity between intake air thermosensor and ECU?

| Intake air thermosensor | ECU |
|---|---|
| Signal side (G) | 2L |
| Ground (BR/B) | 3C |

  - NO → Repair or replace wire harness
  - YES → Replace ECU

**No.12 Code (Throttle sensor (Full range))**

- Are there poor connections at throttle sensor connectors?
  - YES → Repair or replace connector
  - NO ↓
- Is resistance of throttle sensor OK?

| Throttle opening | Specification resistance |
|---|---|
| Idle position | Approx. 1 kΩ |
| Full opening | Approx. 5 kΩ |

  - NO → Replace throttle sensor
  - YES ↓
- Is there continuity between throttle sensor and ECU?

| Throttle sensor | ECU |
|---|---|
| Vref (BR/W) | 2I |
| Ground (BR/B) | 3D |
| Signal (B/G) | 2G |

  - NO → Repair or replace wire harness
  - YES → Replace ECU

## TROUBLE CODE DIAGNOSTIC CHART 1989–90 R7 TURBO

**No.12 Code (Throttle sensor (Full range))**

- Are there poor connections at throttle sensor connectors?
  - YES → Repair or replace connector
  - NO ↓
- Is voltage of ECU 2G terminal OK?

| Throttle opening | Specification voltage |
|---|---|
| Idle position | 0.25—1.25V |
| Full opening | 4.1—4.4V |

  - NO → Adjust or replace throttle sensor
  - YES ↓
- Is there continuity between throttle sensor and ECU?

| Throttle sensor | ECU |
|---|---|
| Vref (BR/W) | 2I |
| Ground (BR/B) | 3D |
| Signal (B/G) | 2G |

  - NO → Repair or replace wire harness
  - YES → Replace ECU

**No.13 Code (Pressure sensor)**

- Are there poor connections at pressure sensor connectors?
  - YES → Repair or replace connector
  - NO ↓
- Is there continuity between pressure sensor and ECU?

| Pressure sensor | ECU |
|---|---|
| Vref (BR/W) | 2I |
| Ground (BR/B) | 3D |
| Signal (G/Y) | 2H |

  - NO → Repair or replace wire harness
  - YES ↓
- Is voltage of pressure sensor OK? When applying 100 mmHg (3.9 inHg) vacuum (Refer to page F2–82)

| Pressure sensor | Voltage |
|---|---|
| Vref (BR/W) | Approx. 5V |
| Ground (BR/B) | Approx. 0V |
| Signal (G/Y) | 2.8—3.2V |

  - NO → Vref or Ground NG: Replace ECU; Signal NG: Replace pressure sensor
  - YES → Replace ECU

## TROUBLE CODE DIAGNOSTIC CHART 1989–90 R7 TURBO

**No.14 Code (Atmospheric pressure (ATP) sensor)**

- Because of the ATP sensor built in ECU, if No.14 Code flash, it's mean malfunction ECU
  - YES → Replace ECU (ATP sensor in ECU)

**No.15 Code (Oxygen sensor)**

**Note**
**If malfunction codes No.15 and 17 are both present, perform the checking procedure for malfunction Code No.17 first.**

- Are there poor connections at connectors?
  - YES → Repair or replace connector
  - NO ↓
- Is oxygen sensor output voltage OK?
  - NO → Replace oxygen sensor
  - YES ↓
- Is oxygen sensor sensitivity OK?
  - NO → Replace oxygen sensor
  - YES ↓
- Is there continuity between oxygen sensor and ECU **2C** terminal?
  - NO → Repair or replace wire harness
  - YES → Replace ECU

**No.18 Code (Throttle sensor (Narrow range))**

- Are there poor connections at throttle sensor connectors?
  - YES → Repair or replace connector
  - NO ↓
- Is voltage of ECU 2F terminal OK?

| Throttle opening | Specification voltage |
|---|---|
| Idle position | 0.75—1.25V |
| Full opening | Approx. 5V |

  - NO → Adjust or replace throttle sensor
  - YES ↓
- Is there continuity between tfrottle sensor and ECU?

| Throttle sensor | ECU |
|---|---|
| Vref (BR/W) | 2I |
| Ground (BR/B) | 3D |
| Signal (G/R) | 2F |

  - NO → Repair or replace wire harness
  - YES → Replace ECU

## TROUBLE CODE DIAGNOSTIC CHART—1990 R7 TURBO

**No.17 Code (Feedback system)**

- Are there poor connections at connectors?
  - YES → Repair or replace connector
  - NO ↓
- Is oxygen sensor sensitivity OK?
  - NO → Possible cause:
    - Air leak in vacuum hoses or intake air system
    - Contaminated oxygen sensor
    - Insufficient fuel injection
  - YES ↓
- Are spark plugs OK?
  - NO → Clean or replace spark plugs
  - YES ↓
- Is there continuity between oxygen sensor and ECU **2C** terminal?
  - NO → Repair or replace wire harness
  - YES → Replace ECU

**No.18 Code (Throttle sensor (Narrow range))**

- Are there poor connections at throttle sensor connectors?
  - YES → Repair or replace connector
  - NO ↓
- Is resistance of throttle sensor OK?

| Throttle opening | Specification resistance |
|---|---|
| Idle position | Approx. 1 kΩ |
| Full opening | Approx. 5 kΩ |

  - NO → Replace throttle sensor
  - YES ↓
- Is there continuity between throttle sensor and ECU?

| Throttle sensor | ECU |
|---|---|
| Vref (BR/W) | 2I |
| Ground (BR/B) | 3D |
| Signal (G/R) | 2F |

  - NO → Repair or replace wire harness
  - YES → Replace ECU

**No.20 Code (Metering oil pump position sensor)**

Troubleshooting of this system in Section D

**No.27 Code (Metering oil pump)**

Troubleshooting of this system in Section D

**No.37 Code (Metering oil pump)**

Troubleshooting of this system in Section D

## TROUBLE CODE DIAGNOSTIC CHART—1989–90 R7 TURBO

**Output device**

**No.25 Code (Solenoid valve (PRC))**

- Are there poor connections at solenoid valve connectors? — YES → Repair or replace connector
- NO ↓ Is there continuity of solenoid valve? — NO → Replace solenoid valve
- YES ↓ Is there continuity between solenoid valve and ECU? — NO → Repair or replace wire harness

| Solenoid valve | ECU |
|---|---|
| Battery (B/W) | 1B |
| Signal side (L/O) | 3M |

- YES → Replace ECU

**No.26 Code (Stepping motor (Metering oil pump))**

Troubleshooting of this system in Section D

**No.30 Code (Split air solenoid valve)**

- Are there poor connections at solenoid valve connectors? — YES → Repair or replace connector
- NO ↓ Is there continuity of solenoid valve? — NO → Replace solenoid valve
- YES ↓ Is there continuity between solenoid valve and ECU? — NO → Repair or replace wire harness

| Solenoid valve | ECU |
|---|---|
| Battery (B/W) | 1B |
| Signal side (L/B) | 3I |

- YES → Replace ECU

**No.31 Code (Solenoid valve (Relief))**

- Are there poor connections at solenoid valve connectors? — YES → Repair or replace connector
- NO ↓ Is there continuity of solenoid valve? — NO → Replace solenoid valve
- YES ↓ Is there continuity between solenoid valve and ECU? — NO → Repair or replace wire harness

| Solenoid valve | ECU |
|---|---|
| Battery (B/W) | 1B |
| Solenoid valve (L/Y) | 2P |

- YES → Replace ECU

**No.32 Code (Solenoid valve (Switch))**

- Are there poor connections at solenoid valve connectors? — YES → Repair or replace connector
- NO ↓ Is there continuity of solenoid valve? — NO → Replace solenoid valve
- YES ↓ Is there continuity between solenoid valve and ECU? — NO → Repair or replace wire harness

| Solenoid valve | ECU |
|---|---|
| Battery (B/W) | 1B |
| Signal side (L/R) | 2Q |

- YES → Replace ECU

**No.33 Code (Port air solenoid valve)**

- Are there poor connections at solenoid valve connectors? — YES → Repair or replace connector
- NO ↓ Is there continuity of solenoid valve? — NO → Replace solenoid valve
- YES ↓ Is there continuity between solenoid valve and ECU? — NO → Repair or replace wire harness

| Solenoid valve | ECU |
|---|---|
| Battery (B/W) | 1B |
| Signal side (L) | 2N |

- YES → Replace ECU

**No.34 Code (Solenoid valve (BAC))**

- Are there poor connections at solenoid valve connectors? — YES → Repair or replace connector
- NO ↓ Is there continuity of solenoid valve? — NO → Replace solenoid valve
- YES ↓ Is there continuity between solenoid valve and ECU? — NO → Repair or replace wire harness

| Solenoid valve | ECU |
|---|---|
| Battery (B/W) | 1B |
| Signal side (L/G) | 3Q |

- YES → Replace ECU

## TROUBLE CODE DIAGNOSTIC CHART—1989–90 R7 TURBO

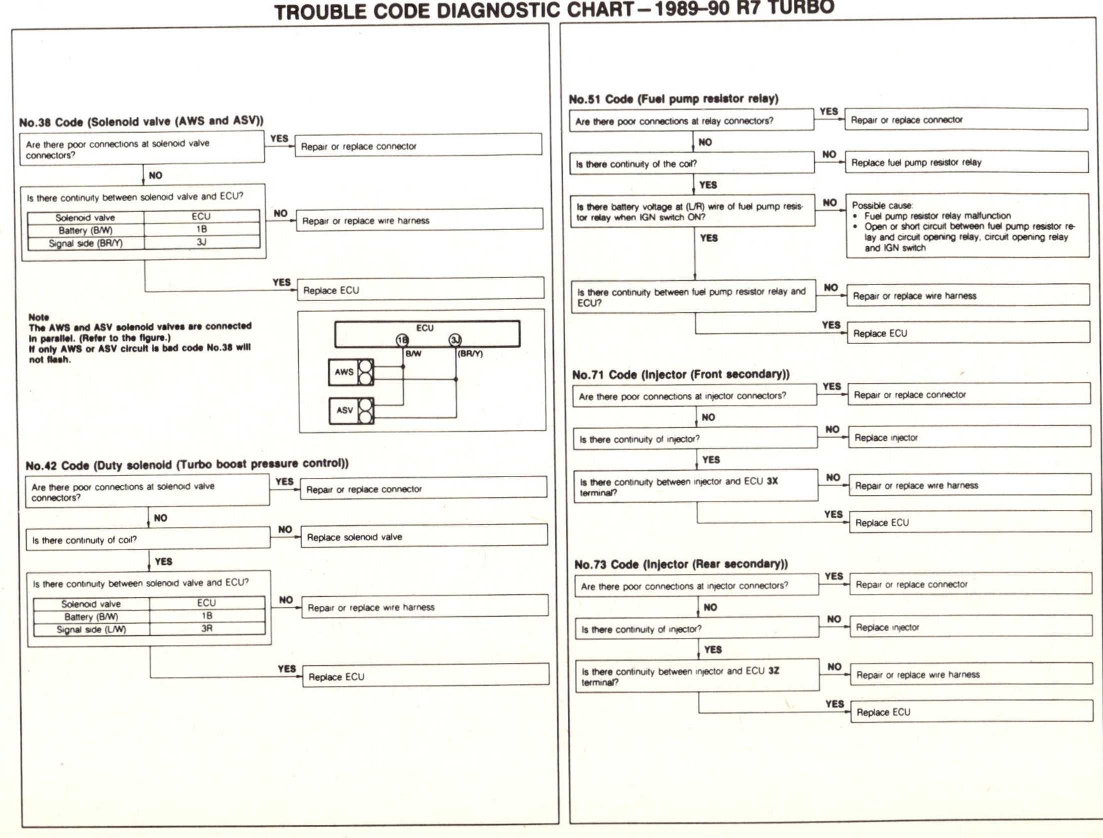

**No.38 Code (Solenoid valve (AWS and ASV))**

- Are there poor connections at solenoid valve connectors? — YES → Repair or replace connector
- NO ↓ Is there continuity between solenoid valve and ECU? — NO → Repair or replace wire harness

| Solenoid valve | ECU |
|---|---|
| Battery (B/W) | 1B |
| Signal side (BR/Y) | 3J |

- YES → Replace ECU

**Note**
**The AWS and ASV solenoid valves are connected in parallel. (Refer to the figure.)**
**If only AWS or ASV circuit is bad code No.38 will not flash.**



**No.42 Code (Duty solenoid (Turbo boost pressure control))**

- Are there poor connections at solenoid valve connectors? — YES → Repair or replace connector
- NO ↓ Is there continuity of coil? — NO → Replace solenoid valve
- YES ↓ Is there continuity between solenoid valve and ECU? — NO → Repair or replace wire harness

| Solenoid valve | ECU |
|---|---|
| Battery (B/W) | 1B |
| Signal side (L/W) | 3R |

- YES → Replace ECU

**No.51 Code (Fuel pump resistor relay)**

- Are there poor connections at relay connectors? — YES → Repair or replace connector
- NO ↓ Is there continuity of the coil? — NO → Replace fuel pump resistor relay
- YES ↓ Is there battery voltage at (L/R) wire of fuel pump resistor relay when IGN switch ON? — NO → Possible cause:
  - Fuel pump resistor relay malfunction
  - Open or short circuit between fuel pump resistor relay and circuit opening relay, circuit opening relay and IGN switch
- YES ↓ Is there continuity between fuel pump resistor relay and ECU? — NO → Repair or replace wire harness
- YES → Replace ECU

**No.71 Code (Injector (Front secondary))**

- Are there poor connections at injector connectors? — YES → Repair or replace connector
- NO ↓ Is there continuity of injector? — NO → Replace injector
- YES ↓ Is there continuity between injector and ECU **3X** terminal? — NO → Repair or replace wire harness
- YES → Replace ECU

**No.73 Code (Injector (Rear secondary))**

- Are there poor connections at injector connectors? — YES → Repair or replace connector
- NO ↓ Is there continuity of injector? — NO → Replace injector
- YES ↓ Is there continuity between injector and ECU **3Z** terminal? — NO → Repair or replace wire harness
- YES → Replace ECU

## CONTROL UNIT PIN LOCATIONS AND TEST VOLTAGES – 1989 – 90 R7 TURBO

**Engine control unit terminal (unit side)**

| 3Y | 3W | 3U | 3S | 3Q | 3O | 3M | 3K | 3I | 3G | 3E | 3C | 3A | 2O | 2M | 2K | 2I | 2G | 2E | 2C | 2A | 1U | 1S | 1Q | 1O | 1M | 1K | 1I | 1G | 1E | 1C | 1A |
|---|---|---|---|---|---|---|---|---|---|---|---|---|---|---|---|---|---|---|---|---|---|---|---|---|---|---|---|---|---|---|---|
| 3Z | 3X | 3V | 3T | 3R | 3P | 3N | 3L | 3J | 3H | 3F | 3D | 3B | 2P | 2N | 2L | 2J | 2H | 2F | 2D | 2B | 1V | 1T | 1R | 1P | 1N | 1L | 1J | 1H | 1F | 1D | 1B |

| Terminal | Input | Output | Connection to | Test condition | Voltage | Remark |
|---|---|---|---|---|---|---|
| 1A | ○ | | Battery | Constant | Approx. 12V | Backup |
| 1B | ○ | | Main relay | Ignition switch ON | Approx. 12V | – |
| | | | | Ignition switch OFF | Approx. 0V | |
| 1C | ○ | | Ignition switch | Ignition switch START (Cranking) | Approx. 12V | – |
| | | | | Ignition switch ON | Approx. 0V | |
| | | | | Ignition switch OFF | Approx. 0V | |
| 1D | | ○ | Self-Diagnosis Checker (Monitor lamp) | Test connector grounded For 3 sec. after ignition switch OFF→ON (Lamp illuminates) | Below 6.2V | With Self-Diagnosis checker |
| | | | | After 3 sec. (Light does not illuminate) | Approx. 12V | |
| | | | | Test connector grounded at idle | Below 6.2V | |
| | | | | Test connector not grounded at idle (Monitor lamp ON) | Below 6.2V | |
| | | | | Test connector not grounded at idle (Monitor lamp OFF) | Approx. 12V | |
| 1E | | ○ | Malfunction indicator light (MIL) lamp | For 3 sec. after ignition switch OFF→ON (Lamp illuminates) | Below 4.8V | Test connector grounded |
| | | | | After 3 sec. (Lamp does not illuminate) | Approx. 12V | |
| | | | | Lamp illuminates | Below 4.8V | |
| | | | | Lamp does not illuminates | Approx. 12V | |
| 1F | | ○ | Self-Diagnosis Checker (Malfunction code number) | For 3 sec. after ignition switch OFF→ON (Buzzer sounds) | Below 6.2V | With Self-Diagnosis Checker and test connector grounded |
| | | | | After 3 sec. (Buzzer does not sound) | Approx. 12V | |
| | | | | Buzzer sounds | Below 6.2V | |
| | | | | Buzzer does not sound | Approx. 12V | |
| 1G | | ○ | Ignition coil (Trailing) | Ignition switch ON | Approx. 0V | IGt-T (Ignition timing signal) |
| | | | | Idle | Approx. 0.8V | |
| 1H | | ○ | Ignition coil (Leading) | Ignition switch ON | Approx. 0V | IGt-L (Ignition timing signal) |
| | | | | Idle | Approx. 0.8V | |
| 1I | ○ | | Test connector (Green: 1-pin) | Test connector grounded | Approx. 0V | Ignition switch ON |
| | | | | Test connector not grounded | Approx. 12V | |
| 1J | | ○ | Ignition coil (Trailing) | Ignition switch ON | Approx. 4.4V | IGs-T (Select signal) |
| | | | | Idle | Approx. 2.2V | |
| 1K | | ○ | Fuel pump resistor relay | Cranking | Approx. 12V | – |
| | | | | Idle (More than 90 sec. after cranking) | Below 2.0V | |
| 1L | | ○ | A/C relay | A/C switch ON | Below 2.5V | Ignition switch ON Blower switch ON |
| | | | | A/C switch OFF | Approx. 12V | |
| 1M | ○ | | Mileage sensor No.1 | Under 20,000 miles (34,000 km) | Approx. 12V | There is an error more or less |
| | | | | Over 20,000 miles (34,000 km) | Below 1.5V | |
| 1N | ○ | | Power steering (P/S) pressure switch | Ignition switch ON | Approx. 12V | P/S ON: Turning P/S OFF: Straight ahead |
| | | | | P/S ON (Idle) | Approx. 0V | |
| | | | | P/S OFF (Idle) | Approx. 12V | |
| 1O | ○ | | A/C switch | A/C switch ON (Idle) | Below 2.5V | Blower switch ON |
| | | | | A/C switch OFF (idle) | Approx. 12V | |
| 1P | ○ | | Heat hazard sensor | Ignition switch ON | Below 1.5V | – |
| | | | | Idle (Floor temp.: Below 110°C (230°F)) | Approx. 12V | |
| | | | | Idle (Floor temp.: Above 110°C (230°F)) | Below 1.5V | |
| 1Q | ○ | | Clutch switch | Clutch pedal: released | Approx. 12V | Ignition switch ON |
| | | | | Clutch pedal: depressed | Below 2.0V | |

| Terminal | Input | Output | Connection to | Test condition | Voltage | Remark |
|---|---|---|---|---|---|---|
| 1R | ○ | | Neutral switch | Neutral | Below 2.0V | Ignition switch ON |
| | | | | In gear | Approx. 12V | |
| 1S | ○ | | Fog light switch | Fog light ON (Idle) | Approx. 12V | If equipped |
| | | | | Fog light OFF (Idle) | Approx. 0V | |
| 1T | ○ | | Back-up light and 5th switch | 5th gear or reverse | Below 2.0V | Ignition switch ON |
| | | | | 1st–4th gear | Approx. 12V | |
| 1U | ○ | | Mileage sensor No.2 | Under 600 miles (1,000 km) | Approx. 12V | Small amount of error acceptable |
| | | | | Over 600 miles (1,000 km) | Below 2.0V | |
| 1V | ○ | | Ignition coil (Trailing) | Ignition switch ON | Below 2.0V | IGf-T (Ignition confirmation signal) |
| | | | | Idle | Approx. 4.0V | |
| 2A | ○ | | Metering oil pump (MOP) position sensor | Ignition switch OFF | 0V | |
| | | | | Idle | Approx. 1.0V | |
| 2B | ○ | | Airflow meter (Vs) | Ignition switch ON | Approx. 4.0V | – |
| | | | | Idle | 2.5V–3.5V | |
| 2C | ○ | | Oxygen sensor | Idle | Below 1.0V | – |
| | | | | Acceleration | 0.5V–1.0V | |
| | | | | Deceleration | 0V–0.4V | |
| 2D | – | – | – | – | – | – |
| 2E | ○ | | Water thermosensor | Idle (Engine cold) | 0.4V–1.8V | – |
| | | | | Water temperature: 20°C (68°F) | Approx. 2.4V | |
| 2F | ○ | | Throttle sensor (Narrow range) | Ignition switch ON (Idle position) | Approx. 1.0V | – |
| | | | | Ignition switch ON (Full throttle) | Approx. 5.0V | |
| | | | | Idle | Approx. 1.0V | |
| 2G | ○ | | Throttle sensor (Full range) | Ignition switch ON (Idle position) | Approx. 0.8V | – |
| | | | | Ignition switch ON (Full throttle) | Approx. 4.3V | |
| | | | | Idle | Approx. 0.8V | |
| 2H | ○ | | Pressure sensor | Vacuum hose disconnected and plugged | 3.4V–3.6V | Ignition switch ON |
| | | | | 100 mmHg (3.9 inHg) vacuum applied to pressure sensor | 2.8V–3.2V | |
| 2I | | ○ | Sensors | Ignition switch ON | 4.5V–5.5V | Vref (Power supply) |
| | | | | Ignition switch OFF | 0V | |
| 2J | ○ | | Ground or open | Canada (Ground) | 0V | – |
| | | | | Except for Canada (Open) | Approx. 12V | |
| 2K | ○ | | Intake air thermosensor (Airflow meter) | Idle (At 20°C (68°F)) | 2V–3V | – |
| 2L | ○ | | Intake air thermosensor (Engine) | Idle (At 80°C (176°F)) | 1V–2V | – |
| 2M | ○ | | Knock sensor | Ignition switch ON | Approx. 0V | Very low voltage at any condition |
| | | | | Idle | Approx. 0V | |
| | | | | Knocking | Approx. 0V | |
| 2N | | ○ | Port air solenoid valve | Idle (Below 20,000 miles (34,000 km)) | Approx. 12V | – |
| | | | | Idle (Above 20,000 miles (34,000 km)) | Below 2.0V | |
| | | | | Engine speed: above 3,500 rpm | Approx. 12V | |
| 2O | | ○ | Solenoid valve (Switch) | Idle | Approx. 12V | Ignition switch ON |
| | | | | Half throttle | Below 2.0V | |
| 2P | | ○ | Solenoid valve (Relief) | Idle | Below 2.0V | – |
| | | | | Engine speed: above 4,000 rpm | Approx. 12V | |

## CONTROL UNIT PIN LOCATIONS AND TEST VOLTAGES (CONT.) – 1989–90 R7 TURBO

| Terminal | Input | Output | Connection to | Test condition | Voltage | Remark |
|---|---|---|---|---|---|---|
| 3A | – | – | Ground | Constant | 0V | Power |
| 3B | – | – | Ground | Constant | 0V | Power |
| 3C | – | – | Ground | Constant | 0V | System |
| 3D | – | – | Ground | Constant | 0V | Analog |
| 3E | ○ | | Crank angle sensor (Ne) | Ignition switch ON | Below 1.0V | Red |
| | | | | Idle | Below 1.0V | |
| 3F | – | – | – | – | – | – |
| 3G | ○ | | Crank angle sensor (G+) | Ignition switch ON | Below 1.0V | Black |
| | | | | Idle | Below 1.0V | |
| 3H | ○ | | Crank angle sensor (G–) | Ignition switch ON | Below 1.0V | White |
| | | | | Idle | Below 1.0V | |
| 3I | | ○ | Split air solenoid valve | 5th gear or reverse | Below 2.0V | Ignition switch ON |
| | | | | Others | Approx. 12V | |
| 3J | | ○ | Solenoid valve (Accelerated warm-up system and air supply valve) | Ignition switch OFF | 0V | Engine coolant temperature: 15°C (59°F)–35°C (95°F) |
| | | | | Ignition switch ON | Approx. 12V | |
| | | | | Idle (Less than 17 sec. after cranking) | Below 2.0V | |
| | | | | Idle (More than 17 sec. after cranking) | Approx. 12V | |
| 3K | ○ | | Circuit opening relay | Ignition switch OFF | 0V | – |
| | | | | Ignition switch ON | Approx. 12V | |
| | | | | Idle | Below 2.0V | |
| 3L | ○ | | Headlight switch | Headlight switch ON | Approx. 12V | – |
| | | | | Headlight switch OFF | 0V | |
| 3M | | ○ | Solenoid valve (Pressure regulator control) | Ignition switch ON | Below 2.0V | – |
| | | | | Cranking | Below 2.0V | Hot condition only |
| | | | | Idle (Less than 20 sec. after cranking) | Below 2.0V | |
| | | | | Idle (More than 90 sec. after cranking) | Approx. 12V | – |
| 3N | – | – | – | – | – | – |
| 3O | ○ | | Blower switch | Blower switch ON | Below 2.0V | – |
| | | | | Blower switch OFF | Approx. 12V | |
| 3P | ○ | | Rear defroster switch | Rear defroster switch ON | Below 2.0V | – |
| | | | | Rear defroster switch OFF | Approx. 12V | |
| 3Q | | ○ | Solenoid valve (Bypass air control) | Ignition switch OFF | 0V | Duty pulse |
| | | | | Ignition switch ON | Approx. 8V | |
| | | | | Idle | Approx. 8V | |
| 3R | | ○ | Duty solenoid valve (Turbo boost pressure control) | Ignition switch OFF | 0V | Duty pulse |
| | | | | Idle | Below 2.0V | |
| 3S<br>3T<br>3U<br>3V | | ○ | Stepping motor (Metering oil pump) | – | – | Can not check with circuit tester (Refer to Section D) |
| 3W | | ○ | Injector (Front primary) | Ignition switch ON | Approx. 12V | Ground time is very short |
| | | | | Idle | Approx. 12V | |
| 3X | | ○ | Injector (Front secondary) | Ignition switch ON | Approx. 12V | Ground time is very short |
| | | | | Idle | Approx. 12V | |
| 3Y | | ○ | Injector (Rear primary) | Ignition switch ON | Approx. 12V | Ground time is very short |
| | | | | Idle | Approx. 12V | |
| 3Z | | ○ | Injector (Rear secondary) | Ignition switch ON | Approx. 12V | Ground time is very short |
| | | | | Idle | Approx. 12V | |

## WIRING SCHEMATIC – 1989–90 RX-7 TURBO

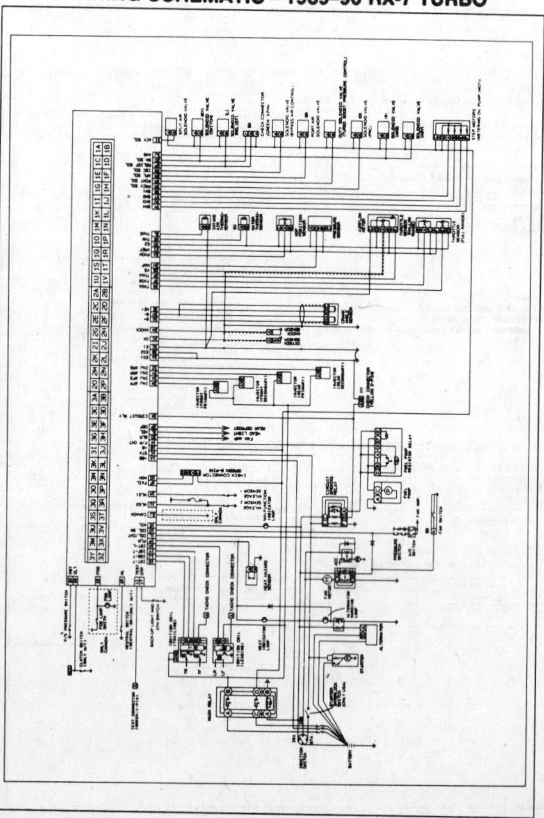

## VACUUM HOSE ROUTING 1989–90 RX-7 EXCEPT TURBO

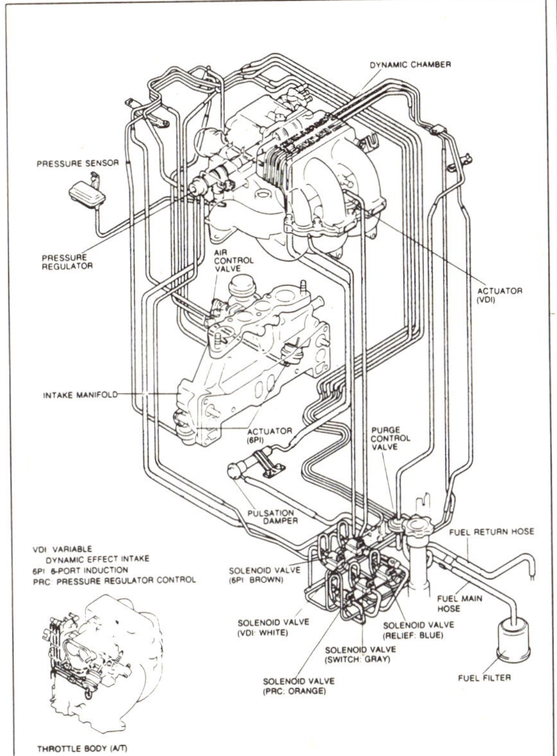


## EXPLODED VIEW OF CONTROL UNIT INPUT COMPONENTS—1989–90 RX-7 EXCEPT TURBO

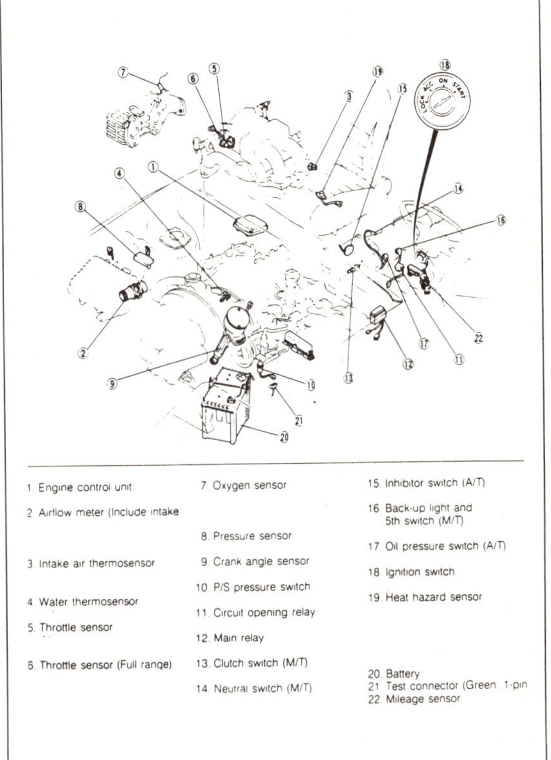

1. Engine control unit
2. Airflow meter (Include intake
3. Intake air thermosensor
4. Water thermosensor
5. Throttle sensor
6. Throttle sensor (Full range)
7. Oxygen sensor
8. Pressure sensor
9. Crank angle sensor
10. P/S pressure switch
11. Circuit opening relay
12. Main relay
13. Clutch switch (M/T)
14. Neutral switch (M/T)
15. Inhibitor switch (A/T)
16. Back-up light and 5th switch (M/T)
17. Oil pressure switch (A/T)
18. Ignition switch
19. Heat hazard sensor
20. Battery
21. Test connector (Green: 1-pin
22. Mileage sensor

## EXPLODED VIEW OF CONTROL UNIT OUTPUT COMPONENTS—1989—90 RX-7 EXCEPT TURBO

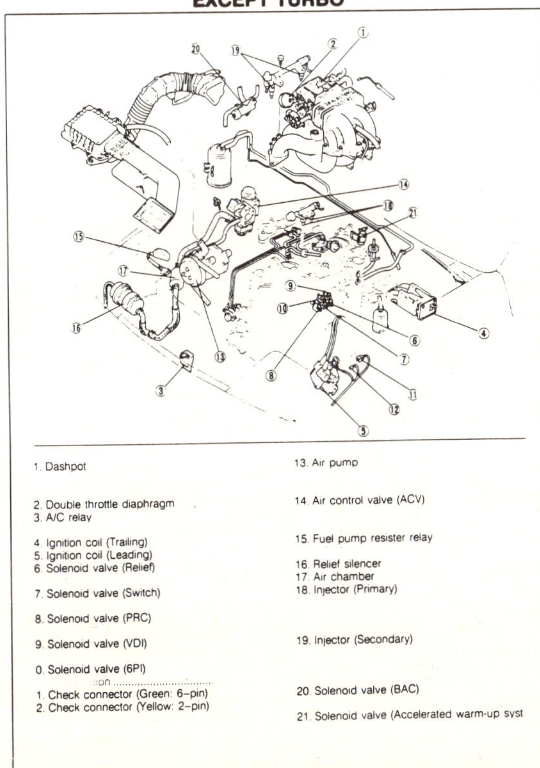

1. Dashpot
2. Double throttle diaphragm
3. A/C relay
4. Ignition coil (Trailing)
5. Ignition coil (Leading)
6. Solenoid valve (Relief)
7. Solenoid valve (Switch)
8. Solenoid valve (PRC)
9. Solenoid valve (VDI)
0. Solenoid valve (6PI)
ion ..............................
1. Check connector (Green: 6–pin)
2. Check connector (Yellow: 2–pin)
13. Air pump
14. Air control valve (ACV)
15. Fuel pump resister relay
16. Relief silencer
17. Air chamber
18. Injector (Primary)
19. Injector (Secondary)
20. Solenoid valve (BAC)
21. Solenoid valve (Accelerated warm-up syst

## EXPLODED VIEW OF FUEL RELATED COMPONENTS 1989–90 RX-7 EXCEPT TURBO

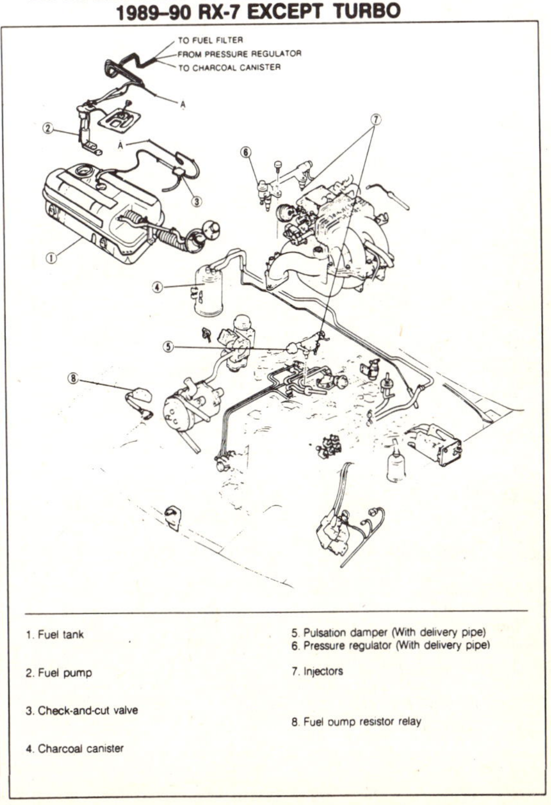


1. Fuel tank
2. Fuel pump
3. Check-and-cut valve
4. Charcoal canister
5. Pulsation damper (With delivery pipe)
6. Pressure regulator (With delivery pipe)
7. Injectors
8. Fuel oump resistor relay

### SPECIFICATIONS CHART – 1989–90 R7 EXCEPT TURBO

| Item | | Model | 13B EGI engine |
|---|---|---|---|
| Idle speed (Test connector grounded) | | rpm | 750 ± 25 (for A/T...N Range) |
| Air cleaner | Element type | | Long life wet |
| Throttle body | Type | | Horizontal — draft (2 stage — 3 barrel) |
| | Throat diameter | Primary mm (in) | 45 (1.772) |
| | | Secondary mm (in) | 45 (1.772) x 2 |
| | Water thermovalve | Operation temp. °C (°F) | M/T: 67–77 (153–171) or more<br>A/T: 60–70 (140–158) or more |
| Dashpot | Adjustment speed | rpm | 2,700–3,100 |
| Fuel tank | Capacity | liters (US gal, Imp gal) | 70 (18.5, 15.4) |
| Fuel filter | Low pressure | | Nylon 6 (164 and 45 mesh) |
| | High pressure | | Filter paper |
| Pressure regulator | Type | | Diaphragm |
| | Regulated pressure | kPa ($kg/cm^2$, psi) | 235–275 (2.4–2.8, 34.1–39.8) |
| Fuel pump | Type | | Impeller (intank) |
| | Outlet pressure | kPa ($kg/cm^2$, psi) | 441–588 (4.5–6.0, 64.0–85.3) |
| Injector (Primary and Secondary) | Drive | | Voltage drive |
| | Injection volume | cc (cu in)/15 sec. | 111–118 (6.8–7.2) |
| Heat hazard sensor | Operation temperature | °C (°F) | 105–115 (221–239) |
| Main silencer | Capacity | cc (cu in) | M/T: 10,300 (628.3) x 2, A/T: 12,000 (732) x 2 |
| Ignition timing (Test connector grounded) | | | Leading: 5° ± 1° ATDC<br>Trailing: 20° ± 2° ATDC |
| Distribution | Type | | Control unit |
| Spark advance | Type | | Control unit |
| Idle-up system | A/C | rpm | M/T: 875 A/T: 800 |
| | "D" range | rpm | 750 (at warm engine) |
| Anti-afterburn valve | Operation time | sec. | M/T: 1.60–2.20<br>A/T: 0.52–0.92 |

### COMPONENT DESCRIPTION CHART 1989–90 R7 EXCEPT TURBO

**COMPONENT DESCRIPTION**

| Component | Function | Remarks |
|---|---|---|
| **Accelerated Warm-up System (AWS)** | Supplies bypass air into dynamic chamber | Controlled by duty signal from control unit |
| **Anti Afterburn Valve** | Supplies fresh air into rear port during deceleration | Included in air control valve |
| **Air Bleed Socket** | Supplies fresh air into injector hole | |
| **Air Cleaner** | Filters air into throttle chamber | |
| **Air Control Valve** | Directs air to one of three locations; exhaust port, main converter or relief air silencer | Consists of 3 valves;<br>Relief valve<br>Switching valve<br>Anti afterburn valve |
| **Air Flow Meter** | Detects amount of intake air; sends signal to control unit | |
| **Atmospheric Pressure Sensor** | Detects atmospheric pressure; sends signal to control unit | Built in ECU |
| **Air Pump** | Supplies secondary air to air control valve | |
| **Catalytic Converter** | Reduces HC, CO and NOx | |
| **Charcoal Canister** | Stores gas tank fumes when engine stoped | Vented to atmosphere through charcoal and filter |
| **Check-and-cut Valve** | Controls pressure in fuel tank | |
| **Coil with Igniter** | Generates high voltage | Leading; ignite simultaneously<br>Trailing; ignite individually |
| **Crank Angle Sensor** | Detects eccentric shaft angle at 30° intervals and front rotor position; sends signal to control unit | |
| **Dashpot** | Gradually closes throttle valve during deceleration | |
| **Double Throttle System** | Gradually opens the No. 2 secondary throttle valve when No. 1 secondary throttle valve suddenly opens | |
| **Dynamic Chamber** | Connects front and rear ports | Primary and secondary separated |
| **Engine Control Unit** | Detects the following:<br>1. Engine speed<br>2. Intake air amount<br>3. Engine coolant temperature<br>4. Throttle opening<br>5. Intake manifold vacuum<br>6. $O_2$ concentration<br>7. In-gear condition<br><br>8. Intake air temperature<br>9. Floor temperature<br>10. A/C operation<br>11. Cranking signal<br>12. Atmospheric pressure<br>13. Initial set signal<br>14. Position of transmission gear<br><br>15. P/S operation<br>16. Metering oil pump (MOP) position signal<br>17. Electric load (E/L) condition<br><br>18. Mileage<br>Controls operation of the following:<br>1. Fuel injection system<br>2. Ignition control system<br>3. ISC system<br>4. Pressure regulator control system<br>5. Secondary air injection control system<br>6. Variable dynamic effect intake (VDI) system<br>7. 6-port induction (6PI) system | <br>Crank angle sensor<br>Air flow meter<br>Water thermosensor<br>Throttle sensors<br>Pressure sensor<br>Oxygen ($O_2$) sensor<br>Neutral switch and clutch switch (Inhibitor switch)<br>Intake air thermosensor<br>Heat hazard sensor<br>A/C switch<br>Starter switch<br>Atmospheric pressure sensor<br>Test connector (Green: 1-pin)<br>Back-up light and 5th switch, oil pressure switch<br>P/S switch<br>MOP position sensor<br>Headlight switch, Blower switch,<br>Rear defroster switch, Fog light switch<br>Mileage sensor |

### COMPONENT DESCRIPTION CHART (CONT.) 1989–90 R7 EXCEPT TURBO

| Component | Function | Remarks |
|---|---|---|
| **Fast Idle System** | Opens primary throttle valve slightly at idle | Only during cold condition |
| **Fuel Filter** | Filters particles from fuel | |
| **Fuel Pump** | Provides fuel to injectors | Operates while engine is running<br>Installed in fuel tank |
| **Heat Hazard Sensor** | Detects floor temperature; sends signal to control unit | Heat hazard sensor turned ON; relieves secondary air |
| **Initial Set Coupler** | Sends initial set signal to control unit | While adjustment of idle speed, idle mixture; coupler is shorted |
| **Injector** | Injects fuel into intermediate housing and secondary intake manifold | Controlled by signals from control unit |
| **Intake Air Thermosensor** | Detects intake air temperature and temperature into the engine; sends signal to control unit | Located on the air flow meter and air intake pipe<br>Thermistor |
| **Mileage Sensor** | Detects vehicle mileage sends signal to control unit | Above 20,000 miles; mileage switch ON |
| **Oxygen ($O_2$) Sensor** | Detects $O_2$ concentration; sends signal to control unit | Zilconia ceramic and platinum coating |
| **Pressure Regulator** | Adjusts fuel pressure supplied to injectors | |
| **Pressure Sensor** | Detects intake manifold pressure sends signal to control unit | |
| **Pulsation Damper** | Absorbs fuel pulsation | |
| **Purge Control Valve** | Regulates evaporative fumes from gas tank and canister to intake manifold | |
| **Solenoid Valve (BAC)** | Supplies bypass air into dynamic chamber | Controlled by duty signal from control unit |
| **Solenoid Valve (Relief)** | Controls relief valve | |
| **Solenoid Valve (Switch)** | Controls switching valve | |
| **Solenoid Valve (6PI)** | Controls auxiliary port valve | |
| **Solenoid Valve (PRC)** | Shuts vacuum passage between dynamic chamber and pressure regulator | Only during hot condition<br>Orange |
| **Solenoid Valve (VDI)** | Controls variable dynamic effect intake valve | |
| **Test Connector (Green: 1-pin)** | Sends initial set signal to control unit | During adjustment of idle speed, ignition timing; connector grounded |
| **Throttle Body** | Controls intake air quantity | |
| **Throttle Sensor** | Detects primary throttle valve opening angle; sends signal to control unit | |
| **VDI Valve** | Change the pressure wave line | |
| **Water Thermosensor** | Detects engine coolant temperature; sends signal to control unit | Thermistor |

### TROUBLESHOOTING DIAGNOSTIC CHART 1989–90 R7 EXCEPT TURBO

This troubleshooting guide shows the malfunction code numbers retrieved by the **SST** and symptoms of various failures. Perform troubleshooting as described below.

**TROUBLESHOOTING PROCEDURE**
**Troubleshooting With SST**
Troubleshooting with the **SST** (Self-Diagnosis Checker 49 H018 9A1) is done to quickly determine what system or unit may be at fault.

**1st:** Check input sensors and output devices with the **SST**.
**2nd:** Check other switches with the **SST**.
**3rd:** Check the following items:

| | | |
|---|---|---|
| Electrical system | : | Battery conditions, fuses |
| Ignition system | : | Ignition spark, ignition timing (with test connector grounded) |
| Fuel system | : | Fuel level, fuel leakage, fuel filter, idle speed (with test connector grounded) |
| Intake air system | : | Air cleaner element, vacuum or air leakage, vacuum hose routing, accelerator cable |
| Engine | : | Compression, overheating |
| Others | : | Clutch slippage, brake dragging |

**4th:** Check fuel and emission control systems.

**Malfunction Code No.**

| Code No. | Input device |
|---|---|
| 01 | Ignition coil (Trailing side) |
| 02 | Crank angle sensor (Ne-signal) |
| 03 | Crank angle sensor (G-signal) |
| 08 | Airflow meter (AFM) |
| 09 | Water thermosensor |
| 10 | Intake air thermosensor (AFM) |
| 11 | Intake air thermosensor (Engine) |
| 12 | Throttle sensor (Full range) |
| 13 | Pressure sensor |
| 14 | Atmospheric pressure sensor (Built in ECU) |
| 15 | Oxygen sensor |
| 17 | Feedback system |
| 18 | Throttle sensor (Narrow range) |
| 20 | Metering oil pump position sensor |
| 27 | Metering oil pump |
| 37 | Metering oil pump |

| Code No. | Output device |
|---|---|
| 25 | Solenoid valve (Pressure regulator control (PRC)) |
| 26 | Stepping motor (Metering oil pump) |
| 30 | Split air solenoid valve |
| 31 | Solenoid valve (Relief) |
| 32 | Solenoid valve (Switch) |
| 33 | Port air solenoid valve |
| 34 | Solenoid valve (Bypass air control (BAC)) |
| 38 | Solenoid valve (Accelerated warm-up system (AWS)) |
| 40 | Auxiliary port valve |
| 41 | Solenoid valve (Variable dynamic effect intake (VDI) control) |
| 51 | Fuel pump resistor relay |
| 71 | Injector (Front secondary) |
| 73 | Injector (Rear secondary) |

## SYSTEM TROUBLESHOOTING CHART 1989–90 R7 EXCEPT TURBO

**Troubleshooting of Each System**

The troubleshooting guide lists the most likely causes to a given symptom. After finding the systems to check, refer to the pages shown for detailed guides.
The numbers of the list show the priorities of inspections from the most probable to that with the lowest probability.
These were determined on the following basis:

- Ease of inspection
- Most possible system
- Most possible point in system

| Possible cause \ Symptom | Hard start or No start (Cranks OK) | Rough idle | High idle speed (At normal operating temperature) | Engine does not run smoothly (At normal operating temperature) | Engine does not run smoothly (Only when engine is cold) | Engine does not run smoothly (Only when engine is hot) | Lack of power, poor acceleration or hesitation | Afterburn | Runs rough on deceleration | Fails emission test | Knocking |
|---|---|---|---|---|---|---|---|---|---|---|---|
| INTAKE AIR SYSTEM | 4 | 1 | 2 | 1 | 3 | | 1 | 2 | | 3 | |
| IDLE SPEED CONTROL (ISC) SYSTEM | | 4 | 1 | | 1 | 4 | | 5 | 2 | 7 | |
| VARIABLE DYNAMIC-EFFECT INTAKE (VDI) SYSTEM | | | | | | | 3 | | | | |
| 6-PORT INDUCTION SYSTEM (6PI) | | 3 | | | | | 2 | | | | |
| FUEL SYSTEM | 2 | 5 | | 4 | 5 | | 5 | 7 | 3 | 6 | |
| PRESSURE REGULATOR CONTROL (PRC) SYSTEM | 3 | | 4 | | | 3 | | | | | |
| EXHAUST SYSTEM | | | | 5 | 6 | | 7 | | | 5 | |
| SECONDARY AIR INJECTION SYSTEM | 5 | | | | 7 | | | 6 | | 4 | |
| CRANKCASE AND EVAPORATIVE EMISSION CONTROL SYSTEM | | | | | | | | | | 2 | |
| DECELERATION CONTROL SYSTEM | | | | | 8 | | | 4 | 1 | 8 | |
| CATALYTIC CONVERTER | | | | | | | | | | 1 | |
| A/C CUT-OUT CONTROL SYSTEM | | | | | | | 8 | | | | |
| CONTROL SYSTEM | 7 | 6 | 3 | 3 | 2 | 2 | 6 | 3 | 5 | 9 | 2 |
| ELECTRONIC SPARK ADVANCE (ESA) SYSTEM | 1 | 2 | | 2 | 4 | 1 | 4 | 1 | 4 | 10 | 1 |

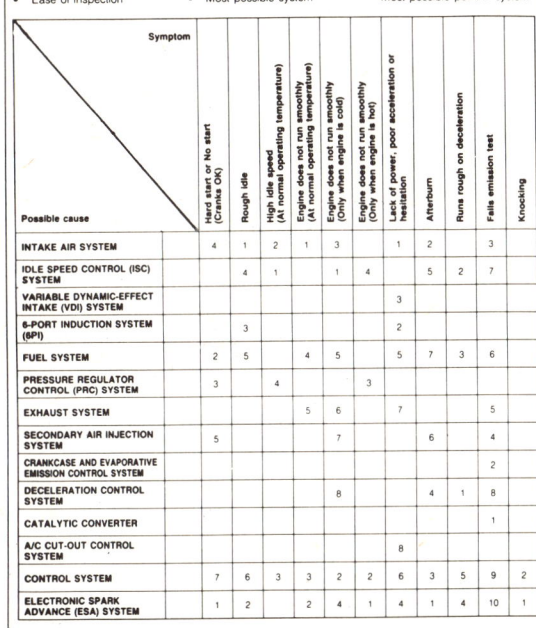

## MALFUNCTION INDIACTOR LAMP (MIL) OPERATION 1989–90 EXCEPT TURBO

**Troubleshooting**
**Principle of code cycle**
Malfunction codes are determined as below by use of the MIL and Self-Diagnosis Checker.

**1. Malfunction code cycle break**
The time between malfunction code cycles is 4.0 sec. (the time the MIL and checker buzzer are off).

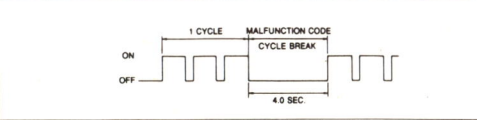

**2. Second digit of malfunction code (ones position)**
The digit in the ones position of the malfunction code represents the number of times the MIL and buzzer are on 0.4 sec. during one cycle.

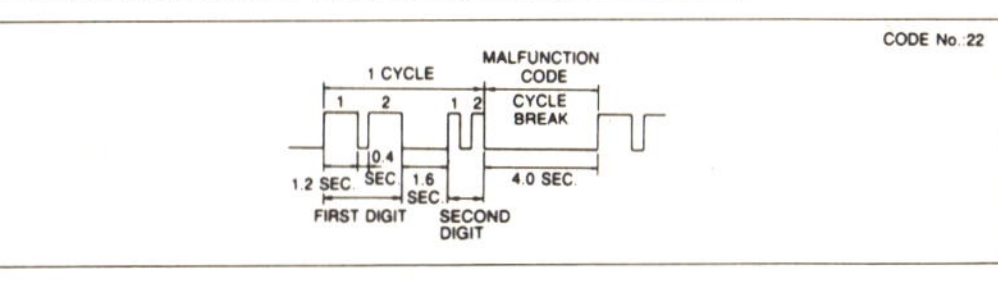


**3. First digit of malfunction code (tens position)**
The digit in the tens position of the malfunction code represents the number of times the MIL and buzzer are on 1.2 sec. during one cycle.

The MIL and buzzer are off for 1.6 sec. between the long and short pulses.



## TROUBLE CODE CHART—1989–90 R7 EXCEPT TRUBO

**Code number**

**Caution**

**a) If there is more than one failure present, the lowest number malfunction code is displayed first, the subsequent malfunction codes appear in order.**
**b) After repairing all failures, turn the ignition switch OFF, disconnect the negative battery cable for at least 5 seconds to erase the malfunction code memory.**

**Input devices**

| Code No. | Input devices | Malfunction | Fail-safe function | Output signal pattern (Self-Diagnosis Checker or MIL) |
|---|---|---|---|---|
| 01 | Ignition coil (Trailing side) | Malfunction of spark plug broken wire, short circuit | Trailing-side ignition pulse cut | |
| 02 | Crank angle sensor (Ne signal) | Broken wire, short circuit | Fuel injection and ignition cut | |
| 03 | Crank angle sensor (G signal) | Broken wire, short circuit | Fuel injection and ignition cut | |
| 08 | Airflow meter (AFM) | Broken wire, short circuit | Basic fuel injection amount and ignition timing fixed | |
| 09 | Water thermosensor | Broken wire, short circuit | Coolant temp. input fixed at 80°C (176°F) | |
| 10 | Intake air thermosensor (AFM) | Broken wire, short circuit | Intake air temp. input fixed at 20°C (68°F) | |
| 11 | Intake air thermosensor (Engine) | Broken wire, short circuit | Intake air temp. input fixed at 20°C (68°F) | |
| 12 | Throttle sensor (Full range) | Broken wire, short circuit | Throttle valve opening angle input signal fixed at 20% open | |
| 13 | Pressure sensor (Intake manifold pressure) | Broken wire, short circuit | Intake manifold pressure input signal fixed at 760 mmHg (29.9 inHg) | |
| 14 | Atmospheric pressure sensor (ATP) | Malfunctioning ECU | Atmospheric pressure input signal fixed at 760 mmHg (29.9 inHg) | |
| 15 | Oxygen sensor | Oxygen sensor output remains below 0.55V 80 sec. after F/B system operation begining | Feedback system canceled (For EGI) | |
| 17 | Feedback system | Oxygen sensor output remains 0.55V 10 sec. after F/B system operation begining | Feedback system canceled (For EGI) | |
| 18 | Throttle sensor (Narrow range) | Broken wire, short circuit | Throttle valve opening angle input signal fixed at full open | |
| 20 | Metering oil pump position sensor | Broken wire, short circuit | MOP fixed smallest open Basic fuel injection amount and ignition timing fixed | |
| 27 | Metering oil pump (MOP) | Malfunctioning MOP, step motors, broken wire, short circuit, or malfunctioning ECU | MOP fixed smallest open Basic fuel injection amount and ignition timing fixed | |
| 37 | Metering oil pump (MOP) | Malfunction MOP, step motors, broken wire, short circuit, malfunctioning ECU, alternator or battery | Basic fuel injection amount and ignition timing fixed | |

**Output devices**

| Code No. | Output devices | Output signal pattern (Self-Diagnosis Checker or MIL) |
|---|---|---|
| 25 | Solenoid valve (Pressure regulator control (PRC)) | |
| 26 | Step motor (Metering oil pump) | |
| 30 | Split air solenoid valve | |
| 31 | Solenid valve (Relief) | |
| 32 | Solenoid valve (Switch) | |
| 33 | Port air solenoid valve | |
| 34 | Solenoid valve (Bypass air control (BAC)) | |
| 38 | Solenoid valve (Accelerated warm-up system (AWS)) | |
| 40 | Auxiliary port valve | |
| 41 | Solenoid valve (Variable dynamic effect intake control (VDI)) | |
| 51 | Fuel pump resistor relay | |
| 71 | Injector (Front secondary) | |
| 73 | Injector (Rear secondary) | |

### TROUBLE CODE DIAGNOSTIC CHART—1989-90 R7 EXCEPT TURBO

**Troubleshooting chart**
If a malfunction code number is shown on the **SST**, check by using the following chart and the wiring diagram.

**Input device**
**No.1 Code (Ignition coil (Trailing side))**

- Are there poor connections at ignition coil connectors? — YES → Repair or replace connector
- NO ↓
- Is (B/W) terminal voltage of coil (Trailing side) OK? **Voltage: Approx. 12V** — NO → Repair or replace wire harness from main relay
- YES ↓
- Is coil (Trailing side) operation OK? — NO → Replace coil (Trailing side)
- YES ↓
- Is there continuity between coil (Trailing side) and ECU **1V** and **1G** terminal? — NO → Repair or replace wire harness
- YES → Replace ECU

**No.2 Code (Crank angle sensor (Ne signal))**

- Are there poor connections at crank angle sensor connectors? — YES → Repair or replace connector
- NO ↓
- Is resistance of Ne pick-up coil OK? — NO → Replace crank angle sensor
- YES ↓
- Is there continuity between crank angle sensor (Ne pick-up coil) and ECU **3E** terminal? — NO → Repair or replace wire harness
- YES → Replace ECU

**No.3 Code (Crank angle sensor (G signal))**

- Are there poor connections, at crank angle sensor connectors? — YES → Repair or replace connector
- NO ↓
- Is resistance of G pick-up coil OK? — NO → Replace crank angle sensor
- YES ↓
- Is there continuity between crank angle sensor (G pick-up coil) and ECU **3G** and **3H** terminals — NO → Repair or replace wire harness
- YES → Replace ECU

**No.8 Code (Airflow meter)**

- Are there poor connections at airflow meter connectors? — YES → Repair or replace connector
- NO ↓
- Is resistance of airflow meter OK? — NO → Replace airflow meter
- YES ↓
- Is there continuity between airflow meter and ECU? — NO → Repair or replace wire harness

| Airflow meter | ECU |
|---|---|
| Vs (GY) | 2B |
| $E_1$ (BR/B) | 3C and 3D |
| VREF (BR/W) | 2I |

- YES → Replace ECU

**No.10 Code (Intake air thermosensor (AFM))**

- Are there poor connections at thermosensor connectors? — YES → Repair or replace connector
- NO ↓
- Is resistance of intake air thermosensor OK? — NO → Replace airflow meter
- YES ↓
- Is there continuity between intake air thermosensor and ECU? — NO → Repair or replace wire harness

| Intake air thermosensor | ECU |
|---|---|
| Signal side (G/O) | 2K |
| Ground (BR/B) | 3C |

- YES → Replace ECU

**No.11 Code (Intake air thermosensor (Engine))**

- Are there poor connections at thermosensor connectors? — YES → Repair or replace connector
- NO ↓
- Is resistance of intake air thermosensor OK? — NO → Replace intake air thermosensor
- YES ↓
- Is there continuity between intake air thermosensor and ECU? — NO → Repair or replace wire harness

| Intake air thermosensor | ECU |
|---|---|
| Signal side (G) | 2L |
| Ground (BR/B) | 3C |

- YES → Replace ECU

### TROUBLE CODE DIAGNOSTIC CHART 1990 R7 EXCEPT TURBO

**No.12 Code (Throttle sensor (Full range))**

- Are there poor connections at throttle sensor connectors? — YES → Repair or replace connector
- NO ↓
- Is voltage of ECU 2G terminal OK? — NO → Adjust or replace throttle sensor

| Throttle opening | Specification voltage |
|---|---|
| Idle position | 0.25—1.25V |
| Full opening | 4.1—4.4V |

- YES ↓
- Is there continuity between throttle sensor and ECU? — NO → Repair or replace wire harness

| Throttle sensor | ECU |
|---|---|
| Vref (BR/W) | 2I |
| Ground (BR/B) | 3D |
| Signal (B/G) | 2G |

- YES → Replace ECU

**No.12 Code (Throttle sensor (Full range))**

- Are there poor connections at throttle sensor connectors? — YES → Repair or replace connector
- NO ↓
- Is resistance of throttle sensor — NO → Replace throttle sensor

| Throttle opening | Specification resistance |
|---|---|
| Idle position | Approx. 1 kΩ |
| Full opening | Approx. 5 kΩ |

- YES ↓
- Is there continuity between throttle sensor and ECU? — NO → Repair or replace wire harness

| Throttle sensor | ECU |
|---|---|
| Vref (BR/W) | 2I |
| Ground (BR/B) | 3D |
| Signal (B/G) | 2G |

- YES → Replace ECU

### TROUBLE CODE DIAGNOSTIC CHART 1989-90 R7 EXCEPT TURBO

**No.13 Code (Pressure sensor)**

- Are there poor connections at pressure sensor connectors? — YES → Repair or replace connector
- NO ↓
- Is there continuity between pressure sensor and ECU? — NO → Repair or replace wire harness

| Pressure sensor | ECU |
|---|---|
| Vref (BR/W) | 2I |
| Ground (BR/B) | 3D |
| Signal (G/Y) | 2H |

- YES ↓
- Is voltage of pressure sensor OK? When applying **100 mmHg (3.9 inHg)** vacuum — NO → Vref or Ground NG: Replace ECU; Signal NG: Replace pressure sensor

| Pressure sensor | Voltage |
|---|---|
| Vref (BR/W) | Approx. 5V |
| Ground (BR/B) | Approx. 0V |
| Signal (G/Y) | 2.8—3.2V |

- YES → Replace ECU

**No.14 Code (Atmospheric pressure (ATP) sensor)**

- Because of the ATP sensor built in ECU, if No.14 Code flash, it's mean malfunction ECU — YES → Replace ECU (ATP sensor in ECU)

**No.15 Code (Oxygen sensor)**

**Note**
**If malfunction codes No.15 and 17 are both present, perform the checking procedure for malfunction Code No.17 first.**

- Are there poor connections at connectors? — YES → Repair or replace connector
- NO ↓
- Is oxygen sensor output voltage OK? — NO → Replace oxygen sensor
- YES ↓
- Is oxygen sensor sensitivity OK? — NO → Replace oxygen sensor
- YES ↓
- Is there continuity between oxygen sensor and ECU **2C** terminal? — NO → Repair or replace wire harness
- YES → Replace ECU

## TROUBLE CODE DIAGNOSTIC CHART 1989–90 R7 EXCEPT TURBO

**No.17 Code (Feedback system)**

Are there poor connections at connectors? — YES → Repair or replace connector
NO
Is oxygen sensor sensivity OK? — NO → Possible cause
• Air leak in vacuum hoses or intake air system
• Contaminated oxygen sensor
• Insufficient fuel injection
YES
Are spark plugs OK? — NO → Clean or replace spark plugs
YES
Is there continuity between oxygen sensor and ECU 2C terminal? — NO → Repair or replace wire harness
YES → Replace ECU

## TROUBLE CODE DIAGNOSTIC CHART 1989 R7 EXCEPT TURBO

**No.18 Code (Throttle sensor (Narrow range))**

Are there poor connections at throttle sensor connectors? — YES → Repair or replace connector
NO
Is resistance of throttle sensor OK? — NO → Replace throttle sensor

| Throttle opening | Specification resistance |
|---|---|
| Idle position | Approx. 1 kΩ |
| Full opening | Approx. 5 kΩ |

YES
Is there continuity between throttle sensor and ECU? — NO → Repair or replace wire harness

| Throttle sensor | ECU |
|---|---|
| Vref (BR/W) | 2I |
| Ground (BR/B) | 3D |
| Signal (G/R) | 2F |

YES → Replace ECU

## TROUBLE CODE DIAGNOSTIC CHART 1990 R7 EXCEPT TRUBO

**No.18 Code (Throttle sensor (Narrow range))**

Are there poor connections at throttle sensor connectors? — YES → Repair or replace connector
NO
Is voltage of ECU 2F terminal OK? — NO → Adjust or replace throttle sensor

| Throttle opening | Specification voltage |
|---|---|
| Idle position | 0.75—1.25V |
| Full opening | Approx. 5V |

YES
Is there continuity between throttle sensor and ECU? — NO → Repair or replace wire harness

| Throttle sensor | ECU |
|---|---|
| Vref (BR/W) | 2I |
| Ground (BR/B) | 3D |
| Signal (G/R) | 2F |

YES → Replace ECU

## TROUBLE CODE DIAGNOSTIC CHART 1989–90 R7 EXCEPT TURBO

**No.20 Code (Metering oil pump position sensor)**

**No.27 Code (Metering oil pump)**

**No.37 Code (Metering oil pump)**

## TROUBLE CODE DIAGNOSTIC CHART—1989–90 R7 EXCEPT TURBO

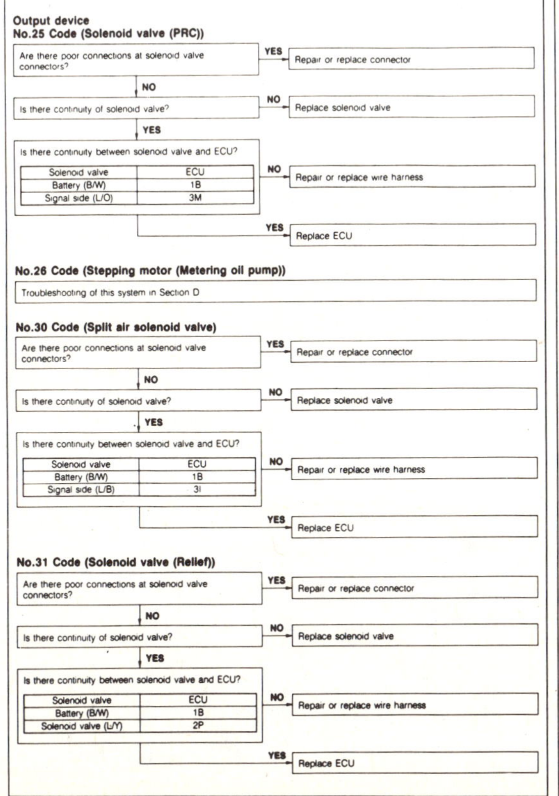

**Output device**
**No.25 Code (Solenoid valve (PRC))**

Are there poor connections at solenoid valve connectors? — YES → Repair or replace connector
NO
Is there continuity of solenoid valve? — NO → Replace solenoid valve
YES
Is there continuity between solenoid valve and ECU? — NO → Repair or replace wire harness

| Solenoid valve | ECU |
|---|---|
| Battery (B/W) | 1B |
| Signal side (L/O) | 3M |

YES → Replace ECU

**No.26 Code (Stepping motor (Metering oil pump))**

Troubleshooting of this system in Section D

**No.30 Code (Split air solenoid valve)**

Are there poor connections at solenoid valve connectors? — YES → Repair or replace connector
NO
Is there continuity of solenoid valve? — NO → Replace solenoid valve
YES
Is there continuity between solenoid valve and ECU? — NO → Repair or replace wire harness

| Solenoid valve | ECU |
|---|---|
| Battery (B/W) | 1B |
| Signal side (L/B) | 3I |

YES → Replace ECU

**No.31 Code (Solenoid valve (Relief))**

Are there poor connections at solenoid valve connectors? — YES → Repair or replace connector
NO
Is there continuity of solenoid valve? — NO → Replace solenoid valve
YES
Is there continuity between solenoid valve and ECU? — NO → Repair or replace wire harness

| Solenoid valve | ECU |
|---|---|
| Battery (B/W) | 1B |
| Solenoid valve (L/Y) | 2P |

YES → Replace ECU

**No.32 Code (Solenoid valve (Switch))**

Are there poor connections at solenoid valve connectors? — YES → Repair or replace connector
NO
Is there continuity of solenoid valve? — NO → Replace solenoid valve
YES
Is there continuity between solenoid valve and ECU? — NO → Repair or replace wire harness

| Solenoid valve | ECU |
|---|---|
| Battery (B/W) | 1B |
| Signal side (L/R) | 2Q |

YES → Replace ECU

**No.33 Code (Port air solenoid valve)**

Are there poor connections at solenoid valve connectors? — YES → Repair or replace connector
NO
Is there continuity of solenoid valve? — NO → Replace solenoid valve
YES
Is there continuity between solenoid valve and ECU? — NO → Repair or replace wire harness

| Solenoid valve | ECU |
|---|---|
| Battery (B/W) | 1B |
| Signal side (L) | 2N |

YES → Replace ECU

**No.34 Code (Solenoid valve (BAC))**

Are there poor connections at solenoid valve connectors? — YES → Repair or replace connector
NO
Is there continuity of solenoid valve? — NO → Replace solenoid valve
YES
Is there continuity between solenoid valve and ECU? — NO → Repair or replace wire harness

| Solenoid valve | ECU |
|---|---|
| Battery (B/W) | 1B |
| Signal side (L/G) | 3Q |

YES → Replace ECU

**No.38 Code (Solenoid valve (AWS))**

Are there poor connections at solenoid valve connectors? — YES → Repair or replace connector
NO
Next page

## TROUBLE CODE DIAGNOSTIC CHART—1989–90 R7 EXCEPT TURBO

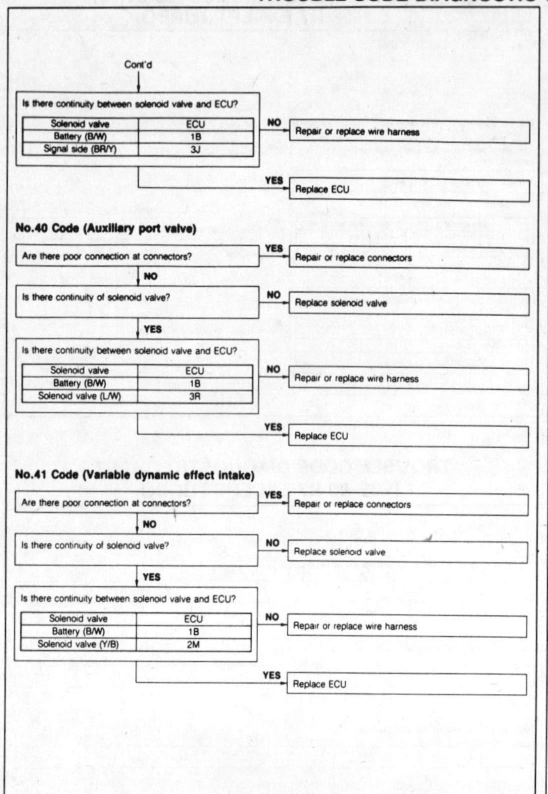


**No.51 Code (Fuel pump resistor relay)**

Are there poor connections at relay connectors? — YES → Repair or replace connector
NO ↓
Is there continuity of the coil? — NO → Replace fuel pump resistor relay
YES ↓
Is there battery voltage at (L/R) wire of fuel pump resistor relay when IGN switch ON? — NO → Possible cause:
- Fuel pump resistor relay malfunction
- Open or short circuit between fuel pump resistor relay and circuit opening relay, circuit opening relay and IGN switch

YES ↓
Is there continuity between fuel pump resistor relay and ECU? — NO → Repair or replace wire harness
YES → Replace ECU

**No.71 Code (Injector (Front secondary))**

Are there poor connections at injector connectors? — YES → Repair or replace connector
NO ↓
Is there continuity of injector? — NO → Replace injector
YES ↓
Is there continuity between injector and ECU **3X** terminal? — NO → Repair or replace wire harness
YES → Replace ECU

**No.73 Code (Injector (Rear secondary))**

Are there poor connections at injector connectors? — YES → Repair or replace connector
NO ↓
Is there continuity of injector? — NO → Replace injector
YES ↓
Is there continuity between injector and ECU **3Z** terminal? — NO → Repair or replace wire harness
YES → Replace ECU

## TROUBLE CODE DIAGNOSTIC CHART 1989–90 R7 EXCEPT TURBO

**RELATIONSHIP CHART**
**Output Devices and Input Devices**

| INPUT DEVICES / OUTPUT DEVICES | FUEL INJECTION AMOUNT | FUEL INJECTION TIMING | FUEL PUMP RESISTOR RELAY | A/C RELAY | PORT AIR SOLENOID VALVE | SPLIT AIR SOLENOID VALVE | SOLENOID VALVE (SWITCH) | SOLENOID VALVE (RELIEF) | SOLENOID VALVE (AWS) | SOLENOID VALVE (PRESSURE REGULATOR) | SOLENOID VALVE (BYPASS AIR CONTROL) | SOLENOID VALVE (VARIABLE DYNAMIC EFFECT INTAKE) | SOLENOID VALVE (6 PORT INJECTION) | IGNITER (TRAILING IGNITION TIMING) | IGNITER (LEADING IGNITION TIMING) |
|---|---|---|---|---|---|---|---|---|---|---|---|---|---|---|---|
| TEST CONNECTOR | ○ | × | × | × | × | × | × | × | × | × | ○ | × | × | ○ | ○ |
| IGNITION SWITCH | ○ | × | ○ | × | × | × | ○ | ○ | ○ | ○ | ○ | × | × | ○ | ○ |
| POWER STEERING PRESSURE SWITCH | × | × | × | × | × | × | × | × | × | × | ○ | × | × | ○ | ○ |
| A/C SWITCH | ○ | × | × | ○ | × | × | × | × | × | × | ○ | × | × | ○ | ○ |
| FOG LIGHT SWITCH (IF EQUIPPED) | ○ | × | × | × | × | × | × | × | × | × | ○ | × | × | ○ | ○ |
| HEAD LIGHT SWITCH | ○ | × | × | × | × | × | × | × | × | × | ○ | × | × | ○ | ○ |
| BLOWER SWITCH | ○ | × | × | × | × | × | × | × | × | × | ○ | × | × | ○ | ○ |
| REAR DEFROSTER SWITCH | ○ | × | × | × | × | × | × | × | × | × | ○ | × | × | ○ | ○ |
| CLUTCH SWITCH (M/T) | ○ | × | ○ | ○ | × | ○ | ○ | ○ | ○ | ○ | ○ | × | ○ | ○ | ○ |
| NEUTRAL SWITCH (M/T) | ○ | × | ○ | ○ | × | ○ | ○ | ○ | ○ | ○ | ○ | × | ○ | ○ | ○ |
| EC-AT CONTROL UNIT (A/T) (INHIBITOR SWITCH) | ○ | × | × | ○ | × | ○ | ○ | ○ | ○ | ○ | ○ | × | ○ | ○ | ○ |
| BACK-UP LIGHT AND 5TH SWITCH (M/T) | × | × | × | × | × | ○ | ○ | ○ | × | × | ○ | × | × | ○ | ○ |
| OIL PRESSURE SWITCH (A/T) | ○ | × | × | × | × | ○ | ○ | ○ | × | × | ○ | × | × | ○ | ○ |
| MILEAGE SENSOR | ○ | × | × | × | ○ | × | ○ | × | × | × | × | × | × | ○ | ○ |
| HEAT HAZARD SENSOR | × | × | × | × | × | × | × | ○ | × | × | × | × | × | × | × |
| OXYGEN SENSOR | ○ | × | × | × | × | × | × | × | × | × | × | × | × | × | × |
| THROTTLE SENSOR (NARROW RANGE) | ○ | × | ○ | × | ○ | × | ○ | ○ | × | ○ | ○ | × | × | ○ | ○ |
| THROTTLE SENSOR (FULL RANGE) | ○ | × | × | ○ | ○ | × | × | ○ | ○ | × | × | × | × | ○ | ○ |
| PRESSURE SENSOR | ○ | × | × | × | ○ | ○ | ○ | ○ | × | × | × | × | × | ○ | ○ |
| CRANK ANGLE SENSOR (NE) | ○ | ○ | ○ | ○ | × | × | ○ | ○ | ○ | × | ○ | ○ | ○ | ○ | ○ |
| CRANK ANGLE SENSOR (G+, G–) | ○ | ○ | ○ | × | × | × | × | × | × | ○ | × | × | × | ○ | ○ |
| WATER THERMOSENSOR | ○ | × | ○ | × | × | × | ○ | ○ | ○ | ○ | ○ | × | ○ | ○ | ○ |
| INTAKE AIR THERMOSENSOR (AFM) | ○ | × | ○ | × | × | × | × | × | × | × | × | × | × | × | × |
| INTAKE AIR THERMOSENSOR (ENGINE) | ○ | × | ○ | × | × | × | × | × | × | ○ | ○ | × | × | ○ | ○ |
| AIRFLOW METER (AFM) | ○ | × | ○ | × | × | × | × | × | × | × | × | ○ | ○ | ○ | ○ |

## OUTPUT COMPONENTS AND ENGINE CONDITION CHART—1989–90 R7 EXCEPT TURBO

<table>
<tr><th colspan="2" rowspan="2">ENGINE CONITION / OUPUT DEVICES</th><th rowspan="2">CRANKING (WHEN ENGINE IS COLD)</th><th rowspan="2">IDLE (DURING WARM-UP)</th><th colspan="2">MEDIUM LOAD</th><th rowspan="2">ACCELERATION</th><th rowspan="2">HEAVY LOAD</th><th rowspan="2">DECELERATION (THROTTLE VALVE FULLY CLOSED)</th><th rowspan="2">IDLE (WHEN ENGINE IS WARM)</th><th rowspan="2">IGNITION: ON (NOT ENGINE RUN)</th><th rowspan="2">REMARKS</th></tr>
<tr><th>COLD</th><th>WARM</th></tr>
<tr><td rowspan="2">FUEL INJECTION AMOUNT</td><td>PRIMARY INJECTOR</td><td colspan="3">Rich</td><td>Rich and Lean</td><td colspan="2" rowspan="2">Rich</td><td>Fuel cut</td><td>Rich</td><td>OFF</td><td rowspan="2">Rear side, Above 8,500 rpm: Fuel cut</td></tr>
<tr><td>SECONDARY INJECTOR</td><td colspan="4">OFF</td><td colspan="3">OFF</td></tr>
<tr><td colspan="2">FUEL PUMP RESISTOR RELAY</td><td>12V drive</td><td colspan="3">9–10V drive</td><td colspan="2">12V drive</td><td colspan="2">9—10V drive</td><td>OFF</td><td>Hot start (at idle) 12V drive</td></tr>
<tr><td colspan="2">A/C RELAY</td><td colspan="4">ON</td><td>*1OFF</td><td colspan="4">ON</td><td>*1Only 7 sec</td></tr>
<tr><td colspan="2">PORT AIR SOLENOID VALVE</td><td colspan="5">Above 20,000 mile (34,000 km), ON<br>Below 20,000 mile (34,000 km), OFF</td><td colspan="4">ON</td><td></td></tr>
<tr><td colspan="2">SPRIT AIR SOLENOID VALVE</td><td colspan="8">O/D switch ON or back-up light and 5th switch ON and above 20,000 mile (34,000 km) ON</td><td>OFF</td><td>Always ON at high altitude</td></tr>
<tr><td colspan="2">SOLENOID VALVE (SWITCH)</td><td colspan="3">OFF (Port air)</td><td colspan="3">ON (Split air)</td><td colspan="2">OFF (Port air)</td><td>OFF</td><td></td></tr>
<tr><td colspan="2">SOLENOID VALVE (RELIEF)</td><td colspan="2">ON (Air injection)</td><td>OFF (Relief)</td><td>ON (Air injection)</td><td colspan="2">OFF (Relief)</td><td colspan="2">ON (Air injection)</td><td>ON</td><td></td></tr>
<tr><td colspan="2">SOLENOID VALVE (AWS)</td><td>ON</td><td>For 17 sec after start, ON</td><td colspan="7">OFF</td><td></td></tr>
<tr><td colspan="2">SOLENOID VALVE (PRESSURE REGULATOR)</td><td colspan="7">OFF</td><td>After start: *2ON</td><td>ON</td><td>*2For 50 sec.—70 sec</td></tr>
<tr><td colspan="2">SOLENOID VALVE (BYPASS AIR CONTROL)</td><td>Large quantity of air</td><td>Feed back duty</td><td colspan="5">Fixed duty</td><td>Feed back duty</td><td>Fixed duty</td><td>Test connector (Green, 1-pin) ground, Stop the feedback control</td></tr>
<tr><td colspan="2">SOLENOID VALVE (VARIABLE DYNAMIC EFFECT INTAKE)</td><td colspan="5">OFF</td><td>ON</td><td colspan="3">OFF</td><td></td></tr>
<tr><td colspan="2">SOLENOID VALVE (6 PORT INJECTION)</td><td colspan="3">OFF</td><td colspan="3">*3ON (Above 3,850 rpm)</td><td colspan="3">OFF</td><td>*3 Only load condition</td></tr>
<tr><td colspan="2">IGNITER (TRAILING IGNITION TIMING)</td><td rowspan="2">BTDC 5°</td><td rowspan="2">Retard</td><td colspan="4" rowspan="2">Advanced</td><td rowspan="2">Retard</td><td>ATDC 20°</td><td rowspan="2">OFF</td><td rowspan="2"></td></tr>
<tr><td colspan="2">IGNITER (LEADING IGNITION TIMING)</td><td>ATDC 5°</td></tr>
</table>

## CONTROL SYSTEM TROUBLESHOOTING CHART 1989–90 R7 EXCEPT TURBO

Check the condition of the wiring harness and connectors before checking the sensors or switches.

| Symptom \ Possible cause | Airflow meter | Intake air thermosensor (Engine) | Throttle sensor (Full and Norrow range) | Oxygen sensor | Pressure sensor | Heat hazard sensor | Main relay | Circuit opening relay | Neutral switch (M/T) | Clutch switch (M/T) | Inhibitor switch (A/T) | Power steering pressure switch | Back-up light and 5th switch (M/T) | Oil pressure switch (A/T) | Water thermosensor | Mileage sensor | Engine control unit terminal |
|---|---|---|---|---|---|---|---|---|---|---|---|---|---|---|---|---|---|
| Hard start or no start (Cranks OK) | 7 | 6 | 4 | | 5 | | 1 | 2 | | | | | | | 3 | | 8 |
| Rough idle | 1 | 4 | 2 | | | | | | 6 | 7 | 6 | | | | 3 | 8 | 9 |
| High idle speed (At nomal operating temperature) | 1 | | 2 | | | | | | | | | 4 | | | 3 | | 5 |
| Engine does not run smoothly (Only when engine is cold) | 1 | 3 | | | | 4 | | | | | | | | | 2 | | 5 |
| Engine does not run smoothly (Only when engine is hot) | 1 | 3 | | | | 4 | | | | | | | | | 2 | | 5 |
| Lack of power, poor acceleration or hesitation | 1 | | 2 | | | 9 | | | 5 | 6 | 5 | 7 | 10 | 10 | 8 | 3 | 11 |
| Afterburn | 1 | | 2 | | 5 | | | | | 3 | | 4 | | | | | 6 |
| Fails emission test | 2 | | 3 | 1 | | 4 | | | | | | | | | 5 | | 6 |
| Knocking | 3 | 5 | 4 | | | | | | 6 | 7 | 6 | | | | 2 | | 8 |

### PREPARATION
### SST

49 H018 9A1

Self-Diagnosis Checker

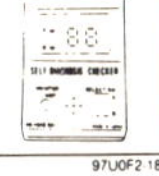

97U0F2-186

97U0F2-187

### ENGINE CONTROL UNIT

**Inspection**

1. Lift up the floor mat in front of the passenger's seat.
2. Remove the protector cover.
3. Turn the ignition switch ON, and measure the voltage of the terminals with circuit tester.

**Caution**

- **If not indicated under remark, warm up the engine to normal operating temperature before checking the control unit.**
- **If the proper voltage is not indicated on the voltmeter, check all wiring, connections, and finally, check the indicated component.**

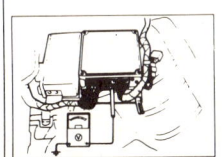

## CONTROL UNIT PIN LOCATIONS AND TEST VOLTAGES 1989–90 R7 EXCEPT TURBO

**Engine control unit terminal (unit side)**

| 3Y | 3W | 3U | 3S | 3Q | 3O | 3M | 3K | 3I | 3G | 3E | 3C | 3A | 2O | 2M | 2K | 2I | 2G | 2E | 2C | 2A | 1U | 1S | 1Q | 1O | 1M | 1K | 1I | 1G | 1E | 1C | 1A |
|---|---|---|---|---|---|---|---|---|---|---|---|---|---|---|---|---|---|---|---|---|---|---|---|---|---|---|---|---|---|---|---|
| 3Z | 3X | 3V | 3T | 3R | 3P | 3N | 3L | 3J | 3H | 3F | 3D | 3B | 2P | 2N | 2L | 2J | 2H | 2F | 2D | 2B | 1V | 1T | 1R | 1P | 1N | 1L | 1J | 1H | 1F | 1D | 1B |

| Terminal | Input | Output | Connection to | Test condition | Voltage | Remark |
|---|---|---|---|---|---|---|
| 1A | ○ | | Battery | Constant | Approx. 12V | Backup |
| 1B | ○ | | Main relay | Ignition switch ON | Approx. 12V | — |
| | | | | Ignition switch OFF | Approx. 0V | |
| 1C | ○ | | Ignition switch | Ignition switch START (Cranking) | Approx. 12V | — |
| | | | | Ignition switch ON | Approx. 0V | |
| | | | | Ignition switch OFF | Approx. 0V | |
| 1D | | ○ | Self-Diagnosis Checker (Monitor lamp) | Test connector grounded For 3 sec. after ignition switch OFF→ON (Lamp illuminates) | Below 6.2V | With Self-Diagnosis checker |
| | | | | After 3 sec. (Light does not illuminate) | Approx. 12V | |
| | | | | Test connector grounded at idle | Below 6.2V | |
| | | | | Test connector not grounded at idle (Monitor lamp ON) | Below 6.2V | |
| | | | | Test connector not grounded at idle (Monitor lamp OFF) | Approx. 12V | |
| 1E | | ○ | Malfunction indicator light (MIL) lamp | For 3 sec. after ignition switch OFF→ON (Lamp illuminates) | Below 4.8V | Test connector grounded |
| | | | | After 3 sec. (Lamp does not illuminate) | Approx. 12V | |
| | | | | Lamp illuminates | Below 4.8V | |
| | | | | Lamp does not illuminates | Approx. 12V | |
| 1F | | ○ | Self-Diagnosis Checker (Malfunction code number) | For 3 sec. after ignition switch OFF→ON (Buzzer sounds) | Below 6.2V | With Self-Diagnosis Checker and test connector grounded |
| | | | | After 3 sec. (Buzzer does not sound) | Approx. 12V | |
| | | | | Buzzer sounds | Below 6.2V | |
| | | | | Buzzer does not sound | Approx. 12V | |
| 1G | | ○ | Ignition coil (Trailing) | Ignition switch ON | Approx. 0V | IGt-T (Ignition timing signal) |
| | | | | Idle | Approx. 0.8V | |
| 1H | | ○ | Ignition coil (Leading) | Ignition switch ON | Approx. 0V | IGt-L (Ignition timing signal) |
| | | | | Idle | Approx. 0.8V | |
| 1I | ○ | | Test connector (Green: 1-pin) | Test connector grounded | Approx. 0V | Ignition switch ON |
| | | | | Test connector not grounded | Approx. 12V | |
| 1J | | ○ | Ignition coil (Trailing) | Ignition switch ON | Approx. 4.4V | IGs-T (Select signal) |
| | | | | Idle | Approx. 2.2V | |
| 1K | | ○ | Fuel pump resistor relay | Cranking | Approx. 12V | — |
| | | | | Idle (More than 90 sec. after cranking) | Below 2.0V | |
| 1L | | ○ | A/C relay | A/C switch ON | Below 2.5V | Ignition switch ON Blower switch ON |
| | | | | A/C switch OFF | Approx. 12V | |
| 1M | ○ | | Mileage sensor | Under 20,000 miles (34,000 km) | Approx. 12V | There is an error more or less |
| | | | | Over 20,000 miles (34,000 km) | Below 1.5V | |
| 1N | ○ | | Power steering (P/S) pressure switch | Ignition switch ON | Approx. 12V | P/S ON: Turning P/S OFF: Straight ahead |
| | | | | P/S ON (idle) | Approx. 0V | |
| | | | | P/S OFF (Idle) | Approx. 12V | |
| 1O | ○ | | A/C switch | A/C switch ON (Idle) | Below 2.5V | Blower switch ON |
| | | | | A/C switch OFF (idle) | Approx. 12V | |
| 1P | ○ | | Heat hazard sensor | Ignition switch ON | Below 1.5V | — |
| | | | | Idle (Floor temp. Below 110°C (230°F)) | Approx. 12V | |
| | | | | Idle (Floor temp. Above 110°C (230°F)) | Below 1.5V | |
| 1Q | ○ | | Clutch switch (M/T) | Clutch pedal released | Approx. 12V | Ignition switch ON |
| | | | | Clutch pedal depressed | Below 2.0V | |

## CONTROL UNIT PIN LOCATIONS AND TEST VOLTAGES—1989–90 R7 EXCEPT TURBO

| Terminal | Input | Output | Connection to | Test condition | Voltage | Remark |
|---|---|---|---|---|---|---|
| 1R | ○ | | Neutral switch (M/T) | Neutral | Below 2.0V | Ignition switch ON |
| | | | | In gear | Approx. 12V | |
| | ○ | | EC-AT control unit (Inhibitor signal) (A/T) | N, P range | Below 2.0V | Ignition switch ON |
| | | | | Others | Approx. 12V | |
| 1S | ○ | | Fog light switch | Fog light ON (Idle) | Approx. 12V | If equipped |
| | | | | Fog light OFF (Idle) | Approx. 0V | |
| 1T | ○ | | Back-up light and 5th switch (M/T) | 5th gear or reverse | Approx. 12V | Ignition switch ON |
| | | | | 1th — 4th gear | Below 2.0V | |
| | ○ | | Oil pressure switch (A/T) | Overdrive | Below 2.0V | Can not check (Load condition only) |
| | | | | Others | Approx. 12V | |
| 1U | ○ | | AT switch | A/T | Approx. 0V | Ignition switch ON |
| | | | | M/T | Approx. 12V | |
| 1V | ○ | | Ignition coil (Trailing) | Ignition switch ON | Below 2.0V | IGf-T (Ignition confirmation signal) |
| | | | | Idle | Approx. 4.0V | |
| 2A | ○ | | Metering oil pump (MOP) position sensor | Ignition switch OFF | 0V | Refer to Section D |
| | | | | Idle | Approx. 1.0V | |
| 2B | ○ | | Airflow meter (Vs) | Ignition switch ON | Approx. 4.0V | — |
| | | | | Idle | 2.5V—3.5V | |
| 2C | ○ | | Oxygen sensor | Idle | Below 1.0V | — |
| | | | | Acceleration | 0.5V—1.0V | |
| | | | | Deceleration | 0V—0.4V | |
| 2D | — | — | — | — | — | — |
| 2E | ○ | | Water thermosensor | Idle (Engine cold) | 0.4V—1.8V | — |
| | | | | Water temperature: 20°C (68°F) | Approx. 2.4V | |
| 2F | ○ | | Throttle sensor (Narrow range) | Ignition switch ON (Idle position) | Approx. 1.0V | — |
| | | | | Ignition switch ON (Full throttle) | Approx. 5.0V | |
| | | | | Idle | Approx. 1.0V | |
| 2G | ○ | | Throttle sensor (Full range) | Ignition switch ON (Idle position) | Approx. 0.8V | — |
| | | | | Ignition switch ON (Full throttle) | Approx. 4.3V | |
| | | | | Idle | Approx. 0.8V | |
| 2H | ○ | | Pressure sensor | Vacuum hose disconnected and plugged | 3.4V—3.6V | Ignition switch ON |
| | | | | 100 mmHg (3.9 inHg) vacuum applied to pressure sensor | 2.8V—3.2V | |
| 2I | | ○ | Sensors | Ignition switch ON | 4.5V—5.5V | Vref (Power supply) |
| | | | | Ignition switch OFF | 0V | |
| 2J | ○ | | Ground or open | Canada (Ground) | 0V | — |
| | | | | Except for Canada (Open) | Approx. 12V | |
| 2K | ○ | | Intake air thermosensor (Airflow meter) | Idle (At 20°C (68°F)) | 2V—3V | — |
| 2L | ○ | | Intake air thermosensor (Engine) | Idle (At 80°C (176°F)) | 1V—2V | — |
| 2M | | ○ | Solenoid valve (Variable dynamic effect intake control) | Above 5,200 rpm | Below 2.0V | Only while driving |
| | | | | Below 5,200 rpm | Approx. 12V | — |
| 2N | | ○ | Port air solenoid valve | Idle (Below 20,000 miles (34,000 km)) | Approx. 12V | — |
| | | | | Idle (Above 20,000 miles (34,000 km)) | Below 2.0V | |
| | | | | Engine speed: above 3,000 rpm | Approx. 12V | |
| 2O | | ○ | Solenoid valve (Switch) | Half throttle | Below 2.0V | Ignition switch ON |
| | | | | Idle | Approx. 12V | |
| 2P | | ○ | Solenoid valve (Relief) | Idle | Below 2.0V | — |
| | | | | Engine speed: above 3,600 rpm | Approx. 12V | |

| Terminal | Input | Output | Connection to | Test condition | Voltage | Remark |
|---|---|---|---|---|---|---|
| 3A | — | — | Ground | Constant | 0V | Power |
| 3B | — | — | Ground | Constant | 0V | Power |
| 3C | — | — | Ground | Constant | 0V | System |
| 3D | — | — | Ground | Constant | 0V | Analog |
| 3E | ○ | | Crank angle sensor (Ne) | Ignition switch ON | Below 1.0V | Red |
| | | | | Idle | Below 1.0V | |
| 3F | — | — | — | — | — | — |
| 3G | ○ | | Crank angle sensor (G+) | Ignition switch ON | Below 1.0V | Black |
| | | | | Idle | Below 1.0V | |
| 3H | ○ | | Crank angle sensor (G−) | Ignition switch ON | Below 1.0V | White |
| | | | | Idle | Below 1.0V | |
| 3I | | ○ | Split air solenoid valve | 5th gear or reverse | Below 2.5V | Refer to page F1–65 |
| | | | | Others | Approx. 12V | |
| 3J | | ○ | Solenoid valve (Accelerated warm-up system) | Ignition switch OFF | 0V | Engine coolant temperature 15°C (59°F)—35°C (95°F) |
| | | | | Ignition switch ON | Approx. 12V | |
| | | | | Idle (Less than 17 sec. after cranking) | Below 2.0V | |
| | | | | Idle (More than 17 sec. after cranking) | Approx. 12V | |
| 3K | | ○ | Circuit opening relay | Ignition switch OFF | 0V | — |
| | | | | Ignition switch ON | Approx. 12V | |
| | | | | Idle | Below 2.0V | |
| 3L | ○ | | Headlight switch | Headlight switch ON | Approx. 12V | — |
| | | | | Headlight switch OFF | 0V | |
| 3M | | ○ | Solenoid valve (Pressure regulator control) | Ignition switch ON | Below [illegible] | — |
| | | | | Cranking | Bel[illegible] | Hot condition only |
| | | | | Idle (Less than 20 sec. after cranking) | B[illegible]V | |
| | | | | Idle (More than 90 sec. after cranking) | [illegible]x. 12V | |
| 3N | | ○ | EC-AT control unit | Idle (After warm-up) | [illegible]low 2.0V | |
| | | | | Others (After warm-up) | Approx. 12V | |
| 3O | ○ | | Blower switch | Blower switch ON | Below 2.0V | — |
| | | | | Blower switch OFF | Approx. 12V | |
| 3P | ○ | | Rear defroster switch | Rear defroster switch ON | Below 2.0V | — |
| | | | | Rear defroster switch OFF | Approx. 12V | |
| 3Q | | ○ | Solenoid valve (Bypass air control) | Ignition switch OFF | 0V | Duty pulse |
| | | | | Ignition switch ON | Approx. 8V | |
| | | | | Idle | Approx. 8V | |
| 3R | | ○ | 6-port induction (6PI) system | Above 3,850 rpm | Below 0V | Cannot check (Warm-up and load condition only) |
| | | | | Below 3,850 rpm | Approx. 12V | |
| 3S 3T 3U 3V | | ○ | Stepping motor (Metering oil pump) | — | — | Can not check with circuit tester (Refer to Section D) |
| 3W | | ○ | Injector (Front primary) | Ignition switch ON | Approx. 12V | Ground time is very short |
| | | | | Idle | Approx. 12V | |
| 3X | | ○ | Injector (Front secondary) | Ignition switch ON | Approx. 12V | Ground time is very short |
| | | | | Idle | Approx. 12V | |
| 3Y | | ○ | Injector (Rear primary) | Ignition switch ON | Approx. 12V | Ground time is very short |
| | | | | Idle | Approx. 12V | |
| 3Z | | ○ | Injector (Rear secondary) | Ignition switch ON | Approx. 12V | Ground time is very short |
| | | | | Idle | Approx. 12V | |

# Component Replacement

## RELIEVING FUEL PRESSURE

The fuel in the fuel system remains under high pressure even when the engine is not running. So before disconnecting any fuel lines, release the fuel pressure to reduce the risk of injury or fire. Relieve the fuel pressure as follows:

1. Start the engine.
2. Disconnect the fuel pump connector with the engine running.
3. After the engine stalls from lack of fuel, turn the ignition switch off.
4. Use a shop rag to cover the fuel lines when disconnecting and plug all fuel lines after they have been disconnected.

## AIR FLOW METER

### Removal and Installation

1. Disconnect the negative battery cable. Remove the high tension lead and connectors.
2. Loosen the hose band and remove the intake hose.
3. Remove the air flow meter attaching bolts.
4. Turn the air cleaner cover upside down and remove the attaching nuts, remove the air flow meter.
5. Installation is the reverse order of the removal procedure.

## THROTTLE BODY

### Removal and Installation

1. Disconnect the negative battery cable. Disconnect the accelerator cable from the throttle linkage.
2. Disconnect the cruise control cable, if equipped.
3. Disconnect the air funnel. Disconnect the hoses and tubes from the throttle body.
4. Disconnect the throttle position sensor connector. Remove the throttle body retaining screws and remove the throttle body assembly.
5. Installation is the reverse order of the removal procedure.

## FUEL PUMP

### Removal and Installation

1. Relieve the fuel pressure from the fuel system.
2. Lift up the rear mat and remove the fuel pump cover.
3. Disconnect the fuel pump wiring connector.
4. Disconnect and plug the fuel main and return lines from the fuel pump.
5. Remove the fuel pump and fuel tank gauge as an assembly.
6. Replace the fuel pump.
7. Installation is the reverse order of the removal procedure.

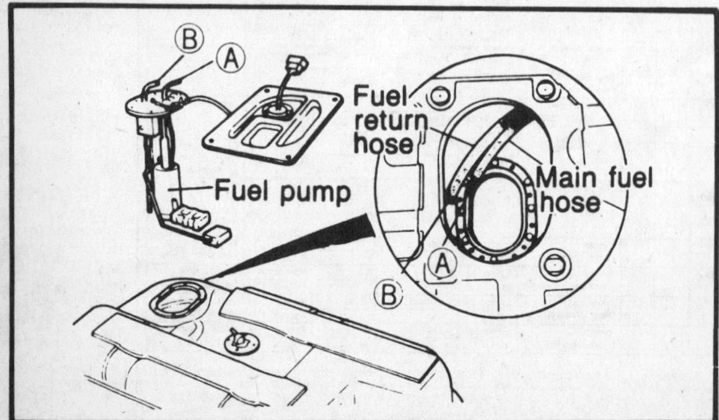


**Fuel pump removal**

## PRESSURE REGULATOR

### Removal and Installation

1. Relieve the fuel pressure from the fuel system. Disconnect the negative battery cable.
2. Remove the throttle chamber (dynamic chamber).
3. Disconnect the fuel return hose.
4. Remove the pressure regulator.
5. Installation is the reverse order of the removal procedure. Be sure to replace the O-ring(s) and after installation check for fuel leaks with fuel pressure applied.

## FUEL FILTER

The fuel filter should be replaced every 30,000 miles. Replace the filter as follows:

1. Relieve the fuel pressure from the fuel system. Disconnect the negative battery cable.
2. Raise the front of the front of the car and support it with jack stands.
2. Disconnect and plug the fuel lines. Remove the fuel filter with bracket as an assembly.
3. Install a new filter and reconnect the fuel lines.

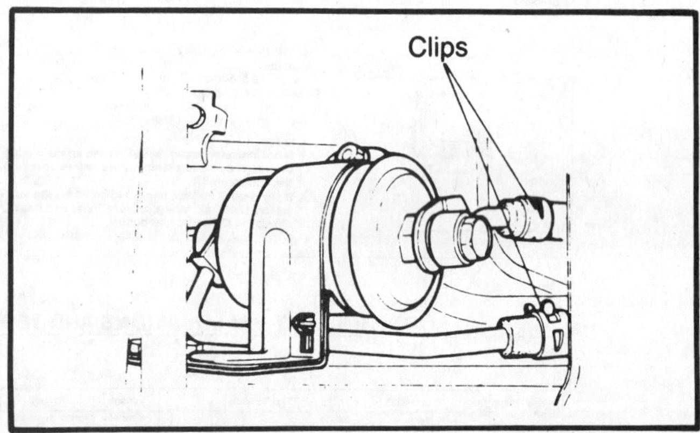


**Fuel filter assembly**

4. Replace the copper washer with a new one and torque it to 18–25 ft.lbs. When installing the fuel filter make sure that the fuel lines are pushed on as far up as possible. Secure the hose with clamps.

## FUEL INJECTORS

### Removal and Installation

1. Relieve the fuel pressure from the fuel system. Disconnect the negative battery cable.

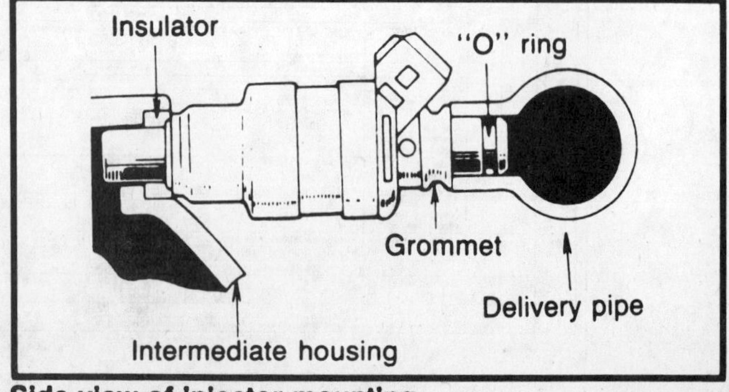


**Side view of injector mounting**

2. Remove the throttle body (dynamic chamber) as described earlier
3. Disconnect the fuel hose and pipe. Disconnect the injector connector.
4. Remove the pressure regulator and distribution pipe.
5. Remove the injector. Once the injector is removed, use an ohmmeter and measure the resistance at the injector connector (on the injector). The resistance should be 12–16 ohms.
6. Installation is the reverse order of the removal procedure. Be sure to replace the O-ring(s) and after installation check for fuel leaks with fuel pressure applied.

## THROTTLE POSITION SENSOR

### Adjustment

#### 1988–90

1. Warm up the engine to operating temperature, then turn it OFF.
2. Disconnect the connector from the throttle position sensor.
3. Connect the throttle position sensor tester (49–F018–001) to the check connector (green).
4. On 1989–90 models, loosen the throttle sensor mounting bolts and move the sensor to set the correct closed position voltage. Tighten the mounting bolts.
5. On all models, turn the ignition switch on and check whether one of the tester lamps illuminates.
6. If both lamps illuminate or if neither does, turn the throttle sensor adjusting screw until one of the lamps illuminates.
   a. If both lamps illuminate turn the adjusting screw counter-clockwise.
   b. If both lamps do not illuminate turn the adjusting screw clockwise.
7. Reinstall the cap on the adjusting screw after adjustment.

**NOTE: Do not apply excessive pressure on the adjusting screw, as it may cause incorrect adjustment.**

## IDLE SYSTEM

### Idle Speed Adjustment

Before adjusting the idle speed and idle mixture, be sure the ignition timing, spark plugs, etc., are all in normal operating condition. Turn off all lights and other unnecessary electrical loads.

1. Connect a suitable tachometer to the check connector (Black), located beside the trailing side coil. If the tachometer does not operate properly, reconnect it to the leading side coil.
2. On 1988 models, install a jumper wire across the set coupler terminals. On 1989–90 models, ground the test connector (green: 1-pin).
3. On all models, after warming up the engine, check that the choke valve has fully opened.
4. **Check the idle speed. If necessary,** turn the throttle adjust screw and set the idle speed to specifications.
   1988–90 – 725–775 rpm

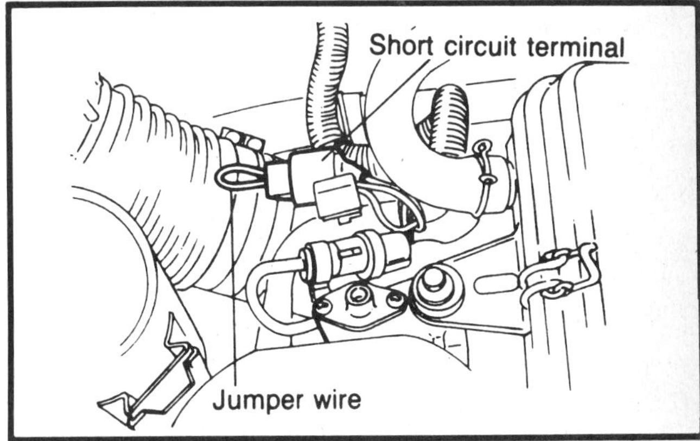

**Fuel pump short-circuit connector**

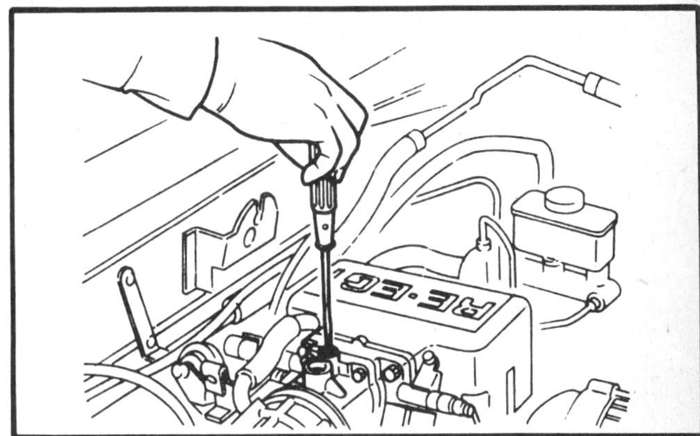

**Idle Speed adjustment – non-turbo engines**

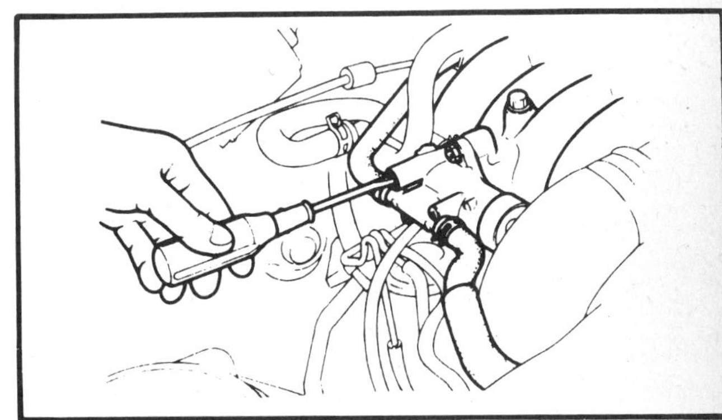
**Idle speed adjustment – turbo**

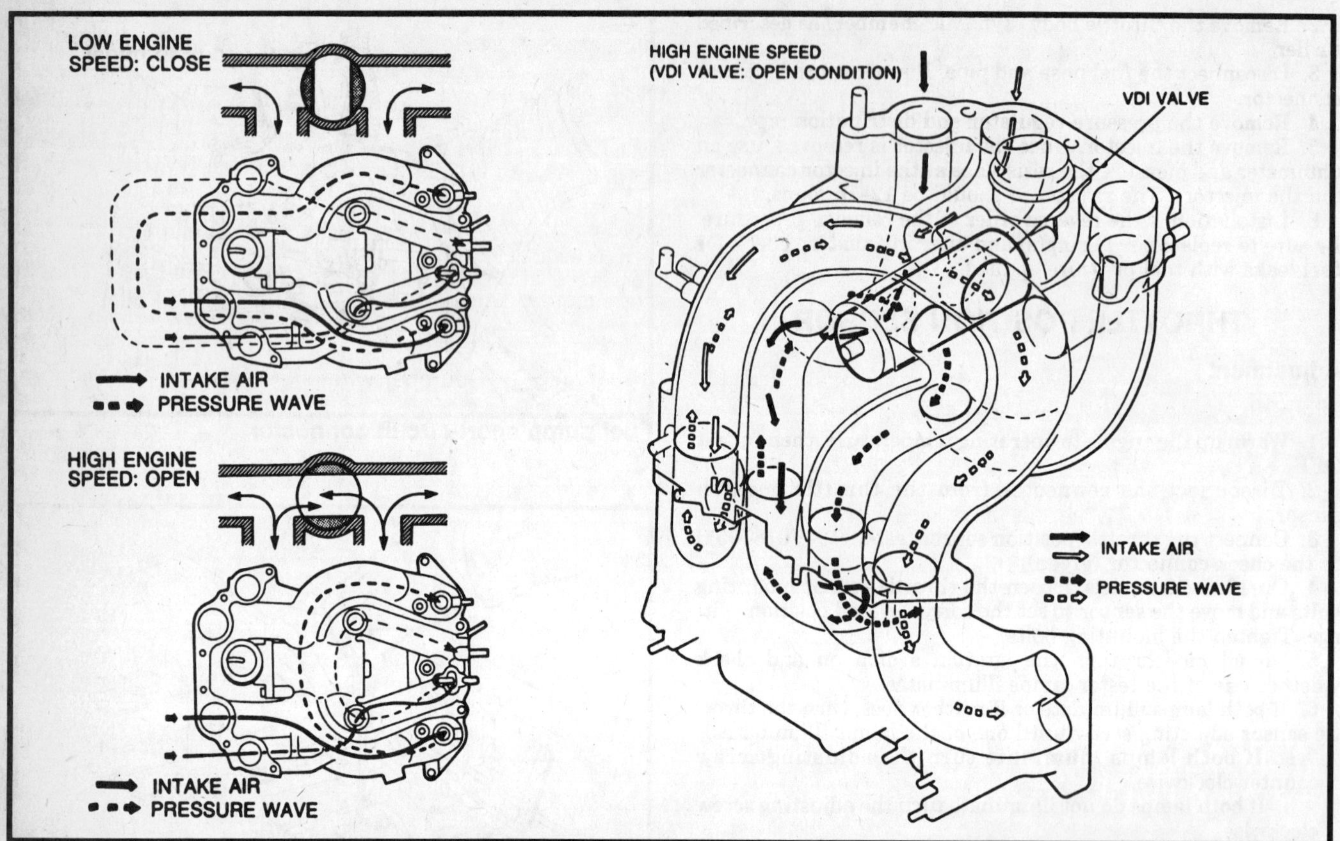


**Variable dynamic effect intake system – 1990 RX-7 shown**

# MITSUBISHI ELECTRONICALLY CONTROLLED INJECTION (ECI) SYSTEM

## MITSUBISHI

| Year | Model | Engine cc (liter) | Family | Fuel System | Ignition System |
|---|---|---|---|---|---|
| 1988 | Tredia | 1997 (2.0) | G63B | FBC | Electronic |
| | | 1795 (1.8) | G62B | ECI | ESC (HEI) |
| | Cordia | 1997 (2.0) | G63B | FBC | Electronic |
| | | 1795 (1.8) | G62B | ECI | ESC (HEI) |
| | Starion | 2555 (2.6) | G54B | ECI | ESC (HEI) |
| | Galant | 2972 (3.0) | 6G72 | MPI | ECIT |
| | Mirage | 1468 (1.5) | G15B | FBC | Electronic |
| | | 1597 (1.6) | G32B | ECI | ECIT |
| | Precis | 1468 (1.5) | G15B | FBC | Electronic |
| | Montero | 2555 (2.6) | G54B | FBC | Electronic |
| | Truck | 1997 (2.0) | G63B | FBC | Electronic |
| | | 2555 (2.6) | G54B | FBC | Electronic |
| | Van/Wagon | 2350 (2.4) | G64B | MPI | |
| 1989 | Starion | 2555 (2.6) | G541B | ECI | ESC (HEI) |
| | Galant | 1997 (2.0) SOHC | 4G63 | MPI | ECIT |
| | | 1997 (2.0) DOHC | 4G63 | MPI | ECIT |
| | Mirage | 1468 (1.5) | 4G15 | MPI | ECIT |
| | | 1597 (1.6) | 4G61 | MPI | ECIT |
| | Precis | 1468 (1.5) | G15B | FBC | Electronic |
| | Montero | 2555 (2.6) | G54B | FBC | Electronic |
| | | 2972 (3.0) | 6G72 | MPI | ECIT |
| | Truck | 1997 (2.0) | G63B | FBC | Electronic |
| | | 2555 (2.6) | G54B | FBC | Electronic |
| | Van/Wagon | 2350 (2.4) | 4G64 | MPI | ECIT |
| 1990 | Galant | 1997 (2.0) SOHC | 4G63 | MPI | ECIT |
| | | 1997 (2.0) DOHC | 4G63 | MPI | ECIT |
| | Mirage | 1468 (1.5) | 4G15 | MPI | ECIT |
| | | 1597 (1.6) | 4G61 | MPI | ECIT |
| | Precis | 1468 (1.5) | G4DJ | MPI | ECIT |
| | Montero | 2972 (3.0) | 6G72 | MPI | ECIT |
| | Truck | 2350 (2.4) | 4G64 | MPI | ECIT |
| | | 2972 (3.0) | 6G72 | MPI | ECIT |
| | Van/Wagon | 2350 (2.4) | 4G64 | MPI | ECIT |
| | Eclipse | 1795 (1.8) | 4G37 | MPI | ECIT |
| | | 1997 (2.0) | 4G63 | MPI | ECIT |

ESC—Electronic spark control
ECIT—Electronic control ignition timing
FBC—Feedback carburetor
ECI—Electronic controlled injection (throttle body inj.)
MPI—Multiport injection
SOHC/DOHC—Single/double overhead cam

## General Information

The Mitsubishi Electronically Controlled Injection System is a standard throttle body system with some variations in operation and components. The fuel control system consists of an electronic control unit (ECU), 1 or 2 solenoid-type fuel injectors, an air flow sensor and several engine sensors. The ECU receives voltage signals from the engine sensors on operating conditions, then sends impulses to the injectors to constantly adjust the fuel mixture. In addition, the ECU controls starting enrichment, warm-up enrichment, fast idle, deceleration fuel cut-off and overboost fuel cut-off on turbocharged models.

One of the primary components is the air flow sensor with its device for generating a Karman vortex or controlled turbulence in the intake air stream. Ultrasonic waves are transmitted across the air flow containing the Karman vortices, which are generated in proportion to the air flow rate. The greater the number of vortices, the more the frequency of the ultrasonic waves is changed (modulated). These modulated ultrasonic waves are picked up by the receiver in the air flow sensor and converted into a voltage signal for the ECU. The ECU uses this signal to measure air flow and control fuel delivery and secondary air management. An intake air temperature sensor is used to provide a signal so that air density changes due to temperature can be calculated.

Other components in the system are common to all throttle body systems. During closed loop operation, the ECU monitors the oxygen sensor to determine the correct fuel mixture according to the oxygen content of the exhaust gases. In the open loop mode, fuel mixture is preprogrammed into the control unit memory.

When the ECI system is activated by the ignition switch, the fuel pump is energized by the ECU. The pump will only operate for about one second unless the engine is running or the starter is cranking. When the engine starts, the fuel pump relay switches to continuous operation and all engine sensors are activated and begin providing input for the ECU. The ISC motor will control idle speed (including fast idle) if the throttle position switch is in the idle position, and the ignition advance shifts from base timing to the preprogrammed ignition advance curve. The fuel pressure regulator maintains system pressure at approximately 14.5 psi (1 bar) by returning excess fuel to the tank.

The ECU provides a ground for the injectors to precisely control the open and closing time (pulse width or driving time) to deliver exact amounts of fuel to the engine, continuously adjusting the air/fuel mixture while monitoring signals from the various engine sensors including:

- Engine coolant temperature
- Intake manifold air temperature and volume
- Barometric pressure
- Intake manifold absolute pressure
- Engine speed (rpm)
- Idle speed
- Detonation
- Boost pressure (turbo models only)
- Throttle position
- Exhaust gas content (oxygen level)

### OPERATION OF ECI SYSTEM

#### Air/Fuel Ratio Control System

The air/fuel ratio control is achieved by controlling the driving time of 1 or 2 injectors installed to the injection mixer. After passing through the in-tank filter, fuel is force-fed by the in-tank fuel pump to the injectors through the main pipe and fuel filter. The fuel pressure applied to the injector is maintained at a fixed level by the fuel pressure regulator. Fuel line pressure is about 36 psi (250 kPa). Care must be used when working around lines and injectors. After pressure regulation, excess fuel is returned to the fuel tank through the return hose. When the injector is energized, the valve inside the injector opens fully to inject the fuel. Since the fuel pressure is kept at a fixed level, supply of fuel injected from the injectors into the injection mixer varies with the energized time.

#### Fuel Injection Control

The amount of fuel injection is determined by the air flow sensor (AFS) frequency corresponding to the amount of intake air. As air flow increases, the amount of injected fuel increases and as the sensor output frequency decreases, the amount of fuel injection decreases. If the air flow sensor fails, the backup within the ECU controls the injectors by means of engine speed sensor signal.

### AIR FLOW SENSOR (AFS)

The AFS measures the intake air volume. It makes use of Karman vortex to detect the air flow rate and sends it to the ECU as the intake air volume signal. The ECU uses this intake air volume signal to decide the basic fuel injection duration.

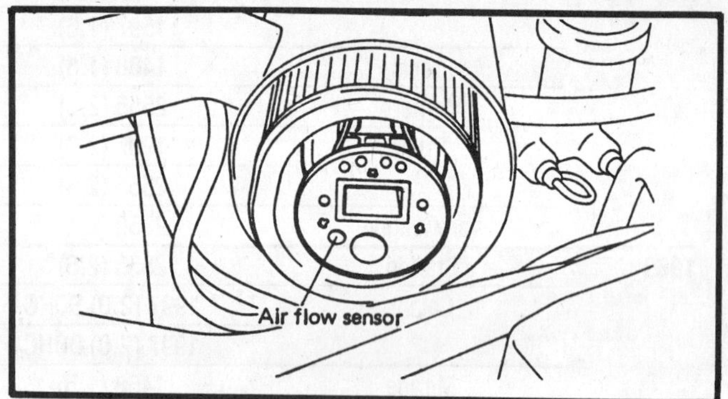


**Air flow sensor (AFS)**

### ATMOSPHERIC PRESSURE SENSOR

The atmospheric pressure sensor installed on the AFS senses the atmospheric pressure and converts it into a voltage which is sent to the ECU. The ECU uses this signal to compute the altitude at which the vehicle is running and adjusts both air/fuel mixture and ignition timing to optimum.

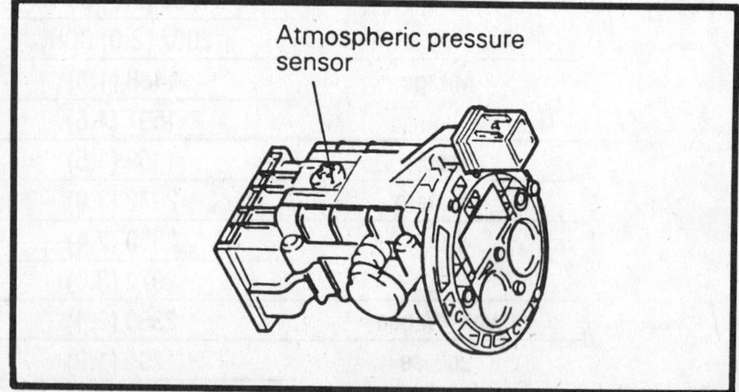


**Atmospheric pressure sensor**

### INTAKE AIR TEMPERATURE SENSOR

The intake air temperature sensor is a resistor-based sensor for detecting the intake air temperature. Since air density is proportional to temperature, the ECU uses this data to adjust the air/fuel ratio.

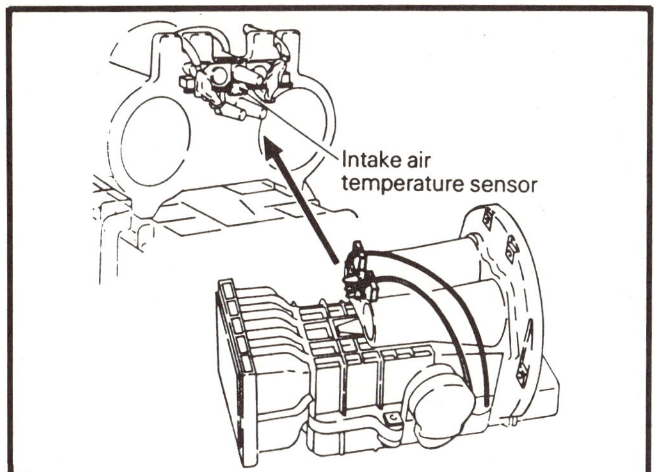


**Intake air temperature sensor**

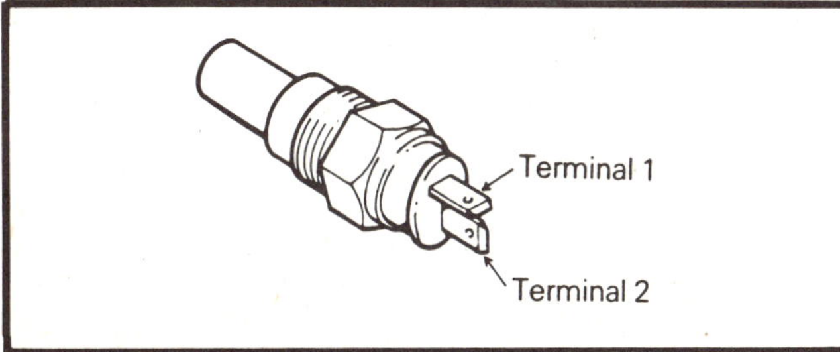


**Engine coolant temperature sensor**

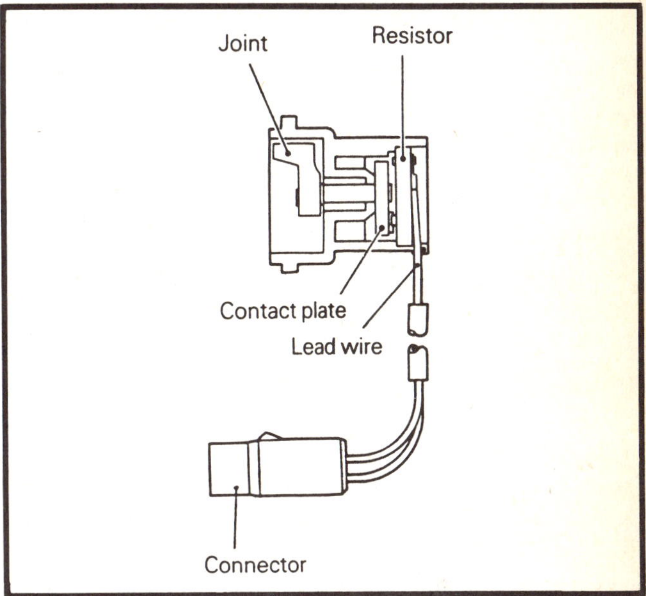


**Throttle position sensor (TPS)**

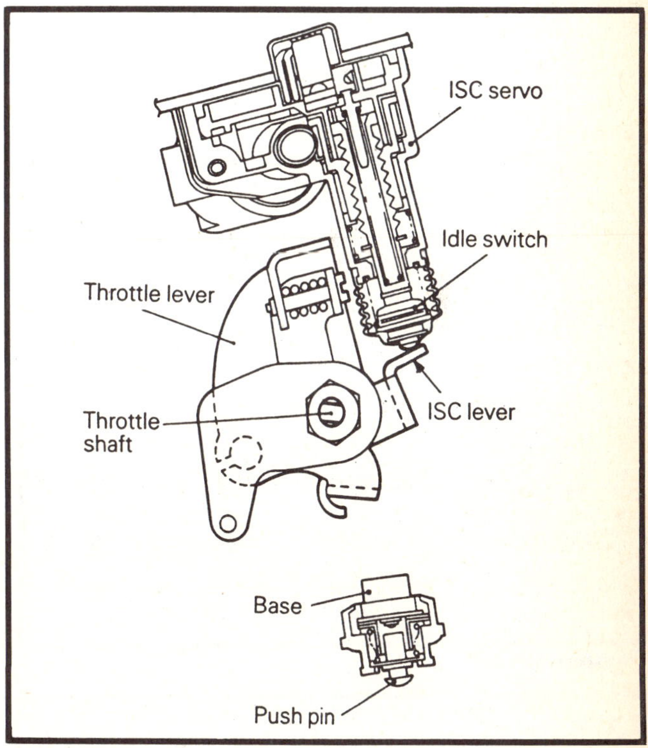


**Idle switch**

## ENGINE COOLANT TEMPERATURE SENSOR

The engine coolant temperature sensor installed in the engine coolant passage of the intake manifold is a resistor-based sensor. This sensor allows ECU to judge cold engine conditions and enrich mixture when necessary.

## THROTTLE POSITION SENSOR (TPS)

The TPS is a rotating variable resistor that moves together with the injection mixer throttle shaft to sense the throttle valve angle. As the throttle shaft rotates, the output voltage of the TPS changes and the ECU detects the throttle valve opening based on the change of the voltage. Based on this output voltage, the ECU computes throttle valve opening change (output voltage change) and judges the engine acceleration/deceleration state accordingly, correcting fuel injection to match.

## IDLE SWITCH

The idle switch, a contact type switch, senses accelerator operation. The switch is installed at the tip of the ISC servo. When the throttle valve is at idle opening, the ISC lever pushes the pin to turn on the switch

## MOTOR POSITION SENSOR (MPS)

The MPS, a variable resistor sensor, is installed in the ISC servo. Its sliding pin is in contact with the plunger end and as the plunger moves, the internal resistance of the MPS changes (the output voltage changes). The MPS senses the ISC servo plunger position and sends the signal to the ECU. The ECU controls the valve opening, and consequently, the idle speed by using the MPS signal, idle signal, engine coolant temperature signal, load signal (automatic transmission and air conditioner) and vehicle speed signal.

## ENGINE SPEED SENSOR (IGNITION NEGATIVE TERMINAL VOLTAGE SENSOR)

The voltage at the ignition coil negative terminal increases sud-

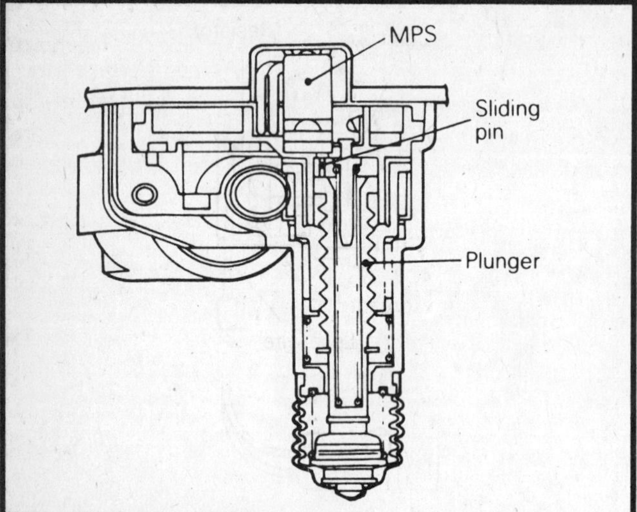


**Motor position sensor**

denly twice per engine revolution with ignition timing. By sensing this ignition coil negative terminal voltage change and measuring the time between peak voltages, the ECU computes the engine speed, judges the engine operating mode and controls the air/fuel ratio and idle speed.

## OXYGEN SENSOR

The oxygen sensor installed in the exhaust pipe makes use of the principles of solid electrolyte oxygen concentration cell. It is characterized by sharp change of output voltage in the vicinity of the 14.7:1 air/fuel ratio.

Using such characteristics, the oxygen sensor senses the oxygen concentration in the exhaust gas and feeds it back (as a voltage signal) to the ECU. The ECU then judges if the air/fuel ratio is richer or leaner than the target 14.7:1 ratio. Fuel metering based on this signal keeps the air/fuel ratio at or close to the ideal (stoichiometric) 14.7:1, allowing the 3-way catalyst to function at maximum efficiency.

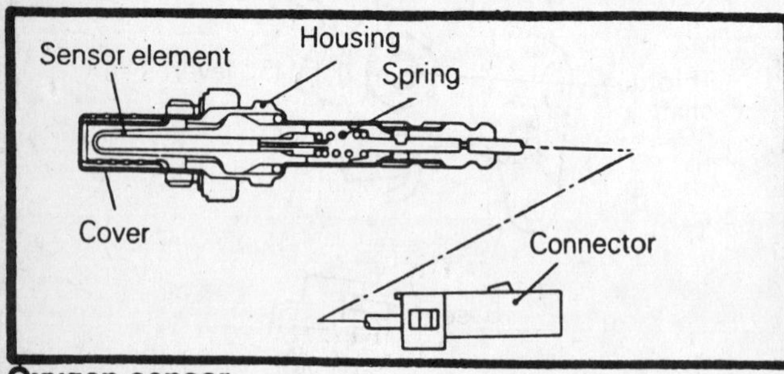


**Oxygen sensor**

**Inhibitor switch (automatic transaxle only)**

## VEHICLE SPEED SENSOR

The vehicle speed sensor uses a reed switch. The speed sensor built in the speedometer converts the transaxle speedometer gear revolution (vehicle speed) into pulse signals, which are sent to the ECU.

## INHIBITOR SWITCH (AUTOMATIC TRANSAXLE)

This switch detects whether the select lever is positioned at **NEUTRAL or PARK**. Based on this signal, the ECU senses the automatic transaxle load and drives the ISC servo to keep optimum idle speed.

## AIR CONDITIONER SWITCH

When the air conditioner is turned on, the air conditioner ON signal is sent to the ECU. Based on this signal, the ECU drives the ISC servo to keep optimum idle speed.

## DETONATION SENSOR

Installed on the cylinder block, the detonation sensor converts vibration into voltage by its piezoelectric element. When detonation occurs, it resonates within the cylinder block, generating voltage which is sent to the igniter as the detonation signal. Based on this signal, the igniter retards the ignition timing to prevent detonation.

## INJECTORS

The injector, an injection nozzle with solenoid valve, injects fuel according to the injection signal from the ECU. The injectors are installed on the injection mixer (throttle body) and inject

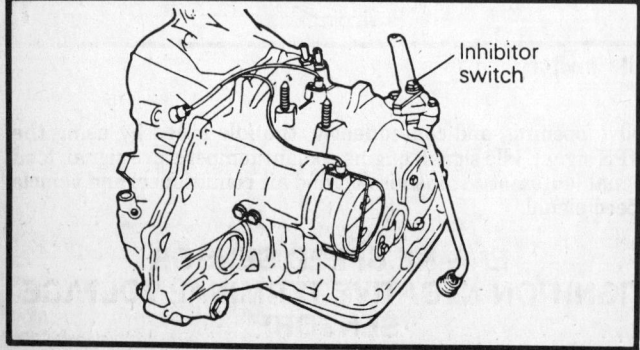

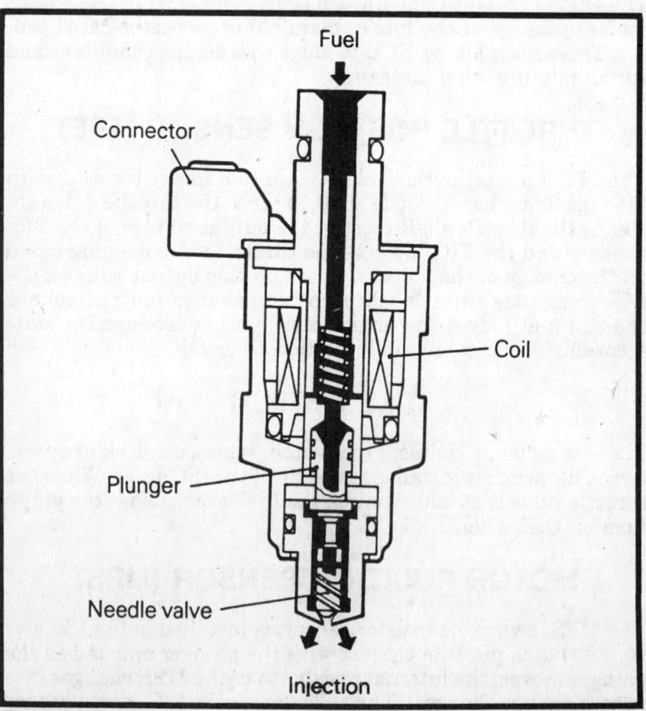


**Fuel injector**

fuel upstream of the throttle valve. When the solenoid coil is energized, the plunger is attracted. The needle valve integral with the plunger is then pulled to the full open position and fuel is injected through the valve. Since the injection nozzle opening is fixed and the fuel pressure is also fixed, the injection amount is determined by the length of time the fuel flows.

## FUEL PRESSURE REGULATOR

The fuel pressure regulator keeps the injector fuel pressure at a level 36 psi (250 kPa) higher than the pressure inside the injection mixer. By doing so, the fuel injection amount is kept constant even when the mixer inside pressure changes. Excess fuel is returned to the tank through the return line.

## RESISTOR

The resistor limits the electric current flowing to the injector coil. The injector is required to respond quickly to the fuel injection signal. This fast response is achieved by reducing the number of turns of the injector coil and thus improving current rise when the coil is energized. This smaller number of turns, however, draws more current and generates more heat. In order to prevent this, a resistor is provided between the power supply (+) and the injector to limit current flowing to the coil.

## IDLE SPEED CONTROL (ISC) SERVO

The ISC servo consists of a motor, worm gear, worm wheel and plunger. The motor position sensor (MPS) detects plunger position and the idle switch detects idle position. The worm gear, installed on the motor shaft, transmits motor rotation to the worm wheel. The worm wheel is meshed with worm on the plunger so that the plunger extends or retracts as the worm wheel rotates. As the motor rotates according to the signal from the ECU, the plunger extends or retracts (depending on the direction of rotation of the motor) to actuate the throttle valve via the ISC lever. In this way, the idle speed is controlled by changing the throttle valve opening.

## IGNITER

The igniter controls timing of the spark to the cylinders. It controls the timing based on signals from the ECU as well as on signals from the detonation sensor. Should the igniter lose the signal from the detonation sensor, the igniter will use its default or "fail-safe" mode to retard ignition timing by a fixed amount (except at idle) to protect the engine.

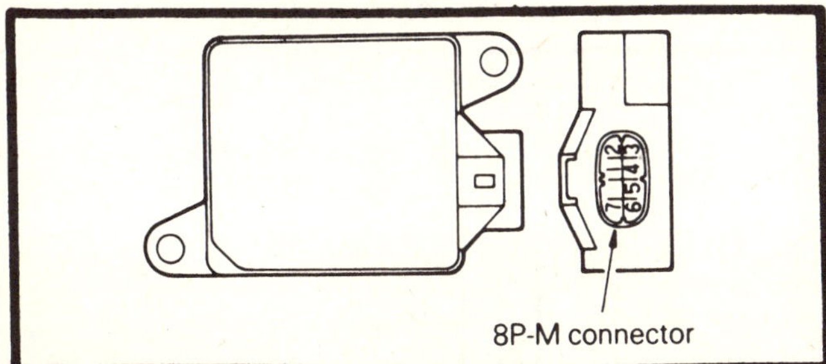


**Igniter**

## ELECTRONIC CONTROL UNIT (ECU)

Based on the information from various sensors, the ECU determines (computes) an optimum control for varying operating conditions and accordingly drives the output actuators. The ECU consists of an microprocessor, random access memory

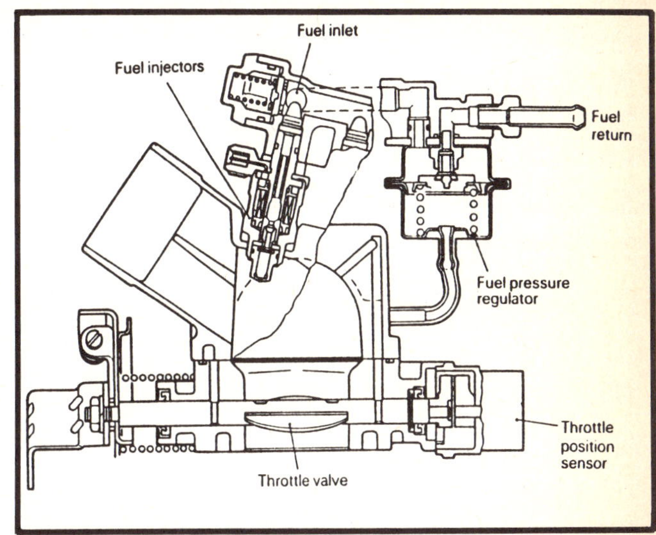


**Injection mixer assembly**

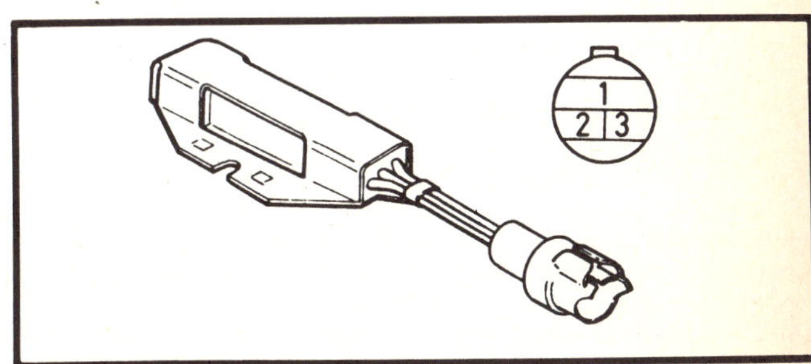


**Injector resistor**

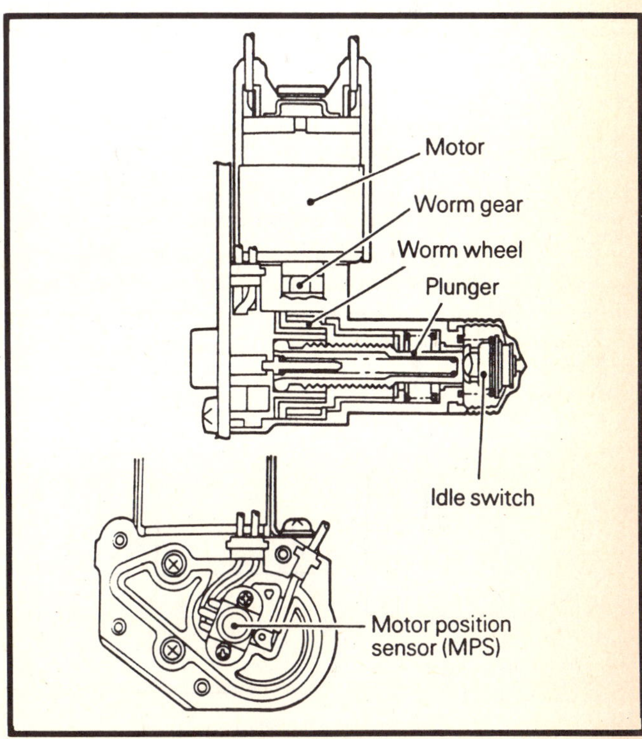


**Idle speed control (ISC) servo**

(RAM), read only memory (ROM) and input/output (I/O) interface.

**NOTE: Antennas or cables from additional radio equipment may affect ECU function. Keep antenna cables at least 8 in. from computers and route radio cables to cross vehicle harnesses at 90 degree angles. Antenna and cable should be matched and adjusted to lowest possible standing wave ratio. (SWR).**

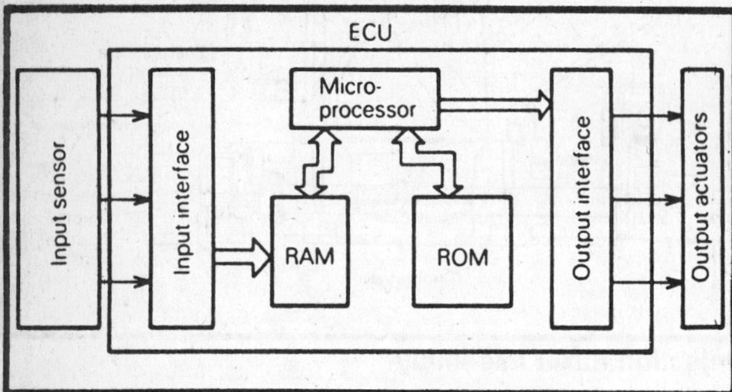


**ECU Schematic**

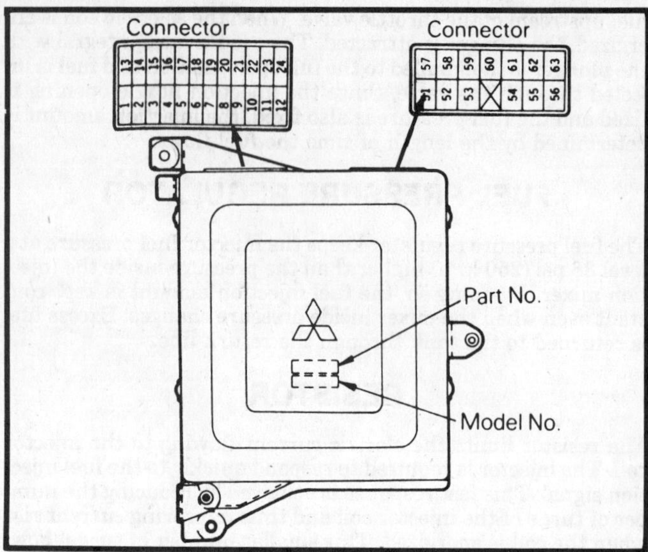


**Typical ECU. Identification numbers are important if replacement is needed**

## Diagnosis and Testing

### SELF-DIAGNOSIS SYSTEM

The Mitsubishi self-diagnosis system monitors the various input signals from the engine sensors and enters a trouble code in the on-board computer memory if a problem is detected. There are nine monitored items, including the "normal operation" code which can be read by using a special ECI tester and adapter. The adapter connects the ECI tester to the diagnosis connector located on the right cowl, next to the control unit. Because the computer memory draws its power directly from the battery, the trouble codes are not erased when the ignition is switched **OFF**. The memory can only be cleared (trouble codes erased) if a battery cable is disconnected or the main ECU wiring harness connector is disconnected from the computer module.

If 2 or more trouble codes are stored in the memory, the computer will read out the codes in numerical order beginning with the lowest numbered code.

**NOTE: The order in which codes are displayed does NOT indicate the order of occurrence.**

The needle of the ECI tester will swing back and forth between 0 and 12 volts to indicate the trouble code stored. There is

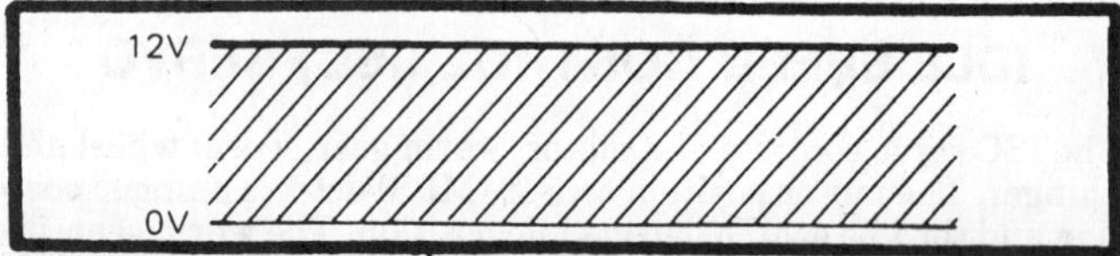


**Constant 12 volt signal indicates no fault codes stored in ECU**

no memory for Code No. 1 (oxygen sensor) once the ignition is switched **OFF**, so it is necessary to perform this diagnosis with theengine running. The oxygen sensor should be allowed to warm up for testing (engine at normal operating temperature) and the trouble code should be read before the ignition is switched **OFF**. All other codes may be read with the engine either on or off. If there are no trouble codes stored in the computer (system is operating normally), the ECI tester will indicate a constant 12 volts on the meter. Consult the instructions supplied with the test equipment to insure proper connections for diagnosis and testing of all components.

The stored codes may also be read on an analog (dial type) voltmeter hooked into the system. Since the output from the ECU is electrical, the meter needle will deflect or sweep as the pulses are generated. Recording the number of sweeps and their time duration will yield the numeric code involved. (For exam-

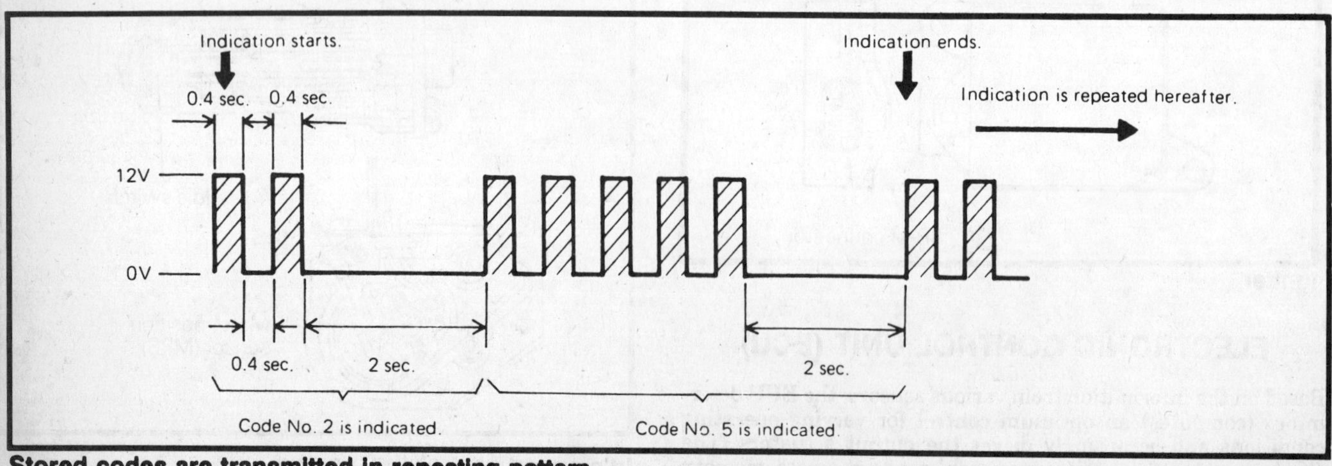


**Stored codes are transmitted in repeating pattern**

ple: 2½–second sweeps followed by a 2 second pause followed by 3½–sweeps might indicate code 23 or Code 2 and Code 3, depending on the system.)

The code, when interpreted, points you to the unit which may be the problem. It must still be checked along with the attendant wiring, connectors and controls. A great number of fault codes are set because of loose or dirty connections in the wiring which fool the ECU into thinking the unit has failed.

Not every component can be tested with a voltmeter or ohmmeter; some require the use of the ECI checker. In each case remember that you are only reading voltage used to transmit a code; the actual voltage running within the system can only be checked with the factory diagnostic unit.

**NOTE: Any resistances given are for 68°F (20°C). Remember that resistance will increase or decrease respectively as the temperature rises or falls. Use common sense in interpreting the readings.**

If the battery voltage is low, the self-diagnosis system will not operate properly; charge should be checked before attempting any self-diagnosis procedures. After completing service procedures, the computer trouble code memory should be erased by disconnecting the battery cable or main harness connector to the control unit for at least 15 seconds.

### INSPECTION OF ECI SYSTEM

**If ECI system components (sensors, ECU, injector, etc.) fail, interruption of fuel supply will result. Therefore, the following situations may be encountered.**

1. Engine is hard to start or does not start at all.
2. Unstable idle.
3. Poor driveability.

If any of above conditions is noted, first perform inspection by self-diagnosis and subsequent basic engine checks (ignition system malfunctions, incorrect engine adjustment, etc.), and then inspect the ECI system components by ECI Checker.
The ECI system can be checked by use of the ECI Checker, Diagnosis Harness Connector and Adapter.

## SERVICE PRECAUTIONS

- Before battery terminals are disconnected, make sure that ignition switch is **OFF**. If battery terminals are disconnected while engine is running or when ignition switch is in **ON** position, ECU damage will result.
- Disconnect battery cables before charging battery.
- When battery is reconnected, be sure not to reverse polarity.
- Make sure that harness connectors are securely connected. Use care not to allow entry of water or oil into connectors.

## INSPECTION PROCEDURE BY SELF-DIAGNOSIS

1. Turn ignition switch to **OFF**.
2. Open glove compartment and pull out connector for diagnosis.
3. Connect a volt meter between terminal for "ECI" and terminal for ground.
4. Turn ignition switch to **ON**, and indication of ECU memory contents will immediately start. If system is in normal condition, pointer of volt meter constantly indicates 12 volts. If any codes are stored in memory, pointer of voltmeter will deflect. Record each code as it is transmitted.
5. Turn ignition switch to **OFF**.
6. Once defective part or circuit has been repaired, disconnect negative battery cable for 15 seconds or more to make sure that abnormal code has been erased.

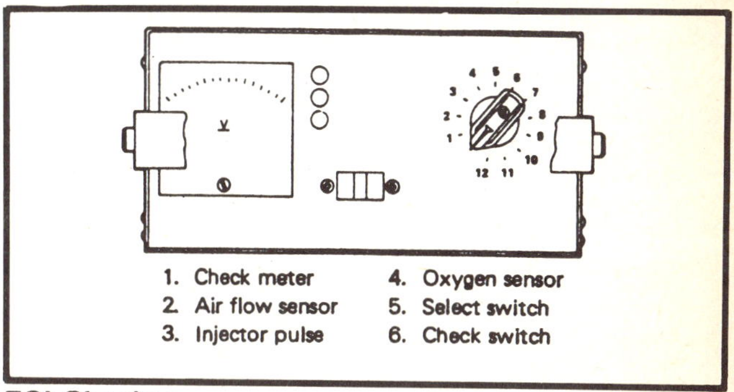


ECI Checker

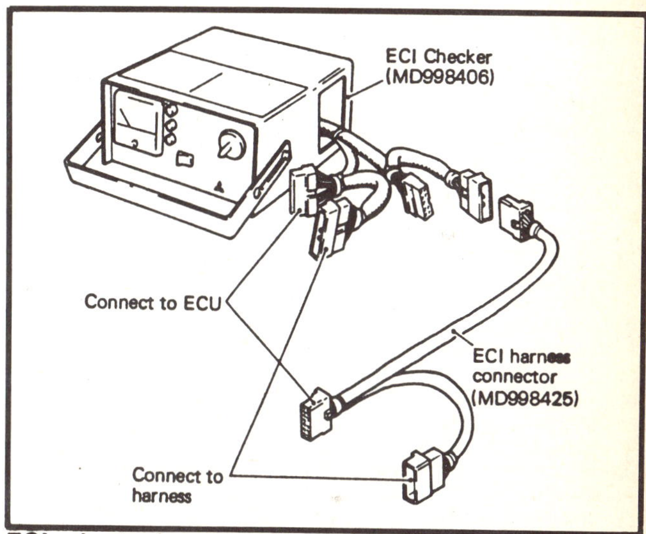


ECI adapter harness. Cordia/Tredia harness shown, Starion harness is MD998437

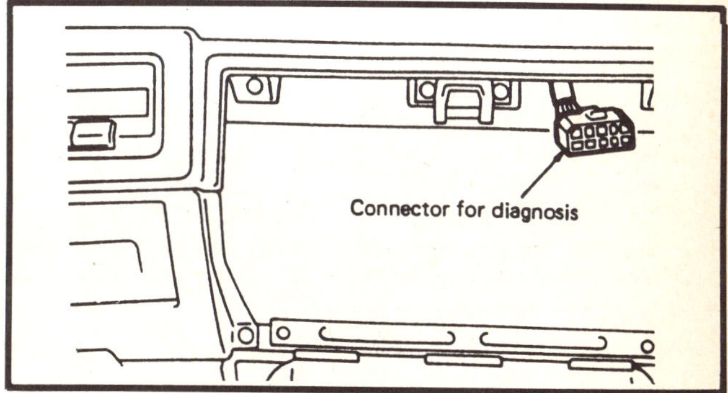


Diagnostic connector location

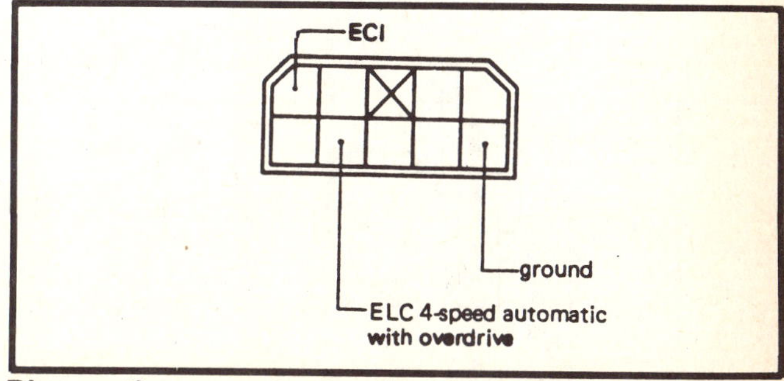


Diagnostic connector terminal identification

## 1988 CORDIA/TREDIA—TESTING WITH ECI CHECKER

| Select switch | Check switch | Check item | Condition | | Check meter reading when normal | Terminal number of computer |
|---|---|---|---|---|---|---|
| A | 1 | Power supply | Ignition switch OFF → ON | | SV | 51 |
| | 2 | Secondary air control solenoid valve | Ignition switch OFF → ST after warming up the engine | | After 15 seconds 0–0.5V ↓ SV | 10 |
| | 3 | Throttle position switch | Ignition switch OFF → ON | Accelerator closed | 0.4–1.5V | 1 |
| | | | | Accelerator wide opened | 4.5–5.0V | |
| | 4 | Coolant temperature sensor | Ignition switch OFF → ON | 0°C (32°F) | 3.5V | 3 |
| | | | | 20°C (68°F) | 2.6V | |
| | | | | 40°C (104°F) | 1.8V | |
| | | | | 80°C (176°F) | 0.6V | |
| | 5 | Intake air temperature | Ignition switch OFF → ON | 0°C (32°F) | 3.5V | 4 |
| | | | | 20°C (68°F) | 2.6V | |
| | | | | 40°C (104°F) | 1.8V | |
| | | | | 80°C (176°F) | 0.6V | |
| | 6 | Idle position switch | Ignition switch OFF → ON | Accelerator closed | 0–0.4V | 5 |
| | | | | Accelerator wide opened | SV | |
| | 7 | ISC motor position switch | Ignition switch OFF → ON | | SV* | 14 |
| | 8 | – | – | | – | – |
| | 9 | – | – | | – | – |
| | 10 | A/C (Air conditioner) relay | Ignition switch OFF → ON | A/C switch OFF | 0–0.5V | 62 |
| | | | | A/C switch ON | SV | |
| | 11 | Lead switch for vehicle speed | Start engine, transaxle in first and operate vehicle slowly | | Flashing 0–0.5V ↕ Over 2V | 15 |
| | 12 | – | – | | – | – |
| B | 1 | Cranking signal | Ignition switch OFF → ST | | Over 8V | 13 |
| | 2 | Control relay | Idling | | 0 – 0.5V | 55 |
| | 3 | – | – | | – | – |

SV: System Voltage

| Select switch | Check switch | Check item | Condition | | Check meter reading when normal | Terminal location of computer |
|---|---|---|---|---|---|---|
| B | 4 | Ignition pulse | Idling | | 12 – 14V | 8 |
| | | | 3000 rpm | | SV | |
| | 5 | Air flow sensor | Idling | | 2.2 – 3.2V | 7 |
| | | | 3000 rpm | | | |
| | 6 | Injector No. 1 | Idling | | SV | 59 |
| | | | 3000 rpm | | 12 – 14V | |
| | 7 | Injector No. 2 | Idling | | SV | 60 |
| | | | 3000 rpm | | 12 – 14V | |
| | 8 | Oxygen sensor | Keep 1300 rpm after warming up the engine | | Flashing 0.4–1V ↕ 2.7V | 6 |
| | 9 | EGR control solenoid valve | Keep idling after warming up the engine | | SV | 54 |
| | | | Raise the engine 3500 rpm | | 0–0.5V | |
| | 10 | Pressure sensor | Ignition switch: OFF → ON | | 1.5–2.6V | 17 |
| | | | Idling | | 0.2–1.2V*2 | |
| | 11 | ISC motor for extension | Idling | | 0–2V | 56 |
| | | | | A/C switch: OFF → ON | Momentarily Over 6V | |
| | 12 | ISC motor for retraction | Idling | | 0–2V | 61 |
| | | | | A/C switch: ON → OFF | Momentarily Over 6V | |

NOTE: *1 : If ignition switch is turned to ON for 15 seconds or more, the reading drops below 5V momentarily.
*2 : The reading rises to 1.5–2.6V every 2 minutes momentarily.

| | Check item | Condition | | | Check meter reading when normal | Terminal location of computer |
|---|---|---|---|---|---|---|
| Checking with voltmeter | Spark advance signal | Idling | Coolant temp. below 35°C (95°F) | | Over 5V | 12 |
| | | | Coolant temp. above 35°C (95°F) | Altitudes below 3,900 ft. | 0 – 0.5V | |
| | | | | Altitudes above 3,900 ft. | Over 5V | |
| | Inhibitor switch | Ignition switch OFF → ON | Select lever in "P" or "N" | | 0 – 0.5V | 11 |
| | | | Select lever in "D" | | SV | 11 |

SV : System Voltage

# 1988–89 STARION – TESTING WITH ECI CHECKER

**STEP 1** (Connect white color connectors labeled "CHECKER" of harness connector to ECI checker).

| ECI Checker Operation | | Check Item | ECU Terminal # Checked | Condition | | Test Specification |
|---|---|---|---|---|---|---|
| Select Switch | Check Switch | | | | | |
| "A" | 1 | Power supply | 51 | Ignition switch "LOCK" → "ON" | | SV |
| | 2 | Ignition pulse | 1 | Ignition switch "LOCK" → "START" | | 4V – 10V |
| | 3 | Intake air temperature sensor | 5 | Ignition switch "LOCK" → "ON" | 0°C (32°F) | 3.4V – 3.6V |
| | | | | | 20°C (68°F) | 2.5V – 2.7V |
| | | | | | 40°C (104°F) | 1.7V – 1.9V |
| | | | | | 80°C (176°F) | 0.6V – 0.8V |
| | 4 | Engine coolant temperature sensor | 6 | Ignition switch "LOCK" → "ON" | 0°C (32°F) | 3.4V – 3.6V |
| | | | | | 20°C (68°F) | 2.5V – 2.7V |
| | | | | | 40°C (104°F) | 1.5V – 1.7V |
| | | | | | 80°C (176°F) | 0.5V – 0.7V |
| | 5 | Power supply for sensor | 10 | Ignition switch "LOCK" → "ON" | | 4.5V – 5.5V |
| | 6 | Throttle position sensor | 15 | Ignition switch "LOCK" → "ON" (Warm engine) | Accelerator fully closed | 0.4V – 0.7V |
| | | | | | Accelerator fully opened | 4.5V – 5.5V |
| | 7 | Motor position sensor | 3 | Ignition switch "LOCK" → "ON" | After 15 seconds | 0.8V – 1.2V |
| | 8 | Idle position switch | 7 | Ignition switch "LOCK" → "ON" | Accelerator fully closed | 0V – 0.6V |
| | | | | | Accelerator fully opened | 8V – 13V |
| | 9 | Cranking signal | 55 | Ignition switch "LOCK" → "START" | | Over 8V |
| | 10 | Vehicle speed sensor reed switch | 19 | Start engine and operate vehicle slowly in 1st or DRIVE range | | 0V – 0.6V ↑ (pulsates) ↓ Over 2V |
| | 11 | Air conditioner switch | 56 | Ignition switch "LOCK" → "ON" | Air conditioner switch "OFF" | 0V – 0.6V |
| | | | | | Air conditioner switch "ON" *1 | SV |
| | 12 | Inhibitor switch | 58 | Ignition switch "LOCK" → "ON" | At "P" or "N" range | 0V – 0.6V |
| | | | | | At "D" range | SV |

NOTE
*1: ON means compressor clutch engaged.
SV = System Voltage

**STEP 1**

| ECI Checker Operation | | Check Item | ECU Terminal # Checked | Condition | | Test Specification |
|---|---|---|---|---|---|---|
| Select Switch | Check Switch | | | | | |
| "B" | 1 | | | | | |
| | 2 | | | | | |
| | 3 | | | | | |
| | 4 | Spark advance signal | 13 | Idling | Engine coolant temperature less than 35°C (95°F) | Over 5V |
| | | | | | Engine coolant temperature 35°C (95°F) or higher, altitude up to approx. 1,200 m (3,900 ft.) | 0 – 0.6V |
| | | | | | Engine coolant temperature 35°C (95°F) or higher, altitude approx. 1,200 m (3,900 ft.) or above | Over 5V |
| | 5 | Air flow sensor | 2 | Idling | | 2.2 – 3.2V |
| | | | | 3,000 rpm | | |
| | 6 | | | | | |
| | 7 | EGR control solenoid valve | 54 | Hold engine at a speed less than 3,500 rpm after warming up | | SV |
| | | | | Hold engine at a speed 3,500 rpm or higher | | 0V – 0.6V |
| | 8 | Oxygen sensor | 11 | Hold engine at a constant speed above 1,300 rpm, after 30 seconds from start of warm engine | | 0V – 0.6V *2 ↑ (pulsates) ↓ 2V – 3V |
| | 9 | | | | | |
| | 10 | | | | | |
| | 11 | | | | | |
| | 12 | | | | | |

NOTE
*2: Failure of parts other than the oxygen sensor can also cause deviation from the specifications. Therefore, check other parts related to air-fuel ratio control.
SV = System Voltage

## 1988–89 STARION – TESTING WITH ECI CHECKER

**STEP 2** (Connect green color connectors labeled "CHECKER" of harness connector to ECI checker)

| ECI Checker Operation | | Check Item | ECU Terminal # Checked | Condition | Test Specification |
|---|---|---|---|---|---|
| Select Switch | Check Switch | | | | |
| "A" | 1 | | | | |
| | 2 | Secondary air control solenoid valve | 20 | Hold engine over 1,500 rpm, 15 seconds after start of warm engine | 0V – 0.6V then SV |
| | 3 | | | | |
| | 4 | | | | |
| | 5 | | | | |
| | 6 | | | | |
| | 7 | | | | |
| | 8 | | | | |
| | 9 | | | | |
| | 10 | | | | |
| | 11 | | | | |
| | 12 | | | | |

ECU terminal

View from front as installed in ECU

SV = System Voltage

| Malfunction No. | Diagnosis item | Self-diagnosis output pattern and output code | Problem | Check item |
|---|---|---|---|---|
| 1 | Oxygen sensor | 12V 0V | Oxygen sensor signal does not change for 20 seconds or more in its feedback range. | • Wire harness and connector • Oxygen sensor • ECU |
| 2 | Ignition pulse | 12V 0V | While cranking the engine, input of ignition signal is not applied to ECU for 3 seconds or more | • Wire harness and connector • Igniter • ECU • Distributor |
| 3 | Air flow sensor (AFS) | 12V 0V | AFS output is 10 Hz or less while engine is idling. | • Wire harness and connector • AFS • ECU |
| 5 | Throttle position sensor (TPS) | 12V 0V | • TPS output is 0.2V or less. • TPS output is 4V or higher for 1 second or more while engine is idling (Idle switch is on). | • Wire harness and connector • TPS • ECU |
| 6 | ISC motor position sensor | 12V 0V | • MPS output voltage is 4.8V or more. • MPS output voltage is 0.2V or less. | • Wire harness and connector • Motor position sensor (MPS) • ECU |
| 7 | Engine coolant temperature sensor | 12V 0V | • Engine coolant temperature sensor output is 4.5V or more. • Engine coolant temperature sensor output is 0.1V or less. | • Wire harness and connector • Engine coolant temperature sensor • ECU |

**Diagnostic code chart – 1988–89 Starion. Note that Code 4 is not used**

| Mal-function No. | Diagnosis item | Self-diagnosis output pattern and output code | Problem | Check item |
|---|---|---|---|---|
| 1 | Oxygen sensor | 12V 0V | When oxygen sensor signal does not change for 20 seconds or more in its feed-back zone. | • Wire harness and connector • Oxygen sensor • ECU |
| 2 | Ignition signal | 12V 0V | While cranking the engine, input of ignition signal is not applied to ECU for 3 seconds or more. | • Wire harness and connector • Igniter • ECU |
| 3 | Air flow sensor | 12V 0V | Air flow sensor output is 10 Hz or less while engine is idling, or it is 100 Hz or more when engine stalls. | • Wire harness and connector • Air flow sensor • ECU |
| 4 | Pressure sensor | 12V 0V | Pressure sensor output is 194.7 kPa (28.2 psi) (4.5V) or more, or it is 8.7 kPa (1.3 psi) (0.2V) or less | • Wire harness and connector • Pressure sensor • ECU |
| 5 | Throttle position sensor | 12V 0V | Throttle position sensor output is 0.2V or less, or it is 4V or more while engine is idling (idle switch is ON). | • Wire harness and connector • Throttle position sensor • ECU |
| 6 | ISC motor position switch | 12V 0V | Throttle sensor output is 0.4V with L switch OFF. | • Wire harness and connector • ISC servo • ECU |
| 7 | Coolant temperature sensor | 12V 0V | Water temperature sensor output is 4.5V or more, or it is 0.1V or less. | • Wire harness and connector • Coolant temperature sensor • ECU |

**Diagnostic code chart – 1988 Cordia and Tredia with ECI**

## AIR FLOW SENSOR

### Check

1. Disconnect air flow sensor connector.
2. Unsnap finger clips and remove air cleaner cover. Make certain sensor is disconnected before removing cover.
3. Remove filter from air cleaner body, the remove airflow sensor.
4. Check following and replace if defective:
   a. Check air cleaner element for contamination and damage.
   b. Check air cleaner case and cover for damage and cracks.
5. Check intake air temperature sensor by measuring resistance. Resistance should be 2.65 kilo–ohms.

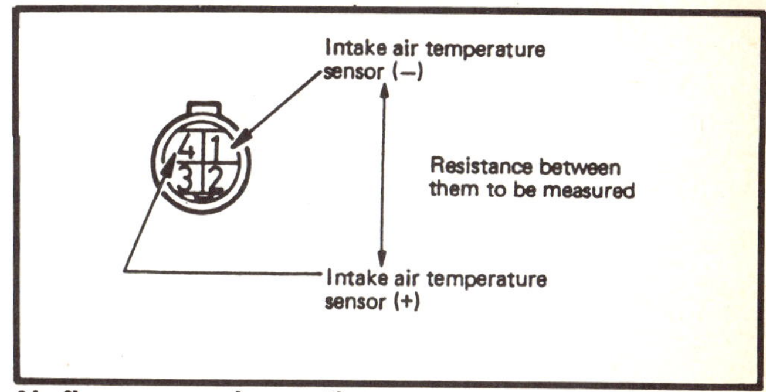


Air flow sensor inspection

## FUEL INJECTOR

### Inspection

#### SOUND TEST

Using a stethoscope or similar tool, check for operating sound as engine idles. Each injector gives a distinct ticking as it opens and closes. Check that the sound occurs at shorter intervals as the engine speed increases. Listen carefully; other injectors may produce similar sound even if the injector being checked is not working.

#### RESISTANCE TEST

Disconnect the injector connector. Measure resistance between the injector terminals. Resistance should be 2–3 ohms. Reconnect injector lead.

#### INJECTION INSPECTION

1. Set ignition switch to **OFF**.
2. Disconnect high tension cable from ignition coil.
3. Remove air intake pipe from injection mixer to make visual inspection of injection condition.
4. When ignition is turned to **START** position, injector should spray in an even pattern.
5. Check to ensure that after ignition switch has been set to **OFF**, there is no fuel leakage or dripping from nozzle of injectors.
6. Reconnect high tension cable to ignition coil and connect air intake hose to injection mixer.

Fuel injection flow test

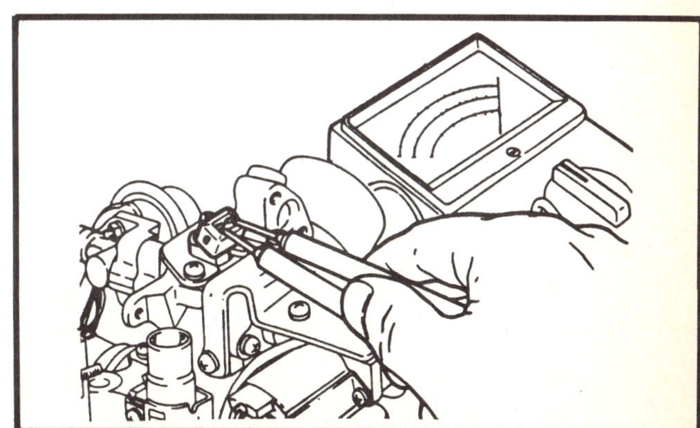

Fuel injector coil continuity test

## IDLE SWITCH

### Test

1. Set ignition switch to **OFF**.
2. Disconnect ISC servo connector.
3. Check for continuity with an ohmmeter circuit tester (ohm range) between terminal No. 2 and good ground.
4. With accelerator released (idle position), meter should show conductivity or 0 ohm resistance. Depress accelerator to position off idle; meter should show no conductivity or infinite resistance.
5. If either condition is not met, replace the ISC servo assembly.
6. Reconnect ISC servo connector.

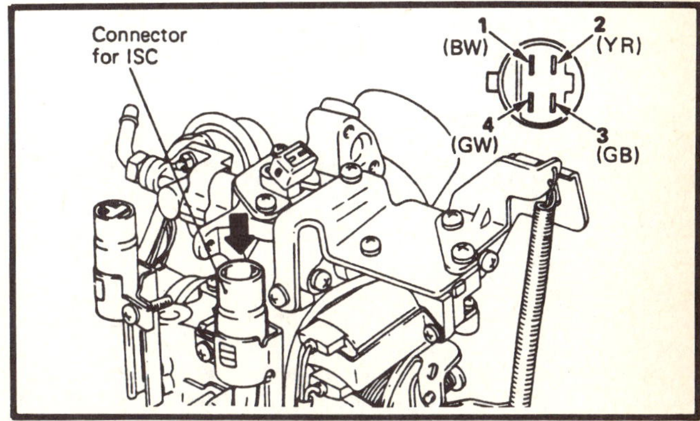


Idle switch test for Cordia/Tredia

## IDLE SPEED CONTROL (ISC) SERVO MOTOR

### Continuity Test

#### CORDIA/TREDIA

1. Set ignition switch to **OFF**.
2. Disconnect connector of ISC servo.

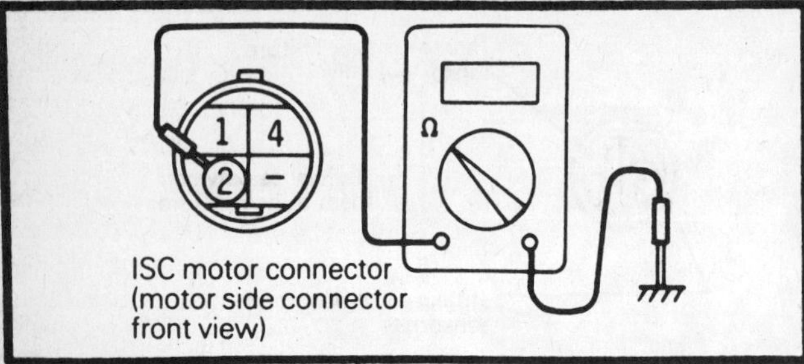


**Testing Starion idle switch**

3. Check motor for continuity with an ohmmeter between terminals 1 and 4. Resistance should be 5–35 ohms. If resistance is 0 or abnormally large, motor coil is open or shorted and must be replaced.

4. There must be no continuity between terminal 1 and ground or terminal 4 and ground. If any continuity exists, the motor coil is shorted and the servo assembly must be replaced.

**Idle speed control (ISC) test**

### Continuity and Function Test

**STARION**

1. Set ignition switch to **OFF**.
2. Disconnect connector of ISC servo.
3. Check motor for continuity with an ohmmeter between terminals 1 and 4. Resistance should be 5–35 ohms. If resistance is 0 or abnormally large, motor coil is open or shorted and must be replaced.
4. Using jumper wires as necessary, connect 6 volt DC or lower voltage between terminals 1 and 4. The ISC motor should operate.

**NOTE: Voltages over 6 volt will cause damage to servo gears. Do NOT use system voltage (12 volt) for this test.**

5. If motor does not operate, replace ISC servo as an assembly.

## THROTTLE POSITION SENSOR (TPS)

### Check

1. Set ignition switch to **OFF**.
2. Disconnect sensor connector.
3. Check resistance across poles 1 and 3. Total resistance of throttle position sensor should be 3500–6500 ohms.
4. Connect a circuit tester across poles 1 and 2; check that when throttle valve is slowly operated from idle to fully opened position, resistance changes smoothly.
5. Fasten connector firmly.

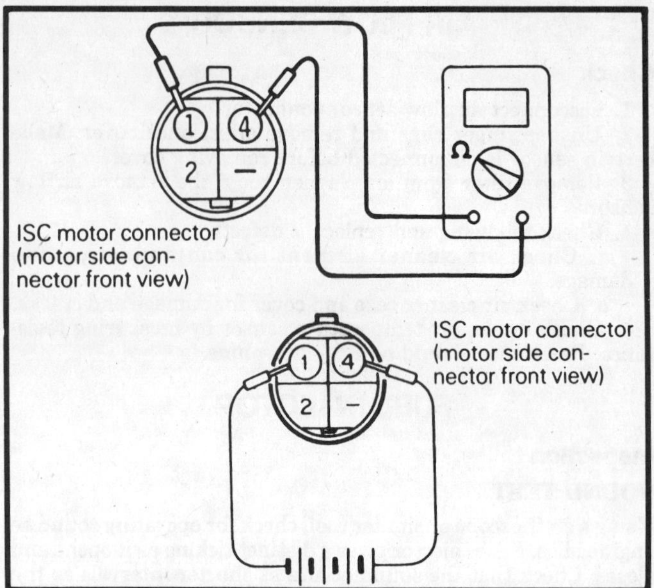


**Testing Starion ISC motors**

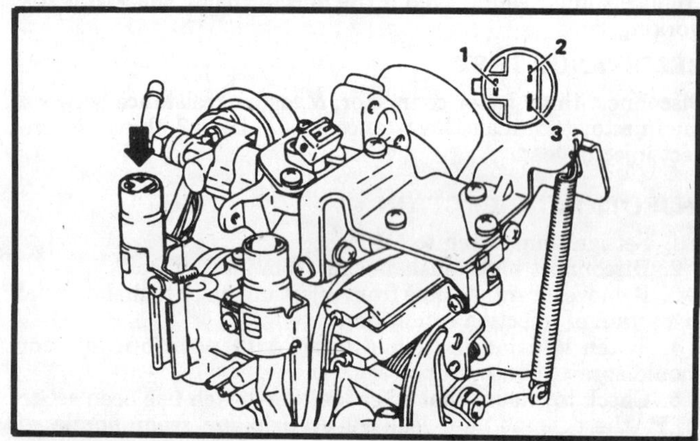


**Throttle position sensor (TPS) test**

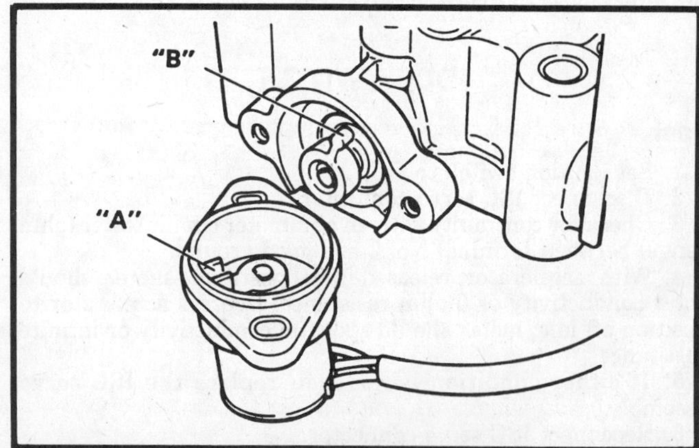


**Throttle position sensor (TPS) servicing**

6. If TPS is to be replaced, carefully remove retaining screws and remove unit from mixer.
7. Install new TPS to mixer body, making certain that the tab within the sensor engages the lever on the mixer.
8. Tighten the retaining screws and connect the wire harness.
9. Adjust the TPS.

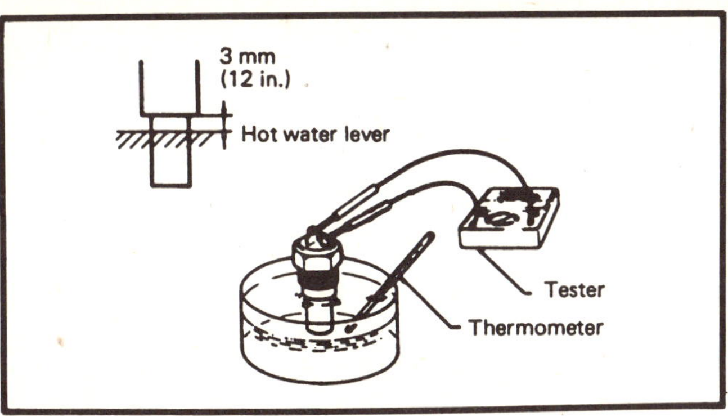

Coolant temperature sensor testing

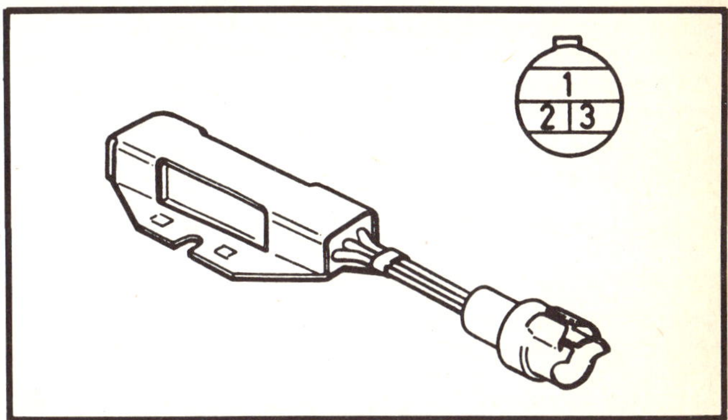

Fuel resistor terminal identification

## COOLANT TEMPERATURE SENSOR

### Inspection

1. Remove coolant temperature sensor from intake manifold.
2. With temperature sensing portion of coolant temperature sensor immersed in hot water, measure resistance. Do not allow temperature sensing portion to directly touch heated portion of hot water container. Stir hot water continuously. Connector terminal portion of sensor should be held 0.12 in. (3mm) above hot water level. Resistance should be:

**Cordia/Tredia:**
2.45 kilo ohms at 20°C (68°F)
296 ohms at 80°C (176°F)

**Starion:**
2.50 kilo ohm at 20°C (68°F)
300 ohm at 80°C (176°F)

3. If resistance meets specifications, sensor is good. If anything abnormal is evident, replace sensor.
4. When reinstalling, coat threads of sensor with sealing compound.
5. Install sensor by hand and tighten to 21 ft. lbs (30 Nm).
6. Connect harness wiring securely

## INJECTOR RESISTOR

### Inspection

1. Disconnect harness connector of resistor.
2. Measure resistance across terminals 1 and 2 and across terminals 1 and 3. Resistance should be 5.5–6.5 ohms.
3. If resistance is within standard value, resistor is good. If resistance is 0 or abnormally large, resistor is shorted or open-circuited. Replace resistor.
4. Fasten connector of resistor firmly.

## OXYGEN SENSOR

### Testing

1. Warm engine until coolant is at normal operating temperature.
2. Use an accurate digital voltmeter.
3. Disconnect oxygen sensor connector; connect voltmeter to sensor.

**CAUTION**

*Components and surfaces will be hot.*

4. Start engine and observe meter. When engine is raced (richening mixture), meter should show approximately 1 volt. Voltage should change as engine returns to idle speed.
5. If replacement is necessary, allow engine to cool before repair. Use correct wrench and remove sensor carefully.
6. Tip of new sensor must be kept clean and free of all grease, oil etc.
7. Install new sensor and tighten to 20 ft. lbs. (27 Nm)
8. Connect wiring securely.

## ECU CONTROL RELAY

### Inspection

**NOTE: To be considered good, relay must pass ALL tests below. Use care to apply voltage to correct terminals during testing; relay may be damaged by incorrect connections.**

1. Test continuity between terminals 1 and 7 and between terminals 3 and 7. There should be NO continuity; if there is continuity, replace control relay.
2. Apply 12 volts to terminal 8 and ground terminal 4. Test continuity between terminals 3 and 7. Continuity should exist.
3. Apply 12 volts to terminal 6 and ground terminal 4. Test continuity between terminals 1 and 7. Continuity should exist.
4. Apply 12 volts across terminals 5 and 2 while testing continuity between terminals 1 and 7. Continuity should exist.

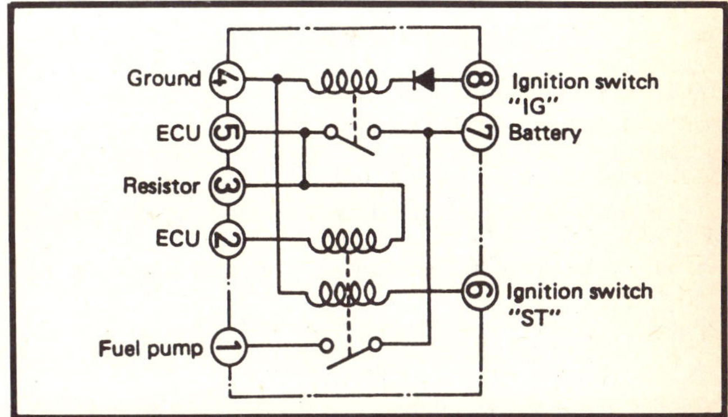

ECI control relay terminal identification

# Component Replacement

## FUEL INJECTOR

### Removal and Installation

**NOTE: The injectors are extremely sensitive to dirt and impact. They must be handled gently and protected at all times. The entire work area must be as clean as possible. Any particle of dirt entering the system can**

**foul an injector or change its operation. Any gaskets or O-rings removed with the injector MUST be replaced with new ones at reassembly. Do not attempt to reuse these seals; high pressure fuel leaks may result.**

1. Remove the injection mixer from the engine and place it on the workbench.

**NOTE: Use a screwdriver which fits the screw head exactly. The screws will be tight; an improper tool can destroy the head of the screw.**

2. Carefully remove the injector holder from the injection mixer. The injector holder is the unit to which the high pressure and fuel return lines connect.
3. Remove the small O-rings and larger gaskets from the top of the injectors.
4. By hand – never with pliers – pull firmly on the injector to remove it. Remove the injector seat from below the injector.

**NOTE: On the G54B engine, the injectors have differently colored electrical connectors. Before removal, note and record which color is in which position (front, rear, left, right). The injectors are not identical and must be reinstalled in their proper locations.**

5. As soon as both injectors have been removed, use tape to seal the injector port against entry of dirt.
6. After the injectors have been tested, checked or replaced, remove the tape over the injector ports and install new seats. Note that the injector seats have a flat side and a round side; the flat side faces up (towards the injector).
7. Each injector should be fitted with a new collar and O-ring. The O-ring may be lightly coated with clean gasoline to make installation easier; do not use oil or grease.
8. Place the injector in its correct location and press it firmly into place.
9. Before the injector holder is reinstalled, examine the small screen filters for any clogging or obstruction.
10. Install the injector holder onto the injection mixer. Install the retaining screws and tighten them evenly to 3 ft. lbs or 36 inch lbs. (4 Nm)
11. Reinstall the injection mixer.

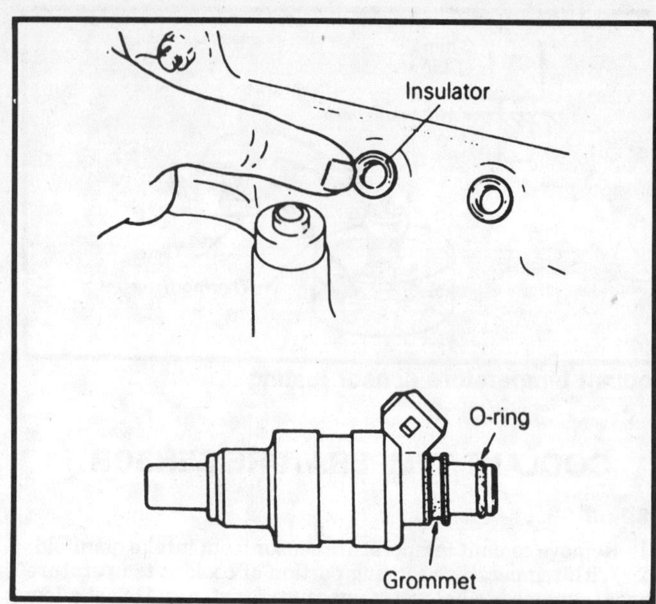


**Always replace the seals when the injectors are removed**

## INJECTION MIXER

### Removal and Installation

1. Disconnect battery ground cable.
2. Drain coolant down to intake manifold level or below.
3. Disconnect air intake hose from injection mixer.
4. Disconnect throttle cable from throttle lever of injection mixer.
5. Disconnect fuel inlet pipe and fuel return hose from injection mixer.
6. Disconnect harness connectors at injectors.
7. Disconnect connectors for ISC servo and throttle position sensor.
8. Disconnect vacuum hoses from nipple of mixer.
9. Remove four bolts and remove injection mixer from intake manifold.
10. Reinstall injection mixer in reverse order of removal, paying special attention to following items:
    a. When replacing throttle position sensor or ISC servo, output voltage or engine speed should be adjusted.
    b. Start and run engine, and check for fuel leakage.

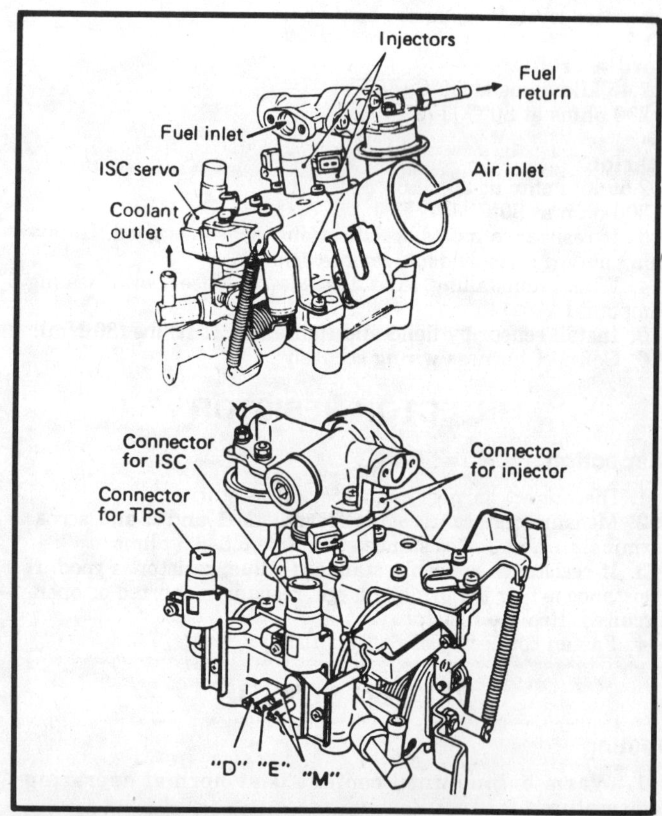


**Injection mixer component identification**

### Overhaul

**NOTE: When a cross-recessed screw is to be loosened, use a cross-recessed screwdriver of proper size. Screw head may be damaged otherwise.**

1. Clamp injection mixer in a vise with soft jaws.
2. Disconnect rubber hose from fuel pressure regulator and mixing body.
3. Remove injector retainer tightening screws and remove retainer.
4. Remove fuel pressure regulator from retainer.
5. Remove pulsation damper cover from retainer, and then take out spring and diaphragm.

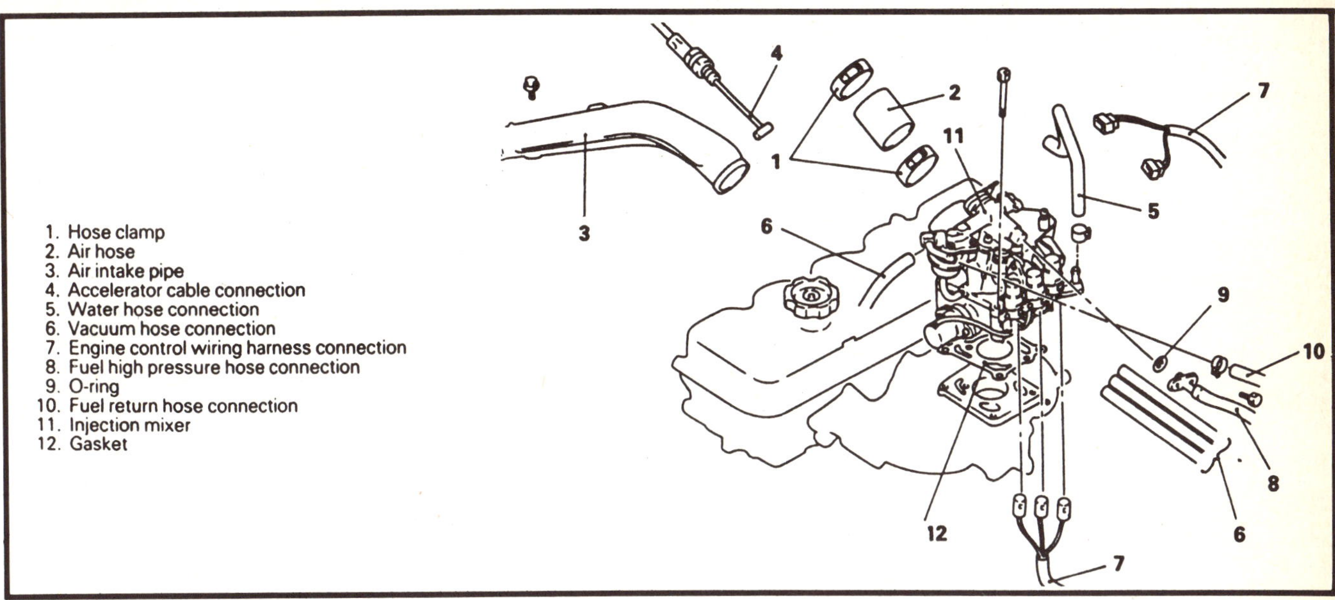


**Starion injection mixer removal**

6. Pull injectors from mixing body. Do not clamp injector by pliers. Then remove gaskets from body.
7. Remove throttle return spring and damper spring.
8. Remove connector bracket.
9. Remove ISC servo mounting bracket retaining bolts and remove ISC servo and bracket.

**NOTE: Do not remove ISC servo except when replacement is required.**

10. Remove 2 screws and then remove mixing body and seal ring from throttle body.
11. Remove throttle position sensor.

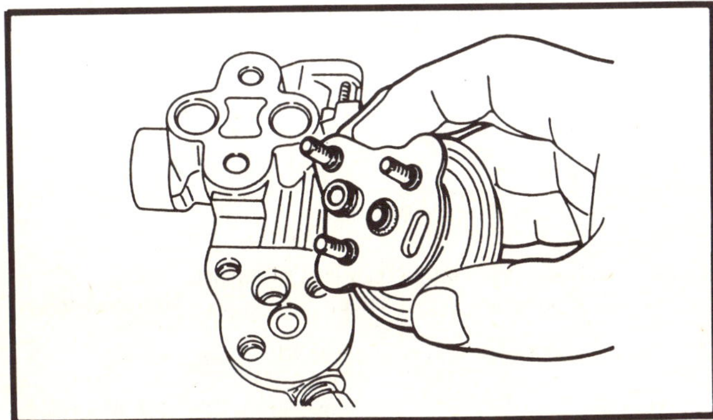

**Fuel pressure regulator**

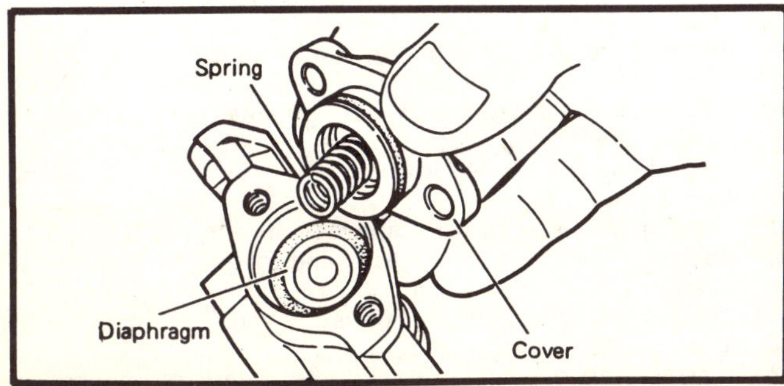


**Pulsation damper servicing**

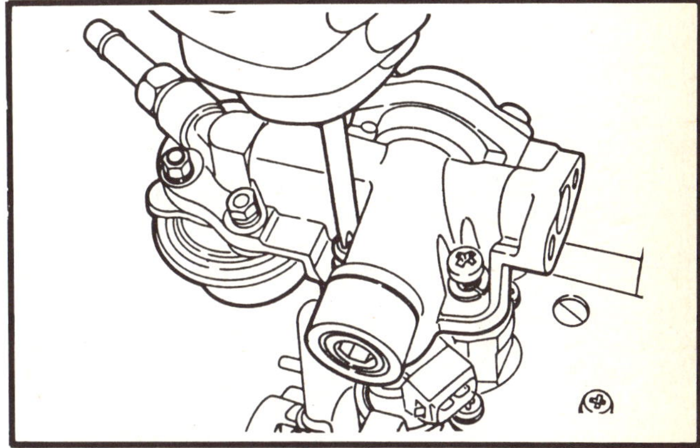

**Injector retainer servicing**

**NOTE: Do not remove throttle position sensor except when replacement is required.**

12. When cleaning injection mixer parts, do not immerse parts in cleaning solvent. Immersing ISC servo, throttle position sensor and injector will damage insulation. Wipe these parts with a cloth only.
13. Do not immerse injectors and fuel pressure regulator in cleaning solvent.
14. Check vacuum parts and passages for clogging. Clean vacuum passage and fuel passage with compressed air.

**CAUTION**

*Always wear eye protection when using compressed air.*

15. To reassemble, insert joint A of throttle position sensor to joint B and install throttle position sensor to throttle body. Then temporarily tighten screws.

**NOTE: After installing injection mixer assembly to engine, adjust throttle position sensor.**

16. Clamp connector into bracket.
17. Check for proper installation of throttle position sensor. Measure resistance value between terminals 1 and 2 or 3 and 4, while moving throttle lever from open to close. If resistance value changes, throttle position sensor is properly installed.

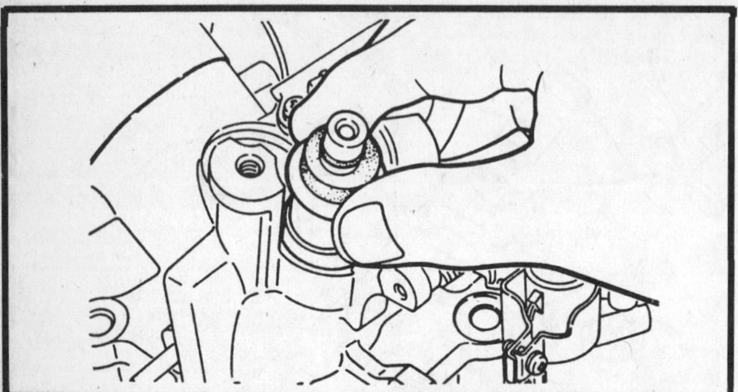

**Fuel injector servicing**

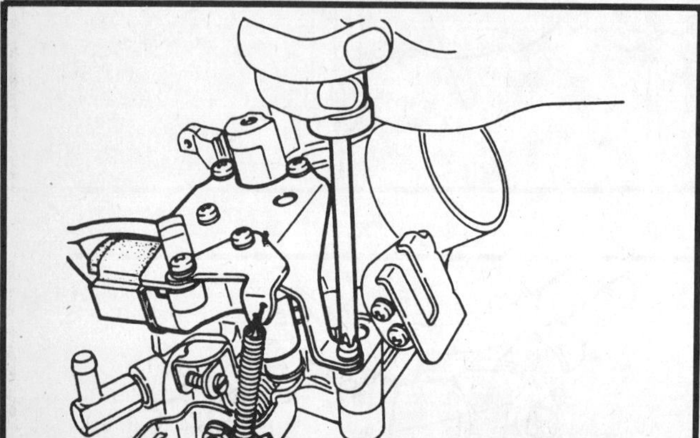

**Idle speed control (ISC) servo and bracket servicing**

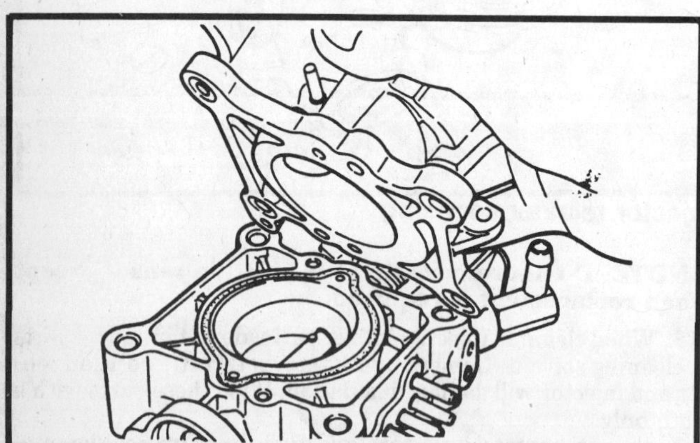

**Injection mixer body servicing**

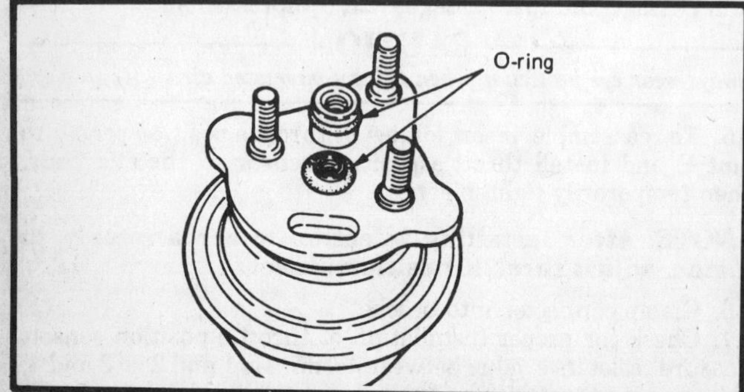


**Fuel pressure regulator O-ring servicing**

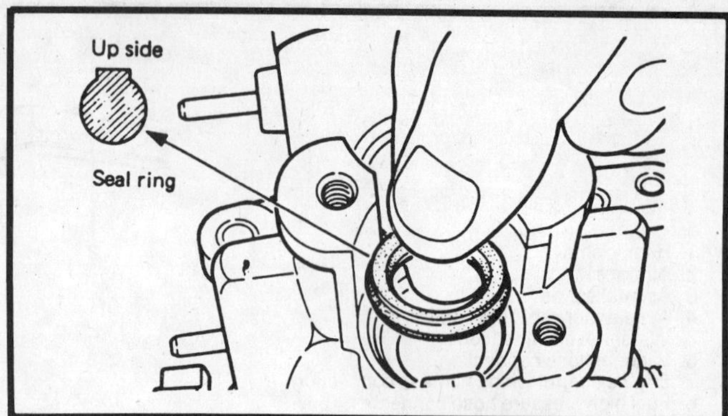


**Injection mixer body seal rings servicing**

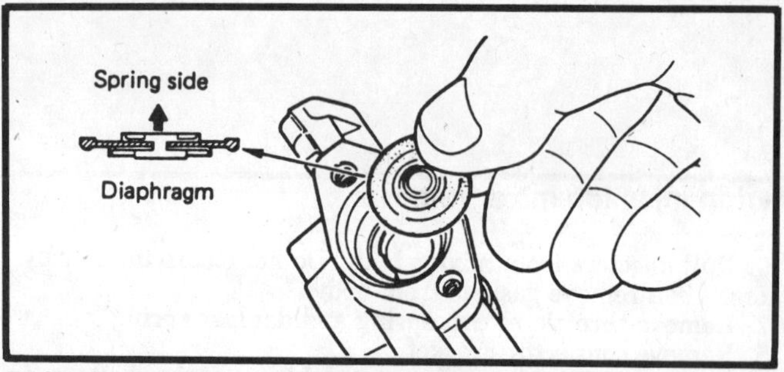


**Pulsation damper diaphragm servicing**

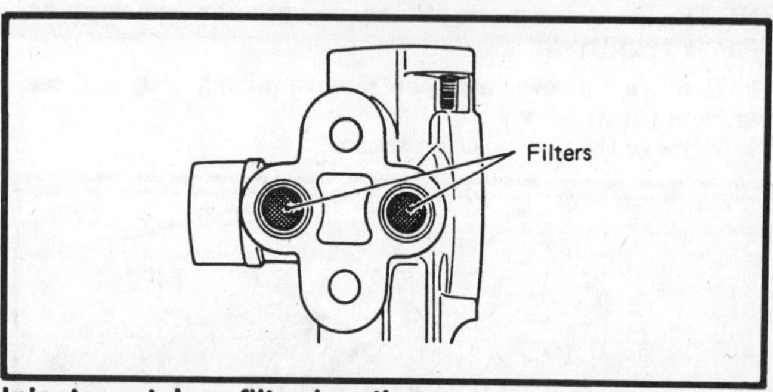


**Injector retainer filter location**

18. Install new seal ring into groove of throttle body.
19. Install mixing body into throttle body and tighten screws firmly.
20. Install ISC servo and bracket, and tighten screws firmly.

**NOTE: If ISC has been replaced, be sure to adjust it correctly.**

21. Install connector bracket to throttle body, and clamp ISC connector into bracket.
22. Install throttle return spring and damper spring.
23. Insert new seal rings into mixing body. When installing seal ring, be sure to install with flat face side up.
24. Install new O-ring onto injectors.
25. Install injectors into mixing body. Push injector down firmly by finger.
26. Insert pulsation damper diaphragm into injector retainer.
27. Install pulsation damper spring and cover, and then tighten screws.
28. Install new O-rings onto regulator, and then install fuel pressure regulator to injector retainer.
29. Check filters in retainer for clogging or damage. Replace as necessary.

1. Throttle position sensor (TPS)
2. Joint
3. Hose
4. Screw
5. Injector holder
6. Fuel pressure regulator
7. O-ring
8. Pulsation damper cover
9. Spring
10. Diaphragm
11. O-ring
12. Injector
13. Injector
14. O-ring
15. Collar
16. Seal ring
17. Damper spring
18. Return spring
19. Connector bracket
20. Connector bracket
21. ISC servo assembly
22. Throttle cable bracket
23. Screw
24. Mixing body
25. Seal ring
26. Throttle valve set screw
27. Return lever
28. Adjusting screw
29. Free lever
30. Ring
31. Throttle lever
32. Spring
33. Return spring
34. Throttle body

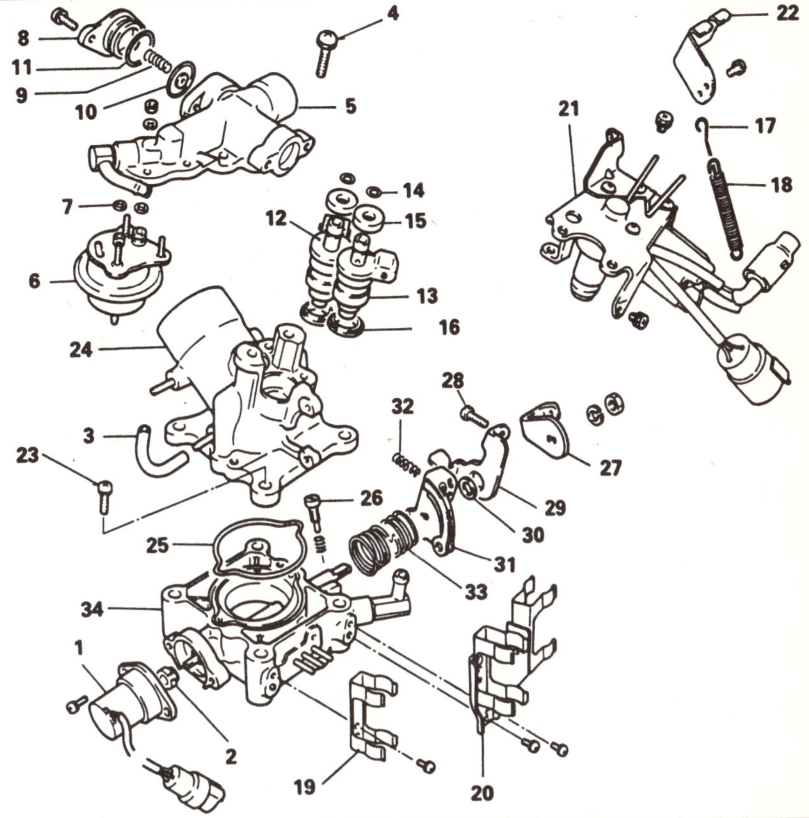

**Starion injection mixer overhaul**

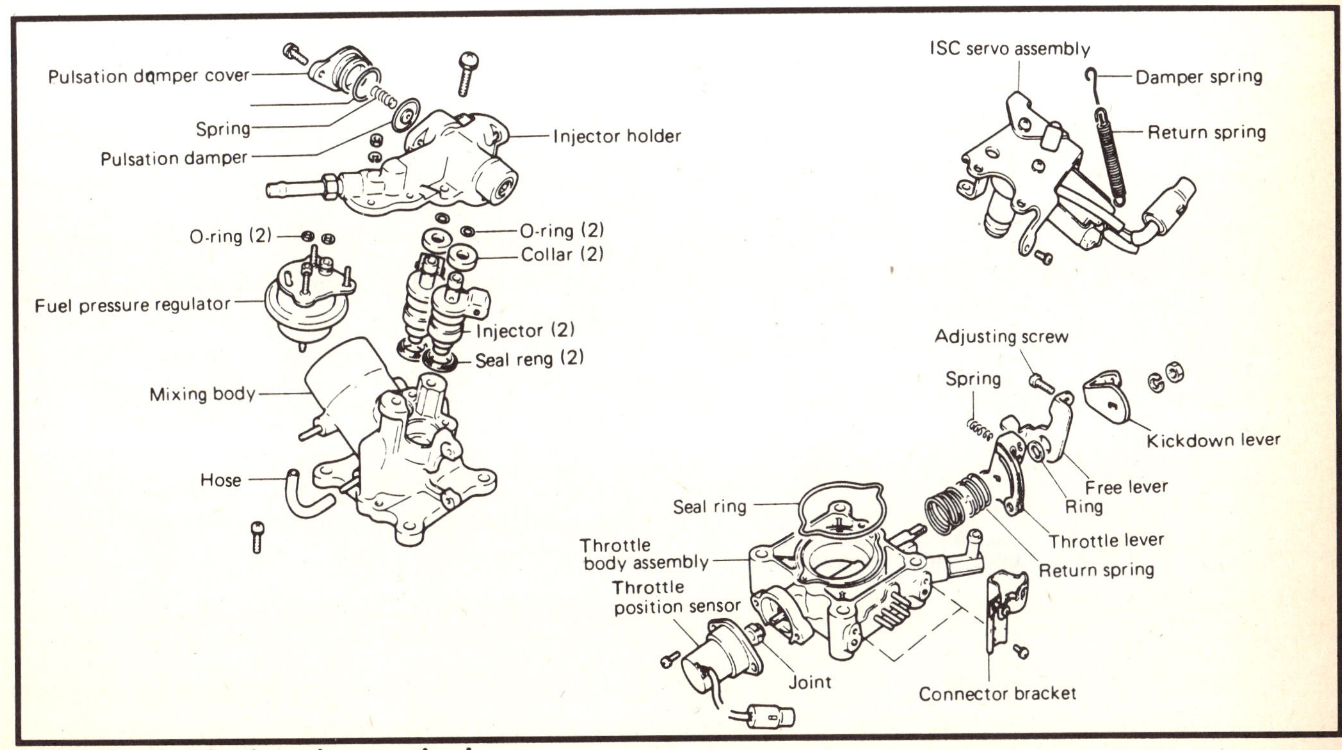


**Cordia/Tredia injection mixer overhaul**

30. Install injector retainer and push down firmly.
31. Tighten screws alternately a little at a time, and then tighten them to specified torque.

## SPEED CONTROL (ISC) SERVO AND THROTTLE POSITION SENSOR (TPS)

### Adjustment

If ISC servo, throttle position sensor, throttle body or injectors have been removed or replaced, following adjustments should be made. These adjustments are important to driveability.

1. Start engine and allow it to reach operating temperature.
2. Loosen or disconnect accelerator cable from throttle lever of injection mixer.
3. Loosen throttle position sensor mounting screws and turn throttle position sensor clockwise as far as it will go, then temporarily tighten screws.
4. Set ISC Servo position by turning ignition **ON** for 15 seconds, then turn ignition **OFF**.
5. Disconnect ISC servo harness connector.
6. Start engine and check idle speed. Idle speed should 600 rpm. If necessary, by turning adjusting screw clockwise to increase rpm or counterclockwise to decrease rpm.
7. Stop engine.
8. In order to read output voltage of TPS, insert digital voltmeter test probe from rubber cap side of TPS connector and bring it into contact with pins in connector. Insert test probes along TPS output and ground lead of body side harness.
9. Turn ignition switch **ON** but do not start engine.
10. Read throttle position sensor output voltage on digital voltmeter. If voltage is not 0.48 ± 0.03 volts, loosen throttle position sensor mounting screws and turn sensor until reading is within specified range, then tighten mounting screws.
11. Open throttle fully and confirm that output voltage is correct when throttle valve is returned to idle position.
12. Remove digital voltmeter connections and recheck idle speed.
13. Turn ignition switch from OFF to ON position, and after lapse of 15 seconds, return it to OFF position.

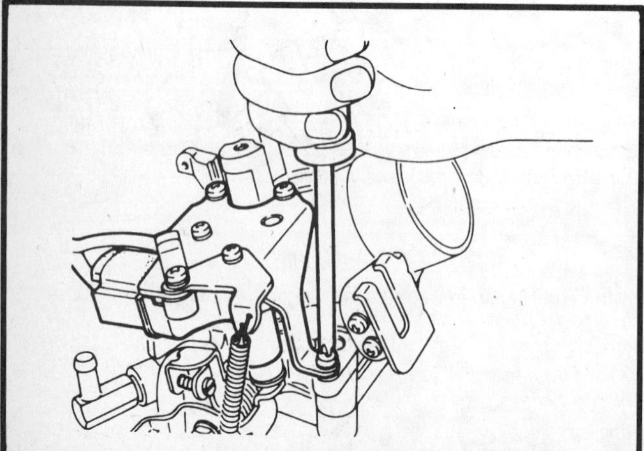

**Do not remove ISC assembly unless it has failed**

14. Connect accelerator cable to throttle lever of injection mixer and adjust accelerator cable.

## IDLE SPEED CONTROL (ISC) SERVO MOTOR

### Removal and Installation

**NOTE: Do not remove ISC assembly unless it is to be replaced.**

**CAUTION**

*Gasoline in either liquid or vapor state is EXTREMELY explosive. Contain spillage; work in an open or well–ventilated area. Observe no smoking/no open flame rules during repairs. Have a dry-chemical fire extinguisher (type B-C) within arm's reach at all times and know how to use it.*

1. Disconnect negative battery cable. Drain the cooling system.
2. Safely release the pressure within the fuel system.
3. Disconnect the breather and air intake hoses.
4. Disconnect the accelerator cable.
5. Disconnect the coolant hose(s) to the mixer.
6. Disconnect the vacuum hoses at the injection mixer.
7. Disconnect the electrical connectors at or near the injection mixer. holding the wire harnesses or connectors.
8. Disconnect high pressure fuel line at the mixer.

**NOTE: Wrap the connection with a clean towel before removal. Some pressure may remain within the fuel system.**

9. Disconnect the fuel return line.
10. Using the proper sized wrench, carefully remove the bolts holding the injection mixer to the manifold. Lift the mixer away from the engine and remove the gasket.
11. Remove the injector holder.
12. Remove the fuel pressure regulator and gaskets.
13. Remove both injectors, with seals and collars.
14. Remove damper spring and return spring.
15. Remove 2 electrical connector clips from side of mixer.
16. Remove the ISC servo motor assembly. Do not disassemble the unit.
17. Install new unit and tighten retaining bolts.
18. Install the 2 connector clips and install the 2 springs.
19. Install the injectors, using new seals and gaskets.
20. Install the pressure regulator, then install the injector holder.
21. Place a new gasket on manifold and set the mixing unit in place.
22. Install four bolts finger tight. Tighten evenly, in small increments to 13.5 ft. lbs. (18 Nm).
23. Connect the fuel return hose.
24. Connect high pressure fuel line. Install a new O-ring and coat surfaces lightly with gasoline. Do not use oil or grease. Install line and tighten to 4 ft. lbs. or 48 inch lbs. (6 Nm).
25. Connect electrical harnesses.
26. Install the vacuum hoses.
27. Connect the water hose(s).
28. Install the accelerator cable.
29. Reassemble the air intake and breather hoses.
30. Refill the coolant.
31. Connect the negative battery cable.

# MITSUBISHI MULTI-POINT INJECTION (MPI) SYSTEM

## General Information

The MPI (Multi-Point Injection) system controls the fuel flow, idle speed, and ignition timing. The basic function of the MPI system is to control the air/fuel ratio according to operational conditions through the Electronic Control Unit (ECU), based on data from various sensors. The MPI System is roughly divided into 3 areas: fuel system, intake system, and control system.

### FUEL SYSTEM

The fuel is pressure-fed through the in-tank filter by the fuel pump and is distributed to the respective injectors via the main pipe and fuel filter. The fuel pressure applied to the injector is a continuous 47.6 psi (335 kPa), higher than the pressure in the intake manifold. The pressure is controlled by the pressure regulator. The excess fuel, after being pressure-adjusted, is returned to the fuel tank through the return pipe.

When an electric current flows in the injector, the injector valve is fully opened to inject the fuel. Since the fuel pressure is a continuous constant, the amount of the fuel injected from the injector into the manifold is increased or decreased in proportion to the time the electric current flows. Based on ECU signals, the injectors inject fuel to the cylinder manifold ports in firing order.

### INTAKE SYSTEM

The flow rate of the air drawn through the air cleaner is measured by the air flow sensor (AFS). The air enters the air intake plenum through the throttle body.

The air is distributed to each cylinder manifold from the air intake plenum. In the manifold, the air is mixed with the fuel from the injectors and is drawn into the cylinder. The air flow rate is controlled according to the degree of the throttle valve and the stepper motor openings.

The amount of air drawn during idling is adjusted by the idle speed control (ISC) servo controlled by the ECU. Further, the amount of air drawn during warm-up and deceleration is also controlled by the ISC servo.

### CONTROL SYSTEM

The control system is composed of a sensor section, which monitors engine conditions, and an Electronic Control Unit (ECU), which calculates the injection timing and rate according to the signals from the sensors. The sensors convert such conditions as the amount of intake air, amount of oxygen in the exhaust gas, coolant temperature, intake air temperature, engine revolution speed, and driving speed into electric signals, which are sent to the ECU.

Certain values and settings are built into the ECU at manufacture; these map or default settings are used under certain extreme operating conditions and/or when the ECU detects loss of signal from certain sensors.

Analyzing these signals, the ECU determines the amount of fuel to inject according to driving conditions and drives the injectors. The fuel injection is by sequential injection type, in which four or six injectors are sequentially driven.

During idling, the ISC Servo is driven according to the load to assure stable idling.

#### Injector Drive Time

Injector activation occurs 1 time per cycle for each cylinder, and the injector activation time (amount of fuel injection), which is the theoretical air/fuel mixture ratio relative to the amount of intake air, is called the basic activation time.

The ECU functions calculate this basic activation time according to air-flow sensor (AFS) signals and crankshaft angle sensor signals.

When starting (cranking) the engine, the map value determined by the engine coolant temperature sensor signal is used as the basic drive time. During the deceleration, the basic driving time is set as zero. The injector driving time is obtained by making the following corrections on the above-mentioned basic driving time.

- Oxygen sensor feedback correction (closed loop control correction) – In normal (warm) operation (including idling), air/fuel ratio is corrected to optimum by using oxygen sensor signals. This allows the 3-way catalytic converter to give the best degree of purification.
- Air/fuel ratio map correction (Open loop control correction) – Correction to the optimum air/fuel ratio is made by the map values set by engine speed and amount of intake air.
- Engine coolant temperature correction – To maintain operability of cold engine, correction is so made that the lower the engine coolant, the greater the amount of fuel injected.
- Intake air temperature correction – Change in air/fuel ratio due to difference in intake air density.
- Barometric pressure correction – Change in air/fuel ratio due to difference in intake air density caused by change in barometric pressure is corrected.
- Acceleration/deceleration correction – In accordance with change in amount of intake air, fuel amount is corrected, improving operability.
- Dead time correction – The injectors opened by the driving signal from ECU have an operation lag which changes according to battery voltage. This means that actual injector opening time becomes less than injector driving signal, failing to provide expected air/fuel ratio. Therefore, battery correction time corresponding to the battery voltage is added by ECU.

characteristic of the conventional type of distributor.

#### Idle Speed Control (ISC) System

The ISC system provides the following 4 modes of control:

**START CONTROL**

The throttle valve opening is controlled to optimum position for start according to the engine coolant temperature and the altitude (atmospheric pressure).

**FAST IDLE CONTROL**

When the idle switch is on, the engine speed is controlled to a target rpm according to the engine coolant temperature (rpm feedback control). When the idle switch is off, the ISC servo is actuated to move the throttle valve to a target opening position (throttle valve opening position) according to the engine coolant temperature (target opening control).

**IDLE CONTROL**

When the air conditioner switch is turned on or when the transmission is shifted from **N** to **D**, the system causes the idle speed to increase to the target rpm according to the load (rpm feedback control).

**DASH POT CONTROL**

The system provides dash pot control according to deceleration conditions to alleviate shock at deceleration

#### Air Conditioner Relay Control

When the air conditioner switch is turned on while the engine is at idle, the ISC servo operates to increase the engine speed. However, there is some delay before the engine speed actually increases. To maintain the engine free from the air conditioner load during that delay period, the ECU keeps the power transistor off for a fixed time (about 0.5 seconds) to open the air conditioner power relay circuit. As a result, even if the air conditioner switch is on, the air compressor is not driven instantly, preventing engine speed drop due to compressor load.

### Fuel Pressure Control System

Ordinarily, the intake manifold negative pressure (vacuum) acts upon the fuel-pressure regulator, the fuel pressure is maintained at a constant fixed level relative to the pressure within the intake manifold, and the amount of fuel injection is regulated to a constant fixed amount. During starting when the temperature of the intake air is high (122°F or higher), and when the temperature of the engine coolant is also high (194°F or higher), the ECU switches ON the power transistor for a certain fixed time (approximately 2 minutes), thus making the fuel-pressure solenoid valve conductive. As a result, the atmospheric pressure acts upon the fuel-pressure regulator. The fuel pressure becomes high relative to the pressure within the intake manifold, and the amount of fuel injection is increased, thus maintaining idling stability immediately after restarting at high temperature.

## System Components

### INJECTORS

Injectors are electromagnetic type injection nozzles which function to inject fuel according to injection signals calculated and provided by the ECU.

#### Injector Operation

When there is continuity at the solenoid coil, the plunger and the needle valve are pulled up and fuel is sprayed from the nozzle. Because the stroke of the needle valve is always constant, the amount of fuel spray (injection) is determined by the time (duration) of the injector opening.

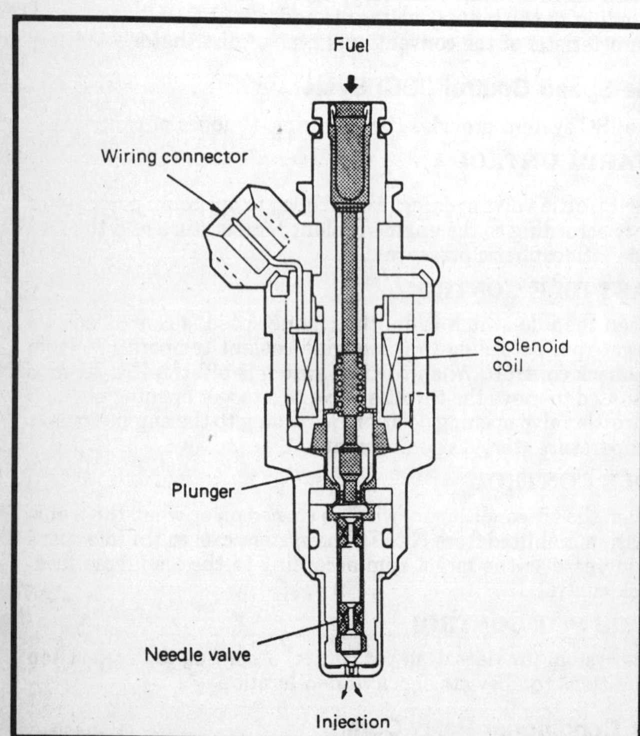

Fuel injector

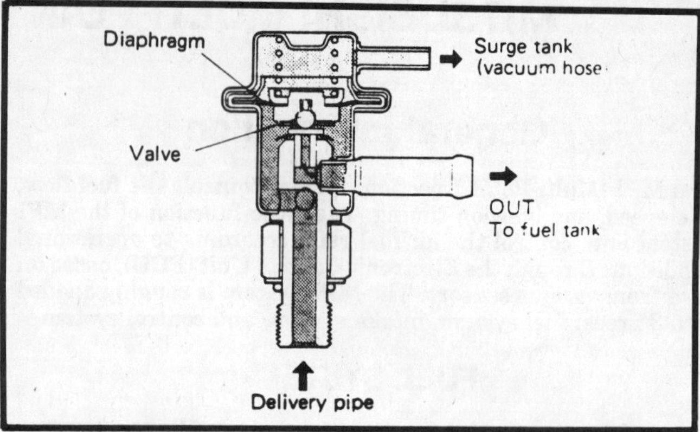

Pressure regulator

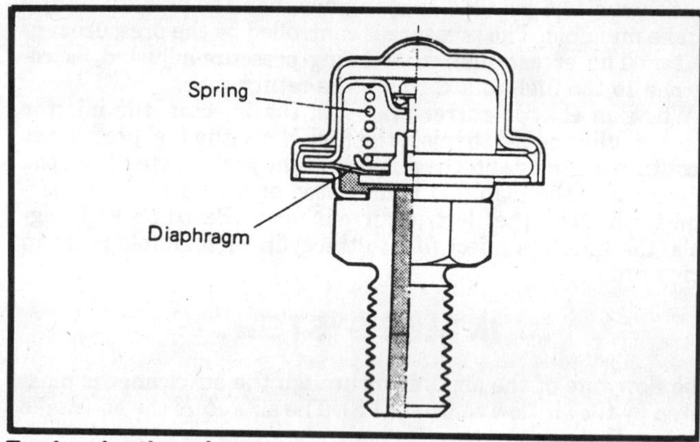

Fuel pulsation damper

### PRESSURE REGULATOR

The pressure regulator functions to maintain the fuel volume at a fixed amount relative to fuel injection time. As a means to improve precision, the pressure regulator controls the fuel pressure applied to the injectors at a constant pressure relative to the surge tank negative pressure.

#### Pressure Regulator Operation

Surge tank negative pressure is applied to the diaphragm chamber of the pressure regulator. When the fuel pressure within the pressure regulator reaches a pressure of approximately 36.26 psi (250 kPa) or higher, extra fuel is by-passed to the return hose and is returned to the fuel tank.

### FUEL PULSATION DAMPER

The fuel pulsation damper absorbs the slight fluctuations of fuel pressure which occur when the injectors inject the fuel.

#### Pulsation Damper Operation

The fuel pressure is constantly maintained by the pressure regulator at a pressure of approximately 36.26 psi (250 kPa) but slight fluctuations of the fuel pressure do occur when the injector sprays the fuel. The fuel pulsation damper, through the action of the diaphragm, absorbs these minor fluctuations of the fuel pressure.

## RESISTOR

The resistor is used to lower the source voltage to a level suitable for the injector. The resistor is connected in series with the injector. It reduces the voltage to approximately ¼ of the source voltage. These resistors protect the injectors from alternator voltage surges and the effects of other components in the vehicle's electrical system.

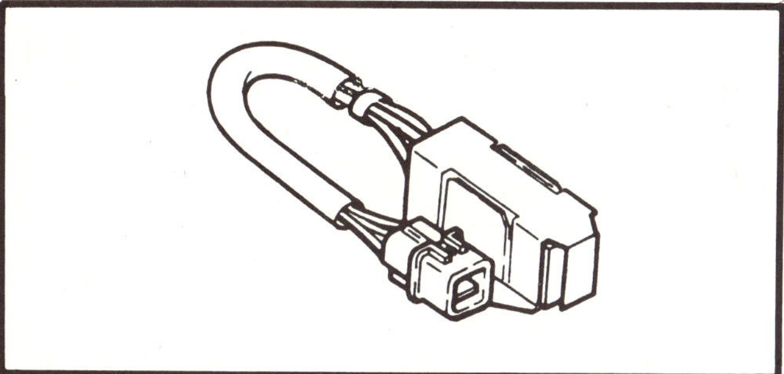

**Injector resistor**

## SPARK-ADVANCE SWITCHING SOLENOID DRIVE

The spark-advance switching solenoid is activated in order to advance the ignition timing at high altitudes where the atmospheric pressure is low or during cold weather.

At a barometric pressure of less than 660mm Hg or a water temperature of less than 95°F (35°C), the spark-advance control solenoid is driven, and, by leading the surge tank negative pressure to the boost control, the ignition timing is advanced 5 degrees.

## AIR CONDITIONER RELAY CONTROL

The air conditioner relay is switched off, even when the air conditioner switch is switched on, according to the engine rpm and the throttle opening.

- The air conditioner relay is switched off at an engine rpm of 400 rpm or less.
- The air conditioner relay is switched off for 5 seconds if the ECU detects full throttle acceleration.

## AIRFLOW SENSOR

This measures the intake airflow; utilizing the Karman vortex phenomenon, it counts the number of vortices and converts this data to electric pulses which are then sent to the ECU. The ECU uses these signals to determine the basic injection time.

## INTAKE AIR TEMPERATURE SENSOR

This sensor detects the temperature of the intake air, and converts this data to voltage which is sent to the ECU. The ECU, based on these signals, performs air/fuel ratio feedback control.

## BAROMETRIC PRESSURE SENSOR

This sensor detects the barometric pressure, and converts this data to voltage which is sent to the ECU. The ECU uses these signals for correct the fuel injection amount and the ignition timing.

## THROTTLE POSITION SENSOR (TPS)

The TPS detects the degree of throttle opening; this data is converted to voltage and is sent to the ECU. The ECU uses these signals to calculate the changes in the degree of throttle opening and to sense the variable speed condition.

## IDLE SWITCH

This switch detects the fact that the throttle valve is in the idle position and sends this data as signals to the ECU. The ECU senses that the engine is idling and then controls the fuel injection amount, the ignition timing, the ISC servo, etc.

## COOLANT TEMPERATURE SENSOR

This sensor detects the temperature of the engine's coolant, and converts this data to voltage which is sent to the ECU. The ECU uses these signals for correction of the fuel injection amount and the ignition spark advance.

## OXYGEN SENSOR

This sensor measures the amount of oxygen in the exhaust gas, and sends these signals to the ECU. The ECU, based on these signals, performs air/fuel ratio feedback control.

## TOP DEAD CENTER (TDC) SENSOR

This sensor identifies the reference signal applicable to each cylinder. The disc slit is read by the light-emitting diode and the photo diode; this data is sent from the unit assembly to the ECU as electrical pulses. The ECU uses these signals and the signals from the crankshaft angle sensor to determine the injection timing.

## CRANKSHAFT ANGLE SENSOR

This sensor detects the position of the crankshaft (and therefore, the piston). The 360 slits in the disc are read by the light-emitting diode and the photo diode, and this data is sent from the unit assembly to the ECU as electrical pulses.

## VEHICLE SPEED SENSOR

This sensor converts the rotation of the transaxle speedometer gear to electrical pulses and sends them to the ECU.

## POWER STEERING OIL PRESSURE SWITCH

Switch detects a large load on power steering system and increases idle to compensate.

## INHIBITOR SWITCH (VEHICLES WITH AN AUTOMATIC TRANSAXLE)

Whether the transaxle is in neutral or drive position is detected, and the result is used as the idle speed control signal.

# Diagnosis and Testing

## SELF-DIAGNOSIS

Self-diagnosis is a system in which the input signal from each sensor is monitored by the ECU and, should any abnormality happen in the input signal, the abnormal item is memorized by the ECU. Many items are diagnosed including that for normal condition and can be confirmed using a volt meter or the Mitsubishi MPI tester.

The abnormality-diagnosis memory is kept by direct power supply from the battery. Therefore the memory is not erased by turning **OFF** the ignition switch. However, it is erased if the back-up power supply is turned off by disconnection of battery cable or ECU connector.

**NOTE: The memory is not erased if the power supply is restored within 15 seconds.**

The ECU checks each sensor for a signal periodically. If a fault is found, the code for that sensor is stored in memory. If 2 or more faults are detected, the codes are stored in numerical order from lowest to highest. The order of codes in the memory does NOT indicate the order of failure.

### Reading the Trouble Codes

The MPI diagnostic connector is located either behind the glove box or by the fusebox depending on model. If using the Multi Use tester, connect it according to the instructions for the unit. The codes may also be read using an analog (dial) voltmeter. Connect the ground probe to the ground terminal and the positive probe to the test terminal within the connector.

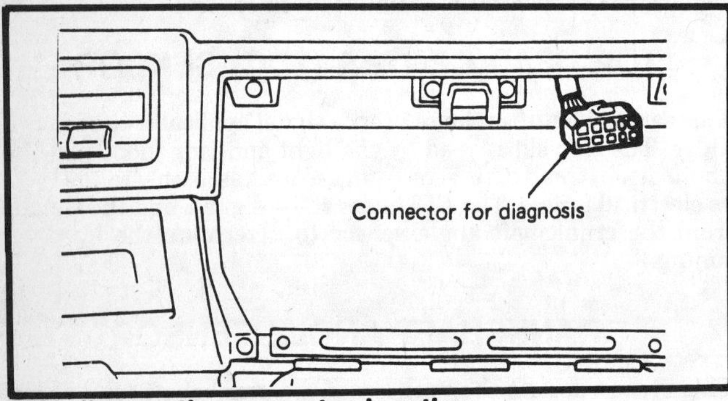

**MPI diagnostic connector location**

**NOTE: When battery voltage is low, fault codes may not store properly. Check battery condition before attempting self-diagnosis procedures**

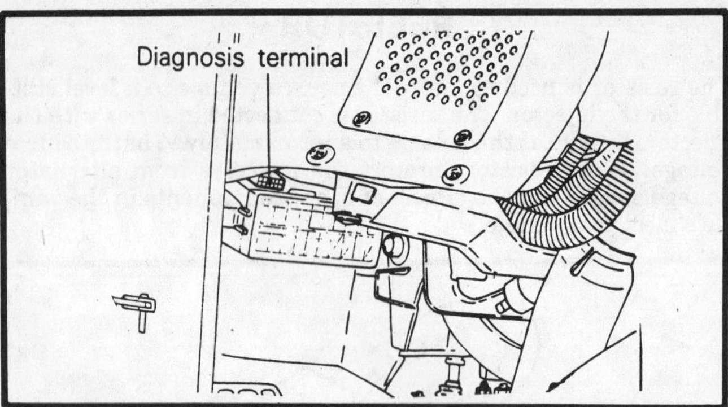

**MPI diagnostic connector location**

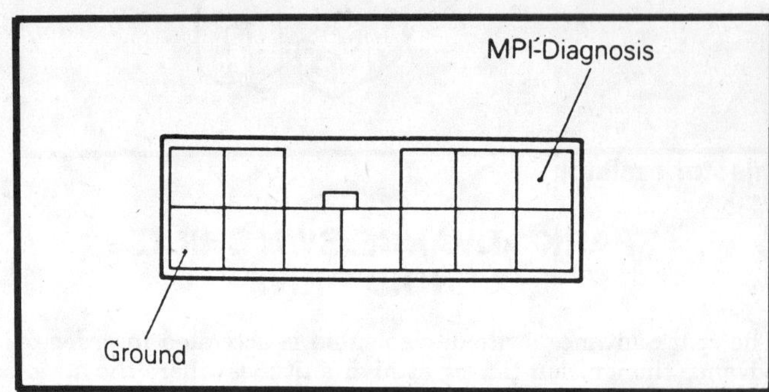

**MPI diagnosis connector terminal identification**

Turn the ignition switch **ON** and use the Multi Use tester to read the codes. If using a voltmeter, the needle will deflect periodically, indicating the codes. If the needle shows a constant 12 volts, the ECU has diagnosed itself as being the failed component. This indication may be due to a failed ECU or (more likely) to loose or faulty wiring running to or from the unit.

Regardless of the method used to read the system, once repairs have been made the negative battery cable must be disconnected for 15 seconds or more. This erases the code from memory. After the engine has been operated, check again for presence of codes; if repair was successful, no code will be present.

## TESTING PROCEDURE WITH MULTI-CHECKER; 1988 VAN/WAGON

| ECI Checker Operation: Select Switch | ECI Checker Operation: Check Switch | Check Item | ECU Terminal # Checked | Condition | | Test Specification |
|---|---|---|---|---|---|---|
| Set to "A" | 1 | Power supply | 51 | Ignition switch "LOCK → ON" | | 5V |
| | 2 | Crank angle sensor | 1 | Ignition switch "LOCK→START" | | 1.8V to 2.5V |
| | | | | 3000 rpm | | |
| | 3 | Intake air temperature sensor | 5 | Ignition switch "LOCK → ON" | 0°C (32°F)<br>20°C (68°F)<br>40°C (104°F)<br>80°C (176°F) | 3.4V to 3.6V<br>2.5V to 2.7V<br>1.7V to 1.9V<br>0.6V to 0.8V |
| | 4 | Engine coolant temperature sensor | 6 | Ignition switch "LOCK → ON" | 0°C (32°F)<br>20°C (68°F)<br>40°C (104°F)<br>80°C (176°F) | 3.4V to 3.6V<br>2.5V to 2.7V<br>1.5V to 1.7V<br>0.5V to 0.7V |
| | 5 | Power supply for sensor | 10 | Ignition switch "LOCK → ON" | | 4.5V to 5.5V |
| | 6 | Throttle position sensor | 15 | Ignition switch "LOCK → ON" (Warm engine) | Accelerator fully closed | 0.4V to 0.7V |
| | | | | | Accelerator fully opened | 4.5V to 5.5V |
| | 7 | Motor position sensor | 3 | Ignition switch "LOCK → ON" | After 15 seconds | 0.8V to 1.2V |
| | 8 | Idle position switch | 7 | Ignition switch "LOCK → ON" | Accelerator fully closed | 0V to 0.6V |
| | | | | | Accelerator fully opened | 8V to 13V |
| | 9 | Cranking signal | 55 | Ignition switch "LOCK → START" | | Over 8V |
| | 10 | Reed switch for vehicle speed | 19 | Start engine, transmission in first or drive and operate vehicle slowly | | 0V to 0.6V<br>↑<br>(pulsates)<br>↓<br>Over 2V |
| | 11 | A/C switch | 56 | Ignition switch "LOCK → ON" | A/C switch OFF | 0V to 0.6V |
| | | | | | A/C switch ON | 5V |
| | 12 | Inhibitor switch | 58 | Ignition switch "LOCK → ON" | Transmission in "P" or "N" | 0V to 0.6V |
| | | | | | Transmission in "D" | 5V |

## FAULT CODES, 1988 VAN/WAGON

| Malfunction No. | Diagnosis item | Self-diagnosis output pattern and output code | Problem | Check item |
|---|---|---|---|---|
| 0 | Normal | 0 0 0 0 0 | None of malfunctions are present. | – |
| 1 | Oxygen sensor | 1 0 0 0 0 | (1) When engine stalls, and for 15 seconds after start<br>(2) Until output voltage reaches 0.6V or higher<br>(3) When oxygen sensor signal doesn't change for 20 seconds or longer during urban driving mode | • Harness and connector<br>• Oxygen sensor |
| 2 | Crank angle sensor | 0 1 0 0 0 | Ignition switch (ST) ON (continuous) and, moreover, there is no crank angle signal input for three seconds or longer | • Harness and connector<br>• Crank angle sensor |
| 3 | AFS | 1 1 0 0 0 | (1) AFS output of 10 Hz or less, with engine rpm 500 rpm or higher<br>(2) AFS output 100 Hz or more, at time of engine stall | • Harness and connector<br>• AFS |
| 4 | Barometric pressure sensor | 0 0 1 0 0 | (1) Barometric pressure sensor output voltage 4.5V (equivalent to 855 mmHg) or higher<br>(2) Barometric pressure sensor output voltage 0.2V or lower | • Harness and connector<br>• Barometric pressure sensor |
| 5 | TPS | 1 0 1 0 0 | (1) TPS output voltage 4V or higher continuously for one seconds or longer, with idling switch ON<br>(2) TPS output voltage 0.2V or lower | • Harness and connector<br>• TPS |
| 6 | MPS | 0 1 1 0 0 | (1) MPS output voltage 4.8V or higher<br>(2) MPS output voltage 0.2V or lower | • Harness and connector<br>• MPS |
| 7 | Engine coolant temperature sensor | 1 1 1 0 0 | (1) Engine coolant temperature sensor thermistor resistance value 45kΩ or higher<br>(2) Engine coolant temperature sensor thermistor resistance value 50Ω or lower | • Harness and connector<br>• Engine coolant temperature sensor |
| 8 | No. 1 cylinder TDC sensor | 0 0 0 1 0 | Absolutely no input of No. 1 cylinder TDC sensor signal during eight ignitions after ignition switch turned to ON or after input of No. 1 cylinder TDC signal | • Harness and connector<br>• No. 1 cylinder TDC sensor |

| ECI Checker Operation: Select Switch | ECI Checker Operation: Check Switch | Check Item | ECU Terminal # Checked | Condition | | Test Specification |
|---|---|---|---|---|---|---|
| Set to "A" | 1 | | | | | |
| | 2 | Barometric pressure sensor | 20 | Ignition switch "LOCK → ON" | at sea level | 3.8V to 4.2V |
| | | | | Idling | | |
| | 3 | | | | | |
| | 4 | | | | | |
| | 5 | | | | | |
| | 6 | | | | | |
| | 7 | | | | | |
| | 8 | | | | | |
| | 9 | | | | | |
| | 10 | | | | | |
| | 11 | | | | | |
| | 12 | | | | | |

ECU Terminal

View from front as installed in ECU

| ECI Checker Operation: Select Switch | ECI Checker Operation: Check Switch | Check Item | ECU Terminal # Checked | Condition | | Test Specification |
|---|---|---|---|---|---|---|
| Set to "B" | 1 | Fuel pressure exhange solenoid valve | 8 | Ignition switch "LOCK → START" | Coolant temp. less than 90°C (194°F) or air temp. less than 50° (122°F) | Over 8V |
| | | | | | Coolant temp. more than 90°C (194°F) or air temp. more than 50° (122°F) | 0V to 0.6V |
| | 2 | | | | | |
| | 3 | | | | | |
| | 4 | No. 1 cylinder sensor | 13 | Ignition switch "LOCK→ START" | | 0.2V to 1.5V (oscillatiing) |
| | | | | 3000 rpm | | 0.8V to 1.2V |
| | 5 | Air-flow sensor | 2 | Idling | | 2.2V to 3.2V |
| | | | | 3000 rpm | | |
| | 6 | | | | | |
| | 7 | Ignition control signal | 54 | Idling | | 0.3V to 0.8V |
| | | | | 3000 rpm | | 1.0V to 2V |
| | 8 | Oxygen senser | 11 | Hold rpm constant above 1300, after 30 seconds from start of warm engine | | 0V to 0.6V ↑ (pulsates) ↓ 2V to 3V |
| | 9 | | | | | |
| | 10 | | | | | |
| | 11 | | | | | |
| | 12 | | | | | |

## TESTING PROCEDURE WITH MULTI-CHECKER; 1988 VAN/WAGON

| ECI Checker Operation: Select Switch | ECI Checker Operation: Check Switch | Check Item | ECU Terminal # Checked | Condition | | Test Specification |
|---|---|---|---|---|---|---|
| Set to "B" | 1 | ISC motor for extension | 23 | Idling | A/C switch OFF → ON | Momentarily over 4V, then 0V to 2V |
| | 2 | ISC motor for retraction | 12 | Idling | A/C switch OFF → ON | Momentarily over 4V, then 0V tc 2V |
| | 3 | A/C cutoff relay | 24 | Idling | A/C switch OFF → ON | Over 12V, then 0V to 0.6V |
| | 4 | Control relay | 22 | Ignition switch "LOCK → ON" | | SV |
| | | | | Idling | | 0V to 0.6V |
| | 5 | | | | | |
| | 6 | Injector No. 1 pulse | 59 | Idling | | 12V to 14V |
| | | | | Quick acceleration from idling to above 2000 rpm with "N" or "P" position | | Slight drop |
| | 7 | Injector No. 2 pulse | 60 | Idling | | 12V to 14V |
| | | | | Quick acceleration from idling to above 2000 rpm with "N" or "P" position | | Slight drop |
| | 8 | | | | | |
| | 9 | Injector No. 3 pulse | 61 | Idling | | 12V to 14V |
| | | | | Quick acceleration from idling to above 2000 rpm with "N" or "P" position | | Slight drop |
| | 10 | | | | | |
| | 11 | Injector No. 4 pulse | 62 | Idling | | 12V to 14V |
| | | | | Quick acceleration from idling to above 2000 rpm with "N" or "P" position | | Slight drop |
| | 12 | Purge control solenoid valve | 17 | Idling (warm engine) | A/C switch OFF | SV |
| | | | | | A/C switch ON | 0V to 0.6V |

## FAULT CODES—1988 GALANT, 1988–89 SIGMA, 1989–90 V-6 MONTERO AND 1990 V-6 TRUCK

| Output preference order | Diagnosis item | Malfunction code: Output signal pattern | Malfunction code: No. | Malfunction code: Memory | Check item (Remedy) |
|---|---|---|---|---|---|
| 1 | Computer | H L | – | – | (Replace electronic control unit) |
| 2 | Oxygen sensor | H L | 11 | Retained | • Harness and connector<br>• Oxygen sensor<br>• Fuel pressure<br>• Injectors (Replace if defective.)<br>• Intake air leaks |
| 3 | Air flow sensor | H L | 12 | Retained | • Harness and connector (If harness and connector are normal, replace air flow sensor assembly.) |
| 4 | Intake air tempera-ture sensor | H L | 13 | Retained | • Harness and connector<br>• Intake air temperature sensor |
| 5 | Throttle position sensor | H L | 14 | Retained | • Harness and connector<br>• Throttle position sensor<br>• Idle switch |
| 6 | Coolant tempera-ture sensor | H L | 21 | Retained | • Harness and connector<br>• Coolant temperature sensor |
| 7 | Crank angle sensor | H L | 22 | Retained | • Harness and connector (If harness and connector are normal, replace distribu-tor assembly.) |

## FAULT CODES—1988 GALANT, 1988–89 SIGMA, 1989–90 V-6 MONTERO AND 1990 V-6 TRUCK

| Output preference order | Diagnosis item | Malfunction code: Output signal pattern | Malfunction code: No. | Malfunction code: Memory | Check item (Remedy) |
|---|---|---|---|---|---|
| 8 | Top dead center sensor (No. 1 cylinder) | H / L | 23 | Retained | • Harness and connector (If harness and connector are normal, replace distributor assembly.) |
| 9 | Vehicle-speed sensor (reed switch) | H / L | 24 | Retained | • Harness and connector<br>• Vehicle-speed sensor (reed switch) |
| 10 | Barometric pressure sensor | H / L | 25 | Retained | • Harness and connector (If harness and connector are normal, replace air flow sensor assembly.) |
| 11 | Injector | H / L | 41 | Retained | • Harness and connector<br>• Injector coil resistance |
| 12 | Fuel pump | H / L | 42 | Retained | • Harness and connector<br>• Control relay |
| 13 | EGR | H / L | 43 | Retained | • Harness and connector<br>• EGR thermo-sensor<br>• EGR valve<br>• Thermo-valve<br>• EGR valve control vacuum |
| 14 | Normal state | H / L | – | – | – |

## TESTING PROCEDURE WITH MULTI-CHECKER, 1989 MONTERO

**Checking the cranking (Check when the engine can't be started or when it is started.)**

| Check items | Check description: Check conditions | | Check description: Normal value | Probable cause of malfunction (or action) |
|---|---|---|---|---|
| Power-supply voltage<br>• Data reading<br>• Item No. 16 | • Ignition switch: ON | | 11–13 V | • Low battery voltage<br>• Power not supplied to the engine control unit<br>(1) Check the power-supply circuit.<br>(2) Check the ignition switch, ignition signal and input circuit.<br>(3) Check the control relay.<br>(4) Check the control relay control circuit.<br>• Malfunction of the engine control unit earth circuit. |
| Throttle position sensor<br>• Data reading<br>• Item No. 14 | • Ignition switch: ON<br>• Throttle valve: idling position (When the Throttle position sensor output voltage is 1200 mV or lower, the engine control unit diagnoses the injector signal.) | | 300–1000 mV | • Maladjustment of the throttle position sensor<br>• Malfunction of the throttle position sensor or related circuitry |
| Self-diagnosis output | • Crank the engine for four seconds or longer.<br>• Ignition switch: ON (Check for injector or top dead center sensor circuitry disconnection or damage.) | | Normal | • Check in accordance with the diagnosis code. (Note that the diagnosis code will be erased if there is disconnection or damage of the engine control unit back-up power-supply circuit.)<br>• If various diagnosis codes are output, the most frequent cause is damage or disconnection of the power-supply or earth circuit. |
| Fuel pump<br>• Actuater test<br>• Item No. 7 | • Make the test with relation to both engine cranking and fuel pump forced actuation. | Pinch closed the return hose. | The pulsations of fuel flow can be felt by the finger. | • Power is not supplied to the fuel pump.<br>(1) Check the ignition switch (IG and ST).<br>(2) Check the control relay.<br>(3) Check related circuitry.<br>• Fuel pump malfunction |
| | | Listen close to the fuel tank. | The pump operation sound can be heard. | |
| Ignition switch – ST<br>• Data reading<br>• Item No. 18 | • Ignition switch: ON | Engine stopped | OFF | • Ignition switch – ST signal circuit check<br>• Ignition switch check |
| | | Cranking | ON | |

# TESTING PROCEDURE WITH MULTI–CHECKER, 1989 MONTERO

| Check items | Check description: Check conditions | | Check description: Normal value | Probable cause of malfunction (or action) |
|---|---|---|---|---|
| Top dead center sensor<br>● Data reading<br>● Item No. 22 | ● Engine cranking<br>● Tachometer connection (Use the tachometer to check the cut-off of the ignition coil's primary currents.) | Cranking rpm | rpm | ● If the tachometer's indicated read-out is 0, there is no cut-off of the ignition coil primary currents.<br>(1) Check the power transistor unit and control circuit.<br>(2) Check the ignition coil and the coil power-supply circuit.<br>● If the multi-use tester rpm. read-out is abnormal<br>(1) Malfunction of the top dead center sensor circuit<br>(2) Malfunction of the top dead center sensor<br>(3) Malfunction of the timing belt |
| | | Approx. 200 | Approx. 200 | |
| Injector<br>● Data reading<br>● Item No. 41 | ● Engine cranking | Listen for operation sound. | Operation sound of injector is audible. | ● Injector malfunction<br>● Improper contact of connector and relay contacts |
| | | Coolant temperature °C (°F) | Actuation time *2 (msec) | ● Engine coolant temperature sensor malfunction<br>● Ignition switch ST malfunction |
| | | 0 (32) *1 | Approx. 14 | |
| | | 20 (68) | Approx. 40 | |
| | | 80 (176) | Approx. 9 | |

**Checking the sensors**

| Check items | Check description: Check conditions | | Check description: Normal value | Probable cause of malfunction (or action) |
|---|---|---|---|---|
| Self-diagnosis output | ● Engine: idling (2 minutes or more after engine start) | | Normal | ● Check in accordance with the diagnosis code. (Note that the diagnosis code will be erased if there is disconnection or damage of the engine control unit back-up power-supply circuit.)<br>● If various diagnosis codes are output, the most frequent cause is damage or disconnection of the power-supply or earth circuit. |
| Oxygen sensor<br>● Data reading<br>● Item No. 11 | ● Engine warm-up (Make the mixture lean by engine speed reduction, and rich by racing.) | Engine condition | Voltage (mV) | ● If the oxygen sensor output voltage is high during sudden deceleration<br>(1) Check for injector leakage.<br>(2) Check the oxygen sensor signal circuit.<br>● If the oxygen sensor output voltage is low during engine racing<br>(1) Check the oxygen sensor and signal circuit. |
| | | When sudden deceleration from 4,000 rpm | 200 or lower | |
| | | When engine is suddenly raced | 600–1,000 | |

| Check items | Check description: Check conditions | | Check description: Normal value | Probable cause of malfunction (or action) |
|---|---|---|---|---|
| Oxygen sensor<br>● Data reading<br>● Item No. 11 | ● Engine warm-up (Using the oxygen sensor signal, check the air/fuel mixture ratio, and also check the condition of control by the engine control unit.) | Engine speed (rpm.) | Voltage (mV) | ● If the oxygen sensor signal is normal, the engine control unit is regulating the air/fuel mixture ratio normally.<br>● If the oxygen sensor output voltage is low at all times, check whether or not there is intake of air.<br>● If the oxygen sensor output voltage is high at all times, check for leakage of the injector. |
| | | 700 (idle) | 400 or lower | |
| | | 2,000 | ↕ (changes)<br>600–1,000 | |
| Air flow sensor<br>● Data reading<br>● Item No. 12 | ● Engine warm-up | Engine condition | Frequency (Hz) | ● If the air flow sensor output frequency suddenly changes greatly, improper contact of the air flow sensor or connector is probable.<br>● If the output frequency of the air flow sensor is unusually high or low, check the air cleaner element.<br>● If the output frequency of the air flow sensor is high, an increase of engine resistance or leakage of compression pressure is probable. |
| | | 700 rpm (Idling) | 25–45 | |
| | | 2,000 rpm | 85–105 | |
| | | Racing | Increase caused by racing | |
| Intake air temperature sensor<br>● Data reading<br>● Item No. 13 | ● Ignition switch: ON, or engine running | Intake-air temperature °C (°F) | Temperature °C (°F) | ● Malfunction of intake air temperature sensor or related circuitry |
| | | – 20 (– 4) | – 20 (– 4) | |
| | | 0 (32) | 0 (32) | |
| | | 20 (68) | 20 (68) | |
| | | 40 (104) | 40 (104) | |
| | | 80 (176) | 80 (176) | |
| | ● Ignition switch: ON | Warm by using hair dryer or other method. | Increases. | |
| Throttle position sensor<br>● Data reading<br>● Item No. 14 | ● Ignition switch: ON | Throttle valve | Voltage (mV) | ● Throttle position sensor maladjustment<br>● Throttle position sensor or related circuitry malfunction<br>● If there is any indication that the fixed SAS has been moved, adjust the fixed SAS. |
| | | Idling position | 300–1000 | |
| | | Opens slowly. | Becomes higher in proportion to valve opening. | |
| | | Fully open | 4,500–5,500 | |
| Ignition switch – ST<br>● Data reading<br>● Item No. 18 | ● Ignition switch: ON | | OFF | ● Ignition switch – ST signal circuit check<br>● Ignition switch check |

# TESTING PROCEDURE WITH MULTI–CHECKER, 1989 MONTERO

| Check items | Check description | | | Probable cause of malfunction (or action) |
|---|---|---|---|---|
| | Check conditions | | Normal value | |
| Engine coolant temperature sensor<br>● Data reading<br>● Item No. 21 | ● Ignition switch: ON, or engine running | Coolant temperature °C (°F) | Temperature °C (°F) | ● Engine coolant temperature sensor or related circuitry malfunction |
| | | −20 (−4) | −20 (−4) | |
| | | 0 (32) | 0 (32) | |
| | | 20 (68) | 20 (68) | |
| | | 40 (104) | 40 (104) | |
| | | 80 (176) | 80 (176) | |
| Top dead center sensor<br>● Data reading<br>● Item No. 22 | ● Engine: idling (Check with the Idle position switch ON.) | Coolant temperature °C (°F) | Idling rpm | ● If the rpm suddenly becomes greater a malfunction of the crank angle sensor or improper contact of the connector is probable.<br>● If the rpm is low when cold, clogging of the fast-idling air valve is probable. |
| | | −20 (−4) | 1,450–1,650 | |
| | | 0 (32) | 1,250–1,450 | |
| | | 20 (68) | 1,050–1,250 | |
| | | 40 (104) | 850–1,050 | |
| | | 80 (176) | 600–800 | |
| Barometric pressure sensor<br>● Data reading<br>● Item No. 25 | ● Ignition switch: ON | Altitude m (ft.) | Pressure mm Hg | ● Barometric pressure sensor or related circuitry malfunction. (If the barometric pressure sensor pressure is low at high speed, clogging of the air cleaner element is probable.) |
| | | 0 (0) | 760 | |
| | | 600 (1,969) | 710 | |
| | | 1,200 (3,937) | 660 | |
| | | 1,800 (5,906) | 610 | |
| | ● Engine: racing at 2,000 rpm | Gradually close the air-intake duct by using a hand. | Decreases. | |
| Idle position switch<br>● Data reading<br>● Item No. 26 | ● Ignition switch: ON (Checking by using the accelerator pedal several times.) | Throttle valve idling position | ON | ● Idle position switch or related circuitry malfunction<br>● Improper adjustment of the accelerator cable or the auto-cruise cable<br>● Idling-position switch: improper adjustment. |
| | | Open the throttle valve slightly. | OFF | |
| Power steering oil pressure switch<br>● Data reading<br>● Item No. 27 | ● Engine: idling | Steering wheel neutral position (wheels straight-ahead direction) | OFF | ● Power steering oil pressure switch or signal circuit malfunction |
| | | Steering wheel half turn | ON | |
| Air conditioner switch<br>● Data reading<br>● Item No. 28 | ● Engine: idling (The air conditioner compressor could be activated when the air conditioner switch is ON.) | Air conditioner switch "OFF" | OFF | ● Check air conditioner system. |
| | | Air conditioner switch "ON" | ON | |
| EGR temperature sensor (California only)<br>● Data reading<br>● Item No. 43 | ● Engine: warmed up (Engine is maintained in a constant state for 2 minutes or more.) | Engine condition | Temperature °C (°F) | ● Check the EGR temperature sensor.<br>● Check the EGR control system.<br>● Check the EGR valve.<br>● Check the thermo valve (A/T models only)<br>● Check the EGR control solenoid valve (M/T models only)<br>● Check the EGR control vacuum. |
| | | 700 (idling) | 100°C (212°F) or lower | |
| | | ● Intake air temperature 10–40°C (50–104°F)<br>● Water temperature: 70°C (158°F) or higher<br>● While driving continuously for more than 30 seconds at a vehicle speed of 80–90 km/h (50–56 mph) | 120°C (248°F) or higher | |

**Checking the actuator**

| Check items | Check description | | | Probable cause of malfunction (or action) |
|---|---|---|---|---|
| | Check conditions | | Normal value | |
| Injectors<br>● Actuater test<br>● Item No. 1-4 | ● Engine: warmed up (Cut off the injectors in sequence during idling after engine warm-up; check the idling condition of a cylinder that doesn't change.) | Injector No. | Engine | ● If the idling condition of one cylinder doesn't change, check that cylinder.<br>(1) Check the injector operation sound.<br>(2) Check the spark plug and high-tension cable.<br>(3) Check the power transistor unit and control circuit. |
| | | 1 | Idling condition changes more. (Becomes more unstable, or engine stalls.) | |
| | | 2 | | |
| | | 3 | | |
| | | 4 | | |
| | | 5 | | |
| | | 6 | | |
| Injector<br>● Data reading<br>● Item No. 41 | ● Engine: warmed up | Engine condition | Actuation time (msec.) | ● If the injector activation times is unusually long or short, there is a malfunction of the air-flow sensor, Engine coolant temperature sensor, intake air temperature sensor, or barometric pressure sensor.<br>● If the injector activation time is long, increased engine resistance or leakage of compression pressure is probable. |
| | | 700 rpm (Idling) | 2,7–3,2 | |
| | | 2,000 rpm | 2,4–2,9 | |
| | | Rapid racing | Increases. | |

## TESTING PROCEDURE WITH MULTI–CHECKER, 1988 GALANT

| Check item (item No.) | Tester mode | Condition | | Test specification | Troubleshooting when outside the test specifications |
|---|---|---|---|---|---|
| Power supply (16) | Data transfer | Ignition switch: ON | | 11–13V | • Measure the battery voltage.<br>• Check the circuit that supplies the ECU power. |
| Throttle position sensor (14) | | Ignition switch: ON<br>Throttle valve: Idle position | | 400–600mV | • Check the throttle position sensor.<br>• Check the sensor circuit.<br>• Adjust the throttle position sensor. |
| Malfunction code read out | Self-diagnosis | Execute cranking for 4 seconds or more.<br>Ignition switch: ON | | Malfunction code is not output. | • Check the check items. (Refer to the self-diagnosis section.) |
| Fuel pump (07) | Actuator forced drive | Ignition switch: ON<br>Actuator drive | Squeeze the return hose. | Feel the fuel pulse with a finger. | • Check the fuel pump.<br>• Check the circuit that supplies power to the fuel pump. |
| | | | Listen near the fuel tank. | Pump driving sound is heard. | |
| TDC sensor (22) | Data transfer | Engine: Cranking<br>Tachometer: Connect | Cranking revolution speed [rpm] | Revolution speed [rpm] | • Check the power transistor and the ignition coil. (The tachometer reading is not proper.)<br>• Check the TDC sensor circuit. If the circuit is proper, replace the distributor assembly and recheck the system. |
| | | | Approx. 200 | Approx. 200 | |
| Ignition switch-ST (18) | | Ignition switch: ON | Engine stop | OFF | • Check the ignition switch-ST circuit.<br>• Check the ignition switch. |
| | | | Cranking | ON | |

## TESTING PROCEDURE WITH MULTI–CHECKER, 1989 MONTERO

| Check items | Check description | | | Probable cause of malfunction (or action) |
|---|---|---|---|---|
| | Check conditions | | Normal value | |
| Ignition advance (power transistor)<br>• Data reading<br>• Item No. 44 | • Engine: warmed up<br>• Timing light: set (The timing light is set so as to check the actual ignition timing.) | Engine rpm | Ignition advance (°BTDC) | • If the ignition advance and actual ignition timing are different, adjust the ignition timing. [The ignition timing may fluctuate during idling, but this is not a problem. The advance is greater (approx. 5°) at high altitude.] |
| | | 700 (Idling) | 13–20 | |
| | | 2,000 | 38–42 | |
| Stepper motor<br>• Data reading<br>• Item No. 45 | • Engine: idling after warm-up (Idle position switch hust be ON.) | Engine condition | Step | • If the number of steps increases to 100 or 120 or decreases to 0, a malfunction of the stepper motor or the activation circuit is probable.<br>• If the number of steps is small, check whether or not air is being sucked in.<br>• If the number of steps is large, either of the following is probable: (1)Deposits adhered to the throttle valve part (2)Increased engine resistance<br>• If the number of steps is abnormal even though the engine is normal, adjust the basic idle speed. |
| | | 700 rpm (Idling) | 2-12 | |
| | (The compressor clutch should be activated when the air conditioner switch is ON.) | Air conditioner switch ON (900 rpm) | 30–70 | • Check the air conditioner system.<br>• If the engine speed does not increase when the air conditioner switch is switched from OFF to ON, check the stepper motor or the activation circuit. |
| | | Air-conditioner switch ON shift lever "D" range (650 rpm) | 20–60 | • Check the inhibitor switch and the signal circuit. |
| Air conditioner relay<br>• Data reading<br>• Item No. 49 | • Engine: idling after warm-up | Air conditioner switch | Air conditioner relay | • If the air conditioner relay output is abnormal, check the air conditioner signal input circuit and the air conditioner system.<br>• If the activation of the air conditioner compressor clutch is not normal, check the compressor clutch and the relay circuit. |
| | | OFF | OFF (compressor clutch non-activation) | |
| | | ON | ON (compressor clutch activation) | |
| Purge control solenoid valve<br>• Actuator test<br>• Item No. 8 | • Ignition switch: ON (Engine Stop) | Actuator forced actuation | Operation sound audible during activation | • Check the purge control solenoid valve<br>• Check the purge control solenoid valve drive circuit. |
| EGR control solenoid valve (M/T models only) | • Ignition switch: ON (Engine stop) | Actuator forced actuation | Operation sound audible during activation | • Check the EGR control solenoid valve<br>• Check the EGR control solenoid valve drive circuit |

## TROUBLESHOOTING

When checking and correcting engine problems, it is important to start with inspection of the basic systems. Such conditions as failure to start, unstable idling or poor acceleration are often caused by items other than the fuel injection or engine management systems. Therefore, first check the following basic systems:

1. Power supply
   a. Battery
   b. Fusible link
   c. Fuse
2. Body ground
3. Fuel supply
   a. Fuel line
   b. Fuel filter
   c. Fuel pump
4. Ignition system
   a. Spark plugs
   b. High tension cable
   c. Distributor
   d. Ignition coil
5. Emission control system
   a. PCV system
   b. EGR system
   c. Vacuum leak
6. Others
   a. Ignition timing
   b. Idle speed

Troubles with the MPI system are often caused by poor contact of harness connectors. It is important to check harness connector contact.

### System Inspection

1. Before removing or installing a part, read diagnosis code and then disconnect the battery (−) terminal cable.
2. Before disconnecting the cable from battery terminal, turn the ignition switch to **OFF**. Removal or connection of battery cable during engine operation or while the ignition switch is **ON** could cause erroneous operation of the ECU or damage to semiconductors.
3. The control harness between the ECU and oxygen sensor are shielded wires with shield grounded to the body in order to prevent ignition noise and radio interference. When the shielded wire is faulty the control harness must be replaced.
4. When Multi–Use or MPI Checker is used, pay attention to the following.

- Avoid rough operation of switches.
- Do not subject test equipment to shock and other external forces, heat, etc.
- Keep the checker away from water and oil.

**NOTE: Follow test procedures below; refer to test specifications chart for exact test values.**

## FUEL PRESSURE

### Check

**NOTE: Threaded adaptors and extra O-rings are required for this test. All line connections must be tight and not leak under pressure.**

1. Release the pressure remaining in fuel pipe line so that fuel will not flow out.
   a. Disconnect the fuel pump connector at the fuel tank.
   b. Start the engine and after it stops by itself, turn the ignition switch to **OFF**.
   c. Disconnect the negative battery terminal.
   d. Connect the fuel pump connector.
2. Disconnect the fuel high pressure hose at the delivery pipe side.

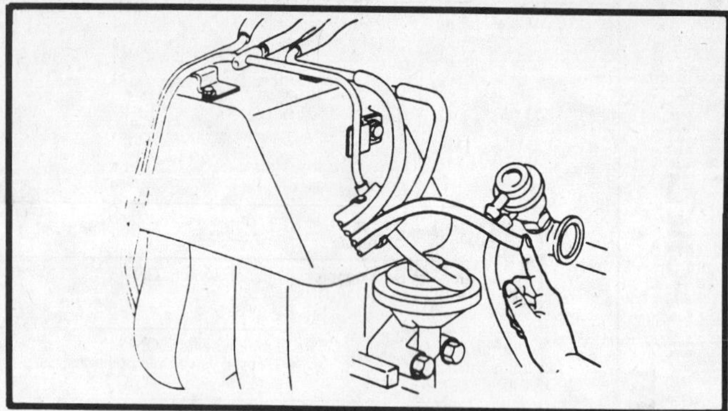

Fuel pressure measurement

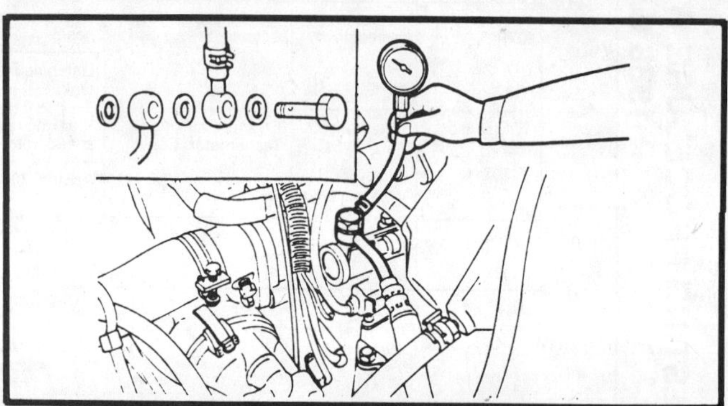

Fuel pressure gauge connection location

**CAUTION**

*Cover the hose connection with rags to prevent splash of fuel that could be caused by residual pressure in the fuel line.*

3. Using the adapters install the fuel-pressure gauge to the delivery pipe. Tighten the bolt to 18–25 ft. lbs. (25–35 Nm).
4. Connect the negative battery cable.
5. Apply battery voltage to the terminal for fuel pump. Check to be sure that there is no fuel leakage from the pressure gauge or the connections.
6. Disconnect the vacuum hose from the pressure regulator, and plug the hose end. Measure the fuel pressure during idling. This is the unregulated pressure.
7. Measure the fuel pressure when the vacuum hose is connected to the pressure regulator. This is regulated pressure.
8. If the results of the measurements made in Steps 6 and 7 above are not within the standard value, determine the probable cause make the necessary repairs.
9. Stop the engine and check fuel pressure gauge; pressure should not drop.
10. Release residual pressure from the fuel line.
11. Disconnect the fuel-pressure gauge from the delivery pipe.

**CAUTION**

*Cover the connection with rags to prevent splash of fuel caused by residual pressure in the fuel line.*

12. Using a new gasket, connect the fuel high-pressure hose.
13. Check for fuel leaks; apply voltage to fuel pump test terminal and check connections for any leak.

**NOTE: The following electrical tests require the use of the various special wiring adapters. These adapters isolate the various conductors within a harness for ease of testing. If testing without the adapters, carefully probe from the rear of the connector. Never insert a test probe into the front of a female harness connector.**

## AIR FLOW SENSOR (AFS)

### Inspection

**NOTE: If the air flow sensor fails, the intake air volume cannot be measured and as a result, normal fuel injection control is no longer available. The engine will run using the default value built into the ECU.**

1. Disconnect the air flow sensor connector.
2. Connect the adapter between the unattached connectors.
3. Warm the engine and bring it to a normal idle.
4. Measure the voltage of terminals as shown on the chart. Most 1988–90 MPI engines require the use of the special tester; the frequency of the AFS is expressed in Hz, not readable with a volt/ohm meter.

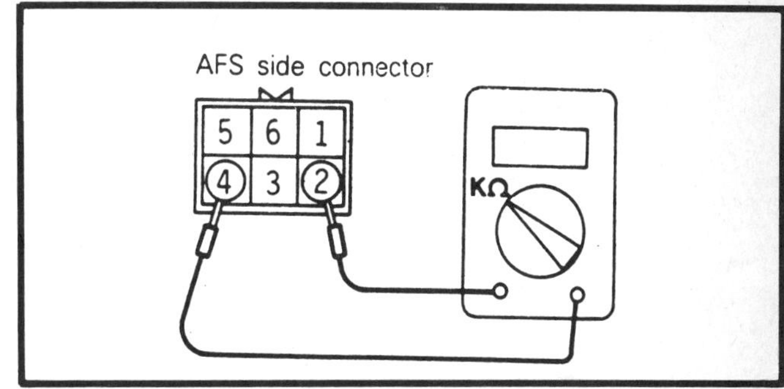


**Testing the intake air temperature sensor**

## INTAKE AIR TEMPERATURE

### Inspection

1. Disconnect the air flow sensor connectors.
2. Measure resistance between terminals. Resistance will be determined by sensor temperature.
3. Measure resistance while heating the sensor using a hair drier. Resistance should become smaller as temperature become higher.
4. If the value deviates from the standard value or the resistance remains unchanged, replace the air flow sensor assembly.

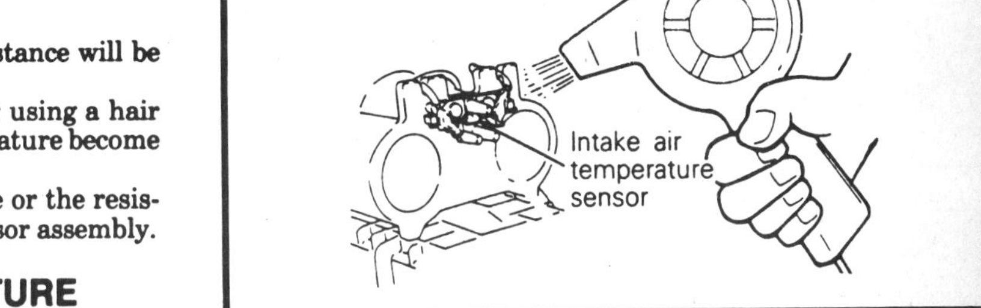


**Intake air temperature sensor testing**

## ENGINE COOLANT TEMPERATURE SENSOR

### Inspection

1. Remove engine coolant temperature sensor from the intake manifold.
2. With temperature sensing portion of engine coolant temperature sensor immersed in hot water, check resistance. The sensor should be held with its housing 0.12 in. (3mm) away from the surface of the container. Resistance will vary according to temperature. Resistance should be close to 0 at boiling point.
3. If the resistance deviates from the standard value greatly, replace the sensor.

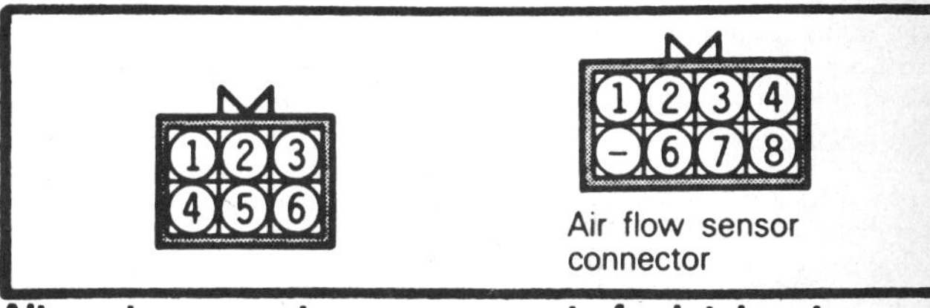


**Alternate connector arrangements for intake air sensor**

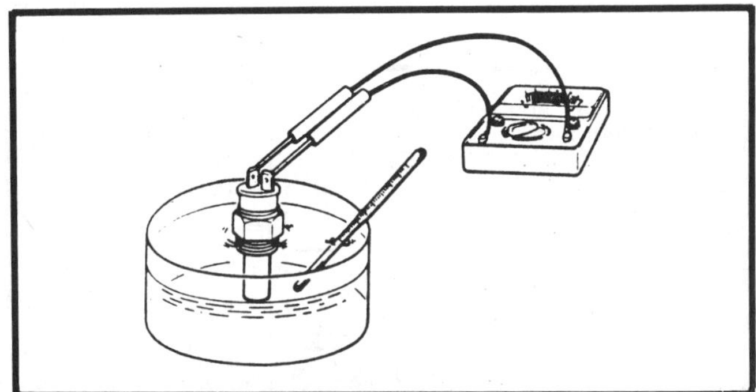

**Engine coolant temperature sensor testing**

## THROTTLE POSITION SENSOR (TPS)

### Inspection

1. Disconnect the throttle position sensor connector; install adapter harness.
2. Measure resistance between terminals as shown on the chart.
3. Connect a pointer type ohmmeter between the ground terminal and the terminal not tested in Step 2.

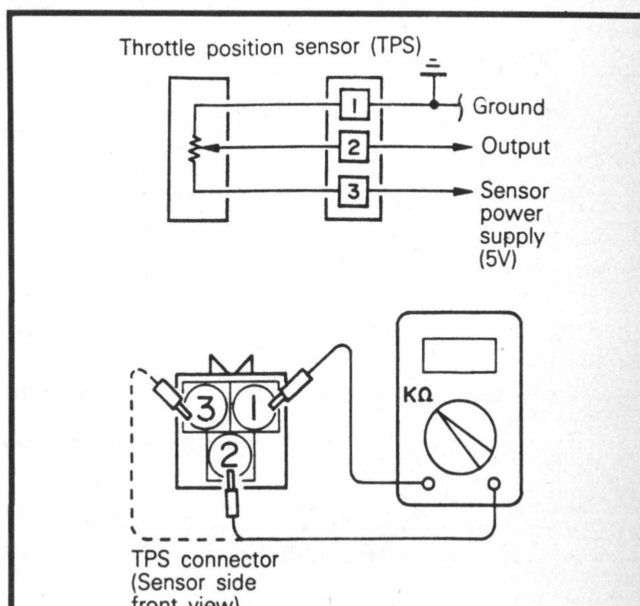


**Throttle position sensor testing**

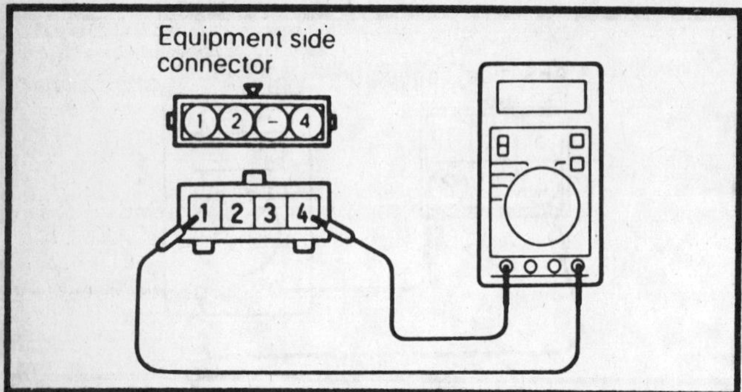

Alternate connector arrangements for TPS connector

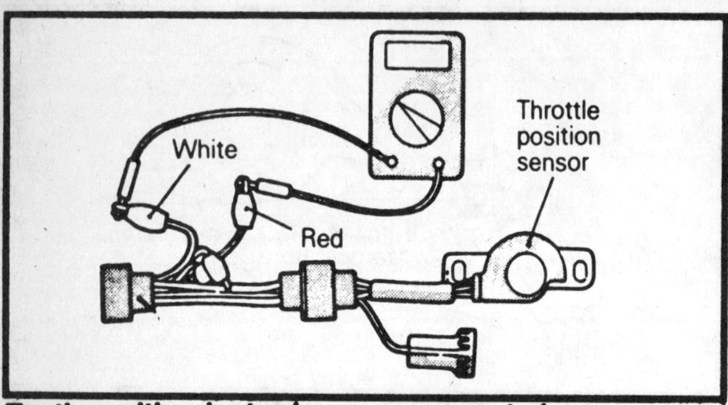

Testing with adapter harness connected

4. Operate the throttle valve slowly from the idle position to the full open position and check that the resistance changes smoothly and in proportion to the throttle valve angle.

5. If the resistance is out of specification, or fails to change smoothly, replace the TPS.

## IDLE SWITCH

### Inspection

1. Disconnect the ISC motor connector.
2. Check continuity between terminal 2 and body ground. Accelerator depressed – no continuity, accelerator released – continuity.
3. If out of specification, replace the ISC servo assembly.

## MOTOR POSITION SENSOR (MPS)

### Inspection

1. Disconnect the motor position sensor connector.
2. Connect the adapter harness to the disconnected MPS connector.
3. Disconnect the ISC motor connector.
4. Connect the adapter harness to the ISC motor end.

**NOTE: Be sure not to connect the harness (ECU side) connector.**

5. Measure the resistance between terminals as shown on the chart.
6. If the standard value is not achieved, replace the ISC servo assembly.

## NO. 1 CYLINDER TDC SENSOR AND CRANKSHAFT ANGLE SENSOR

### Inspection

1. Disconnect the spark plug wires from the ignition coil.

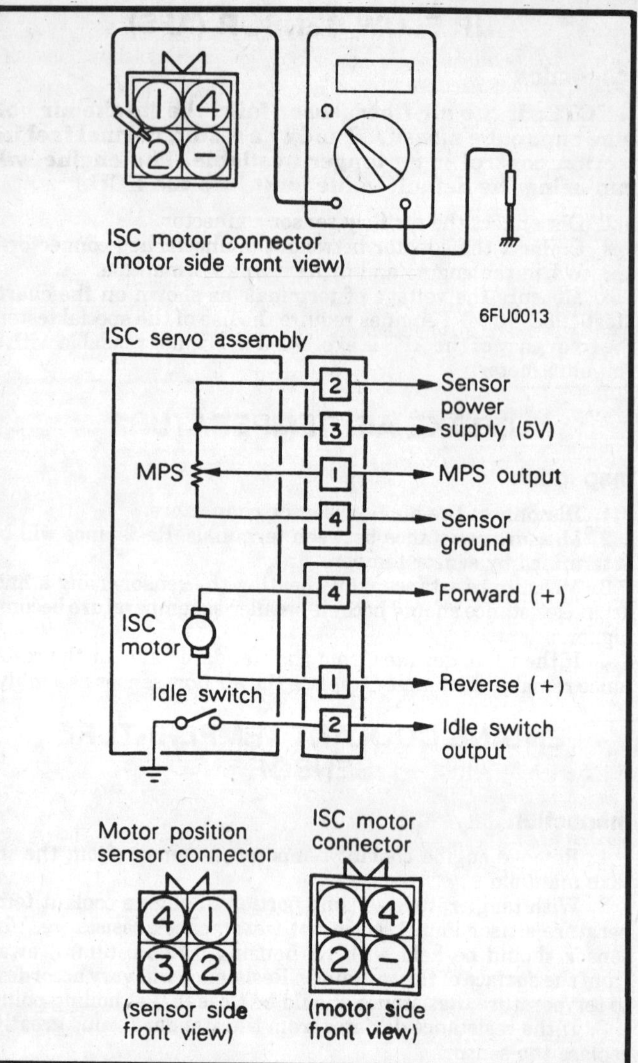

Idle switch testing

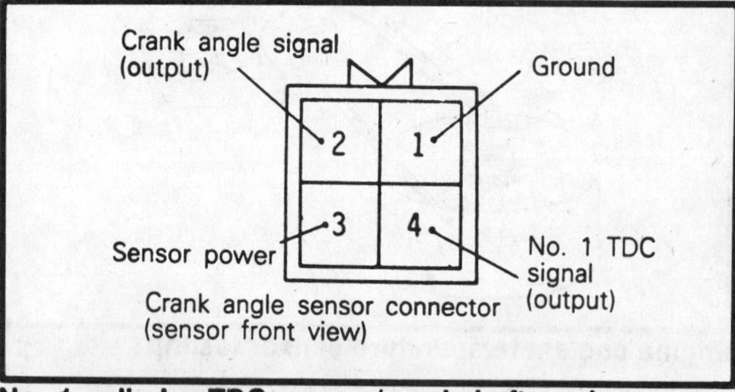

No. 1 cylinder TDC sensor/crankshaft angle sensor connector identification

2. Disconnect the crankshaft angle sensor connector.
3. Connect the adapter harness between the disconnected connectors.
4. Measure the output voltage between terminals 2 and 1 (crank angle signal) and 4 and 1 (No. 1 TDC signal) while cranking the engine. No. 1 cylinder TDC sensor should be 0.5–1.0 volt and crankshaft angle sensor should be 2.0–2.5 volts.

5. When the voltage is abnormal, check the sensor power and ground circuit, and if nothing unusual is found, disassemble the distributor and check it.

## OXYGEN SENSOR

### Inspection

**NOTE: Before checking, warm the engine until engine coolant temperature reaches 185–205°F (85–95°C). Use an accurate digital voltmeter for testing.**

1. Disconnect the oxygen sensor connector and connect a voltmeter to the oxygen sensor connector.

**CAUTION**

*Surfaces and components will be hot.*

2. With engine running, observe voltage at idle, then race engine momentarily. Voltage should increase (when raced) to approximately 1 volt, then drop as engine returns to idle.
3. If voltage is not correct, or no change occurs during rpm increases, replace oxygen sensor.

## POWER STEERING OIL PRESSURE SWITCH

### Inspection

1. Disconnect the power steering oil pressure switch connector.
2. Start the engine and check continuity between the pressure switch terminal and body ground. When the steering wheel is straight ahead, there should be NO continuity. With wheels turned, continuity should exist.
3. If the check result is otherwise than specified, check the oil pump assembly.

## INJECTORS

### Sound Test

Using a stethoscope or similar tool, check for operating sound as engine idles. Each injector gives a distinct ticking as it opens and closes. Check that the sound occurs at shorter intervals as the engine speed increases. Listen carefully; other injectors may produce similar sound even if the injector being checked is not working.

### Resistance Test

Disconnect the injector connector. Measure resistance between the injector terminals. Resistance should be as specified on the chart. Install the injector connector.

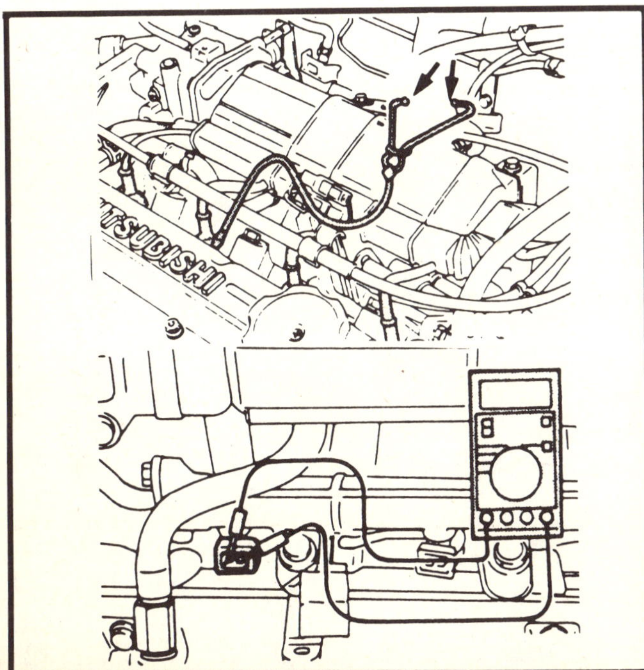


Checking the injectors for sound and resistance

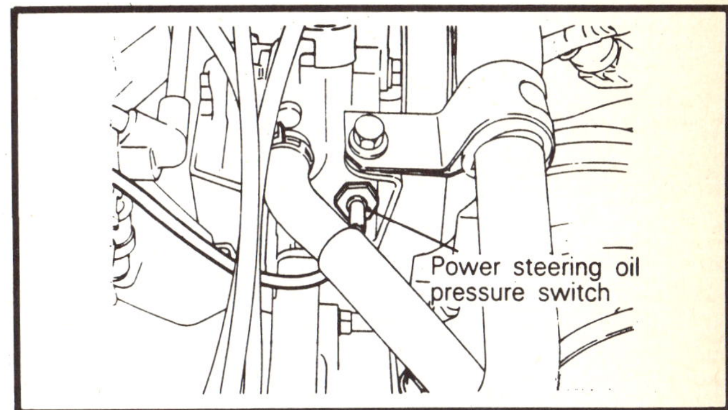


Power steering oil pressure switch location

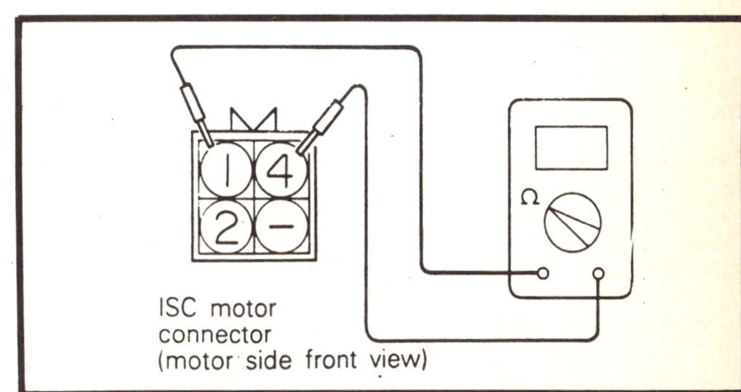


Idle speed control (ISC) motor continuity testing

## IDLE SPEED CONTROL (ISC) MOTOR

### Inspection

1. Disconnect the ISC motor connector.
2. Check continuity of the ISC motor coil.
3. Connect 6 volt DC between terminal 1 and terminal 4 of the ISC motor connector and check to be sure that the ISC servo operates.

**NOTE: Apply only a 6 volts DC or lower voltage. Application of higher voltage could cause locking of the servo gears.**

4. If not, replace ISC servo as an assembly.

# Component Replacement

## THROTTLE BODY

### Removal and Installation

**NOTE: The throttle body for each of the MPI engines is different. The procedure below is general and may require slight alteration of sequence depending on the engine.**

1. Disconnect the negative battery cable.
2. Drain the coolant.

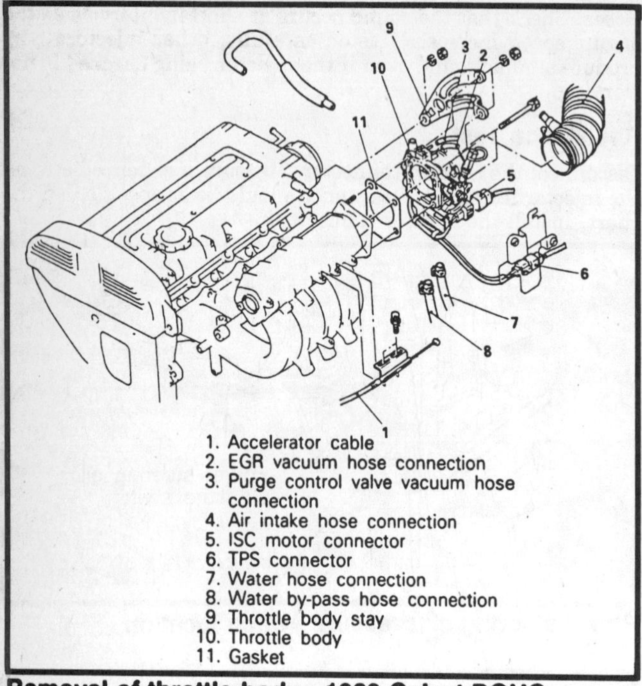


**Removal of throttle body – 1989 Galant DOHC**

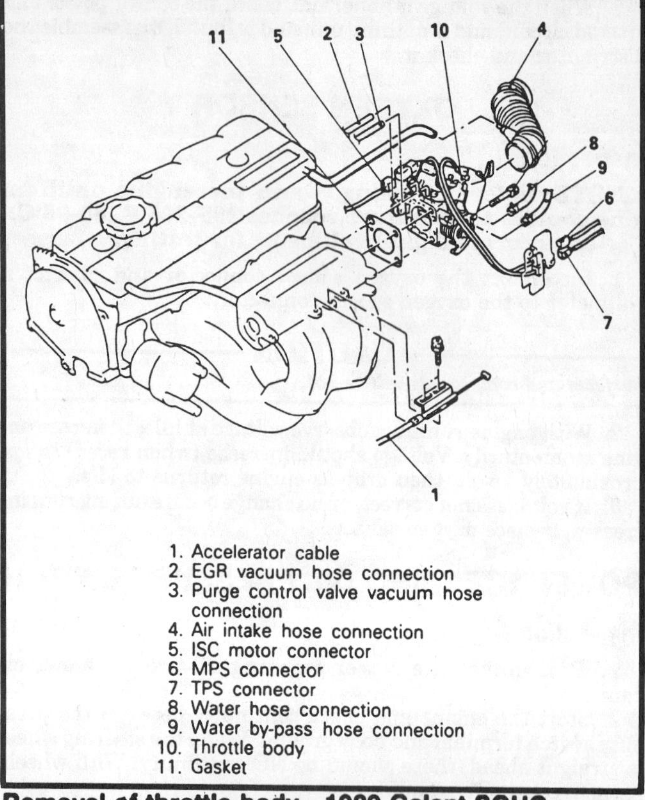


**Removal of throttle body – 1989 Galant SOHC**

3. Disconnect the main air intake duct from the throttle body.
4. Disconnect the vacuum and breather hoses running to the throttle body.
5. Disconnect accelerator cable and cruise control cable if so equipped.
6. Disconnect the coolant hoses running to the throttle body.
7. Disconnect wiring running to throttle body.
8. Remove nuts and bolts holding throttle body to the manifold. Some throttle bodies are supported by a bracket which must be removed before the throttle body can be removed. The 6G72 V6 throttle body uses bolts of 2 different lengths; take note of each bolt's position.

**NOTE: Some throttle bodies have the idle speed control servo assembly mounted on the top of the unit. The bracket may appear to hold the throttle body; it should not be removed from the throttle body.**

9. Lift throttle body away from the manifold; handle it carefully.

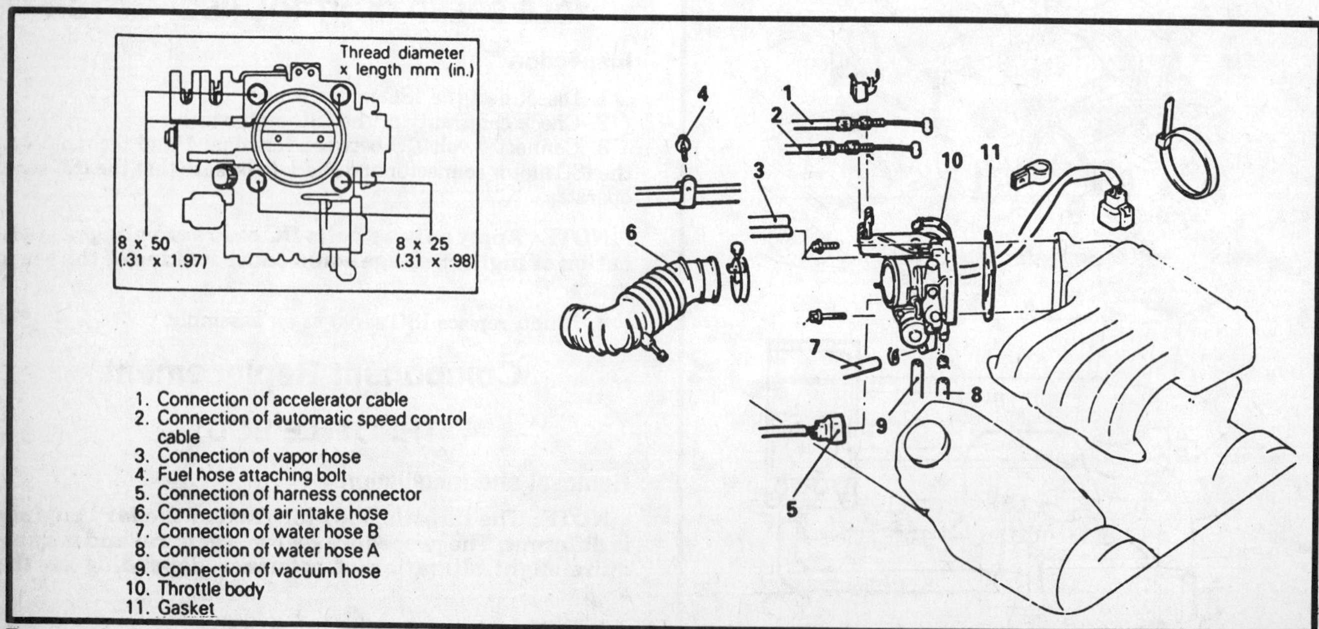


**Removal of throttle body – 1988 Galant and 1988-89 Sigma**

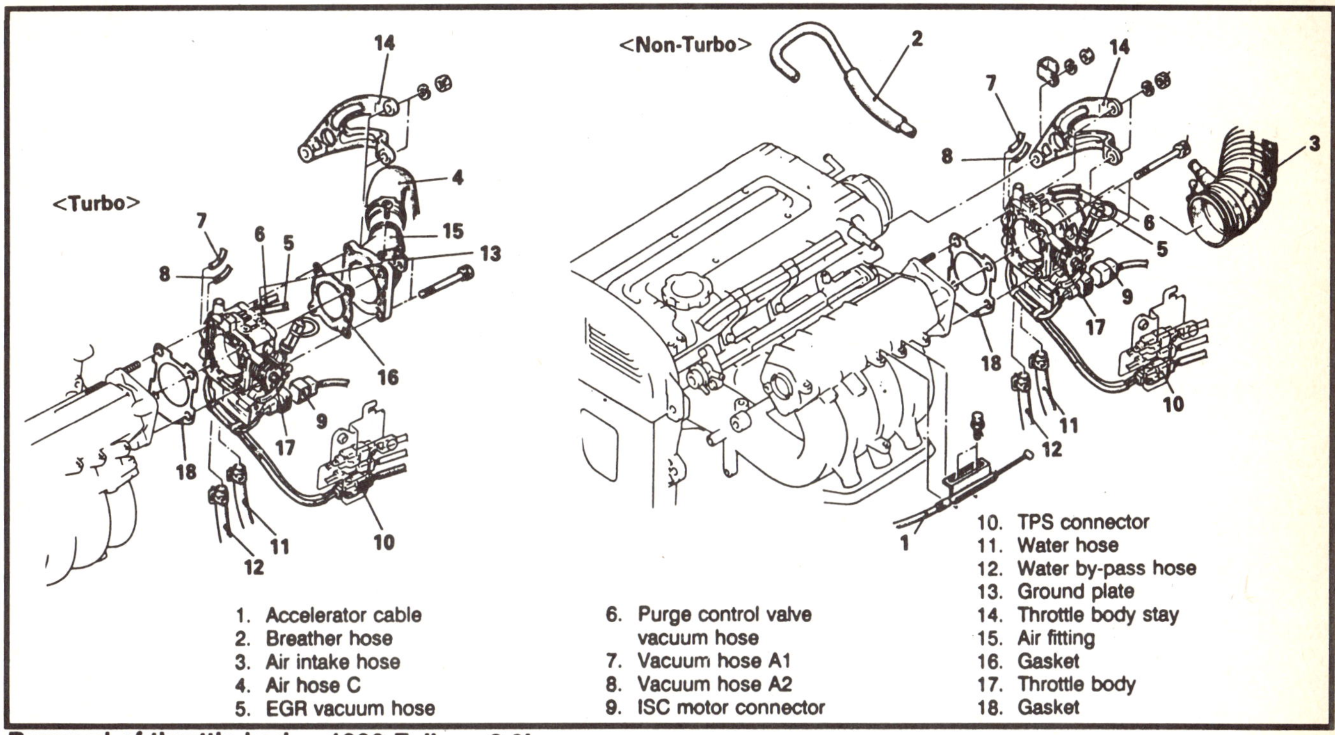


**Removal of throttle body—1990 Eclipse 2.0L**

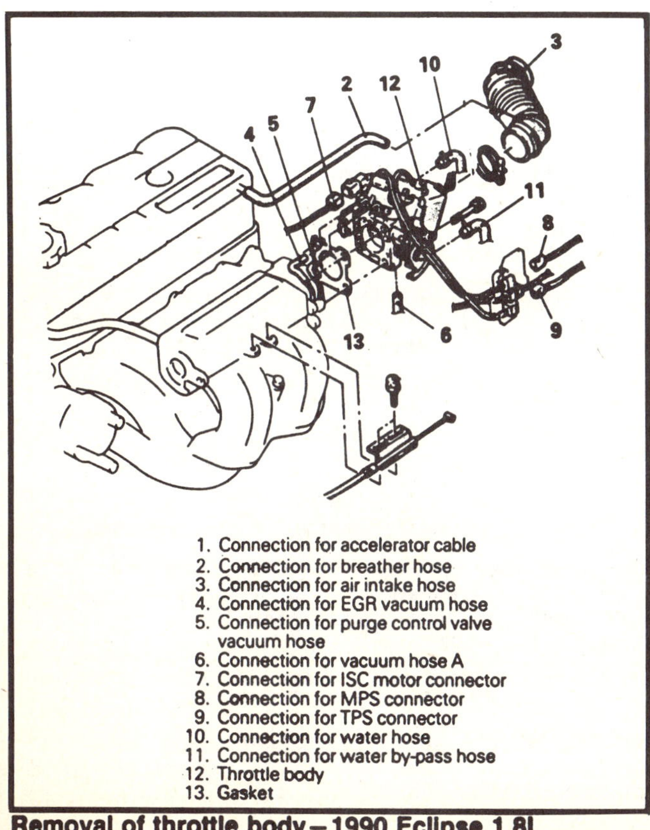


**Removal of throttle body—1990 Eclipse 1.8L**

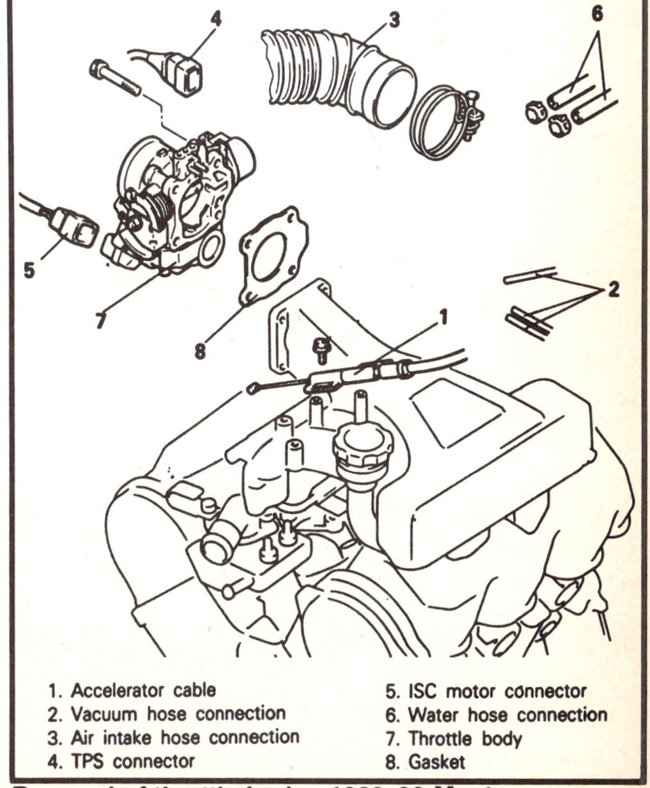


**Removal of throttle body—1989–90 Montero**

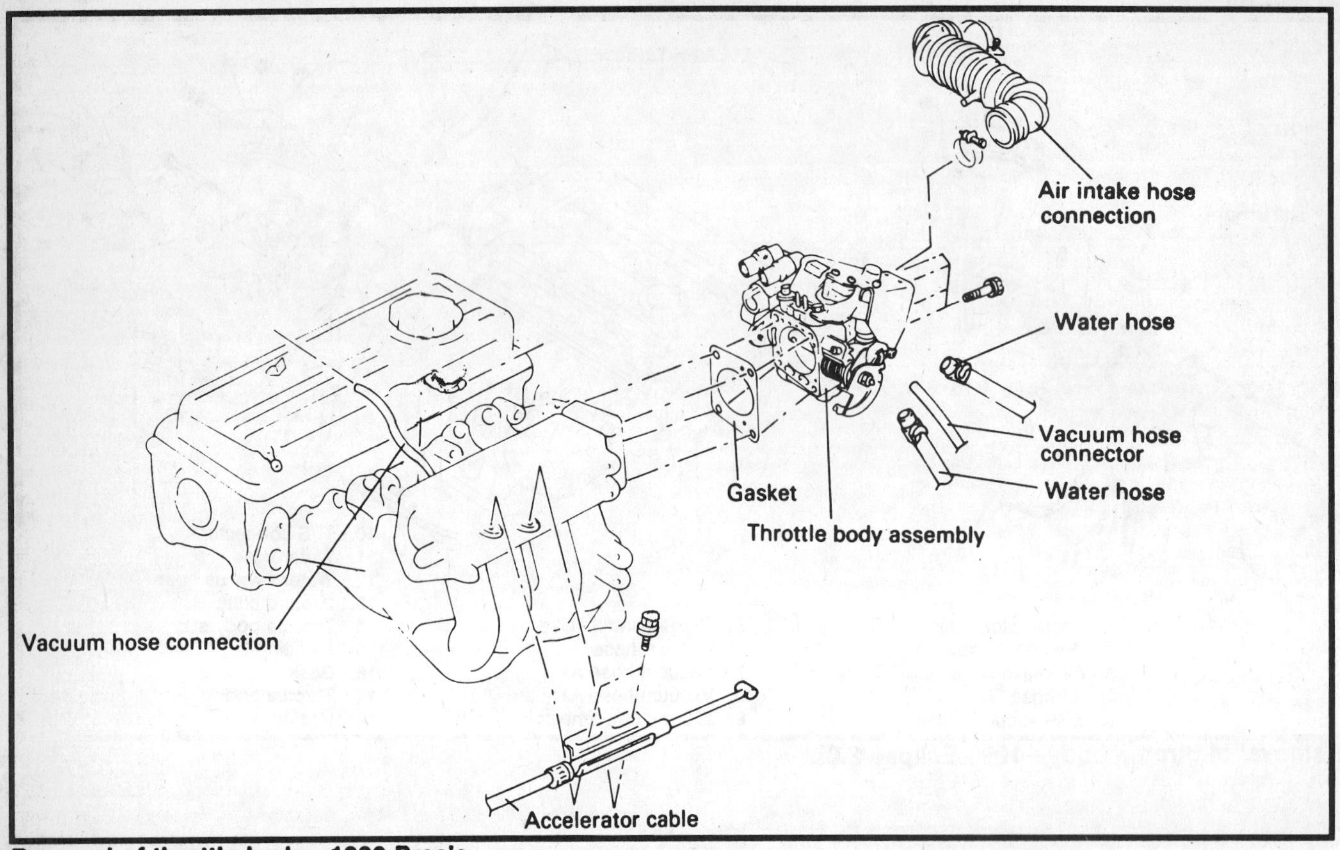


**Removal of throttle body – 1990 Precis**

10. The throttle plate area may be cleaned with a spray cleaner, but must be completely dry before installation. Disassembly of the throttle body is not recommended.
11. Before reinstalling, all remains of old gasket must be removed.
12. Install nuts and bolts finger tight. For V-6 engines, make certain bolts are in correct holes by length. Tighten the bolts evenly creating an airtight seal against the gasket.
13. For all SOHC engines, tighten the nuts and bolts to 8 ft. lbs. (11 Nm). For DOHC, correct torque is 14 ft. lbs. (19 Nm). Do NOT overtighten.
14. Reconnect the wiring connectors.
15. Install the coolant hoses.
16. Install accelerator cable and cruise control cable. Adjust the cables.
17. Connect the vacuum hoses.
18. Install the air duct and breather tubes.
19. Refill the coolant to the proper level.
20. Connect the negative battery cable.

## IDLE SPEED CONTROL (ISC) AND THROTTLE POSITION SENSOR (TPS)

### Adjustment

**NOTE: Before testing, engine coolant temperature must be 185–205°F (85–95°C), all accessories OFF, electric fan off and transaxle in PARK OR NEUTRAL.**

1. Loosen or disconnect the accelerator cable.
2. Disconnect the TPS connector.
3. Connect the adapter harness between the detached connectors.
4. Connect a digital voltmeter between the blue and black terminals of the special tool.
5. Insert a paper clip into the connector of the coil primary wire. Use the paper clip as a terminal for connecting the tachometer.

**NOTE: Wiring connectors must not be separated; the paper clip can be inserted along terminal surface.**

6. Turn the ignition switch to **ON** (do not start the engine) and hold the switch in that position for 15 seconds or more to check that the ISC motor is set at the initial position (idle point).

**NOTE: When the ignition switch is turned to ON, the ISC motor extends to the fast idle position and in 15 seconds retracts, stopping at the initial position.**

7. Disconnect the ISC servo connector and fix the ISC servo at the initial position.
8. In order to prevent binding, open the throttle valve by hand to a half or more opening 2 or 3 times and then release it. Allow it to return with a snap.
9. Start the engine and run at idle.
10. Check that the engine speed is 750 rpm ± 100 rpm.

**NOTE: Engine idle may be low for reasons not cured by adjustment. Deposits may be stuck to the throttle plate, causing binding. Cleaning the throttle plate is indicated in this case.**

11. If the engine speed is not as specified, adjust the ISC adjusting screw for the standard rpm.

**NOTE: When turning the ISC adjusting screw, use hexagon wrench whenever possible. Prevent the screw from becoming loose by turning it only in the tightening direction.**

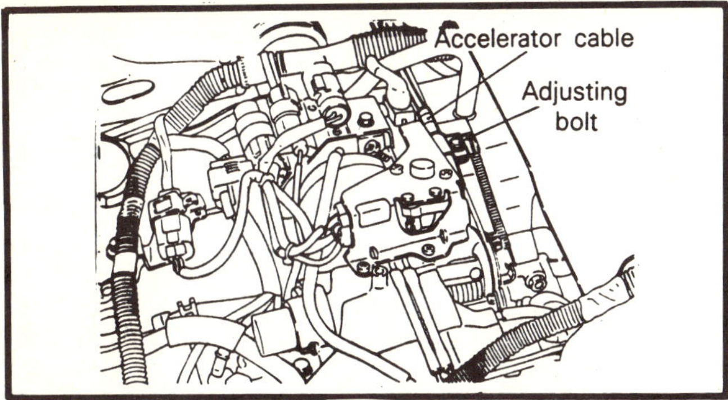


**Idle speed control (ISC) and Throttle position sensor (TPS) adjustment**

**Idle speed check connector location**

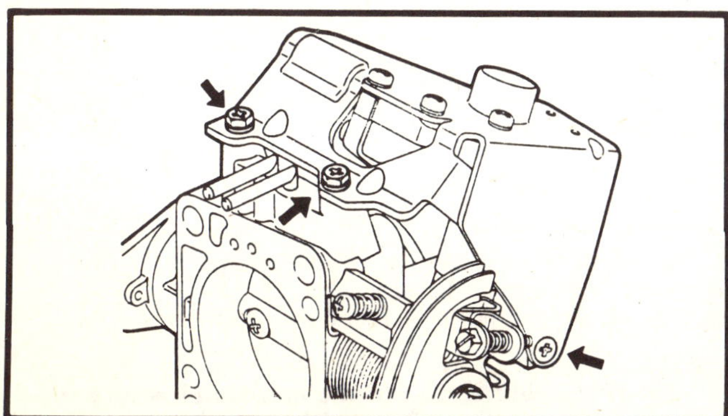

**Idle speed control (ISC) servo servicing**

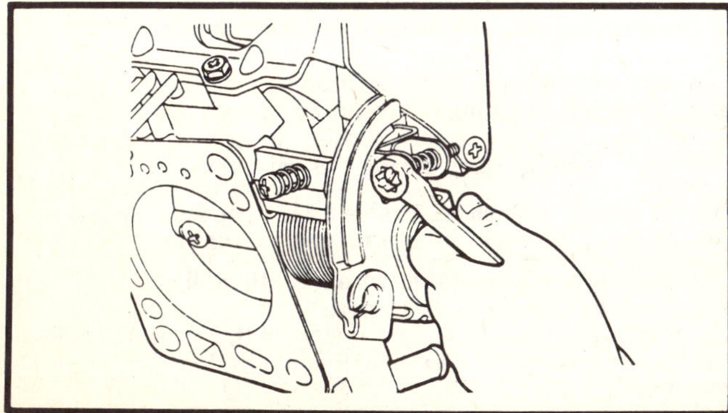

**Idle speed adjustment screw**

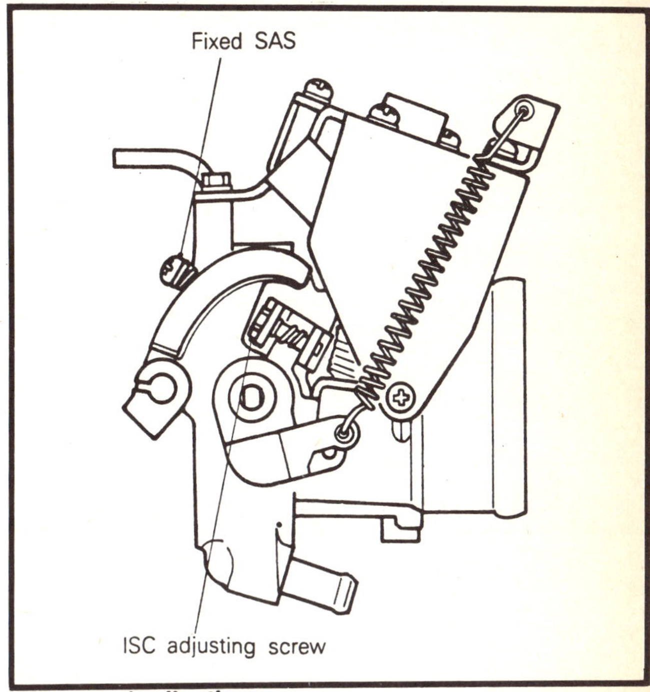


**Idle speed adjusting screws**

12. Tighten the fixed SAS (special adjusting screw) until the engine speed starts to increase. Loosen it until the engine speed ceases to drop (touch point) and then loosen a half turn from the touch point.
13. Stop the engine.
14. Turn the ignition switch to **ON** (engine does not start) and check that the TPS output voltage is correct.
15. If TPS voltage is incorrect, loosen TPS mounting screws and adjust by turning the TPS. Correct output voltage at idle is 0.9 volts.

**NOTE: Turning the TPS clockwise increases the output voltage. Tighten the screws securely after adjustment.**

16. Turn ignition switch to **OFF**.
17. Adjust the accelerator cable play.
18. Connect the ISC servo connector.
19. Disconnect the test harness and voltmeter, and connect the TPS connector.
20. Start the engine and check to be sure that the idling speed is correct. Standard value: 750 ± 100 rpm.
21. Turn the ignition switch to **OFF** and disconnect the battery terminal for 15 seconds. (This erases data stored in memory during the ISC adjustment.)

## FUEL INJECTOR

### Removal and Installation

**NOTE: Injectors are extremely sensitive to dirt. They must be protected at all times. Dirt entering the system can foul an injector or change its operation. Any gaskets or O-rings removed the injector MUST be replaced with new ones at reassembly. Do not attempt to reuse seals; high pressure fuel leaks will result.**

#### 4G15, G4DJ, 4G37 AND 4G63 SOHC ENGINES

1. Safely relieve the pressure within the fuel system.
2. Disconnect the negative battery cable.

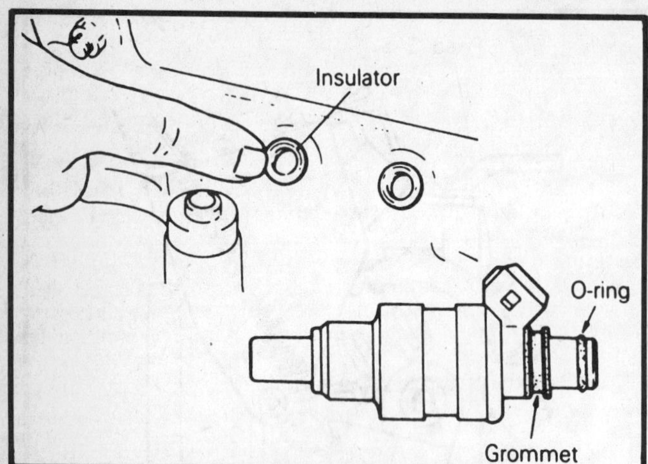

Always replace seats, O-rings and grommets

Injector must turn freely within delivery pipe

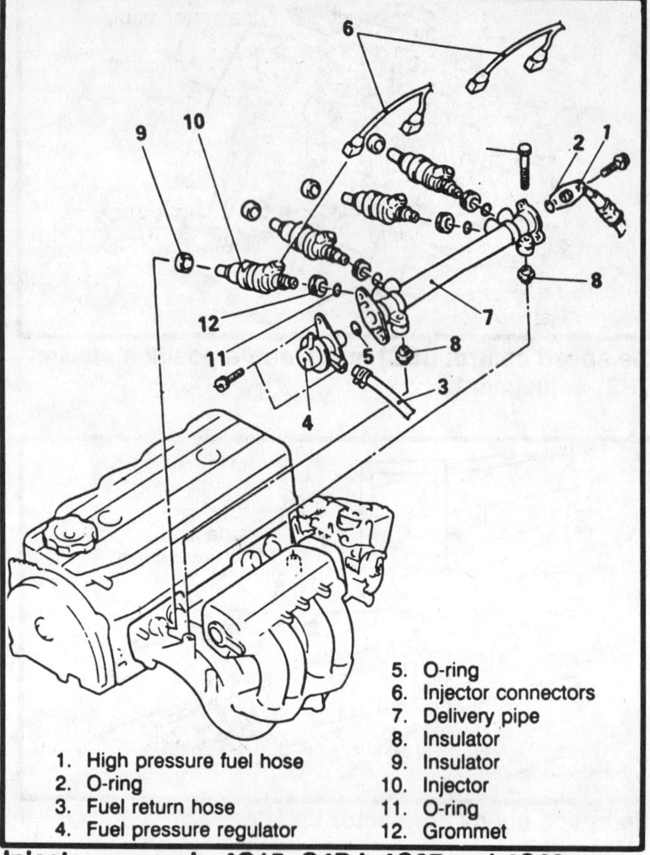

Injector removal – 4G15, G4DJ, 4G37 and 4G63

3. Disconnect the high pressure fuel line at delivery pipe (rail).

**NOTE: Wrap the connection in a clean towel or cloth before disconnecting. Some pressure will remain within the system.**

4. Disconnect fuel return hose.
5. Disconnect electrical connector to each injector.
6. Remove bolts holding the injector rail; remove rubber insulators below the rail mounting points.
7. Lift rail with injectors attached up and away from the engine. Do not drop injectors.

**NOTE: If injector should fall and hit the floor or other hard surface, it must be considered unusable.**

8. Remove injectors from rail with gentle pull. The lower insulator must be removed and replaced.
9. Installing new grommet and O-ring (in that order) onto injector. Coat O-ring with a light coating of gasoline. Do not use grease or oil.
10. Install injector into rail. Injector must turn freely when in place. If it does not, inspect O-ring and reinstall. Injector does not turn during operation but its ability to turn is an indicator of correct installation.
11. Replace seats in intake manifold. Install delivery pipe and injectors onto the manifold. Make certain rubber bushings are in place under the delivery pipe brackets.
12. Tighten fuel rail bolts to 8 ft. lbs or 72 inch lbs. (11 Nm)
13. Connect electrical connector to injectors.
14. Replace O-ring, coat it lightly with gasoline and connect fuel pressure regulator. Tighten connection to 6 ft. lbs or 72 inch lbs. (8 Nm)
15. Connect the fuel return hose.
16. Replace the O-ring, coat lightly with gasoline and install high pressure fuel line. Make certain O-ring is not damaged during installation. Tighten the bolts to 3 ft. lbs. or 36 inch lbs. (4 Nm)
17. Connect the negative battery cable.

### 4G61 AND 4G63 DOHC ENGINES

1. Safely relieve the pressure within fuel system.
2. Disconnect negative battery cable.
3. Disconnect high pressure fuel line at delivery pipe (rail).

**NOTE: Wrap connection in a clean towel before disconnecting. Some pressure will remain within the system.**

4. Disconnect fuel return hose and remove O-ring. Disconnect vacuum hose from the fuel pressure regulator.
5. Remove fuel pressure regulator and its O-ring.
6. Disconnect PCV hose.
7. Remove electrical connector from each injector.
8. Remove bolts holding delivery pipe to the engine.
9. Lift the rail with injectors attached up and away from engine. Do not drop injectors during this removal.

**NOTE: If injector should fall and hit the floor or other hard surface, it must be considered unusable.**

10. Injectors may be removed with a gentle pull. Lower insulator must also be removed.
11. Installing new grommet and O-ring (in that order) onto injector. Coat O-ring with a light coating of gasoline. Do not use grease or oil.
12. Install each injector into rail, making sure that injector turns freely. If it does not turn, inspect O-ring and reinsert.

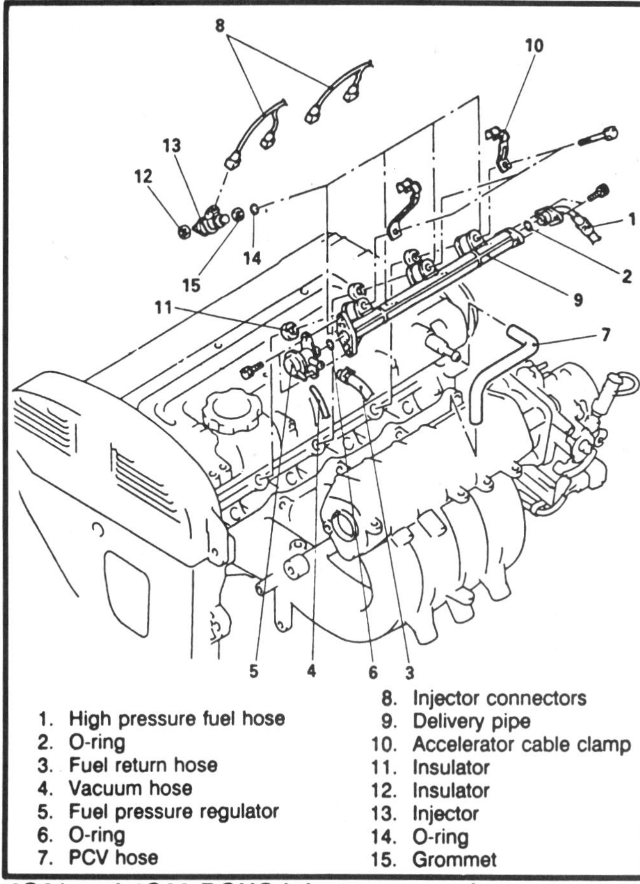


**4G61 and 4G63 DOHC injector removal**

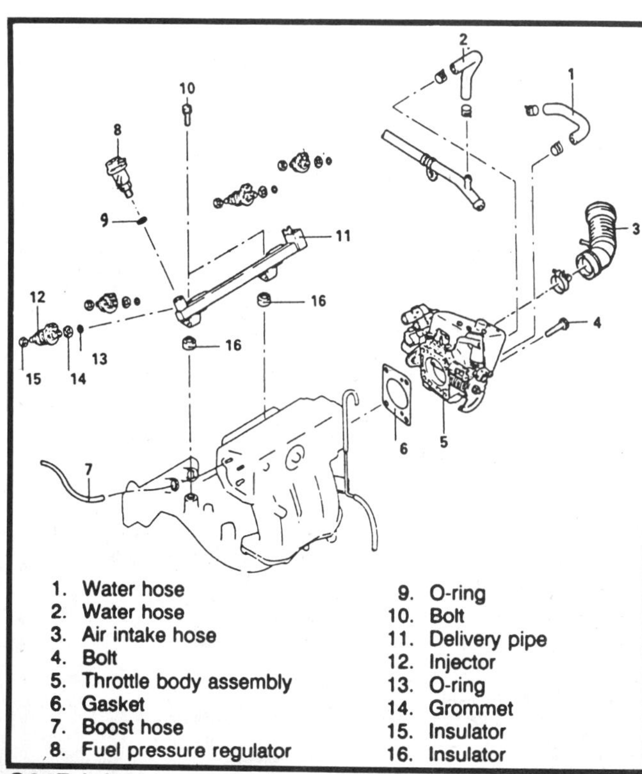


**G64B injector removal**

(Iinjector does not turn during operation, but its ability to turn is indicator of correct installation).

13. Replace seats in the intake manifold. Install delivery pipe and injectors onto the manifold without dropping injector.
14. Tighten fuel rail bolts to 8 ft. lbs or 72 inch lbs. (11 Nm).
15. Connect electrical connectors to the injectors.
16. Connect the PCV hose.
17. Replace O-ring, coat it lightly with gasoline and install fuel pressure regulator. Tighten the fasteners to 6 ft. lbs or 72 inch lbs. (8 Nm)
18. Connect the fuel return hose.
19. Replace the O-ring, coat lightly with gasoline and install high pressure fuel line. Make certain O-ring is not damaged during installation. Tighten the bolts to 3 ft. lbs. or 36 inch lbs. (4 Nm)
20. Connect the negative battery cable.

### G64B ENGINE

1. Safely relieve pressure within the fuel system.
2. Disconnect negative battery cable.
3. Remove the throttle body.
4. Remove the boost hose from opposite end of air plenum.

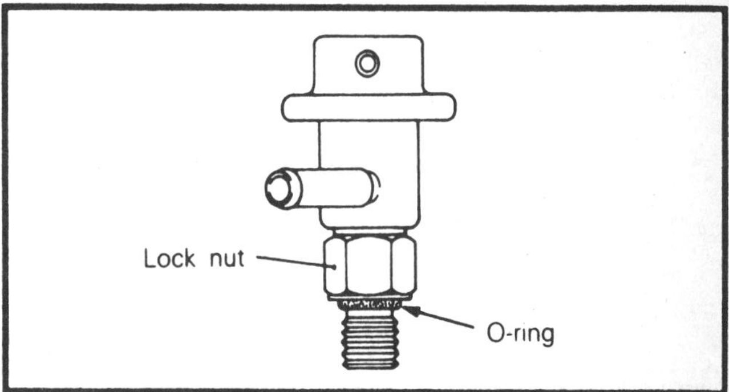


**G64B fuel pressure regulator**

5. Disconnect high pressure fuel line from delivery pipe (fuel rail).

**NOTE: Wrap connection in clean towel or cloth before disconnecting. Some pressure will remain within system.**

6. Remove the fuel return line from fuel pressure regulator. Remove the seal (O-ring) from the bottom of regulator.
7. Disconnect electrical connectors from injectors.
8. Remove the bolts holding fuel delivery pipe.
9. Lift rail with injectors attached up and away from the engine. Do not drop injectors during removal. If injector should fall and hit the floor or other hard surface, it must be considered unusable.
10. Injectors are removed with a gentle pull. The lower insulator must be removed and replaced.
11. Installing new grommet and O-ring (in that order) onto the injector. Coat O-ring with a light coating of gasoline. Do not use grease or oil.
12. Install injector into rail, making sure that injector turns freely. If it does not turn, inspect O-ring and reinsert.
13. Replace seats (insulator) in the intake manifold. Install delivery pipe and injectors onto manifold without dropping an injector. Make certain rubber bushings are correctly seated in installation hole.
14. Tighten fuel rail bolts to 8 ft. lbs or 72 inch lbs. (11 Nm)
15. Install new O-ring on the fuel pressure regulator and coat lightly with clean gasoline. Turn locking nut all the way up the threads.
16. Screw the regulator into fuel rail by hand. When regulator has bottomed on threads, unscrew less than 1 full turn so that

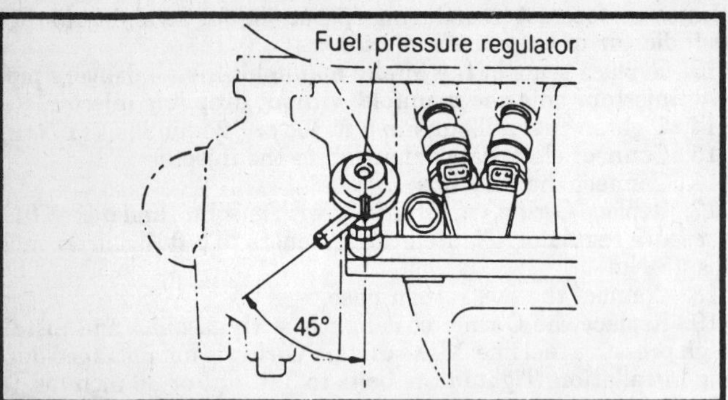


**Pressure regulator must be in correct postition before tightening**

hose port faces away from the injectors and at 45 degree angle to the fuel rail.

17. After position is set, tighten lock nut to 22.5 ft. lbs. (30 Nm)
18. Replace O-ring and connect the high pressure fuel line to delivery pipe. Tighten bolts to 22.5 ft. lbs. (30 Nm)
19. Connect boost hose to air plenum.
20. Reinstall throttle body.
21. Connect the negative battery cable.

### 6G72 V6 ENGINE

1. Safely relieve pressure within the fuel system.
2. Disconnect the negative battery cable.
3. Remove air intake plenum.
4. Disconnect the high pressure fuel line at delivery pipe (rail).

**NOTE: Wrap connection in clean towel or cloth before disconnecting. Some pressure will remain within the system.**

5. Disconnect fuel return hose and remove its O-ring. Disconnect vacuum hose from the fuel pressure regulator.
6. Remove fuel pressure regulator and its O-ring.
7. Remove cover piece from the fuel rail.
8. Remove electrical connector from injectors.
9. Remove bolts holding delivery pipe to the engine. Note that fuel rail is 1 continuous piece serving both banks of cylinders.
10. Lift rail with injectors attached up and away from the engine. Do not drop injectors during removal. If injector should fall and hit the floor or other hard surface, it must be considered unusable.
11. Remove injectors from the fuel rail with a gentle pull. The lower insulator (seat) must be removed and replaced.
12. Install new grommet and O-ring (in that order) onto injector. Coat the O-ring with light coating of gasoline. Do not use grease or oil.
13. Install injectors into rail, making sure that injector turns freely. If it does not, inspect the O-ring.
14. Replace seats (insulators) in the intake manifold. Install new rubber bushings onto mounting points of fuel rail. Install delivery pipe and the injectors onto manifold without dropping an injector.
15. Tighten fuel rail bolts to 8 ft. lbs or 72 inch lbs. (11 Nm)
16. Connect electrical connectors to injectors.
17. Install cover on the fuel rail; tighten bolts to 6 ft. lbs. or 72 inch lbs. (8 Nm)
18. Replace the O-ring, coat lightly with gasoline and install fuel pressure regulator. Tighten fasteners to 6 ft. lbs or 72 inch lbs. (8 Nm)
19. Connect fuel return hose and the vacuum hose.
20. Replace O-ring, coat it lightly with gasoline and install the high pressure fuel line. Make certain O-ring is not damaged during installation. Tighten bolts to 7.5 ft. lbs. or 81 inch lbs. (10 Nm)
21. Reinstall air intake plenum.
22. Connect the negative battery cable.

## IDLE SPEED

### Inspection

**NOTE: Improper adjustment of idle speed will increase exhaust gas temperature and possibly damage catalytic converter. Idle adjustment generally should not be necessary due to computer control.**

Check conditions: Engine coolant temperature: 185–205°F; all lights, electric cooling fan and electric accessories **OFF**; transmission in **NEUTRAL** or **PARK**; steering wheel in straight ahead position (vehicles with power steering).

1. Connect timing light and tachometer.
2. Start engine and let it idle.
3. Check the basic ignition timing and adjust if necessary.

**NOTE: When checking the basic ignition timing at high altitude, stop the engine and disconnect the waterproof female connector from the ignition timing connector. Connect a lead wire with an alligator clip to the ignition timing adjusting terminal to ground it.**

4. Run the engine for more than 5 seconds at 2,000–3,000 rpm.
5. Run the engine at idle for 2 minutes.
6. Take idle speed reading. Curb idle speed as listed on underhood emissions label.

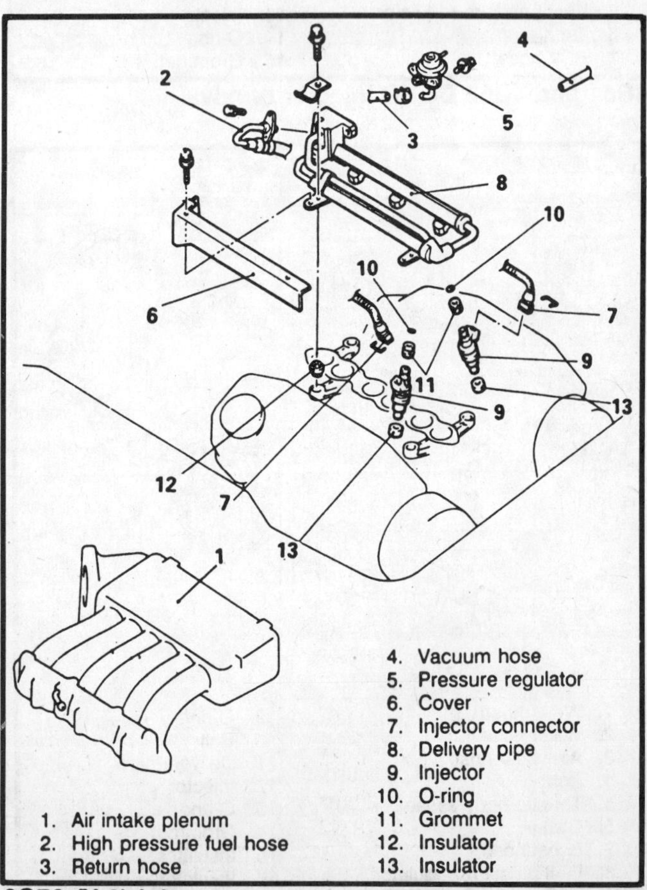


**6G72 (V-6) injector removal**

## MPI TEST VALUES SYSTEM

| | | Fuel Pressure psi | Air Flow Sensor | Intake Air Temp Sensor— Resistance @ Terminals: | Eng. Coolant Temp Sensor | Throttle Position Sensor Resistance 3.5-6.5 Kilo-Ohms @ Terminals: ② | Motor Position Sensor 4-6 KΩ @ Terminals: ② | Injector ② |
|---|---|---|---|---|---|---|---|---|
| 1988 | Van/Wagon | Unregulated 35.6–38.4 psi Regulated 28.4 psi | Term. 4 & 5: 2.2–3.2v @ Idle–3000 rpm | Term. 2 & 4: 32°F=6.0KΩ 68°F=2.7KΩ 176°F=0.4KΩ | 32°F=5.9KΩ 68°F=2.5KΩ 176°F=0.3KΩ | Term. 1 & 3: | Term. 3 & 4: | 2–3Ω |
| | Galant | Unregulated 45.5–48.4 psi Regulated 37.7 psi | ① | Term. 2 & 4: 32°F=6.0KΩ 68°F=2.7KΩ 176°F=0.4KΩ | 32°F=5.9KΩ 68°F=2.5KΩ 176°F=0.3KΩ | Term. 2 & 3: | — | 15–17Ω |
| 1989 | Mirage SOHC Galant SOHC | Unregulated 47–50 psi Regulated Approx. 38 psi | ① | Term. 4 & 6: 32°F=6.0KΩ 68°F=2.7KΩ 176°F=0.4KΩ | 32°F=5.9KΩ 68°F=2.5KΩ 176°F=0.3KΩ | Term. 1 & 2: | Term. 1 & 3: | 13–16Ω |
| | Galant DOHC | Unregulated 47–50 psi Regulated Approx. 38 psi | ① | Term. 4 & 6: 32°F=6.0KΩ 68°F=2.7KΩ 176°F=0.4KΩ | 32°F=5.9KΩ 68°F=2.5KΩ 176°F=0.3KΩ | Term. 2 & 3: | — | 13–16Ω |
| | Mirage DOHC | Unregulated 47–50 psi Regulated Approx. 38 psi | ① | Term. 6 & 8: 32°F=6.0KΩ 68°F=2.7KΩ 176°F=0.4KΩ | 32°F=5.9KΩ 68°F=2.5KΩ 176°F=0.3KΩ | Term. 1 & 2: | Term. 1 & 3: | 2–3Ω |
| | Van/Wagon | Unregulated 47–50 psi Regulated Approx. 38 psi | ① | Term. 4 & 6: 32°F=6.0KΩ 68°F=2.7KΩ 176°F=0.4KΩ | 32°F=5.9KΩ 68°F=2.5KΩ 176°F=0.3KΩ | Term. 1 & 2: | Term. 1 & 3: | 13–16Ω |
| | Montero 3.0 | Unregulated 47–53 psi Regulated 38 psi | ① | Term. 4 & 6: 32°F=6.0KΩ 68°F=2.7KΩ 176°F=0.4KΩ | 32°F=5.9KΩ 68°F=2.5KΩ 176°F=0.3KΩ | Term. 1 & 4: | | 13–16Ω |
| 1990 | Truck 3.0 | Unregulated 47–53 psi Regulated 38 psi | ① | Term. 4 & 6: 32°F=6.0KΩ 68°F=2.7KΩ 176°F=0.4KΩ | 32°F=5.9KΩ 68°F=2.5KΩ 176°F=0.3KΩ | Term. 1 & 4: | — | 13–16Ω |
| | Mirage 1.5L | Unregulated 47–50 psi Regulated Approx. 38 psi | ① | Term. 4 & 6: 32°F=6.0KΩ 68°F=2.7KΩ 176°F=0.4KΩ | 32°F=5.9KΩ 68°F=2.5KΩ 176°F=0.3KΩ | Term. 1 & 4: | Term. 2 & 3: | 13–16Ω |
| | Van/Wagon | Unregulated 47–50 psi Regulated Approx. 38 psi | ① | Term. 4 & 6: 32°F=6.0KΩ 68°F=2.7KΩ 176°F=0.4KΩ | 32°F=5.9KΩ 68°F=2.5KΩ 176°F=0.3KΩ | Term. 1 & 4: | Term. 2 & 3: | 13–16Ω |

## MPI TEST VALUES SYSTEM

| | | Fuel Pressure psi | Air Flow Sensor | Intake Air Temp Sensor—Resistance @ Terminals: | Eng. Coolant Temp Sensor | Throttle Position Sensor Resistance 3.5-6.5 Kilo-Ohms @ Terminals: ② | Motor Position Sensor 4-6 KΩ @ Terminals: ② | Injector ② |
|---|---|---|---|---|---|---|---|---|
| **1990** | Galant SOHC | Unregulated 47-50 psi Regulated Approx. 38 psi | ① | Term. 4 & 6: 32°F=6.0KΩ 68°F=2.7KΩ 176°F=0.4KΩ | 32°F=5.9KΩ 68°F=2.5KΩ 176°F=0.3KΩ | Term. 1 & 4: | Term. 2 & 3: | 13-16Ω |
| | Truck 2.4 | Unregulated 47-50 psi Regulated Approx. 38 psi | ① | Term. 4 & 6: 32°F=6.0KΩ 68°F=2.7KΩ 176°F=0.4KΩ | 32°F=5.9KΩ 68°F=2.5KΩ 176°F=0.3KΩ | Term. 1 & 4: | Term. 2 & 3: | 13-16Ω |
| | Galant DOHC | Unregulated 47-50 psi Regulated Approx. 38 psi | ① | Term. 4 & 6: 32°F=6.0KΩ 68°F=2.7KΩ 176°F=0.4KΩ | 32°F=5.9KΩ 68°F=2.5KΩ 176°F=0.3KΩ | Term. 1 & 4: | — | 13-16Ω |
| | Mirage 1.6L | Unregulated 47-50 psi Regulated Approx. 38 psi | ① | Term. 4 & 6: 32°F=6.0KΩ 68°F=2.7KΩ 176°F=0.4KΩ | 32°F=5.9KΩ 68°F=2.5KΩ 176°F=0.3KΩ | Term. 1 & 4: | Term. 2 & 3: | 13-16Ω |
| | Precis | Unregulated 47-50 psi Regulated Approx. 38 psi | Term. 6 & 3: 2.7-3.2v @ Idle-3000 rpm | Term. 4 & 6: 32°F=6.0KΩ 68°F=2.7KΩ 176°F=0.4KΩ | 32°F=5.9KΩ 68°F=2.5KΩ 176°F=0.3KΩ | Term. 1 & 3: | Term. 1 & 3: | 13-16Ω |
| **1989-1990** | Sigma | Unregulated 47-50 psi Regulated Approx. 38 psi | ① | Term. 4 & 6: 32°F=6.0KΩ 68°F=2.7KΩ 176°F=0.4KΩ | 32°F=5.9KΩ 68°F=2.5KΩ 176°F=0.3KΩ | Term. 3 & 4: | — | 13-16Ω |
| | Eclipse 1.8 | Unregulated 47-50 psi Regulated 38 psi | ① | Term. 4 & 6: 32°F=6.0KΩ 68°F=2.7KΩ 176°F=0.4KΩ | 32°F=5.9KΩ 68°F=2.5KΩ 176°F=0.3KΩ | Adapter Harness Red & Black | Term. 1 & 3: | 13-16Ω |
| | Eclipse 2.0 | Unregulated 47-50 psi Regulated 38 psi | ① | Term. 4 & 6: 32°F=6.0KΩ 68°F=2.7KΩ 176°F=0.4KΩ | 32°F=5.9KΩ 68°F=2.5KΩ 176°F=0.3KΩ | Adapter Harness Red & Black | — | 13-16Ω |
| | Eclipse 2.0 Turbo | Unregulated 36-38 psi Regulated 27 psi | ① | Term. 6 & 8: 32°F=6.0KΩ 68°F=2.7KΩ 176°F=0.4KΩ | 32°F=5.9KΩ 68°F=2.5KΩ 176°F=0.3KΩ | Adapter Harness Red & Black | — | 2-3Ω |

① Must test with multi use tester or BCI checker
② Resistance at 68°F (20°C)